G

NORIE'S
NAUTICAL TABLES

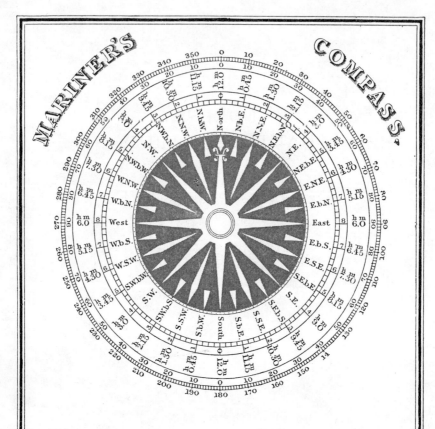

A TABLE OF THE ANGLES

which every Point & Quarter Point of the Compass makes with the Meridian.

NORTH		POINTS	o , ''	POINTS	SOUTH	
		0 – ¼	2 . 48 . 45	0 – ¼		
		0 – ½	5 . 37 . 30	0 – ½		
		0 – ¾	8 . 26 . 15	0 – ¾		
N.b.E.	N.b.W.	1	11 . 15 . 0	1	S.b.E.	S.b.W.
		1 – ¼	14 . 3 . 45	1 – ¼		
		1 – ½	16 . 52 . 30	1 – ½		
		1 – ¾	19 . 41 . 15	1 – ¾		
N.N.E.	N.N.W.	2	22 . 30 . 0	2	S.S.E.	S.S.W.
		2 – ¼	25 . 18 . 45	2 – ¼		
		2 – ½	28 . 7 . 30	2 – ½		
		2 – ¾	30 . 56 . 15	2 – ¾		
N.E.b.N.	N.W.b.N.	3	33 . 45 . 0	3	S.E.b.S.	S.W.b.S.
		3 – ¼	36 . 33 . 45	3 – ¼		
		3 – ½	39 . 22 . 30	3 – ½		
		3 – ¾	42 . 11 . 15	3 – ¾		
N.E.	N.W.	4	45 . 0 . 0	4	S.E.	S.W.
		4 – ¼	47 . 48 . 45	4 – ¼		
		4 – ½	50 . 37 . 30	4 – ½		
		4 – ¾	53 . 26 . 15	4 – ¾		
N.E.b.E.	N.W.b.W.	5	56 . 15 . 0	5	S.E.b.E.	S.W.b.W.
		5 – ¼	59 . 3 . 45	5 – ¼		
		5 – ½	61 . 52 . 30	5 – ½		
		5 – ¾	64 . 41 . 15	5 – ¾		
E.N.E.	W.N.W.	6	67 . 30 . 0	6	E.S.E.	W.S.W.
		6 – ¼	70 . 18 . 45	6 – ¼		
		6 – ½	73 . 7 . 30	6 – ½		
		6 – ¾	75 . 56 . 15	6 – ¾		
E.b.N	W.b.N.	7	78 . 45 . 0	7	E.b.S.	W.b.S.
		7 – ¼	81 . 33 . 45	7 – ¼		
		7 – ½	84 . 22 . 30	7 – ½		
		7 – ¾	87 . 11 . 15	7 – ¾		
East	West	8	90 . 0 . 0	8	East	West

NORIE'S NAUTICAL TABLES

WITH EXPLANATIONS OF THEIR USE

EDITED BY

CAPTAIN F. N. HOPKINS

IMRAY LAURIE NORIE AND WILSON LTD

SAINT IVES, HUNTINGDON

1973

PUBLISHED BY
IMRAY LAURIE NORIE AND WILSON LTD
WYCH HOUSE, SAINT IVES, HUNTINGDON

©

IMRAY LAURIE NORIE AND WILSON LTD
1973

Revised edition   1973

Printed in Great Britain by the
Pendragon Press, Papworth Everard, Cambridgeshire

PREFACE

J. W. Norie was born in London in 1772. He conducted a nautical academy at 157 Leadenhall Street, London, and became well known as a teacher of navigation and nautical astronomy. In 1803 his *Complete Set of Nautical Tables* and *Epitome of Practical Navigation* were first published at the same address by the firm of which he later became head and whose imprint merged during the nineteenth century to become that of the present publishers of this work.

The process of rearranging *Norie's Nautical Tables* which began in the early nineteen-fifties, and which resulted in a number of changes appearing in the 1956 edition, has been continued further so that the present edition brings this long-established and famous set of tables completely up to date and relates them realistically to modern marine navigational techniques and requirements.

Whilst it has never been the policy of the publishers to make unnecessary changes, every opportunity has been taken to introduce such improvements as the circumstances of the times demand. Navigation in these days of increasingly larger, faster and costlier vessels calls for the maximum of accuracy obtained by the minimum of effort. Hence, in designing the tables everything possible has been done to provide navigators with the means of obtaining accurate results quickly. This, it is hoped, has been adequately achieved by preserving wherever possible the familiar layout, by eliminating the need for tedious mental interpolations, and by reducing to a minimum the necessity to use upward reading tables. This should greatly diminish the risk of making mistakes. In short, the ideal which has been aimed at is to produce tables which are accurate, labour-saving and blunder-proof. Students and examination candidates as well as practising navigators should find this a distinct advantage.

The 1963 edition introduced a number of significant changes involving reconstruction of previously existing tables and the addition of new ones. So favourably were these reforms received that the only changes made in the present edition are those rendered necessary by the general demand for metrication and the recalculation of Dip and Distance of the Sea Horizon values to conform with the now recommended formulae. Details of these changes will be found in the Introduction.

F. N. HOPKINS

CONTENTS

CONTENTS

British Isles and Eire p.583.   Arctic and N.E. Europe p.585.   N.W. Europe—Elbe to Gibraltar p.588.   Mediteranean, Black Sea and Sea of Azov p.589. West, South and East Africa p.591.   Asia p.593.   Australasia and Pacific Islands p.596.   North and South America—Pacific Coast p.598.   South America—Atlantic Coast p.600.   West Indies, Gulf of Mexico and Caribbean Sea p.601.   North America—Atlantic Coast, Greenland and Iceland p.604.

INTRODUCTION

EXPLANATION AND USE OF THE TABLES

EXPLANATION AND USE OF THE TABLES

TRAVERSE TABLE

(Pages 2 to 93.)

These Tables afford an easy and expeditious means of solving all problems that resolve themselves into the solution of right-angled plane triangles. They can thus be applied to all the forms of Sailings except Great Circle Sailing; but they are specially useful in resolving a Traverse. On this account they are called Traverse Tables, and the terms *Course, Distance, Difference of Latitude* and *Departure* are used as names of the different parts involved.

The Traverse Table has now been brought into line with the requirements of the modern compass notation by the inclusion, at the top and foot of each page, of the number of degrees of the new (0–360) circular system of reckoning, corresponding to the value printed at centre of title in conformity with the older quadrantal notation, the latter form being retained for its application to the solution of certain problems in the Sailings and for its utility in the conversion of Departure to Difference of Longitude and vice versa, as explained later in this article.

The figures denoting the number of degrees under the new arrangement are placed in the appropriate quadrants of a small diagrammatic symbol, representing the cardinal points of the compass, and in these positions they introduce the equivalents in the new notation corresponding to the number of degrees of the old system, shown at centre of titles, when pertaining to the respective quadrants. The arrangement will be better understood by reference to an example; thus, on page 58—"28 DEGREES"

For old	N28°E	S28°E	S28°W	N28°W
Read new	↑ 028°	↑	↑	332° ↑
	|	| 152°	208° |	|

or vice versa, and, as examples of the reverse process, on page 38, but this time from the foot, with caption "72 DEGREES"—

For new	↑ 072°	↑	↑	288° ↑
	|	| 108°	252° |	|
Read old	N72°E	S72°E	S72°W	N72°W

It will be observed that, in the new notation of the Traverse Table, the three-figure degrees corresponding to Easterly courses are placed in the symbol diagram towards the right-hand side of the page, in contradistinction to those for Westerly equivalents which are printed on the left.

The courses, in both the old and new notation, are displayed at the top and bottom of the pages, while the Distances are arranged in order in the columns marked Dist. The Difference of Latitude and Departure corresponding to any given Run on any given Course will be found in the columns marked D. Lat. and Dep., respectively, of the page for the given Course and opposite the given Distance.   But it must most carefully be observed that when the required Course is found at the *top* of the page, the Difference of Latitude and Departure also are to be taken from the columns as named at the *top* of the page; and when the Course appears at the *foot* of the page; the relevant quantities too must be taken from the columns as named at the *foot* of the page.

When any of the given quantities (except the Course which is never to be changed) exceeds the limits of the tables, any aliquot part, as a half or a third, is to be taken, and the quantities found are to be doubled or trebled; that is, they are to be multiplied by the same figure as the given quantity was divided by. And since the Difference of Latitude and Departure corresponding to any given Course and Distance are to be found opposite the Distance on that page which contains the Course, it follows that if any two of the four parts be given, and these two be found in their proper places in the tables, the other two will be found in their respective places on the same page.

The following examples will illustrate the application of the tables to Plane Sailing:—

Example.—Find the difference of latitude and departure made good by a ship in sailing 84 miles on a course 112°.   (S68°E., Old Style).

Course 112° is found in the Table at foot of page 46.   Opposite 84 in the Distance column on that page we get:—D. Lat. 31·5, Dep. 77·9.

The D. Lat. is named S and Departure E, because it is noted that 112° is shown in the South and East quadrant of the compass symbol.

Example.—Find the course and distance made good by a ship whose difference of latitude was found to be 431 miles S, and departure 132' W.

431 and 132 are not to be found alongside each other, but in the Table on page 37 we find 431·3 and 131·9, and these are sufficiently near to the desired value for all practical purposes.   These give 197°, or 17° old style, as a course, and 451 as a distance.   Hence—

Course S17° W, or 197°, and Distance 451 miles.

These tables may also, as has already been stated, be applied to solving problems in Parallel and Middle Latitude Sailings.   In solving these problems the Course (old notation) at the *top* or *bottom* of page becames the Latitude or Middle Latitude, the Distance column becomes a Diff. Longitude column, and the D. Lat. column becomes a Dep. column.   To facilitate the taking out of these quantities the D. Long. and Dep. are bracketed together, and the words *D. Long.,* and *Dep.* are also printed in italics at the top of their respective columns when the Latitude or Middle Latitude, as course, is at the *top;* but at the bottom of their respective columns when Latitude or Middle Latitude, as course is at the *bottom.*

Example.—In Latitude or Middle Latitude 47° the departure made good was 260'·5; required the difference of Longitude.

With 47° as course at the *bottom* of the page, look in the column with *Dep.* printed in italics at the *bottom,* just over the end of the bracket; and opposite to 260'·5 will be found 382 in the *D. Long.* column, which is the Difference of Longitude required.

Example.—A ship, after sailing East 260'·5, had changed her Longitude 6° 22'. Required the parallel of Latitude on which she sailed.

6° 22' equals 382'. Opposite 382 in *D. Long.* column is 260'·5 in *Dep.* column entered from the *bottom,* and the parallel on which she sailed is Lat. 47°.

RADAR RANGE TABLE
(*Page 94, Explanation with Table*)

RADAR PLOTTERS SPEED AND DISTANCE TABLES
(*Page 95*)

DAY'S RUN AVERAGE SPEED TABLE
(*Pages 96 to 102*)

This table provides a rapid means of finding the average speed directly from the arguments "steaming time" and "distance run". It will be appreciated that there is no necessity to convert minutes into decimals of a day, and that no logarithms or co-logarithms are required. Simple addition is all that is needed.

The scope of the table has been made wide enough to cover cases of high speed vessels (up to 40 knots or so) on easterly or westerly courses in high latitudes where change of longitude between one local noon and the next may amount to some 30°, or 2 hours of time.

Distances are tabulated as multiples of 100 miles. Increments of speed for multiples of 10 miles and multiples of 1 mile are obtained simply by shifting the decimal point one or two places to the left, respectively.

Example.—Given steaming time 23h. 29m., distance 582 miles, find the average speed.

Distance in miles	Speed in knots
500	21·291
80*	3·4066
2†	0·08517
582	24·78277

That is, average speed correct to two places of decimals, which are generally considered sufficient, is 24·78 knots.

* Enter with 800 miles and shift decimal point 1 place to the left
† Enter with 200 miles and shift decimal point 2 places to the left

MEASURED MILE SPEED TABLE
(*Pages* 103 *to* 109)

This table is arranged in "critical table" form and gives speeds correct to the nearest hundredth of a knot without interpolation. If the time argument is an exact tabulated value, the speed immediately above it should be taken.

Examples:—

1. If the time recorded for the measured mile is 9m. 16·2s., the speed is 6 47 knots.
2. If the time is 4m. 55·3s., the speed is 12·19 knots.
3. If the time is 3m. 52·3s., the speed is 15·49 knots.
4. Suppose a ship on trials makes six runs over a measured mile, three against the tide and three with the tide, such that the timings by stop-watch are as follows:—

		m.	s.
First run against tide	....	3	28·8
First run with tide	....	3	18·4
Second run against tide	....	3	30·0
Second run with tide	....	3	17·8
Third run against tide	....	3	31·1
Third run with tide	....	3	16·7
Then total time for 6 miles is	....	20	22·8
∴ Average time for 1 mile is	....	3	23·8

From the table the average speed for the six runs is 17·66 *knots.*

Strictly speaking, the average speed should be computed by finding the "mean of means", in which case the work would be arranged as follows.

Run			Speed knots	1st mean	2nd mean	3rd mean	4th mean	Mean of means
	m.	s.						
1st	3	28·8	17·24					
				17·690				
2nd	3	18·4	18·14		17·6650			
				17·640		17·66000		
3rd	3	30·0	17·14		17·6550		17·655625	
				17·670		17·65125		17·6528125
4th	3	17·8	18·20		17·6475		17·650000	
				17·625		17·64875		
5th	3	31·1	17·05		17·6500			
				17·675				
6th	3	16·7	18·30					
			6)106·07	4)70·6175				
			17·68	17·6544				*True mean speed*
			Ordinary mean speed	*Ordinary mean second means**				

* This is usually regarded as being sufficiently accurate

At speeds greater than about 19½ knots it will be noticed that in certain cases a change of a tenth of a second in the time will make a difference of more than one hundredth of a knot in the tabulated speed. For example, if the time for one mile is between 2m. 38·7s. and 2m. 38·8s. the speed, correct to two places of decimals, could be either 22·68 or 22·67 knots.

In very high speed vessels the recorded time for a measured mile may be so small as to be beyond the scope of the table. Even so, a reasonably accurate speed is easily obtained by entering the table with double the recorded time, and then doubling the speed so obtained. For instance, if a mile is run in 1m. 55·2s., enter with 3m. 50·4s. This gives 15·62 knots which is half the required speed of 31·24 knots (and this will be correct within 0·02 of a knot). By calculation the correct speed is actually 31·250 knots.

Besides its orthodox use for speed trial purposes, the table will be found useful to navigators for other purposes.

For example, suppose it is decided to alter course after the ship has run 6 miles on a certain heading from a position line obtained at 1432, the speed of the ship being 11·75 knots. The table shows that at this speed the ship will run one mile in a little over 5m. 6s., or 6 miles in about 30½ minutes. Therefore, the course should be altered at 1502½.

In certain circumstances it might be considered convenient to plot the radar target of another vessel at regular intervals corresponding to one mile runs of one's own vessel. Suppose the speed to be 9·70 knots, which the table shows to correspond to a mile in about 6m. 11s. Then, if the stop-watch is started from zero at the time of the first observation, successive observations should be taken as nearly as practicable when the watch shows 6m. 11s., 12m. 22s., 18m. 33s., 24m. 44s., and so on.

DISTANCE BY VERTICAL ANGLE
(*Pages* 110 *to* 115)

This table gives the distance of an observer from objects of known height when the angle at the eye of the observer is known. The distances range from 1 cable to 5 miles, and the heights from 40 feet (12·2m.) to 2,000 feet (609·6m.). As the distance is only five miles, the whole of the object from base to summit will be in view of the observer when the eye is elevated about 25 feet, and there will, therefore, be no Dip to apply.

The distances given are from the position of the observer to a point at the base vertically below the summit, and it is to this point that the angle should be measured. In places where there is a big rise and fall of tide it would be necessary to make an allowance for the state of the tide, as heights are always given above high water springs. In the case of light vessels there is no allowance for the state of the tide, as the water plane is always at the same distance with reference to any part of the vessel.

To find the Distance

Measure the angle from summit to base and note the angle; then under the given height find the observed angle, and opposite the angle will be found the distance off in Miles and Cables in the *left* hand column.

Example.—The vertical angle between the summit and base of a light-house situated 200 feet above sea level was 0° 57'. Required the distance.

Under 200 feet and opposite the given angle is 2m. 0 c., the distance.

To find the Angle to place on the Sextant to pass at a given distance from a Point of Known Height

Opposite the given distance and under the known height will be found the required angle to place on the sextant.

Example.—Wishing to pass a point situated 500 feet above sea level at a distance of 4 miles, required the angle to place on the sextant.

Opposite 4 miles in the distance column at the *side*, and under 500 feet at the *top,* is 1° 11', the angle required to place on the sextant.

EXTREME RANGE TABLE

(*Pages* 116 *to* 117)

This table has been compiled for the purpose of determining the maximum distance at which an object may be seen at sea according to its elevation and that of the observer's eye. Heights of eye are now given in feet with equivalents in metres, whilst heights of object are given in metres to conform with the new practice of expressing heights on Admiralty Charts in metres, though equivalents in feet are also given.

The arguments with which the table is ordinarily entered are the height of the observer's eye and the height of the distant object which last, however, need not be a terrestrial one but may be the masthead of a vessel, or some other easily defined detail thereof, provided always that the height of the feature or object observed be definitely known. The arguments, too, can be made interchangeable, thus, should the lookout, stationed at the masthead at an elevation exceeding 100 feet observe a low-lying rock having a height of less than that amount, then the terms can be substituted for each other and the "Height of Eye" can be sought in the "Height of Object" column, and vice versa.

The tables are computed on the basis of normal atmospheric conditions, refraction and visibility, and, in the case of lights, the quantity taken out as "Extreme Range" presupposes that the light possesses sufficient power to be discernible at such a distance. It must be remembered also that the heights of lights and shore objects are usually referred to High Water Level therefore due allowance should be made when the time of observation does not approximate thereto, particularly if the elevation or distance should be small.

Example 1.—At what distance will a tower 200 feet high be visible to an observer whose eye is elevated 15 feet above the water?

Take 200 feet as the "Height of Object" in the marginal column and in the column under 15 feet "Height of Eye" at top of page, will be found the distance 20·66 miles.

Example 2.—The officer of the watch, whose eye is elevated 45 feet above the water, observes a shore light, with an elevation of 150 feet, just dipping; what distance is the ship from the light?

In the column headed 45 feet, "Height of Eye" and abreast of 150 feet, "Height of Object", will be found the distance, 21·72 miles.

CHANGE of HOUR ANGLE per 1' of ALTITUDE

(*Pages* 118 *to* 119)

The formula used in calculating the values tabulated is:—

Change of H.A. (in mins.) due to 1' change of Alt.=cosec. Az. sec. Lat.

The table gives in minutes of arc the error in hour angle resulting from an altitude 1' in error. This is of particular value to those navigators who work their sights by the "Longitude by Chronometer" method. It will be seen that the error is least in the case of a body on the prime vertical and that it increases as the azimuth decreases—very rapidly as the azimuth becomes very small. From the table the observer can readily find the least azimuth on which the altitude of a body should be observed in order that the resulting longitude may not exceed a chosen limit of error. Another use to which this table can be put is to find the correct longitude when a sight has been worked using an altitude in error by a known amount.

Example 1.—In latitude 18° what should be the lowest value of azimuth in order that an error of 1' in the altitude may not produce more than 2' of error in the computed longitude?

Under lat. 18° and against azi. 32°, the error for 1' of alt. is found to be 1'·98. Accordingly, the observation should be taken on a bearing greater than 32°.

(In lat. 36°, it will be seen, an azimuth of about 39° would constitute the limit. In lat. 63° the error would exceed 2' even when the body was on the P.V.)

Example 2.—A sight worked in lat. 54° by the "Longitude Method" resulted in a deduced longitude of 64° 14'·5W. and azimuth N.65°E. Afterwards it was discovered that the sextant index error of 2' 30" off the arc had been applied the wrong way. Find the correct longitude.

Since the longitude is found by comparing the L.H.A. of the body with its G.H.A., it is evident that the error in the L.H.A. will be the error in the computed longitude. The index error of 2'·5, which should have been added, was subtracted, so that the altitude used was 5' too small.

The table shows that in lat. 54°, when the azi. is 65°, the error in H.A. is 1'·88 per 1' of alt. For 5', therefore, the error will be $5 \times 1'·88 = 9'·40$.

As the real altitude was greater than the value used, the observer must be *nearer* to the body than his computed longitude would lead him to suppose.

With an *easterly* azimuth this means that the *westerly* L.H.A. should be greater, and therefore the observer's west longitude should be smaller. Hence:—

> Computed long. .... 64° 14'·5W.
> Error ....   9'·4 to subtract
>
> Correct long. .... 64° 05'·1W

It will be appreciated that this is much quicker than re-working the sight.

CHANGE of ALTITUDE in ONE MINUTE of TIME

(*Pages* 120 *to* 121)

This Table contains the change in the altitude of a celestial body in minutes and tenths of arc in one minute of time. It is useful for finding the correction to be applied to the computed altitude of a heavenly body when the time of observation differs from that used in the computation of the altitude. When the star is East of the Meridian the correction from the Table is subtractive from the computed altitude if the time of observation is earlier than that used in the computation of the altitude; it is additive if the time of observation is later. When the star is West of the meridian the correction is additive if the time of observation is before that used when computing the altitude, it is subtractive if the time of observation is after.

Formula

Change of altitude in one minute of time=15' Sin. Az. Cos. Lat.

The change in 6 seconds of time is found by shifting the decimal point one place to the left.

The change in 1 second of time is found by calling the quantities in the Table seconds instead of minutes.

Example.—In Lat. 51° 30'N. on the Meridian of Greenwich on October 26th, 1925 at 8h. 0m. p.m. the computed altitude of the star *Altair* was 37° 09'·2. Find the true altitude at 8h. 10m. p.m., the Az. being S.49° 37' W. Opposite 52° in the Lat. Col. and under 50° in the Az. Col. is 7'·1 of arc which is the change of altitude in 1 min. of time, and 7'·1 × 10 minutes gives 71' or 1° 11', which is the correction to apply to the computed altitude.

> Computed Alt. .... 37° 09'·2
> Corr. to Subt............. 1° 11'·0
>
> True Alt. required .... 35° 58'·2

EX-MERIDIAN TABLE I

(*Pages* 122 *to* 133)

This table contains the Variation of the Altitude, in seconds of arc and tenths of any heavenly body, for one minute of time from its Meridian Passage; for Latitudes up to 83° and Declinations up to 63°. It is entered with the Latitude at the *side* and Declination at the *top*, care being taken to observe when the Latitude and Declination are of the "Same" or "Different Names".

The values set down in this table may be computed from the Formula—

$$A = \frac{1''{\cdot}9635 \times \cos. \text{ L.} \times \cos. \text{ D.}}{\sin. (\text{L} - \text{D.})}$$

Where A=Variation of Altitude of any heavenly body in one minute from its Meridian Passage.

L=the Latitude.

D=the Declination, positive when of the Same Name as the latitude and negative when of a Different Name to the latitude, at "Upper Transit"; it is negative when of Same Name at "Lower Transit".

The interpolation can be done at sight, but care should be used in taking out the value of A.

EX-MERIDIAN TABLE II

(Pages 134 *to* 137)

This Table contains the Product of the quantity "A", obtained from Table I, by the square of the number of minutes the observed heavenly body is East or West of the Meridian. This Product is the Reduction to be added to the True Altitude at the "Upper Transit", but to be subtracted from the True Altitude at the "Lower Transit".

The Table is entered with the Hour Angle or time from Meridian Passage at the *top* and "A" at the *side*. The value of "A" is given from 1″ to 9″; if the reduction be required for tenths of a second, shift the decimal point one place to the left; if it be required for hundreds of a second the decimal point is moved two places to the left. When the reduction is a whole number, cut off one figure on the right for tenths of a second in the value of "A" and two figures on the right for hundredths of a second in the value of "A".

Example 1.—In D.R. Lat. 48° 13′ N., D.R. Long. 7° 20′ W., the True Altitude of the sun was 19° 52′. Sun's Hour Angle 356° 00′. Declination 21° 39′ S. Determine the Position Line.

Table I	Table II	T. Alt.	19° 52′·0 S.
Different Name	Hour Angle 356° 00′	Reduction	5′·6
Lat. 48° 13′ N.	A=1″·3	T. Mer. Alt.	19° 57′·6 S.
Decl. 21° 39′ S.	Red for. 1″·0=4′·3	T. Mer. Z. Dist.	70° 02′·4 N.
A=1″·3	„ „ ·3=1′·3	Decl.	21° 39′·0S.
	Reduction=5′·6	Lat.	48° 23′·4 N.
		True Azimuth	
		from Az. Tables 176°	

Position Line passes 086° and 266° through Lat. 48° 23′·4 N., Long. 7° 20′ W.

Example 2.—D.R. Lat. 42° 12′ N., D.R. Long. 24° 32′ W., the True Altitude of Antares was 21° 28′. Star's Hour Angle 357° 00′. Declination 26° 18′ ·0S. Determine the Position Line.

Table I	Table II		
Different Name	Hour Angle 357° 00′	T. Alt.	21° 28′·0 S.
Lat. 42° 12′ N.	A=1″·4	Reduction	3′·4
Decl. 26° 18′ S.	Red. for 1″·0=2′·4		
A=1″·4	„ „ ·4=′·96	T. Mer. Alt.	21° 31′·4 S.
		T. Mer. Z. Dist.	68° 28′·6 N.
	Reduction =3′·36	Decl.	26° 18′·0 S.
		Lat	42° 10′·6 N.
		True Azimuth	
		from Az. Tables 177°	

Position Line passes 087° and 267° through Lat. 42° 10′·6 N. Long. 24° 32′ W.

When a celestial object is near its "Lower Transit", and within Ex-Meridian limits, if its Hour Angle is less than 180° it is West of the meridian and the Hour Angle to use in entering the Table is (180°—actual Hour Angle). But if its Hour Angle is greater than 180° it is East of the meridian and the Hour Angle to use in entering the Table is (Actual H.A.—180°).

Example 3.—D.R. Lat. 42° 10′ N., Long. 21° 30′ W., the True Altitude of Dubhe was 14° 20′. Star's Hour Angle 176° 30′. Declination 62° 01′ N. Determine the Position Line.

Table I	Table II		
Same Name	Hour Angle 176° 30′	T. Alt.	14° 20′·0
Lat. 42° 10′ N.	=3° 30′	Reduction	— 2′·3
Decl. 62° 01′ N.	A=0″·7		
A=0″·7	Reduction =2′·29	T. Mer. Alt.	14° 17′·7
	For Lower Transit—2′·3	Polar Dist.	27° 59′·0
		Lat.	42° 16′·7 N.
		True Azimuth	
		from Az. Tables 358°	

Position Line passes 088° and 268° through Lat. 42° 16′·7 N., Long. 21° 30′ W.

Example 4.—D.R. Lat. 50° 02′ S., D.R. Long. 67° 20′ W., the True Altitude of Achernar was 17° 20′. Star's Hour Angle 184° 20′. Declination 57° 29′ S. Determine the Position Line.

Table I	Table II		
Same Name	Hour Angle 184° 20′	T. Alt.	17° 20′·0
Lat. 50° 02′ S.	=4° 20′	Reduction	—3′·5
Decl. 57° 29′ S.	A=0″·7		
A=0″·7	Reduction =3′·5	T. Mer. Alt.	17° 16′·5
	For Lower Transit—3′·5	Polar Dist.	32° 31′·0
		Lat.	49° 47′·5 S.
		True Azimuth	
		from Az. Tables 177·5°	

Position Line passes 087°·5 and 267°·5 through 49° 47′·5 S., Long. 67° 20′ W.

EX-MERIDIAN TABLE III

(*Page* 138)

This Table contains a Second Correction, which, when the amount of the Main Correction is considerable, enables the process of Reduction to Meridian to be applied with advantage on much larger hour angles than could otherwise be the case.

Example.—D.R. Lat. 31° 00' N., D.R. Long. 124° 00' W., the True Altitude of the Sun was 55° 01'. Sun's Hour Angle 347° 30'. Declination 2° 00' S. Determine the Position Line.

Table I	Table II		
Different Name	Hour Angle 347° 30'	T. Alt.	55° 01'·0 S.
Lat. 31° 00' N.	$A=3''·1$	1st Correction	2° 09'·2+
Decl. 2° 00' S.	Red. for 3''·0=125'·0	2nd Correction	3'·6—
$A=3''·1$	,, ,, ·1= 4'·2	T. Mer. Alt.	57° 06'·6 S.
		T. Mer. Z. Dist.	32° 53'·4 N.
	1st Correction=129'·2	Decl.	2° 00'·0 S.
		Lat.	30° 53'·4 N.

Entering Table III with 129' as First Correction and 56° as Altitude we have 3·6' Subtractive for Second Correction.

Lat.   30° 53'·4 N.
True Azimuth from Az. Tables 158°

Position Line passes 068° and 248° through Lat. 30° 53'·4 N., Long. 124° 00' W.

EX-MERIDIAN TABLE IV

(*Page* 139)

This Table gives the limits of Hour Angle or Time before or after the time of the Meridian Passage when an Ex-Meridian observation can be taken. When the observation is taken within the time limit prescribed by this Table the Second Correction from Table III is negligible. The Table is entered with "A" taken from Table I,

Given Lat. 37° N., Declination 18° N., find the limits of Hour Angle for taking an Ex-Meridian observation.

For Lat. 37° and Declination 18°, "Same Name", Table I gives 4''·6 for "A". Entering Table IV with 4''·6 as "A", the time limit abreast is found to be 24 minutes.

MEAN LATITUDE to MIDDLE LATITUDE

(*Page* 139)

The corrections in this table are used to convert mean latitude to middle latitude when finding the course and distance from one position to another by middle latitude sailing.

Corrections taken from the table *above* the heavy black line are *subtractive* from mean latitude to obtain middle latitude. Those *below* are *additive*. The reverse holds good when converting middle latitude to mean latitude.

The corrections are obtained in the following manner. As the corresponding sides of similar plane triangles are proportional, it follows that:—

$$\frac{\text{d. long.}}{\text{dep.}} = \frac{\text{d.m.p.}}{\text{d. lat.}}$$

Hence,
$$\frac{\text{d.m.p.}}{\text{d. lat.}} = \text{sec. mid. lat.}$$

With respect to any two latitudes, the middle latitude can be found from the above equation and, when it is compared with the mean latitude, the difference is the correction required.

It will be noticed that no corrections are given for cases where the mean latitude is less than 14°. The reason for this is that in such cases, owing to the spheroidal shape of the earth, the d.m.p. is a smaller quantity than the d. lat. and secants cannot be less than unity. However, this has no practical significance because when the mean latitude is small d. long. and dep. differ by so little that no appreciable error is introduced by using the mean latitude instead of the middle latitude when converting from one to the other.

MERIDIONAL PARTS (For the Spheroid)

(*Pages* 140 *to* 147)

This table is used in resolving problems by *Mercator's Sailing* and in constructing charts on Mercator's projection. The meridional parts are to be taken out for the degrees answering to the given latitude at the *top* or *bottom*, and for the minutes at either *side* column. Thus, the meridional parts corresponding to the latitude 49° 57′ are 3451·88.

MEAN REFRACTION

(*Page* 148)

This table contains the Refraction of the heavenly bodies, in minutes and decimals at a mean state of the atmosphere, and corresponding to their apparent altitudes. This correction is always to be *subtracted* from the apparent altitude of the object.

Example.—The mean refraction for the apparent altitude 10° 50′, is 4′·85.

Caution.—For low altitudes all refraction tables are more or less inaccurate.

CORRECTION of the MEAN REFRACTION

(Page 148)

The values of mean refraction given in the main table are for an atmosphere of temperature 50° F. (10° C.) and pressure 29·6 mercury inches (1002·4 mb.) If temperature and/or pressure differ from the figures quoted, the mean refraction should be adjusted by applying corrections from the subsidiary table.   To find the correction for temperature the arguments are apparent altitude at the *side* and fahrenheit thermometer reading at the *top*.   To find the correction for pressure the arguments are apparent altitude at the *side* and barometer reading in mercury inches at the *bottom*.

Example.—Find the true refraction for an apparent altitude of 5° 00′ when the thermometer reading is 74° F. (23°·3 C.) and the barometer reading is 30·0 inches (1015·9 mb.).

Mean Refraction for app. alt. 5° 00′	9′·90
Correction for altitude 5°, and temperature 74° F. . . .	—·57
,, ,, ,, ,, ,, pressure 30·0 in.	+·14

True Refraction   9′·47

N.B.—To convert barometer readings from mercury inches to millibars, or vice-versa, see page 152.   To convert temperatures from Fahrenheit to Centigrade, or vice-versa, see page 559.

The adjustment of mean refraction as shown above is important only when the altitude is small.   It should be borne in mind that on account of uncertain refraction position lines obtained from sights taken when the altitude of the body is less than 10° or so should not be relied upon implicitly.   Moreover, due to the effect of atmospheric refraction on dip it is unwise to place too much reliance on sights taken, whatever the altitude, when there is cause for abnormal refraction to be suspected.

DISTANCE of the SEA HORIZON

(Page 149)

The tabulations are derived from the formula—Distance of the sea horizon in nautical miles$=1·17\sqrt{h}$, where h=height of eye in feet.   Thus for example, when h=100 the distance of the sea horizon is 11·7 nautical miles.

Heights are given in feet, ranging from 1 foot to 6000 feet, and also in metres, ranging from 1m. to 1600m.   The following examples show how the table can be used.

Example 1.—At what distance in good visibility should an observer whose height of eye is 47 feet be able to sight a terrestrial object of height 400 feet?

Dist. of horizon for height	47 feet	=	8·0 miles
,, ,, ,, ,, ,,	400 feet	=	23·4 miles

Sum   =   31·4 miles

Hence, the object should be visible at a distance of about 31½ miles.

Similarly, with height of eye 9m. and height of object 150m. the latter should be visible at a distance of 6·4+26·0=32·4 miles.

Example 2.—The range of visibility of a light is stated on a chart to be 21M. At what distance from the light will an observer be at the moment when the light has just dipped below the horizon if his height of eye is 50 feet?

Charted range, i.e. for 15 feet height of eye   21   miles
Subtracting dist. of horizon for height 15 feet .... 4·55 ,,

Range of light at sea level .... 16·45 ,,
Adding dist. of horizon for height 50 feet .... 8·30 ,,

Dipping distance, or maximum range to observer .... 24·75 ,,

N.B.—This method is applicable only in the case of a light of adequate power, and the accuracy of the result will probably be affected by the fact that the charted height of a light never includes a fraction of a mile. Abnormal refraction will also affect the accuracy of distances obtained by using this table.

DIP of the SEA HORIZON
(*Page* 149)

The tabulated values are derived from the formula—Dip (in minutes)= $0·97\sqrt{h}$ where h=height of eye in feet. Thus, for example, when h=100, dip.=9'·7.

Heights of eye are given in feet, ranging from 1 foot to 500 feet, and also in metres, ranging from 1m. to 150m.

SUN'S PARALLAX IN ALTITUDE
(*Page* 149)

This correction is to be taken out opposite the Sun's Altitude, and is always *additive*.

Example.—The sun's parallax corresponding to 51° of altitude is 0'·1.

SUN'S SEMIDIAMETER on the 1st Day of each MONTH
(*Page* 149)

SUN'S TOTAL CORRECTION
(*Page* 150)

The corrections in the main table are combinations of dip, atmospheric refraction, semi-diameter and parallax in altitude, computed on the assumption that the sun's semidiameter is 15'·8. The subsidiary corrections given at the foot of the page should be applied to the main corrections to take into account the difference between the actual and assumed values of semi-diameter. The corrections are *additive* to observed atlitude. The subsidiary (month) correction should always be added to the main correction. No allowance is made for irradiation effects.

Example:—

Obs. alt. sun's L.L.	25° 57′·2
Corr'n. for obs. alt. 26°, H.E. 40 feet ....+7′·9	
Subsidiary corr'n. for March ....+0′·3	
Tot. corr'n. ....+8′·2	+8′·2
True alt. of sun's centre	26° 05′·4

STAR'S TOTAL CORRECTION

(*Page* 151)

This table gives the combined effect of dip and refraction which are the only corrections necessary to apply to the observed altitude of a fixed star. The star's total correction should also be used to correct the observed altitude of a planet, but in the case of Venus or Mars a small additional correction, depending on date and altitude, and given in the Nautical Almanac, may be necessary to incorporate the effects of parallax and phase.

ATMOSPHERIC PRESSURE CONVERSION TABLE

(*Page* 152)

AUGMENTATION of the MOON'S SEMIDIAMETER

(*Page* 152)

REDUCTION of the MOON'S HORIZONTAL EQUATORIAL PARALLAX
for the FIGURE of the EARTH

(*Page* 152)

CORRECTION of MOON'S MERIDIAN PASSAGE

(*Page* 153)

The correction obtained from this table is to be applied to the time of meridian passage given in the Nautical Almanac (i.e. the time of transit at Greenwich) in order to find the time of the local transit according to the observer's longitude.

$$\text{Correction} = \frac{D \times \text{longtitude}}{360}$$

where D is the difference between the times of successive transits. When the observer is in *East* longitude, D is the difference between the time of transit on the day of observation and the time of transit on the *preceding* day. When in *West* longitude it is the difference between the times on the day of observation and the *following* day.

Example.—From Naut. Alm. L.M.T. of moon's upper transit at Greenwich is:—

	h.	m.
1st July	18	44

diff. 48m.

	h.	m.
2nd July	19	32

diff. 53m.

	h.	m.
3rd July	20	25

Find G.M.T. of moon's upper transit on 2nd. July (a) in longitude 156° E., (b) in longitude 63° W.

(a)	July	h.	m.
L.M.T. of transit at Greenwich..................	2	19	32
Corr'n. for D 48m., long. 156° E..................			—20·8
L.M.T. of local transit	2	19	11·2
East longitude in time units		—10	24
G.M.T. of local transit (156° E.)	2	8	47·2

(b)	July	h.	m.
L.M.T. of transit at Greenwich..................	2	19	32
Corr'n. for D 53m., long. 63° W.			+9·2
L.M.T. of local transit	2	19	41·2
West longitude in time units		+4	12
G.M.T. of local transit (63° W.)	2	23	53·2

MOON'S TOTAL CORRECTION

(*Pages* 154 *to* 157)

This table gives the combined effect of dip, atmospheric refraction, augmented semidiameter and parallax in altitude. The dip component of the main correction is a constant 9′·8. Therefore, a subsidiary correction given at the foot of each page must be added to the main correction, the argument for the subsidiary correction being the observer's height of eye. No account has been taken of the reduction of the moon's horizontal parallax for latitude but, in general, this omission will be of no practical significance. In cases where a very high degree of accuracy is justified it will be advisable not to use a total correction table at all, but to apply separate corrections including the adjustment of refraction for the prevailing pressure and temperature of the atmosphere.

Example 1		*Example* 2	
Moon's Hor. Par'x. from		Moon's Hor. Par'x. from	
N. Alm.	57'·5	N. Alm.	59'·0
Obs. alt. moon's lower		Obs. alt. moon's upper	
limb	38° 47'·4	limb	69° 36'·0
Main corr'n......+49'·6		Main corr'n. ......—5'·6	
For H.E. 43 ft.....+ 3'·4		For H.E. 60 ft. ....+2'·2	
Tot. corr'n.	+53'·0	Tot. corr'n.	—3'·4
True alt. of moon's centre	39° 40'·4	True alt. of moon's centre	69° 32'·6

DIP at DIFFERENT DISTANCES from the OBSERVER
Or DIP of the SHORE HORIZON
(*Page* 158)

When that part of the horizon immediately under the sun is obstructed by land, and the observer is nearer the shore than five or six miles, then, if the object be brought down to the line separating the sea and land, the dip will exceed that shown in the Dip of the Sea Horizon Table. In this case, the Dip is to be taken from the present table, with the height of the eye at the *top,* and the distance estimated in miles in the *side* column.

Example.—Distance of water-line, 3 miles, and height of eye 20 feet (6·1 m.), the correction of altitude is 5' to *subtract.*

LOGARITHMS
(*Pages* 159 *to* 174)

This table gives correct to five significant figures the mantissae (or fractional parts) of the common logarithms of numbers, so that the operator must decide for himself the integral or whole number part of the logarithm according to the position of the decimal point in the natural number. It is usual to refer to the integral part of a logarithm as the "index" but, strictly speaking, the entire logarithm is the required index of the base 10, or the power to which the base must be raised, to produce the number. The rules for determining the so-called index, more properly termed the "characteristic" are very simple and are made quite obvious from a consideration of the following statements:—

$$10^4 = 10,000 \qquad \therefore \log_{10} 10,000 = 4$$
$$10^3 = 1,000 \qquad \therefore \log_{10} 1,000 = 3$$
$$10^2 = 100 \qquad \therefore \log_{10} 100 = 2$$
$$10^1 = 10 \qquad \therefore \log_{10} 10 = 1$$
$$10^0 = 1 \qquad \therefore \log_{10} 1 = 0$$
$$10^{-1} = \tfrac{1}{10}=0\cdot1 \qquad \therefore \log_{10} 0\cdot1 = -1$$
$$10^{-2} = \tfrac{1}{100}=0\cdot01 \qquad \therefore \log_{10} 0\cdot01 = -2$$
$$10^{-3} = \tfrac{1}{1000}=0\cdot001 \qquad \therefore \log_{10} 0\cdot001 = -3$$
$$10^{-4} = \tfrac{1}{10000}=0\cdot0001 \qquad \therefore \log_{10} 0\cdot0001 = -4$$

The above, which may be extended infinitely in both directions, shows that the log. of, say, 342 must lie between 2 and 3. Similarly, the log. of 29·64 must be between 1 and 2. From the table it will be found that log. 342=2·53403 and log. 29·64=1·47188. These statements could be expressed as follows :—

$$10^{2 \cdot 53403} = 342$$
$$10^{1 \cdot 47188} = 29 \cdot 64$$

For numbers greater than 1 the rule for finding the characteristic is, therefore, quite simply—The characteristic is the number which is 1 less than the number of figures before the decimal point. If there are five figures before the decimal point the characteristic is 4 ; if there is one figure before the decimal point the characteristic is 0, and so on. Thus :—

$$\log. \ 5378 \quad = 3 \cdot 73062$$
$$\log. \ \ 537 \cdot 8 \quad = 2 \cdot 73062$$
$$\log. \ \ \ 53 \cdot 78 \ = 1 \cdot 73062$$
$$\log. \ \ \ \ 5 \cdot 378 = 0 \cdot 73062$$

For numbers less than 1 the rule is different. Consider the number 0·00237. Since this lies between 0·001 and 0·01 the log. must be a quantity between −2 and −3. Actually it is −2·62525, i.e.

$$10^{-2 \cdot 62525} = 0 \cdot 00237$$

For convenience, however, it is not desirable to have two sets of mantissae, one for numbers greater than 1 and another for numbers less than 1. Accordingly, it is the custom to express the log. of a number less than 1 with a negative characteristic and a positive mantissa

$$\text{Now,} \ -2 \cdot 62525 = -3 + 1 - 0 \cdot 62525$$
$$= -3 + 0 \cdot 37475$$

and this is generally written $\bar{3} \cdot 37475$. The minus sign written over the negative characteristic (in this case read as " Bar 3 ") implies that although the characteristic is negative the mantissa is positive. Thus, log. 0·00237= $\bar{3} \cdot 37475$. Similarly :—

$$\log. \ 0 \cdot 5378 \quad = \bar{1} \cdot 73062$$
$$\log. \ 0 \cdot 05378 \quad = \bar{2} \cdot 73062$$
$$\log. \ 0 \cdot 005378 \quad = \bar{3} \cdot 73062$$
$$\log. \ 0 \cdot 0005378 = \bar{4} \cdot 73062$$

from which it is easy to see that the rule for finding the characteristic is— The negative characteristic of the log. of a number less than 1 is the number which is 1 more than the number of noughts between the decimal point and the first significant figure.

In navigation tables, to avoid having to print columns of negative characteristics, it is customary to display what are called " tabular logs." of various functions. The tabular log. is simply the ordinary log. with 10 added to the characteristic. Thus, in the table of Logs. of Trig. Functions, log. sin. 5° 30′ which is $\bar{2} \cdot 98157$ is printed as 8·98157. Similarly, log. cot. 5° 30′ which is 1·01642 is printed as 11·01642.

Interpolation. When the number whose logarithm is required consists of four significant figures or less, no interpolation is necessary. Where there are five significant figures, the proportional parts can be found from the table on page 174. Thus :—

$$\text{log. } 55 \cdot 687 = 1 \cdot (74570 + 6) = 1 \cdot 74576$$
$$\text{log. } 14936 = 4 \cdot (17406 + 17) = 4 \cdot 17423$$

Where there are more than five significant figures the approximate log. (accurate enough for most purposes) may be found by simple proportion, using the number in the difference column (D.). For instance, suppose log. 140·278 is required. From the table

$$\text{log. } 140 \cdot 2 = 2 \cdot 14675$$

As the difference (D.) is 31, the proportional parts for the two extra digits will be $\dfrac{78 \times 31}{100} = 24$ to nearest integer.

$$\text{Hence, log. } 140 \cdot 278 = 2 \cdot (14675 + 24) = 2 \cdot 14699$$

To find the number, N, whose log. is known. If the number is required to four significant figures or less all that is necessary is to find the series of digits corresponding to the tabulated mantissa which is *nearest* to the one given. The characteristic of the log. will determine the position of the decimal point. Thus :—

Given log. N = 1·87109.
Nearest tabulated mantissa 87111 gives digits 7432.
The characteristic being 1, there are two figures before the decimal point.
The required number, N, is therefore 74·32.

The following examples will serve to illustrate the procedure when more than four significant figures are required. Suppose the number, N, correct to five significant figures is required when log. N is known to be 4·27104. The next less tabulated mantissa 27091 gives the digits 1866 with 23 in the difference column (D.). The excess 13, i.e. (27104−27091), against Difference 23 in the table on page 174 gives 6 as the fifth figure. As there are five figures before the decimal point, the number, N, is therefore 18666. Suppose log. N=2·48946 and N is required to six significant figures. The next less tabulated mantissa is 48940 which gives the digits 3086. The excess 6 when divided by 14 (the number in the difference column) produces a quotient with digits in the order 42857 · · · ·, etc. Hence, to six figures, the required number N is 308·643. It is emphasised, however, that this process should not be carried too far as five-figure logarithms are not really adequate, nor are they intended, for calculations involving numbers containing more than four or five digits.

Arc quantities. It will be seen that the first two columns of the table give arcs or angles in degrees and minutes. This is purely for the convenience of navigators when working sailing problems to avoid the preliminary step of converting d. lat. or d. long. to minutes of arc. For example, if the d. lat. is 4° 37′, log. d. lat., i.e. 2·44248, can be taken out directly.

LOGS. of TRIG. FUNCTIONS

(Pages 176 to 299)

This is an entirely new arrangement of the former table of Log. Sines, Tangents, Secants, &c. Whilst preserving the basic layout which has been a feature of " Norie's ", and " Norie's " alone, since J. W. Norie produced the original edition, changes have been introduced which make the table a much more efficient instrument in conforming with the modern technique of astronomical navigation. For all angles from 0° to 90° the table is now completely *downward* reading and for that reason alone should be practically " blunder-proof ". Modern navigators are no longer interested in working to seconds of arc, so *Parts for "* have been done away with. Instead, the following arrangements have been introduced. In the main part of the table from 4° to 86° the log. functions of angles are tabulated for one minute intervals of the angles and proportional parts for fractions of one minute (from 0'·1 to 0'·9) are given. In the remainder of the table, where that system ceases to be practicable, log. functions are tabulated for intervals of 0'·1 or 0'·2 as necessary and differences between successive tabulations are given. This means that, except in special and rare cases, no mental effort is involved in taking out any log. function of an angle and there is no need to resort to the questionable practice of rounding off angles to the nearest minute in order to " save trouble ". With this table it is no more of an effort to work accurately than it is to work roughly. How far a navigator is justified in working to tenths of a minute is a matter which can be argued about indefinitely, but since the Nautical Almanac gives hour angles and declinations to tenths of a minute and a modern sextant with a decimal vernier enables readings to be taken to tenths of a minute as well, it would seem only logical to use navigation tables which, with the minimum of effort, provide for the same order of precision.

Occasionally, it may be necessary to find the logs. of trigonometrical functions of angles greater than 90°. No difficulty should be experienced in such cases as the second, third and fourth quadrant equivalents of the first quadrant angles are plainly indicated. It should be noted, however, that the table is *upward* reading for angles between 90° and 180° and also for those between 270° and 360°, but *downward* reading for angles, between 180° and 270°. In *all* cases, however, the name of the ratio being used appears at the *top* of the page. When applying proportional parts care should be taken to notice in which direction the log. function is increasing, i.e. upwards or downwards.

Examples :—

$$\text{log. sin.} \quad 1° 38'·7 = 8·(45754 + \tfrac{88}{2}) = 8·45798$$
$$\text{log. tan.} \quad 177° 57'·5 = 8·(55240 - \tfrac{70}{2}) = 8·55205$$
$$\text{log. cosec.} \quad 26° 04'·4 = 10·(35712 - 10) = 10·35702$$
$$\text{log. sec.} \quad 333° 25'·3 = 10·(04852 - 2) = 10·04850$$
$$\text{log. cos.} \quad 138° 17'·6 = 9·(87300 + 7) = 9·87307$$
$$\text{log. sin} \quad 62° 19'·8 = 9·(94720 + 5) = 9·94725$$
$$\text{log. cot.} \quad 117° 53'·0 = \qquad\qquad\qquad 9·72354$$
$$\text{log. cos.} \quad 83° 15'·3 = 9·(07018 - 32) = 9·06986$$

To find the angle whose log. function is given is equally simple. For instance, to find θ when log. sin. $\theta = 9 \cdot 66305$, notice that the next *less* tabulated log. sin. is $9 \cdot 66295$ which gives the angle $27° \, 24' \cdot 0$. The excess 10 gives an additional $0' \cdot 4$. Hence, $\theta = 27° \, 24' \cdot 4$.

In practice, the above processes will, of course, be performed mentally.

ALTITUDE-AZIMUTH TABLE

Supplying a

SHORT METHOD OF SIGHT WORKING INVOLVING THE USE OF A SPECIAL POSITION FOR PLOTTING THE POSITION LINE

(Pages 300 to 372.)

In 1924 Norie's Nautical Tables first published, under the heading of " Short Method for Zenith Distance," a table based on the original by Captain Ogura who gave permission for this to be done. Ogura's method, like other short methods of sight reduction that have been introduced at various times, involves splitting the astronomical triangle PZX into two right angled triangles. In some cases this is done by a perpendicular from X on to PZ (or PZ produced), in others—including Ogura's method—by a perpendicular from Z on to PX (or PX produced).

Many navigators have used Norie's " Short Method " table, popularly known as the A-K table, with acknowledged benefit. This " Altitude-Azimuth " table, which now replaces the A-K table, is Ogura's table redesigned and extended to eliminate certain disadvantages that were associated with the older form. Briefly, the changes are as follows.

1. The new table is arranged to make latitude (l) the primary argument and hour angle (h) the secondary one. This is of considerable advantage when working a group of sights as all the required quantities can be extracted from the same page.

2. K values are given correct to the nearest tenth of a minute instead of being rounded off to the nearest minute as they were for the most part in the older table.

3. The range of latitude of the new table is $0°$ to $72°$, whereas the older table was limited to $65°$. It has not been thought necessary to extend it up to $90°$, but that would be done if there appeared to be a sufficient demand for it.

4. A short method of obtaining the azimuth is now incorporated by extending the table to include the columns headed N and a_1. The latter quantity is a component of the azimuth and is given correct to the nearest minute. When using the former A-K table the azimuth had to be obtained independently, either by calculation or from special azimuth tables (usually the ABC tables). To work a sight completely then required reference to five separate tabulations, viz.:—

    (*i*)   Nautical Almanac for L.H.A. and decl. of body.

   (*ii*)   Altitude correction tables.

  (*iii*) A K table for values of A and K

  (*iv*) Table of logs. of trigonometrical functions for calculated zenith distance.

   (*v*) ABC tables (or other Az. tables) for azimuth and direction of position line.

With the new arrangement the procedure is more direct, especially now that the Nautical Almanac supplies altitude corrections. Using that Almanac it will be now necessary to consult only :—

    (*i*)   Nautical Almanac for L.H.A., decl., and altitude corrections.

   (*ii*)   Altitude-Azimuth table for K, A, N and a_1.

  (*iii*) Table of logs. of trigonometrical functions for calculated altitude and a_2 (the other component of the azimuth).

    5. If the given instructions are followed, the intercept is obtained by comparing the true altitude directly with the computed altitude. This avoids the extra step of converting to zenith distance.

None of these modifications is new, but the general presentation of the table is exclusive to Norie's Nautical Tables and all the tabulated quantities have been calculated afresh. By following the simple instructions given further on in this explanation and summarized at the foot of each pair of pages of the tabulations, there should be no difficulty in using the table correctly.

For the benefit of students and others interested in the construction of the table the following matter is included.

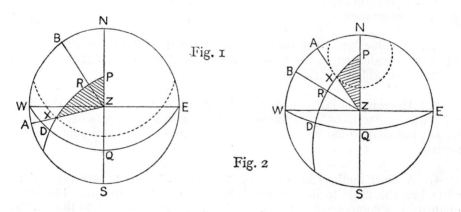

Fig. 1

Fig. 2

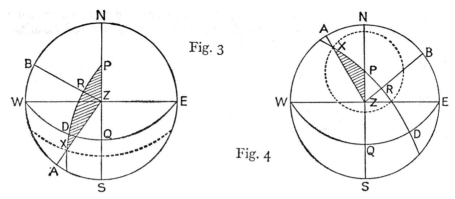

Fig. 3

Fig. 4

In all the above figures the observer is shown in N. latitude and the body W. of the meridian (i.e. H.A. less than 180°) but the same special features arise in S. latitude and with the body E. of the meridian.

The PZX triangle is divided (internally in Figs. 1 and 3, externally in Figs. 2 and 4) into two right angled triangles by drawing ZB to cut the meridian of the body at right angles at the point R. Of the two triangles so formed, triangle PZR is usually called the " time triangle " as it contains the local hour angle (or 360°—L.H.A.) which, in the case of the sun, is directly related to L.A.T. Triangle ZRX is called the " altitude triangle " as it contains ZX (zenith distance) directly related to the altitude of the body.

Special features of the figures are :—

Fig. 1. Lat. and decl. same name. R falls between P and X.
Fig. 2. Lat. and decl. same name. R falls on the side of X remote from P.
Fig. 3. Lat. and decl. different names. R falls between P and X.
Fig. 4. Hour angle greater than 90° and less than 270°. R falls on the side of P remote from X.

Throughout the explanation and in the worked examples of sights the following notation is adhered to.

l = lat. = QZ.

$(90° - l)$ = co-lat. = PZ.

d = decl. = DX.

p = polar distance = PX = $(90° \pm d)$.

h = hour angle or $(360° -$ hour angle$)$ = $Z\hat{P}X$ = $Z\hat{P}R$.

z = zenith distance = ZX.

H_c = computed altitude = AX = $(90° - z)$.

a_1 = azimuth constituent from time triangle = $P\hat{Z}R$.

a_2 = azimuth constituent from alt. triangle = $R\hat{Z}X$.

a = azimuth = $P\hat{Z}X$ = $(a_2 \pm a_1)$.

K = decl. of R, or for convenience in some cases the supplement of that arc. (See reference to Fig. 4 below.)

A = log. secant ZR.

N = log. cosecant ZR.

H_o = corrected observed altitude (i.e. sextant altitude adjusted for some or all of the corrections I.E., dip, refr., semi-diam., and parallax-in-alt., as may be necessary).

In Fig. 1. RX = DR − DX = K − d.

    Azimuth = $P\hat{Z}R + R\hat{Z}K$, i.e. a = $a_1 + a_2$.

In Fig. 2. RX = DX − DR = d − K.

    Azimuth = $P\hat{Z}R − R\hat{Z}X$, i.e. a = $a_1 − a_2$.

In Fig. 3. RX = DR + DX = K + d.

    Azimuth = $P\hat{Z}R + R\hat{Z}X$, i.e. a = $a_1 + a_2$.

In Fig. 4. RX = DX − DR = (180° − d) − K = (180° − K) − d.

    Azimuth = $R\hat{Z}X − P\hat{Z}R$, i.e. a = $a_2 − a_1$.

∴ *In general*, RX = K $\overset{+}{\sim}$ d and a = $a_2 \overset{+}{\sim} a_1$.

The reader should refer to the above when considering the rules for (*i*) combining K and d, (*ii*) combining a_1 and a_2.

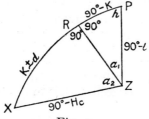

Fig. 5

Fig. 5
This represents the same portion of the celestial sphere as that indicated by the same letters in Fig. 1.

Applying Napier's Rules to triangle PZR,

  sin. (90° − P) = tan. PR tan. (90° − PZ).

 ∴ cos.h   = tan. (90° − K) tan. l.

 ∴ *cos. h cot. l* = *cot. K*

or, *cos. h cot. l* = *cot.* (180° − K) when h is greater than 90° $\left.\right\}$ I

  sin. ZR = cos. (90° − P) cos. (90° − PZ)

 ∴ *sin. ZR* = *sin. h cos. l* II

  sin. (90° − PZ) = tan. (90° − P) tan. (90° − Z)

 ∴ sin. l   = cot. h cot. a_1

or, *cosec. l cot. h* = *tan.* a_1......................................III

Applying Napier's Rules to triangle RZX,

  sin. (90° − ZX) = cos. ZR cos. RX

 ∴ sin. H_c = cos. ZR cos. (K $\overset{+}{\sim}$ d)

or, cosec. H_c = sec. ZR sec. (K $\overset{+}{\sim}$ d)

i.e. *log. cosec.* H_c = *A* + *log. sec.* (K $\overset{+}{\sim}$ *d*) IV

  sin. ZR = tan. (90° − Z) tan. RX

 ∴ sin. ZR = cot. a_2 tan. (K $\overset{+}{\sim}$ d)

 ∴ sin. ZR cot. (K $\overset{+}{\sim}$ d) = cot. a_2

or, cosec. ZR tan. (K $\overset{+}{\sim}$ d) = tan. a_2

i.e. *log. tan.* a_2 = *N* + *log. tan.* (K $\overset{+}{\sim}$ *d*) V

From I values of K or (180° − K) are found.

From II ZR is found, thence log. sec. ZR which is tabulated as A, and log. cosec. ZR which is tabulated as N.

From III values of a_1 are found.

K, A, N and a_1 are tabulated for each integral degree of l from 0° to 72° and each integral degree of h from 0° to 180°.

The procedure for working a sight is as follows :—

(*i*) From the Naut. Alm. find the G.H.A. and decl. of the observed body corresponding to the G.M.T. of observation.

(*ii*) Take the longitude nearest to the D.R. longitude which, when applied to the G.H.A., will make the L.H.A. a whole number of degrees. Use this L.H.A. as h, or (360° − h).

(*iii*) Take the whole degree of latitude nearest to the D.R. latitude and use this as l.

(*iv*) Enter the table with l at the top and h at the side and take out K, A, N and a_1. Should h exceed 90°, the quantity in the K column is to be treated as (180° − K) as shown at the foot of the column.

(*v*) Combine K with d, according to the rule given below, to obtain (K $\underset{\sim}{+}$ d).

RULE $\begin{cases} \text{If l and d have } \textbf{SAME} \text{ name use (K } - \text{ d) when K is greater} \\ \quad \text{than d.} \\ \text{If l and d have } \textbf{SAME} \text{ name use (d } - \text{ K) when d is greater} \\ \quad \text{than K.} \\ \text{If l and d have } \textbf{DIFFERENT} \text{ names use (K } + \text{ d).} \end{cases}$

(*vi*) To A add the log. secant of (K $\underset{\sim}{+}$ d). The sum is the log. cosecant of H_c. Find H_c from table of log. cosecants.

(*vii*) To N add the log. tangent of (K $\underset{\sim}{+}$ d). The sum is the log. tangent of a_2. Find a_2 correct to the nearest minute from the table of log. tangents.

(*viii*) Apply the necessary corrections to the sext. alt. of the body to find the true alt. H_o. Compare H_o with H_c to find the intercept.

RULE $\begin{cases} \text{If } H_o \text{ is } \textbf{greater} \text{ than } H_c, \text{ name the intercept } \textbf{towards.} \\ \text{If } H_o \text{ is } \textbf{less} \text{ than } H_c, \text{ name the intercept } \textbf{away.} \end{cases}$

(*ix*) Combine a_1 with a_2 to find the azimuth.

RULE $\begin{cases} \text{If h is } \textbf{less} \text{ than 90°, use } (a_1 + a_2) \text{ if (K } + \text{ d) or K } - \text{ d)} \\ \quad \text{was used.} \\ \text{If h is } \textbf{less} \text{ than 90°, use } (a_1 - a_2) \text{ if (d } - \text{ K) was used.} \\ \text{If h is } \textbf{greater} \text{ than 90°, use } (a_2 - a_1). \\ \text{Name the azimuth from the elevated pole, E. or W. accord} \\ \quad \text{ing to whether L.H.A. is greater or less than 180°.} \end{cases}$

(*x*) With intercept and azimuth so obtained, plot the position line *FROM THE SPECIAL POSITION* selected as in (*ii*) and (*iii*).

{ **N.B.** Navigators who have been long accustomed to finding the intercept by comparing zenith distances, and who
prefer not to change that practice, should treat A + log.
sec. (K $\pm$ d) as log. sec. C.Z.D. }

Examples to show the recommended lay-out are given below.

Example 1.—This illustrates a case where l and d are of the SAME NAME,
K is greater than d, and h is less than 90°.

D.R. 58° 19′ N., 23° 13′ W. H$_o$ derived from sext. alt. 28° 41′ ·5,
G.H.A. from N. Alm. 333° 23′ ·0, decl. from N. Alm. 10° 01′ ·2 N.

	G.H.A.	333° 23′ ·0	l = 58° N.
	Special long.	23° 23′ ·0 W.	d = 10° 01′ ·2 N.
	L.H.A.	310° 00′ ·0	h = (360° − 310°) = 50°

K	68° 07′·0	A 0·03910	N 0·39154	a$_1$ 44° 42′
d	10° 01′·2			

K $\pm$ d	58° 05′·8 ——> l. sec. 10·27696	l. tan. 10·20585	
H$_c$	28° 52′·9 <—— l. cosec. 10·31606	l. tan. 10·59739 ——> a$_2$ 75° 49′
H$_o$	28° 41′·5		a 120° 31′

Intercept   11′·4 *away*   *i.e. Azi* = N. 120°·5 E.

Example 1 is shown worked in full by the cosine-haversine method and a
plot given to demonstrate that the two methods produce the same position
line. (See below.)

G.H.A.	333° 23′·0		
W. long.	23° 13′·0		
L.H.A.	310° 10′·0	l. hav. 9·24918	From ABC tables using
lat. N.	58° 19′·0	l. cos. 9·72035	lat. 58° and H.A. 310°
decl. N.	10° 01′·2	l. cos. 9·99332	A 1·34 S.
		l. hav. 8·96285	B 0·23 N.
		N. hav. 0·09180	C 1·11 S.
l $\pm$ d	48° 17′·8	N. hav. 0·16736	Azi. S. 59°·5 E.
C.Z.D.	61° 12′·3 <—— N. hav. 0·25916	*or 120°·5*
T.Z.D.	61° 18′·5		
Intercept	6′·2 *away.*		

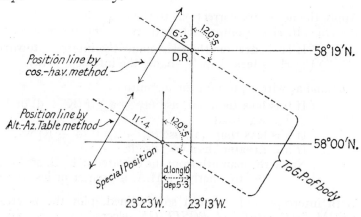

Fig. 6

N.B. If the azimuth is found by careful interpolation for 58° 19′ N. lat. and H.A. 310° 10′ it will be seen to be 120°·8.. For practical purposes the small error may be safely ignored.

Example 2.—This illustrates a case where l and d are of the SAME NAME, d is greater than K, and h is less than 90°.

D.R. 9° 54′ N., 97° 16′ E. H₀ derived from sext. alt. 50° 59′·3, G.H.A. from N. Alm. 301° 14′·6, decl. from N. Alm. 22° 27′·8 N.

	G.H.A.	301° 14′·6	l = 10° N..
	Special long.	97° 45′·4 E.	d = 22° 27′·8 N.
	L.H.A.	39°.00′·0	h = 39°

K	12° 47′·0	A 0·10525	N 0·20778	a₁ 82° 00
d	22° 27′·8			
K + d	9° 40′·8 ——> l. sec. 10·00623		l. tan. 9·23191	
H̃꜀	50° 40′·7 <—— l. cosec. 10·11148		l. tan. 9·43969 ——> a₂ 15° 23′	
H₀	50° 59′·3			a 66° 37′

Intercept 18′·6 *towards.*   *i.e. Azi. = N 66°·6 W.*

Example 3.—This illustrates a case where l and d are of DIFFERENT NAMES, and h is less than 90°.

D.R. 23° 11′ S., 127° 14′ W. H₀ derived from sext. alt. 34° 40′·8, G.H.A. from N. Alm. 078° 30′·6, decl. from N. Alm. 5° 21′·0 N.

	G.H.A.	078° 30′·6	l = 23° S.
	Special long.	127° 30′·6 W.	d = 5° 21′·0 N.
	L.H.A.	311° 00′·0	h = (360° − 311°) = 49°

K	32° 54′·2	A 0·14310	N 0·15819	a₁ 65° 48′
d	5° 21′·0			
K + d	38° 15′·2 ——> l. sec. 10·10498		l. tan. 9·89676	
H̃꜀	34° 23′·5 <—— l. cosec. 10·24808		l. tan. 10·05495 ——> a₂ 48° 37′	
H₀	34° 40′·8			a 114° 25′

Intercept 17′·3 *towards.*   *i.e. Azi. S 114°·4 E.*

Example 4.—This illustrates a case where h is greater than 90°.

D.R. 38° 40′ S., 169° 19′ E. H₀ derived from sext. alt. 13° 09′·7, G.H.A. from N. Alm. 043° 09′·2, decl. from N. Alm. 60° 38′·4 S.

	G.H.A.	043° 09′·2	l = 39° S.
	Special long.	168° 50′·8 E.	d = 60° 38′·4 S.
	L.H.A.	212° 00′·0	h = (360° − 212°) = 148°

180° − K	43° 40′·7			
K	136° 19′·3	A 0·04036	N 0·38529	a₁ 68° 32′
d	60° 38′·4			
K + d	75° 40′·9 ——> l. sec. 10·60676		l. tan. 10·59305	
H̃꜀	13° 01′·5 <—— l. cosec· 10·64712		l. tan. 10·97834 ——> a₂ 84° 00′	
H₀	13° 09′·7			a 15° 28′

Intercept 8′·2 *towards.*   *i.e. Azi. S 15°·5 E.*

ALTITUDE ADJUSTMENT TABLES

(Pages 374 to 379.)

These tables have been constructed on the lines of a smaller table supplied to the publishers by H. C. Kirsopp. Their purpose is to provide a means whereby the navigator can take advantage of the short method of sight reduction, using Norie's Altitude-Azimuth Tables (or other similar tables) and at the same time avoid having to find and employ a " special position". In other words, the navigator can both work the sight and plot the position line from his D.R. position. This is often of considerable advantage, especially in circumstances where the use of the special position results in an inconveniently large intercept, or when that position is not on the plotting chart.

The procedure is as follows :—

From the Nautical Almanac obtain the G.H.A. and declination of the observed body corresponding to the G.M.T. of the observation. Apply the D.R. longitude to the G.H.A. to obtain the L.H.A.

Using the next LESS whole degree of L.H.A. and the next LESS whole degree of latitude to the D.R. latitude, employ the Altitude-Azimuth table to find the calculated altitude and azimuth (as shown in the examples contained in the " explanation " to that table).

Then adjust the calculated altitude by applying corrections for hour angle and latitude from the Altitude Adjustment Tables.

Tables A and B together give the hour angle correction as derived from the formula :—

Hour Angle Correction = d.h.a. (in mins.) cos. lat. sin. az.

Table B gives the latitude correction as derived from the formula :—

Latitude Correction = d. lat. (in mins.) cos. az.

To find the hour angle correction, enter Table A with LAT. at the top and follow the column down until the *AZIMUTH* (or nearest value thereto) is found. Then note the key letters to Table B immediately abreast. Turn to Table B and, using the line having the appropriate key letters at the side, find the correction corresponding to the extra number of minutes of hour angle. Apply this correction to the calculated altitude by the given rule " always *SUBTRACT*".

(Note that when the number of extra minutes exceeds 10, and is not an exact multiple of 10, it will be necessary to add together the corrections for tens and units. Corrections for fractions of a minute can also be made by a shift of decimal place in the corrections given for minutes.)

To find the latitude correction, Table B *alone* is used. Simply find the AZIMUTH in the right-hand column and, using the line against which it is

found, obtain the correction for the additional minutes of latitude. Apply
this correction in accordance with the rules :—

In *NORTH* latitude—
    ADD if Azimuth is *NORTHERLY.*
    SUBTRACT if Azimuth is *SOUTHERLY.*   } *Here, Lat. means*
In *SOUTH* latitude—   *D.R. latitide.*
    ADD if Azimuth is *SOUTHERLY.*
    SUBTRACT if Azimuth is *NORTHERLY.*

For the purpose of these tables always express azimuths in the quadrantal
notation to avoid angles greater than 90°. For a reasonably high standard
of accuracy it may be necessary to interpolate in some cases, but this should
not present any difficulty.

The following "*Example*" is appended to demonstrate the above
procedure.

$$\text{Data} \begin{cases} \text{D.R. position } 23° \ 43' \ \text{N., } 58° \ 16' \ \text{W.} \\ \text{G.H.A. } 350° \ 38'\cdot4, \ \text{Decl. } 18° \ 25'\cdot5 \ \text{N.} \\ \text{Corrected sextant altitude } 27° \ 18'\cdot0. \end{cases}$$

$$\begin{array}{lr} \text{G.H.A.} & 350° \ 38'\cdot4 \\ \text{D.R. long.} & 58° \ 16'\cdot0 \ \text{W.} \\ \hline \text{L.H.A.} & 292° \ 22'\cdot4 \ \text{W.} \\ & 360° \ 00'\cdot0 \\ \hline \text{L.H.A.} & 67° \ 37'\cdot6 \ \text{E.} \\ \hline \end{array}$$

From the Altitude-Azimuth Tables, etc., using LAT. 23° and H.A. 67° :
    K 47° 22'·2   A 0·27485   N 0·07195 a_1 47° 22'
    d 18° 25'·5

K $\overset{+}{\sim}$ d 28° 56'·7 l. sec.   10·05795 l. tan.   9·74277

  H_c 27° 41'·6 l. cosec   10·33280 l. tan.   9·81472 a_2 33° 08'

H.A. Corr.   − 34'·0   a 80° 30'

               27° 07'·6
Lat. Corr.   + 7'·1   Azimuth N80°·5E.

Adj. Alt.   27° 14'·7
True Alt.   27° 18'·0

Intercept   3'·3 Towards.

In the above example the H.A. correction is obtained as follows. Since
23° is not a tabulated latitude, it is necessary to interpolate between 22°

and 24°.   In Table A Lat. 22° and the nearest azimuth (80°) combine to give the key CS. Lat. 24° and azimuth 80° give CU.   So CT is selected which lies between them.   In Table B, from the line keyed as CT, the correction for the extra 37'·6 of H.A. is obtained as shown below.

for 30'   the correction is 27'·2
,,   7'   ,,   ,,   ,,   6'·3
,, 0'·6   ,,   ,,   ,,   0'·5

∴   ,, 37'·6   ,,   ,,   ,, 34'·0 to *SUBTRACT*.

In practice, of course, the above assessment of the total correction can be done mentally.

The latitude correction for the extra 43' of lat. is obtained from Table B for azimuth 80½° by interpolating between 80° and 81°.

80° gives 6'·9 for 40' and 0'·5 for 3'
81° ,,   6'·3 ,, 40' ,, 0'·5 ,, 3'

∴ for 80½° use 6'·6 ,, 40' ,, 0'·5 ,, 3', i.e. total 7'·1

which is *ADDED* because Lat. is North and Azimuth is Northerly.

———————

The same sight when worked by the long method using the cos.-hav. formula and ABC Tables gives " Intercept 3'·5 Towards " and " Azimuth N80°·6E," with which the above is in satisfactorily close agreement.

A SECONDARY USE OF THE ALTITUDE ADJUSTMENT TABLES

These tables can also be used to find the altitude for the D.R. position (or any other position from which it is desired to plot the position line) when the sight *HAS BEEN WORKED* from a " special position".

For this purpose the H.A. correction should be obtained corresponding to the number of minutes' difference between the D.R. longitude and the Special longitude.   Similarly, the Lat. correction should be that for the number of minutes' difference between the D.R. lat. and the Special lat. *BUT* the rules for applying the corrections *WILL NOT BE THE SAME* as those shown at the foot of Table B.   Instead, the following rules will apply.

Let S be the Special Position used for working the sight, and D be the D.R. or other position from which it is desired to plot the position line.

RULES :—

⎧ S East of D, body bearing Easterly ⎫
⎪ S West of D, body bearing Westerly ⎬ H.A. correction *SUBTRACTED*
⎪
⎨ S East of D, body bearing Westerly ⎫
⎪ S West of D, body bearing Easterly ⎬ H.A. correction *ADDED*

⎧ S North of D, body bearing Northerly ⎫ Lat. correction
⎪ S South of D, body bearing Southerly ⎬ *SUBTRACTED*
⎪
⎨ S North of D, body bearing Southerly ⎫
⎪ S South of D, body bearing Northerly ⎬ Lat. correction *ADDED*

NATURAL FUNCTIONS OF ANGLES.

(Pages 380 to 394.)

Previous editions of " Norie's Tables " have included separate tables of natural sines and cosines, natural tangents and cotangents, and natural secants and cosecants. These, which tabulated the various functions to five places of decimals for angles at one minute intervals, are now replaced by this single table, entirely downward reading, giving the six natural ratios and the radian measure of all angles from $0°$ to $90°$ at angular intervals of one-tenth of a degree. This more compact arrangement should be found to be adequate as the employment of natural functions is an advantage only when dealing with calculations of a very simple type. In cases where multiplication or division processes would be awkward and time-absorbing it will always be preferable to use logarithms. This table may be found particularly useful in cases where the approximate evaluation of an expression is considered sufficient.

Examples.

   1. To find the value of 6 tan. $68°$
        6 tan. $68° = 6 \times 2·4751 = 14·8506.$

   2. To find the value of $0·4$ sin. $54°·5.$
        $0·4$ sin. $54°·5 = 0·4 \times 0·8141 = 0·32564.$

   3. To find the approximate value of θ when 7 sec. $\theta = 9·75.$

        Sec. $\theta = \dfrac{9·75}{7} = 1·3929$ (approx.) $\therefore \theta = 44°·1$ (approx.).

   4. To find the radian measure of $15°.$
        $15° = 0·2618$ radians.

   5. To find the radian measure of $163°·4.$
        $90°·0 = 1·5708$ radians.
        $73°·4 = 1·2811$,,

        $163°·4 = 2·8519$,,

FORM OF THE EARTH.

(Page 395.)

This single page table is included mainly for the benefit of those interested in chart construction. The quantities tabulated are obtained from the figures and formulae given below.

Earth's equatorial radius, a $= 3,963·35$ statute miles.
Earth's polar radius,   b $= 3,950·01$ statute miles.

Polar compression,   c $= \dfrac{1}{297}$

Reduction of latitude, r $=$ c sin. $2\,\Phi$, where $\Phi =$ geographical latitude, from which it can be deduced that r in minutes of latitude $= 11'·575$ sin. $2\,\Phi.$

The values of r are not tabulated.

Radius of Earth, R, for any latitude $= a (1 - c \sin^2 \Phi)$ so that

$$R = 3{,}963 \cdot 35 \left(1 - \frac{1}{297} \sin.^2 \text{ lat.}\right).$$

Strictly speaking, the geocentric latitude should be used in this expression when evaluating R, but as the difference is very small no appreciable error is involved if the geographical latitude is used instead.

One minute of latitude $= (6077 \cdot 1 - 30 \cdot 7 \cos. 2 \Phi)$ feet.

One minute of longitude $= \left(6087 \cdot 2 \times \dfrac{R}{3{,}963 \cdot 35} \times \cos. \theta \right)$ feet, where $\theta =$ geocentric latitude.

HAVERSINES

(*Pages 396 to 502.*)

Prior to the introduction of the G.H.A. Almanac, Log. Haversines and Natural Haversines were tabulated in " Norie's Nautical Tables " at angular intervals of 15″. Thereafter the table was rearranged to give Haversines to every 0′·5. This change proved to be not entirely satisfactory as tedious interpolation was frequently found to be necessary to avoid inaccuracies especially when finding the natural haversine corresponding to a given log. haversine or vice versa. To obviate this disadvantage and, moreover, to make interpolation except in the rarest of cases completely unnecessary, Norie's Haversine Table has been redesigned and is now presented in an entirely new form, although the practice of printing Log. Haversines in *darker type* and Natural Haversines in *lighter type* has been retained.

In the range 0° to 90° and 270° to 360°—the range most frequently used—haversines are tabulated at 0′·2 intervals and the proportional parts for 0′·1 are given at the foot of each page. In the remainder of the table haversines are tabulated at 1′·0 intervals and the proportional parts for multiples of 0′·2 are given at the head of each column.

The following examples should suffice to illustrate the manner of reading the table.

Angle	Log. hav.	Nat. hav.
15° 33′·0	8·26249	0·01830
15° 33′·6	8·26304	0·01832
15° 33′·7	8·26313	0·01832
344° 10′·0	8·27807	0·01897
344° 10′·4	8·27771	0·01895
344° 10′·5	8·27762	0·01895
95° 25′·0	9·73815	0·54720
95° 25′·6	9·73822	0·54729
263° 37′·0	9·74475	0·55559
263° 37′·8	9·74466	0·55547

Derivation of Haversine Formulæ :

$$\text{Cos. A} = \frac{\cos. a - \cos. b \cos. c}{\sin. b \sin. c}, \text{ (fundamental formula).}$$

$$\therefore \text{I} - \cos. A = \text{I} - \frac{\cos. a - \cos. b \cos. c}{\sin. b \ \sin. c}$$

i.e. vers A $= \dfrac{\sin. b. \sin. c - \cos. a + \cos. b \cos. c}{\sin. b \sin. c},$

$\therefore \cos. (b \sim c) - \cos. a = \sin. b \sin. c \text{ vers. A,}$
or $- \cos. a = - \cos. (b \sim c) + \sin. b \sin. c \text{ vers. A.}$

By adding unity to each side this becomes—

$\text{I} - \cos. a = \text{I} - \cos. (b \sim c) + \sin. b \sin. c \text{ vers. A,}$
$\therefore \text{vers. a} = \text{vers. } (b \sim c) + \sin. b \sin. c \text{ vers. A,}$
whence hav. a $=$ hav. $(b \sim c) + \sin. b \sin. c \text{ hav. A} \dots (1)$

By transposing we obtain—

hav. A $= [\text{hav. a} - \text{hav. } (b \sim c)] \text{ cosec. b cosec. c} \dots (2)$
and hav. $(b \sim c) = $ hav. a $-$ hav. A sin. b sin. c $\dots (3)$

These three versions of the spherical haversine formula are frequently adapted for navigational purposes as follows.

(1) Hav. z = hav. $(1 \overset{+}{\sim} d)^* +$ hav. h cos. l cos. d.
(2) Hav. h = [hav. z $-$ hav. $(1 \overset{+}{\sim} d)^*]$ sec. l sec. d.
(3) Hav. mer. zen. dist. = hav. z $-$ hav. h cos. l cos. d.

$$\text{where} \begin{cases} z = \text{zenith distance,} \\ l = \text{latitude,} \\ d = \text{declination,} \\ h = \text{hour angle.} \end{cases}$$

$*$ $(1 \sim d)$ when l and d have the same name,
$(1 + d)$ when l and d have different names.

Examples.

(1) Find zenith distance when h $= 66°$ $49'·3$, l $= 31°$ $10'·2$ N., d $= 19°$ $24'·7$ N.

Hav. z = hav. h cos. l cos. d + hav. $(1 \overset{+}{\sim} d)$.

h	.... 66° 49'·3	L. hav. 9·48173	
l	.... 31° 10'·2	L. cos. 9·93228	
d	.... 19° 24'·7	L. cos. 9·97458	
		L. hav. 9·38859	N. hav. 0·24468
(l ~ d)	.... 11° 45'·5	....	N. hav. 0·01049
z	.... 60° 40'·9	....	N. hav. 0·25517

Calculated zenith distance = 60° 40'·9 and is used for comparing with the true zenith distance to find the intercept when establishing the position line by the Marc St. Hilaire method.

(2) Find the hour angle when $l = 41° 21'\cdot6$ N., $d = 9° 34'\cdot1$ S., $z = 63° 45'\cdot8$.

Hav. h = [hav. z — hav. $(l \overset{+}{\sim} d)$] sec. l sec. d.

z	.... 63° 45'·8	N. hav. 0·27896		
(l + d)	.... 50° 55'·7	N. hav. 0·18485		
		N. hav. 0·09411	L. hav.	8·97364
l	.... 41° 21'·6	....	L. sec.	10·12461
d	.... 9° 34'·1	....	L. sec.	10·00608
h	.... 41° 46'·9	....	L. hav.	9·10433

Hour angle = 41° 46'·9 if body is W. of the meridian, or hour angle = 318° 13'·1 if body is E. of the meridian, and is used for finding the computed longitude when establishing the position line by the " chronometer method".

(3) Find the mer. zen. dist. when $h = 355° 57'\cdot2$, $l = 48° 12'\cdot5$ N., $d = 12° 13'\cdot7$ S., $z = 60° 21'\cdot6$.

Hav. mer. zen. dist. =: hav. z — hav. h cos. l cos. d.

h	.... 355° 57'·2	L. hav. 7·09571	
l	.... 48° 12'·5	L. cos. 9·82375	
d	.... 12° 13'·7	L. cos. 9·99003	
		L. hav. 6·90949	N. hav. 0·00081
z	.... 60° 21'·6	....	N. hav. 0·25273
mer. zen. dist.	.... 60° 15'·2	....	N. hav. 0·25192

The mer. zen. dist., 60° 15'·2, when combined with the declination gives the latitude of the point where the position line (at right angles to the direction of the body) cuts the meridian of D.R. longitude used to compute h. This method of working an ex-meridian sight is, of course, an alternative to using ex-meridian tables.

The haversine formulæ and great circle sailing calculations.

Formula (1) is used to find the great circle distance from one point to another and formula (2) is used to find the initial and final courses. The vertex of the track and the latitude of the point where the track cuts any specified meridian can then be found by right angled spherical trigonometry.

Example. Find the great circle distance and the initial course on the track from A (17° 22' N., 25° 28' W.) to B (40° 08' N., 73° 17' W.).

To find the great circle distance.

Hav. AB = hav. (PA ∼ PB) + hav. P sin. PA sin. PB.

P	.... 47° 49'·0	L. hav. 9·21550	
PA	.... 72° 38'·0	L. sin. 9·97974	
PB	.... 49° 52'·0	L. sin. 9·88340	
		L. hav. 9·07864	N. hav. 0·11985
(PA ∼ PB)	.... 22° 46'·0	....	N. hav. 0·03896
AB	.... 46° 58'·2	....	N. hav. 0·15881

∴ Great circle distance = 2818·2 miles.

To find the initial course.

Hav. A = [hav. PB − hav. (PA ∼ AB)] cosec. PA cosec. AB.

PB	.... 49° 52'·0	N. hav. 0·17772	
(PA ∼ AB)	.... 25° 39'·8	N. hav. 0·04932	
		N. hav. 0·12840	L. hav. 9·10856
PA	.... 72° 38'·0	....	L. cosec. 10·02026
AB	.... 46° 58'·2	....	L. cosec. 10·13609
A	.... 50° 48'·5	....	L. hav. 9·26491

∴ Initial course = N. 50° 48'·5 W. or 309° 11'·5.

CORRECTION REQUIRED to CONVERT a RADIO GREAT CIRCLE BEARING to MERCATORIAL BEARING

Page 503.   Explanation with Table.

A, B & C AZIMUTH TABLES

(Pages 504 to 552.)

To conform with the method of presenting data in the " Nautical Almanac " the hour angles in Tables A and B are given in degrees and minutes of arc from 0° 15′ to 359° 45′.

If the H.A. is between 0° and 180° the body is west of the meridian and its hour angle will appear in the upper row of H.A.s at either the top or bottom of the page.   If the H.A. is between 180° and 360° the body is east of the meridian and its hour angle will appear in the lower row.

The A, B and C values and the azimuth are derived by employing the well known formula which connects four adjacent parts of a spherical triangle.   It can be shown, for instance, that in spherical triangle A B C :—

$$\cot a \sin b = \cot A \sin C + \cos b \cos C.$$
$$\therefore \cot a \sin b - \cos C \cos b = \sin C \cot A.$$

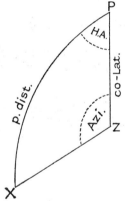

The figure shows the astronomical triangle PZX with the four adjacent parts PX, P, PZ and Z representing, in that order, polar distance, hour angle, co-latitude and azimuth.

Applying the above formula to this particular case, we have :—

$$\cot PX \sin PZ - \cos P \cos PZ = \sin P \cot Z.$$

Dividing by sin. P sin. PZ, this becomes—

$$\frac{\cot PX}{\sin P} \cdot \frac{\sin PZ}{\sin PZ} - \frac{\cos P}{\sin P} \cdot \frac{\cos PZ}{\sin PZ} = \frac{\sin P}{\sin P} \cdot \frac{\cot Z}{\sin PZ}$$

i.e.   cot. PX   cosec. P — cot. P   cot. PZ = cot. Z   cosec. PZ,

or   tan. decl.   cosec. H.A. — cot. H.A. tan. lat.

$$= \cot. \text{ azi. sec. lat.}$$

In the tables :—

cot. H.A. tan. lat. is tabulated as A, and tan. decl. cosec. H. A. is tabulated as B.

Hence (A $\overset{+}{\sim}$ B) cos. lat. = cot. azimuth.

(A $\overset{+}{\sim}$ B), referred to for convenience as " C ", forms the primary argument in Table C with lat. as the secondary argument.   With these two arguments the azimuth is found.

As an example, consider the case where hour angle = 48°, lat. = 52° N., and decl. = 15° N.

lat. 52° 00′ N	L. tan. 0·10719	
H.A. 48° 00′	L. cot. 9·95444	L. cosec. 0·12893
decl. 15° 00′ N	....	L. tan. 9·42805
	Log. A 0·06163	Log. B 9·55698
	∴ A = 1·153 S.	B = 0·361 N.

(A is named opposite to lat. ; B has the same name as decl.) (A $\pm$ B) = " C " = 0·792 S. (Same name as A which is numerically greater than B.).

" C " 0·792 S.	Log. 9·89873
lat. 52° 00′ N.	L. cos. 9·78934
Azi. 64° 00′	L. cot. 9·68807

∴ Azimuth = S. 64° W. or 244°.

(The azimuth takes the names of the " C " factor and hour angle.)

Reference to the tables will show that for the above data A = 1·15 S. and B = 0·36 N. The combination of these is 0·79 S., which in Table C with lat. 52° gives azimuth S. 64°·2 W.

The rules for naming and combining A and B and for naming the azimuth are given on each page of the appropriate table. It is important that they should be applied correctly.

Longitude Correction.

The quantity (A $\pm$ B) or " C ", besides being one of the arguments for finding the azimuth from table C, is also the " longitude correction factor " or the error in longitude due to an error of 1′ of latitude. This can often be very useful to those accustomed to working sights by the longitude method.

A simple sketch showing the direction of the position line will at once make it clear which way the longitude correction should be applied. It will easily be apparent that when working a sight by the longitude method :—

(a) when the position line lies N.E./S.W. (body in N.W. or S.E. quadrant), if the assumed latitude is too far north the computed longitude will be too far east, and if the latitude is too far south the longitude will be too far west ;

(b) when the position line lies N.W./S.E. (body in N.E. or S.W. quadrant) the reverse holds good.

Example. Suppose a sight worked with lat. 49° 06′ N. gives longitude 179° 46′·0 W. and azimuth S. 70°·5 E., the value of " C " being 0·54. If the correct latitude turned out to be 49° 46′ N., i.e. 40′ error, the error in longitude would be 40 x 0′·54 or 21′·6. We should therefore have :—

Computed long. 179° 46′·0 W.
Correction   21′·6 E.

Correct long.   179° 24′·4 W.

This is a case where the latitude being too far south, the computed longitude is too far west.

Examples on the use of the tables.

In each of the following cases find the longitude correction factor and the true azimuth.

Example 1.   H.A. 310°, lat. 48° N., decl. 20° N.

From Table A with H.A. 310°, lat. 48° N., A = 0·93 S.
From Table B with H.A. 310°, decl. 20° N., B = 0·48 N.

Long. corr'n. factor = A — B =   C = 0·45 S.

From Table C with C 0·45 S., lat. 48° N., T. Azi. = S. 73°·2 E.

A is named S. opposite to lat. because H.A. is *not* between 90° and 270°.
B is named N. because the decl. is N.
C = A — B as A and B have different names, and is named S. as the greater quantity is S.
The azimuth is named S. because C is S., and E. because H.A. is between 180° and 360°.

Example 2.   H.A. 244°, lat. 41° S., decl. 5° S.

From Table A with H.A. 244°, lat. 41° S., A = 0·42 S.
From Table B with H.A. 244°, decl. 5° S., B = 0·10 S.

Long. corr'n. factor = A + B =   C = 0·52 S.

From Table C with C 0·52 S., lat. 41° S., T. Azi = S. 68°·6 E.

A is named S. same as lat. because H.A. is between 90° and 270°.
B is named S. because the decl. is S.
C — A + B as A and B have the same name (both S.).
The azimuth is named S. because C is S., and E. because H.A. is between 180° and 360°.

Example 3.   H.A. 108°, lat. 61° N., decl. 20° N.
From Supplementary Table A with H.A. 108°, lat. 61° N.,   A = 0·59 N.
From Table B with H.A. 108°, decl. 20° N.,   B = 0·38 N.

Long. corr'n. factor = A + B =   C = 0·97 N.

From Table C with C o 97 N., lat 61° N,, T. Azi. = N. 64°·8 W.

A is named N. same as lat. because H.A. is between 90° and 270°.
B is named N. because the decl. is N.
C = A + B as A and B have the same name (both N.).

The azimuth is named N. because C is N., and W. because H.A. is between
    0° and 180°.

Use of ABC Tables for Great Circle Sailing.

These tables provide a ready means of finding the initial great circle
course from one point to another.   Suppose, for example, the initial course
from P (49° 30′ N., 5° 00′ W.) to Q (46° 00′ N., 53° 00′ W.) is required.   The
procedure is simply to treat d. long. as hour angle, lat. of P. as latitude,
and lat. of Q as declination.   Thus :—
From Table A with H.A. 48°, lat.   49° 30′ N.,   A = 1·06 S.
From Table B with H.A. 48°, decl. 46° 00′ N.,   B = 1·39 N.

    A − B =   C = 0·33 N.

From Table C with C 0·33 N., lat. 49° 30′ N., T.Azi. = N. 77°·9 W.
    i.e. Initial G.C. Course = N. 77°·9 W. or 282°·1.

The final course, if required, may be obtained in a similar way by finding
the initial course from Q to P and reversing it.

To CONVERT ARC to TIME and TIME to ARC
(Pages 553 to 554.)

HOURS and MINUTES to DECIMAL of a DAY
(Page 555.)

AMPLITUDES and CORRECTIONS
(Pages 556 to 558.   Explanation with Table.)

EQUIVALENTS of THERMOMETER SCALES
(Page 559.)

BRITISH and METRIC WEIGHTS and MEASURES
(Pages 560 to 561.)

TANK TONNAGE TABLE

(Page 562.)

This table is intended for use in calculating the quantity of liquid in tons, at various gravities, from the capacity in cubic feet and will be found serviceable when dealing with petroleum cargoes, oil bunkers, or in any other like case where such information is required.

In the absence of tables, quantities of oil are calculated by dividing the capacity of oil space in cubic feet by the number of cubic feet of fresh water to the ton, then, multiplying the quotient so obtained by the specific gravity of the oil. The product is the quantity in tons.

To simplify the process of calculation a constant has been computed for each specific gravity, as shown in the table. This constant is the logarithm of the quotient of the specific gravity divided by 35·9673 and has been made additive for simplicity of application and consequent lessened risk of error.

In using this table the log. of the capacity in cubic feet is added to the log. constant given for the specific gravity of the oil concerned, subtracting 10 from index if in excess of that amount. The sum is the log. of the number of tons in the tank

Several methods are extensively resorted to on occasions when it is required to determine such information and the work is considerably facilitated by recourse to volumes of tables expressly designed for the purpose but, in the absence of such aids, it will probably be found that the introduction to these pages of this new table, notwithstanding its extreme brevity, provides ready to hand an efficient alternative procedure, as may be judged from a comparison of this with examples of other methods shown hereunder :

Example. Given specific gravity of oil ·814 and capacity of loading space 9,694 cubic feet, to find quantity in tons.

By Norie's New Table, Method " A."		Alternative Tabular Method " B."		
·814	Constant 8·35471(5)	1st Entry 9000	=	203·68
9694	Log. 3·98650	2nd ,, 600	=	13·58
		3rd ,, 90	=	2·04
	Log. 2·34121(5)	4th ,, 4	=	·09
219·4 tons, nearly.		Total 9694	=	**219·39 tons.**

By Log. Calculation Method " C."			Alternative Tabular Method " D."
9694	Log.	3·98650	44·186) 96940 (219·4
·814	Log.	9·91062	88372
	Log.	3·89712	85680
35·97	Log.	1·55594	44186
	Log.	2·34118	414940
			397674

219·4 tons, nearly.

172660
176744
219·4 tons, nearly.

IMPERIAL GALLONS, U.S.A. GALLONS, LITRES

(Page 563.)

NAUTICAL MILES, STATUTE MILES, KILOMETRES

(Page 564.)

PORTS OF THE WORLD. LATITUDES and LONGITUDES

(Pages 565 to 607.)

About three thousand commercial ports are here listed in alphabetical order and, with a few exceptions, each is accompanied by an index number to facilitate reference to a geographical list which gives the approximate latitude and longitude of the port.

No claim is made either that the lists are complete or that the positions are in all cases strictly accurate. The positions, however, will be found accurate enough to enable the port to be located easily on the appropriate chart. Where latitude and longitude are not given the rough location of the port is indicated.

In spite of attempts to standardize the spelling of place names, many variants are still commonly used. Where it has been considered advisable alternative spellings of the same name are given. In some cases the port is known by two or more names, and again where it is thought necessary the different versions are given. In all other cases when faced with alternative spellings the version adopted is that which appears to be most commonly in vogue in the shipping press generally. For the sake of simplicity accents on letters in foreign place names have been dispensed with, but this should not give rise to any difficulty.

TABLES

TRAVERSE TABLE

0°

O DEGREES.

D Lon	Dep.		D Lon	Dep.		D Lon	Dep.		D Lon	Dep.		D Lon	Dep.	
Dist.	D.Lat	Dep.	Dist.	D.Lat	Dep.	Dist.	D.Lat	Dep.	Dist.	D.Lat	Dep.	Dist.	D.Lat	Dep.
1	01·0	00·0	61	61·0	00·0	121	121·0	00·0	181	181·0	00·0	241	241·0	00·0
2	02·0	00·0	62	62·0	00·0	122	122·0	00·0	182	182·0	00·0	242	242·0	00·0
3	03·0	00·0	63	63·0	00·0	123	123·0	00·0	183	183·0	00·0	243	243·0	00·0
4	04·0	00·0	64	64·0	00·0	124	124·0	00·0	184	184·0	00·0	244	244·0	00·0
5	05·0	00·0	65	65·0	00·0	125	125·0	00·0	185	185·0	00·0	245	245·0	00·0
6	06·0	00·0	66	66·0	00·0	126	126·0	00·0	186	186·0	00·0	246	246·0	00·0
7	07·0	00·0	67	67·0	00·0	127	127·0	00·0	187	187·0	00·0	247	247·0	00·0
8	08·0	00·0	68	68·0	00·0	128	128·0	00·0	188	188·0	00·0	248	248·0	00·0
9	09·0	00·0	69	69·0	00·0	129	129·0	00·0	189	189·0	00·0	249	249·0	00·0
10	10·0	00·0	70	70·0	00·0	130	130·0	00·0	190	190·0	00·0	250	250·0	00·0
11	11·0	00·0	71	71·0	00·0	131	131·0	00·0	191	191·0	00·0	251	251·0	00·0
12	12·0	00·0	72	72·0	00·0	132	132·0	00·0	192	192·0	00·0	252	252·0	00·0
13	13·0	00·0	73	73·0	00·0	133	133·0	00·0	193	193·0	00·0	253	253·0	00·0
14	14·0	00·0	74	74·0	00·0	134	134·0	00·0	194	194·0	00·0	254	254·0	00·0
15	15·0	00·0	75	75·0	00·0	135	135·0	00·0	195	195·0	00·0	255	255·0	00·0
16	16·0	00·0	76	76·0	00·0	136	136·0	00·0	196	196·0	00·0	256	256·0	00·0
17	17·0	00·0	77	77·0	00·0	137	137·0	00·0	197	197·0	00·0	257	257·0	00·0
18	18·0	00·0	78	78·0	00·0	138	138·0	00·0	198	198·0	00·0	258	258·0	00·0
19	19·0	00·0	79	79·0	00·0	139	139·0	00·0	199	199·0	00·0	259	259·0	00·0
20	20·0	00·0	80	80·0	00·0	140	140·0	00·0	200	200·0	00·0	260	260·0	00·0
21	21·0	00·0	81	81·0	00·0	141	141·0	00·0	201	201·0	00·0	261	261·0	00·0
22	22·0	00·0	82	82·0	00·0	142	142·0	00·0	202	202·0	00·0	262	262·0	00·0
23	23·0	00·0	83	83·0	00·0	143	143·0	00·0	203	203·0	00·0	263	263·0	00·0
24	24·0	00·0	84	84·0	00·0	144	144·0	00·0	204	204·0	00·0	264	264·0	00·0
25	25·0	00·0	85	85·0	00·0	145	145·0	00·0	205	205·0	00·0	265	265·0	00·0
26	26·0	00·0	86	86·0	00·0	146	146·0	00·0	206	206·0	00·0	266	266·0	00·0
27	27·0	00·0	87	87·0	00·0	147	147·0	00·0	207	207·0	00·0	267	267·0	00·0
28	28·0	00·0	88	88·0	00·0	148	148·0	00·0	208	208·0	00·0	268	268·0	00·0
29	29·0	00·0	89	89·0	00·0	149	149·0	00·0	209	209·0	00·0	269	269·0	00·0
30	30·0	00·0	90	90·0	00·0	150	150·0	00·0	210	210·0	00·0	270	270·0	00·0
31	31·0	00·0	91	91·0	00·0	151	151·0	00·0	211	211·0	00·0	271	271·0	00·0
32	32·0	00·0	92	92·0	00·0	152	152·0	00·0	212	212·0	00·0	272	272·0	00·0
33	33·0	00·0	93	93·0	00·0	153	153·0	00·0	213	213·0	00·0	273	273·0	00·0
34	34·0	00·0	94	94·0	00·0	154	154·0	00·0	214	214·0	00·0	274	274·0	00·0
35	35·0	00·0	95	95·0	00·0	155	155·0	00·0	215	215·0	00·0	275	275·0	00·0
36	36·0	00·0	96	96·0	00·0	156	156·0	00·0	216	216·0	00·0	276	276·0	00·0
37	37·0	00·0	97	97·0	00·0	157	157·0	00·0	217	217·0	00·0	277	277·0	00·0
38	38·0	00·0	98	98·0	00·0	158	158·0	00·0	218	218·0	00·0	278	278·0	00·0
39	39·0	00·0	99	99·0	00·0	159	159·0	00·0	219	219·0	00·0	279	279·0	00·0
40	40·0	00·0	100	100·0	00·0	160	160·0	00·0	220	220·0	00·0	280	280·0	00·0
41	41·0	00·0	101	101·0	00·0	161	161·0	00·0	221	221·0	00·0	281	281·0	00·0
42	42·0	00·0	102	102·0	00·0	162	162·0	00·0	222	222·0	00·0	282	282·0	00·0
43	43·0	00·0	103	103·0	00·0	163	163·0	00·0	223	223·0	00·0	283	283·0	00·0
44	44·0	00·0	104	104·0	00·0	164	164·0	00·0	224	224·0	00·0	284	284·0	00·0
45	45·0	00·0	105	105·0	00·0	165	165·0	00·0	225	225·0	00·0	285	285·0	00·0
46	46·0	00·0	106	106·0	00·0	166	166·0	00·0	226	226·0	00·0	286	286·0	00·0
47	47·0	00·0	107	107·0	00·0	167	167·0	00·0	227	227·0	00·0	287	287·0	00·0
48	48·0	00·0	108	108·0	00·0	168	168·0	00·0	228	228·0	00·0	288	288·0	00·0
49	49·0	00·0	109	109·0	00·0	169	169·0	00·0	229	229·0	00·0	289	289·0	00·0
50	50·0	00·0	110	110·0	00·0	170	170·0	00·0	230	230·0	00·0	290	290·0	00·0
51	51·0	00·0	111	111·0	00·0	171	171·0	00·0	231	231·0	00·0	291	291·0	00·0
52	52·0	00·0	112	112·0	00·0	172	172·0	00·0	232	232·0	00·0	292	292·0	00·0
53	53·0	00·0	113	113·0	00·0	173	173·0	00·0	233	233·0	00·0	293	293·0	00·0
54	54·0	00·0	114	114·0	00·0	174	174·0	00·0	234	234·0	00·0	294	294·0	00·0
55	55·0	00·0	115	115·0	00·0	175	175·0	00·0	235	235·0	00·0	295	295·0	00·0
56	56·0	00·0	116	116·0	00·0	176	176·0	00·0	236	236·0	00·0	296	296·0	00·0
57	57·0	00·0	117	117·0	00·0	177	177·0	00·0	237	237·0	00·0	297	297·0	00·0
58	58·0	00·0	118	118·0	00·0	178	178·0	00·0	238	238·0	00·0	298	298·0	00 0
59	59·0	00·0	119	119·0	00·0	179	179·0	00·0	239	239·0	00·0	299	299·0	00·0
60	60·0	00·0	120	120·0	00·0	180	180·0	00·0	240	240·0	00·0	300	300·0	00·0
Dist.	Dep.	D.Lat	Dist.	Dep.	D.Lat	Dist.	Dep.	D.Lat	Dist.	Dep.	D.Lat	Dist.	Dep.	D.Lat

90°

90 DEGREES.

TRAVERSE TABLE

O DEGREES.

D Lon	Dep.		D Lon	Dep.		D Lon	Dep.		D Lon	Dep.		D Lon	Dep.	
Dist.	D.Lat	Dep.	Dist.	D.Lat	Dep.	Dist.	D.Lat	Dep.	Dist.	D.Lat	Dep.	Dist.	D.Lat	Dep.
301	301·0	00·0	361	361·0	00·0	421	421·0	00·0	481	481·0	00·0	541	541·0	00·0
302	302·0	00·0	362	362·0	00·0	422	422·0	00·0	482	482·0	00·0	542	542·0	00·0
303	303·0	00·0	363	363·0	00·0	423	423·0	00·0	483	483·0	00·0	543	543·0	00·0
304	304·0	00·0	364	364·0	00·0	424	424·0	00·0	484	484·0	00·0	544	544·0	00·0
305	305·0	00·0	365	365·0	00·0	425	425·0	00·0	485	485·0	00·0	545	545·0	00·0
306	306·0	00·0	366	366·0	00·0	426	426·0	00·0	486	486·0	00·0	546	546·0	00·0
307	307·0	00·0	367	367·0	00·0	427	427·0	00·0	487	487·0	00·0	547	547·0	00·0
308	308·0	00·0	368	368·0	00·0	428	428·0	00·0	488	488·0	00·0	548	548·0	00·0
309	309·0	00·0	369	369·0	00·0	429	429·0	00·0	489	489·0	00·0	549	549·0	00·0
310	310·0	00·0	370	370·0	00·0	430	430·0	00·0	490	490·0	00·0	550	550·0	00·0
311	311·0	00·0	371	371·0	00·0	431	431·0	00·0	491	491·0	00·0	551	551·0	00·0
312	312·0	00·0	372	372·0	00·0	432	432·0	00·0	492	492·0	00·0	552	552·0	00·0
313	313·0	00·0	373	373·0	00·0	433	433·0	00·0	493	493·0	00·0	553	553·0	00·0
314	314·0	00·0	374	374·0	00·0	434	434·0	00·0	494	494·0	00·0	554	554·0	00·0
315	315·0	00·0	375	375·0	00·0	435	435·0	00·0	495	495·0	00·0	555	555·0	00·0
316	316·0	00·0	376	376·0	00·0	436	436·0	00·0	496	496·0	00·0	556	556·0	00·0
317	317·0	00·0	377	377·0	00·0	437	437·0	00·0	497	497·0	00·0	557	557·0	00·0
318	318·0	00·0	378	378·0	00·0	438	438·0	00·0	498	498·0	00·0	558	558·0	00·0
319	319·0	00·0	379	379·0	00·0	439	439·0	00·0	499	499·0	00·0	559	559·0	00·0
320	320·0	00·0	380	380·0	00·0	440	440·0	00·0	500	500·0	00·0	560	560·0	00·0
321	321·0	00·0	381	381·0	00·0	441	441·0	00·0	501	501·0	00·0	561	561·0	00·0
322	322·0	00·0	382	382·0	00·0	442	442·0	00·0	502	502·0	00·0	562	562·0	00·0
323	323·0	00·0	383	383·0	00·0	443	443·0	00·0	503	503·0	00·0	563	563·0	00·0
324	324·0	00·0	384	384·0	00·0	444	444·0	00·0	504	504·0	00·0	564	564·0	00·0
325	325·0	00·0	385	385·0	00·0	445	445·0	00·0	505	505·0	00·0	565	565·0	00·0
326	326·0	00·0	386	386·0	00·0	446	446·0	00·0	506	506·C	00·0	566	566·0	00·0
327	327·0	00·0	387	387·0	00·0	447	447·0	00·0	507	507·0	00·0	567	567·0	00·0
328	328·0	00·0	388	388·0	00·0	448	448·0	00·0	508	508·0	00·0	568	568·0	00·0
329	329·0	00·0	389	389·0	00·0	449	449·0	00·0	509	509·0	00·0	569	569·0	00·0
330	330·0	00·0	390	390·0	00·0	450	450·0	00·0	510	510·0	00·0	570	570·0	00·0
331	331·0	00·0	391	391·0	00·0	451	451·0	00·0	511	511·0	00·0	571	571·0	00·0
332	332·0	00·0	392	392·0	00·0	452	452·0	00·0	512	512·0	00·0	572	572·0	00·0
333	333·0	00·0	393	393·0	00·0	453	453·0	00·0	513	513·0	00·0	573	573·0	00·0
334	334·0	00·0	394	394·0	00·0	454	454·0	00·0	514	514·0	00·0	574	574·0	00·0
335	335·0	00·0	395	395·0	00·0	455	455·0	00·0	515	515·0	00·0	575	575·0	00·0
336	336·0	00·0	396	396·0	00·0	456	456·0	00·0	516	516·0	00·0	576	576·0	00·0
337	337·0	00·0	397	397·0	00·0	457	457·0	00·0	517	517·0	00·0	577	577·0	00·0
338	338·0	00·0	398	398·0	00·0	458	458·0	00·0	518	518·0	00·0	578	578·0	00·0
339	339·0	00·0	399	399·0	00·0	459	459·0	00·0	519	519·0	00·0	579	579·0	00·0
340	340·0	00·0	400	400·0	00·0	460	460·0	00·0	520	520·0	00·0	580	580·0	00·0
341	341·0	00·0	401	401·0	00·0	461	461·0	00·0	521	521·0	00·0	581	581·0	00·0
342	342·0	00·0	402	402·0	00·0	462	462·0	00·0	522	522·0	00·0	582	582·0	00·0
343	343·0	00·0	403	403·0	00·0	463	463·0	00·0	523	523·0	00·0	583	583·0	00·0
344	344·0	00·0	404	404·0	00·0	464	464·0	00·0	524	524·0	00·0	584	584·0	00·0
345	345·0	00·0	405	405·0	00·0	465	465·0	00·0	525	525·0	00·0	585	585·0	00·0
346	346·0	00·0	406	406·0	00·0	466	466·0	00·0	526	526·0	00·0	586	586·0	00·0
347	347·0	00·0	407	407·0	00·0	467	467·0	00·0	527	527·0	00·0	587	587·0	00·0
348	348·0	00·0	408	408·0	00·0	468	468·0	00·0	528	528·0	00·0	588	588·0	00·0
349	349·0	00·0	409	409·0	00·0	469	469·0	00·0	529	529·0	00·0	589	589·0	00·0
350	350·0	00·0	410	410·0	00·0	470	470·0	00·0	530	530·0	00·0	590	590·0	00·0
351	351·0	00·0	411	411·0	00·0	471	471·0	00·0	531	531·0	00·0	591	591·0	00·0
352	352·0	00·0	412	412·0	00·0	472	472·0	00·0	532	532·0	00·0	592	592·0	00·0
353	353·0	00·0	413	413·0	00·0	473	473·0	00·0	533	533·0	00·0	593	593·0	00·0
354	354·0	00·0	414	414·0	00·0	474	474·0	00·0	534	534·0	00·0	594	594·0	00·0
355	355·0	00·0	415	415·0	00·0	475	475·0	00·0	535	535·0	00·0	595	595·0	00·0
356	356·0	00·0	416	416·0	00·0	476	476·0	00·0	536	536·0	00·0	596	596·0	00·0
357	357·0	00·0	417	417·0	00·0	477	477·0	00·0	537	537·0	00·0	597	597·0	00·0
358	358·0	00·0	418	418·0	00·0	478	478·0	00·0	538	538·0	00·0	598	598·0	00·0
359	359·0	00·0	419	419·0	00·0	479	479·0	00·0	539	539·0	00·0	599	599·0	00·0
360	360·0	00·0	420	420·0	00·0	480	480·0	00·0	540	540·0	00·0	600	600·0	00·0
Dist.	Dep.	D.Lat	Dist.	Dep.	D.Lat	Dist.	Dep.	D.Lat	Dist.	Dep.	D.Lat	Dist.	Dep.	D.Lat

90°

90 DEGREES.

W.or ↑
270°

↑ E.or   6h 0m
090°

TRAVERSE TABLE
1 DEGREE.

1° 359° / 181°   001° / 179°   0h 4m

Dist	D.Lat	Dep.	Dist	D.Lat	Dep.	Dist	D.Lat	Dep.	Dist	D.Lat	Dep.	Dist	D.Lat	Dep.
1	01·0	00·0	61	61·0	01·1	121	121·0	02·1	181	181·0	03·2	241	241·0	04·2
2	02·0	00·0	62	62·0	01·1	122	122·0	02·1	182	182·0	03·2	242	242·0	04·2
3	03·0	00·1	63	63·0	01·1	123	123·0	02·1	183	183·0	03·2	243	243·0	04·2
4	04·0	00·1	64	64·0	01·1	124	124·0	02·2	184	184·0	03·2	244	244·0	04·3
5	05·0	00·1	65	65·0	01·1	125	125·0	02·2	185	185·0	03·2	245	245·0	04·3
6	06·0	00·1	66	66·0	01·2	126	126·0	02·2	186	186·0	03·2	246	246·0	04·3
7	07·0	00·1	67	67·0	01·2	127	127·0	02·2	187	187·0	03·3	247	247·0	04·3
8	08·0	00·1	68	68·0	01·2	128	128·0	02·2	188	188·0	03·3	248	248·0	04·3
9	09·0	00·2	69	69·0	01·2	129	129·0	02·2	189	189·0	03·3	249	249·0	04·3
10	10·0	00·2	70	70·0	01·2	130	130·0	02·3	190	190·0	03·3	250	250·0	04·4
11	11·0	00·2	71	71·0	01·2	131	131·0	02·3	191	191·0	03·3	251	251·0	04·4
12	12·0	00·2	72	72·0	01·3	132	132·0	02·3	192	192·0	03·4	252	252·0	04·4
13	13·0	00·2	73	73·0	01·3	133	133·0	02·3	193	193·0	03·4	253	253·0	04·4
14	14·0	00·2	74	74·0	01·3	134	134·0	02·3	194	194·0	03·4	254	254·0	04·4
15	15·0	00·3	75	75·0	01·3	135	135·0	02·4	195	195·0	03·4	255	255·0	04·4
16	16·0	00·3	76	76·0	01·3	136	136·0	02·4	196	196·0	03·4	256	256·0	04·5
17	17·0	00·3	77	77·0	01·3	137	137·0	02·4	197	197·0	03·4	257	257·0	04·5
18	18·0	00·3	78	78·0	01·4	138	138·0	02·4	198	198·0	03·5	258	258·0	04·5
19	19·0	00·3	79	79·0	01·4	139	139·0	02·4	199	199·0	03·5	259	259·0	04·5
20	20·0	00·3	80	80·0	01·4	140	140·0	02·4	200	200·0	03·5	260	260·0	04·5
21	21·0	00·4	81	81·0	01·4	141	141·0	02·5	201	201·0	03·5	261	261·0	04·5
22	22·0	00·4	82	82·0	01·4	142	142·0	02·5	202	202·0	03·5	262	262·0	04·6
23	23·0	00·4	83	83·0	01·4	143	143·0	02·5	203	203·0	03·5	263	263·0	04·6
24	24·0	00·4	84	84·0	01·5	144	144·0	02·5	204	204·0	03·6	264	264·0	04·6
25	25·0	00·4	85	85·0	01·5	145	145·0	02·5	205	205·0	03·6	265	265·0	04·6
26	26·0	00·5	86	86·0	01·5	146	146·0	02·5	206	206·0	03·6	266	266·0	04·6
27	27·0	00·5	87	87·0	01·5	147	147·0	02·6	207	207·0	03·6	267	267·0	04·7
28	28·0	00·5	88	88·0	01·5	148	148·0	02·6	208	208·0	03·6	268	268·0	04·7
29	29·0	00·5	89	89·0	01·6	149	149·0	02·6	209	209·0	03·7	269	269·0	04·7
30	30·0	00·5	90	90·0	01·6	150	150·0	02·6	210	210·0	03·7	270	270·0	04·7
31	31·0	00·5	91	91·0	01·6	151	151·0	02·6	211	211·0	03·7	271	271·0	04·7
32	32·0	00·6	92	92·0	01·6	152	152·0	02·7	212	212·0	03·7	272	272·0	04·7
33	33·0	00·6	93	93·0	01·6	153	153·0	02·7	213	213·0	03·7	273	273·0	04·8
34	34·0	00·6	94	94·0	01·6	154	154·0	02·7	214	214·0	03·7	274	274·0	04·8
35	35·0	00·6	95	95·0	01·7	155	155·0	02·7	215	215·0	03·8	275	275·0	04·8
36	36·0	00·6	96	96·0	01·7	156	156·0	02·7	216	216·0	03·8	276	276·0	04·8
37	37·0	00·6	97	97·0	01·7	157	157·0	02·7	217	217·0	03·8	277	277·0	04·8
38	38·0	00·7	98	98·0	01·7	158	158·0	02·8	218	218·0	03·8	278	278·0	04·9
39	39·0	00·7	99	99·0	01·7	159	159·0	02·8	219	219·0	03·8	279	279·0	04·9
40	40·0	00·7	100	100·0	01·7	160	160·0	02·8	220	220·0	03·8	280	280·0	04·9
41	41·0	00·7	101	101·0	01·8	161	161·0	02·8	221	221·0	03·9	281	281·0	04·9
42	42·0	00·7	102	102·0	01·8	162	162·0	02·8	222	222·0	03·9	282	282·0	04·9
43	43·0	00·8	103	103·0	01·8	163	163·0	02·8	223	223·0	03·9	283	283·0	04·9
44	44·0	00·8	104	104·0	01·8	164	164·0	02·9	224	224·0	03·9	284	284·0	05·0
45	45·0	00·8	105	105·0	01·8	165	165·0	02·9	225	225·0	03·9	285	285·0	05·0
46	46·0	00·8	106	106·0	01·8	166	166·0	02·9	226	226·0	03·9	286	286·0	05·0
47	47·0	00·8	107	107·0	01·9	167	167·0	02·9	227	227·0	04·0	287	287·0	05·0
48	48·0	00·8	108	108·0	01·9	168	168·0	02·9	228	228·0	04·0	288	288·0	05·0
49	49·0	00·9	109	109·0	01·9	169	169·0	02·9	229	229·0	04·0	289	289·0	05·0
50	50·0	00·9	110	110·0	01·9	170	170·0	03·0	230	230·0	04·0	290	290·0	05·1
51	51·0	00·9	111	111·0	01·9	171	171·0	03·0	231	231·0	04·0	291	291·0	05·1
52	52·0	00·9	112	112·0	02·0	172	172·0	03·0	232	232·0	04·0	292	292·0	05·1
53	53·0	00·9	113	113·0	02·0	173	173·0	03·0	233	233·0	04·1	293	293·0	05·1
54	54·0	00·9	114	114·0	02·0	174	174·0	03·0	234	234·0	04·1	294	294·0	05·1
55	55·0	01·0	115	115·0	02·0	175	175·0	03·1	235	235·0	04·1	295	295·0	05·1
56	56·0	01·0	116	116·0	02·0	176	176·0	03·1	236	236·0	04·1	296	296·0	05·2
57	57·0	01·0	117	117·0	02·0	177	177·0	03·1	237	237·0	04·1	297	297·0	05·2
58	58·0	01·0	118	118·0	02·1	178	178·0	03·1	238	238·0	04·2	298	298·0	05·2
59	59·0	01·0	119	119·0	02·1	179	179·0	03·1	239	239·0	04·2	299	299·0	05·2
60	60·0	01·0	120	120·0	02·1	180	180·0	03·1	240	240·0	04·2	300	300·0	05·2

Dist	Dep.	D.Lat	Dist	Dep.	D.Lat	Dist	Dep.	D.Lat	Dist	Dep.	D.Lat	Dist	Dep.	D.Lat

89° 271° / 269°   89 DEGREES.   089° / 091°   5h 56m

TRAVERSE TABLE
1 DEGREE.

359° ↑ / 181°   001° ↑ / 179°   0h 4m   1°

D Lon	Dep.		D Lon	Dep.		D Lon	Dep.		D Lon	Dep.		D Lon	Dep.	
Dist	D.Lat	Dep.	Dist	D.Lat	Dep.	Dist	D.Lat	Dep.	Dist	D.Lat	Dep.	Dist	D.Lat	Dep.
301	301·0	05·3	361	360·9	06·3	421	420·9	07·3	481	480·9	08·4	541	540·9	09·4
302	302·0	05·3	362	361·9	06·3	422	421·9	07·4	482	481·9	08·4	542	541·9	09·5
303	303·0	05·3	363	362·9	06·3	423	422·9	07·4	483	482·9	08·4	543	542·9	09·5
304	304·0	05·3	364	363·9	06·4	424	423·9	07·4	484	483·9	08·5	544	543·9	09·5
305	305·0	05·3	365	364·9	06·4	425	424·9	07·4	485	484·9	08·5	545	544·9	09·5
306	306·0	05·3	366	365·9	06·4	426	425·9	07·4	486	485·9	08·5	546	545·9	09·5
307	307·0	05·4	367	366·9	06·4	427	426·9	07·5	487	486·9	08·5	547	546·9	09·5
308	308·0	05·4	368	367·9	06·4	428	427·9	07·5	488	487·9	08·5	548	547·9	09·6
309	309·0	05·4	369	368·9	06·4	429	428·9	07·5	489	488·9	08·5	549	548·9	09·6
310	310·0	05·4	370	369·9	06·5	430	429·9	07·5	490	489·9	08·6	550	549·9	09·6
311	311·0	05·4	371	370·9	06·5	431	430·9	07·5	491	490·9	08·6	551	550·9	09·6
312	312·0	05·4	372	371·9	06·5	432	431·9	07·5	492	491·9	08·6	552	551·9	09·6
313	313·0	05·5	373	372·9	06·5	433	432·9	07·6	493	492·9	08·6	553	552·9	09·7
314	314·0	05·5	374	373·9	06·5	434	433·9	07·6	494	493·9	08·6	554	553·9	09·7
315	315·0	05·5	375	374·9	06·5	435	434·9	07·6	495	494·9	08·7	555	554·9	09·7
316	316·0	05·5	376	375·9	06·6	436	435·9	07·6	496	495·9	08·7	556	555·9	09·7
317	317·0	05·5	377	376·9	06·6	437	436·9	07·6	497	496·9	08·7	557	556·9	09·7
318	318·0	05·5	378	377·9	06·6	438	437·9	07·6	498	497·9	08·7	558	557·9	09·7
319	319·0	05·6	379	378·9	06·6	439	438·9	07·7	499	498·9	08·7	559	558·9	09·8
320	320·0	05·6	380	379·9	06·6	440	439·9	07·7	500	499·9	08·7	560	559·9	09·8
321	321·0	05·6	381	380·9	06·6	441	440·9	07·7	501	500·9	08·7	561	560·9	09·8
322	322·0	05·6	382	381·9	06·7	442	441·9	07·7	502	501·9	08·8	562	561·9	09·8
323	323·0	05·6	383	382·9	06·7	443	442·9	07·7	503	502·9	08·8	563	562·9	09·8
324	324·0	05·7	384	383·9	06·7	444	443·9	07·7	504	503·9	08·8	564	563·9	09·8
325	325·0	05·7	385	384·9	06·7	445	444·9	07·8	505	504·9	08·8	565	564·9	09·9
326	326·0	05·7	386	385·9	06·7	446	445·9	07·8	506	505·9	08·8	566	565·9	09·9
327	327·0	05·7	387	386·9	06·8	447	446·9	07·8	507	506·9	08·8	567	566·9	09·9
328	328·0	05·7	388	387·9	06·8	448	447·9	07·8	508	507·9	08·9	568	567·9	09·9
329	328·9	05·7	389	388·9	06·8	449	448·9	07·8	509	508·9	08·9	569	568·9	09·9
330	329·9	05·8	390	389·9	06·8	450	449·9	07·9	510	509·9	08·9	570	569·9	09·9
331	330·9	05·8	391	390·9	06·8	451	450·9	07·9	511	510·9	08·9	571	570·9	10·0
332	331·9	05·8	392	391·9	06·8	452	451·9	07·9	512	511·9	08·9	572	571·9	10·0
333	332·9	05·8	393	392·9	06·9	453	452·9	07·9	513	512·9	09·0	573	572·9	10·0
334	333·9	05·8	394	393·9	06·9	454	453·9	07·9	514	513·9	09·0	574	573·9	10·0
335	334·9	05·8	395	394·9	06·9	455	454·9	07·9	515	514·9	09·0	575	574·9	10·1
336	335·9	05·9	396	395·9	06·9	456	455·9	08·0	516	515·9	09·0	576	575·9	10·1
337	336·9	05·9	397	396·9	06·9	457	456·9	08·0	517	516·9	09·0	577	576·9	10·1
338	337·9	05·9	398	397·9	06·9	458	457·9	08·0	518	517·9	09·0	578	577·9	10·1
339	338·9	05·9	399	398·9	07·0	459	458·9	08·0	519	518·9	09·1	579	578·9	10·1
340	339·9	05·9	400	399·9	07·0	460	459·9	08·0	520	519·9	09·1	580	579·9	10·1
341	340·9	06·0	401	400·9	07·0	461	460·9	08·0	521	520·9	09·1	581	580·9	10·1
342	341·9	06·0	402	401·9	07·0	462	461·9	08·1	522	521·9	09·1	582	581·9	10·2
343	342·9	06·0	403	402·9	07·0	463	462·9	08·1	523	522·9	09·1	583	582·9	10·2
344	343·9	06·0	404	403·9	07·1	464	463·9	08·1	524	523·9	09·1	584	583·9	10·2
345	344·9	06·0	405	404·9	07·1	465	464·9	08·1	525	524·9	09·2	585	584·9	10·2
346	345·9	06·0	406	405·9	07·1	466	465·9	08·1	526	525·9	09·2	586	585·9	10·2
347	346·9	06·1	407	406·9	07·1	467	466·9	08·2	527	526·9	09·2	587	586·9	10·2
348	347·9	06·1	408	407·9	07·1	468	467·9	08·2	528	527·9	09·2	588	587·9	10·3
349	348·9	06·1	409	408·9	07·1	469	468·9	08·2	529	528·9	09·2	589	588·9	10·3
350	349·9	06·1	410	409·9	07·2	470	469·9	08·2	530	529·9	09·2	590	589·9	10·3
351	350·9	06·1	411	410·9	07·2	471	470·9	08·2	531	530·9	09·3	591	590·9	10·3
352	351·9	06·1	412	411·9	07·2	472	471·9	08·2	532	531·9	09·3	592	591·9	10·3
353	352·9	06·2	413	412·9	07·2	473	472·9	08·3	533	532·9	09·3	593	592·9	10·3
354	353·9	06·2	414	413·9	07·2	474	473·9	08·3	534	533·9	09·3	594	593·9	10·4
355	354·9	06·2	415	414·9	07·2	475	474·9	08·3	535	534·9	09·3	595	594·9	10·4
356	355·9	06·2	416	415·9	07·3	476	475·9	08·3	536	535·9	09·4	596	595·9	10·4
357	356·9	06·2	417	416·9	07·3	477	476·9	08·3	537	536·9	09·4	597	596·9	10·4
358	357·9	06·2	418	417·9	07·3	478	477·9	08·3	538	537·9	09·4	598	597·9	10·4
359	358·9	06·3	419	418·9	07·3	479	478·9	08·4	539	538·9	09·4	599	598·9	10·5
360	359·9	06·3	420	419·9	07·3	480	479·9	08·4	540	539·9	09·4	600	599·9	10·5
Dist	Dep.	D.Lat	Dist	Dep.	D.Lat	Dist	Dep.	D.Lat	Dist	Dep.	D.Lat	Dist	Dep.	D.Lat
D Lon		Dep.	D Lon		Dep.	D Lon		Dep.	D Lon		Dep.	D Lon		Dep.

271° ↑ / 269°   **89 DEGREES.**   089° ↑ / 091°   5h 56m   **89°**

TRAVERSE TABLE
2 DEGREES.

Dist	D.Lat	Dep.	Dist	D.Lat	Dep.	Dist	D.Lat	Dep.	Dist	D.Lat	Dep.	Dist	D.Lat	Dep.
1	01·0	00·0	61	61·0	02·1	121	120·9	04·2	181	180·9	06·3	241	240·9	08·4
2	02·0	00·1	62	62·0	02·2	122	121·9	04·3	182	181·9	06·4	242	241·9	08·4
3	03·0	00·1	63	63·0	02·2	123	122·9	04·3	183	182·9	06·4	243	242·9	08·5
4	04·0	00·1	64	64·0	02·2	124	123·9	04·3	184	183·9	06·4	244	243·9	08·5
5	05·0	00·2	65	65·0	02·3	125	124·9	04·4	185	184·9	06·5	245	244·9	08·6
6	06·0	00·2	66	66·0	02·3	126	125·9	04·4	186	185·9	06·5	246	245·8	08·6
7	07·0	00·2	67	67·0	02·3	127	126·9	04·4	187	186·9	06·5	247	246·8	08·6
8	08·0	00·3	68	68·0	02·4	128	127·9	04·5	188	187·9	06·6	248	247·8	08·7
9	09·0	00·3	69	69·0	02·4	129	128·9	04·5	189	188·9	06·6	249	248·8	08·7
10	10·0	00·3	70	70·0	02·4	130	129·9	04·5	190	189·9	06·6	250	249·8	08·7
11	11·0	00·4	71	71·0	02·5	131	130·9	04·6	191	190·9	06·7	251	250·8	08·8
12	12·0	00·4	72	72·0	02·5	132	131·9	04·6	192	191·9	06·7	252	251·8	08·8
13	13·0	00·5	73	73·0	02·5	133	132·9	04·6	193	192·9	06·7	253	252·8	08·8
14	14·0	00·5	74	74·0	02·6	134	133·9	04·7	194	193·9	06·8	254	253·8	08·9
15	15·0	00·5	75	75·0	02·6	135	134·9	04·7	195	194·9	06·8	255	254·8	08·9
16	16·0	00·6	76	76·0	02·7	136	135·9	04·7	196	195·9	06·9	256	255·8	08·9
17	17·0	00·6	77	77·0	02·7	137	136·9	04·8	197	196·9	06·9	257	256·8	09·0
18	18·0	00·6	78	78·0	02·7	138	137·9	04·8	198	197·9	06·9	258	257·8	09·0
19	19·0	00·7	79	79·0	02·8	139	138·9	04·9	199	198·9	06·9	259	258·8	09·0
20	20·0	00·7	80	80·0	02·8	140	139·9	04·9	200	199·9	07·0	260	259·8	09·1
21	21·0	00·7	81	81·0	02·8	141	140·9	04·9	201	200·9	07·0	261	260·8	09·1
22	22·0	00·8	82	81·9	02·9	142	141·9	05·0	202	201·9	07·0	262	261·8	09·1
23	23·0	00·8	83	82·9	02·9	143	142·9	05·0	203	202·9	07·1	263	262·8	09·2
24	24·0	00·8	84	83·9	02·9	144	143·9	05·0	204	203·9	07·1	264	263·8	09·2
25	25·0	00·9	85	84·9	03·0	145	144·9	05·1	205	204·9	07·2	265	264·8	09·2
26	26·0	00·9	86	85·9	03·0	146	145·9	05·1	206	205·9	07·2	266	265·8	09·3
27	27·0	00·9	87	86·9	03·0	147	146·9	05·1	207	206·9	07·2	267	266·8	09·3
28	28·0	01·0	88	87·9	03·1	148	147·9	05·2	208	207·9	07·3	268	267·8	09·4
29	29·0	01·0	89	88·9	03·1	149	148·9	05·2	209	208·9	07·3	269	268·8	09·4
30	30·0	01·0	90	89·9	03·1	150	149·9	05·2	210	209·9	07·3	270	269·8	09·4
31	31·0	01·1	91	90·9	03·2	151	150·9	05·3	211	210·9	07·4	271	270·8	09·5
32	32·0	01·1	92	91·9	03·2	152	151·9	05·3	212	211·9	07·4	272	271·8	09·5
33	33·0	01·2	93	92·9	03·2	153	152·9	05·3	213	212·9	07·4	273	272·8	09·5
34	34·0	01·2	94	93·9	03·3	154	153·9	05·4	214	213·9	07·5	274	273·8	09·6
35	35·0	01·2	95	94·9	03·3	155	154·9	05·4	215	214·9	07·5	275	274·8	09·6
36	36·0	01·3	96	95·9	03·4	156	155·9	05·4	216	215·9	07·5	276	275·8	09·6
37	37·0	01·3	97	96·9	03·4	157	156·9	05·5	217	216·9	07·6	277	276·8	09·7
38	38·0	01·3	98	97·9	03·4	158	157·9	05·5	218	217·9	07·6	278	277·8	09·7
39	39·0	01·4	99	98·9	03·5	159	158·9	05·5	219	218·9	07·6	279	278·8	09·7
40	40·0	01·4	100	99·9	03·5	160	159·9	05·6	220	219·9	07·7	280	279·8	09·8
41	41·0	01·4	101	100·9	03·5	161	160·9	05·6	221	220·9	07·7	281	280·8	09·8
42	42·0	01·5	102	101·9	03·6	162	161·9	05·7	222	221·9	07·7	282	281·8	09·8
43	43·0	01·5	103	102·9	03·6	163	162·9	05·7	223	222·9	07·8	283	282·8	09·9
44	44·0	01·5	104	103·9	03·6	164	163·9	05·7	224	223·9	07·8	284	283·8	09·9
45	45·0	01·6	105	104·9	03·7	165	164·9	05·8	225	224·9	07·9	285	284·8	09·9
46	46·0	01·6	106	105·9	03·7	166	165·9	05·8	226	225·9	07·9	286	285·8	10·0
47	47·0	01·6	107	106·9	03·7	167	166·9	05·8	227	226·9	07·9	287	286·8	10·0
48	48·0	01·7	108	107·9	03·8	168	167·9	05·9	228	227·9	08·0	288	287·8	10·1
49	49·0	01·7	109	108·9	03·8	169	168·9	05·9	229	228·9	08·0	289	288·8	10·1
50	50·0	01·7	110	109·9	03·8	170	169·9	05·9	230	229·9	08·0	290	289·8	10·1
51	51·0	01·8	111	110·9	03·9	171	170·9	06·0	231	230·9	08·1	291	290·8	10·2
52	52·0	01·8	112	111·9	03·9	172	171·9	06·0	232	231·9	08·1	292	291·8	10·2
53	53·0	01·8	113	112·9	03·9	173	172·9	06·0	233	232·9	08·1	293	292·8	10·2
54	54·0	01·9	114	113·9	04·0	174	173·9	06·1	234	233·9	08·2	294	293·8	10·3
55	55·0	01·9	115	114·9	04·0	175	174·9	06·1	235	234·9	08·2	295	294·8	10·3
56	56·0	02·0	116	115·9	04·0	176	175·9	06·1	236	235·9	08·3	296	295·8	10·3
57	57·0	02·0	117	116·9	04·1	177	176·9	06·2	237	236·9	08·3	297	296·8	10·4
58	58·0	02·0	118	117·9	04·1	178	177·9	06·2	238	237·9	08·3	298	297·8	10·4
59	59·0	02·1	119	118·9	04·2	179	178·9	06·2	239	238·9	08·3	299	298·8	10·4
60	60·0	02·1	120	119·9	04·2	180	179·9	06·3	240	239·9	08·4	300	299·8	10·5

Dist	Dep.	D.Lat	Dist	Dep.	D.Lat	Dist	Dep.	D.Lat	Dist	Dep.	D.Lat	Dist	Dep.	D.Lat

| | 358° ↑ | | | | | | TRAVERSE TABLE | | | | | | ↑ 002° | | **2°** |
| | 182° | | | | | 2 DEGREES. | | | | | | | 178° | 0h 8m | |

D Lon	Dep.		D Lon	Dep.		D Lon	Dep.		D Lon	Dep.		D Lon	Dep.	
Dist	D. Lat	Dep.	Dist	D. Lat	Dep.	Dist	D. Lat	Dep.	Dist	D. Lat	Dep.	Dist	D. Lat	Dep.
301	300·8	10·5	361	360·8	12·6	421	420·7	14·7	481	480·7	16·8	541	540·7	18·9
302	301·8	10·5	362	361·8	12·6	422	421·7	14·7	482	481·7	16·8	542	541·7	18·9
303	302·8	10·6	363	362·8	12·7	423	422·7	14·7	483	482·7	16·9	543	542·7	19·0
304	303·8	10·6	364	363·8	12·7	424	423·7	14·8	484	483·7	16·9	544	543·7	19·0
305	304·8	10·6	365	364·8	12·7	425	424·7	14·8	485	484·7	16·9	545	544·7	19·0
306	305·8	10·7	366	365·8	12·8	426	425·7	14·9	486	485·7	16·9	546	545·7	19·1
307	306·8	10·7	367	366·8	12·8	427	426·7	14·9	487	486·7	17·0	547	546·7	19·1
308	307·8	10·7	368	367·8	12·8	428	427·7	14·9	488	487·7	17·0	548	547·7	19·1
309	308·8	10·8	369	368·8	12·9	429	428·7	15·0	489	488·7	17·0	549	548·7	19·2
310	309·8	10·8	370	369·8	12·9	430	429·7	15·0	490	489·7	17·1	550	549·7	19·2
311	310·8	10·8	371	370·8	12·9	431	430·7	15·0	491	490·7	17·1	551	550·7	19·2
312	311·8	10·9	372	371·8	13·0	432	431·7	15·1	492	491·7	17·1	552	551·7	19·3
313	312·8	10·9	373	372·8	13·0	433	432·7	15·1	493	492·7	17·2	553	552·7	19·3
314	313·8	10·9	374	373·8	13·0	434	433·7	15·1	494	493·7	17·2	554	553·7	19·3
315	314·8	11·0	375	374·8	13·1	435	434·7	15·2	495	494·7	17·2	555	554·7	19·4
316	315·8	11·0	376	375·8	13·1	436	435·7	15·2	496	495·7	17·3	556	555·7	19·4
317	316·8	11·0	377	376·8	13·1	437	436·7	15·2	497	496·7	17·3	557	556·7	19·4
318	317·8	11·1	378	377·8	13·2	438	437·7	15·3	498	497·7	17·3	558	557·7	19·5
319	318·8	11·1	379	378·8	13·2	439	438·7	15·3	499	498·7	17·4	559	558·7	19·5
320	319·8	11·2	380	379·8	13·2	440	439·7	15·3	500	499·7	17·4	560	559·7	19·5
321	320·8	11·2	381	380·8	13·3	441	440·7	15·4	501	500·7	17·5	561	560·7	19·6
322	321·8	11·2	382	381·8	13·3	442	441·7	15·4	502	501·7	17·5	562	561·7	19·6
323	322·8	11·3	383	382·8	13·3	443	442·7	15·4	503	502·7	17·5	563	562·7	19·6
324	323·8	11·3	384	383·8	13·4	444	443·7	15·5	504	503·7	17·6	564	563·7	19·7
325	324·8	11·3	385	384·8	13·4	445	444·7	15·5	505	504·7	17·6	565	564·7	19·7
326	325·8	11·4	386	385·8	13·5	446	445·7	15·6	506	505·7	17·6	566	565·7	19·8
327	326·8	11·4	387	386·8	13·5	447	446·7	15·6	507	506·7	17·7	567	566·7	19·8
328	327·8	11·4	388	387·8	13·5	448	447·7	15·6	508	507·7	17·7	568	567·7	19·8
329	328·8	11·5	389	388·8	13·6	449	448·7	15·7	509	508·7	17·7	569	568·7	19·9
330	329·8	11·5	390	389·8	13·6	450	449·7	15·7	510	509·7	17·8	570	569·7	19·9
331	330·8	11·5	391	390·8	13·6	451	450·7	15·7	511	510·7	17·8	571	570·7	19·9
332	331·8	11·6	392	391·8	13·7	452	451·7	15·8	512	511·7	17·9	572	571·7	20·0
333	332·8	11·6	393	392·8	13·7	453	452·7	15·8	513	512·7	17·9	573	572·7	20·0
334	333·8	11·6	394	393·8	13·7	454	453·7	15·8	514	513·7	17·9	574	573·7	20·0
335	334·8	11·7	395	394·8	13·8	455	454·7	15·9	515	514·7	18·0	575	574·6	20·0
336	335·8	11·7	396	395·8	13·8	456	455·7	15·9	516	515·7	18·0	576	575·6	20·1
337	336·8	11·7	397	396·8	13·8	457	456·7	15·9	517	516·7	18·0	577	576·6	20·1
338	337·8	11·8	398	397·8	13·9	458	457·7	16·0	518	517·7	18·1	578	577·6	20·2
339	338·8	11·8	399	398·8	13·9	459	458·7	16·0	519	518·7	18·1	579	578·6	20·2
340	339·8	11·9	400	399·8	13·9	460	459·7	16·0	520	519·7	18·1	580	579·6	20·2
341	340·8	11·9	401	400·8	14·0	461	460·7	16·1	521	520·7	18·2	581	580·6	20·3
342	341·8	11·9	402	401·8	14·0	462	461·7	16·1	522	521·7	18·2	582	581·6	20·3
343	342·8	12·0	403	402·8	14·0	463	462·7	16·1	523	522·7	18·3	583	582·6	20·3
344	343·8	12·0	404	403·8	14·1	464	463·7	16·2	524	523·7	18·3	584	583·6	20·4
345	344·8	12·0	405	404·8	14·1	465	464·7	16·2	525	524·7	18·3	585	584·6	20·4
346	345·8	12·1	406	405·8	14·2	466	465·7	16·2	526	525·7	18·4	586	585·6	20·5
347	346·8	12·1	407	406·8	14·2	467	466·7	16·3	527	526·7	18·4	587	586·6	20·5
348	347·8	12·1	408	407·8	14·2	468	467·7	16·3	528	527·7	18·4	588	587·6	20·5
349	348·8	12·2	409	408·8	14·3	469	468·7	16·4	529	528·7	18·5	589	588·6	20·6
350	349·8	12·2	410	409·8	14·3	470	469·7	16·4	530	529·7	18·5	590	589·6	20·6
351	350·8	12·2	411	410·7	14·3	471	470·7	16·4	531	530·7	18·5	591	590·6	20·6
352	351·8	12·3	412	411·7	14·4	472	471·7	16·5	532	531·7	18·6	592	591·6	20·7
353	352·8	12·3	413	412·7	14·4	473	472·7	16·5	533	532·7	18·6	593	592·6	20·7
354	353·8	12·3	414	413·7	14·4	474	473·7	16·5	534	533·7	18·6	594	593·6	20·8
355	354·8	12·4	415	414·7	14·5	475	474·7	16·6	535	534·7	18·7	595	594·6	20·8
356	355·8	12·4	416	415·7	14·5	476	475·7	16·6	536	535·7	18·7	596	595·6	20·8
357	356·8	12·4	417	416·7	14·5	477	476·7	16·6	537	536·7	18·7	597	596·6	20·8
358	357·8	12·5	418	417·7	14·6	478	477·7	16·7	538	537·7	18·8	598	597·6	20·9
359	358·8	12·5	419	418·7	14·6	479	478·7	16·7	539	538·7	18·8	599	598·6	20·9
360	359·8	12·5	420	419·7	14·6	480	479·7	16·7	540	539·7	18·8	600	599·6	20·9
Dist	Dep.	D.Lat	Dist	Dep.	D.Lat	Dist	Dep.	D.Lat	Dist	Dep.	D.Lat	Dist	Dep.	D.Lat
D Lon		Dep.	D Lon		Dep.	D Lon		Dep.	D Lon		Dep.	D Lon		Dep.

| | 272° ↑ | | | | 88 DEGREES. | | | | ↑ 088° | | **88°** |
| | 268° | | | | | | | | 092° | 5h 52m | |

TRAVERSE TABLE
3 DEGREES.

357° ↑ | 183°   ↑ 003° | 177°   0h 12m

Dist	D.Lat	Dep.	Dist	D.Lat	Dep.	Dist	D.Lat	Dep.	Dist	D.Lat	Dep.	Dist	D.Lat	Dep.
1	01·0	00·1	61	60·9	03·2	121	120·8	06·3	181	180·8	09·5	241	240·7	12·6
2	02·0	00·1	62	61·9	03·2	122	121·8	06·4	182	181·8	09·5	242	241·7	12·7
3	03·0	00·2	63	62·9	03·3	123	122·8	06·4	183	182·7	09·6	243	242·7	12·7
4	04·0	00·2	64	63·9	03·3	124	123·8	06·5	184	183·7	09·6	244	243·7	12·8
5	05·0	00·3	65	64·9	03·4	125	124·8	06·5	185	184·7	09·7	245	244·7	12·8
6	06·0	00·3	66	65·9	03·5	126	125·8	06·6	186	185·7	09·7	246	245·7	12·9
7	07·0	00·4	67	66·9	03·5	127	126·8	06·6	187	186·7	09·8	247	246·7	12·9
8	08·0	00·4	68	67·9	03·6	128	127·8	06·7	188	187·7	09·8	248	247·7	13·0
9	09·0	00·5	69	68·9	03·6	129	128·8	06·8	189	188·7	09·9	249	248·7	13·0
10	10·0	00·5	70	69·9	03·7	130	129·8	06·8	190	189·7	09·9	250	249·7	13·1
11	11·0	00·6	71	70·9	03·7	131	130·8	06·9	191	190·7	10·0	251	250·7	13·1
12	12·0	00·6	72	71·9	03·8	132	131·8	06·9	192	191·7	10·0	252	251·7	13·2
13	13·0	00·7	73	72·9	03·8	133	132·8	07·0	193	192·7	10·1	253	252·7	13·2
14	14·0	00·7	74	73·9	03·9	134	133·8	07·0	194	193·7	10·2	254	253·7	13·3
15	15·0	00·8	75	74·9	03·9	135	134·8	07·1	195	194·7	10·2	255	254·7	13·3
16	16·0	00·8	76	75·9	04·0	136	135·8	07·1	196	195·7	10·3	256	255·6	13·4
17	17·0	00·9	77	76·9	04·0	137	136·8	07·2	197	196·7	10·3	257	256·6	13·5
18	18·0	00·9	78	77·9	04·1	138	137·8	07·2	198	197·7	10·4	258	257·6	13·5
19	19·0	01·0	79	78·9	04·1	139	138·8	07·3	199	198·7	10·4	259	258·6	13·6
20	20·0	01·0	80	79·9	04·2	140	139·8	07·3	200	199·7	10·5	260	259·6	13·6
21	21·0	01·1	81	80·9	04·2	141	140·8	07·4	201	200·7	10·5	261	260·6	13·7
22	22·0	01·2	82	81·9	04·3	142	141·8	07·4	202	201·7	10·6	262	261·6	13·7
23	23·0	01·2	83	82·9	04·3	143	142·8	07·5	203	202·7	10·6	263	262·6	13·8
24	24·0	01·3	84	83·9	04·4	144	143·8	07·5	204	203·7	10·7	264	263·6	13·8
25	25·0	01·3	85	84·9	04·4	145	144·8	07·6	205	204·7	10·7	265	264·6	13·9
26	26·0	01·4	86	85·9	04·5	146	145·8	07·6	206	205·7	10·8	266	265·6	13·9
27	27·0	01·4	87	86·9	04·6	147	146·8	07·7	207	206·7	10·8	267	266·6	14·0
28	28·0	01·5	88	87·9	04·6	148	147·8	07·7	208	207·7	10·9	268	267·6	14·0
29	29·0	01·5	89	88·9	04·7	149	148·8	07·8	209	208·7	10·9	269	268·6	14·1
30	30·0	01·6	90	89·9	04·7	150	149·8	07·9	210	209·7	11·0	270	269·6	14·1
31	31·0	01·6	91	90·9	04·8	151	150·8	07·9	211	210·7	11·0	271	270·6	14·2
32	32·0	01·7	92	91·9	04·8	152	151·8	08·0	212	211·7	11·1	272	271·6	14·2
33	33·0	01·7	93	92·9	04·9	153	152·8	08·0	213	212·7	11·1	273	272·6	14·3
34	34·0	01·8	94	93·9	04·9	154	153·8	08·1	214	213·7	11·2	274	273·6	14·3
35	35·0	01·8	95	94·9	05·0	155	154·8	08·1	215	214·7	11·3	275	274·6	14·4
36	36·0	01·9	96	95·9	05·0	156	155·8	08·2	216	215·7	11·3	276	275·6	14·4
37	36·9	01·9	97	96·9	05·1	157	156·8	08·2	217	216·7	11·4	277	276·6	14·5
38	37·9	02·0	98	97·9	05·1	158	157·8	08·3	218	217·7	11·4	278	277·6	14·5
39	38·9	02·0	99	98·9	05·2	159	158·8	08·3	219	218·7	11·5	279	278·6	14·6
40	39·9	02·1	100	99·9	05·2	160	159·8	08·4	220	219·7	11·5	280	279·6	14·7
41	40·9	02·1	101	100·9	05·3	161	160·8	08·4	221	220·7	11·6	281	280·6	14·7
42	41·9	02·2	102	101·9	05·3	162	161·8	08·5	222	221·7	11·6	282	281·6	14·8
43	42·9	02·3	103	102·9	05·4	163	162·8	08·5	223	222·7	11·7	283	282·6	14·8
44	43·9	02·3	104	103·9	05·4	164	163·8	08·6	224	223·7	11·7	284	283·6	14·9
45	44·9	02·4	105	104·9	05·5	165	164·8	08·6	225	224·7	11·8	285	284·6	14·9
46	45·9	02·4	106	105·9	05·5	166	165·8	08·7	226	225·7	11·8	286	285·6	15·0
47	46·9	02·5	107	106·9	05·6	167	166·8	08·7	227	226·7	11·9	287	286·6	15·0
48	47·9	02·5	108	107·9	05·7	168	167·8	08·8	228	227·7	11·9	288	287·6	15·1
49	48·9	02·6	109	108·9	05·7	169	168·8	08·8	229	228·7	12·0	289	288·6	15·1
50	49·9	02·6	110	109·8	05·8	170	169·8	08·9	230	229·7	12·0	290	289·6	15·2
51	50·9	02·7	111	110·8	05·8	171	170·8	08·9	231	230·7	12·1	291	290·6	15·2
52	51·9	02·7	112	111·8	05·9	172	171·8	09·0	232	231·7	12·1	292	291·6	15·3
53	52·9	02·8	113	112·8	05·9	173	172·8	09·1	233	232·7	12·2	293	292·6	15·3
54	53·9	02·8	114	113·8	06·0	174	173·8	09·1	234	233·7	12·2	294	293·6	15·4
55	54·9	02·9	115	114·8	06·0	175	174·8	09·2	235	234·7	12·3	295	294·6	15·4
56	55·9	02·9	116	115·8	06·1	176	175·8	09·2	236	235·7	12·4	296	295·6	15·5
57	56·9	03·0	117	116·8	06·1	177	176·8	09·3	237	236·7	12·4	297	296·6	15·5
58	57·9	03·0	118	117·8	06·2	178	177·8	09·3	238	237·7	12·5	298	297·6	15·6
59	58·9	03·1	119	118·8	06·2	179	178·8	09·4	239	238·7	12·5	299	298·6	15·6
60	59·9	03·1	120	119·8	06·3	180	179·8	09·4	240	239·7	12·6	300	299·6	15·7
Dist	Dep.	D.Lat	Dist	Dep.	D.Lat	Dist	Dep.	D.Lat	Dist	Dep.	D.Lat	Dist	Dep.	D.Lat

D Lon | Dep. | D Lon | Dep. | D Lon | Dep. | D Lon | Dep. | D Lon | Dep.

273° | 267°   **87 DEGREES.**   ↑ 087° | 093°   5h 48m

TRAVERSE TABLE
3 DEGREES.

357° / 183° 003° / 177° 0h 12m 3°

D Lon Dep.			D Lon Dep.			D Lon Dep.			D Lon Dep.			D Lon Dep.		
Dist	D.Lat	Dep.	Dist	D.Lat	Dep.	Dist	D.Lat	Dep.	Dist	D.Lat	Dep.	Dist	D.Lat	Dep.
301	300·6	15·8	361	360·5	18·9	421	420·4	22·0	481	480·3	25·2	541	540·3	28·3
302	301·6	15·8	362	361·5	18·9	422	421·4	22·1	482	481·3	25·2	542	541·3	28·4
303	302·6	15·9	363	362·5	19·0	423	422·4	22·1	483	482·3	25·3	543	542·3	28·4
304	303·6	15·9	364	363·5	19·1	424	423·4	22·2	484	483·3	25·3	544	543·3	28·5
305	304·6	16·0	365	364·5	19·1	425	424·4	22·2	485	484·3	25·4	545	544·3	28·5
306	305·6	16·0	366	365·5	19·2	426	425·4	22·3	486	485·3	25·4	546	545·3	28·6
307	306·6	16·1	367	366·5	19·2	427	426·4	22·3	487	486·3	25·5	547	546·3	28·6
308	307·6	16·1	368	367·5	19·3	428	427·4	22·4	488	487·3	25·5	548	547·2	28·7
309	308·6	16·2	369	368·5	19·3	429	428·4	22·5	489	488·3	25·6	549	548·2	28·7
310	309·6	16·2	370	369·5	19·4	430	429·4	22·5	490	489·3	25·6	550	549·2	28·8
311	310·6	16·3	371	370·5	19·4	431	430·4	22·6	491	490·3	25·7	551	550·2	28·8
312	311·6	16·3	372	371·5	19·5	432	431·4	22·6	492	491·3	25·7	552	551·2	28·9
313	312·6	16·4	373	372·5	19·5	433	432·4	22·7	493	492·3	25·8	553	552·2	28·9
314	313·6	16·4	374	373·5	19·6	434	433·4	22·7	494	493·3	25·9	554	553·2	29·0
315	314·6	16·5	375	374·5	19·6	435	434·4	22·8	495	494·3	25·9	555	554·2	29·0
316	315·6	16·5	376	375·5	19·7	436	435·4	22·8	496	495·3	26·0	556	555·2	29·1
317	316·6	16·6	377	376·5	19·7	437	436·4	22·9	497	496·3	26·0	557	556·2	29·2
318	317·6	16·6	378	377·5	19·8	438	437 4	22·9	498	497·3	26·1	558	557·2	29·2
319	318·6	16·7	379	378·5	19·8	439	438·4	23·0	499	498·3	26·1	559	558·2	29·3
320	319·6	16·7	380	379·5	19·9	440	439·4	23·0	500	499·3	26·2	560	559·2	29·3
321	320·6	16·8	381	380·5	19·9	441	440·4	23·1	501	500·3	26·2	561	560·2	29·4
322	321·6	16·9	382	381·5	20·0	442	441·4	23·1	502	501·3	26·3	562	561·2	29·4
323	322·6	16·9	383	382·5	20·0	443	442·4	23·2	503	502·3	26·3	563	562·2	29·5
324	323·6	17·0	384	383·5	20·1	444	443·4	23·2	504	503·3	26·4	564	563·2	29·5
325	324·6	17·0	385	384·5	20·1	445	444·4	23·3	505	504·3	26·4	565	564·2	29·6
326	325·6	17·1	386	385·5	20·2	446	445·4	23·3	506	505·3	26·5	566	565·2	29·6
327	326·6	17·1	387	386·5	20·3	447	446·4	23·4	507	506·3	26·5	567	566·2	29·7
328	327·6	17·2	388	387·5	20·3	448	447·4	23·4	508	507·3	26·6	568	567·2	29·7
329	328·5	17·2	389	388·5	20·4	449	448·4	23·5	509	508·3	26·6	569	568·2	29·8
330	329·5	17·3	390	389·5	20·4	450	449·4	23·6	510	509·3	26·7	570	569·2	29·8
331	330·5	17·3	391	390·5	20·5	451	450·4	23·6	511	510·3	26·7	571	570·2	29·9
332	331·5	17·4	392	391·5	20·5	452	451·4	23·7	512	511·3	26·8	572	571·2	29·9
333	332·5	17·4	393	392·5	20·6	453	452·4	23·7	513	512·3	26·8	573	572·2	30·0
334	333·5	17·5	394	393·5	20·6	454	453·4	23·8	514	513·3	26·9	574	573·2	30·0
335	334·5	17·5	395	394·5	20·7	455	454·4	23·8	515	514·3	27·0	575	574·2	30·1
336	335·5	17·6	396	395·5	20·7	456	455·4	23·9	516	515·3	27·0	576	575·2	30·1
337	336·5	17·6	397	396·5	20·8	457	456·4	23·9	517	516·3	27·1	577	576·2	30·2
338	337·5	17·7	398	397·5	20·8	458	457·4	24·0	518	517·3	27·1	578	577·2	30·3
339	338·5	17·7	399	398·5	20·9	459	458·4	24·0	519	518·3	27·2	579	578·2	30·3
340	339·5	17·8	400	399·5	20·9	460	459·4	24·1	520	519·3	27·2	580	579·2	30·4
341	340·5	17·8	401	400·5	21·0	461	460·4	24·1	521	520·3	27·3	581	580·2	30·4
342	341·5	17·9	402	401·4	21·0	462	461·4	24·2	522	521·3	27·3	582	581·2	30·5
343	342·5	18·0	403	402·4	21·1	463	462·4	24·2	523	522·3	27·4	583	582·2	30·5
344	343·5	18·0	404	403·4	21·1	464	463·4	24·3	524	523·3	27·4	584	583·2	30·6
345	344·5	18·1	405	404·4	21·2	465	464·4	24·3	525	524·3	27·5	585	584·2	30·6
346	345·5	18·1	406	405·4	21·2	466	465·4	24·4	526	525·3	27·5	586	585·2	30·7
347	346·5	18·2	407	406·4	21·3	467	466·4	24·4	527	526·3	27·6	587	586·2	30·7
348	347·5	18·2	408	407·4	21·4	468	467·4	24·5	528	527·3	27·6	588	587·2	30·8
349	348·5	18·3	409	408·4	21·4	469	468·4	24·5	529	528·3	27·7	589	588·2	30·8
350	349·5	18·3	410	409·4	21·5	470	469·4	24·6	530	529·3	27·7	590	589·2	30·9
351	350·5	18·4	411	410·4	21·5	471	470·4	24·7	531	530·3	27·8	591	590·2	30·9
352	351·5	18·4	412	411·4	21·6	472	471·4	24·7	532	531·3	27·8	592	591·2	31·0
353	352·5	18·5	413	412·4	21·6	473	472·4	24·8	533	532·3	27·9	593	592·2	31·0
354	353·5	18·5	414	413·4	21·7	474	473·4	24·8	534	533·3	27·9	594	593·2	31·1
355	354·5	18·6	415	414·4	21·7	475	474·3	24·9	535	534·3	28·0	595	594·2	31·1
356	355·5	18·6	416	415·4	21·8	476	475·3	24·9	536	535·3	28·1	596	595·2	31·2
357	356·5	18·7	417	416·4	21·8	477	476·3	25·0	537	536·3	28·1	597	596·2	31·3
358	357·5	18·7	418	417·4	21·9	478	477·3	25·0	538	537·3	28·2	598	597·2	31·3
359	358·5	18·8	419	418·4	21·9	479	478·3	25·1	539	538·3	28·2	599	598·2	31·3
360	359·5	18·8	420	419·4	22·0	480	479·3	25·1	540	539·3	28·3	600	599·2	31·4
Dist	Dep.	D.Lat	Dist	Dep.	D.Lat	Dist	Dep.	D.Lat	Dist	Dep.	D.Lat	Dist	Dep.	D.Lat
D Lon		Dep.	D Lon		Dep.	D Lon		Dep.	D Lon		Dep.	D Lon		Dep.

273° / 267° 87 DEGREES. 087° / 093° 5h 48m **87°**

TRAVERSE TABLE

4 DEGREES.

Dist	D.Lat	Dep.	Dist	D.Lat	Dep.	Dist	D.Lat	Dep.	Dist	D.Lat	Dep.	Dist	D.Lat	Dep.
1	01·0	00·1	61	60·9	04·3	121	120·7	08·4	181	180·6	12·6	241	240·4	16·8
2	02·0	00·1	62	61·8	04·3	122	121·7	08·5	182	181·6	12·7	242	241·4	16·9
3	03·0	00·2	63	62·8	04·4	123	122·7	08·6	183	182·6	12·8	243	242·4	17·0
4	04·0	00·3	64	63·8	04·5	124	123·7	08·6	184	183·6	12·8	244	243·4	17·0
5	05·0	00·3	65	64·8	04·5	125	124·7	08·7	185	184·5	12·9	245	244·4	17·1
6	06·0	00·4	66	65·8	04·6	126	125·7	08·8	186	185·5	13·0	246	245·4	17·2
7	07·0	00·5	67	66·8	04·7	127	126·7	08·9	187	186·5	13·0	247	246·4	17·2
8	08·0	00·6	68	67·8	04·7	128	127·7	08·9	188	187·5	13·1	248	247·4	17·3
9	09·0	00·6	69	68·8	04·8	129	128·7	09·0	189	188·5	13·2	249	248·4	17·4
10	10·0	00·7	70	69·8	04·9	130	129·7	09·1	190	189·5	13·3	250	249·4	17·4
11	11·0	00·8	71	70·8	05·0	131	130·7	09·1	191	190·5	13·3	251	250·4	17·5
12	12·0	00·8	72	71·8	05·0	132	131·7	09·2	192	191·5	13·4	252	251·4	17·6
13	13·0	00·9	73	72·8	05·1	133	132·7	09·3	193	192·5	13·5	253	252·4	17·6
14	14·0	01·0	74	73·8	05·2	134	133·7	09·3	194	193·5	13·5	254	253·4	17·7
15	15·0	01·0	75	74·8	05·2	135	134·7	09·4	195	194·5	13·6	255	254·4	17·8
16	16·0	01·1	76	75·8	05·3	136	135·7	09·5	196	195·5	13·7	256	255·4	17·9
17	17·0	01·2	77	76·8	05·4	137	136·7	09·6	197	196·5	13·7	257	256·4	17·9
18	18·0	01·3	78	77·8	05·4	138	137·7	09·6	198	197·5	13·8	258	257·4	18·0
19	19·0	01·3	79	78·8	05·5	139	138·7	09·7	199	198·5	13·9	259	258·4	18·1
20	20·0	01·4	80	79·8	05·6	140	139·7	09·8	200	199·5	14·0	260	259·4	18·1
21	20·9	01·5	81	80·8	05·7	141	140·7	09·8	201	200·5	14·0	261	260·4	18·2
22	21·9	01·5	82	81·8	05·7	142	141·7	09·9	202	201·5	14·1	262	261·4	18·3
23	22·9	01·6	83	82·8	05·8	143	142·7	10·0	203	202·5	14·2	263	262·4	18·3
24	23·9	01·7	84	83·8	05·9	144	143·6	10·0	204	203·5	14·2	264	263·4	18·4
25	24·9	01·7	85	84·8	05·9	145	144·6	10·1	205	204·5	14·3	265	264·4	18·5
26	25·9	01·8	86	85·8	06·0	146	145·6	10·2	206	205·5	14·4	266	265·4	18·6
27	26·9	01·9	87	86·8	06·1	147	146·6	10·3	207	206·5	14·4	267	266·3	18·6
28	27·9	02·0	88	87·8	06·1	148	147·6	10·3	208	207·5	14·5	268	267·3	18·7
29	28·9	02·0	89	88·8	06·2	149	148·6	10·4	209	208·5	14·6	269	268·3	18·8
30	29·9	02·1	90	89·8	06·3	150	149·6	10·5	210	209·5	14·6	270	269·3	18·8
31	30·9	02·2	91	90·8	06·3	151	150·6	10·5	211	210·5	14·7	271	270·3	18·9
32	31·9	02·2	92	91·8	06·4	152	151·6	10·6	212	211·5	14·8	272	271·3	19·0
33	32·9	02·3	93	92·8	06·5	153	152·6	10·7	213	212·5	14·9	273	272·3	19·0
34	33·9	02·4	94	93·8	06·6	154	153·6	10·7	214	213·5	14·9	274	273·3	19·1
35	34·9	02·4	95	94·8	06·6	155	154·6	10·8	215	214·5	15·0	275	274·3	19·2
36	35·9	02·5	96	95·8	06·7	156	155·6	10·9	216	215·5	15·1	276	275·3	19·3
37	36·9	02·6	97	96·8	06·8	157	156·6	11·0	217	216·5	15·1	277	276·3	19·3
38	37·9	02·7	98	97·8	06·8	158	157·6	11·0	218	217·5	15·2	278	277·3	19·4
39	38·9	02·7	99	98·8	06·9	159	158·6	11·1	219	218·5	15·3	279	278·3	19·5
40	39·9	02·8	100	99·8	07·0	160	159·6	11·2	220	219·5	15·3	280	279·3	19·5
41	40·9	02·9	101	100·8	07·0	161	160·6	11·2	221	220·5	15·4	281	280·3	19·6
42	41·9	02·9	102	101·8	07·1	162	161·6	11·3	222	221·5	15·5	282	281·3	19·7
43	42·9	03·0	103	102·7	07·2	163	162·6	11·4	223	222·5	15·6	283	282·3	19·7
44	43·9	03·1	104	103·7	07·3	164	163·6	11·4	224	223·5	15·6	284	283·3	19·8
45	44·9	03·1	105	104·7	07·3	165	164·6	11·5	225	224·5	15·7	285	284·3	19·9
46	45·9	03·2	106	105·7	07·4	166	165·6	11·6	226	225·4	15·8	286	285·3	20·0
47	46·9	03·3	107	106·7	07·5	167	166·6	11·6	227	226·4	15·8	287	286·3	20·0
48	47·9	03·3	108	107·7	07·5	168	167·6	11·7	228	227·4	15·9	288	287·3	20·1
49	48·9	03·4	109	108·7	07·6	169	168·6	11·8	229	228·4	16·0	289	288·3	20·2
50	49·9	03·5	110	109·7	07·7	170	169·6	11·9	230	229·4	16·0	290	289·3	20·2
51	50·9	03·6	111	110·7	07·7	171	170·6	11·9	231	230·4	16·1	291	290·3	20·3
52	51·9	03·6	112	111·7	07·8	172	171·6	12·0	232	231·4	16·2	292	291·3	20·4
53	52·9	03·7	113	112·7	07·9	173	172·6	12·1	233	232·4	16·3	293	292·3	20·4
54	53·9	03·8	114	113·7	08·0	174	173·6	12·1	234	233·4	16·3	294	293·3	20·5
55	54·9	03·8	115	114·7	08·0	175	174·6	12·2	235	234·4	16·4	295	294·3	20·6
56	55·9	03·9	116	115·7	08·1	176	175·6	12·3	236	235·4	16·5	296	295·3	20·6
57	56·9	04·0	117	116·7	08·2	177	176·6	12·3	237	236·4	16·5	297	296·3	20·7
58	57·9	04·0	118	117·7	08·2	178	177·6	12·4	238	237·4	16·6	298	297·3	20·8
59	58·9	04·1	119	118·7	08·3	179	178·6	12·5	239	238·4	16·6	299	298·3	20·9
60	59·9	04·2	120	119·7	08·4	180	179·6	12·6	240	239·4	16·7	300	299·3	20·9

Dist	Dep.	D.Lat	Dist	Dep.	D.Lat	Dist	Dep.	D.Lat	Dist	Dep.	D.Lat	Dist	Dep.	D.Lat

D Lon ... Dep. (repeated across groups)

TRAVERSE TABLE
4 DEGREES.

356° / 184°   004° / 176°   0h 16m   **4°**

D Lon	Dep.		D Lon	Dep.		D Lon	Dep.		D Lon	Dep.		D Lon	Dep.	
Dist	D.Lat	Dep.	Dist	D.Lat	Dep.	Dist	D.Lat	Dep.	Dist	D.Lat	Dep.	Dist	D.Lat	Dep.
301	300.3	21.0	361	360.1	25.2	421	420.0	29.4	481	479.8	33.6	541	539.7	37.7
302	301.3	21.1	362	361.1	25.3	422	421.0	29.4	482	480.8	33.6	542	540.7	37.8
303	302.3	21.1	363	362.1	25.3	423	422.0	29.5	483	481.8	33.7	543	541.7	37.9
304	303.3	21.2	364	363.1	25.4	424	423.0	29.6	484	482.8	33.8	544	542.7	37.9
305	304.3	21.3	365	364.1	25.5	425	424.0	29.6	485	483.8	33.8	545	543.7	38.0
306	305.3	21.3	366	365.1	25.5	426	425.0	29.7	486	484.8	33.9	546	544.7	38.1
307	306.3	21.4	367	366.1	25.6	427	426.0	29.8	487	485.8	34.0	547	545.7	38.2
308	307.2	21.5	368	367.1	25.7	428	427.0	29.9	488	486.8	34.0	548	546.7	38.2
309	308.2	21.6	369	368.1	25.7	429	428.0	29.9	489	487.8	34.1	549	547.7	38.3
310	309.2	21.6	370	369.1	25.8	430	429.0	30.0	490	488.8	34.2	550	548.7	38.4
311	310.2	21.7	371	370.1	25.9	431	430.0	30.1	491	489.8	34.3	551	549.7	38.4
312	311.2	21.8	372	371.1	25.9	432	430.9	30.1	492	490.8	34.3	552	550.7	38.5
313	312.2	21.8	373	372.1	26.0	433	431.9	30.2	493	491.8	34.4	553	551.7	38.6
314	313.2	21.9	374	373.1	26.1	434	432.9	30.3	494	492.8	34.5	554	552.7	38.6
315	314.2	22.0	375	374.1	26.2	435	433.9	30.3	495	493.8	34.5	555	553.6	38.7
316	315.2	22.0	376	375.1	26.2	436	434.9	30.4	496	494.8	34.6	556	554.6	38.8
317	316.2	22.1	377	376.1	26.3	437	435.9	30.5	497	495.8	34.7	557	555.6	38.9
318	317.2	22.2	378	377.1	26.4	438	436.9	30.6	498	496.8	34.7	558	556.6	38.9
319	318.2	22.3	379	378.1	26.4	439	437.9	30.6	499	497.8	34.8	559	557.6	39.0
320	319.2	22.3	380	379.1	26.5	440	438.9	30.7	500	498.8	34.9	560	558.6	39.0
321	320.2	22.4	381	380.1	26.6	441	439.9	30.8	501	499.8	34.9	561	559.6	39.1
322	321.2	22.5	382	381.1	26.6	442	440.9	30.8	502	500.8	35.0	562	560.6	39.2
323	322.2	22.5	383	382.1	26.7	443	441.9	30.9	503	501.8	35.1	563	561.6	39.3
324	323.2	22.6	384	383.1	26.8	444	442.9	31.0	504	502.8	35.2	564	562.6	39.3
325	324.2	22.7	385	384.1	26.9	445	443.9	31.0	505	503.8	35.2	565	563.6	39.4
326	325.2	22.7	386	385.1	26.9	446	444.9	31.1	506	504.8	35.3	566	564.6	39.5
327	326.2	22.8	387	386.1	27.0	447	445.9	31.2	507	505.8	35.4	567	565.6	39.5
328	327.2	22.9	388	387.1	27.1	448	446.9	31.3	508	506.8	35.4	568	566.6	39.6
329	328.2	22.9	389	388.1	27.1	449	447.9	31.3	509	507.8	35.5	569	567.6	39.7
330	329.2	23.0	390	389.0	27.2	450	448.9	31.4	510	508.8	35.6	570	568.6	39.8
331	330.2	23.1	391	390.0	27.3	451	449.9	31.5	511	509.8	35.6	571	569.6	39.8
332	331.2	23.2	392	391.0	27.3	452	450.9	31.5	512	510.8	35.7	572	570.6	39.9
333	332.2	23.2	393	392.0	27.4	453	451.9	31.6	513	511.8	35.8	573	571.6	40.0
334	333.2	23.3	394	393.0	27.5	454	452.9	31.7	514	512.7	35.9	574	572.6	40.0
335	334.2	23.4	395	394.0	27.6	455	453.9	31.7	515	513.7	35.9	575	573.6	40.1
336	335.2	23.4	396	395.0	27.6	456	454.9	31.8	516	514.7	36.0	576	574.6	40.2
337	336.2	23.5	397	396.0	27.7	457	455.9	31.9	517	515.7	36.1	577	575.6	40.2
338	337.2	23.6	398	397.0	27.8	458	456.9	31.9	518	516.7	36.1	578	576.6	40.3
339	338.2	23.6	399	398.0	27.8	459	457.9	32.0	519	517.7	36.2	579	577.6	40.4
340	339.2	23.7	400	399.0	27.9	460	458.9	32.1	520	518.7	36.3	580	578.6	40.5
341	340.2	23.8	401	400.0	28.0	461	459.9	32.2	521	519.7	36.3	581	579.6	40.5
342	341.2	23.9	402	401.0	28.0	462	460.9	32.2	522	520.7	36.4	582	580.6	40.6
343	342.2	23.9	403	402.0	28.1	463	461.9	32.3	523	521.7	36.5	583	581.6	40.7
344	343.2	24.0	404	403.0	28.2	464	462.9	32.4	524	522.7	36.6	584	582.6	40.7
345	344.2	24.1	405	404.0	28.3	465	463.9	32.4	525	523.7	36.6	585	583.6	40.8
346	345.2	24.1	406	405.0	28.3	466	464.9	32.5	526	524.7	36.7	586	584.6	40.9
347	346.2	24.2	407	406.0	28.4	467	465.9	32.6	527	525.7	36.8	587	585.6	40.9
348	347.2	24.3	408	407.0	28.5	468	466.9	32.6	528	526.7	36.8	588	586.6	41.0
349	348.1	24.3	409	408.0	28.5	469	467.9	32.7	529	527.7	36.9	589	587.6	41.1
350	349.1	24.4	410	409.0	28.6	470	468.9	32.8	530	528.7	37.0	590	588.6	41.2
351	350.1	24.5	411	410.0	28.7	471	469.9	32.9	531	529.7	37.0	591	589.6	41.2
352	351.1	24.6	412	411.0	28.7	472	470.9	32.9	532	530.7	37.1	592	590.6	41.3
353	352.1	24.6	413	412.0	28.8	473	471.8	33.0	533	531.7	37.2	593	591.6	41.4
354	353.1	24.7	414	413.0	28.9	474	472.8	33.1	534	532.7	37.2	594	592.6	41.4
355	354.1	24.8	415	414.0	28.9	475	473.8	33.1	535	533.7	37.3	595	593.6	41.5
356	355.1	24.8	416	415.0	29.0	476	474.8	33.2	536	534.7	37.4	596	594.5	41.6
357	356.1	24.9	417	416.0	29.1	477	475.8	33.3	537	535.7	37.5	597	595.5	41.6
358	357.1	25.0	418	417.0	29.2	478	476.8	33.3	538	536.7	37.5	598	596.5	41.7
359	358.1	25.0	419	418.0	29.2	479	477.8	33.4	539	537.7	37.6	599	597.5	41.8
360	359.1	25.1	420	419.0	29.3	480	478.8	33.5	540	538.7	37.7	600	598.5	41.9
Dist	Dep.	D.Lat	Dist	Dep.	D.Lat	Dist	Dep.	D.Lat	Dist	Dep.	D.Lat	Dist	Dep.	D.Lat
D Lon		Dep.	D Lon		Dep.	D Lon		Dep.	D Lon		Dep.	D Lon		Dep.

86°

274° / 266°   **86 DEGREES.**   088° / 094°   5h 44m

TRAVERSE TABLE
5 DEGREES.

5° 355° / 185°   005° / 175°   0h 20m

Dist	D.Lat	Dep.	Dist	D.Lat	Dep.	Dist	D.Lat	Dep	Dist	D.Lat	Dep.	Dist	D.Lat	Dep.
1	01.0	00.1	61	60.8	05.3	121	120.5	10.5	181	180.3	15.8	241	240.1	21.0
2	02.0	00.2	62	61.8	05.4	122	121.5	10.6	182	181.3	15.9	242	241.1	21.1
3	03.0	00.3	63	62.8	05.5	123	122.5	10.7	183	182.3	15.9	243	242.1	21.2
4	04.0	00.3	64	63.8	05.6	124	123.5	10.8	184	183.3	16.0	244	243.1	21.3
5	05.0	00.4	65	64.8	05.7	125	124.5	10.9	185	184.3	16.1	245	244.1	21.4
6	06.0	00.5	66	65.7	05.8	126	125.5	11.0	186	185.3	16.2	246	245.1	21.4
7	07.0	00.6	67	66.7	05.8	127	126.5	11.1	187	186.3	16.3	247	246.1	21.5
8	08.0	00.7	68	67.7	05.9	128	127.5	11.2	188	187.3	16.4	248	247.1	21.6
9	09.0	00.8	69	68.7	06.0	129	128.5	11.2	189	188.3	16.5	249	248.1	21.7
10	10.0	00.9	70	69.7	06.1	130	129.5	11.3	190	189.3	16.6	250	249.0	21.8
11	11.0	01.0	71	70.7	06.2	131	130.5	11.4	191	190.3	16.6	251	250.0	21.9
12	12.0	01.0	72	71.7	06.3	132	131.5	11.5	192	191.3	16.7	252	251.0	22.0
13	13.0	01.1	73	72.7	06.4	133	132.5	11.6	193	192.3	16.8	253	252.0	22.1
14	13.9	01.2	74	73.7	06.4	134	133.5	11.7	194	193.3	16.9	254	253.0	22.1
15	14.9	01.3	75	74.7	06.5	135	134.5	11.8	195	194.3	17.0	255	254.0	22.2
16	15.9	01.4	76	75.7	06.6	136	135.5	11.9	196	195.3	17.1	256	255.0	22.3
17	16.9	01.5	77	76.7	06.7	137	136.5	11.9	197	196.3	17.2	257	256.0	22.4
18	17.9	01.6	78	77.7	06.8	138	137.5	12.0	198	197.2	17.3	258	257.0	22.5
19	18.9	01.7	79	78.7	06.9	139	138.5	12.1	199	198.2	17.3	259	258.0	22.6
20	19.9	01.7	80	79.7	07.0	140	139.5	12.2	200	199.2	17.4	260	259.0	22.7
21	20.9	01.8	81	80.7	07.1	141	140.5	12.3	201	200.2	17.5	261	260.0	22.7
22	21.9	01.9	82	81.7	07.1	142	141.5	12.4	202	201.2	17.6	262	261.0	22.8
23	22.9	02.0	83	82.7	07.2	143	142.5	12.5	203	202.2	17.7	263	262.0	22.9
24	23.9	02.1	84	83.7	07.3	144	143.5	12.6	204	203.2	17.8	264	263.0	23.0
25	24.9	02.2	85	84.7	07.4	145	144.4	12.6	205	204.2	17.9	265	264.0	23.1
26	25.9	02.3	86	85.7	07.5	146	145.4	12.7	206	205.2	18.0	266	265.0	23.2
27	26.9	02.4	87	86.7	07.6	147	146.4	12.8	207	206.2	18.0	267	266.0	23.3
28	27.9	02.4	88	87.7	07.7	148	147.4	12.9	208	207.2	18.1	268	267.0	23.4
29	28.9	02.5	89	88.7	07.8	149	148.4	13.0	209	208.2	18.2	269	268.0	23.4
30	29.9	02.6	90	89.7	07.8	150	149.4	13.1	210	209.2	18.3	270	269.0	23.5
31	30.9	02.7	91	90.7	07.9	151	150.4	13.2	211	210.2	18.4	271	270.0	23.6
32	31.9	02.8	92	91.6	08.0	152	151.4	13.2	212	211.2	18.5	272	271.0	23.7
33	32.9	02.9	93	92.6	08.1	153	152.4	13.3	213	212.2	18.6	273	272.0	23.8
34	33.9	03.0	94	93.6	08.2	154	153.4	13.4	214	213.2	18.7	274	273.0	23.9
35	34.9	03.1	95	94.6	08.3	155	154.4	13.5	215	214.2	18.7	275	274.0	24.0
36	35.9	03.1	96	95.6	08.4	156	155.4	13.6	216	215.2	18.8	276	274.9	24.1
37	36.9	03.2	97	96.6	08.5	157	156.4	13.7	217	216.2	18.9	277	275.9	24.1
38	37.9	03.3	98	97.6	08.5	158	157.4	13.8	218	217.2	19.0	278	276.9	24.2
39	38.9	03.4	99	98.6	08.6	159	158.4	13.9	219	218.2	19.1	279	277.9	24.3
40	39.8	03.5	100	99.6	08.7	160	159.4	13.9	220	219.2	19.2	280	278.9	24.4
41	40.8	03.6	101	100.6	08.8	161	160.4	14.0	221	220.2	19.3	281	279.9	24.5
42	41.8	03.7	102	101.6	08.9	162	161.4	14.1	222	221.2	19.3	282	280.9	24.6
43	42.8	03.7	103	102.6	09.0	163	162.4	14.2	223	222.2	19.4	283	281.9	24.7
44	43.8	03.8	104	103.6	09.1	164	163.4	14.3	224	223.1	19.5	284	282.9	24.8
45	44.8	03.9	105	104.6	09.2	165	164.4	14.4	225	224.1	19.6	285	283.9	24.8
46	45.8	04.0	106	105.6	09.2	166	165.4	14.5	226	225.1	19.7	286	284.9	24.9
47	46.8	04.1	107	106.6	09.3	167	166.4	14.6	227	226.1	19.8	287	285.9	25.0
48	47.8	04.2	108	107.6	09.4	168	167.4	14.6	228	227.1	19.9	288	286.9	25.1
49	48.8	04.3	109	108.6	09.5	169	168.4	14.7	229	228.1	20.0	289	287.9	25.2
50	49.8	04.4	110	109.6	09.6	170	169.4	14.8	230	229.1	20.0	290	288.9	25.3
51	50.8	04.4	111	110.6	09.7	171	170.3	14.9	231	230.1	20.1	291	289.9	25.4
52	51.8	04.5	112	111.6	09.8	172	171.3	15.0	232	231.1	20.2	292	290.9	25.4
53	52.8	04.6	113	112.6	09.8	173	172.3	15.1	233	232.1	20.3	293	291.9	25.5
54	53.8	04.7	114	113.6	09.9	174	173.3	15.2	234	233.1	20.4	294	292.9	25.6
55	54.8	04.8	115	114.6	10.0	175	174.3	15.3	235	234.1	20.5	295	293.9	25.7
56	55.8	04.9	116	115.6	10.1	176	175.3	15.3	236	235.1	20.6	296	294.9	25.8
57	56.8	05.0	117	116.6	10.2	177	176.3	15.4	237	236.1	20.7	297	295.9	25.9
58	57.8	05.1	118	117.6	10.3	178	177.3	15.5	238	237.1	20.7	298	296.9	26.0
59	58.8	05.1	119	118.5	10.4	179	178.3	15.6	239	238.1	20.8	299	297.9	26.1
60	59.8	05.2	120	119.5	10.5	180	179.3	15.7	240	239.1	20.9	300	298.9	26.1

Dist	Dep.	D.Lat	Dist	Dep.	D.Lat	Dist	Dep.	D.Lat	Dist	Dep.	D.Lat	Dist	Dep.	D.Lat

85°

275° / 265°   **85 DEGREES.**   085° / 095°   5h 40m

TRAVERSE TABLE
5 DEGREES.

Dist	D.Lat	Dep.	Dist	D.Lat	Dep.	Dist	D.Lat	Dep.	Dist	D.Lat	Dep.	Dist	D.Lat	Dep.
	D Lon / Dep.			D Lon / Dep.			D Lon / Dep.			D Lon / Dep.			D Lon / Dep.	
301	299·9	26·2	361	359·6	31·5	421	419·4	36·7	481	479·2	41·9	541	538·9	47·2
302	300·9	26·3	362	360·6	31·6	422	420·4	36·8	482	480·2	42·0	542	539·9	47·2
303	301·8	26·4	363	361·6	31·6	423	421·4	36·9	483	481·2	42·1	543	540·9	47·3
304	302·8	26·5	364	362·6	31·7	424	422·4	37·0	484	482·2	42·2	544	541·9	47·4
305	303·8	26·6	365	363·6	31·8	425	423·4	37·0	485	483·2	42·3	545	542·9	47·5
306	304·8	26·7	366	364·6	31·9	426	424·4	37·1	486	484·2	42·4	546	543·9	47·6
307	305·8	26·8	367	365·6	32·0	427	425·4	37·2	487	485·1	42·4	547	544·9	47·7
308	306·8	26·8	368	366·6	32·1	428	426·4	37·3	488	486·1	42·5	548	545·9	47·8
309	307·8	26·9	369	367·6	32·2	429	427·4	37·4	489	487·1	42·6	549	546·9	47·8
310	308·8	27·0	370	368·6	32·2	430	428·4	37·5	490	488·1	42·7	550	547·9	47·9
311	309·8	27·1	371	369·6	32·3	431	429·4	37·6	491	489·1	42·8	551	548·9	48·0
312	310·8	27·2	372	370·6	32·4	432	430·4	37·7	492	490·1	42·9	552	549·9	48·1
313	311·8	27·3	373	371·6	32·5	433	431·4	37·7	493	491·1	43·0	553	550·9	48·2
314	312·8	27·4	374	372·6	32·6	434	432·3	37·8	494	492·1	43·1	554	551·9	48·3
315	313·8	27·5	375	373·6	32·7	435	433·3	37·9	495	493·1	43·1	555	552·9	48·4
316	314·8	27·5	376	374·6	32·8	436	434·3	38·0	496	494·1	43·2	556	553·9	48·5
317	315·8	27·6	377	375·6	32·9	437	435·3	38·1	497	495·1	43·3	557	554·9	48·5
318	316·8	27·7	378	376·6	33·0	438	436·3	38·2	498	496·1	43·4	558	555·9	48·6
319	317·8	27·8	379	377·6	33·0	439	437·3	38·3	499	497·1	43·5	559	556·9	48·7
320	318·8	27·9	380	378·6	33·1	440	438·3	38·3	500	498·1	43·6	560	557·9	48·8
321	319·8	28·0	381	379·6	33·2	441	439·3	38·4	501	499·1	43·7	561	558·9	48·9
322	320·8	28·1	382	380·5	33·3	442	440·3	38·5	502	500·1	43·8	562	559·9	49·0
323	321·8	28·2	383	381·5	33·4	443	441·3	38·6	503	501·1	43·8	563	560·9	49·1
324	322·8	28·2	384	382·5	33·5	444	442·3	38·7	504	502·1	43·9	564	561·9	49·2
325	323·8	28·3	385	383·5	33·6	445	443·3	38·8	505	503·1	44·0	565	562·9	49·3
326	324·8	28·4	386	384·5	33·6	446	444·3	38·9	506	504·1	44·1	566	563·8	49·3
327	325·8	28·5	387	385·5	33·7	447	445·3	39·0	507	505·1	44·2	567	564·8	49·4
328	326·8	28·6	388	386·5	33·8	448	446·3	39·0	508	506·1	44·3	568	565·8	49·5
329	327·7	28·7	389	387·5	33·9	449	447·3	39·1	509	507·1	44·4	569	566·8	49·6
330	328·7	28·8	390	388·5	34·0	450	448·3	39·2	510	508·1	44·4	570	567·8	49·7
331	329·7	28·8	391	389·5	34·1	451	449·3	39·3	511	509·1	44·5	571	568·8	49·8
332	330·7	28·9	392	390·5	34·2	452	450·3	39·4	512	510·1	44·6	572	569·8	49·9
333	331·7	29·0	393	391·5	34·3	453	451·3	39·5	513	511·0	44·7	573	570·8	49·9
334	332·7	29·1	394	392·5	34·3	454	452·3	39·6	514	512·0	44·8	574	571·8	50·0
335	333·7	29·2	395	393·5	34·4	455	453·3	39·7	515	513·0	44·9	575	572·8	50·1
336	334·7	29·3	396	394·5	34·5	456	454·3	39·7	516	514·0	45·0	576	573·8	50·2
337	335·7	29·4	397	395·5	34·6	457	455·3	39·8	517	515·0	45·1	577	574·8	50·4
338	336·7	29·5	398	396·5	34·7	458	456·3	39·9	518	516·0	45·1	578	575·8	50·4
339	337·7	29·6	399	397·5	34·8	459	457·3	40·0	519	517·0	45·2	579	576·8	50·5
340	338·7	29·6	400	398·5	34·9	460	458·2	40·1	520	518·0	45·3	580	577·8	50·6
341	339·7	29·7	401	399·5	34·9	461	459·2	40·2	521	519·0	45·4	581	578·8	50·6
342	340·7	29·8	402	400·5	35·0	462	460·2	40·3	522	520·0	45·5	582	579·8	50·7
343	341·7	29·9	403	401·5	35·1	463	461·2	40·4	523	521·0	45·6	583	580·8	50·8
344	342·7	30·0	404	402·5	35·2	464	462·2	40·4	524	522·0	45·7	584	581·8	50·9
345	343·7	30·1	405	403·5	35·3	465	463·2	40·5	525	523·0	45·8	585	582·8	51·0
346	344·7	30·2	406	404·5	35·4	466	464·2	40·6	526	524·0	45·8	586	583·8	51·1
347	345·7	30·2	407	405·5	35·5	467	465·2	40·7	527	525·0	45·9	587	584·8	51·2
348	346·7	30·3	408	406·4	35·6	468	466·2	40·8	528	526·0	46·0	588	585·8	51·2
349	347·7	30·4	409	407·4	35·6	469	467·2	40·9	529	527·0	46·1	589	586·8	51·3
350	348·7	30·5	410	408·4	35·7	470	468·2	41·0	530	528·0	46·2	590	587·8	51·4
351	349·7	30·6	411	409·4	35·8	471	469·2	41·1	531	529·0	46·3	591	588·8	51·5
352	350·7	30·7	412	410·4	35·9	472	470·2	41·1	532	530·0	46·4	592	589·7	51·6
353	351·7	30·8	413	411·4	36·0	473	471·2	41·2	533	531·0	46·5	593	590·7	51·7
354	352·7	30·9	414	412·4	36·1	474	472·2	41·3	534	532·0	46·5	594	591·7	51·8
355	353·6	30·9	415	413·4	36·2	475	473·2	41·4	535	533·0	46·6	595	592·7	51·9
356	354·6	31·0	416	414·4	36·3	476	474·2	41·5	536	534·0	46·8	596	593·7	51·9
357	355·6	31·1	417	415·4	36·3	477	475·2	41·6	537	535·0	46·8	597	594·7	52·0
358	356·6	31·2	418	416·4	36·4	478	476·2	41·7	538	536·0	46·9	598	595·7	52·1
359	357·6	31·3	419	417·4	36·5	479	477·2	41·7	539	536·9	47·0	599	596·7	52·2
360	358·6	31·4	420	418·4	36·6	480	478·2	41·8	540	537·9	47·1	600	597·7	52·3
Dist	Dep.	D.Lat	Dist	Dep.	D.Lat	Dist	Dep.	D.Lat	Dist	Dep.	D.Lat	Dist	Dep.	D.Lat
D Lon		Dep.	D Lon		Dep.	D Lon		Dep.	D Lon		Dep.	D Lon		Dep.

6°

| | 354° / 186° | | **TRAVERSE TABLE** 6 DEGREES. | | 006° / 174° 0h 24m | |

Dist	D.Lat	Dep.	Dist	D.Lat	Dep.	Dist	D.Lat	Dep.	Dist	D.Lat	Dep.	Dist	D.Lat	Dep.
1	01·0	00·1	61	60·7	06·4	121	120·3	12·6	181	180·0	18·9	241	239·7	25·2
2	02·0	00·2	62	61·7	06·5	122	121·3	12·8	182	181·0	19·0	242	240·7	25·3
3	03·0	00·3	63	62·7	06·6	123	122·3	12·9	183	182·0	19·1	243	241·7	25·4
4	04·0	00·4	64	63·6	06·7	124	123·3	13·0	184	183·0	19·2	244	242·7	25·5
5	05·0	00·5	65	64·6	06·8	125	124·3	13·1	185	184·0	19·3	245	243·7	25·6
6	06·0	00·6	66	65·6	06·9	126	125·3	13·2	186	185·0	19·4	246	244·7	25·7
7	07·0	00·7	67	66·6	07·0	127	126·3	13·3	187	186·0	19·5	247	245·6	25·8
8	08·0	00·8	68	67·6	07·1	128	127·3	13·4	188	187·0	19·7	248	246·6	25·9
9	09·0	00·9	69	68·6	07·2	129	128·3	13·5	189	188·0	19·8	249	247·6	26·0
10	09·9	01·0	70	69·6	07·3	130	129·3	13·6	190	189·0	19·9	250	248·6	26·1
11	10·9	01·1	71	70·6	07·4	131	130·3	13·7	191	190·0	20·0	251	249·6	26·2
12	11·9	01·3	72	71·6	07·5	132	131·3	13·8	192	190·9	20·1	252	250·6	26·3
13	12·9	01·4	73	72·6	07·6	133	132·3	13·9	193	191·9	20·2	253	251·6	26·4
14	13·9	01·5	74	73·6	07·7	134	133·3	14·0	194	192·9	20·3	254	252·6	26·6
15	14·9	01·6	75	74·6	07·8	135	134·3	14·1	195	193·9	20·4	255	253·6	26·7
16	15·9	01·7	76	75·6	07·9	136	135·3	14·2	196	194·9	20·5	256	254·6	26·8
17	16·9	01·8	77	76·6	08·0	137	136·2	14·3	197	195·9	20·6	257	255·6	26·9
18	17·9	01·9	78	77·6	08·2	138	137·2	14·4	198	196·9	20·7	258	256·6	27·0
19	18·9	02·0	79	78·6	08·3	139	138·2	14·5	199	197·9	20·8	259	257·6	27·1
20	19·9	02·1	80	79·6	08·4	140	139·2	14·6	200	198·9	20·9	260	258·6	27·2
21	20·9	02·2	81	80·6	08·5	141	140·2	14·7	201	199·9	21·0	261	259·6	27·3
22	21·9	02·3	82	81·6	08·6	142	141·2	14·8	202	200·9	21·1	262	260·6	27·4
23	22·9	02·4	83	82·5	08·7	143	142·2	14·9	203	201·9	21·2	263	261·6	27·5
24	23·9	02·5	84	83·5	08·8	144	143·2	15·1	204	202·9	21·3	264	262·6	27·6
25	24·9	02·6	85	84·5	08·9	145	144·2	15·2	205	203·9	21·4	265	263·5	27·7
26	25·9	02·7	86	85·5	09·0	146	145·2	15·3	206	204·9	21·5	266	264·5	27·8
27	26·9	02·8	87	86·5	09·1	147	146·2	15·4	207	205·9	21·6	267	265·5	27·9
28	27·8	02·9	88	87·5	09·2	148	147·2	15·5	208	206·9	21·7	268	266·5	28·0
29	28·8	03·0	89	88·5	09·3	149	148·2	15·6	209	207·9	21·8	269	267·5	28·1
30	29·8	03·1	90	89·5	09·4	150	149·2	15·7	210	208·8	22·0	270	268·5	28·2
31	30·8	03·2	91	90·5	09·5	151	150·2	15·8	211	209·8	22·1	271	269·5	28·3
32	31·8	03·3	92	91·5	09·6	152	151·2	15·9	212	210·8	22·2	272	270·5	28·4
33	32·8	03·4	93	92·5	09·7	153	152·2	16·0	213	211·8	22·3	273	271·5	28·5
34	33·8	03·6	94	93·5	09·8	154	153·2	16·1	214	212·8	22·4	274	272·5	28·6
35	34·8	03·7	95	94·5	09·9	155	154·2	16·2	215	213·8	22·5	275	273·5	28·7
36	35·8	03·8	96	95·5	10·0	156	155·1	16·3	216	214·8	22·6	276	274·5	28·8
37	36·8	03·9	97	96·5	10·1	157	156·1	16·4	217	215·8	22·7	277	275·5	29·0
38	37·8	04·0	98	97·5	10·2	158	157·1	16·5	218	216·8	22·8	278	276·5	29·1
39	38·8	04·1	99	98·5	10·3	159	158·1	16·6	219	217·8	22·9	279	277·5	29·2
40	39·8	04·2	100	99·5	10·5	160	159·1	16·7	220	218·8	23·0	280	278·5	29·3
41	40·8	04·3	101	100·4	10·6	161	160·1	16·8	221	219·8	23·1	281	279·5	29·4
42	41·8	04·4	102	101·4	10·7	162	161·1	16·9	222	220·8	23·2	282	280·5	29·5
43	42·8	04·5	103	102·4	10·8	163	162·1	17·0	223	221·8	23·3	283	281·4	29·6
44	43·8	04·6	104	103·4	10·9	164	163·1	17·1	224	222·8	23·4	284	282·4	29·7
45	44·8	04·7	105	104·4	11·0	165	164·1	17·2	225	223·8	23·5	285	283·4	29·8
46	45·7	04·8	106	105·4	11·1	166	165·1	17·4	226	224·8	23·6	286	284·4	29·9
47	46·7	04·9	107	106·4	11·2	167	166·1	17·5	227	225·8	23·7	287	285·4	30·0
48	47·7	05·0	108	107·4	11·3	168	167·1	17·6	228	226·8	23·8	288	286·4	30·1
49	48·7	05·1	109	108·4	11·4	169	168·1	17·7	229	227·7	23·9	289	287·4	30·2
50	49·7	05·2	110	109·4	11·5	170	169·1	17·8	230	228·7	24·0	290	288·4	30·3
51	50·7	05·3	111	110·4	11·6	171	170·1	17·9	231	229·7	24·1	291	289·4	30·4
52	51·7	05·4	112	111·4	11·7	172	171·1	18·0	232	230·7	24·3	292	290·4	30·5
53	52·7	05·5	113	112·4	11·8	173	172·1	18·1	233	231·7	24·4	293	291·4	30·6
54	53·7	05·6	114	113·4	11·9	174	173·0	18·2	234	232·7	24·5	294	292·4	30·7
55	54·7	05·7	115	114·4	12·0	175	174·0	18·3	235	233·7	24·6	295	293·4	30·8
56	55·7	05·9	116	115·4	12·1	176	175·0	18·4	236	234·7	24·7	296	294·4	30·9
57	56·7	06·0	117	116·4	12·2	177	176·0	18·5	237	235·7	24·8	297	295·4	31·0
58	57·7	06·1	118	117·4	12·3	178	177·0	18·6	238	236·7	24·9	298	296·4	31·1
59	58·7	06·2	119	118·3	12·4	179	178·0	18·7	239	237·7	25·0	299	297·4	31·3
60	59·7	06·3	120	119·3	12·5	180	179·0	18·8	240	238·7	25·1	300	298·4	31·4

Dist	Dep.	D.Lat	Dist	Dep.	D.Lat	Dist	Dep.	D.Lat	Dist	Dep.	D.Lat	Dist	Dep.	D.Lat
D Lon		Dep.	D Lon		Dep.	D Lon		Dep.	D Lon		Dep.	D Lon		Dep.

84°

276° / 264°   **84 DEGREES.**   084° / 096°   5h 36m

TRAVERSE TABLE

6 DEGREES.

D Lon	Dep.		D Lon	Dep.		D Lon	Dep.		D Lon	Dep.		D Lon	Dep.		D Lon	Dep.	
Dist	D.Lat	Dep.	Dist	D.Lat	Dep.	Dist	D.Lat	Dep.	Dist	D.Lat	Dep.	Dist	D.Lat	Dep.	Dist	D.Lat	Dep.
301	299·4	31·5	361	359·0	37·7	421	418·7	44·0	481	478·4	50·3	541	538·0	56·7			
302	300·3	31·6	362	360·0	37·8	422	419·7	44·1	482	479·4	50·4	542	539·0	56·7			
303	301·3	31·7	363	361·0	37·9	423	420·7	44·2	483	480·4	50·5	543	540·0	56·8			
304	302·3	31·8	364	362·0	38·0	424	421·7	44·3	484	481·3	50·6	544	541·0	56·9			
305	303·3	31·9	365	363·0	38·2	425	422·7	44·4	485	482·3	50·7	545	542·0	57·0			
306	304·3	32·0	366	364·0	38·3	426	423·7	44·5	486	483·3	50·8	546	543·0	57·1			
307	305·3	32·1	367	365·0	38·4	427	424·7	44·6	487	484·3	50·9	547	544·0	57·2			
308	306·3	32·2	368	366·0	38·5	428	425·7	44·7	488	485·3	51·0	548	545·0	57·3			
309	307·3	32·3	369	367·0	38·6	429	426·6	44·8	489	486·3	51·1	549	546·0	57·4			
310	308·3	32·4	370	368·0	38·7	430	427·6	44·9	490	487·3	51·2	550	547·0	57·5			
311	309·3	32·5	371	369·0	38·8	431	428·6	45·1	491	488·3	51·3	551	548·0	57·6			
312	310·3	32·6	372	370·0	38·9	432	429·6	45·2	492	489·3	51·4	552	549·0	57·7			
313	311·3	32·7	373	371·0	39·0	433	430·6	45·3	493	490·3	51·5	553	550·0	57·7			
314	312·3	32·8	374	372·0	39·1	434	431·6	45·4	494	491·3	51·6	554	551·0	57·9			
315	313·3	32·9	375	372·9	39·2	435	432·6	45·5	495	492·3	51·7	555	552·0	58·0			
316	314·3	33·0	376	373·9	39·3	436	433·6	45·6	496	493·3	51·8	556	553·0	58·1			
317	315·3	33·1	377	374·9	39·4	437	434·6	45·7	497	494·3	52·0	557	553·9	58·2			
318	316·3	33·2	378	375·9	39·5	438	435·6	45·8	498	495·3	52·1	558	554·9	58·3			
319	317·3	33·3	379	376·9	39·6	439	436·6	45·9	499	496·3	52·2	559	555·9	58·4			
320	318·2	33·4	380	377·9	39·7	440	437·6	46·0	500	497·3	52·3	560	556·9	58·5			
321	319·2	33·6	381	378·9	39·8	441	438·6	46·1	501	498·3	52·4	561	557·9	58·6			
322	320·2	33·7	382	379·9	39·9	442	439·6	46·2	502	499·2	52·5	562	558·9	58·7			
323	321·2	33·8	383	380·9	40·0	443	440·6	46·3	503	500·2	52·6	563	559·9	58·8			
324	322·2	33·9	384	381·9	40·1	444	441·6	46·4	504	501·2	52·7	564	560·9	59·0			
325	323·2	34·0	385	382·9	40·2	445	442·6	46·5	505	502·2	52·8	565	561·9	59·1			
326	324·2	34·1	386	383·9	40·3	446	443·6	46·6	506	503·2	52·9	566	562·9	59·2			
327	325·2	34·2	387	384·9	40·5	447	444·6	46·7	507	504·2	53·0	567	563·9	59·3			
328	326·2	34·3	388	385·9	40·6	448	445·5	46·8	508	505·2	53·1	568	564·9	59·4			
329	327·2	34·4	389	386·9	40·7	449	446·5	46·9	509	506·2	53·2	569	565·9	59·5			
330	328·2	34·5	390	387·9	40·8	450	447·5	47·0	510	507·2	53·3	570	566·9	59·6			
331	329·2	34·6	391	388·9	40·9	451	448·5	47·1	511	508·2	53·4	571	567·9	59·7			
332	330·2	34·7	392	389·9	41·0	452	449·5	47·2	512	509·2	53·5	572	568·9	59·8			
333	331·2	34·8	393	390·8	41·1	453	450·5	47·4	513	510·2	53·6	573	569·9	59·9			
334	332·2	34·9	394	391·8	41·2	454	451·5	47·5	514	511·2	53·7	574	570·9	60·0			
335	333·2	35·0	395	392·8	41·3	455	452·5	47·6	515	512·2	53·8	575	571·9	60·1			
336	334·2	35·1	396	393·8	41·4	456	453·5	47·7	516	513·2	53·9	576	572·8	60·2			
337	335·2	35·2	397	394·8	41·5	457	454·5	47·8	517	514·2	54·0	577	573·8	60·3			
338	336·1	35·3	398	395·8	41·6	458	455·5	47·9	518	515·2	54·1	578	574·8	60·4			
339	337·1	35·4	399	396·8	41·7	459	456·5	48·0	519	516·2	54·3	579	575·8	60·5			
340	338·1	35·5	400	397·8	41·8	460	457·5	48·1	520	517·2	54·4	580	576·8	60·6			
341	339·1	35·6	401	398·8	41·9	461	458·5	48·2	521	518·1	54·5	581	577·8	60·7			
342	340·1	35·7	402	399·8	42·0	462	459·5	48·3	522	519·1	54·6	582	578·8	60·8			
343	341·1	35·9	403	400·8	42·1	463	460·5	48·4	523	520·1	54·7	583	579·8	60·9			
344	342·1	36·0	404	401·8	42·2	464	461·5	48·5	524	521·1	54·8	584	580·8	61·0			
345	343·1	36·1	405	402·8	42·3	465	462·5	48·6	525	522·1	54·9	585	581·8	61·1			
346	344·1	36·2	406	403·8	42·4	466	463·4	48·7	526	523·1	55·0	586	582·8	61·3			
347	345·1	36·3	407	404·8	42·5	467	464·4	48·8	527	524·1	55·1	587	583·8	61·4			
348	346·1	36·4	408	405·8	42·6	468	465·4	48·9	528	525·1	55·2	588	584·8	61·5			
349	347·1	36·5	409	406·8	42·8	469	466·4	49·0	529	526·1	55·3	589	585·8	61·6			
350	348·1	36·6	410	407·8	42·9	470	467·4	49·1	530	527·1	55·4	590	586·8	61·7			
351	349·1	36·7	411	408·7	43·0	471	468·4	49·2	531	528·1	55·5	591	587·8	61·8			
352	350·1	36·8	412	409·7	43·1	472	469·4	49·3	532	529·1	55·6	592	588·8	61·9			
353	351·1	36·9	413	410·7	43·2	473	470·4	49·5	533	530·1	55·7	593	589·8	62·0			
354	352·1	37·0	414	411·7	43·3	474	471·4	49·5	534	531·1	55·8	594	590·7	62·1			
355	353·1	37·1	415	412·7	43·4	475	472·4	49·7	535	532·1	55·9	595	591·7	62·2			
356	354·0	37·2	416	413·7	43·5	476	473·4	49·8	536	533·1	56·0	596	592·7	62·3			
357	355·0	37·3	417	414·7	43·6	477	474·4	49·9	537	534·1	56·1	597	593·7	62·4			
358	356·0	37·4	418	415·7	43·7	478	475·4	50·0	538	535·1	56·2	598	594·7	62·5			
359	357·0	37·5	419	416·7	43·8	479	476·4	51·1	539	536·0	56·3	599	595·7	62·6			
360	358·0	37·6	420	417·7	43·9	480	477·4	50·2	540	537·0	56·4	600	596·7	62·7			
Dist	Dep.	D.Lat	Dist	Dep.	D.Lat	Dist	Dep.	D.Lat	Dist	Dep.	D.Lat	Dist	Dep.	D.Lat	Dist	Dep.	D.Lat
D Lon		Dep.	D Lon		Dep.	D Lon		Dep.	D Lon		Dep.	D Lon		Dep.	D Lon		Dep.

TRAVERSE TABLE
7 DEGREES.

7° 353° / 187°   007° / 173°   0h 28m

Dist	D.Lat	Dep.	Dist	D.Lat	Dep.	Dist	D.Lat	Dep.	Dist	D.Lat	Dep.	Dist	D.Lat	Dep.
1	01·0	00·1	61	60·5	07·4	121	120·1	14·7	181	179·7	22·1	241	239·2	29·4
2	02·0	00·2	62	61·5	07·6	122	121·1	14·9	182	180·6	22·2	242	240·2	29·5
3	03·0	00·4	63	62·5	07·7	123	122·1	15·0	183	181·6	22·3	243	241·2	29·6
4	04·0	00·5	64	63·5	07·8	124	123·1	15·1	184	182·6	22·4	244	242·2	29·7
5	05·0	00·6	65	64·5	07·9	125	124·1	15·2	185	183·6	22·5	245	243·2	29·9
6	06·0	00·7	66	65·5	08·0	126	125·1	15·4	186	184·6	22·7	246	244·2	30·0
7	06·9	00·9	67	66·5	08·2	127	126·1	15·5	187	185·6	22·8	247	245·2	30·1
8	07·9	01·0	68	67·5	08·3	128	127·0	15·6	188	186·6	22·9	248	246·2	30·2
9	08·9	01·1	69	68·5	08·4	129	128·0	15·7	189	187·6	23·0	249	247·1	30·3
10	09·9	01·2	70	69·5	08·5	130	129·0	15·8	190	188·6	23·2	250	248·1	30·5
11	10·9	01·3	71	70·5	08·7	131	130·0	16·0	191	189·6	23·3	251	249·1	30·6
12	11·9	01·5	72	71·5	08·8	132	131·0	16·1	192	190·6	23·4	252	250·1	30·7
13	12·9	01·6	73	72·5	08·9	133	132·0	16·2	193	191·6	23·5	253	251·1	30·8
14	13·9	01·7	74	73·4	09·0	134	133·0	16·3	194	192·6	23·6	254	252·1	31·0
15	14·9	01·8	75	74·4	09·1	135	134·0	16·5	195	193·5	23·8	255	253·1	31·1
16	15·9	01·9	76	75·4	09·3	136	135·0	16·6	196	194·5	23·9	256	254·1	31·2
17	16·9	02·1	77	76·4	09·4	137	136·0	16·7	197	195·5	24·0	257	255·1	31·3
18	17·9	02·2	78	77·4	09·5	138	137·0	16·8	198	196·5	24·1	258	256·1	31·4
19	18·9	02·3	79	78·4	09·6	139	138·0	16·9	199	197·5	24·3	259	257·1	31·6
20	19·9	02·4	80	79·4	09·7	140	139·0	17·1	200	198·5	24·4	260	258·1	31·7
21	20·8	02·6	81	80·4	09·9	141	139·9	17·2	201	199·5	24·5	261	259·1	31·8
22	21·8	02·7	82	81·4	10·0	142	140·9	17·3	202	200·5	24·6	262	260·0	31·9
23	22·8	02·8	83	82·4	10·1	143	141·9	17·4	203	201·5	24·7	263	261·0	32·1
24	23·8	02·9	84	83·4	10·2	144	142·9	17·5	204	202·5	24·9	264	262·0	32·2
25	24·8	03·0	85	84·4	10·4	145	143·9	17·7	205	203·5	25·0	265	263·0	32·3
26	25·8	03·2	86	85·4	10·5	146	144·9	17·8	206	204·5	25·1	266	264·0	32·4
27	26·8	03·3	87	86·4	10·6	147	145·9	17·9	207	205·5	25·2	267	265·0	32·5
28	27·8	03·4	88	87·3	10·7	148	146·9	18·0	208	206·4	25·3	268	266·0	32·7
29	28·8	03·5	89	88·3	10·8	149	147·9	18·2	209	207·4	25·5	269	267·0	32·8
30	29·8	03·7	90	89·3	11·0	150	148·9	18·3	210	208·4	25·6	270	268·0	32·9
31	30·8	03·8	91	90·3	11·1	151	149·9	18·4	211	209·4	25·7	271	269·0	33·0
32	31·8	03·9	92	91·3	11·2	152	150·9	18·5	212	210·4	25·8	272	270·0	33·1
33	32·8	04·0	93	92·3	11·3	153	151·9	18·6	213	211·4	26·0	273	271·0	33·3
34	33·7	04·1	94	93·3	11·5	154	152·9	18·8	214	212·4	26·1	274	272·0	33·4
35	34·7	04·3	95	94·3	11·6	155	153·8	18·9	215	213·4	26·2	275	273·0	33·5
36	35·7	04·4	96	95·3	11·7	156	154·8	19·0	216	214·4	26·3	276	273·9	33·6
37	36·7	04·5	97	96·3	11·8	157	155·8	19·1	217	215·4	26·4	277	274·9	33·8
38	37·7	04·6	98	97·3	11·9	158	156·8	19·3	218	216·4	26·6	278	275·9	33·9
39	38·7	04·8	99	98·3	12·1	159	157·8	19·4	219	217·4	26·7	279	276·9	34·0
40	39·7	04·9	100	99·3	12·2	160	158·8	19·5	220	218·4	26·8	280	277·9	34·1
41	40·7	05·0	101	100·2	12·3	161	159·8	19·6	221	219·4	26·9	281	278·9	34·2
42	41·7	05·1	102	101·2	12·4	162	160·8	19·7	222	220·3	27·1	282	279·9	34·4
43	42·7	05·2	103	102·2	12·6	163	161·8	19·9	223	221·3	27·2	283	280·9	34·5
44	43·7	05·4	104	103·2	12·7	164	162·8	20·0	224	222·3	27·3	284	281·9	34·6
45	44·7	05·5	105	104·2	12·8	165	163·8	20·1	225	223·3	27·4	285	282·9	34·7
46	45·7	05·7	106	105·2	12·9	166	164·8	20·2	226	224·3	27·5	286	283·9	34·9
47	46·6	05·7	107	106·2	13·0	167	165·8	20·4	227	225·3	27·7	287	284·9	35·0
48	47·6	05·8	108	107·2	13·2	168	166·7	20·5	228	226·3	27·8	288	285·9	35·1
49	48·6	06·0	109	108·2	13·3	169	167·7	20·6	229	227·3	27·9	289	286·8	35·2
50	49·6	06·1	110	109·2	13·4	170	168·7	20·7	230	228·3	28·0	290	287·8	35·3
51	50·6	06·2	111	110·2	13·5	171	169·7	20·8	231	229·3	28·2	291	288·8	35·5
52	51·6	06·3	112	111·2	13·6	172	170·7	21·0	232	230·3	28·3	292	289·8	35·6
53	52·6	06·5	113	112·2	13·8	173	171·7	21·1	233	231·3	28·4	293	290·8	35·7
54	53·6	06·6	114	113·2	13·9	174	172·7	21·2	234	232·3	28·5	294	291·8	35·8
55	54·6	06·7	115	114·1	14·0	175	173·7	21·3	235	233·2	28·6	295	292·8	36·0
56	55·6	06·8	116	115·1	14·1	176	174·7	21·4	236	234·2	28·8	296	293·8	36·1
57	56·6	06·9	117	116·1	14·3	177	175·7	21·6	237	235·2	28·9	297	294·8	36·2
58	57·6	07·1	118	117·1	14·4	178	176·7	21·7	238	236·2	29·0	298	295·8	36·3
59	58·6	07·2	119	118·1	14·5	179	177·7	21·8	239	237·2	29·1	299	296·8	36·4
60	59·6	07·3	120	119·1	14·6	180	178·7	21·9	240	238·2	29·2	300	297·8	36·6
Dist	Dep.	D.Lat	Dist	Dep.	D.Lat	Dist	Dep.	D.Lat	Dist	Dep.	D.Lat	Dist	Dep.	D.Lat

83° 277° / 263°   083° / 097°   5h 32m

83 DEGREES.

Dist	D.Lat	Dep.	Dist	D.Lat	Dep.	Dist	D.Lat	Dep.	Dist	D.Lat	Dep.	Dist	D.Lat	Dep.
301	298·8	36·7	361	358·3	44·0	421	417·9	51·3	481	477·4	58·6	541	537·0	65·9
302	299·7	36·8	362	359·3	44·1	422	418·9	51·4	482	478·4	58·7	542	538·0	66·1
303	300·7	36·9	363	360·3	44·2	423	419·8	51·6	483	479·4	58·9	543	539·0	66·2
304	301·7	37·0	364	361·3	44·4	424	420·8	51·7	484	480·4	59·0	544	539·9	66·3
305	302·7	37·2	365	362·3	44·5	425	421·8	51·8	485	481·4	59·1	545	540·9	66·4
306	303·7	37·3	366	363·3	44·6	426	422·8	51·9	486	482·4	59·2	546	541·9	66·5
307	304·7	37·4	367	364·3	44·7	427	423·8	52·0	487	483·4	59·4	547	542·9	66·7
308	305·7	37·5	368	365·3	44·8	428	424·8	52·2	488	484·4	59·5	548	543·9	66·8
309	306·7	37·7	369	366·2	45·0	429	425·8	52·3	489	485·4	59·6	549	544·9	66·9
310	307·7	37·8	370	367·2	45·1	430	426·8	52·4	490	486·3	59·7	550	545·9	67·0
311	308·7	37·9	371	368·2	45·2	431	427·8	52·5	491	487·3	59·8	551	546·9	67·1
312	309·7	38·0	372	369·2	45·3	432	428·8	52·6	492	488·3	60·0	552	547·9	67·3
313	310·7	38·1	373	370·2	45·5	433	429·8	52·8	493	489·3	60·1	553	548·9	67·4
314	311·7	38·3	374	371·2	45·6	434	430·8	52·9	494	490·3	60·2	554	549·9	67·5
315	312·7	38·4	375	372·2	45·7	435	431·8	53·0	495	491·3	60·3	555	550·9	67·6
316	313·6	38·5	376	373·2	45·8	436	432·8	53·1	496	492·3	60·4	556	551·9	67·8
317	314·6	38·6	377	374·2	45·9	437	433·7	53·3	497	493·3	60·6	557	552·8	67·9
318	315·6	38·8	378	375·2	46·1	438	434·7	53·4	498	494·3	60·7	558	553·8	68·0
319	316·6	38·9	379	376·2	46·2	439	435·7	53·5	499	495·3	60·8	559	554·8	68·1
320	317·6	39·0	380	377·2	46·3	440	436·7	53·6	500	496·3	60·9	560	555·8	68·2
321	318·6	39·1	381	378·2	46·4	441	437·7	53·7	501	497·3	61·1	561	556·8	68·4
322	319·6	39·2	382	379·2	46·6	442	438·7	53·9	502	498·3	61·2	562	557·8	68·5
323	320·6	39·4	383	380·1	46·7	443	439·7	54·0	503	499·3	61·3	563	558·8	68·6
324	321·6	39·5	384	381·1	46·8	444	440·7	54·1	504	500·2	61·4	564	559·8	68·7
325	322·6	39·6	385	382·1	46·9	445	441·7	54·2	505	501·2	61·5	565	560·8	68·9
326	323·6	39·7	386	383·1	47·0	446	442·7	54·4	506	502·2	61·7	566	561·8	69·0
327	324·6	39·8	387	384·1	47·2	447	443·7	54·5	507	503·2	61·8	567	562·8	69·1
328	325·6	40·0	388	385·1	47·3	448	444·7	54·6	508	504·2	61·9	568	563·8	69·2
329	326·5	40·1	389	386·1	47·4	449	445·7	54·7	509	505·2	62·0	569	564·8	69·3
330	327·5	40·2	390	387·1	47·5	450	446·6	54·8	510	506·2	62·2	570	565·8	69·5
331	328·5	40·3	391	388·1	47·7	451	447·6	55·0	511	507·2	62·3	571	566·7	69·6
332	329·5	40·5	392	389·1	47·8	452	448·6	55·1	512	508·2	62·4	572	567·7	69·7
333	330·5	40·6	393	390·1	47·9	453	449·6	55·2	513	509·2	62·5	573	568·7	69·8
334	331·5	40·7	394	391·1	48·0	454	450·6	55·3	514	510·2	62·6	574	569·7	70·0
335	332·5	40·8	395	392·1	48·1	455	451·6	55·5	515	511·2	62·8	575	570·7	70·1
336	333·5	40·9	396	393·0	48·3	456	452·6	55·7	516	512·2	62·9	576	571·7	70·2
337	334·5	41·1	397	394·0	48·4	457	453·6	55·7	517	513·1	63·0	577	572·7	70·3
338	335·5	41·2	398	395·0	48·5	458	454·6	55·8	518	514·1	63·1	578	573·7	70·4
339	336·5	41·3	399	396·0	48·6	459	455·6	55·9	519	515·1	63·3	579	574·7	70·6
340	337·5	41·4	400	397·0	48·7	460	456·6	56·1	520	516·1	63·4	580	575·7	70·7
341	338·5	41·6	401	398·0	48·9	461	457·6	56·2	521	517·1	63·5	581	576·7	70·8
342	339·5	41·7	402	399·0	49·0	462	458·6	56·3	522	518·1	63·6	582	577·7	70·9
343	340·4	41·8	403	400·0	49·1	463	459·5	56·4	523	519·1	63·7	583	578·7	71·0
344	341·4	41·9	404	401·0	49·2	464	460·5	56·5	524	520·1	63·9	584	579·6	71·2
345	342·4	42·0	405	402·0	49·4	465	461·5	56·7	525	521·1	64·0	585	580·6	71·3
346	343·4	42·2	406	403·0	49·5	466	462·5	56·8	526	522·1	64·1	586	581·6	71·4
347	344·4	42·3	407	404·0	49·6	467	463·5	56·9	527	523·1	64·2	587	582·6	71·5
348	345·4	42·4	408	405·0	49·7	468	464·5	57·0	528	524·1	64·3	588	583·6	71·7
349	346·4	42·5	409	406·0	49·8	469	465·5	57·2	529	525·1	64·5	589	584·6	71·8
350	347·4	42·7	410	406·9	50·0	470	466·5	57·3	530	526·0	64·6	590	585·6	71·9
351	348·4	42·8	411	407·9	50·1	471	467·5	57·4	531	527·0	64·7	591	586·6	72·0
352	349·4	42·9	412	408·9	50·2	472	468·5	57·5	532	528·0	64·8	592	587·6	72·1
353	350·4	43·0	413	409·9	50·3	473	469·5	57·6	533	529·0	65·0	593	588·6	72·3
354	351·4	43·1	414	410·9	50·5	474	470·5	57·8	534	530·0	65·1	594	589·6	72·4
355	352·4	43·3	415	411·9	50·6	475	471·5	57·9	535	531·0	65·3	595	590·6	72·5
356	353·3	43·4	416	412·9	50·7	476	472·5	58·0	536	532·0	65·3	596	591·5	72·6
357	354·3	43·5	417	413·9	50·8	477	473·4	58·1	537	533·0	65·4	597	592·5	72·8
358	355·3	43·6	418	414·9	50·9	478	474·4	58·3	538	534·0	65·6	598	593·5	72·9
359	356·3	43·7	419	415·9	51·1	479	475·4	58·4	539	535·0	65·7	599	594·5	73·0
360	357·3	43·8	420	416·9	51·2	480	476·4	58·5	540	536·0	65·8	600	595·5	73·1
Dist	Dep.	D.Lat	Dist	Dep.	D.Lat	Dist	Dep.	D.Lat	Dist	Dep.	D.Lat	Dist	Dep.	D.Lat

8°

TRAVERSE TABLE
8 DEGREES.

D Lon	Dep.		D Lon	Dep.		D Lon	Dep.		D Lon	Dep.		D Lon	Dep.	
Dist	D.Lat	Dep.	Dist	D.Lat	Dep.	Dist	D.Lat	Dep.	Dist	D.Lat	Dep.	Dist	D.Lat	Dep.
1	01·0	00·1	61	60·4	08·5	121	119·8	16·8	181	179·2	25·2	241	238·7	33·5
2	02·0	00·3	62	61·4	08·6	122	120·8	17·0	182	180·2	25·3	242	239·6	33·7
3	03·0	00·4	63	62·4	08·8	123	121·8	17·1	183	181·2	25·5	243	240·6	33·8
4	04·0	00·6	64	63·4	08·9	124	122·8	17·3	184	182·2	25·6	244	241·6	34·0
5	05·0	00·7	65	64·4	09·0	125	123·8	17·4	185	183·2	25·7	245	242·6	34·1
6	05·9	00·8	66	65·4	09·2	126	124·8	17·5	186	184·2	25·9	246	243·6	34·2
7	06·9	01·0	67	66·3	09·3	127	125·8	17·7	187	185·2	26·0	247	244·6	34·4
8	07·9	01·1	68	67·3	09·5	128	126·8	17·8	188	186·2	26·2	248	245·6	34·5
9	08·9	01·3	69	68·3	09·6	129	127·7	18·0	189	187·2	26·3	249	246·6	34·7
10	09·9	01·4	70	69·3	09·7	130	128·7	18·1	190	188·2	26·4	250	247·6	34·8
11	10·9	01·5	71	70·3	09·9	131	129·7	18·2	191	189·1	26·6	251	248·6	34·9
12	11·9	01·7	72	71·3	10·0	132	130·7	18·4	192	190·1	26·7	252	249·5	35·1
13	12·9	01·8	73	72·3	10·2	133	131·7	18·5	193	191·1	26·9	253	250·5	35·2
14	13·9	01·9	74	73·3	10·3	134	132·7	18·6	194	192·1	27·0	254	251·5	35·3
15	14·9	02·1	75	74·3	10·4	135	133·7	18·8	195	193·1	27·1	255	252·5	35·5
16	15·8	02·2	76	75·3	10·6	136	134·7	18·9	196	194·1	27·3	256	253·5	35·6
17	16·8	02·4	77	76·3	10·7	137	135·7	19·1	197	195·1	27·4	257	254·5	35·8
18	17·8	02·5	78	77·2	10·9	138	136·7	19·2	198	196·1	27·6	258	255·5	35·9
19	18·8	02·6	79	78·2	11·0	139	137·7	19·3	199	197·1	27·7	259	256·5	36·0
20	19·8	02·8	80	79·2	11·1	140	138·6	19·5	200	198·1	27·8	260	257·5	36·2
21	20·8	02·9	81	80·2	11·3	141	139·6	19·6	201	199·0	28·0	261	258·5	36·3
22	21·8	03·1	82	81·2	11·4	142	140·6	19·8	202	200·0	28·1	262	259·5	36·5
23	22·8	03·2	83	82·2	11·6	143	141·6	19·9	203	201·0	28·3	263	260·4	36·6
24	23·8	03·3	84	83·2	11·7	144	142·6	20·0	204	202·0	28·4	264	261·4	36·7
25	24·8	03·5	85	84·2	11·8	145	143·6	20·2	205	203·0	28·5	265	262·4	36·9
26	25·7	03·6	86	85·2	12·0	146	144·6	20·3	206	204·0	28·7	266	263·4	37·0
27	26·7	03·8	87	86·2	12·1	147	145·6	20·5	207	205·0	28·8	267	264·4	37·2
28	27·7	03·9	88	87·1	12·2	148	146·6	20·6	208	206·0	28·9	268	265·4	37·3
29	28·7	04·0	89	88·1	12·4	149	147·5	20·7	209	207·0	29·1	269	266·4	37·4
30	29·7	04·2	90	89·1	12·5	150	148·5	20·9	210	208·0	29·2	270	267·4	37·6
31	30·7	04·3	91	90·1	12·7	151	149·5	21·0	211	208·9	29·4	271	268·4	37·7
32	31·7	04·5	92	91·1	12·8	152	150·5	21·2	212	209·9	29·5	272	269·4	37·9
33	32·7	04·6	93	92·1	12·9	153	151·5	21·3	213	210·9	29·6	273	270·3	38·0
34	33·7	04·7	94	93·1	13·1	154	152·5	21·4	214	211·9	29·8	274	271·3	38·1
35	34·7	04·9	95	94·1	13·2	155	153·5	21·6	215	212·9	29·9	275	272·3	38·3
36	35·6	05·0	96	95·1	13·4	156	154·5	21·7	216	213·9	30·1	276	273·3	38·4
37	36·6	05·1	97	96·1	13·5	157	155·5	21·9	217	214·9	30·2	277	274·3	38·6
38	37·6	05·3	98	97·0	13·6	158	156·5	22·0	218	215·9	30·3	278	275·3	38·7
39	38·6	05·4	99	98·0	13·8	159	157·5	22·1	219	216·9	30·5	279	276·3	38·8
40	39·6	05·6	100	99·0	13·9	160	158·4	22·3	220	217·9	30·6	280	277·3	39·0
41	40·6	05·7	101	100·0	14·1	161	159·4	22·4	221	218·8	30·8	281	278·3	39·1
42	41·6	05·8	102	101·0	14·2	162	160·4	22·5	222	219·8	30·9	282	279·3	39·2
43	42·6	06·0	103	102·0	14·3	163	161·4	22·7	223	220·8	31·0	283	280·2	39·4
44	43·6	06·1	104	103·0	14·5	164	162·4	22·8	224	221·8	31·2	284	281·2	39·5
45	44·6	06·3	105	104·0	14·6	165	163·4	23·0	225	222·8	31·3	285	282·2	39·7
46	45·6	06·4	106	105·0	14·8	166	164·4	23·2	226	223·8	31·5	286	283·2	39·8
47	46·5	06·5	107	106·0	14·9	167	165·4	23·2	227	224·8	31·6	287	284·2	39·9
48	47·5	06·7	108	106·9	15·0	168	166·4	23·4	228	225·8	31·7	288	285·2	40·1
49	48·5	06·8	109	107·9	15·2	169	167·4	23·5	229	226·8	31·9	289	286·2	40·2
50	49·5	07·0	110	108·9	15·3	170	168·3	23·7	230	227·8	32·0	290	287·2	40·4
51	50·5	07·1	111	109·9	15·4	171	169·3	23·8	231	228·8	32·1	291	288·2	40·5
52	51·5	07·2	112	110·9	15·6	172	170·3	23·9	232	229·7	32·3	292	289·2	40·6
53	52·5	07·4	113	111·9	15·7	173	171·3	24·1	233	230·7	32·4	293	290·1	40·8
54	53·5	07·5	114	112·9	15·9	174	172·3	24·2	234	231·7	32·6	294	291·1	40·9
55	54·5	07·7	115	113·9	16·0	175	173·3	24·4	235	232·7	32·7	295	292·1	41·1
56	55·5	07·8	116	114·9	16·1	176	174·3	24·5	236	233·7	32·8	296	293·1	41·2
57	56·4	07·9	117	115·9	16·3	177	175·3	24·6	237	234·7	33·0	297	294·1	41·3
58	57·4	08·1	118	116·9	16·4	178	176·3	24·8	238	235·7	33·1	298	295·1	41·5
59	58·4	08·2	119	117·8	16·6	179	177·3	24·9	239	236·7	33·3	299	296·1	41·6
60	59·4	08·4	120	118·8	16·7	180	178·2	25·1	240	237·7	33·4	300	297·1	41·8
Dist	Dep.	D.Lat	Dist	Dep.	D.Lat	Dist	Dep.	D.Lat	Dist	Dep.	D.Lat	Dist	Dep.	D.Lat
D Lon		Dep.	D Lon		Dep.	D Lon		Dep.	D Lon		Dep.	D Lon		Dep.

82°

	352° ↑													
	188°				**TRAVERSE TABLE**					↑ 008°				
					8 DEGREES.					172°		0h 32m		

D Lon	Dep.		D Lon	Dep.		D Lon	Dep.		D Lon	Dep.		D Lon	Dep.	
Dist	D.Lat	Dep.	Dist	D.Lat	Dep.	Dist	D.Lat	Dep.	Dist	D.Lat	Dep.	Dist	D.Lat	Dep.
301	298·1	41·9	361	357·5	50·2	421	416·9	58·6	481	476·3	66·9	541	535·7	75·3
302	299·1	42·0	362	358·5	50·4	422	417·9	58·7	482	477·3	67·1	542	536·7	75·4
303	300·1	42·2	363	359·5	50·5	423	418·9	58·9	483	478·3	67·2	543	537·7	75·6
304	301·0	42·3	364	360·5	50·7	424	419·9	59·0	484	479·3	67·4	544	538·7	75·7
305	302·0	42·4	365	361·4	50·8	425	420·9	59·1	485	480·3	67·5	545	539·7	75·8
306	303·0	42·6	366	362·4	50·9	426	421·9	59·3	486	481·3	67·6	546	540·7	76·0
307	304·0	42·7	367	363·4	51·1	427	422·8	59·4	487	482·3	67·8	547	541·7	76·1
308	305·0	42·9	368	364·4	51·2	428	423·8	59·6	488	483·3	67·9	548	542·7	76·3
309	306·0	43·0	369	365·4	51·4	429	424·8	59·7	489	484·2	68·1	549	543·7	76·4
310	307·0	43·1	370	366·4	51·5	430	425·8	59·8	490	485·2	68·2	550	544·6	76·5
311	308·0	43·3	371	367·4	51·6	431	426·8	60·0	491	486·2	68·3	551	545·6	76·7
312	309·0	43·4	372	368·4	51·8	432	427·8	60·1	492	487·2	68·5	552	546·6	76·8
313	310·0	43·6	373	369·3	51·9	433	428·8	60·3	493	488·2	68·6	553	547·6	77·0
314	310·9	43·7	374	370·4	52·1	434	429·8	60·4	494	489·2	68·8	554	548·6	77·1
315	311·9	43·8	375	371·4	52·2	435	430·8	60·5	495	490·2	68·9	555	549·6	77·2
316	312·9	44·0	376	372·3	52·3	436	431·8	60·7	496	491·2	69·0	556	550·6	77·4
317	313·9	44·1	377	373·3	52·5	437	432·7	60·8	497	492·2	69·2	557	551·6	77·5
318	314·9	44·3	378	374·3	52·6	438	433·7	61·0	498	493·2	69·2	558	552·6	77·7
319	315·9	44·4	379	375·3	52·7	439	434·7	61·1	499	494·1	69·4	559	553·6	77·8
320	316·9	44·5	380	376·3	52·9	440	435·7	61·2	500	495·1	69·6	560	554·6	77·9
321	317·9	44·7	381	377·3	53·0	441	436·7	61·4	501	496·1	69·7	561	555·5	78·1
322	318·9	44·8	382	378·3	53·2	442	437·7	61·5	502	497·1	69·9	562	556·5	78·2
323	319·9	45·0	383	379·3	53·3	443	438·7	61·7	503	498·1	70·0	563	557·5	78·4
324	320·8	45·1	384	380·3	53·4	444	439·7	61·8	504	499·1	70·1	564	558·5	78·5
325	321·8	45·2	385	381·3	53·6	445	440·7	61·9	505	500·1	70·3	565	559·5	78·6
326	322·8	45·4	386	382·2	53·7	446	441·7	62·1	506	501·1	70·4	566	560·5	78·8
327	323·8	45·5	387	383·2	53·9	447	442·6	62·2	507	502·1	70·6	567	561·5	78·9
328	324·8	45·6	388	384·2	54·0	448	443·6	62·3	508	503·1	70·7	568	562·5	79·1
329	325·8	45·8	389	385·2	54·1	449	444·6	62·5	509	504·0	70·8	569	563·5	79·2
330	326·8	45·9	390	386·2	54·3	450	445·6	62·6	510	505·0	71·0	570	564·5	79·3
331	327·8	46·1	391	387·2	54·4	451	446·6	62·8	511	506·0	71·1	571	565·4	79·5
332	328·8	46·2	392	388·2	54·6	452	447·6	62·9	512	507·0	71·3	572	566·4	79·6
333	329·8	46·3	393	389·1	54·7	453	448·6	63·0	513	508·0	71·4	573	567·4	79·7
334	330·7	46·5	394	390·1	54·8	454	449·6	63·2	514	509·0	71·5	574	568·4	79·9
335	331·7	46·6	395	391·1	55·0	455	450·6	63·3	515	510·0	71·7	575	569·4	80·0
336	332·7	46·8	396	392·1	55·1	456	451·6	63·5	516	511·0	71·8	576	570·4	80·2
337	333·7	46·9	397	393·1	55·3	457	452·6	63·6	517	512·0	72·0	577	571·4	80·3
338	334·7	47·0	398	394·1	55·4	458	453·5	63·7	518	513·0	72·1	578	572·4	80·4
339	335·7	47·2	399	395·1	55·5	459	454·5	63·9	519	513·9	72·2	579	573·4	80·6
340	336·7	47·3	400	396·1	55·7	460	455·5	64·0	520	514·9	72·4	580	574·4	80·7
341	337·7	47·5	401	397·1	55·8	461	456·5	64·2	521	515·9	72·5	581	575·3	80·9
342	338·7	47·6	402	398·1	55·9	462	457·5	64·3	522	516·9	72·6	582	576·3	81·0
343	339·7	47·7	403	399·1	56·1	463	458·5	64·4	523	517·9	72·8	583	577·3	81·1
344	340·7	47·9	404	400·1	56·2	464	459·5	64·6	524	518·9	72·9	584	578·3	81·3
345	341·6	48·0	405	401·1	56·4	465	460·5	64·7	525	519·9	73·1	585	579·3	81·4
346	342·6	48·2	406	402·0	56·5	466	461·5	64·9	526	520·9	73·2	586	580·3	81·6
347	343·6	48·3	407	403·0	56·6	467	462·5	65·0	527	521·9	73·3	587	581·3	81·7
348	344·6	48·4	408	404·0	56·8	468	463·4	65·1	528	522·9	73·5	588	582·3	81·8
349	345·6	48·6	409	405·0	56·9	469	464·4	65·3	529	523·9	73·6	589	583·3	82·0
350	346·6	48·7	410	406·0	57·1	470	465·4	65·4	530	524·8	73·8	590	584·3	82·1
351	347·6	48·8	411	407·0	57·2	471	466·4	65·6	531	525·8	73·9	591	585·2	82·3
352	348·6	49·0	412	408·0	57·3	472	467·4	65·7	532	526·8	74·1	592	586·2	82·4
353	349·6	49·1	413	409·0	57·5	473	468·4	65·8	533	527·8	74·2	593	587·2	82·5
354	350·6	49·3	414	410·0	57·6	474	469·4	66·0	534	528·8	74·3	594	588·2	82·7
355	351·5	49·4	415	411·0	57·8	475	470·4	66·1	535	529·8	74·5	595	589·2	82·8
356	352·5	49·5	416	412·0	57·9	476	471·4	66·2	536	530·8	74·6	596	590·2	82·9
357	353·5	49·7	417	412·9	58·0	477	472·4	66·4	537	531·8	74·7	597	591·2	83·1
358	354·5	49·8	418	413·9	58·2	478	473·3	66·5	538	532·8	74·9	598	592·2	83·2
359	355·5	50·0	419	414·9	58·3	479	474·3	66·7	539	533·8	75·0	599	593·2	83·4
360	356·5	50·1	420	415·9	58·5	480	475·3	66·8	540	534·7	75·2	600	694·2	83·5
Dist	Dep.	D.Lat	Dist	Dep.	D.Lat	Dist	Dep.	D.Lat	Dist	Dep.	D.Lat	Dist	Dep.	D.Lat
D Lon		Dep.	D Lon		Dep.	D Lon		Dep.	D Lon		Dep.	D Lon		Dep.

	278° ↑										↑ 082°		
	262°			**82 DEGREES.**							098°		5h 28m

9°

Dist	D.Lat	Dep.	Dist	D.Lat	Dep.	Dist	D.Lat	Dep.	Dist	D.Lat	Dep.	Dist	D.Lat	Dep.
1	01·0	00·2	61	60·2	09·5	121	119·5	18·9	181	178·8	28·3	241	238·0	37·7
2	02·0	00·3	62	61·2	09·7	122	120·5	19·1	182	179·8	28·5	242	239·0	37·9
3	03·0	00·5	63	62·2	09·9	123	121·5	19·2	183	180·7	28·6	243	240·0	38·0
4	04·0	00·6	64	63·2	10·0	124	122·5	19·4	184	181·7	28·8	244	241·0	38·2
5	04·9	00·8	65	64·2	10·2	125	123·5	19·6	185	182·7	28·9	245	242·0	38·3
6	05·9	00·9	66	65·2	10·3	126	124·4	19·7	186	183·7	29·1	246	243·0	38·5
7	06·9	01·1	67	66·2	10·5	127	125·4	19·9	187	184·7	29·3	247	244·0	38·6
8	07·9	01·3	68	67·2	10·6	128	126·4	20·0	188	185·7	29·4	248	244·9	38·8
9	08·9	01·4	69	68·2	10·8	129	127·4	20·2	189	186·7	29·6	249	245·9	39·0
10	09·9	01·6	70	69·1	11·0	130	128·4	20·3	190	187·7	29·7	250	246·9	39·1
11	10·9	01·7	71	70·1	11·1	131	129·4	20·5	191	188·6	29·9	251	247·9	39·3
12	11·9	01·9	72	71·1	11·3	132	130·4	20·6	192	189·6	30·0	252	248·9	39·4
13	12·8	02·0	73	72·1	11·4	133	131·4	20·8	193	190·6	30·2	253	249·9	39·6
14	13·8	02·2	74	73·1	11·6	134	132·4	21·0	194	191·6	30·3	254	250·9	39·7
15	14·8	02·3	75	74·1	11·7	135	133·3	21·1	195	192·6	30·5	255	251·9	39·9
16	15·8	02·5	76	75·1	11·9	136	134·3	21·3	196	193·6	30·7	256	252·8	40·0
17	16·8	02·7	77	76·1	12·0	137	135·3	21·4	197	194·6	30·8	257	253·8	40·2
18	17·8	02·8	78	77·0	12·2	138	136·3	21·6	198	195·6	31·0	258	254·8	40·4
19	18·8	03·0	79	78·0	12·4	139	137·3	21·7	199	196·5	31·1	259	255·8	40·5
20	19·8	03·1	80	79·0	12·5	140	138·3	21·9	200	197·5	31·3	260	256·8	40·7
21	20·7	03·3	81	80·0	12·7	141	139·3	22·1	201	198·5	31·4	261	257·8	40·8
22	21·7	03·4	82	81·0	12·8	142	140·3	22·2	202	199·5	31·6	262	258·8	41·0
23	22·7	03·6	83	82·0	13·0	143	141·2	22·4	203	200·5	31·8	263	259·8	41·1
24	23·7	03·8	84	83·0	13·1	144	142·2	22·5	204	201·5	31·9	264	260·7	41·3
25	24·7	03·9	85	84·0	13·3	145	143·2	22·7	205	202·5	32·1	265	261·7	41·5
26	25·7	04·1	86	84·9	13·5	146	144·2	22·8	206	203·5	32·2	266	262·7	41·6
27	26·7	04·2	87	85·9	13·6	147	145·2	23·0	207	204·5	32·4	267	263·7	41·8
28	27·7	04·4	88	86·9	13·8	148	146·2	23·2	208	205·4	32·5	268	264·7	41·9
29	28·6	04·5	89	87·9	13·9	149	147·2	23·3	209	206·4	32·7	269	265·7	42·1
30	29·6	04·7	90	88·9	14·1	150	148·2	23·5	210	207·4	32·9	270	266·7	42·2
31	30·6	04·8	91	89·9	14·2	151	149·1	23·6	211	208·4	33·0	271	267·7	42·4
32	31·6	05·0	92	90·9	14·4	152	150·1	23·8	212	209·4	33·2	272	268·7	42·6
33	32·6	05·2	93	91·9	14·5	153	151·1	23·9	213	210·4	33·3	273	269·7	42·7
34	33·6	05·3	94	92·8	14·7	154	152·1	24·1	214	211·4	33·5	274	270·6	42·9
35	34·6	05·5	95	93·8	14·9	155	153·1	24·2	215	212·4	33·6	275	271·6	43·0
36	35·6	05·6	96	94·8	15·0	156	154·1	24·4	216	213·3	33·8	276	272·6	43·2
37	36·5	05·8	97	95·8	15·2	157	155·1	24·6	217	214·3	33·9	277	273·6	43·3
38	37·5	05·9	98	96·8	15·3	158	156·1	24·7	218	215·3	34·1	278	274·6	43·5
39	38·5	06·1	99	97·8	15·5	159	157·0	24·9	219	216·3	34·3	279	275·6	43·6
40	39·5	06·3	100	98·8	15·6	160	158·0	25·0	220	217·3	34·4	280	276·6	43·8
41	40·5	06·4	101	99·8	15·8	161	159·0	25·2	221	218·3	34·6	281	277·5	44·0
42	41·5	06·6	102	100·7	16·0	162	160·0	25·3	222	219·3	34·7	282	278·5	44·1
43	42·5	06·7	103	101·7	16·1	163	161·0	25·5	223	220·3	34·9	283	279·5	44·3
44	43·5	06·9	104	102·7	16·3	164	162·0	25·7	224	221·2	35·0	284	280·5	44·4
45	44·4	07·0	105	103·7	16·4	165	163·0	25·8	225	222·2	35·2	285	281·5	44·6
46	45·4	07·2	106	104·7	16·6	166	164·0	26·0	226	223·2	35·4	286	282·5	44·7
47	46·4	07·4	107	105·7	16·7	167	164·9	26·1	227	224·2	35·5	287	283·5	44·9
48	47·4	07·5	108	106·7	16·9	168	165·9	26·3	228	225·2	35·7	288	284·5	45·1
49	48·4	07·7	109	107·7	17·1	169	166·9	26·4	229	226·2	35·8	289	285·4	45·2
50	49·4	07·8	110	108·6	17·2	170	167·9	26·6	230	227·2	36·0	290	286·4	45·4
51	50·4	08·0	111	109·6	17·4	171	168·9	26·8	231	228·2	36·1	291	287·4	45·5
52	51·4	08·1	112	110·6	17·5	172	169·9	26·9	232	229·1	36·3	292	288·4	45·7
53	52·3	08·3	113	111·6	17·7	173	170·9	27·1	233	230·1	36·4	293	289·4	45·8
54	53·3	08·4	114	112·6	17·8	174	171·9	27·2	234	231·1	36·6	294	290·4	46·0
55	54·3	08·6	115	113·6	18·0	175	172·8	27·4	235	232·1	36·8	295	291·4	46·1
56	55·3	08·8	116	114·6	18·1	176	173·8	27·5	236	233·1	36·9	296	292·4	46·3
57	56·3	08·9	117	115·6	18·3	177	174·8	27·7	237	234·1	37·1	297	293·4	46·5
58	57·3	09·1	118	116·5	18·5	178	175·8	27·8	238	235·1	37·2	298	294·3	46·6
59	58·3	09·2	119	117·5	18·6	179	176·8	28·0	239	236·1	37·4	299	295·3	46·8
60	59·3	09·4	120	118·5	18·8	180	177·8	28·2	240	237·0	37·5	300	296·3	46·9

| Dist | Dep. | D.Lat | Dist | Dep. | D.Lat | Dist | Dep. | D.Lat | Dist | Dep. | D.Lat | Dist | Dep. | D.Lat |

81°

TRAVERSE TABLE
9 DEGREES.

D Lon Dep.			D Lon Dep.			D Lon Dep.			D Lon Dep.			D Lon Dep.		
Dist	D.Lat	Dep.	Dist	D.Lat	Dep.	Dist	D.Lat	Dep.	Dist	D.Lat	Dep.	Dist	D.Lat	Dep.
301	297.3	47.1	361	356.6	56.5	421	415.8	65.9	481	475.1	75.2	541	534.3	84.6
302	298.3	47.2	362	357.5	56.6	422	416.8	66.0	482	476.1	75.4	542	535.3	84.8
303	299.3	47.4	363	358.5	56.8	423	417.8	66.2	483	477.1	75.6	543	536.3	84.9
304	300.3	47.6	364	359.5	56.9	424	418.8	66.3	484	478.0	75.7	544	537.3	85.1
305	301.2	47.7	365	360.5	57.1	425	419.8	66.5	485	479.0	75.9	545	538.3	85.3
306	302.2	47.9	366	361.5	57.3	426	420.8	66.6	486	480.0	76.0	546	539.3	85.4
307	303.2	48.0	367	362.5	57.4	427	421.7	66.8	487	481.0	76.2	547	540.3	85.6
308	304.2	48.2	368	363.5	57.6	428	422.7	67.0	488	482.0	76.3	548	541.3	85.7
309	305.2	48.3	369	364.5	57.7	429	423.7	67.1	489	483.0	76.5	549	542.2	85.9
310	306.2	48.5	370	365.4	57.9	430	424.7	67.3	490	484.0	76.7	550	543.2	86.0
311	307.2	48.7	371	366.4	58.0	431	425.7	67.4	491	485.0	76.8	551	544.2	86.2
312	308.2	48.8	372	367.4	58.2	432	426.7	67.6	492	485.9	77.0	552	545.2	86.4
313	309.1	49.0	373	368.4	58.4	433	427.7	67.7	493	486.9	77.1	553	546.2	86.5
314	310.1	49.1	374	369.4	58.5	434	428.7	67.9	494	487.9	77.3	554	547.2	86.7
315	311.1	49.3	375	370.4	58.7	435	429.6	68.0	495	488.9	77.4	555	548.2	86.8
316	312.1	49.4	376	371.4	58.8	436	430.6	68.2	496	489.9	77.6	556	549.2	87.0
317	313.1	49.6	377	372.4	59.0	437	431.6	68.4	497	490.9	77.7	557	550.1	87.1
318	314.1	49.7	378	373.3	59.1	438	432.6	68.5	498	491.9	77.9	558	551.1	87.3
319	315.1	49.9	379	374.3	59.3	439	433.6	68.7	499	492.9	78.1	559	552.1	87.4
320	316.1	50.1	380	375.3	59.4	440	434.6	68.8	500	493.8	78.2	560	553.1	87.6
321	317.0	50.2	381	376.3	59.6	441	435.6	69.0	501	494.8	78.4	561	554.1	87.8
322	318.0	50.4	382	377.3	59.8	442	436.6	69.1	502	495.8	78.5	562	555.1	87.9
323	319.0	50.5	383	378.3	59.9	443	437.5	69.3	503	496.8	78.8	563	556.1	88.1
324	320.0	50.7	384	379.3	60.1	444	438.5	69.5	504	497.8	78.8	564	557.1	88.3
325	321.0	50.8	385	380.3	60.2	445	439.5	69.6	505	498.8	79.0	565	558.0	88.4
326	322.0	51.0	386	381.2	60.3	446	440.5	69.8	506	499.8	79.2	566	559.0	88.5
327	323.0	51.2	387	382.2	60.5	447	441.5	69.9	507	500.8	79.3	567	560.0	88.7
328	324.0	51.3	388	383.2	60.7	448	442.5	70.1	508	501.7	79.5	568	561.0	88.9
329	324.9	51.5	389	384.2	60.9	449	443.5	70.2	509	502.7	79.6	569	562.0	89.0
330	325.9	51.6	390	385.2	61.0	450	444.5	70.4	510	503.7	79.8	570	563.0	89.2
331	326.9	51.8	391	386.2	61.2	451	445.4	70.6	511	504.7	79.9	571	564.0	89.3
332	327.9	51.9	392	387.2	61.3	452	446.4	70.7	512	505.7	80.1	572	565.0	89.5
333	328.9	52.1	393	388.2	61.5	453	447.4	70.9	513	506.7	80.3	573	565.9	89.6
334	329.9	52.2	394	389.1	61.6	454	448.4	71.0	514	507.7	80.4	574	566.9	89.8
335	330.9	52.4	395	390.1	61.8	455	449.4	71.2	515	508.7	80.6	575	567.9	89.9
336	331.9	52.6	396	391.1	61.9	456	450.4	71.3	516	509.6	80.7	576	568.9	90.1
337	332.9	52.7	397	392.1	62.1	457	451.4	71.5	517	510.6	80.9	577	569.9	90.3
338	333.8	52.9	398	393.1	62.3	458	452.4	71.6	518	511.6	81.0	578	570.9	90.4
339	334.8	53.0	399	394.1	62.4	459	453.3	71.8	519	512.6	81.2	579	571.9	90.6
340	335.8	53.2	400	395.1	62.6	460	454.3	72.0	520	513.6	81.3	580	572.9	90.7
341	336.8	53.3	401	396.1	62.7	461	455.3	72.1	521	514.6	81.5	581	573.8	90.9
342	337.8	53.5	402	397.1	62.9	462	456.3	72.3	522	515.6	81.7	582	574.8	91.0
343	338.8	53.7	403	398.0	63.0	463	457.3	72.4	523	516.6	81.8	583	575.8	91.2
344	339.8	53.8	404	399.0	63.2	464	458.3	72.6	524	517.5	82.0	584	576.8	91.4
345	340.8	54.0	405	400.0	63.4	465	459.3	72.7	525	518.5	82.1	585	577.8	91.5
346	341.7	54.1	406	401.0	63.5	466	460.3	72.9	526	519.5	82.3	586	578.8	91.7
347	342.7	54.3	407	402.0	63.7	467	461.3	73.1	527	520.5	82.4	587	579.8	91.8
348	343.7	54.4	408	403.0	63.8	468	462.2	73.2	528	521.5	82.6	588	580.8	92.0
349	344.7	54.6	409	404.0	64.0	469	463.2	73.4	529	522.5	82.8	589	581.7	92.1
350	345.7	54.8	410	405.0	64.1	470	464.2	73.5	530	523.5	82.9	590	582.7	92.3
351	346.7	54.9	411	405.9	64.3	471	465.2	73.7	531	524.5	83.1	591	583.7	92.5
352	347.7	55.1	412	406.9	64.5	472	466.2	73.8	532	525.5	83.2	592	584.7	92.6
353	348.7	55.3	413	407.9	64.6	473	467.2	74.0	533	526.4	83.4	593	585.7	92.7
354	349.6	55.4	414	408.9	64.8	474	468.2	74.1	534	527.4	83.5	594	586.7	92.9
355	350.6	55.5	415	409.9	64.9	475	469.2	74.3	535	528.4	83.7	595	587.7	93.1
356	351.6	55.7	416	410.9	65.1	476	470.1	74.5	536	529.4	83.8	596	588.7	93.2
357	352.6	55.8	417	411.9	65.2	477	471.1	74.6	537	530.4	84.0	597	589.6	93.4
358	353.6	56.0	418	412.9	65.4	478	472.1	74.8	538	531.4	84.2	598	590.6	93.5
359	354.6	56.2	419	413.8	65.5	479	473.1	74.9	539	532.4	84.3	599	591.6	93.7
360	355.6	56.3	420	414.8	65.7	480	474.1	75.1	540	533.4	84.5	600	592.6	93.9
Dist	Dep.	D.Lat	Dist	Dep.	D.Lat	Dist	Dep.	D.Lat	Dist	Dep.	D.Lat	Dist	Dep.	D.Lat
D Lon		Dep.	D Lon		Dep.	D Lon		Dep.	D Lon		Dep.	D Lon		Dep.

TRAVERSE TABLE
10 DEGREES.

10° 350° / 190° ↑010° / 170° 0h 40m

D Lon Dep. / Dist	D.Lat	Dep.	Dist	D.Lat	Dep.	Dist	D.Lat	Dep.	Dist	D.Lat	Dep.	Dist	D.Lat	Dep.
1	01·0	00·2	61	60·1	10·6	121	119·2	21·0	181	178·3	31·4	241	237·3	41·8
2	02·0	00·3	62	61·1	10·8	122	120·1	21·2	182	179·2	31·6	242	238·3	42·0
3	03·0	00·5	63	62·0	10·9	123	121·1	21·4	183	180·2	31·8	243	239·3	42·2
4	03·9	00·7	64	63·0	11·1	124	122·1	21·5	184	181·2	32·0	244	240·3	42·4
5	04·9	00·9	65	64·0	11·3	125	123·1	21·7	185	182·2	32·1	245	241·3	42·5
6	05·9	01·0	66	65·0	11·5	126	124·1	21·9	186	183·2	32·3	246	242·3	42·7
7	06·9	01·2	67	66·0	11·6	127	125·1	22·1	187	184·2	32·5	247	243·2	42·9
8	07·9	01·4	68	67·0	11·8	128	126·1	22·2	188	185·1	32·6	248	244·2	43·1
9	08·9	01·6	69	68·0	12·0	129	127·0	22·4	189	186·1	32·8	249	245·2	43·2
10	09·8	01·7	70	68·9	12·2	130	128·0	22·6	190	187·1	33·0	250	246·2	43·4
11	10·8	01·9	71	69·9	12·3	131	129·0	22·7	191	188·1	33·2	251	247·2	43·6
12	11·8	02·1	72	70·9	12·5	132	130·0	22·9	192	189·1	33·3	252	248·2	43·8
13	12·8	02·3	73	71·9	12·7	133	131·0	23·1	193	190·1	33·5	253	249·2	43·9
14	13·8	02·4	74	72·9	12·8	134	132·0	23·3	194	191·0	33·7	254	250·1	44·1
15	14·8	02·6	75	73·9	13·0	135	132·9	23·4	195	192·0	33·9	255	251·1	44·3
16	15·8	02·8	76	74·8	13·2	136	133·9	23·6	196	193·0	34·0	256	252·1	44·5
17	16·7	03·0	77	75·8	13·4	137	134·9	23·8	197	194·0	34·2	257	253·1	44·6
18	17·7	03·1	78	76·8	13·5	138	135·9	24·0	198	195·0	34·4	258	254·1	44·8
19	18·7	03·3	79	77·8	13·7	139	136·9	24·1	199	196·0	34·6	259	255·1	45·0
20	19·7	03·5	80	78·8	13·9	140	137·9	24·3	200	197·0	34·7	260	256·1	45·1
21	20·7	03·6	81	79·8	14·1	141	138·9	24·5	201	197·9	34·9	261	257·0	45·3
22	21·7	03·8	82	80·8	14·2	142	139·8	24·7	202	198·9	35·1	262	258·0	45·5
23	22·7	04·0	83	81·7	14·4	143	140·8	24·8	203	199·9	35·3	263	259·0	45·7
24	23·6	04·2	84	82·7	14·6	144	141·8	25·0	204	200·9	35·4	264	260·0	45·8
25	24·6	04·3	85	83·7	14·8	145	142·8	25·2	205	201·9	35·6	265	261·0	46·0
26	25·6	04·5	86	84·7	14·9	146	143·8	25·4	206	202·9	35·8	266	262·0	46·2
27	26·6	04·7	87	85·7	15·1	147	144·8	25·5	207	203·9	35·9	267	262·9	46·4
28	27·6	04·9	88	86·7	15·3	148	145·8	25·7	208	204·8	36·1	268	263·9	46·5
29	28·6	05·0	89	87·6	15·5	149	146·7	25·9	209	205·8	36·3	269	264·9	46·7
30	29·5	05·2	90	88·6	15·6	150	147·7	26·0	210	206·8	36·5	270	265·9	46·9
31	30·5	05·4	91	89·6	15·8	151	148·7	26·2	211	207·8	36·6	271	266·9	47·1
32	31·5	05·6	92	90·6	16·0	152	149·7	26·4	212	208·8	36·8	272	267·9	47·2
33	32·5	05·7	93	91·6	16·1	153	150·7	26·6	213	209·8	37·0	273	268·9	47·4
34	33·5	05·9	94	92·6	16·3	154	151·7	26·7	214	210·7	37·2	274	269·8	47·6
35	34·5	06·1	95	93·6	16·5	155	152·6	26·9	215	211·7	37·3	275	270·8	47·8
36	35·5	06·3	96	94·5	16·7	156	153·6	27·1	216	212·7	37·5	276	271·8	47·9
37	36·4	06·4	97	95·5	16·8	157	154·6	27·3	217	213·7	37·7	277	272·8	48·1
38	37·4	06·6	98	96·5	17·0	158	155·6	27·4	218	214·7	37·9	278	273·8	48·3
39	38·4	06·8	99	97·5	17·2	159	156·6	27·6	219	215·7	38·0	279	274·8	48·4
40	39·4	06·9	100	98·5	17·4	160	157·6	27·8	220	216·7	38·2	280	275·7	48·6
41	40·4	07·1	101	99·5	17·5	161	158·6	28·0	221	217·6	38·4	281	276·7	48·8
42	41·4	07·3	102	100·5	17·7	162	159·5	28·1	222	218·6	38·6	282	277·7	49·0
43	42·3	07·5	103	101·4	17·9	163	160·5	28·3	223	219·6	38·7	283	278·7	49·1
44	43·3	07·6	104	102·4	18·1	164	161·5	28·5	224	220·6	38·9	284	279·7	49·3
45	44·3	07·8	105	103·4	18·2	165	162·5	28·7	225	221·6	39·1	285	280·7	49·5
46	45·3	08·0	106	104·4	18·4	166	163·5	28·8	226	222·6	39·2	286	281·7	49·7
47	46·3	08·2	107	105·4	18·6	167	164·5	29·0	227	223·6	39·4	287	282·6	49·8
48	47·3	08·3	108	106·4	18·8	168	165·5	29·2	228	224·5	39·6	288	283·6	50·0
49	48·3	08·5	109	107·3	18·9	169	166·4	29·3	229	225·5	39·8	289	284·6	50·2
50	49·2	08·7	110	108·3	19·1	170	167·4	29·5	230	226·5	39·9	290	285·6	50·4
51	50·2	08·9	111	109·3	19·3	171	168·4	29·7	231	227·5	40·1	291	286·6	50·5
52	51·2	09·0	112	110·3	19·4	172	169·4	29·9	232	228·5	40·3	292	287·6	50·7
53	52·2	09·2	113	111·3	19·6	173	170·4	30·0	233	229·5	40·5	293	288·5	50·9
54	53·2	09·4	114	112·3	19·8	174	171·4	30·2	234	230·4	40·6	294	289·5	51·1
55	54·2	09·6	115	113·3	20·0	175	172·3	30·4	235	231·4	40·8	295	290·5	51·2
56	55·1	09·7	116	114·2	20·1	176	173·3	30·6	236	232·4	41·0	296	291·5	51·4
57	56·1	09·9	117	115·2	20·3	177	174·3	30·7	237	233·4	41·2	297	292·5	51·6
58	57·1	10·1	118	116·2	20·5	178	175·3	30·9	238	234·4	41·3	298	293·5	51·7
59	58·1	10·2	119	117·2	20·7	179	176·3	31·1	239	235·4	41·5	299	294·5	51·9
60	59·1	10·4	120	118·2	20·8	180	177·3	31·3	240	236·4	41·7	300	295·4	52·1

Dist	Dep.	D.Lat	Dist	Dep.	D.Lat	Dist	Dep.	D.Lat	Dist	Dep.	D.Lat	Dist	Dep.	D.Lat
D Lon		Dep.	D Lon		Dep.	D Lon		Dep.	D Lon		Dep.	D Lon		Dep.

80° 280° / 260° ↑080° / 100° 5h 20m

80 DEGREES.

TRAVERSE TABLE
10 DEGREES.

350° ↑ / 190°   010° ↑ 170°   0h 40m   **10°**

D Lon	Dep.		D Lon	Dep.		D Lon	Dep.		D Lon	Dep.		D Lon	Dep.		D Lon	Dep.	
Dist	D.Lat	Dep.	Dist	D.Lat	Dep.	Dist	D.Lat	Dep.	Dist	D.Lat	Dep.	Dist	D.Lat	Dep.	Dist	D.Lat	Dep.
301	296.4	52.3	361	355.5	62.7	421	414.6	73.1	481	473.7	83.5	541	532.8	93.9			
302	297.4	52.4	362	356.5	62.9	422	415.6	73.3	482	474.7	83.7	542	533.8	94.1			
303	298.4	52.6	363	357.5	63.0	423	416.6	73.5	483	475.7	83.9	543	534.8	94.3			
304	299.4	52.8	364	358.5	63.2	424	417.6	73.6	484	476.6	84.1	544	535.7	94.5			
305	300.4	53.0	365	359.5	63.4	425	418.5	73.8	485	477.6	84.2	545	536.7	94.6			
306	301.4	53.1	366	360.4	63.6	426	419.5	74.0	486	478.6	84.4	546	537.7	94.8			
307	302.3	53.3	367	361.4	63.7	427	420.5	74.1	487	479.6	84.6	547	538.7	95.0			
308	303.3	53.5	368	362.4	63.9	428	421.5	74.3	488	480.6	84.7	548	539.7	95.2			
309	304.3	53.7	369	363.4	64.1	429	422.5	74.5	489	481.6	84.9	549	540.7	95.3			
310	305.3	53.8	370	364.4	64.2	430	423.5	74.7	490	482.6	85.1	550	541.6	95.5			
311	306.3	54.0	371	365.4	64.4	431	424.4	74.8	491	483.5	85.2	551	542.6	95.7			
312	307.3	54.2	372	366.4	64.6	432	425.4	75.0	492	484.5	85.4	552	543.6	95.9			
313	308.2	54.4	373	367.3	64.8	433	426.4	75.2	493	485.5	85.6	553	544.6	96.0			
314	309.2	54.5	374	368.3	64.9	434	427.4	75.4	494	486.5	85.8	554	545.6	96.2			
315	310.2	54.7	375	369.3	65.1	435	428.4	75.5	495	487.5	85.9	555	546.6	96.4			
316	311.2	54.9	376	370.3	65.3	436	429.4	75.7	496	488.5	86.1	556	547.6	96.5			
317	312.2	55.0	377	371.3	65.5	437	430.4	75.9	497	489.4	86.3	557	548.5	96.7			
318	313.2	55.2	378	372.3	65.6	438	431.3	76.1	498	490.4	86.5	558	549.5	96.9			
319	314.2	55.4	379	373.2	65.8	439	432.3	76.2	499	491.4	86.7	559	550.5	97.1			
320	315.1	55.6	380	374.2	66.0	440	433.3	76.4	500	492.4	86.8	560	551.5	97.2			
321	316.1	55.7	381	375.2	66.2	441	434.3	76.6	501	493.4	87.0	561	552.5	97.4			
322	317.1	55.9	382	376.2	66.3	442	435.3	76.8	502	494.4	87.2	562	553.5	97.6			
323	318.1	56.1	383	377.2	66.5	443	436.3	76.9	503	495.4	87.3	563	554.4	97.8			
324	319.1	56.3	384	378.2	66.7	444	437.3	77.1	504	496.3	87.5	564	555.4	97.9			
325	320.1	56.4	385	379.2	66.9	445	438.2	77.3	505	497.3	87.7	565	556.4	98.1			
326	321.0	56.6	386	380.1	67.0	446	439.2	77.4	506	498.3	87.9	566	557.4	98.3			
327	322.0	56.8	387	381.1	67.2	447	440.2	77.6	507	499.3	88.0	567	558.4	98.5			
328	323.0	57.0	388	382.1	67.4	448	441.2	77.8	508	500.3	88.2	568	559.4	98.6			
329	324.0	57.1	389	383.1	67.6	449	442.2	78.0	509	501.3	88.4	569	560.4	98.8			
330	325.0	57.3	390	384.1	67.7	450	443.2	78.1	510	502.3	88.6	570	561.3	99.0			
331	326.0	57.5	391	385.1	67.9	451	444.1	78.3	511	503.2	88.7	571	562.3	99.2			
332	327.0	57.7	392	386.0	68.1	452	445.1	78.5	512	504.2	88.9	572	563.3	99.3			
333	327.9	57.8	393	387.0	68.2	453	446.1	78.7	513	505.2	89.1	573	564.3	99.5			
334	328.9	58.0	394	388.0	68.4	454	447.1	78.8	514	506.2	89.2	574	565.3	99.7			
335	329.9	58.2	395	389.0	68.6	455	448.1	79.0	515	507.2	89.4	575	566.3	99.8			
336	330.9	58.3	396	390.0	68.8	456	449.1	79.2	516	508.2	89.6	576	567.2	100.0			
337	331.9	58.5	397	391.0	68.9	457	450.1	79.4	517	509.1	89.8	577	568.2	100.2			
338	332.9	58.7	398	392.0	69.1	458	451.0	79.5	518	510.1	90.0	578	569.2	100.4			
339	333.8	58.9	399	392.9	69.3	459	452.0	79.7	519	511.1	90.1	579	570.2	100.5			
340	334.8	59.0	400	393.9	69.5	460	453.0	79.9	520	512.1	90.3	580	571.2	100.7			
341	335.8	59.2	401	394.9	69.6	461	454.0	80.1	521	513.1	90.5	581	572.2	100.9			
342	336.8	59.4	402	395.9	69.8	462	455.0	80.2	522	514.1	90.6	582	573.2	101.1			
343	337.8	59.6	403	396.9	70.0	463	456.0	80.4	523	515.1	90.8	583	574.1	101.2			
344	338.8	59.7	404	397.9	70.2	464	457.0	80.6	524	516.0	91.0	584	575.1	101.4			
345	339.8	59.9	405	398.9	70.3	465	457.9	80.7	525	517.0	91.2	585	576.1	101.6			
346	340.7	60.1	406	399.8	70.5	466	458.9	80.9	526	518.0	91.3	586	577.1	101.8			
347	341.7	60.3	407	400.8	70.7	467	459.9	81.1	527	519.0	91.5	587	578.1	101.9			
348	342.7	60.4	408	401.8	70.8	468	460.9	81.3	528	520.0	91.7	588	579.1	102.1			
349	343.7	60.6	409	402.8	71.0	469	461.9	81.4	529	521.0	91.9	589	580.1	102.3			
350	344.7	60.8	410	403.8	71.2	470	462.9	81.6	530	521.9	92.0	590	581.0	102.5			
351	345.7	61.0	411	404.8	71.4	471	463.8	81.8	531	522.9	92.2	591	582.0	102.6			
352	346.7	61.1	412	405.7	71.5	472	464.8	82.0	532	523.9	92.4	592	583.0	102.8			
353	347.6	61.3	413	406.7	71.7	473	465.8	82.1	533	524.9	92.6	593	584.0	103.0			
354	348.6	61.5	414	407.7	71.9	474	466.8	82.3	534	525.9	92.7	594	585.0	103.1			
355	349.6	61.6	415	408.7	72.1	475	467.8	82.5	535	526.9	92.9	595	586.0	103.3			
356	350.6	61.8	416	409.7	72.2	476	468.8	82.7	536	527.9	93.1	596	586.9	103.5			
357	351.6	62.0	417	410.7	72.4	477	469.8	82.8	537	528.8	93.2	597	587.8	103.7			
358	352.6	62.2	418	411.6	72.6	478	470.7	83.0	538	529.8	93.4	598	588.9	103.8			
359	353.5	62.3	419	412.6	72.8	479	471.7	83.2	539	530.8	93.6	599	589.9	104.0			
360	354.5	62.5	420	413.6	72.9	480	472.7	83.4	540	531.8	93.8	600	590.9	104.2			
Dist	Dep.	D.Lat	Dist	Dep.	D.Lat	Dist	Dep.	D.Lat	Dist	Dep.	D.Lat	Dist	Dep.	D.Lat	Dist	Dep.	D.Lat
D Lon		Dep.	D Lon		Dep.	D Lon		Dep.	D Lon		Dep.	D Lon		Dep.	D Lon		Dep.

280° ↑ / 260°   **80 DEGREES.**   080° ↑ 100°   5h 20m   **80°**

TRAVERSE TABLE
11 DEGREES.

| 349° | 191° | | | | | | | 011° 169° | 0h 44m |

Dist	D.Lat	Dep.	Dist	D.Lat	Dep.	Dist	D.Lat	Dep.	Dist	D.Lat	Dep.	Dist	D.Lat	Dep.
1	01·0	00·2	61	59·9	11·6	121	118·8	23·1	181	177·7	34·5	241	236·6	46·0
2	02·0	00·4	62	60·9	11·8	122	119·8	23·3	182	178·7	34·7	242	237·6	46·2
3	02·9	00·6	63	61·8	12·0	123	120·7	23·5	183	179·6	34·9	243	238·5	46·4
4	03·9	00·8	64	62·8	12·2	124	121·7	23·7	184	180·6	35·1	244	239·5	46·6
5	04·9	01·0	65	63·8	12·4	125	122·7	23·9	185	181·6	35·3	245	240·5	46·7
6	05·9	01·1	66	64·8	12·6	126	123·7	24·0	186	182·6	35·5	246	241·5	46·9
7	06·9	01·3	67	65·8	12·8	127	124·7	24·2	187	183·6	35·7	247	242·5	47·1
8	07·9	01·5	68	66·8	13·0	128	125·6	24·4	188	184·5	35·9	248	243·4	47·3
9	08·8	01·7	69	67·7	13·2	129	126·6	24·6	189	185·5	36·1	249	244·4	47·5
10	09·8	01·9	70	68·7	13·4	130	127·6	24·8	190	186·5	36·3	250	245·4	47·7
11	10·8	02·1	71	69·7	13·5	131	128·6	25·0	191	187·5	36·4	251	246·4	47·9
12	11·8	02·3	72	70·7	13·7	132	129·6	25·2	192	188·5	36·6	252	247·4	48·1
13	12·8	02·5	73	71·7	13·9	133	130·6	25·4	193	189·5	36·8	253	248·4	48·3
14	13·7	02·7	74	72·6	14·1	134	131·5	25·6	194	190·4	37·0	254	249·3	48·5
15	14·7	02·9	75	73·6	14·3	135	132·5	25·8	195	191·4	37·2	255	250·3	48·7
16	15·7	03·1	76	74·6	14·5	136	133·5	26·0	196	192·4	37·4	256	251·3	48·8
17	16·7	03·2	77	75·6	14·7	137	134·5	26·1	197	193·3	37·6	257	252·3	49·0
18	17·7	03·4	78	76·6	14·9	138	135·5	26·3	198	194·4	37·8	258	253·3	49·2
19	18·7	03·6	79	77·5	15·1	139	136·4	26·5	199	195·3	38·0	259	254·2	49·4
20	19·6	03·8	80	78·5	15·3	140	137·4	26·7	200	196·3	38·2	260	255·2	49·6
21	20·6	04·0	81	79·5	15·5	141	138·4	26·9	201	197·3	38·4	261	256·2	49·8
22	21·6	04·2	82	80·5	15·6	142	139·4	27·1	202	198·3	38·5	262	257·2	50·0
23	22·6	04·4	83	81·5	15·8	143	140·4	27·3	203	199·3	38·7	263	258·2	50·2
24	23·6	04·6	84	82·5	16·0	144	141·4	27·5	204	200·3	38·9	264	259·1	50·4
25	24·5	04·8	85	83·4	16·2	145	142·3	27·7	205	201·2	39·1	265	260·1	50·6
26	25·5	05·0	86	84·4	16·4	146	143·3	27·9	206	202·2	39·3	266	261·1	50·8
27	26·5	05·2	87	85·4	16·6	147	144·3	28·0	207	203·2	39·5	267	262·1	50·9
28	27·5	05·3	88	86·4	16·8	148	145·3	28·2	208	204·2	39·7	268	263·1	51·1
29	28·5	05·5	89	87·4	17·0	149	146·3	28·4	209	205·2	39·9	269	264·1	51·3
30	29·4	05·7	90	88·3	17·2	150	147·2	28·6	210	206·1	40·1	270	265·0	51·5
31	30·4	05·9	91	89·3	17·4	151	148·2	28·8	211	207·1	40·3	271	266·0	51·7
32	31·4	06·1	92	90·3	17·6	152	149·2	29·0	212	208·1	40·5	272	267·0	51·9
33	32·4	06·3	93	91·3	17·7	153	150·2	29·2	213	209·1	40·6	273	268·0	52·1
34	33·4	06·5	94	92·3	17·9	154	151·2	29·4	214	210·1	40·8	274	269·0	52·3
35	34·4	06·7	95	93·3	18·1	155	152·2	29·6	215	211·0	41·0	275	269·9	52·5
36	35·3	06·9	96	94·2	18·3	156	153·1	29·8	216	212·0	41·2	276	270·9	52·7
37	36·3	07·1	97	95·2	18·5	157	154·1	30·0	217	213·0	41·4	277	271·9	52·9
38	37·3	07·3	98	96·2	18·7	158	155·1	30·1	218	214·0	41·6	278	272·9	53·0
39	38·3	07·4	99	97·2	18·9	159	156·1	30·3	219	215·0	41·8	279	273·9	53·2
40	39·3	07·6	100	98·2	19·1	160	157·1	30·5	220	216·0	42·0	280	274·9	53·4
41	40·2	07·8	101	99·1	19·3	161	158·0	30·7	221	216·9	42·2	281	275·8	53·6
42	41·2	08·0	102	100·1	19·5	162	159·0	30·9	222	217·9	42·4	282	276·8	53·8
43	42·2	08·2	103	101·1	19·7	163	160·0	31·1	223	218·9	42·6	283	277·8	54·0
44	43·2	08·4	104	102·1	19·8	164	161·0	31·3	224	219·9	42·7	284	278·8	54·2
45	44·2	08·6	105	103·1	20·0	165	162·0	31·5	225	220·9	42·9	285	279·8	54·4
46	45·2	08·8	106	104·1	20·2	166	163·0	31·7	226	221·8	43·1	286	280·7	54·6
47	46·1	09·0	107	105·0	20·4	167	163·9	31·9	227	222·8	43·3	287	281·7	54·8
48	47·1	09·2	108	106·0	20·6	168	164·9	32·1	228	223·8	43·5	288	282·7	55·0
49	48·1	09·3	109	107·0	20·8	169	165·9	32·2	229	224·8	43·7	289	283·7	55·1
50	49·1	09·5	110	108·0	21·0	170	166·9	32·4	230	225·8	43·9	290	284·7	55·3
51	50·1	09·7	111	109·0	21·2	171	167·9	32·6	231	226·8	44·1	291	285·7	55·5
52	51·0	09·9	112	109·9	21·4	172	168·8	32·8	232	227·7	44·3	292	286·6	55·7
53	52·0	10·1	113	110·9	21·6	173	169·8	33·0	233	228·7	44·5	293	287·6	55·9
54	53·0	10·3	114	111·9	21·8	174	170·8	33·2	234	229·7	44·6	294	288·6	56·1
55	54·0	10·5	115	112·9	21·9	175	171·8	33·4	235	230·7	44·8	295	289·6	56·3
56	55·0	10·7	116	113·9	22·1	176	172·8	33·6	236	231·7	45·0	296	290·6	56·5
57	56·0	10·9	117	114·9	22·3	177	173·7	33·8	237	232·6	45·2	297	291·5	56·7
58	56·9	11·1	118	115·8	22·5	178	174·7	34·0	238	233·6	45·4	298	292·5	56·9
59	57·9	11·3	119	116·8	22·7	179	175·7	34·2	239	234·6	45·6	299	293·5	57·1
60	58·9	11·4	120	117·8	22·9	180	176·7	34·3	240	235·6	45·8	300	294·5	57·2

| Dist | Dep. | D.Lat | Dist | Dep. | D.Lat | Dist | Dep. | D.Lat | Dist | Dep. | D.Lat | Dist | Dep. | D.Lat |

| D Lon | | Dep. | D Lon | | Dep. | D Lon | | Dep. | D Lon | | Dep. | D Lon | | Dep. |

| 281° | 259° | | **79 DEGREES.** | | | 079° 101° | 5h 16m |

TRAVERSE TABLE
11 DEGREES.

349° ↑ / 191°   011° ↑ / 169°   0h 44m   **11°**

D Lon Dep.			D Lon Dep.			D Lon Dep.			D Lon Dep.			D Lon Dep.		
Dist	D.Lat	Dep.	Dist	D.Lat	Dep.	Dist	D.Lat	Dep.	Dist	D.Lat	Dep.	Dist	D.Lat	Dep.
301	295·5	57·4	361	354·4	68·9	421	413·3	80·3	481	472·2	91·8	541	531·1	103·2
302	296·5	57·6	362	355·3	69·1	422	414·2	80·5	482	473·1	92·0	542	532·0	103·4
303	297·4	57·8	363	356·3	69·3	423	415·2	80·7	483	474·1	92·2	543	533·0	103·6
304	298·4	58·0	364	357·3	69·5	424	416·2	80·9	484	475·1	92·4	544	534·0	103·8
305	299·4	58·2	365	358·3	69·8	425	417·2	81·1	485	476·1	92·5	545	535·0	104·0
306	300·4	58·4	366	359·3	69·8	426	418·2	81·3	486	477·1	92·7	546	536·0	104·2
307	301·4	58·6	367	360·3	70·0	427	419·2	81·5	487	478·1	92·9	547	537·0	104·4
308	302·3	58·8	368	361·2	70·2	428	420·1	81·7	488	479·0	93·1	548	537·9	104·6
309	303·3	59·0	369	362·2	70·4	429	421·1	81·9	489	480·0	93·3	549	538·9	104·6
310	304·3	59·2	370	363·2	70·6	430	422·1	82·0	490	481·0	93·5	550	539·9	104·9
311	305·3	59·3	371	364·2	70·8	431	423·0	82·2	491	482·0	93·7	551	540·9	105·1
312	306·3	59·5	372	365·2	71·0	432	424·1	82·4	492	483·0	93·9	552	541·9	105·3
313	307·2	59·7	373	366·1	71·2	433	425·0	82·6	493	483·9	94·1	553	542·8	105·5
314	308·2	59·9	374	367·1	71·4	434	426·0	82·8	494	484·9	94·3	554	543·8	105·7
315	309·2	60·1	375	368·1	71·6	435	427·0	83·0	495	485·9	94·5	555	544·8	105·9
316	310·2	60·3	376	369·1	71·7	436	428·0	83·2	496	486·9	94·6	556	545·8	106·1
317	311·2	60·5	377	370·1	71·9	437	428·9	83·4	497	487·9	94·8	557	546·8	106·3
318	312·2	60·7	378	371·1	72·1	438	430·0	83·6	498	488·9	95·0	558	547·7	106·5
319	313·1	60·9	379	372·0	72·3	439	430·9	83·8	499	489·8	95·2	559	548·7	106·7
320	314·1	61·1	380	373·0	72·5	440	431·9	84·0	500	490·8	95·4	560	549·7	106·9
321	315·1	61·2	381	374·0	72·7	441	432·9	84·1	501	491·8	95·6	561	550·7	107·0
322	316·1	61·4	382	375·0	72·9	442	433·9	84·3	502	492·8	95·8	562	551·7	107·2
323	317·1	61·6	383	376·0	73·1	443	434·9	84·5	503	493·8	96·0	563	552·7	107·4
324	318·0	61·8	384	376·9	73·3	444	435·8	84·7	504	494·7	96·2	564	553·6	107·6
325	319·0	62·0	385	377·9	73·5	445	436·8	84·9	505	495·7	96·4	565	554·6	107·8
326	320·0	62·2	386	378·9	73·7	446	437·8	85·1	506	496·7	96·5	566	555·6	108·0
327	321·0	62·4	387	379·9	73·8	447	438·8	85·3	507	497·7	96·7	567	556·6	108·2
328	322·0	62·6	388	380·9	74·0	448	439·8	85·5	508	498·7	96·9	568	557·6	108·4
329	323·0	62·8	389	381·9	74·2	449	440·8	85·7	509	499·6	97·1	569	558·5	108·6
330	323·9	63·0	390	382·8	74·4	450	441·7	85·9	510	500·6	97·3	570	559·5	108·8
331	324·9	63·2	391	383·8	74·6	451	442·7	86·1	511	501·6	97·5	571	560·5	109·0
332	325·9	63·3	392	384·8	74·8	452	443·7	86·2	512	502·6	97·7	572	561·5	109·1
333	326·9	63·5	393	385·8	75·0	453	444·7	86·4	513	503·6	97·9	573	562·5	109·3
334	327·9	63·7	394	386·8	75·2	454	445·7	86·6	514	504·6	98·1	574	563·5	109·5
335	328·8	63·9	395	387·7	75·4	455	446·6	86·8	515	505·5	98·3	575	564·4	109·7
336	329·8	64·1	396	388·7	75·6	456	447·6	87·0	516	506·5	98·5	576	565·4	109·9
337	330·8	64·3	397	389·7	75·8	457	448·6	87·2	517	507·5	98·6	577	566·4	110·1
338	331·8	64·5	398	390·7	75·9	458	449·6	87·4	518	508·5	98·8	578	567·4	110·3
339	332·7	64·7	399	391·7	76·1	459	450·6	87·6	519	509·5	99·0	579	568·4	110·5
340	333·7	64·9	400	392·7	76·3	460	451·5	87·8	520	510·4	99·2	580	569·3	110·7
341	334·7	65·1	401	393·6	76·5	461	452·5	88·0	521	511·4	99·4	581	570·3	110·9
342	335·7	65·3	402	394·6	76·7	462	453·5	88·2	522	512·4	99·6	582	571·3	111·1
343	336·7	65·4	403	395·6	76·9	463	454·5	88·3	523	513·4	99·8	583	572·3	111·2
344	337·7	65·6	404	396·6	77·1	464	455·5	88·5	524	514·4	100·0	584	573·3	111·4
345	338·7	65·8	405	397·6	77·3	465	456·5	88·7	525	515·4	100·2	585	574·3	111·6
346	339·6	66·0	406	398·5	77·5	466	457·4	88·9	526	516·3	100·4	586	575·2	111·8
347	340·6	66·2	407	399·5	77·7	467	458·4	89·1	527	517·3	100·6	587	576·2	112·1
348	341·6	66·4	408	400·5	77·9	468	459·4	89·3	528	518·3	100·7	588	577·2	112·3
349	342·6	66·6	409	401·5	78·0	469	460·4	89·5	529	519·3	100·9	589	578·2	112·4
350	343·6	66·8	410	402·5	78·2	470	461·4	89·7	530	520·3	101·1	590	579·2	112·6
351	344·6	67·0	411	403·4	78·4	471	462·3	89·9	531	521·2	101·3	591	580·1	112·8
352	345·5	67·2	412	404·4	78·6	472	463·3	90·1	532	522·2	101·5	592	581·1	113·0
353	346·5	67·4	413	405·4	78·8	473	464·3	90·3	533	523·2	101·7	593	582·1	113·2
354	347·5	67·5	414	406·4	79·0	474	465·3	90·4	534	524·2	101·9	594	583·1	113·3
355	348·5	67·7	415	407·4	79·2	475	466·3	90·6	535	525·2	102·1	595	584·1	113·5
356	349·5	67·9	416	408·4	79·4	476	467·3	90·8	536	526·2	102·3	596	585·0	113·7
357	350·4	68·1	417	409·3	79·6	477	468·2	91·0	537	527·1	102·5	597	586·0	113·9
358	351·4	68·3	418	410·3	79·7	478	469·2	91·2	538	528·1	102·7	598	587·0	114·1
359	352·4	68·5	419	411·3	79·9	479	470·2	91·4	539	529·1	102·8	599	588·0	114·3
360	353·4	68·7	420	412·3	80·1	480	471·2	91·6	540	530·1	103·0	600	589·0	114·5
Dist	Dep.	D.Lat	Dist	Dep.	D.Lat	Dist	Dep.	D.Lat	Dist	Dep.	D.Lat	Dist	Dep.	D.Lat
D Lon		Dep.	D Lon		Dep.	D Lon		Dep.	D Lon		Dep.	D Lon		Dep.

281° ↑ / 259°   **79 DEGREES.**   079° ↑ / 101°   5h 16m   **79°**

12°

TRAVERSE TABLE
12 DEGREES.

D Lon	Dep.		D Lon	Dep.		D Lon	Dep.		D Lon	Dep.		D Lon	Dep.	
Dist	D.Lat	Dep.	Dist	D.Lat	Dep.	Dist	D.Lat	Dep.	Dist	D.Lat	Dep.	Dist	D.Lat	Dep.
1	01·0	00·2	61	59·7	12·7	121	118·4	25·2	181	177·0	37·6	241	235·7	50·1
2	02·0	00·4	62	60·6	12·9	122	119·3	25·4	182	178·0	37·8	242	236·7	50·3
3	02·9	00·6	63	61·6	13·1	123	120·3	25·6	183	179·0	38·0	243	237·7	50·5
4	03·9	00·8	64	62·6	13·3	124	121·3	25·8	184	180·0	38·3	244	238·7	50·7
5	04·9	01·0	65	63·6	13·5	125	122·3	26·0	185	181·0	38·6	245	239·6	50·9
6	05·9	01·2	66	64·6	13·7	126	123·2	26·2	186	181·9	38·7	246	240·6	51·1
7	06·8	01·5	67	65·5	13·9	127	124·2	26·4	187	182·9	38·9	247	241·6	51·4
8	07·8	01·7	68	66·5	14·1	128	125·2	26·6	188	183·9	39·1	248	242·6	51·6
9	08·8	01·9	69	67·5	14·3	129	126·2	26·8	189	184·9	39·3	249	243·6	51·8
10	09·8	02·1	70	68·5	14·6	130	127·2	27·0	190	185·8	39·5	250	244·5	52·0
11	10·8	02·3	71	69·4	14·8	131	128·1	27·2	191	186·8	39·7	251	245·5	52·2
12	11·7	02·5	72	70·4	15·0	132	129·1	27·4	192	187·8	39·9	252	246·5	52·4
13	12·7	02·7	73	71·4	15·2	133	130·1	27·7	193	188·8	40·1	253	247·5	52·6
14	13·7	02·9	74	72·4	15·4	134	131·1	27·9	194	189·8	40·3	254	248·4	52·8
15	14·7	03·1	75	73·4	15·6	135	132·0	28·1	195	190·7	40·5	255	249·4	53·0
16	15·7	03·3	76	74·3	15·8	136	133·0	28·3	196	191·7	40·8	256	250·4	53·2
17	16·6	03·5	77	75·3	16·0	137	134·0	28·5	197	192·7	41·0	257	251·4	53·4
18	17·6	03·7	78	76·3	16·2	138	135·0	28·7	198	193·7	41·2	258	252·4	53·6
19	18·6	04·0	79	77·3	16·4	139	136·0	28·9	199	194·7	41·4	259	253·3	53·8
20	19·6	04·2	80	78·3	16·6	140	136·9	29·1	200	195·6	41·6	260	254·3	54·1
21	20·5	04·4	81	79·2	16·8	141	137·9	29·3	201	196·6	41·8	261	255·3	54·3
22	21·5	04·6	82	80·2	17·0	142	138·9	29·5	202	197·6	42·0	262	256·3	54·5
23	22·5	04·8	83	81·2	17·3	143	139·9	29·7	203	198·6	42·2	263	257·3	54·7
24	23·5	05·0	84	82·1	17·5	144	140·9	29·9	204	199·5	42·4	264	258·2	54·9
25	24·5	05·2	85	83·1	17·7	145	141·8	30·1	205	200·5	42·6	265	259·2	55·1
26	25·4	05·4	86	84·1	17·9	146	142·8	30·4	206	201·5	42·8	266	260·2	55·3
27	26·4	05·6	87	85·1	18·1	147	143·8	30·6	207	202·5	43·0	267	261·2	55·5
28	27·4	05·8	88	86·1	18·3	148	144·8	30·8	208	203·5	43·2	268	262·1	55·7
29	28·4	06·0	89	87·1	18·5	149	145·7	31·0	209	204·4	43·4	269	263·1	55·9
30	29·3	06·2	90	88·0	18·7	150	146·7	31·2	210	205·4	43·7	270	264·1	56·1
31	30·3	06·4	91	89·0	18·9	151	147·7	31·4	211	206·4	43·9	271	265·1	56·3
32	31·3	06·7	92	90·0	19·1	152	148·7	31·6	212	207·4	44·1	272	266·1	56·6
33	32·3	06·9	93	91·0	19·3	153	149·7	31·8	213	208·3	44·3	273	267·0	56·8
34	33·3	07·1	94	91·9	19·5	154	150·6	32·0	214	209·3	44·5	274	268·0	57·0
35	34·2	07·3	95	92·9	19·8	155	151·6	32·2	215	210·3	44·7	275	269·0	57·2
36	35·2	07·5	96	93·9	20·0	156	152·6	32·4	216	211·3	44·9	276	270·0	57·4
37	36·2	07·7	97	94·9	20·2	157	153·6	32·6	217	212·3	45·1	277	270·9	57·6
38	37·2	07·9	98	95·9	20·4	158	154·5	32·9	218	213·2	45·3	278	271·9	57·8
39	38·1	08·1	99	96·8	20·6	159	155·5	33·1	219	214·2	45·5	279	272·9	58·0
40	39·1	08·3	100	97·8	20·8	160	156·5	33·3	220	215·2	45·7	280	273·9	58·2
41	40·1	08·5	101	98·8	21·0	161	157·5	33·5	221	216·2	45·9	281	274·9	58·4
42	41·1	08·7	102	99·8	21·2	162	158·5	33·7	222	217·1	46·2	282	275·8	58·6
43	42·1	08·9	103	100·7	21·4	163	159·4	33·9	223	218·1	46·4	283	276·8	58·8
44	43·0	09·1	104	101·7	21·6	164	160·4	34·1	224	219·1	46·6	284	277·8	59·0
45	44·0	09·4	105	102·7	21·8	165	161·4	34·3	225	220·1	46·8	285	278·8	59·3
46	45·0	09·6	106	103·7	22·0	166	162·4	34·5	226	221·1	47·0	286	279·8	59·5
47	46·0	09·8	107	104·7	22·2	167	163·4	34·7	227	222·0	47·2	287	280·7	59·7
48	47·0	10·0	108	105·6	22·5	168	164·3	34·9	228	223·0	47·4	288	281·7	59·9
49	47·9	10·2	109	106·6	22·7	169	165·3	35·1	229	224·0	47·6	289	282·7	60·1
50	48·9	10·4	110	107·6	22·9	170	166·3	35·3	230	225·0	47·8	290	283·7	60·3
51	49·9	10·6	111	108·6	23·1	171	167·3	35·6	231	226·0	48·0	291	284·6	60·5
52	50·9	10·8	112	109·6	23·3	172	168·2	35·8	232	226·9	48·2	292	285·6	60·7
53	51·8	11·0	113	110·5	23·5	173	169·2	36·0	233	227·9	48·4	293	286·6	60·9
54	52·8	11·2	114	111·5	23·7	174	170·2	36·2	234	228·9	48·7	294	287·6	61·1
55	53·8	11·4	115	112·5	23·9	175	171·2	36·4	235	229·9	48·9	295	288·6	61·3
56	54·8	11·6	116	113·5	24·1	176	172·2	36·6	236	230·8	49·1	296	289·5	61·5
57	55·8	11·9	117	114·4	24·3	177	173·1	36·8	237	231·8	49·3	297	290·5	61·7
58	56·7	12·1	118	115·4	24·5	178	174·1	37·0	238	232·8	49·5	298	291·5	62·0
59	57·7	12·3	119	116·4	24·7	179	175·1	37·2	239	233·8	49·7	299	292·5	62·2
60	58·7	12·5	120	117·4	24·9	180	176·1	37·4	240	234·8	49·9	300	293·4	62·4
Dist	Dep.	D.Lat	Dist	Dep.	D.Lat	Dist	Dep.	D.Lat	Dist	Dep.	D.Lat	Dist	Dep.	D.Lat
D Lon		Dep.	D Lon		Dep.	D Lon		Dep.	D Lon		Dep.	D Lon		Dep.

78°

TRAVERSE TABLE
12 DEGREES.

Dist	D.Lat	Dep.	Dist	D.Lat	Dep.	Dist	D.Lat	Dep.	Dist	D.Lat	Dep.	Dist	D.Lat	Dep.
301	294.4	62.6	361	353.1	75.1	421	411.8	87.5	481	470.5	100.0	541	529.2	112.5
302	295.4	62.8	362	354.1	75.3	422	412.8	87.7	482	471.5	100.2	542	530.2	112.7
303	296.4	63.0	363	355.1	75.5	423	413.8	87.9	483	472.4	100.4	543	531.1	112.9
304	297.4	63.2	364	356.0	75.7	424	414.7	88.2	484	473.4	100.6	544	532.1	113.1
305	298.3	63.4	365	357.0	75.9	425	415.7	88.4	485	474.4	100.8	545	533.1	113.3
306	299.3	63.6	366	358.0	76.1	426	416.7	88.6	486	475.4	101.0	546	534.1	113.5
307	300.3	63.8	367	359.0	76.3	427	417.7	88.8	487	476.4	101.3	547	535.0	113.7
308	301.3	64.0	368	360.0	76.5	428	418.6	89.0	488	477.3	101.5	548	536.0	113.9
309	302.2	64.2	369	360.9	76.7	429	419.6	89.2	489	478.3	101.7	549	537.0	114.1
310	303.2	64.5	370	361.9	76.9	430	420.6	89.4	490	479.3	101.9	550	538.0	114.4
311	304.2	64.7	371	362.9	77.1	431	421.6	89.6	491	480.3	102.1	551	539.0	114.6
312	305.2	64.9	372	363.9	77.3	432	422.6	89.8	492	481.2	102.3	552	539.9	114.8
313	306.2	65.1	373	364.8	77.6	433	423.5	90.0	493	482.2	102.5	553	540.9	115.0
314	307.1	65.3	374	365.8	77.8	434	424.5	90.2	494	483.2	102.7	554	541.9	115.2
315	308.1	65.5	375	366.8	78.0	435	425.5	90.4	495	484.2	102.9	555	542.9	115.4
316	309.1	65.7	376	367.8	78.2	436	426.5	90.6	496	485.2	103.1	556	543.9	115.6
317	310.1	65.9	377	368.8	78.4	437	427.5	90.9	497	486.1	103.3	557	544.8	115.8
318	311.1	66.1	378	369.7	78.6	438	428.4	91.1	498	487.1	103.5	558	545.8	116.0
319	312.0	66.3	379	370.7	78.8	439	429.4	91.3	499	488.1	103.7	559	546.8	116.2
320	313.0	66.5	380	371.7	79.0	440	430.4	91.5	500	489.1	104.0	560	547.8	116.4
321	314.0	66.7	381	372.7	79.2	441	431.4	91.7	501	490.1	104.2	561	548.7	116.6
322	315.0	66.9	382	373.7	79.4	442	432.3	91.9	502	491.0	104.4	562	549.7	116.8
323	315.9	67.2	383	374.6	79.6	443	433.3	92.1	503	492.0	104.6	563	550.7	117.1
324	316.9	67.4	384	375.6	79.8	444	434.3	92.3	504	493.0	104.8	564	551.7	117.3
325	317.9	67.6	385	376.6	80.0	445	435.3	92.5	505	494.0	105.0	565	552.7	117.5
326	318.9	67.8	386	377.6	80.3	446	436.3	92.7	506	494.9	105.2	566	553.6	117.7
327	319.9	68.0	387	378.5	80.5	447	437.2	92.9	507	495.9	105.4	567	554.6	117.9
328	320.8	68.2	388	379.5	80.7	448	438.2	93.1	508	496.9	105.6	568	555.6	118.1
329	321.8	68.4	389	380.5	80.9	449	439.2	93.4	509	497.9	105.8	569	556.6	118.3
330	322.8	68.6	390	381.5	81.1	450	440.2	93.6	510	498.9	106.0	570	557.5	118.5
331	323.8	68.8	391	382.5	81.3	451	441.1	93.8	511	499.8	106.2	571	558.5	118.7
332	324.7	69.0	392	383.4	81.5	452	442.1	94.0	512	500.8	106.5	572	559.5	118.9
333	325.7	69.2	393	384.4	81.7	453	443.1	94.2	513	501.8	106.7	573	560.5	119.1
334	326.7	69.4	394	385.4	81.9	454	444.1	94.4	514	502.8	106.9	574	561.5	119.3
335	327.7	69.7	395	386.4	82.1	455	445.1	94.6	515	503.7	107.1	575	562.4	119.5
336	328.7	69.9	396	387.3	82.3	456	446.0	94.8	516	504.7	107.3	576	563.4	119.8
337	329.6	70.1	397	388.3	82.5	457	447.0	95.0	517	505.7	107.5	577	564.4	120.0
338	330.6	70.3	398	389.3	82.7	458	448.0	95.2	518	506.7	107.7	578	565.4	120.2
339	331.6	70.5	399	390.3	83.0	459	449.0	95.4	519	507.7	107.9	579	566.3	120.4
340	332.6	70.7	400	391.3	83.2	460	449.9	95.6	520	508.6	108.1	580	567.3	120.6
341	333.5	70.9	401	392.2	83.4	461	450.9	95.8	521	509.6	108.3	581	568.3	120.8
342	334.5	71.1	402	393.2	83.6	462	451.9	96.1	522	510.6	108.5	582	569.3	121.0
343	335.5	71.3	403	394.2	83.8	463	452.9	96.3	523	511.6	108.7	583	570.3	121.2
344	336.5	71.5	404	395.2	84.0	464	453.9	96.5	524	512.5	108.9	584	571.2	121.4
345	337.5	71.7	405	396.2	84.2	465	454.8	96.7	525	513.5	109.2	585	572.2	121.6
346	338.4	71.9	406	397.1	84.4	466	455.8	96.9	526	514.5	109.4	586	573.2	121.8
347	339.4	72.1	407	398.1	84.6	467	456.8	97.1	527	515.5	109.6	587	574.2	122.0
348	340.4	72.4	408	399.1	84.8	468	457.8	97.3	528	516.5	109.8	588	575.2	122.3
349	341.4	72.6	409	400.1	85.0	469	458.8	97.5	529	517.4	110.0	589	576.1	122.5
350	342.4	72.8	410	401.0	85.2	470	459.7	97.7	530	518.4	110.2	590	577.1	122.7
351	343.3	73.0	411	402.0	85.5	471	460.7	97.9	531	519.4	110.4	591	578.1	122.9
352	344.3	73.2	412	403.0	85.7	472	461.7	98.1	532	520.4	110.6	592	579.1	123.1
353	345.3	73.4	413	404.0	85.9	473	462.7	98.3	533	521.4	110.8	593	580.0	123.3
354	346.3	73.6	414	405.0	86.1	474	463.6	98.6	534	522.3	111.0	594	581.0	123.5
355	347.2	73.8	415	405.9	86.3	475	464.6	98.8	535	523.3	111.2	595	582.0	123.7
356	348.2	74.0	416	406.9	86.5	476	465.6	99.0	536	524.3	111.4	596	583.0	123.9
357	349.2	74.2	417	407.9	86.7	477	466.6	99.2	537	525.3	111.6	597	584.0	124.1
358	350.2	74.4	418	408.9	86.9	478	467.6	99.4	538	526.2	111.9	598	584.9	124.3
359	351.2	74.6	419	409.9	87.1	479	468.5	99.6	539	527.2	112.1	599	585.9	124.5
360	352.1	74.8	420	410.8	87.3	480	469.5	99.8	540	528.2	112.3	600	586.9	124.7

Dist	Dep.	D.Lat	Dist	Dep.	D.Lat	Dist	Dep.	D.Lat	Dist	Dep.	D.Lat	Dist	Dep.	D.Lat

13°

Dist	D.Lat	Dep.	Dist	D.Lat	Dep.	Dist	D.Lat	Dep.	Dist	D.Lat	Dep.	Dist	D.Lat	Dep.
1	01·0	00·2	61	59·4	13·7	121	117·9	27·2	181	176·4	40·7	241	234·8	54·2
2	01 9	00·4	62	60·4	13·9	122	118·9	27·4	182	177·3	40·9	242	235·8	54·4
3	02·9	00·7	63	61·4	14·2	123	119·8	27·7	183	178·3	41·2	243	236·8	54·7
4	03·9	00·9	04	62·4	14·4	124	120·8	27·9	184	179·3	41·4	244	237·7	54·9
5	04·9	01 1	65	63·3	14·6	125	121 8	28·1	185	180·3	41·6	245	238·7	55·1
6	05·8	01·3	66	64·3	14·8	126	122·8	28·3	186	181·2	41·8	246	239·7	55·3
7	06·8	01·6	67	65·3	15·1	127	123·7	28·6	187	182·2	42·1	247	240·7	55 6
8	07·8	01·8	68	66·3	15·3	128	124·7	28·8	188	183·2	42·3	248	241·6	55·8
9	08·8	02·0	69	67·2	15·5	129	125 7	29·0	189	184·2	42·5	249	242·6	56·0
10	09·7	02·2	70	68·2	15·7	130	126·7	29·2	190	185·1	42·7	250	243·6	56·2
11	10·7	02·5	71	69·2	16·0	131	127·6	29·5	191	186·1	43·0	251	244·6	56·5
12	11·7	02·7	72	70·2	16·2	132	128·6	29·7	192	187·1	43·2	252	245·5	56·7
13	12·7	02·9	73	71·1	16·4	133	129·6	29·9	193	188·1	43·4	253	246·5	56·9
14	13 6	03·1	74	72·1	16·6	134	130·6	30·1	194	189·0	43·6	254	247·5	57·1
15	14·6	03·4	75	73·1	16·9	135	131·5	30·4	195	190·0	43·9	255	248·5	57·4
16	15·6	03 6	76	74·1	17·1	136	132·5	30·6	196	191·0	44·1	256	249·4	57·6
17	16·6	03·8	77	75·0	17·3	137	133·5	30·8	197	192·0	44·3	257	250·4	57·8
18	17 5	04·0	78	76·0	17·5	138	134·5	31·0	198	192·9	44·5	258	251·4	58·0
19	18·5	04·3	79	77·0	17·8	139	135·4	31·3	199	193·9	44·8	259	252·4	58·3
20	19·5	04·5	80	77·9	18·0	140	136·4	31·5	200	194·9	45·0	260	253·3	58·5
21	20·5	04·7	81	78·9	18·2	141	137·4	31·7	201	195·8	45·2	261	254·3	58·7
22	21·4	04·9	82	79·9	18·4	142	138·4	31·9	202	196·8	45·4	262	255·3	58·9
23	22·4	05·2	83	80·9	18·7	143	139·3	32·2	203	197·8	45·7	263	256·3	59·2
24	23·4	05·4	84	81·8	18·9	144	140·3	32·4	204	198·8	45·9	264	257·2	59·4
25	24·4	05·6	85	82·8	19·1	145	141·3	32·6	205	199·7	46·1	265	258·2	59·6
26	25·3	05·8	86	83·8	19·3	146	142·3	32·8	206	200·7	46·3	266	259·2	59·8
27	26·3	06·1	87	84·8	19·6	147	143·2	33·1	207	201·7	46·6	267	260·2	60·1
28	27·3	06·3	88	85·7	19·8	148	144·2	33·3	208	202·7	46·8	268	261·1	60·3
29	28·3	06·5	89	86·7	20·0	149	145·2	33·5	209	203·6	47·0	269	262·1	60·5
30	29·2	06·7	90	87·7	20·2	150	146·2	33·7	210	204·6	47·2	270	263·1	60·7
31	30·2	07·0	91	88 7	20·5	151	147·1	34·0	211	205·6	47·5	271	264·1	61·0
32	31·2	07·2	92	89·6	20·7	152	148·1	34·2	212	206·6	47·7	272	265·0	61·2
33	32·2	07·4	93	90·6	20·9	153	149·1	34·4	213	207·5	47·9	273	266 0	61·4
34	33·1	07·6	94	91·6	21·1	154	150·1	34·6	214	208·5	48·1	274	267·0	61·6
35	34·1	07·9	95	92·6	21·4	155	151·0	34·9	215	209·5	48·4	275	268·0	61·9
36	35·1	08·1	96	93·5	21·6	156	152·0	35·1	216	210·5	48·6	276	268·9	62·1
37	36·1	08·3	97	94·5	21·8	157	153·0	35·3	217	211·4	48·8	277	269·9	62·3
38	37·0	08·5	98	95·5	22·0	158	154·0	35·5	218	212·4	49·0	278	270·9	62·5
39	38·0	08·8	99	96·5	22·3	159	154·9	35·8	219	213·4	49·3	279	271·8	62·8
40	39·0	09·0	100	97·4	22·5	160	155·9	36·0	220	214·4	49·5	280	272·8	63·0
41	39·9	09·2	101	98·4	22·7	161	156·9	36·2	221	215·3	49·7	281	273·8	63·2
42	40·9	09·4	102	99·4	22·9	162	157·8	36·4	222	216·3	49·9	282	274·8	63·4
43	41·9	09·7	103	100·4	23·2	163	158·8	36·7	223	217·3	50·2	283	275·7	63·7
44	42·9	09·9	104	101·3	23·4	164	159·8	36·9	224	218·3	50·4	284	276·7	63·9
45	43·8	10·1	105	102·3	23·6	165	160·8	37·1	225	219·2	50·6	285	277·7	64·1
46	44·8	10·3	106	103·3	23·8	166	161·7	37·3	226	220·2	50·8	286	278·7	64·3
47	45·8	10·6	107	104·3	24·1	167	162·7	37·6	227	221·2	51·1	287	279·6	64·6
48	46·8	10·8	108	105·2	24·3	168	163·7	37·8	228	222·2	51·3	288	280·6	64·8
49	47·7	11·0	109	106·2	24·5	169	164·7	38·0	229	223 1	51·5	289	281·6	65·0
50	48·7	11·2	110	107·2	24·7	170	165·6	38·2	230	224·1	51·7	290	282·6	65·2
51	49·7	11·5	111	108·2	25·0	171	166·6	38·5	231	225·1	52·0	291	283·5	65 5
52	50·7	11·7	112	109·1	25·2	172	167·6	38·7	232	226·1	52·2	292	284·5	65·7
53	51·6	11·9	113	110·1	25·4	173	168·6	38·9	233	227·0	52·4	293	285·5	65·9
54	52·6	12·1	114	111·1	25·6	174	169·5	39·1	234	228·0	52·6	294	286·5	66·1
55	53·6	12·4	115	112·1	25·9	175	170·5	39·4	235	229·0	52·9	295	287·4	66·4
56	54·6	12·6	116	113·0	26·1	176	171·5	39·6	236	230·0	53·1	296	288·4	66·6
57	55·5	12·8	117	114·0	26·3	177	172·5	39·8	237	230·9	53·3	297	289·4	66·8
58	56·5	13·0	118	115·0	26 5	178	173·4	40·0	238	231·9	53·5	298	290·4	67·0
59	57·5	13·3	119	116·0	26·8	179	174·4	40·3	239	232·9	53·8	299	291·3	67·3
60	58·5	13·5	120	116·9	27·0	180	175·4	40·5	240	233·8	54·0	300	292·3	67·5

Dist	Dep.	D.Lat	Dist	Dep.	D.Lat	Dist	Dep.	D.Lat	Dist	Dep.	D.Lat	Dist	Dep.	D.Lat

77°

TRAVERSE TABLE
13 DEGREES.

347° ↑ / 193° — ↑013° / 167° 0h 52m   **13°**

Dist	D.Lat	Dep.	Dist	D.Lat	Dep.	Dist	D.Lat	Dep.	Dist	D.Lat	Dep.	Dist	D.Lat	Dep.
301	293·3	67·7	361	351·7	81·2	421	410·2	94·7	481	468·7	108·2	541	527·1	121·7
302	294·3	67·9	362	352·7	81·4	422	411·2	94·9	482	469·6	108·4	542	528·1	121·9
303	295·2	68·2	363	353·7	81·7	423	412·2	95·2	483	470·6	108·7	543	529·1	122·1
304	296·2	68·4	364	354·7	81·9	424	413·1	95·4	484	471·6	108·9	544	530·1	122·4
305	297·2	68·6	365	355·6	82·1	425	414·1	95·6	485	472·6	109·1	545	531·0	122·6
306	298·2	68·8	366	356·6	82·3	426	415·1	95·8	486	473·5	109·3	546	532·0	122·8
307	299·1	69·1	367	357·6	82·6	427	416·1	96·1	487	474·5	109·6	547	533·0	123·0
308	300·1	69·3	368	358·6	82·8	428	417·0	96·3	488	475·5	109·8	548	534·0	123·3
309	301·1	69·5	369	359·5	83·0	429	418·0	96·5	489	476·5	110·0	549	534·9	123·5
310	302·1	69·7	370	360·5	83·2	430	419·0	96·7	490	477·4	110·2	550	535·9	123·7
311	303·0	70·0	371	361·5	83·5	431	420·0	97·0	491	478·4	110·5	551	536·9	123·9
312	304·0	70·2	372	362·5	83·7	432	420·9	97·2	492	479·4	110·7	552	537·9	124·2
313	305·0	70·4	373	363·4	83·9	433	421·9	97·4	493	480·4	110·9	553	538·8	124·4
314	306·0	70·6	374	364·4	84·1	434	422·9	97·6	494	481·3	111·1	554	539·8	124·6
315	306·9	70·9	375	365·4	84·4	435	423·9	97·9	495	482·3	111·4	555	540·8	124·8
316	307·9	71·1	376	366·4	84·6	436	424·8	98·1	496	483·3	111·6	556	541·7	125·1
317	308·9	71·3	377	367·3	84·8	437	425·8	98·3	497	484·3	111·8	557	542·7	125·3
318	309·8	71·5	378	368·3	85·0	438	426·8	98·5	498	485·2	112·0	558	543·7	125·5
319	310·8	71·8	379	369·3	85·3	439	427·7	98·8	499	486·2	112·3	559	544·7	125·7
320	311·8	72·0	380	370·3	85·5	440	428·7	99·0	500	487·2	112·5	560	545·6	126·0
321	312·8	72·2	381	371·2	85·7	441	429·7	99·2	501	488·2	112·7	561	546·6	126·2
322	313·7	72·4	382	372·2	85·9	442	430·7	99·4	502	489·1	112·9	562	547·6	126·4
323	314·7	72·7	383	373·2	86·2	443	431·6	99·7	503	490·1	113·2	563	548·6	126·6
324	315·7	72·9	384	374·2	86·4	444	432·6	99·9	504	491·1	113·4	564	549·5	126·9
325	316·7	73·1	385	375·1	86·6	445	433·6	100·1	505	492·1	113·6	565	550·5	127·1
326	317·6	73·3	386	376·1	86·8	446	434·6	100·3	506	493·0	113·8	566	551·5	127·3
327	318·6	73·6	387	377·1	87·1	447	435·5	100·6	507	494·0	114·1	567	552·5	127·5
328	319·6	73·8	388	378·1	87·3	448	436·5	100·8	508	495·0	114·3	568	553·4	127·8
329	320·6	74·0	389	379·0	87·5	449	437·5	101·0	509	496·0	114·5	569	554·4	128·0
330	321·5	74·2	390	380·0	87·7	450	438·5	101·2	510	496·9	114·7	570	555·4	128·2
331	322·5	74·5	391	381·0	88·0	451	439·4	101·5	511	497·9	115·0	571	556·4	128·4
332	323·5	74·7	392	382·0	88·2	452	440·4	101·7	512	498·9	115·2	572	557·3	128·7
333	324·5	74·9	393	382·9	88·4	453	441·4	101·9	513	499·9	115·4	573	558·3	128·9
334	325·4	75·1	394	383·9	88·6	454	442·4	102·1	514	500·8	115·6	574	559·3	129·1
335	326·4	75·4	395	384·9	88·9	455	443·3	102·4	515	501·8	115·8	575	560·3	129·3
336	327·4	75·6	396	385·9	89·1	456	444·3	102·6	516	502·8	116·1	576	561·2	129·6
337	328·4	75·8	397	386·8	89·3	457	445·3	102·8	517	503·7	116·3	577	562·2	129·8
338	329·3	76·0	398	387·8	89·5	458	446·3	103·0	518	504·7	116·5	578	563·2	130·0
339	330·3	76·3	399	388·8	89·8	459	447·2	103·3	519	505·7	116·7	579	564·2	130·2
340	331·3	76·5	400	389·7	90·0	460	448·2	103·5	520	506·7	117·0	580	565·1	130·5
341	332·3	76·7	401	390·7	90·2	461	449·2	103·7	521	507·6	117·2	581	566·1	130·7
342	333·2	76·9	402	391·7	90·4	462	450·2	103·9	522	508·6	117·5	582	567·1	130·9
343	334·2	77·2	403	392·7	90·7	463	451·1	104·2	523	509·6	117·6	583	568·1	131·1
344	335·2	77·4	404	393·6	90·9	464	452·1	104·4	524	510·6	117·9	584	569·0	131·4
345	336·2	77·6	405	394·6	91·1	465	453·1	104·6	525	511·5	118·1	585	570·1	131·6
346	337·1	77·8	406	395·6	91·3	466	454·1	104·8	526	512·5	118·3	586	571·0	131·8
347	338·1	78·1	407	396·6	91·6	467	455·0	105·1	527	513·5	118·5	587	572·0	132·0
348	339·1	78·3	408	397·5	91·8	468	456·0	105·3	528	514·5	118·8	588	572·9	132·3
349	340·1	78·5	409	398·5	92·0	469	457·0	105·5	529	515·4	119·0	589	573·9	132·5
350	341·0	78·7	410	399·5	92·2	470	458·0	105·7	530	516·4	119·2	590	574·9	132·7
351	342·0	79·0	411	400·5	92·5	471	458·9	106·0	531	517·4	119·4	591	575·9	132·9
352	343·0	79·2	412	401·4	92·7	472	459·9	106·2	532	518·4	119·7	592	576·8	133·2
353	344·0	79·4	413	402·4	92·9	473	460·9	106·4	533	519·3	119·9	593	577·8	133·4
354	344·9	79·6	414	403·4	93·1	474	461·9	106·6	534	520·3	120·1	594	578·8	133·6
355	345·9	79·9	415	404·4	93·4	475	462·8	106·9	535	521·3	120·3	595	579·8	133·8
356	346·9	80·1	416	405·3	93·6	476	463·8	107·1	536	522·3	120·6	596	580·7	134·1
357	347·9	80·3	417	406·3	93·8	477	464·8	107·3	537	523·2	120·8	597	581·7	134·3
358	348·8	80·5	418	407·3	94·0	478	465·7	107·5	538	524·2	121·0	598	582·7	134·5
359	349·8	80·8	419	408·3	94·3	479	466·7	107·8	539	525·2	121·2	599	583·6	134·7
360	350·8	81·0	420	409·2	94·5	480	467·7	108·0	540	526·2	121·5	600	584·6	135·0

| Dist | Dep. | D.Lat | Dist | Dep. | D.Lat | Dist | Dep. | D.Lat | Dist | Dep. | D.Lat | Dist | Dep. | D.Lat |

D Lon … Dep.

77 DEGREES.

283° ↑ / 257° — ↑077° / 103° 5h 8m   **77°**

14°

D Lon	Dep.		D Lon	Dep.		D Lon	Dep.		D Lon	Dep.		D Lon	Dep.	
Dist	D. Lat	Dep.	Dist	D. Lat	Dep.	Dist	D. Lat	Dep.	Dist	D. Lat	Dep.	Dist	D. Lat	Dep.
1	01·0	00·2	61	59·2	14·8	121	117·4	29·3	181	175·6	43·8	241	233·8	58·3
2	01·9	00·5	62	60·2	15·0	122	118·4	29·5	182	176·6	44·0	242	234·8	58·5
3	02·9	00·7	63	61·1	15·2	123	119·3	29·8	183	177·6	44·3	243	235·8	58·8
4	03·9	01·0	64	62·1	15·5	124	120·3	30·0	184	178·5	44·5	244	236·8	59·0
5	04·9	01·2	65	63·1	15·7	125	121·3	30·2	185	179·5	44·8	245	237·7	59·3
6	05·8	01·5	66	64·0	16·0	126	122·3	30·5	186	180·5	45·0	246	238·7	59·5
7	06·8	01·7	67	65·0	16·2	127	123·2	30·7	187	181·4	45·2	247	239·7	59·8
8	07·8	01·9	68	66·0	16·5	128	124·2	31·0	188	182·4	45·5	248	240·6	60·0
9	08·7	02·2	69	67·0	16·7	129	125·2	31·2	189	183·4	45·7	249	241·6	60·2
10	09·7	02·4	70	67·9	16·9	130	126·1	31·4	190	184·4	46·0	250	242·6	60·5
11	10·7	02·7	71	68·9	17·2	131	127·1	31·7	191	185·3	46·2	251	243·5	60·7
12	11·6	02·9	72	69·9	17·4	132	128·1	31·9	192	186·3	46·4	252	244·5	61·0
13	12·6	03·1	73	70·8	17·7	133	129·0	32·2	193	187·3	46·7	253	245·5	61·2
14	13·6	03·4	74	71·8	17·9	134	130·0	32·4	194	188·2	46·9	254	246·5	61·4
15	14·6	03·6	75	72·8	18·1	135	131·0	32·7	195	189·2	47·2	255	247·4	61·7
16	15·5	03·9	76	73·7	18·4	136	132·0	32·9	196	190·2	47·4	256	248·4	61·9
17	16·5	04·1	77	74·7	18·6	137	132·9	33·1	197	191·1	47·7	257	249·4	62·2
18	17·5	04·4	78	75·7	18·9	138	133·9	33·4	198	192·1	47·9	258	250·3	62·4
19	18·4	04·6	79	76·7	19·1	139	134·9	33·6	199	193·1	48·1	259	251·3	62·7
20	19·4	04·8	80	77·6	19·4	140	135·8	33·9	200	194·1	48·4	260	252·3	62·9
21	20·4	05·1	81	78·6	19·6	141	136·8	34·1	201	195·0	48·6	261	253·2	63·1
22	21·3	05·3	82	79·6	19·8	142	137·8	34·4	202	196·0	48·9	262	254·2	63·4
23	22·3	05·6	83	80·5	20·1	143	138·8	34·6	203	197·0	49·1	263	255·2	63·6
24	23·3	05·8	84	81·5	20·3	144	139·7	34·8	204	197·9	49·4	264	256·2	63·9
25	24·3	06·0	85	82·5	20·6	145	140·7	35·1	205	198·9	49·6	265	257·1	64·1
26	25·2	06·3	86	83·4	20·8	146	141·7	35·3	206	199·9	49·8	266	258·1	64·4
27	26·2	06·5	87	84·4	21·0	147	142·6	35·6	207	200·9	50·1	267	259·1	64·6
28	27·2	06·8	88	85·4	21·3	148	143·6	35·8	208	201·8	50·3	268	260·0	64·8
29	28·1	07·0	89	86·4	21·5	149	144·6	36·0	209	202·8	50·6	269	261·0	65·1
30	29·1	07·3	90	87·3	21·8	150	145·5	36·3	210	203·8	50·8	270	262·0	65·3
31	30·1	07·5	91	88·3	22·0	151	146·5	36·5	211	204·7	51·0	271	263·0	65·6
32	31·0	07·7	92	89·3	22·3	152	147·5	36·8	212	205·7	51·3	272	263·9	65·8
33	32·0	08·0	93	90·2	22·5	153	148·5	37·0	213	206·7	51·5	273	264·9	66·0
34	33·0	08·2	94	91·2	22·7	154	149·4	37·3	214	207·6	51·8	274	265·9	66·3
35	34·0	08·5	95	92·2	23·0	155	150·4	37·5	215	208·6	52·0	275	266·8	66·5
36	34·9	08·7	96	93·1	23·2	156	151·4	37·8	216	209·6	52·3	276	267·8	66·8
37	35·9	09·0	97	94·1	23·5	157	152·3	38·0	217	210·6	52·5	277	268·8	67·0
38	36·9	09·2	98	95·1	23·7	158	153·3	38·2	218	211·5	52·7	278	269·7	67·3
39	37·8	09·4	99	96·1	24·0	159	154·3	38·5	219	212·5	53·0	279	270·7	67·5
40	38·8	09·7	100	97·0	24·2	160	155·2	38·7	220	213·5	53·2	280	271·7	67·7
41	39·8	09·9	101	98·0	24·4	161	156·2	38·9	221	214·4	53·5	281	272·7	68·0
42	40·8	10·2	102	99·0	24·7	162	157·2	39·2	222	215·4	53·7	282	273·6	68·2
43	41·7	10·4	103	99·9	24·9	163	158·2	39·4	223	216·4	53·9	283	274·6	68·5
44	42·7	10·6	104	100·9	25·2	164	159·1	39·7	224	217·3	54·2	284	275·6	68·7
45	43·7	10·9	105	101·9	25·4	165	160·1	39·9	225	218·3	54·4	285	276·5	68·9
46	44·6	11·1	106	102·9	25·6	166	161·1	40·2	226	219·3	54·7	286	277·5	69·2
47	45·6	11·4	107	103·8	25·9	167	162·0	40·4	227	220·3	54·9	287	278·5	69·4
48	46·6	11·6	108	104·8	26·1	168	163·0	40·6	228	221·2	55·2	288	279·4	69·7
49	47·5	11·9	109	105·8	26·4	169	164·0	40·9	229	222·2	55·4	289	280·4	69·9
50	48·5	12·1	110	106·7	26·6	170	165·0	41·1	230	223·2	55·6	290	281·4	70·2
51	49·5	12·3	111	107·7	26·9	171	165·9	41·4	231	224·1	55·9	291	282·4	70·4
52	50·5	12·6	112	108·7	27·1	172	166·9	41·6	232	225·1	56·1	292	283·3	70·6
53	51·4	12·8	113	109·6	27·3	173	167·9	41·9	233	226·1	56·4	293	284·3	70·9
54	52·4	13·1	114	110·6	27·6	174	168·8	42·1	234	227·0	56·6	294	285·3	71·1
55	53·4	13·3	115	111·6	27·8	175	169·8	42·3	235	228·0	56·9	295	286·2	71·4
56	54·3	13·5	116	112·6	28·1	176	170·8	42·6	236	229·0	57·1	296	287·2	71·6
57	55·3	13·8	117	113·5	28·3	177	171·7	42·8	237	230·0	57·3	297	288·2	71·9
58	56·3	14·0	118	114·5	28·6	178	172·7	43·1	238	230·9	57·6	298	289·1	72·1
59	57·2	14·3	119	115·5	28·8	179	173·7	43·3	239	231·9	57·8	299	290·1	72·3
60	58·2	14·5	120	116·4	29·0	180	174·7	43·5	240	232·9	58·1	300	291·1	72·6

Dist	Dep.	D. Lat	Dist	Dep.	D. Lat	Dist	Dep.	D. Lat	Dist	Dep.	D. Lat	Dist	Dep.	D. Lat
D Lon		Dep.	D Lon		Dep.	D Lon		Dep.	D Lon		Dep.	D Lon		Dep.

76°

TRAVERSE TABLE
14 DEGREES.

D Lon Dep.			D Lon Dep.			D Lon Dep.			D Lon Dep.			D Lon Dep.		
Dist	D.Lat	Dep.	Dist	D.Lat	Dep.	Dist	D.Lat	Dep.	Dist	D.Lat	Dep.	Dist	D.Lat	Dep.
301	292·1	72·8	361	350·3	87·3	421	408·5	101·8	481	466·7	116·4	541	524·9	130·9
302	293·0	73·1	362	351·2	87·6	422	409·5	102·1	482	467·7	116·6	542	525·9	131·1
303	294·0	73·3	363	352·2	87·8	423	410·4	102·3	483	468·7	116·8	543	526·9	131·4
304	295·0	73·5	364	353·2	88·1	424	411·4	102·6	484	469·6	117·1	544	527·8	131·6
305	295·9	73·8	365	354·2	88·3	425	412·4	102·8	485	470·6	117·3	545	528·8	131·8
306	296·9	74·0	366	355·1	88·5	426	413·3	103·1	486	471·6	117·6	546	529·8	132·1
307	297·9	74·3	367	356·1	88·8	427	414·3	103·3	487	472·5	117·8	547	530·8	132·3
308	298·9	74·5	368	357·1	89·0	428	415·3	103·5	488	473·5	118·1	548	531·7	132·6
309	299·8	74·8	369	358·0	89·3	429	416·3	103·8	489	474·5	118·3	549	532·7	132·8
310	300·8	75·0	370	359·0	89·5	430	417·2	104·0	490	475·4	118·5	550	533·7	133·1
311	301·8	75·2	371	360·0	89·8	431	418·2	104·3	491	476·4	118·8	551	534·6	133·3
312	302·7	75·5	372	361·0	90·0	432	419·0	104·5	492	477·4	119·0	552	535·6	133·5
313	303·7	75·7	373	361·9	90·2	433	420·1	104·8	493	478·4	119·3	553	536·6	133·8
314	304·6	76·0	374	362·9	90·5	434	421·1	105·0	494	479·3	119·5	554	537·5	134·0
315	305·6	76·2	375	363·9	90·7	435	422·0	105·2	495	480·3	119·6	555	538·5	134·3
316	306·6	76·4	376	364·8	91·0	436	423·0	105·5	496	481·3	120·0	556	539·5	134·5
317	307·6	76·7	377	365·8	91·2	437	424·0	105·7	497	482·2	120·2	557	540·5	134·8
318	308·6	76·9	378	366·8	91·4	438	425·0	106·0	498	483·2	120·5	558	541·4	135·0
319	309·5	77·2	379	367·7	91·7	439	426·0	106·2	499	484·2	120·7	559	542·2	135·2
320	310·5	77·4	380	368·7	91·9	440	426·9	106·4	500	485·1	121·0	560	543·4	135·5
321	311·5	77·7	381	369·7	92·2	441	427·9	106·7	501	486·1	121·2	561	544·3	135·7
322	312·4	77·9	382	370·7	92·4	442	428·9	106·9	502	487·1	121·4	562	545·3	136·0
323	313·4	78·1	383	371·6	92·7	443	429·8	107·2	503	488·1	121·7	563	546·3	136·2
324	314·4	78·4	384	372·6	92·9	444	430·8	107·4	504	489·0	121·9	564	547·2	136·4
325	315·3	78·6	385	373·6	93·1	445	431·8	107·7	505	490·0	122·2	565	548·2	136·7
326	316·3	78·9	386	374·5	93·4	446	432·8	107·9	506	491·0	122·4	566	549·2	136·9
327	317·3	79·1	387	375·5	93·6	447	433·7	108·1	507	491·9	122·7	567	550·2	137·2
328	318·3	79·4	388	376·4	93·9	448	434·7	108·4	508	492·9	122·9	568	551·1	137·4
329	319·2	79·6	389	377·4	94·1	449	435·7	108·6	509	493·9	123·1	569	552·1	137·7
330	320·2	79·8	390	378·4	94·3	450	436·6	108·9	510	494·9	123·4	570	553·1	137·9
331	321·2	80·1	391	379·4	94·6	451	437·6	109·1	511	495·8	123·6	571	554·0	138·1
332	322·1	80·3	392	380·3	94·8	452	438·6	109·3	512	496·8	123·9	572	555·0	138·4
333	323·1	80·6	393	381·3	95·1	453	439·5	109·6	513	497·8	124·1	573	556·0	138·6
334	324·1	80·8	394	382·3	95·3	454	440·5	109·8	514	498·7	124·3	574	556·9	138·9
335	325·0	81·0	395	383·3	95·6	455	441·5	110·1	515	499·7	124·6	575	557·9	139·1
336	326·0	81·3	396	384·2	95·8	456	442·5	110·3	516	500·7	124·8	576	558·9	139·3
337	327·0	81·5	397	385·2	96·0	457	443·4	110·6	517	501·6	125·1	577	559·9	139·6
338	328·0	81·8	398	386·2	96·3	458	444·4	110·8	518	502·6	125·3	578	560·8	139·8
339	328·9	82·0	399	387·1	96·5	459	445·4	111·0	519	503·6	125·6	579	561·8	140·1
340	329·9	82·3	400	388·1	96·8	460	446·3	111·3	520	504·6	125·8	580	562·8	140·3
341	330·8	82·5	401	389·1	97·0	461	447·3	111·5	521	505·5	126·0	581	563·7	140·6
342	331·8	82·7	402	390·1	97·3	462	448·3	111·8	522	506·5	126·3	582	564·7	140·8
343	332·8	83·0	403	391·0	97·5	463	449·2	112·0	523	507·5	126·5	583	565·7	141·0
344	333·8	83·2	404	392·0	97·7	464	450·2	112·3	524	508·4	126·8	584	566·7	141·3
345	334·8	83·5	405	393·0	98·0	465	451·2	112·5	525	509·4	127·0	585	567·6	141·5
346	335·7	83·7	406	393·9	98·2	466	452·2	112·7	526	510·4	127·3	586	568·6	141·8
347	336·7	83·9	407	394·9	98·5	467	453·1	113·0	527	511·3	127·5	587	569·6	142·0
348	337·7	84·2	408	395·9	98·7	468	454·1	113·2	528	512·3	127·7	588	570·5	142·3
349	338·6	84·4	409	396·9	98·9	469	455·1	113·5	529	513·3	128·0	589	571·5	142·5
350	339·6	84·7	410	397·8	99·2	470	456·0	113·7	530	514·3	128·2	590	572·5	142·7
351	340·6	84·9	411	398·8	99·4	471	457·0	113·9	531	515·2	128·5	591	573·4	143·0
352	341·5	85·2	412	399·8	99·7	472	458·0	114·2	532	516·2	128·8	592	574·4	143·2
353	342·5	85·4	413	400·7	99·9	473	458·9	114·4	533	517·2	128·9	593	575·4	143·5
354	343·5	85·6	414	401·7	100·2	474	459·9	114·7	534	518·1	129·2	594	576·4	143·7
355	344·5	85·9	415	402·7	100·4	475	460·9	114·9	535	519·1	129·4	595	577·3	143·9
356	345·4	86·1	416	403·6	100·7	476	461·9	115·2	536	520·1	129·7	596	578·3	144·2
357	346·4	86·4	417	404·6	100·9	477	462·8	115·4	537	521·0	129·9	597	579·3	144·4
358	347·4	86·6	418	405·6	101·1	478	463·8	115·6	538	522·0	130·2	598	580·2	144·7
359	348·3	86·8	419	406·6	101·4	479	464·8	115·9	539	523·0	130·4	599	581·2	144·9
360	349·3	87·1	420	407·5	101·6	480	465·7	116·1	540	524·0	130·6	600	582·2	145·2
Dist	Dep.	D.Lat	Dist	Dep.	D.Lat	Dist	Dep.	D.Lat	Dist	Dep.	D.Lat	Dist	Dep.	D.Lat
D Lon		Dep.	D Lon		Dep.	D Lon		Dep.	D Lon		Dep.	D Lon		Dep.

15°

TRAVERSE TABLE
15 DEGREES.

345° ↑ / 195° ↑015° / 165°   1h 0m

D Lon	Dep.		D Lon	Dep.		D Lon	Dep.		D Lon	Dep.		D Lon	Dep.	
Dist	D.Lat	Dep.	Dist	D.Lat	Dep.	Dist	D.Lat	Dep.	Dist	D.Lat	Dep.	Dist	D.Lat	Dep.
1	01·0	00·3	61	58·9	15·8	121	116·9	31·3	181	174·8	46·8	241	232·8	62·4
2	01·9	00·5	62	59·9	16·0	122	117·8	31·6	182	175·8	47·1	242	233·8	62·6
3	02·9	00·8	63	60·9	16·3	123	118·8	31·8	183	176·8	47·4	243	234·7	62·9
4	03·9	01·0	64	61·8	16·6	124	119·8	32·1	184	177·7	47·6	244	235·7	63·2
5	04·8	01·3	65	62·8	16·8	125	120·7	32·4	185	178·7	47·9	245	236·7	63·4
6	05·8	01·6	66	63·8	17·1	126	121·7	32·6	186	179·7	48·1	246	237·6	63·7
7	06·8	01·8	67	64·7	17·3	127	122·7	32·9	187	180·6	48·4	247	238·6	63·9
8	07·7	02·1	68	65·7	17·6	128	123·6	33·1	188	181·6	48·7	248	239·5	64·2
9	08·7	02·3	69	66·6	17·9	129	124·6	33·4	189	182·6	48·9	249	240·5	64·4
10	09·7	02·6	70	67·6	18·1	130	125·6	33·6	190	183·5	49·2	250	241·5	64·7
11	10·6	02·8	71	68·6	18·4	131	126·5	33·9	191	184·5	49·4	251	242·4	65·0
12	11·6	03·1	72	69·5	18·6	132	127·5	34·2	192	185·5	49·7	252	243·4	65·2
13	12·6	03·4	73	70·5	18·9	133	128·5	34·4	193	186·4	50·0	253	244·4	65·5
14	13·5	03·6	74	71·5	19·2	134	129·4	34·7	194	187·4	50·2	254	245·3	65·7
15	14·5	03·9	75	72·4	19·4	135	130·4	34·9	195	188·4	50·5	255	246·3	66·0
16	15·5	04·1	76	73·4	19·7	136	131·4	35·2	196	189·3	50·7	256	247·3	66·3
17	16·4	04·4	77	74·4	19·9	137	132·3	35·5	197	190·3	51·0	257	248·2	66·5
18	17·4	04·7	78	75·3	20·2	138	133·3	35·7	198	191·3	51·2	258	249·2	66·8
19	18·4	04·9	79	76·3	20·4	139	134·3	36·0	199	192·2	51·5	259	250·2	67·0
20	19·3	05·2	80	77·3	20·7	140	135·2	36·2	200	193·2	51·8	260	251·1	67·3
21	20·3	05·4	81	78·2	21·0	141	136·2	36·5	201	194·2	52·0	261	252·1	67·6
22	21·3	05·7	82	79·2	21·2	142	137·2	36·8	202	195·1	52·3	262	253·1	67·8
23	22·2	06·0	83	80·2	21·5	143	138·1	37·0	203	196·1	52·5	263	254·0	68·1
24	23·2	06·2	84	81·1	21·7	144	139·1	37·3	204	197·0	52·8	264	255·0	68·3
25	24·1	06·5	85	82·1	22·0	145	140·1	37·5	205	198·0	53·1	265	256·0	68·6
26	25·1	06·7	86	83·1	22·3	146	141·0	37·8	206	199·0	53·3	266	256·9	68·8
27	26·1	07·0	87	84·0	22·5	147	142·0	38·0	207	199·9	53·6	267	257·9	69·1
28	27·0	07·2	88	85·0	22·8	148	143·0	38·3	208	200·9	53·8	268	258·9	69·4
29	28·0	07·5	89	86·0	23·0	149	143·9	38·6	209	201·9	54·1	269	259·8	69·6
30	29·0	07·8	90	86·9	23·3	150	144·9	38·8	210	202·8	54·4	270	260·8	69·9
31	29·9	08·0	91	87·9	23·6	151	145·9	39·1	211	203·8	54·6	271	261·8	70·1
32	30·9	08·3	92	88·9	23·8	152	146·8	39·3	212	204·8	54·9	272	262·7	70·4
33	31·9	08·5	93	89·8	24·1	153	147·8	39·6	213	205·7	55·1	273	263·7	70·7
34	32·8	08·8	94	90·8	24·3	154	148·8	39·9	214	206·7	55·4	274	264·7	70·9
35	33·8	09·1	95	91·8	24·6	155	149·7	40·1	215	207·7	55·6	275	265·6	71·2
36	34·8	09·3	96	92·7	24·8	156	150·7	40·4	216	208·6	55·9	276	266·6	71·4
37	35·7	09·6	97	93·7	25·1	157	151·7	40·6	217	209·6	56·2	277	267·6	71·7
38	36·7	09·8	98	94·7	25·4	158	152·6	40·9	218	210·6	56·4	278	268·5	72·0
39	37·7	10·1	99	95·6	25·6	159	153·6	41·2	219	211·5	56·7	279	269·5	72·2
40	38·6	10·4	100	96·6	25·9	160	154·5	41·4	220	212·5	56·9	280	270·5	72·5
41	39·6	10·6	101	97·6	26·1	161	155·5	41·7	221	213·5	57·2	281	271·4	72·7
42	40·6	10·9	102	98·5	26·4	162	156·5	41·9	222	214·4	57·5	282	272·4	73·0
43	41·5	11·1	103	99·5	26·7	163	157·4	42·2	223	215·4	57·7	283	273·4	73·2
44	42·5	11·4	104	100·5	26·9	164	158·4	42·4	224	216·4	58·0	284	274·3	73·5
45	43·5	11·6	105	101·4	27·2	165	159·4	42·7	225	217·3	58·2	285	275·3	73·8
46	44·4	11·9	106	102·4	27·4	166	160·3	43·0	226	218·3	58·5	286	276·3	74·0
47	45·4	12·2	107	103·4	27·7	167	161·3	43·2	227	219·3	58·8	287	277·2	74·3
48	46·4	12·4	108	104·3	28·0	168	162·3	43·5	228	220·2	59·0	288	278·2	74·5
49	47·3	12·7	109	105·3	28·2	169	163·2	43·7	229	221·2	59·3	289	279·2	74·8
50	48·3	12·9	110	106·3	28·5	170	164·2	44·0	230	222·2	59·5	290	280·1	75·1
51	49·3	13·2	111	107·2	28·7	171	165·2	44·3	231	223·1	59·8	291	281·1	75·3
52	50·2	13·5	112	108·2	29·0	172	166·1	44·5	232	224·1	60·0	292	282·1	75·6
53	51·2	13·7	113	109·1	29·2	173	167·1	44·8	233	225·1	60·3	293	283·0	75·8
54	52·2	14·0	114	110·1	29·5	174	168·1	45·0	234	226·0	60·6	294	284·0	76·1
55	53·1	14·2	115	111·1	29·8	175	169·0	45·3	235	227·0	60·8	295	284·9	76·4
56	54·1	14·5	116	112·0	30·0	176	170·0	45·6	236	228·0	61·1	296	285·9	76·6
57	55·1	14·8	117	113·0	30·3	177	171·0	45·8	237	228·9	61·3	297	286·9	76·9
58	56·0	15·0	118	114·0	30·5	178	171·9	46·1	238	229·9	61·6	298	287·8	77·1
59	57·0	15·3	119	114·9	30·8	179	172·9	46·3	239	230·9	61·9	299	288·8	77·4
60	58·0	15·5	120	115·9	31·1	180	173·9	46·6	240	231·8	62·1	300	289·8	77·6
Dist	Dep.	D.Lat	Dist	Dep.	D.Lat	Dist	Dep.	D.Lat	Dist	Dep.	D.Lat	Dist	Dep.	D.Lat
D Lon		Dep.	D Lon		Dep.	D Lon		Dep.	D Lon		Dep.	D Lon		Dep.

75°

285° ↑ / 255°   **75 DEGREES.**   ↑075° / 105°   5h 0m

TRAVERSE TABLE
15 DEGREES.

345° ↑		015° ↑		1h 0m
195°		165°		

D Lon	Dep.		D Lon	Dep.		D Lon	Dep.		D Lon	Dep.		D Lon	Dep.	
Dist	D.Lat	Dep.	Dist	D.Lat	Dep.	Dist	D.Lat	Dep.	Dist	D.Lat	Dep.	Dist	D.Lat	Dep.
301	290·7	77·9	361	348·7	93·4	421	406·7	109·0	481	464·6	124·5	541	522·6	140·0
302	291·7	78·2	362	349·7	93·7	422	407·6	109·2	482	465·6	124·8	542	523·5	140·3
303	292·7	78·4	363	350·6	94·0	423	408·6	109·5	483	466·5	125·0	543	524·5	140·5
304	293·6	78·7	364	351·6	94·2	424	409·6	109·7	484	467·5	125·3	544	525·5	140·8
305	294·6	78·9	365	352·6	94·5	425	410·5	110·0	485	468·5	125·5	545	526·4	141·1
306	295·6	79·2	366	353·5	94·7	426	411·5	110·3	486	469·4	125·8	546	527·4	141·3
307	296·5	79·5	367	354·5	95·0	427	412·5	110·5	487	470·4	126·0	547	528·4	141·6
308	297·5	79·7	368	355·5	95·2	428	413·4	110·8	488	471·4	126·3	548	529·3	141·8
309	298·5	80·0	369	356·4	95·5	429	414·4	111·0	489	472·3	126·6	549	530·3	142·1
310	299·4	80·2	370	357·4	95·8	430	415·3	111·3	490	473·3	126·8	550	531·3	142·4
311	300·4	80·5	371	358·4	96·0	431	416·3	111·6	491	474·3	127·1	551	532·2	142·6
312	301·4	80·8	372	359·3	96·3	432	417·3	111·8	492	475·2	127·3	552	533·2	142·9
313	302·3	81·0	373	360·3	96·5	433	418·2	112·1	493	476·2	127·6	553	534·2	143·1
314	303·3	81·3	374	361·3	96·8	434	419·2	112·3	494	477·2	127·9	554	535·1	143·4
315	304·3	81·5	375	362·2	97·1	435	420·2	112·6	495	478·1	128·1	555	536·1	143·6
316	305·2	81·8	376	363·2	97·3	436	421·1	112·8	496	479·1	128·4	556	537·1	143·9
317	306·2	82·0	377	364·2	97·6	437	422·1	113·1	497	480·1	128·6	557	538·0	144·2
318	307·2	82·3	378	365·1	97·8	438	423·1	113·4	498	481·0	128·9	558	539·0	144·4
319	308·1	82·6	379	366·1	98·1	439	424·0	113·6	499	482·0	129·2	559	540·0	144·7
320	309·1	82·8	380	367·1	98·4	440	425·0	113·9	500	483·0	129·4	560	540·9	144·9
321	310·1	83·1	381	368·0	98·6	441	426·0	114·1	501	483·9	129·7	561	541·9	145·2
322	311·0	83·3	382	369·0	98·9	442	426·9	114·4	502	484·9	129·9	562	542·9	145·5
323	312·0	83·6	383	369·9	99·1	443	427·9	114·7	503	485·9	130·2	563	543·8	145·7
324	313·0	83·9	384	370·9	99·4	444	428·9	114·9	504	486·8	130·4	564	544·8	146·0
325	313·9	84·1	385	371·9	99·6	445	429·8	115·2	505	487·8	130·7	565	545·7	146·2
326	314·9	84·4	386	372·8	99·9	446	430·8	115·4	506	488·8	131·0	566	546·7	146·5
327	315·9	84·6	387	373·8	100·2	447	431·8	115·7	507	489·7	131·2	567	547·7	146·8
328	316·8	84·9	388	374·8	100·4	448	432·7	116·0	508	490·7	131·5	568	548·6	147·0
329	317·8	85·2	389	375·7	100·7	449	433·7	116·2	509	491·7	131·7	569	549·6	147·3
330	318·8	85·4	390	376·7	100·9	450	434·7	116·5	510	492·6	132·0	570	550·6	147·5
331	319·7	85·7	391	377·7	101·2	451	435·6	116·7	511	493·6	132·3	571	551·5	147·8
332	320·7	85·9	392	378·6	101·5	452	436·6	117·0	512	494·6	132·5	572	552·5	148·0
333	321·7	86·2	393	379·6	101·7	453	437·6	117·2	513	495·5	132·8	573	553·5	148·3
334	322·6	86·4	394	380·6	102·0	454	438·5	117·5	514	496·5	133·0	574	554·4	148·6
335	323·6	86·7	395	381·5	102·2	455	439·5	117·8	515	497·5	133·3	575	555·4	148·8
336	324·6	87·0	396	382·5	102·5	456	440·5	118·0	516	498·4	133·6	576	556·4	149·1
337	325·5	87·2	397	383·5	102·8	457	441·4	118·3	517	499·4	133·8	577	557·3	149·3
338	326·5	87·5	398	384·4	103·0	458	442·4	118·5	518	500·3	134·1	578	558·3	149·6
339	327·4	87·7	399	385·4	103·3	459	443·4	118·8	519	501·3	134·3	579	559·3	149·8
340	328·4	88·0	400	386·4	103·5	460	444·3	119·1	520	502·3	134·6	580	560·2	150·1
341	329·4	88·3	401	387·3	103·8	461	445·3	119·3	521	503·2	134·8	581	561·2	150·4
342	330·3	88·5	402	388·3	104·0	462	446·3	119·6	522	504·2	135·1	582	562·2	150·6
343	331·3	88·8	403	389·3	104·3	463	447·2	119·8	523	505·2	135·4	583	563·1	150·9
344	332·3	89·0	404	390·2	104·6	464	448·2	120·1	524	506·1	135·6	584	564·1	151·2
345	333·2	89·3	405	391·2	104·8	465	449·2	120·3	525	507·1	135·9	585	565·1	151·4
346	334·2	89·6	406	392·2	105·1	466	450·1	120·6	526	508·1	136·1	586	566·0	151·6
347	335·2	89·8	407	393·1	105·3	467	451·1	120·9	527	509·0	136·4	587	567·0	151·9
348	336·1	90·1	408	394·1	105·6	468	452·1	121·1	528	510·0	136·7	588	568·0	152·2
349	337·1	90·3	409	395·1	105·9	469	453·0	121·4	529	511·0	136·9	589	568·9	152·4
350	338·1	90·6	410	396·0	106·1	470	454·0	121·6	530	511·9	137·2	590	569·9	152·7
351	339·0	90·8	411	397·0	106·4	471	455·0	121·9	531	512·9	137·4	591	570·9	153·0
352	340·0	91·1	412	398·0	106·6	472	455·9	122·2	532	513·9	137·7	592	571·8	153·2
353	341·0	91·4	413	398·9	106·9	473	456·9	122·4	533	514·8	138·0	593	572·8	153·5
354	341·9	91·6	414	399·9	107·2	474	457·8	122·7	534	515·8	138·2	594	573·8	153·7
355	342·9	91·9	415	400·9	107·4	475	458·8	122·9	535	516·8	138·5	595	574·7	154·0
356	343·9	92·1	416	401·8	107·7	476	459·8	123·2	536	517·7	138·7	596	575·7	154·3
357	344·8	92·4	417	402·8	107·9	477	460·7	123·5	537	518·7	139·0	597	576·7	154·5
358	345·8	92·7	418	403·8	108·2	478	461·7	123·7	538	519·7	139·2	598	577·6	154·8
359	346·8	92·9	419	404·7	108·4	479	462·7	124·0	539	520·6	139·5	599	578·6	155·0
360	347·7	93·2	420	405·7	108·7	480	463·6	124·2	540	521·6	139·8	600	579·6	155·3
Dist	Dep.	D.Lat	Dist	Dep.	D.Lat	Dist	Dep.	D.I at	Dist	Dep.	D.Lat	Dist	Dep.	D.Lat
D Lon		Dep.	D Lon		Dep.	D Lon		Dep.	D Lon		Dep.	D Lon		Dep.

285° ↑		075° ↑		5h 0m
255°	**75 DEGREES.**	105°		

TRAVERSE TABLE

344°
196°

16 DEGREES.

016°
164°

1h 4m

D Lon	Dep.		D Lon	Dep.		D Lon	Dep.		D Lon	Dep.		D Lon	Dep.	
Dist	D. Lat	Dep.	Dist	D. Lat	Dep.	Dist	D. Lat	Dep.	Dist	D. Lat	Dep.	Dist	D. Lat	Dep.
1	01·0	00·3	61	58·6	16·8	121	116·3	33·4	181	174·0	49·9	241	231·7	66·4
2	01·9	00·6	62	59·6	17·1	122	117·3	33·6	182	174·9	50·2	242	232·6	66·7
3	02·9	00·8	63	60·6	17·4	123	118·2	33·9	183	175·9	50·4	243	233·6	67·0
4	03·8	01·1	64	61·5	17·6	124	119·2	34·2	184	176·9	50·7	244	234·5	67·3
5	04·8	01·4	65	62·5	17·9	125	120·2	34·5	185	177·8	51·0	245	235·5	67·5
6	05·8	01·7	66	63·4	18·2	126	121·1	34·7	186	178·8	51·3	246	236·5	67·8
7	06·7	01·9	67	64·4	18·5	127	122·1	35·0	187	179·8	51·5	247	237·4	68·1
8	07·7	02·2	68	65·4	18·7	128	123·0	35·3	188	180·7	51·8	248	238·4	68·4
9	08·7	02·5	69	66·3	19·0	129	124·0	35·6	189	181·7	52·1	249	239·4	68·6
10	09·6	02·8	70	67·3	19·3	130	125·0	35·8	190	182·6	52·4	250	240·3	68·9
11	10·6	03·0	71	68·2	19·6	131	125·9	36·1	191	183·6	52·6	251	241·3	69·2
12	11·5	03·3	72	69·2	19·8	132	126·9	36·4	192	184·6	52·9	252	242·2	69·5
13	12·5	03·6	73	70·2	20·1	133	127·8	36·7	193	185·5	53·2	253	243·2	69·7
14	13·5	03·9	74	71·1	20·4	134	128·8	36·9	194	186·5	53·5	254	244·2	70·0
15	14·4	04·1	75	72·1	20·7	135	129·8	37·2	195	187·4	53·7	255	245·1	70·3
16	15·4	04·4	76	73·1	20·9	136	130·7	37·5	196	188·4	54·0	256	246·1	70·6
17	16·3	04·7	77	74·0	21·2	137	131·7	37·8	197	189·4	54·3	257	247·0	70·8
18	17·3	05·0	78	75·0	21·5	138	132·7	38·0	198	190·3	54·6	258	248·0	71·1
19	18·3	05·2	79	75·9	21·8	139	133·6	38·3	199	191·3	54·9	259	249·0	71·4
20	19·2	05·5	80	76·9	22·1	140	134·6	38·6	200	192·3	55·1	260	249·9	71·7
21	20·2	05·8	81	77·9	22·3	141	135·5	38·9	201	193·2	55·4	261	250·9	71·9
22	21·1	06·1	82	78·8	22·6	142	136·5	39·1	202	194·2	55·7	262	251·9	72·2
23	22·1	06·3	83	79·8	22·9	143	137·5	39·4	203	195·1	56·0	263	252·8	72·5
24	23·1	06·6	84	80·7	23·2	144	138·4	39·7	204	196·1	56·2	264	253·8	72·8
25	24·0	06·9	85	81·7	23·4	145	139·4	40·0	205	197·1	56·5	265	254·7	73·0
26	25·0	07·2	86	82·7	23·7	146	140·3	40·2	206	198·0	56·8	266	255·7	73·3
27	26·0	07·4	87	83·6	24·0	147	141·3	40·5	207	199·0	57·1	267	256·7	73·6
28	26·9	07·7	88	84·6	24·3	148	142·3	40·8	208	199·9	57·3	268	257·6	73·9
29	27·9	08·0	89	85·6	24·5	149	143·2	41·1	209	200·9	57·6	269	258·6	74·1
30	28·8	08·3	90	86·5	24·8	150	144·2	41·3	210	201·9	57·9	270	259·5	74·4
31	29·8	08·5	91	87·5	25·1	151	145·2	41·6	211	202·8	58·2	271	260·5	74·7
32	30·8	08·8	92	88·4	25·4	152	146·1	41·9	212	203·8	58·4	272	261·5	75·0
33	31·7	09·1	93	89·4	25·6	153	147·1	42·2	213	204·7	58·7	273	262·4	75·2
34	32·7	09·4	94	90·4	25·9	154	148·0	42·4	214	205·7	59·0	274	263·4	75·5
35	33·6	09·6	95	91·3	26·2	155	149·0	42·7	215	206·7	59·3	275	264·3	75·8
36	34·6	09·9	96	92·3	26·5	156	150·0	43·0	216	207·6	59·5	276	265·3	76·1
37	35·6	10·2	97	93·2	26·7	157	150·9	43·3	217	208·6	59·8	277	266·3	76·4
38	36·5	10·5	98	94·2	27·0	158	151·9	43·6	218	209·6	60·1	278	267·2	76·6
39	37·5	10·7	99	95·2	27·3	159	152·8	43·8	219	210·5	60·4	279	268·2	76·9
40	38·5	11·0	100	96·1	27·6	160	153·8	44·1	220	211·5	60·6	280	269·2	77·2
41	39·4	11·3	101	97·1	27·8	161	154·8	44·4	221	212·4	60·9	281	270·1	77·5
42	40·4	11·6	102	98·0	28·1	162	155·7	44·7	222	213·4	61·2	282	271·1	77·7
43	41·3	11·9	103	99·0	28·4	163	156·7	44·9	223	214·4	61·5	283	272·0	78·0
44	42·3	12·1	104	100·0	28·7	164	157·6	45·2	224	215·3	61·7	284	273·0	78·3
45	43·3	12·4	105	100·9	28·9	165	158·6	45·5	225	216·3	62·0	285	274·0	78·6
46	44·2	12·7	106	101·9	29·2	166	159·6	45·8	226	217·2	62·3	286	274·9	78·8
47	45·2	13·0	107	102·9	29·5	167	160·5	46·0	227	218·2	62·6	287	275·9	79·1
48	46·1	13·2	108	103·8	29·8	168	161·5	46·3	228	219·2	62·8	288	276·8	79·4
49	47·1	13·5	109	104·8	30·0	169	162·5	46·6	229	220·1	63·1	289	277·8	79·7
50	48·1	13·8	110	105·7	30·3	170	163·4	46·9	230	221·1	63·4	290	278·8	79·9
51	49·0	14·1	111	106·7	30·6	171	164·4	47·1	231	222·1	63·7	291	279·7	80·2
52	50·0	14·3	112	107·7	30·9	172	165·3	47·4	232	223·0	63·9	292	280·7	80·5
53	50·9	14·6	113	108·6	31·1	173	166·3	47·7	233	224·0	64·2	293	281·6	80·8
54	51·9	14·9	114	109·6	31·4	174	167·3	48·0	234	224·9	64·5	294	282·6	81·0
55	52·9	15·2	115	110·5	31·7	175	168·2	48·2	235	225·9	64·8	295	283·6	81·3
56	53·8	15·4	116	111·5	32·0	176	169·2	48·5	236	226·9	65·0	296	284·5	81·6
57	54·8	15·7	117	112·5	32·2	177	170·1	48·8	237	227·8	65·3	297	285·5	81·9
58	55·8	16·0	118	113·4	32·5	178	171·1	49·1	238	228·8	65·6	298	286·5	82·1
59	56·7	16·3	119	114·4	32·8	179	172·1	49·3	239	229·7	65·9	299	287·4	82·4
60	57·7	16·5	120	115·4	33·1	180	173·0	49·6	240	230·7	66·2	300	288·4	82·7
Dist	Dep.	D. Lat	Dist	Dep.	D. Lat	Dist	Dep.	D. Lat	Dist	Dep.	D. Lat	Dist	Dep.	D. Lat
D Lon		Dep.	D Lon		Dep.	D Lon		Dep.	D Lon		Dep.	D Lon		Dep.

286°
254°

74 DEGREES.

074°
106°

4h 56m

	344° ↑ 196°		TRAVERSE TABLE 16 DEGREES.		↑016° 164°	1h 4m	16°

D Lon	Dep.		D Lon	Dep.		D Lon	Dep.		D Lon	Dep.		D Lon	Dep.	
Dist	D.Lat	Dep.	Dist	D.Lat	Dep.	Dist	D.Lat	Dep.	Dist	D.Lat	Dep.	Dist	D.Lat	Dep.
301	289.3	83.0	361	347.0	99.5	421	404.7	116.0	481	462.4	132.6	541	520.0	149.1
302	290.3	83.2	362	348.0	99.8	422	405.7	116.3	482	463.3	132.9	542	521.0	149.4
303	291.3	83.5	363	348.9	100.1	423	406.6	116.6	483	464.3	133.1	543	522.0	149.7
304	292.2	83.8	364	349.9	100.3	424	407.6	116.9	484	465.3	133.4	544	522.9	149.9
305	293.2	84.1	365	350.9	100.6	425	408.5	117.0	485	466.2	133.7	545	523.9	150.2
306	294.1	84.3	366	351.8	100.9	426	409.5	117.4	486	467.2	134.0	546	524.8	150.5
307	295.1	84.6	367	352.8	101.2	427	410.5	117.7	487	468.1	134.2	547	525.8	150.8
308	296.1	84.9	368	353.7	101.4	428	411.4	118.0	488	469.1	134.5	548	526.8	151.0
309	297.0	85.2	369	354.7	101.7	429	412.4	118.2	489	470.1	134.8	549	527.7	151.3
310	298.0	85.4	370	355.7	102.0	430	413.3	118.5	490	471.0	135.1	550	528.7	151.6
311	299.0	85.7	371	356.6	102.3	431	414.3	118.8	491	472.0	135.3	551	529.7	151.9
312	299.9	86.0	372	357.6	102.5	432	415.3	119.1	492	472.9	135.6	552	530.6	152.2
313	300.9	86.3	373	358.6	102.8	433	416.2	119.4	493	473.9	135.9	553	531.6	152.4
314	301.8	86.6	374	359.5	103.1	434	417.2	119.6	494	474.9	136.2	554	532.5	152.7
315	302.8	86.8	375	360.5	103.4	435	418.1	119.9	495	475.8	136.4	555	533.5	153.0
316	303.8	87.1	376	361.4	103.6	436	419.1	120.2	496	476.8	136.7	556	534.5	153.3
317	304.7	87.4	377	362.4	103.9	437	420.1	120.5	497	477.7	137.0	557	535.4	153.5
318	305.7	87.7	378	363.4	104.2	438	421.0	120.7	498	478.7	137.3	558	536.4	153.8
319	306.6	87.9	379	364.3	104.5	439	422.0	121.0	499	479.7	137.5	559	537.3	154.1
320	307.6	88.2	380	365.3	104.7	440	423.0	121.3	500	480.6	137.8	560	538.3	154.4
321	308.6	88.5	381	366.2	105.0	441	423.9	121.6	501	481.6	138.1	561	539.3	154.6
322	309.5	88.8	382	367.2	105.3	442	424.9	121.8	502	482.6	138.4	562	540.2	154.9
323	310.5	89.0	383	368.2	105.6	443	425.8	122.1	503	483.5	138.6	563	541.2	155.2
324	311.4	89.3	384	369.1	105.8	444	426.8	122.4	504	484.5	138.9	564	542.2	155.5
325	312.4	89.6	385	370.1	106.1	445	427.8	122.7	505	485.4	139.2	565	543.1	155.7
326	313.4	89.9	386	371.0	106.4	446	428.7	122.9	506	486.4	139.5	566	544.1	156.0
327	314.3	90.1	387	372.0	106.7	447	429.7	123.2	507	487.4	139.7	567	545.0	156.3
328	315.3	90.4	388	373.0	106.9	448	430.6	123.5	508	488.3	140.0	568	546.0	156.6
329	316.3	90.7	389	373.9	107.2	449	431.6	123.8	509	489.3	140.3	569	547.0	156.8
330	317.2	91.0	390	374.9	107.5	450	432.6	124.0	510	490.2	140.6	570	547.9	157.1
331	318.2	91.2	391	375.9	107.8	451	433.5	124.3	511	491.2	140.9	571	548.9	157.4
332	319.1	91.5	392	376.8	108.0	452	434.5	124.6	512	492.2	141.1	572	549.8	157.7
333	320.1	91.8	393	377.8	108.3	453	435.5	124.6	513	493.1	141.4	573	550.8	157.9
334	321.1	92.1	394	378.7	108.6	454	436.4	125.1	514	494.1	141.7	574	551.8	158.2
335	322.0	92.3	395	379.7	108.9	455	437.4	125.4	515	495.0	142.0	575	552.7	158.5
336	323.0	92.6	396	380.7	109.2	456	438.3	125.7	516	496.0	142.2	576	553.7	158.8
337	323.9	92.9	397	381.6	109.4	457	439.3	126.0	517	497.0	142.5	577	554.6	159.0
338	324.9	93.2	398	382.6	109.7	458	440.3	126.2	518	497.9	142.8	578	555.6	159.3
339	325.8	93.4	399	383.5	110.0	459	441.2	126.5	519	498.9	143.1	579	556.6	159.6
340	326.8	93.7	400	384.5	110.3	460	442.2	126.8	520	499.9	143.3	580	557.5	159.9
341	327.8	94.0	401	385.5	110.5	461	443.1	127.1	521	500.8	143.6	581	558.5	160.1
342	328.7	94.3	402	386.4	110.8	462	444.1	127.3	522	501.8	143.9	582	559.5	160.4
343	329.7	94.5	403	387.4	111.1	463	445.1	127.6	523	502.7	144.2	583	560.4	160.6
344	330.7	94.8	404	388.3	111.4	464	446.0	127.9	524	503.7	144.4	584	561.4	161.0
345	331.6	95.1	405	389.3	111.6	465	447.0	128.2	525	504.7	144.7	585	562.3	161.2
346	332.6	95.4	406	390.3	111.9	466	447.9	128.4	526	505.6	145.0	586	563.3	151.5
347	333.6	95.6	407	391.2	112.2	467	448.9	128.7	527	506.6	145.3	587	564.3	161.8
348	334.5	95.9	408	392.2	112.5	468	449.9	129.0	528	507.5	145.5	588	565.2	162.1
349	335.5	96.2	409	393.2	112.7	469	450.8	129.3	529	508.5	145.8	589	566.2	162.4
350	336.4	96.5	410	394.1	113.0	470	451.8	129.5	530	509.5	146.1	590	567.1	162.6
351	337.4	96.7	411	395.1	113.3	471	452.8	129.8	531	510.4	146.4	591	568.1	162.9
352	338.4	97.0	412	396.0	113.6	472	453.7	130.1	532	511.4	146.6	592	569.1	163.2
353	339.3	97.3	413	397.0	113.8	473	454.7	130.4	533	512.4	146.9	593	570.0	163.5
354	340.3	97.6	414	398.0	114.1	474	455.6	130.7	534	513.3	147.2	594	571.0	163.7
355	341.2	97.9	415	398.9	114.4	475	456.6	130.9	535	514.3	147.5	595	572.0	164.0
356	342.2	98.1	416	399.9	114.7	476	457.6	131.2	536	515.2	147.7	596	572.9	164.3
357	343.2	98.4	417	400.8	114.9	477	458.5	131.5	537	516.2	148.0	597	573.0	164.6
358	344.1	98.7	418	401.8	115.2	478	459.5	131.8	538	517.2	148.3	598	574.8	164.8
359	345.1	99.0	419	402.8	115.5	479	460.4	132.0	539	518.1	148.6	599	575.8	165.1
360	346.1	99.2	420	403.7	115.8	480	461.4	132.3	540	519.1	148.8	600	576.8	165.4
Dist	Dep.	D.Lat	Dist	Dep.	D.Lat	Dist	Dep.	D.Lat	Dist	Dep.	D.Lat	Dist	Dep.	D.Lat
D Lon		Dep.	D Lon		Dep.	D Lon		Dep.	D Lon		Dep.	D Lon		Dep.

	286° ↑ 254°		74 DEGREES.		↑074° 106°	4h 56m	74°

TRAVERSE TABLE
17 DEGREES.

17° 343° / 197° | 017° / 163° 1h 8m

D Lon	Dep.		D Lon	Dep.		D Lon	Dep.		D Lon	Dep.		D Lon	Dep.	
Dist	D.Lat	Dep.	Dist	D.Lat	Dep.	Dist	D.Lat	Dep.	Dist	D.Lat	Dep.	Dist	D.Lat	Dep.
1	01·0	00·3	61	58·3	17·8	121	115·7	35·4	181	173·1	52·9	241	230·5	70·5
2	01·9	00·6	62	59·3	18·1	122	116·7	35·7	182	174·0	53·2	242	231·4	70·8
3	02·9	00·9	63	60·2	18·4	123	117·6	36·0	183	175·0	53·5	243	232·4	71·0
4	03·8	01·2	64	61·2	18·7	124	118·6	36·3	184	176·0	53·8	244	233·3	71·3
5	04·8	01·5	65	62·2	19·0	125	119·5	36·5	185	176·9	54·1	245	234·3	71·6
6	05·7	01·8	66	63·1	19·3	126	120·5	36·8	186	177·9	54·4	246	235·3	71·9
7	06·7	02·0	67	64·1	19·6	127	121·5	37·1	187	178·8	54·7	247	236·2	72·2
8	07·7	02·3	68	65·0	19·9	128	122·4	37·4	188	179·8	55·0	248	237·2	72·5
9	08·6	02·6	69	66·0	20·2	129	123·4	37·7	189	180·7	55·3	249	238·1	72·8
10	09·6	02·9	70	66·9	20·5	130	124·3	38·0	190	181·7	55·6	250	239·1	73·1
11	10·5	03·2	71	67·9	20·8	131	125·3	38·3	191	182·7	55·8	251	240·0	73·4
12	11·5	03·5	72	68·9	21·1	132	126·2	38·6	192	183·6	56·1	252	241·0	73·7
13	12·4	03·8	73	69·8	21·3	133	127·2	38·9	193	184·6	56·4	253	241·9	74·0
14	13·4	04·1	74	70·8	21·6	134	128·1	39·2	194	185·5	56·7	254	242·9	74·3
15	14·3	04·4	75	71·7	21·9	135	129·1	39·5	195	186·5	57·0	255	243·9	74·6
16	15·3	04·7	76	72·7	22·2	136	130·1	39·8	196	187·4	57·3	256	244·8	74·8
17	16·3	05·0	77	73·6	22·5	137	131·0	40·1	197	188·4	57·6	257	245·8	75·1
18	17·2	05·3	78	74·6	22·8	138	132·0	40·3	198	189·3	57·9	258	246·7	75·4
19	18·2	05·6	79	75·5	23·1	139	132·9	40·6	199	190·3	58·2	259	247·7	75·7
20	19·1	05·8	80	76·5	23·3	140	133·9	40·9	200	191·3	58·5	260	248·6	76·0
21	20·1	06·1	81	77·5	23·7	141	134·8	41·2	201	192·2	58·8	261	249·6	76·3
22	21·0	06·4	82	78·4	24·0	142	135·8	41·5	202	193·2	59·1	262	250·6	76·6
23	22·0	06·7	83	79·4	24·3	143	136·8	41·8	203	194·1	59·4	263	251·5	76·9
24	23·0	07·0	84	80·3	24·6	144	137·7	42·1	204	195·1	59·6	264	252·5	77·2
25	23·9	07·3	85	81·3	24·9	145	138·7	42·4	205	196·0	59·9	265	253·4	77·5
26	24·9	07·6	86	82·2	25·1	146	139·6	42·7	206	197·0	60·2	266	254·4	77·8
27	25·8	07·9	87	83·2	25·4	147	140·6	43·0	207	198·0	60·5	267	255·3	78·1
28	26·8	08·2	88	84·2	25·7	148	141·5	43·3	208	198·9	60·8	268	256·3	78·4
29	27·7	08·5	89	85·1	26·0	149	142·5	43·6	209	199·9	61·1	269	257·2	78·6
30	28·7	08·8	90	86·1	26·3	150	143·4	43·9	210	200·8	61·4	270	258·2	78·9
31	29·6	09·1	91	87·0	26·6	151	144·4	44·1	211	201·8	61·7	271	259·2	79·2
32	30·6	09·4	92	88·0	26·9	152	145·4	44·4	212	202·7	62·0	272	260·1	79·5
33	31·6	09·6	93	88·9	27·2	153	146·3	44·7	213	203·7	62·3	273	261·1	79·8
34	32·5	09·9	94	89·9	27·5	154	147·3	45·0	214	204·6	62·6	274	262·0	80·1
35	33·5	10·2	95	90·8	27·8	155	148·2	45·3	215	205·6	62·9	275	263·0	80·4
36	34·4	10·5	96	91·8	28·1	156	149·2	45·6	216	206·6	63·2	276	263·9	80·7
37	35·4	10·8	97	92·8	28·4	157	150·1	45·9	217	207·5	63·4	277	264·9	81·0
38	36·3	11·1	98	93·7	28·7	158	151·1	46·2	218	208·5	63·7	278	265·9	81·3
39	37·3	11·4	99	94·7	28·9	159	152·1	46·5	219	209·4	64·0	279	266·8	81·6
40	38·3	11·7	100	95·6	29·2	160	153·0	46·8	220	210·4	64·3	280	267·8	81·9
41	39·2	12·0	101	96·6	29·5	161	154·0	47·1	221	211·3	64·6	281	268·7	82·2
42	40·2	12·3	102	97·5	29·8	162	154·9	47·4	222	212·3	64·9	282	269·7	82·4
43	41·1	12·6	103	98·5	30·1	163	155·9	47·7	223	213·3	65·2	283	270·6	82·7
44	42·1	12·9	104	99·5	30·4	164	156·8	47·9	224	214·2	65·5	284	271·6	83·0
45	43·0	13·2	105	100·4	30·7	165	157·8	48·2	225	215·2	65·8	285	272·5	83·3
46	44·0	13·4	106	101·4	31·0	166	158·7	48·5	226	216·1	66·1	286	273·5	83·6
47	44·9	13·7	107	102·3	31·3	167	159·7	48·8	227	217·1	66·4	287	274·5	83·9
48	45·9	14·0	108	103·3	31·6	168	160·7	49·1	228	218·0	66·7	288	275·4	84·2
49	46·9	14·3	109	104·2	31·9	169	161·6	49·4	229	219·0	67·0	289	276·4	84·5
50	47·8	14·6	110	105·2	32·2	170	162·6	49·7	230	220·0	67·2	290	277·3	84·8
51	48·8	14·9	111	106·1	32·5	171	163·5	50·0	231	220·9	67·5	291	278·3	85·1
52	49·7	15·2	112	107·1	32·7	172	164·5	50·3	232	221·9	67·8	292	279·2	85·4
53	50·7	15·5	113	108·1	33·0	173	165·4	50·6	233	222·8	68·1	293	280·2	85·7
54	51·6	15·8	114	109·0	33·3	174	166·4	50·9	234	223·8	68·4	294	281·2	86·0
55	52·6	16·1	115	110·0	33·6	175	167·4	51·2	235	224·7	68·7	295	282·1	86·2
56	53·6	16·4	116	110·9	33·9	176	168·3	51·5	236	225·7	69·0	296	283·1	86·5
57	54·5	16·7	117	111·9	34·2	177	169·3	51·7	237	226·6	69·3	297	284·0	86·8
58	55·5	17·0	118	112·8	34·5	178	170·2	52·0	238	227·6	69·6	298	285·0	87·1
59	56·4	17·2	119	113·8	34·8	179	171·2	52·3	239	228·6	69·9	299	285·9	87·4
60	57·4	17·5	120	114·8	35·1	180	172·1	52·6	240	229·5	70·2	300	286·9	87·7

Dist	Dep.	D.Lat	Dist	Dep.	D.Lat	Dist	Dep.	D.Lat	Dist	Dep.	D.Lat	Dist	Dep.	D.Lat
D Lon		Dep.	D Lon		Dep.	D Lon		Dep.	D Lon		Dep.	D Lon		Dep.

73° 287° / 253° **73 DEGREES.** 073° / 107° 4h 52m

	343° ↑			017° ↑	
	197°		**TRAVERSE TABLE**	163°	1h 8m

17 DEGREES

DLon	Dep.		DLon	Dep.		DLon	Dep.		DLon	Dep.		DLon	Dep.	
Dist	D.Lat	Dep.	Dist	D.Lat	Dep.	Dist	D.Lat	Dep.	Dist	D.Lat	Dep.	Dist	D.Lat	Dep.
301	287·8	88·0	361	345·2	105·5	421	402·6	123·1	481	460·0	140·6	541	517·4	158·2
302	288·8	88·3	362	346·2	105·8	422	403·6	123·4	482	460·9	140·9	542	518·3	158·5
303	289·8	88·6	363	347·1	106·1	423	404·5	123·7	483	461·9	141·2	543	519·3	158·8
304	290·7	88·9	364	348·1	106·4	424	405·5	124·0	484	462·9	141·5	544	520·2	159·1
305	291·7	89·2	365	349·1	106·7	425	406·4	124·3	485	463·8	141·8	545	521·2	159·3
306	292·6	89·5	366	350·0	107·0	426	407·4	124·6	486	464·8	142·1	546	522·1	159·6
307	293·6	89·8	367	351·0	107·3	427	408·3	124·8	487	465·7	142·4	547	523·1	159·9
308	294·5	90·1	368	351·9	107·6	428	409·3	125·1	488	466·7	142·7	548	524·1	160·2
309	295·5	90·3	369	352·9	107·9	429	410·3	125·4	489	467·6	143·0	549	525·0	160·5
310	296·5	90·6	370	353·8	108·2	430	411·2	125·7	490	468·6	143·3	550	526·0	160·8
311	297·4	90·9	371	354·8	108·5	431	412·2	126·0	491	469·5	143·6	551	526·9	161·1
312	298·4	91·2	372	355·7	108·8	432	413·1	126·3	492	470·5	143·8	552	527·9	161·4
313	299·3	91·5	373	356·7	109·1	433	414·1	126·6	493	471·5	144·1	553	528·8	161·7
314	300·3	91·8	374	357·7	109·3	434	415·0	126·9	494	472·4	144·4	554	529·8	162·0
315	301·2	92·1	375	358·6	109·6	435	416·0	127·2	495	473·4	144·7	555	530·7	162·3
316	302·2	92·4	376	359·6	109·9	436	416·9	127·5	496	474·3	145·0	556	531·7	162·6
317	303·1	92·7	377	360·5	110·2	437	417·9	127·8	497	475·3	145·3	557	532·7	162·9
318	304·1	93·0	378	361·5	110·5	438	418·9	128·1	498	476·2	145·6	558	533·6	163·1
319	305·1	93·3	379	362·4	110·8	439	419·8	128·4	499	477·2	145·9	559	534·6	163·4
320	306·0	93·6	380	363·4	111·1	440	420·8	128·6	500	478·2	146·2	560	535·5	163·7
321	307·0	93·9	381	364·4	111·4	441	421·7	128·9	501	479·1	146·5	561	536·5	164·0
322	307·9	94·1	382	365·3	111·7	442	422·7	129·2	502	480·1	146·8	562	537·4	164·3
323	308·9	94·4	383	366·3	112·0	443	423·6	129·5	503	481·0	147·1	563	538·4	164·6
324	309·8	94·7	384	367·2	112·3	444	424·6	129·8	504	482·0	147·4	564	539·4	164·9
325	310·8	95·0	385	368·2	112·6	445	425·6	130·1	505	482·9	147·6	565	540·3	165·2
326	311·8	95·3	386	369·1	112·9	446	426·5	130·4	506	483·9	147·9	566	541·3	165·5
327	312·7	95·6	387	370·1	113·1	447	427·5	130·7	507	484·8	148·2	567	542·2	165·8
328	313·6	95·9	388	371·0	113·4	448	428·4	131·0	508	485·8	148·5	568	543·2	166·1
329	314·6	96·2	389	372·0	113·7	449	429·4	131·3	509	486·8	148·8	569	544·1	166·4
330	315·5	96·5	390	373·0	114·0	450	430·3	131·6	510	487·7	149·1	570	545·1	166·7
331	316·5	96·8	391	373·9	114·3	451	431·3	131·9	511	488·7	149·4	571	546·1	166·9
332	317·5	97·1	392	374·9	114·6	452	432·2	132·2	512	489·6	149·7	572	547·0	167·2
333	318·4	97·4	393	375·8	114·9	453	433·2	132·4	513	490·6	150·0	573	548·0	167·5
334	319·4	97·7	394	376·8	115·2	454	434·2	132·7	514	491·5	150·3	574	548·9	167·8
335	320·4	97·9	395	377·7	115·5	455	435·1	133·0	515	492·5	150·6	575	549·9	168·1
336	321·3	98·2	396	378·7	115·8	456	436·1	133·3	516	493·5	150·9	576	550·8	168·4
337	322·3	98·5	397	379·7	116·1	457	437·0	133·6	517	494·4	151·2	577	551·8	168·7
338	323·2	98·8	398	380·6	116·4	458	438·0	133·9	518	495·4	151·4	578	552·7	169·0
339	324·2	99·1	399	381·6	116·7	459	438·9	134·2	519	496·3	151·7	579	553·7	169·3
340	325·1	99·4	400	382·5	116·9	460	439·9	134·5	520	497·3	152·0	580	554·7	169·6
341	326·1	99·7	401	383·5	117·2	461	440·9	134·8	521	498·2	152·3	581	555·6	169·9
342	327·1	100·0	402	384·4	117·5	462	441·8	135·1	522	499·2	152·6	582	556·6	170·2
343	328·0	100·3	403	385·4	117·8	463	442·8	135·4	523	500·1	152·9	583	557·5	170·5
344	329·0	100·6	404	386·3	118·1	464	443·7	135·7	524	501·1	153·2	584	558·5	170·7
345	329·9	100·9	405	387·3	118·4	465	444·7	136·0	525	502·1	153·5	585	559·4	171·0
346	330·8	101·2	406	388·3	118·7	466	445·6	136·2	526	503·0	153·8	586	560·4	171·3
347	331·8	101·5	407	389·2	119·0	467	446·6	136·5	527	504·0	154·1	587	561·4	171·6
348	332·8	101·7	408	390·2	119·3	468	447·6	136·8	528	504·9	154·4	588	562·3	171·9
349	333·8	102·0	409	391·1	119·6	469	448·5	137·1	529	505·9	154·7	589	563·3	172·2
350	334·7	102·3	410	392·1	119·9	470	449·5	137·4	530	506·8	155·0	590	564·2	172·5
351	335·7	102·6	411	393·0	120·2	471	450·4	137·7	531	507·8	155·2	591	565·2	172·8
352	336·6	102·9	412	394·0	120·5	472	451·4	138·0	532	508·8	155·5	592	566·1	173·1
353	337·6	103·2	413	394·9	120·7	473	452·3	138·3	533	509·7	155·8	593	567·1	173·4
354	338·5	103·5	414	395·9	121·0	474	453·3	138·6	534	510·7	156·1	594	568·0	173·7
355	339·5	103·8	415	396·8	121·3	475	454·2	138·9	535	511·6	156·4	595	569·0	174·0
356	340·4	104·1	416	397·8	121·6	476	455·2	139·2	536	512·6	156·7	596	570·0	174·3
357	341·4	104·4	417	398·7	121·9	477	456·2	139·5	537	513·5	157·0	597	570·9	174·5
358	342·4	104·7	418	399·7	122·2	478	457·1	139·7	538	514·5	157·3	598	571·9	174·8
359	343·3	105·0	419	400·7	122·5	479	458·1	140·0	539	515·4	157·6	599	572·8	175·1
360	344·3	105·3	420	401·6	122·8	480	459·0	140·3	540	516·4	157·9	600	573·8	175·4
Dist	Dep.	D.Lat	Dist	Dep.	D.Lat	Dist	Dep.	D.Lat	Dist	Dep.	D.Lat	Dist	Dep.	D.Lat
DLon		Dep.	DLon		Dep.	DLon		Dep.	DLon		Dep.	DLon		Dep.

	287° ↑					073° ↑	
	253°		**73 DEGREES.**			107°	4h 52m

TRAVERSE TABLE

| 342° ↑ |
| 198° |

18 DEGREES.

| ↑ 018° |
| 162° | 1h 12m |

D Lon	Dep.		D Lon	Dep.		D Lon	Dep.		D Lon	Dep.		D Lon	Dep.	
Dist	D.Lat	Dep.	Dist	D.Lat	Dep.	Dist	D.Lat	Dep.	Dist	D.Lat	Dep.	Dist	D.Lat	Dep.
1	01·0	00·3	61	58·0	18·9	121	115·1	37·4	181	172·1	55·9	241	229·2	74·5
2	01·9	00·6	62	59·0	19·2	122	116·0	37·7	182	173·1	56·2	242	230·2	74·8
3	02·9	00·9	63	59·9	19·5	123	117·0	38·0	183	174·0	56·6	243	231·1	75·1
4	03·8	01·2	64	60·9	19·8	124	117·9	38·3	184	175·0	56·9	244	232·1	75·4
5	04·8	01·5	65	61·8	20·1	125	118·9	38·6	185	175·9	57·2	245	233·0	75·7
6	05·7	01·9	66	62·8	20·4	126	119·8	38·9	186	176·9	57·5	246	234·0	76·0
7	06·7	02·2	67	63·7	20·7	127	120·8	39·2	187	177·8	57·8	247	234·9	76·3
8	07·6	02·5	68	64·7	21·0	128	121·7	39·6	188	178·8	58·1	248	235·9	76·6
9	08·6	02·8	69	65·6	21·3	129	122·7	39·9	189	179·7	58·4	249	236·8	76·9
10	09·5	03·1	70	66·6	21·6	130	123·6	40·2	190	180·7	58·7	250	237·8	77·3
11	10·5	03·4	71	67·5	21·9	131	124·6	40·5	191	181·7	59·0	251	238·7	77·6
12	11·4	03·7	72	68·5	22·2	132	125·5	40·8	192	182·6	59·3	252	239·7	77·9
13	12·4	04·0	73	69·4	22·6	133	126·5	41·1	193	183·6	59·6	253	240·6	78·2
14	13·3	04·3	74	70·4	22·9	134	127·4	41·4	194	184·5	59·9	254	241·6	78·5
15	14·3	04·6	75	71·3	23·2	135	128·4	41·7	195	185·5	60·3	255	242·5	78·8
16	15·2	04·9	76	72·3	23·5	136	129·3	42·0	196	186·4	60·6	256	243·5	79·1
17	16·2	05·3	77	73·2	23·8	137	130·3	42·3	197	187·4	60·9	257	244·4	79·4
18	17·1	05·6	78	74·2	24·1	138	131·2	42·6	198	188·3	61·2	258	245·4	79·7
19	18·1	05·9	79	75·1	24·4	139	132·2	43·0	199	189·3	61·5	259	246·3	80·0
20	19·0	06·2	80	76·1	24·7	140	133·1	43·3	200	190·2	61·8	260	247·3	80·3
21	20·0	06·5	81	77·0	25·0	141	134·1	43·6	201	191·2	62·1	261	248·2	80·7
22	20·9	06·8	82	78·0	25·3	142	135·1	43·9	202	192·1	62·4	262	249·2	81·0
23	21·9	07·1	83	78·9	25·6	143	136·0	44·2	203	193·1	62·7	263	250·1	81·3
24	22·8	07·4	84	79·9	26·0	144	137·0	44·5	204	194·0	63·0	264	251·1	81·6
25	23·8	07·7	85	80·8	26·3	145	137·9	44·8	205	195·0	63·3	265	252·0	81·9
26	24·7	08·0	86	81·8	26·6	146	138·9	45·1	206	195·9	63·7	266	253·0	82·2
27	25·7	08·3	87	82·7	26·9	147	139·8	45·4	207	196·9	64·0	267	253·9	82·5
28	26·6	08·7	88	83·7	27·2	148	140·8	45·7	208	197·8	64·3	268	254·9	82·8
29	27·6	09·0	89	84·6	27·5	149	141·7	46·0	209	198·8	64·6	269	255·8	83·1
30	28·5	09·3	90	85·6	27·8	150	142·7	46·4	210	199·7	64·9	270	256·8	83·4
31	29·5	09·6	91	86·5	28·1	151	143·6	46·7	211	200·7	65·2	271	257·7	83·7
32	30·4	09·9	92	87·5	28·4	152	144·6	47·0	212	201·6	65·5	272	258·7	84·1
33	31·4	10·2	93	88·4	28·7	153	145·5	47·3	213	202·6	65·8	273	259·6	84·4
34	32·3	10·5	94	89·4	29·0	154	146·5	47·6	214	203·5	66·1	274	260·6	84·7
35	33·3	10·8	95	90·4	29·4	155	147·4	47·9	215	204·5	66·4	275	261·5	85·0
36	34·2	11·1	96	91·3	29·7	156	148·4	48·2	216	205·4	66·7	276	262·5	85·3
37	35·2	11·4	97	92·3	30·0	157	149·3	48·5	217	206·4	67·1	277	263·4	85·6
38	36·1	11·7	98	93·2	30·3	158	150·3	48·8	218	207·3	67·4	278	264·4	85·9
39	37·1	12·1	99	94·2	30·6	159	151·2	49·1	219	208·3	67·7	279	265·3	86·2
40	38·0	12·4	100	95·1	30·9	160	152·2	49·4	220	209·2	68·0	280	266·3	86·5
41	39·0	12·7	101	96·1	31·2	161	153·1	49·8	221	210·2	68·3	281	267·2	86·8
42	39·9	13·0	102	97·0	31·5	162	154·1	50·1	222	211·1	68·6	282	268·2	87·1
43	40·9	13·3	103	98·0	31·8	163	155·0	50·4	223	212·1	68·9	283	269·1	87·5
44	41·8	13·6	104	98·9	32·1	164	156·0	50·7	224	213·0	69·2	284	270·1	87·8
45	42·8	13·9	105	99·9	32·4	165	156·9	51·0	225	214·0	69·5	285	271·1	88·1
46	43·7	14·2	106	100·8	32·8	166	157·9	51·3	226	214·9	69·8	286	272·0	88·4
47	44·7	14·5	107	101·8	33·1	167	158·8	51·6	227	215·9	70·1	287	273·0	88·7
48	45·7	14·8	108	102·7	33·4	168	159·8	51·9	228	216·8	70·5	288	273·9	89·0
49	46·6	15·1	109	103·7	33·7	169	160·7	52·2	229	217·8	70·8	289	274·9	89·3
50	47·6	15·5	110	104·6	34·0	170	161·7	52·5	230	218·7	71·1	290	275·8	89·6
51	48·5	15·8	111	105·6	34·3	171	162·6	52·8	231	219·7	71·4	291	276·8	89·9
52	49·5	16·1	112	106·5	34·6	172	163·6	53·2	232	220·6	71·7	292	277·7	90·2
53	50·4	16·4	113	107·5	34·9	173	164·5	53·5	233	221·6	72·0	293	278·7	90·5
54	51·4	16·7	114	108·4	35·2	174	165·5	53·8	234	222·5	72·3	294	279·6	90·9
55	52·3	17·0	115	109·4	35·5	175	166·4	54·1	235	223·5	72·6	295	280·6	91·2
56	53·3	17·3	116	110·3	35·8	176	167·4	54·4	236	224·4	72·9	296	281·5	91·5
57	54·2	17·6	117	111·3	36·2	177	168·3	54·7	237	225·4	73·2	297	282·5	91·8
58	55·2	17·9	118	112·2	36·5	178	169·3	55·0	238	226·4	73·5	298	283·4	92·1
59	56·1	18·2	119	113·2	36·8	179	170·2	55·3	239	227·3	73·9	299	284·4	92·4
60	57·1	18·5	120	114·1	37·1	180	171·2	55·6	240	228·3	74·2	300	285·3	92·7
Dist	Dep.	D.Lat	Dist	Dep.	D.Lat	Dist	Dep.	D.Lat	Dist	Dep.	D.Lat	Dist	Dep.	D.Lat
D Lon		Dep.	D Lon		Dep.	D Lon		Dep.	D Lon		Dep.	D Lon		Dep.

| 288° ↑ |
| 252° |

72 DEGREES.

| ↑ 072° |
| 108° | 4h 48m |

| | 342° ↑ / 198° | | **TRAVERSE TABLE** — 18 DEGREES. | | ↑018° / 162° 1h 12m | **18°** |

Dist	D.Lat	Dep.	Dist	D.Lat	Dep.	Dist	D.Lat	Dep.	Dist	D.Lat	Dep.	Dist	D.Lat	Dep.
301	286·3	93·0	361	343·3	111·6	421	400·4	130·1	481	457·5	148·6	541	514·5	167·2
302	287·2	93·3	362	344·3	111·9	422	401·3	130·4	482	458·4	148·9	542	515·5	167·5
303	288·2	93·6	363	345·2	112·2	423	402·3	130·7	483	459·4	149·3	543	516·4	167·8
304	289·1	93·9	364	346·2	112·5	424	403·2	131·0	484	460·3	149·6	544	517·4	168·1
305	290·1	94·3	365	347·1	112·8	425	404·2	131·3	485	461·3	149·9	545	518·3	168·4
306	291·0	94·6	366	348·1	113·1	426	405·2	131·6	486	462·2	150·2	546	519·3	168·7
307	292·0	94·9	367	349·0	113·4	427	406·1	132·0	487	463·2	150·5	547	520·2	169·0
308	292·9	95·2	368	350·0	113·7	428	407·1	132·3	488	464·1	150·8	548	521·2	169·3
309	293·9	95·5	369	350·9	114·0	429	408·0	132·6	489	465·1	151·1	549	522·1	169·7
310	294·8	95·8	370	351·9	114·3	430	409·0	132·9	490	466·0	151·4	550	523·1	170·0
311	295·8	96·1	371	352·8	114·6	431	409·9	133·2	491	467·0	151·7	551	524·0	170·3
312	296·7	96·4	372	353·8	115·0	432	410·9	133·5	492	467·9	152·0	552	525·0	170·6
313	297·7	96·7	373	354·7	115·3	433	411·8	133·8	493	468·9	152·3	553	525·9	170·9
314	298·6	97·0	374	355·7	115·6	434	412·8	134·1	494	469·8	152·7	554	526·9	171·2
315	299·6	97·3	375	356·6	115·9	435	413·7	134·4	495	470·8	153·0	555	527·8	171·5
316	300·5	97·6	376	357·6	116·2	436	414·7	134·7	496	471·7	153·3	556	528·8	171·8
317	301·5	98·0	377	358·5	116·5	437	415·6	135·0	497	472·7	153·6	557	529·7	172·1
318	302·4	98·3	378	359·5	116·8	438	416·6	135·3	498	473·6	153·9	558	530·7	172·4
319	303·4	98·6	379	360·5	117·1	439	417·5	135·7	499	474·6	154·2	559	531·6	172·7
320	304·3	98·9	380	361·4	117·4	440	418·5	136·0	500	475·5	154·5	560	532·6	173·0
321	305·3	99·2	381	362·4	117·7	441	419·4	136·3	501	476·5	154·8	561	533·5	173·4
322	306·2	99·5	382	363·3	118·0	442	420·4	136·6	502	477·4	155·1	562	534·5	173·7
323	307·2	99·8	383	364·3	118·4	443	421·3	136·9	503	478·4	155·4	563	535·4	174·0
324	308·2	100·1	384	365·2	118·7	444	422·3	137·2	504	479·3	155·7	564	536·4	174·3
325	309·1	100·4	385	366·2	119·0	445	423·2	137·5	505	480·3	156·1	565	537·3	174·6
326	310·0	100·7	386	367·1	119·3	446	424·2	137·8	506	481·2	156·4	566	538·3	174·9
327	311·0	101·0	387	368·1	119·6	447	425·1	138·1	507	482·2	156·7	567	539·2	175·2
328	311·9	101·4	388	369·0	119·9	448	426·1	138·4	508	483·1	157·0	568	540·2	175·5
329	312·9	101·7	389	370·0	120·2	449	427·0	138·7	509	484·1	157·3	569	541·2	175·8
330	313·8	102·0	390	370·9	120·5	450	428·0	139·1	510	485·0	157·6	570	542·1	176·1
331	314·8	102·3	391	371·9	120·8	451	428·9	139·4	511	486·0	157·9	571	543·1	176·4
332	315·8	102·6	392	372·8	121·1	452	429·9	139·7	512	486·9	158·2	572	544·0	176·8
333	316·7	102·9	393	373·8	121·4	453	430·8	140·0	513	487·9	158·5	573	545·0	177·1
334	317·7	103·2	394	374·7	121·8	454	431·8	140·3	514	488·8	158·8	574	545·9	177·4
335	318·6	103·5	395	375·7	122·1	455	432·7	140·6	515	489·8	159·1	575	546·9	177·7
336	319·6	103·8	396	376·6	122·4	456	433·7	140·9	516	490·7	159·5	576	547·8	178·0
337	320·5	104·1	397	377·6	122·7	457	434·6	141·2	517	491·7	159·8	577	548·8	178·3
338	321·5	104·4	398	378·5	123·0	458	435·6	141·5	518	492·6	160·1	578	549·7	178·6
339	322·4	104·8	399	379·5	123·3	459	436·5	141·8	519	493·6	160·4	579	550·7	178·9
340	323·4	105·1	400	380·4	123·6	460	437·5	142·1	520	494·5	160·7	580	551·6	179·2
341	324·3	105·4	401	381·4	123·9	461	438·4	142·5	521	495·5	161·0	581	552·6	179·5
342	325·3	105·7	402	382·3	124·2	462	439·1	142·8	522	496·5	161·3	582	553·5	179·8
343	326·2	106·0	403	383·3	124·5	463	440·3	143·1	523	497·4	161·6	583	554·5	180·2
344	327·2	106·3	404	384·2	124·8	464	441·3	143·4	524	498·4	161·9	584	555·4	180·5
345	328·1	106·6	405	385·2	125·2	465	442·2	143·7	525	499·3	162·2	585	556·4	180·8
346	329·1	106·9	406	386·1	125·5	466	443·2	144·0	526	500·3	162·5	586	557·3	181·1
347	330·0	107·2	407	387·1	125·8	467	444·1	144·3	527	501·2	162·9	587	558·3	181·4
348	331·0	107·5	408	388·0	126·1	468	445·1	144·6	528	502·2	163·2	588	559·2	181·7
349	331·9	107·8	409	389·0	126·4	469	446·0	144·9	529	503·1	163·5	589	560·2	182·0
350	332·9	108·2	410	389·9	126·7	470	447·0	145·2	530	504·1	163·8	590	561·1	182·3
351	333·8	108·5	411	390·9	127·0	471	447·9	145·5	531	505·0	164·1	591	562·1	182·6
352	334·8	108·8	412	391·8	127·3	472	448·9	145·9	532	506·0	164·4	592	563·0	182·9
353	335·7	109·1	413	392·8	127·6	473	449·8	146·2	533	506·9	164·7	593	564·0	183·2
354	336·7	109·4	414	393·7	127·9	474	450·8	146·5	534	507·9	165·0	594	564·9	183·6
355	337·7	109·7	415	394·7	128·2	475	451·8	146·8	535	508·8	165·3	595	565·9	183·9
356	338·6	110·0	416	395·6	128·6	476	452·7	147·1	536	509·8	165·6	596	566·8	184·2
357	339·5	110·3	417	396·6	128·9	477	453·7	147·4	537	510·7	165·9	597	567·8	184·5
358	340·5	110·6	418	397·5	129·2	478	454·6	147·7	538	511·7	166·3	598	568·7	184·8
359	341·4	110·9	419	398·5	129·5	479	455·6	148·0	539	512·6	166·6	599	569·7	185·1
360	342·4	111·2	420	399·4	129·8	480	456·5	148·3	540	513·6	166·9	600	570·6	185·4

| Dist | Dep. | D.Lat | Dist | Dep. | D.Lat | Dist | Dep. | D.Lat | Dist | Dep. | D.Lat | Dist | Dep. | D.Lat |

341° ↑ / 199°	**TRAVERSE TABLE** 19 DEGREES.	↑019° / 161° 1h 16m

D Lon	Dep.		D Lon	Dep.		D Lon	Dep.		D Lon	Dep.		D Lon	Dep.	
Dist	D.Lat	Dep.	Dist	D.Lat	Dep.	Dist	D.Lat	Dep.	Dist	D.Lat	Dep.	Dist	D.Lat	Dep.
1	00·9	00·3	61	57·7	19·9	121	114·4	39·4	181	171·1	58·9	241	227·9	78·5
2	01·9	00·7	62	58·6	20·2	122	115·4	39·7	182	172·1	59·3	242	228·8	78·8
3	02·8	01·0	63	59·6	20·5	123	116·3	40·0	183	173·0	59·6	243	229·8	79·1
4	03·8	01·3	64	60·5	20·8	124	117·2	40·4	184	174·0	59·9	244	230·7	79·4
5	04·7	01·6	65	61·5	21·2	125	118·2	40·7	185	174·9	60·2	245	231·7	79·8
6	05·7	02·0	66	62·4	21·5	126	119·1	41·0	186	175·9	60·6	246	232·6	80·1
7	06·6	02·3	67	63·3	21·8	127	120·1	41·3	187	176·8	60·9	247	233·5	80·4
8	07·6	02·6	68	64·3	22·1	128	121·0	41·7	188	177·8	61·2	248	234·5	80·7
9	08·5	02·9	69	65·2	22·5	129	122·0	42·0	189	178·7	61·5	249	235·4	81·1
10	09·5	03·3	70	66·2	22·8	130	122·9	42·3	190	179·6	61·9	250	236·4	81·4
11	10·4	03·6	71	67·1	23·1	131	123·9	42·6	191	180·6	62·2	251	237·3	81·7
12	11·3	03·9	72	68·1	23·4	132	124·8	43·0	192	181·5	62·5	252	238·3	82·0
13	12·3	04·2	73	69·0	23·8	133	125·8	43·3	193	182·5	62·8	253	239·2	82·4
14	13·2	04·6	74	70·0	24·1	134	126·7	43·6	194	183·4	63·2	254	240·2	82·7
15	14·2	04·9	75	70·9	24·4	135	127·6	44·0	195	184·4	63·5	255	241·1	83·0
16	15·1	05·2	76	71·9	24·7	136	128·6	44·3	196	185·3	63·8	256	242·1	83·3
17	16·1	05·5	77	72·8	25·1	137	129·5	44·6	197	186·3	64·1	257	243·0	83·7
18	17·0	05·9	78	73·8	25·4	138	130·5	44·9	198	187·2	64·5	258	243·9	84·0
19	18·0	06·2	79	74·7	25·7	139	131·4	45·3	199	188·2	64·8	259	244·9	84·3
20	18·9	06·5	80	75·6	26·0	140	132·4	45·6	200	189·1	65·1	260	245·8	84·6
21	19·9	06·8	81	76·6	26·4	141	133·3	45·9	201	190·0	65·4	261	246·8	85·0
22	20·8	07·2	82	77·5	26·7	142	134·3	46·2	202	191·0	65·8	262	247·7	85·3
23	21·7	07·5	83	78·5	27·0	143	135·2	46·6	203	191·9	66·1	263	248·7	85·6
24	22·7	07·8	84	79·4	27·3	144	136·2	46·9	204	192·9	66·4	264	249·6	86·0
25	23·6	08·1	85	80·4	27·7	145	137·1	47·2	205	193·8	66·7	265	250·6	86·3
26	24·6	08·5	86	81·3	28·0	146	138·0	47·5	206	194·8	67·1	266	251·5	86·6
27	25·5	08·8	87	82·3	28·3	147	139·0	47·9	207	195·7	67·4	267	252·5	86·9
28	26·5	09·1	88	83·2	28·7	148	139·9	48·2	208	196·7	67·7	268	253·4	87·3
29	27·4	09·4	89	84·2	29·0	149	140·9	48·5	209	197·6	68·0	269	254·3	87·6
30	28·4	09·8	90	85·1	29·3	150	141·8	48·8	210	198·6	68·4	270	255·3	87·9
31	29·3	10·1	91	86·0	29·6	151	142·8	49·2	211	199·5	68·7	271	256·2	88·2
32	30·3	10·4	92	87·0	30·0	152	143·7	49·5	212	200·4	69·0	272	257·2	88·6
33	31·2	10·7	93	87·9	30·3	153	144·7	49·8	213	201·4	69·3	273	258·1	88·9
34	32·1	11·1	94	88·9	30·6	154	145·6	50·1	214	202·3	69·7	274	259·1	89·2
35	33·1	11·4	95	89·8	30·9	155	146·6	50·5	215	203·3	70·0	275	260·0	89·5
36	34·0	11·7	96	90·8	31·3	156	147·5	50·8	216	204·2	70·3	276	261·0	89·9
37	35·0	12·0	97	91·7	31·6	157	148·4	51·1	217	205·2	70·6	277	261·9	90·2
38	35·9	12·4	98	92·7	31·9	158	149·4	51·4	218	206·1	71·0	278	262·9	90·5
39	36·9	12·7	99	93·6	32·2	159	150·3	51·8	219	207·1	71·3	279	263·8	90·8
40	37·8	13·0	100	94·6	32·6	160	151·3	52·1	220	208·0	71·6	280	264·7	91·2
41	38·8	13·3	101	95·5	32·9	161	152·2	52·4	221	209·0	72·0	281	265·7	91·5
42	39·7	13·7	102	96·4	33·2	162	153·2	52·7	222	209·9	72·3	282	266·6	91·8
43	40·7	14·0	103	97·4	33·5	163	154·1	53·1	223	210·9	72·6	283	267·6	92·1
44	41·6	14·3	104	98·3	33·9	164	155·1	53·4	224	211·8	72·9	284	268·5	92·5
45	42·5	14·7	105	99·3	34·2	165	156·0	53·7	225	212·7	73·3	285	269·5	92·8
46	43·5	15·0	106	100·2	34·5	166	157·0	54·0	226	213·7	73·6	286	270·4	93·1
47	44·4	15·3	107	101·2	34·8	167	157·9	54·4	227	214·6	73·9	287	271·4	93·5
48	45·4	15·6	108	102·1	35·2	168	158·8	54·7	228	215·6	74·2	288	272·3	93·8
49	46·3	16·0	109	103·1	35·5	169	159·8	55·0	229	216·5	74·6	289	273·3	94·1
50	47·3	16·3	110	104·0	35·8	170	160·7	55·3	230	217·5	74·9	290	274·2	94·4
51	48·2	16·6	111	105·0	36·1	171	161·7	55·7	231	218·4	75·2	291	275·1	94·7
52	49·2	16·9	112	105·9	36·5	172	162·6	56·0	232	219·4	75·5	292	276·1	95·1
53	50·1	17·3	113	106·8	36·8	173	163·6	56·3	233	220·3	75·9	293	277·0	95·4
54	51·1	17·6	114	107·8	37·1	174	164·5	56·6	234	221·3	76·2	294	278·0	95·7
55	52·0	17·9	115	108·7	37·4	175	165·5	57·0	235	222·2	76·5	295	278·9	96·0
56	52·9	18·2	116	109·7	37·8	176	166·4	57·3	236	223·1	76·8	296	279·9	96·4
57	53·9	18·6	117	110·6	38·1	177	167·4	57·6	237	224·1	77·2	297	280·8	96·7
58	54·8	18·9	118	111·6	38·4	178	168·3	58·0	238	225·0	77·5	298	281·8	97·0
59	55·8	19·2	119	112·5	38·7	179	169·2	58·3	239	226·0	77·8	299	282·7	97·3
60	56·7	19·5	120	113·5	39·1	180	170·2	58·6	240	226·9	78·1	300	283·7	97·7
Dist	Dep.	D.Lat	Dist	Dep.	D.Lat	Dist	Dep.	D.Lat	Dist	Dep.	D.Lat	Dist	Dep.	D.Lat
D Lon		Dep.	D Lon		Dep.	D Lon		Dep.	D Lon		Dep.	D Lon		Dep.

289° ↑ / 251°	**71 DEGREES.**	071° / 109° 4h 44m

TRAVERSE TABLE
19 DEGREES.

341° / 199°   ↑ 019° / 161°   1h 16m   **19°**

Dist	D.Lat	Dep.	Dist	D.Lat	Dep.	Dist	D.Lat	Dep.	Dist	D.Lat	Dep.	Dist	D.Lat	Dep.
	D Lon	Dep.		D Lon	Dep.		D Lon	Dep.		D Lon	Dep.		D Lon	Dep.
301	284·6	98·0	361	341·3	117·5	421	398·1	137·1	481	454·8	156·6	541	511·5	176·1
302	285·5	98·3	362	342·3	117·9	422	399·0	137·4	482	455·7	156·9	542	512·5	176·5
303	286·5	98·6	363	343·2	118·2	423	400·0	137·7	483	456·7	157·2	543	513·4	176·8
304	287·4	99·0	364	344·2	118·5	424	400·9	138·0	484	457·6	157·6	544	514·4	177·1
305	288·4	99·3	365	345·1	118·8	425	401·8	138·4	485	458·6	157·9	545	515·3	177·4
306	289·3	99·6	366	346·1	119·2	426	402·8	138·7	486	459·5	158·2	546	516·3	177·8
307	290·3	99·9	367	347·0	119·5	427	403·7	139·0	487	460·5	158·6	547	517·2	178·1
308	291·2	100·3	368	348·0	119·8	428	404·7	139·3	488	461·4	158·9	548	518·1	178·4
309	292·2	100·6	369	348·9	120·1	429	405·6	139·7	489	462·4	159·2	549	519·1	178·7
310	293·1	100·9	370	349·8	120·5	430	406·6	140·0	490	463·3	159·5	550	520·0	179·1
311	294·1	101·3	371	350·8	120·8	431	407·5	140·3	491	464·2	159·9	551	521·0	179·4
312	295·0	101·6	372	351·7	121·1	432	408·5	140·6	492	465·2	160·2	552	521·9	179·7
313	295·9	101·9	373	352·7	121·4	433	409·4	141·0	493	466·1	160·5	553	522·9	180·0
314	296·9	102·2	374	353·6	121·8	434	410·4	141·3	494	467·1	160·8	554	523·8	180·4
315	297·8	102·6	375	354·6	122·1	435	411·3	141·6	495	468·0	161·2	555	524·8	180·7
316	298·8	102·9	376	355·5	122·4	436	412·2	141·9	496	469·0	161·5	556	525·7	181·0
317	299·7	103·2	377	356·5	122·7	437	413·2	142·3	497	469·9	161·8	557	526·7	181·3
318	300·7	103·5	378	357·4	123·1	438	414·1	142·6	498	470·9	162·1	558	527·6	181·7
319	301·6	103·8	379	358·4	123·4	439	415·1	142·9	499	471·8	162·5	559	528·5	182·0
320	302·6	104·2	380	359·3	123·7	440	416·0	143·3	500	472·8	162·8	560	529·5	182·3
321	303·5	104·5	381	360·2	124·0	441	417·0	143·6	501	473·7	163·1	561	530·4	182·6
322	304·5	104·8	382	361·2	124·4	442	417·9	143·9	502	474·7	163·4	562	531·4	183·0
323	305·4	105·2	383	362·1	124·7	443	418·9	144·2	503	475·6	163·8	563	532·3	183·3
324	306·3	105·5	384	363·1	125·0	444	419·8	144·6	504	476·5	164·1	564	533·3	183·6
325	307·3	105·8	385	364·0	125·3	445	420·8	144·9	505	477·5	164·4	565	534·2	183·9
326	308·2	106·1	386	365·1	125·7	446	421·7	145·2	506	478·4	164·7	566	535·2	184·3
327	309·2	106·5	387	365·9	126·0	447	422·6	145·5	507	479·4	165·1	567	536·1	184·6
328	310·1	106·8	388	366·9	126·3	448	423·6	145·8	508	480·3	165·4	568	537·1	184·9
329	311·1	107·1	389	367·8	126·6	449	424·5	146·2	509	481·3	165·7	569	538·0	185·2
330	312·0	107·4	390	368·8	127·0	450	425·5	146·5	510	482·2	166·0	570	538·9	185·6
331	313·0	107·8	391	369·7	127·3	451	426·4	146·8	511	483·2	166·4	571	539·9	185·9
332	313·9	108·1	392	370·6	127·6	452	427·4	147·2	512	484·1	166·7	572	540·8	186·2
333	314·9	108·4	393	371·6	127·9	453	428·3	147·5	513	485·1	167·0	573	541·8	186·6
334	315·8	108·7	394	372·5	128·3	454	429·3	147·8	514	486·0	167·3	574	542·7	186·9
335	316·7	109·1	395	373·5	128·6	455	430·2	148·1	515	486·9	167·7	575	543·7	187·2
336	317·7	109·4	396	374·4	128·9	456	431·2	148·5	516	487·9	168·0	576	544·6	187·5
337	318·6	109·7	397	375·4	129·3	457	432·1	148·8	517	488·8	168·3	577	545·6	187·9
338	319·6	110·0	398	376·3	129·6	458	433·0	149·1	518	489·7	168·6	578	546·5	188·2
339	320·5	110·4	399	377·3	129·9	459	434·0	149·4	519	490·7	169·0	579	547·5	188·5
340	321·5	110·7	400	378·2	130·2	460	434·9	149·8	520	491·6	169·3	580	548·4	188·8
341	322·4	111·0	401	379·2	130·6	461	435·9	150·1	521	492·6	169·6	581	549·3	189·2
342	323·4	111·3	402	380·1	130·9	462	436·8	150·4	522	493·6	169·9	582	550·3	189·5
343	324·3	111·7	403	381·0	131·2	463	437·8	150·7	523	494·5	170·3	583	551·2	189·8
344	325·3	112·0	404	382·0	131·5	464	438·7	151·1	524	495·5	170·6	584	552·2	190·1
345	326·2	112·3	405	382·9	131·9	465	439·7	151·4	525	496·4	170·9	585	553·1	190·5
346	327·1	112·6	406	383·9	132·2	466	440·6	151·7	526	497·3	171·2	586	554·1	190·8
347	328·1	113·0	407	384·8	132·5	467	441·6	152·0	527	498·3	171·6	587	555·0	191·1
348	329·0	113·3	408	385·8	132·8	468	442·5	152·4	528	499·2	171·9	588	556·0	191·4
349	330·0	113·6	409	386·7	133·2	469	443·4	152·7	529	500·2	172·2	589	556·9	191·8
350	330·9	113·9	410	387·7	133·5	470	444·4	153·0	530	501·1	172·6	590	557·9	192·0
351	331·9	114·3	411	388·6	133·8	471	445·3	153·3	531	502·1	172·9	591	558·8	192·4
352	332·8	114·6	412	389·6	134·1	472	446·3	153·7	532	503·0	173·2	592	559·7	192·7
353	333·8	114·9	413	390·5	134·5	473	447·2	154·0	533	504·0	173·5	593	560·7	193·1
354	334·7	115·3	414	391·4	134·8	474	448·2	154·3	534	504·9	173·9	594	561·6	193·4
355	335·7	115·6	415	392·4	135·1	475	449·1	154·6	535	505·9	174·2	595	562·6	193·7
356	336·6	115·9	416	393·3	135·4	476	450·1	155·0	536	506·8	174·5	596	563·5	194·0
357	337·6	116·2	417	394·3	135·8	477	451·0	155·3	537	507·7	174·8	597	564·5	194·4
358	338·5	116·6	418	395·2	136·1	478	452·0	155·6	538	508·7	175·2	598	565·4	194·7
359	339·4	116·9	419	396·2	136·4	479	452·9	155·9	539	509·6	175·5	599	566·4	195·0
360	340·4	117·2	420	397·1	136·7	480	453·8	156·3	540	510·6	175·8	600	567·3	195·3
Dist	Dep.	D.Lat	Dist	Dep.	D.Lat	Dist	Dep.	D.Lat	Dist	Dep.	D.Lat	Dist	Dep.	D.Lat
	D Lon	Dep.		D Lon	Dep.		D Lon	Dep.		D Lon	Dep.		D Lon	Dep.

71°

289° / 251°   **71 DEGREES.**   ↑ 071° / 109°   4h 44m

TRAVERSE TABLE
20 DEGREES.

| 340° ↑ | | 020° ↑ | | 1h 20m |
| 200° | | 160° | | |

D Lon	Dep.		D Lon	Dep.		D Lon	Dep.		D Lon	Dep.		D Lon	Dep.	
Dist	D.Lat	Dep.	Dist	D.Lat	Dep.	Dist	D.Lat	Dep.	Dist	D.Lat	Dep.	Dist	D.Lat	Dep.
1	00·9	00·3	61	57·3	20·9	121	113·7	41·4	181	170·1	61·9	241	226·5	82·4
2	01·9	00·7	62	58·3	21·2	122	114·6	41·7	182	171·0	62·2	242	227·4	82·8
3	02·8	01·0	63	59·2	21·5	123	115·6	42·1	183	172·0	62·6	243	228·3	83·1
4	03·8	01·4	64	60·1	21·9	124	116·5	42·4	184	172·9	62·9	244	229·3	83·5
5	04·7	01·7	65	61·1	22·2	125	117·5	42·8	185	173·8	63·3	245	230·2	83·8
6	05·6	02·1	66	62·0	22·6	126	118·4	43·1	186	174·8	63·6	246	231·2	84·2
7	06·6	02·4	67	63·0	22·9	127	119·3	43·4	187	175·7	64·0	247	232·1	84·5
8	07·5	02·7	68	63·9	23·3	128	120·3	43·8	188	176·7	64·3	248	233·0	84·8
9	08·5	03·1	69	64·8	23·6	129	121·2	44·1	189	177·6	64·6	249	234·0	85·2
10	09·4	03·4	70	65·8	23·9	130	122·2	44·5	190	178·5	65·0	250	234·9	85·5
11	10·3	03·8	71	66·7	24·3	131	123·1	44·8	191	179·5	65·3	251	235·9	85·8
12	11·3	04·1	72	67·7	24·6	132	124·0	45·1	192	180·4	65·7	252	236·8	86·2
13	12·2	04·4	73	68·6	25·0	133	125·0	45·5	193	181·4	66·0	253	237·7	86·5
14	13·2	04·8	74	69·5	25·3	134	125·9	45·8	194	182·3	66·4	254	238·7	86·9
15	14·1	05·1	75	70·5	25·7	135	126·9	46·2	195	183·2	66·7	255	239·6	87·2
16	15·0	05·5	76	71·4	26·0	136	127·8	46·5	196	184·2	67·0	256	240·6	87·6
17	16·0	05·8	77	72·4	26·3	137	128·7	46·9	197	185·1	67·4	257	241·5	87·9
18	16·9	06·2	78	73·3	26·7	138	129·7	47·2	198	186·1	67·7	258	242·4	88·2
19	17·9	06·5	79	74·2	27·0	139	130·6	47·5	199	187·0	68·1	259	243·4	88·6
20	18·8	06·8	80	75·2	27·4	140	131·6	47·9	200	187·9	68·4	260	244·3	88·9
21	19·7	07·2	81	76·1	27·7	141	132·5	48·2	201	188·9	68·7	261	245·3	89·3
22	20·7	07·5	82	77·1	28·0	142	133·4	48·6	202	189·8	69·1	262	246·2	89·6
23	21·6	07·9	83	78·0	28·4	143	134·4	48·9	203	190·8	69·4	263	247·1	90·0
24	22·6	08·2	84	78·9	28·7	144	135·3	49·3	204	191·7	69·8	264	248·1	90·3
25	23·5	08·6	85	79·9	29·1	145	136·3	49·6	205	192·6	70·1	265	249·0	90·6
26	24·4	08·9	86	80·8	29·4	146	137·2	49·9	206	193·6	70·5	266	250·0	91·0
27	25·4	09·2	87	81·8	29·8	147	138·1	50·3	207	194·5	70·8	267	250·9	91·3
28	26·3	09·6	88	82·7	30·1	148	139·1	50·6	208	195·5	71·1	268	251·8	91·7
29	27·3	09·9	89	83·6	30·4	149	140·0	51·0	209	196·4	71·5	269	252·8	92·0
30	28·2	10·3	90	84·6	30·8	150	141·0	51·3	210	197·3	71·8	270	253·7	92·3
31	29·1	10·6	91	85·5	31 1	151	141·9	51·6	211	198·3	72·2	271	254·7	92·7
32	30·1	10·9	92	86·5	31·5	152	142·8	52·0	212	199·2	72·5	272	255·6	93·0
33	31·0	11·3	93	87·4	31·8	153	143·8	52·3	213	200·2	72·9	273	256·5	93·4
34	31·9	11·6	94	88·3	32·1	154	144·7	52·7	214	201·1	73·2	274	257·5	93·7
35	32·9	12·0	95	89·3	32·5	155	145·7	53·0	215	202·0	73·5	275	258·4	94·1
36	33·8	12·3	96	90·2	32·8	156	146·6	53·4	216	203·0	73·9	276	259·4	94·4
37	34·8	12·7	97	91·2	33·2	157	147·5	53·7	217	203·9	74·2	277	260·3	94·7
38	35·7	13·0	98	92·1	33·5	158	148·5	54·0	218	204·9	74·6	278	261·2	95·1
39	36·6	13·3	99	93·0	33·9	159	149·4	54·4	219	205·8	74·9	279	262·2	95·4
40	37·6	13·7	100	94·0	34·2	160	150·4	54·7	220	206·7	75·2	280	263·1	95·8
41	38·5	14·0	101	94·9	34·5	161	151·3	55·1	221	207·7	75·6	281	264·1	96·1
42	39·5	14·4	102	95·8	34·9	162	152·2	55·4	222	208·6	75·9	282	265·0	96·4
43	40·4	14·7	103	96·8	35·2	163	153·2	55·7	223	209·6	76·3	283	265·9	96·8
44	41·3	15·0	104	97·7	35·6	164	154·1	56·1	224	210·5	76·6	284	266·9	97·1
45	42·3	15·4	105	98·7	35·9	165	155·0	56·4	225	211·4	77·0	285	267·8	97·5
46	43·2	15·7	106	99·6	36·3	166	156·0	56·8	226	212·4	77·3	286	268·8	97·8
47	44·2	16·1	107	100·5	36·6	167	156·9	57·1	227	213·3	77·6	287	269·7	98·2
48	45·1	16·4	108	101·5	36·9	168	157·9	57·5	228	214·2	78·0	288	270·6	98·5
49	46·0	16·8	109	102·4	37·3	169	158·8	57·8	229	215·2	78·3	289	271·6	98·8
50	47·0	17·1	110	103·4	37·6	170	159·7	58·1	230	216·1	78·7	290	272·5	99·2
51	47·9	17·4	111	104·3	38·0	171	160·7	58·5	231	217·1	79·0	291	273·5	99·5
52	48·9	17·8	112	105·2	38·3	172	161·6	58·8	232	218·0	79·3	292	274·4	99·9
53	49·8	18·1	113	106·2	38·6	173	162·6	59·2	233	218·9	79·7	293	275·3	100·2
54	50·7	18·5	114	107·1	39·0	174	163·5	59·5	234	219·9	80·0	294	276·3	100·6
55	51·7	18·8	115	108·1	39·3	175	164·4	59·9	235	220·8	80·4	295	277·2	100·9
56	52·6	19·2	116	109·0	39·7	176	165·4	60·2	236	221·8	80·7	296	278·1	101·2
57	53·6	19·5	117	109·9	40·0	177	166·3	60·5	237	222·7	81·1	297	279·1	101·6
58	54·5	19·8	118	110·9	40·4	178	167·3	60·9	238	223·6	81·4	298	280·0	101·9
59	55·4	20·2	119	111·8	40·7	179	168·2	61·2	239	224·6	81·7	299	281·0	102·3
60	56·4	20·5	120	112·8	41·0	180	169·1	61·6	240	225·5	82·1	300	281·9	102·6
Dist	Dep.	D.Lat	Dist	Dep.	D.Lat	Dist	Dep.	D.Lat	Dist	Dep.	D.Lat	Dist	Dep.	D.Lat
D Lon		Dep.	D Lon		Dep.	D Lon		Dep.	D Lon		Dep.	D Lon		Dep.

| 290° ↑ | | 070° ↑ | | 4h 40m |
| 250° | 70 DEGREES. | 110° | | |

TRAVERSE TABLE
20 DEGREES.

340° / 200°   020° / 160°   1h 20m   **20°**

D Lon	Dep.		D Lon	Dep.		D Lon	Dep.		D Lon	Dep.		D Lon	Dep.	
Dist	D.Lat	Dep.	Dist	D.Lat	Dep.	Dist	D.Lat	Dep.	Dist	D.Lat	Dep.	Dist	D.Lat	Dep.
301	282.8	102.9	361	339.2	123.5	421	395.6	144.0	481	452.0	164.5	541	508.4	185.0
302	283.8	103.3	362	340.2	123.8	422	396.6	144.3	482	452.9	164.9	542	509.3	185.4
303	284.7	103.6	363	341.1	124.2	423	397.5	144.7	483	453.9	165.2	543	510.3	185.7
304	285.7	104.0	364	342.0	124.5	424	398.4	145.0	484	454.8	165.5	544	511.2	186.1
305	286.6	104.3	365	343.0	124.8	425	399.4	145.4	485	455.8	165.9	545	512.1	186.4
306	287.5	104.7	366	343.9	125.2	426	400.3	145.7	486	456.7	166.2	546	513.1	186.8
307	288.5	105.0	367	344.9	125.5	427	401.2	146.0	487	457.6	166.6	547	514.0	187.1
308	289.4	105.3	368	345.8	125.9	428	402.2	146.4	488	458.6	166.9	548	515.0	187.4
309	290.4	105.7	369	346.7	126.2	429	403.1	146.7	489	459.5	167.2	549	515.9	187.8
310	291.3	106.0	370	347.7	126.5	430	404.1	147.1	490	460.4	167.6	550	516.8	188.1
311	292.2	106.4	371	348.6	126.9	431	405.0	147.4	491	461.4	167.9	551	517.8	188.5
312	293.2	106.7	372	349.6	127.2	432	405.9	147.8	492	462.3	168.3	552	518.7	188.8
313	294.1	107.1	373	350.5	127.6	433	406.9	148.1	493	463.3	168.6	553	519.7	189.1
314	295.1	107.4	374	351.4	127.9	434	407.8	148.4	494	464.2	169.0	554	520.6	189.5
315	296.0	107.7	375	352.4	128.3	435	408.8	148.8	495	465.1	169.3	555	521.5	189.8
316	296.9	108.1	376	353.3	128.6	436	409.7	149.1	496	466.1	169.6	556	522.5	190.2
317	297.9	108.4	377	354.3	128.9	437	410.6	149.5	497	467.0	170.0	557	523.4	190.5
318	298.8	108.8	378	355.2	129.3	438	411.6	149.8	498	468.0	170.3	558	524.3	190.8
319	299.8	109.1	379	356.1	129.6	439	412.5	150.1	499	468.9	170.7	559	525.3	191.2
320	300.7	109.4	380	357.1	130.0	440	413.5	150.5	500	469.8	171.0	560	526.2	191.5
321	301.6	109.8	381	358.0	130.3	441	414.4	150.8	501	470.8	171.4	561	527.2	191.9
322	302.6	110.1	382	359.0	130.7	442	415.3	151.2	502	471.7	171.7	562	528.1	192.2
323	303.5	110.5	383	359.9	131.0	443	416.3	151.5	503	472.7	172.0	563	529.0	192.6
324	304.5	110.8	384	360.8	131.3	444	417.2	151.9	504	473.6	172.4	564	530.0	192.9
325	305.4	111.2	385	361.8	131.7	445	418.2	152.2	505	474.5	172.7	565	530.9	193.2
326	306.3	111.5	386	362.7	132.0	446	419.1	152.5	506	475.5	173.1	566	531.9	193.6
327	307.3	111.8	387	363.7	132.4	447	420.0	152.9	507	476.4	173.4	567	532.8	193.9
328	308.2	112.2	388	364.6	132.7	448	421.0	153.2	508	477.4	173.7	568	533.7	194.3
329	309.2	112.5	389	365.5	133.0	449	421.9	153.5	509	478.3	174.1	569	534.7	194.6
330	310.1	112.9	390	366.5	133.4	450	422.9	153.9	510	479.2	174.4	570	535.6	195.0
331	311.0	113.2	391	367.4	133.7	451	423.8	154.3	511	480.2	174.8	571	536.6	195.3
332	312.0	113.5	392	368.4	134.1	452	424.7	154.6	512	481.1	175.1	572	537.5	195.6
333	312.9	113.9	393	369.3	134.4	453	425.7	154.9	513	482.1	175.5	573	538.4	196.0
334	313.9	114.2	394	370.2	134.8	454	426.6	155.3	514	483.0	175.8	574	539.4	196.3
335	314.8	114.6	395	371.2	135.1	455	427.6	155.6	515	483.9	176.1	575	540.3	196.7
336	315.7	114.9	396	372.1	135.4	456	428.5	156.0	516	484.9	176.5	576	541.3	197.0
337	316.7	115.3	397	373.1	135.8	457	429.4	156.3	517	485.8	176.8	577	542.2	197.3
338	317.6	115.6	398	374.0	136.1	458	430.4	156.6	518	486.8	177.2	578	543.1	197.7
339	318.6	115.9	399	374.9	136.5	459	431.3	157.0	519	487.7	177.5	579	544.1	198.0
340	319.5	116.3	400	375.9	136.8	460	432.2	157.3	520	488.6	177.9	580	545.0	198.4
341	320.4	116.6	401	376.8	137.2	461	433.2	157.7	521	489.6	178.2	581	546.0	198.7
342	321.4	116.9	402	377.8	137.5	462	434.1	158.0	522	490.5	178.5	582	546.9	199.1
343	322.3	117.3	403	378.7	137.8	463	435.1	158.4	523	491.5	178.9	583	547.8	199.4
344	323.3	117.7	404	379.6	138.2	464	436.0	158.7	524	492.4	179.2	584	548.8	199.7
345	324.2	118.0	405	380.6	138.5	465	437.0	159.0	525	493.3	179.6	585	549.7	200.1
346	325.1	118.3	406	381.5	138.9	466	437.9	159.4	526	494.3	179.9	586	550.7	200.4
347	326.1	118.7	407	382.5	139.2	467	438.8	159.7	527	495.2	180.2	587	551.6	200.8
348	327.0	119.0	408	383.4	139.5	468	439.8	160.1	528	496.2	180.6	588	552.5	201.1
349	328.0	119.4	409	384.3	139.9	469	440.7	160.4	529	497.1	180.9	589	553.5	201.4
350	328.9	119.7	410	385.3	140.2	470	441.7	160.7	530	498.0	181.3	590	554.4	201.8
351	329.8	120.0	411	386.2	140.6	471	442.6	161.1	531	499.0	181.6	591	555.4	202.1
352	330.8	120.4	412	387.2	140.9	472	443.5	161.4	532	499.9	182.0	592	556.3	202.5
353	331.7	120.7	413	388.1	141.3	473	444.5	161.8	533	500.9	182.3	593	557.2	202.8
354	332.7	121.1	414	389.0	141.6	474	445.4	162.1	534	501.8	182.6	594	558.2	203.2
355	333.6	121.4	415	390.0	141.9	475	446.4	162.5	535	502.7	183.0	595	559.1	203.5
356	334.5	121.8	416	390.9	142.3	476	447.3	162.8	536	503.7	183.3	596	560.1	203.8
357	335.5	122.1	417	391.9	142.6	477	448.2	163.1	537	504.6	183.7	597	561.0	204.2
358	336.4	122.4	418	392.8	143.0	478	449.2	163.5	538	505.6	184.0	598	561.9	204.5
359	337.4	122.8	419	393.7	143.3	479	450.1	163.8	539	506.5	184.3	599	562.9	204.9
360	338.3	123.1	420	394.7	143.6	480	451.1	164.2	540	507.4	184.7	600	563.8	205.2
Dist	Dep.	D.Lat	Dist	Dep.	D.Lat	Dist	Dep.	D.Lat	Dist	Dep.	D.Lat	Dist	Dep.	D.Lat
D Lon		Dep.	D Lon		Dep.	D Lon		Dep.	D Lon		Dep.	D Lon		Dep.

290° / 250°   **70 DEGREES.**   070° / 110°   4h 40m   **70°**

TRAVERSE TABLE
21 DEGREES.

21° 339° / 201° 021° / 159° 1h 24m

Dist	D.Lat	Dep.	Dist	D.Lat	Dep.	Dist	D.Lat	Dep.	Dist	D.Lat	Dep.	Dist	D.Lat	Dep.
1	00·9	00·4	61	56·9	21·9	121	113·0	43·4	181	169·0	64·9	241	225·0	86·4
2	01·9	00·7	62	57·9	22·2	122	113·9	43·7	182	169·9	65·2	242	225·9	86·7
3	02·8	01·1	63	58·8	22·6	123	114·8	44·1	183	170·8	65·6	243	226·9	87·1
4	03·7	01·4	64	59·7	22·9	124	115·8	44·4	184	171·8	65·9	244	227·8	87·4
5	04·7	01·8	65	60·7	23·3	125	116·7	44·8	185	172·7	66·3	245	228·7	87·8
6	05·6	02·2	66	61·6	23·7	126	117·6	45·2	186	173·6	66·7	246	229·7	88·2
7	06·5	02·5	67	62·5	24·0	127	118·6	45·5	187	174·6	67·0	247	230·6	88·5
8	07·5	02·9	68	63·5	24·4	128	119·5	45·9	188	175·5	67·4	248	231·5	88·9
9	08·4	03·2	69	64·4	24·7	129	120·4	46·2	189	176·4	67·7	249	232·5	89·2
10	09·3	03·6	70	65·4	25·1	130	121·4	46·6	190	177·4	68·1	250	233·4	89·6
11	10·3	03·9	71	66·3	25·4	131	122·3	46·9	191	178·3	68·4	251	234·3	90·0
12	11·2	04·3	72	67·2	25·8	132	123·2	47·3	192	179·2	68·8	252	235·3	90·3
13	12·1	04·7	73	68·2	26·2	133	124·2	47·7	193	180·2	69·2	253	236·2	90·7
14	13·1	05·0	74	69·1	26·5	134	125·1	48·0	194	181·1	69·5	254	237·1	91·0
15	14·0	05·4	75	70·0	26·9	135	126·0	48·4	195	182·0	69·9	255	238·1	91·4
16	14·9	05·7	76	71·0	27·2	136	127·0	48·7	196	183·0	70·2	256	239·0	91·7
17	15·9	06·1	77	71·9	27·6	137	127·9	49·1	197	183·9	70·6	257	239·9	92·1
18	16·8	06·5	78	72·8	28·0	138	128·8	49·5	198	184·8	71·0	258	240·9	92·5
19	17·7	06·8	79	73·8	28·3	139	129·8	49·8	199	185·8	71·3	259	241·8	92·8
20	18·7	07·2	80	74·7	28·7	140	130·7	50·2	200	186·7	71·7	260	242·7	93·2
21	19·6	07·5	81	75·6	29·0	141	131·6	50·5	201	187·6	72·0	261	243·7	93·5
22	20·5	07·9	82	76·6	29·4	142	132·6	50·9	202	188·6	72·4	262	244·6	93·9
23	21·5	08·2	83	77·5	29·7	143	133·5	51·2	203	189·5	72·7	263	245·5	94·3
24	22·4	08·6	84	78·4	30·1	144	134·4	51·6	204	190·4	73·1	264	246·5	94·6
25	23·3	09·0	85	79·4	30·5	145	135·4	52·0	205	191·4	73·5	265	247·4	95·0
26	24·3	09·3	86	80·3	30·8	146	136·3	52·3	206	192·3	73·8	266	248·3	95·3
27	25·2	09·7	87	81·2	31·2	147	137·2	52·7	207	193·3	74·2	267	249·3	95·7
28	26·1	10·0	88	82·2	31·5	148	138·2	53·0	208	194·2	74·5	268	250·2	96·0
29	27·1	10·4	89	83·1	31·9	149	139·1	53·4	209	195·1	74·9	269	251·1	96·4
30	28·0	10·8	90	84·0	32·3	150	140·0	53·8	210	196·1	75·3	270	252·1	96·8
31	28·9	11·1	91	85·0	32·6	151	141·0	54·1	211	197·0	75·6	271	253·0	97·1
32	29·9	11·5	92	85·9	33·0	152	141·9	54·5	212	197·9	76·0	272	253·9	97·5
33	30·8	11·8	93	86·8	33·3	153	142·8	54·8	213	198·9	76·3	273	254·9	97·8
34	31·7	12·2	94	87·8	33·7	154	143·8	55·2	214	199·8	76·7	274	255·8	98·2
35	32·7	12·5	95	88·7	34·0	155	144·7	55·5	215	200·7	77·0	275	256·7	98·6
36	33·6	12·9	96	89·6	34·4	156	145·6	55·9	216	201·7	77·4	276	257·7	98·9
37	34·5	13·3	97	90·6	34·8	157	146·6	56·3	217	202·6	77·8	277	258·6	99·3
38	35·5	13·6	98	91·5	35·1	158	147·5	56·6	218	203·5	78·1	278	259·5	99·6
39	36·4	14·0	99	92·4	35·5	159	148·4	57·0	219	204·5	78·5	279	260·5	100·0
40	37·3	14·3	100	93·4	35·8	160	149·4	57·3	220	205·4	78·8	280	261·4	100·3
41	38·3	14·7	101	94·3	36·2	161	150·3	57·7	221	206·3	79·2	281	262·3	100·7
42	39·2	15·1	102	95·2	36·6	162	151·2	58·1	222	207·3	79·6	282	263·3	101·1
43	40·1	15·4	103	96·2	36·9	163	152·2	58·4	223	208·2	79·9	283	264·2	101·4
44	41·1	15·8	104	97·1	37·3	164	153·1	58·8	224	209·1	80·3	284	265·1	101·8
45	42·0	16·1	105	98·0	37·6	165	154·0	59·1	225	210·1	80·6	285	266·1	102·1
46	42·9	16·5	106	99·0	38·0	166	155·0	59·5	226	211·0	81·0	286	267·0	102·5
47	43·9	16·8	107	99·9	38·3	167	155·9	59·8	227	211·9	81·3	287	267·9	102·9
48	44·8	17·2	108	100·8	38·7	168	156·8	60·2	228	212·9	81·7	288	268·9	103·2
49	45·7	17·6	109	101·8	39·1	169	157·8	60·6	229	213·8	82·1	289	269·8	103·6
50	46·7	17·9	110	102·7	39·4	170	158·7	60·9	230	214·7	82·4	290	270·7	103·9
51	47·6	18·3	111	103·6	39·8	171	159·6	61·3	231	215·7	82·8	291	271·7	104·3
52	48·5	18·6	112	104·6	40·1	172	160·6	61·6	232	216·6	83·1	292	272·6	104·6
53	49·5	19·0	113	105·5	40·5	173	161·5	62·0	233	217·5	83·5	293	273·5	105·0
54	50·4	19·4	114	106·4	40·9	174	162·4	62·4	234	218·5	83·9	294	274·5	105·4
55	51·3	19·7	115	107·4	41·2	175	163·4	62·7	235	219·4	84·2	295	275·4	105·7
56	52·3	20·1	116	108·3	41·6	176	164·3	63·1	236	220·3	84·6	296	276·3	106·1
57	53·2	20·4	117	109·2	41·9	177	165·2	63·4	237	221·3	84·9	297	277·3	106·4
58	54·1	20·8	118	110·2	42·3	178	166·2	63·8	238	222·2	85·3	298	278·2	106·8
59	55·1	21·1	119	111·1	42·6	179	167·1	64·1	239	223·1	85·6	299	279·1	107·2
60	56·0	21·5	120	112·0	43·0	180	168·0	64·5	240	224·1	86·0	300	280·1	107·5

Dist	Dep.	D.Lat	Dist	Dep.	D.Lat	Dist	Dep.	D.Lat	Dist	Dep.	D.Lat	Dist	Dep.	D.Lat

D Lon | Dep. | D Lon | Dep. | D Lon | Dep. | D Lon | Dep. | D Lon | Dep.

69° 291° / 249° **69 DEGREES.** 069° / 111° 4h 36m

TRAVERSE TABLE
21 DEGREES.

339° / 201° | 021° / 159° 1h 24m | **21°**

Dist	D.Lat	Dep.	Dist	D.Lat	Dep.	Dist	D.Lat	Dep.	Dist	D.Lat	Dep.	Dist	D.Lat	Dep.
301	281.0	107.9	361	337.0	129.4	421	393.0	150.9	481	449.1	172.4	541	505.1	193.9
302	281.9	108.2	362	338.0	129.7	422	394.0	151.2	482	450.0	172.7	542	506.0	194.2
303	282.9	108.6	363	338.9	130.1	423	394.9	151.6	483	450.9	173.1	543	506.9	194.6
304	283.8	108.9	364	339.8	130.4	424	395.8	151.9	484	451.9	173.5	544	507.9	195.0
305	284.7	109.3	365	340.8	130.8	425	396.8	152.3	485	452.8	173.8	545	508.8	195.3
306	285.7	109.7	366	341.7	131.2	426	397.7	152.7	486	453.7	174.2	546	509.7	195.7
307	286.6	110.0	367	342.6	131.5	427	398.6	153.0	487	454.7	174.5	547	510.7	196.0
308	287.5	110.4	368	343.6	131.9	428	399.6	153.4	488	455.6	174.9	548	511.6	196.4
309	288.5	110.7	369	344.5	132.2	429	400.5	153.7	489	456.5	175.2	549	512.5	196.7
310	289.4	111.1	370	345.4	132.6	430	401.4	154.1	490	457.5	175.6	550	513.5	197.1
311	290.3	111.5	371	346.4	133.0	431	402.4	154.5	491	458.4	176.0	551	514.4	197.5
312	291.3	111.8	372	347.3	133.3	432	403.3	154.8	492	459.3	176.3	552	515.3	197.8
313	292.2	112.2	373	348.2	133.7	433	404.2	155.2	493	460.3	176.7	553	516.3	198.2
314	293.1	112.5	374	349.1	134.0	434	405.2	155.5	494	461.2	177.0	554	517.2	198.5
315	294.1	112.9	375	350.1	134.4	435	406.1	155.9	495	462.1	177.4	555	518.1	198.9
316	295.0	113.2	376	351.0	134.7	436	407.0	156.2	496	463.1	177.8	556	519.1	199.3
317	295.9	113.6	377	352.0	135.1	437	408.0	156.6	497	464.0	178.1	557	520.0	199.6
318	296.9	114.0	378	352.9	135.5	438	408.9	157.0	498	464.9	178.5	558	520.9	200.0
319	297.8	114.3	379	353.8	135.8	439	409.8	157.3	499	465.9	178.8	559	521.9	200.3
320	298.7	114.7	380	354.8	136.2	440	410.8	157.7	500	466.8	179.2	560	522.8	200.7
321	299.7	115.0	381	355.7	136.5	441	411.7	158.0	501	467.7	179.5	561	523.7	201.0
322	300.6	115.4	382	356.6	136.9	442	412.6	158.4	502	468.7	179.9	562	524.7	201.4
323	301.5	115.8	383	357.6	137.3	443	413.6	158.8	503	469.6	180.3	563	525.6	201.8
324	302.5	116.1	384	358.5	137.6	444	414.5	159.1	504	470.5	180.6	564	526.5	202.1
325	303.4	116.5	385	359.4	138.0	445	415.4	159.5	505	471.5	181.0	565	527.5	202.5
326	304.3	116.8	386	360.4	138.3	446	416.4	159.8	506	472.4	181.3	566	528.4	202.8
327	305.3	117.2	387	361.3	138.7	447	417.3	160.2	507	473.3	181.7	567	529.3	203.2
328	306.2	117.5	388	362.2	139.0	448	418.2	160.5	508	474.3	182.1	568	530.3	203.6
329	307.1	117.9	389	363.2	139.4	449	419.2	160.9	509	475.2	182.4	569	531.2	203.9
330	308.1	118.3	390	364.1	139.8	450	420.1	161.3	510	476.1	182.8	570	532.1	204.3
331	309.0	118.6	391	365.0	140.1	451	421.0	161.6	511	477.1	183.1	571	533.1	204.6
332	309.9	119.0	392	365.9	140.5	452	422.0	162.0	512	478.0	183.5	572	534.0	205.0
333	310.9	119.3	393	366.9	140.8	453	422.9	162.3	513	478.9	183.8	573	534.9	205.3
334	311.8	119.7	394	367.8	141.2	454	423.8	162.7	514	479.9	184.2	574	535.9	205.7
335	312.7	120.1	395	368.8	141.6	455	424.8	163.1	515	480.8	184.6	575	536.8	206.1
336	313.7	120.4	396	369.7	141.9	456	425.7	163.4	516	481.7	184.9	576	537.7	206.4
337	314.6	120.8	397	370.6	142.3	457	426.6	163.8	517	482.7	185.3	577	538.7	206.8
338	315.6	121.1	398	371.6	142.6	458	427.6	164.1	518	483.6	185.6	578	539.6	207.1
339	316.5	121.5	399	372.5	143.0	459	428.5	164.5	519	484.5	186.0	579	540.5	207.5
340	317.4	121.8	400	373.4	143.3	460	429.4	164.8	520	485.5	186.4	580	541.5	207.9
341	318.4	122.2	401	374.4	143.7	461	430.4	165.2	521	486.4	186.7	581	542.4	208.2
342	319.3	122.6	402	375.3	144.1	462	431.3	165.6	522	487.3	187.1	582	543.3	208.6
343	320.2	122.9	403	376.2	144.4	463	432.2	165.9	523	488.3	187.4	583	544.3	208.9
344	321.2	123.3	404	377.1	144.8	464	433.2	166.3	524	489.2	187.8	584	545.2	209.3
345	322.1	123.6	405	378.1	145.1	465	434.1	166.6	525	490.1	188.1	585	546.1	209.6
346	323.0	124.0	406	379.0	145.5	466	435.0	167.0	526	491.1	188.5	586	547.1	210.0
347	324.0	124.4	407	379.9	145.9	467	436.0	167.4	527	492.0	188.9	587	548.0	210.4
348	324.9	124.7	408	380.9	146.2	468	436.9	167.7	528	492.9	189.2	588	548.9	210.7
349	325.8	125.1	409	381.8	146.6	469	437.8	168.1	529	493.9	189.6	589	549.9	211.1
350	326.8	125.4	410	382.8	146.9	470	438.8	168.4	530	494.8	189.9	590	550.8	211.4
351	327.7	125.8	411	383.7	147.3	471	439.7	168.8	531	495.7	190.3	591	551.7	211.8
352	328.6	126.1	412	384.6	147.6	472	440.6	169.1	532	496.7	190.7	592	552.7	212.2
353	329.6	126.5	413	385.6	148.0	473	441.6	169.5	533	497.6	191.0	593	553.6	212.5
354	330.5	126.9	414	386.5	148.4	474	442.5	169.9	534	498.5	191.4	594	554.5	212.9
355	331.4	127.2	415	387.4	148.7	475	443.4	170.2	535	499.5	191.7	595	555.5	213.2
356	332.4	127.6	416	388.4	149.1	476	444.4	170.6	536	500.4	192.1	596	556.4	213.6
357	333.3	127.9	417	389.3	149.4	477	445.3	170.9	537	501.3	192.4	597	557.3	213.9
358	334.2	128.3	418	390.2	149.8	478	446.3	171.3	538	502.3	192.8	598	558.2	214.3
359	335.2	128.7	419	391.2	150.2	479	447.2	171.7	539	503.2	193.2	599	559.2	214.7
360	336.1	129.0	420	392.1	150.5	480	448.1	172.0	540	504.1	193.5	600	560.1	215.1

| Dist | Dep. | D.Lat | Dist | Dep. | D.Lat | Dist | Dep. | D.Lat | Dist | Dep. | D.Lat | Dist | Dep. | D.Lat |

D Lon / Dep.

291° / 249° | **69 DEGREES.** | 069° / 111° 4h 36m | **69°**

22°

TRAVERSE TABLE
22 DEGREES.

Dist	D.Lat	Dep.	Dist	D.Lat	Dep.	Dist	D.Lat	Dep.	Dist	D.Lat	Dep.	Dist	D.Lat	Dep.
1	00·9	00·4	61	56·6	22·9	121	112·2	45·3	181	167·8	67·8	241	223·5	90·3
2	01·9	00·7	62	57·5	23·2	122	113·1	45·7	182	168·7	68·2	242	224·4	90·7
3	02·8	01·1	63	58·4	23·6	123	114·0	46·1	183	169·7	68·6	243	225·3	91·0
4	03·7	01·6	64	59·3	24·0	124	115·0	46·5	184	170·6	68·9	244	226·2	91·4
5	04·6	01·9	65	60·3	24·3	125	115·9	46·8	185	171·5	69·3	245	227·2	91·8
6	05·6	02·2	66	61·2	24·7	126	116·8	47·2	186	172·5	69·7	246	228·1	92·2
7	06·5	02·6	67	62·1	25·1	127	117·8	47·6	187	173·4	70·1	247	229·0	92·5
8	07·4	03·0	68	63·0	25·5	128	118·7	47·9	188	174·3	70·4	248	229·9	92·9
9	08·3	03·4	69	64·0	25·8	129	119·6	48·3	189	175·2	70·8	249	230·9	93·3
10	09·3	03·7	70	64·9	26·2	130	120·5	48·7	190	176·2	71·2	250	231·8	93·7
11	10·2	04·1	71	65·8	26·6	131	121·5	49·1	191	177·1	71·5	251	232·7	94·0
12	11·1	04·5	72	66·8	27·0	132	122·4	49·4	192	178·0	71·9	252	233·7	94·4
13	12·1	04·9	73	67·7	27·3	133	123·3	49·8	193	178·9	72·3	253	234·6	94·8
14	13·0	05·2	74	68·6	27·7	134	124·2	50·2	194	179·9	72·7	254	235·5	95·2
15	13·9	05·6	75	69·5	28·1	135	125·2	50·6	195	180·8	73·0	255	236·4	95·5
16	14·8	06·0	76	70·5	28·5	136	126·1	50·9	196	181·7	73·4	256	237·4	95·9
17	15·8	06·4	77	71·4	28·8	137	127·0	51·3	197	182·7	73·8	257	238·3	96·3
18	16·7	06·7	78	72·3	29·2	138	128·0	51·7	198	183·6	74·2	258	239·2	96·6
19	17·6	07·1	79	73·2	29·6	139	128·9	52·1	199	184·5	74·5	259	240·1	97·0
20	18·5	07·5	80	74·2	30·0	140	129·8	52·4	200	185·4	74·9	260	241·1	97·4
21	19·5	07·9	81	75·1	30·3	141	130·7	52·8	201	186·4	75·3	261	242·0	97·8
22	20·4	08·2	82	76·0	30·7	142	131·7	53·2	202	187·3	75·7	262	242·9	98·1
23	21·3	08·6	83	77·0	31·1	143	132·6	53·6	203	188·2	76·0	263	243·8	98·5
24	22·3	09·0	84	77·9	31·5	144	133·5	53·9	204	189·1	76·4	264	244·8	98·9
25	23·2	09·4	85	78·8	31·8	145	134·4	54·3	205	190·1	76·8	265	245·7	99·3
26	24·1	09·7	86	79·7	32·2	146	135·4	54·7	206	191·0	77·2	266	246·6	99·6
27	25·0	10·1	87	80·7	32·6	147	136·3	55·1	207	191·9	77·5	267	247·6	100·0
28	26·0	10·5	88	81·6	33·0	148	137·2	55·4	208	192·9	77·9	268	248·5	100·4
29	26·9	10·9	89	82·5	33·3	149	138·2	55·8	209	193·8	78·3	269	249·4	100·8
30	27·8	11·2	90	83·4	33·7	150	139·1	56·2	210	194·7	78·7	270	250·3	101·1
31	28·7	11·6	91	84·4	34·1	151	140·0	56·6	211	195·6	79·0	271	251·3	101·5
32	29·7	12·0	92	85·3	34·5	152	140·9	56·9	212	196·6	79·4	272	252·2	101·9
33	30·6	12·4	93	86·2	34·8	153	141·9	57·3	213	197·5	79·8	273	253·1	102·3
34	31·5	12·7	94	87·2	35·2	154	142·8	57·7	214	198·4	80·2	274	254·0	102·6
35	32·5	13·1	95	88·1	35·6	155	143·7	58·1	215	199·3	80·5	275	255·0	103·0
36	33·4	13·5	96	89·0	36·0	156	144·6	58·4	216	200·3	80·9	276	255·9	103·4
37	34·3	13·9	97	89·9	36·3	157	145·6	58·8	217	201·2	81·3	277	256·8	103·8
38	35·2	14·2	98	90·9	36·7	158	146·5	59·2	218	202·1	81·7	278	257·8	104·1
39	36·2	14·6	99	91·8	37·1	159	147·4	59·6	219	203·1	82·0	279	258·7	104·5
40	37·1	15·0	100	92·7	37·5	160	148·3	59·9	220	204·0	82·4	280	259·6	104·9
41	38·0	15·4	101	93·6	37·8	161	149·3	60·3	221	204·9	82·8	281	260·5	105·3
42	38·9	15·7	102	94·6	38·2	162	150·2	60·7	222	205·8	83·2	282	261·5	105·6
43	39·9	16·1	103	95·5	38·6	163	151·1	61·1	223	206·8	83·5	283	262·4	106·0
44	40·8	16·5	104	96·4	39·0	164	152·1	61·4	224	207·7	83·9	284	263·3	106·4
45	41·7	16·9	105	97·4	39·3	165	153·0	61·8	225	208·6	84·3	285	264·2	106·8
46	42·7	17·2	106	98·3	39·7	166	153·9	62·2	226	209·5	84·7	286	265·2	107·1
47	43·6	17·6	107	99·2	40·1	167	154·8	62·6	227	210·5	85·0	287	266·1	107·5
48	44·5	18·0	108	100·1	40·5	168	155·8	62·9	228	211·4	85·4	288	267·0	107·9
49	45·4	18·4	109	101·1	40·8	169	156·7	63·3	229	212·3	85·8	289	268·0	108·3
50	46·4	18·7	110	102·0	41·2	170	157·6	63·7	230	213·3	86·2	290	268·9	108·6
51	47·3	19·1	111	102·9	41·6	171	158·5	64·1	231	214·2	86·5	291	269·8	109·0
52	48·2	19·5	112	103·8	42·0	172	159·5	64·4	232	215·1	86·9	292	270·7	109·4
53	49·1	19·9	113	104·8	42·3	173	160·4	64·8	233	216·0	87·3	293	271·7	109·8
54	50·1	20·2	114	105·7	42·7	174	161·3	65·2	234	217·0	87·7	294	272·6	110·1
55	51·0	20·6	115	106·6	43·1	175	162·3	65·6	235	217·9	88·0	295	273·5	110·5
56	51·9	21·0	116	107·6	43·5	176	163·2	65·9	236	218·8	88·4	296	274·4	110·9
57	52·8	21·4	117	108·5	43·8	177	164·1	66·3	237	219·7	88·8	297	275·4	111·3
58	53·8	21·7	118	109·4	44·2	178	165·0	66·7	238	220·7	89·2	298	276·3	111·6
59	54·7	22·1	119	110·3	44·6	179	166·0	67·1	239	221·6	89·5	299	277·2	112·0
60	55·6	22·5	120	111·3	45·0	180	166·9	67·4	240	222·5	89·9	300	278·2	112·4
Dist	Dep.	D.Lat	Dist	Dep.	D.Lat	Dist	Dep.	D.Lat	Dist	Dep.	D.Lat	Dist	Dep.	D.Lat

68°

Dist	D.Lat	Dep.	Dist	D.Lat	Dep.	Dist	D.Lat	Dep.	Dist	D.Lat	Dep.	Dist	D.Lat	Dep.	Dist	D.Lat	Dep.
	D Lon	Dep.		D Lon	Dep.		D Lon	Dep.		D Lon	Dep.		D Lon	Dep.		D Lon	Dep.
301	279·1	112·8	361	334·7	135·2	421	390·3	157·7	481	446·0	180·2	541	501·6	202·7			
302	280·0	113·1	362	335·6	135·6	422	391·3	158·1	482	446·9	180·6	542	502·5	203·0			
303	280·9	113·5	363	336·6	136·0	423	392·2	158·5	483	447·8	180·9	543	503·5	203·4			
304	281·9	113·9	364	337·5	136·4	424	393·1	158·8	484	448·8	181·3	544	504·5	203·8			
305	282·8	114·3	365	338·4	136·7	425	394·1	159·2	485	449·7	181·7	545	505·3	204·2			
306	283·7	114·6	366	339·3	137·1	426	395·0	159·6	486	450·6	182·1	546	506·2	204·5			
307	284·6	115·0	367	340·3	137·5	427	395·9	160·0	487	451·5	182·4	547	507·2	204·9			
308	285·6	115·4	368	341·2	137·9	428	396·8	160·3	488	452·5	182·8	548	508·1	205·3			
309	286·5	115·8	369	342·1	138·2	429	397·8	160·7	489	453·4	183·2	549	509·0	205·7			
310	287·4	116·1	370	343·1	138·6	430	398·7	161·1	490	454·3	183·6	550	510·0	206·0			
311	288·4	116·5	371	344·0	139·0	431	399·6	161·5	491	455·2	184·0	551	510·9	206·4			
312	289·3	116·9	372	344·9	139·4	432	400·5	161·8	492	456·2	184·3	552	511·8	206·8			
313	290·2	117·3	373	345·8	139·8	433	401·5	162·2	493	457·1	184·7	553	512·7	207·2			
314	291·1	117·6	374	346·8	140·1	434	402·4	162·6	494	458·0	185·1	554	513·7	207·5			
315	292·1	118·0	375	347·7	140·5	435	403·3	163·0	495	459·0	185·4	555	514·6	207·9			
316	293·0	118·4	376	348·6	140·9	436	404·3	163·3	496	459·9	185·8	556	515·5	208·3			
317	293·9	118·8	377	349·5	141·2	437	405·2	163·7	497	460·8	186·2	557	516·4	208·7			
318	294·8	119·1	378	350·5	141·6	438	406·1	164·1	498	461·7	186·6	558	517·4	209·0			
319	295·8	119·5	379	351·4	142·0	439	407·0	164·5	499	462·7	186·9	559	518·3	209·4			
320	296·7	119·9	380	352·3	142·4	440	408·0	164·8	500	463·6	187·3	560	519·2	209·8			
321	297·6	120·2	381	353·3	142·7	441	408·9	165·2	501	464·5	187·7	561	520·2	210·2			
322	298·6	120·6	382	354·2	143·1	442	409·8	165·6	502	465·4	188·1	562	521·1	210·5			
323	299·5	121·0	383	355·1	143·5	443	410·7	166·0	503	466·4	188·4	563	522·0	210·9			
324	300·4	121·3	384	356·0	143·8	444	411·7	166·3	504	467·3	188·8	564	522·9	211·3			
325	301·3	121·7	385	357·0	144·2	445	412·6	166·7	505	468·2	189·2	565	523·9	211·7			
326	302·3	122·1	386	357·9	144·6	446	413·5	167·1	506	469·2	189·6	566	524·8	212·0			
327	303·2	122·5	387	358·8	145·0	447	414·5	167·4	507	470·1	189·9	567	525·7	212·4			
328	304·1	122·9	388	359·7	145·3	448	415·4	167·8	508	471·0	190·3	568	526·6	212·8			
329	305·0	123·2	389	360·7	145·7	449	416·3	168·2	509	471·9	190·7	569	527·6	213·2			
330	306·0	123·6	390	361·6	146·1	450	417·2	168·6	510	472·9	191·0	570	528·5	213·5			
331	306·9	124·0	391	362·5	146·5	451	418·2	168·9	511	473·8	191·4	571	529·4	213·9			
332	307·8	124·4	392	363·5	146·8	452	419·1	169·3	512	474·7	191·8	572	530·3	214·3			
333	308·8	124·7	393	364·4	147·2	453	420·0	169·7	513	475·6	192·2	573	531·3	214·6			
334	309·7	125·1	394	365·3	147·6	454	420·9	170·1	514	476·6	192·5	574	532·2	215·0			
335	310·6	125·5	395	366·2	148·0	455	421·9	170·4	515	477·5	192·9	575	533·1	215·4			
336	311·5	125·9	396	367·2	148·3	456	422·8	170·8	516	478·4	193·3	576	534·1	215·8			
337	312·5	126·2	397	368·1	148·7	457	423·7	171·2	517	479·4	193·7	577	535·0	216·1			
338	313·4	126·6	398	369·0	149·1	458	424·7	171·6	518	480·3	194·0	578	535·9	216·5			
339	314·3	127·0	399	369·9	149·5	459	425·6	171·9	519	481·2	194·4	579	536·8	216·9			
340	315·2	127·4	400	370·9	149·8	460	426·5	172·3	520	482·1	194·8	580	537·8	217·3			
341	316·2	127·7	401	371·8	150·2	461	427·4	172·7	521	483·1	195·2	581	538·7	217·6			
342	317·1	128·1	402	372·7	150·6	462	428·4	173·1	522	484·0	195·5	582	539·6	218·0			
343	318·0	128·5	403	373·7	151·0	463	429·3	173·4	523	484·9	195·9	583	540·5	218·4			
344	319·0	128·9	404	374·6	151·3	464	430·2	173·8	524	485·8	196·3	584	541·5	218·8			
345	319·9	129·2	405	375·5	151·7	465	431·1	174·2	525	486·8	196·7	585	542·4	219·1			
346	320·8	129·6	406	376·4	152·1	466	432·1	174·6	526	487·7	197·0	586	543·3	219·5			
347	321·7	130·0	407	377·4	152·5	467	433·0	174·9	527	488·6	197·4	587	544·3	219·9			
348	322·7	130·4	408	378·3	152·8	468	433·9	175·3	528	489·6	197·8	588	545·2	220·3			
349	323·6	130·7	409	379·2	153·2	469	434·8	175·7	529	490·5	198·2	589	546·1	220·6			
350	324·5	131·1	410	380·1	153·6	470	435·8	176·1	530	491·4	198·5	590	547·0	221·0			
351	325·4	131·5	411	381·1	154·0	471	436·7	176·4	531	492·3	198·9	591	548·0	221·4			
352	326·4	131·9	412	382·0	154·3	472	437·6	176·8	532	493·3	199·3	592	548·9	221·8			
353	327·3	132·2	413	382·9	154·7	473	438·6	177·2	533	494·2	199·7	593	549·8	222·1			
354	328·2	132·6	414	383·9	155·1	474	439·5	177·6	534	495·1	200·0	594	550·7	222·5			
355	329·2	133·0	415	384·8	155·5	475	440·4	177·9	535	496·0	200·4	595	551·7	222·9			
356	330·1	133·4	416	385·7	155·8	476	441·3	178·3	536	497·0	200·8	596	552·6	223·3			
357	331·0	133·7	417	386·6	156·2	477	442·3	178·7	537	497·9	201·2	597	553·5	223·6			
358	332·0	134·1	418	387·6	156·6	478	443·2	179·1	538	498·8	201·5	598	554·5	224·0			
359	332·9	134·5	419	388·5	157·0	479	444·1	179·4	539	499·8	201·9	599	555·4	224·4			
360	333·8	134·9	420	389·4	157·3	480	445·0	179·8	540	500·7	202·3	600	556·3	224·8			
Dist	Dep.	D.Lat	Dist	Dep.	D.Lat	Dist	Dep.	D.Lat	Dist	Dep.	D.Lat	Dist	Dep.	D.Lat	Dist	Dep.	D.Lat
D Lon		Dep.	D Lon		Dep.	D Lon		Dep.	D Lon		Dep.	D Lon		Dep.	D Lon		Dep.

TRAVERSE TABLE
23 DEGREES.

23° 337° / 203° | 023° / 157° 1h 32m

Dist	D.Lat	Dep.	Dist	D.Lat	Dep.	Dist	D.Lat	Dep.	Dist	D.Lat	Dep.	Dist	D.Lat	Dep.
1	00.9	00.4	61	56.2	23.8	121	111.4	47.3	181	166.6	70.7	241	221.8	94.2
2	01.8	00.8	62	57.1	24.2	122	112.3	47.7	182	167.5	71.1	242	222.8	94.5
3	02.8	01.2	63	58.0	24.6	123	113.2	48.1	183	168.5	71.5	243	223.7	94.9
4	03.7	01.6	64	58.9	25.0	124	114.1	48.5	184	169.4	71.9	244	224.6	95.3
5	04.6	02.0	65	59.8	25.4	125	115.1	48.8	185	170.3	72.3	245	225.5	95.7
6	05.5	02.3	66	60.8	25.8	126	116.0	49.2	186	171.2	72.7	246	226.4	96.1
7	06.4	02.7	67	61.7	26.2	127	116.9	49.6	187	172.1	73.1	247	227.4	96.5
8	07.4	03.1	68	62.6	26.6	128	117.8	50.0	188	173.1	73.5	248	228.3	96.9
9	08.3	03.5	69	63.5	27.0	129	118.7	50.4	189	174.0	73.8	249	229.2	97.3
10	09.2	03.9	70	64.4	27.4	130	119.7	50.8	190	174.9	74.2	250	230.1	97.7
11	10.1	04.3	71	65.4	27.7	131	120.6	51.2	191	175.8	74.6	251	231.0	98.1
12	11.0	04.7	72	66.3	28.1	132	121.5	51.6	192	176.7	75.0	252	232.0	98.5
13	12.0	05.1	73	67.2	28.5	133	122.4	52.0	193	177.7	75.4	253	232.9	98.9
14	12.9	05.5	74	68.1	28.9	134	123.3	52.4	194	178.6	75.8	254	233.8	99.2
15	13.8	05.9	75	69.0	29.3	135	124.3	52.7	195	179.5	76.2	255	234.7	99.6
16	14.7	06.3	76	70.0	29.7	136	125.2	53.1	196	180.4	76.6	256	235.6	100.0
17	15.6	06.6	77	70.9	30.1	137	126.1	53.5	197	181.3	77.0	257	236.6	100.4
18	16.6	07.0	78	71.8	30.5	138	127.0	53.9	198	182.3	77.4	258	237.5	100.8
19	17.5	07.4	79	72.7	30.9	139	128.0	54.3	199	183.2	77.8	259	238.4	101.2
20	18.4	07.8	80	73.6	31.3	140	128.9	54.7	200	184.1	78.1	260	239.3	101.6
21	19.3	08.2	81	74.6	31.6	141	129.8	55.1	201	185.0	78.5	261	240.3	102.0
22	20.3	08.6	82	75.5	32.0	142	130.7	55.5	202	185.9	78.9	262	241.2	102.4
23	21.2	09.0	83	76.4	32.4	143	131.6	55.9	203	186.9	79.3	263	242.1	102.8
24	22.1	09.4	84	77.3	32.8	144	132.6	56.3	204	187.8	79.7	264	243.0	103.2
25	23.0	09.8	85	78.2	33.2	145	133.5	56.7	205	188.7	80.1	265	243.9	103.5
26	23.9	10.2	86	79.2	33.6	146	134.4	57.0	206	189.6	80.5	266	244.9	103.9
27	24.9	10.5	87	80.1	34.0	147	135.3	57.4	207	190.5	80.9	267	245.8	104.3
28	25.8	10.9	88	81.0	34.4	148	136.2	57.8	208	191.5	81.3	268	246.7	104.7
29	26.7	11.3	89	81.9	34.8	149	137.2	58.2	209	192.4	81.7	269	247.6	105.1
30	27.6	11.7	90	82.8	35.2	150	138.1	58.6	210	193.3	82.1	270	248.5	105.5
31	28.5	12.1	91	83.8	35.6	151	139.0	59.0	211	194.2	82.4	271	249.5	105.9
32	29.5	12.5	92	84.7	35.9	152	139.9	59.4	212	195.1	82.8	272	250.4	106.3
33	30.4	12.9	93	85.6	36.3	153	140.8	59.8	213	196.1	83.2	273	251.3	106.7
34	31.3	13.3	94	86.5	36.7	154	141.8	60.2	214	197.0	83.6	274	252.2	107.1
35	32.2	13.7	95	87.4	37.1	155	142.7	60.6	215	197.9	84.0	275	253.1	107.5
36	33.1	14.1	96	88.4	37.5	156	143.6	61.0	216	198.8	84.4	276	254.1	107.9
37	34.1	14.5	97	89.3	37.9	157	144.5	61.3	217	199.7	84.8	277	255.0	108.2
38	35.0	14.8	98	90.2	38.3	158	145.4	61.7	218	200.7	85.2	278	255.9	108.6
39	35.9	15.2	99	91.1	38.7	159	146.4	62.1	219	201.6	85.6	279	256.8	109.0
40	36.8	15.6	100	92.1	39.1	160	147.3	62.5	220	202.5	86.0	280	257.7	109.4
41	37.7	16.0	101	93.0	39.5	161	148.2	62.9	221	203.4	86.4	281	258.7	109.8
42	38.7	16.4	102	93.9	39.9	162	149.1	63.3	222	204.4	86.8	282	259.6	110.2
43	39.6	16.8	103	94.8	40.2	163	150.0	63.7	223	205.3	87.1	283	260.5	110.6
44	40.5	17.2	104	95.7	40.6	164	151.0	64.1	224	206.2	87.5	284	261.4	111.0
45	41.4	17.6	105	96.7	41.0	165	151.9	64.5	225	207.1	87.9	285	262.3	111.4
46	42.3	18.0	106	97.6	41.4	166	152.8	64.9	226	208.0	88.3	286	263.3	111.7
47	43.3	18.4	107	98.5	41.8	167	153.7	65.3	227	209.0	88.7	287	264.2	112.1
48	44.2	18.8	108	99.4	42.2	168	154.6	65.6	228	209.9	89.1	288	265.1	112.5
49	45.1	19.1	109	100.3	42.6	169	155.6	66.0	229	210.8	89.5	289	266.0	112.9
50	46.0	19.5	110	101.3	43.0	170	156.5	66.4	230	211.7	89.9	290	266.9	113.3
51	46.9	19.9	111	102.2	43.4	171	157.4	66.8	231	212.6	90.3	291	267.9	113.7
52	47.9	20.3	112	103.1	43.8	172	158.3	67.2	232	213.6	90.6	292	268.8	114.1
53	48.8	20.7	113	104.0	44.2	173	159.2	67.6	233	214.5	91.0	293	269.7	114.5
54	49.7	21.1	114	104.9	44.5	174	160.2	68.0	234	215.4	91.4	294	270.6	114.9
55	50.6	21.5	115	105.9	44.9	175	161.1	68.4	235	216.3	91.8	295	271.5	115.3
56	51.5	21.9	116	106.8	45.3	176	162.0	68.8	236	217.2	92.2	296	272.5	115.7
57	52.5	22.3	117	107.7	45.7	177	162.9	69.2	237	218.2	92.6	297	273.4	116.0
58	53.4	22.7	118	108.6	46.1	178	163.8	69.6	238	219.1	93.0	298	274.3	116.4
59	54.3	23.1	119	109.5	46.5	179	164.8	69.9	239	220.0	93.4	299	275.2	116.8
60	55.2	23.4	120	110.5	46.9	180	165.7	70.3	240	220.9	93.8	300	276.2	117.2

67° | Dist | Dep. | D.Lat | (repeated across) | D Lon | ... Dep. |

293° / 247° | **67 DEGREES.** | 067° / 113° 4h 28m

TRAVERSE TABLE
23 DEGREES.

337° ↑ / 203°   ↑023° / 157°   1h 32m   **23°**

D Lon	Dep.		D Lon	Dep.		D Lon	Dep.		D Lon	Dep.		D Lon	Dep.	
Dist	D.Lat	Dep.	Dist	D.Lat	Dep.	Dist	D.Lat	Dep.	Dist	D.Lat	Dep.	Dist	D.Lat	Dep.
301	277·1	117·6	361	332·3	141·1	421	387·5	164·5	481	442·8	187·9	541	498·0	211·4
302	278·0	118·0	362	333·2	141·4	422	388·5	164·9	482	443·7	188·3	542	498·9	211·8
303	278·9	118·4	363	334·1	141·8	423	389·4	165·3	483	444·6	188·7	543	499·8	212·2
304	279·8	118·8	364	335·1	142·2	424	390·3	165·7	484	445·5	189·1	544	500·8	212·6
305	280·8	119·2	365	336·0	142·6	425	391·2	166·1	485	446·4	189·5	545	501·7	212·9
306	281·7	119·6	366	336·9	143·0	426	392·1	166·5	486	447·4	189·9	546	502·6	213·3
307	282·6	120·0	367	337·8	143·4	427	393·1	166·8	487	448·3	190·3	547	503·5	213·7
308	283·5	120·3	368	338·7	143·8	428	394·0	167·2	488	449·2	190·7	548	504·4	214·1
309	284·4	120·7	369	339·7	144·2	429	394·9	167·6	489	450·1	191·1	549	505·4	214·5
310	285·4	121·1	370	340·6	144·6	430	395·8	168·0	490	451·0	191·5	550	506·3	214·9
311	286·3	121·5	371	341·5	145·0	431	396·7	168·4	491	452·0	191·8	551	507·2	215·3
312	287·2	121·9	372	342·4	145·4	432	397·7	168·8	492	452·9	192·2	552	508·1	215·7
313	288·1	122·3	373	343·3	145·7	433	398·6	169·2	493	453·8	192·6	553	509·0	216·1
314	289·0	122·7	374	344·3	146·1	434	399·5	169·6	494	454·7	193·0	554	510·0	216·5
315	290·0	123·1	375	345·2	146·5	435	400·4	170·0	495	455·6	193·4	555	510·9	216·9
316	290·9	123·5	376	346·1	146·9	436	401·3	170·4	496	456·6	193·8	556	511·8	217·2
317	291·8	123·9	377	347·0	147·3	437	402·3	170·7	497	457·5	194·2	557	512·7	217·6
318	292·7	124·3	378	348·0	147·7	438	403·2	171·1	498	458·4	194·6	558	513·6	218·0
319	293·6	124·6	379	348·9	148·1	439	404·1	171·5	499	459·3	195·0	559	514·6	218·4
320	294·6	125·0	380	349·8	148·5	440	405·0	171·9	500	460·3	195·4	560	515·5	218·8
321	295·5	125·4	381	350·7	148·9	441	405·9	172·3	501	461·2	195·8	561	516·4	219·2
322	296·4	125·8	382	351·6	149·3	442	406·9	172·7	502	462·1	196·1	562	517·3	219·6
323	297·3	126·2	383	352·6	149·7	443	407·8	173·1	503	463·0	196·5	563	518·2	220·0
324	298·2	126·6	384	353·5	150·0	444	408·7	173·5	504	463·9	196·9	564	519·2	220·4
325	299·2	127·0	385	354·4	150·4	445	409·6	173·9	505	464·9	197·3	565	520·1	220·8
326	300·1	127·4	386	355·3	150·8	446	410·5	174·3	506	465·8	197·7	566	521·0	221·2
327	301·0	127·8	387	356·2	151·2	447	411·5	174·7	507	466·7	198·1	567	521·9	221·5
328	301·9	128·2	388	357·2	151·6	448	412·4	175·0	508	467·6	198·5	568	522·8	221·9
329	302·8	128·6	389	358·1	152·0	449	413·3	175·4	509	468·5	198·9	569	523·8	222·3
330	303·8	128·9	390	359·0	152·4	450	414·2	175·8	510	469·5	199·3	570	524·7	222·7
331	304·7	129·3	391	359·9	152·8	451	415·1	176·2	511	470·4	199·7	571	525·6	223·1
332	305·6	129·7	392	360·8	153·2	452	416·1	176·6	512	471·3	200·1	572	526·5	223·5
333	306·5	130·1	393	361·8	153·6	453	417·0	177·0	513	472·2	200·4	573	527·4	223·9
334	307·4	130·5	394	362·7	153·9	454	417·9	177·4	514	473·1	200·8	574	528·4	224·3
335	308·4	130·9	395	363·6	154·3	455	418·8	177·8	515	474·1	201·2	575	529·3	224·7
336	309·3	131·3	396	364·5	154·7	456	419·8	178·2	516	475·0	201·6	576	530·2	225·1
337	310·2	131·7	397	365·4	155·1	457	420·7	178·6	517	475·9	202·0	577	531·1	225·4
338	311·1	132·1	398	366·4	155·5	458	421·6	179·0	518	476·8	202·4	578	532·1	225·8
339	312·1	132·5	399	367·3	155·9	459	422·5	179·3	519	477·7	202·8	579	533·0	226·2
340	313·0	132·8	400	368·2	156·3	460	423·4	179·7	520	478·7	203·2	580	533·9	226·6
341	313·9	133·2	401	369·1	156·7	461	424·4	180·1	521	479·6	203·6	581	534·8	227·0
342	314·8	133·6	402	370·0	157·1	462	425·3	180·5	522	480·5	204·0	582	535·7	227·4
343	315·7	134·0	403	371·0	157·5	463	426·2	180·9	523	481·4	204·4	583	536·7	227·8
344	316·7	134·4	404	371·9	157·9	464	427·1	181·3	524	482·3	204·7	584	537·6	228·2
345	317·6	134·8	405	372·8	158·2	465	428·0	181·7	525	483·3	205·1	585	538·5	228·6
346	318·5	135·2	406	373·7	158·6	466	429·0	182·1	526	484·2	205·5	586	539·4	229·0
347	319·4	135·6	407	374·6	159·0	467	429·9	182·5	527	485·1	205·9	587	540·3	229·4
348	320·3	136·0	408	375·6	159·4	468	430·8	182·9	528	486·0	206·3	588	541·3	229·7
349	321·3	136·4	409	376·5	159·8	469	431·7	183·3	529	486·9	206·7	589	542·2	230·1
350	322·2	136·8	410	377·4	160·2	470	432·6	183·6	530	487·9	207·1	590	543·1	230·5
351	323·1	137·1	411	378·3	160·6	471	433·6	184·0	531	488·8	207·5	591	544·0	230·9
352	324·0	137·5	412	379·2	161·0	472	434·5	184·4	532	489·7	207·9	592	544·9	231·3
353	324·9	137·9	413	380·2	161·4	473	435·4	184·8	533	490·6	208·3	593	545·9	231·7
354	325·9	138·3	414	381·1	161·8	474	436·3	185·2	534	491·5	208·7	594	546·8	232·1
355	326·8	138·7	415	382·0	162·2	475	437·2	185·6	535	492·5	209·0	595	547·7	232·5
356	327·7	139·1	416	382·9	162·5	476	438·2	186·0	536	493·4	209·4	596	548·6	232·9
357	328·6	139·5	417	383·9	162·9	477	439·1	186·4	537	494·3	209·8	597	549·5	233·3
358	329·5	139·9	418	384·8	163·3	478	440·0	186·8	538	495·2	210·2	598	550·5	233·7
359	330·5	140·3	419	385·7	163·7	479	440·9	187·2	539	496·2	210·6	599	551·4	234·0
360	331·4	140·7	420	386·6	164·1	480	441·8	187·6	540	497·2	211·0	600	552·3	234·4
Dist	Dep.	D.Lat	Dist	Dep.	D.Lat	Dist	Dep.	D.Lat	Dist	Dep.	D.Lat	Dist	Dep.	D.Lat
D Lon		Dep.	D Lon		Dep.	D Lon		Dep.	D Lon		Dep.	D Lon		Dep.

293° ↑ / 247°   **67 DEGREES.**   067° / 113°   4h 28m   **67°**

24°

D Lon	Dep.		D Lon	Dep.		D Lon	Dep.		D Lon	Dep.		D Lon	Dep.	
Dist	D. Lat	Dep.	Dist	D. Lat	Dep.	Dist	D. Lat	Dep.	Dist	D. Lat	Dep.	Dist	D. Lat	Dep.
1	00·9	00·4	61	55·7	24·8	121	110·5	49·2	181	165·4	73·6	241	220·2	98·0
2	01·8	00·8	62	56·6	25·2	122	111·5	49·6	182	166·3	74·0	242	221·1	98·4
3	02·7	01·2	63	57·6	25·6	123	112·4	50·0	183	167·2	74·4	243	222·0	98·8
4	03·7	01·6	64	58·5	26·0	124	113·3	50·4	184	168·1	74·8	244	222·9	99·2
5	04·6	02·0	65	59·4	26·4	125	114·2	50·8	185	169·0	75·2	245	223·8	99·7
6	05·5	02·4	66	60·3	26·8	126	115·1	51·2	186	169·9	75·7	246	224·7	100·1
7	06·4	02·8	67	61·2	27·3	127	116·0	51·7	187	170·8	76·1	247	225·6	100·5
8	07·3	03·3	68	62·1	27·7	128	116·9	52·1	188	171·7	76·5	248	226·6	100·9
9	08·2	03·7	69	63·0	28·1	129	117·8	52·5	189	172·7	76·9	249	227·5	101·3
10	09·1	04·1	70	63·9	28·5	130	118·8	52·9	190	173·6	77·3	250	228·4	101·7
11	10·0	04·5	71	64·9	28·9	131	119·7	53·3	191	174·5	77·7	251	229·3	102·1
12	11·0	04·9	72	65·8	29·3	132	120·6	53·7	192	175·4	78·1	252	230·2	102·5
13	11·9	05·3	73	66·7	29·7	133	121·5	54·1	193	176·3	78·5	253	231·1	102·9
14	12·8	05·7	74	67·6	30·1	134	122·4	54·5	194	177·2	78·9	254	232·0	103·3
15	13·7	06·1	75	68·5	30·5	135	123·3	54·9	195	178·1	79·3	255	233·0	103·7
16	14·6	06·5	76	69·4	30·9	136	124·2	55·3	196	179·1	79·7	256	233·9	104·1
17	15·5	06·9	77	70·3	31·3	137	125·2	55·7	197	180·0	80·1	257	234·8	104·5
18	16·4	07·3	78	71·3	31·7	138	126·1	56·1	198	180·9	80·5	258	235·7	104·9
19	17·4	07·7	79	72·2	32·1	139	127·0	56·5	199	181·8	80·9	259	236·6	105·3
20	18·3	08·1	80	73·1	32·5	140	127·9	56·9	200	182·7	81·3	260	237·5	105·8
21	19·2	08·5	81	74·0	32·9	141	128·8	57·3	201	183·6	81·8	261	238·4	106·2
22	20·1	08·9	82	74·9	33·4	142	129·7	57·8	202	184·5	82·2	262	239·3	106·6
23	21·0	09·4	83	75·8	33·8	143	130·6	58·2	203	185·4	82·6	263	240·3	107·0
24	21·9	09·8	84	76·7	34·2	144	131·6	58·6	204	186·4	83·0	264	241·2	107·4
25	22·8	10·2	85	77·7	34·6	145	132·5	59·0	205	187·3	83·4	265	242·1	107·8
26	23·8	10·6	86	78·6	35·0	146	133·4	59·4	206	188·2	83·8	266	243·0	108·2
27	24·7	11·0	87	79·5	35·4	147	134·3	59·8	207	189·1	84·2	267	243·9	108·6
28	25·6	11·4	88	80·4	35·8	148	135·2	60·2	208	190·0	84·6	268	244·8	109·0
29	26·5	11·8	89	81·3	36·2	149	136·1	60·6	209	190·9	85·0	269	245·7	109·4
30	27·4	12·2	90	82·2	36·6	150	137·0	61·0	210	191·8	85·4	270	246·7	109·8
31	28·3	12·6	91	83·1	37·0	151	137·9	61·4	211	192·8	85·8	271	247·6	110·2
32	29·2	13·0	92	84·0	37·4	152	138·9	61·8	212	193·7	86·2	272	248·5	110·6
33	30·1	13·4	93	85·0	37·8	153	139·8	62·2	213	194·6	86·6	273	249·4	111·0
34	31·1	13·8	94	85·9	38·2	154	140·7	62·6	214	195·5	87·0	274	250·3	111·4
35	32·0	14·2	95	86·8	38·6	155	141·6	63·0	215	196·4	87·4	275	251·2	111·9
36	32·9	14·6	96	87·7	39·0	156	142·5	63·5	216	197·3	87·9	276	252·1	112·3
37	33·8	15·0	97	88·6	39·5	157	143·4	63·9	217	198·2	88·3	277	253·1	112·7
38	34·7	15·5	98	89·5	39·9	158	144·3	64·3	218	199·2	88·7	278	254·0	113·1
39	35·6	15·9	99	90·4	40·3	159	145·3	64·7	219	200·1	89·1	279	254·9	113·5
40	36·5	16·3	100	91·4	40·7	160	146·2	65·1	220	201·0	89·5	280	255·8	113·9
41	37·5	16·7	101	92·3	41·1	161	147·1	65·5	221	201·9	89·9	281	256·7	114·3
42	38·4	17·1	102	93·2	41·5	162	148·0	65·9	222	202·8	90·3	282	257·6	114·7
43	39·3	17·5	103	94·1	41·9	163	148·9	66·3	223	203·7	90·7	283	258·5	115·1
44	40·2	17·9	104	95·0	42·3	164	149·8	66·7	224	204·6	91·1	284	259·4	115·5
45	41·1	18·3	105	95·9	42·7	165	150·7	67·1	225	205·5	91·5	285	260·4	115·9
46	42·0	18·7	106	96·8	43·1	166	151·6	67·5	226	206·5	91·9	286	261·3	116·3
47	42·9	19·1	107	97·7	43·5	167	152·6	67·9	227	207·4	92·3	287	262·2	116·7
48	43·9	19·5	108	98·7	43·9	168	153·5	68·3	228	208·3	92·7	288	263·1	117·1
49	44·8	19·9	109	99·6	44·3	169	154·4	68·7	229	209·2	93·1	289	264·0	117·5
50	45·7	20·3	110	100·5	44·7	170	155·3	69·1	230	210·1	93·5	290	264·9	118·0
51	46·6	20·7	111	101·4	45·1	171	156·2	69·6	231	211·0	94·0	291	265·8	118·4
52	47·5	21·2	112	102·3	45·6	172	157·1	70·0	232	211·9	94·4	292	266·8	118·8
53	48·4	21·6	113	103·2	46·0	173	158·0	70·4	233	212·9	94·8	293	267·7	119·2
54	49·3	22·0	114	104·1	46·4	174	159·0	70·8	234	213·8	95·2	294	268·6	119·6
55	50·2	22·4	115	105·1	46·8	175	159·9	71·2	235	214·7	95·6	295	269·5	120·0
56	51·2	22·8	116	106·0	47·2	176	160·8	71·6	236	215·6	96·0	296	270·4	120·4
57	52·1	23·2	117	106·9	47·6	177	161·7	72·0	237	216·5	96·4	297	271·3	120·8
58	53·0	23·6	118	107·8	48·0	178	162·6	72·4	238	217·4	96·8	298	272·2	121·2
59	53·9	24·0	119	108·7	48·4	179	163·5	72·8	239	218·3	97·2	299	273·2	121·6
60	54·8	24·4	120	109·6	48·8	180	164·4	73·2	240	219·3	97·6	300	274·1	122·0

Dist	Dep.	D. Lat	Dist	Dep.	D. Lat	Dist	Dep.	D. Lat	Dist	Dep.	D. Lat	Dist	Dep.	D. Lat
D Lon		Dep.	D Lon		Dep.	D Lon		Dep.	D Lon		Dep.	D Lon		Dep.

66°

336°↑
204°

TRAVERSE TABLE
24 DEGREES.

↑024°
156°   1h 36m

D Lon	Dep.		D Lon	Dep.		D Lon	Dep.		D Lon	Dep.		D Lon	Dep.	
Dist	D.Lat	Dep.	Dist	D.Lat	Dep.	Dist	D.Lat	Dep.	Dist	D.Lat	Dep.	Dist	D.Lat	Dep.
301	275·0	122·4	361	329·8	146·8	421	384·6	171·2	481	439·4	195·6	541	494·2	220·0
302	275·9	122·8	362	330·7	147·2	422	385·5	171·6	482	440·3	196·0	542	495·1	220·5
303	276·8	123·2	363	331·6	147·6	423	386·4	172·0	483	441·2	196·5	543	496·1	220·9
304	277·7	123·6	364	332·5	148·1	424	387·3	172·5	484	442·2	196·9	544	497·0	221·3
305	278·6	124·1	365	333·4	148·5	425	388·3	172·9	485	443·1	197·3	545	497·9	221·7
306	279·5	124·5	366	334·4	148·9	426	389·2	173·3	486	444·0	197·7	546	498·8	222·1
307	280·5	124·9	367	335·3	149·3	427	390·1	173·7	487	444·9	198·1	547	499·7	222·5
308	281·4	125·3	368	336·2	149·7	428	391·0	174·1	488	445·8	198·5	548	500·6	222·9
309	282·3	125·7	369	337·1	150·1	429	391·9	174·5	489	446·7	198·9	549	501·5	223·3
310	283·2	126·1	370	338·0	150·5	430	392·8	174·9	490	447·6	198·3	550	502·5	223·7
311	284·1	126·5	371	338·9	150·9	431	393·7	175·3	491	448·6	199·7	551	503·4	224·1
312	285·0	126·9	372	339·8	151·3	432	394·7	175·7	492	449·5	200·1	552	504·3	224·5
313	285·9	127·3	373	340·7	151·7	433	395·6	176·1	493	450·4	200·5	553	505·2	224·9
314	286·9	127·7	374	341·7	152·1	434	396·5	176·5	494	451·3	200·9	554	506·1	225·3
315	287·8	128·1	375	342·6	152·5	435	397·4	176·9	495	452·2	201·3	555	507·0	225·7
316	288·7	128·5	376	343·5	152·9	436	398·3	177·3	496	453·1	201·7	556	507·9	226·0
317	289·6	128·9	377	344·4	153·3	437	399·2	177·7	497	454·0	202·1	557	508·8	226·6
318	290·5	129·3	378	345·3	153·7	438	400·1	178·2	498	454·9	202·6	558	509·8	227·0
319	291·4	129·7	379	346·2	154·2	439	401·0	178·6	499	455·9	203·0	559	510·7	227·4
320	292·3	130·2	380	347·1	154·6	440	402·0	179·0	500	456·8	203·4	560	511·6	227·8
321	293·2	130·6	381	348·1	155·0	441	402·9	179·4	501	457·7	203·8	561	512·5	228·2
322	294·2	131·0	382	349·0	155·4	442	403·8	179·8	502	458·6	204·2	562	513·4	228·6
323	295·1	131·4	383	349·9	155·8	443	404·7	180·2	503	459·5	204·6	563	514·3	229·0
324	296·0	131·8	384	350·8	156·2	444	405·6	180·6	504	460·4	205·0	564	515·2	229·4
325	296·9	132·2	385	351·7	156·6	445	406·5	181·0	505	461·3	205·4	565	516·2	229·8
326	297·8	132·6	386	352·6	157·0	446	407·4	181·4	506	462·3	205·8	566	517·1	230·2
327	298·7	133·0	387	353·5	157·4	447	408·4	181·8	507	463·2	206·2	567	518·0	230·6
328	299·6	133·4	388	354·5	157·8	448	409·3	182·2	508	464·1	206·6	568	518·9	231·0
329	300·6	133·8	389	355·4	158·2	449	410·2	182·6	509	465·0	207·0	569	519·8	231·4
330	301·5	134·2	390	356·3	158·6	450	411·1	183·0	510	465·9	207·4	570	520·7	231·8
331	302·4	134·6	391	357·2	159·0	451	412·0	183·4	511	466·8	207·8	571	521·6	232·2
332	303·3	135·0	392	358·1	159·4	452	412·9	183·8	512	467·7	208·2	572	522·5	232·7
333	304·2	135·4	393	359·0	159·8	453	413·8	184·3	513	468·6	208·7	573	523·5	233·1
334	305·1	135·9	394	359·9	160·3	454	414·7	184·7	514	469·6	209·1	574	524·4	233·5
335	306·0	136·3	395	360·9	160·7	455	415·7	185·1	515	470·5	209·5	575	525·3	233·9
336	307·0	136·7	396	361·8	161·1	456	416·6	185·5	516	471·4	209·9	576	526·2	234·3
337	307·9	137·1	397	362·7	161·5	457	417·5	185·9	517	472·3	210·3	577	527·1	234·7
338	308·8	137·5	398	363·6	161·9	458	418·4	186·3	518	473·2	210·7	578	528·0	235·1
339	309·7	137·9	399	364·5	162·3	459	419·3	186·7	519	474·1	211·1	579	528·9	235·5
340	310·6	138·3	400	365·4	162·7	460	420·2	187·1	520	475·0	211·5	580	529·9	235·9
341	311·5	138·7	401	366·3	163·1	461	421·1	187·5	521	476·0	211·9	581	530·8	236·3
342	312·4	139·1	402	367·2	163·5	462	422·1	187·9	522	476·9	212·3	582	531·7	236·7
343	313·3	139·5	403	368·2	163·9	463	423·0	188·3	523	477·8	212·7	583	532·6	237·1
344	314·3	139·9	404	369·1	164·3	464	423·9	188·7	524	478·7	213·1	584	533·5	237·5
345	315·2	140·3	405	370·0	164·7	465	424·8	189·1	525	479·6	213·5	585	534·4	237·9
346	316·1	140·7	406	370·9	165·1	466	425·7	189·5	526	480·5	213·9	586	535·3	238·3
347	317·0	141·1	407	371·8	165·5	467	426·6	189·9	527	481·4	214·4	587	536·3	238·8
348	317·9	141·5	408	372·7	165·9	468	427·5	190·4	528	482·4	214·8	588	537·2	239·2
349	318·8	142·0	409	373·6	166·4	469	428·5	190·8	529	483·3	215·2	589	538·1	239·6
350	319·7	142·4	410	374·6	166·8	470	429·4	191·2	530	484·2	215·6	590	539·0	240·0
351	320·7	142·8	411	375·5	167·2	471	430·3	191·6	531	485·1	216·0	591	539·9	240·4
352	321·6	143·2	412	376·4	167·6	472	431·2	192·0	532	486·0	216·4	592	540·8	240·8
353	322·5	143·6	413	377·3	168·0	473	432·1	192·4	533	486·9	216·8	593	541·7	241·2
354	323·4	144·0	414	378·2	168·4	474	433·0	192·8	534	487·8	217·2	594	542·6	241·6
355	324·3	144·4	415	379·1	168·8	475	433·9	193·2	535	488·7	217·6	595	543·6	242·0
356	325·2	144·8	416	380·0	169·2	476	434·8	193·6	536	489·7	218·0	596	544·5	242·4
357	326·1	145·2	417	380·9	169·6	477	435·8	194·0	537	490·6	218·4	597	545·4	242·8
358	327·0	145·6	418	381·9	170·0	478	436·7	194·4	538	491·5	218·8	598	546·3	243·2
359	328·0	146·0	419	382·8	170·4	479	437·6	194·8	539	492·4	219·2	599	547·2	243·6
360	328·9	146·4	420	383·7	170·8	480	438·5	195·2	540	493·3	219·6	600	548·1	244·0
Dist	Dep.	D.Lat	Dist	Dep.	D.Lat	Dist	Dep.	D.Lat	Dist	Dep.	D.Lat	Dist	Dep.	D.Lat
D Lon		Dep.	D Lon		Dep.	D Lon		Dep.	D Lon		Dep.	D Lon		Dep.

294°↑
246°

66 DEGREES.

↑066°
114°   4h 24m

66°

TRAVERSE TABLE
25 DEGREES.

25° | 335° / 205° | 025° / 155° 1h 40m

D Lon	Dep.		D Lon	Dep.		D Lon	Dep.		D Lon	Dep.		D Lon	Dep.	
Dist	D.Lat	Dep.	Dist	D.Lat	Dep.	Dist	D.Lat	Dep.	Dist	D.Lat	Dep.	Dist	D.Lat	Dep.
1	00·9	00·4	61	55·3	25·8	121	109·7	51·1	181	164·0	76·5	241	218·4	101·9
2	01·8	00·8	62	56·2	26·2	122	110·6	51·6	182	164·9	76·9	242	219·3	102·3
3	02·7	01·3	63	57·1	26·6	123	111·5	52·0	183	165·9	77·3	243	220·2	102·7
4	03·6	01·7	64	58·0	27·0	124	112·4	52·4	184	166·8	77·8	244	221·1	103·1
5	04·5	02·1	65	58·9	27·5	125	113·3	52·8	185	167·7	78·2	245	222·0	103·5
6	05·4	02·5	66	59·8	27·9	126	114·2	53·2	186	168·6	78·6	246	223·0	104·0
7	06·3	03·0	67	60·7	28·3	127	115·1	53·7	187	169·5	79·0	247	223·9	104·4
8	07·3	03·4	68	61·6	28·7	128	116·0	54·1	188	170·4	79·5	248	224·8	104·8
9	08·2	03·8	69	62·5	29·2	129	116·9	54·5	189	171·3	79·9	249	225·7	105·2
10	09·1	04·2	70	63·4	29·6	130	117·8	54·9	190	172·2	80·3	250	226·6	105·7
11	10·0	04·6	71	64·3	30·0	131	118·7	55·4	191	173·1	80·7	251	227·5	106·1
12	10·9	05·1	72	65·3	30·4	132	119·6	55·8	192	174·0	81·1	252	228·4	106·5
13	11·8	05·5	73	66·2	30·9	133	120·5	56·2	193	174·9	81·6	253	229·3	106·9
14	12·7	05·9	74	67·1	31·3	134	121·4	56·6	194	175·8	82·0	254	230·2	107·3
15	13·6	06·3	75	68·0	31·7	135	122·4	57·1	195	176·7	82·4	255	231·1	107·8
16	14·5	06·8	76	68·9	32·1	136	123·3	57·5	196	177·6	82·8	256	232·0	108·2
17	15·4	07·2	77	69·8	32·5	137	124·2	57·9	197	178·5	83·3	257	232·9	108·6
18	16·3	07·6	78	70·7	33·0	138	125·1	58·3	198	179·4	83·7	258	233·8	109·0
19	17·2	08·0	79	71·6	33·4	139	126·0	58·7	199	180·4	84·1	259	234·7	109·5
20	18·1	08·5	80	72·5	33·8	140	126·9	59·2	200	181·3	84·5	260	235·6	109·9
21	19·0	08·9	81	73·4	34·2	141	127·8	59·6	201	182·2	84·9	261	236·5	110·3
22	19·9	09·3	82	74·3	34·7	142	128·7	60·0	202	183·1	85·4	262	237·5	110·7
23	20·8	09·7	83	75·2	35·1	143	129·6	60·4	203	184·0	85·8	263	238·4	111·1
24	21·8	10·1	84	76·1	35·5	144	130·5	60·9	204	184·9	86·2	264	239·3	111·6
25	22·7	10·6	85	77·0	35·9	145	131·4	61·3	205	185·8	86·6	265	240·2	112·0
26	23·6	11·0	86	77·9	36·3	146	132·3	61·7	206	186·7	87·1	266	241·1	112·4
27	24·5	11·4	87	78·8	36·8	147	133·2	62·1	207	187·6	87·5	267	242·0	112·8
28	25·4	11·8	88	79·8	37·2	148	134·1	62·5	208	188·5	87·9	268	242·9	113·3
29	26·3	12·3	89	80·7	37·6	149	135·0	63·0	209	189·4	88·3	269	243·8	113·7
30	27·2	12·7	90	81·6	38·0	150	135·9	63·4	210	190·3	88·7	270	244·7	114·1
31	28·1	13·1	91	82·5	38·5	151	136·9	63·8	211	191·2	89·2	271	245·6	114·5
32	29·0	13·5	92	83·4	38·9	152	137·8	64·2	212	192·1	89·6	272	246·5	115·0
33	29·9	13·9	93	84·3	39·3	153	138·7	64·7	213	193·0	90·0	273	247·4	115·4
34	30·8	14·4	94	85·2	39·7	154	139·6	65·1	214	193·9	90·4	274	248·3	115·8
35	31·7	14·8	95	86·1	40·1	155	140·5	65·5	215	194·9	90·9	275	249·2	116·2
36	32·6	15·2	96	87·0	40·6	156	141·4	65·9	216	195·8	91·3	276	250·1	116·6
37	33·5	15·6	97	87·9	41·0	157	142·3	66·4	217	196·7	91·7	277	251·0	117·1
38	34·4	16·1	98	88·8	41·4	158	143·2	66·8	218	197·6	92·1	278	252·0	117·5
39	35·3	16·5	99	89·7	41·8	159	144·1	67·2	219	198·5	92·6	279	252·9	117·9
40	36·3	16·9	100	90·6	42·3	160	145·0	67·6	220	199·4	93·0	280	253·8	118·3
41	37·2	17·3	101	91·5	42·7	161	145·9	68·0	221	200·3	93·4	281	254·7	118·8
42	38·1	17·7	102	92·4	43·1	162	146·8	68·5	222	201·2	93·8	282	255·6	119·2
43	39·0	18·2	103	93·3	43·5	163	147·7	68·9	223	202·1	94·2	283	256·5	119·6
44	39·9	18·6	104	94·3	44·0	164	148·6	69·3	224	203·0	94·7	284	257·4	120·0
45	40·8	19·0	105	95·2	44·4	165	149·5	69·7	225	203·9	95·1	285	258·3	120·4
46	41·7	19·4	106	96·1	44·8	166	150·4	70·2	226	204·8	95·5	286	259·2	120·9
47	42·6	19·9	107	97·0	45·2	167	151·4	70·6	227	205·7	95·9	287	260·1	121·3
48	43·5	20·3	108	97·9	45·6	168	152·3	71·0	228	206·6	96·4	288	261·0	121·7
49	44·4	20·7	109	98·8	46·1	169	153·2	71·4	229	207·5	96·8	289	261·9	122·1
50	45·3	21·1	110	99·7	46·5	170	154·1	71·8	230	208·5	97·2	290	262·8	122·6
51	46·2	21·6	111	100·6	46·9	171	155·0	72·3	231	209·4	97·6	291	263·7	123·0
52	47·1	22·0	112	101·5	47·3	172	155·9	72·7	232	210·3	98·0	292	264·6	123·4
53	48·0	22·4	113	102·4	47·8	173	156·8	73·1	233	211·2	98·5	293	265·5	123·8
54	48·9	22·8	114	103·3	48·2	174	157·7	73·5	234	212·1	98·9	294	266·5	124·2
55	49·8	23·2	115	104·2	48·6	175	158·6	74·0	235	213·0	99·3	295	267·4	124·7
56	50·8	23·7	116	105·1	49·0	176	159·5	74·4	236	213·9	99·7	296	268·3	125·1
57	51·7	24·1	117	106·0	49·4	177	160·4	74·8	237	214·8	100·2	297	269·2	125·5
58	52·6	24·5	118	106·9	49·9	178	161·3	75·2	238	215·7	100·6	298	270·1	125·9
59	53·5	24·9	119	107·9	50·3	179	162·2	75·6	239	216·6	101·0	299	271·0	126·4
60	54·4	25·4	120	108·8	50·7	180	163·1	76·1	240	217·5	101·4	300	271·9	126·8
Dist	Dep.	D.Lat	Dist	Dep.	D.Lat	Dist	Dep.	D.Lat	Dist	Dep.	D.Lat	Dist	Dep.	D.Lat
D Lon		Dep.	D Lon		Dep.	D Lon		Dep.	D Lon		Dep.	D Lon		Dep.

65° | 295° / 245° | 65 DEGREES. | 065° / 115° 4h 20m

	335° ↑		**TRAVERSE TABLE**				025° ↑	
	205°		**25 DEGREES.**				155°	1h 40m

25°

Dist	D.Lat	Dep.	Dist	D.Lat	Dep.	Dist	D.Lat	Dep.	Dist	D.Lat	Dep.	Dist	D.Lat	Dep.
301	272·8	127·2	361	327·2	152·6	421	381·6	177·9	481	435·9	203·3	541	490·3	228·6
302	273·7	127·6	362	328·0	153·0	422	382·5	178·3	482	436·8	203·7	542	491·2	229·1
303	274·6	128·1	363	329·0	153·4	423	383·4	178·8	483	437·7	204·1	543	492·1	229·5
304	275·5	128·5	364	329·9	153·8	424	384·3	179·2	484	438·7	204·5	544	493·0	229·9
305	276·4	128·9	365	330 8	154·3	425	385·2	179·6	485	439·6	205·0	545	493·9	230·3
306	277·3	129·3	366	331·7	154·7	426	386·1	180·0	486	440·5	205·4	546	494·8	230·7
307	278·2	129·7	367	332·6	155·1	427	387·0	180·5	487	441·4	205·8	547	495·8	231·2
308	279·1	130·2	368	333·5	155·5	428	387·9	180·9	488	442·3	206·2	548	496·7	231·6
309	280·0	130·6	369	334 4	155·9	429	388·8	181·3	489	443·2	206·7	549	497·6	232·0
310	281·0	131·0	370	335·3	156·4	430	389·7	181·7	490	444·1	207·1	550	498·5	232·4
311	281·9	131·4	371	336·2	156·8	431	390·6	182·1	491	445·0	207·5	551	499·4	232·9
312	282·8	131·9	372	337·1	157·2	432	391·5	182·6	492	445·9	207·9	552	500·3	233·3
313	283·7	132·3	373	338·1	157·6	433	392·4	183·0	493	446·8	208·4	553	501·2	233·7
314	284·6	132·7	374	339·0	158·1	434	393·3	183·4	494	447·7	208·8	554	502·1	234·1
315	285·5	133·1	375	339·9	158·5	435	394·2	183·8	495	448·6	209·2	555	503·0	234·6
316	286·4	133·4	376	340·8	158·9	436	395·2	184·3	496	449·5	209·6	556	503·9	235·0
317	287·3	133·9	377	341·7	159·3	437	396·1	184·7	497	450·4	210·0	557	504·8	235·4
318	288·2	134·4	378	342·5	159·7	438	397·0	185·1	498	451·3	210·5	558	505·7	235·8
319	289·1	134·8	379	343·5	160·2	439	397·9	185·5	499	452·2	210·9	559	506·6	236·2
320	290·0	135·2	380	344·4	160·6	440	398·8	186·0	500	453·2	211·3	560	507·5	236·7
321	290·9	135·7	381	345·3	161·0	441	399·6	186·4	501	454·1	211·7	561	508·4	237·1
322	291·8	136·1	382	346·2	161·4	442	400·6	186·8	502	455·0	212·2	562	509·3	237·5
323	292·7	136·5	383	347·1	161·9	443	401·5	187·2	503	455·9	212·6	563	510·3	237·9
324	293·6	136·9	384	348·0	162·3	444	402·4	187·6	504	456·8	213·0	564	511·2	238·4
325	294·6	137·4	385	348·9	162·7	445	403·3	188·1	505	457·7	213·4	565	512·1	238·8
326	295·5	137·8	386	349·8	163·1	446	404·2	188·5	506	458·6	213·8	566	513·0	239·2
327	296·4	138·2	387	350·7	163·6	447	405·1	188·9	507	459·5	214·3	567	513·9	239·6
328	297·3	138·6	388	351·6	164·0	448	406·0	189·3	508	460·4	214·7	568	514·8	240·0
329	298·2	139·0	389	352·6	164·4	449	406·9	189·8	509	461·3	215·1	569	515·7	240·5
330	299·1	139·5	390	353·5	164·8	450	407·8	190·2	510	462·2	215·5	570	516·6	240·9
331	300·0	139·9	391	354·4	165·2	451	408·7	190·6	511	463·1	216·0	571	517·5	241·3
332	300·9	140·3	392	355·3	165·6	452	409·7	191·0	512	464·0	216·4	572	518·4	241·7
333	301·8	140·7	393	356·2	166·1	453	410·6	191·4	513	464·9	216·8	573	519·3	242·2
334	302·7	141·2	394	357·1	166·5	454	411·5	191·9	514	465·8	217·2	574	520·2	242·6
335	303·6	141·6	395	358·0	166·9	455	412·4	192·3	515	466·7	217·6	575	521·1	243·0
336	304·5	142·0	396	358·9	167·4	456	413·3	192·7	516	467·7	218·1	576	522·0	243·4
337	305·4	142·4	397	359·8	167·8	457	414·2	193·1	517	468·6	218·5	577	522·9	243·9
338	306·3	142·8	398	360·7	168·2	458	415·1	193·6	518	469·5	218·9	578	523·8	244·3
339	307·2	143·3	399	361·6	168·6	459	416·0	194·0	519	470·4	219·3	579	524·8	244·7
340	308·1	143·7	400	362·5	169·0	460	416·9	194·4	520	471·3	219·8	580	525·7	245·1
341	309·1	144·1	401	363·4	169·5	461	417·8	194·8	521	472·2	220·2	581	526·6	245·5
342	310·0	144·5	402	364·3	169·9	462	418·7	195·2	522	473·1	220·6	582	527·5	246·0
343	310·9	145·0	403	365·2	170·3	463	419·6	195·7	523	474·0	221·0	583	528·4	246·4
344	311·8	145·4	404	366·1	170·7	464	420·5	196·1	524	474·9	221·5	584	529·3	246·8
345	312·7	145·8	405	367·1	171·2	465	421·4	196·5	525	475·8	221·9	585	530·2	247·2
346	313·6	146·2	406	368·0	171·6	466	422·3	196·9	526	476·7	222·3	586	531·1	247·7
347	314·5	146·5	407	368·9	172·0	467	423·2	197·4	527	477·6	222·7	587	532·0	248·1
348	315·4	147·1	408	369·8	172·4	468	424·2	197·8	528	478·5	223·1	588	532·9	248·5
349	316·3	147·5	409	370·7	172·9	469	425·1	198·2	529	479·4	223·6	589	533·8	248·9
350	317·2	147·9	410	371·6	173·3	470	426·0	198·6	530	480·3	224·0	590	534·7	249·3
351	318·1	148·3	411	372·5	173·7	471	426·9	199·1	531	481·2	224·4	591	535·6	249·8
352	319·0	148·8	412	373·4	174·1	472	427·8	199·5	532	482·2	224·8	592	536·5	250·2
353	319·9	149·2	413	374·3	174·5	473	428·7	199·9	533	483·1	225·3	593	537·4	250·6
354	320·8	149·6	414	375·2	175·0	474	429·6	200·3	534	484·0	225·7	594	538·3	251·0
355	321·7	150·0	415	376·1	175·4	475	430·5	200·7	535	484·9	226·1	595	539·3	251·5
356	322·6	150·5	416	377·0	175·8	476	431·4	201·2	536	485·8	226·5	596	540·2	251·9
357	323·6	150·9	417	377·9	176·2	477	432·3	201·6	537	486·7	226·9	597	541·1	252·3
358	324·5	151·3	418	378·8	176·7	478	433·2	202·0	538	487·6	227·4	598	542·0	252·7
359	325·4	151·7	419	379·7	177·1	479	434·1	202·4	539	488·5	227·8	599	542·9	253·1
360	326·3	152·1	420	380·6	177·5	480	435·0	202·9	540	489·4	228·2	600	543·8	253·6

Dist	Dep.	D.Lat	Dist	Dep.	D.Lat	Dist	Dep.	D.Lat	Dist	Dep.	D.Lat	Dist	Dep.	D.Lat

	295°		**65 DEGREES.**				065°	
	245°						115°	4h 20m

65°

TRAVERSE TABLE
26 DEGREES.

26° 334° ↑ / 206° ↑026° / 154° 1h 44m

64° 296° ↑ / 244° 64 DEGREES ↑064° / 116° 4h 16m

D Lon Dep.			D Lon Dep.			D Lon Dep.			D Lon Dep.			D Lon Dep.		
Dist	D.Lat	Dep.	Dist	D.Lat	Dep.	Dist	D.Lat	Dep.	Dist	D.Lat	Dep.	Dist	D.Lat	Dep.
1	00.9	00.4	61	54.8	26.7	121	108.8	53.0	181	162.7	79.3	241	216.6	105.6
2	01.8	00.9	62	55.7	27.2	122	109.7	53.5	182	163.6	79.8	242	217.5	106.1
3	02.7	01.3	63	56.6	27.6	123	110.6	53.9	183	164.5	80.2	243	218.4	106.5
4	03.6	01.8	64	57.5	28.1	124	111.5	54.4	184	165.4	80.7	244	219.3	107.0
5	04.5	02.2	65	58.4	28.5	125	112.3	54.8	185	166.3	81.1	245	220.2	107.4
6	05.4	02.6	66	59.3	28.9	126	113.2	55.2	186	167.2	81.5	246	221.1	107.8
7	06.3	03.1	67	60.2	29.4	127	114.1	55.7	187	168.1	82.0	247	222.0	108.3
8	07.2	03.5	68	61.1	29.8	128	115.0	56.1	188	169.0	82.4	248	222.9	108.7
9	08.1	03.9	69	62.0	30.2	129	115.9	56.5	189	169.9	82.9	249	223.8	109.2
10	09.0	04.4	70	62.9	30.7	130	116.8	57.0	190	170.8	83.3	250	224.7	109.6
11	09.9	04.8	71	63.8	31.1	131	117.7	57.4	191	171.7	83.7	251	225.6	110.0
12	10.8	05.3	72	64.7	31.6	132	118.6	57.9	192	172.6	84.2	252	226.5	110.5
13	11.7	05.7	73	65.6	32.0	133	119.5	58.3	193	173.5	84.6	253	227.4	110.9
14	12.6	06.1	74	66.5	32.4	134	120.4	58.7	194	174.4	85.0	254	228.3	111.3
15	13.5	06.6	75	67.4	32.9	135	121.3	59.2	195	175.3	85.5	255	229.2	111.8
16	14.4	07.0	76	68.3	33.3	136	122.2	59.6	196	176.2	85.9	256	230.1	112.2
17	15.3	07.5	77	69.2	33.8	137	123.1	60.1	197	177.1	86.4	257	231.0	112.7
18	16.2	07.9	78	70.1	34.2	138	124.0	60.5	198	178.0	86.8	258	231.9	113.1
19	17.1	08.3	79	71.0	34.6	139	124.9	60.9	199	178.9	87.2	259	232.8	113.5
20	18.0	08.8	80	71.9	35.1	140	125.8	61.4	200	179.8	87.7	260	233.7	114.0
21	18.9	09.2	81	72.8	35.5	141	126.7	61.8	201	180.7	88.1	261	234.6	114.4
22	19.8	09.6	82	73.7	35.9	142	127.6	62.2	202	181.6	88.6	262	235.5	114.9
23	20.7	10.1	83	74.6	36.4	143	128.5	62.7	203	182.5	89.0	263	236.4	115.3
24	21.6	10.5	84	75.5	36.8	144	129.4	63.1	204	183.4	89.4	264	237.3	115.7
25	22.5	11.0	85	76.4	37.3	145	130.3	63.6	205	184.3	89.9	265	238.2	116.2
26	23.4	11.4	86	77.3	37.7	146	131.2	64.0	206	185.2	90.3	266	239.1	116.6
27	24.3	11.8	87	78.2	38.1	147	132.1	64.4	207	186.1	90.7	267	240.0	117.0
28	25.2	12.3	88	79.1	38.6	148	133.0	64.9	208	186.9	91.2	268	240.9	117.5
29	26.1	12.7	89	80.0	39.0	149	133.9	65.3	209	187.8	91.6	269	241.8	117.9
30	27.0	13.2	90	80.9	39.5	150	134.8	65.8	210	188.7	92.1	270	242.7	118.4
31	27.9	13.6	91	81.8	39.9	151	135.7	66.2	211	189.6	92.5	271	243.6	118.8
32	28.8	14.0	92	82.7	40.3	152	136.6	66.6	212	190.5	92.9	272	244.5	119.2
33	29.7	14.5	93	83.6	40.8	153	137.5	67.1	213	191.4	93.4	273	245.4	119.7
34	30.6	14.9	94	84.5	41.2	154	138.4	67.5	214	192.3	93.8	274	246.3	120.1
35	31.5	15.3	95	85.4	41.6	155	139.3	67.9	215	193.2	94.2	275	247.2	120.6
36	32.4	15.8	96	86.3	42.1	156	140.2	68.4	216	194.1	94.7	276	248.1	121.0
37	33.3	16.2	97	87.2	42.5	157	141.1	68.8	217	195.0	95.1	277	249.0	121.4
38	34.2	16.7	98	88.1	43.0	158	142.0	69.3	218	195.9	95.6	278	249.9	121.9
39	35.1	17.1	99	89.0	43.4	159	142.9	69.7	219	196.8	96.0	279	250.8	122.3
40	36.0	17.5	100	89.9	43.8	160	143.8	70.1	220	197.7	96.4	280	251.7	122.7
41	36.9	18.0	101	90.8	44.3	161	144.7	70.6	221	198.6	96.9	281	252.6	123.2
42	37.7	18.4	102	91.7	44.7	162	145.6	71.0	222	199.5	97.3	282	253.5	123.6
43	38.6	18.8	103	92.6	45.2	163	146.5	71.5	223	200.4	97.8	283	254.4	124.1
44	39.5	19.3	104	93.5	45.6	164	147.4	71.9	224	201.3	98.2	284	255.3	124.5
45	40.4	19.7	105	94.4	46.0	165	148.3	72.3	225	202.2	98.6	285	256.2	124.9
46	41.3	20.2	106	95.3	46.5	166	149.2	72.8	226	203.1	99.1	286	257.1	125.4
47	42.2	20.6	107	96.2	46.9	167	150.1	73.2	227	204.0	99.5	287	258.0	125.8
48	43.1	21.0	108	97.1	47.3	168	151.0	73.6	228	204.9	99.9	288	258.9	126.3
49	44.0	21.5	109	98.0	47.8	169	151.9	74.1	229	205.8	100.4	289	259.8	126.7
50	44.9	21.9	110	98.9	48.2	170	152.8	74.5	230	206.7	100.8	290	260.7	127.1
51	45.8	22.4	111	99.8	48.7	171	153.7	75.0	231	207.6	101.3	291	261.5	127.6
52	46.7	22.8	112	100.7	49.1	172	154.6	75.4	232	208.5	101.7	292	262.4	128.0
53	47.6	23.2	113	101.6	49.5	173	155.5	75.8	233	209.4	102.1	293	263.3	128.4
54	48.5	23.7	114	102.5	50.0	174	156.4	76.3	234	210.3	102.6	294	264.2	128.9
55	49.4	24.1	115	103.4	50.4	175	157.3	76.7	235	211.2	103.0	295	265.1	129.3
56	50.3	24.5	116	104.3	50.9	176	158.2	77.2	236	212.1	103.5	296	266.0	129.8
57	51.2	25.0	117	105.2	51.3	177	159.1	77.6	237	213.0	103.9	297	266.9	130.2
58	52.1	25.4	118	106.1	51.7	178	160.0	78.0	238	213.9	104.3	298	267.8	130.6
59	53.0	25.9	119	107.0	52.2	179	160.9	78.5	239	214.8	104.8	299	268.7	131.1
60	53.9	26.3	120	107.9	52.6	180	161.8	78.9	240	215.7	105.2	300	269.6	131.5
Dist	Dep.	D.Lat	Dist	Dep.	D.Lat	Dist	Dep.	D.Lat	Dist	Dep.	D.Lat	Dist	Dep.	D.Lat
D Lon		Dep.	D Lon		Dep.	D Lon		Dep.	D Lon		Dep.	D Lon		Dep.

TRAVERSE TABLE
26 DEGREES.

334° ↑ / 206° 026° ↑ / 154° 1h 44m **26°**

DLon	Dep.	Dist	D.Lat	Dep.	Dist	D.Lat	Dep.	Dist	D.Lat	Dep.	Dist	D.Lat	Dep.	Dist	D.Lat	Dep.
Dist	D.Lat	Dep.														
301	270·5	131·9	361	324·5	158·3	421	378·4	184·6	481	432·3	210·9	541	486·2	237·2
302	271·4	132·4	362	325·4	158·7	422	379·3	185·0	482	433·2	211·3	542	487·1	237·6
303	272·3	132·8	363	326·3	159·1	423	380·2	185·4	483	434·1	211·7	543	488·0	238·0
304	273·2	133·3	364	327·2	159·6	424	381·1	185·9	484	435·0	212·2	544	488·9	238·5
305	274·1	133·7	365	328·1	160·0	425	382·0	186·3	485	435·9	212·6	545	489·8	238·9
306	275·0	134·1	366	329·0	160·4	426	382·9	186·7	486	436·8	213·0	546	490·7	239·4
307	275·9	134·6	367	329·9	160·9	427	383·8	187·2	487	437·7	213·5	547	491·6	239·8
308	276·8	135·0	368	330·8	161·3	428	384·7	187·6	488	438·6	213·9	548	492·5	240·2
309	277·7	135·5	369	331·7	161·8	429	385·6	188·1	489	439·5	214·4	549	493·4	240·7
310	278·6	135·9	370	332·6	162·2	430	386·5	188·5	490	440·4	214·8	550	494·3	241·1
311	279·5	136·3	371	333·5	162·6	431	387·4	188·9	491	441·3	215·2	551	495·2	241·5
312	280·4	136·8	372	334·4	163·1	432	388·3	189·4	492	442·2	215·7	552	496·1	242·0
313	281·3	137·2	373	335·3	163·5	433	389·2	189·8	493	443·1	216·1	553	497·0	242·4
314	282·2	137·6	374	336·1	164·0	434	390·1	190·3	494	444·0	216·6	554	497·9	242·9
315	283·1	138·1	375	337·0	164·4	435	391·0	190·7	495	444·9	217·0	555	498·8	243·3
316	284·0	138·5	376	337·9	164·8	436	391·9	191·1	496	445·8	217·4	556	499·7	243·7
317	284·9	139·0	377	338·8	165·3	437	392·8	191·6	497	446·7	217·9	557	500·6	244·2
318	285·8	139·4	378	339·7	165·7	438	393·7	192·0	498	447·6	218·3	558	501·5	244·6
319	286·7	139·8	379	340·6	166·1	439	394·6	192·4	499	448·5	218·7	559	502·4	245·0
320	287·6	140·3	380	341·5	166·6	440	395·5	192·9	500	449·4	219·2	560	503·3	245·5
321	288·5	140·7	381	342·4	167·0	441	396·4	193·3	501	450·3	219·6	561	504·2	245·9
322	289·4	141·2	382	343·3	167·5	442	397·3	193·8	502	451·2	220·1	562	505·1	246·4
323	290·3	141·6	383	344·2	167·9	443	398·2	194·2	503	452·1	220·5	563	506·0	246·8
324	291·2	142·0	384	345·1	168·3	444	399·1	194·6	504	453·0	220·9	564	506·9	247·2
325	292·1	142·5	385	346·0	168·8	445	400·0	195·1	505	453·9	221·4	565	507·8	247·7
326	293·0	142·9	386	346·9	169·2	446	400·9	195·5	506	454·8	221·8	566	508·7	248·1
327	293·9	143·3	387	347·8	169·6	447	401·8	196·0	507	455·7	222·3	567	509·6	248·6
328	294·8	143·8	388	348·7	170·1	448	402·7	196·4	508	456·6	222·7	568	510·5	249·0
329	295·7	144·2	389	349·6	170·5	449	403·6	196·8	509	457·5	223·1	569	511·4	249·4
330	296·6	144·7	390	350·5	171·0	450	404·5	197·3	510	458·4	223·6	570	512·3	249·9
331	297·5	145·1	391	351·4	171·4	451	405·4	197·7	511	459·3	224·0	571	513·2	250·3
332	298·4	145·5	392	352·3	171·8	452	406·3	198·1	512	460·2	224·4	572	514·1	250·7
333	299·3	146·0	393	353·2	172·3	453	407·2	198·6	513	461·1	224·9	573	515·0	251·2
334	300·2	146·4	394	354·1	172·7	454	408·1	199·0	514	462·0	225·3	574	515·9	251·6
335	301·1	146·9	395	355·0	173·2	455	409·0	199·5	515	462·9	225·8	575	516·8	252·1
336	302·0	147·3	396	355·9	173·6	456	409·9	199·9	516	463·8	226·2	576	517·7	252·5
337	302·9	147·7	397	356·8	174·0	457	410·7	200·3	517	464·7	226·6	577	518·6	252·9
338	303·8	148·2	398	357·7	174·5	458	411·6	200·8	518	465·6	227·1	578	519·5	253·4
339	304·7	148·6	399	358·6	174·9	459	412·5	201·2	519	466·5	227·5	579	520·4	253·8
340	305·6	149·0	400	359·5	175·3	460	413·4	201·7	520	467·4	228·0	580	521·3	254·3
341	306·5	149·5	401	360·4	175·8	461	414·3	202·1	521	468·3	228·4	581	522·2	254·7
342	307·4	149·9	402	361·3	176·2	462	415·2	202·5	522	469·2	228·8	582	523·1	255·1
343	308·3	150·4	403	362·2	176·7	463	416·1	203·0	523	470·1	229·3	583	524·0	255·6
344	309·2	150·8	404	363·1	177·1	464	417·0	203·4	524	471·0	229·7	584	524·9	256·0
345	310·1	151·2	405	364·0	177·5	465	417·9	203·8	525	471·9	230·1	585	525·8	256·4
346	311·0	151·7	406	364·9	178·0	466	418·8	204·3	526	472·8	230·6	586	526·7	256·9
347	311·9	152·1	407	365·8	178·4	467	419·7	204·7	527	473·7	231·0	587	527·6	257·3
348	312·8	152·6	408	366·7	178·9	468	420·6	205·2	528	474·6	231·5	588	528·5	257·8
349	313·7	153·0	409	367·6	179·3	469	421·5	205·6	529	475·5	231·9	589	529·4	258·2
350	314·6	153·4	410	368·5	179·7	470	422·4	206·0	530	476·4	232·3	590	530·3	258·6
351	315·5	153·9	411	369·4	180·2	471	423·3	206·5	531	477·3	232·8	591	531·2	259·1
352	316·4	154·3	412	370·3	180·6	472	424·2	206·9	532	478·2	233·2	592	532·1	259·5
353	317·3	154·7	413	371·2	181·0	473	425·1	207·3	533	479·1	233·7	593	533·0	260·0
354	318·2	155·2	414	372·1	181·5	474	426·0	207·8	534	480·0	234·1	594	533·9	260·4
355	319·1	155·6	415	373·0	181·9	475	426·9	208·2	535	480·9	234·5	595	534·8	260·8
356	320·0	156·1	416	373·9	182·4	476	427·8	208·7	536	481·8	235·0	596	535·7	261·3
357	320·9	156·5	417	374·8	182·8	477	428·7	209·1	537	482·7	235·4	597	536·6	261·7
358	321·8	156·9	418	375·7	183·2	478	429·6	209·5	538	483·6	235·8	598	537·5	262·1
359	322·7	157·4	419	376·6	183·7	479	430·5	210·0	539	484·4	236·3	599	538·4	262·6
360	323·6	157·8	420	377·5	184·1	480	431·4	210·4	540	485·3	236·7	600	539·3	263·0

Dist	Dep.	D.Lat	Dist	Dep.	D.Lat	Dist	Dep.	D.Lat	Dist	Dep.	D.Lat	Dist	Dep.	D.Lat
DLon		Dep.	DLon		Dep.	DLon		Dep.	DLon		Dep.	DLon		Dep.

296° ↑ / 244° **64 DEGREES.** 064° ↑ / 116° 4h 16m **64°**

27°

333° ↑
207°

TRAVERSE TABLE
27 Degrees.

↑ 027°
153°   1h 48m

Dist	D.Lat	Dep.	Dist	D.Lat	Dep.	Dist	D.Lat	Dep.	Dist	D.Lat	Dep.	Dist	D.Lat	Dep.
1	00·9	00·5	61	54·4	27·7	121	107·8	54·9	181	161·3	82·2	241	214·7	109·4
2	01·8	00·9	62	55·2	28·1	122	108·7	55·4	182	162·2	82·6	242	215·6	109·9
3	02·7	01·4	63	56·1	28·6	123	109·6	55·8	183	163·1	83·1	243	216·5	110·3
4	03·6	01·8	64	57·0	29·1	124	110·5	66·3	184	163·9	83·5	244	217·4	110·8
5	04·5	02·3	65	57·9	29·5	125	111·4	56·7	185	164·8	84·0	245	218·3	111·2
6	05·3	02·7	66	58·8	30·0	126	112·3	57·2	186	165·7	84·4	246	219·2	111·7
7	06·2	03·2	67	59·7	30·4	127	113·2	57·7	187	166·6	84·9	247	220·1	112·1
8	07·1	03·6	68	60·6	30·9	128	114·0	58·1	188	167·5	85·4	248	221·0	112·6
9	08·0	04·1	69	61·5	31·3	129	114·9	58·6	189	168·4	85·8	249	221·9	113·0
10	08·9	04·5	70	62·4	31·8	130	115·8	59·0	190	169·3	86·3	250	222·8	113·5
11	09·8	05·0	71	63·3	32·2	131	116·7	59·5	191	170·2	86·7	251	223·6	114·0
12	10·7	05·4	72	64·2	32·7	132	117·6	59·9	192	171·1	87·2	252	224·5	114·4
13	11·6	05·9	73	65·0	33·1	133	118·5	60·4	193	172·0	87·6	253	225·4	114·9
14	12·5	06·4	74	65·9	33·6	134	119·4	60·8	194	172·9	88·1	254	226·3	115·3
15	13·4	06·8	75	66·8	34·0	135	120·3	61·3	195	173·7	88·5	255	227·2	115·8
16	14·3	07·3	76	67·7	34·5	136	121·2	61·7	196	174·6	89·0	256	228·1	116·2
17	15·1	07·7	77	68·6	35·0	137	122·1	62·2	197	175·5	89·4	257	229·0	116·7
18	16·0	08·2	78	69·5	35·4	138	123·0	62·7	198	176·4	89·9	258	229·9	117·1
19	16·9	08·6	79	70·4	35·9	139	123·8	63·1	199	177·3	90·3	259	230·8	117·6
20	17·8	09·1	80	71·3	36·3	140	124·7	63·6	200	178·2	90·8	260	231·7	118·0
21	18·7	09·5	81	72·2	36·8	141	125·6	64·0	201	179·1	91·3	261	232·6	118·5
22	19·6	10·0	82	73·1	37·2	142	126·5	64·5	202	180·0	91·7	262	233·4	118·9
23	20·5	10·4	83	74·0	37·7	143	127·4	64·9	203	180·9	92·2	263	234·3	119·4
24	21·4	10·9	84	74·8	38·1	144	128·3	65·4	204	181·8	92·6	264	235·2	119·9
25	22·3	11·3	85	75·7	38·6	145	129·2	65·8	205	182·7	93·1	265	236·1	120·3
26	23·2	11·8	86	76·6	39·0	146	130·1	66·3	206	183·5	93·5	266	237·0	120·8
27	24·1	12·3	87	77·5	39·5	147	131·0	66·7	207	184·4	94·0	267	237·9	121·2
28	24·9	12·7	88	78·4	40·0	148	131·9	67·2	208	185·3	94·4	268	238·8	121·7
29	25·8	13·2	89	79·3	40·4	149	132·8	67·6	209	186·2	94·9	269	239·7	122·1
30	26·7	13·6	90	80·2	40·9	150	133·7	68·1	210	187·1	95·3	270	240·6	122·6
31	27·6	14·1	91	81·1	41·3	151	134·5	68·6	211	188·0	95·8	271	241·5	123·0
32	28·5	14·5	92	82·0	41·8	152	135·4	69·0	212	188·9	96·2	272	242·4	123·5
33	29·4	15·0	93	82·9	42·2	153	136·3	69·5	213	189·8	96·7	273	243·2	123·9
34	30·3	15·4	94	83·8	42·7	154	137·2	69·9	214	190·7	97·2	274	244·1	124·4
35	31·2	15·9	95	84·6	43·1	155	138·1	70·4	215	191·6	97·6	275	245·0	124·8
36	32·1	16·3	96	85·5	43·6	156	139·0	70·8	216	192·5	98·1	276	245·9	125·3
37	33·0	16·8	97	86·4	44·0	157	139·9	71·3	217	193·3	98·5	277	246·8	125·8
38	33·9	17·3	98	87·3	44·5	158	140·8	71·7	218	194·2	99·0	278	247·7	126·2
39	34·7	17·7	99	88·2	44·9	159	141·7	72·2	219	195·1	99·4	279	248·6	126·7
40	35·6	18·2	100	89·1	45·4	160	142·6	72·6	220	196·0	99·9	280	249·5	127·1
41	36·5	18·6	101	90·0	45·9	161	143·5	73·1	221	196·9	100·3	281	250·4	127·6
42	37·4	19·1	102	90·9	46·3	162	144·3	73·5	222	197·8	100·8	282	251·3	128·0
43	38·3	19·5	103	91·8	46·8	163	145·2	74·0	223	198·7	101·2	283	252·2	128·5
44	39·2	20·0	104	92·7	47·2	164	146·1	74·5	224	199·6	101·7	284	253·0	128·9
45	40·1	20·4	105	93·6	47·7	165	147·0	74·9	225	200·5	102·1	285	253·9	129·4
46	41·0	20·9	106	94·4	48·1	166	147·9	75·4	226	201·4	102·6	286	254·8	129·8
47	41·9	21·3	107	95·3	48·6	167	148·8	75·8	227	202·3	103·1	287	255·7	130·3
48	42·8	21·8	108	96·2	49·0	168	149·7	76·3	228	203·1	103·5	288	256·6	130·7
49	43·7	22·2	109	97·1	49·5	169	150·6	76·7	229	204·0	104·0	289	257·5	131·2
50	44·6	22·7	110	98·0	49·9	170	151·5	77·2	230	204·9	104·4	290	258·4	131·7
51	45·4	23·2	111	98·9	50·4	171	152·4	77·6	231	205·8	104·9	291	259·3	132·1
52	46·3	23·6	112	99·8	50·8	172	153·3	78·1	232	206·7	105·3	292	260·2	132·6
53	47·2	24·1	113	100·7	51·3	173	154·1	78·5	233	207·6	105·8	293	261·1	133·0
54	48·1	24·5	114	101·6	51·8	174	155·0	79·0	234	208·5	106·2	294	262·0	133·5
55	49·0	25·0	115	102·5	52·2	175	155·9	79·4	235	209·4	106·7	295	262·8	133·9
56	49·9	25·4	116	103·4	52·7	176	156·8	79·9	236	210·3	107·1	296	263·7	134·4
57	50·8	25·9	117	104·2	53·1	177	157·7	80·4	237	211·2	107·6	297	264·6	134·8
58	51·7	26·3	118	105·1	53·6	178	158·6	80·8	238	212·1	108·0	298	265·5	135·3
59	52·6	26·8	119	106·0	54·0	179	159·5	81·3	239	213·0	108·5	299	266·4	135·7
60	53·5	27·2	120	106·9	54·5	180	160·4	81·7	240	213·8	109·0	300	267·3	136·2

| Dist | Dep. | D.Lat | Dist | Dep. | D.Lat | Dist | Dep. | D.Lat | Dist | Dep. | D.Lat | Dist | Dep. | D.Lat |

63°

TRAVERSE TABLE
27 DEGREES.

1h 48m

D Lon	Dep.		D Lon	Dep.		D Lon	Dep.		D Lon	Dep.		D Lon	Dep.	
Dist	D.Lat	Dep.	Dist	D.Lat	Dep.	Dist	D.Lat	Dep.	Dist	D.Lat	Dep.	Dist	D.Lat	Dep.
301	268·2	136·7	361	321·7	163·9	421	375·1	191·1	481	428·6	218·4	541	482·0	245·6
302	269·1	137·1	362	322·5	164·3	422	376·0	191·6	482	429·5	218·8	542	482·9	246·1
303	270·0	137·6	363	323·4	164·8	423	376·9	192·0	483	430·4	219·3	543	483·8	246·5
304	270·9	138·0	364	324·3	165·3	424	377·8	192·5	484	431·2	219·7	544	484·7	247·0
305	271·8	138·5	365	325·2	165·7	425	378·7	192·9	485	432·1	220·2	545	485·6	247·4
306	272·6	138·9	366	326·1	166·2	426	379·6	193·4	486	433·0	220·6	546	486·5	247·9
307	273·5	139·4	367	327·0	166·6	427	380·5	193·9	487	433·9	221·1	547	487·4	248·3
308	274·4	139·8	368	327·9	167·1	428	381·4	194·3	488	434·8	221·5	548	488·3	248·8
309	275·3	140·3	369	328·8	167·5	429	382·2	194·8	489	435·7	222·0	549	489·2	249·2
310	276·2	140·7	370	329·7	168·0	430	383·1	195·2	490	436·6	222·5	550	490·1	249·7
311	277·1	141·2	371	330·6	168·4	431	384·0	195·7	491	437·5	222·9	551	490·9	250·1
312	278·0	141·6	372	331·5	168·9	432	384·9	196·1	492	438·4	223·4	552	491·8	250·6
313	278·9	142·1	373	332·3	169·3	433	385·8	196·6	493	439·3	223·8	553	492·7	251·1
314	279·8	142·5	374	333·2	169·8	434	386·7	197·0	494	440·2	224·3	554	493·6	251·5
315	280·7	143·0	375	334·1	170·2	435	387·6	197·5	495	441·0	224·7	555	494·5	252·0
316	281·6	143·5	376	335·0	170·7	436	388·5	197·9	496	441·9	225·2	556	495·4	252·4
317	282·4	143·9	377	335·9	171·1	437	389·4	198·4	497	442·8	225·6	557	496·3	252·9
318	283·3	144·4	378	336·8	171·6	438	390·3	198·8	498	443·7	226·1	558	497·2	253·3
319	284·2	144·8	379	337·7	172·1	439	391·2	199·3	499	444·6	226·5	559	498·1	253·8
320	285·1	145·3	380	338·6	172·5	440	392·0	199·8	500	445·5	227·0	560	499·0	254·2
321	286·0	145·7	381	339·5	173·0	441	392·9	200·2	501	446·4	227·4	561	499·9	254·7
322	286·9	146·2	382	340·4	173·4	442	393·8	200·7	502	447·3	227·9	562	500·7	255·1
323	287·8	146·6	383	341·3	173·9	443	394·7	201·1	503	448·2	228·4	563	501·6	255·6
324	288·7	147·1	384	342·1	174·3	444	395·6	201·6	504	449·1	228·8	564	502·5	256·1
325	289·6	147·5	385	343·0	174·8	445	396·5	202·0	505	450·0	229·3	565	503·4	256·5
326	290·5	148·0	386	343·9	175·2	446	397·4	202·5	506	450·8	229·7	566	504·3	257·0
327	291·4	148·5	387	344·8	175·7	447	398·3	202·9	507	451·7	230·2	567	505·2	257·4
328	292·3	148·9	388	345·7	176·1	448	399·2	203·4	508	452·6	230·6	568	506·1	257·9
329	293·1	149·4	389	346·6	176·6	449	400·1	203·8	509	453·5	231·1	569	507·0	258·3
330	294·0	149·8	390	347·5	177·1	450	401·0	204·3	510	454·4	231·5	570	507·9	258·8
331	294·9	150·3	391	348·4	177·5	451	401·8	204·7	511	455·3	232·0	571	508·8	259·2
332	295·8	150·7	392	349·3	178·0	452	402·7	205·2	512	456·2	232·4	572	509·6	259·7
333	296·7	151·2	393	350·2	178·4	453	403·6	205·7	513	457·1	232·9	573	510·5	260·1
334	297·6	151·6	394	351·1	178·9	454	404·5	206·1	514	458·0	233·4	574	511·4	260·6
335	298·5	152·1	395	351·9	179·3	455	405·4	206·6	515	458·9	233·8	575	512·3	261·0
336	299·4	152·5	396	352·8	179·8	456	406·3	207·0	516	459·8	234·3	576	513·2	261·5
337	300·3	153·0	397	353·7	180·2	457	407·2	207·5	517	460·7	234·7	577	514·1	262·0
338	301·2	153·4	398	354·6	180·7	458	408·1	207·9	518	461·5	235·2	578	515·0	262·4
339	302·1	153·9	399	355·5	181·1	459	409·0	208·4	519	462·4	235·6	579	515·9	262·9
340	302·9	154·4	400	356·4	181·6	460	409·9	208·8	520	463·3	236·1	580	516·8	263·4
341	303·8	154·8	401	357·3	182·1	461	410·8	209·3	521	464·2	236·5	581	517·7	263·8
342	304·7	155·3	402	358·2	182·5	462	411·6	209·7	522	465·1	237·0	582	518·6	264·2
343	305·6	155·7	403	359·1	183·0	463	412·5	210·2	523	466·0	237·4	583	519·5	264·7
344	306·5	156·2	404	360·0	183·4	464	413·4	210·7	524	466·9	237·9	584	520·3	265·1
345	307·4	156·6	405	360·9	183·9	465	414·3	211·1	525	467·8	238·3	585	521·2	265·6
346	308·3	157·1	406	361·8	184·3	466	415·2	211·6	526	468·7	238·8	586	522·1	266·0
347	309·2	157·5	407	362·6	184·8	467	416·1	212·0	527	469·6	239·3	587	523·0	266·5
348	310·1	158·0	408	363·5	185·2	468	417·0	212·5	528	470·5	239·7	588	523·9	266·9
349	311·0	158·4	409	364·4	185·7	469	417·9	212·9	529	471·3	240·2	589	524·8	267·4
350	311·9	158·9	410	365·3	186·1	470	418·8	213·4	530	472·2	240·6	590	525·7	267·9
351	312·7	159·4	411	366·2	186·6	471	419·7	213·8	531	473·1	241·1	591	526·6	268·3
352	313·6	159·8	412	367·1	187·0	472	420·6	214·3	532	474·0	241·5	592	527·5	268·8
353	314·5	160·3	413	368·0	187·5	473	421·4	214·7	533	474·9	242·0	593	528·4	269·2
354	315·4	160·7	414	368·9	188·0	474	422·3	215·2	534	475·8	242·4	594	529·3	269·7
355	316·3	161·2	415	369·8	188·4	475	423·2	215·6	535	476·7	242·9	595	530·1	270·1
356	317·2	161·6	416	370·7	188·9	476	424·1	216·1	536	477·6	243·3	596	531·0	270·6
357	318·1	162·1	417	371·5	189·3	477	425·0	216·6	537	478·5	243·8	597	531·9	271·0
358	319·0	162·5	418	372·4	189·8	478	425·9	217·0	538	479·4	244·2	598	532·8	271·5
359	319·9	163·0	419	373·3	190·2	479	426·8	217·5	539	480·3	244·7	599	533·7	271·9
360	320·8	163·4	420	374·2	190·7	480	427·7	217·9	540	481·1	245·2	600	534·6	272·4
Dist	Dep.	D.Lat	Dist	Dep.	D.Lat	Dist	Dep.	D.Lat	Dist	Dep.	D.Lat	Dist	Dep.	D.Lat
D Lon		Dep.	D Lon		Dep.	D Lon		Dep.	D Lon		Dep.	D Lon		Dep.

28°

Dist	D.Lat	Dep.	Dist	D.Lat	Dep.	Dist	D.Lat	Dep.	Dist	D.Lat	Dep.	Dist	D.Lat	Dep.
1	00·9	00·5	61	53·9	28·6	121	106·8	56·8	181	159·8	85·0	241	212·8	113·1
2	01·8	00·9	62	54·7	29·1	122	107·7	57·3	182	160·7	85·4	242	213·7	113·6
3	02·6	01·4	63	55·6	29·6	123	108·6	57·7	183	161·6	85·9	243	214·6	114·1
4	03·5	01·9	64	56·5	30·0	124	109·5	58·2	184	162·5	86·4	244	215·4	114·6
5	04·4	02·3	65	57·4	30·5	125	110·4	58·7	185	163·3	86·9	245	216·3	115·0
6	05·3	02·8	66	58·3	31·0	126	111·3	59·2	186	164·2	87·3	246	217·2	115·5
7	06·2	03·3	67	59·2	31·5	127	112·1	59·6	187	165·1	87·8	247	218·1	116·0
8	07·1	03·8	68	60·0	31·9	128	113·0	60·1	188	166·0	88·3	248	219·0	116·4
9	08·0	04·2	69	60·9	32·4	129	113·9	60·6	189	166·9	88·7	249	219·9	116·9
10	08·8	04·7	70	61·8	32·9	130	114·8	61·0	190	167·8	89·2	250	220·7	117·4
11	09·7	05·2	71	62·7	33·3	131	115·7	61·5	191	168·6	89·7	251	221·6	117·9
12	10·6	05·6	72	63·6	33·8	132	116·5	62·0	192	169·5	90·1	252	222·5	118·3
13	11·5	06·1	73	64·5	34·3	133	117·4	62·4	193	170·4	90·6	253	223·4	118·8
14	12·4	06·6	74	65·3	34·7	134	118·3	62·9	194	171·3	91·1	254	224·3	119·2
15	13·2	07·0	75	66·2	35·2	135	119·2	63·4	195	172·2	91·5	255	225·2	119·7
16	14·1	07·5	76	67·1	35·7	136	120·1	63·8	196	173·1	92·0	256	226·0	120·2
17	15·0	08·0	77	68·0	36·1	137	121·0	64·3	197	173·9	92·5	257	226·9	120·7
18	15·9	08·5	78	68·9	36·6	138	121·8	64·8	198	174·8	93·0	258	227·8	121·1
19	16·8	08·9	79	69·8	37·1	139	122·7	65·3	199	175·7	93·4	259	228·7	121·6
20	17·7	09·4	80	70·6	37·6	140	123·6	65·7	200	176·6	93·9	260	229·6	122·1
21	18·5	09·9	81	71·5	38·0	141	124·5	66·2	201	177·5	94·4	261	230·4	122·5
22	19·4	10·3	82	72·4	38·5	142	125·4	66·7	202	178·4	94·8	262	231·3	123·0
23	20·3	10·8	83	73·3	39·0	143	126·3	67·1	203	179·2	95·3	263	232·2	123·5
24	21·2	11·3	84	74·2	39·4	144	127·1	67·6	204	180·1	95·8	264	233·1	123·9
25	22·1	11·7	85	75·1	39·9	145	128·0	68·1	205	181·0	96·2	265	234·0	124·4
26	23·0	12·2	86	75·9	40·4	146	128·9	68·5	206	181·9	96·7	266	234·9	124·9
27	23·8	12·7	87	76·8	40·8	147	129·8	69·0	207	182·8	97·2	267	235·7	125·3
28	24·7	13·1	88	77·7	41·3	148	130·7	69·5	208	183·7	97·7	268	236·6	125·8
29	25·6	13·6	89	78·6	41·8	149	131·6	70·0	209	184·5	98·1	269	237·5	126·3
30	26·5	14·1	90	79·5	42·3	150	132·4	70·4	210	185·4	98·6	270	238·4	126·8
31	27·4	14·6	91	80·3	42·7	151	133·3	70·9	211	186·3	99·1	271	239·3	127·2
32	28·3	15·0	92	81·2	43·2	152	134·2	71·4	212	187·2	99·5	272	240·2	127·7
33	29·1	15·5	93	82·1	43·7	153	135·1	71·8	213	188·1	100·0	273	241·0	128·2
34	30·0	16·0	94	83·0	44·1	154	136·0	72·3	214	189·0	100·5	274	241·9	128·6
35	30·9	16·4	95	83·9	44·6	155	136·9	72·8	215	189·8	100·9	275	242·8	129·1
36	31·8	16·9	96	84·8	45·1	156	137·7	73·2	216	190·7	101·4	276	243·7	129·6
37	32·7	17·4	97	85·6	45·5	157	138·6	73·7	217	191·6	101·9	277	244·6	130·0
38	33·6	17·8	98	86·5	46·0	158	139·5	74·2	218	192·5	102·3	278	245·5	130·5
39	34·4	18·3	99	87·4	46·5	159	140·4	74·6	219	193·4	102·8	279	246·3	131·0
40	35·3	18·8	100	88·3	46·9	160	141·3	75·1	220	194·2	103·3	280	247·2	131·5
41	36·2	19·2	101	89·2	47·4	161	142·2	75·6	221	195·1	103·8	281	248·1	131·9
42	37·1	19·7	102	90·1	47·9	162	143·0	76·1	222	196·0	104·2	282	249·0	132·4
43	38·0	20·2	103	90·9	48·4	163	143·9	76·5	223	196·9	104·7	283	249·9	132·9
44	38·8	20·7	104	91·8	48·8	164	144·8	77·0	224	197·8	105·2	284	250·8	133·3
45	39·7	21·1	105	92·7	49·3	165	145·7	77·5	225	198·7	105·6	285	251·6	133·8
46	40·6	21·6	106	93·6	49·8	166	146·6	77·9	226	199·5	106·1	286	252·5	134·3
47	41·5	22·1	107	94·5	50·2	167	147·5	78·4	227	200·4	106·6	287	253·4	134·7
48	42·4	22·5	108	95·4	50·7	168	148·3	78·9	228	201·3	107·0	288	254·3	135·2
49	43·3	23·0	109	96·2	51·2	169	149·2	79·3	229	202·2	107·5	289	255·2	135·7
50	44·1	23·5	110	97·1	51·6	170	150·1	79·8	230	203·1	108·0	290	256·1	136·1
51	45·0	23·9	111	98·0	52·1	171	151·0	80·3	231	204·0	108·4	291	256·9	136·6
52	45·9	24·4	112	98·9	52·6	172	151·9	80·7	232	204·8	108·9	292	257·8	137·1
53	46·8	24·9	113	99·8	53·1	173	152·7	81·2	233	205·7	109·4	293	258·7	137·6
54	47·7	25·4	114	100·7	53·5	174	153·6	81·7	234	206·6	109·9	294	259·6	138·0
55	48·6	25·8	115	101·5	54·0	175	154·5	82·2	235	207·5	110·3	295	260·5	138·5
56	49·4	26·3	116	102·4	54·5	176	155·4	82·6	236	208·4	110·8	296	261·3	139·0
57	50·3	26·8	117	103·3	54·9	177	156·3	83·1	237	209·3	111·3	297	262·2	139·4
58	51·2	27·2	118	104·2	55·4	178	157·2	83·6	238	210·1	111·7	298	263·1	139·9
59	52·1	27·7	119	105·1	55·9	179	158·0	84·0	239	211·0	112·2	299	264·0	140·4
60	53·0	28·2	120	106·0	56·3	180	158·9	84·5	240	211·9	112·7	300	264·9	140·8

Dist	Dep.	D.Lat	Dist	Dep.	D.Lat	Dist	Dep.	D.Lat	Dist	Dep.	D.Lat	Dist	Dep.	D.Lat
D Lon		Dep.	D Lon		Dep.	D Lon		Dep.	D Lon		Dep.	D Lon		Dep.

62°

TRAVERSE TABLE
28 DEGREES.

332° / 208°   ↑028° / 152°   1h 52m   **28°**

Dist	D.Lat	Dep.	Dist	D.Lat	Dep.	Dist	D.Lat	Dep.	Dist	D.Lat	Dep.	Dist	D.Lat	Dep.
301	265·8	141·3	361	318·7	169·5	421	371·7	197·6	481	424·7	225·8	541	477·7	254·0
302	266·7	141·8	362	319·6	169·9	422	372·6	198·1	482	425·6	226·3	542	478·6	254·5
303	267·5	142·2	363	320·5	170·4	423	373·5	198·6	483	426·5	226·8	543	479·4	254·9
304	268·4	142·7	364	321·4	170·9	424	374·4	199·1	484	427·3	227·2	544	480·3	255·4
305	269·3	143·2	365	322·3	171·4	425	375·3	199·5	485	428·2	227·7	545	481·2	255·9
306	270·2	143·7	366	323·2	171·8	426	376·1	200·0	486	429·1	228·2	546	482·1	256·3
307	271·1	144·1	367	324·0	172·3	427	377·0	200·5	487	430·0	228·6	547	483·0	256·8
308	271·9	144·6	368	324·9	172·8	428	377·9	200·9	488	430·9	229·1	548	483·9	257·3
309	272·8	145·1	369	325·8	173·2	429	378·8	201·4	489	431·8	229·6	549	484·7	257·7
310	273·7	145·5	370	326·7	173·7	430	379·7	201·9	490	432·6	230·0	550	485·6	258·2
311	274·6	146·0	371	327·6	174·2	431	380·6	202·3	491	433·5	230·5	551	486·5	258·7
312	275·5	146·5	372	328·5	174·6	432	381·4	202·8	492	434·4	231·0	552	487·4	259·1
313	276·4	146·9	373	329·3	175·1	433	382·3	203·3	493	435·3	231·4	553	488·3	259·6
314	277·2	147·4	374	330·2	175·6	434	383·2	203·8	494	436·2	231·9	554	489·2	260·1
315	278·1	147·9	375	331·1	176·1	435	384·1	204·2	495	437·1	232·4	555	490·0	260·6
316	279·0	148·4	376	332·0	176·5	436	385·0	204·7	496	437·9	232·9	556	490·9	261·0
317	279·9	148·8	377	332·9	177·0	437	385·8	205·2	497	438·8	233·3	557	491·8	261·5
318	280·8	149·3	378	333·8	177·5	438	386·7	205·6	498	439·7	233·8	558	492·7	262·0
319	281·7	149·8	379	334·6	177·9	439	387·6	206·1	499	440·6	234·3	559	493·6	262·4
320	282·5	150·2	380	335·5	178·4	440	388·5	206·6	500	441·5	234·7	560	494·5	262·9
321	283·4	150·7	381	336·4	178·9	441	389·4	207·0	501	442·4	235·2	561	495·3	263·4
322	284·3	151·2	382	337·3	179·3	442	390·3	207·5	502	443·2	235·7	562	496·2	263·8
323	285·2	151·6	383	338·2	179·8	443	391·1	208·0	503	444·1	236·1	563	497·1	264·3
324	286·1	152·1	384	339·1	180·3	444	392·0	208·4	504	445·0	236·6	564	498·0	264·8
325	287·0	152·6	385	339·9	180·7	445	392·9	208·9	505	445·9	237·1	565	498·9	265·3
326	287·8	153·0	386	340·8	181·2	446	393·8	209·4	506	446·8	237·6	566	499·7	265·7
327	288·7	153·5	387	341·7	181·7	447	394·7	209·9	507	447·7	238·0	567	500·6	266·2
328	289·6	154·0	388	342·6	182·2	448	395·6	210·3	508	448·5	238·5	568	501·5	266·7
329	290·5	154·5	389	343·5	182·6	449	396·4	210·8	509	449·4	239·0	569	502·4	267·1
330	291·4	154·9	390	344·3	183·1	450	397·3	211·3	510	450·3	239·4	570	503·3	267·6
331	292·3	155·4	391	345·2	183·6	451	398·2	211·7	511	451·2	239·9	571	504·2	268·1
332	293·1	155·9	392	346·1	184·0	452	399·1	212·2	512	452·1	240·4	572	505·0	268·5
333	294·0	156·3	393	347·0	184·5	453	400·0	212·7	513	453·0	240·8	573	505·9	269·0
334	294·9	156·8	394	347·9	185·0	454	400·9	213·1	514	453·8	241·3	574	506·8	269·5
335	295·8	157·3	395	348·8	185·4	455	401·7	213·6	515	454·7	241·8	575	507·7	269·9
336	296·7	157·7	396	349·6	185·9	456	402·6	214·1	516	455·6	242·2	576	508·6	270·4
337	297·6	158·2	397	350·5	186·4	457	403·5	214·5	517	456·5	242·7	577	509·5	270·9
338	298·4	158·7	398	351·4	186·8	458	404·4	215·0	518	457·4	243·2	578	510·3	271·4
339	299·3	159·2	399	352·3	187·3	459	405·3	215·5	519	458·2	243·7	579	511·2	271·8
340	300·2	159·6	400	353·2	187·8	460	406·2	216·0	520	459·1	244·1	580	512·1	272·3
341	301·1	160·1	401	354·1	188·3	461	407·0	216·4	521	460·0	244·6	581	513·0	272·8
342	302·0	160·6	402	354·9	188·7	462	407·9	216·9	522	460·9	245·1	582	513·9	273·2
343	302·9	161·0	403	355·8	189·2	463	408·8	217·4	523	461·8	245·5	583	514·8	273·7
344	303·7	161·5	404	356·7	189·7	464	409·7	217·8	524	462·7	246·0	584	515·6	274·2
345	304·6	162·0	405	357·6	190·1	465	410·6	218·3	525	463·5	246·5	585	516·5	274·6
346	305·5	162·4	406	358·5	190·6	466	411·5	218·8	526	464·4	246·9	586	517·4	275·1
347	306·4	162·9	407	359·4	191·1	467	412·3	219·2	527	465·3	247·4	587	518·3	275·5
348	307·3	163·4	408	360·2	191·5	468	413·2	219·7	528	466·2	247·9	588	519·2	276·0
349	308·1	163·8	409	361·1	192·0	469	414·1	220·2	529	467·1	248·4	589	520·1	276·5
350	309·0	164·3	410	362·0	192·5	470	415·0	220·7	530	468·0	248·8	590	520·9	277·0
351	309·9	164·8	411	362·9	193·0	471	415·9	221·1	531	468·8	249·3	591	521·8	277·5
352	310·8	165·3	412	363·8	193·4	472	416·8	221·6	532	469·7	249·8	592	522·7	277·9
353	311·7	165·7	413	364·7	193·9	473	417·6	222·1	533	470·6	250·2	593	523·6	278·4
354	312·6	166·2	414	365·5	194·4	474	418·5	222·5	534	471·5	250·7	594	524·5	278·9
355	313·4	166·7	415	366·4	194·8	475	419·4	223·0	535	472·4	251·2	595	525·4	279·3
356	314·3	167·1	416	367·3	195·3	476	420·3	223·5	536	473·3	251·6	596	526·2	279·8
357	315·2	167·6	417	368·2	195·8	477	421·2	223·9	537	474·1	252·1	597	527·1	280·3
358	316·1	168·1	418	369·1	196·2	478	422·0	224·4	538	475·0	252·6	598	528·0	280·7
359	317·0	168·5	419	370·0	196·7	479	422·9	224·9	539	475·9	253·0	599	528·9	281·2
360	317·9	169·0	420	370·8	197·2	480	423·8	225·3	540	476·8	253·5	600	529·8	281·7

| Dist | Dep. | D.Lat | Dist | Dep. | D.Lat | Dist | Dep. | D.Lat | Dist | Dep. | D.Lat | Dist | Dep. | D.Lat |

62°

298° / 242°   **62 DEGREES.**   062° / 118°   4h 8m

TRAVERSE TABLE
29 DEGREES.

29° 331° ↑ / 209°   029° ↑ / 151°   1h 56m

Dist	D.Lat	Dep.	Dist	D.Lat	Dep.	Dist	D.Lat	Dep.	Dist	D.Lat	Dep.	Dist	D.Lat	Dep.
1	00·9	00·5	61	53·4	29·6	121	105·8	58·7	181	158·3	87·8	241	210·8	116·8
2	01·7	01·0	62	54·2	30·1	122	106·7	59·1	182	159·2	88·2	242	211·7	117·3
3	02·6	01·5	63	55·1	30·5	123	107·6	59·6	183	160·1	88·7	243	212·5	117·8
4	03·5	01·9	64	56·0	31·0	124	108·5	60·1	184	160·9	89·2	244	213·4	118·3
5	04·4	02·4	65	56·9	31·5	125	109·3	60·6	185	161·8	89·7	245	214·3	118·8
6	05·2	02·9	66	57·7	32·0	126	110·2	61·1	186	162·7	90·2	246	215·2	119·3
7	06·1	03·4	67	58·6	32·5	127	111·1	61·6	187	163·6	90·7	247	216·0	119·7
8	07·0	03·9	68	59·5	33·0	128	112·0	62·1	188	164·4	91·1	248	216·9	120·2
9	07·9	04·4	69	60·3	33·5	129	112·8	62·5	189	165·3	91·6	249	217·8	120·7
10	08·7	04·8	70	61·2	33·9	130	113·7	63·0	190	166·2	92·1	250	218·7	121·2
11	09·6	05·3	71	62·1	34·4	131	114·6	63·5	191	167·1	92·6	251	219·5	121·7
12	10·5	05·8	72	63·0	34·9	132	115·4	64·0	192	167·9	93·1	252	220·4	122·2
13	11·4	06·3	73	63·8	35·4	133	116·3	64·5	193	168·8	93·6	253	221·3	122·7
14	12·2	06·8	74	64·7	35·9	134	117·2	65·0	194	169·7	94·1	254	222·2	123·1
15	13·1	07·3	75	65·6	36·4	135	118·1	65·4	195	170·6	94·5	255	223·0	123·6
16	14·0	07·8	76	66·5	36·8	136	118·9	65·9	196	171·4	95·0	256	223·9	124·1
17	14·9	08·2	77	67·3	37·3	137	119·8	66·4	197	172·3	95·5	257	224·8	124·6
18	15·7	08·7	78	68·2	37·8	138	120·7	66·9	198	173·2	96·0	258	225·7	125·1
19	16·6	09·2	79	69·1	38·3	139	121·6	67·4	199	174·0	96·5	259	226·5	125·6
20	17·5	09·7	80	70·0	38·8	140	122·4	67·9	200	174·9	97·0	260	227·4	126·1
21	18·4	10·2	81	70·8	39·3	141	123·3	68·4	201	175·8	97·4	261	228·3	126·5
22	19·2	10·7	82	71·7	39·8	142	124·2	68·8	202	176·7	97·9	262	229·2	127·0
23	20·1	11·2	83	72·6	40·2	143	125·1	69·3	203	177·5	98·4	263	230·0	127·5
24	21·0	11·6	84	73·5	40·7	144	125·9	69·8	204	178·4	98·9	264	230·9	128·0
25	21·9	12·1	85	74·3	41·2	145	126·8	70·3	205	179·3	99·4	265	231·8	128·5
26	22·7	12·6	86	75·2	41·7	146	127·7	70·8	206	180·2	99·9	266	232·6	129·0
27	23·6	13·1	87	76·1	42·2	147	128·6	71·3	207	181·0	100·4	267	233·5	129·4
28	24·5	13·6	88	77·0	42·7	148	129·4	71·8	208	181·9	100·8	268	234·4	129·9
29	25·4	14·1	89	77·8	43·1	149	130·3	72·2	209	182·8	101·3	269	235·3	130·4
30	26·2	14·5	90	78·7	43·6	150	131·2	72·7	210	183·7	101·8	270	236·1	130·9
31	27·1	15·0	91	79·6	44·1	151	132·1	73·2	211	184·5	102·3	271	237·0	131·4
32	28·0	15·5	92	80·5	44·6	152	132·9	73·7	212	185·4	102·8	272	237·9	131·9
33	28·9	16·0	93	81·3	45·1	153	133·8	74·2	213	186·3	103·3	273	238·8	132·4
34	29·7	16·5	94	82·2	45·6	154	134·7	74·7	214	187·2	103·7	274	239·6	132·8
35	30·6	17·0	95	83·1	46·1	155	135·6	75·1	215	188·0	104·2	275	240·5	133·3
36	31·5	17·5	96	84·0	46·5	156	136·4	75·6	216	188·9	104·7	276	241·4	133·8
37	32·4	17·9	97	84·8	47·0	157	137·3	76·1	217	189·8	105·2	277	242·3	134·3
38	33·2	18·4	98	85·7	47·5	158	138·2	76·6	218	190·7	105·7	278	243·1	134·8
39	34·1	18·9	99	86·6	48·0	159	139·1	77·1	219	191·5	106·2	279	244·0	135·3
40	35·0	19·4	100	87·5	48·5	160	139·9	77·6	220	192·4	106·7	280	244·9	135·7
41	35·9	19·9	101	88·3	49·0	161	140·8	78·1	221	193·3	107·1	281	245·8	136·2
42	36·7	20·4	102	89·2	49·5	162	141·7	78·5	222	194·2	107·6	282	246·6	136·7
43	37·6	20·8	103	90·1	49·9	163	142·6	79·0	223	195·0	108·1	283	247·5	137·2
44	38·5	21·3	104	91·0	50·4	164	143·4	79·5	224	195·9	108·6	284	248·4	137·7
45	39·4	21·8	105	91·8	50·9	165	144·3	80·0	225	196·8	109·1	285	249·3	138·2
46	40·2	22·3	106	92·7	51·4	166	145·2	80·5	226	197·7	109·6	286	250·1	138·7
47	41·1	22·8	107	93·6	51·9	167	146·1	81·0	227	198·5	110·1	287	251·0	139·1
48	42·0	23·3	108	94·5	52·4	168	146·9	81·4	228	199·4	110·6	288	251·9	139·6
49	42·9	23·8	109	95·3	52·8	169	147·8	81·9	229	200·3	111·0	289	252·8	140·1
50	43·7	24·2	110	96·2	53·3	170	148·7	82·4	230	201·2	111·5	290	253·6	140·6
51	44·6	24·7	111	97·1	53·8	171	149·6	82·9	231	202·0	112·0	291	254·5	141·1
52	45·5	25·2	112	98·0	54·3	172	150·4	83·4	232	202·9	112·5	292	255·4	141·6
53	46·4	25·7	113	98·8	54·8	173	151·3	83·9	233	203·8	113·0	293	256·3	142·0
54	47·2	26·2	114	99·7	55·3	174	152·2	84·4	234	204·7	113·4	294	257·1	142·5
55	48·1	26·7	115	100·6	55·8	175	153·1	84·8	235	205·5	113·9	295	258·0	143·0
56	49·0	27·1	116	101·5	56·2	176	153·9	85·3	236	206·4	114·4	296	258·9	143·5
57	49·9	27·6	117	102·3	56·7	177	154·8	85·8	237	207·3	114·9	297	259·8	144·0
58	50·7	28·1	118	103·2	57·2	178	155·7	86·3	238	208·2	115·4	298	260·6	144·5
59	51·6	28·6	119	104·1	57·7	179	156·6	86·8	239	209·0	115·9	299	261·5	145·0
60	52·5	29·1	120	105·0	58·2	180	157·4	87·3	240	209·9	116·4	300	262·4	145·4

Dist	Dep.	D.Lat	Dist	Dep.	D.Lat	Dist	Dep.	D.Lat	Dist	Dep.	D.Lat	Dist	Dep.	D.Lat

61° 299° ↑ / 241°   **61 DEGREES.**   061° ↑ / 119°   4h 4m

TRAVERSE TABLE
29 DEGREES.

| 331° ↑ / 209° | 029° ↑ / 151° 1h 56m | **29°** |

D Lon	Dep.		D Lon	Dep.		D Lon	Dep.		D Lon	Dep.		D Lon	Dep.	
Dist	D.Lat	Dep.	Dist	D.Lat	Dep.	Dist	D.Lat	Dep.	Dist	D.Lat	Dep.	Dist	D.Lat	Dep.
301	263·3	145·9	361	315·7	175·0	421	368·2	204·1	481	420·7	233·2	541	473·2	262·3
302	264·1	146·4	362	316·6	175·5	422	369·1	204·6	482	421·6	233·7	542	474·0	262·8
303	265·0	146·9	363	317·5	176·0	423	370·0	205·1	483	422·4	234·2	543	474·9	263·3
304	265·9	147·4	364	318·4	176·5	424	370·8	205·6	484	423·3	234·6	544	475·8	263·7
305	266·8	147·9	365	319·2	177·0	425	371·7	206·0	485	424·2	235·1	545	476·7	264·2
306	267·6	148·4	366	320·1	177·4	426	372·6	206·5	486	425·1	235·6	546	477·5	264·7
307	268·5	148·8	367	321·0	177·9	427	373·5	207·0	487	425·9	236·1	547	478·4	265·2
308	269·4	149·3	368	321·9	178·4	428	374·3	207·5	488	426·8	236·6	548	479·3	265·7
309	270·3	149·8	369	322·7	178·9	429	375·2	208·0	489	427·7	237·1	549	480·2	266·2
310	271·1	150·3	370	323·6	179·4	430	376·1	208·5	490	428·6	237·6	550	481·0	266·6
311	272·0	150·8	371	324·5	179·9	431	377·0	209·0	491	429·4	238·0	551	481·9	267·1
312	272·9	151·3	372	325·4	180·3	432	377·8	209·4	492	430·3	238·5	552	482·8	267·6
313	273·8	151·7	373	326·2	180·8	433	378·7	209·9	493	431·2	239·0	553	483·7	268·1
314	274·6	152·2	374	327·1	181·3	434	379·6	210·4	494	432·1	239·5	554	484·5	268·6
315	275·5	152·7	375	328·0	181·8	435	380·5	210·9	495	432·9	240·0	555	485·4	269·1
316	276·4	153·2	376	328·9	182·3	436	381·3	211·4	496	433·8	240·5	556	486·3	269·6
317	277·3	153·7	377	329·7	182·8	437	382·2	211·9	497	434·7	241·0	557	487·2	270·0
318	278·1	154·2	378	330·6	183·3	438	383·1	212·3	498	435·6	241·4	558	488·0	270·5
319	279·0	154·7	379	331·5	183·7	439	384·0	212·8	499	436·4	241·9	559	488·9	271·0
320	279·9	155·1	380	332·4	184·2	440	384·8	213·3	500	437·3	242·4	560	489·8	271·5
321	280·8	155·6	381	333·2	184·7	441	385·7	213·8	501	438·2	242·9	561	490·7	272·0
322	281·6	156·1	382	334·1	185·2	442	386·6	214·3	502	439·1	243·4	562	491·5	272·5
323	282·5	156·6	383	335·0	185·7	443	387·5	214·8	503	439·9	243·9	563	492·4	272·9
324	283·4	157·1	384	335·9	186·2	444	388·3	215·3	504	440·8	244·3	564	493·3	273·4
325	284·3	157·6	385	336·7	186·7	445	389·2	215·7	505	441·7	244·8	565	494·2	273·9
326	285·1	158·0	386	337·6	187·1	446	390·1	216·2	506	442·6	245·3	566	495·0	274·4
327	286·0	158·5	387	338·5	187·6	447	391·0	216·7	507	443·4	245·8	567	495·9	274·9
328	286·9	159·0	388	339·4	188·1	448	391·8	217·2	508	444·3	246·3	568	496·8	275·4
329	287·7	159·5	389	340·2	188·6	449	392·7	217·7	509	445·2	246·8	569	497·7	275·9
330	288·6	160·0	390	341·1	189·1	450	393·6	218·2	510	446·1	247·3	570	498·5	276·3
331	289·5	160·5	391	342·0	189·6	451	394·5	218·6	511	446·9	247·7	571	499·4	276·8
332	290·4	161·0	392	342·9	190·0	452	395·3	219·1	512	447·8	248·2	572	500·3	277·3
333	291·2	161·4	393	343·7	190·5	453	396·2	219·6	513	448·7	248·7	573	501·2	277·8
334	292·1	161·9	394	344·6	191·0	454	397·1	220·1	514	449·6	249·2	574	502·0	278·3
335	293·0	162·4	395	345·5	191·5	455	398·0	220·6	515	450·4	249·7	575	502·9	278·8
336	293·9	162·9	396	346·3	192·0	456	398·8	221·1	516	451·3	250·2	576	503·8	279·3
337	294·7	163·4	397	347·2	192·5	457	399·7	221·6	517	452·2	250·6	577	504·7	279·7
338	295·6	163·9	398	348·1	193·0	458	400·6	222·0	518	453·1	251·1	578	505·5	280·2
339	296·5	164·4	399	349·0	193·4	459	401·5	222·5	519	453·9	251·6	579	506·4	280·7
340	297·4	164·8	400	349·8	193·8	460	402·3	223·0	520	454·8	252·1	580	507·3	281·2
341	298·2	165·3	401	350·7	194·4	461	403·2	223·5	521	455·7	252·6	581	508·2	281·7
342	299·1	165·8	402	351·6	194·9	462	404·1	224·0	522	456·6	253·1	582	509·0	282·2
343	300·0	166·3	403	352·5	195·4	463	404·9	224·5	523	457·4	253·6	583	509·9	282·6
344	300·9	166·8	404	353·3	195·9	464	405·8	225·0	524	458·3	254·0	584	510·7	283·1
345	301·7	167·3	405	354·2	196·3	465	406·7	225·4	525	459·2	254·5	585	511·6	283·6
346	302·6	167·7	406	355·1	196·8	466	407·6	225·9	526	460·0	255·0	586	512·5	284·1
347	303·5	168·2	407	356·0	197·3	467	408·4	226·4	527	460·9	255·5	587	513·4	284·6
348	304·4	168·7	408	356·8	197·8	468	409·3	226·9	528	461·8	256·0	588	514·3	285·1
349	305·2	169·2	409	357·7	198·3	469	410·2	227·4	529	462·7	256·5	589	515·2	285·6
350	306·1	169·7	410	358·6	198·8	470	411·1	227·9	530	463·5	256·9	590	516·0	286·0
351	307·0	170·2	411	359·5	199·3	471	411·9	228·3	531	464·4	257·4	591	516·9	286·5
352	307·9	170·7	412	360·3	199·7	472	412·8	228·8	532	465·3	257·9	592	517·8	287·0
353	308·7	171·1	413	361·2	200·2	473	413·7	229·3	533	466·2	258·4	593	518·6	287·5
354	309·6	171·6	414	362·1	200·7	474	414·6	229·8	534	467·0	258·9	594	519·5	288·0
355	310·5	172·1	415	363·0	201·2	475	415·4	230·3	535	467·9	259·4	595	520·4	288·5
356	311·4	172·6	416	363·8	201·7	476	416·3	230·8	536	468·8	259·9	596	521·3	288·9
357	312·2	173·1	417	364·7	202·2	477	417·2	231·3	537	469·6	260·3	597	522·1	289·4
358	313·1	173·6	418	365·6	202·7	478	418·1	231·7	538	470·5	260·8	598	523·0	289·9
359	314·0	174·0	419	366·5	203·1	479	418·9	232·2	539	471·4	261·3	599	523·9	290·4
360	314·9	174·5	420	367·3	203·6	480	419·8	232·7	540	472·3	261·8	600	524·8	290·9
Dist	Dep.	D.Lat	Dist	Dep.	D.Lat	Dist	Dep.	D.Lat	Dist	Dep.	D.Lat	Dist	Dep.	D.Lat
D Lon		Dep.	D Lon		Dep.	D Lon		Dep.	D Lon		Dep.	D Lon		Dep.

| 299° ↑ / 241° | 61 DEGREES. | 061° ↑ / 119° 4h 4m | **61°** |

TRAVERSE TABLE
30 DEGREES.

30° 330° ↑ / 210°   ↑030° / 150°   2h 0m

D Lon Dist	Dep. D.Lat	Dep.	D Lon Dist	Dep. D.Lat	Dep.	D Lon Dist	Dep. D.Lat	Dep.	D Lon Dist	Dep. D.Lat	Dep.	D Lon Dist	Dep. D.Lat	Dep.
1	00·9	00·5	61	52·8	30·5	121	104·8	60·5	181	156·8	90·5	241	208·7	120·5
2	01·7	01·0	62	53·7	31·0	122	105·7	61·0	182	157·6	91·0	242	209·6	121·0
3	02·6	01·5	63	54·6	31·5	123	106·5	61·5	183	158·5	91·5	243	210·4	121·5
4	03·5	02·0	64	55·4	32·0	124	107·4	62·0	184	159·3	92·0	244	211·3	122·0
5	04·3	02·5	65	56·3	32·5	125	108·3	62·5	185	160·2	92·5	245	212·2	122·5
6	05·2	03·0	66	57·2	33·0	126	109·1	63·0	186	161·1	93·0	246	213·0	123·0
7	06·1	03·5	67	58·0	33·5	127	110·0	63·5	187	161·9	93·5	247	213·9	123·5
8	06·9	04·0	68	58·9	34·0	128	110·9	64·0	188	162·8	94·0	248	214·8	124·0
9	07·8	04·5	69	59·8	34·5	129	111·7	64·5	189	163·7	94·5	249	215·6	124·5
10	08·7	05·0	70	60·6	35·0	130	112·6	65·0	190	164·5	95·0	250	216·5	125·0
11	09·5	05·5	71	61·5	35·5	131	113·4	65·5	191	165·4	95·5	251	217·4	125·5
12	10·4	06·0	72	62·4	36·0	132	114·3	66·0	192	166·3	96·0	252	218·2	126·0
13	11·3	06·5	73	63·2	36·5	133	115·2	66·5	193	167·1	96·5	253	219·1	126·5
14	12·1	07·0	74	64·1	37·0	134	116·0	67·0	194	168·0	97·0	254	220·0	127·0
15	13·0	07·5	75	65·0	37·5	135	116·9	67·5	195	168·9	97·5	255	220·8	127·5
16	13·9	08·0	76	65·8	38·0	136	117·8	68·0	196	169·7	98·0	256	221·7	128·0
17	14·7	08·5	77	66·7	38·5	137	118·6	68·5	197	170·6	98·5	257	222·6	128·5
18	15·6	09·0	78	67·5	39·0	138	119·5	69·0	198	171·5	99·0	258	223·4	129·0
19	16·5	09·5	79	68·4	39·5	139	120·4	69·5	199	172·3	99·5	259	224·3	129·5
20	17·3	10·0	80	69·3	40·0	140	121·2	70·0	200	173·2	100·0	260	225·2	130·0
21	18·2	10·5	81	70·1	40·5	141	122·1	70·5	201	174·1	100·5	261	226·0	130·5
22	19·1	11·0	82	71·0	41·0	142	123·0	71·0	202	174·9	101·0	262	226·9	131·0
23	19·9	11·5	83	71·9	41·5	143	123·8	71·5	203	175·8	101·5	263	227·8	131·5
24	20·8	12·0	84	72·7	42·0	144	124·7	72·0	204	176·7	102·0	264	228·6	132·0
25	21·7	12·5	85	73·6	42·5	145	125·6	72·5	205	177·5	102·5	265	229·5	132·5
26	22·5	13·0	86	74·5	43·0	146	126·4	73·0	206	178·4	103·0	266	230·4	133·0
27	23·4	13·5	87	75·3	43·5	147	127·3	73·5	207	179·3	103·5	267	231·2	133·5
28	24·2	14·0	88	76·2	44·0	148	128·2	74·0	208	180·1	104·0	268	232·1	134·0
29	25·1	14·5	89	77·1	44·5	149	129·0	74·5	209	181·0	104·5	269	233·0	134·5
30	26·0	15·0	90	77·9	45·0	150	129·9	75·0	210	181·9	105·0	270	233·8	135·0
31	26·8	15·5	91	78·8	45·5	151	130·8	75·5	211	182·7	105·5	271	234·7	135·5
32	27·7	16·0	92	79·7	46·0	152	131·6	76·0	212	183·6	106·0	272	235·6	136·0
33	28·6	16·5	93	80·5	46·5	153	132·5	76·5	213	184·5	106·5	273	236·4	136·5
34	29·4	17·0	94	81·4	47·0	154	133·4	77·0	214	185·3	107·0	274	237·3	137·0
35	30·3	17·5	95	82·3	47·5	155	134·2	77·5	215	186·2	107·5	275	238·2	137·5
36	31·2	18·0	96	83·1	48·0	156	135·1	78·0	216	187·1	108·0	276	239·0	138·0
37	32·0	18·5	97	84·0	48·5	157	136·0	78·5	217	187·9	108·5	277	239·9	138·5
38	32·9	19·0	98	84·9	49·0	158	136·8	79·0	218	188·8	109·0	278	240·8	139·0
39	33·8	19·5	99	85·7	49·5	159	137·7	79·5	219	189·7	109·5	279	241·6	139·5
40	34·6	20·0	100	86·6	50·0	160	138·6	80·0	220	190·5	110·0	280	242·5	140·0
41	35·5	20·5	101	87·5	50·5	161	139·4	80·5	221	191·4	110·5	281	243·4	140·5
42	36·4	21·0	102	88·3	51·0	162	140·3	81·0	222	192·3	111·0	282	244·2	141·0
43	37·2	21·5	103	89·2	51·5	163	141·2	81·5	223	193·1	111·5	283	245·1	141·5
44	38·1	22·0	104	90·1	52·0	164	142·0	82·0	224	194·0	112·0	284	246·0	142·0
45	39·0	22·5	105	90·9	52·5	165	142·9	82·5	225	194·9	112·5	285	246·8	142·5
46	39·8	23·0	106	91·8	53·0	166	143·8	83·0	226	195·7	113·0	286	247·7	143·0
47	40·7	23·5	107	92·7	53·5	167	144·6	83·5	227	196·6	113·5	287	248·5	143·5
48	41·6	24·0	108	93·5	54·0	168	145·5	84·0	228	197·5	114·0	288	249·4	144·0
49	42·4	24·5	109	94·4	54·5	169	146·4	84·5	229	198·3	114·5	289	250·3	144·5
50	43·3	25·0	110	95·3	55·0	170	147·2	85·0	230	199·2	115·0	290	251·1	145·0
51	44·2	25·5	111	96·1	55·5	171	148·1	85·5	231	200·1	115·5	291	252·0	145·5
52	45·0	26·0	112	97·0	56·0	172	149·0	86·0	232	200·9	116·0	292	252·9	146·0
53	45·9	26·5	113	97·9	56·5	173	149·8	86·5	233	201·8	116·5	293	253·7	146·5
54	46·8	27·0	114	98·7	57·0	174	150·7	87·0	234	202·6	117·0	294	254·6	147·0
55	47·6	27·5	115	99·6	57·5	175	151·6	87·5	235	203·5	117·5	295	255·5	147·5
56	48·5	28·0	116	100·5	58·0	176	152·4	88·0	236	204·4	118·0	296	256·3	148·0
57	49·4	28·5	117	101·3	58·5	177	153·3	88·5	237	205·2	118·5	297	257·2	148·5
58	50·2	29·0	118	102·2	59·0	178	154·2	89·0	238	206·1	119·0	298	258·1	149·0
59	51·1	29·5	119	103·1	59·5	179	155·0	89·5	239	207·0	119·5	299	258·9	149·5
60	52·0	30·0	120	103·9	60·0	180	155·9	90·0	240	207·8	120·0	300	259·8	150·0
Dist	Dep.	D.Lat	Dist	Dep.	D.Lat	Dist	Dep.	D.Lat	Dist	Dep.	D.Lat	Dist	Dep.	D.Lat
D Lon		Dep.	D Lon		Dep.	D Lon		Dep.	D Lon		Dep.	D Lon		Dep.

60° 300° ↑ / 240°   **60 DEGREES.**   060 / 120   4h 0m

TRAVERSE TABLE
30 DEGREES.

330° ↑ / 210° · ↑ 030° / 150° · 2h 0m · **30°**

D Lon / Dist	Dep. / D.Lat	Dep.	D Lon / Dist	Dep. / D.Lat	Dep.	D Lon / Dist	Dep. / D.Lat	Dep.	D Lon / Dist	Dep. / D.Lat	Dep.	D Lon / Dist	Dep. / D.Lat	Dep.
301	260·7	150·5	361	312·6	180·5	421	364·6	210·5	481	416·6	240·5	541	468·5	270·5
302	261·5	151·0	362	313·5	181·0	422	365·5	211·0	482	417·4	241·0	542	469·4	271·0
303	262·4	151·5	363	314·4	181·5	423	366·3	211·5	483	418·3	241·5	543	470·3	271·5
304	263·3	152·0	364	315·2	182·0	424	367·2	212·0	484	419·2	242·0	544	471 1	272·0
305	264·1	152·5	365	316·1	182·5	425	368·1	212·5	485	420·0	242·5	545	472·0	272·5
306	265·0	153·0	366	317·0	183·0	426	368·9	213·0	486	420·9	243·0	546	472·8	273·0
307	265·9	153·5	367	317·8	183·5	427	369·8	213·5	487	421·8	243·5	547	473·7	273·5
308	266·7	154·0	368	318·7	184·0	428	370·7	214·0	488	422·6	244·0	548	474·6	274·0
309	267·6	154·5	369	319·6	184·5	429	371·5	214·5	489	423·5	244·5	549	475·4	274·5
310	268·5	155·0	370	320·4	185·0	430	372·4	215·0	490	424·4	245·0	550	476·3	275·0
311	269·3	155·5	371	321·3	185·5	431	373·3	215·5	491	425·2	245·5	551	477·2	275·5
312	270·2	156·0	372	322·2	186·0	432	374·1	216·0	492	426·1	246·0	552	478·0	276·0
313	271·1	156·5	373	323·0	186·5	433	375·0	216·5	493	427·0	246·5	553	478·9	276·5
314	271·9	157·0	374	323·9	187·0	434	375·9	217·0	494	427·8	247·0	554	479·8	277·0
315	272·8	157·5	375	324·8	187·5	435	376·7	217·5	495	428·7	247·5	555	480·6	277·5
316	273·7	158·0	376	325·6	188·0	436	377·6	218·0	496	429·5	248·0	556	481·5	278·0
317	274·5	158·5	377	326·5	188·5	437	378·5	218·5	497	430·4	248·5	557	482·4	278·5
318	275·4	159·0	378	327·4	189·0	438	379·3	219·0	498	431·3	249·0	558	483·2	279·0
319	276·3	159·5	379	328·2	189·5	439	380·2	219·5	499	432·1	249·5	559	484·1	279·5
320	277·1	160·0	380	329·1	190·0	440	381·1	220·0	500	433·0	250·0	560	485·0	280·0
321	278·0	160·5	381	330·0	190·5	441	381·9	220·5	501	433·9	250·5	561	485·8	280·5
322	278·9	161·0	382	330·8	191·0	442	382·8	221·0	502	434·7	251·0	562	486·7	281·0
323	279·7	161·5	383	331·7	191·5	443	383·6	221·5	503	435·6	251·5	563	487·6	281·5
324	280·6	162·0	384	332·6	192·0	444	384·5	222·0	504	436·5	252·0	564	488·4	282·0
325	281·5	162·5	385	333·4	192·5	445	385·4	222·5	505	437·3	252·5	565	489·3	282·5
326	282·3	163·0	386	334·3	193·0	446	386·3	223·0	506	438·2	253·0	566	490·2	283·0
327	283·2	163·5	387	335·2	193·5	447	387·1	223·5	507	439·1	253·5	567	491·0	283·5
328	284·1	164·0	388	336·0	194·0	448	388·0	224·0	508	439·9	254·0	568	491·9	284·0
329	284·9	164·5	389	336·9	194·5	449	388·8	224·5	509	440·8	254·5	569	492·8	284·5
330	285·8	165·0	390	337·7	195·0	450	389 7	225·0	510	441 7	255·0	570	493·6	285·0
331	286·7	165·5	391	338·6	195·5	451	390·6	225·5	511	442·5	255·5	571	494·5	285·5
332	287·5	166·0	392	339·5	196·0	452	391·4	226·0	512	443·4	256·0	572	495·4	286·0
333	288·4	166·5	393	340·3	196·5	453	392·3	226·5	513	444·3	256·5	573	496·2	286·5
334	289·3	167·0	394	341·2	197·0	454	393·2	227·0	514	445·1	257·0	574	497·1	287·0
335	290·1	167·5	395	342·1	197·5	455	394·0	227·5	515	446·0	257·5	575	498·0	287·5
336	291·0	168·0	396	342·9	198·0	456	394·9	228·0	516	446·9	258·0	576	498·8	288·0
337	291·9	168·5	397	343·8	198·5	457	395·8	228·5	517	447·7	258·5	577	499·7	288·5
338	292·7	169·0	398	344·7	199·0	458	396·6	229·0	518	448·6	259·0	578	500·6	289·0
339	293·6	169·5	399	345·5	199·5	459	397·5	229·5	519	449·5	259·5	579	501·3	289·5
340	294·5	170·0	400	346·4	200·0	460	398·4	230·0	520	450·3	260·0	580	502·3	290·0
341	295·3	170·5	401	347·3	200·5	461	399·2	230·5	521	451·2	260·5	581	503·2	290·5
342	296·2	171·0	402	348·1	201·0	462	400·1	231·0	522	452·1	261·0	582	504·0	291·0
343	297·0	171·5	403	349·0	201·5	463	401·0	231·5	523	452·9	261·5	583	504·9	291·5
344	297·9	172·0	404	349·9	202·0	464	401·8	232·0	524	453·8	262·0	584	505·8	292·0
345	298·8	172·5	405	350·7	202·5	465	402·7	232·5	525	454·7	262·5	585	506·6	292·5
346	299·6	173·0	406	351·6	203·0	466	403·6	233·0	526	455·5	263·0	586	507·5	293·0
347	300·5	173·5	407	352·5	203·5	467	404·4	233·5	527	456·4	263·5	587	508·4	293·5
348	301·4	174·0	408	353·3	204·0	468	405·3	234·0	528	457·3	264·0	588	509·2	294·0
349	302·2	174·5	409	354·2	204·5	469	406·2	234·5	529	458·1	264·5	589	510·1	294·5
350	303·1	175·0	410	355·1	205·0	470	407·0	235·0	530	459·0	265·0	590	511·0	295·0
351	304·0	175·5	411	355·9	205·5	471	407·9	235·5	531	459·9	265·5	591	511·8	295·5
352	304·8	176·0	412	356·8	206·0	472	408·8	236·0	532	460·7	266·0	592	512·7	296·0
353	305·7	176·5	413	357·7	206·5	473	409·6	236·5	533	461·6	266·5	593	513·6	296·5
354	306·6	177·0	414	358·5	207·0	474	410·5	237·0	534	462·5	267·0	594	514·4	297·0
355	307·4	177·5	415	359·4	207·5	475	411·4	237·5	535	463·3	267·5	595	515·3	297·5
356	308·3	178·0	416	360·3	208·0	476	412·2	238·0	536	464·2	268·0	596	516·2	298·0
357	309·2	178·5	417	361·1	208·5	477	413·1	238·5	537	465·1	268·5	597	517·0	298·5
358	310·0	179·0	418	362·0	209·0	478	414·0	239·0	538	465·9	269·0	598	517·9	299·0
359	310·9	179·5	419	362·9	209·5	479	414·8	239·5	539	466·8	269·5	599	518·7	299·5
360	311·8	180·0	420	363·7	210·0	480	415·7	240·0	540	467·7	270·0	600	519·6	300·0

Dist	Dep.	D.Lat	Dist	Dep.	D.Lat	Dist	Dep.	D.Lat	Dist	Dep.	D.Lat	Dist	Dep.	D.Lat
D Lon		Dep.	D Lon		Dep.	D Lon		Dep.	D Lon		Dep.	D Lon		Dep.

300° ↑ / 240° · **60 DEGREES.** · ↑ 060° / 120° · 4h 0m · **60°**

TRAVERSE TABLE

31°

31 DEGREES.

Dist	D.Lat	Dep.	Dist	D.Lat	Dep.	Dist	D.Lat	Dep.	Dist	D.Lat	Dep.	Dist	D.Lat	Dep.
1	00·9	00·5	61	52·3	31·4	121	103·7	62·3	181	155·1	93·2	241	206·6	124·1
2	01·7	01·0	62	53·1	31·9	122	104·6	62·8	182	156·0	93·7	242	207·4	124·6
3	02·6	01·5	63	54·0	32·4	123	105·4	63·3	183	156·9	94·3	243	208·3	125·2
4	03·4	02·1	64	54·9	33·0	124	106·3	63·9	184	157·7	94·8	244	209·1	125·7
5	04·3	02·6	65	55·7	33·5	125	107·1	64·4	185	158·6	95·3	245	210·0	126·2
6	05·1	03·1	66	56·6	34·0	126	108·0	64·9	186	159·4	95·8	246	210·9	126·7
7	06·0	03·6	67	57·4	34·5	127	108·9	65·4	187	160·3	96·3	247	211·7	127·2
8	06·9	04·1	68	58·3	35·0	128	109·7	65·9	188	161·1	96·8	248	212·6	127·7
9	07·7	04·6	69	59·1	35·5	129	110·6	66·4	189	162·0	97·3	249	213·4	128·2
10	08·6	05·2	70	60·0	36·1	130	111·4	67·0	190	162·9	97·9	250	214·3	122·8
11	09·4	05·7	71	60·9	36·6	131	112·3	67·5	191	163·7	98·4	251	215·1	129·3
12	10·3	06·2	72	61·7	37·1	132	113·1	68·0	192	164·6	98·9	252	216·0	129·8
13	11·1	06·7	73	62·6	37·6	133	114·0	68·5	193	165·4	99·4	253	216·9	130·3
14	12·0	07·2	74	63·4	38·1	134	114·9	69·0	194	166·3	99·9	254	217·7	130·8
15	12·9	07·7	75	64·3	38·6	135	115·7	69·5	195	167·1	100·4	255	218·6	131·3
16	13·7	08·2	76	65·1	39·1	136	116·6	70·0	196	168·0	100·9	256	219·4	131·8
17	14·6	08·8	77	66·0	39·7	137	117·4	70·6	197	168·9	101·5	257	220·3	132·4
18	15·4	09·3	78	66·9	40·2	138	118·3	71·1	198	169·7	102·0	258	221·1	132·9
19	16·3	09·8	79	67·7	40·7	139	119·1	71·6	199	170·6	102·5	259	222·0	133·4
20	17·1	10·3	80	68·6	41·2	140	120·0	72·1	200	171·4	103·0	260	222·9	133·9
21	18·0	10·8	81	69·4	41·7	141	120·9	72·6	201	172·3	103·5	261	223·7	134·4
22	18·9	11·3	82	70·3	42·2	142	121·7	73·1	202	173·1	104·0	262	224·6	134·9
23	19·7	11·8	83	71·1	42·7	143	122·6	73·7	203	174·0	104·6	263	225·4	135·5
24	20·6	12·4	84	72·0	43·3	144	123·4	74·2	204	174·9	105·1	264	226·3	136·0
25	21·4	12·9	85	72·9	43·8	145	124·3	74·7	205	175·7	105·6	265	227·1	136·5
26	22·3	13·4	86	73·7	44·3	146	125·1	75·2	206	176·6	106·1	266	228·0	137·0
27	23·1	13·9	87	74·6	44·8	147	126·0	75·7	207	177·4	106·6	267	228·9	137·5
28	24·0	14·4	88	75·4	45·3	148	126·9	76·2	208	178·3	107·1	268	229·7	138·0
29	24·9	14·9	89	76·3	45·8	149	127·7	76·7	209	179·1	107·6	269	230·6	138·5
30	25·7	15·5	90	77·1	46·4	150	128·6	77·3	210	180·0	108·2	270	231·4	139·1
31	26·6	16·0	91	78·0	46·9	151	129·4	77·8	211	180·9	108·7	271	232·3	139·6
32	27·4	16·5	92	78·9	47·4	152	130·3	78·3	212	181·7	109·2	272	233·1	140·1
33	28·3	17·0	93	79·7	47·9	153	131·1	78·8	213	182·6	109·7	273	234·0	140·6
34	29·1	17·5	94	80·6	48·4	154	132·0	79·3	214	183·4	110·2	274	234·9	141·1
35	30·0	18·0	95	81·4	48·9	155	132·9	79·8	215	184·3	110·7	275	235·7	141·6
36	30·9	18·5	96	82·3	49·4	156	133·7	80·3	216	185·1	111·2	276	236·6	142·2
37	31·7	19·1	97	83·1	50·0	157	134·6	80·8	217	186·0	111·8	277	237·4	142·7
38	32·6	19·6	98	84·0	50·5	158	135·4	81·4	218	186·9	112·3	278	238·3	143·2
39	33·4	20·1	99	84·9	51·0	159	136·3	81·9	219	187·7	112·8	279	239·1	143·7
40	34·3	20·6	100	85·7	51·5	160	137·1	82·4	220	188·6	113·3	280	240·0	144·2
41	35·1	21·1	101	86·6	52·0	161	138·0	82·9	221	189·4	113·8	281	240·9	144·7
42	36·0	21·6	102	87·4	52·5	162	138·9	83·4	222	190·3	114·3	282	241·7	145·2
43	36·9	22·1	103	88·3	53·0	163	139·7	84·0	223	191·1	114·9	283	242·6	145·8
44	37·7	22·7	104	89·1	53·6	164	140·6	84·5	224	192·0	115·4	284	243·4	146·3
45	38·6	23·2	105	90·0	54·1	165	141·4	85·0	225	192·9	115·9	285	244·3	146·8
46	39·4	23·7	106	90·9	54·6	166	142·3	85·5	226	193·7	116·4	286	245·1	147·3
47	40·3	24·2	107	91·7	55·1	167	143·1	86·0	227	194·6	116·9	287	246·0	147·8
48	41·1	24·7	108	92·6	55·6	168	144·0	86·5	228	195·4	117·4	288	246·9	148·3
49	42·0	25·2	109	93·4	56·1	169	144·9	87·0	229	196·3	117·9	289	247·7	148·8
50	42·9	25·8	110	94·3	56·7	170	145·7	87·6	230	197·1	118·5	290	248·6	149·4
51	43·7	26·3	111	95·1	57·2	171	146·6	88·1	231	198·0	119·0	291	249·4	149·9
52	44·6	26·8	112	96·0	57·7	172	147·4	88·6	232	198·9	119·5	292	250·3	150·4
53	45·4	27·3	113	96·9	58·2	173	148·3	89·1	233	199·7	120·0	293	251·2	150·9
54	46·3	27·8	114	97·7	58·7	174	149·1	89·6	234	200·6	120·5	294	252·0	151·4
55	47·1	28·3	115	98·6	59·2	175	150·0	90·1	235	201·4	121·0	295	252·9	151·9
56	48·0	28·8	116	99·4	59·7	176	150·9	90·6	236	202·3	121·5	296	253·7	152·5
57	48·9	29·4	117	100·3	60·3	177	151·7	91·2	237	203·1	122·1	297	254·6	153·0
58	49·7	29·9	118	101·1	60·8	178	152·6	91·7	238	204·0	122·6	298	255·4	153·5
59	50·6	30·4	119	102·0	61·3	179	153·4	92·2	239	204·9	123·1	299	256·3	154·0
60	51·4	30·9	120	102·9	61·8	180	154·3	92·7	240	205·7	123·6	300	257·1	154·5

Dist	Dep.	D.Lat	Dist	Dep.	D.Lat	Dist	Dep.	D.Lat	Dist	Dep.	D.Lat	Dist	Dep.	D.Lat

59°

TRAVERSE TABLE
31 DEGREES.

329° / 211°   ↑031° / 149°   2h 4m   **31°**

Dist	D. Lat	Dep.	Dist	D. Lat	Dep.	Dist	D. Lat	Dep.	Dist	D. Lat	Dep.	Dist	D. Lat	Dep.
301	258·0	155·0	361	309·4	185·9	421	360·9	216·8	481	412·3	247·7	541	463·7	278·6
302	258·9	155·5	362	310·3	186·4	422	361·7	217·3	482	413·2	248·2	542	464·6	279·2
303	259·7	156·1	363	311·2	187·0	423	362·6	217 9	483	414·0	248·8	543	465·4	279·7
304	260·6	156·6	364	312·0	187·5	424	363·4	218 4	484	414·9	249·3	544	466·3	280·2
305	261·4	157·1	365	312·9	188·0	425	364·3	218·9	485	415·7	249·8	545	467·2	280·7
306	262·3	157·6	366	313·7	188·5	426	365·2	219·4	486	416·6	250·3	546	468·0	281·2
307	263·2	158·1	367	314·6	189·0	427	366·0	219·9	487	417·4	250·8	547	468·9	281·7
308	264·0	158·6	368	315·4	189·5	428	366·9	220·4	488	418·3	251·3	548	469·7	282·2
309	264·9	159·1	369	316·3	190·0	429	367·7	221·0	489	419·2	251·9	549	470·6	282·8
310	265·7	159·7	370	317·2	190·6	430	368·6	221·5	490	420·0	252·4	550	471·4	283·3
311	266·6	160·2	371	318·0	191·1	431	369·4	222 0	491	420·9	252·9	551	472·3	283·8
312	267·4	160·7	372	318·9	191·6	432	370·3	222·5	492	421·7	253·4	552	473·2	284·3
313	268·3	161·2	373	319·7	192·1	433	371·2	223·0	493	422·6	253 9	553	474·0	284·8
314	269·2	161·7	374	320·6	192·6	434	372·0	223·5	494	423·4	254·4	554	474·9	285·3
315	270·0	162·2	375	321·4	193·1	435	372·9	224·0	495	424·3	254·9	555	475·7	285·8
316	270·9	162·8	376	322·3	193·7	436	373·7	224·6	496	425·2	255·5	556	476·6	286·4
317	271·7	163·3	377	323·2	194·2	437	374·6	225·1	497	426·0	256·0	557	477·4	286·9
318	272·6	163·8	378	324·0	194·7	438	375·4	225·6	498	426·9	256·5	558	478·3	287·4
319	273·4	164·3	379	324·9	195·2	439	376·3	226·1	499	427·7	257·0	559	479·2	287·9
320	274·3	164·8	380	325·7	195·7	440	377·2	226·6	500	428·6	257·5	560	480·0	288·4
321	275·2	165·3	381	326·6	196·2	441	378·0	227·1	501	429·4	258·0	561	480·9	288·9
322	276·0	165·8	382	327·4	196·7	442	378·9	227·6	502	430·3	258·5	562	481·7	289·5
323	276·9	166·4	383	328·3	197·3	443	379·7	228·2	503	431·2	259·1	563	482·6	290·0
324	277·7	166·9	384	329·2	197·8	444	380·6	228·7	504	432·0	259·6	564	483·4	290·5
325	278·6	167·4	385	330·0	198·3	445	381·4	229·2	505	432·9	260·1	565	484·3	291·0
326	279·4	167·9	386	330·9	198·8	446	382·3	229·7	506	433·7	260·6	566	485·2	291·5
327	280·3	168·4	387	331·7	199·3	447	383·2	230·2	507	434·6	261·1	567	486·0	292·0
328	281·2	168·9	388	332·6	199·8	448	384·0	230·7	508	435·4	261·6	568	486·9	292·5
329	282·0	169·4	389	333·4	200·3	449	384·9	231·3	509	436·3	262·2	569	487·7	293·1
330	282·9	170·0	390	334·3	200·9	450	385·7	231·8	510	437·2	262·7	570	488·6	293·6
331	283·7	170·5	391	335·2	201·4	451	386·6	232·3	511	438·0	263·2	571	489·4	294·1
332	284·6	171·0	392	336·0	201·9	452	387·4	232·8	512	438·9	263·7	572	490·3	294·6
333	285·4	171·5	393	336·9	202·4	453	388·3	233·3	513	439·7	264·2	573	491·2	295·1
334	286·3	172·0	394	337·7	202·9	454	389·2	233·8	514	440·6	264·7	574	492·0	295·6
335	287·2	172·5	395	338·6	203·4	455	390·0	234·3	515	441·4	265·2	575	492·9	296·1
336	288·0	173·1	396	339·4	204·0	456	390·9	234·9	516	442·3	265·8	576	493·7	296·7
337	288·9	173·6	397	340·3	204·5	457	391·7	235·4	517	443·2	266·3	577	494·6	297·2
338	289·7	174·1	398	341·2	205·0	458	392·6	235·9	518	444·0	266·8	578	495·4	297·7
339	290·6	174·6	399	342·0	205·5	459	393·4	236·4	519	444·9	267·3	579	496·3	298·2
340	291·4	175·1	400	342·9	206·0	460	394·3	236·9	520	445·7	267·8	580	497·2	298·7
341	292·3	175·6	401	343·7	206·5	461	395·2	237·4	521	446·6	268·3	581	498·0	299·2
342	293·2	176·1	402	344·6	207·0	462	396·0	237·9	522	447·4	268·8	582	498·9	299·8
343	294·0	176·7	403	345·4	207·6	463	396·9	238·5	523	448·3	269·4	583	499·7	300·3
344	294·9	177·2	404	346·3	208·1	464	397·7	239·0	524	449·2	269·9	584	500·6	300·8
345	295·7	177·7	405	347·2	208·6	465	398·6	239·5	525	450·0	270·4	585	501·4	301·3
346	296·6	178·2	406	348·0	209·1	466	399·4	240·0	526	450·9	270·9	586	502·3	301·8
347	297·4	178·7	407	348·9	209·6	467	400·3	240·5	527	451·7	271·4	587	503·2	302·3
348	298·3	179·2	408	349·7	210·1	468	401·2	241·0	528	452·6	271·9	588	504·0	302·8
349	299·2	179·7	409	350·6	210·7	469	402·2	241·6	529	453·4	272·5	589	504·9	303·4
350	300·0	180·3	410	351·4	211·2	470	402·9	242·1	530	454·3	273·0	590	505·7	303·9
351	300·9	180·8	411	352·3	211·7	471	403·7	242·6	531	455·2	273·5	591	506·6	304·4
352	301·7	181·3	412	353·2	212·2	472	404·6	243·1	532	456·0	274·0	592	507·4	304·9
353	302·6	181·8	413	354·0	212·7	473	405·4	243·6	533	456·9	274·5	593	508·3	305·4
354	303·4	182·3	414	354·9	213·2	474	406·3	244·1	534	457·7	275·0	594	509·2	305·9
355	304·3	182·8	415	355·7	213·7	475	407·2	244·6	535	458·6	275·5	595	510·0	306·4
356	305·2	183·4	416	356·6	214·3	476	408·0	245·2	536	459·4	276·1	596	510·9	307·0
357	306·0	183·9	417	357·4	214·8	477	408·9	245·7	537	460·3	276·6	597	511·7	307·5
358	306·9	184·4	418	358·3	215·3	478	409·7	246·2	538	461·2	277·1	598	512·6	308·0
359	307·7	184·9	419	359·2	215·8	479	410·6	246·7	539	462·0	277·6	599	513·4	308·5
360	308·6	185·4	420	360·0	216·3	480	411·4	247·2	540	462·9	278·1	600	514·3	309·0

| Dist | Dep. | D. Lat | Dist | Dep. | D. Lat | Dist | Dep. | D. Lat | Dist | Dep. | D. Lat | Dist | Dep. | D. Lat |

301° / 239°   **59 DEGREES.**   059° / 121°   3h 56m   **59°**

32°

TRAVERSE TABLE
32 DEGREES.

D Lon Dep.			D Lon Dep.			D Lon Dep.			D Lon Dep.			D Lon Dep.			D Lon Dep.		
Dist	D.Lat	Dep.	Dist	D.Lat	Dep.	Dist	D.Lat	Dep.	Dist	D.Lat	Dep.	Dist	D.Lat	Dep.	Dist	D.Lat	Dep.
1	00·8	00·5	61	51·7	32·3	121	102·6	64·1	181	153·5	95·9	241	204·4	127·7			
2	01·7	01·1	62	52·6	32·9	122	103·5	64·7	182	154·3	96·4	242	205·2	128·2			
3	02·5	01·6	63	53·4	33·4	123	104·3	65·2	183	155·2	97·0	243	206·1	128·8			
4	03·4	02·1	64	54·3	33·9	124	105·2	65·7	184	156·0	97·5	244	206·9	129·3			
5	04·2	02·6	65	55·1	34·4	125	106·0	66·2	185	156·9	98·0	245	207·8	129·8			
6	05·1	03·2	66	56·0	35·0	126	106·9	66·8	186	157·7	98·6	246	208·6	130·4			
7	05·9	03·7	67	56·8	35·5	127	107·7	67·3	187	158·6	99·1	247	209·5	130·9			
8	06·8	04·2	68	57·7	36·0	128	108·6	67·8	188	159·4	99·6	248	210·3	131·4			
9	07·6	04·8	69	58·5	36·6	129	109·4	68·4	189	160·3	100·2	249	211·2	131·9			
10	08·5	05·3	70	59·4	37·1	130	110·2	68·9	190	161·1	100·7	250	212·0	132·5			
11	09·3	05·8	71	60·2	37·6	131	111·1	69·4	191	162·0	101·2	251	212·9	133·0			
12	10·2	06·4	72	61·1	38·2	132	111·9	69·9	192	162·8	101·7	252	213·7	133·5			
13	11·0	06·9	73	61·9	38·7	133	112·8	70·5	193	163·7	102·3	253	214·6	134·1			
14	11·9	07·4	74	62·8	39·2	134	113·6	71·0	194	164·5	102·8	254	215·4	134·6			
15	12·7	07·9	75	63·6	39·7	135	114·5	71·5	195	165·4	103·3	255	216·3	135·1			
16	13·6	08·5	76	64·5	40·3	136	115·3	72·1	196	166·2	103·9	256	217·1	135·7			
17	14·4	09·0	77	65·3	40·8	137	116·2	72·6	197	167·1	104·4	257	217·9	136·2			
18	15·3	09·5	78	66·1	41·3	138	117·0	73·1	198	167·9	104·9	258	218·8	136·7			
19	16·1	10·1	79	67·0	41·9	139	117·9	73·7	199	168·8	105·5	259	219·6	137·3			
20	17·0	10·6	80	67·8	42·4	140	118·7	74·2	200	169·6	106·0	260	220·5	137·8			
21	17·8	11·1	81	68·7	42·9	141	119·6	74·7	201	170·5	106·5	261	221·3	138·3			
22	18·7	11·7	82	69·5	43·5	142	120·4	75·2	202	171·3	107·0	262	222·2	138·8			
23	19·5	12·2	83	70·4	44·0	143	121·3	75·8	203	172·2	107·6	263	223·0	139·4			
24	20·4	12·7	84	71·2	44·5	144	122·1	76·3	204	173·0	108·1	264	223·9	139·9			
25	21·2	13·2	85	72·1	45·0	145	123·0	76·8	205	173·8	108·6	265	224·7	140·4			
26	22·0	13·8	86	72·9	45·6	146	123·8	77·4	206	174·7	109·2	266	225·6	141·0			
27	22·9	14·3	87	73·8	46·1	147	124·7	77·9	207	175·5	109·7	267	226·4	141·5			
28	23·7	14·8	88	74·6	46·6	148	125·5	78·4	208	176·4	110·2	268	227·3	142·0			
29	24·6	15·4	89	75·5	47·2	149	126·4	79·0	209	177·2	110·8	269	228·1	142·5			
30	25·4	15·9	90	76·3	47·7	150	127·2	79·5	210	178·1	111·3	270	229·0	143·1			
31	26·3	16·4	91	77·2	48·2	151	128·1	80·0	211	178·9	111·8	271	229·8	143·6			
32	27·1	17·0	92	78·0	48·8	152	128·9	80·5	212	179·8	112·3	272	230·7	144·1			
33	28·0	17·5	93	78·9	49·3	153	129·8	81·1	213	180·6	112·9	273	231·5	144·7			
34	28·8	18·0	94	79·7	49·8	154	130·6	81·6	214	181·5	113·4	274	232·4	145·2			
35	29·7	18·5	95	80·6	50·3	155	131·4	82·1	215	182·3	113·9	275	233·2	145·7			
36	30·5	19·1	96	81·4	50·9	156	132·3	82·7	216	183·2	114·5	276	234·1	146·3			
37	31·4	19·6	97	82·3	51·4	157	133·1	83·2	217	184·0	115·0	277	234·9	146·8			
38	32·2	20·1	98	83·1	51·9	158	134·0	83·7	218	184·9	115·5	278	235·8	147·3			
39	33·1	20·7	99	84·0	52·5	159	134·8	84·3	219	185·7	116·1	279	236·6	147·8			
40	33·9	21·2	100	84·8	53·0	160	135·7	84·8	220	186·6	116·6	280	237·5	148·4			
41	34·8	21·7	101	85·7	53·5	161	136·5	85·3	221	187·4	117·1	281	238·3	148·9			
42	35·6	22·3	102	86·5	54·1	162	137·4	85·8	222	188·3	117·6	282	239·1	149·4			
43	36·5	22·8	103	87·3	54·6	163	138·2	86·4	223	189·1	118·2	283	240·0	150·0			
44	37·3	23·3	104	88·2	55·1	164	139·1	86·9	224	190·0	118·7	284	240·8	150·5			
45	38·2	23·8	105	89·0	55·6	165	139·9	87·4	225	190·8	119·2	285	241·7	151·0			
46	39·0	24·4	106	89·9	56·2	166	140·8	88·0	226	191·7	119·8	286	242·5	151·6			
47	39·9	24·9	107	90·7	56·7	167	141·6	88·5	227	192·5	120·3	287	243·4	152·1			
48	40·7	25·4	108	91·6	57·2	168	142·5	89·0	228	193·4	120·8	288	244·2	152·6			
49	41·6	26·0	109	92·4	57·8	169	143·3	89·6	229	194·2	121·4	289	245·1	153·1			
50	42·4	26·5	110	93·3	58·3	170	144·2	90·1	230	195·1	121·9	290	245·9	153·7			
51	43·3	27·0	111	94·1	58·8	171	145·0	90·6	231	195·9	122·4	291	246·8	154·2			
52	44·1	27·6	112	95·0	59·4	172	145·9	91·1	232	196·7	122·9	292	247·6	154·7			
53	44·9	28·1	113	95·8	59·9	173	146·7	91·7	233	197·6	123·5	293	248·5	155·3			
54	45·8	28·6	114	96·7	60·4	174	147·6	92·2	234	198·4	124·0	294	249·3	155·8			
55	46·6	29·1	115	97·5	60·9	175	148·4	92·7	235	199·3	124·5	295	250·2	156·3			
56	47·5	29·7	116	98·4	61·5	176	149·3	93·3	236	200·1	125·1	296	251·0	156·9			
57	48·3	30·2	117	99·2	62·0	177	150·1	93·8	237	201·0	125·6	297	251·9	157·4			
58	49·2	30·7	118	100·1	62·5	178	151·0	94·3	238	201·8	126·1	298	252·7	157·9			
59	50·0	31·3	119	100·9	63·1	179	151·8	94·9	239	202·7	126·7	299	253·6	158·4			
60	50·9	31·8	120	101·8	63·6	180	152·6	95·4	240	203·5	127·2	300	254·4	159·0			

Dist	Dep.	D.Lat	Dist	Dep.	D.Lat	Dist	Dep.	D.Lat	Dist	Dep.	D.Lat	Dist	Dep.	D.Lat	Dist	Dep.	D.Lat
D Lon		Dep.	D Lon		Dep.	D Lon		Dep.	D Lon		Dep.	D Lon		Dep.	D Lon		Dep.

58°

TRAVERSE TABLE

32 DEGREES.

32°

D Lon	Dep.		D Lon	Dep.		D Lon	Dep.		D Lon	Dep.		D Lon	Dep.	
Dist	D. Lat	Dep.	Dist	D. Lat	Dep.	Dist	D. Lat	Dep.	Dist	D. Lat	Dep.	Dist	D. Lat	Dep.
301	255·3	159·5	361	306·1	191·3	421	357·0	223·1	481	407·9	254·9	541	458·8	286·7
302	256·1	160·0	362	307·0	191·8	422	357·9	223·6	482	408·8	255·4	542	459·6	287·2
303	257·0	160·6	363	307·8	192·4	423	358·7	224·2	483	409·6	256·0	543	460·5	287 7
304	257·8	161·1	364	308·7	192·9	424	359·6	224·7	484	410·5	256·5	544	461·3	288·3
305	258·7	161·6	365	309·5	193·4	425	360·4	225·2	485	411·3	257·0	545	462·2	288·8
306	259·5	162·2	366	310·4	194·0	426	361·3	225·7	486	412·2	257·5	546	463·0	289·3
307	260·4	162·7	367	311·2	194·5	427	362·1	226·3	487	413·0	258·1	547	463·9	289·9
308	261·2	163·2	368	312·1	195·0	428	363·0	226·8	488	413·8	258·6	548	464·7	290·4
309	262·0	163·7	369	312·9	195·5	429	363·8	227·3	489	414·7	259·1	549	465·6	290·9
310	262·9	164·3	370	313·8	196·1	430	364·7	227·9	490	415·5	259·7	550	466·4	291·5
311	263·7	164·8	371	314·6	196·6	431	365·5	228·4	491	416·4	260·2	551	467·3	292·0
312	264·6	165·3	372	315·5	197·1	432	366·4	228·9	492	417·2	260·7	552	468·1	292·5
313	265·4	165·9	373	316·3	197·7	433	367·2	229·5	493	418·1	261·3	553	469·0	293·0
314	266·3	166·4	374	317·2	198·2	434	368·1	230·0	494	418·9	261·8	554	469·8	293·6
315	267·1	166·9	375	318·0	198·7	435	368·9	230·5	495	419·8	262·3	555	470·7	294·1
316	268·0	167·5	376	318·9	199·2	436	369·7	231·0	496	420·6	262·8	556	471·5	294·6
317	268·8	168·0	377	319·7	199·8	437	370·6	231·6	497	421·5	263·4	557	472·4	295·2
318	269·7	168·5	378	320·6	200·3	438	371·4	232·1	498	422·3	263·9	558	473·2	295·7
319	270·5	169·0	379	321·4	200·8	439	372·3	232·6	499	423·2	264·4	559	474·1	296·2
320	271·4	169·6	380	322·3	201·4	440	373·1	233·2	500	424·0	265·0	560	474·9	296·8
321	272·2	170·1	381	323·1	201·9	441	374·0	233·7	501	424·9	265·5	561	475·8	297·3
322	273·1	170·6	382	324·0	202·4	442	374·8	234·2	502	425·7	266·0	562	476·6	297·8
323	273·9	171·2	383	324·8	203·0	443	375·7	234·8	503	426·6	266·5	563	477·5	298·3
324	274·8	171·7	384	325·7	203·5	444	376·5	235·3	504	427·4	267·1	564	478·3	298·9
325	275·6	172·2	385	326·5	204·0	445	377·4	235·8	505	428·3	267·6	565	479·1	299·4
326	276·5	172·8	386	327·3	204·5	446	378·2	236·3	506	429·1	268·1	566	480·0	299·9
327	277·3	173·3	387	328·2	205·1	447	379·1	236·9	507	430·0	268·7	567	480·8	300·5
328	278·2	173·8	388	329·0	205·6	448	379·9	237·4	508	430·8	269·2	568	481·7	301·0
329	279·0	174·3	389	329·9	206·1	449	380·8	237·9	509	431·7	269·7	569	482·5	301·5
330	279·9	174·9	390	330·7	206·7	450	381·6	238·5	510	432·5	270·3	570	483·4	302·1
331	280·7	175·4	391	331·6	207·2	451	382·5	239·0	511	433·4	270·8	571	484·2	302·6
332	281·6	175·9	392	332·4	207·7	452	383·3	239·5	512	434·2	271·3	572	485·1	303·1
333	282·4	176·5	393	333·3	208·3	453	384·2	240·1	513	435·0	271·9	573	485·9	303·6
334	283·2	177·0	394	334·1	208·8	454	385·0	240·6	514	435·9	272·4	574	486·8	304·2
335	284·1	177·5	395	335·0	209·3	455	385·9	241·1	515	436·7	272·9	575	487·6	304·7
336	284·9	178·1	396	335·8	209·8	456	386·7	241·6	516	437·6	273·4	576	488·5	305·2
337	285·8	178·6	397	336·7	210·4	457	387·6	242·2	517	438·4	274·0	577	489·3	305·8
338	286·6	179·1	398	337·5	210·9	458	388·4	242·7	518	439·3	274·5	578	490·2	306·3
339	287·5	179·6	399	338·4	211·4	459	389·3	243·2	519	440·1	275·0	579	491·0	306·8
340	288·3	180·2	400	339·2	212·0	460	390·1	243·8	520	441·0	275·6	580	491·9	307·4
341	289·2	180·7	401	340·1	212·5	461	391·0	244·3	521	441·8	276·1	581	492·7	307·9
342	290·0	181·2	402	340·9	213·0	462	391·8	244·8	522	442·7	276·6	582	493·6	308·4
343	290·9	181·8	403	341·8	213·6	463	392·6	245·4	523	443·5	277·1	583	494·4	308·9
344	291·7	182·3	404	342·6	214·1	464	393·5	245·9	524	444·4	277·7	584	495·3	309·5
345	292·6	182·8	405	343·5	214·6	465	394·3	246·4	525	445·2	278·2	585	496·1	310·0
346	293·4	183·4	406	344·3	215·1	466	395·2	246·9	526	446·1	278·7	586	497·0	310·5
347	294·3	183·9	407	345·2	215·7	467	396·0	247·5	527	446·9	279·3	587	497·8	311·1
348	295·1	184·4	408	346·0	216·2	468	396·9	248·0	528	447·8	279·8	588	498·7	311·6
349	296·0	184·9	409	346·9	216·7	469	397·7	248·5	529	448·6	280·3	589	499·5	312·1
350	296·8	185·5	410	347·7	217·3	470	398·6	249·1	530	449·5	280·9	590	500·3	312·7
351	297·7	186·0	411	348·5	217·8	471	399·4	249·6	531	450·3	281·4	591	501·2	313·2
352	298·5	186·5	412	349·4	218·3	472	400·3	250·1	532	451·2	281·9	592	502·0	313·7
353	299·4	187·1	413	350·2	218·9	473	401·1	250·7	533	452·0	282·4	593	502·9	314·2
354	300·2	187·6	414	351·1	219·4	474	402·0	251·2	534	452·9	283·0	594	503·7	314·8
355	301·1	188·1	415	351·9	219·9	475	402·8	251·7	535	453·7	283·5	595	504·6	315·3
356	301·9	188·7	416	352·8	220·4	476	403·7	252·2	536	454·6	284·0	596	505·4	315·8
357	302·8	189·2	417	353·6	221·0	477	404·5	252·8	537	455·4	284·6	597	506·3	316·4
358	303·6	189·7	418	354·5	221·5	478	405·4	253·3	538	456·2	285·1	598	507·1	316·9
359	304·4	190·2	419	355·3	222·0	479	406·2	253·8	539	457·1	285·6	599	508·0	317·4
360	305·3	190·8	420	356·2	222·6	480	407·1	254·4	540	457·9	286·2	600	508·8	318·0
Dist	Dep.	D. Lat	Dist	Dep.	D. Lat	Dist	Dep.	D. Lat	Dist	Dep.	D. Lat	Dist	Dep.	D. Lat
D Lon		Dep.	D Lon		Dep.	D Lon		Dep.	D Lon		Dep.	D Lon		Dep.

58°

33°

TRAVERSE TABLE
33 DEGREES.

2h 12m

D Lon	Dep.		D Lon	Dep.		D Lon	Dep.		D Lon	Dep.		D Lon	Dep.	
Dist	D.Lat	Dep.	Dist	D.Lat	Dep.	Dist	D.Lat	Dep.	Dist	D.Lat	Dep.	Dist	D.Lat	Dep.
1	00.8	00.5	61	51.2	33.2	121	101.5	65.9	181	151.8	98.6	241	202.1	131.3
2	01.7	01.1	62	52.0	33.8	122	102.3	66.4	182	152.6	99.1	242	203.0	131.8
3	02.5	01.6	63	52.8	34.3	123	103.2	67.0	183	153.5	99.7	243	203.8	132.3
4	03.4	02.2	64	53.7	34.9	124	104.0	67.5	184	154.3	100.2	244	204.6	132.9
5	04.2	02.7	65	54.5	35.4	125	104.8	68.1	185	155.2	100.8	245	205.5	133.4
6	05.0	03.3	66	55.4	35.9	126	105.7	68.6	186	156.0	101.3	246	206.3	134.0
7	05.9	03.8	67	56.2	36.5	127	106.5	69.2	187	156.8	101.8	247	207.2	134.5
8	06.7	04.4	68	57.0	37.0	128	107.3	69.7	188	157.7	102.4	248	208.0	135.1
9	07.5	04.9	69	57.9	37.6	129	108.2	70.3	189	158.5	102.9	249	208.8	135.6
10	08.4	05.4	70	58.7	38.1	130	109.0	70.8	190	159.3	103.5	250	209.7	136.2
11	09.2	06.0	71	59.5	38.7	131	109.9	71.3	191	160.2	104.0	251	210.5	136.7
12	10.1	06.5	72	60.4	39.2	132	110.7	71.9	192	161.0	104.6	252	211.3	137.2
13	10.9	07.1	73	61.2	39.8	133	111.5	72.4	193	161.9	105.1	253	212.2	137.8
14	11.7	07.6	74	62.1	40.3	134	112.4	73.0	194	162.7	105.7	254	213.0	138.3
15	12.6	08.2	75	62.9	40.8	135	113.2	73.5	195	163.5	106.2	255	213.9	138.9
16	13.4	08.7	76	63.7	41.4	136	114.1	74.1	196	164.4	106.7	256	214.7	139.4
17	14.3	09.3	77	64.6	41.9	137	114.9	74.6	197	165.2	107.3	257	215.5	140.0
18	15.1	09.8	78	65.4	42.5	138	115.7	75.2	198	166.1	107.8	258	216.4	140.5
19	15.9	10.3	79	66.3	43.0	139	116.6	75.7	199	166.9	108.4	259	217.2	141.1
20	16.8	10.9	80	67.1	43.6	140	117.4	76.2	200	167.7	108.9	260	218.1	141.6
21	17.6	11.4	81	67.9	44.1	141	118.3	76.8	201	168.6	109.5	261	218.9	142.2
22	18.5	12.0	82	68.8	44.7	142	119.1	77.3	202	169.4	110.0	262	219.7	142.7
23	19.3	12.5	83	69.6	45.2	143	119.9	77.9	203	170.3	110.6	263	220.6	143.2
24	20.1	13.1	84	70.4	45.7	144	120.8	78.4	204	171.1	111.1	264	221.4	143.8
25	21.0	13.6	85	71.3	46.3	145	121.6	79.0	205	171.9	111.7	265	222.2	144.3
26	21.8	14.2	86	72.1	46.8	146	122.4	79.5	206	172.8	112.2	266	223.1	144.9
27	22.6	14.7	87	73.0	47.4	147	123.3	80.1	207	173.6	112.7	267	223.9	145.4
28	23.5	15.2	88	73.8	47.9	148	124.1	80.6	208	174.4	113.3	268	224.8	146.0
29	24.3	15.8	89	74.6	48.5	149	125.0	81.2	209	175.3	113.8	269	225.6	146.5
30	25.2	16.3	90	75.5	49.0	150	125.8	81.7	210	176.1	114.4	270	226.4	147.1
31	26.0	16.9	91	76.3	49.6	151	126.6	82.2	211	177.0	114.9	271	227.3	147.6
32	26.8	17.4	92	77.2	50.1	152	127.5	82.8	212	177.8	115.5	272	228.1	148.1
33	27.7	18.0	93	78.0	50.7	153	128.3	83.3	213	178.6	116.0	273	229.0	148.7
34	28.5	18.5	94	78.8	51.2	154	129.2	83.9	214	179.5	116.6	274	229.8	149.2
35	29.4	19.1	95	79.7	51.7	155	130.0	84.4	215	180.3	117.1	275	230.6	149.8
36	30.2	19.6	96	80.5	52.3	156	130.8	85.0	216	181.2	117.6	276	231.5	150.3
37	31.0	20.2	97	81.4	52.8	157	131.7	85.5	217	182.0	118.2	277	232.3	150.9
38	31.9	20.7	98	82.2	53.4	158	132.5	86.1	218	182.8	118.7	278	233.2	151.4
39	32.7	21.2	99	83.0	53.9	159	133.3	86.6	219	183.7	119.3	279	234.0	152.0
40	33.5	21.8	100	83.9	54.5	160	134.2	87.1	220	184.5	119.8	280	234.8	152.5
41	34.4	22.3	101	84.7	55.0	161	135.0	87.7	221	185.3	120.4	281	235.7	153.0
42	35.2	22.9	102	85.5	55.6	162	135.9	88.2	222	186.2	120.9	282	236.5	153.6
43	36.1	23.4	103	86.4	56.1	163	136.7	88.8	223	187.0	121.5	283	237.3	154.1
44	36.9	24.0	104	87.2	56.6	164	137.5	89.3	224	187.9	122.0	284	238.2	154.7
45	37.7	24.5	105	88.1	57.2	165	138.4	89.9	225	188.7	122.5	285	239.0	155.2
46	38.6	25.1	106	88.9	57.7	166	139.2	90.4	226	189.5	123.1	286	239.9	155.8
47	39.4	25.6	107	89.7	58.3	167	140.1	91.0	227	190.4	123.6	287	240.7	156.3
48	40.3	26.1	108	90.6	58.8	168	140.9	91.5	228	191.2	124.2	288	241.5	156.9
49	41.1	26.7	109	91.4	59.4	169	141.7	92.0	229	192.1	124.7	289	242.4	157.4
50	41.9	27.2	110	92.3	59.9	170	142.6	92.6	230	192.9	125.3	290	243.2	157.9
51	42.8	27.8	111	93.1	60.5	171	143.4	93.1	231	193.7	125.8	291	244.1	158.5
52	43.6	28.3	112	93.9	61.0	172	144.3	93.7	232	194.6	126.4	292	244.9	159.0
53	44.4	28.9	113	94.8	61.5	173	145.1	94.2	233	195.4	126.9	293	245.7	159.6
54	45.3	29.4	114	95.6	62.1	174	145.9	94.8	234	196.2	127.4	294	246.6	160.1
55	46.1	30.0	115	96.4	62.6	175	146.8	95.3	235	197.1	128.0	295	247.4	160.7
56	47.0	30.5	116	97.3	63.2	176	147.6	95.9	236	197.9	128.5	296	248.2	161.2
57	47.8	31.0	117	98.1	63.7	177	148.4	96.4	237	198.8	129.1	297	249.1	161.8
58	48.6	31.6	118	99.0	64.3	178	149.3	96.9	238	199.6	129.6	298	249.9	162.3
59	49.5	32.1	119	99.8	64.8	179	150.1	97.5	239	200.4	130.2	299	250.8	162.8
60	50.3	32.7	120	100.6	65.4	180	151.0	98.0	240	201.3	130.7	300	251.6	163.4
Dist	Dep.	D.Lat	Dist	Dep.	D.Lat	Dist	Dep.	D.Lat	Dist	Dep.	D.Lat	Dist	Dep.	D.Lat
D Lon		Dep.	D Lon		Dep.	D Lon		Dep.	D Lon		Dep.	D Lon		Dep.

57°

	327°	213°	**TRAVERSE TABLE** 33 DEGREES.			033° 147° 2h 12m	**33°**

DLon	Dep.		DLon	Dep.		DLon	Dep.		DLon	Dep.		DLon	Dep.		DLon	Dep.	
Dist	D.Lat	Dep.	Dist	D.Lat	Dep.	Dist	D.Lat	Dep.	Dist	D.Lat	Dep.	Dist	D.Lat	Dep.	Dist	D.Lat	Dep.
301	252·4	163·9	361	302·8	196·6	421	353·1	229·3	481	403·4	262·0	541	453·7	294·6			
302	253·3	164·5	362	303·6	197·2	422	353·9	229·8	482	404·2	262·5	542	454·6	295·2			
303	254·1	165·0	363	304·4	197·7	423	354·8	230·4	483	405·1	263·1	543	455·4	295·7			
304	255·0	165·6	364	305·3	198·2	424	355·6	230·9	484	405·9	263·6	544	456·2	296·3			
305	255·8	166·1	365	306·1	198·8	425	356·4	231·5	485	406·8	264·1	545	457·1	296·8			
306	256·6	166·7	366	307·0	199·3	426	357·3	232·0	486	407·6	264·7	546	457·9	297·4			
307	257·5	167·2	367	307·8	199·9	427	358·1	232·6	487	408·4	265·2	547	458·8	297·9			
308	258·3	167·7	368	308·6	200·4	428	359·0	233·1	488	409·3	265·8	548	459·6	298·5			
309	259·1	168·3	369	309·5	201·0	429	359·8	233·7	489	410·1	266·3	549	460·4	299·0			
310	260·0	168·8	370	310·3	201·5	430	360·6	234·2	490	410·9	266·9	550	461·3	299·6			
311	260·8	169·4	371	311·1	202·1	431	361·5	234·7	491	411·8	267·4	551	462·1	300·1			
312	261·7	169·9	372	312·0	202·6	432	362·3	235·3	492	412·6	268·0	552	462·9	300·6			
313	262·5	170·5	373	312·8	203·2	433	363·1	235·8	493	413·5	268·5	553	463·8	301·2			
314	263·3	171·0	374	313·7	203·7	434	364·0	236·4	494	414·3	269·0	554	464·6	301·7			
315	264·2	171·6	375	314·5	204·2	435	364·8	236·9	495	415·1	269·6	555	465·5	302·3			
316	265·0	172·1	376	315·3	204·7	436	365·7	237·5	496	416·0	270·1	556	466·3	302·8			
317	265·9	172·7	377	316·2	205·3	437	366·5	238·0	497	416·8	270·7	557	467·1	303·4			
318	266·7	173·2	378	317·0	205·9	438	367·3	238·6	498	417·7	271·2	558	468·0	303·9			
319	267·5	173·7	379	317·9	206·4	439	368·2	239·1	499	418·5	271·8	559	468·8	304·5			
320	268·4	174·3	380	318·7	207·0	440	369·2	239·6	500	419·3	272·3	560	469·7	305·0			
321	269·2	174·8	381	319·5	207·5	441	369·9	240·2	501	420·2	272·9	561	470·5	305·5			
322	270·1	175·4	382	320·4	208·1	442	370·7	240·7	502	421·0	273·4	562	471·3	306·1			
323	270·9	175·9	383	321·2	208·6	443	371·5	241·3	503	421·9	274·0	563	472·2	306·6			
324	271·7	176·5	384	322·0	209·1	444	372·4	241·8	504	422·7	274·5	564	473·0	307·2			
325	272·6	177·0	385	322·9	209·7	445	373·2	242·4	505	423·5	275·0	565	473·8	307·7			
326	273·4	177·6	386	323·7	210·2	446	374·0	242·9	506	424·4	275·6	566	474·7	308·3			
327	274·2	178·1	387	324·5	210·8	447	374·9	243·5	507	425·2	276·1	567	475·5	308·8			
328	275·1	178·6	388	325·4	211·3	448	375·7	244·0	508	426·0	276·7	568	476·4	309·4			
329	275·9	179·2	389	326·2	211·9	449	376·6	244·5	509	426·9	277·2	569	477·2	309·9			
330	276·8	179·7	390	327·1	212·4	450	377·4	245·1	510	427·7	277·8	570	478·0	310·4			
331	277·6	180·3	391	327·9	213·0	451	378·2	245·6	511	428·6	278·3	571	478·9	311·0			
332	278·4	180·8	392	328·8	213·5	452	379·1	246·2	512	429·4	278·9	572	479·7	311·5			
333	279·3	181·4	393	329·6	214·0	453	379·9	246·7	513	430·2	279·4	573	480·6	312·1			
334	280·1	181·9	394	330·4	214·6	454	380·8	247·3	514	431·1	279·9	574	481·4	312·6			
335	281·0	182·5	395	331·3	215·1	455	381·6	247·8	515	431·9	280·5	575	482·2	313·2			
336	281·8	183·0	396	332·1	215·7	456	382·4	248·4	516	432·8	281·0	576	483·1	313·7			
337	282·6	183·5	397	333·0	216·2	457	383·3	248·9	517	433·6	281·6	577	483·9	314·3			
338	283·5	184·1	398	333·8	216·8	458	384·1	249·4	518	434·4	282·1	578	484·8	314·8			
339	284·3	184·6	399	334·6	217·3	459	384·9	250·0	519	435·3	282·7	579	485·6	315·3			
340	285·1	185·2	400	335·5	217·9	460	385·8	250·5	520	436·1	283·2	580	486·4	315·9			
341	286·0	185·7	401	336·3	218·4	461	386·6	251·1	521	436·9	283·8	581	487·3	316·4			
342	286·8	186·3	402	337·1	218·9	462	387·5	251·6	522	437·8	284·3	582	488·1	317·0			
343	287·7	186·8	403	338·0	219·5	463	388·3	252·2	523	438·6	284·8	583	488·9	317·5			
344	288·5	187·4	404	338·8	220·0	464	389·1	252·7	524	439·5	285·4	584	489·8	318·1			
345	289·3	187·9	405	339·7	220·6	465	390·0	253·3	525	440·3	285·9	585	490·6	318·6			
346	290·2	188·4	406	340·5	221·1	466	390·8	253·8	526	441·1	286·5	586	491·5	319·2			
347	291·0	189·0	407	341·3	221·7	467	391·7	254·3	527	442·0	287·0	587	492·3	319·7			
348	291·9	189·5	408	342·2	222·2	468	392·5	254·9	528	442·8	287·6	588	493·1	320·2			
349	292·7	190·1	409	343·0	222·8	469	393·3	255·4	529	443·7	288·1	589	494·0	320·8			
350	293·5	190·6	410	343·9	223·3	470	394·2	256·0	530	444·5	288·7	590	494·8	321·3			
351	294·4	191·2	411	344·7	223·8	471	395·0	256·5	531	445·3	289·2	591	495·7	321·9			
352	295·2	191·7	412	345·5	224·4	472	395·9	257·1	532	446·2	289·7	592	496·5	322·4			
353	296·1	192·3	413	346·4	224·9	473	396·7	257·6	533	447·0	290·3	593	497·3	323·0			
354	296·9	192·8	414	347·2	225·5	474	397·5	258·2	534	447·9	290·8	594	498·2	323·5			
355	297·7	193·3	415	348·0	226·0	475	398·4	258·7	535	448·7	291·4	595	499·0	324·1			
356	298·6	193·9	416	348·9	226·6	476	399·2	259·2	536	449·5	291·9	596	499·8	324·6			
357	299·4	194·4	417	349·7	227·1	477	400·0	259·8	537	450·4	292·5	597	500·7	325·1			
358	300·2	195·0	418	350·6	227·7	478	400·9	260·3	538	451·2	293·0	598	501·5	325·7			
359	301·1	195·5	419	351·4	228·2	479	401·7	260·9	539	452·0	293·6	599	502·4	326·2			
360	301·9	196·1	420	352·2	228·7	480	402·6	261·4	540	452·9	294·1	600	503·2	326·8			
Dist	Dep.	D.Lat	Dist	Dep.	D.Lat	Dist	Dep.	D.Lat	Dist	Dep.	D.Lat	Dist	Dep.	D.Lat	Dist	Dep.	D.Lat
DLon		Dep.	DLon		Dep.	DLon		Dep.	DLon		Dep.	DLon		Dep.	DLon		Dep.

303° 237°	**57 DEGREES.**	057° 123° 3h 48m	**57°**

TRAVERSE TABLE

34°

34 DEGREES.

D Lon	Dep.		D Lon	Dep.		D Lon	Dep.		D Lon	Dep.		D Lon	Dep.	
Dist	D.Lat	Dep.	Dist	D.Lat	Dep.	Dist	D.Lat	Dep.	Dist	D.Lat	Dep.	Dist	D.Lat	Dep.
1	00·8	00·6	61	50·6	34·1	121	100·3	67·7	181	150·1	101·2	241	199·8	134·8
2	01·7	01·1	62	51·4	34·7	122	101·1	68·2	182	150·9	101·8	242	200·6	135·3
3	02·5	01·7	63	52·2	35·2	123	102·0	68·8	183	151·7	102·3	243	201·5	135·9
4	03·3	02·2	64	53·1	35·8	124	102·8	69·3	184	152·5	102·9	244	202·3	136·4
5	04·1	02·8	65	53·9	36·3	125	103·6	69·9	185	153·4	103·5	245	203·1	137·0
6	05·0	03·4	66	54·7	36·9	126	104·5	70·5	186	154·2	104·0	246	203·9	137·6
7	05·8	03·9	67	55·5	37·5	127	105·3	71·0	187	155·0	104·6	247	204·8	138·1
8	06·6	04·5	68	56·4	38·0	128	106·1	71·6	188	155·9	105·1	248	205·6	138·7
9	07·5	05·0	69	57·2	38·6	129	106·9	72·1	189	156·7	105·7	249	206·4	139·2
10	08·3	05·6	70	58·0	39·1	130	107·8	72·7	190	157·5	106·2	250	207·3	139·8
11	09·1	06·2	71	58·9	39·7	131	108·6	73·3	191	158·3	106·8	251	208·1	140·4
12	09·9	06·7	72	59·7	40·3	132	109·4	73·8	192	159·2	107·4	252	208·9	140·9
13	10·8	07·3	73	60·5	40·8	133	110·3	74·4	193	160·0	107·9	253	209·7	141·5
14	11·6	07·8	74	61·3	41·4	134	111·1	74·9	194	160·8	108·5	254	210·6	142·0
15	12·4	08·4	75	62·2	41·9	135	111·9	75·5	195	161·7	109·0	255	211·4	142·6
16	13·3	08·9	76	63·0	42·5	136	112·7	76·1	196	162·5	109·6	256	212·2	143·2
17	14·1	09·5	77	63·8	43·1	137	113·6	76·6	197	163·3	110·2	257	213·1	143·7
18	14·9	10·1	78	64·7	43·6	138	114·4	77·2	198	164·1	110·7	258	213·9	144·3
19	15·8	10·6	79	65·5	44·2	139	115·2	77·7	199	165·0	111·3	259	214·7	144·8
20	16·6	11·2	80	66·3	44·7	140	116·1	78·3	200	165·8	111·8	260	215·5	145·4
21	17·4	11·7	81	67·2	45·3	141	116·9	78·8	201	166·6	112·4	261	216·4	145·9
22	18·2	12·3	82	68·0	45·9	142	117·7	79·4	202	167·5	113·0	262	217·2	146·5
23	19·1	12·9	83	68·8	46·4	143	118·6	80·0	203	168·3	113·5	263	218·0	147·1
24	19·9	13·4	84	69·6	47·0	144	119·4	80·5	204	169·1	114·1	264	218·9	147·6
25	20·7	14·0	85	70·5	47·5	145	120·2	81·1	205	170·0	114·6	265	219·7	148·2
26	21·6	14·5	86	71·3	48·1	146	121·0	81·6	206	170·8	115·2	266	220·5	148·7
27	22·4	15·1	87	72·1	48·6	147	121·9	82·2	207	171·6	115·8	267	221·4	149·3
28	23·2	15·7	88	73·0	49·2	148	122·7	82·8	208	172·4	116·3	268	222·2	149·9
29	24·0	16·2	89	73·8	49·8	149	123·5	83·3	209	173·3	116·9	269	223·0	150·4
30	24·9	16·8	90	74·6	50·3	150	124·4	83·9	210	174·1	117·4	270	223·8	151·0
31	25·7	17·3	91	75·4	50·9	151	125·2	84·4	211	174·9	118·0	271	224·7	151·5
32	26·5	17·9	92	76·3	51·4	152	126·0	85·0	212	175·8	118·5	272	225·5	152·1
33	27·4	18·5	93	77·1	52·0	153	126·8	85·6	213	176·6	119·1	273	226·3	152·7
34	28·2	19·0	94	77·9	52·6	154	127·7	86·1	214	177·4	119·7	274	227·2	153·2
35	29·0	19·6	95	78·8	53·1	155	128·5	86·7	215	178·2	120·2	275	228·0	153·8
36	29·8	20·1	96	79·6	53·7	156	129·3	87·2	216	179·1	120·8	276	228·8	154·3
37	30·7	20·7	97	80·4	54·2	157	130·2	87·8	217	179·9	121·3	277	229·6	154·9
38	31·5	21·2	98	81·2	54·8	158	131·0	88·4	218	180·7	121·9	278	230·5	155·5
39	32·3	21·8	99	82·1	55·4	159	131·8	88·9	219	181·6	122·5	279	231·3	156·0
40	33·2	22·4	100	82·9	55·9	160	132·6	89·5	220	182·4	123·0	280	232·1	156·6
41	34·0	22·9	101	83·7	56·5	161	133·5	90·0	221	183·2	123·6	281	233·0	157·1
42	34·8	23·5	102	84·6	57·0	162	134·3	90·6	222	184·0	124·1	282	233·8	157·7
43	35·6	24·0	103	85·4	57·6	163	135·1	91·1	223	184·9	124·7	283	234·6	158·3
44	36·5	24·6	104	86·2	58·2	164	136·0	91·7	224	185·7	125·3	284	235·4	158·8
45	37·3	25·2	105	87·0	58·7	165	136·8	92·3	225	186·5	125·8	285	236·3	159·4
46	38·1	25·7	106	87·9	59·3	166	137·6	92·8	226	187·4	126·4	286	237·1	159·9
47	39·0	26·3	107	88·7	59·8	167	138·4	93·4	227	188·2	126·9	287	237·9	160·5
48	39·8	26·8	108	89·5	60·4	168	139·3	93·9	228	189·0	127·5	288	238·8	161·0
49	40·6	27·4	109	90·4	61·0	169	140·1	94·5	229	189·8	128·1	289	239·6	161·6
50	41·5	28·0	110	91·2	61·5	170	140·9	95·1	230	190·7	128·6	290	240·4	162·2
51	42·3	28·5	111	92·0	62·1	171	141·8	95·6	231	191·5	129·2	291	241·2	162·7
52	43·1	29·1	112	92·9	62·6	172	142·6	96·2	232	192·3	129·7	292	242·1	163·3
53	43·9	29·6	113	93·7	63·2	173	143·4	96·7	233	193·2	130·3	293	242·9	163·8
54	44·8	30·2	114	94·5	63·7	174	144·3	97·3	234	194·0	130·9	294	243·7	164·4
55	45·6	30·8	115	95·3	64·3	175	145·1	97·9	235	194·8	131·4	295	244·6	165·0
56	46·4	31·3	116	96·2	64·9	176	145·9	98·4	236	195·7	132·0	296	245·4	165·5
57	47·3	31·9	117	97·0	65·4	177	146·7	99·0	237	196·5	132·5	297	246·2	166·1
58	48·1	32·4	118	97·8	66·0	178	147·6	99·5	238	197·3	133·1	298	247·1	166·6
59	48·9	33·0	119	98·7	66·5	179	148·4	100·1	239	198·1	133·6	299	247·9	167·2
60	49·7	33·6	120	99·5	67·1	180	149·2	100·7	240	199·0	134·2	300	248·7	167·8

Dist	Dep.	D.Lat	Dist	Dep.	D.Lat	Dist	Dep.	D.Lat	Dist	Dep.	D.Lat	Dist	Dep.	D.Lat
D Lon		Dep.	D Lon		Dep.	D Lon		Dep.	D Lon		Dep.	D Lon		Dep.

56°

56 DEGREES.

TRAVERSE TABLE
34 DEGREES.

326° ↑ / 214°   ↑034° / 146°   2h 16m   **34°**

Dist	D.Lat	Dep.	Dist	D.Lat	Dep.	Dist	D.Lat	Dep.	Dist	D.Lat	Dep.	Dist	D.Lat	Dep.
301	249·5	168·3	361	299·3	201·9	421	349·0	235·4	481	398·8	269·0	541	448·5	302·5
302	250·4	168·9	362	300·1	202·4	422	349·9	236·0	482	399·6	269·5	542	449·3	303·1
303	251·2	169·4	363	300·9	203·0	423	350·7	236·5	483	400·4	270·1	543	450·2	303·6
304	252·0	170·0	364	301·8	203·5	424	351·5	237·1	484	401·3	270·6	544	451·0	304·2
305	252·9	170·6	365	302·6	204·1	425	352·3	237·7	485	402·1	271·2	545	451·8	304·8
306	253·7	171·1	366	303·4	204·7	426	353·2	238·2	486	402·9	271·8	546	452·7	305·3
307	254·5	171·7	367	304·3	205·2	427	354·0	238·8	487	403·7	272·3	547	453·5	305·9
308	255·3	172·2	368	305·1	205·8	428	354·8	239·3	488	404·6	272·9	548	454·3	306·4
309	256·2	172·8	369	305·9	206·3	429	355·7	239·9	489	405·4	273·4	549	455·1	307·0
310	257·0	173·3	370	306·7	206·9	430	356·5	240·5	490	406·2	274·0	550	456·0	307·6
311	257·8	173·9	371	307·6	207·5	431	357·3	241·0	491	407·1	274·6	551	456·8	308·1
312	258·7	174·5	372	308·4	208·0	432	358·1	241·6	492	407·9	275·1	552	457·6	308·7
313	259·5	175·0	373	309·2	208·6	433	359·0	242·1	493	408·7	275·7	553	458·5	309·2
314	260·3	175·6	374	310·1	209·1	434	359·8	242·7	494	409·5	276·2	554	459·3	309·8
315	261·1	176·1	375	310·9	209·7	435	360·6	243·2	495	410·4	276·8	555	460·1	310·4
316	262·0	176·7	376	311·7	210·3	436	361·5	243·8	496	411·2	277·4	556	460·9	310·9
317	262·8	177·3	377	312·5	210·8	437	362·3	244·4	497	412·0	277·9	557	461·8	311·5
318	263·6	177·8	378	313·4	211·4	438	363·1	244·9	498	412·9	278·5	558	462·6	312·0
319	264·5	178·4	379	314·2	211·9	439	363·9	245·5	499	413·7	279·0	559	463·4	312·6
320	265·3	178·9	380	315·0	212·5	440	364·8	246·0	500	414·5	279·6	560	464·3	313·1
321	266·1	179·5	381	315·9	213·1	441	365·6	246·6	501	415·3	280·2	561	465·1	313·7
322	267·0	180·1	382	316·7	213·6	442	366·4	247·2	502	416·2	280·7	562	465·9	314·3
323	267·8	180·6	383	317·5	214·2	443	367·3	247·7	503	417·0	281·3	563	466·7	314·8
324	268·6	181·2	384	318·4	214·7	444	368·1	248·3	504	417·8	281·8	564	467·6	315·5
325	269·4	181·7	385	319·2	215·3	445	368·9	248·8	505	418·7	282·4	565	468·4	315·9
326	270·3	182·3	386	320·0	215·8	446	369·8	249·4	506	419·5	283·0	566	469·2	316·5
327	271·1	182·9	387	320·8	216·4	447	370·6	250·0	507	420·3	283·5	567	470·1	317·1
328	271·9	183·4	388	321·7	217·0	448	371·4	250·5	508	421·2	284·1	568	470·9	317·6
329	272·8	184·0	389	322·5	217·5	449	372·2	251·1	509	422·0	284·6	569	471·7	318·2
330	273·6	184·5	390	323·3	218·1	450	373·1	251·6	510	422·8	285·2	570	472·6	318·7
331	274·4	185·1	391	324·2	218·6	451	373·9	252·2	511	423·6	285·7	571	473·4	319·3
332	275·2	185·7	392	325·0	219·2	452	374·7	252·8	512	424·5	286·3	572	474·2	319·9
333	276·1	186·2	393	325·8	219·8	453	375·6	253·3	513	425·3	287·0	573	475·0	320·4
334	276·9	186·8	394	326·6	220·3	454	376·4	253·9	514	426·1	287·4	574	475·9	321·0
335	277·7	187·3	395	327·5	220·9	455	377·2	254·4	515	427·0	288·0	575	476·7	321·5
336	278·6	187·9	396	328·3	221·4	456	378·0	255·0	516	427·8	288·5	576	477·5	322·1
337	279·4	188·4	397	329·1	222·0	457	378·9	255·6	517	428·6	289·1	577	478·4	322·7
338	280·2	189·0	398	330·0	222·6	458	379·7	256·1	518	429·4	289·7	578	479·2	323·2
339	281·0	189·6	399	330·8	223·1	459	380·5	256·7	519	430·3	290·2	579	480·0	323·8
340	281·9	190·1	400	331·6	223·7	460	381·4	257·2	520	431·1	290·8	580	480·8	324·3
341	282·7	190·7	401	332·4	224·2	461	382·2	257·8	521	431·9	291·3	581	481·7	324·9
342	283·5	191·2	402	333·3	224·8	462	383·0	258·3	522	432·8	291·9	582	482·5	325·4
343	284·4	191·8	403	334·1	225·4	463	383·8	258·9	523	433·6	292·5	583	483·3	326·0
344	285·2	192·4	404	334·9	225·9	464	384·7	259·5	524	434·4	293·0	584	484·2	326·6
345	286·0	192·9	405	335·8	226·5	465	385·5	260·0	525	435·2	293·6	585	485·0	327·1
346	286·8	193·5	406	336·6	227·0	466	386·3	260·6	526	436·1	294·1	586	485·8	327·7
347	287·7	194·0	407	337·4	227·6	467	387·2	261·1	527	436·9	294·7	587	486·6	328·2
348	288·5	194·6	408	338·2	228·2	468	388·0	261·7	528	437·7	295·3	588	487·5	328·8
349	289·3	195·2	409	339·1	228·7	469	388·8	262·3	529	438·6	295·8	589	488·3	329·4
350	290·2	195·7	410	339·9	229·3	470	389·6	262·8	530	439·4	296·4	590	489·1	329·9
351	291·0	196·3	411	340·7	229·8	471	390·5	263·4	531	440·2	296·9	591	490·0	330·5
352	291·8	196·8	412	341·6	230·4	472	391·3	263·9	532	441·0	297·5	592	490·8	331·0
353	292·7	197·4	413	342·4	230·9	473	392·1	264·5	533	441·9	298·0	593	491·6	331·6
354	293·5	198·0	414	343·2	231·5	474	393·0	265·1	534	442·7	298·6	594	492·4	332·2
355	294·3	198·5	415	344·1	232·1	475	393·8	265·6	535	443·5	299·2	595	493·3	332·7
356	295·1	199·1	416	344·9	232·6	476	394·6	266·2	536	444·4	299·7	596	494·1	333·3
357	296·0	199·6	417	345·7	233·2	477	395·5	266·7	537	445·3	300·3	597	494·9	333·8
358	296·8	200·2	418	346·5	233·7	478	396·3	267·3	538	446·0	300·9	598	495·8	334·4
359	297·6	200·8	419	347·4	234·3	479	397·1	267·9	539	446·9	301·4	599	496·6	335·0
360	298·5	201·3	420	348·2	234·9	480	397·9	268·4	540	447·7	302·0	600	497·4	335·5

| Dist | Dep. | D.Lat | Dist | Dep. | D.Lat | Dist | Dep. | D.Lat | Dist | Dep. | D.Lat | Dist | Dep. | D.Lat |

304° ↑ / 236°   **56 DEGREES.**   056° / 124°   3h 44m   **56°**

35°

TRAVERSE TABLE
35 DEGREES.

325° / 215° 035° / 145° 2h 20m

Dist	D.Lat	Dep.	Dist	D.Lat	Dep.	Dist	D.Lat	Dep.	Dist	D.Lat	Dep.	Dist	D.Lat	Dep.
1	00.8	00.6	61	50.0	35.0	121	99.1	69.4	181	148.3	103.8	241	197.4	138.2
2	01.6	01.1	62	50.8	35.6	122	99.9	70.0	182	149.1	104.4	242	198.2	138.8
3	02.5	01.7	63	51.6	36.1	123	100.8	70.5	183	149.9	105.0	243	199.1	139.4
4	03.3	02.3	64	52.4	36.7	124	101.6	71.1	184	150.7	105.5	244	199.9	140.0
5	04.1	02.9	65	53.2	37.3	125	102.4	71.7	185	151.5	106.1	245	200.7	140.5
6	04.9	03.4	66	54.1	37.9	126	103.2	72.3	186	152.4	106.7	246	201.5	141.1
7	05.7	04.0	67	54.9	38.4	127	104.0	72.8	187	153.2	107.3	247	202.3	141.7
8	06.6	04.6	68	55.7	39.0	128	104.9	73.4	188	154.0	107.8	248	203.1	142.2
9	07.4	05.2	69	56.5	39.6	129	105.7	74.0	189	154.8	108.4	249	204.0	142.8
10	08.2	05.7	70	57.3	40.2	130	106.5	74.6	190	155.6	109.0	250	204.8	143.4
11	09.0	06.3	71	58.2	40.7	131	107.3	75.1	191	156.5	109.6	251	205.6	144.0
12	09.8	06.9	72	59.0	41.3	132	108.1	75.7	192	157.3	110.1	252	206.4	144.5
13	10.6	07.5	73	59.8	41.9	133	108.9	76.3	193	158.1	110.7	253	207.2	145.1
14	11.5	08.0	74	60.6	42.4	134	109.8	76.9	194	158.9	111.3	254	208.1	145.7
15	12.3	08.6	75	61.4	43.0	135	110.6	77.4	195	159.7	111.8	255	208.9	146.3
16	13.1	09.2	76	62.3	43.6	136	111.4	78.0	196	160.6	112.4	256	209.7	146.8
17	13.9	09.8	77	63.1	44.2	137	112.2	78.6	197	161.4	113.0	257	210.5	147.4
18	14.7	10.3	78	63.9	44.7	138	113.0	79.2	198	162.2	113.6	258	211.3	148.0
19	15.6	10.9	79	64.7	45.3	139	113.9	79.7	199	163.0	114.1	259	212.2	148.6
20	16.4	11.5	80	65.5	45.9	140	114.7	80.3	200	163.8	114.7	260	213.0	149.1
21	17.2	12.0	81	66.4	46.5	141	115.5	80.9	201	164.6	115.3	261	213.8	149.7
22	18.0	12.6	82	67.2	47.0	142	116.3	81.4	202	165.5	115.9	262	214.6	150.3
23	18.8	13.2	83	68.0	47.6	143	117.1	82.0	203	166.3	116.4	263	215.4	150.9
24	19.7	13.8	84	68.8	48.2	144	118.0	82.6	204	167.1	117.0	264	216.3	151.4
25	20.5	14.3	85	69.6	48.8	145	118.8	83.2	205	167.9	117.6	265	217.1	152.0
26	21.3	14.9	86	70.4	49.3	146	119.6	83.7	206	168.7	118.2	266	217.9	152.6
27	22.1	15.5	87	71.3	49.9	147	120.4	84.3	207	169.6	118.7	267	218.7	153.1
28	22.9	16.1	88	72.1	50.5	148	121.2	84.9	208	170.4	119.3	268	219.5	153.7
29	23.8	16.6	89	72.9	51.0	149	122.1	85.5	209	171.2	119.9	269	220.4	154.3
30	24.6	17.2	90	73.7	51.6	150	122.9	86.0	210	172.0	120.5	270	221.2	154.9
31	25.4	17.8	91	74.5	52.2	151	123.7	86.6	211	172.8	121.0	271	222.0	155.4
32	26.2	18.4	92	75.4	52.8	152	124.5	87.2	212	173.7	121.6	272	222.8	156.0
33	27.0	18.9	93	76.2	53.3	153	125.3	87.8	213	174.5	122.2	273	223.6	156.6
34	27.9	19.5	94	77.0	53.9	154	126.1	88.3	214	175.3	122.7	274	224.4	157.2
35	28.7	20.1	95	77.8	54.5	155	127.0	88.9	215	176.1	123.3	275	225.3	157.7
36	29.5	20.6	96	78.6	55.1	156	127.8	89.5	216	176.9	123.9	276	226.1	158.3
37	30.3	21.2	97	79.5	55.6	157	128.6	90.1	217	177.8	124.5	277	226.9	158.9
38	31.1	21.8	98	80.3	56.2	158	129.4	90.6	218	178.6	125.0	278	227.7	159.5
39	31.9	22.4	99	81.1	56.8	159	130.2	91.2	219	179.4	125.6	279	228.5	160.0
40	32.8	22.9	100	81.9	57.4	160	131.1	91.8	220	180.2	126.2	280	229.4	160.6
41	33.6	23.5	101	82.7	57.9	161	131.9	92.3	221	181.0	126.8	281	230.2	161.2
42	34.4	24.1	102	83.6	58.5	162	132.7	92.9	222	181.9	127.3	282	231.0	161.7
43	35.2	24.7	103	84.4	59.1	163	133.5	93.5	223	182.7	127.9	283	231.8	162.3
44	36.0	25.2	104	85.2	59.7	164	134.3	94.1	224	183.5	128.5	284	232.6	162.9
45	36.9	25.8	105	86.0	60.2	165	135.2	94.6	225	184.3	129.1	285	233.5	163.5
46	37.7	26.4	106	86.8	60.8	166	136.0	95.2	226	185.1	129.6	286	234.3	164.0
47	38.5	27.0	107	87.6	61.4	167	136.8	95.8	227	185.9	130.2	287	235.1	164.6
48	39.3	27.5	108	88.5	61.9	168	137.6	96.4	228	186.8	130.8	288	235.9	165.2
49	40.1	28.1	109	89.3	62.5	169	138.4	96.9	229	187.6	131.3	289	236.7	165.8
50	41.0	28.7	110	90.1	63.1	170	139.3	97.5	230	188.4	131.9	290	237.6	166.3
51	41.8	29.3	111	90.9	63.7	171	140.1	98.1	231	189.2	132.5	291	238.4	166.9
52	42.6	29.8	112	91.7	64.2	172	140.9	98.7	232	190.0	133.1	292	239.2	167.5
53	43.4	30.4	113	92.6	64.8	173	141.7	99.2	233	190.9	133.6	293	240.0	168.0
54	44.2	31.0	114	93.4	65.4	174	142.5	99.8	234	191.7	134.2	294	240.8	168.6
55	45.1	31.5	115	94.2	66.0	175	143.4	100.4	235	192.5	134.8	295	241.6	169.2
56	45.9	32.1	116	95.0	66.5	176	144.2	100.9	236	193.3	135.4	296	242.5	169.8
57	46.7	32.7	117	95.8	67.1	177	145.0	101.5	237	194.1	135.9	297	243.3	170.4
58	47.5	33.3	118	96.7	67.7	178	145.8	102.1	238	195.0	136.5	298	244.1	170.9
59	48.3	33.8	119	97.5	68.3	179	146.6	102.7	239	195.8	137.1	299	244.9	171.5
60	49.1	34.4	120	98.3	68.8	180	147.4	103.2	240	196.6	137.7	300	245.7	172.1
Dist	Dep.	D.Lat	Dist	Dep.	D.Lat	Dist	Dep.	D.Lat	Dist	Dep.	D.Lat	Dist	Dep.	D.Lat

55°

305° / 235° 055° / 125° 3h 40m

55 DEGREES.

	325° ↑												035° ↑				35°
	215°		**TRAVERSE TABLE**										145°		2h 20m		
			35 DEGREES.														

D Lon	Dep.		D Lon	Dep.		D Lon	Dep.		D Lon	Dep.		D Lon	Dep.		D Lon	Dep.	
Dist	D.Lat	Dep.	Dist	D.Lat	Dep.	Dist	D.Lat	Dep.	Dist	D.Lat	Dep.	Dist	D.Lat	Dep.	Dist	D.Lat	Dep.
301	246·6	172·6	361	295·7	207·1	421	344·9	241·5	481	394·0	275 9	541	443·2	310·3			
302	247·4	173·2	362	296·5	207·6	422	345·7	242·0	482	394·8	276·5	542	444·0	310·9			
303	248·2	173·8	363	297·4	208·2	423	346·5	242·6	483	395·7	277·0	543	444·8	311·5			
304	249·0	174·4	364	298·2	208·8	424	347·3	243·2	484	396·5	277·6	544	445·6	312·0			
305	249·8	174·9	365	299·0	209·4	425	348·1	243·8	485	397·3	278·2	545	446·4	312·6			
306	250·7	175·5	366	299·8	209·9	426	349·0	244·3	486	398·1	278·8	546	447·3	313·2			
307	251·5	176·1	367	300·6	210·5	427	349·8	244·9	487	398·9	279·3	547	448·1	313·7			
308	252·3	176·7	368	301·4	211·1	428	350·6	245·5	488	399·7	279·9	548	448·9	314·3			
309	253·1	177·2	369	302·3	211·6	429	351·4	246·1	489	400·6	280·5	549	449·7	314·9			
310	253·9	177·8	370	303·1	212·2	430	352·2	246·6	490	401·4	281·1	550	450·5	315·5			
311	254·8	178·4	371	303·9	212·8	431	353·1	247·2	491	402·2	281·6	551	451·4	316·0			
312	255·6	179·0	372	304·7	213·4	432	353·9	247·8	492	403·0	282·2	552	452·2	316·6			
313	256·4	179·5	373	305·5	213·9	433	354·7	248·4	493	403·8	282·8	553	453·0	317·2			
314	257·2	180·1	374	306·4	214·5	434	355·5	248·9	494	404·7	283·3	554	453·8	317·8			
315	258·0	180·7	375	307·2	215·1	435	356·3	249·5	495	405·5	283·9	555	454·6	318·3			
316	258·9	181·3	376	308·0	215·7	436	357·2	250·1	496	406·3	284·5	556	455·4	318·9			
317	259·7	181·8	377	308·8	216·2	437	358·0	250·7	497	407·1	285·1	557	456·3	319·5			
318	260·5	182·4	378	309·6	216·8	438	358·8	251·2	498	407·9	285·6	558	457·1	320·1			
319	261·3	183·0	379	310·5	217·4	439	359·6	251·8	499	408·8	286·2	559	457·9	320·6			
320	262·1	183·5	380	311·3	218·0	440	360·4	252·4	500	409·6	286·8	560	458·7	321·2			
321	262·9	184·1	381	312·1	218·5	441	361·2	252·9	501	410·4	287·4	561	459·5	321·8			
322	263·8	184·7	382	312·9	219·1	442	362·1	253·5	502	411·2	287·9	562	460·4	322·3			
323	264·6	185·3	383	313·7	219·7	443	362·9	254·1	503	412·0	288·5	563	461·2	322·9			
324	265·4	185·8	384	314·6	220·3	444	363·7	254·7	504	412·9	289·1	564	462·0	323·5			
325	266·2	186·4	385	315·4	220·8	445	364·5	255·2	505	413·7	289·7	565	462·8	324·1			
326	267·0	187·0	386	316·2	221·4	446	365·3	255·8	506	414·5	290·2	566	463·6	324·6			
327	267·9	187·6	387	317·0	222·0	447	366·2	256·4	507	415·3	290·8	567	464·5	325·2			
328	268·7	188·1	388	317·8	222·5	448	367·0	257·0	508	416·1	291·4	568	465·3	325·8			
329	269·5	188·7	389	318·7	223·1	449	367·8	257·5	509	416·9	292·0	569	466·1	326·4			
330	270·3	189·3	390	319·5	223·7	450	368·6	258·1	510	417·8	292·5	570	466·9	326·9			
331	271·1	189·9	391	320·3	224·3	451	369·4	258·7	511	418·6	293·1	571	467·7	327·5			
332	272·0	190·4	392	321·1	224·8	452	370·3	259·3	512	419·4	293·7	572	468·6	328·1			
333	272·8	191·0	393	321·9	225·4	453	371·1	259·8	513	420·2	294·2	573	469·4	328·7			
334	273·6	191·6	394	322·7	226·0	454	371·9	260·4	514	421·0	294·8	574	470·2	329·2			
335	274·4	192·1	395	323·6	226·6	455	372·7	261·0	515	421·9	295·4	575	471·0	329·8			
336	275·2	192·7	396	324·4	227·1	456	373·5	261·6	516	422·7	296·0	576	471·8	330·4			
337	276·1	193·3	397	325·2	227·7	457	374·4	262·1	517	423·5	296·5	577	472·7	331·0			
338	276·9	193·9	398	326·0	228·3	458	375·2	262·7	518	424·3	297·1	578	473·5	331·5			
339	277·7	194·4	399	326·8	228·9	459	376·0	263·3	519	425·1	297·7	579	474·3	332·1			
340	278·5	195·0	400	327·7	229·4	460	376·8	263·8	520	426·0	298·3	580	475·1	332·7			
341	279·3	195·6	401	328·5	230·0	461	377·6	264·4	521	426·8	298·8	581	475·9	333·2			
342	280·1	196·2	402	329·3	230·6	462	378·4	265·0	522	427·6	299·4	582	476·7	333·8			
343	281·0	196·7	403	330·1	231·2	463	379·3	265·6	523	428·4	300·0	583	477·6	334·4			
344	281·8	197·3	404	330·9	231·7	464	380·1	266·1	524	429·2	300·6	584	478·4	335·0			
345	282·6	197·9	405	331·8	232·3	465	380·9	266·7	525	430·1	301·1	585	479·2	335·5			
346	283·4	198·5	406	332·6	232·9	466	381·7	267·3	526	430·9	301·7	586	480·0	336·1			
347	284·2	199·0	407	333·4	233·4	467	382·5	267·9	527	431·7	302·3	587	480·8	336·7			
348	285·1	199·6	408	334·2	234·0	468	383·4	268·4	528	432·5	302·8	588	481·7	337·3			
349	285·9	200·2	409	335·0	234·6	469	384·2	269·0	529	433·3	303·4	589	482·5	337·8			
350	286·7	200·8	410	335·9	235·2	470	385·0	269·6	530	434·2	304·0	590	483·3	338·4			
351	287·5	201·3	411	336·7	235·7	471	385·8	270·2	531	435·0	304·6	591	484·1	339·0			
352	288·3	201·9	412	337·5	236·3	472	386·6	270·7	532	435·8	305·1	592	484·9	339·6			
353	289·2	202·5	413	338·3	236·9	473	387·5	271·3	533	436·6	305·7	593	485·8	340·1			
354	290·0	203·0	414	339·1	237·5	474	388·3	271·9	534	437·4	306·3	594	486·6	340·7			
355	290·8	203·6	415	339·9	238·0	475	389·1	272·4	535	438·2	306·9	595	487·4	341·3			
356	291·6	204·2	416	340·8	238·6	476	389·9	273·0	536	439·1	307·4	596	488·2	341·9			
357	292·4	204·8	417	341·6	239·2	477	390·7	273·6	537	439·9	308·0	597	489·0	342·4			
358	293·3	205·3	418	342·4	239·8	478	391·6	274·2	538	440·7	308·6	598	489·9	343·0			
359	294·1	205·9	419	343·2	240·3	479	392·4	274·7	539	441·5	309·2	599	490·7	343·6			
360	294·9	206·5	420	344·0	240·9	480	393·2	275·3	540	442·3	309·7	600	491·5	344·1			
Dist	Dep.	D.Lat	Dist	Dep.	D.Lat	Dist	Dep.	D.Lat	Dist	Dep.	D.Lat	Dist	Dep.	D.Lat	Dist	Dep.	D.Lat
D Lon		Dep.	D Lon		Dep.	D Lon		Dep.	D Lon		Dep.	D Lon		Dep.	D Lon		Dep.

	305° ↑				**55 DEGREES.**					055° ↑		3h 40m	**55°**
	235°									125°			

36°

TRAVERSE TABLE
36 DEGREES.

Dist	D.Lat	Dep.	Dist	D.Lat	Dep.	Dist	D.Lat	Dep.	Dist	D.Lat	Dep.	Dist	D.Lat	Dep.	Dist	D.Lat	Dep.
1	00.8	00.6	61	49.4	35.9	121	97.9	71.1	181	146.4	106.4	241	195.0	141.7	241	195.0	141.7
2	01.6	01 2	62	50.2	36.4	122	98.7	71.7	182	147.2	107.0	242	195.8	142.2	242	195.8	142.2
3	02.4	01.8	63	51.0	37.0	123	99.5	72.3	183	148.1	107.6	243	196.6	142.8	243	196.6	142.8
4	03.2	02.4	64	51.8	37.6	124	100.3	72.9	184	148.9	108.2	244	197.4	143.4	244	197.4	143.4
5	04.0	02.9	65	52.6	38.2	125	101.1	73.5	185	149.7	108.7	245	198.2	144.0	245	198.2	144.0
6	04.9	03.5	66	53.4	38.8	126	101.9	74.1	186	150.5	109.3	246	199.0	144 6	246	199.0	144 6
7	05.7	04.1	67	54.2	39.4	127	102.7	74.6	187	151.3	109.9	247	199.8	145.2	247	199.8	145.2
8	06.5	04.7	68	55.0	40.0	128	103.6	75.2	188	152.1	110.5	248	200.6	145.8	248	200.6	145.8
9	07.3	05.3	69	55.8	40.6	129	104.4	75.8	189	152.9	111.1	249	201.4	146.4	249	201.4	146.4
10	08.1	05.9	70	56.6	41.1	130	105.2	76.4	190	153.7	111.7	250	202.3	146.9	250	202.3	146.9
11	08.9	06.5	71	57.4	41.7	131	106.0	77.0	191	154.5	112.3	251	203.1	147.5	251	203.1	147.5
12	09.7	07.1	72	58.2	42.3	132	106.8	77.6	192	155.3	112.9	252	203.9	148.1	252	203.9	148.1
13	10.5	07.6	73	59.1	42.9	133	107.6	78.2	193	156.1	113.4	253	204.7	148.7	253	204.7	148.7
14	11.3	08.2	74	59.9	43.5	134	108.4	78.8	194	156.9	114.0	254	205.5	149.3	254	205.5	149.3
15	12.1	08.8	75	60.7	44.1	135	109.2	79.4	195	157.8	114.6	255	206.3	149.9	255	206.3	149.9
16	12.9	09.4	76	61.5	44.7	136	110.0	79.9	196	158.6	115.2	256	207.1	150.5	256	207.1	150.5
17	13.8	10.0	77	62.3	45.3	137	110.8	80.5	197	159.4	115.8	257	207.9	151.1	257	207.9	151.1
18	14.6	10.6	78	63.1	45.8	138	111.6	81.1	198	160.2	116.4	258	208.7	151.6	258	208.7	151.6
19	15.4	11.2	79	63.9	46.4	139	112.5	81.7	199	161.0	117.0	259	209.5	152.2	259	209.5	152.2
20	16.2	11.8	80	64.7	47.0	140	113.3	82.3	200	161.8	117.6	260	210.3	152.8	260	210.3	152.8
21	17.0	12.3	81	65.5	47.6	141	114.1	82.9	201	162.6	118.1	261	211.2	153.4	261	211.2	153.4
22	17.8	12.9	82	66.3	48.2	142	114.9	83.5	202	163.4	118.7	262	212.0	154.0	262	212.0	154.0
23	18.6	13.5	83	67.1	48.8	143	115.7	84.1	203	164.2	119.3	263	212.8	154.6	263	212.8	154.6
24	19.4	14.1	84	68.0	49.4	144	116.5	84.6	204	165.0	119.9	264	213.6	155.2	264	213.6	155.2
25	20.2	14.7	85	68.8	50.0	145	117.3	85.2	205	165.8	120.5	265	214.4	155.8	265	214.4	155.8
26	21.0	15.3	86	69.6	50.5	146	118.1	85.8	206	166.7	121.1	266	215.2	156.4	266	215.2	156.4
27	21 8	15.9	87	70.4	51.1	147	118.9	86.4	207	167.5	121.7	267	216.0	156.9	267	216.0	156.9
28	22.7	16.5	88	71.2	51.7	148	119.7	87.0	208	168.3	122.3	268	216.8	157.5	268	216.8	157.5
29	23.5	17.0	89	72.0	52.3	149	120.5	87.6	209	169.1	122.8	269	217.6	158.1	269	217.6	158.1
30	24.3	17.6	90	72.8	52.9	150	121.4	88.2	210	169.9	123.4	270	218.4	158.7	270	218.4	158.7
31	25.1	18.2	91	73.6	53.5	151	122.2	88.8	211	170.7	124.0	271	219.2	159.3	271	219.2	159.3
32	25.9	18.8	92	74.4	54.1	152	123.0	89.3	212	171.5	124.6	272	220.1	159.9	272	220.1	159.9
33	26.7	19.4	93	75.2	54.7	153	123.8	89.9	213	172.3	125.2	273	220.9	160.5	273	220.9	160.5
34	27.5	20.0	94	76.0	55.3	154	124.6	90.5	214	173.1	125.8	274	221.7	161.1	274	221.7	161.1
35	28.3	20.6	95	76.9	55.8	155	125.4	91.1	215	173.9	126.4	275	222.5	161.6	275	222.5	161.6
36	29.1	21.2	96	77.7	56.4	156	126.2	91.7	216	174.7	127.0	276	223.3	162.2	276	223.3	162.2
37	29.9	21.7	97	78.5	57.0	157	127.0	92.3	217	175.6	127.5	277	224.1	162.8	277	224.1	162.8
38	30.7	22.3	98	79.3	57.6	158	127.8	92.9	218	176.4	128.1	278	224.9	163.4	278	224.9	163.4
39	31.6	22.9	99	80.1	58.2	159	128.6	93.5	219	177.2	128.7	279	225.7	164.0	279	225.7	164.0
40	32.4	23.5	100	80.9	58.8	160	129.4	94.0	220	178.0	129.3	280	226.5	164.6	280	226.5	164.6
41	33.2	24.1	101	81.7	59.4	161	130.3	94.6	221	178.8	129.9	281	227.3	165.2	281	227.3	165.2
42	34.0	24.7	102	82.5	60.0	162	131.1	95.2	222	179.6	130.5	282	228.1	165.8	282	228.1	165.8
43	34.8	25.3	103	83.3	60.5	163	131.9	95.8	223	180.4	131.1	283	229.0	166.3	283	229.0	166.3
44	35.6	25.9	104	84.1	61.1	164	132.7	96.4	224	181.2	131.7	284	229.8	166.9	284	229.8	166.9
45	36.4	26.5	105	84.9	61.7	165	133.5	97.0	225	182.0	132.3	285	230.6	167.5	285	230.6	167.5
46	37.2	27.0	106	85.8	62.3	166	134.4	97.6	226	182.8	132.8	286	231.4	168.1	286	231.4	168.1
47	38.0	27.6	107	86.6	62.9	167	135.1	98.2	227	183.6	133.4	287	232.2	168.7	287	232.2	168.7
48	38.8	28.2	108	87.4	63.5	168	135.9	98.7	228	184.5	134.0	288	233.0	169.3	288	233.0	169.3
49	39.6	28.8	109	88.2	64.1	169	136.7	99.3	229	185.3	134.6	289	233.8	169.9	289	233.8	169.9
50	40.5	29.4	110	89.0	64.7	170	137.5	99.9	230	186.1	135.2	290	234.6	170 5	290	234.6	170 5
51	41.3	30.0	111	89.8	65.2	171	138.3	100.5	231	186.9	135.8	291	235.4	171.0	291	235.4	171.0
52	42.1	30.6	112	90.6	65.8	172	139.2	101.1	232	187.7	136.4	292	236.2	171.6	292	236.2	171.6
53	42.9	31.2	113	91.4	66.4	173	140.0	101.7	233	188.5	137.0	293	237.0	172.2	293	237.0	172.2
54	43.7	31.7	114	92.2	67.0	174	140.8	102.3	234	189.3	137.5	294	237.9	172.8	294	237.9	172.8
55	44.5	32.3	115	93.0	67.6	175	141.6	102.9	235	190.1	138.1	295	238.7	173.4	295	238.7	173.4
56	45.3	32.9	116	93.8	68.2	176	142.4	103.5	236	190.9	138.7	296	239.5	174.0	296	239.5	174.0
57	46.1	33.5	117	94.7	68.8	177	143.2	104.0	237	191.7	139.3	297	240.3	174.6	297	240.3	174.6
58	46.9	34.1	118	95.5	69.4	178	144.0	104.6	238	192.5	139.9	298	241.1	175.2	298	241.1	175.2
59	47.7	34.7	119	96.3	69.9	179	144.8	105.2	239	193.4	140.5	299	241.9	175.7	299	241.9	175.7
60	48.5	35.3	120	97.1	70.5	180	145.6	105.8	240	194.2	141.1	300	242.7	176.3	300	242.7	176.3

Dist	Dep.	D.Lat	Dist	Dep.	D.Lat	Dist	Dep.	D.Lat	Dist	Dep.	D.Lat	Dist	Dep.	D.Lat	Dist	Dep.	D.Lat

54°

TRAVERSE TABLE
36 DEGREES.

324° / 216°   036° / 144°   2h 24m   **36°**

Dist	D.Lat	Dep.	Dist	D.Lat	Dep.	Dist	D.Lat	Dep.	Dist	D.Lat	Dep.	Dist	D.Lat	Dep.
301	243·5	176·9	361	292·1	212·2	421	340·6	247·5	481	389·1	282·7	541	437·7	318·0
302	244·3	177·5	362	292·9	212·8	422	341·4	248·0	482	389·9	283·3	542	438·5	318·6
303	245·1	178·1	363	293·7	213·4	423	342·2	248·6	483	390·8	283·9	543	439·3	319·2
304	245·9	178·7	364	294·5	214·0	424	343·0	249·2	484	391·6	284·5	544	440·2	319·8
305	246·8	179·3	365	295·3	214·5	425	343·8	249·8	485	392·4	285·1	545	440·9	320·3
306	247·6	179·9	366	296·1	215·1	426	344·6	250·4	486	393·2	285·7	546	441·7	320·9
307	248·4	180·5	367	296·9	215·7	427	345·5	251·0	487	394·0	286·3	547	442·5	321·5
308	249·2	181·0	368	297·7	216·3	428	346·3	251·6	488	394·8	286·8	548	443·3	322·1
309	250·0	181·6	369	298·5	216·9	429	347·1	252·2	489	395·6	287·4	549	444·2	322·7
310	250·8	182·2	370	299·3	217·5	430	347·9	252·7	490	396·4	288·0	550	445·0	323·3
311	251·6	182·8	371	300·1	218·1	431	348·7	253·3	491	397·2	288·6	551	445·8	323·9
312	252·4	183·4	372	301·0	218·7	432	349·5	253·9	492	398·0	289·2	552	446·6	324·5
313	253·2	184·0	373	301·8	219·3	433	350·3	254·5	493	398·8	289·8	553	447·4	325·0
314	254·0	184·6	374	302·6	219·8	434	351·1	255·1	494	399·7	290·4	554	448·2	325·6
315	254·8	185·2	375	303·4	220·4	435	351·9	255·7	495	400·5	291·0	555	449·0	326·2
316	255·6	185·7	376	304·2	221·0	436	352·7	256·3	496	401·3	291·5	556	449·8	326·8
317	256·5	186·3	377	305·0	221·6	437	353·5	256·9	497	402·1	292·1	557	450·6	327·4
318	257·3	186·9	378	305·8	222·2	438	354·3	257·4	498	402·9	292·7	558	451·4	328·0
319	258·1	187·5	379	306·6	222·8	439	355·2	258·0	499	403·7	293·3	559	452·2	328·6
320	258·9	188·1	380	307·4	223·4	440	356·0	258·6	500	404·5	293·9	560	453·0	329·2
321	259·7	188·7	381	308·2	223·9	441	356·8	259·2	501	405·3	294·5	561	453·9	329·7
322	260·5	189·3	382	309·0	224·5	442	357·6	259·8	502	406·1	295·1	562	454·7	330·3
323	261·3	189·9	383	309·9	225·1	443	358·4	260·4	503	406·9	295·7	563	455·5	330·9
324	262·1	190·4	384	310·7	225·7	444	359·2	261·0	504	407·7	296·2	564	456·3	331·5
325	262·9	191·0	385	311·5	226·3	445	360·0	261·6	505	408·6	296·8	565	457·1	332·1
326	263·7	191·6	386	312·3	226·9	446	360·8	262·2	506	409·4	297·4	566	457·9	332·7
327	264·5	192·2	387	313·1	227·5	447	361·6	262·7	507	410·2	298·0	567	458·7	333·3
328	265·4	192·8	388	313·9	228·1	448	362·4	263·3	508	411·0	298·6	568	459·5	333·9
329	266·2	193·4	389	314·7	228·6	449	363·2	263·9	509	411·8	299·2	569	460·3	334·4
330	267·0	194·0	390	315·5	229·2	450	364·1	264·5	510	412·6	299·8	570	461·1	335·0
331	267·8	194·6	391	316·3	229·8	451	364·9	265·1	511	413·4	300·4	571	461·9	335·6
332	268·6	195·1	392	317·1	230·4	452	365·7	265·7	512	414·2	300·9	572	462·8	336·2
333	269·4	195·7	393	317·9	231·0	453	366·5	266·3	513	415·0	301·5	573	463·6	336·8
334	270·2	196·3	394	318·8	231·6	454	367·3	266·9	514	415·8	302·1	574	464·4	337·4
335	271·0	196·9	395	319·6	232·2	455	368·1	267·4	515	416·6	302·7	575	465·2	338·0
336	271·8	197·5	396	320·4	232·8	456	368·9	268·0	516	417·5	303·3	576	466·0	338·6
337	272·6	198·1	397	321·2	233·4	457	369·7	268·6	517	418·3	303·9	577	466·8	339·2
338	273·4	198·7	398	322·0	233·9	458	370·5	269·2	518	419·1	304·5	578	467·6	339·7
339	274·3	199·3	399	322·8	234·5	459	371·3	269·8	519	419·9	305·1	579	468·4	340·3
340	275·1	199·8	400	323·6	235·1	460	372·1	270·4	520	420·7	305·6	580	469·2	340·9
341	275·9	200·4	401	324·4	235·7	461	373·0	271·0	521	421·5	306·2	581	470·0	341·5
342	276·7	201·0	402	325·2	236·3	462	373·8	271·6	522	422·3	306·8	582	470·8	342·1
343	277·5	201·6	403	326·0	236·9	463	374·6	272·1	523	423·1	307·4	583	471·7	342·7
344	278·3	202·2	404	326·9	237·5	464	375·4	272·7	524	423·9	308·0	584	472·5	343·3
345	279·1	202·8	405	327·7	238·1	465	376·2	273·3	525	424·7	308·6	585	473·3	343·9
346	279·9	203·4	406	328·5	238·7	466	377·0	273·9	526	425·5	309·2	586	474·1	344·4
347	280·7	204·0	407	329·3	239·2	467	377·8	274·5	527	426·4	309·8	587	474·9	345·0
348	281·5	204·5	408	330·1	239·8	468	378·6	275·1	528	427·2	310·4	588	475·7	345·6
349	282·3	205·1	409	330·9	240·4	469	379·4	275·7	529	428·0	310·9	589	476·5	346·2
350	283·2	205·7	410	331·7	241·0	470	380·2	276·3	530	428·8	311·5	590	477·3	346·8
351	284·0	206·3	411	332·5	241·6	471	381·1	276·8	531	429·6	312·1	591	478·1	347·4
352	284·8	206·9	412	333·3	242·2	472	381·9	277·4	532	430·4	312·7	592	478·9	348·0
353	285·6	207·5	413	334·1	242·8	473	382·7	278·0	533	431·2	313·3	593	479·7	348·6
354	286·4	208·1	414	334·9	243·3	474	383·5	278·6	534	432·0	313·9	594	480·6	349·1
355	287·2	208·7	415	335·7	243·9	475	384·3	279·2	535	432·8	314·5	595	481·4	349·7
356	288·0	209·3	416	336·6	244·5	476	385·1	279·8	536	433·6	315·1	596	482·2	350·3
357	288·8	209·8	417	337·4	245·1	477	385·9	280·4	537	434·4	315·7	597	483·0	350·9
358	289·6	210·4	418	338·2	245·7	478	386·7	281·0	538	435·3	316·2	598	483·8	351·5
359	290·4	211·0	419	339·0	246·3	479	387·5	281·5	539	436·1	316·8	599	484·6	352·1
360	291·2	211·6	420	339·8	246·9	480	388·3	282·1	540	436·9	317·4	600	485·4	352·7

| Dist | Dep. | D.Lat | Dist | Dep. | D.Lat | Dist | Dep. | D.Lat | Dist | Dep. | D.Lat | Dist | Dep. | D.Lat |

306° / 234°   **54 DEGREES.**   054° / 126°   3h 36m   **54°**

E

37°

TRAVERSE TABLE
37 DEGREES.

Dist	D.Lat	Dep.	Dist	D.Lat	Dep.	Dist	D.Lat	Dep.	Dist	D.Lat	Dep.	Dist	D.Lat	Dep.
1	00·8	00·6	61	48·7	36·7	121	96·6	72·8	181	144·6	108·9	241	192·5	145·0
2	01·6	01·2	62	49·5	37·3	122	97·4	73·4	182	145·4	109·5	242	193·3	145·6
3	02·4	01·8	63	50·3	37·9	123	98·2	74·0	183	146·2	110·1	243	194·1	146·2
4	03·2	02·4	64	51·1	38·5	124	99·0	74·6	184	146·9	110·7	244	194·9	146·8
5	04·0	03·0	65	51·9	39·1	125	99·8	75·2	185	147·7	111·3	245	195·7	147·4
6	04·8	03·6	66	52·7	39·7	126	100·6	75·8	186	148·5	111·9	246	196·5	148·0
7	05·6	04·2	67	53·5	40·3	127	101·4	76·4	187	149·3	112·5	247	197·3	148·6
8	06·4	04·8	68	54·3	40·9	128	102·2	77·0	188	150·1	113·1	248	198·1	149·3
9	07·2	05·4	69	55·1	41·5	129	103·0	77·6	189	150·9	113·7	249	198·9	149·9
10	08·0	06·0	70	55·9	42·1	130	103·8	78·2	190	151·7	114·3	250	199·7	150·5
11	08·8	06·6	71	56·7	42·7	131	104·6	78·8	191	152·5	114·9	251	200·5	151·1
12	09·6	07·2	72	57·5	43·3	132	105·4	79·4	192	153·3	115·5	252	201·3	151·7
13	10·4	07·8	73	58·3	43·9	133	106·2	80·0	193	154·1	116·2	253	202·1	152·3
14	11·2	08·4	74	59·1	44·5	134	107·0	80·6	194	154·9	116·8	254	202·9	152·9
15	12·0	09·0	75	59·9	45·1	135	107·8	81·2	195	155·7	117·4	255	203·7	153·5
16	12·8	09·6	76	60·7	45·7	136	108·6	81·8	196	156·5	118·0	256	204·5	154·1
17	13·6	10·2	77	61·5	46·3	137	109·4	82·4	197	157·3	118·6	257	205·2	154·7
18	14·4	10·8	78	62·3	46·9	138	110·2	83·1	198	158·1	119·2	258	206·0	155·3
19	15·2	11·4	79	63·1	47·5	139	111·0	83·7	199	158·9	119·8	259	206·8	155·9
20	16·0	12·0	80	63·9	48·1	140	111·8	84·3	200	159·7	120·4	260	207·6	156·5
21	16·8	12·6	81	64·7	48·7	141	112·6	84·9	201	160·5	121·0	261	208·4	157·1
22	17·6	13·2	82	65·5	49·3	142	113·4	85·5	202	161·3	121·6	262	209·2	157·7
23	18·4	13·8	83	66·3	50·0	143	114·2	86·1	203	162·1	122·2	263	210·0	158·3
24	19·2	14·4	84	67·1	50·6	144	115·0	86·7	204	162·9	122·8	264	210·8	158·9
25	20·0	15·0	85	67·9	51·2	145	115·8	87·3	205	163·7	123·4	265	211·6	159·5
26	20·8	15·6	86	68·7	51·8	146	116·6	87·9	206	164·5	124·0	266	212·4	160·1
27	21·6	16·2	87	69·5	52·4	147	117·4	88·5	207	165·3	124·6	267	213·2	160·7
28	22·4	16·9	88	70·3	53·0	148	118·2	89·1	208	166·1	125·2	268	214·0	161·3
29	23·2	17·5	89	71·1	53·6	149	119·0	89·7	209	166·9	125·8	269	214·8	161·9
30	24·0	18·1	90	71·9	54·2	150	119·8	90·3	210	167·7	126·4	270	215·6	162·5
31	24·8	18·7	91	72·7	54·8	151	120·6	90·9	211	168·5	127·0	271	216·4	163·1
32	25·6	19·3	92	73·5	55·4	152	121·4	91·5	212	169·3	127·6	272	217·2	163·7
33	26·4	19·9	93	74·3	56·0	153	122·2	92·1	213	170·1	128·2	273	218·0	164·3
34	27·2	20·5	94	75·1	56·6	154	123·0	92·7	214	170·9	128·8	274	218·8	164·9
35	28·0	21·1	95	75·9	57·2	155	123·8	93·3	215	171·7	129·4	275	219·6	165·5
36	28·8	21·7	96	76·7	57·8	156	124·6	93·9	216	172·5	130·0	276	220·4	166·1
37	29·5	22·3	97	77·5	58·4	157	125·4	94·5	217	173·3	130·6	277	221·2	166·7
38	30·3	22·9	98	78·3	59·0	158	126·2	95·1	218	174·1	131·2	278	222·0	167·3
39	31·1	23·5	99	79·1	59·6	159	127·0	95·7	219	174·9	131·8	279	222·8	167·9
40	31·9	24·1	100	79·9	60·2	160	127·8	96·3	220	175·7	132·4	280	223·6	168·5
41	32·7	24·7	101	80·7	60·8	161	128·6	96·9	221	176·5	133·0	281	224·4	169·1
42	33·5	25·3	102	81·5	61·4	162	129·4	97·5	222	177·3	133·6	282	225·2	169·7
43	34·3	25·9	103	82·3	62·0	163	130·2	98·1	223	178·1	134·2	283	226·0	170·3
44	35·1	26·5	104	83·1	62·6	164	131·0	98·7	224	178·9	134·8	284	226·8	170·9
45	35·9	27·1	105	83·9	63·2	165	131·8	99·3	225	179·7	135·4	285	227·6	171·5
46	36·7	27·7	106	84·7	63·8	166	132·6	99·9	226	180·5	136·0	286	228·4	172·1
47	37·5	28·3	107	85·5	64·4	167	133·4	100·5	227	181·3	136·6	287	229·2	172·7
48	38·3	28·9	108	86·3	65·0	168	134·2	101·1	228	182·1	137·2	288	230·0	173·3
49	39·1	29·5	109	87·1	65·6	169	135·0	101·7	229	182·9	137·8	289	230·8	173·9
50	39·9	30·1	110	87·8	66·2	170	135·8	102·3	230	183·7	138·4	290	231·6	174·5
51	40·7	30·7	111	88·6	66·8	171	136·6	102·9	231	184·5	139·0	291	232·4	175·1
52	41·5	31·3	112	89·4	67·4	172	137·4	103·5	232	185·3	139·6	292	233·2	175·7
53	42·3	31·9	113	90·2	68·0	173	138·2	104·1	233	186·1	140·2	293	234·0	176·3
54	43·1	32·5	114	91·0	68·6	174	139·0	104·7	234	186·9	140·8	294	234·8	176·9
55	43·9	33·1	115	91·8	69·2	175	139·8	105·3	235	187·7	141·4	295	235·6	177·5
56	44·7	33·7	116	92·6	69·8	176	140·6	105·9	236	188·5	142·0	296	236·4	178·1
57	45·5	34·3	117	93·4	70·4	177	141·4	106·5	237	189·3	142·6	297	237·2	178·7
58	46·3	34·9	118	94·2	71·0	178	142·2	107·1	238	190·1	143·2	298	238·0	179·3
59	47·1	35·5	119	95·0	71·6	179	143·0	107·7	239	190·9	143·8	299	238·8	179·9
60	47·9	36·1	120	95·8	72·2	180	143·8	108·3	240	191·7	144·4	300	239·6	180·5

Dist	Dep.	D.Lat	Dist	Dep.	D.Lat	Dist	Dep.	D.Lat	Dist	Dep.	D.Lat	Dist	Dep.	D.Lat
D Lon		Dep.	D Lon		Dep.	D Lon		Dep.	D Lon		Dep.	D Lon		Dep.

53°

53 DEGREES.

TRAVERSE TABLE
37 DEGREES.

323° ↑ / 217°   ↑ 037° / 143°   2h 28m   **37°**

D Lon	Dep.		D Lon	Dep.		D Lon	Dep.		D Lon	Dep.		D Lon	Dep.		D Lon	Dep.	
Dist	D. Lat	Dep.	Dist	D. Lat	Dep.	Dist	D. Lat	Dep.	Dist	D. Lat	Dep.	Dist	D. Lat	Dep.	Dist	D. Lat	Dep.
301	240·4	181·1	361	288·3	217·3	421	336·2	253·4	481	384·1	289·5	541	432·1	325·6			
302	241·2	181·7	362	289·1	217·9	422	337·0	254·0	482	384·9	290·1	542	432·9	326·2			
303	242·0	182·3	363	289·9	218·5	423	337·8	254·6	483	385·7	290·7	543	433·7	326·8			
304	242·7	183·0	364	290·7	219·1	424	338·6	255·2	484	386·5	291·3	544	434·5	327·4			
305	243·6	183·6	365	291·5	219·7	425	339·4	255·8	485	387·3	291·9	545	435·3	328·0			
306	244·4	184·2	366	292·3	220·3	426	340·2	256·4	486	388·1	292·5	546	436·1	328·6			
307	245·2	184·8	367	293·1	220·9	427	341·0	257·0	487	388·9	293·1	547	436·9	329·2			
308	246·0	185·4	368	293·9	221·5	428	341·8	257·6	488	389·7	293·7	548	437·7	329·8			
309	246·8	186·0	369	294·7	222·1	429	342·6	258·2	489	390·5	294·3	549	438·5	330·4			
310	247·6	186·6	370	295·5	222·7	430	343·4	258·8	490	391·3	294·9	550	439·2	331·0			
311	248·4	187·2	371	296·3	223·3	431	344·2	259·4	491	392·1	295·5	551	440·0	331·6			
312	249·2	187·8	372	297·1	223·9	432	345·0	260·0	492	392·9	296·1	552	440·8	332·2			
313	250·0	188·4	373	297·9	224·5	433	345·8	260·6	493	393·7	296·7	553	441·6	332·8			
314	250·8	189·0	374	298·7	225·1	434	346·6	261·2	494	394·5	297·3	554	442·4	333·4			
315	251·6	189·6	375	299·5	225·7	435	347·4	261·8	495	395·3	297·9	555	443·2	334·0			
316	252·4	190·2	376	300·3	226·3	436	348·2	262·4	496	396·1	298·5	556	444·0	334·6			
317	253·2	190·8	377	301·1	226·9	437	349·0	263·0	497	396·9	299·1	557	444·8	335·2			
318	254·0	191·4	378	301·9	227·5	438	349·8	263·6	498	397·7	299·7	558	445·6	335·8			
319	254·8	192·0	379	302·7	228·1	439	350·6	264·2	499	398·5	300·3	559	446·4	336·4			
320	255·6	192·6	380	303·5	228·7	440	351·4	264·8	500	399·3	300·9	560	447·2	337·0			
321	256·4	193·2	381	304·3	229·3	441	352·2	265·4	501	400·1	301·5	561	448·0	337·6			
322	257·2	193·8	382	305·1	229·9	442	353·0	266·0	502	400·9	302·1	562	448·8	338·2			
323	258·0	194·4	383	305·9	230·5	443	353·8	266·6	503	401·7	302·7	563	449·6	338·8			
324	258·8	195·0	384	306·7	231·1	444	354·6	267·2	504	402·5	303·3	564	450·4	339·4			
325	259·6	195·6	385	307·5	231·7	445	355·4	267·8	505	403·3	303·9	565	451·2	340·0			
326	260·4	196·2	386	308·3	232·3	446	356·2	268·4	506	404·1	304·5	566	452·0	340·6			
327	261·2	196·8	387	309·1	232·9	447	357·0	269·0	507	404·9	305·1	567	452·8	341·2			
328	262·0	197·4	388	309·9	233·5	448	357·8	269·6	508	405·7	305·7	568	453·6	341·8			
329	262·8	198·0	389	310·7	234·1	449	358·6	270·2	509	406·5	306·3	569	454·4	342·4			
330	263·5	198·6	390	311·5	234·7	450	359·4	270·8	510	407·3	306·9	570	455·2	343·0			
331	264·3	199·2	391	312·3	235·3	451	360·2	271·4	511	408·1	307·5	571	456·0	343·6			
332	265·1	199·8	392	313·1	235·9	452	361·0	272·0	512	408·9	308·1	572	456·8	344·2			
333	265·9	200·4	393	313·9	236·5	453	361·8	272·6	513	409·7	308·7	573	457·6	344·8			
334	266·7	201·0	394	314·7	237·1	454	362·6	273·2	514	410·5	309·3	574	458·4	345·4			
335	267·5	201·6	395	315·5	237·7	455	363·4	273·8	515	411·3	309·9	575	459·2	346·0			
336	268·3	202·2	396	316·3	238·3	456	364·2	274·4	516	412·1	310·5	576	460·0	346·6			
337	269·1	202·8	397	317·1	238·9	457	365·0	275·0	517	412·9	311·1	577	460·8	347·2			
338	269·9	203·4	398	317·9	239·5	458	365·8	275·6	518	413·7	311·7	578	461·6	347·8			
339	270·7	204·0	399	318·7	240·1	459	366·6	276·2	519	414·5	312·3	579	462·4	348·5			
340	271·5	204·6	400	319·5	240·7	460	367·4	276·8	520	415·3	312·9	580	463·2	349·1			
341	272·3	205·2	401	320·3	241·3	461	368·2	277·4	521	416·1	313·5	581	464·0	349·7			
342	273·1	205·8	402	321·1	241·9	462	369·0	278·0	522	416·9	314·1	582	464·8	350·3			
343	273·9	206·4	403	321·9	242·5	463	369·8	278·6	523	417·7	314·7	583	465·6	350·9			
344	274·7	207·0	404	322·6	243·1	464	370·6	279·2	524	418·5	315·4	584	466·4	351·5			
345	275·5	207·6	405	323·4	243·7	465	371·4	279·8	525	419·3	316·0	585	467·2	352·1			
346	276·3	208·2	406	324·2	244·3	466	372·2	280·4	526	420·1	316·6	586	468·0	352·7			
347	277·1	208·8	407	325·0	244·9	467	373·0	281·0	527	420·9	317·2	587	468·8	353·3			
348	277·9	209·4	408	325·8	245·5	468	373·8	281·6	528	421·7	317·8	588	469·6	353·9			
349	278·7	210·0	409	326·6	246·1	469	374·6	282·3	529	422·5	318·4	589	470·4	354·5			
350	279·5	210·6	410	327·4	246·7	470	375·4	282·9	530	423·3	319·0	590	471·2	355·1			
351	280·3	211·2	411	328·2	247·3	471	376·2	283·5	531	424·1	319·6	591	472·0	355·7			
352	281·1	211·8	412	329·0	247·9	472	377·0	284·1	532	424·9	320·2	592	472·8	356·3			
353	281·9	212·4	413	329·8	248·5	473	377·8	284·7	533	425·7	320·8	593	473·6	356·9			
354	282·7	213·0	414	330·6	249·2	474	378·6	285·3	534	426·5	321·4	594	474·4	357·5			
355	283·5	213·6	415	331·4	249·8	475	379·4	285·9	535	427·3	322·0	595	475·2	358·1			
356	284·3	214·3	416	332·2	250·4	476	380·2	286·5	536	428·1	322·6	596	476·0	358·7			
357	285·1	214·8	417	333·0	251·0	477	380·9	287·1	537	428·9	323·2	597	476·8	359·3			
358	285·9	215·4	418	333·8	251·6	478	381·7	287·7	538	429·7	323·8	598	477·6	359·9			
359	286·7	216·1	419	334·6	252·2	479	382·5	288·3	539	430·5	324·4	599	478·4	360·5			
360	287·5	216·7	420	335·4	252·8	480	383·3	288·9	540	431·3	325·0	600	479·2	361·1			
Dist	Dep.	D. Lat	Dist	Dep.	D. Lat	Dist	Dep.	D. Lat	Dist	Dep.	D. Lat	Dist	Dep.	D. Lat			
D Lon		Dep.	D Lon		Dep.	D Lon		Dep.	D Lon		Dep.	D Lon		Dep.			

307° ↑ / 233°   ↑ 053° / 127°   3h 32m

53 DEGREES.

53°

38°

TRAVERSE TABLE
38 DEGREES.

D Lon	Dep.		D Lon	Dep.		D Lon	Dep.		D Lon	Dep.		D Lon	Dep.	
Dist	D.Lat	Dep.	Dist	D.Lat	Dep.	Dist	D.Lat	Dep.	Dist	D.Lat	Dep.	Dist	D.Lat	Dep.
1	00·8	00·6	61	48·1	37·6	121	95·3	74·5	181	142·6	111·4	241	189·9	148·4
2	01·6	01·2	62	48·9	38·2	122	96·1	75·1	182	143·4	112·1	242	190·7	149·0
3	02·4	01·8	63	49·6	38·8	123	96·9	75·7	183	144·2	112·7	243	191·5	149·6
4	03·2	02·5	64	50·4	39·4	124	97·7	76·3	184	145·0	113·3	244	192·3	150·2
5	03·9	03·1	65	51·2	40·0	125	98·5	77·0	185	145·8	113·9	245	193·1	150·8
6	04·7	03·7	66	52·0	40·6	126	99·3	77·6	186	146·6	114·5	246	193·9	151·5
7	05·5	04·3	67	52·8	41·2	127	100·1	78·2	187	147·4	115·1	247	194·6	152·1
8	06·3	04·9	68	53·6	41·9	128	100·9	78·8	188	148·1	115·7	248	195·4	152·7
9	07·1	05·5	69	54·4	42·5	129	101·7	79·4	189	148·9	116·4	249	196·2	153·3
10	07·9	06·2	70	55·2	43·1	130	102·4	80·0	190	149·7	117·0	250	197·0	153·9
11	08·7	06·8	71	55·9	43·7	131	103·2	80·7	191	150·5	117·6	251	197·8	154·5
12	09·5	07·4	72	56·7	44·3	132	104·0	81·3	192	151·3	118·2	252	198·6	155·1
13	10·2	08·0	73	57·5	44·9	133	104·8	81·9	193	152·1	118·8	253	199·4	155·8
14	11·0	08·6	74	58·3	45·6	134	105·6	82·5	194	152·9	119·4	254	200·2	156·4
15	11·8	09·2	75	59·1	46·2	135	106·4	83·1	195	153·7	120·1	255	200·9	157·0
16	12·6	09·9	76	59·9	46·8	136	107·2	83·7	196	154·5	120·7	256	201·7	157·6
17	13·4	10·5	77	60·7	47·4	137	108·0	84·3	197	155·2	121·3	257	202·5	158·2
18	14·2	11·1	78	61·5	48·0	138	108·7	85·0	198	156·0	121·9	258	203·3	158·8
19	15·0	11·7	79	62·3	48·6	139	109·5	85·6	199	156·8	122·5	259	204·1	159·5
20	15·8	12·3	80	63·0	49·3	140	110·3	86·2	200	157·6	123·1	260	204·9	160·1
21	16·5	12·9	81	63·8	49·9	141	111·1	86·8	201	158·4	123·7	261	205·7	160·7
22	17·3	13·5	82	64·6	50·5	142	111·9	87·4	202	159·2	124·4	262	206·5	161·3
23	18·1	14·2	83	65·4	51·1	143	112·7	88·0	203	160·0	125·0	263	207·2	161·9
24	18·9	14·8	84	66·2	51·7	144	113·5	88·7	204	160·8	125·6	264	208·0	162·5
25	19·7	15·4	85	67·0	52·3	145	114·3	89·3	205	161·5	126·2	265	208·8	163·2
26	20·5	16·0	86	67·8	52·9	146	115·0	89·9	206	162·3	126·8	266	209·6	163·8
27	21·3	16·6	87	68·6	53·6	147	115·8	90·5	207	163·1	127·4	267	210·4	164·4
28	22·1	17·2	88	69·3	54·2	148	116·6	91·1	208	163·9	128·1	268	211·2	165·0
29	22·9	17·9	89	70·1	54·8	149	117·4	91·7	209	164·7	128·7	269	212·0	165·6
30	23·6	18·5	90	70·9	55·4	150	118·2	92·3	210	165·5	129·3	270	212·8	166·2
31	24·4	19·1	91	71·7	56·0	151	119·0	93·0	211	166·3	129·9	271	213·6	166·8
32	25·2	19·7	92	72·5	56·6	152	119·8	93·6	212	167·1	130·5	272	214·3	167·5
33	26·0	20·3	93	73·3	57·3	153	120·6	94·2	213	167·8	131·1	273	215·1	168·1
34	26·8	20·9	94	74·1	57·9	154	121·4	94·8	214	168·6	131·8	274	215·9	168·7
35	27·6	21·5	95	74·9	58·5	155	122·1	95·4	215	169·4	132·4	275	216·7	169·3
36	28·4	22·2	96	75·6	59·1	156	122·9	96·0	216	170·2	133·0	276	217·5	169·9
37	29·2	22·8	97	76·4	59·7	157	123·7	96·7	217	171·0	133·6	277	218·3	170·5
38	29·9	23·4	98	77·2	60·3	158	124·5	97·3	218	171·8	134·2	278	219·1	171·2
39	30·7	24·0	99	78·0	61·0	159	125·3	97·9	219	172·6	134·8	279	219·9	171·8
40	31·5	24·6	100	78·8	61·6	160	126·1	98·5	220	173·4	135·4	280	220·6	172·4
41	32·3	25·2	101	79·6	62·2	161	126·9	99·1	221	174·2	136·1	281	221·4	173·0
42	33·1	25·9	102	80·4	62·8	162	127·7	99·7	222	174·9	136·7	282	222·2	173·6
43	33·9	26·5	103	81·2	63·4	163	128·4	100·4	223	175·7	137·3	283	223·0	174·2
44	34·7	27·1	104	82·0	64·0	164	129·2	101·0	224	176·5	137·9	284	223·8	174·8
45	35·5	27·7	105	82·7	64·6	165	130·0	101·6	225	177·3	138·5	285	224·6	175·5
46	36·2	28·3	106	83·5	65·3	166	130·8	102·2	226	178·1	139·1	286	225·4	176·1
47	37·0	28·9	107	84·3	65·9	167	131·6	102·8	227	178·9	139·8	287	226·2	176·7
48	37·8	29·6	108	85·1	66·5	168	132·4	103·4	228	179·7	140·4	288	226·9	177·3
49	38·6	30·2	109	85·9	67·1	169	133·2	104·0	229	180·5	141·0	289	227·7	177·9
50	39·4	30·8	110	86·7	67·7	170	134·0	104·7	230	181·2	141·6	290	228·5	178·5
51	40·2	31·4	111	87·5	68·3	171	134·7	105·3	231	182·0	142·2	291	229·3	179·2
52	41·0	32·0	112	88·3	69·0	172	135·5	105·9	232	182·8	142·8	292	230·1	179·8
53	41·8	32·6	113	89·0	69·6	173	136·3	106·5	233	183·6	143·4	293	230·9	180·4
54	42·6	33·2	114	89·8	70·2	174	137·1	107·1	234	184·4	144·1	294	231·7	181·0
55	43·3	33·9	115	90·6	70·8	175	137·9	107·7	235	185·2	144·7	295	232·5	181·6
56	44·1	34·5	116	91·4	71·4	176	138·7	108·4	236	186·0	145·3	296	233·3	182·2
57	44·9	35·1	117	92·2	72·0	177	139·5	109·0	237	186·8	145·9	297	234·0	182·9
58	45·7	35·7	118	93·0	72·6	178	140·3	109·6	238	187·5	146·5	298	234·8	183·5
59	46·5	36·3	119	93·8	73·3	179	141·1	110·2	239	188·3	147·1	299	235·6	184·1
60	47·3	36·9	120	94·6	73·9	180	141·8	110·8	240	189·1	147·8	300	236·4	184·7
Dist	Dep.	D.Lat	Dist	Dep.	D.Lat	Dist	Dep.	D.Lat	Dist	Dep.	D.Lat	Dist	Dep.	D.Lat
D Lon		Dep.	D Lon		Dep.	D Lon		Dep.	D Lon		Dep.	D Lon		Dep.

52°

TRAVERSE TABLE
38 DEGREES.

322° ↑ / 218°   038° ↑ / 142°   2h 32m   **38°**

Dist	D.Lat	Dep.	Dist	D.Lat	Dep.	Dist	D.Lat	Dep.	Dist	D.Lat	Dep.	Dist	D.Lat	Dep.
301	237·2	185·3	361	284·5	222·3	421	331·8	259·2	481	379·0	296·1	541	426·3	333·1
302	238·0	185·9	362	285·3	222·9	422	332·5	259·8	482	379·8	296·7	542	427·1	333·7
303	238·8	186·5	363	286·0	223·5	423	333·3	260·4	483	380·6	297·4	543	427·9	334·3
304	239·6	187·2	364	286·8	224·1	424	334·1	261·0	484	381·4	298·0	544	428·7	334·9
305	240·3	187·8	365	287·6	224·7	425	334·9	261·7	485	382·2	298·6	545	429·5	335·5
306	241·1	188·4	366	288·4	225·3	426	335·7	262·3	486	383·0	299·2	546	430·3	336·2
307	241·9	189·0	367	289·2	225·9	427	336·5	262·9	487	383·8	299·8	547	431·0	336·8
308	242·7	189·6	368	290·0	226·6	428	337·3	263·5	488	384·5	300·4	548	431·8	337·4
309	243·5	190·2	369	290·8	227·2	429	338·1	264·1	489	385·3	301·1	549	432·6	338·0
310	244·3	190·9	370	291·6	227·8	430	338·8	264·7	490	386·1	301·7	550	433·4	338·6
311	245·1	191·5	371	292·4	228·4	431	339·6	265·4	491	386·9	302·3	551	434·2	339·2
312	245·9	192·1	372	293·1	229·0	432	340·4	266·0	492	387·7	302·9	552	435·0	339·8
313	246·6	192·7	373	293·9	229·6	433	341·2	266·6	493	388·5	303·5	553	435·8	340·5
314	247·4	193·3	374	294·7	230·3	434	342·0	267·2	494	389·3	304·1	554	436·6	341·1
315	248·2	193·9	375	295·5	230·9	435	342·8	267·8	495	390·1	304·8	555	437·3	341·7
316	249·0	194·5	376	296·3	231·5	436	343·6	268·4	496	390·9	305·4	556	438·1	342·3
317	249·8	195·2	377	297·1	232·1	437	344·4	269·0	497	391·6	306·0	557	438·9	342·9
318	250·6	195·8	378	297·9	232·7	438	345·1	269·7	498	392·4	306·6	558	439·7	343·5
319	251·4	196·4	379	298·7	233·3	439	345·9	270·3	499	393·2	307·2	559	440·5	344·2
320	252·2	197·0	380	299·4	234·0	440	346·7	270·9	500	394·0	307·8	560	441·3	344·8
321	253·0	197·6	381	300·2	234·6	441	347·5	271·5	501	394·8	308·4	561	442·1	345·4
322	253·8	198·2	382	301·0	235·2	442	348·3	272·1	502	395·6	309·1	562	442·9	346·0
323	254·5	198·9	383	301·8	235·8	443	349·1	272·7	503	396·4	309·7	563	443·7	346·6
324	255·3	199·5	384	302·6	236·4	444	349·9	273·4	504	397·2	310·3	564	444·4	347·2
325	256·1	200·1	385	303·4	237·0	445	350·7	274·0	505	397·9	310·9	565	445·2	347·8
326	256·9	200·7	386	304·2	237·6	446	351·5	274·6	506	398·7	311·5	566	446·0	348·5
327	257·7	201·3	387	305·0	238·3	447	352·2	275·2	507	399·5	312·1	567	446·8	349·1
328	258·5	201·9	388	305·7	238·9	448	353·0	275·8	508	400·3	312·8	568	447·6	349·7
329	259·3	202·6	389	306·5	239·5	449	353·8	276·4	509	401·1	313·4	569	448·4	350·3
330	260·0	203·2	390	307·3	240·1	450	354·6	277·0	510	401·9	314·0	570	449·2	350·9
331	260·8	203·8	391	308·1	240·7	451	355·4	277·7	511	402·7	314·6	571	450·0	351·5
332	261·6	204·4	392	308·9	241·3	452	356·2	278·3	512	403·5	315·2	572	450·7	352·2
333	262·4	205·0	393	309·7	242·0	453	357·0	278·9	513	404·2	315·8	573	451·5	352·8
334	263·2	205·6	394	310·5	242·6	454	357·8	279·5	514	405·0	316·5	574	452·3	353·4
335	264·0	206 2	395	311·3	243·2	455	358·5	280·1	515	405·8	317·1	575	453·1	354·0
336	264·8	206 9	396	312·0	243·8	456	359·3	280·7	516	406·6	317·7	576	453·9	354·6
337	265·6	207·5	397	312·8	244·4	457	360·1	281·4	517	407·4	318·3	577	454·7	355·2
338	266·3	208·1	398	313·6	245·0	458	360·9	282·0	518	408·2	318·9	578	455·5	355·7
339	267·1	208·7	399	314·4	245·6	459	361·7	282·6	519	409·0	319·5	579	456·3	356·5
340	267·9	209·3	400	315·2	246·3	460	362·5	283·2	520	409·8	320·1	580	457·0	357·1
341	268·7	209·9	401	316·0	246·9	461	363·3	283·8	521	410·6	320·8	581	457·8	357·7
342	269·5	210·6	402	316·8	247·5	462	364·1	284·4	522	411·3	321·4	582	458·6	358·3
343	270·3	211·2	403	317·6	248·1	463	364·8	285·1	523	412·1	322·0	583	459·4	358·9
344	271·1	211·8	404	318·4	248·7	464	365·6	285·7	524	412·9	322·6	584	460·2	359·5
345	271·9	212·4	405	319·1	249·3	465	366·4	286·3	525	413·7	323·2	585	461·0	360·2
346	272·7	213·0	406	319·9	250·0	466	367·2	286·9	526	414·5	323·8	586	461·8	360·8
347	273·4	213·6	407	320·7	250·6	467	368·0	287·5	527	415·3	324·5	587	462·6	361·4
348	274·2	214·3	408	321·5	251·2	468	368·8	288·1	528	416·1	325·1	588	463·4	362·0
349	275·0	214·9	409	322·3	251·8	469	369·6	288·7	529	416·9	325·7	589	464·1	362·6
350	275·8	215·5	410	323·1	252·4	470	370·4	289·4	530	417·6	326·3	590	464·9	363·2
351	276·6	216·1	411	323·9	253·0	471	371·2	290·0	531	418·4	326·9	591	465·7	363·9
352	277·4	216·7	412	324·7	253·7	472	371·9	290·6	532	419·2	327·5	592	466·5	364·5
353	278·2	217·3	413	325·4	254·3	473	372·7	291·2	533	420·0	328·1	593	467·3	365·1
354	279·0	217·9	414	326·2	254·9	474	373·5	291·8	534	420·8	328·8	594	468·1	365·7
355	279·7	218·6	415	327·0	255·5	475	374·3	292·4	535	421·6	329·4	595	468·9	366·3
356	280·5	219·2	416	327·8	256·1	476	375·1	293·1	536	422·4	330·0	596	469·7	366·9
357	281·3	219·8	417	328·6	256·7	477	375·9	293·7	537	423·2	330·6	597	470·4	367·5
358	282·1	220·4	418	329·4	257·3	478	376·7	294·3	538	423·9	331·2	598	471·2	368·2
359	282·9	221·0	419	330·2	258·0	479	377·5	294·9	539	424·7	331·8	599	472·0	368·8
360	283·7	221·6	420	331·0	258·6	480	378·2	295·5	540	425·5	332·5	600	472·8	369·4
Dist	Dep.	D.Lat	Dist	Dep.	D.Lat	Dist	Dep.	D.Lat	Dist	Dep.	D.Lat	Dist	Dep.	D.Lat
D Lon		Dep.	D Lon		Dep.	D Lon		Dep.	D Lon		Dep.	D Lon		Dep.

308° ↑ / 232°   **52 DEGREES.**   052° ↑ / 128°   3h 28m   **52°**

39°

TRAVERSE TABLE
39 DEGREES.

Dist	D.Lat	Dep.	Dist	D.Lat	Dep.	Dist	D.Lat	Dep.	Dist	D.Lat	Dep.	Dist	D.Lat	Dep.
1	00.8	00.6	61	47.4	38.4	121	94.0	76.1	181	140.7	113.9	241	187.3	151.7
2	01.6	01.3	62	48.2	39.0	122	94.8	76.8	182	141.4	114.5	242	188.1	152.3
3	02.3	01.9	63	49.0	39.6	123	95.6	77.4	183	142.2	115.2	243	188.8	152.9
4	03.1	02.5	64	49.7	40.3	124	96.4	78.0	184	143.0	115.8	244	189.6	153.6
5	03.9	03.1	65	50.5	40.9	125	97.1	78.7	185	143.8	116.4	245	190.4	154.2
6	04.7	03.8	66	51.3	41.5	126	97.9	79.3	186	144.5	117.1	246	191.2	154.8
7	05.4	04.4	67	52.1	42.2	127	98.7	79.9	187	145.3	117.7	247	192.0	155.4
8	06.2	05.0	68	52.8	42.8	128	99.5	80.6	188	146.1	118.3	248	192.7	156.1
9	07.0	05.7	69	53.6	43.4	129	100.3	81.2	189	146.9	118.9	249	193.5	156.7
10	07.8	06.3	70	54.4	44.1	130	101.0	81.8	190	147.7	119.6	250	194.3	157.3
11	08.5	06.9	71	55.2	44.7	131	101.8	82.4	191	148.4	120.2	251	195.1	158.0
12	09.3	07.6	72	56.0	45.3	132	102.6	83.1	192	149.2	120.8	252	195.8	158.6
13	10.1	08.2	73	56.7	45.9	133	103.4	83.7	193	150.0	121.5	253	196.6	159.2
14	10.9	08.8	74	57.5	46.6	134	104.1	84.3	194	150.8	122.1	254	197.4	159.8
15	11.7	09.4	75	58.3	47.2	135	104.9	85.0	195	151.5	122.7	255	198.2	160.5
16	12.4	10.1	76	59.1	47.8	136	105.7	85.6	196	152.3	123.3	256	198.9	161.1
17	13.2	10.7	77	59.8	48.5	137	106.5	86.2	197	153.1	124.0	257	199.7	161.7
18	14.0	11.3	78	60.6	49.1	138	107.2	86.8	198	153.9	124.6	258	200.5	162.4
19	14.8	12.0	79	61.4	49.7	139	108.0	87.5	199	154.7	125.2	259	201.3	163.0
20	15.5	12.6	80	62.2	50.3	140	108.8	88.1	200	155.4	125.9	260	202.1	163.6
21	16.3	13.2	81	62.9	51.0	141	109.6	88.7	201	156.2	126.5	261	202.8	164.3
22	17.1	13.8	82	63.7	51.6	142	110.4	89.4	202	157.0	127.1	262	203.6	164.9
23	17.9	14.5	83	64.5	52.2	143	111.1	90.0	203	157.8	127.8	263	204.4	165.5
24	18.7	15.1	84	65.3	52.9	144	111.9	90.6	204	158.5	128.4	264	205.2	166.1
25	19.4	15.7	85	66.1	53.5	145	112.7	91.3	205	159.3	129.0	265	205.9	166.8
26	20.2	16.4	86	66.8	54.1	146	113.5	91.9	206	160.1	129.6	266	206.7	167.4
27	21.0	17.0	87	67.6	54.8	147	114.2	92.5	207	160.9	130.3	267	207.5	168.0
28	21.8	17.6	88	68.4	55.4	148	115.0	93.1	208	161.6	130.9	268	208.3	168.7
29	22.5	18.3	89	69.2	56.0	149	115.8	93.8	209	162.4	131.5	269	209.1	169.3
30	23.3	18.9	90	69.9	56.6	150	116.6	94.4	210	163.2	132.2	270	209.8	169.9
31	24.1	19.5	91	70.7	57.3	151	117.3	95.0	211	164.0	132.8	271	210.6	170.5
32	24.9	20.1	92	71.5	57.9	152	118.1	95.7	212	164.8	133.4	272	211.4	171.2
33	25.6	20.8	93	72.3	58.5	153	118.9	96.3	213	165.5	134.0	273	212.2	171.8
34	26.4	21.4	94	73.1	59.2	154	119.7	96.9	214	166.3	134.7	274	212.9	172.4
35	27.2	22.0	95	73.8	59.9	155	120.5	97.5	215	167.1	135.3	275	213.7	173.1
36	28.0	22.7	96	74.6	60.4	156	121.2	98.2	216	167.9	135.9	276	214.5	173.7
37	28.8	23.3	97	75.4	61.0	157	122.0	98.8	217	168.6	136.6	277	215.3	174.3
38	29.5	23.9	98	76.2	61.7	158	122.8	99.4	218	169.4	137.2	278	216.0	175.0
39	30.3	24.5	99	76.9	62.3	159	123.6	100.1	219	170.2	137.8	279	216.8	175.6
40	31.1	25.2	100	77.7	62.9	160	124.3	100.7	220	171.0	138.5	280	217.6	176.2
41	31.9	25.8	101	78.5	63.6	161	125.1	101.3	221	171.7	139.1	281	218.4	176.8
42	32.6	26.4	102	79.3	64.2	162	125.9	101.9	222	172.5	139.7	282	219.2	177.5
43	33.4	27.1	103	80.0	64.8	163	126.7	102.6	223	173.3	140.3	283	219.9	178.1
44	34.2	27.7	104	80.8	65.4	164	127.5	103.2	224	174.1	141.0	284	220.7	178.7
45	35.0	28.3	105	81.6	66.1	165	128.2	103.8	225	174.9	141.6	285	221.5	179.4
46	35.7	28.9	106	82.4	66.7	166	129.0	104.5	226	175.6	142.2	286	222.3	180.0
47	36.5	29.6	107	83.2	67.3	167	129.8	105.1	227	176.4	142.9	287	223.0	180.6
48	37.3	30.2	108	83.9	68.0	168	130.6	105.7	228	177.2	143.5	288	223.8	181.2
49	38.1	30.8	109	84.7	68.6	169	131.3	106.4	229	178.0	144.1	289	224.6	181.9
50	38.9	31.5	110	85.5	69.2	170	132.1	107.0	230	178.7	144.7	290	225.4	182.5
51	39.6	32.1	111	86.3	69.9	171	132.9	107.6	231	179.5	145.4	291	226.1	183.1
52	40.4	32.7	112	87.0	70.5	172	133.7	108.2	232	180.3	146.0	292	226.9	183.8
53	41.2	33.4	113	87.8	71.1	173	134.4	108.9	233	181.1	146.6	293	227.7	184.4
54	42.0	34.0	114	88.6	71.7	174	135.2	109.5	234	181.9	147.3	294	228.5	185.0
55	42.7	34.6	115	89.4	72.4	175	136.0	110.1	235	182.6	147.9	295	229.2	185.6
56	43.5	35.2	116	90.1	73.0	176	136.8	110.8	236	183.4	148.5	296	230.0	186.3
57	44.3	35.9	117	90.9	73.6	177	137.6	111.4	237	184.2	149.1	297	230.8	186.9
58	45.1	36.5	118	91.7	74.3	178	138.3	112.0	238	185.0	149.8	298	231.6	187.5
59	45.9	37.1	119	92.5	74.9	179	139.1	112.6	239	185.7	150.4	299	232.3	188.2
60	46.6	37.8	120	93.3	75.5	180	139.9	113.3	240	186.5	151.0	300	233.1	188.8
Dist	Dep.	D.Lat	Dist	Dep.	D.Lat	Dist	Dep.	D.Lat	Dist	Dep.	D.Lat	Dist	Dep.	D.Lat

51°

TRAVERSE TABLE
39 DEGREES.

321° ↑ / 219°   039° ↑ / 141°   2h 36m   **39°**

D Lon	Dep.		D Lon	Dep.		D Lon	Dep.		D Lon	Dep.		D Lon	Dep.	
Dist	D.Lat	Dep.	Dist	D.Lat	Dep.	Dist	D.Lat	Dep.	Dist	D.Lat	Dep.	Dist	D.Lat	Dep.
301	233·9	189·4	361	280·5	227·2	421	327·2	264·9	481	373·8	302·7	541	420·4	340·5
302	234·7	190·1	362	281·3	227·8	422	328·0	265·6	482	374·6	303·3	542	421·2	341·1
303	235·5	190·7	363	282·1	228·4	423	328·7	266·2	483	375·4	304·0	543	422·0	341·7
304	236·3	191·3	364	282·9	229·1	424	329·5	266·8	484	376·1	304·6	544	422·8	342·2
305	237·0	191·9	365	283·7	229·7	425	330·3	267·5	485	376·9	305·2	545	423·5	343·0
306	237·8	192·6	366	284·4	230·3	426	331·1	268·1	486	377·7	305·8	546	424·3	343·6
307	238·6	193·2	367	285·2	231·0	427	331·8	268·7	487	378·5	306·5	547	425·1	344·2
308	239·4	193·8	368	286·0	231·5	428	332·6	269·3	488	379·2	307·1	548	425·9	344·9
309	240·1	194·5	369	286·8	232·2	429	333·4	270·0	489	380·0	307·7	549	426·7	345·5
310	240·9	195·1	370	287·5	232·8	430	334·2	270·6	490	380·8	308·4	550	427·4	346·1
311	241·7	195·7	371	288·3	233·5	431	334·9	271·2	491	381·6	309·0	551	428·2	346·8
312	242·5	196·3	372	289·1	234·1	432	335·7	271·9	492	382·4	309·6	552	429·0	347·4
313	243·2	197·0	373	289·9	234·7	433	336·5	272·5	493	383·1	310·3	553	429·8	348·0
314	244·0	197·6	374	290·7	235·4	434	337·3	273·1	494	383·9	310·9	554	430·5	348·6
315	244·8	198·2	375	291·4	236·0	435	338·1	273·8	495	384·7	311·5	555	431·3	349·3
316	245·6	198·9	376	292·2	236·6	436	338·8	274·4	496	385·5	312·1	556	432·1	349·9
317	246·4	199·5	377	293·0	237·3	437	339·6	275·0	497	386·2	312·8	557	432·9	350·5
318	247·1	200·1	378	293·8	237·9	438	340·4	275·6	498	387·0	313·4	558	433·6	351·2
319	247·9	200·8	379	294·5	238·5	439	341·2	276·3	499	387·8	314·0	559	434·4	351·8
320	248·7	201·4	380	295·3	239·1	440	341·9	276·9	500	388·6	314·7	560	435·2	352·4
321	249·5	202·0	381	296·1	239·8	441	342·7	277·5	501	389·4	315·3	561	436·0	353·0
322	250·2	202·6	382	296·9	240·4	442	343·5	278·2	502	390·1	315·9	562	436·8	353·7
323	251·0	203·3	383	297·6	241·0	443	344·3	278·8	503	390·9	316·5	563	437·5	354·3
324	251·8	203·9	384	298·4	241·7	444	345·1	279·4	504	391·7	317·2	564	438·3	354·9
325	252·6	204·5	385	299·2	242·3	445	345·8	280·0	505	392·5	317·8	565	439·1	355·6
326	253·3	205·2	386	300·0	242·9	446	346·6	280·7	506	393·2	318·4	566	439·9	356·2
327	254·1	205·8	387	300·8	243·5	447	347·4	281·3	507	394·0	319·1	567	440·6	356·8
328	254·9	206·4	388	301·5	244·2	448	348·2	281·9	508	394·8	319·7	568	441·4	357·5
329	255·7	207·0	389	302·3	244·8	449	348·9	282·6	509	395·6	320·3	569	442·2	358·1
330	256·5	207·7	390	303·1	245·4	450	349·7	283·2	510	396·3	321·0	570	443·0	358·7
331	257·2	208·3	391	303·9	246·1	451	350·5	283·8	511	397·1	321·6	571	443·8	359·3
332	258·0	208·9	392	304·6	246·7	452	351·3	284·5	512	397·9	322·2	572	444·5	360·0
333	258·8	209·6	393	305·4	247·3	453	352·0	285·1	513	398·7	322·8	573	445·3	360·6
334	259·6	210·2	394	306·2	248·0	454	352·8	285·7	514	399·5	323·5	574	446·1	361·2
335	260·3	210·8	395	307·0	248·6	455	353·6	286·3	515	400·2	324·1	575	446·9	361·9
336	261·1	211·5	396	307·7	249·2	456	354·4	287·0	516	401·0	324·7	576	447·7	362·5
337	261·9	212·1	397	308·5	249·8	457	355·2	287·6	517	401·8	325·4	577	448·4	363·1
338	262·7	212·7	398	309·3	250·5	458	355·9	288·2	518	402·6	326·0	578	449·2	363·7
339	263·5	213·3	399	310·1	251·1	459	356·7	288·9	519	403·3	326·6	579	450·0	364·4
340	264·2	214·0	400	310·9	251·7	460	357·5	289·5	520	404·1	327·2	580	450·7	365·0
341	265·0	214·6	401	311·6	252·4	461	358·3	290·1	521	404·9	327·9	581	451·5	365·6
342	265·8	215·2	402	312·4	253·0	462	359·0	290·7	522	405·7	328·5	582	452·3	366·3
343	266·6	215·9	403	313·2	253·6	463	359·8	291·4	523	406·4	329·1	583	453·1	366·9
344	267·3	216·5	404	314·0	254·2	464	360·6	292·0	524	407·2	329·8	584	453·9	367·5
345	268·1	217·1	405	314·7	254·9	465	361·4	292·6	525	408·0	330·4	585	454·6	368·2
346	268·9	217·7	406	315·5	255·5	466	362·2	293·3	526	408·8	331·0	586	455·4	368·8
347	269·7	218·4	407	316·3	256·1	467	362·9	293·9	527	409·6	331·7	587	456·2	369·4
348	270·4	219·0	408	317·1	256·8	468	363·7	294·5	528	410·3	332·3	588	457·0	370·0
349	271·2	219·6	409	317·9	257·4	469	364·5	295·2	529	411·1	332·9	589	457·8	370·7
350	272·0	220·3	410	318·6	258·0	470	365·3	295·8	530	411·9	333·5	590	458·5	371·3
351	272·8	220·9	411	319·4	258·7	471	366·0	296·4	531	412·7	334·2	591	459·3	371·9
352	273·6	221·5	412	320·2	259·3	472	366·8	297·0	532	413·4	334·8	592	460·1	372·6
353	274·3	222·2	413	321·0	259·9	473	367·6	297·7	533	414·2	335·4	593	460·8	373·2
354	275·1	222·7	414	321·7	260·5	474	368·4	298·3	534	415·0	336·1	594	461·6	373·8
355	275·9	223·4	415	322·5	261·2	475	369·1	298·9	535	415·8	336·7	595	462·4	374·4
356	276·7	224·0	416	323·3	261·8	476	369·9	299·6	536	416·6	337·3	596	463·2	375·1
357	277·4	224·7	417	324·1	262·4	477	370·7	300·2	537	417·3	337·9	597	464·0	375·7
358	278·2	225·3	418	324·8	263·1	478	371·5	300·8	538	418·1	338·6	598	464·7	376·3
359	279·0	225·9	419	325·6	263·7	479	372·3	301·4	539	418·9	339·2	599	465·5	377·0
360	279·8	226·6	420	326·4	264·3	480	373·0	302·1	540	419·7	339·8	600	466·3	377·6
Dist	Dep.	D.Lat	Dist	Dep.	D.Lat	Dist	Dep.	D.Lat	Dist	Dep.	D.Lat	Dist	Dep.	D.Lat
D Lon		Dep.	D Lon		Dep.	D Lon		Dep.	D Lon		Dep.	D Lon		Dep.

309° ↑ / 231°   **51 DEGREES.**   051° ↑ / 129°   3h 24m   **51°**

40°

TRAVERSE TABLE
40 DEGREES.

D Lon	Dep.		D Lon	Dep.		D Lon	Dep.		D Lon	Dep.		D Lon	Dep.	
Dist	D.Lat	Dep.	Dist	D.Lat	Dep.	Dist	D.Lat	Dep.	Dist	D.Lat	Dep.	Dist	D.Lat	Dep.
1	00.8	00.6	61	46.7	39.2	121	92.7	77.8	181	138.7	116.3	241	184.6	154.9
2	01.5	01.3	62	47.5	39.9	122	93.5	78.4	182	139.4	117.0	242	185.4	155.6
3	02.3	01.9	63	48.3	40.5	123	94.2	79.1	183	140.2	117.6	243	186.1	156.2
4	03.1	02.6	64	49.0	41.1	124	95.0	79.7	184	141.0	118.3	244	186.9	156.8
5	03.8	03.2	65	49.8	41.8	125	95.8	80.3	185	141.7	118.9	245	187.7	157.5
6	04.6	03.9	66	50.6	42.4	126	96.5	81.0	186	142.5	119.6	246	188.4	158.1
7	05.4	04.5	67	51.3	43.1	127	97.3	81.6	187	143.3	120.2	247	189.2	158.8
8	06.1	05.1	68	52.1	43.7	128	98.1	82.3	188	144.0	120.8	248	190.0	159.4
9	06.9	05.8	69	52.9	44.4	129	98.8	82.9	189	144.8	121.5	249	190.7	160.1
10	07.7	06.4	70	53.6	45.0	130	99.6	83.6	190	145.5	122.1	250	191.5	160.7
11	08.4	07.1	71	54.4	45.6	131	100.4	84.2	191	146.3	122.8	251	192.3	161.3
12	09.2	07.7	72	55.2	46.3	132	101.1	84.8	192	147.1	123.4	252	193.0	162.0
13	10.0	08.4	73	55.9	46.9	133	101.9	85.5	193	147.8	124.1	253	193.8	162.6
14	10.7	09.0	74	56.7	47.6	134	102.6	86.1	194	148.6	124.7	254	194.6	163.3
15	11.5	09.6	75	57.5	48.2	135	103.4	86.8	195	149.4	125.3	255	195.3	163.9
16	12.3	10.3	76	58.2	48.9	136	104.2	87.4	196	150.1	126.0	256	196.1	164.6
17	13.0	10.9	77	59.0	49.5	137	104.9	88.1	197	150.9	126.6	257	196.9	165.2
18	13.8	11.6	78	59.8	50.1	138	105.7	88.7	198	151.7	127.3	258	197.6	165.8
19	14.6	12.2	79	60.5	50.8	139	106.5	89.3	199	152.4	127.9	259	198.4	166.5
20	15.3	12.9	80	61.3	51.4	140	107.2	90.0	200	153.2	128.6	260	199.2	167.1
21	16.1	13.5	81	62.0	52.1	141	108.0	90.6	201	154.0	129.2	261	199.9	167.8
22	16.9	14.1	82	62.8	52.7	142	108.8	91.3	202	154.7	129.8	262	200.7	168.4
23	17.6	14.8	83	63.6	53.4	143	109.5	91.9	203	155.5	130.5	263	201.5	169.1
24	18.4	15.4	84	64.3	54.0	144	110.3	92.6	204	156.3	131.1	264	202.2	169.7
25	19.2	16.1	85	65.1	54.6	145	111.1	93.2	205	157.0	131.8	265	203.0	170.3
26	19.9	16.7	86	65.9	55.3	146	111.8	93.8	206	157.8	132.4	266	203.8	171.0
27	20.7	17.4	87	66.6	55.9	147	112.6	94.5	207	158.6	133.1	267	204.5	171.6
28	21.4	18.0	88	67.4	56.6	148	113.4	95.1	208	159.3	133.7	268	205.3	172.3
29	22.2	18.6	89	68.2	57.2	149	114.1	95.8	209	160.1	134.3	269	206.1	172.9
30	23.0	19.3	90	68.9	57.9	150	114.9	96.4	210	160.9	135.0	270	206.8	173.6
31	23.7	19.9	91	69.7	58.5	151	115.7	97.1	211	161.6	135.6	271	207.6	174.2
32	24.5	20.6	92	70.5	59.1	152	116.4	97.7	212	162.4	136.3	272	208.4	174.8
33	25.3	21.2	93	71.2	59.8	153	117.2	98.3	213	163.2	136.9	273	209.1	175.5
34	26.0	21.9	94	72.0	60.4	154	118.0	99.0	214	163.9	137.6	274	209.9	176.1
35	26.8	22.5	95	72.8	61.1	155	118.7	99.6	215	164.7	138.2	275	210.7	176.8
36	27.6	23.1	96	73.5	61.7	156	119.5	100.3	216	165.5	138.8	276	211.4	177.4
37	28.3	23.8	97	74.3	62.4	157	120.3	100.9	217	166.2	139.5	277	212.2	178.1
38	29.1	24.4	98	75.1	63.0	158	121.0	101.6	218	167.0	140.1	278	213.0	178.7
39	29.9	25.1	99	75.8	63.6	159	121.8	102.2	219	167.8	140.8	279	213.7	179.3
40	30.6	25.7	100	76.6	64.3	160	122.6	102.8	220	168.5	141.4	280	214.5	180.0
41	31.4	26.4	101	77.4	64.9	161	123.3	103.5	221	169.3	142.1	281	215.3	180.6
42	32.2	27.0	102	78.1	65.6	162	124.1	104.1	222	170.1	142.7	282	216.0	181.3
43	32.9	27.6	103	78.9	66.2	163	124.9	104.8	223	170.8	143.3	283	216.8	181.9
44	33.7	28.3	104	79.7	66.8	164	125.6	105.4	224	171.6	144.0	284	217.6	182.6
45	34.5	28.9	105	80.4	67.5	165	126.4	106.1	225	172.4	144.6	285	218.3	183.2
46	35.2	29.6	106	81.2	68.1	166	127.2	106.7	226	173.1	145.3	286	219.1	183.8
47	36.0	30.2	107	82.0	68.8	167	127.9	107.3	227	173.9	145.9	287	219.9	184.5
48	36.8	30.9	108	82.7	69.4	168	128.7	108.0	228	174.7	146.6	288	220.6	185.1
49	37.5	31.5	109	83.5	70.1	169	129.5	108.6	229	175.4	147.2	289	221.4	185.8
50	38.3	32.1	110	84.3	70.7	170	130.2	109.3	230	176.2	147.8	290	222.2	186.4
51	39.1	32.8	111	85.0	71.3	171	131.0	109.9	231	177.0	148.5	291	222.9	187.1
52	39.8	33.4	112	85.8	72.0	172	131.8	110.6	232	177.7	149.1	292	223.7	187.7
53	40.6	34.1	113	86.6	72.6	173	132.5	111.2	233	178.5	149.8	293	224.5	188.3
54	41.4	34.7	114	87.3	73.3	174	133.3	111.8	234	179.3	150.4	294	225.2	189.0
55	42.1	35.4	115	88.1	73.9	175	134.1	112.5	235	180.0	151.1	295	226.0	189.6
56	42.9	36.0	116	88.9	74.6	176	134.8	113.1	236	180.8	151.7	296	226.7	190.3
57	43.7	36.6	117	89.6	75.2	177	135.6	113.8	237	181.6	152.3	297	227.5	190.9
58	44.4	37.3	118	90.4	75.8	178	136.4	114.4	238	182.3	153.0	298	228.3	191.6
59	45.2	37.9	119	91.2	76.5	179	137.1	115.1	239	183.1	153.6	299	229.0	192.2
60	46.0	38.6	120	91.9	77.1	180	137.9	115.7	240	183.9	154.3	300	229.8	192.8
Dist	Dep.	D.Lat	Dist	Dep.	D.Lat	Dist	Dep.	D.Lat	Dist	Dep.	D.Lat	Dist	Dep.	D.Lat
D Lon		Dep.	D Lon		Dep.	D Lon		Dep.	D Lon		Dep.	D Lon		Dep.

50°

50 DEGREES.

D Lon	Dep.		D Lon	Dep.		D Lon	Dep.		D Lon	Dep.		D Lon	Dep.		D Lon	Dep.	
Dist	D.Lat	Dep.	Dist	D.Lat	Dep.	Dist	D.Lat	Dep.	Dist	D.Lat	Dep.	Dist	D.Lat	Dep.	Dist	D.Lat	Dep.
301	230·6	193·5	361	276·5	232·0	421	322·5	270·6	481	368·5	309·2	541	414·4	347·7			
302	231·3	194·1	362	277·3	232·7	422	323·3	271·3	482	369·2	309·8	542	415·2	348·4			
303	232·1	194·8	363	278·1	233·3	423	324·0	271·9	483	370·0	310·5	543	416·0	349·0			
304	232·9	195·4	364	278·8	234·0	424	324·8	272·5	484	370·8	311·1	544	416·7	349·7			
305	233·6	196·1	365	279·6	234·6	425	325·6	273·2	485	371·5	311·8	545	417·5	350·3			
306	234·4	196·7	366	280·4	235·3	426	326·3	273·8	486	372·3	312·4	546	418·3	351·0			
307	235·2	197·3	367	281·1	235·9	427	327·1	274·5	487	373·1	313·0	547	419·0	351·6			
308	235·9	198·0	368	281·9	236·5	428	327·9	275·1	488	373·8	313·7	548	419·8	352·2			
309	236·7	198·6	369	282·7	237·2	429	328·6	275·8	489	374·6	314·3	549	420·6	352·9			
310	237·5	199·3	370	283·4	237·8	430	329·4	276·4	490	375·4	315·0	550	421·3	353·5			
311	238·2	199·9	371	284·2	238·5	431	330·2	277·0	491	376·1	315·6	551	422·1	354·2			
312	239·0	200·5	372	285·0	239·1	432	330·9	277·7	492	376·9	316·3	552	422·9	354·8			
313	239·8	201·2	373	285·7	239·8	433	331·7	278·3	493	377·7	316·9	553	423·6	355·5			
314	240·5	201·8	374	286·5	240·4	434	332·5	279·0	494	378·4	317·5	554	424·4	356·1			
315	241·3	202·5	375	287·3	241·0	435	333·2	279·6	495	379·2	318·2	555	425·2	356·7			
316	242·1	203·1	376	288·0	241·7	436	334·0	280·3	496	380·0	318·8	556	425·9	357·4			
317	242·8	203·8	377	288·8	242·3	437	334·8	280·9	497	380·7	319·5	557	426·7	358·0			
318	243·6	204·4	378	289·6	243·0	438	335·5	281·5	498	381·5	320·1	558	427·5	358·7			
319	244·4	205·0	379	290·3	243·6	439	336·3	282·2	499	382·3	320·8	559	428·2	359·3			
320	245·1	205·7	380	291·1	244·3	440	337·1	282·8	500	383·0	321·4	560	429·0	360·0			
321	245·9	206·3	381	291·9	244·9	441	337·8	283·5	501	383·8	322·0	561	429·8	360·6			
322	246·6	207·0	382	292·6	245·5	442	338·6	284·1	502	384·6	322·7	562	430·5	361·2			
323	247·4	207·6	383	293·4	246·2	443	339·4	284·8	503	385·3	323·3	563	431·3	361·9			
324	248·2	208·3	384	294·2	246·8	444	340·1	285·4	504	386·1	324·0	564	432·0	362·5			
325	349·0	208·9	385	294·9	247·5	445	340·9	286·0	505	386·9	324·6	565	432·8	363·2			
326	249·7	209·5	386	295·7	248·1	446	341·7	286·7	506	387·6	325·3	566	433·6	363·8			
327	250·5	210·2	387	296·5	248·8	447	342·4	287·3	507	388·4	325·9	567	434·3	364·5			
328	251·3	210·8	388	297·2	249·4	448	343·2	288·0	508	389·2	326·5	568	435·1	365·1			
329	252·0	211·5	389	298·0	250·0	449	344·0	288·6	509	389·9	327·2	569	435·9	365·7			
330	252·8	212·1	390	298·8	250·7	450	344·7	289·3	510	390·7	327·8	570	436·6	366·4			
331	253·6	212·8	391	299·5	251·3	451	345·5	289·9	511	391·4	328·5	571	437·4	367·0			
332	254·3	213·4	392	300·3	252·0	452	346·3	290·5	512	392·2	329·1	572	438·2	367·7			
333	255·1	214·0	393	301·1	252·6	453	347·0	291·2	513	393·0	329·8	573	438·9	368·3			
334	255·9	214·7	394	301·8	253·3	454	347·8	291·8	514	393·7	330·4	574	439·7	369·0			
335	256·6	215·3	395	302·6	253·9	455	348·6	292·5	515	394·5	331·0	575	440·5	369·6			
336	257·4	216·0	396	303·4	254·5	456	349·3	293·1	516	395·3	331·7	576	441·2	370·2			
337	258·2	216·6	397	304·1	255·2	457	350·1	293·8	517	396·0	332·3	577	442·0	370·9			
338	258·9	217·3	398	304·9	255·8	458	350·8	294·4	518	396·8	333·0	578	442·8	371·5			
339	259·7	217·9	399	305·7	256·5	459	351·6	295·0	519	397·6	333·6	579	443·5	372·2			
340	260·5	218·5	400	306·4	257·1	460	352·4	295·7	520	398·3	334·2	580	444·3	372·8			
341	261·2	219·2	401	307·2	257·8	461	353·1	296·3	521	399·1	334·9	581	445·1	373·5			
342	262·0	219·8	402	307·9	258·4	462	353·9	297·0	522	399·9	335·5	582	445·8	374·1			
343	262·8	220·5	403	308·7	259·0	463	354·7	297·6	523	400·6	336·2	583	446·6	374·7			
344	263·5	221·1	404	309·5	259·7	464	355·4	298·3	524	401·4	336·8	584	447·4	375·4			
345	264·3	221·8	405	310·2	260·3	465	356·2	298·9	525	402·2	337·5	585	448·1	376·0			
346	265·1	222·4	406	311·0	261·0	466	357·0	299·5	526	402·9	338·1	586	448·9	376·7			
347	265·8	223·0	407	311·8	261·6	467	357·7	300·2	527	403·7	338·7	587	449·7	377·3			
348	266·6	223·7	408	312·5	262·3	468	358·5	300·8	528	404·5	339·4	588	450·4	378·0			
349	267·3	224·3	409	313·3	262·9	469	359·3	301·5	529	405·2	340·0	589	451·2	378·6			
350	268·1	225·0	410	314·1	263·5	470	360·0	302·1	530	406·0	340·7	590	452·0	379·2			
351	268·9	225·6	411	314·8	264·2	471	360·8	302·8	531	406·8	341·3	591	452·7	379·9			
352	269·6	226·3	412	315·6	264·8	472	361·6	303·4	532	407·5	342·0	592	453·5	380·5			
353	270·4	226·9	413	316·4	265·5	473	362·3	304·0	533	408·3	342·6	593	454·3	381·2			
354	271·2	227·5	414	317·1	266·1	474	363·1	304·7	534	409·1	343·2	594	455·0	381·8			
355	271·9	228·2	415	317·9	266·8	475	363·9	305·3	535	409·8	343·9	595	455·8	382·5			
356	272·7	228·8	416	318·7	267·4	476	364·6	306·0	536	410·6	344·5	596	456·6	383·1			
357	273·5	229·5	417	319·4	268·0	477	365·4	306·6	537	411·4	345·2	597	457·3	383·7			
358	274·2	230·1	418	320·2	268·7	478	366·2	307·3	538	412·1	345·8	598	458·1	384·4			
359	275·0	230·8	419	321·0	269·3	479	366·9	307·9	539	412·9	346·5	599	458·9	385·0			
360	275·8	231·4	420	321·7	270·0	480	367·7	308·5	540	413·7	347·1	600	459·6	385·7			
Dist	Dep.	D.Lat	Dist	Dep.	D.Lat	Dist	Dep.	D.Lat	Dist	Dep.	D.Lat	Dist	Dep.	D.Lat	Dist	Dep.	D.Lat
D Lon		Dep.	D Lon		Dep.	D Lon		Dep.	D Lon		Dep.	D Lon		Dep.	D Lon		Dep.

TRAVERSE TABLE

319° ↑				041° ↑	
221°		**41 DEGREES.**		139°	2h 44m

D Lon	Dep.		Dist	D.Lat	Dep.	Dist	D.Lat	Dep.	Dist	D.Lat	Dep.	Dist	D.Lat	Dep.
Dist	D.Lat	Dep.												
1	00·8	00·7	61	46·0	40·0	121	91·3	79·4	181	136·6	118·7	241	181·9	158·1
2	01·5	01·3	62	46·8	40·7	122	92·1	80·0	182	137·4	119·4	242	182·6	158·8
3	02·3	02·0	63	47·5	41·3	123	92·8	80·7	183	138·1	120·1	243	183·4	159·4
4	03·0	02·6	64	48·3	42·0	124	93·6	81·4	184	138·9	120·7	244	184·1	160·1
5	03·8	03·3	65	49·1	42·6	125	94·3	82·0	185	139·6	121·4	245	184·9	160·7
6	04·5	03·9	66	49·8	43·3	126	95·1	82·7	186	140·4	122·0	246	185·7	161·4
7	05·3	04·6	67	50·6	44·0	127	95·8	83·3	187	141·1	122·7	247	186·4	162·0
8	06·0	05·2	68	51·3	44·6	128	96·6	84·0	188	141·9	123·3	248	187·2	162·7
9	06·8	05·9	69	52·1	45·3	129	97·4	84·6	189	142·6	124·0	249	187·9	163·4
10	07·5	06·6	70	52·8	45·9	130	98·1	85·3	190	143·4	124·7	250	188·7	164·0
11	08·3	07·2	71	53·6	46·6	131	98·9	85·9	191	144·1	125·3	251	189·4	164·7
12	09·1	07·9	72	54·3	47·2	132	99·6	86·6	192	144·9	126·0	252	190·2	165·3
13	09·8	08·5	73	55·1	47·9	133	100·4	87·3	193	145·7	126·6	253	190·9	166·0
14	10·6	09·2	74	55·8	48·5	134	101·1	87·9	194	146·4	127·3	254	191·7	166·6
15	11·3	09·8	75	56·6	49·2	135	101·9	88·6	195	147·2	127·9	255	192·5	167·3
16	12·1	10·5	76	57·4	49·9	136	102·6	89·2	196	147·9	128·6	256	193·2	168·0
17	12·8	11·2	77	58·1	50·5	137	103·4	89·9	197	148·7	129·2	257	194·0	168·6
18	13·6	11·8	78	58·9	51·2	138	104·1	90·5	198	149·4	129·9	258	194·7	169·3
19	14·3	12·5	79	59·6	51·8	139	104·9	91·2	199	150·2	130·6	259	195·5	169·9
20	15·1	13·1	80	60·4	52·5	140	105·7	91·8	200	150·9	131·2	260	196·2	170·6
21	15·8	13·8	81	61·1	53·1	141	106·4	92·5	201	151·7	131·9	261	197·0	171·2
22	16·6	14·4	82	61·9	53·8	142	107·2	93·2	202	152·5	132·5	262	197·7	171·9
23	17·4	15·1	83	62·6	54·5	143	107·9	93·8	203	153·2	133·2	263	198·5	172·5
24	18·1	15·7	84	63·4	55·1	144	108·7	94·5	204	154·0	133·8	264	199·2	173·2
25	18·9	16·4	85	64·2	55·8	145	109·4	95·1	205	154·7	134·5	265	200·0	173·9
26	19·6	17·1	86	64·9	56·4	146	110·2	95·8	206	155·5	135·1	266	200·8	174·5
27	20·4	17·7	87	65·7	57·1	147	110·9	96·4	207	156·2	135·8	267	201·5	175·2
28	21·1	18·4	88	66·4	57·7	148	111·7	97·1	208	157·0	136·5	268	202·3	175·8
29	21·9	19·0	89	67·2	58·4	149	112·5	97·8	209	157·7	137·1	269	203·0	176·5
30	22·6	19·7	90	67·9	59·0	150	113·2	98·4	210	158·5	137·8	270	203·8	177·1
31	23·4	20·3	91	68·7	59·7	151	114·0	99·1	211	159·2	138·4	271	204·5	177·8
32	24·2	21·0	92	69·4	60·4	152	114·7	99·7	212	160·0	139·1	272	205·3	178·4
33	24·9	21·6	93	70·2	61·0	153	115·5	100·4	213	160·8	139·7	273	206·0	179·1
34	25·7	22·3	94	70·9	61·7	154	116·2	101·0	214	161·5	140·4	274	206·8	179·8
35	26·4	23·0	95	71·7	62·3	155	117·0	101·7	215	162·3	141·1	275	207·5	180·4
36	27·2	23·6	96	72·5	63·0	156	117·7	102·3	216	163·0	141·7	276	208·3	181·1
37	27·9	24·3	97	73·2	63·6	157	118·5	103·0	217	163·8	142·4	277	209·1	181·7
38	28·7	24·9	98	74·0	64·3	158	119·2	103·7	218	164·5	143·0	278	209·8	182·4
39	29·4	25·6	99	74·7	64·9	159	120·0	104·3	219	165·3	143·7	279	210·6	183·0
40	30·2	26·2	100	75·5	65·6	160	120·8	105·0	220	166·0	144·3	280	211·3	183·7
41	30·9	26·9	101	76·2	66·3	161	121·5	105·6	221	166·8	145·0	281	212·1	184·4
42	31·7	27·6	102	77·0	66·9	162	122·3	106·3	222	167·5	145·6	282	212·8	185·0
43	32·5	28·2	103	77·7	67·6	163	123·0	106·9	223	168·3	146·3	283	213·6	185·7
44	33·2	28·9	104	78·5	68·2	164	123·8	107·6	224	169·1	147·0	284	214·3	186·3
45	34·0	29·5	105	79·2	68·9	165	124·5	108·2	225	169·8	147·6	285	215·1	187·0
46	34·7	30·2	106	80·0	69·5	166	125·3	108·9	226	170·6	148·3	286	215·8	187·6
47	35·5	30·8	107	80·8	70·2	167	126·0	109·6	227	171·3	148·9	287	216·6	188·3
48	36·2	31·5	108	81·5	70·9	168	126·8	110·2	228	172·1	149·6	288	217·4	188·9
49	37·0	32·1	109	82·3	71·5	169	127·5	110·9	229	172·8	150·2	289	218·1	189·6
50	37·7	32·8	110	83·0	72·2	170	128·3	111·5	230	173·6	150·9	290	218·9	190·3
51	38·5	33·5	111	83·8	72·8	171	129·1	112·2	231	174·3	151·5	291	219·6	190·9
52	39·2	34·1	112	84·5	73·5	172	129·8	112·9	232	175·1	152·2	292	220·4	191·6
53	40·0	34·8	113	85·3	74·1	173	130·6	113·5	233	175·8	152·9	293	221·1	192·2
54	40·8	35·4	114	86·0	74·8	174	131·3	114·2	234	176·6	153·5	294	221·9	192·9
55	41·5	36·1	115	86·8	75·4	175	132·1	114·8	235	177·4	154·2	295	222·6	193·5
56	42·3	36·7	116	87·5	76·1	176	132·8	115·5	236	178·1	154·8	296	223·4	194·2
57	43·0	37·4	117	88·3	76·8	177	133·6	116·1	237	178·9	155·5	297	224·1	194·8
58	43·8	38·1	118	89·1	77·4	178	134·3	116·8	238	179·6	156·1	298	224·9	195·5
59	44·5	38·7	119	89·8	78·1	179	135·1	117·4	239	180·4	156·8	299	225·7	196·2
60	45·3	39·4	120	90·6	78·7	180	135·8	118·1	240	181·1	157·5	300	226·4	196·8
Dist	Dep.	D.Lat	Dist	Dep.	D.Lat	Dist	Dep.	D.Lat	Dist	Dep.	D.Lat	Dist	Dep.	D.Lat
D Lon		Dep.	D Lon		Dep.	D Lon		Dep.	D Lon		Dep.	D Lon		Dep.

311° ↑			049° ↑	
229°		**49 DEGREES.**	131°	3h 16m

TRAVERSE TABLE
41 DEGREES.

319° ↑ / 221°   041° ↑ / 139°   2h 44m   **41°**

D Lon Dist	Dep. D.Lat	Dep.	D Lon Dist	Dep. D.Lat	Dep.	D Lon Dist	Dep. D.Lat	Dep.	D Lon Dist	Dep. D.Lat	Dep.	D Lon Dist	Dep. D.Lat	Dep.
301	227.2	197.5	361	272.5	236.8	421	317.7	276.2	481	363.0	315.6	541	408.3	354.9
302	227.9	198.1	362	273.2	237.5	422	318.5	276.9	482	363.8	316.2	542	409.1	355.6
303	228.7	198.8	363	274.0	238.1	423	319.2	277.5	483	364.5	316.9	543	409.8	356.2
304	229.4	199.4	364	274.7	238.8	424	320.0	278.2	484	365.3	317.5	544	410.6	356.9
305	230.2	200.1	365	275.5	239.5	425	320.8	278.8	485	366.0	318.2	545	411.3	357.6
306	230.9	200.8	366	276.2	240.1	426	321.5	279.5	486	366.8	318.8	546	412.1	358.2
307	231.7	201.4	367	277.0	240.8	427	322.3	280.1	487	367.5	319.5	547	412.8	358.9
308	232.5	202.1	368	277.7	241.4	428	323.0	280.8	488	368.3	320.2	548	413.6	359.5
309	233.2	202.7	369	278.5	242.1	429	323.8	281.4	489	369.1	320.8	549	414.3	360.2
310	234.0	203.4	370	279.2	242.7	430	324.5	282.1	490	369.8	321.5	550	415.1	360.8
311	234.7	204.0	371	280.0	243.4	431	325.3	282.8	491	370.6	322.1	551	415.8	361.5
312	235.5	204.7	372	280.8	244.1	432	326.0	283.4	492	371.3	322.8	552	416.6	362.1
313	236.2	205.3	373	281.5	244.7	433	326.8	284.1	493	372.1	323.4	553	417.4	362.8
314	237.0	206.0	374	282.3	245.4	434	327.5	284.7	494	372.8	324.1	554	418.1	363.5
315	237.7	206.7	375	283.0	246.0	435	328.3	285.4	495	373.6	324.7	555	418.9	364.1
316	238.5	207.3	376	283.8	246.7	436	329.1	286.0	496	374.3	325.4	556	419.6	364.8
317	239.2	208.0	377	284.5	247.3	437	329.8	286.7	497	375.1	326.1	557	420.4	365.4
318	240.0	208.6	378	285.3	248.0	438	330.6	287.4	498	375.8	326.7	558	421.1	366.1
319	240.8	209.3	379	286.0	248.6	439	331.3	288.0	499	376.6	327.4	559	421.9	366.7
320	241.5	209.9	380	286.8	249.3	440	332.1	288.7	500	377.4	328.0	560	422.6	367.4
321	242.3	210.6	381	287.5	250.0	441	332.8	289.3	501	378.1	328.7	561	423.4	368.0
322	243.0	211.3	382	288.3	250.6	442	333.6	290.0	502	378.9	329.3	562	424.1	368.7
323	243.8	211.9	383	289.1	251.3	443	334.3	290.6	503	379.6	330.0	563	424.9	369.4
324	244.5	212.6	384	289.8	251.9	444	335.1	291.3	504	380.4	330.7	564	425.7	370.0
325	245.3	213.2	385	290.6	252.5	445	335.8	291.9	505	381.1	331.3	565	426.4	370.7
326	246.0	213.9	386	291.3	253.2	446	336.6	292.6	506	381.9	332.0	566	427.2	371.3
327	246.8	214.5	387	292.1	253.9	447	337.4	293.3	507	382.6	332.6	567	427.9	372.0
328	247.5	215.2	388	292.8	254.6	448	338.1	293.9	508	383.4	333.3	568	428.7	372.6
329	248.3	215.8	389	293.6	255.2	449	338.9	294.6	509	384.1	333.9	569	429.4	373.3
330	249.1	216.5	390	294.3	255.9	450	339.6	295.2	510	384.9	334.6	570	430.2	374.0
331	249.8	217.2	391	295.1	256.5	451	340.4	295.9	511	385.7	335.2	571	430.9	374.6
332	250.6	217.8	392	295.8	257.2	452	341.1	296.5	512	386.4	335.9	572	431.7	375.3
333	251.3	218.5	393	296.6	257.8	453	341.9	297.2	513	387.2	336.6	573	432.4	375.9
334	252.1	219.1	394	297.4	258.5	454	342.6	297.9	514	387.9	337.2	574	433.2	376.6
335	252.8	219.8	395	298.1	259.1	455	343.4	298.5	515	388.7	337.9	575	434.0	377.2
336	253.6	220.4	396	298.9	259.8	456	344.1	299.2	516	389.4	338.5	576	434.7	377.9
337	254.3	221.1	397	299.6	260.5	457	344.9	299.8	517	390.2	339.2	577	435.5	378.5
338	255.1	221.7	398	300.4	261.1	458	345.7	300.5	518	390.9	339.8	578	436.2	379.2
339	255.8	222.4	399	301.1	261.8	459	346.4	301.1	519	391.7	340.5	579	437.0	379.9
340	256.6	223.1	400	301.9	262.4	460	347.2	301.8	520	392.4	341.2	580	437.7	380.5
341	257.4	223.7	401	302.6	263.1	461	347.9	302.4	521	393.2	341.8	581	438.5	381.2
342	258.1	224.4	402	303.4	263.7	462	348.7	303.1	522	394.0	342.5	582	439.2	381.8
343	258.9	225.0	403	304.1	264.4	463	349.4	303.8	523	394.7	343.1	583	440.0	382.5
344	259.6	225.7	404	304.9	265.0	464	350.2	304.4	524	395.5	343.8	584	440.8	383.1
345	260.4	226.3	405	305.7	265.7	465	350.9	305.1	525	396.2	344.4	585	441.5	383.8
346	261.1	227.0	406	306.4	266.4	466	351.7	305.7	526	397.0	345.1	586	442.3	384.5
347	261.9	227.7	407	307.2	267.0	467	352.4	306.4	527	397.7	345.7	587	443.0	385.1
348	262.6	228.3	408	307.9	267.7	468	353.2	307.0	528	398.5	346.4	588	443.8	385.8
349	263.4	229.0	409	308.7	268.3	469	354.0	307.7	529	399.2	347.1	589	444.5	386.4
350	264.1	229.6	410	309.4	269.0	470	354.7	308.5	530	400.0	347.7	590	445.3	387.1
351	264.9	230.3	411	310.2	269.6	471	355.5	309.0	531	400.8	348.4	591	446.0	387.7
352	265.7	230.9	412	310.9	270.3	472	356.2	309.7	532	401.5	349.0	592	446.8	388.4
353	266.4	231.6	413	311.7	271.0	473	357.0	310.3	533	402.3	349.7	593	447.5	389.0
354	267.2	232.2	414	312.4	271.6	474	357.7	311.0	534	403.0	350.3	594	448.3	389.7
355	267.9	232.9	415	313.2	272.3	475	358.5	311.6	535	403.8	351.0	595	449.1	390.4
356	268.7	233.6	416	314.0	272.9	476	359.2	312.3	536	404.5	351.6	596	449.8	391.0
357	269.4	234.2	417	314.7	273.6	477	360.0	312.9	537	405.3	352.3	597	450.6	391.7
358	270.2	234.9	418	315.5	274.2	478	360.8	313.6	538	406.0	353.0	598	451.3	392.3
359	270.9	235.5	419	316.2	274.9	479	361.5	314.3	539	406.8	353.6	599	452.1	393.0
360	271.7	236.2	420	317.0	275.5	480	362.3	314.9	540	407.5	354.3	600	452.8	393.6

| Dist | Dep. | D.Lat | Dist | Dep. | D.Lat | Dist | Dep. | D.Lat | Dist | Dep. | D.Lat | Dist | Dep. | D.Lat |
| D Lon | | Dep. | D Lon | | Dep. | D Lon | | Dep. | D Lon | | Dep. | D Lon | | Dep. |

311° ↑ / 229°   **49 DEGREES.**   049° ↑ / 131°   3h 16m   **49°**

| | 318° ↑ | | | | | | | | | | | | 042° ↑ | |
| | 222° | | | **TRAVERSE TABLE** | | | | 42 DEGREES. | | | | | 138° | 2h 48m |

D Lon	Dep.		D Lon	Dep.		D Lon	Dep.		D Lon	Dep.		D Lon	Dep.	
Dist	D. Lat	Dep.	Dist	D. Lat	Dep.	Dist	D. Lat	Dep.	Dist	D. Lat	Dep.	Dist	D. Lat	Dep.
1	00·7	00·7	61	45·3	40·8	121	89·9	81·0	181	134·5	121·1	241	179·1	161·3
2	01·5	01·3	62	46·1	41·5	122	90·7	81·6	182	135·3	121·8	242	179·8	161·9
3	02·2	02·0	63	46·8	42·2	123	91·4	82·3	183	136·0	122·5	243	180·6	162·6
4	03·0	02·7	64	47·6	42·8	124	92·1	83·0	184	136·7	123·1	244	181·3	163·3
5	03·7	03·3	65	48·3	43·5	125	92·9	83·6	185	137·5	123·8	245	182·1	163·9
6	04·5	04·0	66	49·0	44·2	126	93·6	84·3	186	138·2	124·5	246	182·8	164·6
7	05·2	04·7	67	49·8	44·8	127	94·4	85·0	187	139·0	125·1	247	183·6	165·3
8	05·9	05·4	68	50·5	45·5	128	95·1	85·6	188	139·7	125·8	248	184·3	165·9
9	06·7	06·0	69	51·3	46·2	129	95·9	86·3	189	140·5	126·5	249	185·0	166·6
10	07·4	06·7	70	52·0	46·8	130	96·6	87·0	190	141·2	127·1	250	185·8	167·3
11	08·2	07·4	71	52·8	47·5	131	97·4	87·7	191	141·9	127·8	251	186·5	168·0
12	08·9	08·0	72	53·5	48·2	132	98·1	88·3	192	142·7	128·5	252	187·3	168·6
13	09·7	08·7	73	54·2	48·8	133	98·8	89·0	193	143·4	129·1	253	188·0	169·3
14	10·4	09·4	74	55·0	49·5	134	99·6	89·7	194	144·2	129·8	254	188·8	170·0
15	11·1	10·0	75	55·7	50·2	135	100·3	90·3	195	144·9	130·5	255	189·5	170·6
16	11·9	10·7	76	56·5	50·9	136	101·1	91·0	196	145·7	131·1	256	190·2	171·3
17	12·6	11·4	77	57·2	51·5	137	101·8	91·7	197	146·4	131·8	257	191·0	172·0
18	13·4	12·0	78	58·0	52·2	138	102·6	92·3	198	147·1	132·5	258	191·7	172·6
19	14·1	12·7	79	58·7	52·9	139	103·3	93·0	199	147·9	133·2	259	192·5	173·3
20	14·9	13·4	80	59·5	53·5	140	104·0	93·7	200	148·6	133·8	260	193·2	174·0
21	15·6	14·1	81	60·2	54·2	141	104·8	94·3	201	149·4	134·5	261	194·0	174·6
22	16·3	14·7	82	60·9	54·9	142	105·5	95·0	202	150·1	135·2	262	194·7	175·3
23	17·1	15·4	83	61·7	55·5	143	106·3	95·7	203	150·9	135·8	263	195·4	176·0
24	17·8	16·1	84	62·4	56·2	144	107·0	96·4	204	151·6	136·5	264	196·2	176·7
25	18·6	16·7	85	63·2	56·9	145	107·8	97·0	205	152·3	137·2	265	196·9	177·3
26	19·3	17·4	86	63·9	57·5	146	108·5	97·7	206	153·1	137·8	266	197·7	178·0
27	20·1	18·1	87	64·7	58·2	147	109·2	98·4	207	153·8	138·5	267	198·4	178·7
28	20·8	18·7	88	65·4	58·9	148	110·0	99·0	208	154·6	139·2	268	199·2	179·3
29	21·6	19·4	89	66·1	59·6	149	110·7	99·7	209	155·3	139·8	269	199·9	180·0
30	22·3	20·1	90	66·9	60·2	150	111·5	100·4	210	156·1	140·5	270	200·6	180·7
31	23·0	20·7	91	67·6	60·9	151	112·2	101·0	211	156·8	141·2	271	201·4	181·3
32	23·8	21·4	92	68·4	61·6	152	113·0	101·7	212	157·5	141·9	272	202·1	182·0
33	24·5	22·1	93	69·1	62·2	153	113·7	102·4	213	158·3	142·5	273	202·9	182·7
34	25·3	22·8	94	69·9	62·9	154	114·4	103·0	214	159·0	143·2	274	203·6	183·3
35	26·0	23·4	95	70·6	63·6	155	115·2	103·7	215	159·8	143·9	275	204·4	184·0
36	26·8	24·1	96	71·3	64·2	156	115·9	104·4	216	160·5	144·5	276	205·1	184·7
37	27·5	24·8	97	72·1	64·9	157	116·7	105·1	217	161·3	145·2	277	205·9	185·3
38	28·2	25·4	98	72·8	65·6	158	117·4	105·7	218	162·0	145·9	278	206·6	186·0
39	29·0	26·1	99	73·6	66·2	159	118·2	106·4	219	162·7	146·5	279	207·3	186·7
40	29·7	26·8	100	74·3	66·9	160	118·9	107·1	220	163·5	147·2	280	208·1	187·4
41	30·5	27·4	101	75·1	67·6	161	119·6	107·7	221	164·2	147·9	281	208·8	188·0
42	31·2	28·1	102	75·8	68·3	162	120·4	108·4	222	165·0	148·5	282	209·6	188·7
43	32·0	28·8	103	76·5	68·9	163	121·1	109·1	223	165·7	149·2	283	210·3	189·4
44	32·7	29·4	104	77·3	69·6	164	121·9	109·7	224	166·5	149·9	284	211·1	190·0
45	33·4	30·1	105	78·0	70·3	165	122·6	110·4	225	167·2	150·6	285	211·8	190·7
46	34·2	30·8	106	78·8	70·9	166	123·4	111·1	226	168·0	151·2	286	212·5	191·4
47	34·9	31·4	107	79·5	71·6	167	124·1	111·7	227	168·7	151·9	287	213·3	192·0
48	35·7	32·1	108	80·3	72·3	168	124·8	112·4	228	169·4	152·6	288	214·0	192·7
49	36·4	32·8	109	81·0	72·9	169	125·6	113·1	229	170·2	153·2	289	214·8	193·4
50	37·2	33·5	110	81·7	73·6	170	126·3	113·8	230	170·9	153·9	290	215·5	194·0
51	37·9	34·1	111	82·5	74·3	171	127·1	114·4	231	171·7	154·6	291	216·3	194·7
52	38·6	34·8	112	83·2	74·9	172	127·8	115·1	232	172·4	155·2	292	217·0	195·4
53	39·4	35·5	113	84·0	75·6	173	128·6	115·8	233	173·2	155·9	293	217·7	196·1
54	40·1	36·1	114	84·7	76·3	174	129·3	116·4	234	173·9	156·6	294	218·5	196·7
55	40·9	36·8	115	85·5	77·0	175	130·1	117·1	235	174·6	157·2	295	219·2	197·4
56	41·6	37·5	116	86·2	77·6	176	130·8	117·8	236	175·4	157·9	296	220·0	198·1
57	42·4	38·1	117	86·9	78·3	177	131·5	118·4	237	176·1	158·6	297	220·7	198·7
58	43·1	38·8	118	87·7	79·0	178	132·3	119·1	238	176·9	159·3	298	221·5	199·4
59	43·8	39·5	119	88·4	79·6	179	133·0	119·8	239	177·6	159·9	299	222·2	200·1
60	44·6	40·1	120	89·2	80·3	180	133·8	120·4	240	178·4	160·6	300	222·9	200·7
Dist	Dep.	D. Lat	Dist	Dep.	D. Lat	Dist	Dep.	D. Lat	Dist	Dep.	D. Lat	Dist	Dep.	D. Lat
D Lon		Dep.	D Lon		Dep.	D Lon		Dep.	D Lon		Dep.	D Lon		Dep.

| 312° ↑ | | | | | | | | 048° ↑ | |
| 228° | | | **48 DEGREES.** | | | | | 132° | 3h 12m |

	318° ↑					**TRAVERSE TABLE**					↑ 042°		
	222°					42 DEGREES.					138°	2h 48m	**42°**

D Lon	Dep.		D Lon	Dep.		D Lon	Dep.		D Lon	Dep.		D Lon	Dep.	
Dist	D.Lat	Dep.	Dist	D.Lat	Dep.	Dist	D.Lat	Dep.	Dist	D.Lat	Dep.	Dist	D.Lat	Dep.
301	223·7	201·4	361	268·3	241·6	421	312·9	281·7	481	357·5	321·9	541	402·0	362·0
302	224·4	202·1	362	269·0	242·2	422	313·6	282·4	482	358·2	322·5	542	402·8	362·7
303	225·2	202·7	363	269·8	242·9	423	314·4	283·0	483	358·9	323·2	543	403·5	363·3
304	225·9	203·4	364	270·5	243 6	424	315·1	283·7	484	359·7	323·9	544	404·3	364·0
305	226·7	204·1	365	271·2	244·2	425	315·8	284·4	485	360·4	324·5	545	405·0	364·7
306	227·4	204·8	366	272·0	244·9	426	316·6	285·0	486	361·2	325·2	546	405·8	365·3
307	228·1	205·4	367	272·7	245·6	427	317·3	285·7	487	361·9	325·9	547	406·5	366·0
308	228·9	206·1	368	273·5	246·2	428	318·1	286 4	488	362·7	326·5	548	407·2	366·7
309	229·6	206·8	369	274·2	246·9	429	318·8	287·1	489	363·4	327·2	549	408·0	367·4
310	230·4	207·4	370	275·0	247·6	430	319·6	287·7	490	364·1	327·9	550	408·7	368·0
311	231·1	208·1	371	275·7	248·2	431	320·3	288·4	491	364·9	328·5	551	409·5	368·7
312	231·9	208·8	372	276·4	248·9	432	321·0	289·1	492	365·6	329·2	552	410·2	369·4
313	232·6	209·4	373	277 2	249·6	433	321·8	289·7	493	366·4	329·9	553	411·0	370·0
314	233·3	210·1	374	277·9	250·3	434	322·5	290·4	494	367·1	330·6	554	411·7	370·7
315	234·1	210·8	375	278·7	250·9	435	323·3	291·1	495	367·9	331·2	555	412·4	371·4
316	234·8	211·4	376	279·4	251·6	436	324·0	291·7	496	368·6	331·9	556	413·2	372·0
317	235·6	212·1	377	280·2	252·3	437	324·8	292·4	497	369·3	332·5	557	413·9	372·7
318	236·3	212·8	378	280·9	252·9	438	325·5	293·1	498	370·1	333·2	558	414·7	373·4
319	237·1	213·5	379	281·7	253·6	439	326·2	293·7	499	370·8	333·9	559	415·4	374·0
320	237·8	214·1	380	282·4	254·3	440	327·0	294·4	500	371·6	334·6	560	416·2	374·7
321	238·5	214·8	381	283·1	254·9	441	327·7	295·1	501	372·3	335·2	561	416·9	375·4
322	239·3	215·5	382	283·9	255·6	442	328·5	295·8	502	373·1	335·9	562	417·6	376·1
323	240·0	216·1	383	284·6	256·3	443	329·2	296·4	503	373·8	336·6	563	418·4	376·7
324	240·8	216·8	384	285·4	256·9	444	330·0	297·1	504	374·5	337·2	564	419·1	377·4
325	241·5	217·5	385	286·1	257·6	445	330·7	297·8	505	375·3	337·9	565	419·9	378·1
326	242·3	218·1	386	286·9	258·3	446	331·4	298·4	506	376·0	338·6	566	420·6	378·7
327	243·0	218·8	387	287·6	259·0	447	332·2	299·1	507	376·8	339·2	567	421·4	379·4
328	243·8	219·5	388	288·3	259·6	448	332·9	299·8	508	377·5	339·9	568	422·1	380·1
329	244·5	220·1	389	289·1	260·3	449	333·7	300·4	509	378·3	340·6	569	422·8	380·7
330	245·2	220·8	390	289·8	261·0	450	334·4	301·1	510	379·0	341·3	570	423·6	381·4
331	246·0	221·5	391	290·6	261·6	451	335·2	301·8	511	379·7	341·9	571	424·3	382·1
332	246·7	222·2	392	291·3	262·3	452	335·9	302·4	512	380·5	342·6	572	425·1	382·7
333	247·5	222·8	393	292·1	263·0	453	336·6	303·1	513	381·2	343·3	573	425·8	383·4
334	248·2	223·5	394	292·8	263·6	454	337·4	303·8	514	382·0	343·9	574	426·6	384·1
335	249·0	224·2	395	293·5	264·3	455	338·1	304·5	515	382·7	344·6	575	427·3	384·8
336	249·7	224·8	396	294·3	265·0	456	338·9	305·1	516	383·5	345·3	576	428·1	385·4
337	250·4	225·5	397	295·0	265·6	457	339·6	305·8	517	384·2	345·9	577	428·8	386·1
338	251·2	226·2	398	295·8	266·3	458	340·4	306·5	518	384·9	346·6	578	429·5	386·8
339	251·9	226·8	399	296·5	267·0	459	341·1	307·1	519	385·7	347·3	579	430·3	387·4
340	252·7	227·5	400	297·3	267·7	460	341·8	307·8	520	386·4	347·9	580	431·0	388·1
341	253·4	228·2	401	298·0	268·3	461	342·6	308·5	521	387·2	348·6	581	431·8	388·8
342	254·2	228·8	402	298·7	269·0	462	343·3	309·1	522	387·9	349·3	582	432·5	389·4
343	254·9	229·5	403	299·5	269·7	463	344·1	309·8	523	388·7	350·0	583	433·3	390·1
344	255·6	230·2	404	300·2	270·3	464	344·8	310·5	524	389·4	350·6	584	434·0	390·8
345	256·4	230·9	405	301·0	271·0	465	345·6	311·1	525	390·2	351·3	585	434·7	391·4
346	257·1	231·5	406	301·7	271·7	466	346·3	311·8	526	390·9	352·0	586	435·5	392·1
347	257·9	232·2	407	302·5	272·3	467	347·0	312·5	527	391·6	352·6	587	436·2	392·8
348	258·6	232·9	408	303·2	273·0	468	347·8	313·2	528	392·4	353·3	588	437·0	393·4
349	259·4	233·5	409	303·9	273·7	469	348·5	313·8	529	393·1	354·0	589	437·7	394·1
350	260·1	234·2	410	304·7	274·3	470	349·3	314·5	530	393·9	354·6	590	438·5	394·8
351	260·8	234·9	411	305·4	275·0	471	350·0	315·2	531	394·6	355·3	591	439·2	395·5
352	261·6	235·5	412	306·2	275·7	472	350·8	315·8	532	395·4	356·0	592	439·9	396·1
353	262·3	236·2	413	306·9	276·4	473	351·5	316·5	533	396·1	356·6	593	440·7	396·8
354	263·1	236·9	414	307·7	277·0	474	352·3	317·2	534	396·8	357·3	594	441·4	397·5
355	263·8	237·5	415	308·4	277·7	475	353·0	317·8	535	397·6	358·0	595	442·2	398·1
356	264·6	238·2	416	309·1	278·4	476	353·7	318·5	536	398·3	358·7	596	442·9	398·8
357	265·3	238·9	417	309·9	279·0	477	354·5	319·2	537	399·1	359·3	597	443·7	399·5
358	266·0	239·5	418	310·6	279·7	478	355·2	319·8	538	399·8	360·0	598	444·4	400·1
359	266·8	240·2	419	311·4	280·4	479	356·0	320·5	539	400·6	360·7	599	445·1	400·8
360	267·5	240·9	420	312·1	281·0	480	356·7	321·2	540	401·3	361·3	600	445·9	401·5
Dist	Dep.	D.Lat	Dist	Dep.	D.Lat	Dist	Dep.	D.Lat	Dist	Dep.	D.Lat	Dist	Dep.	D.Lat
D Lon		Dep.	D Lon		Dep.	D Lon		Dep.	D Lon		Dep.	D Lon		Dep.

	312° ↑							↑ 048°		
	228°			**48 DEGREES.**				132°	3h 12m	**48°**

43°

TRAVERSE TABLE
43 DEGREES.

D Lon Dep.			D Lon Dep.			D Lon Dep.			D Lon Dep.			D Lon Dep.		
Dist	D.Lat	Dep.	Dist	D.Lat	Dep.	Dist	D.Lat	Dep.	Dist	D.Lat	Dep.	Dist	D.Lat	Dep.
1	00·7	00·7	61	44·6	41·6	121	88·5	82·5	181	132·4	123·4	241	176·3	164·4
2	01·5	01·4	62	45·3	42·3	122	89·2	83·2	182	133·1	124·1	242	177·0	165·0
3	02·2	02·0	63	46·1	43·0	123	90·0	83·9	183	133·8	124·8	243	177·7	165·7
4	02·9	02·7	64	46·8	43·6	124	90·7	84·6	184	134·6	125·5	244	178·5	166·4
5	03·7	03·4	65	47·5	44·3	125	91·4	85·2	185	135·3	126·2	245	179·2	167·1
6	04·4	04·1	66	48·3	45·0	126	92·2	85·9	186	136·0	126·9	246	179·9	167·8
7	05·1	04·8	67	49·0	45·7	127	92·9	86·6	187	136·8	127·5	247	180·6	168·5
8	05·9	05·5	68	49·7	46·4	128	93·6	87·3	188	137·5	128·2	248	181·4	169·1
9	06·6	06·1	69	50·5	47·1	129	94·3	88·0	189	138·2	128·9	249	182·1	169·8
10	07·3	06·8	70	51·2	47·7	130	95·1	88·7	190	139·0	129·6	250	182·8	170·5
11	08·0	07·5	71	51·9	48·4	131	95·8	89·3	191	139·7	130·3	251	183·6	171·2
12	08·8	08·2	72	52·7	49·1	132	96·5	90·0	192	140·4	130·9	252	184·3	171·9
13	09·5	08·9	73	53·4	49·8	133	97·3	90·7	193	141·2	131·6	253	185·0	172·5
14	10·2	09·5	74	54·1	50·5	134	98·0	91·4	194	141·9	132·3	254	185·8	173·2
15	11·0	10·2	75	54·9	51·1	135	98·7	92·1	195	142·6	133·0	255	186·5	173·9
16	11·7	10·9	76	55·6	51·8	136	99·5	92·8	196	143·3	133·7	256	187·2	174·6
17	12·4	11·6	77	56·3	52·5	137	100·2	93·4	197	144·1	134·4	257	188·0	175·3
18	13·2	12·3	78	57·0	53·2	138	100·9	94·1	198	144·8	135·0	258	188·7	176·0
19	13·9	13·0	79	57·8	53·9	139	101·7	94·8	199	145·5	135·7	259	189·4	176·6
20	14·6	13·6	80	58·5	54·6	140	102·4	95·5	200	146·3	136·4	260	190·2	177·3
21	15·4	14·3	81	59·2	55·2	141	103·1	96·2	201	147·0	137·1	261	190·9	178·0
22	16·1	15·0	82	60·0	55·9	142	103·9	96·8	202	147·7	137·8	262	191·6	178·7
23	16·8	15·7	83	60·7	56·6	143	104·6	97·5	203	148·5	138·4	263	192·3	179·4
24	17·6	16·4	84	61·4	57·3	144	105·3	98·2	204	149·2	139·1	264	193·1	180·0
25	18·3	17·0	85	62·2	58·0	145	106·0	98·9	205	149·9	139·8	265	193·8	180·7
26	19·0	17·7	86	62·9	58·7	146	106·8	99·6	206	150·7	140·5	266	194·5	181·4
27	19·7	18·4	87	63·6	59·3	147	107·5	100·3	207	151·4	141·2	267	195·3	182·1
28	20·5	19·1	88	64·4	60·0	148	108·2	100·9	208	152·1	141·8	268	196·0	182·8
29	21·2	19·8	89	65·1	60·7	149	109·0	101·6	209	152·9	142·5	269	196·7	183·5
30	21·9	20·5	90	65·8	61·4	150	109·7	102·3	210	153·6	143·2	270	197·5	184·1
31	22·7	21·1	91	66·6	62·1	151	110·4	103·0	211	154·3	143·9	271	198·2	184·8
32	23·4	21·8	92	67·3	62·7	152	111·2	103·7	212	155·0	144·6	272	198·9	185·5
33	24·1	22·5	93	68·0	63·4	153	111·9	104·3	213	155·8	145·3	273	199·7	186·2
34	24·9	23·2	94	68·7	64·1	154	112·6	105·0	214	156·5	145·9	274	200·4	186·9
35	25·6	23·9	95	69·5	64·8	155	113·4	105·7	215	157·2	146·6	275	201·1	187·5
36	26·3	24·6	96	70·2	65·5	156	114·1	106·4	216	158·0	147·3	276	201·9	188·2
37	27·1	25·2	97	70·9	66·2	157	114·8	107·1	217	158·7	148·0	277	202·6	188·9
38	27·8	25·9	98	71·7	66·8	158	115·6	107·8	218	159·4	148·7	278	203·3	189·6
39	28·5	26·6	99	72·4	67·5	159	116·3	108·4	219	160·2	149·4	279	204·0	190·3
40	29·3	27·3	100	73·1	68·2	160	117·0	109·1	220	160·9	150·0	280	204·8	191·0
41	30·0	28·0	101	73·9	68·9	161	117·7	109·8	221	161·6	150·7	281	205·5	191·6
42	30·7	28·6	102	74·6	69·6	162	118·5	110·5	222	162·4	151·4	282	206·2	192·3
43	31·4	29·3	103	75·3	70·2	163	119·2	111·2	223	163·1	152·1	283	207·0	193·0
44	32·2	30·0	104	76·1	70·9	164	119·9	111·8	224	163·8	152·8	284	207·7	193·7
45	32·9	30·7	105	76·8	71·6	165	120·7	112·5	225	164·6	153·4	285	208·4	194·4
46	33·6	31·4	106	77·5	72·3	166	121·4	113·2	226	165·3	154·1	286	209·2	195·1
47	34·4	32·1	107	78·3	73·0	167	122·1	113·9	227	166·0	154·8	287	209·9	195·7
48	35·1	32·7	108	79·0	73·7	168	122·9	114·6	228	166·7	155·5	288	210·6	196·4
49	35·8	33·4	109	79·7	74·3	169	123·6	115·3	229	167·5	156·2	289	211·4	197·1
50	36·6	34·1	110	80·4	75·0	170	124·3	115·9	230	168·2	156·9	290	212·1	197·8
51	37·3	34·8	111	81·2	75·7	171	125·1	116·6	231	168·9	157·5	291	212·8	198·5
52	38·0	35·5	112	81·9	76·4	172	125·8	117·3	232	169·7	158·2	292	213·6	199·1
53	38·8	36·1	113	82·6	77·1	173	126·5	118·0	233	170·4	158·9	293	214·3	199·8
54	39·5	36·8	114	83·4	77·7	174	127·3	118·7	234	171·1	159·6	294	215·0	200·5
55	40·2	37·5	115	84·1	78·4	175	128·0	119·3	235	171·9	160·3	295	215·7	201·2
56	41·0	38·2	116	84·8	79·1	176	128·7	120·0	236	172·6	161·0	296	216·5	201·9
57	41·7	38·9	117	85·6	79·8	177	129·4	120·7	237	173·3	161·6	297	217·2	202·6
58	42·4	39·6	118	86·3	80·5	178	130·2	121·4	238	174·1	162·3	298	217·9	203·2
59	43·1	40·2	119	87·0	81·2	179	130·9	122·1	239	174·8	163·0	299	218·7	203·9
60	43·9	40·9	120	87·8	81·8	180	131·6	122·8	240	175·5	163·7	300	219·4	204·6
Dist	Dep.	D.Lat	Dist	Dep.	D.Lat	Dist	Dep.	D.Lat	Dist	Dep.	D.Lat	Dist	Dep.	D.Lat
D Lon		Dep.	D Lon		Dep.	D Lon		Dep.	D Lon		Dep.	D Lon		Dep.

47°

TRAVERSE TABLE
43 DEGREES.

317° ↑ / 223° — 043° ↑ / 137° — 2h 52m **43°**

Dist	D.Lat	Dep.	Dist	D.Lat	Dep.	Dist	D.Lat	Dep.	Dist	D.Lat	Dep.	Dist	D.Lat	Dep.
301	220.1	205.3	361	264.0	246.2	421	307.9	287.1	481	351.8	328.0	541	395.7	369.0
302	220.9	206.0	362	264.8	246.9	422	308.6	287.8	482	352.5	328.7	542	396.4	369.6
303	221.6	206.6	363	265.5	247.6	423	309.4	288.5	483	353.2	329.4	543	397.1	370.3
304	222.3	207.3	364	266.2	248.2	424	310.1	289.2	484	354.0	330.1	544	397.9	371.0
305	223.1	208.0	365	266.9	248.9	425	310.8	289.9	485	354.7	330.8	545	398.6	371.7
306	223.8	208.7	366	267.7	249.6	426	311.6	290.5	486	355.4	331.5	546	399.3	372.4
307	224.5	209.4	367	268.4	250.3	427	312.3	291.2	487	356.2	332.1	547	400.1	373.1
308	225.3	210.1	368	269.1	251.0	428	313.0	291.9	488	356.9	332.8	548	400.8	373.7
309	226.0	210.7	369	269.9	251.7	429	313.8	292.6	489	357.6	333.5	549	401.5	374.4
310	226.7	211.4	370	270.6	252.3	430	314.5	293.3	490	358.4	334.2	550	402.2	375.1
311	227.5	212.1	371	271.3	253.0	431	315.2	293.9	491	359.1	334.9	551	403.0	375.8
312	228.2	212.8	372	272.1	253.7	432	315.9	294.6	492	359.8	335.5	552	403.7	376.5
313	228.9	213.5	373	272.8	254.4	433	316.7	295.3	493	360.6	336.2	553	404.4	377.1
314	229.6	214.1	374	273.5	255.1	434	317.4	296.0	494	361.3	336.9	554	405.2	377.8
315	230.4	214.8	375	274.3	255.7	435	318.1	296.7	495	362.0	337.6	555	405.9	378.5
316	231.1	215.5	376	275.0	256.4	436	318.9	297.4	496	362.8	338.3	556	406.6	379.2
317	231.8	216.2	377	275.7	257.1	437	319.6	298.0	497	363.5	339.0	557	407.4	379.9
318	232.6	216.9	378	276.5	257.8	438	320.3	298.7	498	364.2	339.6	558	408.1	380.6
319	233.3	217.6	379	277.2	258.5	439	321.1	299.4	499	364.9	340.3	559	408.8	381.2
320	234.0	218.2	380	277.9	259.2	440	321.8	300.1	500	365.7	341.0	560	409.6	381.9
321	234.8	218.9	381	278.6	259.8	441	322.5	300.8	501	366.4	341.7	561	410.3	382.6
322	235.5	219.6	382	279.4	260.5	442	323.3	301.4	502	367.1	342.4	562	411.0	383.3
323	236.2	220.3	383	280.1	261.2	443	324.0	302.1	503	367.9	343.0	563	411.8	384.0
324	237.0	221.0	384	280.8	261.9	444	324.7	302.8	504	368.6	343.7	564	412.5	384.6
325	237.7	221.6	385	281.6	262.6	445	325.5	303.5	505	369.3	344.4	565	413.2	385.3
326	238.4	222.3	386	282.3	263.3	446	326.2	304.2	506	370.1	345.1	566	413.9	386.0
327	239.2	223.0	387	283.0	263.9	447	326.9	304.9	507	370.8	345.8	567	414.7	386.7
328	239.9	223.7	388	283.8	264.6	448	327.6	305.5	508	371.5	346.5	568	415.4	387.4
329	240.6	224.4	389	284.5	265.3	449	328.4	306.2	509	372.3	347.1	569	416.1	388.1
330	241.3	225.1	390	285.2	266.0	450	329.1	306.9	510	373.0	347.8	570	416.9	388.8
331	242.1	225.7	391	286.0	266.7	451	329.9	307.6	511	373.7	348.5	571	417.6	389.4
332	242.8	226.4	392	286.7	267.3	452	330.6	308.3	512	374.5	349.2	572	418.3	390.1
333	243.5	227.1	393	287.4	268.0	453	331.3	308.9	513	375.2	349.9	573	419.1	390.8
334	244.3	227.8	394	288.2	268.7	454	332.0	309.6	514	375.9	350.5	574	419.8	391.5
335	245.0	228.5	395	288.9	269.4	455	332.8	310.3	515	376.6	351.2	575	420.5	392.1
336	245.7	229.2	396	289.6	270.1	456	333.5	311.0	516	377.4	351.9	576	421.3	392.8
337	246.5	229.8	397	290.3	270.8	457	334.2	311.7	517	378.1	352.6	577	422.0	393.5
338	247.2	230.5	398	291.1	271.4	458	335.0	312.4	518	378.8	353.3	578	422.7	394.2
339	247.9	231.2	399	291.8	272.1	459	335.7	313.0	519	379.6	354.0	579	423.5	394.9
340	248.7	231.9	400	292.5	272.8	460	336.4	313.7	520	380.3	354.6	580	424.2	395.6
341	249.4	232.6	401	293.3	273.5	461	337.2	314.4	521	381.0	355.3	581	424.9	396.2
342	250.1	233.2	402	294.0	274.2	462	337.9	315.1	522	381.8	356.0	582	425.6	396.9
343	250.9	233.9	403	294.7	274.8	463	338.6	315.8	523	382.5	356.7	583	426.4	397.6
344	251.6	234.6	404	295.5	275.5	464	339.3	316.4	524	383.2	357.4	584	427.1	398.3
345	252.3	235.3	405	296.2	276.2	465	340.1	317.1	525	384.0	358.0	585	427.8	399.0
346	253.0	236.0	406	296.9	276.9	466	340.8	317.8	526	384.7	358.7	586	428.6	399.7
347	253.8	236.7	407	297.7	277.6	467	341.5	318.5	527	385.4	359.4	587	429.3	400.3
348	254.5	237.3	408	298.4	278.3	468	342.3	319.2	528	386.2	360.1	588	430.0	401.0
349	255.2	238.0	409	299.1	278.9	469	343.0	319.9	529	386.9	360.8	589	430.8	401.7
350	256.0	238.7	410	299.9	279.6	470	343.7	320.5	530	387.6	361.5	590	431.5	402.4
351	256.7	239.4	411	300.6	280.3	471	344.5	321.2	531	388.3	362.1	591	432.2	403.1
352	257.4	240.1	412	301.3	281.0	472	345.2	321.9	532	389.1	362.8	592	433.0	403.7
353	258.2	240.7	413	302.0	281.7	473	345.9'	322.6	533	389.8	363.5	593	433.7	404.4
354	258.9	241.4	414	302.8	282.3	474	346.7	323.3	534	390.5	364.2	594	434.4	405.1
355	259.6	242.1	415	303.5	283.0	475	347.4	323.9	535	391.3	364.9	595	435.2	405.8
356	260.4	242.8	416	304.3	283.7	476	348.1	324.6	536	392.0	365.6	596	435.9	406.5
357	261.1	243.5	417	305.0	284.4	477	348.9	325.3	537	392.7	366.2	597	436.6	407.2
358	261.8	244.2	418	305.7	285.1	478	349.6	326.0	538	393.5	366.9	598	437.3	407.8
359	262.6	244.8	419	306.4	285.8	479	350.3	326.7	539	394.2	367.6	599	438.1	408.5
360	263.3	245.5	420	307.2	286.4	480	351.0	327.4	540	394.9	368.3	600	438.8	409.2

Dist	Dep.	D.Lat	Dist	Dep.	D.Lat	Dist	Dep.	D.Lat	Dist	Dep.	D.Lat	Dist	Dep.	D.Lat

313° ↑ / 227° — **47 DEGREES.** — 047° / 133° — 3h 8m **47°**

TRAVERSE TABLE
44 DEGREES.

44°   316° / 224°   044° / 136°   2h 56m

Dist	D.Lat	Dep.	Dist	D.Lat	Dep.	Dist	D.Lat	Dep.	Dist	D.Lat	Dep.	Dist	D.Lat	Dep.
1	00·7	00·7	61	43·9	42·4	121	87·0	84·1	181	130·2	125·7	241	173·4	167·4
2	01·4	01·4	62	44·6	43·1	122	87·8	84·7	182	130·9	126·4	242	174·1	168·1
3	02·2	02·1	63	45·3	43·8	123	88·5	85·4	183	131·6	127·1	243	174·8	168·8
4	02·9	02·8	64	46·0	44·5	124	89·2	86·1	184	132·4	127·8	244	175·5	169·5
5	03·6	03·5	65	46·8	45·2	125	89·9	86·8	185	133·1	128·5	245	176·2	170·2
6	04·3	04·2	66	47·5	45·8	126	90·6	87·5	186	133·8	129·2	246	177·0	170·9
7	05·0	04·9	67	48·2	46·5	127	91·4	88·2	187	134·5	129·9	247	177·7	171·6
8	05·8	05·6	68	48·9	47·2	128	92·1	88·9	188	135·2	130·6	248	178·4	172·3
9	06·5	06·3	69	49·6	47·9	129	92·8	89·6	189	136·0	131·3	249	179·1	173·0
10	07·2	06·9	70	50·4	48·6	130	93·5	90·3	190	136·7	132·0	250	179·8	173·7
11	07·9	07·6	71	51·1	49·3	131	94·2	91·0	191	137·4	132·7	251	180·6	174·4
12	08·6	08·3	72	51·8	50·0	132	95·0	91·7	192	138·1	133·4	252	181·3	175·1
13	09·4	09·0	73	52·5	50·7	133	95·7	92·4	193	138·8	134·1	253	182·0	175·7
14	10·1	09·7	74	53·2	51·4	134	96·4	93·1	194	139·6	134·8	254	182·7	176·4
15	10·8	10·4	75	54·0	52·1	135	97·1	93·8	195	140·3	135·5	255	183·4	177·1
16	11·5	11·1	76	54·7	52·8	136	97·8	94·5	196	141·0	136·2	256	184·2	177·8
17	12·2	11·8	77	55·4	53·5	137	98·5	95·2	197	141·7	136·8	257	184·9	178·5
18	12·9	12·5	78	56·1	54·2	138	99·3	95·9	198	142·4	137·5	258	185·6	179·2
19	13·7	13·2	79	56·8	54·9	139	100·0	96·6	199	143·1	138·2	259	186·3	179·9
20	14·4	13·9	80	57·5	55·6	140	100·7	97·3	200	143·9	138·9	260	187·0	180·6
21	15·1	14·6	81	58·3	56·3	141	101·4	97·9	201	144·6	139·6	261	187·7	181·3
22	15·8	15·3	82	59·0	57·0	142	102·1	98·6	202	145·3	140·3	262	188·5	182·0
23	16·5	16·0	83	59·7	57·7	143	102·9	99·3	203	146·0	141·0	263	189·2	182·7
24	17·3	16·7	84	60·4	58·4	144	103·6	100·0	204	146·7	141·7	264	189·9	183·4
25	18·0	17·4	85	61·1	59·0	145	104·3	100·7	205	147·5	142·4	265	190·6	184·1
26	18·7	18·1	86	61·9	59·7	146	105·0	101·4	206	148·2	143·1	266	191·3	184·8
27	19·4	18·8	87	62·6	60·4	147	105·7	102·1	207	148·9	143·8	267	192·1	185·5
28	20·1	19·5	88	63·3	61·1	148	106·5	102·8	208	149·6	144·5	268	192·8	186·2
29	20·9	20·1	89	64·0	61·8	149	107·2	103·5	209	150·3	145·2	269	193·5	186·9
30	21·6	20·8	90	64·7	62·5	150	107·9	104·2	210	151·1	145·9	270	194·2	187·6
31	22·3	21·5	91	65·5	63·2	151	108·6	104·9	211	151·8	146·6	271	194·9	188·3
32	23·0	22·2	92	66·2	63·9	152	109·3	105·6	212	152·5	147·3	272	195·7	188·9
33	23·7	22·9	93	66·9	64·6	153	110·1	106·3	213	153·2	148·0	273	196·4	189·6
34	24·5	23·6	94	67·6	65·3	154	110·8	107·0	214	153·9	148·7	274	197·1	190·3
35	25·2	24·3	95	68·3	66·0	155	111·5	107·7	215	154·7	149·4	275	197·8	191·0
36	25·9	25·0	96	69·1	66·7	156	112·2	108·4	216	155·4	150·0	276	198·5	191·7
37	26·6	25·7	97	69·8	67·4	157	112·9	109·1	217	156·1	150·7	277	199·3	192·4
38	27·3	26·4	98	70·5	68·1	158	113·7	109·8	218	156·8	151·4	278	200·0	193·1
39	28·1	27·1	99	71·2	68·8	159	114·4	110·5	219	157·5	152·1	279	200·7	193·8
40	28·8	27·8	100	71·9	69·5	160	115·1	111·1	220	158·3	152·8	280	201·4	194·5
41	29·5	28·5	101	72·7	70·2	161	115·8	111·8	221	159·0	153·5	281	202·1	195·2
42	30·2	29·2	102	73·4	70·9	162	116·5	112·5	222	159·7	154·2	282	202·9	195·9
43	30·9	29·9	103	74·1	71·5	163	117·3	113·2	223	160·4	154·9	283	203·6	196·6
44	31·7	30·6	104	74·8	72·2	164	118·0	113·9	224	161·1	155·6	284	204·3	197·3
45	32·4	31·3	105	75·5	72·9	165	118·7	114·6	225	161·9	156·3	285	205·0	198·0
46	33·1	32·0	106	76·3	73·6	166	119·4	115·3	226	162·6	157·0	286	205·7	198·7
47	33·8	32·6	107	77·0	74·3	167	120·1	116·0	227	163·3	157·7	287	206·5	199·4
48	34·5	33·4	108	77·7	75·0	168	120·8	116·7	228	164·0	158·4	288	207·2	200·1
49	35·2	34·0	109	78·4	75·7	169	121·6	117·4	229	164·7	159·1	289	207·9	200·8
50	36·0	34·7	110	79·1	76·4	170	122·3	118·1	230	165·4	159·8	290	208·6	201·5
51	36·7	35·4	111	79·8	77·1	171	123·0	118·8	231	166·2	160·5	291	209·3	202·1
52	37·4	36·1	112	80·6	77·8	172	123·7	119·5	232	166·9	161·2	292	210·0	202·8
53	38·1	36·8	113	81·3	78·5	173	124·4	120·2	233	167·6	161·9	293	210·8	203·5
54	38·8	37·5	114	82·0	79·2	174	125·2	120·9	234	168·3	162·6	294	211·5	204·2
55	39·6	38·2	115	82·7	79·9	175	125·9	121·6	235	169·0	163·2	295	212·2	204·9
56	40·3	38·9	116	83·4	80·6	176	126·6	122·3	236	169·8	163·9	296	212·9	205·6
57	41·0	39·6	117	84·2	81·3	177	127·3	123·0	237	170·5	164·6	297	213·6	206·3
58	41·7	40·3	118	84·9	82·0	178	128·0	123·6	238	171·2	165·3	298	214·4	207·0
59	42·4	41·0	119	85·6	82·7	179	128·8	124·3	239	171·9	166·0	299	215·1	207·7
60	43·2	41·7	120	86·3	83·4	180	129·5	125·0	240	172·6	166·7	300	215·8	208·4
Dist	Dep.	D.Lat	Dist	Dep.	D.Lat	Dist	Dep.	D.Lat	Dist	Dep.	D.Lat	Dist	Dep.	D.Lat

46°   314° / 226°   **46 DEGREES.**   046° / 134°   3h 4m

TRAVERSE TABLE
44 DEGREES.

| 316° ↑ | | | | | ↑ 044° | |
| 224° | | | | | 136° | 2h 56m | **44°** |

Dist	D.Lat	Dep.	Dist	D.Lat	Dep.	Dist	D.Lat	Dep.	Dist	D.Lat	Dep.	Dist	D.Lat	Dep.
301	216·5	209·1	361	259·7	250·8	421	302·8	292·5	481	346·0	334·1	541	389·2	375·8
302	217·2	209·8	362	260·4	251·5	422	303·6	293·1	482	346·7	334·8	542	389·9	376·5
303	218·0	210·5	363	261·1	252·2	423	304·3	293·8	483	347·4	335·5	543	390·6	377·2
304	218·7	211·2	364	261·8	252·9	424	305·0	294·5	484	348·2	336·2	544	391·3	377·9
305	219·4	211·9	365	262·6	253·6	425	305·7	295·2	485	348·9	336·9	545	392·0	378·6
306	220·1	212·6	366	263·3	254·2	426	306·4	295·9	486	349·6	337·6	546	392·8	379·3
307	220·8	213·3	367	264·0	254·9	427	307·2	296·6	487	350·3	338·3	547	393·5	380·0
308	221·6	214·0	368	264·7	255·6	428	307·9	297·3	488	351·0	339·0	548	394·2	380·7
309	222·3	214·6	369	265·4	256·3	429	308·6	298·0	489	351·7	339·7	549	394·9	381·4
310	223·0	215·3	370	266·2	257·0	430	309·3	298·7	490	352·5	340·4	550	395·6	382·1
311	223·7	216·0	371	266·9	257·7	431	310·0	299·4	491	353·2	341·1	551	396·4	382·8
312	224·4	216·7	372	267·6	258·4	432	310·8	300·1	492	353·9	341·8	552	397·1	383·5
313	225·2	217·4	373	268·3	259·1	433	311·5	300·8	493	354·6	342·5	553	397·8	384·1
314	225·9	218·1	374	269·0	259·8	434	312·2	301·5	494	355·4	343·2	554	398·5	384·8
315	226·6	218·8	375	269·8	260·5	435	312·9	302·2	495	356·1	343·9	555	399·2	385·5
316	227·3	219·5	376	270·5	261·2	436	313·6	302·9	496	356·8	344·6	556	400·0	386·2
317	228·0	220·2	377	271·2	261·9	437	314·4	303·6	497	357·5	345·2	557	400·7	386·9
318	228·8	220·9	378	271·9	262·6	438	315·1	304·3	498	358·2	345·9	558	401·4	387·6
319	229·5	221·6	379	272·6	263·3	439	315·8	305·0	499	359·0	346·6	559	402·1	388·3
320	230·2	222·3	380	273·3	264·0	440	316·6	305·6	500	359·7	347·3	560	402·8	389·0
321	230·9	223·0	381	274·1	264·7	441	317·2	306·3	501	360·4	348·0	561	403·5	389·7
322	231·6	223·7	382	274·8	265·4	442	317·9	307·0	502	361·1	348·7	562	404·3	390·4
323	232·3	224·4	383	275·5	266·1	443	318·7	307·7	503	361·8	349·4	563	405·0	391·1
324	233·1	225·1	384	276·2	266·7	444	319·4	308·4	504	362·5	350·1	564	405·7	391·8
325	233·8	225·8	385	276·9	267·4	445	320·1	309·1	505	363·3	350·8	565	406·4	392·5
326	234·5	226·5	386	277·7	268·1	446	320·8	309·8	506	364·0	351·5	566	407·1	393·2
327	235·2	227·2	387	278·4	268·8	447	321·5	310·5	507	364·7	352·2	567	407·9	393·9
328	235·9	227·8	388	279·1	269·5	448	322·3	311·2	508	365·4	352·9	568	408·6	394·6
329	236·7	228·5	389	279·8	270·2	449	323·0	311·9	509	366·1	353·6	569	409·3	395·3
330	237·4	229·2	390	280·5	270·9	450	323·7	312·6	510	366·9	354·3	570	410·0	396·0
331	238·1	229·9	391	281·3	271·6	451	324·4	313·3	511	367·6	355·0	571	410·7	396·6
332	238·8	230·6	392	282·0	272·3	452	325·1	314·0	512	368·3	355·7	572	411·5	397·3
333	239·5	231·3	393	282·7	273·0	453	325·9	314·7	513	369·0	356·4	573	412·2	398·0
334	240·3	232·0	394	283·4	273·7	454	326·6	315·4	514	369·7	357·1	574	412·9	398·7
335	241·0	232·7	395	284·1	274·4	455	327·3	316·1	515	370·5	357·7	575	413·6	399·4
336	241·7	233·4	396	284·9	275·1	456	328·0	316·8	516	371·2	358·4	576	414·3	400·1
337	242·4	234·1	397	285·6	275·8	457	328·7	317·5	517	371·9	359·1	577	415·1	400·8
338	243·1	234·8	398	286·3	276·5	458	329·5	318·2	518	372·6	359·8	578	415·8	401·5
339	243·9	235·5	399	287·0	277·2	459	330·2	318·8	519	373·3	360·5	579	416·5	402·2
340	244·6	236·2	400	287·7	277·9	460	330·9	319·5	520	374·1	361·2	580	417·2	402·9
341	245·3	236·9	401	288·5	278·6	461	331·6	320·2	521	374·8	361·9	581	417·9	403·6
342	246·0	237·6	402	289·2	279·3	462	332·3	320·9	522	375·5	362·6	582	418·7	404·3
343	246·7	238·3	403	289·9	279·9	463	333·1	321·6	523	376·2	363·3	583	419·4	405·0
344	247·5	239·0	404	290·6	280·6	464	333·8	322·3	524	376·9	364·0	584	420·1	405·7
345	248·2	239·7	405	291·3	281·3	465	334·5	323·0	525	377·7	364·7	585	420·8	406·4
346	248·9	240·4	406	292·1	282·0	466	335·2	323·7	526	378·4	365·4	586	421·5	407·1
347	249·6	241·0	407	292·8	282·7	467	335·9	324·4	527	379·1	366·1	587	422·3	407·8
348	250·3	241·7	408	293·5	283·4	468	336·7	325·1	528	379·8	366·8	588	423·0	408·5
349	251·0	242·4	409	294·2	284·1	469	337·4	325·8	529	380·5	367·5	589	423·7	409·2
350	251·8	243·1	410	294·9	284·8	470	338·1	326·5	530	381·3	368·2	590	424·4	409·8
351	252·5	243·8	411	295·6	285·5	471	338·8	327·2	531	382·0	368·9	591	425·1	410·5
352	253·2	244·5	412	296·4	286·2	472	339·5	327·9	532	382·7	369·6	592	425·8	411·2
353	253·9	245·2	413	297·1	286·9	473	340·2	328·6	533	383·4	370·3	593	426·6	411·9
354	254·6	245·9	414	297·8	287·6	474	341·0	329·3	534	384·1	370·9	594	427·3	412·6
355	255·4	246·6	415	298·5	288·3	475	341·7	330·0	535	384·8	371·6	595	428·0	413·3
356	256·1	247·3	416	299·2	289·0	476	342·4	330·7	536	385·6	372·3	596	428·7	414·0
357	256·8	248·0	417	300·0	289·7	477	343·1	331·4	537	386·3	373·0	597	429·4	414·7
358	257·5	248·7	418	300·7	290·4	478	343·8	332·0	538	387·0	373·7	598	430·2	415·4
359	258·2	249·4	419	301·4	291·1	479	344·6	332·7	539	387·7	374·4	599	430·9	416·1
360	259·0	250·1	420	302·1	291·8	480	345·3	333·4	540	388·4	375·1	600	431·6	416·8
Dist	Dep.	D.Lat	Dist	Dep.	D.Lat	Dist	Dep.	D.Lat	Dist	Dep.	D.Lat	Dist	Dep.	D.Lat
D Lon		Dep.	D Lon		Dep.	D Lon		Dep.	D Lon		Dep.	D Lon		Dep.

| 314° ↑ | | | | | 046° | |
| 226° | 46 DEGREES. | | | | 134° | 3h 4m | **46°** |

TRAVERSE TABLE
45 DEGREES.

45°

315° ↑ / 225° 045° ↑ / 135° 3h 0m

Dist	D.Lat	Dep.	Dist	D.Lat	Dep.	Dist	D.Lat	Dep.	Dist	D.Lat	Dep.	Dist	D.Lat	Dep.
1	00·7	00·7	61	43·1	43·1	121	85·6	85·6	181	128·0	128·0	241	170·4	170·4
2	01·4	01·4	62	43·8	43·8	122	86·3	86·3	182	128·7	128·7	242	171·1	171·1
3	02·1	02·1	63	44·5	44·5	123	87·0	87·0	183	129·4	129·4	243	171·8	171·8
4	02·8	02·8	64	45·3	45·3	124	87·7	87·7	184	130·1	130·1	244	172·5	172·5
5	03·5	03·5	65	46·0	46·0	125	88·4	88·4	185	130·8	130·8	245	173·2	173·2
6	04·2	04·2	66	46·7	46·7	126	89·1	89·1	186	131·5	131·5	246	173·9	173·9
7	04·9	04·9	67	47·4	47·4	127	89·8	89·8	187	132·2	132·2	247	174·7	174·7
8	05·7	05·7	68	48·1	48·1	128	90·5	90·5	188	132·9	132·9	248	175·4	175·4
9	06·4	06·4	69	48·8	48·8	129	91·2	91·2	189	133·6	133·6	249	176·1	176·1
10	07·1	07·1	70	49·5	49·5	130	91·9	91·9	190	134·3	134·3	250	176·8	176·8
11	07·8	07·8	71	50·2	50·2	131	92·6	92·6	191	135·1	135·1	251	177·5	177·5
12	08·5	08·5	72	50·9	50·9	132	93·3	93·3	192	135·8	135·8	252	178·2	178·2
13	09·2	09·2	73	51·6	51·6	133	94·0	94·0	193	136·5	136·5	253	178·9	178·9
14	09·9	09·9	74	52·3	52·3	134	94·8	94·8	194	137·2	137·2	254	179·6	179·6
15	10·6	10·6	75	53·0	53·0	135	95·5	95·5	195	137·9	137·9	255	180·3	180·3
16	11·3	11·3	76	53·7	53·7	136	96·2	96·2	196	138·6	138·6	256	181·0	181·0
17	12·0	12·0	77	54·4	54·4	137	96·9	96·9	197	139·3	139·3	257	181·7	181·7
18	12·7	12·7	78	55·2	55·2	138	97·6	97·6	198	140·0	140·0	258	182·4	182·4
19	13·4	13·4	79	55·9	55·9	139	98·3	98·3	199	140·7	140·7	259	183·1	183·1
20	14·1	14·1	80	56·6	56·6	140	99·0	99·0	200	141·4	141·4	260	183·8	183·8
21	14·8	14·8	81	57·3	57·3	141	99·7	99·7	201	142·1	142·1	261	184·6	184·6
22	15·6	15·6	82	58·0	58·0	142	100·4	100·4	202	142·8	142·8	262	185·3	185·3
23	16·3	16·3	83	58·7	58·7	143	101·1	101·1	203	143·5	143·5	263	186·0	186·0
24	17·0	17·0	84	59·4	59·4	144	101·8	101·8	204	144·2	144·2	264	186·7	186·7
25	17·7	17·7	85	60·1	60·1	145	102·5	102·5	205	145·0	145·0	265	187·4	187·4
26	18·4	18·4	86	60·8	60·8	146	103·2	103·2	206	145·7	145·7	266	188·1	188·1
27	19·1	19·1	87	61·5	61·5	147	103·9	103·9	207	146·4	146·4	267	188·8	188·8
28	19·8	19·8	88	62·2	62·2	148	104·7	104·7	208	147·1	147·1	268	189·5	189·5
29	20·5	20·5	89	62·9	62·9	149	105·4	105·4	209	147·8	147·8	269	190·2	190·2
30	21·2	21·2	90	63·6	63·6	150	106·1	106·1	210	148·5	148·5	270	190·9	190·9
31	21·9	21·9	91	64·3	64·3	151	106·8	106·8	211	149·2	149·2	271	191·6	191·6
32	22·6	22·6	92	65·1	65·1	152	107·5	107·5	212	149·9	149·9	272	192·3	192·3
33	23·3	23·3	93	65·8	65·8	153	108·2	108·2	213	150·6	150·6	273	193·0	193·0
34	24·0	24·0	94	66·5	66·5	154	108·9	108·9	214	151·3	151·3	274	193·7	193·7
35	24·7	24·7	95	67·2	67·2	155	109·6	109·6	215	152·0	152·0	275	194·5	194·5
36	25·5	25·5	96	67·9	67·9	156	110·3	110·3	216	152·7	152·7	276	195·2	195·2
37	26·2	26·2	97	68·6	68·6	157	111·0	111·0	217	153·4	153·4	277	195·9	195·9
38	26·9	26·9	98	69·3	69·3	158	111·7	111·7	218	154·1	154·1	278	196·6	196·6
39	27·6	27·6	99	70·0	70·0	159	112·4	112·4	219	154·9	154·9	279	197·3	197·3
40	28·3	28·3	100	70·7	70·7	160	113·1	113·1	220	155·6	155·6	280	198·0	198·0
41	29·0	29·0	101	71·4	71·4	161	113·8	113·8	221	156·3	156·3	281	198·7	198·7
42	29·7	29·7	102	72·1	72·1	162	114·5	114·5	222	157·0	157·0	282	199·4	199·4
43	30·4	30·4	103	72·8	72·8	163	115·3	115·3	223	157·7	157·7	283	200·1	200·1
44	31·1	31·1	104	73·5	73·5	164	116·0	116·0	224	158·4	158·4	284	200·8	200·8
45	31·8	31·8	105	74·2	74·2	165	116·7	116·7	225	159·1	159·1	285	201·5	201·5
46	32·5	32·5	106	75·0	75·0	166	117·4	117·4	226	159·8	159·8	286	202·2	202·2
47	33·2	33·2	107	75·7	75·7	167	118·1	118·1	227	160·5	160·5	287	202·9	202·9
48	33·9	33·9	108	76·4	76·4	168	118·8	118·8	228	161·2	161·2	288	203·6	203·6
49	34·6	34·6	109	77·1	77·1	169	119·5	119·5	229	161·9	161·9	289	204·3	204·3
50	35·4	35·4	110	77·8	77·8	170	120·2	120·2	230	162·6	162·6	290	205·1	205·1
51	36·1	36·1	111	78·5	78·5	171	120·9	120·9	231	163·3	163·3	291	205·8	205·8
52	36·8	36·8	112	79·2	79·2	172	121·6	121·6	232	164·0	164·0	292	206·5	206·5
53	37·5	37·5	113	79·9	79·9	173	122·3	122·3	233	164·8	164·8	293	207·2	207·2
54	38·2	38·2	114	80·6	80·6	174	123·0	123·0	234	165·5	165·5	294	207·9	207·9
55	38·9	38·9	115	81·3	81·3	175	123·7	123·7	235	166·2	166·2	295	208·6	208·6
56	39·6	39·6	116	82·0	82·0	176	124·4	124·4	236	166·9	166·9	296	209·3	209·3
57	40·3	40·3	117	82·7	82·7	177	125·2	125·2	237	167·6	167·6	297	210·0	210·0
58	41·0	41·0	118	83·4	83·4	178	125·9	125·9	238	168·3	168·3	298	210·7	210·7
59	41·7	41·7	119	84·1	84·1	179	126·6	126·6	239	169·0	169·0	299	211·4	211·4
60	42·4	42·4	120	84·9	84·9	180	127·3	127·3	240	169·7	169·7	300	212·1	212·1

| Dist | Dep. | D.Lat | Dist | Dep. | D.Lat | Dist | Dep. | D.Lat | Dist | Dep. | D.Lat | Dist | Dep. | D.Lat |

315° ↑ / 225° **45 DEGREES.** 045° ↑ / 135° 3h 0m

	315° ↑ 225°			TRAVERSE TABLE 45 DEGREES.				↑045° 135°	3h 0m			

D Lon	Dep.		D Lon	Dep.		D Lon	Dep.		D Lon	Dep.		D Lon	Dep.	
Dist	D. Lat	Dep.	Dist	D. Lat	Dep.	Dist	D. Lat	Dep.	Dist	D. Lat	Dep.	Dist	D. Lat	Dep.
301	212·8	212·8	361	255·3	255·3	421	297·7	297·7	481	340·1	340·1	541	382·5	382·5
302	213·5	213·5	362	256·0	256·0	422	298·4	298·4	482	340·8	340·8	542	383·3	383·3
303	214·3	214·3	363	256·7	256·7	423	299·1	299·1	483	341·5	341·5	543	384·0	384·0
304	215·0	215·0	364	257·4	257·4	424	299·8	299·8	484	342·2	342·2	544	384·7	384·7
305	215·7	215·7	365	258·1	258·1	425	300·5	300·5	485	342·9	342·9	545	385·4	385·4
306	216·4	216·4	366	258·8	258·8	426	301·2	301·2	486	343·7	343·7	546	386·1	386·1
307	217·1	217·1	367	259·5	259·5	427	301·9	301·9	487	344·4	344·4	547	386·8	386·8
308	217·8	217·8	368	260·2	260·2	428	302·6	302·6	488	345·1	345·1	548	387·5	387·5
309	218·5	218·5	369	260·9	260·9	429	303·3	303·3	489	345·8	345·8	549	388·2	388·2
310	219·2	219·2	370	261·6	261·6	430	304·1	304·1	490	346·5	346·5	550	388·9	388·9
311	219·9	219·9	371	262·3	262·3	431	304·8	304·8	491	347·2	347·2	551	389·6	389·6
312	220·6	220·6	372	263·0	263·0	432	305·5	305·5	492	347·9	347·9	552	390·3	390·3
313	221·3	221·3	373	263·8	263·8	433	306·2	306·2	493	348·6	348·6	553	391·0	391·0
314	222·0	222·0	374	264·5	264·5	434	306·9	306·9	494	349·3	349·3	554	391·7	391·7
315	222·7	222·7	375	265·2	265·2	435	307·6	307·6	495	350·0	350·0	555	392·4	392·4
316	223·4	223·4	376	265·9	265·9	436	308·3	308·3	496	350·7	350·7	556	393·2	393·2
317	224·2	224·2	377	266·6	266·6	437	309·0	309·0	497	351·4	351·4	557	393·9	393·9
318	224·9	224·9	378	267·3	267·3	438	309·7	309·7	498	352·1	352·1	558	394·6	394·6
319	225·6	225·6	379	268·0	268·0	439	310·4	310·4	499	352·8	352·8	559	395·3	395·3
320	226·3	226·3	380	268·7	268·7	440	311·1	311·1	500	353·6	353·6	560	396·0	396·0
321	227·0	227·0	381	269·4	269·4	441	311·8	311·8	501	354·3	354·3	561	396·7	396·7
322	227·7	227·7	382	270·1	270·1	442	312·5	312·5	502	355·0	355·0	562	397·4	397·4
323	228·4	228·4	383	270·8	270·8	443	313·2	313·2	503	355·7	355·7	563	398·1	398·1
324	229·1	229·1	384	271·5	271·5	444	314·0	314·0	504	356·4	356·4	564	398·8	398·8
325	229·8	229·8	385	272·2	272·2	445	314·7	314·7	505	357·1	357·1	565	399·5	399·5
326	230·5	230·5	386	272·9	272·9	446	315·4	315·4	506	357·8	357·8	566	400·2	400·2
327	231·2	231·2	387	273·7	273·7	447	316·1	316·1	507	358·5	358·5	567	400·9	400·9
328	231·9	231·9	388	274·4	274·4	448	316·8	316·8	508	359·2	359·2	568	401·6	401·6
329	232·6	232·6	389	275·1	275·1	449	317·5	317·5	509	359·9	359·9	569	402·3	402·3
330	233·3	233·3	390	275·8	275·8	450	318·2	318·2	510	360·6	360·6	570	403·1	403·1
331	234·1	234·1	391	276·5	276·5	451	318·9	318·9	511	361·3	361·3	571	403·8	403·8
332	234·8	234·8	392	277·2	277·2	452	319·6	319·6	512	362·0	362·0	572	404·5	404·5
333	235·5	235·5	393	277·9	277·9	453	320·3	320·3	513	362·7	362·7	573	405·2	405·2
334	236·2	236·2	394	278·6	278·6	454	321·0	321·0	514	363·5	363·5	574	405·9	405·9
335	236·9	236·9	395	279·3	279·3	455	321·7	321·7	515	364·2	364·2	575	406·6	406·6
336	237·6	237·6	396	280·0	280·0	456	322·4	322·4	516	364·9	364·9	576	407·3	407·3
337	238·3	238·3	397	280·7	280·7	457	323·1	323·1	517	365·6	365·6	577	408·0	408·0
338	239·0	239·0	398	281·4	281·4	458	323·9	323·9	518	366·3	366·3	578	408·7	408·7
339	239·7	239·7	399	282·1	282·1	459	324·6	324·6	519	367·0	367·0	579	409·4	409·4
340	240·4	240·4	400	282·8	282·8	460	325·3	325·3	520	367·7	367·7	580	410·1	410·1
341	241·1	241·1	401	283·5	283·5	461	326·0	326·0	521	368·4	368·4	581	410·8	410·8
342	241·8	241·8	402	284·3	284·3	462	326·7	326·7	522	369·1	369·1	582	411·5	411·5
343	242·5	242·5	403	285·0	285·0	463	327·4	327·4	523	369·8	369·8	583	412·2	412·2
344	243·2	243·2	404	285·7	285·7	464	328·1	328·1	524	370·5	370·5	584	413·0	413·0
345	244·0	244·0	405	286·4	286·4	465	328·8	328·8	525	371·2	371·2	585	413·7	413·7
346	244·7	244·7	406	287·1	287·1	466	329·5	329·5	526	371·9	371·9	586	414·4	414·4
347	245·4	245·4	407	287·8	287·8	467	330·2	330·2	527	372·6	372·6	587	415·1	415·1
348	246·1	246·1	408	288·5	288·5	468	330·9	330·9	528	373·4	373·4	588	415·8	415·8
349	246·8	246·8	409	289·2	289·2	469	331·6	331·6	529	374·1	374·1	589	416·5	416·5
350	247·5	247·5	410	289·9	289·9	470	332·3	332·3	530	374·8	374·8	590	417·2	417·2
351	248·2	248·2	411	290·6	290·6	471	333·0	333·0	531	375·5	375·5	591	417·9	417·9
352	248·9	248·9	412	291·3	291·3	472	333·8	333·8	532	376·2	376·2	592	418·6	418·6
353	249·6	249·6	413	292·0	292·0	473	334·5	334·5	533	376·9	376·9	593	419·3	419·3
354	250·3	250·3	414	292·7	292·7	474	335·2	335·2	534	377·6	377·6	594	420·0	420·0
355	251·0	251·0	415	293·4	293·4	475	335·9	335·9	535	378·3	378·3	595	420·7	420·7
356	251·7	251·7	416	294·2	294·2	476	336·6	336·6	536	379·0	379·0	596	421·4	421·4
357	252·4	252·4	417	294·9	294·9	477	337·3	337·3	537	379·7	379·7	597	422·1	422·1
358	253·1	253·1	418	295·6	295·6	478	338·0	338·0	538	380·4	380·4	598	422·8	422·8
359	253·9	253·9	419	296·3	296·3	479	338·7	338·7	539	381·1	381·1	599	423·6	423·6
360	254·6	254·6	420	297·0	297·0	480	339·4	339·4	540	381·8	381·8	600	424·3	424·3
Dist	Dep.	D. Lat	Dist	Dep.	D. Lat	Dist	Dep.	D. Lat	Dist	Dep.	D. Lat	Dist	Dep.	D. Lat
D Lon		Dep.	D Lon		Dep.	D Lon		Dep.	D Lon		Dep.	D Lon		Dep.

RADAR RANGE TABLE

h	d		h	d
ft. m.	miles		ft. m.	miles
4·2 1·3			340 103·6	
	3			23
8·2 2·5			371 113·1	
	4			24
13·6 4·1			403 122·8	
	5			25
20·3 6·2			436 132·9	
	6			26
28·3 8·6			471 143·6	
	7			27
37·7 11·5			508 154·8	
	8			28
48·5 14·8			545 166·1	
	9			29
60·6 18·5			584 178·0	
	10			30
74·0 22·6			625 190·5	
	11			31
88·8 27·1			666 203·0	
	12			32
105·0 32·0			709 216·1	
	13			33
122 37·2			754 229·8	
	14			34
141 43·0			799 243·5	
	15			35
161 49·1			846 257·9	
	16			36
182 55·5			895 272·8	
	17			37
205 62·5			944 287·7	
	18			38
229 69·8			995 303·3	
	19			39
255 77·7			1048 319·4	
	20			40
281 85·6			1102 335·9	
	21			41
310 94·5			1157 352·7	
	22			42
340 103·6			1213 369·7	

The accompanying table gives the approximate distance of the "radar horizon" corresponding to different heights of the radar aerial or target from which an echo is returned, and is based on the formula:—

$$\text{horizon dist. in n.mls.} = 1 \cdot 22\sqrt{h}.$$

where h = height of aerial or target in feet.

In similar manner to light waves, radio waves are refracted in passing through the atmosphere. This has the effect of making the distance of the radar horizon for 3 cm. waves, under certain standard conditions of the atmosphere, about 15 per cent. greater than the distance of the geometrical horizon. Hence, taking the latter in nautical miles to be $1 \cdot 06\sqrt{h}$, the distance of the radar horizon becomes $1 \cdot 22\sqrt{h}$. This will be correct only under the standard conditions, and every departure from such standard will cause the distances to vary somewhat. The standard referred to is as follows:—

Atmospheric *pressure* 1013 mb. decreasing with height at the rate of 36 mb. per 1000 feet.

Air *temperature* at sea level 59°F. decreasing with height at the rate of 3·6°F. per 1000 feet.

60 per cent. *relative humidity* remaining constant with height.

Apart from variations from the above standard, the range at which target echoes can be seen on the P.P.I. screen will depend to a considerable extent on the characteristics of the particular radar installation and on the echoing qualities of the target.

Used with discretion, however, the information given in the table can be of much value to the radar observer.

Note:—The sum of the radar horizon distances of aerial and target respectively gives the maximum distance from which that target can return an echo.

Examples:—

1. A target of height 400 feet (122 m.) should begin to appear on the P.P.I. of an installation with an aerial mounted 60 feet (18·3 m.) above sea level at a range of approximately 24+9=33 miles.

2. If an echo first appears on the P.P.I. (aerial 35 feet (10·7 m.) above sea level) at a range of 25 miles, the probable height of the target is of the order of about 215 feet (66·0 m.). This may assist in identifying it.

N.B.—For h =35 ft., d = 7 m.   25—7 = 18.   For d = 18, h = about 215 ft.

CAUTION:—TO BE USED WITH 3 cm. WAVE RADAR ONLY.

RADAR PLOTTER'S SPEED AND DISTANCE TABLE

SPEED IN KNOTS	Miles in 1 min.	Miles in 2 min.	Miles in 2½ min.	Miles in 3 min.	Miles in 4 min.	Miles in 5 min.	Miles in 6 min.	SPEED IN KNOTS
4·0	0·07	0·13	0·17	0·20	0·27	0·33	0·40	4·0
4·5	0·08	0·15	0·19	0·23	0·30	0·38	0·45	4·5
5·0	0·08	0·17	0·21	0·25	0·33	0·42	0·50	5·0
5·5	0·10	0·18	0·23	0·28	0·37	0·46	0·55	5·5
6·0	0·10	0·20	0·25	0·30	0·40	0·50	0·60	6·0
6·5	0·11	0·22	0·27	0·33	0·43	0·54	0·65	6·5
7·0	0·12	0·23	0·29	0·35	0·47	0·58	0·70	7·0
7·5	0·13	0·25	0·31	0·38	0·50	0·63	0·75	7·5
8·0	0·13	0·27	0·33	0·40	0·53	0·67	0·80	8·0
8·5	0·14	0·28	0·35	0·43	0·57	0·71	0·85	8·5
9·0	0·15	0·30	0·38	0·45	0·60	0·75	0·90	9·0
9·5	0·16	0·32	0·40	0·48	0·63	0·80	0·95	9·5
10·0	0·17	0·33	0·42	0·50	0·67	0·83	1·00	10·0
10·5	0·18	0·35	0·44	0·53	0·70	0·88	1·05	10·5
11·0	0·18	0·37	0·46	0·55	0·73	0·91	1·10	11·0
11·5	0·19	0·38	0·48	0·58	0·77	0·96	1·15	11·5
12·0	0·20	0·40	0·50	0·60	0·80	1·00	1·20	12·0
12·5	0·21	0·42	0·52	0·63	0·83	1·04	1·25	12·5
13·0	0·22	0·43	0·54	0·65	0·87	1·08	1·30	13·0
13·5	0·23	0·45	0·56	0·68	0·90	1·13	1·35	13·5
14·0	0·23	0·47	0·58	0·70	0·93	1·17	1·40	14·0
14·5	0·24	0·48	0·60	0·73	0·97	1·21	1·45	14·5
15·0	0·25	0·50	0·63	0·75	1·00	1·25	1·50	15·0
15·5	0·26	0·52	0·65	0·78	1·03	1·29	1·55	15·5
16·0	0·27	0·53	0·67	0·80	1·07	1·33	1·60	16·0
16·5	0·28	0·55	0·69	0·83	1·10	1·38	1·65	16·5
17·0	0·28	0·57	0·71	0·85	1·13	1·42	1·70	17·0
17·5	0·29	0·58	0·73	0·88	1·17	1·46	1·75	17·5
18·0	0·30	0·60	0·75	0·90	1·20	1·50	1·80	18·0
18·5	0·31	0·62	0·77	0·93	1·23	1·54	1·85	18·5
19·0	0·32	0·63	0·79	0·95	1·27	1·58	1·90	19·0
19·5	0·33	0·65	0·81	0·98	1·30	1·63	1·95	19·5
20·0	0·33	0·67	0·83	1·00	1·33	1·67	2·00	20·0
20·5	0·34	0·68	0·85	1·03	1·37	1·71	2·05	20·5
21·0	0·35	0·70	0·88	1·05	1·40	1·75	2·10	21·0
21·5	0·36	0·72	0·90	1·08	1·43	1·79	2·15	21·5
22·0	0·37	0·73	0·92	1·10	1·47	1·83	2·20	22·0
22·5	0·38	0·75	0·94	1·13	1·50	1·88	2·25	22·5
23·0	0·38	0·77	0·96	1·15	1·53	1·92	2·30	23·0
23·5	0·39	0·78	0·98	1·18	1·57	1·96	2·35	23·5
24·0	0·40	0·80	1·00	1·20	1·60	2·00	2·40	24·0
24·5	0·41	0·82	1·02	1·23	1·63	2·04	2·45	24·5
25·0	0·42	0·83	1·04	1·25	1·67	2·08	2·50	25·0
25·5	0·43	0·85	1·06	1·28	1·70	2·13	2·55	25·5
26·0	0·43	0·87	1·08	1·30	1·73	2·17	2·60	26·0
26·5	0·44	0·88	1·10	1·33	1·77	2·21	2·65	26·5
27·0	0·45	0·90	1·13	1·35	1·80	2·25	2·70	27·0
27·5	0·46	0·92	1·15	1·38	1·83	2·29	2·75	27·5
28·0	0·47	0·93	1·17	1·40	1·87	2·33	2·80	28·0
28·5	0·48	0·95	1·19	1·43	1·90	2·38	2·85	28·5
29·0	0·48	0·97	1·21	1·45	1·93	2·42	2·90	29·0
29·5	0·49	0·98	1·23	1·48	1·97	2·46	2·95	29·5
30·0	0·50	1·00	1·25	1·50	2·00	2·50	3·00	30·0
30·5	0·51	1·02	1·27	1·53	2·03	2·54	3·05	30·5
31·0	0·52	1·03	1·29	1·55	2·07	2·58	3·10	31·0
31·5	0·53	1·05	1·31	1·58	2·10	2·63	3·15	31·5
32·0	0·53	1·07	1·33	1·60	2·13	2·67	3·20	32·0
32·5	0·54	1·08	1·35	1·63	2·17	2·71	3·25	32·5
33·0	0·55	1·10	1·38	1·65	2·20	2·75	3·30	33·0
33·5	0·56	1·12	1·40	1·68	2·23	2·79	3·35	33·5
34·0	0·57	1·13	1·42	1·70	2·27	2·83	3·40	34·0

DAY'S RUN—AVERAGE SPEED TABLE

Distance in Nautical Miles	STEAMING TIME								
	h. m. 21 59	h. m. 22 00	h. m. 22 01	h. m. 22 02	h. m. 22 03	h. m. 22 04	h. m. 22 05	h. m. 22 06	h. m. 22 07
	Knots	Knots	Knots	Knots	Knots	Knots	Knots	Knots	Knots
900	40·940	40·909	40·878	40·846	40·816	40·785	40·754	40·724	40·693
800	36·391	36·364	36·336	36·307	36·281	36·254	36·226	36·199	36·172
700	31·842	31·818	31·794	31·769	31·746	31·722	31·698	31·674	31·650
600	27·293	27·274	27·252	27·231	27·211	27·190	27·170	27·149	27·129
500	22·745	22·727	22·710	22·692	22·676	22·659	22·641	22·624	22·607
400	18·196	18·182	18·168	18·154	18·141	18·127	18·113	18·100	18·086
300	13·647	13·636	13·626	13·615	13·605	13·595	13·585	13·575	13·564
200	9·098	9·091	9·084	9·077	9·070	9·063	9·057	9·050	9·043
100	4·549	4·545	4·542	4·538	4·535	4·532	4·528	4·525	4·521
	h. m. 22 08	h. m. 22 09	h. m. 22 10	h. m. 22 11	h. m. 22 12	h. m. 22 13	h. m. 22 14	h. m. 22 15	h. m. 22 16
900	40·663	40·632	40·602	40·571	40·541	40·510	40·480	40·449	40·419
800	36·145	36·117	36·090	36·063	36·036	36·009	35·982	35·955	35·928
700	31·626	31·603	31·579	31·555	31·532	31·508	31·484	31·461	31·437
600	27·108	27·088	27·068	27·047	27·027	27·007	26·987	26·966	26·946
500	22·590	22·573	22·556	22·539	22·523	22·506	22·489	22·472	22·455
400	18·072	18·059	18·045	18·032	18·018	18·005	17·991	17·978	17·964
300	13·554	13·544	13·534	13·524	13·514	13·504	13·493	13·483	13·473
200	9·036	9·029	9·023	9·016	9·009	9·002	8·996	8·989	8·982
100	4·518	4·515	4·511	4·508	4·505	4·501	4·498	4·494	4·491
	h. m. 22 17	h. m. 22 18	h. m. 22 19	h. m. 22 20	h. m. 22 21	h. m. 22 22	h. m. 22 23	h. m. 22 24	h. m. 22 25
900	40·389	40·359	40·329	40·299	40·268	40·238	40·208	40·179	40·149
800	35·901	35·874	35·848	35·821	35·794	35·768	35·741	35·714	35·688
700	31·414	31·390	31·367	31·343	31·320	31·297	31·273	31·250	31·227
600	26·926	26·906	26·886	26·866	26·846	26·826	26·806	26·786	26·766
500	22·438	22·422	22·405	22·388	22·371	22·355	22·338	22·321	22·305
400	17·951	17·937	17·924	17·910	17·898	17·884	17·870	17·857	17·844
300	13·463	13·453	13·443	13·433	13·423	13·413	13·403	13·393	13·383
200	8·975	8·969	8·962	8·955	8·949	8·942	8·935	8·929	8·922
100	4·488	4·484	4·481	4·478	4·474	4·471	4·468	4·464	4·461
	h. m. 22 26	h. m. 22 27	h. m. 22 28	h. m. 22 29	h. m. 22 30	h. m. 22 31	h. m. 22 32	h. m. 22 33	h. m. 22 34
900	40·119	40·089	40·059	40·030	40·000	39·970	39·941	39·911	39·882
800	35·661	35·635	35·608	35·582	35·556	35·529	35·503	35·477	35·450
700	31·204	31·180	31·157	31·134	31·111	31·088	31·065	31·042	31·019
600	26·746	26·726	26·706	26·686	26·667	26·647	26·627	26·608	26·588
500	22·288	22·272	22·255	22·239	22·222	22·206	22·189	22·173	22·157
400	17·831	17·817	17·804	17·791	17·778	17·765	17·751	17·739	17·725
300	13·373	13·363	13·353	13·343	13·333	13·323	13·314	13·304	13·294
200	8·915	8·909	8·902	8·895	8·889	8·882	8·876	8·869	8·863
100	4·458	4·454	4·451	4·448	4·444	4·441	4·438	4·435	4·431

DAY'S RUN—AVERAGE SPEED TABLE

STEAMING TIME

Distance in Nautical Miles	h. m. 22 35	h. m. 22 36	h. m. 22 37	h. m. 22 38	h. m. 22 39	h. m. 22 40	h. m. 22 41	h. m. 22 42	h. m. 22 43
	Knots	Knots	Knots	Knots	Knots	Knots	Knots	Knots	Knots
900	39·852	39·823	29·794	39·764	39·735	29·706	39·677	39·648	39·618
800	35·424	35·398	35·372	35·346	35·320	35·294	35·268	35·242	35·216
700	30·996	30·973	30·951	30·928	30·905	30·882	30·860	30·837	30·814
600	26·568	26·549	26·529	26·510	26·490	26·471	26·451	26·432	26·412
500	22·140	22·124	22·108	22·091	22·075	22·059	22·043	22·026	22·010
400	17·712	17·699	17·686	17·673	17·660	17·647	17·634	17·621	17·608
300	13·284	13·274	13·265	13·255	13·245	13·235	13·226	13·216	13·206
200	8·856	8·850	8·843	8·837	8·830	8·824	8·817	8·811	8·804
100	4·428	4·425	4·422	4·418	4·415	4·412	4·409	4·405	4·402

Distance	h. m. 22 44	h. m. 22 45	h. m. 22 46	h. m. 22 47	h. m. 22 48	h. m. 22 49	h. m. 22 50	h. m. 22 51	h. m. 22 52
900	39·589	39·560	39·532	39·503	39·474	39·445	39·416	39·387	39·359
800	35·191	35·165	35·139	35·113	35·088	35·062	35·036	35·011	34·985
700	30·792	30·769	30·747	30·724	30·702	30·679	30·657	30·635	30·612
600	26·393	26·374	26·354	26·335	26·316	26·297	26·277	26·258	26·239
500	21·994	21·978	21·962	21·946	21·930	21·914	21·898	21·882	21·866
400	17·595	17·582	17·570	17·557	17·544	17·531	17·518	17·505	17·493
300	13·196	13·187	13·177	13·168	13·158	13·148	13·139	13·129	13·120
200	8·798	8·791	8·785	8·778	8·772	8·766	8·759	8·753	8·746
100	4·399	4·396	4·392	4·389	4·386	4·383	4·380	4·376	4·373

Distance	h. m. 22 53	h. m. 22 54	h. m. 22 55	h. m. 22 56	h. m. 22 57	h. m. 22 58	h. m. 22 59	h. m. 23 00	h. m. 23 01
900	39·330	39·301	39·273	39·244	39·216	39·187	39·159	39·130	39·102
800	34·960	34·934	34·909	34·884	34·858	34·833	34·808	34·783	34·757
700	30·590	30·568	30·545	30·523	30·501	30·479	30·457	30·435	30·413
600	26·220	26·201	26·182	26·163	26·144	26·125	26·106	26·087	26·068
500	21·850	21·834	21·818	21·802	21·787	21·771	21·755	21·739	21·723
400	17·480	17·467	17·455	17·442	17·429	17·417	17·404	17·391	17·379
300	13·110	13·100	13·091	13·081	13·072	13·062	13·053	13·043	13·034
200	8·740	8·734	8·727	8·721	8·715	8·708	8·702	8·696	8·689
100	4·370	4·367	4·364	4·360	4·357	4·354	4·351	4·348	4·345

Distance	h. m. 23 02	h. m. 23 03	h. m. 23 04	h. m. 23 05	h. m. 23 06	h. m. 23 07	h. m. 23 08	h. m. 23 09	h. m. 23 10
900	39·074	39·046	39·017	38·989	38·961	38·933	38·905	38·877	38·849
800	34·732	34·707	34·682	34·657	34·632	34·607	34·582	34·557	34·532
700	30·391	30·369	30·347	30·325	30·303	30·281	30·259	30·238	30·216
600	26·049	26·030	26·012	25·993	25·974	25·955	25·937	25·918	25·899
500	21·708	21·692	21·676	21·661	21·645	21·629	21·614	21·598	21·583
400	17·366	17·354	17·341	17·329	17·316	17·304	17·291	17·279	17·266
300	13·025	13·015	13·006	12·996	12·987	12·978	12·968	12·959	12·950
200	8·683	8·677	8·671	8·664	8·658	8·652	8·646	8·639	8·633
100	4·342	4·338	4·335	4·332	4·329	4·326	4·323	4·320	4·317

DAY'S RUN—AVERAGE SPEED TABLE

Distance in Nautical Miles	STEAMING TIME								
	h. m. 23 11	h. m. 23 12	h. m. 23 13	h. m. 23 14	h. m. 23 15	h. m. 23 16	h. m. 23 17	h. m. 23 18	h. m. 23 19
	Knots	Knots	Knots	Knots	Knots	Knots	Knots	Knots	Knots
900	38·821	38·793	38·765	38·737	38·710	38·682	38·654	38·627	38·599
800	34·508	34·483	34·458	34·433	34·409	34·384	34·359	34·335	34·310
700	30·194	30·172	30·151	30·129	30·108	30·086	30·064	30·043	30·021
600	25·881	25·862	25·844	25·825	25·806	25·788	25·770	25·751	25·733
500	21·567	21·552	21·536	21·521	21·505	21·490	21·475	21·459	21·444
400	17·254	17·241	17·229	17·217	17·204	17·192	17·180	17·167	17·155
300	12·940	12·931	12·922	12·912	12·903	12·894	12·885	12·876	12·866
200	8·627	8·621	8·615	8·608	8·602	8·596	8·590	8·584	8·578
100	4·313	4·310	4·307	4·304	4·301	4·298	4·295	4·292	4·289
	h. m. 23 20	h. m. 23 21	h. m. 23 22	h. m. 23 23	h. m. 23 24	h. m. 23 25	h. m. 23 26	h. m. 23 27	h. m. 23 28
900	38·571	38·544	38·516	38·489	38·462	38·434	38·407	38·380	38·352
800	34·296	34·261	34·237	34·212	34·188	34·164	34·139	34·115	34·091
700	30·000	29·979	29·957	29·936	29·915	29·893	29·872	29·851	29·830
600	25·714	25·696	25·678	25·660	25·641	25·623	25·605	25·586	25·568
500	21·429	21·413	21·398	21·383	21·368	21·352	21·337	21·322	21·307
400	17·143	17·131	17·118	17·106	17·094	17·082	17·070	17·058	17·045
300	12·857	12·848	12·839	12·830	12·821	12·811	12·802	12·793	12·784
200	8·571	8·565	8·559	8·553	8·547	8·541	8·535	8·529	8·523
100	4·286	4·283	4·280	4·277	4·274	4·270	4·267	4·264	4·261
	h. m. 23 29	h. m. 23 30	h. m. 23 31	h. m. 23 32	h. m. 23 33	h. m. 23 34	h. m. 23 35	h. m. 23 36	h. m. 23 37
900	38·324	38·298	38·271	38·244	38·217	38·190	38·163	38·136	38·109
800	34·066	34·043	34·018	33·994	33·970	33·946	33·922	33·898	33·874
700	29·808	29·787	29·766	29·745	29·724	29·703	29·682	29·661	29·640
600	25·550	25·532	25·514	25·496	25·478	25·460	25·442	25·424	25·406
500	21·291	21·277	21·262	21·246	21·231	21·216	21·201	21·186	21·172
400	17·033	17·021	17·009	16·997	16·985	16·973	16·961	16·949	16·937
300	12·775	12·766	12·757	12·748	12·739	12·730	12·721	12·712	12·703
200	8·517	8·511	8·505	8·499	8·493	8·487	8·481	8·475	8·469
100	4·258	4·255	4·252	4·249	4·246	4·243	4·240	4·237	4·234
	h. m. 23 38	h. m. 23 39	h. m. 23 40	h. m. 23 41	h. m. 23 42	h. m. 23 43	h. m. 23 44	h. m. 23 45	h. m. 23 46
900	38·082	38·055	38·028	38·001	37·975	37·948	37·921	37·895	37·868
800	33·850	33·827	33·803	33·779	33·755	33·732	33·708	33·684	33·661
700	29·619	29·598	29·577	29·557	29·536	29·515	29·494	29·474	29·453
600	25·388	25·370	25·352	25·334	25·316	25·299	25·281	25·263	25·245
500	21·157	21·142	21·127	21·112	21·097	21·082	21·067	21·053	21·038
400	16·925	16·913	16·901	16·890	16·878	16·866	16·854	16·842	16·830
300	12·694	12·685	12·676	12·667	12·658	12·649	12·640	12·632	12·623
200	8·463	8·457	8·451	8·445	8·439	8·433	8·427	8·421	8·415
100	4·231	4·228	4·225	4·222	4·219	4·216	4·213	4·211	4·208

DAY'S RUN—AVERAGE SPEED TABLE

STEAMING TIME

Distance in Nautical Miles	h. m. 23 47 Knots	h. m. 23 48 Knots	h. m. 23 49 Knots	h. m. 23 50 Knots	h. m. 23 51 Knots	h. m. 23 52 Knots	h. m. 23 53 Knots	h. m. 23 54 Knots	h. m. 23 55 Knots
900	37·842	37·815	37·789	37·762	37·736	37·709	37·683	37·657	37·631
800	33·637	33·613	33·590	33·566	33·543	33·520	33·496	33·473	33·449
700	29·432	29·412	29·391	29·371	29·350	29·330	39·309	29·289	29·268
600	25·228	25·210	25·192	25·175	25·157	25·140	25·122	25·105	25·087
500	21·023	21·008	20·994	20·979	20·964	20·950	20·935	20·921	20·906
400	16·819	16·807	16·795	16·783	16·771	16·760	16·748	16·736	16·725
300	12·614	12·605	12·596	12·587	12·579	12·570	12·561	12·552	12·544
200	8·409	8·403	8·397	8·392	8·386	8·380	8·374	8·368	8·362
100	4·205	4·202	4·199	4·196	4·193	4·190	4·187	4·184	4·181

Distance	h. m. 23 56	h. m. 23 57	h. m. 23 58	h. m. 23 59	h. m. 24 00	h. m. 24 01	h. m. 24 02	h. m. 24 03	h. m. 24 04
900	37·604	37·578	37·552	37·526	37·500	37·474	37·448	37·422	37·396
800	33·426	33·403	33·380	33·356	33·333	33·310	33·287	33·264	33·241
700	29·248	29·228	29·207	29·187	29·167	29·146	29·126	29·106	29·086
600	25·070	25·052	25·035	25·017	25·000	24·983	24·965	24·948	24·931
500	20·891	20·877	20·862	20·848	20·833	20·819	20·804	20·790	20·776
400	16·713	16·701	16·690	16·678	16·667	16·655	16·644	16·632	16·620
300	12·535	12·526	12·517	12·509	12·500	12·491	12·483	12·474	12·465
200	8·357	8·351	8·345	8·339	8·333	8·328	8·322	8·316	8·310
100	4·178	4·175	4·172	4·170	4·167	4·164	4·161	4·158	4·155

Distance	h. m. 24 05	h. m. 24 06	h. m. 24 07	h. m. 24 08	h. m. 24 09	h. m. 24 10	h. m. 24 11	h. m. 24 12	h. m. 24 13
900	37·370	37·344	37·319	37·293	37·267	37·241	37·216	37·190	37·165
800	33·218	33·195	33·172	33·149	33·126	33·103	33·081	33·058	33·035
700	29·066	29·046	29·026	29·006	28·986	28·966	28·946	28·926	28·906
600	24·914	24·896	24·879	24·862	24·845	24·828	24·810	24·793	24·776
500	20·761	20·747	20·733	20·718	20·704	20·690	20·675	20·661	20·647
400	16·609	16·598	16·586	16·575	16·563	16·552	16·540	16·529	16·518
300	12·457	12·448	12·440	12·431	12·422	12·414	12·405	12·397	12·388
200	8·305	8·299	8·293	8·287	8·282	8·276	8·270	8·264	8·259
100	4·152	4·149	4·147	4·144	4·141	4·138	4·135	4·132	4·129

Distance	h. m. 24 14	h. m. 24 15	h. m. 24 16	h. m. 24 17	h. m. 24 18	h. m. 24 19	h. m. 24 20	h. m. 24 21	h. m. 24 22
900	37·139	37·113	37·088	37·062	37·037	37·012	36·986	36·961	36·936
800	33·012	32·990	32·967	32·944	32·922	32·899	32·877	32·854	32·832
700	28·886	28·866	28·846	28·826	28·807	28·787	28·767	28·747	28·728
600	24·759	24·742	24·725	24·708	24·691	24·674	24·658	24·641	24·624
500	20·633	20·619	20·604	20·590	20·576	20·562	20·548	20·534	20·520
400	16·506	16·495	16·484	16·472	16·461	16·450	16·438	16·427	16·416
300	12·380	12·371	12·363	12·354	12·346	12·337	12·329	12·320	12·312
200	8·253	8·247	8·242	8·236	8·230	8·225	8·219	8·214	8·208
100	4·127	4·124	4·121	4·118	4·115	4·112	4·110	4·107	4·104

DAY'S RUN—AVERAGE SPEED TABLE

STEAMING TIME

Distance in Nautical Miles	h. m. 24 23	h. m. 24 24	h. m. 24 25	h. m. 24 26	h. m. 24 27	h. m. 24 28	h. m. 24 29	h. m. 24 30	h. m. 24 31
	Knots	Knots	Knots	Knots	Knots	Knots	Knots	Knots	Knots
900	36·910	36·885	36·860	36·835	36·810	36·785	36·760	36·735	36·710
800	32·809	32·787	32·764	32·742	32·720	32·698	32·675	32·653	32·631
700	28·708	28·689	28·669	28·649	28·630	28·610	28·591	28·571	28·552
600	24·607	24·590	24·573	24·557	24·540	24·523	24·506	24·490	24·473
500	20·506	20·492	20·478	20·464	20·450	20·436	20·422	20·408	20·394
400	16·405	16·393	16·382	16·371	16·360	16·349	16·338	16·327	16·315
300	12·303	12·295	12·287	12·278	12·270	12·262	12·253	12·245	12·237
200	8·202	8·197	8·191	8·186	8·180	8·174	8·169	8·163	8·158
100	4·101	4·098	4·096	4·093	4·090	4·087	4·084	4·082	4·079

Distance	h. m. 24 32	h. m. 24 33	h. m. 24 34	h. m. 24 35	h. m. 24 36	h. m. 24 37	h. m. 24 38	h. m. 24 39	h. m. 24 40
900	36·685	36·660	36·635	36·610	36·585	36·561	36·536	36·511	36·486
800	32·609	32·587	32·564	32·542	32·520	32·498	32·476	32·454	32·432
700	28·533	28·513	28·494	28·475	28·455	28·436	28·417	28·398	28·378
600	24·457	24·440	24·423	24·407	24·390	24·374	24·357	24·341	24·324
500	20·380	20·367	20·353	20·339	20·325	20·312	20·298	20·284	20·270
400	16·304	16·293	16·282	16·271	16·260	16·249	16·238	16·227	16·216
300	12·228	12·220	12·212	12·203	12·195	12·187	12·179	12·170	12·162
200	8·152	8·147	8·141	8·136	8·130	8·125	8·119	8·114	8·108
100	4·076	4·073	4·071	4·068	4·065	4·062	4·060	4·057	4·054

Distance	h. m. 24 41	h. m. 24 42	h. m. 24 43	h. m. 24 44	h. m. 24 45	h. m. 24 46	h. m. 24 47	h. m. 24 48	h. m. 24 49
900	36·462	36·438	36·413	36·388	36·364	36·339	36·315	36·290	36·266
800	32·411	32·389	32·367	32·345	32·323	32·302	32·280	32·258	32·236
700	28·359	28·341	28·321	28·302	28·283	28·264	28·245	28·226	28·207
600	24·308	24·292	24·275	24·259	24·242	24·226	24·210	24·194	24·177
500	20·257	20·243	20·229	20·216	20·202	20·188	20·175	20·161	20·148
400	16·205	16·195	16·183	16·173	16·162	16·151	16·140	16·129	16·118
300	12·154	12·146	12·138	12·129	12·121	12·113	12·105	12·097	12·089
200	8·103	8·097	8·092	8·086	8·081	8·075	8·070	8·065	8·059
100	4·051	4·049	4·046	4·043	4·040	4·038	4·035	4·032	4·030

Distance	h. m. 24 50	h. m. 24 51	h. m. 24 52	h. m. 24 53	h. m. 24 54	h. m. 24 55	h. m. 24 56	h. m. 24 57	h. m. 24 58
900	36·241	36·217	36·193	36·169	36·145	36·120	36·096	36·072	36·048
800	32·215	32·193	32·172	32·150	32·128	32·107	32·086	32·064	32·043
700	28·188	28·169	28·150	28·131	28·112	28·094	28·075	28·056	28·037
600	24·161	24·145	24·129	24·113	24·096	24·080	24·064	24·048	24·032
500	20·134	20·121	20·107	20·094	20·080	20·067	20·054	20·040	20·027
400	16·107	16·097	16·086	16·075	16·064	16·054	16·043	16·032	16·021
300	12·080	12·072	12·064	12·056	12·048	12·040	12·032	12·024	12·016
200	8·054	8·048	8·043	8·038	8·032	8·027	8·021	8·016	8·011
100	4·027	4·024	4·021	4·019	4·016	4·013	4·011	4·008	4·005

DAY'S RUN—AVERAGE SPEED TABLE

Distance in Nautical Miles	STEAMING TIME								
	h. m. 24 59	h. m. 25 00	h. m. 25 01	h. m. 25 02	h. m. 25 03	h. m. 25 04	h. m. 25 05	h. m. 25 06	h. m. 25 07
	Knots	Knots	Knots	Knots	Knots	Knots	Knots	Knots	Knots
900	36·024	36·000	35·976	35·952	35·928	35·904	35·880	35·857	35·833
800	32·021	32·000	31·979	31·957	31·936	31·915	31·894	31·872	31·851
700	28·019	28·000	27·981	27·963	27·944	27·926	27·908	27·888	27·870
600	24·016	24·000	23·984	23·968	23·952	23·936	23·920	23·904	23·889
500	20·013	20·000	19·987	19·973	19·960	19·947	19·934	19·920	19·907
400	16·011	16·000	15·989	15·979	15·968	15·957	15·947	15·936	15·926
300	12·008	12·000	11·992	11·984	11·976	11·968	11·960	11·952	11·944
200	8·005	8·000	7·995	7·989	7·984	7·979	7·973	7·968	7·963
100	4·003	4·000	3·997	3·995	3·992	3·989	3·987	3·984	3·981

Distance	h. m. 25 08	h. m. 25 09	h. m. 25 10	h. m. 25 11	h. m. 25 12	h. m. 25 13	h. m. 25 14	h. m. 25 15	h. m. 25 16
900	35·809	35·785	35·762	35·738	35·714	35·691	35·667	35·644	35·620
800	31·830	31·809	31·788	31·767	31·746	31·725	31·704	31·683	31·662
700	27·851	27·833	27·815	27·796	27·778	27·759	27·741	27·723	27·704
600	23·873	23·857	23·841	23·825	23·810	23·794	23·778	23·762	23·747
500	19·894	19·881	19·868	19·854	19·841	19·828	19·815	19·802	19·789
400	15·915	15·905	15·894	15·884	15·873	15·863	15·852	15·842	15·831
300	11·936	11·928	11·921	11·913	11·905	11·897	11·889	11·881	11·873
200	7·958	7·952	7·947	7·942	7·937	7·931	7·926	7·921	7·916
100	3·979	3·976	3·974	3·971	3·968	3·966	3·963	3·960	3·958

Distance	h. m. 25 17	h. m. 25 18	h. m. 25 19	h. m. 25 20	h. m. 25 21	h. m. 25 22	h. m. 25 23	h. m. 25 24	h. m. 25 25
900	35·597	35·573	35·550	35·526	35·503	35·480	35·456	35·433	35·410
800	31·641	31·621	31·600	31·579	31·558	31·537	31·517	31·496	31·475
700	27·686	27·668	27·650	27·632	27·613	27·595	27·577	27·559	27·541
600	23·731	23·715	23·700	23·684	23·669	23·653	23·638	23·622	23·607
500	19·776	19·763	19·750	19·737	19·724	19·711	19·698	19·685	19·672
400	15·821	15·810	15·800	15·789	15·779	15·769	15·758	15·748	15·738
300	11·866	11·858	11·850	11·842	11·834	11·827	11·819	11·811	11·803
200	7·910	7·905	7·900	7·895	7·890	7·884	7·879	7·874	7·869
100	3·955	3·953	3·950	3·947	3·945	3·942	3·940	3·937	3·934

Distance	h. m. 25 26	h. m. 25 27	h. m. 25 28	h. m. 25 29	h. m. 25 30	h. m. 25 31	h. m. 25 32	h. m. 25 33	h. m. 25 34
900	35·387	35·363	35·340	35·317	35·294	35·271	35·248	35·225	35·202
800	31·455	31·434	31·414	31·393	31·373	31·352	31·332	31·311	31·291
700	27·523	27·505	27·487	27·469	27·451	27·433	27·415	27·397	27·379
600	23·591	23·576	23·560	23·545	23·529	23·514	23·499	23·483	23·468
500	19·659	19·646	19·634	19·621	19·608	19·595	19·582	19·569	19·557
400	15·727	15·717	15·707	15·697	15·686	15·676	15·666	15·656	15·645
300	11·796	11·788	11·780	11·772	11·765	11·757	11·749	11·742	11·734
200	7·864	7·859	7·853	7·848	7·843	7·838	7·833	7·828	7·823
100	3·932	3·929	3·927	3·924	3·922	3·919	3·916	3·914	3·911

DAY'S RUN—AVERAGE SPEED TABLE

STEAMING TIME

Distance in Nautical Miles	h. m. 25 35 Knots	h. m. 25 36 Knots	h. m. 25 37 Knots	h. m. 25 38 Knots	h. m. 25 39 Knots	h. m. 25 40 Knots	h. m. 25 41 Knots	h. m. 25 42 Knots	h. m. 25 43 Knots
900	35·179	35·156	35·133	35·111	35·088	35·065	35·042	35·019	34·997
800	31·270	31·250	31·230	31·209	31·189	31·169	31·149	31·128	31·108
700	27·362	27·344	27·326	27·308	27·290	27·273	27·255	27·237	27·220
600	23·453	23·438	23·422	23·407	23·392	23·377	23·361	23·346	23·331
500	19·544	19·531	19·519	19·506	19·493	19·481	19·468	19·455	19·443
400	15·635	15·625	15·615	15·605	15·595	15·584	15·574	15·564	15·554
300	11·726	11·719	11·711	11·704	11·696	11·688	11·681	11·673	11·666
200	7·818	7·813	7·807	7·802	7·797	7·792	7·787	7·782	7·777
100	3·909	3·906	3·904	3·901	3·899	3·896	3·894	3·891	3·889

Distance	h. m. 25 44	h. m. 25 45	h. m. 25 46	h. m. 25 47	h. m. 25 48	h. m. 25 49	h. m. 25 50	h. m. 25 51	h. m. 25 52
900	34·974	34·952	34·929	34·906	34·884	34·861	34·838	34·816	34·794
800	31·088	31·068	31·048	31·028	31·008	30·988	30·968	30·948	30·928
700	27·202	27·185	27·167	27·149	27·132	27·114	27·097	27·079	27·062
600	23·316	23·301	23·286	23·271	23·256	23·241	23·226	23·211	23·196
500	19·430	19·418	19·405	19·392	19·380	19·367	19·355	19·342	19·330
400	15·544	15·534	15·524	15·514	15·504	15·494	15·484	15·474	15·464
300	11·658	11·651	11·643	11·635	11·628	11·620	11·613	11·605	11·598
200	7·772	7·767	7·762	7·757	7·752	7·747	7·742	7·737	7·732
100	3·886	3·884	3·881	3·878	3·876	3·873	3·871	3·868	3·866

Distance	h. m. 25 53	h. m. 25 54	h. m. 25 55	h. m. 25 56	h. m. 25 57	h. m. 25 58	h. m. 25 59	h. m. 26 00	h. m. 26 01
900	34·771	34·749	34·726	34·704	34·682	34·660	34·638	34·615	34·593
800	30·908	30·888	30·868	30·848	30·828	30·809	30·789	30·769	30·750
700	27·044	27·027	27·010	26·992	26·975	26·958	26·940	26·923	26·906
600	23·181	23·166	23·151	23·136	23·121	23·107	23·092	23·077	23·062
500	19·317	19·305	19·293	19·280	19·268	19·255	19·243	19·231	19·218
400	15·454	15·444	15·434	15·424	15·414	15·404	15·394	15·385	15·375
300	11·590	11·583	11·576	11·568	11·561	11·553	11·546	11·538	11·531
200	7·727	7·722	7·717	7·712	7·707	7·702	7·697	7·692	7·687
100	3·863	3·861	3·858	3·856	3·854	3·851	3·849	3·846	3·844

HALF HOUR INTERVAL SUMMARY

Distance	h. m. 22 00	h. m. 22 30	h. m. 23 00	h. m. 23 30	h. m. 24 00	h. m. 24 30	h. m. 25 00	h. m. 25 30	h. m. 26 00
900	40·909	40·000	39·130	38·298	37·500	36·735	36·000	35·294	34·615
800	36·364	35·556	34·783	34·043	33·333	32·653	32·000	31·373	30·769
700	31·818	31·111	30·435	29·787	29·167	28·571	28·000	27·451	26·923
600	27·274	26·667	26·087	25·532	25·000	24·490	24·000	23·529	23·077
500	22·727	22·222	21·739	21·277	20·833	20·408	20·000	19·608	19·231
400	18·182	17·778	17·391	17·021	16·667	16·327	16·000	15·686	15·385
300	13·636	13·333	13·043	12·766	12·500	12·245	12·000	11·765	11·538
200	9·091	8·889	8·696	8·511	8·333	8·163	8·000	7·843	7·692
100	4·545	4·444	4·348	4·255	4·167	4·082	4·000	3·922	3·846

MEASURED MILE SPEED TABLE

Time to Steam One Mile in Mins. and Secs.

Speed in Knots

12 00·7	11 19·9	10 43·4	10 10·7	9 41·1	9 14·3	8 49·8	8 27·4	8 06·8	7 47·8	7 30·3
5·00	*5·30*	*5·60*	*5·90*	*6·20*	*6·50*	*6·80*	*7·10*	*7·40*	*7·70*	*8·00*
11 59·3	11 18·6	10 42·3	10 09·7	9 40·2	9 13·4	8 49·0	8 26·7	8 06·2	7 47·2	7 29·7
5·01	*5·31*	*5·61*	*5·91*	*6·21*	*6·51*	*6·81*	*7·11*	*7·41*	*7·71*	*8·01*
11 57·9	11 17·3	10 41·1	10 08·6	9 39·2	9 12·6	8 48·2	8 26·0	8 05·5	7 46·6	7 29·2
5·02	*5·32*	*5·62*	*5·92*	*6·22*	*6·52*	*6·82*	*7·12*	*7·42*	*7·72*	*8·02*
11 56·4	11 16·1	10 40·0	10 07·6	9 38·3	9 11·7	8 47·5	8 25·3	8 04·8	7 46·0	7 28·6
5·03	*5·33*	*5·63*	*5·93*	*6·23*	*6·53*	*6·83*	*7·13*	*7·43*	*7·73*	*8·03*
11 55·0	11 14·8	10 38·9	10 06·6	9 37·4	9 10·9	8 46·7	8 24·6	8 04·2	7 45·4	7 28·0
5·04	*5·34*	*5·64*	*5·94*	*6·24*	*6·54*	*6·84*	*7·14*	*7·44*	*7·74*	*8·04*
11 53·6	11 13·5	10 37·7	10 05·6	9 36·5	9 10·0	8 45·9	8 23·8	8 03·5	7 44·8	7 27·5
5·05	*5·35*	*5·65*	*5·95*	*6·25*	*6·55*	*6·85*	*7·15*	*7·45*	*7·75*	*8·05*
11 52·2	11 12·3	10 36·6	10 04·5	9 35·5	9 09·2	8 45·2	8 23·1	8 02·9	7 44·2	7 26·9
5·06	*5·36*	*5·66*	*5·96*	*6·26*	*6·56*	*6·86*	*7·16*	*7·46*	*7·76*	*8·06*
11 50·8	11 11·0	10 35·5	10 03·5	9 34·6	9 08·4	8 44·4	8 22·4	8 02·3	7 43·6	7 26·4
5·07	*5·37*	*5·67*	*5·97*	*6·27*	*6·57*	*6·87*	*7·17*	*7·47*	*7·77*	*8·07*
11 49·4	11 09·8	10 34·4	10 02·5	9 33·7	9 07·5	8 43·6	8 21·7	8 01·6	7 43·0	7 25·8
5·08	*5·38*	*5·68*	*5·98*	*6·28*	*6·58*	*6·88*	*7·18*	*7·48*	*7·78*	*8·08*
11 48·0	11 08·5	10 33·2	10 01·5	9 32·8	9 06·7	8 42·9	8 21·0	8 01·0	7 42·4	7 25·3
5·09	*5·39*	*5·69*	*5·99*	*6·29*	*6·59*	*6·89*	*7·19*	*7·49*	*7·79*	*8·09*
11 46·6	11 07·3	10 32·1	10 00·5	9 31·9	9 05·9	8 42·1	8 20·3	8 00·3	7 41·8	7 24·7
5·10	*5·40*	*5·70*	*6·00*	*6·30*	*6·60*	*6·90*	*7·20*	*7·50*	*7·80*	*8·10*
11 45·2	11 06·1	10 31·0	9 59·5	9 31·0	9 05·0	8 41·4	8 19·7	7 59·7	7 41·2	7 24·2
5·11	*5·41*	*5·71*	*6·01*	*6·31*	*6·61*	*6·91*	*7·21*	*7·51*	*7·81*	*8·11*
11 43·8	11 04·8	10 29·9	9 58·5	9 30·1	9 04·2	8 40·6	8 19·0	7 59·0	7 40·7	7 23·6
5·12	*5·42*	*5·72*	*6·02*	*6·32*	*6·62*	*6·92*	*7·22*	*7·52*	*7·82*	*8·12*
11 42·4	11 03·6	10 28·8	9 57·5	9 29·2	9 03·4	8 39·9	8 18·3	7 58·4	7 40·1	7 23·1
5·13	*5·43*	*5·73*	*6·03*	*6·33*	*6·63*	*6·93*	*7·23*	*7·53*	*7·83*	*8·13*
11 41·1	11 02·4	10 27·7	9 56·5	9 28·3	9 02·6	8 39·1	8 17·6	7 57·8	7 39·5	7 22·5
5·14	*5·44*	*5·74*	*6·04*	*6·34*	*6·64*	*6·94*	*7·24*	*7·54*	*7·84*	*8·14*
11 39·7	11 01·2	10 26·6	9 55·5	9 27·4	9 01·8	8 38·4	8 16·9	7 57·1	7 38·9	7 22·0
5·15	*5·45*	*5·75*	*6·05*	*6·35*	*6·65*	*6·95*	*7·25*	*7·55*	*7·85*	*8·15*
11 38·4	10 59·9	10 25·5	9 54·6	9 26·5	9 00·9	8 37·6	8 16·2	7 56·5	7 38·3	7 21·4
5·16	*5·46*	*5·76*	*6·06*	*6·36*	*6·66*	*6·96*	*7·26*	*7·56*	*7·86*	*8·16*
11 37·0	10 58·7	10 24·5	9 53·6	9 25·6	9 00·1	8 36·9	8 15·5	7 55·9	7 37·7	7 20·9
5·17	*5·47*	*5·77*	*6·07*	*6·37*	*6·67*	*6·97*	*7·27*	*7·57*	*7·87*	*8·17*
11 35·7	10 57·5	10 23·4	9 52·6	9 24·7	8 59·3	8 36·1	8 14·8	7 55·2	7 37·1	7 20·4
5·18	*5·48*	*5·78*	*6·08*	*6·38*	*6·68*	*6·98*	*7·28*	*7·58*	*7·88*	*8·18*
11 34·3	10 56·3	10 22·3	9 51·6	9 23·8	8 58·5	8 35·4	8 14·2	7 54·6	7 36·6	7 19·8
5·19	*5·49*	*5·79*	*6·09*	*6·39*	*6·69*	*6·99*	*7·29*	*7·59*	*7·89*	*8·19*
11 33·0	10 55·1	10 21·2	9 50·7	9 22·9	8 57·7	8 34·7	8 13·5	7 54·0	7 36·0	7 19·3
5·20	*5·50*	*5·80*	*6·10*	*6·40*	*6·70*	*7·00*	*7·30*	*7·60*	*7·90*	*8·20*
11 31·6	10 54·0	10 20·1	9 49·7	9 22·1	8 56·9	8 33·9	8 12·8	7 53·4	7 35·4	7 18·8
5·21	*5·51*	*5·81*	*6·11*	*6·41*	*6·71*	*7·01*	*7·31*	*7·61*	*7·91*	*8·21*
11 30·3	10 52·8	10 19·1	9 48·7	9 21·2	8 56·1	8 33·2	8 12·1	7 52·8	7 34·8	7 18·2
5·22	*5·52*	*5·82*	*6·12*	*6·42*	*6·72*	*7·02*	*7·32*	*7·62*	*7·92*	*8·22*
11 29·0	10 51·6	10 18·0	9 47·8	9 20·3	8 55·3	8 32·5	8 11·5	7 52·1	7 34·3	7 17·7
5·23	*5·53*	*5·83*	*6·13*	*6·43*	*6·73*	*7·03*	*7·33*	*7·63*	*7·93*	*8·23*
11 27·7	10 50·4	10 17·0	9 46·8	9 19·4	8 54·5	8 31·7	8 10·8	7 51·5	7 33·7	7 17·2
5·24	*5·54*	*5·84*	*6·14*	*6·44*	*6·74*	*7·04*	*7·34*	*7·64*	*7·94*	*8·24*
11 26·4	10 49·2	10 15·9	9 45·8	9 18·6	8 53·7	8 31·0	8 10·1	7 50·9	7 33·1	7 16·6
5·25	*5·55*	*5·85*	*6·15*	*6·45*	*6·75*	*7·05*	*7·35*	*7·65*	*7·95*	*8·25*
11 25·1	10 48·1	10 14·8	9 44·9	9 17·7	8 52·9	8 30·3	8 09·5	7 50·3	7 32·5	7 16·1
5·26	*5·56*	*5·86*	*6·16*	*6·46*	*6·76*	*7·06*	*7·36*	*7·66*	*7·96*	*8·26*
11 23·8	10 46·9	10 13·8	9 43·9	9 16·8	8 52·2	8 29·6	8 08·8	7 49·7	7 32·0	7 15·6
5·27	*5·57*	*5·87*	*6·17*	*6·47*	*6·77*	*7·07*	*7·37*	*7·67*	*7·97*	*8·27*
11 22·5	10 45·7	10 12·8	9 43·0	9 16·0	8 51·4	8 28·8	8 08·1	7 49·1	7 31·4	7 15·0
5·28	*5·58*	*5·88*	*6·18*	*6·48*	*6·78*	*7·08*	*7·38*	*7·68*	*7·98*	*8·28*
11 21·2	10 44·6	10 11·8	9 42·1	9 15·1	8 50·6	8 28·1	8 07·5	7 48·4	7 30·8	7 14·5
5·29	*5·59*	*5·89*	*6·19*	*6·49*	*6·79*	*7·09*	*7·39*	*7·69*	*7·99*	*8·29*
11 19·9	10 43·4	10 10·7	9 41·1	9 14·3	8 49·8	8 27·4	8 06·8	7 47·8	7 30·3	7 14·0
5·30	*5·60*	*5·90*	*6·20*	*6·50*	*6·80*	*7·10*	*7·40*	*7·70*	*8·00*	*8·30*
11 18·6	10 42·3	10 09·7	9 40·2	9 13·4	8 49·0	8 26·7	8 06·2	7 47·2	7 29·7	7 13·5

104

MEASURED MILE SPEED TABLE

Time to Steam One Mile in Mins. and Secs.

Speed in Knots

7 14·0	6 58·8	6 44·7	6 31·5	6 19·1	6 07·5	5 56·6	5 46·3	5 36·6	5 27·4	5 18·7
8·30	*8·60*	*8·90*	*9·20*	*9·50*	*9·80*	*10·10*	*10·40*	*10·70*	*11·00*	*11·30*
7 13·5	6 58·4	6 44·3	6 31·1	6 18·7	6 07·2	5 56·3	5 46·0	5 36·3	5 27·1	5 18·4
8·31	*8·61*	*8·91*	*9·21*	*9·51*	*9·81*	*10·11*	*10·41*	*10·71*	*11·01*	*11·31*
7 13·0	6 57·9	6 43·8	6 30·7	6 18·3	6 06·8	5 55·9	5 45·7	5 36·0	5 26·8	5 18·2
8·32	*8·62*	*8·92*	*9·22*	*9·52*	*9·82*	*10·12*	*10·42*	*10·72*	*11·02*	*11·32*
7 12·4	6 57·4	6 43·4	6 30·2	6 18·0	6 06·4	5 55·6	5 45·3	5 35·7	5 26·5	5 17·9
8·33	*8·63*	*8·93*	*9·23*	*9·53*	*9·83*	*10·13*	*10·43*	*10·73*	*11·03*	*11·33*
7 11·9	6 56·9	6 42·9	6 29·8	6 17·6	6 06·0	5 55·2	5 45·0	5 35·4	5 26·2	5 17·6
8·34	*8·64*	*8·94*	*9·24*	*9·54*	*9·84*	*10·14*	*10·44*	*10·74*	*11·04*	*11·34*
7 11·4	6 56·4	6 42·5	6 29·4	6 17·2	6 05·7	5 54·9	5 44·7	5 35·0	5 25·9	5 17·3
8·35	*8·65*	*8·95*	*9·25*	*9·55*	*9·85*	*10·15*	*10·45*	*10·75*	*11·05*	*11·35*
7 10·9	6 55·9	6 42·0	6 29·0	6 16·8	6 05·3	5 54·5	5 44·3	5 34·7	5 25·6	5 17·0
8·36	*8·66*	*8·96*	*9·26*	*9·56*	*9·86*	*10·16*	*10·46*	*10·76*	*11·06*	*11·36*
7 10·4	6 55·4	6 41·6	6 28·6	6 16·4	6 04·9	5 54·2	5 44·0	5 34·4	5 25·4	5 16·8
8·37	*8·67*	*8·97*	*9·27*	*9·57*	*9·87*	*10·17*	*10·47*	*10·77*	*11·07*	*11·37*
7 09·9	6 55·0	6 41·1	6 28·1	6 16·0	6 04·6	5 53·8	5 43·7	5 34·1	5 25·1	5 16·5
8·38	*8·68*	*8·98*	*9·28*	*9·58*	*9·88*	*10·18*	*10·48*	*10·78*	*11·08*	*11·38*
7 09·3	6 54·5	6 40·7	6 27·7	6 15·6	6 04·2	5 53·5	5 43·3	5 33·8	5 24·8	5 16·2
8·39	*8·69*	*8·99*	*9·29*	*9·59*	*9·89*	*10·19*	*10·49*	*10·79*	*11·09*	*11·39*
7 08·8	6 54·0	6 40·2	6 27·3	6 15·2	6 03·8	5 53·1	5 43·0	5 33·5	5 24·5	5 15·9
8·40	*8·70*	*9·00*	*9·30*	*9·60*	*9·90*	*10·20*	*10·50*	*10·80*	*11·10*	*11·40*
7 08·3	6 53·6	6 39·8	6 26·9	6 14·8	6 03·5	5 52·8	5 42·7	5 33·2	5 24·2	5 15·7
8·41	*8·71*	*9·01*	*9·31*	*9·61*	*9·91*	*10·21*	*10·51*	*10·81*	*11·11*	*11·41*
7 07·8	6 53·1	6 39·3	6 26·5	6 14·4	6 03·1	5 52·4	5 42·4	5 32·9	5 23·9	5 15·4
8·42	*8·72*	*9·02*	*9·32*	*9·62*	*9·92*	*10·22*	*10·52*	*10·82*	*11·12*	*11·42*
7 07·3	6 52·6	6 38·8	6 26·1	6 14·0	6 02·7	5 52·1	5 42·0	5 32·6	5 23·6	5 15·1
8·43	*8·73*	*9·03*	*9·33*	*9·63*	*9·93*	*10·23*	*10·53*	*10·83*	*11·13*	*11·43*
7 06·8	6 52·1	6 38·5	6 25·6	6 13·6	6 02·4	5 51·7	5 41·7	5 32·3	5 23·3	5 14·8
8·44	*8·74*	*9·04*	*9·34*	*9·64*	*9·94*	*10·24*	*10·54*	*10·84*	*11·14*	*11·44*
7 06·3	6 51·7	6 38·0	6 25·2	6 13·3	6 02·0	5 51·4	5 41·4	5 32·0	5 23·0	5 14·5
8·45	*8·75*	*9·05*	*9·35*	*9·65*	*9·95*	*10·25*	*10·55*	*10·85*	*11·15*	*11·45*
7 05·8	6 51·2	6 37·6	6 24·8	6 12·9	6 01·6	5 51·0	5 41·1	5 31·6	5 22·7	5 14·3
8·46	*8·76*	*9·06*	*9·36*	*9·66*	*9·96*	*10·26*	*10·56*	*10·86*	*11·16*	*11·46*
7 05·3	6 50·7	6 37·1	6 24·4	6 12·5	6 01·3	5 50·7	5 40·7	5 31·3	5 22·4	5 14·0
8·47	*8·77*	*9·07*	*9·37*	*9·67*	*9·97*	*10·27*	*10·57*	*10·87*	*11·17*	*11·47*
7 04·8	6 50·3	6 36·7	6 24·0	6 12·1	6 00·9	5 50·4	5 40·4	5 31·0	5 22·1	5 13·7
8·48	*8·78*	*9·08*	*9·38*	*9·68*	*9·98*	*10·28*	*10·58*	*10·88*	*11·18*	*11·48*
7 04·3	6 49·8	6 36·3	6 23·6	6 11·7	6 00·5	5 50·0	5 40·1	5 30·7	5 21·9	5 13·5
8·49	*8·79*	*9·09*	*9·39*	*9·69*	*9·99*	*10·29*	*10·59*	*10·89*	*11·19*	*11·49*
7 03·8	6 49·3	6 35·8	6 23·2	6 11·3	6 00·2	5 49·7	5 39·8	5 30·4	5 21·6	5 13·2
8·50	*8·80*	*9·10*	*9·40*	*9·70*	*10·00*	*10·30*	*10·60*	*10·90*	*11·20*	*11·50*
7 03·3	6 48·9	6 35·4	6 22·8	6 10·9	5 59·8	5 49·3	5 39·5	5 30·1	5 21·3	5 12·9
8·51	*8·81*	*9·11*	*9·41*	*9·71*	*10·01*	*10·31*	*10·61*	*10·91*	*11·21*	*11·51*
7 02·8	6 48·4	6 35·0	6 22·4	6 10·6	5 59·5	5 49·0	5 39·1	5 29·8	5 21·0	5 12·6
8·52	*8·82*	*9·12*	*9·42*	*9·72*	*10·02*	*10·32*	*10·62*	*10·92*	*11·22*	*11·52*
7 02·3	6 47·9	6 34·5	6 22·0	6 10·2	5 59·1	5 48·7	5 38·8	5 29·5	5 20·7	5 12·4
8·53	*8·83*	*9·13*	*9·43*	*9·73*	*10·03*	*10·33*	*10·63*	*10·93*	*11·23*	*11·53*
7 01·8	6 47·5	6 34·1	6 21·6	6 09·8	5 58·7	5 48·3	5 38·5	5 29·2	5 20·4	5 12·1
8·54	*8·84*	*9·14*	*9·44*	*9·74*	*10·04*	*10·34*	*10·64*	*10·94*	*11·24*	*11·54*
7 01·3	6 47·0	6 33·7	6 21·2	6 09·4	5 58·4	5 48·0	5 38·2	5 28·9	5 20·1	5 11·8
8·55	*8·85*	*9·15*	*9·45*	*9·75*	*10·05*	*10·35*	*10·65*	*10·95*	*11·25*	*11·55*
7 00·8	6 46·6	6 33·2	6 20·8	6 09·0	5 58·0	5 47·7	5 37·9	5 28·6	5 19·9	5 11·6
8·56	*8·86*	*9·16*	*9·46*	*9·76*	*10·06*	*10·36*	*10·66*	*10·96*	*11·26*	*11·56*
7 00·3	6 46·1	6 32·8	6 20·3	6 08·7	5 57·7	5 47·3	5 37·6	5 28·3	5 19·6	5 11·3
8·57	*8·87*	*9·17*	*9·47*	*9·77*	*10·07*	*10·37*	*10·67*	*10·97*	*11·27*	*11·57*
6 59·8	6 45·6	6 32·4	6 19·9	6 08·3	5 57·3	5 47·0	5 37·2	5 28·0	5 19·3	5 11·0
8·58	*8·88*	*9·18*	*9·48*	*9·78*	*10·08*	*10·38*	*10·68*	*10·98*	*11·28*	*11·58*
6 59·3	6 45·2	6 31·9	6 19·5	6 07·9	5 57·0	5 46·7	5 36·9	5 27·7	5 19·0	5 10·7
8·59	*8·89*	*9·19*	*9·49*	*9·79*	*10·09*	*10·39*	*10·69*	*10·99*	*11·29*	*11·59*
6 58·8	6 44·7	6 31·5	6 19·1	6 07·5	5 56·6	5 46·3	5 36·6	5 27·4	5 18·7	5 10·5
8·60	*8·90*	*9·20*	*9·50*	*9·80*	*10·10*	*10·40*	*10·70*	*11·00*	*11·30*	*11·60*
6 58·4	6 44·3	6 31·1	6 18·7	6 07·2	5 56·3	5 46·0	5 36·3	5 27·1	5 18·4	5 10·2

MEASURED MILE SPEED TABLE

Time to Steam One Mile in Mins. and Secs.

Speed in Knots

5 10·5	5 02·6	4 55·2	4 48·1	4 41·4	4 34·9	4 28·8	4 22·9	4 17·2	4 11·8	4 06·7
11·60	*11·90*	*12·20*	*12·50*	*12·80*	*13·10*	*13·40*	*13·70*	*14·00*	*14·30*	*14·60*
5 10·2	5 02·4	4 55·0	4 47·9	4 41·1	4 34·7	4 28·6	4 22·7	4 17·1	4 11·7	4 06·5
11·61	*11·91*	*12·21*	*12·51*	*12·81*	*13·11*	*13·41*	*13·71*	*14·01*	*14·31*	*14·61*
5 09·9	5 02·1	4 54·7	4 47·7	4 40·9	4 34·5	4 28·4	4 22·5	4 16·9	4 11·5	4 06·3
11·62	*11·92*	*12·22*	*12·52*	*12·82*	*13·12*	*13·42*	*13·72*	*14·02*	*14·32*	*14·62*
5 09·7	5 01·9	4 54·5	4 47·4	4 40·7	4 34·3	4 28·2	4 22·3	4 16·7	4 11·3	4 06·2
11·63	*11·93*	*12·23*	*12·53*	*12·83*	*13·13*	*13·43*	*13·73*	*14·03*	*14·33*	*14·63*
5 09·4	5 01·6	4 54·2	4 47·2	4 40·5	4 34·1	4 28·0	4 22·1	4 16·5	4 11·1	4 06·0
11·64	*11·94*	*12·24*	*12·54*	*12·84*	*13·14*	*13·44*	*13·74*	*14·04*	*14·34*	*14·64*
5 09·1	5 01·4	4 54·0	4 47·0	4 40·3	4 33·9	4 27·8	4 21·9	4 16·3	4 11·0	4 05·8
11·65	*11·95*	*12·25*	*12·55*	*12·85*	*13·15*	*13·45*	*13·75*	*14·05*	*14·35*	*14·65*
5 08·9	5 01·1	4 53·8	4 46·7	4 40·0	4 33·7	4 27·6	4 21·7	4 16·1	4 10·8	4 05·6
11·66	*11·96*	*12·26*	*12·56*	*12·86*	*13·16*	*13·46*	*13·76*	*14·06*	*14·36*	*14·66*
5 08·6	5 00·9	4 53·5	4 46·5	4 39·8	4 33·5	4 27·4	4 21·5	4 16·0	4 10·6	4 05·5
11·67	*11·97*	*12·27*	*12·57*	*12·87*	*13·17*	*13·47*	*13·77*	*14·07*	*14·37*	*14·67*
5 08·4	5 00·6	4 53·3	4 46·3	4 39·6	4 33·2	4 27·2	4 21·3	4 15·8	4 10·4	4 05·3
11·68	*11·98*	*12·28*	*12·58*	*12·88*	*13·18*	*13·48*	*13·78*	*14·08*	*14·38*	*14·68*
5 08·1	5 00·4	4 53·0	4 46·1	4 39·4	4 33·0	4 27·0	4 21·2	4 15·6	4 10·3	4 05·1
11·69	*11·99*	*12·29*	*12·59*	*12·89*	*13·19*	*13·49*	*13·79*	*14·09*	*14·39*	*14·69*
5 07·8	5 00·1	4 52·8	4 45·8	4 39·2	4 32·8	4 26·8	4 21·0	4 15·4	4 10·1	4 05·0
11·70	*12·00*	*12·30*	*12·60*	*12·90*	*13·20*	*13·50*	*13·80*	*14·10*	*14·40*	*14·70*
5 07·6	4 59·9	4 52·6	4 45·6	4 39·0	4 32·6	4 26·6	4 20·8	4 15·2	4 09·9	4 04·8
11·71	*12·01*	*12·31*	*12·61*	*12·91*	*13·21*	*13·51*	*13·81*	*14·11*	*14·41*	*14·71*
5 07·3	4 59·6	4 52·3	4 45·4	4 38·8	4 32·4	4 26·4	4 20·6	4 15·0	4 09·7	4 04·6
11·72	*12·02*	*12·32*	*12·62*	*12·92*	*13·22*	*13·52*	*13·82*	*14·12*	*14·42*	*14·72*
5 07·0	4 59·4	4 52·1	4 45·1	4 38·5	4 32·2	4 26·2	4 20·4	4 14·9	4 09·6	4 04·5
11·73	*12·03*	*12·33*	*12·63*	*12·93*	*13·23*	*13·53*	*13·83*	*14·13*	*14·43*	*14·73*
5 06·8	4 59·1	4 51·9	4 44·9	4 38·3	4 32·0	4 26·0	4 20·2	4 14·7	4 09·4	4 04·3
11·74	*12·04*	*12·34*	*12·64*	*12·94*	*13·24*	*13·54*	*13·84*	*14·14*	*14·44*	*14·74*
5 06·5	4 58·9	4 51·6	4 44·7	4 38·1	4 31·8	4 25·8	4 20·0	4 14·5	4 09·2	4 04·2
11·75	*12·05*	*12·35*	*12·65*	*12·95*	*13·25*	*13·55*	*13·85*	*14·15*	*14·45*	*14·75*
5 06·3	4 58·6	4 51·4	4 44·5	4 37·9	4 31·6	4 25·6	4 19·8	4 14·3	4 09·0	4 04·0
11·76	*12·06*	*12·36*	*12·66*	*12·96*	*13·26*	*13·56*	*13·86*	*14·16*	*14·46*	*14·76*
5 06·0	4 58·4	4 51·1	4 44·2	4 37·7	4 31·4	4 25·4	4 19·6	4 14·1	4 08·9	4 03·8
11·77	*12·07*	*12·37*	*12·67*	*12·97*	*13·27*	*13·57*	*13·87*	*14·17*	*14·47*	*14·77*
5 05·7	4 58·1	4 50·9	4 44·0	4 37·5	4 31·2	4 25·2	4 19·5	4 14·0	4 08·7	4 03·7
11·78	*12·08*	*12·38*	*12·68*	*12·98*	*13·28*	*13·58*	*13·88*	*14·18*	*14·48*	*14·78*
5 05·5	4 57·9	4 50·7	4 43·8	4 37·2	4 31·0	4 25·0	4 19·3	4 13·8	4 08·5	4 03·5
11·79	*12·09*	*12·39*	*12·69*	*12·99*	*13·29*	*13·59*	*13·89*	*14·19*	*14·49*	*14·79*
5 05·2	4 57·6	4 50·4	4 43·6	4 37·0	4 30·8	4 24·8	4 19·1	4 13·6	4 08·4	4 03·3
11·80	*12·10*	*12·40*	*12·70*	*13·00*	*13·30*	*13·60*	*13·90*	*14·20*	*14·50*	*14·80*
5 05·0	4 57·4	4 50·2	4 43·4	4 36·8	4 30·6	4 24·6	4 18·9	4 13·4	4 08·2	4 03·2
11·81	*12·11*	*12·41*	*12·71*	*13·01*	*13·31*	*13·61*	*13·91*	*14·21*	*14·51*	*14·81*
5 04·7	4 57·2	4 50·0	4 43·1	4 36·6	4 30·4	4 24·4	4 18·7	4 13·3	4 08·0	4 03·0
11·82	*12·12*	*12·42*	*12·72*	*13·02*	*13·32*	*13·62*	*13·92*	*14·22*	*14·52*	*14·82*
5 04·4	4 56·9	4 49·7	4 42·9	4 36·4	4 30·2	4 24·2	4 18·5	4 13·1	4 07·8	4 02·8
11·83	*12·13*	*12·43*	*12·73*	*13·03*	*13·33*	*13·63*	*13·93*	*14·23*	*14·53*	*14·83*
5 04·2	4 56·7	4 49·5	4 42·7	4 36·2	4 30·0	4 24·0	4 18·3	4 12·9	4 07·7	4 02·7
11·84	*12·14*	*12·44*	*12·74*	*13·04*	*13·34*	*13·64*	*13·94*	*14·24*	*14·54*	*14·84*
5 03·9	4 56·4	4 49·3	4 42·5	4 36·0	4 29·8	4 23·8	4 18·2	4 12·7	4 07·5	4 02·5
11·85	*12·15*	*12·45*	*12·75*	*13·05*	*13·35*	*13·65*	*13·95*	*14·25*	*14·55*	*14·85*
5 03·7	4 56·2	4 49·0	4 42·2	4 35·8	4 29·6	4 23·6	4 18·0	4 12·5	4 07·3	4 02·3
11·86	*12·16*	*12·46*	*12·76*	*13·06*	*13·36*	*13·66*	*13·96*	*14·26*	*14·56*	*14·86*
5 03·4	4 55·9	4 48·8	4 42·0	4 35·5	4 29·4	4 23·4	4 17·8	4 12·4	4 07·2	4 02·2
11·87	*12·17*	*12·47*	*12·77*	*13·07*	*13·37*	*13·67*	*13·97*	*14·27*	*14·57*	*14·87*
5 03·2	4 55·7	4 48·6	4 41·8	4 35·3	4 29·2	4 23·3	4 17·6	4 12·2	4 07·0	4 02·0
11·88	*12·18*	*12·48*	*12·78*	*13·08*	*13·38*	*13·68*	*13·98*	*14·28*	*14·58*	*14·88*
5 02·9	4 55·4	4 48·3	4 41·6	4 35·1	4 29·0	4 23·1	4 17·4	4 12·0	4 06·8	4 01·9
11·89	*12·19*	*12·49*	*12·79*	*13·09*	*13·39*	*13·69*	*13·99*	*14·29*	*14·59*	*14·89*
5 02·6	4 55·2	4 48·1	4 41·4	4 34·9	4 28·8	4 22·9	4 17·2	4 11·8	4 06·7	4 01·7
11·90	*12·20*	*12·50*	*12·80*	*13·10*	*13·40*	*13·70*	*14·00*	*14·30*	*14·60*	*14·90*
5 02·4	4 55·0	4 47·9	4 41·1	4 34·7	4 28·6	4 22·7	4 17·1	4 11·7	4 06·5	4 01·5

MEASURED MILE SPEED TABLE

Time to Steam One Mile in Mins. and Secs.

Speed in Knots

4 01·7	3 56·9	3 52·3	3 47·9	3 43·7	3 39·6	3 35·0	3 31·8	3 28·2	3 24·6	3 21·2
14·90	*15·20*	*15·50*	*15·80*	*16·10*	*16·40*	*16·70*	*17·00*	*17·30*	*17·60*	*17·90*
4 01·5	3 56·8	3 52·2	3 47·8	3 43·5	3 39·4	3 35·5	3 31·7	3 28·0	3 24·5	3 21·1
14·91	*15·21*	*15·51*	*15·81*	*16·11*	*16·41*	*16·71*	*17·01*	*17·31*	*17·61*	*17·91*
4 01·4	3 56·6	3 52·0	3 47·6	3 43·4	3 39·3	3 35·4	3 31·6	3 27·9	3 24·4	3 20·9
14·92	*15·22*	*15·52*	*15·82*	*16·12*	*16·42*	*16·72*	*17·02*	*17·32*	*17·62*	*17·92*
4 01·2	3 56·5	3 51·9	3 47·5	3 43·3	3 39·2	3 35·2	3 31·5	3 27·8	3 24·3	3 20·8
14·93	*15·23*	*15·53*	*15·83*	*16·13*	*16·43*	*16·73*	*17·03*	*17·33*	*17·63*	*17·93*
4 01·0	3 56·3	3 51·7	3 47·3	3 43·1	3 39·0	3 35·1	3 31·3	3 27·7	3 24·1	3 20·7
14·94	*15·24*	*15·54*	*15·84*	*16·14*	*16·44*	*16·74*	*17·04*	*17·34*	*17·64*	*17·94*
4 00·9	3 56·1	3 51·6	3 47·2	3 43·0	3 38·9	3 35·0	3 31·2	3 27·6	3 24·0	3 20·6
14·95	*15·25*	*15·55*	*15·85*	*16·15*	*16·45*	*16·75*	*17·05*	*17·35*	*17·65*	*17·95*
4 00·7	3 56·0	3 51·4	3 47·1	3 42·8	3 38·8	3 34·9	3 31·1	3 27·4	3 23·9	3 20·5
14·96	*15·26*	*15·56*	*15·86*	*16·16*	*16·46*	*16·76*	*17·06*	*17·36*	*17·66*	*17·96*
4 00·6	3 55·8	3 51·3	3 46·9	3 42·7	3 38·6	3 34·7	3 31·0	3 27·3	3 23·8	3 20·4
14·97	*15·27*	*15·57*	*15·87*	*16·17*	*16·47*	*16·77*	*17·07*	*17·37*	*17·67*	*17·97*
4 00·4	3 55·7	3 51·1	3 46·8	3 42·6	3 38·5	3 34·6	3 30·8	3 27·2	3 23·7	3 20·3
14·98	*15·28*	*15·58*	*15·88*	*16·18*	*16·48*	*16·78*	*17·08*	*17·38*	*17·68*	*17·98*
4 00·2	3 55·5	3 51·0	3 46·6	3 42·4	3 38·4	3 34·5	3 30·7	3 27·1	3 23·6	3 20·2
14·99	*15·29*	*15·59*	*15·89*	*16·19*	*16·49*	*16·79*	*17·09*	*17·39*	*17·69*	*17·99*
4 00·1	3 55·4	3 50·8	3 46·5	3 42·3	3 38·2	3 34·3	3 30·6	3 27·0	3 23·4	3 20·1
15·00	*15·30*	*15·60*	*15·90*	*16·20*	*16·50*	*16·80*	*17·10*	*17·40*	*17·70*	*18·00*
3 59·9	3 55·2	3 50·7	3 46·3	3 42·2	3 38·1	3 34·2	3 30·5	3 26·8	3 23·3	3 19·9
15·01	*15·31*	*15·61*	*15·91*	*16·21*	*16·51*	*16·81*	*17·11*	*17·41*	*17·71*	*18·01*
3 59·8	3 55·1	3 50·5	3 46·2	3 42·0	3 38·0	3 34·1	3 30·3	3 26·7	3 23·2	3 19·8
15·02	*15·32*	*15·62*	*15·92*	*16·22*	*16·52*	*16·82*	*17·12*	*17·42*	*17·72*	*18·02*
3 59·6	3 54·9	3 50·4	3 46·1	3 41·9	3 37·9	3 34·0	3 30·2	3 26·6	3 23·1	3 19·7
15·03	*15·33*	*15·63*	*15·93*	*16·23*	*16·53*	*16·83*	*17·13*	*17·43*	*17·73*	*18·03*
3 59·4	3 54·8	3 50·3	3 45·9	3 41·7	3 37·7	3 33·8	3 30·1	3 26·5	3 23·0	3 19·6
15·04	*15·34*	*15·64*	*15·94*	*16·24*	*16·54*	*16·84*	*17·14*	*17·44*	*17·74*	*18·04*
3 59·3	3 54·6	3 50·1	3 45·8	3 41·6	3 37·6	3 33·7	3 30·0	3 26·4	3 22·9	3 19·5
15·05	*15·35*	*15·65*	*15·95*	*16·25*	*16·55*	*16·85*	*17·15*	*17·45*	*17·75*	*18·05*
3 59·1	3 54·5	3 50·0	3 45·6	3 41·5	3 37·5	3 33·6	3 29·9	3 26·2	3 22·8	3 19·4
15·06	*15·36*	*15·66*	*15·96*	*16·26*	*16·56*	*16·86*	*17·16*	*17·46*	*17·76*	*18·06*
3 59·0	3 54·3	3 49·8	3 45·5	3 41·3	3 37·3	3 33·5	3 29·7	3 26·1	3 22·6	3 19·3
15·07	*15·37*	*15·67*	*15·97*	*16·27*	*16·57*	*16·87*	*17·17*	*17·47*	*17·77*	*18·07*
3 58·8	3 54·1	3 49·7	3 45·4	3 41·2	3 37·2	3 33·3	3 29·6	3 26·0	3 22·5	3 19·2
15·08	*15·38*	*15·68*	*15·98*	*16·28*	*16·58*	*16·88*	*17·18*	*17·48*	*17·78*	*18·08*
3 58·6	3 54·0	3 49·5	3 45·2	3 41·1	3 37·1	3 33·2	3 29·5	3 25·9	3 22·4	3 19·1
15·09	*15·39*	*15·69*	*15·99*	*16·29*	*16·59*	*16·89*	*17·19*	*17·49*	*17·79*	*18·09*
3 58·5	3 53·8	3 49·4	3 45·1	3 40·9	3 36·9	3 33·1	3 29·4	3 25·8	3 22·3	3 18·9
15·10	*15·40*	*15·70*	*16·00*	*16·30*	*16·60*	*16·90*	*17·20*	*17·50*	*17·80*	*18·10*
3 58·3	3 53·7	3 49·2	3 44·9	3 40·8	3 36·8	3 33·0	3 29·2	3 25·7	3 22·2	3 18·8
15·11	*15·41*	*15·71*	*16·01*	*16·31*	*16·61*	*16·91*	*17·21*	*17·51*	*17·81*	*18·11*
3 58·2	3 53·5	3 49·1	3 44·8	3 40·7	3 36·7	3 32·8	3 29·1	3 25·5	3 22·1	3 18·7
15·12	*15·42*	*15·72*	*16·02*	*16·32*	*16·62*	*16·92*	*17·22*	*17·52*	*17·82*	*18·12*
3 58·0	3 53·4	3 48·9	3 44·6	3 40·5	3 36·5	3 32·7	3 29·0	3 25·4	3 22·0	3 18·6
15·13	*15·43*	*15·73*	*16·03*	*16·33*	*16·63*	*16·93*	*17·23*	*17·53*	*17·83*	*18·13*
3 57·9	3 53·2	3 48·8	3 44·5	3 40·5	3 36·4	3 32·6	3 28·9	3 25·3	3 21·9	3 18·5
15·14	*15·44*	*15·74*	*16·04*	*16·34*	*16·64*	*16·94*	*17·24*	*17·54*	*17·84*	*18·14*
3 57·7	3 53·1	3 48·6	3 44·4	3 40·3	3 36·3	3 32·5	3 28·8	3 25·2	3 21·7	3 18·4
15·15	*15·45*	*15·75*	*16·05*	*16·35*	*16·65*	*16·95*	*17·25*	*17·55*	*17·85*	*18·15*
3 57·5	3 52·9	3 48·5	3 44·2	3 40·1	3 36·2	3 32·3	3 28·6	3 25·1	3 21·6	3 18·3
15·16	*15·46*	*15·76*	*16·06*	*16·36*	*16·66*	*16·96*	*17·26*	*17·56*	*17·86*	*18·16*
3 57·4	3 52·8	3 48·4	3 44·1	3 40·0	3 36·0	3 32·2	3 28·5	3 25·0	3 21·5	3 18·2
15·17	*15·47*	*15·77*	*16·07*	*16·37*	*16·67*	*16·97*	*17·27*	*17·57*	*17·87*	*18·17*
3 57·2	3 52·6	3 48·2	3 44·0	3 39·8	3 35·9	3 32·1	3 28·4	3 24·8	3 21·4	3 18·1
15·18	*15·48*	*15·78*	*16·08*	*16·38*	*16·68*	*16·98*	*17·28*	*17·58*	*17·88*	*18·18*
3 57·1	3 52·5	3 48·1	3 43·8	3 39·7	3 35·8	3 32·0	3 28·3	3 24·7	3 21·3	3 18·0
15·19	*15·49*	*15·79*	*16·09*	*16·39*	*16·69*	*16·99*	*17·29*	*17·59*	*17·89*	*18·19*
3 56·9	3 52·3	3 47·9	3 43·7	3 39·6	3 35·6	3 31·8	3 28·2	3 24·6	3 21·2	3 17·9
15·20	*15·50*	*15·80*	*16·10*	*16·40*	*16·70*	*17·00*	*17·30*	*17·60*	*17·90*	*18·20*
3 56·8	3 52·2	3 47·8	3 43·5	3 39·4	3 35·5	3 31·7	3 28·0	3 24·5	3 21·1	3 17·7

MEASURED MILE SPEED TABLE

Time to Steam One Mile in Mins. and Secs.

Speed in Knots

3 17·9	3 14·6	3 11·5	3 08·5	3 05·6	3 02·8	—	2 57·4	2 54·8	2 52·3	2 49·9
18·20	*18·50*	*18·80*	*19·10*	*19·40*	*19·70*	*20·00*	*20·30*	*20·60*	*20·90*	*21·20*
3 17·7	3 14·5	3 11·4	3 08·4	3 05·5	3 02·7	—	2 57·3	2 54·7	2 52·2	2 49·8
18·21	*18·51*	*18·81*	*19·11*	*19·41*	*19·71*	*20·01*	*20·31*	*20·61*	*20·91*	*21·21*
3 17·6	3 14·4	3 11·3	3 08·3	3 05·4	3 02·6	2 59·9	2 57·2	2 54·6	2 52·1	2 49·7
18·22	*18·52*	*18·82*	*19·12*	*19·42*	*19·72*	*20·02*	*20·32*	*20·62*	*20·92*	*21·22*
3 17·5	3 14·3	3 11·2	3 08·2	3 05·3	3 02·5	2 59·8	2 57·1	2 54·5	2 52·0	2 49·6
18·23	*18·53*	*18·83*	*19·13*	*19·43*	*19·73*	*20·03*	*20·33*	*20·63*	*20·93*	*21·23*
3 17·4	3 14·2	3 11·1	3 08·1	3 05·2	3 02·4	2 59·7	2 57·0	—	—	2 49·5
18·24	*18·54*	*18·84*	*19·14*	*19·44*	*19·74*	*20·04*	*20·34*	*20·64*	*20·94*	*21·24*
3 17·3	3 14·1	3 11·0	3 08·0	3 05·1	3 02·3	2 59·6	2 56·9	2 54·4	2 51·9	—
18·25	*18·55*	*18·85*	*19·15*	*19·45*	*19·75*	*20·05*	*20·35*	*20·65*	*20·95*	*21·25*
3 17·2	3 14·0	3 10·9	3 07·9	3 05·0	3 02·2	2 59·5	—	2 54·3	2 51·8	2 49·4
18·26	*18·56*	*18·86*	*19·16*	*19·46*	*19·76*	*20·06*	*20·36*	*20·66*	*20·96*	*21·26*
3 17·1	3 13·9	3 10·8	3 07·8	3 04·9	3 02·1	2 59·4	2 56·8	2 54·2	2 51·7	2 49·3
18·27	*18·57*	*18·87*	*19·17*	*19·47*	*19·77*	*20·07*	*20·37*	*20·67*	*20·97*	*21·27*
3 17·0	3 13·8	3 10·7	3 07·7	—	3 02·0	2 59·3	2 56·7	2 54·1	2 51·6	2 49·2
18·28	*18·58*	*18·88*	*19·18*	*19·48*	*19·78*	*20·08*	*20·38*	*20·68*	*20·98*	*21·28*
3 16·9	3 13·7	3 10·6	3 07·6	3 04·8	—	2 59·2	2 56·6	2 54·0	—	2 49·1
18·29	*18·59*	*18·89*	*19·19*	*19·49*	*19·79*	*20·09*	*20·39*	*20·69*	*20·99*	*21·29*
3 16·8	3 13·6	3 10·5	3 07·5	3 04·7	3 01·9	2 59·1	2 56·5	—	2 51·5	—
18·30	*18·60*	*18·90*	*19·20*	*19·50*	*19·80*	*20·10*	*20·40*	*20·70*	*21·00*	*21·30*
3 16·7	3 13·5	3 10·4	—	3 04·6	3 01·8	—	2 56·4	2 53·9	2 51·4	2 49·0
18·31	*18·61*	*18·91*	*19·21*	*19·51*	*19·81*	*20·11*	*20·41*	*20·71*	*21·01*	*21·31*
3 16·6	3 13·4	3 10·3	3 07·4	3 04·5	3 01·7	2 59·0	2 56·3	2 53·8	2 51·3	2 48·9
18·32	*18·62*	*18·92*	*19·22*	*19·52*	*19·82*	*20·12*	*20·42*	*20·72*	*21·02*	*21·32*
3 16·5	3 13·3	3 10·2	3 07·3	3 04·4	3 01·6	2 58·9	—	2 53·7	2 51·2	2 48·8
18·33	*18·63*	*18·93*	*19·23*	*19·53*	*19·83*	*20·13*	*20·43*	*20·73*	*21·03*	*21·33*
3 16·3	3 13·2	3 10·1	3 07·2	3 04·3	3 01·5	2 58·8	2 56·2	2 53·6	2 51·1	2 48·7
18·34	*18·64*	*18·94*	*19·24*	*19·54*	*19·84*	*20·14*	*20·44*	*20·74*	*21·04*	*21·34*
3 16·2	3 13·1	3 10·0	3 07·1	3 04·2	3 01·4	2 58·7	2 56·1	2 53·5	—	—
18·35	*18·65*	*18·95*	*19·25*	*19·55*	*19·85*	*20·15*	*20·45*	*20·75*	*21·05*	*21·35*
3 16·1	3 13·0	3 09·9	3 07·0	3 04·1	3 01·3	2 58·6	2 56·0	—	2 51·0	2 48·6
18·36	*18·66*	*18·96*	*19·26*	*19·56*	*19·86*	*20·16*	*20·46*	*20·76*	*21·06*	*21·36*
3 16·0	3 12·9	3 09·8	3 06·9	3 04·0	3 01·2	2 58·5	2 55·9	2 53·4	2 50·9	2 48·5
18·37	*18·67*	*18·97*	*19·27*	*19·57*	*19·87*	*20·17*	*20·47*	*20·77*	*21·07*	*21·37*
3 15·9	3 12·8	3 09·7	3 06·8	3 03·9	3 01·1	2 58·4	2 55·8	2 53·3	2 50·8	2 48·4
18·38	*18·68*	*18·98*	*19·28*	*19·58*	*19·88*	*20·18*	*20·48*	*20·78*	*21·08*	*21·38*
3 15·8	3 12·7	3 09·6	3 06·7	3 03·8	3 01·0	—	2 55·7	2 53·2	2 50·7	2 48·3
18·39	*18·69*	*18·99*	*19·29*	*19·59*	*19·89*	*20·19*	*20·49*	*20·79*	*21·09*	*21·39*
3 15·7	3 12·6	3 09·5	3 06·6	3 03·7	3 00·9	2 58·3	—	2 53·1	—	—
18·40	*18·70*	*19·00*	*19·30*	*19·60*	*19·90*	*20·20*	*20·50*	*20·80*	*21·10*	*21·40*
3 15·6	3 12·5	3 09·4	3 06·5	3 03·6	—	2 58·2	2 55·6	2 53·0	2 50·6	2 48·2
18·41	*18·71*	*19·01*	*19·31*	*19·61*	*19·91*	*20·21*	*20·51*	*20·81*	*21·11*	*21·41*
3 15·5	3 12·4	3 09·3	3 06·4	3 03·5	3 00·8	2 58·1	2 55·5	—	2 50·5	2 48·1
18·42	*18·72*	*19·02*	*19·32*	*19·62*	*19·92*	*20·22*	*20·52*	*20·82*	*21·12*	*21·42*
3 15·4	3 12·3	3 09·2	3 06·3	3 03·4	3 00·7	2 58·0	2 55·4	2 52·9	2 50·4	2 48·0
18·43	*18·73*	*19·03*	*19·33*	*19·63*	*19·93*	*20·23*	*20·53*	*20·83*	*21·13*	*21·43*
3 15·3	3 12·2	3 09·1	3 06·2	3 03·3	3 00·6	2 57·9	2 55·3	2 52·8	2 50·3	2 47·9
18·44	*18·74*	*19·04*	*19·34*	*19·64*	*19·94*	*20·24*	*20·54*	*20·84*	*21·14*	*21·44*
3 15·2	3 12·1	3 09·0	3 06·1	—	3 00·5	2 57·8	2 55·2	2 52·7	—	—
18·45	*18·75*	*19·05*	*19·35*	*19·65*	*19·95*	*20·25*	*20·55*	*20·85*	*21·15*	*21·45*
3 15·1	3 11·9	3 08·9	3 06·0	3 03·2	3 00·4	2 57·7	2 55·1	2 52·6	2 50·2	2 47·8
18·46	*18·76*	*19·06*	*19·36*	*19·66*	*19·96*	*20·26*	*20·56*	*20·86*	*21·16*	*21·46*
3 15·0	3 11·8	3 08·8	3 05·9	3 03·1	3 00·3	2 57·6	—	2 52·5	2 50·1	2 47·7
18·47	*18·77*	*19·07*	*19·37*	*19·67*	*19·97*	*20·27*	*20·57*	*20·87*	*21·17*	*21·47*
3 14·9	3 11·7	3 08·7	3 05·8	3 03·0	3 00·2	—	2 55·0	—	2 50·0	2 47·6
18·48	*18·78*	*19·08*	*19·38*	*19·68*	*19·98*	*20·28*	*20·58*	*20·88*	*21·18*	*21·48*
3 14·8	3 11·6	3 08·6	3 05·7	3 02·9	3 00·1	2 57·5	2 54·9	2 52·4	2 49·9	—
18·49	*18·79*	*19·09*	*19·39*	*19·69*	*19·99*	*20·29*	*20·59*	*20·89*	*21·19*	*21·49*
3 14·6	3 11·5	3 08·5	3 05·6	3 02·8	3 00·0	2 57·4	2 54·8	2 52·3	—	2 47·5
18·50	*18·80*	*19·10*	*19·40*	*19·70*	*20·00*	*20·30*	*20·60*	*20·90*	*21·20*	*21·50*
3 14·5	3 11·4	3 08·4	3 05·5	3 02·7	—	2 57·3	2 54·7	2 52·2	2 49·8	2 47·4

MEASURED MILE SPEED TABLE

Time to Steam One Mile in Mins. and Secs.

Speed in Knots

2 47.5	2 45.2	2 42.9	2 40.8	2 38.6	2 36.6	2 34.5	2 32.6	2 30.7	2 28.8	2 27.0
21·50	*21·80*	*22·10*	*22·40*	*22·70*	*23·00*	*23·30*	*23·60*	*23·90*	*24·20*	*24·50*
2 47.4	2 45.1	—	2 40.7	—	2 36.5	—	2 32.5	2 30.6	2 28.7	2 26.9
21·51	*21·81*	*22·11*	*22·41*	*22·71*	*23·01*	*23·31*	*23·61*	*23·91*	*24·21*	*24·51*
2 47.3	2 45.0	2 42.8	2 40.6	2 38.5	2 36.4	2 34.4	2 32.4	2 30.5	—	2 26.8
21·52	*21·82*	*22·12*	*22·42*	*22·72*	*23·02*	*23·32*	*23·62*	*23·92*	*24·22*	*24·52*
2 47.2	2 44.9	2 42.7	2 40.5	2 38.4	—	2 34.3	—	—	2 28.6	—
21·53	*21·83*	*22·13*	*22·43*	*22·73*	*23·03*	*23·33*	*23·63*	*23·93*	*24·23*	*24·53*
—	—	2 42.6	—	2 38.3	2 36.3	—	2 32.3	2 30.4	2 28.5	2 26.7
21·54	*21·84*	*22·14*	*22·44*	*22·74*	*23·04*	*23·34*	*23·64*	*23·94*	*24·24*	*34·54*
2 47.1	2 44.8	—	2 40.4	—	2 36.2	2 34.2	—	2 30.3	—	—
21·55	*21·85*	*22·15*	*22·45*	*22·75*	*23·05*	*23·35*	*23·65*	*23·95*	*24·25*	*24·55*
2 47.0	2 44.7	2 42.5	2 40.3	2 38.2	2 36.1	2 34.1	2 32.2	—	2 28.4	2 26.6
21·56	*21·86*	*22·16*	*22·46*	*22·76*	*23·06*	*23·36*	*23·66*	*23·96*	*24·26*	*24·56*
2 46.9	2 44.6	2 42.4	2 40.2	2 38.1	—	—	2 32.1	2 30.2	—	2 26.5
21·57	*21·87*	*22·17*	*22·47*	*22·77*	*23·07*	*23·37*	*23·67*	*23·97*	*24·27*	*24·57*
—	—	2 42.3	—	—	2 36.0	2 34.0	—	—	2·28.3	—
21·58	*21·88*	*22·18*	*22·48*	*22·78*	*23·08*	*23·38*	*23·68*	*23·98*	*24·28*	*24·58*
2 46.8	2 44.5	—	2 40.1	2 38.0	2 35.9	2 33.9	2 32.0	2 30.1	2 28.2	2 26.4
21·59	*21·89*	*22·19*	*22·49*	*22·79*	*23·09*	*23·39*	*23·69*	*23·99*	*24·29*	*24·59*
2 46.7	2 44.4	2 42.2	2 40.0	2 37.9	2 35.8	—	2 31.9	2 30.0	—	—
21·60	*21·90*	*22·20*	*22·50*	*22·80*	*23·10*	*23·40*	*23·70*	*24·00*	*24·30*	*24·60*
2 46.6	2 44.3	2 42.1	—	—	—	2 33.8	—	—	2 28.1	2 26.3
21·61	*21·91*	*22·21*	*22·51*	*22·81*	*23·11*	*23·41*	*23·71*	*24·01*	*24·31*	*24·61*
—	—	—	2 39.9	2 37.8	2 35.7	2 33.7	2 31.8	2 29.9	—	—
21·62	*21·92*	*22·22*	*22·52*	*22·82*	*23·12*	*23·42*	*23·72*	*24·02*	*24·32*	*24·62*
2 46.5	2 44.2	2 42.0	2 39.8	2 37.7	—	—	2 31.7	2 29.8	2 28.0	2 26.2
21·63	*21·93*	*22·23*	*22·53*	*22·83*	*23·13*	*23·43*	*23·73*	*24·03*	*24·33*	*24·63*
2 46.4	2 44.1	2 41.9	—	—	2 35.6	2 33.6	—	—	2 27.9	2 26.1
21·64	*21·94*	*22·24*	*22·54*	*22·84*	*23·14*	*23·44*	*23·74*	*24·04*	*24·34*	*24·64*
2 46.3	2 44.0	2 41.8	2 39.7	2 37.6	2 35.5	—	2 31.6	2 29.7	—	—
21·65	*21·95*	*22·25*	*22·55*	*22·85*	*23·15*	*23·45*	*23·75*	*24·05*	*24·35*	*24·65*
2 46.2	—	—	2 39.6	2 37.5	—	2 33.5	2 31.5	—	2 27.8	2 26.0
21·66	*21·96*	*22·26*	*22·56*	*22·86*	*23·16*	*23·46*	*23·76*	*24·06*	*24·36*	*24·66*
—	2 43.9	2 41.7	2 39.5	2 37.4	2 35.4	2 33.4	—	2 29.6	—	—
21·67	*21·97*	*22·27*	*22·57*	*22·87*	*23·17*	*23·47*	*23·77*	*24·07*	*24·37*	*24·67*
2 46.1	2 43.8	2 41.6	—	—	2 35.3	—	2 31.4	2 29.5	2 27.7	2 25.9
21·68	*21·98*	*22·28*	*22·58*	*22·88*	*23·18*	*23·48*	*23·78*	*24·08*	*24·38*	*24·68*
2 46.0	2 43.7	2 41.5	2 39.4	2 37.3	—	2 33.3	—	—	2 27.6	2 25.8
21·69	*21·99*	*22·29*	*22·59*	*22·89*	*23·19*	*23·49*	*23·79*	*24·09*	*24·39*	*24·69*
2 45.9	—	—	2 39.3	2 37.2	2 35.2	2 33.2	2 31.3	2 29.4	—	—
21·70	*22·00*	*22·30*	*22·60*	*22·90*	*23·20*	*23·50*	*23·80*	*24·10*	*24·40*	*24·70*
—	2 43.6	2 41.4	—	—	2 35.1	—	2 31.2	2 29.3	2 27.5	2 25.7
21·71	*22·01*	*22·31*	*22·61*	*22·91*	*23·21*	*23·51*	*23·81*	*24·11*	*24·41*	*24·71*
2 45.8	2 43.5	2 41.3	2 39.2	2 37.1	—	2 33.1	—	—	—	—
21·72	*22·02*	*22·32*	*22·62*	*22·92*	*23·22*	*23·52*	*23·82*	*24·12*	*24·42*	*24·72*
2 45.7	—	—	2 39.1	2 37.0	2 35.0	2 33.0	2 31.1	2 29.2	2 27.4	2 25.6
21·73	*22·03*	*22·33*	*22·63*	*22·93*	*23·23*	*23·53*	*23·83*	*24·13*	*24·43*	*24·73*
2 45.6	2 43.4	2 41.2	2 39.0	—	2 34.9	—	2 31.0	—	2 27.3	2 25.5
21·74	*22·04*	*22·34*	*22·64*	*22·94*	*23·24*	*23·54*	*23·84*	*24·14*	*24·44*	*24·74*
—	2 43.3	2 41.1	—	2 36.9	—	2 32.9	—	2 29.1	—	—
21·75	*22·05*	*22·35*	*22·65*	*22·95*	*23·25*	*23·55*	*23·85*	*24·15*	*24·45*	*24·75*
2 45.5	2 43.2	2 41.0	2 38.9	2 36.8	2 34.8	2 32.8	2 30.9	2 29.0	2 27.2	2 25.4
21·76	*22·06*	*22·36*	*22·66*	*22·96*	*23·26*	*23·56*	*23·86*	*24·16*	*24·46*	*24·76*
2 45.4	—	—	2 38.8	—	2 34.7	—	2 30.8	—	2 27.1	—
21·77	*22·07*	*22·37*	*22·67*	*22·97*	*23·27*	*23·57*	*23·87*	*24·17*	*24·47*	*24·77*
2 45.3	2 43.1	2 40.9	—	2 36.7	—	2 32.7	—	2 28.9	—	2 25.3
21·78	*22·08*	*22·38*	*22·68*	*22·98*	*23·28*	*23·58*	*23·88*	*24·18*	*24·48*	*24·78*
—	2 43.0	2 40.8	2 38.7	2 36.6	2 34.6	2 32.6	2 30.7	—	2 27.0	2 25.2
21·79	*22·09*	*22·39*	*22·69*	*22·99*	*23·29*	*23·59*	*23·89*	*24·19*	*24·49*	*24·79*
2 45.2	2 42.9	—	2 38.6	—	2 34.5	—	—	2 28.8	—	—
21·80	*22·10*	*22·40*	*22·70*	*23·00*	*23·30*	*23·60*	*23·90*	*24·20*	*24·50*	*24·80*
2 45.1	—	2 40.7	—	2 36.5	—	2 32.5	2 30.6	2 28.7	2 26.9	2 25.1

MEASURED MILE SPEED TABLE

Time to Steam One Mile in Mins. and Secs.

Speed in Knots

2 25·2	2 23·5	2 21·8	2 20·1	2 18·5	2 16·9	2 15·4	2 13·9	2 12·4	2 10·9	2 09·5
24·80	25·10	25·40	25·70	26·00	26·30	26·60	26·90	27·20	27·50	27·80
2 25·1	2 23·4	2 21·7	—	2 18·4	—	2 15·3	2 13·8	2 12·3	—	—
24·81	25·11	25·41	25·71	26·01	26·31	26·61	26·91	27·21	27·51	27·81
—	2 23·3	2 21·6	2 20·0	—	2 16·8	—	—	—	2 10·8	2 09·4
24·82	25·12	25·42	25·72	26·02	26·32	26·62	26·92	27·22	27·52	27·82
2 25·0	—	—	2 19·9	2 18·3	—	2 15·2	2 13·7	2 12·2	—	—
24·83	25·13	25·43	25·73	26·03	26·33	26·63	26·93	27·23	27·53	27·83
—	2 23·2	2 21·5	—	—	2 16·7	—	—	—	2 10·7	2 09·3
24·84	25·14	25·44	25·74	26·04	26·34	26·64	26·94	27·24	27·54	27·84
2 24·9	—	—	2 19·8	2 18·2	2 16·6	2 15·1	2 13·6	2 12·1	—	—
24·85	25·15	25·45	25·75	26·05	26·35	26·65	26·95	27·25	27·55	27·85
2 24·8	2 23·1	2 21·4	—	—	—	2 15·0	2 13·5	2 12·0	2 10·6	2 09·2
24·86	25·16	25·46	25·76	26·06	26·36	26·66	26·96	27·26	27·56	27·86
—	—	—	2 19·7	2 18·1	2 16·5	—	—	—	—	—
24·87	25·17	25·47	25·77	26·07	26·37	26·67	26·97	27·27	27·57	27·87
2 24·7	2 23·0	2 21·3	—	—	—	—	—	—	—	2 09·1
24·88	25·18	25·48	25·78	26·08	26·38	26·68	26·98	27·28	27·58	27·88
—	2 22·9	2 21·2	2 19·6	2 18·0	2 16·4	2 14·9	2 13·4	2 11·9	2 10·5	—
24·89	25·19	25·49	25·79	26·09	26·39	26·69	26·99	27·29	27·59	27·89
2 24·6	—	—	—	—	—	—	—	—	—	—
24·90	25·20	25·50	25·80	26·10	26·40	26·70	27·00	27·30	27·60	27·90
2 24·5	2 22·8	2 21·1	2 19·5	2 17·9	2 16·3	2 14·8	2 13·3	2 11·8	2 10·4	2 09·0
24·91	25·21	25·51	25·81	26·11	26·41	26·71	27·01	27·31	27·61	27·91
—	—	—	—	—	—	—	—	—	—	—
24·92	25·22	25·52	25·82	26·12	26·42	26·72	27·02	27·32	27·62	27·92
2 24·4	2 22·7	2 21·0	2 19·4	2 17·8	2 16·2	2 14·7	2 13·2	2 11·7	2 10·3	2 08·9
24·93	25·23	25·53	25·83	26·13	26·43	26·73	27·03	27·33	27·63	27·93
—	—	—	2 19·3	2 17·7	—	—	—	—	—	—
24·94	25·24	25·54	25·84	26·14	26·44	26·74	27·04	27·34	27·64	27·94
2 24·3	2 22·6	2 20·9	—	—	2 16·1	2 14·6	2 13·1	—	2 10·2	2 08·8
24·95	25·25	25·55	25·85	26·15	26·45	26·75	27·05	27·35	27·65	27·95
—	2 22·5	2 20·8	2 19·2	2 17·6	2 16·0	2 14·5	2 13·0	2 11·6	—	—
24·96	25·26	25·56	25·86	26·16	26·46	26·76	27·06	27·36	27·66	27·96
2 24·2	—	—	—	—	—	—	—	—	2 10·1	2 08·7
24·97	25·27	25·57	25·87	26·17	26·47	26·77	27·07	27·37	27·67	27·97
2 24·1	2 22·4	—	2 19·1	2 17·5	2 15·9	2 14·4	2 12·9	2 11·5	—	—
24·98	25·28	25·58	25·88	26·18	26·48	26·78	27·08	27·38	27·68	27·98
—	—	2 20·7	—	—	—	—	—	—	2 10·0	2 08·6
24·99	25·29	25·59	25·89	26·19	26·49	26·79	27·09	27·39	27·69	27·99
2 24·0	2 22·3	—	2 19·0	2 17·4	2 15·8	2 14·3	2 12·8	2 11·4	—	—
25·00	25·30	25·60	25·90	26·20	26·50	26·80	27·10	27·40	27·70	28·00
—	—	2 20·6	—	—	—	—	—	—	2 09·9	2 08·5
25·01	25·31	25·61	25·91	26·21	26·51	26·81	27·11	27·41	27·71	28·01
2 23·9	2 22·2	2 20·5	2 18·9	2 17·3	2 15·7	2 14·2	2 12·7	2 11·3	—	—
25·02	25·32	25·62	25·92	26·22	26·52	26·82	27·12	27·42	27·72	28·02
—	—	—	—	—	—	—	—	—	2 09·8	2 08·4
25·03	25·33	25·63	25·93	26·23	26·53	26·83	27·13	27·43	27·73	28·03
2 23·8	2 22·1	2 20·4	2 18·8	2 17·2	2 15·6	2 14·1	2 12·6	2 11·2	—	—
25·04	25·34	25·64	25·94	26·24	26·54	26·84	27·14	27·44	27·74	28·04
2 23·7	2 22·0	—	—	2 17·1	—	—	—	—	2 09·7	2 08·3
25·05	25·35	25·65	25·95	26·25	26·55	26·85	27·15	27·45	27·75	28·05
—	—	2 20·3	2 18·7	—	2 15·5	2 14·0	2 12·5	2 11·1	—	—
25·06	25·36	25·66	25·96	26·26	26·56	26·86	27·16	27·46	27·76	28·06
2 23·6	2 21·9	2 20·2	2 18·6	2 17·0	—	—	—	—	2 09·6	2 08·2
25·07	25·37	25·67	25·97	26·27	26·57	26·87	27·17	27·47	27·77	28·07
—	—	2 20·1	—	—	—	—	—	2 11·0	—	—
25·08	25·38	25·68	25·98	26·28	26·58	26·88	27·18	27·48	27·78	28·08
2 23·5	2 21·8	—	2 18·5	—	2 15·4	2 13·9	2 12·4	2 10·9	2 09·5	2 08·1
25·09	25·39	25·69	25·99	26·29	26·59	26·89	27·19	27·49	27·79	28·09
—	—	—	—	2 16·9	—	—	—	—	—	—
25·10	25·40	25·70	26·00	26·30	26·60	26·90	27·20	27·50	27·80	28·10
2 23·4	2 21·7	—	2 18·4	—	2 15·3	2 13·8	2 12·3	—	—	—

DISTANCE by VERTICAL ANGLE.

$d = h \times \cot \theta$. Where d = distance, h = height of object, and θ = vertical angle.

Height of Object in Feet and Metres. Values are given in degrees (°) and minutes (′).

Distance m.	c.	ft. 40 / m. 12·2	45 / 13·7	50 / 15·2	55 / 16·8	60 / 18·3	65 / 19·8	70 / 21·3	75 / 22·9	80 / 24·4	85 / 25·9	90 / 27·4	95 / 29·0	Distance m.	c.
0	1	3 46	4 14	4 42	5 10	5 38	6 6	6 34	7 2	7 30	7 58	8 25	8 53	0	1
0	2	1 53	2 7	2 21	2 35	2 49	3 4	3 18	3 32	3 46	4 0	4 14	4 28	0	2
0	3	1 15	1 25	1 34	1 44	1 53	2 2	2 12	2 21	2 31	2 40	2 49	2 59	0	3
0	4	0 57	1 4	1 11	1 18	1 25	1 32	1 39	1 46	1 53	2 0	2 7	2 14	0	4
0	5	0 45	0 51	0 57	1 2	1 8	1 14	1 19	1 25	1 30	1 36	1 42	1 47	0	5
0	6	0 38	0 42	0 47	0 52	0 57	1 1	1 6	1 11	1 15	1 20	1 25	1 30	0	6
0	7	0 32	0 36	0 40	0 44	0 48	0 53	0 57	1 1	1 5	1 9	1 13	1 17	0	7
0	8	0 28	0 32	0 35	0 39	0 42	0 46	0 49	0 53	0 57	1 0	1 4	1 7	0	8
0	9	0 25	0 28	0 31	0 35	0 38	0 41	0 44	0 47	0 50	0 53	0 57	1 0	0	9
1	0	0 23	0 25	0 28	0 31	0 34	0 37	0 40	0 42	0 45	0 48	0 51	0 54	1	0
1	1	0 21	0 23	0 26	0 28	0 31	0 33	0 36	0 39	0 41	0 44	0 46	0 49	1	1
1	2	0 19	0 21	0 24	0 26	0 28	0 31	0 33	0 35	0 38	0 40	0 42	0 45	1	2
1	3	0 17	0 20	0 22	0 24	0 26	0 28	0 30	0 33	0 35	0 37	0 39	0 41	1	3
1	4	0 16	0 18	0 20	0 22	0 24	0 26	0 28	0 30	0 32	0 34	0 36	0 38	1	4
1	5	0 15	0 17	0 19	0 21	0 23	0 25	0 26	0 28	0 30	0 32	0 34	0 36	1	5
1	6	0 14	0 16	0 18	0 19	0 21	0 23	0 25	0 27	0 28	0 30	0 32	0 34	1	6
1	7	0 13	0 15	0 17	0 18	0 20	0 22	0 23	0 25	0 27	0 28	0 30	0 32	1	7
1	8	0 13	0 14	0 16	0 17	0 19	0 20	0 22	0 24	0 25	0 27	0 28	0 30	1	8
1	9	0 12	0 13	0 15	0 16	0 18	0 19	0 21	0 22	0 24	0 25	0 27	0 28	1	9
2	0	0 11	0 13	0 14	0 16	0 17	0 18	0 20	0 21	0 23	0 24	0 25	0 27	2	0
2	1	0 11	0 13	0 14	0 15	0 16	0 18	0 19	0 20	0 22	0 23	0 24	0 26	2	1
2	2	0 10	0 12	0 13	0 14	0 15	0 17	0 18	0 19	0 21	0 22	0 23	0 24	2	2
2	3	0 10	0 12	0 12	0 14	0 14	0 16	0 17	0 18	0 20	0 21	0 22	0 23	2	3
2	4	0 10	0 11	0 12	0 13	0 14	0 15	0 17	0 18	0 19	0 20	0 21	0 22	2	4
2	5	0 9	0 11	0 11	0 12	0 13	0 15	0 16	0 17	0 18	0 19	0 20	0 21	2	5
2	6	0 9	0 10	0 11	0 12	0 13	0 14	0 15	0 16	0 17	0 18	0 20	0 21	2	6
2	7	0 9	0 10	0 10	0 11	0 12	0 14	0 15	0 16	0 17	0 18	0 19	0 20	2	7
2	8	0 8	0 9	0 10	0 11	0 12	0 13	0 14	0 15	0 16	0 17	0 18	0 19	2	8
2	9	0 8	0 9	0 10	0 10	0 11	0 13	0 14	0 15	0 16	0 17	0 18	0 19	2	9
3	0	0 8	0 9	0 9	0 10	0 10	0 12	0 13	0 14	0 15	0 16	0 17	0 18	3	0
3	2							0 12	0 13	0 14	0 15	0 16	0 17	3	2
3	4							0 12	0 12	0 13	0 14	0 15	0 16	3	4
3	6							0 11	0 12	0 13	0 13	0 14	0 15	3	6
3	8							0 10	0 11	0 12	0 13	0 13	0 14	3	8
4	0							0 10	0 11	0 11	0 12	0 13	0 13	4	0
4	2										0 11	0 12	0 13	4	2
4	4										0 11	0 12	0 12	4	4
4	6										0 10	0 11	0 12	4	6
4	8										0 10	0 11	0 11	4	8
5	0										0 10	0 10	0 11	5	0

$\text{Tan } \theta = \dfrac{h}{d}$. Where θ = vertical angle, h = height of object, and d = distance.

DISTANCE by VERTICAL ANGLE.

$d = h \times \cot. \theta.$ Where d = distance, h = height of object, and θ = vertical angle.

Height of Object in Feet and Metres.

Distance (m.)	(c.)	ft. 100 / m. 30·5	105 / 32·0	110 / 33·5	115 / 35·1	120 / 36·6	125 / 38·1	130 / 39·6	135 / 41·2	140 / 42·7	145 / 44·2	150 / 45·7	155 / 47·2	(m.)	(c.)
0	1	9 20	9 48	10 15	10 43	11 10	11 37	12 4	12 31	12 58	13 25	13 52	14 18	0	1
0	2	4 42	4 56	5 10	5 24	5 38	5 52	6 6	6 20	6 34	6 48	7 2	7 16	0	2
0	3	3 8	3 18	3 27	3 36	3 46	3 55	4 5	4 14	4 23	4 33	4 42	4 51	0	3
0	4	2 21	2 28	2 35	2 42	2 49	2 57	3 4	3 11	3 18	3 25	3 32	3 39	0	4
0	5	1 53	1 59	2 4	2 10	2 16	2 21	2 27	2 33	2 38	2 44	2 49	2 55	0	5
0	6	1 34	1 39	1 44	1 48	1 53	1 58	2 2	2 7	2 12	2 17	2 21	2 26	0	6
0	7	1 21	1 25	1 29	1 33	1 37	1 41	1 45	1 49	1 53	1 57	2 1	2 5	0	7
0	8	1 11	1 14	1 18	1 21	1 25	1 28	1 32	1 35	1 39	1 42	1 46	1 50	0	8
0	9	1 3	1 6	1 9	1 12	1 15	1 19	1 22	1 25	1 28	1 31	1 34	1 37	0	9
1	0	0 57	0 59	1 2	1 5	1 8	1 11	1 14	1 16	1 19	1 22	1 25	1 26	1	0
1	1	0 51	0 54	0 57	0 59	1 2	1 4	1 7	1 9	1 12	1 15	1 17	1 20	1	1
1	2	0 47	0 49	0 52	0 54	0 57	0 59	1 1	1 4	1 6	1 8	1 11	1 13	1	2
1	3	0 44	0 46	0 48	0 50	0 52	0 54	0 57	0 59	1 1	1 3	1 5	1 7	1	3
1	4	0 40	0 42	0 44	0 46	0 48	0 51	0 53	0 55	0 57	0 59	1 1	1 3	1	4
1	5	0 38	0 40	0 41	0 43	0 45	0 47	0 49	0 51	0 53	0 55	0 57	0 58	1	5
1	6	0 35	0 37	0 39	0 41	0 42	0 44	0 46	0 48	0 49	0 51	0 53	0 55	1	6
1	7	0 33	0 35	0 37	0 38	0 40	0 42	0 43	0 45	0 47	0 48	0 50	0 52	1	7
1	8	0 31	0 33	0 35	0 36	0 38	0 39	0 41	0 42	0 44	0 46	0 47	0 49	1	8
1	9	0 30	0 31	0 33	0 34	0 36	0 37	0 39	0 40	0 42	0 43	0 45	0 46	1	9
2	0	0 28	0 30	0 31	0 33	0 34	0 35	0 37	0 38	0 40	0 41	0 42	0 44	2	0
2	1	0 27	0 28	0 30	0 31	0 32	0 34	0 35	0 36	0 38	0 39	0 40	0 42	2	1
2	2	0 26	0 27	0 28	0 30	0 31	0 32	0 33	0 35	0 36	0 37	0 39	0 40	2	2
2	3	0 25	0 26	0 27	0 28	0 30	0 31	0 32	0 33	0 34	0 36	0 37	0 38	2	3
2	4	0 24	0 25	0 26	0 27	0 28	0 29	0 31	0 32	0 33	0 34	0 35	0 37	2	4
2	5	0 23	0 24	0 25	0 26	0 27	0 28	0 29	0 31	0 32	0 33	0 34	0 35	2	5
2	6	0 22	0 23	0 24	0 25	0 26	0 27	0 28	0 29	0 30	0 32	0 33	0 34	2	6
2	7	0 21	0 22	0 23	0 24	0 25	0 26	0 27	0 28	0 29	0 30	0 31	0 32	2	7
2	8	0 20	0 21	0 22	0 23	0 24	0 25	0 26	0 25	0 28	0 29	0 30	0 31	2	8
2	9	0 20	0 20	0 21	0 22	0 23	0 24	0 25	0 26	0 27	0 28	0 29	0 30	2	9
3	0	0 19	0 20	0 21	0 22	0 23	0 24	0 25	0 25	0 26	0 27	0 28	0 29	3	0
3	2	0 18	0 19	0 19	0 20	0 21	0 22	0 23	0 24	0 25	0 26	0 27	0 27	3	2
3	4	0 17	0 18	0 18	0 19	0 20	0 21	0 22	0 22	0 23	0 24	0 25	0 26	3	4
3	6	0 16	0 16	0 17	0 18	0 19	0 20	0 20	0 21	0 22	0 23	0 24	0 24	3	6
3	8	0 15	0 16	0 16	0 17	0 18	0 19	0 19	0 20	0 21	0 22	0 22	0 23	3	8
4	0	0 14	0 15	0 16	0 16	0 17	0 18	0 18	0 19	0 20	0 21	0 21	0 22	4	0
4	2	0 14	0 15	0 15	0 16	0 16	0 17	0 17	0 18	0 19	0 20	0 20	0 21	4	2
4	4	0 13	0 14	0 14	0 15	0 15	0 16	0 17	0 17	0 18	0 19	0 19	0 20	4	4
4	6	0 13	0 14	0 14	0 15	0 15	0 16	0 16	0 16	0 17	0 18	0 18	0 19	4	6
4	8	0 12	0 13	0 13	0 14	0 14	0 15	0 15	0 15	0 16	0 17	0 18	0 18	4	8
5	0	0 11	0 12	0 12	0 13	0 14	0 14	0 15	0 15	0 16	0 16	0 17	0 17	5	0

$\text{Tan. } \theta = \dfrac{h}{d}.$ Where θ = vertical angle, h = height of object, and d = distance.

DISTANCE by VERTICAL ANGLE.

$d = h \times \cot \theta$. Where d = distance, h = height of object, and θ = vertical angle.

Distance in Miles and Cables	Height of Object in Feet and Metres.												Distance in Miles and Cables
m. c.	ft. 160 / m. 48·8	165 / 50·3	170 / 51·8	175 / 53·3	183 / 54·9	185 / 56·4	190 / 57·9	195 / 59·4	200 / 61·0	210 / 64·0	220 / 67·1	230 / 70·1	m. c.
0 1	14 45	15 11	15 37	16 3	16 29	16 55	17 21	17 47	18 13	19 3	19 54	20 43	0 1
0 2	7 30	7 44	7 58	8 11	8 25	8 39	8 53	9 7	9 20	9 48	10 15	10 43	0 2
0 3	5 1	5 10	5 19	5 29	5 38	5 47	5 57	6 6	6 15	6 34	6 53	7 11	0 3
0 4	3 46	3 53	4 0	4 7	4 14	4 21	4 28	4 35	4 42	4 56	5 10	5 24	0 4
0 5	3 1	3 6	3 12	3 18	3 23	3 29	3 35	3 40	3 46	3 57	4 8	4 20	0 5
0 6	2 31	2 35	2 40	2 45	2 49	2 54	2 59	3 4	3 8	3 18	3 27	3 36	0 6
0 7	2 9	2 13	2 17	2 21	2 25	2 29	2 33	2 37	2 41	2 49	2 58	3 6	0 7
0 8	1 53	1 57	2 0	2 4	2 7	2 11	2 14	2 18	2 21	2 28	2 35	2 42	0 8
0 9	1 40	1 44	1 47	1 50	1 53	1 56	1 59	2 2	2 6	2 12	2 18	2 24	0 9
1 0	1 30	1 33	1 36	1 39	1 42	1 45	1 47	1 50	1 53	1 59	2 4	2 10	1 0
1 1	1 22	1 25	1 27	1 30	1 33	1 35	1 38	1 40	1 43	1 48	1 53	1 58	1 1
1 2	1 15	1 18	1 20	1 22	1 25	1 27	1 30	1 32	1 34	1 39	1 44	1 48	1 2
1 3	1 10	1 12	1 14	1 16	1 18	1 20	1 23	1 25	1 27	1 31	1 36	1 40	1 3
1 4	1 5	1 7	1 9	1 11	1 13	1 15	1 17	1 19	1 21	1 25	1 29	1 33	1 4
1 5	1 0	1 2	1 4	1 6	1 8	1 10	1 12	1 14	1 15	1 19	1 23	1 27	1 5
1 6	0 57	0 58	1 0	1 2	1 4	1 5	1 7	1 9	1 11	1 14	1 18	1 21	1 6
1 7	0 53	0 55	0 57	0 58	1 0	1 2	1 3	1 5	1 7	1 10	1 13	1 16	1 7
1 8	0 50	0 52	0 53	0 55	0 57	0 58	1 0	1 1	1 3	1 6	1 9	1 12	1 8
1 9	0 48	0 49	0 51	0 52	0 54	0 55	0 57	0 58	1 0	1 2	1 5	1 8	1 9
2 0	0 45	0 47	0 48	0 49	0 51	0 52	0 54	0 55	0 57	0 59	1 2	1 5	2 0
2 1	0 43	0 44	0 46	0 47	0 48	0 50	0 51	0 53	0 54	0 57	0 59	1 2	2 1
2 2	0 41	0 42	0 44	0 45	0 46	0 48	0 49	0 50	0 51	0 54	0 57	0 59	2 2
2 3	0 39	0 41	0 42	0 43	0 44	0 45	0 47	0 48	0 49	0 52	0 54	0 57	2 3
2 4	0 38	0 39	0 40	0 41	0 42	0 44	0 45	0 46	0 47	0 49	0 52	0 54	2 4
2 5	0 36	0 37	0 38	0 40	0 41	0 42	0 43	0 44	0 45	0 48	0 50	0 52	2 5
2 6	0 35	0 36	0 37	0 38	0 39	0 40	0 41	0 42	0 44	0 46	0 48	0 50	2 6
2 7	0 34	0 35	0 36	0 37	0 38	0 39	0 40	0 41	0 42	0 44	0 46	0 48	2 7
2 8	0 32	0 33	0 34	0 35	0 36	0 37	0 38	0 39	0 40	0 42	0 44	0 46	2 8
2 9	0 31	0 32	0 33	0 34	0 35	0 36	0 37	0 38	0 39	0 41	0 43	0 45	2 9
3 0	0 30	0 31	0 32	0 33	0 34	0 35	0 36	0 37	0 38	0 40	0 41	0 43	3 0
3 2	0 28	0 29	0 30	0 31	0 32	0 33	0 34	0 34	0 35	0 37	0 39	0 41	3 2
3 4	0 27	0 27	0 28	0 29	0 30	0 31	0 32	0 32	0 33	0 35	0 37	0 38	3 4
3 6	0 25	0 26	0 27	0 27	0 28	0 29	0 30	0 31	0 31	0 33	0 35	0 36	3 6
3 8	0 24	0 25	0 25	0 26	0 27	0 28	0 28	0 29	0 30	0 31	0 33	0 34	3 8
4 0	0 23	0 23	0 24	0 25	0 25	0 26	0 27	0 28	0 28	0 30	0 31	0 33	4 0
4 2	0 22	0 22	0 23	0 24	0 24	0 25	0 26	0 27	0 27	0 28	0 30	0 31	4 2
4 4	0 21	0 21	0 22	0 23	0 23	0 24	0 24	0 26	0 26	0 27	0 28	0 30	4 4
4 6	0 20	0 20	0 21	0 21	0 22	0 23	0 23	0 24	0 25	0 26	0 27	0 28	4 6
4 8	0 19	0 19	0 20	0 20	0 21	0 21	0 22	0 23	0 24	0 25	0 26	0 27	4 8
5 0	0 18	0 18	0 19	0 19	0 20	0 20	0 21	0 22	0 23	0 24	0 25	0 26	5 0

$\tan \theta = \dfrac{h}{d}$. Where θ = vertical angle, h = height of object, and d = distance.

DISTANCE by VERTICAL ANGLE.

$d = h \times \cot \theta$. Where d = distance, h = height of object, and θ = vertical angle.

Distance in Miles and Cables.		Height of Object in Feet and Metres												Distance in Miles and Cables.	
		ft. 240 / m. 73.2	250 / 76.2	260 / 79.2	270 / 82.3	280 / 85.3	290 / 88.4	300 / 91.4	310 / 94.5	320 / 97.5	330 / 100.6	340 / 103.6	350 / 106.7		
m.	c.	° ′	° ′	° ′	° ′	° ′	° ′	° ′	° ′	° ′	° ′	° ′	° ′	m.	c.
0	1	21 32	22 21	23 9	23 57	24 44	25 30	26 16	27 1	27 46	28 29	29 13	29 56	0	1
0	2	11 10	11 37	12 4	12 31	12 58	13 25	13 52	14 18	14 45	15 11	15 37	16 3	0	2
0	3	7 30	7 48	8 7	8 25	8 44	9 2	9 20	9 39	9 57	10 15	10 34	10 52	0	3
0	4	5 38	5 52	6 6	6 20	6 34	6 48	7 2	7 16	7 30	7 44	7 58	8 11	0	4
0	5	4 31	4 42	4 53	5 5	5 16	5 27	5 38	5 49	6 1	6 12	6 23	6 34	0	5
0	6	3 46	3 55	4 5	4 14	4 23	4 33	4 42	4 51	5 1	5 10	5 19	5 29	0	6
0	7	3 14	3 22	3 30	3 38	3 46	3 54	4 2	4 10	4 18	4 26	4 34	4 42	0	7
0	8	2 49	2 57	3 4	3 11	3 18	3 25	3 32	3 39	3 46	3 53	4 0	4 7	0	8
0	9	2 31	2 37	2 43	2 49	2 56	3 2	3 8	3 15	3 21	3 27	3 33	3 40	0	9
1	0	2 16	2 21	2 27	2 33	2 38	2 44	2 49	2 55	3 1	3 6	3 12	3 18	1	0
1	1	2 3	2 8	2 14	2 19	2 24	2 29	2 34	2 39	2 44	2 49	2 55	3 0	1	1
1	2	1 53	1 58	2 2	2 7	2 12	2 17	2 21	2 26	2 31	2 35	2 40	2 45	1	2
1	3	1 44	1 49	1 53	1 57	2 2	2 6	2 10	2 15	2 19	2 23	2 28	2 32	1	3
1	4	1 37	1 41	1 45	1 49	1 53	1 57	2 1	2 5	2 9	2 13	2 17	2 21	1	4
1	5	1 30	1 34	1 38	1 42	1 46	1 49	1 53	1 57	2 1	2 4	2 8	2 12	1	5
1	6	1 25	1 28	1 32	1 35	1 39	1 42	1 46	1 50	1 53	1 57	2 0	2 4	1	6
1	7	1 20	1 23	1 26	1 30	1 33	1 36	1 40	1 43	1 46	1 50	1 53	1 56	1	7
1	8	1 15	1 19	1 22	1 25	1 28	1 31	1 34	1 37	1 40	1 44	1 47	1 50	1	8
1	9	1 11	1 14	1 17	1 20	1 23	1 26	1 29	1 32	1 35	1 38	1 41	1 44	1	9
2	0	1 8	1 11	1 14	1 16	1 19	1 22	1 25	1 28	1 30	1 33	1 36	1 39	2	0
2	1	1 5	1 7	1 10	1 13	1 15	1 18	1 21	1 23	1 26	1 29	1 32	1 34	2	1
2	2	1 2	1 4	1 7	1 9	1 12	1 15	1 17	1 20	1 22	1 25	1 27	1 30	2	2
2	3	0 59	1 1	1 4	1 6	1 9	1 11	1 14	1 16	1 19	1 21	1 24	1 26	2	3
2	4	0 57	0 59	1 1	1 4	1 6	1 8	1 11	1 13	1 15	1 18	1 20	1 22	2	4
2	5	0 54	0 57	0 59	1 1	1 3	1 6	1 8	1 10	1 12	1 15	1 17	1 19	2	5
2	6	0 52	0 54	0 57	0 59	1 1	1 3	1 5	1 7	1 10	1 12	1 14	1 16	2	6
2	7	0 50	0 52	0 54	0 57	0 59	1 1	1 3	1 5	1 7	1 9	1 11	1 13	2	7
2	8	0 48	0 50	0 53	0 55	0 57	0 59	1 1	1 3	1 5	1 7	1 9	1 11	2	8
2	9	0 47	0 49	0 51	0 53	0 55	0 57	0 58	1 0	1 2	1 4	1 6	1 8	2	9
3	0	0 45	0 47	0 49	0 51	0 53	0 55	0 57	0 58	1 0	1 2	1 4	1 6	3	0
3	2	0 42	0 44	0 46	0 48	0 49	0 51	0 53	0 55	0 57	0 58	1 0	1 2	3	2
3	4	0 40	0 42	0 43	0 45	0 47	0 48	0 50	0 52	0 53	0 55	0 57	0 58	3	4
3	6	0 38	0 39	0 41	0 42	0 44	0 46	0 47	0 49	0 50	0 52	0 53	0 55	3	6
3	8	0 36	0 37	0 39	0 40	0 42	0 43	0 45	0 46	0 48	0 49	0 51	0 52	3	8
4	0	0 34	0 35	0 37	0 38	0 40	0 41	0 42	0 44	0 45	0 47	0 48	0 49	4	0
4	2	0 32	0 34	0 35	0 36	0 38	0 39	0 40	0 42	0 43	0 44	0 46	0 47	4	2
4	4	0 31	0 32	0 33	0 35	0 36	0 37	0 39	0 40	0 41	0 42	0 44	0 45	4	4
4	6	0 30	0 31	0 32	0 33	0 34	0 36	0 37	0 38	0 39	0 41	0 42	0 43	4	6
4	8	0 28	0 30	0 31	0 32	0 33	0 34	0 35	0 37	0 38	0 39	0 40	0 41	4	8
5	0	0 27	0 28	0 29	0 31	0 32	0 33	0 34	0 35	0 36	0 37	0 38	0 40	5	0
5	2	0 26	0 27	0 28	0 29	0 30	0 32	0 33	0 34	0 35	0 36	0 37	0 38	5	2
5	4	0 25	0 26	0 27	0 28	0 29	0 30	0 31	0 32	0 34	0 34	0 36	0 37	5	4
5	6	0 24	0 25	0 26	0 27	0 28	0 29	0 30	0 31	0 32	0 33	0 34	0 35	5	6
5	8	0 23	0 24	0 25	0 26	0 27	0 28	0 29	0 30	0 31	0 32	0 33	0 34	5	8
6	0	0 23	0 24	0 25	0 25	0 26	0 27	0 28	0 29	0 30	0 31	0 32	0 33	6	0
6	2	0 22	0 23	0 24	0 25	0 26	0 26	0 27	0 28	0 29	0 30	0 31	0 32	6	2
6	4	0 21	0 22	0 23	0 24	0 25	0 26	0 27	0 27	0 28	0 29	0 30	0 31	6	4
6	6	0 21	0 21	0 22	0 23	0 24	0 25	0 26	0 27	0 27	0 28	0 29	0 30	6	6
6	8	0 20	0 21	0 22	0 22	0 23	0 24	0 25	0 26	0 27	0 27	0 28	0 29	6	8
7	0	0 19	0 20	0 21	0 22	0 23	0 23	0 24	0 25	0 26	0 27	0 27	0 28	7	0
7	2	0 19	0 20	0 20	0 21	0 22	0 23	0 23	0 24	0 25	0 26	0 27	0 27	7	2
7	4	0 18	0 19	0 20	0 21	0 21	0 22	0 23	0 24	0 24	0 25	0 26	0 27	7	4

$\tan \theta = \dfrac{h}{d}$. Where θ = vertical angle, h = height of object, and d = distance.

DISTANCE by VERTICAL ANGLE.

$d = h \times \cot \theta.$ Where d = distance, h = height of object, and θ = vertical angle.

Height of Object in Feet and Metres.

Distance in Miles and Cables (m.)	(c.)	ft. 360 / m. 109·7	370 / 112·8	380 / 115·8	390 / 118·9	400 / 121·9	420 / 128·0	450 / 137·2	480 / 146·3	500 / 152·4	550 / 167·6	600 / 182·9	650 / 198·1	Distance in Miles and Cables (m.)	(c.)
0	1	30 38	31 19	32 0	32 41	33 20	34 38	36 30	38 17	39 26	42 8	44 37		0	1
0	2	16 29	16 55	17 21	17 47	18 13	19 3	20 18	21 32	22 21	24 20	26 16	28 8	0	2
0	3	11 10	11 28	11 46	12 4	12 22	12 58	13 52	14 45	15 20	16 47	18 13	19 37	0	3
0	4	8 25	8 39	8 53	9 7	9 20	9 48	10 29	11 10	11 37	12 45	13 52	14 58	0	4
0	5	6 45	6 56	7 8	7 19	7 30	7 52	8 25	8 58	9 20	10 15	11 10	12 4	0	5
0	6	5 38	5 47	5 57	6 6	6 15	6 34	7 2	7 30	7 48	8 34	9 20	10 6	0	6
0	7	4 50	4 58	5 6	5 14	5 22	5 38	6 2	6 26	6 42	7 22	8 1	8 41	0	7
0	8	4 14	4 21	4 28	4 35	4 42	4 56	5 17	5 38	5 52	6 27	7 2	7 37	0	8
0	9	3 46	3 52	3 58	4 5	4 11	4 23	4 42	5 1	5 13	5 44	6 15	6 46	0	9
1	0	3 23	3 29	3 35	3 40	3 46	3 57	4 14	4 31	4 42	5 10	5 38	6 6	1	0
1	1	3 5	3 10	3 15	3 20	3 25	3 36	3 51	4 6	4 17	4 42	5 8	5 33	1	1
1	2	2 49	2 54	2 59	3 4	3 8	3 18	3 32	3 46	3 55	4 19	4 42	5 5	1	2
1	3	2 36	2 41	2 45	2 49	2 54	3 3	3 16	3 29	3 37	3 59	4 20	4 42	1	3
1	4	2 25	2 29	2 33	2 37	2 41	2 49	3 2	3 14	3 22	3 42	4 2	4 22	1	4
1	5	2 16	2 19	2 23	2 27	2 31	2 38	2 49	3 1	3 8	3 27	3 46	4 5	1	5
1	6	2 7	2 11	2 14	2 18	2 21	2 28	2 39	2 49	2 57	3 14	3 32	3 49	1	6
1	7	2 0	2 3	2 6	2 10	2 13	2 20	2 30	2 40	2 46	3 3	3 19	3 36	1	7
1	8	1 53	1 56	1 59	2 2	2 6	2 12	2 21	2 31	2 37	2 53	3 8	3 24	1	8
1	9	1 47	1 50	1 53	1 56	1 59	2 5	2 14	2 23	2 29	2 44	2 58	3 13	1	9
2	0	1 42	1 45	1 47	1 50	1 53	1 59	2 7	2 16	2 21	2 35	2 49	3 4	2	0
2	1	1 37	1 40	1 42	1 45	1 48	1 53	2 1	2 9	2 15	2 28	2 41	2 55	2	1
2	2	1 33	1 35	1 38	1 40	1 43	1 48	1 56	2 3	2 8	2 21	2 34	2 47	2	2
2	3	1 29	1 31	1 33	1 36	1 38	1 43	1 51	1 58	2 3	2 15	2 27	2 40	2	3
2	4	1 25	1 27	1 30	1 32	1 34	1 39	1 46	1 53	1 58	2 10	2 21	2 33	2	4
2	5	1 21	1 24	1 26	1 28	1 30	1 35	1 42	1 49	1 53	2 4	2 16	2 27	2	5
2	6	1 18	1 20	1 23	1 25	1 27	1 31	1 38	1 44	1 49	2 0	2 10	2 21	2	6
2	7	1 15	1 17	1 20	1 22	1 24	1 28	1 34	1 40	1 45	1 55	2 6	2 16	2	7
2	8	1 13	1 15	1 17	1 19	1 21	1 25	1 31	1 37	1 41	1 51	2 1	2 11	2	8
2	9	1 10	1 12	1 14	1 16	1 18	1 22	1 28	1 34	1 37	1 47	1 57	2 7	2	9
3	0	1 8	1 10	1 12	1 14	1 15	1 19	1 25	1 30	1 34	1 44	1 53	2 2	3	0
3	2	1 4	1 5	1 7	1 9	1 11	1 14	1 20	1 25	1 28	1 37	1 46	1 55	3	2
3	4	1 0	1 2	1 3	1 5	1 7	1 10	1 15	1 20	1 23	1 31	1 40	1 48	3	4
3	6	0 57	0 58	1 0	1 1	1 3	1 6	1 11	1 15	1 19	1 26	1 34	1 42	3	6
3	8	0 54	0 55	0 57	0 58	1 0	1 2	1 7	1 11	1 14	1 22	1 29	1 37	3	8
4	0	0 51	0 52	0 54	0 55	0 57	0 59	1 4	1 8	1 11	1 18	1 25	1 32	4	0
4	2	0 48	0 50	0 51	0 53	0 54	0 57	1 1	1 5	1 7	1 14	1 21	1 28	4	2
4	4	0 46	0 48	0 49	0 50	0 51	0 54	0 58	1 2	1 4	1 11	1 17	1 24	4	4
4	6	0 44	0 45	0 47	0 48	0 49	0 52	0 55	0 59	1 1	1 8	1 14	1 20	4	6
4	8	0 42	0 44	0 45	0 46	0 47	0 49	0 53	0 57	0 59	1 5	1 11	1 17	4	8
5	0	0 41	0 42	0 43	0 44	0 45	0 48	0 51	0 54	0 57	1 2	1 8	1 14	5	0
5	2	0 39	0 40	0 41	0 42	0 43	0 46	0 49	0 52	0 54	1 0	1 5	1 11	5	2
5	4	0 38	0 39	0 40	0 41	0 42	0 44	0 47	0 50	0 52	0 58	1 3	1 8	5	4
5	6	0 36	0 37	0 38	0 39	0 40	0 42	0 45	0 48	0 50	0 56	1 1	1 6	5	6
5	8	0 35	0 36	0 37	0 38	0 39	0 41	0 44	0 47	0 49	0 54	0 58	1 3	5	8
6	0	0 34	0 35	0 36	0 37	0 38	0 40	0 42	0 45	0 47	0 52	0 57	1 1	6	0
6	2	0 33	0 34	0 35	0 36	0 36	0 38	0 41	0 44	0 46	0 50	0 55	0 59	6	2
6	4	0 32	0 33	0 34	0 34	0 35	0 37	0 40	0 42	0 44	0 49	0 53	0 57	6	4
6	6	0 31	0 32	0 33	0 33	0 34	0 36	0 38	0 41	0 43	0 47	0 51	0 56	6	6
6	8	0 30	0 31	0 32	0 32	0 33	0 35	0 37	0 40	0 42	0 46	0 50	0 54	6	8
7	0	0 29	0 30	0 31	0 32	0 32	0 34	0 36	0 39	0 40	0 44	0 48	0 53	7	0
7	2	0 28	0 29	0 30	0 31	0 31	0 33	0 35	0 38	0 39	0 43	0 47	0 51	7	2
7	4	0 27	0 28	0 29	0 30	0 31	0 32	0 34	0 37	0 38	0 42	0 46	0 50	7	4

$\tan \theta = \dfrac{h}{d}.$ Where θ = vertical angle, h = height of object, and d = distance.

DISTANCE by VERTICAL ANGLE.

$d = h \times \cot \theta$. Where d = distance, h = height of object, and θ = vertical angle.

Distance in Miles and Cables.		Height of Object in Feet and Metres.												Distance in Miles and Cables.	
		ft. 700 / m. 213·4	750 / 228·6	800 / 243·8	850 / 259·1	900 / 274·3	950 / 289·6	1,000 / 304·8	1,200 / 365·8	1,400 / 426·7	1,600 / 487·7	1,800 / 548·6	2,000 / 609·6		
m.	c.	° ′	° ′	° ′	° ′	° ′	° ′	° ′	° ′	° ′	° ′	° ′	° ′	m.	c.
0	1													0	1
0	2	29 56	31 40	33 20	34 57	36 30	38 0	39 26						0	2
0	3	21 0	22 21	23 41	24 59	26 16	27 31	28 44						0	3
0	4	16 3	17 8	18 13	19 16	20 18	21 20	22 21	26 16	29 56				0	4
0	5	12 58	13 52	14 45	15 37	16 30	17 21	18 13	21 32	24 44	27 46	30 38		0	5
0	6	10 52	11 37	12 22	13 7	13 52	14 36	15 20	18 13	21 0	23 41	26 16	28 44	0	6
0	7	9 20	10 0	10 39	11 18	11 56	12 35	13 13	15 45	18 13	20 36	22 56	25 10	0	7
0	8	8 11	8 46	9 20	9 55	10 29	11 3	11 37	13 52	16 3	18 13	20 18	22 21	0	8
0	9	7 17	7 48	8 19	8 50	9 20	9 51	10 21	12 22	14 21	16 18	18 13	20 5	0	9
1	0	6 34	7 2	7 30	7 58	8 25	8 53	9 20	11 10	12 58	14 45	16 29	18 13	1	0
1	1	5 59	6 24	6 49	7 15	7 40	8 5	8 30	10 10	11 49	13 27	15 4	16 39	1	1
1	2	5 29	5 52	6 15	6 39	7 2	7 25	7 48	9 20	10 52	12 22	13 52	15 20	1	2
1	3	5 4	5 25	5 47	6 8	6 30	6 51	7 13	8 38	10 3	11 27	12 50	14 12	1	3
1	4	4 42	5 2	5 22	5 42	6 2	6 22	6 42	8 1	9 20	10 39	11 56	13 13	1	4
1	5	4 23	4 42	5 1	5 19	5 38	5 57	6 15	7 30	8 44	9 57	11 10	12 22	1	5
1	6	4 7	4 25	4 42	5 0	5 17	5 35	5 52	7 2	8 11	9 20	10 29	11 37	1	6
1	7	3 52	4 9	4 26	4 42	4 59	5 15	5 32	6 38	7 43	8 48	9 53	10 57	1	7
1	8	3 40	3 55	4 11	4 26	4 42	4 58	5 13	6 15	7 17	8 19	9 20	10 21	1	8
1	9	3 28	3 43	3 58	4 13	4 27	4 42	4 57	5 56	6 55	7 54	8 52	9 50	1	9
2	0	3 18	3 32	3 46	4 0	4 14	4 28	4 42	5 38	6 34	7 30	8 25	9 20	2	0
2	1	3 8	3 22	3 35	3 49	4 2	4 15	4 29	5 22	6 15	7 8	8 1	8 53	2	1
2	2	3 0	3 13	3 25	3 38	3 51	4 4	4 17	5 8	5 59	6 49	7 40	8 30	2	2
2	3	2 52	3 4	3 16	3 29	3 41	3 53	4 5	4 55	5 44	6 32	7 21	8 9	2	3
2	4	2 45	2 57	3 8	3 20	3 32	3 44	3 55	4 42	5 29	6 15	7 2	7 48	2	4
2	5	2 38	2 49	3 1	3 12	3 23	3 35	3 46	4 31	5 16	6 1	6 45	7 30	2	5
2	6	2 32	2 43	2 54	3 5	3 16	3 26	3 37	4 20	5 4	5 47	6 30	7 13	2	6
2	7	2 27	2 37	2 47	2 58	3 8	3 19	3 29	4 11	4 53	5 34	6 15	6 57	2	7
2	8	2 21	2 31	2 41	2 52	3 2	3 12	3 22	4 2	4 42	5 22	6 2	6 42	2	8
2	9	2 16	2 26	2 36	2 46	2 55	3 5	3 15	3 54	4 32	5 11	5 50	6 28	2	9
3	0	2 12	2 21	2 31	2 40	2 49	2 59	3 8	3 46	4 23	5 1	5 38	6 15	3	0
3	2	2 4	2 12	2 21	2 30	2 39	2 48	2 57	3 32	4 7	4 42	5 17	5 52	3	2
3	4	1 56	2 5	2 13	2 21	2 30	2 38	2 46	3 19	3 52	4 26	4 59	5 32	3	4
3	6	1 50	1 58	2 6	2 13	2 21	2 29	2 37	3 8	3 40	4 11	4 42	5 13	3	6
3	8	1 44	1 52	1 59	2 6	2 14	2 21	2 29	2 58	3 28	3 58	4 27	4 57	3	8
4	0	1 39	1 46	1 53	2 0	2 7	2 14	2 21	2 49	3 18	3 46	4 14	4 42	4	0
4	2	1 34	1 41	1 48	1 54	2 1	2 8	2 15	2 41	3 8	3 35	4 2	4 29	4	2
4	4	1 30	1 36	1 43	1 49	1 56	2 2	2 8	2 34	3 0	3 25	3 51	4 17	4	4
4	6	1 26	1 32	1 38	1 44	1 51	1 57	2 3	2 27	2 52	3 16	3 41	4 5	4	6
4	8	1 22	1 28	1 34	1 40	1 46	1 52	1 58	2 21	2 45	3 8	3 32	3 55	4	8
5	0	1 19	1 25	1 30	1 36	1 42	1 47	1 53	2 16	2 38	3 1	3 23	3 46	5	0
5	2	1 16	1 22	1 27	1 33	1 38	1 43	1 49	2 11	2 32	2 54	3 16	3 38	5	2
5	4	1 13	1 19	1 24	1 29	1 34	1 39	1 45	2 6	2 27	2 48	3 9	3 30	5	4
5	6	1 11	1 16	1 21	1 26	1 31	1 36	1 41	2 1	2 21	2 42	3 2	3 22	5	6
5	8	1 8	1 13	1 18	1 23	1 28	1 33	1 37	1 57	2 17	2 36	2 56	3 15	5	8
6	0	1 6	1 11	1 15	1 20	1 25	1 30	1 34	1 53	2 12	2 31	2 50	3 9	6	0
6	2	1 4	1 8	1 13	1 18	1 22	1 27	1 31	1 49	2 8	2 26	2 44	3 2	6	2
6	4	1 2	1 6	1 11	1 15	1 20	1 24	1 28	1 46	2 4	2 21	2 39	2 57	6	4
6	6	1 0	1 4	1 9	1 12	1 17	1 21	1 26	1 43	2 0	2 17	2 34	2 51	6	6
6	8	0 58	1 2	1 7	1 11	1 15	1 19	1 23	1 40	1 56	2 13	2 30	2 46	6	8
7	0	0 57	1 1	1 5	1 9	1 13	1 17	1 21	1 37	1 53	2 9	2 25	2 42	7	0
7	2	0 55	0 59	1 3	1 7	1 11	1 15	1 19	1 34	1 50	2 6	2 21	2 37	7	2
7	4	0 53	0 57	1 1	1 5	1 9	1 13	1 16	1 32	1 47	2 2	2 18	2 33	7	4

$\text{Tan. } \theta = \dfrac{h}{d}$. Where θ = vertical angle, h = height of object, and d = distance.

EXTREME RANGE TABLE.

Heights in feet and metres.   Range in miles.

Height of object		ft. 5 m. 1·5	10 3·0	15 4·6	20 6·1	25 7·6	30 9·1	35 10·7	40 12·2	45 13·7	50 15·2
m.	ft.										
10	32·8	9·32	10·40	11·23	11·93	12·55	13·11	13·65	14·10	14·55	14·98
12	39·4	9·96	11·04	11·87	12·57	13·19	13·75	14·29	14·74	15·19	15·61
14	45·9	10·55	11·63	12·46	13·16	13·78	14·34	14·88	15·33	15·78	16·20
16	52·5	11·09	12·18	13·01	13·71	14·33	14·89	15·43	15·88	16·33	16·75
18	59·1	11·61	12·69	13·52	14·22	14·84	15·40	15·94	16·39	16·84	17·26
20	65·6	12·09	13·18	14·01	14·71	15·33	15·89	16·43	16·88	17·33	17·75
22	72·2	12·56	13·64	14·47	15·17	15·79	16·35	16·89	17·34	17·79	18·21
24	78·7	13·00	14·08	14·91	15·61	16·23	16·79	17·33	17·78	18·23	18·65
26	85·3	13·43	14·51	15·34	16·04	16·66	17·22	17·76	18·21	18·66	19·08
28	91·9	13·83	14·91	15·74	16·44	17·06	17·62	18·16	18·61	19·06	19·48
30	98·4	14·25	15·33	16·16	16·86	17·48	18·04	18·58	19·03	19·48	19·90
32	105·0	14·61	15·69	16·52	17·22	17·84	18·40	18·94	19·39	19·84	20·26
34	111·6	14·98	16·06	16·89	17·59	18·21	18·77	19·31	19·76	20·21	20·63
36	118·1	15·34	16·42	17·25	17·95	18·57	19·13	19·67	20·12	20·57	20·99
38	124·7	15·68	16·76	17·59	18·29	18·91	19·47	20·01	20·46	20·91	21·33
40	131·2	16·02	17·10	17·93	18·63	19·25	19·81	20·35	20·80	21·25	21·67
45	147·6	16·83	17·91	18·74	19·44	20·06	20·62	21·16	21·61	22·06	22·48
50	164·0	17·60	18·68	19·51	20·21	20·83	21·39	21·93	22·38	22·83	23·25
55	180·4	18·34	19·42	20·25	20·95	21·57	22·13	22·67	23·12	23·57	23·99
60	196·9	19·04	20·12	20·95	21·65	22·27	22·83	23·37	23·82	24·27	24·69
65	213·3	19·71	20·79	21·62	22·32	22·94	23·50	24·04	24·49	24·94	25·36
70	229·7	20·35	21·43	22·26	22·96	23·58	24·14	24·68	25·13	25·58	26·00
75	246·1	20·97	22·05	22·88	23·58	24·20	24·76	25·30	25·75	26·20	26·62
80	262·5	21·58	22·66	23·49	24·19	24·81	25·37	25·91	26·36	26·81	27·23
85	278·9	22·16	23·24	24·07	24·77	25·39	25·95	26·49	26·94	27·39	27·81
90	295·3	22·73	23·81	24·64	25·34	25·96	26·52	27·06	27·51	27·96	28·38
95	311·7	23·28	24·36	25·19	25·89	26·51	27·07	27·61	28·06	28·51	28·93
100	328·1	23·81	24·89	25·72	26·42	27·04	27·60	28·14	28·59	29·04	29·46
105	344·5	24·34	25·42	26·25	26·95	27·57	28·13	28·67	29·12	29·57	29·99
110	360·9	24·85	25·93	26·76	27·46	28·08	28·64	29·18	29·63	30·08	30·50
115	377·3	25·35	26·43	27·26	27·96	28·58	29·14	29·68	30·13	30·58	31·00
120	393·7	25·83	26·92	27·75	28·45	29·07	29·62	30·17	30·62	31·07	31·49
125	410·1	26·31	27·39	28·22	28·93	29·54	30·10	30·64	31·09	31·54	31·97
130	426·5	26·78	27·86	28·69	29·39	30·01	30·57	31·11	31·56	32·01	32·43
135	442·9	27·25	28·33	29·16	29·87	30·48	31·04	31·58	32·03	32·48	32·91
140	459·3	27·69	28·78	29·61	30·31	30·93	31·48	32·03	32·48	32·92	33·35
145	475·7	28·14	29·22	30·05	30·75	31·37	31·93	32·47	32·92	33·37	33·79
150	492·1	28·57	29·66	30·49	31·19	31·81	32·36	32·91	33·36	33·80	34·23
155	508·5	29·00	30·08	30·92	31·62	32·23	32·79	33·34	33·78	34·23	34·66
160	524·9	29·42	30·51	31·34	32·04	32·66	33·21	33·76	34·21	34·66	35·08
165	541·3	29·84	30·92	31·75	32·45	33·07	33·63	34·17	34·62	35·07	35·50
170	557·7	30·25	31·33	32·16	32·86	33·48	34·04	34·58	35·03	35·48	35·90
175	574·1	30·65	31·74	32·57	33·27	33·89	34·44	34·99	35·44	35·88	36·31
180	590·6	31·05	32·13	32·96	33·67	34·28	34·84	35·38	35·83	36·28	36·71
185	607·0	31·44	32·53	33·36	34·06	34·68	35·23	35·78	36·23	36·67	37·11
190	623·4	31·83	32·91	33·74	34·44	35·06	35·62	36·16	36·61	37·06	37·49
195	639·8	32·21	33·29	34·12	34·83	35·44	36·00	36·54	36·99	37·44	37·87
200	656·2	32·59	33·67	34·50	35·20	35·82	36·38	36·92	37·37	37·82	38·24
205	672·6	32·96	34·04	34·87	35·58	36·19	36·75	37·29	37·74	38·19	38·62
210	689·0	33·33	34·41	35·25	35·95	36·56	37·12	37·67	38·11	38·56	38·99
215	705·4	33·69	34·77	35·61	36·31	36·92	37·48	38·03	38·47	38·92	39·35
220	721·8	34·05	35·13	35·97	36·67	37·28	37·84	38·39	38·83	39·28	39·71
225	738·2	34·41	35·49	36·32	37·02	37·64	38·20	38·74	39·19	39·64	40·06
230	754·6	34·76	35·84	36·67	37·37	37·99	38·55	39·09	39·54	39·99	40·41
235	771·0	35·10	36·19	37·02	37·73	38·34	38·90	39·44	39·89	40·34	40·76
240	787·4	35·45	36·53	37·36	38·06	38·68	39·24	39·78	40·23	40·68	41·11
245	803·8	35·79	36·87	37·70	38·40	39·02	39·58	40·12	40·57	41·02	41·44
250	820·2	36·12	37·21	38·04	38·74	39·36	39·92	40·46	40·91	41·36	41·78
275	902·2	37·76	38·84	39·68	40·38	40·99	41·55	42·10	42·54	42·99	43·42
300	984·3	39·32	40·40	41·23	41·94	42·55	43·11	43·65	44·10	44·55	44·98

EXTREME RANGE TABLE.

Heights in feet and metres.   Range in miles.

Height of object		ft. 55 m. 16·7	60 18·3	65 19·8	70 21·3	75 22·9	80 24·4	85 25·9	90 27·4	95 29·0	100 30·5
m.	ft.										
10	32·8	15·38	15·77	16·14	16·49	16·83	17·16	17·49	17·80	18·10	18·40
12	39·4	16·02	16·40	16·77	17·13	17·47	17·80	18·13	18·44	18·74	19·04
14	45·9	16·61	16·99	17·36	17·72	18·06	18·39	18·72	19·03	19·33	19·63
16	52·5	17·16	17·54	17·91	18·27	18·61	18·94	19·27	19·58	19·88	20·18
18	59·1	17·67	18·05	18·42	18·78	19·12	19·45	19·78	20·09	20·39	20·69
20	65·6	18·16	18·54	18·91	19·27	19·61	19·94	20·27	20·58	20·88	21·18
22	72·2	18·62	19·00	19·37	19·73	20·07	20·40	20·73	21·04	21·34	21·64
24	78·7	19·06	19·44	19·81	20·17	20·51	20·84	21·17	21·48	21·78	22·08
26	85·3	19·49	19·87	20·24	20·60	20·94	21·27	21·60	21·91	22·21	22·51
28	91·9	19·89	20·27	20·64	21·00	21·34	21·67	22·00	22·31	22·61	22·91
30	98·4	20·31	20·69	21·06	21·42	21·76	22·09	22·42	22·73	23·03	23·33
32	105·0	20·67	21·05	21·42	21·78	22·12	22·45	22·78	23·09	23·39	23·69
34	111·6	21·04	21·42	21·79	22·15	22·49	22·82	23·15	23·46	23·76	24·06
36	118·1	21·40	21·78	22·15	22·51	22·85	23·18	23·51	23·82	24·12	24·42
38	124·7	21·74	22·12	22·49	22·85	23·19	23·52	23·85	24·16	24·46	24·76
40	131·2	22·08	22·46	22·83	23·19	23·53	23·86	24·19	24·50	24·80	25·10
45	147·6	22·89	23·27	23·64	24·00	24·34	24·67	25·00	25·31	25·61	25·91
50	164·0	23·66	24·04	24·41	24·77	25·11	25·44	25·77	26·08	26·38	26·68
55	180·4	24·40	24·78	25·15	25·51	25·85	26·18	26·51	26·82	27·12	27·42
60	196·9	25·10	25·48	25·85	26·21	26·55	26·88	27·21	27·52	27·82	28·12
65	213·3	25·77	26·15	26·52	26·88	27·22	27·55	27·88	28·19	28·49	28·79
70	229·7	26·41	26·79	27·16	27·52	27·86	28·19	28·52	28·83	29·13	29·43
75	246·1	27·03	27·41	27·78	28·14	28·48	28·81	29·14	29·45	29·75	30·05
80	262·5	27·64	28·02	28·39	28·75	29·09	29·42	29·75	30·06	30·36	30·66
85	278·9	28·22	28·60	28·97	29·33	29·67	30·00	30·33	30·64	30·94	31·24
90	295·3	28·79	29·17	29·54	29·90	30·24	30·57	30·90	31·21	31·51	31·81
95	311·7	29·34	29·72	30·09	30·45	30·79	31·12	31·45	31·76	32·06	32·36
100	328·1	29·87	30·25	30·62	30·98	31·32	31·65	31·98	32·29	32·59	32·89
105	344·5	30·40	30·78	31·15	31·51	31·85	32·18	32·51	32·82	33·12	33·42
110	360·9	30·91	31·29	31·66	32·02	32·36	32·69	33·02	33·33	33·63	33·93
115	377·3	31·41	31·79	32·16	32·52	32·86	33·19	33·52	33·83	34·13	34·43
120	393·7	31·89	32·28	32·65	33·00	33·35	33·68	34·01	34·32	34·62	34·92
125	410·1	32·37	32·76	33·13	33·48	33·82	34·15	34·48	34·79	35·09	35·39
130	426·5	32·84	33·22	33·59	33·95	34·29	34·62	34·95	35·36	35·56	35·86
135	442·9	33·31	33·70	34·07	34·42	34·76	35·09	35·42	35·73	36·03	36·33
140	459·3	33·75	34·14	34·51	34·86	35·21	35·54	35·87	36·18	36·48	36·78
145	475·7	34·20	34·58	34·95	35·31	35·65	35·98	36·31	36·62	36·92	37·22
150	492·1	34·63	35·02	35·39	35·74	36·09	36·42	36·75	37·06	37·36	37·66
155	508·5	35·06	35·45	35·82	36·17	36·51	36·84	37·17	37·48	37·78	38·08
160	524·9	35·48	35·87	36·24	36·60	36·94	37·27	37·60	37·91	38·21	38·51
165	541·3	35·90	36·29	36·66	37·01	37·35	37·68	38·01	38·32	38·62	38·92
170	557·7	36·31	36·69	37·06	37·42	37·76	38·09	38·42	38·73	39·03	39·33
175	574·1	36·71	37·10	37·47	37·82	38·17	38·50	38·83	39·14	39·44	39·74
180	590·6	37·11	37·50	37·87	38·22	38·56	38·89	39·22	39·53	39·83	40·13
185	607·0	37·50	37·89	38·26	38·61	38·96	39·29	39·62	39·93	40·23	40·53
190	623·4	37·89	38·28	38·65	39·00	39·34	39·67	40·00	40·31	40·61	40·91
195	639·8	38·27	38·66	39·03	39·38	39·72	40·05	40·38	40·69	40·99	41·29
200	656·2	38·65	39·03	39·40	39·76	40·10	40·43	40·76	41·07	41·37	41·67
205	672·6	39·02	39·41	39·78	40·13	40·47	40·80	41·13	41·44	41·74	42·04
210	689·0	39·39	39·78	40·15	40·50	40·84	41·17	41·50	41·81	42·11	42·41
215	705·4	39·75	40·14	40·51	40·86	41·20	41·53	41·86	42·17	42·47	42·77
220	721·8	40·11	40·50	40·87	41·22	41·56	41·89	42·22	42·53	42·83	43·13
225	738·2	40·47	40·85	41·22	41·58	41·92	42·25	42·58	42·89	43·19	43·49
230	754·6	40·82	41·20	41·57	41·93	42·27	42·60	42·93	43·24	43·54	43·84
235	771·0	41·16	41·55	41·92	42·28	42·62	42·95	43·28	43·59	43·89	44·19
240	787·4	41·51	41·90	42·27	42·62	42·96	43·29	43·62	43·93	44·23	44·53
245	803·8	41·85	42·23	42·60	42·96	43·30	43·63	43·96	44·27	44·57	44·87
250	820·2	42·19	42·57	42·94	43·30	43·64	43·97	44·30	44·61	44·91	45·21
275	902·2	43·82	44·21	44·58	44·93	45·27	45·60	45·93	46·24	46·54	46·84
300	984·3	45·38	45·77	46·14	46·49	46·83	47·16	47·49	47·80	48·10	48·40

118

CHANGE of HOUR ANGLE per 1′ of ALTITUDE.

Latitude.

Azimuth.	0°	3°	6°	9°	12°	15°	18°	21°	24°	27°	30°	33°	36°
°	′	′	′	′	′	′	′	′	′	′	′	′	′
1	57·30	57·38	57·61	58·01	58·58	59·32	60·25	61·38	62·72	64·31	66·16	68·32	70·83
2	28·65	28·69	28·81	29·01	29·29	29·66	30·13	30·69	31·37	32·16	33·09	34·17	35·42
3	19·11	19·13	19·21	19·35	19·53	19·78	20·09	20·47	20·92	21·44	22·06	22·78	23·62
4	14·34	14·36	14·41	14·51	14·66	14·84	15·07	15·36	15·69	16·09	16·55	17·09	17·72
5	11·47	11·49	11·54	11·62	11·73	11·88	12·06	12·29	12·56	12·88	13·25	13·68	14·18
6	9·57	9·58	9·62	9·69	9·78	9·90	10·06	10·25	10·47	10·74	11·05	11·41	11·83
7	8·21	8·22	8·25	8·31	8·39	8·50	8·63	8·79	8·98	9·21	9·48	9·78	10·14
8	7·19	7·20	7·23	7·28	7·35	7·44	7·56	7·70	7·87	8·06	8·30	8·57	8·88
9	6·39	6·40	6·43	6·47	6·54	6·62	6·72	6·85	7·00	7·17	7·38	7·62	7·90
10	5·76	5·77	5·79	5·83	5·89	5·96	6·06	6·17	6·30	6·46	6·65	6·87	7·12
11	5·24	5·25	5·27	5·31	5·36	5·43	5·51	5·61	5·74	5·88	6·05	6·25	6·48
12	4·81	4·82	4·84	4·87	4·92	4·98	5·06	5·15	5·27	5·40	5·55	5·74	5·95
13	4·45	4·45	4·47	4·50	4·55	4·60	4·67	4·76	4·87	4·99	5·13	5·30	5·50
14	4·13	4·14	4·16	4·19	4·23	4·28	4·35	4·43	4·53	4·64	4·77	4·93	5·11
15	3·86	3·87	3·89	3·91	3·95	4·00	4·06	4·14	4·23	4·34	4·46	4·61	4·78
16	3·63	3·63	3·65	3·67	3·71	3·76	3·82	3·89	3·97	4·07	4·19	4·33	4·48
17	3·42	3·43	3·44	3·46	3·50	3·54	3·60	3·66	3·74	3·84	3·95	4·08	4·23
18	3·24	3·24	3·25	3·28	3·31	3·35	3·40	3·47	3·54	3·63	3·74	3·86	4·00
19	3·07	3·08	3·09	3·11	3·14	3·18	3·23	3·29	3·36	3·45	3·55	3·66	3·80
20	2·92	2·93	2·94	2·96	2·99	3·03	3·07	3·13	3·20	3·28	3·38	3·49	3·61
21	2·79	2·79	2·81	2·83	2·85	2·89	2·93	2·99	3·06	3·13	3·22	3·33	3·45
22	2·67	2·67	2·68	2·70	2·73	2·76	2·81	2·86	2·92	3·00	3·08	3·18	3·30
24	2·46	2·46	2·47	2·49	2·51	2·55	2·59	2·63	2·69	2·76	2·84	2·93	3·04
26	2·28	2·28	2·29	2·31	2·33	2·36	2·40	2·44	2·50	2·56	2·63	2·72	2·82
28	2·13	2·13	2·14	2·16	2·18	2·21	2·24	2·28	2·33	2·39	2·46	2·54	2·63
30	2·00	2·00	2·01	2·03	2·05	2·07	2·10	2·14	2·19	2·25	2·31	2·39	2·47
32	1·89	1·89	1·90	1·91	1·93	1·95	1·98	2·02	2·07	2·12	2·18	2·25	2·33
34	1·79	1·79	1·80	1·81	1·83	1·85	1·88	1·92	1·96	2·01	2·07	2·13	2·21
36	1·70	1·70	1·71	1·72	1·74	1·76	1·79	1·82	1·86	1·91	1·96	2·03	2·10
38	1·62	1·63	1·63	1·65	1·66	1·68	1·71	1·74	1·78	1·82	1·88	1·94	2·01
40	1·56	1·56	1·56	1·58	1·59	1·61	1·64	1·67	1·70	1·75	1·80	1·86	1·92
42	1·49	1·50	1·50	1·51	1·53	1·55	1·57	1·60	1·64	1·68	1·73	1·78	1·85
44	1·44	1·44	1·45	1·46	1·47	1·49	1·51	1·54	1·58	1·62	1·66	1·72	1·78
46	1·39	1·39	1·40	1·41	1·42	1·44	1·46	1·49	1·52	1·56	1·61	1·66	1·72
48	1·35	1·35	1·35	1·36	1·38	1·40	1·42	1·44	1·47	1·51	1·55	1·60	1·66
50	1·31	1·31	1·31	1·32	1·34	1·35	1·37	1·40	1·43	1·47	1·51	1·56	1·61
52	1·27	1·27	1·28	1·29	1·30	1·31	1·33	1·36	1·39	1·42	1·47	1·51	1·57
55	1·22	1·22	1·23	1·24	1·25	1·26	1·28	1·31	1·34	1·37	1·41	1·46	1·51
60	1·16	1·16	1·16	1·17	1·18	1·20	1·21	1·24	1·26	1·30	1·33	1·38	1·43
65	1·10	1·11	1·11	1·12	1·13	1·14	1·16	1·18	1·21	1·24	1·27	1·32	1·36
70	1·06	1·07	1·07	1·08	1·09	1·10	1·12	1·14	1·17	1·19	1·23	1·27	1·32
75	1·04	1·04	1·04	1·05	1·06	1·07	1·09	1·11	1·13	1·16	1·19	1·23	1·28
80	1·02	1·02	1·02	1·03	1·04	1·05	1·07	1·09	1·11	1·14	1·17	1·21	1·26
35	1·00	1·01	1·01	1·02	1·03	1·04	1·06	1·08	1·10	1·13	1·16	1·20	1·24
90	1·00	1·00	1·01	1·01	1·02	1·04	1·05	1·07	1·09	1·12	1·15	1·19	1·24

CHANGE of HOUR ANGLE per 1′ of ALTITUDE.

Latitude.

Azimuth.	39°	42°	45°	48°	51°	54°	57°	60°	63°	66°	69°	72°
	′	′	′	′	′	′	′	′	′	′	′	′
1	73·73	77·10	81·03	85·63	91·05	97·48	105·20	114·60	126·21	140·88	159·89	185·43
2	36·87	38·56	40·52	42·82	45·53	48·75	52·61	57·31	63·11	70·45	79·96	92·73
3	24·59	25·71	27·02	28·56	30·36	32·51	35·08	38·21	42·09	46·98	53·32	61·83
4	18·45	19·29	20·27	21·42	22·78	24·39	26·32	28·67	31·58	35·25	40·00	46·39
5	14·76	15·44	16·23	17·15	18·23	19·52	21·07	22·95	25·27	28·21	32·02	37·13
6	12·31	12·87	13·53	14·30	15·20	16·28	17·57	19·13	21·04	23·48	26·65	30·91
7	10·56	11·04	11·60	12·26	13·04	13·96	15·07	16·41	18·07	20·17	22·90	26·55
8	9·25	9·67	10·16	10·74	11·42	12·22	13·19	14·37	15·83	17·67	20·05	23·25
9	8·23	8·60	9·04	9·55	10·16	10·88	11·74	12·78	14·08	15·72	17·84	20·69
10	7·41	7·75	8·14	8·61	9·15	9·80	10·57	11·52	12·68	14·16	16·07	18·64
11	6·74	7·05	7·41	7·83	8·33	8·92	9·62	10·48	11·54	12·89	14·63	16·96
12	6·19	6·47	6·80	7·19	7·64	8·18	8·83	9·62	10·59	11·83	13·42	15·56
13	5·72	5·98	6·29	6·64	7·06	7·56	8·16	8·89	9·79	10·93	12·40	14·39
14	5·32	5·56	5·85	6·18	6·57	7·03	7·59	8·27	9·11	10·16	11·53	13·38
15	4·97	5·20	5·46	5·77	6·14	6·57	7·09	7·73	8·51	9·50	10·78	12·50
16	4·67	4·88	5·13	5·42	5·77	6·17	6·66	7·26	7·99	8·92	10·12	11·74
17	4·40	4·60	4·84	5·11	5·44	5·82	6·28	6·84	7·53	8·41	9·54	11·07
18	4·16	4·36	4·58	4·84	5·14	5·51	5·94	6·47	7·13	7·96	9·03	10·47
19	3·95	4·13	4·34	4·59	4·88	5·23	5·64	6·14	6·77	7·55	8·57	9·94
20	3·76	3·93	4·14	4·37	4·65	4·97	5·37	5·85	6·44	7·19	8·16	9·46
21	3·59	3·76	3·95	4·17	4·43	4·75	5·12	5·58	6·15	6·86	7·79	9·03
22	3·44	3·59	3·78	3·99	4·24	4·54	4·90	5·34	5·88	6·56	7·45	8·64
24	3·16	3·31	3·48	3·67	3·91	4·18	4·51	4·92	5·42	6·04	6·86	7·96
26	2·94	3·07	3·23	3·41	3·63	3·88	4·19	4·56	5·02	5·61	6·37	7·38
28	2·74	2·87	3·01	3·18	3·39	3·62	3·91	4·26	4·69	5·24	5·94	6·89
30	2·57	2·69	2·83	2·99	3·18	3·40	3·67	4·00	4·41	4·92	5·58	6·47
32	2·43	2·54	2·67	2·82	3·00	3·21	3·47	3·77	4·16	4·64	5·27	6·11
34	2·30	2·41	2·53	2·67	2·84	3·04	3·28	3·58	3·94	4·40	4·99	5·79
36	2·19	2·29	2·41	2·54	2·70	2·89	3·12	3·40	3·75	4·18	4·75	5·51
38	2·09	2·19	2·30	2·43	2·58	2·76	2·98	3·25	3·58	3·99	4·53	5·26
40	2·00	2·09	2·20	2·33	2·47	2·65	2·86	3·11	3·43	3·82	4·34	5·03
42	1·92	2·01	2·11	2·23	2·38	2·54	2·74	2·99	3·29	3·67	4·17	4·84
44	1·85	1·94	2·04	2·15	2·29	2·45	2·64	2·88	3·17	3·54	4·02	4·66
46	1·79	1·87	1·97	2·08	2·21	2·37	2·55	2·78	3·06	3·42	3·88	4·50
48	1·73	1·81	1·90	2·01	2·14	2·29	2·47	2·69	2·96	3·31	3·75	4·35
50	1·68	1·76	1·85	1·95	2·07	2·22	2·40	2·61	2·88	3·21	3·64	4·22
52	1·63	1·71	1·80	1·90	2·02	2·16	2·33	2·54	2·80	3·12	3·54	4·11
55	1·57	1·64	1·73	1·82	1·94	2·08	2·24	2·44	2·69	3·00	3·41	3·95
60	1·49	1·55	1·63	1·73	1·84	1·96	2·12	2·31	2·54	2·84	3·22	3·74
65	1·42	1·49	1·56	1·65	1·75	1·88	2·03	2·21	2·43	2·71	3·08	3·57
70	1·37	1·43	1·51	1·59	1·69	1·81	1·95	2·13	2·34	2·62	2·97	3·44
75	1·33	1·39	1·46	1·55	1·65	1·76	1·90	2·07	2·28	2·55	2·89	3·35
80	1·31	1·37	1·44	1·52	1·61	1·73	1·86	2·03	2·24	2·50	2·83	3·29
85	1·29	1·35	1·42	1·50	1·60	1·71	1·84	2·01	2·21	2·47	2·80	3·25
90	1·29	1·35	1·41	1·49	1·59	1·70	1·84	2·00	2·20	2·46	2·79	3·24

CHANGE OF ALTITUDE IN ONE MINUTE OF TIME.

AZIMUTH.

Lat.	0°	2½°	5°	7½°	10°	12½°	15°	17½°	20°	22½°	25°	27½°	30°	32½°	35°	37½°	Lat.
0°	0	0·7	1·3	1·9	2·6	3·2	3·9	4·5	5·1	5·7	6·3	6·9	7·5	8·0	8·6	9·1	0°
4°	0	0·7	1·3	1·9	2·6	3·2	3·9	4·5	5·1	5·7	6·3	6·9	7·5	8·0	8·6	9·1	4°
8°	0	0·7	1·3	1·9	2·6	3·2	3·8	4·4	5·1	5·7	6·3	6·8	7·4	7·9	8·5	9·0	8°
12°	0	0·7	1·3	1·9	2·5	3·1	3·8	4·4	5·0	5·6	6·2	6·7	7·3	7·8	8·4	8·9	12°
16°	0	0·7	1·3	1·9	2·5	3·1	3·7	4·3	4·9	5·5	6·1	6·6	7·2	7·7	8·3	8·8	16°
20°	0	0·6	1·2	1·8	2·4	3·0	3·6	4·2	4·8	5·4	6·0	6·5	7·0	7·5	8·1	8·6	20°
24°	0	0·6	1·2	1·8	2·4	3·0	3·5	4·1	4·7	5·3	5·8	6·3	6·9	7·4	7·9	8·4	24°
26°	0	0·6	1·2	1·7	2·3	2·9	3·5	4·0	4·6	5·1	5·7	6·2	6·7	7·2	7·7	8·2	26°
28°	0	0·6	1·2	1·7	2·3	2·8	3·4	4·0	4·5	5·0	5·6	6·1	6·6	7·1	7·6	8·0	28°
30°	0	0·5	1·1	1·7	2·3	2·8	3·4	3·9	4·4	5·0	5·5	6·0	6·5	7·0	7·4	7·8	30°
32°	0	0·5	1·1	1·6	2·2	2·7	3·3	3·8	4·4	4·9	5·4	5·9	6·4	6·9	7·3	7·7	32°
34°	0	0·5	1·1	1·6	2·2	2·7	3·2	3·7	4·3	4·8	5·3	5·8	6·2	6·6	7·1	7·5	34°
36°	0	0·5	1·1	1·6	2·1	2·6	3·1	3·6	4·2	4·6	5·1	5·6	6·1	6·5	7·0	7·4	36°
38°	0	0·5	1·0	1·5	2·1	2·6	3·1	3·5	4·0	4·5	5·0	5·4	5·9	6·3	6·8	7·2	38°
40°	0	0·5	1·0	1·5	2·0	2·5	3·0	3·5	3·9	4·4	4·9	5·3	5·7	6·1	6·6	7·0	40°
42°	0	0·5	1·0	1·4	1·9	2·4	2·9	3·4	3·8	4·3	4·7	5·1	5·6	6·0	6·4	6·8	42°
44°	0	0·4	0·9	1·4	1·9	2·4	2·8	3·3	3·7	4·2	4·6	5·0	5·4	5·8	6·2	6·5	44°
46°	0	0·4	0·9	1·3	1·8	2·3	2·7	3·2	3·6	4·0	4·4	4·8	5·2	5·6	6·0	6·3	46°
48°	0	0·4	0·9	1·3	1·7	2·2	2·6	3·0	3·4	3·8	4·3	4·6	5·0	5·4	5·8	6·1	48°
49°	0	0·4	0·8	1·3	1·7	2·2	2·5	2·9	3·4	3·8	4·2	4·5	4·9	5·3	5·7	6·0	49°
50°	0	0·4	0·8	1·2	1·7	2·1	2·5	2·9	3·3	3·7	4·1	4·4	4·8	5·1	5·5	5·8	50°
51°	0	0·4	0·8	1·2	1·6	2·0	2·4	2·8	3·2	3·6	4·0	4·3	4·7	5·0	5·4	5·7	51°
52°	0	0·4	0·8	1·2	1·6	2·0	2·4	2·8	3·2	3·5	3·9	4·2	4·6	5·0	5·3	5·6	52°
53°	0	0·4	0·8	1·2	1·6	2·0	2·3	2·7	3·1	3·4	3·8	4·1	4·5	4·9	5·2	5·5	53°
54°	0	0·4	0·9	1·2	1·5	1·9	2·3	2·6	3·0	3·4	3·7	4·0	4·4	4·8	5·1	5·4	54°
55°	0	0·4	0·7	1·1	1·5	1·8	2·2	2·5	2·9	3·2	3·5	3·9	4·3	4·6	4·9	5·2	55°
56°	0	0·4	0·7	1·1	1·5	1·8	2·2	2·5	2·9	3·2	3·5	3·8	4·2	4·5	4·8	5·1	56°
57°	0	0·3	0·7	1·1	1·4	1·7	2·1	2·5	2·8	3·1	3·5	3·8	4·1	4·4	4·7	5·0	57°
58°	0	0·3	0·7	1·0	1·4	1·7	2·0	2·4	2·7	3·0	3·4	3·7	4·0	4·3	4·6	4·8	58°
59°	0	0·3	0·7	1·0	1·3	1·6	2·0	2·3	2·6	3·0	3·3	3·6	3·9	4·2	4·4	4·7	59°
60°	0	0·3	0·7	1·0	1·3	1·6	1·9	2·3	2·6	2·9	3·2	3·6	3·8	4·1	4·3	4·6	60°
61°	0	0·3	0·7	1·0	1·3	1·6	1·9	2·3	2·6	2·9	3·2	3·6	3·8	4·1	4·3	4·6	61°
62°	0	0·3	0·6	0·9	1·2	1·5	1·8	2·1	2·4	2·7	3·0	3·3	3·5	3·8	4·0	4·2	62°
63°	0	0·3	0·6	0·9	1·2	1·5	1·7	2·0	2·3	2·6	2·9	3·1	3·4	3·6	3·8	4·0	63°
64°	0	0·3	0·6	0·9	1·2	1·4	1·7	2·0	2·3	2·5	2·8	3·0	3·3	3·5	3·7	3·9	64°
65°	0	0·3	0·6	0·9	1·1	1·4	1·6	1·9	2·2	2·5	2·7	2·9	3·2	3·4	3·6	3·8	65°
66°	0	0·3	0·5	0·8	1·0	1·3	1·5	1·8	2·1	2·4	2·6	2·8	3·0	3·3	3·5	3·7	66°
67°	0	0·2	0·5	0·8	1·0	1·2	1·5	1·7	2·0	2·3	2·5	2·7	2·9	3·2	3·4	3·6	67°
68°	0	0·2	0·5	0·8	1·0	1·2	1·5	1·7	1·9	2·1	2·4	2·6	2·8	3·0	3·2	3·4	68°
69°	0	0·2	0·5	0·7	0·9	1·1	1·4	1·6	1·8	2·0	2·3	2·5	2·7	2·9	3·1	3·3	69°
70°	0	0·2	0·5	0·7	0·9	1·1	1·4	1·6	1·8	2·0	2·2	2·4	2·6	2·8	3·0	3·1	70°
71°	0	0·2	0·4	0·6	0·9	1·1	1·3	1·5	1·7	1·9	2·1	2·3	2·4	2·6	2·8	2·9	71°
72°	0	0·2	0·4	0·6	0·8	1·0	1·2	1·4	1·6	1·8	2·0	2·1	2·3	2·5	2·6	2·8	72°
73°	0	0·2	0·4	0·6	0·8	0·9	1·1	1·3	1·5	1·7	1·9	2·0	2·2	2·4	2·5	2·7	73°
74°	0	0·2	0·4	0·6	0·8	0·9	1·1	1·2	1·4	1·6	1·8	1·9	2·0	2·2	2·4	2·5	74°
75°	0	0·1	0·3	0·5	0·7	0·8	1·0	1·1	1·3	1·5	1·6	1·8	1·9	2·0	2·2	2·3	75°
Lat.	0°	2½°	5°	7½°	10°	12½°	15°	17½°	20°	22½°	25°	27½°	30°	32½°	35°	37½°	Lat.

CHANGE OF ALTITUDE IN ONE MINUTE OF TIME.

AZIMUTH

Lat.	40°	42½°	45°	47½°	50°	52½°	55°	57½°	60°	62½°	65°	67½°	70°	75°	80°	90°	Lat.
0°	9·6	10·1	10·6	11·0	11·5	11·9	12·3	12·7	13·0	13·3	13·6	13·8	14·1	14·5	14·8	15·0	0°
4°	9·6	10·1	10·6	11·0	11·5	11·9	12·3	12·6	13·0	13·3	13·6	13·8	14·1	14·5	14·7	15·0	4°
8°	9·5	10·0	10·5	11·0	11·4	11·8	12·2	12·5	12·9	13·2	13·5	13·7	14·0	14·4	14·6	14·9	8°
12°	9·4	9·9	10·4	10·8	11·2	11·6	12·0	12·3	12·7	13·0	13·3	13·5	13·8	14·2	14·4	14·7	12°
16°	9·3	9·8	10·2	10·6	11·0	11·4	11·8	12·1	12·5	12·8	13·1	13·3	13·5	13·9	14·2	14·4	16°
20°	9·1	9·5	10·0	10·4	10·8	11·1	11·5	11·8	12·2	12·5	12·8	13·0	13·2	13·6	13·9	14·1	20°
24°	8·8	9·2	9·7	10·2	10·5	10·8	11·2	11·5	11·9	12·2	12·4	12·6	12·9	13·2	13·5	13·7	24°
26°	8·7	9·1	9·5	9·9	10·3	10·6	11·0	11·3	11·7	12·0	12·2	12·5	12·7	13·0	13·3	13·5	26°
28°	8·5	9·0	9·4	9·7	10·1	10·4	10·8	11·1	11·5	11·7	12·0	12·2	12·4	12·8	13·1	13·2	28°
30°	8·3	8·8	9·2	9·6	9·9	10·2	10·6	10·9	11·2	11·5	11·8	12·0	12·2	12·5	12·8	13·0	30°
32°	8·2	8·6	9·0	9·4	9·7	10·0	10·4	10·7	11·0	11·3	11·5	11·8	12·0	12·3	12·5	12·7	32°
34°	8·0	8·4	8·8	9·2	9·5	9·8	10·2	10·5	10·8	11·0	11·3	11·5	11·7	12·0	12·3	12·4	34°
36°	7·8	8·2	8·6	9·0	9·3	9·6	9·9	10·2	10·5	10·7	11·0	11·2	11·4	11·7	12·0	12·1	36°
38°	7·6	8·0	8·4	8·7	9·1	9·4	9·7	10·0	10·2	10·5	10·7	10·9	11·1	11·4	11·6	11·8	38°
40°	7·4	7·7	8·1	8·5	8·8	9·1	9·4	9·7	10·0	10·2	10·4	10·6	10·8	11·1	11·3	11·5	40°
42°	7·2	7·5	7·9	8·2	8·5	8·8	9·1	9·4	9·7	9·9	10·1	10·3	10·5	10·8	11·0	11·1	42°
44°	6·9	7·2	7·6	8·0	8·3	8·6	8·8	9·0	9·3	9·6	9·8	10·0	10·1	10·4	10·6	10·8	44°
46°	6·7	7·0	7·4	7·7	8·0	8·3	8·5	8·7	9·0	9·2	9·4	9·6	9·8	10·1	10·3	10·4	46°
48°	6·5	6·8	7·1	7·4	7·7	8·0	8·2	8·5	8·7	8·9	9·1	9·3	9·4	9·7	9·9	10·0	48°
49°	6·3	6·6	6·9	7·2	7·6	7·8	8·0	8·3	8·5	8·7	8·9	9·1	9·2	9·5	9·7	9·8	49°
50°	6·2	6·5	6·8	7·1	7·4	7·6	7·9	8·1	8·3	8·5	8·7	8·9	9·1	9·3	9·5	9·6	50°
51°	6·0	6·3	6·6	6·8	7·2	7·5	7·7	7·9	8·1	8·3	8·5	8·7	8·9	9·1	9·3	9·4	51°
52°	5·9	6·2	6·5	6·8	7·1	7·4	7·6	7·8	8·0	8·2	8·4	8·6	8·7	8·9	9·1	9·2	52°
53°	5·8	6·1	6·4	6·7	7·0	7·3	7·5	7·7	7·9	8·1	8·3	8·4	8·5	8·7	8·9	9·0	53°
54°	5·7	6·0	6·2	6·5	6·8	7·0	7·2	7·4	7·6	7·8	8·0	8·2	8·3	8·5	8·7	8·8	54°
55°	5·5	5·8	6·1	6·3	6·6	6·8	7·0	7·2	7·5	7·6	7·8	8·0	8·1	8·3	8·5	8·6	55°
56°	5·4	5·7	5·9	6·2	6·4	6·7	6·9	7·1	7·3	7·4	7·6	7·8	7·9	8·1	8·3	8·4	56°
57°	5·2	5·5	5·8	6·0	6·3	6·5	6·7	6·9	7·1	7·2	7·4	7·6	7·7	7·9	8·0	8·2	57°
58°	5·1	5·3	5·6	5·8	6·1	6·3	6·5	6·7	6·9	7·0	7·2	7·4	7·5	7·7	7·8	8·0	58°
59°	5·0	5·2	5·5	5·7	5·9	6·1	6·3	6·5	6·7	6·9	7·0	7·2	7·3	7·5	7·6	7·7	59°
60°	4·8	5·1	5·3	5·5	5·7	5·9	6·1	6·3	6·5	6·7	6·8	6·9	7·0	7·2	7·4	7·5	60°
61°	4·8	5·1	5·3	5·5	5·7	5·9	6·1	6·3	6·5	6·6	6·7	6·8	6·9	7·0	7·2	7·3	61°
62°	4·5	4·7	5·0	5·2	5·4	5·6	5·8	5·9	6·1	6·2	6·4	6·5	6·6	6·7	6·9	7·0	62°
63°	4·3	4·5	4·8	5·0	5·2	5·4	5·6	5·7	5·9	6·0	6·2	6·3	6·4	6·5	6·7	6·8	63°
64°	4·2	4·4	4·7	4·9	5·1	5·2	5·4	5·5	5·7	5·8	6·0	6·1	6·2	6·3	6·4	6·5	64°
65°	4·1	4·3	4·5	4·7	4·9	5·0	5·2	5·3	5·5	5·6	5·8	5·9	6·0	6·1	6·2	6·3	65°
66°	4·0	4·1	4·3	4·5	4·7	4·9	5·0	5·1	5·3	5·4	5·5	5·6	5·7	5·9	6·0	6·1	66°
67°	3·8	4·0	4·1	4·3	4·5	4·7	4·8	5·0	5·1	5·2	5·3	5·4	5·5	5·7	5·8	5·9	67°
68°	3·6	3·8	4·0	4·1	4·3	4·5	4·6	4·8	4·9	5·0	5·1	5·1	5·2	5·4	5·5	5·6	68°
69°	3·5	3·6	3·8	4·0	4·1	4·3	4·4	4·6	4·7	4·8	4·9	5·0	5·0	5·1	5·3	5·4	69°
70°	3·3	3·5	3·6	3·7	3·9	4·1	4·2	4·3	4·5	4·6	4·7	4·7	4·8	4·9	5·0	5·1	70°
71°	3·1	3·3	3·5	3·6	3·7	3·9	4·0	4·1	4·2	4·3	4·4	4·5	4·6	4·7	4·8	4·9	71°
72°	2·9	3·1	3·3	3·4	3·5	3·7	3·8	3·9	4·0	4·1	4·2	4·2	4·3	4·4	4·5	4·6	72°
73°	2·8	3·0	3·1	3·3	3·4	3·5	3·6	3·7	3·8	3·9	4·0	4·1	4·1	4·2	4·3	4·4	73°
74°	2·6	2·8	2·9	3·1	3·2	3·3	3·4	3·5	3·6	3·7	3·8	3·9	3·9	4·0	4·1		74°
75°	2·5	2·6	2·7	2·9	3·0	3·1	3·2	3·3	3·4	3·4	3·5	3·5	3·6	3·7	3·8	3·9	75°
Lat.	40°	42½°	45°	47½°	50°	52½°	55°	57½°	60°	62½°	65°	67½°	70°	75°	80°	90°	Lat.

EX-MERIDIAN TABLE I.

LATITUDE AND DECLINATION Same Name.

Change of Altitude in one minute from Meridian Passage = A.

Lat.	0°	1°	2°	3°	4°	5°	6°	7°	8°	9°	10°	11°	12°	13°	14°	Lat.
	′′	′′	′′	′′	′′	′′	′′	′′	′′	′′	′′	′′	′′	′′	′′	
0					28·1	22·4	18·7	16·0	14·0	12·4	11·1	10 1	0·2	8·5	7·9	0
1						28·0	22·4	18·6	16·0	13·9	12·4	11·1	10·1	9·2	8·5	1
2							28·0	22·3	18·6	15·9	13·9	12·3	11·1	10·0	9·2	2
3								27·9	22·3	18·5	15·8	13·8	12·3	11·0	10·0	3
4	28·1								27·8	22·2	18·5	15·8	13·8	12·2	10·9	4
5	22·4	28·0								27·7	22·1	18·4	15·7	13·7	12·1	5
6	18·7	22·4	28·0								27·6	22·0	18·3	15·6	13·6	6
7	16·0	18·6	22·3	27·9								27·4	21·9	18·2	15·5	7
8	14·0	16·0	18·6	22·3	27·8								27·3	21·7	18·0	8
9	12·4	13·9	15·9	18·5	22·2	27·7								27·1	21·6	9
10	11·1	12·4	13·9	15·8	18·5	22·1	27·6								26·9	10
11	10·1	11·1	12·3	13·8	15·8	18·4	22·0	27·4								11
12	9·2	10·1	11·1	12·3	13·8	15·7	18·3	21·9	27·3							12
13	8·5	9·2	10·0	11·0	12·2	13·7	15·6	18·2	21·7	27·1						13
14	7·9	8·5	9·2	10·0	10·9	12·1	13·6	15·5	18·0	21·6	26·9					14
15	7·3	7·8	8·4	9·1	9·9	10·9	12·1	13·5	15·4	17·9	21·4	26·7				15
16	6·8	7·3	7·8	8·4	9·1	9·8	10·8	12·0	13·4	15·3	17·8	21·3	26·5			16
17	6·4	6·8	7·2	7·8	8·3	9·0	9·8	10·7	11·9	13·3	15·2	17·6	21·1	26·2		17
18	6·0	6·4	6·8	7·2	7·7	8·3	8·9	9·7	10·6	11·8	13·2	15·0	17·5	20·9	26·0	18
19	5·7	6·0	6·3	6·7	7·2	7·6	8·2	8·9	9·6	10·6	11·7	13·1	14·9	17·3	20·7	19
20	5·4	5·7	6·0	6·3	6·7	7·1	7·6	8·1	8·8	9·5	10·5	11·6	13·0	14·8	17·1	20
21	5·1	5·4	5·6	5·9	6·3	6·6	7·0	7·5	8·1	8·7	9·5	10·4	11·5	12·8	14·6	21
22	4·9	5·1	5·3	5·6	5·9	6·2	6·6	7·0	7·5	8·0	8·6	9·4	10·3	11·3	12·7	22
23	4·6	4·8	5·0	5·3	5·5	5·8	6·1	6·5	6·9	7·4	7·9	8·5	9·3	10·1	11·2	23
24	4·4	4·6	4·8	5·0	5·2	5·5	5·8	6·1	6·4	6·8	7·3	7·8	8·4	9·2	10·0	24
25	4·2	4·4	4·6	4·7	5·0	5·2	5·4	5·7	6·0	6·4	6·8	7·2	7·7	8·3	9·0	25
26	4·0	4·2	4·3	4·5	4·7	4·9	5·1	5·4	5·7	6·0	6·3	6·7	7·1	7·6	8·2	26
27	3·9	4·0	4·1	4·3	4·5	4·7	4·9	5·1	5·3	5·6	5·9	6·2	6·6	7·0	7·5	27
28	3·7	3·8	4·0	4·1	4·4	4·4	4·6	4·8	5·0	5·3	5·5	5·8	6·2	6·5	7·0	28
29	3·5	3·7	3·8	3·9	4·1	4·2	4·4	4·6	4·7	5·0	5·2	5·5	5·7	6·1	6·4	29
30	3·4	3·5	3·6	3·7	3·9	4·0	4·2	4·3	4·5	4·7	4·9	5·1	5·4	5·7	6·0	30
31	3·3	3·4	3·5	3·6	3·7	3·8	4·0	4·1	4·3	4·4	4·6	4·8	5·1	5·3	5·6	31
32	3·1	3·2	3·3	3·4	3·5	3·7	3·8	3·9	4·1	4·2	4·4	4·6	4·8	5·0	5·2	32
33	3·0	3·1	3·2	3·3	3·4	3·5	3·6	3·7	3·9	4·0	4·2	4·3	4·5	4·7	4·9	33
34	2·9	3·0	3·1	3·2	3·2	3·3	3·4	3·6	3·7	3·8	3·9	4·1	4·3	4·4	4·6	34
35	2·8	2·9	3·0	3·0	3·1	3·2	3·3	3·4	3·5	3·6	3·7	3·9	4·0	4·2	4·4	35
36	2·7	2·8	2·8	2·9	3·0	3·1	3·2	3·3	3·4	3·5	3·6	3·7	3·8	4·0	4·1	36
37	2·6	2·7	2·7	2·8	2·9	2·9	3·0	3·1	3·2	3·3	3·4	3·5	3·6	3·8	3·9	37
38	2·5	2·6	2·6	2·7	2·8	2·8	2·9	3·0	3·0	3·2	3·2	3·3	3·4	3·6	3·7	38
39	2·4	2·5	2·5	2·6	2·7	2·7	2·8	2·9	2·9	3·0	3·1	3·2	3·3	3·4	3·5	39
40	2·3	2·4	2·4	2·5	2·6	2·6	2·7	2·7	2·8	2·9	3·0	3·0	3·1	3·2	3·3	40
41	2·3	2·3	2·4	2·4	2·5	2·5	2·6	2·6	2·7	2·8	2·8	2·9	3·0	3·1	3·2	41
42	2·2	2·2	2·3	2·3	2·4	2·4	2·5	2·5	2·6	2·6	2·7	2·8	2·9	2·9	3·0	42
43	2·1	2·1	2·2	2·2	2·3	2·3	2·4	2·4	2·5	2·5	2·6	2·7	2·7	2·8	2·9	43
44	2·0	2·1	2·1	2·1	2·2	2·2	2·3	2·3	2·4	2·4	2·5	2·5	2·6	2·7	2·7	44
45	2·0	2·0	2·0	2·1	2·1	2·2	2·2	2·2	2·3	2·3	2·4	2·4	2·5	2·6	2·6	45
46	1·9	1·9	2·0	2·0	2·0	2·1	2·1	2·2	2·2	2·2	2·3	2·3	2·4	2·4	2·5	46
47	1·8	1·9	1·9	1·9	2·0	2·0	2·0	2·1	2·1	2·1	2·2	2·2	2·3	2·3	2·4	47
48	1·8	1·8	1·8	1·9	1·9	1·9	2·0	2·0	2·0	2·1	2·1	2·1	2·2	2·2	2·3	48
49	1·7	1·7	1·8	1·8	1·8	1·8	1·9	1·9	1·9	2·0	2·0	2·1	2·1	2·1	2·2	49
50	1·6	1·7	1·7	1·7	1·8	1·8	1·8	1·8	1·9	1·9	1·9	2·0	2·0	2·0	2·1	50
51	1·6	1·6	1·6	1·7	1·7	1·7	1·7	1·8	1·8	1·8	1·9	1·9	1·9	2·0	2·0	51
52	1·5	1·6	1·6	1·6	1·6	1·6	1·7	1·7	1·7	1·8	1·8	1·8	1·8	1·9	1·9	52
53	1·5	1·5	1·5	1·5	1·6	1·6	1·6	1·6	1·7	1·7	1·7	1·7	1·8	1·8	1·8	53
54	1·4	1·4	1·5	1·5	1·5	1·5	1·5	1·6	1·6	1·6	1·6	1·7	1·7	1·7	1·7	54
55	1·4	1·4	1·4	1·4	1·5	1·5	1·5	1·5	1·5	1·6	1·6	1·6	1·6	1·6	1·7	55
56	1·3	1·3	1·4	1·4	1·4	1·4	1·4	1·4	1·5	1·5	1·5	1·5	1·5	1·6	1·6	56
57	1·3	1·3	1·3	1·3	1·3	1·4	1·4	1·4	1·4	1·4	1·4	1·5	1·5	1·5	1·5	57
58	1·2	1·2	1·3	1·3	1·3	1·3	1·3	1·3	1·3	1·4	1·4	1·4	1·4	1·4	1·5	58
59	1·2	1·2	1·2	1·2	1·2	1·3	1·3	1·3	1·3	1·3	1·3	1·3	1·4	1·4	1·4	59
60	1·1	1·1	1·2	1·2	1·2	1·2	1·2	1·2	1·2	1·2	1·3	1·3	1·3	1·3	1·3	60
	0°	1°	2°	3°	4°	5°	6°	7°	8°	9°	10°	11°	12°	13°	14°	

Declination.

EX-MERIDIAN TABLE I.

Latitude and Declination Same Name.

Change of Altitude in one minute from Meridian Passage = A.

Lat.	15°	16°	17°	18°	19°	20°	21°	22°	23°	24°	25°	26°	27°	28°	29°	30°	Lat.
0	7·3	6·8	6·4	6·0	5·7	5·4	5·1	4·9	4·6	4·4	4·2	4·0	3·9	3·7	3·5	3·4	0
1	7·8	7·3	6·8	6·4	6·0	5·7	5·4	5·1	4·8	4·6	4·4	4·2	4·0	3·8	3·7	3·5	1
2	8·4	7·8	7·2	6·8	6·3	6·0	5·6	5·3	5·0	4·8	4·6	4·3	4·1	4·0	3·8	3·6	2
3	9·1	8·4	7·8	7·2	6·7	6·3	5·9	5·6	5·3	5·0	4·7	4·5	4·3	4·1	3·9	3·7	3
4	9·9	9·1	8·3	7·7	7·2	6·7	6·3	5·9	5·5	5·2	5·0	4·7	4·5	4·3	4·1	3·9	4
5	10·9	9·8	9·0	8·3	7·6	7·1	6·6	6·2	5·8	5·5	5·2	4·9	4·7	4·4	4·2	4·0	5
6	12·1	10·8	9·8	8·9	8·2	7·6	7·0	6·6	6·1	5·8	5·4	5·1	4·9	4·6	4·4	4·2	6
7	13·5	12·0	10·7	9·7	8·9	8·1	7·5	7·0	6·5	6·1	5·7	5·4	5·1	4·8	4·6	4·3	7
8	15·4	13·4	11·9	10·6	9·6	8·8	8·1	7·5	6·9	6·4	6·0	5·7	5·3	5·0	4·8	4·5	8
9	17·9	15·3	13·3	11·8	10·6	9·5	8·7	8·0	7·4	6·8	6·4	6·0	5·6	5·3	5·0	4·7	9
10	21·4	17·8	15·2	13·2	11·7	10·5	9·5	8·6	7·9	7·3	6·8	6·3	5·9	5·5	5·2	4·9	10
11	26·7	21·3	17·6	15·0	13·1	11·6	10·4	9·4	8·5	7·8	7·2	6·7	6·2	5·8	5·5	5·1	11
12		26·5	21·1	17·5	14·9	13·0	11·5	10·3	9·3	8·4	7·7	7·1	6·6	6·2	5·8	5·4	12
13			26·2	20·9	17·3	14·8	12·8	11·3	10·1	9·2	8·3	7·6	7·1	6·5	6·1	5·7	13
14				26·0	20·7	17·1	14·6	12·7	11·2	10·0	9·1	8·2	7·6	7·0	6·4	6·0	14
15					25·7	20·4	16·9	14·4	12·5	11·1	9·9	8·9	8·1	7·4	6·9	6·4	15
16						25·4	20·2	16·7	14·3	12·4	10·9	9·8	8·8	8·0	7·3	6·8	16
17							25·1	20·0	16·5	14·1	12·2	10·8	9·6	8·7	7·9	7·2	17
18								24·8	19·7	16·3	13·9	12·1	10·6	9·5	8·6	7·8	18
19	25·7								24·5	19·5	16·1	13·7	11·9	10·5	9·4	8·4	19
20	20·4	25·4								24·2	19·2	15·9	13·5	11·7	10·3	9·2	20
21	16·9	20·2	25·1								23·8	18·9	15·6	13·3	11·5	10·2	21
22	14·4	16·7	20·0	24·8								23·5	18·6	15·4	13·1	11·3	22
23	12·5	14·3	16·5	19·7	24·5								23·1	18·3	15·1	12·8	23
24	11·1	12·4	14·1	16·3	19·5	24·2								22·7	18·0	14·9	24
25	9·9	10·9	12·2	13·9	16·1	19·2	23·8								22·3	17·7	25
26	8·9	9·8	10·8	12·1	13·7	15·9	18·9	23·5								21·9	26
27	8·1	8·8	9·6	10·6	11·9	13·5	15·6	18·6	23·1								27
28	7·4	8·0	8·7	9·5	10·5	11·7	13·3	15·4	18·3	22·7							28
29	6·9	7·3	7·9	8·6	9·4	10·3	11·5	13·1	15·1	18·0	22·3						29
30	6·4	6·8	7·2	7·8	8·4	9·2	10·1	11·3	12·8	14·9	17·7	21·9					30
31	5·9	6·3	6·7	7·1	7·7	8·3	9·0	10·0	11·1	12·6	14·6	17·4	21·5				31
32	5·5	5·8	6·2	6·5	7·0	7·5	8·1	8·9	9·8	10·9	12·4	14·3	17·0	21·1			32
33	5·1	5·4	5·7	6·1	6·4	6·9	7·4	8·0	8·7	9·6	10·7	12·1	14·0	16·7	20·6		33
34	4·8	5·1	5·3	5·6	5·9	6·3	6·8	7·3	7·8	8·6	9·4	10·5	11·9	13·8	16·3	20·2	34
35	4·5	4·7	5·0	5·2	5·5	5·8	6·2	6·6	7·1	7·7	8·4	9·2	10·3	11·7	13·5	16·0	35
36	4·3	4·5	4·7	4·9	5·1	5·4	5·7	6·1	6·5	7·0	7·5	8·2	9·1	10·1	11·4	13·2	36
37	4·0	4·2	4·4	4·6	4·8	5·0	5·3	5·6	6·0	6·4	6·8	7·4	8·1	8·9	9·9	11·1	37
38	3·8	4·0	4·1	4·3	4·5	4·7	4·9	5·2	5·5	5·8	6·2	6·7	7·2	7·9	8·7	9·6	38
39	3·6	3·8	3·9	4·0	4·2	4·4	4·6	4·8	5·1	5·4	5·7	6·1	6·5	7·1	7·7	8·5	39
40	3·4	3·6	3·7	3·8	4·0	4·1	4·3	4·5	4·7	5·0	5·3	5·6	6·0	6·4	6·9	7·5	40
41	3·3	3·4	3·5	3·6	3·7	3·9	4·0	4·2	4·4	4·6	4·9	5·2	5·5	5·8	6·2	6·7	41
42	3·1	3·2	3·3	3·4	3·5	3·7	3·8	4·0	4·1	4·3	4·5	4·8	5·0	5·3	5·7	6·1	42
43	3·0	3·0	3·1	3·2	3·3	3·5	3·6	3·7	3·9	4·0	4·2	4·4	4·6	4·9	5·2	5·5	43
44	2·8	2·9	3·0	3·1	3·2	3·3	3·4	3·5	3·6	3·8	3·9	4·1	4·3	4·5	4·8	5·1	44
45	2·7	2·8	2·8	2·9	3·0	3·1	3·2	3·3	3·4	3·5	3·7	3·8	4·0	4·2	4·4	4·7	45
46	2·6	2·6	2·7	2·8	2·8	2·9	3·0	3·1	3·2	3·3	3·5	3·6	3·7	3·9	4·1	4·3	46
47	2·4	2·5	2·6	2·6	2·7	2·8	2·9	2·9	3·0	3·1	3·3	3·4	3·5	3·6	3·8	4·0	47
48	2·3	2·4	2·4	2·5	2·6	2·6	2·7	2·8	2·9	3·0	3·1	3·2	3·3	3·4	3·5	3·7	48
49	2·2	2·3	2·3	2·4	2·4	2·5	2·6	2·6	2·7	2·8	2·9	3·0	3·1	3·2	3·3	3·4	49
50	2·1	2·2	2·2	2·3	2·3	2·4	2·4	2·5	2·6	2·6	2·7	2·8	2·9	3·0	3·1	3·2	50
51	2·0	2·1	2·1	2·2	2·2	2·3	2·3	2·4	2·4	2·5	2·6	2·6	2·7	2·8	2·9	3·0	51
52	1·9	2·0	2·0	2·1	2·1	2·1	2·2	2·2	2·3	2·4	2·4	2·5	2·6	2·6	2·7	2·8	52
53	1·9	1·9	1·9	2·0	2·0	2·0	2·1	2·1	2·2	2·2	2·3	2·3	2·4	2·5	2·5	2·6	53
54	1·8	1·8	1·8	1·9	1·9	1·9	2·0	2·0	2·1	2·1	2·2	2·2	2·3	2·3	2·4	2·5	54
55	1·7	1·7	1·8	1·8	1·8	1·9	1·9	1·9	2·0	2·0	2·0	2·1	2·1	2·2	2·3	2·3	55
56	1·6	1·6	1·7	1·7	1·7	1·8	1·8	1·8	1·9	1·9	1·9	2·0	2·0	2·1	2·1	2·2	56
57	1·5	1·6	1·6	1·6	1·6	1·7	1·7	1·7	1·8	1·8	1·8	1·9	1·9	2·0	2·0	2·0	57
58	1·5	1·5	1·5	1·5	1·6	1·6	1·6	1·6	1·7	1·7	1·7	1·8	1·8	1·8	1·9	1·9	58
59	1·4	1·4	1·5	1·5	1·5	1·5	1·5	1·6	1·6	1·6	1·6	1·7	1·7	1·7	1·8	1·8	59
60	1·3	1·4	1·4	1·4	1·4	1·4	1·5	1·5	1·5	1·5	1·6	1·6	1·6	1·6	1·7	1·7	60
	15°	16°	17°	18°	19°	20°	21°	22°	23°	24°	25°	26°	27°	28°	29°	30°	

Declination.

EX-MERIDIAN TABLE I.

LATITUDE AND DECLINATION Same Name.

Change of Altitude in one minute from Meridian Passage = A.

Lat.	31°	32°	33°	34°	35°	36°	37°	38°	39°	40°	41°	42°	43°	44°	45°	46°	Lat.
°	"	"	"	"	"	"	"	"	"	"	"	"	"	"	"	"	°
0	3·3	3·1	3·0	2·9	2·8	2·7	2·6	2·5	2·4	2·3	2·3	2·2	2·1	2·0	2·0	1·9	0
1	3·4	3·2	3·1	3·0	2·9	2·8	2·7	2·6	2·5	2·4	2·3	2·2	2·2	2·1	2·0	1·9	1
2	3·5	3·3	3·2	3·1	3·0	2·8	2·7	2·6	2·5	2·4	2·4	2·3	2·2	2·1	2·0	2·0	2
3	3·6	3·4	3·3	3·2	3·0	2·9	2·8	2·7	2·6	2·5	2·4	2·3	2·2	2·2	2·1	2·0	3
4	3·7	3·5	3·4	3·3	3·1	3·0	2·9	2·8	2·7	2·6	2·5	2·4	2·3	2·2	2·1	2·0	4
5	3·8	3·7	3·5	3·3	3·2	3·1	3·0	2·8	2·7	2·6	2·5	2·4	2·3	2·2	2·2	2·1	5
6	4·0	3·8	3·6	3·5	3·3	3·2	3·0	2·9	2·8	2·7	2·6	2·5	2·4	2·3	2·2	2·1	6
7	4·1	3·9	3·7	3·6	3·4	3·3	3·1	3·0	2·9	2·7	2·6	2·5	2·4	2·3	2·2	2·2	7
8	4·3	4·1	3·9	3·7	3·5	3·4	3·2	3·1	2·9	2·8	2·7	2·6	2·5	2·4	2·3	2·2	8
9	4·4	4·2	4·0	3·8	3·6	3·5	3·3	3·2	3·0	2·9	2·8	2·7	2·5	2·4	2·3	2·2	9
10	4·6	4·4	4·2	3·9	3·8	3·6	3·4	3·3	3·1	3·0	2·8	2·7	2·6	2·5	2·4	2·3	10
11	4·8	4·6	4·3	4·1	3·9	3·7	3·5	3·4	3·2	3·1	2·9	2·8	2·7	2·6	2·4	2·3	11
12	5·1	4·8	4·5	4·3	4·0	3·8	3·6	3·5	3·3	3·1	3·0	2·9	2·7	2·6	2·5	2·4	12
13	5·3	5·0	4·7	4·4	4·2	4·0	3·8	3·6	3·4	3·2	3·1	2·9	2·8	2·7	2·6	2·4	13
14	5·6	5·2	4·9	4·6	4·4	4·1	3·9	3·7	3·5	3·3	3·2	3·0	2·9	2·7	2·6	2·5	14
15	5·9	5·5	5·2	4·8	4·5	4·3	4·0	3·8	3·6	3·4	3·3	3·1	3·0	2·8	2·7	2·6	15
16	6·3	5·8	5·4	5·1	4·8	4·5	4·2	4·0	3·8	3·6	3·4	3·2	3·0	2·9	2·8	2·6	16
17	6·7	6·2	5·7	5·3	5·0	4·7	4·4	4·1	3·9	3·7	3·5	3·3	3·1	3·0	2·8	2·7	17
18	7·1	6·6	6·1	5·6	5·2	4·9	4·6	4·3	4·1	3·8	3·6	3·4	3·2	3·1	2·9	2·8	18
19	7·7	7·0	6·4	6·0	5·5	5·1	4·8	4·5	4·2	4·0	3·7	3·5	3·3	3·2	3·0	2·8	19
20	8·3	7·5	6·9	6·3	5·8	5·4	5·0	4·7	4·4	4·1	3·9	3·7	3·5	3·3	3·1	2·9	20
21	9·1	8·2	7·4	6·8	6·2	5·7	5·3	4·9	4·6	4·3	4·0	3·8	3·6	3·4	3·2	3·0	21
22	10·0	8·9	8·0	7·3	6·6	6·1	5·6	5·2	4·8	4·5	4·2	4·0	3·7	3·5	3·3	3·1	22
23	11·1	9·8	8·7	7·9	7·1	6·5	6·0	5·5	5·1	4·7	4·4	4·1	3·9	3·6	3·4	3·2	23
24	12·6	10·9	9·6	8·6	7·7	7·0	6·4	5·8	5·4	5·0	4·6	4·3	4·0	3·8	3·5	3·3	24
25	14·6	12·4	10·7	9·4	8·4	7·5	6·8	6·2	5·7	5·3	4·9	4·5	4·2	3·9	3·7	3·5	25
26	17·4	14·3	12·1	10·5	9·2	8·2	7·4	6·7	6·1	5·6	5·2	4·8	4·4	4·1	3·8	3·6	26
27	21·5	17·0	14·0	11·9	10·3	9·1	8·1	7·2	6·5	6·0	5·5	5·0	4·6	4·3	4·0	3·7	27
28		21·1	16·7	13·8	11·7	10·1	8·9	7·9	7·1	6·4	5·8	5·3	4·9	4·5	4·2	3·9	28
29			20·6	16·3	13·5	11·4	9·9	8·7	7·7	6·9	6·2	5·7	5·2	4·8	4·4	4·1	29
30				20·2	16·0	13·2	11·1	9·6	8·5	7·5	6·7	6·1	5·5	5·1	4·7	4·3	30
31					19·8	15·6	12·9	10·9	9·4	8·2	7·3	6·6	5·9	5·4	4·9	4·5	31
32						19·3	15·3	12·6	10·6	9·2	8·0	7·1	6·4	5·8	5·2	4·8	32
33							18·9	14·9	12·2	10·4	8·9	7·8	6·9	6·2	5·6	5·1	33
34								18·4	14·5	11·9	10·1	8·7	7·6	6·7	6·0	5·4	34
35	19·8								17·9	14·1	11·6	9·8	8·5	7·4	6·6	5·9	35
36	15·6	19·3								17·4	13·8	11·3	9·5	8·2	7·2	6·4	36
37	12·9	15·3	18·9								17·0	13·4	11·0	9·3	8·0	7·0	37
38	10·9	12·6	14·9	18·4								16·5	13·0	10·7	9·0	7·7	38
39	9·4	10·6	12·2	14·5	17·9								16·0	12·6	10·3	8·7	39
40	8·2	9·2	10·4	11·9	14·1	17·4								15·5	12·2	10·0	40
41	7·3	8·0	8·9	10·1	11·6	13·8	17·0								15·0	11·8	41
42	6·6	7·1	7·8	8·7	9·8	11·3	13·4	16·5								14·5	42
43	5·9	6·4	6·9	7·6	8·5	9·5	11·0	13·0	16·0								43
44	5·4	5·8	6·2	6·7	7·4	8·2	9·3	10·7	12·6	15·5							44
45	4·9	5·2	5·6	6·0	6·6	7·2	8·0	9·0	10·3	12·2	15·0						45
46	4·5	4·8	5·1	5·4	5·9	6·4	7·0	7·7	8·7	10·0	11·8	14·5					46
47	4·2	4·4	4·6	4·9	5·3	5·7	6·2	6·8	7·5	8·4	9·7	11·4	14·0				47
48	3·9	4·0	4·3	4·5	4·8	5·1	5·5	6·0	6·5	7·2	8·1	9·3	11·0	13·6			48
49	3·6	3·7	3·9	4·1	4·4	4·6	5·0	5·3	5·8	6·3	7·0	7·9	9·0	10·6	13·1		49
50	3·3	3·5	3·6	3·8	4·0	4·2	4·5	4·8	5·1	5·6	6·1	6·7	7·6	8·7	10·2	12·6	50
51	3·1	3·2	3·4	3·5	3·7	3·9	4·1	4·3	4·6	5·0	5·4	5·9	6·5	7·3	8·4	9·9	51
52	2·9	3·0	3·1	3·2	3·4	3·6	3·7	3·9	4·2	4·5	4·8	5·2	5·7	6·3	7·0	8·0	52
53	2·7	2·8	2·9	3·0	3·1	3·3	3·4	3·6	3·8	4·0	4·3	4·6	5·0	5·4	6·0	6·7	53
54	2·5	2·6	2·7	2·8	2·9	3·0	3·2	3·3	3·5	3·7	3·9	4·1	4·4	4·8	5·2	5·8	54
55	2·4	2·4	2·5	2·6	2·7	2·8	2·9	3·0	3·2	3·3	3·5	3·7	4·0	4·3	4·6	5·0	55
56	2·2	2·3	2·4	2·4	2·5	2·6	2·7	2·8	2·9	3·1	3·2	3·4	3·6	3·8	4·1	4·4	56
57	2·1	2·2	2·2	2·3	2·3	2·4	2·5	2·6	2·7	2·8	2·9	3·1	3·2	3·4	3·6	3·9	57
58	2·0	2·0	2·1	2·1	2·2	2·3	2·3	2·4	2·5	2·6	2·7	2·8	2·9	3·1	3·3	3·5	58
59	1·9	1·9	1·9	2·0	2·0	2·1	2·2	2·2	2·3	2·4	2·5	2·6	2·7	2·8	3·0	3·1	59
60	1·7	1·8	1·8	1·9	1·9	2·0	2·0	2·1	2·1	2·2	2·3	2·4	2·5	2·6	2·7	2·8	60
	31°	32°	33°	34°	35°	36°	37°	33°	39°	40°	41°	42°	43°	44°	45°	46°	

Declination.

EX-MERIDIAN TABLE I.

Latitude and Declination Same Name.

Change of Altitude in one minute from Meridian Passage=A.

Lat.	47°	48°	49°	50°	51°	52°	53°	54°	55°	56°	57°	58°	59°	60°	61°	62°	63°	Lat.
°	″	″	″	″	″	″	″	″	″	″	″	″	″	″	″	″	″	°
0	1·8	1·8	1·7	1·7	1·6	1·5	1·5	1·4	1·4	1·3	1·3	1·2	1·2	1·1	1·1	1·0	1·0	0
1	1·9	1·8	1·7	1·7	1·6	1·6	1·5	1·4	1·4	1·3	1·3	1·2	1·2	1·2	1·1	1·1	1·0	1
2	1·9	1·8	1·8	1·7	1·6	1·6	1·5	1·5	1·4	1·4	1·3	1·3	1·2	1·2	1·1	1·1	1·0	2
3	1·9	1·9	1·8	1·7	1·7	1·6	1·5	1·5	1·4	1·4	1·3	1·3	1·2	1·2	1·1	1·1	1·0	3
4	2·0	1·9	1·8	1·8	1·7	1·6	1·6	1·5	1·5	1·4	1·4	1·3	1·3	1·2	1·1	1·1	1·0	4
5	2·0	1·9	1·9	1·8	1·7	1·7	1·6	1·5	1·5	1·4	1·4	1·3	1·3	1·2	1·1	1·1	1·1	5
6	2·0	2·0	1·9	1·8	1·7	1·7	1·6	1·5	1·5	1·4	1·4	1·3	1·3	1·2	1·2	1·1	1·1	6
7	2·1	2·0	1·9	1·8	1·8	1·7	1·6	1·6	1·5	1·4	1·4	1·3	1·3	1·2	1·2	1·1	1·1	7
8	2·1	2·0	1·9	1·8	1·8	1·7	1·7	1·6	1·5	1·5	1·4	1·4	1·3	1·2	1·2	1·1	1·1	8
9	2·2	2·1	2·0	1·9	1·8	1·8	1·7	1·6	1·6	1·5	1·4	1·4	1·3	1·3	1·2	1·1	1·1	9
10	2·2	2·1	2·0	1·9	1·9	1·8	1·7	1·6	1·6	1·5	1·4	1·4	1·3	1·3	1·2	1·2	1·1	10
11	2·2	2·1	2·1	2·0	1·9	1·8	1·7	1·7	1·6	1·5	1·5	1·4	1·3	1·3	1·2	1·2	1·1	11
12	2·3	2·2	2·1	2·0	1·9	1·8	1·8	1·7	1·6	1·6	1·5	1·4	1·4	1·3	1·2	1·2	1·1	12
13	2·3	2·2	2·1	2·0	2·0	1·9	1·8	1·7	1·7	1·6	1·6	1·5	1·4	1·4	1·3	1·3	1·1	13
14	2·4	2·3	2·2	2·1	2·0	1·9	1·8	1·7	1·7	1·6	1·5	1·5	1·4	1·3	1·3	1·2	1·2	14
15	2·4	2·3	2·2	2·1	2·0	1·9	1·9	1·8	1·7	1·6	1·5	1·5	1·4	1·3	1·3	1·2	1·2	15
16	2·5	2·4	2·3	2·2	2·1	2·0	1·9	1·8	1·7	1·6	1·6	1·5	1·4	1·4	1·3	1·2	1·2	16
17	2·6	2·4	2·3	2·2	2·1	2·0	1·9	1·8	1·8	1·7	1·6	1·5	1·5	1·3	1·3	1·3	1·2	17
18	2·6	2·5	2·4	2·3	2·2	2·1	2·0	1·9	1·8	1·7	1·6	1·5	1·5	1·4	1·3	1·3	1·2	18
19	2·7	2·6	2·4	2·3	2·2	2·1	2·0	1·9	1·8	1·7	1·6	1·6	1·5	1·4	1·4	1·3	1·2	19
20	2·8	2·6	2·5	2·4	2·3	2·1	2·0	1·9	1·9	1·8	1·7	1·6	1·5	1·4	1·4	1·3	1·2	20
21	2·9	2·7	2·6	2·4	2·3	2·2	2·1	2·0	1·9	1·8	1·7	1·6	1·5	1·5	1·4	1·3	1·2	21
22	2·9	2·8	2·6	2·5	2·4	2·2	2·1	2·0	1·9	1·8	1·7	1·6	1·6	1·5	1·4	1·3	1·3	22
23	3·0	2·9	2·7	2·6	2·4	2·3	2·2	2·1	2·0	1·9	1·8	1·7	1·6	1·5	1·4	1·4	1·3	23
24	3·1	3·0	2·8	2·6	2·5	2·4	2·2	2·1	2·0	1·9	1·8	1·7	1·6	1·5	1·5	1·4	1·3	24
25	3·3	3·1	2·9	2·7	2·6	2·4	2·3	2·2	2·0	1·9	1·8	1·7	1·6	1·6	1·5	1·4	1·3	25
26	3·4	3·2	3·0	2·8	2·6	2·5	2·3	2·2	2·1	2·0	1·9	1·8	1·7	1·6	1·5	1·4	1·3	26
27	3·5	3·3	3·1	2·9	2·7	2·6	2·4	2·3	2·1	2·0	1·9	1·8	1·7	1·6	1·5	1·4	1·4	27
28	3·6	3·4	3·2	3·0	2·8	2·6	2·5	2·3	2·2	2·1	2·0	1·8	1·7	1·6	1·5	1·5	1·4	28
29	3·8	3·5	3·3	3·1	2·9	2·7	2·5	2·4	2·3	2·1	2·0	1·9	1·8	1·7	1·6	1·5	1·4	29
30	4·0	3·7	3·4	3·2	3·0	2·8	2·6	2·5	2·3	2·2	2·0	1·9	1·8	1·7	1·6	1·5	1·4	30
31	4·2	3·9	3·6	3·3	3·1	2·9	2·7	2·5	2·4	2·2	2·1	2·0	1·9	1·7	1·6	1·5	1·4	31
32	4·4	4·0	3·7	3·5	3·2	3·0	2·8	2·6	2·4	2·3	2·2	2·0	1·9	1·8	1·7	1·6	1·5	32
33	4·6	4·3	3·9	3·6	3·4	3·1	2·9	2·7	2·5	2·4	2·2	2·1	1·9	1·8	1·7	1·6	1·5	33
34	4·9	4·5	4·1	3·8	3·5	3·2	3·0	2·8	2·6	2·4	2·3	2·1	2·0	1·9	1·7	1·6	1·5	34
35	5·3	4·8	4·4	4·0	3·7	3·4	3·1	2·9	2·7	2·5	2·3	2·2	2·0	1·9	1·8	1·7	1·6	35
36	5·7	5·1	4·6	4·2	3·9	3·6	3·3	3·0	2·8	2·6	2·4	2·3	2·1	2·0	1·8	1·7	1·6	36
37	6·2	5·5	5·0	4·5	4·1	3·7	3·4	3·2	2·9	2·7	2·5	2·3	2·2	2·0	1·9	1·7	1·6	37
38	6·8	6·0	5·3	4·8	4·3	3·9	3·6	3·3	3·0	2·8	2·6	2·4	2·2	2·1	1·9	1·8	1·7	38
39	7·5	6·5	5·8	5·1	4·6	4·2	3·8	3·5	3·2	2·9	2·7	2·5	2·3	2·1	2·0	1·8	1·7	39
40	8·4	7·2	6·3	5·6	5·0	4·5	4·0	3·7	3·3	3·1	2·8	2·6	2·4	2·2	2·0	1·9	1·8	40
41	9·7	8·1	7·0	6·1	5·4	4·8	4·3	3·9	3·5	3·2	2·9	2·7	2·5	2·3	2·1	1·9	1·8	41
42	11·4	9·3	7·9	6·7	5·9	5·2	4·6	4·1	3·7	3·4	3·1	2·8	2·6	2·4	2·2	2·0	1·9	42
43	14·0	11·0	9·0	7·6	6·5	5·7	5·0	4·4	4·0	3·6	3·2	2·9	2·7	2·5	2·3	2·1	1·9	43
44		13·6	10·6	8·7	7·3	6·3	5·4	4·8	4·3	3·8	3·4	3·1	2·8	2·6	2·3	2·2	2·0	44
45			13·1	10·2	8·4	7·0	6·0	5·2	4·6	4·1	3·6	3·3	3·0	2·7	2·4	2·2	2·0	45
46				12·6	9·9	8·0	6·7	5·8	5·0	4·4	3·9	3·5	3·1	2·8	2·6	2·3	2·1	46
47					12·1	9·5	7·7	6·5	5·5	4·8	4·2	3·7	3·3	3·0	2·7	2·4	2·2	47
48						11·6	9·1	7·4	6·2	5·3	4·6	4·0	3·6	3·2	2·8	2·6	2·3	48
49							11·1	8·7	7·1	5·9	5·0	4·4	3·8	3·4	3·0	2·7	2·4	49
50								10·6	8·3	6·8	5·6	4·8	4·2	3·6	3·2	2·9	2·6	50
51	12·1								10·2	7·9	6·4	5·4	4·6	4·0	3·5	3·0	2·7	51
52	9·5	11·6								9·7	7·6	6·1	5·1	4·3	3·8	3·3	2·9	52
53	7·7	9·1	11·1								9·2	7·2	5·9	4·9	4·1	3·6	3·1	53
54	6·5	7·4	8·7	10·6								8·8	6·8	5·5	4·6	3·9	3·4	54
55	5·5	6·2	7·1	8·3	10·2								8·3	6·5	5·3	4·3	3·7	55
56	4·8	5·3	5·9	6·8	7·9	9·7								7·9	6·1	5·0	4·1	56
57	4·2	4·6	5·0	5·6	6·4	7·6	9·2								7·4	5·8	4·7	57
58	3·7	4·0	4·4	4·8	5·4	6·1	7·2	8·8								7·0	5·4	58
59	3·3	3·6	3·8	4·2	5·1	5·9	6·8	8·3									6·6	59
60	3·0	3·2	3·4	3·6	4·0	4·3	4·9	5·5	6·5	7·9								60
	47°	48°	49°	50°	51°	52°	53°	54°	55°	56°	57°	58°	59°	60°	61°	62°	63°	

Declination.

Supplementary Table to 83°.

EX-MERIDIAN TABLE I.
Latitude and Declination SAME NAME.
Change of Altitude in one minute from Meridian Passage=A.

DECLINATION.

Lat.	0°	1°	2°	3°	4°	5°	6°	7°	8°	9°	10°	11°	12°	13°	14°	Lat.
	″	″	″	″	″	″	″	″	″	″	″	″	″	″	″	
60	1·13	1·15	1·16	1·17	1·18	1·19	1·21	1·22	1·23	1·25	1·26	1·28	1·29	1·31	1·33	60
61	1·09	1·10	1·11	1·12	1·13	1·14	1·16	1·17	1·18	1·20	1·21	1·22	1·23	1·25	1·26	61
62	1·04	1·05	1·06	1·07	1·08	1·09	1·11	1·12	1·13	1·14	1·15	1·16	1·17	1·19	1·20	62
63	1·00	1·01	1·02	1·03	1·04	1·05	1·06	1·07	1·08	1·09	1·10	1·11	1·12	1·13	1·14	63
64	0·96	0·97	0·97	0·98	0·99	1·00	1·01	1·02	1·03	1·04	1·05	1·06	1·07	1·08	1·09	64
65	0·92	0·92	0·93	0·94	0·94	0·95	0·96	0·97	0·98	0·99	1·00	1·01	1·02	1·03	1·04	65
66	0·87	0·88	0·89	0·89	0·90	0·91	0·92	0·92	0·93	0·94	0·95	0·96	0·97	0·97	0·98	66
67	0·83	0·84	0·85	0·85	0·86	0·86	0·87	0·88	0·89	0·90	0·90	0·91	0·92	0·92	0·93	67
68	0·79	0·80	0·80	0·81	0·81	0·82	0·83	0·83	0·84	0·85	0·85	0·86	0·87	0·88	0·88	68
69	0·75	0·76	0·76	0·77	0·77	0·78	0·79	0·79	0·80	0·80	0·81	0·81	0·82	0·83	0·83	69
70	0·71	0·72	0·72	0·73	0·73	0·74	0·74	0·74	0·75	0·75	0·76	0·76	0·77	0·78	0·78	70
71	0·67	0·68	0·68	0·69	0·69	0·70	0·70	0·71	0·71	0·71	0·72	0·72	0·73	0·73	0·74	71
72	0·64	0·64	0·64	0·65	0·65	0·66	0·66	0·66	0·67	0·67	0·68	0·68	0·68	0·68	0·69	72
73	0·60	0·60	0·61	0·61	0·61	0·62	0·62	0·62	0·63	0·63	0·63	0·64	0·64	0·64	0·65	73
74	0·56	0·56	0·57	0·57	0·57	0·58	0·58	0·58	0·59	0·59	0·59	0·59	0·60	0·60	0·60	74
75	0·52	0·53	0·53	0·53	0·54	0·54	0·54	0·54	0·55	0·55	0·55	0·55	0·55	0·56	0·56	75
76	0·49	0·49	0·49	0·50	0·50	0·50	0·50	0·51	0·51	0·51	0·51	0·51	0·52	0·52	0·52	76
77	0·45	0·45	0·46	0·46	0·46	0·46	0·46	0·47	0·47	0·47	0·47	0·47	0·48	0·48	0·48	77
78	0·42	0·42	0·42	0·42	0·42	0·43	0·43	0·43	0·43	0·43	0·43	0·43	0·44	0·44	0·44	78
79	0·38	0·38	0·38	0·39	0·39	0·39	0·39	0·39	0·39	0·39	0·39	0·40	0·40	0·40	0·40	79
80	0·35	0·35	0·35	0·35	0·35	0·35	0·35	0·35	0·35	0·36	0·36	0·36	0·36	0·36	0·36	80
81	0·31	0·31	0·31	0·31	0·31	0·32	0·32	0·32	0·32	0·32	0·32	0·32	0·32	0·32	0·32	81
82	0·28	0·28	0·28	0·28	0·28	0·28	0·28	0·28	0·28	0·28	0·28	0·28	0·28	0·29	0·29	82
83	0·24	0·24	0·24	0·24	0·24	0·24	0·24	0·24	0·25	0·25	0·25	0·25	0·25	0·25	0·25	83
	0°	1°	2°	3°	4°	5°	6°	7°	8°	9°	10°	11°	12°	13°	14°	

DECLINATION.

DECLINATION.

Lat.	15°	16°	17°	18°	19°	20°	21°	22°	23°	24°	25°	26°	27°	28°	29°	30°	Lat.
	″	″	″	″	″	″	″	″	″	″	″	″	″	″	″	″	
60	1·34	1·36	1·38	1·40	1·42	1·44	1·46	1·48	1·50	1·53	1·55	1·58	1·61	1·64	1·67	1·70	60
61	1·28	1·29	1·31	1·33	1·35	1·36	1·38	1·40	1·43	1·45	1·47	1·49	1·52	1·54	1·57	1·60	61
62	1·21	1·23	1·25	1·26	1·28	1·29	1·31	1·33	1·35	1·36	1·38	1·40	1·43	1·46	1·48	1·51	62
63	1·16	1·17	1·19	1·20	1·21	1·22	1·24	1·26	1·27	1·29	1·30	1·32	1·35	1·37	1·39	1·42	63
64	1·10	1·12	1·13	1·14	1·15	1·16	1·18	1·19	1·20	1·21	1·24	1·25	1·27	1·29	1·31	1·33	64
65	1·05	1·06	1·07	1·08	1·09	1·10	1·11	1·13	1·14	1·15	1·17	1·18	1·20	1·21	1·23	1·24	65
66	0·99	1·00	1·01	1·02	1·03	1·04	1·05	1·07	1·08	1·09	1·10	1·11	1·13	1·14	1·16	1·17	66
67	0·93	0·94	0·95	0·96	0·97	0·98	1·00	1·01	1·02	1·03	1·04	1·05	1·06	1·08	1·09	1·10	67
68	0·89	0·89	0·90	0·91	0·92	0·93	0·94	0·95	0·96	0·97	0·98	0·99	1·00	1·01	1·03	1·04	68
69	0·84	0·84	0·85	0·86	0·87	0·88	0·89	0·90	0·90	0·91	0·92	0·93	0·94	0·95	0·96	0·97	69
70	0·79	0·79	0·80	0·81	0·82	0·83	0·83	0·84	0·85	0·85	0·86	0·87	0·88	0·89	0·90	0·91	70
71	0·74	0·74	0·75	0·76	0·77	0·78	0·78	0·79	0·79	0·80	0·80	0·81	0·82	0·83	0·84	0·84	71
72	0·70	0·70	0·71	0·71	0·72	0·73	0·73	0·74	0·74	0·75	0·75	0·75	0·76	0·77	0·78	0·79	72
73	0·65	0·66	0·66	0·67	0·67	0·68	0·68	0·69	0·69	0·70	0·70	0·70	0·71	0·71	0·72	0·73	73
74	0·61	0·61	0·61	0·62	0·62	0·63	0·63	0·64	0·64	0·65	0·65	0·65	0·66	0·66	0·66	0·67	74
75	0·56	0·57	0·57	0·58	0·58	0·58	0·58	0·58	0·59	0·60	0·60	0·61	0·61	0·61	0·61	0·62	75
76	0·52	0·52	0·53	0·53	0·54	0·54	0·54	0·54	0·54	0·55	0·55	0·56	0·56	0·56	0·57	0·57	76
77	0·48	0·48	0·49	0·49	0·49	0·49	0·50	0·50	0·50	0·50	0·50	0·51	0·51	0·52	0·52	0·52	77
78	0·44	0·44	0·45	0·45	0·45	0·45	0·45	0·46	0·46	0·46	0·46	0·46	0·47	0·47	0·47	0·48	78
79	0·40	0·40	0·41	0·41	0·41	0·41	0·41	0·41	0·41	0·42	0·42	0·42	0·42	0·42	0·42	0·43	79
80	0·36	0·36	0·37	0·37	0·37	0·37	0·37	0·37	0·37	0·38	0·38	0·38	0·38	0·38	0·38	0·39	80
81	0·32	0·32	0·33	0·33	0·33	0·33	0·33	0·33	0·33	0·33	0·34	0·34	0·34	0·34	0·34	0·34	81
82	0·29	0·29	0·29	0·29	0·29	0·29	0·29	0·29	0·29	0·29	0·30	0·30	0·30	0·30	0·30	0·30	82
83	0·25	0·25	0·25	0·25	0·25	0·25	0·25	0·25	0·25	0·26	0·26	0·26	0·26	0·26	0·26	0·26	83
	15°	16°	17°	18°	19°	20°	21°	22°	23°	24°	25°	26°	27°	28°	29°	30°	

DECLINATION.

Supplementary Table to 83°.

EX-MERIDIAN TABLE I.

Latitude and Declination SAME NAME.

Change of Altitude in one minute from Meridian Passage=A.

DECLINATION.

Lat.	31°	32°	33°	34°	35°	36°	37°	38°	39°	40°	41°	42°	43°	44°	45°	46°	Lat.
°	″	″	″	″	″	″	″	″	″	″	″	″	″	″	″	″	°
60	1·74	1·77	1·81	1·86	1·90	1·95	2·01	2·07	2·13	2·20	2·28	2·36	2·46	2·56	2·68	2·82	60
61	1·63	1·67	1·70	1·74	1·78	1·82	1·87	1·92	1·97	2·03	2·10	2·17	2·25	2·34	2·44	2·56	61
62	1·53	1·56	1·59	1·63	1·66	1·70	1·74	1·78	1·83	1·88	1·94	2·01	2·08	2·15	2·24	2·32	62
63	1·44	1·47	1·49	1·53	1·56	1·59	1·62	1·66	1·71	1·75	1·79	1·85	1·91	1·97	2·05	2·12	63
64	1·35	1·37	1·40	1·44	1·46	1·49	1·51	1·55	1·59	1·63	1·67	1·71	1·76	1·81	1·87	1·94	64
65	1·26	1·28	1·31	1·34	1·36	1·38	1·41	1·44	1·47	1·50	1·53	1·57	1·62	1·67	1·72	1·77	65
66	1·19	1·20	1·22	1·24	1·27	1·29	1·32	1·34	1·37	1·40	1·42	1·46	1·49	1·53	1·57	1·62	66
67	1·12	1·13	1·14	1·16	1·18	1·20	1·22	1·24	1·26	1·29	1·32	1·34	1·37	1·41	1·45	1·49	67
68	1·05	1·06	1·07	1·09	1·10	1·12	1·14	1·15	1·17	1·19	1·22	1·25	1·27	1·30	1·33	1·36	68
69	0·98	0·99	1·00	1·01	1·02	1·04	1·06	1·08	1·09	1·11	1·13	1·15	1·17	1·19	1·22	1·24	69
70	0·92	0·93	0·94	0·95	0·96	0·97	0·99	1·00	1·01	1·03	1·05	1·06	1·08	1·10	1·12	1·14	70
71	0·85	0·86	0·87	0·88	0·89	0·90	0·92	0·93	0·94	0·95	0·96	0·97	0·99	1·01	1·03	1·05	71
72	0·79	0·80	0·81	0·82	0·83	0·84	0·85	0·86	0·87	0·88	0·89	0·90	0·91	0·92	0·94	0·96	72
73	0·73	0·74	0·75	0·76	0·76	0·77	0·78	0·79	0·80	0·81	0·82	0·83	0·84	0·85	0·86	0·88	73
74	0·67	0·68	0·69	0·70	0·70	0·71	0·72	0·73	0·74	0·75	0·75	0·76	0·77	0·78	0·79	0·80	74
75	0·62	0·63	0·64	0·64	0·64	0·65	0·65	0·66	0·67	0·67	0·68	0·69	0·70	0·71	0·72	0·73	75
76	0·57	0·58	0·59	0·59	0·59	0·59	0·60	0·60	0·61	0·62	0·62	0·63	0·63	0·64	0·65	0·66	76
77	0·53	0·53	0·54	0·54	0·54	0·54	0·55	0·55	0·55	0·56	0·57	0·57	0·57	0·58	0·59	0·60	77
78	0·48	0·48	0·49	0·49	0·49	0·49	0·50	0·50	0·50	0·51	0·51	0·52	0·52	0·53	0·53	0·53	78
79	0·43	0·43	0·44	0·44	0·44	0·44	0·45	0·45	0·45	0·46	0·46	0·46	0·46	0·47	0·47	0·47	79
80	0·39	0·39	0·39	0·39	0·39	0·40	0·40	0·40	0·40	0·41	0·41	0·42	0·42	0·42	0·42	0·42	80
81	0·34	0·35	0·35	0·35	0·35	0·35	0·35	0·35	0·36	0·36	0·36	0·36	0·36	0·37	0·37	0·37	81
82	0·30	0·30	0·30	0·30	0·31	0·31	0·31	0·31	0·31	0·31	0·31	0·32	0·32	0·32	0·32	0·32	82
83	0·26	0·26	0·26	0·26	0·26	0·26	0·27	0·27	0·27	0·27	0·27	0·27	0·27	0·27	0·27	0·28	83
	31°	32°	33°	34°	35°	36°	37°	38°	39°	40°	41°	42°	43°	44°	45°	46°	

DECLINATION.

DECLINATION.

Lat.	47°	48°	49°	50°	51°	52°	53°	54°	55°	56°	57°	58°	59°	60°	61°	62°	63°	Lat.
°	″	″	″	″	″	″	″	″	″	″	″	″	″	″	″	″	″	°
60	2·98	3·16	3·38	3·63	3·95	4·34	4·85	5·52	6·46	7·67								60
61	2·68	2·83	3·00	3·21	3·45	3·75	4·12	4·59	5·22	6·11	7·43							61
62	2·44	2·56	2·70	2·85	3·04	3·27	3·55	3·89	4·34	4·93	5·76	7·00						62
63	2·21	2·29	2·41	2·55	2·70	2·88	3·09	3·35	3·67	4·09	4·65	5·42	6·58					63
64	2·01	2·09	2·19	2·29	2·41	2·55	2·72	2·91	3·16	3·46	3·85	4·36	5·09	6·17				64
65	1·83	1·89	1·98	2·06	2·16	2·27	2·40	2·56	2·74	2·97	3·25	3·61	4·09	4·76	5·77			65
66	1·67	1·73	1·79	1·86	1·94	2·03	2·14	2·26	2·40	2·57	2·78	3·04	3·38	3·82	4·44	5·38		66
67	1·53	1·58	1·63	1·69	1·75	1·83	1·91	2·01	2·12	2·25	2·41	2·60	2·84	3·15	3·56	4·13	4·99	67
68	1·40	1·44	1·48	1·53	1·58	1·64	1·71	1·79	1·88	1·98	2·10	2·24	2·42	2·64	2·93	3·30	3·83	68
69	1·27	1·31	1·35	1·39	1·43	1·48	1·54	1·60	1·67	1·75	1·85	1·95	2·09	2·25	2·45	2·71	3·06	69
70	1·16	1·19	1·23	1·26	1·30	1·34	1·38	1·43	1·49	1·56	1·63	1·71	1·81	1·93	2·08	2·27	2·50	70
71	1·07	1·09	1·12	1·15	1·18	1·21	1·25	1·29	1·33	1·39	1·45	1·52	1·59	1·68	1·79	1·92	2·09	71
72	0·97	0·99	1·01	1·04	1·07	1·10	1·13	1·16	1·19	1·23	1·28	1·33	1·39	1·46	1·54	1·64	1·76	72
73	0·89	0·91	0·93	0·95	0·97	0·99	1·01	1·04	1·07	1·10	1·13	1·17	1·21	1·27	1·32	1·40	1·50	73
74	0·81	0·83	0·84	0·85	0·87	0·89	0·91	0·93	0·95	0·98	1·01	1·04	1·08	1·12	1·16	1·21	1·27	74
75	0·74	0·75	0·76	0·77	0·78	0·80	0·81	0·83	0·86	0·88	0·91	0·93	0·95	0·98	1·02	1·06	1·10	75
76	0·67	0·68	0·69	0·70	0·71	0·72	0·73	0·74	0·76	0·78	0·80	0·82	0·84	0·86	0·89	0·92	0·95	76
77	0·60	0·61	0·62	0·63	0·63	0·64	0·65	0·66	0·67	0·69	0·70	0·72	0·74	0·76	0·78	0·81	0·83	77
78	0·54	0·55	0·55	0·56	0·56	0·57	0·58	0·59	0·60	0·61	0·62	0·63	0·65	0·67	0·68	0·70	0·71	78
79	0·48	0·49	0·49	0·50	0·50	0·51	0·51	0·51	0·52	0·53	0·54	0·55	0·57	0·58	0·59	0·60	0·61	79
80	0·43	0·43	0·43	0·44	0·44	0·45	0·45	0·46	0·46	0·47	0·48	0·48	0·49	0·50	0·51	0·52	0·53	80
81	0·37	0·38	0·38	0·38	0·39	0·39	0·39	0·40	0·40	0·41	0·41	0·42	0·42	0·43	0·44	0·44	0·45	81
82	0·32	0·33	0·33	0·33	0·33	0·34	0·34	0·34	0·35	0·35	0·35	0·36	0·36	0·36	0·37	0·37	0·38	82
83	0·28	0·28	0·28	0·28	0·28	0·29	0·29	0·29	0·29	0·29	0·30	0·30	0·30	0·31	0·31	0·31	0·32	83
	47°	48°	49°	50°	51°	52°	53°	54°	55°	56°	57°	58°	59°	60°	61°	62°	63°	

DECLINATION.

EX-MERIDIAN TABLE I.

LATITUDE AND DECLINATION Different Name.

Change of Altitude in one minute from Meridian Passage = A

Lat.	0°	1°	2°	3°	4°	5°	6°	7°	8°	9°	10°	11°	12°	13°	14°	Lat.
0					28·1	22 4	18·7	16·0	14·0	12·4	11·1	10·1	9·2	8·5	7·9	0
1				28·1	22·4	18·7	16·0	14·0	12·4	11·2	10·1	9 8	8·5	7·9	7·4	1
2			28·1	22·4	18·7	16·0	14·0	12·5	11·2	10·2	9·3	8·6	7·9	7·4	6·9	2
3		28·1	22·4	18·7	16·0	14·0	12·5	11·2	10·2	9·3	8·6	8·0	7·4	6·9	6·5	3
4	28·1	22·4	18·7	16·0	14·0	12·5	11·2	10·2	9·3	8·6	8·0	7·4	7·0	6·5	6·2	4
5	22·4	18·7	16·0	14·0	12·5	11·2	10·2	9·3	8·6	8·0	7·4	7·0	6·5	6·2	5·8	5
6	18·7	16·0	14·0	12·5	11·2	10·2	9·3	8·6	8·0	7·5	7·0	6·6	6·2	5·8	5·5	6
7	16·0	14·0	12·4	11·2	10·2	9·3	8·6	8·0	7·5	7·0	6·6	6·2	5·9	5·6	5·3	7
8	14·0	12·4	11·2	10·2	9·3	8·6	8·0	7·5	7·0	6·6	6·2	5·9	5·6	5·3	5·0	8
9	12·4	11·2	10·2	9·3	8·6	8·0	7·5	7·0	6·6	6·2	5·9	5·6	5·3	5·0	4·8	9
10	11·1	10·1	9·3	8·6	8·0	7·4	7·0	6·6	6·2	5·9	5·6	5·3	5·0	4·8	4·6	10
11	10·1	9·3	8·6	8·0	7·4	7·0	6·6	6·2	5·9	5·6	5·3	5·1	4·8	4·6	4·4	11
12	9·2	8·5	7·9	7·4	7·0	6·5	6·2	5·9	5·6	5·3	5·0	4·8	4·6	4·4	4·3	12
13	8·5	7·9	7·4	6·9	6·5	6·2	5·8	5·6	5·3	5·0	4·8	4·6	4·4	4·3	4·1	13
14	7·9	7·4	6·9	6·5	6·2	5·8	5·5	5·3	5·0	4·8	4·6	4·4	4·2	4·1	3·9	14
15	7·3	6·9	6·5	6·1	5·8	5·5	5·3	5·0	4·8	4·6	4·4	4·2	4·1	3·9	3·8	15
16	6·8	6·5	6·1	5·8	5·5	5·2	5·0	4·8	4·6	4·4	4·2	4·1	3·9	3·8	3·7	16
17	6·4	6·1	5·8	5·5	5·2	5·0	4·8	4·6	4·4	4·2	4·1	3·9	3·8	3·7	3·5	17
18	6·0	5·7	5·5	5·2	5·0	4·8	4·6	4·4	4·2	4·1	3·9	3·8	3·7	3·5	3·4	18
19	5·7	5·4	5·2	4·9	4·7	4·5	4·4	4·2	4·0	3·9	3·8	3·6	3·5	3·4	3·3	19
20	5·4	5·1	4·9	4·7	4·5	4·3	4·2	4·0	3·9	3·8	3·6	3·5	3·4	3·3	3·2	20
21	5·1	4·9	4·7	4·5	4·3	4·2	4·0	3·9	3·7	3·6	3·5	3·4	3·3	3·2	3·1	21
22	4·9	4·7	4·5	4·3	4·1	4 0	3·9	3·7	3·6	3·5	3·4	3·3	3·2	3·1	3·0	22
23	4·6	4·4	4·3	4·1	4·0	3·8	3·7	3·6	3·5	3·4	3·3	3·2	3·1	3·0	2·9	23
24	4·4	4·2	4·1	3·9	3·8	3·7	3·6	3·5	3·4	3·3	3·2	3·1	3·0	2·9	2·8	24
25	4·2	4·1	3·9	3·8	3·7	3·5	3·4	3·3	3·2	3·1	3·1	3·0	2·9	2·8	2·7	25
26	4·0	3·9	3·8	3·6	3·5	3·4	3·3	3·2	3·1	3·0	3·0	2·9	2·8	2·7	2·7	26
27	3·9	3·7	3·6	3·5	3·4	3·3	3·2	3·1	3·0	2·9	2·9	2·8	2·7	2·7	2·6	27
28	3·7	3·6	3·5	3·4	3·3	3·2	3·1	3·0	2·9	2·8	2·8	2·7	2·6	2·6	2·5	28
29	3·5	3·4	3·3	3·2	3·1	3·1	3·0	2·9	2·8	2·8	2·7	2·6	2·6	2·5	2·4	29
30	3·4	3·3	3·2	3·1	3·0	3·0	2·9	2·8	2·7	2·7	2·6	2·5	2·5	2·4	2·4	30
31	3·3	3·2	3·1	3·0	2·9	2·9	2·8	2·7	2·6	2·6	2·5	2·5	2·4	2·4	2·3	31
32	3·2	3·1	3·0	2·9	2·8	2·8	2·7	2·6	2·6	2·5	2·5	2·4	2·3	2·3	2·3	32
33	3·0	2·9	2·9	2·8	2·7	2·7	2·6	2·5	2·5	2·4	2·4	2·3	2·3	2·2	2·2	33
34	2·9	2·8	2·8	2·7	2·6	2·6	2·5	2·5	2·4	2·4	2·3	2·3	2·2	2·2	2·1	34
35	2·8	2·7	2·7	2·6	2·5	2·5	2·4	2·4	2·3	2·3	2·2	2·2	2·2	2·1	2·1	35
36	2·7	2·6	2·6	2·5	2·5	2·4	2·4	2·3	2·3	2·2	2·2	2·1	2·1	2·1	2·0	36
37	2·6	2·5	2·5	2·4	2·4	2·3	2·3	2·2	2·2	2·2	2·1	2·1	2·0	2·0	2·0	37
38	2·5	2·5	2·4	2·4	2·3	2·3	2·2	2·2	2·1	2·1	2·1	2·0	2·0	1·9	1·9	38
39	2·4	2·4	2·3	2·3	2·2	2·2	2·1	2·1	2·1	2·0	2·0	2·0	1·9	1·9	1·9	39
40	2·3	2·3	2·2	2·2	2·2	2·1	2·1	2·0	2·0	2·0	1·9	1·9	1·9	1·8	1·8	40
41	2·3	2·2	2·2	2·1	2·1	2·1	2·0	2·0	1·9	1·9	1·9	1·8	1·8	1·8	1·8	41
42	2·2	2·1	2·1	2·1	2·0	2·0	2·0	1·9	1·9	1·9	1·8	1·8	1·8	1·7	1·7	42
43	2·1	2·1	2·0	2·0	2·0	1·9	1·9	1·9	1·8	1·8	1·8	1·7	1·7	1·7	1·7	43
44	2·0	2·0	2·0	1·9	1·9	1·9	1·8	1·8	1·8	1·7	1·7	1·7	1·7	1·6	1·6	44
45	2·0	1·9	1·9	1·9	1·8	1·8	1·8	1·7	1·7	1·7	1·7	1·6	1·6	1·6	1·6	45
46	1·9	1·9	1·8	1·8	1·8	1·7	1·7	1·7	1·7	1·6	1·6	1·6	1·6	1·6	1·5	46
47	1·8	1·8	1·8	1·7	1·7	1·7	1·7	1·7	1·6	1·6	1·6	1·6	1·5	1·5	1·5	47
48	1·8	1·7	1·7	1·7	1·7	1·6	1·6	1·6	1·6	1·6	1·5	1·5	1·5	1·5	1·4	48
49	1·7	1·7	1·7	1·6	1·6	1·6	1·6	1·5	1·5	1·5	1·5	1·5	1·4	1·4	1·4	49
50	1·6	1·6	1·6	1·6	1·6	1·5	1·5	1·5	1·5	1·5	1·4	1·4	1·4	1·4	1·4	50
51	1·6	1·6	1·6	1·5	1·5	1·5	1·5	1·5	1·4	1·4	1·4	1·4	1·4	1·2	1·3	51
52	1·5	1·5	1·5	1·5	1·5	1·4	1·4	1·4	1·4	1·4	1·4	1·3	1·3	1·3	1·3	52
53	1·5	1·5	1·4	1·4	1·4	1·4	1·4	1·4	1·3	1·3	1·3	1·3	1·3	1·3	1·3	53
54	1·4	1·4	1·4	1·4	1·4	1·3	1·3	1·3	1·3	1·3	1·3	1·3	1·2	1·2	1·2	54
55	1·4	1·4	1·3	1·3	1·3	1·3	1·3	1·3	1·3	1·3	1·2	1·2	1·2	1·2	1·2	55
56	1·3	1·3	1·3	1·3	1·3	1·3	1·2	1·2	1·2	1·2	1·2	1·2	1·2	1·1	1·1	56
57	1·3	1·3	1·3	1·2	1·2	1·2	1·2	1·2	1·2	1·2	1·1	1·1	1·1	1·1	1·1	57
58	1·2	1·2	1·2	1·2	1·2	1·2	1·2	1·1	1·1	1·1	1·1	1·1	1·1	1·1	1·1	58
59	1·2	1·2	1·2	1·2	1·1	1·1	1·1	1·1	1·1	1·1	1·1	1·1	1·1	1·0	1·0	59
60	1·1	1·1	1·1	1·1	1·1	1·1	1·1	1·1	1·0	1·0	1·0	1·0	1·0	1·0	1·0	60
	0°	1°	2°	3°	4°	5°	6°	7°	8°	9°	10°	11°	12°	13°	14°	

Declination.

EX-MERIDIAN TABLE I.

Latitude and Declination Different Name.

Change of Altitude in one minute from Meridian Passage = A.

Lat.	15°	16°	17°	18°	19°	20°	21°	22°	23°	24°	25°	26°	27°	28°	29°	30°	Lat.
°	′′	′′	′′	′′	′′	′′	′′	′′	′′	′′	′′	′′	′′	′′	′′	′′	°
0	7·3	6·8	6·4	6·0	5·7	5·4	5·1	4·9	4·6	4·4	4·2	4·0	3·9	3·7	3·5	3·4	0
1	6·9	6·5	6·1	5·7	5·4	5·1	4·9	4·7	4·4	4·2	4·1	3·9	3·7	3·6	3·4	3·3	1
2	6·5	6·1	5·8	5·5	5·2	4·9	4·7	4·5	4·3	4·1	3·9	3·8	3·6	3·5	3·3	3·2	2
3	6·1	5·8	5·5	5·2	4·9	4·7	4·5	4·3	4·1	3·9	3·8	3·6	3·5	3·4	3·2	8·1	3
4	5·8	5·5	5·2	5·0	4·7	4·5	4·3	4·1	4·0	3·8	3·7	3·5	3·4	8·3	3·2	3·0	4
5	5·5	5·2	5·0	4·8	4·5	4·3	4·2	4·0	3·8	3·7	3·6	3·4	3·3	3·2	3·1	3·0	5
6	5·3	5·0	4·8	4·6	4·4	4·2	4·0	3·9	3·7	3·6	3·4	3·3	3·2	3·1	3·0	2·9	6
7	5·0	4·8	4·6	4·4	4·2	4·0	3·9	3·7	3·6	3·5	3·3	3·2	3·1	3·0	2·9	2·8	7
8	4·8	4·6	4·4	4·2	4·0	3·9	3·7	3·6	3·5	3·4	3·2	3·1	3·0	2·9	2·8	2·7	8
9	4·6	4·4	4·2	4·1	3·9	3·8	3·6	3·5	3·4	3·3	3·1	3·0	2·9	2·9	2·8	2·7	9
10	4·4	4·2	4·1	8·9	3·8	3·6	3·5	3·4	3·3	3·2	3·1	3·0	2·9	2·8	2·7	2·6	10
11	4·2	4·1	3·9	3·8	3·6	3·5	3·4	3·3	3·2	3·1	3·0	2·9	2·8	2·7	2·6	2·5	11
12	4·1	3·9	3·8	8·7	3·5	3·4	3·3	3·2	3·1	3·0	2·9	2·8	2·7	2·6	2·6	2·5	12
13	3·9	3·8	3·7	8·5	3·4	8·3	3·2	3·1	3·0	2·9	2·8	2·7	2·7	2·6	2·5	2·4	13
14	3·8	3·7	3·5	3·4	3·3	3·2	3·1	3·0	2·9	2·8	2·7	2·7	2·6	2·5	2·4	2·4	14
15	3·7	3·5	3·4	3·3	3·2	3·1	3·0	2·9	2·8	2·8	2·7	2·6	2·5	2·5	2·4	2·3	15
16	3·5	3·4	3·3	3·2	3·1	3·0	2·9	2·8	2·8	2·7	2·6	2·5	2·5	2·4	2·3	2·3	16
17	3·4	3·3	8·2	3·1	8·0	2·9	2·8	2·8	2·7	2·6	2·5	2·5	2·4	2·3	2·3	2·2	17
18	3·3	3·2	3·1	3·0	2·9	2·9	2·8	2·7	2·6	2·5	2·5	2·4	2·4	2·3	2·2	2·2	18
19	3·2	3·1	8·0	2·9	2·9	2·8	2·7	2·6	2·6	2·5	2·4	2·4	2·3	2·2	2·2	2·1	19
20	3·1	3·0	2·9	2·9	2·8	2·7	2·6	2·6	2·5	2·4	2·4	2·3	2·3	2·2	2·1	2·1	20
21	3·0	2·9	2·8	2·8	2·7	2·6	2·6	2·5	2·4	2·4	2·3	2·3	2·2	2·1	2·1	2·0	21
22	2·9	2·8	2·8	2·7	2·6	2·6	2·5	2·4	2·4	2·3	2·3	2·2	2·2	2·1	2·1	2·0	22
23	2·8	2·8	2·7	2·6	2·6	2·5	2·4	2·4	2·3	2·3	2·2	2·2	2·1	2·1	2·0	2·0	23
24	2·8	2·7	2·6	2·5	2·5	2·4	2·4	2·3	2·3	2·2	2·2	2·1	2·1	2·0	2·0	1·9	24
25	2·7	2·6	2·5	2·5	2·4	2·4	2·3	2·3	2·2	2·2	2·1	2·1	2·0	2·0	1·9	1·9	25
26	2·6	2·5	2·5	2·4	2·4	2·3	2·3	2·2	2·1	2·1	2·1	2·0	2·0	1·9	1·9	1·9	26
27	2·5	2·5	2·4	2·4	2·3	2·2	2·2	2·1	2·1	2·1	2·0	2·0	1·9	1·9	1·9	1·8	27
28	2·5	2·4	2·3	2·3	2·2	2·2	2·1	2·1	2·1	2·0	2·0	1·9	1·9	1·9	1·8	1·8	28
29	2·4	2·3	2·3	2·2	2·2	2·1	2·1	2·0	2·0	2·0	1·9	1·9	1·9	1·8	1·8	1·7	29
30	2·3	2·3	2·2	2·2	2·1	2·1	2·0	2·0	2·0	2·0	1·9	1·9	1·8	1·8	1·7	1·7	30
31	2·3	2·2	2·2	2·1	2·1	2·0	2·0	2·0	1·9	1·9	1·8	1·8	1·8	1·7	1·7	1·7	31
32	2·2	2·2	2·1	2·1	2·0	2·0	1·9	1·9	1·9	1·8	1·8	1·8	1·7	1·7	1·7	1·6	32
33	2·1	2·1	2·1	2·0	2·0	1·9	1·9	1·9	1·8	1·8	1·8	1·7	1·7	1·7	1·6	1·6	33
34	2·1	2·0	2·0	2·0	1·9	1·9	1·9	1·8	1·8	1·8	1·7	1·7	1·7	1·6	1·6	1·6	34
35	2·0	2·0	2·0	1·9	1·9	1·8	1·8	1·8	1·8	1·7	1·7	1·7	1·6	1·6	1·6	1·5	35
36	2·0	1·9	1·9	1·9	1·8	1·8	1·8	1·7	1·7	1·7	1·6	1·6	1·6	1·6	1·5	1·5	36
37	1·9	1·9	1·9	1·8	1·8	1·8	1·7	1·7	1·7	1·6	1·6	1·6	1·6	1·5	1·5	1·5	37
38	1·9	1·8	1·8	1·8	1·8	1·7	1·7	1·7	1·6	1·6	1·6	1·5	1·5	1·5	1·5	1·5	38
39	1·8	1·8	1·8	1·7	1·7	1·7	1·6	1·6	1·6	1·6	1·5	1·5	1·5	1·5	1·4	1·4	39
40	1·8	1·7	1·7	1·7	1·7	1·6	1·6	1·6	1·6	1·5	1·5	1·5	1·5	1·4	1·4	1·4	40
41	1·7	1·7	1·7	1·6	1·6	1·6	1·6	1·5	1·5	1·5	1·5	1·4	1·4	1·4	1·4	1·4	41
42	1·7	1·7	1·6	1·6	1·6	1·6	1·5	1·5	1·5	1·5	1·5	1·4	1·4	1·4	1·4	1·3	42
43	1·6	1·6	1·6	1·6	1·5	1·5	1·5	1·5	1·4	1·4	1·4	1·4	1·4	1·3	1·3	1·3	43
44	1·6	1·6	1·5	1·5	1·5	1·5	1·5	1·4	1·4	1·4	1·4	1·4	1·3	1·3	1·3	1·3	44
45	1·5	1·5	1·5	1·5	1·5	1·4	1·4	1·4	1·4	1·4	1·3	1·3	1·3	1·3	1·3	1·2	45
46	1·5	1·5	1·5	1·4	1·4	1·4	1·4	1·4	1·3	1·3	1·3	1·3	1·3	1·3	1·2	1·2	46
47	1·5	1·4	1·4	1·4	1·4	1·4	1·3	1·3	1·3	1·3	1·3	1·3	1·2	1·2	1·2	1·2	47
48	1·4	1·4	1·4	1·4	1·4	1·3	1·3	1·3	1·3	1·3	1·2	1·2	1·2	1·2	1·2	1·2	48
49	1·4	1·4	1·3	1·3	1·3	1·3	1·3	1·3	1·2	1·2	1·2	1·2	1·2	1·2	1·2	1·1	49
50	1·3	1·3	1·3	1·3	1·3	1·3	1·3	1·2	1·2	1·2	1·2	1·2	1·2	1·1	1·1	1·1	50
51	1·3	1·3	1·3	1·3	1·2	1·2	1·2	1·2	1·2	1·2	1·2	1·1	1·1	1·1	1·1	1·1	51
52	1·3	1·3	1·3	1·2	1·2	1·2	1·2	1·2	1·1	1·1	1·1	1·1	1·1	1·1	1·1	1·1	52
53	1·2	1·2	1·2	1·2	1·2	1·2	1·2	1·1	1·1	1·1	1·1	1·1	1·1	1·1	1·0	1·0	53
54	1·2	1·2	1·2	1·2	1·1	1·1	1·1	1·1	1·1	1·1	1·1	1·0	1·0	1·0	1·0	1·0	54
55	1·2	1·1	1·1	1·1	1·1	1·1	1·1	1·1	1·1	1·1	1·0	1·0	1·0	1·0	1·0		55
56	1·1	1·1	1·1	1·1	1·1	1·1	1·1	1·0	1·0	1·0	1·0	1·0	1·0	1·0			56
57	1·1	1·1	1·1	1·1	1·0	1·0	1·0	1·0	1·0	1·0	1·0	1·0	1·0				57
58	1·1	1·0	1·0	1·0	1·0	1·0	1·0	1·0	1·0	1·0	1·0	0·9					58
59	1·0	1·0	1·0	1·0	1·0	1·0	1·0	1·0	0·9	0·9	0·9						59
60	1·0	1·0	1·0	1·0	0·9	0·9	0·9	0·9	0·9	0·9	0·9						60
	15°	16°	17°	18°	19°	20°	21°	22°	23°	24°	25°	26°	27°	28°	29°	30°	

Declination.

EX-MERIDIAN TABLE I.

LATITUDE AND DECLINATION Different Name.

Change of Altitude in one minute from Meridian Passage = A.

Lat.	31°	32°	33°	34°	35°	36°	37°	38°	39°	40°	41°	42°	43°	44°	45°	46°	Lat.
	″	″	″	″	″	″	″	″	″	″	″	″	″	″	″	″	
0	3.3	3.1	3.0	2.9	2.8	2.7	2.6	2.5	2.4	2.3	2.3	2.2	2.1	2.0	2.0	1.9	0
1	3.2	3.1	2.9	2.8	2.7	2.6	2.6	2.5	2.4	2.3	2.2	2.1	2.1	2.0	1.9	1.9	1
2	3.1	3.0	2.9	2.8	2.7	2.6	2.5	2.4	2.3	2.3	2.2	2.1	2.0	2.0	1.9	1.8	2
3	3.0	2.9	2.8	2.7	2.6	2.5	2.4	2.4	2.3	2.2	2.2	2.1	2.0	1.9	1.9	1.8	3
4	2.9	2.8	2.7	2.6	2.6	2.5	2.4	2.3	2.2	2.2	2.1	2.0	2.0	1.8	1.8	1.8	4
5	2.9	2.8	2.7	2.6	2.5	2.4	2.3	2.3	2.2	2.1	2.1	2.0	1.9	1.9	1.8	1.8	5
6	2.8	2.7	2.6	2.5	2.4	2.4	2.3	2.2	2.2	2.1	2.0	2.0	1.9	1.8	1.8	1.7	6
7	2.7	2.6	2.5	2.5	2.4	2.3	2.2	2.2	2.1	2.0	2.0	1.9	1.9	1.8	1.8	1.7	7
8	2.7	2.6	2.5	2.4	2.3	2.3	2.2	2.1	2.1	2.0	1.9	1.9	1.8	1.8	1.7	1.7	8
9	2.6	2.5	2.4	2.4	2.3	2.2	2.2	2.1	2.0	2.0	1.9	1.9	1.8	1.8	1.7	1.6	9
10	2.5	2.5	2.4	2.3	2.2	2.2	2.1	2.1	2.0	1.9	1.9	1.8	1.8	1.7	1.7	1.6	10
11	2.5	2.4	2.3	2.3	2.2	2.2	2.1	2.0	2.0	1.9	1.8	1.8	1.7	1.7	1.6	1.6	11
12	2.4	2.3	2.3	2.2	2.2	2.1	2.0	2.0	1.9	1.9	1.8	1.8	1.7	1.7	1.6	1.6	12
13	2.4	2.3	2.2	2.2	2.1	2.1	2.0	1.9	1.9	1.8	1.8	1.7	1.7	1.6	1.6	1.6	13
14	2.3	2.3	2.2	2.1	2.1	2.0	2.0	1.9	1.9	1.8	1.8	1.7	1.7	1.6	1.6	1.5	14
15	2.3	2.2	2.1	2.1	2.0	2.0	1.9	1.9	1.8	1.8	1.7	1.7	1.6	1.6	1.6	1.5	15
16	2.2	2.2	2.1	2.0	2.0	1.9	1.9	1.8	1.8	1.7	1.7	1.6	1.6	1.6	1.5	1.5	16
17	2.2	2.1	2.1	2.0	2.0	1.9	1.9	1.8	1.8	1.7	1.7	1.6	1.6	1.5	1.5	1.5	17
18	2.1	2.1	2.0	2.0	1.9	1.9	1.8	1.8	1.7	1.7	1.6	1.6	1.6	1.5	1.5	1.4	18
19	2.1	2.0	2.0	1.9	1.9	1.8	1.8	1.7	1.7	1.7	1.6	1.6	1.5	1.5	1.5	1.4	19
20	2.0	2.0	1.9	1.9	1.8	1.8	1.7	1.7	1.7	1.6	1.6	1.6	1.5	1.5	1.4	1.4	20
21	2.0	2.0	1.9	1.9	1.8	1.8	1.7	1.7	1.6	1.6	1.6	1.5	1.5	1.5	1.4	1.4	21
22	2.0	1.9	1.9	1.8	1.8	1.7	1.7	1.7	1.6	1.6	1.5	1.5	1.5	1.4	1.4	1.4	22
23	1.9	1.9	1.8	1.8	1.8	1.7	1.7	1.6	1.6	1.6	1.5	1.5	1.4	1.4	1.4	1.3	23
24	1.9	1.8	1.8	1.7	1.7	1.6	1.6	1.6	1.5	1.5	1.5	1.5	1.4	1.4	1.4	1.3	24
25	1.8	1.8	1.8	1.7	1.7	1.6	1.6	1.6	1.5	1.5	1.5	1.4	1.4	1.4	1.3	1.3	25
26	1.8	1.8	1.7	1.7	1.7	1.6	1.6	1.6	1.5	1.5	1.5	1.4	1.4	1.4	1.3	1.3	26
27	1.8	1.7	1.7	1.7	1.6	1.6	1.6	1.5	1.5	1.5	1.4	1.4	1.4	1.3	1.3	1.3	27
28	1.7	1.7	1.6	1.6	1.6	1.5	1.5	1.5	1.5	1.4	1.4	1.4	1.3	1.3	1.3	1.3	28
29	1.7	1.7	1.6	1.6	1.6	1.5	1.5	1.5	1.4	1.4	1.4	1.4	1.3	1.3	1.3	1.2	29
30	1.7	1.6	1.6	1.6	1.5	1.5	1.5	1.5	1.4	1.4	1.4	1.3	1.3	1.3	1.2	1.2	30
31	1.6	1.6	1.6	1.5	1.5	1.5	1.5	1.4	1.4	1.4	1.3	1.3	1.3	1.3	1.2	1.2	31
32	1.6	1.6	1.5	1.5	1.5	1.4	1.4	1.4	1.4	1.3	1.3	1.3	1.3	1.3	1.2	1.2	32
33	1.6	1.5	1.5	1.5	1.5	1.4	1.4	1.4	1.3	1.3	1.3	1.3	1.2	1.2	1.2	1.2	33
34	1.5	1.5	1.5	1.5	1.4	1.4	1.4	1.4	1.3	1.3	1.3	1.3	1.2	1.2	1.2	1.2	34
35	1.5	1.5	1.5	1.4	1.4	1.4	1.4	1.3	1.3	1.3	1.3	1.2	1.2	1.2	1.2	1.1	35
36	1.5	1.5	1.4	1.4	1.4	1.4	1.3	1.3	1.3	1.3	1.2	1.2	1.2	1.2	1.1	1.1	36
37	1.5	1.4	1.4	1.4	1.4	1.3	1.3	1.3	1.3	1.2	1.2	1.2	1.2	1.2	1.1	1.1	37
38	1.4	1.4	1.4	1.4	1.3	1.3	1.3	1.3	1.2	1.2	1.2	1.2	1.2	1.1	1.1	1.1	38
39	1.4	1.4	1.4	1.3	1.3	1.3	1.3	1.2	1.2	1.2	1.2	1.2	1.1	1.1	1.1		39
40	1.4	1.3	1.3	1.3	1.3	1.2	1.2	1.2	1.2	1.2	1.2	1.1	1.1	1.1			40
41	1.3	1.3	1.3	1.3	1.3	1.2	1.2	1.2	1.2	1.2	1.1	1.1	1.1				41
42	1.3	1.3	1.3	1.2	1.2	1.2	1.2	1.2	1.2	1.1	1.1	1.1					42
43	1.3	1.3	1.2	1.2	1.2	1.2	1.2	1.2	1.1	1.1	1.1						43
44	1.3	1.2	1.2	1.2	1.2	1.2	1.1	1.1	1.1								44
45	1.2	1.2	1.2	1.2	1.2	1.1	1.1	1.1									45
46	1.2	1.2	1.2	1.2	1.1	1.1	1.1	1.1									46
47	1.2	1.2	1.1	1.1	1.1	1.1	1.1										47
48	1.1	1.1	1.1	1.1	1.1	1.1											48
49	1.1	1.1	1.1	1.1	1.1												49
50	1.1	1.1	1.1	1.1												0.9	50
51	1.1	1.1	1.0												0.9	0.9	51
52	1.0	1.0												0.9	0.9	0.9	52
53	1.0												0.9	0.9	0.8	0.8	53
54												0.9	0.9	0.8	0.8	0.8	54
55											0.9	0.8	0.8	0.8	0.8	0.8	55
56										0.8	0.8	0.8	0.8	0.8	0.8	0.8	56
57									0.8	0.8	0.8	0.8	0.8	0.8	0.8	0.8	57
58								0.8	0.8	0.8	0.8	0.8	0.8	0.8	0.8	0.8	58
59						0.8	0.8	0.8	0.8	0.8	0.8	0.8	0.8	0.8	0.7	0.7	59
60					0.8	0.8	0.8	0.8	0.8	0.8	0.8	0.8	0.8	0.7	0.7	0.7	60
	31°	32°	33°	34°	35°	36°	37°	38°	39°	40°	41°	42°	43°	44°	45°	46°	

Latitude and Declination SAME NAME. The quantities printed below are to be taken out when the body is circumpolar and at its LOWER TRANSIT.

Latitude & Declination **Same Name.**

EX-MERIDIAN TABLE I.

LATITUDE AND DECLINATION Different Name.

Change of Altitude in one minute from Meridian Passage = A.

Declination.

Lat.	47°	48°	49°	50°	51°	52°	53°	54°	55°	56°	57°	58°	59°	60°	61°	62°	63°	Lat.
0	1.8	1.8	1.7	1.7	1.6	1.5	1.5	1.4	1.4	1.3	1.3	1.2	1.2	1.1	1.1	1.0	1.0	0
1	1.8	1.7	1.7	1.6	1.6	1.5	1.5	1.4	1.4	1.3	1.3	1.2	1.2	1.1	1.1	1.0	1.0	1
2	1.8	1.7	1.7	1.6	1.5	1.5	1.4	1.4	1.3	1.3	1.3	1.2	1.2	1.1	1.1	1.0	1.0	2
3	1.8	1.7	1.6	1.6	1.5	1.5	1.4	1.4	1.3	1.3	1.2	1.2	1.1	1.1	1.1	1.0	1.0	3
4	1.7	1.7	1.6	1.6	1.5	1.5	1.4	1.4	1.3	1.3	1.2	1.2	1.1	1.1	1.1	1.0	1.0	4
5	1.7	1.6	1.6	1.5	1.5	1.4	1.4	1.3	1.3	1.3	1.2	1.2	1.1	1.1	1.0	1.0	1.0	5
6	1.7	1.6	1.6	1.5	1.5	1.4	1.4	1.3	1.3	1.2	1.2	1.2	1.1	1.1	1.0	1.0	1.0	6
7	1.6	1.6	1.5	1.5	1.4	1.4	1.4	1.3	1.3	1.2	1.2	1.1	1.1	1.1	1.0	1.0	0.9	7
8	1.6	1.6	1.5	1.5	1.4	1.4	1.3	1.3	1.3	1.2	1.2	1.1	1.1	1.1	1.0	1.0	0.9	8
9	1.6	1.6	1.5	1.5	1.4	1.4	1.3	1.3	1.2	1.2	1.2	1.1	1.1	1.0	1.0	1.0	0.9	9
10	1.6	1.5	1.5	1.4	1.4	1.4	1.3	1.3	1.2	1.2	1.1	1.1	1.1	1.0	1.0	1.0	0.9	10
11	1.6	1.5	1.5	1.4	1.4	1.3	1.3	1.3	1.2	1.2	1.1	1.1	1.1	1.0	1.0	1.0	0.9	11
12	1.5	1.5	1.4	1.4	1.4	1.3	1.3	1.3	1.2	1.2	1.1	1.1	1.1	1.0	1.0	0.9	0.9	12
13	1.5	1.5	1.4	1.4	1.3	1.3	1.3	1.2	1.2	1.2	1.1	1.1	1.0	1.0	0.9	0.9	0.9	13
14	1.5	1.4	1.4	1.4	1.3	1.3	1.3	1.2	1.2	1.1	1.1	1.1	1.0	1.0	1.0	0.9	0.9	14
15	1.5	1.4	1.4	1.4	1.3	1.3	1.2	1.2	1.2	1.1	1.1	1.1	1.0	1.0	1.0	0.9	0.9	15
16	1.4	1.4	1.4	1.3	1.3	1.3	1.2	1.2	1.1	1.1	1.1	1.1	1.0	1.0	1.0	0.9	0.9	16
17	1.4	1.4	1.4	1.3	1.3	1.2	1.2	1.2	1.1	1.1	1.1	1.1	1.0	1.0	0.9	0.9	0.9	17
18	1.4	1.4	1.3	1.3	1.3	1.2	1.2	1.2	1.1	1.1	1.1	1.0	1.0	1.0	0.9	0.9	0.9	18
19	1.4	1.4	1.3	1.3	1.2	1.2	1.2	1.1	1.1	1.1	1.0	1.0	1.0	1.0	0.9	0.9	0.9	19
20	1.4	1.3	1.3	1.3	1.2	1.2	1.2	1.1	1.1	1.1	1.0	1.0	1.0	1.0	0.9	0.9	0.8	20
21	1.4	1.3	1.3	1.3	1.2	1.2	1.2	1.1	1.1	1.1	1.0	1.0	1.0	1.0	0.9	0.9	0.8	21
22	1.3	1.3	1.3	1.2	1.2	1.2	1.1	1.1	1.1	1.0	1.0	1.0	1.0	0.9	0.9	0.9		22
23	1.3	1.3	1.3	1.2	1.2	1.2	1.1	1.1	1.1	1.0	1.0	1.0	0.9	0.9	0.9			23
24	1.3	1.3	1.2	1.2	1.1	1.1	1.1	1.1	1.0	1.0	1.0	1.0	0.9	0.9				24
25	1.3	1.2	1.2	1.2	1.2	1.1	1.1	1.1	1.0	1.0	1.0	1.0	0.9					25
26	1.3	1.2	1.2	1.2	1.1	1.1	1.1	1.1	1.0	1.0	1.0	0.9						26
27	1.2	1.2	1.2	1.2	1.1	1.1	1.1	1.0	1.0	1.0	1.0							27
28	1.2	1.2	1.2	1.1	1.1	1.1	1.1	1.0	1.0	1.0								28
29	1.2	1.2	1.2	1.1	1.1	1.1	1.0	1.0	1.0									29
30	1.2	1.2	1.1	1.1	1.1	1.1	1.0	1.0										30
31	1.2	1.2	1.1	1.1	1.1	1.0	1.0											31
32	1.2	1.1	1.1	1.1	1.1	1.0												32
33	1.1	1.1	1.1	1.1	1.1												0.8	33
34	1.1	1.1	1.1	1.1												0.8	0.7	34
35	1.1	1.1	1.1												0.8	0.8	0.7	35
36	1.1	1.1												0.8	0.8	0.8	0.7	36
37	1.1												0.8	0.8	0.8	0.7	0.7	37
38												0.8	0.8	0.8	0.8	0.7	0.7	38
39											0.8	0.8	0.8	0.8	0.8	0.7	0.7	39
40										0.8	0.8	0.8	0.8	0.8	0.8	0.7	0.7	40
41									0.9	0.8	0.8	0.8	0.8	0.8	0.7	0.7	0.7	41
42								0.9	0.8	0.8	0.8	0.8	0.8	0.7	0.7	0.7	0.7	42
43							0.9	0.9	0.8	0.8	0.8	0.8	0.8	0.7	0.7	0.7	0.7	43
44						0.9	0.9	0.8	0.8	0.8	0.8	0.8	0.8	0.7	0.7	0.7	0.7	44
45					0.9	0.9	0.8	0.8	0.8	0.8	0.8	0.8	0.7	0.7	0.7	0.7	0.7	45
46				0.9	0.9	0.9	0.8	0.8	0.8	0.8	0.8	0.8	0.7	0.7	0.7	0.7	0.7	46
47			0.9	0.9	0.9	0.8	0.8	0.8	0.8	0.8	0.8	0.7	0.7	0.7	0.7	0.7	0.6	47
48		0.9	0.9	0.9	0.8	0.8	0.8	0.8	0.8	0.8	0.8	0.7	0.7	0.7	0.7	0.7	0.6	48
49	0.0	0.9	0.9	0.8	0.8	0.8	0.8	0.8	0.8	0.7	0.7	0.7	0.7	0.7	0.7	0.6	0.6	49
50	0.9	0.9	0.8	0.8	0.8	0.8	0.8	0.8	0.8	0.7	0.7	0.7	0.7	0.7	0.7	0.6	0.6	50
51	0.9	0.8	0.8	0.8	0.8	0.8	0.8	0.8	0.7	0.7	0.7	0.7	0.7	0.7	0.7	0.6	0.6	51
52	0.8	0.8	0.8	0.8	0.8	0.8	0.8	0.7	0.7	0.7	0.7	0.7	0.7	0.7	0.6	0.6	0.6	52
53	0.8	0.8	0.8	0.8	0.8	0.8	0.7	0.7	0.7	0.7	0.7	0.7	0.7	0.6	0.6	0.6	0.6	53
54	0.8	0.8	0.8	0.8	0.8	0.7	0.7	0.7	0.7	0.7	0.7	0.7	0.6	0.6	0.6	0.6	0.6	54
55	0.8	0.8	0.8	0.7	0.7	0.7	0.7	0.7	0.7	0.7	0.7	0.6	0.6	0.6	0.6	0.6	0.6	55
56	0.8	0.8	0.7	0.7	0.7	0.7	0.7	0.7	0.7	0.7	0.6	0.6	0.6	0.6	0.6	0.6	0.6	56
57	0.8	0.7	0.7	0.7	0.7	0.7	0.7	0.7	0.7	0.6	0.6	0.6	0.6	0.6	0.6	0.6	0.6	57
58	0.7	0.7	0.7	0.7	0.7	0.7	0.7	0.7	0.6	0.6	0.6	0.6	0.6	0.6	0.6	0.6	0.6	58
59	0.7	0.7	0.7	0.7	0.7	0.7	0.6	0.6	0.6	0.6	0.6	0.6	0.6	0.6	0.6	0.6	0.5	59
60	0.7	0.7	0.7	0.7	0.7	0.7	0.6	0.6	0.6	0.6	0.6	0.6	0.6	0.6	0.6	0.6	0.5	60
	47°	48°	49°	50°	51°	52°	53°	54°	55°	56°	57°	58°	59°	60°	61°	62°	63°	

Latitude and Declination SAME NAME. The quantities printed below are to be taken out when the body is circumpolar and at its LOWER TRANSIT.

Latitude & Declination Same Name.

EX-MERIDIAN TABLE I.

Latitude and Declination DIFFERENT NAME.

Change of Altitude in one minute from Meridian Passage=A.

DECLINATION.

Lat.	0°	1°	2°	3°	4°	5°	6°	7°	8°	9°	10°	11°	12°	13°	14°	Lat.
°	*"*	*"*	*"*	*"*	*"*	*"*	*"*	*"*	*"*	*"*	*"*	*"*	*"*	*"*	*"*	°
60	1·13	1·12	1·11	1·10	1·09	1·08	1·07	1·06	1·05	1·04	1·03	1·02	1·01	1·00	0·99	60
61	1·09	1·08	1·07	1·06	1·05	1·04	1·03	1·02	1·01	1·00	0·99	0·98	0·97	0·96	0·96	61
62	1·04	1·03	1·03	1·02	1·01	1·00	0·99	0·98	0·97	0·96	0·95	0·94	0·94	0·93	0·92	62
63	1·00	0·99	0·98	0·97	0·97	0·96	0·95	0·94	0·93	0·92	0·91	0·90	0·90	0·90	0·89	63
64	0·96	0·95	0·94	0·93	0·93	0·92	0·91	0·90	0·89	0·89	0·88	0·87	0·87	0·86	0·85	64
65	0·92	0·91	0·90	0·89	0·89	0·88	0·87	0·86	0·86	0·85	0·85	0·84	0·83	0·82	0·81	65
66	0·87	0·87	0·86	0·85	0·85	0·84	0·84	0·83	0·82	0·81	0·81	0·80	0·80	0·79	0·78	66
67	0·83	0·83	0·82	0·81	0·81	0·80	0·80	0·79	0·79	0·78	0·78	0·77	0·76	0·76	0·75	67
68	0·79	0·79	0·78	0·78	0·77	0·76	0·76	0·75	0·75	0·75	0·74	0·74	0·73	0·73	0·72	68
69	0·75	0·75	0·74	0·74	0·73	0·73	0·72	0·72	0·71	0·71	0·71	0·71	0·70	0·70	0·69	69
70	0·71	0·71	0·70	0·70	0·69	0·69	0·69	0·68	0·68	0·67	0·67	0·67	0·67	0·67	0·66	70
71	0·67	0·67	0·66	0·66	0·66	0·65	0·65	0·64	0·63	0·63	0·63	0·63	0·63	0·63	0·62	71
72	0·64	0·63	0·63	0·62	0·62	0·61	0·61	0·61	0·61	0·61	0·60	0·60	0·60	0·59		72
73	0·60	0·60	0·59	0·59	0·59	0·58	0·58	0·58	0·58	0·57	0·57	0·57	0·56			73
74	0·56	0·56	0·55	0·55	0·55	0·54	0·54	0·54	0·54	0·54	0·53	0·53				74
75	0·52	0·52	0·52	0·52	0·51	0·51	0·51	0·51	0·51	0·50	0·50					75
76	0·49	0·49	0·49	0·48	0·48	0·48	0·48	0·47	0·47	0·47						76
77	0·45	0·45	0·45	0·45	0·44	0·44	0·44	0·44	0·44							77
78	0·42	0·42	0·41	0·41	0·41	0·41	0·41	0·40								78
79	0·38	0·38	0·37	0·37	0·37	0·37	0·37									79
80	0·35	0·35	0·34	0·34	0·34	0·34										80
81	0·31	0·31	0·31	0·31	0·31										0·30	81
82	0·28	0·28	0·27	0·27										0·27	0·27	82
83	0·24	0·24	0·24										0·23	0·23	0·23	83

Latitude and Declination SAME NAME. The quantities printed below are to be taken out when the body is circumpolar and at its LOWER TRANSIT.

| | 0° | 1° | 2° | 3° | 4° | 5° | 6° | 7° | 8° | 9° | 10° | 11° | 12° | 13° | 14° | |

DECLINATION.

(Different Name) **DECLINATION.**

Lat.	15°	16°	17°	18°	19°	20°	21°	22°	23°	24°	25°	26°	27°	28°	29°	30°	Lat.
°	*"*	*"*	*"*	*"*	*"*	*"*	*"*	*"*	*"*	*"*	*"*	*"*	*"*	*"*	*"*	*"*	°
60	0·98	0·97	0·96	0·95	0·95	0·94	0·93	0·92	0·91	0·90	0·89						60
61	0·95	0·94	0·93	0·92	0·91	0·91	0·90	0·89	0·88	0·87							61
62	0·91	0·90	0·90	0·89	0·88	0·87	0·87	0·86	0·85								62
63	0·88	0·87	0·86	0·86	0·85	0·84	0·84	0·83									63
64	0·84	0·83	0·83	0·82	0·82	0·81	0·81										64
65	0·81	0·80	0·80	0·79	0·78	0·78										0·72	65
66	0·78	0·78	0·77	0·76	0·76										0·70	0·70	66
67	0·75	0·74	0·74	0·73										0·68	0·68	0·67	67
68	0·72	0·71	0·71										0·66	0·65	0·65	0·64	68
69	0·68	0·68										0·63	0·63	0·62	0·62	0·61	69
70	0·65										0·61	0·60	0·60	0·60	0·59	0·59	70
71										0·59	0·58	0·57	0·57	0·57	0·57	0·56	71
72									0·56	0·56	0·55	0·54	0·54	0·54	0·53	0·53	72
73								0·53	0·53	0·53	0·52	0·51	0·51	0·51	0·50	0·50	73
74							0·51	0·51	0·50	0·50	0·50	0·49	0·49	0·49	0·48	0·48	74
75						0·48	0·48	0·48	0·47	0·47	0·47	0·47	0·46	0·46	0·46	0·45	75
76					0·45	0·45	0·45	0·44	0·44	0·44	0·44	0·44	0·43	0·43	0·43	0·42	76
77				0·42	0·42	0·42	0·42	0·41	0·41	0·41	0·41	0·41	0·40	0·40	0·40	0·40	77
78			0·39	0·39	0·39	0·39	0·39	0·39	0·38	0·38	0·38	0·38	0·38	0·38	0·37	0·37	78
79		0·36	0·36	0·36	0·36	0·36	0·36	0·36	0·35	0·35	0·35	0·35	0·35	0·35	0·35	0·34	79
80	0·33	0·33	0·33	0·33	0·33	0·33	0·32	0·32	0·32	0·32	0·32	0·32	0·32	0·32	0·32	0·31	80
81	0·30	0·30	0·30	0·30	0·29	0·29	0·29	0·29	0·29	0·29	0·29	0·29	0·29	0·29	0·29	0·28	81
82	0·27	0·27	0·26	0·26	0·26	0·26	0·26	0·26	0·26	0·26	0·26	0·26	0·26	0·26	0·26	0·26	82
83	0·23	0·23	0·23	0·23	0·23	0·23	0·23	0·23	0·23	0·23	0·23	0·23	0·23	0·23	0·23	0·23	83

Latitude and Declination SAME NAME. The quantities printed below are to be taken out when the body is circumpolar and at its LOWER TRANSIT.

| | 15° | 16° | 17° | 18° | 19° | 20° | 21° | 22° | 23° | 24° | 25° | 26° | 27° | 28° | 29° | 30° | |

DECLINATION. (Same Name)

Supplementary Table to 83°.

EX-MERIDIAN TABLE I.
Latitude and Declination SAME NAME.
Lower Transit.
Change of Altitude in one Minute from Meridian Passage=A.

DECLINATION.

Lat.	31°	32°	33°	34°	35°	36°	37°	38°	39°	40°	41°	42°	43°	44°	45°	46°	Lat.
60					0·81	0·80	0·79	0·78	0·77	0·76	0·75	0·75	0·74	0·73	0·72	0·71	60
61				0·79	0·79	0·78	0·77	0·76	0·75	0·74	0·73	0·72	0·72	0·71	0·70	0·69	61
62			0·78	0·77	0·76	0·76	0·75	0·74	0·73	0·72	0·71	0·70	0·70	0·69	0·68	0·67	62
63		0·76	0·75	0·75	0·74	0·74	0·73	0·72	0·71	0·70	0·69	0·68	0·68	0·67	0·66	0·65	63
64	0·74	0·74	0·73	0·73	0·71	0·71	0·70	0·70	0·69	0·68	0·67	0·66	0·66	0·65	0·64	0·63	64
65	0·72	0·71	0·70	0·70	0·69	0·68	0·68	0·67	0·66	0·66	0·65	0·64	0·64	0·63	0·62	0·62	65
66	0·69	0·68	0·68	0·67	0·67	0·66	0·65	0·65	0·64	0·64	0·63	0·62	0·62	0·61	0·60	0·60	66
67	0·66	0·65	0·65	0·64	0·64	0·64	0·63	0·63	0·62	0·62	0·61	0·60	0·60	0·59	0·58	0·58	67
68	0·63	0·63	0·63	0·62	0·62	0·61	0·61	0·60	0·60	0·60	0·59	0·58	0·58	0·57	0·56	0·56	68
69	0·61	0·61	0·61	0·60	0·60	0·59	0·59	0·58	0·58	0·58	0·57	0·56	0·56	0·55	0·54	0·54	69
70	0·59	0·58	0·58	0·57	0·57	0·57	0·56	0·56	0·55	0·55	0·54	0·54	0·53	0·53	0·52	0·52	70
71	0·56	0·56	0·55	0·55	0·54	0·54	0·53	0·53	0·52	0·52	0·51	0·51	0·51	0·50	0·50	0·50	71
72	0·53	0·53	0·52	0·52	0·52	0·51	0·51	0·51	0·50	0·50	0·49	0·49	0·49	0·48	0·48	0·48	72
73	0·50	0·50	0·50	0·50	0·50	0·49	0·49	0·49	0·48	0·48	0·47	0·47	0·47	0·46	0·46	0·46	73
74	0·48	0·47	0·47	0·47	0·47	0·47	0·46	0·46	0·46	0·46	0·45	0·45	0·45	0·44	0·44	0·43	74
75	0·45	0·45	0·45	0·45	0·44	0·44	0·44	0·44	0·43	0·43	0·43	0·42	0·42	0·42	0·42	0·41	75
76	0·42	0·42	0·42	0·42	0·41	0·41	0·41	0·41	0·40	0·40	0·40	0·40	0·40	0·40	0·39	0·39	76
77	0·40	0·39	0·39	0·39	0·39	0·38	0·38	0·38	0·38	0·38	0·38	0·38	0·38	0·38	0·37	0·37	77
78	0·37	0·37	0·37	0·36	0·36	0·36	0·36	0·35	0·35	0·35	0·35	0·35	0·35	0·35	0·35	0·35	78
79	0·34	0·34	0·34	0·34	0·34	0·34	0·33	0·33	0·33	0·33	0·33	0·33	0·33	0·33	0·32	0·32	79
80	0·31	0·31	0·31	0·31	0·31	0·31	0·31	0·30	0·30	0·30	0·30	0·30	0·30	0·30	0·29	0·29	80
81	0·28	0·28	0·28	0·28	0·28	0·28	0·28	0·28	0·28	0·27	0·27	0·27	0·27	0·27	0·27	0·27	81
82	0·25	0·25	0·25	0·25	0·25	0·25	0·25	0·25	0·25	0·25	0·25	0·25	0·24	0·24	0·24	0·24	82
83	0·22	0·22	0·22	0·22	0·22	0·22	0·22	0·22	0·22	0·22	0·22	0·22	0·22	0·22	0·21	0·21	83
	31°	32°	33°	34°	35°	36°	37°	38°	39°	40°	41°	42°	43°	44°	45°	46°	

DECLINATION.

DECLINATION.

Lat.	47°	48°	49°	50°	51°	52°	53°	54°	55°	56°	57°	58°	59°	60°	61°	62°	63°	Lat.
60	0·70	0·69	0·68	0·67	0·66	0·65	0·64	0·63	0·62	0·61	0·60	0·59	0·58	0·57	0·56	0·54	0·53	60
61	0·69	0·68	0·67	0·66	0·65	0·64	0·63	0·62	0·61	0·60	0·59	0·58	0·57	0·56	0·54	0·53	0·52	61
62	0·67	0·66	0·65	0·64	0·63	0·62	0·61	0·61	0·60	0·59	0·58	0·57	0·56	0·55	0·53	0·52	0·51	62
63	0·65	0·64	0·63	0·62	0·61	0·60	0·60	0·59	0·58	0·57	0·56	0·55	0·54	0·53	0·52	0·51	0·50	63
64	0·63	0·62	0·61	0·60	0·59	0·59	0·58	0·57	0·56	0·55	0·54	0·54	0·53	0·52	0·51	0·50	0·49	64
65	0·61	0·60	0·60	0·59	0·58	0·57	0·57	0·56	0·55	0·54	0·53	0·52	0·52	0·51	0·50	0·49	0·48	65
66	0·59	0·58	0·58	0·57	0·56	0·55	0·55	0·54	0·53	0·53	0·52	0·51	0·51	0·50	0·49	0·48	0·47	66
67	0·57	0·56	0·56	0·55	0·55	0·54	0·54	0·53	0·52	0·52	0·51	0·50	0·50	0·49	0·48	0·47	0·46	67
68	0·55	0·54	0·54	0·53	0·53	0·52	0·52	0·51	0·51	0·50	0·50	0·49	0·48	0·47	0·46	0·46	0·44	68
69	0·53	0·52	0·52	0·51	0·51	0·50	0·50	0·49	0·49	0·48	0·48	0·47	0·46	0·45	0·44	0·44	0·43	69
70	0·51	0·51	0·50	0·50	0·49	0·49	0·48	0·48	0·47	0·46	0·46	0·45	0·45	0·44	0·43	0·42	0·42	70
71	0·49	0·49	0·48	0·48	0·47	0·47	0·46	0·46	0·46	0·45	0·45	0·44	0·43	0·43	0·42	0·41	0·40	71
72	0·47	0·47	0·46	0·46	0·45	0·45	0·44	0·44	0·44	0·43	0·43	0·42	0·41	0·41	0·40	0·39	0·38	72
73	0·45	0·45	0·44	0·44	0·43	0·43	0·42	0·42	0·42	0·42	0·41	0·41	0·40	0·39	0·38	0·38	0·37	73
74	0·43	0·43	0·42	0·42	0·41	0·41	0·40	0·40	0·40	0·40	0·39	0·38	0·38	0·37	0·37	0·36	0·35	74
75	0·41	0·41	0·40	0·40	0·40	0·39	0·39	0·38	0·38	0·38	0·37	0·37	0·36	0·36	0·35	0·35	0·34	75
76	0·39	0·39	0·38	0·38	0·38	0·37	0·37	0·36	0·36	0·36	0·35	0·35	0·34	0·34	0·34	0·33	0·33	76
77	0·37	0·37	0·36	0·36	0·36	0·35	0·35	0·34	0·34	0·34	0·33	0·33	0·32	0·32	0·32	0·31	0·31	77
78	0·34	0·34	0·33	0·33	0·33	0·33	0·32	0·32	0·32	0·32	0·31	0·31	0·30	0·30	0·30	0·29	0·29	78
79	0·32	0·32	0·31	0·31	0·31	0·30	0·30	0·30	0·30	0·30	0·30	0·29	0·29	0·28	0·28	0·28	0·27	79
80	0·29	0·29	0·29	0·29	0·29	0·28	0·28	0·28	0·28	0·27	0·27	0·27	0·27	0·27	0·26	0·26	0·26	80
81	0·27	0·26	0·26	0·26	0·26	0·26	0·26	0·26	0·25	0·25	0·25	0·25	0·25	0·24	0·24	0·24	0·24	81
82	0·24	0·24	0·23	0·23	0·23	0·23	0·23	0·23	0·23	0·23	0·23	0·23	0·22	0·22	0·22	0·22	0·22	82
83	0·21	0·21	0·21	0·21	0·21	0·21	0·21	0·21	0·21	0·20	0·20	0·20	0·20	0·20	0·20	0·20	0·19	83
	47°	48°	49°	50°	51°	52°	53°	54°	55°	56°	57°	58°	59°	60°	61°	62°	63°	

DECLINATION.

EX-MERIDIAN TABLE II.

Reduction Plus to True Altitude at Upper Transit.

HOUR ANGLE.

A	0° 5'	0° 10'	0° 15'	0° 20'	0° 25'	0° 30'	0° 35'	0° 40'	0° 45'	0° 50'	0° 55'	1° 0'	A
	359°55'	359°50'	359°45'	359°40'	359°35'	359°30'	359°25'	359°20'	359°15'	359°10'	359° 5'	359° 0'	
1	0.0	0.0	0.0	0.0	0.0	0.1	0.1	0.1	0.2	0.2	0.2	0.3	1
2	0.0	0.0	0.0	0.1	0.1	0.1	0.2	0.2	0.3	0.4	0.4	0.5	2
3	0.0	0.0	0.1	0.1	0.1	0.2	0.3	0.4	0.5	0.6	0.7	0.8	3
4	0.0	0.0	0.1	0.1	0.2	0.3	0.4	0.5	0.6	0.7	0.9	1.1	4
5	0.0	0.0	0.1	0.1	0.2	0.3	0.5	0.6	0.8	0.9	1.1	1.3	5
6	0.0	0.0	0.1	0.2	0.3	0.4	0.5	0.7	0.9	1.1	1.3	1.6	6
7	0.0	0.1	0.1	0.2	0.3	0.5	0.6	0.8	1.1	1.3	1.6	1.9	7
8	0.0	0.1	0.1	0.2	0.4	0.5	0.7	0.9	1.2	1.5	1.8	2.1	8
9	0.0	0.1	0.2	0.3	0.4	0.6	0.8	1.1	1.4	1.7	2.0	2.4	9

A	1° 5'	1° 10'	1° 15'	1° 20'	1° 25'	1° 30'	1° 35'	1° 40'	1° 45'	1° 50'	1° 55'	2° 0'	A
	358°55'	358°50'	358°45'	358°40'	358°35'	358°30'	358°25'	358°20'	358°15'	358° 10'	358° 5'	358° 0'	
1	0.3	0.4	0.4	0.5	0.5	0.6	0.7	0.7	0.8	0.9	1.0	1.1	1
2	0.6	0.7	0.8	0.9	1.0	1.2	1.3	1.5	1.6	1.8	1.9	2.1	2
3	0.9	1.1	1.3	1.5	1.7	1.8	2.0	2.2	2.5	2.7	3.0	3.2	3
4	1.2	1.4	1.7	1.9	2.2	2.4	2.7	3.0	3.3	3.6	3.9	4.3	4
5	1.6	1.8	2.1	2.4	2.7	3.0	3.4	3.7	4.1	4.5	4.9	5.3	5
6	1.9	2.2	2.5	2.9	3.2	3.6	4.0	4.4	4.9	5.4	5.9	6.4	6
7	2.2	2.5	2.9	3.3	3.7	4.2	4.7	5.2	5.7	6.3	6.9	7.5	7
8	2.5	2.9	3.3	3.8	4.3	4.8	5.4	5.9	6.5	7.1	7.8	8.5	8
9	2.8	3.3	3.8	4.3	4.9	5.4	6.0	6.7	7.4	8.1	8.8	9.6	9

A	2° 5'	2° 10'	2° 15'	2° 20'	2° 25'	2° 30'	2° 35'	2° 40'	2° 45'	2° 50'	2° 55'	3° 0'	A
	357°55'	357°50'	357°45'	357°40'	357°35'	357°30'	357°25'	357°20'	357°15'	357°10'	357° 5'	357° 0'	
1	1.2	1.3	1.4	1.5	1.6	1.7	1.8	1.9	2.0	2.1	2.3	2.4	1
2	2.3	2.5	2.7	2.9	3.1	3.3	3.5	3.8	4.0	4.2	4.5	4.8	2
3	3.5	3.8	4.1	4.4	4.7	5.0	5.3	5.7	6.1	6.5	6.9	7.2	3
4	4.7	5.0	5.4	5.8	6.2	6.7	7.1	7.6	8.1	8.6	9.1	9.6	4
5	5.8	6.3	6.8	7.3	7.8	8.3	8.9	9.5	10.1	10.7	11.4	12.0	5
6	6.9	7.5	8.1	8.7	9.3	10.0	10.7	11.4	12.1	12.8	13.6	14.4	6
7	8.1	8.8	9.5	10.2	10.9	11.7	12.5	13.3	14.1	14.9	15.8	16.8	7
8	9.2	10.0	10.8	11.6	12.5	13.3	14.2	15.1	16.1	17.1	18.1	19.2	8
9	10.4	11.3	12.2	13.1	14.1	15.0	16.0	17.1	18.2	19.3	20.5	21.6	9

A	3° 5'	3° 10'	3° 15'	3° 20'	3° 25'	3° 30'	3° 35'	3° 40'	3° 45'	3° 50'	3° 55'	4° 0'	A
	356°55'	356°50'	356°45'	356°40'	356°35'	356°30'	356°25'	356°20'	356°15'	356°10'	356° 5'	356° 0'	
1	2.5	2.7	2.8	2.9	3.1	3.3	3.5	3.6	3.8	4.0	4.1	4.3	1
2	5.1	5.3	5.6	5.9	6.2	6.5	6.8	7.1	7.5	7.8	8.2	8.5	2
3	7.6	8.0	8.5	8.9	9.4	9.8	10.3	10.8	11.3	11.8	12.3	12.8	3
4	10.1	10.7	11.3	11.9	12.5	13.1	13.7	14.4	15.0	15.7	16.4	17.1	4
5	12.7	13.4	14.1	14.8	15.6	16.3	17.1	17.9	18.8	19.6	20.5	21.3	5
6	15.2	16.0	16.9	17.8	18.7	19.6	20.5	21.5	22.5	23.5	24.5	25.6	6
7	17.7	18.7	19.7	20.7	21.8	22.9	24.0	25.1	26.3	27.5	28.7	29.9	7
8	20.3	21.4	22.5	23.7	24.9	26.1	27.4	28.7	30.0	31.3	32.7	34.1	8
9	22.8	24.1	25.4	26.7	28.0	29.4	30.8	32.3	33.8	35.3	36.9	38.4	9

Reduction Minus to True Altitude at Lower Transit.

EX-MERIDIAN TABLE II.

Reduction Plus to True Altitude at Upper Transit.

HOUR ANGLE.

A	4° 5'	4° 10'	4° 15'	4° 20'	4° 25'	4° 30'	4° 35'	4° 40'	4° 45'	4° 50'	4° 55'	5° 0'	A
	355°55'	355°50'	355°45'	355°40'	355°35'	355°30'	355°25'	355°20'	355°15'	355°10'	355° 5'	355° 0'	
"													"
1	4·5	4·7	4·8	5·0	5·2	5·4	5·6	5·8	6·0	6·2	6·4	6·7	1
2	8·9	9·2	9·6	10·0	10·4	10·8	11·2	11·6	12·0	12·4	12·9	13·3	2
3	13·3	13·9	14·5	15·1	15·7	16·2	16·8	17·4	18·1	18·7	19·4	20·0	3
4	17·8	18·5	19·3	20·1	20·8	21·6	22·4	23·2	24·1	24·9	25·7	26·7	4
5	22·2	23·1	24·1	25·1	26·0	27·0	28·0	29·9	30·1	31·2	32·3	33·3	5
6	26·7	27·8	28·9	30·0	31·2	32·4	33·6	34·8	36·1	37·4	38·7	40·0	6
7	31·2	32·4	33·7	35·0	36·4	37·8	39·2	40·6	42·1	43·6	45·1	46·7	7
8	35·5	37·0	38·5	40·0	41·6	43·2	44·8	46·5	48·1	49·8	51·5	53·3	8
9	40·0	41·7	43·4	45·1	46·9	48·6	50·4	52·3	54·2	56·1	58·1	60·0	9

A	5° 5'	5° 10'	5° 15'	5° 20'	5° 25'	5° 30'	5° 35'	5° 40'	5° 45'	5° 50'	5° 55'	6° 0'	A
	354°55'	354°50'	354°45'	354°40'	354°35'	354°30'	354°25'	354°20'	354°15'	354°10'	354° 5'	354° 0'	
"	′	′	′	′	′	′	′	′	′	′	′	′	"
1	6·9	7·2	7·4	7·6	7·9	8·1	8·3	8·6	8·8	9·1	9·3	9·6	1
2	13·7	14·2	14·7	15·2	15·6	16·1	16·6	17·1	17·6	18·1	18·6	19·2	2
3	20·7	21·4	22·1	22·8	23·5	24·2	24·9	25·7	26·5	27·3	28·1	28·8	3
4	27·6	28·5	29·4	30·3	31·3	32·3	33·3	34·3	35·3	36·3	37·4	38·4	4
5	34·4	35·6	36·8	38·0	39·2	40·3	41·5	42·8	44·1	45·4	46·7	48·0	5
6	41·3	42·7	44·1	45·5	46·9	48·4	49·9	51·4	52·9	54·4	56·0	57·6	6
7	48·3	49·9	51·5	53·1	54·8	56·5	58·2	60·0	61·7	63·5	65·3	67·2	7
8	55·1	56·9	58·8	60·7	62·6	64·5	66·5	68·5	70·5	72·6	74·7	76·8	8
9	62·0	64·1	66·2	68·3	70·5	72·6	74·8	77·1	79·4	81·7	84·1	86·4	9

A	6° 5'	6° 10'	6° 15'	6° 20'	6° 25'	6° 30'	6° 35'	6° 40'	6° 45'	6° 50'	6° 55'	7° 0'	A
	353°55'	353°50'	353°45'	353°40'	353°35'	353°30'	353°25'	353°20'	353°15'	353°10'	353° 5'	353° 0'	
"	′	′	′	′	′	′	′	′	′	′	′	′	"
1	9·9	10·1	10·4	10·7	11·0	11·3	11·6	11·9	12·2	12·5	12·8	13·1	1
2	19·7	20·3	20·8	21·4	21·9	22·5	23·1	23·7	24·3	24·9	25·5	26·1	2
3	29·6	30·4	31·3	32·1	33·0	33·8	34·7	35·6	36·5	37·4	38·3	39·2	3
4	39·5	40·6	41·7	42·8	44·0	45·1	46·3	47·4	48·6	49·8	51·0	52·3	4
5	49·3	50·7	52·1	53·5	54·9	56·3	57·8	59·3	60·8	62·3	63·8	65·3	5
6	59·2	60·8	62·5	64·2	65·9	67·6	69·3	71·1	72·9	74·7	76·5	78·4	6
7	69·1	71·0	72·9	74·9	76·9	78·9	80·9	83·0	85·1	87·2	89·4	91·5	7
8	78·9	81·1	83·3	85·5	87·8	90·1	92·4	94·8	97·2	99·6	102	105	8
9	88·8	91·3	93·8	96·3	98·9	101	104	107	109	112	115	118	9

A	7° 5'	7° 10'	7° 15'	7° 20'	7° 25'	7° 30'	7° 35'	7° 40'	7° 45'	7° 50'	7° 55'	8° 0'	A
	352°55'	352°50'	352°45'	352°40'	352°35'	352°30'	352°25'	352°20'	352°15'	352°10'	352° 5'	352° 0'	
"	′	′	′	′	′	′	′	′	′	′	′	′	"
1	13·4	13·7	14·0	14·3	14·7	15·0	15·3	15·7	16·0	16·3	16·7	17·1	1
2	26·7	27·4	28·0	28·6	29·3	30·0	30·7	31·3	32·0	32·7	33·4	34·1	2
3	40·1	41·1	42·1	43·1	44·1	45·0	46·0	47·0	48·1	49·1	50·2	51·2	3
4	53·6	54·8	56·1	57·4	58·7	60·0	61·3	62·7	64·1	65·5	66·9	68·3	4
5	66·9	68·5	70·1	71·7	73·4	75·0	76·7	78·4	80·1	81·8	83·6	85·8	5
6	80·3	82·2	84·1	86·0	88·0	90·0	92·0	94·0	96·1	98·2	100	102	6
7	93·7	95·9	98·1	100	103	105	107	110	112	115	117	120	7
8	107	110	112	115	117	120	123	125	128	131	134	137	8
9	120	123	126	129	132	135	138	141	144	147	151	154	9

Reduction Minus to True Altitude at Lower Transit.

EX-MERIDIAN TABLE II.

Reduction Plus to True Altitude at Upper Transit.

HOUR ANGLE.

A	8° 5′	8° 10′	8° 15′	8° 20′	8° 25′	8° 30′	8° 35′	8° 40′	8° 45′	8° 50′	8° 55′	9° 0′	A
	351°55′	351°50′	351°45′	351°40′	351°35′	351°30′	351°25′	351°20′	351°15′	351°10′	351° 5′	351° 0′	
″	′	′	′	′	′	′	′	′	′	′	′	′	″
1	17·5	17·8	18·2	18·6	18·9	19·3	19·7	20·1	20·4	20·8	21·2	21·6	1
2	34·8	35·5	36·3	37·0	37·8	38·5	39·3	40·0	40·8	41·6	42·4	43·2	2
3	52·3	53·4	54·5	55·6	56·7	57·8	58·9	60·1	61·3	62·5	63·7	64·8	3
4	69·7	71·2	72·6	74·1	75·6	77·1	78·6	80·2	81·7	83·3	84·8	86·4	4
5	87·1	88·9	90·8	92·6	94·5	96·3	98·2	100	102	104	106	108	5
6	105	107	109	111	113	116	118	120	123	125	127	130	6
7	122	125	127	130	132	135	138	140	143	146	148	151	7
8	139	142	145	148	151	154	157	160	163	166	170	173	8
9	157	160	163	167	170	173	177	180	184	187	191	194	9

A	9° 5′	9° 10′	9° 15′	9° 20′	9° 25′	9° 30′	9° 35′	9° 40′	9 45′°	9° 50′	9° 55′	10° 0′	A
	350°55′	350°50′	350°45′	350°40′	350°35′	350°30′	350°25′	350°20′	350°15′	350°10′	350° 5′	350° 0′	
″	′	′	′	′	′	′	′	′	′	′	′	′	″
1	22·0	22·4	22·8	23·2	23·6	24·1	24·5	25·0	25·4	25·8	26·3	26·7	1
2	44·0	44·8	45·6	46·4	47·3	48·1	48·9	49·8	50·7	51·6	52·4	53·3	2
3	66·0	67·2	68·5	69·7	71·0	72·2	73·5	74·8	76·1	77·4	78·7	80·0	3
4	88·0	89·6	91·3	93·0	94·5	96·3	98·0	99·7	101	103	105	107	4
5	110	112	114	116	118	120	122	125	127	129	131	133	5
6	132	134	137	139	142	144	147	150	152	155	157	160	6
7	154	157	160	163	166	169	172	175	178	181	184	187	7
8	176	179	183	186	189	193	196	199	203	206	210	213	8
9	198	202	205	209	213	217	220	224	228	232	236	240	9

A	10° 5′	10° 10′	10° 15′	10° 20′	10° 25′	10° 30′	10° 35′	10° 40′	10° 45′	10° 50′	10° 55′	11° 0′	A
	349°55′	349°50′	349°45′	349°40′	349°35′	349°30′	349°25′	349°20′	349°15′	349°10′	349°5′	349° 0′	
″	′	′	′	′	′	′	′	′	′	′	′	′	″
1	27·1	27·6	28·0	28·5	28·9	29·4	29·9	30·3	30·8	31·3	31·8	32·3	1
2	54·2	55·1	56·0	56·9	57·8	58·8	59·7	60·7	61·6	62·6	63·5	64·5	2
3	81·3	82·7	84·1	85·5	86·9	88·2	89·6	91·0	92·5	93·9	95·4	96·8	3
4	109	110	112	114	116	118	120	121	123	125	127	129	4
5	136	138	140	142	145	147	149	152	154	157	159	161	5
6	163	165	168	171	174	176	179	182	185	188	191	194	6
7	190	193	196	199	203	206	209	212	216	219	222	226	7
8	217	221	224	228	232	235	239	243	247	250	254	258	8
9	244	248	252	256	261	265	269	273	277	282	286	290	9

A	11° 5′	11° 10′	11° 15′	11° 20′	11° 25′	11° 30′	11° 35′	11° 40′	11° 45′	11° 50′	11° 55′	12° 0′	A
	348°55′	348°50′	348°45′	348°40′	348°35′	348°30′	348°25′	348°20′	348°15′	348°10′	348° 5′	348° 0′	
″	′	′	′	′	′	′	′	′	′	′	′	′	″
1	32·8	33·3	33·8	34·3	34·8	35·3	35·8	36·3	36·8	37·3	37·9	38·4	1
2	65·5	66·5	67·5	68·5	69·5	70·5	71·5	72·6	73·6	74·6	75·7	76·8	2
3	98·3	99·8	101	103	104	106	107	109	111	112	114	115	3
4	131	133	135	137	139	141	143	145	147	149	152	154	4
5	164	166	169	171	174	176	179	182	184	187	189	192	5
6	197	200	203	206	209	212	215	218	221	224	227	230	6
7	229	233	236	240	243	247	251	254	258	261	265	269	7
8	262	266	270	274	278	282	286	290	295	299	303	307	8
9	295	299	304	308	313	317	322	327	331	336	341	346	9

Reduction Minus to True Altitude at Lower Transit.

EX-MERIDIAN TABLE II.

Reduction Plus to True Altitude at Upper Transit.

HOUR ANGLE.

A	12° 5'	12° 10'	12° 15'	12° 20'	12° 25'	12° 30'	12° 35'	12° 40'	12° 45'	12° 50'	12° 55'	13° 0'	A
	347°55'	347°50'	347°45'	347°40'	347°35'	347°30'	347°25'	347°20'	347°15'	347°10'	347°5'	347° 0'	
"	'	'	'	'	'	'	'	'	'	'	'	'	"
1	38·9	39·5	40·0	40·5	41·1	41·7	42·3	42·8	43·4	44·0	44·5	45·1	1
2	77·9	78·9	80·0	81·1	82·2	83·3	84·4	85·5	86·7	87·8	89·0	90·1	2
3	117	118	120	122	123	125	127	128	130	132	134	135	3
4	156	158	160	162	165	167	169	171	173	176	178	180	4
5	195	197	200	203	206	208	211	214	217	220	223	225	5
6	234	237	240	243	247	250	253	257	260	264	267	270	6
7	273	276	280	284	288	292	296	300	304	308	312	316	7
8	312	316	320	325	329	333	338	342	347	351	356	361	8
9	350	355	3€0	365	370	375	380	385	390	395	401	406	9

A	13° 5'	13° 10'	13° 15'	13° 20'	13° 25'	13° 30'	13° 35'	13° 40'	13° 45'	13° 50'	13° 55'	14° 0'	A
	346°55'	346°50'	346°45'	346°40'	346°35'	346°30'	346°25'	346°20'	346°15'	346°10'	346° 5'	346° 0'	
"	'	'	'	'	'	'	'	'	'	'	'	'	"
1	45·7	46·3	46·8	47·4	48·0	48·6	49·2	49·8	50·4	51·0	51·6	52·3	1
2	91·3	92·4	93·6	94·8	96·0	97·2	98·4	99·6	101	102	103	105	2
3	137	139	141	142	144	146	148	149	151	153	155	157	3
4	183	185	187	190	192	194	197	199	202	204	207	209	4
5	228	231	234	237	240	243	246	249	252	255	258	261	5
6	274	277	281	284	288	292	295	299	303	306	310	314	6
7	320	324	328	332	336	340	344	349	353	357	362	366	7
8	365	370	375	379	384	389	394	399	404	408	413	418	8
9	411	416	421	427	432	437	443	448	454	459	465	470	9

A	14° 5'	14° 10'	14° 15'	14° 20'	14° 25'	14° 30'	14° 35'	14° 40'	14° 45'	14° 50'	14° 55'	15° 0'	A
	345°55'	345°50'	345°45'	345°40'	345°35'	345°30'	345°25'	345°20'	345°15'	345°10'	345° 5'	345° 0'	
"	'	'	'	'	'	'	'	'	'	'	'	'	"
1	52·9	53·6	54·2	54·8	55·5	56·1	56·7	57·4	58·0	58·7	59·3	60·0	1
2	106	107	108	110	111	112	113	115	116	117	119	120	2
3	159	161	163	164	166	168	170	172	174	176	178	180	3
4	212	214	217	219	222	224	227	230	232	235	237	240	4
5	264	268	271	274	277	280	284	287	290	293	297	300	5
6	317	321	325	329	333	336	340	344	348	352	356	360	6
7	370	375	379	384	388	393	397	402	406	411	415	420	7
8	423	428	433	438	443	449	454	459	464	469	475	480	8
9	476	482	487	493	499	505	510	516	522	528	534	540	9

A	15° 5'	15° 10'	15° 15'	15° 20'	15° 25'	15° 30'	15° 35'	15° 40'	15° 45'	15° 50'	15° 55'	16° 0'	A
	344°55'	344°50'	344°45'	344°40'	344°35'	344°30'	344°25'	344°20'	344°15'	344°10'	344° 5'	344° 0'	
"	'	'	'	'	'	'	'	'	'	'	'	'	"
1	60·7	61·3	62·0	62·7	63·4	64·1	64·8	65·4	66·2	66·9	67·6	68·3	1
2	121	123	124	125	127	128	130	131	132	134	135	137	2
3	182	184	186	188	190	192	194	197	199	201	203	205	3
4	243	245	248	251	254	256	259	262	265	268	270	273	4
5	304	307	310	314	317	320	324	328	331	335	338	341	5
6	364	368	372	376	380	384	389	393	397	401	406	410	6
7	425	429	434	439	444	448	454	459	463	468	473	478	7
8	486	490	496	502	507	513	519	524	530	535	541	546	8
9	546	552	558	564	571	577	583	590	596	602	608	614	9

Reduction Minus to True Altitude at Lower Transit.

EX-MERIDIAN TABLE III.

SECOND CORRECTION. SUBTRACTIVE FROM FIRST CORRECTION.

First Correction	Altitude																First Correction
	15°	30°	35°	40°	45°	50°	53°	56°	59°	62°	65°	68°	71°	74°	77°	80°	
15	·0	·0	·0	·0	·0	·0	·0	·1	·1	·1	·1	·1	·1	·1	·1	·2	15
30	·0	.1	·1	·1	·1	·2	·2	·2	·2	·2	·3	·3	·4	·5	·6	·7	30
35	.1	·1	·1	·2	·2	·2	·2	·3	·3	·3	·4	·4	·5	·6	·8	1·0	35
40	·1	·1	·2	·2	·2	·3	·3	·3	·4	·4	·5	·6	·7	·8	1·0	1·3	40
45	·1	·2	·2	·3	·3	·4	·4	·4	·5	·6	·6	·7	·9	1·0	1·3	1·7	45
50	·1	·2	·3	·3	·4	·4	·5	·5	·6	·7	·8	·9	1·1	1·3	1·6	2·1	50
55	·1	·3	·3	·4	·4	·5	·6	·7	·7	·8	·9	1·1	1·3	1·5	1·9	2·5	55
60	·1	·3	·4	·4	·5	·6	·7	·7	·9	1·0	1·1	1·3	1·5	1·8	2·3	3·0	60
65	·2	·4	·4	·5	·6	·7	·8	·9	1·0	1·2	1·3	1·5	1·8	2·1	2·7	3·5	65
70	·2	·4	·5	·6	·7	·9	·9	1·1	1·2	1·3	1·5	1·8	2·1	2·5	3·1	4·0	70
75	·2	·5	·6	·7	·8	1·0	1·1	1·2	1·4	1·5	1·8	2·0	2·4	2·9	3·5	4·6	75
80	·3	·5	·7	·8	·9	1·1	1·2	1·4	1·6	1·8	2·0	2·3	2·7	3·3	4·0	5·3	80
85	·3	·6	·7	·9	1·1	1·3	1·4	1·6	1·8	2·0	2·3	2·6	3·1	3·7	4·5	6·0	85
90	·3	·7	·8	1·0	1·2	1·4	1·6	1·8	2·0	2·2	2·5	2·9	3·4	4·1	5·1	6·7	90
93	·3	·7	·9	1·1	1·3	1·5	1·7	1·9	2·1	2·4	2·7	3·1	3·7	4·4	5·5	7·1	93
96	·4	·8	·9	1·1	1·3	1·6	1·8	2·0	2·2	2·5	2·9	3·3	3·9	4·7	5·8	7·6	96
99	·4	·8	1·0	1·2	1·4	1·7	1·9	2·1	2·4	2·7	3·1	3·5	4·1	5·0	6·2	8·1	99
102	·4	·9	1·1	1·3	1·5	1·8	2·0	2·2	2·5	2·9	3·2	3·7	4·4	5·3	6·6	8·6	102
105	·4	·9	1·1	1·4	1·6	1·9	2·1	2·4	2·7	3·0	3·4	4·0	4·7	5·6	7·0	9·1	105
108	·5	1·0	1·2	1·4	1·7	2·0	2·3	2·5	2·8	3·2	3·6	4·2	4·9	5·9	7·3	9·6	108
111	·5	1·0	1·3	1·5	1·8	2·1	2·4	2·7	3·0	3·4	3·8	4·4	5·2	6·3	7·8	10·2	111
114	·5	1·1	1·3	1·6	1·9	2·3	2·5	2·8	3·2	3·6	4·1	4·7	5·5	6·6	8·2	10·7	114
117	·5	1·1	1·4	1·7	2·0	2·4	2·6	3·0	3·3	3·7	4·3	4·9	5·8	6·9	8·6	11·3	117
120	·6	1·2	1·5	1·8	2·1	2·5	2·8	3·1	3·5	3·9	4·5	5·2	6·1	7·3	9·1	11·9	120
123	·6	1·3	1·5	1·8	2·2	2·6	2·9	3·3	3·7	4·1	4·7	5·5	6·4	7·6	9·5	12·5	123
126	·6	1·3	1·6	1·9	2·3	2·8	3·1	3·4	3·8	4·3	5·0	5·7	6·7	8·0	10·0	13·1	126
129	·7	1·4	1·7	2·0	2·4	2·9	3·2	3·6	4·0	4·6	5·2	6·0	7·0	8·4	10·5	13·7	129
132	·7	1·5	1·8	2·1	2·5	3·0	3·4	3·8	4·2	4·8	5·4	6·3	7·4	8·8	11·0	14·4	132
135	·7	1·5	1·9	2·2	2·6	3·2	3·5	3·9	4·4	5·0	5·7	6·6	7·7	9·2	11·5	15·0	135
138	·7	1·6	1·9	2·3	2·8	3·3	3·7	4·1	4·6	5·2	5·9	6·9	8·0	9·7	12·0	15·7	138
141	·8	1·7	2·0	2·4	2·9	3·5	3·8	4·3	4·8	5·4	6·2	7·2	8·4	1·0	12·5	16·4	141

EX-MERIDIAN TABLE IV.

Limits of Hour Angle, or Time " before or after " Meridian Passage.

A	Hour Angle	A	Hour Angle	A	Hour Angle	A	Hour Angle	A	Hour Angle
"	m	"	m	"	m	"	m	"	m
52·2	4	7·54	17	3·37	30	1·92	43	1·21	56
40·2	5	6·94	18	3·20	31	1·85	44	1·17	57
31·4	6	6·44	19	3·05	32	1·78	45	1·13	58
25·4	7	6·00	20	2·92	33	1·72	46	1·09	59
21·2	8	5·64	21	2·79	34	1·66	47	1·06	60
18·0	9	5·26	22	2·67	35	1·60	48	1·02	61
15·7	10	4·94	23	2·55	36	1·54	49	0·99	62
13·8	11	4·60	24	2·45	37	1·49	50	0·96	63
12·2	12	4·40	25	2·35	38	1·43	51	0·93	64
10·9	13	4·17	26	2·25	39	1·38	52	0·90	65
9·90	14	3·94	27	2·16	40	1·34	53	0·87	66
9·02	15	3·73	28	2·08	41	1·29	54		
8·22	16	3·54	29	2·00	42	1·25	55		

MEAN LATITUDE to MIDDLE LATITUDE.

SUBTRACT / ADD

DIFFERENCE OF LATITUDE.

Mean Lat.	2°	3°	4°	5°	6°	7°	8°	9°	10°	11°	12°	13°	14°	15°	16°	17°	18°	19°	20°	Mean Lat.
°	'	'	'	'	'	'	'	'	'	'	'	'	'	'	'	'	'	'	'	°
14	93	92	90	89	86	83	80	76	72	67	62	57	51	45	38	31	24	16	9	14
15	85	84	83	81	79	76	73	69	65	61	56	51	46	40	34	27	21	13	6	15
16	79	77	76	74	72	70	66	63	60	56	51	46	41	36	30	24	17	10	4	16
17	72	71	70	68	66	64	61	58	55	51	47	42	37	32	27	21	15	8	2	17
18	67	66	65	63	61	59	56	53	50	46	43	38	34	29	24	18	12	6	1	18
19	62	61	60	59	57	55	52	49	46	43	39	35	30	25	21	15	9	3	3	19
20	58	57	56	54	53	51	48	45	42	39	35	31	27	22	18	13	7	1	5	20
22	50	49	48	47	45	44	41	39	36	33	29	25	22	17	13	8	3	3	9	22
24	44	43	42	41	40	38	36	33	31	28	24	21	17	13	8	4	1	6	12	24
26	39	38	37	36	35	33	31	28	26	23	20	16	13	9	5	0	5	10	15	26
28	34	33	32	31	30	28	26	24	22	19	16	12	9	5	1	3	8	13	18	28
30	30	29	29	28	26	24	22	20	18	15	12	9	6	2	2	6	11	16	21	30
35	22	21	21	19	18	17	15	12	10	7	5	1	2	6	10	14	18	23	28	35
40	16	15	14	13	12	10	8	6	4	1	2	5	8	12	16	20	25	29	34	40
45	11	11	10	8	7	5	3	1	1	4	7	11	14	18	22	27	31	36	41	45
50	8	7	6	5	3	1	1	3	6	9	12	16	20	24	28	33	38	44	49	50
55	5	4	3	2	0	2	5	7	10	14	17	21	25	30	35	40	46	52	58	55
60	3	2	1	1	3	5	8	11	14	18	22	27	32	37	43	49	55	62	69	60
	2°	3°	4°	5°	6°	7°	8°	9°	10°	11°	12°	13°	14°	15°	16°	17°	18°	19°	20'	

G

(Terrestrial Spheroid) MERIDIONAL PARTS. Compression $\frac{1}{293\cdot465}$

M	0°	1°	2°	3°	4°	5°	6°	7°	8°	9°	10°	M
0	0·00	59·60	119·21	178·86	238·56	298·34	358·22	418·20	478·31	538·58	599·01	0
1	0·99	60·59	120·20	179·85	239·56	299·34	359·21	419·20	479·31	539·58	600·02	1
2	1·99	61·58	121·20	180·85	240·55	300·34	360·21	420·20	480·32	540·59	601·03	2
3	2·98	62·58	122·19	181·84	241·55	301·33	361·21	421·20	481·32	541·59	602·04	3
4	3·97	63·57	123·18	182·84	242·55	302·33	362·21	422·20	482·32	542·60	603·04	4
5	4·97	64·56	124·18	183·83	243·54	303·33	363·21	423·20	483·33	543·60	604·05	5
6	5·96	65·56	125·17	184·82	244·54	304·32	364·21	424·20	484·33	544·61	605·06	6
7	6·95	66·55	126·17	185·82	245·53	305·32	365·21	425·20	485·33	545·62	606·07	7
8	7·95	67·54	127·16	186·81	246·53	306·32	366·21	426·20	486·34	546·62	607·08	8
9	8·94	68·54	128·15	187·81	247·53	307·32	367·20	427·21	487·34	547·63	608·09	9
10	9·93	69·53	129·15	188·80	248·52	308·31	368·20	428·21	488·34	548·64	609·10	10
11	10·92	70·52	130·14	189·80	249·52	309·31	369·20	429·21	489·35	549·64	610·11	11
12	11·92	71·52	131·14	190·79	250·51	310·31	370·20	430·21	490·35	550·65	611·12	12
13	12·91	72·51	132·13	191·79	251·51	311·31	371·20	431·21	491·36	551·65	612·13	13
14	13·90	73·50	133·12	192·78	252·50	312·30	372·20	432·21	492·36	552·66	613·14	14
15	14·90	74·50	134·12	193·78	253·50	313·30	373·20	433·21	493·36	553·67	614·15	15
16	15·89	75·49	135·11	194·77	254·50	314·30	374·20	434·21	494·37	554·67	615·15	16
17	16·88	76·48	136·11	195·77	255·49	315·30	375·20	435·22	495·37	555·68	616·16	17
18	17·88	77·48	137·10	196·76	256·49	316·29	376·20	436·22	496·37	556·69	617·17	18
19	18·87	78·47	138·09	197·76	257·48	317·29	377·20	437·22	497·38	557·69	618·18	19
20	19·86	79·46	139·09	198·75	258·48	318·29	378·20	438·22	498·38	558·70	619·19	20
21	20·86	80·46	140·08	199·75	259·48	319·29	379·20	439·22	499·39	559·71	620·20	21
22	21·85	81·45	141·08	200·74	260·47	320·28	380·19	440·22	500·39	560·71	621·21	22
23	22·84	82·44	142·07	201·74	261·47	321·28	381·19	441·23	501·39	561·72	622·22	23
24	23·84	83·44	143·06	202·73	262·46	322·28	382·19	442·23	502·40	562·73	623·23	24
25	24·83	84·43	144·06	203·73	263·46	323·28	383·19	443·23	503·40	563·73	624·24	25
26	25·82	85·42	145·05	204·72	264·46	324·27	384·19	444·23	504·41	564·74	625·25	26
27	26·82	86·42	146·05	205·72	265·45	325·27	385·19	445·23	505·41	565·75	626·26	27
28	27·81	87·41	147·04	206·71	266·45	326·27	386·19	446·23	506·41	566·75	627·27	28
29	28·80	88·40	148·03	207·71	267·45	327·27	387·19	447·24	507·42	567·76	628·28	29
30	29·80	89·40	149·03	208·70	268·44	328·27	388·19	448·24	508·42	568·77	629·29	30
31	30·79	90·39	150·02	209·70	269·44	329·26	389·19	449·24	509·43	569·78	630·30	31
32	31·78	91·39	151·02	210·69	270·43	330·26	390·19	450·24	510·43	570·78	631·31	32
33	32·78	92·38	152·01	211·69	271·43	331·26	391·19	451·24	511·44	571·79	632·33	33
34	33·77	93·37	153·00	212·68	272·43	332·26	392·19	452·25	512·44	572·80	633·34	34
35	34·76	94·37	154·00	213·68	273·42	333·26	393·19	453·25	513·45	573·81	634·35	35
36	35·76	95·36	154·99	214·67	274·42	334·25	394·19	454·25	514·45	574·81	635·36	36
37	36·75	96·35	155·99	215·67	275·42	335·25	395·19	455·25	515·45	575·82	636·37	37
38	37·74	97·35	156·98	216·66	276·41	336·25	396·19	456·25	516·46	576·83	637·38	38
39	38·74	98·34	157·97	217·66	277·41	337·25	397·19	457·25	517·46	577·84	638·39	39
40	39·73	99·33	158·97	218·65	278·41	338·25	398·19	458·26	518·47	578·84	639·40	40
41	40·72	100·33	159·96	219·65	279·40	339·24	399·19	459·26	519·47	579·85	640·41	41
42	41·72	101·32	160·96	220·64	280·40	340·24	400·19	460·26	520·48	580·86	641·42	42
43	42·71	102·31	161·95	221·64	281·40	341·24	401·19	461·26	521·48	581·87	642·43	43
44	43·70	103·31	162·95	222·64	282·39	342·24	402·19	462·27	522·49	582·87	643·44	44
45	44·70	104·30	163·94	223·63	283·39	343·24	403·19	463·27	523·49	583·88	644·46	45
46	45·69	105·30	164·94	224·63	284·39	344·23	404·19	464·27	524·50	584·89	645·47	46
47	46·68	106·29	165·93	225·62	285·38	345·23	405·19	465·27	525·50	585·90	646·48	47
48	47·68	107·28	166·92	226·62	286·38	346·23	406·19	466·28	526·51	586·91	647·49	48
49	48·67	108·28	167·92	227·61	287·38	347·23	407·19	467·28	527·51	587·91	648·50	49
50	49·66	109·27	168·91	228·61	288·37	348·23	408·19	468·28	528·52	588·92	649·51	50
51	50·66	110·26	169·91	229·60	289·37	349·23	409·19	469·29	529·52	589·93	650·52	51
52	51·65	111·26	170·90	230·60	290·37	350·23	410·19	470·29	530·53	590·94	651·53	52
53	52·64	112·25	171·90	231·59	291·36	351·22	411·19	471·29	531·54	591·95	652·55	53
54	53·64	113·24	172·89	232·59	292·36	352·22	412·19	472·29	532·54	592·96	653·56	54
55	54·63	114·24	173·88	233·59	293·36	353·22	413·19	473·30	533·55	593·96	654·57	55
56	55·62	115·23	174·88	234·58	294·35	354·22	414·19	474·30	534·55	594·97	655·58	56
57	56·62	116·23	175·87	235·58	295·35	355·22	415·20	475·30	535·56	595·98	656·59	57
58	57·61	117·22	176·87	236·57	296·35	356·22	416·20	476·30	536·56	596·99	657·60	58
59	58·60	118·21	177·86	237·57	297·35	357·22	417·20	477·31	537·57	598·00	658·62	59
60	59·60	119·21	178·86	238·56	298·34	358·22	418·20	478·31	538·58	599·01	659·63	60
M	0°	1°	2°	3°	4°	5°	6°	7°	8°	9°	10°	M

(Terrestrial Spheroid) **MERIDIONAL PARTS.** *Compression* $\frac{1}{293\cdot465}$

M	11°	12°	13°	14°	15°	16°	17°	18°	19°	20°	M
0	659·63	720·46	781·52	842·83	904·41	966·28	1028·46	1090·99	1153·87	1217·14	0
1	660·64	721·47	782·54	843·85	905·43	967·31	1029·50	1092·03	1154·92	1218·19	1
2	661·65	722·49	783·56	844·88	906·46	968·35	1030·54	1093·08	1155·97	1219·25	2
3	662·67	723·51	784·58	845·90	907·49	969·38	1031·58	1094·12	1157·02	1220·31	3
4	663·68	724·52	785·60	846·92	908·52	970·41	1032·62	1095·17	1158·08	1221·37	4
5	664·69	725·54	786·62	847·95	909·55	971·45	1033·66	1096·21	1159·13	1222·43	5
6	665·70	726·55	787·64	848·97	910·58	972·48	1034·70	1097·26	1160·18	1223·49	6
7	666·72	727·57	788·66	850·00	911·61	973·52	1035·74	1098·30	1161·23	1224·54	7
8	667·73	728·59	789·68	851·02	912·64	974·55	1036·78	1099·35	1162·28	1225·60	8
9	668·74	729·60	790·70	852·05	913·67	975·59	1037·82	1100·40	1163·34	1226·66	9
10	669·75	730·62	791·72	853·07	914·70	976·62	1038·86	1101·44	1164·39	1227·72	10
11	670·77	731·64	792·74	854·10	915·73	977·66	1039·90	1102·49	1165·44	1228·78	11
12	671·78	732·65	793·76	855·12	916·76	978·69	1040·94	1103·53	1166·49	1229·84	12
13	672·79	733·67	794·78	856·14	917·79	979·72	1041·98	1104·58	1167·55	1230·90	13
14	673·80	734·69	795·80	857·17	918·82	980·76	1043·02	1105·63	1168·60	1231·96	14
15	674·82	735·70	796·82	858·19	919·85	981·79	1044·06	1106·67	1169·65	1233·02	15
16	675·83	736·72	797·84	859·22	920·88	982·83	1045·10	1107·72	1170·70	1234·08	16
17	676·84	737·74	798·86	860·25	921·91	983·86	1046·14	1108·77	1171·76	1235·14	17
18	677·86	738·75	799·88	861·27	922·94	984·90	1047·18	1109·81	1172·81	1236·20	18
19	678·87	739·77	800·90	862·30	923·97	985·93	1048·23	1110·86	1173·86	1237·26	19
20	679·88	740·79	801·93	.863·32	925·00	986·97	1049·27	1111·91	1174·92	1238·32	20
21	680·90	741·80	802·95	864·35	926·03	988·01	1050·31	1112·95	1175·97	1239·38	21
22	681·91	742·82	803·97	865·37	927·06	989·04	1051·35	1114·00	1177·02	1240·44	22
23	682·92	743·84	804·99	866·40	928·09	990·08	1052·39	1115·05	1178·08	1241·50	23
24	683·94	744·85	806·01	867·42	929·12	991·11	1053·43	1116·10	1179·13	1242·56	24
25	684·95	745·87	807·03	868·45	930·15	992·15	1054·47	1117·14	1180·18	1243·62	25
26	685·96	746·89	808·05	869·48	931·18	993·19	1055·51	1118·19	1181·24	1244·68	26
27	686·98	747·91	809·07	870·50	932·21	994·22	1056·56	1119·24	1182·29	1245·74	27
28	687·99	748·92	810·10	871·53	933·24	995·26	1057·60	1120·29	1183·35	1246·80	28
29	689·00	749·94	811·12	872·55	934·27	996·29	1058·64	1121·33	1184·40	1247·86	29
30	690·02	750·96	812·14	873·58	935·30	997·33	1059·68	1122·38	1185·45	1248·92	30
31	691·03	751·98	813·16	874·61	936·33	998·37	1060·72	1123·43	1186·51	1249·98	31
32	692·04	752·99	814·18	875·63	937·37	999·40	1061·77	1124·48	1187·56	1251·04	32
33	693·06	754·01	815·21	876·66	938·40	1000·44	1062·81	1125·53	1188·62	1252·11	33
34	694·07	755·03	816·23	877·69	939·43	1001·48	1063·85	1126·58	1189·67	1253·17	34
35	695·09	756·05	817·25	878·71	940·46	1002·51	1064·89	1127·62	1190·73	1254·23	35
36	696·10	757·07	818·27	879·74	941·49	1003·55	1065·94	1128·67	1191·78	1255·29	36
37	697·12	758·08	819·29	880·77	942·52	1004·59	1066·98	1129·72	1192·84	1256·35	37
38	698·13	759·10	820·32	881·79	943·56	1005·62	1068·02	1130·77	1193·89	1257·41	38
39	699·14	760·12	821·34	882·82	944·59	1006·66	1069·06	1131·82	1194·95	1258·48	39
40	700·16	761·14	822·36	883·85	945·62	1007·70	1070·11	1132·87	1196·00	1259·54	40
41	701·17	762·16	823·38	884·88	946·65	1008·74	1071·15	1133·92	1197·06	1260·60	41
42	702·19	763·18	824·41	885·90	947·68	1009·77	1072·19	1134·97	1198·12	1261·66	42
43	703·20	764·19	825·43	886·93	948·72	1010·81	1073·24	1136·02	1199·17	1262·73	43
44	704·22	765·21	826·45	887·96	949·75	1011·85	1074·28	1137·07	1200·23	1263·79	44
45	705·23	766·23	827·47	888·99	950·78	1012·89	1075·32	1138·11	1201·28	1264·85	45
46	706·25	767·25	828·50	890·01	951·81	1013·92	1076·37	1139·16	1202·34	1265·92	46
47	707·26	768·27	829·52	891·04	952·85	1014·96	1077·41	1140·21	1203·40	1266·98	47
48	708·28	769·29	830·54	892·07	953·88	1016·00	1078·46	1141·26	1204·45	1268·04	48
49	709·29	770·31	831·57	893·10	954·91	1017·04	1079·50	1142·31	1205·51	1269·11	49
50	710·31	771·32	832·59	894·12	955·94	1018·08	1080·54	1143·36	1206·57	1270·17	50
51	711·32	772·34	833·61	895·15	956·98	1019·11	1081·59	1144·41	1207·62	1271·23	51
52	712·34	773·36	834·64	896·18	958·01	1020·15	1082·63	1145·46	1208·68	1272·30	52
53	713·35	774·38	835·66	897·21	959·04	1021·19	1083·68	1146·51	1209·74	1273·36	53
54	714·37	775·40	836·68	898·24	960·08	1022·23	1084·72	1147·56	1210·79	1274·43	54
55	715·38	776·42	837·71	899·26	961·11	1023·27	1085·76	1148·62	1211·85	1275·49	55
56	716·40	777·44	838·73	900·29	962·14	1024·31	1086·81	1149·67	1212·91	1276·55	56
57	717·41	778·46	839·76	901·32	963·18	1025·35	1087·85	1150·72	1213·96	1277·62	57
58	718·43	779·48	840·78	902·35	964·21	1026·39	1088·90	1151·77	1215·02	1278·68	58
59	719·44	780·50	841·80	903·38	965·24	1027·42	1089·94	1152·82	1216·08	1279·75	59
60	720·46	781·52	842·83	904·41	966·28	1028·46	1090·99	1153·87	1217·14	1280·81	60
M	11°	12°	13°	14°	15°	16°	17°	18°	19°	20°	M

(Terrestrial Spheroid) **MERIDIONAL PARTS.** *Compression* $\frac{1}{293\cdot465}$

M	21°	22°	23°	24°	25°	26°	27°	28°	29°	30°	M
0	1280·81	1344·92	1409·49	1474·54	1540·11	1606·21	1672·89	1740·18	1808·09	1876·67	0
1	1281·88	1345·99	1410·57	1475·63	1541·20	1607·32	1674·01	1741·30	1809·23	1877·82	1
2	1282·94	1347·07	1411·65	1476·72	1542·30	1608·43	1675·13	1742·43	1810·37	1878·97	2
3	1284·01	1348·14	1412·73	1477·81	1543·40	1609·53	1676·24	1743·56	1011·50	1880·12	3
4	1285·07	1349·21	1413·81	1478·89	1544·50	1610·64	1677·36	1744·68	1812·64	1881·27	4
5	1286·14	1350·28	1414·89	1479·98	1545·59	1611·75	1678·48	1745·81	1813·78	1882·42	5
6	1287·20	1351·36	1415·97	1481·07	1546·69	1612·86	1679·59	1746·94	1814·92	1883·57	6
7	1288·27	1352·43	1417·05	1482·16	1547·79	1613·96	1680·71	1748·07	1816·06	1884·72	7
8	1289·33	1353·50	1418·13	1483·25	1548·89	1615·07	1681·83	1749·19	1817·20	1885·87	8
9	1290·40	1354·58	1419·21	1484·34	1549·99	1616·18	1682·95	1750·32	1818·34	1887·02	9
10	1291·47	1355·65	1420·30	1485·43	1551·09	1617·29	1684·06	1751·45	1819·47	1888·17	10
11	1292·53	1356·72	1421·38	1486·52	1552·18	1618·39	1685·18	1752·58	1820·61	1889·32	11
12	1293·60	1357·80	1422·46	1487·61	1553·28	1619·50	1686·30	1753·71	1821·75	1890·47	12
13	1294·66	1358·87	1423·54	1488·70	1554·38	1620·61	1687·42	1754·84	1822·89	1891·62	13
14	1295·73	1359·94	1424·62	1489·79	1555·48	1621·72	1688·54	1755·96	1824·03	1892·78	14
15	1296·80	1361·02	1425·70	1490·88	1556·58	1622·83	1689·66	1757·09	1825·17	1893·93	15
16	1297·86	1362·09	1426·79	1491·97	1557·68	1623·94	1690·78	1758·22	1826·31	1895·08	16
17	1298·93	1363·17	1427·87	1493·06	1558·78	1625·05	1691·89	1759·35	1827·45	1896·23	17
18	1300·00	1364·24	1428·95	1494·16	1559·88	1626·16	1693·01	1760·48	1828·59	1897·38	18
19	1301·06	1365·32	1430·03	1495·25	1560·98	1627·27	1694·13	1761·61	1829·74	1898·54	19
20	1302·13	1366·39	1431·12	1496·34	1562·08	1628·38	1695·25	1762·74	1830·88	1899·69	20
21	1303·20	1367·47	1432·20	1497·43	1563·18	1629·49	1696·37	1763·87	1832·02	1900·84	21
22	1304·27	1368·54	1433·28	1498·52	1564·28	1630·60	1697·49	1765·00	1833·16	1901·99	22
23	1305·33	1369·62	1434·37	1499·61	1565·38	1631·71	1698·61	1766·13	1834·30	1903·15	23
24	1306·40	1370·69	1435·45	1500·70	1566·48	1632·82	1699·73	1767·26	1835·44	1904·30	24
25	1307·47	1371·77	1436·53	1501·80	1567·58	1633·93	1700·85	1768·39	1836·58	1905·46	25
26	1308·54	1372·84	1437·62	1502·89	1568·68	1635·04	1701·97	1769·53	1837·73	1906·61	26
27	1309·61	1373·92	1438·70	1503·98	1569·79	1636·15	1703·09	1770·66	1838·87	1907·76	27
28	1310·67	1374·99	1439·78	1505·07	1570·89	1637·26	1704·22	1771·79	1840·01	1908·92	28
29	1311·74	1376·07	1440·87	1506·16	1571·99	1638·37	1705·34	1772·92	1841·15	1910·07	29
30	1312·81	1377·14	1441·95	1507·26	1573·09	1639·48	1706·46	1774·05	1842·30	1911·23	30
31	1313·88	1378·22	1443·04	1508·35	1574·19	1640·59	1707·58	1775·18	1843·44	1912·38	31
32	1314·95	1379·30	1444·12	1509·44	1575·29	1641·70	1708·70	1776·32	1844·58	1913·54	32
33	1316·02	1380·37	1445·20	1510·54	1576·40	1642·82	1709·82	1777·45	1845·73	1914·69	33
34	1317·09	1381·45	1446·29	1511·63	1577·50	1643·93	1710·94	1778·58	1846·87	1915·85	34
35	1318·15	1382·53	1447·37	1512·72	1578·60	1645·04	1712·07	1779·71	1848·01	1917·00	35
36	1319·22	1383·60	1448·46	1513·82	1579·70	1646·15	1713·19	1780·85	1849·16	1918·16	36
37	1320·29	1384·68	1449·54	1514·91	1580·81	1647·26	1714·31	1781·98	1850·30	1919·31	37
38	1321·36	1385·76	1450·63	1516·00	1581·91	1648·38	1715·43	1783·11	1851·45	1920·47	38
39	1322·43	1386·84	1451·72	1517·10	1583·01	1649·49	1716·56	1784·25	1852·59	1921·63	39
40	1323·50	1387·91	1452·80	1518·19	1584·12	1650·60	1717·68	1785·38	1853·74	1922·78	40
41	1324·57	1388·99	1453·89	1519·29	1585·22	1651·72	1718·80	1786·51	1854·88	1923·94	41
42	1325·64	1390·07	1454·97	1520·38	1586·32	1652·83	1719·93	1787·65	1856·03	1925·10	42
43	1326·71	1391·15	1456·06	1521·48	1587·43	1653·94	1721·05	1788·78	1857·17	1926·25	43
44	1327·78	1392·22	1457·14	1522·57	1588·53	1655·06	1722·17	1789·92	1858·32	1927·41	44
45	1328·85	1393·30	1458·23	1523·67	1589·63	1656·17	1723·30	1791·05	1859·46	1928·57	45
46	1329·92	1394·38	1459·32	1524·76	1590·74	1657·28	1724·42	1792·19	1860·61	1929·73	46
47	1330·99	1395·46	1460·40	1525·86	1591·84	1658·40	1725·54	1793·32	1861·76	1930·89	47
48	1332·06	1396·54	1461·49	1526·95	1592·95	1659·51	1726·67	1794·46	1862·90	1932·04	48
49	1333·13	1397·61	1462·58	1528·05	1594·05	1660·63	1727·79	1795·59	1864·05	1933·20	49
50	1334·20	1398·69	1463·66	1529·14	1595·16	1661·74	1728·92	1796·73	1865·20	1934·36	50
51	1335·28	1399·77	1464·75	1530·24	1596·26	1662·86	1730·04	1797·86	1866·34	1935·52	51
52	1336·35	1400·85	1465·84	1531·33	1597·37	1663·97	1731·17	1799·00	1867·49	1936·68	52
53	1337·42	1401·93	1466·92	1532·43	1598·47	1665·08	1732·29	1800·13	1868·64	1937·84	53
54	1338·49	1403·01	1468·01	1533·53	1599·58	1666·20	1733·42	1801·27	1869·78	1939·00	54
55	1339·56	1404·09	1469·10	1534·62	1600·68	1667·32	1734·54	1802·41	1870·93	1940·16	55
56	1340·63	1405·17	1470·19	1535·72	1601·79	1668·43	1735·67	1803·54	1872·08	1941·32	56
57	1341·70	1406·25	1471·28	1536·82	1602·90	1669·55	1736·80	1804·68	1873·23	1942·48	57
58	1342·78	1407·33	1472·36	1537·91	1604·00	1670·66	1737·92	1805·82	1874·38	1943·64	58
59	1343·85	1408·41	1473·45	1539·01	1605·11	1671·78	1739·05	1806·95	1875·53	1944·80	59
60	1344·92	1409·49	1474·54	1540·11	1606·21	1672·89	1740·18	1808·09	1876·67	1945·96	60
M	21°	22°	23°	24°	25°	26°	27°	28°	29°	30°	M

(Terrestrial Spheroid) MERIDIONAL PARTS. Compression $\frac{1}{293\cdot465}$

M	31°	32°	33°	34°	35°	36°	37°	38°	39°	40°	M
0	1945·96	2015·98	2086·78	2158·39	2230·86	2304·23	2378·54	2453·85	2530·20	2607·64	0
1	1947·12	2017·15	2087·97	2159·59	2232·08	2305·46	2379·79	2455·11	2531·48	2608·94	1
2	1948·28	2018·33	2089·15	2160·79	2233·29	2306·69	2381·04	2456·38	2532·76	2610·24	2
3	1949·44	2019·50	2090·34	2161·99	2234·51	2307·92	2382·28	2457·64	2534·04	2611·54	3
4	1950·60	2020·68	2091·53	2163·20	2235·72	2309·15	2383·53	2458·91	2535·33	2612·84	4
5	1951·77	2021·85	2092·72	2164·40	2236·94	2310·39	2384·78	2460·17	2536·61	2614·15	5
6	1952·93	2023·03	2093·90	2165·60	2238·16	2311·62	2386·03	2461·44	2537·89	2615·45	6
7	1954·09	2024·20	2095·09	2166·80	2239·37	2312·85	2387·28	2462·70	2539·18	2616·75	7
8	1955·25	2025·38	2096·28	2168·00	2240·59	2314·08	2388·53	2463·97	2540·46	2618·05	8
9	1956·41	2026·55	2097·47	2169·21	2241·81	2315·32	2389·77	2465·23	2541·74	2619·36	9
10	1957·58	2027·73	2098·66	2170·41	2243·03	2316·55	2391·02	2466·50	2543·03	2620·66	10
11	1958·74	2028·90	2099·85	2171·61	2244·24	2317·78	2392·27	2467·77	2544·31	2621·96	11
12	1959·90	2030·08	2101·04	2172·82	2245·46	2319·02	2393·52	2469·03	2545·60	2623·27	12
13	1961·07	2031·25	2102·23	2174·02	2246·68	2320·25	2394·77	2470·30	2546·88	2624·57	13
14	1962·23	2032·43	2103·42	2175·22	2247·90	2321·48	2396·02	2471·57	2548·17	2625·88	14
15	1963·39	2033·61	2104·61	2176·43	2249·12	2322·72	2397·27	2472·84	2549·45	2627·18	15
16	1964·56	2034·78	2105·80	2177·63	2250·34	2323·95	2398·53	2474·10	2550·74	2628·49	16
17	1965·72	2035·96	2106·99	2178·84	2251·56	2325·19	2399·78	2475·37	2552·03	2629·79	17
18	1966·89	2037·14	2108·18	2180·04	2252·78	2326·42	2401·03	2476·64	2553·31	2631·10	18
19	1968·05	2038·31	2109·37	2181·25	2254·00	2327·66	2402·28	2477·91	2554·60	2632·40	19
20	1969·22	2039·49	2110·56	2182·45	2255·22	2328·89	2403·53	2479·18	2555·89	2633·71	20
21	1970·38	2040·67	2111·75	2183·66	2256·44	2330·13	2404·78	2480·45	2557·18	2635·02	21
22	1971·55	2041·85	2112·94	2184·86	2257·66	2331·37	2406·04	2481·72	2558·46	2636·32	22
23	1972·71	2043·03	2114·13	2186·07	2258·88	2332·60	2407·29	2482·99	2559·75	2637·63	23
24	1973·88	2044·20	2115·32	2187·27	2260·10	2333·84	2408·54	2484·26	2561·04	2638·94	24
25	1975·04	2045·38	2116·52	2188·48	2261·32	2335·08	2409·80	2485·53	2562·33	2640·25	25
26	1976·21	2046·56	2117·71	2189·69	2262·54	2336·31	2411·05	2486·80	2563·62	2641·56	26
27	1977·38	2047·74	2118·90	2190·89	2263·76	2337·55	2412·30	2488·07	2564·91	2642·86	27
28	1978·54	2048·92	2120·10	2192·10	2264·98	2338·79	2413·56	2489·35	2566·20	2644·17	28
29	1979·71	2050·10	2121·29	2193·31	2266·21	2340·03	2414·81	2490·62	2567·49	2645·48	29
30	1980·88	2051·28	2122·48	2194·52	2267·43	2341·27	2416·07	2491·89	2568·78	2646·79	30
31	1982·04	2052·46	2123·68	2195·73	2268·65	2342·50	2417·32	2493·16	2570·07	2648·10	31
32	1983·21	2053·64	2124·87	2196·93	2269·88	2343·74	2418·58	2494·44	2571·36	2649·41	32
33	1984·38	2054·82	2126·06	2198·14	2271·10	2344·98	2419·83	2495·71	2572·65	2650·72	33
34	1985·55	2056·00	2127·26	2199·35	2272·32	2346·22	2421·09	2496·98	2573·94	2652·03	34
35	1986·71	2057·18	2128·45	2200·56	2273·55	2347·46	2422·35	2498·26	2575·24	2653·35	35
36	1987·88	2058·36	2129·65	2201·77	2274·77	2348·70	2423·60	2499·53	2576·53	2654·66	36
37	1989·05	2059·55	2130·84	2202·98	2276·00	2349·94	2424·86	2500·80	2577·82	2655·97	37
38	1990·22	2060·73	2132·04	2204·19	2277·22	2351·18	2426·12	2502·08	2579·12	2657·28	38
39	1991·39	2061·91	2133·23	2205·40	2278·45	2352·42	2427·38	2503·35	2580·41	2658·60	39
40	1992·56	2063·09	2134·43	2206·61	2279·67	2353·66	2428·63	2504·63	2581·70	2659·91	40
41	1993·73	2064·27	2135·62	2207·82	2280·90	2354·91	2429·89	2505·90	2583·00	2661·22	41
42	1994·90	2065·46	2136·82	2209·03	2282·12	2356·15	2431·15	2507·18	2584·29	2662·54	42
43	1996·06	2066·64	2138·02	2210·24	2283·35	2357·39	2432·41	2508·46	2585·58	2663·85	43
44	1997·23	2067·82	2139·21	2211·45	2284·57	2358·63	2433·67	2509·73	2586·88	2665·16	44
45	1998·40	2069·01	2140·41	2212·66	2285·80	2359·87	2434·93	2511·01	2588·17	2666·48	45
46	1999·57	2070·19	2141·61	2213·87	2287·03	2361·12	2436·19	2512·29	2589·47	2667·79	46
47	2000·74	2071·37	2142·81	2215·09	2288·25	2362·36	2437·45	2513·56	2590·77	2669·11	47
48	2001·92	2072·56	2144·00	2216·30	2289·48	2363·60	2438·71	2514·84	2592·06	2670·42	48
49	2003·09	2073·74	2145·20	2217·51	2290·71	2364·85	2439·97	2516·12	2593·36	2671·74	49
50	2004·26	2074·92	2146·40	2218·72	2291·94	2366·09	2441·23	2517·40	2594·66	2673·06	50
51	2005·43	2076·11	2147·60	2219·94	2293·17	2367·33	2442·49	2518·68	2595·95	2674·37	51
52	2006·60	2077·29	2148·80	2221·15	2294·39	2368·58	2443·75	2519·96	2597·25	2675·69	52
53	2007·77	2078·48	2149·99	2222·36	2295·62	2369·82	2445·01	2521·23	2598·55	2677·01	53
54	2008·94	2079·66	2151·19	2223·57	2296·85	2371·07	2446·27	2522·51	2599·85	2678·33	54
55	2010·12	2080·85	2152·39	2224·79	2298·08	2372·31	2447·53	2523·79	2601·14	2679·64	55
56	2011·29	2082·03	2153·59	2226·00	2299·31	2373·56	2448·80	2525·07	2602·44	2680·96	56
57	2012·46	2083·22	2154·79	2227·22	2300·54	2374·80	2450·06	2526·35	2603·74	2682·28	57
58	2013·63	2084·41	2155·99	2228·43	2301·77	2376·05	2451·32	2527·63	2605·04	2683·60	58
59	2014·81	2085·59	2157·19	2229·65	2303·00	2377·30	2452·58	2528·92	2606·34	2684·92	59
60	2015·98	2086·78	2158·39	2230·86	2304·23	2378·54	2453·85	2530·20	2607·64	2686·24	60
M	31°	32°	33°	34°	35°	36°	37°	38°	39°	40°	M

(Terrestrial Spheroid) MERIDIONAL PARTS. Compression $\frac{1}{293\cdot465}$

M	41°	42°	43°	44°	45°	46°	47°	48°	49°	50°	M
0	2686·24	2766·05	2847·13	2929·55	3013·38	3098·70	3185·59	3274·13	3364·41	3456·53	0
1	2687·56	2767·39	2848·49	2930·93	3014·79	3100·14	3187·05	3275·62	3365·93	3458·08	1
2	2688·88	2768·73	2849·85	2932·32	3016·20	3101·57	3188·51	3277·11	3367·45	3459·64	2
3	2690·20	2770·07	2851·22	2933·71	3017·61	3103·01	3189·97	3278·60	3368·97	3461·19	3
4	2691·52	2771·41	2852·58	2935·09	3019·02	3104·44	3191·44	3280·09	3370·49	3462·74	4
5	2692·84	2772·75	2853·94	2936·48	3020·43	3105·88	3192·90	3281·58	3372·01	3464·29	5
6	2694·16	2774·10	2855·31	2937·87	3021·85	3107·32	3194·36	3283·07	3373·54	3465·85	6
7	2695·49	2775·44	2856·67	2939·26	3023·26	3108·76	3195·83	3284·57	3375·06	3467·40	7
8	2696·81	2776·78	2858·04	2940·64	3024·67	3110·19	3197·29	3286·06	3376·58	3468·96	8
9	2698·13	2778·13	2859·40	2942·03	3026·08	3111·63	3198·76	3287·55	3378·11	3470·52	9
10	2699·45	2779·47	2860·77	2943·42	3027·50	3113·07	3200·23	3289·05	3379·63	3472·07	10
11	2700·78	2780·81	2862·14	2944·81	3028·91	3114·51	3201·69	3290·54	3381·16	3473·63	11
12	2702·10	2782·16	2863·50	2946·20	3030·32	3115·95	3203·16	3292·04	3382·68	3475·19	12
13	2703·42	2783·50	2864·87	2947·59	3031·74	3117·39	3204·63	3293·54	3384·21	3476·75	13
14	2704·75	2784·85	2866·24	2948·98	3033·15	3118·83	3206·10	3295·03	3385·73	3478·30	14
15	2706·07	2786·19	2867·60	2950·37	3034·57	3120·27	3207·56	3296·53	3387·26	3479·86	15
16	2707·40	2787·54	2868·97	2951·76	3035·99	3121·71	3209·03	3298·03	3388·79	3481·42	16
17	2708·72	2788·89	2870·34	2953·15	3037·40	3123·16	3210·50	3299·52	3390·32	3482·98	17
18	2710·05	2790·23	2871·71	2954·55	3038·82	3124·60	3211·97	3301·02	3391·85	3484·54	18
19	2711·38	2791·58	2873·08	2955·94	3040·23	3126·04	3213·44	3302·52	3393·38	3486·11	19
20	2712·70	2792·93	2874·45	2957·33	3041·65	3127·49	3214·91	3304·02	3394·91	3487·67	20
21	2714·03	2794·28	2875·82	2958·73	3043·07	3128·93	3216·38	3305·52	3396·44	3489·23	21
22	2715·36	2795·62	2877·19	2960·12	3044·49	3130·37	3217·86	3307·02	3397·97	3490·79	22
23	2716·68	2796·97	2878·56	2961·51	3045·91	3131·82	3219·33	3308·52	3399·50	3492·36	23
24	2718·01	2798·32	2879·93	2962·91	3047·33	3133·26	3220·80	3310·02	3401·03	3493·92	24
25	2719·34	2799·67	2881·30	2964·30	3048·75	3134·71	3222·27	3311·53	3402·56	3495·49	25
26	2720·67	2801·02	2882·67	2965·70	3050·17	3136·15	3223·75	3313·03	3404·10	3497·05	26
27	2722·00	2802·37	2884·05	2967·09	3051·59	3137·60	3225·22	3314·53	3405·63	3498·62	27
28	2723·33	2803·72	2885·42	2968·49	3053·01	3139·05	3226·69	3316·03	3407·16	3500·18	28
29	2724·66	2805·07	2886·79	2969·89	3054·43	3140·49	3228·17	3317·54	3408·70	3501·75	29
30	2725·99	2806·42	2888·17	2971·28	3055·85	3141·94	3229·64	3319·04	3410·23	3503·32	30
31	2727·32	2807·77	2889·54	2972·68	3057·27	3143·39	3231·12	3320·55	3411·77	3504·89	31
32	2728·65	2809·13	2890·91	2974·08	3058·70	3144·84	3232·60	3322·05	3413·30	3506·45	32
33	2729·98	2810·48	2892·29	2975·48	3060·12	3146·29	3234·07	3323·56	3414·84	3508·02	33
34	2731·31	2811·83	2893·66	2976·88	3061·54	3147·74	3235·55	3325·07	3416·38	3509·59	34
35	2732·64	2813·18	2895·04	2978·28	3062·97	3149·19	3237·03	3326·57	3417·92	3511·16	35
36	2733·97	2814·54	2896·42	2979·68	3064·39	3150·64	3238·51	3328·08	3419·45	3512·73	36
37	2735·31	2815·89	2897·79	2981·08	3065·81	3152·09	3239·98	3329·59	3420·99	3514·31	37
38	2736·64	2817·25	2899·17	2982·48	3067·24	3153·54	3241·46	3331·10	3422·53	3515·88	38
39	2737·97	2818·60	2900·54	2983·88	3068·66	3154·99	3242·94	3332·60	3424·07	3517·45	39
40	2739·30	2819·95	2901·92	2985·28	3070·09	3156·45	3244·42	3334·11	3425·61	3519·02	40
41	2740·64	2821·31	2903·30	2986·68	3071·52	3157·90	3245·90	3335·62	3427·15	3520·60	41
42	2741·97	2822·67	2904·68	2988·08	3072·94	3159·35	3247·38	3337·13	3428·70	3522·17	42
43	2743·31	2824·02	2906·06	2989·48	3074·37	3160·81	3248·87	3338·65	3430·24	3523·75	43
44	2744·64	2825·38	2907·43	2990·88	3075·80	3162·26	3250·35	3340·16	3431·78	3525·32	44
45	2745·98	2826·73	2908·81	2992·29	3077·23	3163·71	3251·83	3341·67	3433·32	3526·90	45
46	2747·31	2828·09	2910·19	2993·69	3078·66	3165·17	3253·31	3343·18	3434·87	3528·47	46
47	2748·65	2829·45	2911·57	2995·09	3080·09	3166·62	3254·80	3344·69	3436·41	3530·05	47
48	2749·98	2830·81	2912·95	2996·50	3081·52	2168·08	3256·28	3346·21	3437·95	3531·63	48
49	2751·32	2832·16	2914·33	2997·90	3082·95	3169·54	3257·77	3347·72	3439·50	3533·21	49
50	2752·66	2833·52	2915·72	2999·31	3084·38	3170·99	3259·25	3349·24	3441·05	3534·79	50
51	2754·00	2834·88	2917·10	3000·71	3085·81	3172·45	3260·74	3350·75	3442·59	3536·37	51
52	2755·33	2836·24	2918·48	3002·12	3087·24	3173·91	3262·22	3352·27	3444·14	3537·95	52
53	2756·67	2837·60	2919·86	3003·53	3088·67	3175·37	3263·71	3353·78	3445·69	3539·53	53
54	2758·01	2838·96	2921·24	3004·93	3090·10	3176·83	3265·20	3355·30	3447·23	3541·11	54
55	2759·35	2840·32	2922·63	3006·34	3091·53	3178·28	3266·68	3356·82	3448·78	3542·69	55
56	2760·69	2841·68	2924·01	3007·75	3092·97	3179·74	3268·17	3358·33	3450·33	3544·27	56
57	2762·03	2843·04	2925·39	3009·16	3094·40	3181·20	3269·66	3359·85	3451·88	3545·85	57
58	2763·37	2844·40	2926·78	3010·56	3095·83	3182·66	3271·15	3361·37	3453·43	3547·44	58
59	2764·71	2845·77	2928·16	3011·97	3097·27	3184·13	3272·64	3362·89	3454·98	3549·02	59
60	2766·05	2847·13	2929·55	3013·38	3098·70	3185·59	3274·13	3364·41	3456·53	3550·60	60
M	41°	42°	43°	44°	45°	46°	47°	48°	49°	50°	M

(Terrestrial Spheroid) **MERIDIONAL PARTS.** *Compression* $\frac{1}{297.465}$

M	51°	52°	53°	54°	55°	56°	57°	58°	59°	60°	M
0	3550·60	3646·74	3745·05	3845·69	3948·78	4054·48	4162·97	4274·43	4389·06	4507·08	0
1	3552·19	3648·36	3746·71	3847·38	3950·52	4056·27	4164·81	4276·31	4391·00	4509·07	1
2	3553·77	3649·98	3748·37	3849·08	3952·26	4058·05	4166·64	4278·20	4392·94	4511·07	2
3	3555·36	3651·60	3750·03	3850·78	3954·00	4059·84	4168·47	4280·08	4394·88	4513·07	3
4	3556·95	3653·22	3751·69	3852·48	3955·74	4061·63	4170·31	4281·97	4396·82	4515·07	4
5	3558·53	3654·84	3753·35	3854·18	3957·48	4063·41	4172·14	4283·86	4398·76	4517·07	5
6	3560·12	3656·47	3755·01	3855·88	3959·23	4065·20	4173·98	4285·75	4400·70	4519·08	6
7	3561·71	3658·09	3756·67	3857·58	3960·97	4066·99	4175·82	4287·64	4402·65	4521·08	7
8	3563·30	3659·72	3758·33	3859·29	3962·72	4068·78	4177·66	4289·53	4404·59	4523·08	8
9	3564·89	3661·34	3760·00	3860·99	3964·46	4070·57	4179·50	4291·42	4406·54	4525·09	9
10	3566·48	3662·97	3761·66	3862·69	3966·21	4072·37	4181·34	4293·31	4408·49	4527·09	10
11	3568·07	3664·59	3763·33	3864·40	3967·96	4074·16	4183·18	4295·20	4410·43	4529·10	11
12	3569·66	3666·22	3764·99	3866·10	3969·70	4075·95	4185·02	4297·09	4412·38	4531·11	12
13	3571·25	3667·85	3766·66	3867·81	3971·45	4077·75	4186·86	4298·99	4414·33	4533·12	13
14	3572·85	3669·48	3768·32	3869·52	3973·20	4079·54	4188·71	4300·89	4416·28	4535·13	14
15	3574·44	3671·11	3769·99	3871·22	3974·95	4081·34	4190·55	4302·78	4418·24	4537·14	15
16	3576·03	3672·74	3771·66	3872·93	3976·70	4083·13	4192·40	4304·68	4420·19	4539·15	16
17	3577·63	3674·37	3773·33	3874·64	3978·46	4084·93	4194·24	4306·58	4422·14	4541·17	17
18	3579·22	3676·00	3774·99	3876·35	3980·21	4086·73	4196·09	4308·48	4424·10	4543·18	18
19	3580·82	3677·63	3776·66	3878·06	3981·96	4088·53	4197·94	4310·38	4426·05	4545·20	19
20	3582·41	3679·26	3778·33	3879·77	3983·71	4090·33	4199·79	4312·28	4428·01	4547·21	20
21	3584·01	3680·89	3780·00	3881·48	3985·47	4092·13	4201·64	4314·18	4429·97	4549·23	21
22	3585·61	3682·53	3781·68	3883·19	3987·22	4093·93	4203·49	4316·08	4431·93	4551·25	22
23	3587·20	3684·16	3783·35	3884·91	3988·98	4095·73	4205·34	4317·98	4433·89	4553·27	23
24	3588·80	3685·79	3785·02	3886·62	3990·74	4097·54	4207·19	4319·89	4435·85	4555·29	24
25	3590·40	3687·43	3786·69	3888·33	3992·49	4099·34	4209·04	4321·79	4437·81	4557·31	25
26	3592·00	3689·07	3788·37	3890·05	3994·25	4101·14	4210·90	4323·70	4439·77	4559·33	26
27	3593·60	3690·70	3790·04	3891·76	3996·01	4102·95	4212·75	4325·61	4441·73	4561·36	27
28	3595·20	3692·34	3791·72	3893·48	3997·77	4104·75	4214·61	4327·52	4443·70	4563·38	28
29	3596·80	3693·98	3793·40	3895·20	3999·53	4106·56	4216·46	4329·42	4445·66	4565·41	29
30	3598·40	3695·61	3795·07	3896·91	4001·29	4108·37	4218·32	4331·33	4447·63	4567·44	30
31	3600·01	3697·25	3796·75	3898·63	4003·06	4110·18	4220·18	4333·24	4449·60	4569·46	31
32	3601·61	3698·89	3798·43	3900·35	4004·82	4111·99	4222·04	4335·16	4451·56	4571·49	32
33	3603·21	3700·53	3800·11	3902·07	4006·58	4113·80	4223·90	4337·07	4453·53	4573·52	33
34	3604·82	3702·17	3801·79	3903·79	4008·35	4115·61	4225·76	4338·98	4455·50	4575·55	34
35	3606·42	3703·82	3803·47	3905·51	4010·11	4117·42	4227·62	4340·90	4457·48	4577·59	35
36	3608·03	3705·46	3805·15	3907·24	4011·88	4119·23	4229·48	4342·81	4459·45	4579·62	36
37	3609·63	3707·10	3806·83	3908·96	4013·64	4121·05	4231·34	4344·73	4461·42	4581·65	37
38	3611·24	3708·74	3808·51	3910·68	4015·41	4122·86	4233·21	4346·65	4463·40	4583·69	38
39	3612·85	3710·39	3810·19	3912·41	4017·18	4124·67	4235·07	4348·56	4465·37	4585·72	39
40	3614·46	3712·03	3811·88	3914·13	4018·95	4126·49	4236·94	4350·48	4467·35	4587·76	40
41	3616·06	3713·68	3813·56	3915·86	4020·72	4128·31	4238·80	4352·40	4469·32	4589·80	41
42	3617·67	3715·32	3815·25	3917·58	4022·49	4130·12	4240·67	4354·32	4471·30	4591·84	42
43	3619·28	3716·97	3816·93	3919·31	4024·26	4131·94	4242·54	4356·25	4473·28	4593·88	43
44	3620·89	3718·62	3818·62	3921·04	4026·03	4133·76	4244·41	4358·17	4475·26	4595·92	44
45	3622·50	3720·26	3820·30	3922·77	4027·80	4135·58	4246·28	4360·09	4477·24	4597·96	45
46	3624·12	3721·91	3821·99	3924·50	4029·58	4137·40	4248·15	4362·02	4479·22	4600·01	46
47	3625·73	3723·56	3823·68	3926·23	4031·35	4139·22	4250·02	4363·94	4481·21	4602·05	47
48	3627·34	3725·21	3825·37	3927·96	4033·13	4141·04	4251·89	4365·87	4483·19	4604·10	48
49	3628·95	3726·86	3827·06	3929·69	4034·90	4142·87	4253·77	4367·80	4485·18	4606·15	49
50	3630·57	3728·51	3828·75	3931·42	4036·68	4144·69	4255·64	4369·72	4487·16	4608·19	50
51	3632·18	3730·16	3830·44	3933·15	4038·45	4146·52	4257·52	4371·65	4489·15	4610·24	51
52	3633·80	3731·81	3832·13	3934·88	4040·23	4148·34	4259·39	4373·58	4491·14	4612·29	52
53	3635·41	3733·47	3833·82	3936·62	4042·01	4150·17	4261·27	4375·51	4493·13	4614·34	53
54	3637·03	3735·12	3835·52	3938·35	4043·79	4151·99	4263·15	4377·45	4495·12	4616·40	54
55	3638·64	3736·77	3837·21	3940·09	4045·57	4153·82	4265·02	4379·38	4497·11	4618·45	55
56	3640·26	3738·43	3838·90	3941·83	4047·35	4155·65	4266·90	4381·31	4499·10	4620·50	56
57	3641·88	3740·08	3840·60	3943·56	4049·13	4157·48	4268·78	4383·25	4501·09	4622·56	57
58	3643·50	3741·74	3842·29	3945·30	4050·92	4159·31	4270·67	4385·18	4503·09	4624·62	58
59	3645·12	3743·40	3843·99	3947·04	4052·70	4161·14	4272·55	4387·12	4505·08	4626·67	59
60	3646·74	3745·05	3845·69	3948·78	4054·48	4162·97	4274·43	4389·06	4507·08	4628·73	60
M	51°	52°	53°	54°	55°	56°	57°	58°	59°	60°	M

(*Terrestrial Spheroid*) **MERIDIONAL PARTS.** Compression $\frac{1}{293\cdot465}$

M	61°	62°	63°	64°	65°	66°	67°	68°	69°	70°	M
0	4628·73	4754·29	4884·06	5018·36	5157·57	5302·11	5452·43	5609·09	5772·68	5943·89	0
1	4630·79	4756·42	4886·26	5020·64	5159·93	5304·56	5454·99	5611·76	5775·47	5946·82	1
2	4632·85	4758·55	4888·46	5022·92	5162·30	5307·02	5457·55	5614·43	5778·26	5949·74	2
3	4634·92	4760·68	4890·66	5025·20	5164·67	5309·48	5460·11	5617·10	5781·05	5952·67	3
4	4636·98	4762·81	4892·87	5027·48	5167·04	5311·94	5462·67	5619·77	5783·85	5955·60	4
5	4639·04	4764·94	4895·07	5029·77	5169·41	5314·40	5465·24	5622·45	5786·65	5958·53	5
6	4641·11	4767·08	4897·28	5032·05	5171·78	5316·87	5467·81	5625·13	5789·45	5961·47	6
7	4643·17	4769·21	4899·49	5034·34	5174·15	5319·34	5470·37	5627·81	5792·25	5964·40	7
8	4645·24	4771·35	4901·70	5036·63	5176·52	5321·80	5472·94	5630·49	5795·05	5967·34	8
9	4647·31	4773·48	4903·91	5038·92	5178·90	5324·27	5475·51	5633·17	5797·86	5970·28	9
10	4649·38	4775·62	4906·12	5041·21	5181·28	5326·74	5478·09	5635·86	5800·66	5973·23	10
11	4651·45	4777·76	4908·33	5043·50	5183·66	5329·22	5480·66	5638·54	5803·47	5976·17	11
12	4653·52	4779·90	4910·55	5045·80	5186·04	5331·69	5483·24	5641·23	5806·29	5979·12	12
13	4655·60	4782·04	4912·76	5048·09	5188·42	5334·17	5485·82	5643·92	5809·10	5982·07	13
14	4657·67	4784·18	4914·98	5050·39	5190·80	5336·65	5488·40	5646·62	5811·92	5985·03	14
15	4659·75	4786·33	4917·20	5052·69	5193·19	5339·13	5490·98	5649·31	5814·74	5987·98	15
16	4661·82	4788·47	4919·42	5054·99	5195·57	5341·61	5493·57	5652·01	5817·56	5990·94	16
17	4663·90	4790·62	4921·64	5057·29	5197·96	5344·09	5496·15	5654·71	5820·38	5993·90	17
18	4665·98	4792·77	4923·86	5059·59	5200·35	5346·57	5498·74	5657·41	5823·21	5996·86	18
19	4668·06	4794·92	4926·08	5061·89	5202·74	5349·06	5501·33	5660·11	5826·04	5999·83	19
20	4670·14	4797·07	4928·30	5064·20	5205·14	5351·55	5503·92	5662·82	5828·87	6002·80	20
21	4672·22	4799·22	4930·53	5066·51	5207·53	5354·04	5506·51	5665·52	5831·70	6005·77	21
22	4674·30	4801·37	4932·76	5068·81	5209·92	5356·53	5509·11	5668·23	5834·53	6008·74	22
23	4676·39	4803·52	4934·99	5071·12	5212·32	5359·02	5511·71	5670·94	5837·37	6011·72	23
24	4678·47	4805·68	4937·22	5073·43	5214·72	5361·51	5514·31	5673·66	5840·21	6014·69	24
25	4680·56	4807·83	4939·45	5075·75	5217·12	5364·01	5516·91	5676·37	5843·05	6017·67	25
26	4682·65	4809·99	4941·68	5078·06	5219·52	5366·51	5519·51	5679·09	5845·89	6020·66	26
27	4684·74	4812·15	4943·91	5080·37	5221·93	5369·01	5522·11	5681·81	5848·74	6023·64	27
28	4686·83	4814·31	4946·15	5082·69	5224·33	5371·51	5524·72	5684·53	5851·58	6026·63	28
29	4688·92	4816·47	4948·38	5085·01	5226·74	5374·01	5527·33	5687·25	5854·43	6029·62	29
30	4691·01	4818·63	4950·62	5087·33	5229·14	5376·51	5529·94	5689·98	5857·29	6032·61	30
31	4693·10	4820·79	4952·86	5089·65	5231·55	5379·02	5532·55	5692·70	5860·14	6035·60	31
32	4695·19	4822·96	4955·10	5091·97	5233·96	5381·53	5535·16	5695·43	5863·00	6038·60	32
33	4697·29	4825·12	4957·34	5094·29	5236·38	5384·04	5537·78	5698·16	5865·86	6041·60	33
34	4699·39	4827·29	4959·58	5096·62	5238·79	5386·55	5540·39	5700·90	5868·72	6044·60	34
35	4701·48	4829·46	4961·83	5098·94	5241·21	5389·06	5543·01	5703·63	5871·58	6047·61	35
36	4703·58	4831·63	4964·07	5101·27	5243·62	5391·57	5545·63	5706·37	5874·44	6050·62	36
37	4705·68	4833·80	4966·32	5103·60	5246·04	5394·09	5548·25	5709·11	5877·31	6053·63	37
38	4707·78	4835·97	4968·57	5105·93	5248·46	5396·61	5550·88	5711·85	5880·18	6056·64	38
39	4709·89	4838·14	4970·82	5108·26	5250·88	5399·13	5553·50	5714·59	5883·05	6059·65	39
40	4711·99	4840·32	4973·07	5110·60	5253·31	5401·65	5556·13	5717·34	5885·93	6062·67	40
41	4714·09	4842·49	4975·32	5112·93	5255·73	5404·17	5558·76	5720·09	5888·80	6065·69	41
42	4716·20	4844·67	4977·57	5115·27	5258·16	5406·70	5561·39	5722·84	5891·68	6068·71	42
43	4718·30	4846·85	4979·83	5117·61	5260·59	5409·22	5564·03	5725·59	5894·56	6071·73	43
44	4720·41	4849·03	4982·08	5119·95	5263·02	5411·75	5566·66	5728·34	5897·45	6074·76	44
45	4722·52	4851·21	4984·34	5122·29	5265·45	5414·28	5569·30	5731·10	5900·33	6077·79	45
46	4724·63	4853·39	4986·60	5124·63	5267·88	5416·81	5571·94	5733·85	5903·22	6080·82	46
47	4726·74	4855·57	4988·86	5126·97	5270·31	5419·34	5574·58	5736·61	5906·11	6083·86	47
48	4728·86	4857·76	4991·12	5129·32	5272·75	5421·88	5577·23	5739·37	5909·00	6086·90	48
49	4730·97	4859·94	4993·38	5131·66	5275·19	5424·42	5579·87	5742·14	5911·90	6089·94	49
50	4733·08	4862·13	4995·65	5134·01	5277·63	5426·96	5582·52	5744·90	5914·80	6092·98	50
51	4735·20	4864·31	4997·91	5136·36	5280·07	5429·50	5585·17	5747·67	5917·70	6096·02	51
52	4737·32	4866·50	5000·18	5138·71	5282·51	5432·04	5587·82	5750·44	5920·60	6099·07	52
53	4739·43	4868·69	5002·45	5141·06	5284·95	5434·58	5590·47	5753·21	5923·50	6102·12	53
54	4741·55	4870·88	5004·72	5143·42	5287·40	5437·13	5593·12	5755·99	5926·41	6105·17	54
55	4743·67	4873·08	5006·99	5145·77	5289·85	5439·67	5595·78	5758·76	5929·32	6108·23	55
56	4745·80	4875·27	5009·26	5148·13	5292·30	5442·22	5598·44	5761·54	5932·23	6111·28	56
57	4747·92	4877·47	5011·53	5150·49	5294·75	5444·77	5601·10	5764·32	5935·14	6114·34	57
58	4750·04	4879·66	5013·81	5152·85	5297·20	5447·32	5603·76	5767·11	5938·06	6117·41	58
59	4752·17	4881·86	5016·08	5155·21	5299·65	5449·88	5606·42	5769·89	5940·97	6120·47	59
60	4754·29	4884·06	5018·36	5157·57	5302·11	5452·43	5609·09	5772·68	5943·89	6123·54	60
M	61°	62°	63°	64°	65°	66°	67°	68°	69°	70°	M

(Terrestrial Spheroid) **MERIDIONAL PARTS.** Compression $\frac{1}{293.465}$

M	71°	72°	73°	74°	75°	76°	77°	78°	79°	80°	M
0	6123·54	6312·55	6512·01	6723·21	6947·70	7187·33	7444·37	7721·64	8022·70	8352·11	0
1	6126·61	6315·78	6515·43	6726·84	6951·56	7191·46	7448·81	7726·45	8027·94	8357 87	1
2	6129·68	6319·02	6518·85	6730·47	6955·43	7195·60	7453·26	7731·27	8033·19	8363·65	2
3	6132·76	6322·26	6522·28	6734·11	6959·30	7199·74	7457·72	7736·09	8038·45	8369·43	3
4	6135·84	6325·51	6525·71	6737·75	6963·18	7203·89	7462·19	7740·92	8043·72	8375·22	4
5	6138·92	6328·75	6529·14	6741·39	6967·06	7208·05	7466·66	7745·76	8048·99	8381·02	5
6	6142·00	6332·00	6532·58	6745·04	6970·95	7212·21	7471·13	7750·61	8054·27	8386·83	6
7	6145·09	6335·26	6536·02	6748·69	6974·84	7216·37	7475·61	7755·46	8059·56	8392·65	7
8	6148·18	6338·51	6539·46	6752·34	6978·73	7220·54	7480·10	7760·32	8064·86	8398·48	8
9	6151·27	6341·77	6542·91	6756·00	6982·63	7224·71	7484·59	7765·18	8070·17	8404·32	9
10	6154·37	6345·03	6546·36	6759·66	6986·53	7228·89	7489·09	7770·05	8075·49	8410·17	10
11	6157·46	6348·30	6549·81	6763·32	6990·44	7233·07	7493·59	7774·93	8080·81	8416·03	11
12	6160·56	6351·57	6553·27	6766·99	6994·35	7237·26	7498·10	7779·82	8086·14	8421·90	12
13	6163·66	6354·84	6556·73	6770·67	6998·26	7241·45	7502·62	7784·71	8091·48	8427·78	13
14	6166·77	6358·11	6560·19	6774·34	7002·18	7245·65	7507·14	7789·61	8096·83	8433·67	14
15	6169·88	6361·39	6563·65	6778·02	7006·11	7249·85	7511·66	7794·51	8102·18	8439·56	15
16	6172·99	6364·67	6567·12	6781·71	7010·03	7254·06	7516·19	7799·43	8107·55	8445·47	16
17	6176·10	6367·95	6570·60	6785·40	7013·97	7258·27	7520·73	7804·35	8112·92	8451·39	17
18	6179·21	6371·23	6574·07	6789·09	7017·90	7262·49	7525·28	7809·27	8118·30	8457·32	18
19	6182·33	6374·52	6577·55	6792·78	7021·84	7266·72	7529·83	7814·21	8123·69	8463·26	19
20	6185·45	6377·81	6581·04	6796·48	7025·79	7270·95	7534·38	7819·15	8129·09	8469·21	20
21	6188·58	6381·11	6584·52	6800·18	7029·74	7275·18	7538·94	7824·09	8134·49	8475·17	21
22	6191·70	6384·41	6588·01	6803·89	7033·69	7279·42	7543·51	7829·05	8139·91	8481·14	22
23	6194·83	6387·71	6591·50	6807·60	7037·65	7283·66	7548·09	7834·01	8145·33	8487·12	23
24	6197·97	6391·01	6595·00	6811·32	7041·62	7287·91	7552·67	7838·98	8150·76	8493·11	24
25	6201·10	6394·32	6598·50	6815·04	7045·58	7292·16	7557·25	7843·95	8156·20	8499·11	2.
26	6204·24	6397·63	6602·01	6818·76	7049·56	7296·42	7561·84	7848·93	8161·65	8505·12	26
27	6207·38	6400·94	6605·51	6822·49	7053·53	7300·69	7566·44	7853·92	8167·10	8511·14	27
28	6210·52	6404·26	6609·02	6826·22	7057·51	7304·95	7571·05	7858·92	8172·57	8517·17	28
29	6213·66	6407·58	6612·54	6829·95	7061·50	7309·23	7575·66	7863·93	8178·04	8523·21	29
30	6216·81	6410·90	6616·05	6833·69	7065·49	7313·51	7580·27	7868·94	8183·52	8529·26	30
31	6219·96	6414·22	6619·57	6837·43	7069·48	7317·79	7584·89	7873·95	8189·01	8535·33	31
32	6223·12	6417·55	6623·10	6841·18	7073·48	7322·08	7589·52	7878·98	8194·51	8541·40	32
33	6226·27	6420·88	6626·63	6844·93	7077·49	7326·38	7594·15	7884·01	8200·02	8547·49	33
34	6229·43	6424·22	6630·16	6848·68	7081·49	7330·68	7598·79	7889·05	8205·54	8553·58	34
35	6232·59	6427·55	6633·69	6852·44	7085·51	7334·98	7603·44	7894·10	8211·06	8559·69	35
36	6235·76	6430·89	6637·23	6856·20	7089·52	7339·29	7608·09	7899·15	8216·60	8565·80	36
37	6238·92	6434·24	6640·77	6859·97	7093·54	7343·61	7612·75	7904·21	8222·14	8571·93	37
38	6242·09	6437·58	6644·32	6863·74	7097·57	7347·93	7617·42	7909·28	8227·69	8578·07	38
39	6245·27	6440·93	6647·87	6867·51	7101·60	7352·26	7622·09	7914·36	8233·25	8584·22	39
40	6248·44	6444·29	6651·42	6871·29	7105·64	7356·59	7626·76	7919·44	8238·82	8590·38	40
41	6251·62	6447·64	6654·97	6875·07	7109·68	7360·93	7631·45	7924·53	8244·40	8596·55	41
42	6254·80	6451·00	6658·53	6878·86	7113·72	7365·27	7636·14	7929·63	8249·98	8602·73	42
43	6257·99	6454·36	6662·10	6882·65	7117·77	7369·62	7640·83	7934·74	8255·58	8608·92	43
44	6261·17	6457·73	6665·66	6886·45	7121·83	7373·97	7645·53	7939·85	8261·19	8615·12	44
45	6264·36	6461·10	6669·23	6890·24	7125·88	7378·33	7650·24	7944·97	8266·80	8621·34	45
46	6267·56	6464·47	6672·81	6894·04	7129·95	7382·69	7654·96	7950·10	8272·42	8627·56	46
47	6270·75	6467·84	6676·38	6897·85	7134·02	7387·06	7659·68	7955·24	8278·05	8633·80	47
48	6273·95	6471·22	6679·96	6901·66	7138·09	7391·44	7664·41	7960·38	8283·70	8640·05	48
49	6277·15	6474·60	6683·55	6905·47	7142·17	7395·82	7669·14	7965·53	8289·35	8646·31	49
50	6280·35	6477·99	6687·14	6909·29	7146·25	7400·21	7673·88	7970·69	8295·01	8652·58	50
51	6283·56	6481·38	6690·73	6913·11	7150·33	7404·60	7678·63	7975·85	8300·68	8658·86	51
52	6286·77	6484·77	6694·32	6916·94	7154·42	7408·99	7683·38	7981·03	8306·35	8665·15	52
53	6289·98	6488·16	6697·92	6920·77	7158·52	7413·40	7688·14	7986·21	8312·04	8671·46	53
54	6293·20	6491·56	6701·52	6924·60	7162·62	7417·80	7692·90	7991·40	8317·74	8677·77	54
55	6296·42	6494·96	6705·13	6928·44	7166·73	7422·22	7697·68	7996·59	8323·44	8684·10	55
56	6299·64	6498·36	6708·74	6932·28	7170·84	7426·64	7702·46	8001·80	8329·16	8690·44	56
57	6302·86	6501·77	6712·35	6936·13	7174·95	7431·06	7707·24	8007·01	8334·88	8696·79	57
58	6306·09	6505·18	6715·97	6939·98	7179·07	7435·49	7712·03	8012·23	8340·62	8703·15	58
59	6309·32	6508·59	6719·59	6943·84	7183·20	7439·93	7716·83	8017·46	8346·36	8709·52	59
60	6312·55	6512·01	6723·21	6947·70	7187·33	7444·37	7721·64	8022·70	8352·11	8715·91	60
M	**71°**	**72°**	**73°**	**74°**	**75°**	**76°**	**77°**	**78°**	**79°**	**80°**	**M**

MEAN REFRACTION.

BAROMETER 29·6 inches
THERMOMETER (Fahr.) 50°

ALWAY minus

App. Alt.	Refr.	App. Alt.	Refr.	App. Alt.	Refr.	App. Alt.	Refr.	App. Alt.	Refr.
° ′	′	° ′	′	° ′	′	° ′	′	° ′	′
0 00	33·00	5 00	9·90	10 00	5·25	20 00	2·58	34 00	1·40
0 05	32·17	5 05	9·77	10 10	5·17	20 10	2·57	34 30	1·38
0 10	31·37	5 10	9·63	10 20	5·08	20 20	2·53	35 00	1·35
0 15	30·58	5 15	9·50	10 30	5·00	20 30	2·52	35 30	1·33
0 20	29·83	5 20	9·38	10 40	4·93	20 40	2·49	36 00	1·30
0 25	29·10	5 25	9·25	10 50	4·85	20 50	2·47	36 30	1·28
0 30	28·38	5 30	9·13	11 00	4·78	21 00	2·45	37 00	1·27
0 35	27·68	5 35	9·02	11 10	4·72	21 10	2·43	37 30	1·23
0 40	27·00	5 40	8·90	11 20	4·65	21 20	2·42	38 00	1·22
0 45	26·33	5 45	8·78	11 30	4·57	21 30	2·40	38 30	1·18
0 50	25·70	5 50	8·68	11 40	4·52	21 40	2·38	39 00	1·17
0 55	25·83	5 55	8·57	11 50	4·45	21 50	2·35	39 30	1·15
1 00	24·48	6 00	8·47	12 00	4·38	22 00	2·33	40 00	1·13
1 05	23·90	6 05	8·35	12 10	4·33	22 10	2·32	41 00	1·08
1 10	23·33	6 10	8·25	12 20	4·27	22 20	2·30	42 00	1·05
1 15	22·78	6 15	8·15	12 30	4·22	22 30	2·28	43 00	1·02
1 20	22·25	6 20	8·05	12 40	4·15	22 40	2·27	44 00	0·98
1 25	21·73	6 25	7·95	12 50	4·10	22 50	2·25	45 00	0·95
1 30	21·25	6 30	7·85	13 00	4·05	23 00	2·23	46 00	0·92
1 35	20·77	6 35	7·75	13 10	4·00	23 10	2·22	47 00	0·88
1 40	20·30	6 40	7·67	13 20	3·95	23 20	2·20	48 00	0·85
1 45	19·85	6 45	7·58	13 30	3·90	23 30	2·18	49 00	0·82
1 50	19·42	6 50	7·50	13 40	3·85	23 40	2·17	50 00	0·80
1 55	19·00	6 55	7·42	13 50	3·80	23 50	2·15	51 00	0·77
2 00	18·58	7 00	7·33	14 00	3·75	24 00	2·13	52 00	0·73
2 05	18·18	7 05	7·25	14 10	3·72	24 10	2·12	53 00	0·71
2 10	17·80	7 10	7·18	14 20	3·67	24 20	2·10	54 00	0·68
2 15	17·43	7 15	7·10	14 30	3·63	24 30	2·08	55 00	0·67
2 20	17·07	7 20	7·03	14 40	3·58	24 40	2·07	56 00	0·63
2 25	16·73	7 25	6·95	14 50	3·55	24 50	2·05	57 00	0·62
2 30	16·40	7 30	6·88	15 00	3·50	25 00	2·03	58 00	0·58
2 35	16·07	7 35	6·82	15 10	3·47	25 10	2·02	59 00	0·57
2 40	15·75	7 40	6·75	15 20	3·43	25 20	2·00	60 00	0·55
2 45	15·45	7 45	6·68	15 30	3·40	25 30	1·98	61 00	0·53
2 50	15·15	7 50	6·62	15 40	3·35	25 40	1·97	62 00	0·50
2 55	14·87	7 55	6·55	15 50	3·32	25 50	1·95	63 00	0·48
3 00	14·60	8 00	6·48	16 00	3·28	26 00	1·93	64 00	0·47
3 05	14·33	8 05	6·42	16 10	3·25	26 10	1·92	65 00	0·43
3 10	14·07	8 10	6·37	16 20	3·20	26 20	1·91	66 00	0·42
3 15	13·82	8 15	6·30	16 30	3·17	26 30	1·90	67 00	0·40
3 20	13·57	8 20	6·25	16 40	3·13	26 40	1·88	68 00	0·38
3 25	13·33	8 25	6·18	16 50	3·10	26 50	1·87	69 00	0·37
3 30	13·10	8 30	6·13	17 00	3·07	27 00	1·85	70 00	0·35
3 35	12·88	8 35	6·08	17 10	3·05	27 15	1·83	71 00	0·32
3 40	12·67	8 40	6·02	17 20	3·02	27 30	1·82	72 00	0·30
3 45	12·45	8 45	5·97	17 30	2·98	27 45	1·80	73 00	0·28
3 50	12·25	8 50	5·92	17 40	2·95	28 00	1·78	74 00	0·27
3 55	12·05	8 55	5·87	17 50	2·92	28 15	1·77	75 00	0·25
4 00	11·85	9 00	5·80	18 00	2·90	28 30	1·75	76 00	0·23
4 05	11·67	9 05	5·75	18 10	2·87	28 45	1·73	77 00	0·22
4 10	11·48	9 10	5·70	18 20	2·85	29 00	1·70	78 00	0·20
4 15	11·30	9 15	5·65	18 30	2·82	29 30	1·67	79 00	0·18
4 20	11·13	9 20	5·60	18 40	2·78	30 00	1·63	80 00	0·17
4 25	10·97	9 25	5·57	18 50	2·77	30 30	1·62	81 00	0·15
4 30	10·80	9 30	5·52	19 00	2·73	31 00	1·58	82 00	0·13
4 35	10·65	9 35	5·47	19 10	2·72	31 30	1·55	83 00	0·12
4 40	10·48	9 40	5·42	19 20	2·68	32 00	1·52	84 00	0·10
4 45	10·33	9 45	5·38	19 30	2·67	32 30	1·50	86 00	0·07
4 50	10·18	9 50	5·33	19 40	2·64	33 00	1·47	88 00	0·03
4 55	10·03	9 55	5·30	19 50	2·62	33 30	1·43	90 00	0·00

CORRECTION of the MEAN REFRACTION

App. Alt.	HEIGHT OF THE THERMOMETER (FAHR.)										
	20°	26°	32°	38°	44°	50°	56°	62°	68°	74°	80
° ′	+ ′	+ ′	+ ′	+ ′	+ ′	0	− ′	− ′	− ′	′	′
2 00	1·52	1·20	·88	·58	·28	0	·27	·53	·80	1·05	1·3
2 10	1·45	1·13	·83	·55	·27	0	·26	·52	·76	1·00	1·2
2 20	1·38	1·08	·80	·53	·26	0	·25	·50	·74	·96	1·1
2 30	1·33	1·05	·77	·50	·25	0	·24	·48	·71	·93	1·1
2 40	1·28	1·00	·73	·48	·23	0	·23	·46	·69	·90	1·1
2 50	1·23	·97	·71	·47	·23	0	·22	·45	·66	·86	1·0
3 00	1·18	·93	·68	·45	·22	0	·21	·43	·63	·83	1·0
3 10	1·13	·90	·66	·43	·22	0	·20	·41	·61	·80	·9
3 20	1·10	·87	·63	·42	·21	0	·20	·40	·59	·78	·9
3 30	1·07	·83	·62	·40	·20	0	·19	·38	·56	·75	·9
3 40	1·03	·81	·60	·40	·20	0	·18	·36	·54	·72	·8
3 50	1·00	·78	·58	·39	·19	0	·17	·34	·52	·70	·8
4 00	·97	·75	·55	·37	·18	0	·16	·32	·50	·66	·8
4 30	·88	·70	·52	·35	·17	0	·15	·30	·47	·62	·7
5 00	·80	·65	·48	·31	·15	0	·14	·29	·43	·57	·7
5 30	·75	·58	·43	·28	·14	0	·13	·27	·40	·52	·6
6 00	·68	·53	·40	·27	·13	0	·12	·25	·37	·49	·6
6 30	·63	·50	·37	·24	·12	0	·11	·23	·35	·46	·5
7 00	·60	·47	·35	·23	·11	0	·10	·21	·32	·43	·5
7 30	·57	·45	·33	·22	·10	0	·10	·20	·30	·40	·4
8 00	·53	·42	·31	·20	·10	0	·09	·19	·28	·38	·4
8 30	·50	·38	·29	·19	·09	0	·09	·18	·26	·36	·4
9 00	·47	·37	·27	·17	·08	0	·08	·17	·25	·34	·4
9 30	·45	·35	·26	·17	·08	0	·08	·16	·23	·32	·3
10 00	·43	·33	·25	·17	·08	0	·07	·15	·22	·30	·3
10 30	·40	·32	·25	·16	·08	0	·07	·14	·21	·28	·3
11 00	·38	·31	·24	·15	·08	0	·06	·13	·20	·26	·3
11 30	·37	·30	·23	·15	·07	0	·06	·12	·20	·24	·3
12 00	·35	·28	·22	·14	·06	0	·06	·11	·19	·23	·3
13 00	·33	·27	·20	·13	·06	0	·05	·11	·18	·22	·2
14 00	·30	·24	·19	·12	·06	0	·05	·10	·17	·21	·2
15 00	·28	·22	·18	·11	·05	0	·05	·10	·16	·20	·2
16 00	·27	·21	·16	·10	·05	0	·05	·10	·15	·19	·2
17 00	·25	·20	·15	·10	·05	0	·05	·09	·13	·18	·2
18 00	·23	·18	·14	·09	·05	0	·04	·09	·12	·17	·2
19 00	·22	·17	·13	·08	·05	0	·04	·08	·11	·16	·1
20 00	·21	·16	·12	·08	·04	0	·04	·08	·10	·15	·1
21 00	·20	·15	·12	·08	·04	0	·04	·07	·10	·15	·1
22 00	·18	·15	·11	·08	·04	0	·04	·07	·10	·14	·1
23 00	·18	·14	·10	·07	·04	0	·04	·07	·10	·13	·1
24 00	·17	·14	·10	·07	·04	0	·03	·06	·09	·12	·1
25 00	·16	·13	·10	·07	·03	0	·03	·06	·09	·12	·1
26 00	·15	·13	·10	·06	·03	0	·03	·06	·08	·11	·1
27 00	·15	·12	·09	·06	·03	0	·03	·05	·08	·10	·1
28 00	·14	·12	·09	·05	·03	0	·03	·05	·07	·09	·1
29 00	·13	·11	·09	·05	·03	0	·03	·04	·07	·08	·1
30 00	·13	·10	·08	·04	·03	0	·02	·04	·06	·08	·1
32 00	·12	·09	·08	·04	·02	0	·02	·04	·06	·07	·1
34 00	·11	·08	·07	·04	·02	0	·02	·03	·05	·07	·1
36 00	·10	·08	·07	·03	·02	0	·02	·03	·05	·06	·0
38 00	·10	·08	·06	·03	·02	0	·01	·03	·05	·06	·0
40 00	·10	·07	·06	·03	·01	0	·01	·02	·05	·06	·0
45 00	·09	·06	·05	·03	·01	0	·01	·02	·04	·05	·0
50 00	·07	·05	·05	·02	·01	0	·01	·02	·04	·05	·0
55 00	·05	·04	·04	·02	·01	0	·01	·01	·03	·04	·0
60 00	·05	·03	·03	·01	·01	0	0	·01	·02	·03	·0
65 00	·04	·03	·02	·01	0	0	0	0	·01	·02	·0
70 00	·03	·02	·01	0	0	0	0	0	·01	·02	·0
80 00	·02	·01	0	0	0	0	0	0	0	·01	·0
90 00	0	0	0	0	0	0	0	0	0	0	0
	28·3 −	28·7 −	29·2 −	29·6	30·0 +	30·5 +	30·9 +				

HEIGHT OF THE BAROMETER

DISTANCE OF SEA HORIZON

Height (Feet)	Dist. (Miles)	Height (Feet)	Dist. (Miles)	Height (Metres)	Dist. (Miles)
1	1·2	310	20·6	1	2·1
2	1·7	320	20·9	2	3·0
3	2·0	330	21·3	3	3·7
4	2·3	340	21·6	4	4·2
5	2·6	350	21·9	5	4·7
6	2·9	360	22·2	6	5·2
7	3·1	370	22·5	7	5·6
8	3·3	380	22·8	8	6·0
9	3·5	390	23·1	9	6·4
10	3·7	400	23·4	10	6·7
12	4·1	410	23·7	11	7·0
14	4·4	420	24·0	12	7·3
16	4·7	430	24·3	13	7·6
18	5·0	440	24·5	14	7·9
20	5·2	450	24·8	15	8·2
22	5·5	460	25·1	16	8·5
24	5·7	470	25·4	17	8·7
26	6·0	480	25·6	18	9·0
28	6·2	490	25·9	19	9·2
30	6·4	500	26·2	20	9·5
32	6·6	520	26·7	22	9·9
34	6·8	540	27·2	24	10·4
36	7·0	560	27·7	26	10·8
38	7·2	580	28·2	28	11·2
40	7·4	600	28·7	30	11·6
42	7·6	620	29·1	32	12·0
44	7·8	640	29·6	34	12·4
46	7·9	660	30·1	36	12·7
48	8·1	680	30·5	38	13·1
50	8·3	700	31·0	40	13·4
55	8·7	720	31·4	50	15·0
60	9·1	740	31·8	60	16·4
65	9·4	760	32·3	70	17·7
70	9·8	780	32·7	80	19·0
75	10·1	800	33·1	90	20·1
80	10·5	820	33·5	100	21·2
85	10·7	840	33·9	110	22·2
90	11·1	860	34·3	120	23·2
95	11·4	880	34·7	130	24·2
100	11·7	900	35·1	140	25·1
110	12·3	950	36·1	150	26·0
120	12·8	1000	37·0	200	30·0
130	13·3	1050	37·9	250	33·5
140	13·8	1100	38·8	300	36·7
150	14·3	1150	39·7	350	39·7
160	14·8	1200	40·5	400	42·4
170	15·3	1250	41·4	450	45·0
180	15·7	1300	42·2	500	47·4
190	16·1	1350	43·0	550	49·7
200	16·5	1400	43·8	600	51·9
210	17·0	1500	45·3	700	56·1
220	17·4	1600	46·8	800	59·9
230	17·7	1700	48·2	900	63·6
240	18·1	1800	49·6	1000	67·0
250	18·5	1900	51·0	1100	70·3
260	18·9	2000	52·3	1200	73·4
270	19·2	3000	64·1	1300	76·4
280	19·6	4000	74·0	1400	79·3
290	19·9	5000	82·7	1500	82·1
300	20·3	6000	90·6	1600	84·8

DIP OF THE SEA HORIZON
SUBTRACTIVE

H.E. (Feet)	Dip (')	H.E. (Feet)	Dip (')	H.E. (Feet)	Dip (')	H.E. (Metres)	Dip (')
1	1·0	51	6·9	105	9·9	1	1·8
2	1·4	52	7·0	110	10·2	2	2·5
3	1·7	53	7·1	115	10·4	3	3·0
4	1·9	54	7·1	120	10·6	4	3·5
5	2·2	55	7·2	125	10·8	5	3·9
6	2·4	56	7·3	130	11·1	6	4·3
7	2·6	57	7·3	135	11·3	7	4·7
8	2·7	58	7·4	140	11·5	8	5·0
9	2·9	59	7·5	145	11·7	9	5·3
10	3·1	60	7·5	150	11·9	10	5·6
11	3·2	61	7·6	155	12·1	11	5·8
12	3·4	62	7·6	160	12·3	12	6·1
13	3·5	63	7·7	165	12·5	13	6·3
14	3·6	64	7·8	170	12·6	14	6·6
15	3·8	65	7·8	175	12·8	15	6·8
16	3·9	66	7·9	180	13·0	16	7·0
17	4·0	67	7·9	185	13·2	17	7·3
18	4·1	68	8·0	190	13·4	18	7·5
19	4·2	69	8·1	195	13·5	19	7·7
20	4·3	70	8·1	200	13·7	20	7·9
21	4·4	71	8·2	210	14·1	21	8·1
22	4·5	72	8·2	220	14·4	22	8·3
23	4·7	73	8·3	230	14·7	23	8·4
24	4·8	74	8·3	240	15·0	24	8·6
25	4·9	75	8·4	250	15·3	25	8·8
26	4·9	76	8·5	260	15·6	26	9·0
27	5·0	77	8·5	270	15·9	27	9·1
28	5·1	78	8·6	280	16·2	28	9·3
29	5·2	79	8·6	290	16·5	29	9·5
30	5·3	80	8·7	300	16·8	30	9·6
31	5·4	81	8·7	310	17·1	35	10·4
32	5·5	82	8·8	320	17·4	40	11·1
33	5·6	83	8·8	330	17·6	45	11·8
34	5·7	84	8·9	340	17·9	50	12·4
35	5·7	85	8·9	350	18·1	55	13·0
36	5·8	86	9·0	360	18·4	60	13·6
37	5·9	87	9·0	370	18·7	65	14·2
38	6·0	88	9·1	380	18·9	70	14·7
39	6·1	89	9·2	390	19·2	75	15·2
40	6·1	90	9·2	400	19·4	80	15·7
41	6·2	91	9·3	410	19·7	85	16·2
42	6·3	92	9·3	420	19·9	90	16·7
43	6·4	93	9·4	430	20·1	95	17·1
44	6·4	94	9·4	440	20·3	100	17·6
45	6·5	95	9·5	450	20·6	105	18·0
46	6·6	96	9·5	460	20·8	110	18·4
47	6·6	97	9·6	470	21·0	120	19·2
48	6·7	98	9·6	480	21·3	130	20·0
49	6·8	99	9·7	490	21·5	140	20·8
50	6·9	100	9·7	500	21·7	150	21·5

SUN'S PARALLAX in ALTITUDE
ADDITIVE

Alt. (°)	Parlx. (')	Alt. (°)	Parlx. (')
0	0·15	45	0·11
5	0·15	50	0·10
10	0·15	55	0·09
15	0·14	60	0·08
20	0·14	65	0·06
25	0·14	70	0·05
30	0·13	75	0·04
35	0·12	80	0·03
40	0·12	85	0·01

SUN'S SEMI-DIAMETER
on the 1st DAY of EACH MONTH

Month	S.D. (')	Month	S.D. (')
Jan.	16·29	July	15·76
Feb.	16·26	Aug.	15·78
Mar.	16·16	Sept.	15·88
April	16·03	Oct.	16·01
May	15·89	Nov.	16·14
June	15·78	Dec.	16·25

SUN'S TOTAL CORRECTION

To be ADDED to the OBSERVED ALTITUDE of the SUN'S LOWER LIMB.

Obs. Alt.	ft. 6 / m. 1·8	8 / 2·4	10 / 3·0	12 / 3·7	14 / 4·3	16 / 4·9	18 / 5·5	20 / 6·1	22 / 6·7	24 / 7·3	26 / 7·9	28 / 8·5	30 / 9·1	32 / 9·8	36 / 11·0	40 / 12·2	Obs. Alt.
HEIGHT OF EYE IN FEET AND METRES:																	
6	5·0	4·7	4·3	4·0	3·8	3·5	3·3	3·1	2·9	2·6	2·5	2·3	2·1	1·9	1·6	1·3	6
6½	5·6	5·3	4·9	4·6	4·4	4·1	3·9	3·7	3·5	3·2	3·1	2·9	2·7	2·5	2·2	1·9	6½
7	6·2	5·9	5·5	5·2	5·0	4·7	4·5	4·3	4·1	3·8	3·7	3·5	3·3	3·1	2·8	2·5	7
7½	6·6	6·2	5·9	5·6	5·4	5·1	4·9	4·7	4·5	4·2	4·1	3·9	3·7	3·5	3·2	2·9	7½
8	7·0	6·7	6·3	6·0	5·8	5·5	5·3	5·1	4·9	4·6	4·5	4·3	4·1	3·9	3·6	3·3	8
8½	7·4	7·1	6·7	6·4	6·2	5·9	5·7	5·5	5·3	5·0	4·9	4·7	4·5	4·3	4·0	3·7	8½
9	7·7	7·4	7·0	6·7	6·5	6·2	6·0	5·8	5·6	5·3	5·2	5·0	4·8	4·6	4·3	4·0	9
10	8·2	7·9	7·5	7·2	7·0	6·7	6·5	6·3	6·1	5·8	5·7	5·5	5·3	5·1	4·8	4·5	10
11	8·7	8·4	8·0	7·7	7·5	7·2	7·0	6·8	6·6	6·3	6·2	6·0	5·8	5·6	5·3	5·0	11
12	9·1	8·8	8·4	8·1	7·9	7·6	7·4	7·2	7·0	6·7	6·6	6·4	6·2	6·0	5·7	5·4	12
13	9·4	9·1	8·7	8·4	8·2	7·9	7·7	7·5	7·3	7·0	6·9	6·7	6·5	6·3	6·0	5·7	13
14	9·7	9·4	9·0	8·7	8·5	8·2	8·0	7·8	7·6	7·3	7·2	7·0	6·8	6·6	6·3	6·0	14
15	10·0	9·7	9·3	9·0	8·8	8·5	8·3	8·1	7·9	7·6	7·5	7·3	7·1	6·9	6·6	6·3	15
16	10·2	9·9	9·5	9·2	9·0	8·7	8·5	8·3	8·1	7·8	7·7	7·5	7·3	7·1	6·8	6·5	16
18	10·6	10·3	9·9	9·6	9·4	9·1	8·9	8·7	8·5	8·2	8·1	7·9	7·7	7·5	7·2	6·9	18
20	10·9	10·6	10·2	9·9	9·7	9·4	9·2	9·0	8·8	8·5	8·4	8·2	8·0	7·8	7·5	7·2	20
22	11·1	10·9	10·5	10·2	10·0	9·7	9·5	9·3	9·1	8·8	8·7	8·5	8·3	8·1	7·8	7·5	22
26	11·6	11·3	10·9	10·6	10·4	10·1	9·9	9·7	9·5	9·2	9·1	8·9	8·7	8·5	8·2	7·9	26
30	11·9	11·6	11·2	10·9	10·7	10·4	10·2	10·0	9·8	9·5	9·4	9·2	9·0	8·8	8·5	8·2	30
35	12·1	11·8	11·4	11·1	10·9	10·6	10·4	10·2	10·0	9·7	9·6	9·4	9·2	9·0	8·7	8·4	35
40	12·4	12·1	11·7	11·4	11·2	10·9	10·7	10·5	10·3	10·0	9·9	9·7	9·5	9·3	9·0	8·7	40
50	12·7	12·4	12·0	11·7	11·5	11·2	11·0	10·8	10·6	10·3	10·2	10·0	9·8	9·6	9·3	9·0	50
60	12·9	12·6	12·2	11·9	11·7	11·4	11·2	11·0	10·8	10·5	10·4	10·2	10·0	9·8	9·5	9·2	60
70	13·1	12·8	12·4	12·1	11·9	11·6	11·4	11·2	11·0	10·7	10·6	10·4	10·2	10·0	9·7	9·4	70
80	13·2	12·9	12·5	12·2	12·0	11·7	11·5	11·3	11·1	10·8	10·7	10·5	10·3	10·1	9·8	9·5	80
90	13·4	13·1	12·7	12·4	12·2	11·9	11·7	11·5	11·3	11·0	10·9	10·7	10·5	10·3	10·0	9·7	90

Obs. Alt.	ft. 40 / m. 12·2	44 / 13·4	48 / 14·6	52 / 15·9	56 / 17·1	60 / 18·3	64 / 19·5	68 / 20·7	72 / 22·0	76 / 23·2	80 / 24·4	84 / 25·6	88 / 26·8	92 / 28·0	96 / 29·3	100 / 30·5	Obs. Alt.
HEIGHT OF EYE IN FEET AND METRES.																	
6	1·3	1·0	0·7	0·4	0·1												6
6½	1·9	1·6	1·3	1·0	0·7	0·5	0·2	0·0									6½
7	2·5	2·2	1·9	1·6	1·3	1·1	0·8	0·6	0·4	0·1							7
7½	2·9	2·6	2·3	2·0	1·7	1·5	1·2	1·0	0·8	0·5	0·3	0·1					7½
8	3·3	3·0	2·7	2·4	2·2	1·9	1·6	1·4	1·2	0·9	0·7	0·5	0·3	0·1			8
8½	3·7	3·4	3·1	2·8	2·5	2·3	2·0	1·8	1·6	1·3	1·1	0·9	0·7	0·5			8½
9	4·0	3·7	3·4	3·1	2·8	2·6	2·3	2·1	1·9	1·6	1·4	1·2	1·0	0·8	0·6	0·4	9
10	4·5	4·2	3·9	3·6	3·3	3·1	2·8	2·6	2·4	2·1	1·9	1·7	1·5	1·3	1·1	0·9	10
11	5·0	4·7	4·4	4·1	3·8	3·6	3·3	3·1	2·9	2·6	2·4	2·2	2·0	1·8	1·6	1·4	11
12	5·4	5·1	4·8	4·5	4·2	4·0	3·7	3·5	3·3	3·0	2·8	2·6	2·4	2·2	2·0	1·8	12
13	5·7	5·4	5·1	4·8	4·5	4·3	4·0	3·8	3·6	3·3	3·1	2·9	2·7	2·5	2·3	2·1	13
14	6·0	5·7	5·4	5·1	4·8	4·6	4·3	4·1	3·9	3·6	3·4	3·1	2·9	2·7	2·5	2·3	14
15	6·3	6·0	5·7	5·4	5·1	4·9	4·6	4·4	4·2	3·9	3·7	3·5	3·2	3·0	2·8	2·6	15
16	6·5	6·2	5·9	5·6	5·3	5·1	4·8	4·6	4·4	4·1	3·9	3·7	3·5	3·2	3·0	2·8	16
18	6·9	6·6	6·3	6·0	5·7	5·5	5·2	5·0	4·8	4·5	4·3	4·1	3·9	3·7	3·5	3·3	18
20	7·2	6·9	6·6	6·3	6·0	5·8	5·5	5·3	5·1	4·8	4·6	4·4	4·2	4·0	3·8	3·6	20
22	7·5	7·3	7·0	6·7	6·4	6·2	5·9	5·7	5·5	5·2	5·0	4·8	4·6	4·4	4·2	4·0	22
26	7·9	7·6	7·3	7·0	6·7	6·5	6·2	6·0	5·8	5·5	5·3	5·1	4·9	4·7	4·5	4·3	26
30	8·2	7·9	7·6	7·3	7·0	6·8	6·5	6·3	6·1	5·8	5·6	5·4	5·2	5·0	4·8	4·6	30
35	8·4	8·1	7·8	7·5	7·2	7·0	6·7	6·5	6·3	6·0	5·8	5·6	5·4	5·2	5·0	4·8	35
40	8·7	8·4	8·1	7·8	7·5	7·3	7·0	6·8	6·6	6·3	6·1	5·9	5·7	5·5	5·3	5·1	40
50	9·0	8·7	8·4	8·1	7·8	7·6	7·3	7·1	6·9	6·6	6·4	6·2	6·0	5·8	5·6	5·4	50
60	9·2	8·9	8·6	8·3	8·0	7·8	7·5	7·3	7·1	6·8	6·6	6·4	6·2	6·0	5·8	5·6	60
70	9·4	9·1	8·8	8·5	8·2	8·0	7·7	7·5	7·3	7·0	6·8	6·6	6·4	6·2	6·0	5·8	70
80	9·5	9·2	8·9	8·6	8·3	8·1	7·8	7·6	7·4	7·1	6·9	6·7	6·5	6·3	6·1	5·9	80
90	9·7	9·4	9·1	8·8	8·5	8·3	8·0	7·8	7·6	7·3	7·1	6·9	6·7	6·5	6·3	6·1	90

Month	Jan.	Feb.	Mar.	April	May	June	July	Aug.	Sept.	Oct.	Nov.	Dec.
Correction	+0'·5	+0'·4	+0'·3	+0'·2	+0'·1	0'·0	0'·0	0'·0	+0'·1	+0'·3	+0'·4	+0'·5

When UPPER LIMB observed, first SUBTRACT the SUN'S DIAMETER.

STAR'S TOTAL CORRECTION

To be SUBTRACTED from the OBSERVED ALTITUDE of a STAR

Obs. Alt.	ft. 6 m. 1·8	8 2·4	10 3·0	12 3·7	14 4·3	16 4·9	18 5·5	20 6·1	22 6·7	24 7·3	26 7·9	28 8·5	30 9·1	32 9·8	36 11·0	40 12·2	Obs. Alt.
6	10·9	11·2	11·6	11·9	12·1	12·4	12·6	12·8	13·0	13·3	13·4	13·6	13·8	14·0	14·3	14·6	6
6½	10·3	10·6	11·0	11·3	11·5	11·8	12·0	12·2	12·4	12·7	12·8	13·0	13·2	13·4	13·7	14·0	6½
7	9·7	10·0	10·4	10·7	10·9	11·2	11·4	11·6	11·8	12·1	12·2	12·4	12·6	12·8	13·1	13·4	7
7½	9·3	9·6	10·0	10·3	10·5	10·8	11·0	11·2	11·4	11·7	11·8	12·0	12·2	12·4	12·7	13·0	7½
8	8·9	9·2	9·6	9·9	10·1	10·4	10·6	10·8	11·0	11·3	11·4	11·6	11·8	12·0	12·3	12·6	8
8½	8·5	8·8	9·2	9·5	9·7	10·0	10·2	10·4	10·6	10·9	11·0	11·2	11·4	11·6	11·9	12·2	8½
9	8·2	8·5	8·9	9·2	9·4	9·7	9·9	10·1	10·3	10·6	10·7	10·9	11·1	11·3	11·6	11·9	9
10	7·7	8·0	8·4	8·7	8·9	9·2	9·4	9·6	9·8	10·1	10·2	10·4	10·6	10·8	11·1	11·4	10
11	7·2	7·5	7·9	8·2	8·4	8·7	8·9	9·1	9·3	9·6	9·7	9·9	10·1	10·3	10·6	10·9	11
12	6·8	7·1	7·5	7·8	8·0	8·3	8·5	8·7	8·9	9·2	9·3	9·5	9·7	9·9	10·2	10·5	12
13	6·5	6·8	7·2	7·5	7·7	8·0	8·2	8·4	8·6	8·9	9·0	9·2	9·4	9·6	9·9	10·2	13
14	6·2	6·5	6·9	7·2	7·4	7·7	7·9	8·1	8·3	8·6	8·7	8·9	9·1	9·3	9·6	9·9	14
15	5·9	6·2	6·6	6·9	7·1	7·4	7·6	7·8	8·0	8·3	8·4	8·6	8·8	9·0	9·3	9·6	15
16	5·7	6·0	6·4	6·7	6·9	7·2	7·4	7·6	7·8	8·1	8·2	8·4	8·6	8·8	9·1	9·4	16
18	5·3	5·6	6·0	6·3	6·5	6·8	7·0	7·2	7·4	7·7	7·8	8·0	8·2	8·4	8·7	9·0	18
20	5·0	5·3	5·7	6·0	6·2	6·5	6·7	6·9	7·1	7·4	7·5	7·7	7·9	8·1	8·4	8·7	20
22	4·7	5·0	5·4	5·7	5·9	6·2	6·4	6·6	6·8	7·1	7·2	7·4	7·6	7·8	8·1	8·4	22
26	4·3	4·6	5·0	5·3	5·5	5·8	6·0	6·2	6·4	6·7	6·8	7·0	7·2	7·4	7·7	8·0	26
30	4·0	4·3	4·7	5·0	5·2	5·5	5·7	5·9	6·1	6·4	6·5	6·7	6·9	7·1	7·4	7·7	30
35	3·8	4·1	4·5	4·8	5·0	5·3	5·5	5·7	5·9	6·2	6·3	6·5	6·7	6·9	7·2	7·5	35
40	3·5	3·8	4·2	4·5	4·7	5·0	5·2	5·4	5·6	5·9	6·0	6·2	6·4	6·6	6·9	7·2	40
50	3·2	3·5	3·9	4·2	4·4	4·7	4·9	5·1	5·3	5·6	5·7	5·9	6·1	6·3	6·6	6·9	50
60	3·0	3·3	3·7	4·0	4·2	4·5	4·7	4·9	5·1	5·4	5·5	5·7	5·9	6·1	6·4	6·7	60
70	2·8	3·1	3·5	3·8	4·0	4·3	4·5	4·7	4·9	5·2	5·3	5·5	5·7	5·9	6·2	6·5	70
80	2·6	2·9	3·3	3·6	3·8	4·1	4·3	4·5	4·7	5·0	5·1	5·3	5·5	5·7	6·0	6·3	80
90	2·4	2·7	3·1	3·4	3·6	3·9	4·1	4·3	4·5	4·8	4·9	5·1	5·3	5·5	5·8	6·1	90

HEIGHT OF EYE IN FEET AND METRES.

Obs. Alt.	ft. 40 m. 12·2	44 13·4	48 14·6	52 15·9	56 17·1	60 18·3	64 19·5	68 20·7	72 22·0	76 23·2	80 24·4	84 25·6	88 26·8	92 28·0	96 29·3	100 30·5	Obs. Alt.
6	14·6	14·9	15·2	15·5	15·8	16·0	16·3	16·5	16·7	17·0	17·2	17·4	17·6	17·8	18·0	18·2	6
6½	14·0	14·3	14·6	14·9	15·2	15·4	15·7	15·9	16·1	16·4	16·6	16·8	17·0	17·2	17·4	17·6	6½
7	13·4	13·7	14·0	14·3	14·6	14·8	15·1	15·3	15·5	15·8	16·0	16·2	16·4	16·6	16·8	17·0	7
7½	13·0	13·3	13·6	13·9	14·2	14·4	14·7	14·9	15·1	15·4	15·6	15·8	16·0	16·2	16·4	16·6	7½
8	12·6	12·9	13·2	13·5	13·8	14·0	14·3	14·5	14·7	15·0	15·2	15·4	15·6	15·8	16·0	16·2	8
8½	12·2	12·5	12·8	13·1	13·4	13·6	13·9	14·1	14·3	14·6	14·8	15·0	15·2	15·4	15·6	15·8	8½
9	11·9	12·2	12·5	12·8	13·1	13·3	13·6	13·8	14·0	14·3	14·5	14·7	14·9	15·1	15·3	15·5	9
10	11·4	11·7	12·0	12·3	12·6	12·8	13·1	13·3	13·5	13·8	14·0	14·2	14·4	14·6	14·8	15·0	10
11	10·9	11·2	11·5	11·8	12·1	12·3	12·6	12·8	13·0	13·3	13·5	13·7	13·9	14·1	14·3	14·5	11
12	10·5	10·8	11·1	11·4	11·7	11·9	12·2	12·4	12·6	12·9	13·1	13·3	13·5	13·7	13·9	14·1	12
13	10·2	10·5	10·8	11·1	11·4	11·6	11·9	12·1	12·3	12·6	12·8	13·0	13·2	13·4	13·6	13·8	13
14	9·9	10·2	10·5	10·8	11·1	11·3	11·6	11·8	12·0	12·3	12·5	12·7	12·9	13·1	13·3	13·5	14
15	9·6	9·9	10·2	10·5	10·8	11·0	11·3	11·5	11·7	12·0	12·2	12·4	12·6	12·8	13·0	13·2	15
16	9·4	9·7	10·0	10·3	10·6	10·8	11·1	11·3	11·5	11·8	12·0	12·2	12·4	12·6	12·8	13·0	16
18	9·0	9·3	9·6	9·9	10·2	10·4	10·7	10·9	11·1	11·4	11·6	11·8	12·0	12·2	12·4	12·6	18
20	8·7	9·0	9·3	9·6	9·9	10·1	10·4	10·6	10·8	11·1	11·3	11·5	11·7	11·9	12·1	12·3	20
22	8·4	8·7	9·0	9·3	9·6	9·8	10·1	10·3	10·5	10·8	11·0	11·2	11·4	11·6	11·8	12·0	22
26	8·0	8·3	8·6	8·9	9·2	9·4	9·7	9·9	10·1	10·4	10·6	10·8	11·0	11·2	11·4	11·6	26
30	7·7	8·0	8·3	8·6	8·9	9·1	9·4	9·6	9·8	10·1	10·3	10·5	10·7	10·9	11·1	11·1	30
35	7·5	7·8	8·1	8·4	8·7	8·9	9·2	9·4	9·6	9·9	10·1	10·3	10·5	10·7	10·9	11·1	35
40	7·2	7·5	7·8	8·1	8·4	8·6	8·9	9·1	9·3	9·6	9·8	10·0	10·2	10·4	10·6	10·8	40
50	6·9	7·2	7·5	7·8	8·1	8·3	8·6	8·8	9·0	9·3	9·5	9·7	9·9	10·1	10·3	10·5	50
60	6·7	7·0	7·3	7·6	7·9	8·1	8·4	8·6	8·8	9·1	9·3	9·5	9·7	9·9	10·1	10·3	60
70	6·5	6·8	7·1	7·4	7·7	7·9	8·2	8·4	8·6	8·9	9·1	9·3	9·5	9·7	9·9	10·1	70
80	6·3	6·6	6·9	7·2	7·5	7·7	8·0	8·2	8·4	8·7	8·9	9·1	9·3	9·5	9·7	9·9	80
90	6·1	6·4	6·7	7·0	7·3	7·5	7·8	8·0	8·2	8·5	8·7	8·9	9·1	9·3	9·5	9·7	90

REDUCTION of the MOON'S HORIZONTAL PARALLAX.
(SUBTRACTIVE)

Lat.	HORIZONTAL PARALLAX				
	54′	56′	58′	60′	62′
°	′	′	′	′	′
8	·00	·00	·00	·00	·00
16	·01	·01	·01	·02	·02
20	·02	·02	·02	·02	·02
24	·03	·03	·03	·03	·03
28	·04	·04	·04	·04	·05
32	·05	·05	·05	·06	·06
36	·06	·07	·07	·07	·07
40	·07	·08	·08	·08	·09
44	·09	·09	·09	·10	·10
48	·10	·10	·11	·11	·11
52	·11	·12	·12	·12	·13
56	·12	·13	·13	·14	·14
60	·14	·14	·15	·15	·16
64	·15	·15	·16	·16	·17
68	·16	·16	·17	·17	·18
72	·16	·17	·18	·18	·19
76	·17	·18	·18	·19	·20
80	·18	·18	·19	·19	·20

AUGMENTATION of the MOON'S SEMIDIAMETER.

Appt. Alt.	MOON'S SEMI-DIAMETER					
	14·5′	15·0′	15·5′	16·0′	16·5′	17·0′
°	′	′	′	′	′	′
6	·03	·03	·03	·03	·03	·04
8	·03	·04	·04	·04	·04	·05
10	·04	·04	·05	·05	·05	·06
12	·05	·05	·06	·06	·06	·07
14	·06	·06	·06	·07	·07	·08
16	·07	·07	·07	·08	·08	·09
18	·07	·08	·08	·09	·09	·10
21	·08	·09	·09	·10	·10	·11
24	·09	·10	·11	·11	·12	·13
27	·10	·11	·12	·13	·14	·14
30	·12	·12	·13	·14	·15	·16
33	·13	·13	·14	·15	·16	·17
36	·13	·14	·15	·16	·17	·18
39	·14	·15	·17	·18	·19	·20
42	·15	·16	·17	·19	·20	·21
45	·16	·17	·18	·20	·21	·22
48	·17	·18	·19	·21	·22	·23
51	·18	·19	·20	·22	·23	·24
54	·19	·20	·21	·23	·24	·25
57	·19	·21	·22	·23	·25	·26
63	·20	·22	·23	·25	·26	·28
70	·21	·23	·25	·26	·28	·29
78	·22	·24	·26	·27	·29	·30
90	·23	·24	·26	·28	·30	·31

ATMOSPHERIC PRESSURE
CONVERSION TABLES

MERCURY INCHES TO MILLIBARS

In.	Mb.	In.	Mb.	In.	Mb.
28·0	948·2	29·1	985·4	30·2	1022·7
28·1	951·6	29·2	988·8	30·3	1026·1
28·2	954·9	29·3	992·2	30·4	1029·4
28·3	958·3	29·4	995·6	30·5	1032·8
28·4	961·7	29·5	999·0	30·6	1036·2
28·5	965·1	29·6	1002·4	30·7	1039·6
28·6	968·5	29·7	1005·7	30·8	1043·0
28·7	971·9	29·8	1009·1	30·9	1046·4
28·8	975·3	29·9	1012·5	31·0	1049·8
28·9	978·6	30·0	1015·9	31·1	1053·1
29·0	982·0	30·1	1019·3	31·2	1056·5

For { Lat. 45°. Temp. 32° F.

MILLIBARS TO MERCURY INCHES

Mb.	In.	M b.	In.	Mb.	In.
948	28·00	984	29·06	1020	30·12
949	28·02	985	29·09	1021	30·15
950	28·05	986	29·12	1022	30·18
951	28·08	987	29·15	1023	30·21
952	28·11	988	29·18	1024	30·24
953	28·14	989	29·21	1025	30·27
954	28·17	990	29·24	1026	30·30
955	28·20	991	29·26	1027	30·33
956	28·23	992	29·29	1028	30·36
957	28·26	993	29·32	1029	30·39
958	28·29	994	29·35	1030	30·42
959	28·32	995	29·38	1031	30·45
960	28·35	996	29·41	1032	30·48
961	28·38	997	29·44	1033	30·50
962	28·41	998	29·47	1034	30·53
963	28·44	999	29·50	1035	30·56
964	28·47	1000	29·53	1036	30·59
965	28·50	1001	29·56	1037	30·62
966	28·53	1002	29·59	1038	30·65
967	28·56	1003	29·62	1039	30·68
968	28·59	1004	29·65	1040	30·71
969	28·62	1005	29·68	1041	30·74
970	28·64	1006	29·71	1042	30·77
971	28·67	1007	29·74	1043	30·80
972	28·70	1008	29·77	1044	30·83
973	28·73	1009	29·80	1045	30·86
974	28·76	1010	29·83	1046	30·89
975	28·79	1011	29·86	1047	30·92
976	28·82	1012	29·88	1048	30·95
977	28·85	1013	29·91	1049	30·98
978	28·88	1014	29·94	1050	31·01
979	28·91	1015	29·97	1051	31·04
980	28·94	1016	30·00	1052	31·07
981	28·97	1017	30·03	1053	31·10
982	29·00	1018	30·06	1054	31·13
983	29·03	1019	30·09	1055	31·15

CORRECTION OF MOON'S MER. PASS.

CORR. PLUS. to MER. PASS. in WEST LONG.

Long.	Difference between times of successive transits.										Long.
	39 m.	42 m.	45 m.	48 m.	51 m.	54 m.	57 m.	60 m.	63 m.	66 m.	
°	m.	m.	m.	m.	m.	m.	m.	m.	m.	m.	°
3	0·3	0·4	0·4	0·4	0·4	0·4	0·5	0·5	0·5	0·6	3
6	0·6	0·7	0·8	0·8	0·8	0·9	1·0	1·0	1·0	1·1	6
9	1·0	1·0	1·1	1·2	1·3	1·4	1·4	1·5	1·6	1·6	9
12	1·3	1·4	1·5	1·6	1·7	1·8	1·9	2·0	2·1	2·2	12
15	1·6	1·8	1·9	2·0	2·1	2·2	2·4	2·5	2·6	2·8	15
18	2·0	2·1	2·2	2·4	2·6	2·7	2·8	3·0	3·2	3·3	18
21	2·3	2·4	2·6	2·8	3·0	3·2	3·3	3·5	3·7	3·8	21
24	2·6	2·8	3·0	3·2	3·4	3·6	3·8	4·0	4·2	4·4	24
27	2·9	3·2	3·4	3·6	3·8	4·0	4·3	4·5	4·7	5·0	27
30	3·2	3·5	3·8	4·0	4·2	4·5	4·8	5·0	5·2	5·5	30
33	3·6	3·8	4·1	4·4	4·7	5·0	5·2	5·5	5·8	6·0	33
36	3·9	4·2	4·5	4·8	5·1	5·4	5·7	6·0	6·3	6·6	36
39	4·2	4·6	4·9	5·2	5·5	5·8	6·2	6·5	6·8	7·2	39
42	4·6	4·9	5·2	5·6	6·0	6·3	6·6	7·0	7·4	7·7	42
45	4·9	5·2	5·6	6·0	6·4	6·8	7·1	7·5	7·9	8·2	45
48	5·2	5·6	6·0	6·4	6·8	7·2	7·6	8·0	8·4	8·8	48
51	5·5	6·0	6·4	6·8	7·2	7·6	8·1	8·5	8·9	9·4	51
54	5·8	6·3	6·8	7·2	7·6	8·1	8·6	9·0	9·4	9·9	54
57	6·2	6·6	7·1	7·6	8·1	8·6	9·0	9·5	10·0	10·4	57
60	6·5	7·0	7·5	8·0	8·5	9·0	9·5	10·0	10·5	11·0	60
63	6·8	7·4	7·9	8·4	8·9	9·4	10·0	10·5	11·0	11·6	63
66	7·2	7·7	8·2	8·8	9·4	9·9	10·4	11·0	11·6	12·1	66
69	7·5	8·0	8·6	9·2	9·8	10·4	10·9	11·5	12·1	12·6	69
72	7·8	8·4	9·0	9·6	10·2	10·8	11·4	12·0	12·6	13·2	72
75	8·1	8·8	9·4	10·0	10·6	11·2	11·9	12·5	13·2	13·8	75
78	8·4	9·1	9·8	10·4	11·0	11·7	12·4	13·0	13·6	14·3	78
81	8·8	9·4	10·1	10·8	11·5	12·2	12·8	13·5	14·2	14·8	81
84	9·1	9·8	10·5	11·2	11·9	12·6	13·3	14·0	14·7	15·4	84
87	9·4	10·2	10·9	11·6	12·3	13·0	13·8	14·5	15·2	16·0	87
90	9·8	10·5	11·2	12·0	12·8	13·5	14·2	15·0	15·8	16·5	90
93	10·1	10·8	11·6	12·4	13·2	14·0	14·7	15·5	16·3	17·0	93
96	10·4	11·2	12·0	12·8	13·6	14·4	15·2	16·0	16·8	17·6	96
99	10·7	11·6	12·4	13·2	14·0	14·8	15·7	16·5	17·3	18·2	99
102	11·0	11·9	12·8	13·6	14·4	15·3	16·2	17·0	17·8	18·7	102
105	11·4	12·2	13·1	14·0	14·9	15·8	16·6	17·5	18·4	19·2	105
108	11·7	12·6	13·5	14·4	15·3	16·2	17·1	18·0	18·9	19·8	108
111	12·0	13·0	13·9	14·8	15·7	16·6	17·6	18·5	19·4	20·4	111
114	12·4	13·3	14·2	15·2	16·2	17·1	18·0	19·0	20·0	20·9	114
117	12·7	13·6	14·6	15·6	16·6	17·6	18·5	19·5	20·5	21·4	117
120	13·0	14·0	15·0	16·0	17·0	18·0	19·0	20·0	21·0	22·0	120
123	13·3	14·4	15·4	16·4	17·4	18·4	19·5	20·5	21·5	22·6	123
126	13·6	14·7	15·8	16·8	17·8	18·9	20·0	21·0	22·0	23·1	126
129	14·0	15·0	16·1	17·2	18·3	19·4	20·4	21·5	22·6	23·6	129
132	14·3	15·4	16·5	17·6	18·7	19·8	20·9	22·0	23·1	24·2	132
135	14·6	15·8	16·9	18·0	19·1	20·2	21·4	22·5	23·6	24·8	135
138	15·0	16·1	17·2	18·4	19·6	20·7	21·8	23·0	24·2	25·3	138
141	15·3	16·4	17·6	18·8	20·0	21·2	22·3	23·5	24·7	25·8	141
144	15·6	16·8	18·0	19·2	20·4	21·6	22·8	24·0	25·2	26·4	144
147	15·9	17·2	18·4	19·6	20·8	22·0	23·3	24·5	25·7	27·0	147
150	16·2	17·5	18·8	20·0	21·2	22·5	23·8	25·0	26·2	27·5	150
153	16·6	17·8	19·1	20·4	21·7	23·0	24·2	25·5	26·8	28·0	153
156	16·9	18·2	19·5	20·8	22·1	23·4	24·7	26·0	27·3	28·6	156
159	17·2	18·6	19·9	21·2	22·5	23·8	25·2	26·5	27·8	29·2	159
162	17·6	18·9	20·2	21·6	23·0	24·3	25·6	27·0	28·4	29·7	162
165	17·9	19·2	20·6	22·0	23·4	24·8	26·1	27·5	28·9	30·2	165
168	18·2	19·6	21·0	22·4	23·8	25·2	26·6	28·0	29·4	30·8	168
171	18·5	20·0	21·4	22·8	24·2	25·6	27·1	28·5	29·9	31·4	171
174	18·8	20·3	21·8	23·2	24·6	26·1	27·6	29·0	30·4	31·9	174
177	19·2	20·6	22·1	23·6	25·1	26·6	28·0	29·5	31·0	32·4	177
180	19·5	21·0	22·5	24·0	25·5	27·0	28·5	30·0	31·5	33·0	180

CORR. MINUS. to MER. PASS. in EAST LONG.

MOON'S TOTAL CORRECTION

Correction of the MOON'S ALTITUDE for SEMIDIAMETER, PARALLAX and REFRACTION.

LOWER LIMB   **ADDITIVE.**

HORIZONTAL PARALLAX.

Observed Altitude	54′·0	54′·5	55′·0	55′·5	56′·0	56′·5	57′·0	57′·5	58′·0	58′·5	59′·0	59′·5	60′·0	60′·5	61′·0	61′·5	Observed Altitude
°	′	′	′	′	′	′	′	′	′	′	′	′	′	′	′	′	°
6	50·2	50·9	51·5	52·1	52·7	53·4	54·0	54·7	55·3	55·9	56·5	57·2	57·8	58·5	59·1	59·7	6
7	51·0	51·7	52·3	52·9	53·5	54·2	54·8	55·5	56·1	56·8	57·4	58·0	58·6	59·3	59·9	60·5	7
8	51·9	52·5	53·1	53·8	54·4	55·1	55·7	56·3	56·9	57·6	58·2	58·9	59·5	60·1	60·7	61·3	8
9	52·4	53·0	53·6	54·3	54·9	55·6	56·2	56·8	57·4	58·0	58·6	59·3	59·9	60·5	61·1	61·7	9
10	52·9	53·5	54·1	54·8	55·4	56·0	56·6	57·3	57·9	58·5	59·1	59·8	60·4	61·0	61·6	62·2	10
11	53·1	53·8	54·4	55·0	55·6	56·2	56·8	57·5	58·1	58·8	59·4	60·1	60·7	61·3	61·8	62·3	11
12	53·3	54·0	54·6	55·2	55·8	56·5	57·1	57·7	58·3	59·0	59·6	60·3	60·9	61·5	62·1	62·7	12
13	53·4	54·1	54·7	55·3	55·9	56·6	57·2	57·8	58·4	59·1	59·7	60·3	60·9	61·6	62·2	62·8	13
14	53·6	54·2	54·8	55·4	56·0	56·7	57·3	57·9	58·5	59·2	59·8	60·4	61·0	61·7	62·4	63·0	14
15	53·6	54·2	54·8	55·4	56·0	56·6	57·2	57·9	58·5	59·1	59·7	60·4	61·0	61·7	62·3	62·9	15
16	53·6	54·2	54·8	55·4	56·0	56·6	57·2	57·9	58·5	59·1	59·7	60·4	61·0	61·6	62·2	62·8	16
17	53·5	54·1	54·7	55·3	55·9	56·5	57·1	57·7	58·3	59·0	59·6	60·3	60·9	61·5	62·1	62·7	17
18	53·4	54·0	54·6	55·3	55·9	56·5	57·1	57·7	58·3	58·9	59·5	60·2	60·8	61·4	62·0	62·6	18
19	53·2	53·9	54·5	55·1	55·7	56·3	56·9	57·5	58·1	58·7	59·3	60·0	60·6	61·2	61·8	62·4	19
20	53·1	53·7	54·3	54·9	55·5	56·1	56·7	57·3	57·9	58·6	59·2	59·8	60·4	61·0	61·6	62·2	20
21	52·9	53·5	54·1	54·7	55·3	55·9	56·5	57·1	57·7	58·3	58·9	59·5	60·1	60·7	61·3	61·9	21
22	52·7	53·3	53·9	54·5	55·1	55·7	56·3	56·9	57·5	58·1	58·7	59·3	59·9	60·5	61·1	61·7	22
23	52·4	53·0	53·6	54·2	54·8	55·4	56·0	56·6	57·2	57·8	58·4	59·1	59·7	60·3	60·8	61·3	23
24	52·2	52·8	53·4	54·0	54·6	55·2	55·7	56·3	56·9	57·5	58·1	58·8	59·4	59·9	60·5	61·1	24
25	51·9	52·5	53·0	53·7	54·3	54·9	55·4	56·0	56·6	57·2	57·8	58·4	59·0	59·6	60·1	60·6	25
26	51·6	52·2	52·7	53·3	53·9	54·5	55·1	55·7	56·3	56·9	57·5	58·1	58·7	59·3	59·8	60·4	26
27	51·2	51·8	52·4	53·0	53·5	54·1	54·7	55·3	55·9	56·5	57·1	57·7	58·3	58·9	59·4	60·0	27
28	50·9	51·5	52·1	52·6	53·1	53·7	54·3	54·9	55·5	56·1	56·7	57·3	57·9	58·5	59·0	59·6	28
29	50·5	51·1	51·6	52·2	52·7	53·3	53·9	54·5	55·1	55·6	56·2	56·8	57·4	58·0	58·5	59·1	29
30	50·1	50·7	51·2	51·8	52·3	52·9	53·5	54·1	54·7	55·2	55·8	56·4	57·0	57·6	58·1	58·7	30
31	49·7	50·3	50·8	51·3	51·8	52·4	53·0	53·6	54·2	54·8	55·3	55·9	56·5	57·1	57·6	58·2	31
32	49·3	49·9	50·4	50·9	51·4	52·0	52·7	53·3	53·8	54·3	54·9	55·5	56·1	56·7	57·2	57·8	32
33	48·8	49·4	49·9	50·4	50·9	51·5	52·2	52·7	53·2	53·8	54·4	55·0	55·5	56·1	56·6	57·1	33
34	48·4	49·0	49·5	50·0	50·5	51·1	51·7	52·2	52·8	53·4	53·9	54·5	55·0	55·6	56·1	56·6	34
35	47·9	48·5	49·0	49·5	50·0	50·6	51·2	51·8	52·3	52·8	53·4	54·0	54·5	55·0	55·5	56·0	35
36	47·4	48·0	48·5	49·0	49·6	50·1	50·7	51·2	51·8	52·3	52·9	53·5	54·0	54·5	55·0	55·5	36
37	46·9	47·4	47·9	48·5	49·0	49·6	50·1	50·6	51·2	51·8	52·3	52·9	53·4	53·9	54·4	54·9	37
38	46·4	46·9	47·4	48·0	48·5	49·0	49·6	50·1	50·6	51·2	51·7	52·3	52·8	53·3	53·8	54·3	38
39	45·8	46·3	46·8	47·4	47·9	48·5	49·0	49·5	50·0	50·6	51·1	51·7	52·2	52·7	53·2	53·7	39
40	45·3	45·8	46·3	46·9	47·4	47·9	48·4	49·0	49·5	50·0	50·5	51·1	51·6	52·1	52·6	53·1	40
41	44·7	45·2	45·7	46·3	46·8	47·3	47·8	48·3	48·8	49·3	49·8	50·4	50·9	51·4	51·9	52·4	41
42	44·1	44·6	45·1	45·7	46·2	46·7	47·2	47·7	48·2	48·7	49·2	49·8	50·3	50·8	51·3	51·8	42
43	43·5	44·0	44·5	45·0	45·5	46·0	46·5	47·0	47·5	48·0	48·5	49·0	49·6	50·1	50·6	51·1	43
44	42·9	43·4	43·9	44·4	44·9	45·4	45·9	46·4	46·9	47·4	47·9	48·4	48·9	49·4	49·9	50·4	44
45	42·3	42·8	43·3	43·8	44·2	44·7	45·2	45·7	46·2	46·7	47·2	47·7	48·2	48·7	49·2	49·7	45
	54′·0	54′·5	55′·0	55′·5	56′·0	56′·5	57′·0	57′·5	58′·0	58′·5	59′·0	59′·5	60′·0	60′·5	61′·0	61′·5	

ADD the following CORRECTIONS for various HEIGHTS OF EYE.

H.E. in Ft.	10	15	20	25	30	35	40	45	50	55	60	65	70	75	80	85	90	95	100
H.E. in m.	3·0	4·6	6·1	7·6	9·1	10·7	12·2	13·7	15·2	16·8	18·3	19·8	21·3	22·9	24·4	25·9	27·4	29·0	30·5
Corr.	6·7	6·0	5·5	4·9	4·5	4·1	3·7	3·3	2·9	2·6	2·3	2·0	1·7	1·4	1·1	0·9	0·6	0·3	0·1

MOON'S TOTAL CORRECTION

Correction of the MOON'S ALTITUDE for SEMIDIAMETER, PARALLAX and REFRACTION.

Observed Altitude	LOWER LIMB								ADDITIVE.								Observed Altitude
	HORIZONTAL PARALLAX.																
°	54′.0	54′.5	55′.0	55′.5	56′.0	56′.5	57′.0	57′.5	58′.0	58′.5	59′.0	59′.5	60′.0	60′.5	61′.0	61′.5	°
45	42.3	42.8	43.3	43.8	44.2	44.7	45.2	45.7	46.2	46.7	47.2	47.7	48.2	48.7	49.2	49.7	45
46	41.7	42.2	42.6	43.1	43.6	44.1	44.6	45.1	45.6	46.1	46.5	47.1	47.5	48.1	48.5	49.1	46
47	41.0	41.4	41.9	42.4	42.9	43.4	43.9	44.3	44.8	45.3	45.8	46.3	46.8	47.2	47.7	48.2	47
48	40.4	40.9	41.3	41.8	42.3	42.8	43.2	43.7	44.1	44.6	45.1	45.6	46.1	46.6	47.0	47.4	48
49	39.7	40.1	40.6	41.1	41.6	42.0	42.4	42.9	43.3	43.9	44.3	44.8	45.3	45.8	46.2	46.6	49
50	39.0	39.5	39.9	40.4	40.9	41.3	41.7	42.2	42.7	43.2	43.6	44.1	44.5	45.0	45.5	45.9	50
51	38.3	38.8	39.2	39.7	40.1	40.6	41.0	41.5	41.9	42.4	42.8	43.3	43.7	44.2	44.7	45.1	51
52	37.6	38.1	38.5	39.0	39.4	39.9	40.3	40.8	41.2	41.7	42.1	42.6	43.0	43.5	43.9	44.3	52
53	36.9	37.3	37.7	38.2	38.6	39.0	39.4	39.9	40.4	40.9	41.3	41.8	42.2	42.6	43.0	43.4	53
54	36.2	36.6	37.0	37.5	37.9	38.3	38.7	39.2	39.6	40.1	40.5	41.0	41.4	41.8	42.2	42.6	54
55	35.4	35.8	36.2	36.7	37.1	37.5	37.9	38.4	38.8	39.3	39.7	40.1	40.5	40.9	41.3	41.7	55
56	34.7	35.1	35.5	35.9	36.3	36.8	37.2	37.6	38.0	38.5	38.9	39.3	39.7	40.1	40.5	40.9	56
57	33.9	34.3	34.7	35.1	35.5	36.0	36.4	36.8	37.2	37.6	38.0	38.5	38.9	39.3	39.6	40.0	57
58	33.1	33.5	33.9	34.4	34.8	35.2	35.6	36.0	36.4	36.8	37.2	37.6	38.0	38.4	38.8	39.2	58
59	32.3	32.7	33.1	33.5	33.9	34.3	34.7	35.1	35.5	35.9	36.3	36.7	37.1	37.5	37.9	38.3	59
60	31.6	31.9	32.3	32.7	33.1	33.5	33.9	34.3	34.7	35.1	35.5	35.9	36.3	36.7	37.0	37.4	60
61	30.8	31.1	31.5	31.9	32.3	32.7	33.0	33.4	33.8	34.2	34.6	35.0	35.4	35.7	36.1	36.5	61
62	30.0	30.3	30.7	31.1	31.5	31.8	32.2	32.6	32.9	33.3	33.7	34.1	34.5	34.8	35.2	35.6	62
63	29.1	29.4	29.8	30.2	30.6	30.9	31.3	31.7	32.1	32.5	32.8	33.2	33.6	34.0	34.3	34.7	63
64	28.3	28.6	29.0	29.4	29.8	30.1	30.5	30.9	31.2	31.6	31.9	32.3	32.7	33.0	33.4	33.8	64
65	27.5	27.8	28.1	28.5	28.9	29.3	29.6	30.0	30.3	30.7	31.0	31.3	31.7	32.0	32.4	32.8	65
66	26.7	27.0	27.3	27.7	28.1	28.4	28.7	29.0	29.4	29.7	30.1	30.5	30.8	31.2	31.5	31.9	66
67	25.8	26.1	26.4	26.8	27.2	27.5	27.8	28.2	28.5	28.8	29.1	29.4	29.8	30.2	30.6	31.0	67
68	25.0	25.3	25.6	25.9	26.3	26.6	26.9	27.3	27.6	27.9	28.2	28.6	28.9	29.3	29.6	30.0	68
69	24.1	24.4	24.7	25.1	25.4	25.7	26.0	26.3	26.6	27.0	27.3	27.6	27.9	28.3	28.6	29.0	69
70	23.3	23.6	23.9	24.2	24.5	24.8	25.1	25.4	25.7	26.1	26.4	26.7	27.0	27.3	27.6	27.9	70
71	22.4	22.7	23.0	23.3	23.6	23.9	24.2	24.5	24.8	25.1	25.4	25.7	26.0	26.3	26.6	26.9	71
72	21.5	21.8	22.1	22.4	22.7	23.0	23.3	23.6	23.9	24.2	24.5	24.8	25.1	25.4	25.6	25.9	72
73	20.6	20.9	21.2	21.5	21.8	22.1	22.3	22.6	22.9	23.2	23.5	23.8	24.1	24.4	24.6	24.9	73
74	19.7	20.0	20.3	20.6	20.9	21.2	21.4	21.7	22.0	22.3	22.5	22.8	23.1	23.4	23.6	23.9	74
75	18.8	19.1	19.4	19.7	19.9	20.2	20.4	20.7	21.0	21.3	21.5	21.8	22.1	22.4	22.6	22.9	75
76	18.0	18.3	18.5	18.8	19.0	19.3	19.5	19.8	20.0	20.3	20.6	20.9	21.1	21.4	21.6	21.9	76
77	17.1	17.3	17.5	17.8	18.0	18.3	18.5	18.8	19.1	19.4	19.6	19.9	20.1	20.4	20.6	20.9	77
78	16.2	16.4	16.6	16.9	17.1	17.4	17.6	17.9	18.1	18.4	18.6	18.9	19.1	19.4	19.6	19.8	78
79	15.3	15.5	15.7	16.0	16.2	16.4	16.6	16.9	17.2	17.4	17.6	17.9	18.1	18.4	18.6	18.8	79
80	14.4	14.6	14.8	15.1	15.3	15.5	15.7	16.0	16.2	16.4	16.6	16.9	17.1	17.3	17.5	17.7	80
81	13.4	13.6	13.8	14.1	14.3	14.5	14.7	15.0	15.2	15.4	15.6	15.9	16.1	16.3	16.5	16.7	81
82	12.5	12.7	12.9	13.2	13.4	13.6	13.8	14.0	14.2	14.4	14.6	14.9	15.1	15.3	15.5	15.7	82
83	11.6	11.8	12.0	12.2	12.4	12.6	12.8	13.0	13.2	13.4	13.6	13.8	14.0	14.2	14.4	14.6	83
84	10.7	10.9	11.1	11.3	11.5	11.7	11.8	12.0	12.2	12.4	12.6	12.8	13.0	13.2	13.4	13.6	84
85	9.8	10.0	10.1	10.3	10.5	10.7	10.8	11.0	11.2	11.4	11.6	11.8	12.0	12.2	12.3	12.5	85
86	8.9	9.1	9.2	9.4	9.5	9.7	9.9	10.1	10.2	10.4	10.6	10.8	11.0	11.2	11.3	11.5	86
87	8.0	8.1	8.2	8.4	8.5	8.7	8.9	9.1	9.2	9.4	9.6	9.8	10.0	10.1	10.2	10.3	87
88	7.0	7.2	7.3	7.4	7.5	7.7	7.9	8.1	8.2	8.4	8.6	8.8	8.9	9.1	9.2	9.4	88
	54′.0	54′.5	55′.0	55′.5	56′.0	56′.5	57′.0	57′.5	58′.0	58′.5	59′.0	59′.5	60′.0	60′.5	61′.0	61′.5	

ADD the following CORRECTIONS for various HEIGHTS OF EYE.

H.E. in Ft.	10	15	20	25	30	35	40	45	50	55	60	65	70	75	80	85	90	95	100
H.E. in m.	3.0	4.6	6.1	7.6	9.1	10.7	12.2	13.7	15.2	16.8	18.3	19.8	21.3	22.9	24.4	25.9	27.4	29.0	30.5
Corr.	6.7	6.0	5.5	4.9	4.5	4.1	3.7	3.3	2.9	2.6	2.3	2.0	1.7	1.4	1.1	0.9	0.6	0.3	0.1

MOON'S TOTAL CORRECTION

Correction of the MOON'S ALTITUDE for SEMIDIAMETER, PARALLAX and REFRACTION.

Observed Altitude	UPPER LIMB								ADDITIVE.								Observed Altitude
	HORIZONTAL PARALLAX.																
	54'0	54'5	55'0	55'5	56'0	56'5	57'0	57'5	58'0	58'5	59'0	59'5	60'0	60'5	61'0	61'5	
°	'	'	'	'	'	'	'	'	'	'	'	'	'	'	'	'	°
6	20·6	21·0	21·4	21·7	22·0	22·4	22·8	23·2	23·5	23·9	24·2	24·6	24·9	25·3	25·7	26·1	6
7	21·8	22·2	22·5	22·8	23·1	23·4	23·8	24·2	24·5	24·9	25·2	25·6	26·0	26·4	26·8	27·2	7
8	22·5	22·8	23·2	23·5	23·7	24·0	24·2	24·6	25·0	25·4	25·8	26·2	26·6	27·0	27·5	27·9	8
9	23·0	23·3	23·7	24·0	24·2	24·6	24·9	25·3	25·6	26·0	26·4	26·7	27·1	27·6	28·0	28·4	9
10	23·3	23·7	24·0	24·4	24·7	25·1	25·4	25·8	26·1	26·5	26·8	27·2	27·5	27·9	28·3	28·7	10
11	23·6	23·9	24·2	24·6	24·9	25·3	25·6	26·0	26·3	26·7	27·0	27·4	27·7	28·2	28·6	29·0	11
12	23·7	24·1	24·4	24·8	25·1	25·5	25·8	26·2	26·5	26·9	27·2	27·6	27·9	28·3	28·7	29·1	12
13	23·9	24·3	24·6	25·0	25·3	25·7	26·0	26·4	26·7	27·1	27·4	27·8	28·1	28·5	28·8	29·2	13
14	23·9	24·3	24·6	25·0	25·3	25·7	26·0	26·4	26·7	27·1	27·4	27·8	28·1	28·5	28·8	29·2	14
15	23·9	24·3	24·6	25·0	25·3	25·7	26·0	26·4	26·7	27·1	27·4	27·8	28·1	28·5	28·8	29·2	15
16	24·0	24·3	24·6	25·0	25·3	25·7	26·0	26·4	26·7	27·1	27·4	27·8	28·1	28·5	28·8	29·2	16
17	23·9	24·2	24·5	24·9	25·2	25·6	26·0	26·3	26·6	27·0	27·3	27·7	28·0	28·4	28·7	29·1	17
18	23·8	24·2	24·5	24·9	25·2	25·5	25·8	26·2	26·5	26·9	27·2	27·6	27·9	28·3	28·6	29·0	18
19	23·7	24·1	24·4	24·8	25·1	25·4	25·7	26·1	26·4	26·8	27·1	27·4	27·7	28·1	28·4	28·8	19
20	23·5	23·9	24·2	24·6	24·9	25·2	25·5	25·9	26·2	26·6	26·9	27·2	27·5	27·9	28·2	28·5	20
21	23·3	23·7	24·0	24·4	24·7	25·0	25·3	25·7	26·0	26·4	26·7	27·0	27·3	27·7	28·0	28·3	21
22	23·1	23·5	23·8	24·1	24·4	24·8	25·1	25·4	25·7	26·1	26·4	26·7	27·0	27·4	27·7	28·0	22
23	22·8	23·2	23·5	23·8	24·1	24·5	24·8	25·2	25·5	25·8	26·1	26·4	26·7	27·1	27·4	27·7	23
24	22·6	23·0	23·3	23·6	23·9	24·2	24·5	24·9	25·2	25·5	25·8	26·1	26·4	26·8	27·1	27·4	24
25	22·3	22·6	22·9	23·3	23·6	23·9	24·2	24·6	24·9	25·2	25·5	25·8	26·1	26·5	26·8	27·1	25
26	22·0	22·2	22·5	22·9	23·3	23·6	23·9	24·2	24·5	24·8	25·1	25·5	25·8	26·1	26·4	26·7	26
27	21·6	21·9	22·2	22·6	22·9	23·3	23·6	23·9	24·2	24·5	24·8	25·1	25·4	25·7	26·0	26·3	27
28	21·3	21·6	21·9	22·3	22·6	22·9	23·2	23·5	23·8	24·1	24·4	24·7	25·0	25·3	25·6	25·9	28
29	21·0	21·3	21·6	21·9	22·2	22·5	22·8	23·1	23·4	23·7	24·0	24·3	24·6	24·9	25·2	25·5	29
30	20·6	20·9	21·2	21·5	21·8	22·1	22·4	22·7	23·0	23·3	23·6	23·9	24·1	24·4	24·7	25·0	30
31	20·2	20·5	20·8	21·1	21·4	21·7	21·9	22·2	22·5	22·8	23·1	23·4	23·7	24·0	24·2	24·5	31
32	19·7	20·0	20·3	20·6	20·9	21·1	21·4	21·7	22·0	22·3	22·6	22·9	23·2	23·5	23·8	24·1	32
33	19·3	19·5	19·8	20·1	20·4	20·7	21·0	21·3	21·6	21·9	22·1	22·4	22·7	23·0	23·3	23·6	33
34	18·9	19·1	19·4	19·6	19·9	20·2	20·5	20·8	21·1	21·3	21·6	21·9	22·2	22·4	22·7	23·0	34
35	18·4	18·7	18·9	19·2	19·4	19·7	20·0	20·3	20·6	20·9	21·1	21·4	21·7	22·0	22·2	22·5	35
36	17·9	18·2	18·4	18·7	18·9	19·2	19·5	19·8	20·0	20·3	20·6	20·9	21·1	21·4	21·6	21·9	36
37	17·3	17·6	17·9	18·2	18·4	18·7	19·0	19·3	19·5	19·8	20·0	20·3	20·5	20·8	21·0	21·3	37
38	16·8	17·1	17·4	17·7	17·9	18·2	18·4	18·7	18·9	19·2	19·4	19·7	19·9	20·2	20·4	20·7	38
39	16·3	16·6	16·8	17·1	17·3	17·6	17·8	18·1	18·3	18·6	18·8	19·1	19·3	19·6	19·8	20·1	39
40	15·8	16·0	16·2	16·5	16·7	17·0	17·2	17·5	17·7	18·0	18·2	18·5	18·7	19·0	19·2	19·5	40
41	15·2	15·5	15·7	15·9	16·1	16·4	16·6	16·9	17·1	17·4	17·6	17·9	18·1	18·4	18·6	18·9	41
42	14·6	14·9	15·1	15·3	15·5	15·7	15·9	16·2	16·5	16·7	16·9	17·2	17·4	17·7	17·9	18·2	42
43	14·0	14·3	14·5	14·7	14·9	15·1	15·3	15·6	15·8	16·1	16·3	16·6	16·8	17·0	17·2	17·4	43
44	13·4	13·6	13·8	14·0	14·2	14·5	14·7	14·9	15·1	15·4	15·6	15·9	16·1	16·3	16·5	16·7	44
45	12·7	13·0	13·2	13·5	13·7	13·9	14·1	14·3	14·5	14·7	14·9	15·2	15·4	15·6	15·8	16·0	45
	54'0	54'5	55'0	55'5	56'0	56'5	57'0	57'5	58'0	58'5	59'0	59'5	60'0	60'5	61'0	61'5	

ADD the following CORRECTIONS for various HEIGHTS OF EYE.

H.E. in Ft.	10	15	20	25	30	35	40	45	50	55	60	65	70	75	80	85	90	95	100
H.E. in m.	3·0	4·6	6·1	7·6	9·1	10·7	12·2	13·7	15·2	16·8	18·3	19·8	21·3	22·9	24·4	25·9	27·4	29·0	30·5
Corr.	6·7	6·0	5·5	4·9	4·5	4·1	3·7	3·3	2·9	2·6	2·3	2·0	1·7	1·4	1·1	0·9	0·6	0·3	0·1

MOON'S TOTAL CORRECTION

Correction of the MOON'S ALTITUDE for SEMIDIAMETER, PARALLAX and REFRACTION.

Observed Altitude.	UPPER LIMB					ADDITIVE. SUBTRACTIVE.										Observed Altitude.
	HORIZONTAL PARALLAX.															
°	54′·0	54′·5	55′·0	55′·5	56′·0	56′·5	57′·0	57′·5	58′·0	58′·5	59′·0	59′·5	60′·0	60′·5	61′·0	61′·5	°
45	12·7	13·0	13·2	13·5	13·7	13·9	14·1	14·3	14·5	14·7	14·9	15·2	15·4	15·6	15·8	16·0	45
46	12·1	12·3	12·5	12·8	13·0	13·2	13·4	13·6	13·8	14·0	14·2	14·5	14·7	14·9	15·1	15·3	46
47	11·5	11·7	11·9	12·1	12·3	12·5	12·7	12·9	13·1	13·3	13·5	13·8	14·0	14·2	14·4	14·6	47
48	10·8	11·0	11·2	11·4	11·6	11·8	12·0	12·2	12·4	12·6	12·8	13·0	13·2	13·4	13·6	13·8	48
49	10·1	10·3	10·5	10·7	10·9	11·1	11·3	11·5	11·7	11·9	12·1	12·2	12·4	12·6	12·8	13·0	49
50	9·4	9·6	9·8	10·0	10·2	10·3	10·5	10·7	10·9	11·1	11·3	11·4	11·6	11·8	12·0	12·2	50
51	‾8·8	8·9	9·1	9·3	9·5	9·6	9·8	10·0	10·2	10·4	10·6	10·7	10·9	11·1	11·3	11·5	51
52	8·1	8·2	8·4	8·5	8·7	8·9	9·1	9·2	9·4	9·6	9·8	9·9	10·1	10·3	10·5	10·7	52
53	7·4	7·5	7·7	7·9	8·0	8·1	8·3	8·5	8·7	8·8	9·0	9·1	9·3	9·4	9·6	9·8	53
54	6·6	6·7	6·9	7·0	7·2	7·3	7·5	7·7	7·9	8·0	8·2	8·3	8·5	8·6	8·8	9·0	54
55	5·9	6·0	6·2	6·3	6·5	6·6	6·8	6·9	7·1	7·2	7·4	7·5	7·7	7·8	8·0	8·1	55
56	5·1	5·2	5·4	5·5	5·7	5·8	6·0	6·1	6·2	6·3	6·5	6·6	6·8	6·9	7·1	7·2	56
57	4·2	4·4	4·6	4·7	4·9	5·0	5·1	5·2	5·4	5·5	5·7	5·8	6·0	6·1	6·2	6·3	57
58	3·5	3·6	3·8	3·9	4·1	4·2	4·3	4·4	4·5	4·6	4·8	4·9	5·1	5·2	5·3	5·4	58
59	2·7	2·8	3·0	3·1	3·3	3·4	3·5	3·6	3·7	3·8	3·9	4·0	4·2	4·3	4·4	4·5	59
60	2·0	2·1	2·2	2·3	2·5	2·6	2·7	2·8	2·9	3·0	3·1	3·2	3·3	3·4	3·5	3·6	60
61	1·2	1·3	1·4	1·5	1·6	1·7	1·8	1·9	2·0	2·1	2·2	2·3	2·4	2·5	2·6	2·7	61
62	0·4	0·5	0·6	0·7	0·8	0·9	1·0	1·1	1·2	1·3	1·4	1·5	1·6	1·7	1·8	1·9	62
63	0·4	0·3	0·3	0·2	0·1	0·0	0·1	0·2	0·3	0·4	0·5	0·6	0·6	0·7	0·8	0·9	63
64	1·3	1·2	1·1	1·0	0·9	0·8	0·8	0·7	0·6	0·6	0·5	0·4	0·3	0·2	0·0	0·1	64
65	2·1	2·0	1·9	1·9	1·8	1·8	1·7	1·6	1·5	1·5	1·4	1·3	1·2	1·1	1·0	0·9	65
66	2·9	2·9	2·8	2·7	2·6	2·6	2·5	2·5	2·4	2·3	2·2	2·2	2·1	2·1	2·0	2·0	66
67	3·7	3·7	3·6	3·6	3·5	3·5	3·4	3·4	3·3	3·2	3·1	3·1	3·0	3·0	2·9	2·9	67
68	4·6	4·6	4·5	4·5	4·4	4·4	4·3	4·3	4·2	4·2	4·1	4·1	4·0	3·9	3·8	3·8	68
69	5·5	5·5	5·4	5·4	5·3	5·3	5·2	5·2	5·1	5·1	5·1	5·1	5·0	5·0	4·9	4·9	69
70	6·3	6·3	6·3	6·2	6·2	6·2	6·1	6·1	6·1	6·0	6·0	6·0	5·9	5·9	5·9	5·9	70
71	7·2	7·2	7·1	7·1	7·1	7·1	7·1	7·1	7·1	7·0	7·0	7·0	6·9	6·9	6·9	6·9	71
72	8·1	8·1	8·0	8·0	8·0	8·0	8·0	8·0	8·0	7·9	7·9	7·9	7·9	7·8	7·8	7·8	72
73	8·9	8·9	8·9	8·9	8·9	8·9	8·9	8·9	8·9	8·9	8·9	8·9	8·9	8·9	8·9	8·9	73
74	9·8	9·8	9·8	9·8	9·8	9·8	9·8	9·8	9·8	9·8	9·8	9·8	9·8	9·8	9·8	9·8	74
75	10·7	10·7	10·7	10·7	10·7	10·7	10·7	10·7	10·7	10·7	10·8	10·8	10·8	10·8	10·8	10·8	75
76	11·6	11·6	11·6	11·6	11·7	11·7	11·7	11·7	11·7	11·7	11·8	11·8	11·8	11·8	11·8	11·8	76
77	12·5	12·5	12·6	12·6	12·6	12·6	12·6	12·6	12·6	12·7	12·8	12·8	12·8	12·8	12·8	12·8	77
78	13·4	13·4	13·5	13·5	13·5	13·5	13·6	13·6	13·6	13·7	13·7	13·7	13·8	13·8	13·8	13·8	78
79	14·3	14·3	14·4	14·4	14·4	14·4	14·5	14·5	14·6	14·6	14·7	14·7	14·8	14·8	14·9	14·9	79
80	15·2	15·2	15·3	15·3	15·4	15·4	15·5	15·5	15·6	15·6	15·7	15·7	15·8	15·8	15·9	15·9	80
81	16·1	16·1	16·2	16·2	16·3	16·4	16·5	16·5	16·6	16·6	16·7	16·7	16·8	16·8	16·9	16·9	81
82	17·1	17·1	17·2	17·2	17·3	17·4	17·5	17·5	17·6	17·6	17·6	17·7	17·8	17·9	18·0	18·1	82
83	17·9	18·0	18·1	18·1	18·2	18·3	18·4	18·5	18·6	18·6	18·7	18·7	18·8	18·9	19·0	19·1	83
84	18·8	18·9	19·0	19·1	19·2	19·3	19·4	19·5	19·6	19·6	19·6	19·7	19·8	19·9	20·0	20·1	20·2	84
85	19·8	19·9	20·0	20·0	20·1	20·2	20·4	20·5	20·6	20·6	20·7	20·8	20·9	21·0	21·1	21·2	85
86	20·7	20·8	20·9	21·0	21·1	21·2	21·4	21·5	21·6	21·7	21·8	21·9	22·0	22·1	22·2	22·3	86
87	21·7	21·8	21·9	22·0	22·1	22·2	22·3	22·4	22·5	22·6	22·7	22·8	23·0	23·1	23·2	23·3	87
88	22·6	22·7	22·8	22·9	23·1	23·2	23·3	23·4	23·5	23·6	23·8	23·9	24·0	24·1	24·2	24·3	88
	54′·0	54′·5	55′·0	55′·5	56′·0	56′·5	57′·0	57′·5	58′·0	58′·5	59′·0	59′·5	60′·0	60′·5	61′·0	61′·5	

ADD the following CORRECTIONS for various HEIGHTS OF EYE.

H.E. in Ft.	10	15	20	25	30	35	40	45	50	55	60	65	70	75	80	85	90	95	100
H.E. in m.	3·0	4·6	6·1	7·6	9·1	10·7	12·2	13·7	15·2	16·8	18·3	19·8	21·3	22·9	24·4	25·9	27·4	29·0	30·5
Corr.		6·7	6·0	5·5	4·9	4·5	4·1	3·7	3·3	2·9	2·6	2·3	2·0	1·7	1·4	1·1	0·9	0·6	0·3	0·1

DIP at DIFFERENT DISTANCES from the OBSERVER or DIP of the SHORE HORIZON.

Dist. Miles	HEIGHT OF EYE IN FEET AND METRES														
Ft.	5	10	15	20	25	30	35	40	45	50	60	70	80	90	100
m.	1·5	3·0	4·6	6·1	7·6	9·1	10·7	12·2	13·7	15·2	18·3	21·3	24·4	27·4	30·5
	′	′	′	′	′	′	′	′	′	′	′	′	′	′	′
·1	28·3	56·5	84·8	113·0	141·3	169·5	197·8	226·0	254·3	282·5	339·0	395·5	452·0	508·5	565·0
·2	14·1	28·3	42·5	56·6	70·7	84·8	99·0	113·1	127·2	141·3	169·6	197·9	226·1	254·4	282·6
·3	9·5	19·0	28·4	37·8	47·2	56·6	66·0	75·4	84·9	94·3	113·1	131·9	150·7	169·6	188·3
·4	7·2	14·3	21·4	28·4	35·5	42·5	49·6	56·7	63·7	70·8	84·9	99·0	113·1	127·3	141·4
·5	5·9	11·5	17·2	22·8	28·5	34·1	39·8	45·4	51·1	56·7	68·0	79·3	90·6	101·9	113·2
·6	5·0	9·7	14·4	19·1	23·8	28·5	33·2	37·9	42·6	47·3	56·8	66·2	75·6	85·0	94·5
·7	4·3	8·4	12·4	16·4	20·5	24·5	28·5	32·6	36·6	40·7	48·7	56·8	64·9	72·9	81·0
·8	3·9	7·4	10·9	14·5	18·0	21·5	25·1	28·6	32·1	35·7	42·7	49·8	56·8	63·9	70·9
·9	3·5	6·7	9·8	12·9	16·1	19·2	22·4	25·5	28·6	31·8	38·1	44·3	50·6	56·9	63·2
1·0	3·3	6·1	8·9	11·7	14·6	17·4	20·2	23·0	25·9	28·7	34·3	40·0	45·6	51·3	56·9
1·2	2·9	5·2	7·6	9·9	12·3	14·6	17·0	19·4	21·7	24·1	28·8	33·5	38·2	42·9	47·6
1·4	2·6	4·6	6·7	8·7	10·7	12·7	14·7	16·8	18·8	20·8	24·8	28·9	32·9	37·0	41·0
1·6	2·5	4·2	6·0	7·8	9·6	11·3	13·1	14·8	16·6	18·4	21·9	25·4	28·9	32·5	36·0
1·8	2·3	3·9	5·5	7·1	8·6	10·2	11·8	13·3	14·9	16·5	19·6	22·8	25·9	29·0	32·2
2·0	2·2	3·7	5·1	6·5	7·9	9·3	10·8	12·2	13·6	15·0	17·8	20·6	23·5	26·3	29·2
2·5		3·3	4·5	5·6	6·7	7·9	9·0	10·1	11·3	12·4	14·6	16·9	19·2	21·4	23·7
3·0		3·2	4·1	5·0	6·0	6·9	7·9	8·8	9·7	10·7	12·6	14·4	16·3	18·2	20·1
3·5		3·1	3·9	4·7	5·5	6·3	7·1	7·9	8·7	9·6	11·2	12·8	14·4	16·0	17·6
4·0			3·8	4·5	5·2	5·9	6·6	7·3	8·1	8·8	10·2	11·6	13·0	14·4	15·8
5·0				4·4	5·0	5·5	6·1	6·6	7·2	7·8	8·9	10·0	11·2	12·3	13·4
6·0						5·4	5·8	6·3	6·8	7·3	8·2	9·1	10·1	11·0	12·0
7·0								6·2	6·6	7·0	7·8	8·6	9·4	10·2	11·0
8·0										6·9	7·6	8·3	9·0	9·7	10·4
9·0												8·2	8·9	9·5	10·1
10·0													8·8	9·3	9·9

LOGARITHMS.

No. 1——100.   Log. 0.00000——2.00000.

No.	Log.	No.	Log.	No.	Log.	No.	Log.	No.	Log.
1	0·00000	21	1·32222	41	1·61278	61	1·78533	81	1·90849
2	0·30103	22	1·34242	42	1·62325	62	1·79239	82	1·91381
3	0·47712	23	1·36173	43	1·63347	63	1·79934	83	1·91908
4	0·60206	24	1·38021	44	1·64345	64	1·80618	84	1·92428
5	0·69897	25	1·39794	45	1·65321	65	1·81291	85	1·92942
6	0·77815	26	1·41497	46	1·66276	66	1·81954	86	1·93450
7	0·84510	27	1·43136	47	1·67210	67	1·82608	87	1·93952
8	0·90309	28	1·44716	48	1·68124	68	1·83251	88	1·94448
9	0·95424	29	1·46240	49	1·69020	69	1·83885	89	1·94939
10	1·00000	30	1·47712	50	1·69897	70	1·84510	90	1·95424
11	1·04139	31	1·49136	51	1·70757	71	1·85126	91	1·95904
12	1·07918	32	1·50515	52	1·71600	72	1·85733	92	1·96379
13	1·11394	33	1·51851	53	1·72428	73	1·86332	93	1·96848
14	1·14613	34	1·53148	54	1·73239	74	1·86923	94	1·97313
15	1·17609	35	1·54407	55	1·74036	75	1·87506	95	1·97772
16	1·20412	36	1·55630	56	1·74819	76	1·88081	96	1·98227
17	1·23045	37	1·56820	57	1·75588	77	1·88649	97	1·98677
18	1·25527	38	1·57978	58	1·76343	78	1·89210	98	1·99123
19	1·27875	39	1·59107	59	1·77085	79	1·89763	99	1·99564
20	1·30103	40	1·60206	60	1·77815	80	1·90309	100	2·00000

LOGARITHMS

Angles 2·	Angles 3·	No.	0	1	2	3	4	5	6	7	8	9	D
			No. 1000——1599					**Log. 00000——20385**					
° ′	° ′												
1 40	16 40	100	00000	00043	00087	00130	00173	00217	00260	00303	00346	00389	43
1 41	16 50	101	00432	00475	00518	00561	00604	00647	00689	00732	00775	00817	43
1 42	17 00	102	00860	00903	00945	00988	01030	01072	01115	01157	01199	01242	42
1 43	17 10	103	01284	01326	01368	01410	01452	01494	01536	01578	01620	01662	42
1 44	17 20	104	01703	01745	01787	01828	01870	01912	01953	01995	02036	02078	42
1 45	17 30	105	02119	02160	02202	02243	02284	02325	02366	02408	02449	02490	41
1 46	17 40	106	02531	02572	02612	02653	02694	02735	02776	02816	02857	02898	41
1 47	17 50	107	02938	02979	03020	03060	03100	03141	03181	03222	03262	03302	40
1 48	18 00	108	03342	03383	03423	03463	03503	03543	03583	03623	03663	03703	40
1 49	18 10	109	03743	03783	03822	03862	03902	03941	03981	04021	04060	04100	40
1 50	18 20	110	04139	04179	04218	04258	04297	04336	04376	04415	04454	04493	39
1 51	18 30	111	04532	04571	04611	04650	04689	04728	04766	04805	04844	04883	39
1 52	18 40	112	04922	04961	04999	05038	05077	05115	05154	05192	05231	05269	39
1 53	18 50	113	05308	05346	05385	05423	05461	05500	05538	05576	05614	05652	38
1 54	19 00	114	05691	05729	05767	05805	05843	05881	05919	05956	05994	06032	38
1 55	19 10	115	06070	06108	06145	06183	06221	06258	06296	06333	06371	06408	38
1 56	19 20	116	06446	06483	06521	06558	06595	06633	06670	06707	06744	06781	37
1 57	1S 30	117	06819	06856	06893	06930	06967	07004	07041	07078	07115	07151	37
1 58	19 40	118	07188	07225	07262	07299	07335	07372	07409	07445	07482	07518	37
1 59	19 50	119	07555	07591	07628	07664	07700	07737	07773	07809	07846	07882	36
2 00	20 00	120	07918	07954	07990	08027	08063	08099	08135	08171	08207	08243	36
2 01	20 10	121	08279	08314	08350	08386	08422	08458	08493	08529	08565	08600	36
2 02	20 20	122	08636	08672	08707	08743	08778	08814	08849	08885	08920	08955	36
2 03	20 30	123	08991	09026	09061	09096	09132	09167	09202	09237	09272	09307	35
2 04	20 40	124	09342	09377	09412	09447	09482	09517	09552	09587	09622	09656	35
2 05	20 50	125	09691	09726	09760	09795	09830	09864	09899	09934	09968	10003	35
2 06	21 00	126	10037	10072	10106	10140	10175	10209	10243	10278	10312	10346	34
2 07	21 10	127	10380	10415	10449	10483	10517	10551	10585	10619	10653	10687	34
2 08	21 20	128	10721	10755	10789	10823	10857	10890	10924	10958	10992	11025	34
2 09	21 30	129	11059	11093	11126	11160	11193	11227	11261	11294	11328	11361	34
2 10	21 40	130	11394	11428	11461	11494	11528	11561	11594	11628	11661	11694	33
2 11	21 50	131	11727	11760	11793	11827	11860	11893	11926	11959	11992	12025	33
2 12	22 00	132	12057	12090	12123	12156	12189	12222	12254	12287	12320	12353	33
2 13	22 10	133	12385	12418	12450	12483	12516	12548	12581	12613	12646	12678	33
2 14	22 20	134	12711	12743	12775	12808	12840	12872	12905	12937	12969	13001	32
2 15	22 30	135	13033	13066	13098	13130	13162	13194	13226	13258	13290	13322	32
2 16	22 40	136	13354	13386	13418	13450	13481	13513	13545	13577	13609	13640	32
2 17	22 50	137	13672	13704	13735	13767	13799	13830	13862	13893	13925	13956	32
2 18	23 00	138	13988	14019	14051	14082	14114	14145	14176	14208	14239	14270	31
2 19	23 10	139	14302	14333	14364	14395	14426	14457	14489	14520	14551	14582	31
2 20	23 20	140	14613	14644	14675	14706	14737	14768	14799	14829	14860	14891	31
2 21	23 30	141	14922	14953	14984	15014	15045	15076	15106	15137	15168	15198	31
2 22	23 40	142	15229	15259	15290	15321	15351	15382	15412	15442	15473	15503	31
2 23	23 50	143	15534	15564	15594	15625	15655	15685	15715	15746	15776	15806	30
2 24	24 00	144	15836	15866	15897	15927	15957	15987	16017	16047	16077	16107	30
2 25	24 10	145	16137	16167	16197	16227	16256	16286	16316	16346	16376	16406	30
2 26	24 20	146	16435	16465	16495	16524	16554	16584	16613	16643	16673	16702	30
2 27	24 30	147	16732	16761	16791	16820	16850	16879	16909	16938	16967	16997	30
2 28	24 40	148	17026	17056	17085	17114	17143	17173	17202	17231	17260	17290	29
2 29	24 50	149	17319	17348	17377	17406	17435	17464	17493	17522	17551	17580	29
2 30	25 00	150	17609	17638	17667	17696	17725	17754	17783	17811	17840	17869	29
2 31	25 10	151	17898	17926	17955	17984	18013	18041	18070	18099	18127	18156	29
2 32	25 20	152	18184	18213	18242	18270	18299	18327	18355	18384	18412	18441	29
2 33	25 30	153	18469	18498	18526	18554	18583	18611	18639	18667	18696	18724	28
2 34	25 40	154	18752	18780	18808	18837	18865	18893	18921	18949	18977	19005	28
2 35	25 50	155	19033	19061	19089	19117	19145	19173	19201	19229	19257	19285	28
2 36	26 00	156	19313	19340	19368	19396	19424	19451	19479	19507	19535	19562	28
2 37	26 10	157	19590	19618	19645	19673	19701	19728	19756	19783	19811	19838	28
2 38	26 20	158	19866	19893	19921	19948	19976	20003	20030	20058	20085	20112	27
2 39	26 30	159	20140	20167	20194	20222	20249	20276	20303	20331	20358	20385	27
Angles		No.	0	1	2	3	4	5	6	7	8	9	D.

For fifth-figure differences see page 174.

LOGARITHMS

Angles		No.	No. 1600——2199					Log. 20412——34223					D.
2·	3·	No.	0	1	2	3	4	5	6	7	8	9	D.
° ′	° ′												
2 40	26 40	160	20412	20439	20466	20493	20520	20548	20575	20602	20629	20656	27
2 41	26 50	161	20683	20710	20737	20763	20790	20817	20844	20871	20898	20925	27
2 42	27 00	162	20952	20978	21005	21032	21059	21085	21112	21139	21165	21192	27
2 43	27 10	163	21219	21245	21272	21299	21325	21352	21378	21405	21431	21458	27
2 44	27 20	164	21484	21511	21537	21564	21590	21617	21643	21669	21696	21722	26
2 45	27 30	165	21748	21775	21801	21827	21854	21880	21906	21932	21958	21985	26
2 46	27 40	166	22011	22037	22063	22089	22115	22141	22168	22194	22220	22246	26
2 47	27 50	167	22272	22298	22324	22350	22376	22402	22427	22453	22479	22505	26
2 48	28 00	168	22531	22557	22583	22608	22634	22660	22686	22712	22737	22763	26
2 49	28 10	169	22789	22814	22840	22866	22891	22917	22943	22968	22994	23019	26
2 50	28 20	170	23045	23070	23096	23122	23147	23172	23198	23223	23249	23274	26
2 51	28 30	171	23300	23325	23350	23376	23401	23426	23452	23477	23502	23528	25
2 52	28 40	172	23553	23578	23603	23629	23654	23679	23704	23729	23754	23780	25
2 53	28 50	173	23805	23830	23855	23880	23905	23930	23955	23980	24005	24030	25
2 54	29 00	174	24055	24080	24105	24130	24155	24180	24204	24229	24254	24279	25
2 55	29 10	175	24304	24329	24353	24378	24403	24428	24452	24477	24502	24527	25
2 56	29 20	176	24551	24576	24601	24625	24650	24675	24699	24724	24748	24773	25
2 57	29 30	177	24797	24822	24846	24871	24895	24920	24944	24969	24993	25018	25
2 58	29 40	178	25042	25066	25091	25115	25140	25164	25188	25213	25237	25261	24
2 59	29 50	179	25285	25310	25334	25358	25382	25406	25431	25455	25479	25503	24
3 00	30 00	180	25527	25551	25576	25600	25624	25648	25672	25696	25720	25744	24
3 01	30 10	181	25768	25792	25816	25840	25864	25888	25912	25936	25959	25983	24
3 02	30 20	182	26007	26031	26055	26079	26103	26126	26150	26174	26198	26221	24
3 03	30 30	183	26245	26269	26293	26316	26340	26364	26387	26411	26435	26458	24
3 04	30 40	184	26482	26505	26529	26553	26576	26600	26623	26647	26670	26694	24
3 05	30 50	185	26717	26741	26764	26788	26811	26834	26858	26881	26905	26928	23
3 06	31 00	186	26951	26975	26998	27021	27045	27068	27091	27114	27138	27161	23
3 07	31 10	187	27184	27207	27231	27254	27277	27300	27323	27346	27370	27393	23
3 08	31 20	188	27416	27439	27462	27485	27508	27531	27554	27577	27600	27623	23
3 09	31 30	189	27646	27669	27692	27715	27738	27761	27784	27807	27830	27853	23
3 10	31 40	190	27875	27898	27921	27944	27967	27990	28012	28035	28058	28081	23
3 11	31 50	191	28103	28126	28149	28172	28194	28217	28240	28262	28285	28308	23
3 12	32 00	192	28330	28353	28375	28398	28421	28443	28466	28488	28511	28533	23
3 13	32 10	193	28556	28578	28601	28623	28646	28668	28691	28713	28735	28758	23
3 14	32 20	194	28780	28803	28825	28847	28870	28892	28914	28937	28959	28981	22
3 15	32 30	195	29004	29026	29048	29070	29093	29115	29137	29159	29181	29203	22
3 16	32 40	196	29226	29248	29270	29292	29314	29336	29358	29380	29403	29425	22
3 17	32 50	197	29447	29469	29491	29513	29535	29557	29579	29601	29623	29645	22
3 18	33 00	198	29667	29688	29710	29732	29754	29776	29798	29820	29842	29864	22
3 19	33 10	199	29885	29907	29929	29951	29973	29994	30016	30038	30060	30081	22
3 20	33 20	200	30103	30125	30146	30168	30190	30211	30233	30255	30276	30298	22
3 21	33 30	201	30320	30341	30363	30384	30406	30428	30449	30471	30492	30514	22
3 22	33 40	202	30535	30557	30578	30600	30621	30643	30664	30685	30707	30728	22
3 23	33 50	203	30750	30771	30792	30814	30835	30856	30878	30899	30920	30942	21
3 24	34 00	204	30963	30984	31006	31027	31048	31069	31091	31112	31133	31154	21
3 25	34 10	205	31175	31197	31218	31239	31260	31281	31302	31323	31345	31366	21
3 26	34 20	206	31387	31408	31429	31450	31471	31492	31513	31534	31555	31576	21
3 27	34 30	207	31597	31618	31639	31660	31681	31702	31723	31744	31765	31785	21
3 28	34 40	208	31806	31827	31848	31869	31890	31911	31931	31952	31973	31994	21
3 29	34 50	209	32015	32035	32056	32077	32098	32118	32139	32160	32181	32201	21
3 30	35 00	210	32222	32243	32263	32284	32305	32325	32346	32367	32387	32408	21
3 31	35 10	211	32428	32449	32469	32490	32511	32531	32552	32572	32593	32613	21
3 32	35 20	212	32634	32654	32675	32695	32716	32736	32756	32777	32797	32818	20
3 33	35 30	213	32838	32858	32879	32899	32919	32940	32960	32981	33001	33021	20
3 34	35 40	214	33041	33062	33082	33102	33123	33143	33163	33183	33203	33224	20
3 35	35 50	215	33244	33264	33284	33304	33325	33345	33365	33385	33405	33425	20
3 36	36 00	216	33445	33466	33486	33506	33526	33546	33566	33586	33606	33626	20
3 37	36 10	217	33646	33666	33686	33706	33726	33746	33766	33786	33806	33826	20
3 38	36 20	218	33846	33866	33886	33905	33925	33945	33965	33985	34005	34025	20
3 39	36 30	219	34044	34064	34084	34104	34124	34144	34163	34183	34203	34223	20
Angles		No.	0	1	2	3	4	5	6	7	8	9	D.

For fifth-figure differences see page 174.

LOGARITHMS

Angles		No.	No. 2200——2799					Log. 34242——44700					
2·	**3·**	**No.**	**0**	**1**	**2**	**3**	**4**	**5**	**6**	**7**	**8**	**9**	**D.**
° ′	° ′												
3 40	36 40	220	34242	34262	34282	34301	34321	34341	34361	34380	34400	34420	20
3 41	36 50	221	34439	34459	34479	34498	34518	34537	34557	34577	34596	34616	20
3 42	37 00	222	34635	34655	34674	34694	34714	34733	34753	34772	34792	34811	20
3 43	37 10	223	34831	34850	34869	34889	34908	34928	34947	34967	34986	35005	19
3 44	37 20	224	35025	35044	35064	35083	35102	35122	35141	35160	35180	35199	19
3 45	37 30	225	35218	35238	35257	35276	35295	35315	35334	35353	35372	35392	19
3 46	37 40	226	35411	35430	35449	35469	35488	35507	35526	35545	35564	35583	19
3 47	37 50	227	35603	35622	35641	35660	35679	35698	35717	35736	35755	35774	19
3 48	38 00	228	35794	35813	35832	35851	35870	35889	35908	35927	35946	35965	19
3 49	38 10	229	35984	36003	36022	36040	36059	36078	36097	36116	36135	36154	19
3 50	38 20	230	36173	36192	36211	36229	36248	36267	36286	36305	36324	36342	19
3 51	38 30	231	36361	36380	36399	36418	36436	36455	36474	36493	36511	36530	19
3 52	38 40	232	36549	36568	36586	36605	36624	36642	36661	36680	36698	36717	19
3 53	38 50	233	36736	36754	36773	36792	36810	36829	36847	36866	36884	36903	19
3 54	39 00	234	36922	36940	36959	36977	36996	37014	37033	37051	37070	37088	19
3 55	39 10	235	37107	37125	37144	37162	37181	37199	37218	37236	37254	37273	18
3 56	39 20	236	37291	37310	37328	37346	37365	37383	37402	37420	37438	37457	18
3 57	39 30	237	37475	37493	37512	37530	37548	37566	37585	37603	37621	37639	18
3 58	39 40	238	37658	37676	37694	37712	37731	37749	37767	37785	37803	37822	18
3 59	39 50	239	37840	37858	37876	37894	37912	37931	37949	37967	37985	38003	18
4 00	40 00	240	38021	38039	38057	38075	38093	38112	38130	38148	38166	38184	18
4 01	40 10	241	38202	38220	38238	38256	38274	38292	38310	38328	38346	38364	18
4 02	40 20	242	38382	38400	38417	38435	38453	38471	38489	38507	38525	38543	18
4 03	40 30	243	38561	38579	38596	38614	38632	38650	38668	38686	38703	38721	18
4 04	40 40	244	38739	38757	38775	38792	38810	38828	38846	38863	38881	38899	18
4 05	40 50	245	38917	38934	38952	38970	38988	39005	39023	39041	39058	39076	18
4 06	41 00	246	39094	39111	39129	39146	39164	39182	39199	39217	39235	39252	18
4 07	41 10	247	39270	39287	39305	39322	39340	39358	39375	39393	39410	39428	18
4 08	41 20	248	39445	39463	39480	39498	39515	39533	39550	39568	39585	39603	18
4 09	41 30	249	39620	39637	39655	39672	39690	39707	39725	39742	39759	39777	17
4 10	41 40	250	39794	39811	39829	39846	39863	39881	39898	39915	39933	39950	17
4 11	41 50	251	39967	39985	40002	40019	40037	40054	40071	40088	40106	40123	17
4 12	42 00	252	40140	40157	40175	40192	40209	40226	40243	40261	40278	40295	17
4 13	42 10	253	40312	40329	40346	40364	40381	40398	40415	40432	40449	40466	17
4 14	42 20	254	40483	40501	40518	40535	40552	40569	40586	40603	40620	40637	17
4 15	42 30	255	40654	40671	40688	40705	40722	40739	40756	40773	40790	40807	17
4 16	42 40	256	40824	40841	40858	40875	40892	40909	40926	40943	40960	40976	17
4 17	42 50	257	40993	41010	41027	41044	41061	41078	41095	41111	41128	41145	17
4 18	43 00	258	41162	41179	41196	41212	41229	41246	41263	41280	41296	41313	17
4 19	43 10	259	41330	41347	41364	41380	41397	41414	41431	41447	41464	41481	17
4 20	43 20	260	41497	41514	41531	41547	41564	41581	41597	41614	41631	41647	17
4 21	43 30	261	41664	41681	41697	41714	41731	41747	41764	41780	41797	41814	17
4 22	43 40	262	41830	41847	41863	41880	41896	41913	41930	41946	41963	41979	17
4 23	43 50	263	41996	42012	42029	42045	42062	42078	42095	42111	42128	42144	17
4 24	44 00	264	42160	42177	42193	42210	42226	42243	42259	42275	42292	42308	16
4 25	44 10	265	42325	42341	42357	42374	42390	42406	42423	42439	42456	42472	16
4 26	44 20	266	42488	42505	42521	42537	42553	42570	42586	42602	42619	42635	16
4 27	44 30	267	42651	42667	42684	42700	42716	42732	42749	42765	42781	42797	16
4 28	44 40	268	42814	42830	42846	42862	42878	42894	42911	42927	42943	42959	16
4 29	44 50	269	42975	42991	43008	43024	43040	43056	43072	43088	43104	43120	16
4 30	45 00	270	43136	43153	43169	43185	43201	43217	43233	43249	43265	43281	16
4 31	45 10	271	43297	43313	43329	43345	43361	43377	43393	43409	43425	43441	16
4 32	45 20	272	43457	43473	43489	43505	43521	43537	43553	43569	43584	43600	16
4 33	45 30	273	43616	43632	43648	43664	43680	43696	43712	43728	43743	43759	16
4 34	45 40	274	43775	43791	43807	43823	43838	43854	43870	43886	43902	43918	16
4 35	45 50	275	43933	43949	43965	43981	43996	44012	44028	44044	44059	44075	16
4 36	46 00	276	44091	44107	44122	44138	44154	44170	44185	44201	44217	44232	16
4 37	46 10	277	44248	44264	44279	44295	44311	44326	44342	44358	44373	44389	16
4 38	46 20	278	44405	44420	44436	44451	44467	44483	44498	44514	44529	44545	16
4 39	46 30	279	44560	44576	44592	44607	44623	44638	44654	44669	44685	44700	16
Angles		**No.**	**0**	**1**	**2**	**3**	**4**	**5**	**6**	**7**	**8**	**9**	**D.**

For fifth-figure differences see page 174.

LOGARITHMS

Angles		No.	No. 2800———3399				Log. 44716———53135						

2·	3·	No.	0	1	2	3	4	5	6	7	8	9	D.
° ′	° ′												
4 40	46 40	280	44716	44731	44747	44762	44778	44793	44809	44824	44840	44855	16
4 41	46 50	281	44871	44886	44902	44917	44932	44948	44963	44979	44994	45010	15
4 42	47 00	282	45025	45040	45056	45071	45087	45102	45117	45133	45148	45163	15
4 43	47 10	283	45179	45194	45209	45225	45240	45255	45271	45286	45301	45317	15
4 44	47 20	284	45332	45347	45362	45378	45393	45408	45424	45439	45454	45469	15
4 45	47 30	285	45485	45500	45515	45530	45545	45561	45576	45591	45606	45621	15
4 46	47 40	286	45637	45652	45667	45682	45697	45713	45728	45743	45758	45773	15
4 47	47 50	287	45788	45803	45818	45834	45849	45864	45879	45894	45909	45924	15
4 48	48 00	288	45939	45954	45969	45985	46000	46015	46030	46045	46060	46075	15
4 49	48 10	289	46090	46105	46120	46135	46150	46165	46180	46195	46210	46225	15
4 50	48 20	290	46240	46255	46270	46285	46300	46315	46330	46345	46359	46374	15
4 51	48 30	291	46389	46404	46419	46434	46449	46464	46479	46494	46509	46523	15
4 52	48 40	292	46538	46553	46568	46583	46598	46613	46627	46642	46657	46672	15
4 53	48 50	293	46687	46702	46716	46731	46746	46761	46776	46790	46805	46820	15
4 54	49 00	294	46835	46850	46864	46879	46894	46909	46923	46938	46953	46968	15
4 55	49 10	295	46982	46997	47012	47026	47041	47056	47070	47085	47100	47115	15
4 56	49 20	296	47129	47144	47159	47173	47188	47203	47217	47232	47246	47261	15
4 57	49 30	297	47276	47290	47305	47320	47334	47349	47363	47378	47393	47407	15
4 58	49 40	298	47422	47436	47451	47465	47480	47494	47509	47524	47538	47553	15
4 59	49 50	299	47567	47582	47596	47611	47625	47640	47654	47669	47683	47698	15
5 00	50 00	300	47712	47727	47741	47756	47770	47784	47799	47813	47828	47842	15
5 01	50 10	301	47857	47871	47886	47900	47914	47929	47943	47958	47972	47986	14
5 02	50 20	302	48001	48015	48029	48044	48058	48073	48087	48101	48116	48130	14
5 03	50 30	303	48144	48159	48173	48187	48202	48216	48230	48245	48259	48273	14
5 04	50 40	304	48287	48302	48316	48330	48345	48359	48373	48387	48402	48416	14
5 05	50 50	305	48430	48444	48458	48473	48487	48501	48515	48530	48544	48558	14
5 06	51 00	306	48572	48586	48601	48615	48629	48643	48657	48671	48686	48700	14
5 07	51 10	307	48714	48728	48742	48756	48770	48785	48799	48813	48827	48841	14
5 08	51 20	308	48855	48869	48883	48897	48911	48926	48940	48954	48968	48982	14
5 09	51 30	309	48996	49010	49024	49038	49052	49066	49080	49094	49108	49122	14
5 10	51 40	310	49136	49150	49164	49178	49192	49206	49220	49234	49248	49262	14
5 11	51 50	311	49276	49290	49304	49318	49332	49346	49360	49374	49388	49402	14
5 12	52 00	312	49416	49429	49443	49457	49471	49485	49499	49513	49527	49541	14
5 13	52 10	313	49554	49568	49582	49596	49610	49624	49638	49651	49665	49679	14
5 14	52 20	314	49693	49707	49721	49734	49748	49762	49776	49790	49804	49817	14
5 15	52 30	315	49831	49845	49859	49872	49886	49900	49914	49928	49941	49955	14
5 16	52 40	316	49969	49982	49996	50010	50024	50037	50051	50065	50079	50092	14
5 17	52 50	317	50106	50120	50133	50147	50161	50174	50188	50202	50215	50229	14
5 18	53 00	318	50243	50256	50270	50284	50297	50311	50325	50338	50352	50365	14
5 19	53 10	319	50379	50393	50406	50420	50434	50447	50461	50474	50488	50501	14
5 20	53 20	320	50515	50529	50542	50556	50569	50583	50596	50610	50623	50637	14
5 21	53 30	321	50651	50664	50678	50691	50705	50718	50732	50745	50759	50772	14
5 22	53 40	322	50786	50799	50813	50826	50840	50853	50866	50880	50893	50907	14
5 23	53 50	323	50920	50934	50947	50961	50974	50987	51001	51014	51028	51041	13
5 24	54 00	324	51055	51068	51081	51095	51108	51122	51135	51148	51162	51175	13
5 25	54 10	325	51188	51202	51215	51228	51242	51255	51268	51282	51295	51308	13
5 26	54 20	326	51322	51335	51348	51362	51375	51388	51402	51415	51428	51442	13
5 27	54 30	327	51455	51468	51481	51495	51508	51521	51534	51548	51561	51574	13
5 28	54 40	328	51587	51601	51614	51627	51640	51654	51667	51680	51693	51706	13
5 29	54 50	329	51720	51733	51746	51759	51772	51786	51799	51812	51825	51838	13
5 30	55 00	330	51851	51865	51878	51891	51904	51917	51930	51943	51957	51970	13
5 31	55 10	331	51983	51996	52009	52022	52035	52048	52061	52075	52088	52101	13
5 32	55 20	332	52114	52127	52140	52153	52166	52179	52192	52205	52218	52231	13
5 33	55 30	333	52244	52258	52271	52284	52297	52310	52323	52336	52349	52362	13
5 34	55 40	334	52375	52388	52401	52414	52427	52440	52453	52466	52479	52492	13
5 35	55 50	335	52505	52517	52530	52543	52556	52569	52582	52595	52608	52621	13
5 36	56 00	336	52634	52647	52660	52673	52686	52699	52711	52724	52737	52750	13
5 37	56 10	337	52763	52776	52789	52802	52815	52827	52840	52853	52866	52879	13
5 38	56 20	338	52892	52905	52917	52930	52943	52956	52969	52982	52994	53007	13
5 39	56 30	339	53020	53033	53046	53058	53071	53084	53097	53110	53122	53135	13
Angles		No.	0	1	2	3	4	5	6	7	8	9	D.

For fifth-figure differences see page 174.

LOGARITHMS

Angles		No.	No. 3400——3999					Log. 53148——60195					
2·	**3·**	**No.**	**0**	**1**	**2**	**3**	**4**	**5**	**6**	**7**	**8**	**9**	**D.**
° ′	° ′												
5 40	56 40	340	53148	53161	53173	53186	53199	53212	53225	53237	53250	53263	13
5 41	56 50	341	53275	53288	53301	53314	53326	53339	53352	53365	53377	53390	13
5 42	57 00	342	53403	53415	53428	53441	53453	53466	53479	53491	53504	53517	13
5 43	57 10	343	53529	53542	53555	53567	53580	53593	53605	53618	53631	53643	13
5 44	57 20	344	53656	53669	53681	53694	53706	53719	53732	53744	53757	53769	13
5 45	57 30	345	53782	53795	53807	53820	53832	53845	53857	53870	53883	53895	13
5 46	57 40	346	53908	53920	53933	53945	53958	53970	53983	53995	54008	54020	13
5 47	57 50	347	54033	54046	54058	54071	54083	54096	54108	54121	54133	54145	13
5 48	58 00	348	54158	54170	54183	54195	54208	54220	54233	54245	54258	54270	13
5 49	58 10	349	54283	54295	54307	54320	54332	54345	54357	54370	54382	54394	12
5 50	58 20	350	54407	54419	54432	54444	54456	54469	54481	54494	54506	54518	12
5 51	58 30	351	54531	54543	54555	54568	54580	54593	54605	54617	54630	54642	12
5 52	58 40	352	54654	54667	54679	54691	54704	54716	54728	54741	54753	54765	12
5 53	58 50	353	54778	54790	54802	54814	54827	54839	54851	54864	54876	54888	12
5 54	59 00	354	54900	54913	54925	54937	54949	54962	54974	54986	54998	55011	12
5 55	59 10	355	55023	55035	55047	55060	55072	55084	55096	55108	55121	55133	12
5 56	59 20	356	55145	55157	55169	55182	55194	55206	55218	55230	55243	55255	12
5 57	59 30	357	55267	55279	55291	55303	55315	55328	55340	55352	55364	55376	12
5 58	59 40	358	55388	55400	55413	55425	55437	55449	55461	55473	55485	55497	12
5 59	59 50	359	55509	55522	55534	55546	55558	55570	55582	55594	55606	55618	12
6 00	60 00	360	55630	55642	55654	55666	55679	55691	55703	55715	55727	55739	12
6 01	60 10	361	55751	55763	55775	55787	55799	55811	55823	55835	55847	55859	12
6 02	60 20	362	55871	55883	55895	55907	55919	55931	55943	55955	55967	55979	12
6 03	60 30	363	55991	56003	56015	56027	56039	56050	56062	56074	56086	56098	12
6 04	60 40	364	56110	56122	56134	56146	56158	56170	56182	56194	56206	56217	12
6 05	60 50	365	56229	56241	56253	56265	56277	56289	56301	56313	56324	56336	12
6 06	61 00	366	56348	56360	56372	56384	56396	56407	56419	56431	56443	56455	12
6 07	61 10	367	56467	56478	56490	56502	56514	56526	56538	56549	56561	56573	12
6 08	61 20	368	56585	56597	56608	56620	56632	56644	56656	56667	56679	56691	12
6 09	61 30	369	56703	56714	56726	56738	56750	56761	56773	56785	56797	56808	12
6 10	61 40	370	56820	56832	56844	56855	56867	56879	56891	56902	56914	56926	12
6 11	61 50	371	56937	56949	56961	56973	56984	56996	57008	57019	57031	57043	12
6 12	62 00	372	57054	57066	57078	57089	57101	57113	57124	57136	57148	57159	12
6 13	62 10	373	57171	57183	57194	57206	57217	57229	57241	57252	57264	57276	12
6 14	62 20	374	57287	57299	57310	57322	57334	57345	57357	57368	57380	57392	12
6 15	62 30	375	57403	57415	57426	57438	57449	57461	57473	57484	57496	57507	12
6 16	62 40	376	57519	57530	57542	57553	57565	57577	57588	57600	57611	57623	12
6 17	62 50	377	57634	57646	57657	57669	57680	57692	57703	57715	57726	57738	12
6 18	63 00	378	57749	57761	57772	57784	57795	57807	57818	57830	57841	57853	12
6 19	63 10	379	57864	57875	57887	57898	57910	57921	57933	57944	57956	57967	11
6 20	63 20	380	57978	57990	58001	58013	58024	58036	58047	58058	58070	58081	11
6 21	63 30	381	58093	58104	58115	58127	58138	58150	58161	58172	58184	58195	11
6 22	63 40	382	58206	58218	58229	58240	58252	58263	58275	58286	58297	58309	11
6 23	63 50	383	58320	58331	58343	58354	58365	58377	58388	58399	58411	58422	11
6 24	64 00	384	58433	58444	58456	58467	58478	58490	58501	58512	58524	58535	11
6 25	64 10	385	58546	58557	58569	58580	58591	58602	58614	58625	58636	58648	11
6 26	64 20	386	58659	58670	58681	58693	58704	58715	58726	58737	58749	58760	11
6 27	64 30	387	58771	58782	58794	58805	58816	58827	58838	58850	58861	58872	11
6 28	64 40	388	58883	58894	58906	58917	58928	58939	58950	58962	58973	58984	11
6 29	64 50	389	58995	59006	59017	59028	59040	59051	59062	59073	59084	59095	11
6 30	65 00	390	59107	59118	59129	59140	59151	59162	59173	59184	59196	59207	11
6 31	65 10	391	59218	59229	59240	59251	59262	59273	59284	59295	59306	59318	11
6 32	65 20	392	59329	59340	59351	59362	59373	59384	59395	59406	59417	59428	11
6 33	65 30	393	59439	59450	59461	59472	59483	59495	59506	59517	59528	59539	11
6 34	65 40	394	59550	59561	59572	59583	59594	59605	59616	59627	59638	59649	11
6 35	65 50	395	59660	59671	59682	59693	59704	59715	59726	59737	59748	59759	11
6 36	66 00	396	59770	59781	59791	59802	59813	59824	59835	59846	59857	59868	11
6 37	66 10	397	59879	59890	59901	59912	59923	59934	59945	59956	59967	59977	11
6 38	66 20	398	59988	59999	60010	60021	60032	60043	60054	60065	60076	60086	11
6 39	66 30	399	60097	60108	60119	60130	60141	60152	60163	60173	60184	60195	11
Angles		**No.**	**0**	**1**	**2**	**3**	**4**	**5**	**6**	**7**	**8**	**9**	**D.**

For fifth-figure differences see page 174.

LOGARITHMS

Angles		No.	No. 4000———4599								Log. 60206———66266		
2·	3·	No.	0	1	2	3	4	5	6	7	8	9	D.
° ′	° ′												
6 40	66 40	400	60206	60217	60228	60239	60249	60260	60271	60282	60293	60304	11
6 41	66 50	401	60314	60325	60336	60347	60358	60369	60379	60390	60401	60412	11
6 42	67 00	402	60423	60433	60444	60455	60466	60477	60487	60498	60509	60520	11
6 43	67 10	403	60531	60541	60552	60563	60574	60584	60595	60606	60617	60627	11
6 44	67 20	404	60638	60649	60660	60670	60681	60692	60703	60713	60724	60735	11
6 45	67 30	405	60746	60756	60767	60778	60788	60799	60810	60821	60831	60012	11
6 46	67 40	406	60853	60863	60874	60885	60895	60906	60917	60927	60938	60949	11
6 47	67 50	407	60959	60970	60981	60991	61002	61013	61023	61034	61045	61055	11
6 48	68 00	408	61066	61077	61087	61098	61109	61119	61130	61141	61151	61162	11
6 49	68 10	409	61172	61183	61194	61204	61215	61225	61236	61247	61257	61268	11
6 50	68 20	410	61278	61289	61300	61310	61321	61331	61342	61353	61363	61374	11
6 51	68 30	411	61384	61395	61405	61416	61426	61437	61448	61458	61469	61479	11
6 52	68 40	412	61490	61500	61511	61521	61532	61542	61553	61563	61574	61585	11
6 53	68 50	413	61595	61606	61616	61627	61637	61648	61658	61669	61679	61690	11
6 54	69 00	414	61700	61711	61721	61732	61742	61752	61763	61773	61784	61794	11
6 55	69 10	415	61805	61815	61826	61836	61847	61857	61868	61878	61888	61899	11
6 56	69 20	416	61909	61920	61930	61941	61951	61962	61972	61982	61993	62003	10
6 57	69 30	417	62014	62024	62034	62045	62055	62066	62076	62086	62097	62107	10
6 58	69 40	418	62118	62128	62138	62149	62159	62170	62180	62190	62201	62211	10
6 59	69 50	419	62221	62232	62242	62253	62263	62273	62284	62294	62304	62315	10
7 00	70 00	420	62325	62335	62346	62356	62366	62377	62387	62397	62408	62418	10
7 01	70 10	421	62428	62439	62449	62459	62469	62480	62490	62500	62511	62521	10
7 02	70 20	422	62531	62542	62552	62562	62572	62583	62593	62603	62614	62624	10
7 03	70 30	423	62634	62644	62655	62665	62675	62685	62696	62706	62716	62726	10
7 04	70 40	424	62737	62747	62757	62767	62778	62788	62798	62808	62818	62829	10
7 05	70 50	425	62839	62849	62859	62870	62880	62890	62900	62910	62921	62931	10
7 06	71 00	426	62941	62951	62961	62972	62982	62992	63002	63012	63022	63033	10
7 07	71 10	427	63043	63053	63063	63073	63083	63094	63104	63114	63124	63134	10
7 08	71 20	428	63144	63155	63165	63175	63185	63195	63205	63215	63226	63236	10
7 09	71 30	429	63246	63256	63266	63276	63286	63296	63306	63317	63327	63337	10
7 10	71 40	430	63347	63357	63367	63377	63387	63397	63407	63418	63428	63438	10
7 11	71 50	431	63448	63458	63468	63478	63488	63498	63508	63518	63528	63538	10
7 12	72 00	432	63548	63558	63569	63579	63589	63599	63609	63619	63629	63639	10
7 13	72 10	433	63649	63659	63669	63679	63689	63699	63709	63719	63729	63739	10
7 14	72 20	434	63749	63759	63769	63779	63789	63799	63809	63819	63829	63839	10
7 15	72 30	435	63849	63859	63869	63879	63889	63899	63909	63919	63929	63939	10
7 16	72 40	436	63949	63959	63969	63979	63989	63998	64008	64018	64028	64038	10
7 17	72 50	437	64048	64058	64068	64078	64088	64098	64108	64118	64128	64138	10
7 18	73 00	438	64147	64157	64167	64177	64187	64197	64207	64217	64227	64237	10
7 19	73 10	439	64246	64256	64266	64276	64286	64296	64306	64316	64326	64335	10
7 20	73 20	440	64345	64355	64365	64375	64385	64395	64404	64414	64424	64434	10
7 21	73 30	441	64444	64454	64464	64473	64483	64493	64503	64513	64523	64532	10
7 22	73 40	442	64542	64552	64562	64572	64582	64591	64601	64611	64621	64631	10
7 23	73 50	443	64640	64650	64660	64670	64680	64689	64699	64709	64719	64729	10
7 24	74 00	444	64738	64748	64758	64768	64777	64787	64797	64807	64817	64826	10
7 25	74 10	445	64836	64846	64856	64865	64875	64885	64895	64904	64914	64924	10
7 26	74 20	446	64934	64943	64953	64963	64972	64982	64992	65002	65011	65021	10
7 27	74 30	447	65031	65041	65050	65060	65070	65079	65089	65099	65108	65118	10
7 28	74 40	448	65128	65138	65147	65157	65167	65176	65186	65196	65205	65215	10
7 29	74 50	449	65225	65234	65244	65254	65263	65273	65283	65292	65302	65312	10
7 30	75 00	450	65321	65331	65341	65350	65360	65370	65379	65389	65398	65408	10
7 31	75 10	451	65418	65427	65437	65447	65456	65466	65475	65485	65495	65504	10
7 32	75 20	452	65514	65523	65533	65543	65552	65562	65571	65581	65591	65600	10
7 33	75 30	453	65610	65619	65629	65639	65648	65658	65667	65677	65686	65696	10
7 34	75 40	454	65706	65715	65725	65734	65744	65753	65763	65773	65782	65792	10
7 35	75 50	455	65801	65811	65820	65830	65839	65849	65858	65868	65877	65887	10
7 36	76 00	456	65897	65906	65916	65925	65935	65944	65954	65963	65973	65982	10
7 37	76 10	457	65992	66001	66011	66020	66030	66039	66049	66058	66068	66077	10
7 38	76 20	458	66087	66096	66106	66115	66125	66134	66143	66153	66162	66172	10
7 39	76 30	459	66181	66191	66200	66210	66219	66229	66238	66247	66257	66266	10
Angles		No.	0	1	2	3	4	5	6	7	8	9	D.

For fifth-figure differences see page 174.

LOGARITHMS

Angles		No.	No. 4600———5199					Log. 66276———71592					D.
2·	**3·**		**0**	**1**	**2**	**3**	**4**	**5**	**6**	**7**	**8**	**9**	
° ′	° ′												
7 40	76 40	460	66276	66285	66295	66304	66314	66323	66332	66342	66351	66361	9
7 41	76 50	461	66370	66380	66389	66398	66408	66417	66427	66436	66445	66455	9
7 42	77 00	462	66464	66474	66483	66492	66502	66511	66521	66530	66539	66549	9
7 43	77 10	463	66558	66568	66577	66586	66596	66605	66614	66624	66633	66642	9
7 44	77 20	464	66652	66661	66671	66680	66689	66699	66708	66717	66727	66736	9
7 45	77 30	465	66745	66755	66764	66773	66783	66792	66801	66811	66820	66829	9
7 46	77 40	466	66839	66848	66857	66867	66876	66885	66895	66904	66913	66922	9
7 47	77 50	467	66932	66941	66950	66960	66969	66978	66988	66997	67006	67015	9
7 48	78 00	468	67025	67034	67043	67052	67062	67071	67080	67090	67099	67108	9
7 49	78 10	469	67117	67127	67136	67145	67154	67164	67173	67182	67191	67201	9
7 50	78 20	470	67210	67219	67228	67238	67247	67256	67265	67274	67284	67293	9
7 51	78 30	471	67302	67311	67321	67330	67339	67348	67357	67367	67376	67385	9
7 52	78 40	472	67394	67403	67413	67422	67431	67440	67449	67459	67468	67477	9
7 53	78 50	473	67486	67495	67505	67514	67523	67532	67541	67550	67560	67569	9
7 54	79 00	474	67578	67587	67596	67605	67614	67624	67633	67642	67651	67660	9
7 55	79 10	475	67669	67679	67688	67697	67706	67715	67724	67733	67742	67752	9
7 56	79 20	476	67761	67770	67779	67788	67797	67806	67815	67825	67834	67843	9
7 57	79 30	477	67852	67861	67870	67879	67888	67897	67906	67916	67925	67934	9
7 58	79 40	478	67943	67952	67961	67970	67979	67988	67997	68006	68015	68025	9
7 59	79 50	479	68034	68043	68052	68061	68070	68079	68088	68097	68106	68115	9
8 00	80 00	480	68124	68133	68142	68151	68160	68169	68178	68187	68196	68206	9
8 01	80 10	481	68215	68224	68233	68242	68251	68260	68269	68278	68287	68296	9
8 02	80 20	482	68305	68314	68323	68332	68341	68350	68359	68368	68377	68386	9
8 03	80 30	483	68395	68404	68413	68422	68431	68440	68449	68458	68467	68476	9
8 04	80 40	484	68485	68494	68503	68511	68520	68529	68538	68547	68556	68565	9
8 05	80 50	485	68574	68583	68592	68601	68610	68619	68628	68637	68646	68655	9
8 06	81 00	486	68664	68673	68682	68690	68699	68708	68717	68726	68735	68744	9
8 07	81 10	487	68753	68762	68771	68780	68789	68798	68806	68815	68824	68833	9
8 08	81 20	488	68842	68851	68860	68869	68878	68887	68895	68904	68913	68922	9
8 09	81 30	489	68931	68940	68949	68958	68966	68975	68984	68993	69002	69011	9
8 10	81 40	490	69020	69029	69037	69046	69055	69064	69073	69082	69091	69099	9
8 11	81 50	491	69108	69117	69126	69135	69144	69152	69161	69170	69179	69188	9
8 12	82 00	492	69197	69205	69214	69223	69232	69241	69249	69258	69267	69276	9
8 13	82 10	493	69285	69294	69302	69311	69320	69329	69338	69346	69355	69364	9
8 14	82 20	494	69373	69382	69390	69399	69408	69417	69425	69434	69443	69452	9
8 15	82 30	495	69461	69469	69478	69487	69496	69504	69513	69522	69531	69539	9
8 16	82 40	496	69548	69557	69566	69574	69583	69592	69601	69609	69618	69627	9
8 17	82 50	497	69636	69644	69653	69662	69671	69679	69688	69697	69706	69714	9
8 18	83 00	498	69723	69732	69740	69749	69758	69767	69775	69784	69793	69801	9
8 19	83 10	499	69810	69819	69828	69836	69845	69854	69862	69871	69880	69888	9
8 20	83 20	500	69897	69906	69914	69923	69932	69940	69949	69958	69966	69975	9
8 21	83 30	501	69984	69992	70001	70010	70018	70027	70036	70044	70053	70062	9
8 22	83 40	502	70070	70079	70088	70096	70105	70114	70122	70131	70140	70148	9
8 23	83 50	503	70157	70165	70174	70183	70191	70200	70209	70217	70226	70234	9
8 24	84 00	504	70243	70252	70260	70269	70278	70286	70295	70303	70312	70321	9
8 25	84 10	505	70329	70338	70346	70355	70364	70372	70381	70389	70398	70407	9
8 26	84 20	506	70415	70424	70432	70441	70449	70458	70467	70475	70484	70492	9
8 27	84 30	507	70501	70509	70518	70527	70535	70544	70552	70561	70569	70578	9
8 28	84 40	508	70586	70595	70604	70612	70621	70629	70638	70646	70655	70663	9
8 29	84 50	509	70672	70680	70689	70697	70706	70714	70723	70732	70740	70749	9
8 30	85 00	510	70757	70766	70774	70783	70791	70800	70808	70817	70825	70834	9
8 31	85 10	511	70842	70851	70859	70868	70876	70885	70893	70902	70910	70919	9
8 32	85 20	512	70927	70936	70944	70952	70961	70969	70978	70986	70995	71003	9
8 33	85 30	513	71012	71020	71029	71037	71046	71054	71063	71071	71079	71088	9
8 34	85 40	514	71096	71105	71113	71122	71130	71139	71147	71155	71164	71172	8
8 35	85 50	515	71181	71189	71198	71206	71214	71223	71231	71240	71248	71257	8
8 36	86 00	516	71265	71273	71282	71290	71299	71307	71315	71324	71332	71341	8
8 37	86 10	517	71349	71357	71366	71374	71383	71391	71399	71408	71416	71425	8
8 38	86 20	518	71433	71441	71450	71458	71467	71475	71483	71492	71500	71508	8
8 39	86 30	519	71517	71525	71534	71542	71550	71559	71567	71575	71584	71592	8
Angles		No.	0	1	2	3	4	5	6	7	8	9	D.

For fifth-figure differences see page 174.

LOGARITHMS

No. 5200——5799 Log. 71600——76335

2· ° '	3· ° '	No.	0	1	2	3	4	5	6	7	8	9	D.
8 40	86 40	520	71600	71609	71617	71625	71634	71642	71650	71659	71667	71675	8
8 41	86 50	521	71684	71692	71700	71709	71717	71725	71734	71742	71750	71759	8
8 42	87 00	522	71767	71775	71784	71792	71800	71809	71817	71825	71834	71842	8
8 43	87 10	523	71850	71859	71867	71875	71883	71892	71900	71908	71917	71925	8
8 44	87 20	524	71933	71941	71950	71958	71966	71975	71983	71991	71999	72008	8
8 45	87 30	525	72016	72024	72033	72041	72049	72057	72066	72074	72082	72090	8
8 46	87 40	526	72099	72107	72115	72123	72132	72140	72148	72156	72165	72173	8
8 47	87 50	527	72181	72189	72198	72206	72214	72222	72231	72239	72247	72255	8
8 48	88 00	528	72263	72272	72280	72288	72296	72305	72313	72321	72329	72337	8
8 49	88 10	529	72346	72354	72362	72370	72378	72387	72395	72403	72411	72419	8
8 50	88 20	530	72428	72436	72444	72452	72460	72469	72477	72485	72493	72501	8
8 51	88 30	531	72510	72518	72526	72534	72542	72550	72559	72567	72575	72583	8
8 52	88 40	532	72591	72599	72608	72616	72624	72632	72640	72648	72656	72665	8
8 53	88 50	533	72673	72681	72689	72697	72705	72713	72722	72730	72738	72746	8
8 54	89 00	534	72754	72762	72770	72779	72787	72795	72803	72811	72819	72827	8
8 55	89 10	535	72835	72844	72852	72860	72868	72876	72884	72892	72900	72908	8
8 56	89 20	536	72917	72925	72933	72941	72949	72957	72965	72973	72981	72989	8
8 57	89 30	537	72997	73006	73014	73022	73030	73038	73046	73054	73062	73070	8
8 58	89 40	538	73078	73086	73094	73102	73111	73119	73127	73135	73143	73151	8
8 59	89 50	539	73159	73167	73175	73183	73191	73199	73207	73215	73223	73231	8
9 00	90 00	540	73239	73247	73256	73264	73272	73280	73288	73296	73304	73312	8
9 01	90 10	541	73320	73328	73336	73344	73352	73360	73368	73376	73384	73392	8
9 02	90 20	542	73400	73408	73416	73424	73432	73440	73448	73456	73464	73472	8
9 03	90 30	543	73480	73488	73496	73504	73512	73520	73528	73536	73544	73552	8
9 04	90 40	544	73560	73568	73576	73584	73592	73600	73608	73616	73624	73632	8
9 05	90 50	545	73640	73648	73656	73664	73672	73680	73687	73695	73703	73711	8
9 06	91 00	546	73719	73727	73735	73743	73751	73759	73767	73775	73783	73791	8
9 07	91 10	547	73799	73807	73815	73823	73831	73838	73846	73854	73862	73870	8
9 08	91 20	548	73878	73886	73894	73902	73910	73918	73926	73934	73941	73949	8
9 09	91 30	549	73957	73965	73973	73981	73989	73997	74005	74013	74021	74028	8
9 10	91 40	550	74036	74044	74052	74060	74068	74076	74084	74092	74099	74107	8
9 11	91 50	551	74115	74123	74131	74139	74147	74155	74162	74170	74178	74186	8
9 12	92 00	552	74194	74202	74210	74218	74225	74233	74241	74249	74257	74265	8
9 13	92 10	553	74273	74280	74288	74296	74304	74312	74320	74328	74335	74343	8
9 14	92 20	554	74351	74359	74367	74375	74382	74390	74398	74406	74414	74422	8
9 15	92 30	555	74429	74437	74445	74453	74461	74468	74476	74484	74492	74500	8
9 16	92 40	556	74508	74515	74523	74531	74539	74547	74554	74562	74570	74578	8
9 17	92 50	557	74586	74593	74601	74609	74617	74625	74632	74640	74648	74656	8
9 18	93 00	558	74663	74671	74679	74687	74695	74702	74710	74718	74726	74733	8
9 19	93 10	559	74741	74749	74757	74765	74772	74780	74788	74796	74803	74811	8
9 20	93 20	560	74819	74827	74834	74842	74850	74858	74865	74873	74881	74889	8
9 21	93 30	561	74896	74904	74912	74920	74927	74935	74943	74950	74958	74966	8
9 22	93 40	562	74974	74981	74989	74997	75005	75012	75020	75028	75035	75043	8
9 23	93 50	563	75051	75059	75066	75074	75082	75089	75097	75105	75113	75120	8
9 24	94 00	564	75128	75136	75143	75151	75159	75166	75174	75182	75190	75197	8
9 25	94 10	565	75205	75213	75220	75228	75236	75243	75251	75259	75266	75274	8
9 26	94 20	566	75282	75289	75297	75305	75312	75320	75328	75335	75343	75351	8
9 27	94 30	567	75358	75366	75374	75381	75389	75397	75404	75412	75420	75427	8
9 28	94 40	568	75435	75443	75450	75458	75465	75473	75481	75488	75496	75504	8
9 29	94 50	569	75511	75519	75527	75534	75542	75549	75557	75565	75572	75580	8
9 30	95 00	570	75588	75595	75603	75610	75618	75626	75633	75641	75648	75656	8
9 31	95 10	571	75664	75671	75679	75686	75694	75702	75709	75717	75724	75732	8
9 32	95 20	572	75740	75747	75755	75762	75770	75778	75785	75793	75800	75808	8
9 33	95 30	573	75816	75823	75831	75838	75846	75853	75861	75869	75876	75884	8
9 34	95 40	574	75891	75899	75906	75914	75921	75929	75937	75944	75952	75959	8
9 35	95 50	575	75967	75974	75982	75989	75997	76005	76012	76020	76027	76035	8
9 36	96 00	576	76042	76050	76057	76065	76072	76080	76088	76095	76103	76110	8
9 37	96 10	577	76118	76125	76133	76140	76148	76155	76163	76170	76178	76185	8
9 38	96 20	578	76193	76200	76208	76215	76223	76230	76238	76245	76253	76260	8
9 39	96 30	579	76268	76275	76283	76290	76298	76305	76313	76320	76328	76335	8
Angles		No.	0	1	2	3	4	5	6	7	8	9	D.

For fifth-figure differences see page 174.

LOGARITHMS

Angles 2·	Angles 3·	No.	0	1	2	3	4	5	6	7	8	9	D.
° '	° '		No. 5800———6399					Log. 76343———80611					
9 40	96 40	580	76343	76350	76358	76365	76373	76380	76388	76395	76403	76410	8
9 41	96 50	581	76418	76425	76433	76440	76448	76455	76462	76470	76477	76485	8
9 42	97 00	582	76492	76500	76507	76515	76522	76530	76537	76545	76552	76559	8
9 43	97 10	583	76567	76574	76582	76589	76597	76604	76612	76619	76626	76634	7
9 44	97 20	584	76641	76649	76656	76664	76671	76679	76686	76693	76701	76708	7
9 45	97 30	585	76716	76723	76730	76738	76745	76753	76760	76768	76775	76782	7
9 46	97 40	586	76790	76797	76805	76812	76819	76827	76834	76842	76849	76856	7
9 47	97 50	587	76864	76871	76879	76886	76893	76901	76908	76916	76923	76930	7
9 48	98 00	588	76938	76945	76953	76960	76967	76975	76982	76989	76997	77004	7
9 49	98 10	589	77012	77019	77026	77034	77041	77048	77056	77063	77071	77078	7
9 50	98 20	590	77085	77093	77100	77107	77115	77122	77129	77137	77144	77151	7
9 51	98 30	591	77159	77166	77173	77181	77188	77196	77203	77210	77218	77225	7
9 52	98 40	592	77232	77240	77247	77254	77262	77269	77276	77284	77291	77298	7
9 53	98 50	593	77306	77313	77320	77327	77335	77342	77349	77357	77364	77371	7
9 54	99 00	594	77379	77386	77393	77401	77408	77415	77423	77430	77437	77444	7
9 55	99 10	595	77452	77459	77466	77474	77481	77488	77496	77503	77510	77517	7
9 56	99 20	596	77525	77532	77539	77547	77554	77561	77568	77576	77583	77590	7
9 57	99 30	597	77597	77605	77612	77619	77627	77634	77641	77648	77656	77663	7
9 58	99 40	598	77670	77677	77685	77692	77699	77706	77714	77721	77728	77735	7
9 59	99 50	599	77743	77750	77757	77764	77772	77779	77786	77793	77801	77808	7
10 00	100 00	600	77815	77822	77830	77837	77844	77851	77859	77866	77873	77880	7
10 01	100 10	601	77887	77895	77902	77909	77916	77924	77931	77938	77945	77952	7
10 02	100 20	602	77960	77967	77974	77981	77989	77996	78003	78010	78017	78025	7
10 03	100 30	603	78032	78039	78046	78053	78061	78068	78075	78082	78089	78097	7
10 04	100 40	604	78104	78111	78118	78125	78132	78140	78147	78154	78161	78168	7
10 05	100 50	605	78176	78183	78190	78197	78204	78211	78219	78226	78233	78240	7
10 06	101 00	606	78247	78254	78262	78269	78276	78283	78290	78297	78305	78312	7
10 07	101 10	607	78319	78326	78333	78340	78348	78355	78362	78369	78376	78383	7
10 08	101 20	608	78390	78398	78405	78412	78419	78426	78433	78440	78448	78455	7
10 09	101 30	609	78462	78469	78476	78483	78490	78497	78505	78512	78519	78526	7
10 10	101 40	610	78533	78540	78547	78554	78562	78569	78576	78583	78590	78597	7
10 11	101 50	611	78604	78611	78618	78625	78633	78640	78647	78654	78661	78668	7
10 12	102 00	612	78675	78682	78689	78696	78704	78711	78718	78725	78732	78739	7
10 13	102 10	613	78746	78753	78760	78767	78774	78782	78789	78796	78803	78810	7
10 14	102 20	614	78817	78824	78831	78838	78845	78852	78859	78866	78873	78880	7
10 15	102 30	615	78888	78895	78902	78909	78916	78923	78930	78937	78944	78951	7
10 16	102 40	616	78958	78965	78972	78979	78986	78993	79000	79007	79014	79022	7
10 17	102 50	617	79029	79036	79043	79050	79057	79064	79071	79078	79085	79092	7
10 18	103 00	618	79099	79106	79113	79120	79127	79134	79141	79148	79155	79162	7
10 19	103 10	619	79169	79176	79183	79190	79197	79204	79211	79218	79225	79232	7
10 20	103 20	620	79239	79246	79253	79260	79267	79274	79281	79288	79295	79302	7
10 21	103 30	621	79309	79316	79323	79330	79337	79344	79351	79358	79365	79372	7
10 22	103 40	622	79379	79386	79393	79400	79407	79414	79421	79428	79435	79442	7
10 23	103 50	623	79449	79456	79463	79470	79477	79484	79491	79498	79505	79512	7
10 24	104 00	624	79519	79525	79532	79539	79546	79553	79560	79567	79574	79581	7
10 25	104 10	625	79588	79595	79602	79609	79616	79623	79630	79637	79644	79651	7
10 26	104 20	626	79657	79664	79671	79678	79685	79692	79699	79706	79713	79720	7
10 27	104 30	627	79727	79734	79741	79748	79755	79761	79768	79775	79782	79789	7
10 28	104 40	628	79796	79803	79810	79817	79824	79831	79837	79844	79851	79858	7
10 29	104 50	629	79865	79872	79879	79886	79893	79900	79907	79913	79920	79927	7
10 30	105 00	630	79934	79941	79948	79955	79962	79969	79975	79982	79989	79996	7
10 31	105 10	631	80003	80010	80017	80024	80031	80037	80044	80051	80058	80065	7
10 32	105 20	632	80072	80079	80085	80092	80099	80106	80113	80120	80127	80134	7
10 33	105 30	633	80140	80147	80154	80161	80168	80175	80182	80188	80195	80202	7
10 34	105 40	634	80209	80216	80223	80230	80236	80243	80250	80257	80264	80271	7
10 35	105 50	635	80277	80284	80291	80298	80305	80312	80318	80325	80332	80339	7
10 36	106 00	636	80346	80353	80359	80366	80373	80380	80387	80394	80400	80407	7
10 37	106 10	637	80414	80421	80428	80434	80441	80448	80455	80462	80469	80475	7
10 38	106 20	638	80482	80489	80496	80503	80509	80516	80523	80530	80537	80543	7
10 39	106 30	639	80550	80557	80564	80571	80577	80584	80591	80598	80604	80611	7
Angles		No.	0	1	2	3	4	5	6	7	8	9	D.

For fifth-figure differences see page 174.

LOGARITHMS

Angles		No.	No. 6400————6999				Log. 80618————84504						
2·	3·	No.	0	1	2	3	4	5	6	7	8	9	D.
° ′	° ′												
10 40	106 40	640	80618	80625	80632	80638	80645	80652	80659	80666	80672	80679	7
10 41	106 50	641	80686	80693	80699	80706	80713	80720	80726	80733	80740	80747	7
10 42	107 00	642	80754	80760	80767	80774	80781	80787	80794	80801	80808	80814	7
10 43	107 10	643	80821	80028	80835	80841	80848	80855	80862	80868	80875	80882	7
10 44	107 20	644	80889	80895	80902	80909	80916	80922	80929	80936	80943	80949	7
10 45	107 30	645	80956	80963	80969	80976	80983	80990	80996	81003	81010	81017	7
10 46	107 40	646	81023	81030	81037	81043	81050	81057	81064	81070	81077	81084	7
10 47	107 50	647	81090	81097	81104	81111	81117	81124	81131	81137	81144	81151	7
10 48	108 00	648	81158	81164	81171	81178	81184	81191	81198	81204	81211	81218	7
10 49	108 10	649	81225	81231	81238	81245	81251	81258	81265	81271	81278	81285	7
10 50	108 20	650	81291	81298	81305	81311	81318	81325	81331	81338	81345	81351	7
10 51	108 30	651	81358	81365	81371	81378	81385	81391	81398	81405	81411	81418	7
10 52	108 40	652	81425	81431	81438	81445	81451	81458	81465	81471	81478	81485	7
10 53	108 50	653	81491	81498	81505	81511	81518	81525	81531	81538	81545	81551	7
10 54	109 00	654	81558	81564	81571	81578	81584	81591	81598	81604	81611	81618	7
10 55	109 10	655	81624	81631	81637	81644	81651	81657	81664	81671	81677	81684	7
10 56	109 20	656	81690	81697	81704	81710	81717	81724	81730	81737	81743	81750	7
10 57	109 30	657	81757	81763	81770	81776	81783	81790	81796	81803	81809	81816	7
10 58	109 40	658	81823	81829	81836	81842	81849	81856	81862	81869	81875	81882	7
10 59	109 50	659	81889	81895	81902	81908	81915	81922	81928	81935	81941	81948	7
11 00	110 00	660	81954	81961	81968	81974	81981	81987	81994	82000	82007	82014	7
11 01	110 10	661	82020	82027	82033	82040	82046	82053	82060	82066	82073	82079	7
11 02	110 20	662	82086	82092	82099	82106	82112	82119	82125	82132	82138	82145	7
11 03	110 30	663	82151	82158	82164	82171	82178	82184	82191	82197	82204	82210	7
11 04	110 40	664	82217	82223	82230	82236	82243	82250	82256	82263	82269	82276	7
11 05	110 50	665	82282	82289	82295	82302	82308	82315	82321	82328	82334	82341	7
11 06	111 00	666	82347	82354	82361	82367	82374	82380	82387	82393	82400	82406	7
11 07	111 10	667	82413	82419	82426	82432	82439	82445	82452	82458	82465	82471	7
11 08	111 20	668	82478	82484	82491	82497	82504	82510	82517	82523	82530	82536	7
11 09	111 30	669	82543	82549	82556	82562	82569	82575	82582	82588	82595	82601	7
11 10	111 40	670	82608	82614	82620	82627	82633	82640	82646	82653	82659	82666	7
11 11	111 50	671	82672	82679	82685	82692	82698	82705	82711	82718	82724	82731	7
11 12	112 00	672	82737	82743	82750	82756	82763	82769	82776	82782	82789	82795	7
11 13	112 10	673	82802	82808	82814	82821	82827	82834	82840	82847	82853	82860	6
11 14	112 20	674	82866	82872	82879	82885	82892	82898	82905	82911	82918	82924	6
11 15	112 30	675	82930	82937	82943	82950	82956	82963	82969	82975	82982	82988	6
11 16	112 40	676	82995	83001	83008	83014	83020	83027	83033	83040	83046	83053	6
11 17	112 50	677	83059	83065	83072	83078	83085	83091	83097	83104	83110	83117	6
11 18	113 00	678	83123	83129	83136	83142	83149	83155	83161	83168	83174	83181	6
11 19	113 10	679	83187	83193	83200	83206	83213	83219	83225	83232	83238	83245	6
11 20	113 20	680	83251	83257	83264	83270	83276	83283	83289	83296	83302	83308	6
11 21	113 30	681	83315	83321	83328	83334	83340	83347	83353	83359	83366	83372	6
11 22	113 40	682	83378	83385	83391	83398	83404	83410	83417	83423	83429	83436	6
11 23	113 50	683	83442	83448	83455	83461	83468	83474	83480	83487	83493	83499	6
11 24	114 00	684	83506	83512	83518	83525	83531	83537	83544	83550	83556	83563	6
11 25	114 10	685	83569	83575	83582	83588	83594	83601	83607	83613	83620	83626	6
11 26	114 20	686	83632	83639	83645	83651	83658	83664	83670	83677	83683	83689	6
11 27	114 30	687	83696	83702	83708	83715	83721	83727	83734	83740	83746	83753	6
11 28	114 40	688	83759	83765	83772	83778	83784	83790	83797	83803	83809	83816	6
11 29	114 50	689	83822	83828	83835	83841	83847	83853	83860	83866	83872	83879	6
11 30	115 00	690	83885	83891	83898	83904	83910	83916	83923	83929	83935	83942	6
11 31	115 10	691	83948	83954	83960	83967	83973	83979	83986	83992	83998	84004	6
11 32	115 20	692	84011	84017	84023	84029	84036	84042	84048	84055	84061	84067	6
11 33	115 30	693	84073	84080	84086	84092	84098	84105	84111	84117	84123	84130	6
11 34	115 40	694	84136	84142	84149	84155	84161	84167	84174	84180	84186	84192	6
11 35	115 50	695	84199	84205	84211	84217	84224	84230	84236	84242	84248	84255	6
11 36	116 00	696	84261	84267	84273	84280	84286	84292	84298	84305	84311	84317	6
11 37	116 10	697	84323	84330	84336	84342	84348	84354	84361	84367	84373	84379	6
11 38	116 20	698	84386	84392	84398	84404	84410	84417	84423	84429	84435	84442	6
11 39	116 30	699	84448	84454	84460	84466	84473	84479	84485	84491	84497	84504	6
Angles		No.	0	1	2	3	4	5	6	7	8	9	D.

For fifth-figure differences see page 174.

LOGARITHMS

Angles 2.	Angles 3.	No.	0	1	2	3	4	5	6	7	8	9	D.
° ′	° ′		No. 7000——7599					Log. 84510——88076					
11 40	116 40	700	84510	84516	84522	84528	84535	84541	84547	84553	84559	84566	6
11 41	116 50	701	84572	84578	84584	84590	84597	84603	84609	84615	84621	84628	6
11 42	117 00	702	84634	84640	84646	84652	84658	84665	84671	84677	84683	84689	6
11 43	117 10	703	84696	84702	84708	84714	84720	84726	84733	84739	84745	84751	6
11 44	117 20	704	84757	84763	84770	84776	84782	84788	84794	84800	84807	84813	6
11 45	117 30	705	84819	84825	84831	84837	84844	84850	84856	84862	84868	84874	6
11 46	117 40	706	84881	84887	84893	84899	84905	84911	84917	84924	84930	84936	6
11 47	117 50	707	84942	84948	84954	84960	84967	84973	84979	84985	84991	84997	6
11 48	118 00	708	85003	85010	85016	85022	85028	85034	85040	85046	85052	85059	6
11 49	118 10	709	85065	85071	85077	85083	85089	85095	85101	85108	85114	85120	6
11 50	118 20	710	85126	85132	85138	85144	85150	85156	85163	85169	85175	85181	6
11 51	118 30	711	85187	85193	85199	85205	85211	85218	85224	85230	85236	85242	6
11 52	118 40	712	85248	85254	85260	85266	85272	85279	85285	85291	85297	85303	6
11 53	118 50	713	85309	85315	85321	85327	85333	85339	85346	85352	85358	85364	6
11 54	119 00	714	85370	85376	85382	85388	85394	85400	85406	85412	85419	85425	6
11 55	119 10	715	85431	85437	85443	85449	85455	85461	85467	85473	85479	85485	6
11 56	119 20	716	85491	85497	85503	85510	85516	85522	85528	85534	85540	85546	6
11 57	119 30	717	85552	85558	85564	85570	85576	85582	85588	85594	85600	85606	6
11 58	119 40	718	85612	85619	85625	85631	85637	85643	85649	85655	85661	85667	6
11 59	119 50	719	85673	85679	85685	85691	85697	85703	85709	85715	85721	85727	6
12 00	120 00	720	85733	85739	85745	85751	85757	85763	85769	85775	85782	85788	6
12 01	120 10	721	85794	85800	85806	85812	85818	85824	85830	85836	85842	85848	6
12 02	120 20	722	85854	85860	85866	85872	85878	85884	85890	85896	85902	85908	6
12 03	120 30	723	85914	85920	85926	85932	85938	85944	85950	85956	85962	85968	6
12 04	120 40	724	85974	85980	85986	85992	85998	86004	86010	86016	86022	86028	6
12 05	120 50	725	86034	86040	86046	86052	86058	86064	86070	86076	86082	86088	6
12 06	121 00	726	86094	86100	86106	86112	86118	86124	86130	86136	86142	86148	6
12 07	121 10	727	86153	86159	86165	86171	86177	86183	86189	86195	86201	86207	6
12 08	121 20	728	86213	86219	86225	86231	86237	86243	86249	86255	86261	86267	6
12 09	121 30	729	86273	86279	86285	86291	86297	86303	86309	86314	86320	86326	6
12 10	121 40	730	86332	86338	86344	86350	86356	86362	86368	86374	86380	86386	6
12 11	121 50	731	86392	86398	86404	86410	86416	86421	86427	86433	86439	86445	6
12 12	122 00	732	86451	86457	86463	86469	86475	86481	86487	86493	86499	86505	6
12 13	122 10	733	86510	86516	86522	86528	86534	86540	86546	86552	86558	86564	6
12 14	122 20	734	86570	86576	86581	86587	86593	86599	86605	86611	86617	86623	6
12 15	122 30	735	86629	86635	86641	86647	86652	86658	86664	86670	86676	86682	6
12 16	122 40	736	86688	86694	86700	86706	86711	86717	86723	86729	86735	86741	6
12 17	122 50	737	86747	86753	86759	86764	86770	86776	86782	86788	86794	86800	6
12 18	123 00	738	86806	86812	86817	86823	86829	86835	86841	86847	86853	86859	6
12 19	123 10	739	86864	86870	86876	86882	86888	86894	86900	86906	86911	86917	6
12 20	123 20	740	86923	86929	86935	86941	86947	86953	86958	86964	86970	86976	6
12 21	123 30	741	86982	86988	86994	86999	87005	87011	87017	87023	87029	87035	6
12 22	123 40	742	87040	87046	87052	87058	87064	87070	87076	87081	87087	87093	6
12 23	123 50	743	87099	87105	87111	87116	87122	87128	87134	87140	87146	87152	6
12 24	124 00	744	87157	87163	87169	87175	87181	87187	87192	87198	87204	87210	6
12 25	124 10	745	87216	87222	87227	87233	87239	87245	87251	87256	87262	87268	6
12 26	124 20	746	87274	87280	87286	87291	87297	87303	87309	87315	87320	87326	6
12 27	124 30	747	87332	87338	87344	87350	87355	87361	87367	87373	87379	87384	6
12 28	124 40	748	87390	87396	87402	87408	87413	87419	87425	87431	87437	87442	6
12 29	124 50	749	87448	87454	87460	87466	87471	87477	87483	87489	87495	87500	6
12 30	125 00	750	87506	87512	87518	87524	87529	87535	87541	87547	87552	87558	6
12 31	125 10	751	87564	87570	87576	87581	87587	87593	87599	87605	87610	87616	6
12 32	125 20	752	87622	87628	87633	87639	87645	87651	87656	87662	87668	87674	6
12 33	125 30	753	87680	87685	87691	87697	87703	87708	87714	87720	87726	87731	6
12 34	125 40	754	87737	87743	87749	87754	87760	87766	87772	87777	87783	87789	6
12 35	125 50	755	87795	87800	87806	87812	87818	87823	87829	87835	87841	87846	6
12 36	126 00	756	87852	87858	87864	87869	87875	87881	87887	87892	87898	87904	6
12 37	126 10	757	87910	87915	87921	87927	87933	87938	87944	87950	87956	87961	6
12 38	126 20	758	87967	87973	87978	87984	87990	87996	88001	88007	88013	88019	6
12 39	126 30	759	88024	88030	88036	88041	88047	88053	88059	88064	88070	88076	6
Angles		No.	0	1	2	3	4	5	6	7	8	9	D.

For fifth-figure differences see page 174.

LOGARITHMS

Angles 2· (o ')	3· (o ')	No.	0	1	2	3	4	5	6	7	8	9	D.
			No. 7600——8199					Log. 88081——91376					
12 40	126 40	760	88081	88087	88093	88099	88104	88110	88116	88121	88127	88133	6
12 41	126 50	761	88139	88144	88150	88156	88161	88167	88173	88178	88184	88190	6
12 42	127 00	762	88196	88201	88207	88213	88218	88224	88230	88235	88241	88247	6
12 43	127 10	763	88252	88258	88264	00270	88275	88281	88287	88292	88298	88304	6
12 44	127 20	764	88309	88315	88321	88326	88332	88338	88343	88349	88355	88361	6
12 45	127 30	765	88366	88372	88378	88383	88389	88395	88400	88406	88412	88417	6
12 46	127 40	766	88423	88429	88434	88440	88446	88451	88457	88463	88468	88474	6
12 47	127 50	767	88480	88485	88491	88497	88502	88508	88514	88519	88525	88531	6
12 48	128 00	768	88536	88542	88547	88553	88559	88564	88570	88576	88581	88587	6
12 49	128 10	769	88593	88598	88604	88610	88615	88621	88627	88632	88638	88643	6
12 50	128 20	770	88649	88655	88660	88666	88672	88677	88683	88689	88694	88700	6
12 51	128 30	771	88705	88711	88717	88722	88728	88734	88739	88745	88751	88756	6
12 52	128 40	772	88762	88767	88773	88779	88784	88790	88796	88801	88807	88812	6
12 53	128 50	773	88818	88824	88829	88835	88840	88846	88852	88857	88863	88869	6
12 54	129 00	774	88874	88880	88885	88891	88897	88902	88908	88913	88919	88925	6
12 55	129 10	775	88930	88936	88941	88947	88953	88958	88964	88969	88975	88981	6
12 56	129 20	776	88986	88992	88997	89003	89009	89014	89020	89025	89031	89037	6
12 57	129 30	777	89042	89048	89053	89059	89064	89070	89076	89081	89087	89092	6
12 58	129 40	778	89098	89104	89109	89115	89120	89126	89131	89137	89143	89148	6
12 59	129 50	779	89154	89159	89165	89171	89176	89182	89187	89193	89198	89204	6
13 00	130 00	780	89210	89215	89221	89226	89232	89237	89243	89248	89254	89260	6
13 01	130 10	781	89265	89271	89276	89282	89287	89293	89299	89304	89310	89315	6
13 02	130 20	782	89321	89326	89332	89337	89343	89348	89354	89360	89365	89371	6
13 03	130 30	783	89376	89382	89387	89393	89398	89404	89409	89415	89421	89426	6
13 04	130 40	784	89432	89437	89443	89448	89454	89459	89465	89470	89476	89481	6
13 05	130 50	785	89487	89493	89498	89504	89509	89515	89520	89526	89531	89537	6
13 06	131 00	786	89542	89548	89553	89559	89564	89570	89575	89581	89586	89592	6
13 07	131 10	787	89598	89603	89609	89614	89620	89625	89631	89636	89642	89647	6
13 08	131 20	788	89653	89658	89664	89669	89675	89680	89686	89691	89697	89702	6
13 09	131 30	789	89708	89713	89719	89724	89730	89735	89741	89746	89752	89757	6
13 10	131 40	790	89763	89768	89774	89779	89785	89790	89796	89801	89807	89812	6
13 11	131 50	791	89818	89823	89829	89834	89840	89845	89851	89856	89862	89867	6
13 12	132 00	792	89873	89878	89884	89889	89894	89900	89905	89911	89916	89922	6
13 13	132 10	793	89927	89933	89938	89944	89949	89955	89960	89966	89971	89977	6
13 14	132 20	794	89982	89988	89993	89999	90004	90009	90015	90020	90026	90031	6
13 15	132 30	795	90037	90042	90048	90053	90059	90064	90070	90075	90080	90086	6
13 16	132 40	796	90091	90097	90102	90108	90113	90119	90124	90130	90135	90140	6
13 17	132 50	797	90146	90151	90157	90162	90168	90173	90179	90184	90189	90195	5
13 18	133 00	798	90200	90206	90211	90217	90222	90228	90233	90238	90244	90249	5
13 19	133 10	799	90255	90260	90266	90271	90276	90282	90287	90293	90298	90304	5
13 20	133 20	800	90309	90314	90320	90325	90331	90336	90342	90347	90352	90358	5
13 21	133 30	801	90363	90369	90374	90380	90385	90390	90396	90401	90407	90412	5
13 22	133 40	802	90417	90423	90428	90434	90439	90445	90450	90455	90461	90466	5
13 23	133 50	803	90472	90477	90482	90488	90493	90499	90504	90509	90515	90520	5
13 24	134 00	804	90526	90531	90536	90542	90547	90553	90558	90563	90569	90574	5
13 25	134 10	805	90580	90585	90590	90596	90601	90607	90612	90617	90623	90628	5
13 26	134 20	806	90634	90639	90644	90650	90655	90660	90666	90671	90677	90682	5
13 27	134 30	807	90687	90693	90698	90704	90709	90714	90720	90725	90730	90736	5
13 28	134 40	808	90741	90747	90752	90757	90763	90768	90773	90779	90784	90790	5
13 29	134 50	809	90795	90800	90806	90811	90816	90822	90827	90832	90838	90843	5
13 30	135 00	810	90849	90854	90859	90865	90870	90875	90881	90886	90891	90897	5
13 31	135 10	811	90902	90907	90913	90918	90924	90929	90934	90940	90945	90950	5
13 32	135 20	812	90956	90961	90966	90972	90977	90982	90988	90993	90998	91004	5
13 33	135 30	813	91009	91014	91020	91025	91030	91036	91041	91046	91052	91057	5
13 34	135 40	814	91062	91068	91073	91078	91084	91089	91094	91100	91105	91110	5
13 35	135 50	815	91116	91121	91126	91132	91137	91142	91148	91153	91158	91164	5
13 36	136 00	816	91169	91174	91180	91185	91190	91196	91201	91206	91212	91217	5
13 37	136 10	817	91222	91228	91233	91238	91244	91249	91254	91259	91265	91270	5
13 38	136 20	818	91275	91281	91286	91291	91297	91302	91307	91313	91318	91323	5
13 39	136 30	819	91328	91334	91339	91344	91350	91355	91360	91366	91371	91376	5
Angles		No.	0	1	2	3	4	5	6	7	8	9	D.

For fifth-figure differences see page 174.

LOGARITHMS

Angles 2·	Angles 3·	No.	0	1	2	3	4	5	6	7	8	9	D.
			No. 8200——8799					Log. 91381——94443						
° ′	° ′												
13 40	136 40	820	91381	91387	91392	91397	91403	91408	91413	91418	91424	91429	5
13 41	136 50	821	91434	91440	91445	91450	91456	91461	91466	91471	91477	91482	5
13 42	137 00	822	91487	91493	91498	91503	91508	91514	91519	91524	91529	91535	5
13 43	137 10	823	91540	91545	91551	91556	91561	91566	91572	91577	91582	91587	5
13 44	137 20	824	91593	91598	91603	91609	91614	91619	91624	91630	91635	91640	5
13 45	137 30	825	91645	91651	91656	91661	91666	91672	91677	91682	91688	91693	5
13 46	137 40	826	91698	91703	91709	91714	91719	91724	91730	91735	91740	91745	5
13 47	137 50	827	91751	91756	91761	91766	91772	91777	91782	91787	91793	91798	5
13 48	138 00	828	91803	91808	91814	91819	91824	91829	91835	91840	91845	91850	5
13 49	138 10	829	91856	91861	91866	91871	91876	91882	91887	91892	91897	91903	5
13 50	138 20	830	91908	91913	91918	91924	91929	91934	91939	91944	91950	91955	5
13 51	138 30	831	91960	91965	91971	91976	91981	91986	91991	91997	92002	92007	5
13 52	138 40	832	92012	92018	92023	92028	92033	92038	92044	92049	92054	92059	5
13 53	138 50	833	92065	92070	92075	92080	92085	92091	92096	92101	92106	92111	5
13 54	139 00	834	92117	92122	92127	92132	92137	92143	92148	92153	92158	92163	5
13 55	139 10	835	92169	92174	92179	92184	92189	92195	92200	92205	92210	92215	5
13 56	139 20	836	92221	92226	92231	92236	92241	92247	92252	92257	92262	92267	5
13 57	139 30	837	92273	92278	92283	92288	92293	92299	92304	92309	92314	92319	5
13 58	139 40	838	92324	92330	92335	92340	92345	92350	92356	92361	92366	92371	5
13 59	139 50	839	92376	92381	92387	92392	92397	92402	92407	92412	92418	92423	5
14 00	140 00	840	92428	92433	92438	92443	92449	92454	92459	92464	92469	92474	5
14 01	140 10	841	92480	92485	92490	92495	92500	92505	92511	92516	92521	92526	5
14 02	140 20	842	92531	92536	92542	92547	92552	92557	92562	92567	92572	92578	5
14 03	140 30	843	92583	92588	92593	92598	92603	92609	92614	92619	92624	92629	5
14 04	140 40	844	92634	92639	92645	92650	92655	92660	92665	92670	92675	92681	5
14 05	140 50	845	92686	92691	92696	92701	92706	92711	92717	92722	92727	92732	5
14 06	141 00	846	92737	92742	92747	92752	92758	92763	92768	92773	92778	92783	5
14 07	141 10	847	92788	92794	92799	92804	92809	92814	92819	92824	92829	92835	5
14 08	141 20	848	92840	92845	92850	92855	92860	92865	92870	92875	92881	92886	5
14 09	141 30	849	92891	92896	92901	92906	92911	92916	92921	92927	92932	92937	5
14 10	141 40	850	92942	92947	92952	92957	92962	92967	92973	92978	92983	92988	5
14 11	141 50	851	92993	92998	93003	93008	93013	93019	93024	93029	93034	93039	5
14 12	142 00	852	93044	93049	93054	93059	93064	93069	93075	93080	93085	93090	5
14 13	142 10	853	93095	93100	93105	93110	93115	93120	93125	93131	93136	93141	5
14 14	142 20	854	93146	93151	93156	93161	93166	93171	93176	93181	93186	93192	5
14 15	142 30	855	93197	93202	93207	93212	93217	93222	93227	93232	93237	93242	5
14 16	142 40	856	93247	93252	93258	93263	93268	93273	93278	93283	93288	93293	5
14 17	142 50	857	93298	93303	93308	93313	93318	93323	93329	93334	93339	93344	5
14 18	143 00	858	93349	93354	93359	93364	93369	93374	93379	93384	93389	93394	5
14 19	143 10	859	93399	93404	93409	93415	93420	93425	93430	93435	93440	93445	5
14 20	143 20	860	93450	93455	93460	93465	93470	93475	93480	93485	93490	93495	5
14 21	143 30	861	93500	93505	93510	93515	93521	93526	93531	93536	93541	93546	5
14 22	143 40	862	93551	93556	93561	93566	93571	93576	93581	93586	93591	93596	5
14 23	143 50	863	93601	93606	93611	93616	93621	93626	93631	93636	93641	93646	5
14 24	144 00	864	93651	93656	93661	93666	93672	93677	93682	93687	93692	93697	5
14 25	144 10	865	93702	93707	93712	93717	93722	92727	93732	93737	93742	93747	5
14 26	144 20	866	93752	93757	93762	93767	93772	93777	93782	93787	93792	93797	5
14 27	144 30	867	93802	93807	93812	93817	93822	93827	93832	93837	93842	93847	5
14 28	144 40	868	93852	93857	93862	93867	93872	93877	93882	93887	93892	93897	5
14 29	144 50	869	93902	93907	93912	93917	93922	93927	93932	93937	93942	93947	5
14 30	145 00	870	93952	93957	93962	93967	93972	93977	93982	93987	93992	93997	5
14 31	145 10	871	94002	94007	94012	94017	94022	94027	94032	94037	94042	94047	5
14 32	145 20	872	94052	94057	94062	94067	94072	94077	94082	94087	94092	94096	5
14 33	145 30	873	94101	94106	94111	94116	94121	94126	94131	94136	94141	94146	5
14 34	145 40	874	94151	94156	94161	94166	94171	94176	94181	94186	94191	94196	5
14 35	145 50	875	94201	94206	94211	94216	94221	94226	94231	94236	94241	94245	5
14 36	146 00	876	94250	94255	94260	94265	94270	94275	94280	94285	94290	94295	5
14 37	146 10	877	94300	94305	94310	94315	94320	94325	94330	94335	94340	94345	5
14 38	146 20	878	94349	94354	94359	94364	94369	94374	94379	94384	94389	94394	5
14 39	146 30	879	94399	94404	94409	94414	94419	94424	94429	94434	94438	94443	5
Angles		No.	0	1	2	3	4	5	6	7	8	9	D.

For fifth-figure differences see page 174.

H

LOGARITHMS

No. 8800——9399   Log. 94448——97308

Angles 2.	Angles 3.	No.	0	1	2	3	4	5	6	7	8	9	D.
° ′	° ′												
14 40	146 40	880	94448	94453	94458	94463	94468	94473	94478	94483	94488	94493	5
14 41	146 50	881	94498	94503	94507	94512	94517	94522	94527	94532	94537	94542	5
14 42	147 00	882	94547	94552	94557	94562	94567	94572	94576	94581	94586	94591	5
14 43	147 10	883	94596	94601	94606	94611	94616	94621	94626	94631	94636	94640	5
14 44	147 20	884	94645	94650	94655	94660	94665	94670	94675	94680	94685	94689	5
14 45	147 30	885	94694	94699	94704	94709	94714	94719	94724	94729	94734	94739	5
14 46	147 40	886	94743	94748	94753	94758	94763	94768	94773	94778	94783	94788	5
14 47	147 50	887	94792	94797	94802	94807	94812	94817	94822	94827	94832	94836	5
14 48	148 00	888	94841	94846	94851	94856	94861	94866	94871	94876	94880	94885	5
14 49	148 10	889	94890	94895	94900	94905	94910	94915	94920	94924	94929	94934	5
14 50	148 20	890	94939	94944	94949	94954	94959	94963	94968	94973	94978	94983	5
14 51	148 30	891	94988	94993	94998	95002	95007	95012	95017	95022	95027	95032	5
14 52	148 40	892	95037	95041	95046	95051	95056	95061	95066	95071	95075	95080	5
14 53	148 50	893	95085	95090	95095	95100	95105	95110	95114	95119	95124	95129	5
14 54	149 00	894	95134	95139	95144	95148	95153	95158	95163	95168	95173	95177	5
14 55	149 10	895	95182	95187	95192	95197	95202	95207	95211	95216	95221	95226	5
14 56	149 20	896	95231	95236	95241	95245	95250	95255	95260	95265	95270	95274	5
14 57	149 30	897	95279	95284	95289	95294	95299	95303	95308	95313	95318	95323	5
14 58	149 40	898	95328	95333	95337	95342	95347	95352	95357	95362	95366	95371	5
14 59	149 50	899	95376	95381	95386	95391	95395	95400	95405	95410	95415	95419	5
15 00	150 00	900	95424	95429	95434	95439	95444	95448	95453	95458	95463	95468	5
15 01	150 10	901	95473	95477	95482	95487	95492	95497	95501	95506	95511	95516	5
15 02	150 20	902	95521	95526	95530	95535	95540	95545	95550	95554	95559	95564	5
15 03	150 30	903	95569	95574	95578	95583	95588	95593	95598	95602	95607	95612	5
15 04	150 40	904	95617	95622	95626	95631	95636	95641	95646	95651	95655	95660	5
15 05	150 50	905	95665	95670	95675	95679	95684	95689	95694	95698	95703	95708	5
15 06	151 00	906	95713	95718	95722	95727	95732	95737	95742	95746	95751	95756	5
15 07	151 10	907	95761	95766	95770	95775	95780	95785	95789	95794	95799	95804	5
15 08	151 20	908	95809	95813	95818	95823	95828	95833	95837	95842	95847	95852	5
15 09	151 30	909	95856	95861	95866	95871	95876	95880	95885	95890	95895	95899	5
15 10	151 40	910	95904	95909	95914	95918	95923	95928	95933	95938	95942	95947	5
15 11	151 50	911	95952	95957	95961	95966	95971	95976	95980	95985	95990	95995	5
15 12	152 00	912	96000	96004	96009	96014	96019	96023	96028	96033	96038	96042	5
15 13	152 10	913	96047	96052	96057	96061	96066	96071	96076	96080	96085	96090	5
15 14	152 20	914	96095	96099	96104	96109	96114	96118	96123	96128	96133	96137	5
15 15	152 30	915	96142	96147	96152	96156	96161	96166	96171	96175	96180	96185	5
15 16	152 40	916	96190	96194	96199	96204	96209	96213	96218	96223	96228	96232	5
15 17	152 50	917	96237	96242	96246	96251	96256	96261	96265	96270	96275	96280	5
15 18	153 00	918	96284	96289	96294	96299	96303	96308	96313	96317	96322	96327	5
15 19	153 10	919	96332	96336	96341	96346	96350	96355	96360	96365	96369	96374	5
15 20	153 20	920	96379	96384	96388	96393	96398	96402	96407	96412	96417	96421	5
15 21	153 30	921	96426	96431	96435	96440	96445	96450	96454	96459	96464	96468	5
15 22	153 40	922	96473	96478	96483	96487	96492	96497	96501	96506	96511	96516	5
15 23	153 50	923	96520	96525	96530	96534	96539	96544	96548	96553	96558	96563	5
15 24	154 00	924	96567	96572	96577	96581	96586	96591	96595	96600	96605	96610	5
15 25	154 10	925	96614	96619	96624	96628	96633	96638	96642	96647	96652	96656	5
15 26	154 20	926	96661	96666	96671	96675	96680	96685	96689	96694	96699	96703	5
15 27	154 30	927	96708	96713	96717	96722	96727	96731	96736	96741	96745	96750	5
15 28	154 40	928	96755	96760	96764	96769	96774	96778	96783	96788	96792	96797	5
15 29	154 50	929	96802	96806	96811	96816	96820	96825	96830	96834	96839	96844	5
15 30	155 00	930	96848	96853	96858	96862	96867	96872	96876	96881	96886	96890	5
15 31	155 10	931	96895	96900	96904	96909	96914	96918	96923	96928	96932	96937	5
15 32	155 20	932	96942	96946	96951	96956	96960	96965	96970	96974	96979	96984	5
15 33	155 30	933	96988	96993	96998	97002	97007	97011	97016	97021	97025	97030	5
15 34	155 40	934	97035	97039	97044	97049	97053	97058	97063	97067	97072	97077	5
15 35	155 50	935	97081	97086	97090	97095	97100	97104	97109	97114	97118	97123	5
15 36	156 00	936	97128	97132	97137	97142	97146	97151	97155	97160	97165	97169	5
15 37	156 10	937	97174	97179	97183	97188	97193	97197	97202	97206	97211	97216	5
15 38	156 20	938	97220	97225	97230	97234	97239	97243	97248	97253	97257	97262	5
15 39	156 30	939	97267	97271	97276	97280	97285	97290	97294	97299	97304	97308	5
Angles		No.	0	1	2	3	4	5	6	7	8	9	D.

For fifth-figure differences see page 174.

LOGARITHMS

Angles			No. 9400——9999					Log. 97313——99996					
2·	3·	**No.**	**0**	**1**	**2**	**3**	**4**	**5**	**6**	**7**	**8**	**9**	**D.**
° ′	° ′												
15 40	156 40	**940**	97313	97317	97322	97327	97331	97336	97341	97345	97350	97354	5
15 41	156 50	**941**	97359	97364	97368	97373	97377	97382	97387	97391	97396	97401	5
15 42	157 00	**942**	97405	97410	97414	97419	97424	97428	97433	97437	97442	97447	5
15 43	157 10	**943**	97451	97456	97460	97465	97470	97474	97479	97483	97488	97493	5
15 44	157 20	**944**	97497	97502	97506	97511	97516	97520	97525	97529	97534	97539	5
15 45	157 30	**945**	97543	97548	97552	97557	97562	97566	97571	97575	97580	97585	5
15 46	157 40	**946**	97589	97594	97598	97603	97608	97612	97617	97621	97626	97630	5
15 47	157 50	**947**	97635	97640	97644	97649	97653	97658	97663	97667	97672	97676	5
15 48	158 00	**948**	97681	97685	97690	97695	97699	97704	97708	97713	97718	97722	5
15 49	158 10	**949**	97727	97731	97736	97740	97745	97750	97754	97759	97763	97768	5
15 50	158 20	**950**	97772	97777	97782	97786	97791	97795	97800	97804	97809	97814	5
15 51	158 30	**951**	97818	97823	97827	97832	97836	97841	97845	97850	97855	97859	5
15 52	158 40	**952**	97864	97868	97873	97877	97882	97887	97891	97896	97900	97905	5
15 53	158 50	**953**	97909	97914	97918	97923	97928	97932	97937	97941	97946	97950	5
15 54	159 00	**954**	97955	97959	97964	97969	97973	97978	97982	97987	97991	97996	5
15 55	159 10	**955**	98000	98005	98009	98014	98019	98023	98028	98032	98037	98041	5
15 56	159 20	**956**	98046	98050	98055	98059	98064	98069	98073	98078	98082	98087	5
15 57	159 30	**957**	98091	98096	98100	98105	98109	98114	98118	98123	98128	98132	5
15 58	159 40	**958**	98137	98141	98146	98150	98155	98159	98164	98168	98173	98177	5
15 59	159 50	**959**	98182	98186	98191	98195	98200	98205	98209	98214	98218	98223	5
16 00	160 00	**960**	98227	98232	98236	98241	98245	98250	98254	98259	98263	98268	5
16 01	160 10	**961**	98272	98277	98281	98286	98290	98295	98299	98304	98309	98313	5
16 02	160 20	**962**	98318	98322	98327	98331	98336	98340	98345	98349	98354	98358	5
16 03	160 30	**963**	98363	98367	98372	98376	98381	98385	98390	98394	98399	98403	5
16 04	160 40	**964**	98408	98412	98417	98421	98426	98430	98435	98439	98444	98448	5
16 05	160 50	**965**	98453	98457	98462	98466	98471	98475	98480	98484	98489	98493	5
16 06	161 00	**966**	98498	98502	98507	98511	98516	98520	98525	98529	98534	98538	5
16 07	161 10	**967**	98543	98547	98552	98556	98561	98565	98570	98574	98579	98583	5
16 08	161 20	**968**	98588	98592	98597	98601	98606	98610	98614	98619	98623	98628	5
16 09	161 30	**969**	98632	98637	98641	98646	98650	98655	98659	98664	98668	98673	5
16 10	161 40	**970**	98677	98682	98686	98691	98695	98700	98704	98709	98713	98717	5
16 11	161 50	**971**	98722	98726	98731	98735	98740	98744	98749	98753	98758	98762	5
16 12	162 00	**972**	98767	98771	98776	98780	98785	98789	98793	98798	98802	98807	5
16 13	162 10	**973**	98811	98816	98820	98825	98829	98834	98838	98843	98847	98851	5
16 14	162 20	**974**	98856	98860	98865	98869	98874	98878	98883	98887	98892	98896	5
16 15	162 30	**975**	98901	98905	98909	98914	98918	98923	98927	98932	98936	98941	5
16 16	162 40	**976**	98945	98949	98954	98958	98963	98967	98972	98976	98981	98985	4
16 17	162 50	**977**	98990	98994	98998	99003	99007	99012	99016	99021	99025	99029	4
16 18	163 00	**978**	99034	99038	99043	99047	99052	99056	99061	99065	99069	99074	4
16 19	163 10	**979**	99078	99083	99087	99092	99096	99100	99105	99109	99114	99118	4
16 20	163 20	**980**	99123	99127	99132	99136	99140	99145	99149	99154	99158	99163	4
16 21	163 30	**981**	99167	99171	99176	99180	99185	99189	99193	99198	99202	99207	4
16 22	163 40	**982**	99211	99216	99220	99224	99229	99233	99238	99242	99247	99251	4
16 23	163 50	**983**	99255	99260	99264	99269	99273	99277	99282	99286	99291	99295	4
16 24	164 00	**984**	99300	99304	99308	99313	99317	99322	99326	99330	99335	99339	4
16 25	164 10	**985**	99344	99348	99352	99357	99361	99366	99370	99375	99379	99383	4
16 26	164 20	**986**	99388	99392	99397	99401	99405	99410	99414	99419	99423	99427	4
16 27	164 30	**987**	99432	99436	99441	99445	99449	99454	99458	99463	99467	99471	4
16 28	164 40	**988**	99476	99480	99485	99489	99493	99498	99502	99506	99511	99515	4
16 29	164 50	**989**	99520	99524	99528	99533	99537	99542	99546	99550	99555	99559	4
16 30	165 00	**990**	99564	99568	99572	99577	99581	99585	99590	99594	99599	99603	4
16 31	165 10	**991**	99607	99612	99616	99621	99625	99629	99634	99638	99642	99647	4
16 32	165 20	**992**	99651	99656	99660	99664	99669	99673	99677	99682	99686	99690	4
16 33	165 30	**993**	99695	99699	99704	99708	99712	99717	99721	99726	99730	99734	4
16 34	165 40	**994**	99739	99743	99747	99752	99756	99761	99765	99769	99774	99778	4
16 35	165 50	**995**	99782	99787	99791	99795	99800	99804	99809	99813	99817	99822	4
16 36	166 00	**996**	99826	99830	99835	99839	99843	99848	99852	99856	99861	99865	4
16 37	166 10	**997**	99870	99874	99878	99883	99887	99891	99896	99900	99904	99909	4
16 38	166 20	**998**	99913	99917	99922	99926	99931	99935	99939	99944	99948	99952	4
16 39	166 30	**999**	99957	99961	99965	99970	99974	99978	99983	99987	99991	99996	4
Angles		**No.**	**0**	**1**	**2**	**3**	**4**	**5**	**6**	**7**	**8**	**9**	**D.**

For fifth-figure differences see page 174.

LOG DIFFERENCES FOR FIFTH FIGURE

5th fig.	1	2	3	4	5	6	7	8	9	5th fig.
Diff. (D)	p.p.	p.p.	p.p.	p.p.	p.p.	p.p.	p.p.	p.p.	p.p.	**Diff. (D)**
43	4	9	13	17	22	26	30	34	39	**43**
42	4	8	13	17	21	25	29	34	38	**42**
41	4	8	12	16	21	25	29	33	37	**41**
40	4	8	12	16	20	24	28	32	36	**40**
39	4	8	12	16	20	23	27	31	35	**39**
38	4	8	11	15	19	23	27	30	34	**38**
37	4	7	11	15	19	22	26	30	33	**37**
36	4	7	11	14	18	22	25	29	32	**36**
35	4	7	11	14	18	21	25	28	32	**35**
34	3	7	10	14	17	20	24	27	31	**34**
33	3	7	10	13	17	20	23	26	30	**33**
32	3	6	10	13	16	19	22	26	29	**32**
31	3	6	9	12	16	19	22	25	28	**31**
30	3	6	9	12	15	18	21	24	27	**30**
29	3	6	9	12	15	17	20	23	26	**29**
28	3	6	8	11	14	17	20	22	25	**28**
27	3	5	8	11	14	16	19	22	24	**27**
26	3	5	8	10	13	16	18	21	23	**26**
25	3	5	8	10	13	15	18	20	23	**25**
24	2	5	7	10	12	14	17	19	22	**24**
23	2	5	7	9	12	14	16	18	21	**23**
22	2	4	7	9	11	13	15	18	20	**22**
21	2	4	6	8	11	13	15	17	19	**21**
20	2	4	6	8	10	12	14	16	18	**20**
19	2	4	6	8	10	11	13	15	17	**19**
18	2	4	5	7	9	11	13	14	16	**18**
17	2	3	5	7	9	10	12	14	15	**17**
16	2	3	5	6	8	10	11	13	14	**16**
15	2	3	5	6	8	9	11	12	14	**15**
14	1	3	4	6	7	8	10	11	13	**14**
13	1	3	4	5	7	8	9	10	12	**13**
12	1	2	4	5	6	7	8	10	11	**12**
11	1	2	3	4	6	7	8	9	10	**11**
10	1	2	3	4	5	6	7	8	9	**10**
9	1	2	3	4	5	5	6	7	8	**9**
8	1	2	2	3	4	5	6	6	7	**8**
7	1	1	2	3	4	4	5	6	6	**7**
6	1	1	2	2	3	4	4	5	5	**6**
5	1	1	2	2	3	3	4	4	5	**5**
4	0	1	1	2	2	2	3	3	4	**4**
Diff. (D)	p.p.	p.p.	p.p.	p.p.	p.p.	p.p.	p.p.	p.p.	p.p.	**Diff. (D)**
5th fig.	1	2	3	4	5	6	7	8	9	**5th fig.**

LOGARITHMS OF TRIGONOMETRIC FUNCTIONS

This table and the table of Haversines (pages 396 to 502) both conform to the long-established convention of tabular logarithms referred to on page xxviii of the Introduction. (See also Preface).

Those who prefer to use real logarithms will recognise that these tables can be adapted to suit their requirements by mentally subtracting 10 from each of the tabulated quantities. Thus, with respect to characteristics,

$$
\begin{array}{lll}
\text{for } 11 & \text{read} & 1 \\
,, \quad 10 & ,, & 0 \\
,, \quad 9 & ,, & \bar{1} \\
,, \quad 8 & ,, & \bar{2} \\
,, \quad 7 & ,, & \bar{3}
\end{array}
$$

0° *180°*			LOGS. OF TRIG. FUNCTIONS							
′	Sine	Diff.	Cosec.	Tan.	Diff.	Cotan.	Secant	Diff.	Cosine	
00·0	−∞		Infinite	−∞		Infinite	10·00000		10·00000	60′
1	5·46373	30103	14·53627	5·46373	30103	14·53627	10·00000		10·00000	
·2	5·76476	17609	14·23524	5·76476	17609	14·23524	10·00000		10·00000	
·3	5·94085	12494	14·05915	5·94085	12494	14·05915	10·00000		10·00000	
·4	6·06579	9691	13·93421	6·06579	9691	13·93421	10·00000		10·00000	
00·5	6·16270		13·83730	6·16270		13·83730	10·00000		10·00000	
·6	6·24188	7918	13·75812	6·24188	7918	13·75812	10·00000		10·00000	
·7	6·30882	6694	13·69118	6·30882	6694	13·69118	10·00000		10·00000	
·8	6·36682	5800	13·63318	6·36682	5800	13·63318	10·00000		10·00000	
·9	6·41797	5115	13·58203	6·41797	5115	13·58203	10·00000		10·00000	
01·0	6·46373	4576	13·53627	6·46373	4576	13·53627	10·00000		10·00000	59′
·1	6·50512	4139	13·49488	6·50512	4139	13·49488	10·00000		10·00000	
·2	6·54291	3779	13·45709	6·54291	3779	13·45709	10·00000		10·00000	
·3	6·57767	3476	13·42233	6·57767	3476	13·42233	10·00000		10·00000	
·4	6·60985	3218	13·39015	6·60985	3218	13·39015	10·00000		10·00000	
01·5	6·63982	2997	13·36018	6·63982	2997	13·36018	10·00000		10·00000	
·6	6·66785	2803	13·33215	6·66785	2803	13·33215	10·00000		10·00000	
·7	6·69418	2633	13·30582	6·69418	2633	13·30582	10·00000		10·00000	
·8	6·71900	2482	13·28100	6·71900	2482	13·28100	10·00000		10·00000	
·9	6·74248	2348	13·25752	6·74248	2348	13·25752	10·00000		10·00000	
02·0	6·76476	2228	13·23524	6·76476	2228	13·23524	10·00000		10·00000	58′
·1	6·78595	2119	13·21405	6·78595	2119	13·21405	10·00000		10·00000	
·2	6·80615	2020	13·19385	6·80615	2020	13·19385	10·00000		10·00000	
·3	6·82545	1930	13·17455	6·82545	1930	13·17455	10·00000		10·00000	
·4	6·84394	1849	13·15606	6·84394	1849	13·15606	10·00000		10·00000	
02·5	6·86167	1773	13·13833	6·86167	1773	13·13833	10·00000		10·00000	
·6	6·87870	1703	13·12130	6·87870	1703	13·12130	10·00000		10·00000	
·7	6·89509	1639	13·10491	6·89509	1639	13·10491	10·00000		10·00000	
·8	6·91088	1579	13·08912	6·91088	1579	13·08912	10·00000		10·00000	
·9	6·92612	1524	13·07388	6·92612	1524	13·07388	10·00000		10·00000	
03·0	6·94085	1473	13·05915	6·94085	1473	13·05915	10·00000		10·00000	57′
·1	6·95509	1424	13·04491	6·95509	1424	13·04491	10·00000		10·00000	
·2	6·96888	1379	13·03112	6·96888	1379	13·03112	10·00000		10·00000	
·3	6·98224	1336	13·01776	6·98224	1336	13·01776	10·00000		10·00000	
·4	6·99521	1297	13·00479	6·99521	1297	13·00479	10·00000		10·00000	
03·5	7·00779	1258	12·99221	7·00779	1258	12·99221	10·00000		10·00000	
·6	7·02003	1224	12·97997	7·02003	1224	12·97997	10·00000		10·00000	
·7	7·03193	1190	12·96807	7·03193	1190	12·96807	10·00000		10·00000	
·8	7·04351	1158	12·95649	7·04351	1158	12·95649	10·00000		10·00000	
·9	7·05479	1128	12·94521	7·05479	1128	12·94521	10·00000		10·00000	
04·0	7·06579	1100	12·93421	7·06579	1100	12·93421	10·00000		10·00000	56′
·1	7·07651	1072	12·92349	7·07651	1072	12·92349	10·00000		10·00000	
·2	7·08698	1047	12·91302	7·08698	1047	12·91302	10·00000		10·00000	
·3	7·09719	1021	12·90281	7·09719	1021	12·90281	10·00000		10·00000	
·4	7·10718	999	12·89282	7·10718	999	12·89282	10·00000		10·00000	
04·5	7·11694	976	12·88306	7·11694	976	12·88306	10·00000		10·00000	
·6	7·12648	954	12·87352	7·12648	954	12·87352	10·00000		10·00000	
·7	7·13582	934	12·86418	7·13582	934	12·86418	10·00000		10·00000	
·8	7·14497	915	12·85503	7·14497	915	12·85503	10·00000		10·00000	
·9	7·15392	895	12·84608	7·15392	895	12·84608	10·00000		10·00000	
05·0	7·16270	878	12·83730	7·16270	878	12·83730	10·00000		10·00000	55′
·1	7·17130	860	12·82870	7·17130	860	12·82870	10·00000		10·00000	
·2	7·17973	843	12·82027	7·17973	843	12·82027	10·00000		10·00000	
·3	7·18800	827	12·81200	7·18800	827	12·81200	10·00000		10·00000	
·4	7·19612	812	12·80388	7·19612	812	12·80388	10·00000		10·00000	
05·5	7·20409	797	12·79591	7·20409	797	12·79591	10·00000		10·00000	
·6	7·21191	782	12·78809	7·21191	782	12·78809	10·00000		10·00000	
·7	7·21960	769	12·78040	7·21960	769	12·78040	10·00000		10·00000	
·8	7·22715	755	12·77285	7·22715	755	12·77285	10·00000		10·00000	
·9	7·23458	743	12·76542	7·23458	743	12·76542	10·00000		10·00000	
06·0	7·24188	730	12·75812	7·24188	730	12·75812	10·00000		10·00000	54′

LOGS. OF TRIG. FUNCTIONS

,	Sine	Diff.	Cosec.	Tan.	Diff.	Cotan.	Secant	Diff.	Cosine	
06·0	7·24188		12·75812	7·24188		12·75812	10·00000		10·00000	*54'*
·1	7·24906	718	12·75094	7·24906	718	12·75094	10·00000		10·00000	
·2	7·25612	706	12·74388	7·25612	706	12·74388	10·00000		10·00000	
·3	7·26307	695	12·73693	7·26307	695	12·73693	10·00000		10·00000	
·4	7·26991	684	12·73009	7·26991	684	12·73009	10·00000		10·00000	
06·5	7·27664	673	12·72336	7·27664	673	12·72336	10·00000		10·00000	
·6	7·28327	663	12·71673	7·28327	663	12·71673	10·00000		10·00000	
·7	7·28980	653	12·71020	7·28980	653	12·71020	10·00000		10·00000	
·8	7·29624	644	12·70376	7·29624	644	12·70376	10·00000		10·00000	
·9	7·30258	634	12·69742	7·30258	634	12·69742	10·00000		10·00000	
07·0	7·30882	624	12·69118	7·30883	625	12·69118	10·00000		10·00000	*53'*
·1	7·31498	616	12·68502	7·31498	615	12·68502	10·00000		10·00000	
·2	7·32106	608	12·67894	7·32106	608	12·67894	10·00000		10·00000	
·3	7·32705	599	12·67295	7·32705	599	12·67295	10·00000		10·00000	
·4	7·33296	591	12·66704	7·33296	591	12·66704	10·00000		10·00000	
07·5	7·33879	583	12·66121	7·33879	583	12·66121	10·00000		10·00000	
·6	7·34454	575	12·65546	7·34454	575	12·65546	10·00000		10·00000	
·7	7·35022	568	12·64978	7·35022	568	12·64978	10·00000		10·00000	
·8	7·35582	560	12·64418	7·35582	560	12·64418	10·00000		10·00000	
·9	7·36135	553	12·63865	7·36135	553	12·63865	10·00000		10·00000	
08·0	7·36682	547	12·63318	7·36682	547	12·63318	10·00000		10·00000	*52'*
·1	7·37221	539	12·62779	7·37221	539	12·62779	10·00000		10·00000	
·2	7·37754	533	12·62246	7·37754	533	12·62246	10·00000		10·00000	
·3	7·38280	526	12·61720	7·38280	526	12·61720	10·00000		10·00000	
·4	7·38801	521	12·61199	7·38801	521	12·61199	10·00000		10·00000	
08·5	7·39315	514	12·60686	7·39315	514	12·60685	10·00000		10·00000	
·6	7·39822	507	12·60178	7·39822	507	12·60178	10·00000		10·00000	
·7	7·40325	503	12·59675	7·40325	503	12·59675	10·00000		10·00000	
·8	7·40821	496	12·59179	7·40821	496	12·59179	10·00000		10·00000	
·9	7·41312	491	12·58688	7·41312	491	12·58688	10·00000		10·00000	
09·0	7·41797	485	12·58203	7·41797	485	12·58203	10·00000		10·00000	*51'*
·1	7·42277	480	12·57723	7·42277	480	12·57723	10·00000		10·00000	
·2	7·42751	474	12·57249	7·42751	474	12·57249	10·00000		10·00000	
·3	7·43221	470	12·56779	7·43221	470	12·56779	10·00000		10·00000	
·4	7·43685	464	12·56315	7·43685	464	12·56315	10·00000		10·00000	
09·5	7·44145	460	12·55855	7·44145	460	12·55855	10·00000		10·00000	
·6	7·44600	455	12·55400	7·44600	455	12·55400	10·00000		10·00000	
·7	7·45050	450	12·54950	7·45050	450	12·54950	10·00000		10·00000	
·8	7·45495	445	12·54505	7·45495	445	12·54505	10·00000		10·00000	
·9	7·45936	441	12·54064	7·45936	441	12·54064	10·00000		10·00000	
10·0	7·46373	437	12·53628	7·46373	437	12·53627	10·00000		10·00000	*50'*
·1	7·46805	432	12·53195	7·46805	432	12·53195	10·00000		10·00000	
·2	7·47233	428	12·52767	7·47233	428	12·52767	10·00000		10·00000	
·3	7·47656	423	12·52344	7·47656	423	12·52344	10·00000		10·00000	
·4	7·48076	420	12·51924	7·48076	420	12·51924	10·00000		10·00000	
10·5	7·48492	416	12·51509	7·48492	416	12·51508	10·00000		10·00000	
·6	7·48903	411	12·51097	7·48903	411	12·51097	10·00000		10·00000	
·7	7·49311	408	12·50689	7·49311	408	12·50689	10·00000		10·00000	
·8	7·49715	404	12·50285	7·49715	404	12·50285	10·00000		10·00000	
·9	7·50115	400	12·49885	7·50115	400	12·49885	10·00000		10·00000	
11·0	7·50512	397	12·49488	7·50512	397	12·49488	10·00000		10·00000	*49'*
·1	7·50905	393	12·49095	7·50905	393	12·49095	10·00000		10·00000	
·2	7·51294	389	12·48706	7·51294	309	12·48706	10·00000		10·00000	
·3	7·51680	386	12·48320	7·51680	386	12·48320	10·00000		10·00000	
·4	7·52063	383	12·47937	7·52063	383	12·47937	10·00000		10·00000	
11·5	7·52442	379	12·47558	7·52443	380	12·47557	10·00000		10·00000	
·6	7·52818	376	12·47182	7·52818	375	12·47182	10·00000		10·00000	
·7	7·53191	373	12·46809	7·53191	373	12·46809	10·00000		10·00000	
·8	7·53561	370	12·46439	7·53561	370	12·46439	10·00000		10·00000	
·9	7·53927	366	12·46073	7·53927	366	12·46073	10·00000		10·00000	
12·0	7·54291	364	12·45709	7·54291	364	12·45709	10·00000		10·00000	*48'*

LOGS. OF TRIG. FUNCTIONS

| 0°
180° | | | | | | | | | |

′	Sine	Diff.	Cosec.	Tan.	Diff.	Cotan.	Secant	Diff.	Cosine	
12·0	7·54291		12·45709	7·54291		12·45709	10·00000		10·00000	48′
·2	7·55008	717	12·44992	7·55008	717	12·44992	10·00000		10·00000	
·4	7·55715	707	12·44285	7·55715	707	12·44285	10·00000		10·00000	
·6	7·56410	695	12·43590	7·56410	695	12·43590	10·00000		10·00000	
·8	7·57093	683	12·42917	7·57093	683	12·42917	10·00000		10·00000	
13·0	7·57767	674	12·42233	7·57767	674	12·42233	10·00000		10·00000	47′
·2	7·58430	663	12·41570	7·58430	663	12·41570	10·00000		10·00000	
·4	7·59083	653	12·40917	7·59083	653	12·40917	10·00000		10·00000	
·6	7·59726	643	12·40274	7·59726	643	12·40274	10·00000		10·00000	
·8	7·60360	634	12·39640	7·60360	634	12·39640	10·00000		10·00000	
14·0	7·60985	625	12·39015	7·60986	626	12·39014	10·00000		10·00000	46′
·2	7·61601	616	12·38399	7·61601	615	12·38399	10·00000		10·00000	
·4	7·62209	608	12·37791	7·62209	608	12·37791	10·00000		10·00000	
·6	7·62808	599	12·37192	7·62808	599	12·37192	10·00000		10·00000	
·8	7·63399	591	12·36601	7·63399	591	12·36601	10·00000		10·00000	
15·0	7·63982	583	12·36018	7·63982	583	12·36018	10·00000		10·00000	45′
·2	7·64557	575	12·35443	7·64557	575	12·35443	10·00000		10·00000	
·4	7·65124	567	12·34876	7·65124	567	12·34876	10·00000		10·00000	
·6	7·65685	561	12·34315	7·65685	561	12·34315	10·00000		10·00000	
·8	7·66238	553	12·33762	7·66238	553	12·33762	10·00000		10·00000	
16·0	7·66784	546	12·33216	7·66785	547	12·33215	10·00000		10·00000	44′
·2	7·67324	540	12·32676	7·67324	539	12·32676	10·00000		10·00000	
·4	7·67857	533	12·32143	7·67857	533	12·32143	10·00000		10·00000	
·6	7·68383	526	12·31617	7·68384	527	12·31616	10·00001		9·99999	
·8	7·68903	520	12·31097	7·68904	520	12·31096	10·00001		9·99999	
17·0	7·69417	514	12·30583	7·69418	514	12·30582	10·00001		9·99999	43′
·2	7·69925	508	12·30075	7·69926	508	12·30074	10·00001		9·99999	
·4	7·70427	502	12·29573	7·70428	502	12·29572	10·00001		9·99999	
·6	7·70924	497	12·29076	7·70925	497	12·29075	10·00001		9·99999	
·8	7·71414	490	12·28586	7·71415	490	12·28585	10·00001		9·99999	
18·0	7·71900	486	12·28100	7·71900	485	12·28100	10·00001		9·99999	42′
·2	7·72379	479	12·27621	7·72380	480	12·27620	10·00001		9·99999	
·4	7·72854	475	12·27146	7·72855	475	12·27145	10·00001		9·99999	
·6	7·73324	470	12·26676	7·73325	470	12·26675	10·00001		9·99999	
·8	7·73788	464	12·26212	7·73789	464	12·26211	10·00001		9·99999	
19·0	7·74248	460	12·25752	7·74248	459	12·25752	10·00001		9·99999	41′
·2	7·74702	454	12·25298	7·74703	455	12·25297	10·00001		9·99999	
·4	7·75152	450	12·24848	7·75153	450	12·24847	10·00001		9·99999	
·6	7·75598	446	12·24402	7·75599	446	12·24401	10·00001		9·99999	
·8	7·76039	441	12·23961	7·76040	441	12·23960	10·00001		9·99999	
20·0	7·76475	436	12·23525	7·76476	436	12·23524	10·00001		9·99999	40′
·2	7·76907	432	12·23093	7·76908	432	12·23092	10·00001		9·99999	
·4	7·77335	428	12·22665	7·77336	428	12·22664	10·00001		9·99999	
·6	7·77759	424	12·22241	7·77760	424	12·22240	10·00001		9·99999	
·8	7·78179	420	12·21821	7·78180	420	12·21820	10·00001		9·99999	
21·0	7·78594	415	12·21406	7·78595	415	12·21405	10·00001		9·99999	39′
·2	7·79006	412	12·20994	7·79007	412	12·20993	10·00001		9·99999	
·4	7·79414	408	12·20586	7·79415	408	12·20585	10·00001		9·99999	
·6	7·79818	404	12·20182	7·79819	404	12·20181	10·00001		9·99999	
·8	7·80218	400	12·19782	7·80219	400	12·19781	10·00001		9·99999	
22·0	7·80615	397	12·19385	7·80616	396	12·19385	10·00001		9·99999	38′
·2	7·81008	393	12·18992	7 81009	393	12 18991	10 00001		9 99999	
4	7 81397	389	12 18603	7·81398	389	12·18602	10·00001		9·99999	
·6	7·81783	386	12·18217	7·81784	386	12·18216	10·00001		9·99999	
·8	7·82166	383	12·17834	7·82167	383	12·17833	10·00001		9·99999	
23·0	7·82545	379	12·17455	7·82546	379	12·17454	10·00001		9·99999	37′
·2	7·82921	376	12·17079	7·82922	376	12·17078	10·00001		9·99999	
·4	7·83294	373	12·16706	7·83295	373	12·16705	10·00001		9·99999	
·6	7·83663	369	12·16337	7·83664	369	12·16336	10·00001		9·99999	
·8	7·84030	367	12·15970	7·84031	367	12·15969	10·00001		9·99999	
24·0	7·84393	363	12·15607	7·84394	363	12·15606	10·00001		9·99999	36′

LOGS. OF TRIG. FUNCTIONS

0°
180°

,	Sine	Diff.	Cosec.	Tan.	Diff.	Cotan.	Secant	Diff.	Cosine	
24·0	7·84393		12·15607	**7·84394**		12·15606	**10·00001**		9·99999	*36'*
·2	7·84754	361	12·15246	**7·84755**	361	12·15245	**10·00001**		9·99999	
·4	7·85111	357	12·14889	**7·85112**	357	12·14888	**10·00001**		9·99999	
·6	7·85466	355	12·14534	**7·85467**	355	12·14533	**10·00001**		9·99999	
·8	7·85817	351	12·14183	**7·85818**	351	12·14182	**10·00001**		9·99999	
25·0	7·86166	349	12·13834	**7·86167**	349	12·13833	**10·00001**		9·99999	*35'*
·2	7·86512	346	12·13488	**7·86513**	346	12·13487	**10·00001**		9·99999	
·4	7·86856	344	12·13144	**7·86857**	344	12·13143	**10·00001**		9·99999	
·6	7·87196	340	12·12804	**7·87197**	340	12·12803	**10·00001**		9·99999	
·8	7·87534	338	12·12466	**7·87535**	338	12·12465	**10·00001**		9·99999	
26·0	7·87870	336	12·12131	**7·87871**	336	12·12129	**10·00001**		9·99999	*34'*
·2	7·88202	333	12·11798	**7·88203**	332	12·11797	**10·00001**		9·99999	
·4	7·88533	331	12·11467	**7·88534**	331	12·11466	**10·00001**		9·99999	
·6	7·88860	327	12·11140	**7·88861**	327	12·11139	**10·00001**		9·99999	
·8	7·89186	326	12·10814	**7·89187**	326	12·10813	**10·00001**		9·99999	
27·0	7·89509	323	12·10492	**7·89510**	323	12·10490	**10·00001**		9·99999	*33'*
·2	7·89829	321	12·10171	**7·89830**	320	12·10170	**10·00001**		9·99999	
·4	7·90147	318	12·09853	**7·90148**	318	12·09852	**10·00001**		9·99999	
·6	7·90463	316	12·09537	**7·90464**	316	12·09536	**10·00001**		9·99999	
·8	7·90777	314	12·09223	**7·90778**	314	12·09222	**10·00001**		9·99999	
28·0	7·91088	311	12·08912	**7·91089**	311	12·08911	**10·00001**		9·99999	*32'*
·2	7·91397	309	12·08603	**7·91398**	309	12·08602	**10·00001**		9·99999	
·4	7·91704	307	12·08296	**7·91705**	307	12·08295	**10·00001**		9·99999	
·6	7·92009	305	12·07991	**7·92010**	305	12·07990	**10·00001**		9·99999	
·8	7·92311	302	12·07689	**7·92312**	302	12·07688	**10·00001**		9·99999	
29·0	7·92612	301	12·07388	**7·92613**	301	12·07387	**10·00001**		9·99999	*31'*
·2	7·92910	298	12·07090	**7·92912**	299	12·07088	**10·00002**		9·99998	
·4	7·93207	297	12·06793	**7·93209**	297	12·06791	**10·00002**		9·99998	
·6	7·93501	294	12·06499	**7·93503**	294	12·06497	**10·00002**		9·99998	
·8	7·93794	293	12·06206	**7·93796**	293	12·06204	**10·00002**		9·99998	
30·0	7·94084	290	12·05916	**7·94086**	290	12·05914	**10·00002**		9·99998	*30'*
·2	7·94373	289	12·05627	**7·94375**	289	12·05625	**10·00002**		9·99998	
·4	7·94659	286	12·05341	**7·94661**	286	12·05339	**10·00002**		9·99998	
·6	7·94944	285	12·05056	**7·94946**	285	12·05054	**10·00002**		9·99998	
·8	7·95227	283	12·04773	**7·95229**	283	12·04771	**10·00002**		9·99998	
31·0	7·95508	281	12·04492	**7·95510**	281	12·04490	**10·00002**		9·99998	*29'*
·2	7·95787	279	12·04213	**7·95789**	279	12·04211	**10·00002**		9·99998	
·4	7·96065	278	12·03935	**7·96067**	278	12·03933	**10·00002**		9·99998	
·6	7·96341	276	12·03659	**7·96343**	276	12·03657	**10·00002**		9·99998	
·8	7·96615	274	12·03385	**7·96617**	274	12·03383	**10·00002**		9·99998	
32·0	7·96887	272	12·03113	**7·96889**	272	12·03111	**10·00002**		9·99998	*28'*
·2	7·97158	271	12·02842	**7·97160**	271	12·02840	**10·00002**		9·99998	
·4	7·97426	268	12·02574	**7·97428**	268	12·02572	**10·00002**		9·99998	
·6	7·97694	268	12·02306	**7·97696**	268	12·02304	**10·00002**		9·99998	
·8	7·97959	265	12·02041	**7·97961**	265	12·02039	**10·00002**		9·99998	
33·0	7·98223	264	12·01777	**7·98225**	264	12·01775	**10·00002**		9·99998	*27'*
·2	7·98486	263	12·01514	**7·98488**	263	12·01512	**10·00002**		9·99998	
·4	7·98747	261	12·01253	**7·98749**	261	12·01251	**10·00002**		9·99998	
·6	7·99006	259	12·00994	**7·99008**	259	12·00992	**10·00002**		9·99998	
·8	7·99264	258	12·00736	**7·99266**	258	12·00734	**10·00002**		9·99998	
34·0	7·99520	256	12·00480	**7·99522**	256	12·00478	**10·00002**		9·99998	*26'*
·2	7·99774	254	12·00226	**7·99776**	254	12·00224	**10·00002**		9·99998	
·4	8·00028	254	11·99972	**8·00030**	254	11·99970	**10·00002**		9·99998	
·6	8·00279	251	11·99721	**8·00281**	251	11·99719	**10·00002**		9·99998	
·8	8·00530	251	11·99470	**8·00532**	251	11·99468	**10·00002**		9·99998	
35·0	8·00779	249	11·99221	**8·00781**	249	11·99219	**10·00002**		9·99998	*25'*
·2	8·01026	247	11·98974	**8·01028**	247	11·98972	**10·00002**		9·99998	
·4	8·01272	246	11·98728	**8·01274**	246	11·98726	**10·00002**		9·99998	
·6	8·01517	245	11·98483	**8·01519**	245	11·98481	**10·00002**		9·99998	
·8	8·01760	243	11·98240	**8·01762**	243	11·98238	**10·00002**		9·99998	
36·0	8·02002	242	11·97998	**8·02005**	242	11·97996	**10·00002**		9·99998	*24'*

0°										
180°			**LOGS. OF TRIG. FUNCTIONS**							

′	Sine	Diff.	Cosec.	Tan.	Diff.	Cotan.	Secant	Diff.	Cosine	
36·0	8·02002		11·97998	8·02005		11·97996	10·00002		9·99998	24′
		241			241					
·2	8·02243		11·97757	8·02245		11·97755	10·00002		9·99998	
		239			239					
·4	8·02482		11·97518	8·02484		11·97516	10·00002		9·99998	
		238			238					
·6	8·02720		11·97280	8·02722		11·97278	10·00002		9·99998	
		237			237					
·8	8·02957		11·97043	8·02959		11·97041	10·00002		9·99998	
		235			236					
37·0	8·03192		11·96808	8·03195		11·96806	10·00003		9·99997	23′
		234			234					
·2	8·03426		11·96574	8·03429		11·96571	10·00003		9·99997	
		233			233					
·4	8·03659		11·96341	8·03662		11·96338	10·00003		9·99997	
		231			231					
·6	8·03890		11·96110	8·03893		11·96107	10·00003		9·99997	
		231			231					
·8	8·04121		11·95879	8·04124		11·95876	10·00003		9·99997	
		229			229					
38·0	8·04350		11·95650	8·04353		11·95647	10·00003		9·99997	22′
		228			228					
·2	8·04578		11·95422	8·04581		11·95419	10·00003		9·99997	
		227			227					
·4	8·04805		11·95195	8·04808		11·95192	10·00003		9·99997	
		225			225					
·6	8·05030		11·94970	8·05033		11·94967	10·00003		9·99997	
		225			225					
·8	8·05255		11·94745	8·05258		11·94742	10·00003		9·99997	
		223			223					
39·0	8·05478		11·94522	8·05481		11·94519	10·00003		9·99997	21′
		222			222					
·2	8·05700		11·94300	8·05703		11·94297	10·00003		9·99997	
		221			221					
·4	8·05921		11·94079	8·05924		11·94076	10·00003		9·99997	
		220			220					
·6	8·06141		11·93859	8·06144		11·93856	10·00003		9·99997	
		219			219					
·8	8·06360		11·93640	8·06363		11·93637	10·00003		9·99997	
		218			218					
40·0	8·06578		11·93422	8·06581		11·93419	10·00003		9·99997	20′
		216			216					
·2	8·06794		11·93206	8·06797		11·93203	10·00003		9·99997	
		216			216					
·4	8·07010		11·92990	8·07013		11·92987	10·00003		9·99997	
		214			214					
·6	8·07224		11·92776	8·07227		11·92773	10·00003		9·99997	
		214			214					
·8	8·07438		11·92562	8·07441		11·92559	10·00003		9·99997	
		212			212					
41·0	8·07650		11·92350	8·07653		11·92347	10·00003		9·99997	19′
		211			211					
·2	8·07861		11·92139	8·07864		11·92136	10·00003		9·99997	
		211			211					
·4	8·08072		11·91928	8·08075		11·91925	10·00003		9·99997	
		209			209					
·6	8·08281		11·91719	8·08284		11·91716	10·00003		9·99997	
		208			208					
·8	8·08489		11·91511	8·08492		11·91508	10·00003		9·99997	
		207			208					
42·0	8·08696		11·91304	8·08700		11·91300	10·00003		9·99997	18′
		207			206					
·2	8·08903		11·91097	8·08906		11·91094	10·00003		9·99997	
		205			205					
·4	8·09108		11·90892	8·09111		11·90889	10·00003		9·99997	
		204			204					
·6	8·09312		11·90688	8·09315		11·90685	10·00003		9·99997	
		204			204					
·8	8·09516		11·90484	8·09519		11·90481	10·00003		9·99997	
		202			203					
43·0	8·09718		11·90282	8·09722		11·90278	10·00003		9·99997	17′
		202			201					
·2	8·09920		11·90080	8·09923		11·90077	10·00003		9·99997	
		200			200					
·4	8·10120		11·89880	8·10123		11·89877	10·00003		9·99997	
		200			201					
·6	8·10320		11·89680	8·10324		11·89676	10·00004		9·99996	
		199			199					
·8	8·10519		11·89481	8·10523		11·89477	10·00004		9·99996	
		198			197					
44·0	8·10717		11·89283	8·10720		11·89280	10·00004		9·99996	16′
		197			198					
·2	8·10914		11·89086	8·10918		11·89082	10·00004		9·99996	
		196			196					
·4	8·11110		11·88890	8·11114		11·88886	10·00004		9·99996	
		195			195					
·6	8·11305		11·88695	8·11309		11·88691	10·00004		9·99996	
		194			194					
·8	8·11499		11·88501	8·11503		11·88497	10·00004		9·99996	
		194			193					
45·0	8·11693		11·88307	8·11696		11·88304	10·00004		9·99996	15′
		192			193					
·2	8·11885		11·88115	8·11889		11·88111	10·00004		9·99996	
		192			192					
·4	8·12077		11·87923	8·12081		11·87919	10·00004		9·99996	
		191			191					
·6	8·12268		11·87732	8·12272		11·87728	10·00004		9·99996	
		190			190					
·8	8·12458		11·87542	8·12462		11·87538	10·00004		9·99996	
		189			189					
46·0	8·12647		11·87353	8·12651		11·87349	10·00004		9·99996	14′
		188			188					
·2	8·12835		11·87165	8·12839		11·87161	10·00004		9·99996	
		188			188					
·4	8·13023		11·86977	8·13027		11·86973	10·00004		9·99996	
		187			187					
·6	8·13210		11·86790	8·13214		11·87686	10·00004		9·99996	
		186			186					
·8	8·13396		11·86604	8·13400		11·86600	10·00004		9·99996	
		185			185					
47·0	8·13581		11·86419	8·13585		11·86415	10·00004		9·99996	13′
		184			184					
·2	8·13765		11·86235	8·13769		11·86231	10·00004		9·99996	
		184			184					
·4	8·13949		11·86051	8·13953		11·86047	10·00004		9·99996	
		183			183					
·6	8·14132		11·85868	8·14136		11·85864	10·00004		9·99996	
		182			182					
·8	8·14314		11·85686	8·14318		11·85682	10·00004		9·99996	
		181			182					
48·0	8·14495		11·85505	8·14500		11·85500	10·00004		9·99996	12′

0° 180°	LOGS. OF TRIG. FUNCTIONS									
′	Sine	Diff.	Cosec.	Tan.	Diff.	Cotan.	Secant	Diff.	Cosine	
48·0	8·14495		11·85505	8·14500		11·85500	10·00004		9·99996	*12′*
·2	8·14676	181	11·85324	8·14680	180	11·85320	10·00004		9·99996	
·4	8·14856	180	11·85144	8·14860	180	11·85140	10·00004		9·99996	
·6	8·15035	179	11·84965	8·15039	179	11·84961	10·00004		9·99996	
·8	8·15213	178	11·84787	8·15217	178	11·84783	10·00004		9·99996	
		178			178					
49·0	8·15391		11·84609	8·15395		11·84605	10·00004		9·99996	*11′*
·2	8·15568	177	11·84432	8·15572	177	11·84428	10·00004		9·99996	
·4	8·15744	176	11·84256	8·15748	176	11·84252	10·00004		9·99996	
·6	8·15919	175	11·84081	8·15924	176	11·84076	10·00005		9·99995	
·8	8·16094	175	11·83906	8·16099	175	11·83901	10·00005		9·99995	
		174			174					
50·0	8·16268		11·83732	8·16273		11·83727	10·00005		9·99995	*10′*
·2	8·16441	173	11·83559	8·16446	173	11·83554	10·00005		9·99995	
·4	8·16614	173	11·83386	8·16619	173	11·83381	10·00005		9·99995	
·6	8·16786	172	11·83214	8·16791	172	11·83209	10·00005		9·99995	
·8	8·16957	171	11·83043	8·16962	171	11·83038	10·00005		9·99995	
		171			171					
51·0	8·17128		11·82872	8·17133		11·82867	10·00005		9·99995	*9′*
·2	8·17298	170	11·82702	8·17303	170	11·82697	10·00005		9·99995	
·4	8·17467	169	11·82533	8·17472	169	11·82528	10·00005		9·99995	
·6	8·17636	169	11·82364	8·17641	169	11·82359	10·00005		9·99995	
·8	8·17804	168	11·82196	8·17809	168	11·82191	10·00005		9·99995	
		167			167					
52·0	8·17971		11·82029	8·17976		11·82024	10·00005		9·99995	*8′*
·2	8·18138	167	11·81862	8·18143	167	11·81857	10·00005		9·99995	
·4	8·18304	166	11·81696	8·18309	167	11·81691	10·00005		9·99995	
·6	8·18469	165	11·81531	8·18474	165	11·81526	10·00005		9·99995	
·8	8·18634	165	11·81366	8·18639	165	11·81361	10·00005		9·99995	
		164			165					
53·0	8·18798		11·81202	8·18804		11·81196	10·00005		9·99995	*7′*
·2	8·18962	164	11·81038	8·18967	163	11·81033	10·00005		9·99995	
·4	8·19125	163	11·80875	8·19130	163	11·80870	10·00005		9·99995	
·6	8·19287	162	11·80713	8·19292	162	11·80708	10·00005		9·99995	
·8	8·19449	162	11·80551	8·19454	162	11·80546	10·00005		9·99995	
		161			162					
54·0	8·19610		11·80390	8·19616		11·80384	10·00005		9·99995	*6′*
·2	8·19771	161	11·80229	8·19776	160	11·80224	10·00005		9·99995	
·4	8·19931	160	11·80069	8·19936	160	11·80064	10·00005		9·99995	
·6	8·20090	159	11·79910	8·20095	159	11·79905	10·00005		9·99995	
·8	8·20249	159	11·79751	8·20255	160	11·79745	10·00006		9·99994	
		158			158					
55·0	8·20407		11·79593	8·20413		11·79587	10·00006		9·99994	*5′*
·2	8·20565	158	11·79435	8·20571	158	11·79429	10·00006		9·99994	
·4	8·20722	157	11·79278	8·20728	157	11·79272	10·00006		9·99994	
·6	8·20878	156	11·79122	8·20884	156	11·79116	10·00006		9·99994	
·8	8·21034	156	11·78966	8·21040	156	11·78960	10·00006		9·99994	
		155			155					
56·0	8·21189		11·78811	8·21195		11·78805	10·00006		9·99994	*4′*
·2	8·21344	155	11·78656	8·21350	155	11·78650	10·00006		9·99994	
·4	8·21499	155	11·78501	8·21505	155	11·78495	10·00006		9·99994	
·6	8·21652	153	11·78348	8·21658	153	11·78342	10·00006		9·99994	
·8	8·21805	153	11·71895	8·21811	153	11·78189	10·00006		9·99994	
		153			153					
57·0	8·21958		11·78042	8·21964		11·78036	10·00006		9·99994	*3′*
·2	8·22110	152	11·77890	8·22116	152	11·77884	10·00006		9·99994	
·4	8·22262	152	11·77738	8·22268	152	11·77732	10·00006		9·99994	
·6	8·22413	151	11·77587	8·22419	151	11·77581	10·00006		9·99994	
·8	8·22563	150	11·77437	8·22569	150	11·77431	10·00006		9·99994	
		150			150					
58·0	8·22713		11·77287	8·22720		11·77281	10·00006		9·99994	*2′*
·2	8·22863	150	11·77137	8·22869	150	11·77131	10·00006		9·99994	
·4	8·23012	149	11·76988	8·23018	149	11·76982	10·00006		9·99994	
·6	8·23160	148	11·76840	8·23166	148	11·76834	10·00006		9·99994	
·8	8·23308	148	11·76692	8·23314	148	11·76686	10·00006		9·99994	
		148			148					
59·0	8·23456		11·76544	8·23462		11·76538	10·00006		9·99994	*1′*
·2	8·23603	147	11·76397	8·23609	147	11 76391	10·00006		9·99994	
·4	8·23749	146	11·76251	8·23755	147	11·76245	10 00006		9 99994	
6	8 23895	146	11·76105	8·23902	146	11·76098	10·00007		9·99993	
·8	8·24041	146	11·75959	8·24048	146	11·75952	10·00007		9·99993	
		145			146					
60·0	8·24186		11·75815	8·24192		11·75808	10·00007		9·99993	*0′*

LOGS. OF TRIG. FUNCTIONS

1° / **181°**

′	Sine	Diff.	Cosec.	Tan.	Diff.	Cotan.	Secant	Diff.	Cosine	
00·0	8·24186	144	11·75815	8·24192	145	11·75808	10·00007		9·99993	60′
·2	8·24330	144	11·75670	8·24337	144	11·75663	10·00007		9·99993	
·4	8·24474	144	11·75526	8·24481	144	11·75519	10·00007		9·99993	
·6	8·24618	143	11·75382	8·24625	144	11·75375	10·00007		9·99993	
·8	8·24761	142	11·75239	8·24768	143	11·75232	10·00007		9·99993	
01·0	8·24903	143	11·75097	8·24910	142	11·75090	10·00007		9·99993	59′
·2	8·25046	141	11·74954	8·25053	143	11·74947	10·00007		9·99993	
·4	8·25187	141	11·74813	8·25194	141	11·74806	10·00007		9·99993	
·6	8·25328	141	11·74672	8·25335	141	11·74665	10·00007		9·99993	
·8	8·25469	140	11·74531	8·25476	141	11·74524	10·00007		9·99993	
02·0	8·25609	140	11·74391	8·25617	140	11·74384	10·00007		9·99993	58′
·2	8·25749	140	11·74251	8·25756	140	11·74244	10·00007		9·99993	
·4	8·25889	139	11·74111	8·25896	140	11·74104	10·00007		9·99993	
·6	8·26028	138	11·73972	8·26035	139	11·73965	10·00007		9·99993	
·8	8·26166	138	11·73834	8·26173	138	11·73827	10·00007		9·99993	
03·0	8·26304	138	11·73696	8·26312	138	11·73689	10·00007		9·99993	57′
·2	8·26442	137	11·73558	8·26449	137	11·73551	10·00007		9·99993	
·4	8·26579	137	11·73421	8·26586	138	11·73414	10·00007		9·99993	
·6	8·26716	136	11·73284	8·26724	136	11·73276	10·00008		9·99992	
·8	8·26852	136	11·73148	8·26860	136	11·73140	10·00008		9·99992	
04·0	8·26988	136	11·73012	8·26996	136	11·73004	10·00008		9·99992	56′
·2	8·27124	135	11·72876	8·27132	135	11·72868	10·00008		9·99992	
·4	8·27259	134	11·72741	8·27267	134	11·72733	10·00008		9·99992	
·6	8·27393	135	11·72607	8·27401	135	11·72599	10·00008		9·99992	
·8	8·27528	133	11·72472	8·27536	133	11·72464	10·00008		9·99992	
05·0	8·27661	134	11·72339	8·27669	134	11·72331	10·00008		9·99992	55′
·2	8·27795	133	11·72205	8·27803	133	11·72197	10·00008		9·99992	
·4	8·27928	132	11·72072	8·27936	132	11·72064	10·00008		9·99992	
·6	8·28060	133	11·71940	8·28068	133	11·71932	10·00008		9·99992	
·8	8·28193	131	11·71807	8·28201	131	11·71799	10·00008		9·99992	
06·0	8·28324	132	11·71676	8·28332	132	11·71668	10·00008		9·99992	54′
·2	8·28456	131	11·71544	8·28464	131	11·71536	10·00008		9·99992	
·4	8·28587	130	11·71413	8·28595	130	11·71405	10·00008		9·99992	
·6	8·28717	130	11·71283	8·28725	130	11·71275	10·00008		9·99992	
·8	8·28847	130	11·71153	8·28855	131	11·71145	10·00008		9·99992	
07·0	8·28977	130	11·71023	8·28986	129	11·71014	10·00008		9·99992	53′
·2	8·29107	129	11·70893	8·29115	129	11·70885	10·00008		9·99992	
·4	8·29236	128	11·70764	8·29244	128	11·70756	10·00008		9·99992	
·6	8·29364	129	11·70636	8·29372	129	11·70628	10·00008		9·99992	
·8	8·29493	128	11·70507	8·29501	128	11·70499	10·00008		9·99992	
08·0	8·29621	127	11·70379	8·29629	128	11·70371	10·00008		9·99992	52′
·2	8·29748	127	11·70252	8·29757	127	11·70243	10·00009		9·99991	
·4	8·29875	127	11·70125	8·29884	127	11·70116	10·00009		9·99991	
·6	8·30002	127	11·69998	8·30011	127	11·69989	10·00009		9·99991	
·8	8·30129	126	11·69871	8·30138	125	11·69862	10·00009		9·99991	
09·0	8·30255	125	11·69745	8·30263	126	11·69737	10·00009		9·99991	51′
·2	8·30380	126	11·69620	8·30389	126	11·69611	10·00009		9·99991	
·4	8·30506	125	11·69494	8·30515	125	11·69485	10·00009		9·99991	
·6	8·30631	124	11·69369	8·30640	124	11·69360	10·00009		9·99991	
·8	8·30755	124	11·69245	8·30764	124	11·69236	10·00009		9·99991	
10·0	8·30879	124	11·69121	8·30888	124	11·69112	10·00009		9·99991	50′
·2	8·31003	124	11·68997	8·31012	124	11·68988	10·00009		9·99991	
·4	8·31127	123	11·68873	8·31136	123	11·68864	10·00009		9·99991	
·6	8·31250	123	11·68750	8·31259	123	11·68741	10·00009		9·99991	
·8	8·31373	122	11·68627	8·31382	123	11·68618	10·00009		9·99991	
11·0	8·31495	122	11·68505	8·31505	121	11·68495	10·00009		9·99991	49′
·2	8·31617	122	11·68383	8·31626	122	11·68374	10·00009		9·99991	
·4	8·31739	122	11·68261	8·31748	122	11·68252	10·00009		9·99991	
·6	8·31861	121	11·68139	8·31870	121	11·68130	10·00009		9·99991	
·8	8·31982	121	11·68018	8·31991	121	11·68009	10·00009		9·99991	
12·0	8·32103		11·67897	8·32112		11·67888	10·00010		9·99990	48′

178° / **358°**

LOGS. OF TRIG. FUNCTIONS

1° 181°	Sine	Diff.	Cosec.	Tan.	Diff.	Cotan.	Secant	Diff.	Cosine	
12·0	8·32103		11·67897	8·32112		11·67888	**10·00010**		9·99990	*48'*
·2	8·32223	120	11·67777	8·32233	121	11·67767	**10·00010**		9·99990	
·4	8·32343	120	11·67657	8·32353	120	11·67647	**10·00010**		9·99990	
·6	8·32463	120	11·67537	8·32473	120	11·67527	**10·00010**		9·99990	
·8	8·32582	119	11·67418	8·32592	119	11·67408	**10·00010**		9·99990	
		120			119					
13·0	8·32702		11·67298	8·32711		11·67289	**10·00010**		9·99990	*47'*
·2	8·32820	118	11·67180	8·32830	119	11·67170	**10·00010**		9·99990	
·4	8·32939	119	11·67061	8·32949	119	11·67051	**10·00010**		9·99990	
·6	8·33057	118	11·66943	8·33067	118	11·66933	**10·00010**		9·99990	
·8	8·33175	118	11·66825	8·33185	118	11·66815	**10·00010**		9·99990	
		117			117					
14·0	8·33292		11·66708	8·33302		11·66698	**10·00010**		9·99990	*46'*
·2	8·33410	118	11·66590	8·33420	118	11·66580	**10·00010**		9·99990	
·4	8·33527	117	11·66473	8·33537	117	11·66463	**10·00010**		9·99990	
·6	8·33643	116	11·66357	8·33653	116	11·66347	**10·00010**		9·99990	
·8	8·33759	116	11·66241	8·33769	116	11·66231	**10·00010**		9·99990	
		116			117					
15·0	8·33875		11·66125	8·33886		11·66114	**10·00010**		9·99990	*45'*
·2	8·33991	116	11·66009	8·34001	115	11·65999	**10·00010**		9·99990	
·4	8·34106	115	11·65894	8·34116	115	11·65884	**10·00010**		9·99990	
·6	8·34221	115	11·65779	8·34232	116	11·65768	**10·00011**		9·99989	
·8	8·34336	115	11·65664	8·34347	115	11·65653	**10·00011**		9·99989	
		114			114					
16·0	8·34450		11·65550	8·34461		11·65539	**10·00011**		9·99989	*44'*
·2	8·34565	115	11·65435	8·34576	115	11·65424	**10·00011**		9·99989	
·4	8·34678	113	11·65322	8·34689	113	11·65311	**10·00011**		9·99989	
·6	8·34792	114	11·65208	8·34803	114	11·65197	**10·00011**		9·99989	
·8	8·34905	113	11·65095	8·34916	113	11·65084	**10·00011**		9·99989	
		113			113					
17·0	8·35018		11·64982	8·35029		11·64971	**10·00011**		9·99989	*43'*
·2	8·35131	113	11·64869	8·35142	113	11·64858	**10·00011**		9·99989	
·4	8·35243	112	11·64757	8·35254	112	11·64746	**10·00011**		9·99989	
·6	8·35355	112	11·64645	8·35366	112	11·64634	**10·00011**		9·99989	
·8	8·35467	112	11·64533	8·35478	112	11·64522	**10·00011**		9 99989	
		111			112					
18·0	8·35578		11·64422	8·35590		11·64410	**10·00011**		9·99989	*42'*
·2	8·35690	112	11·64310	8·35701	111	11·64299	**10·00011**		9·99989	
·4	8·35801	111	11·64199	8·35812	111	11·64188	**10·00011**		9·99989	
·6	8·35911	110	11·64089	8·35923	111	11·64077	**10·00011**		9·99989	
·8	8·36021	110	11·63979	8·36032	109	11·63968	**10·00011**		9·99989	
		110			111					
19·0	8·36131		11·63869	8·36143		11·63857	**10·00012**		9·99988	*41'*
·2	8·36241	110	11·63759	8·36253	110	11·63747	**10·00012**		9·99988	
·4	8·36351	110	11·63649	8·36363	110	11·63637	**10·00012**		9·99988	
·6	8·36460	109	11·63540	8·36472	109	11·63528	**10·00012**		9·99988	
·8	8·36569	109	11·63431	8·36581	109	11·63419	**10·00012**		9·99988	
		109			109					
20·0	8·36678		11·63322	8·36690		11·63310	**10·00012**		9·99988	*40'*
·2	8·36786	108	11·63214	8·36798	108	11·63202	**10·00012**		9·99988	
·4	8·36894	108	11·63106	8·36906	108	11·63094	**10·00012**		9·99988	
·6	8·37002	108	11·62998	8·37014	108	11·62986	**10·00012**		9·99988	
·8	8·37110	108	11·62890	8·37122	108	11·62878	**10·00012**		9·99988	
		107			107					
21·0	8·37217		11·62783	8·37229		11·62771	**10·00012**		9·99988	*39'*
·2	8·37324	107	11·62676	8·37336	107	11·62664	**10·00012**		9·99988	
·4	8·37431	107	11·62569	8·37443	107	11·62557	**10·00012**		9·99988	
·6	8·37538	107	11·62462	8·37550	107	11·62450	**10·00012**		9·99988	
·8	8·37644	106	11·62356	8·37656	106	11·62344	**10·00012**		9·99988	
		106			106					
22·0	8·37750		11·62250	8·37762		11·62238	**10·00012**		9·99988	*38'*
·2	8·37856	106	11·62144	8·37868	106	11·62132	**10·00012**		9·99988	
·4	0·37961	105	11·62039	8·37973	105	11·62027	**10·00012**		9·99988	
·6	8·38066	105	11·61934	8·38079	106	11·61921	**10·00013**		9·99987	
·8	8·38171	105	11·61829	8·38184	105	11·61816	**10·00013**		9·99987	
		105			105					
23·0	8·38276		11·61724	8·38289		11·61711	**10·00013**		9·99987	*37'*
·2	8·38381	105	11·61619	8·38394	105	11·61606	**10·00013**		9·99987	
·4	8·38485	104	11·61515	8·38498	104	11·61502	**10·00013**		9·99987	
·6	8·38589	104	11·61411	8·38602	104	11·61398	**10·00013**		9·99987	
·8	8·38693	104	11·61307	8·38706	104	11·61294	**10·00013**		9·99987	
		103			103					
24·0	8·38796		11·61204	8·38809		11·61191	**10·00013**		9·99987	*36'*

1° / 181°

LOGS. OF TRIG. FUNCTIONS

,	Sine	Diff.	Cosec.	Tan.	Diff.	Cotan.	Secant	Diff.	Cosine	
24·0	8·38796		11·61204	8·38809		11·61191	10·00013		9·99987	36'
		103			103					
·2	8·38899	103	11·61101	8·38912	103	11·61088	10·00013		9·99987	
·4	8·39002	103	11·60998	8·39015	103	11·60985	10·00013		9·99987	
·6	8·39105	103	11·60895	8·39118	103	11·60882	10·00013		9·99987	
·8	8·39208	102	11·60792	8·39221	102	11·60779	10·00013		9·99987	
25·0	8·39310	102	11·60690	8·39323	102	11·60677	10·00013		9·99987	35'
·2	8·39412	102	11·60588	8·39425	102	11·60575	10·00013		9·99987	
·4	8·39514	101	11·60486	8·39527	102	11·60473	10·00013		9·99987	
·6	8·39615	102	11·60385	8·39628	101	11·60372	10·00013		9·99987	
·8	8·39717	101	11·60283	8·39730	101	11·60270	10·00014		9·99987	
26·0	8·39818	101	11·60182	8·39832	102	11·60169	10·00014		9·99986	34'
·2	8·39919	100	11·60081	8·39933	100	11·60067	10·00014		9·99986	
·4	8·40019	101	11·59981	8·40033	101	11·59967	10·00014		9·99986	
·6	8·40120	100	11·59880	8·40134	100	11·59866	10·00014		9·99986	
·8	8·40220	100	11·59780	8·40234	100	11·59766	10·00014		9·99986	
27·0	8·40320	100	11·59680	8·40334	100	11·59666	10·00014		9·99986	33'
·2	8·40420	99	11·59580	8·40434	99	11·59566	10·00014		9·99986	
·4	8·40519	99	11·59481	8·40533	99	11·59467	10·00014		9·99986	
·6	8·40618	99	11·59382	8·40632	99	11·59368	10·00014		9·99986	
·8	8·40717	99	11·59283	8·40731	99	11·59269	10·00014		9·99986	
28·0	8·40816	99	11·59184	8·40830	99	11·59170	10·00014		9·99986	32'
·2	8·40915	98	11·59085	8·40929	98	11·59071	10·00014		9·99986	
·4	8·41013	98	11·58987	8·41027	98	11·58973	10·00014		9·99986	
·6	8·41111	98	11·58889	8·41125	98	11·58875	10·00014		9·99986	
·8	8·41209	98	11·58791	8·41223	98	11·58777	10·00014		9·99986	
29·0	8·41307	97	11·58693	8·41321	98	11·58679	10·00015		9·99985	31'
·2	8·41404	97	11·58596	8·41419	97	11·58581	10·00015		9·99985	
·4	8·41501	98	11·58499	8·41516	98	11·58484	10·00015		9·99985	
·6	8·41599	96	11·58401	8·41614	96	11·58386	10·00015		9·99985	
·8	8·41695	97	11·58305	8·41710	97	11·58290	10·00015		9·99985	
30·0	8·41792	96	11·58208	8·41807	96	11·58193	10·00015		9·99985	30'
·2	8·41888	96	11·58112	8·41903	96	11·58097	10·00015		9·99985	
·4	8·41984	96	11·58016	8·41999	96	11·58001	10·00015		9·99985	
·6	8·42080	96	11·57920	8·42095	96	11·57905	10·00015		9·99985	
·8	8·42176	96	11·57824	8·42191	96	11·57809	10·00015		9·99985	
31·0	8·42272	95	11·57728	8·42287	95	11·57713	10·00015		9·99985	29'
·2	8·42367	95	11·57633	8·42382	95	11·57618	10·00015		9·99985	
·4	8·42462	95	11·57538	8·42477	95	11·57523	10·00015		9·99985	
·6	8·42557	95	11·57443	8·42572	95	11·57428	10·00015		9·99985	
·8	8·42652	94	11·57348	8·42667	95	11·57333	10·00015		9·99985	
32·0	8·42746	95	11·57254	8·42762	95	11·57238	10·00016		9·99984	28'
·2	8·42841	94	11'57159	8·42857	94	11·57143	10·00016		9·99984	
·4	8·42935	94	11·57065	8·42951	94	11·57049	10·00016		9·99984	
·6	8·43029	93	11·56971	8·43045	93	11·56955	10·00016		9·99984	
·8	8·43122	94	11·56878	8·43138	93	11·56862	10·00016		9·99984	
33·0	8·43216	93	11·56784	8·43232	94	11·56769	10·00016		9·99984	27'
·2	8·43309	93	11·56691	8·43325	93	11·56675	10·00016		9·99984	
·4	8·43402	93	11·56598	8·43418	93	11·56582	10·00016		9·99984	
·6	8·43495	93	11·56505	8·43511	93	11·56489	10·00016		9·99984	
·8	8·43588	92	11·56412	8·43604	92	11·56396	10·00016		9·99984	
34·0	8·43680	92	11·56320	8·43696	92	11·56304	10·00016		9·99984	26'
·2	8·43772	92	11·56228	8·43788	92	11·56212	10·00016		9·99984	
·4	8·43864	92	11·56136	8·43880	92	11·56120	10·00016		9·99984	
·6	8·43956	92	11·56044	8·43972	92	11·56028	10·00016		9·99984	
·8	8·44048	91	11·55952	8·44064	92	11·55936	10·00016		9·99984	
35·0	8·44139	92	11·55861	8·44156	92	11·55844	10·00017		9·99983	25'
·2	8·44231	91	11·55769	8·44248	91	11·55752	10·00017		9·99983	
·4	8·44322	91	11·55678	8·44339	91	11·55661	10·00017		9·99983	
·6	8·44413	91	11·55587	8·44430	91	11·55570	10·00017		9·99983	
·8	8·44504	90	11·55496	8·44521	90	11·55479	10·00017		9·99983	
36·0	8·44594		11·55406	8·44611		11·55389	10·00017		9·99983	24'

LOGS. OF TRIG. FUNCTIONS

1°
181°

′	Sine	Diff.	Cosec.	Tan.	Diff.	Cotan.	Secant	Diff.	Cosine	
36·0	8·44594	90	11·55406	8·44611	90	11·55389	10·00017		9·99983	24′
·2	8·44684	91	11·55316	8·44701	91	11·55299	10·00017		9·99983	
·4	8·44775	90	11·55225	8·44792	90	11·55208	10·00017		9·99983	
·6	8·44865	89	11·55135	8·44882	89	11·55118	10·00017		9·99983	
·8	8·44954	90	11·55046	8·44971	90	11·55029	10·00017		9·99983	
37·0	8·45044	90	11·54956	8·45061	90	11·54939	10·00017		9·99983	23′
·2	8·45134	89	11·54866	8·45151	89	11·54849	10·00017		9·99983	
·4	8·45223	89	11·54777	8·45240	90	11·54760	10·00017		9·99983	
·6	8·45312	89	11·54688	8·45330	89	11·54670	10·00018		9·99982	
·8	8·45401	88	11·54599	8·45419	88	11·54581	10·00018		9·99982	
38·0	8·45489	89	11·54511	8·45507	89	11·54493	10·00018		9·99982	22′
·2	8·45578	88	11·54422	8·45596	88	11·54404	10·00018		9·99982	
·4	8·45666	88	11·54334	8·45684	88	11·54316	10·00018		9·99982	
·6	8·45754	88	11·54246	8·45772	88	11·54228	10·00018		9·99982	
·8	8·45842	88	11·54158	8·45860	88	11·54140	10·00018		9·99982	
39·0	8·45930	88	11·54070	8·45948	88	11·54052	10·00018		9·99982	21′
·2	8·46018	87	11·53982	8·46036	87	11·53964	10·00018		9·99982	
·4	8·46105	88	11·53895	8·46123	88	11·53877	10·00018		9·99982	
·6	8·46193	87	11·53807	8·46211	87	11·53789	10·00018		9·99982	
·8	8·46280	86	11·53720	8·46298	87	11·53702	10·00018		9·99982	
40·0	8·46366	87	11·53634	8·46385	86	11·53615	10·00018		9·99982	20′
·2	8·46453	87	11·53547	8·46471	88	11·53529	10·00018		9·99982	
·4	8·46540	86	11·53460	8·46559	86	11·53441	10·00019		9·99981	
·6	8·46626	86	11·53374	8·46645	86	11·53355	10·00019		9·99981	
·8	8·46712	86	11·53288	8·46731	86	11·53269	10·00019		9·99981	
41·0	8·46799	86	11·53202	8·46817	86	11·53183	10·00019		9·99981	19′
·2	8·46884	86	11·53116	8·46903	86	11·53097	10·00019		9·99981	
·4	8·46970	86	11·53030	8·46989	86	11·53011	10·00019		9·99981	
·6	8·47056	85	11·52944	8·47075	85	11·52925	10·00019		9·99981	
·8	8·47141	85	11·52859	8·47160	85	11·52840	10·00019		9·99981	
42·0	8·47226	85	11·52774	8·47245	85	11·52755	10·00019		9·99981	18′
2	8 47311	85	11 52689	8 47330	85	11 52670	10 00019		9 99981	
4	8 47396	85	11 52604	8 47415	85	11 52585	10 00019		9·99981	
·6	8·47481	85	11·52519	8·47500	85	11·52500	10·00019		9·99981	
·8	8·47566	84	11·52434	8·47585	84	11·52415	10·00019		9·99981	
43·0	8·47650	84	11·52350	8·47669	85	11·52331	10·00019		9·99981	17′
·2	8·47734	84	11·52266	8·47754	84	11·52246	10·00020		9·99980	
·4	8·47818	84	11·52182	8·47838	84	11·52162	10·00020		9·99980	
·6	8·47902	84	11·52098	8·47922	84	11·52078	10·00020		9·99980	
·8	8·47986	83	11·52014	8·48006	84	11·51994	10·00020		9·99980	
44·0	8·48069	84	11·51931	8·48089	83	11·51911	10·00020		9·99980	16′
·2	8·48153	83	11·51847	8·48173	84	11·51827	10·00020		9·99980	
·4	8·48236	83	11·51764	8·48256	83	11·51744	10·00020		9·99980	
·6	8·48319	83	11·51681	8·48339	83	11·51661	10·00020		9·99980	
·8	8·43402	83	11·51598	8·48422	83	11·51578	10·00020		9·99980	
45·0	8·48485	82	11·51515	8·48505	83	11·51495	10·00020		9·99980	15′
·2	8·48567	83	11·51433	8·48587	82	11·51413	10·00020		9·99980	
·4	8·48650	82	11·51350	8·48670	83	11·51330	10·00020		9·99980	
·6	8·48732	82	11·51268	8·48753	83	11·51247	10·00021		9·99979	
·8	8·48814	82	11·51186	8·48835	82	11·51165	10·00021		9·99979	
46·0	8·48896	82	11·51104	8·48917	82	11·51083	10·00021		9·99979	14′
·2	8·48978	82	11·51022	8·48999	82	11·51001	10·00021		9·99979	
·4	8·49060	81	11·50940	8·49081	81	11·50919	10·00021		9·99979	
·6	8·49141	82	11·50859	8·49162	82	11·50838	10·00021		9·99979	
·8	8·49223	81	11·50777	8·49244	81	11·50756	10·00021		9·99979	
47·0	8·49304	81	11·50696	8·49325	81	11·50675	10·00021		9·99979	13′
·2	8·49385	81	11·50615	8·49406	81	11·50594	10·00021		9·99979	
·4	8·49466	81	11·50534	8·49487	81	11·50513	10·00021		9·99979	
·6	8·49547	80	11·50453	8·49568	80	11·50432	10·00021		9·99979	
·8	8·49627	81	11·50373	8·49648	81	11·50352	10·00021		9·99979	
48·0	8·49708		11·50292	8·49729		11·50271	10·00021		9·99979	12′

LOGS. OF TRIG. FUNCTIONS

1° / 181°

′	Sine	Diff.	Cosec.	Tan.	Diff.	Cotan.	Secant	Diff.	Cosine	
48·0	8·49708		11·50292	8·49729		11·50271	10·00021		9·99979	12′
·2	8·49788	80	11·50212	8·49810	81	11·50190	10·00022		9·99978	
·4	8·49868	80	11·50132	8·49890	80	11·50110	10·00022		9·99978	
·6	8·49948	80	11·50052	8·49970	80	11·50030	10·00022		9·99978	
·8	8·50028	80	11·49972	8·50050	80	11·49950	10·00022		9·99978	
49·0	8·50108	80	11·49892	8·50130	80	11·49870	10·00022		9·99978	11′
·2	8·50188	80	11·49812	8·50210	80	11·49790	10·00022		9·99978	
·4	8·50267	79	11·49733	8·50289	79	11·49711	10·00022		9·99978	
·6	8·50346	79	11·49654	8·50368	79	11·49632	10·00022		9·99978	
·8	8·50426	80	11·49574	8·50448	80	11·49552	10·00022		9·99978	
50·0	8·50504	78	11·49496	8·50527	79	11·49473	10·00022		9·99978	10′
·2	8·50583	79	11·49417	8·50605	78	11·49395	10·00022		9·99978	
·4	8·50662	79	11·49338	8·50684	79	11·49316	10·00022		9·99978	
·6	8·50741	79	11·49259	8·50764	80	11·49236	10·00023		9·99977	
·8	8·50819	78	11·49181	8·50842	78	11·49158	10·00023		9·99977	
51·0	8·50897	78	11·49103	8·50920	78	11·49080	10·00023		9·99977	9′
·2	8·50976	79	11·49024	8·50999	79	11·49001	10·00023		9·99977	
·4	8·51054	78	11·48946	8·51077	78	11·48923	10·00023		9·99977	
·6	8·51131	77	11·48869	8·51154	77	11·48846	10·00023		9·99977	
·8	8·51209	78	11·48791	8·51232	78	11·48768	10·00023		9·99977	
52·0	8·51287	78	11·48713	8·51310	78	11·48690	10·00023		9·99977	8′
·2	8·51364	77	11·48636	8·51387	77	11·48613	10·00023		9·99977	
·4	8·51442	78	11·48558	8·51465	78	11·48535	10·00023		9·99977	
·6	8·51519	77	11·48481	8·51542	77	11·48458	10·00023		9·99977	
·8	8·51596	77	11·48404	8·51619	77	11·48381	10·00023		9·99977	
53·0	8·51673	77	11·48327	8·51696	77	11·48304	10·00024		9·99977	7′
·2	8·51749	76	11·48251	8·51773	77	11·48227	10·00024		9·99976	
·4	8·51826	77	11·48174	8·51850	77	11·48150	10·00024		9·99976	
·6	8·51902	76	11·48098	8·51926	76	11·48074	10·00024		9·99976	
·8	8·51979	77	11·48021	8·52003	77	11·47997	10·00024		9·99976	
54·0	8·52055	76	11·47945	8·52079	76	11·47921	10·00024		9·99976	6′
·2	8·52131	76	11·47869	8·52155	76	11·47845	10·00024		9·99976	
·4	8·52207	76	11·47793	8·52231	76	11·47769	10·00024		9·99976	
·6	8·52283	76	11·47717	8·52307	76	11·47693	10·00024		9·99976	
·8	8·52359	76	11·47641	8·52383	76	11·47617	10·00024		9·99976	
55·0	8·52434	75	11·47566	8·52459	76	11·47541	10·00024		9·99976	5′
·2	8·52510	76	11·47490	8·52534	75	11·47466	10·00024		9·99976	
·4	8·52585	75	11·47415	8·52610	76	11·47390	10·00025		9·99975	
·6	8·52660	75	11·47340	8·52685	75	11·47315	10·00025		9·99975	
·8	8·52735	75	11·47265	8·52760	75	11·47240	10·00025		9·99975	
56·0	8·52810	75	11·47190	8·52835	75	11·47165	10·00025		9·99975	4′
·2	8·52885	75	11·47115	8·52910	75	11·47090	10·00025		9·99975	
·4	8·52960	75	11·47040	8·52985	75	11·47015	10·00025		9·99975	
·6	8·53034	74	11·46966	8·53059	74	11·46941	10·00025		9·99975	
·8	8·53109	75	11·46891	8·53134	75	11·46866	10·00025		9·99975	
57·0	8·53183	74	11·46817	8·53208	74	11·46792	10·00025		9·99975	3′
·2	8·53257	74	11·46743	8·53282	74	11·46718	10·00025		9·99975	
·4	8·53331	74	11·46669	8·53356	74	11·46644	10·00025		9·99975	
·6	8·53405	74	11·46595	8·53430	74	11·46570	10·00025		9·99975	
·8	8·53479	74	11·46521	8·53505	75	11·46495	10·00026		9·99974	
58·0	8·53552	73	11·46448	8·53578	73	11·46422	10·00026		9·99974	2′
·2	8·53626	74	11·46374	8·53652	74	11·46348	10·00026		9·99974	
·4	8·53699	73	11·46301	8·53725	73	11·46275	10·00026		9·99974	
·6	8·53773	74	11·46227	8·53799	74	11·46201	10·00026		9·99974	
·8	8·53846	73	11·46154	8·53872	73	11·46128	10·00026		9·99974	
59·0	8·53919	73	11·46081	8·53945	73	11·46055	10·00026		9·99974	1′
·2	8·53991	72	11·46009	8·54017	72	11·45983	10·00026		9·99974	
·4	8·54064	73	11·45936	8·54090	73	11·45910	10·00026		9·99974	
·6	8·54137	73	11·45863	8·54163	73	11·45837	10·00026		9·99974	
·8	8·54210	73	11·45790	8·54236	73	11·45764	10·00026		9·99974	
60·0	8·54282	72	11·45718	8·54308	72	11·45692	10·00027		9·99974	0′

2°
182°

LOGS. OF TRIG. FUNCTIONS

′	Sine	Diff.	Cosec.	Tan.	Diff.	Cotan.	Secant	Diff.	Cosine	
00·0	8·54282		11·45718	8·54308		11·45692	10·00027		9·99974	60′
·2	8·54354	72	11·45646	8·54381	73	11·45619	10·00027		9·99973	
·4	8·54426	72	11·45574	8·54453	72	11·45547	10·00027		9·99973	
·6	8·54498	72	11·45502	8·54525	72	11·45475	10·00027		9·99973	
·8	8·54570	72	11·45430	8·54597	72	11·45403	10·00027		9·99973	
01·0	8·54642	72	11·45358	8·54669	72	11·45331	10·00027		9·99973	59′
·2	8·54714	72	11·45286	8·54741	72	11·45259	10·00027		9·99973	
·4	8·54786	72	11·45214	8·54813	72	11·45187	10·00027		9·99973	
·6	8·54857	71	11·45143	8·54884	71	11·45116	10·00027		9·99973	
·8	8·54928	71	11·45072	8·54955	71	11·45045	10·00027		9·99973	
02·0	8·55000	71	11·45001	8·55027	72	11·44973	10·00027		9·99973	58′
·2	8·55071	72	11·44929	8·55098	71	11·44902	10·00027		9·99973	
·4	8·55142	71	11·44858	8·55170	72	11·44830	10·00028		9·99972	
·6	8·55212	70	11·44788	8·55240	70	11·44760	10·00028		9·99972	
·8	8·55283	71	11·44717	8·55311	71	11·44689	10·00028		9·99972	
03·0	8·55354	71	11·44646	8·55382	71	11·44618	10·00028		9·99972	57′
·2	8·55424	70	11·44576	8·55452	70	11·44548	10·00028		9·99972	
·4	8·55495	71	11·44505	8·55523	71	11·44477	10·00028		9·99972	
·6	8·55565	70	11·44435	8·55593	70	11·44407	10·00028		9·99972	
·8	8·55635	70	11·44365	8·55663	70	11·44337	10·00028		9·99972	
04·0	8·55705	70	11·44295	8·55734	71	11·44266	10·00028		9·99972	56′
·2	8·55775	70	11·44225	8·55803	69	11·44197	10·00028		9·99972	
·4	8·55845	70	11·44155	8·55873	70	11·44127	10·00028		9·99972	
·6	8·55915	70	11·44085	8·55944	71	11·44056	10·00029		9·99971	
·8	8·55984	69	11·44016	8·56013	69	11·43987	10·00029		9·99971	
05·0	8·56054	70	11·43946	8·56083	70	11·43917	10·00029		9·99971	55′
·2	8·56123	69	11·43877	8·56152	69	11·43848	10·00029		9·99971	
·4	8·56193	70	11·43807	8·56222	70	11·43778	10·00029		9·99971	
·6	8·56262	69	11·43738	8·56291	69	11·43709	10·00029		9·99971	
·8	8·56331	69	11·43669	8·56360	69	11·43640	10·00029		9·99971	
06·0	8·56400	69	11·43600	8·56429	69	11·43571	10·00029		9·99971	54′
·2	8·56469	69	11·43531	8·56498	69	11·43502	10·00029		9·99971	
·4	8·56538	69	11·43462	8·56567	69	11·43433	10·00029		9·99971	
·6	8·56606	68	11·43394	8·56636	69	11·43364	10·00030		9·99970	
·8	8·56675	69	11·43325	8·56705	69	11·43295	10·00030		9·99970	
07·0	8·56743	68	11·43257	8·56773	68	11·43227	10·00030		9·99970	53′
·2	8·56811	68	11·43189	8·56841	69	11·43159	10·00030		9·99970	
·4	8·56880	69	11·43120	8·56910	68	11·43090	10·00030		9·99970	
·6	8·56948	68	11·43052	8·56978	68	11·43022	10·00030		9·99970	
·8	8·57016	68	11·42984	8·57046	68	11·42954	10·00030		9·99970	
08·0	8·57084	68	11·42916	8·57114	67	11·42886	10·00030		9·99970	52′
·2	8·57151	67	11·42849	8·57181	68	11·42819	10·00030		9·99970	
·4	8·57219	68	11·42781	8·57249	68	11·42751	10·00030		9·99970	
·6	8·57287	68	11·42713	8·57317	68	11·42683	10·00030		9·99970	
·8	8·57354	67	11·42646	8·57385	67	11·42615	10·00031		9·99969	
09·0	8·57421	67	11·42579	8·57452	68	11·42548	10·00031		9·99969	51′
·2	8·57489	68	11·42511	8·57520	67	11·42480	10·00031		9·99969	
·4	8·57556	67	11·42444	8·57587	67	11·42413	10·00031		9·99969	
·6	8·57623	67	11·42377	8·57654	67	11·42346	10·00031		9·99969	
·8	8·57690	67	11·42310	8·57721	67	11·42279	10·00031		9·99969	
10·0	8·57757	67	11·42243	8·57788	66	11·42212	10·00031		9·99969	50′
·2	8·57823	67	11·42177	8·57854	67	11·42146	10·00031		9·99969	
·4	8·57890	67	11·42110	8·57921	67	11·42079	10·00031		9·99969	
·6	8·57957	66	11·42043	8·57988	66	11·42012	10·00031		9·99969	
·8	8·58023	66	11·41977	8·58054	67	11·41946	10·00031		9·99969	
11·0	8·58089	66	11·41911	8·58121	66	11·41879	10·00032		9·99969	49′
·2	8·58155	67	11·41845	8·58187	67	11·41813	10·00032		9·99968	
·4	8·58222	66	11·41778	8·58254	66	11·41746	10·00032		9·99968	
·6	8·58288	65	11·41712	8·58320	65	11·41680	10·00032		9·99968	
·8	8·58353	66	11·41647	8·58385	66	11·41615	10·00032		9·99968	
12·0	8·58419		11·41581	8·58451		11·41549	10·00032		9·99968	48′

2°   182°   LOGS. OF TRIG. FUNCTIONS

′	Sine	Diff.	Cosec.	Tan.	Diff.	Cotan.	Secant	Diff.	Cosine	
12·0	8·58419	66	11·41581	8·58451	66	11·41549	10·00032		9·99968	48′
·2	8·58485	66	11·41515	8·58517	66	11·41483	10·00032		9·99968	
·4	8·58551	65	11·41449	0·58583	65	11·41417	10·00032		9·99968	
·6	8·58616	66	11·41384	8·58648	66	11·41352	10·00032		9·99968	
·8	8·58682	65	11·41318	8·58714	65	11·41286	10·00032		9·99968	
13·0	8·58747	65	11·41253	8·58780	66	11·41221	10·00033		9·99968	47′
·2	8·58812	65	11·41188	8·58845	65	11·41155	10·00033		9·99967	
·4	8·58877	65	11·41123	8·58910	65	11·41090	10·00033		9·99967	
·6	8·58942	65	11·41058	8·58975	65	11·41025	10·00033		9·99967	
·8	8·59007	65	11·40993	8·59040	65	11·40960	10·00033		9·99967	
14·0	8·59072	65	11·40928	8·59105	65	11·40895	10·00033		9·99967	46′
·2	8·59137	65	11·40863	8·59170	65	11·40830	10·00033		9·99967	
·4	8·59202	64	11·40798	8·59235	64	11·40765	10·00033		9·99967	
·6	8·59266	64	11·40734	8·59299	64	11·40701	10·00033		9·99967	
·8	8·59330	65	11·40670	8·59363	65	11·40637	10·00033		9·99967	
15·0	8·59395	64	11·40605	8·59428	65	11·40572	10·00034		9·99967	45′
·2	8·59459	64	11·40541	8·59493	64	11·40507	10·00034		9·99966	
·4	8·59523	64	11·40477	8·59557	64	11·40443	10·00034		9·99966	
·6	8·59587	64	11·40413	8·59621	64	11·40379	10·00034		9·99966	
·8	8·59651	64	11·40349	8·59685	64	11·40315	10·00034		9·99966	
16·0	8·59715	64	11·40285	8·59749	64	11·40251	10·00034		9·99966	44′
·2	8·59779	64	11·40221	8·59813	64	11·40187	10·00034		9·99966	
·4	8·59843	63	11·40157	8·59877	63	11·40123	10·00034		9·99966	
·6	8·59906	64	11·40094	8·59940	64	11·40060	10·00034		9·99966	
·8	8·59970	63	11·40030	8·60004	64	11·39996	10·00034		9·99966	
17·0	8·60033	64	11·39967	8·60068	64	11·39932	10·00035		9·99966	43′
·2	8·60097	63	11·39903	8·60132	63	11·39868	10·00035		9·99965	
·4	8·60160	63	11·39840	8·60195	63	11·39805	10·00035		9·99965	
·6	8·60223	63	11·39777	8·60258	63	11·39742	10·00035		9·99965	
·8	8·60286	63	11·39714	8·60321	63	11·39679	10·00035		9·99965	
18·0	8·60349	63	11·39651	8·60384	63	11·39616	10·00035		9·99965	42′
·2	8·60412	63	11·39588	8·60447	63	11·39553	10·00035		9·99965	
·4	8·60475	62	11·39525	8·60510	62	11·39490	10·00035		9·99965	
·6	8·60537	63	11·39463	8·60572	63	11·39428	10·00035		9·99965	
·8	8·60600	62	11·39400	8·60635	63	11·39365	10·00035		9·99965	
19·0	8·60662	63	11·39338	8·60698	63	11·39302	10·00036		9·99965	41′
·2	8·60725	62	11·39275	8·60761	62	11·39239	10·00036		9·99964	
·4	8·60787	62	11·39213	8·60823	62	11·39177	10·00036		9·99964	
·6	8·60849	62	11·39151	8·60885	62	11·39115	10·00036		9·99964	
·8	8·60911	62	11·39089	8·60947	62	11·39053	10·00036		9·99964	
20·0	8·60973	62	11·39027	8·61009	62	11·38991	10·00036		9·99964	40′
·2	8·61035	62	11·38965	8·61071	62	11·38929	10·00036		9·99964	
·4	8·61097	62	11·38903	8·61133	62	11·38867	10·00036		9·99964	
·6	8·61159	62	11·38841	8·61195	62	11·38805	10·00036		9·99964	
·8	8·61221	61	11·38779	8·61257	62	11·38743	10·00036		9·99964	
21·0	8·61282	62	11·38718	8·61319	62	11·38681	10·00037		9·99964	39′
·2	8·61344	61	11·38656	8·61381	61	11·38619	10·00037		9·99963	
·4	8·61405	62	11·38595	8·61442	62	11·38558	10·00037		9·99963	
·6	8·61467	61	11·38533	8·61504	61	11·38496	10·00037		9·99963	
·8	8·61528	61	11·38472	8·61565	61	11·38435	10·00037		9·99963	
22·0	8·61589	61	11·38411	8·61626	61	11·38374	10·00037		9·99963	38′
·2	8·61650	61	11·38350	8·61687	61	11·38313	10·00037		9·99963	
·4	8·61711	61	11·38289	8·61748	61	11·38252	10·00037		9·99963	
·6	8·61772	61	11·38228	8·61809	61	11·38191	10·00037		9·99963	
·8	8·61833	61	11·38167	8·61870	61	11·38130	10·00037		9·99963	
23·0	8·61894	60	11·38106	8·61931	61	11·38069	10·00038		9·99962	37′
·2	8·61954	61	11·38046	8·61992	61	11·38008	10·00038		9·99962	
·4	8·62015	60	11·37985	8·62053	60	11·37947	10·00038		9·99962	
·6	8·62075	61	11·37925	8·62113	61	11·37887	10·00038		9·99962	
·8	8·62136	60	11·37864	8·62174	60	11·37826	10·00038		9·99962	
24·0	8·62196		11·37804	8·62234		11·37766	10·00038		9·99962	36′

LOGS. OF TRIG. FUNCTIONS

2° / *182°*

,	Sine	Diff.	Cosec.	Tan.	Diff.	Cotan.	Secant	Diff.	Cosine	
24·0	8·62196		11·37804	8·62234		11·37766	10·00038		9·99962	36'
·2	8·62256	60	11·37744	8·62294	60	11·37706	10·00038		9·99962	
·4	8·62317	61	11·37683	8·62355	61	11·37645	10·00038		9·99962	
·6	8·62377	60	11·37623	8·62415	60	11·37585	10·00038		9·99962	
·8	8·62437	60	11·37563	8·62476	61	11·37524	10·00039		9·99961	
		59			59					
25·0	8·62497		11·37504	8·62535		11·37465	10·00039		9·99961	35'
·2	8·62556	60	11·37444	8·62595	60	11·37405	10·00039		9·99961	
·4	8·62616	60	11·37384	8·62655	60	11·37345	10·00039		9·99961	
·6	8·62676	60	11·37324	8·62715	60	11·37285	10·00039		9·99961	
·8	8·62735	59	11·37265	8·62774	59	11·37226	10·00039		9·99961	
		60			60					
26·0	8·62795		11·37205	8·62834		11·37166	10·00039		9·99961	34'
·2	8·62854	59	11·37146	8·62893	59	11·37107	10·00039		9·99961	
·4	8·62914	60	11·37086	8·62953	60	11·37047	10·00039		9·99961	
·6	8·62973	59	11·37027	8·63012	59	11·36988	10·00039		9·99961	
·8	8·63032	59	11·36968	8·63072	60	11·36928	10·00040		9·99960	
		59			59					
27·0	8·63091		11·36909	8·63131		11·36869	10·00040		9·99960	33'
·2	8·63150	59	11·36850	8·63190	59	11·36810	10·00040		9·99960	
·4	8·63209	59	11·36791	8·63249	59	11·36751	10·00040		9·99960	
·6	8·63268	59	11·36732	8·63308	59	11·36692	10·00040		9·99960	
·8	8·63327	59	11·36673	8·63367	59	11·36633	10·00040		9·99960	
		58			59					
28·0	8·63385		11·36615	8·63426		11·36574	10·00040		9·99960	32'
·2	8·63444	59	11·36556	8·63484	58	11·36516	10·00040		9·99960	
·4	8·63503	59	11·36497	8·63543	59	11·36457	10·00040		9·99960	
·6	8·63561	58	11·36439	8·63602	59	11·36398	10·00041		9·99959	
·8	8·63619	58	11·36381	8·63660	58	11·36340	10·00041		9·99959	
		59			58					
29·0	8·63678		11·36322	8·63718		11·36282	10·00041		9·99959	31'
·2	8·63736	58	11·36264	8·63777	59	11·36223	10·00041		9·99959	
·4	8·63794	58	11·36206	8·63835	58	11·36165	10·00041		9·99959	
·6	8·63852	58	11·36148	8·63893	58	11·36107	10·00041		9·99959	
·8	8·63910	58	11·36090	8·63951	58	11·36049	10·00041		9·99959	
		58			58					
30·0	8·63968		11·36032	8·64009		11·35991	10·00041		9·99959	30'
·2	8·64026	58	11·35974	8·64067	58	11·35933	10·00041		9·99959	
·4	8·64083	57	11·35917	8·64125	58	11·35875	10·00042		9·99958	
·6	8·64141	58	11·35859	8·64183	58	11·35817	10·00042		9·99958	
·8	8·64199	58	11·35801	8·64241	58	11·35759	10·00042		9·99958	
		57			57					
31·0	8·64256		11·35744	8·64298		11·35702	10·00042		9·99958	29'
·2	8·64314	58	11·35686	8·64356	58	11·35644	10·00042		9·99958	
·4	8·64371	57	11·35629	8·64413	57	11·35587	10·00042		9·99958	
·6	8·64429	58	11·35571	8·64471	58	11·35529	10·00042		9·99958	
·8	8·64486	57	11·35514	8·64528	57	11·35472	10·00042		9·99958	
		57			57					
32·0	8·64543		11·35457	8·64585		11·35415	10·00042		9·99958	28'
·2	8·64600	57	11·35400	8·64643	58	11·35357	10·00043		9·99957	
·4	8·64657	57	11·35343	8·64700	57	11·35300	10·00043		9·99957	
·6	8·64714	57	11·35286	8·64757	57	11·35243	10·00043		9·99957	
·8	8·64771	57	11·35229	8·64814	57	11·35186	10·00043		9·99957	
		56			56					
33·0	8·64827		11·35173	8·64870		11·35130	10·00043		9·99957	27'
·2	8·64884	57	11·35116	8·64927	57	11·35073	10·00043		9·99957	
·4	8·64941	57	11·35059	8·64984	57	11·35016	10·00043		9·99957	
·6	8·64997	56	11·35003	8·65040	56	11·34960	10·00043		9·99957	
·8	8·65054	57	11·34946	8·65097	57	11·34903	10·00043		9·99957	
		56			57					
34·0	8·65110		11·34890	8·65154		11·34846	10·00044		9·99956	26'
·2	8·65166	56	11·34834	8·65210	56	11·34790	10·00044		9·99956	
·4	8·65223	57	11·34777	8·65267	57	11·34733	10·00044		9·99956	
·6	8·65279	56	11·34721	8·65323	56	11·34677	10·00044		9·99956	
·8	8·65335	56	11·34665	8·65379	56	11·34621	10·00044		9·99956	
		56			56					
35·0	8·65391		11·34609	8·65435		11·34565	10·00044		9·99956	25'
·2	8·65447	56	11·34553	8·65491	56	11·34509	10·00044		9·99956	
·4	8·65503	56	11·34497	8·65547	56	11·34453	10·00044		9·99956	
·6	8·65559	56	11·34441	8·65604	57	11·34396	10·00045		9·99955	
·8	8·65615	56	11·34385	8·65660	56	11·34340	10·00045		9·99955	
		55			55					
36·0	8·65670		11·34330	8·65715		11·34285	10·00045		9·99955	24'

2°
182°

LOGS. OF TRIG. FUNCTIONS

′	Sine	Diff.	Cosec.	Tan.	Diff.	Cotan.	Secant	Diff.	Cosine	
36·0	8·65670		11·34330	8·65715		11·34285	10·00045		9·99955	24′
·2	8·65726	56	11·34274	8·65771	56	11·34229	10·00045		9·99955	
·4	8·65781	55	11·34219	8·65826	55	11·34174	10·00045		9·99955	
·6	8·65837	56	11·34163	8·65882	56	11·34118	10·00045		9·99955	
·8	8·65892	55	11·34108	8·65937	55	11·34063	10·00045		9·99955	
37·0	8·65948	55	11·34053	8·65993	56	11·34007	10·00045		9·99955	23′
·2	8·66003	56	11·33997	8·66048	55	11·33952	10·00045		9·99955	
·4	8·66058	55	11·33942	8·66104	56	11·33896	10·00046		9·99954	
·6	8·66113	55	11·33887	8·66159	55	11·33841	10·00046		9·99954	
·8	8·66168	55	11·33832	8·66214	55	11·33786	10·00046		9·99954	
38·0	8·66223	55	11·33777	8·66269	55	11·33731	10·00046		9·99954	22′
·2	8·66278	55	11·33722	8·66324	55	11·33676	10·00046		9·99954	
·4	8·66333	55	11·33667	8·66379	55	11·33621	10·00046		9·99954	
·6	8·66388	55	11·33612	8·66434	55	11·33566	10·00046		9·99954	
·8	8·66442	54	11·33558	8·66488	54	11·33512	10·00046		9·99954	
39·0	8·66497	55	11·33503	8·66543	55	11·33457	10·00046		9·99954	21′
·2	8·66551	54	11·33449	8·66598	55	11·33402	10·00047		9·99953	
·4	8·66606	55	11·33394	8·66653	55	11·33347	10·00047		9·99953	
·6	8·66660	54	11·33340	8·66707	54	11·33293	10·00047		9·99953	
·8	8·66715	55	11·33285	8·66762	55	11·33238	10·00047		9·99953	
40·0	8·66769	54	11·33231	8·66816	54	11·33184	10·00047		9·99953	20′
·2	8·66823	54	11·33177	8·66870	54	11·33130	10·00047		9·99953	
·4	8·66877	54	11·33123	8·66924	54	11·33076	10·00047		9·99953	
·6	8·66931	54	11·33069	8·66978	54	11·33022	10·00047		9·99953	
·8	8·66985	54	11·33015	8·67032	54	11·32968	10·00047		9·99953	
41·0	8·67039	54	11·32961	8·67087	55	11·32913	10·00048		9·99952	19′
·2	8·67093	54	11·32907	8·67141	54	11·32859	10·00048		9·99952	
·4	8·67147	54	11·32853	8·67195	54	11·32805	10·00048		9·99952	
·6	8·67201	54	11·32799	8·67249	54	11·32751	10·00048		9·99952	
·8	8·67254	53	11·32746	8·67302	53	11·32698	10·00048		9·99952	
42·0	8·67308	54	11·32692	8·67356	54	11·32644	10·00048		9·99952	18′
·2	8·67362	54	11·32638	8·67410	54	11·32590	10·00048		9·99952	
·4	8·67415	53	11·32585	8·67463	53	11·32537	10·00048		9·99952	
·6	8·67468	53	11·32532	8·67517	54	11·32483	10·00049		9·99951	
·8	8·67522	54	11·32478	8·67571	54	11·32429	10·00049		9·99951	
43·0	8·67575	53	11·32425	8·67624	53	11·32376	10·00049		9·99951	17′
·2	8·67628	53	11·32372	8·67677	53	11·32323	10·00049		9·99951	
·4	8·67682	54	11·32318	8·67731	54	11·32269	10·00049		9·99951	
·6	8·67735	53	11·32265	8·67784	53	11·32216	10·00049		9·99951	
·8	8·67788	53	11·32212	8·67837	53	11·32163	10·00049		9·99951	
44·0	8·67841	52	11·32160	8·67890	53	11·32110	10·00049		9·99951	16′
·2	8·67893	53	11·32107	8·67942	52	11·32058	10·00049		9·99951	
·4	8·67946	53	11·32054	8·67996	54	11·32004	10·00050		9·99950	
·6	8·67999	53	11·32001	8·68049	53	11·31951	10·00050		9·99950	
·8	8·68052	53	11·31948	8·68102	53	11·31898	10·00050		9·99950	
45·0	8·68104	52	11·31896	8·68154	52	11·31846	10·00050		9·99950	15′
·2	8·68157	53	11·31843	8·68207	53	11·31793	10·00050		9·99950	
·4	8·68209	52	11·31791	8·68259	52	11·31741	10·00050		9·99950	
·6	8·68262	53	11·31738	8·68312	53	11·31688	10·00050		9·99950	
·8	8·68314	52	11·31686	8·68364	52	11·31636	10·00050		9·99950	
46·0	8·68367	52	11·31634	8·68417	53	11·31583	10·00051		9·99949	14′
·2	8·68419	53	11·31581	8·68470	53	11·31530	10·00051		9·99949	
·4	8·68471	52	11·31529	8·68522	52	11·31478	10·00051		9·99949	
·6	8·68523	52	11·31477	8·68574	52	11·31426	10·00051		9·99949	
·8	8·68575	52	11·31425	8·68626	52	11·31374	10·00051		9·99949	
47·0	8·68627	52	11·31373	8·68678	52	11·31322	10·00051		9·99949	13′
·2	8·68679	52	11·31321	8·68730	52	11·31270	10·00051		9·99949	
·4	8·68731	52	11·31269	8·68782	52	11·31218	10·00051		9·99949	
·6	8·68783	52	11·31217	8·68835	53	11·31165	10·00052		9·99948	
·8	8·68835	52	11·31165	8·68887	52	11·31113	10·00052		9·99948	
		51			51					
48·0	8·68886		11·31114	8·68938		11·31062	10·00052		9·99948	12′

LOGS. OF TRIG. FUNCTIONS

2°
182°

′	Sine	Diff.	Cosec.	Tan.	Diff.	Cotan.	Secant	Diff.	Cosine	
48·0	8·68886		11·31114	8·68938		11·31062	10·00052		9·99948	12′
·2	8·68938	52	11·31062	8·68990	52	11·31010	10·00052		9·99948	
·4	8·68989	51	11·31011	8·69041	51	11·30959	10·00052		9·99948	
·6	8·68041	52	11·30959	8·69093	52	11·30907	10·00052		9·99948	
·8	8·69092	51	11·30908	8·69144	51	11·30856	10·00052		9·99948	
49·0	8·69144	52	11·30856	8·69196	52	11·30804	10·00053		9·99948	11′
·2	8·69195	51	11·30805	8·69248	52	11·30752	10·00053		9·99947	
·4	8·69246	51	11·30754	8·69299	51	11·30701	10·00053		9·99947	
·6	8·69298	52	11·30702	8·69351	52	11·30649	10·00053		9·99947	
·8	8·69349	51	11·30651	8·69402	51	11·30598	10·00053		9·99947	
50·0	8·69400	51	11·30600	8·69453	51	11·30547	10·00053		9·99947	10′
·2	8·69451	51	11·30549	8·69504	51	11·30496	10·00053		9·99947	
·4	8·69502	51	11·30498	8·69556	52	11·30444	10·00054		9·99947	
·6	8·69553	51	11·30447	8·69607	51	11·30393	10·00054		9·99946	
·8	8·69604	51	11·30396	8·69658	51	11·30342	10·00054		9·99946	
51·0	8·69654	50	11·30346	8·69708	50	11·30292	10·00054		9·99946	9′
·2	8·69705	51	11·30295	8·69759	51	11·30241	10·00054		9·99946	
·4	8·69756	51	11·30244	8·69810	51	11·30190	10·00054		9·99946	
·6	8·69806	50	11·30194	8·69860	50	11·30140	10·00054		9·99946	
·8	8·69857	51	11·30143	8·69911	51	11·30089	10·00054		9·99946	
52·0	8·69907	50	11·30093	8·69962	51	11·30038	10·00054		9·99946	8′
·2	8·69958	51	11·30042	8·70013	51	11·29987	10·00055		9·99945	
·4	8·70008	50	11·29992	8·70063	50	11·29937	10·00055		9·99945	
·6	8·70058	50	11·29942	8·70113	50	11·29887	10·00055		9·99945	
·8	8·70109	51	11·29891	8·70164	51	11·29836	10·00055		9·99945	
53·0	8·70159	50	11·29841	8·70214	50	11·29786	10·00055		9·99945	7′
·2	8·70209	50	11·29791	8·70264	50	11·29736	10·00055		9·99945	
·4	8·70259	50	11·29741	8·70314	50	11·29686	10·00055		9·99945	
·6	8·70309	50	11·29691	8·70364	50	11·29636	10·00055		9·99945	
·8	8·70359	50	11·29641	8·70415	51	11·29585	10·00056		9·99944	
54·0	8·70409	50	11·29591	8·70465	50	11·29535	10·00056		9·99944	6′
·2	8·70459	50	11·29541	8·70515	50	11·29485	10·00056		9·99944	
·4	8·70509	50	11·29491	8·70565	50	11·29435	10·00056		9·99944	
·6	8·70559	50	11·29441	8·70615	50	11·29385	10·00056		9·99944	
·8	8·70608	49	11·29392	8·70664	49	11·29336	10·00056		9·99944	
55·0	8·70658	50	11·29342	8·70714	50	11·29286	10·00056		9·99944	5′
·2	8·70707	49	11·29293	8·70763	49	11·29237	10·00056		9·99944	
·4	8·70757	50	11·29243	8·70814	51	11·29186	10·00057		9·99943	
·6	8·70806	49	11·29194	8·70863	49	11·29137	10·00057		9·99943	
·8	8·70856	50	11·29144	8·70913	50	11·29087	10·00057		9·99943	
56·0	8·70905	49	11·29095	8·70962	49	11·29038	10·00057		9·99943	4′
·2	8·70954	49	11·29046	8·71011	49	11·28989	10·00057		9·99943	
·4	8·71003	50	11·28997	8·71060	50	11·28940	10·00057		9·99943	
·6	8·71053	49	11·28947	8·71110	49	11·28890	10·00057		9·99943	
·8	8·71102	49	11·28898	8·71159	49	11·28841	10·00057		9·99943	
57·0	8·71151	49	11·28849	8·71208	50	11·28791	10·00058		9·99942	3′
·2	8·71200	49	11·28800	8·71258	49	11·28742	10·00058		9·99942	
·4	8·71249	49	11·28751	8·71307	49	11·28693	10·00058		9·99942	
·6	8·71298	48	11·28702	8·71356	48	11·28644	10·00058		9·99942	
·8	8·71346	49	11·28654	8·71404	49	11·28596	10·00058		9·99942	
58·0	8·71395	49	11·28605	8·71453	49	11·28547	10·00058		9·99942	2′
·2	8·71444	49	11·28556	8·71502	49	11·28498	10·00058		9·99942	
·4	8·71493	48	11·28507	8·71551	49	11·28449	10·00058		9·99942	
·6	8·71541	49	11·28459	8·71600	49	11·28400	10·00059		9·99941	
·8	8·71590	48	11·28410	8·71649	48	11·28351	10·00059		9·99941	
59·0	8·71638	49	11·28362	8·71697	49	11·28303	10·00059		9·99941	1′
·2	8·71687	48	11·28313	8·71746	48	11·28254	10·00059		9·99941	
·4	8·71735	49	11·28265	8·71794	49	11·28206	10·00059		9·99941	
·6	8·71784	48	11·28216	8·71843	48	11·28157	10·00059		9·99941	
·8	8·71832	48	11·28168	8·71891	49	11·28109	10·00059		9·99941	
60·0	8·71880		11·28120	8·71940		11·28060	10·00060		9·99940	0′

177°
357°

3°
183°

LOGS. OF TRIG. FUNCTIONS

,	Sine	Diff.	Cosec.	Tan.	Diff.	Cotan.	Secant	Diff.	Cosine	
00·0	8·71880		11·28120	8·71940		11·28060	10·00060		9·99940	60'
·2	8·71928	48	11·28072	8·71988	48	11·28012	10·00060		9·99940	
·4	8·71976	48	11·28024	8·72036	48	11·27964	10·00060		9·99940	
·6	8·72024	48	11·27976	8·72084	48	11·27916	10·00060		9·99940	
·8	8·72072	48	11·27928	8·72132	48	11·27868	10·00060		9·99940	
01·0	8·72120	48	11·27880	8·72181	49	11·27819	10·00060		9·99940	59'
·2	8·72168	48	11·27832	8·72228	47	11·27772	10·00060		9·99940	
·4	8·72216	48	11·27784	8·72276	48	11·27724	10·00060		9·99940	
·6	8·72264	48	11·27736	8·72325	49	11·27675	10·00061		9·99939	
·8	8·72312	48	11·27688	8·72373	48	11·27627	10·00061		9·99939	
02·0	8·72360	48	11·27641	8·72420	47	11·27580	10·00061		9·99939	58'
·2	8·72407	47	11·27593	8·72468	48	11·27532	10·00061		9·99939	
·4	8·72455	48	11·27545	8·72516	48	11·27484	10·00061		9·99939	
·6	8·72502	47	11·27498	8·72563	47	11·27437	10·00061		9·99939	
·8	8·72550	48	11·27450	8·72611	48	11·27389	10·00061		9·99939	
03·0	8·72597	47	11·27403	8·72659	48	11·27341	10·00062		9·99938	57'
·2	8·72645	48	11·27355	8·72707	48	11·27293	10·00062		9·99938	
·4	8·72692	47	11·27308	8·72754	47	11·27246	10·00062		9·99938	
·6	8·72739	47	11·27261	8·72801	47	11·27199	10·00062		9·99938	
·8	8·72786	47	11·27214	8·72848	47	11·27152	10·00062		9·99938	
04·0	8·72834	48	11·27166	8·72896	48	11·27104	10·00062		9·99938	56'
·2	8·71881	47	11·27119	8·72943	47	11·27057	10·00062		9·99938	
·4	8·72928	47	11·27072	8·72990	47	11·27010	10·00062		9·99938	
·6	8·72975	47	11·27025	8·73038	48	11·26962	10·00063		9·99937	
·8	8·73022	47	11·26978	8·73085	47	11·26915	10·00063		9·99937	
05·0	8·73069	47	11·26931	8·73132	47	11·26868	10·00063		9·99937	55'
·2	8·73116	47	11·26884	8·73179	47	11·26821	10·00063		9·99937	
·4	8·73163	47	11·26837	8·73226	46	11·26774	10·00063		9·99937	
·6	8·73209	46	11·26791	8·73272	47	11·26728	10·00063		9·99937	
·8	8·73256	47	11·26744	8·73319	47	11·26681	10·00063		9·99937	
06·0	8·73303	47	11·26697	8·73366	47	11·26634	10·00064		9·99936	54'
·2	8·73349	46	11·26651	8·73413	47	11·26587	10·00064		9·99936	
·4	8·73396	47	11·26604	8·73460	46	11·26540	10·00064		9·99936	
·6	8·73442	46	11·26558	8·73506	47	11·26494	10·00064		9·99936	
·8	8·73489	47	11·26511	8·73553	47	11·26447	10·00064		9·99936	
07·0	8·73535	46	11·26465	8·73600	46	11·26400	10·00064		9·99936	53'
·2	8·73582	47	11·26418	8·73646	47	11·26354	10·00064		9·99936	
·4	8·73628	46	11·26372	8·73693	46	11·26307	10·00065		9·99935	
·6	8·73674	46	11·26326	8·73739	47	11·26261	10·00065		9·99935	
·8	8·73721	47	11·26279	8·73786	46	11·26214	10·00065		9·99935	
08·0	8·73767	46	11·26233	8·73732	46	11·26168	10·00065		9·99935	52'
·2	8·73813	46	11·26187	8·73878	46	11·26122	10·00065		9·99935	
·4	8·73859	46	11·26141	8·73924	46	11·26076	10·00065		9·99935	
·6	8·73905	46	11·26095	8·73970	47	11·26030	10·00065		9·99935	
·8	8·73951	46	11·26049	8·74017	46	11·25983	10·00066		9·99934	
09·0	8·73997	46	11·26003	8·74063	46	11·25937	10·00066		9·99934	51'
·2	8·74043	46	11·25957	8·74109	46	11·25891	10·00066		9·99934	
·4	8·74089	45	11·25911	8·74155	45	11·25845	10·00066		9·99934	
·6	8·74134	46	11·25866	8·74200	46	11·25800	10·00066		9·99934	
·8	8·74180	46	11·25820	8·74246	46	11·25754	10·00066		9·99934	
10·0	8·74226	46	11·25774	8·74292	46	11·25708	10·00066		9·99934	50'
·2	8·74272	45	11·25728	8·74338	46	11·25662	10·00066		9·99934	
·4	8·74317	46	11·25683	8·74384	46	11·25616	10·00067		9·99933	
·6	8·74363	45	11·25637	8·74430	45	11·25570	10·00067		9·99933	
·8	8·74408	46	11·25592	8·74475	46	11·25525	10·00067		9·99933	
11·0	8·74454	45	11·25546	8·74521	45	11·25479	10·00067		9·99933	49'
·2	8·74499	45	11·25501	8·74566	45	11·25434	10·00067		9·99933	
·4	8·74544	46	11·25456	8·74611	46	11·25389	10·00067		9·99933	
·6	8·74590	45	11·25410	8·74657	46	11·25343	10·00067		9·99933	
·8	8·74635	45	11·25365	8·74703	45	11·25297	10·00068		9·99932	
12·0	8·74680		11·25320	8·74748		11·25252	10·00068		9·99932	48'

3°
183°

LOGS. OF TRIG. FUNCTIONS

′	Sine	Diff.	Cosec.	Tan.	Diff.	Cotan.	Secant	Diff.	Cosine	
12·0	8·74680	45	11·25320	8·74748	45	11·25252	10·00068		9·99932	48′
·2	8·74725	45	11·25275	8·74793	45	11·25207	10·00068		9·99932	
·4	8·74770	45	11·25230	8·74838	46	11·25162	10·00068		9·99932	
·6	8·74816	46	11·25184	8·74884	45	11·25116	10·00068		9·99932	
·8	8·74861	45	11·25139	8·74929	45	11·25071	10·00068		9·99932	
13·0	8·74906	45	11·25095	8·74974	45	11·25026	10·00068		9·99932	47′
·2	8·74950	44	11·25050	8·75019	45	11·24981	10·00069		9·99931	
·4	8·74995	45	11·25005	8·75064	45	11·24936	10·00069		9·99931	
·6	8·75040	45	11·24960	8·75109	45	11·24891	10·00069		9·99931	
·8	8·75085	45	11·24915	8·75154	45	11·24846	10·00069		9·99931	
14·0	8·75130	45	11·24870	8·75199	45	11·24801	10·00069		9·99931	46′
·2	8·75174	44	11·24826	8·75243	44	11·24757	10·00069		9·99931	
·4	8·75219	45	11·24781	8·75288	45	11·24712	10·00069		9·99931	
·6	8·75264	45	11·24736	8·75334	46	11·24666	10·00070		9·99930	
·8	8·75308	44	11·24692	8·75378	44	11·24622	10·00070		9·99930	
15·0	8·75353	45	11·24647	8·75423	45	11·24577	10·00070		9·99930	45′
·2	8·75397	44	11·24603	8·75467	44	11·24533	10·00070		9·99930	
·4	8·75442	45	11·24558	8·75512	45	11·24488	10·00070		9·99930	
·6	8·75486	44	11·24514	8·75556	44	11·24444	10·00070		9·99930	
·8	8·75530	44	11·24470	8·75600	44	11·24400	10·00070		9·99930	
16·0	8·75575	45	11·24425	8·75646	46	11·24355	10·00071		9·99929	44′
·2	8·75619	44	11·24381	8·75690	44	11·24310	10·00071		9·99929	
·4	8·75663	44	11·24337	8·75734	44	11·24266	10·00071		9·99929	
·6	8·75707	44	11·24293	8·75778	44	11·24222	10·00071		9·99929	
·8	8·75751	44	11·24249	8·75822	44	11·24178	10·00071		9·99929	
17·0	8·75795	44	11·24205	8·75866	44	11·24133	10·00071		9·99929	43′
·2	8·75839	44	11·24161	8·75910	44	11·24090	10·00071		9·99929	
·4	8·75883	44	11·24117	8·75955	45	11·24045	10·00072		9·99928	
·6	8·75927	44	11·24073	8·75999	44	11·24001	10·00072		9·99928	
·8	8·75971	44	11·24029	8·76043	44	11·23957	10·00072		9·99928	
18·0	8·76015	44	11·23985	8·76087	44	11·23913	10·00072		9·99928	42′
·2	8·76059	44	11·23941	8·76131	44	11·23869	10·00072		9·99928	
·4	8·76103	44	11·23897	8·76175	44	11·23825	10·00072		9·99928	
·6	8·76146	43	11·23854	8·76218	43	11·23782	10·00072		9·99928	
·8	8·76190	44	11·23810	8·76263	45	11·23737	10·00073		9·99927	
19·0	8·76234	44	11·23766	8·76307	44	11·23694	10·00073		9·99927	41′
·2	8·76277	43	11·23723	8·76350	43	11·23650	10·00073		9·99927	
·4	8·76321	44	11·23679	8·76394	44	11·23606	10·00073		9·99927	
·6	8·76364	43	11·23636	8·76437	43	11·23563	10·00073		9·99927	
·8	8·76408	44	11·23592	8·76481	44	11·23519	10·00073		9·99927	
20·0	8·76451	43	11·23549	8·76525	44	11·23475	10·00074		9·99927	40′
·2	8·76494	43	11·23506	8·76568	43	11·23432	10·00074		9·99926	
·4	8·76538	44	11·23462	8·76612	44	11·23388	10·00074		9·99926	
·6	8·76581	43	11·23419	8·76655	43	11·23345	10·00074		9·99926	
·8	8·76624	43	11·23376	8·76698	43	11·23302	10·00074		9·99926	
21·0	8·76667	43	11·23333	8·76742	44	11·23258	10·00074		9·99926	39′
·2	8·76711	44	11·23289	8·76785	43	11·23215	10·00074		9·99926	
·4	8·76754	43	11·23246	8·76829	44	11·23171	10·00075		9·99925	
·6	8·76797	43	11·23203	8·76872	43	11·23128	10·00075		9·99925	
·8	8·76840	43	11·23160	8·76915	43	11·23085	10·00075		9·99925	
22·0	8·76883	43	11·23117	8·76958	43	11·23042	10·00075		9·99925	38′
·2	8·76926	43	11·23074	8·77001	43	11·22999	10·00075		9·99925	
·4	8·76969	42	11·23031	8·77044	43	11·22956	10·00075		9·99925	
·6	8·77011	43	11·22989	8·77087	43	11·22913	10·00076		9·99924	
·8	8·77054	43	11·22946	8·77130	43	11·22870	10·00076		9·99924	
23·0	8·77097	43	11·22903	8·77173	43	11·22827	10·00076		9·99924	37′
·2	8·77140	42	11·22860	8·77216	42	11·22784	10·00076		9·99924	
·4	8·77182	43	11·22818	8·77258	43	11·22742	10·00076		9·99924	
·6	8·77225	43	11·22775	8·77301	43	11·22699	10·00076		9·99924	
·8	8·77268	42	11·22732	8·77344	43	11·22656	10·00076		9·99924	
24·0	8·77310		11·22690	8·77387		11·22613	10·00077		9·99924	36′

3°
183°

LOGS. OF TRIG. FUNCTIONS

′	Sine	Diff.	Cosec.	Tan.	Diff.	Cotan.	Secant	Diff.	Cosine	
24·0	8·77310		11·22690	8·77387		11·22613	10·00077		9·99924	36′
·2	8·77353	43	11·22647	8·77430	43	11·22570	10·00077		9·99923	
·4	8·77395	42	11·22605	8·77472	42	11·22528	10·00077		9·99923	
·6	8·77438	43	11·22562	8·77515	43	11·22485	10·00077		9·99923	
·8	8·77480	42	11·22520	8·77557	42	11·22443	10·00077		9·99923	
25·0	8·77522	42	11·22478	8·77600	43	11·22401	10·00077		9·99923	35′
·2	8·77565	43	11·22435	8·77642	42	11·22358	10·00077		9·99923	
·4	8·77607	42	11·22393	8·77685	43	11·22315	10·00078		9·99922	
·6	8·77649	42	11·22351	8·77727	42	11·22273	10·00078		9·99922	
·8	8·77691	42	11·22309	8·77769	42	11·22231	10·00078		9·99922	
26·0	8·77733	42	11·22267	8·77811	42	11·22189	10·00078		9·99922	34′
·2	8·77775	42	11·22225	8·77853	42	11·22147	10·00078		9·99922	
·4	8·77817	42	11·22183	8·77895	42	11·22105	10·00078		9·99922	
·6	8·77859	42	11·22141	8·77937	42	11·22063	10·00078		9·99922	
·8	8·77901	42	11·22099	8·77980	43	11·22020	10·00079		9·99921	
27·0	8·77943	42	11·22057	8·78022	42	11·21978	10·00079		9·99921	33′
·2	8·77985	42	11·22015	8·78064	42	11·21936	10·00079		9·99921	
·4	8·78027	42	11·21973	8·78106	42	11·21894	10·00079		9·99921	
·6	8·78069	42	11·21931	8·78148	42	11·21852	10·00079		9·99921	
·8	8·78111	42	11·21889	8·78190	42	11·21810	10·00079		9·99921	
28·0	8·78152	41	11·21848	8·78232	42	11·21768	10·00080		9·99921	32′
·2	8·78194	42	11·21806	8·78274	42	11·21726	10·00080		9·99920	
·4	8·78236	42	11·21764	8·78316	42	11·21684	10·00080		9·99920	
·6	8·78277	41	11·21723	8·78357	41	11·21643	10·00080		9·99920	
·8	8·78319	42	11·21681	8·78399	42	11·21601	10·00080		9·99920	
29·0	8·78361	42	11·21640	8·78441	42	11·21559	10·00080		9·99920	31′
·2	8·78402	41	11·21598	8·78482	41	11·21518	10·00080		9·99920	
·4	8·78443	41	11·21557	8·78524	42	11·21476	10·00081		9·99919	
·6	8·78485	42	11·21515	8·78566	42	11·21434	10·00081		9·99919	
·8	8·78526	41	11·21474	8·78607	41	11·21393	10·00081		9·99919	
30·0	8·78568	42	11·21433	8·78649	42	11·21351	10·00081		9·99919	30′
·2	8·78609	41	11·21391	8·78690	41	11·21310	10·00081		9·99919	
·4	8·78650	41	11·21350	8·78731	41	11·21269	10·00081		9·99919	
·6	8·78691	41	11·21309	8·78773	42	11·21227	10·00082		9·99918	
·8	8·78732	41	11·21268	8·78814	41	11·21186	10·00082		9·99918	
31·0	8·78774	42	11·21226	8·78855	41	11·21145	10·00082		9·99918	29′
·2	8·78815	41	11·21185	8·78897	42	11·21103	10·00082		9·99918	
·4	8·78856	41	11·21144	8·78938	41	11·21062	10·00082		9·99918	
·6	8·78897	41	11·21103	8·78979	41	11·21021	10·00082		9·99918	
·8	8·78938	41	11·21062	8·79020	41	11·20980	10·00082		9·99918	
32·0	8·78979	41	11·21021	8·79061	41	11·20939	10·00083		9·99917	28′
·2	8·79020	41	11·20980	8·79103	42	11·20897	10·00083		9·99917	
·4	8·79060	40	11·20940	8·79143	40	11·20857	10·00083		9·99917	
·6	8·79101	41	11·20899	8·79184	41	11·20816	10·00083		9·99917	
·8	8·79142	41	11·20858	8·79225	41	11·20775	10·00083		9·99917	
33·0	8·79183	41	11·20817	8·79266	41	11·20734	10·00083		9·99917	27′
·2	8·79224	41	11·20776	8·79308	42	11·20692	10·00084		9·99916	
·4	8·79264	40	11·20736	8·79348	40	11·20652	10·00084		9·99916	
·6	8·79305	41	11·20695	8·79389	41	11·20611	10·00084		9·99916	
·8	8·79345	40	11·20655	8·79429	40	11·20571	10·00084		9·99916	
34·0	8·79386	41	11·20614	8·79470	41	11·20530	10·00084		9·99916	26′
·2	8·79427	41	11·20573	8·79511	41	11·20489	10·00084		9·99916	
·4	8·79467	40	11·20533	8·79552	41	11·20448	10·00085		9·99915	
·6	8·79507	40	11·20493	8·79592	40	11·20408	10·00085		9·99915	
·8	8·79548	41	11·20452	8·79633	41	11·20367	10·00085		9·99915	
35·0	8·79588	40	11·20412	8·79673	40	11·20327	10·00085		9·99915	25′
·2	8·79629	41	11·20371	8·79714	41	11·20286	10·00085		9·99915	
·4	8·79669	40	11·20331	8·79754	40	11·20246	10·00085		9·99915	
·6	8·79709	40	11·20291	8·79794	40	11·20206	10·00085		9·99915	
·8	8·79749	40	11·20251	8·79835	41	11·20165	10·00086		9·99914	
36·0	8·79789	40	11·20211	8·79875	40	11·20125	10·00086		9·99914	24′

3° 183°

LOGS. OF TRIG. FUNCTIONS

,	Sine	Diff.	Cosec.	Tan.	Diff.	Cotan.	Secant	Diff.	Cosine	
36·0	8·79789		11·20211	8·79875		11·20125	10·00086		9·99914	*24'*
·2	8·79830	41	11·20170	8·79916	41	11·20084	10·00086		9·99914	
·4	8·79870	40	11·20130	8·79956	40	11·20044	10·00086		9·99914	
·6	8·79910	40	11·20090	8·79996	40	11·20004	10·00086		9·99914	
·8	8·79950	40	11·20050	8·80036	40	11·19964	10·00086		9·99914	
37·0	8·79990	40	11·20010	8·80076	40	11·19924	10·00087		9·99913	*23'*
·2	8·80030	40	11·19970	8·80117	40	11·19883	10·00087		9·99913	
·4	8·80070	40	11·19930	8·80157	41	11·19843	10·00087		9·99913	
·6	8·80110	40	11·19890	8·80197	40	11·19803	10·00087		9·99913	
·8	8·80149	39	11·19851	8·80237	40	11·19763	10·00087		9·99913	
38·0	8·80189	40	11·19811	8·80277	40	11·19723	10·00087		9·99913	*22'*
·2	8·80229	40	11·19771	8·80317	40	11·19683	10·00088		9·99912	
·4	8·80269	40	11·19731	8·80357	40	11·19643	10·00088		9·99912	
·6	8·80308	39	11·19692	8·80396	39	11·19604	10·00088		9·99912	
·8	8·80348	40	11·19652	8·80436	40	11·19564	10·00088		9·99912	
39·0	8·80388	40	11·19612	8·80476	40	11·19524	10·00088		9·99912	*21'*
·2	8·80427	39	11·19573	8·80515	39	11·19485	10·00088		9·99912	
·4	8·80467	40	11·19533	8·80555	40	11·19445	10·00088		9·99912	
·6	8·80506	39	11·19494	8·80595	40	11·19405	10·00089		9·99911	
·8	8·80546	40	11·19454	8·80635	40	11·19365	10·00089		9·99911	
40·0	8·80585	39	11·19415	8·80674	39	11·19326	10·00089		9·99911	*20'*
·2	8·80625	40	11·19375	8·80714	40	11·19286	10·00089		9·99911	
·4	8·80664	39	11·19336	8·80753	39	11·19247	10·00089		9·99911	
·6	8·80703	39	11·19297	8·80793	40	11·19207	10·00089		9·99911	
·8	8·80743	40	11·19257	8·80833	40	11·19167	10·00090		9·99910	
41·0	8·80782	39	11·19218	8·80872	39	11·19128	10·00090		9·99910	*19'*
·2	8·80821	39	11·19179	8·80911	39	11·19089	10·00090		9·99910	
·4	8·80860	39	11·19140	8·80950	39	11·19050	10·00090		9·99910	
·6	8·80900	40	11·19100	8·80990	40	11·19010	10·00090		9·99910	
·8	8·80939	39	11·19061	8·81029	39	11·18971	10·00090		9·99910	
42·0	8·80978	39	11·19022	8·81068	39	11·18932	10·00091		9·99909	*18'*
·2	8·81017	39	11·18983	8·81108	40	11·18892	10·00091		9·99909	
·4	8·81056	39	11·18944	8·81147	39	11·18853	10·00091		9·99909	
·6	8·81095	39	11·18905	8·81186	39	11·18814	10·00091		9·99909	
·8	8·81134	39	11·18866	8·81225	39	11·18775	10·00091		9·99909	
43·0	8·81173	39	11·18827	8·81264	39	11·18736	10·00091		9·99909	*17'*
·2	8·81211	38	11·18789	8·81303	39	11·18697	10·00092		9·99908	
·4	8·81250	39	11·18750	8·81342	39	11·18658	10·00092		9·99908	
·6	8·81289	39	11·18711	8·81381	39	11·18619	10·00092		9·99908	
·8	8·81328	39	11·18672	8·81420	39	11·18580	10·00092		9·99908	
44·0	8·81367	39	11·18633	8·81459	39	11·18541	10·00092		9·99908	*16'*
·2	8·81405	38	11·18595	8·81498	39	11·18502	10·00092		9·99908	
·4	8·81444	39	11·18556	8·81537	39	11·18463	10·00093		9·99907	
·6	8·81483	39	11·18517	8·81576	39	11·18424	10·00093		9·99907	
·8	8·81521	38	11·18479	8·81614	38	11·18386	10·00093		9·99907	
45·0	8·81560	39	11·18440	8·81653	39	11·18347	10·00093		9·99907	*15'*
·2	8·81598	38	11·18402	8·81691	38	11·18309	10·00093		9·99907	
·4	8·81637	39	11·18363	8·81730	39	11·18270	10·00093		9·99907	
·6	8·81675	38	11·18325	8·81768	38	11·18232	10·00093		9·99907	
·8	8·81714	39	11·18286	8·81807	39	11·18193	10·00094		9·99906	
46·0	8·81752	38	11·18248	8·81846	39	11·18154	10·00094		9·99906	*14'*
·2	8·81790	38	11·18210	8·81884	38	11·18116	10·00094		9·99906	
·4	8·81829	39	11·18171	8·81923	39	11·18077	10·00094		9·99906	
·6	8·81867	38	11·18133	8·81961	38	11·18039	10·00094		9·99906	
·8	8·81905	38	11·18095	8·81999	38	11·18001	10·00094		9·99906	
47·0	8·81944	39	11·18056	8·82038	39	11·17962	10·00095		9·99905	*13'*
·2	8·81982	38	11·18018	8·82077	39	11·17923	10·00095		9·99905	
·4	8·82020	38	11·17980	8·82115	38	11·17885	10·00095		9·99905	
·6	8·82058	38	11·17942	8·82153	38	11·17847	10·00095		9·99905	
·8	8·82096	38	11·17904	8·82191	38	11·17809	10·00095		9·99905	
48·0	8·82134	38	11·17866	8·82230	39	11·17770	10·00096		9·99904	*12'*

3°
183°

LOGS. OF TRIG. FUNCTIONS

′	Sine	Diff.	Cosec.	Tan.	Diff.	Cotan.	Secant	Diff.	Cosine	
48·0	8·82134		11·17866	8·82230		11·17770	10·00096		9·99904	12′
·2	8·82172	38	11·17828	8·82268	38	11·17732	10·00096		9·99904	
·4	8·82210	38	11·17790	8·82306	38	11·17694	10·00096		9·99904	
·6	8·82248	38	11·17752	8·82344	38	11·17656	10·00096		9·99904	
·8	8·82286	38	11·17714	8·82382	38	11·17618	10·00096		9·99904	
49·0	8·82324	38	11·17676	8·82421	39	11·17580	10·00096		9·99904	11′
·2	8·82362	38	11·17638	8·82459	38	11·17541	10·00097		9·99903	
·4	8·82400	38	11·17600	8·82497	38	11·17503	10·00097		9·99903	
·6	8·82438	38	11·17562	8·82535	38	11·17465	10·00097		9·99903	
·8	8·82475	37	11·17525	8·82572	37	11·17428	10·00097		9·99903	
50·0	8·82513	38	11·17487	8·82610	38	11·17390	10·00097		9·99903	10′
·2	8·82551	38	11·17449	8·82648	38	11·17352	10·00097		9·99903	
·4	8·82588	37	11·17412	8·82686	38	11·17314	10·00098		9·99902	
·6	8·82626	38	11·17374	8·82724	38	11·17276	10·00098		9·99902	
·8	8·82664	38	11·17336	8·82762	38	11·17238	10·00098		9·99902	
51·0	8·82701	37	11·17299	8·82799	37	11·17201	10·00098		9·99902	9′
·2	8·82739	38	11·17261	8·82837	38	11·17163	10·00098		9·99902	
·4	8·82776	37	11·17224	8·82874	37	11·17126	10·00098		9·99902	
·6	8·82814	38	11·17186	8·82912	38	11·17087	10·00099		9·99901	
·8	8·82851	37	11·17149	8·82950	38	11·17050	10·00099		9·99901	
52·0	8·82888	37	11·17112	8·82987	37	11·17013	10·00099		9·99901	8′
·2	8·82926	38	11·17074	8·83025	38	11·16975	10·00099		9·99901	
·4	8·82963	37	11·17037	8·83062	37	11·16938	10·00099		9·99901	
·6	8·83000	37	11·17000	8·83100	38	11·16901	10·00100		9·99901	
·8	8·83038	38	11·16962	8·83138	38	11·16862	10·00100		9·99900	
53·0	8·83075	37	11·16925	8·83175	37	11·16825	10·00100		9·99900	7′
·2	8·83112	37	11·16888	8·83212	37	11·16788	10·00100		9·99900	
·4	8·83149	37	11·16851	8·83249	37	11·16751	10·00100		9·99900	
·6	8·83187	38	11·16813	8·83287	38	11·16713	10·00100		9·99900	
·8	8·83224	37	11·16776	8·83325	38	11·16675	10·00101		9·99899	
54·0	8·83261	37	11·16739	8·83361	36	11·16639	10·00101		9·99899	6′
·2	8·83298	37	11·16702	8·83399	38	11·16601	10·00101		9·99899	
·4	8·83335	37	11·16665	8·83436	37	11·16564	10·00101		9·99899	
·6	8·83372	37	11·16628	8·83473	37	11·16527	10·00101		9·99899	
·8	8·83409	37	11·16591	8·83510	37	11·16490	10·00101		9·99899	
55·0	8·83446	37	11·16554	8·83547	37	11·16453	10·00102		9·99898	5′
·2	8·83482	38	11·16518	8·83584	37	11·16416	10·00102		9·99898	
·4	8·83519	37	11·16481	8·83621	37	11·16379	10·00102		9·99898	
·6	8·83556	37	11·16444	8·83658	37	11·16342	10·00102		9·99898	
·8	8·83593	37	11·16407	8·83695	37	11·16305	10·00102		9·99898	
56·0	8·83630	37	11·16370	8·83732	37	11·16268	10·00102		9·99898	4′
·2	8·83666	36	11·16334	8·83769	37	11·16231	10·00103		9·99897	
·4	8·83703	37	11·16297	8·83806	37	11·16194	10·00103		9·99897	
·6	8·83740	37	11·16260	8·83843	36	11·16157	10·00103		9·99897	
·8	8·83776	36	11·16224	8·83879	37	11·16121	10·00103		9·99897	
57·0	8·83813	37	11·16187	8·83916	37	11·16083	10·00103		9·99897	3′
·2	8·83850	37	11·16150	8·83953	37	11·16047	10·00103		9·99897	
·4	8·83886	36	11·16114	8·83990	37	11·16010	10·00104		9·99896	
·6	8·83923	37	11·16077	8·84027	36	11·15973	10·00104		9·99896	
·8	8·83959	36	11·16041	8·84063	37	11·15937	10·00104		9·99896	
58·0	8·83996	37	11·16004	8·84100	36	11·15900	10·00104		9·99896	2′
·2	8·84032	36	11·15968	8·84136	36	11·15864	10·00104		9·99896	
·4	8·84068	36	11·15932	8·84172	37	11·15828	10·00104		9·99896	
·6	8·84105	37	11·15895	8·84209	37	11·15790	10·00105		9·99895	
·8	8·84141	36	11·15859	8·84246	37	11·15754	10·00105		9·99895	
59·0	8·84177	36	11·15823	8·84283	36	11·15718	10·00105		9·99895	1′
·2	8·84214	37	11·15786	8·84319	36	11·15681	10·00105		9·99895	
·4	8·84250	36	11·15750	8·84355	36	11·15645	10·00105		9·99895	
·6	8·84286	36	11·15714	8·84391	37	11·15609	10·00105		9·99895	
·8	8·84322	37	11·15678	8·84428	36	11·15572	10·00106		9·99894	
60·0	8·84359		11·15642	8·84464		11·15536	10·00106		9·99894	0′

4°

184°

LOGS. OF TRIG. FUNCTIONS

′	Sine	Parts	Cosec.	Tan.	Parts	Cotan.	Secant	Parts	Cosine	
00·0	8·84359	′	11·15642	8·84464	′	11·15536	10·00106		9·99894	60′
01·0	8·84539	·1 18	11·15461	8·84646	·1 18	11·15355	10·00107		9·99893	
02·0	8·84718	·2 35	11·15282	8·84826	·2 36	11·15174	10·00108		9·99892	
03·0	8·84897	·3 53	11·15103	8·85006	·3 53	11·14994	10·00109		9·99891	
04·0	8·85075	·4 71	11·14925	8·85185	·4 71	11·14815	10·00110		9·99891	
05·0	8·85253	·5 88	11·14748	8·85363	·5 89	11·14637	10·00110		9·99890	55′
06·0	8·85429	·6 106	11·14571	8·85540	·6 107	11·14460	10·00111		9·99889	
07·0	8·85605	·7 124	11·14395	8·85717	·7 125	11·14283	10·00112		9·99888	
08·0	8·85780	·8 142	11·14220	8·85893	·8 142	11·14107	10·00113		9·99887	
09·0	8·85955	·9 159	11·14045	8·86069	·9 160	11·13931	10·00114		9·99886	
10·0	8·86128		11·13872	8·86243		11·13757	10·00115		9·99885	50′
11·0	8·86301	·1 17	11·13699	8·86417	·1 17	11·13583	10·00116		9·99884	
12·0	8·86474	·2 34	11·13526	8·86591	·2 34	11·13409	10·00117		9·99883	
13·0	8·86646	·3 51	11·13355	8·86763	·3 51	11·13237	10·00118		9·99882	
14·0	8·86817	·4 68	11·13184	8·86935	·4 68	11·13065	10·00119		9·99881	
15·0	8·86987	·5 85	11·13013	8·87106	·5 86	11·12894	10·00120		9·99880	45′
16·0	8·87157	·6 102	11·12844	8·87277	·6 103	11·12723	10·00121		9·99880	
17·0	8·87326	·7 119	11·12675	8·87447	·7 120	11·12553	10·00122		9·99879	
18·0	8·87494	·8 136	11·12506	8·87616	·8 137	11·12384	10·00122		9·99878	
19·0	8·87662	·9 153	11·12339	8·87785	·9 154	11·12215	10·00123		9·99877	
20·0	8·87829		11·12172	8·87953		11·12047	10·00124		9·99876	40′
21·0	8·87995	·1 16	11·12005	8·88120	·1 16	11·11880	10·00125		9·99875	
22·0	8·88161	·2 33	11·11839	8·88287	·2 33	11·11713	10·00126		9·99874	
23·0	8·88326	·3 49	11·11674	8·88453	·3 49	11·11547	10·00127		9·99873	
24·0	8·88490	·4 65	11·11510	8·88619	·4 66	11·11382	10·00128		9·99872	
25·0	8·88654	·5 82	11·11346	8·88783	·5 82	11·11217	10·00129		9·99871	35′
26·0	8·88817	·6 98	11·11183	8·88948	·6 99	11·11052	10·00130		9·99870	
27·0	8·88980	·7 114	11·11020	8·89111	·7 115	11·10889	10·00131		9·99869	
28·0	8·89142	·8 131	11·10858	8·89274	·8 132	11·10726	10·00132		9·99868	
29·0	8·89304	·9 147	11·10697	8·89437	·9 148	11·10563	10·00133		9·99867	
30·0	8·89464		11·10536	8·89598		11·10402	10·00134		9·99866	30′
31·0	8·89625	·1 16	11·10375	8·89760	·1 16	11·10240	10·00135		9·99865	
32·0	8·89784	·2 32	11·10216	8·89920	·2 32	11·10080	10·00136		9·99864	
33·0	8·89943	·3 47	11·10057	8·90080	·3 48	11·09920	10·00137		9·99863	
34·0	8·90102	·4 63	11·09898	8·90240	·4 63	11·09760	10·00138		9·99862	
35·0	8·90260	·5 79	11·09740	8·90399	·5 79	11·09601	10·00139		9·99861	25′
36·0	8·90417	·6 95	11·09583	8·90557	·6 95	11·09443	10·00140		9·99860	
37·0	8·90574	·7 110	11·09426	8·90715	·7 111	11·09285	10·00141		9·99859	
38·0	8·90730	·8 126	11·09270	8·90872	·8 127	11·09128	10·00142		9·99858	
39·0	8·90885	·9 142	11·09115	8·91029	·9 143	11·08972	10·00143		9·99857	
40·0	8·91040		11·08960	8·91185		11·08815	10·00144		9·99856	20′
41·0	8·91195	·1 15	11·08805	8·91340	·1 15	11·08660	10·00145		9·99855	
42·0	8·91349	·2 30	11·08651	8·91495	·2 31	11·08505	10·00146		9·99854	
43·0	8·91502	·3 46	11·08498	8·91650	·3 46	11·08351	10·00147		9·99853	
44·0	8·91655	·4 61	11·08345	8·91803	·4 61	11·08197	10·00148		9·99852	
45·0	8·91807	·5 76	11·08193	8·91957	·5 77	11·08043	10·00149		9·99851	15′
46·0	8·91959	·6 91	11·08041	8·92110	·6 92	11·07890	10·00151		9·99850	
47·0	8·92110	·7 106	11·07890	8·92262	·7 107	11·07738	10·00152		9·99849	
48·0	8·92261	·8 122	11·07739	8·92414	·8 122	11·07586	10·00153		9·99847	
49·0	8·92411	·9 137	11·07589	8·92565	·9 138	11·07435	10·00154		9·99846	
50·0	8·92561		11·07439	8·92716		11·07284	10·00155		9·99845	10′
51·0	8·92710	·1 15	11·07290	8·92866	·1 15	11·07134	10·00156		9·99844	
52·0	8·92859	·2 29	11·07141	8·93016	·2 30	11·06985	10·00157		9·99843	
53·0	8·93007	·3 44	11·06993	8·93165	·3 44	11·06835	10·00158		9·99842	
54·0	8·93154	·4 59	11·06846	8·93313	·4 59	11·06687	10·00159		9·99841	
55·0	8·93302	·5 73	11·06699	8·93462	·5 74	11·06538	10·00160		9·99840	5′
56·0	8·93438	·6 88	11·06552	8·93609	·6 89	11·06391	10·00161		9·99839	
57·0	8·93594	·7 103	11·06406	8·93757	·7 104	11·06244	10·00162		9·99838	
58·0	8·93740	·8 118	11·06260	8·93903	·8 118	11·06097	10·00163		9·99837	
59·0	8·93885	·9 132	11·06115	8·94049	·9 133	11·05951	10·00165		9·99836	
60·0	8·94030		11·05970	8·94195		11·05805	10·00166		9·99834	0′

175°
355′

5°
185°

LOGS. OF TRIG. FUNCTIONS

′	Sine	Parts	Cosec.	Tan.	Parts	Cotan.	Secant	Parts	Cosine	
00·0	8·94030	′	11·05970	8·94195	′	11·05805	10·00166		9·99834	60"
01·0	8·94174	·1 14	11·05826	8·94340	·1 14	11·05660	10·00167		9·99833	
02·0	8·94317	·2 28	11·05683	8·94485	·2 29	11·05515	10·00168		9·99832	
03·0	8·94461	·3 43	11·05539	8·94630	·3 43	11·05371	10·00169		9·99831	
04·0	8·94603	·4 57	11·05397	8·94773	·4 57	11·05227	10·00170		9·99830	
		·5 71			·5 72					
05·0	8·94746		11·05254	8·94917		11·05083	10·00171		9·99829	55"
06·0	8·94887	·6 85	11·05113	8·95060	·6 86	11·04940	10·00172		9·99828	
07·0	8·95029	·7 99	11·04971	8·95202	·7 100	11·04798	10·00173		9·99827	
08·0	8·95170	·8 114	11·04830	8·95344	·8 115	11·04656	10·00175		9·99826	
09·0	8·95310	·9 128	11·04690	8·95486	·9 129	11·04514	10·00176		9·99824	
10·0	8·95450		11·04550	8·95627		11·04373	10·00177		9·99823	50′
11·0	8·95589	·1 14	11·04411	8·95767	·1 14	11·04233	10·00178		9·99822	
12·0	8·95728	·2 28	11·04272	8·95908	·2 28	11·04093	10·00179		9·99821	
13·0	8·95867	·3 41	11·04133	8·96047	·3 42	11·03953	10·00180		9·99820	
14·0	8·96005	·4 55	11·03995	8·96187	·4 55	11·03813	10·00181		9·99819	
		·5 69			·5 69					
15·0	8·96143		11·03857	8·96326		11·03675	10·00183		9·99817	45′
16·0	8·96280	·6 83	11·03720	8·96464	·6 83	11·03536	10·00184		9·99816	
17·0	8·96417	·7 96	11·03583	8·96602	·7 97	11·03398	10·00185		9·99815	
18·0	8·96553	·8 110	11·03447	8·96739	·8 111	11·03261	10·00186		9·99814	
19·0	8·96689	·9 124	11·03311	8·96877	·9 125	11·03123	10·00187		9·99813	
20·0	8·96825		11·03175	8·97013		11·02987	10·00188		9·99812	40′
21·0	8·96960	·1 13	11·03040	8·97150	·1 13	11·02850	10·00190		9·99810	
22·0	8·97095	·2 27	11·02905	8·97286	·2 27	11·02715	10·00191		9·99809	
23·0	8·97229	·3 40	11·02771	8·97421	·3 40	11·02579	10·00192		9·99808	
24·0	8·97363	·4 53	11·02637	8·97556	·4 54	11·02444	10·00193		9·99807	
		·5 67			·5 67					
25·0	8·97496		11·02504	8·97691		11·02309	10·00194		9·99806	35′
26·0	8·97629	·6 80	11·02371	8·97825	·6 81	11·02175	10·00196		9·99804	
27·0	8·97762	·7 93	11·02238	8·97959	·7 94	11·02041	10·00197		9·99803	
28·0	8·97894	·8 107	11·02106	8·98092	·8 108	11·01908	10·00198		9·99802	
29·0	8·98026	·9 120	11·01974	8·98225	·9 121	11·01775	10·00199		9·99801	
30·0	8·98157		11·01843	8·98358		11·01642	10·00200		9·99800	30′
31·0	8·98288	·1 13	11·01712	8·98490	·1 13	11·01510	10·00202		9·99798	
32·0	8·98419	·2 26	11·01581	8·98622	·2 26	11·01378	10·00203		9·99797	
33·0	8·98549	·3 39	11·01451	8·98753	·3 39	11·01247	10·00204		9·99796	
34·0	8·98679	·4 52	11·01321	8·98884	·4 52	11·01116	10·00205		9·99795	
		·5 65			·5 65					
35·0	8·98808		11·01192	8·99015		11·00985	10·00207		9·99794	25′
36·0	8·98937	·6 78	11·01063	8·99145	·6 78	11·00855	10·00208		9·99792	
37·0	8·99066	·7 91	11·00934	8·99275	·7 91	11·00725	10·00209		9·99791	
38·0	8·99194	·8 103	11·00806	8·99405	·8 104	11·00596	10·00210		9·99790	
39·0	8·99322	·9 116	11·00678	8·99534	·9 117	11·00466	10·00212		9·99789	
40·0	8·99450		11·00550	8·99662		11·00338	10·00213		9·99787	20′
41·0	8·99577	·1 13	11·00423	8·99791	·1 13	11·00209	10·00214		9·99786	
42·0	8·99704	·2 25	11·00296	8·99919	·2 25	11·00081	10·00215		9·99785	
43·0	8·99830	·3 38	11·00170	9·00047	·3 38	10·99954	10·00217		9·99784	
44·0	8·99956	·4 50	11·00044	9·00174	·4 51	10·99826	10·00218		9·99782	
		·5 63			·5 63					
45·0	9·00082		10·99918	9·00301		10·99700	10·00219		9·99781	15′
46·0	9·00207	·6 75	10·99793	9·00427	·6 76	10·99573	10·00220		9·99780	
47·0	9·00332	·7 88	10·99668	9·00553	·7 89	10·99447	10·00222		9·99778	
48·0	9·00456	·8 100	10·99544	9·00679	·8 101	10·99321	10·00223		9·99777	
49·0	9·00581	·9 113	10·99420	9·00805	·9 114	10·99195	10·00224		9·99776	
50·0	9·00704		10·99296	9·00930		10·99070	10·00226		9·99775	10′
51·0	9·00828	·1 12	10·99172	9·01055	·1 12	10·98945	10·00227		9·99773	
52·0	9·00951	·2 24	10·99049	9·01179	·2 25	10·98821	10·00228		9·99772	
53·0	9·01074	·3 37	10·98926	9·01303	·3 37	10·98697	10·00229		9·99771	
54·0	9·01196	·4 49	10·98804	9·01427	·4 49	10·98573	10·00231		9·99769	
		·5 61			·5 62					
55·0	9·01318		10·98682	9·01550		10·98450	10·00232		9·99768	5′
56·0	9·01440	·6 73	10·98560	9·01673	·6 74	10·98327	10·00233		9·99767	
57·0	9·01561	·7 85	10·98439	9·01796	·7 86	10·98204	10·00235		9·99765	
58·0	9·01682	·8 98	10·98318	9·01918	·8 99	10·98082	10·00236		9·99764	
59·0	9·01803	·9 110	10·98197	9·02040	·9 111	10·97960	10·00237		9·99763	
60·0	9·01924		10·98077	9·02162		10·97838	10·00239		9·99761	0′

6°
186°

LOGS. OF TRIG. FUNCTIONS

′	Sine	Parts	Cosec.	Tan.	Parts	Cotan.	Secant	Parts	Cosine	
00·0	9·01924	′	10·98077	9·02162	′	10·97838	10·00239		9·99761	60′
01·0	9·02044	·1 12	10·97957	9·02283	·1 12	10·97717	10·00240		9·99760	
02·0	9·02163	·2 24	10·97837	9·02404	·2 24	10·97596	10·00241		9·99759	
03·0	9·02283	·3 36	10·97718	9·02525	·3 36	10·97475	10·00243		9·99757	
04·0	9·02402	·4 47	10·97598	9·02646	·4 48	10·97355	10·00244		9·99756	
05·0	9·02520	·5 59	10·97480	9·02766	·5 60	10·97235	10·00245		9·99755	55′
06·0	9·02639	·6 71	10·97361	9·02885	·6 72	10·97115	10·00247		9·99753	
07·0	9·02757	·7 83	10·97243	9·03005	·7 84	10·96995	10·00248		9·99752	
08·0	9·02874	·8 95	10·97126	9·03124	·8 96	10·96876	10·00249		9·99751	
09·0	9·02992	·9 107	10·97008	9·03243	·9 108	10·96758	10·00251		9·99749	
10·0	9·03109		10·96891	9·03361		10·96639	10·00252		9·99748	50′
11·0	9·03226	·1 12	10·96774	9·03479	·1 12	10·96521	10·00253		9·99747	
12·0	9·03342	·2 23	10·96658	9·03597	·2 23	10·96403	10·00255		9·99745	
13·0	9·03458	·3 35	10·96542	9·03714	·3 35	10·96286	10·00256		9·99744	
14·0	9·03574	·4 46	10·96426	9·03832	·4 47	10·96168	10·00258		9·99743	
15·0	9·03690	·5 58	10·96310	9·03949	·5 58	10·96052	10·00259		9·99741	45′
16·0	9·03805	·6 69	10·96195	9·04065	·6 70	10·95935	10·00260		9·99740	
17·0	9·03920	·7 81	10·96080	9·04181	·7 82	10·95819	10·00262		9·99738	
18·0	9·04034	·8 92	10·95966	9·04297	·8 93	10·95703	10·00263		9·99737	
19·0	9·04149	·9 104	10·95852	9·04413	·9 105	10·95587	10·00265		9·99736	
20·0	9·04263		10·95738	9·04528		10·95472	10·00266		9·99734	40′
21·0	9·04376	·1 11	10·95624	9·04643	·1 11	10·95357	10·00267		9·99733	
22·0	9·04490	·2 22	10·95511	9·04758	·2 23	10·95242	10·00269		9·99731	
23·0	9·04603	·3 34	10·95397	9·04873	·3 34	10·95127	10·00270		9·99730	
24·0	9·04715	·4 45	10·95285	9·04987	·4 46	10·95013	10·00272		9·99729	
25·0	9·04828	·5 56	10·95172	9·05101	·5 57	10·94899	10·00273		9·99727	35′
26·0	9·04940	·6 67	10·95060	9·05214	·6 68	10·94786	10·00274		9·99726	
27·0	9·05052	·7 79	10·94948	9·05328	·7 80	10·94672	10·00276		9·99724	
28·0	9·05164	·8 90	10·94837	9·05441	·8 91	10·94559	10·00277		9·99723	
29·0	9·05275	·9 101	10·94725	9·05554	·9 102	10·94447	10·00279		9·99721	
30·0	9·05386		10·94614	9·05666		10·94334	10·00280		9·99720	30′
31·0	9·05497	·1 11	10·94503	9·05778	·1 11	10·94222	10·00282		9·99719	
32·0	9·05607	·2 22	10·94393	9·05890	·2 22	10·94110	10·00283		9·99717	
33·0	9·05717	·3 33	10·94283	9·06002	·3 33	10·93998	10·00284		9·99716	
34·0	9·05827	·4 44	10·94173	9·06113	·4 44	10·93887	10·00286		9·99714	
35·0	9·05937	·5 55	10·94063	9·06224	·5 55	10·93776	10·00287		9·99713	25′
36·0	9·06046	·6 66	10·93954	9·06335	·6 67	10·93665	10·00289		9·99711	
37·0	9·06155	·7 77	10·93845	9·06445	·7 78	10·93555	10·00290		9·99710	
38·0	9·06264	·8 88	10·93736	9·06556	·8 89	10·93444	10·00292		9·99708	
39·0	9·06372	·9 99	10·93628	9·06666	·9 100	10·93335	10·00293		9·99707	
40·0	9·06481		10·93519	9·06775		10·93225	10·00295		9·99705	20′
41·0	9·06588	·1 11	10·93412	9·06885	·1 11	10·93115	10·00296		9·99704	
42·0	9·06696	·2 21	10·93304	9·06994	·2 22	10·93006	10·00298		9·99702	
43·0	9·06804	·3 32	10·93196	9·07103	·3 32	10·92897	10·00299		9·99701	
44·0	9·06911	·4 43	10·93089	9·07211	·4 43	10·92789	10·00301		9·99699	
45·0	9·07018	·5 53	10·92982	9·07320	·5 54	10·92680	10·00302		9·99698	15′
46·0	9·07124	·6 64	10·92876	9·07428	·6 65	10·92572	10·00304		9·99696	
47·0	9·07231	·7 75	10·92769	9·07536	·7 76	10·92464	10·00305		9·99695	
48·0	9·07337	·8 85	10·92663	9·07643	·8 87	10·92357	10·00307		9·99693	
49·0	9·07442	·9 96	10·92558	9·07751	·9 97	10·92250	10·00308		9·99692	
50·0	9·07548		10·92452	9·07858		10·92142	10·00310		9·99690	10′
51·0	9·07653	·1 10	10·92347	9·07964	·1 11	10·92036	10·00311		9·99689	
52·0	9·07758	·2 21	10·92242	9·08071	·2 21	10·91929	10·00313		9·99687	
53·0	9·07863	·3 31	10·92137	9·08177	·3 32	10·91823	10·00314		9·99686	
54·0	9·07968	·4 42	10·92032	9·08283	·4 42	10·91717	10·00316		9·99684	
55·0	9·08072	·5 52	10·91928	9·08389	·5 53	10·91611	10·00317		9·99683	5′
56·0	9·08176	·6 62	10·91824	9·08495	·6 63	10·91505	10·00319		9·99681	
57·0	9·08280	·7 73	10·91720	9·08600	·7 74	10·91400	10·C0320		9·99680	
58·0	9·08383	·8 83	10·91617	9·08705	·8 84	10·91295	10·00322		9·99678	
59·0	9·08486	·9 94	10·91514	9·08810	·9 95	10·91190	10·00323		9·99677	
60·0	9·08589		10·91411	9·08914		10·91086	10·00325		9·99675	0′

173°
353°

7°
187°

LOGS. OF TRIG. FUNCTIONS

′	Sine	Parts		Cosec.	Tan.	Parts		Cotan.	Secant	Parts		Cosine	
00·0	9·08589	′		10·91411	9·08914	′		10·91086	10·00325	′		9·99675	60′
01·0	9·08692	·1	10	10·91308	9·09019	·1	10	10·90981	10·00327	·1	0	9·99674	
02·0	9·08795	·2	20	10·91205	9·09123	·2	21	10·90877	10·00328	·2	0	9·99672	
03·0	9·08897	3	30	10·91103	9·09227	·3	31	10·90773	10·00330	·3	0	9·99670	
04·0	9·08999	·4	41	10·91001	9·09330	·4	41	10·90670	10·00331	·4	1	9·99669	
05·0	9·09101	·5	51	10·90899	9·09434	·5	52	10·90566	10·00333	·5	1	9·99667	55′
06·0	9·09202	·6	61	10·90798	9·09537	·6	62	10·90463	10·00334	·6	1	9·99666	
07·0	9·09304	·7	71	10·90696	9·09640	·7	72	10·90361	10·00336	·7	1	9·99664	
08·0	9·09405	·8	81	10·90595	9·09742	·8	83	10·90258	10·00338	·8	1	9·99663	
09·0	9·09506	·9	92	10·90494	9·09845	·9	93	10·90155	10·00339	·9	1	9·99661	
10·0	9·09606			10·90394	9·09947			10·90053	10·00341			9·99659	50′
11·0	9·09707	·1	10	10·90294	9·10049	·1	10	10·89951	10·00342	·1	0	9·99658	
12·0	9·09807	·2	20	10·90193	9·10150	·2	20	10·89850	10·00344	·2	0	9·99656	
13·0	9·09907	·3	30	10·90094	9·10252	·3	30	10·89748	10·00345	·3	0	9·99655	
14·0	9·10006	·4	40	10·89994	9·10353	·4	40	10·89647	10·00347	·4	1	9·99653	
15·0	9·10106	·5	50	10·89894	9·10454	·5	50	10·89546	10·00349	·5	1	9·99651	45′
16·0	9·10205	·6	60	10·89795	9·10555	·6	61	10·89445	10·00350	·6	1	9·99650	
17·0	9·10304	·7	70	10·89696	9·10656	·7	71	10·89344	10·00352	·7	1	9·99648	
18·0	9·10403	·8	79	10·89598	9·10756	·8	81	10·89244	10·00354	·8	1	9·99647	
19·0	9·10501	·9	89	10·89499	9·10856	·9	91	10·89144	10·00355	·9	1	9·99645	
20·0	9·10599			10·89401	9·10956			10·89044	10·00357			9·99643	40′
21·0	9·10697	·1	10	10·89303	9·11056	·1	10	10·88944	10·00358	·1	0	9·99642	
22·0	9·10795	·2	19	10·89205	9·11155	·2	20	10·88845	10·00360	·2	0	9·99640	
23·0	9·10893	·3	29	10·89107	9·11254	·3	30	10·88746	10·00362	·3	0	9·99638	
24·0	9·10990	·4	39	10·89010	9·11353	·4	40	10·88646	10·00363	·4	1	9·99637	
25·0	9·11087	·5	49	10·88913	9·11452	·5	49	10·88548	10·00365	·5	1	9·99635	35′
26·0	9·11184	·6	58	10·88816	9·11551	·6	59	10·88449	10·00367	·6	1	9·99634	
27·0	9·11281	·7	68	10·88719	9·11649	·7	69	10·88351	10·00368	·7	1	9·99632	
28·0	9·11377	·8	78	10·88623	9·11747	·8	79	10·88253	10·00370	·8	1	9·99630	
29·0	9·11474	·9	87	10·88526	9·11845	·9	89	10·88155	10·00372	·9	1	9·99629	
30·0	9·11570			10·88430	9·11943			10·88057	10·00373			9·99627	30′
31·0	9·11666	·1	9	10·88334	9·12040	·1	10	10·87960	10·00375	·1	0	9·99625	
32·0	9·11761	·2	19	10·88239	9·12138	·2	19	10·87862	10·00377	·2	0	9·99624	
33·0	9·11857	·3	28	10·88143	9·12235	·3	29	10·87765	10·00378	·3	1	9·99622	
34·0	9·11952	·4	38	10·88048	9·12332	·4	39	10·87668	10·00380	·4	1	9·99620	
35·0	9·12047	·5	47	10·87953	9·12428	·5	48	10·87572	10·00382	·5	1	9·99619	25′
36·0	9·12142	·6	57	10·87858	9·12525	·6	58	10·87475	10·00383	·6	1	9·99617	
37·0	9·12236	·7	66	10·87764	9·12621	·7	68	10·87379	10·00385	·7	1	9·99615	
38·0	9·12331	·8	76	10·87669	9·12717	·8	77	10·87283	10·00387	·8	1	9·99613	
39·0	9·12425	·9	85	10·87575	9·12813	·9	87	10·87187	10·00388		2	9·99612	
40·0	9·12519			10·87481	9·12909			10·87091	10·00390			9·99610	20′
41·0	9·12613	·1	9	10·87388	9·13004	·1	9	10·86996	10·00392	·1	0	9·99608	
42·0	9·12706	·2	19	10·87294	9·13099	·2	19	10·86901	10·00393	·2	0	9·99607	
43·0	9·12799	·3	28	10·87201	9·13194	·3	28	10·86806	10·00395	·3	1	9·99605	
44·0	9·12893	·4	37	10·87108	9·13289	·4	38	10·86711	10·00397	·4	1	9·99603	
45·0	9·12985	·5	46	10·87015	9·13384	·5	47	10·86616	10·00399	·5	1	9·99602	15′
46·0	9·13078	·6	56	10·86922	9·13478	·6	57	10·86522	10·00400	·6	1	9·99600	
47·0	9·13171	·7	65	10·86829	9·13573	·7	66	10·86427	10·00402	·7	1	9·99598	
48·0	9·13263	·8	74	10·86737	9·13667	·8	76	10·86333	10·00404	·8	1	9·99596	
49·0	9·13355	·9	84	10·86645	9·13761	·9	85	10·86240	10·00405	·9	2	9·99595	
50·0	9·13447			10·86553	9·13854			10·86146	10·00407			9·99593	10′
51·0	9·13539	·1	9	10·86461	9·13948	·1	9	10·86052	10·00409	·1	0	9·99591	
52·0	9·13630	·2	18	10·86370	9·14041	·2	19	10·85959	10·00411	·2	0	9·99589	
53·0	9·13722	·3	27	10·86278	9·14134	·3	28	10·85866	10·00412	·3	1	9·99588	
54·0	9·13813	·4	36	10·86187	9·14227	·4	37	10·85773	10·00414	·4	1	9·99586	
55·0	9·13904	·5	45	10·86096	9·14320	·5	46	10·85680	10·00416	·5	1	9·99584	5′
56·0	9·13994	·6	55	10·86006	9·14412	·6	56	10·85588	10·00418	·6	1	9·99582	
57·0	9·14085	·7	64	10·85915	9·14504	·7	65	10·85496	10·00419	·7	1	9·99581	
58·0	9·14175	·8	73	10·85825	9·14597	·8	74	10·85403	10·00421	·8	1	9·99579	
59·0	9·14266	·9	82	10·85735	9·14689	·9	83	10·85312	10·00423	·9	2	9·99577	
60·0	9·14356			10·85645	9·14780			10·85220	10·00425			9·99575	0′

8°
188°

LOGS. OF TRIG. FUNCTIONS

′	Sine	Parts	Cosec.	Tan.	Parts	Cotan.	Secant	Parts	Cosine	
00·0	9·14356	′	10·85645	9·14780	′	10·85220	10·00425	′	9·99575	60′
01·0	9·14445	·1 9	10·85555	9·14872	·1 9	10·85128	10·00427	·1 0	9·99574	
02·0	9·14535	·2 18	10·85465	9·14963	·2 18	10·85037	10·00428	·2 0	9·99572	
03·0	9·14624	·3 27	10·85376	9·15054	·3 27	10·84946	10·00430	·3 1	9·99570	
04·0	9·14714	·4 36	10·85286	9·15145	·4 36	10·84855	10·00432	·4 1	9·99568	
05·0	9·14803	·5 44	10·85197	9·15236	·5 45	10·84764	10·00434	·5 1	9·99566	55′
06·0	9·14892	·6 53	10·85109	9·15327	·6 54	10·84673	10·00435	·6 1	9·99565	
07·0	9·14980	·7 62	10·85020	9·15417	·7 64	10·84583	10·00437	·7 1	9·99563	
08·0	9·15069	·8 71	10·84931	9·15508	·8 73	10·84492	10·00439	·8 1	9·99561	
09·0	9·15157	·9 80	10·84843	9·15598	·9 82	10·84402	10·00441	·9 2	9·99559	
10·0	9·15245		10·84755	9·15688		10·84312	10·00443		9·99557	50′
11·0	9·15333	·1 9	10·84667	9·15778	·1 9	10·84223	10·00445	·1 0	9·99556	
12·0	9·15421	·2 17	10·84579	9·15867	·2 18	10·84133	10·00446	·2 0	9·99554	
13·0	9·15508	·3 26	10·84492	9·15957	·3 27	10·84044	10·00448	·3 1	9·99552	
14·0	9·15596	·4 35	10·84404	9·16046	·4 36	10·83954	10·00450	·4 1	9·99550	
15·0	9·15683	·5 44	10·84317	9·16135	·5 44	10·83865	10·00452	·5 1	9·99548	45′
16·0	9·15770	·6 52	10·84230	9·16224	·6 53	10·83776	10·00454	·6 1	9·99546	
17·0	9·15857	·7 61	10·84143	9·16312	·7 62	10·83688	10·00455	·7 1	9·99545	
18·0	9·15944	·8 70	10·84057	9·16401	·8 71	10·83599	10·00457	·8 2	9·99543	
19·0	9·16030	·9 78	10·83970	9·16489	·9 80	10·83511	10·00459	·9 2	9·99541	
20·0	9·16116		10·83884	9·16577		10·83423	10·00461		9·99539	40′
21·0	9·16203	·1 9	10·83798	9·16665	·1 9	10·83335	10·00463	·1 0	9·99537	
22·0	9·16289	·2 17	10·83712	9·16753	·2 17	10·83247	10·00465	·2 0	9·99535	
23·0	9·16374	·3 26	10·83626	9·16841	·3 26	10·83159	10·00467	·3 1	9·99533	
24·0	9·16460	·4 34	10·83540	9·16928	·4 35	10·83072	10·00468	·4 1	9·99532	
25·0	9·16545	·5 43	10·83455	9·17016	·5 44	10·82984	10·00470	·5 1	9·99530	35′
26·0	9·16631	·6 51	10·83369	9·17103	·6 52	10·82897	10·00472	·6 1	9·99528	
27·0	9·16716	·7 60	10·83284	9·17190	·7 61	10·82810	10·00474	·7 1	9·99526	
28·0	9·16801	·8 68	10·83199	9·17277	·8 70	10·82723	10·00476	·8 2	9·99524	
29·0	9·16886	·9 77	10·83114	9·17363	·9 79	10·82637	10·00478	·9 2	9·99522	
30·0	9·16970		10·83030	9·17450		10·82550	10·00480		9·99520	30′
31·0	9·17055	·1 8	10·82945	9·17536	·1 9	10·82464	10·00482	·1 0	9·99518	
32·0	9·17139	·2 17	10·82861	9·17622	·2 17	10·82378	10·00484	·2 0	9·99517	
33·0	9·17223	·3 25	10·82777	9·17708	·3 26	10·82292	10·00485	·3 1	9·99515	
34·0	9·17307	·4 33	10·82693	9·17794	·4 34	10·82206	10·00487	·4 1	9·99513	
35·0	9·17391	·5 42	10·82609	9·17880	·5 43	10·82120	10·00489	·5 1	9·99511	25′
36·0	9·17474	·6 50	10·82526	9·17966	·6 51	10·82035	10·00491	·6 1	9·99509	
37·0	9·17558	·7 59	10·82442	9·18051	·7 60	10·81949	10·00493	·7 1	9·99507	
38·0	9·17641	·8 67	10·82359	9·18136	·8 68	10·81864	10·00495	·8 2	9·99505	
39·0	9·17724	·9 75	10·82276	9·18221	·9 77	10·81779	10·00497	·9 2	9·99503	
40·0	9·17807		10·82193	9·18306		10·81694	10·00499		9·99501	20′
41·0	9·17890	·1 8	10·82110	9·18391	·1 8	10·81609	10·00501	·1 0	9·99499	
42·0	9·17973	·2 16	10·82027	9·18475	·2 17	10·81525	10·00503	·2 0	9·99497	
43·0	9·18055	·3 25	10·81945	9·18560	·3 25	10·81440	10·00505	·3 1	9·99496	
44·0	9·18137	·4 33	10·81863	9·18644	·4 34	10·81356	10·00507	·4 1	9·99494	
45·0	9·18220	·5 41	10·81780	9·18728	·5 42	10·81272	10·00508	·5 1	9·99492	15′
46·0	9·18302	·6 49	10·81698	9·18812	·6 50	10·81188	10·00510	·6 1	9·99490	
47·0	9·18383	·7 57	10·81617	9·18896	·7 59	10·81104	10·00512	·7 1	9·99488	
48·0	9·18465	·8 66	10·81535	9·18979	·8 67	10·81021	10·00514	·8 2	9·99486	
49·0	9·18547	·9 74	10·81453	9·19063	·9 76	10·80937	10·00516	·9 2	9·99484	
50·0	9·18628		10·81372	9·19146		10·80854	10·00518		9·99482	10′
51·0	9·18709	·1 8	10·81291	9·19229	·1 8	10·80771	10·00520	·1 0	9·99480	
52·0	9·18790	·2 16	10·81210	9·19312	·2 17	10·80688	10·00522	·2 0	9·99478	
53·0	9·18871	·3 24	10·81129	9·19395	·3 25	10·80605	10·00524	·3 1	9·99476	
54·0	9·18952	·4 32	10·81048	9·19478	·4 33	10·80522	10·00526	·4 1	9·99474	
55·0	9·19033	·5 40	10·80968	9·19561	·5 41	10·80439	10·00528	·5 1	9·99472	5′
56·0	9·19113	·6 48	10·80887	9·19643	·6 50	10·80357	10·00530	·6 1	9·99470	
57·0	9·19193	·7 56	10·80807	9·19725	·7 58	10·80275	10·00532	·7 1	9·99468	
58·0	9·19273	·8 64	10·80727	9·19807	·8 66	10·80193	10·00534	·8 2	9·99466	
59·0	9·19353	·9 72	10·80647	9·19889	·9 74	10·80111	10·00536	·9 2	9·99464	
60·0	9·19433		10·80567	9·19971		10·80029	10·00538		9·99462	0′

171°
351°

9°
LOGS. OF TRIG. FUNCTIONS
189°

,	Sine	Parts		Cosec.	Tan.	Parts		Cotan.	Secant	Parts		Cosine	
00·0	9·19433	′		10·80567	9·19971	′		10·80029	10·00538	′		9·99462	60′
01·0	9·19513	·1	8	10·80487	9·20053	·1	8	10·79947	10·00540	·1	0	9·99460	
02·0	9·19593	·2	16	10·80408	9·20135	·2	16	10·79866	10·00542	·2	0	9·99458	
03·0	9·19672	·3	24	10·80328	9·20216	·3	24	10·79784	10·00544	·3	1	9·99456	
04·0	9·19751	·4	32	10·80249	9·20297	·4	32	10·79703	10·00546	·4	1	9·99454	
05·0	9·19830	·5	40	10·80170	9·20378	·5	41	10·79622	10·00548	·5	1	9·99452	55′
06·0	9·19909	·6	47	10·80091	9·20459	·6	49	10·79541	10·00550	·6	1	9·99450	
07·0	9·19988	·7	55	10·80012	9·20540	·7	57	10·79460	10·00552	·7	1	9·99448	
08·0	9·20067	·8	63	10·79933	9·20621	·8	65	10·79379	10·00554	·8	2	9·99446	
09·0	9·20145	·9	71	10·79855	9·20701	·9	73	10·79299	10·00556	·9	2	9·99444	
10·0	9·20223			10·79777	9·20782			10·79218	10·00558			9·99442	50′
11·0	9·20302	·1	8	10·79698	9·20862	·1	8	10·79138	10·00560	·1	0	9·99440	
12·0	9·20380	·2	16	10·79620	9·20942	·2	16	10·79058	10·00562	·2	0	9·99438	
13·0	9·20458	·3	23	10·79542	9·21022	·3	24	10·78978	10·00564	·3	1	9·99436	
14·0	9·20535	·4	31	10·79465	9·21102	·4	32	10·78898	10·00566	·4	1	9·99434	
15·0	9·20613	·5	39	10·79387	9·21182	·5	40	10·78819	10·00568	·5	1	9·99432	45′
16·0	9·20691	·6	47	10·79309	9·21261	·6	48	10·78739	10·00571	·6	1	9·99430	
17·0	9·20768	·7	54	10·79232	9·21341	·7	56	10·78660	10·00573	·7	1	9·99427	
18·0	9·20845	·8	62	10·79155	9·21420	·8	64	10·78580	10·00575	·8	2	9·99425	
19·0	9·20922	·9	70	10·79078	9·21499	·9	72	10·78501	10·00577	·9	2	9·99423	
20·0	9·20999			10·79001	9·21578			10·78422	10·00579			9·99421	40′
21·0	9·21076	·1	8	10·78924	9·21657	·1	8	10·78343	10·00581	·1	0	9·99419	
22·0	9·21153	·2	15	10·78847	9·21736	·2	16	10·78264	10·00583	·2	0	9·99417	
23·0	9·21229	·3	23	10·78771	9·21814	·3	23	10·78186	10·00585	·3	1	9·99415	
24·0	9·21306	·4	30	10·78695	9·21893	·4	31	10·78107	10·00587	·4	1	9·99413	
25·0	9·21382	·5	38	10·78618	9·21971	·5	39	10·78029	10·00589	·5	1	9·99411	35′
26·0	9·21458	·6	46	10·78542	9·22049	·6	47	10·77951	10·00591	·6	1	9·99409	
27·0	9·21534	·7	53	10·78466	9·22127	·7	55	10·77873	10·00593	·7	1	9·99407	
28·0	9·21610	·8	61	10·78390	9·22205	·8	63	10·77795	10·00596	·8	2	9·99405	
29·0	9·21685	·9	69	10·78315	9·22283	·9	70	10·77717	10·00598	·9	2	9·99402	
30·0	9·21761			10·78239	9·22361			10·77639	10·00600			9·99400	30′
31·0	9·21836	·1	7	10·78164	9·22438	·1	8	10·77562	10·00602	·1	0	9·99398	
32·0	9·21912	·2	15	10·78088	9·22516	·2	15	10·77484	10·00604	·2	0	9·99396	
33·0	9·21987	·3	22	10·78013	9·22593	·3	23	10·77407	10·00606	·3	1	9·99394	
34·0	9·22062	·4	30	10·77938	9·22670	·4	31	10·77330	10·00608	·4	1	9·99392	
35·0	9·22137	·5	37	10·77863	9·22747	·5	38	10·77253	10·00610	·5	1	9·99390	25′
36·0	9·22211	·6	45	10·77789	9·22824	·6	46	10·77176	10·00613	·6	1	9·99388	
37·0	9·22286	·7	52	10·77714	9·22901	·7	54	10·77099	10·00615	·7	1	9·99385	
38·0	9·22361	·8	60	10·77639	9·22977	·8	62	10·77023	10·00617	·8	2	9·99383	
39·0	9·22435	·9	67	10·77565	9·23054	·9	69	10·76946	10·00619	·9	2	9·99381	
40·0	9·22509			10·77491	9·23130			10·76870	10·00621			9·99379	20′
41·0	9·22583	·1	7	10·77417	9·23207	·1	8	10·76794	10·00623	·1	0	9·99377	
42·0	9·22657	·2	15	10·77343	9·23283	·2	15	10·76717	10·00625	·2	0	9·99375	
43·0	9·22731	·3	22	10·77269	9·23359	·3	23	10·76641	10·00628	·3	1	9·99373	
44·0	9·22805	·4	29	10·77195	9·23435	·4	30	10·76566	10·00630	·4	1	9·99370	
45·0	9·22878	·5	37	10·77122	9·23510	·5	38	10·76490	10·00632	·5	1	9·99368	15′
46·0	9·22952	·6	44	10·77048	9·23586	·6	45	10·76414	10·00634	·6	1	9·99366	
47·0	9·23025	·7	51	10·76975	9·23661	·7	53	10·76339	10·00636	·7	2	9·99364	
48·0	9·23098	·8	59	10·76902	9·23737	·8	61	10·76263	10·00638	·8	2	9·99362	
49·0	9·23172	·9	66	10·76829	9·23812	·9	68	10·76188	10·00641	·9	2	9·99359	
50·0	9·23244			10·76756	9·23887			10·76113	10·00643			9·99357	10′
51·0	9·23317	·1	7	10·76683	9·23962	·1	7	10·76038	10·00645	·1	0	9·99355	
52·0	9·23390	·2	14	10·76610	9·24037	·2	15	10·75963	10·00647	·2	0	9·99353	
53·0	9·23463	·3	22	10·76538	9·24112	·3	22	10·75888	10·00649	·3	1	9·99351	
54·0	9·23535	·4	29	10·76465	9·24187	·4	30	10·75814	10·00652	·4	1	9·99348	
55·0	9·23607	·5	36	10·76393	9·24261	·5	37	10·75739	10·00654	·5	1	9·99346	5′
56·0	9·23680	·6	43	10·76321	9·24335	·6	45	10·75665	10·00656	·6	1	9·99344	
57·0	9·23752	·7	51	10·76249	9·24410	·7	52	10·75590	10·00658	·7	2	9·99342	
58·0	9·23824	·8	58	10·76177	9·24484	·8	60	10·75516	10·00660	·8	2	9·99340	
59·0	9·23895	·9	65	10·76105	9·24558	·9	67	10·75442	10·00663	·9	2	9·99337	
60·0	9·23967			10·76033	9·24632			10·75368	10·00665			9·99335	0′

LOGS. OF TRIG. FUNCTIONS

′	Sine	Parts	Cosec.	Tan.	Parts	Cotan.	Secant	Parts	Cosine	
00·0	9·23967	′	10·76033	9·24632	′	10·75368	10·00665	′	9·99335	*60′*
01·0	9·24039	·1 7	10·75961	9·24706	·1 7	10·75294	10·00667	·1 0	9·99333	
02·0	9·24110	·2 14	10·75890	9·24779	·2 15	10·75221	10·00669	·2 0	9·99331	
03·0	9·24181	·3 21	10·75819	9·24853	·3 22	10·75147	10·00672	·3 1	9·99328	
04·0	9·24253	·4 28	10·75747	9·24926	·4 29	10·75074	10·00674	·4 1	9·99326	
05·0	9·24324	·5 36	10·75676	9·25000	·5 37	10·75000	10·00676	·5 1	9·99324	*55′*
06·0	9·24395	·6 43	10·75605	9·25073	·6 44	10·74927	10·00678	·6 1	9·99322	
07·0	9·24466	·7 50	10·75534	9·25146	·7 51	10·74854	10·00681	·7 2	9·99320	
08·0	9·24536	·8 57	10·75464	9·25219	·8 59	10·74781	10·00683	·8 2	9·99317	
09·0	9·24607	·9 64	10·75393	9·25292	·9 66	10·74708	10·00685	·9 2	9·99315	
10·0	9·24678	·1 7	10·75323	9·25365	·1 7	10·74635	10·00687	·1 0	9·99313	*50′*
11·0	9·24748	·2 14	10·75252	9·25437	·2 14	10·74563	10·00690	·2 0	9·99310	
12·0	9·24818	·3 21	10·75182	9·25510	·3 22	10·74490	10·00692	·3 1	9·99308	
13·0	9·24888	·4 28	10·75112	9·25582	·4 29	10·74418	10·00694	·4 1	9·99306	
14·0	9·24958	·5 35	10·75042	9·25655	·5 36	10·74345	10·00696	·5 1	9·99304	
15·0	9·25028	·6 42	10·74972	9·25727	·6 43	10·74273	10·00699	·6 1	9·99301	*45′*
16·0	9·25098	·7 49	10·74902	9·25799	·7 50	10·74201	10·00701	·7 2	9·99299	
17·0	9·25168	·8 56	10·74832	9·25871	·8 58	10·74129	10·00703	·8 2	9·99297	
18·0	9·25237	·9 63	10·74763	9·25943	·9 65	10·74057	10·00706	·9 2	9·99294	
19·0	9·25307		10·74693	9·26015		10·73985	10·00708		9·99292	
20·0	9·25376	·1 7	10·74624	9·26086	·1 7	10·73914	10·00710	·1 0	9·99290	*40′*
21·0	9·25445	·2 14	10·74555	9·26158	·2 14	10·73842	10·00713	·2 0	9·99288	
22·0	9·25514	·3 21	10·74486	9·26229	·3 21	10·73771	10·00715	·3 1	9·99285	
23·0	9·25583	·4 27	10·74417	9·26301	·4 28	10·73700	10·00717	·4 1	9·99283	
24·0	9·25652	·5 34	10·74348	9·26372	·5 36	10·73628	10·00719	·5 1	9·99281	
25·0	9·25721	·6 41	10·74279	9·26443	·6 43	10·73557	10·00722	·6 1	9·99278	*35′*
26·0	9·25790	·7 48	10·74210	9·26514	·7 50	10·73486	10·00724	·7 2	9·99276	
27·0	9·25858	·8 55	10·74142	9·26585	·8 57	10·73415	10·00727	·8 2	9·99274	
28·0	9·25927	·9 62	10·74073	9·26656	·9 64	10·73345	10·00729	·9 2	9·99271	
29·0	9·25995		10·74005	9·26726		10·73274	10·00731		9·99269	
30·0	9·26063	·1 7	10·73937	9·26797	·1 7	10·73203	10·00733	·1 0	9·99267	*30′*
31·0	9·26131	·2 14	10·73869	9·26867	·2 14	10·73133	10·00736	·2 0	9·99264	
32·0	9·26199	·3 20	10·73801	9·26938	·3 21	10·73063	10·00738	·3 1	9·99262	
33·0	9·26267	·4 27	10·73733	9·27008	·4 28	10·72992	10·00740	·4 1	9·99260	
34·0	9·26335	·5 34	10·73665	9·27078	·5 35	10·72922	10·00743	·5 1	9·99257	
35·0	9·26403	·6 41	10·73597	9·27148	·6 42	10·72852	10·00745	·6 1	9·99255	*25′*
36·0	9·26470	·7 47	10·73530	9·27218	·7 49	10·72782	10·00748	·7 2	9·99253	
37·0	9·26538	·8 54	10·73462	9·27288	·8 56	10·72712	10·00750	·8 2	9·99250	
38·0	9·26605	·9 61	10·73395	9·27357	·9 63	10·72643	10·00752	·9 2	9·99248	
39·0	9·26672		10·73328	9·27427		10·72573	10·00755		9·99245	
40·0	9·26740	·1 7	10·73261	9·27496	·1 7	10·72504	10·00757	·1 0	9·99243	*20′*
41·0	9·26807	·2 13	10·73194	9·27566	·2 14	10·72434	10·00759	·2 0	9·99241	
42·0	9·26873	·3 20	10·73127	9·27635	·3 21	10·72365	10·00762	·3 1	9·99238	
43·0	9·26940	·4 27	10·73060	9·27704	·4 28	10·72296	10·00764	·4 1	9·99236	
44·0	9·27007	·5 33	10·72993	9·27773	·5 35	10·72227	10·00767	·5 1	9·99234	
45·0	9·27074	·6 40	10·72927	9·27842	·6 41	10·72158	10·00769	·6 1	9·99231	*15′*
46·0	9·27140	·7 47	10·72860	9·27911	·7 48	10·72089	10·00771	·7 2	9·99229	
47·0	9·27206	·8 53	10·72794	9·27980	·8 55	10·72020	10·00774	·8 2	9·99226	
48·0	9·27273	·9 60	10·72727	9·28049	·9 62	10·71951	10·00776	·9 2	9·99224	
49·0	9·27339		10·72661	9·28117		10·71883	10·00779		9·99221	
50·0	9·27405		10·72595	9·28186		10·71814	10·00781		9·99219	*10′*
51·0	9·27471	·1 7	10·72529	9·28254	·1 7	10·71746	10·00783	·1 0	9·99217	
52·0	9·27537	·2 13	10·72463	9·28323	·2 14	10·71678	10·00786	·2 0	9·99214	
53·0	9·27603	·3 20	10·72398	9·28391	·3 20	10·71609	10·00788	·3 1	9·99212	
54·0	9·27668	·4 26	10·72332	9·28459	·4 27	10·71541	10·00791	·4 1	9·99209	
55·0	9·27734	·5 33	10·72266	9·28527	·5 34	10·71473	10·00793	·5 1	9·99207	*5′*
56·0	9·27799	·6 39	10·72201	9·28595	·6 41	10·71405	10·00796	·6 1	9·99204	
57·0	9·27865	·7 46	10·72136	9·28662	·7 48	10·71338	10·00798	·7 2	9·99202	
58·0	9·27930	·8 52	10·72070	9·28730	·8 54	10·71270	10·00800	·8 2	9·99200	
59·0	9·27995	·9 59	10·72005	9·28798	·9 61	10·71202	10·00803	·9 2	9·99197	
60·0	9·28060		10·71940	9·28865		10·71135	10·00805		9·99195	*0′*

LOGS. OF TRIG. FUNCTIONS

11°
191°

´	Sine	Parts		Cosec.	Tan.	Parts		Cotan.	Secant	Parts		Cosine	
		´				´				´			
00·0	9·28060			10·71940	9·28865			10·71135	10·00805			9·99195	60´
01·0	9·28125			10·71875	9·28933			10·71067	10·00808			9·99192	
02·0	9·28190	·1	6	10·71810	9·29000	·1	7	10·71000	10·00810	·1	0	9·99190	
03·0	9·28254	·2	13	10·71746	9·29067	·2	13	10·70933	10·00813	·2	1	9·99187	
04·0	9·28319	·3	19	10·71681	9·29134	·3	20	10·70866	10·00815	·3	1	9·99185	
05·0	9·28384	·4	26	10·71616	9·29201	·4	27	10·70799	10·00818	·4	1	9·99182	55´
06·0	9·28448			10·71552	9·29268			10·70732	10·00820			9·99180	
07·0	9·28512	·5	32	10·71488	9·29335	·5	33	10·70665	10·00823	·5	1	9·99177	
08·0	9·28577	·6	39	10·71423	9·29402	·6	40	10·70598	10·00825	·6	2	9·99175	
09·0	9·28641			10·71359	9·29468			10·70532	10·00828			9·99172	
10·0	9·28705	·7	45	10·71295	9·29535	·7	47	10·70465	10·00830	·7	2	9·99170	50´
11·0	9·28769	·8	51	10·71231	9·29601	·8	53	10·70399	10·00833	·8	2	9·99167	
12·0	9·28833	·9	58	10·71167	9·29668	·9	60	10·70332	10·00835	·9	2	9·99165	
13·0	9·28896			10·71104	9·29734			10·70266	10·00838			9·99162	
14·0	9·28960			10·71040	9·29800			10·70200	10·00840			9·99160	
15·0	9·29024			10·70976	9·29866			10·70134	10·00843			9·99157	45´
16·0	9·29087			10·70913	9·29932			10·70068	10·00845			9·99155	
17·0	9·29150	·1	6	10·70850	9·29998	·1	7	10·70002	10·00848	·1	0	9·99152	
18·0	9·29214	·2	13	10·70786	9·30064	·2	13	10·69936	10·00850	·2	1	9·99150	
19·0	9·29277	·3	19	10·70723	9·30130	·3	20	10·69871	10·00853	·3	1	9·99147	
20·0	9·29340	·4	25	10·70660	9·30195	·4	26	10·69805	10·00855	·4	1	9·99145	40´
21·0	9·29403	·5	31	10·70597	9·30261	·5	33	10·69739	10·00858	·5	1	9·99142	
22·0	9·29466			10·70534	9·30326			10·69674	10·00860			9·99140	
23·0	9·29529	·6	38	10·70471	9·30391	·6	39	10·69609	10·00863	·6	2	9·99137	
24·0	9·29591	·7	44	10·70409	9·30457	·7	46	10·69543	10·00865	·7	2	9·99135	
25·0	9·29654	·8	50	10·70346	9·30522	·8	52	10·69478	10·00868	·8	2	9·99132	35´
26·0	9·29716	·9	57	10·70284	9·30587	·9	59	10·69413	10·00871	·9	2	9·99130	
27·0	9·29779			10·70221	9·30652			10·69348	10·00873			9·99127	
28·0	9·29841			10·70159	9·30717			10·69283	10·00876			9·99124	
29·0	9·29903			10·70097	9·30781			10·69218	10·00878			9·99122	
30·0	9·29966			10·70035	9·30846			10·69154	10·00881			9·99119	30´
31·0	9·30028			10·69972	9·30911			10·69089	10·00883			9·99117	
32·0	9·30090	·1	6	10·69911	9·30975	·1	6	10·69025	10·00886	·1	0	9·99114	
33·0	9·30151	·2	12	10·69849	9·31040	·2	13	10·68960	10·00889	·2	1	9·99112	
34·0	9·30213	·3	18	10·69787	9·31104	·3	19	10·68896	10·00891	·3	1	9·99109	
35·0	9·30275	·4	25	10·69725	9·31169	·4	26	10·68832	10·00894	·4	1	9·99106	25´
36·0	9·30336	·5	31	10·69664	9·31233	·5	32	10·68767	10·00896	·5	1	9·99104	
37·0	9·30398			10·69602	9·31297			10·68703	10·00899			9·99101	
38·0	9·30459	·6	37	10·69541	9·31361	·6	38	10·68639	10·00901	·6	2	9·99099	
39·0	9·30521	·7	43	10·69479	9·31425	·7	45	10·68575	10·00904	·7	2	9·99096	
40·0	9·30582	·8	49	10·69418	9·31489	·8	51	10·68512	10·00907	·8	2	9·99093	20´
41·0	9·30643	·9	55	10·69357	9·31552	·9	58	10·68448	10·00909	·9	2	9·99091	
42·0	9·30704			10·69296	9·31616			10·68384	10·00912			9·99088	
43·0	9·30765			10·69235	9·31680			10·68321	10·00915			9·99086	
44·0	9·30826			10·69174	9·31743			10·68257	10·00917			9·99083	
45·0	9·30887			10·69113	9·31806			10·68194	10·00920			9·99080	15´
46·0	9·30947			10·69053	9·31870			10·68130	10·00922			9·99078	
47·0	9·31008	·1	6	10·68992	9·31939	·1	6	10·68067	10·00925	·1	0	9·99075	
48·0	9·31069	·2	12	10·68932	9·31996	·2	13	10·68004	10·00928	·2	1	9·99072	
49·0	9·31129	·3	18	10·68871	9·32059	·3	19	10·67941	10·00930	·3	1	9·99070	
50·0	9·31189	·4	24	10·68811	9·32122	·4	25	10·67878	10·00933	·4	1	9·99067	10´
51·0	9·31250	·5	30	10·68751	9·32185	·5	31	10·67815	10·00935	·5	1	9·99065	
52·0	9·31310			10·68690	9·32248			10·67752	10·00938			9·99062	
53·0	9·31370	·6	36	10·68630	9·32311	·6	38	10·67689	10·00941	·6	2	9·99059	
54·0	9·31430	·7	42	10·68570	9·32373	·7	44	10·67627	10·00944	·7	2	9·99057	
55·0	9·31490	·8	48	10·68510	9·32436	·8	50	10·67564	10·00946	·8	2	9·99054	5´
56·0	9·31550	·9	54	10·68451	9·32498	·9	57	10·67502	10·00949	·9	2	9·99051	
57·0	9·31609			10·68391	9·32561			10·67439	10·00952			9·99049	
58·0	9·31669			10·68331	9·32623			10·67377	10·00954			9·99046	
59·0	9·31728			10·68272	9·32685			10·67315	10·00957			9·99043	
60·0	9·31788			10·68212	9·32748			10·67253	10·00960			9·99040	0´

12°
192°

LOGS. OF TRIG. FUNCTIONS

′	Sine	Parts	Cosec.	Tan.	Parts	Cotan.	Secant	Parts	Cosine	
00·0	9·31788		10·68212	9·32748		10·67253	10·00960		9·99040	60′
01·0	9·31847		10·68153	9·32810		10·67191	10·00962		9·99038	
02·0	9·31907	·1 6	10·68093	9·32872	·1 6	10·67129	10·00965	·1 0	9·99035	
03·0	9·31966	·2 12	10·68034	9·32933	·2 12	10·67067	10·00968	·2 1	9·99032	
04·0	9·32025	·3 18	10·67975	9·32995	·3 18	10·67005	10·00970	·3 1	9·99030	
05·0	9·32084	·4 24	10·67916	9·33057	·4 25	10·66943	10·00973	·4 1	9·99027	55′
06·0	9·32143	·5 29	10·67857	9·33119	·5 31	10·66881	10·00976	·5 1	9·99024	
07·0	9·32202	·6 35	10·67798	9·33180	·6 37	10·66820	10·00979	·6 2	9·99022	
08·0	9·32261	·7 41	10·67739	9·33242	·7 43	10·66758	10·00981	·7 2	9·99019	
09·0	9·32319	·8 47	10·67681	9·33303	·8 49	10·66697	10·00984	·8 2	9·99016	
10·0	9·32378	·9 53	10·67622	9·33365	·9 55	10·66635	10·00987	·9 2	9·99013	50′
11·0	9·32437		10·67563	9·33426		10·66574	10·00989		9·99011	
12·0	9·32495		10·67505	9·33487		10·66513	10·00992		9·99008	
13·0	9·32553		10·67447	9·33548		10·66452	10·00995		9·99005	
14·0	9·32612		10·67388	9·33609		10·66391	10·00998		9·99003	
15·0	9·32670		10·67330	9·33670		10·66330	10·01000		9·99000	45′
16·0	9·32728		10·67272	9·33731		10·66269	10·01003		9·98997	
17·0	9·32786	·1 6	10·67214	9·33792	·1 6	10·66208	10·01006	·1 0	9·98994	
18·0	9·32844	·2 12	10·67156	9·33853	·2 12	10·66147	10·01009	·2 1	9·98992	
19·0	9·32902	·3 17	10·67098	9·33913	·3 18	10·66087	10·01011	·3 1	9·98989	
20·0	9·32960	·4 23	10·67040	9·33974	·4 24	10·66026	10·01014	·4 1	9·98986	40′
21·0	9·33018	·5 29	10·66982	9·34034	·5 30	10·65966	10·01017	·5 1	9·98983	
22·0	9·33075	·6 35	10·66925	9·34095	·6 36	10·65905	10·01020	·6 2	9·98980	
23·0	9·33133	·7 40	10·66867	9·34155	·7 42	10·65845	10·01022	·7 2	9·98978	
24·0	9·33190	·8 46	10·66810	9·34216	·8 48	10·65785	10·01025	·8 2	9·98975	
25·0	9·33248	·9 52	10·66752	9·34276	·9 54	10·65724	10·01028	·9 2	9·98972	35′
26·0	9·33305		10·66695	9·34336		10·65664	10·01031		9·98969	
27·0	9·33362		10·66638	9·34396		10·65604	10·01034		9·98967	
28·0	9·33420		10·66581	9·34456		10·65544	10·01036		9·98964	
29·0	9·33477		10·66523	9·34516		10·65484	10·01039		9·98961	
30·0	9·33534		10·66466	9·34576		10·65425	10·01042		9·98958	30′
31·0	9·33591		10·66409	9·34635		10·65365	10·01045		9·98955	
32·0	9·33648	·1 6	10·66353	9·34695	·1 6	10·65305	10·01048	·1 0	9·98953	
33·0	9·33704	·2 11	10·66296	9·34755	·2 12	10·65246	10·01050	·2 1	9·98950	
34·0	9·33761	·3 17	10·66239	9·34814	·3 18	10·65186	10·01053	·3 1	9·98947	
35·0	9·33818	·4 23	10·66182	9·34874	·4 24	10·65127	10·01056	·4 1	9·98944	25′
36·0	9·33874	·5 28	10·66126	9·34933	·5 30	10·65067	10·01059	·5 1	9·98941	
37·0	9·33931	·6 34	10·66069	9·34992	·6 36	10·65008	10·01062	·6 2	9·98939	
38·0	9·33987	·7 39	10·66013	9·35051	·7 41	10·64949	10·01064	·7 2	9·98936	
39·0	9·34043	·8 45	10·65957	9·35111	·8 47	10·64889	10·01067	·8 2	9·98933	
40·0	9·34100	·9 51	10·65900	9·35170	·9 53	10·64830	10·01070	·9 2	9·98930	20′
41·0	9·34156		10·65844	9·35229		10·64771	10·01073		9·98927	
42·0	9·34212		10·65788	9·35288		10·64712	10·01076		9·98924	
43·0	9·34268		10·65732	9·35347		10·64654	10·01079		9·98921	
44·0	9·34324		10·65676	9·35405		10·64595	10·01081		9·98919	
45·0	9·34380		10·65620	9·35464		10·64536	10·01084		9·98916	15′
46·0	9·34436		10·65565	9·35523		10·64477	10·01087		9·98913	
47·0	9·34491	·1 6	10·65509	9·35581	·1 6	10·64419	10·01090	·1 0	9·98910	
48·0	9·34547	·2 11	10·65453	9·35640	·2 12	10·64360	10·01093	·2 1	9·98907	
49·0	9·34602	·3 17	10·65398	9·35698	·3 17	10·64302	10·01096	·3 1	9·98904	
50·0	9·34658	·4 22	10·65342	9·35757	·4 23	10·64243	10·01099	·4 1	9·98901	10′
51·0	9·34713	·5 28	10·65287	9·35815	·5 29	10·64185	10·01102	·5 1	9·98899	
52·0	9·34769	·6 33	10·65231	9·35873	·6 35	10·64127	10·01104	·6 2	9·98896	
53·0	9·34824	·7 39	10·65176	9·35931	·7 41	10·64069	10·01107	·7 2	9·98893	
54·0	9·34879	·8 44	10·65121	9·35989	·8 47	10·64011	10·01110	·8 2	9·98890	
55·0	9·34934	·9 50	10·65066	9·36047	·9 52	10·63953	10·01113	·9 3	9·98887	5′
56·0	9·34989		10·65011	9·36105		10·63895	10·01116		9·98884	
57·0	9·35044		10·64956	9·36163		10·63837	10·01119		9·98881	
58·0	9·35099		10·64901	9·36221		10·63779	10·01122		9·98878	
59·0	9·35154		10·64846	9·36279		10·63721	10·01125		9·98875	
60·0	9·35209		10·64791	9·36336		10·63664	10·01128		9·98872	0′

167°
347°

LOGS. OF TRIG. FUNCTIONS

13°
193°

,	Sine	Parts	Cosec.	Tan.	Parts	Cotan.	Secant	Parts	Cosine	
00·0	9·35209	′	10·64791	9·36336	′	10·63664	10·01128	′	9·98872	60′
01·0	9·35264		10·64737	9·36394		10·63606	10·01131		9·98870	
02·0	9·35318	·1 5	10·64682	9·36452	·1 6	10·63549	10·01133	·1 0	9·98867	
03·0	9·35373	·2 11	10·64627	9·36509	·2 11	10·63491	10·01136	·2 1	9·98864	
04·0	9·35427	·3 16	10·64572	9·36566	·3 17	10·63434	10·01139	·3 1	9·98861	
05·0	9·35482	·4 22	10·64519	9·36624	·4 23	10·63376	10·01142	·4 1	9·98858	55′
06·0	9·35536	·5 27	10·64464	9·36681	·5 29	10·63319	10·01145	·5 1	9·98855	
07·0	9·35590	·6 33	10·64410	9·36738	·6 34	10·63262	10·01148	·6 2	9·98852	
08·0	9·35644	·7 38	10·64356	9·36795	·7 40	10·63205	10·01151	·7 2	9·98849	
09·0	9·35698		10·64302	9·36852		10·63148	10·01154		9·98846	
10·0	9·35752	·8 43	10·64248	9·36909	·8 46	10·63091	10·01157	·8 2	9·98843	50′
11·0	9·35806	·9 49	10·64194	9·36966	·9 51	10·63034	10·01160	·9 3	9·98840	
12·0	9·35860		10·64140	9·37023		10·62977	10·01163		9·98837	
13·0	9·35914		10·64086	9·37080		10·62920	10·01166		9·98834	
14·0	9·35968		10·64032	9·37137		10·62863	10·01169		9·98831	
15·0	9·36022		10·63979	9·37193		10·62807	10·01172		9·98828	45′
16·0	9·36075		10·63925	9·37250		10·62750	10·01175		9·98825	
17·0	9·36129	·1 5	10·63871	9·37306	·1 6	10·62694	10·01178	·1 0	9·98822	
18·0	9·36182	·2 11	10·63818	9·37363	·2 11	10·62637	10·01181	·2 1	9·98819	
19·0	9·36236	·3 16	10·63764	9·37419	·3 17	10·62581	10·01184	·3 1	9·98816	
20·0	9·36289	·4 21	10·63711	9·37476	·4 22	10·62524	10·01187	·4 1	9·98813	40′
21·0	9·36342	·5 27	10·63658	9·37532	·5 28	10·62468	10·01190	·5 2	9·98810	
22·0	9·36395	·6 32	10·63605	9·37595	·6 34	10·62412	10·01193	·6 2	9·98807	
23·0	9·36449	·7 37	10·63552	9·37644	·7 39	10·62356	10·01196	·7 2	9·98804	
24·0	9·36502		10·63498	9·37700		10·62300	10·01199		9·98801	
25·0	9·36555	·8 43	10·63445	9·37756	·8 45	10·62244	10·01202	·8 2	9·98798	35′
26·0	9·36608	·9 48	10·63393	9·37812	·9 51	10·62188	10·01205	·9 3	9·98795	
27·0	9·36660		10·63340	9·37868		10·62132	10·01208		9·98792	
28·0	9·36713		10·63287	9·37924		10·62076	10·01211		9·98789	
29·0	9·36766		10·63234	9·37980		10·62020	10·01214		9·98786	
30·0	9·36819		10·63182	9·38035		10·61965	10·01217		9·98783	30′
31·0	9·36871		10·63129	9·38091		10·61909	10·01220		9·98780	
32·0	9·36924	·1 5	10·63076	9·38147	·1 6	10·61853	10·01223	·1 0	9·98777	
33·0	9·36976	·2 10	10·63024	9·38202	·2 11	10·61798	10·01226	·2 1	9·98774	
34·0	9·37029	·3 16	10·62972	9·38258	·3 17	10·61743	10·01229	·3 1	9·98771	
35·0	9·37081	·4 21	10·62919	9·38313	·4 22	10·61687	10·01232	·4 1	9·98768	25′
36·0	9·37133	·5 26	10·62867	9·38368	·5 28	10·61632	10·01235	·5 2	9·98765	
37·0	9·37185	·6 31	10·62815	9·38423	·6 33	10·61577	10·01238	·6 2	9·98762	
38·0	9·37237	·7 36	10·62763	9·38479	·7 39	10·61521	10·01241	·7 2	9·98759	
39·0	9·37289		10·62711	9·38534		10·61466	10·01244		9·98756	
40·0	9·37341	·8 42	10·62659	9·38589	·8 44	10·61411	10·01247	·8 2	9·98753	20′
41·0	9·37393	·9 47	10·62607	9·38644	·9 50	10·61356	10·01250	·9 3	9·98750	
42·0	9·37445		10·62555	9·38699		10·61301	10·01254		9·98747	
43·0	9·37497		10·62503	9·38754		10·61246	10·01257		9·98743	
44·0	9·37549		10·62451	9·38808		10·61192	10·01260		9·98740	
45·0	9·37600		10·62400	9·38863		10·61137	10·01263		9·98737	15′
46·0	9·37652		10·62348	9·38918		10·61082	10·01266		9·98734	
47·0	9·37704	·1 5	10·62297	9·38972	·1 5	10·61028	10·01269	·1 0	9·98731	
48·0	9·37755	·2 10	10·62245	9·39027	·2 11	10·60973	10·01272	·2 1	9·98728	
49·0	9·37806	·3 15	10·62194	9·39082	·3 16	10·60919	10·01275	·3 1	9·98725	
50·0	9·37858	·4 20	10·62142	9·39136	·4 22	10·60864	10·01278	·4 1	9·98722	10′
51·0	9·37909	·5 26	10·62091	9·39190	·5 27	10·60810	10·01281	·5 2	9·98719	
52·0	9·37960	·6 31	10·62040	9·39245	·6 33	10·60755	10·01285	·6 2	9·98716	
53·0	9·38011	·7 36	10·61989	9·39299	·7 38	10·60701	10·01288	·7 2	9·98712	
54·0	9·38062		10·61938	9·39353		10·60647	10·01291		9·98709	
55·0	9·38113	·8 41	10·61887	9·39407	·8 43	10·60593	10·01294	·8 3	9·98706	5′
56·0	9·38164	·9 46	10·61836	9·39461	·9 49	10·60539	10·01297	·9 3	9·98703	
57·0	9·38215		10·61785	9·39515		10·60485	10·01300		9·98700	
58·0	9·38266		10·61734	9·39569		10·60431	10·01303		9·98697	
59·0	9·38317		10·61683	9·39623		10·60377	10·01306		9·98693	
60·0	9·38368		10·61632	9·39677		10·60323	10·01310		9·98690	0′

14°
194°

LOGS. OF TRIG. FUNCTIONS

′	Sine	Parts	Cosec.	Tan.	Parts	Cotan.	Secant	Parts	Cosine	
00·0	9·38368		10·61632	9·39677		10·60323	10·01310		9·98690	60′
01·0	9·38418		10·61582	9·39731		10·60269	10·01313		9·98687	
02·0	9·38469	·1 5	10·61531	9·39785	·1 5	10·60215	10·01316	·1 0	9·98684	
03·0	9·38519	·2 10	10·61481	9·39838	·2 11	10·60162	10·01319	·2 1	9·98681	
04·0	9·38570	·3 15	10·61430	9·39892	·3 16	10·60108	10·01322	·3 1	9·98678	
05·0	9·38620		10·61380	9·39946		10·60055	10·01325		9·98675	55′
06·0	9·38670	·4 20	10·61330	9·39999	·4 21	10·60001	10·01329	·4 1	9·98671	
07·0	9·38721	·5 25	10·61279	9·40052	·5 27	10·59948	10·01332	·5 2	9·98668	
08·0	9·38771	·6 30	10·61229	9·40106	·6 32	10·59894	10·01335	·6 2	9·98665	
09·0	9·38821		10·61179	9·40159		10·59841	10·01338		9·98662	
10·0	9·38871	·7 35	10·61129	9·40212	·7 37	10·59788	10·01341	·7 2	9·98659	50′
11·0	9·38921	·8 40	10·61079	9·40266	·8 43	10·59734	10·01345	·8 3	9·98656	
12·0	9·38971	·9 45	10·61029	9·40319	·9 48	10·59681	10·01348	·9 3	9·98652	
13·0	9·39021		10·60979	9·40372		10·59628	10·01351		9·98649	
14·0	9·39071		10·60929	9·40425		10·59575	10·01354		9·98646	
15·0	9·39121		10·60879	9·40478		10·59522	10·01358		9·98643	45′
16·0	9·39170		10·60830	9·40531		10·59469	10·01361		9·98640	
17·0	9·39220	·1 5	10·60780	9·40584	·1 5	10·59416	10·01364	·1 0	9·98636	
18·0	9·39270	·2 10	10·60731	9·40636	·2 11	10·59364	10·01367	·2 1	9·98633	
19·0	9·39319	·3 15	10·60681	9·40689	·3 16	10·59311	10·01370	·3 1	9·98630	
20·0	9·39369		10·60632	9·40742		10·59258	10·01373		9·98627	40′
21·0	9·39418	·4 20	10·60582	9·40795	·4 21	10·59206	10·01377	·4 1	9·98623	
22·0	9·39467	·5 25	10·60533	9·40847	·5 26	10·59153	10·01380	·5 2	9·98620	
23·0	9·39517	·6 30	10·60483	9·40900	·6 32	10·59100	10·01383	·6 2	9·98617	
24·0	9·39566		10·60434	9·40952		10·59048	10·01386		9·98614	
25·0	9·39615	·7 34	10·60385	9·41005	·7 37	10·58996	10·01390	·7 2	9·98610	35′
26·0	9·39664	·8 39	10·60336	9·41057	·8 42	10·58943	10·01393	·8 3	9·98607	
27·0	9·39713	·9 44	10·60287	9·41109	·9 47	10·58891	10·01396	·9 3	9·98604	
28·0	9·39762		10·60238	9·41162		10·58839	10·01399		9·98601	
29·0	9·39811		10·60189	9·41214		10·58786	10·01403		9·98597	
30·0	9·39860		10·60140	9·41266		10·58734	10·01406		9·98594	30′
31·0	9·39909		10·60091	9·41318		10·58682	10·01409		9·98591	
32·0	9·39958	·1 5	10·60043	9·41370	·1 5	10·58630	10·01412	·1 0	9·98588	
33·0	9·40006	·2 10	10·59994	9·41422	·2 10	10·58578	10·01416	·2 1	9·98584	
34·0	9·40055	·3 15	10·59945	9·41474	·3 16	10·58526	10·01419	·3 1	9·98581	
35·0	9·40104		10·59897	9·41526		10·58474	10·01422		9·98578	25′
36·0	9·40152	·4 19	10·59848	9·41578	·4 21	10·58423	10·01426	·4 1	9·98575	
37·0	9·40201	·5 24	10·59800	9·41629	·5 26	10·58371	10·01429	·5 2	9·98571	
38·0	9·40249	·6 29	10·59751	9·41681	·6 31	10·58319	10·01432	·6 2	9·98568	
39·0	9·40297		10·59703	9·41733		10·58267	10·01435		9·98565	
40·0	9·40346	·7 34	10·59655	9·41784	·7 36	10·58216	10·01439	·7 2	9·98561	20′
41·0	9·40394	·8 39	10·59606	9·41836	·8 41	10·58164	10·01442	·8 3	9·98558	
42·0	9·40442	·9 44	10·59558	9·41887	·9 47	10·58113	10·01445	·9 3	9·98555	
43·0	9·40490		10·59510	9·41939		10·58061	10·01449		9·98551	
44·0	9·40538		10·59462	9·41990		10·58010	10·01452		9·98548	
45·0	9·40586		10·59414	9·42042		10·57959	10·01455		9·98545	15′
46·0	9·40634		10·59366	9·42093		10·57907	10·01459		9·98541	
47·0	9·40682	·1 5	10·59318	9·42144	·1 5	10·57856	10·01462	·1 0	9·98538	
48·0	9·40730	·2 10	10·59270	9·42195	·2 10	10·57805	10·01465	·2 1	9·98535	
49·0	9·40778	·3 14	10·59222	9·42246	·3 15	10·57754	10·01469	·3 1	9·98531	
50·0	9·40825		10·59175	9·42297		10·57703	10·01472		9·98528	10′
51·0	9·40873	·4 19	10·59127	9·42348	·4 20	10·57652	10·01475	·4 1	9·98525	
52·0	9·40921	·5 24	10·59079	9·42399	·5 25	10·57601	10·01479	·5 2	9·98521	
53·0	9·40968	·6 29	10·59032	9·42450	·6 31	10·57550	10·01482	·6 2	9·98518	
54·0	9·41016		10·58984	9·42501		10·57499	10·01485		9·98515	
55·0	9·41063	·7 33	10·58937	9·42552	·7 36	10·57448	10·01489	·7 2	9·98511	5′
56·0	9·41111	·8 38	10·58889	9·42603	·8 41	10·57397	10·01492	·8 3	9·98508	
57·0	9·41158	·9 43	10·58842	9·42653	·9 46	10·57347	10·01496	·9 3	9·98505	
58·0	9·41205		10·58795	9·42704		10·57296	10·01499		9·98501	
59·0	9·41252		10·58748	9·42755		10·57245	10·01502		9·98498	
60·0	9·41300		10·58700	9·42805		10·57195	10·01506		9·98494	0′

15°
195°

LOGS. OF TRIG. FUNCTIONS

′	Sine	Parts	Cosec.	Tan.	Parts	Cotan.	Secant	Parts	Cosine	
00·0	9·41300	′	10·58700	9·42805	′	10·57195	10·01506	′	9·98494	60′
01·0	9·41347		10·58653	9·42856		10·57144	10·01509		9·98491	
02·0	9·41394		10·58606	9·42906		10·57094	10·01512		9·98488	
03·0	9·41441	·1 5	10·58559	9·42957	·1 5	10·57043	10·01516	·1 0	9·98484	
04·0	9·41488		10·58512	9·43007		10·56993	10·01519		9·98481	
05·0	9·41535	·2 9	10·58465	9·43057	·2 10	10·56943	10·01523	·2 1	9·98477	55′
06·0	9·41582	·3 14	10·58419	9·43108	·3 15	10·56893	10·01526	·3 1	9·98474	
07·0	9·41628		10·58372	9·43158		10·56842	10·01529		9·98471	
08·0	9·41675	·4 19	10·58325	9·43208	·4 20	10·56792	10·01533	·4 1	9·98467	
09·0	9·41722		10·58278	9·43258		10·56742	10·01536		9·98464	
10·0	9·41768	·5 23	10·58232	9·43308	·5 25	10·56692	10·01540	·5 2	9·98460	50′
11·0	9·41815	·6 28	10·58185	9·43358	·6 30	10·56642	10·01543	·6 2	9·98457	
12·0	9·41862		10·58139	9·43408		10·56592	10·01547		9·98454	
13·0	9·41908	·7 33	10·58092	9·43458	·7 35	10·56542	10·01550	·7 2	9·98450	
14·0	9·41954	·8 37	10·58046	9·43508	·8 40	10·56492	10·01553	·8 3	9·98447	
15·0	9·42001		10·57999	9·43558		10·56442	10·01557		9·98443	45′
16·0	9·42047	·9 42	10·57953	9·43607	·9 45	10·56393	10·01560	·9 3	9·98440	
17·0	9·42093		10·57907	9·43657		10·56343	10·01564		9·98436	
18·0	9·42140		10·57861	9·43707		10·56293	10·01567		9·98433	
19·0	9·42186		10·57814	9·43756		10·56244	10·01571		9·98429	
20·0	9·42232		10·57768	9·43806		10·56194	10·01574		9·98426	40′
21·0	9·42278		10·57722	9·43855		10·56145	10·01578		9·98422	
22·0	9·42324		10·57676	9·43905		10·56095	10·01581		9·98419	
23·0	9·42370	·1 5	10·57630	9·43954	·1 5	10·56046	10·01585	·1 0	9·98416	
24·0	9·42416		10·57584	9·44004		10·55996	10·01588		9·98412	
25·0	9·42462	·2 9	10·57539	9·44053	·2 10	10·55947	10·01592	·2 1	9·98409	35′
26·0	9·42507	·3 14	10·57493	9·44102	·3 15	10·55898	10·01595	·3 1	9·98405	
27·0	9·42553		10·57447	9·44151		10·55849	10·01599		9·98402	
28·0	9·42599	·4 18	10·57401	9·44201	·4 20	10·55799	10·01602	4 1	9·98398	
29·0	9·42644		10·57356	9·44250		10·55750	10·01605		9·98395	
30·0	9·42690	·5 23	10·57310	9·44299	·5 25	10·55701	10·01609	5 2	9·98391	30′
31·0	9·42735	·6 27	10·57265	9·44348	·6 29	10·55652	10·01613	·6 2	9·98388	
32·0	9·42781		10·57219	9·44397		10·55603	10·01616		9·98384	
33·0	9·42826	·7 32	10·57174	9·44446	·7 34	10·55554	10·01620	·7 2	9·98381	
34·0	9·42872	·8 36	10·57128	9·44495	·8 39	10·55505	10·01623	·8 3	9·98377	
35·0	9·42917		10·57083	9·44544		10·55457	10·01627		9·98374	25′
36·0	9·42962	·9 41	10·57038	9·44592	·9 44	10·55408	10·01630	·9 3	9·98370	
37·0	9·43008		10·56993	9·44641		10·55359	10·01634		9·98366	
38·0	9·43053		10·56947	9·44690		10·55310	10·01637		9·98363	
39·0	9·43098		10·56902	9·44738		10·55262	10·01641		9·98359	
40·0	9·43143		10·56857	9·44787		10·55213	10·01644		9·98356	20′
41·0	9·43188		10·56812	9·44836		10·55164	10·01648		9·98352	
42·0	9·43233		10·56767	9·44884		10·55116	10·01651		9·98349	
43·0	9·43278	·1 4	10·56722	9·44933	·1 5	10·55067	10·01655	·1 0	9·98345	
44·0	9·43323		10·56677	9·44981		10·55019	10·01658		9·98342	
45·0	9·43368	·2 9	10·56633	9·45029	·2 10	10·54971	10·01662	·2 1	9·98338	15′
46·0	9·43412	·3 13	10·56588	9·45078	·3 14	10·54922	10·01666	·3 1	9·98335	
47·0	9·43457		10·56543	9·45126		10·54874	10·01669		9·98331	
48·0	9·43502	·4 18	10·56498	9·45174	·4 19	10·54826	10·01673	·4 1	9·98327	
49·0	9·43546		10·56454	9·45223		10·54778	10·01676		9·98324	
50·0	9·43591	·5 22	10·56409	9·45271	·5 24	10·54729	10·01680	·5 2	9·98320	10′
51·0	9·43635	·6 27	10·56365	9·45319	·6 29	10·54681	10·01683	·6 2	9·98317	
52·0	9·43680		10·56320	9·45367		10·54633	10·01687		9·98313	
53·0	9·43724	·7 31	10·56276	9·45415	·7 34	10·54585	10·01691	·7 3	9·98309	
54·0	9·43769	·8 36	10·56231	9·45463	·8 39	10·54537	10·01694	·8 3	9·98306	
55·0	9·43813		10·56187	9·45511		10·54489	10·01698		9·98302	5′
56·0	9·43857	·9 40	10·56143	9·45559	·9 43	10·54441	10·01701	·9 3	9·98299	
57·0	9·43901		10·56099	9·45606		10·54394	10·01705		9·98295	
58·0	9·43946		10·56054	9·45654		10·54346	10·01709		9·98291	
59·0	9·43990		10·56010	9·45702		10·54298	10·01712		9·98288	
60·0	9·44034		10·55966	9·45750		10·54250	10·01716		9·98284	0′

16°
196°

LOGS. OF TRIG. FUNCTIONS

′	Sine	Parts	Cosec.	Tan	Parts	Cotan.	Secant	Parts	Cosine	
00·0	9·44034	′	10·55966	9·45750	′	10·54250	10·01716	′	9·98284	60′
01·0	9·44078		10·55922	9·45797		10·54203	10·01720		9·98281	
02·0	9·44122	·1 4	10·55878	9·45845	·1 5	10·54155	10·01723	·1 0	9·98277	
03·0	9·44166		10·55834	9·45893		10·54108	10·01727		9·98273	
04·0	9·44210	·2 9	10·55790	9·45940	·2 9	10·54060	10·01730	·2 1	9·98270	
05·0	9·44254		10·55747	9·45988		10·54013	10·01734		9·98266	55′
06·0	9·44297	·3 13	10·55703	9·46035	·3 14	10·53965	10·01738	·3 1	9·98262	
07·0	9·44341		10·55659	9·46082		10·53918	10·01741		9·98259	
08·0	9·44385	·4 17	10·55615	9·46130	·4 19	10·53870	10·01745	·4 1	9·98255	
09·0	9·44428		10·55572	9·46177		10·53823	10·01749		9·98251	
		·5 22			·5 24			·5 2		
10·0	9·44472		10·55528	9·46224		10·53776	10·01752		9·98248	50′
11·0	9·44516	·6 26	10·55485	9·46272	·6 28	10·53729	10·01756	·6 2	9·98244	
12·0	9·44559		10·55441	9·46319		10·53681	10·01760		9·98240	
13·0	9·44603	·7 30	10·55398	9·46366	·7 33	10·53634	10·01763	·7 3	9·98237	
14·0	9·44646		10·55354	9·46413		10·53587	10·01767		9·98233	
15·0	9·44689	·8 35	10·55311	9·46460	·8 38	10·53540	10·01771	·8 3	9·98229	45′
16·0	9·44733		10·55267	9·46507		10·53493	10·01774		9·98226	
17·0	9·44776	·9 39	10·55224	9·46554	·9 43	10·53446	10·01778	·9 3	9·98222	
18·0	9·44819		10·55181	9·46601		10·53399	10·01782		9·98218	
19·0	9·44862		10·55138	9·46648		10·53352	10·01785		9·98215	
20·0	9·44905		10·55095	9·46695		10·53306	10·01789		9·98211	40′
21·0	9·44949		10·55052	9·46741		10·53259	10·01793		9·98207	
22·0	9·44992	·1 4	10·55009	9·46788	·1 5	10·53212	10·01797	·1 0	9·98204	
23·0	9·45035		10·54966	9·46835		10·53165	10·01800		9·98200	
24·0	9·45078	·2 9	10·54923	9·46881	·2 9	10·53119	10·01804	·2 1	9·98196	
25·0	9·45120		10·54880	9·46928		10·53072	10·01808		9·98192	35′
26·0	9·45163	·3 13	10·54837	9·46975	·3 14	10·53025	10·01811	·3 1	9·98189	
27·0	9·45206		10·54794	9·47021		10·52979	10·01815		9·98185	
28·0	9·45249	·4 17	10·54751	9·47068	·4 19	10·52932	10·01819	·4 2	9·98181	
29·0	9·45292		10·54709	9·47114		10·52886	10·01823		9·98177	
		·5 21			·5 23			·5 2		
30·0	9·45334		10·54666	9·47161		10·52840	10·01826		9·98174	30′
31·0	9·45377	·6 26	10·54623	9·47207	·6 28	10·52793	10·01830	·6 2	9·98170	
32·0	9·45419		10·54581	9·47253		10·52747	10·01834		9·98166	
33·0	9·45462	·7 30	10·54538	9·47300	·7 32	10·52701	10·01838	·7 3	9·98163	
34·0	9·45504		10·54496	9·47346		10·52654	10·01841		9·98159	
35·0	9·45547	·8 34	10·54453	9·47392	·8 37	10·52608	10·01845	·8 3	9·98155	25′
36·0	9·45589		10·54411	9·47438		10·52562	10·01849		9·98151	
37·0	9·45632	·9 38	10·54368	9·47484	·9 42	10·52516	10·01853	·9 3	9·98147	
38·0	9·45674		10·54326	9·47530		10·52470	10·01856		9·98144	
39·0	9·45716		10·54284	9·47576		10·52424	10·01860		9·98140	
40·0	9·45758		10·54242	9·47622		10·52378	10·01864		9·98136	20′
41·0	9·45801		10·54199	9·47668		10·52332	10·01868		9·98132	
42·0	9·45843	·1 4	10·54157	9·47714	·1 5	10·52286	10·01872	·1 0	9·98129	
43·0	9·45885		10·54115	9·47760		10·52240	10·01875		9·98125	
44·0	9·45927	·2 8	10·54073	9·47806	·2 9	10·52194	10·01879	·2 1	9·98121	
45·0	9·45969		10·54031	9·47852		10·52148	10·01883		9·98117	15′
46·0	9·46011	·3 13	10·53989	9·47898	·3 14	10·52103	10·01887	·3 1	9·98113	
47·0	9·46053		10·53947	9·47943		10·52057	10·01891		9·98110	
48·0	9·46095	·4 17	10·53905	9·47989	·4 18	10·52011	10·01894	·4 2	9·98106	
49·0	9·46136		10·53864	9·48035		10·51966	10·01898		9·98102	
		·5 21			·5 23			·5 2		
50·0	9·46178		10·53822	9·48080		10·51920	10·01902		9·98098	10′
51·0	9·46220	·6 25	10·53780	9·48126	·6 27	10·51874	10·01906	·6 2	9·98094	
52·0	9·46262		10·53738	9·48171		10·51829	10·01910		9·98090	
53·0	9·46303	·7 29	10·53697	9·48217	·7 32	10·51783	10·01913	·7 3	9·98087	
54·0	9·46345		10·53655	9·48262		10·51738	10·01917		9·98083	
55·0	9·46386	·8 33	10·53614	9·48308	·8 36	10·51693	10·01921	·8 3	9·98079	5′
56·0	9·46428		10·53572	9·48353		10·51647	10·01925		9·98075	
57·0	9·46469	·9 38	10·53531	9·48398	·9 41	10·51602	10·01929	·9 3	9·98071	
58·0	9·46511		10·53489	9·48444		10·51557	10·01933		9·98067	
59·0	9·46552		10·53448	9·48489		10·51511	10·01937		9·98064	
60·0	9·46594		10·53407	9·48534		10·51466	10·01940		9·98060	0′

163°
343°

17°
197°

LOGS. OF TRIG. FUNCTIONS

,	Sine	Parts		Cosec.	Tan.	Parts		Cotan.	Secant	Parts		Cosine	
00·0	9·46594	′		10·53407	9·48534	′		10·51466	10·01940	′		9·98060	60′
01·0	9·46635			10·53365	9·48579			10·51421	10·01944			9·98056	
02·0	9·46676	·1	4	10·53324	9·48624	·1	4	10·51376	10·01948	·1	0	9·98052	
03·0	9·46717			10·53283	9·48669			10·51331	10·01952			9·98048	
04·0	9·46759	·2	8	10·53242	9·48714	·2	9	10·51286	10·01956	·2	1	9·98044	
05·0	9·46800			10·53200	9·48759			10·51241	10·01960			9·98040	55′
06·0	9·46841	·3	12	10·53159	9·48804	·3	13	10·51196	10·01964	·3	1	9·98036	
07·0	9·46882			10·53118	9·48849			10·51151	10·01968			9·98033	
08·0	9·46923	·4	16	10·53077	9·48894	·4	18	10·51106	10·01971	·4	2	9·98029	
09·0	9·46964			10·53036	9·48939			10·51061	10·01975			9·98025	
10·0	9·47005	·5	20	10·52995	9·48984	·5	22	10·51016	10·01979	·5	2	9·98021	50′
11·0	9·47046			10·52955	9·49029			10·50971	10·01983			9·98017	
12·0	9·47086	·6	25	10·52914	9·49073	·6	27	10·50927	10·01987	·6	2	9·98013	
13·0	9·47127			10·52873	9·49118			10·50882	10·01991			9·98009	
14·0	9·47168	·7	29	10·52832	9·49163	·7	31	10·50837	10·01995	·7	3	9·98005	
15·0	9·47209	·8	33	10·52791	9·49207	·8	36	10·50793	10·01999	·8	3	9·98001	45′
16·0	9·47249			10·52751	9·49252			10·50748	10·02003			9·97997	
17·0	9·47290	·9	37	10·52710	9·49297	·9	40	10·50704	10·02007	·9	4	9·97993	
18·0	9·47330			10·52670	9·49341			10·50659	10·02011			9·97990	
19·0	9·47371			10·52629	9·49385			10·50615	10·02015			9·97986	
20·0	9·47412			10·52589	9·49430			10·50570	10·02018			9·97982	40′
21·0	9·47452			10·52548	9·49474			10·50526	10·02022			9·97978	
22·0	9·47492	·1	4	10·52508	9·49519	·1	4	10·50481	10·02026	·1	0	9·97974	
23·0	9·47533			10·52467	9·49563			10·50437	10·02030			9·97970	
24·0	9·47573	·2	8	10·52427	9·49607	·2	9	10·50393	10·02034	·2	1	9·97966	
25·0	9·47613	·3	12	10·52387	9·49652	·3	13	10·50349	10·02038	·3	1	9·97962	35′
26·0	9·47654			10·52346	9·49696			10·50304	10·02042			9·97958	
27·0	9·47694	·4	16	10·52306	9·49740	·4	18	10·50260	10·02046	·4	2	9·97954	
28·0	9·47734			10·52266	9·49784			10·50216	10·02050			9·97950	
29·0	9·47774	·5	20	10·52226	9·49828	·5	22	10·50172	10·02054	·5	2	9·97946	
30·0	9·47814			10·52186	9·49872			10·50128	10·02058			9·97942	30′
31·0	9·47854	·6	24	10·52146	9·49916	·6	26	10·50084	10·02062	·6	2	9·97938	
32·0	9·47894			10·52106	9·49960			10·50040	10·02066			9·97934	
33·0	9·47934	·7	28	10·52066	9·50004	·7	31	10·49996	10·02070	·7	3	9·97930	
34·0	9·47974			10·52026	9·50048			10·49952	10·02074			9·97926	
35·0	9·48014	·8	32	10·51986	9·50092	·8	35	10·49908	10·02078	·8	3	9·97922	25′
36·0	9·48054			10·51946	9·50136			10·49864	10·02082			9·97918	
37·0	9·48094	·9	36	10·51906	9·50180	·9	40	10·49820	10·02086	·9	4	9·97914	
38·0	9·48133			10·51867	9·50224			10·49777	10·02090			9·97910	
39·0	9·48173			10·51827	9·50267			10·49733	10·02094			9·97906	
40·0	9·48213			10·51787	9·50311			10·49689	10·02098			9·97902	20′
41·0	9·48253			10·51748	9·50355			10·49645	10·02102			9·97898	
42·0	9·48292	·1	4	10·51708	9·50398	·1	4	10·49602	10·02106	·1	0	9·97894	
43·0	9·48332			10·51668	9·50442			10·49558	10·02110			9·97890	
44·0	9·48371	·2	8	10·51629	9·50485	·2	9	10·49515	10·02114	·2	1	9·97886	
45·0	9·48411	·3	12	10·51589	9·50529	·3	13	10·49471	10·02118	·3	1	9·97882	15′
46·0	9·48450			10·51550	9·50572			10·49428	10·02122			9·97878	
47·0	9·48490	·4	16	10·51511	9·50616	·4	17	10·49384	10·02126	·4	2	9·97874	
48·0	9·48529			10·51471	9·50659			10·49341	10·02130			9·97870	
49·0	9·48568	·5	20	10·51432	9·50703	·5	22	10·49297	10·02135	·5	2	9·97866	
50·0	9·48608			10·51393	9·50746			10·49254	10·02139			9·97862	10′
51·0	9·48647	·6	24	10·51353	9·50789	·6	26	10·49211	10·02143	·6	2	9·97857	
52·0	9·48686			10·51314	9·50833			10·49167	10·02147			9·97853	
53·0	9·48725	·7	27	10·51275	9·50876	·7	30	10·49124	10·02151	·7	3	9·97849	
54·0	9·48764			10·51236	9·50919			10·49081	10·02155			9·97845	
55·0	9·48803	·8	31	10·51197	9·50962	·8	35	10·49038	10·02159	·8	3	9·97841	5′
56·0	9·48842			10·51158	9·51005			10·48995	10·02163			9·97837	
57·0	9·48881	·9	35	10·51119	9·51049	·9	39	10·48952	10·02167	·9	4	9·97833	
58·0	9·48920			10·51080	9·51092			10·48908	10·02171			9·97829	
59·0	9·48959			10·51041	9·51135			10·48865	10·02175			9·97825	
60·0	9·48998			10·51002	9·51178			10·48822	10·02179			9·97821	0′

162°
342°

LOGS. OF TRIG. FUNCTIONS

18°
198°

′	Sine	Parts	Cosec.	Tan.	Parts	Cotan.	Secant	Parts	Cosine	
00·0	9·48998	′	10·51002	9·51178	′	10·48822	10·02179	′	9·97821	60′
01·0	9·49037		10·50963	9·51221		10·48779	10·02184		9·97817	
02·0	9·49076	·1 4	10·50924	9·51264	·1 4	10·48737	10·02188	·1 0	9·97812	
03·0	9·49115		10·50885	9·51306		10·48694	10·02192		9·97808	
04·0	9·49154	·2 8	10·50847	9·51349	·2 9	10·48651	10·02196	·2 1	9·97804	
05·0	9·49192		10·50808	9·51392		10·48608	10·02200		9·97800	55′
06·0	9·49231	·3 12	10·50769	9·51435	·3 13	10·48565	10·02204	·3 1	9·97796	
07·0	9·49270		10·50731	9·51478		10·48522	10·02208		9·97792	
08·0	9·49308	·4 15	10·50692	9·51520	·4 17	10·48480	10·02212	·4 2	9·97788	
09·0	9·49347		10·50653	9·51563		10·48437	10·02217		9·97784	
10·0	9·49385	·5 19	10·50615	9·51606	·5 21	10·48394	10·02221	·5 2	9·97779	50′
11·0	9·49424	·6 23	10·50576	9·51648		10·48352	10·02225	·6 2	9·97775	
12·0	9·49462		10·50538	9·51691	·6 26	10·48309	10·02229		9·97771	
13·0	9·49501	·7 27	10·50500	9·51734		10·48267	10·02233	·7 3	9·97767	
14·0	9·49539		10·50461	9·51776	·7 30	10·48224	10·02237		9·97763	
15·0	9·49577	·8 31	10·50423	9·51819	·8 34	10·48181	10·02241	·8 3	9·97759	45′
16·0	9·49615		10·50385	9·51861		10·48139	10·02246		9·97754	
17·0	9·49654	·9 35	10·50346	9·51903	·9 38	10·48097	10·02250	·9 4	9·97750	
18·0	9·49692		10·50308	9·51946		10·48054	10·02254		9·97746	
19·0	9·49730		10·50270	9·51988		10·48012	10·02258		9·97742	
20·0	9·49768		10·50232	9·52031		10·47970	10·02262		9·97738	40′
21·0	9·49806		10·50194	9·52073		10·47927	10·02267		9·97734	
22·0	9·49844	·1 4	10·50156	9·52115	·1 4	10·47885	10·02271	·1 0	9·97729	
23·0	9·49883		10·50118	9·52157		10·47843	10·02275		9·97725	
24·0	9·49920	·2 8	10·50080	9·52200	·2 8	10·47801	10·02279	·2 1	9·97721	
25·0	9·49958		10·50042	9·52242		10·47758	10·02283		9·97717	35′
26·0	9·49996	·3 11	10·50004	9·52284	·3 13	10·47716	10·02288	·3 1	9·97713	
27·0	9·50034		10·49966	9·52326		10·47674	10·02292		9·97708	
28·0	9·50072	·4 15	10·49928	9·52368	·4 17	10·47632	10·02296	·4 2	9·97704	
29·0	9·50110		10·49890	9·52410		10·47590	10·02300		9·97700	
30·0	9·50148	·5 19	10·49852	9·52452	·5 21	10·47548	10·02304	·5 2	9·97696	30′
31·0	9·50185	·6 23	10·49815	9·52494		10·47506	10·02309	·6 3	9·97691	
32·0	9·50223		10·49777	9·52536	·6 25	10·47464	10·02313		9·97687	
33·0	9·50261	·7 26	10·49739	9·52578		10·47422	10·02317	·7 3	9·97683	
34·0	9·50298		10·49702	9·52620	·7 29	10·47380	10·02321		9·97679	
35·0	9·50336	·8 30	10·49664	9·52662	·8 34	10·47339	10·02326	·8 3	9·97675	25′
36·0	9·50374		10·49627	9·52703		10·47297	10·02330		9·97670	
37·0	9·50411	·9 34	10·49589	9·52745	·9 38	10·47255	10·02334	·9 4	9·97666	
38·0	9·50449		10·49552	9·52787		10·47213	10·02338		9·97662	
39·0	9·50486		10·49514	9·52829		10·47172	10·02343		9·97657	
40·0	9·50523		10·49477	9·52870		10·47130	10·02347		9·97653	20′
41·0	9·50561		10·49439	9·52912		10·47088	10·02351		9·97649	
42·0	9·50598	·1 4	10·49402	9·52954	·1 4	10·47047	10·02355	·1 0	9·97645	
43·0	9·50635		10·49365	9·52995		10·47005	10·02360		9·97640	
44·0	9·50673	·2 7	10·49327	9·53037	·2 8	10·46963	10·02364	·2 1	9·97636	
45·0	9·50710		10·49290	9·53078		10·46922	10·02368		9·97632	15′
46·0	9·50747	·3 11	10·49253	9·53120	·3 12	10·46880	10·02373	·3 1	9·97628	
47·0	9·50784		10·49216	9·53161		10·46839	10·02377		9·97623	
48·0	9·50821	·4 15	10·49179	9·53203	·4 17	10·46798	10·02381	·4 2	9·97619	
49·0	9·50859		10·49142	9·53244		10·46756	10·02385		9·97615	
50·0	9·50896	·5 19	10·49104	9·53285	·5 21	10·46715	10·02390	·5 2	9·97610	10′
51·0	9·50933	·6 22	10·49067	9·53327		10·46673	10·02394	·6 3	9·97606	
52·0	9·50970		10·49030	9·53368	·6 25	10·46632	10·02398		9·97602	
53·0	9·51007	·7 26	10·48994	9·53409		10·46591	10·02403	·7 3	9·97597	
54·0	9·51043		10·48957	9·53450	·7 29	10·46550	10·02407		9·97593	
55·0	9·51080	·8 30	10·48920	9·53492	·8 33	10·46508	10·02411	·8 3	9·97589	5′
56·0	9·51117		10·48883	9·53533		10·46467	10·02416		9·97584	
57·0	9·51154	·9 33	10·48846	9·53574	·9 37	10·46426	10·02420	·9 4	9·97580	
58·0	9·51191		10·48809	9·53615		10·46385	10·02424		9·97576	
59·0	9·51228		10·48773	9·53656		10·46344	10·02429		9·97571	
60·0	9·51264		10·48736	9·53697		10·46303	10·02433		9·97567	0′

161°
341°

19°
199°

LOGS. OF TRIG. FUNCTIONS

′	Sine	Parts		Cosec.	Tan.	Parts		Cotan.	Secant	Parts		Cosine	
00·0	9·51264	′		10·48736	9·53697	′		10·46303	10·02433	′		9·97567	60′
01·0	9·51301			10·48699	9·53738			10·46262	10·02437			9·97563	
02·0	9·51338	·1	4	10·48663	9·53779	·1	4	10·46221	10·02442	·1	0	9·97558	
03·0	9·51374			10·48626	9·53820			10·46180	10·02446			9·97554	
04·0	9·51411	·2	7	10·48589	9·53861	·2	8	10·46139	10·02450	·2	1	9·97550	
05·0	9·51447			10·48553	9·53902			10·46098	10·02455			9·97545	55′
06·0	9·51484	·3	11	10·48516	9·53943	·3	12	10·46057	10·02459	·3	1	9·97541	
07·0	9·51520			10·48480	9·53984			10·46016	10·02464			9·97537	
08·0	9·51557	·4	15	10·48443	9·54025	·4	16	10·45976	10·02468	·4	2	9·97532	
09·0	9·51593			10·48407	9·54065			10·45935	10·02472			9·97528	
10·0	9·51629	·5	18	10·48371	9·54106	·5	20	10·45894	10·02477	·5	2	9·97523	50′
11·0	9·51666			10·48334	9·54147			10·45853	10·02481			9·97519	
12·0	9·51702	·6	22	10·48298	9·54188	·6	24	10·45813	10·02486	·6	3	9·97515	
13·0	9·51738			10·48262	9·54228			10·45772	10·02490			9·97510	
14·0	9·51775	·7	25	10·48226	9·54269	·7	29	10·45731	10·02494	·7	3	9·97506	
15·0	9·51811	·8	29	10·48189	9·54309	·8	33	10·45691	10·02499	·8	4	9·97501	45′
16·0	9·51847			10·48153	9·54350			10·45650	10·02503			9·97497	
17·0	9·51883	·9	33	10·48117	9·54391	·9	37	10·45610	10·02508	·9	4	9·97493	
18·0	9·51919			10·48081	9·54431			10·45569	10·02512			9·97488	
19·0	9·51955			10·48045	9·54472			10·45529	10·02516			9·97484	
20·0	9·51991			10·48009	9·54512			10·45488	10·02521			9·97479	40′
21·0	9·52027			10·47973	9·54552			10·45448	10·02525			9·97475	
22·0	9·52063	·1	4	10·47937	9·54593	·1	4	10·45407	10·02530	·1	0	9·97470	
23·0	9·52099			10·47901	9·54633			10·45367	10·02534			9·97466	
24·0	9·52135	·2	7	10·47865	9·54674	·2	8	10·45327	10·02539	·2	1	9·97461	
25·0	9·52171			10·47829	9·54714			10·45286	10·02543			9·97457	35′
26·0	9·52207	·3	11	10·47793	9·54754	·3	12	10·45246	10·02548	·3	1	9·97453	
27·0	9·52242			10·47758	9·54794			10·45206	10·02552			9·97448	
28·0	9·52278	·4	14	10·47722	9·54835	·4	16	10·45166	10·02556	·4	2	9·97444	
29·0	9·52314			10·47686	9·54875			10·45125	10·02561			9·97439	
30·0	9·52350	·5	18	10·47651	9·54915	·5	20	10·45085	10·02565	·5	2	9·97435	30′
31·0	9·52385			10·47615	9·54955			10·45045	10·02570			9·97430	
32·0	9·52421	·6	21	10·47579	9·54995	·6	24	10·45005	10·02574	·6	3	9·97426	
33·0	9·52456			10·47544	9·55035			10·44965	10·02579			9·97421	
34·0	9·52492	·7	25	10·47508	9·55075	·7	28	10·44925	10·02583	·7	3	9·97417	
35·0	9·52528	·8	29	10·47473	9·55115	·8	32	10·44885	10·02588	·8	4	9·97412	25′
36·0	9·52563			10·47437	9·55155			10·44845	10·02592			9·97408	
37·0	9·52598	·9	32	10·47402	9·55195	·9	36	10·44805	10·02597	·9	4	9·97403	
38·0	9·52634			10·47366	9·55235			10·44765	10·02601			9·97399	
39·0	9·52669			10·47331	9·55275			10·44725	10·02606			9·97394	
40·0	9·52705			10·47295	9·55315			10·44685	10·02610			9·97390	20′
41·0	9·52740			10·47260	9·55355			10·44645	10·02615			9·97385	
42·0	9·52775	·1	4	10·47225	9·55395	·1	4	10·44605	10·02619	·1	0	9·97381	
43·0	9·52811			10·47190	9·55434			10·44566	10·02624			9·97376	
44·0	9·52846	·2	7	10·47154	9·55474	·2	8	10·44526	10·02628	·2	1	9·97372	
45·0	9·52881			10·47119	9·55514			10·44486	10·02633			9·97367	15′
46·0	9·52916	·3	11	10·47084	9·55554	·3	12	10·44446	10·02638	·3	1	9·97363	
47·0	9·52951			10·47049	9·55593			10·44407	10·02642			9·97358	
48·0	9·52986	·4	14	10·47014	9·55633	·4	16	10·44367	10·02647	·4	2	9·97354	
49·0	9·53022			10·46979	9·55673			10·44328	10·02651			9·97349	
50·0	9·53057	·5	18	10·46944	9·55712	·5	20	10·44288	10·02656	·5	2	9·97344	10′
51·0	9·53092			10·46909	9·55752			10·44248	10·02660			9·97340	
52·0	9·53127	·6	21	10·46874	9·55791	·6	24	10·44209	10·02665	·6	3	9·97335	
53·0	9·53161			10·46839	9·55831			10·44169	10·02669			9·97331	
54·0	9·53196	·7	25	10·46804	9·55870	·7	28	10·44130	10·02674	·7	3	9·97326	
55·0	9·53231	·8	28	10·46769	9·55910	·8	32	10·44090	10·02679	·8	4	9·97322	5′
56·0	9·53266			10·46734	9·55949			10·44051	10·02683			9·97317	
57·0	9·53301	·9	32	10·46699	9·55989	·9	36	10·44012	10·02688	·9	4	9·97312	
58·0	9·53336			10·46664	9·56028			10·43972	10·02692			9·97308	
59·0	9·53370			10·46630	9·56067			10·43933	10·02697			9·97303	
60·0	9·53405			10·46595	9·56107			10·43893	10·02701			9·97299	0′

20°
200°

LOGS. OF TRIG. FUNCTIONS

′	Sine	Parts	Cosec.	Tan.	Parts	Cotan.	Secant	Parts	Cosine	
00·0	9·53405	′	10·46595	9·56107	′	10·43893	10·02701	′	9·97299	60′
01·0	9·53440		10·46560	9·56146		10·43854	10·02706		9·97294	
02·0	9·53475	·1 3	10·46526	9·56185	·1 4	10·43815	10·02711	·1 0	9·97289	
03·0	9·53509		10·46491	9·56224		10·43776	10·02715		9·97285	
04·0	9·53544		10·46456	9·56264		10·43736	10·02720		9·97280	
05·0	9·53578	·2 7	10·46422	9·56303	·2 8	10·43697	10·02725	·2 1	9·97276	55′
06·0	9·53613		10·46387	9·56342		10·43658	10·02729		9·97271	
07·0	9·53647		10·46353	9·56381		10·43619	10·02734		9·97266	
08·0	9·53682	·3 10	10·46318	9·56420	·3 12	10·43580	10·02738	·3 1	9·97262	
09·0	9·53716		10·46284	9·56459		10·43541	10·02743		9·97257	
10·0	9·53751		10·46249	9·56498		10·43502	10·02748		9·97252	50′
11·0	9·53785	·4 14	10·46215	9·56537	·4 16	10·43463	10·02752	·4 2	9·97248	
12·0	9·53819		10·46181	9·56576		10·43424	10·02757		9·97243	
13·0	9·53854		10·46146	9·56615		10·43385	10·02762		9·97239	
14·0	9·53888	·5 17	10·46112	9·56654	·5 19	10·43346	10·02766	·5 2	9·97234	
15·0	9·53922		10·46078	9·56693		10·43307	10·02771		9·97229	45′
16·0	9·53957		10·46044	9·56732		10·43268	10·02776		9·97225	
17·0	9·53991		10·46009	9·56771		10·43229	10·02780		9·97220	
18·0	9·54025	·6 21	10·45975	9·56810	·6 23	10·43190	10·02785	·6 3	9·97215	
19·0	9·54059		10·45941	9·56849		10·43151	10·02790		9·97211	
20·0	9·54093	·7 24	10·45907	9·56887	·7 27	10·43113	10·02794	·7 3	9·97206	40′
21·0	9·54127		10·45873	9·56926		10·43074	10·02799		9·97201	
22·0	9·54161		10·45839	9·56965		10·43035	10·02804		9·97196	
23·0	9·54195		10·45805	9·57004		10·42997	10·02808		9·97192	
24·0	9·54229	·8 27	10·45771	9·57042	·8 31	10·42958	10·02813	·8 4	9·97187	35′
25·0	9·54263		10·45737	9·57081		10·42919	10·02818		9·97182	
26·0	9·54297		10·45703	9·57120		10·42881	10·02822		9·97178	
27·0	9·54331	·9 31	10·45669	9·57158	·9 35	10·42842	10·02827	·9 4	9·97173	
28·0	9·54365		10·45635	9·57197		10·42803	10·02832		9·97168	
29·0	9·54399		10·45601	9·57235		10·42765	10·02837		9·97164	
30·0	9·54433		10·45568	9·57274		10·42726	10·02841		9·97159	30′
31·0	9·54466		10·45534	9·57312		10·42688	10·02846		9·97154	
32·0	9·54500	·1 3	10·45500	9·57351	·1 4	10·42649	10·02851	·1 0	9·97149	
33·0	9·54534		10·45466	9·57389		10·42611	10·02855		9·97145	
34·0	9·54567		10·45433	9·57428		10·42572	10·02860		9·97140	
35·0	9·54601	·2 7	10·45399	9·57466	·2 8	10·42534	10·02865	·2 1	9·97135	25′
36·0	9·54635		10·45365	9·57504		10·42496	10·02870		9·97130	
37·0	9·54668		10·45332	9·57543		10·42457	10·02874		9·97126	
38·0	9·54702	·3 10	10·45298	9·57581	·3 11	10·42419	10·02879	·3 1	9·97121	
39·0	9·54735		10·45265	9·57619		10·42381	10·02884		9·97116	
40·0	9·54769		10·45231	9·57658		10·42342	10·02889		9·97111	20′
41·0	9·54802	·4 13	10·45198	9·57696	·4 15	10·42304	10·02893	·4 2	9·97107	
42·0	9·54836		10·45164	9·57734		10·42266	10·02898		9·97102	
43·0	9·54869		10·45131	9·57772		10·42228	10·02903		9·97097	
44·0	9·54903	·5 17	10·45097	9·57810	·5 19	10·42190	10·02908	·5 2	9·97092	
45·0	9·54936		10·45064	9·57849		10·42151	10·02913		9·97087	15′
46·0	9·54969		10·45031	9·57887		10·42113	10·02917		9·97083	
47·0	9·55003		10·44997	9·57925		10·42075	10·02922		9·97078	
48·0	9·55036	·6 20	10·44964	9·57963	·6 23	10·42037	10·02927	·6 3	9·97073	
49·0	9·55069		10·44931	9·58001		10·41999	10·02932		9·97068	
50·0	9·55102	·7 23	10·44898	9·58039	·7 27	10·41961	10·02937	·7 3	9·97064	10′
51·0	9·55136		10·44864	9·58077		10·41923	10·02941		9·97059	
52·0	9·55169		10·44831	9·58115		10·41885	10·02946		9·97054	
53·0	9·55202		10·44798	9·58153		10·41847	10·02951		9·97049	
54·0	9·55235	·8 27	10·44765	9·58191	·8 31	10·41809	10·02956	·8 4	9·97044	
55·0	9·55268		10·44732	9·58229		10·41771	10·02961		9·97039	5′
56·0	9·55301		10·44699	9·58267		10·41734	10·02966		9·97035	
57·0	9·55334	·9 30	10·44666	9·58304	·9 34	10·41696	10·02970	·9 4	9·97030	
58·0	9·55367		10·44633	9·58342		10·41658	10·02975		9·97025	
59·0	9·55400		10·44600	9·58380		10·41620	10·02980		9·97020	
60·0	9·55433		10·44567	9·58418		10·41582	10·02985		9·97015	0′

21°
201°

LOGS. OF TRIG. FUNCTIONS

′	Sine	Parts		Cosec.	Tan.	Parts		Cotan.	Secant	Parts		Cosine	
00·0	9·55433	′		10·44567	9·58418	′		10·41582	10·02985	′		9·97015	60′
01·0	9·55466			10·44534	9·58456			10·41545	10·02990			9·97010	
02·0	9·55499	·1	3	10·44501	9·58493	·1	4	10·41507	10·02995	·1	0	9·97006	
03·0	9·55532			10·44469	9·58531			10·41469	10·02999			9·97001	
04·0	9·55564			10·44436	9·58569			10·41431	10·03004			9·96996	
05·0	9·55597	·2	7	10·44403	9·58606	·2	7	10·41394	10·03009	·2	1	9·96991	55′
06·0	9·55630			10·44370	9·58644			10·41356	10·03014			9·96986	
07·0	9·55663			10·44337	9·58682			10·41319	10·03019			9·96981	
08·0	9·55695	·3	10	10·44305	9·58719	·3	11	10·41281	10·03024	·3	1	9·96976	
09·0	9·55728			10·44272	9·58757			10·41243	10·03029			9·96971	
10·0	9·55761			10·44239	9·58794			10·41206	10·03034			9·96967	50′
11·0	9·55793	·4	13	10·44207	9·58832	·4	15	10·41168	10·03038	·4	2	9·96962	
12·0	9·55826			10·44174	9·58869			10·41131	10·03043			9·96957	
13·0	9·55858			10·44141	9·58907			10·41093	10·03048			9·96952	
14·0	9·55891	·5	16	10·44109	9·58944	·5	19	10·41056	10·03053	·5	2	9·96947	
15·0	9·55923			10·44077	9·58981			10·41019	10·03058			9·96942	45′
16·0	9·55956			10·44044	9·59019			10·40981	10·03063			9·96937	
17·0	9·55988			10·44012	9·59056			10·40944	10·03068			9·96932	
18·0	9·56021	·6	20	10·43979	9·59094	·6	22	10·40907	10·03073	·6	3	9·96927	
19·0	9·56053			10·43947	9·59131			10·40869	10·03078			9·96922	
20·0	9·56086			10·43915	9·59168			10·40832	10·03083			9·96917	40′
21·0	9·56118	·7	23	10·43882	9·59205	·7	26	10·40795	10·03088	·7	3	9·96912	
22·0	9·56150			10·43850	9·59243			10·40757	10·03093			9·96908	
23·0	9·56182			10·43818	9·59280			10·40720	10·03098			9·96903	
24·0	9·56215	·8	26	10·43785	9·59317	·8	30	10·40683	10·03102	·8	4	9·96898	
25·0	9·56247			10·43753	9·59354			10·40646	10·03107			9·96893	35′
26·0	9·56279			10·43721	9·59391			10·40609	10·03112			9·96888	
27·0	9·56311	·9	29	10·43689	9·59429	·9	34	10·40572	10·03117	·9	4	9·96883	
28·0	9·56343			10·43657	9·59466			10·40534	10·03122			9·96878	
29·0	9·56376			10·43625	9·59503			10·40497	10·03127			9·96873	
30·0	9·56408			10·43593	9·59540			10·40460	10·03132			9·96868	30′
31·0	9·56440			10·43560	9·59577			10·40423	10·03137			9·96863	
32·0	9·56472	·1	3	10·43528	9·59614	·1	4	10·40386	10·03142	·1	1	9·96858	
33·0	9·56504			10·43496	9·59651			10·40349	10·03147			9·96853	
34·0	9·56536			10·43464	9·59688			10·40312	10·03152			9·96848	
35·0	9·56568	·2	6	10·43432	9·59725	·2	7	10·40275	10·03157	·2	1	9·96843	25′
36·0	9·56600			10·43401	9·59762			10·40238	10·03162			9·96838	
37·0	9·56631			10·43369	9·59799			10·40202	10·03167			9·96833	
38·0	9·56663	·3	10	10·43337	9·59835	·3	11	10·40165	10·03172	·3	2	9·96828	
39·0	9·56695			10·43305	9·59872			10·40128	10·03177			9·96823	
40·0	9·56727			10·43273	9·59909			10·40091	10·03182			9·96818	20′
41·0	9·56759	·4	13	10·43241	9·59946	·4	15	10·40054	10·03187	·4	2	9·96813	
42·0	9·56790			10·43210	9·59983			10·40017	10·03192			9·96808	
43·0	9·56822			10·43178	9·60019			10·39981	10·03197			9·96803	
44·0	9·56854	·5	16	10·43146	9·60056	·5	18	10·39944	10·03202	·5	3	9·96798	
45·0	9·56886			10·43114	9·60093			10·39907	10·03207			9·96793	15′
46·0	9·56917			10·43083	9·60130			10·39870	10·03212			9·96788	
47·0	9·56949			10·43051	9·60166			10·39834	10·03217			9·96783	
48·0	9·56980	·6	19	10·43020	9·60203	·6	22	10·39797	10·03223	·6	3	9·96778	
49·0	9·57012			10·42988	9·60240			10·39761	10·03228			9·96773	
50·0	9·57044			10·42957	9·60276			10·39724	10·03233			9·96767	10′
51·0	9·57075	·7	22	10·42925	9·60313	·7	26	10·39687	10·03238	·7	4	9·96762	
52·0	9·57107			10·42893	9·60349			10·39651	10·03243			9·96757	
53·0	9·57138			10·42862	9·60386			10·39614	10·03248			9·96752	
54·0	9·57170	·8	25	10·42831	9·60422	·8	29	10·39578	10·03253	·8	4	9·96747	
55·0	9·57201			10·42799	9·60459			10·39541	10·03258			9·96742	5′
56·0	9·57232			10·42768	9·60495			10·39505	10·03263			9·96737	
57·0	9·57264	·9	29	10·42736	9·60532	·9	33	10·39468	10·03268	·9	5	9·96732	
58·0	9·57295			10·42705	9·60568			10·39432	10·03273			9·96727	
59·0	9·57326			10·42674	9·60605			10·39395	10·03278			9·96722	
60·0	9·57358			10·42643	9·60641			10·39359	10·03283			9·96717	0′

22°
202°

LOGS. OF TRIG. FUNCTIONS

′	Sine	Parts	Cosec.	Tan.	Parts	Cotan.	Secant	Parts	Cosine	
		′			′			′		
00·0	9·57358		10·42643	9·60641		10·39359	10·03283		9·96717	60′
01·0	9·57389		10·42611	9·60677		10·39323	10·03289		9·96712	
02·0	9·57420	·1 3	10·42580	9·60714	·1 4	10·39286	10·03294	·1 1	9·96706	
03·0	9·57451		10·42549	9·60750		10·39250	10·03299		9·96701	
04·0	9·57482		10·42518	9·60786		10·39214	10·03304		9·96696	
05·0	9·57514	·2 6	10·42486	9·60823	·2 7	10·39178	10·03309	·2 1	9·96691	55′
06·0	9·57545		10·42455	9·60859		10·39141	10·03314		9·96686	
07·0	9·57576		10·42424	9·60895		10·39105	10·03319		9·96681	
08·0	9·57607	·3 9	10·42393	9·60931	·3 11	10·39069	10·03324	·3 2	9·96676	
09·0	9·57638		10·42362	9·60967		10·39033	10·03330		9·96671	
10·0	9·57669		10·42331	9·61004		10·38996	10·03335		9·96665	50′
11·0	9·57700	·4 12	10·42300	9·61040	·4 14	10·38960	10·03340	·4 2	9·96660	
12·0	9·57731		10·42269	9·61076		10·38924	10·03345		9·96655	
13·0	9·57762		10·42238	9·61112		10·38888	10·03350		9·96650	
14·0	9·57793	·5 15	10·42207	9·61148	·5 18	10·38852	10·03355	·5 3	9·96645	
15·0	9·57824		10·42176	9·61184		10·38816	10·03361		9·96640	45′
16·0	9·57855		10·42146	9·61220		10·38780	10·03366		9·96634	
17·0	9·57885		10·42115	9·61256		10·38744	10·03371		9·96629	
18·0	9·57916	·6 19	10·42084	9·61292	·6 22	10·38708	10·03376	·6 3	9·96624	
19·0	9·57947		10·42053	9·61328		10·38672	10·03381		9·96619	
20·0	9·57978	·7 22	10·42022	9·61364	·7 25	10·38636	10·03386	·7 4	9·96614	40′
21·0	9·58009		10·41992	9·61400		10·38600	10·03392		9·96609	
22·0	9·58039		10·41961	9·61436		10·38564	10·03397		9·96603	
23·0	9·58070		10·41930	9·61472		10·38528	10·03402		9·96598	
24·0	9·58101	·8 25	10·41900	9·61508	·8 29	10·38492	10·03407	·8 4	9·96593	
25·0	9·58131		10·41869	9·61544		10·38457	10·03412		9·96588	35′
26·0	9·58162		10·41838	9·61579		10·38421	10·03418		9·96582	
27·0	9·58192	·9 28	10·41808	9·61615	·9 32	10·38385	10·03423	·9 5	9·96577	
28·0	9·58223		10·41777	9·61651		10·38349	10·03428		9·96572	
29·0	9·58254		10·41747	9·61687		10·38313	10·03433		9·96567	
30·0	9·58284		10·41716	9·61722		10·38277	10·03439		9·96562	30′
31·0	9·58314		10·41685	9·61758		10·38242	10·03444		9·96556	
32·0	9·58345	·1 3	10·41655	9·61794	·1 4	10·38206	10·03449	·1 1	9·96551	
33·0	9·58375		10·41625	9·61830		10·38171	10·03454		9·96546	
34·0	9·58406		10·41594	9·61865		10·38135	10·03459		9·96541	
35·0	9·58436	·2 6	10·41564	9·61901	·2 7	10·38099	10·03465	·2 1	9·96535	25′
36·0	9·58467		10·41534	9·61936		10·38064	10·03470		9·96530	
37·0	9·58497		10·41503	9·61972		10·38028	10·03475		9·96525	
38·0	9·58527	·3 9	10·41473	9·62008	·3 11	10·37992	10·03481	·3 2	9·96520	
39·0	9·58557		10·41443	9·62043		10·37957	10·03486		9·96514	
40·0	9·58588		10·41412	9·62079		10·37921	10·03491		9·96509	20′
41·0	9·58618	·4 12	10·41382	9·62114	·4 14	10·37886	10·03496	·4 2	9·96504	
42·0	9·58648		10·41352	9·62150		10·37850	10·03502		9·96498	
43·0	9·58678		10·41322	9·62185		10·37815	10·03507		9·96493	
44·0	9·58709	·5 15	10·41292	9·62221	·5 18	10·37779	10·03512	·5 3	9·96488	
45·0	9·58739		10·41261	9·62256		10·37744	10·03517		9·96483	15′
46·0	9·58769		10·41231	9·62292		10·37709	10·03523		9·96477	
47·0	9·58799		10·41201	9·62327		10·37673	10·03528		9·96472	
48·0	9·58829	·6 18	10·41171	9·62362	·6 21	10·37638	10·03533	·6 3	9·96467	
49·0	9·58859		10·41141	9·62398		10·37602	10·03539		9·96461	
50·0	9·58889		10·41111	9·62433		10·37567	10·03544		9·96456	10′
51·0	9·58919	·7 21	10·41081	9·62468	·7 25	10·37532	10·03549	·7 4	9·96451	
52·0	9·58949		10·41051	9·62504		10·37496	10·03555		9·96445	
53·0	9·58979		10·41021	9·62539		10·37461	10·03560		9·96440	
54·0	9·59009	·8 24	10·40991	9·62574	·8 28	10·37426	10·03565	·8 4	9·96435	
55·0	9·59039		10·40961	9·62609		10·37391	10·03571		9·96429	5′
56·0	9·59069		10·40931	9·62645		10·37356	10·03576		9·96424	
57·0	9·59098	·9 27	10·40902	9·62680	·9 32	10·37320	10·03581	·9 5	9·96419	
58·0	9·59128		10·40872	9·62715		10·37285	10·03587		9·96413	
59·0	9·59158		10·40842	9·62750		10·37250	10·03592		9·96408	
60·0	9·59188		10·40812	9·62785		10·37215	10·03597		9·96403	0′

157°
337°

LOGS. OF TRIG. FUNCTIONS

′	Sine	Parts	Cosec.	Tan.	Parts	Cotan.	Secant	Parts	Cosine	
00·0	9·59188	′	10·40812	9·62785	′	10·37215	10·03597	′	9·96403	*60′*
01·0	9·59218		10·40782	9·62820		10·37180	10·03603		9·96397	
02·0	9·59247	·1 3	10·40753	9·62855	·1 3	10·37145	10·03608	·1 1	9·96392	
03·0	9·59277		10·40723	9·62891		10·37110	10·03614		9·96387	
04·0	9·59307		10·40693	9·62926		10·37075	10·03619		9·96381	
05·0	9·59336	·2 6	10·40664	9·62961	·2 7	10·37039	10·03624	·2 1	9·96376	*55′*
06·0	9·59366		10·40634	9·62996		10·37004	10·03630		9·96370	
07·0	9·59396		10·40605	9·63031		10·36969	10·03635		9·96365	
08·0	9·59425	·3 9	10·40575	9·63066	·3 10	10·36934	10·03640	·3 2	9·96360	
09·0	9·59455		10·40545	9·63101		10·36900	10·03646		9·96354	
10·0	9·59484		10·40516	9·63136		10·36865	10·03651		9·96349	*50′*
11·0	9·59514	·4 12	10·40486	9·63170	·4 14	10·36830	10·03657	·4 2	9·96343	
12·0	9·59543		10·40457	9·63205		10·36795	10·03662		9·96338	
13·0	9·59573		10·40427	9·63240		10·36760	10·03668		9·96333	
14·0	9·59602		10·40398	9·63275		10·36725	10·03673		9·96327	
15·0	9·59632	·5 15	10·40369	9·63310	·5 17	10·36690	10·03678	·5 3	9·96322	*45′*
16·0	9·59661		10·40339	9·63345		10·36655	10·03684		9·96316	
17·0	9·59690		10·40310	9·63380		10·36621	10·03689		9·96311	
18·0	9·59720	·6 18	10·40280	9·63414	·6 21	10·36586	10·03695	·6 3	9·96305	
19·0	9·59749		10·40251	9·63449		10·36551	10·03700		9·96300	
20·0	9·59778	·7 21	10·40222	9·63484	·7 24	10·36516	10·03706	·7 4	9·96295	*40′*
21·0	9·59808		10·40193	9·63519		10·36482	10·03711		9·96289	
22·0	9·59837		10·40163	9·63553		10·36447	10·03716		9·96284	
23·0	9·59866		10·40134	9·63588		10·36412	10·03722		9·96278	
24·0	9·59895	·8 24	10·40105	9·63623	·8 28	10·36377	10·03727	·8 4	9·96273	
25·0	9·59924		10·40076	9·63657		10·36343	10·03733		9·96267	*35′*
26·0	9·59954		10·40046	9·63692		10·36308	10·03738		9·96262	
27·0	9·59983	·9 26	10·40017	9·63727	·9 31	10·36274	10·03744	·9 5	9·96256	
28·0	9·60012		10·39988	9·63761		10·36239	10·03749		9·96251	
29·0	9·60041		10·39959	9·63796		10·36204	10·03755		9·96245	
30·0	9·60070		10·39930	9·63830		10·36170	10·03760		9·96240	*30′*
31·0	9·60099		10·39901	9·63865		10·36135	10·03766		9·96234	
32·0	9·60128	·1 3	10·39872	9·63899	·1 3	10·36101	10·03771	·1 1	9·96229	
33·0	9·60157		10·39843	9·63934		10·36066	10·03777		9·96223	
34·0	9·60186		10·39814	9·63968		10·36032	10·03782		9·96218	
35·0	9·60215	·2 6	10·39785	9·64003	·2 7	10·35997	10·03788	·2 1	9·96212	*25′*
36·0	9·60244		10·39756	9·64037		10·35963	10·03793		9·96207	
37·0	9·60273		10·39727	9·64072		10·35928	10·03799		9·96201	
38·0	9·60302	·3 9	10·39698	9·64106	·3 10	10·35894	10·03804	·3 2	9·96196	
39·0	9·60331		10·39670	9·64140		10·35860	10·03810		9·96190	
40·0	9·60359		10·39641	9·64174		10·35825	10·03815		9·96185	*20′*
41·0	9·60388	·4 11	10·39612	9·64209	·4 14	10·35791	10·03821	·4 2	9·96179	
42·0	9·60417		10·39583	9·64243		10·35757	10·03827		9·96174	
43·0	9·60446		10·39554	9·64278		10·35722	10·03832		9·96168	
44·0	9·60475		10·39526	9·64312		10·35688	10·03838		9·96162	
45·0	9·60503	·5 14	10·39497	9·64346	·5 17	10·35654	10·03843	·5 3	9·96157	*15′*
46·0	9·60532		10·39468	9·64381		10·35619	10·03849		9·96151	
47·0	9·60561		10·39439	9·64415		10·35585	10·03854		9·96146	
48·0	9·60589	·6 17	10·39411	9·64449	·6 21	10·35551	10·03860	·6 3	9·96140	
49·0	9·60618		10·39382	9·64483		10·35517	10·03865		9·96135	
50·0	9·60647		10·39353	9·64517		10·35483	10·03871		9·96129	*10′*
51·0	9·60675	·7 20	10·39325	9·64552	·7 24	10·35448	10·03877	·7 4	9·96124	
52·0	9·60704		10·39296	9·64586		10·35414	10·03882		9·96118	
53·0	9·60732		10·39268	9·64620		10·35380	10·03888		9·96112	
54·0	9·60761	·8 23	10·39239	9·64654	·8 27	10·35346	10·03893	·8 4	9·96107	
55·0	9·60789		10·39211	9·64688		10·35312	10·03899		9·96101	*5′*
56·0	9·60818		10·39182	9·64722		10·35278	10·03905		9·96096	
57·0	9·60846	·9 26	10·39154	9·64756	·9 31	10·35244	10·03910	·9 5	9·96090	
58·0	9·60875		10·39126	9·64790		10·35210	10·03916		9·96084	
59·0	9·60903		10·39097	9·64824		10·35176	10·03921		9·96079	
60·0	9·60931		10·39069	9·64858		10·35142	10·03927		9·96073	*0′*

24°
204°

LOGS. OF TRIG. FUNCTIONS

′	Sine	Parts	Cosec.	Tan.	Parts	Cotan.	Secant	Parts	Cosine	
00·0	9·60931	′	10·39069	9·64858	′	10·35142	10·03927	′	9·96073	60′
01·0	9·60960		10·39040	9·64892		10·35108	10·03933		9·96067	
02·0	9·60988	·1 3	10·39012	9·64926	·1 3	10·35074	10·03938	·1 1	9·96062	
03·0	9·61016		10·38984	9·64960		10·35040	10·03944		9·96056	
04·0	9·61045		10·38955	9·64994		10·35006	10·03950		9·96051	
05·0	9·61073	·2 6	10·38927	9·65028	·2 7	10·34972	10·03955	·2 1	9·96045	55′
06·0	9·61101		10·38899	9·65062		10·34938	10·03961		9·96039	
07·0	9·61129		10·38871	9·65096		10·34904	10·03967		9·96034	
08·0	9·61158	·3 8	10·38842	9·65130	·3 10	10·34870	10·03972	·3 2	9·96028	
09·0	9·61186		10·38814	9·65164		10·34836	10·03978		9·96022	
10·0	9·61214		10·38786	9·65197		10·34803	10·03984		9·96017	50′
11·0	9·61242	·4 11	10·38758	9·65231	·4 13	10·34769	10·03989	·4 2	9·96011	
12·0	9·61270		10·38730	9·65265		10·34735	10·03995		9·96005	
13·0	9·61298		10·38702	9·65299		10·34701	10·04001		9·96000	
14·0	9·61326		10·38674	9·65333		10·34667	10·04006		9·95994	
15·0	9·61355	·5 14	10·38646	9·65366	·5 17	10·34634	10·04012	·5 3	9·95988	45′
16·0	9·61383		10·38618	9·65400		10·34600	10·04018		9·95983	
17·0	9·61411		10·38590	9·65434		10·34566	10·04023		9·95977	
18·0	9·61439	·6 17	10·38562	9·65467	·6 20	10·34533	10·04029	·6 3	9·95971	
19·0	9·61467		10·38534	9·65501		10·34499	10·04035		9·95965	
20·0	9·61494		10·38506	9·65535		10·34465	10·04040		9·95960	40′
21·0	9·61522	·7 20	10·38478	9·65568	·7 24	10·34432	10·04046	·7 4	9·95954	
22·0	9·61550		10·38450	9·65602		10·34398	10·04052		9·95948	
23·0	9·61578		10·38422	9·65636		10·34364	10·04058		9·95943	
24·0	9·61606	·8 22	10·38394	9·65669	·8 27	10·34331	10·04063	·8 5	9·95937	
25·0	9·61634		10·38366	9·65703		10·34297	10·04069		9·95931	35′
26·0	9·61662		10·38338	9·65736		10·34264	10·04075		9·95925	
27·0	9·61689	·9 25	10·38311	9·65770	·9 30	10·34230	10·04081	·9 5	9·95920	
28·0	9·61717		10·38283	9·65803		10·34197	10·04086		9·95914	
29·0	9·61745		10·38255	9·65837		10·34163	10·04092		9·95908	
30·0	9·61773		10·38227	9·65870		10·34130	10·04098		9·95902	30′
31·0	9·61800		10·38200	9·65904		10·34096	10·04104		9·95897	
32·0	9·61828	·1 3	10·38172	9·65937	·1 3	10·34063	10·04109	·1 1	9·95891	
33·0	9·61856		10·38144	9·65971		10·34029	10·04115		9·95885	
34·0	9·61883		10·38117	9·66004		10·33996	10·04121		9·95879	
35·0	9·61911	·2 5	10·38089	9·66038	·2 7	10·33962	10·04127	·2 1	9·95873	25′
36·0	9·61939		10·38061	9·66071		10·33929	10·04132		9·95868	
37·0	9·61966		10·38034	9·66104		10·33896	10·04138		9·95862	
38·0	9·61994	·3 8	10·38006	9·66138	·3 10	10·33862	10·04144	·3 2	9·95856	
39·0	9·62021		10·37979	9·66171		10·33829	10·04150		9·95850	
40·0	9·62049		10·37951	9·66204		10·33796	10·04156		9·95845	20′
41·0	9·62076	·4 11	10·37924	9·66238	·4 13	10·33762	10·04161	·4 2	9·95839	
42·0	9·62104		10·37896	9·66271		10·33729	10·04167		9·95833	
43·0	9·62131		10·37869	9·66304		10·33696	10·04173		9·95827	
44·0	9·62159		10·37841	9·66338		10·33663	10·04179		9·95821	
45·0	9·62186	·5 14	10·37814	9·66371	·5 17	10·33629	10·04185	·5 3	9·95815	15′
46·0	9·62214		10·37787	9·66404		10·33596	10·04190		9·95810	
47·0	9·62241		10·37759	9·66437		10·33563	10·04196		9·95804	
48·0	9·62268	·6 16	10·37732	9·66470	·6 20	10·33530	10·04202	·6 3	9·95798	
49·0	9·62296		10·37704	9·66504		10·33497	10·04208		9·95792	
50·0	9·62323		10·37677	9·66537		10·33463	10·04214		9·95786	10′
51·0	9·62350	·7 19	10·37650	9·66570	·7 23	10·33430	10·04220	·7 4	9·95780	
52·0	9·62377		10·37623	9·66603		10·33397	10·04225		9·95775	
53·0	9·62405		10·37595	9·66636		10·33364	10·04231		9·95769	
54·0	9·62432	·8 22	10·37568	9·66669	·8 27	10·33331	10·04237	·8 5	9·95763	
55·0	9·62459		10·37541	9·66702		10·33298	10·04243		9·95757	5′
56·0	9·62486		10·37514	9·66735		10·33265	10·04249		9·95751	
57·0	9·62514	·9 25	10·37487	9·66768	·9 30	10·33232	10·04255	·9 5	9·95745	
58·0	9·62541		10·37459	9·66801		10·33199	10·04261		9·95739	
59·0	9·62568		10·37432	9·66834		10·33166	10·04267		9·95734	
60·0	9·62595		10·37405	9·66867		10·33133	10·04272		9·95728	0′

155°
335°

25°
205°

LOGS. OF TRIG. FUNCTIONS

′	Sine	Parts	Cosec.	Tan.	Parts	Cotan.	Secant	Parts	Cosine	
00·0	9·62595	′	10·37405	9·66867	′	10·33133	10·04272	′	9·95728	60′
01·0	9·62622		10·37378	9·66900		10·33100	10·04278		9·95722	
02·0	9·62649	·1 3	10·37351	9·66933	·1 3	10·33067	10·04284	·1 1	9·95716	
03·0	9·62676		10·37324	9·66966		10·33034	10·04290		9·95710	
04·0	9·62703		10·37297	9·66999		10·33001	10·04296		9·95704	
05·0	9·62730	·2 5	10·37270	9·67032	·2 7	10·32968	10·04302	·2 1	9·95698	55′
06·0	9·62757		10·37243	9·67065		10·32935	10·04308		9·95692	
07·0	9·62784		10·37216	9·67098		10·32902	10·04314		9·95686	
08·0	9·62811	·3 8	10·37189	9·67131	·3 10	10·32869	10·04320	·3 2	9·95680	
09·0	9·62838		10·37162	9·67164		10·32837	10·04326		9·95674	
10·0	9·62865		10·37135	9·67196		10·32804	10·04332		9·95668	50′
11·0	9·62892	·4 11	10·37108	9·67229	·4 13	10·32771	10·04338	·4 2	9·95663	
12·0	9·62919		10·37082	9·67262		10·32738	10·04343		9·95657	
13·0	9·62945		10·37055	9·67295		10·32705	10·04349		9·95651	
14·0	9·62972		10·37028	9·67327		10·32673	10·04355		9·95645	
15·0	9·62999	·5 13	10·37001	9·67360	·5 16	10·32640	10·04361	·5 3	9·95639	45′
16·0	9·63026		10·36974	9·67393		10·32607	10·04367		9·95633	
17·0	9·63052		10·36948	9·67426		10·32574	10·04373		9·95627	
18·0	9·63079	·6 16	10·36921	9·67458	·6 20	10·32542	10·04379	·6 4	9·95621	
19·0	9·63106		10·36894	9·67491		10·32509	10·04385		9·95615	
20·0	9·63133		10·36867	9·67524		10·32476	10·04391		9·95609	40′
21·0	9·63159	·7 19	10·36841	9·67556	·7 23	10·32444	10·04397	·7 4	9·95603	
22·0	9·63186		10·36814	9·67589		10·32411	10·04403		9·95597	
23·0	9·63213		10·36788	9·67622		10·32378	10·04409		9·95591	
24·0	9·63239	·8 21	10·36761	9·67654	·8 26	10·32346	10·04415	·8 5	9·95585	
25·0	9·63266		10·36734	9·67687		10·32313	10·04421		9·95579	35′
26·0	9·63292		10·36708	9·67719		10·32281	10·04427		9·95573	
27·0	9·63319	·9 24	10·36681	9·67752	·9 29	10·32248	10·04433	·9 5	9·95567	
28·0	9·63345		10·36655	9·67785		10·32215	10·04439		9·95561	
29·0	9·63372		10·36628	9·67817		10·32183	10·04445		9·95555	
30·0	9·63398		10·36602	9·67850		10·32150	10·04451		9·95549	30′
31·0	9·63425		10·36575	9·67882		10·32118	10·04457		9·95543	
32·0	9·63451	·1 3	10·36549	9·67915	·1 3	10·32085	10·04463	·1 1	9·95537	
33·0	9·63478		10·36522	9·67947		10·32053	10·04469		9·95531	
34·0	9·63504		10·36496	9·67980		10·32021	10·04475		9·95525	
35·0	9·63531	·2 5	10·36469	9·68012	·2 6	10·31988	10·04481	·2 1	9·95519	25′
36·0	9·63557		10·36443	9·68044		10·31956	10·04487		9·95513	
37·0	9·63583		10·36417	9·68077		10·31923	10·04494		9·95507	
38·0	9·63610	·3 8	10·36390	9·68109	·3 10	10·31891	10·04500	·3 2	9·95501	
39·0	9·63636		10·36364	9·68142		10·31858	10·04506		9·95494	
40·0	9·63662		10·36338	9·68174		10·31826	10·04512		9·95488	20′
41·0	9·63689	·4 10	10·36311	9·68206	·4 13	10·31794	10·04518	·4 2	9·95482	
42·0	9·63715		10·36285	9·68239		10·31761	10·04524		9·95476	
43·0	9·63741		10·36259	9·68271		10·31729	10·04530		9·95470	
44·0	9·63767		10·36233	9·68303		10·31697	10·04536		9·95464	
45·0	9·63794	·5 13	10·36207	9·68336	·5 16	10·31664	10·04542	·5 3	9·95458	15′
46·0	9·63820		10·36180	9·68368		10·31632	10·04548		9·95452	
47·0	9·63846		10·36154	9·68400		10·31600	10·04554		9·95446	
48·0	9·63872	·6 16	10·36128	9·68432	·6 19	10·31568	10·04560	·6 4	9·95440	
49·0	9·63898		10·36102	9·68465		10·31535	10·04567		9·95434	
50·0	9·63924		10·36076	9·68497		10·31503	10·04573		9·95427	10′
51·0	9·63950	·7 18	10·36050	9·68529	·7 23	10·31471	10·04579	·7 4	9·95421	
52·0	9·63976		10·36024	9·68561		10·31439	10·04585		9·95415	
53·0	9·64002		10·35998	9·68593		10·31407	10·04591		9·95409	
54·0	9·64028	·8 21	10·35972	9·68626	·8 26	10·31375	10·04597	·8 5	9·95403	
55·0	9·64054		10·35946	9·68658		10·31342	10·04603		9·95397	5′
56·0	9·64080		10·35920	9·68690		10·31310	10·04609		9·95391	
57·0	9·64106	·9 24	10·35894	9·68722	·9 29	10·31278	10·04616	·9 5	9·95385	
58·0	9·64132		10·35868	9·68754		10·31246	10·04622		9·95378	
59·0	9·64158		10·35842	9·68786		10·31214	10·04628		9·95372	
60·0	9·64184		10·35816	9·68818		10·31182	10·04634		9·95366	0′

26°
206°

LOGS. OF TRIG. FUNCTIONS

′	Sine	Parts	Cosec.	Tan.	Parts	Cotan.	Secant	Parts	Cosine	
00·0	9·64184	′	10·35816	9·68818	′	10·31182	10·04634	′	9·95366	60′
01·0	9·64210		10·35790	9·68850		10·31150	10·04640		9·95360	
02·0	9·64236	·1 3	10·35764	9·68882	·1 3	10·31118	10·04646	·1 1	9·95354	
03·0	9·64262		10·35738	9·68914		10·31086	10·04653		9·95348	
04·0	9·64288		10·35712	9·68946		10·31054	10·04659		9·95341	
05·0	9·64314	·2 5	10·35687	9·68978	·2 6	10·31022	10·04665	·2 1	9·95335	55′
06·0	9·64339		10·35661	9·69010		10·30990	10·04671		9·95329	
07·0	9·64365		10·35635	9·69042		10·30958	10·04677		9·95323	
08·0	9·64391	·3 8	10·35609	9·69074	·3 10	10·30926	10·04683	·3 2	9·95317	
09·0	9·64417		10·35584	9·69106		10·30894	10·04690		9·95310	
10·0	9·64442		10·35558	9·69138		10·30862	10·04696		9·95304	50′
11·0	9·64468	·4 10	10·35532	9·69170	·4 13	10·30830	10·04702	·4 2	9·95298	
12·0	9·64494		10·35506	9·69202		10·30798	10·04708		9·95292	
13·0	9·64519		10·35481	9·69234		10·30766	10·04715		9·95286	
14·0	9·64545	·5 13	10·35455	9·69266	·5 16	10·30734	10·04721	·5 3	9·95279	
15·0	9·64571		10·35429	9·69298		10·30703	10·04727		9·95273	45′
16·0	9·64596		10·35404	9·69329		10·30671	10·04733		9·95267	
17·0	9·64622		10·35378	9·69361		10·30639	10·04739		9·95261	
18·0	9·64647	·6 15	10·35353	9·69393	·6 19	10·30607	10·04746	·6 4	9·95254	
19·0	9·64673		10·35327	9·69425		10·30575	10·04752		9·95248	
20·0	9·64698		10·35302	9·69457		10·30543	10·04758		9·95242	40′
21·0	9·64724	·7 18	10·35276	9·69488	·7 22	10·30511	10·04764	·7 4	9·95236	
22·0	9·64749		10·35251	9·69520		10·30480	10·04771		9·95229	
23·0	9·64775		10·35225	9·69552		10·30448	10·04777		9·95223	
24·0	9·64800	·8 21	10·35200	9·69584	·8 25	10·30416	10·04783	·8 5	9·95217	
25·0	9·64826		10·35174	9·69615		10·30385	10·04789		9·95211	35′
26·0	9·64851		10·35149	9·69647		10·30353	10·04796		9·95204	
27·0	9·64877	·9 23	10·35123	9·69679	·9 29	10·30321	10·04802	·9 6	9·95198	
28·0	9·64902		10·35098	9·69710		10·30290	10·04808		9·95192	
29·0	9·64927		10·35073	9·69742		10·30258	10·04815		9·95185	
30·0	9·64953		10·35047	9·69774		10·30226	10·04821		9·95179	30′
31·0	9·64978		10·35022	9·69805		10·30195	10·04827		9·95173	
32·0	9·65003	·1 2	10·34997	9·69837	·1 3	10·30163	10·04834	·1 1	9·95167	
33·0	9·65029		10·34971	9·69869		10·30132	10·04840		9·95160	
34·0	9·65054		10·34946	9·69900		10·30100	10·04846		9·95154	
35·0	9·65079	·2 5	10·34921	9·69932	·2 6	10·30068	10·04852	·2 1	9·95148	25′
36·0	9·65104		10·34896	9·69963		10·30037	10·04859		9·95141	
37·0	9·65130		10·34870	9·69995		10·30005	10·04865		9·95135	
38·0	9·65155	·3 7	10·34845	9·70026	·3 9	10·29974	10·04871	·3 2	9·95129	
39·0	9·65180		10·34820	9·70058		10·29942	10·04878		9·95122	
40·0	9·65205		10·34795	9·70089		10·29911	10·04884		9·95116	20′
41·0	9·65230	·4 10	10·34770	9·70121	·4 13	10·29879	10·04890	·4 3	9·95110	
42·0	9·65256		10·34745	9·70152		10·29848	10·04897		9·95103	
43·0	9·65281		10·34719	9·70184		10·29816	10·04903		9·95097	
44·0	9·65306		10·34694	9·70215		10·29785	10·04910		9·95091	
45·0	9·65331	·5 12	10·34669	9·70247	·5 16	10·29753	10·04916	·5 3	9·95084	15′
46·0	9·65356		10·34644	9·70278		10·29722	10·04922		9·95078	
47·0	9·65381		10·34619	9·70310		10·29691	10·04929		9·95071	
48·0	9·65406	·6 15	10·34594	9·70341	·6 19	10·29659	10·04935	·6 4	9·95065	
49·0	9·65431		10·34569	9·70372		10·29628	10·04941		9·95059	
50·0	9·65456		10·34544	9·70404		10·29596	10·04948		9·95052	10′
51·0	9·65481	·7 17	10·34519	9·70435	·7 22	10·29565	10·04954	·7 4	9·95046	
52·0	9·65506		10·34494	9·70466		10·29534	10·04961		9·95039	
53·0	9·65531		10·34469	9·70498		10·29502	10·04967		9·95033	
54·0	9·65556	·8 20	10·34444	9·70529	·8 25	10·29471	10·04973	·8 5	9·95027	
55·0	9·65581		10·34420	9·70560		10·29440	10·04980		9·95020	5′
56·0	9·65605		10·34395	9·70592		10·29408	10·04986		9·95014	
57·0	9·65630	·9 22	10·34370	9·70623	·9 28	10·29377	10·04993	·9 6	9·95007	
58·0	9·65655		10·34345	9·70654		10·29346	10·04999		9·95001	
59·0	9·65680		10·34320	9·70685		10·29315	10·05006		9·94995	
60·0	9·65705		10·34295	9·70717		10·29283	10·05012		9·94988	0′

153°
333°

27°
207°

LOGS. OF TRIG. FUNCTIONS

,	Sine	Parts		Cosec.	Tan.	Parts		Cotan.	Secant	Parts		Cosine	
00·0	9·65705	'		10·34295	9·70717	'		10·29283	10·05012	'		9·94988	60'
01·0	9·65730			10·34271	9·70748			10·29252	10·05018			9·94982	
02·0	9·65754	·1	2	10·34246	9·70779	·1	3	10·29221	10·05025	·1	1	9·94975	
03·0	9·65779			10·34221	9·70810			10·29190	10·05031			9·94969	
04·0	9·65804			10·34196	9·70841			10·29159	10·05038			9·94962	
05·0	9·65828	·2	5	10·34172	9·70873	·2	6	10·29127	10·05044	·2	1	9·94956	55'
06·0	9·65853			10·34147	9·70904			10·29096	10·05051			9·94949	
07·0	9·65878			10·34122	9·70935			10·29065	10·05057			9·94943	
08·0	9·65903	·3	7	10·34098	9·70966	·3	9	10·29034	10·05064	·3	2	9·94930	
09·0	9·65927			10·34073	9·70997			10·29003	10·05070			9·94930	
10·0	9·65952			10·34048	9·71028			10·28972	10·05077			9·94924	50'
11·0	9·65976	·4	10	10·34024	9·71059	·4	12	10·28941	10·05083	·4	3	9·94917	
12·0	9·66001			10·33999	9·71090			10·28910	10·05090			9·94911	
13·0	9·66026			10·33975	9·71122			10·28879	10·05096			9·94904	
14·0	9·66050			10·33950	9·71153			10·28848	10·05103			9·94898	
15·0	9·66075	·5	12	10·33925	9·71184	·5	16	10·28816	10·05109	·5	3	9·94891	45'
16·0	9·66099			10·33901	9·71215			10·28785	10·05116			9·94885	
17·0	9·66124			10·33876	9·71246			10·28754	10·05122			9·94878	
18·0	9·66148	·6	15	10·33852	9·71277	·6	19	10·28723	10·05129	·6	4	9·94872	
19·0	9·66173			10·33827	9·71308			10·28692	10·05135			9·94865	
20·0	9·66197	·7	17	10·33803	9·71339	·7	22	10·28661	10·05142	·7	5	9·94858	40'
21·0	9·66221			10·33779	9·71370			10·28630	10·05148			9·94852	
22·0	9·66246			10·33754	9·71401			10·28600	10·05155			9·94845	
23·0	9·66270			10·33730	9·71431			10·28569	10·05161			9·94839	
24·0	9·66295	·8	20	10·33705	9·71462	·8	25	10·28538	10·05168	·8	5	9·94832	
25·0	9·66319			10·33681	9·71493			10·28507	10·05174			9·94826	35'
26·0	9·66343			10·33657	9·71524			10·28476	10·05181			9·94819	
27·0	9·66368	·9	22	10·33632	9·71555	·9	28	10·28445	10·05187	·9	6	9·94813	
28·0	9·66392			10·33608	9·71586			10·28414	10·05194			9·94806	
29·0	9·66416			10·33584	9·71617			10·28383	10·05201			9·94800	
30·0	9·66441			10·33559	9·71648			10·28352	10·05207			9·94793	30'
31·0	9·66465			10·33535	9·71679			10·28322	10·05214			9·94786	
32·0	9·66489	·1	2	10·33511	9·71709	·1	3	10·28291	10·05220	·1	1	9·94780	
33·0	9·66513			10·33487	9·71740			10·28260	10·05227			9·94773	
34·0	9·66538			10·33463	9·71771			10·28229	10·05234			9·94767	
35·0	9·66562	·2	5	10·33438	9·71802	·2	6	10·28198	10·05240	·2	1	9·94760	25'
36·0	9·66586			10·33414	9·71833			10·28168	10·05247			9·94753	
37·0	9·66610			10·33390	9·71863			10·28137	10·05253			9·94747	
38·0	9·66634	·3	7	10·33366	9·71894	·3	9	10·28106	10·05260	·3	2	9·94740	
39·0	9·66658			10·33342	9·71925			10·28075	10·05267			9·94734	
40·0	9·66682			10·33318	9·71956			10·28045	10·05273			9·94727	20'
41·0	9·66707	·4	10	10·33294	9·71986	·4	12	10·28014	10·05280	·4	3	9·94720	
42·0	9·66731			10·33270	9·72017			10·27983	10·05286			9·94714	
43·0	9·66755			10·33245	9·72048			10·27952	10·05293			9·94707	
44·0	9·66779			10·33221	9·72078			10·27922	10·05300			9·94700	
45·0	9·66803	·5	12	10·33197	9·72109	·5	15	10·27891	10·05306	·5	3	9·94694	15'
46·0	9·66827			10·33173	9·72140			10·27860	10·05313			9·94687	
47·0	9·66851			10·33149	9·72170			10·27830	10·05320			9·94680	
48·0	9·66875	·6	14	10·33125	9·72201	·6	18	10·27799	10·05326	·6	4	9·94674	
49·0	9·66899			10·33101	9·72232			10·27769	10·05333			9·94667	
50·0	9·66923			10·33078	9·72262			10·27738	10·05340			9·94660	10'
51·0	9·66946	·7	17	10·33054	9·72293	·7	21	10·27707	10·05346	·7	5	9·94654	
52·0	9·66970			10·33030	9·72323			10·27677	10·05353			9·94647	
53·0	9·66994			10·33006	9·72354			10·27646	10·05360			9·94640	
54·0	9·67018	·8	19	10·32982	9·72384	·8	25	10·27616	10·05366	·8	5	9·94634	
55·0	9·67042			10·32958	9·72415			10·27585	10·05373			9·94627	5'
56·0	9·67066			10·32934	9·72445			10·27555	10·05380			9·94620	
57·0	9·67090	·9	22	10·32910	9·72476	·9	28	10·27524	10·05386	·9	6	9·94614	
58·0	9·67113			10·32887	9·72507			10·27494	10·05393			9·94607	
59·0	9·67137			10·32863	9·72537			10·27463	10·05400			9·94600	
60·0	9·67161			10·32839	9·72567			10·27433	10·05407			9·94593	0'

28°
208°

LOGS. OF TRIG. FUNCTIONS

′	Sine	Parts	Cosec.	Tan.	Parts	Cotan.	Secant	Parts	Cosine	
00·0	9·67161	′	10·32839	9·72567	′	10·27433	10·05407	′	9·94593	60′
01·0	9·67185		10·32815	9·72598		10·27402	10·05413		9·94587	
02·0	9·67208	·1 2	10·32792	9·72628	·1 3	10·27372	10·05420	·1 1	9·94580	
03·0	9·67232		10·32768	9·72659		10·27341	10·05427		9·94573	
04·0	9·67256		10·32744	9·72689		10·27311	10·05433		9·94567	
05·0	9·67280	·2 5	10·32721	9·72720	·2 6	10·27280	10·05440	·2 1	9·94560	55′
06·0	9·67303		10·32697	9·72750		10·27250	10·05447		9·94553	
07·0	9·67327		10·32673	9·72781		10·27220	10·05454		9·94546	
08·0	9·67351	·3 7	10·32650	9·72811	·3 9	10·27189	10·05460	·3 2	9·94540	
09·0	9·67374		10·32626	9·72841		10·27159	10·05467		9·94533	
10·0	9·67398		10·32602	9·72872		10·27128	10·05474		9·94526	50′
11·0	9·67421	·4 9	10·32579	9·72902	·4 12	10·27098	10·05481	·4 3	9·94519	
12·0	9·67445		10·32555	9·72932		10·27068	10·05488		9·94513	
13·0	9·67468		10·32532	9·72963		10·27037	10·05494		9·94506	
14·0	9·67492	·5 12	10·32508	9·72993	·5 15	10·27007	10·05501	·5 3	9·94499	
15·0	9·67516		10·32485	9·73023		10·26977	10·05508		9·94492	45′
16·0	9·67539		10·32461	9·73054		10·26947	10·05515		9·94485	
17·0	9·67562		10·32438	9·73084		10·26916	10·05521		9·94479	
18·0	9·67586	·6 14	10·32414	9·73114	·6 18	10·26886	10·05528	·6 4	9·94472	
19·0	9·67609		10·32391	9·73144		10·26856	10·05535		9·94465	
20·0	9·67633	·7 16	10·32367	9·73175	·7 21	10·26825	10·05542	·7 5	9·94458	40′
21·0	9·67656		10·32344	9·73205		10·26795	10·05549		9·94451	
22·0	9·67680		10·32320	9·73235		10·26765	10·05555		9·94445	
23·0	9·67703		10·32297	9·73265		10·26735	10·05562		9·94438	
24·0	9·67726	·8 19	10·32274	9·73296	·8 24	10·26705	10·05569	·8 5	9·94431	
25·0	9·67750		10·32250	9·73326		10·26674	10·05576		9·94424	35′
26·0	9·67773		10·32227	9·73356		10·26644	10·05583		9·94417	
27·0	9·67796	·9 21	10·32204	9·73386	·9 27	10·26614	10·05590	·9 6	9·94410	
28·0	9·67820		10·32180	9·73416		10·26584	10·05596		9·94404	
29·0	9·67843		10·32157	9·73446		10·26554	10·05603		9·94397	
30·0	9·67866		10·32134	9·73476		10·26524	10·05610		9·94390	30′
31·0	9·67890		10·32111	9·73507		10·26493	10·05617		9·94383	
32·0	9·67913	·1 2	10·32087	9·73537	·1 3	10·26463	10·05624	·1 1	9·94376	
33·0	9·67936		10·32064	9·73567		10·26433	10·05631		9·94369	
34·0	9·67959		10·32041	9·73597		10·26403	10·05638		9·94362	
35·0	9·67982	·2 5	10·32018	9·73627	·2 6	10·26373	10·05645	·2 1	9·94356	25′
36·0	9·68006		10·31994	9·73657		10·26343	10·05651		9·94349	
37·0	9·68029		10·31971	9·73687		10·26313	10·05658		9·94342	
38·0	9·68052	·3 7	10·31948	9·73717	·3 9	10·26283	10·05665	·3 2	9·94335	
39·0	9·68075		10·31925	9·73747		10·26253	10·05672		9·94328	
40·0	9·68098		10·31902	9·73777		10·26223	10·05679		9·94321	20′
41·0	9·68121		10·31879	9·73807		10·26193	10·05686		9·94314	
42·0	9·68144	·4 9	10·31856	9·73837	·4 12	10·26163	10·05693	·4 3	9·94307	
43·0	9·68167		10·31833	9·73867		10·26133	10·05700		9·94300	
44·0	9·68191		10·31810	9·73897		10·26103	10·05707		9·94293	
45·0	9·68214	·5 12	10·31787	9·73927	·5 15	10·26073	10·05714	·5 3	9·94286	15′
46·0	9·68237		10·31764	9·73957		10·26043	10·05721		9·94280	
47·0	9·68260		10·31741	9·73987		10·26013	10·05727		9·94273	
48·0	9·68283	·6 14	10·31718	9·74017	·6 18	10·25983	10·05734	·6 4	9·94266	
49·0	9·68306		10·31695	9·74047		10·25953	10·05741		9·94259	
50·0	9·68328		10·31672	9·74077		10·25923	10·05748		9·94252	10′
51·0	9·68351	·7 16	10·31649	9·74107	·7 21	10·25893	10·05755	·7 5	9·94245	
52·0	9·68374		10·31626	9·74137		10·25864	10·05762		9·94238	
53·0	9·68397		10·31603	9·74166		10·25834	10·05769		9·94231	
54·0	9·68420	·8 18	10·31580	9·74196	·8 24	10·25804	10·05776	·8 6	9·94224	
55·0	9·68443		10·31557	9·74226		10·25774	10·05783		9·94217	5′
56·0	9·68466		10·31534	9·74256		10·25744	10·05790		9·94210	
57·0	9·68489	·9 21	10·31511	9·74286	·9 27	10·25714	10·05797	·9 6	9·94203	
58·0	9·68512		10·31489	9·74316		10·25684	10·05804		9·94196	
59·0	9·68534		10·31466	9·74345		10·25655	10·05811		9·94189	
60·0	9·68557		10·31443	9·74375		10·25625	10·05818		9·94182	0′

151°
331°

29°
209°

LOGS. OF TRIG. FUNCTIONS

,	Sine	Parts	Cosec.	Tan.	Parts	Cotan.	Secant	Parts	Cosine	
00·0	9·68557	′	10·31443	9·74375	′	10·25625	10·05818	′	9·94182	60′
01·0	9·68580		10·31420	9·74405		10·25595	10·05825		9·94175	
02·0	9·68603	·1 2	10·31397	9·74435	·1 3	10·25565	10·05832	·1 1	9·94168	
03·0	9·68625		10·31375	9·74465		10·25536	10·05839		9·94161	
04·0	9·68648		10·31352	9·74494		10·25506	10·05846		9·94154	
05·0	9·68671	·2 5	10·31329	9·74524	·2 6	10·25476	10·05853	·2 1	9·94147	55′
06·0	9·68694		10·31306	9·74554		10·25446	10·05860		9·94140	
07·0	9·68716		10·31284	9·74584		10·25417	10·05867		9·94133	
08·0	9·68739	·3 7	10·31261	9·74613	·3 9	10·25387	10·05874	·3 2	9·94126	
09·0	9·68762		10·31238	9·74643		10·25357	10·05881		9·94119	
10·0	9·68784		10·31216	9·74673		10·25327	10·05888		9·94112	50′
11·0	9·68807	·4 9	10·31193	9·74702	·4 12	10·25298	10·05895	·4 3	9·94105	
12·0	9·68830		10·31171	9·74732		10·25268	10·05903		9·94098	
13·0	9·68852		10·31148	9·74762		10·25238	10·05910		9·94091	
14·0	9·68875		10·31125	9·74791		10·25209	10·05917		9·94083	
15·0	9·68897	·5 11	10·31103	9·74821	·5 15	10·25179	10·05924	·5 4	9·94076	45′
16·0	9·68920		10·31080	9·74851		10·25150	10·05931		9·94069	
17·0	9·68942		10·31058	9·74880		10·25120	10·05938		9·94062	
18·0	9·68965	·6 14	10·31035	9·74910	·6 18	10·25090	10·05945	·6 4	9·94055	
19·0	9·68987		10·31013	9·74939		10·25061	10·05952		9·94048	
20·0	9·69010		10·30990	9·74969		10·25031	10·05959		9·94041	40′
21·0	9·69032	·7 16	10·30968	9·74999	·7 21	10·25002	10·05966	·7 5	9·94034	
22·0	9·69055		10·30945	9·75028		10·24972	10·05973		9·94027	
23·0	9·69077		10·30923	9·75058		10·24942	10·05980		9·94020	
24·0	9·69100	·8 18	10·30900	9·75087	·8 24	10·24913	10·05988	·8 6	9·94013	
25·0	9·69122		10·30878	9·75117		10·24883	10·05995		9·94005	35′
26·0	9·69144		10·30856	9·75146		10·24854	10·06002		9·93998	
27·0	9·69167	·9 20	10·30833	9·75176	·9 27	10·24824	10·06009	·9 6	9·93991	
28·0	9·69189		10·30811	9·75205		10·24795	10·06016		9·93984	
29·0	9·69212		10·30789	9·75235		10·24765	10·06023		9·93977	
30·0	9·69234		10·30766	9·75264		10·24736	10·06030		9·93970	30′
31·0	9·69256		10·30744	9·75294		10·24706	10·06038		9·93963	
32·0	9·69279	·1 2	10·30722	9·75323	·1 3	10·24677	10·06045	·1 1	9·93955	
33·0	9·69301		10·30699	9·75353		10·24647	10·06052		9·93948	
34·0	9·69323		10·30677	9·75382		10·24618	10·06059		9·93941	
35·0	9·69345	·2 4	10·30655	9·75412	·2 6	10·24589	10·06066	·2 1	9·93934	25′
36·0	9·69368		10·30632	9·75441		10·24559	10·06073		9·93927	
37·0	9·69390		10·30610	9·75470		10·24530	10·06081		9·93920	
38·0	9·69412	·3 7	10·30588	9·75500	·3 9	10·24500	10·06088	·3 2	9·93912	
39·0	9·69434		10·30566	9·75529		10·24471	10·06095		9·93905	
40·0	9·69456		10·30544	9·75559		10·24442	10·06102		9·93898	20′
41·0	9·69479	·4 9	10·30521	9·75588	·4 12	10·24412	10·06109	·4 3	9·93891	
42·0	9·69501		10·30499	9·75617		10·24383	10·06116		9·93884	
43·0	9·69523		10·30477	9·75647		10·24354	10·06124		9·93876	
44·0	9·69545		10·30455	9·75676		10·24324	10·06131		9·93869	
45·0	9·69567	·5 11	10·30433	9·75705	·5 15	10·24295	10·06138	·5 4	9·93862	15′
46·0	9·69589		10·30411	9·75735		10·24266	10·06145		9·93855	
47·0	9·69611		10·30389	9·75764		10·24236	10·06153		9·93848	
48·0	9·69633	·6 13	10·30367	9·75793	·6 18	10·24207	10·06160	·6 4	9·93840	
49·0	9·69655		10·30345	9·75822		10·24178	10·06167		9·93833	
50·0	9·69678		10·30323	9·75852		10·24148	10·06174		9·93826	10′
51·0	9·69700	·7 15	10·30301	9·75881	·7 21	10·24119	10·06182	·7 5	9·93819	
52·0	9·69722		10·30279	9·75910		10·24090	10·06189		9·93811	
53·0	9·69744		10·30257	9·75940		10·24061	10·06196		9·93804	
54·0	9·69765	·8 18	10·30235	9·75969	·8 23	10·24031	10·06203	·8 6	9·93797	
55·0	9·69787		10·30213	9·75998		10·24002	10·06211		9·93790	5′
56·0	9·69809		10·30191	9·76027		10·23973	10·06218		9·93782	
57·0	9·69831	·9 20	10·30169	9·76056	·9 26	10·23944	10·06225	·9 7	9·93775	
58·0	9·69853		10·30147	9·76086		10·23914	10·06232		9·93768	
59·0	9·69875		10·30125	9·76115		10·23885	10·06240		9·93760	
60·0	9·69897		10·30103	9·76144		10·23856	10·06247		9·93753	0′

150°
330°

30°
210°

LOGS. OF TRIG. FUNCTIONS

′	Sine	Parts	Cosec.	Tan.	Parts	Cotan.	Secant	Parts	Cosine	
00·0	9·69897	′	10·30103	9·76144	′	10·23856	10·06247	′	9·93753	60′
01·0	9·69919		10·30081	9·76173		10·23827	10·06254		9·93746	
02·0	9·69941	·1 2	10·30059	9·76202	·1 3	10·23798	10·06262	·1 1	9·93739	
03·0	9·69963		10·30037	9·76231		10·23769	10·06269		9·93731	
04·0	9·69984		10·30016	9·76261		10·23739	10·06276		9·93724	
05·0	9·70006	·2 4	10·29994	9·76290	·2 6	10·23710	10·06284	·2 1	9·93717	55′
06·0	9·70028		10·29972	9·76319		10·23681	10·06291		9·93709	
07·0	9·70050		10·29950	9·76348		10·23652	10·06298		9·93702	
08·0	9·70072	·3 7	10·29928	9·76377	·3 9	10·23623	10·06305	·3 2	9·93695	
09·0	9·70093		10·29907	9·76406		10·23594	10·06313		9·93687	
10·0	9·70115		10·29885	9·76435		10·23565	10·06320		9·93680	50′
11·0	9·70137	·4 9	10·29863	9·76464	·4 12	10·23536	10·06328	·4 3	9·93673	
12·0	9·70159		10·29842	9·76493		10·23507	10·06335		9·93665	
13·0	9·70180		10·29820	9·76522		10·23478	10·06342		9·93658	
14·0	9·70202	·5 11	10·29798	9·76551	·5 15	10·23449	10·06350	·5 4	9·93651	
15·0	9·70224		10·29776	9·76581		10·23420	10·06357		9·93643	45′
16·0	9·70245		10·29755	9·76610		10·23391	10·06364		9·93636	
17·0	9·70267		10·29733	9·76639		10·23362	10·06372		9·93628	
18·0	9·70289	·6 13	10·29712	9·76668	·6 17	10·23333	10·06379	·6 4	9·93621	
19·0	9·70310		10·29690	9·76697		10·23304	10·06386		9·93614	
20·0	9·70332		10·29668	9·76726		10·23275	10·06394		9·93606	40′
21·0	9·70353	·7 15	10·29647	9·76755	·7 20	10·23246	10·06401	·7 5	9·93599	
22·0	9·70375		10·29625	9·76783		10·23217	10·06409		9·93591	
23·0	9·70396		10·29604	9·76812		10·23188	10·06416		9·93584	
24·0	9·70418	·8 17	10·29582	9·76841	·8 23	10·23159	10·06423	·8 6	9·93577	
25·0	9·70440		10·29561	9·76870		10·23130	10·06431		9·93569	35′
26·0	9·70461		10·29539	9·76899		10·23101	10·06438		9·93562	
27·0	9·70483	·9 20	10·29518	9·76928	·9 26	10·23072	10·06446	·9 7	9·93554	
28·0	9·70504		10·29496	9·76957		10·23043	10·06453		9·93547	
29·0	9·70525		10·29475	9·76986		10·23014	10·06461		9·93540	
30·0	9·70547		10·29453	9·77015		10·22985	10·06468		9·93532	30′
31·0	9·70568		10·29432	9·77044		10·22956	10·06475		9·93524	
32·0	9·70590	·1 2	10·29410	9·77073	·1 3	10·22927	10·06483	·1 1	9·93517	
33·0	9·70611		10·29389	9·77102		10·22899	10·06490		9·93510	
34·0	9·70633		10·29367	9·77130		10·22870	10·06498		9·93502	
35·0	9·70654	·2 4	10·29346	9·77159	·2 6	10·22841	10·06505	·2 2	9·93495	25′
36·0	9·70675		10·29325	9·77188		10·22812	10·06513		9·93487	
37·0	9·70697		10·29303	9·77217		10·22783	10·06520		9·93480	
38·0	9·70718	·3 6	10·29282	9·77246	·3 9	10·22754	10·06528	·3 2	9·93472	
39·0	9·70739		10·29261	9·77275		10·22726	10·06535		9·93465	
40·0	9·70761		10·29239	9·77303		10·22697	10·06543		9·93457	20′
41·0	9·70782	·4 8	10·29218	9·77332	·4 11	10·22668	10·06550	·4 3	9·93450	
42·0	9·70803		10·29197	9·77361		10·22639	10·06558		9·93442	
43·0	9·70825		10·29176	9·77390		10·22610	10·06565		9·93435	
44·0	9·70846		10·29154	9·77418		10·22582	10·06573		9·93427	
45·0	9·70867	·5 11	10·29133	9·77447	·5 14	10·22553	10·06580	·5 4	9·93420	15′
46·0	9·70888		10·29112	9·77476		10·22524	10·06588		9·93412	
47·0	9·70909		10·29091	9·77505		10·22495	10·06595		9·93405	
48·0	9·70931	·6 13	10·29069	9·77533	·6 17	10·22467	10·06603	·6 5	9·93397	
49·0	9·70952		10·29048	9·77562		10·22438	10·06610		9·93390	
50·0	9·70973		10·29027	9·77591		10·22409	10·06618		9·93382	10′
51·0	9·70994	·7 15	10·29006	9·77620	·7 20	10·22381	10·06625	·7 5	9·93375	
52·0	9·71015		10·28985	9·77648		10·22352	10·06633		9·93367	
53·0	9·71036		10·28964	9·77677		10·22323	10·06640		9·93360	
54·0	9·71058	·8 17	10·28943	9·77706	·8 23	10·22295	10·06648	·8 6	9·93352	
55·0	9·71079		10·28921	9·77734		10·22266	10·06656		9·93345	5′
56·0	9·71100		10·28900	9·77763		10·22237	10·06663		9·93337	
57·0	9·71121	·9 19	10·28879	9·77792	·9 26	10·22209	10·06671	·9 7	9·93329	
58·0	9·71142		10·28858	9·77820		10·22180	10·06678		9·93322	
59·0	9·71163		10·28837	9·77849		10·22151	10·06686		9·93314	
60·0	9·71184		10·28816	9·77877		10·22123	10·06693		9·93307	0′

149°
329°

31°
211°

LOGS. OF TRIG. FUNCTIONS

,	Sine	Parts		Cosec.	Tan.	Parts		Cotan.	Secant	Parts		Cosine	
00·0	9·71184	′		10·28816	9·77877	′		10·22123	10·06693	′		9·93307	60′
01·0	9·71205			10·28795	9·77906			10·22094	10·06701			9·93299	
02·0	9·71226	·1	2	10·28774	9·77935	·1	3	10·22065	10·06709	·1	1	9·93291	
03·0	9·71247			10·28753	9·77963			10·22037	10·06716			9·93284	
04·0	9·71268			10·28732	9·77992			10·22008	10·06724			9·93276	
05·0	9·71289	·2	4	10·28711	9·78020	·2	6	10·21980	10·06732	·2	2	9·93269	55′
06·0	9·71310			10·28690	9·78049			10·21951	10·06739			9·93261	
07·0	9·71331			10·28669	9·78078			10·21923	10·06747			9·93253	
08·0	9·71352	·3	6	10·28648	9·78106	·3	9	10·21894	10·06754	·3	2	9·93246	
09·0	9·71373			10·28627	9·78135			10·21865	10·06762			9·93238	
10·0	9·71394			10·28607	9·78163			10·21837	10·06770			9·93230	50′
11·0	9·71414	·4	8	10·28586	9·78192	·4	11	10·21808	10·06777	·4	3	9·93223	
12·0	9·71435			10·28565	9·78220			10·21780	10·06785			9·93215	
13·0	9·71456			10·28544	9·78249			10·21751	10·06793			9·93208	
14·0	9·71477	·5	10	10·28523	9·78277	·5	14	10·21723	10·06800	·5	4	9·93200	
15·0	9·71498			10·28502	9·78306			10·21694	10·06808			9·93192	45′
16·0	9·71519			10·28481	9·78334			10·21666	10·06816			9·93185	
17·0	9·71539			10·28461	9·78363			10·21637	10·06823			9·93177	
18·0	9·71560	·6	12	10·28440	9·78391	·6	17	10·21609	10·06831	·6	5	9·93169	
19·0	9·71581			10·28419	9·78420			10·21581	10·06839			9·93161	
20·0	9·71602			10·28398	9·78448			10·21552	10·06846			9·93154	40′
21·0	9·71622	·7	15	10·28378	9·78476	·7	20	10·21524	10·06854	·7	5	9·93146	
22·0	9·71643			10·28357	9·78505			10·21495	10·06862			9·93138	
23·0	9·71664			10·28336	9·78533			10·21467	10·06869			9·93131	
24·0	9·71685	·8	17	10·28315	9·78562	·8	23	10·21438	10·06877	·8	6	9·93123	
25·0	9·71705			10·28295	9·78590			10·21410	10·06885			9·93115	35′
26·0	9·71726			10·28274	9·78618			10·21382	10·06893			9·93108	
27·0	9·71747	·9	19	10·28253	9·78647	·9	26	10·21353	10·06900	·9	7	9·93100	
28·0	9·71767			10·28233	9·78675			10·21325	10·06908			9·93092	
29·0	9·71788			10·28212	9·78704			10·21296	10·06916			9·93084	
30·0	9·71809			10·28192	9·78732			10·21268	10·06923			9·93077	30′
31·0	9·71829			10·28171	9·78760			10·21240	10·06931			9·93069	
32·0	9·71850	·1	2	10·28150	9·78789	·1	3	10·21211	10·06939	·1	1	9·93061	
33·0	9·71870			10·28130	9·78817			10·21183	10·06947			9·93053	
34·0	9·71891			10·28109	9·78845			10·21155	10·06954			9·93046	
35·0	9·71911	·2	4	10·28089	9·78874	·2	6	10·21126	10·06962	·2	2	9·93038	25′
36·0	9·71932			10·28068	9·78902			10·21098	10·06970			9·93030	
37·0	9·71953			10·28048	9·78930			10·21070	10·06978			9·93022	
38·0	9·71973	·3	6	10·28027	9·78959	·3	8	10·21042	10·06986	·3	2	9·93015	
39·0	9·71994			10·28007	9·78987			10·21013	10·06993			9·93007	
40·0	9·72014			10·27986	9·79015			10·20985	10·07001			9·92999	20′
41·0	9·72035	·4	8	10·27966	9·79043	·4	11	10·20957	10·07009	·4	3	9·92991	
42·0	9·72055			10·27945	9·79072			10·20928	10·07017			9·92983	
43·0	9·72075			10·27925	9·79100			10·20900	10·07025			9·92976	
44·0	9·72096			10·27904	9·79128			10·20872	10·07032			9·92968	
45·0	9·72116	·5	10	10·27884	9·79156	·5	14	10·20844	10·07040	·5	4	9·92960	15′
46·0	9·72137			10·27863	9·79185			10·20815	10·07048			9·92952	
47·0	9·72157			10·27843	9·79213			10·20787	10·07056			9·92944	
48·0	9·72177	·6	12	10·27823	9·79241	·6	17	10·20759	10·07064	·6	5	9·92936	
49·0	9·72198			10·27802	9·79269			10·20731	10·07071			9·92929	
50·0	9·72218			10·27782	9·79297			10·20703	10·07079			9·92921	10′
51·0	9·72239	·7	14	10·27762	9·79326	·7	20	10·20674	10·07087	·7	5	9·92913	
52·0	9·72259			10·27741	9·79354			10·20646	10·07095			9·92905	
53·0	9·72279			10·27721	9·79382			10·20618	10·07103			9·92897	
54·0	9·72299	·8	16	10·27701	9·79410	·8	23	10·20590	10·07111	·8	6	9·92889	
55·0	9·72320			10·27680	9·79438			10·20562	10·07119			9·92882	5′
56·0	9·72340			10·27660	9·79466			10·20534	10·07126			9·92874	
57·0	9·72360	·9	18	10·27640	9·79495	·9	25	10·20505	10·07134	·9	7	9·92866	
58·0	9·72381			10·27620	9·79523			10·20477	10·07142			9·92858	
59·0	9·72401			10·27599	9·79551			10·20449	10·07150			9·92850	
60·0	9·72421			10·27579	9·79579			10·20421	10·07158			9·92842	0′

148°
328°

32°
212°

LOGS. OF TRIG. FUNCTIONS

′	Sine	Parts	Cosec.	Tan.	Parts	Cotan·	Secant	Parts	Cosine	
00·0	9·72421	′	10·27579	9·79579	′	10·20421	10·07158	′	9·92842	60′
01·0	9·72441		10·27559	9·79607		10·20393	10·07166		9·92834	
02·0	9·72461	·1 2	10·27539	9·79635	·1 3	10·20365	10·07174	·1 1	9·92826	
03·0	9·72482		10·27518	9·79663		10·20337	10·07182		9·92818	
04·0	9·72502		10·27498	9·79691		10·20309	10·07190		9·92810	
05·0	9·72522	·2 4	10·27478	9·79719	·2 6	10·20281	10·07198	·2 2	9·92803	55′
06·0	9·72542		10·27458	9·79748		10·20253	10·07205		9·92795	
07·0	9·72562		10·27438	9·79776		10·20225	10·07213		9·92787	
08·0	9·72582	·3 6	10·27418	9·79804	·3 8	10·20196	10·07221	·3 2	9·92779	
09·0	9·72602		10·27398	9·79832		10·20168	10·07229		9·92771	
10·0	9·72623		10·27378	9·79860		10·20140	10·07237		9·92763	50′
11·0	9·72643	·4 8	10·27357	9·79888	·4 11	10·20112	10·07245	·4 3	9·92755	
12·0	9·72663		10·27337	9·79916		10·20084	10·07253		9·92747	
13·0	9·72683		10·27317	9·79944		10·20056	10·07261		9·92739	
14·0	9·72703	·5 10	10·27297	9·79972	·5 14	10·20028	10·07269	·5 4	9·92731	
15·0	9·72723		10·27277	9·80000		10·20000	10·07276		9·92723	45′
16·0	9·72743		10·27257	9·80028		10·19972	10·07285		9·92715	
17·0	9·72763		10·27237	9·80056		10·19944	10·07293		9·92707	
18·0	9·72783	·6 12	10·27217	9·80084	·6 17	10·19916	10·07301	·6 5	9·92699	
19·0	9·72803		10·27197	9·80112		10·19888	10·07309		9·92691	
20·0	9·72823		10·27177	9·80140	·7 20	10·19860	10·07317		9·92683	40′
21·0	9·72843	·7 14	10·27157	9·80168		10·19833	10·07325	·7 6	9·92675	
22·0	9·72863		10·27137	9·80196		10·19805	10·07333		9·92667	
23·0	9·72883		10·27118	9·80223		10·19777	10·07341		9·92659	
24·0	9·72902	·8 16	10·27098	9·80251	·8 22	10·19749	10·07349	·8 6	9·92651	
25·0	9·72922		10·27078	9·80279		10·19721	10·07357		9·92643	35′
26·0	9·72942		10·27058	9·80307		10·19693	10·07365		9·92635	
27·0	9·72962	·9 18	10·27038	9·80335	·9 25	10·19665	10·07373	·9 7	9·92627	
28·0	9·72982		10·27018	9·80363		10·19637	10·07381		9·92619	
29·0	9·73002		10·26998	9·80391		10·19609	10·07389		9·92611	
30·0	9·73022		10·26978	9·80419		10·19581	10·07397		9·92603	30′
31·0	9·73042		10·26959	9·80447		10·19553	10·07405		9·92595	
32·0	9·73061	·1 2	10·26939	9·80475	·1 3	10·19526	10·07413	·1 1	9·92587	
33·0	9·73081		10·26919	9·80502		10·19498	10·07421		9·92579	
34·0	9·73101		10·26899	9·80530		10·19470	10·07429		9·92571	
35·0	9·73121	·2 4	10·26879	9·80558	·2 6	10·19442	10·07437	·2 2	9·92563	25′
36·0	9·73140		10·26860	9·80586		10·19414	10·07446		9·92555	
37·0	9·73160		10·26840	9·80614		10·19386	10·07454		9·92547	
38·0	9·73180	·3 6	10·26820	9·80642	·3 8	10·19359	10·07462	·3 2	9·92538	
39·0	9·73200		10·26800	9·80669		10·19331	10·07470		9·92530	
40·0	9·73219		10·26781	9·80697		10·19303	10·07478		9·92522	20′
41·0	9·73239		10·26761	9·80725		10·19275	10·07486		9·92514	
42·0	9·73259	·4 8	10·26741	9·80753	·4 11	10·19247	10·07494	·4 3	9·92506	
43·0	9·73278		10·26722	9·80781		10·19220	10·07502		9·92498	
44·0	9·73298		10·26702	9·80808		10·19192	10·07510		9·92490	
45·0	9·73318	·5 10	10·26682	9·80836	·5 14	10·19164	10·07518	·5 4	9·92482	15′
46·0	9·73337		10·26663	9·80864		10·19136	10·07527		9·92474	
47·0	9·73357		10·26643	9·80892		10·19108	10·07535		9·92465	
48·0	9·73377	·6 12	10·26624	9·80919	·6 17	10·19081	10·07543	·6 5	9·92457	
49·0	9·73396		10·26604	9·80947		10·19053	10·07551		9·92449	
50·0	9·73416		10·26584	9·80975		10·19025	10·07559		9·92441	10′
51·0	9·73435	·7 14	10·26565	9·81003	·7 19	10·18998	10·07567	·7 6	9·92433	
52·0	9·73455		10·26545	9·81030		10·18970	10·07575		9·92425	
53·0	9·73474		10·26526	9·81058		10·18942	10·07584		9·92416	
54·0	9·73494	·8 16	10·26506	9·81086	·8 22	10·18914	10·07592	·8 7	9·92408	
55·0	9·73514		10·26487	9·81113		10·18887	10·07600		9·92400	5′
56·0	9·73533		10·26467	9·81141		10·18859	10·07608		9·92392	
57·0	9·73553	·9 18	10·26448	9·81169	·9 25	10·18831	10·07616	·9 7	9·92384	
58·0	9·73572		10·26428	9·81196		10·18804	10·07625		9·92376	
59·0	9·73591		10·26409	9·81224		10·18776	10·07633		9·92367	
60·0	9·73611		10·26389	9·81252		10·18748	10·07641		9·92359	0′

33°
213°

LOGS. OF TRIG. FUNCTIONS

′	Sine	Parts	Cosec.	Tan.	Parts	Cotan.	Secant	Parts	Cosine	
00·0	9·73611	′	10·26389	9·81252	′	10·18748	10·07641	′	9·92359	60′
01·0	9·73630		10·26370	9·81279		10·18721	10·07649		9·92351	
02·0	9·73650	·1 2	10·26350	9·81307	·1 3	10·18693	10·07657	·1 1	9·92343	
03·0	9·73669		10·26331	9·81335		10·18665	10·07666		9·92335	
04·0	9·73689		10·26311	9·81362		10·18638	10·07674		9·92326	
05·0	9·73708	·2 4	10·26292	9·81390	·2 6	10·18610	10·07682	·2 2	9·92318	55′
06·0	9·73727		10·26273	9·81418		10·18582	10·07690		9·92310	
07·0	9·73747		10·26253	9·81445		10·18555	10·07698		9·92302	
08·0	9·73766	·3 6	10·26234	9·81473	·3 8	10·18527	10·07707	·3 2	9·92293	
09·0	9·73786		10·26215	9·81500		10·18500	10·07715		9·92285	
10·0	9·73805		10·26195	9·81528		10·18472	10·07723		9·92277	50′
11·0	9·73824	·4 8	10·26176	9·81556	·4 11	10·18445	10·07731	·4 3	9·92269	
12·0	9·73843		10·26157	9·81583		10·18417	10·07740		9·92260	
13·0	9·73863		10·26137	9·81611		10·18389	10·07748		9·92252	
14·0	9·73882	·5 10	10·26118	9·81638	·5 14	10·18362	10·07756	·5 4	9·92244	
15·0	9·73901		10·26099	9·81666		10·18334	10·07765		9·92236	45′
16·0	9·73921		10·26079	9·81693		10·18307	10·07773		9·92227	
17·0	9·73940		10·26060	9·81721		10·18279	10·07781		9·92219	
18·0	9·73959	·6 12	10·26041	9·81748	·6 17	10·18252	10·07789	·6 5	9·92211	
19·0	9·73978		10·26022	9·81776		10·18224	10·07798		9·92202	
20·0	9·73998		10·26003	9·81804		10·18197	10·07806		9·92194	40′
21·0	9·74017	·7 13	10·25983	9·81831	·7 19	10·18169	10·07814	·7 6	9·92186	
22·0	9·74036		10·25964	9·81859		10·18142	10·07823		9·92177	
23·0	9·74055		10·25945	9·81886		10·18114	10·07831		9·92169	
24·0	9·74074	·8 15	10·25926	9·81914	·8 22	10·18087	10·07839	·8 7	9·92161	
25·0	9·74093		10·25907	9·81941		10·18059	10·07848		9·92152	35′
26·0	9·74113		10·25888	9·81968		10·18032	10·07856		9·92144	
27·0	9·74132	·9 17	10·25868	9·81996	·9 25	10·18004	10·07864	·9 7	9·92136	
28·0	9·74151		10·25849	9·82023		10·17977	10·07873		9·92127	
29·0	9·74170		10·25830	9·82051		10·17949	10·07881		9·92119	
30·0	9·74189		10·25811	9·82078		10·17922	10·07889		9·92111	30′
31·0	9·74208		10·25792	9·82106		10·17894	10·07898		9·92102	
32·0	9·74227	·1 2	10·25773	9·82133	·1 3	10·17867	10·07906	·1 1	9·92094	
33·0	9·74246		10·25754	9·82161		10·17839	10·07914		9·92086	
34·0	9·74265		10·25735	9·82188		10·17812	10·07923		9·92077	
35·0	9·74284	·2 4	10·25716	9·82215	·2 5	10·17785	10·07931	·2 2	9·92069	25′
36·0	9·74303		10·25697	9·82243		10·17757	10·07940		9·92060	
37·0	9·74322		10·25678	9·82270		10·17730	10·07948		9·92052	
38·0	9·74341	·3 6	10·25659	9·82298	·3 8	10·17702	10·07956	·3 3	9·92044	
39·0	9·74360		10·25640	9·82325		10·17675	10·07965		9·92035	
40·0	9·74379		10·25621	9·82352		10·17648	10·07973		9·92027	20′
41·0	9·74398	·4 8	10·25602	9·82380	·4 11	10·17620	10·07982	·4 3	9·92018	
42·0	9·74417		10·25583	9·82407		10·17593	10·07990		9·92010	
43·0	9·74436		10·25564	9·82435		10·17566	10·07999		9·92002	
44·0	9·74455		10·25545	9·82462		10·17538	10·08007		9·91993	
45·0	9·74474	·5 9	10·25526	9·82489	·5 14	10·17511	10·08015	·5 4	9·91985	15′
46·0	9·74493		10·25507	9·82517		10·17483	10·08024		9·91976	
47·0	9·74512		10·25488	9·82544		10·17456	10·08032		9·91968	
48·0	9·74531	·6 11	10·25469	9·82571	·6 16	10·17429	10·08041	·6 5	9·91959	
49·0	9·74549		10·25451	9·82599		10·17401	10·08049		9·91951	
50·0	9·74568		10·25432	9·82626		10·17374	10·08058		9·91942	10′
51·0	9·74587	·7 13	10·25413	9·82653	·7 19	10·17347	10·08066	·7 6	9·91934	
52·0	9·74606		10·25394	9·82681		10·17320	10·08075		9·91925	
53·0	9·74625		10·25375	9·82708		10·17292	10·08083		9·91917	
54·0	9·74644	·8 15	10·25356	9·82735	·8 22	10·17265	10·08092	·8 7	9·91909	
55·0	9·74662		10·25338	9·82762		10·17238	10·08100		9·91900	5′
56·0	9·74681		10·25319	9·82790		10·17210	10·08109		9·91892	
57·0	9·74700	·9 17	10·25300	9·82817	·9 25	10·17183	10·08117	·9 8	9·91883	
58·0	9·74719		10·25281	9·82844		10·17156	10·08126		9·91875	
59·0	9·74737		10·25263	9·82872		10·17129	10·08134		9·91866	
60·0	9·74756		10·25244	9·82899		10·17101	10·08143		9·91857	0′

146°
326°

34°
214°

LOGS. OF TRIG. FUNCTIONS

′	Sine	Parts		Cosec.	Tan.	Parts		Cotan.	Secant	Parts		Cosine	
00·0	9·74756	′		10·25244	9·82899	′		10·17101	10·08143	′		9·91857	60′
01·0	9·74775			10·25225	9·82926			10·17074	10·08151			9·91849	
02·0	9·74794	·1	2	10·25206	9·82953	·1	3	10·17047	10·08160	·1	1	9·91840	
03·0	9·74812			10·25188	9·82981			10·17020	10·08168			9·91832	
04·0	9·74831			10·25169	9·83008			10·16992	10·08177			9·91823	
05·0	9·74850	·2	4	10·25150	9·83035	·2	5	10·16965	10·08185	·2	2	9·91815	55′
06·0	9·74868			10·25132	9·83062			10·16938	10·08194			9·91806	
07·0	9·74887			10·25113	9·83089			10·16911	10·08202			9·91798	
08·0	9·74906	·3	6	10·25094	9·83117	·3	8	10·16884	10·08211	·3	3	9·91789	
09·0	9·74924			10·25076	9·83144			10·16856	10·08220			9·91781	
10·0	9·74943			10·25057	9·83171			10·16829	10·08228			9·91772	50′
11·0	9·74962	·4	7	10·25039	9·83198	·4	11	10·16802	10·08237	·4	3	9·91763	
12·0	9·74980			10·25020	9·83225			10·16775	10·08245			9·91755	
13·0	9·74999			10·25001	9·83253			10·16748	10·08254			9·91746	
14·0	9·75017	·5	9	10·24983	9·83280	·5	14	10·16720	10·08262	·5	4	9·91738	
15·0	9·75036			10·24964	9·83307			10·16693	10·08271			9·91729	·45′
16·0	9·75054			10·24946	9·83334			10·16666	10·08280			9·91720	
17·0	9·75073			10·24927	9·83361			10·16639	10·08288			9·91712	
18·0	9·75091	·6	11	10·24909	9·83388	·6	16	10·16612	10·08297	·6	5	9·91703	
19·0	9·75110			10·24890	9·83415			10·16585	10·08305			9·91695	
20·0	9·75128			10·24872	9·83443			10·16558	10·08314			9·91686	40′
21·0	9·75147	·7	13	10·24853	9·83470	·7	19	10·16530	10·08323	·7	6	9·91677	
22·0	9·75165			10·24835	9·83497			10·16503	10·08331			9·91669	
23·0	9·75184			10·24816	9·83524			10·16476	10·08340			9·91660	
24·0	9·75202	·8	15	10·24798	9·83551	·8	22	10·16449	10·08349	·8	7	9·91651	
25·0	9·75221			10·24779	9·83578			10·16422	10·08357			9·91643	35′
26·0	9·75239			10·24761	9·83605			10·16395	10·08366			9·91634	
27·0	9·75258	·9	17	10·24742	9·83632	·9	24	10·16368	10·08375	·9	8	9·91625	
28·0	9·75276			10·24724	9·83659			10·16341	10·08383			9·91617	
29·0	9·75294			10·24706	9·83686			10·16314	10·08392			9·91608	
30·0	9·75313			10·24687	9·83713			10·16287	10·08401			9·91599	30′
31·0	9·75331			10·24669	9·83741			10·16260	10·08409			9·91591	
32·0	9·75350	·1	2	10·24651	9·83768	·1	3	10·16233	10·08418	·1	1	9·91582	
33·0	9·75368			10·24632	9·83795			10·16205	10·08427			9·91573	
34·0	9·75386			10·24614	9·83822			10·16178	10·08435			9·91565	
35·0	9·75405	·2	4	10·24595	9·83849	·2	5	10·16151	10·08444	·2	2	9·91556	25′
36·0	9·75423			10·24577	9·83876			10·16124	10·08453			9·91547	
37·0	9·75441			10·24559	9·83903			10·16097	10·08462			9·91539	
38·0	9·75460	·3	5	10·24541	9·83930	·3	8	10·16070	10·08470	·3	3	9·91530	
39·0	9·75478			10·24522	9·83957			10·16043	10·08479			9·91521	
40·0	9·75496			10·24504	9·83984			10·16016	10·08488			9·91512	20′
41·0	9·75514	·4	7	10·24486	9·84011	·4	11	10·15989	10·08497	·4	3	9·91504	
42·0	9·75533			10·24467	9·84038			10·15962	10·08505			9·91495	
43·0	9·75551			10·24449	9·84065			10·15935	10·08514			9·91486	
44·0	9·75569	·5	9	10·24431	9·84092	·5	14	10·15908	10·08523	·5	4	9·91477	
45·0	9·75587			10·24413	9·84119			10·15881	10·08532			9·91469	15′
46·0	9·75605			10·24395	9·84146			10·15854	10·08540			9·91460	
47·0	9·75624			10·24376	9·84173			10·15827	10·08549			9·91451	
48·0	9·75642	·6	11	10·24358	9·84200	·6	16	10·15800	10·08558	·6	5	9·91442	
49·0	9·75660			10·24340	9·84227			10·15773	10·08567			9·91433	
50·0	9·75678			10·24322	9·84254			10·15747	10·08575			9·91425	10′
51·0	9·75696	·7	13	10·24304	9·84281	·7	19	10·15720	10·08584	·7	6	9·91416	
52·0	9·75714			10·24286	9·84307			10·15693	10·08593			9·91407	
53·0	9·75733			10·24267	9·84334			10·15666	10·08602			9·91398	
54·0	9·75751	·8	15	10·24249	9·84361	·8	22	10·15639	10·08611	·8	7	9·91389	
55·0	9·75769			10·24231	9·84388			10·15612	10·08619			9·91381	5′
56·0	9·75787			10·24213	9·84415			10·15585	10·08628			9·91372	
57·0	9·75805	·9	16	10·24195	9·84442	·9	24	10·15558	10·08637	·9	8	9·91363	
58·0	9·75823			10·24177	9·84469			10·15531	10·08646			9·91354	
59·0	9·75841			10·24159	9·84496			10·15504	10·08655			9·91345	
60·0	9·75859			10·24141	9·84523			10·15477	10·08664			9·91337	0′

145°
325°

35°
215°

LOGS. OF TRIG. FUNCTIONS

′	Sine	Parts	Cosec.	Tan.	Parts	Cotan.	Secant	Parts	Cosine	
00·0	9·75859		10·24141	9·84523		10·15477	10·08664		9·91337	60′
01·0	9·75877		10·24123	9·84550		10·15450	10·08672		9·91328	
02·0	9·75895	·1 2	10·24105	9·84576	·1 3	10·15424	10·08681	·1 1	9·91319	
03·0	9·75913		10·24007	9·84603		10·15397	10·08690		9·91310	
04·0	9·75931		10·24069	9·84630		10·15370	10·08699		9·91301	
05·0	9·75949	·2 4	10·24051	9·84657	·2 5	10·15343	10·08708	·2 2	9·91292	55′
06·0	9·75967		10·24033	9·84684		10·15316	10·08717		9·91283	
07·0	9·75985		10·24015	9·84711		10·15289	10·08726		9·91274	
08·0	9·76003	·3 5	10·23997	9·84738	·3 8	10·15262	10·08735	·3 3	9·91266	
09·0	9·76021		10·23979	9·84764		10·15236	10·08743		9·91257	
10·0	9·76039		10·23961	9·84791		10·15209	10·08752		9·91248	50′
11·0	9·76057	·4 7	10·23943	9·84818	·4 11	10·15182	10·08761	·4 4	9·91239	
12·0	9·76075		10·23925	9·84845		10·15155	10·08770		9·91230	
13·0	9·76093		10·23907	9·84872		10·15128	10·08779		9·91221	
14·0	9·76111		10·23889	9·84899		10·15101	10·08788		9·91212	
15·0	9·76129	·5 9	10·23871	9·84925	·5 13	10·15075	10·08797	·5 4	9·91203	45′
16·0	9·76146		10·23854	9·84952		10·15048	10·08806		9·91194	
17·0	9·76164		10·23836	9·84979		10·15021	10·08815		9·91185	
18·0	9·76182	·6 11	10·23818	9·85006	·6 16	10·14994	10·08824	·6 5	9·91176	
19·0	9·76200		10·23800	9·85033		10·14968	10·08833		9·91167	
20·0	9·76218		10·23782	9·85059		10·14941	10·08842		9·91158	40′
21·0	9·76236	·7 13	10·23764	9·85086	·7 19	10·14914	10·08851	·7 6	9·91150	
22·0	9·76253		10·23747	9·85113		10·14887	10·08860		9·91141	
23·0	9·76271		10·23729	9·85140		10·14860	10·08869		9·91132	
24·0	9·76289	·8 14	10·23711	9·85166	·8 21	10·14834	10·08877	·8 7	9·91123	
25·0	9·76307		10·23693	9·85193		10·14807	10·08886		9·91114	35′
26·0	9·76325		10·23676	9·85220		10·14780	10·08895		9·91105	
27·0	9·76342	·9 16	10·23658	9·85247	·9 24	10·14753	10·08904	·9 8	9·91096	
28·0	9·76360		10·23640	9·85273		10·14727	10·08913		9·91087	
29·0	9·76378		10·23622	9·85300		10·14700	10·08922		9·91078	
30·0	9·76395		10·23605	9·85327		10·14673	10·08931		9·91069	30′
31·0	9·76413		10·23587	9·85354		10·14647	10·08940		9·91060	
32·0	9·76431	·1 2	10·23569	9·85380	·1 3	10·14620	10·08949	·1 1	9·91051	
33·0	9·76449		10·23552	9·85407		10·14593	10·08959		9·91042	
34·0	9·76466		10·23534	9·85434		10·14566	10·08968		9·91033	
35·0	9·76484	·2 4	10·23516	9·85460	·2 5	10·14540	10·08977	·2 2	9·91024	25′
36·0	9·76502		10·23499	9·85487		10·14513	10·08986		9·91014	
37·0	9·76519		10·23481	9·85514		10·14486	10·08995		9·91005	
38·0	9·76537	·3 5	10·23463	9·85540	·3 8	10·14460	10·09004	·3 3	9·90996	
39·0	9·76554		10·23446	9·85567		10·14433	10·09013		9·90987	
40·0	9·76572		10·23428	9·85594		10·14406	10·09022		9·90978	20′
41·0	9·76590	·4 7	10·23410	9·85620	·4 11	10·14380	10·09031	·4 4	9·90969	
42·0	9·76607		10·23393	9·85647		10·14353	10·09040		9·90960	
43·0	9·76625		10·23375	9·85674		10·14326	10·09049		9·90951	
44·0	9·76642		10·23358	9·85700		10·14300	10·09058		9·90942	
45·0	9·76660	·5 9	10·23340	9·85727	·5 13	10·14273	10·09067	·5 5	9·90933	15′
46·0	9·76677		10·23323	9·85754		10·14246	10·09076		9·90924	
47·0	9·76695		10·23305	9·85780		10·14220	10·09085		9·90915	
48·0	9·76712	·6 11	10·23288	9·85807	·6 16	10·14193	10·09095	·6 5	9·90906	
49·0	9·76730		10·23270	9·85834		10·14166	10·09104		9·90896	
50·0	9·76748		10·23253	9·85860		10·14140	10·09113		9·90887	10′
51·0	9·76765	·7 12	10·23235	9·85887	·7 19	10·14113	10·09122	·7 6	9·90878	
52·0	9·76782		10·23218	9·85913		10·14087	10·09131		9·90869	
53·0	9·76800		10·23200	9·85940		10·14060	10·09140		9·90860	
54·0	9·76817	·8 14	10·23183	9·85967	·8 21	10·14033	10·09149	·8 7	9·90851	
55·0	9·76835		10·23165	9·85993		10·14007	10·09158		9·90842	5′
56·0	9·76852		10·23148	9·86020		10·13980	10·09168		9·90832	
57·0	9·76870	·9 16	10·23130	9·86046	·9 24	10·13954	10·09177	·9 8	9·90823	
58·0	9·76887		10·23113	9·86073		10·13927	10·09186		9·90814	
59·0	9·76905		10·23096	9·86100		10·13901	10·09195		9·90805	
60·0	9·76922		10·23078	9·86126		10·13874	10·09204		9·90796	0′

36°
216°

LOGS. OF TRIG. FUNCTIONS

,	Sine	Parts	Cosec.	Tan.	Parts	Cotan.	Secant	Parts	Cosine	
00·0	9·76922	'	10·23078	9·86126	'	10·13874	10·09204	'	9·90796	69'
01·0	9·76939		10·23061	9·86153		10·13847	10·09213		9·90787	
02·0	9·76957	·1 2	10·23043	9·86179	·1 3	10·13821	10·09223	·1 1	9·90777	
03·0	9·76974		10·23026	9·86206		10·13794	10·09232		9·90768	
04·0	9·76991		10·23009	9·86232		10·13768	10·09241		9·90759	
05·0	9·77009	·2 3	10·22991	9·86259	·2 5	10·13741	10·09250	·2 2	9·90750	55'
06·0	9·77026		10·22974	9·86285		10·13715	10·09259		9·90741	
07·0	9·77043		10·22957	9·86312		10·13688	10·09269		9·90731	
08·0	9·77061	·3 5	10·22939	9·86339	·3 8	10·13662	10·09278	·3 3	9·90722	
09·0	9·77078		10·22922	9·86365		10·13635	10·09287		9·90713	
10·0	9·77095		10·22905	9·86392		10·13609	10·09296		9·90704	50'
11·0	9·77113	·4 7	10·22888	9·86418	·4 11	10·13582	10·09306	·4 4	9·90695	
12·0	9·77130		10·22870	9·86445		10·13556	10·09315		9·90685	
13·0	9·77147		10·22853	9·86471		10·13529	10·09324		9·90676	
14·0	9·77164		10·22836	9·86498		10·13503	10·09333		9·90667	
15·0	9·77182	·5 9	10·22819	9·86524	·5 13	10·13476	10·09343	·5 5	9·90658	45'
16·0	9·77199		10·22801	9·86551		10·13450	10·09352		9·90648	
17·0	9·77216		10·22784	9·86577		10·13423	10·09361		9·90639	
18·0	9·77233	·6 10	10·22767	9·86604	·6 16	10·13397	10·09370	·6 6	9·90630	
19·0	9·77250		10·22750	9·86630		10·13370	10·09380		9·90620	
20·0	9·77268		10·22733	9·86656		10·13344	10·09389		9·90611	40'
21·0	9·77285	·7 12	10·22715	9·86683	·7 19	10·13317	10·09398	·7 6	9·90602	
22·0	9·77302		10·22698	9·86709		10·13291	10·09408		9·90593	
23·0	9·77319		10·22681	9·86736		10·13264	10·09417		9·90583	
24·0	9·77336	·8 14	10·22664	9·86762	·8 21	10·13238	10·09426	·8 7	9·90574	
25·0	9·77353		10·22647	9·86789		10·13211	10·09436		9·90565	35'
26·0	9·77370		10·22630	9·86815		10·13185	10·09445		9·90555	
27·0	9·77388	·9 16	10·22613	9·86842	·9 24	10·13158	10·09454	·9 8	9·90546	
28·0	9·77405		10·22595	9·86868		10·13132	10·09463		9·90537	
29·0	9·77422		10·22578	9·86895		10·13106	10·09473		9·90527	
30·0	9·77439		10·22561	9·86921		10·13079	10·09482		9·90518	30'
31·0	9·77456		10·22544	9·86947		10·13053	10·09492		9·90509	
32·0	9·77473	·1 2	10·22527	9·86974	·1 3	10·13026	10·09501	·1 1	9·90499	
33·0	9·77490		10·22510	9·87000		10·13000	10·09510		9·90490	
34·0	9·77507		10·22493	9·87027		10·12974	10·09520		9·90480	
35·0	9·77524	·2 3	10·22476	9·87053	·2 5	10·12947	10·09529	·2 2	9·90471	25'
36·0	9·77541		10·22459	9·87079		10·12921	10·09538		9·90462	
37·0	9·77558		10·22442	9·87106		10·12894	10·09548		9·90452	
38·0	9·77575	·3 5	10·22425	9·87132	·3 8	10·12868	10·09557	·3 3	9·90443	
39·0	9·77592		10·22408	9·87159		10·12841	10·09567		9·90434	
40·0	9·77609		10·22391	9·87185		10·12815	10·09576		9·90424	20'
41·0	9·77626	·4 7	10·22374	9·87211	·4 11	10·12789	10·09585	·4 4	9·90415	
42·0	9·77643		10·22357	9·87238		10·12762	10·09595		9·90405	
43·0	9·77660		10·22340	9·87264		10·12736	10·09604		9·90396	
44·0	9·77677		10·22323	9·87290		10·12710	10·09614		9·90386	
45·0	9·77694	·5 8	10·22306	9·87317	·5 13	10·12683	10·09623	·5 5	9·90377	15'
46·0	9·77711		10·22289	9·87343		10·12657	10·09632		9·90368	
47·0	9·77728		10·22273	9·87369		10·12631	10·09642		9·90358	
48·0	9·77744	·6 10	10·22256	9·87396	·6 16	10·12604	10·09651	·6 6	9·90349	
49·0	9·77761		10·22239	9·87422		10·12578	10·09661		9·90339	
50·0	9·77778		10·22222	9·87448		10·12552	10·09670		9·90330	10'
51·0	9·77795	·7 12	10·22205	9·87475	·7 18	10·12525	10·09680	·7 7	9·90320	
52·0	9·77812		10·22188	9·87501		10·12499	10·09689		9·90311	
53·0	9·77829		10·22171	9·87527		10·12473	10·09699		9·90301	
54·0	9·77846	·8 14	10·22155	9·87554	·8 21	10·12446	10·09708	·8 8	9·90292	
55·0	9·77862		10·22138	9·87580		10·12420	10·09718		9·90282	5'
56·0	9·77879		10·22121	9·87606		10·12394	10·09727		9·90273	
57·0	9·77896	·9 15	10·22104	9·87633	·9 24	10·12367	10·09737	·9 8	9·90263	
58·0	9·77913		10·22087	9·87659		10·12341	10·09746		9·90254	
59·0	9·77930		10·22071	9·87685		10·12315	10·09756		9·90244	
60·0	9·77946		10·22054	9·87711		10·12289	10·09765		9·90235	0'

143°
323°

37°
217°

LOGS. OF TRIG. FUNCTIONS

,	Sine	Parts		Cosec.	Tan.	Parts		Cotan.	Secant	Parts		Cosine	
00·0	9·77946	′		10·22054	9·87711	′		10·12289	10·09765	′		9·90235	60′
01·0	9·77963			10·22037	9·87738			10·12262	10·09775			9·90225	
02·0	9·77980	·1	2	10·22020	9·87764	·1	3	10·12236	10·09784	·1	1	9·90216	
03·0	9·77997			10·22003	9·87790			10·12210	10·09794			9·90206	
04·0	9·78013			10·21987	9·87817			10·12184	10·09803			9·90197	
05·0	9·78030	·2	3	10·21970	9·87843	·2	5	10·12157	10·09813	·2	2	9·90187	55′
06·0	9·78047			10·21953	9·87869			10·12131	10·09822			9·90178	
07·0	9·78063			10·21937	9·87895			10·12105	10·09832			9·90168	
08·0	9·78080	·3	5	10·21920	9·87922	·3	8	10·12078	10·09842	·3	3	9·90159	
09·0	9·78097			10·21903	9·87948			10·12052	10·09851			9·90149	
10·0	9·78113			10·21887	9·87974			10·12026	10·09861			9·90139	50′
11·0	9·78130	·4	7	10·21870	9·88000	·4	10	10·12000	10·09870	·4	4	9·90130	
12·0	9·78147			10·21853	9·88027			10·11974	10·09880			9·90120	
13·0	9·78163			10·21837	9·88053			10·11947	10·09889			9·90111	
14·0	9·78180	·5	8	10·21820	9·88079	·5	13	10·11921	10·09899	·5	5	9·90101	
15·0	9·78197			10·21803	9·88105			10·11895	10·09909			9·90091	45′
16·0	9·78213			10·21787	9·88131			10·11869	10·09918			9·90082	
17·0	9·78230			10·21770	9·88158			10·11842	10·09928			9·90072	
18·0	9·78246	·6	10	10·21754	9·88184	·6	16	10·11816	10·09937	·6	6	9·90063	
19·0	9·78263			10·21737	9·88210			10·11790	10·09947			9·90053	
20·0	9·78280			10·21720	9·88236			10·11764	10·09957			9·90043	40′
21·0	9·78296	·7	12	10·21704	9·88263	·7	18	10·11738	10·09966	·7	7	9·90034	
22·0	9·78313			10·21687	9·88289			10·11711	10·09976			9·90024	
23·0	9·78329			10·21671	9·88315			10·11685	10·09986			9·90014	
24·0	9·78346	·8	13	10·21654	9·88341	·8	21	10·11659	10·09995	·8	8	9·90005	
25·0	9·78362			10·21638	9·88367			10·11633	10·10005			9·89995	35′
26·0	9·78379			10·21621	9·88393			10·11607	10·10015			9·89985	
27·0	9·78395	·9	15	10·21605	9·88420	·9	24	10·11580	10·10024	·9	9	9·89976	
28·0	9·78412			10·21588	9·88446			10·11554	10·10034			9·89966	
29·0	9·78428			10·21572	9·88472			10·11528	10·10044			9·89956	
30·0	9·78445			10·21555	9·88498			10·11502	10·10053			9·89947	30′
31·0	9·78461			10·21539	9·88524			10·11476	10·10063			9·89937	
32·0	9·78478	·1	2	10·21522	9·88550	·1	3	10·11450	10·10073	·1	1	9·89927	
33·0	9·78494			10·21506	9·88577			10·11424	10·10082			9·89918	
34·0	9·78511			10·21490	9·88603			10·11397	10·10092			9·89908	
35·0	9·78527	·2	3	10·21473	9·88629	·2	5	10·11371	10·10102	·2	2	9·89898	25′
36·0	9·78543			10·21457	9·88655			10·11345	10·10112			9·89888	
37·0	9·78560			10·21440	9·88681			10·11319	10·10121			9·89879	
38·0	9·78576	·3	5	10·21424	9·88707	·3	8	10·11293	10·10131	·3	3	9·89869	
39·0	9·78593			10·21408	9·88733			10·11267	10·10141			9·89859	
40·0	9·78609			10·21391	9·88759			10·11241	10·10151			9·89849	20′
41·0	9·78625	·4	7	10·21375	9·88786	·4	10	10·11215	10·10160	·4	4	9·89840	
42·0	9·78642			10·21358	9·88812			10·11188	10·10170			9·89830	
43·0	9·78658			10·21342	9·88838			10·11162	10·10180			9·89820	
44·0	9·78674	·5	8	10·21326	9·88864	·5	13	10·11136	10·10190	·5	5	9·89810	
45·0	9·78691			10·21309	9·88890			10·11110	10·10199			9·89801	15′
46·0	9·78707			10·21293	9·88916			10·11084	10·10209			9·89791	
47·0	9·78723			10·21277	9·88942			10·11058	10·10219			9·89781	
48·0	9·78740	·6	10	10·21261	9·88968	·6	16	10·11032	10·10229	·6	6	9·89771	
49·0	9·78756			10·21244	9·88994			10·11006	10·10239			9·89761	
50·0	9·78772			10·21228	9·89020			10·10980	10·10248			9·89752	10′
51·0	9·78788	·7	11	10·21212	9·89047	·7	18	10·10954	10·10258	·7	7	9·89742	
52·0	9·78805			10·21196	9·89073			10·10928	10·10268			9·89732	
53·0	9·78821			10·21179	9·89099			10·10901	10·10278			9·89722	
54·0	9·78837	·8	13	10·21163	9·89125	·8	21	10·10875	10·10288	·8	8	9·89712	
55·0	9·78853			10·21147	9·89151			10·10849	10·10298			9·89703	5′
56·0	9·78869			10·21131	9·89177			10·10823	10·10307			9·89693	
57·0	9·78886	·9	15	10·21114	9·89203	·9	23	10·10797	10·10317	·9	9	9·89683	
58·0	9·78902			10·21098	9·89229			10·10771	10·10327			9·89673	
59·0	9·78918			10·21082	9·89255			10·10745	10·10337			9·89663	
60·0	9·78934			10·21066	9·89281			10·10719	10·10347			9·89653	0′

38°
218°

LOGS. OF TRIG. FUNCTIONS

′	Sine	Parts		Cosec.	Tan.	Parts		Cotan.	Secant	Parts		Cosine	
00·0	9·78934	′		10·21066	9·89281	′		10·10719	10·10347	′		9·89653	60′
01·0	9·78950			10·21050	9·89307			10·10693	10·10357			9·89643	
02·0	9·78967	·1	2	10·21034	9·89333	·1	3	10·10667	10·10367	·1	1	9·89634	
03·0	9·78983			10·21017	9·89359			10·10641	10·10376			9·89624	
04·0	9·78999			10·21001	9·89385			10·10615	10·10386			9·89614	
05·0	9·79015	·2	3	10·20985	9·89411	·2	5	10·10589	10·10396	·2	2	9·89604	55′
06·0	9·79031			10·20969	9·89437			10·10563	10·10406			9·89594	
07·0	9·79047			10·20953	9·89463			10·10537	10·10416			9·89584	
08·0	9·79063	·3	5	10·20937	9·89489	·3	8	10·10511	10·10426	·3	3	9·89574	
09·0	9·79079			10·20921	9·89515			10·10485	10·10436			9·89564	
10·0	9·79095			10·20905	9·89541			10·10459	10·10446			9·89554	50′
11·0	9·79112	·4	6	10·20889	9·89567	·4	10	10·10433	10·10456	·4	4	9·89544	
12·0	9·79128			10·20873	9·89593			10·10407	10·10466			9·89534	
13·0	9·79144			10·20856	9·89619			10·10381	10·10476			9·89524	
14·0	9·79160	·5	8	10·20840	9·89645	·5	13	10·10355	10·10486	·5	5	9·89515	
15·0	9·79176			10·20824	9·89671			10·10329	10·10496			9·89505	45′
16·0	9·79192			10·20808	9·89697			10·10303	10·10506			9·89495	
17·0	9·79208	·6	10	10·20792	9·89723	·6	16	10·10277	10·10515	·6	6	9·89485	
18·0	9·79224			10·20776	9·89749			10·10251	10·10525			9·89475	
19·0	9·79240			10·20760	9·89775			10·10225	10·10535			9·89465	
20·0	9·79256	·7	11	10·20744	9·89801	·7	18	10·10199	10·10545	·7	7	9·89455	40′
21·0	9·79272			10·20728	9·89827			10·10173	10·10555			9·89445	
22·0	9·79288			10·20712	9·89853			10·10147	10·10565			9·89435	
23·0	9·79304			10·20697	9·89879			10·10121	10·10575			9·89425	
24·0	9·79320	·8	13	10·20681	9·89905	·8	21	10·10095	10·10585	·8	8	9·89415	
25·0	9·79335			10·20665	9·89931			10·10069	10·10595			9·89405	35′
26·0	9·79351			10·20649	9·89957			10·10043	10·10605			9·89395	
27·0	9·79367	·9	14	10·20633	9·89983	·9	23	10·10017	10·10615	·9	9	9·89385	
28·0	9·79383			10·20617	9·90009			10·09991	10·10626			9·89375	
29·0	9·79399			10·20601	9·90035			10·09965	10·10636			9·89365	
30·0	9·79415			10·20585	9·90061			10·09940	10·10646			9·89354	30′
31·0	9·79431			10·20569	9·90086			10·09914	10·10656			9·89344	
32·0	9·79447	·1	2	10·20553	9·90112	·1	3	10·09888	10·10666	·1	1	9·89334	
33·0	9·79463			10·20537	9·90138			10·09862	10·10676			9·89324	
34·0	9·79478			10·20522	9·90164			10·09836	10·10686			9·89314	
35·0	9·79494	·2	3	10·20506	9·90190	·2	5	10·09810	10·10696	·2	2	9·89304	25′
36·0	9·79510			10·20490	9·90216			10·09784	10·10706			9·89294	
37·0	9·79526			10·20474	9·90242			10·09758	10·10716			9·89284	
38·0	9·79542	·3	5	10·20458	9·90268	·3	8	10·09732	10·10726	·3	3	9·89274	
39·0	9·79558			10·20443	9·90294			10·09706	10·10736			9·89264	
40·0	9·79573			10·20427	9·90320			10·09680	10·10746			9·89254	20′
41·0	9·79589	·4	6	10·20411	9·90346	·4	10	10·09654	10·10757	·4	4	9·89244	
42·0	9·79605			10·20395	9·90371			10·09629	10·10767			9·89233	
43·0	9·79621			10·20379	9·90397			10·09603	10·10777			9·89223	
44·0	9·79636			10·20364	9·90423			10·09577	10·10787			9·89213	
45·0	9·79652	·5	8	10·20348	9·90449	·5	13	10·09551	10·10797	·5	5	9·89203	15′
46·0	9·79668			10·20332	9·90475			10·09525	10·10807			9·89193	
47·0	9·79684			10·20316	9·90501			10·09499	10·10817			9·89183	
48·0	9·79699	·6	9	10·20301	9·90527	·6	16	10·09473	10·10827	·6	6	9·89173	
49·0	9·79715			10·20285	9·90553			10·09447	10·10838			9·89162	
50·0	9·79731			10·20269	9·90579			10·09422	10·10848			9·89152	10′
51·0	9·79746	·7	11	10·20254	9·90604	·7	18	10·09396	10·10858	·7	7	9·89142	
52·0	9·79762			10·20238	9·90630			10·09370	10·10868			9·89132	
53·0	9·79778			10·20222	9·90656			10·09344	10·10878			9·89122	
54·0	9·79793	·8	13	10·20207	9·90682	·8	21	10·09318	10·10889	·8	8	9·89112	
55·0	9·79809			10·20191	9·90708			10·09292	10·10899			9·89101	5′
56·0	9·79825			10·20175	9·90734			10·09266	10·10909			9·89091	
57·0	9·79840	·9	14	10·20160	9·90759	·9	23	10·09241	10·10919	·9	9	9·89081	
58·0	9·79856			10·20144	9·90785			10·09215	10·10929			9·89071	
59·0	9·79872			10·20128	9·90811			10·09189	10·10940			9·89061	
60·0	9·79887			10·20113	9·90837			10·09163	10·10950			9·89050	0′

39°
219°

LOGS. OF TRIG. FUNCTIONS

'	Sine	Parts	Cosec.	Tan.	Parts	Cotan.	Secant	Parts	Cosine	
00·0	9·79887	'	10·20113	9·90837	'	10·09163	10·10950	'	9·89050	60'
01·0	9·79903		10·20097	9·90863		10·09137	10·10960		9·89040	
02·0	9·79918	·1 2	10·20082	9·90889	·1 3	10·09111	10·10970	·1 1	9·89030	
03·0	9·79934		10·20066	9·90914		10·09086	10·10981		9·89020	
04·0	9·79950		10·20051	9·90940		10·09060	10·10991		9·89009	
05·0	9·79965	·2 3	10·20035	9·90966	·2 5	10·09034	10·11001	·2 2	9·88999	55'
06·0	9·79981		10·20019	9·90992		10·09008	10·11011		9·88989	
07·0	9·79996		10·20004	9·91018		10·08982	10·11022		9·88979	
08·0	9·80012	·3 5	10·19988	9·91044	·3 8	10·08957	10·11032	·3 3	9·88968	
09·0	9·80027		10·19973	9·91069		10·08931	10·11042		9·88958	
10·0	9·80043		10·19957	9·91095		10·08905	10·11052		9·88948	50'
11·0	9·80058	·4 6	10·19942	9·91121	·4 10	10·08879	10·11063	·4 4	9·88937	
12·0	9·80074		10·19926	9·91147		10·08853	10·11073		9·88927	
13·0	9·80089		10·19911	9·91173		10·08827	10·11083		9·88917	
14·0	9·80105	·5 8	10·19895	9·91198	·5 13	10·08802	10·11094	·5 5	9·88906	
15·0	9·80120		10·19880	9·91224		10·08776	10·11104		9·88896	45'
16·0	9·80136		10·19864	9·91250		10·08750	10·11114		9·88886	
17·0	9·80151	·6 9	10·19849	9·91276	·6 15	10·08724	10·11125	·6 6	9·88875	
18·0	9·80167		10·19834	9·91301		10·08699	10·11135		9·88865	
19·0	9·80182		10·19818	9·91327		10·08673	10·11145		9·88855	
20·0	9·80197	·7 11	10·19803	9·91353	·7 18	10·08647	10·11156	·7 7	9·88844	40'
21·0	9·80213		10·19787	9·91379		10·08621	10·11166		9·88834	
22·0	9·80228		10·19772	9·91404		10·08596	10·11176		9·88824	
23·0	9·80244		10·19756	9·91430		10·08570	10·11187		9·88813	
24·0	9·80259	·8 12	10·19741	9·91456	·8 21	10·08544	10·11197	·8 8	9·88803	
25·0	9·80274		10·19726	9·91482		10·08518	10·11207		9·88793	35'
26·0	9·80290		10·19710	9·91508		10·08493	10·11218		9·88782	
27·0	9·80305	·9 14	10·19695	9·91533	·9 23	10·08467	10·11228	·9 9	9·88772	
28·0	9·80320		10·19680	9·91559		10·08441	10·11239		9·88761	
29·0	9·80336		10·19664	9·91585		10·08415	10·11249		9·88751	
30·0	9·80351		10·19649	9·91610		10·08390	10·11259		9·88741	30'
31·0	9·80366		10·19634	9·91636		10·08364	10·11270		9·88730	
32·0	9·80382	·1 2	10·19618	9·91662	·1 3	10·08338	10·11280	·1 1	9·88720	
33·0	9·80397		10·19603	9·91688		10·08312	10·11291		9·88709	
34·0	9·80412		10·19588	9·91713		10·08287	10·11301		9·88699	
35·0	9·80428	·2 3	10·19572	9·91739	·2 5	10·08261	10·11312	·2 2	9·88689	25'
36·0	9·80443		10·19557	9·91765		10·08235	10·11322		9·88678	
37·0	9·80458		10·19542	9·91791		10·08209	10·11332		9·88668	
38·0	9·80473	·3 5	10·19527	9·91816	·3 8	10·08184	10·11343	·3 3	9·88657	
39·0	9·80489		10·19511	9·91842		10·08158	10·11353		9·88647	
40·0	9·80504		10·19496	9·91868		10·08132	10·11364		9·88636	20'
41·0	9·80519	·4 6	10·19481	9·91893	·4 10	10·08107	10·11374	·4 4	9·88626	
42·0	9·80534		10·19466	9·91919		10·08081	10·11385		9·88615	
43·0	9·80550		10·19451	9·91945		10·08055	10·11395		9·88605	
44·0	9·80565		10·19435	9·91971		10·08030	10·11406		9·88594	
45·0	9·80580	·5 8	10·19420	9·91996	·5 13	10·08004	10·11416	·5 5	9·88584	15'
46·0	9·80595		10·19405	9·92022		10·07978	10·11427		9·88573	
47·0	9·80610		10·19390	9·92048		10·07952	10·11437		9·88563	
48·0	9·80625	·6 9	10·19375	9·92073	·6 15	10·07927	10·11448	·6 6	9·88552	
49·0	9·80641		10·19359	9·92099		10·07901	10·11458		9·88542	
50·0	9·80656		10·19344	9·92125		10·07875	10·11469		9·88531	10'
51·0	9·80671	·7 11	10·19329	9·92150	·7 18	10·07850	10·11479	·7 7	9·88521	
52·0	9·80686		10·19314	9·92176		10·07824	10·11490		9·88510	
53·0	9·80701		10·19299	9·92202		10·07798	10·11501		9·88499	
54·0	9·80716	·8 12	10·19284	9·92227	·8 21	10·07773	10·11511	·8 8	9·88489	
55·6	9·80731		10·19269	9·92253		10·07747	10·11522		9·88478	5'
56·0	9·80747		10·19254	9·92279		10·07721	10·11532		9·88468	
57·0	9·80762	·9 14	10·19239	9·92304	·9 23	10·07696	10·11543	·9 9	9·88457	
58·0	9·80777		10·19223	9·92330		10·07670	10·11553		9·08447	
59·0	9·80792		10·19208	9·92356		10·07644	10·11564		9·88436	
60·0	9·80807		10·19193	9·92381		10·07619	10·11575		9·88425	0'

140°
320°

40°
220°

′	Sine	Parts	Cosec.	Tan.	Parts	Cotan.	Secant	Parts	Cosine	
00·0	9·80807	′	10·19193	9·92381	′	10·07619	10·11575	′	9·88425	60′
01·0	9·80822		10·19178	9·92407		10·07593	10·11585		9·88415	
02·0	9·80837	·1 1	10·19163	9·92433	·1 3	10·07567	10·11596	·1 1	9·88404	
03·0	9·80852		10·19148	9·92458		10·07542	10·11606		9·88394	
04·0	9·80867		10·19133	9·92484		10·07516	10·11617		9·88383	
05·0	9·80882	·2 3	10·19118	9·92510	·2 5	10·07490	10·11628	·2 2	9·88372	55′
06·0	9·80897		10·19103	9·92535		10·07465	10·11638		9·88362	
07·0	9·80912		10·19088	9·92561		10·07439	10·11649		9·88351	
08·0	9·80927	·3 4	10·19073	9·92587	·3 8	10·07414	10·11660	·3 3	9·88340	
09·0	9·80942		10·19058	9·92612		10·07388	10·11670		9·88330	
10·0	9·80957		10·19043	9·92638		10·07362	10·11681		9·88319	50′
11·0	9·80972	·4 6	10·19028	9·92663	·4 10	10·07337	10·11692	·4 4	9·88308	
12·0	9·80987		10·19013	9·92689		10·07311	10·11702		9·88298	
13·0	9·81002		10·18998	9·92715		10·07285	10·11713		9·88287	
14·0	9·81017		10·18983	9·92740		10·07260	10·11724		9·88276	
15·0	9·81032	·5 7	10·18968	9·92766	·5 13	10·07234	10·11734	·5 5	9·88266	45′
16·0	9·81047		10·18954	9·92792		10·07209	10·11745		9·88255	
17·0	9·81061		10·18939	9·92817		10·07183	10·11756		9·88244	
18·0	9·81076	·6 9	10·18924	9·92843	·6 15	10·07157	10·11766	·6 6	9·88234	
19·0	9·81091		10·18909	9·92868		10·07132	10·11777		9·88223	
20·0	9·81106	·7 10	10·18894	9·92894	·7 18	10·07106	10·11788	·7 7	9·88212	40′
21·0	9·81121		10·18879	9·92920		10·07080	10·11799		9·88201	
22·0	9·81136		10·18864	9·92945		10·07055	10·11809		9·88191	
23·0	9·81151		10·18849	9·92971		10·07029	10·11820		9·88180	
24·0	9·81166	·8 12	10·18835	9·92996	·8 21	10·07004	10·11831	·8 9	9·88169	
25·0	9·81180		10·18820	9·93022		10·06978	10·11842		9·88158	35′
26·0	9·81195		10·18805	9·93048		10·06953	10·11852		9·88148	
27·0	9·81210	·9 13	10·18790	9·93073	·9 23	10·06927	10·11863	·9 10	9·88137	
28·0	9·81225		10·18775	9·93099		10·06901	10·11874		9·88126	
29·0	9·81240		10·18760	9·93124		10·06876	10·11885		9·88115	
30·0	9·81254		10·18746	9·93150		10·06850	10·11895		9·88105	30′
31·0	9·81269		10·18731	9·93176		10·06825	10·11906		9·88094	
32·0	9·81284	·1 1	10·18716	9·93201	·1 3	10·06799	10·11917	·1 1	9·88083	
33·0	9·81299		10·18701	9·93227		10·06773	10·11928		9·88072	
34·0	9·81314		10·18687	9·93252		10·06748	10·11939		9·88061	
35·0	9·81328	·2 3	10·18672	9·93278	·2 5	10·06722	10·11950	·2 2	9·88051	25′
36·0	9·81343		10·18657	9·93303		10·06697	10·11960		9·88040	
37·0	9·81358		10·18642	9·93329		10·06671	10·11971		9·88029	
38·0	9·81373	·3 4	10·18628	9·93355	·3 8	10·06646	10·11982	·3 3	9·88018	
39·0	9·81387		10·18613	9·93380		10·06620	10·11993		9·88007	
40·0	9·81402		10·18598	9·93406		10·06594	10·12004		9·87996	20′
41·0	9·81417	·4 6	10·18583	9·93431	·4 10	10·06569	10·12015	·4 4	9·87986	
42·0	9·81431		10·18569	9·93457		10·06543	10·12025		9·87975	
43·0	9·81446		10·18554	9·93482		10·06518	10·12036		9·87964	
44·0	9·81461		10·18539	9·93508		10·06492	10·12047		9·87953	
45·0	9·81475	·5 7	10·18525	9·93533	·5 13	10·06467	10·12058	·5 5	9·87942	15′
46·0	9·81490		10·18510	9·93559		10·06441	10·12069		9·87931	
47·0	9·81505		10·18495	9·93584		10·06416	10·12080		9·87920	
48·0	9·81519	·6 9	10·18481	9·93610	·6 15	10·06390	10·12091	·6 7	9·87909	
49·0	9·81534		10·18466	9·93636		10·06365	10·12102		9·87898	
50·0	9·81549		10·18452	9·93661		10·06339	10·12113		9·87888	10′
51·0	9·81563	·7 10	10·18437	9·93687	·7 18	10·06313	10·12123	·7 8	9·87877	
52·0	9·81578		10·18422	9·93712		10·06288	10·12134		9·87866	
53·0	9·81592		10·18408	9·93738		10·06262	10·12145		9·87855	
54·0	9·81607	·8 12	10·18393	9·93763	·8 20	10·06237	10·12156	·8 9	9·87844	
55·0	9·81622		10·18379	9·93789		10·06211	10·12167		9·87833	5′
56·0	9·81636		10·18364	9·93814		10·06186	10·12178		9·87822	
57·0	9·81651	·9 13	10·18349	9·93840	·9 23	10·06160	10·12189	·9 10	9·87811	
58·0	9·81665		10·18335	9·93865		10·06135	10·12200		9·87800	
59·0	9·81680		10·18320	9·93891		10·06109	10·12211		9·87789	
60·0	9·81694		10·18306	9·93916		10·06084	10·12222		9·87778	0′

41°
221°

LOGS. OF TRIG. FUNCTIONS

′	Sine	Parts		Cosec.	Tan.	Parts		Cotan.	Secant	Parts		Cosine	
00·0	9·81694	′		10·18306	9·93916	′		10·06084	10·12222	′		9·87778	60′
01·0	9·81709			10·18291	9·93942			10·06058	10·12233			9·87767	
02·0	9·81723	·1	1	10·18277	9·93967	·1	3	10·06033	10·12244	·1	1	9·87756	
03·0	9·81738			10·18262	9·93993			10·06007	10·12255			9·87745	
04·0	9·81752			10·18248	9·94018			10·05982	10·12266			9·87734	
05·0	9·81767	·2	3	10·18233	9·94044	·2	5	10·05956	10·12277	·2	2	9·87723	55′
06·0	9·81781			10·18219	9·94069			10·05931	10·12288			9·87712	
07·0	9·81796			10·18204	9·94095			10·05905	10·12299			9·87701	
08·0	9·81810	·3	4	10·18190	9·94120	·3	8	10·05880	10·12310	·3	3	9·87690	
09·0	9·81825			10·18175	9·94146			10·05854	10·12321			9·87679	
10·0	9·81839			10·18161	9·94171			10·05829	10·12332			9·87668	50′
11·0	9·81854	·4	6	10·18146	9·94197	·4	10	10·05803	10·12343	·4	4	9·87657	
12·0	9·81868			10·18132	9·94222			10·05778	10·12354			9·87646	
13·0	9·81883			10·18118	9·94248			10·05752	10·12365			9·87635	
14·0	9·81897	·5	7	10·18103	9·94273	·5	13	10·05727	10·12376	·5	6	9·87624	
15·0	9·81911			10·18089	9·94299			10·05701	10·12388			9·87613	45′
16·0	9·81926			10·18074	9·94324			10·05676	10·12399			9·87601	
17·0	9·81940	·6	9	10·18060	9·94350	·6	15	10·05650	10·12410	·6	7	9·87590	
18·0	9·81955			10·18046	9·94375			10·05625	10·12421			9·87579	
19·0	9·81969			10·18031	9·94401			10·05599	10·12432			9·87568	
20·0	9·81983	·7	10	10·18017	9·94426	·7	18	10·05574	10·12443	·7	8	9·87557	40′
21·0	9·81998			10·18002	9·94452			10·05548	10·12454			9·87546	
22·0	9·82012			10·17988	9·94477			10·05523	10·12465			9·87535	
23·0	9·82026			10·17974	9·94503			10·05497	10·12476			9·87524	
24·0	9·82041	·8	12	10·17959	9·94528	·8	20	10·05472	10·12487	·8	9	9·87513	
25·0	9·82055			10·17945	9·94554			10·05447	10·12499			9·87501	35′
26·0	9·82069			10·17931	9·94579			10·05421	10·12510			9·87490	
27·0	9·82084	·9	13	10·17916	9·94605	·9	23	10·05396	10·12521	·9	10	9·87479	
28·0	9·82098			10·17902	9·94630			10·05370	10·12532			9·87468	
29·0	9·82112			10·17888	9·94655			10·05345	10·12543			9·87457	
30·0	9·82127			10·17874	9·94681			10·05319	10·12554			9·87446	30′
31·0	9·82141			10·17859	9·94706			10·05294	10·12566			9·87434	
32·0	9·82155	·1	1	10·17845	9·94732	·1	3	10·05268	10·12577	·1	1	9·87423	
33·0	9·82169			10·17831	9·94757			10·05243	10·12588			9·87412	
34·0	9·82184			10·17817	9·94783			10·05217	10·12599			9·87401	
35·0	9·82198	·2	3	10·17802	9·94808	·2	5	10·05192	10·12610	·2	2	9·87390	25′
36·0	9·82212			10·17788	9·94834			10·05167	10·12622			9·87378	
37·0	9·82226			10·17774	9·94859			10·05141	10·12633			9·87367	
38·0	9·82240	·3	4	10·17760	9·94884	·3	8	10·05116	10·12644	·3	3	9·87356	
39·0	9·82255			10·17745	9·94910			10·05090	10·12655			9·87345	
40·0	9·82269			10·17731	9·94935			10·05065	10·12667			9·87334	20′
41·0	9·82283	·4	6	10·17717	9·94961	·4	10	10·05039	10·12678	·4	5	9·87322	
42·0	9·82297			10·17703	9·94986			10·05014	10·12689			9·87311	
43·0	9·82311			10·17689	9·95012			10·04988	10·12700			9·87300	
44·0	9·82326			10·17675	9·95037			10·04963	10·12712			9·87289	
45·0	9·82340	·5	7	10·17660	9·95063	·5	13	10·04938	10·12723	·5	6	9·87277	15′
46·0	9·82354			10·17646	9·95088			10·04912	10·12734			9·87266	
47·0	9·82368			10·17632	9·95113			10·04887	10·12745			9·87255	
48·0	9·82382	·6	8	10·17618	9·95139	·6	15	10·04861	10·12757	·6	7	9·87243	
49·0	9·82396			10·17604	9·95164			10·04836	10·12768			9·87232	
50·0	9·82410			10·17590	9·95190			10·04810	10·12779			9·87221	10′
51·0	9·82425	·7	10	10·17576	9·95215	·7	18	10·04785	10·12791	·7	8	9·87210	
52·0	9·82439			10·17561	9·95241			10·04760	10·12802			9·87198	
53·0	9·82453			10·17547	9·95266			10·04734	10·12813			9·87187	
54·0	9·82467	·8	11	10·17533	9·95291	·8	20	10·04709	10·12825	·8	9	9·87176	
55·0	9·82481			10·17519	9·95317			10·04683	10·12836			9·87164	5′
56·0	9·82495			10·17505	9·95342			10·04658	10·12847			9·87153	
57·0	9·82509	·9	13	10·17491	9·95368	·9	23	10·04633	10·12859	·9	10	9·87141	
58·0	9·82523			10·17477	9·95393			10·04607	10·12870			9·87130	
59·0	9·82537			10·17463	9·95418			10·04582	10·12881			9·87119	
60·0	9·82551			10·17449	9·95444			10·04556	10·12893			9·87107	0′

42°
222°

LOGS. OF TRIG. FUNCTIONS

′	Sine	Parts	Cosec.	Tan.	Parts	Cotan.	Secant	Parts	Cosine	
00·0	9·82551	′	10·17449	9·95444	′	10·04556	10·12893	′	9·87107	60′
01·0	9·82565		10·17435	9·95469		10·04531	10·12904		9·87096	
02·0	9·82579	·1 1	10·17421	9·95495	·1 3	10·04505	10·12915	·1 1	9·87085	
03·0	9·82593		10·17407	9·95520		10·04480	10·12927		9·87073	
04·0	9·82607		10·17393	9·95545		10·04455	10·12938		9·87062	
05·0	9·82621	·2 3	10·17379	9·95571	·2 5	10·04429	10·12950	·2 2	9·87050	55′
06·0	9·82635		10·17365	9·95596		10·04404	10·12961		9·87039	
07·0	9·82649		10·17351	9·95622		10·04379	10·12972		9·87028	
08·0	9·82663	·3 4	10·17337	9·95647	·3 8	10·04353	10·12984	·3 3	9·87016	
09·0	9·82677		10·17323	9·95672		10·04328	10·12995		9·87005	
10·0	9·82691		10·17309	9·95698		10·04302	10·13007		9·86993	50′
11·0	9·82705	·4 6	10·17295	9·95723	·4 10	10·04277	10·13018	·4 5	9·86982	
12·0	9·82719		10·17281	9·95749		10·04252	10·13030		9·86970	
13·0	9·82733		10·17267	9·95774		10·04226	10·13041		9·86959	
14·0	9·82747		10·17253	9·95799		10·04201	10·13053		9·86947	
15·0	9·82761	·5 7	10·17239	9·95825	·5 13	10·04175	10·13064	·5 6	9·86936	45′
16·0	9·82775		10·17226	9·95850		10·04150	10·13076		9·86925	
17·0	9·82788		10·17212	9·95875		10·04125	10·13087		9·86913	
18·0	9·82802	·6 8	10·17198	9·95901	·6 15	10·04099	10·13099	·6 7	9·86902	
19·0	9·82816		10·17184	9·95926		10·04074	10·13110		9·86890	
20·0	9·82830		10·17170	9·95952		10·04048	10·13122		9·86879	40′
21·0	9·82844	·7 10	10·17156	9·95977	·7 18	10·04023	10·13133	·7 8	9·86867	
22·0	9·82858		10·17142	9·96002		10·03998	10·13145		9·86856	
23·0	9·82872		10·17128	9·96028		10·03972	10·13156		9·86844	
24·0	9·82886	·8 11	10·17115	9·96053	·8 20	10·03947	10·13168	·8 9	9·86832	
25·0	9·82899		10·17101	9·96078		10·03922	10·13179		9·86821	35′
26·0	9·82913		10·17087	9·96104		10·03896	10·13191		9·86809	
27·0	9·82927	·9 13	10·17073	9·96129	·9 23	10·03871	10·13202	·9 10	9·86798	
28·0	9·82941		10·17059	9·96155		10·03846	10·13214		9·86786	
29·0	9·82955		10·17046	9·96180		10·03820	10·13225		9·86775	
30·0	9·82968		10·17032	9·96205		10·03795	10·13237		9·86763	30′
31·0	9·82982		10·17018	9·96231		10·03769	10·13249		9·86752	
32·0	9·82996	·1 1	10·17004	9·96256	·1 3	10·03744	10·13260	·1 1	9·86740	
33·0	9·83010		10·16990	9·96281		10·03719	10·13272		9·86728	
34·0	9·83023		10·16977	9·96307		10·03693	10·13283		9·86717	
35·0	9·83037	·2 3	10·16963	9·96332	·2 5	10·03668	10·13295	·2 2	9·86705	25′
36·0	9·83051		10·16949	9·96357		10·03643	10·13307		9·86694	
37·0	9·83065		10·16935	9·96383		10·03617	10·13318		9·86682	
38·0	9·83078		10·16922	9·96408		10·03592	10·13330		9·86670	
39·0	9·83092	·3 4	10·16908	9·96434	·3 8	10·03567	10·13341	·3 4	9·86659	
40·0	9·83106		10·16894	9·96459		10·03541	10·13353		9·86647	20′
41·0	9·83120		10·16881	9·96484		10·03516	10·13365		9·86635	
42·0	9·83133	·4 5	10·16867	9·96510	·4 10	10·03491	10·13376	·4 5	9·86624	
43·0	9·83147		10·16853	9·96535		10·03465	10·13388		9·86612	
44·0	9·83161		10·16839	9·96560		10·03440	10·13400		9·86600	
45·0	9·83174	·5 7	10·16826	9·96586	·5 13	10·03415	10·13411	·5 6	9·86589	15′
46·0	9·83188		10·16812	9·96611		10·03389	10·13423		9·86577	
47·0	9·83202		10·16799	9·96636		10·03364	10·13435		9·86565	
48·0	9·83215	·6 8	10·16785	9·96662	·6 15	10·03338	10·13446	·6 7	9·86554	
49·0	9·83229		10·16771	9·96687		10·03313	10·13458		9·86542	
50·0	9·83243		10·16758	9·96712		10·03288	10·13470		9·86530	10′
51·0	9·83256	·7 10	10·16744	9·96738	·7 18	10·03262	10·13482	·7 8	9·86519	
52·0	9·83270		10·16730	9·96763		10·03237	10·13493		9·86507	
53·0	9·83283		10·16717	9·96788		10·03212	10·13505		9·86495	
54·0	9·83297	·8 11	10·16703	9·96814	·8 20	10·03186	10·13517	·8 9	9·86483	
55·0	9·83311		10·16690	9·96839		10·03161	10·13528		9·86472	5′
56·0	9·83324		10·16676	9·96864		10·03136	10·13540		9·86460	
57·0	9·83338	·9 12	10·16662	9·96890	·9 23	10·03110	10·13552	·9 11	9·86448	
58·0	9·83351		10·16649	9·96915		10·03085	10·13564		9·86436	
59·0	9·83365		10·16635	9·96940		10·03060	10·13576		9·86425	
60·0	9·83378		10·16622	9·96966		10·03034	10·13587		9·86413	0′

43°
223°

LOGS. OF TRIG. FUNCTIONS

′	Sine	Parts	Cosec.	Tan.	Parts	Cotan.	Secant	Parts	Cosine	
00·0	9·83378	′	10·16622	9·96966	′	10·03034	10·13587	′	9·86413	60′
01·0	9·83392		10·16608	9·96991		10·03009	10·13599		9·86401	
02·0	9·83405	·1 1	10·16595	9·97016	·1 3	10·02984	10·13611	·1 1	9·86389	
03·0	9·83419		10·16581	9·97042		10·02958	10·13623		9·86377	
04·0	9·83433		10·16568	9·97067		10·02933	10·13634		9·86366	
05·0	9·83446	·2 3	10·16554	9·97092	·2 5	10·02908	10·13646	·2 2	9·86354	55′
06·0	9·83460		10·16541	9·97118		10·02883	10·13658		9·86342	
07·0	9·83473		10·16527	9·97143		10·02857	10·13670		9·86330	
08·0	9·83487	·3 4	10·16514	9·97168	·3 8	10·02832	10·13682	·3 4	9·86318	
09·0	9·83500		10·16500	9·97194		10·02807	10·13694		9·86306	
10·0	9·83513		10·16487	9·97219		10·02781	10·13705		9·86295	50′
11·0	9·83527	·4 5	10·16473	9·97244	·4 10	10·02756	10·13717	·4 5	9·86283	
12·0	9·83540		10·16460	9·97270		10·02731	10·13729		9·86271	
13·0	9·83554		10·16446	9·97295		10·02705	10·13741		9·86259	
14·0	9·83567		10·16433	9·97320		10·02680	10·13753		9·86247	
15·0	9·83581	·5 7	10·16419	9·97345	·5 13	10·02655	10·13765	·5 6	9·86235	45′
16·0	9·83594		10·16406	9·97371		10·02629	10·13777		9·86223	
17·0	9·83608		10·16393	9·97396		10·02604	10·13789		9·86212	
18·0	9·83621	·6 8	10·16379	9·97421	·6 15	10·02579	10·13800	·6 7	9·86200	
19·0	9·83634		10·16366	9·97447		10·02553	10·13812		9·86188	
20·0	9·83648		10·16352	9·97472		10·02528	10·13824		9·86176	40′
21·0	9·83661	·7 9	10·16339	9·97497	·7 18	10·02503	10·13836	·7 8	9·86164	
22·0	9·83675		10·16326	9·97523		10·02477	10·13848		9·86152	
23·0	9·83688		10·16312	9·97548		10·02452	10·13860		9·86140	
24·0	9·83701	·8 11	10·16299	9·97573	·8 20	10·02427	10·13872	·8 10	9·86128	
25·0	9·83715		10·16285	9·97599		10·02402	10·13884		9·86116	35′
26·0	9·83728		10·16272	9·97624		10·02376	10·13896		9·86104	
27·0	9·83741	·9 12	10·16259	9·97649	·9 23	10·02351	10·13908	·9 11	9·86092	
28·0	9·83755		10·16245	9·97674		10·02326	10·13920		9·86080	
29·0	9·83768		10·16232	9·97700		10·02300	10·13932		9·86068	
30·0	9·83781		10·16219	9·97725		10·02275	10·13944		9·86056	30′
31·0	9·83795		10·16206	9·97750		10·02250	10·13956		9·86044	
32·0	9·83808	·1 1	10·16192	9·97776	·1 3	10·02224	10·13968	·1 1	9·86032	
33·0	9·83821		10·16179	9·97801		10·02199	10·13980		9·86020	
34·0	9·83834		10·16166	9·97826		10·02174	10·13992		9·86008	
35·0	9·83848	·2 3	10·16152	9·97852	·2 5	10·02149	10·14004	·2 2	9·85996	25′
36·0	9·83861		10·16139	9·97877		10·02123	10·14016		9·85984	
37·0	9·83874		10·16126	9·97902		10·02098	10·14028		9·85972	
38·0	9·83888	·3 4	10·16113	9·97927	·3 8	10·02073	10·14040	·3 4	9·85960	
39·0	9·83901		10·16099	9·97953		10·02047	10·14052		9·85948	
40·0	9·83914		10·16086	9·97978		10·02022	10·14064		9·85936	20′
41·0	9·83927		10·16073	9·98003		10·01997	10·14076		9·85924	
42·0	9·83940	·4 5	10·16060	9·98029	·4 10	10·01971	10·14088	·4 5	9·85912	
43·0	9·83954		10·16046	9·98054		10·01946	10·14100		9·85900	
44·0	9·83967		10·16033	9·98079		10·01921	10·14112		9·85888	
45·0	9·83980	·5 7	10·16020	9·98104	·5 13	10·01896	10·14124	·5 6	9·85876	15′
46·0	9·83993		10·16007	9·98130		10·01870	10·14137		9·85864	
47·0	9·84006		10·15994	9·98155		10·01845	10·14149		9·85851	
48·0	9·84020	·6 8	10·15980	9·98180	·6 15	10·01820	10·14161	·6 7	9·85839	
49·0	9·84033		10·15967	9·98206		10·01794	10·14173		9·85827	
50·0	9·84046		10·15954	9·98231		10·01769	10·14185		9·85815	10′
51·0	9·84059	·7 9	10·15941	9·98256	·7 18	10·01744	10·14197	·7 8	9·85803	
52·0	9·84072		10·15928	9·98281		10·10719	10·14209		9·85791	
53·0	9·84085		10·15915	9·98307		10·01693	10·14221		9·85779	
54·0	9·84099	·8 11	10·15902	9·98332	·8 20	10·01668	10·14234	·8 10	9·85767	
55·0	9·84112		10·15888	9·98357		10·01643	10·14246		9·85754	5′
56·0	9·84125		10·15875	9·98383		10·01617	10·14258		9·85742	
57·0	9·84138	·9 12	10·15862	9·98408	·9 23	10·01592	10·14270	·9 11	9·85730	
58·0	9·84151		10·15849	9·98433		10·01567	10·14282		9·85718	
59·0	9·84164		10·15836	9·98458		10·01542	10·14294		9·85706	
60·0	9·84177		10·15823	9·98484		10·01516	10·14307		9·85693	0′

LOGS. OF TRIG. FUNCTIONS

44°
224°

′	Sine	Parts		Cosec.	Tan.	Parts		Cotan.	Secant	Parts		Cosine	
00·0	9·84177	′		10·15823	9·98484	′		10·01516	10·14307	′		9·85693	60′
01·0	9·84190			10·15810	9·98509			10·01491	10·14319			9·85681	
02·0	9·84203	·1	1	10·15797	9·98534	·1	3	10·01466	10·14331	·1	1	9·85669	
03·0	9·84216			10·15784	9·98560			10·01440	10·14343			9·85657	
04·0	9·84229			10·15771	9·98585			10·01415	10·14355			9·85645	
05·0	9·84242	·2	3	10·15758	9·98610	·2	5	10·01390	10·14368	·2	2	9·85632	55′
06·0	9·84256			10·15745	9·98635			10·01365	10·14380			9·85620	
07·0	9·84269			10·15732	9·98661			10·01339	10·14392			9·85608	
08·0	9·84282	·3	4	10·15719	9·98686	·3	8	10·01314	10·14404	·3	4	9·85596	
09·0	9·84295			10·15705	9·98711			10·01289	10·14417			9·85583	
10·0	9·84308			10·15692	9·98737			10·01264	10·14429			9·85571	50′
11·0	9·84321	·4	5	10·15679	9·98762	·4	10	10·01238	10·14441	·4	5	9·85559	
12·0	9·84334			10·15666	9·98787			10·01213	10·14454			9·85547	
13·0	9·84347			10·15653	9·98812			10·01188	10·14466			9·85534	
14·0	9·84360			10·15641	9·98838			10·01162	10·14478			9·85522	
15·0	9·84373	·5	6	10·15628	9·98863	·5	13	10·01137	10·14490	·5	6	9·85510	45′
16·0	9·84386			10·15615	9·98888			10·01112	10·14503			9·85497	
17·0	9·84398			10·15602	9·98913			10·01087	10·14515			9·85485	
18·0	9·84411	·6	8	10·15589	9·98939	·6	15	10·01061	10·14527	·6	7	9·85473	
19·0	9·84424			10·15576	9·98964			10·01036	10·14540			9·85460	
20·0	9·84437			10·15563	9·98989			10·01011	10·14552			9·85448	40′
21·0	9·84450	·7	9	10·15550	9·99015	·7	18	10·00986	10·14564	·7	9	9·85436	
22·0	9·84463			10·15537	9·99040			10·00960	10·14577			9·85423	
23·0	9·84476			10·15524	9·99065			10·00935	10·14589			9·85411	
24·0	9·84489	·8	10	10·15511	9·99090	·8	20	10·00910	10·14601	·8	10	9·85399	
25·0	9·84502			10·15498	9·99116			10·00884	10·14614			9·85386	35′
26·0	9·84515			10·15485	9·99141			10·00859	10·14626			9·85374	
27·0	9·84528	·9	12	10·15472	9·99166	·9	23	10·00834	10·14639	·9	11	9·85361	
28·0	9·84541			10·15460	9·99191			10·00809	10·14651			9·85349	
29·0	9·84553			10·15447	9·99217			10·00783	10·14663			9·85337	
30·0	9·84566			10·15434	9·99242			10·00758	10·14676			9·85324	30′
31·0	9·84579			10·15421	9·99267			10·00733	10·14688			9·85312	
32·0	9·84592	·1	1	10·15408	9·99293	·1	3	10·00708	10·14701	·1	1	9·85299	
33·0	9·84605			10·15395	9·99318			10·00682	10·14713			9·85287	
34·0	9·84618			10·15383	9·99343			10·00657	10·14726			9·85275	
35·0	9·84630	·2	3	10·15370	9·99368	·2	5	10·00632	10·14738	·2	3	9·85262	25′
36·0	9·84643			10·15357	9·99394			10·00606	10·14750			9·85250	
37·0	9·84656			10·15344	9·99419			10·00581	10·14763			9·85237	
38·0	9·84669	·3	4	10·15331	9·99444	·3	8	10·00556	10·14775	·3	4	9·85225	
39·0	9·84682			10·15318	9·99469			10·00531	10·14788			9·85212	
40·0	9·84694			10·15306	9·99495			10·00505	10·14800			9·85200	20′
41·0	9·84707	·4	5	10·15293	9·99520	·4	10	10·00480	10·14813	·4	5	9·85187	
42·0	9·84720			10·15280	9·99545			10·00455	10·14825			9·85175	
43·0	9·84733			10·15267	9·99571			10·00430	10·14838			9·85162	
44·0	9·84745			10·15255	9·99596			10·00404	10·14850			9·85150	
45·0	9·84758	·5	6	10·15242	9·99621	·5	13	10·00379	10·14863	·5	6	9·85137	15′
46·0	9·84771			10·15229	9·99646			10·00354	10·14875			9·85125	
47·0	9·84784			10·15216	9·99672			10·00329	10·14888			9·85112	
48·0	9·84796	·6	8	10·15204	9·99697	·6	15	10·00303	10·14900	·6	8	9·85100	
49·0	9·84809			10·15191	9·99722			10·00278	10·14913			9·85087	
50·0	9·84822			10·15178	9·99747			10·00253	10·14926			9·85075	10′
51·0	9·84835	·7	9	10·15166	9·99773	·7	18	10·00227	10·14938	·7	9	9·85062	
52·0	9·84847			10·15153	9·99798			10·00202	10·14951			9·85049	
53·0	9·84860			10·15140	9·99823			10·00177	10·14963			9·85037	
54·0	9·84873	·8	10	10·15127	9·99848	·8	20	10·00152	10·14976	·8	10	9·85024	
55·0	9·84885			10·15115	9·99874			10·00126	10·14988			9·85012	5′
56·0	9·84898			10·15102	9·99899			10·00101	10·15001			9·84999	
57·0	9·84911	·9	11	10·15089	9·99924	·9	23	10·00076	10·15014	·9	11	9·84986	
58·0	9·84923			10·15077	9·99950			10·00051	10·15026			9·84974	
59·0	9·84936			10·15064	9·99975			10·00025	10·15039			9·84961	
60·0	9·84949			10·15052	10·00000			10·00000	10·15052			9·84949	0′

135°
315°

45°
225°

LOGS. OF TRIG. FUNCTIONS

′	Sine	Parts		Cosec.	Tan.	Parts		Cotan.	Secant	Parts		Cosine	
00·0	9·84949	′		10·15052	10·00000	′		10·00000	10·15052	′		9·84949	60′
01·0	9·84961			10·15039	10·00025			9·99975	10·15064			9·84936	
02·0	9·84974	·1	1	10·15026	10·00051	·1	3	9·99950	10·15077	·1	1	9·84923	
03·0	9·84986			10·15011	10·00076			9·99924	10·15089			9·84911	
04·0	9·84999			10·15001	10·00101			9·99899	10·15102			9·04098	
05·0	9·85012	·2	3	10·14988	10·00126	·2	5	9·99874	10·15115	·2	3	9·84885	55′
06·0	9·85024			10·14976	10·00152			9·99848	10·15127			9·84873	
07·0	9·85037			10·14963	10·00177			9·99823	10·15140			9·84860	
08·0	9·85049	·3	4	10·14951	10·00202	·3	8	9·99798	10·15153	·3	4	9·84847	
09·0	9·85062			10·14938	10·00227			9·99773	10·15166			9·84835	
10·0	9·85075			10·14926	10·00253			9·99747	10·15178			9·84822	50′
11·0	9·85087	·4	5	10·14913	10·00278	·4	10	9·99722	10·15191	·4	5	9·84809	
12·0	9·85100			10·14900	10·00303			9·99697	10·15204			9·84796	
13·0	9·85112			10·14888	10·00329			9·99672	10·15216			9·84784	
14·0	9·85125	·5	6	10·14875	10·00354	·5	13	9·99646	10·15229	·5	6	9·84771	
15·0	9·85137			10·14863	10·00379			9·99621	10·15242			9·84758	45′
16·0	9·85150			10·14850	10·00404			9·99596	10·15255			9·84745	
17·0	9·85162			10·14838	10·00430			9·99571	10·15267			9·84733	
18·0	9·85175	·6	8	10·14825	10·00455	·6	15	9·99545	10·15280	·6	8	9·84720	
19·0	9·85187			10·14813	10·00480			9·99520	10·15293			9·84707	
20·0	9·85200	·7	9	10·14800	10·00505	·7	18	9·99495	10·15306	·7	9	9·84694	40′
21·0	9·85212			10·14788	10·00531			9·99469	10·15318			9·84682	
22·0	9·85225			10·14775	10·00556			9·99444	10·15331			9·84669	
23·0	9·85237			10·14763	10·00581			9·99419	10·15344			9·84656	
24·0	9·85250	·8	10	10·14750	10·00606	·8	20	9·99394	10·15357	·8	10	9·84643	
25·0	9·85262			10·14738	10·00632			9·99368	10·15370			9·84630	35′
26·0	9·85275			10·14726	10·00657			9·99343	10·15383			9·84618	
27·0	9·85287	·9	11	10·14713	10·00682	·9	23	9·99318	10·15395	·9	11	9·84605	
28·0	9·85299			10·14701	10·00708			9·99293	10·15408			9·84592	
29·0	9·85312			10·14688	10·00733			9·99267	10·15421			9·84579	
30·0	9·85324			10·14676	10·00758			9·99242	10·15434			9·84566	30′
31·0	9·85337			10·14663	10·00783			9·99217	10·15447			9·84553	
32·0	9·85349	·1	1	10·14651	10·00809	·1	3	9·99191	10·15460	·1	1	9·84541	
33·0	9·85361			10·14639	10·00834			9·99166	10·15472			9·84528	
34·0	9·85374			10·14626	10·00859			9·99141	10·15485			9·84515	
35·0	9·85386	·2	2	10·14614	10·00884	·2	5	9·99116	10·15498	·2	3	9·84502	25′
36·0	9·85399			10·14601	10·00910			9·99090	10·15511			9·84489	
37·0	9·85411			10·14589	10·00935			9·99065	10·15524			9·84476	
38·0	9·85423	·3	4	10·14577	10·00960	·3	8	9·99040	10·15537	·3	4	9·84463	
39·0	9·85436			10·14564	10·00986			9·99015	10·15550			9·84450	
40·0	9·85448			10·14552	10·01011			9·98989	10·15563			9·84437	20′
41·0	9·85460	·4	5	10·14540	10·01036	·4	10	9·98964	10·15576	·4	5	9·84424	
42·0	9·85473			10·14527	10·01061			9·98939	10·15589			9·84411	
43·0	9·85485			10·14515	10·01087			9·98913	10·15602			9·84398	
44·0	9·85497			10·14503	10·01112	·5	13	9·98888	10·15615			9·84386	
45·0	9·85510	·5	6	10·14490	10·01137			9·98863	10·15628	·5	6	9·84373	15′
46·0	9·85522			10·14478	10·01162			9·98838	10·15641			9·84360	
47·0	9·85534			10·14466	10·01188			9·98812	10·15653			9·84347	
48·0	9·85547	·6	7	10·14454	10·01213	·6	15	9·98787	10·15666	·6	8	9·84334	
49·0	9·85559			10·14441	10·01238			9·98762	10·15679			9·84321	
50·0	9·85571			10·14429	10·01264			9·98737	10·15692			9·84308	10′
51·0	9·85583	·7	9	10·14417	10·01289	·7	18	9·98711	10·15705	·7	9	9·84295	
52·0	9·85596			10·14404	10·01314			9·98686	10·15719			9·84282	
53·0	9·85608			10·14392	10·01339			9·98661	10·15732			9·84269	
54·0	9·85620	·8	10	10·14380	10·01365	·8	20	9·98635	10·15745	·8	10	9·84256	
55·0	9·85632			10·14368	10·01390			9·98610	10·15758			9·84242	5′
56·0	9·85645			10·14355	10·01415			9·98585	10·15771			9·84229	
57·0	9·85657	·9	11	10·14343	10·01440	·9	23	9·98560	10·15784	·9	12	9·84216	
58·0	9·85669			10·14331	10·01466			9·98534	10·15797			9·84203	
59·0	9·85681			10·14319	10·01491			9·98509	10·15810			9·84190	
60·0	9·85693			10·14307	10·01516			9·98484	10·15823			9·84177	0′

46°
226°

LOGS. OF TRIG. FUNCTIONS

′	Sine	Parts	Cosec.	Tan.	Parts	Cotan.	Secant	Parts	Cosine	
00·0	9·85693	′	10·14307	10·01516	′	9·98484	10·15823	′	9·84177	60′
01·0	9·85706		10·14294	10·01542		9·98458	10·15836		9·84164	
02·0	9·85718	·1 1	10·14282	10·01567	·1 3	9·98433	10·15849	·1 1	9·84151	
03·0	9·85730		10·14270	10·01592		9·98408	10·15862		9·84138	
04·0	9·85742		10·14258	10·01617		9·98383	10 15875		9 84125	
05·0	9·85754	2 2	10·14246	10·01643	2 5	9·98357	10·15888	·2 3	9·84112	55′
06·0	9·85767		10·14234	10·01668		9·98332	10·15902		9·84099	
07·0	9·85779		10·14221	10·01693		9·98307	10·15915		9·84085	
08·0	9·85791	·3 4	10·14209	10·01719	·3 8	9·98281	10·15928	·3 4	9·84072	
09·0	9·85803		10·14197	10·01744		9·98256	10·15941		9·84059	
10·0	9·85815		10·14185	10·01769		9·98231	10·15954		9·84046	50′
11·0	9·85827	·4 5	10·14173	10·01794	·4 10	9·98206	10·15967	·4 5	9·84033	
12·0	9·85839		10·14161	10·01820		9·98180	10·15980		9·84020	
13·0	9·85851		10·14149	10·01845		9·98155	10·15994		9·84006	
14·0	9·85864	·5 6	10·14137	10·01870	·5 13	9·98130	10·16007	·5 7	9·83993	
15·0	9·85876		10·14124	10·01896		9·98104	10·16020		9·83980	45′
16·0	9·85888		10·14112	10·01921		9·98079	10·16033		9·83967	
17·0	9·85900		10·14100	10·01946		9·98054	10·16046		9·83954	
18·0	9·85912	·6 7	10·14088	10·01971	·6 15	9·98029	10·16060	·6 8	9·83940	
19·0	9·85924		10·14076	10·01997		9·98003	10·16073		9·83927	
20·0	9·85936		10·14064	10·02022		9·97978	10·16086		9·83914	40′
21·0	9·85948	·7 8	10·14052	10·02047	·7 18	9·97953	10·16099	·7 9	9·83901	
22·0	9·85960		10·14040	10·02073		9·97927	10·16113		9·83888	
23·0	9·85972		10·14028	10·02098		9·97902	10·16126		9·83874	
24·0	9·85984	·8 10	10·14016	10·02123	·8 20	9·97877	10·16139	·8 11	9·83861	
25·0	9·85996		10·14004	10·02149		9·97852	10·16152		9·83848	35′
26·0	9·86008		10·13992	10·02174		9·97826	10·16166		9·83834	
27·0	9·86020	·9 11	10·13980	10·02199	·9 23	9·97801	10·16179	·9 12	9·83821	
28·0	9·86032		10·13968	10·02224		9·97776	10·16192		9·83808	
29·0	9·86044		10·13956	10·02250		9·97750	10·16206		9·83795	
30·0	9·86056		10·13944	10·02275		9·97725	10·16219		9·83781	30′
31·0	9·86068		10·13932	10·02300		9·97700	10·16232		9·83768	
32·0	9·86080	·1 1	10·13920	10·02326	·1 3	9·97674	10·16245	·1 1	9·83755	
33·0	9·86092		10·13908	10·02351		9·97649	10·16259		9·83741	
34·0	9·86104		10·13896	10·02376		9·97624	10·16272		9·83728	
35·0	9·86116	·2 2	10·13884	10·02402	·2 5	9·97599	10·16285	·2 3	9·83715	25′
36·0	9·86128		10·13872	10·02427		9·97573	10·16299		9·83701	
37·0	9·86140		10·13860	10·02452		9·97548	10·16312		9·83688	
38·0	9·86152		10·13848	10·02477		9·97523	10·16326		9·83675	
39·0	9·86164	·3 4	10·13836	10·02503	·3 8	9·97497	10·16339	·3 4	9·83661	
40·0	9·86176		10·13824	10·02528		9·97472	10·16352		9·83648	20′
41·0	9·86188		10·13812	10·02553		9·97447	10·16366		9·83634	
42·0	9·86200	·4 5	10·13800	10·02579	·4 10	9·97421	10·16379	·4 5	9·83621	
43·0	9·86212		10·13789	10·02604		9·97396	10·16393		9·83608	
44·0	9·86223		10·13777	10·02629		9·97371	10·16406		9·83594	
45·0	9·86235	·5 6	10·13765	10·02655	·5 13	9·97345	10·16419	·5 7	9·83581	15′
46·0	9·86247		10·13753	10·02680		9·97320	10·16433		9·83567	
47·0	9·86259		10·13741	10·02705		9·97295	10·16446		9·83554	
48·0	9·86271	·6 7	10·13729	10·02731	·6 15	9·97270	10·16460	·6 8	9·83540	
49·0	9·86283		10·13717	10·02756		9·97244	10·16473		9·83527	
50·0	9·86295		10·13705	10·02781		9·97219	10·16487		9·83513	10′
51·0	9·86306	·7 8	10·13694	10·02807	·7 18	9·97194	10·16500	·7 9	9·83500	
52·0	9·86318		10·13682	10·02832		9·97168	10·16514		9·83487	
53·0	9·86330		10·13670	10·02857		9·97143	10·16527		9·83473	
54·0	9·86342	·8 10	10·13658	10·02883	·8 20	9·97118	10·16541	·8 11	9·83460	
55·0	9·86354		10·13646	10·02908		9·97092	10·16554		9·83446	5′
56·0	9·86366		10·13634	10·02933		9·97067	10·16568		9·83433	
57·0	9·86377	·9 11	10·13623	10·02958	·9 23	9·97042	10·16581	·9 12	9·83419	
58·0	9·86389		10·13611	10·02984		9·97016	10·16595		9·83405	
59·0	9·86401		10·13599	10·03009		9·96991	10·16608		9·83392	
60·0	9·86413		10·13587	10·03034		9·96966	10·16622		9·83378	0′

47°
227°

LOGS. OF TRIG. FUNCTIONS

′	Sine	Parts	Cosec.	Tan.	Parts	Cotan.	Secant	Parts	Cosine	
00·0	9·86413	′	10·13587	10·03034	′	9·96966	10·16622	′	9·83378	60′
01·0	9·86425		10·13576	10·03060		9·96940	10·16635		9·83365	
02·0	9·86436	·1 1	10·13564	10·03085	·1 3	9·96915	10·16649	·1 1	9·83351	
03·0	9·86448		10·13552	10·03110		9·96890	10·16662		9·83330	
04·0	9·86460		10·13540	10·03136		9·96864	10·16676		9·83324	
05·0	9·86472	·2 2	10·13528	10·03161	·2 5	9·96839	10·16690	·2 3	9·83311	55′
06·0	9·86483		10·13517	10·03186		9·96814	10·16703		9·83297	
07·0	9·86495		10·13505	10·03212		9·96788	10·16717		9·83283	
08·0	9·86507	·3 4	10·13493	10·03237	·3 8	9·96763	10·16730	·3 4	9·83270	
09·0	9·86519		10·13482	10·03262		9·96738	10·16744		9·83256	
10·0	9·86530		10·13470	10·03288		9·96712	10·16758		9·83243	50′
11·0	9·86542	·4 5	10·13458	10·03313	·4 10	9·96687	10·16771	·4 5	9·83229	
12·0	9·86554		10·13446	10·03338		9·96662	10·16785		9·83215	
13·0	9·86565		10·13435	10·03364		9·96636	10·16799		9·83202	
14·0	9·86577	·5 6	10·13423	10·03389	·5 13	9·96611	10·16812	·5 7	9·83188	
15·0	9·86589		10·13411	10·03415		9·96586	10·16826		9·83174	45′
16·0	9·86600		10·13400	10·03440		9·96560	10·16839		9·83161	
17·0	9·86612		10·13388	10·03465		9·96535	10·16853		9·83147	
18·0	9·86624	·6 7	10·13376	10·03491	·6 15	9·96510	10·16867	·6 8	9·83133	
19·0	9·86635		10·13365	10·03516		9·96484	10·16881		9·83120	
20·0	9·86647	·7 8	10·13353	10·03541	·7 18	9·96459	10·16894	·7 10	9·83106	40′
21·0	9·86659		10·13341	10·03567		9·96434	10·16908		9·83092	
22·0	9·86670		10·13330	10·03592		9·96408	10·16922		9·83078	
23·0	9·86682		10·13318	10·03617		9·96383	10·16935		9·83065	
24·0	9·86694	·8 9	10·13307	10·03643	·8 20	9·96357	10·16949	·8 11	9·83051	
25·0	9·86705		10·13295	10·03668		9·96332	10·16963		9·83037	35′
26·0	9·86717		10·13283	10·03693		9·96307	10·16977		9·83023	
27·0	9·86728	·9 11	10·13272	10·03719	·9 23	9·96281	10·16990	·9 12	9·83010	
28·0	9·86740		10·13260	10·03744		9·96256	10·17004		9·82996	
29·0	9·86752		10·13249	10·03769		9·96231	10·17018		9·82982	
30·0	9·86763		10·13237	10·03795		9·96205	10·17032		9·82968	30′
31·0	9·86775		10·13225	10·03820		9·96180	10·17046		9·82955	
32·0	9·86786	·1 1	10·13214	10·03846	·1 3	9·96155	10·17059	·1 1	9·82941	
33·0	9·86798		10·13202	10·03871		9·96129	10·17073		9·82927	
34·0	9·86809		10·13191	10·03896		9·96104	10·17087		9·82913	
35·0	9·86821	·2 2	10·13179	10·03922	·2 5	9·96078	10·17101	·2 3	9·82899	25′
36·0	9·86832		10·13168	10·03947		9·96053	10·17115		9·82886	
37·0	9·86844		10·13156	10·03972		9·96028	10·17128		9·82872	
38·0	9·86856	·3 3	10·13145	10·03998	·3 8	9·96002	10·17142	·3 4	9·82858	
39·0	9·86867		10·13133	10·04023		9·95977	10·17156		9·82844	
40·0	9·86879		10·13122	10·04048		9·95952	10·17170		9·82830	20′
41·0	9·86890	·4 5	10·13110	10·04074	·4 10	9·95926	10·17184	·4 6	9·82816	
42·0	9·86902		10·13099	10·04099		9·95901	10·17198		9·82802	
43·0	9·86913		10·13087	10·04125		9·95875	10·17212		9·82788	
44·0	9·86925	·5 6	10·13076	10·04150	·5 13	9·95850	10·17226	·5 7	9·82775	
45·0	9·86936		10·13064	10·04175		9·95825	10·17239		9·82761	15′
46·0	9·86947		10·13053	10·04201		9·95799	10·17253		9·82747	
47·0	9·86959		10·13041	10·04226		9·95774	10·17267		9·82733	
48·0	9·86970	·6 7	10·13030	10·04252	·6 15	9·95749	10·17281	·6 8	9·82719	
49·0	9·86982		10·13018	10·04277		9·95723	10·17295		9·82705	
50·0	9·86993	·7 8	10·13007	10·04302	·7 18	9·95698	10·17309	·7 10	9·82691	10′
51·0	9·87005		10·12995	10·04328		9·95672	10·17323		9·82677	
52·0	9·87016		10·12984	10·04353		9·95647	10·17337		9·82663	
53·0	9·87028		10·12972	10·04379		9·95622	10·17351		9·82649	
54·0	9·87039	·8 9	10·12961	10·04404	·8 20	9·95596	10·17365	·8 11	9·82635	
55·0	9·87050		10·12950	10·04429		9·95571	10·17379		9·82621	5′
56·0	9·87062		10·12938	10·04455		9·95545	10·17393		9·82607	
57·0	9·87073	·9 10	10·12927	10·04480	·9 23	9·95520	10·17407	·9 13	9·82593	
58·0	9·87085		10·12915	10·04505		9·95495	10·17421		9·82579	
59·0	9·87096		10·12904	10·04531		9·95469	10·17435		9·82565	
60·0	9·87107		10·12893	10·04556		9·95444	10·17449		9·82551	0′

48°
228°

LOGS. OF TRIG. FUNCTIONS

′	Sine	Parts	Cosec.	Tan.	Parts	Cotan.	Secant	Parts	Cosine	
00·0	9·87107	′	10·12893	10·04556	′	9·95444	10·17449	′	9·82551	60′
01·0	9·87119		10·12881	10·04582		9·95418	10·17463		9·82537	
02·0	9·87130	·1 1	10·12870	10·04607	·1 3	9·95393	10·17477	·1 1	9·82523	
03·0	9·87141		10·12859	10·04633		9·95368	10·17491		9·82509	
04·0	9·87153		10·12847	10·04658		9·95342	10·17505		9·82495	
05·0	9·87164	·2 2	10·12836	10·04683	·2 5	9·95317	10·17519	·2 3	9·82481	55′
06·0	9·87176		10·12825	10·04709		9·95291	10·17533		9·82467	
07·0	9·87187		10·12813	10·04734		9·95266	10·17547		9·82453	
08·0	9·87198	·3 3	10·12802	10·04760	·3 8	9·95241	10·17561	·3 4	9·82439	
09·0	9·87210		10·12791	10·04785		9·95215	10·17576		9·82425	
10·0	9·87221		10·12779	10·04810		9·95190	10·17590		9·82410	50′
11·0	9·87232	·4 5	10·12768	10·04836	·4 10	9·95164	10·17604	·4 6	9·82396	
12·0	9·87243		10·12757	10·04861		9·95139	10·17618		9·82382	
13·0	9·87255		10·12745	10·04887		9·95113	10·17632		9·82368	
14·0	9·87266	·5 6	10·12734	10·04912	·5 13	9·95088	10·17646	·5 7	9·82354	
15·0	9·87277		10·12723	10·04938		9·95063	10·17660		9·82340	45′
16·0	9·87289		10·12712	10·04963		9·95037	10·17675		9·82326	
17·0	9·87300	·6 7	10·12700	10·04988	·6 15	9·95012	10·17689	·6 8	9·82311	
18·0	9·87311		10·12689	10·05014		9·94986	10·17703		9·82297	
19·0	9·87322		10·12678	10·05039		9·94961	10·17717		9·82283	
20·0	9·87334	·7 8	10·12667	10·05065	·7 18	9·94935	10·17731	·7 10	9·82269	40′
21·0	9·87345		10·12655	10·05090		9·94910	10·17745		9·82255	
22·0	9·87356		10·12644	10·05116		9·94884	10·17760		9·82240	
23·0	9·87367		10·12633	10·05141		9·94859	10·17774		9·82226	
24·0	9·87378	·8 9	10·12622	10·05167	·8 20	9·94834	10·17788	·8 11	9·82212	
25·0	9·87390		10·12610	10·05192		9·94808	10·17802		9·82198	35′
26·0	9·87401		10·12599	10·05217		9·94783	10·17817		9·82184	
27·0	9·87412	·9 10	10·12588	10·05243	·9 23	9·94757	10·17831	·9 13	9·82169	
28·0	9·87423		10·12577	10·05268		9·94732	10·17845		9·82155	
29·0	9·87434		10·12566	10·05294		9·94706	10·17859		9·82141	
30·0	9·87446		10·12554	10·05319		9·94681	10·17874		9·82127	30′
31·0	9·87457		10·12543	10·05345		9·94655	10·17888		9·82112	
32·0	9·87468	·1 1	10·12532	10·05370	·1 3	9·94630	10·17902	·1 1	9·82098	
33·0	9·87479		10·12521	10·05396		9·94605	10·17916		9·82084	
34·0	9·87490		10·12510	10·05421		9·94579	10·17931		9·82069	
35·0	9·87501	·2 2	10·12499	10·05447	·2 5	9·94554	10·17945	·2 3	9·82055	25′
36·0	9·87513		10·12487	10·05472		9·94528	10·17959		9·82041	
37·0	9·87524		10·12476	10·05497		9·94503	10·17974		9·82026	
38·0	9·87535	·3 3	10·12465	10·05523	·3 8	9·94477	10·17988	·3 4	9·82012	
39·0	9·87546		10·12454	10·05548		9·94452	10·18002		9·81998	
40·0	9·87557		10·12443	10·05574		9·94426	10·18017		9·81983	20′
41·0	9·87568	·4 4	10·12432	10·05599	·4 10	9·94401	10·18031	·4 6	9·81969	
42·0	9·87579		10·12421	10·05625		9·94375	10·18046		9·81955	
43·0	9·87590		10·12410	10·05650		9·94350	10·18060		9·81940	
44·0	9·87601		10·12399	10·05676		9·94324	10·18074		9·81926	
45·0	9·87613	·5 6	10·12388	10·05701	·5 13	9·94299	10·18089	·5 7	9·81911	15′
46·0	9·87624		10·12376	10·05727		9·94273	10·18103		9·81897	
47·0	9·87635		10·12365	10·05752		9·94248	10·18118		9·81883	
48·0	9·87646	·6 7	10·12354	10·05778	·6 15	9·94222	10·18132	·6 9	9·81868	
49·0	9·87657		10·12343	10·05803		9·94197	10·18146		9·81854	
50·0	9·87668		10·12332	10·05829		9·94171	10·18161		9·81839	10′
51·0	9·87679	·7 8	10·12321	10·05854	·7 18	9·94146	10·18175	·7 10	9·81825	
52·0	9·87690		10·12310	10·05880		9·94120	10·18190		9·81810	
53·0	9·87701		10·12299	10·05905		9·94095	10·18204		9·81796	
54·0	9·87712	·8 9	10·12288	10·05931	·8 20	9·94069	10·18219	·8 12	9·81781	
55·0	9·87723		10·12277	10·05956		9·94044	10·18233		9·81767	5′
56·0	9·87734		10·12266	10·05982		9·94018	10·18248		9·81752	
57·0	9·87745	·9 10	10·12255	10·06007	·9 23	9·93993	10·18262	·9 13	9·81738	
58·0	9·87756		10·12244	10·06033		9·93967	10·18277		9·81723	
59·0	9·87767		10·12233	10·06058		9·93942	10·18291		9·81709	
60·0	9·87778		10·12222	10·06084		9·93916	10·18306		9·81694	0′

49°

229°

LOGS. OF TRIG. FUNCTIONS

′	Sine	Parts	Cosec.	Tan.	Parts	Cotan.	Secant	Parts	Cosine	
00·0	9·87778	′	10·12222	10·06084	′	9·93916	10·18306	′	9·81694	60′
01·0	9·87789		10·12211	10·06109		9·93891	10·18320		9·81680	
02·0	9·87800	·1 1	10·12200	10·06135	·1 3	9·93865	10·18335	·1 1	9·81665	
03·0	9·87811		10·12189	10·06160		9·93840	10·18349		9·81651	
04·0	9·87822		10·12178	10·06186		9·93814	10·18364		9·81636	
05·0	9·87833	·2 2	10·12167	10·06211	·2 5	9·93789	10·18379	·2 3	9·81622	55′
06·0	9·87844		10·12156	10·06237		9·93763	10·18393		9·81607	
07·0	9·87855		10·12145	10·06262		9·93738	10·18408		9·81592	
08·0	9·87866	·3 3	10·12134	10·06288	·3 8	9·93712	10·18422	·3 4	9·81578	
09·0	9·87877		10·12123	10·06313		9·93687	10·18437		9·81563	
10·0	9·87888		10·12113	10·06339		9·93661	10·18452		9·81549	50′
11·0	9·87898	·4 4	10·12102	10·06365	·4 10	9·93636	10·18466	·4 6	9·81534	
12·0	9·87909		10·12091	10·06390		9·93610	10·18481		9·81519	
13·0	9·87920		10·12080	10·06416		9·93584	10·18495		9·81505	
14·0	9·87931	·5 5	10·12069	10·06441	·5 13	9·93559	10·18510	·5 7	9·81490	
15·0	9·87942		10·12058	10·06467		9·93533	10·18525		9·81475	45′
16·0	9·87953		10·12047	10·06492		9·93508	10·18539		9·81461	
17·0	9·87964	·6 7	10·12036	10·06518	·6 15	9·93482	10·18554	·6 9	9·81446	
18·0	9·87975		10·12025	10·06543		9·93457	10·18569		9·81431	
19·0	9·87986		10·12015	10·06569		9·93431	10·18583		9·81417	
20·0	9·87996	·7 8	10·12004	10·06594	·7 18	9·93406	10·18598	·7 10	9·81402	40′
21·0	9·88007		10·11993	10·06620		9·93380	10·18613		9·81387	
22·0	9·88018		10·11982	10·06646		9·93355	10·18628		9·81373	
23·0	9·88029		10·11971	10·06671		9·93329	10·18642		9·81358	
24·0	9·88040	·8 9	10·11960	10·06697	·8 20	9·93303	10·18657	·8 12	9·81343	
25·0	9·88051		10·11950	10·06722		9·93278	10·18672		9·81328	35′
26·0	9·88061		10·11939	10·06748		9·93252	10·18687		9·81314	
27·0	9·88072	·9 10	10·11928	10·06773	·9 23	9·93227	10·18701	·9 13	9·81299	
28·0	9·88083		10·11917	10·06799		9·93201	10·18716		9·81284	
29·0	9·88094		10·11906	10·06825		9·93176	10·18731		9·81269	
30·0	9·88105		10·11895	10·06850		9·93150	10·18746		9·81254	30′
31·0	9·88115		10·11885	10·06876		9·93124	10·18760		9·81240	
32·0	9·88126	·1 1	10·11874	10·06901	·1 3	9·93099	10·18775	·1 1	9·81225	
33·0	9·88137		10·11863	10·06927		9·93073	10·18790		9·81210	
34·0	9·88148		10·11852	10·06953		9·93048	10·18805		9·81195	
35·0	9·88158	·2 2	10·11842	10·06978	·2 5	9·93022	10·18820	·2 3	9·81180	25′
36·0	9·88169		10·11831	10·07004		9·92996	10·18835		9·81166	
37·0	9·88180		10·11820	10·07029		9·92971	10·18849		9·81151	
38·0	9·88191	·3 3	10·11809	10·07055	·3 8	9·92945	10·18864	·3 4	9·81136	
39·0	9·88201		10·11799	10·07080		9·92920	10·18879		9·81121	
40·0	9·88212		10·11788	10·07106		9·92894	10·18894		9·81106	20′
41·0	9·88223	·4 4	10·11777	10·07132	·4 10	9·92868	10·18909	·4 6	9·81091	
42·0	9·88234		10·11766	10·07157		9·92843	10·18924		9·81076	
43·0	9·88244		10·11756	10·07183		9·92817	10·18939		9·81061	
44·0	9·88255	·5 5	10·11745	10·07209	·5 13	9·92792	10·18954	·5 7	9·81047	
45·0	9·88266		10·11734	10·07234		9·92766	10·18968		9·81032	15′
46·0	9·88276		10·11724	10·07260		9·92740	10·18983		9·81017	
47·0	9·88287	·6 6	10·11713	10·07285	·6 15	9·92715	10·18998	·6 9	9·81002	
48·0	9·88298		10·11702	10·07311		9·92689	10·19013		9·80987	
49·0	9·88308		10·11692	10·07337		9·92663	10·19028		9·80972	
50·0	9·88319	·7 7	10·11681	10·07362	·7 18	9·92638	10·19043	·7 10	9·80957	10′
51·0	9·88330		10·11670	10·07388		9·92612	10·19058		9·80942	
52·0	9·88340		10·11660	10·07414		9·92587	10·19073		9·80927	
53·0	9·88351		10·11649	10·07439		9·92561	10·19088		9·80912	
54·0	9·88362	·8 9	10·11638	10·07465	·8 21	9·92535	10·19103	·8 12	9·80897	
55·0	9·88372		10·11628	10·07490		9·92510	10·19118		9·80882	5′
56·0	9·88383		10·11617	10·07516		9·92484	10·19133		9·80867	
57·0	9·88394	·9 10	10·11606	10·07542	·9 23	9·92458	10·19148	·9 13	9·80852	
58·0	9·88404		10·11596	10·07567		9·92433	10·19163		9·80837	
59·0	9·88415		10·11585	10·07593		9·92407	10·19178		9·80822	
60·0	9·88425		10·11575	10·07619		9·92381	10·19193		9·80807	0′

50°
230°

LOGS. OF TRIG. FUNCTIONS

′	Sine	Parts	Cosec.	Tan.	Parts	Cotan.	Secant	Parts	Cosine	
00·0	9·88425	′	10·11575	10·07619	′	9·92381	10·19193	′	9·80807	60′
01·0	9·88436		10·11564	10·07644		9·92356	10·19208		9·80792	
02·0	9·88447	·1 1	10·11553	10·07670	·1 3	9·92330	10·19223	·1 2	9·80777	
03·0	9·88457		10·11543	10·07696		9·92304	10·19239		9·80762	
04·0	9·88468		10·11532	10·07721		9·92279	10·19254		9·80747	
05·0	9·88478	·2 2	10·11522	10·07747	·2 5	9·92253	10·19269	·2 3	9·80731	55′
06·0	9·88489		10·11511	10·07773		9·92227	10·19284		9·80716	
07·0	9·88499		10·11501	10·07798		9·92202	10·19299		9·80701	
08·0	9·88510	·3 3	10·11490	10·07824	·3 8	9·92176	10·19314	·3 5	9·80686	
09·0	9·88521		10·11479	10·07850		9·92150	10·19329		9·80671	
10·0	9·88531		10·11469	10·07875		9·92125	10·19344		9·80656	50′
11·0	9·88542	·4 4	10·11458	10·07901	·4 10	9·92099	10·19359	·4 6	9·80641	
12·0	9·88552		10·11448	10·07927		9·92073	10·19375		9·80625	
13·0	9·88563		10·11437	10·07952		9·92048	10·19390		9·80610	
14·0	9·88573	·5 5	10·11427	10·07978	·5 13	9·92022	10·19405	5 8	9·80595	
15·0	9·88584		10·11416	10·08004		9·91996	10·19420		9·80580	45′
16·0	9·88594		10·11406	10·08030		9·91971	10·19435		9·80565	
17·0	9·88605	·6 6	10·11395	10·08055	·6 15	9·91945	10·19451	·6 9	9·80550	
18·0	9·88615		10·11385	10·08081		9·91919	10·19466		9·80534	
19·0	9·88626		10·11374	10·08107		9·91893	10·19481		9·80519	
20·0	9·88636	·7 7	10·11364	10·08132	·7 18	9·91868	10·19496	·7 10	9·80504	40′
21·0	9·88647		10·11353	10·08158		9·91842	10·19511		9·80489	
22·0	9·88657		10·11343	10·08184		9·91816	10·19527		9·80473	
23·0	9·88668		10·11332	10·08209		9·91791	10·19542		9·80458	
24·0	9·88678	·8 8	10·11322	10·08235	·8 21	9·91765	10·19557	·8 12	9·80443	
25·0	9·88689		10·11312	10·08261		9·91739	10·19572		9·80428	35′
26·0	9·88699		10·11301	10·08287		9·91713	10·19588		9·80412	
27·0	9·88709	·9 9	10·11291	10·08312	·9 23	9·91688	10·19603	·9 14	9·80397	
28·0	9·88720		10·11280	10·08338		9·91662	10·19618		9·80382	
29·0	9·88730		10·11270	10·08364		9·91636	10·19634		9·80366	
30·0	9·88741		10·11259	10·08390		9·91610	10·19649		9·80351	30′
31·0	9·88751		10·11249	10·08415		9·91585	10·19664		9·80336	
32·0	9·88761	·1 1	10·11239	10·08441	·1 3	9·91559	10·19680	·1 2	9·80320	
33·0	9·88772		10·11228	10·08467		9·91533	10·19695		9·80305	
34·0	9·88782		10·11218	10·08493		9·91508	10·19710		9·80290	
35·0	9·88793	·2 2	10·11207	10·08518	·2 5	9·91482	10·19726	·2 3	9·80274	25′
36·0	9·88803		10·11197	10·08544		9·91456	10·19741		9·80259	
37·0	9·88813		10·11187	10·08570		9·91430	10·19756		9·80244	
38·0	9·88824	·3 3	10·11176	10·08596	·3 8	9·91404	10·19772	·3 5	9·80228	
39·0	9·88834		10·11166	10·08621		9·91379	10·19787		9·80213	
40·0	9·88844		10·11156	10·08647		9·91353	10·19803		9·80197	20′
41·0	9·88855	·4 4	10·11145	10·08673	·4 10	9·91327	10·19818	·4 6	9·80182	
42·0	9·88865		10·11135	10·08699		9·91301	10·19834		9·80167	
43·0	9·88875		10·11125	10·08724		9·91276	10·19849		9·80151	
44·0	9·88886	·5 5	10·11114	10·08750	·5 13	9·91250	10·19864	·5 8	9·80136	
45·0	9·88896		10·11104	10·08776		9·91224	10·19880		9·80120	15′
46·0	9·88906		10·11094	10·08802		9·91198	10·19895		9·80105	
47·0	9·88917		10·11083	10·08827		9·91173	10·19911		9·80089	
48·0	9·88927	·6 6	10·11073	10·08853	·6 15	9·91147	10·19926	·6 9	9·80074	
49·0	9·88937		10·11063	10·08879		9·91121	10·19942		9·80058	
50·0	9·88948	·7 7	10·11052	10·08905	·7 18	9 91095	10·19957	·7 11	9·80043	10′
51·0	9·88958		10·11042	10·08931		9·91069	10·19973		9·80027	
52·0	9·88968		10·11032	10·08957		9·91044	10·19988		9·80012	
53·0	9·88979		10·11022	10·08982		9·91018	10·20004		9·79996	
54·0	9·88989	·8 8	10·11011	10·09008	·8 21	9·90992	10·20019	·8 12	9·79981	
55·0	9·88999		10·11001	10·09034		9·90966	10·20035		9·79965	5′
56·0	9·89009		10·10991	10·09060		9·90940	10·20051		9·79950	
57·0	9·89020	·9 9	10·10981	10·09086	·9 23	9·90914	10·20066	·9 14	9·79934	
58·0	9·89030		10·10970	10·09111		9·90889	10·20082		9·79918	
59·0	9·89040		10·10960	10·09137		9·90863	10·20097		9·79903	
60·0	9·89050		10·10950	10·09163		9·90837	10·20113		9·79887	0′

51°
231°

LOGS. OF TRIG. FUNCTIONS

′	Sine	Parts		Cosec.	Tan.	Parts		Cotan.	Secant	Parts		Cosine	
00·0	9·89050	′		10·10950	10·09163	′		9·90837	10·20113	′		9·79887	60′
01·0	9·89061			10·10940	10·09189			9·90811	10·20128			9·79872	
02·0	9·09071	·1	1	10·10929	10·09215	·1	3	9·90785	10·20154	·1	2	9·79856	
03·0	9·89081			10·10919	10·09241			9·90759	10·20160			9·79840	
04·0	9·89091			10·10909	10·09266			9·90734	10·20175			9·79825	
05·0	9·89101	·2	2	10·10899	10·09292	·2	5	9·90708	10·20191	·2	3	9·79809	55′
06·0	9·89112			10·10889	10·09318			9·90682	10·20207			9·79793	
07·0	9·89122			10·10878	10·09344			9·90656	10·20222			9·79778	
08·0	9·89132	·3	3	10·10868	10·09370	·3	8	9·90630	10·20238	·3	5	9·79762	
09·0	9·89142			10·10858	10·09396			9·90604	10·20254			9·79746	
10·0	9·89152			10·10848	10·09422			9·90579	10·20269			9·79731	50′
11·0	9·89162	·4	4	10·10838	10·09447	·4	10	9·90553	10·20285	·4	6	9·79715	
12·0	9·89173			10·10827	10·09473			9·90527	10·20301			9·79699	
13·0	9·89183			10·10817	10·09499			9·90501	10·20316			9·79684	
14·0	9·89193	·5	5	10·10807	10·09525	·5	13	9·90475	10·20332	·5	8	9·79668	
15·0	9·89203			10·10797	10·09551			9·90449	10·20348			9·79652	45′
16·0	9·89213			10·10787	10·09577			9·90423	10·20364			9·79636	
17·0	9·89223	·6	6	10·10777	10·09603	·6	16	9·90397	10·20379	·6	9	9·79621	
18·0	9·89233			10·10767	10·09629			9·90371	10·20395			9·79605	
19·0	9·89244			10·10757	10·09654			9·90346	10·20411			9·79589	
20·0	9·89254	·7	7	10·10746	10·09680	·7	18	9·90320	10·20427	·7	11	9·79573	40′
21·0	9·89264			10·10736	10·09706			9·90294	10·20443			9·79558	
22·0	9·89274			10·10726	10·09732			9·90268	10·20458			9·79542	
23·0	9·89284			10·10716	10·09758			9·90242	10·20474			9·79526	
24·0	9·89294	·8	8	10·10706	10·09784	·8	21	9·90216	10·20490	·8	13	9·79510	
25·0	9·89304			10·10696	10·09810			9·90190	10·20506			9·79494	35′
26·0	9·89314			10·10686	10·09836			9·90164	10·20522			9·79478	
27·0	9·89324	·9	9	10·10676	10·09862	·9	23	9·90138	10·20537	·9	14	9·79463	
28·0	9·89334			10·10666	10·09888			9·90112	10·20553			9·79447	
29·0	9·89344			10·10656	10·09914			9·90086	10·20569			9·79431	
30·0	9·89354			10·10646	10·09940			9·90061	10·20585			9·79415	30′
31·0	9·89365			10·10636	10·09965			9·90035	10·20601			9·79399	
32·0	9·89375	·1	1	10·10626	10·09991	·1	3	9·90009	10·20617	·1	2	9·79383	
33·0	9·89385			10·10615	10·10017			9·89983	10·20633			9·79367	
34·0	9·89395			10·10605	10·10043			9·89957	10·20649			9·79351	
35·0	9·89405	·2	2	10·10595	10·10069	·2	5	9·89931	10·20665	·2	3	9·79335	25′
36·0	9·89415			10·10585	10·10095			9·89905	10·20681			9·79320	
37·0	9·89425			10·10575	10·10121			9·89879	10·20697			9·79304	
38·0	9·89435	·3	3	10·10565	10·10147	·3	8	9·89853	10·20712	·3	5	9·79288	
39·0	9·89445			10·10555	10·10173			9·89827	10·20728			9·79272	
40·0	9·89455			10·10545	10·10199			9·89801	10·20744			9·79256	20′
41·0	9·89465	·4	4	10·10535	10·10225	·4	10	9·89775	10·20760	·4	6	9·79240	
42·0	9·89475			10·10525	10·10251			9·89749	10·20776			9·79224	
43·0	9·89485			10·10515	10·10277			9·89723	10·20792			9·79208	
44·0	9·89495	·5	5	10·10506	10·10303	·5	13	9·89697	10·20808	·5	8	9·79192	
45·0	9·89505			10·10496	10·10329			9·89671	10·20824			9·79176	15′
46·0	9·89515			10·10486	10·10355			9·89645	10·20840			9·79160	
47·0	9·89524	·6	6	10·10476	10·10381	·6	16	9·89619	10·20856	·6	10	9·79144	
48·0	9·89534			10·10466	10·10407			9·89593	10·20873			9·79128	
49·0	9·89544			10·10456	10·10433			9·89567	10·20889			9·79112	
50·0	9·89554			10·10446	10·10459			9·89541	10·20905			9·79095	10′
51·0	9·89564	·7	7	10·10436	10·10485	·7	18	9·89515	10·20921	·7	11	9·79079	
52·0	9·89574			10·10426	10·10511			9·89489	10·20937			9·79063	
53·0	9·89584			10·10416	10·10537			9·89463	10·20953			9·79047	
54·0	9·89594	·8	8	10·10406	10·10563	·8	21	9·89437	10·20969	·8	13	9·79031	
55·0	9·89604			10·10396	10·10589			9·89411	10·20985			9·79015	5′
56·0	9·89614			10·10386	10·10615			9·89385	10·21001			9·78999	
57·0	9·89624	·9	9	10·10376	10·10641	·9	23	9·89359	10·21017	·9	14	9·78983	
58·0	9·89634			10·10367	10·10667			9·89333	10·21034			9·78967	
59·0	9·89643			10·10357	10·10693			9·89307	10·21050			9·78950	
60·0	9·89653			10·10347	10·10719			9·89281	10·21066			9·78934	0′

LOGS. OF TRIG. FUNCTIONS

52°
232°

′	Sine	Parts		Cosec.	Tan.	Parts		Cotan.	Secant	Parts		Cosine	
00·0	9·89653	′		10·10347	10·10719	′		9·89281	10·21066	′		9·78934	60′
01·0	9·89663			10·10337	10·10745			9·89255	10·21082			9·78918	
02·0	9·89673	·1	1	10·10327	10·10771	·1	3	9·89229	10·21098	·1	2	9·78902	
03·0	9·89683			10·10317	10·10797			9·89203	10·21114			9·78886	
04·0	9·89693			10·10307	10·10823			9·89177	10·21131			9·78869	
05·0	9·89703	·2	2	10·10298	10·10849	·2	5	9·89151	10·21147	·2	3	9·78853	55′
06·0	9·89712			10·10288	10·10875			9·89125	10·21163			9·78837	
07·0	9·89722			10·10278	10·10901			9·89099	10·21179			9·78821	
08·0	9·89732	·3	3	10·10268	10·10928	·3	8	9·89073	10·21196	·3	5	9·78805	
09·0	9·89742			10·10258	10·10954			9·89047	10·21212			9·78788	
10·0	9·89752			10·10248	10·10980			9·89020	10·21228			9·78772	50′
11·0	9·89761	·4	4	10·10239	10·11006	·4	10	9·88994	10·21244	·4	7	9·78756	
12·0	9·89771			10·10229	10·11032			9·88968	10·21261			9·78740	
13·0	9·89781			10·10219	10·11058			9·88942	10·21277			9·78723	
14·0	9·89791			10·10209	10·11084			9·88916	10·21293			9·78707	
15·0	9·89801	·5	5	10·10199	10·11110	·5	13	9·88890	10·21309	·5	8	9·78691	45′
16·0	9·89810			10·10190	10·11136			9·88864	10·21326			9·78674	
17·0	9·89820			10·10180	10·11162			9·88838	10·21342			9·78658	
18·0	9·89830	·6	6	10·10170	10·11188	·6	16	9·88812	10·21358	·6	10	9·78642	
19·0	9·89840			10·10160	10·11215			9·88786	10·21375			9·78625	
20·0	9·89849			10·10151	10·11241			9·88759	10·21391			9·78609	40′
21·0	9·89859	·7	7	10·10141	10·11267	·7	18	9·88733	10·21408	·7	11	9·78593	
22·0	9·89869			10·10131	10·11293			9·88707	10·21424			9·78576	
23·0	9·89879			10·10121	10·11319			9·88681	10·21440			9·78560	
24·0	9·89888	·8	8	10·10112	10·11345	·8	21	9·88655	10·21457	·8	13	9·78543	
25·0	9·89898			10·10102	10·11371			9·88629	10·21473			9·78527	35′
26·0	9·89908			10·10092	10·11397			9·88603	10·21490			9·78511	
27·0	9·89918	·9	9	10·10082	10·11424	·9	23	9·88577	10·21506	·9	15	9·78494	
28·0	9·89927			10·10073	10·11450			9·88550	10·21522			9·78478	
29·0	9·89937			10·10063	10·11476			9·88524	10·21539			9·78461	
30·0	9·89947			10·10053	10·11502			9·88498	10·21555			9·78445	30′
31·0	9·89956			10·10044	10·11528			9·88472	10·21572			9·78428	
32·0	9·89966	·1	1	10·10034	10·11554	·1	3	9·88446	10·21588	·1	2	9·78412	
33·0	9·89976			10·10024	10·11580			9·88420	10·21605			9·78395	
34·0	9·89985			10·10015	10·11607			9·88393	10·21621			9·78379	
35·0	9·89995	·2	2	10·10005	10·11633	·2	5	9·88367	10·21638	·2	3	9·78362	25′
36·0	9·90005			10·09995	10·11659			9·88341	10·21654			9·78346	
37·0	9·90014			10·09986	10·11685			9·88315	10·21671			9·78329	
38·0	9·90024	·3	3	10·09976	10·11711	·3	8	9·88289	10·21687	·3	5	9·78313	
39·0	9·90034			10·09966	10·11738			9·88263	10·21704			9·78296	
40·0	9·90043			10·09957	10·11764			9·88236	10·21720			9·78280	20′
41·0	9·90053	·4	4	10·09947	10·11790	·4	10	9·88210	10·21737	·4	7	9·78263	
42·0	9·90063			10·09937	10·11816			9·88184	10·21754			9·78246	
43·0	9·90072			10·09928	10·11842			9·88158	10·21770			9·78230	
44·0	9·90082			10·09918	10·11869			9·88131	10·21787			9·78213	
45·0	9·90091	·5	5	10·09909	10·11895	·5	13	9·88105	10·21803	·5	8	9·78197	15′
46·0	9·90101			10·09899	10·11921			9·88079	10·21820			9·78180	
47·0	9·90111			10·09889	10·11947			9·88053	10·21837			9·78163	
48·0	9·90120	·6	6	10·09880	10·11974	·6	16	9·88027	10·21853	·6	10	9·78147	
49·0	9·90130			10·09870	10·12000			9·88000	10·21870			9·78130	
50·0	9·90139			10·09861	10·12026			9·87974	10·21887			9·78113	10′
51·0	9·90149	·7	7	10·09851	10·12052	·7	18	9·87948	10·21903	·7	12	9·78097	
52·0	9·90159			10·09842	10·12078			9·87922	10·21920			9·78080	
53·0	9·90168			10·09832	10·12105			9·87895	10·21937			9·78063	
54·0	9·90178	·8	8	10·09822	10·12131	·8	21	9·87869	10·21953	·8	13	9·78047	
55·0	9·90187			10·09813	10·12157			9·87843	10·21970			9·78030	5′
56·0	9·90197			10·09803	10·12184			9·87817	10·21987			9·78013	
57·0	9·90206	·9	9	10·09794	10·12210	·9	24	9·87790	10·22003	·9	15	9·77997	
58·0	9·90216			10·09784	10·12236			9·87764	10·22020			9·77980	
59·0	9·90225			10·09775	10·12262			9·87738	10·22037			9·77963	
60·0	9·90235			10·09765	10·12289			9·87711	10·22054			9·77946	0′

53°
233°

LOGS. OF TRIG. FUNCTIONS

′	Sine	Parts	Cosec.	Tan.	Parts	Cotan.	Secant	Parts	Cosine	
00·0	9·90235	′	10·09765	10·12289	′	9·87711	10·22054	′	9·77946	60′
01·0	9·90244		10·09756	10·12315		9·87685	10·22071		9·77930	
02·0	9·90254	·1 1	10·09746	10·12341	·1 3	9·87659	10·22087	·1 2	9·77913	
03·0	9·90263		10·09737	10·12367		9·87633	10·22104		9·77896	
04·0	9·90273		10·09727	10·12394		9·87606	10·22121		9·77879	
05·0	9·90282	·2 2	10·09718	10·12420	·2 5	9·87580	10·22138	·2 3	9·77862	55′
06·0	9·90292		10·09708	10·12446		9·87554	10·22155		9·77846	
07·0	9·90301		10·09699	10·12473		9·87527	10·22171		9·77829	
08·0	9·90311	·3 3	10·09689	10·12499	·3 8	9·87501	10·22188	·3 5	9·77812	
09·0	9·90320		10·09680	10·12525		9·87475	10·22205		9·77795	
10·0	9·90330		10·09670	10·12552		9·87448	10·22222		9·77778	50′
11·0	9·90339	·4 4	10·09661	10·12578	·4 11	9·87422	10·22239	·4 7	9·77761	
12·0	9·90349		10·09651	10·12604		9·87396	10·22256		9·77744	
13·0	9·90358		10·09642	10·12631		9·87369	10·22273		9·77728	
14·0	9·90368		10·09632	10·12657		9·87343	10·22289		9·77711	
15·0	9·90377	·5 5	10·09623	10·12683	·5 13	9·87317	10·22306	·5 8	9·77694	45′
16·0	9·90386		10·09614	10·12710		9·87290	10·22323		9·77677	
17·0	9·90396		10·09604	10·12736		9·87264	10·22340		9·77660	
18·0	9·90405	·6 6	10·09595	10·12762	·6 16	9·87238	10·22357	·6 10	9·77643	
19·0	9·90415		10·09585	10·12789		9·87211	10·22374		9·77626	
20·0	9·90424		10·09576	10·12815		9·87185	10·22391		9·77609	40′
21·0	9·90434	·7 7	10·09567	10·12841	·7 18	9·87159	10·22408	·7 12	9·77592	
22·0	9·90443		10·09557	10·12868		9·87132	10·22425		9·77575	
23·0	9·90452		10·09548	10·12894		9·87106	10·22442		9·77558	
24·0	9·90462	·8 8	10·09538	10·12921	·8 21	9·87079	10·22459	·8 14	9·77541	
25·0	9·90471		10·09529	10·12947		9·87053	10·22476		9·77524	35′
26·0	9·90480		10·09520	10·12974		9·87027	10·22493		9·77507	
27·0	9·90490	·9 8	10·09510	10·13000	·9 24	9·87000	10·22510	·9 15	9·77490	
28·0	9·90499		10·09501	10·13026		9·86974	10·22527		9·77473	
29·0	9·90509		10·09492	10·13053		9·86947	10·22544		9·77456	
30·0	9·90518		10·09482	10·13079		9·86921	10·22561		9·77439	30′
31·0	9·90527		10·09473	10·13106		9·86895	10·22578		9·77422	
32·0	9·90537	·1 1	10·09463	10·13132	·1 3	9·86868	10·22595	·1 2	9·77405	
33·0	9·90546		10·09454	10·13158		9·86842	10·22613		9·77388	
34·0	9·90555		10·09445	10·13185		9·86815	10·22630		9·77370	
35·0	9·90565	·2 2	10·09436	10·13211	·2 5	9·86789	10·22647	·2 3	9·77353	25′
36·0	9·90574		10·09426	10·13238		9·86762	10·22664		9·77336	
37·0	9·90583		10·09417	10·13264		9·86736	10·22681		9·77319	
38·0	9·90593	·3 3	10·09408	10·13291	·3 8	9·86709	10·22698	·3 5	9·77302	
39·0	9·90602		10·09398	10·13317		9·86683	10·22715		9·77285	
40·0	9·90611		10·09389	10·13344		9·86656	10·22733		9·77268	20′
41·0	9·90620	·4 4	10·09380	10·13370	·4 11	9·86630	10·22750	·4 7	9·77250	
42·0	9·90630		10·09370	10·13397		9·86604	10·22767		9·77233	
43·0	9·90639		10·09361	10·13423		9·86577	10·22784		9·77216	
44·0	9·90648		10·09352	10·13450		9·86551	10·22801		9·77199	
45·0	9·90658	·5 5	10·09343	10·13476	·5 13	9·86524	10·22819	·5 9	9·77182	15′
46·0	9·90667		10·09333	10·13503		9·86498	10·22836		9·77164	
47·0	9·90676		10·09324	10·13529		9·86471	10·22853		9·77147	
48·0	9·90685	·6 6	10·09315	10·13556	·6 16	9·86445	10·22870	·6 10	9·77130	
49·0	9·90695		10·09306	10·13582		9·86418	10·22888		9·77113	
50·0	9·90704		10·09296	10·13609		9·86392	10·22905		9·77095	10′
51·0	9·90713	·7 6	10·09287	10·13635	·7 19	9·86365	10·22922	·7 12	9·77078	
52·0	9·90722		10·09278	10·13662		9·86339	10·22939		9·77061	
53·0	9·90731		10·09269	10·13688		9·86312	10·22957		9·77043	
54·0	9·90741	·8 7	10·09259	10·13715	·8 21	9·86285	10·22974	·8 14	9·77026	
55·0	9·90750		10·09250	10·13741		9·86259	10·22991		9·77009	5′
56·0	9·90759		10·09241	10·13768		9·86232	10·23009		9·76991	
57·0	9·90768	·9 8	10·09232	10·13794	·9 24	9·86206	10·23026	·9 16	9·76974	
58·0	9·90777		10·09223	10·13821		9·86179	10·23043		9·76957	
59·0	9·90787		10·09213	10·13847		9·86153	10·23061		9·76939	
60·0	9·90796		10·09204	10·13874		9·86126	10·23078		9·76922	0′

54°
234°

LOGS. OF TRIG. FUNCTIONS

'	Sine	Parts	Cosec.	Tan.	Parts	Cotan.	Secant	Parts	Cosine	
00·0	9·90796	'	10·09204	10·13874	'	9·86126	10·23078	'	9·76922	60'
01·0	9·90805		10·09195	10·13901		9·86100	10·23096		9·76905	
02·0	9·90814	·1 1	10·09186	10·13927	·1 3	9·86073	10·23113	·1 2	9·76887	
03·0	9·90823		10·09177	10·13954		9·86046	10·23130		9·76870	
04·0	9·90832		10·09168	10·13980		9·86020	10·23148		9·76852	
05·0	9·90842	·2 2	10·09158	10·14007	·2 5	9·85993	10·23165	·2 4	9·76835	55'
06·0	9·90851		10·09149	10·14033		9·85967	10·23183		9·76817	
07·0	9·90860		10·09140	10·14060		9·85940	10·23200		9·76800	
08·0	9·90869	·3 3	10·09131	10·14087	·3 8	9·85913	10·23218	·3 5	9·76782	
09·0	9·90878		10·09122	10·14113		9·85887	10·23235		9·76765	
10·0	9·90887		10·09113	10·14140		9·85860	10·23253		9·76748	50'
11·0	9·90896	·4 4	10·09104	10·14166	·4 11	9·85834	10·23270	·4 7	9·76730	
12·0	9·90906		10·09095	10·14193		9·85807	10·23288		9·76712	
13·0	9·90915		10·09085	10·14220		9·85780	10·23305		9·76695	
14·0	9·90924	·5 5	10·09076	10·14246		9·85754	10·23323	·5 9	9·76677	
15·0	9·90933		10·09067	10·14273	·5 13	9·85727	10·23340		9·76660	45'
16·0	9·90942		10·09058	10·14300		9·85700	10·23358		9·76642	
17·0	9·90951		10·09049	10·14326		9·85674	10·23375		9·76625	
18·0	9·90960	·6 5	10·09040	10·14353	·6 16	9·85647	10·23393	·6 11	9·76607	
19·0	9·90969		10·09031	10·14380		9·85620	10·23410		9·76590	
20·0	9·90978		10·09022	10·14406		9·85594	10·23428		9·76572	40'
21·0	9·90987	·7 6	10·09013	10·14433	·7 19	9·85567	10·23446	·7 12	9·76554	
22·0	9·90996		10·09004	10·14460		9·85540	10·23463		9·76537	
23·0	9·91005		10·08995	10·14486		9·85514	10·23481		9·76519	
24·0	9·91014	·8 7	10·08986	10·14513	·8 21	9·85487	10·23499	·8 14	9·76502	
25·0	9·91024		10·08977	10·14540		9·85460	10·23516		9·76484	35'
26·0	9·91033		10·08968	10·14566		9·85434	10·23534		9·76466	
27·0	9·91042	·9 8	10·08959	10·14593	·9 24	9·85407	10·23552	·9 16	9·76449	
28·0	9·91051		10·08949	10·14620		9·85380	10·23569		9·76431	
29·0	9·91060		10·08940	10·14647		9·85354	10·23587		9·76413	
30·0	9·91069		10·08931	10·14673		9·85327	10·23605		9·76395	30'
31·0	9·91078		10·08922	10·14700		9·85300	10·23622		9·76378	
32·0	9·91087	·1 1	10·08913	10·14727	·1 3	9·85273	10·23640	·1 2	9·76360	
33·0	9·91096		10·08904	10·14753		9·85247	10·23658		9·76342	
34·0	9·91105		10·08895	10·14780		9·85220	10·23676		9·76325	
35·0	9·91114	·2 2	10·08886	10·14807	·2 5	9·85193	10·23693	·2 4	9·76307	25'
36·0	9·91123		10·08877	10·14834		9·85166	10·23711		9·76289	
37·0	9·91132		10·08869	10·14860		9·85140	10·23729		9·76271	
38·0	9·91141	·3 3	10·08860	10·14887	·3 8	9·85113	10·23747	·3 5	9·76253	
39·0	9·91150		10·08851	10·14914		9·85086	10·23764		9·76236	
40·0	9·91158		10·08842	10·14941		9·85059	10·23782		9·76218	20'
41·0	9·91167	·4 4	10·08833	10·14968	·4 11	9·85033	10·23800	·4 7	9·76200	
42·0	9·91176		10·08824	10·14994		9·85006	10·23818		9·76182	
43·0	9·91185		10·08815	10·15021		9·84979	10·23836		9·76164	
44·0	9·91194	·5 4	10·08806	10·15048		9·84952	10·23854	·5 9	9·76146	
45·0	9·91203		10·08797	10·15075	·5 13	9·84925	10·23871		9·76129	15'
46·0	9·91212		10·08788	10·15101		9·84899	10·23889		9·76111	
47·0	9·91221	·6 5	10·08779	10·15128		9·84872	10·23907	·6 11	9·76093	
48·0	9·91230		10·08770	10·15155	·6 16	9·84845	10·23925		9·76075	
49·0	9·91239		10·08761	10·15182		9·84818	10·23943		9·76057	
50·0	9·91248	·7 6	10·08752	10·15209	·7 19	9·84791	10·23961	·7 13	9·76039	10'
51·0	9·91257		10·08743	10·15236		9·84764	10·23979		9·76021	
52·0	9·91266		10·08735	10·15262		9·84738	10·23997		9·76003	
53·0	9·91274		10·08726	10·15289		9·84711	10·24015		9·75985	
54·0	9·91283	·8 7	10·08717	10·15316	·8 21	9·84684	10·24033	·8 14	9·75967	
55·0	9·91292		10·08708	10·15343		9·84657	10·24051		9·75949	5'
56·0	9·91301		10·08699	10·15370		9·84630	10·24069		9·75931	
57·0	9·91310	·9 8	10·08690	10·15397	·9 24	9·84603	10·24087	·9 16	9·75913	
58·0	9·91319		10·08681	10·15424		9·84576	10·24105		9·75895	
59·0	9·91328		10·08672	10·15450		9·84550	10·24123		9·75877	
60·0	9·91337		10·08664	10·15477		9·84523	10·24141		9·75859	0'

125°
305°

55°
235°

LOGS. OF TRIG. FUNCTIONS

′	Sine	Parts		Cosec.	Tan.	Parts		Cotan.	Secant	Parts		Cosine	
00·0	9·91337	′		10·08663	10·15477	′		9·84523	10·24141	′		9·75859	60′
01·0	9·91345			10·08655	10·15504			9·84496	10·24159			9·75841	
02·0	9·91354	·1	1	10·08646	10·15531	·1	3	9·84469	10·24177	·1	2	9·75823	
03·0	9·91363			10·08637	10·15558			9·84442	10·24195			9·75805	
04·0	9·91372			10·08628	10·15585			9·84415	10·24213			9·75787	
05·0	9·91381	·2	2	10·08619	10·15612	·2	5	9·84388	10·24231	·2	4	9·75769	55′
06·0	9·91389			10·08611	10·15639			9·84361	10·24249			9·75751	
07·0	9·91398			10·08602	10·15666			9·84334	10·24267			9·75733	
08·0	9·91407	·3	3	10·08593	10·15693	·3	8	9·84307	10·24286	·3	5	9·75714	
09·0	9·91416			10·08584	10·15720			9·84281	10·24304			9·75696	
10·0	9·91425			10·08575	10·15747			9·84254	10·24322			9·75678	50′
11·0	9·91433	·4	3	10·08567	10·15773	·4	11	9·84227	10·24340	·4	7	9·75660	
12·0	9·91442			10·08558	10·15800			9·84200	10·24358			9·75642	
13·0	9·91451			10·08549	10·15827			9·84173	10·24376			9·75624	
14·0	9·91460			10·08540	10·15854			9·84146	10·24395			9·75605	
15·0	9·91469	·5	4	10·08532	10·15881	·5	14	9·84119	10·24413	·5	9	9·75587	45′
16·0	9·91477			10·08523	10·15908			9·84092	10·24431			9·75569	
17·0	9·91486			10·08514	10·15935			9·84065	10·24449			9·75551	
18·0	9·91495	·6	5	10·08505	10·15962	·6	16	9·84038	10·24467	·6	11	9·75533	
19·0	9·91504			10·08497	10·15989			9·84011	10·24486			9·75514	
20·0	9·91512			10·08488	10·16016			9·83984	10·24504			9·75496	40′
21·0	9·91521	·7	6	10·08479	10·16043	·7	19	9·83957	10·24522	·7	13	9·75478	
22·0	9·91530			10·08470	10·16070			9·83930	10·24541			9·75460	
23·0	9·91539			10·08462	10·16097			9·83903	10·24559			9·75441	
24·0	9·91547	·8	7	10·08453	10·16124	·8	22	9·83876	10·24577	·8	15	9·75423	
25·0	9·91556			10·08444	10·16151			9·83849	10·24595			9·75405	35′
26·0	9·91565			10·08435	10·16178			9·83822	10·24614			9·75386	
27·0	9·91573	·9	8	10·08427	10·16205	·9	24	9·83795	10·24632	·9	16	9·75368	
28·0	9·91582			10·08418	10·16233			9·83768	10·24651			9·75350	
29·0	9·91591			10·08409	10·16260			9·83741	10·24669			9·75331	
30·0	9·91599			10·08401	10·16287			9·83713	10·24687			9·75313	30′
31·0	9·91608			10·08392	10·16314			9·83686	10·24706			9·75294	
32·0	9·91617	·1	1	10·08383	10·16341	·1	3	9·83659	10·24724	·1	2	9·75276	
33·0	9·91625			10·08375	10·16368			9·83632	10·24742			9·75258	
34·0	9·91634			10·08366	10·16395			9·83605	10·24761			9·75239	
35·0	9·91643	·2	2	10·08357	10·16422	·2	5	9·83578	10·24779	·2	4	9·75221	25′
36·0	9·91651			10·08349	10·16449			9·83551	10·24798			9·75202	
37·0	9·91660			10·08340	10·16476			9·83524	10·24816			9·75184	
38·0	9·91669	·3	3	10·08331	10·16503	·3	8	9·83497	10·24835	·3	6	9·75165	
39·0	9·91677			10·08323	10·16530			9·83470	10·24853			9·75147	
40·0	9·91686			10·08314	10·16558			9·83443	10·24872			9·75128	20′
41·0	9·91695	·4	3	10·08305	10·16585	·4	11	9·83415	10·24890	·4	7	9·75110	
42·0	9·91703			10·08297	10·16612			9·83388	10·24909			9·75091	
43·0	9·91712			10·08288	10·16639			9·83361	10·24927			9·75073	
44·0	9·91720			10·08280	10·16666			9·83334	10·24946			9·75054	
45·0	9·91729	·5	4	10·08271	10·16693	·5	14	9·83307	10·24964	·5	9	9·75036	15′
46·0	9·91738			10·08262	10·16720			9·83280	10·24983			9·75017	
47·0	9·91746			10·08254	10·16748			9·83253	10·25001			9·74999	
48·0	9·91755	·6	5	10·08245	10·16775	·6	16	9·83225	10·25020	·6	11	9·74980	
49·0	9·91763			10·08237	10·16802			9·83198	10·25039			9·74962	
50·0	9·91772			10·08228	10·16829			9·83171	10·25057			9·74943	10′
51·0	9·91781	·7	6	10·08220	10·16856	·7	19	9·83144	10·25076	·7	13	9·74924	
52·0	9·91789			10·08211	10·16884			9·83117	10·25094			9·74906	
53·0	9·91798			10·08202	10·16911			9·83089	10·25113			9·74887	
54·0	9·91806	·8	7	10·08194	10·16938	·8	22	9·83062	10·25132	·8	15	9·74868	
55·0	9·91815			10·08185	10·16965			9·83035	10·25150			9·74850	5′
56·0	9·91823			10·08177	10·16992			9·83008	10·25169			9·74831	
57·0	9·91832	·9	8	10·08168	10·17020	·9	24	9·82981	10·25188	·9	17	9·74812	
58·0	9·91840			10·08160	10·17047			9·82953	10·25206			9·74794	
59·0	9·91849			10·08151	10·17074			9·82926	10·25225			9·74775	
60·0	9·91857			10·08143	10·17101			9·82899	10·25244			9·74756	0′

LOGS. OF TRIG. FUNCTIONS

56°
236°

′	Sine	Parts		Cosec.	Tan.	Parts		Cotan.	Secant	Parts		Cosine	
00·0	9·91857	′		10·08143	10·17101	′		9·82899	10·25244	′		9·74756	60′
01·0	9·91866			10·08134	10·17129			9·82872	10·25263			9·74737	
02·0	9·91875	·1	1	10·08126	10·17156	·1	3	9·82844	10·25281	·1	2	9·74719	
03·0	9·91883			10·08117	10·17183			9·82817	10·25300			9·74700	
04·0	9·91892			10·08109	10·17210			9·82790	10·25319			9·74681	
05·0	9·91900	·2	2	10·08100	10·17238	·2	5	9·82762	10·25338	·2	4	9·74662	55′
06·0	9·91909			10·08092	10·17265			9·82735	10·25356			9·74644	
07·0	9·91917			10·08083	10·17292			9·82708	10·25375			9·74625	
08·0	9·91925	·3	3	10·08075	10·17320	·3	8	9·82681	10·25394	·3	6	9·74606	
09·0	9·91934			10·08066	10·17347			9·82653	10·25413			9·74587	
10·0	9·91942			10·08058	10·17374			9·82626	10·25432			9·74568	50′
11·0	9·91951	·4	3	10·08049	10·17401	·4	11	9·82599	10·25451	·4	8	9·74549	
12·0	9·91959			10·08041	10·17429			9·82571	10·25469			9·74531	
13·0	9·91968			10·08032	10·17456			9·82544	10·25488			9·74512	
14·0	9·91976	·5	4	10·08024	10·17483	·5	14	9·82517	10·25507	·5	9	9·74493	
15·0	9·91985			10·08015	10·17511			9·82489	10·25526			9·74474	45′
16·0	9·91993			10·08007	10·17538			9·82462	10·25545			9·74455	
17·0	9·92002			10·07999	10·17566			9·82435	10·25564			9·74436	
18·0	9·92010	·6	5	10·07990	10·17593	·6	16	9·82407	10·25583	·6	11	9·74417	
19·0	9·92018			10·07982	10·17620			9·82380	10·25602			9·74398	
20·0	9·92027			10·07973	10·17648			9·82352	10·25621			9·74379	40′
21·0	9·92035	·7	6	10·07965	10·17675	·7	19	9·82325	10·25640	·7	13	9·74360	
22·0	9·92044			10·07956	10·17702			9·82298	10·25659			9·74341	
23·0	9·92052			10·07948	10·17730			9·82270	10·25678			9·74322	
24·0	9·92060	·8	7	10·07940	10·17757	·8	22	9·82243	10·25697	·8	15	9·74303	
25·0	9·92069			10·07931	10·17785			9·82215	10·25716			9·74284	35′
26·0	9·92077			10·07923	10·17812			9·82188	10·25735			9·74265	
27·0	9·92086	·9	8	10·07914	10·17839	·9	25	9·82161	10·25754	·9	17	9·74246	
28·0	9·92094			10·07906	10·17867			9·82133	10·25773			9·74227	
29·0	9·92102			10·07898	10·17894			9·82106	10·25792			9·74208	
30·0	9·92111			10·07889	10·17922			9·82078	10·25811			9·74189	30′
31·0	9·92119			10·07881	10·17949			9·82051	10·25830			9·74170	
32·0	9·92127	·1	1	10·07873	10·17977	·1	3	9·82023	10·25849	·1	2	9·74151	
33·0	9·92136			10·07864	10·18004			9·81996	10·25868			9·74132	
34·0	9·92144			10·07856	10·18032			9·81968	10·25888			9·74113	
35·0	9·92152	·2	2	10·07848	10·18059	·2	6	9·81941	10·25907	·2	4	9·74093	25′
36·0	9·92161			10·07839	10·18087			9·81914	10·25926			9·74074	
37·0	9·92169			10·07831	10·18114			9·81886	10·25945			9·74055	
38·0	9·92177	·3	2	10·07823	10·18142	·3	8	9·81859	10·25964	·3	6	9·74036	
39·0	9·92186			10·07814	10·18169			9·81831	10·25983			9·74017	
40·0	9·92194			10·07806	10·18197			9·81804	10·26003			9·73998	20′
41·0	9·92202	·4	3	10·07798	10·18224	·4	11	9·81776	10·26022	·4	8	9·73978	
42·0	9·92211			10·07789	10·18252			9·81748	10·26041			9·73959	
43·0	9·92219			10·07781	10·18279			9·81721	10·26060			9·73940	
44·0	9·92227	·5	4	10·07773	10·18307	·5	14	9·81693	10·26079	·5	10	9·73921	
45·0	9·92236			10·07765	10·18334			9·81666	10·26099			9·73901	15′
46·0	9·92244			10·07756	10·18362			9·81638	10·26118			9·73882	
47·0	9·92252			10·07748	10·18389			9·81611	10·26137			9·73863	
48·0	9·92260	·6	5	10·07740	10·18417	·6	17	9·81583	10·26157	·6	12	9·73843	
49·0	9·92269			10·07731	10·18445			9·81556	10·26176			9·73824	
50·0	9·92277	·7	6	10·07723	10·18472	·7	19	9·81528	10·26195	·7	13	9·73805	10′
51·0	9·92285			10·07715	10·18500			9·81500	10·26215			9·73786	
52·0	9·92293			10·07707	10·18527			9·81473	10·26234			9·73766	
53·0	9·92302			10·07698	10·18555			9·81445	10·26253			9·73747	
54·0	9·92310	·8	7	10·07690	10·18582	·8	22	9·81418	10·26273	·8	15	9·73727	
55·0	9·92318			10·07682	10·18610			9·81390	10·26292			9·73708	5′
56·0	9·92326			10·07674	10·18638			9·81362	10·26311			9·73689	
57·0	9·92335	·9	7	10·07666	10·18665	·9	25	9·81335	10·26331	·9	17	9·73669	
58·0	9·92343			10·07657	10·18693			9·81307	10·26350			9·73650	
59·0	9·92351			10·07649	10·18721			9·81279	10·26370			9·73630	
60·0	9·92359			10·07641	10·18748			9·81252	10·26389			9·73611	0′

123°
303°

57°
237°

LOGS. OF TRIG. FUNCTIONS

′	Sine	Parts		Cosec.	Tan.	Parts		Cotan.	Secant	Parts		Cosine	
00·0	9·92359	′		10·07641	10·18748	′		9·81252	10·26389	′		9·73611	60′
01·0	9·92367			10·07633	10·18776			9·81224	10·26409			9·73591	
02·0	9·92376	·1	1	10·07625	10·18804	·1	3	9·81196	10·26428	·1	2	9·73572	
03·0	9·92384			10 07616	10·18831			9·81169	10·26448			9·73553	
04·0	9·92392			10·07608	10·18859			9·81141	10·26467			9·73533	
05·0	9·92400	·2	2	10·07600	10·18887	·2	6	9·81113	10·26487	·2	4	9·73514	55′
06·0	9·92408			10·07592	10·18914			9·81086	10·26506			9·73494	
07·0	9·92416			10·07584	10·18942			9·81058	10·26526			9·73474	
08·0	9·92425	·3	2	10·07575	10·18970	·3	8	9·81030	10·26545	·3	6	9·73455	
09·0	9·92433			10·07567	10·18998			9·81003	10·26565			9·73435	
10·0	9·92441			10·07559	10·19025			9·80975	10·26584			9·73416	50′
11·0	9·92449	·4	3	10·07551	10·19053	·4	11	9·80947	10·26604	·4	8	9·73396	
12·0	9·92457			10·07543	10·19081			9·80919	10·26624			9·73377	
13·0	9·92465			10·07535	10·19108			9·80892	10·26643			9·73357	
14·0	9·92474			10·07527	10·19136			9·80864	10·26663			9·73337	
15·0	9·92482	·5	4	10·07518	10·19164	·5	14	9·80836	10·26682	·5	10	9·73318	45′
16·0	9·92490			10·07510	10·19192			9·80808	10·26702			9·73298	
17·0	9·92498			10·07502	10·19220			9·80781	10·26722			9·73278	
18·0	9·92506	·6	5	10·07494	10·19247	·6	17	9·80753	10·26741	·6	12	9·73259	
19·0	9·92514			10·07486	10·19275			9·80725	10·26761			9·73239	
20·0	9·92522			10·07478	10·19303			9·80697	10·26781			9·73219	40′
21·0	9·92530	·7	6	10·07470	10·19331	·7	19	9·80669	10·26800	·7	14	9·73200	
22·0	9·92538			10·07462	10·19359			9·80642	10·26820			9·73180	
23·0	9·92547			10·07454	10·19386			9·80614	10·26840			9·73160	
24·0	9·92555	·8	7	10·07446	10·19414	·8	22	9·80586	10·26860	·8	16	9·73140	
25·0	9·92563			10·07437	10·19442			9·80558	10·26879			9·73121	35′
26·0	9·92571			10·07429	10·19470			9·80530	10·26899			9·73101	
27·0	9·92579	·9	7	10·07421	10·19498	·9	25	9·80502	10·26919	·9	18	9·73081	
28·0	9·92587			10·07413	10·19526			9·80475	10·26939			9·73061	
29·0	9·92595			10·07405	10·19553			9·80447	10·26959			9·73042	
30·0	9·92603			10·07397	10·19581			9·80419	10·26978			9·73022	30′
31·0	9·92611			10·07389	10·19609			9·80391	10·26998			9·73002	
32·0	9·92619	·1	1	10·07381	10·19637	·1	3	9·80363	10·27018	·1	2	9·72982	
33·0	9·92627			10·07373	10·19665			9·80335	10·27038			9·72962	
34·0	9·92635			10·07365	10·19693			9·80307	10·27058			9·72942	
35·0	9·92643	·2	2	10·07357	10·19721	·2	6	9·80279	10·27078	·2	4	9·72922	25′
36·0	9·92651			10·07349	10·19749			9·80251	10·27098			9·72902	
37·0	9·92659			10·07341	10·19777			9·80223	10·27118			9·72883	
38·0	9·92667	·3	2	10·07333	10·19805	·3	8	9·80196	10·27137	·3	6	9·72863	
39·0	9·92675			10·07325	10·19833			9·80168	10·27157			9·72843	
40·0	9·92683			10·07317	10·19860			9·80140	10·27177			9·72823	20′
41·0	9·92691	·4	3	10·07309	10·19888	·4	11	9·80112	10·27197	·4	8	9·72803	
42·0	9·92699			10·07301	10·19916			9·80084	10·27217			9·72783	
43·0	9·92707			10·07293	10·19944			9·80056	10·27237			9·72763	
44·0	9·92715			10·07285	10·19972			9·80028	10·27257			9·72743	
45·0	9·92723	·5	4	10·07276	10·20000	·5	14	9·80000	10·27277	·5	10	9·72723	15′
46·0	9·92731			10·07269	10·20028			9·79972	10·27297			9·72703	
47·0	9·92739			10·07261	10·20056			9·79944	10·27317			9·72683	
48·0	9·92747	·6	5	10·07253	10·20084	·6	17	9·79916	10·27337	·6	12	9·72663	
49·0	9·92755			10·07245	10·20112			9·79888	10·27357			9·72643	
50·0	9·92763			10·07237	10·20140			9·79860	10·27378			9·72623	10′
51·0	9·92771	·7	6	10·07229	10·20168	·7	20	9·79832	10·27398	·7	14	9·72602	
52·0	9·92779			10·07221	10·20196			9·79804	10·27418			9·72582	
53·0	9·92787			10·07213	10·20225			9·79776	10·27438			9·72562	
54·0	9·92795	·8	6	10·07205	10·20253	·8	22	9·79748	10·27458	·8	16	9·72542	
55·0	9·92803			10·07198	10·20281			9·79719	10·27478			9·72522	5′
56·0	9·92810			10·07190	10·20309			9·79691	10·27498			9·72502	
57·0	9·92818	·9	7	10·07182	10·20337	·9	25	9·79663	10·27518	·9	18	9·72482	
58·0	9·92826			10·07174	10·20365			9·79635	10·27539			9·72461	
59·0	9·92834			10·07166	10·20393			9·79607	10·27559			9·72441	
60·0	9·92842			10·07158	10·20421			9·79579	10·27579			9·72421	0′

58°
238°

LOGS. OF TRIG. FUNCTIONS

′	Sine	Parts		Cosec.	Tan.	Parts		Cotan.	Secant	Parts		Cosine	
00·0	9·92842	′		10·07158	10·20421	′		9·79579	10·27579	′		9·72421	60′
01·0	9·92850			10·07150	10·20449			9·79551	10·27599			9·72401	
02·0	9·92858	·1	1	10·07142	10·20477	·1	3	9·79523	10·27620	·1	2	9·72381	
03·0	9·92866			10·07134	10·20505			9·79495	10·27640			9·72360	
04·0	9·92874			10·07126	10·20534			9·79466	10·27660			9·72340	
05·0	9·92882	·2	2	10·07119	10·20562	·2	6	9·79438	10·27680	·2	4	9·72320	55′
06·0	9·92889			10·07111	10·20590			9·79410	10·27701			9·72299	
07·0	9·92897			10·07103	10·20618			9·79382	10·27721			9·72279	
08·0	9·92905	·3	2	10·07095	10·20646	·3	8	9·79354	10·27741	·3	6	9·72259	
09·0	9·92913			10·07087	10·20674			9·79326	10·27762			9·72239	
10·0	9·92921			10·07079	10·20703			9·79297	10·27782			9·72218	50′
11·0	9·92929	·4	3	10·07071	10·20731	·4	11	9·79269	10·27802	·4	8	9·72198	
12·0	9·92936			10·07064	10·20759			9·79241	10·27823			9·72177	
13·0	9·92944			10·07056	10·20787			9·79213	10·27843			9·72157	
14·0	9·92952			10·07048	10·20815			9·79185	10·27863			9·72137	
15·0	9·92960	·5	4	10·07040	10·20844	·5	14	9·79156	10·27884	·5	10	9·72116	45′
16·0	9·92968			10·07032	10·20872			9·79128	10·27904			9·72096	
17·0	9·92976			10·07025	10·20900			9·79100	10·27925			9·72075	
18·0	9·92983	·6	5	10·07017	10·20928	·6	17	9·79072	10·27945	·6	12	9·72055	
19·0	9·92991			10·07009	10·20957			9·79043	10·27966			9·72035	
20·0	9·92999			10·07001	10·20985			9·79015	10·27986			9·72014	40′
21·0	9·93007	·7	5	10·06993	10·21013	·7	20	9·78987	10·28007	·7	14	9·71994	
22·0	9·93015			10·06986	10·21042	·		9·78959	10·28027			9·71973	
23·0	9·93022			10·06978	10·21070			9·78930	10·28040			9·71953	
24·0	9·93030	·8	6	10·06970	10·21098	·8	23	9·78902	10·28068	·8	16	9·71932	
25·0	9·93038			10·06962	10·21126			9·78874	10·28089			9·71911	35′
26·0	9·93046			10·06954	10·21155			9·78845	10·28109			9·71891	
27·0	9·93053	·9	7	10·06947	10·21183	·9	25	9·78817	10·28130	·9	18	9·71870	
28·0	9·93061			10·06939	10·21211			9·78789	10·28150			9·71850	
29·0	9·93069			10·06931	10·21240			9·78760	10·28171			9·71829	
30·0	9·93077			10·06923	10·21268			9·78732	10·28192			9·71809	30′
31·0	9·93084			10·06916	10·21296			9·78704	10·28212			9·71788	
32·0	9·93092	·1	1	10·06908	10·21325	·1	3	9·78675	10·28233	·1	2	9·71767	
33·0	9·93100			10·06900	10·21353			9·78647	10·28253			9·71747	
34·0	9·93108			10·06893	10·21382			9·78618	10·28274			9·71726	
35·0	9·93115	·2	2	10·06885	10·21410	·2	6	9·78590	10·28295	·2	4	9·71705	25′
36·0	9·93123			10·06877	10·21438			9·78562	10·28315			9·71685	
37·0	9·93131			10·06869	10·21467			9·78533	10·28336			9·71664	
38·0	9·93138	·3	2	10·06862	10·21495	·3	9	9·78505	10·28357	·3	6	9·71643	
39·0	9·93146			10·06854	10·21524			9·78476	10·28378			9·71622	
40·0	9·93154			10·06846	10·21552			9·78448	10·28398			9·71602	20′
41·0	9·93161	·4	3	10·06839	10·21581	·4	11	9·78420	10·28419	·4	8	9·71581	
42·0	9·93169			10·06831	10·21609			9·78391	10·28440			9·71560	
43·0	9·93177			10·06823	10·21637			9·78363	10·28461			9·71539	
44·0	9·93185			10·06816	10·21666			9·78334	10·28481			9·71519	
45·0	9·93192	·5	4	10·06808	10·21694	·5	14	9·78306	10·28502	·5	10	9·71498	15′
46·0	9·93200			10·06800	10·21723			9·78277	10·28523			9·71477	
47·0	9·93208			10·06793	10·21751			9·78249	10·28544			9·71456	
48·0	9·93215	·6	5	10·06785	10·21780	·6	17	9·78220	10·28565	·6	12	9·71435	
49·0	9·93223			10·06777	10·21808			9·78192	10·28586			9·71414	
50·0	9·93230			10·06770	10·21837			9·78163	10·28607			9·71394	10′
51·0	9·93238	·7	5	10·06762	10·21865	·7	20	9·78135	10·28627	·7	15	9·71373	
52·0	9·93246			10·06754	10·21894			9·78106	10·28648			9·71352	
53·0	9·93253			10·06747	10·21923			9·78078	10·28669			9·71331	
54·0	9·93261	·8	6	10·06739	10·21951	·8	23	9·78049	10·28690	·8	17	9·71310	
55·0	9·93269			10·06732	10·21980			9·78020	10·28711			9·71289	5′
56·0	9·93276			10·06724	10·22008			9·77992	10·28732			9·71268	
57·0	9·93284	·9	7	10·06716	10·22037	·9	26	9·77963	10·28753	·9	19	9·71247	
58·0	9·93291			10·06709	10·22065			9·77935	10·28774			9·71226	
59·0	9·93299			10·06701	10·22094			9·77906	10·28795			9·71205	
60·0	9·93307			10·06693	10·22123			9·77877	10·28816			9·71184	0′

59°
239°

′	Sine	Parts		Cosec.	Tan.	Parts		Cotan.	Secant	Parts		Cosine	
00·0	9·93307	′		10·06693	10·22123	′		9·77877	10·28816	′		9·71184	60′
01·0	9·93314			10·06686	10·22151			9·77849	10·28837			9·71163	
02·0	9·93322	·1	1	10·06678	10·22180	·1	3	9·77820	10·28858	·1	2	9·71142	
03·0	9·93329			10·06671	10·22209			9·77792	10·28879			9·71121	
04·0	9·93337			10·06663	10·22237			9·77763	10·28900			9·71100	
05·0	9·93345	·2	2	10·06656	10·22266	·2	6	9·77734	10·28921	·2	4	9·71079	55′
06·0	9·93352			10·06648	10·22295			9·77706	10·28943			9·71058	
07·0	9·93360			10·06640	10·22323			9·77677	10·28964			9·71036	
08·0	9·93367	·3	2	10·06633	10·22352	·3	9	9·77648	10·28985	·3	6	9·71015	
09·0	9·93375			10·06625	10·22381			9·77620	10·29006			9·70994	
10·0	9·93382			10·06618	10·22409			9·77591	10·29027			9·70973	50′
11·0	9·93390	·4	3	10·06610	10·22438	·4	11	9·77562	10·29048	·4	8	9·70952	
12·0	9·93397			10·06603	10·22467			9·77533	10·29069			9·70931	
13·0	9·93405			10·06595	10·22495			9·77505	10·29091			9·70909	
14·0	9·93412	·5	4	10·06588	10·22524	·5	14	9·77476	10·29112	·5	11	9·70888	
15·0	9·93420			10·06580	10·22553			9·77447	10·29133			9·70867	45′
16·0	9·93427			10·06573	10·22582			9·77418	10·29154			9·70846	
17·0	9·93435	·6	5	10·06565	10·22610	·6	17	9·77390	10·29176	·6	13	9·70825	
18·0	9·93442			10·06558	10·22639			9·77361	10·29197			9·70803	
19·0	9·93450			10·06550	10·22668			9·77332	10·29218			9·70782	
20·0	9·93457			10·06543	10·22697			9·77303	10·29239			9·70761	40′
21·0	9·93465	·7	5	10·06535	10·22726	·7	20	9·77275	10·29261	·7	15	9·70739	
22·0	9·93472			10·06528	10·22754			9·77246	10·29282			9·70718	
23·0	9·93480			10·06520	10·22783			9·77217	10·29303			9·70697	
24·0	9·93487	·8	6	10·06513	10·22812	·8	23	9·77188	10·29325	·8	17	9·70675	
25·0	9·93495			10·06505	10·22841			9·77159	10·29346			9·70654	35′
26·0	9·93502			10·06498	10·22870			9·77130	10·29367			9·70633	
27·0	9·93510	·9	7	10·06490	10·22899	·9	26	9·77102	10·29389	·9	19	9·70611	
28·0	9·93517			10·06483	10·22927			9·77073	10·29410			9·70590	
29·0	9·93524			10·06475	10·22956			9·77044	10·29432			9·70568	
30·0	9·93532	′		10·06468	10·22985			9·77015	10·29453	′		9·70547	30′
31·0	9·93540			10·06461	10·23014			9·76986	10·29475			9·70525	
32·0	9·93547	·1	1	10·06453	10·23043	·1	3	9·76957	10·29496	·1	2	9·70504	
33·0	9·93554			10·06446	10·23072			9·76928	10·29518			9·70483	
34·0	9·93562			10·06438	10·23101			9·76899	10·29539			9·70461	
35·0	9·93569	·2	1	10·06431	10·23130	·2	6	9·76870	10·29561	·2	4	9·70440	25′
36·0	9·93577			10·06423	10·23159			9·76841	10·29582			9·70418	
37·0	9·93584			10·06416	10·23188			9·76812	10·29604			9·70396	
38·0	9·93591	·3	2	10·06409	10·23217	·3	9	9·76783	10·29625	·3	7	9·70375	
39·0	9·93599			10·06401	10·23246			9·76755	10·29647			9·70353	
40·0	9·93606			10·06394	10·23275			9·76726	10·29668			9·70332	20′
41·0	9·93614	·4	3	10·06386	10·23304	·4	12	9·76697	10·29690	·4	9	9·70310	
42·0	9·93621			10·06379	10·23333			9·76668	10·29712			9·70289	
43·0	9·93628			10·06372	10·23362			9·76639	10·29733			9·70267	
44·0	9·93636	·5	4	10·06364	10·23391	·5	15	9·76610	10·29755	·5	11	9·70245	
45·0	9·93643			10·06357	10·23420			9·76581	10·29776			9·70224	15′
46·0	9·93651			10·06350	10·23449			9·76551	10·29798			9·70202	
47·0	9·93658	·6	4	10·06342	10·23478	·6	17	9·76522	10·29820	·6	13	9·70180	
48·0	9·93665			10·06335	10·23507			9·76493	10·29842			9·70159	
49·0	9·93673			10·06328	10·23536			9·76464	10·29863			9·70137	
50·0	9·93680			10·06320	10·23565			9·76435	10·29885			9·70115	10′
51·0	9·93687	·7	5	10·06313	10·23594	·7	20	9·76406	10·29907	·7	15	9·70093	
52·0	9·93695			10·06305	10·23623			9·76377	10·29928			9·70072	
53·0	9·93702			10·06298	10·23652			9·76348	10·29950			9·70050	
54·0	9·93709	·8	6	10·06291	10·23681	·8	23	9·76319	10·29972	·8	17	9·70028	
55·0	9·93717			10·06284	10·23710			9·76290	10·29994			9·70006	5′
56·0	9·93724			10·06276	10·23739			9·76261	10·30016			9·69984	
57·0	9·93731	·9	7	10·06269	10·23769	·9	26	9·76231	10·30037	·9	20	9·69963	
58·0	9·93739			10·06262	10·23798			9·76202	10·30059			9·69941	
59·0	9·93746			10·06254	10·23827			9·76173	10·30081			9·69919	
60·0	9·93753			10·06247	10·23856			9·76144	10·30103			9·69897	0′

60°
240°

LOGS. OF TRIG. FUNCTIONS

′	Sine	Parts		Cosec.	Tan.	Parts		Cotan.	Secant	Parts		Cosine	
00·0	9·93753	′		10·06247	10·23856	′		9·76144	10·30103	′		9·69897	60′
01·0	9·93760			10·06240	10·23885			9·76115	10·30125			9·69875	
02·0	9·93768	·1	1	10·06232	10·23914	·1	3	9·76086	10·30147	·1	2	9·69853	
03·0	9·93775			10·06225	10·23944			9·76056	10·30169			9·69831	
04·0	9·93782			10·06218	10·23973			9·76027	10·30191			9·69809	
05·0	9·93790	·2	1	10·06211	10·24002	·2	6	9·75998	10·30213	·2	4	9·69787	55′
06·0	9·93797			10·06203	10·24031			9·75969	10·30235			9·69765	
07·0	9·93804			10·06196	10·24061			9·75940	10·30257			9·69744	
08·0	9·93811	·3	2	10·06189	10·24090	·3	9	9·75910	10·30279	·3	7	9·69722	
09·0	9·93819			10·06182	10·24119			9·75881	10·30301			9·69700	
10·0	9·93826			10·06174	10·24148			9·75852	10·30323			9·69678	50′
11·0	9·93833	·4	3	10·06167	10·24178	·4	12	9·75822	10·30345	·4	9	9·69655	
12·0	9·93840			10·06160	10·24207			9·75793	10·30367			9·69633	
13·0	9·93848			10·06153	10·24236			9·75764	10·30389			9·69611	
14·0	9·93855			10·06145	10·24266			9·75735	10·30411			9·69589	
15·0	9·93862	·5	4	10·06138	10·24295	·5	15	9·75705	10·30433	·5	11	9·69567	45′
16·0	9·93869			10·06131	10·24324			9·75676	10·30455			9·69545	
17·0	9·93876			10·06124	10·24354			9·75647	10·30477			9·69523	
18·0	9·93884	·6	4	10·06116	10·24383	·6	18	9·75617	10·30499	·6	13	9·69501	
19·0	9·93891			10·06109	10·24412			9·75588	10·30521			9·69479	
20·0	9·93898			10·06102	10·24442			9·75559	10·30544			9·69456	40′
21·0	9·93905	·7	5	10·06095	10·24471	·7	21	9·75529	10·30566	·7	15	9·69434	
22·0	9·93912			10·06088	10·24500			9·75500	10·30588			9·69412	
23·0	9·93920			10·06081	10·24530			9·75470	10·30610			9·69390	
24·0	9·93927	·8	6	10·06073	10·24559	·8	23	9·75441	10·30632	·8	18	9·69368	
25·0	9·93934			10·06066	10·24589			9·75412	10·30655			9·69345	35′
26·0	9·93941			10·06059	10·24618			9·75382	10·30677			9·69323	
27·0	9·93948	·9	7	10·06052	10·24647	·9	26	9·75353	10·30699	·9	20	9·69301	
28·0	9·93955			10·06045	10·24677			9·75323	10·30722			9·69279	
29·0	9·93963			10·06038	10·24706			9·75294	10·30744			9·69256	
30·0	9·93970			10·06030	10·24736			9·75264	10·30766			9·69234	30′
31·0	9·93977			10·06023	10·24765			9·75235	10·30789			9·69212	
32·0	9·93984	·1	1	10·06016	10·24795	·1	3	9·75205	10·30811	·1	2	9·69189	
33·0	9·93991			10·06009	10·24824			9·75176	10·30833			9·69167	
34·0	9·93998			10·06002	10·24854			9·75146	10·30856			9·69144	
35·0	9·94005	·2	1	10·05995	10·24883	·2	6	9·75117	10·30878	·2	5	9·69122	25′
36·0	9·94013			10·05988	10·24913			9·75087	10·30900			9·69100	
37·0	9·94020			10·05980	10·24942			9·75058	10·30923			9·69077	
38·0	9·94027	·3	2	10·05973	10·24972	·3	9	9·75028	10·30945	·3	7	9·69055	
39·0	9·94034			10·05966	10·25002			9·74999	10·30968			9·69032	
40·0	9·94041			10·05959	10·25031			9·74969	10·30990			9·69010	20′
41·0	9·94048			10·05952	10·25061			9·74939	10·31013			9·68987	
42·0	9·94055	·4	3	10·05945	10·25090	·4	12	9·74910	10·31035	·4	9	9·68965	
43·0	9·94062			10·05938	10·25120			9·74880	10·31058			9·68942	
44·0	9·94069			10·05931	10·25150			9·74851	10·31080			9·68920	
45·0	9·94076	·5	4	10·05924	10·25179	·5	15	9·74821	10·31103	·5	11	9·68897	15′
46·0	9·94083			10·05917	10·25209			9·74791	10·31125			9·68875	
47·0	9·94091			10·05910	10·25238			9·74762	10·31148			9·68852	
48·0	9·94098	·6	4	10·05903	10·25268	·6	18	9·74732	10·31171	·6	14	9·68830	
49·0	9·94105			10·05895	10·25298			9·74702	10·31193			9·68807	
50·0	9·94112			10·05888	10·25327			9·74673	10·31216			9·68784	10′
51·0	9·94119	·7	5	10·05881	10·25357	·7	21	9·74643	10·31238	·7	16	9·68762	
52·0	9·94126			10·05874	10·25387			9·74613	10·31261			9·68739	
53·0	9·94133			10·05867	10·25417			9·74584	10·31284			9·68716	
54·0	9·94140	·8	6	10·05860	10·25446	·8	24	9·74554	10·31306	·8	18	9·68694	
55·0	9·94147			10·05853	10·25476			9·74524	10·31329			9·68671	5′
56·0	9·94154			10·05846	10·25506			9·74494	10·31352			9·68648	
57·0	9·94161	·9	6	10·05839	10·25536	·9	27	9·74465	10·31375	·9	20	9·68625	
58·0	9·94168			10·05832	10·25565			9·74435	10·31397			9·68603	
59·0	9·94175			10·05825	10·25595			9·74405	10·31420			9·68580	
60·0	9·94182			10·05818	10·25625			9·74375	10·31443			9·68557	0′

61°
241°

LOGS. OF TRIG. FUNCTIONS

′	Sine	Parts	Cosec.	Tan.	Parts	Cotan.	Secant	Parts	Cosine	
00·0	9·94182	′	10·05818	10·25625	′	9·74375	10·31443	′	9·68557	60′
01·0	9·94189		10·05811	10·25655		9·74345	10·31466		9·68534	
02·0	9·94196	·1 1	10·05804	10·25684	·1 3	9·74316	10·31489	·1 2	9·68512	
03·0	9·94203		10·05797	10·25714		9·74286	10·31511		9·68489	
04·0	9·94210		10·05790	10·25744		9·74256	10·31534		9·68466	
05·0	9·94217	·2 1	10·05783	10·25774	·2 6	9·74226	10·31557	·2 5	9·68443	55′
06·0	9·94224		10·05776	10·25804		9·74196	10·31580		9·68420	
07·0	9·94231		10·05769	10·25834		9·74166	10·31603		9·68397	
08·0	9·94238	·3 2	10·05762	10·25864	·3 9	9·74137	10·31626	·3 7	9·68374	
09·0	9·94245		10·05755	10·25893		9·74107	10·31649		9·68351	
10·0	9·94252		10·05748	10·25923		9·74077	10·31672		9·68328	50′
11·0	9·94259	·4 3	10·05741	10·25953	·4 12	9·74047	10·31695	·4 9	9·68306	
12·0	9·94266		10·05734	10·25983		9·74017	10·31718		9·68283	
13·0	9·94273		10·05727	10·26013		9·73987	10·31741		9·68260	
14·0	9·94280		10·05721	10·26043		9·73957	10·31764		9·68237	
15·0	9·94286	·5 3	10·05714	10·26073	·5 15	9·73927	10·31787	·5 12	9·68214	45′
16·0	9·94293		10·05707	10·26103		9·73897	10·31810		9·68191	
17·0	9·94300		10·05700	10·26133		9·73867	10·31833		9·68167	
18·0	9·94307	·6 4	10·05693	10·26163	·6 18	9·73837	10·31856	·6 14	9·68144	
19·0	9·94314		10·05686	10·26193		9·73807	10·31879		9·68121	
20·0	9·94321		10·05679	10·26223		9·73777	10·31902		9·68098	40′
21·0	9·94328	·7 5	10·05672	10·26253	·7 21	9·73747	10·31925	·7 16	9·68075	
22·0	9·94335		10·05665	10·26283		9·73717	10·31948		9·68052	
23·0	9·94342		10·05658	10·26313		9·73687	10·31971		9·68029	
24·0	9·94349	·8 6	10·05651	10·26343	·8 24	9·73657	10·31994	·8 18	9·68006	
25·0	9·94356		10·05645	10·26373		9·73627	10·32018		9·67982	35′
26·0	9·94362		10·05638	10·26403		9·73597	10·32041		9·67959	
27·0	9·94369	·9 6	10·05631	10·26433	·9 27	9·73567	10·32064	·9 21	9·67936	
28·0	9·94376		10·05624	10·26463		9·73537	10·32087		9·67913	
29·0	9·94383		10·05617	10·26493		9·73507	10·32111		9·67890	
30·0	9·94390		10·05610	10·26524		9·73476	10·32134		9·67866	30′
31·0	9·94397		10·05603	10·26554		9·73446	10·32157		9·67843	
32·0	9·94404	·1 1	10·05596	10·26584	·1 3	9·73416	10·32180	·1 2	9·67820	
33·0	9·94410		10·05590	10·26614		9·73386	10·32204		9·67796	
34·0	9·94417		10·05583	10·26644		9·73356	10·32227		9·67773	
35·0	9·94424	·2 1	10·05576	10·26674	·2 6	9·73326	10·32250	·2 5	9·67750	25′
36·0	9·94431		10·05569	10·26705		9·73296	10·32274		9·67726	
37·0	9·94438		10·05562	10·26735		9·73265	10·32297		9·67703	
38·0	9·94445	·3 2	10·05555	10·26765	·3 9	9·73235	10·32320	·3 7	9·67680	
39·0	9·94451		10·05549	10·26795		9·73205	10·32344		9·67656	
40·0	9·94458		10·05542	10·26825		9·73175	10·32367		9·67633	20′
41·0	9·94465	·4 3	10·05535	10·26856	·4 12	9·73144	10·32391	·4 9	9·67609	
42·0	9·94472		10·05528	10·26886		9·73114	10·32414		9·67586	
43·0	9·94479		10·05521	10·26916		9·73084	10·32438		9·67562	
44·0	9·94485		10·05515	10·26947		9·73054	10·32461		9·67539	
45·0	9·94492	·5 3	10·05508	10·26977	·5 15	9·73023	10·32485	·5 12	9·67516	15′
46·0	9·94499		10·05501	10·27007		9·72993	10·32508		9·67492	
47·0	9·94506		10·05494	10·27037		9·72963	10·32532		9·67468	
48·0	9·94513	·6 4	10·05488	10·27068	·6 18	9·72932	10·32555	·6 14	9·67445	
49·0	9·94519		10·05481	10·27098		9·72902	10·32579		9·67421	
50·0	9·94526		10·05474	10·27128		9·72872	10·32602		9·67398	10′
51·0	9·94533	·7 5	10·05467	10·27159	·7 21	9·72841	10·32626	·7 16	9·67374	
52·0	9·94540		10·05460	10·27189		9·72811	10·32650		9·67351	
53·0	9·94546		10·05454	10·27220		9·72781	10·32673		9·67327	
54·0	9·94553	·8 5	10·05447	10·27250	·8 24	9·72750	10·32697	·8 19	9·67303	
55·0	9·94560		10·05440	10·27290		9·72720	10·32721		9·67280	5′
56·0	9·94567		10·05433	10·27311		9·72689	10·32744		9·67256	
57·0	9·94573	·9 6	10·05427	10·27341	·9 27	9·72659	10·32768	·9 21	9·67232	
58·0	9·94580		10·05420	10·27372		9·72628	10·32792		9·67208	
59·0	9·94587		10·05413	10·27402		9·72598	10·32815		9·67185	
60·0	9·94593		10·05407	10·27433		9·72567	10·32839		9·67161	0′

118°
298°

62°
242°

LOGS. OF TRIG. FUNCTIONS

′	Sine	Parts	Cosec.	Tan.	Parts	Cotan.	Secant	Parts	Cosine	
00·0	9·94593	′	10·05407	10·27433	′	9·72567	10·32839	′	9·67161	60′
01·0	9·94600		10·05400	10·27463		9·72537	10·32863		9·67137	
02·0	9·94607	·1 1	10·05393	10·27494	·1 3	9·72507	10·32887	·1 2	9·67113	
03·0	9·94614		10·05386	10·27524		9·72476	10·32910		9·67090	
04·0	9·94620		10·05380	10·27555		9·72445	10·32934		9·67066	
05·0	9·94627	·2 1	10·05373	10·27585	·2 6	9·72415	10·32958	·2 5	9·67042	55′
06·0	9·94634		10·05366	10·27616		9·72384	10·32982		9·67018	
07·0	9·94640		10·05360	10·27646		9·72354	10·33006		9·66994	
08·0	9·94647	·3 2	10·05353	10·27677	·3 9	9·72323	10·33030	·3 7	9·66970	
09·0	9·94654		10·05346	10·27707		9·72293	10·33054		9·66946	
10·0	9·94660		10·05340	10·27738		9·72262	10·33078		9·66923	50′
11·0	9·94667	·4 3	10·05333	10·27769	·4 12	9·72232	10·33101	·4 10	9·66899	
12·0	9·94674		10·05326	10·27799		9·72201	10·33125		9·66875	
13·0	9·94680		10·05320	10·27830		9·72170	10·33149		9·66851	
14·0	9·94687		10·05313	10·27860		9·72140	10·33173		9·66827	
15·0	9·94694	·5 3	10·05306	10·27891	·5 15	9·72109	10·33197	·5 12	9·66803	45′
16·0	9·94700		10·05300	10·27922		9·72078	10·33221		9·66779	
17·0	9·94707		10·05293	10·27952		9·72048	10·33245		9·66755	
18·0	9·94714	·6 4	10·05286	10·27983	·6 18	9·72017	10·33270	·6 14	9·66731	
19·0	9·94720		10·05280	10·28014		9·71986	10·33294		9·66707	
20·0	9·94727		10·05273	10·28045		9·71956	10·33318		9·66682	40′
21·0	9·94734	·7 5	10·05267	10·28075	·7 21	9·71925	10·33342	·7 17	9·66658	
22·0	9·94740		10·05260	10·28106		9·71894	10·33366		9·66634	
23·0	9·94747		10·05253	10·28137		9·71863	10·33390		9·66610	
24·0	9·94753	·8 5	10·05247	10·28168	·8 25	9·71833	10·33414	·8 19	9·66586	
25·0	9·94760		10·05240	10·28198		9·71802	10·33438		9·66562	35′
26·0	9·94767		10·05234	10·28229		9·71771	10·33463		9·66538	
27·0	9·94773	·9 6	10·05227	10·28260	·9 28	9·71740	10·33487	·9 22	9·66513	
28·0	9·94780		10·05220	10·28291		9·71709	10·33511		9·66489	
29·0	9·94786		10·05214	10·28322		9·71679	10·33535		9·66465	
30·0	9·94793		10·05207	10·28352		9·71648	10·33559		9·66441	30′
31·0	9·94800		10·05201	10·28383		9·71617	10·33584		9·66416	
32·0	9·94806	·1 1	10·05194	10·28414	·1 3	9·71586	10·33608	·1 2	9·66392	
33·0	9·94813		10·05187	10·28445		9·71555	10·33632		9·66368	
34·0	9·94819		10·05181	10·28476		9·71524	10·33657		9·66343	
35·0	9·94826	·2 1	10·05174	10·28507	·2 6	9·71493	10·33681	·2 5	9·66319	25′
36·0	9·94832		10·05168	10·28538		9·71462	10·33705		9·66295	
37·0	9·94839		10·05161	10·28569		9·71431	10·33730		9·66270	
38·0	9·94845	·3 2	10·05155	10·28600	·3 9	9·71401	10·33754	·3 7	9·66246	
39·0	9·94852		10·05148	10·28630		9·71370	10·33779		9·66221	
40·0	9·94858		10·05142	10·28661		9·71339	10·33803		9·66197	20′
41·0	9·94865	·4 3	10·05135	10·28692	·4 12	9·71308	10·33827	·4 10	9·66173	
42·0	9·94872		10·05129	10·28723		9·71277	10·33852		9·66148	
43·0	9·94878		10·05122	10·28754		9·71246	10·33876		9·66124	
44·0	9·94885		10·05116	10·28785		9·71215	10·33901		9·66099	
45·0	9·94891	·5 3	10·05109	10·28816	·5 16	9·71184	10·33925	·5 12	9·66075	15′
46·0	9·94898		10·05103	10·28848		9·71153	10·33950		9·66050	
47·0	9·94904		10·05096	10·28879		9·71122	10·33975		9·66026	
48·0	9·94911	·6 4	10·05090	10·28910	·6 19	9·71090	10·33999	·6 15	9·66001	
49·0	9·94917		10·05083	10·28941		9·71059	10·34024		9·65976	
50·0	9·94924		10·05077	10·28972		9·71028	10·34048		9·65952	10′
51·0	9·94930	·7 5	10·05070	10·29003	·7 22	9·70997	10·34073	·7 17	9·65927	
52·0	9·94936		10·05064	10·29034		9·70966	10·34098		9·65903	
53·0	9·94943		10·05057	10·29065		9·70935	10·34122		9·65878	
54·0	9·94949	·8 5	10·05051	10·29096	·8 25	9·70904	10·34147	·8 20	9·65853	
55·0	9·94956		10·05044	10·29127		9·70873	10·34172		9·65828	5′
56·0	9·94962		10·05038	10·29159		9·70841	10·34196		9·65804	
57·0	9·94969	·9 6	10·05031	10·29190	·9 28	9·70810	10·34221	·9 22	9·65779	
58·0	9·94975		10·05025	10·29221		9·70779	10·34246		9·65754	
59·0	9·94982		10·05018	10·29252		9·70748	10·34271		9·65730	
60·0	9·94988		10·05012	10·29283		9·70717	10·34295		9·65705	0′

117°
297°

63°
243°

LOGS. OF TRIG. FUNCTIONS

′	Sine	Parts		Cosec.	Tan.	Parts		Cotan.	Secant	Parts		Cosine	
00·0	9·94988	′		10·05012	10·29283	′		9·70717	10·34295	′		9·65705	60′
01·0	9·94995			10·05006	10·29315			9·70685	10·34320			9·65680	
02·0	9·95001	·1	1	10·04999	10·29346	·1	3	9·70654	10·34345	·1	2	9·65655	
03·0	9·95007			10·04993	10·29377			9·70623	10·34370			9·65630	
04·0	9·95014			10·04986	10·29408			9·70592	10·34395			9·65605	
05·0	9·95020	·2	1	10·04980	10·29440	·2	6	9·70560	10·34420	·2	5	9·65581	55′
06·0	9·95027			10·04973	10·29471			9·70529	10·34444			9·65556	
07·0	9·95033			10·04967	10·29502			9·70498	10·34469			9·65531	
08·0	9·95039	·3	2	10·04961	10·29534	·3	9	9·70466	10·34494	·3	7	9·65506	
09·0	9·95046			10·04954	10·29565			9·70435	10·34519			9·65481	
10·0	9·95052			10·04948	10·29596			9·70404	10·34544			9·65456	50′
11·0	9·95059	·4	3	10·04941	10·29628	·4	13	9·70372	10·34569	·4	10	9·65431	
12·0	9·95065			10·04935	10·29659			9·70341	10·34594			9·65406	
13·0	9·95071			10·04929	10·29691			9·70310	10·34619			9·65381	
14·0	9·95078			10·04922	10·29722			9·70278	10·34644			9·65356	
15·0	9·95084	·5	3	10·04916	10·29753	·5	16	9·70247	10·34669	·5	12	9·65331	45′
16·0	9·95091			10·04910	10·29785			9·70215	10·34694			9·65306	
17·0	9·95097			10·04903	10·29816			9·70184	10·34719			9·65281	
18·0	9·95103	·6	4	10·04897	10·29848	·6	19	9·70152	10·34745	·6	15	9·65256	
19·0	9·95110			10·04890	10·29879			9·70121	10·34770			9·65230	
20·0	9·95116			10·04884	10·29911			9·70089	10·34795			9·65205	40′
21·0	9·95122	·7	4	10·04878	10·29942	·7	22	9·70058	10·34820	·7	17	9·65180	
22·0	9·95129			10·04871	10·29974			9·70026	10·34845			9·65155	
23·0	9·95135			10·04865	10·30005			9·69995	10·34870			9·65130	
24·0	9·95141	·8	5	10·04859	10·30037	·8	25	9·69963	10·34896	·8	20	9·65104	
25·0	9·95148			10·04852	10·30068			9·69932	10·34921			9·65079	35′
26·0	9·95154			10·04846	10·30100			9·69900	10·34946			9·65054	
27·0	9·95160	·9	6	10·04840	10·30132	·9	28	9·69868	10·34971	·9	22	9·65029	
28·0	9·95167			10·04834	10·30163			9·69837	10·34997			9·65003	
29·0	9·95173			10·04827	10·30195			9·69805	10·35022			9·64978	
30·0	9·95179			10·04821	10·30226			9·69774	10·35047			9·64953	30′
31·0	9·95185			10·04815	10·30258			9·69742	10·35073			9·64927	
32·0	9·95192	·1	1	10·04808	10·30290	·1	3	9·69710	10·35098	·1	3	9·64902	
33·0	9·95198			10·04802	10·30321			9·69679	10·35123			9·64877	
34·0	9·95204			10·04796	10·30353			9·69647	10·35149			9·64851	
35·0	9·95211	·2	1	10·04789	10·30385	·2	6	9·69615	10·35174	·2	5	9·64826	25′
36·0	9·95217			10·04783	10·30416			9·69584	10·35200			9·64800	
37·0	9·95223			10·04777	10·30448			9·69552	10·35225			9·64775	
38·0	9·95229	·3	2	10·04771	10·30480	·3	10	9·69520	10·35251	·3	8	9·64749	
39·0	9·95236			10·04764	10·30511			9·69488	10·35276			9·64724	
40·0	9·95242			10·04758	10·30543			9·69457	10·35302			9·64698	20′
41·0	9·95248	·4	2	10·04752	10·30575	·4	13	9·69425	10·35327	·4	10	9·64673	
42·0	9·95254			10·04746	10·30607			9·69393	10·35353			9·64647	
43·0	9·95261			10·04739	10·30639			9·69361	10·35378			9·64622	
44·0	9·95267			10·04733	10·30671			9·69329	10·35404			9·64596	
45·0	9·95273	·5	3	10·04727	10·30703	·5	16	9·69298	10·35429	·5	13	9·64571	15′
46·0	9·95279			10·04721	10·30734			9·69266	10·35455			9·64545	
47·0	9·95286			10·04715	10·30766			9·69234	10·35481			9·64519	
48·0	9·95292	·6	4	10·04708	10·30798	·6	19	9·69202	10·35506	·6	15	9·64494	
49·0	9·95298			10·04702	10·30830			9·69170	10·35532			9·64468	
50·0	9·95304			10·04696	10·30862			9·69138	10·35558			9·64442	10′
51·0	9·95310	·7	4	10·04690	10·30894	·7	22	9·69106	10·35584	·7	18	9·64417	
52·0	9·95317			10·04683	10·30926			9·69074	10·35609			9·64391	
53·0	9·95323			10·04677	10·30958			9·69042	10·35635			9·64365	
54·0	9·95329	·8	5	10·04671	10·30990	·8	25	9·69010	10·35661	·8	21	9·64339	
55·0	9·95335			10·04665	10·31022			9·68978	10·35687			9·64314	5′
56·0	9·95341			10·04659	10·31054			9·68946	10·35712			9·64288	
57·0	9·95348	·9	6	10·04653	10·31086	·9	29	9·68914	10·35738	·9	23	9·64262	
58·0	9·95354			10·04646	10·31118			9·68882	10·35764			9·64236	
59·0	9·95360			10·04640	10·31150			9·68850	10·35790			9·64210	
60·0	9·95366			10·04634	10·31182			9·68818	10·35816			9·64184	0′

64°
244°

LOGS. OF TRIG. FUNCTIONS

′	Sine	Parts		Cosec.	Tan.	Parts		Cotan.	Secant	Parts		Cosine	
00·0	9·95366	′		10·04634	10·31182	′		9·68818	10·35816	′		9·64184	60′
01·0	9·95372			10·04628	10·31214			9·68786	10·35842			9·64158	
02·0	9·95378	·1	1	10·04622	10·31246	·1	3	9·68754	10·35868	·1	3	9·64132	
03·0	9·95385			10·04616	10·31278			9·68722	10·35894			9·64106	
04·0	9·95391			10·04609	10·31310			9·68690	10·35920			9·64080	
05·0	9·95397	·2	1	10·04603	10·31342	·2	6	9·68658	10·35946	·2	5	9·64054	55′
06·0	9·95403			10·04597	10·31375			9·68626	10·35972			9·64028	
07·0	9·95409			10·04591	10·31407			9·68593	10·35998			9·64002	
08·0	9·95415	·3	2	10·04585	10·31439	·3	10	9·68561	10·36024	·3	8	9·63976	
09·0	9·95421			10·04579	10·31471			9·68529	10·36050			9·63950	
10·0	9·95427			10·04573	10·31503			9·68497	10·36076			9·63924	50′
11·0	9·95434	·4	2	10·04567	10·31535	·4	13	9·68465	10·36102	·4	10	9·63898	
12·0	9·95440			10·04560	10·31568			9·68432	10·36128			9·63872	
13·0	9·95446			10·04554	10·31600			9·68400	10·36154			9·63846	
14·0	9·95452	·5	3	10·04548	10·31632	·5	16	9·68368	10·36180	·5	13	9·63820	
15·0	9·95458			10·04542	10·31664			9·68336	10·36207			9·63794	45′
16·0	9·95464			10·04536	10·31697			9·68303	10·36233			9·63767	
17·0	9·95470			10·04530	10·31729			9·68271	10·36259			9·63741	
18·0	9·95476	·6	4	10·04524	10·31761	·6	19	9·68239	10·36285	·6	16	9·63715	
19·0	9·95482			10·04518	10·31794			9·68206	10·36311			9·63689	
20·0	9·95488	·7	4	10·04512	10·31826	·7	23	9·68174	10·36338	·7	18	9·63662	40′
21·0	9·95494			10·04506	10·31858			9·68142	10·36364			9·63636	
22·0	9·95501			10·04500	10·31891			9·68109	10·36390			9·63610	
23·0	9·95507			10·04494	10·31923			9·68077	10·36417			9·63583	
24·0	9·95513	·8	5	10·04487	10·31956	·8	26	9·68044	10·36443	·8	21	9·63557	
25·0	9·95519			10·04481	10·31988			9·68012	10·36469			9·63531	35′
26·0	9·95525			10·04475	10·32021			9·67980	10·36496			9·63504	
27·0	9·95531	·9	5	10·04469	10·32053	·9	29	9·67947	10·36522	·9	24	9·63478	
28·0	9·95537			10·04463	10·32085			9·67915	10·36549			9·63451	
29·0	9·95543			10·04457	10·32118			9·67882	10·36575			9·63425	
30·0	9·95549			10·04451	10·32150			9·67850	10·36602			9·63398	30′
31·0	9·95555			10·04445	10·32183			9·67817	10·36628			9·63372	
32·0	9·95561	·1	1	10·04439	10·32215	·1	3	9·67785	10·36655	·1	3	9·63345	
33·0	9·95567			10·04433	10·32248			9·67752	10·36681			9·63319	
34·0	9·95573			10·04427	10·32281			9·67719	10·36708			9·63292	
35·0	9·95579	·2	1	10·04421	10·32313	·2	7	9·67687	10·36734	·2	5	9·63266	25′
36·0	9·95585			10·04415	10·32346			9·67654	10·36761			9·63239	
37·0	9·95591			10·04409	10·32378			9·67622	10·36788			9·63213	
38·0	9·95597	·3	2	10·04403	10·32411	·3	10	9·67589	10·36814	·3	8	9·63186	
39·0	9·95603			10·04397	10·32444			9·67556	10·36841			9·63159	
40·0	9·95609			10·04391	10·32476			9·67524	10·36867			9·63133	20′
41·0	9·95615			10·04385	10·32509			9·67491	10·36894			9·63106	
42·0	9·95621	·4	2	10·04379	10·32542	·4	13	9·67458	10·36921	·4	11	9·63079	
43·0	9·95627			10·04373	10·32574			9·67426	10·36948			9·63052	
44·0	9·95633			10·04367	10·32607			9·67393	10·36974			9·63026	
45·0	9·95639	·5	3	10·04361	10·32640	·5	16	9·67360	10·37001	·5	13	9·62999	15′
46·0	9·95645			10·04355	10·32673			9·67327	10·37028			9·62972	
47·0	9·95651			10·04349	10·32705			9·67295	10·37055			9·62945	
48·0	9·95657	·6	4	10·04343	10·32738	·6	20	9·67262	10·37082	·6	16	9·62919	
49·0	9·95663			10·04338	10·32771			9·67229	10·37108			9·62892	
50·0	9·95668			10·04332	10·32804			9·67196	10·37135			9·62865	10′
51·0	9·95674	·7	4	10·04326	10·32837	·7	23	9·67164	10·37162	·7	19	9·62838	
52·0	9·95680			10·04320	10·32869			9·67131	10·37189			9·62811	
53·0	9·95686			10·04314	10·32902			9·67098	10·37216			9·62784	
54·0	9·95692	·8	5	10·04308	10·32935	·8	26	9·67065	10·37243	·8	21	9·62757	
55·0	9·95698			10·04302	10·32968			9·67032	10·37270			9·62730	5′
56·0	9·95704			10·04296	10·33001			9·66999	10·37297			9·62703	
57·0	9·95710	·9	5	10·04290	10·33034	·9	29	9·66966	10·37324	·9	24	9·62676	
58·0	9·95716			10·04284	10·33067			9·66933	10·37351			9·62649	
59·0	9·95722			10·04278	10·33100			9·66900	10·37378			9·62622	
60·0	9·95728			10·04272	10·33133			9·66867	10·37405			9·62595	0′

65°
245°

LOGS. OF TRIG. FUNCTIONS

′	Sine	Parts	Cosec.	Tan.	Parts	Cotan.	Secant	Parts	Cosine	
00·0	9·95728	′	10·04272	10·33133	′	9·66867	10·37405	′	9·62595	60′
01·0	9·95734		10·04267	10·33166		9·66834	10·37432		9·62568	
02·0	9·95739	·1 1	10·04261	10·33199	·1 3	9·66801	10·37459	·1 3	9·62541	
03·0	9·95745		10·04255	10·33232		9·66760	10·37487		9·62514	
04·0	9·95751		10·04249	10·33265		9·66735	10·37514		9·62486	
05·0	9·95757	·2 1	10·04243	10·33298	·2 7	9·66702	10·37541	·2 5	9·62459	55′
06·0	9·95763		10·04237	10·33331		9·66669	10·37568		9·62432	
07·0	9·95769		10·04231	10·33364		9·66636	10·37595		9·62405	
08·0	9·95775	·3 2	10·04225	10·33397	·3 10	9·66603	10·37623	·3 8	9·62377	
09·0	9·95780		10·04220	10·33430		9·66570	10·37650		9·62350	
10·0	9·95786		10·04214	10·33463		9·66537	10·37677		9·62323	50′
11·0	9·95792	·4 2	10·04208	10·33497	·4 13	9·66504	10·37704	·4 11	9·62296	
12·0	9·95798		10·04202	10·33530		9·66470	10·37732		9·62268	
13·0	9·95804		10·04196	10·33563		9·66437	10·37759		9·62241	
14·0	9·95810	·5 3	10·04190	10·33596	·5 17	9·66404	10·37787	·5 14	9·62214	
15·0	9·95815		10·04185	10·33629		9·66371	10·37814		9·62186	45′
16·0	9·95821		10·04179	10·33663		9·66338	10·37841		9·62159	
17·0	9·95827		10·04173	10·33696		9·66304	10·37869		9·62131	
18·0	9·95833	·6 3	10·04167	10·33729	·6 20	9·66271	10·37896	·6 16	9·62104	
19·0	9·95839		10·04161	10·33762		9·66238	10·37924		9·62076	
20·0	9·95845		10·04156	10·33796		9·66204	10·37951		9·62049	40′
21·0	9·95850	·7 4	10·04150	10·33829	·7 23	9·66171	10·37979	·7 19	9·62021	
22·0	9·95856		10·04144	10·33862		9·66138	10·38006		9·61994	
23·0	9·95862		10·04138	10·33896		9·66104	10·38034		9·61966	
24·0	9·95868	·8 5	10·04132	10·33929	·8 27	9·66071	10·38061	·8 22	9·61939	
25·0	9·95873		10·04127	10·33962		9·66038	10·38089		9·61911	35′
26·0	9·95879		10·04121	10·33996		9·66004	10·38117		9·61883	
27·0	9·95885	·9 5	10·04115	10·34029	·9 30	9·65971	10·38144	·9 25	9·61856	
28·0	9·95891		10·04109	10·34063		9·65937	10·38172		9·61828	
29·0	9·95897		10·04104	10·34096		9·65904	10·38200		9·61800	
30·0	9·95902		10·04098	10·34130		9·65870	10·38227		9·61773	30′
31·0	9·95908		10·04092	10·34163		9·65837	10·38255		9·61745	
32·0	9·95914	·1 1	10·04086	10·34197	·1 3	9·65803	10·38283	·1 3	9·61717	
33·0	9·95920		10·04081	10·34230		9·65770	10·38311		9·61689	
34·0	9·95925		10·04075	10·34264		9·65736	10·38338		9·61662	
35·0	9·95931	·2 1	10·04069	10·34297	·2 7	9·65703	10·38366	·2 6	9·61634	25′
36·0	9·95937		10·04063	10·34331		9·65669	10·38394		9·61606	
37·0	9·95943		10·04058	10·34364		9·65636	10·38422		9·61578	
38·0	9·95948	·3 2	10·04052	10·34398	·3 10	9·65602	10·38450	·3 8	9·61550	
39·0	9·95954		10·04046	10·34432		9·65568	10·38478		9·61522	
40·0	9·95960		10·04040	10·34465		9·65535	10·38506		9·61494	20′
41·0	9·95965		10·04035	10·34499		9·65501	10·38534		9·61467	
42·0	9·95971	·4 2	10·04029	10·34533	·4 13	9·65467	10·38562	·4 11	9·61439	
43·0	9·95977		10·04023	10·34566		9·65434	10·38590		9·61411	
44·0	9·95983		10·04018	10·34600		9·65400	10·38618		9·61383	
45·0	9·95988	·5 3	10·04012	10·34634	·5 17	9·65366	10·38646	·5 14	9·61355	15′
46·0	9·95994		10·04006	10·34667		9·65333	10·38674		9·61326	
47·0	9·96000		10·04001	10·34701		9·65299	10·38702		9·61298	
48·0	9·96005	·6 3	10·03995	10·34735	·6 20	9·65265	10·38730	·6 17	9·61270	
49·0	9·96011		10·03989	10·34769		9·65231	10·38758		9·61242	
50·0	9·96017		10·03984	10·34803		9·65197	10·38786		9·61214	10′
51·0	9·96022	·7 4	10·03978	10·34836	·7 24	9·65164	10·38814	·7 20	9·61186	
52·0	9·96028		10·03972	10·34870		9·65130	10·38842		9·61158	
53·0	9·96034		10·03967	10·34904		9·65096	10·38871		9·61129	
54·0	9·96039	·8 5	10·03961	10·34938	·8 27	9·65062	10·38899	·8 22	9·61101	
55·0	9·96045		10·03955	10·34972		9·65028	10·38927		9·61073	5′
56·0	9·96051		10·03950	10·35006		9·64994	10·38955		9·61045	
57·0	9·96056	·9 5	10·03944	10·35040	·9 30	9·64960	10·38984	·9 25	9·61016	
58·0	9·96062		10·03938	10·35074		9·64926	10·39012		9·60988	
59·0	9·96067		10·03933	10·35108		9·64892	10·39040		9·60960	
60·0	9·96073		10·03927	10·35142		9·64858	10·39069		9·60931	0′

66°
246°

LOGS. OF TRIG. FUNCTIONS

′	Sine	Parts		Cosec.	Tan.	Parts		Cotan.	Secant	Parts		Cosine	
00·0	9·96073	′		10·03927	10·35142	′		9·64858	10·39069	′		9·60931	60′
01·0	9·96079			10·03921	10·35176			9·64824	10·39097			9·60903	
02·0	9·96084	·1	1	10·03916	10·35210	·1	3	9·64790	10·39126	·1	3	9·60875	
03·0	9·96090			10·03910	10·35244			9·64756	10·39154			9·60846	
04·0	9·96096			10·03905	10·35278			9·64722	10·39182			9·60818	
05·0	9·96101	·2	1	10·03899	10·35312	·2	7	9·64688	10·39211	·2	6	9·60789	55′
06·0	9·96107			10·03893	10·35346			9·64654	10·39239			9·60761	
07·0	9·96112			10·03888	10·35380			9·64620	10·39268			9·60732	
08·0	9·96118	·3	2	10·03882	10·35414	·3	10	9·64586	10·39296	·3	9	9·60704	
09·0	9·96124			10·03877	10·35448			9·64552	10·39325			9·60675	
10·0	9·96129			10·03871	10·35483			9·64517	10·39354			9·60647	50′
11·0	9·96135	·4	2	10·03865	10·35517	·4	14	9·64483	10·39382	·4	11	9·60618	
12·0	9·96140			10·03860	10·35551			9·64449	10·39411			9·60589	
13·0	9·96146			10·03854	10·35585			9·64415	10·39439			9·60561	
14·0	9·96151	·5	3	10·03849	10·35619	·5	17	9·64381	10·39468	·5	14	9·60532	
15·0	9·96157			10·03843	10·35654			9·64346	10·39497			9·60503	45′
16·0	9·96162			10·03838	10·35688			9·64312	10·39526			9·60475	
17·0	9·96168			10·03832	10·35722			9·64278	10·39554			9·60446	
18·0	9·96174	·6	3	10·03827	10·35757	·6	21	9·64243	10·39583	·6	17	9·60417	
19·0	9·96179			10·03821	10·35791			9·64209	10·39612			9·60388	
20·0	9·96185	·7	4	10·03815	10·35825	·7	24	9·64174	10·39641	·7	20	9·60359	40′
21·0	9·96190			10·03810	10·35860			9·64140	10·39670			9·60331	
22·0	9·96196			10·03804	10·35894			9·64106	10·39698			9·60302	
23·0	9·96201			10·03799	10·35928			9·64072	10·39727			9·60273	
24·0	9·96207	·8	4	10·03793	10·35963	·8	27	9·64037	10·39756	·8	23	9·60244	
25·0	9·96212			10·03788	10·35997			9·64003	10·39785			9·60215	35′
26·0	9·96218			10·03782	10·36032			9·63968	10·39814			9·60186	
27·0	9·96223	·9	5	10·03777	10·36066	·9	31	9·63934	10·39843	·9	26	9·60157	
28·0	9·96229			10·03771	10·36101			9·63899	10·39872			9·60128	
29·0	9·96234			10·03766	10·36135			9·63865	10·39901			9·60099	
30·0	9·96240			10·03760	10·36170			9·63830	10·39930			9·60070	30′
31·0	9·96245			10·03755	10·36204			9·63796	10·39959			9·60041	
32·0	9·96251	·1	1	10·03749	10·36239	·1	3	9·63761	10·39988	·1	3	9·60012	
33·0	9·96256			10·03744	10·36274			9·63727	10·40017			9·59983	
34·0	9·96262			10·03738	10·36308			9·63692	10·40046			9·59954	
35·0	9·96267	·2	1	10·03733	10·36343	·2	7	9·63657	10·40076	·2	6	9·59924	25′
36·0	9·96273			10·03727	10·36377			9·63623	10·40105			9·59895	
37·0	9·96278			10·03722	10·36412			9·63588	10·40134			9·59866	
38·0	9·96284	·3	2	10·03716	10·36447	·3	10	9·63553	10·40163	·3	9	9·59837	
39·0	9·96289			10·03711	10·36482			9·63519	10·40193			9·59808	
40·0	9·96295			10·03706	10·36516			9·63484	10·40222			9·59778	20′
41·0	9·96300	·4	2	10·03700	10·36551	·4	14	9·63449	10·40251	·4	12	9·59749	
42·0	9·96305			10·03695	10·36586			9·63414	10·40280			9·59720	
43·0	9·96311			10·03689	10·36621			9·63380	10·40310			9·59690	
44·0	9·96316	·5	3	10·03684	10·36655	·5	17	9·63345	10·40339	·5	15	9·59661	
45·0	9·96322			10·03678	10·36690			9·63310	10·40369			9·59632	15′
46·0	9·96327			10·03673	10·36725			9·63275	10·40398			9·59602	
47·0	9·96333			10·03668	10·36760			9·63240	10·40427			9·59573	
48·0	9·96338	·6	3	10·03662	10·36795	·6	21	9·63205	10·40457	·6	18	9·59543	
49·0	9·96343			10·03657	10·36830			9·63170	10·40486			9·59514	
50·0	9·96349			10·03651	10·36865			9·63136	10·40516			9·59484	10′
51·0	9·96354	·7	4	10·03646	10·36900	·7	24	9·63101	10·40545	·7	21	9·59455	
52·0	9·96360			10·03640	10·36934			9·63066	10·40575			9·59425	
53·0	9·96365			10·03635	10·36969			9·63031	10·40605			9·59396	
54·0	9·96370	·8	4	10·03630	10·37004	·8	28	9·62996	10·40634	·8	24	9·59366	
55·0	9·96376			10·03624	10·37039			9·62961	10·40664			9·59336	5′
56·0	9·96381			10·03619	10·37075			9·62926	10·40693			9·59307	
57·0	9·96387	·9	5	10·03614	10·37110	·9	31	9·62891	10·40723	·9	26	9·59277	
58·0	9·96392			10·03608	10·37145			9·62855	10·40753			9·59247	
59·0	9·96397			10·03603	10·37180			9·62820	10·40782			9·59218	
60·0	9·96403			10·03597	10·37215			9·62785	10·40812			9·59188	0′

113°
293°

LOGS. OF TRIG. FUNCTIONS

67°
247°

′	Sine	Parts		Cosec.	Tan.	Parts		Cotan.	Secant	Parts		Cosine	
00·0	9·96403	′		10·03597	10·37215	′		9·62785	10·40812	′		9·59188	60′
01·0	9·96408			10·03592	10·37250			9·62750	10·40842			9·59158	
02·0	9·96413	·1	1	10·03587	10·37285	·1	4	9·62715	10·40872	·1	3	9·59128	
03·0	9·96419			10·03581	10·37320			9·62680	10·40902			9·59098	
04·0	9·96424			10·03576	10·37356			9·62645	10·40931			9·59069	
05·0	9·96429	·2	1	10·03571	10·37391	·2	7	9·62609	10·40961	·2	6	9·59039	55′
06·0	9·96435			10·03565	10·37426			9·62574	10·40991			9·59009	
07·0	9·96440			10·03560	10·37461			9·62539	10·41021			9·58979	
08·0	9·96445	·3	2	10·03555	10·37496	·3	11	9·62504	10·41051	·3	9	9·58949	
09·0	9·96451			10·03549	10·37532			9·62468	10·41081			9·58919	
10·0	9·96456			10·03544	10·37567			9·62433	10·41111			9·58889	50′
11·0	9·96461	·4	2	10·03539	10·37602	·4	14	9·62398	10·41141	·4	12	9·58859	
12·0	9·96467			10·03533	10·37638			9·62362	10·41171			9·58829	
13·0	9·96472			10·03528	10·37673			9·62327	10·41201			9·58799	
14·0	9·96477			10·03523	10·37709			9·62292	10·41231			9·58769	
15·0	9·96483	·5	3	10·03517	10·37744	·5	18	9·62256	10·41261	·5	15	9·58739	45′
16·0	9·96488			10·03512	10·37779			9·62221	10·41292			9·58709	
17·0	9·96493			10·03507	10·37815			9·62185	10·41322			9·58678	
18·0	9·96498	·6	3	10·03502	10·37850	·6	21	9·62150	10·41352	·6	18	9·58648	
19·0	9·96504			10·03496	10·37886			9·62114	10·41382			9·58618	
20·0	9·96509			10·03491	10·37921			9·62079	10·41412			9·58588	40′
21·0	9·96514	·7	4	10·03486	10·37957	·7	25	9·62043	10·41443	·7	21	9·58557	
22·0	9·96520			10·03481	10·37992			9·62008	10·41473			9·58527	
23·0	9·96525			10·03475	10·38028			9·61972	10·41503			9·58497	
24·0	9·96530	·8	4	10·03470	10·38064	·8	28	9·61936	10·41534	·8	24	9·58467	
25·0	9·96535			10·03465	10·38099			9·61901	10·41564			9·58436	35′
26·0	9·96541			10·03459	10·38135			9·61865	10·41594			9·58406	
27·0	9·96546	·9	5	10·03454	10·38171	·9	32	9·61830	10·41625	·9	27	9·58375	
28·0	9·96551			10·03449	10·38206			9·61794	10·41655			9·58345	
29·0	9·96556			10·03444	10·38242			9·61758	10·41685			9·58314	
30·0	9·96562			10·03439	10·38277			9·61722	10·41716			9·58284	30′
31·0	9·96567			10·03433	10·38313			9·61687	10·41747			9·58254	
32·0	9·96572	·1	1	10·03428	10·38349	·1	4	9·61651	10·41777	·1	3	9·58223	
33·0	9·96577			10·03423	10·38385			9·61615	10·41808			9·58192	
34·0	9·96582			10·03418	10·38421			9·61579	10·41838			9·58162	
35·0	9·96588	·2	1	10·03412	10·38457	·2	7	9·61544	10·41869	·2	6	9·58131	25′
36·0	9·96593			10·03407	10·38492			9·61508	10·41900			9·58101	
37·0	9·96598			10·03402	10·38528			9·61472	10·41930			9·58070	
38·0	9·96603	·3	2	10·03397	10·38564	·3	11	9·61436	10·41961	·3	9	9·58039	
39·0	9·96609			10·03392	10·38600			9·61400	10·41992			9·58009	
40·0	9·96614			10·03386	10·38636			9·61364	10·42022			9·57978	20′
41·0	9·96619	·4	2	10·03381	10·38672	·4	14	9·61328	10·42053	·4	12	9·57947	
42·0	9·96624			10·03376	10·38708			9·61292	10·42084			9·57916	
43·0	9·96629			10·03371	10·38744			9·61256	10·42115			9·57885	
44·0	9·96634			10·03366	10·38780			9·61220	10·42146			9·57855	
45·0	9·96640	·5	3	10·03361	10·38816	·5	18	9·61184	10·42176	·5	15	9·57824	15′
46·0	9·96645			10·03355	10·38852			9·61148	10·42207			9·57793	
47·0	9·96650			10·03350	10·38888			9·61112	10·42238			9·57762	
48·0	9·96655	·6	3	10·03345	10·38924	·6	22	9·61076	10·42269	·6	19	9·57731	
49·0	9·96660			10·03340	10·38960			9·61040	10·42300			9·57700	
50·0	9·96665			10·03335	10·38996			9·61004	10·42331			9·57669	10′
51·0	9·96671	·7	4	10·03330	10·39033	·7	25	9·60967	10·42362	·7	22	9·57638	
52·0	9·96676			10·03324	10·39069			9·60931	10·42393			9·57607	
53·0	9·96681			10·03319	10·39105			9·60895	10·42424			9·57576	
54·0	9·96686	·8	4	10·03314	10·39141	·8	29	9·60859	10·42455	·8	25	9·57545	
55·0	9·96691			10·03309	10·39178			9·60823	10·42486			9·57514	5′
56·0	9·96696			10·03304	10·39214			9·60786	10·42518			9·57482	
57·0	9·96701	·9	5	10·03299	10·39250	·9	32	9·60750	10·42549	·9	28	9·57451	
58·0	9·96706			10·03294	10·39286			9·60714	10·42580			9·57420	
59·0	9·96712			10·03289	10·39323			9·60677	10·42611			9·57389	
60·0	9·96717			10·03283	10·39359			9·60641	10·42643			9·57358	0′

68°
248°

LOGS. OF TRIG. FUNCTIONS

′	Sine	Parts	Cosec.	Tan.	Parts	Cotan.	Secant	Parts	Cosine	
00·0	9·96717	′	10·03283	10·39359	′	9·60641	10·42643	′	9·57358	60′
01·0	9·96722		10·03278	10·39395		9·60605	10·42674		9·57326	
02·0	9·96727	·1 1	10·03273	10·39432	·1 4	9·60568	10·42705	·1 3	9·57295	
03·0	9·96732		10·03268	10·39468		9·60532	10·42736		9·57264	
04·0	9·96737		10·03263	10·39505		9·60495	10·42768		9·57232	
05·0	9·96742	·2 1	10·03258	10·39541	·2 7	9·60459	10·42799	·2 6	9·57201	55′
06·0	9·96747		10·03253	10·39578		9·60422	10·42831		9·57170	
07·0	9·96752		10·03248	10·39614		9·60386	10·42862		9·57138	
08·0	9·96757	·3 2	10·03243	10·39651	·3 11	9·60349	10·42893	·3 10	9·57107	
09·0	9·96762		10·03238	10·39687		9·60313	10·42925		9·57075	
10·0	9·96767		10·03233	10·39724		9·60276	10·42957		9·57044	50′
11·0	9·96773	·4 2	10·03228	10·39761	·4 15	9·60240	10·42988	·4 13	9·57012	
12·0	9·96778		10·03223	10·39797		9·60203	10·43020		9·56980	
13·0	9·96783		10·03217	10·39834		9·60166	10·43051		9·56949	
14·0	9·96788	·5 3	10·03212	10·39870	·5 18	9·60130	10·43083	·5 16	9·56917	
15·0	9·96793		10·03207	10·39907		9·60093	10·43114		9·56886	45′
16·0	9·96798		10·03202	10·39944		9·60056	10·43146		9·56854	
17·0	9·96803		10·03197	10·39981		9·60019	10·43178		9·56822	
18·0	9·96808	·6 3	10·03192	10·40017	·6 22	9·59983	10·43210	·6 19	9·56790	
19·0	9·96813		10·03187	10·40054		9·59946	10·43241		9·56759	
20·0	9·96818		10·03182	10·40091		9·59909	10·43273		9·56727	40′
21·0	9·96823	·7 4	10·03177	10·40128	·7 26	9·59872	10·43305	·7 22	9·56695	
22·0	9·96828		10·03172	10·40165		9·59835	10·43337		9·56663	
23·0	9·96833		10·03167	10·40202		9·59799	10·43369		9·56631	
24·0	9·96838	·8 4	10·03162	10·40238	·8 29	9·59762	10·43401	·8 25	9·56600	
25·0	9·96843		10·03157	10·40275		9·59725	10·43432		9·56568	35′
26·0	9·96848		10·03152	10·40312		9·59688	10·43464		9·56536	
27·0	9·96853	·9 5	10·03147	10·40349	·9 33	9·59651	10·43496	·9 29	9·56504	
28·0	9·96858		10·03142	10·40386		9·59614	10·43528		9·56472	
29·0	9·96863		10·03137	10·40423		9·59577	10·43560		9·56440	
30·0	9·96868		10·03132	10·40460		9·59540	10·43593		9·56408	30′
31·0	9·96873		10·03127	10·40497		9·59503	10·43625		9·56376	
32·0	9·96878	·1 0	10·03122	10·40534	·1 4	9·59466	10·43657	·1 3	9·56343	
33·0	9·96883		10·03117	10·40572		9·59429	10·43689		9·56311	
34·0	9·96888		10·03112	10·40609		9·59391	10·43721		9·56279	
35·0	9·96893	·2 1	10·03107	10·40646	·2 7	9·59354	10·43753	·2 7	9·56247	25′
36·0	9·96898		10·03102	10·40683		9·59317	10·43785		9·56215	
37·0	9·96903		10·03098	10·40720		9·59280	10·43818		9·56182	
38·0	9·96908	·3 1	10·03093	10·40757	·3 11	9·59243	10·43850	·3 10	9·56150	
39·0	9·96912		10·03088	10·40795		9·59205	10·43882		9·56118	
40·0	9·96917		10·03083	10·40832		9·59168	10·43915		9·56086	20′
41·0	9·96922		10·03078	10·40869		9·59131	10·43947		9·56053	
42·0	9·96927	·4 2	10·03073	10·40907	·4 15	9·59094	10·43979	·4 13	9·56021	
43·0	9·96932		10·03068	10·40944		9·59056	10·44012		9·55988	
44·0	9·96937		10·03063	10·40981		9·59019	10·44044		9·55956	
45·0	9·96942	·5 2	10·03058	10·41019	·5 19	9·58981	10·44077	·5 16	9·55923	15′
46·0	9·96947		10·03053	10·41056		9·58944	10·44109		9·55891	
47·0	9·96952		10·03048	10·41093		9·58907	10·44141		9·55858	
48·0	9·96957	·6 3	10·03043	10·41131	·6 22	9·58869	10·44174	·6 20	9·55826	
49·0	9·96962		10·03038	10·41168		9·58832	10·44207		9·55793	
50·0	9·96967		10·03034	10·41206		9·58794	10·44239		9·55761	10′
51·0	9·96971	·7 3	10·03029	10·41243	·7 26	9·58757	10·44272	·7 23	9·55728	
52·0	9·96976		10·03024	10·41281		9·58719	10·44305		9·55695	
53·0	9·96981		10·03019	10·41319		9·58682	10·44337		9·55663	
54·0	9·96986	·8 4	10·03014	10·41356	·8 30	9·58644	10·44370	·8 26	9·55630	
55·0	9·96991		10·03009	10·41394		9·58606	10·44403		9·55597	5′
56·0	9·96996		10·03004	10·41431		9·58569	10·44436		9·55564	
57·0	9·97001	·9 4	10·02999	10·41469	·9 34	9·58531	10·44469	·9 29	9·55532	
58·0	9·97006		10·02995	10·41507		9·58493	10·44501		9·55499	
59·0	9·97010		10·02990	10·41545		9·58456	10·44534		9·55466	
60·0	9·97015		10·02985	10·41582		9·58418	10·44567		9·55433	0′

69°
249°

LOGS. OF TRIG. FUNCTIONS

′	Sine	Parts	Cosec.	Tan.	Parts	Cotan.	Secant	Parts	Cosine	
00·0	9·97015	′	10·02985	10·41582	′	9·58418	10·44567	′	9·55433	60′
01·0	9·97020		10·02980	10·41620		9·58380	10·44600		9·55400	
02·0	9·97025	·1 0	10·02975	10·41658	·1 4	9·58342	10·44633	·1 3	9·55367	
03·0	9·97030		10·02970	10·41696		9·58304	10·44666		9·55334	
04·0	9·97035		10·02966	10·41734		9·58267	10·44699		9·55301	
05·0	9·97039	·2 1	10·02961	10·41771	·2 8	9·58229	10·44732	·2 7	9·55268	55′
06·0	9·97044		10·02956	10·41809		9·58191	10·44765		9·55235	
07·0	9·97049		10·02951	10·41847		9·58153	10·44798		9·55202	
08·0	9·97054	·3 1	10·02946	10·41885	·3 11	9·58115	10·44831	·3 10	9·55169	
09·0	9·97059		10·02941	10·41923		9·58077	10·44864		9·55136	
10·0	9·97064		10·02937	10·41961		9·58039	10·44898		9·55102	50′
11·0	9·97068	·4 2	10·02932	10·41999	·4 15	9·58001	10·44931	·4 13	9·55069	
12·0	9·97073		10·02927	10·42037		9·57963	10·44964		9·55036	
13·0	9·97078		10·02922	10·42075		9·57925	10·44997		9·55003	
14·0	9·97083	·5 2	10·02917	10·42113	·5 19	9·57887	10·45031	·5 17	9·54969	
15·0	9·97087		10·02913	10·42151		9·57849	10·45064		9·54936	45′
16·0	9·97092		10·02908	10·42190		9·57810	10·45097		9·54903	
17·0	9·97097		10·02903	10·42228		9·57772	10·45131		9·54869	
18·0	9·97102	·6 3	10·02898	10·42266	·6 23	9·57734	10·45164	·6 20	9·54836	
19·0	9·97107		10·02893	10·42304		9·57696	10·45198		9·54802	
20·0	9·97111		10·02889	10·42342		9·57658	10·45231		9·54769	40′
21·0	9·97116	·7 3	10·02884	10·42381	·7 27	9·57619	10·45265	·7 23	9·54735	
22·0	9·97121		10·02879	10·42419		9·57581	10·45298		9·54702	
23·0	9·97126		10·02874	10·42457		9·57543	10·45332		9·54668	
24·0	9·97130	·8 4	10·02870	10·42496	·8 31	9·57504	10·45365	·8 27	9·54635	
25·0	9·97135		10·02865	10·42534		9·57466	10·45399		9·54601	35′
26·0	9·97140		10·02860	10·42572		9·57428	10·45433		9·54567	
27·0	9·97145	·9 4	10·02855	10·42611	·9 34	9·57389	10·45466	·9 30	9·54534	
28·0	9·97149		10·02851	10·42649		9·57351	10·45500		9·54500	
29·0	9·97154		10·02846	10·42688		9·57312	10·45534		9·54466	
30·0	9·97159		10·02841	10·42726		9·57274	10·45568		9·54433	30′
31·0	9·97164		10·02837	10·42765		9·57235	10·45601		9·54399	
32·0	9·97168	·1 0	10·02832	10·42803	·1 4	9·57197	10·45635	·1 3	9·54365	
33·0	9·97173		10·02827	10·42842		9·57158	10·45669		9·54331	
34·0	9·97178		10·02822	10·42881		9·57120	10·45703		9·54297	
35·0	9·97182	·2 1	10·02818	10·42919	·2 8	9·57081	10·45737	·2 7	9·54263	25′
36·0	9·97187		10·02813	10·42958		9·57042	10·45771		9·54229	
37·0	9·97192		10·02808	10·42997		9·57004	10·45805		9·54195	
38·0	9·97196	·3 1	10·02804	10·43035	·3 12	9·56965	10·45839	·3 10	9·54161	
39·0	9·97201		10·02799	10·43074		9·56926	10·45873		9·54127	
40·0	9·97206		10·02794	10·43113		9·56887	10·45907		9·54093	20′
41·0	9·97211		10·02790	10·43151		9·56849	10·45941		9·54059	
42·0	9·97215	·4 2	10·02785	10·43190	·4 16	9·56810	10·45975	·4 14	9·54025	
43·0	9·97220		10·02780	10·43229		9·56771	10·46009		9·53991	
44·0	9·97225		10·02776	10·43268		9·56732	10·46044		9·53957	
45·0	9·97229	·5 2	10·02771	10·43307	·5 19	9·56693	10·46078	·5 17	9·53922	15′
46·0	9·97234		10·02766	10·43346		9·56654	10·46112		9·53888	
47·0	9·97239		10·02762	10·43385		9·56615	10·46146		9·53854	
48·0	9·97243	·6 3	10·02757	10·43424	·6 23	9·56576	10·46181	·6 21	9·53819	
49·0	9·97248		10·02752	10·43463		9·56537	10·46215		9·53785	
50·0	9·97252		10·02748	10·43502		9·56498	10·46249		9·53751	10′
51·0	9·97257	·7 3	10·02743	10·43541	·7 27	9·56459	10·46284	·7 24	9·53716	
52·0	9·97262		10·02738	10·43580		9·56420	10·46318		9·53682	
53·0	9·97266		10·02734	10·43619		9·56381	10·46353		9·53647	
54·0	9·97271	·8 4	10·02729	10·43658	·8 31	9·56342	10·46387	·8 27	9·53613	
55·0	9·97276		10·02725	10·43697		9·56303	10·46422		9·53578	5′
56·0	9·97280		10·02720	10·43736		9·56264	10·46456		9·53544	
57·0	9·97285	·9 4	10·02715	10·43776	·9 35	9·56224	10·46491	·9 31	9·53509	
58·0	9·97289		10·02711	10·43815		9·56185	10·46526		9·53475	
59·0	9·97294		10·02706	10·43854		9·56146	10·46560		9·53440	0′
60·0	9·97299		10·02701	10·43893		9·56107	10·46595		9·53405	

70°
250°

LOGS. OF TRIG. FUNCTIONS

′	Sine	Parts		Cosec.	Tan.	Parts		Cotan.	Secant	Parts		Cosine	
00·0	9·97299			10·02701	10·43893			9·56107	10·46595			9·53405	60′
01·0	9·97303			10·02697	10·43933			9·56067	10·46630			9·53370	
02·0	9·97308	·1	0	10·02692	10·43972	·1	4	9·56028	10·46664	·1	4	9·53336	
03·0	9·97312			10·02688	10·44012			9·55989	10·46699			9·53301	
04·0	9·97317	·2	1	10·02683	10·44051	·2	8	9·55949	10·46734	·2	7	9·53266	
05·0	9·97322			10·02679	10·44090			9·55910	10·46769			9·53231	55′
06·0	9·97326	·3	1	10·02674	10·44130	·3	12	9·55870	10·46804	·3	11	9·53196	
07·0	9·97331			10·02669	10·44169			9·55831	10·46839			9·53161	
08·0	9·97335	·4	2	10·02665	10·44209	·4	16	9·55791	10·46874	·4	14	9·53127	
09·0	9·97340			10·02660	10·44248			9·55752	10·46909			9·53092	
10·0	9·97344	·5	2	10·02656	10·44288	·5	20	9·55712	10·46944	·5	18	9·53057	50′
11·0	9·97349	·6	3	10·02651	10·44328	·6	24	9·55673	10·46979	·6	21	9·53022	
12·0	9·97354			10·02647	10·44367			9·55633	10·47014			9·52986	
13·0	9·97358	·7	3	10·02642	10·44407	·7	28	9·55593	10·47049	·7	25	9·52951	
14·0	9·97363			10·02638	10·44446			9·55554	10·47084			9·52916	
15·0	9·97367	·8	4	10·02633	10·44486	·8	32	9·55514	10·47119	·8	28	9·52881	45′
16·0	9·97372			10·02628	10·44526			9·55474	10·47154			9·52846	
17·0	9·97376	·9	4	10·02624	10·44566	·9	36	9·55434	10·47190	·9	32	9·52811	
18·0	9·97381			10·02619	10·44605			9·55395	10·47225			9·52775	
19·0	9·97385			10·02615	10·44645			9·55355	10·47260			9·52740	
20·0	9·97390			10·02610	10·44685			9·55315	10·47295			9·52705	40′
21·0	9·97394			10·02606	10·44725			9·55275	10·47331			9·52669	
22·0	9·97399	·1	0	10·02601	10·44765	·1	4	9·55235	10·47366	·1	4	9·52634	
23·0	9·97403			10·02597	10·44805			9·55195	10·47402			9·52598	
24·0	9·97408	·2	1	10·02592	10·44845	·2	8	9·55155	10·47437	·2	7	9·52563	
25·0	9·97412			10·02588	10·44885			9·55115	10·47473			9·52528	35′
26·0	9·97417	·3	1	10·02583	10·44925	·3	12	9·55075	10·47508	·3	11	9·52492	
27·0	9·97421			10·02579	10·44965			9·55035	10·47544			9·52456	
28·0	9·97426	·4	2	10·02574	10·45005	·4	16	9·54995	10·47579	·4	14	9·52421	
29·0	9·97430			10·02570	10·45045			9·54955	10·47615			9·52385	
30·0	9·97435	·5	2	10·02565	10·45085	·5	20	9·54915	10·47651	·5	18	9·52350	30′
31·0	9·97439	·6	3	10·02561	10·45125	·6	24	9·54875	10·47686	·6	21	9·52314	
32·0	9·97444			10·02556	10·45166			9·54835	10·47722			9·52278	
33·0	9·97448	·7	3	10·02552	10·45206	·7	28	9·54794	10·47758	·7	25	9·52242	
34·0	9·97453			10·02548	10·45246			9·54754	10·47793			9·52207	
35·0	9·97457	·8	4	10·02543	10·45286	·8	32	9·54714	10·47829	·8	29	9·52171	25′
36·0	9·97461			10·02539	10·45327			9·54674	10·47865			9·52135	
37·0	9·97466	·9	4	10·02534	10·45367	·9	36	9·54633	10·47901	·9	32	9·52099	
38·0	9·97470			10·02530	10·45407			9·54593	10·47937			9·52063	
39·0	9·97475			10·02525	10·45448			9·54552	10·47973			9·52027	
40·0	9·97479			10·02521	10·45488			9·54512	10·48009			9·51991	20′
41·0	9·97484			10·02516	10·45529			9·54472	10·48045			9·51955	
42·0	9·97488	·1	0	10·02512	10·45569	·1	4	9·54431	10·48081	·1	4	9·51919	
43·0	9·97493			10·02508	10·45610			9·54391	10·48117			9·51883	
44·0	9·97497	·2	1	10·02503	10·45650	·2	8	9·54350	10·48153	·2	7	9·51847	
45·0	9·97501	·3	1	10·02499	10·45691	·3	12	9·54309	10·48189	·3	11	9·51811	15′
46·0	9·97506			10·02494	10·45731			9·54269	10·48226			9·51775	
47·0	9·97510	·4	2	10·02490	10·45772	·4	16	9·54228	10·48262	·4	15	9·51738	
48·0	9·97515			10·02486	10·45813			9·54188	10·48298			9·51702	
49·0	9·97519	·5	2	10·02481	10·45853	·5	20	9·54147	10·48334	·5	18	9·51666	
50·0	9·97523			10·02477	10·45894			9·54106	10·48371			9·51629	10′
51·0	9·97528	·6	3	10·02472	10·45935	·6	24	9·54065	10·48407	·6	22	9·51593	
52·0	9·97532			10·02468	10·45976			9·54025	10·48443			9·51557	
53·0	9·97537	·7	3	10·02464	10·46016	·7	29	9·53984	10·48480	·7	25	9·51520	
54·0	9·97541			10·02459	10·46057			9·53943	10·48516			9·51484	
55·0	9·97545	·8	4	10·02455	10·46098	·8	33	9·53902	10·48553	·8	29	9·51447	5′
56·0	9·97550			10·02450	10·46139			9·53861	10·48589			9·51411	
57·0	9·97554	·9	4	10·02446	10·46180	·9	37	9·53820	10·48626	·9	33	9·51374	
58·0	9·97558			10·02442	10·46221			9·53779	10·48663			9·51338	
59·0	9·97563			10·02437	10·46262			9·53738	10·48699			9·51301	
60·0	9·97567			10·02433	10·46303			9·53697	10·48736			9·51264	0′

71°
251°

LOGS. OF TRIG. FUNCTIONS

′	Sine	Parts	Cosec.	Tan.	Parts	Cotan.	Secant	Parts	Cosine	
00·0	9·97567	′	10·02433	10·46303	′	9·53697	10·48736	′	9·51264	60′
01·0	9·97571		10·02429	10·46344		9·53656	10·48773		9·51228	
02·0	9·97576	·1 0	10·02424	10·46385	·1 4	9·53615	10·48809	·1 4	9·51191	
03·0	9·97580		10·02420	10·46426		9·53574	10·48846		9·51154	
04·0	9·97584	·2 1	10·02416	10·46467	·2 8	9·53533	10·48883	·2 7	9·51117	
05·0	9·97589		10·02411	10·46508		9·53492	10·48920		9·51080	55′
06·0	9·97593	·3 1	10·02407	10·46550	·3 12	9·53450	10·48957	·3 11	9·51043	
07·0	9·97597		10·02403	10·46591		9·53409	10·48994		9·51007	
08·0	9·97602	·4 2	10·02398	10·46632	·4 17	9·53368	10·49030	·4 15	9·50970	
09·0	9·97606		10·02394	10·46673		9·53327	10·49067		9·50933	
10·0	9·97610	·5 2	10·02390	10·46715	·5 21	9·53285	10·49104	·5 19	9·50896	50′
11·0	9·97615		10·02385	10·46756		9·53244	10·49142		9·50859	
12·0	9·97619	·6 3	10·02381	10·46798	·6 25	9·53203	10·49179	·6 22	9·50821	
13·0	9·97623		10·02377	10·46839		9·53161	10·49216		9·50784	
14·0	9·97628	·7 3	10·02373	10·46880	·7 29	9·53120	10·49253	·7 26	9·50747	
15·0	9·97632	·8 3	10·02368	10·46922	·8 33	9·53078	10·49290	·8 30	9·50710	45′
16·0	9·97636		10·02364	10·46963		9·53037	10·49327		9·50673	
17·0	9·97640	·9 4	10·02360	10·47005	·9 37	9·52995	10·49365	·9 33	9·50635	
18·0	9·97645		10·02355	10·47047		9·52954	10·49402		9·50598	
19·0	9·97649		10·02351	10·47088		9·52912	10·49439		9·50561	
20·0	9·97653		10·02347	10·47130		9·52870	10·49477		9·50523	40′
21·0	9·97657		10·02343	10·47172		9·52829	10·49514		9·50486	
22·0	9·97662	·1 0	10·02338	10·47213	·1 4	9·52787	10·49552	·1 4	9·50449	
23·0	9·97666		10·02334	10·47255		9·52745	10·49589		9·50411	
24·0	9·97670	·2 1	10·02330	10·47297	·2 8	9·52703	10·49627	·2 8	9·50374	
25·0	9·97675		10·02326	10·47339		9·52662	10·49664		9·50336	35′
26·0	9·97679	·3 1	10·02321	10·47380	·3 13	9·52620	10·49702	·3 11	9·50298	
27·0	9·97683		10·02317	10·47422		9·52578	10·49739		9·50261	
28·0	9·97687	·4 2	10·02313	10·47464	·4 17	9·52536	10·49777	·4 15	9·50223	
29·0	9·97691		10·02309	10·47506		9·52494	10·49815		9·50185	
30·0	9·97696	·5 2	10·02304	10·47548	·5 21	9·52452	10·49852	·5 19	9·50148	30′
31·0	9·97700		10·02300	10·47590		9·52410	10·49890		9·50110	
32·0	9·97704	·6 3	10·02296	10·47632	·6 25	9·52368	10·49928	·6 23	9·50072	
33·0	9·97708		10·02292	10·47674		9·52326	10·49966		9·50034	
34·0	9·97713	·7 3	10·02288	10·47716	·7 29	9·52284	10·50004	·7 26	9·49996	
35·0	9·97717	·8 3	10·02283	10·47758	·8 34	9·52242	10·50042	·8 30	9·49958	25′
36·0	9·97721		10·02279	10·47801		9·52200	10·50080		9·49920	
37·0	9·97725	·9 4	10·02275	10·47843	·9 38	9·52157	10·50118	·9 34	9·49883	
38·0	9·97729		10·02271	10·47885		9·52115	10·50156		9·49844	
39·0	9·97734		10·02267	10·47927		9·52073	10·50194		9·49806	
40·0	9·97738		10·02262	10·47970		9·52031	10·50232		9·49768	20′
41·0	9·97742		10·02258	10·48012		9·51988	10·50270		9·49730	
42·0	9·97746	·1 0	10·02254	10·48054	·1 4	9·51946	10·50308	·1 4	9·49692	
43·0	9·97750		10·02250	10·48097		9·51903	10·50346		9·49654	
44·0	9·97754	·2 1	10·02246	10·48139	·2 9	9·51861	10·50385	·2 8	9·49615	
45·0	9·97759		10·02241	10·48181		9·51819	10·50423		9·49577	15′
46·0	9·97763	·3 1	10·02237	10·48224	·3 13	9·51776	10·50461	·3 12	9·49539	
47·0	9·97767		10·02233	10·48267		9·51734	10·50500		9·49501	
48·0	9·97771	·4 2	10·02229	10·48309	·4 17	9·51691	10·50538	·4 15	9·49462	
49·0	9·97775		10·02225	10·48352		9·51648	10·50576		9·49424	
50·0	9·97779	·5 2	10·02221	10·48394	·5 21	9·51606	10·50615	·5 19	9·49385	10′
51·0	9·97784		10·02217	10·48437		9·51563	10·50653		9·49347	
52·0	9·97788	·6 2	10·02212	10·48480	·6 26	9·51520	10·50692	·6 23	9·49308	
53·0	9·97792		10·02208	10·48522		9·51478	10·50731		9·49270	
54·0	9·97796	·7 3	10·02204	10·48565	·7 30	9·51435	10·50769	·7 27	9·49231	
55·0	9·97800	·8 3	10·02200	10·48608	·8 34	9·51392	10·50008	·8 31	9·49192	5′
56·0	9·97804		10·02196	10·48651		9·51349	10·50847		9·49154	
57·0	9·97808	·9 4	10·02192	10·48694	·9 38	9·51306	10·50885	·9 35	9·49115	
58·0	9·97812		10·02188	10·48737		9·51264	10·50924		9·49076	
59·0	9·97817		10·02184	10·48779		9·51221	10·50963		9·49037	
60·0	9·97821		10·02179	10·48822		9·51178	10·51002		9·48998	0′

72°
252°

LOGS. OF TRIG. FUNCTIONS

′	Sine	Parts	Cosec	Tan.	Parts	Cotan.	Secant	Parts	Cosine	
00·0	9·97821	′	10·02179	10·48822	′	9·51178	10·51002	′	9·48998	60′
01·0	9·97825		10·02175	10·48865		9·51135	10·51041		9·48959	
02·0	9·97829	·1 0	10·02171	10·48908	·1 4	9·51092	10·51080	·1 4	9·48920	
03·0	9·97833		10·02167	10·48952		9·51049	10·51119		9·48881	
04·0	9·97837	·2 1	10·02163	10·48995	·2 9	9·51005	10·51158	·2 8	9·48842	
05·0	9·97841		10·02159	10·49038		9·50962	10·51197		9·48803	55′
06·0	9·97845	·3 1	10·02155	10·49081	·3 13	9·50919	10·51236	·3 12	9·48764	
07·0	9·97849		10·02151	10·49124		9·50876	10·51275		9·48725	
08·0	9·97853	·4 2	10·02147	10·49167	·4 17	9·50833	10·51314	·4 16	9·48686	
09·0	9·97857		10·02143	10·49211		9·50789	10·51353		9·48647	
10·0	9·97862	·5 2	10·02139	10·49254	·5 22	9·50746	10·51393	·5 20	9·48608	50′
11·0	9·97866	·6 2	10·02135	10·49297	·6 26	9·50703	10·51432	·6 24	9·48568	
12·0	9·97870		10·02130	10·49341		9·50659	10·51471		9·48529	
13·0	9·97874	·7 3	10·02126	10·49384	·7 30	9·50616	10·51511	·7 27	9·48490	
14·0	9·97878		10·02122	10·49428		9·50572	10·51550		9·48450	
15·0	9·97882	·8 3	10·02118	10·49471	·8 35	9·50529	10·51589	·8 31	9·48411	45′
16·0	9·97886		10·02114	10·49515		9·50485	10·51629		9·48371	
17·0	9·97890	·9 4	10·02110	10·49558	·9 39	9·50442	10·51668	·9 35	9·48332	
18·0	9·97894		10·02106	10·49602		9·50398	10·51708		9·48292	
19·0	9·97898		10·02102	10·49645		9·50355	10·51748		9·48253	
20·0	9·97902		10·02098	10·49689		9·50311	10·51787		9·48213	40′
21·0	9·97906		10·02094	10·49733		9·50267	10·51827		9·48173	
22·0	9·97910	·1 0	10·02090	10·49777	·1 4	9·50224	10·51867	·1 4	9·48133	
23·0	9·97914		10·02086	10·49820		9·50180	10·51906		9·48094	
24·0	9·97918	·2 1	10·02082	10·49864	·2 9	9·50136	10·51946	·2 8	9·48054	
25·0	9·97922		10·02078	10·49908		9·50092	10·51986		9·48014	35′
26·0	9·97926	·3 1	10·02074	10·49952	·3 13	9·50048	10·52026	·3 12	9·47974	
27·0	9·97930		10·02070	10·49996		9·50004	10·52066		9·47934	
28·0	9·97934	·4 2	10·02066	10·50040	·4 18	9·49960	10·52106	·4 16	9·47894	
29·0	9·97938		10·02062	10·50084		9·49916	10·52146		9·47854	
30·0	9·97942	·5 2	10·02058	10·50128	·5 22	9·49872	10·52186	·5 20	9·47814	30′
31·0	9·97946	·6 2	10·02054	10·50172	·6 26	9·49828	10·52226	·6 24	9·47774	
32·0	9·97950		10·02050	10·50216		9·49784	10·52266		9·47734	
33·0	9·97954	·7 3	10·02046	10·50260	·7 31	9·49740	10·52306	·7 28	9·47694	
34·0	9·97958		10·02042	10·50304		9·49696	10·52346		9·47654	
35·0	9·97962	·8 3	10·02038	10·50349	·8 35	9·49652	10·52387	·8 32	9·47613	25′
36·0	9·97966		10·02034	10·50393		9·49607	10·52427		9·47573	
37·0	9·97970	·9 4	10·02030	10·50437	·9 40	9·49563	10·52467	·9 36	9·47533	
38·0	9·97974		10·02026	10·50481		9·49519	10·52508		9·47492	
39·0	9·97978		10·02022	10·50526		9·49474	10·52548		9·47452	
40·0	9·97982		10·02018	10·50570		9·49430	10·52589		9·47412	20′
41·0	9·97986		10·02015	10·50615		9·49385	10·52629		9·47371	
42·0	9·97990	·1 0	10·02011	10·50659	·1 4	9·49341	10·52670	·1 4	9·47330	
43·0	9·97993		10·02007	10·50704		9·49297	10·52710		9·47290	
44·0	9·97997	·2 1	10·02003	10·50748	·2 9	9·49252	10·52751	·2 8	9·47249	
45·0	9·98001		10·01999	10·50793		9·49207	10·52791		9·47209	15′
46·0	9·98005	·3 1	10·01995	10·50837	·3 13	9·49163	10·52832	·3 12	9·47168	
47·0	9·98009		10·01991	10·50882		9·49118	10·52873		9·47127	
48·0	9·98013	·4 2	10·01987	10·50927	·4 18	9·49073	10·52914	·4 16	9·47086	
49·0	9·98017		10·01983	10·50971		9·49029	10·52955		9·47046	
50·0	9·98021	·5 2	10·01979	10·51016	·5 22	9·48984	10·52995	·5 20	9·47005	10′
51·0	9·98025	·6 2	10·01975	10·51061	·6 27	9·48939	10·53036	·6 25	9·46964	
52·0	9·98029		10·01971	10·51106		9·48894	10·53077		9·46923	
53·0	9·98033	·7 3	10·01968	10·51151	·7 31	9·48849	10·53118	·7 29	9·46882	
54·0	9·98036		10·01964	10·51196		9·48804	10·53159		9·46841	
55·0	9·98040	·8 3	10·01960	10·51241	·8 36	9·48759	10·53200	·8 33	9·46800	5′
56·0	9·98044		10·01956	10·51286		9·48714	10·53242		9·46759	
57·0	9·98048	·9 4	10·01952	10·51331	·9 40	9·48669	10·53283	·9 37	9·46717	
58·0	9·98052		10·01948	10·51376		9·48624	10·53324		9·46676	
59·0	9·98056		10·01944	10·51421		9·48579	10·53365		9·46635	
60·0	9·98060		10·01940	10·51466		9·48534	10·53407		9·46594	0′

73°
253°

LOGS. OF TRIG. FUNCTIONS

′	Sine	Parts		Cosec.	Tan.	Parts		Cotan.	Secant	Parts		Cosine	
00·0	9·98060	′		10·01940	10·51466	′		9·48534	10·53407	′		9·46594	60′
01·0	9·98064			10·01937	10·51511			9·48489	10·53448			9·46552	
02·0	9·98067	·1	0	10·01933	10·51557	·1	5	9·48444	10·53489	·1	4	9·46511	
03·0	9·98071			10·01929	10·51602			9·48398	10·53531			9·46469	
04·0	9·98075	·2	1	10·01925	10·51647	·2	9	9·48353	10·53572	·2	8	9·46428	
05·0	9·98079			10·01921	10·51693			9·48308	10·53614			9·46386	55′
06·0	9·98083	·3	1	10·01917	10·51738	·3	14	9·48262	10·53655	·3	13	9·46345	
07·0	9·98087			10·01913	10·51783			9·48217	10·53697			9·46303	
08·0	9·98090	·4	2	10·01910	10·51829	·4	18	9·48171	10·53738	·4	17	9·46262	
09·0	9·98094			10·01906	10·51874			9·48126	10·53780			9·46220	
10·0	9·98098	·5	2	10·01902	10·51920	·5	23	9·48080	10·53822	·5	21	9·46178	50′
11·0	9·98102	·6	2	10·01898	10·51966	·6	27	9·48035	10·53864	·6	25	9·46136	
12·0	9·98106			10·01894	10·52011			9·47989	10·53905			9·46095	
13·0	9·98110	·7	3	10·01891	10·52057	·7	32	9·47943	10·53947	·7	29	9·46053	
14·0	9·98113			10·01887	10·52103			9·47898	10·53989			9·46011	
15·0	9·98117	·8	3	10·01883	10·52148	·8	36	9·47852	10·54031	·8	33	9·45969	45′
16·0	9·98121			10·01879	10·52194			9·47806	10·54073			9·45927	
17·0	9·98125	·9	3	10·01875	10·52240	·9	41	9·47760	10·54115	·9	38	9·45885	
18·0	9·98129			10·01872	10·52286			9·47714	10·54157			9·45843	
19·0	9·98132			10·01868	10·52332			9·47668	10·54199			9·45801	
20·0	9·98136			10·01864	10·52378			9·47622	10·54242			9·45758	40′
21·0	9·98140			10·01860	10·52424			9·47576	10·54284			9·45716	
22·0	9·98144	·1	0	10·01856	10·52470	·1	5	9·47530	10·54326	·1	4	9·45674	
23·0	9·98147			10·01853	10·52516			9·47484	10·54368			9·45632	
24·0	9·98151	·2	1	10·01849	10·52562	·2	9	9·47438	10·54411	·2	9	9·45589	
25·0	9·98155	·3	1	10·01845	10·52608	·3	14	9·47392	10·54453	·3	13	9·45547	35′
26·0	9·98159			10·01841	10·52654			9·47346	10·54496			9·45504	
27·0	9·98163	·4	2	10·01838	10·52701	·4	19	9·47300	10·54538	·4	17	9·45462	
28·0	9·98166			10·01834	10·52747			9·47253	10·54581			9·45419	
29·0	9·98170	·5	2	10·01830	10·52793	·5	23	9·47207	10·54623	·5	21	9·45377	
30·0	9·98174			10·01826	10·52840			9·47161	10·54666			9·45334	30′
31·0	9·98177	·6	2	10·01823	10·52886	·6	28	9·47114	10·54709	·6	26	9·45292	
32·0	9·98181			10·01819	10·52932			9·47068	10·54751			9·45249	
33·0	9·98185	·7	3	10·01815	10·52979	·7	32	9·47021	10·54794	·7	30	9·45206	
34·0	9·98189			10·01811	10·53025			9·46975	10·54837			9·45163	
35·0	9·98192	·8	3	10·01808	10·53072	·8	37	9·46928	10·54880	·8	34	9·45120	25′
36·0	9·98196			10·01804	10·53119			9·46881	10·54923			9·45078	
37·0	9·98200	·9	3	10·01800	10·53165	·9	42	9·46835	10·54966	·9	38	9·45035	
38·0	9·98204			10·01797	10·53212			9·46788	10·55009			9·44992	
39·0	9·98207			10·01793	10·53259			9·46741	10·55052			9·44949	
40·0	9·98211			10·01789	10·53306			9·46695	10·55095			9·44905	20′
41·0	9·98215			10·01785	10·53352			9·46648	10·55138			9·44862	
42·0	9·98218	·1	0	10·01782	10·53399	·1	5	9·46601	10·55181	·1	4	9·44819	
43·0	9·98222			10·01778	10·53446			9·46554	10·55224			9·44776	
44·0	9·98226	·2	1	10·01774	10·53493	·2	9	9·46507	10·55267	·2	9	9·44733	
45·0	9·98229			10·01771	10·53540			9·46460	10·55311			9·44689	15′
46·0	9·98233	·3	1	10·01767	10·53587	·3	14	9·46413	10·55354	·3	13	9·44646	
47·0	9·98237			10·01763	10·53634			9·46366	10·55398			9·44603	
48·0	9·98240	·4	1	10·01760	10·53681	·4	19	9·46319	10·55441	·4	17	9·44559	
49·0	9·98244			10·01756	10·53729			9·46272	10·55485			9·44516	
50·0	9·98248	·5	2	10·01752	10·53776	·5	24	9·46224	10·55528	·5	22	9·44472	10′
51·0	9·98251	·6	2	10·01749	10·53823	·6	28	9·46177	10·55572	·6	26	9·44428	
52·0	9·98255			10·01745	10·53870			9·46130	10·55615			9·44385	
53·0	9·98258	·7	3	10·01741	10·53918	·7	33	9·46082	10·55659	·7	30	9·44341	
54·0	9·98262			10·01738	10·53965			9·46035	10·55703			9·44297	
55·0	9·98266	·8	3	10·01734	10·54013	·8	38	9·45988	10·55747	·8	35	9·44254	5′
56·0	9·98270			10·01730	10·54060			9·45940	10·55790			9·44210	
57·0	9·98273	·9	3	10·01727	10·54108	·9	43	9·45893	10·55834	·9	39	9·44166	
58·0	9·98277			10·01723	10·54155			9·45845	10·55878			9·44122	
59·0	9·98281			10·01720	10·54203			9·45797	10·55922			9·44078	
60·0	9·98284			10·01716	10·54250			9·45750	10·55966			9·44034	0′

LOGS. OF TRIG. FUNCTIONS

74°
254°

′	Sine	Parts	Cosec	Tan.	Parts	Cotan.	Secant	Parts	Cosine	
00·0	9·98284	′	10·01716	10·54250	′	9·45750	10·55966	′	9·44034	60′
01·0	9·98288		10·01712	10·54298		9·45702	10·56010		9·43990	
02·0	9·98291	·1 0	10·01709	10·54346	·1 5	9·45654	10·56054	·1 4	9·43946	
03·0	9·98295		10·01705	10·54394		9·45606	10·56099		9·43901	
04·0	9·98299	·2 1	10·01701	10·54441	·2 10	9·45559	10·56143	·2 9	9·43857	
05·0	9·98302	·3 1	10·01698	10·54489	·3 14	9·45511	10·56187	·3 13	9·43813	55′
06·0	9·98306		10·01694	10·54537		9·45463	10·56231		9·43769	
07·0	9·98309		10·01691	10·54585		9·45415	10·56276		9·43724	
08·0	9·98313	·4 1	10·01687	10·54633	·4 19	9·45367	10·56320	·4 18	9·43680	
09·0	9·98317	·5 2	10·01683	10·54681	·5 24	9·45319	10·56365	·5 22	9·43635	
10·0	9·98320		10·01680	10·54729		9·45271	10·56409		9·43591	50′
11·0	9·98324	·6 2	10·01676	10·54778	·6 29	9·45223	10·56454	·6 27	9·43546	
12·0	9·98327		10·01673	10·54826		9·45174	10·56498		9·43502	
13·0	9·98331	·7 3	10·01669	10·54874	·7 34	9·45126	10·56543	·7 31	9·43457	
14·0	9·98335		10·01666	10·54922		9·45078	10·56588		9·43412	
15·0	9·98338	·8 3	10·01662	10·54971	·8 39	9·45029	10·56633	·8 36	9·43368	45′
16·0	9·98342		10·01658	10·55019		9·44981	10·56677		9·43323	
17·0	9·98345	·9 3	10·01655	10·55067	·9 43	9·44933	10·56722	·9 40	9·43278	
18·0	9·98349		10·01651	10·55116		9·44884	10·56767		9·43233	
19·0	9·98352		10·01648	10·55164		9·44836	10·56812		9·43188	
20·0	9·98356		10·01644	10·55213		9·44787	10·56857		9·43143	40′
21·0	9·98359		10·01641	10·55262		9·44738	10·56902		9·43098	
22·0	9·98363	·1 0	10·01637	10·55310	·1 5	9·44690	10·56947	·1 5	9·43053	
23·0	9·98366		10·01634	10·55359		9·44641	10·56993		9·43008	
24·0	9·98370	·2 1	10·01630	10·55408	·2 10	9·44592	10·57038	·2 9	9·42962	
25·0	9·98374	·3 1	10·01627	10·55457	·3 15	9·44544	10·57083	·3 14	9·42917	35′
26·0	9·98377		10·01623	10·55505		9·44495	10·57128		9·42872	
27·0	9·98381		10·01620	10·55554		9·44446	10·57174		9·42826	
28·0	9·98384	·4 1	10·01616	10·55603	·4 20	9·44397	10·57219	·4 18	9·42781	
29·0	9·98388	·5 2	10·01613	10·55652	·5 25	9·44348	10·57265	·5 23	9·42735	
30·0	9·98391		10·01609	10·55701		9·44299	10·57310		9·42690	30′
31·0	9·98395	·6 2	10·01605	10·55750	·6 29	9·44250	10·57356	·6 27	9·42644	
32·0	9·98398		10·01602	10·55799		9·44201	10·57401		9·42599	
33·0	9·98402	·7 2	10·01599	10·55849	·7 34	9·44151	10·57447	·7 32	9·42553	
34·0	9·98405		10·01595	10·55898		9·44102	10·57493		9·42507	
35·0	9·98409	·8 3	10·01592	10·55947	·8 39	9·44053	10·57539	·8 36	9·42462	25′
36·0	9·98412		10·01588	10·55996		9·44004	10·57584		9·42416	
37·0	9·98416	·9 3	10·01585	10·56046	·9 44	9·43954	10·57630	·9 41	9·42370	
38·0	9·98419		10·01581	10·56095		9·43905	10·57676		9·42324	
39·0	9·98422		10·01578	10·56145		9·43855	10·57722		9·42278	
40·0	9·98426		10·01574	10·56194		9·43806	10·57768		9·42232	20′
41·0	9·98429		10·01571	10·56244		9·43756	10·57814		9·42186	
42·0	9·98433	·1 0	10·01567	10·56293	·1 5	9·43707	10·57861	·1 5	9·42140	
43·0	9·98436		10·01564	10·56343		9·43657	10·57907		9·42093	
44·0	9·98440	·2 1	10·01560	10·56393	·2 10	9·43607	10·57953	·2 9	9·42047	
45·0	9·98443	·3 1	10·01557	10·56442	·3 15	9·43558	10·57999	·3 14	9·42001	15′
46·0	9·98447		10·01553	10·56492		9·43508	10·58046		9·41954	
47·0	9·98450		10·01550	10·56542		9·43458	10·58092		9·41908	
48·0	9·98454	·4 1	10·01547	10·56592	·4 20	9·43408	10·58139	·4 19	9·41862	
49·0	9·98457	·5 2	10·01543	10·56642	·5 25	9·43358	10·58185	·5 23	9·41815	
50·0	9·98460		10·01540	10·56692		9·43308	10·58232		9·41768	10′
51·0	9·98464	·6 2	10·01536	10·56742	·6 30	9·43258	10·58278	·6 28	9·41722	
52·0	9·98467		10·01533	10·56792		9·43208	10·58325		9·41675	
53·0	9·98471	·7 2	10·01529	10·56842	·7 35	9·43158	10·58372	·7 33	9·41628	
54·0	9·98474		10·01526	10·56893		9·43108	10·58419		9·41582	
55·0	9·98477	·8 3	10·01523	10·56943	·8 40	9·43057	10·58465	·8 37	9·41535	5′
56·0	9·98481		10·01519	10·56993		9·43007	10·58512		9·41488	
57·0	9·98484	·9 3	10·01516	10·57043	·9 45	9·42957	10·58559	·9 42	9·41441	
58·0	9·98488		10·01512	10·57094		9·42906	10·58606		9·41394	
59·0	9·98491		10·01509	10·57144		9·42856	10·58653		9·41347	
60·0	9·98494		10·01506	10·57195		9·42805	10·58700		9·41300	0′

105°
285°

K

LOGS. OF TRIG. FUNCTIONS

75°
255°

′	Sine	Parts		Cosec.	Tan.	Parts		Cotan.	Secant	Parts		Cosine	
00·0	9·98494	′		10·01506	10·57195	′		9·42805	10·58700	′		9·41300	60′
01·0	9·98498			10·01502	10·57245			9·42755	10·58748			9·41252	
02·0	9·98501	·1	0	10·01499	10·57296	·1	5	9·42704	10·58795	·1	5	9·41205	
03·0	9·98505	·2	1	10·01496	10·57347	·2	10	9·42653	10·58842	·2	10	9·41158	
04·0	9·98508	·3	1	10·01492	10·57397	·3	15	9·42603	10·58889	·3	14	9·41111	
05·0	9·98511	·4	1	10·01489	10·57448	·4	20	9·42552	10·58937	·4	19	9·41063	55′
06·0	9·98515			10·01485	10·57499			9·42501	10·58984			9·41016	
07·0	9·98518	·5	2	10·01482	10·57550	·5	25	9·42450	10·59032	·5	24	9·40968	
08·0	9·98521	·6	2	10·01479	10·57601	·6	31	9·42399	10·59079	·6	29	9·40921	
09·0	9·98525	·7	2	10·01475	10·57652	·7	36	9·42348	10·59127	·7	33	9·40873	
10·0	9·98528	·8	3	10·01472	10·57703	·8	41	9·42297	10·59175	·8	38	9·40825	50′
11·0	9·98531	·9	3	10·01469	10·57754	·9	46	9·42246	10·59222	·9	43	9·40778	
12·0	9·98535			10·01465	10·57805			9·42195	10·59270			9·40730	
13·0	9·98538			10·01462	10·57856			9·42144	10·59318			9·40682	
14·0	9·98541			10·01459	10·57907			9·42093	10·59366			9·40634	
15·0	9·98545			10·01455	10·57959			9·42042	10·59414			9·40586	45′
16·0	9·98548			10·01452	10·58010			9·41990	10·59462			9·40538	
17·0	9·98551	·1	0	10·01449	10·58061	·1	5	9·41939	10·59510	·1	5	9·40490	
18·0	9·98555	·2	1	10·01445	10·58113	·2	10	9·41887	10·59558	·2	10	9·40442	
19·0	9·98558	·3	1	10·01442	10·58164	·3	16	9·41836	10·59606	·3	15	9·40394	
20·0	9·98561	·4	1	10·01439	10·58216	·4	21	9·41784	10·59655	·4	19	9·40346	40′
21·0	9·98565			10·01435	10·58267			9·41733	10·59703			9·40297	
22·0	9·98568	·5	2	10·01432	10·58319	·5	26	9·41681	10·59751	·5	24	9·40249	
23·0	9·98571	·6	2	10·01429	10·58371	·6	31	9·41629	10·59800	·6	29	9·40201	
24·0	9·98575	·7	2	10·01426	10·58423	·7	36	9·41578	10·59848	·7	34	9·40152	
25·0	9·98578	·8	3	10·01422	10·58474	·8	41	9·41526	10·59897	·8	39	9·40104	35′
26·0	9·98581	·9	3	10·01419	10·58526	·9	47	9·41474	10·59945	·9	44	9·40055	
27·0	9·98584			10·01416	10·58578			9·41422	10·59994			9·40006	
28·0	9·98588			10·01412	10·58630			9·41370	10·60043			9·39958	
29·0	9·98591			10·01409	10·58682			9·41318	10·60091			9·39909	
30·0	9·98594			10·01406	10·58734			9·41266	10·60140			9·39860	30′
31·0	9·98597			10·01403	10·58786			9·41214	10·60189			9·39811	
32·0	9·98601	·1	0	10·01399	10·58839	·1	5	9·41162	10·60238	·1	5	9·39762	
33·0	9·98604	·2	1	10·01396	10·58891	·2	11	9·41109	10·60287	·2	10	9·39713	
34·0	9·98607	·3	1	10·01393	10·58943	·3	16	9·41057	10·60336	·3	15	9·39664	
35·0	9·98610	·4	1	10·01390	10·58996	·4	21	9·41005	10·60385	·4	20	9·39615	25′
36·0	9·98614			10·01386	10·59048			9·40952	10·60434			9·39566	
37·0	9·98617	·5	2	10·01383	10·59100	·5	26	9·40900	10·60483	·5	25	9·39517	
38·0	9·98620	·6	2	10·01380	10·59153	·6	32	9·40847	10·60533	·6	30	9·39467	
39·0	9·98623	·7	2	10·01377	10·59206	·7	37	9·40795	10·60582	·7	34	9·39418	
40·0	9·98627	·8	3	10·01373	10·59258	·8	42	9·40742	10·60632	·8	39	9·39369	20′
41·0	9·98630	·9	3	10·01370	10·59311	·9	47	9·40689	10·60681	·9	44	9·39319	
42·0	9·98633			10·01367	10·59364			9·40636	10·60731			9·39270	
43·0	9·98636			10·01364	10·59416			9·40584	10·60780			9·39220	
44·0	9·98640			10·01361	10·59469			9·40531	10·60830			9·39170	
45·0	9·98643			10·01358	10·59522			9·40478	10·60879			9·39121	15′
46·0	9·98646			10·01354	10·59575			9·40425	10·60929			9·39071	
47·0	9·98649	·1	0	10·01351	10·59628	·1	5	9·40372	10·60979	·1	5	9·39021	
48·0	9·98652	·2	1	10·01348	10·59681	·2	11	9·40319	10·61029	·2	10	9·38971	
49·0	9·98656	·3	1	10·01345	10·59734	·3	16	9·40266	10·61079	·3	15	9·38921	
50·0	9·98659	·4	1	10·01341	10·59788	·4	21	9·40212	10·61129	·4	20	9·38871	10′
51·0	9·98662			10·01338	10·59841			9·40159	10·61179			9·38821	
52·0	9·98665	·5	2	10·01335	10·59894	·5	27	9·40106	10·61229	·5	25	9·38771	
53·0	9·98668	·6	2	10·01332	10·59948	·6	32	9·40052	10·61279	·6	30	9·38721	
54·0	9·98671	·7	2	10·01329	10·60001	·7	37	9·39999	10·61330	·7	35	9·38670	
55·0	9·98675	·8	3	10·01325	10·60055	·8	43	9·39946	10·61380	·8	40	9·38620	5′
56·0	9·98678	·9	3	10·01322	10·60108	·9	48	9·39892	10·61430	·9	45	9·38570	
57·0	9·98681			10·01319	10·60162			9·39838	10·61481			9·38519	
58·0	9·98684			10·01316	10·60215			9·39785	10·61531			9·38469	
59·0	9·98687			10·01313	10·60269			9·39731	10·61582			9·38418	
60·0	9·98690			10·01310	10·60323			9·39677	10·61632			9·38368	0′

76°
256°

LOGS. OF TRIG. FUNCTIONS

′	Sine	Parts		Cosec.	Tan.	Parts		Cotan.	Secant	Parts		Cosine	
00·0	9·98690			10·01310	10·60323			9·39677	10·61632			9·38368	60′
01·0	9·98693			10·01306	10·60377			9·39623	10·61683			9·38317	
02·0	9·98697	·1	0	10·01303	10·60431	·1	5	9·39569	10·61734	·1	5	9·38266	
03·0	9·98700	·2	1	10·01300	10·60485	·2	11	9·39515	10·61785	·2	10	9·38215	
04·0	9·98703	·3	1	10·01297	10·60539	·3	16	9·39461	10·61836	·3	15	9·38164	55′
05·0	9·98706	·4	1	10·01294	10·60593	·4	22	9·39407	10·61887	·4	20	9·38113	
06·0	9·98709			10·01291	10·60647			9·39353	10·61938			9·38062	
07·0	9·98712	·5	2	10·01288	10·60701	·5	27	9·39299	10·61989	·5	26	9·38011	
08·0	9·98716	·6	2	10·01285	10·60755	·6	33	9·39245	10·62040	·6	31	9·37960	
09·0	9·98719			10·01281	10·60810			9·39190	10·62091			9·37909	50′
10·0	9·98722	·7	2	10·01278	10·60864	·7	38	9·39136	10·62142	·7	36	9·37858	
11·0	9·98725	·8	3	10·01275	10·60919	·8	43	9·39082	10·62194	·8	41	9·37806	
12·0	9·98728	·9	3	10·01272	10·60973	·9	49	9·39027	10·62245	·9	46	9·37755	
13·0	9·98731			10·01269	10·61028			9·38972	10·62297			9·37704	
14·0	9·98734			10·01266	10·61082			9·38918	10·62348			9·37652	
15·0	9·98737			10·01263	10·61137			9·38863	10·62400			9·37600	45′
16·0	9·98740			10·01260	10·61192			9·38808	10·62451			9·37549	
17·0	9·98743	·1	0	10·01257	10·61246	·1	6	9·38754	10·62503	·1	5	9·37497	
18·0	9·98747	·2	1	10·01254	10·61301	·2	11	9·38699	10·62555	·2	10	9·37445	
19·0	9·98750	·3	1	10·01250	10·61356	·3	17	9·38644	10·62607	·3	16	9·37393	
20·0	9·98753	·4	1	10·01247	10·61411	·4	22	9·38589	10·62659	·4	21	9·37341	40′
21·0	9·98756			10·01244	10·61466			9·38534	10·62711			9·37289	
22·0	9·98759	·5	2	10·01241	10·61521	·5	28	9·38479	10·62763	·5	26	9·37237	
23·0	9·98762	·6	2	10·01238	10·61577	·6	33	9·38423	10·62815	·6	31	9·37185	
24·0	9·98765			10·01235	10·61632			9·38368	10·62867			9·37133	
25·0	9·98768	·7	2	10·01232	10·61687	·7	39	9·38313	10·62919	·7	36	9·37081	35′
26·0	9·98771	·8	2	10·01229	10·61743	·8	44	9·38258	10·62972	·8	42	9·37029	
27·0	9·98774	·9	3	10·01226	10·61798	·9	50	9·38202	10·63024	·9	47	9·36976	
28·0	9·98777			10·01223	10·61853			9·38147	10·63076			9·36924	
29·0	9·98780			10·01220	10·61909			9·38091	10·63129			9·36871	
30·0	9·98783			10·01217	10·61965			9·38035	10·63182			9·36819	30′
31·0	9·98786			10·01214	10·62020			9·37980	10·63234			9·36766	
32·0	9·98789	·1	0	10·01211	10·62076	·1	6	9·37924	10·63287	·1	5	9·36713	
33·0	9·98792	·2	1	10·01208	10·62132	·2	11	9·37868	10·63340	·2	11	9·36660	
34·0	9·98795	·3	1	10·01205	10·62188	·3	17	9·37812	10·63393	·3	16	9·36608	
35·0	9·98798	·4	1	10·01202	10·62244	·4	22	9·37756	10·63445	·4	21	9·36555	25′
36·0	9·98801			10·01199	10·62300			9·37700	10·63498			9·36502	
37·0	9·98804	·5	2	10·01196	10·62356	·5	28	9·37644	10·63552	·5	27	9·36449	
38·0	9·98807	·6	2	10·01193	10·62412	·6	34	9·37588	10·63605	·6	32	9·36395	
39·0	9·98810			10·01190	10·62468			9·37532	10·63658			9·36342	
40·0	9·98813	·7	2	10·01187	10·62524	·7	39	9·37476	10·63711	·7	37	9·36289	20′
41·0	9·98816	·8	2	10·01184	10·62581	·8	45	9·37419	10·63764	·8	43	9·36236	
42·0	9·98819	·9	3	10·01181	10·62637	·9	51	9·37363	10·63818	·9	48	9·36182	
43·0	9·98822			10·01178	10·62694			9·37306	10·63871			9·36129	
44·0	9·98825			10·01175	10·62750			9·37250	10·63925			9·36075	
45·0	9·98828			10·01172	10·62807			9·37193	10·63979			9·36022	15′
46·0	9·98831			10·01169	10·62863			9·37137	10·64032			9·35968	
47·0	9·98834	·1	0	10·01166	10·62920	·1	6	9·37080	10·64086	·1	5	9·35914	
48·0	9·98837	·2	1	10·01163	10·62977	·2	11	9·37023	10·64140	·2	11	9·35860	
49·0	9·98840	·3	1	10·01160	10·63034	·3	17	9·36966	10·64194	·3	16	9·35806	
50·0	9·98843	·4	1	10·01157	10·63091	·4	23	9·36909	10·64248	·4	22	9·35752	10′
51·0	9·98846			10·01154	10·63148			9·36852	10·64302			9·35698	
52·0	9·98849	·5	1	10·01151	10·63205	·5	29	9·36795	10·64356	·5	27	9·35644	
53·0	9·98852	·6	2	10·01148	10·63262	·6	34	9·36738	10·64410	·6	33	9·35590	
54·0	9·98855			10·01145	10·63319			9·36681	10·64464			9·35536	
55·0	9·98858	·7	2	10·01142	10·63376	·7	40	9·36624	10·64519	·7	38	9·35482	5′
56·0	9·98861	·8	2	10·01139	10·63434	·8	46	9·36566	10·64572	·8	43	9·35427	
57·0	9·98864	·9	3	10·01136	10·63491	·9	51	9·36509	10·64627	·9	49	9·35373	
58·0	9·98867			10·01133	10·63549			9·36452	10·64682			9·35318	
59·0	9·98870			10·01131	10·63606			9·36394	10·64737			9·35264	0′
60·0	9·98872			10·01128	10·63664			9·36336	10·64791			9·35209	

77°
257°

LOGS. OF TRIG. FUNCTIONS

′	Sine	Parts		Cosec.	Tan.	Parts		Cotan.	Secant	Parts		Cosine	
00·0	9·98872	′		10·01128	10·63664	′		9·36336	10·64791	′		9·35209	60′
01·0	9·98875			10·01125	10·63721			9·36279	10·64846			9·35154	
02·0	9·98878	·1	0	10·01122	10·63779	·1	6	9·36221	10·64901	·1	6	9·35099	
03·0	9·98881	·2	1	10·01119	10·63837	·2	12	9·36163	10·64956	·2	11	9·35044	
04·0	9·98884	·3	1	10·01116	10·63895	·3	17	9·36105	10·65011	·3	17	9·34989	
05·0	9·98887	·4	1	10·01113	10·63953	·4	23	9·36047	10·65066	·4	22	9·34934	55′
06·0	9·98890	·5	1	10·01110	10·64011	·5	29	9·35989	10·65121	·5	28	9·34879	
07·0	9·98893	·6	2	10·01107	10·64069	·6	35	9·35931	10·65176	·6	33	9·34824	
08·0	9·98896	·7	2	10·01104	10·64127	·7	41	9·35873	10·65231	·7	39	9·34769	
09·0	9·98899	·8	2	10·01102	10·64185	·8	47	9·35815	10·65287	·8	44	9·34713	
10·0	9·98901	·9	3	10·01099	10·64243	·9	52	9·35757	10·65342	·9	50	9·34658	50′
11·0	9·98904			10·01096	10·64302			9·35698	10·65398			9·34602	
12·0	9·98907			10·01093	10·64360			9·35640	10·65453			9·34547	
13·0	9·98910			10·01090	10·64419			9·35581	10·65509			9·34491	
14·0	9·98913			10·01087	10·64477			9·35523	10·65565			9·34436	
15·0	9·98916			10·01084	10·64536			9·35464	10·65620			9·34380	45′
16·0	9·98919			10·01081	10·64595			9·35405	10·65676			9·34324	
17·0	9·98921	·1	0	10·01079	10·64654	·1	6	9·35347	10·65732	·1	6	9·34268	
18·0	9·98924	·2	1	10·01076	10·64712	·2	12	9·35288	10·65788	·2	11	9·34212	
19·0	9·98927	·3	1	10·01073	10·64771	·3	18	9·35229	10·65844	·3	17	9·34156	
20·0	9·98930	·4	1	10·01070	10·64830	·4	24	9·35170	10·65900	·4	23	9·34100	40′
21·0	9·98933	·5	1	10·01067	10·64889	·5	30	9·35111	10·65957	·5	28	9·34043	
22·0	9·98936	·6	2	10·01064	10·64949	·6	36	9·35051	10·66013	·6	34	9·33987	
23·0	9·98939	·7	2	10·01062	10·65008	·7	41	9·34992	10·66069	·7	39	9·33931	
24·0	9·98941	·8	2	10·01059	10·65067	·8	47	9·34933	10·66126	·8	45	9·33874	
25·0	9·98944	·9	2	10·01056	10·65127	·9	53	9·34874	10·66182	·9	51	9·33818	35′
26·0	9·98947			10·01053	10·65186			9·34814	10·66239			9·33761	
27·0	9·98950			10·01050	10·65246			9·34755	10·66296			9·33704	
28·0	9·98953			10·01048	10·65305			9·34695	10·66353			9·33648	
29·0	9·98955			10·01045	10·65365			9·34635	10·66409			9·33591	
30·0	9·98958			10·01042	10·65425			9·34576	10·66466			9·33534	30′
31·0	9·98961			10·01039	10·65484			9·34516	10·66523			9·33477	
32·0	9·98964	·1	0	10·01036	10·65544	·1	6	9·34456	10·66581	·1	6	9·33420	
33·0	9·98967	·2	1	10·01034	10·65604	·2	12	9·34396	10·66638	·2	12	9·33362	
34·0	9·98969	·3	1	10·01031	10·65664	·3	18	9·34336	10·66695	·3	17	9·33305	
35·0	9·98972	·4	1	10·01028	10·65724	·4	24	9·34276	10·66752	·4	23	9·33248	25′
36·0	9·98975	·5	1	10·01025	10·65785	·5	30	9·34216	10·66810	·5	29	9·33190	
37·0	9·98978	·6	2	10·01022	10·65845	·6	36	9·34155	10·66867	·6	35	9·33133	
38·0	9·98980	·7	2	10·01020	10·65905	·7	42	9·34095	10·66925	·7	40	9·33075	
39·0	9·98983	·8	2	10·01017	10·65966	·8	48	9·34034	10·66982	·8	46	9·33018	
40·0	9·98986	·9	2	10·01014	10·66026	·9	54	9·33974	10·67040	·9	52	9·32960	20′
41·0	9·98989			10·01011	10·66087			9·33913	10·67098			9·32902	
42·0	9·98992			10·01009	10·66147			9·33853	10·67156			9·32844	
43·0	9·98994			10·01006	10·66208			9·33792	10·67214			9·32786	
44·0	9·98997			10·01003	10·66269			9·33731	10·67272			9·32728	
45·0	9·99000			10·01000	10·66330			9·33670	10·67330			9·32670	15′
46·0	9·99003			10·00998	10·66391			9·33609	10·67388			9·32612	
47·0	9·99005	·1	0	10·00995	10·66452	·1	6	9·33548	10·67447	·1	6	9·32553	
48·0	9·99008	·2	1	10·00992	10·66513	·2	12	9·33487	10·67505	·2	12	9·32495	
49·0	9·99011	·3	1	10·00989	10·66574	·3	18	9·33426	10·67563	·3	18	9·32437	
50·0	9·99013	·4	1	10·00987	10·66635	·4	25	9·33365	10·67622	·4	24	9·32378	10′
51·0	9·99016	·5	1	10·00984	10·66697	·5	31	9·33303	10·67681	·5	29	9·32319	
52·0	9·99019	·6	2	10·00981	10·66758	·6	37	9·33242	10·67739	·6	35	9·32261	
53·0	9·99022	·7	2	10·00979	10·66820	·7	43	9·33180	10·67798	·7	41	9·32202	
54·0	9·99024	·8	2	10·00976	10·66881	·8	49	9·33119	10·67857	·8	47	9·32143	
55·0	9·99027	·9	2	10·00973	10·66943	·9	55	9·33057	10·67916	·9	53	9·32084	5′
56·0	9·99030			10·00970	10·67005			9·32995	10·67975			9·32025	
57·0	9·99032			10·00968	10·67067			9·32933	10·68034			9·31966	
58·0	9·99035			10·00965	10·67129			9·32872	10·68093			9·31907	
59·0	9·99038			10·00962	10·67191			9·32810	10·68153			9·31847	
60·0	9·99040			10·00960	10·67253			9·32748	10·68212			9·31788	0′

78°
258°

LOGS. OF TRIG. FUNCTIONS

′	Sine	Parts	Cosec.	Tan.	Parts	Cotan.	Secant	Parts	Cosine	
00·0	9·99040	′	10·00960	10·67253	′	9·32748	10·68212	′	9·31788	60′
01·0	9·99043		10·00957	10·67315		9·32685	10·68272		9·31728	
02·0	9·99046	·1 0	10·00954	10·67377	·1 6	9·32623	10·68331	·1 6	9·31669	
03·0	9·99049	·2 1	10·00952	10·67439	·2 13	9·32561	10·68391	·2 12	9·31609	
04·0	9·99051	·3 1	10·00949	10·67502	·3 19	9·32498	10·68451	·3 18	9·31550	
05·0	9·99054	·4 1	10·00946	10·67564	·4 25	9·32436	10·68510	·4 24	9·31490	55′
06·0	9·99057		10·00944	10·67627		9·32373	10·68570		9·31430	
07·0	9·99059	·5 1	10·00941	10·67689	·5 31	9·32311	10·68630	·5 30	9·31370	
08·0	9·99062	·6 2	10·00938	10·67752	·6 38	9·32248	10·68690	·6 36	9·31310	
09·0	9·99065	·7 2	10·00935	10·67815	·7 44	9·32185	10·68751	·7 42	9·31250	
10·0	9·99067		10·00933	10·67878		9·32122	10·68811		9·31189	50′
11·0	9·99070	·8 2	10·00930	10·67941	·8 50	9·32059	10·68871	·8 48	9·31129	
12·0	9·99072	·9 2	10·00928	10·68004	·9 57	9·31996	10·68932	·9 54	9′31069	
13·0	9·99075		10·00925	10·68067		9·31933	10·68992		9·31008	
14·0	9·99078		10·00922	10·68130		9·31870	10·69053		9·30947	
15·0	9·99080		10·00920	10·68194		9·31806	10·69113		9·30887	45′
16·0	9·99083		10·00917	10·68257		9·31743	10·69174		9·30826	
17·0	9·99086	·1 0	10·00915	10·68321	·1 6	9·31680	10·69235	·1 6	9·30765	
18·0	9·99088	·2 1	10·00912	10·68384	·2 13	9·31616	10·69296	·2 12	9·30704	
19·0	9·99091	·3 1	10·00909	10·68448	·3 19	9·31552	10·69357	·3 18	9·30643	
20·0	9·99093	·4 1	10·00907	10·68512	·4 26	9·31489	10·69418	·4 25	9·30582	40′
21·0	9·99096		10·00904	10·68575		9·31425	10·69479		9·30521	
22·0	9·99099	·5 1	10·00901	10·68639	·5 32	9·31361	10·69541	·5 31	9·30459	
23·0	9·99101	·6 2	10·00899	10·68703	·6 38	9·31297	10·69602	·6 37	9·30398	
24·0	9·99104	·7 2	10·00896	10·68767	·7 45	9·31233	10·69664	·7 43	9·30336	
25·0	9·99106	·8 2	10·00894	10·68832	·8 51	9·31169	10·69725	·8 49	9·30275	35′
26·0	9·99109	·9 2	10·00891	10·68896	·9 58	9·31104	10·69787	·9 55	9·30213	
27·0	9·99112		10·00889	10·68960		9·31040	10·69849		9·30151	
28·0	9·99114		10·00886	10·69025		9·30975	10·69911		9·30090	
29·0	9·99117		10·00883	10·69089		9·30911	10·69972		9·30028	
30·0	9·99119		10·00881	10·69154		9·30846	10·70035		9·29966	30′
31·0	9·99122		10·00878	10·69218		9·30781	10·70097		9·29903	
32·0	9·99124	·1 0	10·00876	10·69283	·1 7	9·30717	10·70159	·1 6	9·29841	
33·0	9·99127	·2 1	10·00873	10·69348	·2 13	9·30652	10·70221	·2 13	9·29779	
34·0	9·99130	·3 1	10·00871	10·69413	·3 20	9·30587	10·70284	·3 19	9·29716	
35·0	9·99132	·4 1	10·00868	10·69478	·4 26	9·30522	10·70346	·4 25	9·29654	25′
36·0	9·99135		10·00865	10·69543		9·30457	10·70409		9·29591	
37·0	9·99137	·5 1	10·00863	10·69609	·5 33	9·30391	10·70471	·5 31	9·29529	
38·0	9·99140	·6 2	10·00860	10·69674	·6 39	9·30326	10·70534	·6 38	9·29466	
39·0	9·99142	·7 2	10·00858	10·69739	·7 46	9·30261	10·70597	·7 44	9·29403	
40·0	9·99145	·8 2	10·00855	10·69805	·8 52	9·30195	10·70660	·8 50	9·29340	20′
41·0	9·99147	·9 2	10·00853	10·69871	·9 59	9·30130	10·70723	·9 57	9·29277	
42·0	9·99150		10·00850	10·69936		9·30064	10·70786		9·29214	
43·0	9·99152		10·00848	10·70002		9·29998	10·70850		9·29150	
44·0	9·99155		10·00845	10·70068		9·29932	10·70913		9·29087	
45·0	9·99157		10·00843	10·70134		9·29866	10·70976		9·29024	15′
46·0	9·99160		10·00840	10·70200		9·29800	10·71040		9·28960	
47·0	9·99162	·1 0	10·00838	10·70266	·1 7	9·29734	10·71104	·1 6	9·28896	
48·0	9·99165	·2 1	10·00835	10·70332	·2 13	9·29668	10·71167	·2 13	9·28833	
49·0	9·99167	·3 1	10·00833	10·70399	·3 20	9·29601	1C·71231	·3 19	9·28769	
50·0	9·99170	·4 1	10·00830	10·70465	·4 27	9·29535	10·71295	·4 26	9·28705	10′
51·0	9·99172		10·00828	10·70532		9·29468	10·71359		9·28641	
52·0	9·99175	·5 1	10·00825	10·70598	·5 33	9·29402	10·71423	·5 32	9·28577	
53·0	9·99177	·6 2	10·00823	10·70665	·6 40	9·29335	10·71488	·6 39	9·28512	
54·0	9·99180	·7 2	10·00820	10·70732	·7 47	9·29268	10·71552	·7 45	9·28448	
55·0	9·99182	·8 2	10·00818	10·70799	·8 53	9·29201	10·71616	·8 51	9·28384	5′
56·0	9·99185	·9 2	10·00815	10·70866	·9 60	9·29134	10·71681	·9 58	9·28319	
57·0	9·99187		10·00813	10·70933		9·29067	10·71746		9·28254	
58·0	9·99190		10·00810	10·71000		9·29000	10·71810		9·28190	
59·0	9·99192		10·00808	10·71067		9·28933	10·71875		9·28125	
60·0	9·99195		10·00805	10·71135		9·28865	10·71940		9·28060	0′

79°
259°

LOGS. OF TRIG. FUNCTIONS

′	Sine	Parts		Cosec.	Tan.	Parts		Cotan.	Secant	Parts		Cosine	
00·0	9·99195	′		10·00805	10·71135	′		9·28865	10·71940	′		9·28060	60′
01·0	9·99197	·1	0	10·00803	10·71202	·1	7	9·28798	10·72005	·1	7	9·27995	
02·0	9·99200	·2	0	10·00800	10·71270	·2	14	9·28730	10·72070	·2	13	9·27930	
03·0	9·99202	·3	1	10·00798	10·71338	·3	20	9·28662	10·72136	·3	20	9·27865	
04·0	9·99204	·4	1	10·00796	10·71405	·4	27	9·28595	10·72201	·4	26	9·27799	
05·0	9·99207	·5	1	10·00793	10·71473	·5	34	9·28527	10·72266	·5	33	9·27734	55′
06·0	9·99209	·6	1	10·00791	10·71541	·6	41	9·28459	10·72332	·6	39	9·27668	
07·0	9·99212	·7	2	10·00788	10·71609	·7	48	9·28391	10·72398	·7	46	9·27603	
08·0	9·99214	·8	2	10·00786	10·71678	·8	54	9·28323	10·72463	·8	52	9·27537	
09·0	9·99217	·9	2	10·00783	10·71746	·9	61	9·28254	10·72529	·9	59	9·27471	
10·0	9·99219			10·00781	10·71814			9·28186	10·72595			9·27405	50′
11·0	9·99221	·1	0	10·00779	10·71883	·1	7	9·28117	10·72661	·1	7	9·27339	
12·0	9·99224	·2	0	10·00776	10·71951	·2	14	9·28049	10·72727	·2	13	9·27273	
13·0	9·99226	·3	1	10·00774	10·72020	·3	21	9·27980	10·72794	·3	20	9·27206	
14·0	9·99229	·4	1	10·00771	10·72089	·4	28	9·27911	10·72860	·4	27	9·27140	
15·0	9·99231	·5	1	10·00769	10·72158	·5	35	9·27842	10·72927	·5	33	9·27074	45′
16·0	9·99234	·6	1	10·00767	10·72227	·6	41	9·27773	10·72993	·6	40	9·27007	
17·0	9·99236	·7	2	10·00764	10·72296	·7	48	9·27704	10·73060	·7	47	9·26940	
18·0	9·99238	·8	2	10·00762	10·72365	·8	55	9·27635	10·73127	·8	53	9·26873	
19·0	9·99241	·9	2	10·00759	10·72434	·9	62	9·27566	10·73194	·9	60	9·26807	
20·0	9·99243			10·00757	10·72504			9·27496	10·73261			9·26740	40′
21·0	9·99245	·1	0	10·00755	10·72573	·1	7	9·27427	10·73328	·1	7	9·26672	
22·0	9·99248	·2	0	10·00752	10·72643	·2	14	9·27357	10·73395	·2	14	9·26605	
23·0	9·99250	·3	1	10·00750	10·72712	·3	21	9·27288	10·73462	·3	20	9·26538	
24·0	9·99253	·4	1	10·00748	10·72782	·4	28	9·27218	10·73530	·4	27	9·26470	
25·0	9·99255	·5	1	10·00745	10·72852	·5	35	9·27148	10·73597	·5	34	9·26403	35′
26·0	9·99257	·6	1	10·00743	10·72922	·6	42	9·27078	10·73665	·6	41	9·26335	
27·0	9·99260	·7	2	10·00740	10·72992	·7	49	9·27008	10·73733	·7	47	9·26267	
28·0	9·99262	·8	2	10·00738	10·73063	·8	56	9·26938	10·73801	·8	54	9·26199	
29·0	9·99264	·9	2	10·00736	10·73133	·9	63	9·26867	10·73869	·9	61	9·26131	
30·0	9·99267			10·00733	10·73203			9·26797	10·73937			9·26063	30′
31·0	9·99269	·1	0	10·00731	10·73274	·1	7	9·26726	10·74005	·1	7	9·25995	
32·0	9·99271	·2	0	10·00729	10·73345	·2	14	9·26656	10·74073	·2	14	9·25927	
33·0	9·99274	·3	1	10·00727	10·73415	·3	21	9·26585	10·74142	·3	21	9·25858	
34·0	9·99276	·4	1	10·00724	10·73486	·4	28	9·26514	10·74210	·4	27	9·25790	
35·0	9·99278	·5	1	10·00722	10·73557	·5	36	9·26443	10·74279	·5	34	9·25721	25′
36·0	9·99281	·6	1	10·00719	10·73628	·6	43	9·26372	10·74348	·6	41	9·25652	
37·0	9·99283	·7	2	10·00717	10·73700	·7	50	9·26301	10·74417	·7	48	9·25583	
38·0	9·99285	·8	2	10·00715	10·73771	·8	57	9·26229	10·74486	·8	55	9·25514	
39·0	9·99288	·9	2	10·00713	10·73842	·9	64	9·26158	10·74555	·9	62	9·25445	
40·0	9·99290			10·00710	10·73914			9·26086	10·74624			9·25376	20′
41·0	9·99292	·1	0	10·00708	10·73985	·1	7	9·26015	10·74693	·1	7	9·25307	
42·0	9·99294	·2	0	10·00706	10·74057	·2	14	9·25943	10·74763	·2	14	9·25237	
43·0	9·99297	·3	1	10·00703	10·74129	·3	22	9·25871	10·74832	·3	21	9·25168	
44·0	9·99299	·4	1	10·00701	10·74201	·4	29	9·25799	10·74902	·4	28	9·25098	
45·0	9·99301	·5	1	10·00699	10·74273	·5	36	9·25727	10·74972	·5	35	9·25028	15′
46·0	9·99304	·6	1	10·00696	10·74345	·6	43	9·25655	10·75042	·6	42	9·24958	
47·0	9·99306	·7	2	10·00694	10·74418	·7	50	9·25582	10·75112	·7	49	9·24888	
48·0	9·99308	·8	2	10·00692	10·74490	·8	58	9·25510	10·75182	·8	56	9·24818	
49·0	9·99310	·9	2	10·00690	10·74563	·9	65	9·25437	10·75252	·9	63	9·24748	
50·0	9·99313			10·00687	10·74635			9·25365	10·75323			9·24678	10′
51·0	9·99315	·1	0	10·00685	10·74708	·1	7	9·25292	10·75393	·1	7	9·24607	
52·0	9·99317	·2	0	10·00683	10·74781	·2	15	9·25219	10·75464	·2	14	9·24536	
53·0	9·99320	·3	1	10·00681	10·74854	·3	22	9·25146	10·75534	·3	21	9·24466	
54·0	9·99322	·4	1	10·00678	10·74927	·4	29	9·25073	10·75605	·4	28	9·24395	
55·0	9·99324	·5	1	10·00676	10·75000	·5	37	9·25000	10·75676	·5	35	9·24324	5′
56·0	9·99326	·6	1	10·00674	10·75074	·6	44	9·24926	10·75747	·6	43	9·24253	
57·0	9·99328	·7	2	10·00672	10·75147	·7	51	9·24853	10·75819	·7	50	9·24181	
58·0	9·99331	·8	2	10·00669	10·75221	·8	59	9·24779	10·75890	·8	57	9·24110	
59·0	9·99333	·9	2	10·00667	10·75294	·9	66	9·24706	10·75961	·9	64	9·24039	
60·0	9·99335			10·00665	10·75368			9·24632	10·76033			9·23967	0′

100°
280°

LOGS. OF TRIG. FUNCTIONS

30°
260°

′	Sine	Parts		Cosec.	Tan.	Parts		Cotan.	Secant	Parts		Cosine	
00·0	9·99335	′		10·00665	10·75368	′		9·24632	10·76033	′		9·23967	60′
01·0	9·99337	·1	0	10·00663	10·75442	·1	7	9·24558	10·76105	·1	7	9·23895	
02·0	9·99340	·2	0	10·00660	10·75516	·2	15	9·24484	10·76177	·2	14	9·23824	
03·0	9·99342	·3	1	10·00658	10·75590	·3	22	9·24410	10·76249	·3	22	9·23752	
04·0	9·99344	·4	1	10·00656	10·75665	·4	30	9·24335	10·76321	·4	29	9·23680	
05·0	9·99346	·5	1	10·00654	10·75739	·5	37	9·24261	10·76393	·5	36	9·23607	55′
06·0	9·99348	·6	1	10·00652	10·75814	·6	45	9·24187	10·76465	·6	43	9·23535	
07·0	9·99351	·7	2	10·00649	10·75888	·7	52	9·24112	10·76538	·7	51	9·23463	
08·0	9·99353	·8	2	10·00647	10·75963	·8	60	9·24037	10·76610	·8	58	9·23390	
09·0	9·99355	·9	2	10·00645	10·76038	·9	67	9·23962	10·76683	·9	65	9·23317	
10·0	9·99357			10·00643	10·76113			9·23887	10·76756			9·23244	50′
11·0	9·99359	·1	0	10·00641	10·76188	·1	8	9·23812	10·76829	·1	7	9·23172	
12·0	9·99362	·2	0	10·00638	10·76263	·2	15	9·23737	10·76902	·2	15	9·23098	
13·0	9·99364	·3	1	10·00636	10·76339	·3	23	9·23661	10·76975	·3	22	9·23025	
14·0	9·99366	·4	1	10·00634	10·76414	·4	30	9·23586	10·77048	·4	29	9·22952	
15·0	9·99368	·5	1	10·00632	10·76490	·5	38	9·23510	10·77122	·5	37	9·22878	45′
16·0	9·99370	·6	1	10·00630	10·76566	·6	45	9·23435	10·77195	·6	44	9·22805	
17·0	9·99373	·7	2	10·00628	10·76641	·7	53	9·23359	10·77269	·7	51	9·22731	
18·0	9·99375	·8	2	10·00625	10·76717	·8	61	9·23283	10·77343	·8	59	9·22657	
19·0	9·99377	·9	2	10·00623	10·76794	·9	68	9·23207	10·77417	·9	66	9·22583	
20·0	9·99379			10·00621	10·76870			9·23130	10·77491			9·22509	40′
21·0	9·99381	·1	0	10·00619	10·76946	·1	8	9·23054	10·77565	·1	7	9·22435	
22·0	9·99383	·2	0	10·00617	10·77023	·2	15	9·22977	10·77639	·2	15	9·22361	
23·0	9·99385	·3	1	10·00615	10·77099	·3	23	9·22901	10·77714	·3	22	9·22286	
24·0	9·99388	·4	1	10·00613	10·77176	·4	31	9·22824	10·77789	·4	30	9·22211	
25·0	9·99390	·5	1	10·00610	10·77253	·5	38	9·22747	10·77863	·5	37	9·22137	35′
26·0	9·99392	·6	1	10·00608	10·77330	·6	46	9·22670	10·77938	·6	45	9·22062	
27·0	9·99394	·7	1	10·00606	10·77407	·7	54	9·22593	10·78013	·7	52	9·21987	
28·0	9·99396	·8	2	10·00604	10·77484	·8	62	9·22516	10·78088	·8	60	9·21912	
29·0	9·99398	·9	2	10·00602	10·77562	·9	69	9·22438	10·78164	·9	67	9·21836	
30·0	9·99400			10·00600	10·77639			9·22361	10·78239			9·21761	30′
31·0	9·99402	·1	0	10·00598	10·77717	·1	8	9·22283	10·78315	·1	8	9·21685	
32·0	9·99405	·2	0	10·00596	10·77795	·2	16	9·22205	10·78390	·2	15	9·21610	
33·0	9·99407	·3	1	10·00593	10·77873	·3	23	9·22127	10·78466	·3	23	9·21534	
34·0	9·99409	·4	1	10·00591	10·77951	·4	31	9·22049	10·78542	·4	30	9·21458	
35·0	9·99411	·5	1	10·00589	10·78029	·5	39	9·21971	10·78618	·5	38	9·21382	25′
36·0	9·99413	·6	1	10·00587	10·78107	·6	47	9·21893	10·78695	·6	46	9·21306	
37·0	9·99415	·7	1	10·00585	10·78186	·7	55	9·21814	10·78771	·7	53	9·21229	
38·0	9·99417	·8	2	10·00583	10·78264	·8	63	9·21736	10·78847	·8	61	9·21153	
39·0	9·99419	·9	2	10·00581	10·78343	·9	70	9·21657	10·78924	·9	69	9·21076	
40·0	9·99421			10·00579	10·78422			9·21578	10·79001			9·20999	20′
41·0	9·99423	·1	0	10·00577	10·78501	·1	8	9·21499	10·79078	·1	8	9·20922	
42·0	9·99425	·2	0	10·00575	10·78580	·2	16	9·21420	10·79155	·2	16	9·20845	
43·0	9·99427	·3	1	10·00573	10·78660	·3	24	9·21341	10·79232	·3	23	9·20768	
44·0	9·99430	·4	1	10·00571	10·78739	·4	32	9·21261	10·79309	·4	31	9·20691	
45·0	9·99432	·5	1	10·00568	10·78819	·5	40	9·21182	10·79387	·5	39	9·20613	15′
46·0	9·99434	·6	1	10·00566	10·78898	·6	48	9·21102	10·79465	·6	47	9·20535	
47·0	9·99436	·7	1	10·00564	10·78978	·7	56	9·21022	10·79542	·7	54	9·20458	
48·0	9·99438	·8	2	10·00562	10·79058	·8	64	9·20942	10·79620	·8	62	9·20380	
49·0	9·99440	·9	2	10·00560	10·79138	·9	72	9·20862	10·79698	·9	70	9·20302	
50·0	9·99442			10·00558	10·79218			9·20782	10·79777			9·20223	10′
51·0	9·99444	·1	0	10·00556	10·79299	·1	8	9·20701	10·79855	·1	8	9·20145	
52·0	9·99446	·2	0	10·00554	10·79379	·2	16	9·20621	10·79933	·2	16	9·20067	
53·0	9·99448	·3	1	10·00552	10·79460	·3	24	9·20540	10·80012	·3	24	9·19988	
54·0	9·99450	·4	1	10·00550	10·79541	·4	32	9·20459	10·80091	·4	32	9·19909	
55·0	9·99452	·5	1	10·00548	10·79622	·5	41	9·20378	10·80170	·5	40	9·19830	5′
56·0	9·99454	·6	1	10·00546	10·79703	·6	49	9·20297	10·80249	·6	47	9·19751	
57·0	9·99456	·7	1	10·00544	10·79784	·7	57	9·20216	10·80328	·7	55	9·19672	
58·0	9·99458	·8	2	10·00542	10·79866	·8	65	9·20135	10·80408	·8	63	9·19593	
59·0	9·99460	·9	2	10·00540	10·79947	·9	73	9·20053	10·80487	·9	71	9·19513	
60·0	9·99462			10·00538	10·80029			9·19971	10·80567			9·19433	0′

81°
261°

LOGS. OF TRIG. FUNCTIONS

′	Sine	Parts		Cosec.	Tan.	Parts		Cotan.	Secant	Parts		Cosine	
00·0	9·99462	′		10·00538	10·80029	′		9·19971	10·80567	′		9·19433	60′
01·0	9·99464	·1	0	10·00536	10·80111	·1	8	9·19889	10·80647	·1	8	9·19353	
02·0	9·99466	·2	0	10·00534	10·80193	·2	17	9·19807	10·80727	·2	16	9·19273	
03·0	9·99468	·3	1	10·00532	10·00275	·3	25	9·19725	10·80807	·3	24	9·19193	
04·0	9·99470	·4	1	10·00530	10·80357	·4	33	9·19643	10·80887	·4	32	9·19113	
05·0	9·99472	·5	1	10·00528	10·80439	·5	41	9·19561	10·80968	·5	40	9·19033	55′
06·0	9·99474	·6	1	10·00526	10·80522	·6	50	9·19478	10·81048	·6	48	9·18952	
07·0	9·99476	·7	1	10·00524	10·80605	·7	58	9·19395	10·81129	·7	56	9·18871	
08·0	9·99478	·8	2	10·00522	10·80688	·8	66	9·19312	10·81210	·8	64	9·18790	
09·0	9·99480	·9	2	10·00520	10·80771	·9	74	9·19229	10·81291	·9	72	9·18709	
10·0	9·99482			10·00518	10·80854			9·19146	10·81372			9·18628	50′
11·0	9·99484	·1	0	10·00516	10·80937	·1	8	9·19063	10·81453	·1	8	9·18547	
12·0	9·99486	·2	0	10·00514	10·81021	·2	17	9·18979	10·81535	·2	16	9·18465	
13·0	9·99488	·3	1	10·00512	10·81104	·3	25	9·18896	10·81617	·3	25	9·18383	
14·0	9·99490	·4	1	10·00510	10·81188	·4	34	9·18812	10·81698	·4	33	9·18302	
15·0	9·99492	·5	1	10·00508	10·81272	·5	42	9·18728	10·81780	·5	41	9·18220	45′
16·0	9·99494	·6	1	10·00507	10·81356	·6	50	9·18644	10·81863	·6	49	9·18137	
17·0	9·99496	·7	1	10·00505	10·81440	·7	59	9·18560	10·81945	·7	57	9·18055	
18·0	9·99497	·8	2	10·00503	10·81525	·8	67	9·18475	10·82027	·8	66	9·17973	
19·0	9·99499	·9	2	10·00501	10·81609	·9	76	9·18391	10·82110	·9	74	9·17890	
20·0	9·99501			10·00499	10·81694			9·18306	10·82193			9·17807	40′
21·0	9·99503	·1	0	10·00497	10·81779	·1	9	9·18221	10·82276	·1	8	9·17724	
22·0	9·99505	·2	0	10·00495	10·81864	·2	17	9·18136	10·82359	·2	17	9·17641	
23·0	9·99507	·3	1	10·00493	10·81949	·3	26	9·18051	10·82442	·3	25	9·17558	
24·0	9·99509	·4	1	10·00491	10·82035	·4	34	9·17966	10·82526	·4	33	9·17474	
25·0	9·99511	·5	1	10·00489	10·82120	·5	43	9·17880	10·82609	·5	42	9·17391	35′
26·0	9·99513	·6	1	10·00487	10·82206	·6	51	9·17794	10·82693	·6	50	9·17307	
27·0	9·99515	·7	1	10·00485	10·82292	·7	60	9·17708	10·82777	·7	59	9·17223	
28·0	9·99517	·8	2	10·00484	10·82378	·8	68	9·17622	10·82861	·8	67	9·17139	
29·0	9·99518	·9	2	10·00482	10·82464	·9	77	9·17536	10·82945	·9	75	9·17055	
30·0	9·99520			10·00480	10·82550			9·17450	10·83030			9·16970	30′
31·0	9·99522	·1	0	10·00478	10·82637	·1	9	9·17363	10·83114	·1	9	9·16886	
32·0	9·99524	·2	0	10·00476	10·82723	·2	17	9·17277	10·83199	·2	17	9·16801	
33·0	9·99526	·3	1	10·00474	10·82810	·3	26	9·17190	10·83284	·3	26	9·16716	
34·0	9·99528	·4	1	10·00472	10·82897	·4	35	9·17103	10·83369	·4	34	9·16631	
35·0	9·99530	·5	1	10·00470	10·82984	·5	44	9·17016	10·83455	·5	43	9·16545	25′
36·0	9·99532	·6	1	10·00468	10·83072	·6	52	9·16928	10·83540	·6	51	9·16460	
37·0	9·99533	·7	1	10·00467	10·83159	·7	61	9·16841	10·83626	·7	60	9·16374	
38·0	9·99535	·8	2	10·00465	10·83247	·8	70	9·16753	10·83712	·8	68	9·16289	
39·0	9·99537	·9	2	10·00463	10·83335	·9	79	9·16665	10·83798	·9	77	9·16203	
40·0	9·99539			10·00461	10·83423			9·16577	10·83884			9·16116	20′
41·0	9·99541	·1	0	10·00459	10·83511	·1	9	9·16489	10·83970	·1	9	9·16030	
42·0	9·99543	·2	0	10·00457	10·83599	·2	18	9·16401	10·84057	·2	17	9·15944	
43·0	9·99545	·3	1	10·00455	10·83688	·3	27	9·16312	10·84143	·3	26	9·15857	
44·0	9·99546	·4	1	10·00454	10·83776	·4	36	9·16224	10·84230	·4	35	9·15770	
45·0	9·99548	·5	1	10·00452	10·83865	·5	44	9·16135	10·84317	·5	44	9·15683	15′
46·0	9·99550	·6	1	10·00450	10·83954	·6	53	9·16046	10·84404	·6	52	9·15596	
47·0	9·99552	·7	1	10·00448	10·84044	·7	62	9·15957	10·84492	·7	61	9·15508	
48·0	9·99554	·8	1	10·00446	10·84133	·8	71	9·15867	10·84579	·8	70	9·15421	
49·0	9·99556	·9	2	10·00445	10·84223	·9	80	9·15778	10·84667	·9	78	9·15333	
50·0	9·99557			10·00443	10·84312			9·15688	10·84755			9·15245	10′
51·0	9·99559	·1	0	10·00441	10·84402	·1	9	9·15598	10·84843	·1	9	9·15157	
52·0	9·99561	·2	0	10·00439	10·84492	·2	18	9·15508	10·84931	·2	18	9·15069	
53·0	9·99563	·3	1	10·00437	10·84583	·3	27	9·15417	10·85020	·3	27	9·14980	
54·0	9·99565	·4	1	10·00435	10·84673	·4	36	9·15327	10·85109	·4	36	9·14892	
55·0	9·99566	·5	1	10·00434	10·84764	·5	45	9·15236	10·85197	·5	44	9·14803	5′
56·0	9·99568	·6	1	10·00432	10·84855	·6	54	9·15145	10·85286	·6	53	9·14714	
57·0	9·99570	·7	1	10·00430	10·84946	·7	64	9·15054	10·85376	·7	62	9·14624	
58·0	9·99572	·8	1	10·00428	10·85037	·8	73	9·14963	10·85465	·8	71	9·14535	
59·0	9·99574	·9	2	10·00427	10·85128	·9	82	9·14872	10·85555	·9	80	9·14445	
60·0	9·99575			10·00425	10·85220			9·14780	10·85645			9·14356	0′

LOGS. OF TRIG. FUNCTIONS

82°
262°

′	Sine	Parts		Cosec.	Tan.	Parts		Cotan.	Secant	Parts		Cosine	
00·0	9·99575	′		10·00425	10·85220	′		9·14780	10·85645	′		9·14356	60′
01·0	9·99577	·1	0	10·00423	10·85312	·1	9	9·14689	10·85735	·1	9	9·14266	
02·0	9·99579	·2	0	10·00421	10·85403	·2	19	9·14597	10·85825	·2	18	9·14175	
03·0	9·99581	·3	1	10·00419	10·85496	·3	28	9·14504	10·85915	·3	27	9·14085	
04·0	9·99582	·4	1	10·00418	10·85588	·4	37	9·14412	10·86006	·4	36	9·13994	
05·0	9·99584	·5	1	10·00416	10·85680	·5	46	9·14320	10·86096	·5	45	9·13904	55′
06·0	9·99586	·6	1	10·00414	10·85773	·6	56	9·14227	10·86187	·6	55	9·13813	
07·0	9·99588	·7	1	10·00412	10·85866	·7	65	9·14134	10·86278	·7	64	9·13722	
08·0	9·99589	·8	1	10·00411	10·85959	·8	74	9·14041	10·86370	·8	73	9·13630	
09·0	9·99591	·9	2	10·00409	10·86052	·9	83	9·13948	10·86461	·9	82	9·13539	
10·0	9·99593			10·00407	10·86146			9·13854	10·86553			9·13447	50′
11·0	9·99595	·1	0	10·00405	10·86240	·1	9	9·13761	10·86645	·1	9	9·13355	
12·0	9·99596	·2	0	10·00404	10·86333	·2	19	9·13667	10·86737	·2	19	9.13263	
13·0	9·99598	·3	1	10·00402	10·86427	·3	28	9·13573	10·86829	·3	28	9·13171	
14·0	9·99600	·4	1	10·00400	10·86522	·4	38	9·13478	10·86922	·4	37	9·13078	
15·0	9·99602	·5	1	10·00399	10·86616	·5	47	9·13384	10·87015	·5	46	9·12985	45′
16·0	9·99603	·6	1	10·00397	10·86711	·6	57	9·13289	10·87108	·6	56	9·12893	
17·0	9·99605	·7	1	10·00395	10·86806	·7	66	9·13194	10·87201	·7	65	9·12799	
18·0	9·99607	·8	1	10·00393	10·86901	·8	76	9·13099	10·87294	·8	74	9·12706	
19·0	9·99608	·9	2	10·00392	10·86996	·9	85	9·13004	10·87388	·9	84	9·12613	
20·0	9·99610			10·00390	10·87091			9·12909	10·87481			9·12519	40′
21·0	9·99612	·1	0	10·00388	10·87187	·1	10	9·12813	10·87575	·1	9	9·12425	
22·0	9·99613	·2	0	10·00387	10·87283	·2	19	9·12717	10·87669	·2	19	9·12331	
23·0	9·99615	·3	1	10·00385	10·87379	·3	29	9·12621	10·87764	·3	28	9·12236	
24·0	9·99617	·4	1	10·00383	10·87475	·4	39	9·12525	10·87858	·4	38	9·12142	
25·0	9·99619	·5	1	10·00382	10·87572	·5	48	9·12428	10·87953	·5	47	9·12047	35′
26·0	9·99620	·6	1	10·00380	10·87668	·6	58	9·12332	10·88048	·6	57	9·11952	
27·0	9·99622	·7	1	10·00378	10·87765	·7	68	9·12235	10·88143	·7	66	9·11857	
28·0	9·99624	·8	1	10·00377	10·87862	·8	77	9·12138	10·88239	·8	76	9·11761	
29·0	9·99625	·9	2	10·00375	10·87960	·9	87	9·12040	10·88334	·9	85	9·11666	
30·0	9·99627			10·00373	10·88057			9·11943	10·88430			9·11570	30′
31·0	9·99629	·1	0	10·00372	10·88155	·1	10	9·11845	10·88526	·1	10	9·11474	
32·0	9·99630	·2	0	10·00370	10·88253	·2	20	9·11747	10·88623	·2	19	9·11377	
33·0	9·99632	·3	0	10·00368	10·88351	·3	30	9·11649	10·88719	·3	29	9·11281	
34·0	9·99634	·4	1	10·00367	10·88449	·4	40	9·11551	10·88816	·4	39	9·11184	
35·0	9·99635	·5	1	10·00365	10·88548	·5	49	9·11452	10·88913	·5	49	9·11087	25′
36·0	9·99637	·6	1	10·00363	10·88646	·6	59	9·11353	10·89010	·6	58	9·10990	
37·0	9·99638	·7	1	10·00362	10·88746	·7	69	9·11254	10·89107	·7	68	9·10893	
38·0	9·99640	·8	1	10·00360	10·88845	·8	79	9·11155	10·89205	·8	78	9·10795	
39·0	9·99642	·9	1	10·00358	10·88944	·9	89	9·11056	10·89303	·9	87	9·10697	
40·0	9·99643			10·00357	10·89044			9·10956	10·89401			9·10599	20′
41·0	9·99645	·1	0	10·00355	10·89144	·1	10	9·10856	10·89499	·1	10	9·10501	
42·0	9·99647	·2	0	10·00354	10·89244	·2	20	9·10756	10·89598	·2	20	9·10403	
43·0	9·99648	·3	0	10·00352	10·89344	·3	30	9·10656	10·89696	·3	30	9·10304	
44·0	9·99650	·4	1	10·00350	10·89445	·4	40	9·10555	10·89795	·4	40	9·10205	
45·0	9·99651	·5	1	10·00349	10·89546	·5	50	9·10454	10·89894	·5	50	9·10106	15′
46·0	9·99653	·6	1	10·00347	10·89647	·6	61	9·10353	10·89994	·6	60	9·10006	
47·0	9·99655	·7	1	10·00345	10·89748	·7	71	9·10252	10·90094	·7	70	9·09907	
48·0	9·99656	·8	1	10·00344	10·89850	·8	81	9·10150	10·90193	·8	79	9·09807	
49·0	9·99658	·9	1	10·00342	10·89951	·9	91	9·10049	10·90294	·9	89	9·09707	
50·0	9·99659			10·00341	10·90053			9·09947	10·90394			9·09606	10′
51·0	9·99661	·1	0	10·00339	10·90155	·1	10	9·09845	10·90494	·1	10	9·09506	
52·0	9·99663	·2	0	10·00338	10·90258	·2	21	9·09742	10·90595	·2	20	9·09405	
53·0	9·99664	·3	0	10·00336	10·90361	·3	31	9·09640	10·90696	·3	30	9·09304	
54·0	9·99666	·4	1	10·00334	10·90463	·4	41	9·09537	10·90798	·4	41	9·09202	
55·0	9·99667	·5	1	10·00333	10·90566	·5	52	9·09434	10·90899	·5	51	9·09101	5′
56·0	9·99669	·6	1	10·00331	10·90670	·6	62	9·09330	10·91001	·6	61	9·08999	
57·0	9·99670	·7	1	10·00330	10·90773	·7	72	9·09227	10·91103	·7	71	9·08897	
58·0	9·99672	·8	1	10·00328	10·90877	·8	83	9·09123	10·91205	·8	81	9·08795	
59·0	9·99674	·9	1	10·00327	10·90981	·9	93	9·09019	10·91308	·9	92	9·08692	
60·0	9·99675			10·00325	10·91086			9·08914	10·91411			9·08589	0′

83°
263°

LOGS. OF TRIG. FUNCTIONS

′	Sine	Parts	Cosec.	Tan.	Parts	Cotan.	Secant	Parts	Cosine	
00·0	9·99675		10·00325	10·91086	′	9·08914	10·91411	′	9·08589	60′
01·0	9·99677		10·00323	10·91190	·1 11	9·08810	10·91514	·1 10	9·08486	
02·0	9·99678		10·00322	10·91295	·2 21	9·08705	10·91617	·2 21	9·08383	
03·0	9·99680		10·00320	10·91400	·3 32	9·08600	10·91720	·3 31	9·08280	
04·0	9·99681		10·00319	10·91505	·4 42	9·08495	10·91824	·4 42	9·08176	
05·0	9·99683		10·00317	10·91611	·5 53	9·08389	10·91928	·5 52	9·08072	55′
06·0	9·99684		10·00316	10·91717	·6 63	9·08283	10·92032	·6 62	9·07968	
07·0	9·99686		10·00314	10·91823	·7 74	9·08177	10·92137	·7 73	9·07863	
08·0	9·99687		10·00313	10·91929	·8 84	9·08071	10·92242	·8 83	9·07758	
09·0	9·99689		10·00311	10·92036	·9 95	9·07964	10·92347	·9 94	9·07653	
10·0	9·99690		10·00310	10·92142		9·07858	10·92452		9·07548	50′
11·0	9·99692		10·00308	10·92250	·1 11	9·07751	10·92558	·1 11	9·07442	
12·0	9·99693		10·00307	10·92357	·2 22	9·07643	10·92663	·2 21	9·07337	
13·0	9·99695		10·00305	10·92464	·3 32	9·07536	10·92769	·3 32	9·07231	
14·0	9·99696		10·00304	10·92572	·4 43	9·07428	10·92876	·4 43	9·07124	
15·0	9·99698		10·00302	10·92680	·5 54	9·07320	10·92982	·5 53	9·07018	45′
16·0	9·99699		10·00301	10·92789	·6 65	9·07211	10·93089	·6 64	9·06911	
17·0	9·99701		10·00299	10·92897	·7 76	9·07103	10·93196	·7 75	9·06804	
18·0	9·99702		10·00298	10·93006	·8 87	9·06994	10·93304	·8 85	9·06696	
19·0	9·99704		10·00296	10·93115	·9 97	9·06885	10·93412	·9 96	9·06589	
20·0	9·99705		10·00295	10·93225		9·06775	10·93519		9·06481	40′
21·0	9·99707		10·00293	10·93335	·1 11	9·06666	10·93628	·1 11	9·06372	
22·0	9·99708		10·00292	10·93444	·2 22	9·06556	10·93736	·2 22	9·06264	
23·0	9·99710		10·00290	10·93555	·3 33	9·06445	10·93845	·3 33	9·06155	
24·0	9·99711		10·00289	10·93665	·4 44	9·06335	10·93954	·4 44	9·06046	
25·0	9·99713		10·00287	10·93776	·5 55	9·06224	10·94063	·5 55	9·05937	35′
26·0	9·99714		10·00286	10·93887	·6 67	9·06113	10·94173	·6 66	9·05827	
27·0	9·99716		10·00284	10·93998	·7 78	9·06002	10·94283	·7 77	9·05717	
28·0	9·99717		10·00283	10·94110	·8 89	9·05890	10·94393	·8 88	9·05607	
29·0	9·99719		10·00282	10·94222	·9 100	9·05778	10·94503	·9 99	9·05497	
30·0	9·99720		10·00280	10·94334		9·05666	10·94614		9·05386	30′
31·0	9·99721		10·00279	10·94447	·1 11	9·05554	10·94725	·1 11	9·05275	
32·0	9·99723		10·00277	10·94559	·2 23	9·05441	10·94837	·2 22	9·05164	
33·0	9·99724		10·00276	10·94672	·3 34	9·05328	10·94948	·3 34	9·05052	
34·0	9·99726		10·00274	10·94786	·4 46	9·05214	10·95060	·4 45	9·04940	
35·0	9·99727		10·00273	10·94899	·5 57	9·05101	10·95172	·5 56	9·04828	25′
36·0	9·99729		10·00272	10·95013	·6 68	9·04987	10·95285	·6 67	9·04715	
37·0	9·99730		10·00270	10·95127	·7 80	9·04873	10·95397	·7 79	9·04603	
38·0	9·99731		10·00269	10·95242	·8 91	9·04758	10·95511	·8 90	9·04490	
39·0	9·99733		10·00267	10·95357	·9 102	9·04643	10·95624	·9 101	9·04376	
40·0	9·99734		10·00266	10·95472		9·04528	10·95738		9·04263	20′
41·0	9·99736		10·00265	10·95587	·1 12	9·04413	10·95852	·1 12	9·04149	
42·0	9·99737		10·00263	10·95703	·2 23	9·04297	10·95966	·2 23	9·04034	
43·0	9·99738		10·00262	10·95819	·3 35	9·04181	10·96080	·3 35	9·03920	
44·0	9·99740		10·00260	10·95935	·4 47	9·04065	10·96195	·4 46	9·03805	
45·0	9·99741		10·00259	10·96052	·5 58	9·03949	10·96310	·5 58	9·03690	15′
46·0	9·99743		10·00258	10·96168	·6 70	9·03832	10·96426	·6 69	9·03574	
47·0	9·99744		10·00256	10·96286	·7 82	9·03714	10·96542	·7 81	9·03458	
48·0	9·99745		10·00255	10·96403	·8 93	9·03597	10·96658	·8 92	9·03342	
49·0	9·99747		10·00253	10·96521	·9 105	9·03479	10·96774	·9 104	9·03226	
50·0	9·99748		10·00252	10·96639		9·03361	10·96891		9·03109	10′
51·0	9·99749		10·00251	10·96758	·1 12	9·03243	10·97008	·1 12	9·02992	
52·0	9·99751		10·00249	10·96876	·2 24	9·03124	10·97126	·2 24	9·02874	
53·0	9·99752		10·00248	10·96995	·3 36	9·03005	10·97243	·3 36	9·02757	
54·0	9·99753		10·00247	10·97115	·4 48	9·02885	10·97361	·4 47	9·02639	
55·0	9·99755		10·00245	10·97235	·5 60	9·02766	10·97480	·5 59	9·02520	5′
56·0	9·99756		10·00244	10·97355	·6 72	9·02646	10·97598	·6 71	9·02402	
57·0	9·99757		10·00243	10·97475	·7 84	9·02525	10·97718	·7 83	9·02283	
58·0	9·99759		10·00241	10·97596	·8 96	9·02404	10·97837	·8 95	9·02163	
59·0	9·99760		10·00240	10·97717	·9 108	9·02283	10·97957	·9 107	9·02044	
60·0	9·99761		10·00239	10·97838		9·02162	10·98077		9·01924	0′

84°
264°

LOGS. OF TRIG. FUNCTIONS

′	Sine	Parts	Cosec.	Tan.	Parts	Cotan.	Secant	Parts	Cosine	
00·0	9·99761		10·00239	10·97838	′	9·02162	10·98077	′	9·01924	60′
01·0	9·99763		10·00237	10·97960	·1 12	9·02040	10·98197	·1 12	9·01803	
02·0	9·99764		10·00236	10·98082	·2 25	9·01918	10·98318	·2 24	9·01682	
03·0	9·99765		10·00235	10·98204	·3 37	9·01796	10·98439	·3 37	9·01561	
04·0	9·99767		10·00233	10·98327	·4 49	9·01673	10·98560	·4 49	9·01440	
05·0	9·99768		10·00232	10·98450	·5 62	9·01550	10·98682	·5 61	9·01318	55′
06·0	9·99769		10·00231	10·98573	·6 74	9·01427	10·98804	·6 73	9·01196	
07·0	9·99771		10·00229	10·98697	·7 86	9·01303	10·98926	·7 85	9·01074	
08·0	9·99772		10·00228	10·98821	·8 99	9·01179	10·99049	·8 98	9·00951	
09·0	9·99773		10·00227	10·98945	·9 111	9·01055	10·99172	·9 110	9·00828	
10·0	9·99775		10·00226	10·99070		9·00930	10·99296		9·00704	50′
11·0	9·99776		10·00224	10·99195	·1 13	9·00805	10·99420	·1 13	9·00581	
12·0	9·99777		10·00223	10·99321	·2 25	9·00679	10·99544	·2 25	9·00456	
13·0	9·99778		10·00222	10·99447	·3 38	9·00553	10·99668	·3 38	9·00332	
14·0	9·99780		10·00220	10·99573	·4 51	9·00427	10·99793	·4 50	9·00207	
15·0	9·99781		10·00219	10·99700	·5 63	9·00301	10·99918	·5 63	9·00082	45′
16·0	9·99782		10·00218	10·99826	·6 76	9·00174	11·00044	·6 75	8·99956	
17·0	9·99784		10·00217	10·99954	·7 89	9·00047	11·00170	·7 88	8·99830	
18·0	9·99785		10·00215	11·00081	·8 101	8·99919	11·00296	·8 100	8·99704	
19·0	9·99786		10·00214	11·00209	·9 114	8·99791	11·00423	·9 113	8·99577	
20·0	9·99787		10·00213	11·00338		8·99662	11·00550		8·99450	40′
21·0	9·99789		10·00212	11·00466	·1 13	8·99534	11·00678	·1 13	8·99322	
22·0	9·99790		10·00210	11·00596	·2 26	8·99405	11·00806	·2 26	8·99194	
23·0	9·99791		10·00209	11·00725	·3 39	8·99275	11·00934	·3 39	8·99066	
24·0	9·99792		10·00208	11·00855	·4 52	8·99145	11·01063	·4 52	8·98937	
25·0	9·99794		10·00207	11·00985	·5 65	8·99015	11·01192	·5 65	8·98808	35′
26·0	9·99795		10·00205	11·01116	·6 78	8·98884	11·01321	·6 78	8·98679	
27·0	9·99796		10·00204	11·01247	·7 91	8·98753	11·01451	·7 91	8·98549	
28·0	9·99797		10·00203	11·01378	·8 104	8·98622	11·01581	·8 103	8·98419	
29·0	9·99798		10·00202	11·01510	·9 117	8·98490	11·01712	·9 116	8·98288	
30·0	9·99800		10·00200	11·01642		8·98358	11·01843		8·98157	30′
31·0	9·99801		10·00199	11·01775	·1 13	8·98225	11·01974	·1 13	8·98026	
32·0	9·99802		10·00198	11·01908	·2 27	8·98092	11·02106	·2 27	8·97894	
33·0	9·99803		10·00197	11·02041	·3 40	8·97959	11·02238	·3 40	8·97762	
34·0	9·99804		10·00196	11·02175	·4 54	8·97825	11·02371	·4 53	8·97629	
35·0	9·99806		10·00194	11·02309	·5 67	8·97691	11·02504	·5 67	8·97496	25′
36·0	9·99807		10·00193	11·02444	·6 81	8·97556	11·02637	·6 80	8·97363	
37·0	9·99808		10·00192	11·02579	·7 94	8·97421	11·02771	·7 93	8·97229	
38·0	9·99809		10·00191	11·02715	·8 108	8·97286	11·02905	·8 107	8·97095	
39·0	9·99810		10·00190	11·02850	·9 121	8·97150	11·03040	·9 120	8·96960	
40·0	9·99812		10·00188	11·02987		8·97013	11·03175		8·96825	20′
41·0	9·99813		10·00187	11·03123	·1 14	8·96887	11·03311	·1 14	8·96689	
42·0	9·99814		10·00186	11·03261	·2 28	8·96739	11·03447	·2 28	8·96553	
43·0	9·99815		10·00185	11·03398	·3 42	8·96602	11·03583	·3 41	8·96417	
44·0	9·99816		10·00184	11·03536	·4 55	8·96464	11·03720	·4 55	8·96280	
45·0	9·99817		10·00183	11·03675	·5 69	8·96326	11·03857	·5 69	8·96143	15′
46·0	9·99819		10·00181	11·03813	·6 83	8·96187	11·03995	·6 83	8·96005	
47·0	9·99820		10·00180	11·03953	·7 97	8·96047	11·04133	·7 96	8·95867	
48·0	9·99821		10·00179	11·04093	·8 111	8·95908	11·04272	·8 110	8·95728	
49·0	9·99822		10·00178	11·04233	·9 125	8·95767	11·04411	·9 124	8·95589	
50·0	9·99823		10·00177	11·04373		8·95627	11·04550		8·95450	10′
51·0	9·99824		10·00176	11·04514	·1 14	8·95486	11·04690	·1 14	8·95310	
52·0	9·99826		10·00175	11·04656	·2 29	8·95344	11·04830	·2 28	8·95170	
53·0	9·99827		10·00173	11·04798	·3 43	8·95202	11·04971	·3 43	8·95029	
54·0	9·99828		10·00172	11·04940	·4 57	8·95060	11·05113	·4 57	8·94887	
55·0	9·99829		10·00171	11·05083	·5 72	8·94917	11·05254	·5 71	8·94746	5′
56·0	9·99830		10·00170	11·05227	·6 86	8·94773	11·05397	·6 85	8·94603	
57·0	9·99831		10·00169	11·05371	·7 100	8·94630	11·05539	·7 99	8·94461	
58·0	9·99832		10·00168	11·05515	·8 115	8·94485	11·05683	·8 114	8·94317	
59·0	9·99833		10·00167	11·05660	·9 129	8·94340	11·05826	·9 128	8·94174	
60·0	9·99834		10·00166	11·05805		8·94195	11·05970		8·94030	0′

85°
265°

LOGS. OF TRIG. FUNCTIONS

′	Sine	Parts	Cosec.	Tan.	Parts	Cotan.	Secant	Parts	Cosine	
00·0	9·99834		10·00166	11·05805	′	8·94195	11·05970	′	8·94030	60′
01·0	9·99836		10·00165	11·05951	·1 15	8·94049	11·06115	·1 15	8·93885	
02·0	9·99837		10·00163	11·06097	·2 30	8·93903	11·06260	·2 29	8·93740	
03·0	9·99030		10·00162	11·06244	·3 44	8·93757	11·06406	·3 44	8·93594	
04·0	9·99839		10·00161	11·06391	·4 59	8·93609	11·06552	·4 59	8·93448	
05·0	9·99840		10·00160	11·06538	·5 74	8·93462	11·06699	·5 73	8·93302	55′
06·0	9·99841		10·00159	11·06687	·6 89	8·93313	11·06846	·6 88	8·93154	
07·0	9·99842		10·00158	11·06835	·7 104	8·93165	11·06993	·7 103	8·93007	
08·0	9·99843		10·00157	11·06985	·8 118	8·93016	11·07141	·8 118	8·92859	
09·0	9·99844		10·00156	11·07134	·9 133	8·92866	11·07290	·9 132	8·92710	
10·0	9·99845		10·00155	11·07284		8·92716	11·07439		8·92561	50′
11·0	9·99846		10·00154	11·07435	·1 15	8·92565	11·07589	·1 15	8·92411	
12·0	9·99847		10·00153	11·07586	·2 31	8·92414	11·07739	·2 30	8·92261	
13·0	9·99849		10·00152	11·07738	·3 46	8·92262	11·07890	·3 46	8·92110	
14·0	9·99850		10·00151	11·07890	·4 61	8·92110	11·08041	·4 61	8·91959	
15·0	9·99851		10·00149	11·08043	·5 77	8·91957	11·08193	·5 76	8·91807	45′
16·0	9·99852		10·00148	11·08197	·6 92	8·91803	11·08345	·6 91	8·91655	
17·0	9·99853		10·00147	11·08351	·7 107	8·91650	11·08498	·7 106	8·91502	
18·0	9·99854		10·00146	11·08505	·8 122	8·91495	11·08651	·8 122	8·91349	
19·0	9·99855		10·00145	11·08660	·9 138	8·91340	11·08805	·9 137	8·91195	
20·0	9·99856		10·00144	11·08815		8·91185	11·08960		8·91040	40′
21·0	9·99857		10·00143	11·08972	·1 16	8·91029	11·09115	·1 16	8·90885	
22·0	9·99858		10·00142	11·09128	·2 32	8·90872	11·09270	·2 32	8·90730	
23·0	9·99859		10·00141	11·09285	·3 48	8·90715	11·09426	·3 47	8·90574	
24·0	9·99860		10·00140	11·09443	·4 63	8·90557	11·09583	·4 63	8·90417	
25·0	9·99861		10·00139	11·09601	·5 79	8·90399	11·09740	·5 79	8·90260	35′
26·0	9·99862		10·00138	11·09760	·6 95	8·90240	11·09898	·6 95	8·90102	
27·0	9·99863		10·00137	11·09920	·7 111	8·90080	11·10057	·7 110	8·89943	
28·0	9·99864		10·00136	11·10080	·8 127	8·89920	11·10216	·8 126	8·89784	
29·0	9·99865		10·00135	11·10240	·9 143	8·89760	11·10375	·9 142	8·89625	
30·0	9·99866		10·00134	11·10402		8·89598	11·10536		8·89464	30′
31·0	9·99867		10·00133	11·10563	·1 16	8·89437	11·10697	·1 16	8·89304	
32·0	9·99868		10·00132	11·10726	·2 33	8·89274	11·10858	·2 33	8·89142	
33·0	9·99869		10·00131	11·10889	·3 49	8·89111	11·11020	·3 49	8·88980	
34·0	9·99870		10·00130	11·11052	·4 66	8·88948	11·11183	·4 65	8·88817	
35·0	9·99871		10·00129	11·11217	·5 82	8·88783	11·11346	·5 82	8·88654	25′
36·0	9·99872		10·00128	11·11382	·6 99	8·88619	11·11510	·6 98	8·88490	
37·0	9·99873		10·00127	11·11547	·7 115	8·88453	11·11674	·7 114	8·88326	
38·0	9·99874		10·00126	11·11713	·8 132	8·88287	11·11839	·8 131	8·88161	
39·0	9·99875		10·00125	11·11880	·9 148	8·88120	11·12005	·9 147	8·87995	
40·0	9·99876		10·00124	11·12047		8·87953	11·12172		8·87829	20′
41·0	9·99877		10·00123	11·12215	·1 17	8·87785	11·12333	·1 17	8·87662	
42·0	9·99878		10·00122	11·12384	·2 34	8·87616	11·12506	·2 34	8·87494	
43·0	9·99879		10·00122	11·12553	·3 51	8·87447	11·12675	·3 51	8·87326	
44·0	9·99880		10·00121	11·12723	·4 68	8·87277	11·12844	·4 68	8·87157	
45·0	9·99880		10·00120	11·12894	·5 86	8·87106	11·13013	·5 85	8·86987	15′
46·0	9·99881		10·00119	11·13065	·6 103	8·86935	11·13184	·6 102	8·86817	
47·0	9·99882		10·00118	11·13237	·7 120	8·86763	11·13355	·7 119	8·86646	
48·0	9·99883		10·00117	11·13409	·8 137	8·86591	11·13526	·8 136	8·86474	
49·0	9·99884		10·00116	11·13583	·9 154	8·86417	11·13699	·9 153	8·86301	
50·0	9·99885		10·00115	11·13757		8·86243	11·13872		8·86128	10′
51·0	9·99886		10·00114	11·13931	·1 18	8·86069	11·14045	·1 18	8·85955	
52·0	9·99887		10·00113	11·14107	·2 36	8·85893	11·14220	·2 35	8·85780	
53·0	9·99888		10·00112	11·14283	·3 53	8·85717	11·14395	·3 53	8·85605	
54·0	9·99889		10·00111	11·14460	·4 71	8·85540	11·14571	·4 71	8·85429	
55·0	9·99890		10·00110	11·14637	·5 89	8·85363	11·14748	·5 88	8·85253	5′
56·0	9·99891		10·00110	11·14815	·6 107	8·85185	11·14925	·6 106	8·85075	
57·0	9·99891		10·00109	11·14994	·7 125	8·85006	11·15103	·7 124	8·84897	
58·0	9·99892		10·00108	11·15174	·8 142	8·84826	11·15282	·8 142	8·84718	
59·0	9·99893		10·00107	11·15355	·9 160	8·84646	11·15461	·9 159	8·84539	
60·0	9·99894		10·00106	11·15536		8·84464	11·15642		8·84359	0′

LOGS. OF TRIG. FUNCTIONS

′	Sine	Diff.	Cosec.	Tan.	Diff.	Cotan.	Secant	Diff.	Cosine	
00·0	9·99894		10·00106	11·15536	36	8·84464	11·15642	36	8·84359	60′
00·2	9·99894		10·00106	11·15572	37	8·84428	11·15678	36	8·84322	
00·4	9·99895		10·00105	11·15609	36	8·84391	11·15714	36	8·84286	
00·6	9·99895		10·00105	11·15645	36	8·84355	11·15750	36	8·84250	
00·8	9·99895		10·00105	11·15681	37	8·84319	11·15786	37	8·84214	
01·0	9·99895		10·00105	11·15718	36	8·84283	11·15823	36	8·84177	59′
01·2	9·99895		10·00105	11·15754	36	8·84246	11·15859	36	8·84141	
01·4	9·99895		10·00105	11·15790	38	8·84209	11·15895	37	8·84105	
01·6	9·99896		10·00104	11·15828	36	8·84172	11·15932	36	8·84068	
01·8	9·99896		10·00104	11·15864	36	8·84136	11·15968	36	8·84032	
02·0	9·99896		10·00104	11·15900	37	8·84100	11·16004	37	8·83996	58′
02·2	9·99896		10·00104	11·15937	36	8·84063	11·16041	36	8·83959	
02·4	9·99896		10·00104	11·15973	37	8·84027	11·16077	37	8·83923	
02·6	9·99896		10·00104	11·16010	37	8·83990	11·16114	36	8·83886	
02·8	9·99897		10·00103	11·16047	36	8·83953	11·16150	37	8·83850	
03·0	9·99897		10·00103	11·16083	38	8·83916	11·16187	37	8·83813	57′
03·2	9·99897		10·00103	11·16121	36	8·83879	11·16224	36	8·83776	
03·4	9·99897		10·00103	11·16157	37	8·83843	11·16260	37	8·83740	
03·6	9·99897		10·00103	11·16194	37	8·83806	11·16297	37	8·83703	
03·8	9·99897		10·00103	11·16231	37	8·83769	11·16334	36	8·83666	
04·0	9·99898		10·00102	11·16268	37	8·83732	11·16370	37	8·83630	56′
04·2	9·99898		10·00102	11·16305	37	8·83695	11·16407	37	8·83593	
04·4	9·99898		10·00102	11·16342	37	8·83658	11·16444	37	8·83556	
04·6	9·99898		10·00102	11·16379	37	8·83621	11·16481	37	8·83519	
04·8	9·99898		10·00102	11·16416	37	8·83584	11·16518	36	8·83482	
05·0	9·99898		10·00102	11·16453	37	8·83547	11·16554	37	8·83446	55′
05·2	9·99899		10·00101	11·16490	37	8·83510	11·16591	37	8·83409	
05·4	9·99899		10·00101	11·16527	37	8·83473	11·16628	37	8·83372	
05·6	9·99899		10·00101	11·16564	37	8·83436	11·16665	37	8·83335	
05·8	9·99899		10·00101	11·16601	38	8·83399	11·16702	37	8·83298	
06·0	9·99899		10·00101	11·16639	36	8·83361	11·16739	37	8·83261	54′
06·2	9·99899		10·00101	11·16675	38	8·83325	11·16776	37	8·83224	
06·4	9·99900		10·00100	11·16713	38	8·83287	11·16813	38	8·83187	
06·6	9·99900		10·00100	11·16751	37	8·83249	11·16851	37	8·83149	
06·8	9·99900		10·00100	11·16788	37	8·83212	11·16888	37	8·83112	
07·0	9·99900		10·00100	11·16825	37	8·83175	11·16925	37	8·83075	53′
07·2	9·99900		10·00100	11·16862	39	8·83138	11·16962	38	8·83038	
07·4	9·99901		10·00100	11·16901	37	8·83100	11·17000	37	8·83000	
07·6	9·99901		10·00099	11·16938	37	8·83062	11·17037	37	8·82963	
07·8	9·99901		10·00099	11·16975	38	8·83025	11·17074	38	8·82926	
08·0	9·99901		10·00099	11·17013	37	8·82987	11·17112	37	8·82888	52′
08·2	9·99901		10·00099	11·17050	37	8·82950	11·17149	37	8·82851	
08·4	9·99901		10·00099	11·17087	39	8·82912	11·17186	38	8·82814	
08·6	9·99902		10·00098	11·17126	37	8·82874	11·17224	37	8·82776	
08·8	9·99902		10·00098	11·17163	38	8·82837	11·17261	38	8·82739	
09·0	9·99902		10·00098	11·17201	37	8·82799	11·17299	37	8·82701	51′
09·2	9·99902		10·00098	11·17238	38	8·82762	11·17336	38	8·82664	
09·4	9·99902		10·00098	11·17276	38	8·82724	11·17374	38	8·82626	
09·6	9·99902		10·00098	11·17314	38	8·82686	11·17412	37	8·82588	
09·8	9·99903		10·00097	11·17352	38	8·82648	11·17449	38	8·82551	
10·0	9·99903		10·00097	11·17390	38	8·82610	11·17487	38	8·82513	50′
10·2	9·99903		10·00097	11·17428	37	8·82572	11·17525	37	8·82475	
10·4	9·99903		10·00097	11·17465	38	8·82535	11·17562	38	8·82438	
10·6	9·99903		10·00097	11·17503	38	8·82497	11·17600	38	8·82400	
10·8	9·99903		10·00097	11·17541	39	8·82459	11·17638	38	8·82362	
11·0	9·99904		10·00096	11·17580	38	8·82421	11·17676	38	8·82324	49′
11·2	9·99904		10·00096	11·17618	38	8·82382	11·17714	38	8·82286	
11·4	9·99904		10·00096	11·17656	38	8·82344	11·17752	38	8·82248	
11·6	9·99904		10·00096	11·17694	38	8·82306	11·17790	38	8·82210	
11·8	9·99904		10·00096	11·17732	38	8·82268	11·17828	38	8·82172	
12·0	9·99904		10·00096	11·17770		8·82230	11·17866		8·82134	48′

LOGS. OF TRIG. FUNCTIONS

′	Sine	Diff.	Cosec.	Tan.	Diff.	Cotan.	Secant	Diff.	Cosine	
12·0	9·99904		10·00096	11·17770		8·82230	11·17866		8·82134	48′
·2	9·99905		10·00095	11·17809	39	8·82191	11·17904	38	8·82096	
4	9·99905		10·00095	11·17847	38	8·82153	11·17942	38	8·82058	
·6	9·99905		10·00095	11·17885	38	8·82115	11·17980	38	8·82020	
·8	9·99905		10·00095	11·17923	38	8·02077	11·18018	38	8·81982	
13·0	9·99905		10·00095	11·17962	39	8·82038	11·18056	38	8·81944	47′
·2	9·99906		10·00094	11·18001	39	8·81999	11·18095	39	8·81905	
·4	9·99906		10·00094	11·18039	38	8·81961	11·18133	38	8·81867	
·6	9·99906		10·00094	11·18077	38	8·81923	11·18171	38	8·81829	
·8	9·99906		10·00094	11·18116	39	8·81884	11·18210	39	8·81790	
14·0	9·99906		10·00094	11·18154	38	8·81846	11·18248	38	8·81752	46′
·2	9·99906		10·00094	11·18193	39	8·81807	11·18286	38	8·81714	
·4	9·99907		10·00093	11·18232	39	8·81768	11·18325	39	8·81675	
·6	9·99907		10·00093	11·18270	38	8·81730	11·18363	38	8·81637	
·8	9·99907		10·00093	11·18309	39	8·81691	11·18402	39	8·81598	
15·0	9·99907		10·00093	11·18347	38	8·81653	11·18440	38	8·81560	45′
·2	9·99907		10·00093	11·18386	39	8·81614	11·18479	39	8·81521	
·4	9·99907		10·00093	11·18424	38	8·81576	11·18517	38	8·81483	
·6	9·99907		10·00093	11·18463	39	8·81537	11·18556	39	8·81444	
·8	9·99908		10·00092	11·18502	39	8·81498	11·18595	39	8·81405	
16·0	9·99908		10·00092	11·18541	39	8·81459	11·18633	38	8·81367	44′
·2	9·99908		10·00092	11·18580	39	8·81420	11·18672	39	8·81328	
·4	9·99908		10·00092	11·18619	39	8·81381	11·18711	39	8·81289	
·6	9·99908		10·00092	11·18658	39	8·81342	11·18750	39	8·81250	
·8	9·99908		10·00092	11·18697	39	8·81303	11·18789	39	8·81211	
17·0	9·99909		10·00091	11·18736	39	8·81264	11·18827	38	8·81173	43′
·2	9·99909		10·00091	11·18775	39	8·81225	11·18866	39	8·81134	
·4	9·99909		10·00091	11·18814	39	8·81186	11·18905	39	8·81095	
·6	9·99909		10·00091	11·18853	39	8·81147	11·18944	39	8·81056	
·8	9·99909		10·00091	11·18892	40	8·81108	11·18983	39	8·81017	
18·0	9·99909		10·00091	11·18932	39	8·81068	11·19022	39	8·80978	42′
·2	9·99910		10·00090	11·18971	39	8·81029	11·19061	39	8·80939	
·4	9·99910		10·00090	11·19010	40	8·80990	11·19100	40	8·80900	
·6	9·99910		10·00090	11·19050	39	8·80950	11·19140	39	8·80860	
·8	9·99910		10·00090	11·19089	39	8·80911	11·19179	39	8·80821	
19·0	9·99910		10·00090	11·19128	39	8·80872	11·19218	39	8·80782	41′
·2	9·99910		10·00090	11·19167	40	8·80833	11·19257	40	8·80743	
·4	9·99911		10·00089	11·19207	40	8·80793	11·19297	39	8·80703	
·6	9·99911		10·00089	11·19247	39	8·80753	11·19336	39	8·80664	
·8	9·99911		10·00089	11·19286	40	8·80714	11·19375	40	8·80625	
20·0	9·99911		10·00089	11·19326	39	8·80674	11·19415	39	8·80585	40′
·2	9·99911		10·00089	11·19365	40	8·80635	11·19454	40	8·80546	
·4	9·99911		10·00089	11·19405	40	8·80595	11·19494	39	8·80506	
·6	9·99912		10·00088	11·19445	40	8·80555	11·19533	40	8·80467	
·8	9·99912		10·00088	11·19485	39	8·80515	11·19573	39	8·80427	
21·0	9·99912		10·00088	11·19524	40	8·80476	11·19612	40	8·80388	39′
·2	9·99912		10·00088	11·19564	40	8·80436	11·19652	40	8·80348	
·4	9·99912		10·00088	11·19604	39	8·80396	11·19692	39	8·80308	
·6	9·99912		10·00088	11·19643	40	8·80357	11·19731	40	8·80269	
·8	9·99912		10·00088	11·19683	40	8·80317	11·19771	40	8·80229	
22·0	9·99913		10·00087	11·19723	40	8·80277	11·19811	40	8·80189	38′
·2	9·99913		10·00087	11·19763	40	8·80237	11·19851	39	8·80149	
·4	9·99913		10·00087	11·19803	40	8·80197	11·19890	40	8·80110	
·6	9·99913		10·00087	11·19843	40	8·80157	11·19930	40	8·80070	
·8	9·99913		10·00087	11·19883	41	8·80117	11·19970	40	8·80030	
23·0	9·99913		10·00087	11·19924	40	8·80076	11·20010	40	8·79990	37′
·2	9·99914		10·00086	11·19964	40	8·80036	11·20050	40	8·79950	
·4	9·99914		10·00086	11·20004	40	8·79996	11·20090	40	8·79910	
·6	9·99914		10·00086	11·20044	40	8·79956	11·20130	40	8·79870	
·8	9·99914		10·00086	11·20084	41	8·79916	11·20170	41	8·79830	
24·0	9·99914		10·00086	11·20125		8·79875	11·20211		8·79789	36′

LOGS. OF TRIG. FUNCTIONS

86°
266°

′	Sine	Diff.	Cosec.	Tan.	Diff.	Cotan.	Secant	Diff.	Cosine	
24·0	9·99914		10·00086	11·20125	40	8·79875	11·20211	40	8·79789	36′
·2	9·99914		10·00086	11·20165	41	8·79835	11·20251	40	8·79749	
·4	9·99915		10·00085	11·20206	40	8·79794	11·20291	40	8·79709	
·6	9·99915		10·00085	11·20246	40	8·79754	11·20331	40	8·79669	
·8	9·99915		10·00085	11·20286	41	8·79714	11·20371	41	8·79629	
25·0	9·99915		10·00085	11·20327	40	8·79673	11·20412	40	8·79588	35′
·2	9·99915		10·00085	11·20367	41	8·79633	11·20452	41	8·79548	
·4	9·99915		10·00085	11·20408	40	8·79592	11·20493	40	8·79507	
·6	9·99915		10·00085	11·20448	41	8·79552	11·20533	40	8·79467	
·8	9·99916		10·00084	11·20489	41	8·79511	11·20573	41	8·79427	
26·0	9·99916		10·00084	11·20530	41	8·79470	11·20614	41	8·79386	34′
·2	9·99916		10·00084	11·20571	40	8·79429	11·20655	40	8·79345	
·4	9·99916		10·00084	11·20611	41	8·79389	11·20695	41	8·79305	
·6	9·99916		10·00084	11·20652	40	8·79348	11·20736	40	8·79264	
·8	9·99916		10·00084	11·20692	42	8·79308	11·20776	41	8·79224	
27·0	9·99917		10·00083	11·20734	41	8·79266	11·20817	41	8·79183	33′
·2	9·99917		10·00083	11·20775	41	8·79225	11·20858	41	8·79142	
·4	9·99917		10·00083	11·20816	41	8·79184	11·20899	41	8·79101	
·6	9·99917		10·00083	11·20857	40	8·79143	11·20940	40	8·79060	
·8	9·99917		10·00083	11·20897	42	8·79103	11·20980	41	8·79020	
28·0	9·99917		10·00083	11·20939	41	8·79061	11·21021	41	8·78979	32′
·2	9·99918		10·00082	11·20980	41	8·79020	11·21062	41	8·78938	
·4	9·99918		10·00082	11·21021	41	8·78979	11·21103	41	8·78897	
·6	9·99918		10·00082	11·21062	41	8·78938	11·21144	41	8·78856	
·8	9·99918		10·00082	11·21103	42	8·78897	11·21185	41	8·78815	
29·0	9·99918		10·00082	11·21145	41	8·78855	11·21226	42	8·78774	31′
·2	9·99918		10·00082	11·21186	41	8·78814	11·21268	41	8·78732	
·4	9·99918		10·00082	11·21227	42	8·78773	11·21309	41	8·78691	
·6	9·99919		10·00081	11·21269	41	8·78731	11·21350	41	8·78650	
·8	9·99919		10·00081	11·21310	41	8·78690	11·21391	42	8·78609	
30·0	9·99919		10·00081	11·21351	42	8·78649	11·21433	41	8·78568	30′
·2	9·99919		10·00081	11·21393	41	8·78607	11·21474	41	8·78526	
·4	9·99919		10·00081	11·21434	42	8·78566	11·21515	42	8·78485	
·6	9·99919		10·00081	11·21476	42	8·78524	11·21557	41	8·78443	
·8	9·99920		10·00080	11·21518	41	8·78482	11·21598	42	8·78402	
31·0	9·99920		10·00080	11·21559	42	8·78441	11·21640	41	8·78361	29′
·2	9·99920		10·00080	11·21601	42	8·78399	11·21681	42	8·78319	
·4	9·99920		10·00080	11·21643	41	8·78357	11·21723	41	8·78277	
·6	9·99920		10·00080	11·21684	42	8·78316	11·21764	42	8·78236	
·8	9·99920		10·00080	11·21726	42	8·78274	11·21806	42	8·78194	
32·0	9·99921		10·00080	11·21768	42	8·78232	11·21848	41	8·78152	28′
·2	9·99921		10·00079	11·21810	42	8·78190	11·21889	42	8·78111	
·4	9·99921		10·00079	11·21852	42	8·78148	11·21931	42	8·78069	
·6	9·99921		10·00079	11·21894	42	8·78106	11·21973	42	8·78027	
·8	9·99921		10·00079	11·21936	42	8·78064	11·22015	42	8·77985	
33·0	9·99921		10·00079	11·21978	42	8·78022	11·22057	42	8·77943	27′
·2	9·99921		10·00079	11·22020	43	8·77980	11·22099	42	8·77901	
·4	9·99922		10·00078	11·22063	42	8·77937	11·22141	42	8·77859	
·6	9·99922		10·00078	11·22105	42	8·77895	11·22183	42	8·77817	
·8	9·99922		10·00078	11·22147	42	8·77853	11·22225	42	8·77775	
34·0	9·99922		10·00078	11·22189	42	8·77811	11·22267	42	8·77733	26′
·2	9·99922		10·00078	11·22231	42	8·77769	11·22309	42	8·77691	
·4	9·99922		10·00078	11·22273	42	8·77727	11·22351	42	8·77649	
·6	9·99922		10·00078	11·22315	43	8·77685	11·22393	42	8·77607	
·8	9·99923		10·00077	11·22358	43	8·77642	11·22435	43	8·77565	
35·0	9·99923		10·00077	11·22401	42	8·77600	11·22478	42	8·77522	25′
·2	9·99923		10·00077	11·22443	42	8·77557	11·22520	42	8·77480	
·4	9·99923		10·00077	11·22485	43	8·77515	11·22562	43	8·77438	
·6	9·99923		10·00077	11·22528	42	8·77472	11·22605	42	8·77395	
·8	9·99923		10·00077	11·22570	43	8·77430	11·22647	43	8·77353	
36·0	9·99924		10·00077	11·22613		8·77387	11·22690		8·77310	24′

86° 266° — LOGS. OF TRIG. FUNCTIONS

′	Sine	Diff.	Cosec.	Tan.	Diff.	Cotan.	Secant	Diff.	Cosine	
36·0	9·99924		10·00077	11·22613	43	8·77387	11·22690	42	8·77310	24′
·2	9·99924		10·00076	11·22656	43	8·77344	11·22732	43	8·77268	
·4	9·99924		10·00076	11·22699	43	8·77301	11·22775	43	8·77225	
·6	9·99924		10 00076	11·22742	42	8·77258	11·22818	43	8·77182	
·8	9·99924		10·00076	11·22784	43	8·77216	11·22860	42	8·77140	
37·0	9·99924		10·00076	11·22827	43	8·77173	11·22903	43	8·77097	23′
·2	9·99924		10·00076	11·22870	43	8·77130	11·22946	43	8·77054	
·4	9·99924		10·00076	11·22913	43	8·77087	11·22989	43	8·77011	
·6	9·99925		10·00075	11·22956	43	8·77044	11·23031	42	8·76969	
·8	9·99925		10·00075	11·22999	43	8·77001	11·23074	43	8·76926	
38·0	9·99925		10·00075	11·23042	43	8·76958	11·23117	43	8·76883	22′
·2	9·99925		10·00075	11·23085	43	8·76915	11·23160	43	8·76840	
·4	9·99925		10·00075	11·23128	43	8·76872	11·23203	43	8·76797	
·6	9·99925		10·00075	11·23171	44	8·76829	11·23246	43	8·76754	
·8	9·99926		10·00074	11·23215	43	8·76785	11·23289	44	8·76711	
39·0	9·99926		10·00074	11·23258	44	8·76742	11·23333	43	8·76667	21′
·2	9·99926		10·00074	11·23302	43	8·76698	11·23376	43	8·76624	
·4	9·99926		10·00074	11·23345	43	8·76655	11·23419	43	8·76581	
·6	9·99926		10·00074	11·23388	44	8·76612	11·23462	44	8·76538	
·8	9·99926		10·00074	11·23432	43	8·76568	11·23506	43	8·76494	
40·0	9·99927		10·00074	11·23475	44	8·76525	11·23549	43	8·76451	20′
·2	9·99927		10·00073	11·23519	44	8·76481	11·23592	44	8·76408	
·4	9·99927		10·00073	11·23563	43	8·76437	11·23636	43	8·76364	
·6	9·99927		10·00073	11·23606	44	8·76394	11·23679	44	8·76321	
·8	9·99927		10·00073	11·23650	44	8·76350	11·23723	43	8·76277	
41·0	9·99927		10·00073	11·23694	43	8·76307	11·23766	44	8·76234	19′
·2	9·99927		10·00073	11·23737	45	8·76263	11·23810	44	8·76190	
·4	9·99928		10·00072	11·23782	43	8·76218	11·23854	43	8·76146	
·6	9·99928		10·00072	11·23825	44	8·76175	11·23897	44	8·76103	
·8	9·99928		10·00072	11·23869	44	8·76131	11·23941	44	8·76059	
42·0	9·99928		10·00072	11·23913	44	8·76087	11·23985	44	8·76015	18′
·2	9·99928		10·00072	11·23957	44	8·76043	11·24029	44	8·75971	
·4	9·99928		10·00072	11·24001	44	8·75999	11·24073	44	8·75927	
·6	9·99928		10·00072	11·24045	45	8·75955	11·24117	44	8·75883	
·8	9·99928		10·00072	11·24090	43	8·75910	11·24161	44	8·75839	
43·0	9·99929		10·00071	11·24133	45	8·75866	11·24205	44	8·75795	17′
·2	9·99929		10·00071	11·24178	44	8·75822	11·24249	44	8·75751	
·4	9·99929		10·00071	11·24222	44	8·75778	11·24293	44	8·75707	
·6	9·99929		10·00071	11·24266	44	8·75734	11·24337	44	8·75663	
·8	9·99929		10·00071	11·24310	45	8·75690	11·24381	44	8·75619	
44·0	9·99929		10·00071	11·24355	45	8·75646	11·24425	45	8·75575	16′
·2	9·99930		10·00070	11·24400	44	8·75600	11·24470	44	8·75530	
·4	9·99930		10·00070	11·24444	44	8·75556	11·24514	44	8·75486	
·6	9·99930		10·00070	11·24488	45	8·75512	11·24558	45	8·75442	
·8	9·99930		10·00070	11·24533	44	8·75467	11·24603	44	8·75397	
45·0	9·99930		10·00070	11·24577	45	8·75423	11·24647	45	8·75353	15′
·2	9·99930		10·00070	11·24622	44	8·75378	11·24692	44	8·75308	
·4	9·99930		10·00070	11·24666	46	8·75334	11·24736	45	8·75264	
·6	9·99931		10·00069	11·24712	45	8·75288	11·24781	45	8·75219	
·8	9·99931		10·00069	11·24757	44	8·75243	11·24826	44	8·75174	
46·0	9·99931		10·00069	11·24801	45	8·75199	11·24870	45	8·75130	14′
·2	9·99931		10·00069	11·24846	45	8·75154	11·24915	45	8·75085	
·4	9·99931		10·00069	11·24891	45	8·75109	11·24960	45	8·75040	
·6	9·99931		10·00069	11·24936	45	8·75064	11·25005	45	8·74995	
·8	9·99931		10·00069	11·24981	45	8·75019	11·25050	45	8·74950	
47·0	9·99932		10·00068	11·25026	45	8·74974	11·25095	44	8·74906	13′
·2	9·99932		10·00068	11·25071	45	8·74929	11·25139	45	8·74861	
·4	9·99932		10·00068	11·25116	46	8·74884	11·25184	46	8·74816	
·6	9·99932		10·00068	11·25162	45	8·74838	11·25230	45	8·74770	
·8	9·99932		10·00068	11·25207	45	8·74793	11·25275	45	8·74725	
48·0	9·99932		10·00068	11·25252		8·74748	11·25320		8·74680	12′

86°
266°

LOGS. OF TRIG. FUNCTIONS

′	Sine	Diff.	Cosec.	Tan.	Diff.	Cotan.	Secant	Diff.	Cosine	
48·0	9·99932		10·00068	11·25252	45	8·74748	11·25320	45	8·74680	12′
·2	9·99932		10·00068	11·25297	46	8·74703	11·25365	45	8·74635	
·4	9·99933		10·00067	11·25343	46	8·74657	11·25410	46	8·74590	
·6	9·99933		10·00067	11·25389	45	8·74611	11·25456	45	8·74544	
·8	9·99933		10·00067	11·25434	45	8·74566	11·25501	45	8·74499	
49·0	9·99933		10·00067	11·25479	46	8·74521	11·25546	46	8·74454	11′
·2	9·99933		10·00067	11·25525	45	8·74475	11·25592	45	8·74408	
·4	9·99933		10·00067	11·25570	46	8·74430	11·25637	46	8·74363	
·6	9·99933		10·00067	11·25616	46	8·74384	11·25683	45	8·74317	
·8	9·99934		10·00066	11·25662	46	8·74338	11·25728	46	8·74272	
50·0	9·99934		10·00066	11·25708	46	8·74292	11·25774	46	8·74226	10′
·2	9·99934		10·00066	11·25754	46	8·74246	11·25820	46	8·74180	
·4	9·99934		10·00066	11·25800	45	8·74200	11·25866	45	8·74134	
·6	9·99934		10·00066	11·25845	46	8·74155	11·25911	46	8·74089	
·8	9·99934		10·00066	11·25891	46	8·74109	11·25957	46	8·74043	
51·0	9·99934		10·00066	11·25937	46	8·74063	11·26003	46	8·73997	9′
·2	9·99934		10·00066	11·25983	47	8·74017	11·26049	46	8·73951	
·4	9·99935		10·00065	11·26030	46	8·73970	11·26095	46	8·73905	
·6	9·99935		10·00065	11·26076	46	8·73924	11·26141	46	8·73859	
·8	9·99935		10·00065	11·26122	46	8·73878	11·26187	46	8·73813	
52·0	9·99935		10·00065	11·26168	46	8·73832	11·26233	46	8·73767	8′
·2	9·99935		10·00065	11·26214	47	8·73786	11·26279	47	8·73721	
·4	9·99935		10·00065	11·26261	46	8·73739	11·26326	46	8·73674	
·6	9·99935		10·00065	11·26307	47	8·73693	11·26372	46	8·73628	
·8	9·99936		10·00064	11·26354	46	8·73646	11·26418	47	8·73582	
53·0	9·99936		10·00064	11·26400	47	8·73600	11·26465	46	8·73535	7′
·2	9·99936		10·00064	11·26447	47	8·73553	11·26511	47	8·73489	
·4	9·99936		10·00064	11·26494	46	8·73506	11·26558	46	8·73442	
·6	9·99936		10·00064	11·26540	47	8·73460	11·26604	47	8·73396	
·8	9·99936		10·00064	11·26587	47	8·73413	11·26651	46	8·73349	
54·0	9·99936		10·00064	11·26634	47	8·73366	11·26697	47	8·73303	6′
·2	9·99937		10·00063	11·26681	47	8·73319	11·26744	47	8·73256	
·4	9·99937		10·00063	11·26728	46	8·73272	11·26791	46	8·73209	
·6	9·99937		10·00063	11·26774	47	8·73226	11·26837	47	8·73163	
·8	9·99937		10·00063	11·26821	47	8·73179	11·26884	47	8·73116	
55·0	9·99937		10·00063	11·26868	47	8·73132	11·26931	47	8·73069	5′
·2	9·99937		10·00063	11·26915	47	8·73085	11·26978	47	8·73022	
·4	9·99937		10·00063	11·26962	48	8·73038	11·27025	47	8·72975	
·6	9·99938		10·00062	11·27010	47	8·72990	11·27072	47	8·72928	
·8	9·99938		10·00062	11·27057	47	8·72943	11·27119	47	8·72881	
56·0	9·99938		10·00062	11·27104	48	8·72896	11·27166	48	8·72834	4′
·2	9·99938		10·00062	11·27152	47	8·72848	11·27214	47	8·72786	
·4	9·99938		10·00062	11·27199	47	8·72801	11·27261	47	8·72739	
·6	9·99938		10·00062	11·27246	47	8·72754	11·27308	47	8·72692	
·8	9·99938		10·00062	11·27293	48	8·72707	11·27355	48	8·72645	
57·0	9·99938		10·00062	11·27341	48	8·72659	11·27403	47	8·72597	3′
·2	9·99939		10·00061	11·27389	48	8·72611	11·27450	48	8·72550	
·4	9·99939		10·00061	11·27437	47	8·72563	11·27498	47	8·72502	
·6	9·99939		10·00061	11·27484	48	8·72516	11·27545	48	8·72455	
·8	9·99939		10·00061	11·27532	48	8·72468	11·27593	48	8·72407	
58·0	9·99939		10·00061	11·27580	47	8·72420	11·27641	47	8·72360	2′
·2	9·99939		10·00061	11·27627	48	8·72373	11·27688	48	8·72312	
·4	9·99939		10·00061	11·27675	49	8·72325	11·27736	48	8·72264	
·6	9·99940		10·00060	11·27724	48	8·72276	11·27784	48	8·72216	
·8	9·99940		10·00060	11·27772	47	8·72228	11·27832	48	8·72168	
59·0	9·99940		10·00060	11·27819	49	8·72181	11·27880	48	8·72120	1′
·2	9·99940		10·00060	11·27868	48	8·72132	11·27928	48	8·72072	
·4	9·99940		10·00060	11·27916	48	8·72084	11·27976	48	8·72024	
·6	9·99940		10·00060	11·27964	48	8·72036	11·28024	48	8·71976	
·8	9·99940		10·00060	11·28012	48	8·71988	11·28072	48	8·71928	
60·0	9·99940		10·00060	11·28060		8·71940	11·28120		8·71880	0′

87°
267°

LOGS. OF TRIG. FUNCTIONS

′	Sine	Diff.	Cosec.	Tan.	Diff.	Cotan.	Secant	Diff.	Cosine	
00·0	9·99940		10·00060	11·28060	49	8·71940	11·28120	48	8·71880	60′
·2	9·99941		10·00059	11·28109	48	8·71891	11·28168	48	8·71832	
·4	9·99941		10·00059	11·28157	49	8·71843	11·28216	49	8·71784	
·6	9·99941		10·00059	11·28206	48	8·71794	11·28265	48	8·71735	
·8	9·99941		10·00059	11·28254	49	8·71746	11·28313	49	8·71687	
01·0	9·99941		10·00059	11·28303	48	8·71697	11·28362	48	8·71638	59′
·2	9·99941		10·00059	11·28351	49	8·71649	11·28410	49	8·71590	
·4	9·99941		10·00059	11·28400	49	8·71600	11·28459	48	8·71541	
·6	9·99942		10·00058	11·28449	49	8·71551	11·28507	49	8·71493	
·8	9·99942		10·00058	11·28498	49	8·71502	11·28556	49	8·71444	
02·0	9·99942		10·00058	11·28547	49	8·71453	11·28605	49	8·71395	58′
·2	9·99942		10·00058	11·28596	48	8·71404	11·28654	48	8·71346	
·4	9·99942		10·00058	11·28644	49	8·71356	11·28702	49	8·71298	
·6	9·99942		10·00058	11·28693	49	8·71307	11·28751	49	8·71249	
·8	9·99942		10·00058	11·28742	49	8·71258	11·28800	49	8·71200	
03·0	9·99942		10·00058	11·28791	50	8·71208	11·28849	49	8·71151	57′
·2	9·99943		10·00057	11·28841	49	8·71159	11·28898	49	8·71102	
·4	9·99943		10·00057	11·28890	50	8·71110	11·28947	50	8·71053	
·6	9·99943		10·00057	11·28940	49	8·71060	11·28997	49	8·71003	
·8	9·99943		10·00057	11·28989	49	8·71011	11·29046	49	8·70954	
04·0	9·99943		10·00057	11·29038	49	8·70962	11·29095	49	8·70905	56′
·2	9·99943		10·00057	11·29087	50	8·70913	11·29144	50	8·70856	
·4	9·99943		10·00057	11·29137	49	8·70863	11·29194	49	8·70806	
·6	9·99943		10·00057	11·29186	51	8·70814	11·29243	50	8·70757	
·8	9·99944		10·00056	11·29237	49	8·70763	11·29293	49	8·70707	
05·0	9·99944		10·00056	11·29286	50	8·70714	11·29342	50	8·70658	55′
·2	9·99944		10·00056	11·29336	49	8·70664	11·29392	49	8·70608	
·4	9·99944		10·00056	11·29385	50	8·70615	11·29441	50	8·70559	
·6	9·99944		10·00056	11·29435	50	8·70565	11·29491	50	8·70509	
·8	9·99944		10·00056	11·29485	50	8·70515	11·29541	50	8·70459	
06·0	9·99944		10·00056	11·29535	50	8·70465	11·29591	50	8·70409	54′
·2	9·99944		10·00056	11·29585	51	8·70415	11·29641	50	8·70359	
·4	9·99945		10·00055	11·29636	50	8·70364	11·29691	50	8·70309	
·6	9·99945		10·00055	11·29686	50	8·70314	11·29741	50	8·70259	
·8	9·99945		10·00055	11·29736	50	8·70264	11·29791	50	8·70209	
07·0	9·99945		10·00055	11·29786	50	8·70214	11·29841	50	8·70159	53′
·2	9·99945		10·00055	11·29836	51	8·70164	11·29891	51	8·70109	
·4	9·99945		10·00055	11·29887	50	8·70113	11·29942	50	8·70058	
·6	9·99945		10·00055	11·29937	50	8·70063	11·29992	50	8·70008	
·8	9·99945		10·00055	11·29987	51	8·70013	11·30042	51	8·69958	
08·0	9·99946		10·00054	11·30038	51	8·69962	11·30093	50	8·69907	52′
·2	9·99946		10·00054	11·30089	51	8·69911	11·30143	51	8·69857	
·4	9·99946		10·00054	11·30140	50	8·69860	11·30194	50	8·69806	
·6	9·99946		10·00054	11·30190	51	8·69810	11·30244	51	8·69756	
·8	9·99946		10·00054	11·30241	51	8·69759	11·30295	51	8·69705	
09·0	9·99946		10·00054	11·30292	50	8·69708	11·30346	50	8·69654	51′
·2	9·99946		10·00054	11·30342	51	8·69658	11·30396	51	8·69604	
·4	9·99946		10·00054	11·30393	51	8·69607	11·30447	51	8·69553	
·6	9·99947		10·00054	11·30444	52	8·69556	11·30498	51	8·69502	
·8	9·99947		10·00053	11·30496	51	8·69504	11·30549	51	8·69451	
10·0	9·99947		10·00053	11·30547	51	8·69453	11·30600	51	8·69400	50′
·2	9·99947		10·00053	11·30598	51	8·69402	11·30651	51	8·69349	
·4	9·99947		10·00053	11·30649	52	8·69351	11·30702	52	8·69298	
·6	9·99947		10·00053	11·30701	51	8·69299	11·30754	51	8·69246	
·8	9·99947		10·00053	11·30752	52	8·69248	11·30805	51	8·69195	
11·0	9·99948		10·00053	11·30804	52	8·69196	11·30856	52	8·69144	49′
·2	9·99948		10·00052	11·30856	51	8·69144	11·30908	51	8·69092	
·4	9·99948		10·00052	11·30907	52	8·69093	11·30959	52	8·69041	
·6	9·99948		10·00052	11·30959	51	8·69041	11·31011	51	8·68989	
·8	9·99948		10·00052	11·31010	52	8·68990	11·31062	52	8·68938	
12·0	9·99948		10·00052	11·31062		8·68938	11·31114		8·68886	48′

87° 267°

LOGS. OF TRIG. FUNCTIONS

′	Sine	Diff.	Cosec.	Tan.	Diff.	Cotan.	Secant	Diff.	Cosine	
12·0	9·99948		10·00052	11·31062	51	8·68938	11·31114	51	8·68886	48′
·2	9·99948		10·00052	11·31113	52	8·68887	11·31165	52	8·68835	
·4	9·99948		10·00052	11·31165	53	8·68835	11·31217	52	8·68783	
·6	9·99949		10·00051	11·31218	52	8·68782	11·31269	52	8·68731	
·8	9·99949		10·00051	11·31270	52	8·68730	11·31321	52	8·68679	
13·0	9·99949		10·00051	11·31322	52	8·68678	11·31373	52	8·68627	47′
·2	9·99949		10·00051	11·31374	52	8·68626	11·31425	52	8·68575	
·4	9·99949		10·00051	11·31426	52	8·68574	11·31477	52	8·68523	
·6	9·99949		10·00051	11·31478	52	8·68522	11·31529	52	8·68471	
·8	9·99949		10·00051	11·31530	53	8·68470	11·31581	53	8·68419	
14·0	9·99949		10·00051	11·31583	53	8·68417	11·31634	52	8·68367	46′
·2	9·99950		10·00050	11·31636	52	8·68364	11·31686	52	8·68314	
·4	9·99950		10·00050	11·31688	53	8·68312	11·31738	53	8·68262	
·6	9·99950		10·00050	11·31741	52	8·68259	11·31791	52	8·68209	
·8	9·99950		10·00050	11·31793	53	8·68207	11·31843	53	8·68157	
15·0	9·99950		10·00050	11·31846	52	8·68154	11·31896	52	8·68104	45′
·2	9·99950		10·00050	11·31898	53	8·68102	11·31948	53	8·68052	
·4	9·99950		10·00050	11·31951	53	8·68049	11·32001	53	8·67999	
·6	9·99950		10·00050	11·32004	54	8·67996	11·32054	53	8·67946	
·8	9·99951		10·00049	11·32058	52	8·67942	11·32107	53	8·67893	
16·0	9·99951		10·00049	11·32110	53	8·67890	11·32160	52	8·67841	44′
2	9·99951		10·00049	11·32163	53	8·67837	11·32212	53	8·67788	
·4	9·99951		10·00049	11·32216	53	8·67784	11·32265	53	8·67735	
·6	9·99951		10·00049	11·32269	54	8·67731	11·32318	54	8·67682	
·8	9·99951		10·00049	11·32323	53	8·67677	11·32372	53	8·67628	
17·0	9·99951		10·00049	11·32376	53	8·67624	11·32425	53	8·67575	43′
·2	9·99951		10·00049	11·32429	54	8·67571	11·32478	54	8·67522	
·4	9·99951		10·00049	11·32483	54	8·67517	11·32532	53	8·67468	
·6	9·99952		10·00048	11·32537	53	8·67463	11·32585	53	8·67415	
·8	9·99952		10·00048	11·32590	54	8·67410	11·32638	54	8·67362	
18·0	9·99952		10·00048	11·32644	54	8·67356	11·32692	54	8·67308	42′
·2	9·99952		10·00048	11·32698	53	8·67302	11·32746	53	8·67254	
·4	9·99952		10·00048	11·32751	54	8·67249	11·32799	54	8·67201	
·6	9·99952		10·00048	11·32805	54	8·67195	11·32853	54	8·67147	
·8	9·99952		10·00048	11·32859	54	8·67141	11·32907	54	8·67093	
19·0	9·99952		10·00048	11·32913	55	8·67087	11·32961	54	8·67039	41′
·2	9·99953		10·00047	11·32968	54	8·67032	11·33015	54	8·66985	
·4	9·99953		10·00047	11·33022	54	8·66978	11·33069	54	8·66931	
·6	9·99953		10·00047	11·33076	54	8·66924	11·33123	54	8·66877	
·8	9·99953		10·00047	11·33130	54	8·66870	11·33177	54	8·66823	
20·0	9·99953		10·00047	11·33184	54	8·66816	11·33231	54	8·66769	40′
·2	9·99953		10·00047	11·33238	55	8·66762	11·33285	55	8·66715	
·4	9·99953		10·00047	11·33293	54	8·66707	11·33340	54	8·66660	
·6	9·99953		10·00047	11·33347	55	8·66653	11·33394	55	8·66606	
·8	9·99953		10·00047	11·33402	55	8·66598	11·33449	54	8·66551	
21·0	9·99954		10·00046	11·33457	55	8·66543	11·33503	55	8·66497	39′
·2	9·99954		10·00046	11·33512	54	8·66488	11·33558	54	8·66442	
·4	9·99954		10·00046	11·33566	55	8·66434	11·33612	55	8·66388	
·6	9·99954		10·00046	11·33621	55	8·66379	11·33667	55	8·66333	
·8	9·99954		10·00046	11·33676	55	8·66324	11·33722	55	8·66278	
22·0	9·99954		10·00046	11·33731	55	8·66269	11·33777	55	8·66223	38′
·2	9·99954		10·00046	11·33786	55	8·66214	11·33832	55	8·66168	
·4	9·99954		10·00046	11·33841	55	8·66159	11·33887	55	8·66113	
·6	9·99954		10·00046	11·33896	56	8·66104	11·33942	55	8·66058	
·8	9·99955		10·00045	11·33952	55	8·66048	11·33997	56	8·66003	
23·0	9·99955		10·00045	11·34007	56	8·65993	11·34053	55	8·65948	37′
·2	9·99955		10·00045	11·34063	55	8·65937	11·34108	55	8·65892	
·4	9·99955		10·00045	11·34118	56	8·65882	11·34163	56	8·65837	
·6	9·99955		10·00045	11·34174	55	8·65826	11·34219	55	8·65781	
·8	9·99955		10·00045	11·34229	56	8·65771	11·34274	56	8·65726	
24·0	9·99955		10·00045	11·34285		8·65715	11·34330		8·65670	36′

87°
267°

LOGS. OF TRIG. FUNCTIONS

′	Sine	Diff.	Cosec.	Tan.	Diff.	Cotan.	Secant	Diff.	Cosine	
24.0	9.99955		10.00045	11.34285		8.65715	11.34330		8.65670	36′
.2	9.99955		10.00045	11.34340	55	8.65660	11.34385	55	8.65615	
.4	9.99955		10.00045	11.34396	56	8.65604	11.34441	56	8.65559	
.6	9.99956		10.00014	11.34453	57	8.65547	11.34497	56	8.65503	
.8	9.99956		10.00044	11.34509	56	8.65491	11.34553	56	8.65447	
25.0	9.99956		10.00044	11.34565	56	8.65435	11.34609	56	8.65391	35′
.2	9.99956		10.00044	11.34621	56	8.65379	11.34665	56	8.65335	
.4	9.99956		10.00044	11.34677	56	8.65323	11.34721	56	8.65279	
.6	9.99956		10.00044	11.34733	56	8.65267	11.34777	56	8.65223	
.8	9.99956		10.00044	11.34790	57	8.65210	11.34834	57	8.65166	
26.0	9.99956		10.00044	11.34846	56	8.65154	11.34890	56	8.65110	34′
.2	9.99957		10.00043	11.34903	57	8.65097	11.34946	56	8.65054	
.4	9.99957		10.00043	11.34960	57	8.65040	11.35003	57	8.64997	
.6	9.99957		10.00043	11.35016	56	8.64984	11.35059	56	8.64941	
.8	9.99957		10.00043	11.35073	57	8.64927	11.35116	57	8.64884	
27.0	9.99957		10.00043	11.35130	57	8.64870	11.35173	57	8.64827	33′
.2	9.99957		10.00043	11.35186	56	8.64814	11.35229	56	8.64771	
.4	9.99957		10.00043	11.35243	57	8.64757	11.35286	57	8.64714	
.6	9.99957		10.00043	11.35300	57	8.64700	11.35343	57	8.64657	
.8	9.99957		10.00043	11.35357	57	8.64643	11.35400	57	8.64600	
28.0	9.99958		10.00042	11.35415	58	8.64585	11.35457	57	8.64543	32′
.2	9.99958		10.00042	11.35472	57	8.64528	11.35514	57	8.64486	
.4	9.99958		10.00042	11.35529	57	8.64471	11.35571	57	8.64429	
.6	9.99958		10.00042	11.35587	58	8.64413	11.35629	58	8.64371	
.8	9.99958		10.00042	11.35644	57	8.64356	11.35686	57	8.64314	
29.0	9.99958		10.00042	11.35702	58	8.64298	11.35744	58	8.64256	31′
.2	9.99958		10.00042	11.35759	57	8.64241	11.35801	57	8.64199	
.4	9.99958		10.00042	11.35817	58	8.64183	11.35859	58	8.64141	
.6	9.99958		10.00042	11.35875	58	8.64125	11.35917	58	8.64083	
.8	9.99959		10.00041	11.35933	58	8.64067	11.35974	57	8.64026	
30.0	9.99959		10.00041	11.35991	58	8.64009	11.36032	58	8.63968	30′
.2	9.99959		10.00041	11.36049	58	8.63951	11.36090	58	8.63910	
.4	9.99959		10.00041	11.36107	58	8.63893	11.36148	58	8.63852	
.6	9.99959		10.00041	11.36165	58	8.63835	11.36206	58	8.63794	
.8	9.99959		10.00041	11.36223	58	8.63777	11.36264	58	8.63736	
31.0	9.99959		10.00041	11.36282	59	8.63718	11.36322	58	8.63678	29′
.2	9.99959		10.00041	11.36340	58	8.63660	11.36381	59	8.63619	
.4	9.99959		10.00041	11.36398	58	8.63602	11.36439	58	8.63561	
.6	9.99960		10.00040	11.36457	59	8.63543	11.36497	58	8.63503	
.8	9.99960		10.00040	11.36516	59	8.63484	11.36556	59	8.63444	
32.0	9.99960		10.00040	11.36574	58	8.63426	11.36615	59	8.63385	28′
.2	9.99960		10.00040	11.36633	59	8.63367	11.36673	58	8.63327	
.4	9.99960		10.00040	11.36692	59	8.63308	11.36732	59	8.63268	
.6	9.99960		10.00040	11.36751	59	8.63249	11.36791	59	8.63209	
.8	9.99960		10.00040	11.36810	59	8.63190	11.36850	59	8.63150	
33.0	9.99960		10.00040	11.36869	59	8.63131	11.36909	59	8.63091	27′
.2	9.99960		10.00040	11.36928	59	8.63072	11.36968	59	8.63032	
.4	9.99961		10.00039	11.36988	60	8.63012	11.37027	59	8.62973	
.6	9.99961		10.00039	11.37047	59	8.62953	11.37086	59	8.62914	
.8	9.99961		10.00039	11.37107	60	8.62893	11.37146	60	8.62854	
34.0	9.99961		10.00039	11.37166	59	8.62834	11.37205	59	8.62795	26′
.2	9.99961		10.00039	11.37226	60	8.62774	11.37265	60	8.62735	
.4	9.99961		10.00039	11.37285	59	8.62715	11.37324	59	8.62676	
.6	9.99961		10.00039	11.37345	60	8.62655	11.37384	60	8.62616	
.8	9.99961		10.00039	11.37405	60	8.62595	11.37444	60	8.62556	
35.0	9.99961		10.00039	11.37465	60	8.62535	11.37504	60	8.62497	25′
.2	9.99961		10.00039	11.37524	59	8.62476	11.37563	59	8.62437	
.4	9.99962		10.00038	11.37585	61	8.62415	11.37623	60	8.62377	
.6	9.99962		10.00038	11.37645	60	8.62355	11.37683	60	8.62317	
.8	9.99962		10.00038	11.37706	61	8.62294	11.37744	61	8.62256	
36.0	9.99962		10.00038	11.37766	60	8.62234	11.37804	60	8.62196	24′

87°
267°

LOGS. OF TRIG. FUNCTIONS

′	Sine	Diff.	Cosec.	Tan.	Diff.	Cotan.	Secant	Diff.	Cosine	
36·0	9·99962		10·00038	11·37766	60	8·62234	11·37804	60	8·62196	24′
·2	9·99962		10·00038	11·37826	61	8·62174	11·37864	61	8·62136	
·4	9·99962		10·00038	11·37887	60	8·62113	11·37925	60	8·62075	
·6	9·99962		10·00038	11·37947	61	8·62053	11·37985	61	8·62015	
·8	9·99962		10·00038	11·38008	61	8·61992	11·38046	60	8·61954	
37·0	9·99962		10·00038	11·38069	61	8·61931	11·38106	61	8·61894	23′
·2	9·99963		10·00037	11·38130	61	8·61870	11·38167	61	8·61833	
·4	9·99963		10·00037	11·38191	61	8·61809	11·38228	61	8·61772	
·6	9·99963		10·00037	11·38252	61	8·61748	11·38289	61	8·61711	
·8	9·99963		10·00037	11·38313	61	8·61687	11·38350	61	8·61650	
38·0	9·99963		10·00037	11·38374	61	8·61626	11·38411	61	8·61589	22′
·2	9·99963		10·00037	11·38435	61	8·61565	11·38472	61	8·61528	
·4	9·99963		10·00037	11·38496	62	8·61504	11·38533	62	8·61467	
·6	9·99963		10·00037	11·38558	61	8·61442	11·38595	61	8·61405	
·8	9·99963		10·00037	11·38619	62	8·61381	11·38656	62	8·61344	
39·0	9·99964		10·00037	11·38681	62	8·61319	11·38718	61	8·61282	21′
·2	9·99964		10·00036	11·38743	62	8·61257	11·38779	62	8·61221	
·4	9·99964		10·00036	11·38805	62	8·61195	11·38841	62	8·61159	
·6	9·99964		10·00036	11·38867	62	8·61133	11·38903	62	8·61097	
·8	9·99964		10·00036	11·38929	62	8·61071	11·38965	62	8·61035	
40·0	9·99964		10·00036	11·38991	62	8·61009	11·39027	62	8·60973	20′
·2	9·99964		10·00036	11·39053	62	8·60947	11·39089	62	8·60911	
·4	9·99964		10·00036	11·39115	62	8·60885	11·39151	62	8·60849	
·6	9·99964		10·00036	11·39177	62	8·60823	11·39213	62	8·60787	
·8	9·99964		10·00036	11·39239	63	8·60761	11·39275	63	8·60725	
41·0	9·99965		10·00036	11·39302	63	8·60698	11·39338	62	8·60662	19′
2	9·99965		10·00035	11·39365	63	8·60635	11·39400	63	8·60600	
·4	9·99965		10·00035	11·39428	62	8·60572	11·39463	62	8·60537	
·6	9·99965		10·00035	11·39490	63	8·60510	11·39525	63	8·60475	
·8	9·99965		10·00035	11·39553	63	8·60447	11·39588	63	8·60412	
42·0	9·99965		10·00035	11·39616	63	8·60384	11·39651	63	8·60349	18′
·2	9·99965		10·00035	11·39679	63	8·60321	11·39714	63	8·60286	
·4	9·99965		10·00035	11·39742	63	8·60258	11·39777	63	8·60223	
·6	9·99965		10·00035	11·39805	63	8·60195	11·39840	63	8·60160	
·8	9·99965		10·00035	11·39868	64	8·60132	11·39903	64	8·60097	
43·0	9·99966		10·00035	11·39932	64	8·60068	11·39967	63	8·60033	17′
·2	9·99966		10·00034	11·39996	64	8·60004	11·40030	64	8·59970	
·4	9·99966		10·00034	11·40060	63	8·59940	11·40094	63	8·59906	
·6	9·99966		10·00034	11·40123	64	8·59877	11·40157	64	8·59843	
·8	9·99966		10·00034	11·40187	64	8·59813	11·40221	64	8·59779	
44·0	9·99966		10·00034	11·40251	64	8·59749	11·40285	64	8·59715	16′
·2	9·99966		10·00034	11·40315	64	8·59685	11·40349	64	8·59651	
·4	9·99966		10·00034	11·40379	64	8·59621	11·40413	64	8·59587	
·6	9·99966		10·00034	11·40443	64	8·59557	11·40477	64	8·59523	
·8	9·99966		10·00034	11·40507	65	8·59493	11·40541	64	8·59459	
45·0	9·99967		10·00034	11·40572	65	8·59428	11·40605	65	8·59395	15′
·2	9·99967		10·00033	11·40637	64	8·59363	11·40670	64	8·59330	
·4	9·99967		10·00033	11·40701	64	8·59299	11·40734	64	8·59266	
·6	9·99967		10·00033	11·40765	65	8·59235	11·40798	65	8·59202	
·8	9·99967		10·00033	11·40830	65	8·59170	11·40863	65	8·59137	
46·0	9·99967		10·00033	11·40895	65	8·59105	11·40928	65	8·59072	14′
·2	9·99967		10·00033	11·40960	65	8·59040	11·40993	65	8·59007	
·4	9·99967		10·00033	11·41025	65	8·58975	11·41058	65	8·58942	
·6	9·99967		10·00033	11·41090	65	8·58910	11·41123	65	8·58877	
·8	9·99967		10·00033	11·41155	65	8·58845	11·41188	65	8·58812	
47·0	9·99968		10·00033	11·41221	66	8·58780	11·41253	65	8·58747	13′
·2	9·99968		10·00032	11·41286	66	8·58714	11·41318	66	8·58682	
·4	9·99968		10·00032	11·41352	65	8·58648	11·41384	65	8·58616	
·6	9·99968		10·00032	11·41417	66	8·58583	11·41449	66	8·58551	
·8	9·99968		10·00032	11·41483	66	8·58517	11·41515	66	8·58485	
48·0	9·99968		10·00032	11·41549		8·58451	11·41581		8·58419	12′

87°
267°

LOGS. OF TRIG. FUNCTIONS

′	Sine	Diff.	Cosec.	Tan.	Diff.	Cotan.	Secant	Diff.	Cosine	
48·0	9·99968		10·00032	11·41549	66	8·58451	11·41581	66	8·58419	*12′*
·2	9·99968		10·00032	11·41615	65	8·58385	11·41647	65	8·58353	
·4	9·99968		10·00032	11·41680	66	8·58320	11·41712	66	8·58288	
·6	9·99968		10·00032	11·41746	67	8·58254	11·41778	67	8·58222	
·8	9·99968		10·00032	11·41813	66	8·58187	11·41845	66	8·58155	
49·0	9·99969		10·00032	11·41879	67	8·58121	11·41911	66	8·58089	*11′*
·2	9·99969		10·00031	11·41946	66	8·58054	11·41977	66	8·58023	
·4	9·99969		10·00031	11·42012	67	8·57988	11·42043	67	8·57957	
·6	9·99969		10·00031	11·42079	67	8·57921	11·42110	67	8·57890	
·8	9·99969		10·00031	11·42146	66	8·57854	11·42177	66	8·57823	
50·0	9·99969		10·00031	11·42212	67	8·57788	11·42243	67	8·57757	*10′*
·2	9·99969		10·00031	11·42279	67	8·57721	11·42310	67	8·57690	
·4	9·99969		10·00031	11·42346	67	8·57654	11·42377	67	8·57623	
·6	9·99969		10·00031	11·42413	67	8·57587	11·42444	67	8·57556	
·8	9·99969		10·00031	11·42480	68	8·57520	11·42511	68	8·57489	
51·0	9·99969		10·00031	11·42548	67	8·57452	11·42579	67	8·57421	*9′*
·2	9·99969		10·00031	11·42615	68	8·57385	11·42646	67	8·57354	
·4	9·99970		10·00030	11·42683	68	8·57317	11·42713	68	8·57287	
·6	9·99970		10·00030	11·42751	68	8·57249	11·42781	68	8·57219	
·8	9·99970		10·00030	11·42819	67	8·57181	11·42849	67	8·57151	
52·0	9·99970		10·00030	11·42886	68	8·57114	11·42916	68	8·57084	*8′*
·2	9·99970		10·00030	11·42954	68	8·57046	11·42984	68	8·57016	
·4	9·99970		10·00030	11·43022	68	8·56978	11·43052	68	8·56948	
·6	9·99970		10·00030	11·43090	69	8·56910	11·43120	69	8·56880	
·8	9·99970		10·00030	11·43159	68	8·56841	11·43189	68	8·56811	
53·0	9·99970		10·00030	11·43227	68	8·56773	11·43257	68	8·56743	*7′*
·2	9·99970		10·00030	11·43295	69	8·56705	11·43325	69	8·56675	
·4	9·99970		10·00030	11·43364	69	8·56636	11·43394	68	8·56606	
·6	9·99971		10·00029	11·43433	69	8·56567	11·43462	69	8·56538	
·8	9·99971		10·00029	11·43502	69	8·56498	11·43531	69	8·56469	
54·0	9·99971		10·00029	11·43571	69	8·56429	11·43600	69	8·56400	*6′*
·2	9·99971		10·00029	11·43640	69	8·56360	11·43669	69	8·56331	
·4	9·99971		10·00029	11·43709	69	8·56291	11·43738	69	8·56262	
·6	9·99971		10·00029	11·43778	70	8·56222	11·43807	70	8·56193	
·8	9·99971		10·00029	11·43848	69	8·56152	11·43877	69	8·56123	
55·0	9·99971		10·00029	11·43917	70	8·56083	11·43946	70	8·56054	*5′*
·2	9·99971		10·00029	11·43987	69	8·56013	11·44016	69	8·55984	
·4	9·99971		10·00029	11·44056	71	8·55944	11·44085	70	8·55915	
·6	9·99972		10·00028	11·44127	70	8·55873	11·44155	70	8·55845	
·8	9·99972		10·00028	11·44197	69	8·55803	11·44225	70	8·55775	
56·0	9·99972		10·00028	11·44266	71	8·55734	11·44295	70	8·55705	*4′*
·2	9·99972		10·00028	11·44337	70	8·55663	11·44365	70	8·55635	
·4	9·99972		10·00028	11·44407	70	8·55593	11·44435	70	8·55565	
·6	9·99972		10·00028	11·44477	71	8·55523	11·44505	71	8·55495	
·8	9·99972		10·00028	11·44548	70	8·55452	11·44576	70	8·55424	
57·0	9·99972		10·00028	11·44618	71	8·55382	11·44646	71	8·55354	*3′*
·2	9·99972		10·00028	11·44689	71	8·55311	11·44717	71	8·55283	
·4	9·99972		10·00028	11·44760	70	8·55240	11·44788	70	8·55212	
·6	9·99972		10·00028	11·44830	72	8·55170	11·44858	71	8·55142	
·8	9·99973		10·00027	11·44902	71	8·55098	11·44929	71	8·55071	
58·0	9·99973		10·00027	11·44973	72	8·55027	11·45001	72	8·55000	*2′*
·2	9·99973		10·00027	11·45045	71	8·54955	11·45072	71	8·54928	
·4	9·99973		10·00027	11·45116	71	8·54884	11·45143	71	8·54857	
·6	9·99973		10·00027	11·45187	72	8·54813	11·45214	72	8·54786	
·8	9·99973		10·00027	11·45259	72	8·54741	11·45286	72	8·54714	
59·0	9·99973		10·00027	11·45331	72	8·54669	11·45358	72	8·54642	*1′*
·2	9·99973		10·00027	11·45403	72	8·54597	11·45430	72	8·54570	
·4	9·99973		10·00027	11·45475	72	8·54525	11·45502	72	8·54498	
·6	9·99973		10·00027	11·45547	72	8·54453	11·45574	72	8·54426	
·8	9·99973		10·00027	11·45619	73	8·54381	11·45646	72	8·54354	
60·0	9·99974		10·00027	11·45692		8·54308	11·45718		8·54282	*0′*

88° 268°

LOGS. OF TRIG. FUNCTIONS

′	Sine	Diff.	Cosec.	Tan.	Diff.	Cotan.	Secant	Diff.	Cosine	
00·0	9·99974		10·00027	11·45692	72	8·54308	11·45718	72	8·54282	60′
·2	9·99974		10·00026	11·45764	73	8·54236	11·45790	73	8·54210	
·4	9·99974		10·00026	11·45837	73	8·54163	11·45863	73	8·54137	
·6	9·99974		10·00026	11·45910	73	8·54090	11·45936	73	8·54064	
·8	9·99974		10·00026	11·45983	72	8·54017	11·46009	72	8·53991	
01·0	9·99974		10·00026	11·46055	73	8·53945	11·46081	73	8·53919	59′
·2	9·99974		10·00026	11·46128	73	8·53872	11·46154	73	8·53846	
·4	9·99974		10·00026	11·46201	74	8·53799	11·46227	73	8·53773	
·6	9·99974		10·00026	11·46275	73	8·53725	11·46301	74	8·53699	
·8	9·99974		10·00026	11·46348	74	8·53652	11·46374	73	8·53626	
02·0	9·99974		10·00026	11·46422	73	8·53578	11·46448	74	8·53552	58′
·2	9·99974		10·00026	11·46495	75	8·53505	11·46521	73	8·53479	
·4	9·99975		10·00025	11·46570	74	8·53430	11·46595	74	8·53405	
·6	9·99975		10·00025	11·46644	74	8·53356	11·46669	74	8·53331	
·8	9·99975		10·00025	11·46718	74	8·53282	11·46743	74	8·53257	
03·0	9·99975		10·00025	11·46792	74	8·53208	11·46817	74	8·53183	57′
·2	9·99975		10·00025	11·46866	75	8·53134	11·46891	75	8·53109	
·4	9·99975		10·00025	11·46941	74	8·53059	11·46966	74	8·53034	
·6	9·99975		10·00025	11·47015	75	8·52985	11·47040	75	8·52960	
·8	9·99975		10·00025	11·47090	75	8·52910	11·47115	75	8·52885	
04·0	9·99975		10·00025	11·47165	75	8·52835	11·47190	75	8·52810	56′
·2	9·99975		10·00025	11·47240	75	8·52760	11·47265	75	8·52735	
·4	9·99975		10·00025	11·47315	75	8·52685	11·47340	75	8·52660	
·6	9·99975		10·00025	11·47390	76	8·52610	11·47415	75	8·52585	
·8	9·99976		10·00024	11·47466	75	8·52534	11·47490	76	8·52510	
05·0	9·99976		10·00024	11·47541	76	8·52459	11·47566	75	8·52434	55′
·2	9·99976		10·00024	11·47617	76	8·52383	11·47641	76	8·52359	
·4	9·99976		10·00024	11·47693	76	8·52307	11·47717	76	8·52283	
·6	9·99976		10·00024	11·47769	76	8·52231	11·47793	76	8·52207	
·8	9·99976		10·00024	11·47845	76	8·52155	11·47869	76	8·52131	
06·0	9·99976		10·00024	11·47921	76	8·52079	11·47945	76	8·52055	54′
·2	9·99976		10·00024	11·47997	77	8·52003	11·48021	77	8·51979	
·4	9·99976		10·00024	11·48074	76	8·51926	11·48098	76	8·51902	
·6	9·99976		10·00024	11·48150	77	8·51850	11·48174	77	8·51826	
·8	9·99976		10·00024	11·48227	77	8·51773	11·48251	76	8·51749	
07·0	9·99977		10·00024	11·48304	77	8·51696	11·48327	77	8·51673	53′
·2	9·99977		10·00023	11·48381	77	8·51619	11·48404	77	8·51596	
·4	9·99977		10·00023	11·48458	77	8·51542	11·48481	77	8·51519	
·6	9·99977		10·00023	11·48535	78	8·51465	11·48558	78	8·51442	
·8	9·99977		10·00023	11·48613	77	8·51387	11·48636	77	8·51364	
08·0	9·99977		10·00023	11·48690	78	8·51310	11·48713	78	8·51287	52′
·2	9·99977		10·00023	11·48768	78	8·51232	11·48791	78	8·51209	
·4	9·99977		10·00023	11·48846	77	8·51154	11·48869	77	8·51131	
·6	9·99977		10·00023	11·48923	78	8·51077	11·48946	78	8·51054	
·8	9·99977		10·00023	11·49001	79	8·50999	11·49024	79	8·50976	
09·0	9·99977		10·00023	11·49080	78	8·50920	11·49103	78	8·50897	51′
·2	9·99977		10·00023	11·49158	78	8·50842	11·49181	78	8·50819	
·4	9·99977		10·00023	11·49236	80	8·50764	11·49259	79	8·50741	
·6	9·99978		10·00022	11·49316	79	8·50684	11·49338	79	8·50662	
·8	9·99978		10·00022	11·49395	78	8·50605	11·49417	79	8·50583	
10·0	9·99978		10·00022	11·49473	79	8·50527	11·49496	78	8·50504	50′
·2	9·99978		10·00022	11·49552	80	8·50448	11·49574	80	8·50426	
·4	9·99978		10·00022	11·49632	79	8·50368	11·49654	79	8·50346	
·6	9·99978		10·00022	11·49711	79	8·50289	11·49733	79	8·50267	
·8	9·99978		10·00022	11·49790	80	8·50210	11·49812	80	8·50188	
11·0	9·99978		10·00022	11·49870	80	8·50130	11·49892	80	8·50108	49′
·2	9·99978		10·00022	11·49950	80	8·50050	11·49972	80	8·50028	
·4	9·99978		10·00022	11·50030	80	8·49970	11·50052	80	8·49948	
·6	9·99978		10·00022	11·50110	80	8·49890	11·50132	80	8·49868	
·8	9·99978		10·00022	11·50190	81	8·49810	11·50212	80	8·49788	
12·0	9·99979		10·00021	11·50271		8·49729	11·50292		8·49708	48′

88°
268°

LOGS. OF TRIG. FUNCTIONS

′	Sine	Diff.	Cosec.	Tan.	Diff.	Cotan.	Secant	Diff.	Cosine	
12·0	9·99979		10·00021	11·50271		8·49729	11·50292		8·49708	48′
·2	9·99979		10·00021	11·50352	81	8·49648	11·50373	81	8·49627	
·4	9·99979		10·00021	11·50432	80	8·49568	11·50453	80	8·49547	
·6	9·99979		10·00021	11·50513	81	8·49487	11·50534	81	8·49466	
·8	9·99979		10·00021	11·50594	81	8·49406	11·50615	81	8·49385	
13·0	9·99979		10·00021	11·50675	81	8·49325	11·50696	81	8·49304	47′
·2	9·99979		10·00021	11·50756	81	8·49244	11·50777	81	8·49223	
·4	9·99979		10·00021	11·50838	82	8·49162	11·50859	82	8·49141	
·6	9·99979		10·00021	11·50919	81	8·49081	11·50940	81	8·49060	
·8	9·99979		10·00021	11·51001	82	8·48999	11·51022	82	8·48978	
14·0	9·99979		10·00021	11·51083	82	8·48917	11·51104	82	8·48896	46′
·2	9·99979		10·00021	11·51165	82	8·48835	11·51186	82	8·48814	
·4	9·99979		10·00021	11·51247	82	8·48753	11·51268	82	8·48732	
·6	9·99979		10·00021	11·51330	83	8·48670	11·51350	82	8·48650	
·8	9·99980		10·00020	11·51413	83	8·48587	11·51433	83	8·48567	
15·0	9·99980		10·00020	11·51495	82	8·48505	11·51515	82	8·48485	45′
·2	9·99980		10·00020	11·51578	83	8·48422	11·51598	83	8·48402	
·4	9·99980		10·00020	11·51661	83	8·48339	11·51681	83	8·48319	
·6	9·99980		10·00020	11·51744	83	8·48256	11·51764	83	8·48236	
·8	9·99980		10·00020	11·51827	83	8·48173	11·51847	83	8·48153	
16·0	9·99980		10·00020	11·51911	84	8·48089	11·51931	84	8·48069	44′
·2	9·99980		10·00020	11·51994	83	8·48006	11·52014	83	8·47986	
·4	9·99980		10·00020	11·52078	84	8·47922	11·52098	84	8·47902	
·6	9·99980		10·00020	11·52162	84	8·47838	11·52182	84	8·47818	
·8	9·99980		10·00020	11·52246	84	8·47754	11·52266	84	8·47734	
17·0	9·99981		10·00019	11·52331	85	8·47669	11·52350	84	8·47650	43′
·2	9·99981		10·00019	11·52415	84	8·47585	11·52434	84	8·47566	
·4	9·99981		10·00019	11·52500	85	8·47500	11·52519	85	8·47481	
·6	9·99981		10·00019	11·52585	85	8·47415	11·52604	85	8·47396	
·8	9·99981		10·00019	11·52670	85	8·47330	11·52689	85	8·47311	
18·0	9·99981		10·00019	11·52755	85	8·47245	11·52774	85	8·47226	42′
·2	9·99981		10·00019	11·52840	85	8·47160	11·52859	85	8·47141	
·4	9·99981		10·00019	11·52925	85	8·47075	11·52944	85	8·47056	
·6	9·99981		10·00019	11·53011	86	8·46989	11·53030	86	8·46970	
·8	9·99981		10·00019	11·53097	86	8·46903	11·53116	86	8·46884	
19·0	9·99981		10·00019	11·53183	86	8·46817	11·53202	85	8·46799	41′
·2	9·99981		10·00019	11·53269	86	8·46731	11·53288	87	8·46712	
·4	9·99981		10·00019	11·53355	86	8·46645	11·53374	86	8·46626	
·6	9·99981		10·00019	11·53441	88	8·46559	11·53460	86	8·46540	
·8	9·99982		10·00018	11·53529	86	8·46471	11·53547	87	8·46453	
20·0	9·99982		10·00018	11·53615	87	8·46385	11·53634	87	8·46366	40′
·2	9·99982		10·00018	11·53702	87	8·46298	11·53720	86	8·46280	
·4	9·99982		10·00018	11·53789	88	8·46211	11·53807	87	8·46193	
·6	9·99982		10·00018	11·53877	87	8·46123	11·53895	88	8·46105	
·8	9·99982		10·00018	11·53964	88	8·46036	11·53982	87	8·46018	
21·0	9·99982		10·00018	11·54052	88	8·45948	11·54070	88	8·45930	39′
·2	9·99982		10·00018	11·54140	88	8·45860	11·54158	88	8·45842	
·4	9·99982		10·00018	11·54228	88	8·45772	11·54246	88	8·45754	
·6	9·99982		10·00018	11·54316	88	8·45684	11·54334	88	8·45666	
·8	9·99982		10·00018	11·54404	89	8·45596	11·54422	88	8·45578	
22·0	9·99982		10·00018	11·54493	88	8·45507	11·54511	89	8·45489	38′
·2	9·99982		10·00018	11·54581	89	8·45419	11·54599	88	8·45401	
·4	9·99982		10·00018	11·54670	90	8·45330	11·54688	89	8·45312	
·6	9·99983		10·00017	11·54760	89	8·45240	11·54777	89	8·45223	
·8	9·99983		10·00017	11·54849	90	8·45151	11·54866	89	8·45134	
23·0	9·99983		10·00017	11·54939	90	8·45061	11·54956	90	8·45044	37′
·2	9·99983		10·00017	11·55029	89	8·44971	11·55046	90	8·44954	
·4	9·99983		10·00017	11·55118	90	8·44882	11·55135	89	8·44865	
·6	9·99983		10·00017	11·55208	91	8·44792	11·55225	90	8·44775	
·8	9·99983		10·00017	11·55299	90	8·44701	11·55316	91	8·44684	
24·0	9·99983		10·00017	11·55389		8·44611	11·55406	90	8·44594	36′

91°
271°

88°
268°

LOGS. OF TRIG. FUNCTIONS

′	Sine	Diff.	Cosec.	Tan.	Diff.	Cotan.	Secant	Diff.	Cosine	
24·0	9·99983		10·00017	11·55389		8·44611	11·55406		8·44594	36′
·2	9·99983		10·00017	11·55479	90	8·44521	11·55496	90	8·44504	
·4	9·99983		10·00017	11·55570	91	8·44430	11·55587	91	8·44413	
·6	9·99983		10·00017	11·55661	91	8·44339	11·55678	91	8·44322	
·8	9·99983		10·00017	11·55752	91	8·44248	11·55769	91	8·44231	
25·0	9·99983		10·00017	11·55844	92	8·44156	11·55861	92	8·44139	35′
·2	9·99984		10·00016	11·55936	92	8·44064	11·55952	91	8·44048	
·4	9·99984		10·00016	11·56028	92	8·43972	11·56044	92	8·43956	
·6	9·99984		10·00016	11·56120	92	8·43880	11·56136	92	8·43864	
·8	9·99984		10·00016	11·56212	92	8·43788	11·56228	92	8·43772	
26·0	9·99984		10·00016	11·56304	92	8·43696	11·56320	92	8·43680	34′
·2	9·99984		10·00016	11·56396	92	8·43604	11·56412	92	8·43588	
·4	9·99984		10·00016	11·56489	93	8·43511	11·56505	93	8·43495	
·6	9·99984		10·00016	11·56582	93	8·43418	11·56598	93	8·43402	
·8	9·99984		10·00016	11·56675	93	8·43325	11·56691	93	8·43309	
27·0	9·99984		10·00016	11·56769	93	8·43232	11·56784	93	8·43216	33′
·2	9·99984		10·00016	11·56862	94	8·43138	11·56878	94	8·43122	
·4	9·99984		10·00016	11·56955	93	8·43045	11·56971	93	8·43029	
·6	9·99984		10·00016	11·57049	94	8·42951	11·57065	94	8·42935	
·8	9·99984		10·00016	11·57143	94	8·42857	11·57159	94	8·42841	
28·0	9·99984		10·00016	11·57238	95	8·42762	11·57254	95	8·42746	32′
·2	9·99985		10·00015	11·57333	95	8·42667	11·57348	94	8·42652	
·4	9·99985		10·00015	11·57428	95	8·42572	11·57443	95	8·42557	
·6	9·99985		10·00015	11·57523	95	8·42477	11·57538	95	8·42462	
·8	9·99985		10·00015	11·57618	95	8·42382	11·57633	95	8·42367	
29·0	9·99985		10·00015	11·57713	95	8·42287	11·57728	95	8·42272	31′
·2	9·99985		10·00015	11·57809	96	8·42191	11·57824	96	8·42176	
·4	9·99985		10·00015	11·57905	96	8·42095	11·57920	96	8·42080	
·6	9·99985		10·00015	11·58001	96	8·41999	11·58016	96	8·41984	
·8	9·99985		10·00015	11·58097	96	8·41903	11·58112	96	8·41888	
30·0	9·99985		10·00015	11·58193	96	8·41807	11·58208	96	8·41792	30′
·2	9·99985		10·00015	11·58290	97	8·41710	11·58305	97	8·41695	
·4	9·99985		10·00015	11·58386	96	8·41614	11·58401	96	8·41599	
·6	9·99985		10·00015	11·58484	98	8·41516	11·58499	98	8·41501	
·8	9·99985		10·00015	11·58581	97	8·41419	11·58596	97	8·41404	
31·0	9·99985		10·00015	11·58679	98	8·41321	11·58693	97	8·41307	29′
·2	9·99986		10·00014	11·58777	98	8·41223	11·58791	98	8·41209	
·4	9·99986		10·00014	11·58875	98	8·41125	11·58889	98	8·41111	
·6	9·99986		10·00014	11·58973	98	8·41027	11·58987	98	8·41013	
·8	9·99986		10·00014	11·59071	98	8·40929	11·59085	98	8·40915	
32·0	9·99986		10·00014	11·59170	99	8·40830	11·59184	99	8·40816	28′
·2	9·99986		10·00014	11·59269	99	8·40731	11·59283	99	8·40717	
·4	9·99986		10·00014	11·59368	99	8·40632	11·59382	99	8·40618	
·6	9·99986		10·00014	11·59467	99	8·40533	11·59481	99	8·40519	
·8	9·99986		10·00014	11·59566	99	8·40434	11·59580	99	8·40420	
33·0	9·99986		10·00014	11·59666	100	8·40334	11·59680	100	8·40320	27′
·2	9·99986		10·00014	11·59766	100	8·40234	11·59780	100	8·40220	
·4	9·99986		10·00014	11·59866	100	8·40134	11·59880	100	8·40120	
·6	9·99986		10·00014	11·59967	101	8·40033	11·59981	101	8·40019	
·8	9·99986		10·00014	11·60067	100	8·39933	11·60081	100	8·39919	
34·0	9·99986		10·00014	11·60169	101	8·39832	11·60182	101	8·39818	26′
·2	9·99987		10·00014	11·60270	102	8·39730	11·60283	101	8·39717	
·4	9·99987		10·00013	11·60372	102	8·39628	11·60385	102	8·39615	
·6	9·99987		10·00013	11·60473	101	8·39527	11·60486	101	8·39514	
·8	9·99987		10·00013	11·60575	102	8·39425	11·60588	102	8·39412	
35·0	9·99987		10·00013	11·60677	102	8·39323	11·60690	102	8·39310	25′
·2	9·99987		10·00013	11·60779	102	8·39221	11·60792	102	8·39208	
·4	9·99987		10·00013	11·60882	103	8·39118	11·60895	103	8·39105	
·6	9·99987		10·00013	11·60985	103	8·39015	11·60998	103	8·39002	
·8	9·99987		10·00013	11·61088	103	8·38912	11·61101	103	8·38899	
36·0	9·99987		10·00013	11·61191	103	8·38809	11·61204	103	8·38796	24′

88°
268°

LOGS. OF TRIG. FUNCTIONS

′	Sine	Diff.	Cosec.	Tan.	Diff.	Cotan.	Secant	Diff.	Cosine	
36·0	9·99987		10·00013	11·61191		8·38809	11·61204		8·38796	24′
·2	9·99987		10·00013	11·61294	103	8·38706	11·61307	103	8·38693	
·4	9·99987		10 00013	11·61398	104	8·38602	11·61411	104	8·38589	
·6	9·99987		10·00013	11·61502	104	8·38498	11·61515	104	8·38485	
·8	9·99987		10·00013	11·61606	104	8·38394	11·61619	104	8·38381	
					105			105		
37·0	9·99987		10·00013	11·61711	105	8·38289	11·61724	105	8·38276	23′
·2	9·99987		10·00013	11·61816	105	8·38184	11·61829	105	8·38171	
·4	9·99987		10·00013	11·61921	106	8·38079	11·61934	105	8·38066	
·6	9·99988		10·00012	11·62027	105	8·37973	11·62039	105	8·37961	
·8	9·99988		10·00012	11·62132	106	8·37868	11·62144	106	8·37856	
38·0	9·99988		10·00012	11·62238	106	8·37762	11·62250	106	8·37750	22′
·2	9·99988		10·00012	11·62344	106	8·37656	11·62356	106	8·37644	
·4	9·99988		10·00012	11·62450	107	8·37550	11·62462	107	8·37538	
·6	9·99988		10·00012	11·62557	107	8·37443	11·62569	107	8·37431	
·8	9·99988		10·00012	11·62664	107	8·37336	11·62676	107	8·37324	
39·0	9·99988		10·00012	11·62771	107	8·37229	11·62783	107	8·37217	21′
·2	9·99988		10·00012	11·62878	108	8·37122	11·62890	108	8·37110	
·4	9·99988		10·00012	11·62986	108	8·37014	11·62998	108	8·37002	
·6	9·99988		10·00012	11·63094	108	8·36906	11·63106	108	8·36894	
·8	9·99988		10·00012	11·63202	108	8·36798	11·63214	108	8·36786	
40·0	9·99988		10·00012	11·63310	109	8·36690	11·63322	109	8·36678	20′
·2	9·99988		10·00012	11·63419	109	8·36581	11·63431	109	8·36569	
·4	9·99988		10·00012	11·63528	109	8·36472	11·63540	109	8·36460	
·6	9·99988		10·00012	11·63637	110	8·36363	11·63649	110	8·36351	
·8	9·99988		10·00012	11·63747	110	8·36253	11·63759	110	8·36241	
41·0	9·99988		10·00012	11·63857	111	8·36143	11·63869	110	8·36131	19′
·2	9·99988		10·00012	11·63968	109	8·36032	11·63979	110	8·36021	
·4	9·99989		10·00011	11·64077	111	8·35923	11·64089	110	8·35911	
·6	9·99989		10·00011	11·64188	111	8·35812	11·64199	111	8·35801	
·8	9·99989		10·00011	11·64299	111	8·35701	11·64310	112	8·35690	
42·0	9·99989		10·00011	11·64410	112	8·35590	11·64422	111	8·35578	18′
·2	9·99989		10·00011	11·64522	112	8·35478	11·64533	112	8·35467	
·4	9·99989		10·00011	11·64634	112	8·35366	11·64645	112	8·35355	
·6	9·99989		10·00011	11·64746	112	8·35254	11·64757	112	8·35243	
·8	9·99989		10·00011	11·64858	113	8·35142	11·64869	113	8·35131	
43·0	9·99989		10·00011	11·64971	113	8·35029	11·64982	113	8·35018	17′
·2	9·99989		10·00011	11·65084	113	8·34916	11·65095	113	8·34905	
·4	9·99989		10·00011	11·65197	114	8·34803	11·65208	114	8·34792	
·6	9·99989		10·00011	11·65311	113	8·34689	11·65322	113	8·34678	
·8	9·99989		10·00011	11·65424	115	8·34576	11·65435	115	8·34565	
44·0	9·99989		10·00011	11·65539	114	8·34461	11·65550	114	8·34450	16′
·2	9·99989		10·00011	11·65653	115	8·34347	11·65664	115	8·34336	
·4	9·99989		10·00011	11·65768	116	8·34232	11·65779	115	8·34221	
·6	9·99990		10·00010	11·65884	115	8·34116	11·65894	115	8·34106	
·8	9·99990		10·00010	11·65999	115	8·34001	11·66009	116	8·33991	
45·0	9·99990		10·00010	11·66114	117	8·33886	11·66125	116	8·33875	15′
·2	9·99990		10·00010	11·66231	116	8·33769	11·66241	116	8·33759	
·4	9·99990		10·00010	11·66347	116	8·33653	11·66357	116	8·33643	
·6	9·99990		10·00010	11·66463	117	8·33537	11·66473	117	8·33527	
·8	9·99990		10·00010	11·66580	118	8·33420	11·66590	118	8·33410	
46·0	9·99990		10·00010	11·66698	117	8·33302	11·66708	117	8·33292	14′
·2	9·99990		10·00010	11·66815	118	8·33185	11·66825	118	8·33175	
·4	9·99990		10·00010	11·66933	118	8·33067	11·66943	118	8·33057	
·6	9·99990		10·00010	11·67051	119	8·32949	11·67061	119	8·32939	
·8	9·99990		10·00010	11·67170	119	8·32830	11·67180	118	8·32820	
47·0	9·99990		10·00010	11·67289	119	8·32711	11·67298	120	8·32702	13′
·2	9·99990		10·00010	11·67408	119	8·32592	11·67418	119	8·32582	
·4	9·99990		10·00010	11·67527	120	8·32473	11·67537	120	8·32463	
·6	9·99990		10·00010	11·67647	120	8·32353	11·67657	120	8·32343	
·8	9·99990		10·00010	11·67767	121	8·32233	11·67777	120	8·32223	
48·0	9·99990		10·00010	11·67888		8·32112	11·67897		8·32103	12′

LOGS. OF TRIG. FUNCTIONS

88°
268°

'	Sine	Diff.	Cosec.	Tan.	Diff.	Cotan.	Secant	Diff.	Cosine	
48·0	9·99990		10·00010	11·67888		8·32112	11·67897		8·32103	12'
·2	9·99991		10·00009	11·68009	121	8·31991	11·68018	121	8·31982	
·4	9·99991		10·00009	11·68130	121	8·31870	11·68139	121	8·31861	
·6	9·99991		10·00009	11·68252	122	8·31748	11·68261	122	8·31739	
·8	9·99991		10·00009	11·68374	122	8·31626	11·68383	122	8·31617	
49·0	9·99991		10·00009	11·68495	121	8·31505	11·68505	122	8·31495	11'
·2	9·99991		10·00009	11·68618	123	8·31382	11·68627	122	8·31373	
·4	9·99991		10·00009	11·68741	123	8·31259	11·68750	123	8·31250	
·6	9·99991		10·00009	11·68864	123	8·31136	11·68873	123	8·31127	
·8	9·99991		10·00009	11·68988	124	8·31012	11·68997	124	8·31003	
50·0	9·99991		10·00009	11·69112	124	8·30888	11·69121	124	8·30879	10'
·2	9·99991		10·00009	11·69236	124	8·30764	11·69245	124	8·30755	
·4	9·99991		10·00009	11·69360	124	8·30604	11·69369	124	8·30631	
·6	9·99991		10·00009	11·69485	125	8·30515	11·69494	125	8·30506	
·8	9·99991		10·00009	11·69611	126	8·30389	11·69620	126	8·30380	
51·0	9·99991		10·00009	11·69737	126	8·30263	11·69745	125	8·30255	9'
·2	9·99991		10·00009	11·69862	125	8·30138	11·69871	126	8·30129	
·4	9·99991		10·00009	11·69989	127	8·30011	11·69998	127	8·30002	
·6	9·99991		10·00009	11·70116	127	8·29884	11·70125	127	8·29875	
·8	9·99991		10·00009	11·70243	127	8·29757	11·70252	127	8·29748	
52·0	9·99992		10·00008	11·70371	128	8·29629	11·70379	127	8·29621	8'
·2	9·99992		10·00008	11·70499	128	8·29501	11·70507	128	8·29493	
·4	9·99992		10·00008	11·70628	129	8·29372	11·70636	129	8·29364	
·6	9·99992		10·00008	11·70756	128	8·29244	11·70764	128	8·29236	
·8	9·99992		10·00008	11·70885	129	8·29115	11·70893	129	8·29107	
53·0	9·99992		10·00008	11·71014	129	8·28986	11·71023	130	8·28977	7'
·2	9·99992		10·00008	11·71145	131	8·28855	11·71153	130	8·28847	
·4	9·99992		10·00008	11·71275	130	8·28725	11·71283	130	8·28717	
·6	9·99992		10·00008	11·71405	130	8·28595	11·71413	130	8·28587	
·8	9·99992		10·00008	11·71536	131	8·28464	11·71544	131	8·28456	
54·0	9·99992		10·00008	11·71668	132	8·28332	11·71676	132	8·28324	6'
·2	9·99992		10·00008	11·71799	131	8·28201	11·71807	131	8·28193	
·4	9·99992		10·00008	11·71932	133	8·28068	11·71940	133	8·28060	
·6	9·99992		10·00008	11·72064	132	8·27936	11·72072	132	8·27928	
·8	9·99992		10·00008	11·72197	133	8·27803	11·72205	133	8·27795	
55·0	9·99992		10·00008	11·72331	134	8·27669	11·72339	134	8·27661	5'
·2	9·99992		10·00008	11·72464	133	8·27536	11·72472	133	8·27528	
·4	9·99992		10·00008	11·72599	135	8·27401	11·72607	135	8·27393	
·6	9·99992		10·00008	11·72733	134	8·27267	11·72741	134	8·27259	
·8	9·99992		10·00008	11·72868	135	8·27132	11·72876	135	8·27124	
56·0	9·99992		10·00008	11·73004	136	8·26996	11·73012	136	8·26988	4'
·2	9·99992		10·00008	11·73140	136	8·26860	11·73148	136	8·26852	
·4	9·99992		10·00008	11·73276	136	8·26724	11·73284	136	8·26716	
·6	9·99993		10·00007	11·73414	138	8·26586	11·73421	137	8·26579	
·8	9·99993		10·00007	11·73551	137	8·26449	11·73558	137	8·26442	
57·0	9·99993		10·00007	11·73689	137	8·26312	11·73696	138	8·26304	3'
·2	9·99993		10·00007	11·73827	139	8·26173	11·73834	138	8·26166	
·4	9·99993		10·00007	11·73965	138	8·26035	11·73972	138	8·26028	
·6	9·99993		10·00007	11·74104	139	8·25896	11·74111	139	8·25889	
·8	9·99993		10·00007	11·74244	140	8·25756	11·74251	140	8·25749	
58·0	9·99993		10·00007	11·74384	139	8·25617	11·74391	140	8·25609	2'
·2	9·99993		10·00007	11·74524	141	8·25476	11·74531	140	8·25469	
·4	9·99993		10·00007	11·74665	141	8·25335	11·74672	141	8·25328	
·6	9·99993		10·00007	11·74806	141	8·25194	11·74813	141	8·25187	
·8	9·99993		10·00007	11·74947	141	8·25053	11·74954	141	8·25046	
59·0	9·99993		10·00007	11·75090	143	8·24910	11·75097	143	8·24903	1'
·2	9·99993		10·00007	11·75232	142	8·24768	11·75239	142	8·24761	
·4	9·99993		10·00007	11·75375	143	8·24625	11·75382	143	8·24618	
·6	9·99993		10·00007	11·75519	144	8·24481	11·75526	144	8·24474	
·8	9·99993		10·00007	11·75663	144	8·24337	11·75670	144	8·24330	
60·0	9·99993		10·00007	11·75808	145	8·24192	11·75815	144	8·24186	0'

91°
271°

89°
269°

LOGS. OF TRIG. FUNCTIONS

'	Sine	Diff.	Cosec.	Tan.	Diff.	Cotan.	Secant	Diff.	Cosine	
00·0	9·99993		10·00007	11·75808		8·24192	11·75815		8·24186	60'
·2	9·99993		10·00007	11·75952	146	8·24048	11·75959	145	8·24041	
·4	9·99993		10·00007	11·76098	146	8·23902	11·76105	146	8·23895	
·6	9·99994		10·00006	11·76245	147	0·23755	11·76251	146	8·23749	
·8	9·99994		10·00006	11·76391	146	8·23609	11·76397	146	8·23603	
01·0	9·99994		10·00006	11·76538	147	8·23462	11·76544	147	8·23456	59'
·2	9·99994		10·00006	11·76686	148	8·23314	11·76692	148	8·23308	
·4	9·99994		10·00006	11·76834	148	8·23166	11·76840	148	8·23160	
·6	9·99994		10·00006	11·76982	148	8·23018	11·76988	148	8·23012	
·8	9·99994		10·00006	11·77131	149	8·22869	11·77137	149	8·22863	
02·0	9·99994		10·00006	11·77281	149	8·22720	11·77287	150	8·22713	58'
·2	9·99994		10·00006	11·77431	151	8·22569	11·77437	150	8·22563	
·4	9·99994		10·00006	11·77581	150	8·22419	11·77587	150	8·22413	
·6	9·99994		10·00006	11·77732	151	8·22268	11·77738	151	8·22262	
·8	9·99994		10·00006	11·77884	152	8·22116	11·77890	152	8·22110	
03·0	9·99994		10·00006	11·78036	152	8·21964	11·78042	152	8·21958	57'
·2	9·99994		10·00006	11·78189	153	8·21811	11·78195	153	8·21805	
·4	9·99994		10·00006	11·78342	153	8·21658	11·78348	153	8·21652	
·6	9·99994		10·00006	11·78495	153	8·21505	11·78501	153	8·21499	
·8	9·99994		10·00006	11·78650	155	8·21350	11·78656	155	8·21344	
04·0	9·99994		10·00006	11·78805	155	8·21195	11·78811	155	8·21189	56'
·4	9·99994		10·00006	11·78960	155	8·21040	11·78966	156	8·21034	
·4	9·99994		10·00006	11·79116	156	8·20884	11·79122	156	8·20878	
·6	9·99994		10·00006	11·79272	156	8·20728	11·79278	157	8·20722	
·8	9·99994		10·00006	11·79429	157	8·20571	11·79435	158	8·20565	
05·0	9·99994		10·00006	11·79587	158	8·20413	11·79593	158	8·20407	55'
·2	9·99994		10·00006	11·79745	158	8·20255	11·79751	159	8·20249	
·4	9·99995		10·00005	11·79905	160	8·20095	11·79910	159	8·20090	
·6	9·99995		10·00005	11·80064	159	8·19936	11·80069	160	8·19931	
·8	9·99995		10·00005	11·80224	160	8·19776	11·80229	161	8·19771	
06·0	9·99995		10·00005	11·80384	160	8·19616	11·80390	161	8·19610	54'
·2	9·99995		10·00005	11·80546	162	8·19454	11·80551	162	8·19449	
·4	9·99995		10·00005	11·80708	162	8·19292	11·80713	162	8·19287	
·6	9·99995		10·00005	11·80870	162	8·19130	11·80875	163	8·19125	
·8	9·99995		10·00005	11·81033	163	8·18967	11·81038	164	8·18962	
07·0	9·99995		10·00005	11·81196	163	8·18804	11·81202	164	8·18798	53'
·2	9·99995		10·00005	11·81361	165	8·18639	11·81366	165	8·18634	
·4	9·99995		10·00005	11·81526	165	8·18474	11·81531	165	8·18469	
·6	9·99995		10·00005	11·81691	165	8·18309	11·81696	166	8·18304	
·8	9·99995		10·00005	11·81857	166	8·18143	11·81862	167	8·18138	
08·0	9·99995		10·00005	11·82024	167	8·17976	11·82029	167	8·17971	52'
·2	9·99995		10·00005	11·82191	167	8·17809	11·82196	168	8·17804	
·4	9·99995		10·00005	11·82359	168	8·17641	11·82364	169	8·17636	
·6	9·99995		10·00005	11·82528	169	8·17472	11·82533	169	8·17467	
·8	9·99995		10·00005	11·82697	169	8·17303	11·82702	170	8·17298	
09·0	9·99995		10·00005	11·82867	170	8·17133	11·82872	171	8·17128	51'
·2	9·99995		10·00005	11·83038	171	8·16962	11·83043	171	8·16957	
·4	9·99995		10·00005	11·83209	171	8·16791	11·83214	172	8·16786	
·6	9·99995		10·00005	11·83381	172	8·16619	11·83386	173	8·16614	
·8	9·99995		10·00005	11·83554	173	8·16446	11·83559	173	8·16441	
10·0	9·99995		10·00005	11·83727	173	8·16273	11·83732	174	8·16268	50'
·2	9·99995		10·00005	11·83901	174	8·16099	11·83906	175	8·16094	
·4	9·99995		10·00005	11·84076	175	8·15924	11·84081	175	8·15919	
·6	9·99996		10·00004	11·84252	176	8·15748	11·84256	176	8·15744	
·8	9·99996		10·00004	11·84428	176	8·15572	11·84432	177	8·15568	
11·0	9·99996		10·00004	11·84605	177	8·15395	11·84609	177	8·15391	49'
·2	9·99996		10·00004	11·84783	178	8·15217	11·84787	178	8·15213	
·4	9·99996		10·00004	11·84961	178	8·15039	11·84965	178	8·15035	
·6	9·99996		10·00004	11·85140	179	8·14860	11·85144	179	8·14856	
·8	9·99996		10·00004	11·85320	180	8·14680	11·85324	180	8·14676	
12·0	9·99996		10·00004	11·85500	180	8·14500	11·85505	181	8·14495	48'

90°
270°

89°
269°

LOGS. OF TRIG. FUNCTIONS

′	Sine	Diff.	Cosec.	Tan.	Diff.	Cotan.	Secant	Diff.	Cosine	
12·0	9·99996		10·00004	11·85500		8·14500	11·85505		8·14495	48′
·2	9·99996		10·00004	11·85682	182	8·14318	11·85686	181	8·14314	
·4	9·99996		10·00004	11·85864	182	8·14136	11·85868	182	8·14132	
·6	9·99996		10·00004	11·86047	183	8·13953	11·86051	183	8·13949	
·8	9·99996		10·00004	11·86231	184	8·13769	11·86235	184	8·13765	
13·0	9·99996		10·00004	11·86415	184	8·13585	11·86419	184	8·13581	47′
·2	9·99996		10·00004	11·86600	185	8·13400	11·86604	185	8·13396	
·4	9·99996		10·00004	11·86786	186	8·13214	11·86790	186	8·13210	
·6	9·99996		10·00004	11·86973	187	8·13027	11·86977	187	8·13023	
·8	9·99996		10·00004	11·87161	188	8·12839	11·87165	188	8·12835	
14·0	9·99996		10·00004	11·87349	188	8·12651	11·87353	188	8·12647	46′
·2	9·99996		10·00004	11·87538	189	8·12462	11·87542	189	8·12458	
·4	9·99996		10·00004	11·87728	190	8·12272	11·87732	190	8·12268	
·6	9·99996		10·00004	11·87919	191	8·12081	11·87923	191	8·12077	
·8	9·99996		10·00004	11·88111	192	8·11889	11·88115	192	8·11885	
15·0	9·99996		10·00004	11·88304	193	8·11696	11·88307	192	8·11693	45′
·2	9·99996		10·00004	11·88497	193	8·11503	11·88501	194	8·11499	
·4	9·99996		10·00004	11·88691	194	8·11309	11·88695	194	8·11305	
·6	9·99996		10·00004	11·88886	195	8·11114	11·88890	195	8·11110	
·8	9·99996		10·00004	11·89082	196	8·10918	11·89086	196	8·10914	
16·0	9·99996		10·00004	11·89280	198	8·10720	11·89283	197	8·10717	44′
·2	9·99996		10·00004	11·89477	197	8·10523	11·89481	198	8·10519	
·4	9·99996		10·00004	11·89676	199	8·10324	11·89680	199	8·10320	
·6	9·99997		10·00003	11·89877	201	8·10123	11·89880	200	8·10120	
·8	9·99997		10·00003	11·90077	200	8·09923	11·90080	200	8·09920	
17·0	9·99997		10·00003	11·90278	201	8·09722	11·90282	202	8·09718	43′
·2	9·99997		10·00003	11·90481	203	8·09519	11·90484	202	8·09516	
·4	9·99997		10·00003	11·90685	204	8·09315	11·90688	204	8·09312	
·6	9·99997		10·00003	11·90889	204	8·09111	11·90892	204	8·09108	
·8	9·99997		10·00003	11·91094	205	8·08906	11·91097	205	8·08903	
18·0	9·99997		10·00003	11·91300	206	8·08700	11·91304	207	8·08696	42′
·2	9·99997		10·00003	11·91508	208	8·08492	11·91511	207	8·08489	
·4	9·99997		10·00003	11·91716	208	8·08284	11·91719	208	8·08281	
·6	9·99997		10·00003	11·91925	209	8·08075	11·91928	209	8·08072	
·8	9·99997		10·00003	11·92136	211	8·07864	11·92139	211	8·07861	
19·0	9·99997		10·00003	11·92347	211	8·07653	11·92350	211	8·07650	41′
·2	9·99997		10·00003	11·92559	212	8·07441	11·92562	212	8·07438	
·4	9·99997		10·00003	11·92773	214	8·07227	11·92776	214	8·07224	
·6	9·99997		10·00003	11·92987	214	8·07013	11·92990	214	8·07010	
·8	9·99997		10·00003	11·93203	216	8·06797	11·93206	216	8·06794	
20·0	9·99997		10·00003	11·93419	216	8·06581	11·93422	216	8·06578	40′
·2	9·99997		10·00003	11·93637	218	8·06363	11·93640	218	8·06360	
·4	9·99997		10·00003	11·93856	219	8·06144	11·93859	219	8·06141	
·6	9·99997		10·00003	11·94076	220	8·05924	11·94079	220	8·05921	
·8	9·99997		10·00003	11·94297	221	8·05703	11·94300	221	8·05700	
21·0	9·99997		10·00003	11·94519	222	8·05481	11·94522	222	8·05478	39′
·2	9·99997		10·00003	11·94742	223	8·05258	11·94745	223	8·05255	
·4	9·99997		10·00003	11·94967	225	8·05033	11·94970	225	8·05030	
·6	9·99997		10·00003	11·95192	225	8·04808	11·95195	225	8·04805	
·8	9·99997		10·00003	11·95419	227	8·04581	11·95422	227	8·04578	
22·0	9·99997		10·00003	11·95647	228	8·04353	11·95650	228	8·04350	38′
·2	9·99997		10·00003	11·95876	229	8·04124	11·95879	229	8·04121	
·4	9·99997		10·00003	11·96107	231	8·03893	11·96110	231	8·03890	
·6	9·99997		10·00003	11·96338	231	8·03662	11·96341	231	8·03659	
·8	9·99997		10·00003	11·96571	233	8·03429	11·96574	233	8·03426	
23·0	9·99998		10·00002	11·96806	234	8·03195	11·96808	234	8·03192	37′
·2	9·99998		10·00002	11·97041	236	8·02959	11·97043	235	8·02957	
·4	9·99998		10·00002	11·97278	237	8·02722	11·97280	237	8·02720	
·6	9·99998		10·00002	11·97516	238	8·02484	11·97518	238	8·02482	
·8	9·99998		10·00002	11·97755	239	8·02245	11·97757	239	8·02243	
24·0	9·99998		10·00002	11·97996	240	8·02005	11·97998	241	8·02002	36′

89°
269°

LOGS. OF TRIG. FUNCTIONS

′	Sine	Diff.	Cosec.	Tan.	Diff.	Cotan.	Secant	Diff.	Cosine	
24·0	9·99998		10·00002	11·97996	242	8·02005	11·97998	242	8·02002	36′
·2	9·99998		10·00002	11·98238	243	8·01762	11·98240	243	8·01760	
·4	9·99998		10·00002	11·98481	245	8·01519	11·98483	245	8·01517	
·6	9·99998		10·00002	11·98728	246	8·01274	11·98728	246	8·01272	
·8	9·99998		10·00002	11·98972	247	8·01028	11·98974	247	8·01026	
25·0	9·99998		10·00002	11·99219	249	8·00781	11·99221	249	8·00779	35′
·2	9·99998		10·00002	11·99468	251	8·00532	11·99470	251	8·00530	
·4	9·99998		10·00002	11·99719	251	8·00281	11·99721	251	8·00279	
·6	9·99998		10·00002	11·99970	254	8·00030	11·99972	254	8·00028	
·8	9·99998		10·00002	12·00224	254	7·99776	12·00226	254	7·99774	
26·0	9·99998		10·00002	12·00478	256	7·99522	12·00480	256	7·99520	34′
·2	9·99998		10·00002	12·00734	258	7·99266	12·00736	258	7·99264	
·4	9·99998		10·00002	12·00992	259	7·99008	12·00994	259	7·99006	
·6	9·99998		10·00002	12·01251	261	7·98749	12·01253	261	7·98747	
·8	9·99998		10·00002	12·01512	263	7·98488	12·01514	263	7·98486	
27·0	9·99998		10·00002	12·01775	264	7·98225	12·01777	264	7·98223	33′
·2	9·99998		10·00002	12·02039	265	7·97961	12·02041	265	7·97959	
·4	9·99998		10·00002	12·02304	268	7·97696	12·02306	268	7·97694	
·6	9·99998		10·00002	12·02572	268	7·97428	12·02574	268	7·97426	
·8	9·99998		10·00002	12·02840	271	7·97160	12·02842	271	7·97158	
28·0	9·99998		10·00002	12·03111	272	7·96889	12·03113	272	7·96887	32′
·2	9·99998		10·00002	12·03383	274	7·96617	12·03385	274	7·96615	
·4	9·99998		10·00002	12·03657	276	7·96343	12·03659	276	7·96341	
·6	9·99998		10·00002	12·03933	278	7·96067	12·03935	278	7·96065	
·8	9·99998		10·00002	12·04211	279	7·95789	12·04213	279	7·95787	
29·0	9·99998		10·00002	12·04490	281	7·95510	12·04492	281	7·95508	31′
·2	9·99998		10·00002	12·04771	283	7·95229	12·04773	283	7·95227	
·4	9·99998		10·00002	12·05054	285	7·94946	12·05056	285	7·94944	
·6	9·99998		10·00002	12·05339	286	7·94661	12·05341	286	7·94659	
·8	9·99998		10·00002	12·05625	289	7·94375	12·05627	289	7·94373	
30·0	9·99998		10·00002	12·05914	290	7·94086	12·05916	290	7·94084	30′
·2	9·99998		10·00002	12·06204	293	7·93796	12·06206	293	7·93794	
·4	9·99998		10·00002	12·06497	294	7·93503	12·06499	294	7·93501	
·6	9·99998		10·00002	12·06791	297	7·93209	12·06793	297	7·93207	
·8	9·99998		10·00002	12·07088	299	7·92912	12·07090	298	7·92910	
31·0	9·99999		10·00001	12·07387	301	7·92613	12·07388	301	7·92612	29′
·2	9·99999		10·00001	12·07688	302	7·92312	12·07689	302	7·92311	
·4	9·99999		10·00001	12·07990	305	7·92010	12·07991	305	7·92009	
·6	9·99999		10·00001	12·08295	307	7·91705	12·08296	307	7·91704	
·8	9·99999		10·00001	12·08602	309	7·91398	12·08603	309	7·91397	
32·0	9·99999		10·00001	12·08911	311	7·91089	12·08912	311	7·91088	28′
·2	9·99999		10·00001	12·09222	314	7·90778	12·09223	314	7·90777	
·4	9·99999		10·00001	12·09536	316	7·90464	12·09537	316	7·90463	
·6	9·99999		10·00001	12·09852	318	7·90148	12·09853	318	7·90147	
·8	9·99999		10·00001	12·10170	320	7·89830	12·10171	320	7·89829	
33·0	9·99999		10·00001	12·10490	323	7·89510	12·10492	323	7·89509	27′
·2	9·99999		10·00001	12·10813	326	7·89187	12·10814	326	7·89186	
·4	9·99999		10·00001	12·11139	327	7·88861	12·11140	327	7·88860	
·6	9·99999		10·00001	12·11466	331	7·88534	12·11467	331	7·88533	
·8	9·99999		10·00001	12·11797	332	7·88203	12·11798	332	7·88202	
34·0	9·99999		10·00001	12·12129	336	7·87871	12·12131	336	7·87870	26′
·2	9·99999		10·00001	12·12465	338	7·87535	12·12466	338	7·87534	
·4	9·99999		10·00001	12·12803	340	7·87197	12·12804	340	7·87196	
·6	9·99999		10·00001	12·13143	344	7·86857	12·13144	344	7·86856	
·8	9·99999		10·00001	12·13487	346	7·86513	12·13488	346	7·86512	
35·0	9·99999		10·00001	12·13833	349	7·86167	12·13834	349	7·86166	25′
·2	9·99999		10·00001	12·14182	351	7·85818	12·14183	351	7·85817	
·4	9·99999		10·00001	12·14533	355	7·85467	12·14534	355	7·85466	
·6	9·99999		10·00001	12·14888	357	7·85112	12·14889	357	7·85111	
·8	9·99999		10·00001	12·15245	361	7·84755	12·15246	361	7·84754	
36·0	9·99999		10·00001	12·15606		7·84394	12·15607		7·84393	24′

90°
270°

89°
269°

LOGS. OF TRIG. FUNCTIONS

′	Sine	Diff.	Cosec.	Tan.	Diff.	Cotan.	Secant	Diff.	Cosine	
36·0	9·99999		10·00001	12·15606		7·84394	12·15607		7·84393	24′
·2	9·99999		10·00001	12·15969	363	7·84031	12·15970	363	7·84030	
·4	9·99999		10·00001	12·16336	367	7·83664	12·16337	367	7·83663	
·6	9·99999		10·00001	12·16705	369	7·83295	12·16706	369	7·83294	
·8	9·99999		10·00001	12·17078	373	7·82922	12·17079	373	7·82921	
37·0	9·99999		10·00001	12·17454	376	7·82546	12·17455	376	7·82545	23′
·2	9·99999		10·00001	12·17833	379	7·82167	12·17834	379	7·82166	
·4	9·99999		10·00001	12·18216	383	7·81784	12·18217	383	7·81783	
·6	9·99999		10·00001	12·18602	386	7·81398	12·18603	386	7·81397	
·8	9·99999		10·00001	12·18991	389	7·81009	12·18992	389	7·81008	
38·0	9·99999		10·00001	12·19385	393	7·80616	12·19385	393	7·80615	22′
·2	9·99999		10·00001	12·19781	396	7·80219	12·19782	397	7·80218	
·4	9·99999		10·00001	12·20181	400	7·79819	12·20182	400	7·79818	
·6	9·99999		10·00001	12·20585	404	7·79415	12·20586	404	7·79414	
·8	9·99999		10·00001	12·20993	408	7·79007	12·20994	408	7·79006	
39·0	9·99999		10·00001	12·21405	412	7·78595	12·21406	412	7·78594	21′
·2	9·99999		10·00001	12·21820	415	7·78180	12·21821	415	7·78179	
·4	9·99999		10·00001	12·22240	420	7·77760	12·22241	420	7·77759	
·6	9·99999		10·00001	12·22664	424	7·77336	12·22665	424	7·77335	
·8	9·99999		10·00001	12·23092	428	7·76908	12·23093	428	7·76907	
40·0	9·99999		10·00001	12·23524	432	7·76476	12·23525	432	7·76475	20′
·2	9·99999		10·00001	12·23960	436	7·76040	12·23961	436	7·76039	
·4	9·99999		10·00001	12·24401	441	7·75599	12·24402	441	7·75598	
·6	9·99999		10·00001	12·24847	446	7·75153	12·24848	446	7·75152	
·8	9·99999		10·00001	12·25297	450	7·74703	12·25298	450	7·74702	
41·0	9·99999		10·00001	12·25752	455	7·74248	12·25752	454	7·74248	19′
·2	9·99999		10·00001	12·26211	459	7·73789	12·26212	460	7·73788	
·4	9·99999		10·00001	12·26675	464	7·73325	12·26676	464	7·73324	
·6	9·99999		10·00001	12·27145	470	7·72855	12·27146	470	7·72854	
·8	9·99999		10·00001	12·27620	475	7·72380	12·27621	475	7·72379	
42·0	9·99999		10·00001	12·28100	480	7·71900	12·28100	479	7·71900	18′
·2	9·99999		10·00001	12·28585	485	7·71415	12·28586	486	7·71414	
·4	9·99999		10·00001	12·29075	490	7·70925	12·29076	490	7·70924	
·6	9·99999		10·00001	12·29572	497	7·70428	12·29573	497	7·70427	
·8	9·99999		10·00001	12·30074	502	7·69926	12·30075	502	7·69925	
43·0	9·99999		10·00001	12·30582	508	7·69418	12·30583	508	7·69417	17′
·2	9·99999		10·00001	12·31096	514	7·68904	12·31097	514	7·68903	
·4	9·99999		10·00001	12·31616	520	7·68384	12·31617	520	7·68383	
·6	10·00000		10·00000	12·32143	527	7·67857	12·32143	526	7·67857	
·8	10·00000		10·00000	12·32676	533	7·67324	12·32676	533	7·67324	
44·0	10·00000		10·00000	12·33215	539	7·66785	12·33216	540	7·66784	16′
·2	10·00000		10·00000	12·33762	547	7·66238	12·33762	546	7·66238	
·4	10·00000		10·00000	12·34315	553	7·65685	12·34315	553	7·65685	
·6	10·00000		10·00000	12·34876	561	7·65124	12·34876	561	7·65124	
·8	10·00000		10·00000	12·35443	567	7·64557	12·35443	567	7·64557	
45·0	10·00000		10·00000	12·36018	575	7·63982	12·36018	575	7·63982	15′
·2	10·00000		10·00000	12·36601	583	7·63399	12·36601	583	7·63399	
·4	10·00000		10·00000	12·37192	591	7·62808	12·37192	591	7·62808	
·6	10·00000		10·00000	12·37791	599	7·62209	12·37791	599	7·62209	
·8	10·00000		10·00000	12·38399	608	7·61601	12·38399	608	7·61601	
46·0	10·00000		10·00000	12·39014	615	7·60986	12·39015	616	7·60985	14′
·2	10·00000		10·00000	12·39640	626	7·60360	12·39640	625	7·60360	
·4	10·00000		10·00000	12·40274	634	7·59726	12·40274	634	7·59726	
·6	10·00000		10·00000	12·40917	643	7·59083	12·40917	643	7·59083	
·8	10·00000		10·00000	12·41570	653	7·58430	12·41570	653	7·58430	
47·0	10·00000		10·00000	12·42233	663	7·57767	12·42233	663	7·57767	13′
·2	10·00000		10·00000	12·42917	674	7·57093	12·42917	674	7·57093	
·4	10·00000		10·00000	12·43590	683	7·56410	12·43590	683	7·56410	
·6	10·00000		10·00000	12·44285	695	7·55715	12·44285	695	7·55715	
·8	10·00000		10·00000	12·44992	707	7·55008	12·44992	707	7·55008	
48·0	10·00000		10·00000	12·45709	717	7·54291	12·45709	717	7·54291	12′

LOGS. OF TRIG. FUNCTIONS

′	Sine	Diff.	Cosec.	Tan.	Diff.	Cotan.	Secant	Diff.	Cosine	
48·0	10·00000		10·00000	12·45709		7·54291	12·45709		7·54291	12′
·1	10·00000		10·00000	12·46073	364	7·53927	12·46073	364	7·53927	
·2	10·00000		10·0c000	12·46439	366	7·53561	12·46439	366	7·53561	
·3	10·00000		10·00000	12·46809	370	7 53191	12·46809	370	7·53191	
·4	10·00000		10·00000	12·47182	373	7·52818	12·47182	373	7·52818	
48·5	10·00000		10·00000	12·47557	375	7·52443	12·47558	376	7·52442	
·6	10·00000		10·00000	12·47937	380	7·52063	12·47937	379	7·52063	
·7	10·00000		10·00000	12·48320	383	7·51680	12·48320	383	7·51680	
·8	10·00000		10·00000	12·48706	386	7·51294	12·48706	386	7·51294	
·9	10·00000		10·00000	12·49095	389	7·50905	12·49095	389	7·50905	
49·0	10·00000		10·00000	12·49488	393	7·50512	12·49488	393	7·50512	11′
·1	10·00000		10·00000	12·49885	397	7·50115	12·49885	397	7·50115	
·2	10·00000		10·00000	12·50285	400	7·49715	12·50285	400	7·49715	
·3	10·00000		10·00000	12·50689	404	7·49311	12·50689	404	7·49311	
·4	10·00000		10·00000	12·51097	408	7·48903	12·51097	408	7·48903	
49·5	10·00000		10·00000	12·51508	411	7·48492	12·50509	411	7·48492	
·6	10·00000		10·00000	12·51924	416	7·48076	12·51924	416	7·48076	
·7	10·00000		10·00000	12·52344	420	7·47656	12·52344	420	7·47656	
·8	10·00000		10·00000	12·52767	423	7·47233	12·52767	423	7·47233	
·9	10·00000		10·00000	12·53195	428	7·46805	12·53195	428	7·46805	
50·0	10·00000		10·00000	12·53627	432	7·46373	12·53628	432	7·46373	10′
·1	10·00000		10·00000	12·54064	437	7·45936	12·54064	437	7·45936	
·2	10·00000		10·00000	12·54505	441	7·45495	12·54505	441	7·45495	
·3	10·00000		10·00000	12·54950	445	7·45050	12·54950	445	7·45050	
·4	10·00000		10·00000	12·55400	450	7·44600	12·55400	450	7·44600	
50·5	10·00000		10·00000	12·55855	455	7·44145	12·55855	455	7·44145	
·6	10·00000		10·00000	12·56315	460	7·43685	12·56315	460	7·43685	
·7	10·00000		10·00000	12·56779	464	7·43221	12·56779	464	7·43221	
·8	10·00000		10·00000	12·57249	470	7·42751	12·57249	470	7·42751	
·9	10·00000		10·00000	12·57723	474	7·42277	12·57723	474	7·42277	
51·0	10·00000		10·00000	12·58203	480	7·41797	12·58203	480	7·41797	9′
·1	10·00000		10·00000	12·58688	485	7·41312	12·58688	485	7·41312	
·2	10·00000		10·00000	12·59179	491	7·40821	12·59179	491	7·40821	
·3	10·00000		10·00000	12·59675	496	7·40325	12·59675	496	7·40325	
·4	10·00000		10·00000	12·60178	503	7·39822	12·60178	503	7·39822	
51·5	10·00000		10·00000	12·60685	507	7·39315	12·60686	507	7·39315	
·6	10·00000		10·00000	12·61199	514	7·38801	12·61199	514	7·38801	
·7	10·00000		10·00000	12·61720	521	7·38280	12·61720	521	7·38280	
·8	10·00000		10·00000	12·62246	526	7·37754	12·62246	526	7·37754	
·9	10·00000		10·00000	12·62779	533	7·37221	12·62779	533	7·37221	
52·0	10·00000		10·00000	12·63318	539	7·36682	12·63318	539	7·36682	8′
·1	10·00000		10·00000	12·63865	547	7·36135	12·63865	547	7·36135	
·2	10·00000		10·00000	12·64418	553	7·35582	12·64418	553	7·35582	
·3	10·00000		10·00000	12·64978	560	7·35022	12·64978	560	7·35022	
·4	10·00000		10·00000	12·65546	568	7·34454	12·65546	568	7·34454	
52·5	10·00000		10·00000	12·66121	575	7·33879	12·66121	575	7·33879	
·6	10·00000		10·00000	12·66704	583	7·33296	12·66704	583	7·33296	
·7	10·00000		10·00000	12·67295	591	7·32705	12·67295	591	7·32705	
·8	10·00000		10·00000	12·67894	599	7·32106	12·67894	599	7·32106	
·9	10·00000		10·00000	12·68502	608	7·31498	12·68502	608	7·31498	
53·0	10·00000		10·00000	12·69118	616	7·30883	12·69118	616	7·30882	7′
·1	10·00000		10·00000	12·69742	624	7·30258	12·69742	624	7·30258	
·2	10·00000		10·00000	12·70376	634	7·29624	12·70376	634	7·29624	
·3	10·00000		10·00000	12·71020	644	7·28980	12·71020	644	7·28980	
·4	10·00000		10·00000	12·71673	653	7·28327	12·71673	653	7·28327	
53·5	10·00000		10·00000	12·72336	663	7·27664	12·72336	663	7·27664	
·6	10·00000		10·00000	12·73009	673	7·26991	12·73009	673	7·26991	
·7	10·00000		10·00000	12·73693	684	7·26307	12·73693	684	7·26307	
·8	10·00000		10·00000	12·74388	695	7·25612	12·74388	695	7·25612	
·9	10·00000		10·00000	12·75094	706	7·24906	12·75094	706	7·24906	
54·0	10·00000		10·00000	12·75812	718	7·24188	12·75812	718	7·24188	6′

LOGS. OF TRIG. FUNCTIONS

89°
269°

′	Sine	Diff.	Cosec.	Tan.	Diff.	Cotan.	Secant	Diff.	Cosine	
54·0	10·00000		10·00000	12·75812	730	7·24188	12·75812	730	7·24188	6′
·1	10·00000		10·00000	12·76542	743	7·23458	12·76542	743	7·23458	
·2	10·00000		10·00000	12·77285	755	7·22715	12·77285	755	7·22715	
·3	10·00000		10·00000	12·78040	769	7·21960	12·78040	769	7·21960	
·4	10·00000		10·00000	12·78809	782	7·21191	12·78809	782	7·21191	
54·5	10·00000		10·00000	12·79591	797	7·20409	12·79591	797	7·20409	
·6	10·00000		10·00000	12·80388	812	7·19612	12·80388	812	7·19612	
·7	10·00000		10·00000	12·81200	827	7·18800	12·81200	827	7·18800	
·8	10·00000		10·00000	12·82027	843	7·17973	12·82027	843	7·17973	
·9	10·00000		10·00000	12·82870	860	7·17130	12·82870	860	7·17130	
55·0	10·00000		10·00000	12·83730	878	7·16270	12·83730	878	7·16270	5′
·1	10·00000		10·00000	12·84608	895	7·15392	12·84608	895	7·15392	
·2	10·00000		10·00000	12·85503	915	7·14497	12·85503	915	7·14497	
·3	10·00000		10·00000	12·86418	934	7·13582	12·86418	934	7·13582	
·4	10·00000		10·00000	12·87352	954	7·12648	12·87352	954	7·12648	
55·5	10·00000		10·00000	12·88306	976	7·11694	12·88306	976	7·11694	
·6	10·00000		10·00000	12·89282	999	7·10718	12·89282	999	7·10718	
·7	10·00000		10·00000	12·90281	1021	7·09719	12·90281	1021	7·09719	
·8	10·00000		10·00000	12·91302	1047	7·08698	12·91302	1047	7·08698	
·9	10·00000		10·00000	12·92349	1072	7·07651	12·92349	1072	7·07651	
56·0	10·00000		10·00000	12·93421	1100	7·06579	12·93421	1100	7·06579	4′
·1	10·00000		10·00000	12·94521	1128	7·05479	12·94521	1128	7·05479	
·2	10·00000		10·00000	12·95649	1158	7·04351	12·95649	1158	7·04351	
·3	10·00000		10·00000	12·96807	1190	7·03193	12·96807	1190	7·03193	
·4	10·00000		10·00000	12·97997	1224	7·02003	12·97997	1224	7·02003	
56·5	10·00000		10·00000	12·99221	1258	7·00779	12·99221	1258	7·00779	
·6	10·00000		10·00000	13·00479	1297	6·99521	13·00479	1297	6·99521	
·7	10·00000		10·00000	13·01776	1336	6·98224	13·01776	1336	6·98224	
·8	10·00000		10·00000	13·03112	1379	6·96888	13·03112	1379	6·96888	
·9	10·00000		10·00000	13·04491	1424	6·95509	13·04491	1424	6·95509	
57·0	10·00000		10·00000	13·05915	1473	6·94085	13·05915	1473	6·94085	3′
·1	10·00000		10·00000	13·07388	1524	6·92612	13·07388	1524	6·92612	
·2	10·00000		10·00000	13·08912	1579	6·91088	13·08912	1579	6·91088	
·3	10·00000		10·00000	13·10491	1639	6·89509	13·10491	1639	6·89509	
·4	10·00000		10·00000	13·12130	1703	6·87870	13·12130	1703	6·87870	
57·5	10·00000		10·00000	13·13833	1773	6·86167	13·13833	1773	6·86167	
·6	10·00000		10·00000	13·15606	1849	6·84394	13·15606	1849	6·84394	
·7	10·00000		10·00000	13·17455	1930	6·82545	13·17455	1930	6·82545	
·8	10·00000		10·00000	13·19385	2020	6·80615	13·19385	2020	6·80615	
·9	10·00000		10·00000	13·21405	2119	6·78595	13·21405	2119	6·78595	
58·0	10·00000		10·00000	13·23524	2228	6·76476	13·23524	2228	6·76476	2′
·1	10·00000		10·00000	13·25752	2348	6·74248	13·25752	2348	6·74248	
·2	10·00000		10·00000	13·28100	2482	6·71900	13·28100	2482	6·71900	
·3	10·00000		10·00000	13·30582	2633	6·69418	13·30582	2633	6·69418	
·4	10·00000		10·00000	13·33215	2803	6·66785	13·33215	2803	6·66785	
58·5	10·00000		10·00000	13·36018	2997	6·63982	13·36018	2997	6·63982	
·6	10·00000		10·00000	13·39015	3218	6·60985	13·39015	3218	6·60985	
·7	10·00000		10·00000	13·42233	3476	6·57767	13·42233	3476	6·57767	
·8	10·00000		10·00000	13·45709	3779	6·54291	13·45709	3779	6·54291	
·9	10·00000		10·00000	13·49488	4139	6·50512	13·49488	4139	6·50512	
59·0	10·00000		10·00000	13·53627	4576	6·46373	13·53627	4576	6·46373	1′
·1	10·00000		10·00000	13·58203	5115	6·41797	13·58203	5115	6·41797	
·2	10·00000		10·00000	13·63318	5800	6·36682	13·63318	5800	6·36682	
·3	10·00000		10·00000	13·69118	6694	6·30882	13·69118	6694	6·30882	
·4	10·00000		10·00000	13·75812	7918	6·24188	13·75812	7918	6·24188	
59·5	10·00000		10·00000	13·83730	9691	6·16270	13·83730	9691	6·16270	
·6	10·00000		10·00000	13·93421	12494	6·06579	13·93421	12494	6·06579	
·7	10·00000		10·00000	14·05915	17609	5·94085	14·05915	17609	5·94085	
·8	10·00000		10·00000	14·23524	30103	5·76476	14·23524	30103	5·76476	
·9	10·00000		10·00000	14·53627		5·46373	14·53627		5·46373	
60·0	10·00000		10·00000	Infinite		— ∞	Infinite		— ∞	0′

LAT. 0° ALT.–AZ. TABLE

h	K	A	N	a₁	h	K	A	N	a₁
°	° ′			° ′	°	° ′			° ′
0 or 180	0 00·0	0·00000	—	90 00	45 or 135	0 00·0	0·15052	0·15052	90 00
1 „ 179	0 00·0	·00007	1·75815	90 00	46 „ 134	0 00·0	·15823	·14307	90 00
2 „ 178	0 00·0	·00027	1·45718	90 00	47 „ 133	0 00·0	·16622	·13507	90 00
3 „ 177	0 00·0	·00060	1·28120	90 00	48 „ 132	0 00·0	·17449	·12893	90 00
4 „ 176	0 00·0	·00106	1·15642	90 00	49 „ 131	0 00·0	·18306	·12222	90 00
5 „ 175	0 00·0	0·00166	1·05970	90 00	50 „ 130	0 00·0	0·19193	0·11575	90 00
6 „ 174	0 00·0	·00239	0·98077	90 00	51 „ 129	0 00·0	·20113	·10950	90 00
7 „ 173	0 00·0	·00325	·91411	90 00	52 „ 128	0 00·0	·21066	·10347	90 00
8 „ 172	0 00·0	·00425	·85645	90 00	53 „ 127	0 00·0	·22054	·09765	90 00
9 „ 171	0 00·0	·00538	·80567	90 00	54 „ 126	0 00·0	·23078	·09204	90 00
10 „ 170	0 00·0	0·00665	0·76033	90 00	55 „ 125	0 00·0	0·24141	0·08664	90 00
11 „ 169	0 00·0	·00805	·71940	90 00	56 „ 124	0 00·0	·25244	·08143	90 00
12 „ 168	0 00·0	·00960	·68212	90 00	57 „ 123	0 00·0	·26389	·07641	90 00
13 „ 167	0 00·0	·01128	·64791	90 00	58 „ 122	0 00·0	·27579	·07158	90 00
14 „ 166	0 00·0	·01310	·61633	90 00	59 „ 121	0 00·0	·28816	·06693	90 00
15 „ 165	0 00·0	0·01506	0·58700	90 00	60 „ 120	0 00·0	0·30103	0·06247	90 00
16 „ 164	0 00·0	·01716	·55966	90 00	61 „ 119	0 00·0	·31443	·05818	90 00
17 „ 163	0 00·0	·01940	·53407	90 00	62 „ 118	0 00·0	·32839	·05407	90 00
18 „ 162	0 00·0	·02179	·51002	90 00	63 „ 117	0 00·0	·34295	·05012	90 00
19 „ 161	0 00·0	·02433	·48736	90 00	64 „ 116	0 00·0	·35816	·04634	90 00
20 „ 160	0 00·0	0·02701	0·46595	90 00	65 „ 115	0 00·0	0·37405	0·04272	90 00
21 „ 159	0 00·0	·02985	·44567	90 00	66 „ 114	0 00·0	·39069	·03927	90 00
22 „ 158	0 00·0	·03283	·42643	90 00	67 „ 113	0 00·0	·40812	·03597	90 00
23 „ 157	0 00·0	·03597	·40812	90 00	68 „ 112	0 00·0	·42643	·03283	90 00
24 „ 156	0 00·0	·03927	·39069	90 00	69 „ 111	0 00·0	·44567	·02985	90 00
25 „ 155	0 00·0	0·04272	0·37405	90 00	70 „ 110	0 00·0	0·46595	0·02701	90 00
26 „ 154	0 00·0	·04634	·35816	90 00	71 „ 109	0 00·0	·48736	·02433	90 00
27 „ 153	0 00·0	·05012	·34295	90 00	72 „ 108	0 00·0	·51002	·02179	90 00
28 „ 152	0 00·0	·05407	·32839	90 00	73 „ 107	0 00·0	·53407	·01940	90 00
29 „ 151	0 00·0	·05818	·31443	90 00	74 „ 106	0 00·0	·55966	·01716	90 00
30 „ 150	0 00·0	0·06247	0·30103	90 00	75 „ 105	0 00·0	0·58700	0·01506	90 00
31 „ 149	0 00·0	·06693	·28816	90 00	76 „ 104	0 00·0	·61633	·01310	90 00
32 „ 148	0 00·0	·07158	·27579	90 00	77 „ 103	0 00·0	·64791	·01128	90 00
33 „ 147	0 00·0	·07641	·26389	90 00	78 „ 102	0 00·0	·68212	·00960	90 00
34 „ 146	0 00·0	·08143	·25244	90 00	79 „ 101	0 00·0	·71940	·00805	90 00
35 „ 145	0 00·0	0·08664	0·24141	90 00	80 „ 100	0 00·0	0·76033	0·00665	90 00
36 „ 144	0 00·0	·09204	·23078	90 00	81 „ 99	0 00·0	·80567	·00538	90 00
37 „ 143	0 00·0	·09765	·22054	90 00	82 „ 98	0 00·0	·85645	·00425	90 00
38 „ 142	0 00·0	·10347	·21066	90 00	83 „ 97	0 00·0	·91411	·00325	90 00
39 „ 141	0 00·0	·10950	·20113	90 00	84 „ 96	0 00·0	0·98077	·00239	90 00
40 „ 140	0 00·0	0·11575	0·19193	90 00	85 „ 95	0 00·0	1·05970	0·00166	90 00
41 „ 139	0 00·0	·12222	·18306	90 00	86 „ 94	0 00·0	1·15642	·00106	90 00
42 „ 138	0 00·0	·12893	·17449	90 00	87 „ 93	0 00·0	1·28120	·00060	90 00
43 „ 137	0 00·0	·13587	·16622	90 00	88 „ 92	0 00·0	1·45718	·00027	90 00
44 „ 136	0 00·0	·14307	·15823	90 00	89 „ 91	0 00·0	1·75815	·00007	90 00
45 „ 135	0 00·0	0·15052	0·15052	90 00	90	0 00·0	—	0·00000	90 00
h	180°-K	A	N	a₁	h	180°-K	A	N	a₁

Directions with regard to (K ⌒ d) :—

If l and d have **Same Name,** use (K ⌒ d), i.e. (K — d) or (d — K).

If l and d have **Different Names,** use (K + d).

To name Intercept :—

If H₀ is greater than H_c, intercept is " **Towards.**"

If H₀ is less than H_c, intercept is " **Away.**"

Log. cosec. H_c = A + log. sec. (K ⌒ d).

N.B. Whenever h is greater than 90°, subtract the quantity in the K column from 180°.

ALT.--AZ. TABLE. LAT. 1°

h	K	A	N	a₁	h	K	A	N	a₁
°	° ′			° ′	°	° ′			° ′
0 or 180	1 00·0	0·00000	—	90 00	45 or 135	1 24·8	0·15045	0·15058	89 00
1 „ 179	1 00·0	·00007	1·75821	90 00	46 „ 134	1 26·4	·15816	·14313	88 58
2 „ 178	1 00·0	·00027	1·45725	89 58	47 „ 133	1 28·0	·16614	·13594	88 56
3 „ 177	1 00·1	·00060	1·28127	89 57	48 „ 132	1 29·7	·17441	·12899	88 53
4 „ 176	1 00·1	·00106	1·15648	89 56	49 „ 131	1 31·4	·18297	·12229	88 51
5 „ 175	1 00·2	0·00166	1·05977	89 55	50 „ 130	1 33·3	0·19184	0·11581	88 49
6 „ 174	1 00·3	·00239	0·98083	89 54	51 „ 129	1 35·3	·20103	·10956	88 46
7 „ 173	1 00·5	·00325	·91417	89 53	52 „ 128	1 37·4	·21055	·10353	88 43
8 „ 172	1 00·6	·00425	·85651	89 52	53 „ 127	1 39·7	·22042	·09772	88 40
9 „ 171	1 00·7	·00538	·80573	89 50	54 „ 126	1 42·1	·23065	·09211	88 37
10 „ 170	1 00·9	0·00665	0·76040	89 49	55 „ 125	1 44·6	0·24127	0·08670	88 34
11 „ 169	1 01·1	·00805	·71947	89 48	56 „ 124	1 47·3	·25229	·08149	88 31
12 „ 168	1 01·3	·00959	·68219	89 47	57 „ 123	1 50·1	·26373	·07648	88 28
13 „ 167	1 01·6	·01127	·64798	89 46	58 „ 122	1 53·2	·27562	·07165	88 24
14 „ 166	1 01·8	·01309	·61639	89 45	59 „ 121	1 56·5	·28797	·06700	88 20
15 „ 165	1 02·1	0·01505	0·58707	89 44	60 „ 120	2 00·0	0·30083	0·06254	88 16
16 „ 164	1 02·4	·01715	·55825	89 43	61 „ 119	2 03·7	·31421	·05825	88 12
17 „ 163	1 02·7	·01940	·53413	89 42	62 „ 118	2 07·8	·32815	·05413	88 07
18 „ 162	1 03·1	·02179	·51008	89 41	63 „ 117	2 12·1	·34270	·05019	88 02
19 „ 161	1 03·5	·02432	·48742	89 39	64 „ 116	2 16·8	·35787	·04641	87 57
20 „ 160	1 03·8	0·02701	0·46601	89 38	65 „ 115	2 21·9	0·37375	0·04279	87 51
21 „ 159	1 04·3	·02984	·44574	89 37	66 „ 114	2 27·4	·39035	·03934	87 45
22 „ 158	1 04·7	·03283	·42649	89 36	67 „ 113	2 33·5	·40776	·03604	87 39
23 „ 157	1 05·2	·03596	·40819	89 35	68 „ 112	2 40·1	·42602	·03290	87 32
24 „ 156	1 05·7	·03926	·39075	89 33	69 „ 111	2 47 3	44522	02991	87 24
25 „ 155	1 06·2	0·04271	0·37412	89 32	70 „ 110	2 55·3	0·46545	0·02708	87 15
26 „ 154	1 06·8	·04632	·35822	89 31	71 „ 109	3 04·1	·48680	·02440	87 06
27 „ 153	1 07·3	·05010	·34302	89 29	72 „ 108	3 14·0	·50939	·02186	86 56
28 „ 152	1 08·0	·05405	·32846	89 28	73 „ 107	3 25·0	·53336	·01947	86 44
29 „ 151	1 08·6	·05816	·31450	89 27	74 „ 106	3 37·4	·55886	·01722	86 31
30 „ 150	1 09·3	0·06245	0·30110	89 25	75 „ 105	3 51·5	0·58608	0·01512	86 17
31 „ 149	1 10·0	·06691	·28823	89 24	76 „ 104	4 07·6	·61526	·01316	86 00
32 „ 148	1 10·7	·07155	·27587	89 23	77 „ 103	4 26·2	·64667	·01134	85 41
33 „ 147	1 11·5	·07638	·26396	89 21	78 „ 102	4 47·9	·68066	·00966	85 19
34 „ 146	1 12·4	·08140	·25250	89 20	79 „ 101	5 13·6	·71766	·00812	84 52
35 „ 145	1 13·2	0·08660	0·24148	89 18	80 „ 100	5 44·4	0·75821	0·00672	84 21
36 „ 144	1 14·2	·09200	·23085	89 16	81 „ 99	6 22·0	·80305	·00545	83 43
37 „ 143	1 15·1	·09761	·22060	89 15	82 „ 98	7 08·9	·85312	·00431	82 56
38 „ 142	1 16·1	·10343	·21072	89 13	83 „ 97	8 09·1	·90976	·00332	81 55
39 „ 141	1 17·2	·10946	·20119	89 11	84 „ 96	9 28·8	0·97486	·00245	80 35
40 „ 140	1 18·3	0·11570	0·19200	89 10	85 „ 95	11 19·5	1·05123	0·00173	78 44
41 „ 139	1 19·5	·12217	·18312	89 08	86 „ 94	14 02·9	1·14329	·00113	76 00
42 „ 138	1 20·7	·12887	·17456	89 06	87 „ 93	18 26·7	1·25836	·00066	71 36
43 „ 137	1 22·0	·13581	·16628	89 04	88 „ 92	26 34·3	1·40877	·00033	63 28
44 „ 136	1 23·4	·14300	·15830	89 02	89 „ 91	45 00·3	1·60766	·00013	45 01
45 „ 135	1 24·8	0·15045	0·15058	89 00	90	90 00·0	1·75815	·00007	0 00
h	180°-K	A	N	a₁	h	180°-K	A	N	a₁

Directions with regard to (a₂ $\overset{+}{\underset{\sim}{}}$ a₁) :—

Whenever h is greater than 90°, use (a₂ − a₁).

If h is less than 90°, use :—

(a₁ + a₂) when (K ± d) has been used,

(a₁ − a₂) when (d − K) has been used.

Always **name azimuth** from elevated pole.

Note that {h = L.H.A. of body **West** of the meridian.
{h = (360° − L.H.A.) of body **East** of the meridian.

Log. tan. a₂ = N + log. tan. (K $\overset{+}{\underset{\sim}{}}$ *d'*

LAT. 2°   ALT.—AZ. TABLE

h	K	A	N	a₁	h	K	A	N	a₁
°	° ′			° ′	°	° ′			^ ′
0 or 180	2 00·0	0·00000	—	90 00	45 or 135	2 49·6	0·15025	0·15076	88 00
1 „ 179	2 00·0	·00007	1·75841	89 58	46 „ 134	2 52·7	·15794	·14333	87 56
2 „ 178	2 00·1	·00027	1·45745	89 56	47 „ 133	2 55·9	·16591	·13614	87 51
3 „ 177	2 00·2	·00060	1·28147	89 54	48 „ 132	2 59·2	·17416	·12919	87 47
4 „ 176	2 00·3	·00106	1·15668	89 52	49 „ 131	3 02·8	·18271	·12249	87 42
5 „ 175	2 00·5	0·00165	1·05977	89 50	50 „ 130	3 06·6	0·19156	0·11601	87 37
6 „ 174	2 00·7	·00238	0·98103	89 47	51 „ 129	3 10·6	·20072	·10976	87 32
7 „ 173	2 00·9	·00325	·91437	89 45	52 „ 128	3 14·8	·21023	·10373	87 27
8 „ 172	2 01·2	·00424	·85671	89 43	53 „ 127	3 19·3	·22007	·09792	87 21
9 „ 171	2 01·5	·00537	·80593	89 41	54 „ 126	3 24·0	·23028	·09231	87 15
10 „ 170	2 01·8	0·00664	0·76060	89 39	55 „ 125	3 29·0	0·24087	0·08690	87 09
11 „ 169	2 02·2	·00804	·71967	89 37	56 „ 124	3 34·4	·25186	·08169	87 02
12 „ 168	2 02·7	·00958	·68239	89 34	57 „ 123	3 40·1	·26326	·07667	86 55
13 „ 167	2 03·2	·01126	·64818	89 32	58 „ 122	3 46·2	·27511	·07185	86 48
14 „ 166	2 03·7	·01308	·61659	89 30	59 „ 121	3 52·7	·28743	·06720	86 41
15 „ 165	2 04·2	0·01504	0·58727	89 28	60 „ 120	3 59·7	0·30024	0·06273	86 32
16 „ 164	2 04·8	·01714	·55993	89 26	61 „ 119	4 07·2	·31357	·05845	86 24
17 „ 163	2 05·5	·01938	·53433	89 23	62 „ 118	4 15·2	·32746	·05433	86 15
18 „ 162	2 06·2	·02177	·51028	89 21	63 „ 117	4 23·9	·34194	·05038	86 05
19 „ 161	2 06·9	·02430	·48762	89 19	64 „ 116	4 33·3	·35705	·04661	85 54
20 „ 160	2 07·7	0·02698	0·46621	89 16	65 „ 115	4 43·4	0·37284	0·04299	85 43
21 „ 159	2 08·5	·02981	·44594	89 14	66 „ 114	4 54·4	·38935	·03954	85 31
22 „ 158	2 09·4	·03279	·42669	89 12	67 „ 113	5 06·4	·40666	·03624	85 18
23 „ 157	2 10·4	·03593	·40839	89 09	68 „ 112	5 19·5	·42481	·03310	85 04
24 „ 156	2 11·3	·03922	·39095	89 07	69 „ 111	5 33·9	·44388	·03011	84 48
25 „ 155	2 12·4	0·04267	0·37432	89 04	70 „ 110	5 49·8	0·46396	0·02728	84 31
26 „ 154	2 13·5	·04628	·35841	89 01	71 „ 109	6 07·3	·48514	·02460	84 13
27 „ 153	2 14·7	·05005	·34322	88 59	72 „ 108	6 26·8	·50753	·02206	83 52
28 „ 152	2 15·9	·05399	·32866	88 56	73 „ 107	6 48·7	·53125	·01967	83 30
29 „ 151	2 17·2	·05810	·31469	88 54	74 „ 106	7 13·2	·55647	·01742	83 04
30 „ 150	2 18·5	0·06238	0·30130	88 51	75 „ 105	7 41·0	0·58335	0·01532	82 35
31 „ 149	2 20·0	·06683	·28843	88 48	76 „ 104	8 12·8	·61211	·01336	82 02
32 „ 148	2 21·5	·07148	·27606	88 45	77 „ 103	8 49·4	·64301	·01154	81 24
33 „ 147	2 23·1	·07630	·26416	88 42	78 „ 102	9 32·1	·67634	·00986	80 41
34 „ 146	2 24·7	·08131	·25270	88 39	79 „ 101	10 22·3	·71250	·00832	79 49
35 „ 145	2 26·5	0·08650	0·24167	88 36	80 „ 100	11 22·2	0·75199	0·00691	78 48
36 „ 144	2 28·3	·09190	·23104	88 33	81 „ 99	12 35·0	·79537	·00565	77 34
37 „ 143	2 30·2	·09750	·22080	88 30	82 „ 98	14 05·1	·84345	·00451	76 03
38 „ 142	2 32·2	·10331	·21092	88 26	83 „ 97	15 59·4	·89724	·00351	74 08
39 „ 141	2 34·4	·10932	·20139	88 23	84 „ 96	18 28·4	·95805	·00265	71 38
40 „ 140	2 36·6	0·11556	0·19220	88 19	85 „ 95	21 50·1	1·02764	0·00192	68 15
41 „ 139	2 39·0	·12202	·18332	88 16	86 „ 94	26 35·6	1·10812	·00132	63 29
42 „ 138	2 41·4	·12871	·17475	88 12	87 „ 93	33 42·8	1·20150	·00086	56 20
43 „ 137	2 44·0	·13564	·16648	88 08	88 „ 92	45 01·0	1·30680	·00053	45 01
44 „ 136	2 46·8	·14282	·15849	88 04	89 „ 91	63 26·7	1·40877	·00033	26 34
45 „ 135	2 49·6	0·15025	0·15076	88 00	90	90 00·0	1·45718	0·00027	0 00
h	180°-K	A	N	a₁	h	180°-K	A	N	a₁

Directions with regard to (K $\overset{+}{\underset{\sim}{}}$ d) :—

If l and d have **Same Name**, use (K ⌣ d), i.e. (K — d) or (d — K).
If l and d have **Different Names**, use (K + d).

To name Intercept :—

If H_o is greater than H_c, intercept is " **Towards.** "
If H_o is less than H_c, intercept is " **Away.** "

Log. cosec. H_c = A + log. sec. (K $\overset{\sim}{+}$ d).

N.B.   Whenever h is greater than 90°, subtract the quantity in the K column from 180°.

ALT.—AZ. TABLE LAT. 3°

h	K	A	N	a₁	h	K	A	N	a₁
°	° ′			° ′	°	° ′			° ′
0 or 180	3 00·0	0·00000	—	90 00	**45 or 135**	4 14·3	0·14992	0·15111	87 00
1 „ 179	3 00·0	·00007	1·75874	89 57	**46 „ 134**	4 18·9	·15759	·14366	86 54
2 „ 178	3 00·1	·00027	1·45778	89 54	**47 „ 133**	4 23·7	·16553	·13647	86 47
3 „ 177	3 00·2	·00059	1·28180	89 51	**48 „ 132**	4 28·7	·17376	·12952	86 40
4 „ 176	3 00·4	·00106	1·15701	89 47	**49 „ 131**	4 34·0	·18227	·12282	86 33
5 „ 175	3 00·7	0·00165	1·06030	89 44	**50 „ 130**	4 39·7	0·19109	0·11634	86 26
6 „ 174	3 01·0	·00238	0·98136	89 41	**51 „ 129**	4 45·6	·20022	·11009	86 18
7 „ 173	3 01·3	·00324	·91470	89 38	**52 „ 128**	4 51·9	·20969	·10406	86 10
8 „ 172	3 01·8	·00424	·85704	89 35	**53 „ 127**	4 58·6	·21950	·09825	86 02
9 „ 171	3 02·2	·00537	·80626	89 32	**54 „ 126**	5 05·7	·22966	·09264	85 53
10 „ 170	3 02·8	0·00663	0·76093	89 28	**55 „ 125**	5 13·2	0·24020	0·08723	85 44
11 „ 169	3 03·4	·00803	·71946	89 25	**56 „ 124**	5 21·2	·25113	·08203	85 34
12 „ 168	3 04·0	·00957	·68272	89 22	**57 „ 123**	5 29·8	·26249	·07701	85 24
13 „ 167	3 04·7	·01124	·64851	89 18	**58 „ 122**	5 38·9	·27428	·07218	85 13
14 „ 166	3 05·5	·01306	·61692	89 15	**59 „ 121**	5 48·6	·28652	·06753	85 01
15 „ 165	3 06·3	0·01501	0·58760	89 12	**60 „ 120**	5 59·0	0·29926	0·06307	84 49
16 „ 164	3 07·2	·01711	·56026	89 08	**61 „ 119**	6 10·2	·31250	·05878	84 36
17 „ 163	3 08·2	·01935	·53466	89 05	**62 „ 118**	6 22·2	·32630	·05466	84 23
18 „ 162	3 09·2	·02173	·51061	89 02	**63 „ 117**	6 35·1	·34067	·05072	84 08
19 „ 161	3 10·4	·02426	·48795	88 58	**64 „ 116**	6 49·0	·35567	·04694	83 53
20 „ 160	3 11·5	0·02694	0·46654	88 55	**65 „ 115**	7 04·1	0·37133	0·04332	83 36
21 „ 159	3 12·8	·02976	·44627	88 51	**66 „ 114**	7 20·5	·38771	·03987	83 18
22 „ 158	3 14·1	·03274	·42702	88 47	**67 „ 113**	7 38·4	·40485	·03657	82 58
23 „ 157	3 15·5	·03587	·40872	88 44	**68 „ 112**	7 57·8	·42281	·03343	82 37
24 „ 156	3 17·0	·03915	·39128	88 40	**69 „ 111**	8 19·2	·44167	·03044	82 14
25 „ 155	3 18·6	0·04260	0·37465	88 36	**70 „ 110**	8 42·7	0·46150	0·02761	81 49
26 „ 154	3 20·2	·04620	·35875	88 32	**71 „ 109**	9 08·7	·48240	·02493	81 21
27 „ 153	3 22·0	·04996	·34355	88 28	**72 „ 108**	9 37·5	·50446	·02239	80 51
28 „ 152	3 23·8	·05390	·32899	88 24	**73 „ 107**	10 09·7	·52779	·02000	80 17
29 „ 151	3 25·7	·05800	·31503	88 20	**74 „ 106**	10 45·9	·55255	·01775	79 39
30 „ 150	3 27·8	0·06227	0·30163	88 16	**75 „ 105**	11 26·8	0·57887	0·01565	78 57
31 „ 149	3 29·9	·06672	·28876	88 12	**76 „ 104**	12 13·4	·60696	·01369	78 09
32 „ 148	3 32·2	·07135	·27639	88 08	**77 „ 103**	13 06·9	·63703	·01187	77 14
33 „ 147	3 34·5	·07616	·26449	88 03	**78 „ 102**	14 08·9	·66934	·01019	76 10
34 „ 146	3 37·0	·08116	·25303	87 59	**79 „ 101**	15 21·5	·70420	·00865	74 56
35 „ 145	3 39·6	0·08634	0·24201	87 54	**80 „ 100**	16 47·6	0·74200	0·00725	73 28
36 „ 144	3 42·4	·09173	·23138	87 49	**81 „ 99**	18 31·3	·78315	·00598	71 43
37 „ 143	3 45·3	·09732	·22113	87 44	**82 „ 98**	20 38·1	·82824	·00485	69 35
38 „ 142	3 48·3	·10311	·21125	87 40	**83 „ 97**	23 16·2	·87786	·00385	66 55
39 „ 141	3 51·5	·10911	·20178	87 34	**84 „ 96**	26 37·7	·93267	·00298	63 32
40 „ 140	3 54·8	0·11533	0·19253	87 29	**85 „ 95**	31 01·1	0·99328	0·00225	59 07
41 „ 139	3 58·3	·12177	·18365	87 24	**86 „ 94**	36 55·0	1·05983	·00166	53 11
42 „ 138	4 02·0	·12845	·17509	87 18	**87 „ 93**	45 02·4	1·13098	·00119	45 02
43 „ 137	4 05·9	·13536	·16681	87 12	**88 „ 92**	56 20·4	1·20150	·00086	33 43
44 „ 136	4 10·0	·14251	·15883	87 06	**89 „ 91**	71 34·9	1·25836	·00066	18 27
45 „ 135	4 14·3	0·14992	0·15111	87 00	**90**	90 00·0	1·28120	0·00060	0 00
h	180°-K	A	N	a₁	h	180°-K	A	N	a₁

Directions with regard to $(a_2 \overset{+}{\sim} a_1)$:—

Whenever h is greater than 90°, use $(a_2 - a_1)$.

If h is less than 90°, use :—

$(a_1 + a_2)$ when $(K \pm d)$ has been used,
$(a_1 - a_2)$ when $(d - K)$ has been used.

Always **name** **azimuth** from elevated pole.

Note that { h = L.H.A. of body **West** of the meridian.
{ h = (360° — L.H.A.) of body **East** of the meridian.

Log. tan. a_2 = N + log. tan. $(K \overset{+}{\sim} d)$.

LAT. 4°   ALT.—AZ. TABLE

h	K	A	N	a₁	h	K	A	N	a₁
°	° ′			° ′	°	° ′			° ′
0 or 180	4 00·0	0·00000	—	90 00	45 or 135	5 38·9	0·14946	0·15157	86 01
1 „ 179	4 00·0	·00007	1·75920	80 66	46 „ 134	5 44·9	·15709	·14413	85 52
2 „ 178	4 00·1	·00026	1·45824	89 52	47 „ 133	5 51·3	·16500	·13692	85 43
3 „ 177	4 00·3	·00059	1·28226	89 47	48 „ 132	5 58·0	·17319	·12999	85 34
4 „ 176	4 00·6	·00105	1·15748	89 43	49 „ 131	6 05·0	·18166	·12328	85 25
5 „ 175	4 00·9	0·00165	1·06076	89 39	50 „ 130	6 12·5	0·19043	0·11681	85 15
6 „ 174	4 01·3	·00237	0·98182	89 35	51 „ 129	6 20·4	·19953	·11056	85 05
7 „ 173	4 01·8	·00323	·91517	89 31	52 „ 128	6 28·8	·20894	·10453	84 54
8 „ 172	4 02·4	·00423	·85750	89 26	53 „ 127	6 37·7	·21869	·09871	84 43
9 „ 171	4 03·0	·00535	·80673	89 22	54 „ 126	6 47·1	·22879	·09310	84 31
10 „ 170	4 03·7	0·00662	0·76139	89 18	55 „ 125	6 57·1	0·23927	0·08766	84 19
11 „ 169	4 04·5	·00801	·72046	89 13	56 „ 124	7 07·7	·25013	·08249	84 06
12 „ 168	4 05·3	·00955	·68318	89 09	57 „ 123	7 19·0	·26140	·07747	83 52
13 „ 167	4 06·3	·01122	·64897	89 05	58 „ 122	7 31·0	·27310	·07264	83 38
14 „ 166	4 07·3	·01303	·61738	89 00	59 „ 121	7 43·9	·28525	·06799	83 23
15 „ 165	4 08·4	0·01498	0·58806	88 56	60 „ 120	7 57·7	0·29788	0·06353	83 07
16 „ 164	4 09·6	·01707	·56072	88 51	61 „ 119	8 12·4	·31102	·05924	82 50
17 „ 163	4 10·9	·01930	·53512	88 47	62 „ 118	8 28·3	·32469	·05512	82 32
18 „ 162	4 12·3	·02168	·51108	88 42	63 „ 117	8 45·4	·33892	·05118	82 12
19 „ 161	4 13·8	·02420	·48842	88 37	64 „ 116	9 03·8	·35376	·04740	81 52
20 „ 160	4 15·3	0·02687	0·46701	88 33	65 „ 115	9 23·7	0·36925	0·04378	81 30
21 „ 159	4 17·0	·02969	·44673	88 28	66 „ 114	9 45·3	·38542	·04033	81 06
22 „ 158	4 18·8	·03266	·42748	88 23	67 „ 113	10 08·8	·40234	·03703	80 40
23 „ 157	4 20·7	·03578	·40918	88 18	68 „ 112	10 34·4	·42004	·03389	80 12
24 „ 156	4 22·6	·03906	·39175	88 13	69 „ 111	11 02·5	·43862	·03091	79 42
25 „ 155	4 24·7	0·04249	0·37511	88 08	70 „ 110	11 33·3	0·45812	0·02807	79 09
26 „ 154	4 26·9	·04609	·35922	88 03	71 „ 109	12 07·3	·47862	·02539	78 33
27 „ 153	4 29·2	·04984	·34401	87 58	72 „ 108	12 45·0	·50023	·02285	77 53
28 „ 152	4 31·7	·05377	·32945	87 53	73 „ 107	13 27·0	·52304	·02046	77 09
29 „ 151	4 34·3	·05786	·31549	87 47	74 „ 106	14 14·1	·54718	·01822	76 20
30 „ 150	4 37·0	0·06212	0·30209	87 42	75 „ 105	15 07·1	0·57276	0·01612	75 25
31 „ 149	4 39·8	·06655	·28922	87 36	76 „ 104	16 07·3	·59996	·01416	74 22
32 „ 148	4 42·8	·07117	·27685	87 30	77 „ 103	17 16·1	·62894	·01233	73 11
33 „ 147	4 46·0	·07597	·26495	87 24	78 „ 102	18 35·4	·65991	·01066	71 50
34 „ 146	4 49·3	·08095	·25350	87 18	79 „ 101	20 07·6	·69310	·00911	70 16
35 „ 145	4 52·8	0·08612	0·24247	87 12	80 „ 100	21 56·1	0·72876	0·00771	68 25
36 „ 144	4 56·4	·09148	·23184	87 06	81 „ 99	24 05·1	·76719	·00644	66 14
37 „ 143	5 00·2	·09706	·22160	86 59	82 „ 98	26 40·6	·80862	·00531	63 36
38 „ 142	5 04·3	·10283	·21172	86 53	83 „ 97	29 50·8	·85336	·00431	60 24
39 „ 141	5 08·5	·10881	·20216	86 46	84 „ 96	33 46·9	·90150	·00345	56 26
40 „ 140	5 12·9	0·11501	0·19299	86 39	85 „ 95	38 44·4	0·95285	0·00272	51 26
41 „ 139	5 17·6	·12142	·18412	86 32	86 „ 94	45 04·2	1·00643	·00212	45 04
42 „ 138	5 22·5	·12807	·17555	86 24	87 „ 93	53 11·2	1·05983	·00166	36 55
43 „ 137	5 27·7	·13496	·16728	86 17	88 „ 92	63 28·6	1·10812	·00132	26 36
44 „ 136	5 33·1	·14209	·15929	86 09	89 „ 91	75 59·2	1·14329	·00113	14 03
45 „ 135	5 38·9	0·14946	0·15157	86 01	90	90 00·0	1·15642	0·00106	0 00
h	180°-K	A	N	a₁	h	180°-K	A	N	a₁

Directions with regard to (K ± d) :—

If l and d have **Same Name,** use (K ⌣ d), i.e. (K — d) or (d — K).

If l and d have **Different Names,** use (K + d).

To name Intercept :—

If H_o is greater than H_c, intercept is " **Towards.**"

If H_o is less than H_c, intercept is " **Away.**"

Log. cosec. H_c = A + log. sec. (K ± d).

N.B.   Whenever h is greater than 90°, subtract the quantity in the K column from 180°.

ALT.—AZ. TABLE　　LAT. 5°

h	K	A	N	a₁	h	K	A	N	a₁
°	° ′			° ′	°	° ′			° ′
0 or 180	5 00·0	0·00000	—	90 00	45 or 135	7 03·2	0·14887	0·15217	85 01
1 ,, 179	5 00·0	·00007	1·75980	89 55	46 ,, 134	7 10·7	·15646	·14472	84 51
2 ,, 178	5 00·2	·00026	1·45884	89 50	47 ,, 133	7 18·6	·16433	·13753	84 40
3 ,, 177	5 00·4	·00059	1·28286	89 44	48 ,, 132	7 26·9	·17246	·13058	84 28
4 ,, 176	5 00·7	·00105	1·15807	89 39	49 ,, 131	7 35·8	·18088	·12387	84 16
5 ,, 175	5 01·1	0·00164	1·06136	89 34	50 ,, 130	7 45·0	0·18959	0·11740	84 04
6 ,, 174	5 01·6	·00237	0·98272	89 29	51 ,, 129	7 54·9	·19863	·11115	83 51
7 ,, 173	5 02·2	·00322	·91576	89 23	52 ,, 128	8 05·3	·20797	·10512	83 38
8 ,, 172	5 02·9	·00422	·85810	89 18	53 ,, 127	8 16·3	·21765	·09931	83 24
9 ,, 171	5 03·7	·00534	·80732	89 13	54 ,, 126	8 28·0	·22768	·09370	83 10
10 ,, 170	5 04·6	0·00660	0·76198	89 07	55 ,, 125	8 40·4	0·23807	0·08829	82 54
11 ,, 169	5 05·6	·00799	·72106	89 02	56 ,, 124	8 53·5	·24884	·08308	82 38
12 ,, 168	5 06·7	·00952	·68378	88 56	57 ,, 123	9 07·5	·26001	·07806	82 21
13 ,, 167	5 07·8	·01119	·64957	88 51	58 ,, 122	9 22·5	·27161	·07324	82 04
14 ,, 166	5 09·1	·01299	·61798	88 45	59 ,, 121	9 38·4	·28364	·06859	81 45
15 ,, 165	5 10·5	0·01494	0·58866	88 40	60 ,, 120	9 55·5	0·29614	0·06413	81 25
16 ,, 164	5 12·0	·01702	·56131	88 34	61 ,, 119	10 13·8	·30913	·05984	81 04
17 ,, 163	5 13·6	·01925	·53572	88 28	62 ,, 118	10 33·4	·32264	·05572	80 41
18 ,, 162	5 15·4	·02162	·51167	88 23	63 ,, 117	10 54·5	·33669	·05178	80 18
19 ,, 161	5 17·2	·02413	·48901	88 17	64 ,, 116	11 17·2	·35133	·04800	79 53
20 ,, 160	5 19·1	0·02680	0·46760	88 11	65 ,, 115	11 41·8	0·36660	0·04438	79 25
21 ,, 159	5 21·2	·02961	·44733	88 05	66 ,, 114	12 08·4	·38252	·04093	78 55
22 ,, 158	5 23·4	·03257	·42808	87 59	67 ,, 113	12 37·3	·39916	·03763	78 24
23 ,, 157	5 25·8	·03568	·40978	87 53	68 ,, 112	13 08·7	·41654	·03449	77 50
24 ,, 156	5 28·2	·03894	·39234	87 47	69 ,, 111	13 43·2	·43476	·03150	77 12
25 ,, 155	5 30·8	0·04237	0·37571	87 40	70 ,, 110	14 20·9	0·45384	0·02867	76 32
26 ,, 154	5 33·6	·04595	·35981	87 34	71 ,, 109	15 02·5	·47387	·02599	75 48
27 ,, 153	5 36·5	·04969	·34461	87 27	72 ,, 108	15 48·5	·49493	·02345	74 59
28 ,, 152	5 39·5	·05360	·33005	87 21	73 ,, 107	16 39·6	·51709	·02106	74 06
29 ,, 151	5 42·7	·05767	·31609	87 14	74 ,, 106	17 36·6	·54047	·01881	73 06
30 ,, 150	5 46·1	0·06192	0·30269	87 07	75 ,, 105	18 40·6	0·56516	0·01671	71 59
31 ,, 149	5 49·7	·06634	·28982	87 00	76 ,, 104	19 52·9	·59129	·01475	70 44
32 ,, 148	5 53·4	·07094	·27745	86 53	77 ,, 103	21 15·1	·61898	·01293	69 19
33 ,, 147	5 57·3	·07571	·26555	86 46	78 ,, 102	22 49·3	·64838	·01125	67 42
34 ,, 146	6 01·4	·08068	·25409	86 38	79 ,, 101	24 37·9	·67962	·00971	65 51
35 ,, 145	6 05·8	0·08583	0·24307	86 30	80 ,, 100	26 44·4	0·71286	0·00831	63 42
36 ,, 144	6 10·3	·09117	·23244	86 33	81 ,, 99	29 13·0	·74823	·00704	61 11
37 ,, 143	6 15·1	·09672	·22219	86 15	82 ,, 98	32 09·3	·78580	·00590	58 12
38 ,, 142	6 20·1	·10246	·21231	86 06	83 ,, 97	35 40·4	·82550	·00491	54 38
39 ,, 141	6 25·4	·10842	·20278	85 58	84 ,, 96	39 55·7	·86713	·00404	50 20
40 ,, 140	6 30·9	0·11459	0·19359	85 49	85 ,, 95	45 06·6	0·91001	0·00331	45 07
41 ,, 139	6 36·7	·12097	·18471	85 40	86 ,, 94	51 26·0	·95285	·00272	38 41
42 ,, 138	6 42·9	·12759	·17615	85 31	87 ,, 93	59 06·7	·99328	·00225	31 01
43 ,, 137	6 49·3	·13444	·16787	85 21	88 ,, 92	68 15·2	1·02764	·00192	21 50
44 ,, 136	6 56·1	·14154	·15989	85 11	89 ,, 91	78 43·1	1·05123	·00172	11 20
45 ,, 135	7 03·2	0·14887	0·15217	85 01	90	90 00·0	1·05970	0·00166	0 00
h	180°-K	A	N	a₁	h	180°-K	A	N	a₁

Directions with regard to $(a_2 \overset{+}{\sim} a_1)$:—

Whenever h is greater than 90°, use $(a_2 - a_1)$.

If h is less than 90°, use :—

Always **name**
azimuth from
elevated pole.

$(a_1 + a_2)$ when $(K \pm d)$ has been used,

$(a_1 - a_2)$ when $(d - K)$ has been used.

Note that { h = L.H.A. of body **West** of the meridian.
{ h = (360° — L.H.A.) of body **East** of the meridian.

Log. tan. a_2 = N + log. tan. $(K \overset{+}{\sim} d)$.

LAT. 6°　　ALT.—AZ. TABLE

h	K	A	N	a₁	h	K	A	N	a₁
°	° ′			° ′	°	° ′			° ′
0 or 180	6 00·0	0·00000	—	90 00	45 or 135	8 27·3	0·14815	0·15290	84 02
1 „ 179	6 00·1	·00007	1·76053	89 54	46 „ 134	8 36·2	·15570	·14545	83 49
2 „ 178	6 00·2	·00026	1·45957	89 47	47 „ 133	8 45·7	·16351	·13826	83 36
3 „ 177	6 00·5	·00059	1·28359	89 41	48 „ 132	8 55·6	·17158	·13131	83 23
4 „ 176	6 00·9	·00105	1·15881	89 35	49 „ 131	9 06·1	·17994	·12461	83 09
5 „ 175	6 01·4	0·00164	1·06209	89 29	50 „ 130	9 17·2	0·18858	0·11813	82 54
6 „ 174	6 02·0	·00236	0·98315	89 22	51 „ 129	9 28·9	·19754	·11188	82 39
7 „ 173	6 02·7	·00321	·91649	89 16	52 „ 128	9 41·3	·20681	·10585	82 23
8 „ 172	6 03·5	·00420	·85883	89 10	53 „ 127	9 54·4	·21640	·10004	82 06
9 „ 171	6 04·5	·00532	·80805	89 03	54 „ 126	10 08·3	·22633	·09443	81 49
10 „ 170	6 05·5	0·00658	0·76272	88 57	55 „ 125	10 23·0	0·23662	0·08902	81 31
11 „ 169	6 06·7	·00796	·72179	88 50	56 „ 124	10 38·7	·24729	·08381	81 11
12 „ 168	6 08·0	·00949	·68451	88 44	57 „ 123	10 55·4	·25834	·07880	80 51
13 „ 167	6 09·4	·01115	·65030	88 37	58 „ 122	11 13·1	·26980	·07397	80 30
14 „ 166	6 10·9	·01295	·61871	88 30	59 „ 121	11 32·0	·28169	·06932	80 08
15 „ 165	6 12·6	0·01489	0·58939	88 24	60 „ 120	11 52·3	0·29403	0·06486	79 44
16 „ 164	6 14·4	·01696	·56205	88 17	61 „ 119	12 13·9	·30685	·06057	79 19
17 „ 163	6 16·3	·01918	·53645	88 10	62 „ 118	12 37·2	·32016	·05645	78 53
18 „ 162	6 18·4	·02154	·51240	88 03	63 „ 117	13 02·1	·33400	·05251	78 24
19 „ 161	6 20·6	·02405	·48974	87 56	64 „ 116	13 29·0	·34841	·04873	77 54
20 „ 160	6 22·9	0·02670	0·46883	87 49	65 „ 115	13 58·0	0·36341	0·04511	77 22
21 „ 159	6 25·4	·02950	·44806	87 42	66 „ 114	14 29·3	·37904	·04167	76 47
22 „ 158	6 28·0	·03245	·42881	87 35	67 „ 113	15 03·3	·39534	·03836	76 10
23 „ 157	6 30·8	·03555	·41051	87 28	68 „ 112	15 40·4	·41236	·03522	75 30
24 „ 156	6 33·8	·03880	·39307	87 20	69 „ 111	16 20·7	·43014	·03223	74 46
25 „ 155	6 36·9	0·04221	0·37644	87 13	70 „ 110	17 04·9	0·44874	0·02940	73 59
26 „ 154	6 40·2	·04578	·36054	87 05	71 „ 109	17 53·5	·46821	·02672	73 07
27 „ 153	6 43·7	·04950	·34534	86 57	72 „ 108	18 47·1	·48863	·02418	72 10
28 „ 152	6 47·3	·05340	·33078	86 49	73 „ 107	19 46·4	·51006	·02179	71 07
29 „ 151	6 51·1	·05745	·31682	86 41	74 „ 106	20 52·4	·53257	·01954	69 58
30 „ 150	6 55·2	0·06168	0·30342	86 33	75 „ 105	22 06·1	0·55624	0·01744	68 41
31 „ 149	6 59·4	·06608	·29055	86 24	76 „ 104	23 29·0	·58115	·01548	67 18
32 „ 148	7 03·9	·07066	·27818	86 16	77 „ 103	25 02·6	·60742	·01366	65 38
33 „ 147	7 08·6	·07541	·26628	86 07	78 „ 102	26 49·1	·63509	·01198	63 49
34 „ 146	7 13·5	·08035	·25482	85 58	79 „ 101	28 50·9	·66424	·01044	61 44
35 „ 145	7 18·7	0·08548	0·24380	85 49	80 „ 100	31 11·1	0·69493	0·00904	59 20
36 „ 144	7 24·1	·09079	·23317	85 39	81 „ 99	33 53·8	·72716	·00777	56 35
37 „ 143	7 29·8	·09631	·22292	85 30	82 „ 98	37 03·6	·76083	·00663	53 22
38 „ 142	7 35·8	·10202	·21304	85 20	83 „ 97	40 46·5	·79574	·00564	49 36
39 „ 141	7 42·1	·10795	·20351	85 10	84 „ 96	45 09·4	·83144	·00477	45 09
40 „ 140	7 48·7	0·11408	0·19432	84 59	85 „ 95	50 20·0	0·86713	0·00404	39 56
41 „ 139	7 55·7	·12043	·18542	84 48	86 „ 94	56 25·7	·90151	·00344	33 47
42 „ 138	8 03·0	·12701	·17688	84 37	87 „ 93	63 31·8	·93267	·00298	26 38
43 „ 137	8 10·7	·13382	·16860	84 26	88 „ 92	71 37·9	·95805	·00265	18 28
44 „ 136	8 18·8	·14087	·16062	84 14	89 „ 91	80 34·3	·97486	·00245	9 29
45 „ 135	8 27·3	0·14815	0·15290	84 02	90	90 00·0	0·98077	0·00239	0 00
h	180°-K	A	N	a₁	h	180°-K	A	N	a₁

Directions with regard to (K ⁓ d) :—

If l and d have **Same Name,** use (K ⁓ d), i.e. (K — d) or (d — K).

If l and d have **Different Names,** use (K + d).

To name Intercept :—

If H_o is greater than H_c, intercept is "**Towards.**"

If H_o is less than H_c, intercept is "**Away.**"

Log. cosec. H_c = A + log. sec. (K ± d).

N.B. *Whenever h is greater than 90°, subtract the quantity in the K column from 180°.*

ALT.—AZ. TABLE LAT. 7°

h	K	A	N	a_1	h	K	A	N	a_1
°	° ′			° ′	°	° ′			° ′
0 or 180	7 00·0	0·00000	—	90 00	**45 or 135**	9 51·0	0·14731	0·15376	83 03
1 „ 179	7 00·1	·00007	1·76139	89 53	**46 „ 134**	10 01·4	·15480	·14632	82 48
2 „ 178	7 00·3	·00026	1·46043	89 45	**47 „ 133**	10 12·4	·16254	·13912	82 33
3 „ 177	7 00·6	·00059	1·28445	89 38	**48 „ 132**	10 23·9	·17055	·13218	82 18
4 „ 176	7 01·0	·00104	1·15966	89 31	**49 „ 131**	10 36·0	·17883	·12547	82 01
5 „ 175	7 01·6	0·00163	1·06296	89 23	**50 „ 130**	10 48·9	0·18740	0·11900	81 44
6 „ 174	7 02·3	·00235	0·98401	89 16	**51 „ 129**	11 02·4	·19626	·11275	81 26
7 „ 173	7 03·1	·00320	·91736	89 09	**52 „ 128**	11 16·7	·20544	·10672	81 08
8 „ 172	7 04·1	·00418	·85969	89 01	**53 „ 127**	11 31·9	·21493	·10090	80 49
9 „ 171	7 05·2	·00530	·80892	88 54	**54 „ 126**	11 47·9	·22476	·09529	80 29
10 „ 170	7 06·4	0·00655	0·76358	88 46	**55 „ 125**	12 05·0	0·23493	0·08988	80 08
11 „ 169	7 07·8	·00793	·72265	88 39	**56 „ 124**	12 23·1	·24547	·08468	79 46
12 „ 168	7 09·3	·00945	·68537	88 31	**57 „ 123**	12 42·3	·25638	·07966	79 22
13 „ 167	7 10·9	·01110	·65126	88 23	**58 „ 122**	13 02·7	·26769	·07483	78 58
14 „ 166	7 12·7	·01290	·61957	88 16	**59 „ 121**	13 24·5	·27941	·07018	78 32
15 „ 165	7 14·7	0·01482	0·59026	88 08	**60 „ 120**	13 47·8	0·29157	0·06572	78 05
16 „ 164	7 16·7	·01689	·56291	88 00	**61 „ 119**	14 12·7	·30418	·06143	77 36
17 „ 163	7 19·0	·01910	·53731	87 52	**62 „ 118**	14 39·4	·31727	·05731	77 05
18 „ 162	7 21·4	·02145	·51327	87 44	**63 „ 117**	15 08·0	·33087	·05337	76 33
19 „ 161	7 23·9	·02395	·49061	87 36	**64 „ 116**	15 38·8	·34501	·04959	75 58
20 „ 160	7 26·7	0·02659	0·46920	87 28	**65 „ 115**	16 12·0	0·35970	0·04597	75 21
21 „ 159	7 29·6	·02937	·44892	87 19	**66 „ 114**	16 47·9	·37500	·04252	74 42
22 „ 158	7 32·6	·03231	·42967	87 11	**67 „ 113**	17 26·7	·39092	·03922	73 59
23 „ 157	7 35·9	·03539	·41137	87 02	**68 „ 112**	18 08·9	·40752	·03608	73 13
24 „ 156	7 39·3	·03863	·39394	86 54	**69 „ 111**	18 54·8	·42482	·03320	72 23
25 „ 155	7 42·9	0·04202	0·37730	86 45	**70 „ 110**	19 44·9	0·44288	0·03026	71 29
26 „ 154	7 46·7	·04557	·36141	86 36	**71 „ 109**	20 39·8	·46173	·02756	70 31
27 „ 153	7 50·8	·04928	·34620	86 27	**72 „ 108**	21 40·2	·48144	·02504	69 26
28 „ 152	7 55·0	·05316	·33164	86 18	**73 „ 107**	22 46·8	·50204	·02265	68 16
29 „ 151	7 59·5	·05719	·31768	86 08	**74 „ 106**	24 00·7	·52360	·02041	66 58
30 „ 150	8 04·2	0·06140	0·30428	85 59	**75 „ 105**	25 22·8	0·54617	0·01831	65 33
31 „ 149	8 09·1	·06577	·29141	85 49	**76 „ 104**	26 54·6	·56980	·01635	63 57
32 „ 148	8 14·3	·07032	·27904	85 39	**77 „ 103**	28 37·6	·59454	·01453	62 10
33 „ 147	8 19·7	·07505	·26714	85 28	**78 „ 102**	30 33·9	·62040	·01285	60 14
34 „ 146	8 25·5	·07996	·25569	85 18	**79 „ 101**	32 45·7	·64742	·01130	57 55
35 „ 145	8 31·5	0·08506	0·24466	85 07	**80 „ 100**	35 15·8	0·67554	0·00990	55 21
36 „ 144	8 37·8	·09034	·23403	84 56	**81 „ 99**	38 07·7	·70469	·00863	52 25
37 „ 143	8 44·4	·09583	·22379	84 45	**82 „ 98**	41 25·2	·73468	·00750	49 04
38 „ 142	8 51·4	·10151	·21392	84 34	**83 „ 34**	45 12·9	·76521	·00650	45 13
39 „ 141	8 58·7	·10740	·20437	84 22	**84 „ 96**	49 35·5	·79574	·00564	40 47
40 „ 140	9 06·4	0·11349	0·19518	84 10	**85 „ 95**	54 37·9	0·82550	0·00491	35 40
41 „ 139	9 14·4	·11980	·18631	83 57	**86 „ 94**	60 23·9	·85336	·00431	29 51
42 „ 138	9 22·9	·12633	·17774	83 44	**87 „ 93**	66 54·9	·87786	·00385	23 16
43 „ 137	9 31·8	·13309	·16947	83 31	**88 „ 92**	74 08·0	·89724	·00351	15 59
44 „ 136	9 41·2	·14008	·16148	83 17	**89 „ 91**	81 54·6	·90976	·00332	8 09
45 „ 135	9 51·0	0·14731	0·15376	83 03	**90**	90 00·0	0·91411	0·00325	0 00
h	180°-K	A	N	a_1	h	180°-K	A	N	a_1

Directions with regard to ($a_2 \stackrel{+}{\smile} a_1$) :—

Whenever h is greater than 90°, use ($a_2 - a_1$).

If h is less than 90°, use :—

($a_1 + a_2$) when (K ± d) has been used,

($a_1 - a_2$) when (d — K) has been used.

Always **name azimuth** from elevated pole.

Note that { h = L.H.A. of body **West** of the meridian.
{ h = (360° — L.H.A.) of body **East** of the meridian.

Log. tan. a_2 = N + log. tan. (K $\stackrel{+}{\smile}$ d).

LAT. 8°   ALT.—AZ. TABLE

h	K	A	N	a₁	h	K	A	N	a₁
°	° ′			° ′	°	° ′			° ′
0 or 180	8 00·0	0·00000	—	90 00	**45 or 135**	11 14·5	0·14635	0·15476	82 05
1 „ 179	8 00·1	·00007	1·76239	89 52	46 „ 134	11 26·3	·15376	·14731	81 48
2 „ 178	8 00·3	·00026	1·46143	89 43	47 „ 133	11 38·6	·16143	·14012	81 31
3 „ 177	8 00·7	·00058	1·28545	89 35	48 „ 132	11 51·7	·16936	·13317	81 13
4 „ 176	8 01·2	·00104	1·16066	89 27	49 „ 131	12 05·5	·17756	·12647	80 54
5 „ 175	8 01·8	0·00162	1·06395	89 18	**50 „ 130**	12 20·0	0·18604	0·11999	80 35
6 „ 174	8 02·6	·00234	0·98501	89 10	51 „ 129	12 35·3	·19481	·11374	80 15
7 „ 173	8 03·6	·00319	·91835	89 01	52 „ 128	12 51·5	·20387	·10772	79 54
8 „ 172	8 04·7	·00417	·86069	88 53	53 „ 127	13 08·7	·21325	·10190	79 32
9 „ 171	8 05·9	·00528	·80992	88 44	54 „ 126	13 26·8	·22296	·09629	79 09
10 „ 170	8 07·3	0·00652	0·76458	88 36	**55 „ 125**	13 46·1	0·23300	0·09088	78 46
11 „ 169	8 08·9	·00789	·72365	88 27	56 „ 124	14 06·5	·24339	·08567	78 20
12 „ 168	8 10·6	·00941	·68637	88 18	57 „ 123	14 28·2	·25414	·08066	77 54
13 „ 167	8 12·5	·01105	·65216	88 10	58 „ 122	14 51·2	·26528	·07583	77 27
14 „ 166	8 14·5	·01283	·62057	88 01	59 „ 121	15 15·8	·27681	·07118	76 58
15 „ 165	8 16·7	0·01475	0·59126	87 52	**60 „ 120**	15 42·0	0·28877	0·06672	76 27
16 „ 164	8 19·1	·01681	·56391	87 43	61 „ 119	16 10·0	·30115	·06243	75 54
17 „ 163	8 21·6	·01901	·53831	87 34	62 „ 118	16 39·9	·31400	·05831	75 20
18 „ 162	8 24·4	·02135	·51426	87 25	63 „ 117	17 12·1	·32732	·05436	74 43
19 „ 161	8 27·3	·02383	·49161	87 15	64 „ 116	17 46·5	·34116	·05059	74 04
20 „ 160	8 30·4	0·02646	0·47020	87 06	**65 „ 115**	18 23·7	0·35552	0·04697	73 23
21 „ 159	8 33·7	·02923	·44992	86 57	66 „ 114	19 03·7	·37044	·04352	72 38
22 „ 158	8 37·2	·03215	·43067	86 47	67 „ 113	19 47·0	·38595	·04022	71 51
23 „ 157	8 40·8	·03522	·41237	86 37	68 „ 112	20 33·9	·40208	·03708	71 00
24 „ 156	8 44·8	·03844	·39493	86 27	69 „ 111	21 24·8	·41885	·03410	70 04
25 „ 155	8 48·9	0·04181	0·37830	86 17	**70 „ 110**	22 20·3	0·43631	0·03126	69 04
26 „ 154	8 53·2	·04534	·36241	86 07	71 „ 109	23 20·9	·45450	·02858	68 00
27 „ 153	8 57·8	·04903	·34720	85 57	72 „ 108	24 27·4	·47344	·02604	66 49
28 „ 152	9 02·6	·05288	·33264	85 46	73 „ 107	25 40·4	·49317	·02365	65 31
29 „ 151	9 07·7	·05689	·31868	85 35	74 „ 106	27 01·0	·51373	·02141	64 07
30 „ 150	9 13·1	0·06107	0·30528	85 24	**75 „ 105**	28 30·1	0·53514	0·01930	62 33
31 „ 149	9 18·7	·06542	·29241	85 13	76 „ 104	30 09·2	·55743	·01734	60 50
32 „ 148	9 24·6	·06994	·28004	85 02	77 „ 103	31 59·7	·58060	·01552	58 55
33 „ 147	9 30·8	·07464	·26814	84 50	78 „ 102	34 03·4	·60465	·01384	56 47
34 „ 146	9 37·3	·07952	·25669	84 38	79 „ 101	36 22·4	·62953	·01230	54 24
35 „ 145	9 44·1	0·08458	0·24566	84 26	**80 „ 100**	38 59·1	0·65517	0·01090	51 43
36 „ 144	9 51·3	·08983	·23503	84 14	81 „ 99	41 56·2	·68142	·00963	48 42
37 „ 143	9 58·8	·09528	·22478	84 01	82 „ 98	45 16·8	·70804	·00849	45 17
38 „ 142	10 06·7	·10092	·21491	83 48	83 „ 97	49 04·2	·73468	·00750	41 25
39 „ 141	10 15·0	·10676	·20538	83 34	84 „ 96	53 21·6	·76083	·00663	37 04
40 „ 140	10 23·8	0·11281	0·19618	83 20	**85 „ 95**	58 11·7	0·78580	0·00590	32 09
41 „ 139	10 32·9	·11907	·18730	83 06	86 „ 94	63 36·2	·80862	·00531	26 41
42 „ 138	10 42·5	·12554	·17874	82 51	87 „ 93	69 34·5	·82824	·00484	20 38
43 „ 137	10 52·7	·13225	·17046	82 36	88 „ 92	76 03·3	·84345	·00451	14 05
44 „ 136	11 03·3	·13918	·16248	82 21	89 „ 91	82 55·3	·85312	·00431	7 09
45 „ 135	11 14·5	0·14635	0·15476	82 05	**90**	90 00·0	0·85645	0·00425	0 00
h	180°-K	A	N	a₁	h	180°-K	A	N	a₁

Directions with regard to (K $\overset{+}{\sim}$ d) :—

If l and d have **Same Name,** use (K $\frown$ d), i.e. (K — d) or (d — K).
If l and d have **Different Names,** use (K + d).

To name Intercept :—

If H₀ is greater than H_c, intercept is **" Towards."**
If H₀ is less than H_c, intercept is **" Away."**

Log. cosec. H_c = A + log. sec. (K $\overset{+}{\sim}$ d).

N.B.   Whenever h is greater than 90°, subtract the quantity in the K column from 180°.

ALT.—AZ. TABLE   LAT. 9°

h	K	A	N	a₁	h	K	A	N	a₁
°	° ′			° ′	°	° ′			° ′
0 or 180	9 00·0	0·00000	—	90 00	45 or 135	12 37·5	0·14526	0·15590	81 07
1 „ 179	9 00·1	·00006	1·76353	89 51	46 „ 134	12 50·6	·15261	·14845	80 48
2 „ 178	9 00·3	·00026	1·46256	89 41	47 „ 133	13 04·5	·16019	·14125	80 29
3 „ 177	9 00·7	·00058	1·28658	89 32	48 „ 132	13 19·0	·16803	·13431	80 09
4 „ 176	9 01·3	·00103	1·16180	89 22	49 „ 131	13 34·4	·17614	·12760	79 48
5 „ 175	9 02·0	0·00162	1·06508	89 13	50 „ 130	13 50·5	0·18451	0·12113	79 26
6 „ 174	9 02·9	·00233	0·98621	89 03	51 „ 129	14 07·6	·19317	·11488	79 04
7 „ 173	9 04·0	·00317	·91949	88 54	52 „ 128	14 25·6	·20212	·10885	78 41
8 „ 172	9 05·2	·00414	·86183	88 44	53 „ 127	14 44·7	·21137	·10303	78 16
9 „ 171	9 06·6	·00525	·81095	88 35	54 „ 126	15 04·8	·22094	·09742	77 51
10 „ 170	9 08·2	0·00648	0·76571	88 25	55 „ 125	15 26·2	0·23083	0·09202	77 24
11 „ 169	9 09·9	·00785	·72478	88 15	56 „ 124	15 48·8	·24107	·08681	76 57
12 „ 168	9 11·9	·00936	·68750	88 06	57 „ 123	16 12·9	·25164	·08179	76 27
13 „ 167	9 14·0	·01099	·65329	87 56	58 „ 122	16 38·4	·26259	·07696	75 57
14 „ 166	9 16·2	·01276	·62171	87 46	59 „ 121	17 05·6	·27392	·07231	75 24
15 „ 165	9 18·7	0·01468	0·59238	87 36	60 „ 120	17 34·6	0·28565	0·06785	74 50
16 „ 164	9 21·4	·01672	·56500	87 26	61 „ 119	18 05·5	·29779	·06356	74 12
17 „ 163	9 24·2	·01891	·53944	87 16	62 „ 118	18 38·6	·31036	·05945	73 36
18 „ 162	9 27·3	·02123	·51540	87 05	63 „ 117	19 13·9	·32339	·05550	72 56
19 „ 161	9 30·6	·02370	·49274	86 55	64 „ 116	19 51·9	·33690	·05172	72 13
20 „ 160	9 34·0	0·02631	0·47133	86 45	65 „ 115	20 32·7	0·35089	0·04810	71 27
21 „ 159	9 37·7	·02907	·45105	86 34	66 „ 114	21 16·6	·36541	·04465	70 38
22 „ 158	9 41·6	·03197	·43181	86 23	67 „ 113	22 03·9	·38047	·04135	69 46
23 „ 157	9 45·8	·03502	·41350	86 12	68 „ 112	22 55·1	·39609	·03821	68 50
24 „ 156	9 50·1	·03822	·39607	86 01	69 „ 111	23 50·6	·41231	·03523	67 50
25 „ 155	9 54·8	0·04157	0·37943	85 50	70 „ 110	24 50·9	0·42914	0·03239	66 45
26 „ 154	9 59·6	·04508	·36354	85 38	71 „ 109	25 56·5	·44661	·02971	65 34
27 „ 153	10 04·8	·04874	·34833	85 27	72 „ 108	27 08·2	·46475	·02718	64 17
28 „ 152	10 10·2	·05257	·33377	85 15	73 „ 107	28 26·7	·48357	·02478	62 54
29 „ 151	10 15·9	·05655	·31981	85 03	74 „ 106	29 52·9	·50309	·02254	61 23
30 „ 150	10 21·8	0·06070	0·30641	84 50	75 „ 105	31 27·9	0·52331	0·02044	59 43
31 „ 149	10 28·1	·06502	·29354	84 38	76 „ 104	33 12·7	·54425	·01848	57 54
32 „ 148	10 34·7	·06951	·28117	84 25	77 „ 103	35 08·9	·56586	·01666	55 53
33 „ 147	10 41·7	·07418	·26927	84 12	78 „ 102	37 18·0	·58813	·01498	53 39
34 „ 146	10 48·9	·07902	·25782	83 59	79 „ 101	39 41·7	·61096	·01343	51 10
35 „ 145	10 56·6	0·08405	0·24679	83 45	80 , 100	42 22·1	0·63426	0·01203	48 25
36 „ 144	11 04·6	·08926	·23616	83 31	81 „ 99	45 21·3	·65783	·01076	45 21
37 „ 143	11 13·0	·09465	·22592	83 17	82 „ 98	48 41·6	·68142	·00963	41 56
38 „ 142	11 21·9	·10025	·21604	83 02	83 „ 97	52 25·4	·70469	·00863	38 08
39 „ 141	11 31·2	·10605	·20651	82 47	84 „ 96	56 34·6	·72716	·00777	33 54
40 „ 140	11 40·9	0·11204	0·19731	82 31	85 „ 95	61 10·6	0·74823	0·00704	29 13
41 „ 139	11 51·1	·11824	·18844	82 15	86 „ 94	66 13·8	·76719	·00644	24 05
42 „ 138	12 01·9	·12466	·17987	81 59	87 „ 93	71 42·9	·78315	·00598	18 31
43 „ 137	12 13·2	·13130	·17160	81 42	88 „ 92	77 34·4	·79536	·00564	12 35
44 „ 136	12 25·0	·13817	·16361	81 25	89 „ 91	83 42·7	·80304	·00545	6 22
45 „ 135	12 37·5	0·14526	0·15590	81 07	90	90 00·0	0·80567	0·00538	0 00
h	180°-K	A	N	a₁	h	180°-K	A	N	a₁

Directions with regard to $(a_2 \pm a_1)$:—

Whenever h is greater than 90°, use $(a_2 - a_1)$.

If h is less than 90°, use :—

$(a_1 + a_2)$ when $(K \pm d)$ has been used,

$(a_1 - a_2)$ when $(d - K)$ has been used.

Always **name azimuth** from elevated pole.

Note that $\begin{cases} \text{h = L.H.A. of body } \textbf{West} \text{ of the meridian.} \\ \text{h = (360° — L.H.A.) of body } \textbf{East} \text{ of the meridian.} \end{cases}$

Log. tan. a_2 = N + log. tan. $(K \pm d)$.

LAT. 10°   ALT.—AZ. TABLE

h	K	A	N	a₁	h	K	A	N	a₁
°	° ′			° ′	°	° ′			° ′
0 or 180	10 00·0	0·00000	—	90 00	45 or 135	14 00·1	0·14406	0·15717	80 09
1 „ 179	10 00·1	·00006	1·76499	89 50	46 „ 134	14 14·6	·15132	·14972	79 48
2 „ 178	10 00·4	·00026	1·46380	89 39	47 „ 133	14 30·8	·15882	·14252	79 27
3 „ 177	10 00·8	·00058	1·28785	89 29	48 „ 132	14 45·8	·16656	·13558	79 05
4 „ 176	10 01·4	·00103	1·16306	89 18	49 „ 131	15 02·6	·17456	·12887	78 42
5 „ 175	10 02·2	0·00161	1·06635	89 08	50 „ 130	15 20·4	0·18283	0·12240	78 18
6 „ 174	10 03·2	·00231	0·98741	88 57	51 „ 129	15 39·1	·19137	·11615	77 54
7 „ 173	10 04·4	·00315	·92076	88 47	52 „ 128	15 58·9	·20019	·11012	77 28
8 „ 172	10 05·8	·00412	·86339	88 36	53 „ 127	16 19·8	·20930	·10430	77 01
9 „ 171	10 07·3	·00522	·81232	88 25	54 „ 126	16 41·9	·21872	·09869	76 33
10 „ 170	10 09·1	0·00645	0·76698	88 15	55 „ 125	17 05·3	0·22845	0·09328	76 04
11 „ 169	10 11·0	·00781	·72651	88 04	56 „ 124	17 30·1	·23850	·08808	75 34
12 „ 168	10 13·1	·00930	·68877	87 53	57 „ 123	17 56·4	·24890	·08306	75 02
13 „ 167	10 15·5	·01093	·65456	87 42	58 „ 122	18 24·3	·25964	·07823	74 28
14 „ 166	10 18·0	·01269	·62297	87 31	59 „ 121	18 53·9	·27074	·07358	73 53
15 „ 165	10 20·7	0·01459	0·59367	87 20	60 „ 120	19 25·5	0·28222	0·06912	73 16
16 „ 164	10 23·7	·01662	·56631	87 09	61 „ 119	19 59·2	·29410	·06483	72 36
17 „ 163	10 26·8	·01879	·54071	86 58	62 „ 118	20 35·1	·30638	·06071	71 55
18 „ 162	10 30·2	·02110	·51667	86 46	63 „ 117	21 13·6	·31909	·05677	71 11
19 „ 161	10 33·8	·02355	·49401	86 35	64 „ 116	21 54·7	·33224	·05299	70 24
20 „ 160	10 37·7	0·02615	0·47260	86 23	65 „ 115	22 38·8	0·34585	0·04937	69 35
21 „ 159	10 41·7	·02889	·45232	86 11	66 „ 114	23 26·3	·35994	·04592	68 42
22 „ 158	10 46·1	·03177	·43307	85 59	67 „ 113	24 17·3	·37452	·04262	67 45
23 „ 157	10 50·6	·03480	·41476	85 47	68 „ 112	25 12·4	·38962	·03948	66 45
24 „ 156	10 55·5	·03798	·39750	85 35	69 „ 111	26 11·9	·40526	·03650	65 40
25 „ 155	11 00·6	0·04131	0·38070	85 22	70 „ 110	27 16·4	0·42142	0·03366	64 30
26 „ 154	11 06·0	·04479	·36481	85 10	71 „ 109	28 26·4	·43815	·03098	63 14
27 „ 153	11 11·6	·04843	·34960	84 57	72 „ 108	29 42·6	·45546	·02844	61 53
28 „ 152	11 17·6	·05222	·33504	84 43	73 „ 107	31 05·6	·47335	·02605	60 24
29 „ 151	11 23·9	·05618	·32108	84 30	74 „ 106	32 36·4	·49182	·02381	58 48
30 „ 150	11 30·5	0·06030	0·30768	84 16	75 „ 105	34 15·9	0·51086	0·02171	57 03
31 „ 149	11 37·4	·06458	·29481	84 03	76 „ 104	36 05·2	·53045	·01975	55 09
32 „ 148	11 44·7	·06904	·28244	83 48	77 „ 103	38 05·5	·55055	·01793	53 03
33 „ 147	11 52·4	·07367	·27054	83 34	78 „ 102	40 18·0	·57110	·01625	50 45
34 „ 146	12 00·4	·07847	·25909	83 19	79 „ 101	42 44·5	·59200	·01470	48 13
35 „ 145	12 08·9	0·08345	0·24806	83 04	80 „ 100	45 26·3	0·61311	0·01330	45 26
36 „ 144	12 17·7	·08861	·23743	82 49	81 „ 99	48 25·3	·63425	·01203	42 22
37 „ 143	12 27·0	·09396	·22719	82 33	82 „ 98	51 43·0	·65517	·01090	38 59
38 „ 142	12 36·8	·09951	·21731	82 16	83 „ 97	55 21·0	·67553	·00990	35 16
39 „ 141	12 47·0	·10525	·20778	82 00	84 „ 96	59 20·4	·69492	·00909	31 11
40 „ 140	12 57·7	0·11119	0·19858	81 43	85 „ 95	63 41·9	0·71286	0·00830	26 44
41 „ 139	13 09·0	·11733	·18971	81 25	86 „ 94	68 24·9	·72876	·00771	21 56
42 „ 138	13 20·9	·12368	·18114	81 07	87 „ 93	73 28·1	·74200	·00725	16 48
43 „ 137	13 33·3	·13025	·17287	80 48	88 „ 92	78 48·3	·75199	·00691	11 22
44 „ 136	13 46·4	·13704	·16488	80 29	89 „ 91	84 20·8	·75821	·00672	5 44
45 „ 135	14 00·1	0·14406	0·15717	80 09	90	90 00·0	0·76033	0·00665	0 00
h	180°-K	A	N	a₁	h	180°-K	A	N	a₁

Directions with regard to (K ± d) :—

If l and d have **Same Name,** use (K ⌣ d), i.e. (K — d) or (d — K).

If l and d have **Different Names,** use (K + d).

To name Intercept :—

If H_o is greater than H_c, intercept is **" Towards."**

If H_o is less than H_c, intercept is **" Away."**

Log. cosec. H_c = A + log. sec. (K ± d).

N.B. *Whenever h is greater than 90°, subtract the quantity in the K column from 180°.*

ALT.—AZ. TABLE LAT. 11°

h	K		A	N	a₁		h	K		A	N	a₁	
°	°	′			°	′	°	°	′			°	′
0 or 180	11	00·0	0·00000	—	90	00	45 or 135	15	22·2	0·14275	0·15857	79	12
1 „ 179	11	00·1	·00006	1·76620	89	49	46 „ 134	15	38·0	·14991	·15112	78	49
2 „ 178	11	00·4	·00026	1·46533	89	37	47 „ 133	15	54·5	·15731	·14393	78	26
3 „ 177	11	00·9	·00057	1·29019	89	26	48 „ 132	16	11·9	·16495	·13698	78	02
4 „ 176	11	01·6	·00102	1·16478	89	14	49 „ 131	16	30·2	·17284	·13027	77	37
5 „ 175	11	02·5	0·00160	1·06778	89	03	50 „ 130	16	49·5	0·18099	0·12380	77	11
6 „ 174	11	03·5	·00230	0·98884	88	51	51 „ 129	17	09·9	·18940	·11755	76	44
7 „ 173	11	04·8	·00313	·92268	88	39	52 „ 128	17	31·3	·19808	·11152	76	17
8 „ 172	11	06·3	·00409	·86450	88	28	53 „ 127	17	54·0	·20704	·10570	75	47
9 „ 171	11	08·0	·00518	·81372	88	16	54 „ 126	18	17·9	·21629	·10010	75	17
10 „ 170	11	09·9	0·00640	0·76838	88	04	55 „ 125	18	43·3	0·22585	0·09469	74	45
11 „ 169	11	12·0	·00776	·72745	87	53	56 „ 124	19	10·1	·23572	·08948	74	12
12 „ 168	11	14·4	·00924	·69017	87	41	57 „ 123	19	38·5	·24591	·08446	73	38
13 „ 167	11	16·9	·01086	·65588	87	29	58 „ 122	20	08·6	·25643	·07963	73	01
14 „ 166	11	19·7	·01260	·62439	87	17	59 „ 121	20	40·6	·26730	·07499	72	23
15 „ 165	11	22·7	0·01449	0·59506	87	05	60 „ 120	21	14·6	0·27852	0·07052	71	43
16 „ 164	11	25·9	·01651	·56770	86	52	61 „ 119	21	50·9	·29012	·06623	71	00
17 „ 163	11	29·4	·01867	·54218	86	40	62 „ 118	22	29·5	·30209	·06212	70	16
18 „ 162	11	33·1	·02096	·51807	86	27	63 „ 117	23	10·7	·31445	·05817	69	28
19 „ 161	11	37·0	·02339	·49541	86	14	64 „ 116	23	54·8	·32724	·05439	68	38
20 „ 160	11	41·2	0·02597	0·47400	86	02	65 „ 115	24	42·0	0·34044	0·05078	67	45
21 „ 159	11	45·7	·02869	·45372	85	49	66 „ 114	25	32·6	·35407	·04732	66	48
22 „ 158	11	50·4	·03155	·43448	85	35	67 „ 113	26	27·0	·36816	·04403	65	48
23 „ 157	11	55·4	·03455	·41618	85	22	68 „ 112	27	25·5	·38271	·04089	64	43
24 „ 156	12	00·7	·03771	·39874	85	09	69 „ 111	28	28·5	·39773	·03790	63	34
25 „ 155	12	06·3	0·04101	0·38211	84	55	70 „ 110	29	36·7	0·41323	0·03507	62	20
26 „ 154	12	12·2	·04447	·36621	84	41	71 „ 109	30	50·4	·42921	·03238	61	00
27 „ 153	12	18·4	·04808	·35101	84	27	72 „ 108	32	10·3	·44568	·02985	59	35
28 „ 152	12	24·9	·05184	·33644	84	12	73 „ 107	33	37·1	·46263	·02746	58	02
29 „ 151	12	31·8	·05577	·32248	83	58	74 „ 106	35	11·5	·48006	·02521	56	22
30 „ 150	12	39·0	0·05985	0·30908	83	43	75 „ 105	36	54·5	0·49793	0·02311	54	33
31 „ 149	12	46·6	·06410	·29621	83	28	76 „ 104	38	46·9	·51622	·02115	52	34
32 „ 148	12	54·6	·06851	·28384	83	12	77 „ 103	40	49·8	·53486	·01933	50	26
33 „ 147	13	02·9	·07310	·27194	82	56	78 „ 102	43	04·4	·55378	·01765	48	05
34 „ 146	13	11·7	·07786	·26049	82	40	79 „ 101	45	31·9	·57288	·01611	45	32
35 „ 145	13	20·9	0·08280	0·24946	82	23	80 „ 100	48	13·5	0·59200	0·01470	42	44
36 „ 144	13	30·6	·08791	·23888	82	06	81 „ 99	51	10·4	·61096	·01343	39	42
37 „ 143	13	40·8	·09321	·22859	81	49	82 „ 98	54	23·9	·62953	·01230	36	22
38 „ 142	13	51·4	·09870	·21871	81	31	83 „ 97	57	54·8	·64742	·01130	32	46
39 „ 141	14	02·6	·10438	·20918	81	13	84 „ 96	61	43·8	·66424	·01044	28	51
40 „ 140	14	14·3	0·11025	0·19999	80	54	85 „ 95	65	51·0	0·67962	0·00971	24	38
41 „ 139	14	26·6	·11633	·19111	80	35	86 „ 94	70	16·5	·69310	·00911	20	08
42 „ 138	14	39·5	·12261	·18254	80	15	87 „ 93	74	55·8	·70420	·00865	15	21
43 „ 137	14	53·0	·12911	·17427	79	55	88 „ 92	79	49·3	·71250	·00832	10	22
44 „ 136	15	07·3	·13582	·16628	79	34	89 „ 91	84	52·2	·71767	·00812	5	14
45 „ 135	15	22·2	0·14275	0·15857	79	12	90	90	00·0	0·71940	0·00805	0	00
h	180°-K		A	N	a₁		h	180°-K		A	N	a₁	

Directions with regard to (a₂ ~ a₁) :—

Whenever h is greater than 90°, use (a₂ − a₁).

If h is less than 90°, use :—

$\quad$ (a₁ + a₂) when (K ± d) has been used,

$\quad$ (a₁ − a₂) when (d − K) has been used.

Always **name azimuth** from elevated pole.

Note that { h = L.H.A. of body **West** of the meridian.
{ h = (360° − L.H.A.) of body **East** of the meridian.

Log. tan. a₂ = N + log. tan. (K ~ d).

LAT. 12°　　ALT.—AZ. TABLE

h	K		A	N	a₁		h	K		A	N	a₁	
°	°	′			°	′		°	′			°	′
0 or 180	12	00·0	0·00000	—	90	00	45 or 135	16	43·8	0·14132	0·16011	78	15
1 „ 179	12	00·1	·00006	1·70774	09	48	46 „ 134	17	00·8	·14838	·15267	77	51
2 „ 178	12	00·4	·00025	1·46678	89	35	47 „ 133	17	18·6	·15568	·14547	77	26
3 „ 177	12	01·0	·00057	1·29080	89	23	48 „ 132	17	37·4	·16320	·13853	77	00
4 „ 176	12	01·7	·00101	1·16601	89	10	49 „ 131	17	57·1	·17097	·13182	76	33
5 „ 175	12	02·7	0·00159	1·06930	88	58	50 „ 130	18	17·9	0·17899	0·12535	76	05
6 „ 174	12	03·8	·00228	0·99036	88	45	51 „ 129	18	39·8	·18726	·11910	75	36
7 „ 173	12	05·2	·00310	·92371	88	32	52 „ 128	19	02·8	·19580	·11307	75	06
8 „ 172	12	06·9	·00406	·86604	88	20	53 „ 127	19	27·2	·20460	·10725	74	35
9 „ 171	12	08·7	·00514	·81527	88	07	54 „ 126	19	52·8	·21369	·10164	74	02
10 „ 170	12	10·8	0·00636	0·76993	87	54	55 „ 125	20	20·0	0·22306	0·09623	73	28
11 „ 169	12	13·1	·00770	·72900	87	41	56 „ 124	20	48·7	·23272	·09103	72	52
12 „ 168	12	15·6	·00917	·69176	87	28	57 „ 123	21	19·1	·24269	·08601	72	15
13 „ 167	12	18·4	·01078	·65751	87	15	58 „ 122	21	51·4	·25298	·08118	71	36
14 „ 166	12	21·4	·01251	·62570	87	02	59 „ 121	22	25·6	·26360	·07653	70	55
15 „ 165	12	24·6	0·01439	0·59661	86	48	60 „ 120	23	01·8	0·27455	0·07207	70	12
16 „ 164	12	28·1	·01639	·56926	86	35	61 „ 119	23	40·4	·28583	·06778	69	26
17 „ 163	12	31·9	·01853	·54366	86	22	62 „ 118	24	21·5	·29749	·06366	68	39
18 „ 162	12	35·9	·02080	·51962	86	08	63 „ 117	25	05·3	·30951	·05972	67	48
19 „ 161	12	40·2	·02322	·49696	85	54	64 „ 116	25	52·1	·32190	·05594	66	55
20 „ 160	12	44·7	0·02577	0·47556	85	40	65 „ 115	26	42·0	0·33467	0·05233	65	58
21 „ 159	12	49·6	·02847	·45527	85	26	66 „ 114	27	35·5	·34785	·04887	64	58
22 „ 158	12	54·7	·03131	·43603	85	12	67 „ 113	28	32·7	·36143	·04557	63	54
23 „ 157	13	00·2	·03429	·41782	84	57	68 „ 112	29	34·2	·37541	·04243	62	46
24 „ 156	13	05·9	·03742	·40029	84	43	69 „ 111	30	40·4	·38981	·03945	61	34
25 „ 155	13	11·9	0·04069	0·38365	84	28	70 „ 110	31	51·6	0·40462	0·03661	60	16
26 „ 154	13	18·3	·04412	·36776	84	13	71 „ 109	33	08·4	·41985	·03393	58	53
27 „ 153	13	25·0	·04770	·35255	83	57	72 „ 108	34	31·3	·43549	·03139	57	23
28 „ 152	13	32·1	·05143	·33799	83	42	73 „ 107	36	01·0	·45153	·02900	55	47
29 „ 151	13	39·6	·05532	·32403	83	26	74 „ 106	37	38·2	·46792	·02676	54	03
30 „ 150	13	47·4	0·05936	0·31063	83	09	75 „ 105	39	23·7	0·48466	0·02466	52	11
31 „ 149	13	55·6	·06357	·29776	82	53	76 „ 104	41	18·2	·50169	·02269	50	11
32 „ 148	14	04·3	·06795	·28539	82	36	77 „ 103	43	22·6	·51894	·02087	48	00
33 „ 147	14	13·3	·07248	·27349	82	19	78 „ 102	45	38·0	·53634	·01919	45	38
34 „ 146	14	22·8	·07719	·26204	82	01	79 „ 101	48	05·2	·55378	·01765	43	04
35 „ 145	14	32·8	0·08208	0·25101	81	43	80 „ 100	50	45·2	0·57109	0·01625	40	18
36 „ 144	14	43·2	·08714	·24038	81	25	81 „ 99	53	38·9	·58812	·01497	37	18
37 „ 143	14	54·2	·09238	·23014	81	06	82 „ 98	56	47·1	·60465	·01384	34	03
38 „ 142	15	05·7	·09781	·22026	80	46	83 „ 97	60	10·3	·62040	·01285	30	34
39 „ 141	15	17·8	·10343	·21073	80	27	84 „ 96	63	48·8	·63509	·01198	26	49
40 „ 140	15	30·5	0·10923	0·20153	80	06	85 „ 95	67	42·3	0·64838	0·01125	22	49
41 „ 139	15	43·7	·11524	·19266	79	45	86 „ 94	71	49·9	·65991	·01066	18	35
42 „ 138	15	57·7	·12144	·18409	79	24	87 „ 93	76	10·1	·66933	·01019	14	09
43 „ 137	16	12·3	·12786	·17582	79	02	88 „ 92	80	40·6	·67633	·00986	9	32
44 „ 136	16	27·7	·13448	·16783	78	39	89 „ 91	85	18·4	·68066	·00966	4	48
45 „ 135	16	43·8	0·14132	0·16011	78	15	90	90	00·0	0·68212	·00960	0	00
h	180°-K		A	N	a₁		h	180°-K		A	N	a₁	

Directions with regard to (K ⌣̲ d) :—

If l and d have **Same Name,** use (K ⌣ d), i.e. (K — d) or (d — K).

If l and d have **Different Names,** use (K + d).

To name Intercept :—

If H₀ is greater than H_c, intercept is " **Towards.**"

If H₀ is less than H_c, intercept is " **Away.**"

Log. cosec. H_c = A + log. sec. (K ⌣̲ d).

N.B. Whenever h is greater than 90°, subtract the quantity in the K column from 180°.

ALT.—AZ. TABLE   LAT. 13°

h	K	A	N	a₁	h	K	A	N	a₁
°	° ′			° ′		° ′			° ′
0 or 180	13 00·0	0·00000	—	90 00	45 or 135	18 04·9	0·13980	0·16179	77 19
1 „ 179	13 00·1	·00006	1·76942	89 47	46 „ 134	18 23·0	·14676	·15434	76 53
2 „ 178	13 00·5	·00025	1·46846	89 33	47 „ 133	18 42·1	·15394	·14715	76 26
3 „ 177	13 01·0	·00057	1·29248	89 19	48 „ 132	19 02·1	·16134	·14020	75 58
4 „ 176	13 01·8	·00101	1·16769	89 06	49 „ 131	19 23·2	·16898	·13350	75 29
5 „ 175	13 02·9	0·00157	1·07098	88 52	50 „ 130	19 45·4	0·17685	0·12702	75 00
6 „ 174	13 04·1	·00227	0·99204	88 39	51 „ 129	20 08·7	·18499	·12077	74 29
7 „ 173	13 05·7	·00308	·92538	88 25	52 „ 128	20 33·3	·19337	·11474	73 56
8 „ 172	13 07·4	·00403	·86772	88 11	53 „ 127	20 59·3	·20200	·10893	73 23
9 „ 171	13 09·4	·00510	·81684	87 58	54 „ 126	21 26·6	·21090	·10332	72 48
10 „ 170	13 11·6	0·00631	0·77161	87 44	55 „ 125	21 55·5	0·22008	0·09791	72 11
11 „ 169	13 14·1	·00764	·73068	87 30	56 „ 124	22 26·0	·22954	·09270	71 33
12 „ 168	13 16·8	·00910	·69340	87 16	57 „ 123	22 58·3	·23928	·08769	70 54
13 „ 167	13 19·8	·01069	·65938	87 02	58 „ 122	23 32·5	·24933	·08286	70 12
14 „ 166	13 23·0	·01241	·62760	86 47	59 „ 121	24 08·7	·25968	·07821	69 29
15 „ 165	13 26·5	0·01427	0·59828	86 33	60 „ 120	24 47·1	0·27034	0·07375	68 43
16 „ 164	13 30·3	·01626	·57084	86 19	61 „ 119	25 27·8	·28132	·06946	67 55
17 „ 163	13 34·3	·01838	·54534	86 04	62 „ 118	26 11·2	·29263	·06534	67 04
18 „ 162	13 38·7	·02064	·52129	85 49	63 „ 117	26 57·3	·30428	·06140	66 11
19 „ 161	13 43·3	·02303	·49863	85 34	64 „ 116	27 46·4	·31628	·05762	65 14
20 „ 160	13 48·2	0·02556	0·47722	85 19	65 „ 115	28 38·8	0·32862	0·05400	64 15
21 „ 159	13 53·4	·02823	·45695	85 04	66 „ 114	29 34·8	·34132	·05055	63 12
22 „ 158	13 58·9	·03105	·43769	84 48	67 „ 113	30 34·6	·35437	·04725	62 05
23 „ 157	14 04·8	·03400	·41940	84 33	68 „ 112	31 38·7	·36779	·04411	60 54
24 „ 156	14 11·0	·03710	·40196	84 17	69 „ 111	32 47·4	·38157	·04112	59 38
25 „ 155	14 17·5	0·04035	0·38533	84 01	70 „ 110	34 01·2	0·39570	0·03829	58 17
26 „ 154	14 24·3	·04374	·36943	83 44	71 „ 109	35 20·5	·41018	·03561	56 51
27 „ 153	14 31·6	·04728	·35423	83 28	72 „ 108	36 45·8	·42499	·03307	55 18
28 „ 152	14 39·2	·05098	·33967	83 11	73 „ 107	38 17·8	·44011	·03068	53 39
29 „ 151	14 47·2	·05483	·32571	82 54	74 „ 106	39 56·9	·45552	·02843	51 53
30 „ 150	14 55·6	0·05883	0·31231	82 36	75 „ 105	41 44·0	0·47116	0·02633	49 59
31 „ 149	15 04·5	·06300	·29944	82 18	76 „ 104	43 39·6	·48700	·02437	47 57
32 „ 148	15 13·7	·06733	·28707	82 00	77 „ 103	45 44·6	·50296	·02255	45 05
33 „ 147	15 23·5	·07182	·27517	81 41	78 „ 102	47 59·7	·51895	·02087	43 23
34 „ 146	15 33·7	·07648	·26371	81 22	79 „ 101	50 25·6	·53486	·01933	40 50
35 „ 145	15 44·4	0·08132	0·25269	81 03	80 „ 100	53 03·1	0·55055	0·01793	38 05
36 „ 144	15 55·6	·08632	·24206	80 43	81 „ 99	55 52·7	·56586	·01666	35 09
37 „ 143	16 07·4	·09150	·23181	80 23	82 „ 98	58 55·0	·58060	·01552	32 00
38 „ 142	16 19·8	·09686	·22193	80 02	83 „ 97	62 10·3	·59454	·01453	28 38
39 „ 141	16 32·7	·10241	·21240	79 41	84 „ 96	65 38·4	·60742	·01366	25 03
40 „ 140	16 46·3	0·10814	0·20321	79 19	85 „ 95	69 19·1	0·61898	0·01293	21 15
41 „ 139	17 00·5	·11407	·19433	78 56	86 „ 94	73 11·3	·62894	·01234	17 16
42 „ 138	17 15·5	·12020	·18577	78 33	87 „ 93	77 13·6	·63703	·01187	13 07
43 „ 137	17 31·2	·12652	·17749	78 09	88 „ 92	81 24·2	·64301	·01154	8 49
44 „ 136	17 47·6	·13305	·16951	77 45	89 „ 91	85 40·6	·64667	·01134	4 26
45 „ 135	18 04·9	0·13980	0·16179	77 19	90	90 00·0	0·64791	0·01128	0 00
h	180°-K	A	N	a₁	h	180°-K	A	N	a₁

Directions with regard to $(a_2 \overset{+}{\sim} a_1)$:—

Whenever h is greater than 90°, use $(a_2 - a_1)$.

If h is less than 90°, use :—

    $(a_1 + a_2)$ when $(K \pm d)$ has been used,

    $(a_1 - a_2)$ when $(d - K)$ has been used.

Always **name azimuth** from elevated pole.

Note that $\begin{cases} \text{h = L.H.A. of body } \textbf{West} \text{ of the meridian.} \\ \text{h = (360° — L.H.A.) of body } \textbf{East} \text{ of the meridian.} \end{cases}$

Log. tan. $a_2 = N + \log. \tan. (K \overset{+}{\sim} d)$.

LAT. 14° ALT.—AZ. TABLE

h	K		A	N	a₁		h	K		A	N	a₁	
°	°	′			°	′		°	′			°	′
0 or 180	14	00·0	0·00000	—	90	00	45 or 135	19	25·4	0·13816	0·16361	76	24
1 „ 179	14	00·1	·00006	1·77116	89	45	46 „ 134	19	44·7	·14501	·15616	75	56
2 „ 178	14	00·5	·00025	1·47028	89	31	47 „ 133	20	04·9	·15207	·14897	75	27
3 „ 177	14	01·1	·00056	1·29430	89	16	48 „ 132	20	26·2	·15936	·14202	74	58
4 „ 176	14	02·0	·00100	1·16951	89	02	49 „ 131	20	48·5	·16686	·13532	74	27
5 „ 175	14	03·1	0·00156	1·07280	88	47	50 „ 130	21	12·0	0·17459	·012884	73	55
6 „ 174	14	04·4	·00225	0·99386	88	32	51 „ 129	21	36·8	·18256	·12259	73	22
7 „ 173	14	06·1	·00306	·92720	88	17	52 „ 128	22	02·8	·19077	·11656	72	48
8 „ 172	14	07·9	·00400	·86954	88	02	53 „ 127	22	30·2	·19924	·11075	72	12
9 „ 171	14	10·1	·00506	·81876	87	47	54 „ 126	22	59·1	·20795	·10514	71	35
10 „ 170	14	12·4	0·00625	0·77343	87	33	55 „ 125	23	29·6	0·21692	0·09974	70	56
11 „ 169	14	15·1	·00757	·73250	87	18	56 „ 124	24	01·9	·22616	·09452	70	16
12 „ 168	14	18·0	·00902	·69522	87	03	57 „ 123	24	35·9	·23567	·08951	69	34
13 „ 167	14	21·2	·01060	·66101	86	48	58 „ 122	25	11·8	·24546	·08467	68	50
14 „ 166	14	24·7	·01231	·62949	86	33	59 „ 121	25	49·9	·25554	·08003	68	04
15 „ 165	14	28·4	0·01415	0·60010	86	17	60 „ 120	26	30·2	0·26590	·07557	67	16
16 „ 164	14	32·4	·01612	·57276	86	02	61 „ 119	27	13·0	·27656	·07128	66	25
17 „ 163	14	36·8	·01822	·54716	85	46	62 „ 118	27	58·3	·28753	·06716	65	32
18 „ 162	14	41·4	·02046	·52311	85	30	63 „ 117	28	46·5	·29881	·06322	64	36
19 „ 161	14	46·3	·02283	·50045	85	14	64 „ 116	29	37·8	·31039	·05944	63	37
20 „ 160	14	51·6	0·02534	0·47904	84	58	65 „ 115	30	32·3	0·32229	0·05582	62	35
21 „ 159	14	57·2	·02798	·45877	84	42	66 „ 114	31	30·5	·33451	·05237	61	29
22 „ 158	15	03·1	·03077	·43952	84	35	67 „ 113	32	32·5	·34704	·04907	60	19
23 „ 157	15	09·3	·03370	·42122	84	08	68 „ 112	33	38·8	·35989	·04593	59	05
24 „ 156	15	15·9	·03677	·40378	83	51	69 „ 111	34	49·7	·37304	·04294	57	47
25 „ 155	15	22·9	0·03998	0·38715	83	34	70 „ 110	36	05·5	0·38650	0·04011	56	23
26 „ 154	15	30·2	·04334	·37125	83	16	71 „ 109	37	26·7	·40024	·03743	54	54
27 „ 153	15	38·0	·04684	·35605	82	58	72 „ 108	38	53·9	·41424	·03489	53	20
28 „ 152	15	46·1	·05050	·34149	82	40	73 „ 107	40	27·4	·42848	·03250	51	39
29 „ 151	15	54·7	·05431	·32753	82	22	74 „ 106	42	07·9	·44294	·03025	49	51
30 „ 150	16	03·7	0·05827	0·31413	82	03	75 „ 105	43	55·8	0·45754	0·02815	47	55
31 „ 149	16	13·1	·06239	·30126	81	44	76 „ 104	45	51·8	·47225	·02619	45	52
32 „ 148	16	23·0	·06667	·28889	81	24	77 „ 103	47	56·5	·48700	·02437	43	40
33 „ 147	16	33·4	·07111	·27699	81	04	78 „ 102	50	10·5	·50169	·02269	41	18
34 „ 146	16	44·3	·07572	·26553	80	44	79 „ 101	52	34·4	·51622	·02115	38	47
35 „ 145	16	55·7	0·08049	0·25451	80	23	80 „ 100	55	08·6	0·53045	0·01975	36	05
36 „ 144	17	07·7	·08543	·24388	80	02	81 „ 99	57	53·7	·54425	·01848	33	13
37 „ 143	17	20·3	·09055	·23363	79	40	82 „ 98	60	49·8	·55743	·01734	30	09
38 „ 142	17	33·4	·09584	·22375	79	18	83 „ 97	63	57·1	·56980	·01635	26	55
39 „ 141	17	47·3	·10132	·21422	78	55	84 „ 96	67	15·3	·58115	·01548	23	29
40 „ 140	18	01·7	0·10698	0·20503	78	32	85 „ 95	70	43·9	0·59128	0·01475	19	53
41 „ 139	18	16·9	·11282	·19615	78	07	86 „ 94	74	22·2	·59997	·01416	16	07
42 „ 138	18	32·8	·11886	·18759	77	43	87 „ 93	78	08·7	·60696	·01369	12	13
43 „ 137	18	49·5	·12509	·17931	77	17	88 „ 92	82	01·9	·61211	·01336	8	13
44 „ 136	19	07·0	·13153	·17132	76	51	89 „ 91	85	59·8	·61526	·01316	4	08
45 „ 135	19	25·4	0·13816	0·16361	76	24	90	90	00·0	0·61633	0·01310	0	00
h	180°-K		A	N	a₁		h	180°-K		A	N	a₁	

Directions with regard to (K ⁓ d) :—

If l and d have **Same Name,** use (K ⁓ d), i.e. (K — d) or (d — K).

If l and d have **Different Names,** use (K + d).

To name Intercept :—

If H_0 is greater than H_c, intercept is " **Towards.** "

If H_0 is less than H_c, intercept is " **Away.** "

Log. cosec. H_c = A + log. sec. (K ⁓ d).

N.B. Whenever h is greater than 90°, subtract the quantity in the K column from 180°

ALT.—AZ. TABLE LAT. 15°

h	K		A	N	a₁		h	K		A	N	a₁	
°	°	′			°	′		°	′			°	′
0 or 180	15	00·0	0·00000	—	90	00	45 or 135	20	45·2	0·13643	0·16557	75	29
1 „ 179	15	00·1	·00006	1·77320	89	44	46 „ 134	21	05·6	·14317	·15812	75	00
2 „ 178	15	00·5	·00025	1·47224	89	29	47 „ 133	21	27·0	·15010	·15093	74	29
3 „ 177	15	01·2	·00056	1·29626	89	13	48 „ 132	21	49·4	·15725	·14398	73	58
4 „ 176	15	02·1	·00099	1·17147	88	58	49 „ 131	22	13·0	·16461	·13728	73	25
5 „ 175	15	03·3	0·00155	1·07476	88	42	50 „ 130	22	37·7	0·17220	0·13080	72	51
6 „ 174	15	04·7	·00223	0·99582	88	27	51 „ 129	23	03·8	·18001	·12455	72	17
7 „ 173	15	06·5	·00303	·92916	88	11	52 „ 128	23	31·2	·18805	·11852	71	40
8 „ 172	15	08·4	·00396	·87151	87	55	53 „ 127	24	00·0	·19632	·11271	71	03
9 „ 171	15	10·7	·00502	·82072	87	39	54 „ 126	24	30·4	·20484	·10710	70	24
10 „ 170	15	13·2	0·00620	0·77539	87	23	55 „ 125	25	02·4	0·21360	0·10169	69	43
11 „ 169	15	16·1	·00750	·73446	87	07	56 „ 124	25	36·1	·22261	·09648	69	00
12 „ 168	15	19·2	·00894	·69733	86	51	57 „ 123	26	11·8	·23188	·09147	68	16
13 „ 167	15	22·6	·01050	·66297	86	35	58 „ 122	26	49·4	·24141	·08664	67	30
14 „ 166	15	26·3	·01219	·63138	86	18	59 „ 121	27	29·1	·25120	·08199	66	42
15 „ 165	15	30·2	0·01401	0·60206	86	02	60 „ 120	28	11·2	0·26126	0·07753	65	51
16 „ 164	15	34·5	·01597	·57472	85	45	61 „ 119	28	55·7	·27160	·07324	64	58
17 „ 163	15	39·1	·01805	·54912	85	29	62 „ 118	29	42·9	·28222	·06912	64	03
18 „ 162	15	44·1	·02026	·52507	85	12	63 „ 117	30	33·0	·29311	·06518	63	04
19 „ 161	15	49·3	·02261	·50241	84	54	64 „ 116	31	26·1	·30428	·06140	62	03
20 „ 160	15	54·9	0·02510	0·48100	84	37	65 „ 115	32	22·5	0·31573	0·05778	60	58
21 „ 159	16	00·8	·02772	·46073	84	20	66 „ 114	33	22·6	·32747	·05433	59	50
22 „ 158	16	07·1	·03047	·44148	84	02	67 „ 113	34	26·5	·33948	·05103	58	38
23 „ 157	16	13·8	·03337	·42318	83	44	68 „ 112	35	34·5	·35176	·04789	57	21
24 „ 156	16	20·8	·03641	·40574	83	26	69 „ 111	36	47·1	·36430	·04490	56	01
25 „ 155	16	28·2	0·03958	0·38911	83	07	70 „ 110	38	04·6	0·37708	0·04207	54	35
26 „ 154	16	36·0	·04291	·37321	82	48	71 „ 109	39	27·3	·39010	·03939	53	04
27 „ 153	16	44·2	·04638	·35801	82	29	72 „ 108	40	55·7	·40332	·03685	51	28
28 „ 152	16	52·9	·04999	·34345	82	10	73 „ 107	42	30·3	·41672	·03446	49	45
29 „ 151	17	02·0	·05375	·32949	81	50	74 „ 106	44	11·4	·43026	·03221	47	56
30 „ 150	17	11·5	0·05767	0·31609	81	30	75 „ 105	45	59·6	0·44389	0·03011	46	00
31 „ 149	17	21·5	·06174	·30322	81	10	76 „ 104	47	55·3	·45754	·02815	43	56
32 „ 148	17	32·1	·06597	·29085	80	49	77 „ 103	49	59·1	·47116	·02633	41	44
33 „ 147	17	43·1	·07036	·27895	80	28	78 „ 102	52	11·4	·48466	·02465	39	24
34 „ 146	17	54·7	·07491	·26749	80	06	79 „ 101	54	32·7	·49793	·02311	36	54
35 „ 145	18	06·8	0·07961	0·25647	79	44	80 „ 100	57	03·3	0·51086	0·02171	34	16
36 „ 144	18	19·5	·08449	·24584	79	21	81 „ 99	59	43·4	·52331	·02044	31	28
37 „ 143	18	32·8	·08954	·23559	78	58	82 „ 98	62	33·2	·53514	·01930	28	30
38 „ 142	18	46·8	·09477	·22571	78	34	83 „ 97	65	32·6	·54617	·01831	25	23
39 „ 141	19	01·4	·10016	·21618	78	10	84 „ 96	68	41·3	·55624	·01744	22	06
40 „ 140	19	16·7	0·10574	0·20699	77	45	85 „ 95	71	58·9	0·56516	0·01671	18	41
41 „ 139	19	32·8	·11150	·19811	77	19	86 „ 94	75	24·5	·57276	·01612	15	07
42 „ 138	19	49·6	·11744	·18955	76	53	87 „ 93	78	56·9	·57887	·01565	11	27
43 „ 137	20	07·3	·12357	·18127	76	26	88 „ 92	82	34·8	·58335	·01532	7	41
44 „ 136	20	25·8	·12991	·17329	75	58	89 „ 91	86	16·4	·58608	·01512	3	51
45 „ 135	20	45·2	0·13643	0·16557	75	29	90	90	00·0	0·58700	0·01506	0	00
h	180°-K		A	N	a₁		h	180°-K		A	N	a₁	

Directions with regard to (a₂ $\overset{+}{\sim}$ a₁) :—

Whenever h is greater than 90°, use (a₂ — a₁).

If h is less than 90°, use :—

  (a₁ + a₂) when (K ± d) has been used,
  (a₁ — a₂) when (d — K) has been used.

Always **name azimuth** from elevated pole.

Note that $\begin{cases} \text{h = L.H.A. of body \textbf{West} of the meridian.} \\ \text{h = (360° — L.H.A.) of body \textbf{East} of the meridian.} \end{cases}$

Log. tan. a₂ = N + log. tan. (K $\overset{+}{\sim}$ d).

LAT. 16° ALT.—AZ. TABLE

h	K	A	N	a₁	h	K	A	N	a₁
°	° ′			° ′	°	° ′			° ′
0 or 180	16 00·0	0·00000	—	90 00	**45 or 135**	22 04·4	0·13461	0·16767	74 35
1 „ 179	16 00·1	·00006	1·77530	89 44	**46 „ 134**	22 25·8	·14122	·16022	74 04
2 „ 178	16 00·6	·00024	1·47434	89 27	**47 „ 133**	22 48·3	·14803	·15303	73 32
3 „ 177	16 01·2	·00055	1·29836	89 10	**48 „ 132**	23 11·8	·15504	·14609	72 59
4 „ 176	16 02·2	·00098	1·17357	88 54	**49 „ 131**	23 36·5	·16225	·13938	72 24
5 „ 175	16 03·5	0·00153	1·07686	88 37	**50 „ 130**	24 02·5	0·16968	0·13290	71 49
6 „ 174	16 05·0	·00220	0·99792	88 20	**51 „ 129**	24 29·8	·17733	·12666	71 12
7 „ 173	16 06·8	·00300	·93126	88 04	**52 „ 128**	24 58·4	·18519	·12064	70 34
8 „ 172	16 08·9	·00392	·87367	87 47	**53 „ 127**	25 28·6	·19327	·11481	69 54
9 „ 171	16 11·3	·00497	·82283	87 30	**54 „ 126**	26 00·3	·20158	·10920	69 13
10 „ 170	16 14·0	0·00614	0·77749	87 13	**55 „ 125**	26 33·7	0·21012	0·10379	68 31
11 „ 169	16 17·0	·00743	·73656	86 56	**56 „ 124**	27 08·9	·21890	·09859	67 46
12 „ 168	16 20·3	·00885	·69928	86 39	**57 „ 123**	27 46·0	·22792	·09357	67 00
13 „ 167	16 23·9	·01040	·66508	86 22	**58 „ 122**	28 25·1	·23718	·08874	66 12
14 „ 166	16 27·8	·01207	·63328	86 04	**59 „ 121**	29 06·4	·24669	·08409	65 21
15 „ 165	16 32·0	0·01387	0·60416	85 47	**60 „ 120**	29 50·0	0·25644	0·07963	64 29
16 „ 164	16 36·6	·01581	·57682	85 29	**61 „ 119**	30 36·2	·26645	·07534	63 34
17 „ 163	16 41·5	·01787	·55122	85 11	**62 „ 118**	31 25·0	·27670	·07122	62 36
18 „ 162	16 46·7	·02006	·52718	84 53	**63 „ 117**	32 16·6	·28721	·06728	61 35
19 „ 161	16 52·3	·02238	·50452	84 35	**64 „ 116**	33 11·4	·29797	·06350	60 32
20 „ 160	16 58·2	0·02484	0·48301	84 16	**65 „ 115**	34 09·4	0·30898	0·05989	59 25
21 „ 159	17 04·4	·02743	·46283	83 58	**66 „ 114**	35 11·0	·32023	·05643	58 44
22 „ 158	17 11·1	·03016	·44358	83 39	**67 „ 113**	36 16·4	·33172	·05313	57 00
23 „ 157	17 18·1	·03302	·42528	83 20	**68 „ 112**	37 26·0	·34344	·04999	55 42
24 „ 156	17 25·6	·03602	·40785	83 00	**69 „ 111**	38 39·9	·35537	·04701	54 19
25 „ 155	17 33·4	0·03917	0·39121	82 41	**70 „ 110**	39 58·6	0·36752	0·04417	52 52
26 „ 154	17 41·7	·04245	·37532	82 21	**71 „ 109**	41 22·3	·37983	·04149	51 19
27 „ 153	17 50·4	·04588	·36011	82 00	**72 „ 108**	42 51·5	·39230	·03895	49 41
28 „ 152	17 59·5	·04945	·34555	81 40	**73 „ 107**	44 26·6	·40489	·03656	47 58
29 „ 151	18 09·1	·05317	·33159	81 19	**74 „ 106**	46 07·9	·41756	·03432	46 08
30 „ 150	18 19·2	0·05704	0·31819	80 57	**75 „ 105**	47 55·8	0·43026	0·03221	44 11
31 „ 149	18 29·8	·06106	·30532	80 36	**76 „ 104**	49 50·8	·44294	·03025	42 08
32 „ 148	18 40·9	·06523	·29295	80 14	**77 „ 103**	51 53·2	·45552	·02843	39 57
33 „ 147	18 52·5	·06956	·28105	79 51	**78 „ 102**	54 03·3	·46792	·02675	37 38
34 „ 146	19 04·8	·07405	·26960	79 28	**79 „ 101**	56 21·5	·48006	·02521	35 11
35 „ 145	19 17·6	0·07869	0·25857	79 05	**80 „ 100**	58 48·1	0·49182	0·02381	32 36
36 „ 144	19 31·0	·08350	·24794	78 41	**81 „ 99**	61 23·1	·50309	·02254	29 53
37 „ 143	19 45·0	·08848	·23770	78 16	**82 „ 98**	64 06·6	·51373	·02141	27 01
38 „ 142	19 59·7	·09362	·22782	77 51	**83 „ 97**	66 58·4	·52360	·02041	24 01
39 „ 141	20 15·2	·09894	·21828	77 25	**84 „ 96**	69 58·3	·53257	·01954	20 52
40 „ 140	20 31·3	0·10443	0·20909	76 59	**85 „ 95**	73 05·6	0·54047	0·01881	17 37
41 „ 139	20 48·2	·11010	·20022	76 32	**86 „ 94**	76 19·6	·54718	·01822	14 14
42 „ 138	21 06·0	·11595	·19165	76 04	**87 „ 93**	79 39·4	·55255	·01775	10 46
43 „ 137	21 24·5	·12198	·18338	75 35	**88 „ 92**	83 03·6	·55647	·01742	7 13
44 „ 136	21 44·0	·12820	·17539	75 06	**89 „ 91**	86 31·0	·55886	·01722	3 37
45 „ 135	22 04·4	0·13461	0·16767	74 35	**90**	90 00·0	0·55966	0·01716	0 00
h	180°-K	A	N	a₁	h	180°-K	A	N	a₁

Directions with regard to (K $\overset{+}{\sim}$ d) :—

If l and d have **Same Name,** use (K $\frown$ d), i.e. (K — d) or (d — K).
If l and d have **Different Names,** use (K + d).

To name Intercept :—

If H_o is greater than H_c, intercept is " **Towards.**"
If H_o is less than H_c, intercept is " **Away.**"

Log. cosec. H_c = A + log. sec. (K $\overset{+}{\sim}$ d).

N.B. Whenever h is greater than 90°, subtract the quantity in the K column from 180°.

ALT.—AZ. TABLE LAT. 17°

h	K		A	N	a₁		h	K		A	N	a₁	
°	°	′			°	′		°	′			°	′
0 or 180	17	00·0	0·00000	—	90	00	45 or 135	23	22·9	0·13270	0·16992	73	42
1 „ 179	17	00·1	·00006	1·77755	89	42	46 „ 134	23	45·3	·13919	·16247	73	09
2 „ 178	17	00·6	·00024	1·47659	89	25	47 „ 133	24	08·8	·14586	·15528	72	36
3 „ 177	17	01·3	·00055	1·30060	89	07	48 „ 132	24	33·4	·15272	·14833	72	01
4 „ 176	17	02·3	·00097	1·17590	88	50	49 „ 131	24	59·2	·15979	·14162	71	25
5 „ 175	17	03·7	0·00152	1·07911	88	32	50 „ 130	25	26·2	0·16705	0·13515	70	47
6 „ 174	17	05·3	·00218	1·00017	88	14	51 „ 129	25	54·7	·17452	·12890	70	09
7 „ 173	17	07·2	·00297	0·93351	87	57	52 „ 128	26	24·5	·18220	·12287	69	29
8 „ 172	17	09·4	·00388	·87585	87	39	53 „ 127	26	55·9	·19009	·11706	68	48
9 „ 171	17	12·0	·00492	·82507	87	21	54 „ 126	27	28·8	·19819	·11145	68	05
10 „ 170	17	14·8	0·00607	0·77973	87	03	55 „ 125	28	03·5	0·20651	0·10604	67	20
11 „ 169	17	18·0	·00735	·73881	86	45	56 „ 124	28	40·0	·21505	·10083	66	34
12 „ 168	17	21·4	·00876	·70153	86	27	57 „ 123	29	18·4	·22381	·09581	65	46
13 „ 167	17	25·2	·01029	·66732	86	08	58 „ 122	29	58·9	·23280	·09098	64	56
14 „ 166	17	29·3	·01195	·63573	85	50	59 „ 121	30	41·6	·24202	·08634	64	03
15 „ 165	17	33·8	0·01373	0·60641	85	31	60 „ 120	31	26·7	0·25146	0·08187	63	09
16 „ 164	17	38·6	·01564	·57907	85	12	61 „ 119	32	14·2	·26113	·07759	62	11
17 „ 163	17	43·7	·01768	·55347	84	54	62 „ 118	33	04·4	·27103	·07347	61	12
18 „ 162	17	49·2	·01984	·52942	84	34	63 „ 117	33	57·4	·28115	·06952	60	09
19 „ 161	17	55·1	·02214	·50676	84	15	64 „ 116	34	53·6	·29150	·06574	59	04
20 „ 160	18	01·3	0·02457	0·48535	83	56	65 „ 115	35	53·0	0·30206	0·06213	57	55
21 „ 159	18	08·0	·02713	·46508	83	36	66 „ 114	36	55·9	·31283	·05867	56	42
22 „ 158	18	15·0	·02982	·44583	83	16	67 „ 113	38	02·5	·32381	·05538	55	26
23 „ 157	18	22·4	·03265	·42752	82	56	68 „ 112	39	13·2	·33498	·05224	54	07
24 „ 156	18	30·2	·03562	·41009	82	35	69 „ 111	40	28·1	·34632	·04925	52	42
25 „ 155	18	38·5	0·03873	0·39346	82	14	70 „ 110	41	47·6	0·35783	0·04642	51	14
26 „ 154	18	47·2	·04197	·37756	81	53	71 „ 109	43	12·0	·36947	·04373	49	40
27 „ 153	18	56·3	·04535	·36236	81	32	72 „ 108	44	41·6	·38122	·04120	48	01
28 „ 152	19	05·9	·04888	·34780	81	10	73 „ 107	46	16·8	·39303	·03881	46	17
29 „ 151	19	16·1	·05255	·33383	80	48	74 „ 106	47	57·8	·40489	·03656	44	27
30 „ 150	19	26·7	0·05637	0·32043	80	25	75 „ 105	49	45·0	0·41672	0·03446	42	30
31 „ 149	19	37·8	·06033	·30757	80	02	76 „ 104	51	38·7	·42849	·03250	40	27
32 „ 148	19	49·5	·06445	·29519	79	39	77 „ 103	53	39·3	·44011	·03068	38	18
33 „ 147	20	01·7	·06872	·28330	79	15	78 „ 102	55	46·9	·45152	·02900	36	01
34 „ 146	20	14·6	·07314	·27184	78	51	79 „ 101	58	01·9	·46263	·02746	33	37
35 „ 145	20	28·0	0·07772	0·26081	78	26	80 „ 100	60	24·3	0·47335	0·02605	31	06
36 „ 144	20	42·1	·08246	·25019	78	00	81 „ 99	62	54·1	·48357	·02478	28	27
37 „ 143	20	56·9	·08736	·23994	77	35	82 „ 98	65	31·5	·49317	·02365	25	40
38 „ 142	21	12·3	·09242	·23006	77	08	83 „ 97	68	16·0	·50204	·02265	22	47
39 „ 141	21	28·5	·09765	·22054	76	41	84 „ 96	71	07·5	·51006	·02179	19	46
40 „ 140	21	45·4	0·10305	0·21134	76	13	85 „ 95	74	05·3	0·51710	0·02106	16	40
41 „ 139	22	03·2	·10863	·20246	75	44	86 „ 94	77	08·8	·52304	·02046	13	27
42 „ 138	22	21·7	·11438	·19389	75	15	87 „ 93	80	17·2	·52779	·02000	10	10
43 „ 137	22	41·2	·12030	·18562	74	45	88 „ 92	83	29·3	·53125	·01967	6	49
44 „ 136	23	01·6	·12641	·17763	74	14	89 „ 91	86	44·0	·53336	·01947	3	25
45 „ 135	23	22·9	0·13270	0·16992	73	42	90	90	00·0	0·53407	0·01940	0	00
h	180°-K		A	N	a₁		h	180°-K		A	N	a₁	

Directions with regard to ($a_2 \overset{+}{\sim} a_1$) :—

Whenever h is greater than 90°, use ($a_2 - a_1$).

If h is less than 90°, use :—

($a_1 + a_2$) when (K ± d) has been used,
($a_1 - a_2$) when (d — K) has been used.

Always **name azimuth** from elevated pole.

Note that $\begin{cases} \text{h = L.H.A. of body \textbf{West} of the meridian.} \\ \text{h = (360° — L.H.A.) of body \textbf{East} of the meridian.} \end{cases}$

Log. tan. a_2 = N + log. tan. (K $\overset{+}{\sim}$ d).

318

LAT. 18° — ALT.—AZ. TABLE

h	K	A	N	a₁	h	K	A	N	a₁
°	° ′			° ′	°	° ′			° ′
0 or 180	18 00·0	0·00000	—	90 00	**45 or 135**	24 40·7	0·13071	0·17231	72 50
1 „ 179	18 00·2	·00006	1·77994	89 41	**46 „ 134**	25 04·0	·13706	·16486	72 15
2 „ 178	18 00·6	·00024	1·47898	89 23	**47 „ 133**	25 28·5	·14359	·15767	71 40
3 „ 177	18 01·4	·00054	1·30299	89 04	**48 „ 132**	25 54·0	·15031	·15072	71 03
4 „ 176	18 02·5	·00096	1·17821	88 46	**49 „ 131**	26 20·8	·15722	·14401	70 26
5 „ 175	18 03·9	0·00150	1·08150	88 27	**50 „ 130**	26 49·0	0·16432	0·13754	69 47
6 „ 174	18 05·6	·00216	1·00256	88 08	**51 „ 129**	27 18·4	·17161	·13129	69 07
7 „ 173	18 07·6	·00294	0·93590	87 50	**52 „ 128**	27 49·4	·17910	·12526	68 25
8 „ 172	18 09·9	·00384	·87824	87 31	**53 „ 127**	28 21·9	·18679	·11945	67 42
9 „ 171	18 12·6	·00486	·82746	87 12	**54 „ 126**	28 56·0	·19468	·11384	66 58
10 „ 170	18 15·6	0·00601	0·78212	86 53	**55 „ 125**	29 31·8	0·20277	0·10843	66 11
11 „ 169	18 18·9	·00727	·74118	86 34	**56 „ 124**	30 09·5	·21107	·10322	65 23
12 „ 168	18 22·5	·00866	·70392	86 15	**57 „ 123**	30 49·2	·21957	·09820	64 33
13 „ 167	18 26·5	·01017	·66971	85 55	**58 „ 122**	31 30·9	·22828	·09337	63 41
14 „ 166	18 30·8	·01181	·63812	85 36	**59 „ 121**	32 14·8	·23720	·08873	62 47
15 „ 165	18 35·5	0·01357	0·60880	85 16	**60 „ 120**	33 01·0	0·24633	0·08427	61 51
16 „ 164	18 40·6	·01546	·58146	84 56	**61 „ 119**	33 49·8	·25566	·07998	60 52
17 „ 163	18 46·0	·01747	·55586	84 36	**62 „ 118**	34 41·2	·26520	·07586	59 50
18 „ 162	18 51·7	·01962	·53181	84 16	**63 „ 117**	35 35·5	·27494	·07191	58 46
19 „ 161	18 57·9	·02189	·50915	83 56	**64 „ 116**	36 32·7	·28487	·06813	57 39
20 „ 160	19 04·4	0·02428	0·48774	83 35	**65 „ 115**	37 33·2	0·29500	0·06542	56 28
21 „ 159	19 11·4	·02681	·46747	83 14	**66 „ 114**	38 37·2	·30531	·06106	55 14
22 „ 158	19 18·7	·02948	·44822	82 53	**67 „ 113**	39 44·8	·31578	·05778	53 57
23 „ 157	19 26·5	·03227	·42992	82 32	**68 „ 112**	40 56·2	·32641	·05463	52 35
24 „ 156	19 34·7	·03520	·41248	82 10	**69 „ 111**	42 11·8	·33719	·05164	51 10
25 „ 155	19 43·4	0·03826	0·39585	81 48	**70 „ 110**	43 31·9	0·34808	0·04881	49 40
26 „ 154	19 52·5	·04146	·37995	81 26	**71 „ 109**	44 56·6	·35907	·04612	48 06
27 „ 153	20 02·1	·04480	·36475	81 03	**72 „ 108**	46 26·2	·37012	·04359	46 26
28 „ 152	20 12·2	·04828	·35019	80 40	**73 „ 107**	48 01·1	·38121	·04120	44 42
29 „ 151	20 22·8	·05190	·33622	80 17	**74 „ 106**	49 41·5	·39230	·03895	42 52
30 „ 150	20 33·9	0·05566	0·32282	79 53	**75 „ 105**	51 27·6	0·40332	0·03683	40 54
31 „ 149	20 45·6	·05957	·30995	79 29	**76 „ 104**	53 19·8	·41424	·03489	38 54
32 „ 148	20 57·8	·06363	·29758	79 04	**77 „ 103**	55 18·2	·42499	·03307	36 46
33 „ 147	21 10·6	·06784	·28569	78 39	**73 „ 102**	57 23·1	·43549	·03139	34 31
34 „ 146	21 24·1	·07219	·27423	78 14	**79 „ 101**	59 34·6	·44568	·02985	32 10
35 „ 145	21 38·2	0·07670	0·26320	77 47	**80 „ 100**	61 52·7	0·45546	0·02844	29 43
36 „ 144	21 52·9	·08136	·25258	77 21	**81 „ 99**	64 17·5	·46474	·02717	27 08
37 „ 143	22 08·3	·08619	·24233	76 54	**82 „ 98**	66 48·8	·47344	·02604	24 27
38 „ 142	22 24·5	·09117	·23245	76 26	**83 „ 97**	69 26·4	·48144	·02504	21 40
39 „ 141	22 41·4	·09631	·22292	75 57	**84 „ 96**	72 10·0	·48863	·02418	18 47
40 „ 140	22 59·1	0·10162	0·21373	75 28	**85 „ 95**	74 59·1	0·49493	0·02345	15 48
41 „ 139	23 17·6	·10709	·20485	74 58	**86 „ 94**	77 53·0	·50023	·02285	12 45
42 „ 138	23 37·0	·11273	·19628	74 27	**87 „ 93**	80 51·0	·50446	·02239	9 38
43 „ 137	23 57·3	·11855	·18801	73 56	**88 „ 92**	83 52·2	·50753	·02206	6 27
44 „ 136	24 18·5	·12454	·18002	73 23	**89 „ 91**	86 55·5	·50939	·02186	3 14
45 „ 135	24 40·7	0·13071	0·17231	72 50	**90**	90 00·0	0·51002	0·02179	0 00
h	180°-K	A	N	a₁	h	180°-K	A	N	a₁

Directions with regard to (K ⁓ d) :—

If l and d have **Same Name,** use (K ⁓ d), i.e. (K — d) or (d — K).
If l and d have **Different Names,** use (K + d).

To name Intercept :—

If H₀ is greater than H_c, intercept is " **Towards.** "
If H₀ is less than H_c, intercept is " **Away.** "

Log. cosec. H_c = A + log. sec. (K ⁓ d).

N.B.   Whenever h is greater than 90°, subtract the quantity in the K column from 180°.

ALT.—AZ. TABLE — LAT. 19°

h	K	A	N	a₁	h	K	A	N	a₁
°	° ′			° ′	°	° ′			° ′
0 or 180	19 00·0	0·00000	—	90 00	45 or 135	25 57·8	0·12864	0·17484	71 58
1 „ 179	19 00·2	·00006	1·78248	89 40	46 „ 134	26 22·0	·13485	·16740	71 22
2 „ 178	19 00·6	·00024	1·48151	89 21	47 „ 133	26 47·3	·14124	·16020	70 45
3 „ 177	19 01·5	·00053	1·30553	89 01	48 „ 132	27 13·8	·14781	·15326	70 07
4 „ 176	19 02·6	·00095	1·18075	88 42	49 „ 131	27 41·5	·15456	·14655	69 28
5 „ 175	19 04·0	0·00148	1·08403	88 22	50 „ 130	28 10·6	0·16148	0·14008	68 48
6 „ 174	19 05·8	·00213	1·00510	88 02	51 „ 129	28 41·1	·16859	·13383	68 06
7 „ 173	19 07·9	·00290	0·93844	87 43	52 „ 128	29 13·0	·17589	·12780	67 23
8 „ 172	19 10·4	·00379	·88077	87 23	53 „ 127	29 46·6	·18338	·12198	66 38
9 „ 171	19 13·2	·00480	·83000	87 03	54 „ 126	30 21·7	·19105	·11637	65 52
10 „ 170	19 16·3	0·00593	0·78466	86 43	55 „ 125	30 58·6	0·19891	0·11097	65 04
11 „ 169	19 19·8	·00719	·74373	86 23	56 „ 124	31 37·4	·20696	·10576	64 14
12 „ 168	19 23·6	·00856	·70645	86 02	57 „ 123	32 18·1	·21520	·10074	63 22
13 „ 167	19 27·8	·01005	·67224	85 42	58 „ 122	33 00·9	·22364	·09591	62 29
14 „ 166	19 32·3	·01167	·64066	85 21	59 „ 121	33 45·9	·23226	·09126	61 33
15 „ 165	19 37·2	0·01341	0·61133	85 01	60 „ 120	34 33·2	0·24107	0·08680	60 35
16 „ 164	19 42·5	·01527	·58399	84 40	61 „ 119	35 23·0	·25007	·08251	59 34
17 „ 163	19 48·1	·01726	·55840	84 19	62 „ 118	36 15·5	·25925	·07840	58 31
18 „ 162	19 54·2	·01938	·53435	83 58	63 „ 117	37 10·7	·26861	·07445	57 25
19 „ 161	20 00·6	·02162	·51170	83 36	64 „ 116	38 08·9	·27814	·07067	56 17
20 „ 160	20 07·4	0·02399	0·49028	83 15	65 „ 115	39 10·3	0·28783	0·06705	55 05
21 „ 159	20 14·7	·02648	·47000	82 53	66 „ 114	40 15·0	·29767	·06360	53 49
22 „ 158	20 22·4	·02911	·45076	82 30	67 „ 113	41 23·3	·30766	·06030	52 31
23 „ 157	20 30·5	·03187	·43245	82 08	68 „ 112	42 35·3	·31777	·05716	51 08
24 „ 156	20 39·1	·03475	·41502	81 45	69 „ 111	43 51·3	·32799	·05418	49 42
25 „ 155	20 48·2	0·03778	0·39838	81 22	70 „ 110	45 11·6	0·33830	0·05134	48 11
26 „ 154	20 57·7	·04093	·38249	80 59	71 „ 109	46 36·2	·34867	·04866	46 36
27 „ 153	21 07·7	·04422	·36728	80 35	72 „ 108	48 05·6	·35907	·04612	44 57
28 „ 152	21 18·3	·04765	·35272	80 11	73 „ 107	49 39·9	·36947	·04373	43 12
29 „ 151	21 29·3	·05122	·33876	79 46	74 „ 106	51 19·3	·37983	·04149	41 22
30 „ 150	21 41·0	0·05493	0·32536	79 21	75 „ 105	53 04·1	0·39010	0·03939	39 27
31 „ 149	21 53·1	·05878	·31249	78 56	76 „ 104	54 54·5	·40024	·03743	37 27
32 „ 148	22 05·9	·06277	·30012	78 30	77 „ 103	56 50·6	·41018	·03561	35 20
33 „ 147	22 19·3	·06691	·28822	78 04	78 „ 102	58 52·5	·41986	·03393	33 08
34 „ 146	22 33·3	·07120	·27677	77 37	79 „ 101	61 00·4	·42921	·03238	30 50
35 „ 145	22 48·0	0·07563	0·26574	77 09	80 „ 100	63 14·3	0·43815	0·03098	28 26
36 „ 144	23 03·3	·08022	·25511	76 42	81 „ 99	65 34·0	·44661	·02971	25 57
37 „ 143	23 19·4	·08496	·24487	76 13	82 „ 98	67 59·5	·45450	·02858	23 21
38 „ 142	23 36·2	·08986	·23499	75 44	83 „ 97	70 30·6	·46173	·02758	20 40
39 „ 141	23 53·8	·09491	·22546	75 14	84 „ 96	73 06·8	·46821	·02672	17 53
40 „ 140	24 12·2	0·10012	0·21626	74 43	85 „ 95	75 47·7	0·47387	0·02599	15 03
41 „ 139	24 31·5	·10549	·20739	74 12	86 „ 94	78 32·9	·47862	·02539	12 07
42 „ 138	24 51·6	·11103	·19882	73 40	87 „ 93	81 21·4	·48240	·02493	9 09
43 „ 137	25 12·7	·11672	·19055	73 07	88 „ 92	84 12·8	·48514	·02460	6 07
44 „ 136	25 34·7	·12260	·18256	72 33	89 „ 91	87 05·9	·48680	·02440	3 04
45 „ 135	25 57·8	0·12864	0·17484	71 58	90	90 00·0	0·48736	0·02433	0 00
h	180°-K	A	N	a₁	h	180°-K	A	N	a₁

Directions with regard to (a₂ $\overset{+}{\sim}$ a₁) :—

Whenever h is greater than 90°, use (a₂ — a₁).

If h is less than 90°, use :—

(a₁ + a₂) when (K ± d) has been used,

(a₁ — a₂) when (d — K) has been used.

Always **name azimuth** from elevated pole.

Note that $\begin{cases} h = \text{L.H.A. of body } \textbf{West} \text{ of the meridian.} \\ h = (360° - \text{L.H.A.}) \text{ of body } \textbf{East} \text{ of the meridian.} \end{cases}$

Log. tan. a₂ = N + log. tan. (K $\overset{+}{\sim}$ d).

LAT. 20°   ALT.—AZ. TABLE

h	K		A	N	a₁		h	K		A	N	a₁	
°	°	′			°	′	°	°	′			°	′
0 or 180	20	00·0	0·00000	—	90	00	**45 or 135**	27	14·2	0·12649	0·17753	71	07
1 ,, 179	20	00·2	·00006	1·78516	89	39	**46 ,, 134**	27	39·2	·13257	·17008	70	30
2 ,, 178	20	00·7	·00023	1·48420	89	19	**47 ,, 133**	28	05·3	·13881	·16289	69	52
3 ,, 177	20	01·5	·00053	1·30821	88	58	**48 ,, 132**	28	32·6	·14522	·15594	69	12
4 ,, 176	20	02·7	·00094	1·18345	88	38	**49 ,, 131**	29	01·2	·15181	·14923	68	31
5 ,, 175	20	04·2	0·00146	1·08672	88	17	**50 ,, 130**	29	31·2	0·15856	0·14276	67	49
6 ,, 174	20	06·1	·00211	1·00778	87	56	**51 ,, 129**	30	02·6	·16549	·13651	67	06
7 ,, 173	20	08·3	·00287	0·94112	87	36	**52 ,, 128**	30	35·5	·17259	·13048	66	21
8 ,, 172	20	10·8	·00375	·88346	87	15	**53 ,, 127**	31	09·9	·17987	·12467	65	35
9 ,, 171	20	13·8	·00474	·83268	86	54	**54 ,, 126**	31	46·0	·18732	·11906	64	47
10 ,, 170	20	17·0	0·00586	0·78734	86	33	**55 ,, 125**	32	23·9	0·19495	0·11365	63	58
11 ,, 169	20	20·6	·00710	·74642	86	12	**56 ,, 124**	33	03·6	·20275	·10844	63	07
12 ,, 168	20	24·6	·00845	·70914	85	51	**57 ,, 123**	33	45·2	·21073	·10342	62	14
13 ,, 167	20	29·0	·00993	·67493	85	29	**58 ,, 122**	34	29·0	·21889	·09859	61	18
14 ,, 166	20	33·7	·01152	·64334	85	08	**59 ,, 121**	35	14·9	·22722	·09395	60	21
15 ,, 165	20	38·8	0·01324	0·61402	84	46	**60 ,, 120**	36	03·1	0·23571	0·08948	59	21
16 ,, 164	20	44·3	·01508	·58668	84	24	**61 ,, 119**	36	53·8	·24438	·08520	58	19
17 ,, 163	20	50·2	·01704	·56130	84	02	**62 ,, 118**	37	47·1	·25320	·08108	57	15
18 ,, 162	20	56·5	·01913	·53703	83	40	**63 ,, 117**	38	43·2	·26218	·07713	56	08
19 ,, 161	21	03·2	·02134	·51437	83	17	**64 ,, 116**	39	42·1	·27131	·07335	54	58
20 ,, 160	21	10·4	0·02367	0·49296	82	54	**65 ,, 115**	40	44·2	0·28058	0·06974	53	44
21 ,, 159	21	17·9	·02614	·47269	82	31	**66 ,, 114**	41	49·4	·28998	·06628	52	28
22 ,, 158	21	26·0	·02873	·45344	82	08	**67 ,, 113**	42	58·2	·29948	·06299	51	08
23 ,, 157	21	34·4	·03145	·43514	81	44	**68 ,, 112**	44	10·5	·30909	·05985	49	45
24 ,, 156	21	43·4	·03429	·41770	81	20	**69 ,, 111**	45	26·7	·31878	·05686	48	18
25 ,, 155	21	52·8	0·03727	0·40107	80	56	**70 ,, 110**	46	46·8	0·32852	0·05403	46	47
26 ,, 154	22	02·7	·04038	·38517	80	32	**71 ,, 109**	48	11·3	·33830	·05134	45	12
27 ,, 153	22	13·2	·04362	·36997	80	07	**72 ,, 108**	49	40·1	·34808	·04881	43	32
28 ,, 152	22	24·2	·04700	·35541	79	42	**73 ,, 107**	51	13·5	·35783	·04642	41	48
29 ,, 151	22	35·7	·05052	·34144	79	16	**74 ,, 106**	52	51·8	·36752	·04417	39	59
30 ,, 150	22	47·8	0·05416	0·32804	78	50	**75 ,, 105**	54	35·0	0·37708	0·04207	38	05
31 ,, 149	23	00·4	·05795	·31518	78	23	**76 ,, 104**	56	23·3	·38650	·04011	36	05
32 ,, 148	23	13·7	·06188	·30280	77	56	**77 ,, 103**	58	16·9	·39570	·03829	34	01
33 ,, 147	23	27·6	·06595	·29091	77	29	**78 ,, 102**	60	15·8	·40463	·03661	31	52
34 ,, 146	23	42·2	·07017	·27945	77	00	**79 ,, 101**	62	20·1	·41322	·03507	29	37
35 ,, 145	23	57·4	0·07452	0·26842	76	32	**80 ,, 100**	64	29·7	0·42142	0·03366	27	16
36 ,, 144	24	13·4	·07903	·25780	76	03	**81 ,, 99**	66	44·5	·42914	·03239	24	51
37 ,, 143	24	30·0	·08369	·24755	75	33	**82 ,, 98**	69	04·5	·43631	·03126	22	20
38 ,, 142	24	47·5	·08849	·23767	75	02	**83 ,, 97**	71	29·3	·44288	·03026	19	45
39 ,, 141	25	05·7	·09346	·22814	74	31	**84 ,, 96**	73	58·6	·44874	·02940	17	05
40 ,, 140	25	24·8	0·09856	0·21895	73	59	**85 ,, 95**	76	32·0	0·45384	0·02867	14	21
41 ,, 139	25	44·8	·10383	·21007	73	27	**86 ,, 94**	79	09·0	·45812	·02807	11	33
42 ,, 138	26	05·7	·10925	·20150	72	53	**87 ,, 93**	81	49·0	·46150	·02761	8	43
43 ,, 137	26	27·5	·11484	·19323	72	19	**88 ,, 92**	84	31·4	·46396	·02728	5	50
44 ,, 136	26	50·3	·12058	·18524	71	43	**89 ,, 91**	87	15·3	·46545	·02708	2	55
45 ,, 135	27	14·2	0·12649	0·17753	71	07	**90**	90	00·0	0·46595	0·02701	0	00
h	180°-K		A	N	a₁		h	180°-K		A	N	a₁	

Directions with regard to (K $\stackrel{+}{\smile}$ d) :—

If l and d have **Same Name,** use (K $\smile$ d), i.e. (K — d) or (d — K).
If l and d have **Different Names,** use (K + d).

To name Intercept :—

If H_o is greater than H_c, intercept is **" Towards."**
If H_o is less than H_c, intercept is **" Away."**

Log. cosec. H_c = A + log. sec. (K $\stackrel{+}{\smile}$ d).

N.B.   Whenever h is greater than 90°, subtract the quantity in the K column from 180°.

ALT.—AZ. TABLE LAT. 21°

h	K	A	N	a₁	h	K	A	N	a₁
°	° ′			° ′	°	° ′			° ′
0 or 180	21 00·0	0·00000	—	90 00	45 or 135	28 29·8	0·12428	0·18036	70 17
1 „ 179	21 00·2	·00006	1·78799	89 38	46 „ 134	28 55·5	·13021	·17291	69 38
2 „ 178	21 00·7	·00023	1·48703	89 17	47 „ 133	29 22·4	·13631	·16572	68 59
3 „ 177	21 01·6	·00052	1·31105	88 55	48 „ 132	29 50·5	·14256	·15878	68 18
4 „ 176	21 02·8	·00092	1·18626	88 34	49 „ 131	30 19·9	·14897	·15207	67 36
5 „ 175	21 04·4	0·00144	1·08955	88 12	50 „ 130	30 50·7	0·15555	0·14559	66 52
6 „ 174	21 06·3	·00208	1·01061	87 51	51 „ 129	31 22·9	·16229	·13935	66 08
7 „ 173	21 08·6	·00283	0·94395	87 29	52 „ 128	31 56·6	·16920	·13332	65 22
8 „ 172	21 11·3	·00370	·88629	87 07	53 „ 127	32 31·9	·17626	·12750	64 34
9 „ 171	21 14·3	·00468	·83552	86 45	54 „ 126	33 08·8	·18349	·12189	63 45
10 „ 170	21 17·7	0·00578	0·79018	86 23	55 „ 125	33 47·5	0·19089	0·11648	62 54
11 „ 169	21 21·5	·00700	·74925	86 01	56 „ 124	34 28·1	·19845	·11127	62 01
12 „ 168	21 25·6	·00834	·71197	85 39	57 „ 123	35 10·6	·20616	·10626	61 06
13 „ 167	21 30·2	·00979	·67776	85 16	58 „ 122	35 55·1	·21405	·10143	60 10
14 „ 166	21 35·1	·01137	·64617	84 54	59 „ 121	36 41·9	·22208	·09678	59 11
15 „ 165	21 40·4	0·01306	0·61685	84 31	60 „ 120	37 30·9	0·23026	0·09232	58 10
16 „ 164	21 46·1	·01488	·58951	84 08	61 „ 119	38 22·3	·23859	·08803	57 07
17 „ 163	21 52·2	·01681	·56391	83 45	62 „ 118	39 16·3	·24707	·08391	56 01
18 „ 162	21 58·8	·01887	·53987	83 22	63 „ 117	40 12·9	·25568	·07997	54 53
19 „ 161	22 05·8	·02105	·51721	82 58	64 „ 116	41 12·4	·26442	·07619	53 42
20 „ 160	22 13·2	0·02335	0·49580	82 34	65 „ 115	42 14·9	0·27327	0·07257	52 27
21 „ 159	22 21·1	·02578	·47552	82 10	66 „ 114	43 20·6	·28222	·06912	51 10
22 „ 158	22 29·4	·02833	·45627	81 46	67 „ 113	44 29·5	·29217	·06582	49 50
23 „ 157	22 38·2	·03101	·43797	81 21	68 „ 112	45 42·0	·30039	·06268	48 26
24 „ 156	22 47·5	·03381	·42054	80 56	69 „ 111	46 58·0	·30957	·05970	46 58
25 „ 155	22 57·3	0·03674	0·40390	80 31	70 „ 110	48 18·0	0·31878	0·05686	45 27
26 „ 154	23 07·6	·03981	·38801	80 05	71 „ 109	49 41·9	·32799	·05418	43 51
27 „ 153	23 18·4	·04300	·37280	79 39	72 „ 108	51 09·9	·33718	·05164	42 12
28 „ 152	23 29·8	·04632	·35824	79 13	73 „ 107	52 42·3	·34632	·04925	40 28
29 „ 151	23 41·8	·04978	·34428	78 46	74 „ 106	54 19·2	·35537	·04701	38 40
30 „ 150	23 54·3	0·05337	0·33089	78 19	75 „ 105	56 00·6	0·36430	0·04490	36 47
31 „ 149	24 07·5	·05709	·31801	77 51	76 „ 104	57 46·8	·37304	·04294	34 50
32 „ 148	24 21·2	·06096	·30564	77 23	77 „ 103	59 37·7	·38157	·04112	32 47
33 „ 147	24 35·6	·06496	·29374	76 54	78 „ 102	61 33·5	·38981	·03944	30 40
34 „ 146	24 50·7	·06910	·28229	76 25	79 „ 101	63 34·2	·39773	·03790	28 29
35 „ 145	25 06·5	0·07338	0·27126	75 55	80 „ 100	65 39·6	0·40525	0·03650	26 12
36 „ 144	25 23·0	·07780	·26063	75 24	81 „ 99	67 49·7	·41231	·03523	23 51
37 „ 143	25 40·3	·08237	·25039	74 53	82 „ 98	70 04·3	·41885	·03410	21 25
38 „ 142	25 58·3	·08708	·24051	74 22	83 „ 97	72 23·2	·42482	·03310	18 55
39 „ 141	26 17·2	·09194	·23098	73 49	84 „ 96	74 46·0	·43014	·03223	16 21
40 „ 140	26 36·9	0·09695	0·22178	73 16	85 „ 95	77 12·5	0·43476	0·03150	13 43
41 „ 139	26 57·5	·10211	·21291	72 42	86 „ 94	79 42·0	·43862	·03091	11 02
42 „ 138	27 19·1	·10742	·20434	72 07	87 „ 93	82 14·2	·44167	·03044	8 19
43 „ 137	27 41·6	·11288	·19606	71 31	88 „ 92	84 48·3	·44388	·03011	5 34
44 „ 136	28 05·2	·11850	·18808	70 55	89 „ 91	87 23·8	·44522	·02991	2 47
45 „ 135	28 29·8	0·12428	0·18036	70 17	90	90 00·0	0·44567	0·02985	0 00
h	180°-K	A	N	a₁	h	180°-K	A	N	a₁

Directions with regard to (a₂ $\overset{+}{\sim}$ a₁) :—

Whenever h is greater than 90°, use (a₂ — a₁).

If h is less than 90°, use :—

(a₁ + a₂) when (K ± d) has been used,

(a₁ — a₂) when (d — K) has been used.

Always **name azimuth** from elevated pole.

Note that { h = L.H.A. of body **West** of the meridian.
{ h = (360° — L.H.A.) of body **East** of the meridian.

Log. tan. a₂ = N + log. tan. (K $\overset{+}{\sim}$ d).

LAT. 22°   ALT.—AZ. TABLE

h	K	A	N	a₁	h	K	A	N	a₁
°	° ′			° ′		° ′			° ′
0 or 180	22 00·0	0·00000	—	90 00	45 or 135	29 44·6	0·12200	0·18335	69 28
1 „ 179	22 00·2	·00006	1·79098	89 38	46 „ 134	30 11·0	·12779	·17590	68 48
2 „ 178	22 00·7	·00023	1·40002	89 15	47 „ 133	30 38·6	·13373	·16871	68 07
3 „ 177	22 01·6	·00051	1·31403	88 53	48 „ 132	31 07·4	·13982	·16176	67 25
4 „ 176	22 02·9	·00091	1·18925	88 30	49 „ 131	31 37·6	·14607	0·15505	66 41
5 „ 175	22 04·6	0·00142	1·09254	88 07	50 „ 130	32 09·1	0·15247	·14857	65 57
6 „ 174	22 06·6	·00205	1·01360	87 45	51 „ 129	32 42·0	·15902	·14233	65 10
7 „ 173	22 09·0	·00279	0·94694	87 22	52 „ 128	33 16·5	·16573	·13630	64 23
8 „ 172	22 11·7	·00365	·88928	86 59	53 „ 127	33 52·5	·17258	·13049	63 34
9 „ 171	22 14·9	·00462	·83850	86 36	54 „ 126	34 30·2	·17959	·12488	62 43
10 „ 170	22 18·4	0·00570	0·79316	86 13	55 „ 125	35 09·6	0·18675	0·11947	61 51
11 „ 169	22 22·3	·00691	·75224	85 50	56 „ 124	35 50·9	·19406	·11426	60 57
12 „ 168	22 26·6	·00822	·71495	85 27	57 „ 123	36 34·1	·20152	·10924	60 01
13 „ 167	22 31·3	·00966	·68075	85 03	58 „ 122	37 19·4	·20912	·10441	59 03
14 „ 166	22 36·4	·01121	·64916	84 40	59 „ 121	38 06·8	·21686	·09977	58 03
15 „ 165	22 41·9	0·01288	0·61984	84 16	60 „ 120	38 56·4	0·22473	0·09530	57 01
16 „ 164	22 47·8	·01467	·59250	83 52	61 „ 119	39 48·4	·23274	·09102	55 57
17 „ 163	22 54·2	·01657	·56690	83 28	62 „ 118	40 42·9	·24087	·08690	54 50
18 „ 162	23 01·0	·01860	·54285	83 04	63 „ 117	41 40·0	·24912	·08295	53 41
19 „ 161	23 08·2	·02075	·52019	82 39	64 „ 116	42 39·9	·25747	·07917	52 28
20 „ 160	23 15·9	0·02301	0·49878	82 14	65 „ 115	43 42·7	0·26592	0·07556	51 13
21 „ 159	23 24·1	·02540	·47851	81 49	66 „ 114	44 48·5	·27446	·07210	49 55
22 „ 158	23 32·7	·02792	·45926	81 24	67 „ 113	45 57·5	·28305	·06881	48 38
23 „ 157	23 41·9	·03055	·44096	80 58	68 „ 112	47 09·8	·29171	·06567	47 10
24 „ 156	23 51·5	·03331	·42352	80 32	69 „ 111	48 25·6	·30039	·06268	45 42
25 „ 155	24 01·6	0·03620	0·40689	80 05	70 „ 110	49 45·1	0·30909	0·05985	44 11
26 „ 154	24 12·3	·03921	·39099	79 39	71 „ 109	51 08·3	·31777	·05716	42 35
27 „ 153	24 23·5	·04235	·37579	79 12	72 „ 108	52 35·4	·32641	·05463	40 56
28 „ 152	24 35·3	·04562	·36123	78 44	73 „ 107	54 06·5	·33498	·05224	39 13
29 „ 151	24 47·7	·04901	·34726	78 16	74 „ 106	55 41·8	·34344	·04999	37 26
30 „ 150	25 00·6	0·05254	0·33386	77 48	75 „ 105	57 21·4	0·35176	0·04789	35 35
31 „ 149	25 14·2	·05620	·32100	77 19	76 „ 104	59 05·3	·35989	·04593	33 39
32 „ 148	25 28·4	·06000	·30862	76 50	77 „ 103	60 53·5	·36779	·04411	31 39
33 „ 147	25 43·3	·06392	·29673	76 20	78 „ 102	62 46·2	·37541	·04243	29 34
34 „ 146	25 58·9	·06799	·28527	75 49	79 „ 101	64 43·2	·38270	·04089	27 25
35 „ 145	26 15·2	0·07219	0·27424	75 18	80 „ 100	66 44·5	0·38962	0·03948	25 12
36 „ 144	26 32·3	·07652	·26362	74 46	81 „ 99	68 50·0	·39609	·03821	22 55
37 „ 143	26 50·1	·08100	·25337	74 14	82 „ 98	70 59·6	·40208	·03708	20 34
38 „ 142	27 08·7	·08562	·24349	73 41	83 „ 97	73 12·9	·40752	·03608	18 09
39 „ 141	27 28·2	·09038	·23396	73 07	84 „ 96	75 29·7	·41236	·03522	15 40
40 „ 140	27 48·5	0·09529	0·22477	72 33	85 „ 95	77 49·6	0·41654	0·03449	13 09
41 „ 139	28 09·7	·10033	·21589	71 58	86 „ 94	80 12·3	·42004	·03389	10 34
42 „ 138	28 31·9	·10553	·20732	71 22	87 „ 93	82 37·2	·42281	·03343	7 58
43 „ 137	28 55·1	·11087	·19905	70 45	88 „ 92	85 03·8	·42481	·03310	5 20
44 „ 136	29 19·3	·11636	·19106	70 07	89 „ 91	87 31·6	·42602	·03290	2 40
45 „ 135	29 44·6	0·12200	0·18335	69 28	90	90 00·0	0·42643	0·03283	0 00
h	180°-K	A	N	a₁	h	180°-K	A	N	a₁

Directions with regard to (K ⸪ d) :—

If l and d have **Same Name,** use (K ⌣ d), i.e. (K — d) or (d — K).

If l and d have **Different Names,** use (K + d).

To name Intercept :—

If H₀ is greater than H_c, intercept is " **Towards.**"

If H₀ is less than H_c, intercept is " **Away.**"

Log. cosec. H_c = A + log. sec. (K ⸪ d).

N.B.   Whenever h is greater than 90°, subtract the quantity in the K column from 180°.

ALT.—AZ. TABLE — LAT. 23°

h	K	A	N	a_1	h	K	A	N	a_1
°	° ′			° ′		° ′			° ′
0 or 180	23 00·0	0·00000	—	90 00	45 or 135	30 58·6	0·11966	0·18649	68 39
1 „ 179	23 00·2	·00006	1·79412	89 37	46 „ 134	31 25·6	·12531	·17904	67 58
2 „ 178	23 00·8	·00022	1·49316	89 13	47 „ 133	31 53·9	·13109	·17185	67 16
3 „ 177	23 01·7	·00051	1·31717	88 50	48 „ 132	32 23·4	·13702	·16490	66 32
4 „ 176	23 03·0	·00090	1·19239	88 26	49 „ 131	32 54·2	·14310	·15819	65 48
5 „ 175	23 04·7	0·00140	1·09568	88 03	50 „ 130	33 26·4	0·14932	0·15172	65 02
6 „ 174	23 06·8	·00202	1·01674	87 39	51 „ 129	34 00·0	·15568	·14547	64 15
7 „ 173	23 09·3	·00275	0·95008	87 15	52 „ 128	34 35·1	·16219	·13944	63 26
8 „ 172	23 12·1	·00359	·89242	86 51	53 „ 127	35 11·8	·16883	·13363	62 36
9 „ 171	23 15·4	·00455	·84164	86 28	54 „ 126	35 50·1	·17562	·12802	61 44
10 „ 170	23 19·0	0·00562	0·79630	86 04	55 „ 125	36 30·2	0·18255	0·12261	60 50
11 „ 169	23 23·1	·00680	·75538	85 39	56 „ 124	37 12·1	·18960	·11740	59 55
12 „ 168	23 27·5	·00810	·71810	85 15	57 „ 123	37 55·9	·19680	·11238	58 58
13 „ 167	23 32·4	·00952	·68389	84 51	58 „ 122	38 41·7	·20413	·10755	57 59
14 „ 166	23 37·7	·01104	·65230	84 26	59 „ 121	39 29·6	·21158	·10291	56 58
15 „ 165	23 43·4	0·01269	0·62298	84 01	60 „ 120	40 19·8	0·21915	0·09844	55 55
16 „ 164	23 49·5	·01445	·59564	83 36	61 „ 119	41 12·2	·22683	·09416	54 49
17 „ 163	23 56·1	·01633	·57004	83 11	62 „ 118	42 07·1	·23463	·09004	53 41
18 „ 162	24 03·1	·01832	·54599	82 46	63 „ 117	43 04·5	·24251	·08609	52 31
19 „ 161	24 10·6	·02043	·52333	82 20	64 „ 116	44 04·6	·25050	·08232	51 18
20 „ 160	24 18·6	0·02267	0·50192	81 54	65 „ 115	45 07·5	0·25855	0·07870	50 02
21 „ 159	24 27·0	·02502	·48164	81 28	66 „ 114	46 13·4	·26668	·07524	48 44
22 „ 158	24 35·9	·02749	·46240	81 02	67 „ 113	47 22·2	·27485	·07195	47 22
23 „ 157	24 45·4	·03008	·44410	80 35	68 „ 112	48 34·3	·28305	·06881	45 58
24 „ 156	24 55·3	·03280	·42666	80 08	69 „ 111	49 49·6	·29127	·06582	44 30
25 „ 155	25 05·8	0·03563	0·41003	79 40	70 „ 110	51 08·4	0·29948	0·06299	42 58
26 „ 154	25 16·8	·03859	·39413	79 13	71 „ 109	52 30·7	·30766	·06030	41 23
27 „ 153	25 28·4	·04168	·37893	78 44	72 „ 108	53 56·7	·31578	·05777	39 45
28 „ 152	25 40·5	·04489	·36437	78 16	73 „ 107	55 26·5	·32381	·05538	38 03
29 „ 151	25 53·3	·04823	·35040	77 47	74 „ 106	57 00·1	·33172	·05313	36 16
30 „ 150	26 06·7	0·05169	0·33700	77 17	58 „ 105	58 37·7	0·33948	0·05103	34 26
31 „ 149	26 20·7	·05528	·32414	76 47	76 „ 104	60 19·2	·34704	·04907	32 33
32 „ 148	26 35·4	·05900	·31176	76 17	77 „ 103	62 04·7	·35437	·04725	30 35
33 „ 147	26 50·7	·06286	·29987	75 46	78 „ 102	63 54·2	·36143	·04557	28 33
34 „ 146	27 06·8	·06684	·28841	75 14	79 „ 101	65 47·7	·36816	·04403	26 27
35 „ 145	27 23·6	0·07097	0·27738	74 42	80 „ 100	67 45·1	0·37452	0·04262	24 17
36 „ 144	27 41·1	·07521	·26676	74 09	81 „ 99	69 46·2	·38047	·04135	22 04
37 „ 143	27 59·4	·07959	·25651	73 36	82 „ 98	71 50·8	·38595	·04022	19 47
38 „ 142	28 18·6	·08412	·24663	73 01	83 „ 97	73 58·9	·39092	·03922	17 27
39 „ 141	28 38·6	·08878	·23710	72 27	84 „ 96	76 10·0	·39534	·03836	15 03
40 „ 140	28 59·5	0·09358	0·22791	71 51	85 „ 95	78 23·8	0·39916	0·03763	12 37
41 „ 139	29 21·3	·09851	·21903	71 14	86 „ 94	80 40·1	·40234	·03703	10 09
42 „ 138	29 44·1	·10359	·21046	70 37	87 „ 93	82 58·3	·40485	·03657	7 38
43 „ 137	30 07·8	·10880	·20219	69 59	88 „ 92	85 18·0	·40666	·03624	5 06
44 „ 136	30 32·7	·11416	·19420	69 20	89 „ 91	87 38·7	·40776	·03604	2 33
45 „ 135	30 58·6	0·11966	0·18649	68 39	90	90 00·0	0·40812	0·03597	0 00
h	180°-K	A	N	a_1	h	180°-K	A	N	a_1

Directions with regard to $(a_2 \overset{+}{\sim} a_1)$:—

Whenever h is greater than 90°, use $(a_2 - a_1)$.

If h is less than 90°, use :—

$(a_1 + a_2)$ when $(K \pm d)$ has been used,
$(a_1 - a_2)$ when $(d - K)$ has been used.

Always **name azimuth** from elevated pole.

Note that $\begin{cases} \text{h = L.H.A. of body } \textbf{West} \text{ of the meridian.} \\ \text{h = (360° — L.H.A.) of body } \textbf{East} \text{ of the meridian.} \end{cases}$

Log. tan. a_2 = N + log. tan. $(K \overset{+}{\sim} d)$.

LAT. 24°   ALT.—AZ. TABLE

h	K	A	N	a₁	h	K	A	N	a₁
°	° ′			° ′		° ′			° ′
0 or 180	24 00·0	0·00000	—	90 00	45 or 135	32 11·8	0·11727	0·18979	67 52
1 „ 179	24 00·2	·00006	1·79742	89 36	46 „ 134	32 39·4	·12277	·18234	67 10
2 „ 178	24 00·8	·00022	1·49645	89 11	47 „ 133	33 08·3	·12840	·17514	66 26
3 „ 177	24 01·8	·00050	1·32047	88 47	48 „ 132	33 38·3	·13416	·16820	65 41
4 „ 176	24 03·1	·00088	1·19569	88 22	49 „ 131	34 09·7	·14007	·16149	64 56
5 „ 175	24 04·9	0·00138	1·09899	87 58	50 „ 130	34 42·5	0·14611	0·15502	64 08
6 „ 174	24 07·0	·00199	1·02004	87 33	51 „ 129	35 16·7	·15228	·14877	63 20
7 „ 173	24 09·6	·00271	0·95338	87 08	52 „ 128	35 52·4	·15859	·14274	62 30
8 „ 172	24 12·5	·00354	·89572	86 44	53 „ 127	36 29·7	·16502	·13692	61 39
9 „ 171	24 15·9	·00448	·84494	86 19	54 „ 126	37 08·6	·17158	·13131	60 46
10 „ 170	24 19·7	0·00554	0·79960	85 54	55 „ 125	37 49·2	0·17827	0·12591	59 51
11 „ 169	24 23·8	·00670	·75867	85 29	56 „ 124	38 31·6	·18509	·12070	58 55
12 „ 168	24 28·4	·00798	·72139	85 04	57 „ 123	39 15·9	·19203	·11568	57 56
13 „ 167	24 33·5	·00937	·68718	84 38	58 „ 122	40 02·2	·19908	·11085	56 56
14 „ 166	24 38·9	·01087	·65559	84 13	59 „ 121	40 50·5	·20625	·10620	55 54
15 „ 165	24 44·8	0·01249	0·62627	83 47	60 „ 120	41 41·0	0·21352	0·10174	54 50
16 „ 164	24 51·1	·01422	·59893	83 21	61 „ 119	42 33·8	·22089	·09745	53 44
17 „ 163	24 57·9	·01607	·57334	82 55	62 „ 118	43 28·9	·22835	·09334	52 35
18 „ 162	25 05·2	·01803	·54929	82 28	63 „ 117	44 26·5	·23590	·08939	51 24
19 „ 161	25 12·9	·02011	·52663	82 02	64 „ 116	45 26·7	·24351	·08561	50 10
20 „ 160	25 21·1	0·02231	0·50522	81 35	65 „ 115	46 29·5	0·25119	0·08199	48 54
21 „ 159	25 29·8	·02462	·48494	81 08	66 „ 114	47 35·2	·25892	·07854	47 35
22 „ 158	25 39·0	·02705	·46570	80 40	67 „ 113	48 43·8	·26668	·07524	46 13
23 „ 157	25 48·7	·02960	·44739	80 12	68 „ 112	49 55·4	·27446	·07210	44 49
24 „ 156	25 59·0	·03226	·42996	79 44	69 „ 111	51 10·1	·28222	·06912	43 21
25 „ 155	26 09·8	0·03505	0·41332	79 16	70 „ 110	52 28·1	0·28998	0·06628	41 49
26 „ 154	26 21·1	·03796	·39743	78 47	71 „ 109	53 49·5	·29768	·06360	40 15
27 „ 153	26 33·1	·04099	·38222	78 18	72 „ 108	55 14·2	·30531	·06106	38 37
28 „ 152	26 45·6	·04414	·36766	77 48	73 „ 107	56 42·5	·31283	·05867	36 56
29 „ 151	26 58·7	·04741	·35370	77 18	74 „ 106	58 14·3	·32023	·05643	35 11
30 „ 150	27 12·5	0·05081	0·34030	76 47	75 „ 105	59 49·8	0·32747	0·05433	33 23
31 „ 149	27 26·9	·05434	·32743	76 16	76 „ 104	61 28·9	·33451	·05237	31 30
32 „ 148	27 42·0	·05799	·31506	75 44	77 „ 103	63 11·7	·34132	·05055	29 35
33 „ 147	27 57·8	·06177	·30316	75 12	78 „ 102	64 58·1	·34785	·04887	27 35
34 „ 146	28 14·3	·06567	·29171	74 40	79 „ 101	66 48·1	·35407	·04732	25 33
35 „ 145	28 31·5	0·06970	0·28068	74 06	80 „ 100	68 41·6	0·35994	0·04592	23 26
36 „ 144	28 49·5	·07386	·27005	73 32	81 „ 99	70 38·4	·36541	·04465	21 17
37 „ 143	29 08·3	·07815	·25981	72 58	82 „ 98	72 38·5	·37044	·04352	19 04
38 „ 142	29 28·0	·08258	·24993	72 22	83 „ 97	74 41·5	·37500	·04252	16 48
39 „ 141	29 48·5	·08713	·24040	71 46	84 „ 96	76 47·3	·37904	·04166	14 29
40 „ 140	30 09·9	0·09182	0·23120	71 09	85 „ 95	78 55·4	0·38252	0·04093	12 08
41 „ 139	30 32·3	·09664	·22233	70 32	86 „ 94	81 05·7	·38542	·04033	9 45
42 „ 138	30 55·6	·10160	·21376	69 53	87 „ 93	83 17·7	·38771	·03987	7 21
43 „ 137	31 19·9	·10669	·20549	69 14	88 „ 92	85 31·1	·38935	·03954	4 54
44 „ 136	31 45·3	·11191	·19750	68 33	89 „ 91	87 45·3	·39035	·03934	2 27
45 „ 135	32 11·8	0·11727	0·18979	67 52	90	90 00·0	0·39069	0·03927	0 00
h	180°-K	A	N	a₁	h	180°-K	A	N	a₁

Directions with regard to (K $\overset{+}{\underset{-}{}}$ d) :—

If l and d have **Same Name**, use (K ⌣ d), i.e. (K — d) or (d — K).
If l and d have **Different Names**, use (K + d).

To name Intercept :—

If H_o is greater than H_c, intercept is " **Towards.** "
If H_o is less than H_c, intercept is " **Away.** "

Log. cosec. H_c = A + log. sec. (K $\overset{+}{\underset{-}{}}$ d).

N.B.   *Whenever h is greater than 90°, subtract the quantity in the K column from 180°.*

ALT.—AZ. TABLE — LAT. 25°

h	K		A	N	a₁		h	K		A	N	a₁	
°	°	′			°	′	°	°	′			°	′
0 or 180	25	00·0	0·00000	—	90	00	**45 or 135**	33	24·2	0·11483	0·19324	67	05
1 „ 179	25	00·2	·00005	1·80087	89	35	**46 „ 134**	33	52·4	·12018	·18579	66	22
2 „ 178	25	00·8	·00022	1·49991	89	09	**47 „ 133**	34	21·7	·12565	·17860	65	37
3 „ 177	25	01·8	·00049	1·32392	88	44	**48 „ 132**	34	52·3	·13125	·17165	64	51
4 „ 176	25	03·2	·00087	1·19914	88	18	**49 „ 131**	35	24·2	·13699	·16494	64	04
5 „ 175	25	05·0	0·00136	1·10243	87	53	**50 „ 130**	35	57·5	0·14284	0·15847	63	16
6 „ 174	25	07·2	·00196	1·02349	87	27	**51 „ 129**	36	32·2	·14882	·15222	62	26
7 „ 173	25	09·9	·00267	0·95683	87	02	**52 „ 128**	37	08·4	·15493	·14619	61	35
8 „ 172	25	12·9	·00348	·89917	86	36	**53 „ 127**	37	46·2	·16115	·14038	60	43
9 „ 171	25	16·4	·00441	·84839	86	10	**54 „ 126**	38	25·6	·16750	·13477	59	49
10 „ 170	25	20·3	0·00545	0·80305	85	44	**55 „ 125**	39	06·6	0·17396	0·12936	58	53
11 „ 169	25	24·6	·00659	·76213	85	18	**56 „ 124**	39	49·5	·18053	·12415	57	56
12 „ 168	25	29·3	·00785	·72485	84	52	**57 „ 123**	40	34·2	·18721	·11913	56	57
13 „ 167	25	34·5	·00922	·69064	84	26	**58 „ 122**	41	20·8	·19400	·11430	55	56
14 „ 166	25	40·1	·01070	·65905	83	59	**59 „ 121**	42	09·4	·20088	·10966	54	53
15 „ 165	25	46·2	0·01229	0·62973	83	32	**60 „ 120**	43	00·2	0·20786	0·10519	53	48
16 „ 164	25	52·7	·01399	·60239	83	05	**61 „ 119**	43	53·1	·21492	·10091	52	41
17 „ 163	25	59·7	·01581	·57679	82	38	**62 „ 118**	44	48·4	·22206	·09679	51	31
18 „ 162	26	07·1	·01774	·55273	82	11	**63 „ 117**	45	46·0	·22927	·09284	50	20
19 „ 161	26	15·1	·01978	·53008	81	43	**64 „ 116**	46	46·1	·23654	·08906	49	05
20 „ 160	26	23·5	0·02194	0·50867	81	15	**65 „ 115**	47	48·8	0·24385	0·08545	47	49
21 „ 159	26	32·5	·02421	·48840	80	47	**66 „ 114**	48	54·2	·25119	·08199	46	30
22 „ 158	26	41·9	·02659	·46915	80	19	**67 „ 113**	50	02·4	·25855	·07870	45	08
23 „ 157	26	51·9	·02910	·45085	79	50	**68 „ 112**	51	13·4	·26592	·07556	43	43
24 „ 156	27	02·5	·03171	·43341	79	21	**69 „ 111**	52	27·4	·27327	·07257	42	15
25 „ 155	27	13·6	0·03445	0·41678	78	51	**70 „ 110**	53	44·5	0·28058	0·06974	40	44
26 „ 154	27	25·3	·03730	·40088	78	21	**71 „ 109**	55	04·7	·28783	·06705	39	10
27 „ 153	27	37·5	·04028	·38568	77	51	**72 „ 108**	56	28·1	·29500	·06452	37	33
28 „ 152	27	50·4	·04337	·37112	77	20	**73 „ 107**	57	54·8	·30206	·06213	35	53
29 „ 151	28	03·9	·04658	·35715	76	49	**74 „ 106**	59	24·7	·30898	·05988	34	09
30 „ 150	28	18·0	0·04991	0·34375	76	17	**75 „ 105**	60	58·1	0·31573	0·05778	32	23
31 „ 149	28	32·8	·05336	·33089	75	45	**76 „ 104**	62	34·8	·32229	·05582	30	32
32 „ 148	28	48·3	·05694	·31851	75	12	**77 „ 103**	64	14·8	·32862	·05400	28	39
33 „ 147	29	04·5	·06064	·30662	74	39	**78 „ 102**	65	58·2	·33463	·05232	26	42
34 „ 146	29	21·4	·06446	·29516	74	05	**79 „ 101**	67	44·8	·34043	·05078	24	42
35 „ 145	29	39·1	0·06841	0·28413	73	31	**80 „ 100**	69	34·5	0·34585	0·04937	22	39
36 „ 144	29	57·5	·07247	·27351	72	56	**81 „ 99**	71	27·3	·35089	·04810	20	33
37 „ 143	30	16·8	·07667	·26326	72	20	**82 „ 98**	73	22·9	·35552	·04697	18	24
38 „ 142	30	36·9	·08100	·25338	71	44	**83 „ 97**	75	21·2	·35970	·04597	16	12
39 „ 141	30	57·9	·08545	·24385	71	06	**84 „ 96**	77	21·9	·36341	·04511	13	58
40 „ 140	31	19·8	0·09003	0·23466	70	28	**85 „ 95**	79	24·8	0·36660	0·04438	11	42
41 „ 139	31	42·6	·09473	·22578	69	50	**86 „ 94**	81	29·5	·36925	·04378	9	24
42 „ 138	32	06·4	·09956	·21721	69	10	**87 „ 93**	83	35·8	·37133	·04332	7	04
43 „ 137	32	31·3	·10453	·20894	68	29	**88 „ 92**	85	43·2	·37284	·04299	4	43
44 „ 136	32	57·2	·10961	·20095	67	48	**89 „ 91**	87	51·4	·37375	·04279	2	22
45 „ 135	33	24·2	0·11483	0·19324	67	05	**90**	90	00·0	0·37405	0·04272	0	00
h	180°-K		A	N	a₁		h	180°-K		A	N	a₁	

Directions with regard to $(a_2 \overset{+}{\sim} a_1)$:—

Whenever h is greater than 90°, use $(a_2 - a_1)$.

If h is less than 90°, use :—

$(a_1 + a_2)$ when $(K \pm d)$ has been used,
$(a_1 - a_2)$ when $(d - K)$ has been used.

Always **name**
azimuth from
elevated pole.

Note that $\begin{cases} \text{h} = \text{L.H.A. of body \textbf{West} of the meridian.} \\ \text{h} = (360° - \text{L.H.A.}) \text{ of body \textbf{East} of the meridian.} \end{cases}$

Log. tan. a_2 = N + log. tan. $(K \overset{+}{\sim} d)$.

LAT. 26°   ALT.—AZ. TABLE

h	K		A	N	a₁		h	K		A	N	a₁	
°	°	′			°	′	°	°	′			°	′
0 or 180	26	00·0	0·00000	—	90	00	45 or 135	34	35·8	0·11235	0·19686	66	20
1 „ 179	26	00·2	·00005	1·80449	89	34	46 „ 134	35	04·4	·11755	·18941	65	35
2 „ 178	26	00·8	·00021	1·50352	89	07	47 „ 133	35	34·2	·12286	·18221	64	49
3 „ 177	26	01·9	·00048	1·32754	88	41	48 „ 132	36	05·3	·12830	·17527	64	02
4 „ 176	26	03·3	·00086	1·20276	88	15	49 „ 131	36	37·7	·13386	·16856	63	14
5 „ 175	26	05·2	0·00134	1·10604	87	48	50 „ 130	37	11·4	0·13953	0·16209	62	25
6 „ 174	26	07·5	·00193	1·02711	87	22	51 „ 129	37	46·6	·14532	·15584	61	34
7 „ 173	26	10·2	·00262	0·96045	86	55	52 „ 128	38	23·2	·15123	·14981	60	42
8 „ 172	26	13·3	·00343	·90279	86	28	53 „ 127	39	01·4	·15724	·14399	59	49
9 „ 171	26	16·8	·00434	·85201	86	02	54 „ 126	39	41·1	·16337	·13838	58	54
10 „ 170	26	20·8	0·00536	0·80667	85	35	55 „ 125	40	22·5	0·16960	0·13298	57	57
11 „ 169	26	25·3	·00648	·76574	85	08	56 „ 124	41	05·7	·17593	·12777	56	59
12 „ 168	26	30·1	·00772	·72846	84	41	57 „ 123	41	50·7	·18236	·12275	55	59
13 „ 167	26	35·4	·00906	·69425	84	13	58 „ 122	42	37·6	·18888	·11792	54	57
14 „ 166	26	41·2	·01052	·66267	83	46	59 „ 121	43	26·4	·19549	·11327	53	53
15 „ 165	26	47·5	0·01208	0·63334	83	18	60 „ 120	44	17·3	0·20218	0·10881	52	47
16 „ 164	26	54·2	·01375	·60600	82	50	61 „ 119	45	10·3	·20894	·10452	51	40
17 „ 163	27	01·3	·01554	·58041	82	22	62 „ 118	46	05·6	·21577	·10041	50	30
18 „ 162	27	09·0	·01743	·55636	81	54	63 „ 117	47	03·1	·22266	·09646	49	18
19 „ 161	27	17·2	·01944	·53370	81	25	64 „ 116	48	03·1	·22958	·09268	48	03
20 „ 160	27	25·9	0·02156	0·51229	80	56	65 „ 115	49	05·5	0·23654	0·08906	46	46
21 „ 159	27	35·0	·02378	·49201	80	27	66 „ 114	50	10·4	·24351	·08561	45	27
22 „ 158	27	44·8	·02613	·47277	79	57	67 „ 113	51	18·1	·25050	·08232	44	05
23 „ 157	27	55·0	·02858	·45446	79	28	68 „ 112	52	28·4	·25747	·07917	42	40
24 „ 156	28	05·8	·03115	·43703	78	57	69 „ 111	53	41·6	·26442	·07619	41	12
25 „ 155	28	17·2	0·03383	0·42039	78	27	70 „ 110	54	57·6	0·27131	0·07335	39	42
26 „ 154	28	29·2	·03663	·40450	77	56	71 „ 109	56	16·6	·27814	·07067	38	09
27 „ 153	28	41·8	·03955	·38929	77	25	72 „ 108	57	38·6	·28487	·06813	36	33
28 „ 152	28	55·0	·04258	·37473	76	53	73 „ 107	59	03·6	·29150	·06574	34	54
29 „ 151	29	08·8	·04573	·36077	76	21	74 „ 106	60	31·6	·29797	·06350	33	11
30 „ 150	29	23·3	0·04899	0·34737	75	48	75 „ 105	62	02·8	0·30428	0·06140	31	26
31 „ 149	29	38·4	·05237	·33450	75	15	76 „ 104	63	37·1	·31039	·05944	29	38
32 „ 148	29	54·3	·05587	·32213	74	41	77 „ 103	65	14·4	·31628	·05762	27	46
33 „ 147	30	10·8	·05949	·31023	74	07	78 „ 102	66	54·7	·32190	·05594	25	52
34 „ 146	30	28·1	·06323	·29878	73	32	79 „ 101	68	38·0	·32723	·05439	23	55
35 „ 145	30	46·2	0·06709	0·28775	72	56	80 „ 100	70	24·2	0·33224	0·05299	21	55
36 „ 144	31	05·1	·07106	·27712	72	20	81 „ 99	72	13·0	·33690	·05172	19	52
37 „ 143	31	24·8	·07516	·26688	71	43	82 „ 98	74	04·4	·34116	·05059	17	47
38 „ 142	31	45·3	·07938	·25700	71	06	83 „ 97	75	58·2	·34501	·04959	15	39
39 „ 141	32	06·7	·08373	·24747	70	27	84 „ 96	77	54·2	·34841	·04873	13	29
40 „ 140	32	29·1	0·08819	0·23827	69	48	85 „ 95	79	52·1	0·35133	0·04800	11	17
41 „ 139	32	52·4	·09278	·22940	69	08	86 „ 94	81	51·6	·35376	·04740	9	04
42 „ 138	33	16·6	·09749	·22083	68	28	87 „ 93	83	52·5	·35567	·04694	6	49
43 „ 137	33	41·9	·10232	·21256	67	46	88 „ 92	85	54·4	·35705	·04661	4	34
44 „ 136	34	08·3	·10727	·20457	67	03	89 „ 91	87	57·0	·35787	·04641	2	17
45 „ 135	34	35·8	0·11235	0·19686	66	20	90	90	00·0	0·35816	0·04634	0	00
h	180°-K		A	N	a₁		h	180°-K		A	N	a₁	

Directions with regard to (K ⌣ d) :—

If l and d have **Same Name,** use (K ⌣ d), i.e. (K — d) or (d — K).

If l and d have **Different Names,** use (K ⊢ d).

To name Intercept :—

If H₀ is greater than H𝒸, intercept is **" Towards."**

If H₀ is less than H𝒸, intercept is **" Away."**

Log. cosec. H𝒸 = A + log. sec. (K ⌣ d).

N.B.   Whenever h is greater than 90°, subtract the quantity in the K column from 180°.

ALT.—AZ. TABLE   LAT. 27°

h	K	A	N	a₁	h	K	A	N	a₁
°	° ′			° ′	°	° ′			° ′
0 or 180	27 00·0	0·00000	—	90 00	45 or 135	35 46·5	0·10982	0·20063	65 35
1 „ 179	27 00·2	·00005	1·80826	89 33	46 „ 134	36 15·6	·11487	·19319	64 49
2 „ 178	27 00·8	·00021	1·50730	89 06	47 „ 133	36 45·8	·12003	·18599	64 02
3 „ 177	27 01·9	·00047	1·33132	88 38	48 „ 132	37 17·3	·12530	·17905	63 15
4 „ 176	27 03·4	·00084	1·20653	88 11	49 „ 131	37 50·1	·13069	·17234	62 25
5 „ 175	27 05·3	0·00131	1·10982	87 44	50 „ 130	38 24·2	0·13618	0·16587	61 35
6 „ 174	27 07·7	·00189	1·03088	87 16	51 „ 129	38 59·7	·14178	·15962	60 43
7 „ 173	27 10·4	·00257	0·96423	86 49	52 „ 128	39 36·7	·14749	·15359	59 50
8 „ 172	27 13·6	·00336	·90656	86 21	53 „ 127	40 15·2	·15329	·14777	58 56
9 „ 171	27 17·3	·00426	·85579	85 53	54 „ 126	40 55·2	·15920	·14216	58 00
10 „ 170	27 21·4	0·00526	0·81045	85 25	55 „ 125	41 36·9	0·16521	0·13675	57 03
11 „ 169	27 25·9	·00637	·76952	84 57	56 „ 124	42 20·3	·17130	·13155	56 03
12 „ 168	27 30·9	·00758	·73224	84 29	57 „ 123	43 05·5	·17748	·12653	55 03
13 „ 167	27 36·4	·00890	·69803	84 01	58 „ 122	43 52·6	·18375	·12170	54 00
14 „ 166	27 42·3	·01033	·66644	83 33	59 „ 121	44 41·5	·19009	·11705	52 56
15 „ 165	27 48·7	0·01187	0·63712	83 04	60 „ 120	45 32·4	0·19650	0·11259	51 50
16 „ 164	27 55·6	·01351	·60978	82 35	61 „ 119	46 25·4	·20296	·10830	50 41
17 „ 163	28 02·9	·01526	·58418	82 06	62 „ 118	47 20·6	·20949	·10418	49 30
18 „ 162	28 10·8	·01712	·56014	81 37	63 „ 117	48 17·9	·21606	·10024	48 18
19 „ 161	28 19·2	·01909	·53748	81 07	64 „ 116	49 17·6	·22266	·09646	47 03
20 „ 160	28 28·1	0·02116	0·51607	80 37	65 „ 115	50 19·6	0·22927	0·09284	45 46
21 „ 159	28 37·5	·02335	·49579	80 07	66 „ 114	51 24·1	·23590	·08939	44 26
22 „ 158	28 47·4	·02565	·47654	79 36	67 „ 113	52 31·0	·24251	·08609	43 05
23 „ 157	28 57·9	·02806	·45824	79 06	68 „ 112	53 40·6	·24912	·08295	41 40
24 „ 156	29 09·0	·03057	·44081	78 34	69 „ 111	54 52·8	·25568	·07997	40 13
25 „ 155	29 20·7	0·03320	0·42417	78 03	70 „ 110	56 07·7	0·26218	0·07713	38 43
26 „ 154	29 32·9	·03595	·40828	77 31	71 „ 109	57 25·4	·26861	·07445	37 11
27 „ 153	29 45·8	·03880	·39307	76 59	72 „ 108	58 45·8	·27494	·07191	35 35
28 „ 152	29 59·3	·04177	·37851	76 26	73 „ 107	60 09·1	·28115	·06952	33 57
29 „ 151	30 13·4	·04485	·36455	75 52	74 „ 106	61 35·3	·28721	·06728	32 17
30 „ 150	30 28·2	0·04804	0·35115	75 19	75 „ 105	63 04·3	0·29311	0·06518	30 33
31 „ 149	30 43·7	·05135	·33828	74 44	76 „ 104	64 36·1	·29881	·06322	28 47
32 „ 148	30 59·9	·05477	·32591	74 10	77 „ 103	66 10·7	·30428	·06140	26 57
33 „ 147	31 16·8	·05831	·31401	73 34	78 „ 102	67 48·1	·30951	·05972	25 05
34 „ 146	31 34·5	·06196	·30256	72 58	79 „ 101	69 28·2	·31445	·05817	23 11
35 „ 145	31 52·9	0·06573	0·29153	72 22	80 „ 100	71 10·8	0·31909	0·05677	21 14
36 „ 144	32 12·2	·06962	·28090	71 45	81 „ 99	72 55·9	·32339	·05550	19 14
37 „ 143	32 32·3	·07362	·27066	71 07	82 „ 98	74 43·4	·32732	·05437	17 12
38 „ 142	32 53·2	·07773	·26078	70 28	83 „ 97	76 32·9	·33087	·05337	15 08
39 „ 141	33 15·0	·08197	·25125	69 49	84 „ 96	78 24·4	·33400	·05251	13 02
40 „ 140	33 37·8	0·08632	0·24205	69 09	85 „ 95	80 17·6	0·33669	0·05178	10 54
41 „ 139	34 01·5	·09079	·23318	68 28	86 „ 94	82 12·3	·33892	·05118	8 45
42 „ 138	34 26·2	·09537	·22461	67 46	87 „ 93	84 08·1	·34067	·05072	6 35
43 „ 137	34 51·9	·10007	·21634	67 03	88 „ 92	86 04·9	·34194	·05038	4 24
44 „ 136	35 18·7	·10489	·20835	66 20	89 „ 91	88 02·3	·34270	·05019	2 12
45 „ 135	35 46·5	0·10982	0·20063	65 35	90	90 00·0	0·34295	0·05012	0 00
h	180°-K	A	N	a₁	h	180°-K	A	N	a₁

Directions with regard to $(a_2 \overset{+}{\sim} a_1)$:—

Whenever h is greater than 90°, use $(a_2 - a_1)$.

If h is less than 90°, use :—

   $(a_1 + a_2)$ when $(K \pm d)$ has been used,

   $(a_1 - a_2)$ when $(d - K)$ has been used.

Always **name azimuth** from elevated pole.

Note that { h = L.H.A. of body **West** of the meridian.
{ h = (360° — L.H.A.) of body **East** of the meridian.

Log. tan. a_2 = N + log. tan. $(K \overset{+}{\sim} d)$.

LAT. 28°   ALT.—AZ. TABLE

h	K	A	N	a₁	h	K	A	N	a₁
°	° ′			° ′	°	° ′			° ′
0 or 180	28 00·0	0·00000	—	90 00	45 „ 135	36 56·5	0·10726	0·20458	64 51
1 „ 179	28 00·2	·00005	1·81221	89 32	46 „ 134	37 25·9	·11216	·19713	64 04
2 „ 178	28 00·9	·00021	1·51121	89 04	47 „ 133	37 56·5	·11716	·18994	63 17
3 „ 177	28 02·0	·00046	1·33527	88 35	48 „ 132	38 28·3	·12227	·18299	62 28
4 „ 176	28 03·5	·00083	1·21048	88 07	49 „ 131	39 01·4	·12748	·17629	61 28
5 „ 175	28 05·4	0·00129	1·11377	87 39	50 „ 130	39 35·8	0·13280	0·16981	60 46
6 „ 174	28 07·8	·00186	1·03483	87 11	51 „ 129	40 11·7	·13821	·16356	59 54
7 „ 173	28 10·7	·00253	0·96817	86 42	52 „ 128	40 48·9	·14372	·15753	59 00
8 „ 172	28 14·0	·00330	·91051	86 14	53 „ 127	41 27·7	·14932	·15172	58 05
9 „ 171	28 17·7	·00418	·85973	85 45	54 „ 126	42 07·9	·15501	·14611	57 08
10 „ 170	28 21·9	0·00517	0·81440	85 16	55 „ 125	42 49·8	0·16079	0·14070	56 10
11 „ 169	28 26·6	·00625	·77347	84 47	56 „ 124	43 33·4	·16665	·13549	55 10
12 „ 168	28 31·7	·00744	·73619	84 18	57 „ 123	44 18·7	·17259	·13047	54 08
13 „ 167	28 37·3	·00874	·70198	83 49	58 „ 122	45 05·8	·17860	·12565	53 05
14 „ 166	28 43·3	·01014	·67039	83 19	59 „ 121	45 54·7	·18468	·12100	52 00
15 „ 165	28 49·9	0·01165	0·64107	82 50	60 „ 120	46 45·6	0·19082	0·11653	50 53
16 „ 164	28 56·9	·01326	·61373	82 20	61 „ 119	47 38·5	·19700	·11225	49 44
17 „ 163	29 04·5	·01498	·58813	81 50	62 „ 118	48 33·4	·20323	·10813	48 33
18 „ 162	29 12·5	·01680	·56408	81 20	63 „ 117	49 30·5	·20949	·10418	47 21
19 „ 161	29 21·1	·01873	·54142	80 49	64 „ 116	50 29·8	·21577	·10041	46 06
20 „ 160	29 30·2	0·02076	0·52001	80 18	65 „ 115	51 31·3	0·22206	0·09679	44 48
21 „ 159	29 39·8	·02291	·49974	79 47	66 „ 114	52 35·1	·22835	·09334	43 29
22 „ 158	29 50·0	·02516	·48049	79 16	67 „ 113	53 41·4	·23462	·09004	42 07
23 „ 157	30 00·7	·02752	·46219	78 44	68 „ 112	54 50·0	·24087	·08690	40 43
24 „ 156	30 12·0	·02998	·44475	78 12	69 „ 111	56 01·2	·24707	·08391	39 16
25 „ 155	30 23·9	0·03256	0·42812	77 39	70 „ 110	57 14·9	0·25320	0·08108	37 47
26 „ 154	30 36·5	·03524	·41222	77 06	71 „ 109	58 31·2	·25925	·07840	36 15
27 „ 153	30 49·6	·03804	·39702	76 33	72 „ 108	59 50·2	·26520	·07586	34 41
28 „ 152	31 03·4	·04094	·38246	75 59	73 „ 107	61 11·7	·27103	·07347	33 04
29 „ 151	31 17·8	·04395	·36849	75 25	74 „ 106	62 35·9	·27671	·07122	31 25
30 „ 150	31 32·9	0·04707	0·35510	74 50	75 „ 105	64 02·7	0·28222	0·06912	29 43
31 „ 149	31 48·7	·05031	·34223	74 15	76 „ 104	65 32·1	·28753	·06716	27 58
32 „ 148	32 05·2	·05365	·32986	73 39	77 „ 103	67 04·1	·29263	·06534	26 11
33 „ 147	32 22·5	·05711	·31796	73 03	78 „ 102	68 38·6	·29749	·06366	24 22
34 „ 146	32 40·5	·06067	·30650	72 26	79 „ 101	70 15·5	·30209	·06212	22 29
35 „ 145	32 59·3	0·06435	0·29547	71 48	80 „ 100	71 54·8	0·30638	0·06071	20 35
36 „ 144	33 18·8	·06814	·28485	71 10	81 „ 99	73 36·3	·31036	·05945	18 39
37 „ 143	33 39·3	·07204	·27460	70 31	82 „ 98	75 19·9	·31400	·05831	16 40
38 „ 142	34 00·6	·07606	·26472	69 51	83 „ 97	77 05·4	·31727	·05731	14 39
39 „ 141	34 22·8	·08018	·25519	69 11	84 „ 96	78 52·7	·32016	·05645	12 37
40 „ 140	34 45·9	0·08442	0·24600	68 30	85 „ 95	80 41·5	0·32264	0·05572	10 33
41 „ 139	35 09·9	·08877	·23712	67 48	86 „ 94	82 31·6	·32469	·05512	8 28
42 „ 138	35 35·0	·09323	·22855	67 05	87 „ 93	84 22·7	·32630	·05466	6 22
43 „ 137	36 01·1	·09780	·22028	66 21	88 „ 92	86 14·7	·32746	·05433	4 15
44 „ 136	36 28·2	·10248	·21229	65 37	89 „ 91	88 07·2	·32815	·05413	2 08
45 „ 135	36 56·5	0·10726	0·20458	64 51	90	90 00·0	0·32839	0·05407	0 00
h	180°-K	A	N	a₁	h	180°-K	A	N	a₁

Directions with regard to (K ∓ d) :—

If l and d have **Same Name,** use (K ⌣ d), i.e. (K — d) or (d — K).
If l and d have **Different Names,** use (K + d).

To name Intercept :—

If H₀ is greater than H_c, intercept is **" Towards."**
If H₀ is less than H_c, intercept is **" Away."**

Log. cosec. H_c = A + log. sec. (K ∓ d).

N.B.   Whenever h is greater than 90°, subtract the quantity in the K column from 180°.

ALT.—AZ. TABLE LAT. 29°

h	K	A	N	a₁	h	K	A	N	a₁
0 or 180	29 00·0	0·00000	—	90 00	**45 or 135**	38 05·6	0·10467	0·20870	64 08
1 „ 179	29 00·2	·00005	1·81633	89 31	46 „ 134	38 35·3	·10942	·20125	63 21
2 „ 178	29 00·9	·00020	1·51536	89 02	47 „ 133	39 06·2	·11427	·19405	62 32
3 „ 177	29 02·0	·00046	1·33938	88 33	48 „ 132	39 38·3	·11921	·18711	61 42
4 „ 176	29 03·6	·00081	1·21460	88 04	49 „ 131	40 11·7	·12425	·18040	60 51
5 „ 175	29 05·6	0·00127	1·11789	87 34	50 „ 130	40 46·4	0·12938	0·17393	59 59
6 „ 174	29 08·0	·00182	1·03895	87 05	51 „ 129	41 22·4	·13461	·16768	59 05
7 „ 173	29 10·9	·00248	0·97229	86 36	52 „ 128	41 59·9	·13992	·16165	58 11
8 „ 172	29 14·3	·00324	·91463	86 06	53 „ 127	42 38·8	·14533	·15583	57 15
9 „ 171	29 18·1	·00410	·86385	85 37	54 „ 126	43 19·3	·15081	·15022	56 17
10 „ 170	29 22·4	0·00507	0·81851	85 07	55 „ 125	44 01·3	0·15637	0·14482	55 18
11 „ 169	29 27·2	·00613	·77758	84 37	56 „ 124	44 44·9	·16200	·13961	54 18
12 „ 168	29 32·4	·00730	·74030	84 07	57 „ 123	45 30·2	·16770	·13459	53 15
13 „ 167	29 38·1	·00857	·70609	83 37	58 „ 122	46 17·3	·17347	·12976	52 12
14 „ 166	29 44·3	·00995	·67451	83 06	59 „ 121	47 06·2	·17928	·12512	51 06
15 „ 165	29 51·0	0·01142	0·64519	82 36	60 „ 120	47 56·9	0·18515	0·12065	49 59
16 „ 164	29 58·2	·01300	·61784	82 05	61 „ 119	48 49·6	·19106	·11636	48 50
17 „ 163	30 05·9	·01468	·59225	81 34	62 „ 118	49 44·2	·19700	·11225	47 39
18 „ 162	30 14·1	·01647	·56820	81 03	63 „ 117	50 40·9	·20296	·10830	46 25
19 „ 161	30 22·9	·01836	·54554	80 31	64 „ 116	51 39·7	·20894	·10452	45 10
20 „ 160	30 32·1	0·02036	0·52413	80 00	65 „ 115	52 40·6	0·21492	0·10091	43 53
21 „ 159	30 42·0	·02245	·50385	79 27	66 „ 114	53 43·8	·22089	·09745	42 34
22 „ 158	30 52·4	·02466	·48461	78 55	67 „ 113	54 49·2	·22683	·09416	41 13
23 „ 157	31 03·3	·02697	·46630	78 22	68 „ 112	55 56·9	·23274	·09102	39 48
24 „ 156	31 14·9	·02938	·44887	77 49	69 „ 111	57 07·0	·23859	·08803	38 22
25 „ 155	31 27·0	0·03190	0·43223	77 16	70 „ 110	58 19·5	0·24438	0·08520	36 54
26 „ 154	31 39·8	·03453	·41634	76 42	71 „ 109	59 34·4	·25007	·08251	35 23
27 „ 153	31 53·2	·03726	·40113	76 07	72 „ 108	60 51·7	·25566	·07998	33 50
28 „ 152	32 07·2	·04009	·38657	75 33	73 „ 107	62 11·4	·26113	·07759	32 14
29 „ 151	32 21·9	·04304	·37261	74 57	74 „ 106	63 33·6	·26645	·07534	30 36
30 „ 150	32 37·3	0·04609	0·35921	74 22	75 „ 105	64 58·3	0·27160	0·07324	28 56
31 „ 149	32 53·4	·04925	·34634	73 46	76 „ 104	66 25·3	·27656	·07128	27 13
32 „ 148	33 10·2	·05252	·33397	73 09	77 „ 103	67 54·7	·28132	·06946	25 28
33 „ 147	33 27·7	·05589	·32207	72 31	78 „ 102	69 26·4	·28584	·06778	23 40
34 „ 146	33 46·0	·05937	·31062	71 54	79 „ 101	71 00·3	·29011	·06623	21 51
35 „ 145	34 05·1	0·06295	0·29959	71 15	80 „ 100	72 36·3	0·29410	0·06483	19 59
36 „ 144	34 25·1	·06665	·28896	70 36	81 „ 99	74 14·4	·29779	·06356	18 06
37 „ 143	34 45·8	·07045	·27872	69 56	82 „ 98	75 54·3	·30115	·06243	16 10
38 „ 142	35 07·4	·07436	·26884	69 15	83 „ 97	77 36·0	·30418	·06143	14 13
39 „ 141	35 29·9	·07837	·25931	68 34	84 „ 96	79 19·3	·30685	·06057	12 14
40 „ 140	35 53·4	0·08249	0·25011	67 52	85 „ 95	81 03·9	0·30913	0·05984	10 14
41 „ 139	36 17·8	·08672	·24124	67 09	86 „ 94	82 49·6	·31102	·05924	8 12
42 „ 138	36 43·1	·09105	·23267	66 25	87 „ 93	84 36·4	·31250	·05878	6 10
43 „ 137	37 09·6	·09549	·22440	65 40	88 „ 92	86 23·8	·31357	·05845	4 07
44 „ 136	37 37·0	·10003	·21641	64 55	89 „ 91	88 11·8	·31421	·05825	2 04
45 „ 135	38 05·6	0·10467	0·20870	64 08	**90**	90 00·0	0·31443	0·05818	0 00
h	180°-K	A	N	a₁	h	180°-K	A	N	a₁

Directions with regard to $(a_2 \overset{+}{-} a_1)$:—

Whenever h is greater than 90°, use $(a_2 - a_1)$.

If h is less than 90°, use :—

$(a_1 + a_2)$ when $(K \pm d)$ has been used,
$(a_1 - a_2)$ when $(d - K)$ has been used.

Always **name**
azimuth from
elevated pole.

Note that $\begin{cases} h = \text{L.H.A. of body } \textbf{West} \text{ of the meridian.} \\ h = (360° - \text{L.H.A.}) \text{ of body } \textbf{East} \text{ of the meridian.} \end{cases}$

Log. tan. $a_2 = N + \log. \tan. (K \overset{+}{-} d).$

LAT. 30°   ALT.—AZ. TABLE

h	K	A	N	a₁	h	K	A	N	a₁
°	° ′			° ′	°	° ′			° ′
0 or 180	30 00·0	0·00000	—	90 00	**45 „ 135**	39 13·9	0·10206	0·21298	63 26
1 „ 179	30 00·2	·00005	1·82061	89 30	**46 „ 134**	39 43·9	·10666	·20554	62 38
2 „ 178	30 00·9	·00020	1·51965	89 00	**47 „ 133**	40 15·0	·11135	·19834	61 48
3 „ 177	30 02·0	·00045	1·34367	88 30	**48 „ 132**	40 47·3	·11612	·19140	60 57
4 „ 176	30 03·6	·00079	1·21888	88 00	**49 „ 131**	41 20·9	·12100	·18469	60 06
5 „ 175	30 05·7	0·00124	1·12217	87 30	**50 „ 130**	41 55·8	0·12595	0·17822	59 13
6 „ 174	30 08·2	·00179	1·04323	87 00	**51 „ 129**	42 32·0	·13099	·17197	58 18
7 „ 173	30 11·2	·00243	0·97658	86 29	**52 „ 128**	43 09·6	·13612	·16594	57 23
8 „ 172	30 14·6	·00318	·91891	85 59	**53 „ 127**	43 48·7	·14131	·16012	56 26
9 „ 171	30 18·5	·00402	·86814	85 28	**54 „ 126**	44 29·2	·14659	·15451	55 28
10 „ 170	30 22·9	0·00497	0·82280	84 58	**55 „ 125**	45 11·3	0·15194	0·14910	54 28
11 „ 169	30 27·7	·00601	·78187	84 27	**56 „ 124**	45 54·9	·15734	·14390	53 27
12 „ 168	30 33·1	·00716	·74459	83 56	**57 „ 123**	46 40·2	·16281	·13888	52 24
13 „ 167	30 38·9	·00840	·71038	83 25	**58 „ 122**	47 27·2	·16834	·13405	51 20
14 „ 166	30 45·2	·00975	·67879	82 54	**59 „ 121**	48 15·9	·17390	·12940	50 14
15 „ 165	30 52·0	0·01119	0·64947	82 22	**60 „ 120**	49 06·4	0·17591	0·12494	49 06
16 „ 164	30 59·4	·01274	·62213	81 50	**61 „ 119**	49 58·8	·18515	·12065	47 57
17 „ 163	31 07·2	·01439	·59653	81 19	**62 „ 118**	50 53·0	·19082	·11653	46 16
18 „ 162	31 15·6	·01614	·57249	80 46	**63 „ 117**	51 49·2	·19650	·11259	45 32
19 „ 161	31 24·5	·01799	·54983	80 14	**64 „ 116**	52 47·5	·20218	·10881	44 17
20 „ 160	31 34·0	0·01994	0·52842	79 41	**65 „ 115**	53 47·8	0·20786	0·10519	43 00
21 „ 159	31 44·0	·02199	·50814	79 08	**66 „ 114**	54 50·1	·21352	·10174	41 41
22 „ 158	31 54·6	·02415	·48889	78 35	**67 „ 113**	55 54·7	·21915	·09844	40 20
23 „ 157	32 05·8	·02641	·47059	78 01	**68 „ 112**	57 01·4	·22473	·09530	38 56
24 „ 156	32 17·5	·02877	·45316	77 27	**69 „ 111**	58 10·3	·23026	·09232	37 31
25 „ 155	32 29·9	0·03123	0·43652	76 53	**70 „ 110**	59 21·5	0·23571	0·08948	36 03
26 „ 154	32 42·9	·03380	·42063	76 18	**71 „ 109**	60 34·9	·24107	·08680	34 33
27 „ 153	32 56·5	·03646	·40542	75 42	**72 „ 108**	61 50·6	·24633	·08426	33 01
28 „ 152	33 10·8	·03924	·39086	75 07	**73 „ 107**	63 08·5	·25146	·08187	31 27
29 „ 151	33 25·8	·04211	·37690	74 31	**74 „ 106**	64 28·8	·25644	·07963	29 50
30 „ 150	33 41·4	0·04509	0·36350	73 54	**75 „ 105**	65 51·2	0·26126	0·07753	28 11
31 „ 149	33 57·7	·04817	·35063	73 17	**76 „ 104**	67 15·9	·26590	·07557	26 30
32 „ 148	34 14·8	·05136	·33826	72 39	**77 „ 103**	68 42·8	·27034	·07375	24 47
33 „ 147	34 32·6	·05464	·32636	72 01	**78 „ 102**	70 11·7	·27455	·07207	23 02
34 „ 146	34 51·2	·05803	·31491	71 22	**79 „ 101**	71 42·7	·27852	·07052	21 15
35 „ 145	35 10·6	0·06153	0·30388	70 42	**80 „ 100**	73 15·6	0·28222	0·06912	19 26
36 „ 144	35 30·8	·06513	·29325	70 02	**81 „ 99**	74 50·4	·28565	·06785	17 35
37 „ 143	35 51·8	·06882	·28301	69 21	**82 „ 98**	76 26·8	·28877	·06672	15 42
38 „ 142	36 13·7	·07263	·27313	68 40	**83 „ 97**	78 04·8	·29157	·06572	13 48
39 „ 141	36 36·5	·07653	·26360	67 57	**84 „ 96**	79 44·3	·29403	·06486	11 52
40 „ 140	37 00·3	0·08054	0·25440	67 14	**85 „ 95**	81 24·9	0·29614	0·06413	9 56
41 „ 139	37 25·0	·08465	·24553	66 30	**86 „ 94**	83 06·6	·29788	·06353	7 58
42 „ 138	37 50·6	·08885	·23696	65 46	**87 „ 93**	84 49·2	·29926	·06307	5 59
43 „ 137	38 17·3	·09316	·22869	65 00	**88 „ 92**	86 32·4	·30024	·06273	4 00
44 „ 136	38 45·1	·09756	·22070	64 14	**89 „ 91**	88 16·1	·30083	·06254	2 00
45 „ 135	39 13·9	0·10206	0·21298	63 26	**90**	90 00·0	0·30103	0·06247	0 00
h	**180°-K**	**A**	**N**	**a₁**	**h**	**180°-K**	**A**	**N**	**a₁**

Directions with regard to (K $\pm$ d) :—

If l and d have **Same Name,** use (K $\smile$ d), i.e. (K — d) or (d — K).

If l and d have **Different Names,** use (K + d).

To name Intercept :—

If H_o is greater than H_c, intercept is **" Towards."**

If H_o is less than H_c, intercept is **" Away."**

Log. cosec. H_c = A + log. sec. (K $\pm$ d).

N.B. Whenever h is greater than 90°, subtract the quantity in the K column from 180°.

ALT.—AZ. TABLE LAT. 31°

h	K	A	N	a₁	h	K	A	N	a₁
°	° ′			° ′	°	° ′			° ′
0 or 180	31 00·0	0·00000	—	90 00	45 „ 135	40 21·4	0·09942	0·21745	62 45
1 „ 179	31 00·2	·00005	1·82508	89 29	46 „ 134	40 51·5	·10387	·21000	61 56
2 „ 178	31 00·9	·00019	1·52412	88 58	47 „ 133	41 22·9	·10841	·20281	61 05
3 „ 177	31 02·1	·00044	1·34813	88 27	48 „ 132	41 55·4	·11302	·19586	60 14
4 „ 176	31 03·7	·00078	1·22335	87 56	49 „ 131	42 29·1	·11773	·18915	59 21
5 „ 175	31 05·8	0·00122	1·12664	87 25	50 „ 130	43 04·1	0·12251	0·18268	58 28
6 „ 174	31 08·4	·00175	1·04770	86 54	51 „ 129	43 40·5	·12736	·17643	57 33
7 „ 173	31 11·4	·00238	0·98104	86 23	52 „ 128	44 18·2	·13230	·17040	56 36
8 „ 172	31 14·9	·00311	·92338	85 52	53 „ 127	44 57·3	·13730	·16458	55 39
9 „ 171	31 18·9	·00394	·87260	85 20	54 „ 126	45 37·8	·14237	·15898	54 40
10 „ 170	31 23·3	0·00487	0·82726	84 49	55 „ 125	46 19·9	0·14750	0·15357	53 40
11 „ 169	31 28·3	·00589	·78634	84 17	56 „ 124	47 03·4	·15269	·14836	52 38
12 „ 168	31 33·7	·00701	·74906	83 45	57 „ 123	47 48·6	·15793	·14334	51 35
13 „ 167	31 39·6	·00823	·71485	83 13	58 „ 122	48 35·4	·16322	·13851	50 30
14 „ 166	31 46·1	·00954	·68326	82 41	59 „ 121	49 23·9	·16854	·13387	49 24
15 „ 165	31 53·0	0·01096	0·65394	82 09	60 „ 120	50 14·1	0·17390	0·12940	48 16
16 „ 164	32 00·5	·01247	·62660	81 36	61 „ 119	51 06·1	·17928	·12511	47 06
17 „ 163	32 08·5	·01409	·60100	81 03	62 „ 118	51 59·9	·18468	·12100	45 55
18 „ 162	32 17·0	·01580	·57695	80 30	63 „ 117	52 55·6	·19009	·11705	44 41
19 „ 161	32 26·1	·01761	·55429	79 57	64 „ 116	53 53·2	·19549	·11327	43 26
20 „ 160	32 35·7	0·01951	0·53288	79 23	65 „ 115	54 52·8	0·20088	0·10966	42 09
21 „ 159	32 45·9	·02152	·51261	78 49	66 „ 114	55 54·3	·20625	·10620	40 51
22 „ 158	32 56·7	·02363	·49336	78 15	67 „ 113	56 57·9	·21158	·10291	39 30
23 „ 157	33 08·1	·02584	·47506	77 40	68 „ 112	58 03·5	·21686	·09977	38 07
24 „ 156	33 20·0	·02814	·45762	77 05	69 „ 111	59 11·2	·22208	·09678	36 42
25 „ 155	33 32·6	0·03055	0·44099	76 30	70 „ 110	60 21·0	0·22722	0·09395	35 15
26 „ 154	33 45·8	·03305	·42510	75 54	71 „ 109	61 33·0	·23227	·09126	33 46
27 „ 153	33 59·7	·03566	·40989	75 18	72 „ 108	62 47·0	·23720	·08873	32 15
28 „ 152	34 14·2	·03836	·39533	74 41	73 „ 107	64 03·2	·24202	·08634	30 42
29 „ 151	34 29·3	·04117	·38136	74 04	74 „ 106	65 21·4	·24669	·08409	29 06
30 „ 150	34 45·2	0·04407	0·36796	73 26	75 „ 105	66 41·8	0·25120	0·08199	27 29
31 „ 149	35 01·8	·04707	·35510	72 48	76 „ 104	68 04·1	·25554	·08003	25 50
32 „ 148	35 19·1	·05018	·34272	72 10	77 „ 103	69 28·5	·25968	·07821	24 09
33 „ 147	35 37·2	·05338	·33083	71 30	78 „ 102	70 54·8	·26360	·07653	22 26
34 „ 146	35 56·0	·05668	·31937	70 51	79 „ 101	72 22·9	·26730	·07499	20 41
35 „ 145	36 15·6	0·06009	0·30834	70 10	80 „ 100	73 52·8	0·27074	0·07358	18 54
36 „ 144	36 36·1	·06358	·29771	69 29	81 „ 99	75 24·4	·27392	·07231	17 06
37 „ 143	36 57·4	·06718	·28747	68 47	82 „ 98	76 57·5	·27681	·07118	15 16
38 „ 142	37 19·5	·07088	·27759	68 05	83 „ 97	78 32·1	·27941	·07018	13 25
39 „ 141	37 42·6	·07467	·26806	67 22	84 „ 96	80 07·9	·28169	·06932	11 32
40 „ 140	38 06·6	0·07856	0·25887	66 38	85 „ 95	81 44·8	0·28364	0·06859	9 38
41 „ 139	38 31·5	·08255	·24999	65 53	86 „ 94	83 22·7	·28525	·06799	7 44
42 „ 138	38 57·4	·08663	·24142	65 07	87 „ 93	85 01·3	·28652	·06753	5 49
43 „ 137	39 24·3	·09080	·23315	64 21	88 „ 92	86 40·6	·28743	·06720	3 53
44 „ 136	39 52·3	·09507	·22516	63 33	89 „ 91	88 20·2	·28797	·06700	1 56
45 „ 135	40 21·4	0·09942	0·21745	62 45	90	90 00·0	0·28816	0·06693	0 00
h	180°−K	A	N	a₁	h	180°−K	A	N	a₁

Directions with regard to (a₂ ± a₁) : –

Whenever h is greater than 90°, use (a₂ − a₁).

If h is less than 90°, use :—

(a₁ + a₂) when (K ± d) has been used,

(a₁ − a₂) when (d − K) has been used.

Always **name** **azimuth** from elevated pole.

Note that { h = L.H.A. of body **West** of the meridian.
{ h = (360° − L.H.A.) of body **East** of the meridian.

Log. tan. a₂ = N + log. tan. (K ± d).

M

LAT. 32°   ALT.—AZ. TABLE

h	K	A	N	a₁	h	K	A	N	a₁
°	° ′			° ′	°	° ′			° ′
0 or 180	32 00.0	0·00000	—	90 00	45 or 135	41 28.0	0·09677	0·22209	62 05
1 „ 179	32 00.2	·00005	1·82972	89 28	46 „ 134	41 58.3	·10107	·21465	61 15
2 „ 178	32 00.9	·00019	1·52876	88 56	47 „ 133	42 29.8	·10545	·20745	60 24
3 „ 177	32 02.1	·00043	1·35278	88 25	48 „ 132	43 02.5	·10991	·20051	59 31
4 „ 176	32 03.8	·00076	1·22799	87 53	49 „ 131	43 36.3	·11444	·19380	58 30
5 „ 175	32 05.9	0·00119	1·13128	87 21	50 „ 130	44 11.4	0·11905	0·18733	57 44
6 „ 174	32 08.5	·00171	1·05234	86 49	51 „ 129	44 47.8	·12373	·18108	56 48
7 „ 173	32 11.6	·00233	0·98569	86 17	52 „ 128	45 25.5	·12847	·17505	55 51
8 „ 172	32 15.1	·00305	·92802	85 44	53 „ 127	46 04.6	·13328	·16923	54 53
9 „ 171	32 19.2	·00386	·87725	85 12	54 „ 126	46 45.1	·13815	·16362	53 54
10 „ 170	32 23.7	0·00476	0·83191	84 40	55 „ 125	47 27.0	0·14308	0·15821	52 53
11 „ 169	32 28.8	·00576	·79098	84 07	56 „ 124	48 10.5	·14805	·15301	51 51
12 „ 168	32 34.3	·00686	·75370	83 34	57 „ 123	48 55.5	·15307	·14799	50 47
13 „ 167	32 40.3	·00805	·71949	83 01	58 „ 122	49 42.0	·15813	·14316	49 42
14 „ 166	32 46.9	·00934	·68790	82 28	59 „ 121	50 30.2	·16322	·13851	48 35
15 „ 165	32 54.0	0·01072	0·65858	81 55	60 „ 120	51 20.1	0·16834	0·13405	47 27
16 „ 164	33 01.6	·01220	·63124	81 22	61 „ 119	52 11.6	·17347	·12976	46 17
17 „ 163	33 09.7	·01378	·60564	80 48	62 „ 118	53 04.9	·17860	·12564	45 06
18 „ 162	33 18.4	·01545	·58160	80 14	63 „ 117	54 00.0	·18375	·12170	43 53
19 „ 161	33 27.6	·01722	·55894	79 40	64 „ 116	54 56.9	·18888	·11792	42 38
20 „ 160	33 37.4	0·01908	0·53753	79 05	65 „ 115	55 55.7	0·19400	0·11430	41 21
21 „ 159	33 47.7	·02104	·51725	78 30	66 „ 114	56 56.4	·19908	·11085	40 02
22 „ 158	33 58.7	·02310	·49800	77 55	67 „ 113	57 58.9	·20413	·10755	38 42
23 „ 157	34 10.2	·02526	·47970	77 19	68 „ 112	59 03.5	·20912	·10441	37 19
24 „ 156	34 22.3	·02751	·46227	76 43	69 „ 111	60 09.9	·21405	·10143	35 55
25 „ 155	34 35.1	0·02985	0·44563	76 07	70 „ 110	61 18.4	0·21889	0·09859	34 29
26 „ 154	34 48.5	·03230	·42974	75 31	71 „ 109	62 28.8	·22364	·09591	33 01
27 „ 153	35 02.5	·03484	·41453	74 53	72 „ 108	63 41.2	·22828	·09337	31 31
28 „ 152	35 17.2	·03748	·39997	74 16	73 „ 107	64 55.5	·23280	·09098	29 59
29 „ 151	35 32.6	·04021	·38601	73 38	74 „ 106	66 11.8	·23718	·08874	28 25
30 „ 150	35 48.7	0·04304	0·37261	72 59	75 „ 105	67 30.0	0·24141	0·08664	26 49
31 „ 149	36 05.5	·04596	·35974	72 20	76 „ 104	68 50.1	·24546	·08468	25 12
32 „ 148	36 23.0	·04899	·34737	71 41	77 „ 103	70 12.1	·24933	·08286	23 32
33 „ 147	36 41.3	·05211	·33547	71 01	78 „ 102	71 35.8	·25299	·08118	21 51
34 „ 146	37 00.4	·05532	·32402	70 20	79 „ 101	73 01.2	·25643	·07963	20 09
35 „ 145	37 20.2	0·05863	0·31299	69 39	80 „ 100	74 28.2	0·25964	0·07823	18 24
36 „ 144	37 40.9	·06203	·30236	68 57	81 „ 99	75 56.7	·26259	·07696	16 38
37 „ 143	38 02.4	·06553	·29212	68 14	82 „ 98	77 26.6	·26528	·07583	14 51
38 „ 142	38 24.8	·06911	·28224	67 31	83 „ 97	78 57.8	·26769	·07483	13 03
39 „ 141	38 48.1	·07279	·27271	66 47	84 „ 96	80 30.2	·26980	·07397	11 13
40 „ 140	39 12.3	0·07657	0·26351	66 02	85 „ 95	82 03.6	0·27161	0·07324	9 22
41 „ 139	39 37.4	·08043	·25464	65 16	86 „ 94	83 37.8	·27310	·07264	7 31
42 „ 138	40 03.5	·08439	·24607	64 30	87 „ 93	85 12.7	·27428	·07218	5 39
43 „ 137	40 30.6	·08843	·23780	63 42	88 „ 92	86 48.2	·27511	·07184	3 46
44 „ 136	40 58.8	·09256	·22981	62 54	89 „ 91	88 24.0	·27562	·07165	1 53
45 „ 135	41 28.0	0·09677	0·22209	62 05	90	90 00.0	0·27579	0·07158	0 00
h	180°-K	A	N	a₁	h	180°-K	A	N	a₁

Directions with regard to (K $\overset{+}{\underset{-}{\ }}$ d) :—

If l and d have **Same Name,** use (K ⌣ d), i.e. (K — d) or (d — K).
If l and d have **Different Names,** use (K + d).

To name Intercept :—

If H_o is greater than H_c, intercept is **" Towards."**
If H_o is less than H_c, intercept is **" Away."**

Log. cosec. H_c = A + log. sec. (K $\overset{+}{\underset{-}{\ }}$ d).

N.B.   Whenever h is greater than 90°, subtract the quantity in the K column from 180°.

ALT.—AZ. TABLE   LAT. 33°

h	K	A	N	a₁	h	K	A	N	a₁
°	° ′			° ′	°	° ′			° ′
0 or 180	33 00.0	0.00000	—	90 00	**45 or 135**	42 33.9	0.09411	0.22692	61 26
1 „ 179	33 00.2	.00005	1.83455	89 27	**46 „ 134**	43 04.3	.09826	.21947	60 35
2 „ 178	33 01.0	.00019	1.53359	88 55	**47 „ 133**	43 35.9	.10248	.21228	59 43
3 „ 177	33 02.2	.00042	1.35761	88 22	**48 „ 132**	44 08.6	.10678	.20534	58 50
4 „ 176	33 03.8	.00074	1.23282	87 49	**49 „ 131**	44 42.5	.11115	.19863	57 56
5 „ 175	33 06.0	0.00116	1.13611	87 16	**50 „ 130**	45 17.6	0.11559	0.19215	57 01
6 „ 174	33 08.6	.00168	1.05717	86 43	**51 „ 129**	45 54.0	.12009	.18591	56 05
7 „ 173	33 11.8	.00228	0.99051	86 10	**52 „ 128**	46 31.7	.12465	.17988	55 07
8 „ 172	33 15.4	.00298	.93285	85 37	**53 „ 127**	47 10.7	.12928	.17406	54 09
9 „ 171	33 19.5	.00377	.88208	85 04	**54 „ 126**	47 51.1	.13395	.16845	53 09
10 „ 170	33 24.1	0.00466	0.83674	84 31	**55 „ 125**	48 32.9	0.13867	0.16304	52 07
11 „ 169	33 29.2	.00563	.79581	83 57	**56 „ 124**	49 16.1	.14344	.15783	51 05
12 „ 168	33 34.8	.00670	.75853	83 24	**57 „ 123**	50 00.9	.14824	.15282	50 01
13 „ 167	33 41.0	.00787	.72432	82 50	**58 „ 122**	50 47.1	.15307	.14799	48 55
14 „ 166	33 47.6	.00913	.69273	82 16	**59 „ 121**	51 34.9	.15793	.14334	47 49
15 „ 165	33 54.8	0.01048	0.66341	81 42	**60 „ 120**	52 24.4	0.16281	0.13888	46 40
16 „ 164	34 02.5	.01193	.63607	81 07	**61 „ 119**	53 15.4	.16770	.13459	45 30
17 „ 163	34 10.8	.01346	.61047	80 33	**62 „ 118**	54 08.2	.17259	.13043	44 19
18 „ 162	34 19.6	.01510	.58643	79 58	**63 „ 117**	55 02.6	.17748	.12653	43 06
19 „ 161	34 28.9	.01682	.56377	79 23	**64 „ 116**	55 58.8	.18236	.12275	41 51
20 „ 160	34 38.9	0.01864	0.54236	78 47	**65 „ 115**	56 56.7	0.18721	0.11913	40 34
21 „ 159	34 49.4	.02056	.52208	78 11	**66 „ 114**	57 56.4	.19203	.11568	39 16
22 „ 158	35 00.5	.02257	.50283	77 35	**67 „ 113**	58 57.9	.19680	.11238	37 56
23 „ 157	35 12.2	.02467	.48453	76 59	**68 „ 112**	60 01.3	.20152	.10924	36 34
24 „ 156	35 24.5	.02686	.46710	76 22	**69 „ 111**	61 06.5	.20616	.10626	35 11
25 „ 155	35 37.4	0.02915	0.45046	75 45	**70 „ 110**	62 13.5	0.21073	0.10342	33 45
26 „ 154	35 51.0	.03153	.43457	75 07	**71 „ 109**	63 22.4	.21521	.10074	32 18
27 „ 153	36 05.2	.03401	.41936	74 29	**72 „ 108**	64 33.2	.21957	.09820	30 49
28 „ 152	36 20.1	.03658	.40480	73 51	**73 „ 107**	65 45.7	.22381	.09581	29 18
29 „ 151	36 35.6	.03924	.39084	73 12	**74 „ 106**	67 00.1	.22792	.09357	27 46
30 „ 150	36 51.9	0.04199	0.37744	72 33	**75 „ 105**	68 16.2	0.23188	0.09146	26 12
31 „ 149	37 08.9	.04484	.36457	71 53	**76 „ 104**	69 34.1	.23567	.08950	24 36
32 „ 148	37 26.6	.04778	.35220	71 12	**77 „ 103**	70 53.7	.23928	.08768	22 58
33 „ 147	37 45.1	.05082	.34030	70 31	**78 „ 102**	72 14.8	.24270	.08600	21 19
34 „ 146	38 04.4	.05394	.32885	69 50	**79 „ 101**	73 37.6	.24591	.08446	19 38
35 „ 145	38 24.4	0.05715	0.31782	69 08	**80 „ 100**	75 01.8	0.24890	0.08306	17 56
36 „ 144	38 45.3	.06046	.30719	68 25	**81 „ 99**	76 27.4	.25164	.08179	16 13
37 „ 143	39 07.0	.06385	.29695	67 41	**82 „ 98**	77 54.2	.25414	.08066	14 28
38 „ 142	39 29.5	.06733	.28707	66 57	**83 „ 97**	79 22.3	.25638	.07966	12 42
39 „ 141	39 53.0	.07090	.27754	66 12	**84 „ 96**	80 51.4	.25834	.07879	10 55
40 „ 140	40 17.4	0.07456	0.26834	65 26	**85 „ 95**	82 21.4	0.26001	0.07806	9 08
41 „ 139	40 42.7	.07830	.25947	64 40	**86 „ 94**	83 52.1	.26140	.07747	7 19
42 „ 138	41 08.9	.08213	.25090	63 53	**87 „ 93**	85 23.5	.26249	.07700	5 30
43 „ 137	41 36.2	.08604	.24263	63 04	**88 „ 92**	86 55.4	.26326	.07667	3 40
44 „ 136	42 04.5	.09004	.23464	62 15	**89 „ 91**	88 27.6	.26373	.07647	1 50
45 „ 135	42 33.9	0.09411	0.22692	61 26	**90**	90 00.0	0.26389	0.07641	0 00
h	180°-K	A	N	a₁	h	180°-K	A	N	a₁

Directions with regard to $(a_2 \overset{+}{\underset{-}{\sim}} a_1)$:—

Whenever h is greater than 90°, use $(a_2 - a_1)$.

If h is less than 90°, use :—

$(a_1 + a_2)$ when $(K \pm d)$ has been used,

$(a_1 - a_2)$ when $(d - K)$ has been used.

Always **name azimuth** from elevated pole.

Note that $\begin{cases} \text{h = L.H.A. of body } \textbf{West} \text{ of the meridian.} \\ \text{h = } (360° - \text{L.H.A.}) \text{ of body } \textbf{East} \text{ of the meridian.} \end{cases}$

Log. tan. a_2 = N + log. tan. $(K \overset{+}{\underset{-}{\sim}} d)$.

LAT. 34°   ALT.—AZ. TABLE

h	K		A	N	a₁		h	K		A	N	a₁	
°	°	′			°	′	°	°	′			°	′
0 or 180	34	00·0	0·00000	—	90	00	**45 or 135**	43	38·9	0·09143	0·23194	60	47
1 „ 179	34	00·2	·00005	1·83957	89	26	**46 „ 134**	44	09·4	·09544	·22449	59	56
2 „ 178	34	01·0	·00018	1·53861	88	53	**47 „ 133**	44	41·0	·09951	·21730	59	03
3 „ 177	34	02·2	·00041	1·36263	88	19	**48 „ 132**	45	13·8	·10366	·21035	58	09
4 „ 176	34	03·9	·00073	1·23784	87	46	**49 „ 131**	45	47·7	·10787	·20365	57	15
5 „ 175	34	06·1	0·00114	1·14113	87	12	**50 „ 130**	46	22·8	0·11213	0·19717	56	19
6 „ 174	34	08·8	·00164	1·06219	86	38	**51 „ 129**	46	59·1	·11646	·19092	55	22
7 „ 173	34	11·9	·00223	0·99553	86	04	**52 „ 128**	47	36·7	·12084	·18489	54	24
8 „ 172	34	15·6	·00291	·93787	85	30	**53 „ 127**	48	15·6	·12528	·17908	53	25
9 „ 171	34	19·8	·00368	·88709	84	56	**54 „ 126**	48	55·8	·12976	·17347	52	25
10 „ 170	34	24·5	0·00455	0·84176	84	22	**55 „ 125**	49	37·4	0·13428	0·16806	51	23
11 „ 169	34	29·7	·00550	·80083	83	48	**56 „ 124**	50	20·4	·13884	·16285	50	20
12 „ 168	34	35·4	·00655	·76355	83	13	**57 „ 123**	51	04·8	·14344	·15783	49	16
13 „ 167	34	41·6	·00769	·72934	82	39	**58 „ 122**	51	50·7	·14805	·15301	48	10
14 „ 166	34	48·3	·00892	·69775	82	04	**59 „ 121**	52	38·1	·15269	·14836	47	03
15 „ 165	34	55·6	0·01024	0·66843	81	29	**60 „ 120**	53	27·1	0·15734	0·14389	45	55
16 „ 164	35	03·4	·01165	·64109	80	53	**61 „ 119**	54	17·6	·16200	·13961	44	45
17 „ 163	35	11·8	·01315	·61541	80	18	**62 „ 118**	55	09·7	·16665	·13549	43	33
18 „ 162	35	20·7	·01474	·59144	79	42	**63 „ 117**	56	03·4	·17130	·13154	42	20
19 „ 161	35	30·2	·01643	·56878	79	06	**64 „ 116**	56	58·8	·17593	·12777	41	06
20 „ 160	35	40·2	0·01820	0·54737	78	30	**65 „ 115**	57	55·8	0·18053	0·12415	39	49
21 „ 159	35	50·9	·02007	·52710	77	53	**66 „ 114**	58	54·6	·18509	·12070	38	32
22 „ 158	36	02·1	·02202	·50785	77	16	**67 „ 113**	59	55·0	·18960	·11740	37	12
23 „ 157	36	13·9	·02407	·48955	76	39	**68 „ 112**	60	57·2	·19406	·11426	35	51
24 „ 156	36	26·4	·02621	·47211	76	01	**69 „ 111**	62	01·0	·19845	·11127	34	28
25 „ 155	36	39·5	0·02844	0·45548	75	23	**70 „ 110**	63	06·7	0·20275	0·10844	33	04
26 „ 154	36	53·2	·03076	·43958	74	45	**71 „ 109**	64	14·1	·20696	·10576	31	37
27 „ 153	37	07·6	·03317	·42438	74	06	**72 „ 108**	65	23·1	·21107	·10322	30	10
28 „ 152	37	22·6	·03567	·40982	73	26	**73 „ 107**	66	33·9	·21505	·10083	28	40
29 „ 151	37	38·4	·03826	·39585	72	47	**74 „ 106**	67	46·4	·21890	·09858	27	09
30 „ 150	37	54·8	0·04094	0·38246	72	06	**75 „ 105**	69	00·4	0·22261	0·09648	25	36
31 „ 149	38	12·0	·04371	·36959	71	26	**76 „ 104**	70	16·1	·22616	·09452	24	02
32 „ 148	38	29·9	·04656	·35722	70	44	**77 „ 103**	71	33·4	·22954	·09270	22	26
33 „ 147	38	48·5	·04951	·34532	70	03	**78 „ 102**	72	52·1	·23273	·09102	20	49
34 „ 146	39	07·9	·05254	·33386	69	20	**79 „ 101**	74	12·3	·23572	·08948	19	10
35 „ 145	39	28·1	0·05566	0·32283	68	37	**80 „ 100**	75	33·8	0·23850	0·08807	17	30
36 „ 144	39	49·2	·05887	·31221	67	53	**81 „ 99**	76	56·6	·24107	·08681	15	49
37 „ 143	40	11·0	·06216	·30196	67	09	**82 „ 98**	78	20·5	·24339	·08567	14	06
38 „ 142	40	33·7	·06554	·29208	66	24	**83 „ 97**	79	45·5	·24547	·08467	12	23
39 „ 141	40	57·3	·06900	·28255	65	38	**84 „ 96**	81	11·5	·24729	·08381	10	39
40 „ 140	41	21·9	0·07254	0·27336	64	52	**85 „ 95**	82	38·2	0·24884	0·08308	8	54
41 „ 139	41	47·3	·07616	·26442	64	05	**86 „ 94**	84	05·7	·25013	·08247	7	08
42 „ 138	42	13·7	·07986	·25591	63	16	**87 „ 93**	85	33·8	·25113	·08202	5	21
43 „ 137	42	41·1	·08364	·24764	62	28	**88 „ 92**	87	02·3	·25186	·08169	3	34
44 „ 136	43	09·5	·08750	·23965	61	38	**89 „ 91**	88	31·1	·25229	·08149	1	47
45 „ 135	43	38·9	0·09143	0·23194	60	47	**90**	90	00·0	0·25244	0·08143	0	00
h	180°-K		A	N	a₁		**h**	180°-K		A	N	a₁	

Directions with regard to (K $\overset{+}{\sim}$ d) :—

If l and d have **Same Name,** use (K $\smile$ d), i.e. (K — d) or (d — K).

If l and d have **Different Names,** use (K + d).

To name Intercept :—

If H_o is greater than H_c, intercept is " **Towards.** "

If H_o is less than H_c, intercept is " **Away.** "

Log. cosec. H_c = A + log. sec. (K $\overset{+}{\sim}$ d).

N.B.   Whenever h is greater than 90°, subtract the quantity in the K column from 180°.

ALT.—AZ. TABLE LAT. 35°

h	K	A	N	a₁	h	K	A	N	a₁
°	° ′			° ′	°	° ′			° ′
0 or 180	35 00·0	0·00000	—	90 00	45 or 135	44 43·1	0·08875	0·23715	60 10
1 ,, 179	35 00·2	·00004	1·84478	89 26	46 ,, 134	45 13·7	·09261	·22970	59 17
2 ,, 178	35 01·0	·00018	1·54382	88 51	47 ,, 133	45 45·3	·09654	·22251	58 24
3 ,, 177	35 02·2	·00040	1·36783	88 17	48 ,, 132	46 18·0	·10053	·21556	57 30
4 ,, 176	35 03·9	·00071	1·24305	87 43	49 ,, 131	46 51·9	·10458	·20885	56 35
5 ,, 175	35 06·2	0·00111	1·14634	87 08	50 ,, 130	47 26·9	0·10868	0·20238	55 39
6 ,, 174	35 08·9	·00160	1·06740	86 33	51 ,, 129	48 03·1	·11284	·19613	54 41
7 ,, 173	35 12·1	·00217	1·00074	85 58	52 ,, 128	48 40·6	·11705	·19010	53 43
8 ,, 172	35 15·8	·00284	0·94308	85 23	53 ,, 127	49 19·3	·12130	·18429	52 43
9 ,, 171	35 20·1	·00360	·89230	84 49	54 ,, 126	49 59·3	·12559	·17868	51 43
10 ,, 170	35 24·8	0·00444	0·84696	84 13	55 ,, 125	50 40·6	0·12992	0·17327	50 41
11 ,, 169	35 30·0	·00537	·80604	83 38	56 ,, 124	51 23·3	·13428	·16806	49 37
12 ,, 168	35 35·8	·00639	·76876	83 03	57 ,, 123	52 07·4	·13867	·16304	48 33
13 ,, 167	35 42·1	·00750	·73455	82 27	58 ,, 122	52 52·9	·14308	·15821	47 27
14 ,, 166	35 49·0	·00870	·70296	81 52	59 ,, 121	53 39·8	·14750	·15357	46 20
15 ,, 165	35 56·3	0·00999	0·67364	81 16	60 ,, 120	54 28·2	0·15194	0·14910	45 11
16 ,, 164	36 04·2	·01136	·64630	80 40	61 ,, 119	55 18·1	·15637	·14482	44 01
17 ,, 163	36 12·7	·01283	·62070	80 03	62 ,, 118	56 09·6	·16079	·14070	42 50
18 ,, 162	36 21·7	·01438	·59665	79 27	63 ,, 117	57 02·5	·16521	·13675	41 37
19 ,, 161	36 31·3	·01602	·57399	78 50	64 ,, 116	57 57·1	·16960	·13297	40 23
20 ,, 160	36 41·5	0·01775	0·55258	78 12	65 ,, 115	58 53·2	0·17396	0·12936	39 07
21 ,, 159	36 52·2	·01957	·53231	77 35	66 ,, 114	59 50·9	·17827	·12590	37 49
22 ,, 158	37 03·6	·02147	·51306	76 57	67 ,, 113	60 50·3	·18255	·12261	36 30
23 ,, 157	37 15·6	·02347	·49476	76 19	68 ,, 112	61 51·2	·18675	·11947	35 10
24 ,, 156	37 28·2	·02555	·47732	75 40	69 ,, 111	62 53·8	·19089	·11642	33 48
25 ,, 155	37 41·4	0·02772	0·46069	75 02	70 ,, 110	63 58·0	0·19495	0·11365	32 24
26 ,, 154	37 55·2	·02998	·44479	74 22	71 ,, 109	65 03·8	·19891	·11096	30 59
27 ,, 153	38 09·7	·03232	·42959	73 43	72 ,, 108	66 11·2	·20277	·10843	29 32
28 ,, 152	38 24·9	·03475	·41503	73 02	73 ,, 107	67 20·2	·20651	·10604	28 04
29 ,, 151	38 40·8	·03727	·40106	72 22	74 ,, 106	68 30·8	·21012	·10379	26 34
30 ,, 150	38 57·4	0·03987	0·38766	71 41	75 ,, 105	69 42·8	0·21360	0·10169	25 02
31 ,, 149	39 14·7	·04256	·37480	70 59	76 ,, 104	70 56·4	·21692	·09973	23 30
32 ,, 148	39 32·7	·04534	·36242	70 17	77 ,, 103	72 11·4	·22008	·09791	21 56
33 ,, 147	39 51·5	·04820	·35053	69 34	78 ,, 102	73 27·7	·22306	·09623	20 20
34 ,, 146	40 11·1	·05114	·33907	68 51	79 ,, 101	74 45·4	·22585	·09469	18 43
35 ,, 145	40 31·4	0·05417	0·32804	68 07	80 ,, 100	76 04·3	0·22845	0·09328	17 05
36 ,, 144	40 52·6	·05727	·31742	67 23	81 ,, 99	77 24·4	·23083	·09201	15 26
37 ,, 143	41 14·6	·06046	·30717	66 38	82 ,, 98	78 45·5	·23300	·09088	13 46
38 ,, 142	41 37·4	·06373	·29729	65 52	83 ,, 97	80 07·6	·23493	·08988	12 05
39 ,, 141	42 01·1	·06708	·28776	65 05	84 ,, 96	81 30·6	·23662	·08902	10 23
40 ,, 140	42 25·7	0·07051	0·27857	64 18	85 ,, 95	82 54·3	0·23807	0·08829	8 40
41 ,, 139	42 51·3	·07401	·26969	63 30	86 ,, 94	84 18·6	·23927	·08769	6 57
42 ,, 138	43 17·8	·07759	·26112	62 41	87 ,, 93	85 43·5	·24020	·08723	5 13
43 ,, 137	43 45·2	·08124	·25285	61 52	88 ,, 92	87 08·8	·24087	·08690	3 29
44 ,, 136	44 13·7	·08497	·24486	61 01	89 ,, 91	88 34·3	·24127	·08670	1 45
45 ,, 135	44 43·1	0·08875	0·23715	60 10	90	90 00·0	0·24141	0·08663	0 00
h	180°-K	A	N	a₁	h	180°-K	A	N	a₁

Directions with regard to $(a_2 \overset{+}{\underset{\sim}{}} a_1)$:—

    Whenever h is greater than 90°, use $(a_2 - a_1)$.   Always **name**

    If h is less than 90°, use :—   **azimuth** from

        $(a_1 + a_2)$ when $(K \pm d)$ has been used,   elevated pole.

        $(a_1 - a_2)$ when $(d - K)$ has been nsed.

Note that { h = L.H.A. of body **West** of the meridian.
{ h = (360° − L.H.A.) of body **East** of the meridian.

Log. tan. a_2 = N + log. tan. $(K \overset{+}{\underset{\sim}{}} d)$.

LAT. 36°   ALT.—AZ. TABLE

h	K	A	N	a₁	h	K	A	N	a₁
°	° ′			° ′	°	° ′			° ′
0 or 180	36 00·0	0·00000	—	90 00	45 or 135	45 46·6	0·08607	0·24256	59 33
1 „ 179	36 00·2	·00004	1·85019	89 25	46 „ 134	46 17·1	·08979	·23511	58 40
2 „ 178	36 01·0	·00017	1·54922	88 40	47 „ 133	46 48·7	·09357	·22791	57 47
3 „ 177	36 02·2	·00039	1·37324	88 14	48 „ 132	47 21·3	·09741	·22097	56 52
4 „ 176	36 04·0	·00069	1·24846	87 39	49 „ 131	47 55·1	·10130	·21426	55 56
5 „ 175	36 06·2	0·00108	1·15175	87 03	50 „ 130	48 30·0	0·10524	0·20779	54 59
6 „ 174	36 09·0	·00156	1·07281	86 28	51 „ 129	49 06·1	·10923	·20154	54 02
7 „ 173	36 12·2	·00212	1·00615	85 52	52 „ 128	49 43·4	·11326	·19551	53 03
8 „ 172	36 16·0	·00277	0·94849	85 17	53 „ 127	50 21·8	·11734	·18969	52 03
9 „ 171	36 20·3	·00351	·89771	84 41	54 „ 126	51 01·6	·12145	·18408	51 02
10 „ 170	36 25·1	0·00433	0·85237	84 05	55 „ 125	51 42·6	0·12559	0·17868	49 59
11 „ 169	36 30·4	·00524	·81144	83 29	56 „ 124	52 25·0	·12976	·17347	48 56
12 „ 168	36 36·2	·00623	·77416	82 53	57 „ 123	53 08·6	·13395	·16845	47 51
13 „ 167	36 42·6	·00731	·73995	82 16	58 „ 122	53 53·6	·13815	·16362	46 45
14 „ 166	36 49·5	·00848	·70837	81 40	59 „ 121	54 40·1	·14237	·15898	45 38
15 „ 165	36 57·0	0·00974	0·67905	81 03	60 „ 120	55 27·9	0·14659	0·15451	44 29
16 „ 164	37 05·0	·01108	·65170	80 26	61 „ 119	56 17·1	·15081	·15022	43 19
17 „ 163	37 13·5	·01250	·62611	79 49	62 „ 118	57 07·8	·15501	·14611	42 08
18 „ 162	37 22·6	·01401	·60206	79 11	63 „ 117	58 00·0	·15920	·14216	40 55
19 „ 161	37 32·3	·01561	·57940	78 33	64 „ 116	58 57·3	·16337	·13838	39 41
20 „ 160	37 42·6	0·01730	0·55799	77 55	65 „ 115	59 48·8	0·16750	0·13477	38 26
21 „ 159	37 53·5	·01907	·53771	77 17	66 „ 114	60 45·5	·17158	·13131	37 09
22 „ 158	38 04·9	·02092	·51847	76 38	67 „ 113	61 43·7	·17562	·12802	35 50
23 „ 157	38 17·0	·02286	·50016	75 59	68 „ 112	62 43·5	·17959	·12488	34 30
24 „ 156	38 29·7	·02489	·48273	75 20	69 „ 111	63 44·7	·18349	·12189	33 09
25 „ 155	38 43·0	0·02699	0·46609	74 40	70 „ 110	64 47·5	0·18732	0·11906	31 46
26 „ 154	38 57·0	·02919	·45020	74 00	71 „ 109	65 51·8	·19105	·11637	30 22
27 „ 153	39 11·7	·03147	·43499	73 20	72 „ 108	66 57·5	·19468	·11384	28 56
28 „ 152	39 27·0	·03383	·42043	72 39	73 „ 107	68 04·8	·19819	·11145	27 29
29 „ 151	39 43·0	·03627	·40647	71 57	74 „ 106	69 13·4	·20158	·10920	26 00
30 „ 150	39 59·7	0·03880	0·39307	71 15	75 „ 105	70 23·5	0·20484	0·10710	24 30
31 „ 149	40 17·1	·04141	·38020	70 33	76 „ 104	71 35·0	·20795	·10514	22 59
32 „ 148	40 35·2	·04410	·36783	69 50	77 „ 103	72 47·8	·21090	·10332	21 27
33 „ 147	40 54·1	·04687	·35593	69 06	78 „ 102	74 01·8	·21368	·10164	19 53
34 „ 146	41 13·8	·04973	·34448	68 22	79 „ 101	75 17·1	·21629	·10009	18 18
35 „ 145	41 34·3	0·05266	0·33345	67 38	80 „ 100	76 33·5	0·21872	0·09869	16 42
36 „ 144	41 55·5	·05566	·32282	66 53	81 „ 99	77 50·9	·22094	·09742	15 05
37 „ 143	42 17·6	·05875	·31258	66 07	82 „ 98	79 09·4	·22296	·09629	13 27
38 „ 142	42 40·6	·06192	·30270	65 20	83 „ 97	80 28·7	·22476	·09529	11 48
39 „ 141	43 04·4	·06516	·29317	64 33	84 „ 96	81 48·8	·22633	·09443	10 08
40 „ 140	43 29·0	0·06847	0·28397	63 45	85 „ 95	83 09·6	0·22768	0·09370	8 28
41 „ 139	43 54·6	·07185	·27510	62 56	86 „ 94	84 30·9	·22879	·09310	6 47
42 „ 138	44 21·2	·07531	·26653	62 07	87 „ 93	85 52·8	·22966	·09264	5 06
43 „ 137	44 48·7	·07883	·25826	61 16	88 „ 92	87 15·0	·23028	·09231	3 24
44 „ 136	45 17·1	·08242	·25027	60 25	89 „ 91	88 37·4	·23065	·09211	1 42
45 „ 135	45 46·6	0·08607	0·24256	59 33	90	90 00·0	0·23078	0·09204	0 00
h	180°-K	A	N	a₁	h	180°-K	A	N	a₁

Directions with regard to (K ± d) :—

If l and d have **Same Name,** use (K ⌣ d), i.e. (K — d) or (d — K).
If l and d have **Different Names,** use (K + d).

To name Intercept :—

If Hₒ is greater than Hc, intercept is "**Towards.**
If Hₒ is less than Hc, intercept is "**Away.**"

Log. cosec. Hc = A + log. sec. (K ± d).

N.B.   Whenever h is greater than 90°, subtract the quantity in the K column from 180°.

ALT.—AZ. TABLE   LAT. 37°

h	K	A	N	a₁	h	K	A	N	a₁
°	° ′			° ′	°	° ′			° ′
0 or 180	37 00·0	0·00000	—	90 00	**45 or 135**	46 49·3	0·08340	0·24817	58 58
1 ,, 179	37 00·3	·00004	1·85580	89 24	**46 ,, 134**	47 19·7	·08698	·24072	58 04
2 ,, 178	37 01·0	·00017	1·55483	88 48	**47 ,, 133**	47 51·2	·09061	·23352	57 10
3 ,, 177	37 02·3	·00038	1·37885	88 12	**48 ,, 132**	48 23·8	·09430	·22658	56 15
4 ,, 176	37 04·0	·00067	1·25407	87 35	**49 ,, 131**	48 57·4	·09803	·21987	55 18
5 ,, 175	37 06·3	0·00105	1·15735	86 59	**50 ,, 130**	49 32·1	0·10181	0·21340	54 21
6 ,, 174	37 09·1	·00152	1·07832	86 23	**51 ,, 129**	50 08·0	·10564	·20715	53 23
7 ,, 173	37 12·4	·00207	1·01176	85 46	**52 ,, 128**	50 45·0	·10951	·20112	52 24
8 ,, 172	37 16·2	·00270	0·95410	85 10	**53 ,, 127**	51 23·3	·11340	·19530	51 23
9 ,, 171	37 20·5	·00342	·90332	84 33	**54 ,, 126**	52 02·7	·11734	·18969	50 22
10 ,, 170	37 25·3	0·00422	0·85798	83 57	**55 ,, 125**	52 43·4	0·12130	0·18429	49 19
11 ,, 169	37 30·7	·00510	·81705	83 20	**56 ,, 124**	53 25·3	·12528	·17908	48 16
12 ,, 168	37 36·6	·00607	·77977	82 43	**57 ,, 123**	54 08·5	·12928	·17406	47 11
13 ,, 167	37 43·0	·00712	·74556	82 05	**58 ,, 122**	54 53·0	·13328	·16923	46 05
14 ,, 166	37 50·0	·00826	·71398	81 28	**59 ,, 121**	55 38·9	·13730	·16458	44 57
15 ,, 165	37 57·5	0·00948	0·68465	80 50	**60 ,, 120**	56 26·1	0·14131	0·16012	43 49
16 ,, 164	38 05·6	·01079	·65731	80 13	**61 ,, 119**	57 14·7	·14533	·15583	42 39
17 ,, 163	38 14·3	·01217	·63172	79 34	**62 ,, 118**	58 04·6	·14932	·15172	41 28
18 ,, 162	38 23·5	·01365	·60767	78 56	**63 ,, 117**	58 55·9	·15329	·14777	40 15
19 ,, 161	38 33·2	·01520	·58501	78 18	**64 ,, 116**	59 48·7	·15724	·14399	39 01
20 ,, 160	38 43·6	0·01684	0·56360	77 39	**65 ,, 115**	60 42·9	0·16115	0·14037	37 46
21 ,, 159	38 54·6	·01856	·54332	77 00	**66 ,, 114**	61 38·5	·16502	·13692	36 30
22 ,, 158	39 06·1	·02036	·52408	76 20	**67 ,, 113**	62 35·5	·16883	·13362	35 12
23 ,, 157	39 18·3	·02225	·50577	75 40	**68 ,, 112**	63 33·9	·17258	·13048	33 53
24 ,, 156	39 31·1	·02421	·48834	75 00	**69 ,, 111**	64 33·9	·17627	·12750	32 32
25 ,, 155	39 44·5	0·02626	0·47170	74 19	**70 ,, 110**	65 35·3	0·17987	0·12466	31 10
26 ,, 154	39 58·6	·02839	·45581	73 39	**71 ,, 109**	66 38·0	·18338	·12198	29 47
27 ,, 153	40 13·3	·03060	·44060	72 57	**72 ,, 108**	67 42·1	·18679	·11944	28 22
28 ,, 152	40 28·8	·03290	·42604	72 15	**73 ,, 107**	68 47·7	·19009	·11705	26 56
29 ,, 151	40 44·8	·03527	·41208	71 33	**74 ,, 106**	69 54·5	·19327	·11481	25 29
30 ,, 150	41 01·6	0·03772	0·39868	70 50	**75 ,, 105**	71 02·7	0·19632	0·11271	24 00
31 ,, 149	41 19·2	·04025	·38581	70 07	**76 ,, 104**	72 12·1	·19924	·11075	22 30
32 ,, 148	41 37·4	·04286	·37344	69 23	**77 ,, 103**	73 22·7	·20200	·10893	20 59
33 ,, 147	41 56·4	·04554	·36154	68 39	**78 ,, 102**	74 34·5	·20460	·10725	19 27
34 ,, 146	42 16·2	·04831	·35009	67 54	**79 ,, 101**	75 47·4	·20704	·10570	17 54
35 ,, 145	42 36·7	0·05115	0·33906	67 09	**80 ,, 100**	77 01·4	0·20930	0·10430	16 20
36 ,, 144	42 58·0	·05405	·32843	66 23	**81 ,, 99**	78 16·3	·21137	·10303	14 45
37 ,, 143	43 20·2	·05704	·31819	65 36	**82 ,, 98**	79 32·2	·21325	·10190	13 09
38 ,, 142	43 43·2	·06010	·30831	64 49	**83 ,, 97**	80 48·8	·21493	·10090	11 32
39 ,, 141	44 07·0	·06323	·29878	64 01	**84 ,, 96**	82 06·2	·21640	·10004	9 54
40 ,, 140	44 31·7	0·06643	0·28958	63 12	**85 ,, 95**	83 24·2	0·21765	0·09931	8 16
41 ,, 139	44 57·4	·06969	·28071	62 23	**86 ,, 94**	84 42·7	·21869	·09871	6 38
42 ,, 138	45 23·9	·07302	·27214	61 33	**87 ,, 93**	86 01·6	·21950	·09825	4 59
43 ,, 137	45 51·4	·07642	·26387	60 42	**88 ,, 92**	87 20·9	·22007	·09792	3 19
44 ,, 136	46 19·8	·07988	·25588	59 50	**89 ,, 91**	88 40·4	·22042	·09772	1 40
45 ,, 135	46 49·3	0·08340	0·24817	58 58	**90**	90 00·0	0·22054	0·09765	0 00
h	180°-K	A	N	a₁	h	180°-K	A	N	a₁

Directions with regard to $(a_2 \overset{+}{\sim} a_1)$:—

Whenever h is greater than 90°, use $(a_2 - a_1)$.

If h is less than 90°, use :—

$(a_1 + a_2)$ when $(K \pm d)$ has been used,

$(a_1 - a_2)$ when $(d - K)$ has been used.

Always **name azimuth** from elevated pole.

Note that $\begin{cases} \text{h} = \text{L.H.A. of body } \textbf{West} \text{ of the meridian.} \\ \text{h} = (360° - \text{L.H.A.}) \text{ of body } \textbf{East} \text{ of the meridian.} \end{cases}$

Log. tan. $a_2 = N + \log. \tan. (K \overset{+}{\sim} d)$.

LAT. 38°   ALT.—AZ. TABLE

h	K	A	N	a₁	h	K	A	N	a₁
°	° ′			° ′	°	° ′			° ′
0 or 180	38 00.0	0.00000	—	90 00	45 or 135	47 51.2	0.08073	0.25392	58 23
1 ,, 179	38 00.3	.00004	1.86161	89 23	46 ,, 134	48 21.5	.08417	.24653	57 29
2 ,, 178	38 01.0	.00016	1.56065	88 46	47 ,, 133	48 50.9	.08766	.23934	56 34
3 ,, 177	38 02.3	.00037	1.38467	88 09	48 ,, 132	49 25.3	.09120	.23239	55 38
4 ,, 176	38 04.1	.00066	1.25988	87 32	49 ,, 131	49 58.8	.09478	.22569	54 42
5 ,, 175	38 06.4	0.00103	1.16317	86 55	50 ,, 130	50 33.3	0.09841	0.21921	53 44
6 ,, 174	38 09.2	.00148	1.08423	86 18	51 ,, 129	51 08.9	.10207	.21296	52 45
7 ,, 173	38 12.5	.00201	1.01757	85 41	52 ,, 128	51 45.7	.10577	.20694	51 46
8 ,, 172	38 16.3	.00263	0.95991	85 03	53 ,, 127	52 23.6	.10951	.20112	50 45
9 ,, 171	38 20.7	.00333	.90914	84 26	54 ,, 126	53 02.7	.11326	.19551	49 43
10 ,, 170	38 25.6	0.00410	0.86380	83 48	55 ,, 125	53 43.0	0.11705	0.19010	48 41
11 ,, 169	38 31.0	.00497	.82287	83 11	56 ,, 124	54 24.4	.12084	.18489	47 37
12 ,, 168	38 36.9	.00591	.78559	82 33	57 ,, 123	55 07.2	.12465	.17988	46 32
13 ,, 167	38 43.4	.00693	.75138	81 55	58 ,, 122	55 51.1	.12847	.17505	45 26
14 ,, 166	38 50.5	.00804	.71979	81 16	59 ,, 121	56 36.4	.13230	.17040	44 18
15 ,, 165	38 58.1	0.00923	0.69047	80 38	60 ,, 120	57 22.9	0.13612	0.16594	43 10
16 ,, 164	39 06.2	.01049	.66313	79 59	61 ,, 119	58 10.8	.13993	.16165	42 00
17 ,, 163	39 14.9	.01184	.63753	79 20	62 ,, 118	58 59.9	.14372	.15753	40 49
18 ,, 162	39 24.2	.01327	.61349	78 41	63 ,, 117	59 50.4	.14749	.15359	39 37
19 ,, 161	39 34.0	.01478	.59083	78 02	64 ,, 116	60 42.2	.15123	.14981	38 23
20 ,, 160	39 44.5	0.01638	0.56942	77 22	65 ,, 115	61 35.4	0.15493	0.14619	37 08
21 ,, 159	39 55.5	.01805	.54914	76 42	66 ,, 114	62 29.9	.15859	.14274	35 52
22 ,, 158	40 07.1	.01980	.52989	76 02	67 ,, 113	63 25.8	.16219	.13944	34 35
23 ,, 157	40 19.4	.02163	.51159	75 21	68 ,, 112	64 23.0	.16573	.13630	33 16
24 ,, 156	40 32.3	.02354	.49415	74 40	69 ,, 111	65 21.6	.16920	.13332	31 53
25 ,, 155	40 45.8	0.02553	0.47752	73 59	70 ,, 110	66 21.5	0.17259	0.13048	30 35
26 ,, 154	40 59.9	.02759	.46163	73 17	71 ,, 109	67 22.7	.17589	.12780	29 13
27 ,, 153	41 14.8	.02974	.44642	72 35	72 ,, 108	68 25.2	.17910	.12526	27 49
28 ,, 152	41 30.3	.03196	.43186	71 52	73 ,, 107	69 29.0	.18220	.12287	26 25
29 ,, 151	41 46.4	.03426	.41790	71 09	74 ,, 106	70 34.0	.18519	.12063	24 58
30 ,, 150	42 03.3	0.03663	0.40450	70 26	75 ,, 105	71 40.3	0.18805	0.11852	23 31
31 ,, 149	42 20.9	.03908	.39163	69 42	76 ,, 104	72 47.7	.19077	.11656	22 03
32 ,, 148	42 39.2	.04161	.37926	68 57	77 ,, 103	73 56.3	.19337	.11474	20 33
33 ,, 147	42 58.3	.04421	.36736	68 12	78 ,, 102	75 05.9	.19580	.11306	19 03
34 ,, 146	43 18.1	.04688	.35591	67 27	79 ,, 101	76 16.5	.19808	.11152	17 31
35 ,, 145	43 38.7	0.04963	0.34488	66 41	80 ,, 100	77 28.1	0.20019	0.11012	15 59
36 ,, 144	44 00.1	.05244	.33425	65 54	81 ,, 99	78 40.7	.20212	.10885	14 26
37 ,, 143	44 22.2	.05532	.32400	65 07	82 ,, 98	79 54.0	.20387	.10771	12 52
38 ,, 142	44 45.3	.05827	.31413	64 19	83 ,, 97	81 08.0	.20544	.10672	11 17
39 ,, 141	45 09.1	.06130	.30460	63 30	84 ,, 96	82 22.8	.20681	.10585	9 41
40 ,, 140	45 33.9	0.06438	0.29540	62 41	85 ,, 95	83 38.1	0.20797	0.10512	8 05
41 ,, 139	45 59.5	.06753	.28652	61 51	86 ,, 94	84 53.9	.20894	.10452	6 29
42 ,, 138	46 26.0	.07074	.27796	61 00	87 ,, 93	86 10.1	.20969	.10406	4 52
43 ,, 137	46 53.4	.07401	.26968	60 08	88 ,, 92	87 26.5	.21023	.10373	3 15
44 ,, 136	47 21.8	.07734	.26170	59 16	89 ,, 91	88 43.2	.21055	.10353	1 37
45 ,, 135	47 51.2	0.08073	0.25392	58 23	90	90 00.0	0.21066	0.10347	0 00
h	180°-K	A	N	a₁	h	180°-K	A	N	a₁

Directions with regard to (K ± d) :—

If l and d have **Same Name,** use (K ⌣ d), i.e. (K — d) or (d — K).

If l and d have **Different Names,** use (K + d).

To name Intercept :—

If H₀ is greater than H_c, intercept is **" Towards."**

If H₀ is less than H_c, intercept is **" Away."**

Log. cosec. H₀ = A + iog. sec. (K ± d).

N.B. Whenever h is greater than 90°, subtract the quantity in the K column from 180°.

ALT.—AZ. TABLE — LAT. 39°

h	K	A	N	a_1	h	K	A	N	a_1
°	° ′			° ′	°	° ′			° ′
0 or 180	39 00·0	0·00000	—	90 00	45 or 135	48 52·3	0·07806	0·26001	57 49
1 „ 179	39 00·3	·00004	1·86764	89 22	46 „ 134	49 22·6	·08137	·25256	56 55
2 „ 178	39 01·0	·00016	1·56668	88 44	47 „ 133	49 53·8	·08472	·24537	55 59
3 „ 177	39 02·3	·00036	1·39070	88 07	48 „ 132	50 26·0	·08812	·23842	55 03
4 „ 176	39 04·1	·00064	1·26591	87 29	49 „ 131	50 59·2	·09155	·23172	54 06
5 „ 175	39 06·4	0·00100	1·16920	86 51	50 „ 130	51 33·5	0·09503	0·22524	53 08
6 „ 174	39 09·2	·00144	1·09026	86 13	51 „ 129	52 08·9	·09853	·21899	52 09
7 „ 173	39 12·6	·00196	1·02360	85 35	52 „ 128	52 45·3	·10207	·21296	51 09
8 „ 172	39 16·5	·00256	0·96594	84 57	53 „ 127	53 22·9	·10564	·20715	50 08
9 „ 171	39 20·9	·00323	·91516	84 18	54 „ 126	54 01·5	·10923	·20154	49 06
10 „ 170	39 25·8	0·00399	0·86983	83 40	55 „ 125	54 41·4	0·11284	0·19613	48 03
11 „ 169	39 31·2	·00483	·82890	83 02	56 „ 124	55 22·4	·11646	·19092	46 59
12 „ 168	39 37·2	·00574	·79162	82 23	57 „ 123	56 04·6	·12009	·18591	45 54
13 „ 167	39 43·8	·00674	·75741	81 44	58 „ 122	56 48·0	·12373	·18107	44 48
14 „ 166	39 50·9	·00781	·72582	81 05	59 „ 121	57 32·6	·12736	·17643	43 40
15 „ 165	39 58·5	0·00897	0·69650	80 26	60 „ 120	58 18·4	0·13099	0·17197	42 32
16 „ 164	40 06·7	·01020	·66916	79 46	61 „ 119	59 05·5	·13461	·16768	41 22
17 „ 163	40 15·4	·01151	·64356	79 07	62 „ 118	59 53·8	·13821	·16356	40 12
18 „ 162	40 24·8	·01290	·61951	78 27	63 „ 117	60 43·4	·14178	·15962	39 00
19 „ 161	40 34·7	·01437	·59685	77 46	64 „ 116	61 34·3	·14532	·15584	37 47
20 „ 160	40 45·2	0·01591	0·57544	77 06	65 „ 115	62 26·4	0·14882	0·15222	36 32
21 „ 159	40 56·3	·01753	·55517	76 25	66 „ 114	63 19·8	·15228	·14877	35 17
22 „ 158	41 08·0	·01923	·53592	75 44	67 „ 113	64 14·5	·15568	·14547	34 00
23 „ 157	41 20·3	·02101	·51762	75 03	68 „ 112	65 10·5	·15902	·14233	32 42
24 „ 156	41 33·3	·02286	·50018	74 21	69 „ 111	66 07·7	·16229	·13934	31 23
25 „ 155	41 46·8	0·02479	0·48355	73 39	70 „ 110	67 06·2	0·16549	0·13651	30 03
26 „ 154	42 01·1	·02679	·46765	72 56	71 „ 109	68 05·9	·16859	·13383	28 41
27 „ 153	42 16·0	·02887	·45245	72 13	72 „ 108	69 06·8	·17161	·13129	27 18
28 „ 152	42 31·5	·03102	·43789	71 30	73 „ 107	70 08·9	·17452	·12890	25 55
29 „ 151	42 47·7	·03325	·42393	70 46	74 „ 106	71 12·1	·17733	·12665	24 30
30 „ 150	43 04·7	0·03554	0·41053	70 02	75 „ 105	72 16·5	0·18001	0·12455	23 04
31 „ 149	43 22·3	·03791	·39766	69 17	76 „ 104	73 22·0	·18256	·12259	21 37
32 „ 148	43 40·7	·04036	·38529	68 32	77 „ 103	74 28·5	·18499	·12077	20 09
33 „ 147	43 59·8	·04287	·37339	67 46	78 „ 102	75 36·0	·18727	·11909	18 40
34 „ 146	44 19·6	·04545	·36193	67 00	79 „ 101	76 44·5	·18940	·11755	17 10
35 „ 145	44 40·2	0·04810	0·35091	66 13	80 „ 100	77 53·8	0·19137	0·11615	15 39
36 „ 144	45 01·6	·05082	·34028	65 26	81 „ 99	79 04·0	·19317	·11488	14 08
37 „ 143	45 23·8	·05360	·33003	64 38	82 „ 98	80 14·9	·19481	·11374	12 35
38 „ 142	45 46·8	·05645	·32015	63 49	83 „ 97	81 26·5	·19626	·11275	11 02
39 „ 141	46 10·7	·05937	·31062	63 00	84 „ 96	82 38·7	·19754	·11188	9 29
40 „ 140	46 35·4	0·06234	0·30133	62 10	85 „ 95	83 51·4	0·19863	0·11115	7 55
41 „ 139	47 01·0	·06537	·29255	61 19	86 „ 94	85 04·6	·19953	·11056	6 20
42 „ 138	47 27·4	·06846	·28399	60 28	87 „ 93	86 18·1	·20022	·11009	4 46
43 „ 137	47 54·8	·07161	·27571	59 36	88 „ 92	87 31·9	·20072	·10976	3 11
44 „ 136	48 23·1	·07481	·26773	58 43	89 „ 91	88 45·9	·20103	·10956	1 35
45 „ 135	48 52·3	0·07806	0·26001	57 49	90	90 00·0	0·20113	0·10950	0 00
h	180°-K	A	N	a_1	h	180°-K	A	N	a_1

Directions with regard to ($a_2 \overset{+}{\underset{-}{\sim}} a_1$) :—

Whenever h is greater than 90°, use ($a_2 - a_1$).

If h is less than 90°, use :—

($a_1 + a_2$) when ($K \pm d$) has been used,

($a_1 - a_2$) when ($d - K$) has been used.

Always **name** **azimuth** from elevated pole.

Note that $\begin{cases} h = \text{L.H.A. of body } \textbf{West} \text{ of the meridian.} \\ h = (360° - \text{L.H.A.}) \text{ of body } \textbf{East} \text{ of the meridian} \end{cases}$

Log. tan. a_2 = N + log. tan. ($K \overset{+}{\underset{-}{\sim}} d$).

LAT. 40° ALT.—AZ. TABLE

h	K	A	N	a₁	h	K	A	N	a₁
°	° ′			° ′	°	° ′			° ′
0 or 180	40 00·0	0·00000	—	90 00	45 or 135	49 52·8	0·07542	0·26626	57 16
1 „ 179	40 00·3	·00004	1·87389	89 21	46 „ 134	50 22·8	·07859	·25881	56 21
2 „ 178	40 01·0	·00016	1·57293	88 43	47 „ 133	50 53·8	·08180	·25162	55 22
3 „ 177	40 02·3	·00035	1·39695	88 04	48 „ 132	51 25·8	·08500	·24467	54 29
4 „ 176	40 04·1	·00062	1·27216	87 26	49 „ 131	51 58·8	·08834	·23797	53 31
5 „ 175	40 06·5	0·00097	1·17545	86 47	50 „ 130	52 32·8	0·09167	0·23149	52 33
6 „ 174	40 09·3	·00140	1·09651	86 08	51 „ 129	53 07·8	·09503	·22524	51 34
7 „ 173	40 12·7	·00190	1·02985	85 29	52 „ 128	53 43·9	·09841	·21921	50 33
8 „ 172	40 16·6	·00248	0·97219	84 50	53 „ 127	54 21·1	·10181	·21340	49 32
9 „ 171	40 21·0	·00314	·92141	84 11	54 „ 126	54 59·3	·10524	·20779	48 30
10 „ 170	40 25·9	0·00388	0·87608	83 32	55 „ 125	55 38·7	0·10868	0·20238	47 27
11 „ 169	40 31·4	·00469	·83515	82 53	56 „ 124	56 19·2	·11213	·19717	46 23
12 „ 168	40 37·5	·00558	·79787	82 13	57 „ 123	57 00·8	·11559	·19215	45 18
13 „ 167	40 44·0	·00655	·76366	81 34	58 „ 122	57 43·6	·11905	·18733	44 11
14 „ 166	40 51·2	·00759	·73207	80 54	59 „ 121	58 27·5	·12251	·18268	43 04
15 „ 165	40 58·9	0·00871	0·70275	80 14	60 „ 120	59 12·6	0·12595	0·17821	41 56
16 „ 164	41 07·1	·00990	·67541	79 33	61 „ 119	59 58·9	·12938	·17393	40 46
17 „ 163	41 15·9	·01118	·64981	78 53	62 „ 118	60 46·4	·13280	·16981	39 36
18 „ 162	41 25·3	·01252	·62576	78 12	63 „ 117	61 35·1	·13618	·16586	38 24
19 „ 161	41 35·2	·01394	·60310	77 31	64 „ 116	62 25·0	·13953	·16209	37 11
20 „ 160	41 45·8	0·01544	0·58169	76 50	65 „ 115	63 16·1	0·14284	0·15847	35 58
21 „ 159	41 56·9	·01701	·56142	76 08	66 „ 114	64 08·4	·14611	·15502	34 43
22 „ 158	42 08·7	·01866	·54217	75 27	67 „ 113	65 01·8	·14932	·15172	33 26
23 „ 157	42 21·1	·02038	·52387	74 44	68 „ 112	65 56·5	·15247	·14858	32 09
24 „ 156	42 34·1	·02218	·50643	74 02	69 „ 111	66 52·4	·15555	·14559	30 51
25 „ 155	42 47·7	0·02404	0·48980	73 19	70 „ 110	67 49·4	0·15856	0·14276	29 31
26 „ 154	43 02·0	·02598	·47390	72 36	71 „ 109	68 47·6	·16148	·14008	28 11
27 „ 153	43 16·9	·02799	·45870	71 52	72 „ 108	69 47·0	·16432	·13754	26 49
28 „ 152	43 32·5	·03008	·44414	71 08	73 „ 107	70 47·4	·16705	·13515	25 26
29 „ 151	43 48·8	·03223	·43017	70 23	74 „ 106	71 48·9	·16968	·13290	24 02
30 „ 150	44 05·7	0·03445	0·41678	69 38	75 „ 105	72 51·5	0·17220	0·13080	22 38
31 „ 149	44 23·4	·03674	·40391	68 53	76 „ 104	73 55·0	·17459	·12884	21 12
32 „ 148	44 41·8	·03910	·39154	68 07	77 „ 103	74 59·6	·17685	·12702	19 45
33 „ 147	45 00·9	·04153	·37964	67 21	78 „ 102	76 05·0	·17899	·12534	18 18
34 „ 146	45 20·7	·04402	·36818	66 34	79 „ 101	77 11·3	·18099	·12380	16 50
35 „ 145	45 41·4	0·04658	0·35715	65 46	80 „ 100	78 18·5	0·18283	0·12239	15 20
36 „ 144	46 02·7	·04920	·34653	64 58	81 „ 99	79 26·4	·18451	·12113	13 51
37 „ 143	46 24·9	·05188	·33628	64 09	82 „ 98	80 35·0	·18604	·11999	12 20
38 „ 142	46 47·9	·05463	·32640	63 20	83 „ 97	81 44·2	·18740	·11899	10 49
39 „ 141	47 11·7	·05744	·31687	62 30	84 „ 96	82 53·9	·18858	·11813	9 17
40 „ 140	47 36·4	0·06030	0·30768	61 40	85 „ 95	84 04·2	0·18959	0·11740	7 45
41 „ 139	48 01·9	·06321	·29880	60 48	86 „ 94	85 14·9	·19043	·11680	6 13
42 „ 138	48 28·2	·06619	·29023	59 56	87 „ 93	86 25·9	·19109	·11634	4 40
43 „ 137	48 55·5	·06922	·28196	59 04	88 „ 92	87 37·1	·19155	·11601	3 07
44 „ 136	49 23·7	·07229	·27397	58 10	89 „ 91	88 48·5	·19184	·11581	1 33
45 „ 135	49 52·8	0·07542	0·26626	57 16	90	90 00·0	0·19193	0·11575	0 00
h	180°-K	A	N	a₁	h	180°-K	A	N	a₁

Directions with regard to (K $\overset{+}{\underset{-}{\sim}}$ d) :—

If l and d have **Same Name**, use (K ⌣ d), i.e. (K — d) or (d — K).

If l and d have **Different Names,** use (K + d).

To name Intercept :—

If H₀ is greater than H꜀, intercept is **" Towards."**

If H₀ is less than H꜀, intercept is **" Away."**

Log. cosec. H꜀ = A + log. sec. (K $\overset{+}{\underset{-}{\sim}}$ d).

N.B.   Whenever h is greater than 90°, subtract the quantity in the K column from 180°.

ALT.—AZ. TABLE — LAT. 41°

h	K	A	N	a₁	h	K	A	N	a₁
°	° ′			° ′	°	° ′			° ′
0 or 180	41 00·0	0·00000	—	90 00	45 or 135	50 52·4	0·07278	0·27273	56 44
1 „ 179	41 00·3	·00004	1·88036	89 21	46 „ 134	51 22·3	·07582	·26529	55 49
2 „ 178	41 01·0	·00015	1·57940	88 41	47 „ 133	51 53·0	·07890	·25809	54 52
3 „ 177	41 02·3	·00034	1·40342	88 02	48 „ 132	52 24·8	·08202	·25115	53 55
4 „ 176	41 04·2	·00060	1·27863	87 22	49 „ 131	52 57·5	·08516	·24444	52 57
5 „ 175	41 06·5	0·00094	1·18192	86 43	50 „ 130	53 31·2	0·08834	0·23797	51 59
6 „ 174	41 09·4	·00136	1·10298	86 03	51 „ 129	54 05·8	·09155	·23172	50 59
7 „ 173	41 12·7	·00184	1·03633	85 24	52 „ 128	54 41·5	·09478	·22569	49 59
8 „ 172	41 16·7	·00241	0·97866	84 44	53 „ 127	55 18·3	·09803	·21987	48 57
9 „ 171	41 21·1	·00305	·92789	84 04	54 „ 126	55 56·1	·10130	·21426	47 55
10 „ 170	41 26·1	0·00376	0·88255	83 24	55 „ 125	56 34·9	0·10458	0·20885	46 52
11 „ 169	41 31·6	·00455	·84162	82 44	56 „ 124	57 14·9	·10787	·20365	45 48
12 „ 168	41 37·7	·00541	·80434	82 40	57 „ 123	57 55·9	·11115	·19863	44 42
13 „ 167	41 44·3	·00635	·77013	81 23	58 „ 122	58 38·0	·11444	·19380	43 36
14 „ 166	41 51·4	·00736	·73854	80 43	59 „ 121	59 21·2	·11773	·18915	42 29
15 „ 165	41 59·1	0·00845	0·70922	80 02	60 „ 120	60 05·6	0·12100	0·18469	41 21
16 „ 164	42 07·4	·00961	·68188	79 21	61 „ 119	60 51·1	·12425	·18040	40 12
17 „ 163	42 16·3	·01084	·65628	78 39	62 „ 118	61 37·7	·12748	·17628	39 01
18 „ 162	42 25·7	·01214	·63224	77 58	63 „ 117	62 25·4	·13069	·17234	37 50
19 „ 161	42 35·7	·01352	·60958	77 16	64 „ 116	63 14·3	·13386	·16856	36 38
20 „ 160	42 46·3	0·01497	0·58817	76 34	65 „ 115	64 04·4	0·13699	0·16494	35 24
21 „ 159	42 57·5	·01650	·56789	75 52	66 „ 114	64 55·5	·14007	·16149	34 10
22 „ 158	43 09·2	·01809	·54864	75 09	67 „ 113	65 47·8	·14310	·15819	32 54
23 „ 157	43 21·7	·01975	·53034	74 26	68 „ 112	66 41·2	·14607	·15505	31 38
24 „ 156	43 34·7	·02149	·51291	73 43	69 „ 111	67 35·8	·14897	·15207	30 20
25 „ 155	43 48·3	0·02330	0·49627	72 59	70 „ 110	68 31·4	0·15181	0·14923	29 01
26 „ 154	44 02·6	·02517	·48038	72 15	71 „ 109	69 28·1	·15456	·14655	27 42
27 „ 153	44 17·6	·02712	·46517	71 31	72 „ 108	70 25·8	·15722	·14401	26 21
28 „ 152	44 33·2	·02913	·45061	70 46	73 „ 107	71 24·6	·15979	·14162	24 59
29 „ 151	44 49·5	·03121	·43665	70 01	74 „ 106	72 24·4	·16226	·13938	23 37
30 „ 150	45 06·5	0·03336	0·42325	69 15	75 „ 105	73 25·2	0·16461	0·13728	22 13
31 „ 149	45 24·1	·03557	·41038	68 29	76 „ 104	74 26·9	·16686	·13532	20 49
32 „ 148	45 42·5	·03785	·39801	67 43	77 „ 103	75 29·5	·16898	·13350	19 23
33 „ 147	46 01·6	·04019	·38611	66 55	78 „ 102	76 32·9	·17098	·13182	17 57
34 „ 146	46 21·5	·04260	·37466	66 08	79 „ 101	77 37·2	·17284	·13027	16 30
35 „ 145	46 42·0	0·04506	0·36363	65 20	80 „ 100	78 42·2	0·17456	0·12887	15 03
36 „ 144	47 03·4	·04758	·35300	64 31	81 „ 99	79 47·9	·17614	·12760	13 34
37 „ 143	47 25·5	·05017	·34276	63 42	82 „ 98	80 54·2	·17756	·12647	12 05
38 „ 142	47 48·5	·05281	·33288	62 52	83 „ 97	82 01·2	·17883	·12547	10 36
39 „ 141	48 12·2	·05551	·32335	62 01	84 „ 96	83 08·6	·17994	·12461	9 06
40 „ 140	48 36·7	0·05827	0·31415	61 10	85 „ 95	84 16·5	0·18088	0·12388	7 36
41 „ 139	49 02·1	·06107	·30528	60 18	86 „ 94	85 24·7	·18166	·12328	6 05
42 „ 138	49 28·4	·06393	·29671	59 26	87 „ 93	86 33·3	·18227	·12282	4 34
43 „ 137	49 55·5	·06683	·28844	58 33	88 „ 92	87 42·1	·18271	·12248	3 03
44 „ 136	50 23·5	·06979	·28045	57 39	89 „ 91	88 51·0	·18297	·12229	1 31
45 „ 135	50 52·4	0·07278	0·27273	56 44	90	90 00·0	0·18306	0·12222	0 00

h	180°-K	A	N	a₁	h	180°-K	A	N	a₁

Directions with regard to (a₂ ∼ a₁) :—

Whenever h is greater than 90°, use (a₂ − a₁).

If h is less than 90°, use :—

    (a₁ + a₂) when (K ± d) has been used,

    (a₁ − a₂) when (d − K) has been used.

Always **name azimuth** from elevated pole.

Note that { h = L.H.A. of body **West** of the meridian.

{ h = (360° − L.H.A.) of body **East** of the meridian

Log. tan. a₂ = N + log. tan. (K ∼ d).

LAT. 42° ALT.—AZ. TABLE

h	K	A	N	a₁
°	° ′			° ′
0 or 180	42 00·0	0·00000	—	90 00
1 ,, 179	42 00·3	·00004	1·88707	89 20
2 ,, 178	42 01·0	·00015	1·58611	88 40
3 ,, 177	42 02·3	·00033	1·41013	87 59
4 ,, 176	42 04·2	·00058	1·28534	87 19
5 ,, 175	42 06·5	0·00091	1·18863	86 39
6 ,, 174	42 09·4	·00131	1·10969	85 19
7 ,, 173	42 12·8	·00179	1·04303	85 18
8 ,, 172	42 16·7	·00234	0·98537	84 38
9 ,, 171	42 21·2	·00295	·93459	83 57
10 ,, 170	42 26·2	0·00365	0·88926	83 16
11 ,, 169	42 31·7	·00441	·84833	82 35
12 ,, 168	42 37·8	·00525	·81105	81 54
13 ,, 167	42 44·4	·00615	·77684	81 07
14 ,, 166	42 51·6	·00713	·74525	80 32
15 ,, 165	42 59·4	0·00819	0·71593	79 50
16 ,, 164	43 07·7	·00931	·68859	79 06
17 ,, 163	43 16·5	·01050	·66299	78 26
18 ,, 162	43 26·0	·01176	·63894	77 44
19 ,, 161	43 36·0	·01310	·61628	77 02
20 ,, 160	43 46·6	0·01450	0·59487	76 19
21 ,, 159	43 57·8	·01597	·57460	75 36
22 ,, 158	44 09·6	·01752	·55535	74 52
23 ,, 157	44 22·1	·01913	·53705	74 09
24 ,, 156	44 35·1	·02080	·51961	73 25
25 ,, 155	44 48·8	0·02255	0·50298	72 40
26 ,, 154	45 03·1	·02436	·48708	71 56
27 ,, 153	45 18·0	·02624	·47188	71 10
28 ,, 152	45 33·6	·02818	·45732	70 25
29 ,, 151	45 49·9	·03019	·44336	69 39
30 ,, 150	46 06·9	0·03226	0·42996	68 53
31 ,, 149	46 24·6	·03439	·41709	68 06
32 ,, 148	46 42·9	·03659	·40472	67 19
33 ,, 147	47 02·0	·03885	·39282	66 31
34 ,, 146	47 21·8	·04117	·38136	65 43
35 ,, 145	47 42·3	0·04354	0·37034	64 54
36 ,, 144	48 03·6	·04597	·35971	64 04
37 ,, 143	48 25·7	·04846	·34946	63 14
38 ,, 142	48 48·5	·05100	·33958	62 24
39 ,, 141	49 12·1	·05360	·33005	61 33
40 ,, 140	49 36·6	0·05625	0·32086	60 41
41 ,, 139	50 01·8	·05894	·31198	59 49
42 ,, 138	50 27·9	·06168	·30342	58 56
43 ,, 137	50 54·9	·06447	·29514	58 02
44 ,, 136	51 22·7	·06730	·28716	57 08
45 ,, 135	51 51·4	0·07017	0·27944	56 13

h	K	A	N	a₁
°	° ′			° ′
45 or 135	51 51·4	0·07017	0·27944	56 13
46 ,, 134	52 21·0	·07308	·27199	55 17
47 ,, 133	52 51·5	·07603	·26480	54 20
48 ,, 132	53 22·9	·07901	·25785	53 23
49 ,, 131	53 55·3	·08202	·25115	52 25
50 ,, 130	54 28·6	0·08506	0·24467	51 26
51 ,, 129	55 02·9	·08812	·23842	50 26
52 ,, 128	55 38·2	·09120	·23239	49 25
53 ,, 127	56 14·5	·09430	·22658	48 24
54 ,, 126	56 51·8	·09741	·22097	47 21
55 ,, 125	57 30·1	0·10053	0·21556	46 18
56 ,, 124	58 09·5	·10366	·21035	45 14
57 ,, 123	58 49·9	·10678	·20534	44 09
58 ,, 122	59 31·3	·10991	·20051	43 02
59 ,, 121	60 13·8	·11302	·19586	41 55
60 ,, 120	60 57·4	0·11612	0·19140	40 47
61 ,, 119	61 42·0	·11921	·18711	39 38
62 ,, 118	62 27·7	·12227	·18299	38 28
63 ,, 117	63 14·5	·12530	·17905	37 17
64 ,, 116	64 02·4	·12830	·17527	36 05
65 ,, 115	64 51·4	0·13125	0·17165	34 52
66 ,, 114	65 41·4	·13416	·16820	33 38
67 ,, 113	66 32·5	·13702	·16490	32 25
68 ,, 112	67 24·6	·13982	·16176	31 07
69 ,, 111	68 17·8	·14256	·15877	29 51
70 ,, 110	69 12·0	0·14522	0·15594	28 33
71 ,, 109	70 07·3	·14781	·15326	27 14
72 ,, 108	71 03·5	·15031	·15072	25 54
73 ,, 107	72 00·6	·15272	·14833	24 33
74 ,, 106	72 58·8	·15504	·14608	23 12
75 ,, 105	73 57·8	0·15725	0·14398	21 49
76 ,, 104	74 57·7	·15936	·14202	20 26
77 ,, 103	75 58·4	·16134	·14020	19 02
78 ,, 102	76 59·9	·16321	·13852	17 37
79 ,, 101	78 02·1	·16495	·13698	16 12
80 ,, 100	79 05·1	0·16656	0·13558	14 46
81 ,, 99	80 08·6	·16803	·13431	13 19
82 ,, 98	81 12·8	·16936	·13317	11 52
83 ,, 97	82 17·5	·17055	·13218	10 24
84 ,, 96	83 22·7	·17158	·13131	8 56
85 ,, 95	84 28·3	0·17246	0·13058	7 27
86 ,, 94	85 34·2	·17319	·12999	5 58
87 ,, 93	86 40·4	·17376	·12952	4 29
88 ,, 92	87 46·8	·17416	·12919	2 58
89 ,, 91	88 53·4	·17411	·12899	1 30
90	90 00·0	0·17449	0·12892	0 00

h	180°-K	A	N	a₁

Directions with regard to $(K \overset{+}{\sim} d)$:—

If l and d have **Same Name,** use $(K \smile d)$, i.e. $(K - d)$ or $(d - K)$.
If l and d have **Different Names,** use $(K + d)$.

To name Intercept :—

If H_o is greater than H_c, intercept is " **Towards.** "
If H_o is less than H_c, intercept is " **Away.** "

Log. cosec. H_c = A + log. sec. $(K \overset{+}{\sim} d)$.

N.B. *Whenever h is greater than 90°, subtract the quantity in the K column from 180°.*

ALT.—AZ. TABLE   LAT. 43°

h	K	A	N	a_1	h	K	A	N	a_1
°	° ′			° ′	°	° ′			° ′
0 or 180	43 00·0	0·00000	—	90 00	45 or 135	52 49·7	0·06758	0·28639	55 42
1 „ 179	43 00·3	·00004	1·89402	89 19	46 „ 134	53 19·0	·07036	·27894	54 46
2 „ 178	43 01·0	·00014	1·59305	88 38	47 „ 133	53 49·2	·07318	·27175	53 49
3 „ 177	43 02·4	·00032	1·41707	87 57	48 „ 132	54 20·3	·07603	·26480	52 52
4 „ 176	43 04·2	·00057	1·29229	87 16	49 „ 131	54 52·3	·07890	·25809	51 53
5 „ 175	43 06·5	0·00088	1·19558	86 35	50 „ 130	55 25·3	0·08180	0·25162	50 54
6 „ 174	43 09·4	·00127	1·11664	85 54	51 „ 129	55 59·2	·08472	·24537	49 54
7 „ 173	43 12·8	·00173	1·04998	85 13	52 „ 128	56 34·0	·08766	·23934	48 53
8 „ 172	43 16·8	·00226	0·99232	84 32	53 „ 127	57 09·8	·09061	·23352	47 51
9 „ 171	43 21·3	·00286	·94154	83 48	54 „ 126	57 46·6	·09357	·22791	46 49
10 „ 170	43 26·3	0·00353	0·89620	83 09	55 „ 125	58 24·3	0·09654	0·22251	43 45
11 „ 169	43 31·8	·00427	·85527	82 27	56 „ 124	59 03·0	·09951	·21730	44 41
12 „ 168	43 37·9	·00508	·81709	81 45	57 „ 123	59 42·8	·10248	·21228	43 36
13 „ 167	43 44·6	·00596	·78378	81 03	58 „ 122	60 23·5	·10545	·20745	42 30
14 „ 166	43 51·8	·00691	·75220	80 21	59 „ 121	61 05·3	·10841	·20281	41 23
15 „ 165	43 59·5	0·00792	0·72288	79 39	60 „ 120	61 48·0	0·11135	0·19834	40 15
16 „ 164	44 07·8	·00901	·69553	78 56	61 „ 119	62 31·8	·11427	·19405	39 06
17 „ 163	44 16·7	·01016	·66994	78 13	62 „ 118	63 16·6	·11716	·18994	37 56
18 „ 162	44 26·2	·01138	·64589	77 30	63 „ 117	64 02·5	·12003	·18599	36 46
19 „ 161	44 36·2	·01267	·62323	76 47	64 „ 116	64 49·3	·12286	·18221	35 34
20 „ 160	44 46·8	0·01403	0·60182	76 04	65 „ 115	65 37·2	0·12565	0·17860	34 22
21 „ 159	44 58·0	·01545	·58154	75 20	66 „ 114	66 26·1	·12840	·17514	33 08
22 „ 158	45 09·9	·01694	·56230	74 36	67 „ 113	67 16·0	·13109	·17185	31 54
23 „ 157	45 22·3	·01850	·54399	73 51	68 „ 112	68 06·8	·13373	·16871	30 39
24 „ 156	45 35·3	·02012	·52656	73 07	69 „ 111	68 58·7	·13631	·16572	29 22
25 „ 155	45 49·0	0·02180	0·50992	72 21	70 „ 110	69 51·5	0·13881	0·16289	28 05
26 „ 154	46 03·3	·02355	·49403	71 36	71 „ 109	70 45·3	·14124	·16020	26 47
27 „ 153	46 18·2	·02536	·47883	70 50	72 „ 108	71 40·0	·14359	·15767	25 28
28 „ 152	46 33·8	·02724	·46426	70 04	73 „ 107	72 35·5	·14586	·15528	24 09
29 „ 151	46 50·1	·02918	·45030	69 17	74 „ 106	73 32·0	·14803	·15303	22 48
30 „ 150	47 07·0	0·03117	0·43690	68 30	75 „ 105	74 29·3	0·15010	0·15093	21 27
31 „ 149	47 24·6	·03323	·42403	67 43	76 „ 104	75 27·4	·15207	·14897	20 05
32 „ 148	47 43·0	·03534	·41166	66 55	77 „ 103	76 26·3	·15394	·14715	18 43
33 „ 147	48 02·0	·03752	·39976	66 07	78 „ 102	77 25·9	·15568	·14547	17 19
34 „ 146	48 21·7	·03975	·38831	65 18	79 „ 101	78 26·2	·15731	·14393	15 54
35 „ 145	48 42·2	0·04203	0·37728	64 28	80 „ 100	79 27·1	0·15882	0·14252	14 30
36 „ 144	49 03·4	·04437	·36665	63 38	81 „ 99	80 28·6	·16019	·14125	13 04
37 „ 143	49 25·3	·04676	·35641	62 48	82 „ 98	81 30·7	·16143	·14012	11 39
38 „ 142	49 48·1	·04920	·34653	61 57	83 „ 97	82 33·3	·16254	·13912	10 12
39 „ 141	50 11·6	·05169	·33700	61 05	84 „ 96	83 36·3	·16351	·13826	8 46
40 „ 140	50 35·9	0·05423	0·32781	60 13	85 „ 95	84 39·6	0·16433	0·13753	7 19
41 „ 139	51 00·9	·05682	·31893	59 20	86 „ 94	85 43·3	·16500	·13693	5 51
42 „ 138	51 26·9	·05945	·31036	58 27	87 „ 93	86 47·3	·16553	·13647	4 24
43 „ 137	51 53·6	·06212	·30209	57 33	88 „ 92	87 51·4	·16591	·13614	2 56
44 „ 136	52 21·2	·06483	·29410	56 38	89 „ 91	88 55·7	·16614	·13594	1 28
45 „ 135	52 49·7	0·06758	0·28639	55 42	90	90 00·0	0·16622	0·13587	0 00
h	180°-K	A	N	a_1	h	180°-K	A	N	a_1

Directions with regard to $(a_2 \overset{+}{\underset{\sim}{}} a_1)$:—

Whenever h is greater than 90°, use $(a_2 - a_1)$.

If h is less than 90°, use :—

$(a_1 + a_2)$ when $(K \pm d)$ has been used,

$(a_1 - a_2)$ when $(d - K)$ has been used.

Always **name azimuth** from elevated pole.

Note that { h = L.H.A. of body **West** of the meridian.
{ h = (360° — L.H.A.) of body **East** of the meridian.

Log. tan. a_2 = N + log. tan. $(K \overset{+}{\underset{\sim}{}} d)$.

LAT. 44° ALT.—AZ. TABLE

h	K	A	N	a₁	h	K	A	N	a₁
°	° '			° '	°	° '			° '
0 or 180	44 00·0	0·00000	—	90 00	**45 or 135**	53 47·2	0·06501	0·29358	55 13
1 „ 179	44 00·3	·00003	1·90121	89 18	46 „ 134	54 16·3	·06767	·28613	54 16
2 „ 178	44 01·0	·00014	1·60025	88 37	47 „ 133	54 46·1	·07036	·27894	53 19
3 „ 177	44 02·4	·00031	1·42427	87 55	48 „ 132	55 16·9	·07308	·27100	52 21
4 „ 176	44 04·2	·00055	1·29948	87 13	49 „ 131	55 48·5	·07582	·26529	51 22
5 „ 175	44 06·6	0·00086	1·20277	86 31	**50 „ 130**	56 21·1	0·07859	0·25881	50 23
6 „ 174	44 09·4	·00123	1·12383	85 49	51 „ 129	56 54·5	·08137	·25256	49 23
7 „ 173	44 12·9	·00168	1·05717	85 07	52 „ 128	57 28·9	·08417	·24653	48 22
8 „ 172	44 16·8	·00219	0·99951	84 25	53 „ 127	58 04·1	·08698	·24072	47 20
9 „ 171	44 21·3	·00277	·94873	83 43	54 „ 126	58 40·3	·08979	·23511	46 17
10 „ 170	44 26·3	0·00341	0·90340	83 01	**55 „ 125**	59 17·5	0·09261	0·22970	45 14
11 „ 169	44 31·9	·00413	·86247	82 19	56 „ 124	59 55·6	·09544	·22449	44 09
12 „ 168	44 38·0	·00491	·82519	81 36	57 „ 123	60 34·6	·09826	·21947	43 04
13 „ 167	44 44·6	·00576	·79098	80 53	58 „ 122	61 14·7	·10107	·21465	41 58
14 „ 166	44 51·8	·00668	·75939	80 10	59 „ 121	61 55·6	·10387	·21000	40 52
15 „ 165	44 59·6	0·00766	0·73007	79 27	**60 „ 120**	62 37·6	0·10666	0·20553	39 44
16 „ 164	45 07·9	·00871	·70273	78 44	61 „ 119	63 20·5	·10942	·20125	38 35
17 „ 163	45 16·8	·00982	·67713	78 01	62 „ 118	64 04·4	·11216	·19713	37 26
18 „ 162	45 26·2	·01100	·65308	77 17	63 „ 117	64 49·2	·11487	·19318	36 16
19 „ 161	45 36·3	·01225	·63042	76 33	64 „ 116	65 35·1	·11755	·18941	35 04
20 „ 160	45 46·9	0·01356	0·60901	75 49	**65 „ 115**	66 21·9	0·12018	0·18579	33 52
21 „ 159	45 58·1	·01493	·58874	75 04	66 „ 114	67 09·6	·12277	·18234	32 39
22 „ 158	46 09·9	·01637	·56949	74 19	67 „ 113	67 58·3	·12531	·17904	31 26
23 „ 157	46 22·3	·01787	·55119	73 34	68 „ 112	68 47·9	·12779	·17590	30 11
24 „ 156	46 35·4	·01943	·53375	72 49	69 „ 111	69 38·4	·13021	·17291	28 55
25 „ 155	46 49·0	0·02106	0·51712	72 03	**70 „ 110**	70 29·8	0·13257	0·17008	27 39
26 „ 154	47 03·3	·02274	·50122	71 17	71 „ 109	71 22·2	·13485	·16740	26 22
27 „ 153	47 18·2	·02449	·48602	70 31	72 „ 108	72 15·3	·13706	·16486	25 04
28 „ 152	47 33·8	·02629	·47146	69 44	73 „ 107	73 09·3	·13919	·16247	23 45
29 „ 151	47 50·0	·02816	·45749	68 56	74 „ 106	74 04·2	·14122	·16022	22 26
30 „ 150	48 06·9	0·03008	0·44410	68 09	**75 „ 105**	74 59·8	0·14317	0·15812	21 06
31 „ 149	48 24·4	·03206	·43123	67 21	76 „ 104	75 56·2	·14501	·15616	19 45
32 „ 148	48 42·7	·03410	·41886	66 32	77 „ 103	76 53·2	·14676	·15434	18 23
33 „ 147	49 01·6	·03619	·40696	65 43	78 „ 102	77 51·0	·14839	·15266	17 01
34 „ 146	49 21·2	·03833	·39550	64 54	79 „ 101	78 49·4	·14991	·15112	15 38
35 „ 145	49 41·6	0·04052	0·38447	64 04	**80 „ 100**	79 48·4	0·15132	0·14971	14 15
36 „ 144	50 02·7	·04277	·37385	63 13	81 „ 99	80 47·9	·15261	·14845	12 51
37 „ 143	50 24·5	·04507	·36360	62 22	82 „ 98	81 47·9	·15376	·14731	11 26
38 „ 142	50 47·1	·04741	·35372	61 31	83 „ 97	82 48·4	·15480	·14631	10 01
39 „ 141	51 10·5	·04980	·34419	60 38	84 „ 96	83 49·3	·15570	·14545	8 36
40 „ 140	51 34·6	0·05223	0·33500	59 46	**85 „ 95**	84 50·6	0·15646	0·14472	7 11
41 „ 139	51 59·5	·05471	·32612	58 52	86 „ 94	85 52·1	·15709	·14412	5 45
42 „ 138	52 25·2	·05723	·31755	57 59	87 „ 93	86 53·9	·15759	·14366	4 19
43 „ 137	52 51·7	·05979	·30928	57 04	88 „ 92	87 55·8	·15794	·14333	2 53
44 „ 136	53 19·1	·06238	·30129	56 09	89 „ 91	88 57·9	·15816	·14313	1 26
45 „ 135	53 47·2	0·06501	0·29358	55 13	**90**	90 00·0	0·15823	0·14307	0 00
h	180°-K	A	N	a₁	h	180°-K	A	N	a₁

Directions with regard to (K ⁺⁻ d) :—

If l and d have **Same Name,** use (K ⌣ d), i.e. (K — d) or (d — K).

If l and d have **Different Names,** use (K | d).

To name Intercept :—

If H_o is greater than H_c, intercept is " **Towards.** "

If H_o is less than H_c, intercept is " **Away.** "

Log. cosec. H_c = A + log. sec. (K ⁺⁻ d).

N.B. Whenever h is greater than 90°, subtract the quantity in the K column from 180°.

ALT.—AZ. TABLE   LAT. 45°

h	K	A	N	a₁	h	K	A	N	a₁
°	° ′			° ′	°	° ′			° ′
0 or 180	45 00·0	0·00000	—	90 00	**45 or 135**	54 44·1	0·06247	0·30103	54 44
1 „ 179	45 00·3	·00003	1·90866	89 18	**46 „ 134**	55 12·8	·06501	·29358	53 47
2 „ 178	45 01·0	·00013	1·60770	88 35	**47 „ 133**	55 42·4	·06758	·28639	52 50
3 „ 177	45 02·4	·00030	1·43171	87 53	**48 „ 132**	56 12·7	·07017	·27944	51 51
4 „ 176	45 04·2	·00053	1·30693	87 10	**49 „ 131**	56 44·0	·07278	·27273	50 52
5 „ 175	45 06·6	0·00083	1·21022	86 28	**50 „ 130**	57 16·1	0·07542	0·26626	49 53
6 „ 174	45 09·4	·00119	1·13128	85 45	**51 „ 129**	57 49·0	·07806	·26001	48 52
7 „ 173	45 12·9	·00162	1·06462	85 02	**52 „ 128**	58 22·9	·08073	·25398	47 51
8 „ 172	45 16·8	·00211	1·00696	84 19	**53 „ 127**	58 57·6	·08340	·24817	46 49
9 „ 171	45 21·3	·00267	0·95618	83 37	**54 „ 126**	59 33·2	·08607	·24256	45 47
10 „ 170	45 26·3	0·00330	0·91084	82 54	**55 „ 125**	60 09·7	0·08875	0·23715	44 43
11 „ 169	45 31·9	·00399	·86992	82 10	**56 „ 124**	60 47·2	·09143	·23194	43 39
12 „ 168	45 38·0	·00474	·83264	81 27	**57 „ 123**	61 25·5	·09411	·22692	42 34
13 „ 167	45 44·6	·00556	·79843	80 44	**58 „ 122**	62 04·8	·09677	·22209	41 28
14 „ 166	45 51·8	·00645	·76684	80 00	**59 „ 121**	62 45·0	·09942	·21745	40 21
15 „ 165	45 59·6	0·00740	0·73752	79 16	**60 „ 120**	63 26·1	0·10206	0·21298	39 14
16 „ 164	46 07·9	·00841	·71018	78 32	**61 „ 119**	64 08·1	·10467	·20870	38 06
17 „ 163	46 16·8	·00949	·68458	77 48	**62 „ 118**	64 51·1	·10726	·20458	36 56
18 „ 162	46 26·2	·01062	·66053	77 04	**63 „ 117**	65 34·9	·10982	·20063	35 47
19 „ 161	46 36·2	·01182	·63787	76 19	**64 „ 116**	66 19·7	·11235	·19685	34 36
20 „ 160	46 46·8	0·01309	0·61646	75 34	**65 „ 115**	67 05·4	0·11483	0·19324	33 24
21 „ 159	46 58·0	·01441	·59619	74 49	**66 „ 114**	67 52·0	·11727	·18978	32 12
22 „ 158	47 09·8	·01580	·57694	74 03	**67 „ 113**	68 39·5	·11966	·18649	30 59
23 „ 157	47 22·2	·01724	·55864	73 18	**68 „ 112**	69 27·8	·12200	·18335	29 45
24 „ 156	47 35·2	·01875	·54120	72 31	**69 „ 111**	70 17·0	·12428	·18036	28 30
25 „ 155	47 48·8	0·02031	0·52457	71 45	**70 „ 110**	71 07·1	0·12649	0·17753	27 14
26 „ 154	48 03·1	·02194	·50867	70 58	**71 „ 109**	71 58·0	·12864	·17484	25 58
27 „ 153	48 17·9	·02362	·49347	70 11	**72 „ 108**	72 49·7	·13071	·17231	24 41
28 „ 152	48 33·4	·02535	·47891	69 24	**73 „ 107**	73 42·2	·13270	·16992	23 23
29 „ 151	48 49·6	·02715	·46494	68 36	**74 „ 106**	74 35·4	·13461	·16767	22 04
30 „ 150	49 06·4	0·02900	0·45154	67 48	**75 „ 105**	75 29·4	0·13643	0·16557	20 45
31 „ 149	49 23·9	·03090	·43868	66 59	**76 „ 104**	76 24·0	·13816	·16361	19 25
32 „ 148	49 42·0	·03285	·42630	66 10	**77 „ 103**	77 19·3	·13980	·16179	18 05
33 „ 147	50 00·9	·03486	·41441	65 20	**78 „ 102**	78 15·3	·14132	·16011	16 44
34 „ 146	50 20·4	·03692	·40295	64 30	**79 „ 101**	79 11·8	·14275	·15857	15 22
35 „ 145	50 40·6	0·03902	0·39192	63 40	**80 „ 100**	80 08·9	0·14406	0·15716	14 00
36 „ 144	51 01·6	·04118	·38130	62 49	**81 „ 99**	81 06·5	·14526	·15589	12 38
37 „ 143	51 23·3	·04338	·37105	61 57	**82 „ 98**	82 04·6	·14635	·15476	11 14
38 „ 142	51 45·7	·04563	·36117	61 05	**83 „ 97**	83 03·1	·14731	·15376	9 51
39 „ 141	52 08·9	·04792	·35164	60 12	**84 „ 96**	84 02·0	·14815	·15290	8 27
40 „ 140	52 32·8	0·05025	0·34245	59 19	**85 „ 95**	85 01·1	0·14887	0·15127	7 03
41 „ 139	52 57·5	·05263	·33357	58 25	**86 „ 94**	86 00·6	·14946	·15157	5 39
42 „ 138	53 22·9	·05503	·32500	57 31	**87 „ 93**	87 00·2	·14992	·15111	4 14
43 „ 137	53 49·2	·05748	·31673	56 36	**88 „ 92**	88 00·1	·15025	·15078	2 50
44 „ 136	54 16·3	·05996	·30874	55 40	**89 „ 91**	89 00·0	·15045	·15058	1 25
45 „ 135	54 44·1	0·06247	0·30103	54 44	**90**	90 00·0	0·15051	0·15051	0 00
h	180°-K	A	N	a₁	h	180°-K	A	N	a₁

Directions with regard to (a₂ $\overset{+}{\sim}$ a₁) :—

Whenever h is greater than 90°, use (a₂ — a₁).

If h is less than 90°, use :—

(a₁ + a₂) when (K ± d) has been used,

(a₁ — a₂) when (d — K) has been used.

Always **name** **azimuth** from elevated pole.

Note that { h = L.H.A. of body **West** of the meridian.
{ h = (360° — L.H.A.) of body **East** of the meridian.

Log. tan. a₂ = N + log. tan. (K $\overset{+}{\sim}$ d).

LAT. 46°   ALT.—AZ. TABLE

h	K	A	N	a₁	h	K	A	N	a₁
°	° ′			° ′	°	° ′			° ′
0 or 180	46 00·0	0·00000	—	90 00	**45 or 135**	55 40·4	0·05996	0·30874	54 16
1 „ 179	46 00·3	·00003	1·91637	89 17	**46 „ 134**	56 08·7	·06238	·30129	53 19
2 „ 178	46 01·0	·00013	1 61541	88 34	**47 „ 133**	56 37·9	·06483	·29410	52 21
3 „ 177	46 02·4	·00029	1·43943	87 50	**48 „ 132**	57 07·8	·06730	·28716	51 23
4 „ 176	46 04·2	·00051	1·31464	87 07	**49 „ 131**	57 38·6	·06979	·28045	50 24
5 „ 175	46 06·6	0·00080	1·21793	86 24	**50 „ 130**	58 10·2	0·07229	0·27397	49 24
6 „ 174	46 09·4	·00115	1·13899	85 41	**51 „ 129**	58 42·7	·07481	·26773	48 23
7 „ 173	46 12·9	·00156	1·07233	84 57	**52 „ 128**	59 16·0	·07734	·26170	47 22
8 „ 172	46 16·8	·00204	1·01467	84 14	**53 „ 127**	59 50·2	·07988	·25588	46 20
9 „ 171	46 21·3	·00258	0·96390	83 30	**54 „ 126**	60 25·2	·08242	·25027	45 17
10 „ 170	46 26·3	0·00318	0·91856	82 46	**55 „ 125**	61 01·1	0·08497	0·24486	44 14
11 „ 169	46 31·8	·00385	·87763	82 02	**56 „ 124**	61 37·8	·08750	·23965	43 09
12 „ 168	46 37·9	·00458	·84035	81 18	**57 „ 123**	62 15·5	·09004	·23464	42 05
13 „ 167	46 44·6	·00537	·80614	80 34	**58 „ 122**	62 54·0	·09256	·22981	40 59
14 „ 166	46 51·8	·00622	·77455	79 50	**59 „ 121**	63 33·3	·09507	·22516	39 52
15 „ 165	46 59·5	0·00714	0·74523	79 05	**60 „ 120**	64 13·6	0·09756	0·22070	38 45
16 „ 164	47 07·8	·00811	·71789	78 21	**61 „ 119**	64 54·7	·10003	·21641	37 37
17 „ 163	47 16·7	·00915	·69229	77 36	**62 „ 118**	65 36·7	·10248	·21229	36 28
18 „ 162	47 26·1	·01024	·66825	76 51	**63 „ 117**	66 19·6	·10489	·20835	35 19
19 „ 161	47 36·1	·01140	·64559	76 05	**64 „ 116**	67 03·3	·10727	·20457	34 08
20 „ 160	47 46·7	0·01262	0·62418	75 20	**65 „ 115**	67 47·9	0·10961	0·20095	32 57
21 „ 159	47 57·8	·01389	·60390	74 34	**66 „ 114**	68 33·4	·11191	·19750	31 45
22 „ 158	48 09·6	·01523	·58465	73 48	**67 „ 113**	69 19·6	·11416	·19420	30 33
23 „ 157	48 21·9	·01662	·56635	73 01	**68 „ 112**	70 06·7	·11636	·19106	29 19
24 „ 156	48 34·9	·01807	·54892	72 14	**69 „ 111**	70 54·6	·11850	·18808	28 05
25 „ 155	48 48·4	0·01957	0·53228	71 27	**70 „ 110**	71 43·3	0·12058	0·18524	26 50
26 „ 154	49 02·6	·02113	·51639	70 40	**71 „ 109**	72 32·8	·12260	·18256	25 35
27 „ 153	49 17·4	·02275	·50118	69 52	**72 „ 108**	73 23·0	·12454	·18002	24 19
28 „ 152	49 32·8	·02442	·48662	69 04	**73 „ 107**	74 14·0	·12641	·17763	23 02
29 „ 151	49 48·9	·02614	·47266	68 16	**74 „ 106**	75 05·7	·12820	·17539	21 44
30 „ 150	50 05·6	0·02792	0·45926	67 27	**75 „ 105**	75 58·0	0·12991	0·17328	20 26
31 „ 149	50 23·0	·02974	·44639	66 38	**76 „ 104**	76 51·0	·13153	·17132	19 07
32 „ 148	50 41·1	·03162	·43402	65 48	**77 „ 103**	77 44·6	·13305	·16950	17 48
33 „ 147	50 59·8	·03354	·42212	64 58	**78 „ 102**	78 38·8	·13448	·16782	16 28
34 „ 146	51 19·2	·03552	·41067	64 07	**79 „ 101**	79 33·6	·13581	·16628	15 07
35 „ 145	51 39·3	0·03754	0·39964	63 16	**80 „ 100**	80 28·8	0·13704	0·16488	13 46
36 „ 144	52 00·1	·03960	·38901	62 24	**81 „ 99**	81 24·6	·13817	·16361	12 25
37 „ 143	52 21·6	·04171	·37877	61 32	**82 „ 98**	82 20·7	·13918	·16248	11 03
38 „ 142	52 43·8	·04386	·36889	60 40	**83 „ 97**	83 17·3	·14008	·16148	9 41
39 „ 141	53 06·7	·04606	·35936	59 47	**84 „ 96**	84 14·2	·14087	·16061	8 19
40 „ 140	53 30·4	0·04829	0·35016	58 53	**85 „ 95**	85 11·3	0·14154	0·15988	6 56
41 „ 139	53 54·9	·05056	·34129	57 59	**86 „ 94**	86 08·8	·14209	·15929	5 33
42 „ 138	54 20·1	·05286	·33272	57 04	**87 „ 93**	87 06·4	·14251	·15882	4 10
43 „ 137	54 46·1	·05519	·32445	56 09	**88 „ 92**	88 04·2	·14282	·15849	2 47
44 „ 136	55 12·8	·05756	·31646	55 13	**89 „ 91**	89 02·1	·14300	·15829	1 23
45 „ 135	55 40·4	0·05996	0·30874	54 16	**90**	90 00·0	0·14307	0·15823	0 00
h	180°-K	A	N	a₁	h	180°-K	A	N	a₁

Directions with regard to (K ⌃ d) :—

If l and d have **Same Name,** use (K ⌣ d), i.e. (K — d) or (d — K).
If l and d have **Different Names,** use (K + d).

To name Intercept :—

If H_o is greater than H_c, intercept is " **Towards.** "
If H_o is less than H_c, intercept is " **Away.** "

Log. cosec. H_c = A + log. sec. (K ⌃ d).

N.B.   Whenever h is greater than 90°, subtract the quantity in the K column from 180°.

ALT.—AZ. TABLE — LAT. 47°

h	K	A	N	a₁	h	K	A	N	a₁
°	° ′			° ′	°	° ′			° ′
0 or 180	47 00·0	0·00000	—	90 00	45 or 135	56 36·0	0·05748	0·31673	53 49
1 „ 179	47 00·3	·00003	1·92436	89 16	46 „ 134	57 03·9	·05979	·30928	52 52
2 „ 178	47 01·0	·00012	1·62340	88 32	47 „ 133	57 32·7	·06212	·30209	51 54
3 „ 177	47 02·4	·00028	1·44742	87 48	48 „ 132	58 02·2	·06447	·29514	50 55
4 „ 176	47 04·2	·00049	1·32263	87 04	49 „ 131	58 32·5	·06683	·28844	49 56
5 „ 175	47 06·5	0·00077	1·22592	86 20	50 „ 130	59 03·7	0·06922	0·28196	48 55
6 „ 174	47 09·4	·00111	1·14698	85 36	51 „ 129	59 35·6	·07161	·27571	47 55
7 „ 173	47 12·8	·00151	1·08032	84 52	52 „ 128	60 08·4	·07401	·26968	46 53
8 „ 172	47 16·8	·00197	1·02266	84 08	53 „ 127	60 41·9	·07642	·26387	45 51
9 „ 171	47 21·2	·00249	0·97188	83 24	54 „ 126	61 16·3	·07883	·25826	44 49
10 „ 170	47 26·2	0·00307	0·92655	82 39	55 „ 125	61 51·5	0·08124	0·25285	43 45
11 „ 169	47 31·8	·00371	·88562	81 55	56 „ 124	62 27·6	·08364	·24764	42 41
12 „ 168	47 37·9	·00441	·84834	81 10	57 „ 123	63 04·5	·08604	·24263	41 36
13 „ 167	47 44·5	·00517	·81413	80 25	58 „ 122	63 42·2	·08843	·23780	40 31
14 „ 166	47 51·6	·00599	·78254	79 40	59 „ 121	64 20·8	·09080	·23315	39 24
15 „ 165	47 59·4	0·00687	0·75322	78 55	60 „ 120	65 00·1	0·09316	0·22869	38 17
16 „ 164	48 07·6	·00781	·72588	78 09	61 „ 119	65 40·4	·09549	·22440	37 10
17 „ 163	48 16·5	·00881	·70028	77 24	62 „ 118	66 21·4	·09780	·22028	36 01
18 „ 162	48 25·9	·00987	·67623	76 38	63 „ 117	67 03·3	·10007	·21634	34 52
19 „ 161	48 35·8	·01098	·65357	75 52	64 „ 116	67 46·0	·10232	·21256	33 42
20 „ 160	48 46·4	0·01215	0·63216	75 06	65 „ 115	68 29·4	0·10453	0·20894	32 31
21 „ 159	48 57·5	·01337	·61189	74 19	66 „ 114	69 13·7	·10669	·20549	31 20
22 „ 158	49 09·2	·01466	·59264	73 32	67 „ 113	69 58·8	·10880	·20219	30 08
23 „ 157	49 21·5	·01599	·57434	72 45	68 „ 112	70 44·7	·11087	·19905	28 55
24 „ 156	49 34·3	·01739	·55690	71 58	69 „ 111	71 31·3	·11288	·19606	27 42
25 „ 155	49 47·8	0·01883	0·54027	71 10	70 „ 110	72 18·6	0·11484	0·19323	26 27
26 „ 154	50 01·9	·02033	·52437	70 22	71 „ 109	73 06·7	·11672	·19055	25 13
27 „ 153	50 16·7	·02188	·50917	69 34	72 „ 108	73 55·5	·11855	·18801	23 57
28 „ 152	50 32·0	·02349	·49461	68 45	73 „ 107	74 45·0	·12030	·18562	22 41
29 „ 151	50 48·0	·02514	·48065	67 56	74 „ 106	75 35·1	·12198	·18337	21 25
30 „ 150	51 04·6	0·02684	0·46725	67 06	75 „ 105	76 25·9	0·12357	0·18127	20 07
31 „ 149	51 21·8	·02859	·45438	66 17	76 „ 104	77 17·2	·12509	·17931	18 49
32 „ 148	51 39·7	·03039	·44201	65 26	77 „ 103	78 09·2	·12652	·17749	17 32
33 „ 147	51 58·3	·03224	·43011	64 36	78 „ 102	79 01·7	·12786	·17581	16 12
34 „ 146	52 17·6	·03413	·41865	63 45	79 „ 101	79 54·6	·12911	·17427	14 53
35 „ 145	52 37·5	0·03607	0·40763	62 53	80 „ 100	80 48·1	0·13025	0·17287	13 33
36 „ 144	52 58·1	·03804	·39700	62 01	81 „ 99	81 42·0	·13130	·17160	12 13
37 „ 143	53 19·4	·04005	·38675	61 08	82 „ 98	82 36·3	·13225	·17046	10 53
38 „ 142	53 41·4	·04211	·37687	60 15	83 „ 97	83 31·0	·13309	·16947	9 32
39 „ 141	54 04·1	·04421	·36734	59 22	84 „ 96	84 26·0	·13382	·16860	8 11
40 „ 140	54 27·6	0·04634	0·35815	58 28	85 „ 95	85 21·2	0·13444	0·16787	6 49
41 „ 139	54 51·8	·04851	·34927	57 33	86 „ 94	86 16·7	·13496	·16728	5 28
42 „ 138	55 16·7	·05071	·34071	56 38	87 „ 93	87 12·4	·13536	·16681	4 06
43 „ 137	55 42·4	·05294	·33243	55 42	88 „ 92	88 08·2	·13564	·16648	2 44
44 „ 136	56 08·8	·05519	·32445	54 46	89 „ 91	89 04·1	·13581	·16628	1 22
45 „ 135	56 36·0	0·05748	0·31673	53 49	90	90 00·0	0·13587	0·16622	0 00
h	180°-K	A	N	a₁	h	180°-K	A	N	a₁

Directions with regard to $(a_2 \stackrel{+}{\sim} a_1)$:—

Whenever h is greater than 90°, use $(a_2 - a_1)$.

If h is less than 90°, use :—

$(a_1 + a_2)$ when $(K \pm d)$ has been used,

$(a_1 - a_2)$ when $(d - K)$ has been used.

Always **name azimuth** from elevated pole.

Note that $\begin{cases} h = \text{L.H.A. of body } \textbf{West} \text{ of the meridian.} \\ h = (360° - \text{L.H.A.}) \text{ of body } \textbf{East} \text{ of the meridian.} \end{cases}$

Log. tan. $a_2 = N + \log. \tan. (K \stackrel{+}{\sim} d)$.

LAT. 48°   ALT.—AZ. TABLE

h	K	A	N	a₁	h	K	A	N	a₁
°	° ′			° ′	°	° ′			° ′
0 or 180	48 00·0	0·00000	—	90 00	45 or 135	57 31·0	0·05503	0·32500	53 23
1 „ 179	48 00·3	·00003	1·93263	89 15	46 „ 134	57 58·5	·05723	·31755	52 25
2 „ 178	48 01·0	·00012	1·65107	88 31	47 „ 133	58 26·8	·05945	·31036	51 27
3 „ 177	48 02·3	·00027	1·45569	87 46	48 „ 132	58 55·9	·06168	·30342	50 28
4 „ 176	48 04·2	·00047	1·33090	87 02	49 „ 131	59 25·7	·06393	·29071	49 28
5 „ 175	48 06·5	0·00074	1·23419	86 17	50 „ 130	59 56·3	0·06619	0·29023	48 28
6 „ 174	48 09·4	·00106	1·15525	85 32	51 „ 129	60 27·7	·06846	·28399	47 27
7 „ 173	48 12·8	·00145	1·08859	84 47	52 „ 128	60 59·9	·07074	·27796	46 26
8 „ 172	48 16·7	·00189	1·03093	84 02	53 „ 127	61 32·9	·07302	·27214	45 24
9 „ 171	48 21·2	·00239	0·98016	83 17	54 „ 126	62 06·6	·07531	·26653	44 21
10 „ 170	48 26·2	0·00295	0·93482	82 32	55 „ 125	62 41·2	0·07759	0·26112	43 18
11 „ 169	48 31·7	·00357	·89389	81 47	56 „ 124	63 16·5	·07986	·25591	42 14
12 „ 168	48 37·7	·00424	·85661	81 01	57 „ 123	63 52·6	·08213	·25090	41 09
13 „ 167	48 44·3	·00498	·82240	80 16	58 „ 122	64 29·5	·08439	·24607	40 04
14 „ 166	48 51·5	·00577	·79081	79 30	59 „ 121	65 07·3	·08663	·24142	38 57
15 „ 165	48 59·1	0·00661	0·76149	78 44	60 „ 120	65 45·8	0·08885	0·23696	37 51
16 „ 164	49 07·4	·00752	·73415	77 58	61 „ 119	66 25·1	·09105	·23267	36 43
17 „ 163	49 16·2	·00847	·70855	77 12	62 „ 118	67 05·1	·09323	·22855	35 35
18 „ 162	49 25·5	·00949	·68451	76 26	63 „ 117	67 46·0	·09537	·22461	34 26
19 „ 161	49 35·4	·01056	·66185	75 39	64 „ 116	68 27·6	·09749	·22083	33 17
20 „ 160	49 45·9	0·01168	0·64044	74 52	65 „ 115	69 10·0	0·09956	0·21721	32 03
21 „ 159	49 57·0	·01286	·62016	74 05	66 „ 114	69 53·2	·10160	·21376	30 56
22 „ 158	50 08·6	·01409	·60091	73 17	67 „ 113	70 37·0	·10359	·21046	29 44
23 „ 157	50 20·8	·01538	·58261	72 30	68 „ 112	71 21·7	·10553	·20732	28 32
24 „ 156	50 33·6	·01671	·56518	71 42	69 „ 111	72 07·0	·10742	·20434	27 19
25 „ 155	50 47·0	0·01810	0·54854	70 53	70 „ 110	72 53·0	0·10925	0·20150	26 03
26 „ 154	51 01·1	·01954	·53265	70 05	71 „ 109	73 39·7	·11103	·19882	24 52
27 „ 153	51 15·7	·02102	·51744	69 16	72 „ 108	74 27·1	·11273	·19628	23 37
28 „ 152	51 30·9	·02256	·50288	68 26	73 „ 107	75 15·1	·11438	·19389	22 22
29 „ 151	51 46·8	·02415	·48892	67 37	74 „ 106	76 03·7	·11595	·19165	21 06
30 „ 150	52 03·2	0·02578	0·47552	66 47	75 „ 105	76 52·9	0·11744	0·18954	19 50
31 „ 149	52 20·3	·02745	·46265	65 56	76 „ 104	77 42·7	·11886	·18758	18 33
32 „ 148	52 38·1	·02918	·45028	65 05	77 „ 103	78 33·0	·12020	·18576	17 15
33 „ 147	52 56·5	·03095	·43838	64 14	78 „ 102	79 23·8	·12145	·18408	15 58
34 „ 146	53 15·6	·03275	·42693	63 23	79 „ 101	80 15·1	·12261	·18254	14 39
35 „ 145	53 35·3	0·03461	0·41590	62 31	80 „ 100	81 06·8	0·12368	0·18114	13 21
36 „ 144	53 55·7	·03649	·40527	61 38	81 „ 99	81 58·9	·12466	·17987	12 02
37 „ 143	54 16·8	·03842	·39503	60 45	82 „ 98	82 51·4	·12554	·17874	10 43
38 „ 142	54 38·6	·04039	·38515	59 52	83 „ 97	83 44·3	·12633	·17774	9 23
39 „ 141	55 01·1	·04239	·37562	58 58	84 „ 96	84 37·4	·12701	·17687	8 03
40 „ 140	55 24·2	0·04442	0·36642	58 03	85 „ 95	85 30·8	0·12759	0·17614	6 43
41 „ 139	55 48·1	·04649	·35755	57 08	86 „ 94	86 24·4	·12807	·17555	5 23
42 „ 138	56 12·7	·04858	·34898	56 13	87 „ 93	87 18·1	·12845	·17508	4 02
43 „ 137	56 38·1	·05071	·34071	55 17	88 „ 92	88 12·0	·12871	·17475	2 41
44 „ 136	57 04·1	·05286	·33272	54 20	89 „ 91	89 06·0	·12887	·17455	1 21
45 „ 135	57 31·0	0·05503	0·32500	53 23	90	90 00·0	0·12893	0·17449	0 00
h	180°-K	A	N	a₁	h	180°-K	A	N	a₁

Directions with regard to (K ⁺⁄₋ d) :—

If l and d have **Same Name,** use (K ⌣ d), i.e. (K — d) or (d — K).

If l and d have **Different Names,** use (K + d).

To name Intercept :—

If H₀ is greater than Hᴄ, intercept is " **Towards.**"

If H₀ is less than Hₒ, intercept is " **Away.**"

Log. cosec. Hᴄ = A + log. sec. (K ⁺⁄₋ d).

N.B.   Whenever h is greater than 90°, subtract the quantity in the K column from 180°.

ALT.—AZ. TABLE $\qquad$ LAT. 49°

h	K	A	N	a₁	h	K	A	N	a₁
°	° ′			° ′	°	° ′			° ′
0 or 180	49 00·0	0·00000	—	90 00	**45 or 135**	58 25·3	0·05263	0·33357	52 57
1 „ 179	49 00·3	·00003	1·94120	89 15	**46 „ 134**	58 52·4	·05471	·32612	51 59
2 „ 178	49 01·0	·00011	1·64024	88 29	**47 „ 133**	59 20·3	·05682	·31893	51 01
3 „ 177	49 02·3	·00026	1·46426	87 44	**48 „ 132**	59 48·9	·05894	·31198	50 02
4 „ 176	49 04·2	·00046	1·33947	86 59	**49 „ 131**	60 18·2	·06107	·30528	49 02
5 „ 175	49 06·5	0·00071	1·24276	86 13	**50 „ 130**	60 48·3	0·06321	0·29880	48 02
6 „ 174	49 09·3	·00102	1·16382	85 28	**51 „ 129**	61 19·1	·06537	·29255	47 01
7 „ 173	49 12·7	·00139	1·09716	84 42	**52 „ 128**	61 50·7	·06753	·28652	45 59
8 „ 172	49 16·6	·00182	1·03950	83 57	**53 „ 127**	62 23·0	·06969	·28071	44 57
9 „ 171	49 21·1	·00230	0·98872	83 11	**54 „ 126**	62 56·1	·07185	·27510	43 55
10 „ 170	49 26·0	0·00284	0·94339	82 25	**55 „ 125**	63 29·9	0·07401	0·26969	42 51
11 „ 169	49 31·5	·00343	·90246	81 39	**56 „ 124**	64 04·5	·07616	·26448	41 47
12 „ 168	49 37·6	·00408	·86518	80 53	**57 „ 123**	64 39·9	·07830	·25947	40 43
13 „ 167	49 44·1	·00478	·83097	80 07	**58 „ 122**	65 16·0	·08043	·25464	39 37
14 „ 166	49 51·2	·00554	·79938	79 21	**59 „ 121**	65 52·9	·08255	·24999	38 32
15 „ 165	49 58·9	0·00635	0·77006	78 34	**60 „ 120**	66 30·5	0·08465	0·24553	37 25
16 „ 164	50 07·0	·00722	·74272	77 47	**61 „ 119**	67 08·9	·08672	·24124	36 18
17 „ 163	50 15·8	·00814	·71712	77 00	**62 „ 118**	67 48·0	·08877	·23712	35 10
18 „ 162	50 25·1	·00911	·69307	76 13	**63 „ 117**	68 27·5	·09079	·23318	34 01
19 „ 161	50 34·9	·01014	·67041	75 26	**64 „ 116**	69 08·4	·09278	·22940	32 52
20 „ 160	50 45·4	0·01122	0·64900	74 38	**65 „ 115**	69 49·7	0·09473	0·22578	31 43
21 „ 159	50 56·3	·01235	·62873	73 51	**66 „ 114**	70 31·7	·09664	·22233	30 32
22 „ 158	51 07·9	·01353	·60948	73 03	**67 „ 113**	71 14·4	·09851	·21903	29 21
23 „ 157	51 20·0	·01476	·59118	72 14	**68 „ 112**	71 57·8	·10033	·21589	28 10
24 „ 156	51 32·7	·01604	·57350	71 26	**69 „ 111**	72 41·8	·10211	·21290	26 58
25 „ 155	51 46·0	0·01737	0·55711	70 37	**70 „ 110**	73 26·5	0·10383	0·21007	25 45
26 „ 154	51 59·9	·01875	·54121	69 47	**71 „ 109**	74 11·9	·10549	·20739	24 31
27 „ 153	52 14·4	·02017	·52601	68 58	**72 „ 108**	74 57·8	·10709	·20485	23 18
28 „ 152	52 29·6	·02164	·51145	68 08	**73 „ 107**	75 44·4	·10863	·20246	22 03
29 „ 151	52 45·3	·02316	·49749	67 18	**74 „ 106**	76 31·5	·11010	·20021	20 48
30 „ 150	53 01·6	0·02472	0·48409	66 27	**75 „ 105**	77 19·2	0·11150	0·19811	19 33
31 „ 149	53 18·6	·02633	·47122	65 36	**76 „ 104**	78 07·4	·11282	·19615	18 17
32 „ 148	53 36·1	·02797	·45885	64 45	**77 „ 103**	78 56·1	·11407	·19433	17 01
33 „ 147	53 54·4	·02966	·44695	63 53	**78 „ 102**	79 45·3	·11524	·19265	15 44
34 „ 146	54 13·2	·03139	·43549	63 01	**79 „ 101**	80 34·9	·11633	·19111	14 27
35 „ 145	54 32·8	0·03316	0·42447	62 09	**80 „ 100**	81 25·0	0·11733	0·18971	13 09
36 „ 144	54 53·0	·03496	·41384	61 16	**81 „ 99**	82 15·4	·11824	·18844	11 51
37 „ 143	55 13·8	·03680	·40359	60 22	**82 „ 98**	83 06·1	·11907	·18730	10 33
38 „ 142	55 35·3	·03868	·39371	59 28	**83 „ 97**	83 57·2	·11980	·18631	9 14
39 „ 141	55 57·5	·04059	·38418	58 34	**84 „ 96**	84 48·5	·12043	·18544	7 56
40 „ 140	56 20·4	0·04252	0·37499	57 39	**85 „ 95**	85 40·0	0·12097	0·18471	6 37
41 „ 139	56 44·0	·04449	·36611	56 44	**86 „ 94**	86 31·8	·12142	·18412	5 18
42 „ 138	57 08·2	·04649	·35755	55 48	**87 „ 93**	87 23·7	·12177	·18365	3 58
43 „ 137	57 33·2	·04851	·34927	54 52	**88 „ 92**	88 15·7	·12202	·18332	2 39
44 „ 136	57 58·9	·05056	·34129	53 55	**89 „ 91**	89 07·8	·12216	·18312	1 19
45 „ 135	58 25·3	0·05263	0·33357	52 57	**90**	90 00·0	0·12222	0·18306	0 00
h	180°-K	A	N	a₁	h	180°-K	A	N	a₁

Directions with regard to $(a_2 \overset{+}{\sim} a_1)$:—

Whenever h is greater than 90°, use $(a_2 - a_1)$.

If h is less than 90°, use :—

$\qquad$ $(a_1 + a_2)$ when $(K \pm d)$ has been used,

$\qquad$ $(a_1 - a_2)$ when $(d - K)$ has been used.

Always **name** **azimuth** from elevated pole.

Note that $\begin{cases} \text{h = L.H.A. of body \textbf{West} of the meridian.} \\ \text{h = (360° — L.H.A.) of body \textbf{East} of the meridian} \end{cases}$

Log. tan. a_2 = N + log. tan. $(K \overset{+}{\sim} d)$.

LAT. 50°   ALT.—AZ. TABLE

h	K	A	N	a₁	h	K	A	N	a₁
°	° ′			° ′	°	° ′			° ′
0 or 180	50 00·0	0·00000	—	90 00	45 or 135	59 19·1	0·05025	0·34245	52 33
1 ,, 179	50 00·3	·00003	1·95008	89 14	46 ,, 134	59 45·8	·05223	·33500	51 35
2 ,, 178	50 01·0	·00011	1·64911	88 28	47 ,, 133	60 13·1	·05423	·32781	50 36
3 ,, 177	50 02·3	·00025	1·47313	87 42	48 ,, 132	60 41·2	·05625	·32086	49 37
4 ,, 176	50 04·1	·00044	1·34835	86 56	49 ,, 131	61 10·0	·05827	·31415	48 37
5 ,, 175	50 06·5	0·00068	1·25164	86 10	50 ,, 130	61 39·6	0·06030	0·30768	47 36
6 ,, 174	50 09·3	·00098	1·17270	85 24	51 ,, 129	62 09·8	·06234	·30143	46 35
7 ,, 173	50 12·7	·00134	1·10604	84 38	52 ,, 128	62 40·7	·06438	·29540	45 34
8 ,, 172	50 16·5	·00174	1·04838	83 51	53 ,, 127	63 12·4	·06643	·28958	44 32
9 ,, 171	50 20·9	·00221	0·99760	83 05	54 ,, 126	63 44·8	·06847	·28397	43 29
10 ,, 170	50 25·9	0·00272	0·95226	82 18	55 ,, 125	64 17·9	0·07051	0·27857	42 26
11 ,, 169	50 31·3	·00329	·91133	81 32	56 ,, 124	64 51·8	·07254	·27336	41 22
12 ,, 168	50 37·3	·00391	·87405	80 45	57 ,, 123	65 26·4	·07456	·26834	40 17
13 ,, 167	50 43·8	·00459	·83984	79 58	58 ,, 122	66 01·6	·07657	·26351	39 12
14 ,, 166	50 50·9	·00532	·80826	79 11	59 ,, 121	66 37·7	·07856	·25887	38 07
15 ,, 165	50 58·5	0·00609	0·77894	78 24	60 ,, 120	67 14·4	0·08054	0·25440	37 00
16 ,, 164	51 06·6	·00693	·75159	77 37	61 ,, 119	67 51·8	·08249	·25011	35 53
17 ,, 163	51 15·3	·00781	·72600	76 49	62 ,, 118	68 29·9	·08442	·24600	34 46
18 ,, 162	51 24·5	·00874	·70195	76 01	63 ,, 117	69 08·8	·08632	·24205	33 38
19 ,, 161	51 34·3	·00972	·67921	75 13	64 ,, 116	69 48·3	·08819	·23827	32 29
20 ,, 160	51 44·7	0·01076	0·65788	74 25	65 ,, 115	70 28·5	0·09003	0·23466	31 20
21 ,, 159	51 55·6	·01184	·63760	73 37	66 ,, 114	71 09·3	·09182	·23120	30 10
22 ,, 158	52 07·0	·01297	·61836	72 48	67 ,, 113	71 50·9	·09358	·22791	28 59
23 ,, 157	52 19·1	·01415	·60005	71 59	68 ,, 112	72 33·0	·09529	·22477	27 48
24 ,, 156	52 31·7	·01537	·58262	71 10	69 ,, 111	73 15·8	·09695	·22178	26 37
25 ,, 155	52 44·9	0·01665	0·56598	70 21	70 ,, 110	73 59·2	0·09856	0·21895	25 25
26 ,, 154	52 58·6	·01796	·55009	69 31	71 ,, 109	74 43·2	·10012	·21626	24 12
27 ,, 153	53 13·0	·01933	·53489	68 41	72 ,, 108	75 27·8	·10162	·21373	22 59
28 ,, 152	53 28·0	·02073	·52032	67 50	73 ,, 107	76 13·0	·10305	·21134	21 45
29 ,, 151	53 43·5	·02218	·50636	67 00	74 ,, 106	76 58·6	·10443	·20909	20 31
30 ,, 150	53 59·7	0·02367	0·49296	66 08	75 ,, 105	77 44·8	0·10574	0·20699	19 17
31 ,, 149	54 16·5	·02521	·48009	65 17	76 ,, 104	78 31·5	·10698	·20503	18 02
32 ,, 148	54 33·9	·02678	·46772	64 25	77 ,, 103	79 18·6	·10814	·20321	16 46
33 ,, 147	54 51·9	·02839	·45582	63 33	78 ,, 102	80 06·2	·10923	·20153	15 30
34 ,, 146	55 10·5	·03004	·44437	62 40	79 ,, 101	80 54·2	·11025	·19999	14 14
35 ,, 145	55 29·8	0·03173	0·43334	61 47	80 ,, 100	81 42·6	0·11119	0·19858	12 58
36 ,, 144	55 49·8	·03345	·42271	60 54	81 ,, 99	82 31·3	·11204	·19731	11 41
37 ,, 143	56 10·4	·03520	·41247	60 00	82 ,, 98	83 20·3	·11281	·19618	10 24
38 ,, 142	56 31·6	·03699	·40259	59 06	83 ,, 97	84 09·7	·11349	·19518	9 06
39 ,, 141	56 53·5	·03881	·39306	58 11	84 ,, 96	84 59·2	·11408	·19432	7 49
40 ,, 140	57 16·1	0·04065	0·38387	57 16	85 ,, 95	85 49·0	0·11459	0·19359	6 31
41 ,, 139	57 39·3	·04252	·37499	56 20	86 ,, 94	86 39·0	·11501	·19299	5 13
42 ,, 138	58 03·2	·04442	·36642	55 24	87 ,, 93	87 29·1	·11533	·19253	3 55
43 ,, 137	58 27·8	·04634	·35815	54 28	88 ,, 92	88 19·4	·11556	·19220	2 37
44 ,, 136	58 53·1	·04829	·35016	53 30	89 ,, 91	89 09·7	·11570	·19200	1 18
45 ,, 135	59 19·1	0·05025	0·34245	52 33	90	90 00·0	0·11575	0·19193	0 00
h	180°-K	A	N	a₁	h	180°-K	A	N	a₁

Directions with regard to (K $\pm$ d) :—

    If l and d have **Same Name,** use (K $\frown$ d), i.e. (K — d) or (d — K).

    If l and d have **Different Names,** use (K + d).

To name Intercept :—

    If H₀ is greater than H₍, intercept is " **Towards.**"

    If H₀ is less than H₍, intercept is " **Away.**"

Log₀ cosec. H₍ = A + log. sec. (K $\pm$ d).

N.B. Whenever h is greater than 90°, subtract the quantity in the K column from 180°.

ALT.—AZ. TABLE   LAT. 51°

h	K	A	N	a₁	h	K	A	N	a₁
°	° ′			° ′	°	° ′			° ′
0 or 180	51 00·0	0·00000	—	90 00	45 or 135	60 12·3	0·04792	0·35164	52 09
1 „ 179	51 00·3	·00003	1·95927	89 13	46 „ 134	60 38·5	·04980	·34419	51 10
2 „ 178	51 01·0	·00010	1·65831	88 27	47 „ 133	61 05·4	·05169	·33700	50 12
3 „ 177	51 02·3	·00024	1·48233	87 40	48 „ 132	61 32·9	·05360	·33005	49 12
4 „ 176	51 04·1	·00042	1·35754	86 53	49 „ 131	62 01·2	·05551	·32335	48 12
5 „ 175	51 06·4	0·00065	1·26083	86 07	50 „ 130	62 30·1	0·05744	0·31687	47 12
6 „ 174	51 09·2	·00094	1·18189	85 20	51 „ 129	62 59·8	·05937	·31062	46 11
7 „ 173	51 12·6	·00128	1·11523	84 33	52 „ 128	63 30·1	·06130	·30460	45 09
8 „ 172	51 16·4	·00167	1·05757	83 46	53 „ 127	64 01·1	·06323	·29878	44 07
9 „ 171	51 20·8	·00211	1·00680	82 59	54 „ 126	64 32·8	·06516	·29317	43 04
10 „ 170	51 25·7	0·00261	0·96146	82 12	55 „ 125	65 05·2	0·06708	0·28776	42 01
11 „ 169	51 31·1	·00315	·92053	81 25	56 „ 124	65 38·3	·06900	·28255	40 57
12 „ 168	51 37·1	·00375	·88325	80 37	57 „ 123	66 12·0	·07090	·27754	39 53
13 „ 167	51 43·5	·00440	·84904	79 50	58 „ 122	66 46·5	·07279	·27271	38 48
14 „ 166	51 50·5	·00509	·81745	79 02	59 „ 121	67 21·6	·07467	·26806	37 43
15 „ 165	51 58·1	0·00584	0·78813	78 14	60 „ 120	67 57·4	0·07653	0·26360	36 37
16 „ 164	52 06·1	·00663	·76079	77 26	61 „ 119	68 33·9	·07837	·25931	35 30
17 „ 163	52 14·7	·00748	·73519	76 38	62 „ 118	69 11·1	·08018	·25519	34 23
18 „ 162	52 23·9	·00837	·71115	75 50	63 „ 117	69 48·9	·08197	·25125	33 15
19 „ 161	52 33·6	·00931	·68849	75 01	64 „ 116	70 27·4	·08373	·24747	32 07
20 „ 160	52 43·8	0·01030	0·66708	74 12	65 „ 115	71 06·5	0·08545	0·24385	30 58
21 „ 159	52 54·6	·01134	·64680	73 23	66 „ 114	71 46·2	·08713	·24040	29 49
22 „ 158	53 06·0	·01242	·62755	72 34	67 „ 113	72 26·5	·08878	·23710	28 39
23 „ 157	53 17·9	·01354	·60925	71 45	68 „ 112	73 07·5	·09038	·23396	27 28
24 „ 156	53 30·4	·01471	·59181	70 55	69 „ 111	73 49·0	·09194	·23098	26 17
25 „ 155	53 43·5	0·01593	0·57518	70 05	70 „ 110	74 31·2	0·09346	0·22814	25 06
26 „ 154	53 57·1	·01719	·55929	69 14	71 „ 109	75 13·8	·09491	·22546	23 54
27 „ 153	54 11·3	·01849	·54408	68 24	72 „ 108	75 57·1	·09631	·22292	22 41
28 „ 152	54 26·1	·01983	·52952	67 53	73 „ 107	76 40·8	·09765	·22053	21 28
29 „ 151	54 41·5	·02122	·51556	66 42	74 „ 106	77 25·0	·09894	·21829	20 15
30 „ 150	54 57·5	0·02264	0·50216	65 50	75 „ 105	78 09·8	0·10016	0·21618	19 01
31 „ 149	55 14·1	·02410	·48929	64 58	76 „ 104	78 55·0	·10132	·21422	17 47
32 „ 148	55 31·3	·02560	·47692	64 06	77 „ 103	79 40·6	·10241	·21240	16 33
33 „ 147	55 49·1	·02714	·46502	63 13	78 „ 102	80 26·6	·10343	·21072	15 18
34 „ 146	56 07·5	·02871	·45357	62 20	79 „ 101	81 13·0	·10438	·20918	14 03
35 „ 145	56 26·5	0·03032	0·44254	61 27	80 „ 100	81 59·7	0·10525	0·20778	12 47
36 „ 144	56 46·2	·03195	·43191	60 33	81 „ 99	82 46·8	·10605	·20651	11 31
37 „ 143	57 06·5	·03362	·42166	59 39	82 „ 98	83 34·2	·10676	·20537	10 15
38 „ 142	57 27·4	·03532	·41179	58 44	83 „ 97	84 21·8	·10740	·20438	8 59
39 „ 141	57 49·0	·03705	·40226	57 49	84 „ 96	85 09·7	·10795	·20351	7 42
40 „ 140	58 11·2	0·03881	0·39306	56 53	85 „ 95	85 57·8	0·10842	0·20278	6 25
41 „ 139	58 34·1	·04059	·38418	55 58	86 „ 94	86 46·0	·10881	·20219	5 08
42 „ 138	58 57·7	·04239	·37562	55 01	87 „ 93	87 34·4	·10911	·20172	3 51
43 „ 137	59 21·9	·04421	·36734	54 04	88 „ 92	88 22·9	·10932	·20139	2 34
44 „ 136	59 46·7	·04606	·35936	53 07	89 „ 91	89 11·4	·10946	·20119	1 17
45 „ 135	60 12·3	0·04792	0·35164	52 09	90	90 00·0	0·10950	0·20113	0 00
h	180°-K	A	N	a₁	h	180°-K	A	N	a₁

Directions with regard to (a₂ ± a₁) :—

Whenever h is greater than 90°, use (a₂ — a₁).

If h is less than 90°, use :—

(a₁ + a₂) when (K ± d) has been used,

(a₁ — a₂) when (d — K) has been used.

Always **name azimuth** from elevated pole.

Note that { h = L.H.A. of body **West** of the meridian.
{ h = (360° — L.H.A.) of body **East** of the meridian.

Log. tan. a₂ = N + log. tan. (K ± d).

LAT. 52° ALT.—AZ. TABLE

h	K		A	N	a₁		h	K		A	N	a₁	
°	°	′			°	′	°	°	′			°	′
0 or 180	52	00·0	0·00000	—	90	00	45 or 135	61	04·9	0·04563	0·36117	51	46
1 „ 179	52	00·3	·00003	1·96880	89	13	46 „ 134	61	30·6	·04741	·35372	50	47
2 „ 178	52	01·0	·00010	1·66784	88	25	47 „ 133	61	57·0	·04920	·34653	49	48
3 „ 177	52	02·3	·00023	1·49186	87	38	48 „ 132	62	24·0	·05100	·33958	48	49
4 „ 176	52	04·1	·00040	1·36707	86	51	49 „ 131	62	51·7	·05281	·33288	47	48
5 „ 175	52	06·4	0·00063	1·27036	86	03	50 „ 130	63	20·0	0·05463	0·32640	46	48
6 „ 174	52	09·2	·00090	1·19142	85	15	51 „ 129	63	49·1	·05645	·32015	45	47
7 „ 173	52	12·5	·00123	1·12476	84	28	52 „ 128	64	18·7	·05827	·31413	44	45
8 „ 172	52	16·3	·00160	1·06710	83	41	53 „ 127	64	49·1	·06010	·30831	43	43
9 „ 171	52	20·6	·00202	1·01633	82	53	54 „ 126	65	20·0	·06192	·30270	42	41
10 „ 170	52	25·5	0·00250	0·97099	82	05	55 „ 125	65	51·7	0·06373	0·29729	41	37
11 „ 169	52	30·9	·00302	·93006	81	17	56 „ 124	66	24·0	·06554	·29208	40	34
12 „ 168	52	36·7	·00359	·89278	80	29	57 „ 123	66	57·0	·06733	·28707	39	30
13 „ 167	52	43·2	·00421	·85857	79	41	58 „ 122	67	30·6	·06911	·28224	38	25
14 „ 166	52	50·1	·00487	·82698	78	53	59 „ 121	68	04·8	·07088	·27759	37	20
15 „ 165	52	57·6	0·00558	0·79766	78	05	60 „ 120	68	39·7	0·07263	0·27313	36	14
16 „ 164	53	05·6	·00635	·77032	77	16	61 „ 119	69	15·3	·07436	·26884	35	07
17 „ 163	53	14·1	·00715	·74472	76	27	62 „ 118	69	51·4	·07606	·26472	34	01
18 „ 162	53	23·2	·00801	·72068	75	38	63 „ 117	70	28·2	·07773	·26078	32	53
19 „ 161	53	32·8	·00890	·69802	74	49	64 „ 116	71	05·6	·07938	·25700	31	45
20 „ 160	53	42·9	0·00985	0·67661	74	00	65 „ 115	71	43·7	0·08100	0·25338	30	57
21 „ 159	53	53·6	·01084	·65633	73	10	66 „ 114	72	22·3	·08258	·24993	29	28
22 „ 158	54	04·8	·01187	·63708	72	20	67 „ 113	73	01·4	·08412	·24663	28	19
23 „ 157	54	16·6	·01294	·61878	71	30	68 „ 112	73	41·2	·08562	·24349	27	09
24 „ 156	54	29·0	·01406	·60134	70	40	69 „ 111	74	21·5	·08708	·24051	25	58
25 „ 155	54	41·9	0·01522	0·58472	69	49	70 „ 110	75	02·4	0·08849	0·23767	24	47
26 „ 154	54	55·4	·01642	·56882	68	59	71 „ 109	75	43·7	·08986	·23499	23	36
27 „ 153	55	09·4	·01766	·55361	68	07	72 „ 108	76	25·6	·09117	·23245	22	24
28 „ 152	55	24·0	·01894	·53905	67	16	73 „ 107	77	08·0	·09242	·23006	21	12
29 „ 151	55	39·2	·02026	·52509	66	24	74 „ 106	77	50·8	·09362	·22782	20	00
30 „ 150	55	55·0	0·02162	0·51169	65	32	75 „ 105	78	34·1	0·09477	0·22571	18	47
31 „ 149	56	11·4	·02301	·49882	64	40	76 „ 104	79	17·8	·09584	·22375	17	33
32 „ 148	56	28·4	·02444	·48645	63	47	77 „ 103	80	01·9	·09686	·22193	16	20
33 „ 147	56	45·9	·02590	·47455	62	54	78 „ 102	80	46·4	·09781	·22025	15	06
34 „ 146	57	04·1	·02739	·46310	62	01	79 „ 101	81	31·3	·09870	·21871	13	51
35 „ 145	57	22·9	0·02892	0·45207	61	07	80 „ 100	82	16·4	0·09951	0·21731	12	37
36 „ 144	57	42·2	·03047	·44144	60	12	81 „ 99	83	01·9	·10025	·21604	11	22
37 „ 143	58	02·2	·03206	·43119	59	18	82 „ 98	83	47·7	·10092	·21490	10	07
38 „ 142	58	22·9	·03368	·42132	58	23	83 „ 97	84	33·7	·10151	·21391	8	51
39 „ 141	58	44·1	·03532	·41179	57	27	84 „ 96	85	19·9	·10202	·21304	7	36
40 „ 140	59	06·0	0·03699	0·40251	56	32	85 „ 95	86	06·3	0·10246	0·21231	6	20
41 „ 139	59	28·5	·03868	·39371	55	35	86 „ 94	86	52·8	·10283	·21172	5	04
42 „ 138	59	51·6	·04039	·38515	54	39	87 „ 93	87	39·5	·10311	·21125	3	48
43 „ 137	60	15·4	·04211	·37687	53	41	88 „ 92	88	26·3	·10331	·21092	2	32
44 „ 136	60	39·8	·04386	·36889	52	44	89 „ 91	89	13·1	·10343	·21072	1	16
45 „ 135	61	04·9	0·04563	0·36117	51	46	90	90	00·0	0·10347	0·21066	0	00
h	180°-K		A	N	a₁		h	180°-K		A	N	a₁	

Directions with regard to $(K \overset{+}{\underset{-}{\sim}} d)$:—

    If l and d have **Same Name,** use $(K \frown d)$, i.e. $(K — d)$ or $(d — K)$.
    If l and d have **Different Names,** use $(K + d)$.

To name Intercept :—

    If H_o is greater than H_c, intercept is " **Towards.** "
    If H_o is less than H_c, intercept is " **Away.** "

Log. cosec. $H_c = A + $ log. sec. $(K \overset{+}{\underset{-}{\sim}} d)$.

N.B.   Whenever h is greater than 90°, subtract the quantity in the K column from 180°.

ALT.—AZ. TABLE LAT. 53°

h	K		A	N	a₁		h	K		A	N	a₁	
°	°	′			°	′	°	°	′			°	′
0 or 180	53	00.0	0.00000	—	90	00	45 or 135	61	57.0	0.04338	0.37105	51	23
1 „ 179	53	00.3	.00002	1.97868	89	12	46 „ 134	62	22.2	.04507	.36360	50	25
2 „ 178	53	01.0	.00010	1.67772	88	24	47 „ 133	62	48.0	.04676	.35641	49	25
3 „ 177	53	02.3	.00022	1.50174	87	36	48 „ 132	63	14.5	.04846	.34946	48	26
4 „ 176	53	04.0	.00038	1.37695	86	48	49 „ 131	63	41.6	.05017	.34286	47	26
5 „ 175	53	06.3	0.00060	1.28024	86	00	50 „ 130	64	09.3	0.05188	0.33628	46	25
6 „ 174	53	09.1	.00086	1.20130	85	12	51 „ 129	64	37.7	.05360	.33003	45	24
7 „ 173	53	12.4	.00117	1.13464	84	24	52 „ 128	65	06.7	.05532	.32400	44	22
8 „ 172	53	16.1	.00153	1.07698	83	36	53 „ 127	65	36.3	.05704	.31819	43	20
9 „ 171	53	20.4	.00193	1.02620	82	47	54 „ 126	66	06.6	.05875	.31258	42	18
10 „ 170	53	25.2	0.00238	0.98087	81	59	55 „ 125	66	37.5	0.06046	0.30717	41	15
11 „ 169	53	30.6	.00288	.93994	81	11	56 „ 124	67	09.0	.06216	.30196	40	11
12 „ 168	53	36.4	.00343	.90266	80	22	57 „ 123	67	41.2	.06385	.29695	39	07
13 „ 167	53	42.7	.00402	.86845	79	33	58 „ 122	68	13.9	.06553	.29212	38	02
14 „ 166	53	49.6	.00465	.83686	78	44	59 „ 121	68	47.3	.06718	.28747	36	57
15 „ 165	53	57.0	0.00533	0.80754	77	55	60 „ 120	69	21.3	0.06882	0.28301	35	52
16 „ 164	54	04.9	.00606	.78020	77	06	61 „ 119	69	55.9	.07045	.27872	34	46
17 „ 163	54	13.3	.00683	.75460	76	17	62 „ 118	70	31.1	.07204	.27460	33	39
18 „ 162	54	22.3	.00764	.73055	75	27	63 „ 117	71	06.8	.07362	.27066	32	32
19 „ 161	54	31.8	.00850	.70789	74	37	64 „ 116	71	43.2	.07516	.26688	31	25
20 „ 160	54	41.8	0.00940	0.68648	73	48	65 „ 115	72	20.1	0.07667	0.26326	30	17
21 „ 159	54	52.4	.01034	.66621	72	57	66 „ 114	72	57.6	.07815	.25981	29	08
22 „ 158	55	03.5	.01133	.64696	72	07	67 „ 113	73	35.6	.07959	.25651	27	59
23 „ 157	55	15.2	.01235	.62866	71	16	68 „ 112	74	14.2	.08100	.25337	26	50
24 „ 156	55	27.4	.01342	.61122	70	26	69 „ 111	74	53.3	.08237	.25038	25	40
25 „ 155	55	40.1	0.01452	0.59459	69	34	70 „ 110	75	32.9	0.08369	0.24755	24	30
26 „ 154	55	53.4	.01567	.57869	68	43	71 „ 109	76	12.9	.08496	.24487	23	19
27 „ 153	56	07.3	.01685	.56349	67	51	72 „ 108	76	53.5	.08619	.24233	22	08
28 „ 152	56	21.7	.01807	.54893	67	00	73 „ 107	77	34.5	.08736	.23994	20	57
29 „ 151	56	36.7	.01932	.53497	66	07	74 „ 106	78	16.0	.08848	.23769	19	45
30 „ 150	56	52.3	0.02061	0.52157	65	15	75 „ 105	78	57.8	0.08954	0.23559	18	33
31 „ 149	57	08.4	.02193	.50870	64	22	76 „ 104	79	40.1	.09055	.23363	17	20
32 „ 148	57	25.2	.02329	.49633	63	29	77 „ 103	80	22.7	.09150	.23181	16	07
33 „ 147	57	42.5	.02468	.48443	62	35	78 „ 102	81	05.7	.09238	.23013	14	54
34 „ 146	58	00.4	.02610	.47297	61	41	79 „ 101	81	49.1	.09321	.22859	13	41
35 „ 145	58	18.8	0.02755	0.46195	60	47	80 „ 100	82	32.7	0.09396	0.22719	12	27
36 „ 144	58	37.9	.02903	.45132	59	53	81 „ 99	83	16.6	.09465	.22592	11	13
37 „ 143	58	57.6	.03053	.44107	58	58	82 „ 98	84	00.8	.09528	.22478	9	59
38 „ 142	59	17.9	.03206	.43119	58	02	83 „ 97	84	45.2	.09583	.22379	8	44
39 „ 141	59	38.8	.03362	.42166	57	07	84 „ 96	85	29.8	.09631	.22292	7	30
40 „ 140	60	00.2	0.03520	0.41247	56	10	85 „ 95	86	14.5	0.09672	0.22219	6	15
41 „ 139	60	22.4	.03680	.40359	55	14	86 „ 94	86	59.5	.09706	.22160	5	00
42 „ 138	60	45.1	.03842	.39503	54	17	87 „ 93	87	44.7	.09732	.22113	3	45
43 „ 137	61	08.4	.04005	.38675	53	19	88 „ 92	88	29.6	.09750	.22080	2	30
44 „ 136	61	32.4	.04171	.37877	52	22	89 „ 91	89	14.8	.09761	.22060	1	15
45 „ 135	61	57.0	0.04338	0.37105	51	23	90	90	00.0	0.09765	0.22054	0	00
h	180°-K		A	N	a₁		h	180°-K		A	N	a₁	

Directions with regard to $(a_2 \overset{+}{\underset{-}{\sim}} a_1)$:—

Whenever h is greater than 90°, use $(a_2 - a_1)$.

If h is less than 90°, use :—

    $(a_1 + a_2)$ when $(K \pm d)$ has been used,

    $(a_1 - a_2)$ when $(d - K)$ has been used.

Always **name azimuth** from elevated pole.

Note that $\begin{cases} \text{h = L.H.A. of body } \textbf{West} \text{ of the meridian.} \\ \text{h = } (360° - \text{L.H.A.}) \text{ of body } \textbf{East} \text{ of the meridian.} \end{cases}$

Log. tan. a_2 = N + log. tan. $(K \overset{+}{\underset{-}{\sim}} d)$.

LAT. 54°   ALT.—AZ. TABLE

h	K	A	N	a₁	h	K	A	N	a₁
°	° ′			° ′	°	° ′			° ′
0 or 180	54 00·0	0·00000	—	90 00	45 or 135	62 48·5	0·04118	0·38130	51 02
1 „ 179	54 00·2	·00002	1·98893	89 11	46 „ 134	63 13·2	·04277	·37385	50 03
2 „ 178	54 01·0	·00009	1·68796	88 23	47 „ 133	63 38·5	·04437	·36665	49 03
3 „ 177	54 02·2	·00021	1·51198	87 34	48 „ 132	64 04·4	·04597	·35971	48 04
4 „ 176	54 04·0	·00037	1·38720	86 46	49 „ 131	64 30·9	·04758	·35300	47 03
5 „ 175	54 06·2	0·00057	1·29048	85 57	50 „ 130	64 58·0	0·04920	0·34653	46 03
6 „ 174	54 09·0	·00082	1·21155	85 08	51 „ 129	65 25·7	·05082	·34028	45 02
7 „ 173	54 12·2	·00112	1·14489	84 20	52 „ 128	65 54·0	·05244	·33425	44 00
8 „ 172	54 16·0	·00146	1·08723	83 31	53 „ 127	66 23·0	·05405	·32843	42 58
9 „ 171	54 20·2	·00184	1·03645	82 42	54 „ 126	66 52·5	·05566	·32282	41 56
10 „ 170	54 25·0	0·00227	0·99111	81 53	55 „ 125	67 22·6	0·05727	0·31742	40 53
11 „ 169	54 30·2	·00275	·95018	81 04	56 „ 124	67 53·4	·05887	·31221	39 49
12 „ 168	54 36·0	·00327	·91290	80 15	57 „ 123	68 24·7	·06046	·30719	38 45
13 „ 167	54 42·3	·00383	·87869	79 25	58 „ 122	68 56·6	·06203	·30236	37 41
14 „ 166	54 49·1	·00444	·84711	78 36	59 „ 121	69 29·1	·06358	·29771	36 36
15 „ 165	54 56·4	0·00508	0·81778	77 45	60 „ 120	70 02·1	0·06513	0·29325	35 31
16 „ 164	55 04·2	·00578	·79044	76 56	61 „ 119	70 35·8	·06665	·28896	34 25
17 „ 163	55 12·5	·00651	·76485	76 06	62 „ 118	71 10·0	·06814	·28485	33 19
18 „ 162	55 21·4	·00728	·74080	75 16	63 „ 117	71 44·7	·06962	·28090	32 12
19 „ 161	55 30·8	·00810	·71814	74 26	64 „ 116	72 20·0	·07106	·27712	31 05
20 „ 160	55 40·7	0·00896	0·69673	73 36	65 „ 115	72 55·9	0·07247	0·27350	29 58
21 „ 159	55 51·1	·00986	·67645	72 45	66 „ 114	73 32·2	·07386	·27005	28 50
22 „ 158	56 02·1	·01079	·65721	71 54	67 „ 113	74 09·1	·07521	·26675	27 41
23 „ 157	56 13·6	·01177	·63890	71 03	68 „ 112	74 46·5	·07652	·26361	26 32
24 „ 156	56 25·6	·01278	·62147	70 11	69 „ 111	75 24·4	·07780	·26063	25 23
25 „ 155	56 38·2	0·01383	0·60483	69 20	70 „ 110	76 02·7	0·07903	0·25779	24 13
26 „ 154	56 51·3	·01492	·58894	68 28	71 „ 109	76 41·5	·08022	·25511	23 03
27 „ 153	57 05·0	·01604	·57373	67 36	72 „ 108	77 20·8	·08136	·25257	21 53
28 „ 152	57 19·2	·01720	·55917	66 43	73 „ 107	78 00·4	·08246	·25018	20 42
29 „ 151	57 34·0	·01839	·54521	65 51	74 „ 106	78 40·5	·08350	·24794	19 31
30 „ 150	57 49·3	0·01962	0·53181	64 58	75 „ 105	79 21·0	0·08449	0·24584	18 20
31 „ 149	58 05·2	·02087	·51894	64 05	76 „ 104	80 01·9	·08543	·24388	17 08
32 „ 148	58 21·7	·02216	·50657	63 11	77 „ 103	80 43·1	·08632	·24206	15 56
33 „ 147	58 38·7	·02348	·49467	62 17	78 „ 102	81 24·6	·08714	·24038	14 43
34 „ 146	58 56·3	·02483	·48322	61 23	79 „ 101	82 06·4	·08791	·23883	13 31
35 „ 145	59 14·5	0·02620	0·47219	60 28	80 „ 100	82 48·6	0·08861	0·23743	12 18
36 „ 144	59 33·2	·02760	·46156	59 33	81 „ 99	83 30·9	·08926	·23616	11 05
37 „ 143	59 52·6	·02903	·45132	58 38	82 „ 98	84 13·6	·08983	·23503	9 51
38 „ 142	60 12·5	·03047	·44144	57 42	83 „ 97	84 56·5	·09034	·23403	8 38
39 „ 141	60 33·0	·03195	·43191	56 46	84 „ 96	85 39·4	·09079	·23317	7 24
40 „ 140	60 54·1	0·03345	0·42271	55 50	85 „ 95	86 22·6	0·09117	0·23244	6 10
41 „ 139	61 15·8	·03496	·41384	54 53	86 „ 94	87 05·9	·09148	·23184	4 56
42 „ 138	61 38·1	·03649	·40527	53 56	87 „ 93	87 49·3	·09173	·23138	3 42
43 „ 137	62 00·9	·03804	·39700	52 58	88 „ 92	88 32·9	·09190	·23105	2 28
44 „ 136	62 24·4	·03960	·38901	52 00	89 „ 91	89 16·4	·09200	·23085	1 15
45 „ 135	62 48·5	0·04118	0·38130	51 02	90	90 00·0	0·09204	0·23078	0 00
h	180°-K	A	N	a₁	h	180°-K	A	N	a₁

Directions with regard to (K ⁓ d) :—

If l and d have **Same Name**, use (K ⁓ d), i.e. (K — d) or (d — K).

If l and d have **Different Names**, use (K + d).

To name Intercept :—

If H_o is greater than H_c, intercept is " **Towards.** "
If H_o is less than H_c, intercept is " **Away.** "

Log. cosec. H_c = A + log. sec. (K ⁓ d).

N.B. *Whenever h is greater than 90°, subtract the quantity in the K column from 180°.*

ALT.—AZ. TABLE LAT. 55°

h	K	A	N	a₁	h	K	A	N	a₁
°	° ′			° ′	°	° ′			° ′
0 or 180	55 00·0	0·00000	—	90 00	45 or 135	63 39·5	0·03902	0·39192	50 41
1 „ 179	55 00·2	·00002	1·99955	89 11	46 „ 134	64 03·7	·04052	·38447	49 42
2 „ 178	55 01·0	·00009	1·69859	88 22	47 „ 133	64 28·4	·04203	·37728	48 42
3 „ 177	55 02·2	·00020	1·52261	87 33	48 „ 132	64 53·7	·04354	·37034	47 42
4 „ 176	55 03·9	·00035	1·39782	86 43	49 „ 131	65 19·6	·04506	·36363	46 42
5 „ 175	55 06·2	0·00054	1·30111	85 54	50 „ 130	65 46·1	0·04658	0·35715	45 41
6 „ 174	55 08·9	·00078	1·22217	85 05	51 „ 129	66 13·1	·04810	·35091	44 40
7 „ 173	55 12·1	·00106	1·15551	84 15	52 „ 128	66 40·8	·04963	·34488	43 39
8 „ 172	55 15·8	·00139	1·09785	83 26	53 „ 127	67 09·0	·05115	·33906	42 37
9 „ 171	55 20·0	·00176	1·04708	82 36	54 „ 126	67 37·8	·05266	·33345	41 34
10 „ 170	55 24·7	0·00216	1·00174	81 47	55 „ 125	68 07·1	0·05417	0·32804	40 31
11 „ 169	55 29·9	·00262	0·96081	80 57	56 „ 124	68 37·0	·05566	·32283	39 28
12 „ 168	55 35·6	·00311	·92353	80 07	57 „ 123	69 07·5	·05715	·31782	38 24
13 „ 167	55 41·7	·00365	·88932	79 17	58 „ 122	69 38·5	·05863	·31299	37 20
14 „ 166	55 48·4	·00422	·85773	78 27	59 „ 121	70 10·1	·06009	·30834	36 16
15 „ 165	55 55·7	0·00484	0·82841	77 37	60 „ 120	70 42·3	0·06153	0·30388	35 11
16 „ 164	56 03·4	·00550	·80107	76 47	61 „ 119	71 15·0	·06295	·29959	34 05
17 „ 163	56 11·6	·00619	·77547	75 56	62 „ 118	71 48·2	·06435	·29547	32 59
18 „ 162	56 20·3	·00693	·75143	75 06	63 „ 117	72 21·9	·06573	·29153	31 53
19 „ 161	56 29·6	·00771	·72877	74 15	64 „ 116	72 56·1	·06709	·28775	30 46
20 „ 160	56 39·4	0·00852	0·70736	73 24	65 „ 115	73 30·9	0·06841	0·28413	29 39
21 „ 159	56 49·6	·00937	·68708	72 33	66 „ 114	74 06·2	·06970	·28068	28 31
22 „ 158	57 00·5	·01026	·66783	71 41	67 „ 113	74 41·9	·07097	·27738	27 24
23 „ 157	57 11·8	·01119	·64953	70 50	68 „ 112	75 18·1	·07219	·27424	26 15
24 „ 156	57 23·6	·01215	·63210	69 58	69 „ 111	75 54·8	·07338	·27126	25 07
25 „ 155	57 36·0	0·01315	0·61546	69 06	70 „ 110	76 31·9	0·07453	0·26842	23 57
26 „ 154	57 49·0	·01418	·59957	68 13	71 „ 109	77 09·5	·07563	·26574	22 48
27 „ 153	58 02·4	·10525	·58436	67 21	72 „ 108	77 47·4	·07670	·26320	21 38
28 „ 152	58 16·4	·01635	·56980	66 28	73 „ 107	78 25·8	·07772	·26081	20 28
29 „ 151	58 31·0	·01748	·55584	65 35	74 „ 106	79 04·6	·07869	·25857	19 18
30 „ 150	58 46·0	0·01864	0·54244	64 41	75 „ 105	79 43·7	0·07961	0·25646	18 07
31 „ 149	59 01·7	·01983	·52957	63 48	76 „ 104	80 23·1	·08049	·25450	16 56
32 „ 148	59 17·9	·02105	·51720	62 54	77 „ 103	81 02·9	·08132	·25268	15 44
33 „ 147	59 34·6	·02230	·50530	61 59	78 „ 102	81 43·0	·08208	·25100	14 33
34 „ 146	59 51·9	·02358	·49385	61 05	79 „ 101	82 23·4	·08280	·24946	13 21
35 „ 145	60 09·7	0·02487	0·48282	60 10	80 „ 100	83 04·0	0·08345	0·24806	12 09
36 „ 144	60 28·2	·02620	·47219	59 14	81 „ 99	83 44·9	·08405	·24679	10 57
37 „ 143	60 47·1	·02755	·46195	58 19	82 „ 98	84 26·0	·08458	·24566	9 44
38 „ 142	61 06·7	·02892	·45207	57 23	83 „ 97	85 07·4	·08506	·24466	8 31
39 „ 141	61 26·8	·03032	·44254	56 27	84 „ 96	85 48·8	·08548	·24379	7 19
40 „ 140	61 47·5	0·03173	0·43334	55 30	85 „ 95	86 30·5	0·08583	0·24306	6 06
41 „ 139	62 08·7	·03316	·42447	54 33	86 „ 94	87 12·2	·08612	·24247	4 53
42 „ 138	62 30·6	·03461	·41590	53 35	87 „ 93	87 54·1	·08634	·24200	3 40
43 „ 137	62 53·0	·03607	·40763	52 37	88 „ 92	88 36·0	·08650	·24167	2 26
44 „ 136	63 16·0	·03754	·39964	51 39	89 „ 91	89 18·0	·08660	·24147	1 13
45 „ 135	63 39·5	0·03902	0·39192	50 41	90	90 00·0	0·08664	0·24141	0 00
h	180°-K	A	N	a₁	h	180°-K	A	N	a₁

Directions with regard to (a₂ ± a₁) :—

Whenever h is greater than 90°, use (a₂ − a₁).

If h is less than 90°, use :—

(a₁ + a₂) when (K ± d) has been used,

(a₁ − a₂) when (d − K) has been used.

Always **name** azimuth from elevated pole.

Note that { h = L.H.A. of body **West** of the meridian.
{ h = (360° − L.H.A.) of body **East** of the meridian.

Log. tan. a₂ = N + log. tan. (K ± d).

LAT. 56°   ALT.—AZ. TABLE

h	K	A	N	a₁	h	K	A	N	a₁
°	° ′			° ′	°	° ′			° ′
0 or 180	56 00·0	0·00000	—	90 00	45 or 135	64 30·1	0·03692	0·40295	50 20
1 „ 179	56 00·2	·00002	2·01058	89 10	46 „ 134	64 53·7	·03833	·39550	49 21
2 „ 178	56 01·0	·00008	1·70902	88 21	47 „ 133	65 17·8	·03975	·38831	48 22
3 „ 177	56 02·2	·00019	1·53364	87 31	48 „ 132	65 42·5	·04117	·38136	47 22
4 „ 176	56 03·9	·00033	1·40885	86 41	49 „ 131	66 07·8	·04260	·37466	46 21
5 „ 175	56 06·1	0·00052	1·31214	85 51	50 „ 130	66 33·6	0·04402	0·36818	45 21
6 „ 174	56 08·7	·00074	1·23320	85 01	51 „ 129	67 00·0	·04545	·36193	44 20
7 „ 173	56 11·9	·00101	1·16654	84 11	52 „ 128	67 26·9	·04688	·35591	43 18
8 „ 172	56 15·6	·00132	1·10888	83 21	53 „ 127	67 54·4	·04831	·35009	42 16
9 „ 171	56 19·7	·00167	1·05811	82 31	54 „ 126	68 22·4	·04973	·34448	41 14
10 „ 170	56 24·3	0·00206	1·01277	81 41	55 „ 125	68 51·0	0·05114	0·33907	40 11
11 „ 169	56 29·5	·00249	0·97184	80 51	56 „ 124	69 20·1	·05254	·33386	39 08
12 „ 168	56 35·1	·00296	·93456	80 00	57 „ 123	69 49·7	·05394	·32885	38 04
13 „ 167	56 41·2	·00346	·90035	79 10	58 „ 122	70 19·9	·05532	·32402	37 00
14 „ 166	56 47·8	·00401	·86876	78 19	59 „ 121	70 50·6	·05668	·31937	35 56
15 „ 165	56 54·9	0·00460	0·83944	77 29	60 „ 120	71 21·8	0·05803	0·31491	34 51
16 „ 164	57 02·5	·00522	·81210	76 38	61 „ 119	71 53·5	·05937	·31062	33 46
17 „ 163	57 10·6	·00588	·78650	75 47	62 „ 118	72 25·7	·06067	·30650	32 40
18 „ 162	57 19·2	·00658	·76246	74 55	63 „ 117	72 58·5	·06196	·30256	31 34
19 „ 161	57 28·3	·00732	·73980	74 04	64 „ 116	73 31·7	·06323	·29878	30 28
20 „ 160	57 37·9	0·00809	0·71839	73 13	65 „ 115	74 05·4	0·06446	0·29516	29 21
21 „ 159	57 48·1	·00890	·69811	72 21	66 „ 114	74 39·5	·06567	·29171	28 14
22 „ 158	57 58·7	·00974	·67886	71 29	67 „ 113	75 14·1	·06684	·28841	27 07
23 „ 157	58 09·9	·01062	·66056	70 37	68 „ 112	75 49·2	·06799	·28527	25 59
24 „ 156	58 21·5	·01153	·64312	69 44	69 „ 111	76 24·7	·06910	·28229	24 51
25 „ 155	58 33·7	0·01248	0·62649	68 52	70 „ 110	77 00·6	0·07017	0·27945	23 42
26 „ 154	58 46·4	·01346	·61060	67 59	71 „ 109	77 36·9	·07120	·27677	22 33
27 „ 153	58 59·7	·01447	·59539	67 06	72 „ 108	78 13·6	·07219	·27423	21 24
28 „ 152	59 13·4	·01551	·58083	66 13	73 „ 107	78 50·6	·07314	·27184	20 15
29 „ 151	59 27·7	·01658	·56687	65 19	74 „ 106	79 28·1	·07405	·26960	19 05
30 „ 150	59 42·5	0·01768	0·55347	64 25	75 „ 105	80 05·8	0·07491	0·26749	17 55
31 „ 149	59 57·9	·01880	·54060	63 31	76 „ 104	80 43·9	·07572	·26553	16 44
32 „ 148	60 13·8	·01996	·52823	62 37	77 „ 103	81 22·3	·07648	·26371	15 34
33 „ 147	60 30·2	·02114	·51633	61 42	78 „ 102	82 01·0	·07720	·26203	14 23
34 „ 146	60 47·2	·02234	·50488	60 47	79 „ 101	82 40·0	·07786	·26049	13 12
35 „ 145	61 04·7	0·02358	0·49385	59 52	80 „ 100	83 19·2	0·07847	0·25909	12 00
36 „ 144	61 22·8	·02483	·48322	58 56	81 „ 99	83 58·6	·07902	·25782	10 49
37 „ 143	61 41·4	·02610	·47297	58 00	82 „ 98	84 38·2	·07952	·25668	9 37
38 „ 142	62 00·5	·02739	·46310	57 04	83 „ 97	85 18·0	·07996	·25569	8 25
39 „ 141	62 20·2	·02871	·45367	56 07	84 „ 96	85 58·0	·08035	·25482	7 14
40 „ 140	62 40·5	0·03004	0·44437	55 11	85 „ 95	86 38·1	0·08068	0·25409	6 01
41 „ 139	63 01·3	·03139	·43549	54 13	86 „ 94	87 18·4	·08095	·25350	4 49
42 „ 138	63 22·6	·03275	·42693	53 16	87 „ 93	87 58·7	·08116	·25303	3 37
43 „ 137	63 44·6	·03413	·41865	52 18	88 „ 92	88 39·1	·08131	·25270	2 25
44 „ 136	64 07·0	·03552	·41067	51 19	89 „ 91	89 19·5	·08140	·25250	1 12
45 „ 135	64 30·1	0·03692	0·40295	50 20	90	90 00·0	0·08143	0·25244	0 00
h	180°-K	A	N	a₁	h	180°-K	A	N	a₁

Directions with regard to (K ⌢ d) :—

If l and d have **Same Name,** use (K ⌢ d), i.e. (K — d) or (d — K).

If l and d have **Different Names,** use (K + d).

To name Intercept :—

If H₀ is greater than H_c, intercept is " **Towards.** "

If H₀ is less than H_c, intercept is " **Away.** "

Log. cosec. H_c = A + log. sec. (K ⌢ d).

N.B.   Whenever h is greater than 90°, subtract the quantity in the K column from 180°.

ALT.—AZ. TABLE   LAT. 57°

h	K	A	N	a_1	h	K	A	N	a_1
°	° ′			° ′	°	° ′			° ′
0 or 180	57 00·0	0·00000	—	90 00	45 or 135	65 20·1	0·03486	0·41441	50 01
1 „ 179	57 00·2	·00002	2·02204	89 10	46 „ 134	65 43·1	·03619	·40696	49 02
2 „ 178	57 01·0	·00008	1·72107	88 19	47 „ 133	66 06·7	·03752	·39976	48 02
3 „ 177	57 02·2	·00018	1·54509	87 29	48 „ 132	66 30·8	·03885	·39282	47 02
4 „ 176	57 03·8	·00031	1·42031	86 39	49 „ 131	66 55·4	·04019	·38611	46 02
5 „ 175	57 06·0	0·00049	1·32359	85 48	50 „ 130	67 20·6	0·04153	0·37964	45 01
6 „ 174	57 08·6	·00070	1·24466	84 58	51 „ 129	67 46·3	·04287	·37339	44 00
7 „ 173	57 11·7	·00096	1·17800	84 07	52 „ 128	68 12·5	·04421	·36736	42 58
8 „ 172	57 15·3	·00125	1·12034	83 17	53 „ 127	68 39·2	·04554	·36154	41 56
9 „ 171	57 19·4	·00158	1·06956	82 26	54 „ 126	69 06·5	·04687	·35593	40 54
10 „ 170	57 24·0	0·00195	1·02422	81 35	55 „ 125	69 34·2	0·04820	0·35053	39 52
11 „ 169	57 29·0	·00236	0·98329	80 44	56 „ 124	70 02·5	·04951	·34532	38 48
12 „ 168	57 34·5	·00280	·94601	79 54	57 „ 123	70 31·3	·05082	·34020	37 45
13 „ 167	57 40·6	·00328	·91180	79 03	58 „ 122	71 00·6	·05211	·33547	36 41
14 „ 166	57 47·1	·00380	·88022	78 11	59 „ 121	71 30·4	·05338	·33082	35 37
15 „ 165	57 54·0	0·00436	0·85089	77 20	60 „ 120	72 00·7	0·05464	0·32636	34 33
16 „ 164	58 01·5	·00495	·82355	76 29	61 „ 119	72 31·4	·05589	·32207	33 28
17 „ 163	58 09·5	·00558	·79796	75 37	62 „ 118	73 02·7	·05711	·31796	32 22
18 „ 162	58 18·0	·00624	·77391	74 45	63 „ 117	73 34·4	·05831	·31401	31 17
19 „ 161	58 26·9	·00694	·75125	73 54	64 „ 116	74 06·6	·05949	·31023	30 11
20 „ 160	58 36·4	0·00767	0·72984	73 02	65 „ 115	74 39·2	0·06064	0·30661	29 04
21 „ 159	58 46·4	·00843	·70956	72 09	66 „ 114	75 12·2	·06177	·30316	27 58
22 „ 158	58 56·8	·00923	·69032	71 17	67 „ 113	75 45·7	·06286	·29986	26 51
23 „ 157	59 07·8	·01006	·67201	70 24	68 „ 112	76 19·6	·06392	·29672	25 43
24 „ 156	59 19·3	·01093	·65458	69 31	69 „ 111	76 53·9	·06496	·29374	24 36
25 „ 155	59 31·2	0·01182	0·63794	68 38	70 „ 110	77 28·6	0·06595	0·29090	23 28
26 „ 154	59 43·7	·01274	·62205	67 45	71 „ 109	78 03·7	·06691	·28822	22 19
27 „ 153	59 56·7	·01370	·60684	66 52	72 „ 108	78 39·2	·06784	·28568	21 11
28 „ 152	60 10·2	·01468	·59228	65 58	73 „ 107	79 15·0	·06872	·28329	20 02
29 „ 151	60 24·2	·01569	·57832	65 04	74 „ 106	79 51·1	·06956	·28105	18 53
30 „ 150	60 38·8	0·01673	0·56492	64 10	75 „ 105	80 27·5	0·07036	0·27895	17 43
31 „ 149	60 53·8	·01780	·55205	63 15	76 „ 104	81 04·3	·07111	·27699	16 33
32 „ 148	61 09·4	·01889	·53968	62 21	77 „ 103	81 41·3	·07182	·27517	15 23
33 „ 147	61 25·5	·02000	·52778	61 26	78 „ 102	82 18·6	·07248	·27349	14 13
34 „ 146	61 42·2	·02114	·51633	60 30	79 „ 101	82 56·2	·07310	·27194	13 03
35 „ 145	61 59·3	0·02230	0·50530	59 35	80 „ 100	83 34·0	0·07367	0·27054	11 52
36 „ 144	62 17·0	·02348	·49467	58 39	81 „ 99	84 12·0	·07418	·26927	10 42
37 „ 143	62 35·2	·02468	·48443	57 42	82 „ 98	84 50·1	·07464	·26814	9 31
38 „ 142	62 54·0	·02590	·47455	56 46	83 „ 97	85 28·5	·07505	·26714	8 20
39 „ 141	63 13·2	·02714	·46502	55 49	84 „ 96	86 07·0	·07541	·26628	7 09
40 „ 140	63 33·1	0·02839	0·45582	54 52	85 „ 95	86 45·6	0·07571	0·26555	5 57
41 „ 139	63 53·4	·02966	·44695	53 54	86 „ 94	87 24·4	·07597	·26495	4 46
42 „ 138	64 14·3	·03095	·43838	52 57	87 „ 93	88 03·2	·07616	·26449	3 35
43 „ 137	64 35·7	·03224	·43011	51 58	88 „ 92	88 42·1	·07630	·26416	2 23
44 „ 136	64 57·6	·03354	·42212	51 00	89 „ 91	89 21·0	·07638	·26396	1 12
45 „ 135	65 20·1	0·03486	0·41441	50 01	90	90 00·0	0·07641	0·26389	0 00
h	180°-K	A	N	a_1	h	180°-K	A	N	a_1

Directions with regard to $(a_2 \overset{+}{\underset{-}{\sim}} a_1)$:—

Whenever h is greater than 90°, use $(a_2 - a_1)$.   Always **name**

If h is less than 90°, use :—   **azimuth** from

    $(a_1 + a_2)$ when $(K \pm d)$ has been used,   elevated pole.

    $(a_1 - a_2)$ when $(d - K)$ has been used.

Note that $\begin{cases} h = \text{L.H.A. of body } \textbf{West} \text{ of the meridian.} \\ h = (360° - \text{L.H.A.}) \text{ of body } \textbf{East} \text{ of the meridian.} \end{cases}$

Log. tan. a_2 = N + log. tan. $(K \overset{+}{\underset{-}{\sim}} d)$.

LAT. 58°   ALT.—AZ. TABLE

h	K	A	N	a_1	h	K	A	N	a_1
°	° ′			° ′	°	° ′			° ′
0 or 180	58 00·0	0·00000	—	90 00	45 or 135	66 09·7	0·03285	0·42630	49 42
1 ,, 179	58 00·2	·00002	2·03393	89 09	46 ,, 134	66 32·1	·03410	·41886	48 43
2 ,, 178	58 00·9	·00007	1·73297	88 10	47 ,, 133	66 55·1	·03534	·41166	47 43
3 ,, 177	58 02·1	·00017	1·55699	87 27	48 ,, 132	67 18·8	·03659	·40472	46 43
4 ,, 176	58 03·8	·00030	1·43220	86 36	49 ,, 131	67 42·5	·03785	·39801	45 43
5 ,, 175	58 05·9	0·00046	1·33549	85 45	50 ,, 130	68 07·0	0·03910	0·39154	44 42
6 ,, 174	58 08·5	·00067	1·25655	84 54	51 ,, 129	68 32·0	·04036	·38529	43 41
7 ,, 173	58 11·5	·00091	1·18990	84 03	52 ,, 128	68 57·5	·04161	·37926	42 39
8 ,, 172	58 15·1	·00118	1·13223	83 12	53 ,, 127	69 23·5	·04286	·37344	41 37
9 ,, 171	58 19·1	·00150	1·08146	82 21	54 ,, 126	69 49·9	·04410	·36783	40 35
10 ,, 170	58 23·6	0·00185	1·03612	81 30	55 ,, 125	70 16·9	0·04534	0·36242	39 33
11 ,, 169	58 28·5	·00223	0·99519	80 38	56 ,, 124	70 44·4	·04656	·35722	38 30
12 ,, 168	58 34·0	·00265	·95791	79 47	57 ,, 123	71 12·3	·04778	·35220	37 27
13 ,, 167	58 39·9	·00311	·92370	78 55	58 ,, 122	71 40·7	·04899	·34737	36 23
14 ,, 166	58 46·3	·00360	·89211	78 04	59 ,, 121	72 09·6	·05018	·34272	35 19
15 ,, 165	58 53·1	0·00412	0·86279	77 12	60 ,, 120	72 39·0	0·05136	0·33826	34 15
16 ,, 164	59 00·5	·00468	·83545	76 19	61 ,, 119	73 08·8	·05252	·33397	33 10
17 ,, 163	59 08·3	·00528	·80985	75 28	62 ,, 118	73 39·0	·05365	·32985	32 05
18 ,, 162	59 16·7	·00590	·78581	74 36	63 ,, 117	74 09·7	·05477	·32591	31 00
19 ,, 161	59 25·5	·00656	·76315	73 43	64 ,, 116	74 40·9	·05587	·32213	29 54
20 ,, 160	59 34·8	0·00725	0·74174	72 51	65 ,, 115	75 12·4	0·05694	0·31851	28 48
21 ,, 159	59 44·5	·00798	·72146	71 58	66 ,, 114	75 44·4	·05799	·31506	27 42
22 ,, 158	59 54·8	·00873	·70221	71 05	67 ,, 113	76 16·8	·05900	·31176	26 35
23 ,, 157	60 05·6	·00952	·68391	70 12	68 ,, 112	76 49·5	·06000	·30862	25 28
24 ,, 156	60 16·8	·01033	·66648	69 19	69 ,, 111	77 22·7	·06096	·30564	24 21
25 ,, 155	60 28·6	0·01117	0·64984	68 25	70 ,, 110	77 56·2	0·06188	0·30280	23 14
26 ,, 154	60 40·8	·01205	·63395	67 32	71 ,, 109	78 30·0	·06277	·30012	22 06
27 ,, 153	60 53·6	·01295	·61874	66 38	72 ,, 108	79 04·3	·06363	·29758	20 58
28 ,, 152	61 06·8	·01387	·60418	65 44	73 ,, 107	79 38·8	·06445	·29519	19 49
29 ,, 151	61 20·5	·01483	·59022	64 49	74 ,, 106	80 13·6	·06523	·29295	18 41
30 ,, 150	61 34·8	0·01581	0·57682	63 55	75 ,, 105	80 48·8	0·06597	0·29085	17 32
31 ,, 149	61 49·5	·01681	·56395	63 00	76 ,, 104	81 24·2	·06667	·28889	16 23
32 ,, 148	62 04·8	·01784	·55158	62 05	77 ,, 103	81 59·9	·06733	·28707	15 14
33 ,, 147	62 20·6	·01889	·53968	61 09	78 ,, 102	82 25·9	·06795	·28539	14 04
34 ,, 146	62 36·8	·01996	·52823	60 14	79 ,, 101	83 12·0	·06851	·28384	12 55
35 ,, 145	62 53·6	0·02105	0·51720	59 18	80 ,, 100	83 48·4	0·06904	0·28244	11 45
36 ,, 144	63 10·9	·02216	·50657	58 22	81 ,, 99	84 25·0	·06951	·28117	10 35
37 ,, 143	63 28·7	·02329	·49633	57 25	82 ,, 98	85 01·8	·06994	·28004	9 25
38 ,, 142	63 47·1	·02444	·48645	56 28	83 ,, 97	85 38·7	·07032	·27904	8 14
39 ,, 141	64 05·9	·02560	·47692	55 31	84 ,, 96	86 15·8	·07066	·27818	7 04
40 ,, 140	64 25·2	0·02678	0·46772	54 34	85 ,, 95	86 53·0	0·07094	0·27745	5 53
41 ,, 139	64 45·1	·02797	·45885	53 36	86 ,, 94	87 30·2	·07117	·27685	4 43
42 ,, 138	65 05·5	·02918	·45028	52 38	87 ,, 93	88 07·6	·07135	·27639	3 32
43 ,, 137	65 26·4	·03039	·44201	51 40	88 ,, 92	88 45·0	·07148	·27605	2 21
44 ,, 136	65 47·8	·03162	·43402	50 41	89 ,, 91	89 22·5	·07155	·27586	1 11
45 ,, 135	66 09·7	0·03285	0·42630	49 42	90	90 00·0	0·07158	0·27579	0 00
h	180°-K	A	N	a_1	h	180°-K	A	N	a_1

Directions with regard to (K ± d) :—

If l and d have **Same Name,** use (K ⌣ d), i.e. (K — d) or (d — K).
If l and d have **Different Names,** use (K + d).

To name Intercept :—

If H_o is greater than H_c, intercept is " **Towards.** "
If H_o is less than H_c, intercept is " **Away.** "

Log. cosec. H_c = A + log. sec. (K ± d).

N.B.   Whenever h is greater than 90°, subtract the quantity in the K column from 180°.

ALT.—AZ. TABLE   LAT. 59°

h	K	A	N	a₁	h	K	A	N	a₁
°	° ′			° ′	°	° ′			° ′
0 or 180	59 00·0	0·00000	—	90 00	**45 or 135**	66 58·8	0·03090	0·43868	49 24
1 „ 179	59 00·2	·00002	2·04631	89 09	**46 „ 134**	67 20·7	·03206	·43123	48 24
2 „ 178	59 00·9	·00007	1·74534	88 17	**47 „ 133**	67 43·0	·03323	·42403	47 25
3 „ 177	59 02·1	·00016	1·56936	87 26	**48 „ 132**	68 05·8	·03439	·41709	46 25
4 „ 176	59 03·7	·00028	1·44458	86 54	**49 „ 131**	68 29·1	·03557	·41038	45 24
5 „ 175	59 05·8	0·00044	1·34786	85 43	**50 „ 130**	68 52·9	0·03674	0·40391	44 23
6 „ 174	59 08·3	·00063	1·26893	84 51	**51 „ 129**	69 17·2	·03791	·39766	43 22
7 „ 173	59 11·3	·00086	1·20227	84 00	**52 „ 128**	69 42·0	·03908	·39163	42 21
8 „ 172	59 14·8	·00112	1·14461	83 08	**53 „ 127**	70 07·2	·04025	·38581	41 19
9 „ 171	59 18·7	·00141	1·09383	82 16	**54 „ 126**	70 32·9	·04141	·38020	40 17
10 „ 170	59 23·2	0·00174	1·04849	81 24	**55 „ 125**	70 59·0	0·04256	0·37480	39 15
11 „ 169	59 28·0	·00211	1·00756	80 32	**56 „ 124**	71 25·7	·04371	·36959	38 12
12 „ 168	59 33·4	·00250	0·97028	79 40	**57 „ 123**	71 52·8	·04484	·36457	37 09
13 „ 167	59 39·2	·00293	·93607	78 48	**58 „ 122**	72 20·3	·04596	·35974	36 06
14 „ 166	59 45·4	·00340	·90449	77 56	**59 „ 121**	72 48·3	·04707	·35509	35 02
15 „ 165	59 52·2	0·00389	0·87516	77 04	**60 „ 120**	73 16·7	0·04817	0·35063	33 58
16 „ 164	59 59·4	·00442	·84782	76 11	**61 „ 119**	73 45·5	·04925	·34634	32 53
17 „ 163	60 07·1	·00498	·82223	75 19	**62 „ 118**	74 14·8	·05031	·34223	31 49
18 „ 162	60 15·2	·00557	·79818	74 26	**63 „ 117**	74 44·5	·05135	·33828	30 44
19 „ 161	60 23·9	·00619	·77552	73 33	**64 „ 116**	75 14·6	·05236	·33450	29 38
20 „ 160	60 33·0	0·00684	0·75411	72 40	**65 „ 115**	75 45·1	0·05336	0·33088	28 33
21 „ 159	60 42·6	·00753	·73383	71 47	**66 „ 114**	76 16·0	·05434	·32743	27 27
22 „ 158	60 52·6	·00824	·71459	70 54	**67 „ 113**	76 47·3	·05528	·32403	26 21
23 „ 157	61 03·2	·00898	·69628	70 00	**68 „ 112**	77 18·9	·05620	·32099	25 14
24 „ 156	61 14·2	·00974	·67885	69 07	**69 „ 111**	77 50·9	·05709	·31801	24 07
25 „ 155	61 25·7	0·01054	0·66221	68 13	**70 „ 110**	78 23·2	0·05795	0·31517	23 00
26 „ 154	61 37·7	·01136	·64632	67 19	**71 „ 109**	78 55·9	·05878	·31249	21 53
27 „ 153	61 50·2	·01221	·63111	66 24	**72 „ 108**	79 28·9	·05957	·30995	20 46
28 „ 152	62 03·2	·01308	·61655	65 30	**73 „ 107**	80 02·2	·06033	·30756	19 38
29 „ 151	62 16·6	·01398	·60259	64 35	**74 „ 106**	80 35·8	·06106	·30532	18 30
30 „ 150	62 30·6	0·01490	0·58919	63 40	**75 „ 105**	81 09·6	0·06174	0·30322	17 22
31 „ 149	62 45·0	·01584	·57632	62 45	**76 „ 104**	81 43·8	·06239	·30126	16 13
32 „ 148	62 59·9	·01681	·56395	61 50	**77 „ 103**	82 18·1	·06300	·29944	15 05
33 „ 147	63 15·3	·01780	·55205	60 54	**78 „ 102**	82 52·7	·06357	·29776	13 56
34 „ 146	63 31·2	·01880	·54060	59 58	**79 „ 101**	83 27·6	·06410	·29621	12 47
35 „ 145	63 47·6	0·01983	0·52957	59 02	**80 „ 100**	84 02·6	0·06458	0·29481	11 37
36 „ 144	64 04·5	·02087	·51894	58 05	**81 „ 99**	84 37·8	·06502	·29354	10 28
37 „ 143	64 21·9	·02193	·50870	57 08	**82 „ 98**	85 13·2	·06542	·29241	9 19
38 „ 142	64 39·8	·02301	·49882	56 11	**83 „ 97**	85 48·7	·06577	·29141	8 09
39 „ 141	64 58·2	·02410	·48929	55 14	**84 „ 96**	86 24·4	·06608	·29055	6 59
40 „ 140	65 17·0	0·02521	0·48009	54 16	**85 „ 95**	87 00·1	0·06634	0·28982	5 50
41 „ 139	65 36·4	·02633	·47122	53 19	**86 „ 94**	87 36·0	·06655	·28922	4 40
42 „ 138	65 56·3	·02745	·46265	52 20	**87 „ 93**	88 11·9	·06672	·28876	3 30
43 „ 137	66 16·6	·02859	·45438	51 22	**88 „ 92**	88 47·9	·06683	·28843	2 20
44 „ 136	66 37·5	·02974	·44639	50 23	**89 „ 91**	89 24·0	·06691	·28823	1 10
45 „ 135	66 58·8	0·03090	0·43868	49 24	**90**	90 00·0	0·06693	0·28816	0 00
h	**180°-K**	**A**	**N**	**a₁**	**h**	**180°-K**	**A**	**N**	**a₁**

Directions with regard to ($a_2 \overset{+}{\sim} a_1$):—

    Whenever h is greater than 90°, use $(a_2 - a_1)$.   Always **name**

    If h is less than 90°, use :—   **azimuth** from

        $(a_1 + a_2)$ when $(K \pm d)$ has been used,   elevated pole.

        $(a_1 - a_2)$ when $(d - K)$ has been used.

Note that $\begin{cases} \text{h = L.H.A. of body \textbf{West} of the meridian.} \\ \text{h = (360° — L.H.A.) of body \textbf{East} of the meridian.} \end{cases}$

Log. tan. a_2 = N + log. tan. $(K \overset{+}{\sim} d)$.

LAT. 60°   ALT.—AZ. TABLE

h	K	A	N	a₁	h	K	A	N	a₁
°	° ′			° ′	°	° ′			° ′
0 or 180	60 00·0	0·00000	—	90 00	45 or 135	67 47·5	0·02900	0·45154	49 06
1 „ 179	60 00·2	·00002	2·05917	89 08	46 „ 134	68 08·8	·03008	·44410	48 07
2 „ 178	60 00·9	·00007	1·75821	00 16	47 „ 133	68 30·5	·03117	·43690	47 07
3 „ 177	60 02·0	·00015	1·58223	87 24	48 „ 132	68 52·6	·03226	·42966	46 07
4 „ 176	60 03·6	·00026	1·45744	86 32	49 „ 131	69 15·3	·03336	·42305	45 06
5 „ 175	60 05·7	0·00041	1·36073	85 40	50 „ 130	69 38·4	0·03445	0·41678	44 06
6 „ 174	60 08·2	·00059	1·28179	84 48	51 „ 129	70 01·9	·03554	·41053	43 05
7 „ 173	60 11·1	·00081	1·21514	83 56	52 „ 128	70 25·9	·03663	·40450	42 03
8 „ 172	60 14·5	·00105	1·15747	83 04	53 „ 127	70 50·4	·03772	·39868	41 02
9 „ 171	60 18·4	·00133	1·10670	82 11	54 „ 126	71 15·3	·03880	·39307	40 00
10 „ 170	60 22·7	·00164	1·06136	81 19	55 „ 125	71 40·7	0·03987	0·38766	38 57
11 „ 169	60 27·5	·00199	1·02043	80 27	56 „ 124	72 06·4	·04094	·38246	37 55
12 „ 168	60 32·7	·00236	0·98315	79 34	57 „ 123	72 32·7	·04199	·37744	36 52
13 „ 167	60 38·4	·00276	·94894	78 42	58 „ 122	72 59·3	·04304	·37261	35 49
14 „ 166	60 44·5	·00320	·91735	77 49	59 „ 121	73 26·4	·04407	·36796	34 45
15 „ 165	60 51·2	0·00367	0·88803	76 56	60 „ 120	73 53·9	0·04509	0·36350	33 41
16 „ 164	60 58·2	·00416	·86069	76 03	61 „ 119	74 21·8	·04609	·35921	32 37
17 „ 163	61 05·8	·00469	·83509	75 10	62 „ 118	74 50·1	·04707	·35509	31 33
18 „ 162	61 13·7	·00525	·81105	74 17	63 „ 117	75 18·8	·04804	·35115	30 28
19 „ 161	61 22·2	·00583	·78839	73 24	64 „ 116	75 47·8	·04899	·34737	29 23
20 „ 160	61 31·1	0·00645	0·76698	72 30	65 „ 115	76 17·3	0·04991	0·34375	28 18
21 „ 159	61 40·5	·00709	·74670	71 37	66 „ 114	76 47·1	·05081	·34030	27 12
22 „ 158	61 50·4	·00775	·72745	70 43	67 „ 113	77 17·3	·05169	·33700	26 07
23 „ 157	62 00·7	·00845	·70915	69 49	68 „ 112	77 47·8	·05254	·33386	25 01
24 „ 156	62 11·5	·00917	·69172	68 55	69 „ 111	78 18·6	·05337	·33088	23 54
25 „ 155	62 22·7	0·00992	0·67508	68 01	70 „ 110	78 49·8	0·05416	0·32804	22 48
26 „ 154	62 34·5	·01069	·65919	67 06	71 „ 109	79 21·3	·05493	·32536	21 41
27 „ 153	62 46·7	·01149	·64398	66 11	72 „ 108	79 53·1	·05566	·32282	20 34
28 „ 152	62 59·3	·01231	·62942	65 17	73 „ 107	80 25·1	·05637	·32043	19 27
29 „ 151	63 12·5	·01315	·61546	64 21	74 „ 106	80 57·5	·05704	·31819	18 19
30 „ 150	63 26·1	0·01402	0·60206	63 26	75 „ 105	81 30·1	0·05767	0·31609	17 12
31 „ 149	63 40·2	·01490	·58919	62 31	76 „ 104	82 02·9	·05827	·31413	16 04
32 „ 148	63 54·8	·01581	·57682	61 35	77 „ 103	82 36·0	·05883	·31231	14 56
33 „ 147	64 09·8	·01673	·56492	60 39	78 „ 102	83 09·3	·05936	·31063	13 47
34 „ 146	64 25·3	·01768	·55347	59 43	79 „ 101	83 42·8	·05985	·30908	12 39
35 „ 145	64 41·3	0·01864	0·54244	58 46	80 „ 100	84 16·5	0·06030	0·30768	11 30
36 „ 144	64 57·8	·01962	·53181	57 49	81 „ 99	84 50·4	·06070	·30641	10 22
37 „ 143	65 14·8	·02061	·52157	56 52	82 „ 98	85 24·4	·06107	·30528	9 13
38 „ 142	65 32·2	·02162	·51169	55 55	83 „ 97	85 58·5	·06140	·30428	8 04
39 „ 141	65 50·1	·02264	·50216	54 58	84 „ 96	86 32·8	·06168	·30342	6 55
40 „ 140	66 08·5	0·02367	0·49296	54 00	85 „ 95	87 07·2	0·06192	0·30269	5 46
41 „ 139	66 27·3	·02472	·48409	53 02	86 „ 94	87 41·6	·06212	·30209	4 37
42 „ 138	66 46·7	·02578	·47552	52 03	87 „ 93	88 16·2	·06227	·30163	3 28
43 „ 137	67 06·5	·02684	·46725	51 05	88 „ 92	88 50·7	·06238	·30129	2 19
44 „ 136	67 26·8	·02792	·45926	50 06	89 „ 91	89 25·4	·06245	·30110	1 09
45 „ 135	67 47·5	0·02900	0·45154	49 06	90	90 00·0	0·06247	0·30103	0 00
h	180°-K	A	N	a₁	h	180°-K	A	N	a₁

Directions with regard to (K ± d) :—

If l and d have **Same Name,** use (K ⌣ d), i.e. (K — d) or (d — K).

If l and d have **Different Names,** use (K + d).

To name Intercept :—

If H$_o$ is greater than H$_c$, intercept is **" Towards."**

If H$_o$ is less than H$_c$, intercept is **" Away."**

Log. cosec. H$_c$ = A + log. sec. (K ± d).

N.B.   Whenever h is greater than 90°, subtract the quantity in the K column from 180°.

ALT.—AZ. TABLE — LAT. 61°

h	K	A	N	a₁	h	K	A	N	a₁
0 or 180	61 00·0	0·00000	—	90 00	**45 or 135**	68 35·8	0·02715	0·46494	48 50
1 „ 179	61 00·2	·00002	2·07257	89 08	**46 „ 134**	68 56·4	·02816	·45749	47 50
2 „ 178	61 00·9	·00006	1·77161	88 15	**47 „ 133**	69 17·5	·02918	·45030	46 50
3 „ 177	61 02·0	·00014	1·59563	87 23	**48 „ 132**	69 39·0	·03019	·44336	45 50
4 „ 176	61 03·6	·00025	1·47084	86 30	**49 „ 131**	70 00·9	·03121	·43665	44 49
5 „ 175	61 05·6	0·00039	1·37413	85 37	**50 „ 130**	70 23·3	0·03223	0·43017	43 49
6 „ 174	61 08·0	·00056	1·29519	84 45	**51 „ 129**	70 46·2	·03325	·42393	42 48
7 „ 173	61 10·9	·00076	1·22853	83 52	**52 „ 128**	71 09·4	·03426	·41790	41 46
8 „ 172	61 14·2	·00099	1·17087	83 00	**53 „ 127**	71 33·1	·03527	·41208	40 45
9 „ 171	61 18·0	·00125	1·12010	82 07	**54 „ 126**	71 57·2	·03627	·40647	39 43
10 „ 170	61 22·2	0·00154	1·07476	81 14	**55 „ 125**	72 21·8	0·03727	0·40106	38 41
11 „ 169	61 26·9	·00187	1·03383	80 21	**56 „ 124**	72 46·7	·03826	·39585	37 38
12 „ 168	61 32·0	·00222	0·99655	79 28	**57 „ 123**	73 12·1	·03924	·39084	36 36
13 „ 167	61 37·6	·00260	·96234	78 35	**58 „ 122**	73 37·8	·04021	·38601	35 33
14 „ 166	61 43·6	·00301	·93075	77 42	**59 „ 121**	74 04·0	·04117	·38136	34 29
15 „ 165	61 50·1	0·00345	0·90143	76 49	**60 „ 120**	74 30·5	0·04211	0·37690	33 26
16 „ 164	61 57·0	·00391	·87409	75 55	**61 „ 119**	74 57·5	·04304	·37261	32 22
17 „ 163	62 04·3	·00441	·84849	75 02	**62 „ 118**	75 24·8	·04395	·36849	31 18
18 „ 162	62 12·2	·00493	·82445	74 08	**63 „ 117**	75 52·5	·04485	·36455	30 13
19 „ 161	62 20·4	·00548	·80179	73 14	**64 „ 116**	76 20·5	·04573	·36077	29 09
20 „ 160	62 29·2	0·00605	0·78038	72 21	**65 „ 115**	76 48·9	0·04658	0·35715	28 04
21 „ 159	62 38·3	·00666	·76010	71 26	**66 „ 114**	77 17·7	·04741	·35370	26 59
22 „ 158	62 48·0	·00728	·74085	70 32	**67 „ 113**	77 46·8	·04823	·35040	25 43
23 „ 157	62 58·0	·00794	·72255	69 38	**68 „ 112**	78 16·2	·04901	·34726	24 48
24 „ 156	63 08·6	·00861	·70512	68 43	**69 „ 111**	78 45·9	·04978	·34428	23 42
25 „ 155	63 19·6	0·00931	0·68848	67 49	**70 „ 110**	79 15·9	0·05052	0·34144	22 36
26 „ 154	63 31·0	·01004	·67259	66 54	**71 „ 109**	79 46·2	·05122	·33876	21 29
27 „ 153	63 42·9	·01078	·65738	65 59	**72 „ 108**	80 16·8	·05190	·33622	20 23
28 „ 152	63 55·3	·01155	·64282	65 04	**73 „ 107**	80 47·7	·05255	·33383	19 16
29 „ 151	64 08·1	·01234	·62886	64 08	**74 „ 106**	81 18·8	·05317	·33159	18 09
30 „ 150	64 21·4	0·01315	0·61546	63 12	**75 „ 105**	81 50·1	0·05375	0·32948	17 02
31 „ 149	64 35·2	·01398	·60259	62 17	**76 „ 104**	82 21·7	·05431	·32752	15 55
32 „ 148	64 49·4	·01483	·59022	61 21	**77 „ 103**	82 53·5	·05483	·32570	14 47
33 „ 147	65 04·0	·01569	·57832	60 24	**78 „ 102**	83 25·5	·05532	·32402	13 40
34 „ 146	65 19·2	·01658	·56687	59 28	**79 „ 101**	83 57·7	·05576	·32248	12 32
35 „ 145	65 34·7	0·01748	0·55584	58 31	**80 „ 100**	84 30·1	0·05618	0·32108	11 24
36 „ 144	65 50·8	·01839	·54521	57 34	**81 „ 99**	85 02·6	·05655	·31981	10 16
37 „ 143	66 07·3	·01932	·53497	56 37	**82 „ 98**	85 35·3	·05689	·31868	9 08
38 „ 142	66 24·3	·02026	·52509	55 39	**83 „ 97**	86 08·1	·05719	·31768	7 59
39 „ 141	66 41·7	·02122	·51556	54 42	**84 „ 96**	86 41·0	·05745	·31681	6 51
40 „ 140	66 59·6	0·02218	0·50636	53 44	**85 „ 95**	87 14·0	0·05767	0·31608	5 43
41 „ 139	67 17·9	·02316	·49749	52 45	**86 „ 94**	87 47·1	·05786	·31549	4 34
42 „ 138	67 36·7	·02415	·48892	51 47	**87 „ 93**	88 20·3	·05800	·31502	3 26
43 „ 137	67 56·0	·02514	·48065	50 48	**88 „ 92**	88 53·5	·05810	·31469	2 17
44 „ 136	68 15·7	·02614	·47266	49 49	**89 „ 91**	89 26·7	·05816	·31449	1 09
45 „ 135	68 35·8	0·02715	0·46494	48 50	**90**	90 00·0	0·05818	0·31443	0 00
h	**180°-K**	**A**	**N**	**a₁**	**h**	**180°-K**	**A**	**N**	**a₁**

Directions with regard to $(a_2 \pm a_1)$:—

Whenever h is greater than 90°, use $(a_2 - a_1)$.

If h is less than 90°, use :—

$(a_1 + a_2)$ when $(K \pm d)$ has been used,

$(a_1 - a_2)$ when $(d - K)$ has been used.

Always **name azimuth** from elevated pole.

Note that { h = L.H.A. of body **West** of the meridian.
{ h = (360° — L.H.A.) of body **East** of the meridian.

Log. tan. a_2 = N + log. tan. $(K \pm d)$.

LAT. 62° ALT.—AZ. TABLE

h	K	A	N	a₁	h	K	A	N	a₁
°	° ′			° ′	°	° ′			° ′
0 or 180	62 00.0	0.00000	—	90 00	45 or 135	69 23.7	0.02535	0.47891	48 33
1 „ 179	62 00.2	.00001	2.08654	89 07	46 „ 134	69 43.7	.02629	.47146	47 34
2 „ 178	62 00.9	.00006	1.78557	88 14	47 „ 133	70 04.1	.02724	.46126	46 34
3 „ 177	62 02.0	.00013	1.60959	87 21	48 „ 132	70 24.9	.02818	.45732	45 34
4 „ 176	62 03.5	.00023	1.48471	86 28	49 „ 131	70 46.2	.02913	.45061	44 33
5 „ 175	62 05.4	0.00036	1.38809	85 35	50 „ 130	71 07.8	0.03008	0.44414	43 32
6 „ 174	62 07.8	.00052	1.30916	84 42	51 „ 129	71 29.9	.03102	.43789	42 32
7 „ 173	62 10.6	.00071	1.24250	83 49	52 „ 128	71 52.4	.03196	.43186	41 30
8 „ 172	62 13.9	.00093	1.18484	82 56	53 „ 127	72 15.3	.03290	.42604	40 29
9 „ 171	62 17.6	.00117	1.13406	82 02	54 „ 126	72 38.7	.03383	.42043	39 27
10 „ 170	62 21.7	0.00145	1.08872	81 09	55 „ 125	73 02.4	0.03475	0.41503	38 25
11 „ 169	62 26.3	.00175	1.04779	80 16	56 „ 124	73 26.5	.03567	.40982	37 23
12 „ 168	62 31.3	.00208	1.01051	79 22	57 „ 123	73 51.0	.03658	.40480	36 20
13 „ 167	62 36.7	.00244	0.97630	78 29	58 „ 122	74 15.8	.03748	.39997	35 17
14 „ 166	62 42.6	.00282	.94472	77 35	59 „ 121	74 41.1	.03836	.39532	34 14
15 „ 165	62 48.9	0.00323	0.91539	76 41	60 „ 120	75 06.7	0.03924	0.39086	33 11
16 „ 164	62 55.7	.00367	.88805	75 48	61 „ 119	75 32.7	.04009	.38657	32 07
17 „ 163	63 02.9	.00413	.86246	74 54	62 „ 118	75 59.0	.04094	.38246	31 03
18 „ 162	63 10.5	.00462	.83841	74 00	63 „ 117	76 25.7	.04177	.37851	29 59
19 „ 161	63 18.6	.00513	.81575	73 05	64 „ 116	76 52.8	.04258	.37473	28 55
20 „ 160	63 27.1	0.00567	0.79434	72 11	65 „ 115	77 20.1	0.04337	0.37111	27 50
21 „ 159	63 36.0	.00624	.77406	71 17	66 „ 114	77 47.8	.04414	.36766	26 46
22 „ 158	63 45.4	.00682	.75482	70 22	67 „ 113	78 15.8	.04489	.36436	25 41
23 „ 157	63 55.3	.00743	.73651	69 27	68 „ 112	78 44.1	.04562	.36122	24 35
24 „ 156	64 05.5	.00807	.71908	68 32	69 „ 111	79 12.7	.04632	.35824	23 30
25 „ 155	64 16.3	0.00872	0.70244	67 37	70 „ 110	79 41.6	0.04700	0.35540	22 24
26 „ 154	64 27.4	.00940	.68655	66 42	71 „ 109	80 10.7	.04765	.35272	21 18
27 „ 153	64 39.0	.01010	.67134	65 47	72 „ 108	80 40.2	.04828	.35018	20 12
28 „ 152	64 51.1	.01081	.65678	64 51	73 „ 107	81 09.8	.04888	.34779	19 06
29 „ 151	65 03.6	.01155	.64282	63 55	74 „ 106	81 39.7	.04945	.34555	18 00
30 „ 150	65 16.5	0.01231	0.62942	62 59	75 „ 105	82 09.9	0.04999	0.34345	16 53
31 „ 149	65 29.9	.01308	.61655	62 03	76 „ 104	82 40.2	.05050	.34149	15 46
32 „ 148	65 43.7	.01387	.60418	61 07	77 „ 103	83 10.8	.05098	.33967	14 40
33 „ 147	65 58.0	.01468	.59228	60 10	78 „ 102	83 41.5	.05143	.33799	13 32
34 „ 146	66 12.7	.01551	.58083	59 13	79 „ 101	84 12.4	.05184	.33644	12 25
35 „ 145	66 27.9	0.01635	0.56980	58 16	80 „ 100	84 43.5	0.05222	0.33504	11 18
36 „ 144	66 43.5	.01720	.55917	57 19	81 „ 99	85 14.7	.05257	.33377	10 10
37 „ 143	66 59.5	.01807	.54893	56 22	82 „ 98	85 46.1	.05283	.33264	9 03
38 „ 142	67 16.0	.01894	.53905	55 24	83 „ 97	86 17.5	.05316	.33164	7 55
39 „ 141	67 32.9	.01983	.52952	54 26	84 „ 96	86 49.1	.05340	.33078	6 47
40 „ 140	67 50.3	0.02073	0.52032	53 28	85 „ 95	87 20.8	0.05360	0.33005	5 40
41 „ 139	68 08.1	.02164	.51145	52 30	86 „ 94	87 52.6	.05377	.32945	4 32
42 „ 138	68 26.4	.02256	.50288	51 31	87 „ 93	88 24.4	.05390	.32899	3 24
43 „ 137	68 45.0	.02349	.49461	50 32	88 „ 92	88 56.2	.05399	.32866	2 16
44 „ 136	69 04.1	.02442	.48662	49 33	89 „ 91	89 28.1	.05405	.32846	1 08
45 „ 135	69 23.7	0.02535	0.47891	48 33	90	90 00.0	0.05407	0.32839	0 00
h	180°-K	A	N	a₁	h	180°-K	A	N	a₁

Directions with regard to (K ± d) :—

If l and d have **Same Name**, use (K ⁓ d), i.e. (K — d) or (d — K).

If l and d have **Different Names**, use (K + d).

To name Intercept :—

If H_o is greater than H_c, intercept is " **Towards.** "

If H_o is less than H_c, intercept is " **Away.** "

Log. cosec. $H_o = A + $ log. sec. (K ± d).

N.B. Whenever h is greater than 90°, subtract the quantity in the K column from 180°.

ALT.—AZ. TABLE LAT. 63°

h	K		A	N	a₁		h	K		A	N	a₁	
°	°	′			°	′	°	°	′			°	′
0 or 180	63	00·0	0·00000	—	90	00	45 or 135	70	11·2	0·02362	0·49347	48	18
1 „ 179	63	00·2	·00001	2·10110	89	07	46 „ 134	70	30·5	·02449	·48602	47	18
2 „ 178	63	00·8	·00005	1·80013	88	13	47 „ 133	70	50·3	·02536	·47883	46	18
3 „ 177	63	01·9	·00012	1·62415	87	20	48 „ 132	71	10·4	·02624	·47188	45	18
4 „ 176	63	03·4	·00022	1·49937	86	26	49 „ 131	71	31·0	·02712	·46517	44	18
5 „ 175	63	05·3	0·00034	1·40266	85	33	50 „ 130	71	51·9	·02799	0·45870	43	17
6 „ 174	63	07·6	·00049	1·32372	84	39	51 „ 129	72	13·3	·02887	·45245	42	16
7 „ 173	63	10·4	·00067	1·25706	83	45	52 „ 128	72	35·0	·02974	·44642	41	15
8 „ 172	63	13·6	·00087	1·19940	82	52	53 „ 127	72	57·1	·03060	·44060	40	13
9 „ 171	63	17·2	·00110	1·14862	81	58	54 „ 126	73	19·6	·03147	·43499	39	12
10 „ 170	63	21·2	0·00135	1·10328	81	04	55 „ 125	73	42·5	0·03232	0·42959	38	10
11 „ 169	63	25·6	·00164	1·06235	80	10	56 „ 124	74	05·8	·03317	·42438	37	08
12 „ 168	63	30·5	·00194	1·02507	79	17	57 „ 123	74	29·4	·03401	·41936	36	05
13 „ 167	63	35·8	·00228	0·99086	78	23	58 „ 122	74	53·4	·03484	·41453	35	03
14 „ 166	63	41·6	·00264	·95928	77	28	59 „ 121	75	17·7	·03566	·40989	34	00
15 „ 165	63	47·7	0·00302	0·92996	76	34	60 „ 120	75	42·4	0·03646	0·40542	32	57
16 „ 164	63	54·3	·00343	·90261	75	40	61 „ 119	76	07·5	·03726	·40113	31	53
17 „ 163	64	01·3	·00386	·87702	74	46	62 „ 118	76	32·8	·03804	·39702	30	50
18 „ 162	64	08·7	·00432	·85297	73	51	63 „ 117	76	58·5	·03880	·39307	29	46
19 „ 161	64	16·6	·00480	·83031	72	57	64 „ 116	77	24·5	·03955	·38929	28	42
20 „ 160	64	24·9	0·00530	0·80890	72	02	65 „ 115	77	50·9	0·04028	0·38568	27	38
21 „ 159	64	33·6	·00583	·78862	71	07	66 „ 114	78	17·5	·04099	·38222	26	33
22 „ 158	64	42·8	·00637	·76938	70	12	67 „ 113	78	44·4	·04168	·37893	25	28
23 „ 157	64	52·3	·00694	·75107	69	17	68 „ 112	79	11·6	·04235	·37579	24	24
24 „ 156	65	02·4	·00753	·73364	68	22	69 „ 111	79	39·1	·04300	·37280	23	18
25 „ 155	65	12·8	0·00814	0·71700	67	26	70 „ 110	80	06·9	0·04362	0·36997	22	13
26 „ 154	65	23·7	·00878	·70111	66	31	71 „ 109	80	34·9	·04422	·36728	21	08
27 „ 153	65	34·9	·00943	·68591	65	35	72 „ 108	81	03·1	·04480	·36475	20	02
28 „ 152	65	46·7	·01010	·67134	64	39	73 „ 107	81	31·6	·04535	·36236	18	56
29 „ 151	65	58·8	·01078	·65738	63	43	74 „ 106	82	00·3	·04588	·36011	17	50
30 „ 150	66	11·4	0·01149	0·64398	62	47	75 „ 105	82	29·2	0·04638	0·35801	16	44
31 „ 149	66	24·4	·01221	·63111	61	50	76 „ 104	82	58·4	·04684	·35605	15	38
32 „ 148	66	37·8	·01295	·61874	60	54	77 „ 103	83	27·7	·04728	·35423	14	32
33 „ 147	66	51·7	·01370	·60684	59	57	78 „ 102	83	57·2	·04770	·35255	13	25
34 „ 146	67	06·0	·01447	·59539	59	00	79 „ 101	84	26·8	·04808	·35101	12	18
35 „ 145	67	20·7	0·01525	0·58436	58	02	80 „ 100	84	56·6	0·04843	0·34960	11	12
36 „ 144	67	35·9	·01604	·57373	57	05	81 „ 99	85	26·6	·04874	·34833	10	05
37 „ 143	67	51·4	·01685	·56349	56	07	82 „ 98	85	56·6	·04903	·34720	8	58
38 „ 142	68	07·4	·01766	·55361	55	09	83 „ 97	86	26·8	·04928	·34620	7	51
39 „ 141	68	23·9	·01849	·54408	54	11	84 „ 96	86	57·1	·04950	·34534	6	44
40 „ 140	68	40·7	0·01933	0·53489	53	13	85 „ 95	87	27·4	0·04969	0·34461	5	36
41 „ 139	68	58·0	·02017	·52601	52	14	86 „ 94	87	57·9	·04984	·34401	4	29
42 „ 138	69	15·6	·02102	·51744	51	16	87 „ 93	88	28·3	·04996	·34355	3	22
43 „ 137	69	33·7	·02188	·50917	50	17	88 „ 92	88	58·9	·05005	·34322	2	15
44 „ 136	69	52·3	·02275	·50118	49	17	89 „ 91	89	29·4	·05010	·34302	1	07
45 „ 135	70	11·2	0·02362	0·49347	48	18	90	90	00·0	0·05012	0·34295	0	00
h	180°-K		A	N	a₁		h	180°-K		A	N	a₁	

Directions with regard to $(a_2 \overset{+}{\sim} a_1)$:—

Whenever h is greater than 90°, use $(a_2 - a_1)$.

If h is less than 90°, use :—

$(a_1 + a_2)$ when $(K \pm d)$ has been used,

$(a_1 - a_2)$ when $(d - K)$ has been used.

Always **name azimuth** from elevated pole.

Note that $\begin{cases} h = \text{L.H.A. of body } \textbf{West} \text{ of the meridian.} \\ h = (360° - \text{L.H.A.}) \text{ of body } \textbf{East} \text{ of the meridian} \end{cases}$

Log. tan. $a_2 = N + \log.$ tan. $(K \overset{+}{\sim} d)$.

LAT. 64° ALT.—AZ. TABLE

h	K	A	N	a₁	h	K	A	N	a₁
°	° ′			° ′	°	° ′			° ′
0 or 180	64 00·0	0·00000	—	90 00	**45 or 135**	70 58·3	0·02194	0·50867	48 03
1 „ 179	64 00·2	·00001	2·11630	89 06	**46 „ 134**	71 17·0	·02274	·50122	47 03
2 „ 178	64 00·8	·00005	1·81534	88 12	**47 „ 133**	71 36·1	·02355	·49403	46 03
3 „ 177	64 01·9	·00011	1·63936	87 18	**48 „ 132**	71 55·5	·02436	·48708	45 03
4 „ 176	64 03·3	·00020	1·51457	86 24	**49 „ 131**	72 15·4	·02517	·48038	44 03
5 „ 175	64 05·2	0·00032	1·41786	85 30	**50 „ 130**	72 35·6	0·02598	0·47390	43 02
6 „ 174	64 07·4	·00046	1·33892	84 36	**51 „ 129**	72 56·2	·02679	·46765	42 01
7 „ 173	64 10·1	·00062	1·27226	83 42	**52 „ 128**	73 17·2	·02759	·46163	41 00
8 „ 172	64 13·2	·00081	1·21460	82 48	**53 „ 127**	73 38·5	·02839	·45581	39 59
9 „ 171	64 16·7	·00102	1·16383	81 54	**54 „ 126**	74 00·2	·02919	·45020	38 57
10 „ 170	64 20·6	0·00126	1·11849	81 00	**55 „ 125**	74 22·3	0·02998	0·44479	37 55
11 „ 169	64 25·0	·00152	1·07756	80 05	**56 „ 124**	74 44·7	·03076	·43958	36 53
12 „ 168	64 29·7	·00181	1·04028	79 11	**57 „ 123**	75 07·4	·03153	·43457	35 51
13 „ 167	64 34·9	·00212	1·00607	78 17	**58 „ 122**	75 30·5	·03230	·42974	34 48
14 „ 166	64 40·5	·00246	0·97448	77 22	**59 „ 121**	75 53·9	·03305	·42509	33 46
15 „ 165	64 46·5	0·00281	0·94516	76 28	**60 „ 120**	76 17·7	0·03380	0·42063	32 43
16 „ 164	64 52·9	·00319	·91782	75 33	**61 „ 119**	76 41·8	·03453	·41634	31 40
17 „ 163	64 59·7	·00360	·89222	74 38	**62 „ 118**	77 06·2	·03524	·41222	30 36
18 „ 162	65 06·9	·00402	·86818	73 43	**63 „ 117**	77 30·9	·03595	·40828	29 33
19 „ 161	65 14·6	·00447	·84552	72 48	**64 „ 116**	77 55·9	·03663	·40450	28 29
20 „ 160	65 22·6	0·00494	0·82411	71 53	**65 „ 115**	78 21·2	0·03730	0·40088	27 25
21 „ 159	65 31·1	·00543	·80383	70 58	**66 „ 114**	78 46·8	·03796	·39743	26 21
22 „ 158	65 40·0	·00594	·78458	70 03	**67 „ 113**	79 12·6	·03859	·39413	25 17
23 „ 157	65 49·3	·00647	·76628	69 07	**68 „ 112**	79 38·7	·03921	·39091	24 12
24 „ 156	65 59·0	·00702	·74884	68 11	**69 „ 111**	80 05·1	·03981	·38801	23 08
25 „ 155	66 09·2	0·00758	0·73221	67 16	**70 „ 110**	80 31·8	0·04038	0·38517	22 03
26 „ 154	66 19·7	·00817	·71632	66 20	**71 „ 109**	80 58·6	·04093	·38249	20 58
27 „ 153	66 30·7	·00878	·70111	65 24	**72 „ 108**	81 25·7	·04146	·37995	19 53
28 „ 152	66 42·1	·00940	·68655	64 27	**73 „ 107**	81 53·1	·04197	·37756	18 47
29 „ 151	66 53·9	·01004	·67259	63 31	**74 „ 106**	82 20·6	·04245	·37532	17 42
30 „ 150	67 06·1	0·01069	0·65919	62 34	**75 „ 105**	82 48·3	0·04291	0·37321	16 36
31 „ 149	67 18·7	·01136	·64632	61 38	**76 „ 104**	83 16·2	·04334	·37125	15 30
32 „ 148	67 31·7	·01205	·63395	60 41	**77 „ 103**	83 44·3	·04374	·36943	14 24
33 „ 147	67 45·2	·01274	·62205	59 44	**78 „ 102**	84 12·6	·04412	·36775	13 18
34 „ 146	67 59·0	·01346	·61060	58 46	**79 „ 101**	84 41·0	·04447	·36621	12 12
35 „ 145	68 13·3	0·01418	0·59957	57 49	**80 „ 100**	85 09·5	0·04479	0·36481	11 06
36 „ 144	68 28·0	·01492	·58894	56 51	**81 „ 99**	85 38·2	·04508	·36354	10 00
37 „ 143	68 43·1	·01567	·57869	55 53	**82 „ 98**	86 07·0	·04534	·36240	8 53
38 „ 142	68 58·6	·01642	·56882	54 55	**83 „ 97**	86 35·9	·04557	·36141	7 47
39 „ 141	69 14·5	·01719	·55929	53 57	**84 „ 96**	87 04·9	·04578	·36054	6 40
40 „ 140	69 30·8	0·01796	0·55009	52 59	**85 „ 95**	87 34·0	0·04595	0·35981	5 34
41 „ 139	69 47·5	·01875	·54121	52 00	**86 „ 94**	88 03·1	·04609	·35922	4 27
42 „ 138	70 04·6	·01954	·53265	51 01	**87 „ 93**	88 32·3	·04620	·35875	3 20
43 „ 137	70 22·1	·02033	·52437	50 02	**88 „ 92**	89 01·5	·04628	·35842	2 13
44 „ 136	70 40·0	·02113	·51639	49 03	**89 „ 91**	89 30·7	·04632	·35822	1 07
45 „ 135	70 58·3	0·02194	0·50867	48 03	**90**	90 00·0	0·04634	0·35816	0 00
h	180°-K	A	N	a₁	**h**	180°-K	A	N	a₁

Directions with regard to (K $\pm$ d) :—

If l and d have **Same Name,** use (K ⌣ d), i.e. (K — d) or (d — K).
If l and d have **Different Names,** use (K + d).

To name Intercept :—

If H_o is greater than H_c, intercept is " **Towards.** "
If H_o is less than H_c, intercept is " **Away.** "

Log. cosec. H_o = A + log. sec. (K $\pm$ d).

N.B. Whenever h is greater than 90°, subtract the quantity in the K column from 180°.

ALT.—AZ. TABLE # LAT. 65°

h	K	A	N	a₁	h	K	A	N	a₁
°	° ′	° ′		° ′	°	° ′	° ′		° ′
0 or 180	65 00·0	0·00000	—	90 00	45 or 135	71 45·1	0·02031	0·52457	47 49
1 „ 179	65 00·2	·00001	2·13220	89 06	46 „ 134	72 03·1	·02106	·51712	46 49
2 „ 178	65 00·8	·00005	1·83123	88 11	47 „ 133	72 21·5	·02180	·50902	45 49
3 „ 177	65 01·8	·00011	1·65525	87 17	48 „ 132	72 40·3	·02255	·50298	44 49
4 „ 176	65 03·2	·00019	1·53047	86 22	49 „ 131	72 59·4	·02330	·49627	43 48
5 „ 175	65 05·0	0·00029	1·43376	85 28	50 „ 130	73 18·9	0·02404	0·48980	42 48
6 „ 174	65 07·2	·00042	1·35482	84 34	51 „ 129	73 38·7	·02479	·48355	41 47
7 „ 173	65 09·8	·00058	1·28816	83 39	52 „ 128	73 58·9	·02553	·47752	40 46
8 „ 172	65 12·8	·00075	1·23050	82 44	53 „ 127	74 19·5	·02626	·47170	39 45
9 „ 171	65 16·2	·00095	1·17972	81 50	54 „ 126	74 40·3	·02699	·46609	38 43
10 „ 170	65 20·1	0·00117	1·13438	80 55	55 „ 125	75 01·6	0·02772	0·46069	37 41
11 „ 169	65 24·3	·00142	1·09345	80 01	56 „ 124	75 23·1	·02844	·45548	36 39
12 „ 168	65 28·9	·00168	1·05617	79 06	57 „ 123	75 45·0	·02915	·45046	35 37
13 „ 167	65 33·9	·00197	1·02196	78 11	58 „ 122	76 07·2	·02985	·44563	34 35
14 „ 166	65 39·3	·00228	0·99038	77 16	59 „ 121	76 29·7	·03055	·44099	33 33
15 „ 165	65 45·1	0·00261	0·96106	76 21	60 „ 120	76 52·5	0·03123	0·43652	32 30
16 „ 164	65 51·4	·00297	·93371	75 26	61 „ 119	77 15·7	·03190	·43223	31 27
17 „ 163	65 58·0	·00334	·90812	74 31	62 „ 118	77 39·1	·03256	·42812	30 24
18 „ 162	66 05·0	·00374	·88407	73 35	63 „ 117	78 02·8	·03320	·42417	29 21
19 „ 161	66 12·4	·00415	·86141	72 40	64 „ 116	78 26·8	·03384	·42039	28 17
20 „ 160	66 20·3	0·00458	0·84000	71 45	65 „ 115	78 51·1	0·03445	0·41678	27 14
21 „ 159	66 28·5	·00504	·81972	70 49	66 „ 114	79 15·6	·03505	·41332	26 10
22 „ 158	66 37·1	·00551	·80048	69 53	67 „ 113	79 40·4	·03563	·41003	25 06
23 „ 157	66 46·1	·00600	·78217	68 57	68 „ 112	80 05·5	·03620	·40689	24 02
24 „ 156	66 55·6	·00651	·76474	68 02	69 „ 111	80 30·8	·03674	·40390	22 57
25 „ 155	67 05·4	0·00704	0·74810	67 05	70 „ 110	80 56·3	0·03727	0·40107	21 53
26 „ 154	67 15·6	·00758	·73221	66 09	71 „ 109	81 22·1	·03778	·39838	20 48
27 „ 153	67 26·3	·00814	·71700	65 13	72 „ 108	81 48·0	·03826	·39585	19 43
28 „ 152	67 37·3	·00872	·70244	64 16	73 „ 107	82 14·2	·03873	·39346	18 38
29 „ 151	67 48·7	·00931	·68848	63 20	74 „ 106	82 40·5	·03917	·39121	17 33
30 „ 150	68 00·6	0·00992	0·67508	62 23	75 „ 105	83 07·1	0·03958	0·38911	16 28
31 „ 149	68 12·8	·01054	·66221	61 26	76 „ 104	83 33·8	·03998	·38715	15 23
32 „ 148	68 25·4	·01117	·64984	60 29	77 „ 103	84 00·7	·04035	·38533	14 17
33 „ 147	68 38·4	·01182	·63794	59 31	78 „ 102	84 27·7	·04069	·38365	13 12
34 „ 146	68 51·9	·01248	·62649	58 34	79 „ 101	84 54·9	·04101	·38210	12 06
35 „ 145	69 05·7	0·01315	0·61546	57 36	80 „ 100	85 22·2	0·04131	0·38070	11 01
36 „ 144	69 19·9	·01383	·60483	56 38	81 „ 99	85 49·7	·04157	·37943	9 55
37 „ 143	69 34·4	·01452	·59459	55 40	82 „ 98	86 17·2	·04181	·37830	8 49
38 „ 142	69 49·4	·01522	·58471	54 42	83 „ 97	86 44·8	·04202	·37730	7 43
39 „ 141	70 04·8	·01593	·57518	53 43	84 „ 96	87 12·6	·04221	·37644	6 37
40 „ 140	70 20·6	0·01665	0·56598	52 45	85 „ 95	87 40·4	0·04237	0·37571	5 31
41 „ 139	70 36·7	·01737	·55711	51 46	86 „ 94	88 08·2	·04249	·37511	4 25
42 „ 138	70 53·2	·01810	·54854	50 47	87 „ 93	88 36·1	·04260	·37465	3 19
43 „ 137	71 10·1	·01883	·54027	49 48	88 „ 92	89 04·1	·04267	·37432	2 12
44 „ 136	71 27·4	·01957	·53228	48 48	89 „ 91	89 32·0	·04271	·37412	1 06
45 „ 135	71 45·1	0·02031	0·52457	47 49	90	90 00·0	0·04272	0·37405	0 00
h	180°-K	A	N	a₁	h	180°-K	A	N	a₁

Directions with regard to (a₂ $\overset{+}{\sim}$ a₁) :—

Whenever h is greater than 90°, use (a₂ — a₁).

If h is less than 90°, use :—

(a₁ + a₂) when (K ± d) has been used,

(a₁ — a₂) when (d — K) has been used.

Always **name azimuth** from elevated pole.

Note that { h = L.H.A. of body **West** of the meridian.
{ h = (360° — L.H.A.) of body **East** of the meridian.

Log. tan. a₂ = N + log. tan. (K $\overset{+}{\sim}$ d).

LAT. 66°   ALT.—AZ. TABLE

h	K	A	N	a₁		h	K	A	N	a₁
°	° ′			° ′		°	° ′			° ′
0 or 180	66 00·0	0·00000	—	90 00		**45 or 135**	72 31·5	0·01875	0·54120	47 35
1 „ 179	66 00·2	·00001	2·14883	89 05		**46 „ 134**	72 48·8	·01943	·53375	46 35
2 „ 178	66 00·8	·00004	1·84787	88 10		**47 „ 133**	73 06·6	·02012	·52656	45 35
3 „ 177	66 01·8	·00010	1·67189	87 16		**48 „ 132**	73 24·6	·02080	·51961	44 35
4 „ 176	66 03·1	·00017	1·54710	86 21		**49 „ 131**	73 43·0	·02149	·51291	43 35
5 „ 175	66 04·9	0·00027	1·45039	85 26		**50 „ 130**	74 01·8	0·02218	0·50643	42 34
6 „ 174	66 07·0	·00039	1·37145	84 31		**51 „ 129**	74 20·9	·02286	·50018	41 33
7 „ 173	66 09·5	·00053	1·30479	83 36		**52 „ 128**	74 40·3	·02354	·49415	40 32
8 „ 172	66 12·5	·00070	1·24713	82 41		**53 „ 127**	75 00·0	·02421	·48834	39 31
9 „ 171	66 15·8	·00088	1·19635	81 46		**54 „ 126**	75 20·1	·02489	·48273	38 30
10 „ 170	66 19·5	0·00109	1·15102	80 51		**55 „ 125**	75 40·5	0·02555	0·47732	37 28
11 „ 169	66 23·5	·00131	1·11009	79 56		**56 „ 124**	76 01·2	·02621	·47211	36 26
12 „ 168	66 28·0	·00156	1·07281	79 01		**57 „ 123**	76 22·2	·02686	·46710	35 24
13 „ 167	66 32·9	·00183	1·03860	78 05		**58 „ 122**	76 43·5	·02751	·46227	34 22
14 „ 166	66 38·1	·00211	1·00701	77 10		**59 „ 121**	77 05·1	·02814	·45762	33 20
15 „ 165	66 43·8	0·00242	0·97769	76 15		**60 „ 120**	77 27·0	0·02877	0·45316	32 18
16 „ 164	66 49·8	·00275	·95035	75 19		**61 „ 119**	77 49·2	·02938	·44887	31 15
17 „ 163	66 56·2	·00309	·92475	74 24		**62 „ 118**	78 11·6	·02998	·44475	30 12
18 „ 162	67 03·0	·00346	·90070	73 28		**63 „ 117**	78 34·4	·03057	·44081	29 09
19 „ 161	67 10·2	·00384	·87804	72 32		**64 „ 116**	78 57·4	·03115	·43703	28 06
20 „ 160	67 17·8	0·00424	0·85663	71 36		**65 „ 115**	79 20·6	0·03171	0·43341	27 02
21 „ 159	67 25·8	·00466	·83636	70 41		**66 „ 114**	79 44·1	·03226	·42996	25 59
22 „ 158	67 34·1	·00510	·81711	69 44		**67 „ 113**	80 07·9	·03280	·42666	24 55
23 „ 157	67 42·9	·00555	·79881	68 48		**68 „ 112**	80 31·9	·03331	·42352	23 51
24 „ 156	67 52·0	·00603	·78137	67 52		**69 „ 111**	80 56·1	·03381	·42053	22 47
25 „ 155	68 01·5	0·00651	0·76474	66 56		**70 „ 110**	81 20·5	0·03429	0·41770	21 43
26 „ 154	68 11·4	·00702	·74884	65 59		**71 „ 109**	81 45·1	·03475	·41502	20 39
27 „ 153	68 21·7	·00753	·73364	65 02		**72 „ 108**	82 10·0	·03520	·41248	19 35
28 „ 152	68 32·4	·00807	·71908	64 06		**73 „ 107**	82 35·0	·03562	·41009	18 30
29 „ 151	68 43·4	·00861	·70512	63 09		**74 „ 106**	83 00·2	·03602	·40784	17 26
30 „ 150	68 54·9	0·00917	0·69172	62 11		**75 „ 105**	83 25·6	0·03641	0·40574	16 21
31 „ 149	69 06·7	·00974	·67885	61 14		**76 „ 104**	83 51·1	·03677	·40378	15 16
32 „ 148	69 18·9	·01033	·66648	60 17		**77 „ 103**	84 16·8	·03710	·40196	14 11
33 „ 147	69 31·5	·01093	·65458	59 19		**78 „ 102**	84 42·7	·03742	·40028	13 06
34 „ 146	69 44·4	·01153	·64312	58 22		**79 „ 101**	85 08·7	·03771	·39874	12 01
35 „ 145	69 57·8	0·01215	0·63210	57 24		**80 „ 100**	85 34·7	0·03798	0·39734	10 55
36 „ 144	70 11·5	·01278	·62147	56 26		**81 „ 99**	86 01·0	·03822	·39607	9 50
37 „ 143	70 25·6	·01342	·61122	55 27		**82 „ 98**	86 27·3	·03844	·39493	8 45
38 „ 142	70 40·0	·01406	·60134	54 29		**83 „ 97**	86 53·7	·03863	·39394	7 39
39 „ 141	70 54·8	·01471	·59181	53 30		**84 „ 96**	87 20·1	·03880	·39307	6 34
40 „ 140	71 10·0	0·01537	0·58262	52 32		**85 „ 95**	87 46·7	0·03894	0·39234	5 28
41 „ 139	71 25·6	·01604	·57374	51 33		**86 „ 94**	88 13·3	·03906	·39175	4 23
42 „ 138	71 41·5	·01671	·56518	50 34		**87 „ 93**	88 39·9	·03915	·39128	3 17
43 „ 137	71 57·8	·01739	·55690	49 34		**88 „ 92**	89 06·6	·03922	·39095	2 11
44 „ 136	72 14·5	·01807	·54892	48 35		**89 „ 91**	89 33·3	·03926	·39075	1 06
45 „ 135	72 31·5	0·01875	0·54120	47 35		**90**	90 00·0	0·03927	0·39069	0 00
h	180°-K	A	N	a₁		h	180°-K	A	N	a₁

Directions with regard to (K $\overset{+}{-}$ d) :—

If l and d have **Same Name,** use (K ⌣ d), i.e. (K — d) or (d — K).

If l and d have **Different Names,** use (K + d).

To name Intercept :—

If H_o is greater than H_c, intercept is " **Towards.** "

If H_o is less than H_c, intercept is " **Away.** "

Log. cosec. H_c = A + log. sec. (K $\overset{+}{-}$ d).

N.B.   Whenever h is greater than 90°, subtract the quantity in the K column from 180°.

ALT.—AZ. TABLE LAT. 67°

h	K		A	N	a₁		h	K		A	N	a₁	
°	°	′			°	′	°	°	′			°	′
0 or 180	67	00·0	0·00000	—	90	00	**45 or 135**	73	17·6	0·01724	0·55864	47	22
1 „ 179	67	00·2	·00001	2·16627	89	05	**46 „ 134**	73	34·3	·01787	·55119	46	22
2 „ 178	67	00·8	·00004	1·86530	88	10	**47 „ 133**	73	51·3	·01850	·54399	45	22
3 „ 177	67	01·7	·00009	1·68932	87	14	**48 „ 132**	74	08·6	·01913	·53705	44	22
4 „ 176	67	03·0	·00016	1·56454	86	19	**49 „ 131**	74	26·3	·01975	·53034	43	22
5 „ 175	67	04·7	0·00025	1·46783	85	24	**50 „ 130**	74	44·3	0·02038	0·52387	42	21
6 „ 174	67	06·8	·00036	1·38889	84	28	**51 „ 129**	75	02·6	·02101	·51762	41	20
7 „ 173	67	09·2	·00049	1·32223	83	33	**52 „ 128**	75	21·3	·02163	·51159	40	19
8 „ 172	67	12·1	·00064	1·26457	82	38	**53 „ 127**	75	40·2	·02225	·50577	39	18
9 „ 171	67	15·3	·00081	1·21379	81	42	**54 „ 126**	75	59·4	·02286	·50016	38	17
10 „ 170	67	18·8	0·00100	1·16845	80	47	**55 „ 125**	76	19·0	0·02347	0·49476	37	16
11 „ 169	67	22·8	·00121	1·12752	79	51	**56 „ 124**	76	38·8	·02407	·48955	36	14
12 „ 168	67	27·1	·00144	1·09024	78	56	**57 „ 123**	76	59·0	·02467	·48453	35	12
13 „ 167	67	31·8	·00168	1·05603	78	00	**58 „ 122**	77	19·4	·02526	·47970	34	10
14 „ 166	67	36·9	·00195	1·02445	77	04	**59 „ 121**	77	40·1	·02584	·47506	33	08
15 „ 165	67	42·3	0·00223	0·99513	76	09	**60 „ 120**	78	01·1	0·02641	0·47059	32	06
16 „ 164	67	48·2	·00253	·96778	75	13	**61 „ 119**	78	22·3	·02697	·46630	31	03
17 „ 163	67	54·4	·00285	·94219	74	17	**62 „ 118**	78	43·8	·02752	·46218	30	01
18 „ 162	68	00·9	·00319	·91814	73	21	**63 „ 117**	79	05·5	·02806	·45824	28	58
19 „ 161	68	07·9	·00354	·89548	72	25	**64 „ 116**	79	27·5	·02858	·45446	27	55
20 „ 160	68	15·2	0·00391	0·87407	71	29	**65 „ 115**	79	49·8	0·02910	0·45085	26	52
21 „ 159	68	22·9	·00430	·85379	70	32	**66 „ 114**	80	12·3	·02960	·44739	25	49
22 „ 158	68	31·0	·00470	·83455	69	36	**67 „ 113**	80	35·0	·03008	·44410	24	45
23 „ 157	68	39·5	·00512	·81624	68	39	**68 „ 112**	80	57·9	·03055	·44096	23	42
24 „ 156	68	48·3	·00555	·79881	67	43	**69 „ 111**	81	21·0	·03101	·43797	22	38
25 „ 155	68	57·5	0·00600	0·78217	66	46	**70 „ 110**	81	44·4	0·03144	0·43514	21	34
26 „ 154	69	07·0	·00647	·76628	65	49	**71 „ 109**	82	07·9	·03187	·43245	20	31
27 „ 153	69	17·0	·00694	·75107	64	52	**72 „ 108**	82	31·6	·03227	·42992	19	27
28 „ 152	69	27·3	·00743	·73651	63	55	**73 „ 107**	82	55·5	·03265	·42753	18	22
29 „ 151	69	37·9	·00794	·72255	62	58	**74 „ 106**	83	19·6	·03302	·42528	17	18
30 „ 150	69	49·0	0·00845	0·70915	62	01	**75 „ 105**	83	43·8	0·03337	0·42318	16	14
31 „ 149	70	00·4	·00898	·69628	61	03	**76 „ 104**	84	08·2	·03370	·42122	15	09
32 „ 148	70	12·1	·00952	·68391	60	06	**77 „ 103**	84	32·7	·03400	·41940	14	05
33 „ 147	70	24·3	·01006	·67201	59	08	**78 „ 102**	84	57·4	·03429	·41772	13	00
34 „ 146	70	36·8	·01062	·66056	58	10	**79 „ 101**	85	22·2	·03455	·41617	11	55
35 „ 145	70	49·6	0·01119	0·64953	57	12	**80 „ 100**	85	47·1	0·03480	0·41477	10	51
36 „ 144	71	02·8	·01177	·63890	56	14	**81 „ 99**	86	12·1	·03502	·41350	9	46
37 „ 143	71	16·4	·01235	·62866	55	15	**82 „ 98**	86	37·1	·03522	·41237	8	41
38 „ 142	71	30·3	·01294	·61878	54	17	**83 „ 97**	87	02·3	·03539	·41137	7	36
39 „ 141	71	44·6	·01354	·60925	53	18	**84 „ 96**	87	27·6	·03555	·41051	6	31
40 „ 140	71	59·2	0·01415	0·60005	52	19	**85 „ 95**	87	52·9	0·03568	0·40978	5	26
41 „ 139	72	14·2	·01476	·59118	51	20	**86 „ 94**	88	18·2	·03578	·40918	4	21
42 „ 138	72	29·5	·01538	·58261	50	21	**87 „ 93**	88	43·6	·03587	·40872	3	16
43 „ 137	72	45·2	·01599	·57434	49	21	**88 „ 92**	89	09·1	·03593	·40839	2	10
44 „ 136	73	01·2	·01662	·56635	48	22	**89 „ 91**	89	34·5	·03596	·40819	1	05
45 „ 135	73	17·6	0·01724	0·55864	47	22	**90**	90	00·0	0·03597	0·40812	0	00
h	180°-K		A	N	a₁		h	180°-K		A	N	a₁	

Directions with regard to ($a_2 \overset{+}{\sim} a_1$) :—

Whenever h is greater than 90°, use ($a_2 - a_1$). Always **name**

If h is less than 90°, use :— **azimuth** from

elevated pole.

($a_1 + a_2$) when (K ± d) has been used,

($a_1 - a_2$) when (d — K) has been used.

Note that $\begin{cases} \text{h = L.H.A. of body \textbf{West} of the meridian.} \\ \text{h = (360° — L.H.A.) of body \textbf{East} of the meridian.} \end{cases}$

Log. tan. a_2 = N + log. tan. (K $\overset{+}{\sim}$ d).

LAT. 68° ALT.—AZ. TABLE

h	K	A	N	a₁	h	K	A	N	a₁
°	° ′			° ′	°	° ′			° ′
0 or 180	68 00·0	0·00000	—	90 00	**45 or 135**	74 03·4	0·01580	0·57694	47 10
1 „ 170	68 00·2	·00001	2·18457	89 04	**46 „ 134**	74 19·4	·01637	·56949	46 10
2 „ 178	68 00·7	·00004	1·88361	88 09	**47 „ 133**	74 35·7	·01694	·50230	45 10
3 „ 177	68 01·6	·00008	1·70762	87 13	**48 „ 132**	74 52·3	·01752	·55535	44 10
4 „ 176	68 02·9	·00015	1·58284	86 17	**49 „ 131**	75 09·3	·01809	·54864	43 09
5 „ 175	68 04·5	0·00023	1·48613	85 22	**50 „ 130**	75 26·5	0·01866	0·54217	42 09
6 „ 174	68 06·5	·00033	1·40719	84 26	**51 „ 129**	75 44·0	·01923	·53592	41 08
7 „ 173	68 08·9	·00045	1·34053	83 30	**52 „ 128**	76 01·9	·01980	·52989	40 07
8 „ 172	68 11·6	·00059	1·28287	82 35	**53 „ 127**	76 20·0	·02036	·52407	39 06
9 „ 171	68 14·7	·00075	1·23209	81 39	**54 „ 126**	76 38·4	·02092	·51847	38 05
10 „ 170	68 18·2	0·00092	1·18675	80 43	**55 „ 125**	76 57·2	0·02147	0·51306	37 04
11 „ 169	68 22·0	·00111	1·14583	79 47	**56 „ 124**	77 16·1	·02202	·50785	36 02
12 „ 168	68 26·2	·00132	1·10855	78 51	**57 „ 123**	77 35·4	·02257	·50283	35 00
13 „ 167	68 30·7	·00155	1·07434	77 55	**58 „ 122**	77 54·9	·02310	·49800	33 59
14 „ 166	68 ·35·6	·00179	1·04275	76 59	**59 „ 121**	78 14·7	·02363	·49336	32 57
15 „ 165	68 40·9	0·00205	1·01343	76 03	**60 „ 120**	78 34·8	0·02415	·048889	31 55
16 „ 164	68 46·5	·00233	0·98609	75 07	**61 „ 119**	78 55·0	·02466	·48461	30 52
17 „ 163	68 52·5	·00262	·96049	74 10	**62 „ 118**	79 15·6	·02516	·48048	29 50
18 „ 162	68 58·9	·00293	·93644	73 14	**63 „ 117**	79 36·4	·02565	·47654	28 47
19 „ 161	69 05·5	·00325	·91378	72 18	**64 „ 116**	79 57·4	·02613	·47276	27 45
20 „ 160	69 12·6	0·00359	0·89237	71 2J	**65 „ 115**	80 18·6	0·02659	0·46915	26 42
21 „ 159	69 20·0	·00395	·87210	70 25	**66 „ 114**	80 40·1	·02705	·46569	25 39
22 „ 158	69 27·8	·00432	·85285	69 28	**67 „ 113**	81 01·7	·02749	·46240	24 36
23 „ 157	69 36·0	·00470	·83455	68 31	**68 „ 112**	81 23·6	·02792	·45926	23 33
24 „ 156	69 44·5	·00510	·81711	67 34	**69 „ 111**	81 45·7	·02833	·45627	22 29
25 „ 155	69 53·3	0·00551	0·80048	66 37	**70 „ 110**	82 07·9	0·02873	0·45344	21 26
26 „ 154	70 02·5	·00594	·78458	65 40	**71 „ 109**	82 30·4	·02911	·45075	20 22
27 „ 153	70 12·1	·00637	·76939	64 43	**72 „ 108**	82 53·0	·02948	·44822	19 19
28 „ 152	70 22·0	·00682	·75482	63 45	**73 „ 107**	83 15·8	·02982	·44583	18 15
29 „ 151	70 32·3	·00728	·74085	62 48	**74 „ 106**	83 38·7	·03016	·44358	17 11
30 „ 150	70 42·9	0·00775	0·72745	61 50	**75 „ 105**	84 01·8	0·03047	0·44148	16 07
31 „ 149	70 53·9	·00824	·71459	60 53	**76 „ 104**	84 25·0	·03077	·43952	15 03
32 „ 148	71 05·2	·00873	·70221	59 55	**77 „ 103**	84 48·4	·03105	·43770	13 59
33 „ 147	71 16·9	·00923	·69032	58 57	**78 „ 102**	85 11·9	·03131	·43602	12 55
34 „ 146	71 28·9	·00974	·67886	57 59	**79 „ 101**	85 35·5	·03155	·43448	11 50
35 „ 145	71 41·3	0·01026	0·66783	57 00	**80 „ 100**	85 59·2	0·03177	0·43307	10 46
36 „ 144	71 54·0	·01079	·65721	56 02	**81 „ 99**	86 23·0	·03197	·43180	9 42
37 „ 143	72 07·0	·01133	·64696	55 04	**82 „ 98**	86 46·9	·03215	·43067	8 37
38 „ 142	72 20·4	·01187	·63708	54 05	**83 „ 97**	87 10·9	·03231	·42967	7 33
39 „ 141	72 34·1	·01242	·62755	53 06	**84 „ 96**	87 34·9	·03245	·42881	6 28
40 „ 140	72 48·2	0·01297	0·61836	52 07	**85 „ 95**	87 59·0	0·03257	0·42808	5 23
41 „ 139	73 02·5	·01353	·60948	51 08	**86 „ 94**	88 23·1	·03266	·42748	4 19
42 „ 138	73 17·3	·01409	·60091	50 09	**87 „ 93**	88 47·3	·03274	·42702	3 14
43 „ 137	73 32·3	·01466	·59264	49 09	**88 „ 92**	89 11·5	·03279	·42669	2 09
44 „ 136	73 47·7	·01523	·58465	48 10	**89 „ 91**	89 35·8	·03282	·42649	1 05
45 „ 135	74 03·4	0·01580	0·57694	47 10	**90**	90 00·0	0·03283	0·42642	0 00
h	180°-K	A	N	a₁	h	180°-K	A	N	a₁

Directions with regard to (K ⌣ d) :—

If l and d have **Same Name,** use (K ⌣ d), i.e. (K — d) or (d — K).

If l and d have **Different Names,** use (K + d).

To name Intercept :—

If H_o is greater than H_c, intercept is "**Towards.**"

If H_o is less than H_c, intercept is "**Away.**"

Log. cosec. H_c = A + log. sec. (K ⌣ d).

N.B. Whenever h is greater than 90°, subtract the quantity in the K column from 180°.

ALT.—AZ. TABLE   LAT. 69°

h	K		A	N	a₁		h	K		A	N	a₁	
°	°	′			°	′	°	°	′			°	′
0 or 180	69	00·0	0·00000	—	90	00	45 or 135	74	48·8	0·01441	0·59619	46	58
1 „ 179	69	00·2	·00001	2·20382	89	04	46 „ 134	75	04·2	·01493	·58874	45	58
2 „ 178	69	00·7	·00003	1·90285	88	08	47 „ 133	75	19·8	·01545	·58154	44	58
3 „ 177	69	01·6	·00008	1·72687	87	12	48 „ 132	75	35·7	·01597	·57460	43	58
4 „ 176	69	02·8	·00014	1·60209	86	16	49 „ 131	75	51·9	·01650	·56789	42	57
5 „ 175	69	04·4	0·00021	1·50537	85	20	50 „ 130	76	08·4	0·01701	0·56142	41	57
6 „ 174	69	06·3	·00030	1·42644	84	24	51 „ 129	76	25·1	·01753	·55517	40	56
7 „ 173	69	08·6	·00041	1·35978	83	28	52 „ 128	76	42·2	·01805	·54914	39	55
8 „ 172	69	11·2	·00054	1·30212	82	32	53 „ 127	76	59·5	·01856	·54332	38	55
9 „ 171	69	14·2	·00068	1·25134	81	35	54 „ 126	77	17·1	·01907	·53771	37	53
10 „ 170	69	17·5	0·00084	1·20600	80	39	55 „ 125	77	35·0	0·01957	0·53231	36	52
11 „ 169	69	21·2	·00102	1·16507	79	43	56 „ 124	77	53·1	·02007	·52710	35	51
12 „ 168	69	25·2	·00121	1·12779	78	47	57 „ 123	78	11·5	·02056	·52208	34	49
13 „ 167	69	29·6	·00142	1·09358	77	50	58 „ 122	78	30·1	·02104	·51725	33	48
14 „ 166	69	34·3	·00164	1·06200	76	54	59 „ 121	78	49·0	·02152	·51260	32	46
15 „ 165	69	39·4	0·00188	1·03267	75	57	60 „ 120	79	08·1	0·02199	0·50814	31	44
16 „ 164	69	44·8	·00213	1·00533	75	01	61 „ 119	79	27·5	·02245	·50385	30	42
17 „ 163	69	50·5	·00240	0·97974	74	04	62 „ 118	79	47·1	·02291	·49974	29	40
18 „ 162	69	56·6	·00268	·95569	73	08	63 „ 117	80	06·9	·02335	·49579	28	37
19 „ 161	70	03·1	·00298	·93303	72	11	64 „ 116	80	26·9	·02378	·49201	27	35
20 „ 160	70	09·9	0·00329	0·91162	71	14	65 „ 115	80	47·1	0·02421	0·48839	26	32
21 „ 159	70	17·0	·00361	·89134	70	17	66 „ 114	81	07·6	·02462	·48494	25	30
22 „ 158	70	24·5	·00395	·87210	69	20	67 „ 113	81	28·2	·02502	·48164	24	27
23 „ 157	70	32·4	·00430	·85379	68	23	68 „ 112	81	49·0	·02540	·47850	23	24
24 „ 156	70	40·5	·00466	·83636	67	26	69 „ 111	82	10·0	·02578	·47552	22	21
25 „ 155	70	49·0	0·00504	0·81972	66	28	70 „ 110	82	31·2	0·02614	0·47268	21	18
26 „ 154	70	57·9	·00543	·80383	65	31	71 „ 109	82	52·6	·02648	·47000	20	15
27 „ 153	71	07·1	·00583	·78862	64	34	72 „ 108	83	14·1	·02681	·46746	19	11
28 „ 152	71	16·6	·00624	·77406	63	36	73 „ 107	83	35·8	·02713	·46507	18	08
29 „ 151	71	26·5	·00666	·76010	62	38	74 „ 106	83	57·6	·02743	·46283	17	04
30 „ 150	71	36·7	0·00709	0·74670	61	40	75 „ 105	84	19·6	0·02772	0·46073	16	01
31 „ 149	71	47·2	·00753	·73383	60	43	76 „ 104	84	41·7	·02798	·45877	14	57
32 „ 148	71	58·1	·00798	·72146	59	45	77 „ 103	85	03·9	·02823	·45695	13	53
33 „ 147	72	09·3	·00843	·70956	58	46	78 „ 102	85	26·2	·02847	·45527	12	50
34 „ 146	72	20·8	·00890	·69811	57	48	79 „ 101	85	48·7	·02869	·45372	11	46
35 „ 145	72	32·7	0·00937	0·68708	56	50	80 „ 100	86	11·2	0·02889	0·45232	10	42
36 „ 144	72	44·9	·00986	·67645	55	51	81 „ 99	86	33·8	·02907	·45105	9	38
37 „ 143	72	57·4	·01034	·66621	54	52	82 „ 98	86	56·5	·02923	·44992	8	34
38 „ 142	73	10·2	·01084	·65633	53	54	83 „ 97	87	19·3	·02937	·44892	7	30
39 „ 141	73	23·4	·01134	·64680	52	55	84 „ 96	87	42·1	·02950	·44806	6	25
40 „ 140	73	36·8	0·01184	0·63760	51	56	85 „ 95	88	05·0	0·02961	0·44733	5	21
41 „ 139	73	50·6	·01235	·62873	50	56	86 „ 94	88	28·0	·02969	·44673	4	17
42 „ 138	74	04·7	·01286	·62016	49	57	87 „ 93	88	50·9	·02976	·44627	3	13
43 „ 137	74	19·1	·01337	·61189	48	57	88 „ 92	89	13·9	·02981	·44594	2	09
44 „ 136	74	33·8	·01389	·60390	47	58	89 „ 91	89	37·0	·02984	·44574	1	04
45 „ 135	74	48·8	0·01441	0·59619	46	58	90	90	00·0	0·02985	0·44567	0	00
h	180°-K		A	N	a₁		h	180°-K		A	N	a₁	

Directions with regard to ($a_2 \overset{+}{\sim} a_1$) :—

Whenever h is greater than 90°, use ($a_2 - a_1$).

If h is less than 90°, use :—

    ($a_1 + a_2$) when (K ± d) has been used,

    ($a_1 - a_2$) when (d — K) has been used.

Always **name azimuth** from elevated pole.

Note that $\begin{cases} \text{h} = \text{L.H.A. of body } \textbf{West} \text{ of the meridian.} \\ \text{h} = (360° - \text{L.H.A.}) \text{ of body } \textbf{East} \text{ of the meridian.} \end{cases}$

Log. tan. $a_2 = \text{N} + \text{log. tan. (K} \overset{+}{\sim} \text{d)}.$

LAT. 70°   ALT.—AZ. TABLE

h	K	A	N	a$_1$	h	K	A	N	a$_1$
°	° ′			° ′	°	° ′			° ′
0 or 180	70 00·0	0·00000	—	90 00	**45 „ 135**	75 34·0	0·01309	0·61646	46 47
1 „ 179	70 00·2	·00001	2·22409	89 04	**46 „ 134**	75 48·7	·01356	·60901	45 47
2 „ 178	70 00·7	·00003	1·92313	88 07	**47 „ 133**	76 03·6	·01403	·60182	44 47
3 „ 177	70 01·5	·00007	1·74715	87 11	**48 „ 132**	76 18·7	·01450	·59487	43 47
4 „ 176	70 02·7	·00012	1·62236	86 14	**49 „ 131**	76 34·2	·01497	·58817	42 46
5 „ 175	70 04·2	0·00019	1·52565	85 18	**50 „ 130**	76 49·9	0·01544	0·58169	41 46
6 „ 174	70 06·1	·00028	1·44671	84 22	**51 „ 129**	77 05·9	·01591	·57544	40 45
7 „ 173	70 08·2	·00038	1·38005	83 25	**52 „ 128**	77 22·2	·01638	·56942	39 44
8 „ 172	70 10·8	·00049	1·32239	82 29	**53 „ 127**	77 38·7	·01684	·56360	38 44
9 „ 171	70 13·6	·00062	1·27162	81 32	**54 „ 126**	77 55·5	·01730	·55799	37 43
10 „ 170	70 16·8	0·00077	1·22628	80 36	**55 „ 125**	78 12·5	0·01775	0·55258	36 42
11 „ 169	70 20·3	·00093	1·18535	79 39	**56 „ 124**	78 29·7	·01820	·54737	35 40
12 „ 168	70 24·2	·00110	1·14807	78 42	**57 „ 123**	78 47·2	·01864	·54236	34 39
13 „ 167	70 28·4	·00129	1·11386	77 46	**58 „ 122**	79 05·0	·01908	·53753	33 37
14 „ 166	70 32·9	·00149	1·08227	76 49	**59 „ 121**	79 23·0	·01951	·53288	32 36
15 „ 165	70 37·8	0·00171	1·05295	75 52	**60 „ 120**	79 41·2	0·01994	0·52842	31 34
16 „ 164	70 43·0	·00194	1·02561	74 55	**61 „ 119**	79 59·6	·02036	·52413	30 32
17 „ 163	70 48·5	·00218	1·00001	73 58	**62 „ 118**	80 18·2	·02076	·52001	29 30
18 „ 162	70 54·4	·00244	0·97597	73 01	**63 „ 117**	80 37·0	·02116	·51607	28 28
19 „ 161	71 00·6	·00271	·95331	72 04	**64 „ 116**	80 56·1	·02156	·51229	27 26
20 „ 160	71 07·1	0·00299	0·93190	71 07	**65 „ 115**	81 15·3	0·02194	0·50867	26 24
21 „ 159	71 13·9	·00329	·91162	70 10	**66 „ 114**	81 34·7	·02231	·50522	25 21
22 „ 158	71 21·1	·00359	·89237	69 13	**67 „ 113**	81 54·4	·02267	·50192	24 19
23 „ 157	71 28·6	·00391	·87407	68 15	**68 „ 112**	82 14·2	·02301	·49878	23 16
24 „ 156	71 36·5	·00424	·85663	67 18	**69 „ 111**	82 34·1	·02335	·49580	22 13
25 „ 155	71 44·6	0·00458	0·84000	66 20	**70 „ 110**	82 54·2	0·02367	0·49296	21 10
26 „ 154	71 53·1	·00494	·82411	65 23	**71 „ 109**	83 14·5	·02399	·49028	20 07
27 „ 153	72 01·9	·00530	·80890	64 25	**72 „ 108**	83 35·0	·02428	·48774	19 04
28 „ 152	72 11·1	·00567	·79434	63 27	**73 „ 107**	83 55·5	·02457	·48535	18 01
29 „ 151	72 20·5	·00605	·78038	62 29	**74 „ 106**	84 16·3	·02484	·48311	16 58
30 „ 150	72 30·3	0·00645	0·76698	61 31	**75 „ 105**	84 37·1	0·02510	0·48100	15 55
31 „ 149	72 40·4	·00684	·75411	60 33	**76 „ 104**	84 58·1	·02534	·47904	14 52
32 „ 148	72 50·8	·00725	·74174	59 35	**77 „ 103**	85 19·2	·02556	·47722	13 48
33 „ 147	73 01·5	·00767	·72984	58 36	**78 „ 102**	85 40·3	·02577	·47554	12 45
34 „ 146	73 12·5	·00809	·71839	57 38	**79 „ 101**	86 01·6	·02597	·47400	11 41
35 „ 145	73 23·9	0·00852	0·70736	56 39	**80 „ 100**	86 23·0	0·02615	0·47260	10 38
36 „ 144	73 35·6	·00896	·69673	55 41	**81 „ 99**	86 44·5	·02631	·47133	9 34
37 „ 143	73 47·5	·00940	·68648	54 42	**82 „ 98**	87 06·0	·02646	·47019	8 30
38 „ 142	73 59·8	·00985	·67661	53 43	**83 „ 97**	87 27·6	·02659	·46920	7 27
39 „ 141	74 12·4	·01030	·66708	52 44	**84 „ 96**	87 49·3	·02670	·46833	6 23
40 „ 140	74 25·2	0·01076	0·65788	51 45	**85 „ 95**	88 11·0	0·02680	0·46760	5 19
41 „ 139	74 38·4	·01122	·64900	50 45	**86 „ 94**	88 32·7	·02687	·46701	4 15
42 „ 138	74 51·9	·01168	·64044	49 46	**87 „ 93**	88 54·5	·02694	·46654	3 12
43 „ 137	75 05·6	·01215	·63216	48 46	**88 „ 92**	89 16·3	·02698	·46621	2 08
44 „ 136	75 19·7	·01262	·62418	47 47	**89 „ 91**	89 38·2	·02701	·46601	1 04
45 „ 135	75 34·0	0·01309	0·61646	46 47	**90**	90 00·0	0·02701	0·46595	0 00
h	180°-K	A	N	a$_1$	h	180°-K	A	N	a$_1$

Directions with regard to (K $\overset{+}{\sim}$ d) :—

If l and d have **Same Name,** use (K $\smile$ d), i.e. (K — d) or (d — K).
If l and d have **Different Names,** use (K + d).

To name Intercept :—

If H$_o$ is greater than H$_c$, intercept is " **Towards.** "
If H$_o$ is less than H$_c$, intercept is " **Away.** "

Log. cosec. H$_c$ = A + log. sec. (K $\overset{+}{\sim}$ d'.

N.B. Whenever h is greater than 90°. subtract the quantity in the K column from 180°.

ALT.—AZ. TABLE   LAT. 71°

h	K	A	N	a₁	h	K	A	N	a₁
°	° ′		° ′	° ′	°	° ′		° ′	° ′
0 or 180	71 00·0	0·00000	—	90 00	45 or 135	76 19·0	0·01182	0·63787	46 36
1 „ 179	71 00·2	·00001	2·24550	89 03	46 „ 134	76 32·9	·01225	·63042	45 36
2 „ 178	71 00·6	·00003	1·94454	88 07	47 „ 133	76 47·1	·01267	·62323	44 36
3 „ 177	71 01·5	·00006	1·76856	87 10	48 „ 132	77 01·5	·01310	·61628	43 36
4 „ 176	71 02·6	·00011	1·64377	86 13	49 „ 131	77 16·2	·01352	·60958	42 36
5 „ 175	71 04·0	0·00017	1·54706	85 16	50 „ 130	77 31·2	0·01394	0·60310	41 35
6 „ 174	71 05·8	·00025	1·46812	84 19	51 „ 129	77 46·4	·01437	·59685	40 35
7 „ 173	71 07·9	·00034	1·40146	83 23	52 „ 128	78 01·9	·01478	·59083	39 34
8 „ 172	71 10·3	·00045	1·34380	82 26	53 „ 127	78 17·6	·01520	·58501	38 33
9 „ 171	71 13·0	·00056	1·29303	81 29	54 „ 126	78 33·5	·01561	·57940	37 32
10 „ 170	71 16·0	0·00070	1·24769	80 32	55 „ 125	78 49·7	0·01602	0·57399	36 31
11 „ 169	71 19·5	·00084	1·20676	79 35	56 „ 124	79 06·1	·01643	·56878	35 30
12 „ 168	71 23·2	·00100	1·16948	78 38	57 „ 123	79 22·7	·01682	·56377	34 29
13 „ 167	71 27·2	·00117	1·13527	77 41	58 „ 122	79 39·6	·01722	·55894	33 28
14 „ 166	71 31·5	·00135	1·10368	76 44	59 „ 121	79 56·6	·01761	·55429	32 26
15 „ 165	71 36·2	0·00155	1·07436	75 47	60 „ 120	80 13·9	0·01799	0·54983	31 25
16 „ 164	71 41·2	·00176	1·04702	74 50	61 „ 119	80 31·4	·01836	·54554	30 23
17 „ 163	71 46·5	·00198	1·02142	73 53	62 „ 118	80 49·0	·01873	·54142	29 21
18 „ 162	71 52·1	·00221	0·99738	72 55	63 „ 117	81 06·9	·01909	·53748	28 19
19 „ 161	71 58·0	·00245	·97472	71 58	64 „ 116	81 25·0	·01944	·53370	27 17
20 „ 160	72 04·2	0·00271	0·95331	71 01	65 „ 115	81 43·2	0·01978	0·53008	26 15
21 „ 159	72 10·8	·00298	·93303	70 03	66 „ 114	82 01·7	·02011	·52663	25 13
22 „ 158	72 17·6	·00325	·91378	69 06	67 „ 113	82 20·2	·02043	·52333	24 11
23 „ 157	72 24·8	·00354	·89548	68 08	68 „ 112	82 39·0	·02075	·52019	23 08
24 „ 156	72 32·3	·00384	·87804	67 10	69 „ 111	82 57·9	·02105	·51721	22 06
25 „ 155	72 40·1	0·00415	0·86141	66 12	70 „ 110	83 17·0	0·02134	0·51437	21 03
26 „ 154	72 48·2	·00447	·84552	65 15	71 „ 109	83 36·2	·02162	·51169	20 01
27 „ 153	72 56·6	·00480	·83031	64 17	72 „ 108	83 55·6	·02189	·50915	18 58
28 „ 152	73 05·4	·00513	·81575	63 19	73 „ 107	84 15·1	·02214	·50676	17 55
29 „ 151	73 14·4	·00548	·80179	62 20	74 „ 106	84 34·7	·02238	·50452	16 52
30 „ 150	73 23·7	0·00583	0·78839	61 22	75 „ 105	84 54·4	0·02261	0·50241	15 49
31 „ 149	73 33·4	·00619	·77552	60 24	76 „ 104	85 14·3	·02283	·50045	14 46
32 „ 148	73 43·3	·00656	·76315	59 25	77 „ 103	85 34·3	·02303	·49863	13 43
33 „ 147	73 53·5	·00694	·75125	58 27	78 „ 102	85 54·3	·02322	·49695	12 40
34 „ 146	74 04·1	·00732	·73980	57 28	79 „ 101	86 14·5	·02339	·49541	11 37
35 „ 145	74 14·9	0·00771	0·72877	56 30	80 „ 100	86 34·7	0·02355	0·49401	10 34
36 „ 144	74 26·0	·00810	·71814	55 31	81 „ 99	86 55·0	·02370	·49274	9 31
37 „ 143	74 37·4	·00850	·70789	54 32	82 „ 98	87 15·4	·02383	·49160	8 27
38 „ 142	74 49·2	·00890	·69802	53 33	83 „ 97	87 35·8	·02395	·49061	7 24
39 „ 141	75 01·1	·00931	·68849	52 34	84 „ 96	87 56·3	·02405	·48974	6 21
40 „ 140	75 13·4	0·00972	0·67929	51 34	85 „ 95	88 16·9	0·02413	0·48901	5 17
41 „ 139	75 26·0	·01014	·67041	50 35	86 „ 94	88 37·4	·02420	·48842	4 14
42 „ 138	75 38·8	·01056	·66185	49 35	87 „ 93	88 58·1	·02426	·48795	3 10
43 „ 137	75 51·9	·01098	·65357	48 36	88 „ 92	89 18·7	·02430	·48762	2 07
44 „ 136	76 05·3	·01140	·64559	47 36	89 „ 91	89 39·3	·02432	·48742	1 03
45 „ 135	76 19·0	0·01182	0·63787	46 36	90	90 00·0	0·02433	0·48736	0 00
h	180°-K	A	N	a₁	h	180°-K	A	N	a₁

Directions with regard to (a₂ $\overset{+}{\underset{-}{\sim}}$ a₁) :—

Whenever h is greater than 90°, use (a₂ — a₁).

If h is less than 90°, use :—

(a₁ + a₂) when (K ± d) has been used,

(a₁ — a₂) when (d — K) has been used.

Always **name azimuth** from elevated pole.

Note that { h = L.H.A. of body **West** of the meridian.
{ h = (360° — L.H.A.) of body **East** of the meridian.

Log. tan. a₂ = N + log. tan. (K $\overset{+}{\underset{-}{\sim}}$ d).

LAT. 72° ALT.—AZ. TABLE

h	K		A	N	a_1		h	K		A	N	a_1	
°	°	′			°	′		°	°	′			°	′
0 or 180	72	00·0	0·00000	—	90	00	45 or 135	77	03·6	0·01062	0·66053	46	26
1 „ 179	72	00·2	·00001	2·26816	89	03	46 „ 134	77	16·9	·01100	·65308	45	26
2 „ 178	72	00·6	·00003	1·96693	88	06	47 „ 133	77	30·3	·01138	·64589	44	26
3 „ 177	72	01·4	·00006	1·79122	87	09	48 „ 132	77	44·0	·01176	·63894	43	26
4 „ 176	72	02·5	·00010	1·66643	86	12	49 „ 131	77	58·0	·01214	·63224	42	26
5 „ 175	72	03·8	0·00016	1·56972	85	15	50 „ 130	78	12·2	0·01252	0·62576	41	25
6 „ 174	72	05·5	·00023	1·49078	84	18	51 „ 129	78	26·6	·01290	·61951	40	25
7 „ 173	72	07·5	·00031	1·42412	83	20	52 „ 128	78	41·3	·01327	·61349	39	24
8 „ 172	72	09·8	·00040	1·36646	82	23	53 „ 127	78	56·2	·01365	·60767	38	23
9 „ 171	72	12·5	·00051	1·31569	81	26	54 „ 126	79	11·3	·01401	·60206	37	23
10 „ 170	72	15·4	0·00063	1·27035	80	29	55 „ 125	79	26·6	0·01438	0·59665	36	22
11 „ 169	72	18·6	·00076	1·22942	79	32	56 „ 124	79	42·1	·01474	·59144	35	21
12 „ 168	72	22·1	·00090	1·19214	78	34	57 „ 123	79	57·9	·01510	·58643	34	20
13 „ 167	72	26·0	·00105	1·15793	77	37	58 „ 122	80	13·8	·01545	·58160	33	18
14 „ 166	72	30·1	·00122	1·12634	76	40	59 „ 121	80	30·0	·01580	·57695	32	17
15 „ 165	72	34·5	0·00139	1·09702	75	42	60 „ 120	80	46·3	0·01614	0·57249	31	16
16 „ 164	72	39·3	·00158	1·06968	74	45	61 „ 119	81	02·9	·01647	·56820	30	14
17 „ 163	72	44·3	·00178	1·04408	73	47	62 „ 118	81	19·6	·01680	·56408	29	12
18 „ 162	72	49·7	·00199	1·02004	72	50	63 „ 117	81	36·5	·01712	·56014	28	11
19 „ 161	72	55·3	·00221	0·99738	71	52	64 „ 116	81	53·6	·01743	·55636	27	09
20 „ 160	73	01·3	0·00244	0·97597	70	54	65 „ 115	82	10·9	0·01774	0·55274	26	07
21 „ 159	73	07·5	·00268	·95569	69	57	66 „ 114	82	28·3	·01803	·54929	25	05
22 „ 158	73	14·1	·00293	·93644	68	59	67 „ 113	82	45·9	·01832	·54599	24	03
23 „ 157	73	20·9	·00319	·91814	68	01	68 „ 112	83	03·6	·01860	·54285	23	01
24 „ 156	73	28·1	·00346	·90070	67	03	69 „ 111	83	21·5	·01887	·53987	21	59
25 „ 155	73	35·5	0·00374	0·88407	66	05	70 „ 110	83	39·5	0·01913	0·53703	20	57
26 „ 154	73	43·2	·00402	·86818	65	07	71 „ 109	83	57·7	·01938	·53435	19	54
27 „ 153	73	51·2	·00432	·85297	64	09	72 „ 108	84	16·0	·01962	·53181	18	52
28 „ 152	73	59·6	·00462	·83841	63	11	73 „ 107	84	34·4	·01984	·52942	17	49
29 „ 151	74	08·2	·00493	·82445	62	12	74 „ 106	84	52·9	·02006	·52718	16	47
30 „ 150	74	17·0	0·00525	0·81105	61	14	75 „ 105	85	11·6	0·02026	0·52507	15	44
31 „ 149	74	26·2	·00557	·79818	60	15	76 „ 104	85	30·3	·02046	·52311	14	41
32 „ 148	74	35·7	·00590	·78581	59	17	77 „ 103	85	49·2	·02064	·52129	13	39
33 „ 147	74	45·4	·00624	·77391	58	18	78 „ 102	86	08·1	·02080	·51961	12	36
34 „ 146	74	55·4	·00658	·76246	57	19	79 „ 101	86	27·1	·02096	·51807	11	33
35 „ 145	75	05·7	0·00693	0·75143	56	20	80 „ 100	86	46·2	0·02110	0·51667	10	30
36 „ 144	75	16·3	·00728	·74080	55	21	81 „ 99	87	05·4	·02123	·51540	9	27
37 „ 143	75	27·2	·00764	·73055	54	22	82 „ 98	87	24·7	·02135	·51426	8	24
38 „ 142	75	38·3	·00801	·72068	53	23	83 „ 97	87	43·9	·02145	·51327	7	21
39 „ 141	75	49·7	·00837	·71115	52	24	84 „ 96	88	03·3	·02154	·51240	6	18
40 „ 140	76	01·4	0·00874	0·70195	51	25	85 „ 95	88	22·7	0·02162	0·51167	5	15
41 „ 139	76	13·3	·00911	·69307	50	25	86 „ 94	88	42·1	·02168	·51100	4	12
42 „ 138	76	25·5	·00949	·68451	49	26	87 „ 93	89	01·5	·02173	·51061	3	09
43 „ 137	76	38·0	·00987	·67623	48	26	88 „ 92	89	21·0	·02177	·51028	2	06
44 „ 136	76	50·7	·01024	·66825	47	26	89 „ 91	89	40·5	·02179	·51008	1	03
45 „ 135	77	03·6	0·01062	0·66053	46	26	90	90	00·0	0·02179	0·51002	0	00
h	180°-K		A	N	a_1		h	180°-K		A	N	a_1	

Directions with regard to ($K \overset{+}{\sim} d$) :—

If l and d have **Same Name**, use ($K \frown d$), i.e. ($K - d$) or ($d - K$).

If l and d have **Different Names**, use ($K + d$).

To name Intercept :—

If H_o is less than H_c, intercept is " **Away.** "

If H_o is greater than H_c, intercept is " **Towards.** "

Log. cosec. $H_c = A +$ log. sec. ($K \overset{+}{\sim} d$).

N.B. *Whenever h is greater than 90°, subtract the quantity in the K column from 180°.*

ALTITUDE ADJUSTMENT TABLES

ALTITUDE ADJUSTMENT TABLES

TABLE A. Latitudes 0° to 34°.

AZIMUTHS IN DEGREES.

Lat.	0°	3°	6°	9°	12°	14°	16°	18°	20°	22°	24°	26°	28°	29°	30°	31°	32°	33°	34°	Key to Table B
	0	0	0	0	0	0	0	0	0	0	0	0	0	0	0	0	0	0	0	AA
	1	1	1	1	1	1	1	1	1	1	1	1	1	1	1	1	1	1	1	AB
	2	2	2	2	2	2	2	2	2	2	2	2	2	2	2	2	2	2	2	AC
	3	3	3	3	3	3	3	3	3	3	3	3	3	3	3	4	4	4	4	AD
	4	4	4	4	4	4	4	4	4	4	4	4	4	5	5	5	5	5	5	AE
	5	5	5	5	5	5	5	5	5	5	5	6	6	6	6	6	6	6	6	AF
	6	6	6	6	6	6	6	6	6	6	7	7	7	7	7	7	7	7	7	AG
	7	7	7	7	7	7	7	7	7	8	8	8	8	8	8	8	8	8	8	AH
	8	8	8	8	8	8	8	8	9	9	9	9	9	9	9	9	9	10	10	AK
	9	9	9	9	9	9	9	9	10	10	10	10	10	10	10	11	11	11	11	AL
	10	10	10	10	10	10	10	11	11	11	11	11	11	11	11	12	12	12	12	AM
	11	11	11	11	11	11	11	12	12	12	12	12	12	12	13	13	13	13	13	AN
	12	12	12	12	12	12	13	13	13	13	13	14	14	14	14	14	14	14	15	AO
	13	13	13	13	13	13	14	14	14	14	14	15	15	15	15	15	15	16	16	AP
	14	14	14	14	14	14	15	15	15	15	15	16	16	16	16	16	17	17	17	AQ
	15	15	15	15	15	15	16	16	16	16	16	17	17	17	17	18	18	18	18	AR
	16	16	16	16	16	17	17	17	17	17	18	18	18	18	19	19	19	19	19	AS
	17	17	17	17	17	18	18	18	18	18	19	19	19	20	20	20	20	20	21	AT
	18	18	18	18	18	19	19	19	19	19	20	20	20	21	21	21	21	22	22	AU
	19	19	19	19	19	20	20	20	20	21	21	21	22	22	22	22	23	23	23	AV
	20	20	20	20	20	21	21	21	21	22	22	22	23	23	23	24	24	24	24	AW
	21	21	21	21	21	22	22	22	23	23	23	24	24	24	25	25	25	26	26	AX
	22	22	22	22	23	23	23	24	24	24	25	25	25	26	26	26	27	27	27	AY
	23	23	23	23	24	24	24	24	25	25	25	26	26	26	27	27	27	28	28	AZ
	24	24	24	24	25	25	25	25	26	26	26	27	27	28	28	28	29	29	29	BA
	25	25	25	25	26	26	26	26	27	27	27	28	28	29	29	29	30	30	31	BB
	26	26	26	26	27	27	27	27	28	28	29	29	30	30	30	31	31	32	32	BC
	27	27	27	27	28	28	28	29	29	29	30	30	31	31	32	32	32	33	33	BD
	28	28	28	28	29	29	29	30	30	30	31	31	32	32	33	33	34	34	34	BE
	29	29	29	29	30	30	30	31	31	32	32	32	33	34	34	34	35	35	36	BF
	30	30	30	30	31	31	31	32	32	33	33	34	35	35	35	36	36	37	37	BG
	31	31	31	31	32	32	32	33	33	34	34	35	36	36	36	37	37	38	39	BH
	32	32	32	32	33	33	33	34	34	35	35	36	37	37	38	38	39	39	40	BK
	33	33	33	33	34	34	34	35	35	36	37	37	38	39	39	39	40	40	41	BL
	34	34	34	34	35	35	35	36	37	37	38	38	39	40	40	41	41	42	42	BM
	35	35	35	36	36	36	36	37	38	38	39	40	41	41	41	42	43	43	44	BN
	36	36	36	37	37	37	37	38	39	39	40	41	42	42	43	43	44	44	45	BO
	37	37	37	38	38	38	38	39	40	40	41	42	43	43	44	45	45	46	47	BP
	38	38	38	39	39	39	40	40	41	42	42	43	44	45	45	46	47	47	48	BQ
	39	39	39	40	40	40	41	41	42	43	44	44	45	46	47	47	48	49	49	BR
	40	40	40	41	41	41	42	43	43	44	45	46	47	47	48	49	49	50	51	BS
	41	41	41	42	42	43	43	44	44	45	46	47	48	49	49	50	51	51	52	BT
	42	42	42	43	43	44	44	45	45	46	47	48	49	50	51	51	52	53	54	BU
	43	43	43	44	44	45	45	46	47	47	48	49	51	51	52	53	54	54	55	BV

Continued on next page

ALTITUDE ADJUSTMENT TABLES

TABLE A.   Latitudes 0° to 34° (continued).

Lat.	0°	3°	6°	9°	12°	14°	16°	18°	20°	22°	24°	26°	28°	29°	30°	31°	32°	33°	34°	Key to Table B
	44	44	44	45	45	46	46	47	48	49	49	51	52	53	53	54	55	56	57	**BW**
	45	45	45	46	46	47	47	48	49	50	51	52	53	54	55	56	56	57	59	**BX**
	46	46	46	47	47	48	48	49	50	51	52	53	55	55	56	57	58	59	60	**BY**
	47	47	47	48	48	49	50	50	51	52	53	54	56	57	58	59	60	61	62	**BZ**
	48	48	48	49	49	50	51	51	52	53	54	56	57	58	59	60	61	62	64	**CA**
	49	49	49	50	50	51	52	53	53	54	56	57	59	60	61	62	63	64	66	**CB**
	50	50	50	51	51	52	53	54	55	56	57	58	60	61	62	63	65	66	68	**CC**
	51	51	51	52	52	53	54	55	56	57	58	60	62	63	64	65	66	68	70	**CD**
	52	52	52	53	53	54	55	56	57	58	60	61	63	64	65	67	68	70	72	**CE**
	53	53	53	54	54	55	56	57	58	59	61	63	65	66	67	68	70	72	74	**CF**
	54	54	54	55	55	56	57	58	59	61	62	64	66	68	69	71	73	75	77	**CG**
	55	55	55	56	56	58	58	59	61	62	64	66	68	69	71	73	75	78	81	**CH**
	56	56	56	57	57	59	60	61	62	63	65	67	70	71	73	75	78	81	90	**CK**
	57	57	57	58	58	60	61	62	63	65	67	69	72	74	76	78	81	90		**CL**
	58	58	59	59	59	61	62	63	64	66	68	71	74	76	78	82	90			**CM**
	59	59	60	60	61	62	63	64	66	68	70	72	76	79	82	90				**CN**
	60	60	61	61	62	63	64	66	67	69	71	74	79	82	90					**CO**
	61	61	62	62	63	64	65	67	69	71	73	77	82	90						**CP**
	62	62	63	63	65	66	67	68	70	72	75	79	90							**CQ**
	63	63	64	64	66	67	68	70	71	74	77	82								**CR**
	64	64	65	66	67	68	69	71	73	76	80	90								**CS**
	65	65	66	67	68	69	71	72	75	78	83									**CT**
	66	66	67	68	69	70	72	74	76	80	90									**CU**
	67	67	68	69	70	72	73	75	78	83										**CV**
	68	68	69	70	71	73	75	77	81	90										**CW**
	69	69	70	71	73	74	76	79	83											**CX**
	70	70	71	72	74	76	78	81	90											**CY**
	71	71	72	73	75	77	80	84												**CZ**
	72	72	73	74	76	79	82	90												**DA**
	73	73	74	76	78	80	84													**DB**
	74	74	75	77	79	82	90													**DC**
	75	75	76	78	81	85														**DD**
	76	76	77	79	83	90														**DE**
	77	77	78	81	85															**DF**
	78	78	80	82	90															**DG**
	79	79	81	84																**DH**
	80	80	82	86																**DK**
	81	82	83	90																**DL**
	82	83	85																	**DM**
	83	84	86																	**DN**
	84	85	90																	**DO**
	86	87																		**DP**
	88																			**DQ**
	90																			**DR**

AZIMUTHS IN DEGREES.

ALTITUDE ADJUSTMENT TABLES

TABLE A.   Latitudes 35° to 53°.

AZIMUTHS IN DEGREES.

Lat.	35°	36°	37°	38°	39°	40°	41°	42°	43°	44°	45°	46°	47°	48°	49°	50°	51°	52°	53°	Key to Table B
	0	0	0	0	0	0	0	0	0	0	0	0	0	0	0	0	0	0	0	AA
	1	1	1	1	1	1	1	1	1	1	1	1	1	1	2	2	2	2	2	AB
	2	2	3	3	3	3	3	3	3	3	3	3	3	3	3	3	3	3	3	AC
	4	4	4	4	4	4	4	4	4	4	4	4	4	4	5	5	5	5	5	AD
	5	5	5	5	5	5	5	5	5	6	6	6	6	6	6	6	6	7	7	AE
	6	6	6	6	6	7	7	7	7	7	7	7	7	7	8	8	8	8	8	AF
	7	7	8	8	8	8	8	8	8	8	9	9	9	9	9	9	10	10	10	AG
	9	9	9	9	9	9	9	9	10	10	10	10	10	10	11	11	11	11	12	AH
	10	10	10	10	10	10	11	11	11	11	11	12	12	12	12	13	13	13	13	AK
	11	11	11	11	12	12	12	12	12	13	13	13	13	14	14	14	14	15	15	AL
	12	12	13	13	13	13	13	14	14	14	14	14	15	15	15	15	16	16	16	AM
	13	14	14	14	14	14	15	15	15	15	16	16	16	17	17	17	18	18	18	AN
	15	15	15	15	16	16	16	16	17	17	17	17	18	18	18	19	19	20	20	AO
	16	16	16	17	17	17	17	18	18	18	19	19	19	20	20	20	21	21	22	AP
	17	17	18	18	18	18	19	19	19	20	20	20	21	21	22	22	23	23	24	AQ
	18	19	19	19	19	20	20	20	21	21	21	22	22	23	23	24	24	25	25	AR
	20	20	20	20	21	21	21	22	22	23	23	23	24	24	25	25	26	27	27	AS
	21	21	21	22	22	22	23	23	24	24	24	25	25	26	26	27	28	28	29	AT
	22	22	23	23	23	24	24	25	25	25	26	26	27	28	28	29	29	30	31	AU
	23	24	24	24	25	25	26	26	26	27	27	28	29	29	30	30	31	32	33	AV
	25	25	25	26	26	27	27	27	28	28	29	29	30	31	31	32	33	34	35	AW
	26	26	27	27	27	28	28	29	29	30	30	31	32	32	33	34	35	36	37	AX
	27	28	28	28	29	29	30	30	31	31	32	33	33	34	35	36	37	37	38	AY
	28	29	29	30	30	31	31	32	32	33	34	34	35	36	37	37	38	39	40	AZ
	30	30	31	31	32	32	33	33	34	34	35	36	37	37	38	39	40	41	43	BA
	31	31	32	32	33	33	34	35	35	36	36	37	37	39	40	41	42	43	45	BB
	32	33	33	34	34	35	36	36	37	38	38	39	40	41	42	43	44	45	47	BC
	34	34	35	35	36	36	37	38	38	39	40	41	42	43	44	45	46	48	49	BD
	35	35	36	37	37	38	38	39	40	41	42	43	44	45	46	47	48	50	51	BE
	36	37	37	38	39	39	40	41	42	42	43	44	45	46	48	49	50	52	54	BF
	38	38	39	39	40	41	41	42	43	44	44	45	46	47	48	51	53	54	56	BG
	39	40	40	41	42	42	43	44	45	46	47	48	49	50	52	53	55	57	59	BH
	40	41	42	42	43	44	45	45	46	47	49	50	51	52	54	56	57	59	62	BK
	42	42	43	44	44	45	46	47	48	49	50	52	53	54	56	58	60	62	65	BL
	43	44	44	45	46	47	48	49	50	51	52	54	55	57	58	60	63	65	68	BM
	44	45	46	47	48	48	49	51	52	53	54	56	57	59	61	63	66	69	72	BN
	46	47	47	48	49	50	51	52	53	55	56	58	59	61	64	66	69	73	78	BO
	47	48	49	50	51	52	53	54	55	57	58	60	62	64	67	69	73	78	90	BP
	49	50	50	51	52	53	55	56	57	59	61	62	65	67	70	73	78	90		BQ
	50	51	52	53	54	55	56	58	59	61	63	65	67	70	74	78	90			BR
	52	53	54	55	56	57	58	60	62	63	65	68	70	74	78	90				BS
	53	54	55	56	58	59	60	62	64	66	68	71	74	79	90					BT
	55	56	57	58	59	61	62	64	66	68	71	74	79	90						BU
	56	57	59	60	61	63	65	67	69	71	75	79	90							BV

Continued on next page

ALTITUDE ADJUSTMENT TABLES

TABLE A.   Latitudes 35° to 53° (continued).

AZIMUTHS

Lat.	35°	36°	37°	38°	39°	40°	41°	42°	43°	44°	45°	46°	47°	48°	49°	50°	51°	52°	53°	Key to Table B
	58	59	60	63	63	65	67	69	72	75	79	90								BW
	60	61	62	64	65	67	70	72	75	79	90									BX
	61	63	64	66	68	70	72	75	80	90										BY
	63	65	66	68	70	73	76	80	90											BZ
	65	67	69	71	73	76	80	90												CA
	67	69	71	73	76	80	90													CB
	69	71	74	76	80	90														CC
	72	74	77	80	90															CD
	74	77	81	90																CE
	77	81	90																	CF
	81	90																		CG
	90																			CH

TABLE A.   Latitudes 54° to 72°.

AZIMUTHS IN DEGREES.

Lat.	54°	55°	56°	57°	58°	59°	60°	61°	62°	63°	64°	65°	66°	67°	68°	69°	70°	71°	72°	Key to Table B
	0	0	0	0	0	0	0	0	0	0	0	0	0	0	0	0	0	0	0	AA
	2	2	2	2	2	2	2	2	2	2	2	2	2	3	3	3	3	3	3	AB
	3	3	4	4	4	4	4	4	4	4	5	5	5	5	6	6	6	6	6	AC
	5	5	5	6	6	6	6	6	6	7	7	7	7	8	8	8	9	9	10	AD
	7	7	7	7	8	8	8	8	9	9	9	10	10	10	11	11	12	12	13	AE
	9	9	9	9	9	10	10	10	11	11	11	12	12	13	13	14	15	16	16	AF
	10	11	11	11	11	12	12	12	13	13	14	14	15	16	16	17	18	19	20	AG
	12	12	13	13	13	14	14	15	15	16	16	17	17	18	19	20	21	22	23	AH
	14	14	14	15	15	16	16	17	17	18	19	19	20	21	22	23	24	25	27	AK
	15	16	16	17	17	18	18	19	19	20	21	22	23	24	25	26	27	29	30	AL
	17	17	18	18	19	19	20	20	21	22	23	24	25	26	27	28	30	31	33	AM
	19	19	20	21	21	22	22	23	24	25	26	27	28	29	31	32	34	36	38	AN
	21	21	22	22	23	24	25	25	26	27	28	29	31	32	34	35	37	40	42	AO
	23	23	24	24	25	26	27	28	29	30	31	32	34	35	37	39	41	44	47	AP
	24	25	26	26	27	28	29	30	31	32	33	35	36	38	40	42	45	48	52	AQ
	26	27	28	28	29	30	31	32	33	35	36	38	40	41	44	46	49	53	57	AR
	28	29	30	30	31	32	33	35	36	37	39	41	43	45	47	50	54	58	63	AS
	30	31	32	32	33	35	36	37	39	40	42	44	46	48	51	55	59	64	71	AT
	32	33	34	35	36	37	38	40	41	43	45	47	49	52	56	60	65	72	90	AU
	34	35	36	37	38	39	41	42	44	46	48	50	53	56	60	65	72	90		AV
	36	37	38	39	40	42	43	45	47	49	51	54	57	61	66	73	90			AW
	38	39	40	41	43	44	46	48	50	52	55	58	62	67	73	90				AX
	40	41	42	43	45	47	49	51	53	56	59	62	67	73	90					AY
	42	43	44	46	48	49	51	54	56	59	63	68	74	90						AZ
	44	45	47	48	50	52	54	57	60	64	68	74	90							BA
	46	47	49	51	53	55	58	61	64	68	75	90								BB
	48	50	52	54	56	58	61	65	69	75	90									BC
	51	52	54	56	59	62	65	69	75	90										BD
	53	55	57	60	62	66	70	76	90											BE
	56	58	60	63	66	70	76	90												BF
	58	61	63	67	71	76	90													BG
	61	64	67	71	76	90														BH
	64	68	71	77	90															BK
	68	72	77	90																BL
	72	77	90																	BM
	77	90																		BN
	90																			BO

ALTITUDE ADJUSTMENT TABLES

TABLE B.   Azimuths 90° to 45°.

Key	1′	2′	3′	4′	5′	6′	7′	8′	9′	10′	20′	30′	40′	50′	Azimuth for latitude Correction.°
AA	0·0	0·0	0·0	0·0	0·0	0·0	0·0	0·0	0·0	0·0	0·0	0·0	0·0	0·0	90
AB	0·0	0·0	0·1	0·1	0·1	0·1	0·1	0·1	0·2	0·2	0·4	0·5	0·7	0·9	89
AC	0·0	0·1	0·1	0·1	0·2	0·2	0·2	0·3	0·3	0·3	0·7	1·0	1·4	1·7	88
AD	0·1	0·1	0·2	0·2	0·3	0·3	0·4	0·4	0·5	0·5	1·0	1·6	2·1	2·6	87
AE	0·1	0·1	0·2	0·3	0·3	0·4	0·5	0·6	0·6	0·7	1·4	2·1	2·8	3·5	86
AF	0·1	0·2	0·3	0·3	0·4	0·5	0·6	0·7	0·8	0·9	1·7	2·6	3·5	4·4	85
AG	0·1	0·2	0·3	0·4	0·5	0·6	0·7	0·8	0·9	1·0	2·1	3·1	4·2	5·2	84
AH	0·1	0·2	0·4	0·5	0·6	0·7	0·9	1·0	1·1	1·2	2·4	3·7	4·9	6·1	83
AK	0·1	0·3	0·4	0·6	0·7	0·8	1·0	1·1	1·3	1·4	2·8	4·2	5·6	7·0	82
AL	0·2	0·3	0·5	0·6	0·8	0·9	1·1	1·3	1·4	1·6	3·1	4·7	6·3	7·8	81
AM	0·2	0·3	0·5	0·7	0·9	1·0	1·2	1·4	1·6	1·7	3·5	5·2	6·9	8·7	80
AN	0·2	0·4	0·6	0·8	1·0	1·1	1·3	1·5	1·7	1·9	3·8	5·7	7·6	9·5	79
AO	0·2	0·4	0·6	0·8	1·0	1·2	1·5	1·7	1·9	2·1	4·2	6·2	8·3	10·4	78
AP	0·2	0·4	0·7	0·9	1·1	1·3	1·6	1·8	2·0	2·2	4·5	6·7	9·0	11·2	77
AQ	0·2	0·5	0·7	1·0	1·2	1·5	1·7	1·9	2·2	2·4	4·8	7·3	9·7	12·1	76
AR	0·3	0·5	0·8	1·0	1·3	1·6	1·8	2·1	2·3	2·6	5·2	7·8	10·4	12·9	75
AS	0·3	0·6	0·8	1·1	1·4	1·7	1·9	2·2	2·5	2·8	5·5	8·3	11·0	13·8	74
AT	0·3	0·6	0·9	1·2	1·5	1·8	2·0	2·3	2·6	2·9	5·8	8·8	11·7	14·6	73
AU	0·3	0·6	0·9	1·2	1·5	1·9	2·2	2·5	2·8	3·1	6·2	9·3	12·4	15·5	72
AV	0·3	0·7	1·0	1·3	1·6	2·0	2·3	2·6	2·9	3·3	6·5	9·8	13·0	16·3	71
AW	0·3	0·7	1·0	1·4	1·7	2·1	2·4	2·7	3·1	3·4	6·8	10·3	13·7	17·1	70
AX	0·4	0·7	1·1	1·4	1·8	2·2	2·5	2·9	3·2	3·6	7·2	10·8	14·3	17·9	69
AY	0·4	0·7	1·1	1·5	1·9	2·2	2·6	3·0	3·4	3·7	7·5	11·2	15·0	18·7	68
AZ	0·4	0·8	1·2	1·6	2·0	2·3	2·7	3·1	3·5	3·9	7·8	11·7	15·6	19·5	67
BA	0·4	0·8	1·2	1·6	2·0	2·4	2·8	3·3	3·7	4·1	8·1	12·2	16·3	20·3	66
BB	0·4	0·8	1·3	1·7	2·1	2·5	3·0	3·4	3·8	4·2	8·5	12·7	16·9	21·1	65
BC	0·4	0·9	1·3	1·8	2·2	2·6	3·1	3·5	3·9	4·4	8·8	13·2	17·5	21·9	64
BD	0·5	0·9	1·4	1·8	2·3	2·7	3·2	3·6	4·1	4·5	9·1	13·6	18·2	22·7	63
BE	0·5	0·9	1·4	1·9	2·3	2·8	3·3	3·8	4·2	4·7	9·4	14·1	18·8	23·5	62
BF	0·5	1·0	1·5	1·9	2·4	2·9	3·4	3·9	4·4	4·8	9·7	14·5	19·4	24·2	61
BG	0·5	1·0	1·5	2·0	2·5	3·0	3·5	4·0	4·5	5·0	10·0	15·0	20·0	25·0	60
BH	0·5	1·0	1·5	2·1	2·6	3·1	3·6	4·1	4·6	5·2	10·3	15·5	20·6	25·8	59
BK	0·5	1·1	1·6	2·1	2·6	3·2	3·7	4·2	4·8	5·3	10·6	15·9	21·2	26·5	58
BL	0·5	1·1	1·6	2·2	2·7	3·3	3·8	4·4	4·9	5·4	10·9	16·3	21·8	27·2	57
BM	0·6	1·1	1·7	2·2	2·8	3·4	3·9	4·5	5·0	5·6	11·2	16·8	22·4	28·0	56
BN	0·6	1·1	1·7	2·3	2·9	3·4	4·0	4·6	5·2	5·7	11·5	17·2	22·9	28·7	55
BO	0·6	1·2	1·8	2·4	2·9	3·5	4·1	4·7	5·3	5·9	11·8	17·6	23·5	29·4	54
BP	0·6	1·2	1·8	2·4	3·0	3·6	4·2	4·8	5·4	6·0	12·0	18·1	24·1	30·1	53
BQ	0·6	1·2	1·8	2·5	3·1	3·7	4·3	4·9	5·5	6·2	12·3	18·5	24·6	30·8	52
BR	0·6	1·3	1·9	2·5	3·1	3·8	4·4	5·0	5·7	6·3	12·6	18·9	25·2	31·5	51
BS	0·6	1·3	1·9	2·6	3·2	3·9	4·5	5·1	5·8	6·4	12·9	19·3	25·7	32·1	50
BT	0·7	1·3	2·0	2·6	3·3	3·9	4·6	5·2	5·9	6·6	13·1	19·7	26·2	32·8	49
BU	0·7	1·3	2·0	2·7	3·3	4·0	4·7	5·4	6·0	6·7	13·4	20·1	26·8	33·5	48
BV	0·7	1·4	2·0	2·7	3·4	4·1	4·8	5·5	6·1	6·8	13·6	20·5	27·3	34·1	47
BW	0·7	1·4	2·1	2·8	3·5	4·2	4·9	5·6	6·3	6·9	13·9	20·8	27·8	34·7	46
BX	0·7	1·4	2·1	2·8	3·5	4·2	4·9	5·7	6·4	7·1	14·1	21·2	28·3	35·4	45

Minutes of Hour Angle and Latitude.

RULES for applying corrections to Calculated Altitude :—

HOUR ANGLE CORR'N   Always **SUBTRACT**.

LATITUDE CORR'N   { In N. lat. **ADD** if Azimuth is N'ly ; **SUBTRACT** if S'ly.
{ In S. lat. **ADD** if Azimuth is S'ly ; **SUBTRACT** if N'ly.

ALTITUDE ADJUSTMENT TABLES

TABLE B.	Azimuths 45° to 0°.

| Key | | | | | | | | | | | | | | | Azimuth for latitude Correction. |

						Minutes of Hour Angle and Latitude.									Azimuth for latitude Correction. °
Key	1′	2′	3′	4′	5′	6′	7′	8′	9′	10′	20′	30′	40′	50′	
BX	0·7	1·4	2·1	2·8	3·5	4·2	4·9	5·7	6·4	7·1	14·1	21·2	28·3	35·4	45
BY	0·7	1·4	2·2	2·9	3·6	4·3	5·0	5·8	6·5	7·2	14·4	21·6	28·8	36·0	44
BZ	0·7	1·5	2·2	2·9	3·7	4·4	5·1	5·9	6·6	7·3	14·6	21·9	29·3	36·6	43
CA	0·7	1·5	2·2	3·0	3·7	4·5	5·2	5·9	6·7	7·4	14·9	22·3	29·7	37·2	42
CB	0·8	1·5	2·3	3·0	3·8	4·5	5·3	6·0	6·8	7·5	15·1	22·6	30·2	37·7	41
CC	0·8	1·5	2·3	3·1	3·8	4·6	5·4	6·1	6·9	7·7	15·3	23·0	30·6	38·3	40
CD	0·8	1·6	2·3	3·1	3·9	4·7	5·4	6·2	7·0	7·8	15·5	23·3	31·1	38·9	39
CE	0·8	1·6	2·4	3·2	3·9	4·7	5·5	6·3	7·1	7·9	15·8	23·6	31·5	39·4	38
CF	0·8	1·6	2·4	3·2	4·0	4·8	5·6	6·4	7·2	8·0	16·0	24·0	31·9	39·9	37
CG	0·8	1·6	2·4	3·2	4·0	4·9	5·7	6·5	7·3	8·1	16·2	24·3	32·4	40·5	36
CH	0·8	1·6	2·5	3·3	4·1	4·9	5·7	6·6	7·4	8·2	16 4	24·6	32·8	41·0	35
CK	0·8	1·7	2·5	3·3	4·1	5·0	5·8	6·6	7·5	8·3	16·6	24·9	33·2	41·5	34
CL	0·8	1·7	2·5	3·4	4·2	5·0	5·9	6·7	7·5	8·4	16·8	25·2	33·5	41·9	33
CM	0·8	1·7	2·5	3·4	4·2	5·1	5·9	6·8	7·6	8·5	17·0	25·4	33·9	42·4	32
CN	0·9	1·7	2·6	3·4	4·3	5·1	6·0	6·9	7·7	8·6	17·1	25·7	34·3	42·9	31
CO	0·9	1·7	2·6	3·5	4·3	5·2	6·1	6·9	7·8	8·7	17·3	26·0	34·6	43·3	30
CP	0·9	1·7	2·6	3·5	4·4	5·2	6·1	7·0	7·9	8·7	17·5	26·2	35·0	43·7	29
CQ	0·9	1·8	2·6	3·5	4·4	5·3	6·2	7·1	8·0	8·8	17·7	26·5	35·3	44·1	28
CR	0·9	1·8	2·7	3·6	4·5	5·3	6·2	7·1	8·0	8·9	17·8	26·7	35·6	44·6	27
CS	0·9	1·8	2·7	3·6	4·5	5·4	6·3	7·2	8·1	9·0	18·0	27·0	36·0	44·9	26
CT	0·9	1·8	2·7	3·6	4·5	5·4	6·3	7·3	8·2	9·1	18·1	27·2	36·3	45·3	25
CU	0·9	1·8	2·7	3·7	4·6	5·5	6·4	7·3	8·2	9·1	18·3	27·4	36·5	45·7	24
CV	0·9	1·8	2·8	3·7	4·6	5·5	6·4	7·4	8·3	9·2	18·4	27·6	36·8	46·0	23
CW	0·9	1·9	2·8	3·7	4·6	5·6	6·5	7·4	8·3	9·3	18·5	27·8	37·1	46·4	22
CX	0·9	1·9	2·8	3·7	4·7	5·6	6·5	7·5	8·4	9·3	18·7	28·0	37·3	46·7	21
CY	0·9	1·9	2·8	3·8	4·7	5·6	6·6	7·5	8·5	9·4	18·8	28·2	37·6	47·0	20
CZ	0·9	1·9	2·8	3·8	4·7	5·7	6·6	7·6	8·5	9·5	18·9	28·4	37·8	47·3	19
DA	1·0	1·9	2·9	3·8	4·8	5·7	6·7	7·6	8·6	9·5	19·0	28·5	38·0	47·6	18
DB	1·0	1·9	2·9	3·8	4·8	5·7	6·7	7·7	8·6	9·6	19·1	28·7	38·3	47·8	17
DC	1·0	1·9	2·9	3·8	4·8	5·8	6·7	7·7	8·7	9·6	19·2	28·8	38·5	48·1	16
DD	1·0	1·9	2·9	3·9	4·8	5·8	6·8	7·7	8·7	9·7	19·3	29·0	38·6	48·3	15
DE	1·0	1·9	2·9	3·9	4·9	5·8	6·8	7·8	8·7	9·7	19·4	29·1	38·8	48·5	14
DF	1·0	1·9	2·9	3·9	4·9	5·8	6·8	7·8	8·8	9·7	19·5	29·2	39·0	48·7	13
DG	1·0	2·0	2·9	3·9	4·9	5·9	6·8	7·8	8·8	9·8	19·6	29·3	39·1	48·9	12
DH	1·0	2·0	2·9	3·9	4·9	5·9	6·9	7·9	8·8	9·8	19·6	29·4	39·3	49·1	11
DK	1·0	2·0	3·0	3·9	4·9	5·9	6·9	7·9	8·9	9·8	19·7	29·5	39·4	49·2	10
DL	1·0	2·0	3·0	4·0	4·9	5·9	6·9	7·9	8·9	9·9	19·8	29·6	39·5	49·4	9
DM	1·0	2·0	3·0	4·0	5·0	5·9	6·9	7·9	8·9	9·9	19·8	29·7	39·6	49·5	8
DN	1·0	2·0	3·0	4·0	5·0	6·0	6·9	7·9	8·9	9·9	19·9	29·8	39·7	49·6	7
DO	1·0	2·0	3·0	4·0	5·0	6·0	7·0	8·0	9·0	9·9	19·9	29·8	39·8	49·7	6
DP	1·0	2·0	3·0	4·0	5·0	6·0	7·0	8·0	9·0	10·0	20·0	29·9	39·9	49·9	4
DQ	1·0	2·0	3·0	4·0	5·0	6·0	7·0	8·0	9·0	10·0	20·0	30·0	40·0	50·0	2
DR	1·0	2·0	3·0	4·0	5·0	6·0	7·0	8·0	9·0	10·0	20·0	30·0	40·0	50·0	0

RULES for applying corrections to Calculated Altitude :—

HOUR ANGLE CORR'N   Always **SUBTRACT**.

LATITUDE CORR'N   { In N. lat. **ADD** if Azimuth is N'ly ; **SUBTRACT** if S'ly.
{ In S. lat. **ADD** if Azimuth is S'ly ; **SUBTRACT** if N'ly.

0° TO 6°		NATURAL FUNCTIONS OF ANGLES						
	Sine	Cosec.	Tan.	Cotan.	Secant	Cosine	Radians
° 0·0	0·0000	Infinite	0·0000	Infinite	1·0000	1·0000	0·0000
·1	0·0017	572·96	0·0017	572·96	1·0000	1·0000	0·0017
·2	0·0035	286·48	0·0035	286·48	1·0000	1·0000	0·0035
·3	0·0052	190·99	0·0052	190·98	1·0000	1·0000	0·0052
·4	0·0070	143·24	0·0070	143·24	1·0000	1·0000	0·0070
0·5	0·0087	114·59	0·0087	114·59	1·0000	1·0000	0·0087
·6	0·0105	95·495	0·0105	95·490	1·0001	0·9999	0·0105
·7	0·0122	81·853	0·0122	81·847	1·0001	0·9999	0·0122
·8	0·0140	71·622	0·0140	71·615	1·0001	0·9999	0·0140
·9	0·0157	63·665	0·0157	63·657	1·0001	0·9999	0·0157
1·0	0·0175	57·299	0·0175	57·290	1·0002	0·9998	0·0175
·1	0·0192	52·090	0·0192	52·081	1·0002	0·9998	0·0192
·2	0·0209	47·750	0·0209	47·740	1·0002	0·9998	0·0209
·3	0·0227	44·078	0·0227	44·066	1·0003	0·9997	0·0227
·4	0·0244	40·930	0·0244	40·917	1·0003	0·9997	0·0244
1·5	0·0262	38·201	0·0262	38·188	1·0003	0·9997	0·0262
·6	0·0279	35·814	0·0279	35·801	1·0004	0·9996	0·0279
·7	0·0297	33·708	0·0297	33·694	1·0004	0·9996	0·0297
·8	0·0314	31·836	0·0314	31·820	1·0005	0·9995	0·0314
·9	0·0332	30·161	0·0332	30·145	1·0006	0·9995	0·0332
2·0	0·0349	28·654	0·0349	28·636	1·0006	0·9994	0·0349
·1	0·0366	27·290	0·0367	27·271	1·0007	0·9993	0·0367
·2	0·0384	26·050	0·0384	26·031	1·0007	0·9993	0·0384
·3	0·0401	24·918	0·0402	24·898	1·0008	0·9992	0·0401
·4	0·0419	23·880	0·0419	23·859	1·0009	0·9991	0·0419
2·5	0·0436	22·926	0·0437	22·904	1·0010	0·9990	0·0436
·6	0·0454	22·044	0·0454	22·022	1·0010	0·9990	0·0454
·7	0·0471	21·229	0·0472	21·205	1·0011	0·9989	0·0471
·8	0·0489	20·471	0·0489	20·446	1·0012	0·9988	0·0489
·9	0·0506	19·766	0·0507	19·740	1·0013	0·9987	0·0506
3·0	0·0523	19·107	0·0524	19·081	1·0014	0·9986	0·0524
·1	0·0541	18·491	0·0542	18·464	1·0015	0·9985	0·0541
·2	0·0558	17·914	0·0559	17·886	1·0016	0·9984	0·0559
·3	0·0576	17·372	0·0577	17·343	1·0017	0·9983	0·0576
·4	0·0593	16·862	0·0594	16·832	1·0018	0·9982	0·0593
3·5	0·0610	16·380	0·0612	16·350	1·0019	0·9981	0·0611
·6	0·0628	15·926	0·0629	15·894	1·0020	0·9980	0·0628
·7	0·0645	15·496	0·0647	15·464	1·0021	0·9979	0·0646
·8	0·0663	15·089	0·0664	15·056	1·0022	0·9978	0·0663
·9	0·0680	14·703	0·0682	14·669	1·0023	0·9977	0·0681
4·0	0·0698	14·336	0·0699	14·301	1·0024	0·9976	0·0698
·1	0·0715	13·987	0·0717	13·951	1·0026	0·9974	0·0716
·2	0·0732	13·654	0·0734	13·617	1·0027	0·9973	0·0733
·3	0·0750	13·337	0·0752	13·300	1·0028	0·9972	0·0750
·4	0·0767	13·035	0·0769	12·996	1·0030	0·9971	0·0768
4·5	0·0785	12·746	0·0787	12 706	1·0031	0·9969	0·0785
·6	0·0802	12·469	0·0805	12·429	1·0032	0·9968	0·0803
·7	0·0819	12·204	0·0822	12·163	1·0034	0·9967	0·0820
·8	0·0837	11·951	0·0840	11·909	1·0035	0·9965	0·0838
·9	0·0854	11·707	0·0857	11·664	1·0037	0·9963	0·0855
5·0	0·0872	11·474	0·0875	11·430	1·0038	0·9962	0·0873
·1	0·0889	11·249	0·0892	11·205	1·0040	0·9960	0·0890
·2	0·0906	11·034	0·0910	10·988	1·0041	0·9959	0·0908
·3	0·0924	10·826	0·0928	10·780	1·0043	0·9957	0·0925
·4	0·0941	10·626	0·0945	10·579	1·0045	0·9956	0·0942
5·5	0·0958	10·433	0·0963	10·385	1·0046	0·9954	0·0960
·6	0·0976	10·248	0·0981	10·199	1·0048	0·9952	0·0977
·7	0·0993	10·068	0·0998	10·019	1·0050	0·9951	0·0995
·8	0·1011	9·8955	0·1016	9·8448	1·0051	0·9949	0·1012
·9	0·1028	9·7283	0·1033	9·6768	1·0053	0·9947	0·1030
6·0	0·1045	9·5668	0·1051	9·5144	1·0055	0·9945	0·1047

6° TO 12°	NATURAL FUNCTIONS OF ANGLES						
	Sine	Cosec.	Tan.	Cotan.	Secant	Cosine	Radians
6·0	0·1045	9·5668	0·1051	9·5144	1·0055	0·9945	0·1047
·1	0·1063	9·4105	0·1069	9·3572	1·0057	0·9944	0·1065
·2	0·1080	9·2593	0·1086	9·2052	1·0059	0·9942	0·1082
·3	0·1097	9·1129	0·1104	9·0579	1·0061	0·9940	0·1100
·4	0·1115	8·9711	0·1122	8·9152	1·0063	0·9938	0·1117
6·5	0·1132	8·8337	0·1139	8·7769	1·0065	0·9936	0·1134
·6	0·1149	8·7004	0·1157	8·6428	1·0067	0·9934	0·1152
·7	0·1167	8·5711	0·1175	8·5126	1·0069	0·9932	0·1169
·8	0·1184	8·4457	0·1192	8·3863	1·0071	0·9930	0·1187
·9	0·1201	8·3238	0·1210	8·2636	1·0073	0·9928	0·1204
7·0	0·1219	8·2055	0·1228	8·1444	1·0075	0·9925	0·1222
·1	0·1236	8·0905	0·1246	8·0285	1·0077	0·9923	0·1239
·2	0·1253	7·9787	0·1263	7·9158	1·0079	0·9921	0·1257
·3	0·1271	7·8700	0·1281	7·8062	1·0082	0·9919	0·1274
·4	0·1288	7·7642	0·1299	7·6996	1·0084	0·9917	0·1292
7·5	0·1305	7·6613	0·1317	7·5958	1·0086	0·9914	0·1309
·6	0·1323	7·5611	0·1334	7·4947	1·0089	0·9912	0·1326
·7	0·1340	7·4635	0·1352	7·3962	1·0091	0·9910	0·1344
·8	0·1357	7·3684	0·1370	7·3002	1·0093	0·9907	0·1361
·9	0·1374	7·2757	0·1388	7·2066	1·0096	0·9905	0·1379
8·0	0·1392	7·1853	0·1405	7·1154	1·0098	0·9903	0·1396
·1	0·1409	7·0972	0·1423	7·0264	1·0101	0·9900	0·1414
·2	0·1426	7·0112	0·1441	6·9395	1·0103	0·9898	0·1431
·3	0·1444	6·9273	0·1459	6·8548	1·0106	0·9895	0·1449
·4	0·1461	6·8454	0·1477	6·7720	1·0108	0·9893	0·1466
8·5	0·1478	6·7655	0·1495	6·6912	1·0111	0·9890	0·1484
·6	0·1495	6·6874	0·1512	6·6122	1·0114	0·9888	0·1501
·7	0·1513	6·6111	0·1530	6·5350	1·0116	0·9885	0·1518
·8	0·1530	6·5366	0·1548	6·4596	1·0119	0·9882	0·1536
·9	0·1547	6·4637	0·1566	6·3859	1·0122	0·9880	0·1553
9·0	0·1564	6·3925	0·1584	6·3138	1·0125	0·9877	0·1571
·1	0·1582	6·3228	0·1602	6·2432	1·0128	0·9874	0·1588
·2	0·1599	6·2546	0·1620	6·1742	1·0130	0·9871	0·1606
·3	0·1616	6·1880	0·1638	6·1066	1·0133	0·9869	0·1623
·4	0·1633	6·1227	0·1656	6·0405	1·0136	0·9866	0·1641
9·5	0·1650	6·0589	0·1673	5·9758	1·0139	0·9863	0·1658
·6	0·1668	5·9963	0·1691	5·9124	1·0142	0·9860	0·1676
·7	0·1685	5·9351	0·1709	5·8502	1·0145	0·9857	0·1693
·8	0·1702	5·8751	0·1727	5·7894	1·0148	0·9854	0·1710
·9	0·1719	5·8164	0·1745	5·7297	1·0151	0·9851	0·1728
10·0	0·1736	5·7588	0·1763	5·6713	1·0154	0·9848	0·1745
·1	0·1754	5·7023	0·1781	5·6140	1·0157	0·9845	0·1763
·2	0·1771	5·6470	0·1799	5·5578	1·0161	0·9842	0·1780
·3	0·1788	5·5928	0·1817	5·5026	1·0164	0·9839	0·1798
·4	0·1805	5·5396	0·1835	5·4486	1·0167	0·9836	0·1815
10·5	0·1822	5·4874	0·1853	5·3955	1·0170	0·9833	0·1833
·6	0·1840	5·4362	0·1871	5·3435	1·0174	0·9829	0·1850
·7	0·1857	5·3860	0·1889	5·2924	1·0177	0·9826	0·1868
·8	0·1874	5·3367	0·1908	5·2422	1·0180	0·9823	0·1885
·9	0·1891	5·2883	0·1926	5·1929	1·0184	0·9820	0·1902
11·0	0·1908	5·2408	0·1944	5·1446	1·0187	0·9816	0·1920
·1	0·1925	5·1942	0·1962	5·0970	1·0191	0·9813	0·1937
·2	0·1942	5·1484	0·1980	5·0504	1·0194	0·9810	0·1955
·3	0·1959	5·1034	0·1998	5·0045	1·0198	0·9806	0·1972
·4	0·1977	5·0593	0·2016	4·9595	1·0201	0·9803	0·1990
11·5	0·1994	5·0159	0·2034	4·9152	1·0205	0·9799	0·2007
·6	0·2011	4·9732	0·2053	4·8716	1·0209	0·9796	0·2025
·7	0·2028	4·9313	0·2071	4·8288	1·0212	0·9792	0·2042
·8	0·2045	4·8901	0·2089	4·7867	1·0216	0·9789	0·2069
·9	0·2062	4·8496	0·2107	4·7453	1·0220	0·9785	0·2077
12·0	0·2079	4·8097	0·2126	4·7046	1·0223	0·9781	0·2094

12° TO 18°	Sine	Cosec.	Tan.	Cotan.	Secant	Cosine	Radians
°							
12·0	0·2079	4·8097	0·2126	4·7046	1·0223	0·9781	0·2094
·1	0·2096	4·7706	0·2144	4·6646	1·0227	0·9778	0·2112
·2	0·2113	4·7320	0·2162	4·6252	1·0231	0·9774	0·2130
·3	0·2130	4·6942	0·2180	4·5864	1·0235	0·9770	0·2147
·4	0·2147	4·6569	0·2199	4·5483	1·0239	0·9767	0·2164
12·5	0·2164	4·6202	0·2217	4·5107	1·0243	0·9763	0·2182
·6	0·2181	4·5841	0·2235	4·4737	1·0247	0·9759	0·2199
·7	0·2198	4·5486	0·2254	4·4373	1·0251	0·9755	0·2217
·8	0·2215	4·5137	0·2272	4·4015	1·0255	0·9751	0·2234
·9	0·2233	4·4793	0·2290	4·3662	1·0259	0·9748	0·2251
13·0	0·2250	4·4454	0·2309	4·3315	1·0263	0·9744	0·2269
·1	0·2267	4·4121	0·2327	4·2972	1·0267	0·9740	0·2286
·2	0·2284	4·3792	0·2346	4·2635	1·0271	0·9736	0·2304
·3	0·2301	4·3469	0·2364	4·2303	1·0276	0·9732	0·2321
·4	0·2317	4·3150	0·2382	4·1976	1·0280	0·9728	0·2339
13·5	0·2334	4·2837	0·2401	4·1653	1·0284	0·9724	0·2356
·6	0·2351	4·2527	0·2419	4·1335	1·0289	0·9720	0·2374
·7	0·2368	4·2223	0·2438	4·1022	1·0293	0·9715	0·2391
·8	0·2385	4·1923	0·2456	4·0713	1·0297	0·9711	0·2409
·9	0·2402	4·1627	0·2475	4·0408	1·0302	0·9707	0·2426
14·0	0·2419	4·1336	0·2493	4·0108	1·0306	0·9703	0·2443
·1	0·2436	4·1048	0·2512	3·9812	1·0311	0·9699	0·2461
·2	0·2453	4·0765	0·2530	3·9520	1·0315	0·9694	0·2478
·3	0·2470	4·0486	0·2549	3·9232	1·0320	0·9690	0·2496
·4	0·2487	4·0211	0·2568	3·8947	1·0324	0·9686	0·2513
14·5	0·2504	3·9939	0·2586	3·8667	1·0329	0·9681	0·2531
·6	0·2521	3·9672	0·2605	3·8391	1·0334	0·9677	0·2548
·7	0·2538	3·9408	0·2624	3·8118	1·0338	0·9673	0·2566
·8	0·2554	3·9147	0·2642	3·7848	1·0343	0·9668	0·2583
·9	0·2571	3·8890	0·2661	3·7583	1·0348	0·9664	0·2601
15·0	0·2588	3·8637	0·2680	3·7321	1·0353	0·9659	0·2618
·1	0·2605	3·8387	0·2698	3·7062	1·0358	0·9655	0·2635
·2	0·2622	3·8140	0·2717	3·6806	1·0363	0·9650	0·2653
·3	0·2639	3·7897	0·2736	3·6554	1·0367	0·9646	0·2670
·4	0·2656	3·7657	0·2755	3·6305	1·0372	0·9641	0·2688
15·5	0·2672	3·7420	0·2773	3·6059	1·0377	0·9636	0·2705
·6	0·2689	3·7186	0·2792	3·5816	1·0382	0·9632	0·2723
·7	0·2706	3·6955	0·2811	3·5576	1·0387	0·9627	0·2740
·8	0·2723	3·6727	0·2830	3·5339	1·0393	0·9622	0·2758
·9	0·2740	3·6502	0·2849	3·5105	1·0398	0·9617	0·2775
16·0	0·2756	3·6280	0·2868	3·4874	1·0403	0·9613	0·2792
·1	0·2773	3·6060	0·2886	3·4646	1·0408	0·9608	0·2810
·2	0·2790	3·5843	0·2905	3·4420	1·0414	0·9603	0·2827
·3	0·2807	3·5629	0·2924	3·4197	1·0419	0·9598	0·2845
·4	0·2823	3·5418	0·2943	3·3977	1·0424	0·9593	0·2862
16·5	0·2840	3·5209	0·2962	3·3759	1·0430	0·9588	0·2880
·6	0·2857	3·5003	0·2981	3·3544	1·0435	0·9583	0·2897
·7	0·2874	3·4799	0·3000	3·3332	1·0440	0·9578	0·2915
·8	0·2890	3·4598	0·3019	3·3122	1·0446	0·9573	0·2932
·9	0·2907	3·4399	0·3038	3·2914	1·0451	0·9568	0·2950
17·0	0·2924	3·4203	0·3057	3·2709	1·0457	0·9563	0·2967
·1	0·2940	3·4009	0·3076	3·2506	1·0462	0·9558	0·2985
·2	0·2957	3·3817	0·3095	3·2305	1·0468	0·9553	0·3002
·3	0·2974	3·3628	0·3115	3·2106	1·0474	0·9548	0·3019
·4	0·2990	3·3440	0·3134	3·1910	1·0480	0·9542	0·3037
17·5	0·3007	3·3255	0·3153	3·1716	1·0485	0·9537	0·3054
·6	0·3024	3·3072	0·3172	3·1524	1·0491	0·9532	0·3072
·7	0·3040	3·2891	0·3191	3·1334	1·0497	0·9527	0·3089
·8	0·3057	3·2712	0·3211	3·1146	1·0503	0·9521	0·3107
·9	0·3074	3·2535	0·3230	3·0961	1·0509	0·9516	0·3124
18·0	0·3090	3·2361	0·3249	3·0777	1·0515	0·9511	0·3142

18° TO 24°	**NATURAL FUNCTIONS OF ANGLES**						
°	Sine	Cosec.	Tan.	Cotan.	Secant	Cosine	Radians
18·0	0·3090	3·2361	0·3249	3·0777	1·0515	0·9511	0·3142
·1	0·3107	3·2188	0·3268	3·0595	1·0521	0·9505	0·3159
·2	0·3123	3·2017	0·3288	3·0415	1·0527	0·9500	0·3176
·3	0·3140	3·1848	0·3307	3·0237	1·0533	0·9494	0·3194
·4	0·3156	3·1681	0·3327	3·0061	1·0539	0·9489	0·3211
18·5	0·3173	3·1516	0·3346	2·9887	1·0545	0·9483	0·3229
·6	0·3190	3·1352	0·3365	2·9714	1·0551	0·9478	0·3246
·7	0·3206	3·1190	0·3385	2·9544	1·0557	0·9472	0·3264
·8	0·3223	3·1030	0·3404	2·9375	1·0564	0·9466	0·3281
·9	0·3239	3·0872	0·3424	2·9208	1·0570	0·9461	0·3299
19·0	0·3256	3·0716	0·3443	2·9042	1·0576	0·9455	0·3316
·1	0·3272	3·0561	0·3463	2·8878	1·0583	0·9449	0·3334
·2	0·3289	3·0407	0·3482	2·8716	1·0589	0·9444	0·3351
·3	0·3305	3·0256	0·3502	2·8555	1·0595	0·9438	0·3368
·4	0·3322	3·0106	0·3522	2·8397	1·0602	0·9432	0·3386
19·5	0·3338	2·9957	0·3541	2·8239	1·0608	0·9426	0·3403
·6	0·3355	2·9811	0·3561	2·8083	1·0615	0·9421	0·3421
·7	0·3371	2·9665	0·3581	2·7929	1·0622	0·9415	0·3438
·8	0·3387	2·9521	0·3600	2·7776	1·0628	0·9409	0·3456
·9	0·3404	2·9379	0·3620	2·7625	1·0635	0·9403	0·3473
20·0	0·3420	2·9238	0·3640	2·7475	1·0642	0·9397	0·3491
·1	0·3437	2·9099	0·3659	2·7326	1·0649	0·9391	0·3508
·2	0·3453	2·8960	0·3679	2·7179	1·0655	0·9385	0·3526
·3	0·3469	2·8824	0·3699	2·7033	1·0662	0·9379	0·3543
·4	0·3486	2·8688	0·3719	2·6889	1·0669	0·9373	0·3560
20·5	0·3502	2·8555	0·3739	2·6746	1·0676	0·9367	0·3578
·6	0·3518	2·8422	0·3759	2·6605	1·0683	0·9361	0·3595
·7	0·3535	2·8291	0·3779	2·6464	1·0690	0·9354	0·3613
·8	0·3551	2·8161	0·3799	2·6325	1·0697	0·9348	0·3630
·9	0·3567	2·8032	0·3819	2·6187	1·0704	0·9342	0·3648
21·0	0·3584	2·7904	0·3839	2·6051	1·0711	0·9336	0·3665
·1	0·3600	2·7778	0·3859	2·5916	1·0719	0·9330	0·3683
·2	0·3616	2·7653	0·3879	2·5782	1·0726	0·9323	0·3700
·3	0·3633	2·7529	0·3899	2·5649	1·0733	0·9317	0·3718
·4	0·3649	2·7407	0·3919	2·5517	1·0740	0·9311	0·3735
21·5	0·3665	2·7285	0·3939	2·5387	1·0748	0·9304	0·3752
·6	0·3681	2·7165	0·3959	2·5257	1·0755	0·9298	0·3770
·7	0·3697	2·7046	0·3979	2·5129	1·0763	0·9291	0 3787
8	0 3714	2 6928	0·4000	2·5002	1·0770	0·9285	0·3805
·9	0·3730	2·6810	0·4020	2·4876	1·0778	0·9278	0 3822
22·0	0·3746	2·6695	0·4040	2·4751	1·0785	0·9272	0·3840
·1	0·3762	2·6580	0·4061	2·4627	1·0793	0·9265	0·3857
·2	0·3778	2·6466	0·4081	2·4504	1·0801	0·9259	0·3874
·3	0·3795	2·6353	0·4101	2·4383	1·0808	0·9252	0·3892
·4	0·3811	2·6242	0·4122	2·4232	1·0816	0·9245	0·3909
22·5	0·3827	2·6131	0·4142	2·4142	1·0824	0·9239	0·3927
·6	0·3843	2·6022	0·4163	2·4023	1·0832	0·9232	0·3944
·7	0·3859	2·5913	0·4183	2·3906	1·0840	0·9225	0·3962
·8	0·3875	2·5805	0 4204	2·3789	1·0848	0·9219	0·3980
·9	0·3891	2·5699	0·4224	2·3673	1·0856	0·9212	0·3997
23·0	0·3907	2·5593	0·4245	2·3559	1·0864	0·9205	0·4014
·1	0·3923	2·5488	0·4265	2·3445	1·0872	0·9198	0·4032
·2	0·3939	2·5384	0·4286	2·3332	1·0880	0·9191	0·4049
·3	0·3955	2·5282	0·4307	2·3220	1·0888	0·9184	0·4067
·4	0·3971	2·5180	0·4327	2·3109	1·0896	0·9178	0·4084
23·5	0·3987	2·5078	0·4348	2·2999	1·0904	0·9171	0·4102
·6	0·4003	2·4978	0·4369	2·2889	1·0913	0·9164	0·4119
·7	0·4019	2·4879	0·4390	2·2781	1·0921	0·9157	0·4136
·8	0·4035	2·4780	0·4411	2·2673	1·0929	0 9150	0·4154
·9	0·4051	2·4683	0·4431	2·2566	1·0938	0·9143	0·4171
24·0	0·4067	2·4586	0·4452	2·2460	1·0946	0·9135	0·4189

NATURAL FUNCTIONS OF ANGLES

°	Sine	Cosec.	Tan.	Cotan.	Secant	Cosine	Radians
24·0	0·4067	2·4586	0·4452	2·2460	1·0946	0·9135	0·4189
·1	0·4083	2·4490	0·4473	2·2355	1·0955	0·9128	0·4206
·2	0·4099	2·4395	0·4494	2·2251	1·0963	0·9121	0·4224
·3	0·4115	2·4301	0·4515	2·2148	1·0972	0·9114	0·4241
·4	0·4131	2·4207	0·4536	2·2045	1·0981	0·9107	0·4259
24·5	0·4147	2·4114	0·4557	2·1943	1·0989	0·9100	0·4276
·6	0·4163	2·4022	0·4578	2·1842	1·0998	0·9092	0·4293
·7	0·4179	2·3931	0·4599	2·1742	1·1007	0·9085	0·4311
·8	0·4195	2·3841	0·4621	2·1642	1·1016	0·9078	0·4328
·9	0·4210	2·3751	0·4642	2·1543	1·1025	0·9070	0·4346
25·0	0·4226	2·3662	0·4663	2·1445	1·1034	0·9063	0·4363
·1	0·4242	2·3574	0·4684	2·1348	1·1043	0·9056	0·4381
·2	0·4258	2·3486	0·4706	2·1251	1·1052	0·9048	0·4398
·3	0·4274	2·3400	0·4727	2·1155	1·1061	0·9041	0·4416
·4	0·4289	2·3314	0·4748	2·1060	1·1070	0·9033	0·4433
25·5	0·4305	2·3228	0·4770	2·0965	1·1079	0·9026	0·4451
·6	0·4321	2·3144	0·4791	2·0872	1·1089	0·9018	0·4468
·7	0·4337	2·3060	0·4813	2·0778	1·1098	0·9011	0·4485
·8	0·4352	2·2976	0·4834	2·0686	1·1107	0·9003	0·4503
·9	0·4368	2·2894	0·4856	2·0594	1·1117	0·8996	0·4520
26·0	0·4384	2·2812	0·4877	2·0503	1·1126	0·8988	0·4538
·1	0·4399	2·2730	0·4899	2·0413	1·1136	0·8980	0·4555
·2	0·4415	2·2650	0·4921	2·0323	1·1145	0·8973	0·4573
·3	0·4431	2·2570	0·4942	2·0233	1·1155	0·8965	0·4590
·4	0·4446	2·2490	0·4964	2·0145	1·1164	0·8957	0·4608
26·5	0·4462	2·2412	0·4986	2·0057	1·1174	0·8949	0·4625
·6	0·4478	2·2333	0·5008	1·9970	1·1184	0·8942	0·4643
·7	0·4493	2·2256	0·5029	1·9883	1·1194	0·8934	0·4660
·8	0·4509	2·2179	0·5051	1·9797	1·1203	0·8926	0·4677
·9	0·4524	2·2103	0·5073	1·9711	1·1213	0·8918	0·4695
27·0	0·4540	2·2027	0·5095	1·9626	1·1223	0·8910	0·4712
·1	0·4555	2·1952	0·5117	1·9542	1·1233	0·8902	0·4730
·2	0·4571	2·1877	0·5139	1·9458	1·1243	0·8894	0·4747
·3	0·4587	2·1803	0·5161	1·9375	1·1253	0·8886	0·4765
·4	0·4602	2·1730	0·5184	1·9292	1·1264	0·8878	0·4782
27·5	0·4617	2·1657	0·5206	1·9210	1·1274	0·8870	0·4800
·6	0·4633	2·1584	0·5228	1·9128	1·1284	0·8862	0·4817
·7	0·4648	2·1513	0·5250	1·9047	1·1294	0·8854	0·4835
·8	0·4664	2·1441	0·5272	1·8967	1·1305	0·8846	0·4852
·9	0·4679	2·1371	0·5295	1·8887	1·1315	0·8838	0·4869
28·0	0·4695	2·1301	0·5317	1·8807	1·1326	0·8829	0·4887
·1	0·4710	2·1231	0·5340	1·8728	1·1336	0·8821	0·4904
·2	0·4726	2·1162	0·5362	1·8650	1·1347	0·8813	0·4922
·3	0·4741	2·1093	0·5384	1·8572	1·1357	0·8805	0·4939
·4	0·4756	2·1025	0·5407	1·8495	1·1368	0·8796	0·4957
28·5	0·4772	2·0957	0·5430	1·8418	1·1379	0·8788	0·4974
·6	0·4787	2·0890	0·5452	1·8341	1·1390	0·8780	0·4992
·7	0·4802	2·0824	0·5475	1·8265	1·1401	0·8771	0·5009
·8	0·4818	2·0757	0·5498	1·8190	1·1412	0·8763	0·5026
·9	0·4833	2·0692	0·5520	1·8115	1·1423	0·8755	0·5044
29·0	0·4848	2·0627	0·5543	1·8040	1·1434	0·8746	0·5061
·1	0·4863	2·0562	0·5566	1·7966	1·1445	0·8738	0·5079
·2	0·4879	2·0498	0·5589	1·7893	1·1456	0·8729	0·5096
·3	0·4894	2·0434	0·5612	1·7820	1·1467	0·8721	0·5114
·4	0·4909	2·0371	0·5635	1·7747	1·1478	0·8712	0·5131
29·5	0·4924	2·0308	0·5658	1·7675	1·1490	0·8704	0·5149
·6	0·4939	2·0245	0·5681	1·7603	1·1501	0·8695	0·5166
·7	0·4955	2·0183	0·5704	1·7532	1·1512	0·8686	0·5184
·8	0·4970	2·0122	0·5727	1·7461	1·1524	0·8678	0·5201
·9	0·4985	2·0061	0·5750	1·7391	1·1535	0·8669	0·5219
30·0	0·5000	2·0000	0·5774	1·7321	1·1547	0·8660	0·5236

30° TO 36°	Sine	Cosec.	Tan.	Cotan.	Secant	Cosine	Radians
30·0	**0·5000**	2·0000	**0·5774**	1·7321	**1·1547**	0·8660	0·5236
·1	0·5015	1·9940	0·5797	1·7251	1·1559	0·8652	0·5253
·2	0·5030	1·9880	0·5820	1·7182	1·1570	0·8643	0·5271
·3	0·5045	1·9821	0·5844	1·7113	1·1582	0·8634	0·5288
·4	0·5060	1·9762	0·5867	1·7045	1·1594	0·8625	0·5306
30·5	**0·5075**	1·9703	**0·5890**	1·6977	**1·1606**	0·8616	0·5323
·6	0·5090	1·9645	0·5914	1·6909	1·1618	0·8607	0·5341
·7	0·5105	1·9587	0·5938	1·6842	1·1630	0·8599	0·5358
·8	0·5120	1·9530	0·5961	1·6775	1·1642	0·8590	0·5376
·9	0·5135	1·9473	0·5985	1·6709	1·1654	0·8581	0·5393
31·0	**0·5150**	1·9416	**0·6009**	1·6643	**1·1666**	0·8572	0·5411
·1	0·5165	1·9360	0·6032	1·6577	1·1679	0·8563	0·5428
·2	0·5180	1·9304	0·6056	1·6512	1·1691	0·8554	0·5445
·3	0·5195	1·9249	0·6080	1·6447	1·1703	0·8545	0·5463
·4	0·5210	1·9193	0·6104	1·6383	1·1716	0·8536	0·5480
31·5	**0·5225**	1·9139	**0·6128**	1·6319	**1·1728**	0·8526	0·5498
·6	0·5240	1·9084	0·6152	1·6255	1·1741	0·8517	0·5515
·7	0·5255	1·9031	0·6176	1·6191	1·1753	0·8508	0·5533
·8	0·5270	1·8977	0·6200	1·6128	1·1766	0·8499	0·5550
·9	0·5284	1·8924	0·6224	1·6066	1·1779	0·8490	0·5568
32·0	**0·5299**	1·8871	**0·6249**	1·6003	**1·1792**	0·8480	0·5585
·1	0·5314	1·8818	0·6273	1·5941	1·1805	0·8471	0·5602
·2	0·5329	1·8766	0·6297	1·5880	1·1818	0·8462	0·5620
·3	0·5344	1·8714	0·6322	1·5818	1·1831	0·8453	0·5637
·4	0·5358	1·8663	0·6346	1·5757	1·1844	0·8443	0·5655
32·5	**0·5373**	1·8612	**0·6371**	1·5697	**1·1857**	0·8434	0·5672
·6	0·5388	1·8561	0·6395	1·5637	1·1870	0·8425	0·5690
·7	0·5402	1·8510	0·6420	1·5577	1·1883	0·8415	0·5707
·8	0·5417	1·8460	0·6445	1·5517	1·1897	0·8406	0·5725
·9	0·5432	1·8410	0·6469	1·5458	1·1910	0·8396	0·5742
33·0	**0·5446**	1·8361	**0·6494**	1·5399	**1·1924**	0·8387	0·5760
·1	0·5461	1·8312	0·6519	1·5340	1·1937	0·8377	0·5777
·2	0·5476	1·8263	0·6544	1·5282	1·1951	0·8368	0·5794
·3	0·5490	1·8214	0·6569	1·5224	1·1964	0·8358	0·5812
·4	0·5505	1·8166	0·6594	1·5166	1·1978	0·8348	0·5829
33·5	**0·5519**	1·8118	**0·6619**	1·5108	**1·1992**	0·8339	0·5847
·6	0·5534	1·8070	0·6644	1·5051	1·2006	0·8329	0·5864
·7	0·5548	1·8023	0·6669	1·4994	1·2020	0·8320	0·5882
·8	0·5563	1·7976	0·6694	1·4938	1·2034	0·8310	0·5899
·9	0·5577	1·7929	0·6720	1·4882	1·2048	0·8300	0·5917
34·0	**0·5592**	1·7883	**0·6745**	1·4826	**1·2062**	0·8290	0·5934
·1	0·5606	1·7837	0·6771	1·4770	1·2076	0·8281	0·5952
·2	0·5621	1·7791	0·6796	1·4715	1·2091	0·8271	0·5969
·3	0·5635	1·7745	0·6822	1·4659	1·2105	0·8261	0·5987
·4	0·5650	1·7700	0·6847	1·4605	1·2119	0·8251	0·6004
34·5	**0·5664**	1·7655	**0·6873**	1·4550	**1·2134**	0·8241	0·6021
·6	0·5678	1·7610	0·6899	1·4496	1·2149	0·8231	0·6039
·7	0·5693	1·7566	0·6924	1·4442	1·2163	0·8221	0·6056
·8	0·5707	1·7522	0·6950	1·4388	1·2178	0·8211	0·6074
·9	0·5721	1·7478	0·6976	1·4335	1·2193	0·8202	0·6091
35·0	**0·5736**	1·7434	**0·7002**	1·4281	**1·2208**	0·8192	0·6109
·1	0·5750	1·7391	0·7028	1·4229	1·2223	0·8182	0·6126
·2	0·5764	1·7348	0·7054	1·4176	1·2238	0·8171	0·6144
·3	0·5779	1·7305	0·7080	1·4124	1·2253	0·8161	0·6161
·4	0·5793	1·7263	0·7107	1·4071	1·2268	0·8151	0·6178
35·5	**0·5807**	1·7221	**0·7133**	1·4019	**1·2283**	0·8141	0·6196
·6	0·5821	1·7179	0·7159	1·3968	1·2299	0·8131	0·6213
·7	0·5835	1·7137	0·7186	1·3916	1·2314	0·8121	0·6231
·8	0·5850	1·7095	0·7212	1·3865	1·2329	0·8111	0·6248
·9	0·5864	1·7054	0·7239	1·3814	1·2345	0·8100	0·6266
36·0	**0·5878**	1·7013	**0·7265**	1·3764	**1·2361**	0·8090	0·6283

NATURAL FUNCTIONS OF ANGLES

36° TO 42°	NATURAL FUNCTIONS OF ANGLES						
°	Sine	Cosec.	Tan.	Cotan.	Secant	Cosine	Radians
36·0	0·5878	1·7013	0·7265	1·3764	1·2361	0·8090	0·6283
·1	0·5892	1·6972	0·7292	1·3713	1·2376	0·8080	0·6301
·2	0·5906	1·6932	0·7319	1·3663	1·2392	0·0070	0·6318
·3	0·5920	1·6892	0·7346	1·3613	1·2408	0·8059	0·6336
·4	0·5934	1·6851	0·7373	1·3564	1·2424	0·8049	0·6353
36·5	0·5948	1·6812	0·7400	1·3514	1·2440	0·8039	0·6370
·6	0·5962	1·6772	0·7427	1·3465	1·2456	0·8028	0·6388
·7	0·5976	1·6733	0·7454	1·3416	1·2472	0·8018	0·6405
·8	0·5990	1·6694	0·7481	1·3367	1·2489	0·8007	0·6423
·9	0·6004	1·6655	0·7508	1·3319	1·2505	0·7997	0·6440
37·0	0·6018	1·6616	0·7536	1·3270	1·2521	0·7986	0·6458
·1	0·6032	1·6578	0·7563	1·3222	1·2538	0·7976	0·6475
·2	0·6046	1·6540	0·7590	1·3175	1·2554	0·7965	0·6493
·3	0·6060	1·6502	0·7618	1·3127	1·2571	0·7955	0·6510
·4	0·6074	1·6464	0·7646	1·3079	1·2588	0·7944	0·6528
37·5	0·6088	1·6427	0·7673	1·3032	1·2605	0·7934	0·6545
·6	0·6101	1·6390	0·7701	1·2985	1·2622	0·7923	0·6562
·7	0·6115	1·6353	0·7729	1·2938	1·2639	0·7912	0·6580
·8	0·6129	1·6316	0·7757	1·2892	1·2656	0·7902	0·6597
·9	0·6143	1·6279	0·7785	1·2846	1·2673	0·7891	0·6615
38·0	0·6157	1·6243	0·7813	1·2799	1·2690	0·7880	0·6632
·1	0·6170	1·6207	0·7841	1·2753	1·2708	0·7869	0·6650
·2	0·6184	1·6171	0·7869	1·2708	1·2725	0·7859	0·6667
·3	0·6198	1·6135	0·7898	1·2662	1·2742	0·7848	0·6685
·4	0·6211	1·6099	0·7926	1·2617	1·2760	0·7837	0·6702
38·5	0·6225	1·6064	0·7954	1·2572	1·2778	0·7826	0·6719
·6	0·6239	1·6029	0·7983	1·2527	1·2796	0·7815	0·6737
·7	0·6252	1·5994	0·8012	1·2482	1·2813	0·7804	0·6754
·8	0·6266	1·5959	0·8040	1·2437	1·2831	0·7793	0·6772
·9	0·6280	1·5925	0·8069	1·2393	1·2849	0·7782	0·6789
39·0	0·6293	1·5890	0·8098	1·2349	1·2868	0·7771	0·6807
·1	0·6307	1·5856	0·8127	1·2305	1·2886	0·7760	0·6824
·2	0·6320	1·5822	0·8156	1·2261	1·2904	0·7749	0·6842
·3	0·6334	1·5788	0·8185	1·2218	1·2923	0·7738	0·6859
·4	0·6347	1·5755	0·8214	1·2174	1·2941	0·7727	0·6877
39·5	0·6361	1·5721	0·8243	1·2131	1·2960	0·7716	0·6894
·6	0·6374	1·5688	0·8273	1·2088	1·2978	0·7705	0·6911
·7	0·6388	1·5655	0·8302	1·2045	1·2997	0·7694	0·6929
·8	0·6401	1·5622	0·8332	1·2002	1·3016	0·7683	0·6946
·9	0·6415	1·5590	0·8361	1·1960	1·3035	0·7672	0·6964
40·0	0·6428	1·5557	0·8391	1·1918	1·3054	0·7660	0·6981
·1	0·6441	1·5525	0·8421	1·1875	1·3073	0·7649	0·6999
·2	0·6455	1·5493	0·8451	1·1833	1·3093	0·7638	0·7016
·3	0·6468	1·5461	0·8481	1·1792	1·3112	0·7627	0·7034
·4	0·6481	1·5429	0·8511	1·1750	1·3131	0·7615	0·7051
40·5	0·6494	1·5398	0·8541	1·1708	1·3151	0·7604	0·7069
·6	0·6508	1·5366	0·8571	1·1667	1·3171	0·7593	0·7086
·7	0·6521	1·5335	0·8601	1·1626	1·3190	0·7581	0·7103
·8	0·6534	1·5304	0·8632	1·1585	1·3210	0·7570	0·7121
·9	0·6547	1·5273	0·8662	1·1544	1·3230	0·7559	0·7138
41·0	0·6561	1·5243	0·8693	1·1504	1·3250	0·7547	0·7156
·1	0·6574	1·5212	0·8724	1·1463	1·3270	0·7536	0·7173
·2	0·6587	1·5182	0·8754	1·1423	1·3291	0·7524	0·7191
·3	0·6600	1·5151	0·8785	1·1383	1·3311	0·7513	0·7208
·4	0·6613	1·5121	0·8816	1·1343	1·3331	0·7501	0·7226
41·5	0·6626	1·5092	0·8847	1·1303	1·3352	0·7490	0·7243
·6	0·6639	1·5062	0·8878	1·1263	1·3373	0·7478	0·7261
·7	0·6652	1·5032	0·8910	1·1224	1·3393	0·7466	0·7278
·8	0·6665	1·5003	0·8941	1·1184	1·3414	0·7455	0·7295
·9	0·6678	1·4974	0·8972	1·1145	1·3435	0·7443	0·7313
42·0	0·6691	1·4945	0·9004	1·1106	1·3456	0·7431	0·7330

42° TO 48°	Sine	Cosec.	Tan.	Cotan.	Secant	Cosine	Radians
NATURAL FUNCTIONS OF ANGLES							

°	Sine	Cosec.	Tan.	Cotan.	Secant	Cosine	Radians
42·0	0·6691	1·4945	0·9004	1·1106	1·3456	0·7431	0·7330
·1	0·6704	1·4916	0·9036	1·1067	1·3478	0·7420	0·7348
·2	0·6717	1·4887	0·9067	1·1028	1·3499	0·7408	0·7365
·3	0·6730	1·4859	0·9099	1·0990	1·3520	0·7396	0·7383
·4	0·6743	1·4830	0·9131	1·0951	1·3542	0·7385	0·7400
42·5	0·6756	1·4802	0·9163	1·0913	1·3563	0·7373	0·7418
·6	0·6769	1·4774	0·9195	1·0875	1·3585	0·7361	0·7435
·7	0·6782	1·4746	0·9228	1·0837	1·3607	0·7349	0·7453
·8	0·6794	1·4718	0·9260	1·0799	1·3629	0·7337	0·7470
·9	0·6807	1·4690	0·9293	1·0761	1·3651	0·7325	0·7487
43·0	0·6820	1·4663	0·9325	1·0724	1·3673	0·7314	0·7505
·1	0·6833	1·4635	0·9358	1·0686	1·3696	0·7302	0·7522
·2	0·6846	1·4608	0·9391	1·0649	1·3718	0·7290	0·7540
·3	0·6858	1·4581	0·9424	1·0612	1·3741	0·7278	0·7557
·4	0·6871	1·4554	0·9457	1·0575	1·3763	0·7266	0·7575
43·5	0·6884	1·4527	0·9490	1·0538	1·3786	0·7254	0·7592
·6	0·6896	1·4501	0·9523	1·0501	1·3809	0·7242	0·7610
·7	0·6909	1·4474	0·9556	1·0464	1·3832	0·7230	0·7627
·8	0·6921	1·4448	0·9590	1·0428	1·3855	0·7218	0·7645
·9	0·6934	1·4422	0·9623	1·0392	1·3878	0·7206	0·7662
44·0	0·6947	1·4396	0·9657	1·0355	1·3902	0·7193	0·7679
·1	0·6959	1·4370	0·9691	1·0319	1·3925	0·7181	0·7697
·2	0·6972	1·4344	0·9725	1·0283	1·3949	0·7169	0·7714
·3	0·6984	1·4318	0·9759	1·0247	1·3972	0·7157	0·7732
·4	0·6997	1·4293	0·9793	1·0212	1·3996	0·7145	0·7749
44·5	0·7009	1·4267	0·9827	1·0176	1·4020	0·7133	0·7767
·6	0·7022	1·4242	0·9861	1·0141	1·4044	0·7120	0·7784
·7	0·7034	1·4217	0·9896	1·0105	1·4069	0·7108	0·7802
·8	0·7046	1·4192	0·9930	1·0070	1·4093	0·7096	0·7819
·9	0·7059	1·4167	0·9965	1·0035	1·4118	0·7083	0·7837
45·0	0·7071	1·4142	1·0000	1·0000	1·4142	0·7071	0·7854
·1	0·7083	1·4118	1·0035	0·9965	1·4167	0·7059	0·7871
·2	0·7096	1·4093	1·0070	0·9930	1·4192	0·7046	0·7889
·3	0·7108	1·4069	1·0105	0·9896	1·4217	0·7034	0·7906
·4	0·7120	1·4044	1·0141	0·9861	1·4242	0·7022	0·7924
45·5	0·7133	1·4020	1·0176	0·9827	1·4267	0·7009	0·7941
·6	0·7145	1·3996	1·0212	0·9793	1·4293	0·6997	0·7959
·7	0·7157	1·3972	1·0247	0·9759	1·4318	0·6984	0·7976
·8	0·7169	1·3949	1·0283	0·9725	1·4344	0·6972	0·7994
·9	0·7181	1·3925	1·0319	0·9691	1·4370	0·6959	0·8011
46·0	0·7193	1·3902	1·0355	0·9657	1·4396	0·6947	0·8029
·1	0·7206	1·3878	1·0392	0·9623	1·4422	0·6934	0·8046
·2	0·7218	1·3855	1·0428	0·9590	1·4448	0·6921	0·8063
·3	0·7230	1·3832	1·0464	0·9556	1·4474	0·6909	0·8081
·4	0·7242	1·3809	1·0501	0·9523	1·4501	0·6896	0·8098
46·5	0·7254	1·3786	1·0538	0·9490	1·4527	0·6884	0·8116
·6	0·7266	1·3763	1·0575	0·9457	1·4554	0·6871	0·8133
·7	0·7278	1·3741	1·0612	0·9424	1·4581	0·6858	0·8151
·8	0·7290	1·3718	1·0649	0·9391	1·4608	0·6846	0·8168
·9	0·7302	1·3696	1·0686	0·9358	1·4635	0·6833	0·8186
47·0	0·7314	1·3673	1·0724	0·9325	1·4663	0·6820	0·8203
·1	0·7325	1·3651	1·0761	0·9293	1·4690	0·6807	0·8220
·2	0·7337	1·3629	1·0799	0·9260	1·4718	0·6794	0·8238
·3	0·7349	1·3607	1·0837	0·9228	1·4746	0·6782	0·8255
·4	0·7361	1·3585	1·0875	0·9195	1·4774	0·6769	0·8273
47·5	0·7373	1·3563	1·0913	0·9163	1·4802	0·6756	0·8290
·6	0·7385	1·3542	1·0951	0·9131	1·4830	0·6743	0·8308
·7	0·7396	1·3520	1·0990	0·9099	1·4859	0·6730	0·8325
·8	0·7408	1·3499	1·1028	0·9067	1·4887	0·6717	0·8343
·9	0·7420	1·3478	1·1067	0·9036	1·4916	0·6704	0·8360
48·0	0·7431	1·3456	1·1106	0·9004	1·4945	0·6691	0·8378

48° TO 54°

NATURAL FUNCTIONS OF ANGLES

	Sine	Cosec.	Tan.	Cotan.	Secant	Cosine	Radians
48·0	0·7431	1·3456	1·1106	0·9004	1·4945	0·6691	0·8378
·1	0·7443	1·3435	1·1145	0·0972	1·4974	0·6678	0·8395
·2	0·7455	1·3414	1·1184	0·8941	1·5003	0·6665	0·8413
·3	0·7466	1·3393	1·1224	0·8910	1·5032	0·6652	0·8430
·4	0·7478	1·3373	1·1263	0·8878	1·5062	0·6639	0·8447
48·5	0·7490	1·3352	1·1303	0·8847	1·5092	0·6626	0·8465
·6	0·7501	1·3331	1·1343	0·8816	1·5121	0·6613	0·8482
·7	0·7513	1·3311	1·1383	0·8785	1·5151	0·6600	0·8500
·8	0·7524	1·3291	1·1423	0·8754	1·5182	0·6587	0·8517
·9	0·7536	1·3270	1·1463	0·8724	1·5212	0·6574	0·8535
49·0	0·7547	1·3250	1·1504	0·8693	1·5243	0·6561	0·8552
·1	0·7559	1·3230	1·1544	0·8662	1·5273	0·6547	0·8569
·2	0·7570	1·3210	1·1585	0·8632	1·5304	0·6534	0·8587
·3	0·7581	1·3190	1·1626	0·8601	1·5335	0·6521	0·8604
·4	0·7593	1·3171	1·1667	0·8571	1·5366	0·6508	0·8622
49·5	0·7604	1·3151	1·1708	0·8541	1·5398	0·6494	0·8639
·6	0·7615	1·3131	1·1750	0·8511	1·5429	0·6481	0·8657
·7	0·7627	1·3112	1·1792	0·8481	1·5461	0·6468	0·8674
·8	0·7638	1·3093	1·1833	0·8451	1·5493	0·6455	0·8692
·9	0·7649	1·3073	1·1875	0·8421	1·5525	0·6441	0·8709
50·0	0·7660	1·3054	1·1918	0·8391	1·5557	0·6428	0·8727
·1	0·7672	1·3035	1·1960	0·8361	1·5590	0·6415	0·8744
·2	0·7683	1·3016	1·2002	0·8332	1·5622	0·6401	0·8762
·3	0·7694	1·2997	1·2045	0·8302	1·5655	0·6388	0·8779
·4	0·7705	1·2978	1·2088	0·8273	1·5688	0·6374	0·8796
50·5	0·7716	1·2960	1·2131	0·8243	1·5721	0·6361	0·8814
·6	0·7727	1·2941	1·2174	0·8214	1·5755	0·6347	0·8831
·7	0·7738	1·2923	1·2218	0·8185	1·5788	0·6334	0·8849
·8	0·7749	1·2904	1·2261	0·8156	1·5822	0·6320	0·8866
·9	0·7760	1·2886	1·2305	0·8127	1·5856	0·6307	0·8884
51·0	0·7771	1·2868	1·2349	0·8098	1·5890	0·6293	0·8901
·1	0·7782	1·2849	1·2393	0·8069	1·5925	0·6280	0·8919
·2	0·7793	1·2831	1·2437	0·8040	1·5959	0·6266	0·8936
·3	0·7804	1·2813	1·2482	0·8012	1·5994	0·6252	0·8954
·4	0·7815	1·2796	1·2527	0·7983	1·6029	0·6239	0·8971
51·5	0·7826	1·2778	1·2572	0·7954	1·6064	0·6225	0·8988
·6	0·7837	1·2760	1·2617	0·7926	1·6099	0·6211	0·9006
·7	0·7848	1·2742	1·2662	0·7898	1·6135	0·6198	0·9023
·8	0·7859	1·2725	1·2708	0·7869	1·6171	0·6184	0·9041
·9	0·7869	1·2708	1·2753	0·7841	1·6207	0·6170	0·9058
52·0	0·7880	1·2690	1·2799	0·7813	1·6243	0·6157	0·9076
·1	0·7891	1·2673	1·2846	0·7785	1·6279	0·6143	0·9093
·2	0·7902	1·2656	1·2892	0·7757	1·6316	0·6129	0·9111
·3	0·7912	1·2639	1·2938	0·7729	1·6353	0·6115	0·9128
·4	0·7923	1·2622	1·2985	0·7701	1·6390	0·6101	0·9146
52·5	0·7934	1·2605	1·3032	0·7673	1·6427	0·6088	0·9163
·6	0·7944	1·2588	1·3079	0·7646	1·6464	0·6074	0·9180
·7	0·7955	1·2571	1·3127	0·7618	1·6502	0·6060	0·9198
·8	0·7965	1·2554	1·3175	0·7590	1·6540	0·6046	0·9215
·9	0·7976	1·2538	1·3222	0·7563	1·6578	0·6032	0·9233
53·0	0·7986	1·2521	1·3270	0·7536	1·6616	0·6018	0·9250
·1	0·7997	1·2505	1·3319	0·7508	1·6655	0·6004	0·9268
·2	0·8007	1·2489	1·3367	0·7481	1·6694	0·5990	0·9285
·3	0·8018	1·2472	1·3416	0·7454	1·6733	0·5976	0·9303
·4	0·8028	1·2456	1·3465	0·7427	1·6772	0·5962	0·9320
53·5	0·8039	1·2440	1·3514	0·7400	1·6812	0·5948	0·9338
·6	0·8049	1·2424	1·3564	0·7373	1·6851	0·5934	0·9355
·7	0·8059	1·2408	1·3613	0·7346	1·6892	0·5920	0·9372
·8	0·8070	1·2392	1·3663	0·7319	1·6932	0·5906	0·9390
·9	0·8080	1·2376	1·3713	0·7292	1·6972	0·5892	0·9407
54·0	0·8090	1·2361	1·3764	0·7265	1·7013	0·5878	0·9425

54° TO 60°	NATURAL FUNCTIONS OF ANGLES						
	Sine	Cosec.	Tan.	Cotan.	Secant	Cosine	Radians
54·0	0·8090	1·2361	1·3764	0·7265	1·7013	0·5878	0·9425
·1	0·8100	1·2345	1·3814	0·7239	1·7054	0·5864	0·9442
·2	0·8111	1·2329	1·3865	0·7212	1·7095	0·5850	0·9460
·3	0·8121	1·2314	1·3916	0·7186	1·7137	0·5835	0·9477
·4	0·8131	1·2299	1·3968	0·7159	1·7179	0·5821	0·9495
54·5	0·8141	1·2283	1·4019	0·7133	1·7221	0·5807	0·9512
·6	0·8151	1·2268	1·4071	0·7107	1·7263	0·5793	0·9530
·7	0·8161	1·2253	1·4124	0·7080	1·7305	0·5779	0·9547
·8	0·8171	1·2238	1·4176	0·7054	1·7348	0·5764	0·9564
·9	0·8182	1·2223	1·4229	0·7028	1·7391	0·5750	0·9582
55·0	0·8192	1·2208	1·4281	0·7002	1·7434	0·5736	0·9599
·1	0·8202	1·2193	1·4335	0·6976	1·7478	0·5721	0·9617
·2	0·8211	1·2178	1·4388	0·6950	1·7522	0·5707	0·9634
·3	0·8221	1·2163	1·4442	0·6924	1·7566	0·5693	0·9652
·4	0·8231	1·2149	1·4496	0·6899	1·7610	0·5678	0·9669
55·5	0·8241	1·2134	1·4550	0·6873	1·7655	0·5664	0·9687
·6	0·8251	1·2119	1·4605	0·6847	1·7700	0·5650	0·9704
·7	0·8261	1·2105	1·4659	0·6822	1·7745	0·5635	0·9721
·8	0·8271	1·2091	1·4715	0·6796	1·7791	0·5621	0·9739
·9	0·8281	1·2076	1·4770	0·6771	1·7837	0·5606	0·9756
56·0	0·8290	1·2062	1·4826	0·6745	1·7883	0·5592	0·9774
·1	0·8300	1·2048	1·4882	0·6720	1·7929	0·5577	0·9791
·2	0·8310	1·2034	1·4938	0·6694	1·7976	0·5563	0·9809
·3	0·8320	1·2020	1·4994	0·6669	1·8023	0·5548	0·9826
·4	0·8329	1·2006	1·5051	0·6644	1·8070	0·5534	0·9844
56·5	0·8339	1·1992	1·5108	0·6619	1·8118	0·5519	0·9861
·6	0·8348	1·1978	1·5166	0·6594	1·8166	0·5505	0·9879
·7	0·8358	1·1964	1·5224	0·6569	1·8214	0·5490	0·9896
·8	0·8368	1·1951	1·5282	0·6544	1·8263	0·5476	0·9913
·9	0·8377	1·1937	1·5340	0·6519	1·8312	0·5461	0·9931
57·0	0·8387	1·1924	1·5399	0·6494	1·8361	0·5446	0·9948
·1	0·8396	1·1910	1·5458	0·6469	1·8410	0·5432	0·9966
·2	0·8406	1·1897	1·5517	0·6445	1·8460	0·5417	0·9983
·3	0·8415	1·1883	1·5577	0·6420	1·8510	0·5402	1·0000
·4	0·8425	1·1870	1·5637	0·6395	1·8561	0·5388	1·0018
57·5	0·8434	1·1857	1·5697	0·6371	1·8612	0·5373	1·0035
·6	0·8443	1·1844	1·5757	0·6346	1·8663	0·5358	1·0053
·7	0·8453	1·1831	1·5818	0·6322	1·8714	0·5344	1·0071
·8	0·8462	1·1818	1·5880	0·6297	1·8766	0·5329	1·0088
·9	0·8471	1·1805	1·5941	0·6273	1·8818	0·5314	1·0105
58·0	0·8480	1·1792	1·6003	0·6249	1·8871	0·5299	1·0123
·1	0·8490	1·1779	1·6066	0·6224	1·8924	0·5284	1·0140
·2	0·8499	1·1766	1·6128	0·6200	1·8977	0·5270	1·0168
·3	0·8508	1·1753	1·6191	0·6176	1·9031	0·5255	1·0175
·4	0·8517	1·1741	1·6255	0·6152	1·9084	0·5240	1·0193
58·5	0·8526	1·1728	1·6319	0·6128	1·9139	0·5225	1·0210
·6	0·8536	1·1716	1·6383	0·6104	1·9193	0·5210	1·0228
·7	0·8545	1·1703	1·6447	0·6080	1·9249	0·5195	1·0245
·8	0·8554	1·1691	1·6512	0·6056	1·9304	0·5180	1·0262
·9	0·8563	1·1679	1·6577	0·6032	1·9360	0·5165	1·0280
59·0	0·8572	1·1666	1·6643	0·6009	1·9416	0·5150	1·0297
·1	0·8581	1·1654	1·6709	0·5985	1·9473	0·5135	1·0315
·2	0·8590	1·1642	1·6775	0·5961	1·9530	0·5120	1·0332
·3	0·8599	1·1630	1·6842	0·5938	1·9587	0·5105	1·0350
·4	0·8607	1·1618	1·6909	0·5914	1·9645	0·5090	1·0367
59·5	0·8616	1·1606	1·6977	0·5890	1·9703	0·5075	1·0385
·6	0·8625	1·1594	1·7045	0·5867	1·9762	0·5060	1·0402
·7	0·8634	1·1582	1·7113	0·5844	1·9821	0·5045	1·0420
·8	0·8643	1·1570	1·7182	0·5820	1·9880	0·5030	1·0437
·9	0·8652	1·1559	1·7251	0·5797	1·9940	0·5015	1·0455
60·0	0·8660	1·1547	1·7321	0·5774	2·0000	0·5000	1·0472

60° TO 66°	Sine	Cosec.	Tan.	Cotan.	Secant	Cosine	Radians
°							
60·0	0·8660	1·1547	1·7321	0·5774	2·0000	0·5000	1·0472
·1	0·8669	1·1535	1·7391	0·5750	2·0061	0·4985	1·0489
·2	0·8678	1·1524	1·7461	0·5727	2·0122	0·4970	1·0507
·3	0·8686	1·1512	1·7532	0·5704	2·0183	0·4955	1·0524
·4	0·8695	1·1501	1·7603	0·5681	2·0245	0·4939	1·0542
60·5	0·8704	1·1490	1·7675	0·5658	2·0308	0·4924	1·0559
·6	0·8712	1·1478	1·7747	0·5635	2·0371	0·4909	1·0577
·7	0·8721	1·1467	1·7820	0·5612	2·0434	0·4894	1·0594
·8	0·8729	1·1456	1·7893	0·5589	2·0498	0·4879	1·0612
·9	0·8738	1·1445	1·7966	0·5566	2·0562	0·4863	1·0629
61·0	0·8746	1·1434	1·8040	0·5543	2·0627	0·4848	1·0646
·1	0·8755	1·1423	1·8115	0·5520	2·0692	0·4833	1·0664
·2	0·8763	1·1412	1·8190	0·5498	2·0757	0·4818	1·0681
·3	0·8771	1·1401	1·8265	0·5475	2·0824	0·4802	1·0699
·4	0·8780	1·1390	1·8341	0·5452	2·0890	0·4787	1·0716
61·5	0·8788	1·1379	1·8418	0·5430	2·0957	0·4772	1·0734
·6	0·8796	1·1368	1·8495	0·5407	2·1025	0·4756	1·0751
·7	0·8805	1·1357	1·8572	0·5384	2·1093	0·4741	1·0769
·8	0·8813	1·1347	1·8650	0·5362	2·1162	0·4726	1·0786
·9	0·8821	1·1336	1·8728	0·5340	2·1231	0·4710	1·0804
62·0	0·8829	1·1326	1·8807	0·5317	2·1301	0·4695	1·0821
·1	0·8838	1·1315	1·8887	0·5295	2·1371	0·4679	1·0839
·2	0·8846	1·1305	1·8967	0·5272	2·1441	0·4664	1·0856
·3	0·8854	1·1294	1·9047	0·5250	2·1513	0·4648	1·0873
·4	0·8862	1·1284	1·9128	0·5228	2·1584	0·4633	1·0891
62·5	0·8870	1·1274	1·9210	0·5206	2·1657	0·4617	1·0908
·6	0·8878	1·1264	1·9292	0·5184	2·1730	0·4602	1·0926
·7	0·8886	1·1253	1·9375	0·5161	2·1803	0·4587	1·0943
·8	0·8894	1·1243	1·9458	0·5139	2·1877	0·4571	1·0961
·9	0·8902	1·1233	1·9542	0·5117	2·1952	0·4555	1·0978
63·0	0·8910	1·1223	1·9626	0·5095	2·2027	0·4540	1·0996
·1	0·8918	1·1213	1·9711	0·5073	2·2103	0·4524	1·1013
·2	0·8926	1·1203	1·9797	0·5051	2·2179	0·4509	1·1030
·3	0·8934	1·1194	1·9883	0·5029	2·2256	0·4493	1·1048
·4	0·8942	1·1184	1·9970	0·5008	2·2333	0·4478	1·1065
63·5	0·8949	1·1174	2·0057	0·4986	2·2412	0·4462	1·1083
·6	0·8957	1·1164	2·0145	0·4964	2·2490	0·4446	1·1100
·7	0·8965	1·1155	2·0233	0·4942	2·2570	0·4431	1·1118
·8	0·8973	1·1145	2·0323	0·4921	2·2650	0·4415	1·1135
·9	0·8980	1·1136	2·0413	0·4899	2·2730	0·4399	1·1153
64·0	0·8988	1·1126	2·0503	0·4877	2·2812	0·4384	1·1170
·1	0·8996	1·1117	2·0594	0·4856	2·2894	0·4368	1·1187
·2	0·9003	1·1107	2·0686	0·4834	2·2976	0·4352	1·1205
·3	0·9011	1·1098	2·0778	0·4813	2·3060	0·4337	1·1223
·4	0·9018	1·1089	2·0872	0·4791	2·3144	0·4321	1·1240
64·5	0·9026	1·1079	2·0965	0·4770	2·3228	0·4305	1·1257
·6	0·9033	1·1070	2·1060	0·4748	2·3314	0·4289	1·1275
·7	0·9041	1·1061	2·1155	0·4727	2·3400	0·4274	1·1292
·8	0·9048	1·1052	2·1251	0·4706	2·3486	0·4258	1·1310
·9	0·9056	1·1043	2·1348	0·4684	2·3574	0·4242	1·1327
65·0	0·9063	1·1034	2·1445	0·4663	2·3662	0·4226	1·1345
·1	0·9070	1·1025	2·1543	0·4642	2·3751	0·4210	1·1362
·2	0·9078	1·1016	2·1642	0·4621	2·3841	0·4195	1·1380
·3	0·9085	1·1007	2·1742	0·4599	2·3931	0·4179	1·1397
·4	0·9092	1·0998	2·1842	0·4578	2·4022	0·4163	1·1414
65·5	0·9100	1·0989	2·1943	0·4557	2·4114	0·4147	1·1432
·6	0·9107	1·0981	2·2045	0·4536	2·4207	0·4131	1·1449
·7	0·9114	1·0972	2·2148	0·4515	2·4301	0·4115	1·1467
·8	0·9121	1·0963	2·2251	0·4494	2·4395	0·4099	1·1484
·9	0·9128	1·0955	2·2355	0·4473	2·4490	0·4083	1·1502
66·0	0·9135	1·0946	2·2460	0·4452	2·4586	0·4067	1·1519

NATURAL FUNCTIONS OF ANGLES

66° TO 72°	Sine	Cosec.	Tan.	Cotan.	Secant	Cosine	Radians
°							
66·0	0·9135	1·0946	2·2460	0·4452	2·4586	0·4067	1·1519
·1	0·9143	1·0938	2·2566	0·4431	2·4683	0·4051	1·1537
·2	0·9150	1·0929	2·2673	0·4411	2·4780	0·4035	1·1554
·3	0·9157	1·0921	2·2781	0·4390	2·4879	0·4019	1·1572
·4	0·9164	1·0913	2·2889	0·4369	2·4978	0·4003	1·1589
66·5	0·9171	1·0904	2·2999	0·4348	2·5078	0·3987	1·1606
·6	0·9178	1·0896	2·3109	0·4327	2·5180	0·3971	1·1624
·7	0·9184	1·0888	2·3220	0·4307	2·5282	0·3955	1·1641
·8	0·9191	1·0880	2·3332	0·4286	2·5384	0·3939	1·1659
·9	0·9198	1·0872	2·3445	0·4265	2·5488	0·3923	1·1676
67·0	0·9205	1·0864	2·3559	0·4245	2·5593	0·3907	1·1694
·1	0·9212	1·0856	2·3673	0·4224	2·5699	0·3891	1·1711
·2	0·9219	1·0848	2·3789	0·4204	2·5805	0·3875	1·1729
·3	0·9225	1·0840	2·3906	0·4183	2·5913	0·3859	1·1746
·4	0·9232	1·0832	2·4023	0·4163	2·6022	0·3843	1·1764
67·5	0·9239	1·0824	2·4142	0·4142	2·6131	0·3827	1·1781
·6	0·9245	1·0816	2·4232	0·4122	2·6242	0·3811	1·1798
·7	0·9252	1·0808	2·4383	0·4101	2·6353	0·3795	1·1816
·8	0·9259	1·0801	2·4504	0·4081	2·6466	0·3778	1·1833
·9	0·9265	1·0793	2·4627	0·4061	2·6580	0·3762	1·1851
68·0	0·9272	1·0785	2·4751	0·4040	2·6695	0·3746	1·1868
·1	0·9278	1·0778	2·4876	0·4020	2·6810	0·3730	1·1886
·2	0·9285	1·0770	2·5002	0·4000	2·6928	0·3714	1·1903
·3	0·9291	1·0763	2·5129	0·3979	2·7046	0·3697	1·1921
·4	0·9298	1·0755	2·5257	0·3959	2·7165	0·3681	1·1938
68·5	0·9304	1·0748	2·5387	0·3939	2·7285	0·3665	1·1956
·6	0·9311	1·0740	2·5517	0·3919	2·7407	0·3649	1·1973
·7	0·9317	1·0733	2·5649	0·3899	2·7529	0·3633	1·1990
·8	0·9323	1·0726	2·5782	0·3879	2·7653	0·3616	1·2008
·9	0·9330	1·0719	2·5916	0·3859	2·7778	0·3600	1·2025
69·0	0·9336	1·0711	2·6051	0·3839	2·7904	0·3584	1·2043
·1	0·9342	1·0704	2·6187	0·3819	2·8032	0·3567	1·2060
·2	0·9348	1·0697	2·6325	0·3799	2·8161	0·3551	1·2078
·3	0·9354	1·0690	2·6464	0·3779	2·8292	0·3535	1·2095
·4	0·9361	1·0683	2·6605	0·3759	2·8422	0·3518	1·2113
69·5	0·9367	1·0676	2·6746	0·3739	2·8555	0·3502	1·2130
·6	0·9373	1·0669	2·6889	0·3719	2·8688	0·3486	1·2147
·7	0·9379	1·0662	2·7033	0·3699	2·8824	0·3469	1·2165
·8	0·9385	1·0655	2·7179	0·3679	2·8960	0·3453	1·2182
·9	0·9391	1·0649	2·7326	0·3659	2·9099	0·3437	1·2200
70·0	0·9397	1·0642	2·7475	0·3640	2·9238	0·3420	1·2217
·1	0·9403	1·0635	2·7625	0·3620	2·9379	0·3404	1·2235
·2	0·9409	1·0628	2·7776	0·3600	2·9521	0·3387	1·2252
·3	0·9415	1·0622	2·7929	0·3581	2·9665	0·3371	1·2270
·4	0·9421	1·0615	2·8083	0·3561	2·9811	0·3355	1·2287
70·5	0·9426	1·0608	2·8239	0·3541	2·9957	0·3338	1·2305
·6	0·9432	1·0602	2·8397	0·3522	3·0106	0·3322	1·2322
·7	0·9438	1·0595	2·8555	0·3502	3·0256	0·3305	1·2339
·8	0·9444	1·0589	2·8716	0·3482	3·0407	0·3289	1·2357
·9	0·9449	1·0583	2·8878	0·3463	3·0561	0·3272	1·2374
71·0	0·9455	1·0576	2·9042	0·3443	3·0716	0·3256	1·2391
·1	0·9461	1·0570	2·9208	0·3424	3·0872	0·3239	1·2409
·2	0·9466	1·0564	2·9375	0·3404	3·1030	0·3223	1·2427
·3	0·9472	1·0557	2·9544	0·3385	3·1190	0·3206	1·2444
·4	0·9478	1·0551	2·9714	0·3365	3·1352	0·3190	1·2462
71·5	0·9483	1·0545	2·9887	0·3346	3·1516	0·3173	1·2479
·6	0·9489	1·0539	3·0061	0·3327	3·1681	0·3156	1·2497
·7	0·9494	1·0533	3·0237	0·3307	3·1848	0·3140	1·2514
·8	0·9500	1·0527	3·0415	0·3288	3·2017	0·3123	1·2531
·9	0·9505	1·0521	3·0595	0·3268	3·2188	0·3107	1·2549
72·0	0·9511	1·0515	3·0777	0·3249	3·2361	0·3090	1·2566

NATURAL FUNCTIONS OF ANGLES

72° TO 78°	Sine	Cosec.	Tan.	Cotan.	Secant	Cosine	Radians
72·0	0·9511	1·0515	3·0777	0·3249	3·2361	0·3090	1·2566
·1	0·9516	1·0509	3·0961	0·3230	3·2535	0·3074	1·2584
·2	0·9521	1·0503	3·1146	0·3211	3·2712	0·3057	1·2601
·3	0·9527	1·0497	3·1334	0·3191	3·2891	0·3040	1·2619
·4	0·9532	1·0491	3·1524	0·3172	3·3072	0·3024	1·2636
72·5	0·9537	1·0485	3·1716	0·3153	3·3255	0·3007	1·2654
·6	0·9542	1·0480	3·1910	0·3134	3·3440	0·2990	1·2671
·7	0·9548	1·0474	3·2106	0·3115	3·3628	0·2974	1·2689
·8	0·9553	1·0468	3·2305	0·3095	3·3817	0·2957	1·2706
·9	0·9558	1·0462	3·2506	0·3076	3·4009	0·2940	1·2723
73·0	0·9563	1·0457	3·2709	0·3057	3·4203	0·2924	1·2741
·1	0·9568	1·0451	3·2914	0·3038	3·4399	0·2907	1·2758
·2	0·9573	1·0446	3·3122	0·3019	3·4598	0·2890	1·2776
·3	0·9578	1·0440	3·3332	0·3000	3·4799	0·2874	1·2793
·4	0·9583	1·0435	3·3544	0·2981	3·5003	0·2857	1·2811
73·5	0·9588	1·0430	3·3759	0·2962	3·5209	0·2840	1·2828
·6	0·9593	1·0424	3·3977	0·2943	3·5418	0·2823	1·2846
·7	0·9598	1·0419	3·4197	0·2924	3·5629	0·2807	1·2863
·8	0·9603	1·0414	3·4420	0·2905	3·5843	0·2790	1·2881
·9	0·9608	1·0408	3·4646	0·2886	3·6060	0·2773	1·2898
74·0	0·9613	1·0403	3·4874	0·2868	3·6280	0·2756	1·2915
·1	0·9617	1·0398	3·5105	0·2849	3·6502	0·2740	1·2933
·2	0·9622	1·0393	3·5339	0·2830	3·6727	0·2723	1·2950
·3	0·9627	1·0387	3·5576	0·2811	3·6955	0·2706	1·2968
·4	0·9632	1·0382	3·5816	0·2792	3·7186	0·2689	1·2985
74·5	0·9636	1·0377	3·6059	0·2773	3·7420	0·2672	1·3003
·6	0·9641	1·0372	3·6305	0·2755	3·7657	0·2656	1·3020
·7	0·9646	1·0367	3·6554	0·2736	3·7897	0·2639	1·3038
·8	0·9650	1·0363	3·6806	0·2717	3·8140	0·2622	1·3055
·9	0·9655	1·0358	3·7062	0·2698	3·8387	0·2605	1·3072
75·0	0·9659	1·0353	3·7321	0·2680	3·8637	0·2588	1·3090
·1	0·9664	1·0348	3·7583	0·2661	3·8890	0·2571	1·3107
·2	0·9668	1·0343	3·7848	0·2642	3·9147	0·2554	1·3125
·3	0·9673	1·0338	3·8118	0·2624	3·9408	0·2538	1·3142
·4	0·9677	1·0334	3·8391	0·2605	3·9672	0·2521	1·3160
75·5	0·9681	1·0329	3·8667	0·2586	3·9939	0·2504	1·3177
·6	0·9686	1·0324	3·8947	0·2568	4·0211	0·2487	1·3195
·7	0·9690	1·0320	3·9232	0·2549	4·0486	0·2470	1·3212
·8	0·9694	1·0315	3·9520	0·2530	4·0765	0·2453	1·3230
·9	0·9699	1·0311	3·9812	0·2512	4·1048	0·2436	1·3247
76·0	0·9703	1·0306	4·0108	0·2493	4·1336	0·2419	1·3265
·1	0·9707	1·0302	4·0408	0·2475	4·1627	0·2402	1·3282
·2	0·9711	1·0297	4·0713	0·2456	4·1923	0·2385	1·3300
·3	0·9715	1·0293	4·1022	0·2438	4·2223	0·2368	1·3317
·4	0·9720	1·0289	4·1335	0·2419	4·2527	0·2351	1·3334
76·5	0·9724	1·0284	4·1653	0·2401	4·2837	0·2334	1·3352
·6	0·9728	1·0280	4·1976	0·2382	4·3150	0·2317	1·3369
·7	0·9732	1·0276	4·2303	0·2364	4·3469	0·2301	1·3387
·8	0·9736	1·0271	4·2635	0·2346	4·3792	0·2284	1·3404
·9	0·9740	1·0267	4·2972	0·2327	4·4121	0·2267	1·3422
77·0	0·9744	1·0263	4·3315	0·2309	4·4454	0·2250	1·3439
·1	0·9748	1·0259	4·3662	0·2290	4·4793	0·2233	1·3457
·2	0·9751	1·0255	4·4015	0·2272	4·5137	0·2215	1·3474
·3	0·9755	1·0251	4·4373	0·2254	4·5486	0·2198	1·3491
·4	0·9759	1·0247	4·4737	0·2235	4·5841	0·2181	1·3509
77·5	0·9763	1·0243	4·5107	0·2217	4·6202	0·2164	1·3527
·6	0·9767	1·0239	4·5483	0·2199	4·6569	0·2147	1·3544
·7	0·9770	1·0235	4·5864	0·2180	4·6942	0·2130	1·3561
·8	0·9774	1·0231	4·6252	0·2162	4·7320	0·2113	1·3579
·9	0·9778	1·0227	4·6646	0·2144	4·7706	0·2096	1·3596
78·0	0·9781	1·0223	4·7046	0·2126	4·8097	0·2079	1·3614

78° TO 84°	NATURAL FUNCTIONS OF ANGLES						
°	Sine	Cosec.	Tan.	Cotan.	Secant	Cosine	Radians
78·0	0·9781	1·0223	4·7046	0·2126	4·8097	0·2079	1·3614
·1	0·9785	1·0220	4·7453	0·2107	4·8496	0·2062	1·3631
·2	0·9789	1·0216	4·7867	0·2089	4·8901	0·2045	1·3648
·3	0·9792	1·0212	4·8288	0·2071	4·9313	0·2028	1·3666
·4	0·9796	1·0209	4·8716	0·2053	4·9732	0·2011	1·3683
78·5	0·9799	1·0205	4·9152	0·2034	5·0159	0·1994	1·3701
·6	0·9803	1·0201	4·9595	0·2016	5·0593	0·1977	1·3718
·7	0·9806	1·0198	5·0045	0·1998	5·1034	0·1959	1·3736
·8	0·9810	1·0194	5·0504	0·1980	5·1484	0·1942	1·3753
·9	0·9813	1·0191	5·0970	0·1962	5·1942	0·1925	1·3771
79·0	0·9816	1·0187	5·1446	0·1944	5·2408	0·1908	1·3788
·1	0·9820	1·0184	5·1929	0·1926	5·2883	0·1891	1·3806
·2	0·9823	1·0180	5·2422	0·1908	5·3367	0·1874	1·3823
·3	0·9826	1·0177	5·2924	0·1889	5·3860	0·1857	1·3840
·4	0·9829	1·0174	5·3435	0·1871	5·4362	0·1840	1·3858
79·5	0·9833	1·0170	5·3955	0·1853	5·4874	0·1822	1·3875
·6	0·9836	1·0167	5·4486	0·1835	5·5396	0·1805	1·3893
·7	0·9839	1·0164	5·5026	0·1817	5·5928	0·1788	1·3910
·8	0·9842	1·0161	5·5578	0·1799	5·6470	0·1771	1·3928
·9	0·9845	1·0157	5·6140	0·1781	5·7023	0·1754	1·3945
80·0	0·9848	1·0154	5·6713	0·1763	5·7588	0·1736	1·3963
·1	0·9851	1·0151	5·7297	0·1745	5·8164	0·1719	1·3980
·2	0·9854	1·0148	5·7894	0·1727	5·8751	0·1702	1·3997
·3	0·9857	1·0145	5·8502	0·1709	5·9351	0·1685	1·4015
·4	0·9860	1·0142	5·9124	0·1691	5·9963	0·1668	1·4032
80·5	0·9863	1·0139	5·9758	0·1673	6·0589	0·1650	1·4050
·6	0·9866	1·0136	6·0405	0·1656	6·1227	0·1633	1·4067
·7	0·9869	1·0133	6·1066	0·1638	6·1880	0·1616	1·4085
·8	0·9871	1·0130	6·1742	0·1620	6·2546	0·1599	1·4102
·9	0·9874	1·0128	6·2432	0·1602	6·3228	0·1582	1·4120
81·0	0·9877	1·0125	6·3138	0·1584	6·3925	0·1564	1·4137
·1	0·9880	1·0122	6·3859	0·1566	6·4637	0·1547	1·4155
·2	0·9882	1·0119	6·4596	0·1548	6·5366	0·1530	1·4172
·3	0·9885	1·0116	6·5350	0·1530	6·6111	0·1513	1·4189
·4	0·9888	1·0114	6·6122	0·1512	6·6874	0·1495	1·4207
81·5	0·9890	1·0111	6·6912	0·1495	6·7655	0·1478	1·4224
·6	0·9893	1·0108	6·7720	0·1477	6·8454	0·1461	1·4242
·7	0·9895	1·0106	6·8548	0·1459	6·9273	0·1444	1·4259
·8	0·9898	1·0103	6·9395	0·1441	7·0112	0·1426	1·4277
·9	0·9900	1·0101	7·0264	0·1423	7·0972	0·1409	1·4294
82·0	0·9903	1·0098	7·1154	0·1405	7·1853	0·1392	1·4312
·1	0·9905	1·0096	7·2066	0·1388	7·2757	0·1374	1·4329
·2	0·9907	1·0093	7·3002	0·1370	7·3684	0·1357	1·4347
·3	0·9910	1·0091	7·3962	0·1352	7·4635	0·1340	1·4364
·4	0·9912	1·0089	7·4947	0·1334	7·5611	0·1323	1·4381
82·5	0·9914	1·0086	7·5958	0·1317	7·6613	0·1305	1·4399
·6	0·9917	1·0084	7·6996	0·1299	7·7642	0·1288	1·4416
·7	0·9919	1·0082	7·8062	0·1281	7·8700	0·1271	1·4434
·8	0·9921	1·0079	7·9158	0·1263	7·9787	0·1253	1·4451
·9	0·9923	1·0077	8·0285	0·1246	8·0905	0·1236	1·4469
83·0	0·9925	1·0075	8·1444	0·1228	8·2055	0·1219	1·4486
·1	0·9928	1·0073	8·2636	0·1210	8·3238	0·1201	1·4504
·2	0·9930	1·0071	8·3863	0·1192	8·4457	0·1184	1·4521
·3	0·9932	1·0069	8·5126	0·1175	8·5711	0·1167	1·4538
·4	0·9934	1·0067	8·6428	0·1157	8·7004	0·1149	1·4556
83·5	0·9936	1·0065	8·7769	0·1139	8·8337	0·1132	1·4573
·6	0·9938	1·0063	8·9152	0·1122	8·9711	0·1115	1·4591
·7	0·9940	1·0061	9·0579	0·1104	9·1129	0·1097	1·4608
·8	0·9942	1·0059	9·2052	0·1086	9·2593	0·1080	1·4626
·9	0·9944	1·0057	9·3572	0·1069	9·4105	0·1063	1·4643
84·0	0·9945	1·0055	9·5144	0·1051	9·5668	0·1045	1·4661

84° TO 90°	**NATURAL FUNCTIONS OF ANGLES**						
°	Sine	Cosec.	Tan.	Cotan.	Secant	Cosine	Radians
84·0	0·9945	1·0055	9·5144	0·1051	9·5668	0·1045	1·4661
·1	0·9947	1·0053	9·6768	0·1033	9·7283	0·1028	1·4678
·2	0·9949	1·0051	9·8448	0·1016	9·8955	0·1011	1·4695
·3	0·9951	1·0050	10·019	0·0998	10·068	0·0993	1·4713
·4	0·9952	1·0048	10·199	0·0981	10·248	0·0976	1·4731
84·5	0·9954	1·0046	10·385	0·0963	10·433	0·0958	1·4748
·6	0·9956	1·0045	10·579	0·0945	10·626	0·0941	1·4765
·7	0·9957	1·0043	10·780	0·0928	10·826	0·0924	1·4783
·8	0·9959	1·0041	10·988	0·0910	11·034	0·0906	1·4800
·9	0·9960	1·0040	11·205	0·0892	11·249	0·0889	1·4818
85·0	0·9962	1·0038	11·430	0·0875	11·474	0·0872	1·4835
·1	0·9963	1·0037	11·664	0·0857	11·707	0·0854	1·4853
·2	0·9965	1·0035	11·909	0·0840	11·951	0·0837	1·4870
·3	0·9967	1·0034	12·163	0·0822	12·204	0·0819	1·4887
·4	0·9968	1·0032	12·429	0·0805	12·469	0·0802	1·4905
85·5	0·9969	1·0031	12·706	0·0787	12·746	0·0785	1·4923
·6	0·9971	1·0030	12·996	0·0769	13·035	0·0767	1·4940
·7	0·9972	1·0028	13·300	0·0752	13·337	0·0750	1·4957
·8	0·9973	1·0027	13·617	0·0734	13·654	0·0732	1·4975
·9	0·9974	1·0026	13·951	0·0717	13·987	0·0715	1·4992
86·0	0·9976	1·0024	14·301	0·0699	14·336	0·0698	1·5010
·1	0·9977	1·0023	14·669	0·0682	14·703	0·0680	1·5027
·2	0·9978	1·0022	15·056	0·0664	15·089	0·0663	1·5045
·3	0·9979	1·0021	15·464	0·0647	15·496	0·0645	1·5062
·4	0·9980	1·0020	15·894	0·0629	15·926	0·0628	1·5080
86·5	0·9981	1·0019	16·350	0·0612	16·380	0·0610	1·5097
·6	0·9982	1·0018	16·832	0·0594	16·862	0·0593	1·5115
·7	0·9983	1·0017	17·343	0·0577	17·372	0·0576	1·5132
·8	0·9984	1·0016	17·886	0·0559	17·914	0·0588	1·5149
·9	0·9985	1·0015	18·464	0·0542	18·491	0·0541	1·5167
87·0	0·9986	1·0014	19·081	0·0524	19·107	0·0523	1·5184
·1	0·9987	1·0013	19·740	0·0507	19·766	0·0506	1·5202
·2	0·9988	1·0012	20·446	0·0489	20·471	0·0489	1·5219
·3	0·9989	1·0011	21·205	0·0472	21·229	0·0471	1·5237
·4	0·9990	1·0010	22·022	0·0454	22·044	0·0454	1·5254
87·5	0·9990	1·0010	22·904	0·0437	22·926	0·0436	1·5272
·6	0·9991	1·0009	23·859	0·0419	23·880	0·0419	1·5289
·7	0·9992	1·0008	24·898	0·0402	24·918	0·0401	1·5307
·8	0·9993	1·0007	26·031	0·0384	26·050	0·0384	1·5324
·9	0·9993	1·0007	27·271	0·0367	27·290	0·0366	1·5342
88·0	0·9994	1·0006	28·636	0·0349	28·654	0·0349	1·5359
·1	0·9995	1·0006	30·145	0·0332	30·161	0·0332	1·5376
·2	0·9995	1·0005	31·820	0·0314	31·836	0·0314	1·5394
·3	0·9996	1·0004	33·694	0·0297	33·708	0·0297	1·5411
·4	0·9996	1·0004	35·801	0·0279	35·814	0·0279	1·5429
88·5	0·9997	1·0003	38·188	0·0262	38·201	0·0262	1·5446
·6	0·9997	1·0003	40·917	0·0244	40·930	0·0244	1·5464
·7	0·9997	1·0003	44·066	0·0227	44·078	0·0227	1·5481
·8	0·9998	1·0002	47·740	0·0209	47·750	0·0209	1·5498
·9	0·9998	1·0002	52·081	0·0192	52·090	0·0192	1·5516
89·0	0·9998	1·0002	57·290	0·0175	57·299	0·0175	1·5533
·1	0·9999	1·0001	63·657	0·0157	63·665	0·0157	1·5551
·2	0·9999	1·0001	71·615	0·0140	71·622	0·0140	1·5568
·3	0·9999	1·0001	81·847	0·0122	81·853	0·0122	1·5586
·4	0·9999	1·0001	95·490	0·0105	95·495	0·0105	1·5603
89·5	1·0000	1·0000	114·59	0·0087	114·59	0·0087	1·5621
·6	1·0000	1·0000	143·24	0·0070	143·24	0·0070	1·5638
·7	1·0000	1·0000	190·98	0·0052	190·99	0·0052	1·5656
·8	1·0000	1·0000	286·48	0·0035	286·48	0·0035	1·5673
·9	1·0000	1·0000	572·96	0·0017	572·96	0·0017	1·5691
90·0	1·0000	1·0000	Infinite	0·0000	Infinite	0·0000	1·5708

FORM OF THE EARTH

Lat.	Radius of Earth		Length of 1 min. of Lat.		Length of 1 min. of Long.		Lat.	Radius of Earth		Length of 1 min. of Lat.		Length of 1 min. of Long.	
°	Statute Miles	Km.	Feet	Metres	Feet	Metres	°	Statute Miles	Km.	Feet	Metres	Feet	Metres
0	3963·35	6378·38	6046·4	1842·9	6087·2	1855·4	45	3956·71	6367·70	6077·1	1852·3	4311·5	1314·1
1	3963·35	6378·38	6046·4	1842·9	6086·3	1855·1	46	3956·48	6367·33	6078·2	1852·6	4235·9	1291·1
2	3963·34	6378·36	6046·5	1843·0	6083·5	1854·3	47	3956·25	6366·96	6079·2	1852·9	4158·9	1267·6
3	3963·32	6378·33	6046·6	1843·0	6078·9	1852·8	48	3956·02	6366·59	6080·3	1853·3	4080·7	1243·8
4	3963·29	6378·28	6046·7	1843·0	6072·5	1850·9	49	3955·78	6366·20	6081·4	1853·6	4001·2	1219·6
5	3963·26	6378·23	6046·9	1843·1	6064·2	1848·4	50	3955·55	6365·83	6082·4	1853·9	3920·5	1195·0
6	3963·21	6378·15	6047·1	1843·2	6054·1	1845·3	51	3955·33	6365·48	6083·5	1854·2	3838·6	1170·0
7	3963·16	6378·07	6047·3	1843·2	6041·2	1841·4	52	3955·10	6365·11	6084·5	1854·6	3755·5	1144·7
8	3963·10	6377·97	6047·6	1843·3	6028·2	1837·4	53	3954·87	6364·74	6085·6	1854·9	3671·2	1119·0
9	3963·03	6377·86	6047·9	1843·4	6012·8	1832·7	54	3954·64	6364·37	6086·6	1855·2	3585·8	1093·0
10	3962·96	6377·75	6048·3	1843·5	5995·3	1827·4	55	3954·42	6364·01	6087·6	1855·5	3499·3	1066·6
11	3962·88	6377·62	6048·6	1843·6	5976·1	1821·5	56	3954·20	6363·66	6088·6	1855·8	3411·8	1039·9
12	3962·79	6377·48	6049·1	1843·8	5955·0	1815·1	57	3953·99	6363·32	6089·6	1856·1	3323·1	1012·9
13	3962·69	6377·32	6049·5	1843·9	5932·2	1808·1	58	3953·78	6362·98	6090·5	1856·4	3233·5	985·6
14	3962·58	6377·14	6050·0	1844·0	5907·5	1800·6	59	3953·57	6362·65	6091·5	1856·7	3142·9	958·0
15	3962·47	6376·96	6050·5	1844·2	5881·1	1792·6	60	3953·37	6362·32	6092·5	1857·0	3051·3	930·0
16	3962·35	6376·77	6051·1	1844·4	5852·9	1784·0	61	3953·17	6362·00	6093·4	1857·3	2958·7	901·8
17	3962·23	6376·58	6051·6	1844·5	5822·9	1774·6	62	3952·97	6361·68	6094·3	1857·5	2865·2	873·3
18	3962·09	6376·35	6052·3	1844·7	5791·1	1765·1	63	3952·78	6361·37	6095·1	1857·8	2770·9	844·6
19	3961·95	6376·13	6052·9	1844·9	5757·6	1754·9	64	3952·59	6361·07	6096·0	1858·1	2675·7	815·6
20	3961·80	6375·89	6053·6	1845·1	5722·3	1744·2	65	3952·41	6360·78	6096·8	1858·3	2579·6	786·3
21	3961·65	6375·65	6054·3	1845·3	5685·3	1732·9	66	3952·23	6360·49	6097·6	1858·5	2482·8	756·8
22	3961·49	6375·39	6055·0	1845·6	5646·6	1721·1	67	3952·06	6360·22	6098·4	1858·8	2385·2	727·0
23	3961·33	6375·13	6055·8	1845·8	5606·2	1708·8	68	3951·90	6359·96	6099·2	1859·0	2286·9	697·0
24	3961·16	6374·86	6056·6	1846·0	5564·0	1695·9	69	3951·74	6359·70	6099·9	1859·2	2187·8	666·8
25	3960·98	6374·57	6057·4	1846·3	5520·2	1682·6	70	3951·58	6359·44	6100·6	1859·5	2088·1	636·5
26	3960·80	6374·28	6058·2	1846·5	5474·7	1668·7	71	3951·44	6359·22	6101·3	1859·7	1987·7	605·9
27	3960·62	6373·99	6059·1	1846·8	5427·5	1654·3	72	3951·29	6358·98	6101·9	1859·9	1886·7	575·1
28	3960·43	6373·68	6059·9	1847·1	5378·7	1639·4	73	3951·16	6358·77	6102·6	1860·1	1785·2	544·1
29	3960·24	6373·38	6060·8	1847·3	5328·2	1624·0	74	3951·03	6358·56	6103·1	1860·2	1683·1	513·0
30	3960·04	6373·06	6061·7	1847·6	5276·1	1608·2	75	3950·91	6368·37	6103·7	1860·4	1580·4	481·7
31	3959·84	6372·73	6062·7	1847·9	5222·4	1591·8	76	3950·80	6358·19	6104·2	1860·6	1477·3	450·3
32	3959·63	6372·40	6063·6	1848·2	5167·1	1574·9	77	3950·69	6358·01	6104·7	1860·7	1373·7	418·7
33	3959·42	6372·06	6064·6	1848·5	5110·2	1557·6	78	3950·59	6357·85	6105·1	1860·8	1269·7	387·0
34	3959·21	6371·72	6065·6	1848·8	5051·8	1539·8	79	3950·50	6357·71	6105·6	1861·0	1165·2	355·2
35	3958·99	6371·37	6066·6	1849·1	4991·9	1521·5	80	3950·42	6357·58	6105·9	1861·1	1060·5	323·2
36	3958·77	6371·01	6067·6	1849·4	4930·4	1502·8	81	3950·34	6357·45	6106·3	1861·2	955·4	291·2
37	3958·55	6370·66	6068·6	1849·7	4867·4	1483·6	82	3950·27	6357·33	6106·6	1861·3	849·9	259·0
38	3958·32	6370·29	6069·7	1850·0	4802·9	1463·9	83	3950·21	6357·24	6106·9	1861·4	744·3	226·9
39	3958·09	6369·92	6070·8	1850·4	4736·9	1443·8	84	3950·16	6357·16	6107·1	1861·4	638·4	194·6
40	3957·87	6369·56	6071·8	1850·7	4669·5	1423·3	85	3950·11	6357·08	6107·3	1861·5	532·3	162·2
41	3957·64	6369·19	6072·8	1851·0	4600·7	1402·3	86	3950·08	6357·03	6107·5	1861·6	426·0	129·8
42	3957·41	6368·82	6073·9	1851·3	4530·5	1380·9	87	3950·05	6356·98	6107·6	1861·6	319·7	97·4
43	3957·18	6368·45	6075·0	1851·7	4458·8	1359·0	88	3950·03	6356·95	6107·7	1861·6	213·2	65·0
44	3956·94	6368·07	6076·0	1852·0	4385·8	1336·8	89	3950·02	6356·93	6107·8	1861·7	106·6	32·5
45	3956·71	6367·70	6077·1	1852·3	4311·5	1314·1	90	3950·01	6356·92	6107·8	1861·7	—	—

o

0°   HAVERSINES

′	·0 Log.	·0 Nat.	·2 Log.	·2 Nat.	·4 Log.	·4 Nat.	′
00	—∞	0·00000	0·92745	0·00000	1·52951	0·00000	59
01	2·32539	0·00000	2·48375	0·00000	2·61765	0·00000	58
02	2·92745	0·00000	3·01024	0·00000	3·08581	0·00000	57
03	3·27963	0·00000	3·33569	0·00000	3·38835	0·00000	56
04	3·52951	0·00000	3·57189	0·00000	3·61230	0·00000	55
05	3·72333	0·00000	3·75740	0·00000	3·79018	0·00000	54
06	3·88169	0·00000	3·91018	0·00000	3·93775	0·00000	53
07	4·01559	0·00000	4·04006	0·00000	4·06386	0·00000	52
08	4·13157	0·00000	4·15302	0·00000	4·17395	0·00000	51
09	4·23388	0·00000	4·25297	0·00000	4·27165	0·00000	50
10	4·32539	0·00000	4·34259	0·00000	4·35946	0·00000	49
11	4·40818	0·00000	4·42383	0·00000	4·43920	0·00000	48
12	4·48375	0·00000	4·49811	0·00000	4·51224	0·00000	47
13	4·55328	0·00000	4·56654	0·00000	4·57960	0·00000	46
14	4·61765	0·00000	4·62997	0·00000	4·64212	0·00000	45
15	4·67757	0·00000	4·68908	0·00000	4·70043	0·00000	44
16	4·73363	0·00001	4·74442	0·00001	4·75508	0·00001	43
17	4·78629	0·00001	4·79645	0·00001	4·80649	0·00001	42
18	4·83594	0·00001	4·84553	0·00001	4·85503	0·00001	41
19	4·88290	0·00001	4·89199	0·00001	4·90099	0·00001	40
20	4·92745	0·00001	4·93609	0·00001	4·94465	0·00001	39
21	4·96983	0·00001	4·97806	0·00001	4·98622	0·00001	38
22	5·01024	0·00001	5·01810	0·00001	5·02589	0·00001	37
23	5·04885	0·00001	5·05637	0·00001	5·06382	0·00001	36
24	5·08581	0·00001	5·09302	0·00001	5·10017	0·00001	35
25	5·12127	0·00001	5·12819	0·00001	5·13506	0·00001	34
26	5·15534	0·00001	5·16199	0·00001	5·16860	0·00001	33
27	5·18812	0·00002	5·19453	0·00002	5·20089	0·00002	32
28	5·21971	0·00002	5·22589	0·00002	5·23203	0·00002	31
29	5·25019	0·00002	5·25616	0·00002	5·26208	0·00002	30
30	5·27963	0·00002	5·28540	0·00002	5·29114	0·00002	29
31	5·30811	0·00002	5·31370	0·00002	5·31925	0·00002	28
32	5·33569	0·00002	5·34110	0·00002	5·34648	0·00002	27
33	5·36242	0·00002	5·36766	0·00002	5·37288	0·00002	26
34	5·38835	0·00002	5·39344	0·00002	5·39851	0·00003	25
35	5·41352	0·00003	5·41847	0·00003	5·42339	0·00003	24
36	5·43799	0·00003	5·44281	0·00003	5·44759	0·00003	23
37	5·46179	0·00003	5·46647	0·00003	5·47113	0·00003	22
38	5·48495	0·00003	5·48951	0·00003	5·49405	0·00003	21
39	5·50752	0·00003	5·51196	0·00003	5·51638	0·00003	20
40	5·52951	0·00003	5·53384	0·00003	5·53815	0·00003	19
41	5·55095	0·00004	5·55518	0·00004	5·55939	0·00004	18
42	5·57189	0·00004	5·57601	0·00004	5·58012	0·00004	17
43	5·59232	0·00004	5·59635	0·00004	5·60037	0·00004	16
44	5·61229	0·00004	5·61623	0·00004	5·62015	0·00004	15
45	5·63181	0·00004	5·63566	0·00004	5·63950	0·00004	14
46	5·65090	0·00004	5·65467	0·00005	5·65842	0·00005	13
47	5·66958	0·00005	5·67327	0·00005	5·67694	0·00005	12
48	5·68787	0·00005	5·69148	0·00005	5·69508	0·00005	11
49	5·70578	0·00005	5·70931	0·00005	5·71284	0·00005	10
50	5·72332	0·00005	5·72679	0·00005	5·73025	0·00005	09
51	5·74052	0·00006	5·74392	0·00006	5·74731	0·00006	08
52	5·75739	0·00006	5·76072	0·00006	5·76405	0·00006	07
53	5·77394	0·00006	5·77721	0·00006	5·78101	0·00006	06
54	5·79017	0·00006	5·79338	0·00006	5·79658	0·00006	05
55	5·80611	0·00006	5·80926	0·00007	5·81240	0·00007	04
56	5·82176	0·00007	5·82486	0·00007	5·82794	0·00007	03
57	5·83713	0·00007	5·84017	0·00007	5·84321	0·00007	02
58	5·85224	0·00007	5·85523	0·00007	5·85821	0·00007	01
59	5·86709	0·00007	5·87002	0·00007	5·87295	0·00007	00
			·8		·6		

PARTS for 0′·1 :—   LOGS. variable ;   NATURALS, negligible.

| 0° | HAVERSINES | | | | | | |

	·6		·8				
′	Log.	Nat.	Log.	Nat.	Log.	Nat.	′
00	1·88169	0·00000	2·13157	0·00000	2·32539	0·00000	59
01	2·73363	0·00000	2·83594	0·00000	2·92745	0·00000	58
02	3·15534	0·00000	3·21971	0·00000	3·27963	0·00000	57
03	3·43800	0·00000	3·48496	0·00000	3·52951	0·00000	56
04	3·65091	0·00000	3·68787	0·00000	3·72333	0·00000	55
05	3·82177	0·00000	3·85225	0·00000	3·88169	0·00000	54
06	3·96448	0·00000	3·99041	0·00000	4·01559	0·00000	53
07	4·08702	0·00000	4·10958	0·00000	4·13157	0·00000	52
08	4·19439	0·00000	4·21436	0·00000	4·23388	0·00000	51
09	4·28993	0·00000	4·30784	0·00000	4·32539	0·00000	50
10	4·37600	0·00000	4·39224	0·00000	4·40818	0·00000	49
11	4·45431	0·00000	4·46916	0·00000	4·48375	0·00000	48
12	4·52613	0·00000	4·53981	0·00000	4·55328	0·00000	47
13	4·59247	0·00000	4·60515	0·00000	4·61765	0·00000	46
14	4·65410	0·00000	4·66591	0·00000	4·67757	0·00000	45
15	4·71164	0·00001	4·72271	0·00001	4·73363	0·00001	44
16	4·76561	0·00001	4·77601	0·00001	4·78629	0·00001	43
17	4·81642	0·00001	4·82623	0·00001	4·83594	0·00001	42
18	4·86442	0·00001	4·87371	0·00001	4·88290	0·00001	41
19	4·90990	0·00001	4·91872	0·00001	4·92745	0·00001	40
20	4·95313	0·00001	4·96152	0·00001	4·96983	0·00001	39
21	4·99430	0·00001	5·00230	0·00001	5·01024	0·00001	38
22	5·03361	0·00001	5·04126	0·00001	5·04885	0·00001	37
23	5·07121	0·00001	5·07854	0·00001	5·08581	0·00001	36
24	5·10726	0·00001	5·11429	0·00001	5·12127	0·00001	35
25	5·14187	0·00001	5·14863	0·00001	5·15534	0·00001	34
26	5·17515	0·00002	5·18166	0·00002	5·18812	0·00002	33
27	5·20721	0·00002	5·21348	0·00002	5·21971	0·00002	32
28	5·23812	0·00002	5·24417	0·00002	5·25019	0·00002	31
29	5·26797	0·00002	5·27382	0·00002	5·27963	0·00002	30
30	5·29683	0·00002	5·30249	0·00002	5·30811	0·00002	29
31	5·32476	0·00002	5·33024	0·00002	5·33569	0·00002	28
32	5·35182	0·00002	5·35714	0·00002	5·36242	0·00002	27
33	5·37807	0·00002	5·38322	0·00002	5·38835	0·00002	26
34	5·40354	0·00003	5·40855	0·00003	5·41352	0·00003	25
35	5·42829	0·00003	5·43315	0·00003	5·43799	0·00003	24
36	5·45235	0·00003	5·45708	0·00003	5·46179	0·00003	23
37	5·47576	0·00003	5·48037	0·00003	5·48495	0·00003	22
38	5·49856	0·00003	5·50305	0·00003	5·50752	0·00003	21
39	5·52078	0·00003	5·52515	0·00003	5·52951	0·00003	20
40	5·54244	0·00004	5·54671	0·00004	5·55095	0·00004	19
41	5·56357	0·00004	5·56774	0·00004	5·57189	0·00004	18
42	5·58421	0·00004	5·58827	0·00004	5·59232	0·00004	17
43	5·60436	0·00004	5·60833	0·00004	5·61229	0·00004	16
44	5·62406	0·00004	5·62794	0·00004	5·63181	0·00004	15
45	5·64332	0·00004	5·64712	0·00004	5·65090	0·00004	14
46	5·66216	0·00005	5·66588	0·00005	5·66958	0·00005	13
47	5·68060	0·00005	5·68424	0·00005	5·68787	0·00005	12
48	5·69866	0·00005	5·70222	0·00005	5·70578	0·00005	11
49	5·71635	0·00005	5·71984	0·00005	5·72332	0·00005	10
50	5·73369	0·00005	5·73711	0·00005	5·74052	0·00006	09
51	5·75068	0·00006	5·75404	0·00006	5·75739	0·00006	08
52	5·76736	0·00006	5·77065	0·00006	5·77394	0·00006	07
53	5·78371	0·00006	5·78695	0·00006	5·79017	0·00006	06
54	5·79977	0·00006	5·80294	0·00006	5·80611	0·00006	05
55	5·81553	0·00007	5·81865	0·00007	5·82176	0·00007	04
56	5·83102	0·00007	5·83408	0·00007	5·83713	0·00007	03
57	5·84623	0·00007	5·84924	0·00007	5·85224	0·00007	02
58	5·86118	0·00007	5·86414	0·00007	5·86709	0·00007	01
59	5·87587	0·00008	5·87878	0·00008	5·88168	0·00008	00

| | ·4 | | ·2 | | ·0 | | |

PARTS for 0′·1 :— LOGS. variable ; NATURALS, negligible.

1° HAVERSINES

′	·0 Log. 5·/6·	·0 Nat. 0·	·2 Log. 5·/6·	·2 Nat. 0·	·4 Log. 5·/6·	·4 Nat. 0·	·6 Log. 5·/6·	·6 Nat. 0·	·8 Log. 5·/6·	·8 Nat. 0·	Log. 5·/6·	Nat. 0·	′
00	88168	00008	88457	00008	88745	00008	89033	00008	89319	00008	89604	00008	59
01	89604	00008	89888	00008	90172	00008	90454	00008	90736	00008	91016	00008	58
02	91016	00008	91296	00008	91575	00008	91853	00008	92130	00008	92406	00008	57
03	92406	00008	92681	00008	92956	00009	93229	00009	93502	00009	93774	00009	56
04	93774	00009	94045	00009	94315	00009	94584	00009	94853	00009	95121	00009	55
05	95121	00009	95387	00009	95653	00009	95919	00009	96183	00009	96447	00009	54
06	96447	00009	96709	00009	96971	00009	97233	00009	97493	00009	97753	00010	53
07	97753	00010	98012	00010	98270	00010	98527	00010	98784	00010	99040	00010	52
08	99040	00010	99295	00010	99549	00010	99803	00010	00055	00010	00308	00010	51
09	00308	00010	00559	00010	00810	00010	01060	00010	01309	00010	01557	00010	50
10	01557	00010	01805	00010	02052	00010	02299	00011	02544	00011	02789	00011	49
11	02789	00011	03034	00011	03277	00011	03520	00011	03763	00011	04004	00011	48
12	04004	00011	04245	00011	04485	00011	04725	00011	04964	00011	05202	00011	47
13	05202	00011	05440	00011	05677	00011	05913	00011	06149	00012	06384	00012	46
14	06384	00012	06618	00012	06852	00012	07085	00012	07318	00012	07550	00012	45
15	07550	00012	07781	00012	08012	00012	08242	00012	08471	00012	08700	00012	44
16	08700	00012	08928	00012	09156	00012	09383	00012	09610	00012	09836	00013	43
17	09836	00013	10061	00013	10286	00013	10510	00013	10733	00013	10956	00013	42
18	10956	00013	11180	00013	11401	00013	11622	00013	11843	00013	12063	00013	41
19	12063	00013	12282	00013	12501	00013	12720	00013	12938	00013	13155	00014	40
20	13155	00014	13372	00014	13588	00014	13804	00014	14019	00014	14234	00014	39
21	14234	00014	14448	00014	14662	00014	14875	00014	15088	00014	15300	00014	38
22	15300	00014	15512	00014	15723	00014	15933	00014	16143	00014	16353	00015	37
23	16353	00015	16562	00015	16770	00015	16978	00015	17186	00015	17393	00015	36
24	17393	00015	17599	00015	17806	00015	18011	00015	18216	00015	18421	00015	35
25	18421	00015	18625	00015	18829	00015	19032	00016	19234	00016	19437	00016	34
26	19437	00016	19638	00016	19840	00016	20040	00016	20241	00016	20441	00016	33
27	20441	00016	20640	00016	20839	00016	21038	00016	21236	00016	21433	00016	32
28	21433	00016	21631	00016	21827	00017	22024	00017	22219	00017	22415	00017	31
29	22415	00017	22610	00017	22804	00017	22998	00017	23192	00017	23385	00017	30
30	23385	00017	23578	00017	23770	00017	23962	00017	24154	00017	24345	00018	29
31	24345	00018	24536	00018	24726	00018	24916	00018	25105	00018	25294	00018	28
32	25294	00018	25483	00018	25671	00018	25859	00018	26046	00019	26233	00019	27
33	26233	00018	26420	00018	26606	00018	26792	00019	26977	00019	27162	00019	26
34	27162	00019	27347	00019	27531	00019	27715	00019	27898	00019	28081	00019	25
35	28081	00019	28264	00019	28446	00019	28628	00019	28810	00019	28991	00019	24
36	28991	00019	29171	00020	29352	00020	29532	00020	29711	00020	29891	00020	23
37	29891	00020	30070	00020	30248	00020	30426	00020	30604	00020	30781	00020	22
38	30781	00020	30959	00020	31135	00020	31312	00021	31470	00021	31663	00021	21
39	31663	00021	31839	00021	32013	00021	32188	00021	32362	00021	32536	00021	20
40	32536	00021	32710	00021	32883	00021	33056	00021	33228	00022	33400	00022	19
41	33400	00022	33572	00022	33744	00022	33915	00022	34086	00022	34256	00022	18
42	34256	00022	34426	00022	34596	00022	34765	00022	34935	00022	35103	00023	17
43	35103	00022	35272	00023	35440	00023	35608	00023	35775	00023	35943	00023	16
44	35943	00023	36109	00023	36276	00023	36442	00023	36608	00023	36774	00023	15
45	36774	00023	36939	00023	37104	00023	37269	00024	37433	00024	37597	00024	14
46	37597	00024	37761	00024	37924	00024	38087	00024	38250	00024	38412	00024	13
47	38412	00024	38575	00024	38737	00024	38898	00024	39059	00025	39220	00025	12
48	39220	00025	39381	00025	39541	00025	39702	00025	39861	00025	40021	00025	11
49	40021	00025	40180	00025	40339	00025	40498	00025	40656	00026	40814	00026	10
50	40814	00026	40972	00026	41129	00026	41287	00026	41443	00026	41600	00026	09
51	41600	00026	41756	00026	41912	00026	42068	00026	42224	00026	42379	00027	08
52	42379	00027	42534	00027	42689	00027	42843	00027	42997	00027	43151	00027	07
53	43151	00027	43305	00027	43458	00027	43611	00027	43764	00027	43916	00027	06
54	43916	00027	44068	00028	44220	00028	44372	00028	44524	00028	44675	00028	05
55	44675	00028	44826	00028	44976	00028	45127	00028	45277	00028	45427	00028	04
56	45427	00028	45576	00029	45726	00029	45875	00029	46024	00029	46172	00029	03
57	46172	00029	46321	00029	46469	00029	46616	00029	46764	00029	46911	00029	02
58	46911	00029	47058	00030	47205	00030	47352	00030	47498	00030	47644	00030	01
59	47644	00030	47790	00030	47936	00030	48081	00030	48226	00030	48371	00030	00

| ·8 | | ·6 | | ·4 | | ·2 | | ·0 | | |

PARTS for 0′·1 :— LOGS. variable; NATURALS, negligible.

2°   HAVERSINES

′	·0 Log. 6·	·0 Nat. 0·	·2 Log. 6·	·2 Nat. 0·	·4 Log. 6·	·4 Nat. 0·	·6 Log. 6·	·6 Nat. 0·	·8 Log. 6·	·8 Nat. 0·	′
00	48371	00030	48516	00031	48660	00031	48804	00031	48948	00031	49092	00031	59
01	49092	00031	49235	00031	49378	00031	49521	00031	49664	00031	49807	00031	58
02	49807	00031	49949	00032	50091	00032	50233	00032	50374	00032	50516	00032	57
03	50516	00032	50657	00032	50798	00032	50938	00032	51079	00032	51219	00033	56
04	51219	00033	51359	00033	51499	00033	51638	00033	51777	00033	51916	00033	55
05	51916	00033	52055	00033	52194	00033	52332	00033	52471	00033	52608	00034	54
06	52608	00034	52746	00034	52884	00034	53021	00034	53158	00034	53295	00034	53
07	53295	00034	53432	00034	53568	00034	53704	00034	53840	00035	53976	00035	52
08	53976	00035	54112	00035	54247	00035	54382	00035	54517	00035	54652	00035	51
09	54652	00035	54787	00035	54921	00035	55055	00036	55189	00036	55323	00036	50
10	55323	00036	55456	00036	55590	00036	55723	00036	55856	00036	55988	00036	49
11	55988	00036	56121	00036	56253	00037	56385	00037	56517	00037	56649	00037	48
12	56649	00037	56780	00037	56911	00037	57043	00037	57173	00037	57304	00037	47
13	57304	00037	57435	00038	57565	00038	57695	00038	57825	00038	57955	00038	46
14	57955	00038	58084	00038	58214	00038	58343	00038	58472	00038	58600	00039	45
15	58600	00039	58729	00039	58857	00039	58986	00039	59114	00039	59241	00039	44
16	59241	00039	59369	00039	59496	00039	59624	00039	59751	00040	59878	00040	43
17	59878	00040	60004	00040	60131	00040	60257	00040	60383	00040	60509	00040	42
18	60509	00040	60635	00040	60761	00041	60886	00041	61011	00041	61136	00041	41
19	61136	00041	61261	00041	61386	00041	61510	00041	61635	00041	61759	00041	40
20	61759	00041	61883	00042	62007	00042	62130	00042	62254	00042	62377	00042	39
21	62377	00042	62500	00042	62623	00042	62746	00042	62868	00043	62991	00043	38
22	62991	00043	63113	00043	63235	00043	63357	00043	63479	00043	63600	00043	37
23	63600	00043	63722	00043	63843	00043	63964	00044	64085	00044	64205	00044	36
24	64205	00044	64326	00044	64446	00044	64566	00044	64687	00044	64806	00044	35
25	64806	00044	64926	00045	65046	00045	65165	00045	65284	00045	65403	00045	34
26	65403	00045	65522	00045	65641	00045	65759	00045	65878	00046	65996	00046	33
27	65996	00046	66114	00046	66232	00046	66350	00046	66467	00046	66585	00046	32
28	66585	00046	66702	00046	66819	00047	66936	00047	67053	00047	67170	00047	31
29	67170	00047	67286	00047	67403	00047	67519	00047	67635	00047	67751	00048	30
30	67751	00048	67866	00048	67982	00048	68097	00048	68213	00048	68328	00048	29
31	68328	00048	68443	00048	68557	00049	68672	00049	68787	00049	68901	00049	28
32	68901	00049	69015	00049	69129	00049	69243	00049	69356	00050	69470	00050	27
33	69470	00050	69584	00050	69697	00050	69810	00050	69923	00050	70036	00050	26
34	70036	00050	70149	00050	70261	00050	70374	00051	70486	00051	70598	00051	25
35	70598	00051	70710	00051	70822	00051	70934	00051	71045	00051	71157	00051	24
36	71157	00051	71268	00052	71379	00052	71490	00052	71601	00052	71712	00052	23
37	71712	00052	71822	00052	71933	00052	72043	00053	72153	00053	72263	00053	22
38	72263	00053	72373	00053	72483	00053	72592	00053	72702	00053	72811	00053	21
39	72811	00053	72920	00054	73029	00054	73138	00054	73247	00054	73355	00054	20
40	73355	00054	73464	00054	73572	00054	73680	00055	73789	00055	73896	00055	19
41	73896	00055	74004	00055	74112	00055	74220	00055	74327	00055	74434	00056	18
42	74434	00056	74541	00056	74648	00056	74755	00056	74862	00056	74969	00056	17
43	74969	00056	75075	00056	75181	00056	75288	00057	75394	00057	75500	00057	16
44	75500	00057	75606	00057	75711	00057	75817	00057	75922	00057	76028	00058	15
45	76028	00058	76133	00058	76238	00058	76343	00058	76448	00058	76552	00058	14
46	76552	00058	76657	00058	76761	00059	76866	00059	76970	00059	77074	00059	13
47	77074	00059	77178	00059	77282	00059	77385	00059	77489	00060	77592	00060	12
48	77592	00060	77696	00060	77799	00060	77902	00060	78005	00060	78108	00060	11
49	78108	00060	78211	00061	78313	00061	78416	00061	78518	00061	78620	00061	10
50	78620	00061	78722	00061	78824	00061	78926	00062	79028	00062	79129	00062	09
51	79129	00062	79231	00062	79332	00062	79434	00062	79535	00062	79636	00063	08
52	79636	00063	79737	00063	79838	00063	79938	00063	80039	00063	80139	00063	07
53	80139	00063	80240	00063	80340	00064	80440	00064	80540	00064	80640	00064	06
54	80640	00064	80740	00064	80839	00064	80939	00064	81038	00065	81137	00065	05
55	81137	00065	81237	00065	81336	00065	81435	00065	81534	00065	81632	00066	04
56	81632	00066	81731	00066	81829	00066	81928	00066	82026	00066	82124	00066	03
57	82124	00066	82222	00066	82320	00067	82418	00067	82516	00067	82614	00067	02
58	82614	00067	82711	00067	82808	00067	82906	00067	83003	00068	83100	00068	01
59	83100	00068	83197	00068	83294	00068	83391	00068	83487	00068	83584	00069	00

·8	·6	·4	·2	·0	← ↑

357°

PARTS for 0′·1 :—   LOGS. variable.   NATURALS, negligible;

3°   HAVERSINES

′	·0 Log. 6·/7·	·0 Nat. 0·	·2 Log. 6·/7·	·2 Nat. 0·	·4 Log. 6·/7·	·4 Nat. 0·	·6 Log. 6·/7·	·6 Nat. 0·	·8 Log. 6·/7·	·8 Nat. 0·	′
00	83584	00069	83680	00069	83777	00069	83873	00069	83969	00069	59
00	84065	00069	84161	00069	83777	00069	83873	00069	83969	00069	

′	·0 Log.	·0 Nat.	·2 Log.	·2 Nat.	·4 Log.	·4 Nat.	·6 Log.	·6 Nat.	·8 Log.	·8 Nat.	′
00	83584	00069	83680	00069	83777	00069	83873	00069	83969	00069	59
01	84065	00069	84161	00069	84257	00070	84352	00070	84448	00070	58
02	84543	00070	84639	00070	84734	00070	84829	00071	84924	00071	57
03	85019	00071	85114	00071	85209	00071	85303	00071	85398	00071	56
04	85492	00072	85587	00072	85681	00072	85774	00072	85869	00072	55
05	85963	00072	86057	00073	86151	00073	86244	00073	86338	00073	54
06	86431	00073	86525	00073	86618	00073	86711	00074	86804	00074	53
07	86897	00074	86990	00074	87082	00074	87175	00074	87268	00075	52
08	87360	00075	87452	00075	87545	00075	87637	00075	87729	00075	51
09	87821	00076	87912	00076	88004	00076	88096	00076	88187	00076	50
10	88279	00076	88370	00077	88462	00077	88553	00077	88644	00077	49
11	88735	00077	88826	00077	88916	00077	89007	00078	89098	00078	48
12	89188	00078	89279	00078	89369	00078	89459	00078	89549	00079	47
13	89639	00079	89729	00079	89819	00079	89909	00079	89998	00079	46
14	90088	00080	90178	00080	90267	00080	90356	00080	90445	00080	45
15	90535	00080	90624	00081	90712	00081	90801	00081	90890	00081	44
16	90979	00081	91067	00081	91156	00082	91244	00082	91332	00082	43
17	91421	00082	91509	00082	91597	00082	91685	00083	91773	00083	42
18	91860	00083	91948	00083	92036	00083	92123	00083	92210	00084	41
19	92298	00084	92385	00084	92472	00084	92559	00084	92646	00084	40
20	92733	00085	92820	00085	92906	00085	92993	00085	93080	00085	39
21	93166	00085	93252	00086	93339	00086	93425	00086	93511	00086	38
22	93597	00086	93683	00087	93769	00087	93855	00087	93940	00087	37
23	94026	00087	94111	00087	94197	00087	94282	00088	94367	00088	36
24	94453	00088	94538	00088	94623	00088	94708	00089	94792	00089	35
25	94877	00089	94962	00089	95046	00089	95131	00089	95215	00090	34
26	95300	00090	95384	00090	95468	00090	95552	00090	95636	00090	33
27	95720	00091	95804	00091	95888	00091	95972	00091	96055	00091	32
28	96139	00091	96222	00092	96305	00092	96389	00092	96472	00092	31
29	96555	00092	96638	00093	96721	00093	96804	00093	96888	00093	30
30	96970	00093	97052	00093	97135	00094	97217	00094	97300	00094	29
31	97382	00094	97464	00094	97547	00095	97629	00095	97711	00095	28
32	97793	00095	97875	00095	97956	00095	98038	00096	98120	00096	27
33	98201	00096	98283	00096	98364	00096	98446	00096	98527	00097	26
34	98608	00097	98689	00097	98770	00097	98851	00097	98932	00098	25
35	99013	00098	99094	00098	99174	00098	99255	00098	99334	00098	24
36	99416	00099	99496	00099	99576	00099	99657	00099	99737	00099	23
37	99817	00100	99897	00100	99977	00100	00057	00100	00136	00100	22
38	00216	00101	00296	00101	00375	00101	00455	00101	00534	00101	21
39	00613	00101	00693	00102	00772	00102	00851	00102	00930	00102	20
40	01009	00102	01088	00103	01167	00103	01245	00103	01324	00103	19
41	01403	00103	01481	00103	01560	00104	01638	00104	01716	00104	18
42	01795	00104	01873	00104	01951	00105	02029	00105	02107	00105	17
43	02185	00105	02263	00105	02341	00106	02418	00106	02496	00106	16
44	02573	00106	02651	00106	02728	00106	02806	00107	02883	00107	15
45	02960	00107	03037	00107	03114	00107	03191	00108	03268	00108	14
46	03345	00108	03422	00108	03499	00108	03575	00109	03652	00109	13
47	03729	00109	03805	00109	03881	00109	03958	00110	04034	00110	12
48	04110	00110	04186	00110	04262	00110	04338	00111	04414	00111	11
49	04490	00111	04566	00111	04642	00111	04717	00111	04793	00112	10
50	04869	00112	04944	00112	05019	00112	05095	00112	05170	00113	09
51	05245	00113	05320	00113	05396	00113	05471	00113	05545	00114	08
52	05620	00114	05695	00114	05770	00114	05845	00114	05919	00115	07
53	05994	00115	06068	00115	06143	00115	06217	00115	06291	00116	06
54	06366	00116	06440	00116	06514	00116	06588	00116	06662	00117	05
55	06736	00117	06810	00117	06884	00117	06957	00117	07031	00118	04
56	07105	00118	07178	00118	07252	00118	07325	00118	07398	00119	03
57	07472	00119	07545	00119	07618	00119	07691	00119	07764	00120	02
58	07837	00120	07910	00120	07983	00120	08056	00120	08129	00121	01
59	08201	00121	08274	00121	08347	00121	08419	00121	08491	00122	00

·8   ·6   ·4   ·2   ·0

PARTS for 0′·1 :—   LOGS. variable ;   NATURALS, negligible.

4°　　HAVERSINES

′	·0 Log. 7·	·0 Nat. 0·	·2 Log. 7·	·2 Nat. 0·	·4 Log. 7·	·4 Nat. 0·	·6 Log. 7·	·6 Nat. 0·	·8 Log. 7·	·8 Nat. 0·	Log. 7·	Nat. 0·	′
00	08564	00122	08636	00122	08708	00122	08781	00122	08852	00123	08925	00123	59
01	08925	00123	08997	00123	09069	00123	09141	00123	09213	00124	09284	00124	58
02	09284	00124	09356	00124	09428	00124	09499	00124	09571	00125	09642	00125	57
03	09642	00125	09714	00125	09785	00125	09857	00125	09928	00126	09999	00126	56
04	09999	00126	10070	00126	10141	00126	10212	00127	10283	00127	10354	00127	55
05	10354	00127	10425	00127	10496	00127	10566	00128	10637	00128	10708	00128	54
06	10708	00128	10778	00128	10849	00128	10919	00129	10990	00129	11060	00129	53
07	11060	00129	11130	00129	11200	00129	11271	00130	11341	00130	11411	00130	52
08	11411	00130	11481	00130	11551	00130	11621	00131	11690	00131	11760	00131	51
09	11760	00131	11830	00131	11899	00132	11969	00132	12039	00132	12108	00132	50
10	12108	00132	12178	00132	12247	00133	12316	00133	12385	00133	12455	00133	49
11	12455	00133	12524	00133	12593	00134	12662	00134	12731	00134	12800	00134	48
12	12800	00134	12869	00135	12938	00135	13006	00135	13075	00135	13144	00135	47
13	13144	00135	13212	00136	13281	00136	13349	00136	13418	00136	13486	00136	46
14	13486	00136	13555	00137	13623	00137	13691	00137	13759	00137	13827	00137	45
15	13827	00137	13895	00138	13963	00138	14031	00138	14099	00138	14167	00139	44
16	14167	00139	14235	00139	14303	00139	14370	00139	14438	00139	14506	00140	43
17	14506	00140	14573	00140	14641	00140	14708	00140	14775	00141	14843	00141	42
18	14843	00141	14910	00141	14977	00141	15044	00141	15112	00142	15179	00142	41
19	15179	00142	15246	00142	15313	00142	15380	00143	15446	00143	15513	00143	40
20	15513	00143	15580	00143	15647	00143	15713	00144	15780	00144	15846	00144	39
21	15846	00144	15913	00144	15979	00144	16046	00145	16112	00145	16178	00145	38
22	16178	00145	16245	00145	16311	00146	16377	00146	16443	00146	16509	00146	37
23	16509	00146	16575	00146	16641	00147	16707	00147	16773	00147	16839	00147	36
24	16839	00147	16904	00148	16970	00148	17036	00148	17101	00148	17167	00148	35
25	17167	00148	17232	00149	17298	00149	17363	00149	17429	00149	17494	00150	34
26	17494	00150	17559	00150	17624	00150	17689	00150	17755	00151	17820	00151	33
27	17820	00151	17885	00151	17950	00151	18015	00151	18079	00152	18144	00152	32
28	18144	00152	18209	00152	18274	00152	18338	00153	18403	00153	18468	00153	31
29	18468	00153	18532	00153	18597	00153	18661	00154	18725	00154	18790	00154	30
30	18790	00154	18854	00154	18918	00155	18982	00155	19047	00155	19111	00155	29
31	19111	00155	19175	00156	19239	00156	19303	00156	19367	00156	19430	00156	28
32	19430	00156	19494	00157	19558	00157	19622	00157	19685	00157	19749	00158	27
33	19749	00158	19813	00158	19876	00158	19940	00158	20003	00159	20066	00159	26
34	20066	00159	20130	00159	20193	00159	20256	00159	20319	00160	20383	00160	25
35	20383	00160	20446	00160	20509	00160	20572	00161	20635	00161	20698	00161	24
36	20698	00161	20761	00161	20823	00162	20886	00162	20949	00162	21012	00162	23
37	21012	00162	21074	00162	21137	00163	21200	00163	21262	00163	21325	00163	22
38	21325	00163	21387	00164	21449	00164	21512	00164	21574	00164	21636	00165	21
39	21636	00165	21698	00165	21761	00165	21823	00165	21885	00166	21947	00166	20
40	21947	00166	22009	00166	22071	00166	22133	00166	22195	00167	22256	00167	19
41	22256	00167	22318	00167	22380	00167	22441	00168	22503	00168	22565	00168	18
42	22565	00168	22626	00168	22688	00169	22749	00169	22811	00169	22872	00169	17
43	22872	00169	22933	00170	22995	00170	23056	00170	23117	00170	23178	00171	16
44	23178	00171	23239	00171	23300	00171	23361	00171	23422	00171	23483	00172	15
45	23483	00172	23542	00172	23605	00172	23666	00172	23727	00173	23787	00173	14
46	23787	00173	23848	00173	23909	00173	23969	00174	24030	00174	24090	00174	13
47	24090	00174	24151	00174	24211	00175	24272	00175	24332	00175	24392	00175	12
48	24392	00175	24453	00176	24513	00176	24573	00176	24633	00176	24693	00177	11
49	24693	00177	24753	00177	24813	00177	24873	00177	24933	00178	24993	00178	10
50	24993	00178	25053	00178	25113	00178	25172	00179	25232	00179	25292	00179	09
51	25292	00179	25352	00179	25411	00180	25471	00180	25530	00180	25590	00180	08
52	25590	00180	25649	00181	25709	00181	25768	00181	25827	00181	25886	00181	07
53	25886	00181	25946	00182	26005	00182	26064	00182	26123	00182	26182	00183	06
54	26182	00183	26241	00183	26300	00183	26359	00183	26418	00184	26477	00184	05
55	26477	00184	26536	00184	26595	00184	26653	00185	26712	00185	26771	00185	04
56	26771	00185	26829	00185	26888	00186	26947	00186	27005	00186	27063	00186	03
57	27063	00186	27122	00187	27180	00187	27239	00187	27297	00187	27355	00188	02
58	27355	00188	27414	00188	27472	00188	27530	00188	27588	00189	27646	00189	01
59	27646	00189	27704	00189	27762	00189	27820	00190	27878	00190	27936	00190	00

	·8	·6	·4	·2	·0	

355°

PARTS for 0′·1 :—　　　LOGS. variable ;　　　NATURALS, negligible.

5° HAVERSINES

′	·0 Log. 7·	·0 Nat. 0·	·2 Log. 7·	·2 Nat. 0·	·4 Log. 7·	·4 Nat. 0·	·6 Log. 7·	·6 Nat. 0·	·8 Log. 7·	·8 Nat. 0·	Log. 7·	Nat. 0·	′
00	27936	00190	27994	00191	28052	00191	28109	00191	28167	00191	28225	00192	59
01	28225	00192	28282	00192	28340	00192	28390	00192	28455	00193	28513	00193	58
02	28513	00193	28570	00193	28628	00193	28685	00194	28742	00194	28800	00194	57
03	28800	00194	28859	00194	28914	00195	28971	00195	29029	00195	29086	00195	56
04	29086	00195	29143	00196	29200	00196	29257	00196	29314	00196	29371	00197	55
05	29371	00197	29428	00197	29484	00197	29541	00197	29598	00198	29655	00198	54
06	29655	00198	29712	00198	29768	00198	29825	00199	29881	00199	29938	00199	53
07	29938	00199	29995	00200	30051	00200	30108	00200	30164	00200	30220	00201	52
08	30220	00201	30277	00201	30333	00201	30389	00201	30445	00202	30502	00202	51
09	30502	00202	30558	00202	30614	00202	30670	00203	30726	00203	30782	00203	50
10	30782	00203	30838	00203	30894	00204	30950	00204	31006	00204	31062	00204	49
11	31062	00204	31117	00205	31173	00205	31229	00205	31285	00206	31340	00206	48
12	31340	00206	31396	00206	31452	00206	31507	00207	31563	00207	31618	00207	47
13	31618	00207	31674	00207	31729	00208	31784	00208	31840	00208	31895	00208	46
14	31895	00208	31950	00209	32005	00209	32061	00209	32116	00209	32171	00210	45
15	32171	00210	32226	00210	32281	00210	32336	00211	32391	00211	32446	00211	44
16	32446	00211	32501	00211	32556	00212	32611	00212	32666	00212	32720	00212	43
17	32720	00212	32775	00213	32830	00213	32884	00213	32939	00213	32994	00214	42
18	32994	00214	33048	00214	33103	00214	33157	00215	33212	00215	33266	00215	41
19	33266	00215	33321	00215	33375	00216	33429	00216	33484	00216	33538	00216	40
20	33538	00216	33592	00217	33646	00217	33700	00217	33755	00218	33809	00218	39
21	33809	00218	33863	00218	33917	00218	33971	00219	34025	00219	34079	00219	38
22	34079	00219	34133	00219	34186	00220	34240	00220	34294	00220	34348	00221	37
23	34348	00221	34402	00221	34455	00221	34509	00221	34562	00222	34616	00222	36
24	34616	00222	34670	00222	34723	00222	34777	00223	34830	00223	34884	00223	35
25	34884	00223	34937	00224	34990	00224	35044	00224	35097	00224	35150	00225	34
26	35150	00225	35203	00225	35257	00225	35310	00225	35363	00226	35416	00226	33
27	35416	00226	35469	00226	35522	00227	35575	00227	35628	00227	35681	00227	32
28	35681	00227	35734	00228	35787	00228	35840	00228	35892	00229	35945	00229	31
29	35945	00229	35998	00229	36051	00229	36103	00230	36156	00230	36209	00230	30
30	36209	00230	36261	00230	36314	00231	36366	00231	36419	00231	36471	00232	29
31	36471	00232	36524	00232	36576	00232	36628	00232	36681	00233	36733	00233	28
32	36733	00233	36785	00233	36838	00234	36890	00234	36942	00234	36994	00234	27
33	36994	00234	37046	00235	37098	00235	37150	00235	37202	00236	37254	00236	26
34	37254	00236	37306	00236	37358	00236	37410	00237	37462	00237	37514	00237	25
35	37514	00237	37566	00237	37617	00238	37669	00238	37721	00238	37772	00239	24
36	37772	00239	37824	00239	37876	00239	37927	00239	37979	00240	38030	00240	23
37	38030	00240	38082	00240	38133	00241	38185	00241	38236	00241	38288	00241	22
38	38288	00241	38339	00242	38390	00242	38442	00242	38493	00243	38544	00243	21
39	38544	00243	38595	00243	38646	00243	38697	00244	38749	00244	38800	00244	20
40	38800	00244	38851	00245	38902	00245	38953	00245	39004	00246	39054	00246	19
41	39054	00246	39105	00246	39156	00246	39207	00247	39258	00247	39309	00247	18
42	39309	00247	39359	00247	39410	00248	39461	00248	39511	00248	39562	00249	17
43	39562	00249	39613	00249	39663	00249	39714	00250	39764	00250	39815	00250	16
44	39815	00250	39865	00250	39916	00251	39966	00251	40016	00251	40067	00252	15
45	40067	00252	40117	00252	40167	00252	40217	00252	40268	00253	40318	00253	14
46	40318	00253	40368	00253	40418	00254	40468	00254	40518	00254	40568	00255	13
47	40568	00255	40618	00255	40668	00255	40718	00255	40768	00256	40818	00256	12
48	40818	00256	40868	00256	40918	00257	40967	00257	41017	00257	41067	00257	11
49	41067	00257	41117	00258	41166	00258	41216	00258	41266	00259	41315	00259	10
50	41315	00259	41365	00259	41414	00259	41464	00260	41513	00260	41563	00260	09
51	41563	00260	41612	00261	41662	00261	41711	00261	41760	00262	41810	00262	08
52	41810	00262	41858	00262	41908	00262	41958	00263	42007	00263	42056	00263	07
53	42056	00263	42105	00264	42154	00264	42203	00264	42252	00265	42301	00265	06
54	42301	00265	42351	00265	42400	00265	42448	00266	42497	00266	42546	00266	05
55	42546	00266	42595	00267	42644	00267	42693	00267	42742	00268	42790	00268	04
56	42790	00268	42839	00268	42888	00268	42937	00269	42985	00269	43034	00269	03
57	43034	00269	43082	00270	43131	00270	43180	00270	43228	00271	43277	00271	02
58	43277	00271	43325	00271	43373	00271	43422	00272	43470	00272	43519	00272	01
59	43519	00272	43567	00273	43615	00273	43664	00273	43712	00274	43760	00274	00
	·8		6·		·4		·2		0·				

PARTS for 0′·1 :— LOGS. variable from 29 to 24 ; NATURALS, negligible.

6°   HAVERSINES

	·0		·2		·4		·6		·8		·0		
	Log.	Nat.	Log.	Nat.	Log.	Nat.	Log.	Nat.	Log.	Nat.	Log.	Nat.	
	7·	0·	7·	0·	7·	0·	7·	0·	7·	0·	7·	0·	
00	43760	00274	43808	00274	43856	00275	43905	00275	43953	00275	44001	00275	59
01	44001	00275	44049	00276	44097	00276	44145	00276	44193	00277	44241	00277	58
02	44241	00277	44289	00277	44337	00278	44385	00278	44432	00278	44480	00278	57
03	44480	00278	44528	00279	44576	00279	44624	00279	44671	00280	44719	00280	56
04	44719	00280	44767	00280	44814	00281	44862	00281	44909	00281	44957	00282	55
05	44957	00282	45005	00282	45052	00282	45100	00282	45147	00283	45194	00283	54
06	45194	00283	45242	00283	45289	00284	45337	00284	45384	00284	45431	00285	53
07	45431	00285	45478	00285	45526	00285	45573	00286	45620	00286	45667	00286	52
08	45667	00286	45714	00287	45762	00287	45809	00287	45856	00287	45903	00288	51
09	45903	00288	45950	00288	45997	00288	46044	00289	46091	00289	46138	00289	50
10	46138	00289	46185	00290	46231	00290	46278	00290	46325	00291	46372	00291	49
11	46372	00291	46419	00291	46465	00292	46512	00292	46559	00292	46605	00292	48
12	46605	00292	46652	00293	46699	00293	46745	00293	46792	00294	46838	00294	47
13	46838	00294	46885	00294	46931	00295	46978	00295	47024	00295	47071	00296	46
14	47071	00296	47117	00296	47163	00296	47210	00297	47256	00297	47302	00297	45
15	47302	00297	47349	00298	47395	00298	47441	00298	47487	00298	47533	00299	44
16	47533	00299	47580	00299	47626	00299	47672	00300	47718	00300	47764	00300	43
17	47764	00300	47810	00301	47856	00301	47902	00301	47948	00302	47994	00302	42
18	47994	00302	48040	00302	48086	00303	48131	00303	48177	00303	48223	00304	41
19	48223	00304	48269	00304	48315	00304	48360	00305	48406	00305	48452	00305	40
20	48452	00305	48497	00305	48543	00306	48589	00306	48634	00306	48680	00307	39
21	48680	00307	48725	00307	48771	00307	48816	00308	48862	00308	48907	00308	38
22	48907	00308	48953	00309	48998	00309	49043	00309	49089	00310	49134	00310	37
23	49134	00310	49179	00310	49225	00311	49270	00311	49315	00311	49360	00312	36
24	49360	00312	49405	00312	49451	00312	49496	00313	49541	00313	49586	00313	35
25	49586	00313	49631	00314	49676	00314	49721	00314	49766	00315	49811	00315	34
26	49811	00315	49856	00315	49901	00316	49946	00316	49991	00316	50036	00316	33
27	50036	00316	50080	00317	50125	00317	50170	00317	50215	00318	50259	00318	32
28	50259	00318	50304	00318	50349	00319	50394	00319	50438	00319	50483	00320	31
29	50483	00320	50527	00320	50572	00320	50617	00321	50661	00321	50706	00321	30
30	50706	00321	50750	00322	50795	00322	50839	00322	50883	00323	50928	00323	29
31	50928	00323	50972	00323	51016	00324	51061	00324	51105	00324	51149	00325	28
32	51149	00325	51194	00325	51238	00325	51282	00326	51326	00326	51370	00326	27
33	51370	00326	51415	00327	51459	00327	51503	00327	51547	00328	51591	00328	26
34	51591	00328	51635	00328	51679	00329	51723	00329	51767	00329	51811	00330	25
35	51811	00330	51855	00330	51899	00330	51943	00331	51986	00331	52030	00331	24
36	52030	00331	52074	00332	52118	00332	52162	00332	52205	00333	52249	00333	23
37	52249	00333	52293	00333	52336	00334	52380	00334	52424	00334	52467	00335	22
38	52467	00335	52511	00335	52554	00335	52598	00336	52642	00336	52685	00336	21
39	52685	00336	52729	00337	52772	00337	52815	00337	52859	00338	52902	00338	20
40	52902	00338	52946	00338	52989	00339	53032	00339	53076	00339	53119	00340	19
41	53119	00340	53162	00340	53205	00340	53249	00341	53292	00341	53335	00341	18
42	53335	00341	53378	00342	53421	00342	53464	00342	53507	00343	53550	00343	17
43	53550	00343	53594	00343	53637	00344	53680	00344	53723	00345	53765	00345	16
44	53765	00345	53808	00345	53851	00346	53894	00346	53939	00346	53980	00347	15
45	53980	00347	54023	00347	54066	00347	54108	00348	54151	00348	54194	00348	14
46	54194	00348	54237	00349	54279	00349	54322	00349	54365	00350	54407	00350	13
47	54407	00350	54450	00350	54493	00351	54535	00351	54578	00351	54620	00352	12
48	54620	00352	54663	00352	54705	00352	54748	00353	54790	00353	54833	00353	11
49	54833	00353	54875	00354	54917	00354	54960	00354	55002	00355	55045	00355	10
50	55045	00355	55087	00356	55129	00356	55171	00356	55214	00357	55256	00357	09
51	55256	00357	55298	00357	55340	00358	55382	00358	55425	00358	55467	00359	08
52	55467	00359	55509	00359	55551	00359	55593	00360	55635	00360	55677	00360	07
53	55677	00360	55719	00361	55761	00361	55803	00361	55845	00362	55887	00362	06
54	55887	00362	55929	00362	55971	00363	56012	00363	56054	00364	56096	00364	05
55	56096	00364	56138	00364	56180	00365	56221	00365	56263	00365	56305	00366	04
56	56305	00366	56347	00366	56388	00366	56430	00367	56472	00367	56513	00367	03
57	56513	00367	56555	00368	56596	00368	56638	00368	56679	00369	56721	00369	02
58	56721	00369	56762	00369	56804	00370	56845	00370	56887	00371	56928	00371	01
59	56928	00371	56970	00371	57011	00372	57052	00372	57094	00372	57135	00373	00
		·8		·6		·4		·2		·0			

PARTS for 0'·1 :—   LOGS. variable from 24 to 21 ;   NATURALS, negligible.

353°

7° HAVERSINES

′	·0 Log. 7·	·0 Nat. 0·	·2 Log. 7·	·2 Nat. 0·	·4 Log. 7·	·4 Nat. 0·	·6 Log. 7·	·6 Nat. 0·	·8 Log. 7·	·8 Nat. 0·	′
00	57135	00373	57176	00373	57218	00373	57259	00374	57300	00374	57341	00374	59
01	57341	00374	57383	00375	57424	00375	57465	00376	57506	00376	57547	00376	58
02	57547	00376	57588	00377	57629	00377	57670	00377	57711	00378	57752	00378	57
03	57752	00378	57794	00378	57835	00379	57875	00379	57916	00379	57957	00380	56
04	57957	00380	57998	00380	58039	00381	58080	00381	58121	00381	58162	00382	55
05	58162	00382	58203	00382	58243	00382	58284	00383	58325	00383	58366	00383	54
06	58366	00383	58406	00384	58447	00384	58488	00384	58528	00385	58569	00385	53
07	58569	00385	58609	00386	58650	00386	58691	00386	58731	00387	58772	00387	52
08	58772	00387	58812	00387	58853	00388	58893	00388	58934	00388	58974	00389	51
09	58974	00389	59015	00389	59055	00390	59096	00390	59136	00390	59176	00391	50
10	59176	00391	59217	00391	59257	00391	59297	00392	59338	00392	59378	00392	49
11	59378	00392	59418	00393	59458	00393	59498	00394	59539	00394	59579	00394	48
12	59579	00394	59619	00395	59659	00395	59698	00395	59739	00396	59779	00396	47
13	59779	00396	59819	00396	59859	00397	59900	00397	59940	00398	59979	00398	46
14	59979	00398	60019	00398	60059	00399	60099	00399	60139	00399	60179	00400	45
15	60179	00400	60219	00400	60259	00400	60299	00401	60339	00401	60378	00402	44
16	60378	00402	60418	00402	60458	00402	60498	00403	60537	00403	60577	00403	43
17	60577	00403	60617	00404	60656	00404	60696	00405	60736	00405	60775	00405	42
18	60775	00405	60815	00406	60854	00406	60894	00406	60934	00407	60973	00407	41
19	60973	00407	61013	00408	61052	00408	61092	00408	61131	00409	61170	00409	40
20	61170	00409	61210	00409	61249	00410	61289	00410	61328	00410	61367	00411	39
21	61367	00411	61407	00411	61446	00412	61485	00412	61525	00412	61564	00413	38
22	61564	00413	61603	00413	61642	00413	61682	00414	61721	00414	61760	00415	37
23	61760	00415	61799	00415	61838	00415	61877	00416	61916	00416	61955	00416	36
24	61955	00416	61995	00417	62034	00417	62073	00418	62112	00418	62151	00418	35
25	62151	00418	62190	00419	62229	00419	62267	00419	62306	00420	62345	00420	34
26	62345	00420	62384	00421	62423	00421	62462	00421	62501	00422	62540	00422	33
27	62540	00422	62578	00422	62617	00423	62656	00423	62695	00424	62733	00424	32
28	62733	00424	62772	00424	62811	00425	62849	00425	62888	00425	62927	00426	31
29	62927	00426	62965	00426	63004	00427	63043	00427	63081	00427	63120	00428	30
30	63120	00428	63158	00428	63197	00429	63235	00429	63274	00429	63312	00430	29
31	63312	00430	63351	00430	63389	00430	63428	00431	63466	00431	63504	00432	28
32	63504	00432	63543	00432	63581	00432	63619	00433	63658	00433	63696	00433	27
33	63696	00433	63734	00434	63773	00434	63811	00435	63849	00435	63887	00435	26
34	63887	00435	63925	00436	63964	00436	64002	00437	64040	00437	64078	00437	25
35	64078	00437	64116	00438	64154	00438	64192	00438	64230	00439	64269	00439	24
36	64269	00439	64307	00440	64345	00440	64383	00440	64421	00441	64458	00441	23
37	64458	00441	64496	00442	64534	00442	64572	00442	64610	00443	64648	00443	22
38	64648	00443	64686	00443	64724	00444	64762	00444	64799	00445	64837	00445	21
39	64837	00445	64875	00445	64913	00446	64951	00446	64988	00447	65026	00447	20
40	65026	00447	65064	00447	65101	00448	65139	00448	65177	00449	65214	00449	19
41	65214	00449	65252	00449	65290	00450	65327	00450	65365	00450	65402	00451	18
42	65402	00451	65440	00451	65477	00452	65515	00452	65552	00452	65590	00453	17
43	65590	00453	65627	00453	65665	00454	65702	00454	65739	00454	65777	00455	16
44	65777	00455	65814	00455	65852	00456	65889	00456	65926	00456	65964	00457	15
45	65964	00457	66001	00457	66038	00457	66075	00458	66113	00458	66150	00459	14
46	66150	00459	66187	00459	66224	00459	66261	00460	66299	00460	66336	00461	13
47	66336	00461	66373	00461	66410	00461	66447	00462	66484	00462	66521	00463	12
48	66521	00463	66558	00463	66595	00463	66632	00464	66669	00464	66706	00465	11
49	66706	00465	66743	00465	66780	00465	66817	00466	66854	00466	66891	00467	10
50	66891	00467	66928	00467	66965	00467	67002	00468	67039	00468	67075	00469	09
51	67075	00469	67112	00469	67149	00469	67186	00470	67223	00470	67259	00471	08
52	67259	00471	67296	00471	67333	00471	67370	00472	67406	00472	67443	00473	07
53	67443	00473	67480	00473	67516	00473	67553	00474	67589	00474	67626	00475	06
54	67626	00475	67663	00475	67699	00475	67736	00476	67772	00476	67809	00477	05
55	67809	00477	67845	00477	67882	00477	67918	00478	67955	00478	67991	00479	04
56	67991	00479	68028	00479	68064	00479	68100	00480	68137	00480	68173	00481	03
57	68173	00481	68210	00481	68246	00481	68282	00482	68319	00482	68355	00483	02
58	68355	00483	68391	00483	68427	00483	68464	00484	68500	00484	68536	00485	01
59	68536	00485	68572	00485	68608	00485	68645	00486	68681	00486	68717	00487	00

| | ·8 | ·6 | ·4 | ·2 | ·0 | ← |

352°

PARTS for 0′·1 :— LOGS. variable from 21 to 18 ; NATURALS, negligible.

8° HAVERSINES

′	.0 Log. 7·	.0 Nat. 0·	.2 Log. 7·	.2 Nat. 0·	.4 Log. 7·	.4 Nat. 0·	.6 Log. 7·	.6 Nat. 0·	.8 Log. 7·	.8 Nat. 0·	′
00	68717	00487	68753	00487	68789	00487	68825	00488	68861	00488	68897	00489	59
01	68897	00489	68933	00489	68969	00489	69005	00490	69041	00490	69077	00491	58
02	69077	00491	69113	00491	69149	00491	69185	00492	69221	00492	69257	00493	57
03	69257	00493	69293	00493	69329	00494	69365	00494	69401	00494	69437	00495	56
04	69437	00495	69472	00495	69508	00496	69544	00496	69580	00496	69616	00497	55
05	69616	00497	69651	00497	69687	00498	69723	00498	69758	00498	69794	00499	54
06	69794	00499	69830	00499	69865	00500	69901	00500	69937	00500	69972	00501	53
07	69972	00501	70008	00501	70044	00502	70079	00502	70115	00502	70150	00503	52
08	70150	00503	70186	00503	70221	00504	70257	00504	70292	00505	70328	00505	51
09	70328	00505	70363	00505	70399	00506	70434	00506	70470	00507	70505	00507	50
10	70505	00507	70540	00507	70576	00508	70611	00508	70646	00509	70682	00509	49
11	70682	00509	70717	00510	70752	00510	70788	00510	70823	00511	70858	00511	48
12	70858	00511	70893	00512	70929	00512	70964	00512	70999	00513	71034	00513	47
13	71034	00513	71069	00514	71104	00514	71140	00515	71175	00515	71210	00515	46
14	71210	00515	71245	00516	71280	00516	71315	00517	71350	00517	71385	00517	45
15	71385	00517	71420	00518	71455	00518	71490	00519	71525	00519	71560	00520	44
16	71560	00520	71595	00520	71630	00520	71665	00521	71700	00521	71735	00522	43
17	71735	00522	71770	00522	71805	00522	71839	00523	71874	00523	71909	00524	42
18	71909	00524	71944	00524	71979	00525	72013	00525	72048	00525	72083	00526	41
19	72083	00526	72118	00526	72152	00527	72187	00527	72222	00527	72257	00528	40
20	72257	00528	72291	00528	72326	00529	72361	00529	72395	00530	72430	00530	39
21	72430	00530	72464	00530	72499	00531	72534	00531	72568	00532	72603	00532	38
22	72603	00532	72637	00533	72672	00533	72706	00533	72741	00534	72775	00534	37
23	72775	00534	72810	00535	72844	00535	72879	00536	72913	00536	72948	00536	36
24	72948	00536	72982	00537	73016	00537	73051	00538	73085	00538	73119	00539	35
25	73119	00539	73154	00539	73188	00539	73222	00540	73257	00540	73291	00541	34
26	73291	00541	73325	00541	73359	00541	73394	00542	73428	00542	73462	00543	33
27	73462	00543	73496	00543	73530	00544	73564	00544	73598	00544	73633	00545	32
28	73633	00545	73667	00545	73701	00546	73735	00546	73769	00547	73803	00547	31
29	73803	00547	73837	00547	73871	00548	73906	00548	73940	00549	73974	00549	30
30	73974	00549	74008	00550	74042	00550	74075	00550	74109	00551	74143	00551	29
31	74143	00551	74177	00552	74211	00552	74245	00553	74279	00553	74313	00554	28
32	74313	00554	74347	00554	74381	00554	74414	00555	74448	00555	74482	00556	27
33	74482	00556	74516	00556	74550	00557	74583	00557	74617	00557	74651	00558	26
34	74651	00558	74685	00558	74718	00559	74752	00559	74786	00560	74819	00560	25
35	74819	00560	74853	00560	74887	00561	74920	00561	74954	00562	74988	00562	24
36	74988	00562	75021	00563	75055	00563	75088	00563	75122	00564	75155	00564	23
37	75155	00564	75189	00565	75222	00565	75256	00566	75289	00566	75323	00567	22
38	75323	00567	75356	00567	75390	00567	75423	00568	75457	00568	75490	00569	21
39	75490	00569	75524	00569	75557	00570	75590	00570	75624	00570	75657	00571	20
40	75657	00571	75690	00571	75724	00572	75757	00572	75790	00573	75824	00573	19
41	75824	00573	75857	00574	75890	00574	75923	00574	75957	00575	75990	00575	18
42	75990	00575	76023	00576	76056	00576	76089	00577	76123	00577	76156	00578	17
43	76156	00578	76189	00578	76222	00578	76255	00579	76288	00579	76321	00580	16
44	76321	00530	76354	00580	76387	00581	76421	00531	76454	00581	76487	00582	15
45	76487	00532	76520	00582	76553	00583	76586	00583	76619	00584	76652	00584	14
46	76652	00534	76685	00585	76717	00585	76750	00585	76783	00586	76816	00586	13
47	76816	00536	76849	00587	76882	00587	76915	00588	76948	00588	76981	00589	12
48	76981	00589	77013	00589	77046	00589	77079	00590	77112	00590	77145	00591	11
49	77145	00591	77177	00591	77210	00592	77243	00592	77276	00593	77308	00593	10
50	77308	00593	77341	00594	77374	00594	77406	00594	77439	00595	77472	00595	09
51	77472	00595	77504	00596	77537	00596	77570	00597	77602	00597	77635	00598	08
52	77635	00598	77667	00598	77700	00598	77732	00599	77765	00599	77798	00600	07
53	77798	00600	77830	00600	77863	00601	77895	00601	77928	00602	77960	00602	06
54	77960	00602	77993	00602	78025	00603	78058	00603	78090	00604	78122	00604	05
55	78122	00604	78155	00605	78187	00605	78219	00606	78252	00606	78284	00607	04
56	78284	00607	78317	00607	78349	00607	78381	00608	78413	00608	78446	00609	03
57	78446	00609	78478	00609	78510	00610	78543	00610	78575	00611	78607	00611	02
58	78607	00611	78639	00612	78671	00612	78704	00612	78736	00613	78768	00613	01
59	78768	00613	78800	00614	78832	00614	78864	00615	78896	00615	78929	00616	00

.8   .6   .4   .2   .0 ← 351°

PARTS for 0′·1 :— LOGS. variable from 18 to 16 ; NATURALS, negligible.

9° HAVERSINES

′	·0 Log. 7·	·0 Nat. 0·	·2 Log. 7·	·2 Nat. 0·	·4 Log. 7·	·4 Nat. 0·	·6 Log. 7·	·6 Nat. 0·	·8 Log. 7·	·8 Nat. 0·	Log. 7·	Nat. 0·	′
00	78929	00616	78961	00616	78993	00617	79025	00617	79057	00617	79089	00618	59
01	79089	00618	79121	00618	79153	00619	79185	00619	79217	00620	79249	00620	58
02	79249	00620	79281	00621	79313	00621	79345	00622	79377	00622	79409	00622	57
03	79409	00622	79441	00623	79473	00623	79505	00624	79536	00624	79568	00625	56
04	79568	00625	79600	00625	79632	00626	79664	00626	79696	00627	79728	00627	55
05	79728	00627	79759	00627	79791	00628	79823	00628	79855	00629	79886	00629	54
06	79886	00629	79918	00630	79950	00630	79982	00631	80013	00631	80045	00632	53
07	80045	00632	80077	00632	80108	00633	80140	00633	80172	00633	80203	00634	52
08	80203	00634	80235	00634	80267	00635	80298	00635	80330	00636	80361	00636	51
09	80361	00636	80393	00637	80425	00637	80456	00638	80488	00638	80519	00639	50
10	80519	00639	80551	00639	80582	00639	80614	00640	80645	00640	80677	00641	49
11	80677	00641	80708	00641	80739	00642	80771	00642	80802	00643	80834	00643	48
12	80834	00643	80865	00644	80897	00644	80928	00645	80959	00645	80991	00646	47
13	80991	00646	81022	00646	81053	00646	81085	00647	81116	00647	81147	00648	46
14	81147	00648	81178	00648	81210	00649	81241	00649	81272	00650	81303	00650	45
15	81303	00650	81335	00651	81366	00651	81397	00652	81428	00652	81459	00653	44
16	81459	00653	81491	00653	81522	00653	81553	00654	81584	00654	81615	00655	43
17	81615	00655	81646	00655	81677	00656	81708	00656	81740	00657	81771	00657	42
18	81771	00657	81802	00658	81833	00658	81864	00659	81895	00659	81926	00660	41
19	81926	00660	81957	00660	81988	00660	82019	00661	82050	00661	82081	00662	40
20	82081	00662	82112	00662	82143	00663	82174	00663	82204	00664	82235	00664	39
21	82235	00664	82266	00665	82297	00665	82328	00666	82359	00666	82390	00667	38
22	82390	00667	82421	00667	82451	00668	82482	00668	82513	00669	82544	00669	37
23	82544	00669	82575	00669	82605	00670	82636	00670	82667	00671	82698	00671	36
24	82698	00671	82728	00672	82759	00672	82790	00673	82820	00673	82851	00674	35
25	82851	00674	82882	00674	82912	00675	82943	00675	82974	00676	83004	00676	34
26	83004	00676	83035	00677	83066	00677	83096	00678	83127	00678	83157	00679	33
27	83157	00679	83188	00679	83218	00680	83249	00680	83279	00680	83310	00681	32
28	83310	00681	83341	00681	83371	00682	83402	00682	83432	00683	83463	00683	31
29	83463	00683	83493	00684	83523	00684	83554	00685	83584	00685	83615	00686	30
30	83615	00686	83645	00686	83675	00687	83706	00687	83736	00688	83767	00688	29
31	83767	00688	83797	00689	83827	00689	83858	00690	83888	00690	83918	00691	28
32	83918	00691	83948	00691	83979	00692	84009	00692	84039	00692	84070	00693	27
33	84070	00693	84100	00693	84130	00694	84160	00694	84190	00695	84221	00695	26
34	84221	00695	84251	00696	84281	00696	84311	00697	84341	00697	84372	00698	25
35	84372	00698	84402	00698	84432	00699	84462	00699	84492	00700	84522	00700	24
36	84522	00700	84452	00701	84582	00701	84612	00702	84642	00702	84672	00703	23
37	84672	00703	84702	00703	84732	00704	84762	00704	84792	00705	84822	00705	22
38	84822	00705	84852	00706	84882	00706	84912	00707	84942	00707	84972	00707	21
39	84972	00707	85002	00708	85032	00708	85062	00709	85092	00709	85122	00710	20
40	85122	00710	85152	00710	85181	00711	85211	00711	85241	00712	85271	00712	19
41	85271	00712	85301	00713	85331	00713	85360	00714	85390	00714	85420	00715	18
42	85420	00715	85450	00715	85480	00716	85509	00716	85539	00717	85569	00717	17
43	85569	00717	85599	00718	85628	00718	85658	00719	85688	00719	85717	00720	16
44	85717	00720	85747	00720	85777	00721	85806	00721	85836	00722	85866	00722	15
45	85866	00722	85895	00723	85925	00723	85954	00724	85984	00724	86014	00725	14
46	86014	00725	86043	00725	86073	00726	86102	00726	86132	00727	86161	00727	13
47	86161	00727	86191	00728	86220	00728	86250	00729	86279	00729	86309	00730	12
48	86309	00730	86338	00730	86368	00731	86397	00731	86427	00732	86456	00732	11
49	86456	00732	86485	00733	86515	00733	86544	00734	86574	00734	86603	00735	10
50	86603	00735	86632	00735	86662	00736	86691	00736	86720	00737	86750	00737	09
51	86750	00737	86779	00738	86808	00738	86838	00739	86867	00739	86896	00740	08
52	86896	00740	86925	00740	86955	00741	86984	00741	87013	00742	87042	00742	07
53	87042	00742	87072	00743	87101	00743	87130	00744	87159	00744	87188	00745	06
54	87188	00745	87218	00745	87247	00746	87276	00746	87305	00747	87334	00747	05
55	87334	00747	87363	00748	87392	00748	87421	00749	87451	00749	87480	00750	04
56	87480	00750	87509	00750	87538	00751	87567	00751	87596	00752	87625	00752	03
57	87625	00752	87654	00753	87683	00753	87712	00754	87741	00754	87770	00755	02
58	87770	00755	87799	00755	87828	00756	87857	00756	87886	00757	87915	00757	01
59	87915	00757	87944	00758	87972	00758	88001	00759	88030	00759	88059	00760	00

·8 ·6 ·4 ·2 ·0 ← 350°

PARTS for 0′·1 :— LOGS. variable from 16 to 14 ; NATURALS negligible.,

10°   HAVERSINES

′	·0 Log. 7·	·0 Nat. 0·	·2 Log. 7·	·2 Nat. 0·	·4 Log. 7·	·4 Nat. 0·	·6 Log. 7·	·6 Nat. 0·	·8 Log. 7·	·8 Nat. 0·	Log. 7·	Nat. 0·	′
00	88059	00760	88088	00760	88117	00761	88146	00761	88175	00762	88203	00762	59
01	88203	00762	88232	00763	88261	00763	88290	00764	88319	00764	88348	00765	58
02	88348	00765	88376	00765	88405	00766	88434	00766	88463	00767	88491	00767	57
03	88491	00767	88520	00768	88549	00768	88577	00769	88606	00769	88635	00770	56
04	88635	00770	88663	00770	88692	00771	88721	00771	88750	00772	88778	00772	55
05	88778	00772	88807	00773	88835	00773	88864	00774	88893	00774	88921	00775	54
06	88921	00775	88950	00775	88978	00776	89007	00776	89036	00777	89064	00777	53
07	89064	00777	89093	00778	89121	00778	89150	00779	89178	00779	89207	00780	52
08	89207	00780	89235	00780	89264	00781	89292	00781	89321	00782	89349	00783	51
09	89349	00783	89377	00783	89406	00784	89434	00784	89463	00785	89491	00785	50
10	89491	00785	89520	00786	89548	00786	89576	00787	89605	00787	89633	00788	49
11	89633	00788	89662	00788	89690	00789	89718	00789	89746	00790	89775	00790	48
12	89775	00790	89803	00791	89831	00791	89860	00792	89888	00792	89916	00793	47
13	89916	00793	89944	00793	89973	00794	90001	00794	90029	00795	90057	00795	46
14	90057	00795	90086	00796	90114	00796	90142	00797	90170	00797	90198	00798	45
15	90198	00798	90227	00798	90255	00799	90283	00800	90311	00800	90339	00801	44
16	90339	00801	90367	00801	90395	00802	90423	00802	90452	00803	90480	00803	43
17	90480	00803	90508	00804	90536	00804	90564	00805	90592	00805	90620	00806	42
18	90620	00806	90648	00806	90676	00807	90704	00807	90732	00808	90760	00808	41
19	90760	00808	90788	00809	90816	00809	90844	00810	90872	00810	90900	00811	40
20	90900	00811	90928	00811	90956	00812	90984	00813	91011	00813	91039	00814	39
21	91039	00814	91067	00814	91095	00815	91123	00815	91151	00816	91179	00816	38
22	91179	00816	91207	00817	91234	00817	91262	00818	91290	00818	91318	00819	37
23	91318	00819	91346	00819	91374	00820	91401	00820	91429	00821	91457	00821	36
24	91457	00821	91485	00822	91512	00822	91540	00823	91568	00824	91596	00824	35
25	91596	00824	91623	00825	91651	00825	91679	00826	91706	00826	91734	00827	34
26	91734	00827	91762	00827	91789	00828	91817	00828	91845	00829	91872	00829	33
27	91872	00829	91900	00830	91928	00830	91955	00831	91983	00831	92010	00832	32
28	92010	00832	92038	00832	92066	00833	92093	00834	92121	00834	92148	00835	31
29	92148	00835	92176	00835	92203	00836	92231	00836	92258	00837	92286	00837	30
30	92286	00837	92313	00838	92341	00838	92368	00839	92396	00839	92423	00840	29
31	92423	00840	92451	00840	92478	00841	92505	00841	92533	00842	92560	00843	28
32	92560	00843	92588	00843	92615	00844	92642	00844	92670	00845	92697	00845	27
33	92697	00845	92725	00846	92752	00846	92779	00847	92807	00847	92834	00848	26
34	92834	00848	92861	00848	92889	00849	92916	00849	92943	00850	92970	00851	25
35	92970	00851	92998	00851	93025	00852	93052	00852	93080	00853	93107	00853	24
36	93107	00853	93134	00854	93161	00854	93188	00855	93216	00855	93243	00856	23
37	93243	00856	93270	00856	93297	00857	93324	00858	93351	00858	93379	00859	22
38	93379	00859	93406	00859	93433	00860	93460	00860	93487	00861	93514	00861	21
39	93514	00861	93541	00862	93569	00862	93596	00863	93623	00863	93650	00864	20
40	93650	00864	93677	00864	93704	00865	93731	00866	93758	00866	93785	00867	19
41	93785	00867	93812	00867	93839	00868	93866	00868	93893	00869	93920	00869	18
42	93920	00869	93947	00870	93974	00870	94001	00871	94028	00872	94055	00872	17
43	94055	00872	94082	00873	94109	00873	94136	00874	94162	00874	94189	00875	16
44	94189	00875	94216	00875	94243	00876	94270	00876	94297	00877	94324	00877	15
45	94324	00877	94351	00878	94377	00879	94404	00879	94431	00880	94458	00880	14
46	94458	00880	94485	00881	94511	00881	94538	00882	94565	00882	94592	00883	13
47	94592	00883	94619	00883	94645	00384	94672	00885	94699	00885	94726	00886	12
48	94726	00886	94752	00886	94779	00887	94806	00887	94832	00888	94859	00888	11
49	94859	00888	94886	00889	94912	00889	94939	00890	94966	00891	94992	00891	10
50	94992	00891	95019	00892	95046	00892	95072	00893	95099	00893	95126	00894	09
51	95126	00894	95152	00894	95179	00895	95205	00895	95232	00896	95259	00897	08
52	95259	00897	95285	00897	95312	00898	95338	00898	95365	00899	95391	00899	07
53	95391	00899	95418	00900	95444	00900	95471	00901	95497	00901	95524	00902	06
54	95524	00902	95550	00903	95577	00903	95603	00904	95630	00904	95656	00905	05
55	95656	00905	95682	00905	95709	00906	95735	00906	95762	00907	95788	00908	04
56	95788	00908	95815	00908	95841	00909	95867	00909	95894	00910	95920	00910	03
57	95920	00910	95946	00911	95973	00911	95999	00912	96025	00913	96052	00913	02
58	96052	00913	96078	00914	96104	00914	96131	00915	96157	00915	96183	00916	01
59	96183	00916	96209	00916	96236	00917	96262	00918	96288	00918	96315	00919	00

   ·8   ·6   ·4   ·2   ·0 ← ↑

349°

PARTS for 0′·1 :—   LOGS. 14 ;   NATURALS, negligible.

11° HAVERSINES

′	.0 Log. 7./8.	.0 Nat. 0.	.2 Log. 7./8.	.2 Nat. 0.	.4 Log. 7./8.	.4 Nat. 0.	.6 Log. 7./8.	.6 Nat. 0.	.8 Log. 7./8.	.8 Nat. 0.	Log. 7./8.	Nat. 0.	′
00	96315	00919	96341	00919	96367	00920	96393	00920	96419	00921	96446	00921	59
01	96446	00921	96472	00922	96498	00923	96524	00923	96550	00924	96577	00924	58
02	96577	00924	96603	00925	96629	00925	96655	00926	96681	00926	96707	00927	57
03	96707	00927	96733	00928	96759	00928	96786	00929	96812	00929	96838	00930	56
04	96838	00930	96864	00930	96890	00931	96916	00931	96942	00932	96968	00933	55
05	96968	00933	96994	00933	97020	00934	97046	00934	97072	00935	97098	00935	54
06	97098	00935	97124	00936	97150	00936	97176	00937	97202	00938	97228	00938	53
07	97228	00938	97254	00939	97280	00939	97306	00940	97332	00940	97358	00941	52
08	97358	00941	97384	00942	97410	00942	97436	00943	97461	00943	97487	00944	51
09	97487	00944	97513	00944	97539	00945	97565	00945	97591	00946	97617	00947	50
10	97617	00947	97642	00947	97668	00948	97694	00948	97720	00949	97746	00949	49
11	97746	00949	97772	00950	97797	00951	97823	00951	97849	00952	97875	00952	48
12	97875	00952	97901	00953	97926	00953	97952	00954	97978	00954	98003	00955	47
13	98003	00955	98029	00956	98055	00956	98081	00957	98106	00957	98132	00958	46
14	98132	00958	98158	00958	98183	00959	98209	00960	98235	00960	98260	00961	45
15	98260	00961	98286	00961	98312	00962	98337	00962	98363	00963	98389	00964	44
16	98389	00964	98414	00964	98440	00965	98465	00965	98491	00966	98517	00966	43
17	98517	00966	98542	00967	98568	00968	98593	00968	98619	00969	98644	00969	42
18	98644	00969	98670	00970	98695	00970	98721	00971	98746	00972	98772	00972	41
19	98772	00972	98797	00973	98823	00973	98848	00974	98874	00974	98899	00975	40
20	98899	00975	98925	00976	98950	00976	98976	00977	99001	00977	99027	00978	39
21	99027	00978	99052	00978	99077	00979	99103	00980	99128	00980	99154	00981	38
22	99154	00981	99179	00981	99204	00982	99230	00982	99255	00983	99280	00984	37
23	99280	00984	99306	00984	99331	00985	99356	00985	99382	00986	99407	00986	36
24	99407	00986	99432	00987	99458	00988	99483	00988	99508	00989	99534	00989	35
25	99534	00989	99559	00990	99584	00990	99609	00991	99635	00992	99660	00992	34
26	99660	00992	99685	00993	99710	00993	99736	00994	99761	00995	99786	00995	33
27	99786	00995	99811	00996	99836	00996	99862	00997	99887	00997	99912	00998	32
28	99912	00998	99937	00999	99962	00999	99987	01000	00012	01000	00038	01001	31
29	00038	01001	00063	01001	00088	01002	00113	01003	00138	01003	00163	01004	30
30	00163	01004	00188	01004	00213	01005	00238	01005	00264	01006	00289	01007	29
31	00289	01007	00314	01007	00339	01008	00364	01008	00389	01009	00414	01010	28
32	00414	01010	00439	01010	00464	01011	00489	01011	00514	01012	00539	01012	27
33	00539	01012	00564	01013	00589	01014	00614	01014	00639	01015	00664	01015	26
34	00664	01015	00689	01016	00713	01017	00738	01017	00763	01018	00788	01018	25
35	00788	01018	00813	01019	00838	01019	00863	01020	00888	01021	00913	01021	24
36	00913	01021	00937	01022	00962	01022	00987	01023	01012	01024	01037	01024	23
37	01037	01024	01062	01025	01087	01025	01111	01026	01136	01027	01161	01027	22
38	01161	01027	01186	01028	01211	01028	01235	01029	01260	01029	01285	01030	21
39	01285	01030	01310	01031	01335	01031	01359	01032	01384	01032	01409	01033	20
40	01409	01033	01433	01034	01458	01034	01483	01035	01508	01035	01532	01036	19
41	01532	01036	01557	01036	01582	01037	01606	01038	01631	01038	01656	01039	18
42	01656	01039	01680	01039	01705	01040	01730	01041	01754	01041	01779	01042	17
43	01779	01042	01803	01042	01828	01043	01853	01044	01877	01044	01902	01045	16
44	01902	01045	01927	01045	01951	01046	01976	01047	02000	01047	02025	01048	15
45	02025	01048	02049	01048	02074	01049	02098	01050	02123	01050	02147	01051	14
46	02147	01051	02172	01051	02196	01052	02221	01052	02245	01053	02270	01054	13
47	02270	01054	02294	01054	02319	01055	02343	01055	02368	01056	02392	01057	12
48	02392	01057	02417	01057	02441	01058	02466	01058	02490	01059	02514	01060	11
49	02514	01060	02539	01060	02563	01061	02588	01061	02612	01062	02636	01063	10
50	02636	01063	02661	01063	02685	01064	02709	01064	02734	01065	02758	01066	09
51	02758	01066	02783	01066	02807	01067	02831	01067	02856	01068	02880	01069	08
52	02880	01069	02904	01069	02929	01070	02953	01070	02977	01071	03001	01072	07
53	03001	01072	03026	01072	03050	01073	03074	01073	03098	01074	03123	01075	06
54	03123	01075	03147	01075	03171	01076	03195	01076	03220	01077	03244	01078	05
55	03244	01078	03268	01078	03292	01079	03316	01079	03341	01080	03365	01081	04
56	03365	01081	03389	01081	03413	01082	03437	01082	03461	01083	03486	01084	03
57	03486	01084	03510	01084	03534	01085	03558	01085	03582	01086	03606	01087	02
58	03606	01087	03630	01087	03654	01088	03678	01088	03703	01089	03727	01090	01
59	03727	01090	03751	01090	03775	01091	03799	01091	03823	01092	03847	01093	00

.8   .6   .4   .2   .0 ←

348°

PARTS for 0′·1 :—   LOGS. 13 ;   NATURALS, negligible.

12° HAVERSINES

Logs. prefixed by **8·** ; Naturals prefixed by **0·**

′	·0 Log	·0 Nat	·2 Log	·2 Nat	·4 Log	·4 Nat	·6 Log	·6 Nat	·8 Log	·8 Nat	Log	Nat	′
00	03847	01093	03871	01093	03895	01094	03919	01094	03943	01095	03967	01096	59
01	03967	01096	03991	01096	04015	01097	04039	01097	04063	01098	04087	01099	58
02	04087	01099	04111	01099	04135	01100	04159	01100	04183	01101	04207	01102	57
03	04207	01102	04231	01102	04255	01103	04279	01104	04302	01104	04326	01105	56
04	04326	01105	04350	01105	04374	01106	04398	01107	04422	01107	04446	01108	55
05	04446	01108	04470	01108	04493	01109	04517	01110	04541	01110	04565	01111	54
06	04565	01111	04589	01111	04613	01112	04637	01113	04660	01113	04684	01114	53
07	04684	01114	04708	01115	04732	01115	04756	01116	04779	01116	04803	01117	52
08	04803	01117	04827	01118	04851	01118	04874	01119	04898	01119	04922	01120	51
09	04922	01120	04946	01121	04969	01121	04993	01122	05017	01122	05041	01123	50
10	05041	01123	05064	01124	05088	01124	05112	01125	05135	01126	05159	01126	49
11	05159	01126	05183	01127	05206	01127	05230	01128	05254	01129	05277	01129	48
12	05277	01129	05301	01130	05324	01130	05348	01131	05372	01132	05395	01132	47
13	05395	01132	05419	01133	05443	01134	05466	01134	05490	01135	05513	01135	46
14	05513	01135	05537	01136	05561	01137	05584	01137	05608	01138	05631	01138	45
15	05631	01138	05655	01139	05678	01140	05702	01140	05725	01141	05749	01142	44
16	05749	01142	05772	01142	05796	01143	05819	01143	05843	01144	05866	01145	43
17	05866	01145	05890	01145	05913	01146	05937	01146	05960	01147	05984	01148	42
18	05984	01148	06007	01148	06030	01149	06054	01150	06077	01150	06101	01151	41
19	06101	01151	06124	01151	06148	01152	06171	01153	06194	01153	06218	01154	40
20	06218	01154	06241	01155	06265	01155	06288	01156	06311	01156	06335	01157	39
21	06335	01157	06358	01158	06381	01158	06405	01159	06428	01160	06451	01160	38
22	06451	01160	06475	01161	06498	01161	06521	01162	06545	01163	06568	01163	37
23	06568	01163	06591	01164	06614	01165	06638	01165	06661	01166	06684	01166	36
24	06684	01166	06707	01167	06731	01168	06754	01168	06777	01169	06800	01170	35
25	06800	01170	06824	01170	06847	01171	16870	01171	06893	01172	06916	01173	34
26	06916	01173	06940	01173	06963	01174	06986	01175	07009	01175	07032	01176	33
27	07032	01176	07056	01176	07079	01177	07102	01178	07125	01178	07148	01179	32
28	07148	01179	07171	01180	07194	01180	07218	01181	07241	01181	07264	01182	31
29	07264	01182	07287	01183	07310	01183	07333	01184	07356	01185	07379	01185	30
30	07379	01185	07402	01186	07425	01186	07448	01187	07471	01188	07494	01188	29
31	07494	01188	07517	01189	07540	01190	07564	01190	07587	01191	07610	01192	28
32	07610	01192	07633	01192	07656	01193	07679	01193	07702	01194	07725	01195	27
33	07725	01195	07748	01195	07771	01196	07793	01197	07816	01197	07839	01198	26
34	07839	01198	07862	01198	07885	01199	07908	01200	07931	01200	07954	01201	25
35	07954	01201	07977	01202	08000	01202	08023	01203	08046	01204	08068	01204	24
36	08068	01204	08091	01205	08114	01205	08137	01206	08160	01207	08183	01207	23
37	08183	01207	08206	01208	08228	01209	08251	01209	08274	01210	08297	01211	22
38	08297	01211	08320	01211	08343	01212	08365	01212	08388	01213	08411	01214	21
39	08411	01214	08434	01214	08457	01215	08479	01216	08502	01216	08525	01217	20
40	08525	01217	08548	01218	08571	01218	08593	01219	08616	01219	08639	01220	19
41	08639	01220	08662	01221	08684	01221	08707	01222	08730	01223	08752	01223	18
42	08752	01223	08775	01224	08798	01225	08820	01225	08843	01226	08866	01226	17
43	08866	01226	08888	01227	08911	01228	08934	01228	08956	01229	08979	01230	16
44	08979	01230	09002	01230	09024	01231	09047	01232	09070	01232	09092	01233	15
45	09092	01233	09115	01234	09137	01234	09160	01235	09183	01235	09205	01236	14
46	09205	01236	09228	01237	09250	01237	09273	01238	09295	01239	09318	01239	13
47	09318	01239	09341	01240	09363	01241	09386	01241	09408	01242	09431	01243	12
48	09431	01243	09453	01243	09476	01244	09498	01244	09521	01245	09543	01246	11
49	09543	01246	09566	01246	09588	01247	09611	01248	09633	01248	09656	01249	10
50	09656	01249	09678	01250	09701	01250	09723	01251	09746	01252	09768	01252	09
51	09768	01252	09790	01253	09813	01254	09835	01254	09858	01255	09880	01255	08
52	09880	01255	09902	01256	09925	01257	09947	01257	09970	01258	09992	01259	07
53	09992	01259	10014	01259	10037	01260	10059	01261	10081	01261	10104	01262	06
54	10104	01262	10126	01263	10149	01263	10171	01264	10193	01265	10216	01265	05
55	10216	01265	10238	01266	10260	01267	10282	01267	10305	01268	10327	01268	04
56	10327	01268	10349	01269	10372	01270	10394	01270	10416	01271	10438	01272	03
57	10438	01272	10461	01272	10483	01273	10505	01274	10528	01274	10550	01275	02
58	10550	01275	10572	01276	10594	01276	10616	01277	10639	01278	10661	01278	01
59	10661	01278	10683	01279	10705	01280	10727	01280	10750	01281	10772	01282	00

Bottom column labels: ·8 · ·6 · ·4 · ·2 · ·0 ←

347°

PARTS for 0′·1 :—   LOGS. 12 ;   NATURALS, negligible.

13°   HAVERSINES

′	·0 Log.	·0 Nat.	·2 Log.	·2 Nat.	·4 Log.	·4 Nat.	·6 Log.	·6 Nat.	·8 Log.	·8 Nat.	·8 Log.	·0 Nat.	′
	8·	0·	8·	0·	8·	0·	8·	0·	8·	0·	·8	·0	
00	10772	01282	10794	01282	10816	01283	10838	01283	10860	01284	10883	01285	59
01	10883	01285	10905	01285	10927	01286	10949	01287	10971	01287	10993	01288	58
02	10993	01288	11015	01289	11037	01289	11060	01290	11082	01291	11104	01291	57
03	11104	01291	11126	01292	11148	01293	11170	01293	11192	01294	11214	01295	56
04	11214	01295	11236	01295	11258	01296	11280	01297	11302	01297	11324	01298	55
05	11324	01298	11346	01299	11368	01299	11390	01300	11412	01301	11434	01301	54
06	11434	01301	11456	01302	11478	01303	11500	01303	11522	01304	11544	01305	53
07	11544	01305	11566	01305	11588	01306	11610	01306	11632	01307	11654	01308	52
08	11654	01308	11676	01308	11698	01309	11720	01310	11742	01310	11764	01311	51
09	11764	01311	11786	01312	11808	01312	11830	01313	11852	01314	11873	01314	50
10	11873	01314	11895	01315	11917	01316	11939	01316	11961	01317	11983	01318	49
11	11983	01318	12005	01318	12026	01319	12048	01320	12070	01320	12092	01321	48
12	12092	01321	12114	01322	12136	01322	12158	01323	12179	01324	12201	01324	47
13	12201	01324	12223	01325	12245	01326	12267	01326	12288	01327	12310	01328	46
14	12310	01328	12332	01328	12354	01329	12375	01330	12397	01330	12419	01331	45
15	12419	01331	12441	01332	12463	01332	12484	01333	12506	01334	12528	01334	44
16	12528	01334	12549	01335	12571	01336	12593	01336	12615	01337	12636	01338	43
17	12636	01338	12658	01338	12680	01339	12701	01340	12723	01340	12745	01341	42
18	12745	01341	12766	01342	12788	01342	12810	01343	12831	01344	12853	01344	41
19	12853	01344	12875	01345	12896	01346	12918	01346	12940	01347	12961	01348	40
20	12961	01348	12983	01348	13004	01349	13026	01350	13048	01350	13069	01351	39
21	13069	01351	13091	01352	13112	01352	13134	01353	13155	01354	13177	01354	38
22	13177	01354	13199	01355	13220	01356	13242	01356	13263	01357	13285	01358	37
23	13285	01358	13306	01359	13328	01359	13349	01360	13371	01361	13392	01361	36
24	13392	01361	13414	01362	13435	01363	13457	01363	13478	01364	13500	01365	35
25	13500	01365	13521	01365	13543	01366	13564	01367	13586	01367	13607	01368	34
26	13607	01368	13629	01369	13650	01369	13672	01370	13693	01371	13714	01371	33
27	13714	01371	13736	01372	13757	01373	13779	01373	13800	01374	13821	01375	32
28	13821	01375	13843	01375	13864	01376	13886	01377	13907	01377	13928	01378	31
29	13928	01378	13950	01379	13971	01379	13992	01380	14014	01381	14035	01382	30
30	14035	01382	14057	01382	14078	01383	14099	01384	14120	01384	14142	01385	29
31	14142	01385	14163	01386	14184	01386	14206	01387	14227	01388	14248	01388	28
32	14248	01388	14270	01389	14291	01390	14312	01390	14334	01391	14355	01392	27
33	14355	01392	14376	01392	14397	01393	14419	01394	14440	01394	14461	01395	26
34	14461	01395	14482	01396	14504	01396	14525	01397	14546	01398	14567	01399	25
35	14567	01399	14588	01399	14610	01400	14631	01401	14652	01401	14673	01402	24
36	14673	01402	14694	01403	14716	01403	14737	01404	14758	01405	14779	01405	23
37	14779	01405	14800	01406	14821	01407	14843	01407	14864	01408	14885	01409	22
38	14885	01409	14906	01409	14927	01410	14948	01411	14969	01412	14990	01412	21
39	14990	01412	15012	01413	15033	01414	15054	01414	15075	01415	15096	01416	20
40	15096	01416	15117	01416	15138	01417	15159	01418	15180	01418	15201	01419	19
41	15201	01419	15222	01420	15243	01420	15265	01421	15286	01422	15307	01423	18
42	15307	01423	15328	01423	15349	01424	15370	01425	15391	01425	15412	01426	17
43	15412	01426	15433	01427	15454	01427	15475	01428	15496	01429	15517	01429	16
44	15517	01429	15538	01430	15559	01431	15579	01432	15600	01432	15621	01433	15
45	15621	01433	15642	01434	15663	01434	15684	01435	15705	01436	15726	01436	14
46	15726	01436	15747	01437	15768	01438	15789	01438	15810	01439	15831	01440	13
47	15831	01440	15852	01441	15873	01441	15893	01442	15914	01443	15935	01443	12
48	15935	01443	15956	01444	15977	01445	15998	01445	16019	01446	16040	01447	11
49	16040	01447	16060	01447	16081	01448	16102	01449	16123	01450	16144	01450	10
50	16144	01450	16165	01451	16185	01452	16206	01452	16227	01453	16248	01454	09
51	16248	01454	16269	01454	16289	01455	16310	01456	16331	01457	16352	01457	08
52	16352	01457	16373	01458	16393	01459	16414	01459	16435	01460	16456	01461	07
53	16456	01461	16476	01461	16497	01462	16518	01463	16539	01464	16559	01464	06
54	16559	01464	16580	01465	16601	01466	16621	01466	16642	01467	16663	01468	05
55	16663	01468	16684	01468	16704	01469	16725	01470	16746	01470	16766	01471	04
56	16766	01471	16787	01472	16808	01473	16828	01473	16849	01474	16870	01475	03
57	16870	01475	16890	01475	16911	01476	16932	01477	16952	01478	16973	01478	02
58	16973	01478	16993	01479	17014	01480	17035	01480	17055	01481	17076	01482	01
59	17076	01482	17097	01482	17117	01483	17138	01484	17158	01485	17179	01485	00
		·8		·6		·4		·2		·0 ←			

PARTS for 0′·1 ;—   LOGS. 11 ;   NATURALS, negligible.

14°	HAVERSINES

′	Log. 8·	Nat. 0·	Log. 8·	Nat. 0·	Log. 8·	Nat. 0·	Log. 8·	Nat. 0·	Log. 8·	Nat. 0·	Log. 8·	Nat. 0·	′
00	17179	01485	17199	01486	17220	01487	17241	01487	17261	01488	17282	01489	59
01	17282	01489	17302	01489	17323	01490	17343	01491	17364	01492	17384	01492	58
02	17384	01492	17405	01493	17425	01494	17446	01494	17466	01495	17487	01496	57
03	17487	01496	17507	01497	17528	01497	17548	01498	17569	01499	17589	01499	56
04	17589	01499	17610	01500	17630	01501	17651	01501	17671	01502	17692	01503	55
05	17692	01503	17712	01504	17733	01505	17753	01505	17774	01506	17794	01506	54
06	17794	01506	17814	01507	17835	01508	17855	01509	17876	01509	17896	01510	53
07	17896	01510	17916	01511	17937	01511	17957	01512	17978	01513	17998	01513	52
08	17998	01513	18018	01514	18039	01515	18059	01516	18080	01516	18100	01517	51
09	18100	01517	18121	01518	18141	01518	18161	01519	18181	01520	18202	01521	50
10	18202	01521	18222	01521	18242	01522	18263	01523	18283	01523	18303	01524	49
11	18303	01524	18324	01525	18344	01526	18364	01526	18384	01527	18405	01528	48
12	18405	01528	18425	01528	18445	01529	18466	01530	18486	01531	18506	01531	47
13	18506	01531	18526	01532	18547	01533	18567	01533	18587	01534	18607	01535	46
14	18607	01535	18628	01536	18648	01536	18668	01537	18688	01538	18708	01538	45
15	18708	01539	18729	01539	18749	01540	18769	01541	18789	01541	18809	01542	44
16	18809	01542	18830	01543	18850	01543	18870	01544	18890	01545	18910	01546	43
17	18910	01546	18931	01546	18951	01547	18971	01548	18991	01548	19011	01549	42
18	19011	01549	19031	01550	19051	01551	19072	01551	19092	01552	19112	01553	41
19	19112	01553	19132	01554	19152	01554	19172	01555	19192	01556	19212	01556	40
20	19212	01556	19232	01557	19253	01558	19273	01559	19293	01559	19313	01560	39
21	19313	01560	19333	01561	19353	01561	19373	01562	19393	01563	19413	01564	38
22	19413	01564	19433	01564	19453	01565	19473	01566	19493	01566	19513	01567	37
23	19513	01567	19533	01568	19553	01569	19573	01569	19593	01570	19613	01571	36
24	19613	01571	19633	01572	19653	01572	19673	01573	19693	01574	19713	01574	35
25	19713	01574	19733	01575	19753	01576	19773	01577	19793	01577	19813	01578	34
26	19813	01578	19833	01579	19853	01580	19873	01580	19893	01581	19913	01582	33
27	19913	01582	19933	01582	19953	01583	19973	01584	19992	01585	20012	01585	32
28	20012	01585	20032	01586	20052	01587	20072	01588	20092	01588	20112	01589	31
29	20112	01589	20132	01590	20152	01590	20171	01591	20191	01592	20211	01593	30
30	20211	01593	20231	01593	20251	01594	20271	01595	20291	01596	20310	01596	29
31	20310	01596	20330	01597	20350	01598	20370	01598	20390	01599	20410	01600	28
32	20410	01600	20429	01601	20449	01601	20469	01602	20489	01603	20509	01604	27
33	20509	01604	20528	01604	20548	01605	20568	01606	20588	01606	20607	01607	26
34	20607	01607	20627	01608	20647	01609	20667	01609	20686	01610	20706	01611	25
35	20706	01611	20726	01612	20746	01612	20765	01613	20785	01614	20805	01615	24
36	20805	01615	20825	01615	20844	01616	20864	01617	20884	01617	20903	01618	23
37	20903	01618	20923	01619	20943	01620	20963	01620	20982	01621	21002	01622	22
38	21002	01622	21022	01623	21041	01623	21061	01624	21081	01625	21100	01626	21
39	21100	01626	21120	01626	21139	01627	21159	01628	21179	01629	21198	01629	20
40	21198	01629	21218	01630	21238	01631	21257	01631	21277	01632	21297	01633	19
41	21297	01633	21316	01634	21336	01634	21355	01635	21375	01636	21395	01637	18
42	21395	01637	21414	01637	21434	01638	21453	01639	21473	01640	21492	01640	17
43	21492	01640	21512	01641	21532	01642	21551	01643	21571	01643	21590	01644	16
44	21590	01644	21610	01645	21629	01645	21649	01646	21668	01647	21688	01648	15
45	21688	01648	21707	01648	21727	01649	21746	01650	21766	01651	21785	01651	14
46	21785	01651	21805	01652	21824	01653	21844	01654	21863	01654	21883	01655	13
47	21883	01655	21902	01656	21922	01657	21941	01657	21961	01658	21980	01659	12
48	21980	01659	22000	01660	22019	01660	22039	01661	22058	01662	22077	01663	11
49	22077	01663	22097	01663	22116	01664	22136	01665	22155	01666	22175	01666	10
50	22175	01666	22194	01667	22213	01668	22233	01669	22252	01669	22272	01670	09
51	22272	01670	22291	01671	22310	01671	22330	01672	22349	01673	22368	01674	08
52	22368	01674	22388	01674	22407	01675	22426	01676	22446	01677	22465	01677	07
53	22465	01677	22485	01678	22504	01679	22523	01680	22542	01680	22562	01681	06
54	22562	01681	22581	01682	22600	01683	22620	01683	22639	01684	22658	01685	05
55	22658	01685	22678	01686	22697	01686	22716	01687	22736	01688	22755	01689	04
56	22755	01689	22774	01689	22793	01690	22813	01691	22832	01692	22851	01692	03
57	22851	01692	22870	01693	22890	01694	22909	01695	22928	01695	22947	01696	02
58	22947	01696	22967	01697	22986	01698	23005	01698	23024	01699	23044	01700	01
59	23044	01700	23063	01701	23082	01701	23101	01702	23120	01703	23140	01704	00

PARTS for 0′·1 :—   LOGS. 10 ;   NATURALS, negligible.

15° HAVERSINES

′	·0 Log. 8·	·0 Nat. 0·	·2 Log. 8·	·2 Nat. 0·	·4 Log. 8·	·4 Nat. 0·	·6 Log. 8·	·6 Nat. 0·	·8 Log. 8·	·8 Nat. 0·	′
00	23140	01704	23159	01704	23178	01705	23197	01706	23216	01707	23235	01707	59
01	23235	01707	23255	01708	23274	01709	23293	01710	23312	01710	23331	01711	58
02	23331	01711	23350	01712	23370	01713	23389	01713	23408	01714	23427	01715	57
03	23427	01715	23446	01716	23465	01717	23484	01717	23503	01718	23523	01719	56
04	23523	01719	23542	01720	23561	01720	23580	01721	23599	01722	23618	01723	55
05	23618	01723	23637	01723	23656	01724	23675	01725	23694	01726	23713	01726	54
06	23713	01726	23732	01727	23751	01728	23771	01729	23790	01729	23809	01730	53
07	23809	01730	23828	01731	23847	01732	23866	01732	23885	01733	23904	01734	52
08	23904	01734	23923	01735	23942	01735	23961	01736	23980	01737	23999	01738	51
09	23999	01738	24018	01739	24037	01739	24056	01740	24075	01741	24094	01742	50
10	24094	01742	24113	01742	24132	01743	24151	01744	24170	01745	24189	01745	49
11	24189	01745	24208	01746	24227	01747	24245	01748	24264	01748	24283	01749	48
12	24283	01749	24302	01750	24321	01751	24340	01751	24359	01752	24378	01753	47
13	24378	01753	24397	01754	24416	01755	24435	01755	24454	01756	24472	01757	46
14	24472	01757	24491	01758	24510	01758	24529	01759	24548	01760	24567	01761	45
15	24567	01761	24586	01761	24605	01762	24624	01763	24642	01764	24661	01764	44
16	24661	01764	24680	01765	24699	01766	24718	01767	24737	01768	24755	01768	43
17	24755	01768	24774	01769	24793	01770	24812	01771	24831	01771	24850	01772	42
18	24850	01772	24868	01773	24887	01774	24906	01774	24925	01775	24944	01776	41
19	24944	01776	24962	01777	24981	01777	25000	01778	25019	01779	25037	01780	40
20	25037	01780	25056	01781	25075	01781	25094	01782	25112	01783	25131	01784	39
21	25131	01784	25150	01784	25169	01785	25187	01786	25206	01787	25225	01788	38
22	25225	01788	25244	01788	25262	01789	25281	01790	25300	01791	25319	01791	37
23	25319	01791	25337	01792	25356	01793	25375	01794	25393	01794	25412	01795	36
24	25412	01795	25431	01796	25449	01797	25468	01798	25487	01798	25505	01799	35
25	25505	01799	25524	01800	25543	01801	25561	01801	25580	01802	25599	01803	34
26	25599	01803	25617	01804	25636	01805	25655	01805	25673	01806	25692	01807	33
27	25692	01807	25710	01808	25729	01808	25748	01809	25766	01810	25785	01811	32
28	25785	01811	25804	01811	25822	01812	25841	01813	25859	01814	25878	01815	31
29	25878	01815	25897	01815	25915	01816	25934	01817	25952	01818	25971	01818	30
30	25971	01818	25989	01819	26008	01820	26026	01821	26045	01822	26064	01822	29
31	26064	01822	26082	01823	26101	01824	26119	01825	26138	01825	26156	01826	28
32	26156	01826	26175	01827	26193	01828	26212	01829	26230	01829	26249	01830	27
33	26249	01830	26267	01831	26286	01832	26304	01832	26323	01833	26341	01834	26
34	26341	01834	26360	01835	26378	01836	26397	01836	26415	01837	26434	01838	25
35	26434	01838	26452	01839	26471	01840	26489	01840	26508	01841	26526	01842	24
36	26526	01842	26544	01843	26563	01843	26581	01844	26600	01845	26618	01846	23
37	26618	01846	26637	01847	26655	01847	26673	01848	26692	01849	26710	01850	22
38	26710	01850	26729	01850	26747	01851	26765	01852	26784	01853	26802	01854	21
39	26802	01854	26821	01854	26839	01855	26857	01856	26876	01857	26894	01858	20
40	26894	01858	26912	01858	26931	01859	26949	01860	26967	01861	26986	01861	19
41	26986	01861	27004	01862	27022	01863	27041	01864	27059	01865	27077	01865	18
42	27077	01865	27096	01866	27114	01867	27132	01868	27151	01869	27169	01869	17
43	27169	01869	27187	01870	27206	01871	27224	01872	27242	01873	27261	01873	16
44	27261	01873	27279	01874	27297	01875	27315	01876	27334	01876	27352	01877	15
45	27352	01877	27370	01878	27388	01879	27407	01880	27425	01880	27443	01881	14
46	27443	01881	27461	01882	27480	01883	27498	01884	27516	01884	27534	01885	13
47	27534	01885	27553	01886	27571	01887	27589	01888	27607	01888	27626	01889	12
48	27626	01889	27644	01890	27662	01891	27680	01891	27698	01892	27717	01893	11
49	27717	01893	27735	01894	27753	01895	27771	01895	27789	01896	27807	01897	10
50	27807	01897	27826	01898	27844	01899	27862	01899	27880	01900	27898	01901	09
51	27898	01901	27916	01902	27934	01903	27953	01903	27971	01904	27989	01905	08
52	27989	01905	28007	01906	28025	01907	28043	01907	28061	01908	28080	01909	07
53	28080	01909	28098	01910	28116	01911	28134	01911	28152	01912	28170	01913	06
54	28170	01913	28188	01914	28206	01915	28224	01915	28242	01916	28260	01917	05
55	28260	01917	28278	01918	28297	01919	28315	01919	28333	01920	28351	01921	04
56	28351	01921	28369	01922	28387	01923	28405	01923	28423	01924	28441	01925	03
57	28441	01925	28459	01926	28477	01927	28495	01927	28513	01928	28531	01929	02
58	28531	01929	28549	01930	28567	01931	28585	01931	28603	01932	28621	01933	01
59	28621	01933	28639	01934	28657	01935	28675	01935	28693	01936	28711	01937	00

·8	·6	·4	·2	·0 ←

344°

PARTS for 0′·1 :— LOGS. 9 ; NATURALS, negligible.

16° HAVERSINES

′	·0 Log. 8·	Nat. 0·	·2 Log. 8·	Nat. 0·	·4 Log. 8·	Nat. 0·	·6 Log. 8·	Nat. 0·	·8 Log. 8·	Nat. 0·	Log. 8·	Nat. 0·	′
00	28711	01937	28729	01938	28747	01939	28765	01939	28783	01940	28801	01941	59
01	28801	01941	28819	01942	28837	01943	28855	01943	28873	01944	28891	01945	58
02	28891	01945	28909	01946	28927	01947	28944	01947	28962	01948	28980	01949	57
03	28980	01949	28998	01950	29016	01951	29034	01951	29052	01952	29070	01953	56
04	29070	01953	29088	01954	29106	01955	29124	01955	29141	01956	29159	01957	55
05	29159	01957	29177	01958	29195	01959	29213	01959	29231	01960	29249	01961	54
06	29249	01961	29266	01962	29284	01963	29302	01963	29320	01964	29338	01965	53
07	29338	01965	29356	01966	29374	01967	29392	01968	29409	01968	29427	01969	52
08	29427	01969	29445	01970	29463	01971	29481	01972	29498	01972	29516	01973	51
09	29516	01973	29534	01974	29552	01975	29570	01976	29587	01976	29605	01977	50
10	29605	01977	29623	01978	29641	01979	29659	01980	29676	01980	29694	01981	49
11	29694	01981	29712	01982	29730	01983	29747	01984	29765	01984	29783	01985	48
12	29783	01985	29801	01986	29818	01987	29836	01988	29854	01989	29872	01989	47
13	29872	01989	29889	01990	29907	01991	29925	01992	29943	01993	29960	01993	46
14	29960	01993	29978	01994	29996	01995	30013	01996	30031	01997	30049	01998	45
15	30049	01998	30066	01998	30084	01999	30102	02000	30120	02001	30137	02002	44
16	30137	02002	30155	02002	30173	02003	30190	02004	30208	02005	30226	02006	43
17	30226	02006	30243	02006	30261	02007	30279	02008	30296	02009	30314	02010	42
18	30314	02010	30331	02011	30349	02011	30367	02012	30384	02013	30402	02014	41
19	30402	02014	30420	02015	30437	02015	30455	02016	30473	02017	30490	02018	40
20	30490	02018	30508	02019	30525	02020	30543	02020	30561	02021	30578	02022	39
21	30578	02022	30596	02023	30613	02024	30631	02024	30648	02025	30666	02026	38
22	30666	02026	30684	02027	30701	02028	30719	02029	30736	02029	30754	02030	37
23	30754	02030	30771	02031	30789	02032	30806	02033	30824	02033	30842	02034	36
24	30842	02034	30859	02035	30877	02036	30894	02037	30912	02038	30929	02038	35
25	30929	02038	30947	02039	30964	02040	30982	02041	30999	02042	31017	02043	34
26	31017	02043	31034	02043	31052	02044	31069	02045	31087	02046	31104	02047	33
27	31104	02047	31122	02047	31139	02048	31157	02049	31174	02050	31191	02051	32
28	31191	02051	31209	02052	31226	02052	31244	02053	31261	02054	31279	02055	31
29	31279	02055	31296	02056	31314	02057	31331	02057	31349	02058	31366	02059	30
30	31366	02059	31383	02060	31401	02061	31418	02061	31436	02062	31453	02063	29
31	31453	02063	31470	02064	31488	02065	31505	02066	31523	02066	31540	02067	28
32	31540	02067	31557	02068	31575	02069	31592	02070	31610	02071	31627	02071	27
33	31627	02071	31644	02072	31662	02073	31679	02074	31696	02075	31714	02076	26
34	31714	02076	31731	02076	31748	02077	31766	02078	31783	02079	31800	02080	25
35	31800	02080	31818	02081	31835	02081	31852	02082	31870	02083	31887	02084	24
36	31887	02084	31904	02085	31922	02086	31939	02086	31956	02087	31974	02088	23
37	31974	02088	31991	02089	32008	02090	32026	02091	32043	02091	32060	02092	22
38	32060	02092	32077	02093	32095	02094	32112	02095	32129	02096	32146	02096	21
39	32146	02096	32164	02097	32181	02098	32198	02099	32216	02100	32233	02101	20
40	32233	02101	32250	02101	32267	02102	32285	02103	32302	02104	32319	02105	19
41	32319	02105	32336	02106	32353	02106	32371	02107	32388	02108	32405	02109	18
42	32405	02109	32422	02110	32439	02111	32457	02111	32474	02112	32491	02113	17
43	32491	02113	32508	02114	32526	02115	32543	02116	32560	02116	32577	02117	16
44	32577	02117	32594	02118	32611	02119	32629	02120	32646	02121	32663	02121	15
45	32663	02121	32680	02122	32697	02123	32714	02124	32731	02125	32749	02126	14
46	32749	02126	32766	02126	32783	02127	32800	02128	32817	02129	32834	02130	13
47	32834	02130	32852	02131	32869	02132	32886	02132	32903	02133	32920	02134	12
48	32920	02134	32937	02135	32954	02136	32971	02137	32988	02137	33005	02138	11
49	33005	02138	33022	02139	33040	02140	33057	02141	33074	02142	33091	02142	10
50	33091	02142	33108	02143	33125	02144	33142	02145	33159	02146	33176	02147	09
51	33176	02147	33193	02147	33210	02148	33227	02149	33244	02150	33261	02151	08
52	33261	02151	33278	02152	33295	02153	33313	02153	33330	02154	33347	02155	07
53	33347	02155	33364	02156	33381	02157	33398	02158	33415	02158	33432	02159	06
54	33432	02159	33449	02160	33466	02161	33483	02162	33500	02163	33517	02164	05
55	33517	02164	33534	02164	33551	02165	33568	02166	33585	02167	33602	02168	04
56	33602	02168	33619	02169	33636	02169	33652	02170	33669	02171	33686	02172	03
57	33686	02172	33703	02173	33720	02174	33737	02175	33754	02175	33771	02176	02
58	33771	02176	33788	02177	33805	02178	33822	02179	33839	02180	33856	02181	01
59	33856	02181	33873	02181	33890	02182	33907	02183	33923	02184	33940	02185	00

·8 ·6 ·4 ·2 ·0 ←

343°

PARTS for 0′·1 :— LOGS. 9 ; NATURALS, negligible.

17° HAVERSINES

′	·0 Log.	·0 Nat.	·2 Log.	·2 Nat.	·4 Log.	·4 Nat.	·6 Log.	·6 Nat.	·8 Log.	·8 Nat.	Log.	Nat.	′
	8·	0·	8·	0·	8·	0·	8·	0·	8·	0·	8·	0·	
00	33940	02185	33957	02186	33974	02186	33991	02187	34008	02188	34025	02189	59
01	34025	02189	34042	02190	34059	02191	34076	02192	34093	02192	34109	02193	58
02	34109	02193	34126	02194	34143	02195	34160	02196	34177	02197	34194	02198	57
03	34194	02198	34210	02198	34227	02199	34244	02200	34261	02201	34278	02202	56
04	34278	02202	34295	02203	34311	02204	34328	02204	34345	02205	34362	02206	55
05	34362	02206	34379	02207	34396	02208	34412	02209	34429	02209	34446	02210	54
06	34446	02210	34463	02211	34480	02212	34496	02213	34513	02214	34530	02215	53
07	34530	02215	34547	02216	34564	02216	34580	02217	34597	02218	34614	02219	52
08	34614	02219	34631	02220	34648	02221	34664	02221	34681	02222	34698	02223	51
09	34698	02223	34715	02224	34731	02225	34748	02226	34765	02227	34782	02227	50
10	34782	02227	34798	02228	34815	02229	34832	02230	34848	02231	34865	02232	49
11	34865	02232	34882	02233	34899	02234	34915	02234	34932	02235	34949	02236	48
12	34949	02236	34965	02237	34982	02238	34999	02239	35016	02240	35032	02240	47
13	35032	02240	35049	02241	35066	02242	35082	02243	35099	02244	35116	02245	46
14	35116	02245	35132	02246	35149	02246	35166	02247	35182	02248	35199	02249	45
15	35199	02249	35216	02250	35232	02251	35249	02252	35266	02252	35282	02253	44
16	35282	02253	35299	02254	35315	02255	35332	02256	35349	02257	35365	02258	43
17	35365	02258	35382	02259	35399	02259	35415	02260	35432	02261	35448	02262	42
18	35448	02262	35465	02263	35482	02264	35498	02265	35515	02265	35531	02266	41
19	35531	02266	35548	02267	35565	02268	35581	02269	35598	02270	35614	02271	40
20	35614	02271	35631	02271	35648	02272	35664	02273	35681	02274	35697	02275	39
21	35697	02275	35714	02276	35730	02277	35747	02278	35763	02278	35780	02279	38
22	35780	02279	35797	02280	35813	02281	35830	02282	35846	02283	35863	02284	37
23	35863	02284	35879	02285	35896	02285	35912	02286	35929	02287	35945	02288	36
24	35945	02288	35962	02289	35978	02290	35995	02291	36011	02291	36028	02292	35
25	36028	02292	36044	02293	36061	02294	36077	02295	36094	02296	36110	02297	34
26	36110	02297	36127	02298	36143	02298	36160	02299	36176	02300	36193	02301	33
27	36193	02301	36209	02302	36225	02303	36242	02304	36258	02305	36275	02305	32
28	36275	02305	36291	02306	36308	02307	36324	02308	36341	02309	36357	02310	31
29	36357	02310	36373	02311	36390	02312	36406	02312	36423	02313	36439	02314	30
30	36439	02314	36456	02315	36472	02316	36488	02317	36505	02318	36521	02319	29
31	36521	02319	36538	02319	36554	02320	36570	02321	36587	02322	36603	02323	28
32	36603	02323	36620	02324	36636	02325	36652	02326	36669	02326	36685	02327	27
33	36685	02327	36701	02328	36718	02329	36734	02330	36750	02331	36767	02332	26
34	36767	02332	36783	02333	36800	02333	36816	02334	36832	02335	36849	02336	25
35	36849	02336	36865	02337	36881	02338	36898	02339	36914	02340	36930	02340	24
36	36930	02340	36947	02341	36963	02342	36979	02343	36995	02344	37012	02345	23
37	37012	02345	37028	02346	37044	02347	37061	02347	37077	02348	37093	02349	22
38	37093	02349	37110	02350	37126	02351	37142	02352	37158	02353	37175	02354	21
39	37175	02354	37191	02355	37207	02355	37224	02356	37240	02357	37256	02358	20
40	37256	02358	37272	02359	37288	02360	37305	02361	37321	02362	37337	02363	19
41	37337	02363	37353	02363	37370	02364	37386	02365	37402	02366	37418	02367	18
42	37418	02367	37435	02368	37451	02369	37467	02370	37483	02370	37500	02371	17
43	37500	02371	37516	02372	37532	02373	37548	02374	37564	02375	37581	02376	16
44	37581	02376	37597	02377	37613	02378	37629	02378	37645	02379	37662	02380	15
45	37662	02380	37678	02381	37694	02382	37710	02383	37726	02384	37742	02385	14
46	37742	02385	37759	02386	37775	02386	37791	02387	37807	02388	37823	02389	13
47	37823	02389	37839	02390	37855	02391	37872	02392	37888	02393	37904	02394	12
48	37904	02394	37920	02394	37936	02395	37952	02396	37968	02397	37985	02398	11
49	37985	02398	38001	02399	38017	02400	38033	02401	38049	02402	38065	02402	10
50	38065	02402	38081	02403	38097	02404	38113	02405	38129	02406	38146	02407	09
51	38146	02407	38162	02408	38178	02409	38194	02410	38210	02411	38226	02411	08
52	38226	02411	38242	02412	38258	02413	38274	02414	38290	02415	38306	02416	07
53	38306	02416	38322	02417	38338	02418	38355	02419	38371	02419	38387	02420	06
54	38387	02420	38403	02421	38419	02422	38435	02423	38451	02424	38467	02425	05
55	38467	02425	38483	02426	38499	02427	38515	02427	38531	02428	38547	02429	04
56	38547	02429	38563	02430	38579	02431	38595	02432	38611	02433	38627	02434	03
57	38627	02434	38643	02435	38659	02436	38675	02436	38691	02437	38707	02438	02
58	38707	02438	38723	02439	38739	02440	38755	02441	38771	02442	38787	02443	01
59	38787	02443	38803	02444	38819	02444	38835	02445	38851	02446	38867	02447	00
	·8		·6		·4		·2		·0				

PARTS for 0′·1 :—   LOGS. 8 ;   NATURALS, negligible.

18° HAVERSINES

′	·0 Log. 8·	·0 Nat. 0·	·2 Log. 8·	·2 Nat. 0·	·4 Log. 8·	·4 Nat. 0·	·6 Log. 8·	·6 Nat. 0·	·8 Log. 8·	·8 Nat. 0·	′
00	38867	02447	38882	02448	38898	02449	38914	02450	38930	02451	38946 02452	59
01	38946	02452	38962	02453	38978	02453	38994	02454	39010	02455	39026 02456	58
02	39026	02456	39042	02457	39058	02458	39074	02459	39089	02460	39105 02461	57
03	39105	02461	39121	02462	39137	02463	39153	02463	39169	02464	39185 02465	56
04	39185	02465	39201	02466	39217	02467	39233	02468	39249	02469	39264 02470	55
05	39264	02470	39280	02471	39296	02472	39312	02472	39328	02473	39344 02474	54
06	39344	02474	39360	02475	39375	02476	39391	02477	39407	02478	39423 02479	53
07	39423	02479	39439	02480	39455	02481	39471	02481	39486	02482	39502 02483	52
08	39502	02483	39518	02484	39534	02485	39550	02486	39566	02487	39581 02488	51
09	39581	02488	39597	02489	39613	02490	39629	02491	39645	02491	39660 02492	50
10	39660	02492	39676	02493	39692	02494	39708	02495	39724	02496	39739 02497	49
11	39739	02497	39755	02498	39771	02499	39787	02500	39802	02500	39818 02501	48
12	39818	02501	39834	02502	39850	02503	39866	02504	39881	02505	39897 02506	47
13	39897	02506	39913	02507	39929	02508	39944	02509	39960	02510	39976 02510	46
14	39976	02510	39992	02511	40007	02512	40023	02513	40039	02514	40055 02515	45
15	40055	02515	40070	02516	40086	02517	40102	02518	40117	02519	40133 02520	44
16	40133	02520	40149	02521	40165	02521	40180	02522	40196	02523	40212 02524	43
17	40212	02524	40227	02525	40243	02526	40259	02527	40275	02528	40290 02529	42
18	40290	02529	40306	02530	40322	02531	40337	02531	40353	02532	40369 02533	41
19	40369	02533	40384	02534	40400	02535	40416	02536	40431	02537	40447 02538	40
20	40447	02538	40462	02539	40478	02540	40494	02541	40510	02542	40525 02542	39
21	40525	02542	40541	02543	40556	02544	40572	02545	40588	02546	40603 02547	38
22	40603	02547	40619	02548	40635	02549	40650	02550	40666	02551	40681 02552	37
23	40681	02552	40697	02553	40713	02553	40728	02554	40744	02555	40759 02556	36
24	40759	02556	40775	02557	40791	02558	40806	02559	40822	02560	40837 02561	35
25	40837	02561	40853	02562	40869	02563	40884	02564	40900	02564	40915 02565	34
26	40915	02565	40931	02566	40947	02567	40962	02568	40978	02569	40993 02570	33
27	40993	02570	41009	02571	41024	02572	41040	02573	41055	02574	41071 02575	32
28	41071	02575	41086	02576	41102	02576	41117	02577	41133	02578	41149 02579	31
29	41149	02579	41164	02580	41180	02581	41195	02582	41211	02583	41226 02584	30
30	41226	02584	41242	02585	41257	02586	41273	02587	41288	02588	41304 02588	29
31	41304	02588	41319	02589	41335	02590	41350	02591	41366	02592	41381 02593	28
32	41381	02593	41397	02594	41412	02595	41428	02596	41443	02597	41459 02598	27
33	41459	02598	41474	02599	41489	02600	41505	02600	41520	02601	41536 02602	26
34	41536	02602	41551	02603	41567	02604	41582	02605	41598	02606	41613 02607	25
35	41613	02607	41629	02608	41644	02609	41660	02610	41675	02611	41690 02612	24
36	41690	02612	41706	02613	41721	02613	41737	02614	41752	02615	41767 02616	23
37	41767	02616	41783	02617	41798	02618	41814	02619	41829	02620	41844 02621	22
38	41844	02621	41860	02622	41875	02623	41891	02624	41906	02625	41921 02626	21
39	41921	02626	41937	02626	41952	02627	41968	02628	41983	02629	41998 02630	20
40	41998	02630	42014	02631	42029	02632	42044	02633	42060	02634	42075 02635	19
41	42075	02635	42091	02636	42106	02637	42121	02638	42137	02639	42152 02639	18
42	42152	02639	42167	02640	42183	02641	42198	02642	42213	02643	42229 02644	17
43	42229	02644	42244	02645	42259	02646	42275	02647	42290	02648	42305 02649	16
44	42305	02649	42321	02650	42336	02651	42351	02652	42366	02653	42382 02653	15
45	42382	02653	42397	02654	42412	02655	42428	02656	42443	02657	42458 02658	14
46	42458	02658	42474	02659	42489	02660	42504	02661	42519	02662	42535 02663	13
47	42535	02663	42550	02664	42565	02665	42581	02666	42596	02667	42611 02668	12
48	42611	02668	42626	02668	42642	02669	42657	02670	42672	02671	42687 02672	11
49	42687	02672	42703	02673	42718	02674	42733	02675	42748	02676	42764 02677	10
50	42764	02677	42779	02678	42794	02679	42809	02680	42824	02681	42840 02682	09
51	42840	02682	42855	02683	42870	02683	42885	02684	42901	02685	42916 02686	08
52	42916	02686	42931	02687	42946	02688	42961	02689	42977	02690	42992 02691	07
53	42992	02691	43007	02692	43022	02693	43037	02694	43052	02695	43068 02696	06
54	43068	02696	43083	02697	43098	02698	43113	02699	43128	02700	43144 02700	05
55	43144	02700	43159	02701	43174	02702	43189	02703	43204	02704	43219 02705	04
56	43219	02705	43235	02706	43250	02707	43265	02708	43280	02709	43295 02710	03
57	43295	02710	43310	02711	43325	02712	43340	02713	43356	02714	43371 02715	02
58	43371	02715	43386	02716	43401	02717	43416	02717	43431	02718	43446 02719	01
59	43446	02719	43461	02720	43477	02721	43492	02722	43507	02723	43522 02724	00

·8 ·6 ·4 ·2 ·0 ←

341°

PARTS for 0′·1 :— LOGS. 8 ; NATURALS, negligible.

HAVERSINES

19°												

	·0		·2		·4		·6		·8			
	Log.	Nat.	Log.	Nat.	Log.	Nat.	Log.	Nat.	Log.	Nat.	Log.	Nat.	′
′	8·	0·	8·	0·	8·	0·	8·	0·	8·	0·	8·	0·	
00	43522	02724	43537	02725	43552	02726	43567	02727	43582	02728	43597	02729	59
01	43597	02729	43612	02730	43627	02731	43643	02732	43658	02733	43673	02734	58
02	43673	02734	43688	02735	43703	02735	43718	02736	43733	02737	43748	02738	57
03	43748	02738	43763	02739	43778	02740	43793	02741	43808	02742	43823	02743	56
04	43823	02743	43838	02744	43853	02745	43868	02746	43883	02747	43898	02748	55
05	43898	02748	43913	02749	43928	02750	43944	02751	43959	02752	43974	02753	54
06	43974	02753	43989	02754	44004	02754	44019	02755	44034	02756	44049	02757	53
07	44049	02757	44064	02758	44079	02759	44094	02760	44109	02761	44124	02762	52
08	44124	02762	44139	02763	44154	02764	44169	02765	44185	02766	44199	02767	51
09	44199	02767	44214	02768	44229	02769	44243	02770	44258	02771	44273	02772	50
10	44273	02772	44288	02773	44303	02774	44318	02774	44333	02775	44348	02776	49
11	44348	02776	44363	02777	44378	02778	44393	02779	44408	02780	44423	02781	48
12	44423	02781	44438	02782	44453	02783	44468	02784	44483	02785	44498	02786	47
13	44498	02786	44513	02787	44527	02788	44542	02789	44557	02790	44572	02791	46
14	44572	02791	44587	02792	44602	02793	44617	02794	44632	02795	44647	02796	45
15	44647	02796	44662	02797	44677	02797	44691	02798	44706	02799	44721	02800	44
16	44721	02800	44736	02801	44751	02802	44766	02803	44781	02804	44796	02805	43
17	44796	02805	44810	02806	44825	02807	44840	02808	44855	02809	44870	02810	42
18	44870	02810	44885	02811	44900	02812	44914	02813	44929	02814	44944	02815	41
19	44944	02815	44959	02816	44974	02817	44989	02818	45004	02819	45018	02820	40
20	45018	02820	45033	02821	45048	02821	45063	02822	45078	02823	45092	02824	39
21	45092	02824	45107	02825	45122	02826	45137	02827	45152	02828	45167	02829	38
22	45167	02829	45181	02830	45196	02831	45211	02832	45226	02833	45241	02834	37
23	45241	02834	45255	02835	45270	02836	45285	02837	45300	02838	45315	02839	36
24	45315	02839	45329	02840	45344	02841	45359	02842	45374	02843	45388	02844	35
25	45388	02844	45403	02845	45418	02846	45433	02847	45447	02848	45462	02849	34
26	45462	02849	45477	02850	45492	02850	45506	02851	45521	02852	45536	02853	33
27	45536	02853	45551	02854	45565	02855	45580	02856	45595	02857	45610	02858	32
28	45610	02858	45624	02859	45639	02860	45654	02861	45668	02862	45683	02863	31
29	45683	02863	45698	02864	45713	02865	45727	02866	45742	02867	45757	02868	30
30	45757	02868	45771	02869	45786	02870	45801	02871	45816	02872	45830	02873	29
31	45830	02873	45845	02874	45860	02875	45874	02876	45889	02877	45904	02878	28
32	45904	02878	45918	02879	45933	02880	45948	02881	45962	02882	45977	02883	27
33	45977	02883	45992	02883	46006	02884	46021	02885	46036	02886	46050	02887	26
34	46050	02887	46065	02888	46080	02889	46094	02890	46109	02891	46124	02892	25
35	46124	02892	46138	02893	46153	02894	46168	02895	46182	02896	46197	02897	24
36	46197	02897	46211	02898	46226	02899	46241	02900	46255	02901	46270	02902	23
37	46270	02902	46284	02903	46299	02904	46314	02905	46328	02906	46343	02907	22
38	46343	02907	46358	02908	46372	02909	46387	02910	46401	02911	46416	02912	21
39	46416	02912	46430	02913	46445	02914	46460	02915	46474	02916	46489	02917	20
40	46489	02917	46503	02918	46518	02919	46532	02920	46547	02921	46562	02922	19
41	46562	02922	46576	02923	46591	02924	46605	02924	46620	02925	46634	02926	18
42	46634	02926	46649	02927	46664	02928	46678	02929	46693	02930	46707	02931	17
43	46707	02931	46722	02932	46736	02933	46751	02934	46765	02935	46780	02936	16
44	46780	02936	46794	02937	46809	02938	46823	02939	46838	02940	46852	02941	15
45	46852	02941	46867	02942	46881	02943	46896	02944	46910	02945	46925	02946	14
46	46925	02946	46939	02947	46954	02948	46968	02949	46983	02950	46997	02951	13
47	46997	02951	47012	02952	47026	02953	47041	02954	47055	02955	47070	02956	12
48	47070	02956	47084	02957	47099	02958	47113	02959	47128	02960	47142	02961	11
49	47142	02961	47157	02962	47171	02963	47186	02964	47200	02965	47215	02966	10
50	47215	02966	47229	02967	47243	02968	47258	02969	47272	02970	47287	02971	09
51	47287	02971	47301	02972	47316	02973	47330	02974	47345	02975	47359	02976	08
52	47359	02976	47373	02977	47388	02978	47402	02979	47417	02980	47431	02981	07
53	47431	02981	47445	02982	47460	02983	47474	02984	47489	02985	47503	02986	06
54	47503	02986	47517	02987	47532	02988	47546	02989	47561	02990	47575	02991	05
55	47575	02991	47589	02992	47604	02993	47618	02994	47633	02995	47647	02996	04
56	47647	02996	47661	02996	47676	02997	47690	02998	47704	02999	47719	03000	03
57	47719	03000	47733	03001	47748	03002	47762	03003	47776	03004	47791	03005	02
58	47791	03005	47805	03006	47819	03007	47834	03008	47848	03009	47862	03010	01
59	47862	03010	47877	03011	47891	03012	47905	03013	47920	03014	47934	03015	00

	·8		·6		·4		·2		·0		

PARTS for 0′·1 :—   LOGS. 7 ;   NATURALS, negliigble.

HAVERSINES

20° | **339°**

All Log. values are prefixed **8·** and all Nat. values are prefixed **0·**

′	·0 Log.	·0 Nat.	·2 Log.	·2 Nat.	·4 Log.	·4 Nat.	·6 Log.	·6 Nat.	·8 Log.	·8 Nat.	Log.	Nat.	′
00	47934	03015	47948	03016	47963	03017	47977	03018	47991	03019	48006	03020	59
01	48006	03020	48020	03021	48034	03022	48049	03023	48063	03024	48077	03025	58
02	48077	03025	48092	03026	48106	03027	48120	03028	48134	03029	48149	03030	57
03	48149	03030	48163	03031	48177	03032	48192	03033	48206	03034	48220	03035	56
04	48220	03035	48234	03036	48249	03037	48263	03038	48277	03039	48292	03040	55
05	48292	03040	48306	03041	48320	03042	48334	03043	48349	03044	48363	03045	54
06	48363	03045	48377	03046	48391	03047	48406	03048	48420	03049	48434	03050	53
07	48434	03050	48448	03051	48462	03052	48477	03053	48491	03054	48505	03055	52
08	48505	03055	48519	03056	48534	03057	48548	03058	48562	03059	48576	03060	51
09	48576	03060	48591	03061	48605	03062	48619	03063	48633	03064	48647	03065	50
10	48647	03065	48662	03066	48676	03067	48690	03068	48704	03069	48718	03070	49
11	48718	03070	48733	03071	48747	03072	48761	03073	48775	03074	48789	03075	48
12	48789	03075	48804	03076	48818	03077	48832	03078	48846	03079	48860	03080	47
13	48860	03080	48875	03081	48889	03082	48903	03083	48917	03084	48931	03085	46
14	48931	03085	48945	03086	48960	03087	48974	03088	48988	03089	49002	03090	45
15	49002	03090	49016	03091	49030	03092	49044	03093	49058	03094	49073	03095	44
16	49073	03095	49087	03096	49101	03097	49115	03098	49129	03099	49143	03101	43
17	49143	03101	49157	03102	49172	03103	49186	03104	49200	03105	49214	03106	42
18	49214	03106	49228	03107	49242	03108	49256	03109	49270	03110	49284	03111	41
19	49284	03111	49299	03112	49313	03113	49327	03114	49341	03115	49355	03116	40
20	49355	03116	49369	03117	49383	03118	49397	03119	49411	03120	49425	03121	39
21	49425	03121	49439	03122	49453	03123	49468	03124	49482	03125	49496	03126	38
22	49496	03126	49510	03127	49524	03128	49538	03129	49552	03130	49566	03131	37
23	49566	03131	49580	03132	49594	03133	49608	03134	49622	03135	49636	03136	36
24	49636	03136	49650	03137	49664	03138	49678	03139	49692	03140	49706	03141	35
25	49706	03141	49720	03142	49734	03143	49749	03144	49763	03145	49777	03146	34
26	49777	03146	49791	03147	49805	03148	49819	03149	49833	03150	49847	03151	33
27	49847	03151	49861	03152	49875	03153	49889	03154	49903	03155	49917	03156	32
28	49917	03156	49931	03157	49945	03158	49959	03159	49973	03160	49987	03161	31
29	49987	03161	50001	03162	50015	03163	50028	03164	50042	03165	50056	03166	30
30	50056	03166	50070	03167	50084	03168	50098	03169	50112	03170	50126	03171	29
31	50126	03171	50140	03173	50154	03174	50168	03175	50182	03176	50196	03177	28
32	50196	03177	50210	03178	50224	03179	50238	03180	50252	03181	50266	03182	27
33	50266	03182	50280	03183	50294	03184	50308	03185	50321	03186	50335	03187	26
34	50335	03187	50349	03188	50363	03189	50377	03190	50391	03191	50405	03192	25
35	50405	03192	50419	03193	50433	03194	50447	03195	50461	03196	50475	03197	24
36	50475	03197	50488	03198	50502	03199	50516	03200	50530	03201	50544	03202	23
37	50544	03202	50558	03203	50572	03204	50586	03205	50600	03206	50613	03207	22
38	50613	03207	50627	03208	50641	03209	50655	03210	50669	03211	50683	03212	21
39	50683	03212	50697	03213	50711	03214	50724	03215	50738	03216	50752	03218	20
40	50752	03218	50766	03219	50780	03220	50794	03221	50808	03222	50821	03223	19
41	50821	03223	50835	03224	50849	03225	50863	03226	50877	03227	50891	03228	18
42	50891	03228	50904	03229	50918	03230	50932	03231	50946	03232	50960	03233	17
43	50960	03233	50974	03234	50987	03235	51001	03236	51015	03237	51029	03238	16
44	51029	03238	51043	03239	51056	03240	51070	03241	51084	03242	51098	03243	15
45	51098	03243	51112	03244	51126	03245	51139	03246	51153	03247	51167	03248	14
46	51167	03248	51181	03249	51194	03250	51208	03251	51222	03253	51236	03254	13
47	51236	03254	51250	03255	51263	03256	51277	03257	51291	03258	51305	03259	12
48	51305	03259	51318	03260	51332	03261	51346	03262	51360	03263	51373	03264	11
49	51373	03264	51387	03265	51401	03266	51415	03267	51428	03268	51442	03269	10
50	51442	03269	51456	03270	51470	03271	51483	03272	51497	03273	51511	03274	09
51	51511	03274	51525	03275	51538	03276	51552	03277	51566	03278	51580	03279	08
52	51580	03279	51593	03280	51607	03281	51621	03283	51634	03284	51648	03285	07
53	51648	03285	51662	03286	51676	03287	51689	03288	51703	03289	51717	03290	06
54	51717	03290	51730	03291	51744	03292	51758	03293	51771	03294	51785	03295	05
55	51785	03295	51799	03296	51813	03297	51826	03298	51840	03299	51854	03300	04
56	51854	03300	51867	03301	51881	03302	51895	03303	51908	03304	51922	03305	03
57	51922	03305	51935	03306	51949	03307	51963	03308	51977	03310	51990	03311	02
58	51990	03311	52004	03312	52017	03313	52031	03314	52045	03315	52058	03316	01
59	52058	03316	52072	03317	52086	03318	52099	03319	52113	03320	52127	03321	00

Bottom decimal headings (for 339°, read upward): ·8   ·6   ·4   ·2   ·0

PARTS for 0′·1 :—   LOGS. 7 ;   NATURALS, negligible.

21° — HAVERSINES

′	·0 Log	·0 Nat	·2 Log	·2 Nat	·4 Log	·4 Nat	·6 Log	·6 Nat	·8 Log	·8 Nat	Log	Nat	′
	8·	0·	8·	0·	8·	0·	8·	0·	8·	0·	8·	0·	
00	52127	03321	52140	03322	52154	03323	52168	03324	52181	03325	52195	03326	59
01	52195	03326	52208	03327	52222	03328	52236	03329	52249	03330	52263	03331	58
02	52263	03331	52276	03332	52290	03334	52304	03335	52317	03336	52331	03337	57
03	52331	03337	52344	03338	52358	03339	52372	03340	52385	03341	52399	03342	56
04	52399	03342	52412	03343	52426	03344	52440	03345	52453	03346	52467	03347	55
05	52467	03347	52480	03348	52494	03349	52507	03350	52521	03351	52535	03352	54
06	52535	03352	52548	03353	52562	03354	52575	03355	52589	03357	52602	03358	53
07	52602	03358	52616	03359	52630	03360	52643	03361	52657	03362	52670	03363	52
08	52670	03363	52684	03364	52697	03365	52711	03366	52724	03367	52738	03368	51
09	52738	03368	52751	03369	52765	03370	52778	03371	52792	03372	52805	03373	50
10	52805	03373	52819	03374	52832	03375	52846	03376	52859	03378	52873	03379	49
11	52873	03379	52887	03380	52900	03381	52914	03382	52927	03383	52941	03384	48
12	52941	03384	52954	03385	52968	03386	52981	03387	52995	03388	53008	03389	47
13	53008	03389	53021	03390	53035	03391	53048	03392	53062	03393	53075	03394	46
14	53075	03394	53089	03395	53102	03396	53116	03398	53129	03399	53143	03400	45
15	53143	03400	53156	03401	53170	03402	53183	03403	53197	03404	53210	03405	44
16	53210	03405	53224	03406	53237	03407	53251	03408	53264	03409	53277	03410	43
17	53277	03410	53291	03411	53304	03412	53318	03413	53331	03414	53345	03415	42
18	53345	03415	53358	03417	53371	03418	53385	03419	53398	03420	53412	03421	41
19	53412	03421	53425	03422	53439	03423	53452	03424	53466	03425	53479	03426	40
20	53479	03426	53492	03427	53506	03428	53519	03429	53533	03430	53546	03431	39
21	53546	03431	53559	03432	53573	03433	53586	03434	53600	03436	53613	03437	38
22	53613	03437	53626	03438	53640	03439	53653	03440	53666	03441	53680	03442	37
23	53680	03442	53693	03443	53707	03444	53720	03445	53733	03446	53747	03447	36
24	53747	03447	53760	03448	53774	03449	53787	03450	53800	03451	53814	03453	35
25	53814	03453	53827	03454	53840	03455	53854	03456	53867	03457	53880	03458	34
26	53880	03458	53894	03459	53907	03460	53920	03461	53934	03462	53947	03463	33
27	53947	03463	53961	03464	53974	03465	53987	03466	54000	03467	54014	03468	32
28	54014	03468	54027	03470	54040	03471	54054	03472	54067	03473	54080	03474	31
29	54080	03474	54094	03475	54107	03476	54120	03477	54134	03478	54147	03479	30
30	54147	03479	54160	03480	54174	03481	54187	03482	54200	03483	54213	03484	29
31	54213	03484	54227	03486	54240	03487	54253	03488	54267	03489	54280	03490	28
32	54280	03490	54293	03491	54307	03492	54320	03493	54333	03494	54346	03495	27
33	54346	03495	54360	03496	54373	03497	54386	03498	54400	03499	54413	03500	26
34	54413	03500	54426	03502	54439	03503	54453	03504	54466	03505	54479	03506	25
35	54479	03506	54492	03507	54505	03508	54519	03509	54532	03510	54545	03511	24
36	54545	03511	54558	03512	54572	03513	54585	03514	54598	03515	54611	03517	23
37	54611	03517	54625	03518	54638	03519	54651	03520	54664	03521	54678	03522	22
38	54678	03522	54691	03523	54704	03524	54717	03525	54730	03526	54744	03527	21
39	54744	03527	54757	03529	54770	03530	54783	03531	54797	03532	54810	03533	20
40	54810	03533	54823	03534	54836	03535	54849	03536	54863	03537	54876	03538	19
41	54876	03538	54889	03539	54902	03540	54915	03541	54928	03542	54942	03543	18
42	54942	03543	54955	03544	54968	03546	54981	03547	54994	03548	55008	03549	17
43	55008	03549	55021	03550	55034	03551	55047	03552	55060	03553	55073	03554	16
44	55073	03554	55086	03555	55100	03556	55113	03558	55126	03560	55139	03560	15
45	55139	03560	55152	03561	55166	03562	55179	03563	55192	03564	55205	03565	14
46	55205	03565	55218	03566	55231	03567	55244	03568	55257	03569	55271	03570	13
47	55271	03570	55284	03571	55297	03572	55310	03574	55323	03575	55336	03576	12
48	55336	03576	55349	03577	55362	03578	55376	03579	55389	03580	55402	03581	11
49	55402	03581	55415	03582	55428	03583	55441	03584	55454	03585	55467	03587	10
50	55467	03587	55480	03588	55494	03589	55507	03590	55520	03591	55533	03592	09
51	55533	03592	55546	03593	55559	03594	55572	03595	55585	03596	55598	03597	08
52	55598	03597	55611	03598	55624	03600	55637	03601	55651	03602	55664	03603	07
53	55664	03603	55677	03604	55690	03605	55703	03606	55716	03607	55729	03608	06
54	55729	03608	55742	03609	55755	03610	55768	03611	55781	03613	55794	03614	05
55	55794	03614	55807	03615	55820	03616	55833	03617	55846	03618	55859	03619	04
56	55859	03619	55872	03620	55885	03621	55899	03622	55912	03623	55925	03624	03
57	55925	03624	55938	03626	55951	03627	55964	03628	55977	03629	55990	03630	02
58	55990	03630	56003	03631	56016	03632	56029	03633	56042	03634	56055	03635	01
59	56055	03635	56068	03636	56081	03638	56094	03639	56107	03640	56120	03641	00
			·8		·6		·4		·2		·0		

PARTS for 0′·1 :—   LOGS. 7 ;   NATURALS, negligible.

HAVERSINES

22°

′	.0 Log. 8·	.0 Nat. 0·	.2 Log. 8·	.2 Nat. 0·	.4 Log. 8·	.4 Nat. 0·	.6 Log. 8·	.6 Nat. 0·	.8 Log. 8·	.8 Nat. 0·	′
00	56120	03641	56133	03642	56146	03643	56159	03644	56172	03645	59
01	56185	03646	56198	03647	56211	03648	56224	03650	56237	03651	58
02	56250	03652	56263	03653	56276	03654	56289	03655	56302	03656	57
03	56315	03657	56328	03658	56341	03659	56353	03660	56366	03662	56
04	56379	03663	56392	03664	56405	03665	56418	03666	56431	03667	55
05	56444	03668	56457	03669	56470	03670	56483	03671	56496	03672	54
06	56509	03674	56522	03675	56535	03676	56548	03677	56560	03678	53
07	56573	03679	56586	03680	56599	03681	56612	03682	56625	03683	52
08	56638	03685	56651	03686	56664	03687	56677	03688	56690	03689	51
09	56703	03690	56716	03691	56728	03692	56741	03693	56754	03694	50
10	56767	03695	56780	03697	56793	03698	56806	03699	56819	03700	49
11	56832	03701	56844	03702	56857	03703	56870	03704	56883	03705	48
12	56896	03706	56909	03708	56922	03709	56935	03710	56948	03711	47
13	56960	03712	56973	03713	56986	03714	56999	03715	57012	03716	46
14	57025	03717	57038	03719	57050	03720	57063	03721	57076	03722	45
15	57089	03723	57102	03724	57115	03725	57128	03726	57140	03727	44
16	57153	03728	57166	03730	57179	03731	57192	03732	57205	03733	43
17	57217	03734	57230	03735	57243	03736	57256	03737	57269	03738	42
18	57282	03740	57294	03741	57307	03742	57320	03743	57333	03744	41
19	57346	03745	57358	03746	57371	03747	57384	03748	57397	03749	40
20	57410	03751	57422	03752	57435	03753	57448	03754	57461	03755	39
21	57474	03756	57486	03757	57499	03758	57512	03759	57525	03761	38
22	57538	03762	57550	03763	57563	03764	57576	03765	57589	03766	37
23	57601	03767	57614	03768	57627	03769	57640	03770	57652	03772	36
24	57665	03773	57678	03774	57691	03775	57703	03776	57716	03777	35
25	57729	03778	57742	03779	57755	03780	57767	03782	57780	03783	34
26	57793	03784	57806	03785	57818	03786	57831	03787	57844	03788	33
27	57856	03789	57869	03790	57882	03792	57895	03793	57907	03794	32
28	57920	03795	57933	03796	57945	03797	57958	03798	57971	03799	31
29	57984	03800	57996	03802	58009	03803	58022	03804	58034	03805	30
30	58047	03806	58060	03807	58073	03808	58085	03809	58098	03810	29
31	58111	03812	58123	03813	58136	03814	58149	03815	58161	03816	28
32	58174	03817	58187	03818	58199	03819	58212	03821	58225	03822	27
33	58237	03823	58250	03824	58263	03825	58275	03826	58288	03827	26
34	58301	03828	58313	03829	58326	03831	58339	03832	58351	03833	25
35	58364	03834	58377	03835	58389	03836	58402	03837	58415	03838	24
36	58427	03839	58440	03841	58453	03842	58465	03843	58478	03844	23
37	58491	03845	58503	03846	58516	03847	58528	03848	58541	03850	22
38	58554	03851	58566	03852	58579	03853	58592	03854	58604	03855	21
39	58617	03856	58629	03857	58642	03859	58655	03860	58667	03861	20
40	58680	03862	58692	03863	58705	03864	58718	03865	58730	03866	19
41	58743	03867	58755	03869	58768	03870	58780	03871	58793	03872	18
42	58806	03873	58818	03874	58831	03875	58844	03876	58856	03878	17
43	58869	03879	58881	03880	58894	03881	58906	03882	58919	03883	16
44	58932	03884	58944	03885	58957	03887	58969	03888	58982	03889	15
45	58994	03890	59007	03891	59020	03892	59032	03893	59045	03894	14
46	59057	03896	59070	03897	59082	03898	59095	03899	59107	03900	13
47	59120	03901	59133	03902	59145	03903	59158	03905	59170	03906	12
48	59183	03907	59195	03908	59208	03909	59220	03910	59233	03911	11
49	59245	03912	59258	03914	59270	03915	59283	03916	59295	03917	10
50	59308	03918	59320	03919	59333	03920	59345	03921	59358	03923	09
51	59370	03924	59383	03925	59395	03926	59408	03927	59420	03928	08
52	59433	03929	59445	03931	59458	03932	59470	03933	59483	03934	07
53	59495	03935	59508	03936	59520	03937	59533	03938	59545	03940	06
54	59558	03941	59570	03942	59583	03943	59595	03944	59608	03945	05
55	59620	03946	59632	03948	59645	03949	59657	03950	59670	03951	04
56	59682	03952	59695	03953	59707	03954	59720	03955	59732	03957	03
57	59745	03958	59757	03959	59769	03960	59782	03961	59794	03962	02
58	59807	03963	59819	03965	59832	03966	59844	03967	59857	03968	01
59	59869	03969	59881	03970	59894	03971	59906	03972	59919	03974	00

.8   .6   .4   .2   .0

337°

PARTS for 0′·1 :—   LOGS. 6;   NATURALS, 0 or 1 (by inspection).

23° HAVERSINES

′	·0 Log. 8·	·0 Nat. 0·	·2 Log. 8·	·2 Nat. 0·	·4 Log. 8·	·4 Nat. 0·	·6 Log. 8·	·6 Nat. 0·	·8 Log. 8·	·8 Nat. 0·	Log. 8·	Nat. 0·	′
00	59931	03975	59943	03976	59956	03977	59968	03978	59981	03979	59993	03980	59
01	59993	03980	60006	03982	60010	03983	60030	03984	60043	03985	60055	03986	58
02	60055	03986	60068	03987	60080	03988	60092	03990	60105	03991	60117	03992	57
03	60117	03992	60130	03993	60142	03994	60154	03995	60166	03996	60179	03998	56
04	60179	03998	60191	03999	60204	04000	60216	04001	60229	04002	60241	04003	55
05	60241	04003	60253	04004	60266	04006	60278	04007	60290	04008	60303	04009	54
06	60303	04009	60315	04010	60328	04011	60340	04012	60352	04014	60365	04015	53
07	60365	04015	60377	04016	60389	04017	60402	04018	60414	04019	60426	04020	52
08	60426	04020	60439	04021	60451	04023	60463	04024	60476	04025	60488	04026	51
09	60488	04026	60500	04027	60513	04028	60525	04030	60537	04031	60550	04032	50
10	60550	04032	60562	04033	60574	04034	60587	04035	60599	04036	60611	04038	49
11	60611	04038	60624	04039	60636	04040	60648	04041	60661	04042	60673	04043	48
12	60673	04043	60685	04044	60697	04046	60710	04047	60722	04048	60734	04049	47
13	60734	04049	60747	04050	60759	04051	60771	04052	60784	04054	60796	04055	46
14	60796	04055	60808	04056	60820	04057	60833	04058	60845	04059	60857	04060	45
15	60857	04060	60870	04062	60882	04063	60894	04064	60906	04065	60919	04066	44
16	60919	04066	60931	04067	60943	04069	60955	04070	60968	04071	60980	04072	43
17	60980	04072	60992	04073	61005	04074	61017	04075	61029	04077	61041	04078	42
18	61041	04078	61054	04079	61066	04080	61078	04081	61090	04082	61103	04083	41
19	61103	04083	61115	04085	61127	04086	61139	04087	61152	04088	61164	04089	40
20	61164	04089	61176	04090	61188	04092	61201	04093	61213	04094	61225	04095	39
21	61225	04095	61237	04096	61249	04097	61262	04098	61274	04100	61286	04101	38
22	61286	04101	61298	04102	61310	04103	61323	04104	61335	04105	61347	04106	37
23	61347	04106	61359	04108	61372	04109	61384	04110	61396	04111	61408	04112	36
24	61408	04112	61420	04113	61433	04115	61445	04116	61457	04117	61469	04118	35
25	61469	04118	61481	04119	61494	04120	61506	04122	61518	04123	61530	04124	34
26	61530	04124	61542	04125	61554	04126	61567	04127	61579	04128	61591	04130	33
27	61591	04130	61603	04131	61615	04132	61627	04133	61640	04134	61652	04135	32
28	61652	04135	61664	04137	61676	04138	61688	04139	61700	04140	61713	04141	31
29	61713	04141	61725	04142	61737	04144	61749	04145	61761	04146	61773	04147	30
30	61773	04147	61786	04148	61798	04149	61810	04150	61822	04152	61834	04153	29
31	61834	04153	61846	04154	61858	04155	61870	04156	61883	04157	61895	04159	28
32	61895	04159	61907	04160	61919	04161	61931	04162	61943	04163	61955	04164	27
33	61955	04164	61968	04166	61980	04167	61992	04168	62004	04169	62016	04170	26
34	62016	04170	62028	04171	62040	04173	62052	04174	62064	04175	62076	04176	25
35	62076	04176	62089	04177	62101	04178	62113	04180	62125	04181	62137	04182	24
36	62137	04182	62149	04183	62161	04184	62173	04185	62185	04187	62197	04188	23
37	62197	04188	62209	04189	62222	04190	62234	04191	62246	04192	62258	04194	22
38	62258	04194	62270	04195	62282	04196	62294	04197	62306	04198	62318	04199	21
39	62318	04199	62330	04201	62342	04202	62354	04203	62367	04204	62379	04205	20
40	62379	04205	62391	04206	62403	04208	62415	04209	62427	04210	62439	04211	19
41	62439	04211	62451	04212	62463	04213	62475	04215	62487	04216	62499	04217	18
42	62499	04217	62511	04218	62523	04219	62535	04220	62547	04222	62559	04223	17
43	62559	04223	62571	04224	62583	04225	62595	04226	62607	04227	62619	04229	16
44	62619	04229	62631	04230	62643	04231	62655	04232	62667	04233	62679	04234	15
45	62679	04234	62691	04236	62703	04237	62716	04238	62728	04239	62740	04240	14
46	62740	04240	62752	04241	62764	04243	62776	04244	62788	04245	62800	04246	13
47	62800	04246	62812	04247	62824	04248	62835	04250	62847	04251	62859	04252	12
48	62859	04252	62871	04253	62883	04254	62895	04256	62907	04257	62919	04258	11
49	62919	04258	62931	04260	62943	04260	62955	04261	62967	04263	62979	04264	10
50	62979	04264	62991	04265	63003	04266	63015	04267	63027	04268	63039	04270	09
51	63039	04270	63051	04271	63063	04272	63075	04273	63087	04274	63099	04276	08
52	63099	04276	63111	04277	63123	04278	63135	04279	63147	04280	63159	04281	07
53	63159	04281	63171	04283	63183	04284	63194	04285	63206	04286	63218	04287	06
54	63218	04287	63230	04288	63242	04290	63254	04291	63266	04292	63278	04293	05
55	63278	04293	63290	04294	63302	04296	63314	04297	63326	04298	63338	04299	04
56	63338	04299	63350	04300	63362	04301	63373	04303	63385	04304	63397	04305	03
57	63397	04305	63409	04306	63421	04307	63433	04309	63445	04310	63457	04311	02
58	63457	04311	63469	04312	63481	04313	63493	04314	63504	04316	63516	04317	01
59	63516	04317	63528	04318	63540	04319	63552	04320	63564	04322	63576	04323	00

·8 ·6 ·4 ·2 ·0 ← ↑

PARTS for 0′·1 :—   LOGS. 6 ;   NATURALS, 0 or 1 (by inspection).

24° HAVERSINES

Log. values prefixed by **8·** ; Nat. values prefixed by **0·**

′	·0 Log.	·0 Nat.	·2 Log.	·2 Nat.	·4 Log.	·4 Nat.	·6 Log.	·6 Nat.	·8 Log.	·8 Nat.	Log.	Nat.	′
00	63576	04323	63588	04324	63600	04325	63611	04326	63623	04327	63635	04329	59
01	63635	04329	63647	04330	63659	04331	63671	04332	63683	04333	63695	04335	58
02	63695	04335	63706	04336	63718	04337	63730	04338	63742	04339	63754	04340	57
03	63754	04340	63766	04342	63778	04343	63789	04344	63801	04345	63813	04346	56
04	63813	04346	63825	04348	63837	04349	63849	04350	63861	04351	63872	04352	55
05	63872	04352	63884	04354	63896	04355	63908	04356	63920	04357	63932	04358	54
06	63932	04358	63943	04359	63955	04361	63967	04362	63979	04363	63991	04364	53
07	63991	04364	64003	04365	64014	04367	64026	04368	64038	04369	64050	04370	52
08	64050	04370	64062	04371	64073	04373	64085	04374	64097	04375	64109	04376	51
09	64109	04376	64121	04377	64133	04379	64144	04380	64156	04381	64168	04382	50
10	64168	04382	64180	04383	64192	04384	64203	04386	64215	04387	64227	04388	49
11	64227	04388	64239	04389	64251	04390	64262	04392	64274	04393	64286	04394	48
12	64286	04394	64298	04395	64310	04396	64321	04398	64333	04399	64345	04400	47
13	64345	04400	64357	04401	64368	04402	64380	04404	64392	04405	64404	04406	46
14	64404	04406	64416	04407	64427	04408	64439	04410	64451	04411	64463	04412	45
15	64463	04412	64474	04413	64486	04414	64498	04415	64510	04417	64521	04418	44
16	64521	04418	64533	04419	64545	04420	64557	04421	64568	04423	64580	04424	43
17	64580	04424	64592	04425	64604	04426	64615	04427	64627	04429	64639	04430	42
18	64639	04430	64651	04431	64662	04432	64674	04433	64686	04435	64697	04436	41
19	64697	04436	64709	04437	64721	04438	64733	04439	64744	04441	64756	04442	40
20	64756	04442	64768	04443	64779	04444	64791	04445	64803	04447	64815	04448	39
21	64815	04448	64826	04449	64838	04450	64850	04451	64861	04453	64873	04454	38
22	64873	04454	64885	04455	64897	04456	64908	04457	64920	04459	64932	04460	37
23	64932	04460	64943	04461	64955	04462	64967	04463	64978	04465	64990	04466	36
24	64990	04466	65002	04467	65013	04468	65025	04469	65037	04471	65049	04472	35
25	65049	04472	65060	04473	65072	04474	65084	04475	65095	04477	65107	04478	34
26	65107	04478	65118	04479	65130	04480	65142	04481	65154	04483	65165	04484	33
27	65165	04484	65177	04485	65188	04486	65200	04487	65212	04489	65223	04490	32
28	65223	04490	65235	04491	65247	04492	65258	04494	65270	04495	65282	04496	31
29	65282	04496	65293	04497	65305	04498	65317	04500	65328	04501	65340	04502	30
30	65340	04502	65352	04503	65363	04504	65375	04506	65387	04507	65398	04508	29
31	65398	04508	65410	04509	65421	04510	65433	04512	65445	04513	65456	04514	28
32	65456	04514	65468	04515	65479	04516	65491	04518	65503	04519	65514	04520	27
33	65514	04520	65526	04521	65538	04522	65549	04524	65561	04525	65572	04526	26
34	65572	04526	65584	04527	65596	04529	65607	04530	65619	04531	65630	04532	25
35	65630	04532	65642	04533	65654	04535	65665	04536	65677	04537	65688	04538	24
36	65688	04538	65700	04539	65712	04541	65723	04542	65735	04543	65746	04544	23
37	65746	04544	65758	04545	65769	04547	65781	04548	65793	04549	65804	04550	22
38	65804	04550	65816	04552	65827	04553	65839	04554	65850	04555	65862	04556	21
39	65862	04556	65874	04558	65885	04559	65897	04560	65908	04561	65920	04562	20
40	65920	04562	65931	04564	65943	04565	65954	04566	65966	04567	65978	04569	19
41	65978	04569	65989	04570	66001	04571	66012	04572	66024	04573	66035	04575	18
42	66035	04575	66047	04576	66058	04577	66070	04578	66081	04579	66093	04581	17
43	66093	04581	66105	04582	66116	04583	66128	04584	66139	04586	66151	04587	16
44	66151	04587	66162	04588	66174	04589	66185	04590	66197	04592	66208	04593	15
45	66208	04593	66220	04594	66231	04595	66243	04597	66254	04598	66266	04599	14
46	66266	04599	66277	04600	66289	04601	66300	04603	66312	04604	66323	04605	13
47	66323	04605	66335	04606	66346	04607	66358	04609	66369	04610	66381	04611	12
48	66381	04611	66392	04612	66404	04614	66415	04615	66427	04616	66438	04617	11
49	66438	04617	66450	04618	66461	04620	66473	04621	66484	04622	66496	04623	10
50	66496	04623	66507	04625	66518	04626	66530	04627	66541	04628	66553	04629	09
51	66553	04629	66564	04631	66576	04632	66587	04633	66599	04634	66610	04636	08
52	66610	04636	66622	04637	66633	04638	66645	04639	66656	04640	66667	04642	07
53	66667	04642	66679	04643	66690	04644	66702	04645	66713	04647	66725	04648	06
54	66725	04648	66736	04649	66748	04650	66759	04651	66771	04653	66782	04654	05
55	66782	04654	66793	04655	66805	04656	66816	04658	66828	04659	66839	04660	04
56	66839	04660	66850	04661	66862	04663	66873	04664	66885	04665	66896	04666	03
57	66896	04666	66908	04667	66919	04669	66931	04670	66942	04671	66953	04672	02
58	66953	04672	66965	04674	66976	04675	66988	04676	66999	04677	67010	04678	01
59	67010	04678	67022	04680	67033	04681	67045	04682	67056	04683	67067	04685	00

Bottom column labels (for 335°): ·8 ·6 ·4 ·2 ·0 ←

335°

PARTS for 0′·1 :— LOGS. 6 ; NATURALS, 0 or 1 (by inspection).

HAVERSINES

25°

′	·0		·2		·4		·6		·8			′
	Log.	Nat.	Log.	Nat.	Log.	Nat.	Log.	Nat.	Log.	Nat.	Log.	Nat.	
	8·	0·	8·	0·	8·	0·	8·	0·	8·	0·	8·	0·	
00	67067	04685	67079	04686	67090	04687	67102	04688	67113	04690	67124	04691	59
01	67124	04691	67136	04692	67147	04693	67158	04694	67170	04696	67181	04697	58
02	67181	04697	67193	04698	67204	04699	67215	04701	67227	04702	67238	04703	57
03	67238	04703	67250	04704	67261	04706	67272	04707	67284	04708	67295	04709	56
04	67295	04709	67306	04710	67318	04712	67329	04713	67340	04714	67352	04715	55
05	67352	04715	67363	04717	67375	04718	67386	04719	67397	04720	67409	04722	54
06	67409	04722	67420	04723	67431	04724	67443	04725	67454	04727	67465	04728	53
07	67465	04728	67477	04729	67488	04730	67499	04731	67511	04733	67522	04734	52
08	67522	04734	67533	04735	67545	04736	67556	04738	67567	04739	67579	04740	51
09	67579	04740	67590	04741	67601	04743	67613	04744	67624	04745	67635	04746	50
10	67635	04746	67647	04747	67658	04749	67669	04750	67680	04751	67692	04752	49
11	67692	04752	67703	04754	67714	04755	67726	04756	67737	04757	67748	04759	48
12	67748	04759	67760	04760	67771	04761	67782	04762	67794	04764	67805	04765	47
13	67805	04765	67816	04766	67827	04767	67839	04769	67850	04770	67861	04771	46
14	67861	04771	67873	04772	67884	04774	67895	04775	67907	04776	67918	04777	45
15	67918	04777	67929	04778	67940	04780	67952	04781	67963	04782	67974	04783	44
16	67974	04783	67985	04785	67997	04786	68008	04787	68019	04788	68030	04790	43
17	68030	04790	68042	04791	68053	04792	68064	04793	68076	04795	68087	04796	42
18	68087	04796	68098	04797	68109	04798	68121	04800	68132	04801	68143	04802	41
19	68143	04802	68154	04803	68165	04805	68177	04806	68188	04807	68199	04808	40
20	68199	04808	68210	04810	68222	04811	68233	04812	68244	04813	68255	04815	39
21	68255	04815	68267	04816	68278	04817	68289	04818	68300	04820	68312	04821	38
22	68312	04821	68323	04822	68334	04823	68345	04825	68357	04826	68368	04827	37
23	68368	04827	68379	04828	68390	04830	68401	04831	68413	04832	68424	04833	36
24	68424	04833	68435	04835	68446	04836	68457	04837	68469	04838	68480	04840	35
25	68480	04840	68491	04841	68502	04842	68513	04843	68525	04844	68536	04846	34
26	68536	04846	68547	04847	68558	04848	68569	04849	68581	04851	68592	04852	33
27	68592	04852	68603	04853	68614	04854	68625	04856	68537	04857	68648	04858	32
28	68648	04858	68659	04859	68670	04861	68681	04862	68692	04863	68704	04864	31
29	68704	04864	68715	04866	68726	04867	68737	04868	68748	04870	68759	04871	30
30	68759	04871	68771	04872	68782	04873	68793	04875	68804	04876	68815	04877	29
31	68815	04877	68826	04878	68838	04880	68849	04881	68860	04882	68871	04883	28
32	68871	04883	68882	04885	68893	04886	68904	04887	68916	04888	68927	04890	27
33	68927	04890	68938	04891	68949	04892	68960	04893	68971	04895	68982	04896	26
34	68982	04896	68994	04897	69005	04898	69016	04900	69027	04901	69038	04902	25
35	69038	04902	69049	04903	69060	04905	69072	04906	69083	04907	69094	04908	24
36	69094	04908	69105	04910	69116	04911	69127	04912	69138	04913	69149	04915	23
37	69149	04915	69160	04916	69172	04917	69183	04918	69194	04920	69205	04921	22
38	69205	04921	69216	04922	69227	04923	69238	04925	69249	04926	69260	04927	21
39	69260	04927	69271	04929	69283	04930	69294	04931	69305	04932	69316	04934	20
40	69316	04934	69327	04935	69338	04936	69349	04937	69360	04939	69371	04940	19
41	69371	04940	69382	04941	69394	04942	69405	04944	69416	04945	69427	04946	18
42	69427	04946	69438	04947	69449	04949	69460	04950	69471	04951	69482	04952	17
43	69482	04952	69493	04954	69504	04955	69515	04956	69526	04957	69537	04959	16
44	69537	04959	69548	04960	69559	04961	69571	04963	69582	04964	69593	04965	15
45	69593	04965	69604	04966	69615	04968	69626	04969	69637	04970	69648	04971	14
46	69648	04971	69659	04973	69670	04974	69681	04975	69692	04976	69703	04978	13
47	69703	04978	69714	04979	69725	04980	69736	04982	69747	04983	69758	04984	12
48	69758	04984	69769	04985	69780	04987	69791	04988	69802	04989	69813	04990	11
49	69813	04990	69824	04992	69835	04993	69847	04994	69858	04995	69869	04997	10
50	69869	04997	69880	04998	69891	04999	69902	05001	69913	05002	69924	05003	09
51	69924	05003	69935	05004	69946	05006	69957	05007	69968	05008	69979	05009	08
52	69979	05009	69990	05011	70001	05012	70012	05013	70023	05014	70034	05016	07
53	70034	05016	70045	05017	70056	05018	70067	05020	70078	05021	70089	05022	06
54	70089	05022	70100	05023	70111	05025	70122	05026	70133	05027	70144	05028	05
55	70144	05028	70155	05030	70166	05031	70176	05032	70187	05034	70198	05035	04
56	70198	05035	70209	05036	70220	05037	70231	05039	70242	05040	70253	05041	03
57	70253	05041	70264	05042	70275	05044	70286	05045	70297	05046	70308	05048	02
58	70308	05048	70319	05049	70330	05050	70341	05051	70352	05053	70363	05054	01
59	70363	05054	70374	05055	70385	05056	70396	05058	70407	05059	70418	05060	00
	·8		·6		·4		·2		·0				

PARTS for 0′·1 :—   LOGS. 6 ;   NATURALS, 1.

HAVERSINES

′	·0 Log. 8·	·0 Nat. 0·	·2 Log. 8·	·2 Nat. 0·	·4 Log. 8·	·4 Nat. 0·	·6 Log. 8·	·6 Nat· 0·	·8 Log. 8·	·8 Nat· 0·	·8 Log. ·8	·0 Nat. 0·	′
00	70418	05060	70429	05062	70440	05063	70450	05064	70461	05065	70472	05067	59
01	70472	05067	70483	05068	70494	05069	70505	05071	70516	05072	70527	05073	58
02	70527	05073	70538	05074	70549	05076	70560	05077	70571	05078	70582	05079	57
03	70582	05079	70593	05081	70603	05082	70614	05083	70625	05085	70636	05086	56
04	70636	05086	70647	05087	70658	05088	70669	05090	70680	05091	70691	05092	55
05	70691	05092	70702	05093	70713	05095	70723	05096	70734	05097	70745	05099	54
06	70745	05099	70756	05100	70767	05101	70778	05102	70789	05104	70800	05105	53
07	70800	05105	70811	05106	70822	05108	70832	05109	70843	05110	70854	05111	52
08	70854	05111	70865	05113	70876	05114	70887	05115	70898	05117	70909	05118	51
09	70909	05118	70919	05119	70930	05120	70941	05122	70952	05123	70963	05124	50
10	70963	05124	70974	05126	70985	05127	70996	05128	71007	05129	71017	05131	49
11	71017	05131	71028	05132	71039	05133	71050	05135	71061	05136	71072	05137	48
12	71072	05137	71082	05138	71093	05140	71104	05141	71115	05142	71126	05144	47
13	71126	05144	71137	05145	71148	05146	71159	05147	71169	05149	71180	05150	46
14	71180	05150	71191	05151	71202	05153	71213	05154	71224	05155	71234	05156	45
15	71234	05156	71245	05158	71256	05159	71267	05160	71278	05162	71289	05163	44
16	71289	05163	71299	05164	71310	05165	71321	05167	71332	05168	71343	05169	43
17	71343	05169	71353	05171	71364	05172	71375	05173	71386	05174	71397	05176	42
18	71397	05176	71408	05177	71418	05178	71429	05180	71440	05181	71451	05182	41
19	71451	05182	71462	05183	71472	05185	71483	05186	71494	05187	71505	05189	40
20	71505	05189	71516	05190	71526	05191	71537	05192	71548	05194	71559	05195	39
21	71559	05195	71570	05196	71580	05198	71591	05199	71602	05200	71613	05201	38
22	71613	05201	71624	05203	71634	05204	71645	05205	71656	05207	71667	05208	37
23	71667	05208	71677	05209	71688	05211	71699	05212	71710	05213	71721	05214	36
24	71721	05214	71731	05216	71742	05217	71753	05218	71764	05220	71774	05221	35
25	71774	05221	71785	05222	71796	05223	71807	05225	71817	05226	71828	05227	34
26	71828	05227	71839	05229	71850	05230	71860	05231	71871	05233	71882	05234	33
27	71882	05234	71893	05235	71903	05236	71914	05238	71925	05239	71936	05240	32
28	71936	05240	71946	05242	71957	05243	71968	05244	71979	05246	71989	05247	31
29	71989	05247	72000	05248	72011	05249	72022	05251	72032	05252	72043	05253	30
30	72043	05253	72054	05255	72065	05256	72075	05257	72086	05258	72097	05260	29
31	72097	05260	72108	05261	72118	05262	72129	05264	72140	05265	72150	05266	28
32	72150	05266	72161	05268	72172	05269	72182	05270	72193	05271	72204	05273	27
33	72204	05273	72215	05274	72225	05275	72236	05277	72247	05278	72257	05279	26
34	72257	05279	72268	05281	72279	05282	72289	05283	72300	05284	72311	05286	25
35	72311	05286	72322	05287	72332	05288	72343	05290	72354	05291	72364	05292	24
36	72364	05292	72375	05294	72386	05295	72396	05296	72407	05298	72418	05299	23
37	72418	05299	72428	05300	72439	05301	72450	05303	72461	05304	72471	05305	22
38	72471	05305	72482	05307	72493	05308	72503	05309	72514	05311	72525	05312	21
39	72525	05312	72535	05313	72546	05314	72557	05316	72567	05317	72578	05318	20
40	72578	05318	72588	05320	72599	05321	72610	05322	72621	05324	72631	05325	19
41	72631	05325	72642	05326	72652	05328	72663	05329	72674	05330	72684	05331	18
42	72684	05331	72695	05333	72706	05334	72716	05335	72727	05337	72738	05338	17
43	72738	05338	72748	05339	72759	05341	72770	05342	72780	05343	72791	05345	16
44	72791	05345	72801	05346	72812	05347	72823	05348	72833	05350	72844	05351	15
45	72844	05351	72855	05352	72865	05354	72876	05355	72886	05356	72897	05358	14
46	72897	05358	72908	05359	72918	05360	72929	05362	72940	05363	72950	05364	13
47	72950	05364	72961	05365	72971	05367	72982	05368	72993	05369	73003	05371	12
48	73003	05371	73014	05372	73024	05373	73035	05375	73046	05376	73056	05377	11
49	73056	05377	73067	05379	73077	05380	73088	05381	73099	05383	73109	05384	10
50	73109	05384	73120	05385	73130	05386	73141	05388	73151	05389	73162	05390	09
51	73162	05390	73173	05392	73183	05393	73194	05394	73204	05396	73215	05397	08
52	73215	05397	73226	05398	73236	05400	73247	05401	73257	05402	73268	05404	07
53	73268	05404	73278	05405	73289	05406	73300	05407	73310	05409	73321	05410	06
54	73321	05410	73342	05411	73342	05413	73352	05414	73363	05415	73374	05417	05
55	73374	05417	73384	05418	73395	05419	73405	05421	73416	05422	73426	05423	04
56	73426	05423	73437	05425	73447	05426	73458	05427	73468	05429	73479	05430	03
57	73479	05430	73490	05431	73500	05433	73511	05434	73521	05435	73532	05436	02
58	73532	05436	73542	05438	73553	05439	73563	05440	73574	05442	73584	05443	01
59	73584	05443	73595	05444	73605	05446	73616	05447	73626	05448	73637	05450	00
			·8		·6		·4		·2		·0 ←		

PARTS for 0′·1 :—   LOGS. 5 ;   NATURALS, 1.

27° HAVERSINES

′	.0 Log. 8·	.0 Nat. 0·	.2 Log. 8·	.2 Nat. 0·	.4 Log. 8·	.4 Nat. 0·	.6 Log. 8·	.6 Nat. 0·	.8 Log. 8·	.8 Nat. 0·	Log. 8·	Nat. 0·	′
00	73637	05450	73648	05451	73658	05452	73669	05454	73679	05455	73690	05456	59
01	73690	05456	73700	05458	73711	05459	73721	05460	73732	05462	73742	05463	58
02	73742	05463	73753	05464	73763	05466	73774	05467	73784	05468	73795	05470	57
03	73795	05470	73805	05471	73816	05472	73826	05473	73837	05475	73847	05476	56
04	73847	05476	73858	05477	73868	05479	73879	05480	73889	05481	73900	05483	55
05	73900	05483	73910	05484	73921	05485	73931	05487	73942	05488	73952	05489	54
06	73952	05489	73963	05491	73973	05492	73984	05493	73994	05495	74005	05496	53
07	74005	05496	74015	05497	74026	05499	74036	05500	74047	05501	74057	05503	52
08	74057	05503	74067	05504	74078	05505	74088	05507	74099	05508	74109	05509	51
09	74109	05509	74120	05511	74130	05512	74141	05513	74151	05515	74162	05516	50
10	74162	05516	74172	05517	74182	05519	74193	05520	74203	05521	74214	05523	49
11	74214	05523	74224	05524	74235	05525	74245	05527	74256	05528	74266	05529	48
12	74266	05529	74276	05531	74287	05532	74297	05533	74308	05535	74318	05536	47
13	74318	05536	74329	05537	74339	05539	74350	05540	74360	05541	74370	05542	46
14	74370	05542	74381	05544	74391	05545	74402	05546	74412	05548	74423	05549	45
15	74423	05549	74433	05550	74443	05552	74454	05553	74464	05554	74475	05556	44
16	74475	05556	74485	05557	74495	05558	74506	05560	74516	05561	74527	05562	43
17	74527	05562	74537	05564	74548	05565	74558	05566	74568	05568	74579	05569	42
18	74579	05569	74589	05570	74600	05572	74610	05573	74620	05574	74631	05576	41
19	74631	05576	74641	05577	74652	05578	74662	05580	74672	05581	74683	05582	40
20	74683	05582	74693	05584	74704	05585	74714	05586	74724	05588	74735	05589	39
21	74735	05589	74745	05590	74756	05592	74766	05593	74776	05595	74787	05596	38
22	74787	05596	74797	05597	74807	05599	74818	05600	74828	05601	74838	05603	37
23	74838	05603	74849	05604	74859	05605	74870	05607	74880	05608	74890	05609	36
24	74890	05609	74901	05611	74911	05612	74921	05613	74932	05615	74942	05616	35
25	74942	05616	74953	05617	74963	05619	74973	05620	74984	05621	74994	05623	34
26	74994	05623	75004	05624	75015	05625	75025	05627	75035	05628	75046	05629	33
27	75046	05629	75056	05631	75066	05632	75077	05633	75087	05635	75097	05636	32
28	75097	05636	75108	05637	75118	05639	75128	05640	75139	05641	75149	05643	31
29	75149	05643	75159	05644	75170	05645	75180	05647	75190	05648	75201	05649	30
30	75201	05649	75211	05651	75221	05652	75232	05654	75242	05655	75252	05656	29
31	75252	05656	75263	05658	75273	05659	75283	05660	75293	05662	75304	05663	28
32	75304	05663	75314	05664	75324	05666	75335	05667	75345	05668	75355	05670	27
33	75355	05670	75366	05671	75376	05672	75386	05674	75397	05675	75407	05676	26
34	75407	05676	75417	05678	75428	05679	75438	05680	75448	05682	75458	05683	25
35	75458	05683	75469	05684	75479	05686	75489	05687	75500	05688	75510	05690	24
36	75510	05690	75520	05691	75530	05693	75541	05694	75551	05695	75561	05697	23
37	75561	05697	75571	05698	75582	05699	75592	05701	75602	05702	75613	05703	22
38	75613	05703	75623	05705	75633	05706	75643	05707	75654	05709	75664	05710	21
39	75664	05710	75674	05711	75685	05713	75695	05714	75705	05715	75715	05717	20
40	75715	05717	75726	05718	75736	05720	75746	05721	75756	05722	75767	05724	19
41	75767	05724	75777	05725	75787	05726	75797	05728	75808	05729	75818	05730	18
42	75818	05730	75828	05732	75838	05733	75849	05734	75859	05736	75869	05737	17
43	75869	05737	75879	05738	75890	05740	75900	05741	75910	05743	75920	05744	16
44	75920	05744	75930	05745	75941	05747	75951	05748	75961	05749	75971	05751	15
45	75971	05751	75982	05752	75992	05753	76002	05755	76012	05756	76023	05757	14
46	76023	05757	76033	05759	76043	05760	76053	05761	76063	05763	76074	05764	13
47	76074	05764	76084	05766	76094	05767	76104	05768	76115	05770	76125	05771	12
48	76125	05771	76135	05772	76145	05774	76155	05775	76166	05776	76176	05778	11
49	76176	05778	76186	05779	76196	05780	76206	05782	76217	05783	76227	05785	10
50	76227	05785	76237	05786	76247	05787	76257	05789	76268	05790	76278	05791	09
51	76278	05791	76288	05793	76298	05794	76308	05795	76318	05797	76329	05798	08
52	76329	05798	76339	05799	76349	05801	76359	05802	76369	05804	76380	05805	07
53	76380	05805	76390	05806	76400	05808	76410	05809	76420	05810	76430	05812	06
54	76430	05812	76441	05813	76451	05814	76461	05816	76471	05817	76481	05819	05
55	76481	05819	76491	05820	76502	05821	76512	05823	76522	05824	76532	05825	04
56	76532	05825	76542	05827	76552	05828	76563	05829	76573	05831	76583	05832	03
57	76583	05832	76593	05834	76603	05835	76613	05836	76623	05838	76634	05839	02
58	76634	05839	76644	05840	76654	05842	76664	05843	76674	05844	76684	05846	01
59	76684	05846	76695	05847	76705	05849	76715	05850	76725	05851	76735	05853	00
	.8		.6		.4		.2		.0				

PARTS for 0′·1 :—   LOGS. 5 ;   NATURALS, 1.

28° HAVERSINES

′	·0 Log.	·0 Nat.	·2 Log.	·2 Nat.	·4 Log.	·4 Nat.	·6 Log.	·6 Nat.	·8 Log.	·8 Nat.	·0 Log.	·0 Nat.	′
	8·	0·	8·	0·	8·	0·	8·	0·	8·	0·	8·	0·	
00	76735	05853	76745	05854	76755	05855	76765	05857	76775	05858	76786	05859	59
01	76786	05859	76796	05861	76806	05862	76816	05864	76826	05865	76836	05866	58
02	76836	05866	76847	05868	76857	05869	76867	05870	76877	05872	76887	05873	57
03	76887	05873	76897	05874	76907	05876	76917	05877	76927	05879	76937	05880	56
04	76937	05880	76948	05881	76958	05883	76968	05884	76978	05885	76988	05887	55
05	76988	05887	76998	05888	77008	05890	77018	05891	77028	05892	77038	05894	54
06	77038	05894	77049	05895	77059	05896	77069	05898	77079	05899	77089	05901	53
07	77089	05901	77099	05902	77109	05903	77119	05905	77129	05906	77139	05907	52
08	77139	05907	77149	05909	77160	05910	77170	05911	77180	05913	77190	05914	51
09	77190	05914	77200	05916	77210	05917	77220	05918	77230	05920	77240	05921	50
10	77240	05921	77250	05922	77260	05924	77270	05925	77280	05927	77290	05928	49
11	77290	05928	77300	05929	77311	05931	77321	05932	77331	05933	77341	05935	48
12	77341	05935	77351	05936	77361	05938	77371	05939	77381	05940	77391	05942	47
13	77391	05942	77401	05943	77411	05944	77421	05946	77431	05947	77441	05949	46
14	77441	05949	77451	05950	77461	05951	77472	05953	77482	05954	77492	05955	45
15	77492	05955	77502	05957	77512	05958	77522	05960	77532	05961	77542	05962	44
16	77542	05962	77552	05964	77562	05965	77572	05966	77582	05968	77592	05969	43
17	77592	05969	77602	05971	77612	05972	77622	05973	77632	05975	77642	05976	42
18	77642	05976	77652	05978	77662	05979	77672	05980	77682	05982	77692	05983	41
19	77692	05983	77702	05984	77712	05986	77722	05987	77732	05989	77742	05990	40
20	77742	05990	77752	05991	77762	05993	77772	05994	77782	05995	77792	05997	39
21	77792	05997	77802	05998	77812	06000	77822	06001	77832	06002	77842	06004	38
22	77842	06004	77852	06005	77862	06007	77872	06008	77882	06009	77892	06011	37
23	77892	06011	77902	06012	77912	06013	77922	06015	77932	06016	77942	06018	36
24	77942	06018	77952	06019	77962	06020	77972	06022	77982	06023	77992	06024	35
25	77992	06024	78002	06026	78012	06027	78022	06029	78032	06030	78042	06031	34
26	78042	06031	78052	06033	78062	06034	78072	06036	78082	06037	78092	06038	33
27	78092	06038	78102	06040	78112	06041	78122	06042	78132	06044	78142	06045	32
28	78142	06045	78152	06047	78162	06048	78171	06049	78181	06051	78191	06052	31
29	78191	06052	78201	06054	78211	06055	78221	06056	78231	06058	78241	06059	30
30	78241	06059	78251	06061	78261	06062	78271	06063	78281	06065	78291	06066	29
31	78291	06066	78301	06067	78311	06069	78321	06070	78331	06072	78341	06073	28
32	78341	06073	78351	06074	78360	06076	78370	06077	78380	06079	78390	06080	27
33	78390	06080	78400	06081	78410	06083	78420	06084	78430	06086	78440	06087	26
34	78440	06087	78450	06088	78460	06090	78470	06091	78479	06093	78489	06094	25
35	78489	06094	78499	06095	78509	06097	78519	06098	78529	06099	78539	06101	24
36	78539	06101	78549	06102	78559	06104	78569	06105	78579	06106	78589	06108	23
37	78589	06108	78599	06109	78608	06111	78618	06112	78628	06113	78638	06115	22
38	78638	06115	78648	06116	78658	06118	78668	06119	78678	06120	78688	06122	21
39	78688	06122	78697	06123	78707	06125	78717	06126	78727	06127	78737	06129	20
40	78737	06129	78747	06130	78757	06132	78767	06133	78777	06134	78786	06136	19
41	78786	06136	78796	06137	78806	06139	78816	06140	78826	06141	78836	06143	18
42	78836	06143	78846	06144	78856	06146	78865	06147	78875	06148	78885	06150	17
43	78885	06150	78895	06151	78905	06152	78915	06154	78925	06155	78935	06157	16
44	78935	06157	78944	06158	78954	06159	78964	06161	78974	06162	78984	06164	15
45	78984	06164	78994	06165	79004	06166	79013	06168	79023	06169	79033	06171	14
46	79033	06171	79043	06172	79053	06173	79063	06175	79073	06176	79082	06178	13
47	79082	06178	79092	06179	79102	06180	79112	06182	79122	06183	79132	06185	12
48	79132	06185	79141	06186	79151	06187	79161	06189	79171	06190	79181	06192	11
49	79181	06192	79191	06193	79200	06194	79210	06196	79220	06197	79230	06199	10
50	79230	06199	79240	06200	79250	06202	79259	06203	79269	06204	79279	06206	09
51	79279	06206	79289	06207	79299	06209	79309	06210	79318	06211	79328	06213	08
52	79328	06213	79338	06214	79348	06216	79358	06217	79367	06218	79377	06220	07
53	79377	06220	79387	06221	79397	06223	79407	06224	79417	06225	79426	06227	06
54	79426	06227	79436	06228	79446	06230	79456	06231	79466	06232	79475	06234	05
55	79475	06234	79485	06235	79495	06237	79505	06238	79514	06239	79524	06241	04
56	79524	06241	79534	06242	79544	06244	79554	06245	79563	06246	79573	06248	03
57	79573	06248	79583	06249	79593	06251	79603	06252	79612	06254	79622	06255	02
58	79622	06255	79632	06256	79642	06258	79651	06259	79661	06261	79671	06262	01
59	79671	06262	79681	06263	79691	06265	79700	06266	79710	06268	79720	06269	00

| ·8 | ·6 | ·4 | ·2 | ·0 ← |

331°

PARTS for 0′·1 :—   LOGS. 5 ;   NATURALS, 1.

29° HAVERSINES

'	·0 Log.	·0 Nat.	·2 Log.	·2 Nat.	·4 Log.	·4 Nat.	·6 Log.	·6 Nat.	·8 Log.	·8 Nat.	Log.	Nat.	'
	8·	0·	8·	0·	8·	0·	8·	0·	8·	0·	8·	0·	
00	79720	06269	79730	06270	79740	06272	79749	06273	79759	06275	79769	06276	59
01	79769	06276	79779	06277	79788	06279	79790	06280	79808	06282	79818	06283	58
02	79818	06283	79827	06285	79837	06286	79847	06287	79857	06289	79866	06290	57
03	79866	06290	79876	06292	79886	06293	79896	06294	79905	06296	79915	06297	56
04	79915	06297	79925	06299	79935	06300	79944	06301	79954	06303	79964	06304	55
05	79964	06304	79974	06306	79983	06307	79993	06309	80003	06310	80012	06311	54
06	80012	06311	80022	06313	80032	06314	80042	06316	80051	06317	80061	06318	53
07	80061	06318	80071	06320	80081	06321	80090	06323	80100	06324	80110	06326	52
08	80110	06326	80120	06327	80129	06328	80139	06330	80149	06331	80158	06333	51
09	80158	06333	80168	06334	80178	06335	80188	06337	80197	06338	80207	06340	50
10	80207	06340	80217	06341	80226	06343	80236	06344	80246	06345	80255	06347	49
11	80255	06347	80265	06348	80275	06350	80285	06351	80294	06352	80304	06354	48
12	80304	06354	80314	06355	80323	06357	80333	06358	80343	06360	80352	06361	47
13	80352	06361	80362	06362	80372	06364	80382	06365	80391	06367	80401	06368	46
14	80401	06368	80411	06370	80420	06371	80430	06372	80440	06374	80449	06375	45
15	80449	06375	80459	06377	80469	06378	80478	06379	80488	06381	80498	06382	44
16	80498	06382	80507	06384	80517	06385	80527	06387	80537	06388	80546	06389	43
17	80546	06389	80556	06391	80565	06392	80575	06394	80585	06395	80594	06397	42
18	80594	06397	80604	06398	80614	06399	80623	06401	80633	06402	80643	06404	41
19	80643	06404	80652	06405	80662	06407	80672	06408	80681	06409	80691	06411	40
20	80691	06411	80701	06412	80710	06414	80720	06415	80730	06416	80739	06418	39
21	80739	06418	80749	06419	80759	06421	80768	06422	80778	06424	80788	06425	38
22	80788	06425	80797	06426	80807	06428	80817	06429	80826	06431	80836	06432	37
23	80836	06432	80845	06434	80855	06435	80865	06436	80874	06438	80884	06439	36
24	80884	06439	80894	06441	80903	06442	80913	06444	80922	06445	80932	06446	35
25	80932	06446	80942	06448	80951	06449	80961	06451	80971	06452	80980	06454	34
26	80980	06454	80990	06455	80999	06456	81009	06458	81019	06459	81028	06461	33
27	81028	06461	81038	06462	81047	06464	81057	06465	81067	06466	81076	06468	32
28	81076	06468	81086	06469	81096	06471	81105	06472	81115	06474	81124	06475	31
29	81124	06475	81134	06476	81144	06478	81153	06479	81163	06481	81172	06482	30
30	81172	06482	81182	06484	81192	06485	81201	06487	81211	06488	81220	06489	29
31	81220	06489	81230	06491	81240	06492	81249	06494	81259	06495	81268	06497	28
32	81268	06497	81278	06498	81287	06499	81297	06501	81307	06502	81316	06504	27
33	81316	06504	81326	06505	81335	06507	81345	06508	81354	06509	81364	06511	26
34	81364	06511	81374	06512	81383	06514	81393	06515	81402	06517	81412	06518	25
35	81412	06518	81422	06519	81431	06521	81441	06522	81450	06524	81460	06525	24
36	81460	06525	81469	06527	81479	06528	81488	06530	81498	06531	81508	06532	23
37	81508	06532	81517	06534	81527	06535	81536	06537	81546	06538	81555	06540	22
38	81555	06540	81565	06541	81574	06543	81584	06544	81593	06545	81603	06547	21
39	81603	06547	81613	06548	81622	06550	81632	06551	81641	06553	81651	06554	20
40	81651	06554	81660	06555	81670	06557	81679	06558	81689	06560	81698	06561	19
41	81698	06561	81708	06563	81718	06564	81727	06566	81737	06567	81746	06568	18
42	81746	06568	81756	06570	81765	06571	81775	06573	81784	06574	81794	06576	17
43	81794	06576	81803	06577	81813	06579	81822	06580	81832	06581	81841	06583	16
44	81841	06583	81851	06584	81860	06586	81870	06587	81879	06589	81889	06590	15
45	81889	06590	81898	06592	81908	06593	81917	06594	81927	06596	81936	06597	14
46	81936	06597	81946	06599	81955	06600	81965	06602	81974	06603	81984	06605	13
47	81984	06605	81993	06606	82003	06607	82012	06609	82022	06610	82031	06612	12
48	82031	06612	82041	06613	82050	06615	82060	06616	82069	06618	82079	06619	11
49	82079	06619	82088	06620	82098	06622	82107	06623	82117	06625	82126	06626	10
50	82126	06626	82136	06628	82145	06629	82155	06631	82164	06632	82174	06633	09
51	82174	06633	82183	06635	82193	06636	82202	06638	82212	06639	82221	06641	08
52	82221	06641	82231	06642	82240	06644	82250	06645	82259	06646	82269	06648	07
53	82269	06648	82278	06649	82287	06651	82297	06652	82306	06654	82316	06655	06
54	82316	06655	82325	06657	82335	06658	82344	06660	82354	06661	82363	06662	05
55	82363	06662	82373	06664	82382	06665	82392	06667	82401	06668	82410	06670	04
56	82410	06670	82420	06671	82429	06673	82439	06674	82448	06675	82458	06677	03
57	82458	06677	82467	06678	82476	06680	82486	06681	82495	06683	82505	06684	02
58	82505	06684	82514	06686	82524	06687	82533	06689	82543	06690	82552	06691	01
59	82552	06691	82561	06693	82571	06694	82580	06696	82590	06697	82599	06699	00
	·8		·6		·4		·2		·0				

30° HAVERSINES

′	·0 Log. 8·	·0 Nat. 0·	·2 Log. 8·	·2 Nat. 0·	·4 Log. 8·	·4 Nat. 0·	·6 Log. 8·	·6 Nat. 0·	·8 Log. 8·	·8 Nat. 0·	′
00	82599	06699	82609	06700	82618	06702	82628	06703	82637	06705	82646	06706	59
01	82646	06706	82656	06707	82665	06709	82675	06710	82684	06712	82693	06713	58
02	82693	06713	82703	06715	82712	06716	82722	06718	82731	06719	82741	06721	57
03	82741	06721	82750	06722	82759	06723	82769	06725	82778	06726	82788	06728	56
04	82788	06728	82797	06729	82806	06731	82816	06732	82825	06734	82835	06735	55
05	82835	06735	82844	06737	82853	06738	82863	06740	82872	06741	82882	06742	54
06	82882	06742	82891	06744	82900	06745	82910	06747	82919	06748	82929	06750	53
07	82929	06750	82938	06751	82947	06753	82957	06754	82966	06756	82976	06757	52
08	82976	06757	82985	06758	82994	06760	83004	06761	83013	06763	83022	06764	51
09	83022	06764	83032	06766	83041	06767	83051	06769	83060	06770	83069	06772	50
10	83069	06772	83079	06773	83088	06775	83097	06776	83107	06777	83116	06779	49
11	83116	06779	83126	06780	83135	06782	83144	06783	83154	06785	83163	06786	48
12	83163	06786	83172	06788	83182	06789	83191	06791	83200	06792	83210	06794	47
13	83210	06794	83219	06795	83229	06797	83238	06798	83247	06799	83257	06801	46
14	83257	06801	83266	06802	83275	06804	83285	06805	83294	06807	83303	06808	45
15	83303	06808	83313	06810	83322	06811	83331	06813	83341	06814	83350	06816	44
16	83350	06816	83360	06817	83369	06818	83378	06820	83387	06821	83397	06823	43
17	83397	06823	83406	06824	83415	06826	83425	06827	83434	06829	83443	06830	42
18	83443	06830	83453	06832	83462	06833	83471	06835	83481	06836	83490	06838	41
19	83490	06838	83500	06839	83509	06840	83518	06842	83527	06843	83537	06845	40
20	83537	06845	83546	06846	83555	06848	83565	06849	83574	06851	83583	06852	39
21	83583	06852	83593	06854	83602	06855	83611	06857	83621	06858	83630	06860	38
22	83630	06860	83639	06861	83649	06863	83658	06864	83667	06865	83676	06867	37
23	83676	06867	83686	06868	83695	06870	83704	06871	83714	06873	83723	06874	36
24	83723	06874	83732	06876	83742	06877	83751	06879	83760	06880	83769	06882	35
25	83769	06882	83779	06883	83788	06885	83797	06886	83807	06888	83816	06889	34
26	83816	06889	83825	06891	83834	06892	83844	06893	83853	06895	83862	06896	33
27	83862	06896	83872	06898	83881	06899	83890	06901	83900	06902	83909	06904	32
28	83909	06904	83918	06905	83927	06907	83937	06908	83946	06910	83955	06911	31
29	83955	06911	83964	06913	83974	06914	83983	06916	83992	06917	84001	06919	30
30	84001	06919	84011	06920	84020	06922	84029	06923	84039	06924	84048	06926	29
31	84048	06926	84057	06927	84066	06929	84076	06930	84085	06932	84094	06933	28
32	84094	06933	84103	06935	84113	06936	84122	06938	84131	06939	84140	06941	27
33	84140	06941	84150	06942	84159	06944	84168	06945	84177	06947	84187	06948	26
34	84187	06948	84196	06950	84205	06951	84214	06953	84224	06954	84233	06956	25
35	84233	06956	84242	06957	84251	06958	84261	06960	84270	06961	84279	06963	24
36	84279	06963	84288	06964	84297	06966	84307	06967	84316	06969	84325	06970	23
37	84325	06970	84334	06972	84344	06973	84353	06975	84362	06976	84371	06978	22
38	84371	06978	84381	06979	84390	06981	84399	06982	84408	06984	84417	06985	21
39	84417	06985	84427	06987	84436	06988	84445	06990	84454	06991	84464	06993	20
40	84464	06993	84473	06994	84482	06996	84491	06997	84500	06998	84510	07000	19
41	84510	07000	84519	07001	84528	07003	84537	07004	84546	07006	84556	07007	18
42	84556	07007	84565	07009	84574	07010	84583	07012	84592	07013	84602	07015	17
43	84602	07015	84611	07016	84620	07018	84629	07019	84638	07021	84648	07022	16
44	84648	07022	84657	07024	84666	07025	84675	07027	84684	07028	84694	07030	15
45	84694	07030	84703	07031	84712	07033	84721	07034	84730	07036	84739	07037	14
46	84739	07037	84749	07039	84758	07040	84767	07042	84776	07043	84785	07045	13
47	84785	07045	84795	07046	84804	07048	84813	07049	84822	07051	84831	07052	12
48	84831	07052	84840	07053	84850	07055	84859	07056	84868	07058	84877	07059	11
49	84877	07059	84886	07061	84896	07062	84905	07064	84914	07065	84923	07067	10
50	84923	07067	84932	07068	84941	07070	84950	07071	84960	07073	84969	07074	09
51	84969	07074	84978	07076	84987	07077	84996	07079	85005	07080	85015	07082	08
52	85015	07082	85024	07083	85033	07085	85042	07086	85051	07088	85060	07089	07
53	85060	07089	85069	07091	85079	07092	85088	07094	85097	07095	85106	07097	06
54	85106	07097	85115	07098	85124	07100	85133	07101	85142	07103	85152	07104	05
55	85152	07104	85161	07106	85170	07107	85179	07109	85188	07110	85197	07112	04
56	85197	07112	85207	07113	85216	07115	85225	07116	85234	07118	85243	07119	03
57	85243	07119	85252	07121	85261	07122	85270	07124	85279	07125	85289	07127	02
58	85289	07127	85298	07128	85307	07130	85316	07131	85325	07133	85334	07134	01
59	85334	07134	85343	07136	85352	07137	85362	07139	85371	07140	85380	07142	00

·8	·6	·4	·2	·0

329°

PARTS for 0′·1 :— LOGS. 5 ; NATURALS, 1.

P

HAVERSINES

31°

′	·0 Log.	·0 Nat.	·2 Log.	·2 Nat.	·4 Log.	·4 Nat.	·6 Log.	·6 Nat.	·8 Log.	·8 Nat.	·8 Log.	·0 Nat.	′
	8·	0·	8·	0·	8·	0·	8·	0·	8·	0·	8·	0·	
00	85380	07142	85389	07143	85398	07145	85407	07146	85416	07148	85425	07149	59
01	85425	07149	85434	07151	85444	07152	85453	07154	85462	07155	85471	07157	58
02	85471	07157	85480	07158	85489	07160	85498	07161	85507	07163	85516	07164	57
03	85516	07164	85525	07166	85535	07167	85544	07169	85553	07170	85562	07172	56
04	85562	07172	85571	07173	85580	07175	85589	07176	85598	07178	85607	07179	55
05	85607	07179	85616	07181	85625	07182	85634	07184	85644	07185	85653	07187	54
06	85653	07187	85662	07188	85671	07190	85680	07191	85689	07193	85698	07194	53
07	85698	07194	85707	07196	85716	07197	85725	07199	85734	07200	85743	07202	52
08	85743	07202	85752	07203	85762	07205	85771	07206	85780	07208	85789	07209	51
09	85789	07209	85798	07211	85807	07212	85816	07214	85825	07215	85834	07217	50
10	85834	07217	85843	07218	85852	07220	85861	07221	85870	07223	85879	07224	49
11	85879	07224	85888	07226	85897	07227	85907	07229	85916	07230	85925	07232	48
12	85925	07232	85934	07233	85943	07235	85952	07236	85961	07238	85970	07239	47
13	85970	07239	85979	07241	85988	07242	85997	07244	86006	07245	86015	07247	46
14	86015	07247	86024	07248	86033	07250	86042	07251	86051	07253	86060	07254	45
15	86060	07254	86069	07256	86078	07257	86087	07259	86096	07260	86105	07262	44
16	86105	07262	86114	07263	86123	07265	86132	07266	86141	07268	86150	07270	43
17	86150	07270	86159	07271	86168	07273	86178	07274	86187	07276	86196	07277	42
18	86196	07277	86205	07279	86214	07280	86223	07282	86232	07283	86241	07285	41
19	86241	07285	86250	07286	86259	07288	86268	07289	86277	07291	86286	07292	40
20	86286	07292	86295	07294	86304	07295	86313	07297	86322	07298	86331	07300	39
21	86331	07300	86340	07301	86349	07303	86358	07304	86367	07306	86376	07307	38
22	86376	07307	86385	07309	86394	07310	86403	07312	86412	07313	86421	07315	37
23	86421	07315	86430	07316	86439	07318	86448	07319	86457	07321	86466	07322	36
24	86466	07322	86475	07324	86484	07325	86493	07327	86502	07329	86511	07330	35
25	86511	07330	86520	07332	86529	07333	86538	07335	86547	07336	86556	07338	34
26	86556	07338	86565	07339	86574	07341	86582	07342	86591	07344	86600	07345	33
27	86600	07345	86609	07347	86618	07348	86627	07350	86636	07351	86645	07353	32
28	86645	07353	86654	07354	86663	07356	86672	07357	86681	07359	86690	07360	31
29	86690	07360	86699	07362	86708	07363	86717	07365	86726	07366	86735	07368	30
30	86735	07368	86744	07370	86753	07371	86762	07373	86771	07374	86780	07376	29
31	86780	07376	86789	07377	86798	07379	86806	07380	86815	07382	86824	07383	28
32	86824	07383	86833	07385	86842	07386	86851	07388	86860	07389	86869	07391	27
33	86869	07391	86878	07392	86887	07394	86896	07395	86905	07397	86914	07398	26
34	86914	07398	86923	07400	86932	07401	86941	07403	86950	07405	86959	07406	25
35	86959	07406	86968	07408	86976	07409	86985	07411	86994	07412	87003	07414	24
36	87003	07414	87012	07415	87021	07417	87030	07418	87039	07420	87048	07421	23
37	87048	07421	87057	07423	87066	07424	87075	07426	87083	07427	87092	07429	22
38	87092	07429	87101	07430	87110	07432	87119	07433	87128	07435	87137	07437	21
39	87137	07437	87146	07438	87155	07440	87164	07441	87173	07443	87182	07444	20
40	87182	07444	87191	07446	87199	07447	87208	07449	87217	07450	87226	07452	19
41	87226	07452	87235	07453	87244	07455	87253	07456	87262	07458	87271	07459	18
42	87271	07459	87280	07461	87288	07463	87297	07464	87306	07466	87315	07467	17
43	87315	07467	87324	07469	87333	07470	87342	07472	87351	07473	87360	07475	16
44	87360	07475	87368	07476	87377	07478	87386	07479	87395	07481	87404	07482	15
45	87404	07482	87413	07484	87422	07485	87431	07487	87440	07489	87448	07490	14
46	87448	07490	87457	07492	87466	07493	87475	07495	87484	07496	87493	07498	13
47	87493	07498	87502	07499	87511	07501	87519	07502	87528	07504	87537	07505	12
48	87537	07505	87546	07507	87555	07508	87564	07510	87573	07512	87582	07513	11
49	87582	07513	87590	07515	87599	07516	87608	07518	87617	07519	87626	07521	10
50	87626	07521	87635	07522	87644	07524	87652	07525	87661	07527	87670	07528	09
51	87670	07528	87679	07530	87688	07531	87697	07533	87706	07535	87714	07536	08
52	87714	07536	87723	07538	87732	07539	87741	07541	87750	07542	87759	07544	07
53	87759	07544	87767	07545	87776	07547	87785	07548	87794	07550	87803	07551	06
54	87803	07551	87812	07553	87820	07554	87829	07556	87838	07558	87847	07559	05
55	87847	07559	87856	07561	87865	07562	87874	07564	87882	07565	87891	07567	04
56	87891	07567	87900	07568	87909	07570	87918	07571	87927	07573	87935	07575	03
57	87935	07575	87944	07576	87953	07578	87962	07579	87971	07581	87979	07582	02
58	87979	07582	87988	07584	87997	07585	88006	07587	88015	07588	88024	07590	01
59	88024	07590	88032	07591	88041	07593	88050	07595	88059	07596	88068	07598	00
	·8		·6		·4		·2		·0				

328°

PARTS for 0′·1 :— LOGS. 4 ; NATURALS, 1,

32° HAVERSINES

′	·0 Log. 8·	·0 Nat. 0·	·2 Log. 8·	·2 Nat. 0·	·4 Log. 8·	·4 Nat. 0·	·6 Log. 8·	·6 Nat. 0·	·8 Log. 8·	·8 Nat. 0·	Log. 8·	Nat. 0·	′
00	88068	07598	88076	07599	88085	07601	88094	07602	88103	07604	88112	07605	59
01	88112	07605	88120	07607	88129	07608	88138	07610	88147	07611	88156	07613	58
02	88156	07613	88164	07615	88173	07616	88182	07618	88191	07619	88200	07621	57
03	88200	07621	88208	07622	88217	07624	88226	07625	88235	07627	88244	07628	56
04	88244	07628	88252	07630	88261	07632	88270	07633	88279	07635	88288	07636	55
05	88288	07636	88296	07638	88305	07639	88314	07641	88323	07642	88332	07644	54
06	88332	07644	88340	07645	88349	07647	88358	07649	88367	07650	88375	07652	53
07	88375	07652	88384	07653	88393	07655	88402	07656	88410	07658	88419	07659	52
08	88419	07659	88428	07661	88437	07662	88446	07664	88454	07666	88463	07667	51
09	88463	07667	88472	07669	88481	07670	88489	07672	88498	07673	88507	07675	50
10	88507	07675	88516	07676	88525	07678	88533	07680	88542	07681	88551	07683	49
11	88551	07683	88560	07684	88568	07686	88577	07687	88586	07689	88595	07690	48
12	88595	07690	88603	07692	88612	07693	88621	07695	88630	07697	88638	07698	47
13	88638	07698	88647	07700	88656	07701	88665	07703	88673	07704	88682	07706	46
14	88682	07706	88691	07707	88700	07709	88708	07711	88717	07712	88726	07714	45
15	88726	07714	88735	07715	88743	07717	88752	07718	88761	07720	88769	07721	44
16	88769	07721	88778	07723	88787	07724	88796	07726	88804	07728	88813	07729	43
17	88813	07729	88822	07731	88831	07732	88839	07734	88848	07735	88857	07737	42
18	88857	07737	88866	07738	88874	07740	88883	07742	88892	07743	88900	07745	41
19	88900	07745	88909	07746	88918	07748	88927	07749	88935	07751	88944	07752	40
20	88944	07752	88953	07754	88961	07756	88970	07757	88979	07759	88988	07760	39
21	88988	07760	88996	07762	89005	07763	89014	07765	89022	07766	89031	07768	38
22	89031	07768	89040	07770	89048	07771	89057	07773	89066	07774	89075	07776	37
23	89075	07776	89083	07777	89092	07779	89101	07780	89109	07782	89118	07784	36
24	89118	07784	89127	07785	89135	07787	89144	07788	89153	07790	89162	07791	35
25	89162	07791	89170	07793	89179	07795	89188	07796	89196	07798	89205	07799	34
26	89205	07799	89214	07801	89222	07802	89231	07804	89240	07805	89248	07807	33
27	89248	07807	89257	07809	89266	07810	89274	07812	89283	07813	89292	07815	32
28	89292	07815	89300	07816	89309	07818	89318	07819	89327	07821	89335	07823	31
29	89335	07823	89344	07824	89353	07826	89361	07827	89370	07829	89379	07830	30
30	89379	07830	89387	07832	89396	07834	89405	07835	89413	07837	89422	07838	29
31	89422	07838	89430	07840	89439	07841	89448	07843	89457	07844	89465	07846	28
32	89465	07846	89474	07848	89482	07849	89491	07851	89500	07852	89508	07854	27
33	89508	07854	89517	07855	89526	07857	89534	07859	89543	07860	89552	07862	26
34	89552	07862	89560	07863	89569	07865	89578	07866	89586	07868	89595	07870	25
35	89595	07870	89604	07871	89612	07873	89621	07874	89630	07876	89638	07877	24
36	89638	07877	89647	07879	89655	07881	89664	07882	89673	07884	89681	07885	23
37	89681	07885	89690	07887	89699	07888	89707	07890	89716	07891	89725	07893	22
38	89725	07893	89733	07895	89742	07896	89750	07898	89759	07899	89768	07901	21
39	89768	07901	89776	07902	89785	07904	89794	07906	89802	07907	89811	07909	20
40	89811	07909	89819	07910	89828	07912	89837	07913	89845	07915	89854	07917	19
41	89854	07917	89862	07918	89871	07920	89880	07921	89888	07923	89897	07924	18
42	89897	07924	89906	07926	89914	07928	89923	07929	89931	07931	89940	07932	17
43	89940	07932	89949	07934	89957	07935	89966	07937	89974	07939	89983	07940	16
44	89983	07940	89992	07942	90000	07943	90009	07945	90017	07946	90026	07948	15
45	90026	07948	90035	07950	90043	07951	90052	07953	90060	07954	90069	07956	14
46	90069	07956	90078	07957	90086	07959	90095	07961	90103	07962	90112	07964	13
47	90112	07964	90121	07965	90129	07967	90138	07969	90146	07970	90155	07972	12
48	90155	07972	90164	07973	90172	07975	90181	07976	90189	07978	90198	07980	11
49	90198	07980	90206	07981	90215	07983	90224	07984	90232	07986	90241	07987	10
50	90241	07987	90249	07989	90258	07991	90266	07992	90275	07994	90284	07995	09
51	90284	07995	90292	07997	90301	07998	90309	08000	90318	08002	90326	08003	08
52	90326	08003	90335	08005	90344	08006	90352	08008	90361	08010	90369	08011	07
53	90369	08011	90378	08013	90386	08014	90395	08016	90403	08017	90412	08019	06
54	90412	08019	90421	08021	90429	08022	90438	08024	90446	08025	90455	08027	05
55	90455	08027	90463	08028	90472	08030	90480	08032	90489	08033	90498	08035	04
56	90498	08035	90506	08036	90515	08038	90523	08040	90532	08041	90540	08043	03
57	90540	08043	90549	08044	90557	08046	90566	08047	90574	08049	90583	08051	02
58	90583	08051	90592	08052	90600	08054	90609	08055	90617	08057	90626	08059	01
59	90626	08059	90634	08060	90643	08062	90651	08063	90660	08065	90668	08066	00

| ·8 | ·6 | ·4 | ·2 | ·0 |

PARTS for 0′·1 :— LOGS. 4 ; NATURALS, 1.

33° HAVERSINES

'	·0 Log.	·0 Nat.	·2 Log.	·2 Nat.	·4 Log.	·4 Nat.	·6 Log.	·6 Nat.	·8 Log.	·8 Nat.	'
	8·	0·	8·	0·	8·	0·	8·	0·	8·	0·	
00	90668	08066	90677	08068	90685	08070	90694	08071	90702	08073	59
01	90711	08074	90720	08076	90728	08078	90737	08079	90745	08081	58
02	90754	08082	90762	08084	90771	08085	90779	08087	90788	08089	57
03	90796	08090	90805	08092	90813	08093	90822	08095	90830	08097	56
04	90839	08098	90847	08100	90856	08101	90864	08103	90873	08105	55
05	90881	08106	90890	08108	90898	08109	90907	08111	90915	08112	54
06	90924	08114	90932	08116	90941	08117	90949	08119	90958	08120	53
07	90966	08122	90975	08124	90983	08125	90992	08127	91000	08128	52
08	91009	08130	91017	08132	91026	08133	91034	08135	91043	08136	51
09	91051	08138	91060	08139	91068	08141	91077	08143	91085	08144	50
10	91094	08146	91102	08147	91111	08149	91119	08151	91128	08152	49
11	91136	08154	91145	08155	91153	08157	91162	08159	91170	08160	48
12	91179	08162	91187	08163	91195	08165	91204	08167	91212	08168	47
13	91221	08170	91229	08171	91238	08173	91246	08175	91255	08176	46
14	91263	08178	91272	08179	91280	08181	91289	08183	91297	08184	45
15	91306	08186	91314	08187	91322	08189	91331	08190	91339	08192	44
16	91348	08194	91356	08195	91365	08197	91373	08198	91382	08200	43
17	91390	08202	91399	08203	91407	08205	91416	08206	91424	08208	42
18	91432	08210	91441	08211	91449	08213	91458	08214	91466	08216	41
19	91475	08218	91483	08219	91491	08221	91500	08222	91508	08224	40
20	91517	08226	91525	08227	91534	08229	91542	08230	91551	08232	39
21	91559	08234	91567	08235	91576	08237	91584	08238	91593	08240	38
22	91601	08242	91610	08243	91618	08245	91627	08246	91635	08248	37
23	91643	08250	91652	08251	91660	08253	91669	08254	91677	08256	36
24	91685	08258	91694	08259	91702	08261	91711	08262	91719	08264	35
25	91728	08266	91736	08267	91744	08269	91753	08270	91761	08272	34
26	91770	08274	91778	08275	91786	08277	91795	08278	91803	08280	33
27	91812	08282	91820	08283	91828	08285	91837	08286	91845	08288	32
28	91854	08290	91862	08291	91870	08293	91879	08294	91887	08296	31
29	91896	08298	91904	08299	91912	08301	91921	08303	91929	08304	30
30	91938	08306	91946	08307	91954	08309	91963	08311	91971	08312	29
31	91980	08314	91988	08315	91996	08317	92005	08319	92013	08320	28
32	92022	08322	92030	08323	92038	08325	92047	08327	92055	08328	27
33	92063	08330	92072	08331	92080	08333	92089	08335	92097	08336	26
34	92105	08338	92114	08339	92122	08341	92130	08343	92139	08344	25
35	92147	08346	92156	08348	92164	08349	92172	08351	92181	08352	24
36	92189	08354	92198	08356	92206	08357	92214	08359	92223	08360	23
37	92231	08362	92239	08364	92248	08365	92256	08367	92264	08368	22
38	92273	08370	92281	08372	92290	08373	92298	08375	92306	08376	21
39	92315	08378	92323	08380	92331	08381	92340	08383	92348	08385	20
40	92356	08386	92365	08388	92373	08389	92381	08391	92390	08393	19
41	92398	08394	92406	08396	92415	08397	92423	08399	92431	08401	18
42	92440	08402	92448	08404	92456	08406	92465	08407	92473	08409	17
43	92481	08410	92490	08412	92498	08414	92506	08415	92515	08417	16
44	92523	08418	92532	08420	92540	08422	92548	08423	92556	08425	15
45	92565	08427	92573	08428	92581	08430	92590	08431	92598	08433	14
46	92606	08435	92615	08436	92623	08438	92631	08439	92640	08441	13
47	92648	08443	92656	08444	92665	08446	92673	08448	92681	08449	12
48	92690	08451	92698	08452	92706	08454	92715	08456	92723	08457	11
49	92731	08459	92740	08460	92748	08462	92756	08464	92764	08465	10
50	92773	08467	92781	08469	92789	08470	92798	08472	92806	08473	09
51	92814	08475	92823	08477	92831	08478	92839	08480	92848	08482	08
52	92856	08483	92864	08485	92872	08486	92881	08488	92889	08490	07
53	92897	08491	92906	08493	92914	08495	92922	08496	92930	08498	06
54	92939	08499	92947	08501	92955	08503	92964	08504	92972	08506	05
55	92980	08508	92988	08509	92997	08511	93005	08512	93013	08514	04
56	93022	08516	93030	08517	93038	08519	93046	08520	93055	08522	03
57	93063	08524	93071	08525	93080	08527	93088	08529	93096	08530	02
58	93104	08532	93113	08533	93121	08535	93129	08537	93138	08538	01
59	93146	08540	93154	08542	93162	08543	93171	08545	93179	08546	00

·8   ·6   ·4   ·2   ·0 ←

PARTS for 0'·1 :—   LOGS. 4 ;   NATURALS, 1.

34°   HAVERSINES

Bottom reversed heading: **325°**

'	·0 Log 8·	·0 Nat 0·	·2 Log 8·	·2 Nat 0·	·4 Log 8·	·4 Nat 0·	·6 Log 8·	·6 Nat 0·	·8 Log 8·	·8 Nat 0·	Log 8·	Nat 0·	'
00	93187	08548	93195	08550	93204	08551	93212	08553	93220	08555	93228	08556	59
01	93228	08556	93237	08558	93245	08560	93253	08561	93261	08563	93270	08564	58
02	93270	08564	93278	08566	93286	08568	93294	08569	93303	08571	93311	08573	57
03	93311	08573	93319	08574	93327	08576	93336	08577	93344	08579	93352	08581	56
04	93352	08581	93360	08582	93369	08584	93377	08586	93385	08587	93393	08589	55
05	93393	08589	93402	08590	93410	08592	93418	08594	93426	08595	93435	08597	54
06	93435	08597	93443	08599	93451	08600	93459	08602	93468	08604	93476	08605	53
07	93476	08605	93484	08607	93492	08608	93501	08610	93509	08612	93517	08613	52
08	93517	08613	93525	08615	93533	08617	93542	08618	93550	08620	93558	08621	51
09	93558	08621	93566	08623	93574	08625	93583	08626	93591	08628	93599	08630	50
10	93599	08630	93607	08631	93616	08633	93624	08635	93632	08636	93640	08638	49
11	93640	08638	93649	08639	93657	08641	93665	08643	93673	08644	93681	08646	48
12	93681	08646	93690	08648	93698	08649	93706	08651	93714	08653	93722	08654	47
13	93722	08654	93731	08656	93739	08657	93747	08659	93755	08661	93763	08662	46
14	93763	08662	93772	08664	93780	08666	93788	08667	93796	08669	93804	08671	45
15	93804	08671	93813	08672	93821	08674	93829	08675	93837	08677	93845	08679	44
16	93845	08679	93854	08680	93862	08682	93870	08684	93878	08685	93886	08687	43
17	93886	08687	93895	08689	93903	08690	93911	08692	93919	08693	93927	08695	42
18	93927	08695	93936	08697	93944	08698	93952	08700	93960	08702	93968	08703	41
19	93968	08703	93977	08705	93985	08707	93993	08708	94001	08710	94009	08712	40
20	94009	08712	94017	08713	94026	08715	94034	08716	94042	08718	94050	08720	39
21	94050	08720	94058	08721	94067	08723	94075	08725	94083	08726	94091	08728	38
22	94091	08728	94099	08730	94107	08731	94115	08733	94124	08734	94132	08736	37
23	94132	08736	94140	08738	94148	08739	94156	08741	94164	08743	94173	08744	36
24	94173	08744	94181	08746	94189	08748	94197	08749	94205	08751	94213	08753	35
25	94213	08753	94222	08754	94230	08756	94238	08757	94246	08759	94254	08761	34
26	94254	08761	94262	08762	94271	08764	94279	08766	94287	08767	94295	08769	33
27	94295	08769	94303	08771	94311	08772	94319	08774	94328	08776	94336	08777	32
28	94336	08777	94344	08779	94352	08781	94360	08782	94368	08784	94376	08785	31
29	94376	08785	94385	08787	94393	08789	94401	08790	94409	08792	94417	08794	30
30	94417	08794	94425	08795	94433	08797	94442	08799	94450	08800	94458	08802	29
31	94458	08802	94466	08804	94474	08805	94482	08807	94490	08809	94498	08810	28
32	94498	08810	94507	08812	94515	08813	94523	08815	94531	08817	94539	08818	27
33	94539	08818	94547	08820	94556	08822	94563	08823	94571	08825	94580	08827	26
34	94580	08827	94588	08828	94596	08830	94604	08832	94612	08833	94620	08835	25
35	94620	08835	94628	08837	94636	08838	94645	08840	94653	08842	94661	08843	24
36	94661	08843	94669	08845	94677	08847	94685	08848	94693	08850	94701	08851	23
37	94701	08851	94710	08853	94718	08855	94726	08856	94734	08858	94742	08860	22
38	94742	08860	94750	08861	94758	08863	94766	08865	94774	08866	94782	08868	21
39	94782	08868	94791	08870	94799	08871	94807	08873	94815	08875	94823	08876	20
40	94823	08876	94831	08878	94839	08880	94847	08881	94855	08883	94863	08885	19
41	94863	08885	94871	08886	94880	08888	94888	08889	94896	08891	94904	08893	18
42	94904	08893	94912	08894	94920	08896	94928	08898	94936	08899	94944	08901	17
43	94944	08901	94952	08903	94960	08904	94968	08906	94977	08908	94985	08909	16
44	94985	08909	94993	08911	95001	08913	95009	08914	95017	08916	95025	08918	15
45	95025	08918	95033	08919	95041	08921	95049	08923	95057	08924	95065	08926	14
46	95065	08926	95073	08928	95082	08929	95090	08931	95098	08933	95106	08934	13
47	95106	08934	95114	08936	95122	08938	95130	08939	95138	08941	95146	08943	12
48	95146	08943	95154	08944	95162	08946	95170	08948	95178	08949	95186	08951	11
49	95186	08951	95194	08953	95202	08954	95211	08956	95219	08957	95227	08959	10
50	95227	08959	95235	08961	95243	08962	95251	08964	95259	08966	95267	08967	09
51	95267	08967	95275	08969	95283	08971	95291	08972	95299	08974	95307	08976	08
52	95307	08976	95315	08977	95323	08979	95331	08981	95339	08982	95347	08984	07
53	95347	08984	95355	08986	95363	08987	95372	08989	95380	08991	95388	08992	06
54	95388	08992	95396	08994	95404	08996	95412	08997	95420	08999	95428	09001	05
55	95428	09001	95436	09002	95444	09004	95452	09006	95460	09007	95468	09009	04
56	95468	09009	95476	09011	95484	09012	95492	09014	95500	09016	95508	09017	03
57	95508	09017	95516	09019	95524	09021	95532	09022	95540	09024	95548	09026	02
58	95548	09026	95556	09027	95564	09029	95572	09031	95580	09032	95588	09034	01
59	95588	09034	95596	09036	95604	09037	95612	09039	95620	09041	95628	09042	00

Bottom column labels (for 325°):   ·8   ·6   ·4   ·2   ·0

PARTS for 0'·1 :—   LOGS. 4 ;   NATURALS, 1.

35° HAVERSINES

′	·0 Log. 8·	·0 Nat. 0·	·2 Log. 8·	·2 Nat. 0·	·4 Log. 8·	·4 Nat. 0·	·6 Log. 8·	·6 Nat. 0·	·8 Log. 8·	·8 Nat. 0·	·8 Log. 8·	·8 Nat. 0·	′
00	95628	09042	95636	09044	95644	09046	95652	09047	95660	09049	95668	09051	59
01	95668	09051	95676	09052	95684	09054	95692	09056	95700	09057	95708	09059	58
02	95708	09059	95716	09061	95724	09062	95732	09064	95740	09066	95740	09067	57
03	95748	09067	95756	09069	95764	09071	95772	09072	95780	09074	95788	09076	56
04	95788	09076	95796	09077	95804	09079	95812	09081	95820	09082	95828	09084	55
05	95828	09084	95836	09086	95844	09087	95852	09089	95860	09091	95868	09093	54
06	95868	09093	95876	09094	95884	09096	95892	09098	95900	09099	95908	09101	53
07	95908	09101	95916	09103	95924	09104	95932	09106	95940	09108	95948	09109	52
08	95948	09109	95956	09111	95964	09113	95972	09114	95980	09116	95988	09118	51
09	95988	09118	95996	09119	96004	09121	96012	09123	96020	09124	96028	09126	50
10	96028	09126	96036	09128	96044	09129	96052	09131	96060	09133	96068	09134	49
11	96068	09134	96076	09136	96084	09138	96092	09139	96100	09141	96108	09143	48
12	96108	09143	96116	09144	96124	09146	96132	09148	96140	09149	96148	09151	47
13	96148	09151	96156	09153	96164	09155	96171	09156	96179	09158	96187	09160	46
14	96187	09160	96195	09161	96203	09163	96211	09165	96219	09166	96227	09168	45
15	96227	09168	96235	09170	96243	09171	96251	09173	96259	09175	96267	09176	44
16	96267	09176	96275	09178	96283	09180	96291	09181	96299	09183	96307	09185	43
17	96307	09185	96315	09186	96322	09188	96330	09190	96338	09191	96346	09193	42
18	96346	09193	96354	09195	96362	09196	96370	09198	96378	09200	96386	09202	41
19	96386	09202	96394	09203	96402	09205	96410	09207	96418	09208	96426	09210	40
20	96426	09210	96434	09212	96442	09213	96449	09215	96457	09217	96465	09218	39
21	96465	09218	96473	09220	96481	09222	96489	09223	96497	09225	96505	09227	38
22	96505	09227	96513	09228	96521	09230	96529	09232	96537	09234	96545	09235	37
23	96545	09235	96553	09237	96560	09239	96568	09240	96576	09242	96584	09244	36
24	96584	09244	96592	09245	96600	09247	96608	09249	96616	09250	96624	09252	35
25	96624	09252	96632	09254	96640	09255	96647	09257	96655	09259	96663	09260	34
26	96663	09260	96671	09262	96679	09264	96687	09266	96695	09267	96703	09269	33
27	96703	09269	96711	09271	96719	09272	96727	09274	96734	09276	96742	09277	32
28	96742	09277	96750	09279	96758	09281	96766	09282	96774	09284	96782	09286	31
29	96782	09286	96790	09287	96798	09289	96806	09291	96813	09293	96821	09294	30
30	96821	09294	96829	09296	96837	09298	96845	09299	96853	09301	96861	09303	29
31	96861	09303	96869	09304	96877	09306	96884	09308	96892	09309	96900	09311	28
32	96900	09311	96908	09313	96916	09315	96924	09316	96932	09318	96940	09320	27
33	96940	09320	96947	09321	96955	09323	96963	09325	96971	09326	96979	09328	26
34	96979	09328	96987	09330	96995	09331	97003	09333	97011	09335	97018	09337	25
35	97018	09337	97026	09338	97034	09340	97042	09342	97050	09343	97058	09345	24
36	97058	09345	97066	09347	97073	09348	97081	09350	97089	09352	97097	09353	23
37	97097	09353	97105	09355	97113	09357	97121	09359	97129	09360	97136	09362	22
38	97136	09362	97144	09364	97152	09365	97160	09367	97168	09369	97176	09370	21
39	97176	09370	97184	09372	97191	09374	97199	09375	97207	09377	97215	09379	20
40	97215	09379	97223	09381	97231	09382	97239	09384	97246	09386	97254	09387	19
41	97254	09387	97262	09389	97270	09391	97278	09392	97286	09394	97293	09396	18
42	97293	09396	97301	09398	97309	09399	97317	09401	97325	09403	97333	09404	17
43	97333	09404	97341	09406	97348	09408	97356	09409	97364	09411	97372	09413	16
44	97372	09413	97380	09415	97388	09416	97395	09418	97403	09420	97411	09421	15
45	97411	09421	97419	09423	97427	09425	97435	09426	97442	09428	97450	09430	14
46	97450	09430	97458	09432	97466	09433	97474	09435	97482	09437	97489	09438	13
47	97489	09438	97497	09440	97505	09442	97513	09443	97521	09445	97529	09447	12
48	97529	09447	97536	09449	97544	09450	97552	09452	97560	09454	97568	09455	11
49	97568	09455	97575	09457	97583	09459	97591	09460	97599	09462	97607	09464	10
50	97607	09464	97615	09466	97622	09467	97630	09469	97638	09471	97646	09472	09
51	97646	09472	97654	09474	97661	09476	97669	09477	97677	09479	97685	09481	08
52	97685	09481	97693	09483	97700	09484	97708	09486	97716	09488	97724	09489	07
53	97724	09489	97732	09491	97739	09493	97747	09495	97755	09496	97763	09498	06
54	97763	09498	97771	09500	97778	09501	97786	09503	97794	09505	97802	09506	05
55	97802	09506	97810	09508	97817	09510	97825	09512	97833	09513	97841	09515	04
56	97841	09515	97849	09517	97856	09518	97864	09520	97872	09522	97880	09524	03
57	97880	09524	97888	09525	97895	09527	97903	09529	97911	09530	97919	09532	02
58	97919	09532	97926	09534	97934	09535	97942	09537	97950	09539	97958	09541	01
59	97958	09541	97965	09542	97973	09544	97981	09546	97989	09547	97996	09549	00

| | ·8 | | ·6 | | ·4 | | ·2 | | ·0 | | | |

PARTS for 0′·1 :—   LOGS. 4 ;   NATURALS, 1.

36° HAVERSINES

	·0		·2		·4		·6		·8		
	Log.	Nat.	Log.	Nat.	Log.	Nat.	Log.	Nat.	Log.	Nat.	
′	8·/9·	0·	8·/9·	0·	8·/9·	0·	8·/9·	0·	8·/9·	0·	′
00	97997	09549	98004	09551	98012	09553	98020	09554	98028	09556	98035	09558	59
01	98035	09558	98043	09559	98051	09561	98059	09563	98066	09565	98074	09566	58
02	98074	09566	98082	09568	98090	09570	98097	09571	98105	09573	98113	09575	57
03	98113	09575	98121	09577	98129	09578	98136	09580	98144	09582	98152	09583	56
04	98152	09583	98160	09585	98167	09587	98175	09589	98183	09590	98191	09592	55
05	98191	09592	98198	09594	98206	09595	98214	09597	98222	09599	98229	09601	54
06	98229	09601	98237	09602	98245	09604	98253	09606	98260	09607	98268	09609	53
07	98268	09609	98276	09611	98284	09613	98291	09614	98299	09616	98307	09618	52
08	98307	09618	98315	09619	98322	09621	98330	09623	98338	09625	98346	09626	51
09	98346	09626	98353	09628	98361	09630	98369	09631	98377	09633	98384	09635	50
10	98384	09635	98392	09637	98400	09638	98408	09640	98415	09642	98423	09643	49
11	98423	09643	98431	09645	98438	09647	98446	09649	98454	09650	98462	09652	48
12	98462	09652	98469	09654	98477	09655	98485	09657	98493	09659	98500	09661	47
13	98500	09661	98508	09662	98516	09664	98524	09666	98531	09667	98539	09669	46
14	98539	09669	98547	09671	98554	09673	98562	09674	98570	09676	98578	09678	45
15	98578	09678	98585	09679	98593	09681	98601	09683	98608	09685	98616	09686	44
16	98616	09686	98624	09688	98632	09690	98639	09692	98647	09693	98655	09695	43
17	98655	09695	98662	09697	98670	09698	98678	09700	98685	09702	98693	09704	42
18	98693	09704	98701	09705	98709	09707	98716	09709	98724	09710	98732	09712	41
19	98732	09712	98739	09714	98747	09716	98755	09717	98763	09719	98770	09721	40
20	98770	09721	98778	09723	98786	09724	98793	09726	98801	09728	98809	09729	39
21	98809	09729	98816	09731	98824	09733	98832	09735	98840	09736	98847	09738	38
22	98847	09738	98855	09740	98863	09742	98870	09743	98878	09745	98886	09747	37
23	98886	09747	98893	09748	98901	09750	98909	09752	98916	09754	98924	09755	36
24	98924	09755	98932	09757	98940	09759	98947	09760	98955	09762	98963	09764	35
25	98963	09764	98970	09766	98978	09767	98986	09769	98993	09771	99001	09773	34
26	99001	09773	99009	09774	99016	09776	99024	09778	99032	09779	99039	09781	33
27	99039	09781	99047	09783	99055	09785	99062	09786	99070	09788	99078	09790	32
28	99078	09790	99085	09792	99093	09793	99101	09795	99108	09797	99116	09799	31
29	99116	09799	99124	09800	99131	09802	99139	09804	99147	09805	99154	09807	30
30	99154	09807	99162	09809	99170	09811	99177	09812	99185	09814	99193	09816	29
31	99193	09816	99200	09818	99208	09819	99216	09821	99223	09823	99231	09824	28
32	99231	09824	99238	09826	99246	09828	99254	09830	99262	09831	99269	09833	27
33	99269	09833	99277	09835	99284	09837	99292	09838	99300	09840	99307	09842	26
34	99307	09842	99315	09844	99323	09845	99330	09847	99338	09849	99346	09850	25
35	99346	09850	99353	09852	99361	09854	99369	09856	99376	09857	99384	09859	24
36	99384	09859	99391	09861	99399	09863	99407	09864	99414	09866	99422	09868	23
37	99422	09868	99430	09870	99437	09871	99445	09873	99453	09875	99460	09877	22
38	99460	09877	99468	09878	99475	09880	99483	09882	99491	09883	99498	09885	21
39	99498	09885	99506	09887	99514	09889	99521	09890	99529	09892	99536	09894	20
40	99536	09894	99544	09896	99552	09897	99559	09899	99567	09901	99575	09903	19
41	99575	09903	99582	09904	99590	09906	99598	09908	99605	09909	99613	09911	18
42	99613	09911	99620	09913	99628	09915	99636	09916	99643	09918	99651	09920	17
43	99651	09920	99658	09922	99666	09923	99674	09925	99681	09927	99689	09929	16
44	99689	09929	99696	09930	99704	09932	99712	09934	99719	09936	99727	09937	15
45	99727	09937	99734	09939	99742	09941	99750	09943	99757	09944	99765	09946	14
46	99765	09946	99773	09948	99780	09949	99788	09951	99795	09953	99803	09955	13
47	99803	09955	99811	09956	99818	09958	99826	09960	99833	09962	99841	09963	12
48	99841	09963	99848	09965	99856	09967	99864	09969	99871	09970	99879	09972	11
49	99879	09972	99886	09974	99894	09976	99902	09977	99909	09979	99917	09981	10
50	99917	09981	99924	09983	99932	09984	99940	09986	99947	09988	99955	09990	09
51	99955	09990	99962	09991	99970	09993	99977	09995	99985	09997	99993	09998	08
52	99993	09998	00000	10000	00008	10002	00015	10004	00023	10005	00031	10007	07
53	00031	10007	00038	10009	00046	10011	00053	10012	00061	10014	00068	10016	06
54	00068	10016	00076	10018	00084	10019	00091	10021	00099	10023	00106	10025	05
55	00106	10025	00114	10026	00121	10028	00129	10030	00137	10032	00144	10033	04
56	00144	10033	00152	10035	00159	10037	00167	10039	00174	10040	00182	10042	03
57	00182	10042	00190	10044	00197	10046	00205	10047	00212	10049	00220	10051	02
58	00220	10051	00227	10052	00235	10054	00242	10056	00250	10058	00258	10059	01
59	00258	10059	00265	10061	00273	10063	00280	10065	00288	10066	00295	10068	00

| | ·8 | ·6 | ·4 | ·2 | ·0 | |

323°

PARTS for 0′·1 :—   LOGS. 4 ;   NATURALS, 1.

37° HAVERSINES

′	·0 Log. 9·	·0 Nat. 0·	·2 Log. 9·	·2 Nat. 0·	·4 Log. 9·	·4 Nat. 0·	·6 Log. 9·	·6 Nat. 0·	·8 Log. 9·	·8 Nat. 0·	Log. 9·	Nat. 0·	′
00	00295	10068	00303	10070	00310	10072	00318	10073	00325	10075	00333	10077	59
01	00333	10077	00341	10079	00348	10081	00356	10082	00363	10084	00371	10086	58
02	00371	10086	00378	10088	00386	10089	00393	10091	00401	10093	00408	10095	57
03	00408	10095	00416	10096	00424	10098	00431	10100	00439	10102	00446	10103	56
04	00446	10103	00454	10105	00461	10107	00469	10109	00476	10110	00484	10112	55
05	00484	10112	00491	10114	00499	10116	00506	10117	00514	10119	00521	10121	54
06	00521	10121	00529	10123	00537	10124	00544	10126	00552	10128	00559	10130	53
07	00559	10130	00567	10131	00574	10133	00582	10135	00589	10137	00597	10138	52
08	00597	10138	00604	10140	00612	10142	00619	10144	00627	10145	00634	10147	51
09	00634	10147	00642	10149	00649	10151	00657	10152	00664	10154	00672	10156	50
10	00672	10156	00680	10158	00687	10159	00695	10161	00702	10163	00710	10165	49
11	00710	10165	00717	10166	00725	10168	00732	10170	00740	10172	00747	10174	48
12	00747	10174	00755	10175	00762	10177	00770	10179	00777	10181	00785	10182	47
13	00785	10182	00792	10184	00800	10186	00807	10188	00815	10189	00822	10191	46
14	00822	10191	00830	10193	00837	10195	00845	10196	00852	10198	00860	10200	45
15	00860	10200	00867	10202	00875	10203	00882	10205	00890	10207	00897	10209	44
16	00897	10209	00905	10210	00912	10212	00920	10214	00927	10216	00935	10218	43
17	00935	10218	00942	10219	00950	10221	00957	10223	00965	10225	00972	10226	42
18	00972	10226	00979	10228	00987	10230	00994	10232	01002	10233	01009	10235	41
19	01009	10235	01017	10237	01024	10239	01032	10240	01039	10242	01047	10244	40
20	01047	10244	01054	10246	01062	10247	01069	10249	01077	10251	01084	10253	39
21	01084	10253	01092	10255	01099	10256	01107	10258	01114	10260	01122	10262	38
22	01122	10262	01129	10263	01136	10265	01144	10267	01151	10269	01159	10270	37
23	01159	10270	01166	10272	01174	10274	01181	10276	01189	10277	01196	10279	36
24	01196	10279	01204	10281	01211	10283	01219	10285	01226	10286	01234	10288	35
25	01234	10288	01241	10290	01248	10292	01256	10293	01263	10295	01271	10297	34
26	01271	10297	01278	10299	01286	10300	01293	10302	01301	10304	01308	10306	33
27	01308	10306	01316	10308	01323	10309	01331	10311	01338	10313	01345	10315	32
28	01345	10315	01353	10316	01360	10318	01368	10320	01375	10322	01383	10323	31
29	01383	10323	01390	10325	01397	10327	01405	10329	01412	10331	01420	10332	30
30	01420	10332	01427	10334	01435	10336	01442	10338	01450	10339	01457	10341	29
31	01457	10341	01464	10343	01472	10345	01479	10347	01487	10348	01494	10350	28
32	01494	10350	01502	10352	01509	10354	01517	10355	01524	10357	01531	10359	27
33	01531	10359	01539	10361	01546	10362	01554	10364	01561	10366	01569	10368	26
34	01569	10368	01576	10370	01583	10371	01591	10373	01598	10375	01606	10377	25
35	01606	10377	01613	10378	01620	10380	01628	10382	01635	10384	01643	10386	24
36	01643	10386	01650	10387	01658	10389	01665	10391	01673	10393	01680	10394	23
37	01680	10394	01687	10396	01695	10398	01702	10400	01710	10401	01717	10403	22
38	01717	10403	01724	10405	01732	10407	01739	10409	01747	10410	01754	10412	21
39	01754	10412	01761	10414	01769	10416	01776	10417	01784	10419	01791	10421	20
40	01791	10421	01799	10423	01806	10425	01813	10426	01821	10428	01828	10430	19
41	01828	10430	01836	10432	01843	10433	01850	10435	01858	10437	01865	10439	18
42	01865	10439	01873	10441	01880	10442	01887	10444	01895	10446	01902	10448	17
43	01902	10448	01910	10449	01917	10451	01924	10453	01932	10455	01939	10457	16
44	01939	10457	01947	10458	01954	10460	01961	10462	01969	10464	01976	10466	15
45	01976	10466	01983	10467	01991	10469	01998	10471	02006	10473	02013	10474	14
46	02013	10474	02020	10476	02028	10478	02035	10480	02043	10482	02050	10483	13
47	02050	10483	02057	10485	02065	10487	02072	10489	02079	10490	02087	10492	12
48	02087	10492	02094	10494	02102	10496	02109	10498	02116	10499	02124	10501	11
49	02124	10501	02131	10503	02139	10505	02146	10507	02153	10508	02161	10510	10
50	02161	10510	02168	10512	02175	10514	02183	10515	02190	10517	02197	10519	09
51	02197	10519	02205	10521	02212	10523	02220	10524	02227	10526	02234	10528	08
52	02234	10528	02242	10530	02249	10532	02256	10533	02264	10535	02271	10537	07
53	02271	10537	02279	10539	02286	10540	02293	10542	02301	10544	02308	10546	06
54	02308	10546	02315	10548	02323	10549	02330	10551	02337	10553	02345	10555	05
55	02345	10555	02352	10557	02359	10558	02367	10560	02374	10562	02381	10564	04
56	02381	10564	02389	10565	02396	10567	02403	10569	02411	10571	02418	10573	03
57	02418	10573	02426	10574	02433	10576	02440	10578	02448	10580	02455	10582	02
58	02455	10582	02462	10583	02470	10585	02477	10587	02484	10589	02492	10591	01
59	02492	10591	02499	10592	02506	10594	02514	10596	02521	10598	02528	10599	00

·8   ·6   ·4   ·2   ·0

PARTS for 0′·1 :—   LOGS. 4 ;   NATURALS, 1.

38° HAVERSINES

	·0		·2		·4		·6		·8		
	Log.	Nat.	Log.	Nat.	Log.	Nat.	Log.	Nat.	Log.	Nat.	
′	9·	0·	9·	0·	9·	0·	9·	0·	9·	0·	′
00	02528	10599	02536	10601	02543	10603	02550	10605	02558	10607	02565	10608	59
01	02565	10608	02572	10610	02580	10612	02587	10614	02594	10616	02602	10617	58
02	02602	10617	02609	10619	02616	10621	02624	10623	02631	10625	02638	10626	57
03	02638	10626	02646	10628	02653	10630	02660	10632	02668	10634	02675	10635	56
04	02675	10635	02682	10637	02690	10639	02697	10641	02704	10643	02712	10644	55
05	02712	10644	02719	10646	02726	10648	02734	10650	02741	10651	02748	10653	54
06	02748	10653	02756	10655	02763	10657	02770	10659	02777	10660	02785	10662	53
07	02785	10662	02792	10664	02799	10666	02807	10668	02814	10669	02821	10671	52
08	02821	10671	02829	10673	02836	10675	02843	10677	02851	10678	02858	10680	51
09	02858	10680	02865	10682	02872	10684	02880	10686	02887	10687	02894	10689	50
10	02894	10689	02902	10691	02909	10693	02916	10695	02924	10696	02931	10698	49
11	02931	10698	02938	10700	02946	10702	02953	10704	02960	10705	02967	10707	48
12	02967	10707	02975	10709	02982	10711	02989	10713	02997	10714	03004	10716	47
13	03004	10716	03011	10718	03018	10720	03026	10722	03033	10723	03040	10725	46
14	03040	10725	03048	10727	03055	10729	03062	10731	03070	10732	03077	10734	45
15	03077	10734	03084	10736	03091	10738	03099	10740	03106	10741	03113	10743	44
16	03113	10743	03120	10745	03128	10747	03135	10749	03142	10750	03150	10752	43
17	03150	10752	03157	10754	03164	10756	03171	10758	03179	10759	03186	10761	42
18	03186	10761	03193	10763	03201	10765	03208	10767	03215	10768	03222	10770	41
19	03222	10770	03230	10772	03237	10774	03244	10776	03252	10777	03259	10779	40
20	03259	10779	03266	10781	03273	10783	03281	10785	03288	10786	03295	10788	39
21	03295	10788	03302	10790	03310	10792	03317	10794	03324	10795	03331	10797	38
22	03331	10797	03339	10799	03346	10801	03353	10803	03360	10804	03368	10806	37
23	03368	10806	03375	10808	03382	10810	03389	10812	03397	10814	03404	10815	36
24	03404	10815	03411	10817	03418	10819	03426	10821	03433	10823	03440	10824	35
25	03440	10824	03447	10826	03455	10828	03462	10830	03469	10832	03476	10833	34
26	03476	10833	03484	10835	03491	10837	03498	10839	03506	10841	03513	10842	33
27	03513	10842	03520	10844	03527	10846	03535	10848	03542	10850	03549	10851	32
28	03549	10851	03556	10853	03563	10855	03571	10857	03578	10859	03585	10861	31
29	03585	10861	03592	10862	03600	10864	03607	10866	03614	10868	03621	10870	30
30	03621	10870	03629	10871	03636	10873	03643	10875	03650	10877	03657	10879	29
31	03657	10879	03665	10880	03672	10882	03679	10884	03686	10886	03694	10888	28
32	03730	10888	03701	10890	03708	10891	03715	10893	03723	10895	03730	10897	27
33	03730	10897	03737	10899	03744	10900	03751	10902	03759	10904	03766	10906	26
34	03766	10906	03773	10908	03780	10909	03788	10911	03795	10913	03802	10915	25
35	03802	10915	03809	10917	03816	10919	03824	10920	03831	10922	03838	10924	24
36	03838	10924	03845	10926	03852	10928	03860	10929	03867	10931	03874	10933	23
37	03874	10933	03881	10935	03889	10937	03896	10939	03903	10940	03910	10942	22
38	03910	10942	03917	10944	03925	10946	03932	10948	03939	10949	03946	10951	21
39	03946	10951	03953	10953	03961	10955	03968	10957	03975	10958	03982	10960	20
40	03982	10960	03989	10962	03997	10964	04004	10966	04011	10968	04018	10969	19
41	04018	10969	04025	10971	04033	10973	04040	10975	04047	10977	04054	10978	18
42	04054	10978	04061	10980	04069	10982	04076	10984	04083	10986	04090	10988	17
43	04090	10988	04097	10989	04105	10991	04112	10993	04119	10995	04126	10997	16
44	04126	10997	04133	10998	04141	11000	04148	11002	04155	11004	04162	11006	15
45	04162	11006	04169	11008	04176	11009	04184	11011	04191	11013	04198	11015	14
46	04198	11015	04205	11017	04212	11019	04219	11020	04227	11022	04234	11024	13
47	04234	11024	04241	11026	04248	11028	04255	11029	04263	11031	04270	11033	12
48	04270	11033	04277	11035	04284	11037	04291	11039	04298	11040	04306	11042	11
49	04306	11042	04313	11044	04320	11046	04327	11048	04334	11050	04341	11051	10
50	04341	11051	04349	11053	04356	11055	04363	11057	04370	11059	04377	11060	09
51	04377	11060	04385	11062	04392	11064	04399	11066	04406	11068	04413	11070	08
52	04413	11070	04420	11071	04428	11073	04435	11075	04442	11077	04449	11079	07
53	04449	11079	04456	11081	04463	11082	04470	11084	04478	11086	04485	11088	06
54	04485	11088	04492	11090	04499	11092	04506	11093	04513	11095	04520	11097	05
55	04520	11097	04528	11099	04535	11101	04542	11102	04549	11104	04556	11106	04
56	04556	11106	04563	11108	04571	11110	04578	11112	04585	11113	04592	11115	03
57	04592	11115	04599	11117	04606	11119	04613	11121	04620	11123	04628	11124	02
58	04628	11124	04635	11126	04642	11128	04649	11130	04656	11132	04663	11134	01
59	04663	11134	04671	11135	04678	11137	04685	11139	04692	11141	04699	11143	00
	·8		·6		·4		·2		·0		

PARTS for 0′·1 :—   LOGS. 4 ;   NATURALS, 1.

321°

39°   HAVERSINES

′	·0 Log. 9·	·0 Nat. 0·	·2 Log. 9·	·2 Nat. 0·	·4 Log. 9·	·4 Nat. 0·	·6 Log. 9·	·6 Nat. 0·	·8 Log. 9·	·8 Nat. 0·	′
00	04699	11113	04706	11145	04713	11146	04720	11148	04728	11150	59
01	04735	11152	04742	11154	04749	11156	04756	11157	04763	11159	58
02	04770	11161	04778	11163	04785	11165	04792	11167	04799	11168	57
03	04806	11170	04813	11172	04820	11174	04827	11176	04834	11178	56
04	04842	11179	04849	11181	04856	11183	04863	11185	04870	11187	55
05	04877	11189	04884	11190	04891	11192	04899	11194	04906	11196	54
06	04913	11198	04920	11200	04927	11201	04934	11203	04941	11205	53
07	04948	11207	04956	11209	04963	11211	04970	11212	04977	11214	52
08	04984	11216	04991	11218	04998	11220	05005	11222	05012	11223	51
09	05019	11225	05027	11227	05034	11229	05041	11231	05048	11233	50
10	05055	11234	05062	11236	05069	11238	05076	11240	05083	11242	49
11	05090	11244	05098	11245	05105	11247	05112	11249	05119	11251	48
12	05126	11253	05133	11255	05140	11256	05147	11258	05154	11260	47
13	05161	11262	05168	11264	05176	11266	05183	11267	05190	11269	46
14	05197	11271	05204	11273	05211	11275	05218	11277	05225	11279	45
15	05232	11280	05239	11282	05247	11284	05254	11286	05261	11288	44
16	05268	11290	05275	11291	05282	11293	05289	11295	05296	11297	43
17	05303	11299	05310	11301	05317	11302	05324	11304	05332	11306	42
18	05339	11308	05346	11310	05353	11312	05360	11314	05367	11315	41
19	05374	11317	05381	11319	05388	11321	05395	11323	05402	11325	40
20	05409	11326	05416	11328	05423	11330	05430	11332	05438	11334	39
21	05445	11336	05452	11337	05459	11339	05466	11341	05473	11343	38
22	05480	11345	05487	11347	05494	11349	05501	11350	05508	11352	37
23	05515	11354	05522	11356	05529	11358	05536	11360	05544	11361	36
24	05551	11363	05558	11365	05565	11367	05572	11369	05579	11371	35
25	05586	11373	05593	11374	05600	11376	05607	11378	05614	11380	34
26	05621	11382	05628	11384	05635	11386	05642	11387	05649	11389	33
27	05656	11391	05663	11393	05670	11395	05678	11397	05685	11398	32
28	05692	11400	05699	11402	05706	11404	05713	11406	05720	11408	31
29	05727	11410	05734	11411	05741	11413	05748	11415	05755	11417	30
30	05762	11419	05769	11421	05776	11422	05783	11424	05790	11426	29
31	05797	11428	05804	11430	05811	11432	05818	11434	05825	11435	28
32	05832	11437	05839	11439	05846	11441	05853	11443	05860	11445	27
33	05867	11447	05874	11448	05881	11450	05889	11452	05896	11454	26
34	05903	11456	05910	11458	05917	11460	05924	11461	05931	11463	25
35	05938	11465	05945	11467	05952	11469	05959	11471	05966	11472	24
36	05973	11474	05980	11476	05987	11478	05994	11480	06001	11482	23
37	06008	11484	06015	11485	06022	11487	06029	11489	06036	11491	22
38	06043	11493	06050	11495	06057	11497	06064	11498	06071	11500	21
39	06078	11502	06085	11504	06092	11506	06099	11508	06106	11510	20
40	06113	11511	06120	11513	06127	11515	06134	11517	06141	11519	19
41	06148	11521	06155	11523	06162	11524	06169	11526	06176	11528	18
42	06183	11530	06190	11532	06197	11534	06204	11536	06211	11537	17
43	06218	11539	06225	11541	06232	11543	06239	11545	06246	11547	16
44	06253	11549	06260	11550	06267	11552	06274	11554	06281	11556	15
45	06288	11558	06295	11560	06302	11562	06309	11563	06316	11565	14
46	06323	11567	06330	11569	06337	11571	06344	11573	06351	11575	13
47	06358	11577	06365	11578	06372	11580	06379	11582	06386	11584	12
48	06393	11586	06400	11588	06407	11590	06414	11591	06421	11593	11
49	06428	11595	06435	11597	06442	11599	06448	11601	06455	11603	10
50	06462	11604	06469	11606	06476	11608	06483	11610	06490	11612	09
51	06497	11614	06504	11616	06511	11617	06518	11619	06525	11621	08
52	06532	11623	06539	11625	06546	11627	06553	11629	06560	11631	07
53	06567	11632	06574	11634	06581	11636	06588	11638	06595	11640	06
54	06602	11642	06609	11644	06616	11645	06623	11647	06630	11649	05
55	06637	11651	06644	11653	06651	11655	06657	11657	06664	11659	04
56	06671	11660	06678	11662	06685	11664	06692	11666	06699	11668	03
57	06706	11670	06713	11672	06720	11673	06727	11675	06734	11677	02
58	06741	11679	06748	11681	06755	11683	06762	11685	06769	11687	01
59	06776	11688	06783	11690	06790	11692	06796	11694	06803	11696	00

·8   ·6   ·4   ·2   ·0

320°

PARTS for 0′·1 :—   LOGS. 4 ;   NATURALS, 1.

40° HAVERSINES

′	.0 Log. 9·	.0 Nat. 0·	.2 Log. 9·	.2 Nat. 0·	.4 Log. 9·	.4 Nat. 0·	.6 Log. 9·	.6 Nat. 0·	.8 Log. 9·	.8 Nat. 0·	′
00	06810	11698	06817	11700	06824	11702	06831	11703	06838	11705	06845	11707	59
01	06845	11707	06852	11709	06859	11711	06866	11713	06873	11715	06880	11716	58
02	06880	11716	06887	11718	06894	11720	06901	11722	06907	11724	06914	11726	57
03	06914	11726	06921	11728	06928	11730	06935	11731	06942	11733	06949	11735	56
04	06949	11735	06956	11737	06963	11739	06970	11741	06977	11743	06984	11745	55
05	06984	11745	06991	11746	06998	11748	07004	11750	07011	11752	07018	11754	54
06	07018	11754	07025	11756	07032	11758	07039	11760	07046	11761	07053	11763	53
07	07053	11763	07060	11765	07067	11767	07074	11769	07081	11771	07088	11773	52
08	07088	11773	07094	11775	07101	11776	07108	11778	07115	11780	07122	11782	51
09	07122	11782	07129	11784	07136	11786	07143	11788	07150	11790	07157	11791	50
10	07157	11791	07164	11793	07170	11795	07177	11797	07184	11799	07191	11801	49
11	07191	11801	07198	11803	07205	11805	07212	11806	07219	11808	07226	11810	48
12	07226	11810	07233	11812	07240	11814	07246	11816	07253	11818	07260	11820	47
13	07260	11820	07267	11821	07274	11823	07281	11825	07288	11827	07295	11829	46
14	07295	11829	07302	11831	07309	11833	07315	11835	07322	11837	07329	11838	45
15	07329	11838	07336	11840	07343	11842	07350	11844	07357	11846	07364	11848	44
16	07364	11848	07371	11850	07377	11852	07384	11853	07391	11855	07398	11857	43
17	07398	11857	07405	11859	07412	11861	07419	11863	07426	11865	07433	11867	42
18	07433	11867	07439	11868	07446	11870	07453	11872	07460	11874	07467	11876	41
19	07467	11876	07474	11878	07481	11880	07488	11882	07495	11884	07501	11885	40
20	07501	11885	07508	11887	07515	11889	07522	11891	07529	11893	07536	11895	39
21	07536	11895	07543	11897	07550	11899	07556	11900	07563	11902	07570	11904	38
22	07570	11904	07577	11906	07584	11908	07591	11910	07598	11912	07605	11914	37
23	07605	11914	07611	11916	07618	11917	07625	11919	07632	11921	07639	11923	36
24	07639	11923	07646	11925	07653	11927	07659	11929	07666	11931	07673	11933	35
25	07673	11933	07680	11934	07687	11936	07694	11938	07701	11940	07708	11942	34
26	07708	11942	07714	11944	07721	11946	07728	11948	07735	11950	07742	11951	33
27	07742	11951	07749	11953	07755	11955	07762	11957	07769	11959	07776	11961	32
28	07776	11961	07783	11963	07790	11965	07797	11966	07804	11968	07810	11970	31
29	07810	11970	07817	11972	07824	11974	07831	11976	07838	11978	07845	11980	30
30	07845	11980	07851	11982	07858	11983	07865	11985	07872	11987	07879	11989	29
31	07879	11989	07886	11991	07892	11993	07899	11995	07906	11997	07913	11999	28
32	07913	11999	07920	12000	07927	12002	07934	12004	07940	12006	07947	12008	27
33	07947	12008	07954	12010	07961	12012	07968	12014	07975	12016	07981	12018	26
34	07981	12018	07988	12019	07995	12021	08002	12023	08009	12025	08016	12027	25
35	08016	12027	08022	12029	08029	12031	08036	12033	08043	12035	08050	12036	24
36	08050	12036	08057	12038	08053	12040	08070	12042	08077	12044	08084	12046	23
37	08084	12046	08091	12048	08098	12050	08104	12052	08111	12053	08118	12055	22
38	08118	12055	08125	12057	08132	12059	08139	12061	08145	12063	08152	12065	21
39	08152	12065	08159	12067	08166	12069	08173	12071	08179	12072	08186	12074	20
40	08186	12074	08193	12076	08200	12078	08207	12080	08214	12082	08220	12084	19
41	08220	12084	08227	12086	08234	12088	08241	12090	08248	12091	08254	12093	18
42	08254	12093	08261	12095	08268	12097	08275	12099	08282	12101	08288	12103	17
43	08288	12103	08295	12105	08302	12107	08309	12108	08316	12110	08323	12112	16
44	08323	12112	08329	12114	08336	12116	08343	12118	08350	12120	08357	12122	15
45	08357	12122	08363	12124	08370	12126	08377	12127	08384	12129	08391	12131	14
46	08391	12131	08397	12133	08404	12135	08411	12137	08418	12139	08425	12141	13
47	08425	12141	08431	12143	08438	12145	08445	12146	08452	12148	08459	12150	12
48	08459	12150	08465	12152	08472	12154	08479	12156	08486	12158	08492	12160	11
49	08492	12160	08499	12162	08506	12164	08513	12165	08520	12167	03526	12169	10
50	08526	12169	08533	12171	08540	12173	08547	12175	08554	12177	08560	12179	09
51	08560	12179	08567	12181	08574	12183	08581	12184	08587	12186	08594	12188	08
52	08594	12188	08601	12190	08608	12192	08615	12194	08621	12196	08628	12198	07
53	08628	12198	08635	12200	08642	12202	08648	12204	08655	12205	08662	12207	06
54	08662	12207	08669	12209	08676	12211	08682	12213	08689	12215	08696	12217	05
55	08696	12217	08703	12219	08710	12221	08716	12223	08723	12224	08730	12226	04
56	08730	12226	08737	12228	08743	12230	08750	12232	08757	12234	08764	12236	03
57	08764	12236	08770	12238	08777	12240	08784	12242	08791	12244	08797	12245	02
58	08797	12245	08804	12247	08811	12249	08818	12251	08824	12253	08831	12255	01
59	08831	12255	08838	12257	08845	12259	08851	12261	08858	12263	08865	12265	00

| | .8 | | .6 | | .4 | | .2 | | .0 | |

PARTS for 0′·1 :— LOGS. 3 ; NATURALS, 1.

319°

41° HAVERSINES

′	·0 Log.	·0 Nat.	·2 Log.	·2 Nat.	·4 Log.	·4 Nat.	·6 Log.	·6 Nat.	·8 Log.	·8 Nat.	Log.	Nat.	′
	9·	0·	9·	0·	9·	0·	9·	0·	9·	0·	9·	0·	
00	08865	12265	08872	12266	08879	12268	08885	12270	08892	12272	08899	12274	59
01	08899	12274	08906	12276	08912	12270	08919	12280	08926	12282	08933	12284	58
02	08933	12284	08939	12286	08946	12287	08953	12289	08960	12291	08966	12293	57
03	08966	12293	08973	12295	08980	12297	08987	12299	08993	12301	09000	12303	56
04	09000	12303	09007	12305	09014	12307	09020	12308	09027	12310	09034	12312	55
05	09034	12312	09041	12314	09047	12316	09054	12318	09061	12320	09068	12322	54
06	09068	12322	09074	12324	09081	12326	09088	12328	09094	12329	09101	12331	53
07	09101	12331	09108	12333	09115	12335	09121	12337	09128	12339	09135	12341	52
08	09135	12341	09142	12343	09148	12345	09155	12347	09162	12349	09169	12351	51
09	09169	12351	09175	12352	09182	12354	09189	12356	09195	12358	09202	12360	50
10	09202	12360	09209	12362	09216	12364	09222	12366	09229	12368	09236	12370	49
11	09236	12370	09243	12372	09249	12374	09256	12375	09263	12377	09269	12379	48
12	09269	12379	09276	12381	09283	12383	09290	12385	09296	12387	09303	12389	47
13	09303	12389	09310	12391	09316	12393	09323	12395	09330	12396	09337	12398	46
14	09337	12398	09343	12400	09350	12402	09357	12404	09363	12406	09370	12408	45
15	09370	12408	09377	12410	09384	12412	09390	12414	09397	12416	09404	12418	44
16	09404	12418	09411	12420	09417	12421	09424	12423	09431	12425	09437	12427	43
17	09437	12427	09444	12429	09451	12431	09457	12433	09464	12435	09471	12437	42
18	09471	12437	09478	12439	09484	12441	09491	12443	09498	12444	09504	12446	41
19	09504	12446	09511	12448	09518	12450	09524	12452	09531	12454	09538	12456	40
20	09538	12456	09545	12458	09551	12460	09558	12462	09565	12464	09571	12466	39
21	09571	12466	09578	12468	09585	12469	09591	12471	09598	12473	09605	12475	38
22	09605	12475	09611	12477	09618	12479	09625	12481	09632	12483	09638	12485	37
23	09638	12485	09645	12487	09652	12489	09658	12491	09665	12493	09672	12494	36
24	09672	12494	09678	12496	09685	12498	09692	12500	09699	12502	09705	12504	35
25	09705	12504	09712	12506	09719	12508	09725	12510	09732	12512	09739	12514	34
26	09739	12514	09745	12516	09752	12518	09759	12519	09765	12521	09772	12523	33
27	09772	12523	09779	12525	09785	12527	09792	12529	09799	12531	09805	12533	32
28	09805	12533	09812	12535	09819	12537	09825	12539	09832	12541	09839	12543	31
29	09839	12543	09845	12545	09852	12546	09859	12548	09865	12550	09872	12552	30
30	09872	12552	09879	12554	09885	12556	09892	12558	09899	12560	09905	12562	29
31	09905	12562	09912	12564	09919	12566	09925	12568	09932	12570	09939	12572	28
32	09939	12572	09945	12573	09952	12575	09959	12577	09965	12579	09972	12581	27
33	09972	12581	09979	12583	09985	12585	09992	12587	09999	12589	10005	12591	26
34	10005	12591	10012	12593	10019	12595	10025	12597	10032	12599	10039	12600	25
35	10039	12600	10045	12602	10052	12604	10059	12606	10065	12608	10072	12610	24
36	10072	12610	10078	12612	10085	12614	10092	12616	10098	12618	10105	12620	23
37	10105	12620	10112	12622	10118	12624	10125	12626	10132	12627	10138	12629	22
38	10138	12629	10145	12631	10152	12633	10158	12635	10165	12637	10172	12639	21
39	10172	12639	10178	12641	10185	12643	10192	12645	10198	12647	10205	12649	20
40	10205	12649	10211	12651	10218	12653	10225	12655	10231	12656	10238	12658	19
41	10238	12658	10245	12660	10251	12662	10258	12664	10265	12666	10271	12668	18
42	10271	12668	10278	12670	10284	12672	10291	12674	10298	12676	10304	12678	17
43	10304	12678	10311	12680	10317	12682	10324	12684	10331	12686	10337	12687	16
44	10337	12687	10344	12689	10351	12691	10357	12693	10364	12695	10371	12697	15
45	10371	12697	10377	12699	10384	12701	10390	12703	10397	12705	10404	12707	14
46	10404	12707	10410	12709	10417	12711	10424	12713	10430	12715	10437	12717	13
47	10437	12717	10443	12718	10450	12720	10457	12722	10463	12724	10470	12726	12
48	10470	12726	10476	12728	10483	12730	10490	12732	10496	12734	10503	12736	11
49	10503	12736	10510	12738	10516	12740	10523	12742	10529	12744	10536	12746	10
50	10536	12746	10543	12748	10549	12749	10556	12751	10562	12753	10569	12755	09
51	10569	12755	10576	12757	10582	12759	10589	12761	10595	12763	10602	12765	08
52	10602	12765	10609	12767	10615	12769	10622	12771	10629	12773	10635	12775	07
53	10635	12775	10642	12777	10648	12779	10655	12781	10662	12782	10668	12784	06
54	10668	12784	10675	12786	10681	12788	10688	12790	10695	12792	10701	12794	05
55	10701	12794	10708	12796	10714	12798	10721	12800	10727	12802	10734	12804	04
56	10734	12804	10741	12806	10747	12808	10754	12810	10760	12812	10767	12814	03
57	10767	12814	10774	12816	10780	12817	10787	12819	10793	12821	10800	12823	02
58	10800	12823	10807	12825	10813	12827	10820	12829	10826	12831	10833	12833	01
59	10833	12833	10840	12835	10846	12837	10853	12839	10859	12841	10866	12843	00

·8	·6	·4	·2	·0

318°

PARTS for 0′·1 :—   LOGS. 3 ;   NATURALS, 1.

HAVERSINES

42°

′	·0 Log.	·0 Nat.	·2 Log.	·2 Nat.	·4 Log.	·4 Nat.	·6 Log.	·6 Nat.	·8 Log.	·8 Nat.	′
	9·	0·	9·	0·	9·	0·	9·	0·	9·	0·	
00	10866	12843	10872	12845	10879	12847	10886	12849	10892	12851	10899 12852 · 59
00	10866	12843	10872	12845	10879	12847	10886	12849	10892	12851	59
01	10899	12852	10905	12854	10912	12856	10918	12858	10925	12860	58
02	10932	12862	10938	12864	10945	12866	10951	12868	10958	12870	57
03	10964	12872	10971	12874	10978	12876	10984	12878	10991	12880	56
04	10997	12882	11004	12884	11011	12886	11017	12888	11024	12890	55
05	11030	12891	11037	12893	11043	12895	11050	12897	11056	12899	54
06	11063	12901	11070	12903	11076	12905	11083	12907	11089	12909	53
07	11096	12911	11102	12913	11109	12915	11115	12917	11122	12919	52
08	11129	12921	11135	12923	11142	12925	11148	12927	11155	12929	51
09	11161	12930	11168	12932	11175	12934	11181	12936	11188	12938	50
10	11194	12940	11201	12942	11207	12944	11214	12946	11220	12948	49
11	11227	12950	11234	12952	11240	12954	11247	12956	11253	12958	48
12	11260	12960	11266	12962	11273	12964	11279	12966	11286	12968	47
13	11292	12970	11299	12972	11306	12973	11312	12975	11319	12977	46
14	11325	12979	11332	12981	11338	12983	11345	12985	11351	12987	45
15	11358	12989	11364	12991	11371	12993	11377	12995	11384	12997	44
16	11391	12999	11397	13001	11404	13003	11410	13005	11417	13007	43
17	11423	13009	11430	13011	11436	13013	11443	13015	11449	13016	42
18	11456	13018	11463	13020	11469	13022	11476	13024	11482	13026	41
19	11489	13028	11495	13030	11502	13032	11508	13034	11515	13036	40
20	11521	13038	11528	13040	11534	13042	11541	13044	11547	13046	39
21	11554	13048	11560	13050	11567	13052	11573	13054	11580	13056	38
22	11586	13058	11593	13060	11599	13062	11606	13064	11612	13065	37
23	11619	13067	11626	13069	11632	13071	11639	13073	11645	13075	36
24	11652	13077	11658	13079	11665	13081	11671	13083	11678	13085	35
25	11684	13087	11691	13089	11697	13091	11704	13093	11710	13095	34
26	11717	13097	11723	13099	11730	13101	11736	13103	11743	13105	33
27	11749	13107	11756	13109	11762	13111	11769	13113	11775	13115	32
28	11782	13116	11788	13118	11795	13120	11801	13122	11808	13124	31
29	11814	13126	11821	13128	11827	13130	11834	13132	11840	13134	30
30	11847	13136	11853	13138	11860	13140	11866	13142	11873	13144	29
31	11879	13146	11886	13148	11892	13150	11899	13152	11905	13154	28
32	11912	13156	11918	13158	11925	13160	11931	13162	11938	13164	27
33	11944	13166	11951	13168	11957	13170	11964	13172	11970	13173	26
34	11977	13175	11983	13177	11990	13179	11996	13181	12003	13183	25
35	12009	13185	12015	13187	12022	13189	12028	13191	12035	13193	24
36	12041	13195	12048	13197	12054	13199	12061	13201	12067	13203	23
37	12074	13205	12080	13207	12087	13209	12093	13211	12100	13213	22
38	12106	13215	12113	13217	12119	13219	12126	13221	12132	13223	21
39	12139	13225	12145	13227	12152	13229	12158	13231	12165	13233	20
40	12171	13235	12177	13237	12184	13238	12190	13240	12197	13242	19
41	12203	13244	12210	13246	12216	13248	12223	13250	12229	13252	18
42	12236	13254	12242	13256	12248	13258	12255	13260	12261	13262	17
43	12268	13264	12274	13266	12281	13268	12287	13270	12294	13272	16
44	12300	13274	12307	13276	12313	13278	12320	13280	12326	13282	15
45	12332	13284	12339	13286	12345	13288	12352	13290	12358	13292	14
46	12365	13294	12371	13296	12378	13298	12384	13300	12391	13302	13
47	12397	13304	12403	13306	12410	13308	12416	13310	12423	13312	12
48	12429	13314	12436	13315	12442	13317	12449	13319	12455	13321	11
49	12461	13323	12468	13325	12474	13327	12481	13329	12487	13331	10
50	12494	13333	12500	13335	12506	13337	12513	13339	12519	13341	09
51	12526	13343	12532	13345	12539	13347	12545	13349	12552	13351	08
52	12558	13353	12564	13355	12571	13357	12577	13359	12584	13361	07
53	12590	13363	12597	13365	12603	13367	12610	13369	12616	13371	06
54	12622	13373	12629	13375	12635	13377	12642	13379	12648	13381	05
55	12655	13383	12661	13385	12667	13387	12674	13389	12680	13391	04
56	12687	13393	12693	13395	12699	13397	12706	13399	12712	13401	03
57	12719	13403	12725	13405	12732	13407	12738	13409	12745	13411	02
58	12751	13412	12757	13414	12764	13416	12770	13418	12777	13420	01
59	12783	13422	12789	13424	12796	13426	12802	13428	12809	13430	00

(·8)   (·6)   (·4)   (·2)   (·0) ←

317°

PARTS for 0′·1 :—   LOGS, 3 ;   NATURALS, 1.

43° HAVERSINES

′	·0 Log. 9·	·0 Nat. 0·	·2 Log. 9·	·2 Nat. 0·	·4 Log. 9·	·4 Nat. 0·	·6 Log. 9·	·6 Nat. 0·	·8 Log. 9·	·8 Nat. 0·	Log. 9·	Nat. 0·	′
00	12815	13432	12821	13434	12828	13436	12834	13438	12841	13440	12847	13442	59
01	12847	13442	12854	13444	12860	13446	12866	13448	12873	13450	12879	13452	58
02	12879	13452	12886	13454	12892	13456	12898	13458	12905	13460	12911	13462	57
03	12911	13462	12918	13464	12924	13466	12930	13468	12937	13470	12943	13472	56
04	12943	13472	12950	13474	12956	13476	12962	13478	12969	13480	12975	13482	55
05	12975	13482	12982	13484	12988	13486	12994	13488	13001	13490	13007	13492	54
06	13007	13492	13014	13494	13020	13496	13026	13498	13033	13500	13039	13502	53
07	13039	13502	13046	13504	13052	13506	13058	13508	13065	13510	13071	13512	52
08	13071	13512	13078	13514	13084	13516	13090	13518	13097	13520	13103	13522	51
09	13103	13522	13110	13524	13116	13526	13122	13528	13129	13530	13135	13532	50
10	13135	13532	13142	13534	13148	13536	13154	13538	13161	13540	13167	13542	49
11	13167	13542	13173	13544	13180	13546	13186	13548	13193	13550	13199	13552	48
12	13199	13552	13205	13554	13212	13556	13218	13558	13224	13560	13231	13562	47
13	13231	13562	13237	13564	13244	13566	13250	13568	13256	13570	13263	13571	46
14	13263	13571	13269	13573	13276	13575	13282	13577	13288	13579	13295	13581	45
15	13295	13581	13301	13583	13307	13585	13314	13587	13320	13589	13326	13591	44
16	13326	13591	13333	13593	13339	13595	13346	13597	13352	13599	13358	13601	43
17	13358	13601	13365	13603	13371	13605	13377	13607	13384	13609	13390	13611	42
18	13390	13611	13397	13613	13403	13615	13409	13617	13416	13619	13422	13621	41
19	13422	13621	13428	13623	13435	13625	13441	13627	13447	13629	13454	13631	40
20	13454	13631	13460	13633	13467	13635	13473	13637	13479	13639	13486	13641	39
21	13486	13641	13492	13643	13498	13645	13505	13647	13511	13649	13517	13651	38
22	13517	13651	13524	13653	13530	13655	13536	13657	13543	13659	13549	13661	37
23	13549	13661	13556	13663	13562	13665	13568	13667	13574	13669	13581	13671	36
24	13581	13671	13587	13673	13594	13675	13600	13677	13606	13679	13613	13681	35
25	13613	13681	13619	13683	13625	13685	13632	13687	13638	13689	13644	13691	34
26	13644	13691	13651	13693	13657	13695	13663	13697	13670	13699	13676	13701	33
27	13676	13701	13682	13703	13689	13705	13695	13707	13701	13709	13708	13711	32
28	13708	13711	13714	13713	13720	13715	13727	13717	13733	13719	13739	13721	31
29	13739	13721	13746	13723	13752	13725	13758	13727	13765	13729	13771	13731	30
30	13771	13731	13778	13733	13784	13735	13790	13737	13796	13739	13803	13741	29
31	13803	13741	13809	13743	13815	13745	13822	13747	13828	13749	13834	13751	28
32	13834	13751	13841	13753	13847	13755	13853	13757	13860	13759	13866	13761	27
33	13866	13761	13872	13763	13879	13765	13885	13767	13891	13769	13898	13771	26
34	13898	13771	13904	13773	13910	13775	13917	13777	13923	13779	13929	13781	25
35	13929	13781	13936	13783	13942	13785	13948	13787	13954	13789	13961	13791	24
36	13961	13791	13967	13793	13973	13795	13980	13797	13986	13799	13992	13801	23
37	13992	13801	13999	13803	14005	13805	14011	13807	14018	13809	14024	13811	22
38	14024	13811	14030	13813	14037	13815	14043	13817	14049	13819	14056	13822	21
39	14056	13822	14062	13824	14068	13826	14074	13828	14081	13830	14087	13832	20
40	14087	13832	14093	13834	14100	13836	14106	13838	14112	13840	14119	13842	19
41	14119	13842	14125	13844	14131	13846	14138	13848	14144	13850	14150	13852	18
42	14150	13852	14156	13854	14163	13856	14169	13858	14175	13860	14182	13862	17
43	14182	13862	14188	13864	14194	13866	14201	13868	14207	13870	14213	13872	16
44	14213	13872	14219	13874	14226	13876	14232	13878	14238	13880	14245	13882	15
45	14245	13882	14251	13884	14257	13886	14263	13888	14270	13890	14276	13892	14
46	14276	13892	14282	13894	14289	13896	14295	13898	14301	13900	14307	13902	13
47	14307	13902	14314	13904	14320	13906	14326	13908	14333	13910	14339	13912	12
48	14339	13912	14345	13914	14352	13916	14358	13918	14364	13920	14370	13922	11
49	14370	13922	14377	13924	14383	13926	14389	13928	14396	13930	14402	13932	10
50	14402	13932	14408	13934	14414	13936	14421	13938	14427	13940	14433	13942	09
51	14433	13942	14439	13944	14446	13946	14452	13948	14458	13950	14465	13952	08
52	14465	13952	14471	13954	14477	13956	14483	13958	14490	13960	14496	13962	07
53	14496	13962	14502	13964	14508	13966	14515	13968	14521	13970	14527	13972	06
54	14527	13972	14533	13974	14540	13976	14546	13979	14552	13981	14559	13983	05
55	14559	13983	14565	13985	14571	13987	14577	13989	14584	13991	14590	13993	04
56	14590	13993	14596	13995	14602	13997	14609	13999	14615	14001	14621	14003	03
57	14621	14003	14627	14005	14634	14007	14640	14009	14646	14011	14653	14013	02
58	14653	14013	14659	14015	14665	14017	14671	14019	14678	14021	14684	14023	01
59	14684	14023	14690	14025	14696	14027	14703	14029	14709	14031	14715	14033	00

| | ·8 | | ·6 | | ·4 | | ·2 | | ·0 | |

316°

PARTS for 0′·1 :— LOGS, 3 ; NATURALS, 1.

44° HAVERSINES

′	.0 Log. 9·	.0 Nat. 0·	.2 Log. 9·	.2 Nat. ·0	.4 Log. 9·	.4 Nat. 0·	.6 Log. 9·	.6 Nat. 0·	.8 Log. 9·	.8 Nat. 0·	Log. 9·	Nat. 0·	′
00	14715	14033	14721	14035	14728	14037	14734	14039	14740	14041	14746	14043	59
01	14746	14043	14753	14045	14759	14047	14765	14049	14771	14051	14778	14053	58
02	14778	14053	14784	14055	14790	14057	14796	14059	14803	14061	14809	14063	57
03	14809	14063	14815	14065	14821	14067	14828	14069	14834	14071	14840	14073	56
04	14840	14073	14846	14075	14852	14077	14859	14080	14865	14082	14871	14084	55
05	14871	14084	14877	14086	14884	14088	14890	14090	14896	14092	14902	14094	54
06	14902	14094	14909	14096	14915	14098	14921	14100	14927	14102	14934	14104	53
07	14934	14104	14940	14106	14946	14108	14952	14110	14959	14112	14965	14114	52
08	14965	14114	14971	14116	14977	14118	14984	14120	14990	14122	14996	14124	51
09	14996	14124	15002	14126	15008	14128	15015	14130	15021	14132	15027	14134	50
10	15027	14134	15033	14136	15040	14138	15046	14140	15052	14142	15058	14144	49
11	15058	14144	15064	14146	15071	14148	15077	14150	15083	14152	15089	14154	48
12	15089	14154	15096	14156	15102	14159	15108	14161	15114	14163	15120	14165	47
13	15120	14165	15127	14167	15133	14169	15139	14171	15145	14173	15152	14175	46
14	15152	14175	15158	14177	15164	14179	15171	14181	15176	14183	15183	14185	45
15	15183	14185	15189	14187	15195	14189	15201	14191	15208	14193	15214	14195	44
16	15214	14195	15220	14197	15226	14199	15232	14201	15239	14203	15245	14205	43
17	15245	14205	15251	14207	15257	14209	15263	14211	15270	14213	15276	14215	42
18	15276	14215	15282	14217	15288	14219	15294	14221	15301	14223	15307	14226	41
19	15307	14226	15313	14228	15319	14230	15325	14232	15332	14234	15338	14236	40
20	15338	14236	15344	14238	15350	14240	15356	14242	15363	14244	15369	14246	39
21	15369	14246	15375	14248	15381	14250	15387	14252	15394	14254	15400	14256	38
22	15400	14256	15406	14258	15412	14260	15418	14262	15425	14264	15431	14266	37
23	15431	14266	15437	14268	15443	14270	15449	14272	15456	14274	15462	14276	36
24	15462	14276	15468	14278	15474	14280	15480	14282	15487	14285	15493	14287	35
25	15493	14287	15499	14289	15505	14291	15511	14293	15517	14295	15524	14297	34
26	15524	14297	15530	14299	15536	14301	15542	14303	15548	14305	15555	14307	33
27	15555	14307	15561	14309	15567	14311	15573	14313	15579	14315	15585	14317	32
28	15585	14317	15592	14319	15598	14321	15604	14323	15610	14325	15616	14327	31
29	15616	14327	15623	14329	15629	14331	15635	14333	15641	14335	15647	14337	30
30	15647	14337	15653	14340	15660	14342	15666	14344	15672	14346	15678	14348	29
31	15678	14348	15684	14350	15691	14352	15697	14354	15703	14356	15709	14358	28
32	15709	14358	15715	14360	15721	14362	15727	14364	15734	14366	15740	14368	27
33	15740	14368	15746	14370	15752	14372	15758	14374	15764	14376	15771	14378	26
34	15771	14378	15777	14380	15783	14382	15789	14384	15795	14386	15802	14388	25
35	15802	14388	15808	14391	15814	14393	15820	14395	15826	14397	15832	14399	24
36	15832	14399	15839	14401	15845	14403	15851	14405	15857	14407	15863	14409	23
37	15863	14409	15869	14411	15876	14413	15882	14415	15888	14417	15894	14419	22
38	15894	14419	15900	14421	15906	14423	15912	14425	15918	14427	15925	14429	21
39	15925	14429	15931	14431	15937	14433	15943	14435	15949	14438	15955	14440	20
40	15955	14440	15962	14442	15968	14444	15974	14446	15980	14448	15986	14450	19
41	15986	14450	15992	14452	15999	14454	16005	14456	16011	14458	16017	14460	18
42	16017	14460	16023	14462	16029	14464	16035	14466	16041	14468	16048	14470	17
43	16048	14470	16054	14472	16060	14474	16066	14476	16072	14478	16078	14480	16
44	16078	14480	16085	14483	16091	14485	16097	14487	16103	14489	16109	14491	15
45	16109	14491	16115	14493	16121	14495	16127	14497	16134	14499	16140	14501	14
46	16140	14501	16146	14503	16152	14505	16158	14507	16164	14509	16170	14511	13
47	16170	14511	16177	14513	16183	14515	16189	14517	16195	14519	16201	14521	12
48	16201	14521	16207	14524	16213	14526	16219	14528	16225	14530	16232	14532	11
49	16232	14532	16238	14534	16244	14536	16250	14538	16256	14540	16262	14542	10
50	16262	14542	16269	14544	16275	14546	16281	14548	16287	14550	16293	14552	09
51	16293	14552	16299	14554	16305	14556	16311	14558	16317	14560	16324	14562	08
52	16324	14562	16330	14565	16336	14567	16342	14569	16348	14571	16354	14573	07
53	16354	14573	16360	14575	16366	14577	16373	14579	16379	14581	16385	14583	06
54	16385	14583	16391	14585	16397	14587	16403	14589	16409	14591	16415	14593	05
55	16415	14593	16421	14595	16427	14597	16434	14599	16440	14602	16446	14604	04
56	16446	14604	16452	14606	16458	14608	16464	14610	16470	14612	16476	14614	03
57	16476	14614	16483	14616	16489	14618	16495	14620	16501	14622	16507	14624	02
58	16507	14624	16513	14626	16519	14628	16525	14630	16531	14632	16537	14634	01
59	16537	14634	16544	14636	16550	14639	16556	14641	16562	14643	16568	14645	00

·8 ·6 ·4 ·2 0·

PARTS for 0′.1 :—   LOGS, 3 ;   NATURALS, 1.

315°

45°   HAVERSINES

′	·0 Log. 9·	·0 Nat. 0·	·2 Log. 9·	·2 Nat. 0·	·4 Log. 9·	·4 Nat. 0·	·6 Log. 9·	·6 Nat. 0·	·8 Log. 9·	·8 Nat. 0·	Log. 9·	Nat. 0·	′
00	16568	14645	16574	14647	16580	14649	16586	14651	16592	14653	16598	14655	59
01	16598	14655	16605	14657	16611	14659	16617	14661	16623	14663	16629	14665	58
02	16629	14665	16635	14667	16641	14669	16647	14671	16653	14673	16659	14676	57
03	16659	14676	16665	14678	16672	14680	16678	14682	16684	14684	16690	14686	56
04	16690	14686	16696	14688	16702	14690	16708	14692	16714	14694	16720	14696	55
05	16720	14696	16726	14698	16732	14700	16739	14702	16745	14704	16751	14706	54
06	16751	14706	16757	14708	16763	14711	16769	14713	16775	14715	16781	14717	53
07	16781	14717	16787	14719	16793	14721	16799	14723	16806	14725	16812	14727	52
08	16812	14727	16818	14729	16824	14731	16830	14733	16836	14735	16842	14737	51
09	16842	14737	16848	14739	16854	14741	16860	14744	16866	14746	16872	14748	50
10	16872	14748	16878	14750	16884	14752	16890	14754	16897	14756	16903	14758	49
11	16903	14758	16909	14760	16915	14762	16921	14764	16927	14766	16933	14768	48
12	16933	14768	16939	14770	16945	14772	16951	14774	16957	14777	16963	14779	47
13	16963	14779	16969	14781	16975	14783	16982	14785	16988	14787	16994	14789	46
14	16994	14789	17000	14791	17006	14793	17012	14795	17018	14797	17024	14799	45
15	17024	14799	17030	14801	17036	14803	17042	14805	17048	14808	17054	14810	44
16	17054	14810	17060	14812	17066	14814	17073	14816	17079	14818	17085	14820	43
17	17085	14820	17091	14822	17097	14824	17103	14826	17109	14828	17115	14830	42
18	17115	14830	17121	14832	17127	14834	17133	14836	17139	14839	17145	14841	41
19	17145	14841	17151	14843	17157	14845	17163	14847	17169	14849	17175	14851	40
20	17175	14851	17181	14853	17187	14855	17194	14857	17200	14859	17206	14861	39
21	17206	14861	17212	14863	17218	14865	17224	14868	17230	14870	17236	14872	38
22	17236	14872	17242	14874	17248	14876	17254	14878	17260	14880	17266	14882	37
23	17266	14882	17272	14884	17278	14886	17284	14888	17290	14890	17296	14892	36
24	17296	14892	17302	14894	17308	14896	17315	14899	17321	14901	17327	14903	35
25	17327	14903	17333	14905	17339	14907	17345	14909	17351	14911	17357	14913	34
26	17357	14913	17363	14915	17369	14917	17375	14919	17381	14921	17387	14923	33
27	17387	14923	17393	14926	17399	14928	17405	14930	17411	14932	17417	14934	32
28	17417	14934	17423	14936	17429	14938	17435	14940	17441	14942	17447	14944	31
29	17447	14944	17453	14946	17459	14948	17465	14950	17471	14952	17477	14955	30
30	17477	14955	17483	14957	17489	14959	17495	14961	17501	14963	17507	14965	29
31	17507	14965	17513	14967	17519	14969	17526	14971	17532	14973	17538	14975	28
32	17538	14975	17544	14977	17550	14979	17556	14982	17562	14984	17568	14986	27
33	17568	14986	17574	14988	17580	14990	17586	14992	17592	14994	17598	14996	26
34	17598	14996	17604	14998	17610	15000	17616	15002	17622	15004	17628	15006	25
35	17628	15006	17634	15009	17640	15011	17646	15013	17652	15015	17658	15017	24
36	17658	15017	17664	15019	17670	15021	17676	15023	17682	15025	17688	15027	23
37	17688	15027	17694	15029	17700	15031	17706	15033	17712	15036	17718	15038	22
38	17718	15038	17724	15040	17730	15042	17736	15044	17742	15046	17748	15048	21
39	17748	15048	17754	15050	17760	15052	17766	15054	17772	15056	17778	15058	20
40	17778	15058	17784	15060	17790	15063	17796	15065	17802	15067	17808	15069	19
41	17808	15069	17814	15071	17820	15073	17826	15075	17832	15077	17838	15079	18
42	17838	15079	17844	15081	17850	15083	17856	15085	17862	15088	17868	15090	17
43	17868	15090	17874	15092	17880	15094	17886	15096	17892	15098	17898	15100	16
44	17898	15100	17904	15102	17910	15104	17916	15106	17922	15108	17928	15110	15
45	17928	15110	17934	15113	17940	15115	17946	15117	17952	15119	17958	15121	14
46	17958	15121	17964	15123	17970	15125	17976	15127	17982	15129	17988	15131	13
47	17988	15131	17994	15133	18000	15135	18006	15138	18012	15140	18018	15142	12
48	18018	15142	18024	15144	18030	15146	18036	15148	18042	15150	18048	15152	11
49	18048	15152	18053	15154	18059	15156	18065	15158	18071	15161	18077	15163	10
50	18077	15163	18083	15165	18089	15167	18095	15169	18101	15171	18107	15173	09
51	18107	15173	18113	15175	18119	15177	18125	15179	18131	15181	18137	15183	08
52	18137	15183	18143	15186	18149	15188	18155	15190	18161	15192	18167	15194	07
53	18167	15194	18173	15196	18179	15198	18185	15200	18191	15202	18197	15204	06
54	18197	15204	18203	15206	18209	15209	18215	15211	18221	15213	18227	15215	05
55	18227	15215	18233	15217	18239	15219	18244	15221	18250	15223	18256	15225	04
56	18256	15225	18262	15227	18268	15229	18274	15232	18280	15234	18286	15236	03
57	18286	15236	18292	15238	18298	15240	18304	15242	18310	15244	18316	15246	02
58	18316	15246	18322	15248	18328	15250	18334	15252	18340	15255	18346	15257	01
59	18346	15257	18352	15259	18358	15261	18364	15263	18370	15265	18376	15267	00

·8   ·6   ·4   ·2   ·0 ←

PARTS for 0′·1 :—   LOGS. 3 ;   NATURALS, 1.

314°

46° HAVERSINES

′	·0 Log.	·0 Nat.	·2 Log.	·2 Nat.	·4 Log.	·4 Nat.	·6 Log.	·6 Nat.	·8 Log.	·8 Nat.	′
	9·	0·	9·	0·	9·	0·	9·	0·	9·	0·	
00	18376	15267	18382	15269	18388	15271	18393	15273	18399	15275	18405 15278	59
01	18405	15278	18411	15280	18417	15282	18423	15284	18429	15286	18435 15288	58
02	18435	15288	18441	15290	18447	15292	18453	15294	18459	15296	18465 15298	57
03	18465	15298	18471	15301	18477	15303	18483	15305	18489	15307	18495 15309	56
04	18495	15309	18501	15311	18506	15313	18512	15315	18518	15317	18524 15319	55
05	18524	15319	18530	15322	18536	15324	18542	15326	18548	15328	18554 15330	54
06	18554	15330	18560	15332	18566	15334	18572	15336	18578	15338	18584 15340	53
07	18584	15340	18590	15342	18596	15345	18601	15347	18607	15349	18613 15351	52
08	18613	15351	18619	15353	18625	15355	18631	15357	18637	15359	18643 15361	51
09	18643	15361	18649	15363	18655	15366	18661	15368	18667	15370	18673 15372	50
10	18673	15372	18679	15374	18684	15376	18690	15378	18696	15380	18702 15382	49
11	18702	15382	18708	15384	18714	15387	18720	15389	18726	15391	18732 15393	48
12	18732	15393	18738	15395	18744	15397	18750	15399	18756	15401	18762 15403	47
13	18762	15403	18767	15405	18773	15408	18779	15410	18785	15412	18791 15414	46
14	18791	15414	18797	15416	18803	15418	18809	15420	18815	15422	18821 15424	45
15	18821	15424	18827	15426	18832	15429	18838	15431	18844	15433	18850 15435	44
16	18850	15435	18856	15437	18862	15439	18868	15441	18874	15443	18880 15445	43
17	18880	15445	18886	15447	18892	15450	18898	15452	18903	15454	18909 15456	42
18	18909	15456	18915	15458	18921	15460	18927	15462	18933	15464	18939 15466	41
19	18939	15466	18945	15469	18951	15471	18957	15473	18963	15475	18968 15477	40
20	18968	15477	18974	15479	18980	15481	18986	15483	18992	15485	18998 15487	39
21	18998	15487	19004	15490	19010	15492	19016	15494	19022	15496	19027 15498	38
22	19027	15498	19033	15500	19039	15502	19045	15504	19051	15506	19057 15509	37
23	19057	15509	19063	15511	19069	15513	19075	15515	19081	15517	19086 15519	36
24	19086	15519	19092	15521	19098	15523	19104	15525	19110	15527	19116 15530	35
25	19116	15530	19122	15532	19128	15534	19134	15536	19140	15538	19145 15540	34
26	19145	15540	19151	15542	19157	15544	19163	15546	19169	15549	19175 15551	33
27	19175	15551	19181	15553	19187	15555	19192	15557	19198	15559	19204 15561	32
28	19204	15561	19210	15563	19216	15565	19222	15568	19228	15570	19234 15572	31
29	19234	15572	19239	15574	19245	15576	19251	15578	19257	15580	19263 15582	30
30	19263	15582	19269	15584	19275	15586	19281	15589	19287	15591	19292 15593	29
31	19292	15593	19298	15595	19304	15597	19310	15599	19316	15601	19322 15603	28
32	19322	15603	19328	15606	19334	15608	19339	15610	19345	15612	19351 15614	27
33	19351	15614	19357	15616	19363	15618	19369	15620	19375	15622	19381 15625	26
34	19381	15625	19386	15627	19392	15629	19398	15631	19404	15633	19410 15635	25
35	19410	15635	19416	15637	19422	15639	19428	15641	19433	15643	19439 15646	24
36	19439	15646	19445	15648	19451	15650	19457	15652	19463	15654	19469 15656	23
37	19469	15656	19474	15658	19480	15660	19486	15663	19492	15665	19498 15667	22
38	19498	15667	19504	15669	19510	15671	19516	15673	19521	15675	19527 15677	21
39	19527	15677	19533	15679	19539	15682	19545	15684	19551	15686	19557 15688	20
40	19557	15688	19562	15690	19568	15692	19574	15694	19580	15696	19586 15699	19
41	19586	15699	19592	15701	19597	15703	19603	15705	19609	15707	19615 15709	18
42	19615	15709	19621	15711	19627	15713	19633	15715	19639	15718	19644 15720	17
43	19644	15720	19650	15722	19656	15724	19662	15726	19668	15728	19674 15730	16
44	19674	15730	19679	15732	19685	15734	19691	15737	19697	15739	19703 15741	15
45	19703	15741	19709	15743	19714	15745	19720	15747	19726	15749	19732 15751	14
46	19732	15751	19738	15754	19744	15756	19750	15758	19755	15760	19761 15762	13
47	19761	15762	19767	15764	19773	15766	19779	15768	19785	15771	19790 15773	12
48	19790	15773	19796	15775	19802	15777	19808	15779	19814	15781	19820 15783	11
49	19820	15783	19825	15785	19831	15787	19837	15790	19843	15792	19849 15794	10
50	19849	15794	19855	15796	19860	15798	19866	15800	19872	15802	19878 15804	09
51	19878	15804	19884	15807	19890	15809	19896	15811	19901	15813	19907 15815	08
52	19907	15815	19913	15817	19919	15819	19925	15821	19930	15824	19936 15826	07
53	19936	15826	19942	15828	19948	15830	19954	15832	19960	15834	19965 15836	06
54	19965	15836	19971	15838	19977	15841	19983	15843	19989	15845	19995 15847	05
55	19995	15847	20000	15849	20006	15851	20012	15853	20018	15855	20024 15858	04
56	20024	15858	20029	15860	20035	15862	20041	15864	20047	15866	20053 15868	03
57	20053	15868	20059	15870	20064	15872	20070	15875	20076	15877	20082 15879	02
58	20082	15879	20088	15881	20093	15883	20099	15885	20105	15887	20111 15889	01
59	20111	15889	20117	15892	20122	15894	20128	15896	20134	15898	20140 15900	00

| | ·8 | | ·6 | | ·4 | | ·2 | | ·0 | |

PARTS for 0′·1 :— LOGS. 3 ; NATURALS, 1.

313°

47° HAVERSINES

	·0		·2		·4		·6		·8			
	Log.	Nat.	Log.	Nat.	Log.	Nat.	Log.	Nat.	Log.	Nat.	Log.	Nat.
′	9·	0·	9·	0·	9·	0·	9·	0·	9·	0·	9·	0·	′
00	20140	15900	20146	15902	20152	15904	20157	15906	20163	15909	20169	15911	59
01	20169	15911	20175	15913	20181	15915	20186	15917	20192	15919	20198	15921	58
02	20198	15921	20204	15923	20210	15926	20215	15928	20221	15930	20227	15932	57
03	20227	15932	20233	15934	20239	15936	20244	15938	20250	15941	20256	15943	56
04	20256	15943	20262	15945	20268	15947	20273	15949	20279	15951	20285	15953	55
05	20285	15953	20291	15955	20297	15958	20302	15960	20308	15962	20314	15964	54
06	20314	15964	20320	15966	20326	15968	20331	15970	20337	15972	20343	15975	53
07	20343	15975	20349	15977	20355	15979	20360	15981	20366	15983	20372	15985	52
08	20372	15985	20378	15987	20384	15990	20389	15992	20395	15994	20401	15996	51
09	20401	15996	20407	15998	20413	16000	20418	16002	20424	16004	20430	16007	50
10	20430	16007	20436	16009	20442	16011	20447	16013	20453	16015	20459	16017	49
11	20459	16017	20465	16019	20470	16022	20476	16024	20482	16026	20488	16028	48
12	20488	16028	20494	16030	20499	16032	20505	16034	20511	16036	20517	16039	47
13	20517	16039	20522	16041	20528	16043	20534	16045	20540	16047	20546	16049	46
14	20546	16049	20551	16051	20557	16054	20563	16056	20569	16058	20574	16060	45
15	20574	16060	20580	16062	20586	16064	20592	16066	20598	16069	20603	16071	44
16	20603	16071	20609	16073	20615	16075	20621	16077	20626	16079	20632	16081	43
17	20632	16081	20638	16083	20644	16086	20649	16088	20655	16090	20661	16092	42
18	20661	16092	20667	16094	20673	16096	20678	16098	20684	16101	20690	16103	41
19	20690	16103	20696	16105	20701	16107	20707	16109	20713	16111	20719	16113	40
20	20719	16113	20725	16116	20730	16118	20736	16120	20742	16122	20748	16124	39
21	20748	16124	20753	16126	20759	16128	20765	16131	20771	16133	20776	16135	38
22	20776	16135	20782	16137	20788	16139	20794	16141	20799	16143	20805	16146	37
23	20805	16146	20811	16148	20817	16150	20822	16152	20828	16154	20834	16156	36
24	20834	16156	20840	16158	20845	16160	20851	16163	20857	16165	20863	16167	35
25	20863	16167	20868	16169	20874	16171	20880	16173	20886	16175	20891	16178	34
26	20891	16178	20897	16180	20903	16182	20909	16184	20914	16186	20920	16188	33
27	20920	16188	20926	16190	20932	16193	20937	16195	20943	16197	20949	16199	32
28	20949	16199	20955	16201	20960	16203	20966	16205	20972	16208	20978	16210	31
29	20978	16210	20983	16212	20989	16214	20995	16216	21001	16218	21006	16220	30
30	21006	16220	21012	16223	21018	16225	21024	16227	21029	16229	21035	16231	29
31	21035	16231	21041	16233	21047	16236	21052	16238	21058	16240	21064	16242	28
32	21064	16242	21070	16244	21075	16246	21081	16248	21087	16251	21092	16253	27
33	21092	16253	21098	16255	21104	16257	21110	16259	21115	16261	21121	16263	26
34	21121	16263	21127	16266	21133	16268	21138	16270	21144	16272	21150	16274	25
35	21150	16274	21156	16276	21161	16278	21167	16281	21173	16283	21178	16285	24
36	21178	16285	21184	16287	21190	16289	21196	16291	21201	16293	21207	16296	23
37	21207	16296	21213	16298	21219	16300	21224	16302	21230	16304	21236	16306	22
38	21236	16306	21242	16309	21247	16311	21253	16313	21259	16315	21264	16317	21
39	21264	16317	21270	16319	21276	16321	21282	16324	21287	16326	21293	16328	20
40	21293	16328	21299	16330	21304	16332	21310	16334	21316	16336	21322	16339	19
41	21322	16339	21327	16341	21333	16343	21339	16345	21344	16347	21350	16349	18
42	21350	16349	21356	16352	21362	16354	21367	16356	21373	16358	21379	16360	17
43	21379	16360	21384	16362	21390	16364	21396	16367	21401	16369	21407	16371	16
44	21407	16371	21413	16373	21419	16375	21424	16377	21430	16380	21436	16382	15
45	21436	16382	21442	16384	21447	16386	21453	16388	21459	16390	21464	16392	14
46	21464	16392	21470	16395	21476	16397	21481	16399	21487	16401	21493	16403	13
47	21493	16403	21499	16405	21504	16408	21510	16410	21516	16412	21521	16414	12
48	21521	16414	21527	16416	21533	16418	21538	16420	21544	16423	21550	16425	11
49	21550	16425	21556	16427	21561	16429	21567	16431	21573	16433	21578	16436	10
50	21578	16436	21584	16438	21590	16440	21595	16442	21601	16444	21607	16446	09
51	21607	16446	21613	16448	21618	16451	21624	16453	21630	16455	21635	16457	08
52	21635	16457	21641	16459	21647	16461	21652	16464	21658	16466	21664	16468	07
53	21664	16468	21669	16470	21675	16472	21681	16474	21687	16477	21692	16479	06
54	21692	16479	21698	16481	21704	16483	21709	16485	21715	16487	21721	16489	05
55	21721	16489	21726	16492	21732	16494	21738	16496	21743	16498	21749	16500	04
56	21749	16500	21755	16502	21760	16505	21766	16507	21772	16509	21777	16511	03
57	21777	16511	21783	16513	21789	16515	21794	16518	21800	16520	21806	16522	02
58	21806	16522	21811	16524	21817	16526	21823	16528	21829	16530	21834	16533	01
59	21834	16533	21840	16535	21846	16537	21851	16539	21857	16541	21863	16543	00
			·8		·6		·4		·2		·0		

PARTS for 0′·1 :— LOGS. 3; NATURALS, 1.

312°

HAVERSINES

48°

′	·0 Log.	·0 Nat.	·2 Log.	·2 Nat.	·4 Log.	·4 Nat.	·6 Log.	·6 Nat.	·8 Log.	·8 Nat.	Log.	Nat.	′
	9·	0·	9·	0·	9·	0·	9·	0·	9·	0·	9·	0·	
00	21863	16543	21868	16546	21874	16548	21880	16550	21885	16552	21891	16554	59
01	21891	16554	21897	16556	21902	16559	21908	16561	21914	16563	21919	16565	58
02	21919	16565	21925	16567	21931	16569	21936	16572	21942	16574	21948	16576	57
03	21948	16576	21953	16578	21959	16580	21965	16582	21970	16585	21976	16587	56
04	21976	16587	21982	16589	21987	16591	21993	16593	21999	16595	22004	16598	55
05	22004	16598	22010	16600	22016	16602	22021	16604	22027	16606	22033	16608	54
06	22033	16608	22038	16611	22044	16613	22050	16615	22055	16617	22061	16619	53
07	22061	16619	22067	16621	22072	16624	22078	16626	22084	16628	22089	16630	52
08	22089	16630	22095	16632	22101	16634	22106	16637	22112	16639	22118	16641	51
09	22118	16641	22123	16643	22129	16645	22135	16647	22140	16650	22146	16652	50
10	22146	16652	22151	16654	22157	16656	22163	16658	22169	16660	22174	16663	49
11	22174	16663	22180	16665	22185	16667	22191	16669	22197	16671	22202	16673	48
12	22202	16673	22208	16676	22214	16678	22219	16680	22225	16682	22231	16684	47
13	22231	16684	22236	16686	22242	16689	22248	16691	22253	16693	22259	16695	46
14	22259	16695	22264	16697	22270	16699	22276	16702	22281	16704	22287	16706	45
15	22287	16706	22293	16708	22298	16710	22304	16712	22310	16715	22315	16717	44
16	22315	16717	22321	16719	22326	16721	22332	16723	22338	16725	22343	16728	43
17	22343	16728	22349	16730	22355	16732	22360	16734	22366	16736	22372	16738	42
18	22372	16738	22377	16741	22383	16743	22389	16745	22394	16747	22400	16749	41
19	22400	16749	22405	16752	22411	16754	22417	16756	22422	16758	22428	16760	40
20	22428	16760	22434	16762	22439	16765	22445	16767	22450	16769	22456	16771	39
21	22456	16771	22462	16773	22467	16775	22473	16778	22479	16780	22484	16782	38
22	22484	16782	22490	16784	22495	16786	22501	16788	22507	16791	22512	16793	37
23	22512	16793	22518	16795	22524	16797	22529	16799	22535	16802	22540	16804	36
24	22540	16804	22546	16806	22552	16808	22557	16810	22563	16812	22569	16815	35
25	22569	16815	22574	16817	22580	16819	22585	16821	22591	16823	22597	16825	34
26	22597	16825	22602	16828	22608	16830	22614	16832	22619	16834	22625	16836	33
27	22625	16836	22630	16839	22636	16841	22642	16843	22647	16845	22653	16847	32
28	22653	16847	22658	16849	22664	16852	22670	16854	22675	16856	22681	16858	31
29	22681	16858	22686	16860	22692	16862	22698	16865	22703	16867	22709	16869	30
30	22709	16869	22715	16871	22720	16873	22726	16876	22731	16878	22737	16880	29
31	22737	16880	22743	16882	22748	16884	22754	16886	22759	16889	22765	16891	28
32	22765	16891	22771	16893	22776	16895	22782	16897	22787	16900	22793	16902	27
33	22793	16902	22799	16904	22804	16906	22810	16908	22815	16910	22821	16913	26
34	22821	16913	22827	16915	22832	16917	22838	16919	22843	16921	22849	16924	25
35	22849	16924	22854	16926	22860	16928	22866	16930	22871	16932	22877	16934	24
36	22877	16934	22883	16937	22888	16939	22894	16941	22899	16943	22905	16945	23
37	22905	16945	22911	16947	22916	16950	22922	16952	22927	16954	22933	16956	22
38	22933	16956	22939	16958	22944	16961	22950	16963	22955	16965	22961	16967	21
39	22961	16967	22966	16969	22972	16972	22978	16974	22983	16976	22989	16978	20
40	22989	16978	22994	16980	23000	16982	23006	16985	23011	16987	23017	16989	19
41	23017	16989	23022	16991	23028	16993	23033	16996	23039	16998	23045	17000	18
42	23045	17000	23050	17002	23056	17004	23061	17006	23067	17009	23073	17011	17
43	23073	17011	23078	17013	23084	17015	23089	17017	23095	17020	23100	17022	16
44	23100	17022	23106	17024	23112	17026	23117	17028	23123	17030	23128	17033	15
45	23128	17033	23134	17035	23140	17037	23145	17039	23151	17041	23156	17044	14
46	23156	17044	23162	17046	23167	17048	23173	17050	23179	17052	23184	17055	13
47	23184	17055	23190	17057	23195	17059	23201	17061	23206	17063	23212	17066	12
48	23212	17066	23218	17068	23223	17070	23229	17072	23234	17074	23240	17076	11
49	23240	17076	23245	17079	23251	17081	23256	17083	23262	17085	23268	17087	10
50	23268	17087	23273	17090	23279	17092	23284	17094	23290	17096	23295	17098	09
51	23295	17098	23301	17101	23307	17103	23312	17105	23318	17107	23323	17109	08
52	23323	17109	23329	17112	23334	17114	23340	17116	23345	17118	23351	17120	07
53	23351	17120	23357	17122	23362	17125	23368	17127	23373	17129	23379	17131	06
54	23379	17131	23384	17133	23390	17136	23395	17138	23401	17140	23407	17142	05
55	23407	17142	23412	17144	23418	17147	23423	17149	23429	17151	23434	17153	04
56	23434	17153	23440	17155	23446	17158	23451	17160	23457	17162	23462	17164	03
57	23462	17164	23468	17166	23473	17169	23479	17171	23484	17173	23490	17175	02
58	23490	17175	23496	17177	23501	17179	23507	17182	23512	17184	23518	17186	01
59	23518	17186	23523	17188	23529	17190	23534	17193	23540	17195	23545	17197	00

| | ·8 | | ·6 | | ·4 | | ·2 | | ·0 | | | | |

311°

PARTS for 0′·1 :—   LOGS. 3;   NATURALS, 1.

49° HAVERSINES

′	·0 Log. 9·	·0 Nat. 0·	·2 Log. 9·	·2 Nat. 0·	·4 Log. 9·	·4 Nat. 0·	·6 Log. 9·	·6 Nat. 0·	·8 Log. 9·	·8 Nat. 0·	Log. 9·	Nat. 0·	′
00	23545	17197	23551	17199	23557	17201	23562	17204	23568	17206	23573	17208	59
01	23573	17208	23579	17210	23584	17212	23590	17215	23595	17217	23601	17219	58
02	23601	17219	23606	17221	23612	17223	23617	17226	23623	17228	23629	17230	57
03	23629	17230	23634	17232	23640	17234	23645	17237	23651	17239	23656	17241	56
04	23656	17241	23662	17243	23667	17245	23673	17248	23678	17250	23684	17252	55
05	23684	17252	23689	17254	23695	17256	23700	17259	23706	17261	23712	17263	54
06	23712	17263	23717	17265	23723	17267	23728	17270	23734	17272	23739	17274	53
07	23739	17274	23745	17276	23750	17278	23756	17281	23761	17283	23767	17285	52
08	23767	17285	23772	17287	23778	17289	23783	17292	23789	17294	23794	17296	51
09	23794	17296	23800	17298	23805	17300	23811	17303	23816	17305	23822	17307	50
10	23822	17307	23828	17309	23833	17311	23839	17314	23844	17316	23850	17318	49
11	23850	17318	23855	17320	23861	17322	23866	17325	23872	17327	23877	17329	48
12	23877	17329	23883	17331	23888	17333	23894	17336	23899	17338	23905	17340	47
13	23905	17340	23910	17342	23916	17344	23921	17347	23927	17349	23932	17351	46
14	23932	17351	23938	17353	23943	17355	23949	17358	23954	17360	23960	17362	45
15	23960	17362	23966	17364	23971	17366	23977	17369	23982	17371	23988	17373	44
16	23988	17373	23993	17375	23999	17377	24004	17380	24010	17382	24015	17384	43
17	24015	17384	24021	17386	24026	17388	24032	17391	24037	17393	24043	17395	42
18	24043	17395	24048	17397	24054	17400	24059	17402	24065	17404	24070	17406	41
19	24070	17406	24076	17408	24081	17411	24087	17413	24092	17415	24098	17417	40
20	24098	17417	24103	17419	24109	17422	24114	17424	24120	17426	24125	17428	39
21	24125	17428	24131	17430	24136	17433	24142	17435	24147	17437	24153	17439	38
22	24153	17439	24158	17441	24164	17444	24169	17446	24175	17448	24180	17450	37
23	24180	17450	24186	17452	24191	17455	24197	17457	24202	17459	24208	17461	36
24	24208	17461	24213	17464	24219	17466	24224	17468	24230	17470	24235	17472	35
25	24235	17472	24241	17475	24246	17477	24252	17479	24257	17481	24263	17483	34
26	24263	17483	24268	17486	24274	17488	24279	17490	24285	17492	24290	17494	33
27	24290	17494	24295	17497	24301	17499	24306	17501	24312	17503	24317	17505	32
28	24317	17505	24323	17508	24328	17510	24334	17512	24339	17514	24345	17517	31
29	24345	17517	24350	17519	24356	17521	24361	17523	24367	17525	24372	17528	30
30	24372	17528	24378	17530	24383	17532	24389	17534	24394	17536	24400	17539	29
31	24400	17539	24405	17541	24411	17543	24416	17545	24422	17547	24427	17550	28
32	24427	17550	24432	17552	24438	17554	24443	17556	24449	17559	24454	17561	27
33	24454	17561	24460	17563	24465	17565	24471	17568	24476	17570	24482	17572	26
34	24482	17572	24487	17574	24493	17576	24498	17579	24504	17581	24509	17583	25
35	24509	17583	24515	17585	24520	17587	24526	17590	24531	17592	24536	17594	24
36	24536	17594	24542	17596	24547	17598	24553	17601	24558	17603	24564	17605	23
37	24564	17605	24569	17607	24575	17610	24580	17612	24586	17614	24591	17616	22
38	24591	17616	24597	17618	24602	17621	24608	17623	24613	17625	24618	17627	21
39	24618	17627	24624	17629	24629	17632	24635	17634	24640	17636	24646	17638	20
40	24646	17638	24651	17641	24657	17643	24662	17645	24668	17647	24673	17649	19
41	24673	17649	24678	17652	24684	17654	24689	17656	24695	17658	24700	17661	18
42	24700	17661	24706	17663	24711	17665	24717	17667	24722	17669	24728	17672	17
43	24728	17672	24733	17674	24738	17676	24744	17678	24749	17680	24755	17683	16
44	24755	17683	24760	17685	24766	17687	24771	17689	24777	17692	24782	17694	15
45	24782	17694	24788	17696	24793	17698	24799	17700	24804	17703	24809	17705	14
46	24809	17705	24815	17707	24820	17709	24826	17712	24831	17714	24837	17716	13
47	24837	17716	24842	17718	24847	17720	24853	17723	24858	17725	24864	17727	12
48	24864	17727	24869	17729	24875	17732	24880	17734	24886	17736	24891	17738	11
49	24891	17738	24896	17740	24902	17743	24907	17745	24913	17747	24918	17749	10
50	24918	17749	24924	17752	24929	17754	24935	17756	24940	17758	24945	17760	09
51	24945	17760	24951	17763	24956	17765	24962	17767	24967	17769	24973	17772	08
52	24973	17772	24978	17774	24983	17776	24989	17778	24994	17780	25000	17783	07
53	25000	17783	25005	17785	25011	17787	25016	17789	25022	17792	25027	17794	06
54	25027	17794	25032	17796	25038	17798	25043	17800	25049	17803	25054	17805	05
55	25054	17805	25059	17807	25065	17809	25070	17812	25076	17814	25081	17816	04
56	25081	17816	25087	17818	25092	17821	25098	17823	25103	17825	25108	17827	03
57	25108	17827	25114	17829	25119	17832	25125	17834	25130	17836	25135	17838	02
58	25135	17838	25141	17841	25146	17843	25152	17845	25157	17847	25163	17849	01
59	25163	17849	25168	17852	25173	17854	25179	17856	25184	17858	25190	17861	00

·8	·6	·4	·2	·0 ←

310°

PARTS for 0′·1 :—   LOGS. 3 ;   NATURALS, 1.

50° HAVERSINES

′	·0 Log.	·0 Nat.	·2 Log.	·2 Nat.	·4 Log.	·4 Nat.	·6 Log.	·6 Nat.	·8 Log.	·8 Nat.	Log.	Nat.	′
	9·	0·	9·	0·	9·	0·	9·	0·	9·	0·	9·	0·	
00	25190	17861	25195	17863	25200	17865	25206	17867	25211	17870	25217	17872	59
01	25217	17872	25222	17874	25228	17876	25233	17878	25238	17881	25244	17883	58
02	25244	17883	25249	17885	25255	17887	25260	17890	25265	17892	25271	17894	57
03	25271	17894	25276	17896	25282	17899	25287	17901	25293	17903	25298	17905	56
04	25298	17905	25303	17907	25309	17910	25314	17912	25320	17914	25325	17916	55
05	25325	17916	25330	17919	25336	17921	25341	17923	25347	17925	25352	17928	54
06	25352	17928	25357	17930	25363	17932	25368	17934	25374	17936	25379	17939	53
07	25379	17939	25384	17941	25390	17943	25395	17945	25401	17948	25406	17950	52
08	25406	17950	25411	17952	25417	17954	25422	17957	25428	17959	25433	17961	51
09	25433	17961	25438	17963	25444	17965	25449	17968	25455	17970	25460	17972	50
10	25460	17972	25465	17974	25471	17977	25476	17979	25482	17981	25487	17983	49
11	25487	17983	25492	17986	25498	17988	25503	17990	25509	17992	25514	17995	48
12	25514	17995	25519	17997	25525	17999	25530	18001	25536	18003	25541	18006	47
13	25541	18006	25546	18008	25552	18010	25557	18012	25563	18015	25568	18017	46
14	25568	18017	25573	18019	25579	18021	25584	18024	25589	18026	25595	18028	45
15	25595	18028	25600	18030	25606	18033	25611	18035	25616	18037	25622	18039	44
16	25622	18039	25627	18041	25633	18044	25638	18046	25643	18048	25649	18050	43
17	25649	18050	25654	18053	25660	18055	25665	18057	25670	18059	25676	18062	42
18	25676	18062	25681	18064	25686	18066	25692	18068	25697	18071	25703	18073	41
19	25703	18073	25708	18075	25713	18077	25719	18080	25724	18082	25729	18084	40
20	25729	18084	25735	18086	25740	18088	25746	18091	25751	18093	25756	18095	39
21	25756	18095	25762	18097	25767	18100	25773	18102	25778	18104	25783	18106	38
22	25783	18106	25789	18109	25794	18111	25799	18113	25805	18115	25810	18118	37
23	25810	18118	25815	18120	25821	18122	25826	18124	25832	18127	25837	18129	36
24	25837	18129	25842	18131	25848	18133	25853	18136	25858	18138	25864	18140	35
25	25864	18140	25869	18142	25875	18144	25880	18147	25885	18149	25891	18151	34
26	25891	18151	25896	18153	25901	18156	25907	18158	25912	18160	25917	18162	33
27	25917	18162	25923	18165	25928	18167	25933	18169	25939	18171	25944	18174	32
28	25944	18174	25950	18176	25955	18178	25960	18180	25966	18183	25971	18185	31
29	25971	18185	25976	18187	25982	18189	25987	18192	25992	18194	25998	18196	30
30	25998	18196	26003	18198	26009	18201	26014	18203	26019	18205	26025	18207	29
31	26025	18207	26030	18210	26035	18212	26041	18214	26046	18216	26051	18219	28
32	26051	18219	26057	18221	26062	18223	26067	18225	26073	18228	26078	18230	27
33	26078	18230	26084	18232	26089	18234	26094	18237	26099	18239	26105	18241	26
34	26105	18241	26110	18243	26116	18246	26121	18248	26126	18250	26132	18252	25
35	26132	18252	26137	18254	26142	18257	26148	18259	26153	18261	26158	18263	24
36	26158	18263	26164	18266	26169	18268	26174	18270	26180	18272	26185	18275	23
37	26185	18275	26190	18277	26196	18279	26201	18281	26206	18284	26212	18286	22
38	26212	18286	26217	18288	26222	18290	26228	18293	26233	18295	26238	18297	21
39	26238	18297	26244	18299	26249	18302	26254	18304	26260	18306	26265	18308	20
40	26265	18308	26271	18311	26276	18313	26281	18315	26286	18317	26292	18320	19
41	26292	18320	26297	18322	26303	18324	26308	18326	26313	18329	26319	18331	18
42	26319	18331	26324	18333	26329	18335	26335	18338	26340	18340	26345	18342	17
43	26345	18342	26351	18344	26356	18347	26361	18349	26366	18351	26372	18353	16
44	26372	18353	26377	18356	26382	18358	26388	18360	26393	18362	26398	18365	15
45	26398	18365	26404	18367	26409	18369	26414	18372	26420	18374	26425	18376	14
46	26425	18376	26430	18378	26436	18381	26441	18383	26446	18385	26452	18387	13
47	26452	18387	26457	18390	26462	18392	26468	18394	26473	18396	26478	18399	12
48	26478	18399	26484	18401	26489	18403	26494	18405	26500	18408	26505	18410	11
49	26505	18410	26510	18412	26516	18414	26521	18417	26526	18419	26532	18421	10
50	26532	18421	26537	18423	26542	18426	26547	18428	26553	18430	26558	18432	09
51	26558	18432	26563	18435	26569	18437	26574	18439	26579	18441	26585	18444	08
52	26585	18444	26590	18446	26595	18448	26601	18450	26606	18453	26611	18455	07
53	26611	18455	26617	18457	26622	18459	26627	18462	26632	18464	26638	18466	06
54	26638	18466	26643	18468	26648	18471	26654	18473	26659	18475	26664	18478	05
55	26664	18478	26670	18480	26675	18482	26680	18484	26686	18487	26691	18489	04
56	26691	18489	26696	18491	26701	18493	26707	18496	26712	18498	26717	18500	03
57	26717	18500	26723	18502	26728	18505	26733	18507	26739	18509	26744	18511	02
58	26744	18511	26749	18514	26754	18516	26760	18518	26765	18520	26770	18523	01
59	26770	18523	26776	18525	26781	18527	26786	18529	26792	18532	26797	18534	00

| | ·8 | | ·6 | | ·4 | | ·2 | | ·0 | | | | |

309°

PARTS for 0′·1 :— LOGS. 3 ; NATURALS, 1.

51°   HAVERSINES

′	·0 Log	·0 Nat	·2 Log	·2 Nat	·4 Log	·4 Nat	·6 Log	·6 Nat	·8 Log	·8 Nat	Log	Nat	′
	9·	0·	9·	0·	9·	0·	9·	0·	9·	0·	9·	0·	
00	26797	18534	26802	18536	26807	18539	26813	18541	26818	18543	26823	18545	59
01	26823	18545	26829	18548	26834	18550	26839	18552	26845	18554	26850	18557	58
02	26850	18557	26855	18559	26860	18561	26866	18563	26871	18566	26876	18568	57
03	26876	18568	26882	18570	26887	18572	26892	18575	26897	18577	26903	18579	56
04	26903	18579	26908	18581	26913	18584	26919	18586	26924	18588	26929	18591	55
05	26929	18591	26934	18593	26940	18595	26945	18597	26950	18600	26956	18602	54
06	26956	18602	26961	18604	26966	18606	26971	18609	26977	18611	26982	18613	53
07	26982	18613	26987	18615	26993	18618	26998	18620	27003	18622	27008	18624	52
08	27008	18624	27014	18627	27019	18629	27024	18631	27030	18634	27035	18636	51
09	27035	18636	27040	18638	27045	18640	27051	18643	27056	18645	27061	18647	50
10	27061	18647	27066	18649	27072	18652	27077	18654	27082	18656	27088	18658	49
11	27088	18658	27093	18661	27098	18663	27103	18665	27109	18668	27114	18670	48
12	27114	18670	27119	18672	27125	18674	27130	18677	27135	18679	27140	18681	47
13	27140	18681	27146	18683	27151	18686	27156	18688	27162	18690	27167	18692	46
14	27167	18692	27172	18695	27177	18697	27183	18699	27188	18702	27193	18704	45
15	27193	18704	27198	18706	27204	18708	27209	18711	27214	18713	27219	18715	44
16	27219	18715	27225	18717	27230	18720	27235	18722	27241	18724	27246	18727	43
17	27246	18727	27251	18729	27256	18731	27262	18733	27267	18736	27272	18738	42
18	27272	18738	27277	18740	27283	18742	27288	18745	27293	18747	27298	18749	41
19	27298	18749	27304	18751	27309	18754	27314	18756	27319	18758	27325	18761	40
20	27325	18761	27330	18763	27335	18765	27340	18767	27346	18770	27351	18772	39
21	27351	18772	27356	18774	27361	18776	27367	18779	27372	18781	27377	18783	38
22	27377	18783	27382	18786	27388	18788	27393	18790	27398	18792	27403	18795	37
23	27403	18795	27409	18797	27414	18799	27419	18801	27424	18804	27430	18806	36
24	27430	18806	27435	18808	27440	18811	27445	18813	27451	18815	27456	18817	35
25	27456	18817	27461	18820	27466	18822	27472	18824	27477	18826	27482	18829	34
26	27482	18829	27487	18831	27493	18833	27498	18836	27503	18838	27508	18840	33
27	27508	18840	27514	18842	27519	18845	27524	18847	27529	18849	27535	18852	32
28	27535	18852	27540	18854	27545	18856	27550	18858	27556	18861	27561	18863	31
29	27561	18863	27566	18865	27571	18867	27577	18870	27582	18872	27587	18874	30
30	27587	18874	27592	18877	27597	18879	27603	18881	27608	18883	27613	18886	29
31	27613	18886	27618	18888	27624	18890	27629	18892	27634	18895	27639	18897	28
32	27639	18897	27645	18899	27650	18902	27655	18904	27660	18906	27666	18908	27
33	27666	18908	27671	18911	27676	18913	27681	18915	27686	18918	27692	18920	26
34	27692	18920	27697	18922	27702	18924	27707	18927	27713	18929	27718	18931	25
35	27718	18931	27723	18933	27728	18936	27734	18938	27739	18940	27744	18943	24
36	27744	18943	27749	18945	27754	18947	27760	18949	27765	18952	27770	18954	23
37	27770	18954	27775	18956	27781	18959	27786	18961	27791	18963	27796	18965	22
38	27796	18965	27801	18968	27807	18970	27812	18972	27817	18975	27822	18977	21
39	27822	18977	27828	18979	27833	18981	27838	18984	27843	18986	27848	18988	20
40	27848	18988	27854	18991	27859	18993	27864	18995	27869	18997	27875	19000	19
41	27875	19000	27880	19002	27885	19004	27890	19006	27895	19009	27901	19011	18
42	27901	19011	27906	19013	27911	19016	27916	19018	27921	19020	27927	19022	17
43	27927	19022	27932	19025	27937	19027	27942	19029	27948	19032	27953	19034	16
44	27953	19034	27958	19036	27963	19038	27968	19041	27974	19043	27979	19045	15
45	27979	19045	27984	19048	27989	19050	27994	19052	28000	19054	28005	19057	14
46	28005	19057	28010	19059	28015	19061	28020	19064	28026	19066	28031	19068	13
47	28031	19068	28036	19070	28041	19073	28046	19075	28052	19077	28057	19080	12
48	28057	19080	28062	19082	28067	19084	28072	19086	28078	19089	28083	19091	11
49	28083	19091	28088	19093	28093	19096	28098	19098	28104	19100	28109	19102	10
50	28109	19102	28114	19105	28119	19107	28124	19109	28130	19112	28135	19114	09
51	28135	19114	28140	19116	28145	19118	28150	19121	28156	19123	28161	19125	08
52	28161	19125	28166	19128	28171	19130	28176	19132	28182	19134	28187	19137	07
53	28187	19137	28193	19139	28197	19141	28202	19144	28208	19146	28213	19148	06
54	28213	19148	28218	19150	28223	19153	28228	19155	28234	19157	28239	19160	05
55	28239	19160	28244	19162	28249	19164	28254	19167	28260	19169	28265	19171	04
56	28265	19171	28270	19173	28275	19176	28280	19178	28285	19180	28291	19183	03
57	28291	19183	28296	19185	28301	19187	28306	19189	28311	19192	28317	19194	02
58	28317	19194	28322	19196	28327	19199	28332	19201	28337	19203	28342	19205	01
59	28342	19205	28348	19208	28353	19210	28358	19212	28363	19215	28368	19217	00

·8   ·6   ·4   ·2   ·0

PARTS for 0′·1 :—   LOGS. 3 ;   NATURALS, 1.

308°

52° HAVERSINES

′	·0 Log. 9·	·0 Nat. 0·	·2 Log. 9·	·2 Nat. 0·	·4 Log. 9·	·4 Nat. 0·	·6 Log. 9·	·6 Nat. 0·	·8 Log. 9·	·8 Nat. 0·	′
00	28368	19217	28374	19219	28379	19221	28384	19224	28389	19226	28394 19228	59
01	28394	19228	28399	19231	28405	19233	28410	19235	28415	19238	28420 19240	58
02	28420	19240	28425	19242	28431	19244	28436	19247	28441	19249	28446 19251	57
03	28446	19251	28451	19254	28456	19256	28462	19258	28467	19261	28472 19263	56
04	28472	19263	28477	19265	28482	19267	28487	19270	28493	19272	28498 19274	55
05	28498	19274	28503	19277	28508	19279	28513	19281	28518	19283	28524 19286	54
06	28524	19286	28529	19288	28534	19290	28539	19293	28544	19295	28549 19297	53
07	28549	19297	28555	19300	28560	19302	28565	19304	28570	19306	28575 19309	52
08	28575	19309	28581	19311	28586	19313	28591	19316	28596	19318	28601 19320	51
09	28601	19320	28606	19322	28612	19325	28617	19327	28622	19329	28627 19332	50
10	28627	19332	28632	19334	28637	19336	28642	19339	28648	19341	28653 19343	49
11	28653	19343	28658	19345	28663	19348	28668	19350	28673	19352	28679 19355	48
12	28679	19355	28684	19357	28689	19359	28694	19362	28699	19364	28704 19366	47
13	28704	19366	28710	19368	28715	19371	28720	19373	28725	19375	28730 19378	46
14	28730	19378	28735	19380	28740	19382	28745	19385	28751	19387	28756 19389	45
15	28756	19389	28761	19391	28766	19394	28771	19396	28776	19398	28782 19401	44
16	28782	19401	28787	19403	28792	19405	28797	19408	28802	19410	28807 19412	43
17	28807	19412	28813	19414	28818	19417	28823	19419	28828	19421	28833 19424	42
18	28833	19424	28838	19426	28843	19428	28848	19431	28854	19433	28859 19435	41
19	28859	19435	28864	19437	28869	19440	28874	19442	28879	19444	28885 19447	40
20	28885	19447	28890	19449	28895	19451	28900	19454	28905	19456	28910 19458	39
21	28910	19458	28915	19461	28921	19463	28926	19465	28931	19467	28936 19470	38
22	28936	19470	28941	19472	28946	19474	28951	19477	28956	19479	28962 19481	37
23	28962	19481	28967	19484	28972	19486	28977	19488	28982	19490	28987 19493	36
24	28987	19493	28992	19495	28998	19497	29003	19500	29008	19502	29013 19504	35
25	29013	19504	29018	19507	29023	19509	29028	19511	29033	19513	29039 19516	34
26	29039	19516	29044	19518	29049	19520	29054	19523	29059	19525	29064 19527	33
27	29064	19527	29069	19530	29075	19532	29080	19534	29085	19537	29090 19539	32
28	29090	19539	29095	19541	29100	19543	29105	19546	29110	19548	29116 19550	31
29	29116	19550	29121	19553	29126	19555	29131	19557	29136	19560	29141 19562	30
30	29141	19562	29146	19564	29151	19567	29157	19569	29162	19571	29167 19573	29
31	29167	19573	29172	19576	29177	19578	29182	19580	29187	19583	29192 19585	28
32	29192	19585	29198	19587	29203	19590	29208	19592	29213	19594	29218 19597	27
33	29218	19597	29223	19599	29228	19601	29233	19603	29238	19606	29244 19608	26
34	29244	19608	29249	19610	29254	19613	29259	19615	29264	19617	29269 19620	25
35	29269	19620	29274	19622	29279	19624	29285	19627	29290	19629	29295 19631	24
36	29295	19631	29300	19634	29305	19636	29310	19638	29315	19640	29320 19643	23
37	29320	19643	29325	19645	29330	19647	29336	19650	29341	19652	29346 19654	22
38	29346	19654	29351	19657	29356	19659	29361	19661	29366	19664	29371 19666	21
39	29371	19666	29376	19668	29382	19671	29387	19673	29392	19675	29397 19677	20
40	29397	19677	29402	19680	29407	19682	29412	19684	29417	19687	29422 19689	19
41	29422	19689	29428	19691	29433	19694	29438	19696	29443	19698	29448 19701	18
42	29448	19701	29453	19703	29458	19705	29463	19708	29468	19710	29473 19712	17
43	29473	19712	29478	19714	29484	19717	29489	19719	29494	19721	29499 19724	16
44	29499	19724	29504	19726	29509	19728	29514	19731	29519	19733	29524 19735	15
45	29524	19735	29529	19738	29535	19740	29540	19742	29545	19745	29550 19747	14
46	29550	19747	29555	19749	29560	19752	29565	19754	29570	19756	29575 19758	13
47	29575	19758	29580	19761	29586	19763	29591	19765	29596	19768	29601 19770	12
48	29601	19770	29606	19772	29611	19775	29616	19777	29621	19779	29626 19782	11
49	29626	19782	29631	19784	29636	19786	29641	19789	29647	19791	29652 19793	10
50	29652	19793	29657	19796	29662	19798	29667	19800	29672	19802	29677 19805	09
51	29677	19805	29682	19807	29687	19809	29692	19812	29697	19814	29702 19816	08
52	29702	19816	29707	19819	29713	19821	29718	19823	29723	19826	29728 19828	07
53	29728	19828	29733	19830	29738	19833	29743	19835	29748	19837	29753 19840	06
54	29753	19840	29758	19842	29763	19844	29768	19847	29774	19849	29779 19851	05
55	29779	19851	29784	19854	29789	19856	29794	19858	29799	19860	29804 19863	04
56	29804	19863	29809	19865	29814	19867	29819	19870	29824	19872	29829 19874	03
57	29829	19874	29834	19877	29840	19879	29845	19881	29850	19884	29855 19886	02
58	29855	19886	29860	19888	29865	19891	29870	19893	29875	19895	29880 19898	01
59	29880	19898	29885	19900	29890	19902	29895	19905	29900	19907	29905 19909	00

| | ·8 | | ·6 | | ·4 | | ·2 | | ·0 | |

307°

PARTS for 0′·1:—   LOGS. 3 ;   NATURALS, 1.

53° HAVERSINES

	·0		·2		·4		·6		·8				
	Log.	Nat.	Log.	Nat.	Log.	Nat.	Log.	Nat.	Log.	Nat.	Log.	Nat.	
	9·	0·	9·	0·	9·	0·	9·	0·	9·	0·	9·	0·	
00	29905	19909	29910	19912	29916	19914	29921	19916	29926	19919	29931	19921	59
01	29931	19921	29936	19923	29941	19926	29946	19928	29951	19930	29956	19932	58
02	29956	19932	29961	19935	29966	19937	29971	19939	29976	19942	29981	19944	57
03	29981	19944	29986	19946	29992	19949	29997	19951	30002	19953	30007	19956	56
04	30007	19956	30012	19958	30017	19960	30022	19963	30027	19965	30032	19967	55
05	30032	19967	30037	19970	30042	19972	30047	19974	30052	19977	30057	19979	54
06	30057	19979	30062	19981	30067	19984	30073	19986	30078	19988	30083	19991	53
07	30083	19991	30088	19993	30093	19995	30098	19998	30103	20000	30108	20002	52
08	30108	20002	30113	20005	30118	20007	30123	20009	30128	20012	30133	20014	51
09	30133	20014	30138	20016	30143	20019	30148	20021	30153	20023	30158	20026	50
10	30158	20026	30163	20028	30168	20030	30174	20033	30179	20035	30184	20037	49
11	30184	20037	30189	20040	30194	20042	30199	20044	30204	20046	30209	20049	48
12	30209	20049	30214	20051	30219	20053	30224	20056	30229	20058	30234	20060	47
13	30234	20060	30239	20063	30244	20065	30249	20067	30254	20070	30259	20072	46
14	30259	20072	30264	20074	30269	20077	30275	20079	30280	20081	30285	20084	45
15	30285	20084	30290	20086	30295	20088	30300	20091	30305	20093	30310	20095	44
16	30310	20095	30315	20098	30320	20100	30325	20102	30330	20105	30335	20107	43
17	30335	20107	30340	20109	30345	20112	30350	20114	30355	20116	30360	20119	42
18	30360	20119	30365	20121	30370	20123	30375	20126	30380	20128	30385	20130	41
19	30385	20130	30390	20133	30395	20135	30400	20137	30405	20140	30410	20142	40
20	30410	20142	30415	20144	30420	20147	30426	20149	30431	20151	30436	20154	39
21	30436	20154	30441	20156	30446	20158	30451	20161	30456	20163	30461	20165	38
22	30461	20165	30466	20168	30471	20170	30476	20172	30481	20175	30486	20177	37
23	30486	20177	30491	20179	30496	20182	30501	20184	30506	20186	30511	20189	36
24	30511	20189	30516	20191	30521	20193	30526	20196	30531	20198	30536	20200	35
25	30536	20200	30541	20203	30546	20205	30551	20207	30556	20210	30561	20212	34
26	30561	20212	30566	20214	30571	20217	30576	20219	30581	20221	30586	20224	33
27	30586	20224	30591	20226	30596	20228	30601	20231	30606	20233	30611	20235	32
28	30611	20235	30616	20238	30621	20240	30626	20242	30631	20245	30636	20247	31
29	30636	20247	30641	20249	30646	20252	30652	20254	30657	20257	30662	20259	30
30	30662	20259	30667	20261	30672	20264	30677	20266	30682	20268	30687	20271	29
31	30687	20271	30692	20273	30697	20275	30702	20278	30707	20280	30712	20282	28
32	30712	20282	30717	20285	30722	20287	30727	20289	30732	20292	30737	20294	27
33	30737	20294	30742	20296	30747	20299	30752	20301	30757	20303	30762	20306	26
34	30762	20306	30767	20308	30772	20310	30777	20313	30782	20315	30787	20317	25
35	30787	20317	30792	20320	30797	20322	30802	20324	30807	20327	30812	20329	24
36	30812	20329	30817	20331	30822	20334	30827	20336	30832	20338	30837	20341	23
37	30837	20341	30842	20343	30847	20345	30852	20348	30857	20350	30862	20352	22
38	30862	20352	30867	20355	30872	20357	30877	20360	30882	20362	30887	20364	21
39	30887	20364	30892	20367	30897	20369	30902	20371	30907	20374	30912	20376	20
40	30912	20376	30917	20378	30922	20381	30927	20383	30932	20385	30937	20388	19
41	30937	20388	30942	20390	30947	20392	30952	20395	30957	20397	30962	20399	18
42	30962	20399	30967	20402	30972	20404	30977	20406	30982	20409	30987	20411	17
43	30987	20411	30992	20413	30997	20416	31002	20418	31007	20420	31012	20423	16
44	31012	20423	31017	20425	31022	20427	31026	20430	31031	20432	31036	20435	15
45	31036	20435	31041	20437	31046	20439	31051	20442	31056	20444	31061	20446	14
46	31061	20446	31066	20449	31071	20451	31076	20453	31081	20456	31086	20458	13
47	31086	20458	31091	20460	31096	20463	31101	20465	31106	20467	31111	20470	12
48	31111	20470	31116	20472	31121	20474	31126	20477	31131	20479	31136	20481	11
49	31136	20481	31141	20484	31146	20486	31151	20488	31156	20491	31161	20493	10
50	31161	20493	31166	20496	31171	20498	31176	20500	31181	20503	31186	20505	09
51	31186	20505	31191	20507	31196	20510	31201	20512	31206	20514	31211	20517	08
52	31211	20517	31216	20519	31221	20521	31226	20524	31231	20526	31236	20528	07
53	31236	20528	31241	20531	31246	20533	31250	20536	31255	20538	31260	20540	06
54	31260	20540	31265	20543	31270	20545	31275	20547	31280	20550	31285	20552	05
55	31285	20552	31290	20554	31295	20557	31300	20559	31305	20561	31310	20564	04
56	31310	20564	31315	20566	31320	20568	31325	20571	31330	20573	31335	20575	03
57	31335	20575	31340	20578	31345	20580	31350	20582	31355	20585	31360	20587	02
58	31360	20587	31365	20590	31370	20592	31375	20594	31380	20597	31385	20599	01
59	31385	20599	31390	20601	31395	20604	31399	20606	31404	20608	31409	20611	00
	·8		·6		·4		·2		·0 ←				

PARTS for 0'·1 :—   LOGS. 3 (2 after 53° 50');   NATURALS, 1.

HAVERSINES

'	·0 Log.	·0 Nat.	·2 Log.	·2 Nat.	·4 Log.	·4 Nat.	·6 Log.	·6 Nat.	·8 Log.	·8 Nat.	Log.	Nat.	'
	9·	0·	9·	0·	9·	0·	9·	0·	9·	0·	9·	0·	
00	31409	20611	31414	20613	31419	20615	31424	20618	31429	20620	31434	20623	59
01	31434	20623	31439	20625	31444	20627	31449	20630	31454	20632	31459	20634	58
02	31459	20634	31464	20637	31469	20639	31474	20641	31479	20644	31484	20646	57
03	31484	20646	31489	20648	31494	20651	31498	20653	31503	20655	31508	20658	56
04	31508	20658	31513	20660	31518	20663	31523	20665	31528	20667	31533	20670	55
05	31533	20670	31538	20672	31543	20674	31548	20677	31553	20679	31558	20681	54
06	31558	20681	31563	20684	31568	20686	31573	20688	31578	20691	31583	20693	53
07	31583	20693	31588	20696	31593	20698	31597	20700	31602	20703	31607	20705	52
08	31607	20705	31612	20707	31617	20710	31622	20712	31627	20714	31632	20717	51
09	31632	20717	31637	20719	31642	20721	31647	20724	31652	20726	31657	20729	50
10	31657	20729	31662	20731	31667	20733	31672	20736	31677	20738	31682	20740	49
11	31682	20740	31687	20743	31691	20745	31696	20747	31701	20750	31706	20752	48
12	31706	20752	31711	20754	31716	20757	31721	20759	31726	20762	31731	20764	47
13	31731	20764	31736	20766	31741	20769	31746	20771	31751	20773	31756	20776	46
14	31756	20776	31761	20778	31765	20780	31770	20783	31775	20785	31780	20788	45
15	31780	20788	31785	20790	31790	20792	31795	20795	31800	20797	31805	20799	44
16	31805	20799	31810	20802	31815	20804	31820	20806	31825	20809	31830	20811	43
17	31830	20811	31835	20814	31839	20816	31844	20818	31849	20821	31854	20823	42
18	31854	20823	31859	20825	31864	20828	31869	20830	31874	20832	31879	20835	41
19	31879	20835	31884	20837	31889	20839	31894	20842	31898	20844	31903	20847	40
20	31903	20847	31908	20849	31913	20851	31918	20854	31923	20856	31928	20858	39
21	31928	20858	31933	20861	31938	20863	31943	20865	31948	20868	31953	20870	38
22	31953	20870	31958	20873	31962	20875	31967	20877	31972	20880	31977	20882	37
23	31977	20882	31982	20884	31987	20887	31992	20889	31997	20891	32002	20894	36
24	32002	20894	32007	20896	32012	20899	32017	20901	32021	20903	32026	20906	35
25	32026	20906	32031	20908	32036	20910	32041	20913	32046	20915	32051	20918	34
26	32051	20918	32056	20920	32061	20922	32066	20925	32071	20927	32076	20929	33
27	32076	20929	32081	20932	32085	20934	32090	20936	32095	20939	32100	20941	32
28	32100	20941	32105	20944	32110	20946	32115	20948	32120	20951	32125	20953	31
29	32125	20953	32130	20955	32134	20958	32139	20960	32144	20962	32149	20965	30
30	32149	20965	32154	20967	32159	20970	32164	20972	32169	20974	32174	20977	29
31	32174	20977	32179	20979	32184	20981	32188	20984	32193	20986	32198	20989	28
32	32198	20989	32203	20991	32208	20993	32213	20996	32218	20998	32223	21000	27
33	32223	21000	32228	21003	32233	21005	32237	21008	32242	21010	32247	21012	26
34	32247	21012	32252	21015	32257	21017	32262	21019	32267	21022	32272	21024	25
35	32272	21024	32277	21026	32282	21029	32286	21031	32291	21034	32296	21036	24
36	32296	21036	32301	21038	32306	21041	32311	21043	32316	21045	32321	21048	23
37	32321	21048	32326	21050	32330	21053	32335	21055	32340	21057	32345	21060	22
38	32345	21060	32350	21062	32355	21064	32360	21067	32365	21069	32370	21072	21
39	32370	21072	32374	21074	32379	21076	32384	21079	32389	21081	32394	21083	20
40	32394	21083	32399	21086	32404	21088	32409	21091	32414	21093	32418	21095	19
41	32418	21095	32423	21098	32428	21100	32433	21102	32438	21105	32443	21107	18
42	32443	21107	32448	21109	32453	21112	32458	21114	32462	21117	32467	21119	17
43	32467	21119	32472	21121	32477	21124	32482	21126	32487	21128	32492	21131	16
44	32492	21131	32497	21133	32502	21136	32506	21138	32511	21140	32516	21143	15
45	32516	21143	32521	21145	32526	21147	32531	21150	32536	21152	32541	21155	14
46	32541	21155	32545	21157	32550	21159	32555	21162	32560	21164	32565	21167	13
47	32565	21167	32570	21169	32575	21171	32580	21174	32584	21176	32589	21178	12
48	32589	21178	32594	21181	32599	21183	32604	21186	32608	21188	32614	21190	11
49	32614	21190	32618	21193	32623	21195	32628	21197	32633	21200	32638	21202	10
50	32638	21202	32643	21205	32648	21207	32653	21209	32658	21212	32662	21214	09
51	32662	21214	32667	21216	32672	21219	32677	21221	32682	21224	32687	21226	08
52	32687	21226	32691	21228	32696	21231	32701	21233	32706	21235	32711	21238	07
53	32711	21238	32716	21240	32721	21243	32726	21245	32731	21247	32735	21250	06
54	32735	21250	32740	21252	32745	21255	32750	21257	32755	21259	32760	21262	05
55	32760	21262	32764	21264	32769	21266	32774	21269	32779	21271	32784	21274	04
56	32784	21274	32789	21276	32794	21278	32799	21281	32803	21283	32808	21285	03
57	32808	21285	32813	21288	32818	21290	32823	21293	32828	21295	32833	21297	02
58	32833	21297	32837	21300	32842	21302	32847	21304	32852	21307	32857	21309	01
59	32857	21309	32862	21312	32867	21314	32871	21316	32876	21319	32881	21321	00
	·8		·6		·4		·2		·0 ←				

PARTS for 0'·1 :— LOGS. 2 ; NATURALS, 1.

55° HAVERSINES

′	·0 Log. 9·	·0 Nat. 0·	·2 Log. 9·	·2 Nat. 0·	·4 Log. 9·	·4 Nat. 0·	·6 Log. 9·	·6 Nat. 0·	·8 Log. 9·	·8 Nat. 0·	′
00	32881	21321	32886	21324	32891	21326	32896	21328	32901	21331	59
	32905	21333									
01	32905	21333	32910	21335	32915	21330	32920	21340	32925	21343	58
	32930	21345									
02	32930	21345	32934	21347	32939	21350	32944	21352	32949	21355	57
	32954	21357									
03	32954	21357	32959	21359	32964	21362	32968	21364	32973	21366	56
	32978	21369									
04	32978	21369	32983	21371	32988	21374	32993	21376	32998	21378	55
	33002	21381									
05	33002	21381	33007	21383	33012	21386	33017	21388	33022	21390	54
	33027	21393									
06	33027	21393	33031	21395	33036	21397	33041	21400	33046	21402	53
	33051	21405									
07	33051	21405	33056	21407	33060	21409	33065	21412	33070	21414	52
	33075	21417									
08	33075	21417	33080	21419	33085	21421	33090	21424	33094	21426	51
	33099	21429									
09	33099	21429	33104	21431	33109	21433	33114	21436	33119	21438	50
	33123	21440									
10	33123	21440	33128	21443	33133	21445	33138	21448	33143	21450	49
	33148	21452									
11	33148	21452	33152	21455	33157	21457	33162	21460	33167	21462	48
	33172	21464									
12	33172	21464	33177	21467	33181	21469	33186	21471	33191	21474	47
	33196	21476									
13	33196	21476	33201	21479	33205	21481	33210	21483	33215	21486	46
	33220	21488									
14	33220	21488	33225	21491	33230	21493	33235	21495	33239	21498	45
	33244	21500									
15	33244	21500	33249	21503	33254	21505	33259	21507	33264	21510	44
	33268	21512									
16	33268	21512	33273	21514	33278	21517	33283	21519	33288	21522	43
	33292	21524									
17	33292	21524	33297	21526	33302	21529	33307	21531	33312	21534	42
	33317	21536									
18	33317	21536	33321	21538	33326	21541	33331	21543	33336	21546	41
	33341	21548									
19	33341	21548	33345	21550	33350	21553	33355	21555	33360	21558	40
	33365	21560									
20	33365	21560	33370	21562	33374	21565	33379	21567	33384	21570	39
	33389	21572									
21	33389	21572	33394	21574	33398	21577	33403	21579	33408	21581	38
	33413	21584									
22	33413	21584	33418	21586	33423	21589	33427	21591	33432	21593	37
	33437	21596									
23	33437	21596	33442	21598	33447	21601	33451	21603	33456	21605	36
	33461	21608									
24	33461	21608	33466	21610	33471	21613	33476	21615	33480	21617	35
	33485	21620									
25	33485	21620	33490	21622	33495	21625	33500	21627	33504	21629	34
	33509	21632									
26	33509	21632	33514	21634	33519	21637	33524	21639	33528	21641	33
	33533	21644									
27	33533	21644	33538	21646	33543	21649	33548	21651	33552	21653	32
	33557	21656									
28	33557	21656	33562	21658	33567	21661	33572	21663	33576	21665	31
	33581	21668									
29	33581	21668	33586	21670	33591	21673	33596	21675	33601	21677	30
	33605	21680									
30	33605	21680	33610	21682	33615	21685	33620	21687	33625	21689	29
	33629	21692									
31	33629	21692	33634	21694	33639	21696	33644	21699	33649	21701	28
	33653	21704									
32	33653	21704	33658	21706	33663	21708	33668	21711	33672	21713	27
	33677	21716									
33	33677	21716	33682	21718	33687	21720	33692	21723	33696	21725	26
	33701	21728									
34	33701	21728	33706	21730	33711	21732	33716	21735	33720	21737	25
	33725	21740									
35	33725	21740	33730	21742	33735	21744	33740	21747	33744	21749	24
	33749	21752									
36	33749	21752	33754	21754	33759	21756	33764	21759	33768	21761	23
	33773	21764									
37	33773	21764	33778	21766	33783	21768	33788	21771	33792	21773	22
	33797	21776									
38	33797	21776	33802	21778	33807	21780	33811	21783	33816	21785	21
	33821	21788									
39	33821	21788	33826	21790	33831	21792	33835	21795	33840	21797	20
	33845	21800									
40	33845	21800	33850	21802	33855	21804	33859	21807	33864	21809	19
	33869	21812									
41	33869	21812	33874	21814	33879	21817	33883	21819	33888	21821	18
	33893	21824									
42	33893	21824	33898	21826	33902	21829	33907	21831	33912	21833	17
	33917	21836									
43	33917	21836	33922	21838	33926	21841	33931	21843	33936	21845	16
	33941	21848									
44	33941	21848	33945	21850	33950	21853	33955	21855	33960	21857	15
	33965	21860									
45	33965	21860	33969	21862	33974	21865	33979	21867	33984	21869	14
	33988	21872									
46	33988	21872	33993	21874	33998	21877	34003	21879	34007	21881	13
	34012	21884									
47	34012	21884	34017	21886	34022	21889	34027	21891	34031	21893	12
	34036	21896									
48	34036	21896	34041	21898	34046	21901	34050	21903	34055	21905	11
	34060	21908									
49	34060	21908	34065	21910	34070	21913	34074	21915	34079	21917	10
	34084	21920									
50	34084	21920	34089	21922	34093	21925	34098	21927	34103	21930	09
	34108	21932									
51	34108	21932	34112	21934	34117	21937	34122	21939	34127	21942	08
	34132	21944									
52	34132	21944	34136	21946	34141	21949	34146	21951	34151	21954	07
	34155	21956									
53	34155	21956	34160	21958	34165	21961	34170	21963	34174	21966	06
	34179	21968									
54	34179	21968	34184	21970	34189	21973	34193	21975	34198	21978	05
	34203	21980									
55	34203	21980	34208	21983	34213	21985	34217	21987	34222	21990	04
	34227	21992									
56	34227	21992	34232	21995	34236	21997	34241	21999	34246	22002	03
	34251	22004									
57	34251	22004	34255	22007	34260	22009	34265	22011	34270	22014	02
	34274	22016									
58	34274	22016	34279	22019	34284	22021	34289	22023	34292	22026	01
	34298	22028									
59	34298	22028	34303	22031	34308	22033	34312	22036	34317	22038	00
	34322	22040									

·8 ·6 ·4 ·2 ·0

PARTS for 0′·1 :— LOGS. 2 ; NATURALS, 1.

304°

56° HAVERSINES

	·0		·2		·4		·6		·8			
	Log.	Nat.	Log.	Nat.	Log.	Nat.	Log.	Nat.	Log.	Nat.	Log.	Nat	′
′	9·	0·	9·	0·	9·	0·	9·	0·	9·	0·	9·	0·	
00	34322	22040	34327	22043	34331	22045	34336	22048	34341	22050	34346	22052	59
01	34346	22052	34350	22055	34355	22057	34360	22060	34365	22062	34369	22064	58
02	34369	22064	34374	22067	34379	22069	34384	22072	34388	22074	34393	22077	57
03	34393	22077	34398	22079	34403	22081	34407	22084	34412	22086	34417	22089	56
04	34417	22089	34422	22091	34426	22093	34431	22096	34436	22098	34441	22101	55
05	34441	22101	34445	22103	34450	22105	34455	22108	34459	22110	34464	22113	54
06	34464	22113	34469	22115	34474	22118	34478	22120	34483	22122	34488	22125	53
07	34488	22125	34493	22127	34497	22130	34502	22132	34507	22134	34512	22137	52
08	34512	22137	34516	22139	34521	22142	34526	22144	34531	22147	34535	22149	51
09	34535	22149	34540	22151	34545	22154	34550	22156	34554	22159	34559	22161	50
10	34559	22161	34564	22163	34569	22166	34573	22168	34578	22171	34583	22173	49
11	34583	22173	34588	22176	34592	22178	34597	22180	34602	22183	34606	22185	48
12	34606	22185	34611	22188	34616	22190	34621	22192	34625	22195	34630	22197	47
13	34630	22197	34635	22200	34639	22202	34644	22205	34649	22207	34654	22209	46
14	34654	22209	34658	22212	34663	22214	34668	22217	34673	22219	34677	22221	45
15	34677	22221	34682	22224	34687	22226	34692	22229	34696	22231	34701	22234	44
16	34701	22234	34706	22236	34710	22238	34715	22241	34720	22243	34725	22246	43
17	34725	22246	34729	22248	34734	22251	34739	22253	34743	22255	34748	22258	42
18	34748	22258	34753	22260	34758	22263	34762	22265	34767	22267	34772	22270	41
19	34772	22270	34777	22272	34781	22275	34786	22277	34791	22280	34795	22282	40
20	34795	22282	34800	22284	34805	22287	34810	22289	34814	22292	34819	22294	39
21	34819	22294	34824	22297	34828	22299	34833	22301	34838	22304	34843	22306	38
22	34843	22306	34847	22309	34852	22311	34857	22313	34861	22316	34866	22318	37
23	34866	22318	34871	22321	34876	22323	34880	22326	34885	22328	34890	22330	36
24	34890	22330	34894	22333	34899	22335	34904	22338	34908	22340	34913	22343	35
25	34913	22343	34918	22345	34923	22347	34927	22350	34932	22352	34937	22355	34
26	34937	22355	34942	22357	34946	22360	34951	22362	34956	22364	34960	22367	33
27	34960	22367	34965	22369	34970	22372	34974	22374	34979	22376	34984	22379	32
28	34984	22379	34989	22381	34993	22384	34998	22386	35003	22389	35007	22391	31
29	35007	22391	35012	22393	35017	22396	35022	22398	35026	22401	35031	22403	30
30	35031	22403	35036	22406	35040	22408	35045	22410	35050	22413	35054	22415	29
31	35054	22415	35059	22418	35064	22420	35069	22423	35073	22425	35078	22427	28
32	35078	22427	35083	22430	35087	22432	35092	22435	35097	22437	35101	22440	27
33	35101	22440	35106	22442	35111	22444	35115	22447	35120	22449	35125	22452	26
34	35125	22452	35130	22454	35134	22457	35139	22459	35144	22461	35148	22464	25
35	35148	22464	35153	22466	35158	22469	35162	22471	35167	22472	35172	22476	24
36	35172	22476	35176	22478	35181	22481	35186	22483	35191	22486	35195	22488	23
37	35195	22488	35200	22491	35205	22493	35209	22495	35214	22498	35219	22500	22
38	35219	22500	35223	22503	35228	22505	35233	22508	35238	22510	35242	22512	21
39	35242	22512	35247	22515	35252	22517	35256	22520	35261	22522	35266	22525	20
40	35266	22525	35270	22527	35275	22529	35280	22532	35284	22534	35289	22537	19
41	35289	22537	35294	22539	35298	22542	35303	22544	35308	22546	35312	22549	18
42	35312	22549	35317	22551	35322	22554	35326	22556	35331	22559	35336	22561	17
43	35336	22561	35340	22563	35345	22566	35350	22568	35355	22571	35359	22573	16
44	35359	22573	35364	22576	35369	22578	35373	22580	35378	22583	35383	22585	15
45	35383	22585	35387	22588	35392	22590	35397	22593	35401	22595	35406	22598	14
46	35406	22598	35411	22600	35415	22602	35420	22605	35425	22607	35429	22610	13
47	35429	22610	35434	22612	35439	22615	35443	22617	35448	22619	35453	22622	12
48	35453	22622	35457	22624	35462	22627	35467	22629	35472	22632	35476	22634	11
49	35476	22634	35481	22636	35486	22639	35490	22641	35495	22644	35500	22646	10
50	35500	22646	35504	22649	35509	22651	35514	22653	35518	22656	35523	22658	09
51	35523	22658	35527	22661	35532	22663	35537	22666	35542	22668	35546	22671	08
52	35546	22671	35551	22672	35556	22675	35560	22678	35565	22680	35570	22683	07
53	35570	22683	35574	22685	35579	22688	35584	22690	35588	22692	35593	22695	06
54	35593	22695	35597	22697	35602	22700	35607	22702	35611	22705	35616	22707	05
55	35616	22707	35621	22710	35625	22712	35630	22714	35635	22717	35639	22719	04
56	35639	22719	35644	22722	35649	22724	35653	22727	35658	22729	35663	22731	03
57	35663	22731	35667	22734	35672	22736	35677	22739	35681	22741	35686	22744	02
58	35686	22744	35691	22746	35695	22749	35700	22751	35705	22753	35709	22756	01
59	35709	22756	35714	22758	35719	22761	35723	22763	35728	22766	35733	22768	00
	·8		·6		·4		·2		·0				

PARTS for 0′·1 :—   LOGS. 2 ;   NATURALS, 1.

57° HAVERSINES

	·0		·2		·4		·6		·8		
	Log.	Nat.	Log.	Nat.	Log.	Nat.	Log.	Nat.	Log.	Nat.	
′	9·	0·	9·	0·	9·	0·	9·	0·	9·	0·	′
00	35733	22768	35737	22770	35742	22773	35747	22775	35751	22778	35756	22780	59
01	35756	22780	35760	22783	35765	22785	35770	22788	35774	22790	35779	22792	58
02	35779	22792	35784	22795	35788	22797	35793	22800	35798	22802	35802	22805	57
03	35802	22805	35807	22807	35812	22810	35816	22812	35821	22814	35826	22817	56
04	35826	22817	35830	22819	35835	22822	35840	22824	35844	22827	35849	22829	55
05	35849	22829	35853	22832	35858	22834	35863	22836	35867	22839	35872	22841	54
06	35872	22841	35877	22844	35881	22846	35886	22849	35891	22851	35895	22853	53
07	35895	22853	35900	22856	35904	22858	35909	22861	35914	22863	35918	22866	52
08	35918	22866	35923	22868	35928	22871	35932	22873	35937	22875	35942	22878	51
09	35942	22878	35946	22880	35951	22883	35956	22885	35960	22888	35965	22890	50
10	35965	22890	35969	22893	35974	22895	35979	22897	35983	22900	35988	22902	49
11	35988	22902	35993	22905	35997	22907	36002	22910	36007	22912	36011	22915	48
12	36011	22915	36016	22917	36020	22919	36025	22922	36030	22924	36034	22927	47
13	36034	22927	36039	22929	36044	22932	36048	22934	36053	22937	36058	22939	46
14	36058	22939	36062	22942	36067	22944	36071	22946	36076	22949	36081	22951	45
15	36081	22951	36085	22954	36090	22956	36095	22959	36099	22961	36104	22964	44
16	36104	22964	36108	22966	36113	22968	36118	22971	36122	22973	36127	22976	43
17	36127	22976	36132	22978	36136	22981	36141	22983	36145	22986	36150	22988	42
18	36150	22988	36155	22990	36159	22993	36164	22995	36169	22998	36173	23000	41
19	36173	23000	36178	23003	36182	23005	36187	23008	36192	23010	36196	23012	40
20	36196	23012	36201	23015	36206	23017	36210	23020	36215	23022	36219	23025	39
21	36219	23025	36224	23027	36229	23030	36233	23032	36238	23035	36243	23037	38
22	36243	23037	36247	23039	36252	23042	36256	23044	36261	23047	36266	23049	37
23	36266	23049	36270	23052	36275	23054	36279	23057	36284	23059	36289	23061	36
24	36289	23061	36293	23064	36298	23066	36303	23069	36307	23071	36312	23074	35
25	36312	23074	36316	23076	36321	23079	36326	23081	36330	23084	36335	23086	34
26	36335	23086	36339	23088	36344	23091	36349	23093	36353	23096	36358	23098	33
27	36358	23098	36362	23101	36367	23103	36372	23106	36376	23108	36381	23110	32
28	36381	23110	36386	23113	36390	23115	36395	23118	36399	23120	36404	23123	31
29	36404	23123	36409	23125	36413	23128	36418	23130	36422	23133	36427	23135	30
30	36427	23135	36432	23137	36436	23140	36441	23142	36445	23145	36450	23147	29
31	36450	23147	36455	23150	36459	23152	36464	23155	36468	23157	36473	23160	28
32	36473	23160	36478	23162	36482	23164	36487	23167	36491	23169	36496	23172	27
33	36496	23172	36501	23174	36505	23177	36510	23179	36514	23182	36519	23184	26
34	36519	23184	36524	23187	36528	23189	36533	23191	36537	23194	36542	23196	25
35	36542	23196	36547	23199	36551	23201	36556	23204	36560	23206	36565	23209	24
36	36565	23209	36570	23211	36574	23214	36579	23216	36583	23218	36588	23221	23
37	36588	23221	36593	23223	36597	23226	36602	23228	36606	23231	36611	23233	22
38	36611	23233	36616	23236	36620	23238	36625	23241	36629	23243	36634	23246	21
39	36634	23246	36639	23248	36643	23250	36648	23253	36652	23255	36657	23258	20
40	36657	23258	36661	23260	36666	23263	36671	23265	36675	23268	36680	23270	19
41	36680	23270	36684	23273	36689	23275	36694	23277	36698	23280	36703	23282	18
42	36703	23282	36707	23285	36712	23287	36716	23290	36721	23292	36726	23295	17
43	36726	23295	36730	23297	36735	23300	36739	23302	36744	23305	36749	23307	16
44	36749	23307	36753	23309	36758	23312	36762	23314	36767	23317	36772	23319	15
45	36772	23319	36776	23322	36781	23324	36785	23327	36790	23329	36794	23332	14
46	36794	23332	36799	23334	36804	23337	36808	23339	36813	23341	36817	23344	13
47	36817	23344	36822	23346	36827	23349	36831	23351	36836	23354	36840	23356	12
48	36840	23356	36845	23359	36849	23361	36854	23364	36858	23366	36863	23368	11
49	36863	23368	36868	23371	36872	23373	36877	23376	36881	23378	36886	23381	10
50	36886	23381	36891	23383	36895	23386	36900	23388	36904	23391	36909	23393	09
51	36909	23393	36913	23396	36918	23398	36922	23401	36927	23403	36932	23405	08
52	36932	23405	36936	23408	36941	23410	36945	23413	36950	23415	36955	23418	07
53	36955	23418	36959	23420	36964	23423	36968	23425	36973	23428	36977	23430	06
54	36977	23430	36982	23433	36987	23435	36991	23437	36996	23440	37000	23442	05
55	37000	23442	37005	23445	37009	23447	37014	23450	37018	23452	37023	23455	04
56	37023	23455	37028	23457	37032	23460	37037	23462	37041	23465	37046	23467	03
57	37046	23467	37050	23470	37055	23472	37059	23474	37064	23477	37069	23479	02
58	37069	23479	37073	23482	37078	23484	37082	23487	37087	23489	37091	23492	01
59	37091	23492	37096	23494	37101	23497	37105	23499	37110	23502	37114	23504	00

| | ·8 | | ·6 | | ·4 | | ·2 | | ·0 | | |

PARTS for 0′·1 :— LOGS. 2 ; NATURALS, 1.

58° HAVERSINES

′	·0 Log. 9·	·0 Nat. 0·	·2 Log. 9·	·2 Nat. 0·	·4 Log. 9·	·4 Nat. 0·	·6 Log. 9·	·6 Nat. 0·	·8 Log. 9·	·8 Nat. 0·	Log. 9·	Nat. 0·	′
00	37114	23504	37119	23507	37123	23509	37128	23511	37132	23514	37137	23516	59
01	37137	23516	37142	23519	37146	23521	37151	23524	37155	23526	37160	23529	58
02	37160	23529	37164	23531	37169	23534	37173	23536	37178	23539	37183	23541	57
03	37183	23541	37187	23544	37192	23546	37196	23548	37201	23551	37205	23553	56
04	37205	23553	37210	23556	37215	23558	37219	23561	37224	23563	37228	23566	55
05	37228	23566	37233	23568	37237	23571	37242	23573	37246	23576	37251	23578	54
06	37251	23578	37255	23581	37260	23583	37264	23586	37269	23588	37274	23590	53
07	37274	23590	37278	23593	37283	23595	37287	23598	37292	23600	37296	23603	52
08	37296	23603	37301	23605	37305	23608	37310	23610	37314	23613	37319	23615	51
09	37319	23615	37324	23618	37328	23620	37333	23623	37337	23625	37342	23627	50
10	37342	23627	37346	23630	37351	23632	37355	23635	37360	23637	37364	23640	49
11	37364	23640	37369	23642	37374	23645	37378	23647	37383	23650	37387	23652	48
12	37387	23652	37392	23655	37396	23657	37401	23660	37405	23662	37410	23665	47
13	37410	23665	37414	23667	37419	23670	37423	23672	37428	23674	37433	23677	46
14	37433	23677	37437	23679	37442	23682	37446	23684	37451	23687	37455	23689	45
15	37455	23689	37460	23692	37464	23694	37469	23697	37473	23699	37478	23702	44
16	37478	23702	37482	23704	37487	23707	37491	23709	37496	23712	37501	23714	43
17	37501	23714	37505	23717	37510	23719	37514	23721	37519	23724	37523	23726	42
18	37523	23726	37528	23729	37532	23731	37537	23734	37541	23736	37546	23739	41
19	37546	23739	37550	23741	37555	23744	37559	23746	37564	23749	37569	23751	40
20	37569	23751	37573	23754	37578	23756	37582	23759	37587	23761	37591	23764	39
21	37591	23764	37596	23766	37600	23769	37605	23771	37609	23773	37614	23776	38
22	37614	23776	37618	23778	37623	23781	37627	23783	37632	23786	37636	23788	37
23	37636	23788	37641	23791	37645	23793	37650	23796	37654	23798	37659	23801	36
24	37659	23801	37664	23803	37668	23806	37673	23808	37677	23811	37682	23813	35
25	37682	23813	37686	23816	37691	23818	37695	23821	37700	23823	37704	23825	34
26	37704	23825	37709	23828	37713	23830	37718	23833	37722	23835	37727	23838	33
27	37727	23838	37731	23840	37736	23843	37740	23845	37745	23848	37749	23850	32
28	37749	23850	37754	23853	37758	23855	37763	23858	37767	23860	37772	23863	31
29	37772	23863	37776	23865	37781	23868	37785	23870	37790	23873	37794	23875	30
30	37794	23875	37799	23878	37803	23880	37808	23882	37812	23885	37817	23887	29
31	37817	23887	37822	23890	37826	23892	37831	23895	37835	23897	37840	23900	28
32	37840	23900	37844	23902	37849	23905	37853	23907	37858	23910	37862	23912	27
33	37862	23912	37867	23915	37871	23917	37876	23920	37880	23922	37885	23925	26
34	37885	23925	37889	23927	37894	23930	37898	23932	37903	23935	37907	23937	25
35	37907	23937	37912	23940	37916	23942	37921	23945	37925	23947	37930	23950	24
36	37930	23950	37934	23952	37939	23954	37943	23957	37948	23959	37952	23962	23
37	37952	23962	37957	23964	37961	23967	37966	23969	37970	23972	37975	23974	22
38	37975	23974	37979	23977	37984	23979	37988	23982	37993	23984	37997	23987	21
39	37997	23987	38002	23989	38006	23992	38011	23994	38015	23997	38020	23999	20
40	38020	23999	38024	24002	38029	24004	38033	24007	38038	24009	38042	24012	19
41	38042	24012	38047	24014	38051	24017	38056	24019	38060	24022	38065	24024	18
42	38065	24024	38069	24027	38074	24029	38078	24031	38083	24034	38087	24036	17
43	38087	24036	38092	24039	38096	24041	38101	24044	38105	24046	38110	24049	16
44	38110	24049	38114	24051	38119	24054	38123	24056	38128	24059	38132	24061	15
45	38132	24061	38136	24064	38141	24066	38145	24069	38150	24071	38154	24074	14
46	38154	24074	38159	24076	38163	24079	38168	24081	38172	24084	38177	24086	13
47	38177	24086	38181	24089	38186	24091	38190	24094	38195	24096	38199	24099	12
48	38199	24099	38204	24101	38208	24104	38213	24106	38217	24109	38222	24111	11
49	38222	24111	38226	24114	38231	24116	38235	24119	38240	24121	38244	24124	10
50	38244	24124	38248	24126	38253	24129	38257	24131	38262	24134	38267	24136	09
51	38267	24136	38271	24138	38275	24141	38280	24143	38284	24146	38289	24148	08
52	38289	24148	38293	24151	38298	24153	38302	24156	38307	24158	38311	24161	07
53	38311	24161	38316	24163	38320	24166	38325	24168	38329	24171	38334	24173	06
54	38334	24173	38338	24176	38343	24178	38347	24181	38352	24183	38356	24186	05
55	38356	24186	38360	24188	38365	24191	38369	24193	38374	24196	38378	24198	04
56	38378	24198	38383	24201	38387	24203	38392	24206	38396	24208	38401	24211	03
57	38401	24211	38405	24213	38410	24216	38414	24218	38419	24221	38423	24223	02
58	38423	24223	38427	24226	38432	24228	38436	24231	38441	24233	38445	24236	01
59	38445	24236	38450	24238	38454	24241	38459	24243	38463	24246	38468	24248	00

| ·8 | ·6 | ·4 | ·2 | ·0 |

PARTS for 0′·1 :—   LOGS. 2 ;   NATURALS, 1.

59° HAVERSINES

′	·0 Log. 9·	·0 Nat. 0·	·2 Log. 9·	·2 Nat. 0·	·4 Log. 9·	·4 Nat. 0·	·6 Log. 9·	·6 Nat. 0·	·8 Log. 9·	·8 Nat. 0·	Log. 9·	Nat. 0·	′
00	38468	24248	38472	24251	38477	24253	38481	24256	38486	24258	38490	24261	59
01	38490	24261	38494	24263	38499	24266	38503	24268	38508	24271	38512	24273	58
02	38512	24273	38517	24276	38521	24278	38526	24281	38530	24283	38535	24286	57
03	38535	24286	38539	24288	38544	24291	38548	24293	38553	24296	38557	24298	56
04	38557	24298	38561	24300	38566	24303	38570	24305	38575	24308	38579	24310	55
05	38579	24310	38584	24313	38588	24315	38593	24318	38597	24320	38602	24323	54
06	38602	24323	38606	24325	38610	24328	38615	24330	38619	24333	38624	24335	53
07	38624	24335	38628	24338	38633	24340	38637	24343	38642	24345	38646	24348	52
08	38646	24348	38651	24350	38655	24353	38660	24355	38664	24358	38668	24360	51
09	38668	24360	38673	24363	38677	24365	38682	24368	38686	24370	38691	24373	50
10	38691	24373	38695	24375	38700	24378	38704	24380	38709	24383	38713	24385	49
11	38713	24385	38717	24388	38722	24390	38726	24393	38731	24395	38735	24398	48
12	38735	24398	38740	24400	38744	24403	38749	24405	38753	24408	38757	24410	47
13	38757	24410	38762	24413	38766	24415	38771	24418	38775	24420	38780	24423	46
14	38780	24423	38784	24425	38788	24428	38793	24430	38797	24433	38802	24435	45
15	38802	24435	38806	24438	38811	24440	38815	24443	38820	24445	38824	24448	44
16	38824	24448	38828	24450	38833	24453	38837	24455	38842	24458	38846	24460	43
17	38846	24460	38851	24463	38855	24465	38860	24468	38864	24470	38868	24473	42
18	38868	24473	38873	24475	38877	24478	38882	24480	38886	24483	38891	24485	41
19	38891	24485	38895	24488	38899	24490	38904	24493	38908	24495	38913	24498	40
20	38913	24498	38917	24500	38922	24503	38926	24505	38931	24508	38935	24510	39
21	38935	24510	38939	24513	38944	24515	38948	24518	38953	24520	38957	24523	38
22	38957	24523	38962	24525	38966	24528	38971	24530	38975	24533	38979	24535	37
23	38979	24535	38984	24538	38988	24540	38993	24543	38997	24545	39001	24548	36
24	39001	24548	39006	24550	39010	24553	39015	24555	39019	24558	39024	24560	35
25	39024	24560	39028	24563	39032	24565	39037	24568	39041	24570	39046	24573	34
26	39046	24573	39050	24575	39055	24578	39059	24580	39063	24583	39068	24586	33
27	39068	24586	39072	24588	39077	24591	39081	24593	39086	24596	39090	24598	32
28	39090	24598	39094	24601	39099	24603	39103	24606	39108	24608	39112	24611	31
29	39112	24611	39117	24613	39121	24616	39125	24618	39130	24621	39134	24623	30
30	39134	24623	39139	24626	39143	24628	39148	24631	39152	24633	39156	24636	29
31	39156	24636	39161	24638	39165	24641	39170	24643	39174	24646	39178	24648	28
32	39178	24648	39183	24651	39187	24653	39192	24656	39196	24658	39201	24661	27
33	39201	24661	39205	24663	39209	24666	39214	24668	39218	24671	39223	24673	26
34	39223	24673	39227	24676	39231	24678	39236	24681	39240	24683	39245	24686	25
35	39245	24686	39249	24688	39253	24691	39258	24693	39262	24696	39267	24698	24
36	39267	24698	39271	24701	39276	24703	39280	24706	39284	24708	39289	24711	23
37	39289	24711	39293	24713	39298	24716	39302	24718	39306	24721	39311	24723	22
38	39311	24723	39315	24726	39320	24728	39324	24731	39328	24733	39333	24736	21
39	39333	24736	39337	24738	39342	24741	39346	24743	39351	24746	39355	24749	20
40	39355	24749	39359	24751	39364	24754	39368	24756	39373	24759	39377	24761	19
41	39377	24761	39381	24764	39386	24766	39390	24769	39395	24771	39399	24774	18
42	39399	24774	39403	24776	39408	24779	39412	24781	39417	24784	39421	24786	17
43	39421	24786	39425	24789	39430	24791	39434	24794	39439	24796	39443	24799	16
44	39443	24799	39447	24801	39452	24804	39456	24806	39461	24809	39465	24811	15
45	39465	24811	39469	24814	39474	24816	39478	24819	39483	24821	39487	24824	14
46	39487	24824	39491	24826	39496	24829	39500	24831	39505	24834	39509	24836	13
47	39509	24836	39513	24839	39518	24841	39522	24844	39526	24846	39531	24849	12
48	39531	24849	39535	24852	39540	24854	39544	24857	39548	24859	39553	24862	11
49	39553	24862	39557	24864	39562	24867	39566	24869	39570	24872	39575	24874	10
50	39575	24874	39579	24877	39584	24879	39588	24882	39592	24884	39597	24887	09
51	39597	24887	39601	24889	39606	24892	39610	24894	39614	24897	39619	24899	08
52	39619	24899	39623	24902	39628	24904	39632	24907	39636	24909	39641	24912	07
53	39641	24912	39645	24914	39649	24917	39654	24919	39658	24922	39663	24924	06
54	39663	24924	39667	24927	39671	24929	39676	24932	39680	24934	39685	24937	05
55	39685	24937	39689	24940	39693	24942	39698	24945	39702	24947	39706	24950	04
56	39706	24950	39711	24952	39715	24955	39720	24957	39724	24960	39728	24962	03
57	39728	24962	39733	24965	39737	24967	39741	24970	39746	24972	39750	24975	02
58	39750	24975	39755	24977	39759	24980	39763	24982	39768	24985	39772	24987	01
59	39772	24987	39777	24990	39781	24992	39785	24995	39790	24997	39794	25000	00

·8 ·6 ·4 ·2 ·0 ←

300°

PARTS for 0′·1 :— LOGS. 2 ; NATURALS, 1.

HAVERSINES

60°

′	·0 Log. 9·	·0 Nat. 0·	·2 Log. 9·	·2 Nat. 0·	·4 Log. 9·	·4 Nat. 0·	·6 Log. 9·	·6 Nat. 0·	·8 Log. 9·	·8 Nat. 0·	Log. 9·	Nat. 0·	′
00	39794	25000	39798	25003	39803	25005	39807	25008	39811	25010	39816	25013	59
01	39816	25013	39820	25015	39825	25018	39829	25020	39833	25023	39838	25025	58
02	39838	25025	39842	25028	39847	25030	39851	25033	39855	25035	39860	25038	57
03	39860	25038	39864	25040	39868	25043	39873	25045	39877	25048	39881	25050	56
04	39881	25050	39886	25053	39890	25055	39895	25058	39899	25060	39903	25063	55
05	39903	25063	39908	25066	39912	25068	39916	25071	39921	25073	39925	25076	54
06	39925	25076	39930	25078	39934	25081	39938	25083	39943	25086	39947	25088	53
07	39947	25088	39951	25091	39956	25093	39960	25096	39964	25098	39969	25101	52
08	39969	25101	39973	25103	39978	25106	39982	25108	39986	25111	39991	25113	51
09	39991	25113	39995	25116	39999	25118	40004	25121	40008	25124	40012	25126	50
10	40012	25126	40017	25129	40021	25131	40025	25134	40030	25136	40034	25139	49
11	40034	25139	40039	25141	40043	25144	40047	25146	40052	25149	40056	25151	48
12	40056	25151	40060	25154	40065	25156	40069	25159	40073	25161	40078	25164	47
13	40078	25164	40082	25166	40087	25169	40091	25171	40095	25174	40100	25177	46
14	40100	25177	40104	25179	40108	25182	40113	25184	40117	25187	40121	25189	45
15	40121	25189	40126	25192	40130	25194	40134	25197	40139	25199	40143	25202	44
16	40143	25202	40148	25204	40152	25207	40156	25209	40161	25212	40165	25214	43
17	40165	25214	40169	25217	40174	25219	40178	25222	40182	25225	40187	25227	42
18	40187	25227	40191	25230	40195	25232	40200	25235	40204	25237	40208	25240	41
19	40208	25240	40213	25242	40217	25245	40221	25247	40226	25250	40230	25252	40
20	40230	25252	40235	25255	40239	25257	40243	25260	40247	25262	40252	25265	39
21	40252	25265	40256	25268	40261	25270	40265	25273	40269	25275	40274	25278	38
22	40274	25278	40278	25280	40282	25283	40287	25285	40291	25288	40295	25290	37
23	40295	25290	40300	25293	40304	25295	40308	25298	40313	25300	40317	25303	36
24	40317	25303	40321	25305	40326	25308	40330	25310	40334	25313	40339	25316	35
25	40339	25316	40343	25318	40347	25321	40352	25323	40356	25326	40360	25328	34
26	40360	25328	40365	25331	40369	25333	40373	25336	40378	25338	40382	25341	33
27	40382	25341	40387	25343	40391	25346	40395	25348	40399	25351	40404	25354	32
28	40404	25354	40408	25356	40412	25359	40417	25361	40421	25364	40425	25366	31
29	40425	25366	40430	25369	40434	25371	40438	25374	40443	25376	40447	25379	30
30	40447	25379	40452	25381	40456	25384	40460	25386	40464	25389	40469	25391	29
31	40469	25391	40473	25394	40477	25397	40482	25399	40486	25402	40490	25404	28
32	40490	25404	40495	25407	40499	25409	40503	25412	40508	25414	40512	25417	27
33	40512	25417	40517	25419	40521	25422	40525	25424	40529	25427	40534	25429	26
34	40534	25429	40538	25432	40542	25435	40547	25437	40551	25440	40555	25442	25
35	40555	25442	40560	25445	40564	25447	40568	25450	40573	25452	40577	25455	24
36	40577	25455	40581	25457	40586	25460	40590	25462	40594	25465	40599	25467	23
37	40599	25467	40603	25470	40607	25473	40612	25475	40616	25478	40620	25480	22
38	40620	25480	40625	25483	40629	25485	40633	25488	40637	25490	40642	25493	21
39	40642	25493	40646	25495	40650	25498	40655	25500	40659	25503	40663	25506	20
40	40663	25506	40668	25508	40672	25511	40676	25513	40681	25516	40685	25519	19
41	40685	25518	40689	25521	40694	25523	40698	25526	40702	25528	40707	25531	18
42	40707	25531	40711	25533	40715	25536	40720	25538	40724	25541	40728	25544	17
43	40728	25544	40733	25546	40737	25549	40741	25551	40745	25554	40750	25556	16
44	40750	25556	40754	25559	40758	25561	40763	25564	40767	25566	40771	25569	15
45	40771	25569	40776	25571	40780	25574	40784	25577	40788	25579	40793	25582	14
46	40793	25582	40797	25584	40801	25587	40806	25589	40810	25592	40814	25594	13
47	40814	25594	40819	25597	40823	25599	40827	25602	40832	25604	40836	25607	12
48	40836	25607	40840	25610	40844	25612	40849	25615	40853	25617	40858	25620	11
49	40858	25620	40862	25622	40866	25625	40870	25627	40875	25630	40879	25632	10
50	40879	25632	40883	25635	40888	25637	40892	25640	40896	25643	40900	25645	09
51	40900	25645	40905	25648	40909	25650	40913	25653	40918	25655	40922	25658	08
52	40922	25658	40926	25660	40931	25663	40935	25665	40939	25668	40943	25671	07
53	40943	25671	40948	25673	40952	25676	40956	25678	40961	25681	40965	25683	06
54	40965	25683	40969	25686	40974	25688	40978	25691	40982	25693	40986	25696	05
55	40986	25696	40991	25698	40995	25701	40999	25704	41004	25706	41008	25709	04
56	41008	25709	41012	25711	41017	25714	41021	25716	41025	25719	41029	25721	03
57	41029	25721	41034	25724	41038	25726	41042	25729	41047	25732	41051	25734	02
58	41051	25734	41055	25737	41059	25739	41064	25742	41068	25744	41072	25747	01
59	41072	25747	41077	25749	41081	25752	41085	25754	41090	25757	41094	25760	00

| ·8 | ·6 | ·4 | ·2 | ·0 ← |

299°

PARTS for 0′·1 :—   LOGS. 2 ;   NATURALS, 1.

61° HAVERSINES

′	·0 Log. 9·	·0 Nat. 0·	·2 Log. 9·	·2 Nat. 0·	·4 Log. 9·	·4 Nat. 0·	·6 Log. 9·	·6 Nat. 0·	·8 Log. 9·	·8 Nat. 0·	Log. 9·	Nat. 0·	′
00	41094	25760	41098	25762	41102	25765	41107	25767	41111	25770	41115	25772	59
01	41115	25772	41119	25775	41124	25777	41128	25780	41132	25782	41137	25785	58
02	41137	25785	41141	25788	41145	25790	41149	25793	41154	25795	41158	25798	57
03	41158	25798	41162	25800	41167	25803	41171	25805	41175	25808	41180	25810	56
04	41180	25810	41184	25813	41188	25816	41192	25818	41197	25821	41201	25823	55
05	41201	25823	41205	25826	41210	25828	41214	25831	41218	25833	41222	25836	54
06	41222	25836	41227	25838	41231	25841	41235	25844	41240	25846	41244	25849	53
07	41244	25849	41248	25851	41252	25854	41257	25856	41261	25859	41265	25861	52
08	41265	25861	41269	25864	41274	25866	41278	25869	41282	25872	41287	25874	51
09	41287	25874	41291	25877	41295	25879	41299	25882	41304	25884	41308	25887	50
10	41308	25887	41312	25889	41316	25892	41321	25894	41325	25897	41329	25900	49
11	41329	25900	41333	25902	41338	25905	41342	25907	41346	25910	41351	25912	48
12	41351	25912	41355	25915	41359	25917	41363	25920	41368	25923	41372	25925	47
13	41372	25925	41376	25928	41381	25930	41385	25933	41389	25935	41393	25938	46
14	41393	25938	41398	25940	41402	25943	41406	25945	41410	25948	41415	25951	45
15	41415	25951	41419	25953	41423	25956	41427	25958	41432	25961	41436	25963	44
16	41436	25963	41440	25966	41445	25968	41449	25971	41453	25974	41457	25976	43
17	41457	25976	41462	25979	41466	25981	41470	25984	41474	25986	41479	25989	42
18	41479	25989	41483	25991	41487	25994	41491	25997	41496	25999	41500	26002	41
19	41500	26002	41504	26004	41508	26007	41513	26009	41517	26012	41521	26014	40
20	41521	26014	41525	26017	41530	26019	41534	26022	41538	26025	41543	26027	39
21	41543	26027	41547	26030	41551	26032	41555	26035	41560	26037	41564	26040	38
22	41564	26040	41568	26042	41572	26045	41577	26048	41581	26050	41585	26053	37
23	41585	26053	41589	26055	41594	26058	41598	26060	41602	26063	41606	26065	36
24	41606	26065	41611	26068	41615	26071	41619	26073	41624	26076	41628	26078	35
25	41628	26078	41632	26081	41636	26083	41641	26086	41645	26088	41649	26091	34
26	41649	26091	41653	26094	41657	26096	41662	26099	41666	26101	41670	26104	33
27	41670	26104	41674	26106	41679	26109	41683	26111	41687	26114	41692	26117	32
28	41692	26117	41696	26119	41700	26122	41704	26124	41709	26127	41713	26129	31
29	41713	26129	41717	26132	41721	26134	41726	26137	41730	26140	41734	26142	30
30	41734	26142	41738	26145	41742	26147	41747	26150	41751	26152	41755	26155	29
31	41755	26155	41759	26157	41764	26160	41768	26163	41772	26165	41776	26168	28
32	41776	26168	41781	26170	41785	26173	41789	26175	41793	26178	41798	26180	27
33	41798	26180	41802	26183	41806	26186	41810	26188	41815	26191	41819	26193	26
34	41819	26193	41823	26196	41827	26198	41832	26201	41836	26203	41840	26206	25
35	41840	26206	41844	26209	41848	26211	41853	26214	41857	26216	41861	26219	24
36	41861	26219	41865	26221	41870	26224	41874	26226	41878	26229	41882	26232	23
37	41882	26232	41887	26234	41891	26237	41895	26239	41899	26242	41904	26244	22
38	41904	26244	41908	26247	41912	26250	41916	26252	41921	26255	41925	26257	21
39	41925	26257	41929	26260	41933	26262	41938	26265	41942	26267	41946	26270	20
40	41946	26270	41950	26273	41954	26275	41959	26278	41963	26280	41967	26283	19
41	41967	26283	41971	26285	41976	26288	41980	26290	41984	26293	41988	26296	18
42	41988	26296	41992	26298	41997	26301	42001	26303	42005	26306	42009	26308	17
43	42009	26308	42014	26311	42018	26314	42022	26316	42026	26319	42031	26321	16
44	42031	26321	42035	26324	42039	26326	42043	26329	42048	26331	42052	26334	15
45	42052	26334	42056	26337	42060	26339	42064	26342	42069	26344	42073	26347	14
46	42073	26347	42077	26349	42081	26352	42086	26355	42090	26357	42094	26360	13
47	42094	26360	42098	26362	42102	26365	42107	26367	42111	26370	42115	26372	12
48	42115	26372	42119	26375	42123	26378	42128	26380	42132	26383	42136	26385	11
49	42136	26385	42140	26388	42145	26390	42149	26393	42153	26396	42157	26398	10
50	42157	26398	42161	26401	42166	26403	42170	26406	42174	26408	42178	26411	09
51	42178	26411	42183	26413	42187	26416	42191	26419	42195	26421	42199	26424	08
52	42199	26424	42204	26426	42208	26429	42212	26431	42216	26434	42221	26437	07
53	42221	26437	42225	26439	42229	26442	42233	26444	42237	26447	42242	26449	06
54	42242	26449	42246	26452	42250	26455	42254	26457	42258	26460	42263	26462	05
55	42263	26462	42267	26465	42271	26467	42275	26470	42280	26472	42284	26475	04
56	42284	26475	42288	26478	42292	26480	42296	26483	42301	26485	42305	26488	03
57	42305	26488	42309	26490	42313	26493	42317	26496	42322	26498	42326	26501	02
58	42326	26501	42330	26503	42334	26506	42338	26508	42343	26511	42347	26514	01
59	42347	26514	42351	26516	42355	26519	42359	26521	42364	26524	42368	26526	00
	·8		·6		·4		·2		·0 ←				

PARTS for 0′·1 :— LOGS. 2 ; NATURALS, 1.

298°

62°

HAVERSINES

′	·0 Log.	·0 Nat.	·2 Log.	·2 Nat.	·4 Log.	·4 Nat.	·6 Log.	·6 Nat.	·8 Log.	·8 Nat.	′
	9·	0·	9·	0·	9·	0·	9·	0·	9·	0·	
00	42368	26526	42372	26529	42376	26532	42380	26534	42385	26537	59
01	42389	26539	42393	26542	42397	26544	42402	26547	42406	26550	58
02	42410	26552	42414	26555	42418	26557	42423	26560	42427	26562	57
03	42431	26565	42435	26568	42439	26570	42444	26573	42448	26575	56
04	42452	26578	42456	26580	42460	26583	42465	26586	42469	26588	55
05	42473	26591	42477	26593	42481	26596	42485	26598	42490	26601	54
06	42494	26604	42498	26606	42502	26609	42506	26611	42511	26614	53
07	42515	26616	42519	26619	42523	26621	42527	26624	42532	26627	52
08	42536	26629	42540	26632	42544	26634	42548	26637	42553	26640	51
09	42557	26642	42561	26645	42565	26647	42569	26650	42574	26652	50
10	42578	26655	42582	26658	42586	26660	42590	26663	42595	26665	49
11	42599	26668	42603	26670	42607	26673	42611	26676	42615	26678	48
12	42620	26681	42624	26683	42628	26686	42632	26688	42636	26691	47
13	42641	26694	42645	26696	42649	26699	42653	26701	42657	26704	46
14	42662	26706	42666	26709	42670	26712	42674	26714	42678	26717	45
15	42682	26719	42687	26722	42691	26724	42695	26727	42699	26730	44
16	42703	26732	42708	26735	42712	26737	42716	26740	42720	26742	43
17	42724	26745	42728	26748	42733	26750	42737	26753	42741	26755	42
18	42745	26758	42749	26760	42754	26763	42758	26766	42762	26768	41
19	42766	26771	42770	26773	42775	26776	42779	26779	42783	26781	40
20	42787	26784	42791	26786	42795	26789	42799	26791	42804	26794	39
21	42808	26797	42812	26799	42816	26802	42820	26804	42825	26807	38
22	42829	26809	42833	26812	42837	26815	42841	26817	42845	26820	37
23	42850	26822	42854	26825	42858	26827	42862	26830	42866	26833	36
24	42870	26835	42875	26838	42879	26840	42883	26843	42887	26846	35
25	42891	26848	42896	26851	42900	26853	42904	26856	42908	26858	34
26	42912	26861	42916	26864	42921	26866	42925	26869	42929	26871	33
27	42933	26874	42937	26876	42941	26879	42945	26882	42950	26884	32
28	42954	26887	42958	26889	42962	26892	42966	26894	42970	26897	31
29	42975	26900	42979	26902	42983	26905	42987	26907	42991	26910	30
30	42996	26913	43000	26915	43004	26918	43008	26920	43012	26923	29
31	43016	26925	43021	26928	43025	26931	43029	26933	43033	26936	28
32	43037	26938	43041	26941	43046	26944	43050	26946	43054	26949	27
33	43058	26951	43062	26954	43066	26956	43070	26959	43075	26962	26
34	43079	26964	43083	26967	43087	26969	43091	26972	43095	26975	25
35	43100	26977	43104	26980	43108	26982	43112	26985	43116	26987	24
36	43120	26990	43125	26993	43129	26995	43133	26998	43137	27000	23
37	43141	27003	43145	27005	43149	27008	43153	27011	43158	27013	22
38	43162	27016	43166	27018	43170	27021	43174	27024	43178	27026	21
39	43183	27029	43187	27031	43191	27034	43195	27037	43199	27039	20
40	43203	27042	43208	27044	43212	27047	43216	27049	43220	27052	19
41	43224	27055	43228	27057	43232	27060	43237	27062	43241	27065	18
42	43245	27068	43249	27070	43253	27073	43257	27075	43261	27078	17
43	43266	27080	43270	27083	43274	27086	43278	27088	43282	27091	16
44	43286	27093	43291	27096	43295	27099	43299	27101	43303	27104	15
45	43307	27106	43311	27109	43315	27111	43319	27114	43324	27117	14
46	43328	27119	43332	27122	43336	27124	43340	27127	43344	27130	13
47	43348	27132	43353	27135	43357	27137	43361	27140	43365	27143	12
48	43369	27145	43373	27148	43377	27150	43382	27153	43386	27155	11
49	43390	27158	43394	27161	43398	27163	43402	27166	43406	27168	10
50	43411	27171	43415	27174	43419	27176	43423	27179	43427	27181	09
51	43431	27184	43435	27186	43439	27189	43444	27192	43448	27194	08
52	43452	27197	43456	27199	43460	27202	43464	27205	43468	27207	07
53	43473	27210	43477	27212	43481	27215	43485	27218	43489	27220	06
54	43493	27223	43497	27225	43501	27228	43506	27231	43510	27233	05
55	43514	27236	43518	27238	43522	27241	43526	27243	43530	27246	04
56	43535	27249	43539	27251	43543	27254	43547	27256	43551	27259	03
57	43555	27262	43559	27264	43563	27267	43568	27269	43572	27272	02
58	43576	27275	43580	27277	43584	27280	43588	27282	43592	27285	01
59	43596	27288	43601	27290	43605	27293	43609	27295	43613	27298	00
	·8		·6		·4		·2		·0		

297°

PARTS for 0′·1 :—   LOGS. 2 ;   NATURALS, 1.

Q

63°   HAVERSINES

↓ ′	·0 Log. 9·	·0 Nat. 0·	·2 Log. 9·	·2 Nat. 0·	·4 Log. 9·	·4 Nat. 0·	·6 Log. 9·	·6 Nat. 0·	·8 Log. 9·	·8 Nat. 0·	Log. 9·	Nat. 0·	
00	43617	27300	43621	27303	43625	27306	43629	27308	43633	27311	43638	27313	59
01	43638	27313	43642	27316	43646	27319	43650	27321	43654	27324	43658	27326	58
02	43658	27326	43662	27329	43666	27332	43671	27334	43675	27337	43679	27339	57
03	43679	27339	43683	27342	43687	27345	43691	27347	43695	27350	43699	27352	56
04	43699	27352	43704	27355	43708	27358	43712	27360	43716	27363	43720	27365	55
05	43720	27365	43724	27368	43728	27370	43732	27373	43736	27376	43741	27378	54
06	43741	27378	43745	27381	43749	27383	43753	27386	43757	27389	43761	27391	53
07	43761	27391	43765	27394	43769	27396	43774	27399	43778	27402	43782	27404	52
08	43782	27404	43786	27407	43790	27409	43794	27412	43798	27415	43802	27417	51
09	43802	27417	43806	27420	43810	27422	43815	27425	43819	27428	43823	27430	50
10	43823	27430	43827	27433	43831	27435	43835	27438	43839	27441	43843	27443	49
11	43843	27443	43848	27446	43852	27448	43856	27451	43860	27454	43864	27456	48
12	43864	27456	43868	27459	43872	27461	43876	27464	43880	27467	43884	27469	47
13	43884	27469	43889	27472	43893	27474	43897	27477	43901	27480	43905	27482	46
14	43905	27482	43909	27485	43913	27487	43917	27490	43921	27493	43925	27495	45
15	43925	27495	43930	27498	43934	27500	43938	27503	43942	27505	43946	27508	44
16	43946	27508	43950	27511	43954	27513	43958	27516	43962	27518	43967	27521	43
17	43967	27521	43971	27524	43975	27526	43979	27529	43983	27531	43987	27534	42
18	43987	27534	43991	27537	43995	27539	43999	27542	44003	27544	44008	27547	41
19	44008	27547	44012	27550	44016	27552	44020	27555	44024	27557	44028	27560	40
20	44028	27560	44032	27563	44036	27565	44040	27568	44044	27570	44048	27573	39
21	44048	27573	44053	27576	44057	27578	44061	27581	44065	27583	44069	27586	38
22	44069	27586	44073	27589	44077	27591	44081	27594	44085	27596	44089	27599	37
23	44089	27599	44093	27602	44098	27604	44102	27607	44106	27609	44110	27612	36
24	44110	27612	44114	27615	44118	27617	44122	27620	44126	27622	44130	27625	35
25	44130	27625	44134	27628	44139	27630	44143	27633	44147	27635	44151	27638	34
26	44151	27638	44155	27641	44159	27643	44163	27646	44167	27648	44171	27651	33
27	44171	27651	44175	27654	44179	27656	44183	27659	44188	27661	44192	27664	32
28	44192	27664	44196	27667	44200	27669	44204	27672	44208	27675	44212	27677	31
29	44212	27677	44216	27680	44220	27682	44224	27685	44228	27688	44232	27690	30
30	44232	27690	44236	27693	44241	27695	44245	27698	44249	27701	44253	27703	29
31	44253	27703	44257	27706	44261	27708	44265	27711	44269	27714	44273	27716	28
32	44273	27716	44277	27719	44281	27721	44285	27724	44290	27727	44294	27729	27
33	44294	27729	44298	27732	44302	27734	44306	27737	44310	27740	44314	27742	26
34	44314	27742	44318	27745	44322	27747	44326	27750	44330	27753	44334	27755	25
35	44334	27755	44338	27758	44343	27760	44347	27763	44351	27766	44355	27768	24
36	44355	27768	44359	27771	44363	27773	44367	27776	44371	27779	44375	27781	23
37	44375	27781	44379	27784	44383	27786	44387	27789	44392	27792	44396	27794	22
38	44396	27794	44400	27797	44404	27800	44408	27802	44412	27805	44416	27807	21
39	44416	27807	44420	27810	44424	27813	44428	27815	44432	27818	44436	27820	20
40	44436	27820	44440	27823	44444	27826	44448	27828	44453	27831	44457	27833	19
41	44457	27833	44461	27836	44465	27839	44469	27841	44473	27844	44477	27846	18
42	44477	27846	44481	27849	44485	27852	44489	27854	44493	27857	44497	27859	17
43	44497	27859	44501	27862	44505	27865	44510	27867	44514	27870	44518	27873	16
44	44518	27873	44522	27875	44526	27878	44530	27880	44534	27883	44538	27886	15
45	44538	27886	44542	27888	44546	27891	44550	27893	44554	27896	44558	27899	14
46	44558	27899	44562	27901	44566	27904	44570	27906	44575	27909	44579	27912	13
47	44579	27912	44583	27914	44587	27917	44591	27919	44595	27922	44599	27925	12
48	44599	27925	44603	27927	44607	27930	44611	27933	44615	27935	44619	27938	11
49	44619	27938	44623	27940	44627	27943	44631	27946	44635	27948	44639	27951	10
50	44639	27951	44643	27953	44648	27956	44652	27959	44656	27961	44660	27964	09
51	44660	27964	44664	27966	44668	27969	44672	27972	44676	27974	44680	27977	08
52	44680	27977	44684	27980	44688	27982	44692	27985	44696	27987	44700	27990	07
53	44700	27990	44704	27993	44708	27995	44712	27998	44717	28000	44721	28003	06
54	44721	28003	44725	28006	44729	28008	44733	28011	44737	28013	44741	28016	05
55	44741	28016	44745	28019	44749	28021	44753	28024	44757	28027	44761	28029	04
56	44761	28029	44765	28032	44769	28034	44773	28037	44777	28040	44781	28042	03
57	44781	28042	44785	28045	44789	28047	44793	28050	44797	28053	44801	28055	02
58	44801	28055	44805	28058	44810	28061	44814	28063	44818	28066	44822	28068	01
59	44822	28068	44826	28071	44830	28074	44834	28076	44838	28079	44842	28081	00

·8   ·6   ·4   ·2   ·0 ← ↑

PARTS for 0′·1 :—   LOGS. 2 ;   NATURALS, 1.

64° HAVERSINES

	·0		·2		·4		·6		·8			
	Log.	Nat.	Log.	Nat.	Log.	Nat.	Log.	Nat.	Log.	Nat.	Log.	Nat.	′
′	9·	0·	9·	0·	9·	0·	9·	0·	9·	0·	9·	0·	
00	44842	28081	44846	28084	44850	28087	44854	28089	44858	28092	44862	28095	59
01	44862	28095	44866	28097	44870	28100	44874	28102	44878	28105	44882	28108	58
02	44882	28108	44886	28110	44890	28113	44895	28115	44899	28118	44903	28121	57
03	44903	28121	44907	28123	44911	28126	44915	28129	44919	28131	44923	28134	56
04	44923	28134	44927	28136	44931	28139	44935	28142	44939	28144	44943	28147	55
05	44943	28147	44947	28149	44951	28152	44955	28155	44959	28157	44963	28160	54
06	44963	28160	44967	28163	44971	28165	44975	28168	44979	28170	44983	28173	53
07	44983	28173	44987	28176	44991	28178	44995	28181	44999	28183	45003	28186	52
08	45003	28186	45007	28189	45011	28191	45016	28194	45020	28197	45024	28199	51
09	45024	28199	45028	28202	45032	28204	45036	28207	45040	28210	45044	28212	50
10	45044	28212	45048	28215	45052	28217	45056	28220	45060	28223	45064	28225	49
11	45064	28225	45068	28228	45072	28231	45076	28233	45080	28236	45084	28238	48
12	45084	28238	45088	28241	45092	28244	45096	28246	45100	28249	45104	28252	47
13	45104	28252	45108	28254	45112	28257	45116	28259	45120	28262	45124	28265	46
14	45124	28265	45128	28267	45132	28270	45136	28273	45140	28275	45144	28278	45
15	45144	28278	45148	28280	45152	28283	45157	28286	45161	28288	45165	28291	44
16	45165	28291	45169	28293	45173	28296	45177	28299	45181	28301	45185	28304	43
17	45185	28304	45189	28307	45193	28309	45197	28312	45201	28314	45205	28317	42
18	45205	28317	45209	28320	45213	28322	45217	28325	45221	28328	45225	28330	41
19	45225	28330	45229	28333	45233	28335	45237	28338	45241	28341	45245	28343	40
20	45245	28343	45249	28346	45253	28348	45257	28351	45261	28354	45265	28356	39
21	45265	28356	45269	28359	45273	28362	45277	28364	45281	28367	45285	28369	38
22	45285	28369	45289	28372	45293	28375	45297	28377	45301	28380	45305	28383	37
23	45305	28383	45309	28385	45313	28388	45317	28390	45321	28393	45325	28396	36
24	45325	28396	45329	28398	45333	28401	45337	28404	45341	28406	45345	28409	35
25	45345	28409	45349	28411	45353	28414	45357	28417	45361	28419	45365	28422	34
26	45365	28422	45369	28425	45373	28427	45377	28430	45381	28432	45385	28435	33
27	45385	28435	45389	28438	45393	28440	45397	28443	45401	28446	45405	28448	32
28	45405	28448	45409	28451	45413	28453	45418	28456	45422	28459	45426	28461	31
29	45426	28461	45430	28464	45434	28467	45438	28469	45442	28472	45446	28474	30
30	45446	28474	45450	28477	45454	28480	45458	28482	45462	28485	45466	28488	29
31	45466	28488	45470	28490	45474	28493	45478	28495	45482	28498	45486	28501	28
32	45486	28501	45490	28503	45494	28506	45498	28509	45502	28511	45506	28514	27
33	45506	28514	45510	28516	45514	28519	45518	28522	45522	28524	45526	28527	26
34	45526	28527	45530	28530	45534	28532	45538	28535	45542	28537	45546	28540	25
35	45546	28540	45550	28543	45554	28545	45558	28548	45562	28551	45566	28553	24
36	45566	28553	45570	28556	45574	28559	45578	28561	45582	28564	45586	28566	23
37	45586	28566	45590	28569	45594	28572	45597	28574	45601	28577	45605	28580	22
38	45605	28580	45609	28582	45613	28585	45617	28587	45621	28590	45625	28593	21
39	45625	28593	45629	28595	45633	28598	45637	28601	45641	28603	45645	28606	20
40	45645	28606	45649	28608	45653	28611	45657	28614	45661	28616	45665	28619	19
41	45665	28619	45669	28622	45673	28624	45677	28627	45681	28629	45685	28632	18
42	45685	28632	45689	28635	45693	28637	45697	28640	45701	28643	45705	28645	17
43	45705	28645	45709	28648	45713	28651	45717	28653	45721	28656	45725	28658	16
44	45725	28658	45729	28661	45733	28664	45737	28666	45741	28669	45745	28672	15
45	45745	28672	45749	28674	45753	28677	45757	28679	45761	28682	45765	28685	14
46	45765	28685	45769	28687	45773	28690	45777	28693	45781	28695	45785	28698	13
47	45785	28698	45789	28701	45793	28703	45797	28706	45801	28708	45805	28711	12
48	45805	28711	45809	28714	45813	28716	45817	28719	45821	28722	45825	28724	11
49	45825	28724	45829	28727	45833	28729	45837	28732	45841	28735	45845	28737	10
50	45845	28737	45849	28740	45853	28743	45857	28745	45861	28748	45865	28751	09
51	45865	28751	45869	28753	45873	28756	45876	28758	45880	28761	45884	28764	08
52	45884	28764	45888	28766	45892	28769	45896	28772	45900	28774	45904	28777	07
53	45904	28777	45908	28779	45912	28782	45916	28785	45920	28787	45924	28790	06
54	45924	28790	45928	28793	45932	28795	45936	28798	45940	28801	45944	28803	05
55	45944	28803	45948	28806	45952	28808	45956	28811	45960	28814	45964	28816	04
56	45964	28816	45968	28819	45972	28822	45976	28824	45980	28827	45984	28830	03
57	45984	28830	45988	28832	45992	28835	45996	28837	46000	28840	46004	28843	02
58	46004	28843	46008	28845	46012	28848	46015	28851	46019	28853	46023	28856	01
59	46023	28856	46027	28859	46031	28861	46035	28864	46039	28866	46043	28869	00
		·8		·6		·4		·2		·0	←		

PARTS for 0′·1 :—   LOGS. 2 ;   NATURALS, 1.

295°

65°   HAVERSINES

′	.0 Log.	.0 Nat.	.2 Log.	.2 Nat.	.4 Log.	.4 Nat.	.6 Log.	.6 Nat.	.8 Log.	.8 Nat.	Log.	Nat.	
	9·	0·	9·	0·	9·	0·	9·	0·	9·	0·	9·	0·	
00	46043	28869	46047	28872	46051	28874	46055	28877	46059	28880	46063	28882	59
01	46063	28882	46067	28885	46071	28888	46075	28890	46079	28893	46083	28895	58
02	46083	28895	46087	28898	46091	28901	46095	28903	46099	28906	46103	28909	57
03	46103	28909	46107	28911	46111	28914	46115	28917	46119	28919	46123	28922	56
04	46123	28922	46127	28924	46131	28927	46134	28930	46138	28932	46142	28935	55
05	46142	28935	46146	28938	46150	28940	46154	28943	46158	28946	46162	28948	54
06	46162	28948	46166	28951	46170	28953	46174	28956	46178	28959	46182	28961	53
07	46182	28961	46186	28964	46190	28967	46194	28969	46198	28972	46202	28975	52
08	46202	28975	46206	28977	46210	28980	46214	28983	46218	28985	46222	28988	51
09	46222	28988	46226	28990	46229	28993	46233	28996	46237	28998	46241	29001	50
10	46241	29001	46245	29004	46249	29006	46253	29009	46257	29012	46261	29014	49
11	46261	29014	46265	29017	46269	29019	46273	29022	46277	29025	46281	29027	48
12	46281	29027	46285	29030	46289	29033	46293	29035	46297	29038	46301	29041	47
13	46301	29041	46305	29043	46309	29046	46312	29049	46316	29051	46320	29054	46
14	46320	29054	46324	29056	46328	29059	46332	29062	46336	29064	46340	29067	45
15	46340	29067	46344	29070	46348	29072	46352	29075	46356	29078	46360	29080	44
16	46360	29080	46364	29083	46368	29086	46372	29088	46375	29091	46379	29093	43
17	46379	29093	46383	29096	46387	29099	46391	29101	46395	29104	46399	29107	42
18	46399	29107	46403	29109	46407	29112	46411	29115	46415	29117	46419	29120	41
19	46419	29120	46423	29123	46427	29125	46431	29128	46435	29130	46439	29133	40
20	46439	29133	46443	29136	46447	29138	46450	29141	46454	29144	46458	29146	39
21	46458	29146	46462	29149	46466	29152	46470	29154	46474	29157	46478	29160	38
22	46478	29160	46482	29162	46486	29165	46490	29167	46494	29170	46498	29173	37
23	46498	29173	46502	29175	46506	29178	46510	29181	46513	29183	46517	29186	36
24	46517	29186	46521	29189	46525	29191	46529	29194	46533	29197	46537	29199	35
25	46537	29199	46541	29202	46545	29204	46549	29207	46553	29210	46557	29212	34
26	46557	29212	46561	29215	46565	29218	46569	29220	46572	29223	46576	29226	33
27	46576	29226	46580	29228	46584	29231	46588	29234	46592	29236	46596	29239	32
28	46596	29239	46600	29242	46604	29244	46608	29247	46612	29249	46616	29252	31
29	46616	29252	46620	29255	46624	29257	46628	29260	46631	29263	46635	29265	30
30	46635	29265	46639	29268	46643	29271	46647	29273	46651	29276	46655	29279	29
31	46655	29279	46659	29281	46663	29284	46667	29287	46671	29289	46675	29292	28
32	46675	29292	46679	29294	46682	29297	46686	29300	46690	29302	46694	29305	27
33	46694	29305	46698	29308	46702	29310	46706	29313	46710	29316	46714	29318	26
34	46714	29318	46718	29321	46722	29324	46726	29326	46729	29329	46733	29332	25
35	46733	29332	46737	29334	46741	29337	46745	29340	46749	29342	46753	29345	24
36	46753	29345	46757	29347	46761	29350	46765	29353	46769	29355	46773	29358	23
37	46773	29358	46777	29361	46780	29363	46784	29366	46788	29369	46792	29371	22
38	46792	29371	46796	29374	46800	29377	46804	29379	46808	29382	46812	29385	21
39	46812	29385	46816	29387	46820	29390	46824	29392	46827	29395	46831	29398	20
40	46831	29398	46835	29400	46839	29403	46843	29406	46847	29408	46851	29411	19
41	46851	29411	46855	29414	46859	29416	46863	29419	46867	29422	46871	29424	18
42	46871	29424	46875	29427	46878	29430	46882	29432	46886	29435	46890	29438	17
43	46890	29438	46894	29440	46898	29443	46902	29446	46906	29448	46910	29451	16
44	46910	29451	46914	29453	46918	29456	46921	29459	46925	29461	46929	29464	15
45	46929	29464	46933	29467	46937	29469	46941	29472	46945	29475	46949	29477	14
46	46949	29477	46953	29480	46957	29483	46961	29485	46964	29488	46968	29491	13
47	46968	29491	46972	29493	46976	29496	46980	29499	46984	29501	46988	29504	12
48	46988	29504	46992	29507	46996	29509	47000	29512	47003	29514	47007	29517	11
49	47007	29517	47011	29520	47015	29522	47019	29525	47023	29528	47027	29530	10
50	47027	29530	47031	29533	47035	29536	47039	29538	47043	29541	47046	29544	09
51	47046	29544	47050	29546	47054	29549	47058	29552	47062	29554	47066	29557	08
52	47066	29557	47070	29560	47074	29562	47078	29565	47082	29568	47085	29570	07
53	47085	29570	47089	29573	47093	29576	47097	29578	47101	29581	47105	29583	06
54	47105	29583	47109	29586	47113	29589	47117	29591	47121	29594	47124	29597	05
55	47124	29597	47128	29599	47132	29602	47136	29605	47140	29607	47144	29610	04
56	47144	29610	47148	29613	47152	29615	47156	29618	47159	29621	47163	29623	03
57	47163	29623	47167	29626	47171	29629	47175	29631	47179	29634	47183	29637	02
58	47183	29637	47187	29639	47191	29642	47194	29645	47198	29647	47202	29650	01
59	47202	29650	47206	29653	47210	29655	47214	29658	47218	29661	47222	29663	00

| | .8 | | .6 | | .4 | | .2 | | .0 | ← |

PARTS for 0′.1 :—   LOGS. 2 ;   NATURALS, 1.

66° HAVERSINES

	.0		.2		.4		.6		.8				
′	Log. 9·	Nat. 0·	Log. 9·	Nat. 0·	Log. 9·	Nat. 0·	Log. 9·	Nat. 0·	Log. 9·	Nat. 0·	Log. 9·	Nat. 0·	′
00	47222	29663	47226	29666	47230	29668	47233	29671	47237	29674	47241	29676	59
01	47241	29676	47245	29679	47249	29682	47253	29684	47257	29687	47261	29690	58
02	47261	29690	47264	29692	47268	29695	47272	29698	47276	29700	47280	29703	57
03	47280	29703	47284	29706	47288	29708	47292	29711	47296	29714	47300	29716	56
04	47300	29716	47303	29719	47307	29722	47311	29724	47315	29727	47319	29730	55
05	47319	29730	47323	29732	47327	29735	47331	29738	47335	29740	47338	29743	54
06	47338	29743	47342	29746	47346	29748	47350	29751	47354	29754	47358	29756	53
07	47358	29756	47362	29759	47366	29762	47369	29764	47373	29767	47377	29770	52
08	47377	29770	47381	29772	47385	29775	47389	29777	47393	29780	47397	29783	51
09	47397	29783	47400	29785	47404	29788	47408	29791	47412	29793	47416	29796	50
10	47416	29796	47420	29799	47424	29801	47428	29804	47432	29807	47435	29809	49
11	47435	29809	47439	29812	47443	29815	47447	29817	47451	29820	47455	29823	48
12	47455	29823	47459	29825	47463	29828	47466	29831	47470	29833	47474	29836	47
13	47474	29836	47478	29839	47482	29841	47486	29844	47490	29847	47493	29849	46
14	47493	29849	47497	29852	47501	29855	47505	29857	47509	29860	47513	29863	45
15	47513	29863	47517	29865	47521	29868	47524	29871	47528	29873	47532	29876	44
16	47532	29876	47536	29879	47540	29881	47544	29884	47548	29887	47552	29889	43
17	47552	29889	47556	29892	47559	29895	47563	29897	47567	29900	47571	29903	42
18	47571	29903	47575	29905	47579	29908	47583	29911	47586	29913	47590	29916	41
19	47590	29916	47594	29919	47598	29921	47602	29924	47606	29927	47610	29929	40
20	47610	29929	47613	29932	47617	29935	47621	29937	47625	29940	47629	29943	39
21	47629	29943	47633	29945	47637	29948	47641	29951	47644	29953	47648	29956	38
22	47648	29956	47652	29959	47656	29961	47660	29964	47664	29967	47668	29969	37
23	47668	29969	47671	29972	47675	29975	47679	29977	47683	29980	47687	29983	36
24	47687	29983	47691	29985	47695	29988	47698	29991	47702	29993	47706	29996	35
25	47706	29996	47710	29999	47714	30001	47718	30004	47722	30007	47725	30009	34
26	47725	30009	47729	30012	47733	30015	47737	30017	47741	30020	47745	30023	33
27	47745	30023	47749	30025	47752	30028	47756	30031	47760	30033	47764	30036	32
28	47764	30036	47768	30039	47772	30041	47776	30044	47780	30047	47783	30049	31
29	47783	30049	47787	30052	47791	30055	47795	30057	47799	30060	47803	30063	30
30	47803	30063	47806	30065	47810	30068	47814	30071	47818	30073	47822	30076	29
31	47822	30076	47826	30079	47830	30081	47833	30084	47837	30087	47841	30089	28
32	47841	30089	47845	30092	47849	30095	47853	30097	47857	30100	47860	30103	27
33	47860	30103	47864	30105	47868	30108	47872	30111	47876	30113	47880	30116	26
34	47880	30116	47883	30119	47887	30121	47891	30124	47895	30127	47899	30129	25
35	47899	30129	47903	30132	47906	30135	47910	30137	47914	30140	47918	30143	24
36	47918	30143	47922	30145	47926	30148	47930	30151	47934	30153	47937	30156	23
37	47937	30156	47941	30159	47945	30161	47949	30164	47953	30167	47957	30169	22
38	47957	30169	47960	30172	47964	30175	47968	30177	47972	30180	47976	30183	21
39	47976	30183	47980	30185	47983	30188	47987	30191	47991	30193	47995	30196	20
40	47995	30196	47999	30199	48003	30201	48007	30204	48010	30207	48014	30209	19
41	48014	30209	48018	30212	48022	30215	48026	30217	48030	30220	48033	30223	18
42	48033	30223	48037	30225	48041	30228	48045	30231	48049	30233	48053	30236	17
43	48053	30236	48056	30239	48060	30241	48064	30244	48068	30247	48072	30249	16
44	48072	30249	48076	30252	48079	30255	48083	30257	48087	30261	48091	30263	15
45	48091	30263	48095	30265	48099	30268	48102	30271	48106	30273	48110	30276	14
46	48110	30276	48114	30279	48118	30281	48122	30284	48125	30287	48129	30290	13
47	48129	30290	48133	30292	48137	30295	48141	30298	48145	30300	48148	30303	12
48	48148	30303	48152	30306	48156	30308	48160	30311	48164	30314	48168	30316	11
49	48168	30316	48171	30319	48175	30322	48179	30324	48183	30327	48187	30330	10
50	48187	30330	48191	30332	48194	30335	48198	30338	48202	30340	48206	30343	09
51	48206	30343	48210	30346	48213	30348	48217	30351	48221	30354	48225	30356	08
52	48225	30356	48229	30359	48233	30362	48237	30364	48240	30367	48244	30370	07
53	48244	30370	48248	30372	48252	30375	48256	30378	48259	30380	48263	30383	06
54	48263	30383	48267	30386	48271	30388	48275	30391	48279	30394	48282	30397	05
55	48282	30397	48286	30399	48290	30402	48294	30405	48298	30407	48302	30410	04
56	48302	30410	48305	30413	48309	30415	48313	30418	48317	30421	48321	30423	03
57	48321	30423	48324	30426	48328	30429	48332	30431	48336	30434	48340	30437	02
58	48340	30437	48344	30439	48347	30442	48351	30445	48355	30447	48359	30450	01
59	48359	30450	48363	30453	48366	30455	48370	30458	48374	30461	48378	30463	00

·8		·6		·4		·2		·0	

293°

PARTS for 0′·1 :—   LOGS. 2 ;   NATURALS, 1.

HAVERSINES

67°

	·0		·2		·4		·6		·8		
	Log.	Nat.	Log.	Nat.	Log.	Nat.	Log.	Nat.	Log.	Nat.	
′	9·	0·	9·	0·	9·	0·	9·	0·	9·	0·	′
00	48378	30463	48382	30166	48385	30169	48389	30171	40393	30474	40397	30477	59
01	48397	30477	48401	30480	48405	30482	48408	30485	48412	30488	48416	30490	58
02	48416	30490	48420	30493	48424	30496	48428	30498	48431	30501	48435	30504	57
03	48435	30504	48439	30506	48443	30509	48447	30512	48450	30514	48454	30517	56
04	48454	30517	48458	30520	48462	30522	48466	30525	48469	30528	48473	30530	55
05	48473	30530	48477	30533	48481	30536	48485	30538	48489	30541	48492	30544	54
06	48492	30544	48496	30546	48500	30549	48504	30552	48508	30555	48511	30557	53
07	48511	30557	48515	30560	48519	30563	48523	30565	48527	30568	48530	30571	52
08	48530	30571	48534	30573	48538	30576	48542	30579	48546	30581	48549	30584	51
09	48549	30584	48553	30587	48557	30589	48561	30592	48565	30595	48568	30597	50
10	48568	30597	48572	30600	48576	30603	48580	30605	48584	30608	48587	30611	49
11	48587	30611	48591	30613	48595	30616	48599	30619	48603	30622	48607	30624	48
12	48607	30624	48610	30627	48614	30630	48618	30632	48622	30635	48626	30638	47
13	48626	30638	48629	30640	48633	30643	48637	30646	48641	30648	48645	30651	46
14	48645	30651	48648	30654	48652	30656	48656	30659	48660	30662	48664	30664	45
15	48664	30664	48667	30667	48671	30670	48675	30672	48679	30675	48683	30678	44
16	48683	30678	48686	30681	48690	30683	48694	30686	48698	30689	48702	30691	43
17	48702	30691	48705	30694	48709	30697	48713	30699	48717	30702	48720	30705	42
18	48720	30705	48724	30707	48728	30710	48732	30713	48736	30715	48739	30718	41
19	48739	30718	48743	30721	48747	30723	48751	30726	48755	30729	48758	30732	40
20	48758	30732	48762	30734	48766	30737	48770	30740	48774	30742	48777	30745	39
21	48777	30745	48781	30748	48785	30750	48789	30753	48793	30756	48796	30758	38
22	48796	30758	48800	30761	48804	30764	48808	30766	48812	30769	48815	30772	37
23	48815	30772	48819	30774	48823	30777	48827	30780	48830	30783	48834	30785	36
24	48834	30785	48838	30788	48842	30791	48846	30793	48849	30796	48853	30799	35
25	48853	30799	48857	30801	48861	30804	48865	30807	48868	30809	48872	30812	34
26	48872	30812	48876	30815	48880	30817	48883	30820	48887	30823	48891	30826	33
27	48891	30826	48895	30828	48899	30831	48902	30834	48906	30836	48910	30839	32
28	48910	30839	48914	30842	48918	30844	48921	30847	48925	30850	48929	30852	31
29	48929	30852	48933	30855	48936	30858	48940	30860	48944	30863	48948	30866	30
30	48948	30866	48952	30869	48955	30871	48959	30874	48963	30877	48967	30879	29
31	48967	30879	48970	30882	48974	30885	48978	30887	48982	30890	48986	30893	28
32	48986	30893	48989	30895	48993	30898	48997	30901	49001	30904	49004	30906	27
33	49004	30906	49008	30909	49012	30912	49016	30914	49020	30917	49023	30920	26
34	49023	30920	49027	30922	49031	30925	49035	30928	49038	30930	49042	30933	25
35	49042	30933	49046	30936	49050	30938	49054	30941	49057	30944	49061	30946	24
36	49061	30946	49065	30949	49069	30952	49072	30955	49076	30957	49080	30960	23
37	49080	30960	49084	30963	49088	30965	49091	30968	49095	30971	49099	30973	22
38	49099	30973	49103	30976	49106	30979	49110	30981	49114	30984	49118	30987	21
39	49118	30987	49122	30990	49125	30992	49129	30995	49133	30998	49137	31000	20
40	49137	31000	49140	31003	49144	31006	49148	31008	49152	31011	49155	31014	19
41	49155	31014	49159	31016	49163	31019	49167	31022	49170	31025	49174	31027	18
42	49174	31027	49178	31030	49182	31033	49185	31035	49189	31038	49193	31041	17
43	49193	31041	49197	31043	49201	31046	49204	31049	49208	31051	49212	31054	16
44	49212	31054	49216	31057	49220	31059	49223	31062	49227	31065	49231	31068	15
45	49231	31068	49235	31070	49238	31073	49242	31076	49246	31078	49250	31081	14
46	49250	31081	49253	31084	49257	31086	49261	31089	49265	31092	49268	31095	13
47	49268	31095	49272	31097	49276	31100	49280	31103	49283	31105	49287	31108	12
48	49287	31108	49291	31111	49295	31113	49298	31116	49302	31119	49306	31121	11
49	49306	31121	49310	31124	49314	31127	49317	31130	49321	31132	49325	31135	10
50	49325	31135	49329	31138	49332	31140	49336	31143	49340	31146	49344	31148	09
51	49344	31148	49347	31151	49351	31154	49355	31156	49359	31159	49362	31162	08
52	49362	31162	49366	31165	49370	31167	49374	31170	49377	31173	49381	31175	07
53	49381	31175	49385	31178	49389	31181	49392	31183	49396	31186	49400	31189	06
54	49400	31189	49404	31191	49407	31194	49411	31197	49415	31200	49419	31202	05
55	49419	31202	49422	31205	49426	31208	49430	31210	49434	31213	49437	31216	04
56	49437	31216	49441	31218	49445	31221	49449	31224	49452	31227	49456	31229	03
57	49456	31229	49460	31232	49464	31235	49467	31237	49471	31240	49475	31243	02
58	49475	31243	49479	31245	49482	31248	49486	31251	49490	31254	49494	31256	01
59	49494	31256	49497	31259	49501	31262	49505	31264	49509	31267	49512	31270	00

	·8	·6	·4	·2	·0	← ↑

292°

PARTS for 0′·1 :—   LOGS. 2 ;   NATURALS, 1.

68° HAVERSINES

′	·0 Log.	·0 Nat.	·2 Log.	·2 Nat.	·4 Log.	·4 Nat.	·6 Log.	·6 Nat.	·8 Log.	·8 Nat.	Log.	Nat.	′
	9·	0·	9·	0·	9·	0·	9·	0·	9·	0·	9·	0·	
00	49512	31270	49516	31272	49520	31275	49524	31278	49527	31280	49531	31283	59
01	49531	31283	49535	31286	49539	31289	49542	31291	49546	31294	49550	31297	58
02	49550	31297	49554	31299	49557	31302	49561	31305	49565	31307	49568	31310	57
03	49568	31310	49572	31313	49576	31316	49580	31318	49583	31321	49587	31324	56
04	49587	31324	49591	31326	49595	31329	49598	31332	49602	31334	49606	31337	55
05	49606	31337	49610	31340	49613	31343	49617	31345	49621	31348	49625	31351	54
06	49625	31351	49628	31353	49632	31356	49636	31359	49639	31361	49643	31364	53
07	49643	31364	49647	31367	49651	31370	49654	31372	49658	31375	49662	31378	52
08	49662	31378	49666	31380	49669	31383	49673	31386	49677	31388	49681	31391	51
09	49681	31391	49684	31394	49688	31397	49692	31399	49696	31402	49699	31405	50
10	49699	31405	49703	31407	49707	31410	49711	31413	49714	31415	49718	31418	49
11	49718	31418	49722	31421	49725	31424	49729	31426	49733	31429	49737	31432	48
12	49737	31432	49740	31434	49744	31437	49748	31440	49752	31442	49755	31445	47
13	49755	31445	49759	31448	49763	31451	49767	31453	49770	31456	49774	31459	46
14	49774	31459	49778	31461	49781	31464	49785	31467	49789	31469	49793	31472	45
15	49793	31472	49796	31475	49800	31478	49804	31480	49807	31483	49811	31486	44
16	49811	31486	49815	31488	49819	31491	49822	31494	49826	31496	49830	31499	43
17	49830	31499	49834	31502	49837	31505	49841	31507	49845	31510	49849	31513	42
18	49849	31513	49852	31515	49856	31518	49860	31521	49863	31523	49867	31526	41
19	49867	31526	49871	31529	49875	31532	49878	31534	49882	31537	49886	31540	40
20	49886	31540	49890	31542	49893	31545	49897	31548	49901	31551	49904	31553	39
21	49904	31553	49908	31556	49912	31559	49916	31561	49919	31564	49923	31567	38
22	49923	31567	49927	31569	49930	31572	49934	31575	49938	31578	49942	31580	37
23	49942	31580	49945	31583	49949	31586	49953	31588	49956	31591	49960	31594	36
24	49960	31594	49964	31596	49968	31599	49971	31602	49975	31605	49979	31607	35
25	49979	31607	49983	31610	49986	31613	49990	31615	49994	31618	49997	31621	34
26	49997	31621	50001	31624	50005	31626	50008	31629	50012	31632	50016	31634	33
27	50016	31634	50020	31637	50023	31640	50027	31642	50031	31645	50034	31648	32
28	50034	31648	50038	31651	50042	31653	50046	31656	50049	31659	50053	31661	31
29	50053	31661	50057	31664	50060	31667	50064	31670	50068	31672	50072	31675	30
30	50072	31675	50075	31678	50079	31680	50083	31683	50086	31686	50090	31688	29
31	50090	31688	50094	31691	50098	31694	50101	31697	50105	31699	50109	31702	28
32	50109	31702	50112	31705	50116	31707	50120	31710	50123	31713	50127	31716	27
33	50127	31716	50131	31718	50135	31721	50138	31724	50142	31726	50146	31729	26
34	50146	31729	50149	31732	50153	31735	50157	31737	50161	31740	50164	31743	25
35	50164	31743	50168	31745	50172	31748	50175	31751	50179	31753	50183	31756	24
36	50183	31756	50187	31759	50190	31762	50194	31764	50198	31767	50201	31770	23
37	50201	31770	50205	31772	50209	31775	50212	31778	50216	31781	50220	31783	22
38	50220	31783	50224	31786	50227	31789	50231	31791	50235	31794	50238	31797	21
39	50238	31797	50242	31800	50246	31802	50249	31805	50253	31808	50257	31810	20
40	50257	31810	50261	31813	50264	31816	50268	31818	50272	31821	50275	31824	19
41	50275	31824	50279	31827	50283	31829	50286	31832	50290	31835	50294	31837	18
42	50294	31837	50297	31840	50301	31843	50305	31846	50309	31848	50312	31851	17
43	50312	31851	50316	31854	50320	31856	50323	31859	50327	31862	50331	31865	16
44	50331	31865	50334	31867	50338	31870	50342	31873	50346	31875	50349	31878	15
45	50349	31878	50353	31881	50357	31884	50360	31886	50364	31889	50368	31892	14
46	50368	31892	50371	31894	50375	31897	50379	31900	50383	31902	50386	31905	13
47	50386	31905	50390	31908	50394	31911	50397	31913	50401	31916	50405	31919	12
48	50405	31919	50408	31921	50412	31924	50416	31927	50419	31930	50423	31932	11
49	50423	31932	50427	31935	50430	31938	50434	31940	50438	31943	50442	31946	10
50	50442	31946	50445	31949	50449	31951	50453	31954	50456	31957	50460	31959	09
51	50460	31959	50464	31962	50467	31965	50471	31968	50475	31970	50478	31973	08
52	50478	31973	50482	31976	50486	31978	50489	31981	50493	31984	50497	31987	07
53	50497	31987	50500	31989	50504	31992	50508	31995	50512	31997	50515	32000	06
54	50515	32000	50519	32003	50523	32006	50526	32008	50530	32011	50534	32014	05
55	50534	32014	50537	32016	50541	32019	50545	32022	50548	32025	50552	32027	04
56	50552	32027	50556	32030	50559	32033	50563	32035	50567	32038	50570	32041	03
57	50570	32041	50574	32044	50578	32046	50581	32049	50585	32052	50589	32054	02
58	50589	32054	50592	32057	50596	32060	50600	32063	50604	32065	50607	32068	01
59	50607	32068	50611	32071	50615	32073	50618	32076	50622	32079	50626	32082	00

| ·8 | ·6 | ·4 | ·2 | ·0 ← |

PARTS for 0′·1 :— LOGS. 2 ; NATURALS 1.

291°

HAVERSINES

69°

′	·0 Log. 9·	·0 Nat. 0·	·2 Log. 9·	·2 Nat. 0·	·4 Log. 9·	·4 Nat. 0·	·6 Log. 9·	·6 Nat. 0·	·8 Log. 9·	·8 Nat. 0·	Log. 9·	Nat. 0·	′
00	50626	32082	50629	32084	50633	32087	50637	32090	50640	32092	50644	32095	59
01	50644	32095	50648	32098	50651	32101	50655	32103	50659	32106	50662	32109	58
02	50662	32109	50666	32111	50670	32114	50673	32117	50677	32120	50681	32122	57
03	50681	32122	50684	32125	50688	32128	50692	32131	50695	32133	50699	32136	56
04	50699	32136	50703	32139	50706	32141	50710	32144	50714	32147	50717	32150	55
05	50717	32150	50721	32152	50725	32155	50728	32158	50732	32160	50736	32163	54
06	50736	32163	50739	32166	50743	32169	50747	32171	50751	32174	50754	32177	53
07	50754	32177	50758	32179	50761	32182	50765	32185	50769	32188	50772	32190	52
08	50772	32190	50776	32193	50780	32196	50783	32198	50787	32201	50791	32204	51
09	50791	32204	50794	32207	50798	32209	50802	32212	50806	32215	50809	32217	50
10	50809	32217	50813	32220	50816	32223	50820	32226	50824	32228	50827	32231	49
11	50827	32231	50831	32234	50835	32236	50838	32239	50842	32242	50846	32245	48
12	50846	32245	50849	32247	50853	32250	50857	32253	50860	32256	50864	32258	47
13	50864	32258	50868	32261	50871	32264	50875	32266	50879	32269	50882	32272	46
14	50882	32272	50886	32275	50890	32277	50893	32280	50897	32283	50901	32285	45
15	50901	32285	50904	32288	50908	32291	50912	32294	50915	32296	50919	32299	44
16	50919	32299	50923	32302	50926	32304	50930	32307	50934	32310	50937	32313	43
17	50937	32313	50941	32315	50945	32318	50948	32321	50952	32324	50956	32326	42
18	50956	32326	50959	32329	50963	32332	50967	32334	50970	32337	50974	32340	41
19	50974	32340	50977	32343	50981	32345	50985	32348	50988	32351	50992	32353	40
20	50992	32353	50996	32356	50999	32359	51003	32362	51007	32364	51010	32367	39
21	51010	32367	51014	32370	51018	32373	51021	32375	51025	32378	51029	32381	38
22	51029	32381	51032	32383	51036	32386	51040	32389	51043	32392	51047	32394	37
23	51047	32394	51050	32397	51054	32400	51058	32402	51062	32405	51065	32408	36
24	51065	32408	51069	32411	51072	32413	51076	32416	51080	32419	51083	32422	35
25	51083	32422	51087	32424	51091	32427	51094	32430	51098	32432	51102	32435	34
26	51102	32435	51105	32438	51109	32441	51113	32443	51116	32446	51120	32449	33
27	51120	32449	51123	32451	51127	32454	51131	32457	51134	32460	51138	32462	32
28	51138	32462	51142	32465	51145	32468	51149	32471	51153	32473	51156	32476	31
29	51156	32476	51160	32479	51163	32481	51167	32484	51171	32487	51174	32490	30
30	51174	32490	51178	32492	51182	32495	51185	32498	51189	32501	51193	32503	29
31	51193	32503	51196	32506	51200	32509	51204	32511	51207	32514	51211	32517	28
32	51211	32517	51214	32520	51218	32522	51222	32525	51225	32528	51229	32531	27
33	51229	32531	51233	32533	51236	32536	51240	32539	51244	32541	51247	32544	26
34	51247	32544	51251	32547	51254	32550	51258	32552	51262	32555	51265	32558	25
35	51265	32558	51269	32560	51273	32563	51276	32566	51280	32569	51284	32571	24
36	51284	32571	51287	32574	51291	32577	51295	32580	51298	32582	51302	32585	23
37	51302	32585	51305	32588	51309	32590	51313	32593	51316	32596	51320	32599	22
38	51320	32599	51324	32601	51327	32604	51331	32607	51335	32610	51338	32612	21
39	51338	32612	51342	32615	51345	32618	51349	32620	51353	32623	51356	32626	20
40	51356	32626	51360	32629	51363	32631	51367	32634	51371	32637	51374	32640	19
41	51374	32640	51378	32642	51382	32645	51385	32648	51389	32650	51393	32653	18
42	51393	32653	51396	32656	51400	32659	51404	32661	51407	32664	51411	32667	17
43	51411	32667	51414	32670	51418	32672	51422	32675	51425	32678	51429	32681	16
44	51429	32681	51432	32683	51436	32686	51440	32689	51443	32691	51447	32694	15
45	51447	32694	51451	32697	51454	32700	51458	32702	51462	32705	51465	32708	14
46	51465	32708	51469	32711	51472	32713	51476	32716	51480	32719	51483	32721	13
47	51483	32721	51487	32724	51490	32727	51494	32730	51498	32732	51501	32735	12
48	51501	32735	51505	32738	51509	32741	51512	32743	51516	32746	51519	32749	11
49	51519	32749	51523	32751	51527	32754	51530	32757	51534	32760	51538	32762	10
50	51538	32762	51541	32765	51545	32768	51548	32771	51552	32773	51556	32776	09
51	51556	32776	51559	32779	51563	32782	51567	32784	51570	32787	51574	32790	08
52	51574	32790	51577	32792	51581	32795	51585	32798	51588	32801	51592	32803	07
53	51592	32803	51595	32806	51599	32809	51603	32812	51606	32814	51610	32817	06
54	51610	32817	51614	32820	51617	32822	51621	32825	51624	32828	51628	32831	05
55	51628	32831	51632	32833	51635	32836	51639	32839	51642	32842	51646	32844	04
56	51646	32844	51650	32847	51653	32850	51657	32853	51661	32855	51664	32858	03
57	51664	32858	51668	32861	51671	32863	51675	32866	51679	32869	51682	32872	02
58	51682	32872	51686	32874	51689	32877	51693	32880	51697	32883	51700	32885	01
59	51700	32885	51704	32888	51707	32891	51711	32894	51715	32896	51718	32899	00

·8	·6	·4	·2	·0

290°

PARTS for 0′·1 :— LOGS, 2 ; NATURALS, 1.

HAVERSINES

70°

′	·0 Log. 9·	·0 Nat. 0·	·2 Log. 9·	·2 Nat. 0·	·4 Log. 9·	·4 Nat. 0·	·6 Log. 9·	·6 Nat· 0·	·8 Log. 9·	·8 Nat· 0·	·0 Log. 9·	·0 Nat. 0·	′
00	51718	32899	51722	32902	51725	32904	51729	32907	51733	32910	51736	32913	59
01	51736	32913	51740	32915	51743	32918	51747	32921	51751	32924	51754	32926	58
02	51754	32926	51758	32929	51762	32932	51765	32935	51769	32937	51772	32940	57
03	51772	32940	51776	32943	51780	32945	51783	32948	51787	32951	51790	32954	56
04	51790	32954	51794	32956	51798	32959	51801	32962	51805	32965	51808	32967	55
05	51808	32967	51812	32970	51816	32973	51819	32976	51823	32978	51826	32981	54
06	51826	32981	51830	32984	51834	32986	51837	32989	51841	32992	51844	32995	53
07	51844	32995	51848	32997	51852	33000	51855	33003	51859	33006	51862	33008	52
08	51862	33008	51866	33011	51870	33014	51873	33017	51877	33019	51880	33022	51
09	51880	33022	51884	33025	51888	33028	51891	33030	51895	33033	51898	33036	50
10	51898	33036	51902	33038	51906	33041	51909	33044	51913	33047	51916	33049	49
11	51916	33049	51920	33052	51924	33055	51927	33058	51931	33060	51934	33063	48
12	51934	33063	51938	33066	51942	33069	51945	33071	51949	33074	51952	33077	47
13	51952	33077	51956	33080	51960	33082	51963	33085	51967	33088	51970	33090	46
14	51970	33090	51974	33093	51978	33096	51981	33099	51985	33101	51988	33104	45
15	51988	33104	51992	33107	51996	33110	51999	33112	52003	33115	52006	33118	44
16	52006	33118	52010	33121	52014	33123	52017	33126	52021	33129	52024	33132	43
17	52024	33132	52028	33134	52032	33137	52035	33140	52039	33143	52042	33145	42
18	52042	33145	52046	33148	52050	33151	52053	33153	52057	33156	52060	33159	41
19	52060	33159	52064	33162	52068	33164	52071	33167	52075	33170	52078	33173	40
20	52078	33173	52082	33175	52086	33178	52089	33181	52093	33184	52096	33186	39
21	52096	33186	52100	33189	52104	33192	52107	33195	52111	33197	52114	33200	38
22	52114	33200	52118	33203	52122	33205	52125	33208	52129	33211	52132	33214	37
23	52132	33214	52136	33216	52140	33219	52143	33222	52147	33225	52150	33227	36
24	52150	33227	52154	33230	52158	33233	52161	33236	52165	33238	52168	33241	35
25	52168	33241	52171	33244	52175	33247	52178	33249	52182	33252	52185	33255	34
26	52185	33255	52189	33258	52193	33260	52196	33263	52200	33266	52203	33269	33
27	52203	33269	52207	33271	52211	33274	52214	33277	52218	33280	52221	33282	32
28	52221	33282	52225	33285	52228	33288	52232	33290	52236	33293	52239	33296	31
29	52239	33296	52243	33299	52246	33301	52250	33304	52253	33307	52257	33310	30
30	52257	33310	52261	33312	52264	33315	52268	33318	52271	33321	52275	33323	29
31	52275	33323	52278	33326	52282	33329	52286	33332	52289	33334	52293	33337	28
32	52293	33337	52296	33340	52300	33343	52303	33345	52307	33348	52311	33351	27
33	52311	33351	52314	33354	52318	33356	52321	33359	52325	33362	52328	33365	26
34	52328	33365	52332	33367	52336	33370	52339	33373	52343	33376	52346	33378	25
35	52346	33378	52350	33381	52354	33384	52357	33386	52361	33389	52364	33392	24
36	52364	33392	52368	33395	52371	33397	52375	33400	52378	33403	52382	33406	23
37	52382	33406	52386	33408	52389	33411	52393	33414	52396	33417	52400	33419	22
38	52400	33419	52403	33422	52407	33425	52410	33428	52414	33430	52418	33433	21
39	52418	33433	52421	33436	52425	33439	52428	33441	52432	33444	52436	33447	20
40	52436	33447	52439	33450	52443	33452	52446	33455	52450	33458	52453	33461	19
41	52453	33461	52457	33463	52461	33466	52464	33469	52468	33472	52471	33474	18
42	52471	33474	52475	33477	52478	33480	52482	33483	52485	33485	52489	33488	17
43	52489	33488	52493	33491	52496	33494	52500	33496	52503	33499	52507	33502	16
44	52507	33502	52510	33504	52514	33507	52517	33510	52521	33513	52525	33515	15
45	52525	33515	52528	33518	52532	33521	52535	33524	52539	33526	52542	33529	14
46	52542	33529	52546	33532	52550	33535	52553	33537	52557	33540	52560	33543	13
47	52560	33543	52564	33546	52567	33548	52571	33551	52574	33554	52578	33557	12
48	52578	33557	52581	33559	52585	33562	52588	33565	52592	33568	52596	33570	11
49	52596	33570	52599	33573	52603	33576	52606	33579	52610	33581	52613	33584	10
50	52613	33584	52617	33587	52621	33590	52624	33592	52628	33595	52631	33598	09
51	52631	33598	52635	33601	52638	33603	52642	33606	52645	33609	52649	33612	08
52	52649	33612	52653	33614	52656	33617	52660	33620	52663	33623	52667	33625	07
53	52667	33625	52670	33628	52674	33631	52677	33634	52681	33636	52684	33639	06
54	52684	33639	52688	33642	52692	33645	52695	33647	52699	33650	52702	33653	05
55	52702	33653	52706	33656	52709	33658	52713	33661	52716	33664	52720	33667	04
56	52720	33667	52724	33669	52727	33672	52731	33675	52734	33677	52738	33680	03
57	52738	33680	52741	33683	52745	33686	52748	33689	52752	33691	52755	33694	02
58	52755	33694	52759	33697	52762	33700	52766	33702	52769	33705	52773	33708	01
59	52773	33708	52777	33711	52780	33713	52784	33716	52787	33719	52791	33722	00

| ·8 | ·6 | ·4 | ·2 | ·0 |

289°

PARTS for 0′·1 :—   LOGS. 2 ;   NATURALS, 1.

71° HAVERSINES

′	·0 Log. 9·	·0 Nat. 0·	·2 Log. 9·	·2 Nat. 0·	·4 Log. 9·	·4 Nat. 0·	·6 Log. 9·	·6 Nat. 0·	·8 Log. 9·	·8 Nat. 0·		Nat. 0·	′
00	52791	33722	52794	33724	52798	33727	52801	33730	52805	33733	52809	33735	59
01	52809	33735	52812	33738	52816	33741	52819	33744	52823	33746	52826	33749	58
02	52826	33749	52830	33752	52833	33755	52837	33757	52840	33760	52844	33763	57
03	52844	33763	52848	33766	52851	33768	52855	33771	52858	33774	52862	33777	56
04	52862	33777	52865	33779	52869	33782	52872	33785	52876	33788	52879	33790	55
05	52879	33790	52883	33793	52886	33796	52890	33799	52893	33801	52897	33804	54
06	52897	33804	52901	33807	52904	33810	52908	33812	52911	33815	52915	33818	53
07	52915	33818	52918	33821	52922	33823	52925	33826	52929	33829	52932	33832	52
08	52932	33832	52936	33834	52939	33837	52943	33840	52946	33843	52950	33845	51
09	52950	33845	52954	33848	52957	33851	52961	33854	52964	33856	52968	33859	50
10	52968	33859	52971	33862	52975	33865	52978	33867	52982	33870	52985	33873	49
11	52985	33873	52989	33876	52992	33878	52996	33881	52999	33884	53003	33887	48
12	53003	33887	53007	33889	53010	33892	53014	33895	53017	33898	53021	33900	47
13	53021	33900	53024	33903	53028	33906	53031	33909	53035	33912	53038	33914	46
14	53038	33914	53042	33917	53045	33920	53049	33923	53052	33925	53056	33928	45
15	53056	33928	53059	33931	53063	33934	53066	33936	53070	33939	53073	33942	44
16	53073	33942	53077	33945	53081	33947	53084	33950	53088	33953	53091	33956	43
17	53091	33956	53095	33958	53098	33961	53102	33964	53105	33967	53109	33969	42
18	53109	33969	53112	33972	53116	33975	53119	33978	53123	33980	53126	33983	41
19	53126	33983	53130	33986	53133	33989	53137	33991	53140	33994	53144	33997	40
20	53144	33997	53148	34000	53151	34002	53155	34005	53158	34008	53162	34011	39
21	53162	34011	53165	34013	53169	34016	53172	34019	53176	34022	53179	34024	38
22	53179	34024	53183	34027	53186	34030	53190	34033	53193	34035	53197	34038	37
23	53197	34038	53200	34041	53204	34044	53207	34047	53211	34049	53214	34052	36
24	53214	34052	53218	34055	53221	34058	53225	34060	53228	34063	53232	34066	35
25	53232	34066	53235	34069	53239	34071	53242	34074	53246	34077	53249	34080	34
26	53249	34080	53253	34082	53256	34085	53260	34088	53263	34091	53267	34093	33
27	53267	34093	53271	34096	53274	34099	53278	34102	53281	34104	53285	34107	32
28	53285	34107	53288	34110	53292	34113	53295	34115	53299	34118	53302	34121	31
29	53302	34121	53306	34124	53309	34127	53313	34129	53316	34132	53320	34135	30
30	53320	34135	53323	34138	53327	34140	53330	34143	53334	34146	53337	34149	29
31	53337	34149	53341	34151	53344	34154	53348	34157	53351	34160	53355	34162	28
32	53355	34162	53358	34165	53362	34168	53365	34171	53369	34173	53372	34176	27
33	53372	34176	53376	34179	53379	34182	53383	34184	53386	34187	53390	34190	26
34	53390	34190	53393	34193	53397	34195	53400	34198	53404	34201	53407	34204	25
35	53407	34204	53411	34207	53414	34209	53418	34212	53421	34215	53425	34218	24
36	53425	34218	53428	34220	53432	34223	53435	34226	53439	34229	53442	34231	23
37	53442	34231	53446	34234	53449	34237	53453	34240	53456	34242	53460	34245	22
38	53460	34245	53463	34248	53467	34251	53470	34253	53474	34256	53477	34259	21
39	53477	34259	53481	34262	53484	34264	53488	34267	53491	34270	53495	34273	20
40	53495	34273	53498	34276	53502	34278	53505	34281	53509	34284	53512	34287	19
41	53512	34287	53516	34289	53519	34292	53523	34295	53526	34298	53530	34300	18
42	53530	34300	53533	34303	53537	34306	53540	34309	53544	34311	53547	34314	17
43	53547	34314	53551	34317	53554	34320	53558	34322	53561	34325	53565	34328	16
44	53565	34328	53568	34331	53572	34334	53575	34336	53579	34339	53582	34342	15
45	53582	34342	53586	34345	53589	34347	53593	34350	53596	34353	53600	34356	14
46	53600	34356	53603	34358	53607	34361	53610	34364	53614	34367	53617	34369	13
47	53617	34369	53621	34372	53624	34375	53628	34378	53631	34381	53635	34383	12
48	53635	34383	53638	34386	53642	34389	53645	34392	53649	34394	53652	34397	11
49	53652	34397	53656	34400	53659	34403	53663	34405	53666	34408	53670	34411	10
50	53670	34411	53673	34414	53677	34416	53680	34419	53684	34422	53687	34425	09
51	53687	34425	53690	34427	53694	34430	53697	34433	53701	34436	53704	34439	08
52	53704	34439	53708	34441	53711	34444	53715	34447	53718	34450	53722	34452	07
53	53722	34452	53725	34455	53729	34458	53732	34461	53736	34463	53739	34466	06
54	53739	34466	53743	34469	53746	34472	53750	34474	53753	34477	53757	34480	05
55	53757	34480	53760	34483	53764	34486	53767	34488	53771	34491	53774	34494	04
56	53774	34494	53778	34497	53781	34499	53785	34502	53788	34505	53792	34508	03
57	53792	34508	53795	34510	53799	34513	53802	34516	53806	34519	53809	34521	02
58	53809	34521	53812	34524	53816	34527	53819	34530	53823	34533	53826	34535	01
59	53826	34535	53830	34538	53833	34541	53837	34544	53840	34546	53844	34549	00

·8 ·6 ·4 ·2 ·0 ←

PARTS for 0′·1 :— LOGS. 2 ; NATURALS, 1.

72° HAVERSINES

′	·0 Log. 9·	·0 Nat. 0·	·2 Log. 9·	·2 Nat. 0·	·4 Log. 9·	·4 Nat. 0·	·6 Log. 9·	·6 Nat. 0·	·8 Log. 9·	·8 Nat. 0·	Log. 9·	Nat. 0·	′
00	53844	34549	53847	34552	53851	34555	53854	34557	53858	34560	53861	34563	59
01	53861	34563	53865	34566	53868	34569	53872	34571	53875	34574	53879	34577	58
02	53879	34577	53882	34580	53885	34582	53889	34585	53892	34588	53896	34591	57
03	53896	34591	53899	34593	53903	34596	53906	34599	53910	34602	53913	34604	56
04	53913	34604	53917	34607	53920	34610	53924	34613	53927	34616	53931	34618	55
05	53931	34618	53934	34621	53938	34624	53941	34627	53945	34629	53948	34632	54
06	53948	34632	53951	34635	53955	34638	53958	34640	53962	34643	53965	34646	53
07	53965	34646	53969	34649	53972	34652	53976	34654	53979	34657	53983	34660	52
08	53983	34660	53986	34663	53990	34665	53993	34668	53997	34671	54000	34674	51
09	54000	34674	54003	34676	54007	34679	54010	34682	54014	34685	54017	34688	50
10	54017	34688	54021	34690	54024	34693	54028	34696	54031	34699	54035	34701	49
11	54035	34701	54038	34704	54042	34707	54045	34710	54049	34712	54052	34715	48
12	54052	34715	54055	34718	54059	34721	54062	34724	54066	34726	54069	34729	47
13	54069	34729	54073	34732	54076	34735	54080	34737	54083	34740	54087	34743	46
14	54087	34743	54090	34746	54094	34748	54097	34751	54101	34754	54104	34757	45
15	54104	34757	54107	34760	54111	34762	54114	34765	54118	34768	54121	34771	44
16	54121	34771	54125	34773	54128	34776	54132	34779	54135	34782	54139	34784	43
17	54139	34784	54142	34787	54145	34790	54149	34793	54152	34796	54156	34798	42
18	54156	34798	54159	34801	54163	34804	54166	34807	54170	34809	54173	34812	41
19	54173	34812	54177	34815	54180	34818	54184	34821	54187	34823	54190	34826	40
20	54190	34826	54194	34829	54197	34832	54201	34834	54204	34837	54208	34840	39
21	54208	34840	54211	34843	54215	34845	54218	34848	54222	34851	54225	34854	38
22	54225	34854	54228	34857	54232	34859	54235	34862	54239	34865	54242	34868	37
23	54242	34868	54246	34870	54249	34873	54253	34876	54256	34879	54260	34882	36
24	54260	34882	54263	34884	54266	34887	54270	34890	54273	34893	54277	34895	35
25	54277	34895	54280	34898	54284	34901	54287	34904	54291	34906	54294	34909	34
26	54294	34909	54297	34912	54301	34915	54304	34918	54308	34920	54311	34923	33
27	54311	34923	54315	34926	54318	34929	54322	34931	54325	34934	54329	34937	32
28	54329	34937	54332	34940	54335	34943	54339	34945	54342	34948	54346	34951	31
29	54346	34951	54349	34954	54353	34956	54356	34959	54360	34962	54363	34965	30
30	54363	34965	54366	34967	54370	34970	54373	34973	54377	34976	54380	34979	29
31	54380	34979	54384	34981	54387	34984	54391	34987	54394	34990	54397	34992	28
32	54397	34992	54401	34995	54404	34998	54408	35001	54411	35004	54415	35006	27
33	54415	35006	54418	35009	54421	35012	54425	35015	54428	35017	54432	35020	26
34	54432	35020	54435	35023	54439	35026	54442	35029	54446	35031	54449	35034	25
35	54449	35034	54452	35037	54456	35040	54459	35042	54463	35045	54466	35048	24
36	54466	35048	54470	35051	54473	35054	54477	35056	54480	35059	54483	35062	23
37	54483	35062	54487	35065	54490	35067	54494	35070	54497	35073	54501	35076	22
38	54501	35076	54504	35078	54507	35081	54511	35084	54514	35087	54518	35090	21
39	54518	35090	54521	35092	54525	35095	54528	35098	54532	35101	54535	35103	20
40	54535	35103	54538	35106	54542	35109	54545	35112	54549	35115	54552	35117	19
41	54552	35117	54556	35120	54559	35123	54563	35126	54566	35128	54569	35131	18
42	54569	35131	54573	35134	54576	35137	54580	35140	54583	35142	54587	35145	17
43	54587	35145	54590	35148	54593	35151	54597	35153	54600	35156	54604	35159	16
44	54604	35159	54607	35162	54610	35165	54614	35167	54617	35170	54621	35173	15
45	54621	35173	54624	35176	54628	35178	54631	35181	54635	35184	54638	35187	14
46	54638	35187	54641	35190	54645	35192	54648	35195	54652	35198	54655	35201	13
47	54655	35201	54659	35203	54662	35206	54666	35209	54669	35212	54672	35215	12
48	54672	35215	54676	35217	54679	35220	54683	35223	54686	35226	54689	35228	11
49	54689	35228	54693	35231	54696	35234	54700	35237	54703	35240	54707	35242	10
50	54707	35242	54710	35245	54713	35248	54717	35251	54720	35254	54724	35256	09
51	54724	35256	54727	35259	54730	35262	54734	35265	54737	35267	54741	35270	08
52	54741	35270	54744	35273	54748	35276	54751	35279	54754	35281	54758	35284	07
53	54758	35284	54761	35287	54765	35290	54768	35292	54772	35295	54775	35298	06
54	54775	35298	54778	35301	54782	35304	54785	35306	54789	35309	54792	35312	05
55	54792	35312	54795	35315	54799	35317	54802	35320	54806	35323	54809	35326	04
56	54809	35326	54813	35329	54816	35331	54819	35334	54823	35337	54826	35340	03
57	54826	35340	54830	35342	54833	35345	54837	35348	54840	35351	54843	35354	02
58	54843	35354	54847	35356	54850	35359	54854	35362	54857	35365	54860	35368	01
59	54860	35368	54864	35370	54867	35373	54871	35376	54874	35379	54878	35381	00

·8 ·6 ·4 ·2 ·0

287°

PARTS for 0′·1 :— LOGS. 2 ; NATURALS, 1.

73° HAVERSINES

′	·0 Log. 9·	·0 Nat. 0·	·2 Log. 9·	·2 Nat. 0·	·4 Log. 9·	·4 Nat. 0·	·6 Log. 9·	·6 Nat. 0·	·8 Log. 9·	·8 Nat. 0·	Log. 9·	Nat. 0·	′
00	54878	35381	54881	35384	54884	35387	54888	35390	54891	35393	54895	35395	59
01	54895	35395	54898	35398	54901	35401	54905	35404	54908	35406	54912	35409	58
02	54912	35409	54915	35412	54918	35415	54922	35418	54925	35420	54929	35423	57
03	54929	35423	54932	35426	54936	35429	54939	35431	54942	35434	54946	35437	56
04	54946	35437	54949	35440	54953	35443	54956	35445	54959	35448	54963	35451	55
05	54963	35451	54966	35454	54970	35457	54973	35459	54976	35462	54980	35465	54
06	54980	35465	54983	35468	54987	35470	54990	35473	54994	35476	54997	35479	53
07	54997	35479	55000	35482	55004	35484	55007	35487	55011	35490	55014	35493	52
08	55014	35493	55017	35496	55021	35498	55024	35501	55028	35504	55031	35507	51
09	55031	35507	55034	35509	55038	35512	55041	35515	55045	35518	55048	35521	50
10	55048	35521	55051	35523	55055	35526	55058	35529	55062	35532	55065	35534	49
11	55065	35534	55068	35537	55072	35540	55075	35543	55079	35546	55082	35548	48
12	55082	35548	55085	35551	55089	35554	55092	35557	55096	35560	55099	35562	47
13	55099	35562	55102	35565	55106	35568	55109	35571	55113	35573	55116	35576	46
14	55116	35576	55119	35579	55123	35582	55126	35585	55130	35587	55133	35590	45
15	55133	35590	55136	35593	55140	35596	55143	35599	55147	35601	55150	35604	44
16	55150	35604	55153	35607	55157	35610	55160	35612	55164	35615	55167	35618	43
17	55167	35618	55170	35621	55174	35624	55177	35626	55181	35629	55184	35632	42
18	55184	35632	55187	35635	55191	35638	55194	35640	55198	35643	55201	35646	41
19	55201	35646	55204	35649	55208	35651	55211	35654	55215	35657	55218	35660	40
20	55218	35660	55221	35663	55225	35665	55228	35668	55232	35671	55235	35674	39
21	55235	35674	55238	35677	55242	35679	55245	35682	55248	35685	55252	35688	38
22	55252	35688	55255	35690	55259	35693	55262	35696	55265	35699	55269	35702	37
23	55269	35702	55272	35704	55276	35707	55279	35710	55282	35713	55286	35716	36
24	55286	35716	55289	35718	55293	35721	55296	35724	55299	35727	55303	35730	35
25	55303	35730	55306	35732	55310	35735	55313	35738	55316	35741	55320	35743	34
26	55320	35743	55323	35746	55326	35749	55330	35752	55333	35755	55337	35757	33
27	55337	35757	55340	35760	55343	35763	55347	35766	55350	35769	55354	35771	32
28	55354	35771	55357	35774	55360	35777	55364	35780	55367	35783	55370	35785	31
29	55370	35785	55374	35788	55377	35791	55381	35794	55384	35796	55387	35799	30
30	55387	35799	55391	35802	55394	35805	55397	35808	55401	35810	55404	35813	29
31	55404	35813	55408	35816	55411	35819	55414	35822	55418	35824	55421	35827	28
32	55421	35827	55425	35830	55428	35833	55431	35835	55435	35838	55438	35841	27
33	55438	35841	55442	35844	55445	35847	55448	35849	55452	35852	55455	35855	26
34	55455	35855	55458	35858	55462	35861	55465	35863	55468	35866	55472	35869	25
35	55472	35869	55475	35872	55479	35875	55482	35877	55485	35880	55489	35883	24
36	55489	35883	55492	35886	55496	35889	55499	35891	55502	35894	55506	35897	23
37	55506	35897	55509	35900	55512	35902	55516	35905	55519	35908	55523	35911	22
38	55523	35911	55526	35914	55529	35916	55533	35919	55536	35922	55539	35925	21
39	55530	35925	55543	35928	55546	35930	55549	35933	55553	35936	55556	35939	20
40	55556	35939	55560	35942	55563	35944	55566	35947	55570	35950	55573	35953	19
41	55573	35953	55577	35955	55580	35958	55583	35961	55587	35964	55590	35967	18
42	55590	35967	55593	35969	55597	35972	55600	35975	55603	35978	55607	35981	17
43	55607	35981	55610	35983	55614	35986	55617	35989	55620	35992	55624	35995	16
44	55624	35995	55627	35997	55631	36000	55634	36003	55637	36006	55641	36009	15
45	55641	36009	55644	36011	55647	36014	55651	36017	55654	36020	55657	36023	14
46	55657	36023	55661	36025	55664	36028	55667	36031	55671	36034	55674	36036	13
47	55674	36036	55678	36039	55681	36042	55684	36045	55688	36048	55691	36050	12
48	55691	36050	55694	36053	55698	36056	55701	36059	55704	36062	55708	36064	11
49	55708	36064	55711	36067	55715	36070	55718	36073	55721	36076	55725	36078	10
50	55725	36078	55728	36081	55732	36084	55735	36087	55738	36090	55742	36092	09
51	55742	36092	55745	36095	55748	36098	55752	36101	55755	36104	55758	36106	08
52	55758	36106	55762	36109	55765	36112	55768	36115	55772	36118	55775	36120	07
53	55775	36120	55779	36123	55782	36126	55785	36129	55789	36131	55792	36134	06
54	55792	36134	55795	36137	55799	36140	55802	36143	55805	36145	55809	36148	05
55	55809	36148	55812	36151	55816	36154	55819	36157	55822	36159	55826	36162	04
56	55826	36162	55829	36165	55832	36168	55836	36171	55839	36173	55842	36176	03
57	55842	36176	55846	36179	55849	36182	55852	36185	55856	36187	55859	36190	02
58	55859	36190	55862	36193	55866	36196	55869	36199	55872	36201	55876	36204	01
59	55876	36204	55879	36207	55883	36210	55886	36213	55889	36215	55893	36218	00

·8 ·6 ·4 ·2 ·0

286°

PARTS for 0′·1 :—   LOGS. 2 ;   NATURALS, 1.

HAVERSINES

74°

′	·0 Log. 9·	·0 Nat. 0·	·2 Log. 9·	·2 Nat. 0·	·4 Log. 9·	·4 Nat. 0·	·6 Log. 9·	·6 Nat. 0·	·8 Log. 9·	·8 Nat. 0·	Log. 9·	Nat. 0·	′
00	55893	36218	55896	36221	55899	36224	55903	36227	55906	36229	55909	36232	59
01	55909	36232	55913	36235	55916	36238	55919	36241	55923	36243	55926	36246	58
02	55926	36246	55930	36249	55933	36252	55936	36255	55939	36257	55943	36260	57
03	55943	36260	55946	36263	55950	36266	55953	36268	55956	36271	55960	36274	56
04	55960	36274	55963	36277	55966	36280	55970	36282	55973	36285	55976	36288	55
05	55976	36288	55980	36291	55983	36294	55986	36296	55990	36299	55993	36302	54
06	55993	36302	55997	36305	56000	36308	56003	36310	56006	36313	56010	36316	53
07	56010	36316	56013	36319	56017	36322	56020	36324	56023	36327	56027	36330	52
08	56027	36330	56030	36333	56033	36336	56037	36338	56040	36341	56043	36344	51
09	56043	36344	56047	36347	56050	36350	56053	36352	56057	36355	56060	36358	50
10	56060	36358	56063	36361	56067	36364	56070	36366	56073	36369	56077	36372	49
11	56077	36372	56080	36375	56083	36378	56087	36380	56090	36383	56093	36386	48
12	56093	36386	56097	36389	56100	36392	56103	36394	56107	36397	56110	36400	47
13	56110	36400	56114	36403	56117	36406	56120	36408	56123	36411	56127	36414	46
14	56127	36414	56130	36417	56134	36420	56137	36422	56140	36425	56144	36428	45
15	56144	36428	56147	36431	56150	36434	56154	36436	56157	36439	56160	36442	44
16	56160	36442	56164	36445	56167	36448	56170	36450	56173	36453	56177	36456	43
17	56177	36456	56180	36459	56184	36462	56187	36464	56190	36467	56194	36470	42
18	56194	36470	56197	36473	56200	36476	56204	36478	56207	36481	56210	36484	41
19	56210	36484	56214	36487	56217	36490	56220	36492	56223	36495	56227	36498	40
20	56227	36498	56230	36501	56234	36504	56237	36506	56240	36509	56244	36512	39
21	56244	36512	56247	36515	56250	36518	56254	36520	56257	36523	56260	36526	38
22	56260	36526	56264	36529	56267	36532	56270	36534	56273	36537	56277	36540	37
23	56277	36540	56280	36543	56284	36546	56287	36548	56290	36551	56294	36554	36
24	56294	36554	56297	36557	56300	36560	56304	36562	56307	36565	56310	36568	35
25	56310	36568	56314	36571	56317	36574	56320	36576	56323	36579	56327	36582	34
26	56327	36582	56330	36585	56333	36588	56337	36590	56340	36593	56343	36596	33
27	56343	36596	56347	36599	56350	36602	56353	36604	56357	36607	56360	36610	32
28	56360	36610	56363	36613	56367	36616	56370	36618	56373	36621	56377	36624	31
29	56377	36624	56380	36627	56383	36630	56387	36632	56390	36635	56393	36638	30
30	56393	36638	56397	36641	56400	36644	56403	36647	56406	36649	56410	36652	29
31	56410	36652	56413	36655	56416	36658	56420	36661	56423	36663	56426	36666	28
32	56426	36666	56430	36669	56433	36672	56436	36675	56440	36677	56443	36680	27
33	56443	36680	56446	36683	56450	36686	56453	36689	56456	36691	56460	36694	26
34	56460	36694	56463	36697	56466	36700	56470	36703	56473	36705	56476	36708	25
35	56476	36708	56480	36711	56483	36714	56486	36717	56489	36719	56493	36722	24
36	56493	36722	56496	36725	56499	36728	56503	36731	56506	36733	56509	36736	23
37	56509	36736	56513	36739	56516	36742	56519	36745	56523	36747	56526	36750	22
38	56526	36750	56529	36753	56533	36756	56536	36759	56539	36761	56543	36764	21
39	56543	36764	56546	36767	56549	36770	56553	36773	56556	36775	56559	36778	20
40	56559	36778	56563	36781	56566	36784	56569	36787	56572	36790	56576	36792	19
41	56576	36792	56579	36795	56582	36798	56586	36801	56589	36804	56592	36806	18
42	56592	36806	56596	36809	56599	36812	56602	36815	56605	36818	56609	36820	17
43	56609	36820	56612	36823	56615	36826	56619	36829	56622	36832	56625	36834	16
44	56625	36834	56629	36837	56632	36840	56635	36843	56639	36846	56642	36848	15
45	56642	36848	56645	36851	56649	36854	56652	36857	56655	36860	56658	36862	14
46	56658	36862	56662	36865	56665	36868	56668	36871	56672	36874	56675	36877	13
47	56675	36877	56678	36879	56682	36882	56685	36885	56688	36888	56692	36891	12
48	56692	36891	56695	36893	56698	36896	56701	36899	56705	36902	56708	36905	11
49	56708	36905	56711	36907	56715	36910	56718	36913	56721	36916	56725	36919	10
50	56725	36919	56728	36921	56731	36924	56734	36927	56738	36930	56741	36933	09
51	56741	36933	56744	36935	56748	36938	56751	36941	56754	36944	56758	36947	08
52	56758	36947	56761	36950	56764	36952	56767	36955	56771	36958	56774	36961	07
53	56774	36961	56777	36964	56781	36966	56784	36969	56787	36972	56791	36975	06
54	56791	36975	56794	36978	56797	36980	56800	36983	56804	36986	56807	36989	05
55	56807	36989	56810	36992	56814	36994	56817	36997	56820	37000	56824	37003	04
56	56824	37003	56827	37006	56830	37008	56833	37011	56837	37014	56840	37017	03
57	56840	37017	56843	37020	56847	37023	56850	37025	56853	37028	56856	37031	02
58	56856	37031	56860	37034	56863	37037	56866	37039	56870	37042	56873	37045	01
59	56873	37045	56876	37048	56880	37051	56883	37053	56886	37056	56889	37059	00

| | ·8 | ·6 | ·4 | ·2 | ·0 | ← |

285°

PARTS for 0′·1 :—   LOGS. 2 ;   NATURALS, 1.

75° HAVERSINES

′	·0 Log. 9·	·0 Nat. 0·	·2 Log. 9·	·2 Nat. 0·	·4 Log. 9·	·4 Nat. 0·	·6 Log. 9·	·6 Nat. 0·	·8 Log. 9·	·8 Nat. 0·	′
00	56889	37059	56893	37062	56896	37065	56899	37067	56903	37070	56906 37073	59
01	56906	37073	56909	37076	56912	37079	56916	37082	56919	37084	56922 37087	58
02	56922	37087	56926	37090	56929	37093	56932	37096	56936	37098	56939 37101	57
03	56939	37101	56942	37104	56945	37107	56949	37110	56952	37112	56955 37115	56
04	56955	37115	56958	37118	56962	37121	56965	37124	56968	37126	56972 37129	55
05	56972	37129	56975	37132	56978	37135	56982	37138	56985	37141	56988 37143	54
06	56988	37143	56991	37146	56995	37149	56998	37152	57001	37155	57005 37157	53
07	57005	37157	57008	37160	57011	37163	57014	37166	57018	37169	57021 37171	52
08	57021	37171	57024	37174	57028	37177	57031	37180	57034	37183	57037 37186	51
09	57037	37186	57041	37188	57044	37191	57047	37194	57051	37197	57054 37200	50
10	57054	37200	57057	37202	57060	37205	57064	37208	57067	37211	57070 37214	49
11	57070	37214	57073	37216	57077	37219	57080	37222	57083	37225	57087 37228	48
12	57087	37228	57090	37231	57093	37233	57096	37236	57100	37239	57103 37242	47
13	57103	37242	57106	37245	57110	37247	57113	37250	57116	37253	57119 37256	46
14	57119	37256	57123	37259	57126	37261	57129	37264	57133	37267	57136 37270	45
15	57136	37270	57139	37273	57142	37276	57146	37278	57149	37281	57152 37284	44
16	57152	37284	57155	37287	57159	37290	57162	37292	57165	37295	57169 37298	43
17	57169	37298	57172	37301	57175	37304	57178	37306	57182	37309	57185 37312	42
18	57185	37312	57188	37315	57192	37318	57195	37321	57198	37323	57201 37326	41
19	57201	37326	57205	37329	57208	37332	57211	37335	57215	37337	57218 37340	40
20	57218	37340	57221	37343	57224	37346	57228	37349	57231	37351	57234 37354	39
21	57234	37354	57237	37357	57241	37360	57244	37363	57247	37366	57250 37368	38
22	57250	37368	57254	37371	57257	37374	57260	37377	57264	37380	57267 37382	37
23	57267	37382	57270	37385	57273	37388	57277	37391	57280	37394	57283 37397	36
24	57283	37397	57286	37399	57290	37402	57293	37405	57296	37408	57299 37411	35
25	57299	37411	57303	37413	57306	37416	57309	37419	57313	37422	57316 37425	34
26	57316	37425	57319	37428	57322	37430	57326	37433	57329	37436	57332 37439	33
27	57332	37439	57335	37442	57339	37444	57342	37447	57345	37450	57348 37453	32
28	57348	37453	57352	37456	57355	37458	57358	37461	57362	37464	57365 37467	31
29	57365	37467	57368	37470	57371	37473	57375	37475	57378	37478	57381 37481	30
30	57381	37481	57384	37484	57388	37487	57391	37489	57394	37492	57397 37495	29
31	57397	37495	57401	37498	57404	37501	57407	37504	57411	37506	57414 37509	28
32	57414	37509	57417	37512	57420	37515	57424	37518	57427	37520	57430 37523	27
33	57430	37523	57433	37526	57437	37529	57440	37532	57443	37535	57446 37537	26
34	57446	37537	57450	37540	57453	37543	57456	37546	57459	37549	57463 37551	25
35	57463	37551	57466	37554	57469	37557	57472	37560	57476	37563	57479 37566	24
36	57479	37566	57482	37568	57485	37571	57489	37574	57492	37577	57495 37580	23
37	57495	37580	57498	37582	57502	37585	57505	37588	57508	37591	57511 37594	22
38	57511	37594	57515	37597	57518	37599	57521	37602	57525	37605	57528 37608	21
39	57528	37608	57531	37611	57534	37613	57538	37616	57541	37619	57544 37622	20
40	57544	37622	57547	37625	57550	37627	57554	37630	57557	37633	57560 37636	19
41	57560	37636	57563	37639	57567	37642	57570	37644	57573	37647	57577 37650	18
42	57577	37650	57580	37653	57583	37656	57586	37659	57590	37661	57593 37664	17
43	57593	37664	57596	37667	57599	37670	57603	37673	57606	37675	57609 37678	16
44	57609	37678	57612	37681	57616	37684	57619	37687	57622	37690	57625 37692	15
45	57625	37692	57629	37695	57632	37698	57635	37701	57638	37704	57642 37706	14
46	57642	37706	57645	37709	57648	37712	57651	37715	57655	37718	57658 37721	13
47	57658	37721	57661	37723	57664	37726	57668	37729	57671	37732	57674 37735	12
48	57674	37735	57677	37737	57680	37740	57684	37743	57687	37746	57690 37749	11
49	57690	37749	57693	37752	57697	37754	57700	37757	57703	37760	57706 37763	10
50	57706	37763	57710	37766	57713	37768	57716	37771	57719	37774	57723 37777	09
51	57723	37777	57726	37780	57729	37783	57732	37785	57736	37788	57739 37791	08
52	57739	37791	57742	37794	57745	37797	57749	37800	57752	37802	57755 37805	07
53	57755	37805	57758	37808	57761	37811	57765	37814	57768	37816	57771 37819	06
54	57771	37819	57774	37822	57778	37825	57781	37828	57784	37831	57787 37833	05
55	57787	37833	57791	37836	57794	37839	57797	37842	57800	37845	57804 37847	04
56	57804	37847	57807	37850	57810	37853	57813	37856	57817	37859	57820 37862	03
57	57820	37862	57823	37864	57826	37867	57830	37870	57833	37873	57836 37876	02
58	57836	37876	57839	37879	57842	37881	57846	37884	57849	37887	57852 37890	01
59	57852	37890	57855	37893	57859	37895	57862	37898	57865	37901	57868 37904	00

·8   ·6   ·4   ·2   ·0

284°

PARTS for 0′·1 :— LOGS. 2 ; NATURALS, 1.

76° HAVERSINES

′	·0 Log. 9·	·0 Nat. 0·	·2 Log. 9·	·2 Nat. 0·	·4 Log. 9·	·4 Nat. 0·	·6 Log. 9·	·6 Nat. 0·	·8 Log. 9·	·8 Nat. 0·	Log. 9·	Nat. 0·	′
00	57868	37904	57872	37907	57875	37910	57878	37912	57881	37915	57885	37918	59
01	57885	37918	57888	37921	57891	37924	57894	37926	57898	37929	57901	37932	58
02	57901	37932	57904	37935	57907	37938	57910	37941	57914	37943	57917	37946	57
03	57917	37946	57920	37949	57923	37952	57927	37955	57930	37958	57933	37960	56
04	57933	37960	57936	37963	57939	37966	57943	37969	57946	37972	57949	37974	55
05	57949	37974	57952	37977	57956	37980	57959	37983	57962	37986	57965	37989	54
06	57965	37989	57969	37991	57972	37994	57975	37997	57978	38000	57981	38003	53
07	57981	38003	57985	38006	57988	38008	57991	38011	57994	38014	57998	38017	52
08	57998	38017	58001	38020	58004	38022	58007	38025	58011	38028	58014	38031	51
09	58014	38031	58017	38034	58020	38037	58023	38039	58027	38042	58030	38045	50
10	58030	38045	58033	38048	58036	38051	58040	38054	58043	38056	58046	38059	49
11	58046	38059	58049	38062	58052	38065	58056	38068	58059	38071	58062	38073	48
12	58062	38073	58065	38076	58068	38079	58072	38082	58075	38085	58078	38087	47
13	58078	38087	58081	38090	58085	38093	58088	38096	58091	38099	58094	38102	46
14	58094	38102	58097	38104	58101	38107	58104	38110	58107	38113	58110	38116	45
15	58110	38116	58114	38119	58117	38121	58120	38124	58123	38127	58126	38130	44
16	58126	38130	58130	38133	58133	38135	58136	38138	58139	38141	58143	38144	43
17	58143	38144	58146	38147	58149	38150	58152	38152	58155	38155	58159	38158	42
18	58159	38158	58162	38161	58165	38164	58168	38167	58172	38169	58175	38172	41
19	58175	38172	58178	38175	58181	38178	58184	38181	58188	38184	58191	38186	40
20	58191	38186	58194	38189	58197	38192	58200	38195	58204	38198	58207	38200	39
21	58207	38200	58210	38203	58213	38206	58217	38209	58220	38212	58223	38215	38
22	58223	38215	58226	38217	58229	38220	58233	38223	58236	38226	58239	38229	37
23	58239	38229	58242	38232	58245	38234	58249	38237	58252	38240	58255	38243	36
24	58255	38243	58258	38246	58261	38249	58265	38251	58268	38254	58271	38257	35
25	58271	38257	58274	38260	58278	38263	58281	38266	58284	38268	58287	38271	34
26	58287	38271	58290	38274	58294	38277	58297	38280	58300	38282	58303	38285	33
27	58303	38285	58306	38288	58310	38291	58313	38294	58316	38297	58319	38299	32
28	58319	38299	58322	38302	58326	38305	58329	38308	58332	38311	58335	38314	31
29	58335	38314	58338	38316	58342	38319	58345	38322	58348	38325	58351	38328	30
30	58351	38328	58355	38331	58358	38333	58361	38336	58364	38339	58367	38342	29
31	58367	38342	58371	38345	58374	38348	58377	38350	58380	38353	58383	38356	28
32	58383	38356	58387	38359	58390	38362	58393	38364	58396	38367	58399	38370	27
33	58399	38370	58403	38373	58406	38376	58409	38379	58412	38381	58415	38384	26
34	58415	38384	58419	38387	58422	38390	58425	38393	58428	38396	58431	38398	25
35	58431	38398	58435	38401	58438	38404	58441	38407	58444	38410	58447	38413	24
36	58447	38413	58451	38415	58454	38418	58457	38421	58460	38424	58463	38427	23
37	58463	38427	58467	38430	58470	38432	58473	38435	58476	38438	58479	38441	22
38	58479	38441	58483	38444	58486	38447	58489	38449	58492	38452	58495	38455	21
39	58495	38455	58499	38458	58502	38461	58505	38464	58508	38466	58511	38469	20
40	58511	38469	58515	38472	58518	38475	58521	38478	58524	38481	58527	38483	19
41	58527	38483	58530	38486	58534	38489	58537	38492	58540	38495	58543	38498	18
42	58543	38498	58546	38500	58550	38503	58553	38506	58556	38509	58559	38512	17
43	58559	38512	58562	38514	58566	38517	58569	38520	58572	38523	58575	38526	16
44	58575	38526	58578	38529	58582	38531	58585	38534	58588	38537	58591	38540	15
45	58591	38540	58594	38543	58598	38546	58601	38548	58604	38551	58607	38554	14
46	58607	38554	58610	38557	58613	38560	58617	38563	58620	38565	58623	38568	13
47	58623	38568	58626	38571	58629	38574	58633	38577	58636	38580	58639	38582	12
48	58639	38582	58642	38585	58645	38588	58649	38591	58652	38594	58655	38597	11
49	58655	38597	58658	38599	58661	38602	58664	38605	58668	38608	58671	38611	10
50	58671	38611	58674	38614	58677	38616	58680	38619	58684	38622	58687	38625	09
51	58687	38625	58690	38628	58693	38631	58696	38633	58700	38636	58703	38639	08
52	58703	38639	58706	38642	58709	38645	58712	38648	58715	38650	58719	38653	07
53	58719	38653	58722	38656	58725	38659	58728	38662	58731	38665	58735	38667	06
54	58735	38667	58738	38670	58741	38673	58744	38676	58747	38679	58750	38682	05
55	58750	38682	58754	38684	58757	38687	58760	38690	58763	38693	58766	38696	04
56	58766	38696	58770	38699	58773	38701	58776	38704	58779	38707	58782	38710	03
57	58782	38710	58785	38713	58789	38716	58792	38718	58795	38721	58798	38724	02
58	58798	38724	58801	38727	58805	38730	58808	38733	58811	38735	58814	38738	01
59	58814	38738	58817	38741	58820	38744	58824	38747	58827	38750	58830	38752	00
		·8		·6		·4		·2		·0	←		

PARTS for 0′·1 :—   LOGS. 2 ;   NATURALS, 1.

283°

77° HAVERSINES

′	.0 Log.	.0 Nat.	.2 Log.	.2 Nat.	.4 Log.	.4 Nat.	.6 Log.	.6 Nat.	.8 Log.	.8 Nat.	Log.	Nat.	
	9·	0·	9·	0·	9·	0·	9·	0·	9·	0·	9·	0·	
00	58830	38752	58833	38755	58836	38758	58839	38761	58843	38764	58846	38767	59
01	58846	38767	58849	38769	58852	38772	58855	38775	58858	38778	58862	38781	58
02	58862	38781	58865	38784	58868	38786	58871	38789	58874	38792	58878	38795	57
03	58878	38795	58881	38798	58884	38801	58887	38803	58890	38806	58893	38809	56
04	58893	38809	58897	38812	58900	38815	58903	38818	58906	38820	58909	38823	55
05	58909	38823	58912	38826	58916	38829	58919	38832	58922	38835	58925	38837	54
06	58925	38837	58928	38840	58932	38843	58935	38846	58938	38849	58941	38852	53
07	58941	38852	58944	38854	58947	38857	58950	38860	58954	38863	58957	38866	52
08	58957	38866	58960	38869	58963	38872	58966	38874	58969	38877	58973	38880	51
09	58973	38880	58976	38883	58979	38886	58982	38889	58985	38891	58988	38894	50
10	58988	38894	58992	38897	58995	38900	58998	38903	59001	38906	59004	38908	49
11	59004	38908	59008	38911	59011	38914	59014	38917	59017	38920	59020	38923	48
12	59020	38923	59023	38925	59027	38928	59030	38931	59033	38934	59036	38937	47
13	59036	38937	59039	38940	59042	38942	59045	38945	59049	38948	59052	38951	46
14	59052	38951	59055	38954	59058	38957	59061	38959	59064	38962	59068	38965	45
15	59068	38965	59071	38968	59074	38971	59077	38974	59080	38976	59083	38979	44
16	59083	38979	59087	38982	59090	38985	59093	38988	59096	38991	59099	38994	43
17	59099	38994	59102	38996	59106	38999	59109	39002	59112	39005	59115	39008	42
18	59115	39008	59118	39011	59121	39013	59124	39016	59128	39019	59131	39022	41
19	59131	39022	59134	39025	59137	39028	59140	39030	59143	39033	59147	39036	40
20	59147	39036	59150	39039	59153	39042	59156	39045	59159	39047	59162	39050	39
21	59162	39050	59166	39053	59169	39056	59172	39059	59175	39062	59178	39064	38
22	59178	39064	59181	39067	59185	39070	59188	39073	59191	39076	59194	39079	37
23	59194	39079	59197	39081	59200	39084	59203	39087	59207	39090	59210	39093	36
24	59210	39093	59213	39096	59216	39099	59219	39101	59222	39104	59225	39107	35
25	59225	39107	59229	39110	59232	39113	59235	39116	59238	39118	59241	39121	34
26	59241	39121	59244	39124	59248	39127	59251	39130	59254	39133	59257	39135	33
27	59257	39135	59260	39138	59263	39141	59266	39144	59270	39147	59273	39150	32
28	59273	39150	59276	39152	59279	39155	59282	39158	59285	39161	59289	39164	31
29	59289	39164	59292	39167	59295	39169	59298	39172	59301	39175	59304	39178	30
30	59304	39178	59307	39181	59311	39184	59314	39187	59317	39189	59320	39192	29
31	59320	39192	59323	39195	59326	39198	59329	39201	59333	39204	59336	39206	28
32	59336	39206	59339	39209	59342	39212	59345	39215	59348	39218	59351	39221	27
33	59351	39221	59355	39223	59358	39226	59361	39229	59364	39232	59367	39235	26
34	59367	39235	59370	39238	59373	39241	59377	39243	59380	39246	59383	39249	25
35	59383	39249	59386	39252	59389	39255	59392	39258	59395	39260	59399	39263	24
36	59399	39263	59402	39266	59405	39269	59408	39272	59411	39275	59414	39277	23
37	59414	39277	59418	39280	59421	39283	59424	39286	59427	39289	59430	39292	22
38	59430	39292	59433	39294	59436	39297	59439	39300	59443	39303	59446	39306	21
39	59446	39306	59449	39309	59452	39312	59455	39314	59458	39317	59461	39320	20
40	59461	39320	59465	39323	59468	39326	59471	39329	59474	39331	59477	39334	19
41	59477	39334	59480	39337	59483	39340	59487	39343	59490	39346	59493	39348	18
42	59493	39348	59496	39351	59499	39354	59502	39357	59505	39360	59508	39363	17
43	59508	39363	59512	39366	59515	39368	59518	39371	59521	39374	59524	39377	16
44	59524	39377	59527	39380	59530	39383	59534	39385	59537	39388	59540	39391	15
45	59540	39391	59543	39394	59546	39397	59549	39400	59552	39403	59556	39405	14
46	59556	39405	59559	39408	59562	39411	59565	39414	59568	39417	59571	39420	13
47	59571	39420	59574	39422	59577	39425	59581	39428	59584	39431	59587	39434	12
48	59587	39434	59590	39437	59593	39439	59596	39442	59599	39445	59602	39448	11
49	59602	39448	59606	39451	59609	39454	59612	39457	59615	39459	59618	39462	10
50	59618	39462	59621	39465	59624	39468	59628	39471	59631	39474	59634	39476	09
51	59634	39476	59637	39479	59640	39482	59643	39485	59646	39488	59649	39491	08
52	59649	39491	59653	39493	59656	39496	59659	39499	59662	39502	59665	39505	07
53	59665	39505	59668	39508	59671	39511	59674	39513	59677	39516	59681	39519	06
54	59681	39519	59684	39522	59687	39525	59690	39528	59693	39530	59696	39533	05
55	59696	39533	59700	39536	59703	39539	59706	39542	59709	39545	59712	39548	04
56	59712	39548	59715	39550	59718	39553	59721	39556	59724	39559	59728	39562	03
57	59728	39562	59731	39565	59734	39567	59737	39570	59740	39573	59743	39576	02
58	59743	39576	59746	39579	59749	39582	59753	39585	59756	39588	59759	39590	01
59	59759	39590	59762	39593	59765	39596	59768	39599	59771	39602	59774	39604	00

| | .8 | | .6 | | .4 | | .2 | | .0 | |

PARTS for 0′·1 :—   LOGS. 2 ;   NATURALS, 1.

78° HAVERSINES

′	·0 Log.	·0 Nat.	·2 Log.	·2 Nat.	·4 Log.	·4 Nat.	·6 Log.	·6 Nat.	·8 Log.	·8 Nat.	·0 Log.	·0 Nat.	′
	9·	0·	9·	0·	9·	0·	9·	0·	9·	0·	9·	0·	
00	59774	39604	59778	39607	59781	39610	59784	39613	59787	39616	59790	39619	59
01	59790	39619	59793	39621	59796	39624	59799	39627	59802	39630	59806	39633	58
02	59806	39633	59809	39636	59812	39639	59815	39641	59818	39644	59821	39647	57
03	59821	39647	59824	39650	59827	39653	59831	39656	59834	39659	59837	39661	56
04	59837	39661	59840	39664	59843	39667	59846	39670	59849	39673	59852	39676	55
05	59852	39676	59855	39678	59859	39681	59862	39684	59865	39687	59868	39690	54
06	59868	39690	59871	39693	59874	39695	59877	39698	59880	39701	59883	39704	53
07	59883	39704	59887	39707	59890	39710	59893	39713	59896	39715	59899	39718	52
08	59899	39718	59902	39721	59905	39724	59908	39727	59911	39730	59915	39732	51
09	59915	39732	59918	39735	59921	39738	59924	39741	59927	39744	59930	39747	50
10	59930	39747	59933	39750	59936	39752	59939	39755	59943	39758	59946	39761	49
11	59946	39761	59949	39764	59952	39767	59955	39770	59958	39772	59961	39775	48
12	59961	39775	59964	39778	59967	39781	59971	39784	59974	39787	59977	39789	47
13	59977	39789	59980	39792	59983	39795	59986	39798	59989	39801	59992	39804	46
14	59992	39804	59995	39807	59999	39809	60002	39812	60005	39815	60008	39818	45
15	60008	39818	60011	39821	60014	39824	60017	39826	60020	39829	60023	39832	44
16	60023	39832	60027	39835	60030	39838	60033	39841	60036	39844	60039	39846	43
17	60039	39846	60042	39849	60045	39852	60048	39855	60051	39858	60054	39861	42
18	60054	39861	60058	39864	60061	39866	60064	39869	60067	39872	60070	39875	41
19	60070	39875	60073	39878	60076	39881	60079	39883	60082	39886	60085	39889	40
20	60085	39889	60089	39892	60092	39895	60095	39898	60098	39901	60101	39903	39
21	60101	39903	60104	39906	60107	39909	60110	39912	60113	39915	60116	39918	38
22	60116	39918	60120	39920	60123	39923	60126	39926	60129	39929	60132	39932	37
23	60132	39932	60135	39935	60138	39938	60141	39940	60144	39943	60147	39946	36
24	60147	39946	60151	39949	60154	39952	60157	39955	60160	39958	60163	39960	35
25	60163	39960	60166	39963	60169	39966	60172	39969	60175	39972	60178	39975	34
26	60178	39975	60181	39977	60185	39980	60188	39983	60191	39986	60194	39989	33
27	60194	39989	60197	39992	60200	39995	60203	39997	60206	40000	60209	40003	32
28	60209	40003	60212	40006	60216	40009	60219	40012	60222	40015	60225	40017	31
29	60225	40017	60228	40020	60231	40023	60234	40026	60237	40029	60240	40032	30
30	60240	40032	60243	40034	60246	40037	60250	40040	60253	40043	60256	40046	29
31	60256	40046	60259	40049	60262	40052	60265	40054	60268	40057	60271	40060	28
32	60271	40060	60274	40063	60277	40066	60280	40069	60284	40072	60287	40074	27
33	60287	40074	60290	40077	60293	40080	60296	40083	60299	40086	60302	40089	26
34	60302	40089	60305	40091	60308	40094	60311	40097	60315	40100	60318	40103	25
35	60318	40103	60321	40106	60324	40109	60327	40111	60330	40114	60333	40117	24
36	60333	40117	60336	40120	60339	40123	60342	40126	60345	40129	60348	40131	23
37	60348	40131	60351	40134	60355	40137	60358	40140	60361	40143	60364	40146	22
38	60364	40146	60367	40149	60370	40151	60373	40154	60376	40157	60379	40160	21
39	60379	40160	60382	40163	60385	40166	60388	40168	60392	40171	60395	40174	20
40	60395	40174	60398	40177	60401	40180	60404	40183	60407	40186	60410	40188	19
41	60410	40188	60413	40191	60416	40194	60419	40197	60423	40200	60426	40203	18
42	60426	40203	60429	40206	60432	40208	60435	40211	60438	40214	60441	40217	17
43	60441	40217	60444	40220	60447	40223	60450	40226	60453	40228	60456	40231	16
44	60456	40231	60459	40234	60463	40237	60466	40240	60469	40243	60472	40245	15
45	60472	40245	60475	40248	60478	40251	60481	40254	60484	40257	60487	40260	14
46	60487	40260	60490	40263	60493	40265	60496	40268	60499	40271	60502	40274	13
47	60502	40274	60505	40277	60509	40280	60512	40283	60515	40285	60518	40288	12
48	60518	40288	60521	40291	60524	40294	60527	40297	60530	40300	60533	40303	11
49	60533	40303	60536	40305	60539	40308	60542	40311	60546	40314	60549	40317	10
50	60549	40317	60552	40320	60555	40323	60558	40325	60561	40328	60564	40331	09
51	60564	40331	60567	40334	60570	40337	60573	40340	60576	40343	60579	40345	08
52	60579	40345	60582	40348	60586	40351	60589	40354	60592	40357	60595	40360	07
53	60595	40360	60598	40362	60601	40365	60604	40368	60607	40371	60610	40374	06
54	60610	40374	60613	40377	60616	40380	60619	40382	60622	40385	60625	40388	05
55	60625	40388	60628	40391	60632	40394	60635	40397	60638	40400	60641	40402	04
56	60641	40402	60644	40405	60647	40408	60650	40411	60653	40414	60656	40417	03
57	60656	40417	60659	40420	60662	40422	60665	40425	60668	40428	60671	40431	02
58	60671	40431	60674	40434	60678	40437	60681	40440	60684	40442	60687	40445	01
59	60687	40445	60690	40448	60693	40451	60696	40454	60699	40457	60702	40460	00

| | ·8 | | ·6 | | ·4 | | ·2 | | ·0 | | | |

PARTS for 0′·1 :— LOGS. 2 ; NATURALS, 1.

79° HAVERSINES

	·0		·2		·4		·6		·8			
	Log.	Nat.	Log.	Nat.	Log.	Nat.	Log.	Nat.	Log.	Nat.	Log.	Nat.	
′	9·	0·	9·	0·	9·	0·	9·	0·	9·	0·	9·	0·	′
00	60702	40460	60705	40462	60708	40465	60711	40468	60714	40471	60717	40474	59
01	60717	40474	60720	40477	60724	40480	60727	40482	60730	40485	60733	40488	58
02	60733	40488	60736	40491	60739	40494	60742	40497	60745	40500	60748	40502	57
03	60748	40502	60751	40505	60754	40508	60757	40511	60760	40514	60763	40517	56
04	60763	40517	60766	40520	60769	40522	60773	40525	60776	40528	60779	40531	55
05	60779	40531	60782	40534	60785	40537	60788	40540	60791	40542	60794	40545	54
06	60794	40545	60797	40548	60800	40551	60803	40554	60806	40557	60809	40560	53
07	60809	40560	60812	40562	60815	40565	60818	40568	60822	40571	60825	40574	52
08	60825	40574	60828	40577	60831	40580	60834	40582	60837	40585	60840	40588	51
09	60840	40588	60843	40591	60846	40594	60849	40597	60852	40600	60855	40602	50
10	60855	40602	60858	40605	60861	40608	60864	40611	60867	40614	60870	40617	49
11	60870	40617	60873	40620	60876	40622	60880	40625	60883	40628	60886	40631	48
12	60886	40631	60889	40634	60892	40637	60895	40640	60898	40642	60901	40645	47
13	60901	40645	60904	40648	60907	40651	60910	40654	60913	40657	60916	40660	46
14	60916	40660	60919	40662	60922	40665	60925	40668	60928	40671	60931	40674	45
15	60931	40674	60934	40677	60938	40680	60941	40682	60944	40685	60947	40688	44
16	60947	40688	60950	40691	60953	40694	60956	40697	60959	40700	60962	40702	43
17	60962	40702	60965	40705	60968	40708	60971	40711	60974	40714	60977	40717	42
18	60977	40717	60980	40720	60983	40722	60986	40725	60989	40728	60992	40731	41
19	60992	40731	60995	40734	60999	40737	61002	40740	61005	40742	61008	40745	40
20	61008	40745	61011	40748	61014	40751	61017	40754	61020	40757	61023	40760	39
21	61023	40760	61026	40762	61029	40765	61032	40768	61035	40771	61038	40774	38
22	61038	40774	61041	40777	61044	40780	61047	40782	61050	40785	61053	40788	37
23	61053	40788	61056	40791	61059	40794	61063	40797	61066	40800	61069	40802	36
24	61069	40802	61072	40805	61075	40808	61078	40811	61081	40814	61084	40817	35
25	61084	40817	61087	40820	61090	40822	61093	40825	61096	40828	61099	40831	34
26	61099	40831	61102	40834	61105	40837	61108	40840	61111	40842	61114	40845	33
27	61114	40845	61117	40848	61120	40851	61123	40854	61126	40857	61129	40860	32
28	61129	40860	61132	40862	61135	40865	61139	40868	61142	40871	61145	40874	31
29	61145	40874	61148	40877	61151	40880	61154	40882	61157	40885	61160	40888	30
30	61160	40888	61163	40891	61166	40894	61169	40897	61172	40900	61175	40903	29
31	61175	40903	61178	40905	61181	40908	61184	40911	61187	40914	61190	40917	28
32	61190	40917	61193	40920	61196	40923	61199	40925	61202	40928	61205	40931	27
33	61205	40931	61208	40934	61211	40937	61215	40940	61218	40943	61221	40945	26
34	61221	40945	61224	40948	61227	40951	61230	40954	61233	40957	61236	40960	25
35	61236	40960	61239	40963	61242	40965	61245	40968	61248	40971	61251	40974	24
36	61251	40974	61254	40977	61257	40980	61260	40983	61263	40985	61266	40988	23
37	61266	40988	61269	40991	61272	40994	61275	40997	61278	41000	61281	41003	22
38	61281	41003	61284	41006	61287	41008	61290	41011	61293	41014	61296	41017	21
39	61296	41017	61299	41020	61302	41023	61305	41026	61308	41028	61311	41031	20
40	61311	41031	61314	41034	61317	41037	61321	41040	61324	41043	61327	41046	19
41	61327	41046	61330	41048	61333	41051	61336	41054	61339	41057	61342	41060	18
42	61342	41060	61345	41063	61348	41066	61351	41068	61354	41071	61357	41074	17
43	61357	41074	61363	41077	61363	41080	61366	41083	61369	41086	61372	41089	16
44	61372	41089	61375	41091	61378	41094	61381	41097	61384	41100	61387	41103	15
45	61387	41103	61390	41106	61393	41109	61396	41111	61399	41114	61402	41117	14
46	61402	41117	61405	41120	61408	41123	61411	41126	61414	41129	61417	41131	13
47	61417	41131	61420	41134	61423	41137	61427	41140	61430	41143	61433	41146	12
48	61433	41146	61436	41149	61439	41151	61442	41154	61445	41157	61448	41160	11
49	61448	41160	61451	41163	61454	41166	61457	41169	61460	41172	61463	41174	10
50	61463	41174	61466	41177	61469	41180	61472	41183	61475	41186	61478	41189	09
51	61478	41189	61481	41192	61484	41194	61487	41197	61490	41200	61493	41203	08
52	61493	41203	61496	41206	61499	41209	61502	41212	61505	41214	61508	41217	07
53	61508	41217	61511	41220	61514	41223	61517	41226	61520	41229	61523	41232	06
54	61523	41232	61526	41235	61529	11237	61532	41240	61535	41243	61538	41246	05
55	61538	41246	61541	41249	61544	41252	61547	41255	61550	41257	61553	41260	04
56	61553	41260	61556	41263	61559	41266	61562	41269	61565	41272	61568	41275	03
57	61568	41275	61571	41277	61574	41280	61577	41283	61580	41286	61583	41289	02
58	61583	41289	61586	41292	61589	41295	61592	41298	61595	41300	61598	41303	01
59	61598	41303	61601	41306	61604	41309	61607	41312	61610	41315	61613	41318	00
	·8		·6		·4		·2		·0				

PARTS for 0′·1 :—   LOGS. 2 ;   NATURALS, 1.

HAVERSINES

80°												
	·0		·2		·4		·6		·8			
	Log.	Nat.	Log.	Nat.	Log.	Nat.	Log.	Nat.	Log.	Nat.	Log.	Nat.
′	9·	0·	9·	0·	9·	0·	9·	0·	9·	0·	9·	0·
00	61613	41318	61616	41320	61619	41323	61623	41326	61626	41329	61629	41332	59
01	61629	41332	61632	41335	61635	41338	61638	41341	61641	41343	61644	41346	58
02	61644	41346	61647	41349	61650	41352	61653	41355	61656	41358	61659	41361	57
03	61659	41361	61662	41363	61665	41366	61668	41369	61671	41372	61674	41375	56
04	61674	41375	61677	41378	61680	41381	61683	41383	61686	41386	61689	41389	55
05	61689	41389	61692	41392	61695	41395	61698	41398	61701	41401	61704	41404	54
06	61704	41404	61707	41406	61710	41409	61713	41412	61716	41415	61719	41418	53
07	61719	41418	61722	41421	61725	41424	61728	41426	61731	41429	61734	41432	52
08	61734	41432	61737	41435	61740	41438	61743	41441	61746	41444	61749	41447	51
09	61749	41447	61752	41449	61755	41452	61758	41455	61761	41458	61764	41461	50
10	61764	41461	61767	41464	61770	41467	61773	41469	61776	41472	61779	41475	49
11	61779	41475	61782	41478	61785	41481	61788	41484	61791	41487	61794	41490	48
12	61794	41490	61797	41492	61800	41495	61803	41498	61806	41501	61809	41504	47
13	61809	41504	61812	41507	61815	41510	61818	41512	61821	41515	61824	41518	46
14	61824	41518	61827	41521	61830	41524	61833	41527	61836	41530	61839	41533	45
15	61839	41533	61842	41535	61845	41538	61848	41541	61851	41544	61854	41547	44
16	61854	41547	61857	41550	61860	41553	61863	41555	61866	41558	61869	41561	43
17	61869	41561	61872	41564	61875	41567	61878	41570	61881	41573	61884	41576	42
18	61884	41576	61887	41578	61890	41581	61893	41584	61896	41587	61899	41590	41
19	61899	41590	61902	41593	61905	41596	61908	41598	61911	41601	61914	41604	40
20	61914	41604	61917	41607	61920	41610	61923	41613	61926	41616	61929	41619	39
21	61929	41619	61932	41621	61935	41624	61938	41627	61941	41630	61944	41633	38
22	61944	41633	61947	41636	61950	41639	61953	41641	61956	41644	61959	41647	37
23	61959	41647	61962	41650	61965	41653	61968	41656	61971	41659	61974	41662	36
24	61974	41662	61977	41664	61980	41667	61983	41670	61986	41673	61989	41676	35
25	61989	41676	61992	41679	61995	41682	61997	41685	62000	41687	62003	41690	34
26	62003	41690	62006	41693	62009	41696	62012	41699	62015	41702	62018	41705	33
27	62018	41705	62021	41707	62024	41710	62027	41713	62030	41716	62033	41719	32
28	62033	41719	62036	41722	62039	41725	62042	41728	62045	41730	62048	41733	31
29	62048	41733	62051	41736	62054	41739	62057	41742	62060	41745	62063	41748	30
30	62063	41748	62066	41750	62069	41753	62072	41756	62075	41759	62078	41762	29
31	62078	41762	62081	41765	62084	41768	62087	41771	62090	41773	62093	41776	28
32	62093	41776	62096	41779	62099	41782	62102	41785	62105	41788	62108	41791	27
33	62108	41791	62111	41794	62114	41796	62117	41799	62120	41802	62123	41805	26
34	62123	41805	62126	41808	62129	41811	62132	41814	62135	41816	62138	41819	25
35	62138	41819	62141	41822	62144	41825	62147	41828	62150	41831	62153	41834	24
36	62153	41834	62156	41837	62159	41839	62162	41842	62165	41845	62168	41848	23
37	62168	41848	62171	41851	62174	41854	62176	41857	62179	41860	62182	41862	22
38	62182	41862	62185	41865	62188	41868	62191	41871	62194	41874	62197	41877	21
39	62197	41877	62200	41880	62203	41882	62206	41885	62209	41888	62212	41891	20
40	62212	41891	62215	41894	62218	41897	62221	41900	62224	41903	62227	41905	19
41	62227	41905	62230	41908	62233	41911	62236	41914	62239	41917	62242	41920	18
42	62242	41920	62245	41923	62248	41926	62251	41928	62254	41931	62257	41934	17
43	62257	41934	62260	41937	62263	41940	62266	41943	62269	41946	62272	41949	16
44	62272	41949	62275	41951	62278	41954	62281	41957	62284	41960	62287	41963	15
45	62287	41963	62290	41966	62293	41969	62295	41971	62298	41974	62301	41977	14
46	62301	41977	62304	41980	62307	41983	62310	41986	62313	41989	62316	41992	13
47	62316	41992	62319	41994	62322	41997	62325	42000	62328	42003	62331	42006	12
48	62331	42006	62334	42009	62337	42012	62340	42015	62343	42017	62346	42020	11
49	62346	42020	62349	42023	62352	42026	62355	42029	62358	42032	62361	42035	10
50	62361	42035	62364	42038	62367	42040	62370	42043	62373	42046	62376	42049	09
51	62376	42049	62379	42052	62382	42055	62384	42058	62387	42061	62390	42063	08
52	62390	42063	62393	42066	62396	42069	62399	42072	62402	42075	62405	42078	07
53	62405	42078	62408	42081	62411	42083	62414	42086	62417	42089	62420	42092	06
54	62420	42092	62423	42095	62426	42098	62429	42101	62432	42104	62435	42106	05
55	62435	42106	62438	42109	62441	42112	62444	42115	62447	42118	62450	42121	04
56	62450	42121	62453	42124	62456	42127	62458	42129	62461	42132	62464	42135	03
57	62464	42135	62467	42138	62470	42141	62473	42144	62476	42147	62479	42150	02
58	62479	42150	62482	42152	62485	42155	62488	42158	62491	42161	62494	42164	01
59	62494	42164	62497	42167	62500	42170	62503	42173	62506	42175	62509	42178	00
	·8		·6		·4		·2		·0			

PARTS for 0′·1 :—   LOGS. 1 ;   NATURALS, 1.

81°   HAVERSINES

′	·0 Log	·0 Nat	·2 Log	·2 Nat	·4 Log	·4 Nat	·6 Log	·6 Nat	·8 Log	·8 Nat	Log	Nat	′
	9·	0·	9·	0·	9·	0·	9·	0·	9·	0·	9·	0·	
00	62509	42178	62512	42181	62515	42184	62518	42187	62521	42190	62524	42193	59
01	62524	42193	62527	42196	62530	42198	62532	42201	62535	42204	62538	42207	58
02	62538	42207	62541	42210	62544	42213	62547	42216	62550	42219	62553	42221	57
03	62553	42221	62556	42224	62559	42227	62562	42230	62565	42233	62568	42236	56
04	62568	42236	62571	42239	62574	42241	62577	42244	62580	42247	62583	42250	55
05	62583	42250	62586	42253	62589	42256	62592	42259	62595	42262	62598	42264	54
06	62598	42264	62601	42267	62604	42270	62606	42273	62609	42276	62612	42279	53
07	62612	42279	62615	42282	62618	42285	62621	42287	62624	42290	62627	42293	52
08	62627	42293	62630	42296	62633	42299	62636	42302	62639	42305	62642	42308	51
09	62642	42308	62645	42310	62648	42313	62651	42316	62654	42319	62657	42322	50
10	62657	42322	62660	42325	62663	42328	62665	42331	62668	42333	62671	42336	49
11	62671	42336	62674	42339	62677	42342	62680	42345	62683	42348	62686	42351	48
12	62686	42351	62689	42354	62692	42356	62695	42359	62698	42362	62701	42365	47
13	62701	42365	62704	42368	62707	42371	62710	42374	62713	42377	62716	42379	46
14	62716	42379	62719	42382	62721	42385	62724	42388	62727	42391	62730	42394	45
15	62730	42394	62733	42397	62736	42400	62739	42402	62742	42405	62745	42408	44
16	62745	42408	62748	42411	62751	42414	62754	42417	62757	42420	62760	42423	43
17	62760	42423	62763	42425	62766	42428	62769	42431	62771	42434	62774	42437	42
18	62774	42437	62777	42440	62780	42443	62783	42446	62786	42448	62789	42451	41
19	62789	42451	62792	42454	62795	42457	62798	42460	62801	42463	62804	42466	40
20	62804	42466	62807	42469	62810	42471	62813	42474	62816	42477	62819	42480	39
21	62819	42480	62822	42483	62824	42486	62827	42489	62830	42492	62833	42494	38
22	62833	42494	62836	42497	62839	42500	62842	42503	62845	42506	62848	42509	37
23	62848	42509	62851	42512	62854	42515	62857	42517	62860	42520	62863	42523	36
24	62863	42523	62866	42526	62869	42529	62871	42532	62874	42535	62877	42538	35
25	62877	42538	62880	42540	62883	42543	62886	42546	62889	42549	62892	42552	34
26	62892	42552	62895	42555	62898	42558	62901	42561	62904	42564	62907	42566	33
27	62907	42566	62910	42569	62913	42572	62915	42575	62918	42578	62921	42581	32
28	62921	42581	62924	42584	62927	42587	62930	42589	62933	42592	62936	42595	31
29	62936	42595	62939	42598	62942	42601	62945	42604	62948	42607	62951	42610	30
30	62951	42610	62954	42612	62957	42615	62959	42618	62962	42621	62965	42624	29
31	62965	42624	62968	42627	62971	42630	62974	42633	62977	42635	62980	42638	28
32	62980	42638	62983	42641	62986	42644	62989	42647	62992	42650	62995	42653	27
33	62995	42653	62998	42656	63000	42658	63003	42661	63006	42664	63009	42667	26
34	63009	42667	63012	42670	63015	42673	63018	42676	63021	42679	63024	42681	25
35	63024	42681	63027	42684	63030	42687	63033	42690	63036	42693	63039	42696	24
36	63039	42696	63042	42699	63044	42702	63047	42704	63050	42707	63053	42710	23
37	63053	42710	63056	42713	63059	42716	63062	42719	63065	42722	63068	42725	22
38	63068	42725	63071	42728	63074	42730	63077	42733	63079	42736	63082	42739	21
39	63082	42739	63085	42742	63088	42745	63091	42748	63094	42751	63097	42753	20
40	63097	42753	63100	42756	63103	42759	63106	42762	63109	42765	63112	42768	19
41	63112	42768	63115	42771	63118	42774	63120	42776	63123	42779	63126	42782	18
42	63126	42782	63129	42785	63132	42788	63135	42791	63138	42794	63141	42797	17
43	63141	42797	63144	42799	63147	42802	63150	42805	63153	42808	63156	42811	16
44	63156	42811	63159	42814	63161	42817	63164	42820	63167	42822	63170	42825	15
45	63170	42825	63173	42828	63176	42831	63179	42834	63182	42837	63185	42840	14
46	63185	42840	63188	42843	63191	42846	63193	42848	63196	42851	63199	42854	13
47	63199	42854	63202	42857	63205	42860	63208	42863	63211	42866	63214	42869	12
48	63214	42869	63217	42871	63220	42874	63223	42877	63225	42880	63228	42883	11
49	63228	42883	63231	42886	63234	42889	63237	42892	63240	42894	63243	42897	10
50	63243	42897	63246	42900	63249	42903	63252	42906	63255	42909	63258	42912	09
51	63258	42912	63261	42915	63263	42918	63266	42920	63269	42923	63272	42926	08
52	63272	42926	63275	42929	63278	42932	63281	42935	63284	42938	63287	42941	07
53	63287	42941	63290	42943	63293	42946	63296	42949	63298	42952	63301	42955	06
54	63301	42955	63304	42958	63307	42961	63310	42964	63313	42966	63316	42969	05
55	63316	42969	63319	42972	63322	42975	63325	42978	63327	42981	63330	42984	04
56	63330	42984	63333	42987	63336	42990	63339	42992	63342	42995	63345	42998	03
57	63345	42998	63348	43001	63351	43004	63354	43007	63357	43010	63360	43013	02
58	63360	43013	63362	43015	63365	43018	63368	43021	63371	43024	63374	43027	01
59	63374	43027	63377	43030	63380	43033	63383	43036	63386	43038	63389	43041	00

·8   ·6   ·4   ·2   ·0 ←

278°

PARTS for 0′·1 :—   LOGS. 1 ;   NATURALS, 1.

HAVERSINES

82°

′	·0 Log. 9·	·0 Nat. 0·	·2 Log. 9·	·2 Nat. 0·	·4 Log. 9·	·4 Nat. 0·	·6 Log. 9·	·6 Nat. 0·	·8 Log. 9·	·8 Nat. 0·	Log. 9·	Nat. 0·	′
00	63389	43041	63392	43044	63394	43047	63397	43050	63400	43053	63403	43056	59
01	63403	43056	63406	43059	63409	43062	63412	43064	63415	43067	63418	43070	58
02	63418	43070	63421	43073	63423	43076	63426	43079	63429	43082	63432	43085	57
03	63432	43085	63435	43087	63438	43090	63441	43093	63444	43096	63447	43099	56
04	63447	43099	63450	43102	63452	43105	63455	43108	63458	43110	63461	43113	55
05	63461	43113	63464	43116	63467	43119	63470	43122	63473	43125	63476	43128	54
06	63476	43128	63479	43131	63482	43134	63484	43136	63487	43139	63490	43142	53
07	63490	43142	63493	43145	63496	43148	63499	43151	63502	43154	63505	43157	52
08	63505	43157	63508	43159	63511	43162	63513	43165	63516	43168	63519	43171	51
09	63519	43171	63522	43174	63525	43177	63528	43180	63531	43183	63534	43185	50
10	63534	43185	63537	43188	63539	43191	63542	43194	63545	43197	63548	43200	49
11	63548	43200	63551	43203	63554	43206	63557	43208	63560	43211	63563	43214	48
12	63563	43214	63566	43217	63568	43220	63571	43223	63574	43226	63577	43229	47
13	63577	43229	63580	43232	63583	43234	63586	43237	63589	43240	63592	43243	46
14	63592	43243	63594	43246	63597	43249	63600	43252	63603	43255	63606	43257	45
15	63606	43257	63609	43260	63612	43263	63615	43266	63618	43269	63621	43272	44
16	63621	43272	63623	43275	63626	43278	63629	43281	63632	43283	63635	43286	43
17	63635	43286	63638	43289	63641	43292	63644	43295	63647	43298	63649	43301	42
18	63649	43301	63652	43304	63655	43306	63658	43309	63661	43312	63664	43315	41
19	63664	43315	63667	43318	63670	43321	63673	43324	63676	43327	63678	43330	40
20	63678	43330	63681	43332	63684	43335	63687	43338	63690	43341	63693	43344	39
21	63693	43344	63696	43347	63699	43350	63701	43353	63704	43355	63707	43358	38
22	63707	43358	63710	43361	63713	43364	63716	43367	63719	43370	63722	43373	37
23	63722	43373	63725	43376	63728	43379	63730	43381	63733	43384	63736	43387	36
24	63736	43387	63739	43390	63742	43393	63745	43396	63748	43399	63751	43402	35
25	63751	43402	63753	43404	63756	43407	63759	43410	63762	43413	63765	43416	34
26	63765	43416	63768	43419	63771	43422	63774	43425	63777	43428	63779	43430	33
27	63779	43430	63782	43433	63785	43436	63788	43439	63791	43442	63794	43445	32
28	63794	43445	63797	43448	63800	43451	63802	43453	63805	43456	63808	43459	31
29	63808	43459	63811	43462	63814	43465	63817	43468	63820	43471	63823	43474	30
30	63823	43474	63825	43477	63828	43479	63831	43482	63834	43485	63837	43488	29
31	63837	43488	63840	43491	63843	43494	63846	43497	63849	43500	63851	43503	28
32	63851	43503	63854	43505	63857	43508	63860	43511	63863	43514	63866	43517	27
33	63866	43517	63869	43520	63872	43523	63874	43526	63877	43528	63880	43531	26
34	63880	43531	63883	43534	63886	43537	63889	43540	63892	43543	63895	43546	25
35	63895	43546	63897	43549	63900	43552	63903	43554	63906	43557	63909	43560	24
36	63909	43560	63912	43563	63915	43566	63918	43569	63921	43572	63923	43575	23
37	63923	43575	63926	43578	63929	43580	63932	43583	63935	43586	63938	43589	22
38	63938	43589	63941	43592	63944	43595	63946	43598	63949	43601	63952	43603	21
39	63952	43603	63955	43606	63958	43609	63961	43612	63964	43615	63966	43618	20
40	63966	43618	63969	43621	63972	43624	63975	43627	63978	43629	63981	43632	19
41	63981	43632	63984	43635	63987	43638	63989	43641	63992	43644	63995	43647	18
42	63995	43647	63998	43650	64001	43653	64004	43655	64007	43658	64010	43661	17
43	64010	43661	64012	43664	64015	43667	64018	43670	64021	43673	64024	43676	16
44	64024	43676	64027	43678	64030	43681	64033	43684	64035	43687	64038	43690	15
45	64038	43690	64041	43693	64044	43696	64047	43699	64050	43702	64053	43704	14
46	64053	43704	64055	43707	64058	43710	64061	43713	64064	43716	64067	43719	13
47	64067	43719	64070	43722	64073	43725	64076	43728	64078	43730	64081	43733	12
48	64081	43733	64084	43736	64087	43739	64090	43742	64093	43745	64096	43748	11
49	64096	43748	64098	43751	64101	43754	64104	43756	64107	43759	64110	43762	10
50	64110	43762	64113	43765	64116	43768	64119	43771	64121	43774	64124	43777	09
51	64124	43777	64127	43780	64130	43782	64133	43785	64136	43788	64139	43791	08
52	64139	43791	64141	43794	64144	43797	64147	43800	64150	43803	64153	43805	07
53	64153	43805	64156	43808	64159	43811	64161	43814	64164	43817	64167	43820	06
54	64167	43820	64170	43823	64173	43826	64176	43829	64179	43831	64181	43834	05
55	64181	43834	64184	43837	64187	43840	64190	43843	64193	43846	64196	43849	04
56	64196	43849	64199	43852	64201	43855	64204	43857	64207	43860	64210	43863	03
57	64210	43863	64213	43866	64216	43869	64219	43872	64222	43875	64224	43878	02
58	64224	43878	64227	43881	64230	43883	64233	43886	64236	43889	64239	43892	01
59	64239	43892	64241	43895	64244	43898	64247	43901	64250	43904	64253	43907	00

| | ·8 | | ·6 | | ·4 | | ·2 | | ·0 | |

277°

PARTS for 0′·1 :—   LOGS. 1 ;   NATURALS, 1.

83°   HAVERSINES

′	·0 Log.	·0 Nat.	·2 Log.	·2 Nat.	·4 Log.	·4 Nat.	·6 Log.	·6 Nat.	·8 Log.	·8 Nat.	·0 Log.	·0 Nat.	′
	9·	0·	9·	0·	9·	0·	9·	0·	9·	0·	9·	0·	
00	64253	43907	64256	43909	64259	43912	64262	43915	64264	43918	64267	43921	59
01	64267	43921	64270	43924	64273	43927	64276	43930	64279	43933	64281	43935	58
02	64281	43935	64284	43938	64287	43941	64290	43944	64293	43947	64296	43950	57
03	64296	43950	64299	43953	64301	43956	64304	43959	64307	43961	64310	43964	56
04	64310	43964	64313	43967	64316	43970	64319	43973	64321	43976	64324	43979	55
05	64324	43979	64327	43982	64330	43984	64333	43987	64336	43990	64339	43993	54
06	64339	43993	64341	43996	64344	43999	64347	44002	64350	44005	64353	44008	53
07	64353	44008	64356	44010	64358	44013	64361	44016	64364	44019	64367	44022	52
08	64367	44022	64370	44025	64373	44028	64376	44031	64378	44034	64381	44036	51
09	64381	44036	64384	44039	64387	44042	64390	44045	64393	44048	64396	44051	50
10	64396	44051	64398	44054	64401	44057	64404	44060	64407	44062	64410	44065	49
11	64410	44065	64413	44068	64415	44071	64418	44074	64421	44077	64424	44080	48
12	64424	44080	64427	44083	64430	44086	64433	44088	64435	44091	64438	44094	47
13	64438	44094	64441	44097	64444	44100	64447	44103	64450	44106	64452	44109	46
14	64452	44109	64455	44112	64458	44114	64461	44117	64464	44120	64467	44123	45
15	64467	44123	64469	44126	64472	44129	64475	44132	64478	44135	64481	44138	44
16	64481	44138	64484	44140	64486	44143	64489	44146	64492	44149	64495	44152	43
17	64495	44152	64498	44155	64501	44158	64504	44161	64506	44164	64509	44166	42
18	64509	44166	64512	44169	64515	44172	64518	44175	64521	44178	64523	44181	41
19	64523	44181	64526	44184	64529	44187	64532	44190	64535	44192	64538	44195	40
20	64538	44195	64540	44198	64543	44201	64546	44204	64549	44207	64552	44210	39
21	64552	44210	64555	44213	64557	44216	64560	44218	64563	44221	64566	44224	38
22	64566	44224	64569	44227	64572	44230	64575	44233	64577	44236	64580	44239	37
23	64580	44239	64583	44242	64586	44244	64589	44247	64592	44250	64594	44253	36
24	64594	44253	64597	44256	64600	44259	64603	44262	64606	44265	64609	44268	35
25	64609	44268	64611	44270	64614	44273	64617	44276	64620	44279	64623	44282	34
26	64623	44282	64626	44285	64628	44288	64631	44291	64634	44294	64637	44296	33
27	64637	44296	64640	44299	64643	44302	64645	44305	64648	44308	64651	44311	32
28	64651	44311	64654	44314	64657	44317	64660	44320	64662	44323	64665	44325	31
29	64665	44325	64668	44328	64671	44331	64674	44334	64677	44337	64679	44340	30
30	64679	44340	64682	44343	64685	44346	64688	44349	64691	44351	64694	44354	29
31	64694	44354	64696	44357	64699	44360	64702	44363	64705	44366	64708	44369	28
32	64708	44369	64711	44372	64713	44375	64716	44377	64719	44380	64722	44383	27
33	64722	44383	64725	44386	64727	44389	64730	44392	64733	44395	64736	44398	26
34	64736	44398	64739	44401	64742	44403	64745	44406	64747	44409	64750	44412	25
35	64750	44412	64753	44415	64756	44418	64759	44421	64761	44424	64764	44427	24
36	64764	44427	64767	44429	64770	44432	64773	44435	64776	44438	64778	44441	23
37	64778	44441	64781	44444	64784	44447	64787	44450	64790	44453	64793	44455	22
38	64793	44455	64795	44458	64798	44461	64801	44464	64804	44467	64807	44470	21
39	64807	44470	64809	44473	64812	44476	64815	44479	64818	44481	64821	44484	20
40	64821	44484	64824	44487	64826	44490	64829	44493	64832	44496	64835	44499	19
41	64835	44499	64838	44502	64840	44505	64843	44508	64846	44510	64849	44513	18
42	64849	44513	64852	44516	64855	44519	64857	44522	64860	44525	64863	44528	17
43	64863	44528	64866	44531	64869	44534	64872	44536	64874	44539	64877	44542	16
44	64877	44542	64880	44545	64883	44548	64886	44551	64888	44554	64891	44557	15
45	64891	44557	64894	44560	64897	44562	64900	44565	64903	44568	64905	44571	14
46	64905	44571	64908	44574	64911	44577	64914	44580	64917	44583	64919	44586	13
47	64919	44586	64922	44588	64925	44591	64928	44594	64931	44597	64934	44600	12
48	64934	44600	64936	44603	64939	44606	64942	44609	64945	44612	64948	44614	11
49	64948	44614	64950	44617	64953	44620	64956	44623	64959	44626	64962	44629	10
50	64962	44629	64964	44632	64967	44635	64970	44638	64973	44641	64976	44643	09
51	64976	44643	64979	44646	64981	44649	64984	44652	64987	44655	64990	44658	08
52	64990	44658	64993	44661	64995	44664	64998	44667	65001	44669	65004	44672	07
53	65004	44672	65021	44675	65009	44678	65012	44681	65015	44684	65018	44687	06
54	65018	44687	65021	44690	65024	44693	65026	44695	65029	44698	65032	44701	05
55	65032	44701	65035	44704	65038	44707	65040	44710	65043	44713	65046	44716	04
56	65046	44716	65049	44719	65052	44722	65054	44724	65057	44727	65060	44730	03
57	65060	44730	65063	44733	65066	44736	65069	44739	65071	44742	65074	44745	02
58	65074	44745	65077	44748	65080	44750	65083	44753	65085	44756	65088	44759	01
59	65088	44759	65091	44762	65094	44765	65097	44768	65099	44771	65102	44774	00
	·8		·6		·4		·2		·0				

276°

PARTS for 0′·1:—   LOGS. 1;   NATURALS, 1.

84° — HAVERSINES

′	·0 Log 9.	·0 Nat 0.	·2 Log 9.	·2 Nat 0.	·4 Log 9.	·4 Nat 0.	·6 Log 9.	·6 Nat 0.	·8 Log 9.	·8 Nat 0.	Log 9.	Nat 0.	′
00	65102	44774	65105	44776	65108	44779	65111	44782	65113	44785	65116	44788	59
01	65116	44788	65119	44791	65122	44794	65125	44797	65127	44800	65130	44803	58
02	65130	44803	65133	44805	65136	44808	65139	44811	65141	44814	65144	44817	57
03	65144	44817	65147	44820	65150	44823	65153	44826	65155	44829	65158	44831	56
04	65158	44831	65161	44834	65164	44837	65167	44840	65169	44843	65172	44846	55
05	65172	44846	65175	44849	65178	44852	65181	44855	65183	44858	65186	44860	54
06	65186	44860	65189	44863	65192	44866	65195	44869	65197	44872	65200	44875	53
07	65200	44875	65203	44878	65206	44881	65209	44884	65211	44886	65214	44889	52
08	65214	44889	65217	44892	65220	44895	65223	44898	65225	44901	65228	44904	51
09	65228	44907	65231	44910	65234	44910	65237	44912	65239	44915	65242	44918	50
10	65242	44918	65245	44921	65248	44924	65251	44927	65253	44930	65256	44933	49
11	65256	44933	65259	44936	65262	44939	65265	44941	65267	44944	65270	44947	48
12	65270	44947	65273	44950	65276	44953	65279	44956	65281	44959	65284	44962	47
13	65284	44962	65287	44965	65290	44967	65293	44970	65295	44973	65298	44976	46
14	65298	44976	65301	44979	65304	44982	65307	44985	65309	44988	65312	44991	45
15	65312	44991	65315	44993	65318	44996	65321	44999	65323	45002	65326	45005	44
16	65326	45005	65329	45008	65332	45011	65335	45014	65337	45017	65340	45020	43
17	65340	45020	65343	45022	65346	45025	65348	45028	65351	45031	65354	45034	42
18	65354	45034	65357	45037	65360	45040	65362	45043	65365	45046	65368	45048	41
19	65368	45051	65371	45051	65374	45054	65376	45057	65379	45060	65382	45063	40
20	65382	45063	65385	45066	65388	45069	65390	45072	65393	45075	65396	45077	39
21	65396	45077	65399	45080	65402	45083	65404	45086	65407	45089	65410	45092	38
22	65410	45092	65413	45095	65415	45098	65418	45101	65421	45104	65424	45106	37
23	65424	45106	65427	45109	65429	45112	65432	45115	65435	45118	65438	45121	36
24	65438	45124	65441	45124	65443	45127	65446	45130	65449	45132	65452	45135	35
25	65452	45135	65454	45138	65457	45141	65460	45144	65463	45147	65466	45150	34
26	65466	45150	65468	45153	65471	45156	65474	45159	65477	45161	65480	45164	33
27	65480	45164	65482	45167	65485	45170	65488	45173	65491	45176	65493	45179	32
28	65493	45179	65496	45182	65499	45185	65502	45187	65505	45190	65507	45193	31
29	65507	45193	65510	45196	65513	45199	65516	45202	65518	45205	65521	45208	30
30	65521	45208	65524	45211	65527	45214	65530	45216	65532	45219	65535	45222	29
31	65535	45222	65538	45225	65541	45228	65543	45231	65546	45234	65549	45237	28
32	65549	45237	65552	45240	65555	45242	65557	45245	65560	45248	65563	45251	27
33	65563	45251	65566	45254	65569	45257	65571	45260	65574	45263	65577	45266	26
34	65577	45266	65580	45269	65582	45271	65585	45274	65588	45277	65591	45280	25
35	65591	45280	65594	45283	65596	45286	65599	45289	65602	45292	65605	45295	24
36	65605	45295	65607	45297	65610	45300	65613	45303	65616	45306	65619	45309	23
37	65619	45309	65621	45312	65624	45315	65627	45318	65630	45321	65632	45324	22
38	65632	45324	65635	45326	65638	45329	65641	45332	65643	45335	65646	45338	21
39	65646	45338	65649	45341	65652	45344	65655	45347	65657	45350	65660	45353	20
40	65660	45353	65663	45355	65666	45358	65668	45361	65671	45364	65674	45367	19
41	65674	45367	65677	45370	65680	45373	65682	45376	65685	45379	65688	45381	18
42	65688	45381	65691	45384	65693	45387	65696	45390	65699	45393	65702	45396	17
43	65702	45396	65705	45399	65707	45402	65710	45405	65713	45408	65716	45410	16
44	65716	45410	65718	45413	65721	45416	65724	45419	65727	45422	65729	45425	15
45	65729	45425	65732	45428	65735	45431	65738	45434	65740	45437	65743	45439	14
46	65743	45439	65746	45442	65749	45445	65752	45448	65754	45451	65757	45454	13
47	65757	45454	65760	45457	65763	45460	65765	45463	65768	45465	65771	45468	12
48	65771	45468	65774	45471	65777	45474	65779	45477	65782	45480	65785	45483	11
49	65785	45483	65788	45486	65790	45489	65793	45492	65796	45494	65799	45497	10
50	65799	45497	65801	45500	65804	45503	65807	45506	65810	45509	65812	45512	09
51	65812	45512	65815	45515	65818	45518	65821	45521	65823	45523	65826	45526	08
52	65826	45526	65829	45529	65832	45532	65834	45535	65837	45538	65840	45541	07
53	65840	45541	65843	45544	65846	45547	65848	45550	65851	45552	65854	45555	06
54	65854	45555	65857	45558	65859	45561	65862	45564	65865	45567	65868	45570	05
55	65868	45570	65870	45573	65873	45576	65876	45578	65879	45581	65881	45584	04
56	65881	45584	65884	45587	65887	45590	65890	45593	65892	45596	65895	45599	03
57	65895	45599	65898	45602	65901	45605	65903	45607	65906	45610	65909	45613	02
58	65909	45613	65912	45616	65915	45619	65917	45622	65920	45625	65923	45628	01
59	65923	45628	65926	45631	65928	45634	65931	45636	65934	45639	65937	45642	00

Bottom decimal headings (for 275°): ·8 ·6 ·4 ·2 ·0

PARTS for 0′·1 :— LOGS. 1 ; NATURALS, 1.

HAVERSINES

85°		·0		·2		·4		·6		·8		
↓		Log.	Nat.	Log.	Nat.	Log.	Nat.	Log.	Nat.	Log.	Nat.	↑
′		9·	0·	9·	0·	9·	0·	9·	0·	9·	0·	′
00		65937	45642	65939	45645	65942	45648	65945	45651	65948	45654	65950	45657	59
01		65950	45657	65953	45660	65956	45663	65959	45665	65961	45668	65964	45671	58
02		65964	45671	65967	45674	65970	45677	65972	45680	65975	45683	65978	45686	57
03		65978	45686	65981	45689	65984	45691	65986	45694	65989	45697	65992	45700	56
04		65992	45700	65995	45703	65997	45706	66000	45709	66003	45712	66006	45715	55
05		66006	45715	66008	45718	66011	45720	66014	45723	66017	45726	66019	45729	54
06		66019	45729	66022	45732	66025	45735	66028	45738	66030	45741	66033	45744	53
07		66033	45744	66036	45747	66039	45749	66041	45752	66044	45755	66047	45758	52
08		66047	45758	66050	45761	66052	45764	66055	45767	66058	45770	66061	45773	51
09		66061	45773	66063	45776	66066	45778	66069	45781	66072	45784	66074	45787	50
10		66074	45787	66077	45790	66080	45793	66083	45796	66085	45799	66088	45802	49
11		66088	45802	66091	45805	66094	45807	66096	45810	66099	45813	66102	45816	48
12		66102	45816	66105	45819	66107	45822	66110	45825	66113	45828	66116	45831	47
13		66116	45831	66118	45834	66121	45836	66124	45839	66126	45842	66129	45845	46
14		66129	45845	66132	45848	66135	45851	66137	45854	66140	45857	66143	45860	45
15		66143	45860	66146	45863	66149	45865	66151	45868	66154	45871	66157	45874	44
16		66157	45874	66160	45877	66162	45880	66165	45883	66168	45886	66170	45889	43
17		66170	45889	66173	45891	66176	45894	66179	45897	66181	45900	66184	45903	42
18		66184	45903	66187	45906	66190	45909	66192	45912	66195	45915	66198	45918	41
19		66198	45918	66201	45920	66203	45923	66206	45926	66209	45929	66212	45932	40
20		66212	45932	66214	45935	66217	45938	66220	45941	66223	45944	66225	45947	39
21		66225	45947	66228	45949	66231	45952	66234	45955	66236	45958	66239	45961	38
22		66239	45961	66242	45964	66244	45967	66247	45970	66250	45973	66253	45976	37
23		66253	45976	66255	45978	66258	45981	66261	45984	66264	45987	66266	45990	36
24		66266	45990	66269	45993	66272	45999	66275	45999	66277	46002	66280	46005	35
25		66280	46005	66283	46007	66286	46010	66288	46013	66291	46016	66294	46019	34
26		66294	46019	66297	46022	66299	46025	66302	46028	66305	46031	66307	46034	33
27		66307	46034	66310	46036	66313	46039	66316	46042	66318	46045	66321	46048	32
28		66321	46048	66324	46051	66327	46054	66329	46057	66332	46060	66335	46063	31
29		66335	46063	66338	46065	66340	46068	66343	46071	66346	46074	66348	46077	30
30		66348	46077	66351	46080	66354	46083	66357	46086	66359	46089	66362	46092	29
31		66362	46092	66365	46094	66368	46097	66370	46100	66373	46103	66376	46106	28
32		66376	46106	66379	46109	66381	46112	66384	46115	66387	46118	66389	46121	27
33		66389	46121	66392	46123	66395	46126	66398	46129	66400	46132	66403	46135	26
34		66403	46135	66406	46138	66409	46141	66411	46144	66414	46147	66417	46150	25
35		66417	46150	66420	46152	66422	46155	66425	46158	66428	46161	66430	46164	24
36		66430	46164	66433	46167	66436	46170	66439	46173	66441	46176	66444	46179	23
37		66444	46179	66447	46181	66449	46184	66452	46187	66455	46190	66458	46193	22
38		66458	46193	66460	46196	66463	46199	66466	46202	66468	46205	66471	46208	21
39		66471	46208	66474	46210	66477	46213	66479	46216	66482	46219	66485	46222	20
40		66485	46222	66488	46225	66490	46228	66493	46231	66496	46234	66499	46237	19
41		66499	46237	66501	46239	66504	46242	66507	46245	66509	46248	66512	46251	18
42		66512	46251	66515	46254	66518	46257	66520	46260	66523	46263	66526	46266	17
43		66526	46266	66529	46268	66531	46271	66534	46274	66537	46277	66539	46280	16
44		66539	46280	66542	46283	66545	46286	66548	46289	66550	46292	66553	46295	15
45		66553	46295	66556	46298	66558	46300	66561	46303	66564	46306	66567	46309	14
46		66567	46309	66569	46312	66572	46315	66575	46318	66577	46321	66580	46324	13
47		66580	46324	66583	46327	66586	46329	66588	46332	66591	46335	66594	46338	12
48		66594	46338	66597	46341	66599	46344	66602	46347	66605	46350	66607	46353	11
49		66607	46353	66610	46356	66613	46358	66616	46361	66618	46364	66621	46367	10
50		66621	46367	66624	46370	66626	46373	66629	46376	66632	46379	66635	46382	09
51		66635	46382	66637	46385	66640	46387	66643	46390	66645	46393	66648	46396	08
52		66648	46396	66651	46399	66654	46402	66656	46405	66659	46408	66662	46411	07
53		66662	46411	66665	46414	66667	46416	66670	46419	66673	46422	66675	46425	06
54		66675	46425	66678	46428	66681	46431	66683	46434	66686	46437	66689	46440	05
55		66689	46440	66692	46443	66694	46445	66697	46448	66700	46451	66702	46454	04
56		66702	46454	66705	46457	66708	46460	66711	46463	66713	46466	66716	46469	03
57		66716	46469	66719	46472	66721	46474	66724	46477	66727	46480	66730	46483	02
58		66730	46483	66732	46486	66735	46489	66738	46492	66740	46495	66743	46498	01
59		66743	46498	66746	46501	66749	46503	66751	46506	66754	46509	66757	46512	00
		·8		·6		·4		·2		·0		←	

PARTS for 0′·1 :—   LOGS. 1 ;   NATURALS, 1.

274°

86° HAVERSINES

↓	.0 Log.	.0 Nat.	.2 Log.	.2 Nat.	.4 Log.	.4 Nat.	.6 Log.	.6 Nat.	.8 Log.	.8 Nat.	.0 Log.	.0 Nat.	
′	9·	0·	9·	0·	9·	0·	9·	0·	9·	0·	9·	0·	
00	66757	46512	66759	46515	66762	46518	66765	46521	66767	46524	66770	46527	59
01	66770	46527	66773	46530	66776	46533	66778	46535	66781	46538	66784	46541	58
02	66784	46541	66786	46544	66789	46547	66792	46550	66795	46553	66797	46556	57
03	66797	46556	66800	46559	66803	46562	66805	46564	66808	46567	66811	46570	56
04	66811	46570	66814	46573	66816	46576	66819	46579	66822	46582	66824	46585	55
05	66824	46585	66827	46588	66830	46591	66832	46593	66835	46596	66838	46599	54
06	66838	46599	66841	46602	66843	46605	66846	46608	66849	46611	66851	46614	53
07	66851	46614	66854	46617	66857	46620	66860	46622	66862	46625	66865	46628	52
08	66865	46628	66868	46631	66870	46634	66873	46637	66876	46640	66878	46643	51
09	66878	46643	66881	46646	66884	46649	66887	46651	66889	46654	66892	46657	50
10	66892	46657	66895	46660	66897	46663	66900	46666	66903	46669	66905	46672	49
11	66905	46672	66908	46675	66911	46678	66914	46681	66916	46683	66919	46686	48
12	66919	46686	66922	46689	66924	46692	66927	46695	66930	46698	66932	46701	47
13	66932	46701	66935	46704	66938	46707	66941	46710	66943	46712	66946	46715	46
14	66946	46715	66949	46718	66951	46721	66954	46724	66957	46727	66959	46730	45
15	66959	46730	66962	46733	66965	46736	66968	46739	66970	46741	66973	46744	44
16	66973	46744	66976	46747	66978	46750	66981	46753	66984	46756	66986	46759	43
17	66986	46759	66989	46762	66992	46765	66994	46767	66997	46770	67000	46773	42
18	67000	46773	67003	46776	67005	46779	67008	46782	67011	46785	67013	46788	41
19	67013	46788	67016	46791	67019	46794	67021	46797	67024	46800	67027	46802	40
20	67027	46802	67029	46805	67032	46808	67035	46811	67038	46814	67040	46817	39
21	67040	46817	67043	46820	67046	46823	67048	46826	67051	46829	67054	46831	38
22	67054	46831	67056	46834	67059	46837	67062	46840	67065	46843	67067	46846	37
23	67067	46846	67070	46849	67073	46852	67075	46855	67078	46858	67081	46860	36
24	67081	46860	67083	46863	67086	46866	67089	46869	67091	46872	67094	46875	35
25	67094	46875	67097	46878	67100	46881	67102	46884	67105	46887	67108	46890	34
26	67108	46890	67110	46892	67113	46895	67116	46898	67118	46901	67121	46904	33
27	67121	46904	67124	46907	67126	46910	67129	46913	67132	46916	67134	46919	32
28	67134	46919	67137	46921	67140	46924	67142	46927	67145	46930	67148	46933	31
29	67148	46933	67151	46936	67153	46939	67156	46942	67159	46945	67161	46948	30
30	67161	46948	67164	46950	67167	46953	67169	46956	67172	46959	67175	46962	29
31	67175	46962	67177	46965	67180	46968	67183	46971	67186	46974	67188	46977	28
32	67188	46977	67191	46980	67194	46982	67196	46985	67199	46988	67202	46991	27
33	67202	46991	67204	46994	67207	46997	67210	47000	67212	47003	67215	47006	26
34	67215	47006	67218	47009	67220	47011	67223	47014	67226	47017	67228	47020	25
35	67228	47020	67231	47023	67234	47026	67236	47029	67239	47032	67242	47035	24
36	67242	47035	67244	47038	67247	47041	67250	47043	67253	47046	67255	47049	23
37	67255	47049	67258	47052	67261	47055	67263	47058	67266	47061	67269	47064	22
38	67269	47064	67271	47067	67274	47070	67277	47072	67279	47075	67282	47078	21
39	67282	47078	67285	47081	67287	47084	67290	47087	67293	47090	67295	47093	20
40	67295	47093	67298	47096	67301	47099	67303	47101	67306	47104	67309	47107	19
41	67309	47107	67311	47110	67314	47113	67317	47116	67320	47119	67322	47122	18
42	67322	47122	67325	47125	67328	47128	67330	47131	67333	47133	67336	47136	17
43	67336	47139	67338	47139	67341	47142	67344	47145	67346	47148	67349	47151	16
44	67349	47151	67352	47154	67354	47157	67357	47160	67360	47162	67362	47165	15
45	67362	47165	67365	47168	67368	47171	67370	47174	67373	47177	67376	47180	14
46	67376	47180	67378	47183	67381	47186	67384	47189	67386	47192	67389	47194	13
47	67389	47194	67392	47197	67394	47200	67397	47203	67400	47206	67402	47209	12
48	67402	47209	67405	47212	67408	47215	67410	47218	67413	47221	67416	47223	11
49	67416	47223	67418	47226	67421	47229	67424	47232	67427	47235	67429	47238	10
50	67429	47238	67432	47241	67434	47244	67437	47247	67440	47250	67442	47252	09
51	67442	47252	67445	47255	67448	47258	67450	47261	67453	47264	67456	47267	08
52	67456	47267	67458	47270	67461	47273	67464	47276	67467	47279	67469	47282	07
53	67469	47282	67472	47284	67474	47287	67477	47290	67480	47293	67482	47296	06
54	67482	47296	67485	47299	67488	47302	67490	47305	67493	47308	67496	47311	05
55	67496	47311	67498	47314	67501	47316	67504	47319	67507	47322	67509	47325	04
56	67509	47325	67512	47328	67514	47331	67517	47334	67520	47337	67522	47340	03
57	67522	47340	67525	47343	67528	47345	67530	47348	67533	47351	67536	47354	02
58	67536	47354	67538	47357	67541	47360	67544	47363	67547	47366	67549	47369	01
59	67549	47369	67552	47372	67554	47375	67557	47377	67560	47380	67562	47383	00
			.8		.6		.4		.2		.0 ←		↑

PARTS for 0′·1 :— LOGS. 1 ; NATURALS, 1.

273°

87° HAVERSINES

′	·0 Log.	·0 Nat.	·2 Log.	·2 Nat.	·4 Log.	·4 Nat.	·6 Log.	·6 Nat.	·8 Log.	·8 Nat.	Log.	Nat.	′
	9·	0·	9·	0·	9·	0·	9·	0·	9·	0·	9·	0·	
00	67562	47383	67565	47386	67568	47389	67570	47392	67573	47395	67576	47398	60
01	67576	47398	67578	47401	67581	47404	67584	47406	67586	47409	67589	47412	58
02	67589	47412	67592	47415	67594	47418	67597	47421	67600	47424	67602	47427	57
03	67602	47427	67605	47430	67608	47433	67610	47436	67613	47438	67616	47441	56
04	67616	47441	67618	47444	67621	47447	67624	47450	67626	47453	67629	47456	55
05	67629	47456	67632	47459	67634	47462	67637	47465	67640	47467	67642	47470	54
06	67642	47470	67645	47473	67648	47476	67650	47479	67653	47482	67656	47485	53
07	67656	47485	67658	47488	67661	47491	67664	47494	67666	47497	67669	47499	52
08	67669	47499	67671	47502	67674	47505	67677	47508	67679	47511	67682	47514	51
09	67682	47514	67685	47517	67687	47520	67690	47523	67693	47526	67695	47528	50
10	67695	47528	67698	47531	67701	47534	67703	47537	67706	47540	67709	47543	49
11	67709	47543	67711	47546	67714	47549	67717	47552	67719	47555	67722	47558	48
12	67722	47558	67725	47560	67727	47563	67730	47566	67733	47569	67735	47572	47
13	67735	47572	67738	47575	67740	47578	67743	47581	67746	47584	67748	47587	46
14	67748	47587	67751	47589	67754	47592	67756	47595	67759	47598	67762	47601	45
15	67762	47601	67764	47604	67767	47607	67770	47610	67772	47613	67775	47616	44
16	67775	47616	67778	47619	67780	47621	67783	47624	67786	47627	67788	47630	43
17	67788	47630	67791	47633	67793	47636	67796	47639	67799	47642	67801	47645	42
18	67801	47645	67804	47648	67807	47651	67809	47653	67812	47656	67815	47659	41
19	67815	47659	67817	47662	67820	47665	67823	47668	67825	47671	67828	47674	40
20	67828	47674	67831	47677	67833	47680	67836	47682	67839	47685	67841	47688	39
21	67841	47688	67844	47691	67846	47694	67849	47697	67852	47700	67854	47703	38
22	67854	47703	67857	47706	67860	47709	67862	47712	67865	47714	67868	47717	37
23	67868	47717	67870	47720	67873	47723	67876	47726	67878	47719	67881	47732	36
24	67881	47732	67883	47735	67886	47738	67889	47741	67891	47743	67894	47746	35
25	67894	47746	67897	47749	67899	47752	67902	47755	67905	47758	67907	47761	34
26	67907	47761	67910	47764	67912	47767	67915	47770	67918	47773	67920	47775	33
27	67920	47775	67923	47778	67926	47781	67928	47784	67931	47787	67934	47790	32
28	67934	47790	67936	47793	67939	47796	67942	47799	67944	47802	67947	47805	31
29	67947	47805	67949	47807	67952	47810	67955	47813	67957	47816	67960	47819	30
30	67960	47819	67963	47822	67965	47825	67968	47828	67971	47831	67973	47834	29
31	67973	47834	67976	47836	67978	47839	67981	47842	67984	47845	67986	47848	28
32	67986	47848	67989	47851	67992	47854	67994	47857	67997	47860	68000	47863	27
33	68000	47863	68002	47866	68005	47868	68008	47871	68010	47874	68013	47877	26
34	68013	47877	68015	47880	68018	47883	68021	47886	68023	47889	68026	47892	25
35	68026	47892	68029	47895	68031	47898	68034	47900	68037	47903	68039	47906	24
36	68039	47906	68042	47909	68044	47912	68047	47915	68050	47918	68052	47921	23
37	68052	47921	68055	47924	68058	47927	68060	47929	68063	47932	68066	47935	22
38	68066	47935	68068	47938	68071	47941	68073	47944	68076	47947	68079	47950	21
39	68079	47950	68081	47953	68084	47956	68087	47959	68089	47961	68092	47964	20
40	68092	47964	68094	47967	68097	47970	68100	47973	68102	47976	68105	47979	19
41	68105	47979	68108	47982	68110	47985	68113	47988	68116	47991	68118	47993	18
42	68118	47993	68121	47996	68123	47999	68126	48002	68129	48005	68131	48008	17
43	68131	48008	68134	48011	68137	48014	68139	48017	68142	48020	68144	48022	16
44	68144	48022	68147	48025	68150	48028	68152	48031	68155	48034	68158	48037	15
45	68158	48037	68160	48040	68163	48043	68166	48046	68168	48049	68171	48052	14
46	68171	48052	68173	48054	68176	48057	68179	48060	68181	48063	68184	48066	13
47	68184	48066	68186	48069	68189	48072	68192	48075	68194	48078	68197	48081	12
48	68197	48081	68200	48084	68202	48086	68205	48089	68208	48092	68210	48095	11
49	68210	48095	68213	48098	68215	48101	68218	48104	68221	48107	68223	48110	10
50	68223	48110	68226	48113	68228	48116	68231	48118	68234	48121	68236	48124	09
51	68236	48124	68239	48127	68242	48130	68244	48133	68247	48136	68249	48139	08
52	68249	48139	68252	48142	68255	48145	68257	48147	68260	48150	68263	48153	07
53	68263	48153	68265	48156	68268	48159	68270	48162	68273	48165	68276	48168	06
54	68276	48168	60278	48171	68281	48174	68284	48177	68286	48179	68289	48182	05
55	68289	48182	68291	48185	68294	48188	68297	48191	68299	48194	68302	48197	04
56	68302	48197	68304	48200	68307	48203	68310	48206	68312	48209	68315	48211	03
57	68315	48211	68318	48214	68320	48217	68323	48220	68325	48223	68328	48226	02
58	68328	48226	68331	48229	68333	48232	68336	48235	68339	48238	68341	48241	01
59	68341	48241	68344	48243	68346	48246	68349	48249	68352	48252	68354	48255	00
			·8		·6		·4		·2		·0 ←		↑

PARTS for 0′·1 :— LOGS. 1 NATURALS, 1.

272°

88° HAVERSINES

′	·0 Log. 9·	·0 Nat. 0·	·2 Log. 9·	·2 Nat. 0·	·4 Log. 9·	·4 Nat. 0·	·6 Log. 9·	·6 Nat. 0·	·8 Log. 9·	·8 Nat. 0·	′
00	68354	48255	68357	48258	68359	48261	68362	48264	68365	48267	68367	48270	59
01	68367	48270	68370	48272	68373	48275	68375	48278	68378	48281	68380	48284	58
02	68380	48284	68383	48287	68386	48290	68388	48293	68391	48296	68393	48299	57
03	68393	48299	68396	48302	68399	48304	68401	48307	68404	48310	68407	48313	56
04	68407	48313	68409	48316	68412	48319	68414	48322	68417	48325	68420	48328	55
05	68420	48328	68422	48331	68425	48334	68427	48336	68430	48339	68433	48342	54
06	68433	48342	68435	48345	68438	48348	68441	48351	68443	48354	68446	48357	53
07	68446	48357	68448	48360	68451	48363	68454	48366	68456	48368	68459	48371	52
08	68459	48371	68461	48374	68464	48377	68467	48380	68469	48383	68472	48386	51
09	68472	48386	68474	48389	68477	48392	68480	48395	68482	48397	68485	48400	50
10	68485	48400	68487	48403	68490	48406	68493	48409	68495	48412	68498	48415	49
11	68498	48415	68501	48418	68503	48421	68506	48424	68508	48427	68511	48429	48
12	68511	48429	68514	48432	68516	48435	68519	48438	68521	48441	68524	48444	47
13	68524	48444	68527	48447	68529	48450	68532	48453	68534	48456	68537	48459	46
14	68537	48459	68540	48461	68542	48464	68545	48467	68547	48470	68550	48473	45
15	68550	48473	68553	48476	68555	48479	68558	48482	68560	48485	68563	48488	44
16	68563	48488	68566	48491	68568	48493	68571	48496	68573	48499	68576	48502	43
17	68576	48502	68579	48505	68581	48508	68584	48511	68587	48514	68589	48517	42
18	68589	48517	68592	48520	68594	48523	68597	48525	68600	48528	68602	48531	41
19	68602	48531	68605	48534	68607	48537	68610	48540	68613	48543	68615	48546	40
20	68615	48546	68618	48549	68620	48552	68623	48554	68626	48557	68628	48560	39
21	68628	48560	68631	48563	68633	48566	68636	48569	68639	48572	68641	48575	38
22	68641	48575	68644	48578	68646	48581	68649	48584	68652	48586	68654	48589	37
23	68654	48589	68657	48592	68659	48595	68662	48598	68665	48601	68667	48604	36
24	68667	48604	68670	48607	68672	48610	68675	48613	68678	48616	68680	48618	35
25	68680	48618	68683	48621	68685	48624	68688	48627	68691	48630	68693	48633	34
26	68693	48633	68696	48636	68698	48639	68701	48642	68703	48645	68706	48648	33
27	68706	48648	68709	48650	68711	48653	68714	48656	68716	48659	68719	48662	32
28	68719	48662	68722	48665	68724	48668	68727	48671	68729	48674	68732	48677	31
29	68732	48677	68735	48680	68737	48682	68740	48685	68742	48688	68745	48691	30
30	68745	48691	68748	48694	68750	48697	68753	48700	68755	48703	68758	48706	29
31	68758	48706	68761	48709	68763	48712	68766	48714	68768	48717	68771	48720	28
32	68771	48720	68774	48723	68776	48726	68779	48729	68781	48732	68784	48735	27
33	68784	48735	68786	48738	68789	48741	68792	48743	68794	48746	68797	48749	26
34	68797	48749	68799	48752	68802	48755	68805	48758	68807	48761	68810	48764	25
35	68810	48764	68812	48767	68815	48770	68818	48773	68820	48775	68823	48778	24
36	68823	48778	68825	48781	68828	48784	68830	48787	68833	48790	68836	48793	23
37	68836	48793	68838	48796	68841	48799	68843	48802	68846	48805	68849	48807	22
38	68849	48807	68851	48810	68854	48813	68856	48816	68859	48819	68862	48822	21
39	68862	48822	68864	48825	68867	48828	68869	48831	68872	48834	68874	48837	20
40	68874	48837	68877	48839	68880	48842	68882	48845	68885	48848	68887	48851	19
41	68887	48851	68890	48854	68893	48857	68895	48860	68898	48863	68900	48866	18
42	68900	48866	68903	48869	68906	48871	68908	48874	68911	48877	68913	48880	17
43	68913	48880	68916	48883	68918	48886	68921	48889	68924	48892	68926	48895	16
44	68926	48895	68929	48898	68931	48901	68934	48903	68937	48906	68939	48909	15
45	68939	48909	68942	48912	68944	48915	68947	48918	68949	48921	68952	48924	14
46	68952	48924	68955	48927	68957	48930	68960	48933	68962	48935	68965	48938	13
47	68965	48938	68968	48941	68970	48944	68973	48947	68975	48950	68978	48953	12
48	68978	48953	68980	48956	68983	48959	68986	48962	68988	48965	68991	48967	11
49	68991	48967	68993	48970	68996	48973	68998	48976	69001	48979	69004	48982	10
50	69004	48982	69006	48985	69009	48988	69011	48991	69014	48994	69017	48997	09
51	69017	48997	69019	48999	69022	49002	69024	49005	69027	49008	69029	49011	08
52	69029	49011	69032	49014	69035	49017	69037	49020	69040	49023	69042	49026	07
53	69042	49026	69045	49029	69047	49031	69050	49034	69053	49037	69055	49040	06
54	69055	49040	69058	49043	69060	49046	69063	49049	69065	49052	69068	49055	05
55	69068	49055	69071	49058	69073	49060	69076	49063	69078	49066	69081	49069	04
56	69081	49069	69084	49072	69086	49075	69089	49078	69091	49081	69094	49084	03
57	69094	49084	69096	49087	69099	49090	69101	49092	69104	49095	69107	49098	02
58	69107	49098	69109	49101	69112	49104	69114	49107	69117	49110	69120	49113	01
59	69120	49113	69122	49116	69125	49119	69127	49122	69130	49124	69132	49127	00

| | ·8 | | ·6 | | ·4 | | ·2 | | ·0 | |

271°

PARTS for 0′·1 :—   LOGS. 1 ;   NATURALS, 1.

89° HAVERSINES

Log. = 9·xxxxx Nat. = 0·xxxxx

'	·0 Log	·0 Nat	·2 Log	·2 Nat	·4 Log	·4 Nat	·6 Log	·6 Nat	·8 Log	·8 Nat	Log	Nat	'
00	69132	49127	69135	49130	69138	49133	69140	49136	69143	49139	69145	49142	59
01	69145	49142	69148	49145	69150	49148	69153	49151	69155	49154	69158	49156	58
02	69158	49156	69161	49159	69163	49162	69166	49165	69168	49168	69171	49171	57
03	69171	49171	69174	49174	69176	49177	69179	49180	69181	49183	69184	49186	56
04	69184	49186	69186	49188	69189	49191	69191	49194	69194	49197	69197	49200	55
05	69197	49200	69199	49203	69202	49206	69204	49209	69207	49212	69209	49215	54
06	69209	49215	69212	49218	69215	49220	69217	49223	69220	49226	69222	49229	53
07	69222	49229	69225	49232	69227	49235	69230	49238	69232	49241	69235	49244	52
08	69235	49244	69238	49247	69240	49250	69243	49252	69245	49255	69248	49258	51
09	69248	49258	69251	49261	69253	49264	69256	49267	69258	49270	69261	49273	50
10	69261	49273	69263	49276	69266	49279	69268	49282	69271	49284	69274	49287	49
11	69274	49287	69276	49290	69279	49293	69281	49296	69284	49299	69286	49302	48
12	69286	49302	69289	49305	69292	49308	69294	49311	69297	49314	69299	49316	47
13	69299	49316	69302	49319	69304	49322	69307	49325	69309	49328	69312	49331	46
14	69312	49331	69315	49334	69317	49337	69320	49340	69322	49343	69325	49346	45
15	69325	49346	69327	49348	69330	49351	69332	49354	69335	49357	69338	49360	44
16	69338	49360	69340	49363	69343	49366	69345	49369	69348	49372	69350	49375	43
17	69350	49375	69353	49378	69356	49380	69358	49383	69361	49386	69363	49389	42
18	69363	49389	69366	49392	69368	49395	69371	49398	69373	49401	69376	49404	41
19	69376	49404	69379	49407	69381	49410	69384	49412	69386	49415	69389	49418	40
20	69389	49418	69391	49421	69394	49424	69396	49427	69399	49430	69402	49433	39
21	69402	49433	69404	49436	69407	49439	69409	49442	69412	49444	69414	49447	38
22	69414	49447	69417	49450	69419	49453	69422	49456	69424	49459	69427	49462	37
23	69427	49462	69430	49465	69432	49468	69435	49471	69437	49473	69440	49476	36
24	69440	49476	69442	49479	69445	49482	69447	49485	69450	49488	69453	49491	35
25	69453	49491	69455	49494	69458	49497	69460	49500	69463	49503	69465	49506	34
26	69465	49506	69468	49508	69471	49511	69473	49514	69476	49517	69478	49520	33
27	69478	49520	69481	49523	69483	49526	69486	49529	69488	49532	69491	49535	32
28	69491	49535	69493	49538	69496	49540	69498	49543	69501	49546	69504	49549	31
29	69504	49549	69506	49552	69509	49555	69511	49558	69514	49561	69516	49564	30
30	69516	49564	69519	49567	69522	49569	69524	49572	69527	49575	69529	49578	29
31	69529	49578	69532	49581	69534	49584	69537	49587	69539	49590	69542	49593	28
32	69542	49593	69544	49596	69547	49599	69549	49601	69552	49604	69555	49607	27
33	69555	49607	69557	49610	69560	49613	69562	49616	69565	49619	69567	49622	26
34	69567	49622	69570	49625	69572	49628	69575	49631	69577	49633	69580	49636	25
35	69580	49636	69583	49639	69585	49642	69588	49645	69590	49648	69593	49651	24
36	69593	49651	69595	49654	69598	49657	69600	49660	69603	49663	69605	49665	23
37	69605	49665	69608	49668	69611	49671	69613	49674	69616	49677	69618	49680	22
38	69618	49680	69621	49683	69623	49686	69626	49689	69628	49692	69631	49695	21
39	69631	49695	69634	49697	69636	49700	69639	49703	69641	49706	69644	49709	20
40	69644	49709	69646	49712	69649	49715	69651	49718	69654	49721	69656	49724	19
41	69656	49724	69659	49727	69661	49729	69664	49732	69666	49735	69669	49738	18
42	69669	49738	69672	49741	69674	49744	69677	49747	69679	49750	69682	49753	17
43	69682	49753	69684	49756	69687	49759	69689	49761	69692	49764	69694	49767	16
44	69694	49767	69697	49770	69699	49773	69702	49776	69704	49779	69707	49782	15
45	69707	49782	69710	49785	69712	49788	69715	49791	69717	49793	69720	49796	14
46	69720	49796	69722	49799	69725	49802	69727	49805	69730	49808	69732	49811	13
47	69732	49811	69735	49814	69738	49817	69740	49820	69743	49823	69745	49825	12
48	69745	49825	69748	49828	69750	49831	69753	49834	69755	49837	69758	49840	11
49	69758	49840	69760	49843	69763	49846	69765	49849	69768	49852	69770	49855	10
50	69770	49855	69773	49857	69776	49860	69778	49863	69781	49866	69783	49869	09
51	69783	49869	69786	49872	69788	49875	69791	49878	69793	49881	69796	49884	08
52	69796	49884	69798	49887	69801	49889	69803	49892	69806	49895	69808	49898	07
53	69808	49898	69811	49901	69814	49904	69816	49907	69819	49910	69821	49913	06
54	69821	49913	69824	49916	69826	49919	69829	49921	69831	49924	69834	49927	05
55	69834	49927	69836	49930	69839	49933	69841	49936	69844	49939	69846	49942	04
56	69846	49942	69849	49945	69851	49948	69854	49951	69856	49953	69859	49956	03
57	69859	49956	69862	49959	69864	49962	69867	49965	69869	49968	69872	49971	02
58	69872	49971	69874	49974	69877	49977	69879	49980	69882	49983	69884	49985	01
59	69884	49985	69887	49988	69889	49991	69892	49994	69894	49997	69897	50000	00

Bottom scale: ·8 ·6 ·4 ·2 ·0 → 270°

PARTS for 0'·1 :— LOGS. 1 ; NATURALS, 1.

HAVERSINES

90° (top left) — **269°** (bottom right)

PARTS for 0′·1 :— LOGS. 1 ; NATURALS, 1.

′	.0 Log	.0 Nat	.2 Log	.2 Nat	.4 Log	.4 Nat	.6 Log	.6 Nat	.8 Log	.8 Nat	Log	Nat	′
	9·	0·	9·	0·	9·	0·	9·	0·	9·	0·	9·	0·	
00	69897	50000	69900	50003	69902	50006	69905	50009	69907	50012	69910	50015	59
01	69910	50015	69912	50017	69915	50020	69917	50023	69920	50026	69922	50029	58
02	69922	50029	69925	50032	69927	50035	69930	50038	69932	50041	69935	50044	57
03	69935	50044	69937	50047	69940	50049	69942	50052	69945	50055	69948	50058	56
04	69948	50058	69950	50061	69953	50064	69955	50067	69958	50070	69960	50073	55
05	69960	50073	69963	50076	69965	50079	69968	50081	69970	50084	69973	50087	54
06	69973	50087	69975	50090	69978	50093	69980	50096	69983	50099	69985	50102	53
07	69985	50102	69988	50105	69990	50108	69993	50111	69995	50113	69998	50116	52
08	69998	50116	70001	50119	70003	50122	70006	50125	70008	50128	70011	50131	51
09	70011	50131	70013	50134	70016	50137	70018	50140	70021	50143	70023	50145	50
10	70023	50145	70026	50148	70028	50151	70031	50154	70033	50157	70036	50160	49
11	70036	50160	70038	50163	70041	50166	70043	50169	70046	50172	70048	50175	48
12	70048	50175	70051	50177	70053	50180	70056	50183	70058	50186	70061	50189	47
13	70061	50189	70064	50192	70066	50195	70069	50198	70071	50201	70074	50204	46
14	70074	50204	70076	50207	70079	50209	70081	50212	70084	50215	70086	50218	45
15	70086	50218	70089	50221	70091	50224	70094	50227	70096	50230	70099	50233	44
16	70099	50233	70101	50236	70104	50239	70106	50241	70109	50244	70111	50247	43
17	70111	50247	70114	50250	70116	50253	70119	50256	70121	50259	70124	50262	42
18	70124	50262	70126	50265	70129	50268	70131	50271	70134	50273	70136	50276	41
19	70136	50276	70139	50279	70141	50282	70144	50285	70146	50288	70149	50291	40
20	70149	50291	70151	50294	70154	50297	70156	50300	70159	50303	70161	50305	39
21	70161	50305	70164	50308	70166	50311	70169	50314	70171	50317	70174	50320	38
22	70174	50320	70177	50323	70179	50326	70182	50329	70184	50332	70187	50335	37
23	70187	50335	70189	50337	70192	50340	70194	50343	70197	50346	70199	50349	36
24	70199	50349	70202	50352	70204	50355	70207	50358	70209	50361	70212	50364	35
25	70212	50364	70214	50367	70217	50369	70219	50372	70222	50375	70224	50378	34
26	70224	50378	70227	50381	70229	50384	70232	50387	70234	50390	70237	50393	33
27	70237	50393	70239	50396	70242	50399	70244	50401	70247	50404	70249	50407	32
28	70249	50407	70252	50410	70254	50413	70257	50416	70259	50419	70262	50422	31
29	70262	50422	70264	50425	70267	50428	70269	50431	70272	50433	70274	50436	30
30	70274	50436	70277	50439	70279	50442	70282	50445	70284	50448	70287	50451	29
31	70287	50451	70289	50454	70292	50457	70294	50460	70297	50462	70299	50465	28
32	70299	50465	70302	50468	70304	50471	70307	50474	70309	50477	70312	50480	27
33	70312	50480	70314	50483	70317	50486	70319	50489	70322	50492	70324	50495	26
34	70324	50495	70327	50497	70329	50500	70332	50503	70334	50506	70337	50509	25
35	70337	50509	70339	50512	70342	50515	70344	50518	70347	50521	70349	50524	24
36	70349	50524	70352	50527	70354	50529	70357	50532	70359	50535	70362	50538	23
37	70362	50538	70364	50541	70367	50544	70369	50547	70372	50550	70374	50553	22
38	70374	50553	70377	50556	70379	50559	70382	50561	70384	50564	70387	50567	21
39	70387	50567	70389	50570	70392	50573	70394	50576	70397	50579	70399	50582	20
40	70399	50582	70402	50585	70404	50588	70407	50590	70409	50593	70412	50596	19
41	70412	50596	70414	50599	70417	50602	70419	50605	70422	50608	70424	50611	18
42	70424	50611	70427	50614	70429	50617	70432	50620	70434	50622	70437	50625	17
43	70437	50625	70439	50628	70442	50631	70444	50634	70447	50637	70449	50640	16
44	70449	50640	70452	50643	70454	50646	70457	50649	70459	50652	70462	50654	15
45	70462	50654	70464	50657	70467	50660	70469	50663	70472	50666	70474	50669	14
46	70474	50669	70477	50672	70479	50675	70482	50678	70484	50681	70487	50684	13
47	70487	50684	70489	50686	70492	50689	70494	50692	70497	50695	70499	50698	12
48	70499	50698	70502	50701	70504	50704	70507	50707	70509	50710	70512	50713	11
49	70512	50713	70514	50716	70517	50718	70519	50721	70522	50724	70524	50727	10
50	70524	50727	70527	50730	70529	50733	70532	50736	70534	50739	70537	50742	09
51	70537	50742	70539	50745	70542	50748	70544	50750	70547	50753	70549	50756	08
52	70549	50756	70551	50759	70554	50762	70556	50765	70559	50768	70561	50771	07
53	70561	50771	70564	50774	70566	50777	70569	50780	70571	50782	70574	50785	06
54	70574	50785	70576	50788	70579	50791	70581	50794	70584	50797	70586	50800	05
55	70586	50800	70589	50803	70591	50806	70594	50809	70596	50812	70599	50814	04
56	70599	50814	70601	50817	70604	50820	70606	50823	70609	50826	70611	50829	03
57	70611	50829	70614	50832	70616	50835	70619	50838	70621	50841	70624	50844	02
58	70624	50844	70626	50846	70629	50849	70631	50852	70634	50855	70636	50858	01
59	70636	50858	70638	50861	70641	50864	70643	50867	70646	50870	70648	50873	00
		·8		·6		·4		·2		·0			

488

HAVERSINES

'	91° Log.	91° Nat.	92° Log.	92° Nat.	93° Log.	93° Nat.	94° Log.	94° Nat.	95° Log.	95° Nat.	96° Log.	96° Nat.	'
	9·	0·	9·	0·	9·	0·	9·	0·	9·	0·	9·	0·	
00	70648	50873	71387	51745	72112	52617	72825	53488	73526	54358	74215	55226	60
01	70661	50887	71399	51760	72124	52631	72837	53502	73538	54372	74226	55241	59
02	70673	50902	71411	51774	72136	52646	72849	53517	73549	54387	74237	55255	58
03	70686	50916	71423	51789	72148	52660	72861	53531	73561	54401	74249	55270	57
04	70698	50931	71436	51803	72160	52675	72873	53546	73572	54416	74260	55284	56
05	70710	50945	71448	51818	72172	52689	72884	53560	73584	54430	74272	55299	55
06	70723	50960	71460	51832	72184	52704	72896	53575	73596	54445	74283	55313	54
07	70735	50974	71472	51847	72196	52718	72908	53589	73607	54459	74294	55328	53
08	70748	50989	71484	51861	72208	52733	72920	53604	73619	54474	74306	55342	52
09	70760	51004	71496	51876	72220	52748	72931	53618	73630	54488	74317	55357	51
10	70772	51018	71509	51890	72232	52762	72943	53633	73642	54503	74328	55371	50
11	70785	51033	71521	51905	72244	52776	72955	53647	73653	54517	74340	55386	49
12	70797	51047	71533	51919	72256	52791	72967	53662	73665	54532	74351	55400	48
13	70809	51062	71545	51934	72268	52806	72978	53676	73676	54546	74362	55414	47
14	70822	51076	71557	51948	72280	52820	72990	53691	73688	54561	74374	55429	46
15	70834	51091	71569	51963	72292	52835	73002	53705	73699	54575	74385	55443	45
16	70847	51105	71582	51978	72304	52849	73014	53720	73711	54590	74396	55458	44
17	70859	51120	71594	51992	72316	52864	73025	53734	73722	54604	74408	55472	43
18	70871	51134	71606	52007	72328	52878	73037	53749	73734	54619	74419	55487	42
19	70884	51149	71618	52021	72340	52893	73049	53763	73746	54633	74430	55501	41
20	70896	51163	71630	52036	72352	52907	73060	53778	73757	54647	74442	55516	40
21	70908	51178	71642	52050	72363	52922	73072	53792	73769	54662	74453	55530	39
22	70921	51193	71654	52065	72375	52936	73084	53807	73780	54676	74464	55545	38
23	70933	51207	71666	52079	72387	52951	73096	53821	73792	54691	74475	55559	37
24	70945	51222	71679	52094	72399	52965	73107	53836	73803	54705	74487	55573	36
25	70958	51236	71691	52108	72411	52980	73119	53850	73815	54720	74498	55588	35
26	70970	51251	71703	52123	72423	52994	73131	53865	73826	54734	74509	55602	34
27	70982	51265	71715	52137	72435	53009	73142	53879	73838	54749	74521	55617	33
28	70995	51280	71727	52152	72447	53023	73154	53894	73849	54763	74532	55631	32
29	71007	51294	71739	52166	72459	53038	73166	53908	73860	54778	74543	55646	31
30	71019	51309	71751	52181	72471	53052	73177	53923	73872	54792	74554	55660	30
31	71032	51323	71763	52196	72482	53067	73189	53937	73883	54807	74566	55675	29
32	71044	51338	71775	52210	72494	53081	73201	53952	73895	54821	74577	55689	28
33	71056	51352	71787	52225	72506	53096	73212	53966	73906	54836	74588	55704	27
34	71068	51367	71800	52239	72518	53110	73224	53981	73918	54850	74600	55718	26
35	71081	51382	71812	52254	72530	53125	73236	53995	73929	54865	74611	55732	25
36	71093	51396	71824	52268	72542	53140	73247	54010	73941	54879	74622	55747	24
37	71105	51411	71836	52283	72554	53154	73259	54024	73952	54894	74633	55761	23
38	71118	51425	71848	52297	72565	53169	73271	54039	73964	54908	74645	55776	22
39	71130	51440	71860	52312	72577	53183	73282	54053	73975	54923	74656	55790	21
40	71142	51454	71872	52326	72589	53198	73294	54068	73987	54937	74667	55805	20
41	71154	51469	71884	52341	72601	53212	73306	54082	73998	54952	74678	55819	19
42	71167	51483	71896	52355	72613	53227	73317	54097	74009	54966	74690	55834	18
43	71179	51498	71908	52370	72625	53241	73329	54111	74021	54980	74701	55848	17
44	71191	51512	71920	52384	72637	53256	73341	54126	74032	54995	74712	55862	16
45	71203	51527	71932	52399	72648	53270	73352	54140	74044	55009	74723	55877	15
46	71216	51541	71944	52413	72660	53285	73364	54155	74055	55024	74734	55891	14
47	71228	51556	71956	52428	72672	53299	73375	54169	74067	55038	74746	55906	13
48	71240	51571	71968	52442	72684	53314	73387	54184	74078	55053	74757	55920	12
49	71252	51585	71980	52457	72696	53328	73399	54198	74089	55067	74768	55935	11
50	71265	51600	71992	52472	72708	53343	73410	54213	74101	55082	74779	55949	10
51	71277	51614	72004	52486	72719	53357	73422	54227	74112	55096	74791	55964	09
52	71289	51629	72016	52501	72731	53372	73433	54242	74124	55111	74802	55978	08
53	71301	51643	72028	52515	72743	53386	73445	54256	74135	55125	74813	55992	07
54	71314	51658	72040	52530	72755	53401	73457	54271	74146	55140	74824	56007	06
55	71326	51672	72052	52544	72767	53415	73468	54285	74158	55154	74835	56021	05
56	71338	51687	72064	52559	72778	53430	73480	54300	74169	55169	74846	56036	04
57	71350	51701	72076	52573	72790	53444	73491	54314	74181	55183	74858	56050	03
58	71362	51716	72088	52588	72802	53459	73503	54329	74192	55197	74869	56065	02
59	71375	51730	72100	52602	72814	53473	73515	54343	74203	55212	74880	56079	01
60	71387	51745	72112	52617	72825	53488	73526	54358	74215	55226	74891	56093	00

| 268° | 267° | 266° | 265° | 264° | 263° |

HAVERSINES

′	97° Log.	97° Nat.	98° Log.	98° Nat.	99° Log.	99° Nat.	100° Log.	100° Nat.	101° Log.	101° Nat.	102° Log.	102° Nat.	′
	9·	0·	9·	0·	9·	0·	9·	0·	9·	0·	9·	0·	
00	74891	56093	75556	56959	76209	57822	76851	58682	77481	59540	78101	60396	60
01	74902	56108	75567	56973	76220	57836	76861	58697	77492	59555	78111	60410	59
02	74914	56122	75578	56987	76231	57850	76872	58711	77502	59569	78121	60424	58
03	74925	56137	75589	57002	76241	57865	76883	58725	77512	59583	78131	60438	57
04	74936	56151	75600	57016	76252	57879	76893	58740	77523	59598	78141	60452	56
05	74947	56166	75611	57031	76263	57894	76904	58754	77533	59612	78152	60467	55
06	74958	56180	75622	57045	76274	57908	76914	58768	77544	59626	78162	60481	54
07	74969	56195	75633	57059	76285	57922	76925	58783	77554	59640	78172	60495	53
08	74981	56209	75644	57074	76296	57937	76936	58797	77564	59655	78182	60509	52
09	74992	56223	75655	57088	76306	57951	76946	58811	77575	59669	78192	60524	51
10	75003	56238	75666	57103	76317	57965	76957	58826	77585	59683	78203	60538	50
11	75014	56252	75677	57117	76328	57980	76967	58840	77596	59697	78213	60552	49
12	75025	56267	75688	57131	76338	57994	76978	58854	77606	59712	78223	60566	48
13	75036	56281	75698	57146	76349	58008	76988	58869	77616	59726	78233	60580	47
14	75047	56296	75709	57160	76360	58023	76999	58883	77627	59740	78243	60595	46
15	75059	56310	75720	57175	76371	58037	77009	58897	77637	59755	78254	60609	45
16	75070	56324	75731	57189	76381	58051	77020	58911	77647	59769	78264	60623	44
17	75081	56339	75742	57203	76392	58066	77031	58926	77658	59783	78274	60637	43
18	75092	56353	75753	57218	76403	58080	77041	58940	77668	59797	78284	60652	42
19	75103	56368	75764	57232	76414	58095	77052	58954	77679	59812	78294	60666	41
20	75114	56382	75775	57247	76424	58109	77062	58969	77689	59826	78305	60680	40
21	75125	56397	75786	57261	76435	58123	77073	58983	77699	59840	78315	60694	39
22	75136	56411	75797	57275	76446	58138	77083	58997	77710	59854	78325	60708	38
23	75147	56425	75808	57290	76456	58152	77094	59012	77720	59869	78335	60723	37
24	75159	56440	75819	57304	76467	58166	77104	59026	77730	59883	78345	60737	36
25	75170	56454	75830	57318	76478	58181	77115	59040	77741	59897	78355	60751	35
26	75181	56469	75840	57333	76489	58195	77125	59055	77751	59911	78365	60765	34
27	75192	56483	75851	57347	76499	58209	77136	59069	77761	59926	78376	60779	33
28	75203	56497	75862	57362	76510	58224	77146	59083	77772	59940	78386	60794	32
29	75214	56512	75873	57376	76521	58238	77157	59097	77782	59954	78396	60808	31
30	75225	56526	75884	57390	76531	58252	77167	59112	77792	59968	78406	60822	30
31	75236	56541	75895	57405	76542	58267	77178	59126	77803	59983	78416	60836	29
32	75247	56555	75906	57419	76553	58281	77188	59140	77813	59997	78426	60850	28
33	75258	56570	75917	57434	76563	58295	77199	59155	77823	60011	78436	60865	27
34	75269	56584	75927	57448	76574	58310	77209	59169	77834	60025	78447	60879	26
35	75280	56598	75938	57462	76585	58324	77220	59183	77844	60040	78457	60893	25
36	75291	56613	75949	57477	76595	58338	77230	59198	77854	60054	78467	60907	24
37	75303	56627	75960	57491	76606	58353	77241	59212	77864	60068	78477	60921	23
38	75314	56642	75971	57506	76617	58367	77251	59226	77875	60082	78487	60936	22
39	75325	56656	75982	57520	76627	58381	77262	59240	77885	60097	78497	60950	21
40	75336	56670	75993	57534	76638	58396	77272	59255	77895	60111	78507	60964	20
41	75347	56685	76004	57549	76649	58410	77283	59269	77906	60125	78517	60978	19
42	75358	56699	76014	57563	76659	58424	77293	59283	77916	60139	78528	60992	18
43	75369	56714	76025	57577	76670	58439	77304	59298	77926	60154	78538	61007	17
44	75380	56728	76036	57592	76681	58453	77314	59312	77936	60168	78548	61021	16
45	75391	56743	76047	57606	76691	58467	77325	59326	77947	60182	78558	61035	15
46	75402	56757	76058	57621	76702	58482	77335	59340	77957	60196	78568	61049	14
47	75413	56771	76069	57635	76713	58496	77346	59355	77967	60211	78578	61063	13
48	75424	56786	76079	57649	76723	58510	77356	59369	77978	60225	78588	61077	12
49	75435	56800	76090	57664	76734	58525	77366	59383	77988	60239	78598	61092	11
50	75446	56815	76101	57678	76745	58539	77377	59398	77998	60253	78608	61106	10
51	75457	56829	76112	57692	76755	58553	77387	59412	78008	60268	78618	61120	09
52	75468	56843	76123	57707	76766	58568	77398	59426	78019	60282	78628	61134	08
53	75479	56858	76134	57721	76777	58582	77408	59440	78029	60296	78638	61148	07
54	75490	56872	76144	57736	76787	58596	77419	59455	78039	60310	78649	61163	06
55	75501	56887	76155	57750	76798	58611	77429	59469	78049	60324	78659	61177	05
56	75512	56901	76166	57764	76808	58625	77440	59483	78060	60339	78669	61191	04
57	75523	56915	76177	57779	76819	58639	77450	59498	78070	60353	78679	61205	03
58	75534	56930	76188	57793	76830	58654	77460	59512	78080	60367	78689	61219	02
59	75545	56944	76198	57807	76840	58668	77471	59526	78090	60381	78699	61233	01
60	75556	56959	76209	57822	76851	58682	77481	59540	78101	60396	78709	61248	00

Parts for 0′·2, etc.

97°		98°		99°		100°		101°		102°	
·2 2	·3 3	·2 2	·3 3	·2 2	·3 3	·2 2	·3 3	·2 2	·3 3	·2 2	·3 3
·4 4	·6 6	·4 4	·6 6	·4 4	·6 6	·4 4	·6 6	·4 4	·6 6	·4 4	·6 6
·6 7	·9 9	·6 7	·9 9	·6 6	·9 9	·6 6	·9 9	·6 6	·9 9	·6 6	·9 9
·8 9	·8 12	·8 9	·8 12	·8 9	·8 11	·8 8	·8 11	·8 8	·8 11	·8 8	·8 11

| 262° | 261° | 260° | 259° | 258° | 257° |

HAVERSINES

Parts for 0'·2, etc.	103° Log.	103° Nat.	104° Log.	104° Nat.	105° Log.	105° Nat.	106° Log.	106° Nat.	107° Log.	107° Nat.	108° Log.	108° Nat.	Parts for 0'·2, etc.
·2 2 ·3 3 / ·4 4 ·4 6 / ·6 6 ·6 8 / ·8 8 ·8 11			·2 2 ·3 3 / ·4 4 ·4 6 / ·6 6 ·6 8 / ·8 8 ·8 11		·2 2 ·3 3 / ·4 4 ·4 6 / ·6 6 ·6 8 / ·8 8 ·8 11		·2 2 ·3 3 / ·4 4 ·4 6 / ·6 6 ·6 8 / ·8 8 ·8 11		·2 2 ·3 3 / ·4 4 ·4 6 / ·6 6 ·6 8 / ·8 7 ·8 11		·2 2 ·3 3 / ·4 4 ·4 6 / ·6 5 ·6 6 / ·8 7 ·8 11		
′	9·	0·	9·	0·	9·	0·	9·	0·	9·	0·	9·	0·	′
00	78709	61248	79306	62096	79893	62941	80470	63782	81036	64619	81592	65451	60
01	78719	61262	79316	62110	79903	62955	80479	63796	81045	64632	81601	65465	59
02	78729	61276	79326	62124	79913	62969	80489	63810	81054	64646	81610	65479	58
03	78739	61290	79336	62138	79922	62983	80498	63824	81064	64660	81619	65492	57
04	78749	61304	79346	62153	79932	62997	80508	63838	81073	64674	81628	65506	56
05	78759	61318	79356	62167	79942	63011	80517	63852	81082	64688	81637	65520	55
06	78769	61333	79366	62181	79951	63025	80527	63866	81092	64702	81647	65534	54
07	78779	61347	79376	62195	79961	63039	80536	63880	81101	64716	81656	65548	53
08	78789	61361	79385	62209	79971	63053	80546	63894	81110	64730	81665	65561	52
09	78799	61375	79395	62223	79980	63067	80555	63908	81120	64744	81674	65575	51
10	78809	61389	79405	62237	79990	63081	80565	63922	81129	64758	81683	65589	50
11	78819	61403	79415	62251	80000	63095	80574	63936	81138	64772	81692	65603	49
12	78829	61418	79425	62265	80009	63109	80584	63950	81148	64785	81701	65617	48
13	78839	61432	79434	62279	80019	63123	80593	63964	81157	64799	81711	65630	47
14	78849	61446	79444	62294	80029	63138	80603	63977	81166	64813	81720	65644	46
15	78859	61460	79454	62308	80038	63152	80612	63991	81176	64827	81729	65658	45
16	78869	61474	79464	62322	80048	63166	80622	64005	81185	64841	81738	65672	44
17	78879	61488	79474	62336	80058	63180	80631	64019	81194	64855	81747	65686	43
18	78889	61502	79484	62350	80067	63194	80641	64033	81204	64869	81756	65700	42
19	78899	61517	79493	62364	80077	63208	80650	64047	81213	64883	81765	65713	41
20	78909	61531	79503	62378	80087	63222	80660	64061	81222	64897	81775	65727	40
21	78919	61545	79513	62392	80096	63236	80669	64075	81231	64910	81784	65741	39
22	78929	61559	79523	62406	80106	63250	80678	64089	81241	64924	81793	65755	38
23	78939	61573	79533	62420	80116	63264	80688	64103	81250	64938	81802	65769	37
24	78949	61587	79542	62434	80125	63278	80697	64117	81259	64952	81811	65782	36
25	78959	61602	79552	62449	80135	63292	80707	64131	81269	64966	81820	65796	35
26	78969	61616	79562	62463	80144	63306	80716	64145	81278	64980	81829	65810	34
27	78979	61630	79572	62477	80154	63320	80726	64159	81287	64994	81838	65824	33
28	78989	61644	79582	62491	80164	63334	80735	64173	81296	65008	81847	65838	32
29	78999	61658	79591	62505	80173	63348	80745	64187	81306	65021	81857	65851	31
30	79009	61672	79601	62519	80183	63362	80754	64201	81315	65035	81866	65865	30
31	79019	61686	79611	62533	80192	63376	80763	64215	81324	65049	81875	65879	29
32	79029	61701	79621	62547	80202	63390	80773	64229	81333	65063	81884	65893	28
33	79039	61715	79631	62561	80212	63404	80782	64243	81343	65077	81893	65907	27
34	79049	61729	79640	62575	80221	63418	80792	64257	81352	65091	81902	65920	26
35	79059	61743	79650	62589	80231	63432	80801	64270	81361	65105	81911	65934	25
36	79069	61757	79660	62603	80240	63446	80811	64284	81370	65118	81920	65948	24
37	79079	61771	79670	62618	80250	63460	80820	64298	81380	65132	81929	65962	23
38	79089	61785	79679	62632	80260	63474	80829	64312	81389	65146	81938	65976	22
39	79099	61800	79689	62646	80269	63488	80839	64326	81398	65160	81947	65989	21
40	79108	61814	79699	62660	80279	63502	80848	64340	81407	65174	81956	66003	20
41	79118	61828	79709	62674	80288	63516	80858	64354	81417	65188	81965	66017	19
42	79128	61842	79718	62688	80298	63530	80867	64368	81426	65202	81975	66031	18
43	79138	61856	79728	62702	80307	63544	80876	64382	81435	65216	81984	66044	17
44	79148	61870	79738	62716	80317	63558	80886	64396	81444	65229	81993	66058	16
45	79158	61884	79748	62730	80327	63572	80895	64410	81454	65243	82002	66072	15
46	79168	61898	79757	62744	80336	63586	80905	64424	81463	65257	82011	66086	14
47	79178	61913	79767	62758	80346	63600	80914	64438	81472	65271	82020	66100	13
48	79188	61927	79777	62772	80355	63614	80923	64452	81481	65285	82029	66113	12
49	79198	61941	79787	62786	80365	63628	80933	64466	81490	65299	82038	66127	11
50	79208	61955	79796	62800	80374	63642	80942	64479	81500	65312	82047	66141	10
51	79217	61969	79806	62814	80384	63656	80952	64493	81509	65326	82056	66155	09
52	79227	61983	79816	62829	80393	63670	80961	64507	81518	65340	82065	66168	08
53	79237	61997	79825	62843	80403	63684	80970	64521	81527	65354	82074	66182	07
54	79247	62011	79835	62857	80413	63698	80980	64535	81536	65368	82083	66196	06
55	79257	62026	79845	62871	80422	63712	80989	64549	81546	65382	82092	66210	05
56	79267	62040	79855	62885	80432	63726	80998	64563	81555	65396	82101	66223	04
57	79277	62054	79864	62899	80441	63740	81008	64577	81564	65409	82110	66237	03
58	79287	62068	79874	62913	80451	63754	81017	64591	81573	65423	82119	66251	02
59	79297	62082	79884	62927	80460	63768	81026	64605	81582	65437	82128	66265	01
60	79306	62096	79893	62941	80470	63782	81036	64619	81592	65451	82137	66278	00

256°	255°	254°	253°	252°	251°

HAVERSINES

Parts for 0'·2, etc. (each degree column):
- 109°: ·2 2 ·3 3 / ·4 4 ·5 5 / ·6 5 ·6 8 / ·8 7 ·8 11
- 110°: ·2 2 ·3 3 / ·4 4 ·5 5 / ·6 5 ·6 8 / ·8 7 ·8 11
- 111°: ·2 2 ·3 3 / ·4 4 ·5 5 / ·6 5 ·6 8 / ·8 7 ·8 11
- 112°: ·2 2 ·3 3 / ·4 3 ·5 5 / ·6 5 ·6 8 / ·8 7 ·8 11
- 113°: ·2 2 ·3 3 / ·4 3 ·5 5 / ·6 5 ·6 8 / ·8 7 ·8 11
- 114°: ·2 2 ·3 3 / ·4 3 ·5 5 / ·6 5 ·6 8 / ·8 7 ·8 11

'	109° Log. 9·	109° Nat. 0·	110° Log. 9·	110° Nat. 0·	111° Log. 9·	111° Nat. 0·	112° Log. 9·	112° Nat. 0·	113° Log. 9·	113° Nat. 0·	114° Log. 9·	114° Nat. 0·	'
00	82137	66278	82673	67101	83199	67918	83715	68730	84221	69537	84718	70337	60
01	82146	66292	82682	67115	83207	67932	83723	68744	84230	69550	84726	70350	59
02	82155	66306	82691	67128	83216	67946	83732	68757	84238	69563	84735	70363	58
03	82164	66320	82699	67142	83225	67959	83740	68771	84246	69577	84743	70377	57
04	82173	66333	82708	67156	83233	67973	83749	68784	84255	69590	84751	70390	56
05	82182	66347	82717	67169	83242	67986	83757	68798	84263	69603	84759	70403	55
06	82191	66361	82726	67183	83251	68000	83766	68811	84271	69617	84767	70417	54
07	82200	66375	82735	67197	83259	68013	83774	68825	84280	69630	84776	70430	53
08	82209	66388	82744	67210	83268	68027	83783	68838	84288	69644	84784	70443	52
09	82218	66402	82752	67224	83277	68041	83791	68852	84296	69657	84792	70456	51
10	82227	66416	82761	67238	83285	68054	83800	68865	84305	69670	84800	70470	50
11	82236	66430	82770	67251	83294	68068	83808	68879	84313	69684	84808	70483	49
12	82245	66443	82779	67265	83303	68081	83817	68892	84321	69697	84817	70496	48
13	82254	66457	82788	67279	83311	68095	83825	68906	84330	69710	84825	70509	47
14	82263	66471	82796	67292	83320	68108	83834	68919	84338	69724	84833	70523	46
15	82272	66485	82805	67306	83329	68122	83842	68932	84346	69737	84841	70536	45
16	82281	66498	82814	67320	83337	68135	83851	68946	84355	69751	84849	70549	44
17	82290	66512	82823	67333	83346	68149	83859	68959	84363	69764	84857	70562	43
18	82299	66526	82832	67347	83355	68163	83868	68973	84371	69777	84866	70576	42
19	82308	66539	82840	67360	83363	68176	83876	68986	84380	69791	84874	70589	41
20	82317	66553	82849	67374	83372	68190	83885	69000	84388	69804	84882	70602	40
21	82326	66567	82858	67388	83380	68203	83893	69013	84396	69817	84890	70615	39
22	82335	66581	82867	67401	83389	68217	83902	69027	84405	69831	84898	70629	38
23	82344	66594	82876	67415	83398	68230	83910	69040	84413	69844	84906	70642	37
24	82353	66608	82884	67429	83406	68244	83919	69054	84421	69857	84914	70655	36
25	82362	66622	82893	67442	83415	68257	83927	69067	84430	69871	84923	70668	35
26	82371	66635	82902	67456	83424	68271	83935	69080	84438	69884	84931	70682	34
27	82380	66649	82911	67469	83432	68284	83944	69094	84446	69897	84939	70695	33
28	82388	66663	82920	67483	83441	68298	83952	69107	84454	69911	84947	70708	32
29	82397	66677	82928	67497	83449	68312	83961	69121	84463	69924	84955	70721	31
30	82406	66690	82937	67510	83458	68325	83969	69134	84471	69937	84963	70735	30
31	82415	66704	82946	67524	83467	68339	83978	69148	84479	69951	84971	70748	29
32	82424	66718	82955	67538	83475	68352	83986	69161	84488	69964	84979	70761	28
33	82433	66731	82963	67551	83484	68366	83995	69174	84496	69977	84988	70774	27
34	82442	66745	82972	67565	83492	68379	84003	69188	84504	69991	84996	70788	26
35	82451	66759	82981	67578	83501	68393	84011	69201	84512	70004	85004	70801	25
36	82460	66773	82990	67592	83510	68406	84020	69215	84521	70017	85012	70814	24
37	82469	66786	82998	67606	83518	68420	84028	69228	84529	70031	85020	70827	23
38	82478	66800	83007	67619	83527	68433	84037	69242	84537	70044	85028	70840	22
39	82487	66814	83016	67633	83535	68447	84045	69255	84545	70057	85036	70854	21
40	82495	66827	83025	67647	83544	68460	84054	69268	84554	70071	85044	70867	20
41	82504	66841	83033	67660	83552	68474	84062	69282	84562	70084	85052	70880	19
42	82513	66855	83042	67674	83561	68487	84070	69295	84570	70097	85061	70893	18
43	82522	66868	83051	67687	83570	68501	84079	69309	84578	70111	85069	70907	17
44	82531	66882	83059	67701	83578	68515	84087	69322	84587	70124	85077	70920	16
45	82540	66896	83068	67715	83587	68528	84096	69336	84595	70137	85085	70933	15
46	82549	66910	83077	67728	83595	68541	84104	69349	84603	70151	85093	70946	14
47	82558	66923	83086	67742	83604	68555	84112	69362	84611	70164	85101	70959	13
48	82567	66937	83094	67755	83612	68568	84121	69376	84620	70177	85109	70973	12
49	82575	66951	83103	67769	83621	68582	84129	69389	84628	70191	85117	70986	11
50	82584	66964	83112	67783	83630	68595	84138	69403	84636	70204	85125	70999	10
51	82593	66978	83120	67796	83638	68609	84146	69416	84644	70217	85133	71012	09
52	82602	66992	83129	67810	83647	68622	84154	69429	84653	70230	85141	71025	08
53	82611	67005	83138	67823	83655	68636	84163	69443	84661	70244	85149	71039	07
54	82620	67019	83147	67837	83664	68649	84171	69456	84669	70257	85158	71052	06
55	82629	67033	83155	67850	83672	68663	84179	69470	84677	70270	85166	71065	05
56	82638	67046	83164	67864	83681	68676	84188	69483	84685	70284	85174	71078	04
57	82646	67060	83173	67878	83689	68690	84196	69496	84694	70297	85182	71091	03
58	82655	67074	83181	67891	83698	68703	84205	69510	84702	70310	85190	71105	02
59	82664	67087	83190	67905	83706	68717	84213	69523	84710	70324	85198	71118	01
60	82673	67101	83199	67918	83715	68730	84221	69537	84718	70337	85206	71131	00

250° | 249° | 248° | 247° | 246° | 245°

HAVERSINES

Parts for 0'·2, etc.:

	115° Log	115° Nat	116° Log	116° Nat	117° Log	117° Nat	118° Log	118° Nat	119° Log	119° Nat	120° Log	120° Nat
·2	2	3	2	3	2	3	2	3	1	3	1	3
·4	3	5	3	5	3	5	3	5	3	5	3	5
·6	5	8	5	8	5	8	5	8	4	8	4	8
·8	6	11	6	10	6	10	6	10	6	10	6	10

′	115° Log 9·	Nat 0·	116° Log 9·	Nat 0·	117° Log 9·	Nat 0·	118° Log 9·	Nat 0·	119° Log 9·	Nat 0·	120° Log 9·	Nat 0·	′
00	85206	71131	85684	71919	86153	72700	86613	73474	87064	74240	87506	75000	60
01	85214	71144	85692	71932	86161	72712	86621	73486	87072	74253	87513	75013	59
02	85222	71157	85700	71945	86169	72725	86628	73499	87079	74266	87521	75025	58
03	85230	71170	85708	71958	86176	72738	86636	73512	87086	74279	87528	75038	57
04	85238	71184	85716	71971	86184	72751	86643	73525	87094	74291	87535	75050	56
05	85246	71197	85724	71984	86192	72764	86651	73538	87101	74304	87543	75063	55
06	85254	71210	85731	71997	86200	72777	86659	73551	87109	74317	87550	75075	54
07	85262	71223	85739	72010	86207	72790	86666	73563	87116	74229	87557	75088	53
08	85270	71236	85747	72023	86215	72803	86674	73576	87124	74342	87564	75101	52
09	85278	71249	85755	72036	86223	72816	86681	73589	87131	74355	87572	75113	51
10	85286	71263	85763	72049	86230	72829	86689	73602	87138	74368	87579	75126	50
11	85294	71276	85771	72062	86238	72842	86696	73615	87146	74380	87586	75138	49
12	85302	71289	85779	72075	86246	72855	86704	73628	87153	74393	87593	75151	48
13	85310	71302	85787	72088	86254	72868	86712	73640	87161	74406	87601	75164	47
14	85318	71315	85794	72101	86261	72881	86719	73653	87168	74418	87608	75176	46
15	85326	71328	85802	72114	86269	72894	86727	73666	87175	74431	87615	75189	45
16	85334	71342	85810	72127	86277	72907	86734	73679	87183	74444	87623	75201	44
17	85342	71355	85818	72141	86284	72920	86742	73692	87190	74456	87630	75214	43
18	85350	71368	85826	72154	86292	72932	86749	73704	87198	74469	87637	75226	42
19	85358	71381	85834	72167	86300	72945	86757	73717	87205	74482	87644	75239	41
20	85366	71394	85841	72180	86307	72958	86764	73730	87212	74494	87652	75251	40
21	85374	71407	85849	72193	86315	72971	86772	73743	87220	74507	87659	75264	39
22	85382	71420	85857	72206	86323	72984	86780	73756	87227	74520	87666	75277	38
23	85390	71434	85865	72219	86331	72997	86787	73768	87235	74533	87673	75289	37
24	85398	71447	85873	72232	86338	73010	86795	73781	87242	74545	87680	75302	36
25	85406	71460	85881	72245	86346	73023	86802	73794	87249	74558	87688	75314	35
26	85414	71473	85888	72258	86354	73036	86810	73807	87257	74571	87695	75327	34
27	85422	71486	85896	72271	86361	73049	86817	73820	87264	74583	87702	75339	33
28	85430	71499	85904	72284	86369	73062	86825	73832	87271	74596	87709	75352	32
29	85438	71512	85912	72297	86377	73075	86832	73845	87279	74609	87717	75364	31
30	85446	71526	85920	72310	86384	73087	86840	73858	87286	74621	87724	75377	30
31	85454	71539	85928	72323	86392	73100	86847	73871	87294	74634	87731	75389	29
32	85462	71552	85935	72336	86400	73113	86855	73884	87301	74646	87738	75402	28
33	85470	71565	85943	72349	86407	73126	86862	73896	87308	74659	87745	75415	27
34	85478	71578	85951	72362	86415	73139	86870	73909	87316	74672	87753	75427	26
35	85486	71591	85959	72375	86423	73152	86877	73922	87323	74684	87760	75440	25
36	85494	71604	85967	72388	86430	73165	86885	73935	87330	74697	87767	75452	24
37	85502	71617	85974	72401	86438	73178	86892	73947	87338	74710	87774	75465	23
38	85510	71631	85982	72414	86446	73191	86900	73960	87345	74722	87782	75477	22
39	85518	71644	85990	72427	86453	73203	86907	73973	87352	74735	87789	75490	21
40	85526	71657	85998	72440	86461	73216	86915	73986	87360	74748	87796	75502	20
41	85534	71670	86006	72453	86468	73229	86922	73998	87367	74760	87803	75515	19
42	85542	71683	86013	72466	86476	73242	86930	74011	87374	74773	87810	75527	18
43	85550	71696	86021	72479	86484	73255	86937	74024	87382	74786	87818	75540	17
44	85557	71709	86029	72492	86491	73268	86945	74037	87389	74798	87825	75552	16
45	85565	71722	86037	72505	86499	73281	86952	74049	87396	74811	87832	75565	15
46	85573	71735	86045	72518	86507	73294	86960	74062	87404	74823	87839	75577	14
47	85581	71748	86052	72531	86514	73306	86967	74075	87411	74836	87846	75590	13
48	85589	71762	86060	72544	86522	73319	86975	74088	87418	74849	87853	75602	12
49	85597	71775	86068	72557	86529	73332	86982	74100	87426	74861	87861	75615	11
50	85605	71788	86076	72570	86537	73345	86990	74113	87433	74874	87868	75627	10
51	85613	71801	86083	72583	86545	73358	86997	74126	87440	74887	87875	75640	09
52	85621	71814	86091	72596	86552	73371	87004	74139	87448	74899	87882	75652	08
53	85629	71827	86099	72609	86560	73384	87012	74151	87455	74912	87889	75665	07
54	85637	71840	86107	72622	86568	73396	87019	74164	87462	74924	87896	75677	06
55	85645	71853	86114	72635	86575	73409	87027	74177	87470	74937	87904	75690	05
56	85653	71866	86122	72648	86583	73422	87034	74190	87477	74950	87911	75702	04
57	85660	71879	86130	72661	86590	73435	87042	74202	87484	74962	87918	75714	03
58	85668	71892	86138	72674	86598	73448	87049	74215	87492	74975	87925	75727	02
59	85676	71905	86145	72687	86606	73461	87057	74228	87499	74987	87932	75739	01
60	85684	71919	86153	72700	86613	73474	87064	74240	87506	75000	87939	75752	00

244°	243°	242°	241°	240°	239°

HAVERSINES

	121°		122°		123°		124°		125°		126°		
	Log.	Nat.	Log.	Nat.	Log.	Nat.	Log.	Nat.	Log.	Nat.	Log.	Nat.	
	9·	0·	9·	0·	9·	0·	9·	0·	9·	0·	9·	0·	′
00	87939	75752	88364	76496	88780	77232	89187	77960	89586	78679	89976	79389	60
01	87947	75764	88371	76508	88787	77244	89194	77972	89592	78691	89983	79401	59
02	87954	75777	88378	76521	88793	77256	89200	77984	89599	78703	89989	79413	58
03	87961	75789	88385	76533	88800	77269	89207	77996	89606	78715	89995	79425	57
04	87968	75802	88392	76545	88807	77281	89214	78008	89612	78726	90002	79436	56
05	87975	75814	88399	76558	88814	77293	89221	78020	89619	78738	90008	79448	55
06	87982	75827	88406	76570	88821	77305	89227	78032	89625	78750	90015	79460	54
07	87989	75839	88413	76582	88828	77317	89234	78044	89632	78762	90021	79471	53
08	87996	75852	88420	76595	88835	77329	89241	78056	89638	78774	90028	79483	52
09	88004	75864	88427	76607	88841	77342	89247	78068	89645	78786	90034	79495	51
10	88011	75876	88434	76619	88848	77354	89254	78080	89651	78798	90040	79507	50
11	88018	75889	88441	76632	88855	77366	89261	78092	89658	78810	90047	79519	49
12	88025	75901	88448	76644	88862	77378	89267	78104	89665	78822	90053	79530	48
13	88032	75914	88455	76656	88869	77390	89274	78116	89671	78834	90060	79542	47
14	88039	75926	88462	76668	88876	77403	89281	78128	89678	78845	90066	79554	46
15	88046	75939	88469	76681	88882	77415	89287	78140	89684	78857	90072	79565	45
16	88053	75951	88476	76693	88889	77427	89294	78152	89691	78869	90079	79577	44
17	88061	75964	88483	76705	88896	77439	89301	78164	89697	78881	90085	79589	43
18	88068	75976	88490	76718	88903	77451	89308	78176	89704	78893	90092	79601	42
19	88075	75988	88496	76730	88910	77463	89314	78188	89710	78905	90098	79612	41
20	88082	76001	88503	76742	88916	77475	89321	78200	89717	78917	90104	79624	40
21	88089	76013	88510	76754	88923	77488	89328	78212	89723	78928	90111	79636	39
22	88096	76026	88517	76767	88930	77500	89334	78224	89730	78940	90117	79648	38
23	88103	76038	88524	76779	88937	77512	89341	78236	89736	78952	90124	79659	37
24	88110	76050	88531	76791	88944	77524	89348	78248	89743	78964	90130	79671	36
25	88117	76063	88538	76804	88950	77536	89354	78260	89749	78976	90136	79683	35
26	88124	76075	88545	76816	88957	77548	89361	78272	89756	78988	90143	79694	34
27	88131	76088	88552	76828	88964	77560	89368	78284	89763	79000	90149	79706	33
28	88139	76100	88559	76840	88971	77573	89374	78296	89769	79011	90156	79718	32
29	88146	76113	88566	76853	88978	77585	89381	78308	89776	79023	90162	79729	31
30	88153	76125	88573	76865	88984	77597	89387	78320	89782	79035	90168	79741	30
31	88160	76137	88580	76877	88991	77609	89394	78332	89789	79047	90175	79753	29
32	88167	76150	88587	76890	88998	77621	89400	78344	89795	79059	90181	79765	28
33	88174	76162	88594	76902	89005	77633	89407	78356	89802	79071	90187	79776	27
34	88181	76175	88600	76914	89012	77645	89414	78368	89808	79082	90194	79788	26
35	88188	76187	88607	76926	89018	77657	89421	78380	89815	79094	90200	79800	25
36	88195	76199	88614	76939	89025	77670	89427	78392	89821	79106	90206	79811	24
37	88202	76212	88621	76951	89032	77682	89434	78404	89828	79118	90213	79823	23
38	88209	76224	88628	76963	89039	77694	89441	78416	89834	79130	90219	79835	22
39	88216	76236	88635	76975	89045	77706	89447	78428	89840	79142	90225	79846	21
40	88223	76249	88642	76988	89052	77718	89454	78440	89847	79153	90232	79858	20
41	88230	76261	88649	77000	89059	77730	89460	78452	89853	79165	90238	79870	19
42	88237	76274	88656	77012	89066	77742	89467	78464	89860	79177	90244	79881	18
43	88244	76286	88663	77024	89072	77754	89474	78476	89866	79189	90251	79893	17
44	88252	76298	88670	77036	89079	77766	89480	78488	89873	79201	90257	79905	16
45	88259	76311	88677	77049	89086	77779	89487	78500	89879	79212	90264	79916	15
46	88266	76323	88683	77061	89093	77791	89493	78512	89886	79224	90270	79928	14
47	88273	76335	88690	77073	89099	77803	89500	78524	89892	79236	90276	79940	13
48	88280	76348	88697	77085	89106	77815	89507	78536	89899	79248	90282	79951	12
49	88287	76360	88704	77098	89113	77827	89513	78548	89905	79260	90289	79963	11
50	88294	76373	88711	77110	89120	77839	89520	78560	89912	79271	90295	79974	10
51	88301	76385	88718	77122	89126	77851	89527	78572	89918	79283	90301	79986	09
52	88308	76397	88725	77134	89133	77863	89533	78583	89925	79295	90308	79998	08
53	88315	76410	88732	77147	89140	77875	89540	78595	89931	79307	90314	80009	07
54	88322	76422	88739	77159	89147	77887	89546	78607	89938	79319	90320	80021	06
55	88329	76434	88745	77171	89153	77899	89553	78619	89944	79330	90327	80033	05
56	88336	76447	88752	77183	89160	77911	89559	78631	89950	79342	90333	80044	04
57	88343	76459	88759	77195	89167	77923	89566	78643	89957	79354	90339	80056	03
58	88350	76471	88766	77208	89174	77936	89573	78655	89963	79366	90346	80068	02
59	88357	76484	88773	77220	89180	77948	89579	78667	89970	79377	90352	80079	01
60	88364	76496	88780	77232	89187	77960	89586	78679	89976	79389	90358	80091	00

Parts for 0′·2, etc.

Left column parts (121°–123°):
·2 1 · ·2 2 · ·4 3 · ·4 5 · ·6 4 · ·6 7 · ·8 6 · ·8 10

Right column parts (124°–126°):
·2 1 · ·2 2 · ·4 3 · ·4 5 · ·6 4 · ·6 7 · ·8 5 · ·8 9

| 238° | 237° | 236° | 235° | 234° | 233° |

HAVERSINES

	127°		128°		129°		130°		131°		132°	
	Log.	Nat.	Log.	Nat.	Log.	Nat.	Log.	Nat.	Log.	Nat.	Log.	Nat.	

Parts for 0′·2, etc.

127°: ·2 1 / ·2 2 · ·4 2 / ·4 5 · ·6 4 / ·6 7 · ·8 5 / ·8 9
128°: ·2 1 / ·2 2 · ·4 2 / ·4 5 · ·6 4 / ·6 7 · ·8 5 / ·8 9
129°: ·2 1 / ·2 2 · ·4 2 / ·4 5 · ·6 4 / ·6 7 · ·8 5 / ·8 9
130°: ·2 1 / ·2 2 · ·4 2 / ·4 4 · ·6 3 / ·6 7 · ·8 5 / ·8 9
131°: ·2 1 / ·2 2 · ·4 2 / ·4 4 · ·6 3 / ·6 7 · ·8 5 / ·8 9
132°: ·2 1 / ·2 2 · ·4 2 / ·4 4 · ·6 3 / ·6 6 · ·8 4 / ·8 9

′	9·	0·	9·	0·	9·	0·	9·	0·	9·	0·	9·	0·	′
00	90358	80091	90732	80783	91098	81466	91455	82139	91805	82803	92146	83457	60
01	90365	80102	90738	80795	91104	81477	91461	82151	91810	82814	92152	83467	59
02	90371	80114	90744	80806	91110	81489	91467	82162	91816	82825	92157	83478	58
03	90377	80126	90751	80817	91116	81500	91473	82173	91822	82836	92163	83489	57
04	90383	80137	90757	80829	91122	81511	91479	82184	91828	82847	92169	83500	56
05	90390	80149	90763	80840	91128	81523	91485	82195	91833	82858	92174	83511	55
06	90396	80160	90769	80852	91134	81534	91490	82206	91839	82869	92180	83521	54
07	90402	80172	90775	80863	91140	81545	91496	82217	91845	82880	92185	83532	53
08	90409	80184	90781	80875	91146	81556	91502	82228	91851	82891	92191	83543	52
09	90415	80195	90787	80886	91152	81568	91508	82240	91856	82902	92197	83554	51
10	90421	80207	90794	80898	91158	81579	91514	82251	91862	82913	92202	83564	50
11	90428	80218	90800	80909	91164	81590	91520	82262	91868	82924	92208	83575	49
12	90434	80230	90806	80920	91170	81601	91526	82273	91874	82934	92213	83586	48
13	90440	80242	90812	80932	91176	81613	91532	82284	91879	82945	92219	83597	47
14	90446	80253	90818	80943	91182	81624	91537	82295	91885	82956	92225	83608	46
15	90452	80265	90824	80955	91188	81635	91543	82306	91891	82967	92230	83618	45
16	90459	80276	90830	80966	91194	81647	91549	82317	91896	82978	92236	83629	44
17	90465	80288	90836	80978	91200	81658	91555	82328	91902	82989	92241	83640	43
18	90471	80299	90843	80989	91206	81669	91561	82339	91908	83000	92247	83651	42
19	90478	80311	90849	81000	91212	81680	91567	82351	91914	83011	92253	83661	41
20	90484	80323	90855	81012	91218	81692	91573	82362	91919	83022	92258	83672	40
21	90490	80334	90861	81023	91224	81703	91578	82373	91925	83033	92264	83683	39
22	90496	80346	90867	81035	91230	81714	91584	82384	91931	83044	92269	83694	38
23	90503	80357	90873	81046	91236	81725	91590	82395	91936	83055	92275	83704	37
24	90509	80369	90879	81057	91242	81737	91596	82406	91942	83066	92280	83715	36
25	90515	80380	90885	81068	91248	81748	91602	82417	91948	83077	92286	83726	35
26	90521	80392	90892	81080	91254	81759	91608	82428	91954	83087	92292	83737	34
27	90527	80403	90898	81092	91260	81770	91613	82439	91959	83098	92297	83747	33
28	90534	80415	90904	81103	91265	81781	91619	82450	91965	83109	92303	83758	32
29	90540	80427	90910	81114	91271	81793	91625	82461	91971	83120	92308	83769	31
30	90546	80438	90916	81126	91277	81804	91631	82472	91976	83131	92314	83780	30
31	90552	80450	90922	81137	91283	81815	91637	82483	91982	83142	92319	83790	29
32	90559	80461	90928	81148	91289	81826	91643	82495	91988	83153	92325	83801	28
33	90565	80473	90934	81160	91295	81838	91648	82506	91993	83164	92330	83812	27
34	90571	80484	90940	81171	91301	81849	91654	82517	91999	83175	92336	83822	26
35	90577	80496	90946	81183	91307	81860	91660	82528	92005	83185	92342	83833	25
36	90584	80507	90952	81194	91313	81871	91666	82539	92010	83196	92347	83844	24
37	90590	80519	90958	81205	91319	81882	91672	82550	92016	83207	92353	83855	23
38	90596	80530	90965	81217	91325	81894	91677	82561	92022	83218	92358	83865	22
39	90602	80542	90971	81228	91331	81905	91683	82572	92027	83229	92364	83876	21
40	90608	80553	90977	81239	91337	81916	91689	82583	92033	83240	92369	83887	20
41	90615	80565	90983	81251	91343	81927	91695	82594	92039	83251	92375	83897	19
42	90621	80576	90989	81262	91349	81938	91701	82605	92044	83262	92380	83908	18
43	90627	80588	90995	81273	91355	81950	91706	82616	92050	83272	92386	83919	17
44	90633	80599	91001	81285	91361	81961	91712	82627	92056	83283	92391	83929	16
45	90639	80611	91007	81296	91367	81972	91718	82638	92061	83294	92397	83940	15
46	90645	80622	91013	81308	91372	81983	91724	82649	92067	83305	92402	83951	14
47	90652	80634	91019	81319	91378	81994	91730	82660	92073	83316	92408	83961	13
48	90658	80645	91025	81330	91384	82005	91735	82671	92078	83327	92413	83972	12
49	90664	80657	91031	81342	91390	82017	91741	82682	92084	83337	92419	83983	11
50	90670	80668	91037	81353	91396	82028	91747	82693	92090	83348	92425	83993	10
51	90676	80680	91043	81364	91402	82039	91753	82704	92095	83359	92430	84004	09
52	90683	80691	91049	81376	91408	82050	91758	82715	92101	83370	92436	84015	08
53	90689	80703	91055	81387	91414	82061	91764	82726	92107	83381	92441	84025	07
54	90695	80714	91061	81398	91420	82072	91770	82737	92112	83392	92447	84036	06
55	90701	80726	91067	81409	91426	82084	91776	82748	92118	83402	92452	84047	05
56	90707	80737	91074	81421	91432	82095	91782	82759	92124	83413	92458	84057	04
57	90714	80749	91080	81432	91437	82106	91787	82770	92129	83424	92463	84068	03
58	90720	80760	91086	81443	91443	82117	91793	82781	92135	83435	92469	84079	02
59	90726	80772	91092	81455	91449	82128	91799	82792	92140	83446	92474	84089	01
60	90732	80783	91098	81466	91455	82139	91805	82803	92146	83457	92480	84100	00

| 232° | | 231° | | 230° | | 229° | | 228° | | 227° | |

HAVERSINES

Parts for 0'·2, etc.	133° Log.	133° Nat.	134° Log.	134° Nat.	135° Log.	135° Nat.	136° Log.	136° Nat.	137° Log.	137° Nat.	138° Log.	138° Nat.	Parts for 0'·2, etc.
·2 1 ·4 2 ·6 3 ·8 4	·2 2 ·4 4 ·6 6 ·8 8		·2 1 ·4 2 ·6 3 ·8 4	·2 2 ·4 4 ·6 6 ·8 8									
'	9·	0·	9·	0·	9·	0·	9·	0·	9·	0·	9·	0·	'
00	92480	84100	92805	84733	93123	85355	93433	85967	93736	86568	94030	87157	60
01	92485	84111	92811	84743	93128	85366	93438	85977	93741	86578	94035	87167	59
02	92491	84121	92816	84754	93134	85376	93443	85987	93746	86588	94040	87177	58
03	92496	84132	92821	84764	93139	85386	93448	85997	93751	86597	94045	87186	57
04	92502	84142	92827	84775	93144	85396	93454	86007	93755	86607	94050	87196	56
05	92507	84153	92832	84785	93149	85407	93459	86017	93760	86617	94055	87206	55
06	92512	84164	92837	84796	93154	85417	93464	86028	93765	86627	94059	87216	54
07	92518	84174	92843	84806	93160	85427	93469	86038	93770	86637	94064	87225	53
08	92523	84185	92848	84817	93165	85438	93474	86048	93775	86647	94069	87235	52
09	92529	84196	92853	84827	93170	85448	93479	86058	93780	86657	94074	87245	51
10	92534	84206	92859	84837	93175	85458	93484	86068	93785	86667	94079	87254	50
11	92540	84217	92864	84848	93181	85468	93489	86078	93790	86677	94084	87264	49
12	92545	84227	92869	84858	93186	85479	93494	86088	93795	86686	94088	87274	48
13	92551	84238	92875	84869	93191	85489	93499	86098	93800	86696	94093	87283	47
14	92556	84249	92880	84879	93196	85499	93504	86108	93805	86706	94098	87293	46
15	92562	84259	92885	84890	93201	85509	93509	86118	93810	86716	94103	87303	45
16	92567	84270	92891	84900	93207	85520	93515	86128	93815	86726	94108	87313	44
17	92573	84280	92896	84910	93212	85530	93520	86138	93820	86736	94112	87322	43
18	92578	84291	92901	84921	93217	85540	93525	86148	93825	86746	94117	87332	42
19	92584	84302	92907	84931	93222	85550	93530	86158	93830	86756	94122	87342	41
20	92589	84312	92912	84942	93227	85560	93535	86168	93835	86765	94127	87351	40
21	92594	84323	92917	84952	93232	85571	93540	86178	93840	86775	94132	87361	39
22	92600	84333	92923	84962	93238	85581	93545	86189	93845	86785	94137	87371	38
23	92605	84344	92928	84973	93243	85591	93550	86199	93849	86795	94141	87380	37
24	92611	84354	92933	84983	93248	85601	93555	86209	93854	86805	94146	87390	36
25	92616	84365	92939	84994	93253	85612	93560	86219	93859	86815	94151	87400	35
26	92622	84376	92944	85004	93258	85622	93565	86229	93864	86825	94156	87409	34
27	92627	84386	92949	85014	93264	85632	93570	86239	93869	86834	94161	87419	33
28	92633	84397	92955	85025	93269	85642	93575	86249	93874	86844	94165	87428	32
29	92638	84407	92960	85035	93274	85652	93580	86259	93879	86854	94170	87438	31
30	92643	84418	92965	85045	93279	85663	93585	86269	93884	86864	94175	87448	30
31	92649	84428	92970	85056	93284	85673	93590	86279	93889	86874	94180	87457	29
32	92654	84439	92975	85066	93289	85683	93595	86289	93894	86884	94184	87467	28
33	92660	84449	92981	85077	93295	85693	93600	86299	93899	86893	94189	87477	27
34	92665	84460	92986	85087	93300	85703	93605	86309	93904	86903	94194	87486	26
35	92670	84470	92992	85097	93305	85713	93611	86319	93908	86913	94199	87496	25
36	92676	84481	92997	85108	93310	85724	93616	86329	93913	86923	94204	87505	24
37	92681	84492	93002	85118	93315	85734	93621	86339	93918	86933	94208	87515	23
38	92687	84502	93007	85128	93320	85744	93626	86349	93923	86942	94213	87525	22
39	92692	84513	93013	85139	93326	85754	93631	86359	93928	86952	94218	87534	21
40	92698	84523	93018	85149	93331	85764	93636	86369	93933	86962	94223	87544	20
41	92703	84534	93023	85159	93336	85774	93641	86379	93938	86972	94227	87554	19
42	92708	84544	93029	85170	93341	85785	93646	86389	93943	86982	94232	87563	18
43	92714	84555	93034	85180	93346	85795	93651	86399	93948	86991	94237	87573	17
44	92719	84565	93039	85190	93351	85805	93656	86409	93952	87001	94242	87582	16
45	92725	84576	93044	85201	93356	85815	93661	86419	93957	87011	94246	87592	15
46	92730	84586	93050	85211	93362	85825	93666	86429	93962	87021	94251	87602	14
47	92735	84597	93055	85221	93367	85835	93671	86438	93967	87030	94256	87611	13
48	92741	84607	93060	85232	93372	85846	93676	86448	93972	87040	94261	87621	12
49	92746	84618	93065	85242	93377	85856	93681	86458	93977	87050	94265	87630	11
50	92751	84628	93071	85252	93382	85866	93686	86468	93982	87060	94270	87640	10
51	92757	84639	93076	85263	93387	85876	93691	86478	93987	87070	94275	87649	09
52	92762	84649	93081	85273	93392	85886	93696	86488	93991	87079	94280	87659	08
53	92768	84660	93086	85283	93397	85896	93701	86498	93996	87089	94284	87669	07
54	92773	84670	93092	85294	93403	85906	93706	86508	94001	87099	94289	87678	06
55	92778	84681	93097	85304	93408	85916	93711	86518	94006	87109	94294	87688	05
56	92784	84691	93102	85314	93413	85926	93716	86528	94011	87118	94299	87697	04
57	92789	84702	93107	85324	93418	85937	93721	86538	94016	87128	94303	87707	03
58	92794	84712	93113	85335	93423	85947	93726	86548	94021	87138	94308	87716	02
59	92800	84722	93118	85345	93428	85957	93731	86558	94026	87148	94313	87726	01
60	92805	84733	93123	85355	93433	85967	93736	86568	94030	87157	94318	87735	00

| 226° | 225° | 224° | 223° | 222° | 221° |

HAVERSINES

′	139° Log.	139° Nat.	140° Log.	140° Nat.	141° Log.	141° Nat.	142° Log.	142° Nat.	143° Log.	143° Nat.	144° Log.	144° Nat.	′
	9·	0·	9·	0·	9·	0·	9·	0·	9·	0·	9·	0·	
00	94318	87735	94597	88302	94869	88857	95134	89401	95391	89932	95641	90451	60
01	94322	87745	94602	88312	94874	88866	95138	89409	95396	89941	95645	90459	59
02	94327	87755	94606	88321	94878	88876	95143	89418	95400	89949	95649	90468	58
03	94332	87764	94611	88330	94883	88885	95147	89427	95404	89958	95654	90476	57
04	94336	87774	94616	88340	94887	88894	95151	89436	95408	89967	95658	90485	56
05	94341	87783	94620	88349	94892	88903	95156	89445	95412	89976	95662	90494	55
06	94346	87793	94625	88358	94896	88912	95160	89454	95417	89984	95666	90502	54
07	94351	87802	94629	88368	94901	88921	95164	89463	95421	89993	95670	90511	53
08	94355	87812	94634	88377	94905	88930	95169	89472	95425	90002	95674	90519	52
09	94360	87821	94638	88386	94909	88940	95173	89481	95429	90010	95678	90528	51
10	94365	87831	94643	88396	94914	88949	95177	89490	95433	90019	95682	90537	50
11	94369	87840	94648	88405	94918	88958	95182	89499	95438	90028	95686	90545	49
12	94374	87850	94652	88414	94923	88967	95186	89508	95442	90037	95690	90553	48
13	94379	87859	94657	88423	94927	88976	95190	89517	95446	90045	95694	90562	47
14	94383	87869	94661	88433	94932	88985	95195	89526	95450	90054	95699	90570	46
15	94388	87878	94666	88442	94936	88994	95199	89534	95454	90063	95703	90579	45
16	94393	87888	94670	88451	94941	89003	95203	89543	95459	90071	95707	90588	44
17	94398	87897	94675	88461	94945	89012	95208	89552	95463	90080	95711	90596	43
18	94402	87907	94680	88470	94950	89022	95212	89561	95467	90089	95715	90604	42
19	94407	87916	94684	88479	94954	89031	95216	89570	95471	90097	95719	90613	41
20	94412	87926	94689	88489	94958	89040	95221	89579	95475	90106	95723	90621	40
21	94416	87935	94693	88498	94963	89049	95225	89588	95480	90115	95727	90630	39
22	94421	87945	94698	88507	94967	89058	95229	89597	95484	90124	95731	90638	38
23	94426	87954	94702	88516	94972	89067	95234	89606	95488	90132	95735	90647	37
24	94430	87964	94707	88526	94976	89076	95238	89614	95492	90141	95739	90655	36
25	94435	87973	94711	88535	94981	89085	95242	89623	95496	90150	95743	90664	35
26	94440	87983	94716	88544	94985	89094	95246	89632	95501	90158	95747	90672	34
27	94444	87992	94721	88553	94989	89103	95251	89641	95505	90167	95751	90680	33
28	94449	88001	94725	88563	94994	89112	95255	89650	95509	90176	95755	90689	32
29	94454	88011	94730	88572	94998	89121	95259	89659	95513	90184	95759	90697	31
30	94458	88020	94734	88581	95003	89130	95264	89668	95517	90193	95763	90706	30
31	94463	88030	94739	88590	95007	89139	95268	89677	95521	90201	95768	90714	29
32	94468	88039	94743	88600	95011	89149	95272	89685	95526	90210	95772	90723	28
33	94472	88049	94748	88609	95016	89158	95276	89694	95530	90219	95776	90731	27
34	94477	88058	94752	88618	95020	89167	95281	89703	95534	90227	95780	90740	26
35	94482	88068	94757	88627	95025	89176	95285	89712	95538	90236	95784	90748	25
36	94486	88078	94761	88637	95029	89185	95289	89721	95542	90245	95788	90756	24
37	94491	88086	94766	88646	95033	89194	95294	89730	95546	90253	95792	90765	23
38	94496	88096	94770	88655	95038	89203	95298	89738	95550	90262	95796	90773	22
39	94500	88105	94774	88664	95042	89212	95302	89747	95555	90271	95800	90782	21
40	94505	88115	94779	88674	95047	89221	95306	89756	95559	90279	95804	90790	20
41	94509	88124	94784	88683	95051	89230	95311	89765	95563	90288	95808	90798	19
42	94514	88133	94788	88692	95055	89239	95315	89774	95567	90296	95812	90807	18
43	94519	88143	94793	88701	95060	89248	95319	89783	95571	90305	95816	90815	17
44	94523	88152	94797	88710	95064	89257	95323	89791	95575	90314	95820	90824	16
45	94528	88162	94802	88720	95069	89266	95328	89800	95579	90322	95824	90832	15
46	94533	88171	94806	88729	95073	89275	95332	89809	95584	90331	95828	90840	14
47	94537	88180	94811	88738	95077	89284	95336	89818	95588	90339	95832	90849	13
48	94542	88190	94815	88747	95082	89293	95340	89827	95592	90348	95836	90857	12
49	94546	88199	94820	88756	95086	89302	95345	89835	95596	90357	95840	90866	11
50	94551	88209	94824	88766	95090	89311	95349	89844	95600	90365	95844	90874	10
51	94556	88218	94829	88775	95095	89320	95353	89853	95604	90374	95848	90882	09
52	94560	88227	94833	88784	95099	89329	95357	89862	95608	90382	95852	90891	08
53	94565	88237	94838	88793	95104	89338	95362	89870	95613	90391	95856	90899	07
54	94570	88246	94842	88802	95108	89347	95366	89879	95617	90399	95860	90907	06
55	94574	88255	94847	88811	95112	89356	95370	89888	95621	90408	95864	90916	05
56	94579	88265	94851	88821	95117	89365	95374	89897	95625	90417	95868	90924	04
57	94583	88274	94856	88830	95121	89374	95379	89906	95629	90425	95872	90933	03
58	94588	88284	94860	88839	95125	89383	95383	89914	95633	90434	95876	90941	02
59	94593	88293	94865	88848	95130	89392	95387	89923	95637	90442	95880	90949	01
60	94597	88302	94869	88857	95134	89401	95391	89932	95641	90451	95884	90958	00

Parts for 0′·2, etc.

	139°	140°	141°	142°	143°	144°
Log.	·2 1 ·4 2 ·6 3 ·8 4	·2 1 ·4 2 ·6 3 ·8 4	·2 1 ·4 2 ·6 3 ·8 4	·2 1 ·4 2 ·6 3 ·8 3	·2 1 ·4 2 ·6 3 ·8 3	·2 1 ·4 2 ·6 2 ·8 3
Nat.	·2 2 ·4 4 ·6 6 ·8 8	·2 2 ·4 4 ·6 6 ·8 7	·2 2 ·4 4 ·6 5 ·8 7	·2 2 ·4 4 ·6 5 ·8 7	·2 2 ·4 3 ·6 5 ·8 7	·2 2 ·4 3 ·6 5 ·8 7

220°	219°	218°	217°	216°	215°

HAVERSINES

′	145° Log.	145° Nat.	146° Log.	146° Nat.	147° Log.	147° Nat.	148° Log.	148° Nat.	149° Log.	149° Nat.	150° Log.	150° Nat.	′
	9·	0·	9·	0·	9·	0·	9·	0·	9·	0·	9·	0·	
00	95884	90958	96119	91452	96347	91934	96568	92402	96782	92858	96989	93301	60
01	95888	90966	96123	91460	96351	91941	96572	92410	96786	92866	96992	93309	59
02	95892	90974	96127	91468	96355	91949	96576	92418	96789	92873	96996	93316	58
03	95896	90983	96131	91476	96359	91957	96579	92426	96793	92881	96999	93323	57
04	95900	90991	96135	91484	96362	91965	96583	92433	96796	92888	97002	93330	56
05	95904	90999	96139	91493	96366	91973	96586	92441	96800	92896	97006	93338	55
06	95908	91008	96142	91501	96370	91981	96590	92449	96803	92903	97009	93345	54
07	95912	91016	96146	91509	96374	91989	96594	92456	96807	92911	97012	93352	53
08	95916	91024	96150	91517	96377	91997	96597	92464	96810	92918	97016	93359	52
09	95920	91033	96154	91525	96381	92005	96601	92472	96814	92926	97019	93367	51
10	95924	91041	96158	91533	96385	92013	96604	92479	96817	92933	97022	93374	50
11	95928	91049	96162	91541	96388	92020	96608	92487	96821	92941	97026	93381	49
12	95932	91057	96165	91549	96392	92028	96612	92495	96824	92948	97029	93388	48
13	95936	91066	96169	91557	96396	92036	96615	92502	96827	92955	97033	93395	47
14	95939	91074	96173	91565	96400	92044	96619	92510	96831	92963	97036	93403	46
15	95943	91082	96177	91574	96403	92052	96622	92518	96834	92970	97039	93410	45
16	95947	91091	96181	91582	96407	92060	96626	92525	96837	92978	97043	93417	44
17	95951	91099	96185	91590	96411	92068	96630	92533	96841	92985	97046	93424	43
18	95955	91107	96188	91598	96415	92076	96633	92541	96845	92993	97049	93432	42
19	95959	91115	96192	91606	96418	92083	96637	92548	96848	93000	97052	93439	41
20	95963	91124	96196	91614	96422	92091	96640	92556	96852	93007	97056	93446	40
21	95967	91132	96200	91622	96426	92099	96644	92563	96855	93015	97059	93453	39
22	95971	91140	96204	91630	96429	92107	96648	92571	96859	93022	97063	93460	38
23	95975	91149	96208	91638	96433	92115	96651	92579	96862	93030	97066	93468	37
24	95979	91157	96211	91646	96437	92123	96655	92586	96866	93037	97069	93475	36
25	95983	91165	96215	91654	96440	92130	96658	92594	96869	93045	97073	93482	35
26	95987	91173	96219	91662	96444	92138	96662	92602	96873	93052	97076	93489	34
27	95991	91182	96223	91670	96448	92146	96665	92609	96876	93059	97079	93496	33
28	95995	91190	96227	91678	96451	92154	96669	92617	96879	93067	97083	93503	32
29	95999	91198	96230	91686	96455	92162	96673	92624	96883	93074	97086	93511	31
30	96002	91206	96234	91694	96459	92170	96676	92632	96886	93081	97089	93518	30
31	96006	91215	96238	91702	96462	92177	96680	92640	96890	93089	97093	93525	29
32	96010	91223	96242	91710	96466	92185	96683	92647	96894	93096	97096	93532	28
33	96014	91231	96246	91718	96470	92193	96687	92655	96897	93104	97099	93539	27
34	96018	91239	96249	91726	96473	92201	96690	92662	96900	93111	97103	93546	26
35	96022	91247	96253	91734	96477	92209	96694	92670	96904	93118	97106	93554	25
36	96026	91256	96257	91742	96481	92216	96697	92678	96907	93126	97109	93561	24
37	96030	91264	96261	91750	96484	92224	96701	92685	96910	93133	97113	93568	23
38	96034	91272	96265	91758	96488	92232	96705	92693	96914	93140	97116	93575	22
39	96038	91280	96268	91766	96492	92240	96708	92700	96917	93148	97119	93582	21
40	96042	91289	96272	91774	96495	92248	96712	92708	96921	93155	97123	93589	20
41	96046	91297	96276	91782	96499	92255	96715	92715	96924	93162	97126	93596	19
42	96049	91305	96280	91790	96503	92263	96719	92723	96928	93170	97129	93603	18
43	96053	91313	96283	91798	96506	92271	96722	92731	96931	93177	97132	93611	17
44	96057	91321	96287	91806	96510	92279	96726	92738	96934	93184	97136	93618	16
45	96061	91329	96291	91814	96514	92286	96729	92746	96938	93192	97139	93625	15
46	96065	91338	96295	91822	96517	92294	96733	92753	96941	93199	97142	93632	14
47	96069	91346	96299	91830	96521	92302	96736	92761	96945	93206	97146	93639	13
48	96073	91354	96302	91838	96525	92310	96740	92768	96948	93214	97149	93646	12
49	96077	91362	96306	91846	96528	92317	96743	92776	96951	93221	97152	93653	11
50	96081	91370	96310	91854	96532	92325	96747	92783	96955	93228	97156	93660	10
51	96084	91379	96314	91862	96536	92333	96750	92791	96958	93236	97159	93667	09
52	96088	91387	96317	91870	96539	92341	96754	92798	96962	93243	97162	93674	08
53	96092	91395	96321	91878	96543	92348	96758	92806	96965	93250	97165	93682	07
54	96096	91403	96325	91886	96547	92356	96761	92813	96968	93258	97169	93689	06
55	96100	91411	96329	91894	96550	92364	96765	92821	96972	93265	97172	93696	05
56	96104	91419	96332	91902	96554	92372	96768	92828	96975	93272	97175	93703	04
57	96108	91427	96336	91910	96557	92379	96772	92836	96979	93279	97179	93710	03
58	96112	91436	96340	91918	96561	92387	96775	92843	96982	93287	97182	93717	02
59	96115	91444	96344	91926	96565	92394	96779	92851	96985	93294	97185	93724	01
60	96119	91452	96347	91934	96568	92402	96782	92858	96989	93301	97188	93731	00

Parts for 0′·2, etc.

145°		146°		147°		148°		149°		150°	
·2 1	·2 2	·2 1	·2 2	·2 1	·2 2	·2 1	·2 2	·2 1	·2 1	·2 1	·2 1
·4 2	·4 3	·4 2	·4 3	·4 1	·4 3	·4 1	·4 3	·4 1	·4 3	·4 1	·4 3
·6 2	·6 5	·6 2	·6 5	·6 2	·6 5	·6 2	·6 5	·6 2	·6 4	·6 2	·6 4
·8 3	·8 7	·8 3	·8 6	·8 3	·8 6	·8 3	·8 6	·8 3	·8 6	·8 3	·8 6

| 214° | 213° | 212° | 211° | 210° | 209° |

HAVERSINES

Parts for 0'·2, etc.	151°		152°		153°		154°		155°		156°		Parts for 0'·2, etc.
	Log.	Nat.	Log.	Nat.	Log.	Nat.	Log.	Nat.	Log.	Nat.	Log.	Nat.	
·2 1 ·2 1 / ·4 1 ·4 3 / ·6 2 ·6 4 / ·8 3 ·8 6			·2 1 ·2 1 / ·4 1 ·4 3 / ·6 2 ·6 4 / ·8 2 ·8 5			·2 1 ·2 1 / ·4 1 ·4 3 / ·6 2 ·6 4 / ·8 2 ·8 5			·2 1 ·2 1 / ·4 1 ·4 3 / ·6 2 ·6 4 / ·8 2 ·8 5			·2 1 ·2 1 / ·4 1 ·4 2 / ·6 2 ·6 4 / ·8 2 ·8 5			·2 1 ·2 1 / ·4 1 ·4 2 / ·6 2 ·6 3 / ·8 2 ·8 5		
'	9·	0·	9·	0·	9·	0·	9·	0·	9·	0·	9.	0·	'
00	97188	93731	97381	94147	97566	94550	97745	94940	97916	95315	98081	95677	60
01	97192	93738	97384	94154	97569	94557	97748	94946	97919	95322	98084	95683	59
02	97195	93745	97387	94161	97572	94564	97751	94952	97922	95328	98086	95689	58
03	97198	92752	97390	94168	97570	94570	97754	94959	97925	95334	98089	95695	57
04	97201	93759	97393	94175	97578	94577	97756	94965	97927	95340	98092	95701	56
05	97205	93766	97397	94181	97581	94583	97759	94972	97930	95346	98094	95707	55
06	97208	93773	97400	94188	97584	94590	97762	94978	97933	95352	98097	95713	54
07	97211	93780	97403	94195	97587	94596	97765	94984	97936	95358	98100	95719	53
08	97214	93787	97406	94202	97591	94603	97768	94991	97939	95364	98102	95724	52
09	97218	93794	97409	94209	97594	94610	97771	94997	97941	95371	98105	95730	51
10	97221	93801	97412	94215	97597	94616	97774	95003	97944	95377	98108	95736	50
11	97224	93808	97415	94222	97600	94623	97777	95010	97947	95383	98110	95742	49
12	97227	93815	97418	94229	97603	94629	97780	95016	97950	95389	98113	95748	48
13	97231	93822	97422	94236	97606	94636	97783	95022	97953	95395	98116	95754	47
14	97234	93829	97425	94243	97609	94642	97785	95029	97955	95401	98118	95760	46
15	97237	93836	97428	94249	97612	94649	97788	95035	97958	95407	98121	95766	45
16	97240	93843	97431	94256	97615	94655	97791	95041	97961	95413	98124	95771	44
17	97244	93850	97434	94263	97618	94662	97794	95048	97964	95419	98126	95777	43
18	97247	93857	97437	94270	97621	94669	97797	95054	97966	95425	98129	95783	42
19	97250	93864	97440	94276	97624	94675	97800	95060	97969	95431	98132	95789	41
20	97253	93871	97443	94283	97627	94682	97803	95066	97972	95438	98134	95795	40
21	97257	93878	97447	94290	97630	94688	97806	95073	97975	95444	98137	95801	39
22	97260	93885	97450	94297	97633	94695	97808	95079	97977	95450	98139	95806	38
23	97263	93892	97453	94303	97636	94701	97811	95085	97980	95456	98142	95812	37
24	97266	93899	97456	94310	97639	94708	97814	95092	97983	95462	98145	95818	36
25	97269	93906	97459	94317	97642	94714	97817	95098	97986	95468	98147	95824	35
26	97273	93913	97462	94324	97645	94721	97820	95104	97988	95474	98150	95830	34
27	97276	93920	97465	94330	97647	94727	97823	95111	97991	95480	98153	95836	33
28	97279	93927	97468	94337	97650	94734	97826	95117	97994	95486	98155	95841	32
29	97282	93934	97471	94344	97653	94740	97829	95123	97997	95492	98158	95847	31
30	97285	93941	97474	94351	97656	94747	97831	95129	97999	95498	98161	95853	30
31	97289	93948	97478	94357	97659	94753	97834	95136	98002	95504	98163	95859	29
32	97292	93955	97481	94364	97662	94760	97837	95142	98005	95510	98166	95865	28
33	97295	93962	97484	94371	97665	94766	97840	95148	98008	95516	98168	95870	27
34	97298	93969	97487	94377	97668	94773	97843	95154	98010	95522	98171	95876	26
35	97301	93976	97490	94384	97671	94779	97846	95161	98013	95528	98174	95882	25
36	97305	93982	97493	94391	97674	94786	97849	95167	98016	95534	98176	95888	24
37	97308	93989	97496	94397	97677	94792	97851	95173	98019	95540	98179	95894	23
38	97311	93996	97499	94404	97680	94799	97854	95179	98021	95546	98182	95899	22
39	97314	94003	97502	94411	97683	94805	97857	95185	98024	95552	98184	95905	21
40	97317	94010	97505	94418	97686	94811	97860	95192	98027	95558	98187	95911	20
41	97321	94017	97508	94424	97689	94818	97863	95198	98030	95564	98189	95917	19
42	97324	94024	97511	94431	97692	94824	97866	95204	98032	95570	98192	95922	18
43	97327	94031	97514	94438	97695	94831	97868	95210	98035	95576	98195	95928	17
44	97330	94038	97518	94444	97698	94837	97871	95217	98038	95582	98197	95934	16
45	97333	94045	97521	94451	97701	94844	97874	95223	98040	95588	98200	95940	15
46	97337	94051	97524	94458	97704	94850	97877	95229	98042	95594	98202	95945	14
47	97340	94058	97527	94464	97707	94857	97880	95235	98046	95600	98205	95951	13
48	97343	94065	97530	94471	97710	94863	97883	95241	98049	95606	98208	95957	12
49	97346	94072	97533	94477	97713	94869	97885	95248	98051	95612	98210	95962	11
50	97349	94079	97536	94484	97716	94876	97888	95254	98054	95618	98213	95968	10
51	97352	94086	97539	94491	97718	94882	97891	95260	98057	95624	98215	95974	09
52	97356	94093	97542	94497	97721	94889	97894	95266	98059	95630	98218	95980	08
53	97359	94099	97545	94504	97724	94895	97897	95272	98062	95636	98221	95985	07
54	97362	94106	97548	94511	97727	94901	97899	95278	98065	95642	98223	95991	06
55	96365	94113	97551	94517	97730	94908	97902	95285	98067	95648	98226	95997	05
56	97368	94120	97554	94524	97733	94914	97905	95291	98070	95654	98228	96002	04
57	97371	94127	97557	94531	97736	94921	97908	95297	98073	95660	98231	96008	03
58	97375	94134	97560	94537	97739	94927	97911	95303	98076	95665	98233	96014	02
59	97378	94141	97563	94544	97742	94933	97914	95309	98078	95671	98236	96020	01
60	97381	94147	97566	94550	97745	94940	97916	95315	98081	95677	98239	96025	00

208°	207°	206°	205°	204°	203°

HAVERSINES

	157°		158°		159°		160°		161°		162°		
	Log.	Nat.	Log.	Nat.	Log.	Nat.	Log.	Nat.	Log.	Nat.	Log.	Nat.	
Parts for 0'·2, etc.	·2 1 / ·4 1 / ·6 2 / ·8 2	·2 1 / ·4 2 / ·6 3 / ·8 4	·2 0 / ·4 1 / ·6 1 / ·8 2	·2 1 / ·4 2 / ·6 3 / ·8 4	·2 0 / ·4 1 / ·6 1 / ·8 2	·2 1 / ·4 2 / ·6 3 / ·8 4	·2 0 / ·4 1 / ·6 1 / ·8 2	·2 1 / ·4 2 / ·6 3 / ·8 4	·2 0 / ·4 1 / ·6 1 / ·8 2	·2 1 / ·4 2 / ·6 3 / ·8 4	·2 0 / ·4 1 / ·6 1 / ·8 2	·2 1 / ·4 2 / ·6 3 / ·8 3	Parts for 0'·2, etc.
′	9·	0·	9·	0·	9·	0·	9·	0·	9.	0·	9·	0·	′
00	98239	96025	98389	96359	98533	96679	98670	96985	98801	97276	98924	97553	60
01	98241	96031	98392	96365	98536	96684	98673	96990	98803	97281	98926	97557	59
02	98244	96037	98394	96370	98538	96689	98675	96995	98805	97285	98928	97562	58
03	98246	96042	98397	96376	98540	96695	98677	97000	98807	97290	98930	97566	57
04	98249	96048	98399	96381	98543	96700	98679	97005	98809	97295	98932	97571	56
05	98251	96054	98402	96386	98545	96705	98681	97009	98811	97300	98934	97575	55
06	98254	96059	98404	96392	98547	96710	98684	97014	98813	97304	98936	97580	54
07	98256	96065	98406	96397	98550	96715	98686	97019	98815	97309	98938	97584	53
08	98259	96071	98409	96403	98552	96721	98688	97024	98817	97314	98940	97589	52
09	98262	96076	98411	96408	98554	96726	98690	97029	98819	97318	98942	97593	51
10	98264	96082	98414	96413	98557	96731	98692	97034	98822	97323	98944	97598	50
11	98267	96088	98416	96419	98559	96736	98695	97039	98824	97328	98946	97602	49
12	98269	96093	98419	96424	98561	96741	98697	97044	98826	97332	98948	97606	48
13	98272	96099	98421	96430	98564	96746	98699	97049	98828	97337	98950	97611	47
14	98274	96104	98424	96435	98566	96752	98701	97054	98830	97342	98952	97615	46
15	98277	96110	98426	96440	98568	96757	98703	97059	98832	97347	98954	97620	45
16	98279	96116	98428	96446	98570	96762	98706	97064	98834	97351	98956	97624	44
17	98282	96121	98431	96451	98573	96767	98708	97069	98836	97356	98958	97629	43
18	98285	96127	98433	96457	98575	96772	98710	97074	98838	97361	98960	97633	42
19	98287	96133	98436	96462	98577	96777	98712	97078	98840	97365	98962	97637	41
20	98290	96138	98438	96467	98580	96782	98714	97083	98842	97370	98964	97642	40
21	98292	96144	98440	96473	98582	96788	98717	97088	98845	97374	98966	97646	39
22	98295	96149	98443	96478	98584	96793	98719	97093	98847	97379	98968	97651	38
23	98297	96155	98445	96483	98587	96798	98721	97098	98849	97384	98970	97655	37
24	98300	96161	98448	96489	98589	96803	98723	97103	98851	97388	98971	97660	36
25	98302	96166	98450	96494	98591	96808	98725	97108	98853	97393	98973	97664	35
26	98305	96172	98453	96500	98593	96813	98728	97113	98855	97398	98975	97668	34
27	98307	96177	98455	96505	98596	96818	98730	97117	98857	97402	98977	97673	33
28	98310	96183	98457	96510	98598	96823	98732	97122	98859	97407	98979	97677	32
29	98312	96188	98460	96516	98600	96829	98734	97127	98861	97412	98981	97681	31
30	98315	96194	98462	96521	98603	96834	98736	97132	98863	97416	98983	97686	30
31	98317	96200	98465	96526	98605	96839	98738	97137	98865	97421	98985	97690	29
32	98320	96205	98467	96532	98607	96844	98741	97142	98867	97425	98987	97695	28
33	98322	96211	98469	96537	98609	96849	98743	97147	98869	97430	98989	97699	27
34	98325	96216	98472	96542	98612	96854	98745	97151	98871	97435	98991	97703	26
35	98327	96222	98474	96547	98614	96859	98747	97156	98873	97439	98993	97708	25
36	98330	96227	98476	96553	98616	96864	98749	97161	98875	97444	98995	97712	24
37	98332	96233	98479	96558	98619	96869	98751	97166	98877	97448	98997	97716	23
38	98335	96238	98481	96563	98621	96874	98754	97171	98880	97453	98999	97721	22
39	98337	96244	98484	96569	98623	96879	98756	97176	98882	97458	99001	97725	21
40	98340	96249	98486	96574	98625	96884	98758	97180	98884	97462	99003	97729	20
41	98342	96255	98488	96579	98628	96889	98760	97185	98886	97467	99004	97734	19
42	98345	96260	98491	96585	98630	96894	98762	97190	98888	97471	99006	97738	18
43	98347	96266	98493	96590	98632	96899	98764	97195	98890	97476	99008	97742	17
44	98350	96272	98496	96595	98634	96905	98766	97200	98892	97480	99010	97747	16
45	98352	96277	98498	96600	98637	96910	98769	97204	98894	97485	99012	97751	15
46	98355	96283	98500	96606	98639	96915	98771	97209	98896	97490	99014	97755	14
47	98357	96288	98503	96611	98641	96920	98773	97214	98898	97494	99016	97760	13
48	98360	96294	98505	96616	98643	96925	98775	97219	98900	97499	99018	97764	12
49	98362	96299	98507	96621	98646	96930	98777	97224	98902	97503	99020	97768	11
50	98365	96305	98510	96627	98648	96935	98779	97228	98904	97508	99022	97773	10
51	98367	96310	98512	96632	98650	96940	98781	97233	98906	97512	99024	97777	09
52	98370	96315	98514	96637	98652	96945	98784	97238	98908	97517	99026	97781	08
53	98372	96321	98517	96642	98655	96950	98786	97243	98910	97521	99027	97785	07
54	98375	96326	98519	96648	98657	96955	98788	97247	98912	97526	99029	97790	06
55	98377	96332	98521	96653	98659	96960	98790	97252	98914	97530	99031	97794	05
56	98379	96337	98524	96658	98661	96965	98792	97257	98916	97535	99033	97798	04
57	98382	96343	98526	96663	98664	96970	98794	97262	98918	97539	99035	97802	03
58	98384	96348	98529	96669	98666	96975	98796	97266	98920	97544	99037	97807	02
59	98387	96354	98531	96674	98668	96980	98798	97271	98922	97548	99039	97811	01
60	98389	96359	98533	96679	98670	96985	98801	97276	98924	97553	99041	97815	00

202°	201°	200°	199°	198°	197°

500

HAVERSINES

| Parts for 0′·2, etc. | 163° Log. | 163° Nat. | 164° Log. | 164° Nat. | 165° Log. | 165° Nat. | 166° Log. | 166° Nat. | 167° Log. | 167° Nat. | 168° Log. | 168° Nat. | Parts for 0′·2, etc. |

Parts for 0′·2, etc.

163°	164°	165°	166°	167°	168°
·2 0 / ·2 1	·2 0 / ·2 1	·2 0 / ·2 1	·2 0 / ·2 1	·2 0 / ·2 1	·2 0 / ·2 1
·4 1 / ·4 2	·4 1 / ·4 2	·4 1 / ·4 1	·4 1 / ·4 1	·4 1 / ·4 1	·4 1 / ·4 1
·6 1 / ·6 2	·6 1 / ·6 2	·6 1 / ·6 2	·6 1 / ·6 2	·6 1 / ·6 2	·6 1 / ·6 2
·8 1 / ·8 3	·8 1 / ·8 3	·8 1 / ·8 3	·8 1 / ·8 3	·8 1 / ·8 3	·8 1 / ·8 2

′	9·	0·	9·	0·	9·	0·	9·	0·	9·	0·	9·	0·	′
00	99041	97815	99151	98063	99254	98296	99350	98515	99440	98719	99523	98907	60
01	99043	97819	99152	98067	99255	98300	99352	98518	99441	98722	99524	98910	59
02	99044	97824	99154	98071	99257	98304	99353	98522	99443	98725	99526	98913	58
03	99046	97828	99156	98075	99259	98308	99355	98525	99444	98728	99527	98916	57
04	99048	97832	99158	98079	99260	98311	99356	98529	99446	98732	99528	98919	56
05	99050	97836	99159	98083	99262	98315	99358	98532	99447	98735	99529	98922	55
06	99052	97841	99161	98087	99264	98319	99359	98536	99448	98738	99531	98925	54
07	99054	97845	99163	98091	99265	98323	99361	98539	99450	98741	99532	98928	53
08	99056	97849	99165	98095	99267	98326	99362	98543	99451	98745	99533	98931	52
09	99058	97853	99166	98099	99269	98330	99364	98546	99453	98748	99535	98934	51
10	99059	97858	99168	98103	99270	98334	99366	98550	99454	98751	99536	98937	50
11	99061	97862	99170	98107	99272	98337	99367	98553	99456	98754	99537	98940	49
12	99063	97866	99172	98111	99274	98341	99369	98557	99457	98757	99539	98943	48
13	99065	97870	99173	98115	99275	98345	99370	98560	99458	98761	99540	98946	47
14	99067	97874	99175	98119	99277	98349	99372	98564	99460	98764	99541	98949	46
15	99069	97879	99177	98123	99278	98352	99373	98567	99461	98767	99543	98952	45
16	99071	97883	99179	98127	99280	98356	99375	98571	99463	98770	99544	98955	44
17	99072	97887	99180	98131	99282	98360	99376	98574	99464	98774	99545	98958	43
18	99074	97891	99182	98135	99283	98363	99378	98577	99465	98777	99546	98961	42
19	99076	97895	99184	98139	99285	98367	99379	98581	99467	98780	99548	98964	41
20	99078	97899	99186	98142	99287	98371	99381	98584	99468	98783	99549	98967	40
21	99080	97904	99187	98146	99288	98374	99382	98588	99470	98786	99550	98970	39
22	99082	97908	99189	98150	99290	98378	99384	98591	99471	98789	99552	98973	38
23	99084	97912	99191	98154	99291	98382	99385	98595	99472	98793	99553	98976	37
24	99085	97916	99193	98158	99293	98385	99387	98598	99474	98796	99554	98979	36
25	99087	97920	99194	98162	99295	98389	99388	98601	99475	98799	99555	98982	35
26	99089	97924	99196	98166	99296	98393	99390	98605	99477	98802	99557	98985	34
27	99091	97929	99198	98170	99298	98396	99391	98608	99478	98805	99558	98987	33
28	99093	97933	99200	98174	99300	98400	99393	98611	99479	98809	99559	98990	32
29	99095	97937	99201	98178	99301	98404	99394	98615	99481	98812	99561	98993	31
30	99096	97941	99203	98182	99303	98407	99396	98619	99482	98815	99562	98996	30
31	99098	97945	99205	98185	99304	98411	99397	98622	99484	98818	99563	98999	29
32	99100	97949	99206	98189	99306	98415	99399	98625	99485	98821	99564	99002	28
33	99102	97953	99208	98193	99308	98418	99400	98629	99486	98824	99566	99005	27
34	99104	97957	99210	98197	99309	98422	99402	98632	99488	98827	99567	99008	26
35	99106	97962	99212	98201	99311	98426	99403	98635	99489	98830	99568	99011	25
36	99107	97966	99213	98205	99312	98429	99405	98639	99490	98834	99569	99014	24
37	99109	97970	99215	98209	99314	98433	99406	98642	99492	98837	99571	99016	23
38	99111	97974	99217	98212	99316	98436	99408	98646	99493	98840	99572	99019	22
39	99113	97978	99218	98216	99317	98440	99409	98649	99495	98843	99573	99022	21
40	99115	97982	99220	98220	99319	98444	99411	98652	99496	98846	99575	99025	20
41	99116	97986	99222	98224	99320	98447	99412	98656	99497	98849	99576	99028	19
42	99118	97990	99223	98228	99322	98451	99414	98659	99499	98852	99577	99031	18
43	99120	97994	99225	98232	99324	98454	99415	98662	99500	98855	99578	99034	17
44	99122	97998	99227	98236	99325	98458	99417	98666	99501	98858	99580	99036	16
45	99124	98002	99229	98239	99327	98462	99418	98669	99503	98862	99581	99039	15
46	99126	98007	99230	98243	99328	98465	99420	98672	99504	98865	99582	99042	14
47	99127	98011	99232	98247	99330	98469	99421	98676	99505	98868	99583	99045	13
48	99129	98015	99234	98251	99331	98472	99422	98679	99507	98871	99584	99048	12
49	99131	98019	99235	98255	99333	98476	99424	98682	99508	98874	99586	99051	11
50	99133	98023	99237	98258	99335	98479	99425	98686	99510	98877	99587	99053	10
51	99135	98027	99239	98262	99336	98483	99427	98689	99511	98880	99588	99056	09
52	99136	98031	99240	98266	99338	98487	99429	98692	99512	98883	99589	99059	08
53	99138	98035	99242	98270	99339	98490	99430	98696	99514	98886	99591	99062	07
54	99140	98039	99244	98274	99341	98494	99431	98699	99515	98889	99592	99065	06
55	99142	98043	99245	98277	99342	98497	99433	98702	99516	98892	99593	99067	05
56	99143	98047	99247	98281	99344	98501	99434	98705	99518	98895	99594	99070	04
57	99145	98051	99249	98285	99345	98504	99436	98709	99519	98898	99596	99073	03
58	99147	98055	99250	98289	99347	98508	99437	98712	99520	98901	99597	99076	02
59	99149	98059	99252	98293	99349	98511	99438	98715	99522	98904	99598	99079	01
60	99151	98063	99254	98296	99350	98515	99440	98719	99523	98907	99599	99081	00

| 196° | 195° | 194° | 193° | 192° | 191° |

HAVERSINES

Parts for 0'·2, etc.	169° Log.	169° Nat.	170° Log.	170° Nat.	171° Log.	171° Nat.	172° Log.	172° Nat.	173° Log.	173° Nat.	174° Log.	174° Nat.	Parts for 0'·2, etc.
·2 0 ·4 0 ·6 1 ·8 1	·2 1 ·4 1 ·6 2 ·8 2	·2 0 ·4 0 ·6 1 ·8 1	·2 0 ·4 1 ·6 1 ·8 2	·2 0 ·4 0 ·6 1 ·8 1	·2 0 ·4 1 ·6 1 ·8 2	·2 0 ·4 0 ·6 0 ·8 1	·2 0 ·4 1 ·6 1 ·8 2	·2 0 ·4 0 ·6 0 ·8 1	·2 0 ·4 1 ·6 1 ·8 1	·2 0 ·4 0 ·6 0 ·8 0	·2 0 ·4 1 ·6 1 ·8 1	
′	9·	0·	9·	0·	9·	0·	9·	0·	9·	0·	9.	0·	′
00	99599	99081	99669	99240	99732	99384	99788	99513	99838	99627	99881	99726	60
01	99600	99084	99670	99243	99733	99387	99789	99515	99839	99629	99882	99728	59
02	99602	99087	99671	99245	99734	99389	99790	99517	99839	99631	99882	99729	58
03	99603	99090	99672	99248	99735	99391	99790	99519	99840	99633	99883	99731	57
04	99604	99092	99673	99250	99736	99393	99792	99521	99841	99634	99884	99732	56
05	99605	99095	99674	99253	99737	99396	99793	99523	99842	99636	99884	99734	55
06	99606	99098	99675	99255	99738	99398	99793	99525	99842	99638	99885	99735	54
07	99608	99101	99677	99258	99739	99400	99794	99527	99843	99640	99885	99737	53
08	99609	99103	99678	99260	99740	99402	99795	99529	99844	99641	99886	99738	52
09	99610	99106	99679	99263	99741	99405	99796	99531	99845	99643	99887	99740	51
10	99611	99109	99680	99265	99742	99407	99797	99533	99845	99645	99887	99741	50
11	99612	99112	99681	99268	99743	99409	99798	99535	99846	99647	99888	99743	49
12	99614	99114	99682	99270	99744	99411	99799	99537	99847	99648	99889	99744	48
13	99615	99117	99683	99273	99745	99414	99800	99539	99848	99650	99889	99746	47
14	99616	99120	99684	99275	99746	99416	99800	99541	99848	99652	99890	99747	46
15	99617	99123	99685	99278	99747	99418	99801	99543	99849	99653	99891	99748	45
16	99618	99125	99686	99280	99748	99420	99802	99545	99850	99655	99891	99750	44
17	99620	99128	99687	99283	99748	99422	99803	99547	99851	99657	99892	99751	43
18	99621	99131	99688	99285	99749	99425	99804	99549	99851	99659	99893	99753	42
19	99622	99133	99690	99288	99750	99427	99805	99551	99852	99660	99893	99754	41
20	99623	99136	99691	99290	99751	99429	99805	99553	99853	99662	99894	99756	40
21	99624	99139	99692	99293	99752	99431	99806	99555	99854	99664	99894	99757	39
22	99626	99141	99693	99295	99753	99433	99807	99557	99854	99665	99895	99759	38
23	99627	99144	99694	99297	99754	99436	99808	99559	99855	99667	99896	99760	37
24	99628	99147	99695	99300	99755	99438	99809	99561	99856	99669	99896	99761	36
25	99629	99149	99696	99302	99756	99440	99810	99563	99857	99670	99897	99763	35
26	99630	99152	99697	99305	99757	99442	99811	99565	99857	99672	99897	99764	34
27	99631	99155	99698	99307	99758	99444	99811	99567	99858	99674	99898	99766	33
28	99633	99157	99699	99309	99759	99446	99812	99568	99859	99675	99899	99767	32
29	99634	99160	99700	99312	99760	99449	99813	99570	99859	99677	99899	99768	31
30	99635	99163	99701	99314	99761	99451	99814	99572	99860	99679	99900	99770	30
31	99636	99165	99702	99317	99762	99453	99815	99574	99861	99680	99901	99771	29
32	99637	99168	99703	99319	99763	99455	99815	99576	99862	99682	99901	99773	28
33	99638	99171	99704	99321	99764	99457	99816	99578	99862	99684	99902	99774	27
34	99639	99173	99705	99324	99765	99459	99817	99580	99863	99685	99902	99775	26
35	99641	99176	99706	99326	99766	99461	99818	99582	99864	99687	99903	99777	25
36	99642	99179	99707	99329	99766	99464	99819	99584	99864	99688	99904	99778	24
37	99643	99181	99708	99331	99767	99466	99820	99585	99865	99690	99904	99780	23
38	99644	99184	99710	99333	99768	99468	99820	99587	99866	99692	99905	99781	22
39	99645	99186	99711	99336	99769	99470	99821	99589	99867	99693	99905	99782	21
40	99646	99189	99712	99338	99770	99472	99822	99591	99867	99695	99906	99784	20
41	99648	99192	99713	99340	99771	99474	99823	99593	99868	99696	99906	99785	19
42	99649	99194	99714	99343	99772	99476	99824	99595	99869	99698	99907	99786	18
43	99650	99197	99715	99345	99773	99478	99824	99597	99869	99700	99908	99788	17
44	99651	99199	99716	99347	99774	99480	99825	99598	99870	99701	99908	99789	16
45	99652	99202	99717	99350	99774	99483	99826	99600	99871	99703	99909	99790	15
46	99653	99205	99718	99352	99775	99485	99827	99602	99871	99704	99909	99792	14
47	99654	99207	99719	99354	99776	99487	99828	99604	99872	99706	99910	99793	13
48	99655	99210	99720	99357	99777	99489	99828	99606	99873	99708	99911	99794	12
49	99657	99212	99721	99359	99778	99491	99829	99608	99874	99709	99911	99796	11
50	99658	99215	99722	99361	99779	99493	99830	99609	99874	99711	99912	99797	10
51	99659	99217	99723	99364	99780	99495	99831	99611	99875	99712	99912	99798	09
52	99660	99220	99724	99366	99781	99497	99832	99613	99876	99714	99913	99799	08
53	99661	99223	99725	99368	99782	99499	99832	99615	99876	99715	99913	99801	07
54	99662	99225	99726	99371	99783	99501	99833	99617	99877	99717	99914	99802	06
55	99664	99228	99727	99373	99784	99503	99834	99618	99878	99719	99915	99803	05
56	99664	99230	99728	99375	99785	99505	99835	99620	99878	99720	99915	99805	04
57	99666	99233	99729	99378	99786	99507	99836	99622	99879	99722	99916	99806	03
58	99667	99235	99730	99380	99786	99509	99836	99624	99880	99723	99916	99807	02
59	99668	99238	99731	99382	99787	99511	99837	99626	99880	99725	99917	99808	01
60	99669	99240	99732	99384	99788	99513	99838	99627	99881	99726	99917	99810	00
	190°		189°		188°		187°		186°		185°		

HAVERSINES

Parts for 0'·2, etc.	175° Log.	175° Nat.	176° Log.	176° Nat.	177° Log.	177° Nat.	178° Log.	178° Nat.	179° Log.	179° Nat.	Log.	Nat.	Parts for 0'·2, etc.
·2 0 ·4 0 ·6 0 ·8 1	·2 0 ·4 0 ·6 0 ·8 1	·2 0 ·4 0 ·6 1 ·8 1	·2 0 ·4 0 ·6 0 ·8 1	·2 0 ·4 0 ·6 1 ·8 1	·2 0 ·4 0 ·6 0 ·8 1	·2 0 ·4 0 ·6 1 ·8 1	·2 0 ·4 0 ·6 0 ·8 0	·2 0 ·4 0 ·6 0 ·8 0	·2 0 ·4 0 ·6 0 ·8 0	·2 0 ·4 0 ·6 0 ·8 0			
′	9·	0·	9·	0·	9·	0.	9·	0·	9·/10·	0·/1·			′
00	99917	99810	99947	99878	99970	99931	99987	99970	99997	99992			60
01	99918	99811	99948	99879	99971	99932	99987	99971	99997	99993			59
02	99918	99812	99948	99880	99971	99933	99987	99971	99997	99993			58
03	99919	99814	99948	99881	99971	99934	99987	99971	99997	99993			57
04	99919	99815	99949	99882	99972	99934	99988	99972	99997	99994			56
05	99920	99816	99949	99883	99972	99935	99988	99972	99997	99994			55
06	99921	99817	99950	99884	99972	99936	99988	99973	99997	99994			54
07	99921	99819	99950	99885	99973	99937	99988	99973	99997	99994			53
08	99922	99820	99951	99886	99973	99937	99988	99973	99998	99994			52
09	99922	99821	99951	99887	99973	99938	99989	99974	99998	99995			51
10	99923	99822	99951	99888	99973	99939	99989	99974	99998	99995			50
11	99923	99823	99952	99889	99974	99940	99989	99975	99998	99995			49
12	99924	99825	99952	99890	99974	99940	99989	99975	99998	99995			48
13	99924	99826	99953	99891	99974	99941	99989	99976	99998	99995			47
14	99925	99827	99953	99892	99975	99942	99990	99976	99998	99996			46
15	99925	99828	99953	99893	99975	99942	99990	99977	99998	99996			45
16	99926	99829	99954	99894	99975	99943	99990	99977	99998	99996			44
17	99926	99831	99954	99895	99976	99944	99990	99978	99998	99996			43
18	99927	99832	99954	99896	99976	99944	99990	99978	99998	99996			42
19	99927	99833	99955	99897	99976	99945	99991	99978	99998	99996			41
20	99928	99834	99955	99898	99976	99946	99991	99979	99999	99997			40
21	99928	99835	99956	99899	99977	99947	99991	99979	99999	99997			39
22	99929	99837	99956	99900	99977	99947	99991	99980	99999	99997			38
23	99929	99838	99957	99900	99977	99948	99991	99980	99999	99997			37
24	99930	99839	99957	99901	99978	99949	99992	99981	99999	99997			36
25	99931	99840	99958	99902	99978	99949	99992	99981	99999	99997			35
26	99931	99841	99958	99903	99978	99950	99992	99981	99999	99998			34
27	99932	99842	99958	99904	99978	99950	99992	99982	99999	99998			33
28	99932	99844	99959	99905	99979	99951	99992	99982	99999	99998			32
29	99933	99845	99959	99906	99979	99952	99992	99982	99999	99998			31
30	99933	99846	99959	99907	99979	99952	99993	99983	99999	99998			30
31	99934	99847	99960	99908	99980	99953	99993	99983	99999	99998			29
32	99934	99848	99960	99909	99980	99954	99993	99984	99999	99998			28
33	99935	99849	99961	99909	99980	99954	99993	99984	99999	99998			27
34	99935	99850	99961	99910	99980	99955	99993	99984	99999	99999			26
35	99935	99851	99961	99911	99981	99956	99993	99985	99999	99999			25
36	99936	99853	99962	99912	99981	99956	99994	99985	99999	99999			24
37	99936	99854	99962	99913	99981	99957	99994	99985	00000	99999			23
38	99937	99855	99963	99914	99981	99957	99994	99986	00000	99999			22
39	99937	99856	99963	99915	99982	99958	99994	99986	00000	99999			21
40	99938	99857	99963	99915	99982	99959	99994	99986	00000	99999			20
41	99938	99858	99964	99916	99982	99959	99994	99987	00000	99999			19
42	99939	99859	99964	99917	99983	99960	99994	99987	00000	99999			18
43	99939	99860	99964	99918	99983	99960	99995	99987	00000	99999			17
44	99940	99861	99965	99919	99983	99961	99995	99988	00000	99999			16
45	99940	99863	99965	99920	99983	99961	99995	99988	00000	00000			15
46	99941	99864	99965	99920	99983	99962	99995	99988	00000	00000			14
47	99941	99865	99966	99921	99984	99963	99995	99989	00000	00000			13
48	99942	99866	99966	99922	99984	99963	99995	99989	00000	00000			12
49	99942	99867	99966	99923	99984	99964	99995	99989	00000	00000			11
50	99943	99868	99967	99924	99984	99964	99996	99990	00000	00000			10
51	99943	99869	99967	99924	99985	99965	99996	99990	00000	00000			09
52	99943	99870	99968	99925	99985	99965	99996	99990	00000	00000			08
53	99944	99871	99968	99926	99985	99966	99996	99991	00000	00000			07
54	99944	99872	99968	99927	99985	99966	99996	99991	00000	00000			06
55	99945	99873	99969	99928	99986	99967	99996	99991	00000	00000			05
56	99945	99874	99969	99928	99986	99967	99996	99991	00000	00000			04
57	99946	99875	99969	99929	99986	99968	99996	99992	00000	00000			03
58	99946	99876	99970	99930	99986	99969	99996	99992	00000	00000			02
59	99947	99877	99970	99931	99987	99969	99997	99992	00000	00000			01
60	99947	99878	99970	99931	99987	99970	99997	99992	00000	00000			00

| 184° | 183° | 182° | 181° | 180° |

CORRECTION REQUIRED to CONVERT a RADIO GREAT CIRCLE BEARING to MERCATORIAL BEARING.

Mean Lat.	DIFFERENCE OF LONGITUDE OF SHIP AND RADIO STATION.															Mean Lat.
	2°	4°	6°	8°	10°	12°	14°	16°	18°	20°	22°	24°	26°	28°	30°	
°	°	°	°	°	°	°	°	°	°	°	°	°	°	°	°	°
84	1.0	2.0	3.0	4.0	5.0	6.0	7.0	8.0	9.0	9.9	10.9	11.9	12.9	13.9	14.9	84
81	1.0	2.0	2.9	4.0	4.9	5.9	6.9	7.9	8.9	9.9	10.9	11.9	12.8	13.8	14.8	81
78	1.0	2.0	2.9	3.9	4.9	5.9	6.8	7.8	8.8	9.8	10.8	11.7	12.7	13.7	14.7	78
75	1.0	1.9	2.9	3.9	4.8	5.8	6.8	7.7	8.7	9.7	10.7	11.6	12.6	13.5	14.4	75
72	1.0	1.9	2.9	3.8	4.8	5.7	6.7	7.6	8.6	9.5	10.5	11.4	12.4	13.3	14.3	72
69	0.9	1.9	2.8	3.7	4.7	5.6	6.5	7.5	8.4	9.3	10.3	11.2	12.1	13.1	14.0	69
66	0.9	1.8	2.8	3.7	4.6	5.5	6.4	7.3	8.2	9.1	10.0	11.0	11.9	12.8	13.7	66
63	0.9	1.8	2.7	3.6	4.5	5.4	6.3	7.1	8.0	8.9	9.8	10.7	11.6	12.5	13.3	63
60	0.9	1.7	2.6	3.5	4.3	5.2	6.1	6.9	7.8	8.6	9.5	10.4	11.2	12.1	12.9	60
57	0.8	1.7	2.5	3.4	4.2	5.0	5.9	6.7	7.5	8.4	9.2	10.0	10.9	11.7	12.5	57
54	0.8	1.6	2.4	3.3	4.1	4.9	5.7	6.5	7.3	8.1	8.9	9.7	10.5	11.3	12.1	54
51	0.8	1.6	2.3	3.1	3.9	4.7	5.5	6.2	7.0	7.8	8.5	9.3	10.1	10.8	11.6	51
48	0.8	1.5	2.2	3.0	3.7	4.5	5.2	5.9	6.7	7.4	8.2	8.9	9.6	10.4	11.1	48
45	0.7	1.4	2.1	2.8	3.5	4.2	4.9	5.6	6.3	7.1	7.8	8.5	9.2	9.9	10.6	45
42	0.7	1.4	2.0	2.7	3.4	4.0	4.7	5.4	6.0	6.7	7.4	8.0	8.7	9.4	10.0	42
39	0.6	1.3	1.9	2.5	3.2	3.8	4.4	5.0	5.7	6.3	6.9	7.5	8.1	8.8	9.4	39
36	0.6	1.2	1.8	2.4	3.0	3.5	4.1	4.7	5.3	5.9	6.4	7.0	7.6	8.2	8.7	36
33	0.5	1.1	1.6	2.2	2.7	3.3	3.8	4.4	4.9	5.4	6.0	6.5	7.1	7.6	8.1	33
30	0.5	1.0	1.5	2.0	2.5	3.0	3.5	4.0	4.5	5.0	5.5	6.0	6.5	7.0	7.4	30
27	0.5	0.9	1.4	1.8	2.3	2.7	3.2	3.6	4.1	4.5	5.0	5.4	5.9	6.3	6.8	27
24	0.4	0.8	1.2	1.6	2.1	2.4	2.9	3.3	3.6	4.0	4.4	4.8	5.2	5.6	6.0	24
21	0.3	0.7	1.1	1.4	1.8	2.2	2.5	2.9	3.2	3.6	3.9	4.3	4.6	5.0	5.3	21
18	0.3	0.6	0.9	1.2	1.6	1.9	2.2	2.5	2.8	3.1	3.4	3.7	4.0	4.3	4.6	18
15	0.3	0.5	0.8	1.0	1.3	1.6	1.8	2.1	2.3	2.6	2.8	3.1	3.3	3.6	3.8	15
12	0.2	0.4	0.6	0.8	1.0	1.3	1.5	1.7	1.9	2.1	2.3	2.5	2.7	2.9	3.1	12
9	0.2	0.3	0.5	0.6	0.8	1.0	1.1	1.2	1.4	1.6	1.7	1.9	2.0	2.2	2.3	9
6	0.1	0.2	0.3	0.4	0.5	0.6	0.7	0.8	0.9	1.0	1.1	1.2	1.3	1.5	1.6	6
3	0.1	0.1	0.2	0.2	0.3	0.3	0.4	0.4	0.5	0.5	0.6	0.6	0.7	0.7	0.8	3
	2°	4°	6°	8°	10°	12°	14°	16°	18°	20°	22°	24°	26°	28°	30°	

In both North and South latitudes always allow the above corrections towards the Equator from the Radio Great Circle bearing to obtain the corresponding Mercatorial line of bearing.

N.B.—The Bearings must always be laid off, on the chart, from the Radio Station.

EXAMPLE I. A ship in D.R. position Lat. 39° 37′ N., Long. 56° 25′ W., receives from a Radio Station in Lat. 35° 14′ N., Long. 75° 32′ W., the Radio bearing 074°. Find the correction and the corresponding Mercatorial bearing.

Mean Lat. is ½ (39° 37′+35° 14′) or ½ (74° 51′) =37° 4.

D. Long. is 75° 32′ —56° 25′ =19° I.

For Mean Lat. 37°·4 and D. Long. 19°·I the Table gives a correction of 6° (approx.).

Allowing this correction towards the Equator the corresponding Mercatorial bearing is found to be 074° + 6° =080°.

EXAMPLE 2. A ship in D.R. position Lat. 37° 26′ S., Long. 84° 35′ W., finds, with her own apparatus, the Radio bearing of a station in Lat. 36° 37′ S., Long. 73° 03′ W., to be 089°. Find the correction and the corresponding Mercatorial bearing.

For Mean Lat. 37° and D. Long. 11°·5 the Table gives a correction of 3°·5 (approx.).

Allowing this correction towards the Equator the corresponding Mercatorial bearing is found to be 089° – 3°·5 =085°·5.

TABLE A — HOUR ANGLE

A — HOUR ANGLE — A

Lat.	Diff.	0°15' 359°45'	0°30' 359°30'	0°45' 359°15'	1°00' 359°00'	1°15' 358°45'	1°30' 358°30'	1°45' 358°15'	2°00' 358°00'	2°15' 357°45'	2°30' 357°30'	2°45' 357°15'	3°00' 357°00'	3°15' 356°45'	3°30' 356°30'	3°45' 356°15'	Lat.
0	4·00	·00	·00	·00	·00	·00	·00	·00	·00	·00	·00	·00	·00	·00	·00	·00	0
1	4·00	4·00	2·00	1·33	1·00	·80	·67	·57	·50	·44	·40	·36	·33	·31	·29	·27	1
2	4·01	8·00	4·00	2·67	2·00	1·60	1·33	1·14	1·00	·89	·80	·73	·67	·61	·57	·53	2
3	4·01	12·0	6·01	4·00	3·00	2·40	2·00	1·72	1·50	1·33	1·20	1·09	1·00	·92	·86	·80	3
4	4·02	16·0	8·01	5·34	4·01	3·21	2·67	2·29	2·00	1·78	1·60	1·46	1·33	1·23	1·14	1·07	4
5	4·03	20·1	10·0	6·68	5·01	4·01	3·34	2·86	2·51	2·23	2·00	1·82	1·67	1·54	1·43	1·33	5
6	4·04	24·1	12·0	8·03	6·02	4·82	4·01	3·44	3·01	2·68	2·41	2·19	2·01	1·85	1·72	1·60	6
7	4·06	28·1	14·1	9·38	7·03	5·63	4·69	4·02	3·52	3·13	2·81	2·56	2·34	2·16	2·01	1·87	7
8	4·08	32·2	16·1	10·7	8·05	6·44	5·37	4·60	4·02	3·58	3·22	2·93	2·68	2·48	2·30	2·14	8
9	4·10	36·3	18·1	12·1	9·07	7·26	6·05	5·18	4·54	4·03	3·63	3·30	3·02	2·79	2·59	2·42	9
10	4·12	40·4	20·2	13·5	10·1	8·08	6·73	5·77	5·05	4·49	4·04	3·67	3·36	3·10	2·88	2·69	10
11	4·15	44·5	22·3	14·9	11·1	8·91	7·42	6·36	5·57	4·95	4·45	4·05	3·71	3·42	3·18	2·97	11
12	4·18	48·7	24·4	16·2	12·2	9·74	8·12	6·96	6·09	5·41	4·87	4·43	4·06	3·74	3·48	3·24	12
13	4·21	52·9	26·5	17·6	13·2	10·6	8·82	7·56	6·61	5·88	5·29	4·81	4·41	4·07	3·78	3·52	13
14	4·25	57·1	28·6	19·1	14·3	11·4	9·52	8·16	7·14	6·35	5·71	5·19	4·76	4·39	4·08	3·80	14
15	4·29	61·4	30·7	20·5	15·4	12·3	10·2	8·77	7·67	6·82	6·14	5·58	5·11	4·72	4·38	4·09	15
16	4·33	65·7	32·9	21·9	16·4	13·1	11·0	9·39	8·21	7·30	6·57	5·97	5·47	5·05	4·69	4·37	16
17	4·37	70·1	35·0	23·4	17·5	14·0	11·7	10·0	8·75	7·78	7·00	6·36	5·83	5·38	5·00	4·66	17
18	4·42	74·5	37·2	24·8	18·6	14·9	12·4	10·6	9·30	8·27	7·44	6·76	6·20	5·72	5·31	4·96	18
19	4·47	78·9	39·5	26·3	19·7	15·8	13·1	11·3	9·86	8·76	7·89	7·17	6·57	6·06	5·63	5·25	19
20	4·53	83·4	41·7	27·8	20·9	16·7	13·9	11·9	10·4	9·26	8·34	7·58	6·94	6·41	5·95	5·55	20
21	4·59	88·0	44·0	29·3	22·0	17·6	14·7	12·6	11·0	9·77	8·79	7·99	7·32	6·76	6·28	5·86	21
22	4·65	92·6	46·3	30·9	23·1	18·5	15·4	13·2	11·6	10·3	9·25	8·41	7·71	7·12	6·61	6·16	22
23	4·72	97·3	48·6	32·4	24·3	19·5	16·2	13·9	12·2	10·8	9·72	8·84	8·10	7·48	6·94	6·48	23
24	4·79	102	51·0	34·0	25·5	20·4	17·0	14·6	12·7	11·3	10·2	9·27	8·50	7·84	7·28	6·79	24
25	4·87	107	53·4	35·6	26·7	21·4	17·8	15·3	13·4	11·9	10·7	9·71	8·90	8·21	7·62	7·11	25
26	4·95	112	55·9	37·3	27·9	22·4	18·6	16·0	14·0	12·4	11·2	10·2	9·31	8·59	7·97	7·44	26
27	5·04	117	58·4	38·9	29·2	23·4	19·5	16·7	14·6	13·0	11·7	10·6	9·72	8·97	8·33	7·77	27
28	5·13	122	60·9	40·6	30·5	24·4	20·3	17·4	15·2	13·5	12·2	11·1	10·1	9·36	8·69	8·11	28
29	5·23	127	63·5	42·3	31·8	25·4	21·2	18·1	15·9	14·1	12·7	11·5	10·6	9·76	9·06	8·46	29
30	5·33	132	66·2	44·1	33·1	26·5	22·0	18·9	16·5	14·7	13·2	12·0	11·0	10·2	9·44	8·81	30
31	5·44	138	68·9	45·9	34·4	27·5	22·9	19·7	17·2	15·3	13·8	12·5	11·5	10·6	9·82	9·17	31
32	5·56	143	71·6	47·7	35·8	28·6	23·9	20·5	17·9	15·9	14·3	13·0	11·9	11·0	10·2	9·53	32
33	5·69	149	74·4	49·6	37·2	29·8	24·8	21·3	18·6	16·5	14·9	13·5	12·4	11·4	10·6	9·91	33
34	5·82	155	77·3	51·5	38·6	30·9	25·8	22·1	19·3	17·2	15·4	14·0	12·9	11·9	11·0	10·3	34
35	5·96	160	80·2	53·5	40·1	32·1	26·7	22·9	20·1	17·8	16·0	14·6	13·4	12·3	11·4	10·7	35
36	6·11	167	83·3	55·5	41·6	33·3	27·7	23·8	20·8	18·5	16·6	15·1	13·9	12·8	11·9	11·1	36
37	6·27	173	86·3	57·6	43·2	34·5	28·8	24·7	21·6	19·2	17·3	15·7	14·4	13·3	12·3	11·5	37
38	6·44	179	89·5	59·7	44·8	35·8	29·8	25·6	22·4	19·9	17·9	16·3	14·9	13·8	12·8	11·9	38
39	6·62	186	92·8	61·9	46·4	37·1	30·9	26·5	23·2	20·6	18·6	16·9	15·5	14·3	13·2	12·4	39
40	6·82	192	96·2	64·1	48·1	38·5	32·0	27·5	24·0	21·4	19·2	17·5	16·0	14·8	13·7	12·8	40
41	7·03	199	99·6	66·4	49·8	39·8	33·2	28·5	24·9	22·1	19·9	18·1	16·6	15·3	14·2	13·3	41
42	7·25	206	103	68·8	51·6	41·3	34·4	29·5	25·8	22·9	20·6	18·8	17·2	15·9	14·7	13·7	42
43	7·48	214	107	71·2	53·4	42·7	35·6	30·5	26·7	23·7	21·4	19·4	17·8	16·4	15·2	14·2	43
44	7·73	221	111	73·8	55·3	44·3	36·9	31·6	27·7	24·6	22·1	20·1	18·4	17·0	15·8	14·7	44
45	8·00	229	115	76·4	57·3	45·8	38·2	32·7	28·6	25·5	22·9	20·8	19·1	17·6	16·4	15·3	45
46	8·29	237	119	79·1	59·3	47·5	39·6	33·9	29·7	26·4	23·7	21·6	19·8	18·2	16·9	15·8	46
47	8·60	246	123	81·9	61·4	49·2	41·0	35·1	30·7	27·3	24·6	22·3	20·5	18·9	17·5	16·4	47
48	8·93	255	127	84·8	63·6	50·9	42·4	36·4	31·8	28·3	25·4	23·1	21·2	19·6	18·2	16·9	48
49	9·29	264	132	87·9	65·9	52·7	43·9	37·7	32·9	29·3	26·4	24·0	22·0	20·3	18·8	17·6	49
50	9·68	273	137	91·0	68·3	54·6	45·5	39·0	34·1	30·3	27·3	24·8	22·7	21·0	19·5	18·2	50
51	10·1	283	142	94·3	70·8	56·6	47·2	40·4	35·4	31·4	28·3	25·7	23·6	21·8	20·2	18·8	51
52	10·6	293	147	97·8	73·3	58·7	48·9	41·9	36·7	32·6	29·3	26·7	24·4	22·5	20·9	19·5	52
53	11·1	304	152	101	76·0	60·8	50·7	43·4	38·0	33·8	30·4	27·6	25·3	23·4	21·7	20·3	53
54	11·6	315	158	105	78·9	63·1	52·6	45·1	39·4	35·0	31·5	28·7	26·3	24·2	22·5	21·0	54
55	12·2	327	164	109	81·8	65·5	54·5	46·7	40·9	36·4	32·7	29·7	27·3	25·2	23·4	21·8	55
56	12·8	340	170	113	84·9	67·9	56·6	48·5	42·5	37·7	34·0	30·9	28·3	26·1	24·2	22·6	56
57	13·5	353	176	118	88·2	70·6	58·8	50·4	44·1	39·2	35·3	32·1	29·4	27·1	25·2	23·5	57
58	14·3	367	183	122	91·7	73·3	61·1	52·4	45·8	40·7	36·7	33·3	30·5	28·2	26·2	24·4	58
59	15·1	381	191	127	95·4	76·3	63·6	54·5	47·7	42·4	38·1	34·7	31·8	29·3	27·2	25·4	59
60	16·0	397	198	132	99·2	79·4	66·1	56·7	49·6	44·1	39·7	36·1	33·0	30·5	28·3	26·4	60
Lat.	Diff.	179°45'	179°30'	179°15'	179°00'	178°45'	178°30'	178°15'	178°00'	177°45'	177°30'	177°15'	177°00'	176°45'	176°30'	176°15'	Lat.
		180°15'	180°30'	180°45'	181°00'	181°15'	181°30'	181°45'	182°00'	182°15'	182°30'	182°45'	183°00'	183°15'	183°30'	183°45'	

A — HOUR ANGLE — **A**

A—Named opposite to Latitude, **except** when Hour Angle is between 90° and 270°

TABLE B — HOUR ANGLE

B — DLONG LAT T · DLON LAT F — B

Dec.°	Diff.	0° 15' / 359° 45'	0° 30' / 359° 30'	0° 45' / 359° 15'	1° 00' / 359° 00'	1° 15' / 358° 45'	1° 30' / 358° 30'	1° 45' / 358° 15'	2° 00' / 358° 00'	2° 15' / 357° 45'	2° 30' / 357° 30'	2° 45' / 357° 15'	3° 00' / 357° 00'	3° 15' / 356° 45'	3° 30' / 356° 30'	3° 45' / 356° 15'	Dec.°
0	4·00	·00	·00	·00	·00	·00	·00	·00	·00	·00	·00	·00	·00	·00	·00	·00	0
1	4·00	4.00	2.00	1.33	1.00	·80	·67	·57	·50	·45	·40	·36	·33	·31	·29	·27	1
2	4·01	8.00	4.00	2.67	2.00	1.60	1.33	1.14	1.00	·89	·80	·73	·67	·62	·57	·53	2
3	4·01	12.0	6.01	4.00	3.00	2.40	2.00	1.72	1.50	1.34	1.20	1.09	1.00	·92	·86	·80	3
4	4·02	16.0	8.01	5.34	4.01	3.21	2.67	2.29	2.00	1.78	1.60	1.46	1.34	1.23	1.15	1.07	4
5	4·03	20.1	10.0	6.68	5.01	4.01	3.34	2.87	2.51	2.23	2.01	1.82	1.67	1.54	1.43	1.34	5
6	4·04	24.1	12.0	8.03	6.02	4.82	4.02	3.44	3.01	2.68	2.41	2.19	2.01	1.85	1.72	1.61	6
7	4·06	28.1	14.1	9.38	7.04	5.63	4.69	4.02	3.52	3.13	2.81	2.56	2.35	2.17	2.01	1.88	7
8	4·08	32.2	16.1	10.7	8.05	6.44	5.37	4.60	4.03	3.58	3.22	2.93	2.69	2.48	2.30	2.15	8
9	4·10	36.3	18.1	12.1	9.08	7.26	6.05	5.19	4.54	4.03	3.63	3.30	3.03	2.79	2.59	2.42	9
10	4·12	40.4	20.2	13.5	10.1	8.08	6.74	5.77	5.05	4.49	4.04	3.68	3.37	3.11	2.89	2.70	10
11	4·15	44.5	22.3	14.9	11.1	8.91	7.43	6.37	5.57	4.95	4.46	4.05	3.71	3.43	3.18	2.97	11
12	4·18	48.7	24.4	16.2	12.2	9.74	8.12	6.96	6.09	5.41	4.87	4.43	4.06	3.75	3.48	3.25	12
13	4·21	52.9	26.5	17.6	13.2	10.6	8.82	7.56	6.62	5.88	5.29	4.81	4.41	4.07	3.78	3.53	13
14	4·25	57.1	28.6	19.1	14.3	11.4	9.53	8.16	7.14	6.35	5.72	5.20	4.76	4.40	4.08	3.81	14
15	4·29	61.4	30.7	20.5	15.4	12.3	10.2	8.77	7.68	6.83	6.14	5.58	5.12	4.73	4.39	4.10	15
16	4·33	65.7	32.9	21.9	16.4	13.1	11.0	9.39	8.22	7.30	6.57	5.98	5.48	5.06	4.70	4.38	16
17	4·37	70.1	35.0	23.4	17.5	14.0	11.7	10.0	8.76	7.79	7.01	6.37	5.84	5.39	5.01	4.67	17
18	4·42	74.5	37.2	24.8	18.6	14.9	12.4	10.6	9.31	8.28	7.45	6.77	6.21	5.73	5.32	4.97	18
19	4·47	78.9	39.5	26.3	19.7	15.8	13.2	11.3	9.87	8.77	7.89	7.18	6.58	6.07	5.64	5.26	19
20	4·53	83.4	41.7	27.8	20.9	16.7	13.9	11.9	10.4	9.27	8.34	7.59	6.95	6.42	5.96	5.57	20
21	4·59	88.0	44.0	29.3	22.0	17.6	14.7	12.6	11.0	9.78	8.80	8.00	7.33	6.77	6.29	5.87	21
22	4·65	92.6	46.3	30.9	23.2	18.5	15.4	13.2	11.6	10.3	9.26	8.42	7.72	7.13	6.62	6.18	22
23	4·72	97.3	48.6	32.4	24.3	19.5	16.2	13.9	12.2	10.8	9.73	8.85	8.11	7.49	6.95	6.49	23
24	4·79	102	51.0	34.0	25.5	20.4	17.0	14.6	12.8	11.3	10.2	9.28	8.51	7.85	7.29	6.81	24
25	4·87	107	53.4	35.6	26.7	21.4	17.8	15.3	13.4	11.9	10.7	9.72	8.91	8.23	7.64	7.13	25
26	4·95	112	55.9	37.3	27.9	22.4	18.6	16.0	14.0	12.4	11.2	10.2	9.32	8.60	7.99	7.46	26
27	5·04	117	58.4	38.9	29.2	23.4	19.5	16.7	14.6	13.0	11.7	10.6	9.74	8.99	8.35	7.79	27
28	5·13	122	60.9	40.6	30.5	24.4	20.3	17.4	15.2	13.5	12.2	11.1	10.2	9.38	8.71	8.13	28
29	5·23	127	63.5	42.4	31.8	25.4	21.2	18.2	15.9	14.1	12.7	11.6	10.6	9.78	9.08	8.48	29
30	5·33	132	66.2	44.1	33.1	26.5	22.1	18.9	16.5	14.7	13.2	12.0	11.0	10.2	9.46	8.83	30
31	5·44	138	68.9	45.9	34.4	27.5	23.0	19.7	17.2	15.3	13.8	12.5	11.5	10.6	9.84	9.19	31
32	5·56	143	71.6	47.7	35.8	28.6	23.9	20.5	17.9	15.9	14.3	13.0	11.9	11.0	10.2	9.55	32
33	5·69	149	74.4	49.6	37.2	29.8	24.8	21.3	18.6	16.5	14.9	13.5	12.4	11.5	10.6	9.93	33
34	5·82	155	77.3	51.5	38.6	30.9	25.8	22.1	19.3	17.2	15.5	14.1	12.9	11.9	11.0	10.3	34
35	5·96	160	80.2	53.5	40.1	32.1	26.7	22.9	20.1	17.8	16.1	14.6	13.4	12.4	11.5	10.7	35
36	6·11	167	83.3	55.5	41.6	33.3	27.8	23.8	20.8	18.5	16.7	15.1	13.9	12.8	11.9	11.1	36
37	6·27	173	86.4	57.6	43.2	34.5	28.8	24.7	21.6	19.2	17.3	15.7	14.4	13.3	12.3	11.5	37
38	6·44	179	89.5	59.7	44.8	35.8	29.8	25.6	22.4	19.9	17.9	16.3	14.9	13.8	12.8	12.0	38
39	6·62	186	92.8	61.9	46.4	37.1	30.9	26.5	23.2	20.6	18.6	16.9	15.5	14.3	13.3	12.4	39
40	6·82	192	96.2	64.1	48.1	38.5	32.1	27.5	24.0	21.4	19.2	17.5	16.0	14.8	13.7	12.8	40
41	7·03	199	99.6	66.4	49.8	39.9	33.2	28.5	24.9	22.1	19.9	18.1	16.6	15.3	14.2	13.3	41
42	7·25	206	103	68.8	51.6	41.3	34.4	29.5	25.8	22.9	20.6	18.8	17.2	15.9	14.7	13.8	42
43	7·48	214	107	71.2	53.4	42.8	35.6	30.5	26.7	23.8	21.4	19.4	17.8	16.5	15.3	14.3	43
44	7·73	221	111	73.8	55.3	44.3	36.9	31.6	27.7	24.6	22.1	20.1	18.5	17.0	15.8	14.8	44
45	8·00	229	115	76.4	57.3	45.8	38.2	32.8	28.7	25.5	22.9	20.8	19.1	17.6	16.4	15.3	45
46	8·29	237	119	79.1	59.3	47.5	39.6	33.9	29.7	26.4	23.7	21.6	19.8	18.3	17.0	15.8	46
47	8·60	246	123	81.9	61.4	49.2	41.0	35.1	30.7	27.3	24.6	22.4	20.5	18.9	17.6	16.4	47
48	8·93	255	127	84.9	63.6	50.9	42.4	36.4	31.8	28.3	25.5	23.2	21.2	19.6	18.2	17.0	48
49	9·29	264	132	87.9	65.9	52.7	43.9	37.7	33.0	29.3	26.4	24.0	22.0	20.3	18.8	17.6	49
50	9·68	273	137	91.1	68.3	54.6	45.5	39.0	34.1	30.4	27.3	24.8	22.8	21.0	19.5	18.2	50
51	10·1	283	142	94.3	70.8	56.6	47.2	40.4	35.4	31.5	28.3	25.7	23.6	21.8	20.2	18.9	51
52	10·6	293	147	97.8	73.3	58.7	48.9	41.9	36.7	32.6	29.3	26.7	24.5	22.6	21.0	19.6	52
53	11·1	304	152	101	76.0	60.8	50.7	43.5	38.0	33.8	30.4	27.7	25.4	23.4	21.7	20.3	53
54	11·6	315	158	105	78.9	63.1	52.6	45.1	39.4	35.1	31.6	28.7	26.3	24.3	22.5	21.0	54
55	12·2	327	164	109	81.8	65.5	54.6	46.8	40.9	36.4	32.7	29.8	27.3	25.2	23.4	21.8	55
56	12·8	340	170	113	84.9	68.0	56.6	48.6	42.5	37.8	34.0	30.9	28.3	26.2	24.3	22.7	56
57	13·5	353	176	118	88.2	70.6	58.8	50.4	44.1	39.2	35.3	32.1	29.4	27.2	25.2	23.5	57
58	14·3	367	183	122	91.7	73.4	61.1	52.4	45.9	40.8	36.7	33.4	30.6	28.2	26.2	24.5	58
59	15·1	381	191	127	95.4	76.3	63.6	54.5	47.7	42.4	38.2	34.7	31.8	29.4	27.3	25.5	59
60	16·0	397	198	132	99.2	79.4	66.2	56.7	49.6	44.1	39.7	36.1	33.1	30.6	28.4	26.5	60
Dec.	Diff.	179° 45' / 180° 15'	179° 30' / 180° 30'	179° 15' / 180° 45'	179° 00' / 181° 00'	178° 45' / 181° 15'	178° 30' / 181° 30'	178° 15' / 181° 45'	178° 00' / 182° 00'	177° 45' / 182° 15'	177° 30' / 182° 30'	177° 15' / 182° 45'	177° 00' / 183° 00'	176° 45' / 183° 15'	176° 30' / 183° 30'	176° 15' / 183° 45'	Dec.

B—Always named the **same** as Declination.

B — HOUR ANGLE — B

TABLE A — HOUR ANGLE

A   **A**

Lat.	3° 45' 356° 15'	4° 00' 356° 00'	4° 15' 355° 45'	4° 30' 355° 30'	4° 45' 355° 15'	5° 00' 355° 00'	5° 15' 354° 45'	5° 30' 354° 30'	5° 45' 354° 15'	6° 00' 354° 00'	6° 15' 353° 45'	6° 30' 353° 30'	6° 45' 353° 15'	7° 00' 353° 00'	7° 15' 352° 45'	7° 30' 352° 30'	Lat.
0	·00	·00	·00	·00	·00	·00	·00	·00	·00	·00	·00	·00	·00	·00	·00	·00	0
1	·27	·25	·23	·22	·21	·20	·19	·18	·17	·17	·16	·15	·15	·14	·14	·13	1
2	·53	·50	·47	·44	·42	·40	·38	·36	·35	·33	·32	·31	·30	·28	·27	·27	2
3	·80	·75	·71	·67	·63	·60	·57	·54	·52	·50	·48	·46	·44	·43	·41	·40	3
4	1·07	1·00	·94	·89	·84	·80	·76	·73	·69	·67	·64	·61	·59	·57	·55	·53	4
5	1·33	1·25	1·18	1·11	1·05	1·00	·95	·91	·87	·83	·80	·77	·74	·71	·69	·66	5
6	1·60	1·50	1·41	1·34	1·26	1·20	1·14	1·09	1·04	1·00	·96	·92	·89	·86	·83	·80	6
7	1·87	1·76	1·65	1·56	1·48	1·40	1·34	1·28	1·22	1·17	1·12	1·08	1·04	1·00	·97	·93	7
8	2·14	2·01	1·89	1·79	1·69	1·61	1·53	1·46	1·40	1·34	1·28	1·23	1·19	1·14	1·10	1·07	8
9	2·42	2·27	2·13	2·01	1·91	1·81	1·72	1·65	1·57	1·51	1·45	1·39	1·34	1·29	1·25	1·20	9
10	2·69	2·52	2·37	2·24	2·12	2·02	1·92	1·83	1·75	1·68	1·61	1·55	1·49	1·44	1·39	1·34	10
11	2·97	2·78	2·62	2·47	2·34	2·22	2·12	2·02	1·93	1·85	1·77	1·71	1·64	1·58	1·53	1·48	11
12	3·24	3·04	2·86	2·70	2·56	2·43	2·31	2·21	2·11	2·02	1·94	1·87	1·80	1·73	1·67	1·61	12
13	3·52	3·30	3·11	2·93	2·78	2·64	2·51	2·40	2·29	2·20	2·11	2·03	1·95	1·88	1·81	1·75	13
14	3·80	3·57	3·35	3·17	3·00	2·85	2·71	2·59	2·48	2·37	2·28	2·19	2·11	2·03	1·96	1·89	14
15	4·09	3·83	3·61	3·40	3·22	3·06	2·92	2·78	2·66	2·55	2·45	2·35	2·26	2·18	2·11	2·04	15
16	4·37	4·10	3·86	3·64	3·45	3·28	3·12	2·98	2·85	2·73	2·62	2·52	2·42	2·34	2·25	2·18	16
17	4·66	4·37	4·11	3·88	3·68	3·49	3·33	3·18	3·04	2·91	2·79	2·68	2·58	2·49	2·40	2·32	17
18	4·96	4·65	4·37	4·13	3·91	3·71	3·54	3·38	3·23	3·09	2·97	2·85	2·75	2·65	2·55	2·47	18
19	5·25	4·92	4·63	4·38	4·14	3·94	3·75	3·58	3·42	3·28	3·14	3·02	2·91	2·80	2·71	2·62	19
20	5·55	5·21	4·90	4·62	4·38	4·16	3·96	3·78	3·61	3·46	3·32	3·19	3·08	2·96	2·86	2·76	20
21	5·86	5·49	5·17	4·88	4·62	4·39	4·18	3·99	3·81	3·65	3·51	3·37	3·24	3·13	3·02	2·92	21
22	6·16	5·78	5·44	5·13	4·86	4·62	4·40	4·20	4·01	3·84	3·69	3·55	3·41	3·29	3·18	3·07	22
23	6·48	6·07	5·71	5·39	5·11	4·85	4·62	4·41	4·22	4·04	3·88	3·73	3·59	3·46	3·34	3·22	23
24	6·79	6·37	5·99	5·66	5·36	5·09	4·85	4·62	4·42	4·24	4·07	3·91	3·76	3·63	3·50	3·38	24
25	7·11	6·67	6·27	5·92	5·61	5·33	5·08	4·84	4·63	4·44	4·26	4·09	3·94	3·80	3·67	3·54	25
26	7·44	6·98	6·56	6·20	5·87	5·57	5·31	5·07	4·84	4·64	4·45	4·28	4·12	3·97	3·83	3·70	26
27	7·77	7·29	6·86	6·47	6·13	5·82	5·55	5·29	5·06	4·85	4·65	4·47	4·30	4·15	4·01	3·87	27
28	8·11	7·60	7·15	6·76	6·40	6·08	5·79	5·52	5·28	5·06	4·85	4·67	4·49	4·33	4·18	4·04	28
29	8·46	7·93	7·46	7·04	6·67	6·34	6·03	5·76	5·50	5·27	5·06	4·87	4·68	4·52	4·36	4·21	29
30	8·81	8·26	7·77	7·34	6·95	6·60	6·28	6·00	5·73	5·49	5·27	5·07	4·88	4·70	4·54	4·39	30
31	9·17	8·59	8·09	7·63	7·23	6·87	6·54	6·24	5·97	5·72	5·49	5·27	5·08	4·89	4·72	4·56	31
32	9·53	8·94	8·41	7·94	7·52	7·14	6·80	6·49	6·21	5·95	5·71	5·48	5·28	5·09	4·91	4·75	32
33	9·91	9·29	8·74	8·25	7·82	7·42	7·07	6·74	6·45	6·18	5·93	5·70	5·49	5·29	5·10	4·93	33
34	10·3	9·65	9·08	8·57	8·12	7·71	7·34	7·01	6·70	6·42	6·16	5·92	5·70	5·49	5·30	5·12	34
35	10·7	10·0	9·42	8·90	8·43	8·00	7·62	7·27	6·95	6·66	6·39	6·15	5·92	5·70	5·50	5·32	35
36	11·1	10·4	9·78	9·23	8·74	8·30	7·91	7·55	7·22	6·91	6·63	6·38	6·14	5·92	5·71	5·52	36
37	11·5	10·8	10·1	9·57	9·07	8·61	8·20	7·83	7·48	7·17	6·88	6·61	6·37	6·14	5·92	5·72	37
38	11·9	11·2	10·5	9·93	9·40	8·93	8·50	8·11	7·76	7·43	7·13	6·86	6·60	6·36	6·14	5·93	38
39	12·4	11·6	10·9	10·3	9·75	9·26	8·81	8·41	8·04	7·70	7·39	7·11	6·84	6·60	6·37	6·15	39
40	12·8	12·0	11·3	10·7	10·1	9·59	9·13	8·71	8·33	7·98	7·66	7·36	7·09	6·83	6·60	6·37	40
41	13·3	12·4	11·7	11·0	10·5	9·94	9·46	9·03	8·63	8·27	7·94	7·63	7·34	7·08	6·83	6·60	41
42	13·7	12·9	12·1	11·4	10·8	10·3	9·80	9·35	8·94	8·57	8·22	7·90	7·61	7·33	7·08	6·84	42
43	14·2	13·3	12·5	11·8	11·2	10·7	10·1	9·68	9·26	8·87	8·51	8·18	7·88	7·59	7·33	7·08	43
44	14·7	13·8	13·0	12·3	11·6	11·0	10·5	10·0	9·59	9·19	8·82	8·48	8·16	7·86	7·59	7·34	44
45	15·3	14·3	13·5	12·7	12·0	11·4	10·9	10·4	9·93	9·51	9·13	8·78	8·45	8·14	7·86	7·60	45
46	15·8	14·8	13·9	13·2	12·5	11·8	11·3	10·8	10·3	9·85	9·46	9·09	8·75	8·43	8·14	7·87	46
47	16·4	15·3	14·4	13·6	12·9	12·3	11·7	11·1	10·6	10·2	9·79	9·41	9·06	8·73	8·43	8·15	47
48	16·9	15·9	14·9	14·1	13·4	12·7	12·1	11·5	11·0	10·6	10·1	9·75	9·38	9·05	8·73	8·44	48
49	17·6	16·5	15·5	14·6	13·8	13·1	12·5	11·9	11·4	10·9	10·5	10·1	9·72	9·37	9·04	8·74	49
50	18·2	17·0	16·0	15·1	14·3	13·6	13·0	12·4	11·8	11·3	10·9	10·5	10·1	9·71	9·37	9·05	50
51	18·8	17·7	16·6	15·7	14·9	14·1	13·4	12·8	12·3	11·7	11·3	10·8	10·4	10·1	9·71	9·38	51
52	19·5	18·3	17·2	16·3	15·4	14·6	13·9	13·3	12·7	12·2	11·7	11·2	10·8	10·4	10·1	9·72	52
53	20·2	19·0	17·9	16·9	16·0	15·2	14·4	13·8	13·2	12·6	12·1	11·6	11·2	10·8	10·4	10·1	53
54	21·0	19·7	18·5	17·5	16·6	15·7	15·0	14·3	13·7	13·1	12·6	12·1	11·6	11·2	10·8	10·5	54
55	21·8	20·4	19·2	18·1	17·2	16·3	15·5	14·8	14·2	13·6	13·0	12·5	12·1	11·6	11·2	10·8	55
56	22·6	21·2	20·0	18·8	17·8	16·9	16·1	15·4	14·7	14·1	13·5	13·0	12·5	12·1	11·7	11·3	56
57	23·5	22·0	20·7	19·5	18·5	17·6	16·8	16·0	15·3	14·7	14·1	13·5	13·0	12·5	12·1	11·7	57
58	24·4	22·9	21·5	20·3	19·3	18·3	17·4	16·6	15·9	15·2	14·6	14·0	13·5	13·0	12·6	12·2	58
59	25·4	23·8	22·4	21·1	20·0	19·0	18·1	17·3	16·5	15·8	15·2	14·6	14·1	13·6	13·1	12·6	59
60	26·4	24·8	23·3	22·0	20·8	19·8	18·8	18·0	17·2	16·5	15·8	15·2	14·6	14·1	13·6	13·2	60
Lat.	176° 15' 183° 45'	176° 00' 184° 00'	175° 45' 184° 15'	175° 30' 184° 30'	175° 15' 184° 45'	175° 00' 185° 00'	174° 45' 185° 15'	174° 30' 185° 30'	174° 15' 185° 45'	174° 00' 186° 00'	173° 45' 186° 15'	173° 30' 186° 30'	173° 15' 186° 45'	173° 00' 187° 00'	172° 45' 187° 15'	172° 30' 187° 30'	Lat.

A   **HOUR ANGLE**   **A**

TABLE B — HOUR ANGLE

Dec °	3° 45' 356° 15'	4° 00' 356° 00'	4° 15' 355° 45'	4° 30' 355° 30'	4° 45' 355° 15'	5° 00' 355° 00'	5° 15' 354° 45'	5° 30' 354° 30'	5° 45' 354° 15'	6° 00' 354° 00'	6° 15' 353° 45'	6° 30' 353° 30'	6° 45' 353° 15'	7° 00' 353° 00'	7° 15' 352° 45'	7° 30' 352° 30'	Dec.
0	·00	·00	·00	·00	·00	·00	·00	·00	·00	·00	·00	·00	·00	·00	·00	·00	0
1	·27	·25	·24	·22	·21	·20	·19	·18	·17	·17	·16	·15	·15	·14	·14	·13	1
2	·53	·50	·47	·45	·42	·40	·38	·36	·35	·33	·32	·31	·30	·29	·28	·27	2
3	·80	·75	·71	·67	·63	·60	·57	·55	·52	·50	·48	·46	·45	·43	·42	·40	3
4	1·07	1·00	·94	·89	·84	·80	·76	·73	·70	·67	·64	·62	·60	·57	·55	·54	4
5	1·34	1·25	1·18	1·12	1·06	1·00	·96	·91	·87	·84	·80	·77	·74	·72	·69	·67	5
6	1·61	1·51	1·42	1·34	1·27	1·21	1·15	1·10	1·05	1·01	·97	·93	·89	·86	·83	·81	6
7	1·88	1·76	1·66	1·56	1·48	1·41	1·34	1·28	1·23	1·17	1·13	1·08	1·04	1·01	·97	·94	7
8	2·15	2·01	1·90	1·79	1·70	1·61	1·54	1·47	1·40	1·34	1·29	1·24	1·20	1·15	1·11	1·08	8
9	2·42	2·27	2·14	2·02	1·91	1·82	1·73	1·65	1·58	1·52	1·45	1·40	1·35	1·30	1·26	1·21	9
10	2·70	2·53	2·38	2·25	2·13	2·02	1·93	1·84	1·76	1·69	1·62	1·56	1·50	1·45	1·40	1·35	10
11	2·97	2·79	2·62	2·48	2·35	2·23	2·12	2·03	1·94	1·86	1·79	1·72	1·65	1·59	1·54	1·49	11
12	3·25	3·05	2·87	2·71	2·57	2·44	2·32	2·22	2·12	2·03	1·95	1·88	1·81	1·74	1·68	1·63	12
13	3·53	3·31	3·12	2·94	2·79	2·65	2·52	2·41	2·30	2·21	2·12	2·04	1·96	1·89	1·83	1·77	13
14	3·81	3·57	3·36	3·18	3·01	2·86	2·72	2·60	2·49	2·39	2·29	2·20	2·12	2·05	1·98	1·91	14
15	4·10	3·84	3·62	3·41	3·24	3·07	2·93	2·80	2·67	2·56	2·46	2·37	2·28	2·20	2·12	2·05	15
16	4·38	4·11	3·87	3·65	3·46	3·29	3·13	2·99	2·86	2·74	2·63	2·53	2·44	2·35	2·27	2·20	16
17	4·67	4·38	4·13	3·90	3·69	3·51	3·34	3·19	3·05	2·92	2·81	2·70	2·60	2·51	2·42	2·34	17
18	4·97	4·66	4·38	4·14	3·92	3·73	3·55	3·39	3·24	3·11	2·98	2·87	2·76	2·67	2·57	2·49	18
19	5·26	4·94	4·65	4·39	4·16	3·95	3·76	3·59	3·44	3·29	3·16	3·04	2·93	2·83	2·73	2·64	19
20	5·57	5·22	4·91	4·64	4·40	4·18	3·98	3·80	3·63	3·48	3·34	3·22	3·10	2·99	2·88	2·79	20
21	5·87	5·50	5·18	4·89	4·64	4·40	4·20	4·01	3·83	3·67	3·53	3·39	3·27	3·15	3·04	2·94	21
22	6·18	5·79	5·45	5·15	4·88	4·64	4·42	4·22	4·03	3·87	3·71	3·57	3·44	3·32	3·20	3·10	22
23	6·49	6·09	5·73	5·41	5·13	4·87	4·64	4·43	4·24	4·06	3·90	3·75	3·61	3·48	3·36	3·25	23
24	6·81	6·38	6·01	5·67	5·38	5·11	4·87	4·65	4·44	4·26	4·09	3·93	3·79	3·65	3·53	3·41	24
25	7·13	6·68	6·29	5·94	5·63	5·35	5·10	4·87	4·65	4·46	4·28	4·12	3·97	3·83	3·70	3·57	25
26	7·46	6·99	6·58	6·22	5·89	5·60	5·33	5·09	4·87	4·67	4·48	4·31	4·15	4·00	3·86	3·74	26
27	7·79	7·30	6·88	6·49	6·15	5·85	5·57	5·32	5·09	4·87	4·68	4·50	4·34	4·18	4·04	3·90	27
28	8·13	7·62	7·17	6·78	6·42	6·10	5·81	5·55	5·31	5·09	4·88	4·70	4·52	4·36	4·21	4·07	28
29	8·48	7·95	7·48	7·06	6·69	6·36	6·06	5·78	5·53	5·30	5·09	4·90	4·72	4·55	4·39	4·25	29
30	8·83	8·28	7·79	7·36	6·97	6·62	6·31	6·02	5·76	5·52	5·30	5·10	4·91	4·74	4·57	4·42	30
31	9·19	8·61	8·11	7·66	7·26	6·89	6·57	6·27	6·00	5·75	5·52	5·31	5·11	4·93	4·76	4·60	31
32	9·55	8·96	8·43	7·96	7·55	7·17	6·83	6·52	6·24	5·98	5·74	5·52	5·32	5·13	4·95	4·79	32
33	9·93	9·31	8·76	8·28	7·84	7·45	7·10	6·78	6·48	6·21	5·97	5·74	5·53	5·33	5·15	4·98	33
34	10·3	9·67	9·10	8·60	8·15	7·74	7·37	7·04	6·73	6·45	6·20	5·96	5·74	5·54	5·34	5·17	34
35	10·7	10·0	9·45	8·92	8·46	8·03	7·65	7·31	6·99	6·70	6·43	6·19	5·96	5·75	5·55	5·36	35
36	11·1	10·4	9·80	9·26	8·77	8·34	7·94	7·58	7·25	6·95	6·67	6·42	6·18	5·96	5·76	5·57	36
37	11·5	10·8	10·2	9·60	9·10	8·65	8·24	7·86	7·52	7·21	6·92	6·66	6·41	6·18	5·97	5·77	37
38	11·9	11·2	10·5	9·96	9·43	8·96	8·54	8·15	7·80	7·47	7·18	6·90	6·65	6·41	6·19	5·99	38
39	12·4	11·6	10·9	10·3	9·78	9·29	8·85	8·45	8·08	7·75	7·44	7·15	6·89	6·65	6·42	6·20	39
40	12·8	12·0	11·3	10·7	10·1	9·63	9·17	8·75	8·38	8·03	7·71	7·41	7·14	6·89	6·65	6·43	40
41	13·3	12·5	11·7	11·1	10·5	9·97	9·50	9·07	8·68	8·32	7·98	7·68	7·40	7·13	6·89	6·66	41
42	13·8	12·9	12·1	11·5	10·9	10·3	9·84	9·39	8·99	8·61	8·27	7·95	7·66	7·39	7·13	6·90	42
43	14·3	13·4	12·6	11·9	11·3	10·7	10·2	9·73	9·31	8·92	8·57	8·24	7·93	7·65	7·39	7·14	43
44	14·8	13·8	13·0	12·3	11·7	11·1	10·6	10·1	9·64	9·24	8·87	8·53	8·22	7·92	7·65	7·40	44
45	15·3	14·3	13·5	12·7	12·1	11·5	10·9	10·4	9·98	9·57	9·19	8·83	8·51	8·21	7·92	7·66	45
46	15·8	14·8	14·0	13·2	12·5	11·9	11·3	10·8	10·3	9·91	9·51	9·15	8·81	8·50	8·21	7·93	46
47	16·4	15·4	14·5	13·7	13·0	12·3	11·7	11·2	10·7	10·3	9·85	9·47	9·12	8·80	8·50	8·22	47
48	17·0	15·9	15·0	14·2	13·4	12·7	12·1	11·6	11·1	10·6	10·2	9·81	9·45	9·11	8·80	8·51	48
49	17·6	16·5	15·5	14·7	13·9	13·2	12·6	12·0	11·5	11·0	10·6	10·2	9·79	9·44	9·12	8·81	49
50	18·2	17·1	16·1	15·2	14·4	13·7	13·0	12·4	11·9	11·4	10·9	10·5	10·1	9·73	9·44	9·13	50
51	18·9	17·7	16·7	15·7	14·9	14·2	13·5	12·9	12·3	11·8	11·3	10·9	10·5	10·1	9·79	9·46	51
52	19·6	18·3	17·3	16·3	15·5	14·7	14·0	13·4	12·8	12·2	11·8	11·3	10·9	10·5	10·1	9·81	52
53	20·3	19·0	17·9	16·9	16·0	15·2	14·5	13·8	13·2	12·7	12·2	11·7	11·3	10·9	10·5	10·2	53
54	21·0	19·7	18·6	17·5	16·6	15·8	15·0	14·4	13·7	13·2	12·6	12·2	11·7	11·3	10·9	10·5	54
55	21·8	20·5	19·3	18·2	17·2	16·4	15·6	14·9	14·3	13·7	13·1	12·6	12·2	11·7	11·3	10·9	55
56	22·7	21·3	20·0	18·9	17·9	17·0	16·2	15·5	14·8	14·2	13·6	13·1	12·6	12·2	11·7	11·4	56
57	23·5	22·1	20·8	19·6	18·6	17·7	16·8	16·1	15·4	14·7	14·1	13·6	13·1	12·6	12·2	11·8	57
58	24·5	22·9	21·6	20·4	19·4	18·5	17·6	16·8	16·0	15·3	14·7	14·1	13·6	13·1	12·7	12·3	58
59	25·4	23·9	22·5	21·2	20·1	19·1	18·2	17·4	16·6	15·9	15·3	14·7	14·2	13·7	13·2	12·8	59
60	26·5	24·8	23·4	22·1	20·9	19·9	18·9	18·1	17·3	16·6	15·9	15·3	14·7	14·2	13·7	13·3	60
Dec.	176° 15' 183° 45'	176° 00' 184° 00'	175° 45' 184° 15'	175° 30' 184° 30'	175° 15' 184° 45'	175° 00' 185° 00'	174° 45' 185° 15'	174° 30' 185° 30'	174° 15' 185° 45'	174° 00' 186° 00'	173° 45' 186° 15'	173° 30' 186° 30'	173° 15' 186° 45'	173° 00' 187° 00'	172° 45' 187° 15'	172° 30' 187° 30'	Dec.

B — HOUR ANGLE — **B**

B—Always named the **same** as Declination.

| A | TABLE A — HOUR ANGLE | | | | | | | | | | | | | | | A |

Lat. °	7°30'	7°45'	8°00'	8°15'	8°30'	8°45'	9°00'	9°15'	9°30'	9°45'	10°00'	10°15'	10°30'	10°45'	11°00'	11°15'	Lat. °
	352°30'	352°15'	352°00'	351°45'	351°30'	351°15'	351°00'	350°45'	350°30'	350°15'	350°00'	349°45'	349°30'	349°15'	349°00'	348°45'	
0	·00	·00	·00	·00	·00	·00	·00	·00	·00	·00	·00	·00	·00	·00	·00	·00	0
1	·13	·13	·12	·12	·12	·11	·11	·11	·10	·10	·10	·10	·09	·09	·09	·09	1
2	·27	·26	·25	·24	·23	·23	·22	·21	·21	·20	·20	·19	·19	·18	·18	·18	2
3	·40	·39	·37	·36	·35	·34	·33	·32	·31	·30	·29	·28	·28	·27	·26	·26	3
4	·53	·51	·50	·48	·47	·45	·44	·43	·42	·41	·40	·39	·38	·37	·36	·35	4
5	·66	·64	·62	·60	·59	·57	·55	·54	·52	·51	·50	·48	·47	·46	·45	·44	5
6	·80	·77	·75	·72	·70	·68	·66	·65	·63	·61	·60	·58	·57	·55	·54	·53	6
7	·93	·90	·87	·85	·82	·80	·78	·75	·73	·72	·70	·68	·66	·65	·63	·62	7
8	1·07	1·03	1·00	·97	·94	·91	·89	·86	·84	·82	·80	·78	·76	·74	·72	·71	8
9	1·20	1·16	1·13	1·09	1·06	1·03	1·00	·97	·95	·92	·90	·88	·85	·83	·81	·80	9
10	1·34	1·30	1·25	1·22	1·18	1·15	1·11	1·08	1·05	1·03	1·00	·98	·95	·93	·91	·89	10
11	1·48	1·43	1·38	1·34	1·30	1·26	1·23	1·19	1·16	1·13	1·10	1·07	1·05	1·02	1·00	·98	11
12	1·61	1·56	1·51	1·47	1·42	1·38	1·34	1·31	1·27	1·24	1·21	1·18	1·15	1·12	1·09	1·07	12
13	1·75	1·70	1·64	1·59	1·54	1·50	1·46	1·42	1·38	1·34	1·31	1·28	1·25	1·22	1·19	1·16	13
14	1·89	1·83	1·77	1·72	1·67	1·62	1·57	1·53	1·49	1·45	1·41	1·38	1·35	1·31	1·28	1·25	14
15	2·04	1·97	1·91	1·85	1·79	1·74	1·69	1·64	1·60	1·56	1·52	1·48	1·45	1·41	1·38	1·35	15
16	2·18	2·11	2·04	1·98	1·92	1·86	1·81	1·76	1·71	1·67	1·63	1·59	1·55	1·51	1·48	1·44	16
17	2·32	2·25	2·18	2·11	2·05	1·99	1·93	1·88	1·83	1·78	1·73	1·69	1·65	1·61	1·57	1·54	17
18	2·47	2·39	2·31	2·24	2·17	2·11	2·05	2·00	1·94	1·89	1·84	1·80	1·75	1·71	1·67	1·63	18
19	2·62	2·53	2·45	2·37	2·30	2·24	2·17	2·11	2·06	2·00	1·95	1·90	1·86	1·81	1·77	1·73	19
20	2·76	2·67	2·59	2·51	2·44	2·36	2·30	2·23	2·18	2·12	2·06	2·01	1·96	1·92	1·87	1·83	20
21	2·92	2·82	2·73	2·65	2·57	2·49	2·42	2·36	2·29	2·23	2·18	2·12	2·07	2·02	1·97	1·93	21
22	3·07	2·97	2·87	2·79	2·70	2·63	2·55	2·48	2·41	2·35	2·29	2·23	2·18	2·13	2·08	2·03	22
23	3·22	3·12	3·02	2·93	2·84	2·76	2·68	2·61	2·54	2·47	2·41	2·35	2·29	2·24	2·18	2·13	23
24	3·38	3·27	3·17	3·07	2·98	2·89	2·81	2·73	2·66	2·59	2·53	2·46	2·40	2·35	2·29	2·24	24
25	3·54	3·43	3·32	3·22	3·12	3·03	2·94	2·86	2·79	2·71	2·64	2·58	2·52	2·46	2·40	2·34	25
26	3·70	3·58	3·47	3·36	3·26	3·17	3·08	2·99	2·91	2·84	2·77	2·70	2·63	2·57	2·51	2·45	26
27	3·87	3·74	3·63	3·51	3·41	3·31	3·22	3·13	3·04	2·97	2·89	2·82	2·75	2·68	2·62	2·56	27
28	4·04	3·91	3·78	3·67	3·56	3·45	3·36	3·26	3·18	3·09	3·02	2·94	2·87	2·80	2·74	2·67	28
29	4·21	4·07	3·94	3·82	3·71	3·60	3·50	3·40	3·31	3·23	3·14	3·07	2·99	2·92	2·85	2·79	29
30	4·39	4·24	4·11	3·98	3·86	3·75	3·65	3·55	3·45	3·36	3·27	3·19	3·12	3·04	2·97	2·90	30
31	4·56	4·42	4·28	4·14	4·02	3·90	3·79	3·69	3·59	3·50	3·41	3·32	3·24	3·16	3·09	3·02	31
32	4·75	4·59	4·45	4·31	4·18	4·06	3·95	3·84	3·73	3·64	3·54	3·46	3·37	3·29	3·21	3·14	32
33	4·93	4·77	4·62	4·48	4·35	4·22	4·10	3·99	3·88	3·78	3·68	3·59	3·50	3·42	3·34	3·26	33
34	5·12	4·96	4·80	4·65	4·51	4·38	4·26	4·14	4·03	3·93	3·83	3·73	3·64	3·55	3·47	3·39	34
35	5·32	5·15	4·98	4·83	4·69	4·55	4·42	4·30	4·18	4·08	3·97	3·87	3·78	3·69	3·60	3·52	35
36	5·52	5·34	5·17	5·01	4·86	4·72	4·59	4·46	4·34	4·23	4·12	4·02	3·92	3·83	3·74	3·65	36
37	5·72	5·54	5·36	5·20	5·04	4·90	4·76	4·63	4·50	4·39	4·27	4·17	4·07	3·97	3·88	3·79	37
38	5·93	5·74	5·56	5·39	5·23	5·08	4·93	4·80	4·67	4·55	4·43	4·32	4·22	4·12	4·02	3·93	38
39	6·15	5·95	5·76	5·58	5·42	5·26	5·11	4·97	4·84	4·71	4·59	4·48	4·37	4·27	4·17	4·07	39
40	6·37	6·17	5·97	5·79	5·61	5·45	5·30	5·15	5·01	4·88	4·76	4·64	4·53	4·42	4·32	4·22	40
41	6·60	6·39	6·19	6·00	5·82	5·65	5·49	5·34	5·19	5·06	4·93	4·81	4·69	4·58	4·47	4·37	41
42	6·84	6·62	6·41	6·21	6·02	5·85	5·69	5·53	5·38	5·24	5·11	4·98	4·86	4·74	4·63	4·53	42
43	7·08	6·85	6·63	6·43	6·24	6·06	5·89	5·73	5·57	5·43	5·29	5·16	5·03	4·91	4·80	4·69	43
44	7·34	7·10	6·87	6·66	6·46	6·27	6·10	5·93	5·77	5·62	5·48	5·34	5·21	5·09	4·97	4·85	44
45	7·60	7·35	7·12	6·90	6·69	6·50	6·31	6·14	5·98	5·82	5·67	5·53	5·40	5·27	5·15	5·03	45
46	7·87	7·61	7·37	7·14	6·93	6·73	6·54	6·36	6·19	6·03	5·87	5·73	5·59	5·46	5·33	5·21	46
47	8·15	7·88	7·63	7·40	7·18	6·97	6·77	6·59	6·41	6·24	6·08	5·93	5·79	5·65	5·52	5·39	47
48	8·44	8·16	7·90	7·66	7·43	7·22	7·01	6·82	6·64	6·46	6·30	6·14	5·99	5·85	5·71	5·58	48
49	8·74	8·45	8·19	7·93	7·70	7·47	7·26	7·06	6·87	6·69	6·52	6·36	6·21	6·06	5·92	5·78	49
50	9·05	8·76	8·48	8·22	7·97	7·74	7·52	7·32	7·12	6·94	6·76	6·59	6·43	6·28	6·13	5·99	50
51	9·38	9·07	8·79	8·52	8·26	8·02	7·80	7·58	7·38	7·19	7·00	6·83	6·66	6·50	6·35	6·21	51
52	9·72	9·40	9·11	8·83	8·56	8·32	8·08	7·86	7·65	7·45	7·26	7·08	6·91	6·74	6·58	6·43	52
53	10·1	9·75	9·44	9·15	8·88	8·62	8·38	8·15	7·93	7·72	7·53	7·34	7·16	6·99	6·83	6·67	53
54	10·5	10·1	9·79	9·49	9·21	8·94	8·69	8·45	8·23	8·01	7·81	7·61	7·43	7·25	7·08	6·92	54
55	10·8	10·5	10·2	9·85	9·56	9·28	9·02	8·77	8·53	8·31	8·10	7·90	7·71	7·52	7·35	7·18	55
56	11·3	10·9	10·5	10·2	9·92	9·63	9·36	9·10	8·86	8·63	8·41	8·20	8·00	7·81	7·63	7·45	56
57	11·7	11·3	11·0	10·6	10·3	10·0	9·72	9·46	9·20	8·96	8·73	8·52	8·31	8·11	7·92	7·74	57
58	12·2	11·8	11·4	11·0	10·7	10·4	10·1	9·83	9·56	9·31	9·08	8·85	8·64	8·43	8·23	8·05	58
59	12·6	12·2	11·8	11·5	11·1	10·8	10·5	10·2	9·95	9·69	9·44	9·20	8·98	8·77	8·56	8·37	59
60	13·2	12·7	12·3	11·9	11·6	11·3	10·9	10·6	10·4	10·1	9·82	9·58	9·35	9·12	8·91	8·71	60
Lat.	172°30'	172°15'	172°00'	171°45'	171°30'	171°15'	171°00'	170°45'	170°30'	170°15'	170°00'	169°45'	169°30'	169°15'	169°00'	168°45'	Lat.
	187°30'	187°45'	188°00'	188°15'	188°30'	188°45'	189°00'	189°15'	189°30'	189°45'	190°00'	190°15'	190°30'	190°45'	191°00'	191°15'	

| A | HOUR ANGLE | A |

TABLE B — HOUR ANGLE

B | ... | B

Dec.	7°30' 352°30'	7°45' 352°15'	8°00' 352°00'	8°15' 351°45'	8°30' 351°30'	8°45' 351°15'	9°00' 351°00'	9°15' 350°45'	9°30' 350°30'	9°45' 350°15'	10°00' 350°00'	10°15' 349°45'	10°30' 349°30'	10°45' 349°15'	11°00' 349°00'	11°15' 348°45'	Dec.
0	·00	·00	·00	·00	·00	·00	·00	·00	·00	·00	·00	·00	·00	·00	·00	·00	0
1	·13	·13	·13	·12	·12	·12	·11	·11	·11	·10	·10	·10	·10	·09	·09	·09	1
2	·27	·26	·25	·24	·24	·23	·22	·22	·21	·21	·20	·20	·19	·19	·18	·18	2
3	·40	·39	·38	·37	·36	·35	·34	·33	·32	·31	·30	·30	·29	·28	·27	·27	3
4	·54	·52	·50	·49	·47	·46	·45	·44	·42	·41	·40	·39	·38	·38	·37	·36	4
5	·67	·65	·63	·61	·59	·58	·56	·54	·53	·52	·50	·49	·48	·47	·46	·45	5
6	·81	·78	·76	·73	·71	·69	·67	·65	·64	·62	·61	·59	·58	·56	·55	·54	6
7	·94	·91	·88	·86	·83	·81	·79	·76	·74	·73	·71	·69	·67	·66	·64	·63	7
8	1·08	1·04	1·01	·98	·95	·92	·90	·87	·85	·83	·81	·79	·77	·75	·74	·72	8
9	1·21	1·17	1·14	1·10	1·07	1·04	1·01	·99	·96	·94	·91	·89	·87	·85	·83	·81	9
10	1·35	1·31	1·27	1·23	1·19	1·16	1·13	1·10	1·07	1·04	1·02	·99	·97	·95	·92	·90	10
11	1·49	1·44	1·40	1·35	1·32	1·28	1·24	1·21	1·18	1·15	1·12	1·09	1·07	1·04	1·02	1·00	11
12	1·63	1·58	1·53	1·48	1·44	1·40	1·36	1·32	1·29	1·26	1·22	1·19	1·17	1·14	1·11	1·09	12
13	1·77	1·71	1·66	1·61	1·56	1·52	1·48	1·44	1·40	1·36	1·33	1·30	1·27	1·24	1·21	1·18	13
14	1·91	1·85	1·79	1·74	1·69	1·64	1·59	1·55	1·51	1·47	1·44	1·40	1·37	1·34	1·31	1·28	14
15	2·05	1·99	1·93	1·87	1·81	1·76	1·71	1·67	1·62	1·58	1·54	1·51	1·47	1·44	1·40	1·37	15
16	2·20	2·13	2·06	2·00	1·94	1·88	1·83	1·78	1·74	1·69	1·65	1·61	1·57	1·54	1·50	1·47	16
17	2·34	2·27	2·20	2·13	2·07	2·01	1·95	1·90	1·85	1·80	1·76	1·72	1·68	1·64	1·60	1·57	17
18	2·49	2·41	2·33	2·26	2·20	2·14	2·08	2·02	1·97	1·92	1·87	1·83	1·78	1·74	1·70	1·67	18
19	2·64	2·55	2·47	2·40	2·33	2·26	2·20	2·14	2·09	2·03	1·98	1·94	1·89	1·85	1·80	1·76	19
20	2·79	2·70	2·62	2·54	2·46	2·39	2·33	2·26	2·21	2·15	2·10	2·05	2·00	1·95	1·91	1·87	20
21	2·94	2·85	2·76	2·68	2·60	2·52	2·45	2·39	2·33	2·27	2·21	2·16	2·11	2·06	2·01	1·97	21
22	3·10	3·00	2·90	2·82	2·73	2·66	2·58	2·51	2·45	2·39	2·33	2·27	2·22	2·17	2·12	2·07	22
23	3·25	3·15	3·05	2·96	2·87	2·79	2·71	2·64	2·57	2·51	2·44	2·39	2·33	2·28	2·22	2·18	23
24	3·41	3·30	3·20	3·10	3·01	2·93	2·85	2·77	2·70	2·63	2·56	2·50	2·44	2·39	2·33	2·28	24
25	3·57	3·46	3·35	3·25	3·15	3·07	2·98	2·90	2·83	2·75	2·69	2·62	2·56	2·50	2·44	2·39	25
26	3·74	3·62	3·50	3·40	3·30	3·21	3·12	3·03	2·96	2·88	2·81	2·74	2·68	2·61	2·56	2·50	26
27	3·90	3·78	3·66	3·55	3·45	3·35	3·26	3·17	3·09	3·01	2·93	2·86	2·80	2·73	2·67	2·61	27
28	4·07	3·94	3·82	3·71	3·60	3·50	3·40	3·31·	3·22	3·14	3·06	2·99	2·92	2·85	2·79	2·73	28
29	4·25	4·11	3·98	3·86	3·75	3·64	3·54	3·45	3·36	3·27	3·19	3·12	3·04	2·97	2·91	2·84	29
30	4·42	4·28	4·15	4·02	3·91	3·80	3·69	3·59	3·50	3·41	3·32	3·24	3·17	3·10	3·03	2·96	30
31	4·60	4·46	4·32	4·19	4·07	3·95	3·84	3·74	3·64	3·55	3·46	3·38	3·30	3·22	3·15	3·08	31
32	4·79	4·63	4·49	4·35	4·23	4·11	3·99	3·89	3·79	3·69	3·60	3·51	3·43	3·35	3·27	3·20	32
33	4·98	4·82	4·67	4·53	4·39	4·27	4·15	4·04	3·93	3·83	3·74	3·65	3·56	3·48	3·40	3·33	33
34	5·17	5·00	4·85	4·70	4·56	4·43	4·31	4·20	4·09	3·98	3·88	3·79	3·70	3·62	3·53	3·46	34
35	5·36	5·19	5·03	4·88	4·74	4·60	4·48	4·36	4·24	4·13	4·03	3·93	3·84	3·75	3·67	3·59	35
36	5·57	5·39	5·22	5·06	4·92	4·78	4·64	4·52	4·40	4·29	4·18	4·08	3·99	3·90	3·81	3·72	36
37	5·77	5·59	5·41	5·25	5·10	4·95	4·82	4·69	4·57	4·45	4·34	4·23	4·14	4·04	3·95	3·86	37
38	5·99	5·79	5·61	5·44	5·29	5·14	5·00	4·86	4·73	4·61	4·50	4·39	4·29	4·19	4·09	4·00	38
39	6·20	6·01	5·82	5·64	5·48	5·32	5·18	5·04	4·91	4·78	4·66	4·55	4·44	4·34	4·24	4·15	39
40	6·43	6·22	6·03	5·85	5·68	5·52	5·36	5·22	5·08	4·95	4·83	4·72	4·60	4·50	4·40	4·30	40
41	6·66	6·45	6·25	6·06	5·88	5·71	5·56	5·41	5·27	5·13	5·01	4·89	4·77	4·66	4·56	4·46	41
42	6·90	6·68	6·47	6·27	6·09	5·92	5·76	5·60	5·46	5·32	5·19	5·06	4·94	4·83	4·72	4·62	42
43	7·14	6·92	6·70	6·50	6·31	6·13	5·96	5·80	5·65	5·51	5·37	5·24	5·12	5·00	4·89	4·78	43
44	7·40	7·16	6·94	6·73	6·53	6·35	6·17	6·01	5·85	5·70	5·56	5·43	5·30	5·18	5·06	4·95	44
45	7·66	7·42	7·19	6·97	6·77	6·57	6·39	6·22	6·06	5·90	5·76	5·62	5·49	5·36	5·24	5·13	45
46	7·93	7·68	7·44	7·22	7·01	6·81	6·62	6·44	6·27	6·11	5·96	5·82	5·68	5·55	5·43	5·31	46
47	8·22	7·95	7·71	7·47	7·26	7·05	6·86	6·67	6·50	6·33	6·18	6·03	5·88	5·75	5·62	5·50	47
48	8·51	8·24	7·98	7·74	7·51	7·30	7·10	6·91	6·73	6·56	6·40	6·24	6·09	5·95	5·82	5·69	48
49	8·81	8·53	8·27	8·02	7·78	7·56	7·35	7·16	6·97	6·79	6·62	6·46	6·31	6·17	6·03	5·90	49
50	9·13	8·84	8·56	8·31	8·06	7·83	7·62	7·41	7·22	7·04	6·86	6·70	6·54	6·39	6·25	6·11	50
51	9·46	9·16	8·87	8·61	8·35	8·12	7·89	7·68	7·48	7·29	7·11	6·94	6·78	6·62	6·47	6·33	51
52	9·81	9·49	9·20	8·92	8·66	8·41	8·18	7·96	7·75	7·56	7·37	7·19	7·02	6·86	6·71	6·56	52
53	10·2	9·84	9·54	9·25	8·98	8·72	8·48	8·26	8·04	7·84	7·64	7·46	7·28	7·11	6·95	6·80	53
54	10·5	10·2	9·89	9·59	9·31	9·05	8·80	8·56	8·34	8·13	7·93	7·73	7·55	7·38	7·21	7·06	54
55	10·9	10·6	10·3	9·95	9·66	9·39	9·13	8·88	8·65	8·43	8·22	8·03	7·84	7·66	7·48	7·32	55
56	11·4	11·0	10·7	10·3	10·0	9·75	9·48	9·22	8·98	8·75	8·54	8·33	8·14	7·95	7·77	7·60	56
57	11·8	11·4	11·1	10·7	10·4	10·1	9·84	9·58	9·33	9·09	8·87	8·65	8·45	8·26	8·07	7·89	57
58	12·3	11·9	11·5	11·2	10·8	10·5	10·2	9·96	9·70	9·45	9·22	8·99	8·78	8·58	8·39	8·20	58
59	12·8	12·3	12·0	11·6	11·3	10·9	10·6	10·4	10·1	9·83	9·58	9·35	9·13	8·92	8·72	8·53	59
60	13·3	12·8	12·4	12·1	11·7	11·4	11·1	10·8	10·5	10·2	9·97	9·73	9·50	9·29	9·08	8·88	60
Dec.	172°30' 187°30'	172°15' 187°45'	172°00' 188°00'	171°45' 188°15'	171°30' 188°30'	171°15' 188°45'	171°00' 189°00'	170°45' 189°15'	170°30' 189°30'	170°15' 189°45'	170°00' 190°00'	169°45' 190°15'	169°30' 190°30'	169°15' 190°45'	169°00' 191°00'	168°45' 191°15'	Dec.

B — Always named the same as Declination.

HOUR ANGLE

A — TABLE A — HOUR ANGLE — A

A—Named opposite to Latitude, **except** when Hour Angle is between 90° and 270°

Lat.°	11°15' / 348°45'	11°30' / 348°30'	11°45' / 348°15'	12°00' / 348°00'	12°15' / 347°45'	12°30' / 347°30'	12°45' / 347°15'	13°00' / 347°00'	13°15' / 346°45'	13°30' / 346°30'	13°45' / 346°15'	14°00' / 346°00'	14°15' / 345°45'	14°30' / 345°30'	14°45' / 345°15'	15°00' / 345°00'	Lat.°
0	·00	·00	·00	·00	·00	·00	·00	·00	·00	·00	·00	·00	·00	·00	·00	·00	0
1	·09	·09	·08	·08	·08	·08	·08	·08	·07	·07	·07	·07	·07	·07	·07	·07	1
2	·18	·17	·17	·16	·16	·16	·15	·15	·15	·15	·14	·14	·14	·14	·13	·13	2
3	·26	·26	·25	·24	·24	·24	·23	·23	·22	·22	·21	·21	·21	·20	·20	·20	3
4	·35	·34	·31	·33	·32	·32	·31	·30	·30	·29	·29	·28	·28	·27	·27	·26	4
5	·44	·43	·42	·41	·40	·39	·39	·38	·37	·36	·36	·35	·34	·34	·33	·33	5
6	·53	·52	·51	·49	·48	·47	·46	·46	·45	·44	·43	·42	·41	·41	·40	·39	6
7	·62	·60	·59	·58	·57	·55	·54	·53	·52	·51	·50	·49	·48	·47	·47	·46	7
8	·71	·69	·68	·66	·65	·63	·62	·61	·60	·59	·57	·56	·55	·54	·53	·52	8
9	·80	·78	·76	·75	·73	·71	·70	·69	·67	·66	·65	·64	·62	·61	·60	·59	9
10	·89	·87	·85	·83	·81	·80	·78	·76	·75	·73	·72	·71	·69	·68	·67	·66	10
11	·98	·96	·94	·91	·90	·88	·86	·84	·83	·81	·79	·78	·77	·75	·74	·73	11
12	1·07	1·04	1·02	1·00	·98	·96	·94	·92	·90	·89	·87	·85	·84	·82	·81	·79	12
13	1·16	1·13	1·11	1·09	1·06	1·04	1·02	1·00	·98	·96	·94	·93	·91	·89	·88	·86	13
14	1·25	1·23	1·20	1·17	1·15	1·12	1·10	1·08	1·06	1·04	1·02	1·00	·98	·96	·95	·93	14
15	1·35	1·32	1·29	1·26	1·23	1·21	1·18	1·16	1·14	1·12	1·10	1·07	1·06	1·04	1·02	1·00	15
16	1·44	1·41	1·38	1·35	1·32	1·29	1·27	1·24	1·22	1·19	1·17	1·15	1·13	1·11	1·09	1·07	16
17	1·54	1·50	1·47	1·44	1·41	1·38	1·35	1·32	1·30	1·27	1·25	1·23	1·20	1·18	1·16	1·14	17
18	1·63	1·60	1·56	1·53	1·50	1·47	1·44	1·41	1·38	1·35	1·33	1·30	1·28	1·26	1·23	1·21	18
19	1·73	1·69	1·66	1·62	1·59	1·55	1·52	1·49	1·46	1·43	1·41	1·38	1·36	1·33	1·31	1·29	19
20	1·83	1·79	1·75	1·71	1·68	1·64	1·61	1·58	1·55	1·52	1·49	1·46	1·43	1·41	1·38	1·36	20
21	1·93	1·89	1·85	1·81	1·77	1·73	1·70	1·66	1·63	1·60	1·57	1·54	1·51	1·48	1·46	1·43	21
22	2·03	1·99	1·94	1·90	1·86	1·82	1·79	1·75	1·72	1·68	1·65	1·62	1·59	1·56	1·53	1·51	22
23	2·13	2·09	2·04	2·00	1·96	1·91	1·88	1·84	1·80	1·77	1·73	1·70	1·67	1·64	1·61	1·58	23
24	2·24	2·19	2·14	2·09	2·05	2·01	1·97	1·93	1·89	1·85	1·82	1·79	1·75	1·72	1·69	1·66	24
25	2·34	2·29	2·24	2·19	2·15	2·10	2·06	2·02	1·98	1·94	1·91	1·87	1·84	1·80	1·77	1·74	25
26	2·45	2·40	2·34	2·29	2·25	2·20	2·16	2·11	2·07	2·03	1·99	1·96	1·92	1·89	1·85	1·82	26
27	2·56	2·50	2·45	2·40	2·35	2·30	2·25	2·21	2·16	2·12	2·08	2·04	2·01	1·97	1·94	1·90	27
28	2·67	2·61	2·56	2·50	2·45	2·40	2·35	2·30	2·26	2·21	2·17	2·13	2·09	2·06	2·02	1·98	28
29	2·79	2·73	2·66	2·61	2·55	2·50	2·45	2·40	2·35	2·31	2·27	2·22	2·18	2·14	2·11	2·07	29
30	2·90	2·84	2·78	2·72	2·66	2·60	2·55	2·50	2·45	2·40	2·36	2·32	2·27	2·23	2·19	2·15	30
31	3·02	2·95	2·89	2·83	2·77	2·71	2·66	2·60	2·55	2·50	2·46	2·41	2·37	2·32	2·28	2·24	31
32	3·14	3·07	3·00	2·94	2·88	2·82	2·76	2·71	2·65	2·60	2·55	2·51	2·46	2·42	2·37	2·33	32
33	3·26	3·19	3·12	3·06	2·99	2·93	2·87	2·81	2·76	2·71	2·66	2·61	2·56	2·51	2·47	2·42	33
34	3·39	3·32	3·24	3·17	3·11	3·04	2·98	2·92	2·86	2·81	2·76	2·71	2·66	2·61	2·56	2·52	34
35	3·52	3·44	3·37	3·29	3·22	3·16	3·09	3·03	2·97	2·92	2·86	2·81	2·76	2·71	2·66	2·61	35
36	3·65	3·57	3·49	3·42	3·35	3·28	3·21	3·15	3·09	3·03	2·97	2·91	2·86	2·81	2·76	2·71	36
37	3·79	3·70	3·62	3·55	3·47	3·40	3·33	3·26	3·20	3·14	3·08	3·02	2·97	2·91	2·86	2·81	37
38	3·93	3·84	3·76	3·68	3·60	3·52	3·45	3·38	3·32	3·25	3·19	3·13	3·08	3·02	2·97	2·92	38
39	4·07	3·98	3·89	3·81	3·73	3·65	3·58	3·51	3·44	3·37	3·31	3·25	3·19	3·13	3·08	3·02	39
40	4·22	4·12	4·03	3·95	3·86	3·78	3·71	3·63	3·56	3·50	3·43	3·37	3·30	3·24	3·19	3·13	40
41	4·37	4·27	4·18	4·09	4·00	3·92	3·84	3·77	3·69	3·62	3·55	3·49	3·42	3·36	3·30	3·24	41
42	4·53	4·43	4·33	4·24	4·15	4·06	3·98	3·90	3·82	3·75	3·68	3·61	3·55	3·48	3·42	3·36	42
43	4·69	4·58	4·48	4·39	4·30	4·21	4·12	4·04	3·96	3·88	3·81	3·74	3·67	3·61	3·54	3·48	43
44	4·85	4·75	4·64	4·54	4·45	4·36	4·27	4·18	4·10	4·02	3·95	3·87	3·80	3·73	3·67	3·60	44
45	5·03	4·92	4·81	4·70	4·61	4·51	4·42	4·33	4·25	4·17	4·09	4·01	3·94	3·87	3·80	3·73	45
46	5·21	5·09	4·98	4·87	4·77	4·67	4·58	4·49	4·40	4·31	4·23	4·15	4·08	4·00	3·93	3·86	46
47	5·39	5·27	5·16	5·05	4·94	4·84	4·74	4·65	4·56	4·47	4·38	4·30	4·22	4·15	4·07	4·00	47
48	5·58	5·46	5·34	5·23	5·12	5·01	4·91	4·81	4·72	4·63	4·54	4·45	4·37	4·29	4·22	4·14	48
49	5·78	5·65	5·53	5·41	5·30	5·19	5·08	4·98	4·89	4·79	4·70	4·61	4·53	4·45	4·37	4·29	49
50	5·99	5·86	5·73	5·61	5·49	5·38	5·27	5·16	5·06	4·96	4·87	4·78	4·69	4·61	4·53	4·45	50
51	6·21	6·07	5·94	5·81	5·69	5·57	5·46	5·35	5·24	5·14	5·05	4·95	4·86	4·77	4·69	4·61	51
52	6·43	6·29	6·15	6·02	5·90	5·77	5·66	5·55	5·44	5·33	5·23	5·13	5·04	4·95	4·86	4·78	52
53	6·67	6·52	6·38	6·24	6·11	5·99	5·86	5·75	5·64	5·53	5·42	5·32	5·23	5·13	5·04	4·95	53
54	6·92	6·77	6·62	6·48	6·34	6·21	6·08	5·96	5·85	5·73	5·62	5·52	5·42	5·32	5·23	5·14	54
55	7·18	7·02	6·87	6·72	6·58	6·44	6·31	6·19	6·07	5·95	5·84	5·73	5·62	5·52	5·42	5·33	55
56	7·45	7·29	7·13	6·97	6·83	6·69	6·55	6·42	6·30	6·18	6·06	5·95	5·84	5·73	5·63	5·53	56
57	7·74	7·57	7·40	7·24	7·09	6·95	6·81	6·67	6·54	6·41	6·29	6·18	6·06	5·95	5·85	5·75	57
58	8·05	7·87	7·69	7·53	7·37	7·22	7·07	6·93	6·80	6·67	6·54	6·42	6·30	6·19	6·08	5·97	58
59	8·37	8·18	8·00	7·83	7·67	7·51	7·36	7·21	7·07	6·93	6·80	6·68	6·55	6·44	6·32	6·21	59
60	8·71	8·51	8·33	8·15	7·98	7·81	7·65	7·50	7·36	7·21	7·08	6·95	6·82	6·70	6·58	6·46	60
Lat.	168°45'	168°30'	168°15'	168°00'	167°45'	167°30'	167°15'	167°00'	166°45'	166°30'	166°15'	166°00'	165°45'	165°30'	165°15'	165°00'	Lat.
	191°15'	191°30'	191°45'	192°00'	192°15'	192°30'	192°45'	193°00'	193°15'	193°30'	193°45'	194°00'	194°15'	194°30'	194°45'	195°00'	

A — HOUR ANGLE — A

TABLE B — HOUR ANGLE

Dec.	11°15'	11°30'	11°45'	12°00'	12°15'	12°30'	12°45'	13°00'	13°15'	13°30'	13°45'	14°00'	14°15'	14°30'	14°45'	15°00'	Dec.
°	348°45'	348°30'	348°15'	348°00'	347°45'	347°30'	347°15'	347°00'	346°45'	346°30'	346°15'	346°00'	345°45'	345°30'	345°15'	345°00'	°
0	·00	·00	·00	·00	·00	·00	·00	·00	·00	·00	·00	·00	·00	·00	·00	·00	0
1	·09	·09	·09	·08	·08	·08	·08	·08	·08	·08	·07	·07	·07	·07	·07	·07	1
2	·18	·18	·17	·17	·17	·16	·16	·16	·15	·15	·15	·14	·14	·14	·14	·13	2
3	·27	·26	·26	·25	·25	·24	·24	·23	·23	·22	·22	·22	·21	·21	·21	·20	3
4	·36	·35	·34	·34	·33	·32	·32	·31	·31	·30	·29	·29	·28	·28	·28	·27	4
5	·45	·44	·43	·42	·41	·40	·40	·39	·38	·38	·37	·36	·36	·35	·34	·34	5
6	·54	·53	·52	·51	·50	·49	·48	·47	·46	·45	·44	·43	·43	·42	·41	·41	6
7	·63	·62	·60	·59	·58	·57	·56	·55	·54	·53	·52	·51	·50	·49	·48	·47	7
8	·72	·71	·69	·68	·66	·65	·64	·62	·61	·60	·59	·58	·57	·56	·55	·54	8
9	·81	·79	·78	·76	·75	·73	·72	·70	·69	·68	·67	·65	·64	·63	·62	·61	9
10	·90	·88	·87	·85	·83	·82	·80	·78	·77	·76	·74	·73	·72	·70	·69	·68	10
11	1·00	·98	·96	·93	·92	·90	·88	·86	·85	·83	·82	·80	·79	·78	·76	·75	11
12	1·09	1·07	1·04	1·02	1·00	·98	·96	·94	·93	·91	·89	·88	·86	·85	·84	·82	12
13	1·18	1·16	1·13	1·11	1·09	1·07	1·05	1·03	1·01	·99	·97	·95	·94	·92	·91	·89	13
14	1·28	1·25	1·22	1·20	1·18	1·15	1·13	1·11	1·09	1·07	1·05	1·03	1·01	1·00	·98	·96	14
15	1·37	1·34	1·32	1·29	1·26	1·24	1·21	1·19	1·17	1·15	1·13	1·11	1·09	1·07	1·05	1·04	15
16	1·47	1·44	1·41	1·38	1·35	1·32	1·30	1·27	1·25	1·23	1·21	1·19	1·16	1·15	1·13	1·11	16
17	1·57	1·53	1·50	1·47	1·44	1·41	1·39	1·36	1·33	1·31	1·29	1·26	1·24	1·22	1·20	1·18	17
18	1·67	1·63	1·60	1·56	1·53	1·50	1·47	1·44	1·42	1·39	1·37	1·34	1·32	1·30	1·28	1·26	18
19	1·76	1·73	1·69	1·66	1·62	1·59	1·56	1·53	1·50	1·47	1·45	1·42	1·40	1·38	1·35	1·33	19
20	1·87	1·83	1·79	1·75	1·72	1·68	1·65	1·62	1·59	1·56	1·53	1·50	1·48	1·45	1·43	1·41	20
21	1·97	1·93	1·89	1·85	1·81	1·77	1·74	1·71	1·67	1·64	1·62	1·59	1·56	1·53	1·51	1·48	21
22	2·07	2·03	1·98	1·94	1·90	1·87	1·83	1·80	1·76	1·73	1·70	1·67	1·64	1·61	1·59	1·56	22
23	2·18	2·13	2·08	2·04	2·00	1·96	1·92	1·89	1·85	1·82	1·79	1·75	1·72	1·70	1·67	1·64	23
24	2·28	2·23	2·19	2·14	2·10	2·06	2·02	1·98	1·94	1·91	1·87	1·84	1·81	1·78	1·75	1·72	24
25	2·39	2·34	2·29	2·24	2·20	2·15	2·11	2·07	2·03	2·00	1·96	1·93	1·89	1·86	1·83	1·80	25
26	2·50	2·45	2·40	2·35	2·30	2·25	2·21	2·17	2·13	2·09	2·05	2·02	1·98	1·95	1·92	1·88	26
27	2·61	2·56	2·50	2·45	2·40	2·35	2·31	2·27	2·22	2·18	2·14	2·11	2·07	2·04	2·00	1·97	27
28	2·73	2·67	2·61	2·56	2·51	2·46	2·41	2·36	2·32	2·28	2·24	2·20	2·16	2·12	2·09	2·05	28
29	2·84	2·78	2·72	2·67	2·61	2·56	2·51	2·46	2·42	2·37	2·33	2·29	2·25	2·21	2·18	2·14	29
30	2·96	2·90	2·84	2·78	2·72	2·67	2·62	2·57	2·52	2·47	2·43	2·39	2·35	2·31	2·27	2·23	30
31	3·08	3·01	2·95	2·89	2·83	2·78	2·72	2·67	2·62	2·57	2·53	2·48	2·44	2·40	2·36	2·32	31
32	3·20	3·13	3·07	3·01	2·95	2·89	2·83	2·78	2·73	2·68	2·63	2·58	2·54	2·50	2·45	2·41	32
33	3·33	3·26	3·19	3·12	3·06	3·00	2·94	2·89	2·83	2·78	2·73	2·68	2·64	2·59	2·55	2·51	33
34	3·46	3·38	3·31	3·24	3·18	3·12	3·06	3·00	2·94	2·89	2·84	2·79	2·74	2·69	2·65	2·61	34
35	3·59	3·51	3·44	3·37	3·30	3·24	3·17	3·11	3·06	3·00	2·95	2·89	2·84	2·80	2·75	2·71	35
36	3·72	3·64	3·57	3·49	3·42	3·36	3·29	3·23	3·17	3·11	3·06	3·00	2·95	2·90	2·85	2·81	36
37	3·86	3·78	3·70	3·62	3·55	3·48	3·41	3·35	3·29	3·23	3·17	3·11	3·06	3·01	2·96	2·91	37
38	4·00	3·92	3·84	3·76	3·68	3·61	3·54	3·47	3·41	3·35	3·29	3·23	3·17	3·12	3·07	3·02	38
39	4·15	4·06	3·98	3·90	3·82	3·74	3·67	3·60	3·53	3·47	3·41	3·35	3·29	3·23	3·18	3·13	39
40	4·30	4·21	4·12	4·04	3·95	3·88	3·80	3·73	3·66	3·59	3·53	3·47	3·41	3·35	3·30	3·24	40
41	4·46	4·36	4·27	4·18	4·10	4·02	3·94	3·86	3·79	3·72	3·66	3·59	3·53	3·47	3·41	3·36	41
42	4·62	4·52	4·42	4·33	4·24	4·16	4·08	4·00	3·93	3·86	3·79	3·72	3·66	3·60	3·54	3·48	42
43	4·78	4·68	4·58	4·49	4·39	4·31	4·23	4·15	4·07	3·99	3·92	3·85	3·79	3·72	3·66	3·60	43
44	4·95	4·84	4·74	4·64	4·55	4·46	4·38	4·29	4·21	4·14	4·06	3·99	3·92	3·86	3·79	3·73	44
45	5·13	5·02	4·91	4·81	4·71	4·62	4·53	4·45	4·36	4·28	4·21	4·13	4·06	3·99	3·93	3·86	45
46	5·31	5·19	5·09	4·98	4·88	4·78	4·69	4·60	4·52	4·44	4·36	4·28	4·21	4·14	4·07	4·00	46
47	5·50	5·38	5·27	5·16	5·05	4·96	4·86	4·77	4·68	4·59	4·51	4·43	4·36	4·28	4·21	4·14	47
48	5·69	5·57	5·45	5·34	5·23	5·13	5·03	4·94	4·85	4·76	4·67	4·59	4·51	4·44	4·36	4·29	48
49	5·90	5·77	5·65	5·53	5·42	5·31	5·21	5·11	5·02	4·93	4·84	4·76	4·67	4·59	4·52	4·44	49
50	6·11	5·98	5·85	5·73	5·62	5·51	5·40	5·30	5·20	5·11	5·01	4·93	4·84	4·76	4·68	4·60	50
51	6·33	6·19	6·06	5·94	5·82	5·71	5·60	5·49	5·39	5·29	5·20	5·10	5·02	4·93	4·85	4·77	51
52	6·56	6·42	6·29	6·16	6·03	5·91	5·80	5·69	5·58	5·48	5·39	5·29	5·20	5·11	5·03	4·95	52
53	6·80	6·66	6·52	6·38	6·25	6·13	6·01	5·90	5·79	5·68	5·58	5·49	5·39	5·30	5·21	5·13	53
54	7·06	6·90	6·76	6·62	6·49	6·36	6·24	6·12	6·01	5·90	5·79	5·69	5·59	5·50	5·41	5·32	54
55	7·32	7·16	7·01	6·87	6·73	6·60	6·47	6·35	6·23	6·12	6·01	5·90	5·80	5·70	5·61	5·52	55
56	7·60	7·44	7·28	7·13	6·99	6·85	6·72	6·59	6·47	6·35	6·24	6·13	6·02	5·92	5·82	5·73	56
57	7·89	7·72	7·56	7·41	7·26	7·12	6·98	6·85	6·72	6·60	6·48	6·37	6·26	6·15	6·05	5·95	57
58	8·20	8·03	7·86	7·70	7·54	7·39	7·25	7·11	6·98	6·86	6·73	6·62	6·50	6·39	6·29	6·18	58
59	8·53	8·35	8·17	8·00	7·84	7·69	7·54	7·40	7·26	7·13	7·00	6·88	6·76	6·65	6·54	6·43	59
60	8·88	8·69	8·51	8·33	8·16	8·00	7·85	7·70	7·56	7·42	7·29	7·16	7·04	6·92	6·80	6·69	60
Dec.	168°45'	168°30'	168°15'	168°00'	167°45'	167°30'	167°15'	167°00'	166°45'	166°30'	166°15'	166°00'	165°45'	165°30'	165°15'	165°00'	Dec.
	191°15'	191°30'	191°45'	192°00'	192°15'	192°30'	192°45'	193°00'	193°15'	193°30'	193°45'	194°00'	194°15'	194°30'	194°45'	195°00'	

B — HOUR ANGLE — **B**

B — Always named the **same** as Declination.

TABLE A — HOUR ANGLE

A—Named opposite to Latitude, **except** when Hour Angle is between 90° and 270°

Lat.	15°00' / 345°00'	15°30' / 344°30'	16°00' / 344°00'	16°30' / 343°30'	17°00' / 343°00'	17°30' / 342°30'	18°00' / 342°00'	18°30' / 341°30'	19°00' / 341°00'	19°30' / 340°30'	20°00' / 340°00'	20°30' / 339°30'	21°00' / 339°00'	21°30' / 338°30'	22°00' / 338°00'	22°30' / 337°30'	Lat.
0	·00	·00	·00	·00	·00	·00	·00	·00	·00	·00	·00	·00	·00	·00	·00	·00	0
1	·07	·06	·06	·06	·06	·06	·05	·05	·05	·05	·05	·05	·05	·05	·04	·04	1
2	·13	·13	·12	·12	·11	·11	·11	·10	·10	·10	·10	·09	·09	·09	·09	·08	2
3	·20	·19	·18	·18	·17	·17	·16	·16	·15	·15	·14	·14	·14	·13	·13	·13	3
4	·26	·25	·24	·24	·23	·22	·22	·21	·20	·20	·19	·19	·10	·10	·17	17	4
5	·33	·32	·31	·30	·29	·28	·27	·26	·25	·25	·24	·23	·23	·22	·22	·21	5
6	·39	·38	·37	·35	·34	·33	·32	·31	·31	·30	·29	·28	·27	·27	·26	·25	6
7	·46	·44	·43	·41	·40	·39	·38	·37	·36	·35	·34	·33	·32	·31	·30	·30	7
8	·52	·51	·49	·47	·46	·45	·43	·42	·41	·40	·39	·38	·37	·36	·35	·34	8
9	·59	·57	·55	·53	·52	·50	·49	·47	·46	·45	·44	·42	·41	·40	·39	·38	9
10	·66	·64	·61	·60	·58	·56	·54	·53	·51	·50	·48	·47	·46	·45	·44	·43	10
11	·73	·70	·68	·66	·64	·62	·60	·58	·56	·55	·53	·52	·51	·49	·48	·47	11
12	·79	·77	·74	·72	·70	·67	·65	·64	·62	·60	·58	·57	·55	·54	·53	·51	12
13	·86	·83	·81	·78	·76	·73	·71	·69	·67	·65	·63	·62	·60	·59	·57	·56	13
14	·93	·90	·87	·84	·82	·79	·77	·75	·72	·70	·69	·67	·65	·63	·62	·60	14
15	1·00	·97	·93	·90	·88	·85	·82	·80	·78	·76	·74	·72	·70	·68	·66	·65	15
16	1·07	1·03	1·00	·97	·94	·91	·88	·86	·83	·81	·79	·77	·75	·73	·71	·69	16
17	1·14	1·10	1·07	1·03	1·00	·97	·94	·91	·89	·86	·84	·82	·80	·78	·76	·74	17
18	1·21	1·17	1·13	1·10	1·06	1·03	1·00	·97	·94	·92	·89	·87	·85	·82	·80	·78	18
19	1·29	1·24	1·20	1·16	1·13	1·09	1·06	1 03	1·00	·97	·95	·92	·90	·87	·85	·83	19
20	1·36	1·31	1·27	1·23	1·19	1·15	1·12	1·09	1·06	1·03	1·00	·97	·95	·92	·90	·88	20
21	1·43	1·38	1·34	1·30	1·26	1·22	1·18	1·15	1·11	1·08	1·05	1·03	1·00	·97	·95	·93	21
22	1·51	1·46	1·41	1·36	1·32	1·28	1·24	1·21	1·17	1·14	1·11	1·08	1·05	1·03	1·00	·97	22
23	1·58	1·53	1·48	1·43	1·39	1·35	1·31	1·27	1·23	1·20	1·17	1·14	1·11	1·08	1·05	1·02	23
24	1·66	1·61	1·55	1·50	1·46	1·41	1·37	1·33	1·29	1·26	1·22	1·19	1·16	1·13	1·10	1·07	24
25	1·74	1·68	1·63	1·57	1·53	1·48	1·44	1·39	1·35	1·32	1·28	1·25	1·21	1·18	1·15	1·13	25
26	1·82	1·76	1·70	1·65	1·60	1·55	1·50	1·46	1·42	1·38	1·34	1·30	1·27	1·24	1·21	1·18	26
27	1·90	1·84	1·78	1·72	1·67	1·62	1·57	1·52	1·48	1·44	1·40	1·36	1·33	1·29	1·26	1·23	27
28	1·98	1·92	1·85	1·79	1·74	1·69	1·64	1·59	1·54	1·50	1·46	1·42	1·39	1·35	1·32	1·28	28
29	2·07	2·00	1·93	1·87	1·81	1·76	1·71	1·66	1·61	1·57	1·52	1·48	1·44	1·41	1·37	1·34	29
30	2·15	2·08	2·01	1·95	1·89	1·83	1·78	1·73	1·68	1·63	1·59	1·54	1·50	1·47	1·43	1·39	30
31	2·24	2·17	2·10	2·03	1·97	1·91	1·85	1·80	1·75	1·70	1·65	1·61	1·57	1·53	1·49	1·45	31
32	2·33	2·25	2·18	2·11	2·04	1·98	1·92	1·87	1·81	1·76	1·72	1·67	1·63	1·59	1·55	1·51	32
33	2·42	2·34	2·26	2·19	2·12	2·06	2·00	1·94	1·89	1·83	1·78	1·74	1·69	1·65	1·61	1·57	33
34	2·52	2·43	2·35	2·28	2·21	2·14	2·08	2·02	1·96	1·90	1·85	1·80	1·76	1·71	1·67	1·63	34
35	2·61	2·52	2·44	2·36	2·29	2·22	2·16	2·09	2·03	1·98	1·92	1·87	1·82	1·78	1·73	1·69	35
36	2·71	2·62	2·53	2·45	2·38	2·30	2·24	2·17	2·11	2·05	2·00	1·94	1·89	1·84	1·80	1·75	36
37	2·81	2·72	2·63	2·54	2·46	2·39	2·32	2·25	2·19	2·13	2·07	2·02	1·96	1·91	1·87	1·82	37
38	2·92	2·82	2·72	2·64	2·56	2·48	2·40	2·34	2·27	2·21	2·15	2·09	2·04	1·98	1·93	1·89	38
39	3·02	2·92	2·82	2·73	2·65	2·57	2·49	2·42	2·35	2·29	2·23	2·17	2·11	2·06	2·00	1·95	39
40	3·13	3·03	2·93	2·83	2·74	2·66	2·58	2·51	2·44	2·37	2·31	2·25	2·19	2·13	2·08	2·03	40
41	3·24	3·13	3·03	2·93	2·84	2·76	2·68	2·60	2·52	2·45	2·39	2·33	2·27	2·21	2·15	2·10	41
42	3·36	3·25	3·14	3·04	2·95	2·86	2·77	2·69	2·61	2·54	2·47	2·41	2·35	2·29	2·23	2·17	42
43	3·48	3·36	3·25	3·15	3·05	2·96	2·87	2·79	2·71	2·63	2·56	2·49	2·43	2·37	2·31	2·25	43
44	3·60	3·48	3·37	3·26	3·16	3·06	2·97	2·89	2·80	2·73	2·65	2·58	2·52	2·45	2·39	2·33	44
45	3·73	3·61	3·49	3·38	3·27	3·17	3·08	2·99	2·90	2·82	2·75	2·67	2·61	2·54	2·48	2·41	45
46	3·86	3·73	3·61	3·50	3·39	3·28	3·19	3·09	3·01	2·92	2·85	2·77	2·70	2·63	2·56	2·50	46
47	4·00	3·87	3·74	3·62	3·51	3·40	3·30	3·21	3·11	3·03	2·95	2·87	2·79	2·72	2·65	2·59	47
48	4·14	4·00	3·87	3·75	3·63	3·52	3·42	3·32	3·23	3·14	3·05	2·97	2·89	2·82	2·75	2·68	48
49	4·29	4·15	4·01	3·88	3·76	3·65	3·54	3·44	3·34	3·25	3·16	3·08	3·00	2·92	2·85	2·78	49
50	4·45	4·30	4·16	4·02	3·90	3·78	3·67	3·56	3·46	3·37	3·27	3·19	3·10	3·03	2·95	2·88	50
51	4·61	4·45	4·31	4·17	4·04	3·92	3·80	3·69	3·59	3·49	3·39	3·30	3·22	3·14	3·06	2·98	51
52	4·78	4·62	4·46	4·32	4·19	4·06	3·94	3·83	3·72	3·62	3·52	3·42	3·33	3·25	3·17	3·09	52
53	4·95	4·79	4·63	4·48	4·34	4·21	4·08	3·97	3·86	3·75	3·65	3·55	3·46	3·37	3·28	3·20	53
54	5·14	4·96	4·80	4·65	4·50	4·37	4·24	4·11	4·00	3·89	3·78	3·68	3·59	3·49	3·41	3·32	54
55	5·33	5·15	4·98	4·82	4·67	4·53	4·40	4·27	4·15	4·03	3·92	3·82	3·72	3·63	3·53	3·45	55
56	5·53	5·35	5·17	5·01	4·85	4·70	4·56	4·43	4·31	4·19	4·07	3·97	3·86	3·76	3·67	3·58	56
57	5·75	5·55	5·37	5·20	5·04	4·88	4·74	4·60	4·47	4·35	4·23	4·12	4 01	3 91	3·81	3·72	57
58	5·97	5·77	5·58	5·40	5·23	5·08	4·93	4·78	4·65	4·52	4·40	4·28	4·17	4·06	3·96	3·86	58
59	6·21	6·00	5·80	5·62	5·44	5·28	5·12	4·97	4·83	4·70	4·57	4·45	4·34	4·23	4·12	4·02	59
60	6·46	6·25	6·04	5·85	5·67	5·49	5·33	5·18	5·03	4·89	4·76	4·63	4·51	4·40	4·29	4·18	60
Lat.	165°00' / 195°00'	164°30' / 195°30'	164°00' / 196°00'	163°30' / 196°30'	163°00' / 197°00'	162°30' / 197°30'	162°00' / 198°00'	161°30' / 198°30'	161°00' / 199°00'	160°30' / 199°30'	160°00' / 200°00'	159°30' / 200°30'	159°00' / 201°00'	158°30' / 201°30'	158°00' / 202°00'	157°30' / 202°30'	Lat.

A — HOUR ANGLE — A

B																	B

TABLE B — HOUR ANGLE

Dec.°	15°00'	15°30'	16°00'	16°30'	17°00'	17°30'	18°00'	18°30'	19°00'	19°30'	20°00'	20°30'	21°00'	21°30'	22°00'	22°30'	Dec.°
	345°00'	344°30'	344°00'	343°30'	343°00'	342°30'	342°00'	341°30'	341°00'	340°30'	340°00'	339°30'	339°00'	338°30'	338°00'	337°30'	
0	·00	·00	·00	·00	·00	·00	·00	·00	·00	·00	·00	·00	·00	·00	·00	·00	0
1	·07	·07	·06	·06	·06	·06	·06	·06	·05	·05	·05	·05	·05	·05	·05	·05	1
2	·13	·13	·13	·12	·12	·12	·11	·11	·11	·10	·10	·10	·10	·10	·09	·09	2
3	·20	·20	·19	·19	·18	·17	·17	·17	·16	·16	·15	·15	·15	·14	·14	·14	3
4	·27	·26	·25	·25	·24	·23	·23	·22	·21	·21	·20	·20	·20	·19	·19	·18	4
5	·34	·33	·32	·31	·30	·29	·28	·28	·27	·26	·26	·25	·24	·24	·23	·23	5
6	·41	·39	·38	·37	·36	·35	·34	·33	·32	·31	·31	·30	·29	·29	·28	·27	6
7	·47	·46	·45	·43	·42	·41	·40	·39	·38	·37	·36	·35	·34	·34	·33	·32	7
8	·54	·53	·51	·49	·48	·47	·45	·44	·43	·42	·41	·40	·39	·38	·38	·37	8
9	·61	·59	·57	·56	·54	·53	·51	·50	·49	·47	·46	·45	·44	·43	·42	·41	9
10	·68	·66	·64	·62	·60	·59	·57	·56	·54	·53	·52	·50	·49	·48	·47	·46	10
11	·75	·73	·71	·68	·66	·65	·63	·61	·60	·58	·57	·55	·54	·53	·52	·51	11
12	·82	·80	·77	·75	·73	·71	·69	·67	·65	·64	·62	·61	·59	·58	·57	·56	12
13	·89	·86	·84	·81	·79	·77	·75	·73	·71	·69	·68	·66	·64	·63	·62	·60	13
14	·96	·93	·90	·88	·85	·83	·81	·79	·77	·75	·73	·71	·70	·68	·67	·65	14
15	1·04	1·00	·97	·94	·92	·89	·87	·84	·82	·80	·78	·77	·75	·73	·72	·70	15
16	1·11	1·07	1·04	1·01	·98	·95	·93	·90	·88	·86	·84	·82	·80	·78	·77	·75	16
17	1·18	1·14	1·11	1·08	1·05	1·02	·99	·96	·94	·92	·89	·87	·85	·83	·82	·80	17
18	1·26	1·22	1·18	1·14	1·11	1·08	1·05	1·02	1·00	·97	·95	·93	·91	·89	·87	·85	18
19	1·33	1·29	1·25	1·21	1·18	1·15	1·11	1·09	1·06	1·03	1·01	·98	·96	·94	·92	·90	19
20	1·41	1·36	1·32	1·28	1·24	1·21	1·18	1·15	1·12	1·09	1·06	1·04	1·02	·99	·97	·95	20
21	1·48	1·44	1·39	1·35	1·31	1·28	1·24	1·21	1·18	1·15	1·12	1·10	1·07	1·05	1·02	1·00	21
22	1·56	1·51	1·47	1·42	1·38	1·34	1·31	1·27	1·24	1·21	1·18	1·15	1·13	1·10	1·08	1·06	22
23	1·64	1·59	1·54	1·49	1·45	1·41	1·37	1·34	1·30	1·27	1·24	1·21	1·18	1·16	1·13	1·11	23
24	1·72	1·67	1·62	1·57	1·52	1·48	1·44	1·40	1·37	1·33	1·30	1·27	1·24	1·21	1·19	1·16	24
25	1·80	1·74	1·69	1·64	1·59	1·55	1·51	1·47	1·43	1·40	1·36	1·33	1·30	1·27	1·24	1·22	25
26	1·88	1·83	1·77	1·72	1·67	1·62	1·58	1·54	1·50	1·46	1·43	1·39	1·36	1·33	1·30	1·27	26
27	1·97	1·91	1·85	1·79	1·74	1·69	1·65	1·61	1·57	1·53	1·49	1·45	1·42	1·39	1·36	1·33	27
28	2·05	1·99	1·93	1·87	1·82	1·77	1·72	1·68	1·63	1·59	1·55	1·52	1·48	1·45	1·42	1·39	28
29	2·14	2·07	2·01	1·95	1·90	1·84	1·79	1·75	1·70	1·66	1·62	1·58	1·55	1·51	1·48	1·45	29
30	2·23	2·16	2·09	2·03	1·97	1·92	1·87	1·82	1·77	1·73	1·69	1·65	1·61	1·58	1·54	1·51	30
31	2·32	2·25	2·18	2·12	2·06	2·00	1·94	1·89	1·85	1·80	1·76	1·72	1·68	1·64	1·60	1·57	31
32	2·41	2·34	2·27	2·20	2·14	2·08	2·02	1·97	1·92	1·87	1·83	1·78	1·74	1·70	1·67	1·63	32
33	2·51	2·43	2·36	2·29	2·22	2·16	2·10	2·05	1·99	1·95	1·90	1·85	1·81	1·77	1·73	1·70	33
34	2·61	2·52	2·45	2·37	2·31	2·24	2·18	2·13	2·07	2·02	1·97	1·93	1·88	1·84	1·80	1·76	34
35	2·71	2·62	2·54	2·46	2·39	2·33	2·27	2·21	2·15	2·10	2·05	2·00	1·95	1·91	1·87	1·83	35
36	2·81	2·72	2·64	2·56	2·48	2·42	2·35	2·29	2·23	2·18	2·12	2·07	2·03	1·98	1·94	1·90	36
37	2·91	2·82	2·73	2·65	2·58	2·51	2·44	2·37	2·31	2·26	2·20	2·15	2·10	2·06	2·01	1·97	37
38	3·02	2·92	2·83	2·75	2·67	2·60	2·53	2·46	2·40	2·34	2·28	2·23	2·18	2·13	2·09	2·04	38
39	3·13	3·03	2·94	2·85	2·77	2·69	2·62	2·55	2·49	2·43	2·37	2·31	2·26	2·21	2·16	2·12	39
40	3·24	3·14	3·04	2·95	2·87	2·79	2·72	2·64	2·58	2·51	2·45	2·40	2·34	2·29	2·24	2·19	40
41	3·36	3·25	3·15	3·06	2·97	2·89	2·81	2·74	2·67	2·60	2·54	2·48	2·43	2·37	2·32	2·27	41
42	3·48	3·37	3·27	3·17	3·08	2·99	2·91	2·84	2·77	2·70	2·63	2·57	2·51	2·46	2·40	2·35	42
43	3·60	3·49	3·38	3·28	3·19	3·10	3·02	2·94	2·86	2·79	2·73	2·66	2·60	2·54	2·49	2·44	43
44	3·73	3·61	3·50	3·40	3·30	3·21	3·13	3·04	2·97	2·89	2·82	2·76	2·69	2·63	2·58	2·52	44
45	3·86	3·74	3·63	3·52	3·42	3·33	3·24	3·15	3·07	3·00	2·92	2·86	2·79	2·73	2·67	2·61	45
46	4·00	3·87	3·76	3·65	3·54	3·44	3·35	3·26	3·18	3·10	3·03	2·96	2·89	2·83	2·76	2·71	46
47	4·14	4·01	3·89	3·78	3·67	3·57	3·47	3·38	3·29	3·21	3·14	3·06	2·99	2·93	2·86	2·80	47
48	4·29	4·16	4·03	3·91	3·80	3·69	3·59	3·50	3·41	3·33	3·25	3·17	3·10	3·03	2·96	2·90	48
49	4·44	4·30	4·17	4·05	3·93	3·83	3·72	3·63	3·53	3·45	3·36	3·28	3·21	3·14	3·07	3·01	49
50	4·60	4·46	4·32	4·20	4·08	3·96	3·86	3·76	3·66	3·57	3·48	3·40	3·33	3·25	3·18	3·11	50
51	4·77	4·62	4·48	4·35	4·22	4·11	4·00	3·89	3·79	3·70	3·61	3·53	3·45	3·37	3·30	3·23	51
52	4·95	4·79	4·64	4·51	4·38	4·26	4·14	4·03	3·93	3·83	3·74	3·65	3·57	3·49	3·42	3·35	52
53	5·13	4·97	4·81	4·67	4·54	4·42	4·29	4·18	4·08	3·98	3·88	3·79	3·70	3·62	3·54	3·47	53
54	5·32	5·15	4·99	4·85	4·71	4·58	4·45	4·34	4·23	4·12	4·02	3·93	3·84	3·76	3·67	3·60	54
55	5·52	5·34	5·18	5·03	4·88	4·75	4·62	4·50	4·39	4·28	4·18	4·08	3·99	3·90	3·81	3·73	55
56	5·73	5·55	5·38	5·22	5·07	4·93	4·80	4·67	4·55	4·44	4·34	4·23	4·14	4·05	3·96	3·87	56
57	5·95	5·76	5·59	5·42	5·27	5·12	4·98	4·85	4·73	4·61	4·50	4·40	4·30	4·20	4·11	4·02	57
58	6·18	5·99	5·81	5·63	5·47	5·32	5·18	5·04	4·92	4·79	4·68	4·57	4·47	4·37	4·27	4·18	58
59	6·43	6·23	6·04	5·86	5·69	5·54	5·39	5·25	5·11	4·99	4·87	4·75	4·64	4·54	4·44	4·35	59
60	6·69	6·48	6·28	6·10	5·92	5·76	5·61	5·46	5·32	5·19	5·06	4·95	4·83	4·73	4·62	4·53	60
Dec.	165°00'	164°30'	164°00'	163°30'	163°00'	162°30'	162°00'	161°30'	161°00'	160°30'	160°00'	159°30'	159°00'	158°30'	158°00'	157°30'	Dec.
	195°00'	195°30'	196°00'	196°30'	197°00'	197°30'	198°00'	198°30'	199°00'	199°30'	200°00'	200°30'	201°00'	201°30'	202°00'	202°30'	

B								HOUR ANGLE									B

| A | | | | | | | TABLE A — HOUR ANGLE | | | | | | | | | A |

A — Named opposite to Latitude, **except** when Hour Angle is between 90° and 270°

Lat.	22° 30′	23° 00′	23° 30′	24° 00′	24° 30′	25° 00′	25° 30′	26° 00′	26° 30′	27° 00′	27° 30′	28° 00′	28° 30′	29° 00′	29° 30′	30° 00′	Lat.
°	337° 30′	337° 00′	336° 30′	336° 00′	335° 30′	335° 00′	334° 30′	334° 00′	333° 30′	333° 00′	332° 30′	332° 00′	331° 30′	331° 00′	330° 30′	330° 00′	°
0	·00	·00	·00	·00	·00	·00	·00	·00	·00	·00	·00	·00	·00	·00	·00	·00	0
1	·04	·04	·04	·04	·04	·04	·04	·04	·04	·03	·03	·03	·03	·03	·03	·03	1
2	·08	·08	·08	·08	·08	·08	·07	·07	·07	·07	·07	·07	·06	·06	·06	·06	2
3	·13	·12	·12	·12	·12	·11	·11	·11	·11	·10	·10	·10	·10	·09	·09	·09	3
4	·17	·16	·16	·16	·15	·15	·15	·14	·14	·14	·13	·13	·13	·13	·12	·12	4
5	·21	·21	·20	·20	·19	·19	·18	·18	·18	·17	·17	·16	·16	·16	·16	·15	5
6	·25	·25	·24	·24	·23	·23	·22	·22	·21	·21	·20	·20	·19	·19	·19	·18	6
7	·30	·29	·28	·28	·27	·26	·26	·25	·25	·24	·24	·23	·23	·22	·22	·21	7
8	·34	·33	·32	·32	·31	·30	·29	·29	·28	·28	·27	·26	·26	·25	·25	·24	8
9	·38	·37	·36	·36	·35	·34	·33	·32	·32	·31	·30	·30	·29	·29	·28	·27	9
10	·43	·42	·41	·40	·39	·38	·37	·36	·35	·35	·34	·33	·33	·32	·31	·31	10
11	·47	·46	·45	·44	·43	·42	·41	·40	·39	·38	·37	·37	·36	·35	·34	·34	11
12	·51	·50	·49	·48	·47	·46	·45	·44	·43	·42	·41	·40	·39	·38	·38	·37	12
13	·56	·54	·53	·52	·51	·50	·48	·47	·46	·45	·44	·43	·43	·42	·41	·40	13
14	·60	·59	·57	·56	·55	·53	·52	·51	·50	·49	·48	·47	·46	·45	·44	·43	14
15	·65	·63	·62	·60	·59	·57	·56	·55	·54	·53	·51	·50	·49	·48	·47	·46	15
16	·69	·68	·66	·64	·63	·61	·60	·59	·58	·56	·55	·54	·53	·52	·51	·50	16
17	·74	·72	·70	·69	·67	·66	·64	·63	·61	·60	·59	·58	·56	·55	·54	·53	17
18	·78	·77	·75	·73	·71	·70	·68	·67	·65	·64	·62	·61	·60	·59	·57	·56	18
19	·83	·81	·79	·77	·76	·74	·72	·71	·69	·68	·66	·65	·63	·62	·61	·60	19
20	·88	·86	·84	·82	·80	·78	·76	·75	·73	·71	·70	·68	·67	·66	·64	·63	20
21	·93	·90	·88	·86	·84	·82	·81	·79	·77	·75	·74	·72	·71	·69	·68	·66	21
22	·97	·95	·93	·91	·89	·87	·85	·83	·81	·79	·78	·76	·74	·73	·71	·70	22
23	1·02	1·00	·98	·95	·93	·91	·89	·87	·85	·83	·82	·80	·78	·77	·75	·74	23
24	1·07	1·05	1·02	1·00	·98	·95	·93	·91	·89	·87	·86	·84	·82	·80	·79	·77	24
25	1·13	1·10	1·07	1·05	1·02	1·00	·98	·96	·94	·92	·90	·88	·86	·84	·82	·81	25
26	1·18	1·15	1·12	1·10	1·07	1·05	1·02	1·00	·98	·96	·94	·92	·90	·88	·86	·84	26
27	1·23	1·20	1·17	1·14	1·12	1·09	1·07	1·04	1·02	1·00	·98	·96	·94	·92	·90	·88	27
28	1·28	1·25	1·22	1·19	1·17	1·14	1·11	1·09	1·07	1·04	1·02	1·00	·98	·96	·94	·92	28
29	1·34	1·31	1·27	1·25	1·22	1·19	1·16	1·14	1·11	1·09	1·06	1·04	1·02	1·00	·98	·96	29
30	1·39	1·36	1·33	1·30	1·27	1·24	1·21	1·18	1·16	1·13	1·11	1·09	1·06	1·04	1·02	1·00	30
31	1·45	1·42	1·38	1·35	1·32	1·29	1·26	1·23	1·21	1·18	1·15	1·13	1·11	1·08	1·06	1·04	31
32	1·51	1·47	1·44	1·40	1·37	1·34	1·31	1·28	1·25	1·23	1·20	1·18	1·15	1·13	1·10	1·08	32
33	1·57	1·53	1·49	1·46	1·42	1·39	1·36	1·33	1·30	1·27	1·25	1·22	1·20	1·17	1·15	1·12	33
34	1·63	1·59	1·55	1·51	1·48	1·45	1·41	1·38	1·35	1·32	1·30	1·27	1·24	1·22	1·19	1·17	34
35	1·69	1·65	1·61	1·57	1·54	1·50	1·47	1·44	1·40	1·37	1·35	1·32	1·29	1·26	1·24	1·21	35
36	1·75	1·71	1·67	1·63	1·59	1·56	1·52	1·49	1·46	1·43	1·40	1·37	1·34	1·31	1·28	1·26	36
37	1·82	1·78	1·73	1·69	1·65	1·62	1·58	1·55	1·51	1·48	1·45	1·42	1·39	1·36	1·33	1·31	37
38	1·89	1·84	1·80	1·75	1·71	1·68	1·64	1·60	1·57	1·53	1·50	1·47	1·44	1·41	1·38	1·35	38
39	1·95	1·91	1·86	1·82	1·78	1·74	1·70	1·66	1·62	1·59	1·56	1·52	1·49	1·46	1·43	1·40	39
40	2·03	1·98	1·93	1·88	1·84	1·80	1·76	1·72	1·68	1·65	1·61	1·58	1·55	1·51	1·48	1·45	40
41	2·10	2·05	2·00	1·95	1·91	1·86	1·82	1·78	1·74	1·71	1·67	1·63	1·60	1·57	1·54	1·51	41
42	2·17	2·12	2·07	2·02	1·98	1·93	1·89	1·85	1·81	1·77	1·73	1·69	1·66	1·62	1·59	1·56	42
43	2·25	2·20	2·14	2·09	2·05	2·00	1·96	1·91	1·87	1·83	1·79	1·75	1·72	1·68	1·65	1·62	43
44	2·33	2·28	2·22	2·17	2·12	2·07	2·02	1·98	1·94	1·90	1·86	1·82	1·78	1·74	1·71	1·67	44
45	2·41	2·36	2·30	2·25	2·19	2·14	2·10	2·05	2·01	1·96	1·92	1·88	1·84	1·80	1·77	1·73	45
46	2·50	2·44	2·38	2·33	2·27	2·22	2·17	2·12	2·08	2·03	1·99	1·95	1·91	1·87	1·83	1·79	46
47	2·59	2·53	2·47	2·41	2·35	2·30	2·25	2·20	2·15	2·10	2·06	2·02	1·98	1·93	1·90	1·86	47
48	2·68	2·62	2·55	2·49	2·44	2·38	2·33	2·28	2·23	2·18	2·13	2·09	2·05	2·00	1·96	1·92	48
49	2·78	2·71	2·65	2·58	2·52	2·47	2·41	2·36	2·31	2·26	2·21	2·16	2·12	2·08	2·03	1·99	49
50	2·89	2·81	2·74	2·68	2·62	2·56	2·50	2·44	2·39	2·34	2·29	2·24	2·19	2·15	2·11	2·06	50
51	2·98	2·91	2·84	2·77	2·71	2·65	2·59	2·53	2·48	2·42	2·37	2·32	2·27	2·23	2·18	2·14	51
52	3·09	3·02	2·94	2·87	2·81	2·74	2·68	2·62	2·57	2·51	2·46	2·41	2·36	2·31	2·26	2·22	52
53	3·20	3·13	3·05	2·98	2·91	2·85	2·78	2·72	2·66	2·60	2·55	2·50	2·44	2·39	2·35	2·30	53
54	3·32	3·24	3·17	3·09	3·02	2·95	2·89	2·82	2·76	2·70	2·64	2·59	2·53	2·48	2·43	2·38	54
55	3·45	3·36	3·28	3·21	3·13	3·06	2·99	2·93	2·86	2·80	2·74	2·69	2·63	2·58	2·52	2·47	55
56	3·58	3·49	3·41	3·33	3·25	3·18	3·11	3·04	2·97	2·91	2·85	2·79	2·73	2·67	2·62	2·57	56
57	3·72	3·63	3·54	3·46	3·38	3·30	3·23	3·16	3·09	3·02	2·96	2·90	2·84	2·78	2·72	2·67	57
58	3·86	3·77	3·68	3·59	3·51	3·43	3·36	3·28	3·21	3·14	3·07	3·01	2·95	2·89	2·83	2·77	58
59	4·02	3·92	3·83	3·74	3·65	3·57	3·49	3·41	3·34	3·27	3·20	3·13	3·07	3·00	2·94	2·88	59
60	4·18	4·08	3·98	3·89	3·80	3·71	3·63	3·55	3·47	3·40	3·33	3·26	3·19	3·12	3·06	3·00	60
Lat.	157° 30′	157° 00′	156° 30′	156° 00′	155° 30′	155° 00′	154° 30′	154° 00′	153° 30′	153° 00′	152° 30′	152° 00′	151° 30′	151° 00′	150° 30′	150° 00′	Lat.
	202° 30′	203° 00′	203° 30′	204° 00′	204° 30′	205° 00′	205° 30′	206° 00′	206° 30′	207° 00′	207° 30′	208° 00′	208° 30′	209° 00′	209° 30′	210° 00′	

| A | | | | | | | HOUR ANGLE | | | | | | | | | A |

| B | TABLE B — HOUR ANGLE | | | | | | | | | | | | | | | | B |

Dec.°	22°30' 337°30'	23°00' 337°00'	23°30' 336°30'	24°00' 336°00'	24°30' 335°30'	25°00' 335°00'	25°30' 334°30'	26°00' 334°00'	26°30' 333°30'	27°00' 333°00'	27°30' 332°30'	28°00' 332°00'	28°30' 331°30'	29°00' 331°00'	29°30' 330°30'	30°00' 330°00'	Dec.°
0	·00	·00	·00	·00	·00	·00	·00	·00	·00	·00	·00	·00	·00	·00	·00	·00	0
1	·05	·04	·04	·04	·04	·04	·04	·04	·04	·04	·04	·04	·04	·04	·04	·03	1
2	·09	·09	·09	·09	·08	·08	·08	·08	·08	·08	·08	·07	·07	·07	·07	·07	2
3	·14	·13	·13	·13	·13	·12	·12	·12	·12	·12	·11	·11	·11	·11	·11	·10	3
4	·18	·18	·18	·17	·17	·16	·16	·16	·15	·15	·15	·15	·14	·14	·14	·14	4
5	·23	·22	·22	·22	·21	·21	·20	·20	·20	·19	·19	·19	·18	·18	·18	·17	5
6	·27	·27	·26	·26	·25	·25	·24	·24	·24	·23	·23	·22	·22	·22	·21	·21	6
7	·32	·31	·31	·30	·30	·29	·29	·28	·28	·27	·27	·26	·26	·25	·25	·25	7
8	·37	·36	·35	·35	·34	·33	·33	·32	·31	·31	·30	·30	·29	·29	·29	·28	8
9	·41	·41	·40	·39	·38	·37	·37	·35	·35	·35	·34	·34	·33	·33	·32	·32	9
10	·46	·45	·44	·43	·43	·42	·41	·40	·40	·39	·38	·38	·37	·36	·36	·35	10
11	·51	·50	·49	·48	·47	·46	·45	·44	·44	·43	·42	·41	·41	·40	·39	·39	11
12	·56	·54	·53	·52	·51	·50	·49	·48	·48	·47	·46	·45	·45	·44	·43	·43	12
13	·60	·59	·58	·57	·56	·55	·54	·53	·52	·51	·50	·49	·48	·48	·47	·46	13
14	·65	·64	·63	·61	·60	·59	·58	·57	·57	·55	·54	·53	·52	·51	·51	·50	14
15	·70	·69	·67	·66	·65	·63	·62	·61	·60	·59	·58	·57	·56	·55	·54	·54	15
16	·75	·73	·72	·71	·69	·68	·67	·65	·64	·63	·62	·61	·60	·59	·58	·57	16
17	·80	·78	·77	·75	·74	·72	·71	·70	·69	·67	·66	·65	·64	·63	·62	·61	17
18	·85	·83	·81	·80	·78	·77	·75	·74	·73	·72	·70	·69	·68	·67	·66	·65	18
19	·90	·88	·86	·85	·83	·81	·80	·79	·77	·76	·75	·73	·72	·71	·70	·69	19
20	·95	·93	·91	·89	·88	·86	·85	·83	·82	·80	·79	·78	·76	·75	·74	·73	20
21	1·00	·98	·96	·94	·93	·91	·89	·88	·86	·85	·83	·82	·80	·79	·78	·77	21
22	1·06	1·03	1·01	·99	·97	·96	·94	·92	·91	·89	·87	·86	·85	·83	·82	·81	22
23	1·11	1·09	1·06	1·04	1·02	1·00	·99	·97	·95	·93	·92	·90	·89	·88	·86	·85	23
24	1·16	1·14	1·12	1·09	1·07	1·05	1·03	1·02	1·00	·98	·96	·95	·93	·92	·90	·89	24
25	1·22	1·19	1·17	1·15	1·12	1·10	1·08	1·06	1·05	1·03	1·01	·99	·98	·96	·95	·93	25
26	1·27	1·25	1·22	1·20	1·18	1·15	1·13	1·11	1·09	1·07	1·06	1·04	1·02	1·01	·99	·98	26
27	1·33	1·30	1·28	1·25	1·23	1·21	1·18	1·16	1·14	1·12	1·10	1·09	1·07	1·05	1·03	1·02	27
28	1·39	1·36	1·33	1·31	1·28	1·26	1·24	1·21	1·19	1·17	1·15	1·13	1·11	1·10	1·08	1·06	28
29	1·45	1·42	1·39	1·36	1·34	1·31	1·29	1·26	1·24	1·22	1·20	1·18	1·16	1·14	1·13	1·11	29
30	1·51	1·48	1·45	1·42	1·39	1·37	1·34	1·32	1·29	1·27	1·25	1·23	1·21	1·19	1·17	1·15	30
31	1·57	1·54	1·51	1·48	1·45	1·42	1·40	1·37	1·35	1·32	1·30	1·28	1·26	1·24	1·22	1·20	31
32	1·63	1·60	1·57	1·54	1·51	1·48	1·45	1·43	1·40	1·38	1·35	1·33	1·31	1·29	1·27	1·25	32
33	1·70	1·66	1·63	1·60	1·57	1·54	1·51	1·48	1·46	1·43	1·41	1·38	1·36	1·34	1·32	1·30	33
34	1·76	1·73	1·69	1·66	1·63	1·60	1·57	1·54	1·51	1·49	1·46	1·44	1·41	1·39	1·37	1·35	34
35	1·83	1·79	1·76	1·72	1·69	1·66	1·63	1·60	1·57	1·54	1·52	1·49	1·47	1·44	1·42	1·40	35
36	1·90	1·86	1·82	1·79	1·75	1·72	1·69	1·66	1·63	1·60	1·57	1·55	1·52	1·50	1·48	1·45	36
37	1·97	1·93	1·89	1·85	1·82	1·78	1·75	1·72	1·69	1·66	1·63	1·61	1·58	1·55	1·53	1·51	37
38	2·04	2·00	1·96	1·92	1·88	1·85	1·81	1·78	1·75	1·72	1·69	1·66	1·64	1·61	1·59	1·56	38
39	2·12	2·07	2·03	1·99	1·95	1·92	1·88	1·85	1·81	1·78	1·75	1·72	1·70	1·67	1·64	1·62	39
40	2·19	2·15	2·10	2·06	2·02	1·99	1·95	1·91	1·88	1·85	1·82	1·79	1·76	1·73	1·70	1·68	40
41	2·27	2·22	2·18	2·14	2·10	2·06	2·02	1·98	1·95	1·91	1·88	1·85	1·82	1·79	1·77	1·74	41
42	2·35	2·30	2·26	2·21	2·17	2·13	2·09	2·05	2·02	1·98	1·95	1·92	1·89	1·86	1·83	1·80	42
43	2·44	2·39	2·34	2·29	2·25	2·21	2·17	2·13	2·09	2·05	2·02	1·99	1·95	1·92	1·89	1·87	43
44	2·52	2·47	2·42	2·37	2·33	2·29	2·24	2·20	2·16	2·13	2·09	2·06	2·02	1·99	1·96	1·93	44
45	2·61	2·56	2·51	2·46	2·41	2·37	2·32	2·28	2·24	2·20	2·17	2·13	2·10	2·06	2·03	2·00	45
46	2·71	2·65	2·60	2·55	2·50	2·45	2·41	2·36	2·32	2·28	2·24	2·21	2·17	2·14	2·10	2·07	46
47	2·80	2·74	2·69	2·64	2·59	2·54	2·49	2·45	2·40	2·36	2·32	2·28	2·25	2·21	2·18	2·14	47
48	2·90	2·84	2·79	2·73	2·68	2·63	2·58	2·53	2·49	2·45	2·41	2·37	2·33	2·29	2·26	2·22	48
49	3·01	2·94	2·88	2·83	2·77	2·72	2·67	2·62	2·58	2·53	2·49	2·45	2·41	2·37	2·34	2·30	49
50	3·11	3·05	2·99	2·93	2·87	2·82	2·77	2·72	2·67	2·63	2·58	2·54	2·50	2·46	2·42	2·38	50
51	3·23	3·16	3·10	3·04	2·98	2·92	2·87	2·82	2·77	2·72	2·67	2·63	2·59	2·55	2·51	2·47	51
52	3·35	3·28	3·21	3·15	3·09	3·03	2·97	2·92	2·87	2·82	2·77	2·73	2·68	2·64	2·60	2·56	52
53	3·47	3·40	3·33	3·26	3·20	3·14	3·08	3·03	2·97	2·92	2·87	2·83	2·78	2·74	2·69	2·65	53
54	3·60	3·52	3·45	3·38	3·32	3·26	3·20	3·14	3·08	3·03	2·98	2·93	2·88	2·84	2·80	2·75	54
55	3·73	3·66	3·58	3·51	3·44	3·38	3·32	3·26	3·20	3·15	3·09	3·04	2·99	2·95	2·90	2·86	55
56	3·87	3·79	3·72	3·65	3·58	3·51	3·44	3·38	3·32	3·27	3·21	3·16	3·11	3·06	3·01	2·97	56
57	4·02	3·94	3·86	3·79	3·71	3·64	3·58	3·51	3·45	3·39	3·34	3·28	3·23	3·18	3·13	3·08	57
58	4·18	4·10	4·01	3·93	3·86	3·79	3·72	3·65	3·59	3·53	3·47	3·41	3·35	3·30	3·25	3·20	58
59	4·35	4·26	4·17	4·09	4·01	3·94	3·87	3·80	3·73	3·67	3·60	3·55	3·49	3·43	3·38	3·33	59
60	4·53	4·43	4·34	4·26	4·18	4·10	4·02	3·95	3·88	3·82	3·75	3·69	3·63	3·57	3·52	3·46	60
Dec.	157°30'	157°00'	156°30'	156°00'	155°30'	155°00'	154°30'	154°00'	153°30'	153°00'	152°30'	152°00'	151°30'	151°00'	150°30'	150°00'	Dec.
	202°30'	203°00'	203°30'	204°00'	204°30'	205°00'	205°30'	206°00'	206°30'	207°00'	207°30'	208°00'	208°30'	209°00'	209°30'	210°00'	

| B | HOUR ANGLE | | | | | | | | | | | | | | | | B |

B—Always named the same as Declination.

| A | TABLE A — HOUR ANGLE | A |

A—Named opposite to Latitude, **except** when Hour Angle is between 90° and 270°

Lat.°	30° / 330°	31° / 329°	32° / 328°	33° / 327°	34° / 326°	35° / 325°	36° / 324°	37° / 323°	38° / 322°	39° / 321°	40° / 320°	41° / 319°	42° / 318°	43° / 317°	44° / 316°	45° / 315°	Lat.°
0	·00	·00	·00	·00	·00	·00	·00	·00	·00	·00	·00	·00	·00	·00	·00	·00	0
1	·03	·03	·03	·03	·03	·02	·02	·02	·02	·02	·02	·02	·02	·02	·02	·02	1
2	·06	·06	·06	·05	·05	·05	·05	·05	·05	·04	·04	·04	·04	·04	·04	·04	2
3	·09	·09	·08	·08	·08	·07	·07	·07	·07	·06	·06	·06	·06	·06	·06	·05	3
4	·12	·12	·11	·11	·10	·10	·10	·10	·09	·09	·08	·08	·08	·07	·07	·07	4
5	·15	·15	·14	·13	·13	·12	·12	·12	·11	·11	·10	·10	·10	·09	·09	·09	5
6	·18	·17	·17	·16	·16	·15	·14	·14	·13	·13	·13	·12	·12	·11	·11	·11	6
7	·21	·20	·20	·19	·18	·18	·17	·16	·16	·15	·15	·14	·14	·13	·13	·12	7
8	·24	·23	·22	·22	·21	·21	·19	·19	·18	·18	·17	·16	·16	·15	·15	·14	8
9	·27	·26	·25	·24	·23	·23	·22	·21	·20	·20	·19	·18	·18	·17	·16	·16	9
10	·31	·29	·28	·27	·26	·25	·24	·23	·23	·22	·21	·20	·20	·19	·18	·18	10
11	·34	·32	·31	·30	·29	·28	·27	·26	·25	·24	·23	·22	·22	·21	·20	·19	11
12	·37	·35	·34	·33	·32	·30	·29	·28	·27	·26	·25	·24	·24	·23	·22	·21	12
13	·40	·38	·37	·36	·34	·33	·32	·31	·30	·29	·28	·27	·26	·25	·24	·23	13
14	·43	·41	·40	·38	·37	·36	·34	·33	·32	·31	·30	·29	·27	·27	·26	·25	14
15	·46	·45	·43	·41	·40	·38	·37	·36	·34	·33	·32	·31	·30	·29	·28	·27	15
16	·50	·48	·46	·44	·43	·41	·39	·38	·37	·35	·34	·33	·32	·31	·30	·29	16
17	·53	·51	·49	·47	·45	·44	·42	·41	·39	·38	·36	·35	·34	·33	·32	·31	17
18	·56	·54	·52	·50	·48	·46	·45	·43	·42	·40	·39	·37	·36	·35	·34	·32	18
19	·60	·57	·55	·53	·51	·49	·47	·46	·44	·43	·41	·40	·38	·37	·36	·34	19
20	·63	·61	·58	·56	·54	·52	·50	·48	·47	·45	·43	·42	·40	·39	·38	·36	20
21	·66	·64	·61	·59	·57	·55	·53	·51	·49	·47	·46	·44	·43	·41	·40	·38	21
22	·70	·67	·65	·62	·60	·58	·56	·54	·52	·50	·48	·46	·45	·43	·42	·40	22
23	·74	·71	·68	·65	·63	·61	·58	·56	·54	·52	·51	·49	·47	·45	·44	·42	23
24	·77	·74	·71	·69	·66	·64	·61	·59	·57	·55	·53	·51	·49	·48	·46	·45	24
25	·81	·78	·75	·72	·69	·67	·64	·62	·60	·58	·56	·54	·52	·50	·48	·47	25
26	·84	·81	·78	·75	·72	·70	·67	·65	·62	·60	·58	·56	·54	·52	·51	·49	26
27	·88	·85	·82	·78	·76	·73	·70	·68	·65	·63	·61	·59	·57	·55	·53	·51	27
28	·92	·88	·85	·82	·79	·76	·73	·71	·68	·66	·63	·61	·59	·57	·55	·53	28
29	·96	·92	·89	·85	·82	·79	·76	·74	·71	·68	·66	·64	·62	·59	·57	·55	29
30	1·00	·96	·92	·89	·86	·82	·79	·77	·74	·71	·69	·66	·64	·62	·60	·58	30
31	1·04	1·00	·96	·93	·89	·86	·83	·80	·77	·74	·72	·69	·67	·64	·62	·60	31
32	1·08	1·04	1·00	·96	·93	·89	·86	·83	·80	·77	·74	·72	·69	·67	·65	·62	32
33	1·12	1·08	1·04	1·00	·96	·93	·89	·86	·83	·80	·77	·75	·72	·70	·67	·65	33
34	1·17	1·12	1·08	1·04	1·00	·96	·93	·90	·86	·83	·80	·78	·75	·72	·70	·67	34
35	1·21	1·17	1·12	1·08	1·04	1·00	·96	·93	·90	·86	·83	·81	·78	·75	·73	·70	35
36	1·26	1·21	1·16	1·12	1·08	1·04	1·00	·96	·93	·90	·87	·84	·81	·78	·75	·73	36
37	1·31	1·25	1·21	1·16	1·12	1·08	1·04	1·00	·96	·93	·90	·87	·84	·81	·78	·75	37
38	1·35	1·30	1·25	1·20	1·16	1·12	1·08	1·04	1·00	·96	·93	·90	·87	·84	·81	·78	38
39	1·40	1·35	1·30	1·25	1·20	1·16	1·11	1·07	1·04	1·00	·97	·93	·90	·87	·84	·81	39
40	1·45	1·40	1·34	1·29	1·24	1·20	1·15	1·11	1·07	1·04	1·00	·97	·93	·90	·87	·84	40
41	1·51	1·45	1·39	1·34	1·29	1·24	1·20	1·15	1·11	1·07	1·04	1·00	·97	·93	·90	·87	41
42	1·56	1·50	1·44	1·39	1·33	1·29	1·24	1·19	1·15	1·11	1·07	1·04	1·00	·97	·93	·90	42
43	1·62	1·55	1·49	1·44	1·38	1·33	1·28	1·24	1·19	1·15	1·11	1·07	1·04	1·00	·97	·93	43
44	1·67	1·61	1·55	1·49	1·43	1·38	1·33	1·28	1·24	1·19	1·15	1·11	1·07	1·04	1·00	·97	44
45	1·73	1·66	1·60	1·54	1·48	1·43	1·38	1·33	1·28	1·23	1·19	1·15	1·11	1·07	1·04	1·00	45
46	1·79	1·72	1·66	1·59	1·54	1·48	1·43	1·37	1·33	1·28	1·23	1·19	1·15	1·11	1·07	1·04	46
47	1·86	1·78	1·72	1·65	1·59	1·53	1·48	1·42	1·37	1·32	1·28	1·23	1·19	1·15	1·11	1·07	47
48	1·92	1·85	1·78	1·71	1·65	1·59	1·53	1·47	1·42	1·37	1·32	1·28	1·23	1·19	1·15	1·11	48
49	1·99	1·91	1·84	1·77	1·71	1·64	1·58	1·53	1·47	1·42	1·37	1·32	1·28	1·23	1·19	1·15	49
50	2·06	1·98	1·91	1·84	1·77	1·70	1·64	1·58	1·53	1·47	1·42	1·37	1·32	1·28	1·23	1·19	50
51	2·14	2·06	1·98	1·90	1·83	1·76	1·70	1·64	1·58	1·52	1·47	1·42	1·37	1·32	1·28	1·23	51
52	2·22	2·13	2·05	1·97	1·90	1·83	1·76	1·70	1·64	1·58	1·52	1·47	1·42	1·37	1·33	1·28	52
53	2·30	2·21	2·12	2·04	1·97	1·90	1·83	1·76	1·70	1·64	1·58	1·52	1·47	1·42	1·37	1·33	53
54	2·38	2·29	2·20	2·12	2·04	1·97	1·89	1·83	1·76	1·70	1·64	1·58	1·53	1·48	1·43	1·38	54
55	2·47	2·38	2·29	2·20	2·12	2·04	1·97	1·90	1·83	1·76	1·70	1·64	1·59	1·53	1·48	1·43	55
56	2·57	2·47	2·37	2·28	2·20	2·12	2·04	1·97	1·90	1·83	1·77	1·71	1·65	1·59	1·54	1·48	56
57	2·67	2·56	2·46	2·37	2·28	2·20	2·12	2·04	1·97	1·90	1·84	1·77	1·71	1·65	1·59	1·54	57
58	2·77	2·66	2·56	2·46	2·37	2·29	2·20	2·12	2·05	1·98	1·91	1·84	1·78	1·72	1·66	1·60	58
59	2·88	2·77	2·66	2·56	2·47	2·38	2·29	2·21	2·13	2·06	1·98	1·91	1·85	1·78	1·72	1·66	59
60	3·00	2·88	2·77	2·67	2·57	2·47	2·38	2·30	2·22	2·14	2·06	1·99	1·92	1·86	1·79	1·73	60
Lat.	150° / 210°	149° / 211°	148° / 212°	147° / 213°	146° / 214°	145° / 215°	144° / 216°	143° / 217°	142° / 218°	141° / 219°	140° / 220°	139° / 221°	138° / 222°	137° / 223°	136° / 224°	135° / 225°	Lat.

| A | HOUR ANGLE | A |

TABLE B — HOUR ANGLE

Dec.°	30° 330°	31° 329°	32° 328°	33° 327°	34° 326°	35° 325°	36° 324°	37° 323°	38° 322°	39° 321°	40° 320°	41° 319°	42° 318°	43° 317°	44° 316°	45° 315°	Dec.°
0	·00	·00	·00	·00	·00	·00	·00	·00	·00	·00	·00	·00	·00	·00	·00	·00	0
1	·03	·03	·03	·03	·03	·03	·03	·03	·03	·03	·03	·03	·03	·03	·03	·02	1
2	·07	·07	·07	·06	·06	·06	·06	·06	·06	·06	·05	·05	·05	·05	·05	·05	2
3	·10	·10	·10	·10	·09	·09	·09	·09	·09	·09	·08	·08	·08	·08	·08	·07	3
4	·14	·14	·13	·13	·13	·12	·12	·12	·11	·11	·11	·11	·10	·10	·10	·10	4
5	·17	·17	·17	·16	·16	·15	·15	·15	·14	·14	·14	·13	·13	·13	·13	·12	5
6	·21	·20	·20	·19	·19	·18	·18	·17	·17	·17	·16	·16	·16	·15	·15	·15	6
7	·25	·24	·23	·23	·22	·21	·21	·20	·20	·20	·19	·19	·18	·18	·18	·17	7
8	·28	·27	·27	·26	·25	·25	·24	·23	·23	·22	·22	·21	·21	·21	·20	·20	8
9	·32	·31	·30	·29	·28	·28	·27	·26	·26	·25	·25	·24	·24	·23	·23	·22	9
10	·35	·34	·33	·32	·32	·31	·30	·29	·29	·28	·27	·27	·26	·26	·25	·25	10
11	·39	·38	·37	·36	·35	·34	·33	·32	·32	·31	·30	·30	·29	·29	·28	·27	11
12	·43	·41	·40	·39	·38	·37	·36	·35	·35	·34	·33	·32	·32	·31	·31	·30	12
13	·46	·45	·44	·42	·41	·40	·39	·38	·37	·37	·36	·35	·35	·34	·33	·33	13
14	·50	·48	·47	·46	·45	·43	·42	·41	·40	·40	·39	·38	·37	·37	·36	·35	14
15	·54	·52	·51	·49	·48	·47	·46	·45	·44	·43	·42	·41	·40	·39	·39	·38	15
16	·57	·56	·54	·53	·51	·50	·49	·48	·47	·46	·45	·44	·43	·42	·41	·41	16
17	·61	·59	·58	·56	·55	·53	·52	·51	·50	·49	·48	·47	·46	·45	·44	·43	17
18	·65	·63	·61	·60	·58	·57	·55	·54	·53	·52	·51	·50	·49	·48	·47	·46	18
19	·69	·67	·65	·63	·62	·60	·59	·57	·56	·55	·54	·52	·51	·50	·50	·49	19
20	·73	·71	·69	·67	·65	·63	·62	·60	·59	·58	·57	·55	·54	·53	·52	·51	20
21	·77	·75	·72	·70	·69	·67	·65	·64	·62	·61	·60	·59	·57	·56	·55	·54	21
22	·81	·78	·76	·74	·72	·70	·69	·67	·66	·64	·63	·62	·60	·59	·58	·57	22
23	·85	·82	·80	·78	·76	·74	·72	·71	·69	·67	·66	·65	·63	·62	·61	·60	23
24	·89	·86	·84	·82	·80	·78	·76	·74	·72	·71	·69	·68	·67	·65	·64	·63	24
25	·93	·91	·88	·86	·83	·81	·79	·77	·76	·74	·73	·71	·70	·68	·67	·66	25
26	·98	·95	·92	·90	·87	·85	·83	·81	·79	·78	·76	·74	·73	·72	·70	·69	26
27	1·02	·99	·96	·94	·91	·89	·87	·85	·83	·81	·79	·78	·76	·75	·73	·72	27
28	1·06	1·03	1·00	·98	·95	·93	·90	·88	·86	·84	·83	·81	·79	·78	·77	·75	28
29	1·11	1·08	1·05	1·02	·99	·97	·94	·92	·90	·88	·86	·84	·83	·81	·80	·78	29
30	1·15	1·12	1·09	1·06	1·03	1·01	·98	·96	·94	·92	·90	·88	·86	·85	·83	·82	30
31	1·20	1·17	1·13	1·10	1·07	1·05	1·02	1·00	·98	·95	·93	·92	·90	·88	·87	·85	31
32	1·25	1·21	1·18	1·15	1·12	1·09	1·06	1·04	1·01	·99	·97	·95	·93	·92	·90	·88	32
33	1·30	1·26	1·23	1·19	1·16	1·13	1·11	1·08	1·05	1·03	1·01	·99	·97	·95	·93	·92	33
34	1·35	1·31	1·27	1·24	1·21	1·18	1·15	1·12	1·10	1·07	1·05	1·03	1·01	·99	·97	·96	34
35	1·40	1·36	1·32	1·29	1·25	1·22	1·19	1·16	1·14	1·11	1·09	1·07	1·05	1·03	1·01	·99	35
36	1·45	1·41	1·37	1·33	1·30	1·27	1·24	1·21	1·18	1·15	1·13	1·11	1·09	1·07	1·05	1·03	36
37	1·51	1·46	1·42	1·38	1·35	1·31	1·28	1·25	1·22	1·20	1·17	1·15	1·13	1·10	1·08	1·07	37
38	1·56	1·52	1·47	1·43	1·40	1·36	1·33	1·30	1·27	1·24	1·22	1·19	1·17	1·15	1·12	1·11	38
39	1·62	1·57	1·53	1·49	1·45	1·41	1·38	1·35	1·32	1·29	1·26	1·23	1·21	1·19	1·17	1·15	39
40	1·68	1·63	1·58	1·54	1·50	1·46	1·43	1·39	1·36	1·33	1·31	1·28	1·25	1·23	1·21	1·19	40
41	1·74	1·69	1·64	1·60	1·55	1·52	1·48	1·44	1·41	1·38	1·35	1·33	1·30	1·27	1·25	1·23	41
42	1·80	1·75	1·70	1·65	1·61	1·57	1·53	1·50	1·46	1·43	1·40	1·37	1·35	1·32	1·30	1·28	42
43	1·87	1·81	1·76	1·71	1·67	1·63	1·59	1·55	1·51	1·48	1·45	1·42	1·39	1·37	1·34	1·32	43
44	1·93	1·87	1·82	1·77	1·73	1·68	1·64	1·60	1·57	1·53	1·50	1·47	1·44	1·42	1·39	1·37	44
45	2·00	1·94	1·89	1·84	1·79	1·74	1·70	1·66	1·62	1·59	1·56	1·52	1·49	1·47	1·44	1·41	45
46	2·07	2·01	1·95	1·90	1·85	1·81	1·76	1·72	1·68	1·65	1·61	1·58	1·55	1·52	1·49	1·47	46
47	2·14	2·08	2·02	1·97	1·92	1·87	1·82	1·78	1·74	1·70	1·67	1·63	1·60	1·57	1·54	1·52	47
48	2·22	2·16	2·10	2·04	1·99	1·94	1·89	1·85	1·80	1·76	1·73	1·69	1·66	1·63	1·60	1·57	48
49	2·30	2·23	2·17	2·11	2·06	2·01	1·96	1·91	1·87	1·83	1·79	1·75	1·72	1·69	1·66	1·63	49
50	2·38	2·31	2·25	2·19	2·13	2·08	2·03	1·98	1·94	1·89	1·85	1·82	1·78	1·75	1·72	1·69	50
51	2·47	2·40	2·33	2·27	2·21	2·15	2·10	2·05	2·01	1·96	1·92	1·88	1·85	1·81	1·78	1·75	51
52	2·56	2·49	2·42	2·35	2·29	2·23	2·18	2·13	2·08	2·03	1·99	1·95	1·91	1·88	1·84	1·81	52
53	2·65	2·58	2·50	2·44	2·37	2·31	2·26	2·21	2·16	2·11	2·06	2·02	1·98	1·95	1·91	1·88	53
54	2·75	2·67	2·60	2·53	2·46	2·40	2·34	2·29	2·24	2·19	2·14	2·10	2·06	2·02	1·98	1·95	54
55	2·86	2·77	2·70	2·62	2·55	2·49	2·43	2·37	2·32	2·27	2·22	2·18	2·13	2·09	2·06	2·02	55
56	2·97	2·88	2·80	2·72	2·65	2·58	2·52	2·46	2·41	2·36	2·31	2·26	2·22	2·17	2·13	2·10	56
57	3·08	2·99	2·91	2·83	2·75	2·68	2·62	2·56	2·50	2·45	2·40	2·35	2·30	2·26	2·22	2·18	57
58	3·20	3·11	3·02	2·94	2·86	2·79	2·72	2·66	2·60	2·54	2·49	2·44	2·39	2·35	2·30	2·27	58
59	3·33	3·23	3·14	3·06	2·98	2·90	2·83	2·77	2·70	2·64	2·59	2·54	2·49	2·44	2·40	2·36	59
60	3·46	3·36	3·27	3·18	3·10	3·02	2·95	2·88	2·81	2·75	2·69	2·64	2·59	2·54	2·49	2·45	60
Dec.	150° 210°	149° 211°	148° 212°	147° 213°	146° 214°	145° 215°	144° 216°	143° 217°	142° 218°	141° 219°	140° 220°	139° 221°	138° 222°	137° 223°	136° 224°	135° 225°	Dec.

B—Always named the **same** as Declination.

B   HOUR ANGLE   **B**

A	TABLE A — HOUR ANGLE														A

Lat. °	45° 315°	46° 314°	47° 313°	48° 312°	49° 311°	50° 310°	51° 309°	52° 308°	53° 307°	54° 306°	55° 305°	56° 304°	57° 303°	58° 302°	59° 301°	60° 300°	Lat. °
0	·00	·00	·00	·00	·00	·00	·00	·00	·00	·00	·00	·00	·00	·00	·00	·00	0
1	·02	·02	·02	·02	·02	·01	·01	·01	·01	·01	·01	·01	·01	·01	·01	·01	1
2	·03	·03	·03	·03	·03	·03	·03	·03	·03	·03	·02	·02	·02	·02	·02	·02	2
3	·05	·05	·05	·05	·05	·04	·04	·04	·04	·04	·04	·04	·03	·03	·03	·03	3
4	·07	·07	·07	·06	·06	·06	·06	·05	·05	·05	·05	·05	·05	·04	·04	·04	4
5	·09	·08	·08	·08	·08	·07	·07	·07	·07	·06	·06	·06	·06	·05	·05	·05	5
6	·11	·10	·10	·09	·09	·09	·09	·08	·08	·08	·07	·07	·07	·07	·06	·06	6
7	·12	·12	·11	·11	·11	·10	·10	·10	·09	·09	·09	·08	·08	·08	·07	·07	7
8	·14	·14	·13	·13	·12	·12	·11	·11	·11	·10	·10	·09	·09	·09	·08	·08	8
9	·16	·15	·15	·14	·14	·13	·13	·12	·12	·12	·11	·11	·10	·10	·10	·09	9
10	·18	·17	·16	·16	·15	·15	·14	·14	·13	·13	·12	·12	·11	·11	·11	·10	10
11	·19	·19	·18	·18	·17	·16	·16	·15	·15	·14	·14	·13	·13	·12	·12	·11	11
12	·21	·21	·20	·19	·18	·18	·17	·17	·16	·15	·15	·14	·14	·13	·13	·12	12
13	·23	·22	·22	·21	·20	·19	·19	·18	·17	·17	·16	·16	·15	·14	·14	·13	13
14	·25	·24	·23	·22	·22	·21	·20	·19	·19	·18	·17	·17	·16	·16	·15	·14	14
15	·27	·26	·25	·24	·23	·22	·22	·21	·20	·19	·19	·18	·17	·17	·16	·15	15
16	·29	·28	·27	·26	·25	·24	·23	·22	·22	·21	·20	·19	·19	·18	·17	·17	16
17	·31	·30	·29	·28	·27	·26	·25	·24	·23	·22	·21	·21	·20	·19	·18	·18	17
18	·32	·31	·30	·29	·28	·27	·26	·25	·24	·24	·23	·22	·21	·20	·20	·19	18
19	·34	·33	·32	·31	·30	·29	·28	·27	·26	·25	·24	·23	·22	·22	·21	·20	19
20	·36	·35	·34	·33	·32	·31	·29	·28	·27	·26	·25	·25	·24	·23	·22	·21	20
21	·38	·37	·36	·35	·33	·32	·31	·30	·29	·28	·27	·26	·25	·24	·23	·22	21
22	·40	·39	·38	·36	·35	·34	·33	·32	·30	·29	·28	·27	·26	·25	·24	·23	22
23	·42	·41	·40	·38	·37	·36	·34	·33	·32	·31	·30	·29	·28	·27	·26	·25	23
24	·45	·43	·42	·40	·39	·37	·36	·35	·34	·32	·31	·30	·29	·28	·27	·26	24
25	·47	·45	·44	·42	·41	·39	·38	·36	·35	·34	·33	·31	·30	·29	·28	·27	25
26	·49	·47	·46	·44	·42	·41	·39	·38	·37	·35	·34	·33	·32	·30	·29	·28	26
27	·51	·49	·48	·46	·44	·43	·41	·40	·38	·37	·36	·34	·33	·32	·31	·29	27
28	·53	·51	·50	·48	·46	·45	·43	·42	·40	·39	·37	·36	·35	·33	·32	·31	28
29	·55	·54	·52	·50	·48	·47	·45	·43	·42	·40	·39	·37	·36	·35	·33	·32	29
30	·58	·56	·54	·52	·50	·48	·47	·45	·44	·42	·40	·39	·37	·36	·35	·33	30
31	·60	·58	·56	·54	·52	·50	·49	·47	·45	·44	·42	·40	·39	·38	·36	·35	31
32	·62	·60	·58	·56	·54	·52	·51	·49	·47	·45	·44	·42	·41	·39	·38	·36	32
33	·65	·63	·61	·58	·56	·55	·53	·51	·49	·47	·45	·44	·42	·41	·39	·37	33
34	·67	·65	·63	·61	·59	·57	·55	·53	·51	·49	·47	·46	·44	·42	·41	·39	34
35	·70	·68	·65	·63	·61	·59	·57	·55	·53	·51	·49	·47	·45	·44	·42	·40	35
36	·73	·70	·68	·65	·63	·61	·59	·57	·55	·53	·51	·49	·47	·45	·44	·42	36
37	·75	·73	·70	·68	·66	·63	·61	·59	·57	·55	·53	·51	·49	·47	·45	·44	37
38	·78	·75	·73	·70	·68	·66	·63	·61	·59	·57	·55	·53	·51	·49	·47	·45	38
39	·81	·78	·76	·73	·70	·68	·66	·63	·61	·59	·57	·55	·53	·51	·49	·47	39
40	·84	·81	·78	·76	·73	·70	·68	·66	·63	·61	·59	·57	·55	·52	·50	·48	40
41	·87	·84	·81	·78	·76	·73	·70	·68	·66	·63	·61	·59	·56	·54	·52	·50	41
42	·90	·87	·84	·81	·78	·76	·73	·70	·68	·65	·63	·61	·58	·56	·54	·52	42
43	·93	·90	·87	·84	·81	·78	·76	·73	·70	·68	·65	·63	·61	·58	·56	·54	43
44	·97	·93	·90	·87	·84	·81	·78	·75	·73	·70	·68	·65	·63	·60	·58	·56	44
45	1·00	·97	·93	·90	·87	·84	·81	·78	·75	·73	·70	·68	·65	·63	·60	·58	45
46	1·04	1·00	·97	·93	·90	·87	·84	·81	·78	·75	·73	·70	·67	·65	·62	·60	46
47	1·07	1·04	1·00	·97	·93	·90	·87	·84	·81	·78	·75	·72	·70	·67	·64	·62	47
48	1·11	1·07	1·04	1·00	·97	·93	·90	·87	·84	·81	·78	·75	·72	·69	·67	·64	48
49	1·15	1·11	1·07	1·04	1·00	·97	·93	·90	·87	·84	·81	·78	·75	·72	·69	·66	49
50	1·19	1 15	1·11	1·07	1·04	1·00	·97	·93	·90	·87	·83	·80	·77	·75	·72	·69	50
51	1·23	1·19	1·15	1·11	1·07	1·04	1·00	·97	·93	·90	·86	·83	·80	·77	·74	·71	51
52	1·28	1·24	1·19	1·15	1·11	1·07	1·04	1·00	·96	·93	·90	·86	·83	·80	·77	·74	52
53	1·33	1·28	1·24	1·19	1·15	1·11	1·07	1·04	1·00	·96	·93	·90	·86	·83	·80	·77	53
54	1·38	1·33	1·28	1·24	1·20	1·15	1·11	1·08	1·04	1·00	·96	·93	·89	·86	·83	·79	54
55	1·43	1·38	1·33	1·29	1·24	1·20	1·16	1·12	1·08	1·04	1·00	·96	·93	·89	·86	·82	55
56	1·48	1·43	1·38	1·34	1·29	1·24	1·20	1·16	1·12	1·08	1·04	1·00	·96	·93	·89	·86	56
57	1·54	1·49	1·44	1·39	1·34	1·29	1·25	1·20	1·16	1·12	1·08	1·04	1·00	·96	·93	·89	57
58	1·60	1·55	1·49	1·44	1·39	1·34	1·30	1·25	1·21	1·16	1·12	1·08	1·04	1·00	·96	·92	58
59	1·66	1·61	1·55	1·50	1·45	1·40	1·35	1·30	1·25	1·21	1·17	1·12	1·08	1·04	1·00	·96	59
60	1·73	1·67	1·62	1·56	1·51	1·45	1·40	1·35	1·31	1·26	1·21	1·17	1·12	1·08	1·04	1·00	60

Lat.	135° 225°	134° 226°	133° 227°	132° 228°	131° 229°	130° 230°	129° 231°	128° 232°	127° 233°	126° 234°	125° 235°	124° 236°	123° 237°	122° 238°	121° 239°	120° 240°	Lat.

A	HOUR ANGLE	A

A—Named opposite to Latitude, **except** when Hour Angle is between 90° and 270°

B TABLE B — HOUR ANGLE. **B**

Dec.°	45° 315°	46° 314°	47° 313°	48° 312°	49° 311°	50° 310°	51° 309°	52° 308°	53° 307°	54° 306°	55° 305°	56° 304°	57° 303°	58° 302°	59° 301°	60° 300°	Dec.°
0	·00	·00	·00	·00	·00	·00	·00	·00	·00	·00	·00	·00	·00	·00	·00	·00	0
1	·02	·02	·02	·02	·02	·02	·02	·02	·02	·02	·02	·02	·02	·02	·02	·02	1
2	·05	·05	·05	·05	·05	·05	·04	·04	·04	·04	·04	·04	·04	·04	·04	·04	2
3	·07	·07	·07	·07	·67	·07	·07	·07	·07	·06	·06	·06	·06	·06	·06	·06	3
4	·10	·10	·10	·09	·09	·09	·09	·09	·09	·09	·09	·08	·08	·08	·08	·08	4
5	·12	·12	·12	·12	·12	·11	·11	·11	·11	·11	·11	·11	·10	·10	·10	·10	5
6	·15	·15	·14	·14	·14	·14	·14	·13	·13	·13	·13	·13	·13	·12	·12	·12	6
7	·17	·17	·17	·17	·16	·16	·16	·16	·15	·15	·15	·15	·15	·14	·14	·14	7
8	·20	·20	·19	·19	·19	·18	·18	·18	·18	·17	·17	·17	·17	·17	·16	·16	8
9	·22	·22	·22	·21	·21	·21	·20	·20	·20	·20	·19	·19	·19	·19	·18	·18	9
10	·25	·25	·24	·24	·23	·23	·23	·22	·22	·22	·22	·21	·21	·21	·21	·20	10
11	·27	·27	·27	·26	•26	·25	·25	·25	·24	·24	·24	·23	·23	·23	·23	·22	11
12	·30	·30	·29	·29	·28	·28	·27	·27	·27	·26	·26	·26	·25	·25	·25	·25	12
13	·33	·32	·32	·31	·31	·30	·30	·29	·29	·29	·28	·28	·28	·27	·27	·27	13
14	·35	·35	·34	·34	·33	·33	·32	·32	·31	·31	·30	·30	·30	·29	·29	·29	14
15	·38	·37	·37	·37	·36	·36	·35	·34	·34	·34	·33	·33	·32	·32	·32	·31	15
16	·41	·40	·39	·39	·38	·37	·37	·36	·36	·35	·35	·35	·34	·34	·33	·33	16
17	·43	·43	·42	·41	·41	·40	·39	·39	·38	·38	·37	·37	·36	·36	·36	·35	17
18	·46	·45	·44	·44	·43	·42	·42	·41	·41	·40	·40	·39	·39	·38	·38	·38	18
19	·49	·48	·47	·46	·46	·45	·44	·44	·43	·43	·42	·42	·41	·41	·40	·40	19
20	·51	·51	·50	·49	·48	·48	·47	·46	·46	·45	·44	·44	·43	·43	·42	·42	20
21	·54	·53	·52	·52	·51	·50	·49	·49	·48	·47	·47	·46	·46	·45	·45	·44	21
22	·57	·56	·55	·54	·54	·53	·52	·51	·51	·50	·49	·49	·48	·48	·47	·47	22
23	·60	·59	·58	·57	·56	·55	·55	·54	·53	·52	·52	·51	·51	·50	·50	·49	23
24	·63	·62	·61	·60	·59	·58	·57	·57	·56	·55	·55	·54	·54	·53	·53	·52	24
25	·66	·65	·64	·63	·62	·61	·60	·59	·58	·58	·57	·56	·56	·55	·54	·54	25
26	·69	·68	·67	·66	·65	·64	·63	·62	·61	·60	·60	·59	·58	·58	·57	·56	26
27	·72	·71	·70	·69	·68	·67	·66	·65	·64	·63	·62	·61	·61	·60	·59	·59	27
28	·75	·74	·73	·72	·70	·69	·68	·67	·67	·66	·65	·64	·63	·63	·62	·61	28
29	·78	·77	·76	·75	·73	·72	·71	·70	·69	·69	·68	·67	·66	·65	·65	·64	29
30	·82	·80	·79	·78	·76	·75	·74	·73	·72	·71	·70	·70	·69	·68	·67	·67	30
31	·85	·84	·82	·81	·80	·78	·77	·76	·75	·74	·73	·72	·72	·71	·70	·69	31
32	·88	·87	·85	·84	·83	·82	·80	·79	·78	·77	·76	·75	·75	·74	·73	·72	32
33	·92	·90	·89	·87	·86	·85	·84	·82	·81	·80	·79	·78	·77	·77	·76	·75	33
34	·96	·94	·92	·91	·89	·88	·87	·86	·84	·83	·82	·81	·80	·80	·79	·78	34
35	·99	·97	·96	·94	·93	·91	·90	·89	·88	·87	·85	·84	·83	·83	·82	·81	35
36	1·03	1·01	·99	·98	·96	·95	·93	·92	·91	·90	·89	·88	·87	·86	·85	·84	36
37	1·07	1·05	1·03	1·01	1·00	·98	·97	·96	·94	·93	·92	·91	·90	·89	·88	·87	37
38	1·11	1·09	1·07	1·05	1·04	1·02	1·00	·99	·98	·97	·95	·94	·93	·92	·91	·90	38
39	1·15	1·13	1·11	1·09	1·07	1·06	1·04	1·03	1·01	1·00	·99	·98	·97	·95	·94	·94	39
40	1·19	1·17	1·15	1·13	1·11	1·10	1·08	1·06	1·05	1·04	1·02	1·01	1·00	·99	·98	·97	40
41	1·23	1·21	1·19	1·17	1·15	1·13	1·12	1·10	1·09	1·07	1·06	1·05	1·04	1·03	1·01	1·00	41
42	1·28	1·25	1·23	1·21	1·19	1·18	1·16	1·14	1·13	1·11	1·10	1·09	1·07	1·06	1·05	1·04	42
43	1·32	1·30	1·28	1·25	1·24	1·22	1·20	1·18	1·17	1·15	1·14	1·12	1·11	1·10	1·09	1·08	43
44	1·37	1·34	1·32	1·30	1·28	1·26	1·24	1·23	1·21	1·19	1·18	1·16	1·15	1·14	1·13	1·12	44
45	1·41	1·39	1·37	1·35	1·33	1·31	1·29	1·27	1·25	1·24	1·22	1·21	1·19	1·18	1·17	1·15	45
46	1·47	1·44	1·42	1·39	1·37	1·35	1·33	1·31	1·30	1·28	1·26	1·25	1·23	1·22	1·21	1·20	46
47	1·52	1·49	1·47	1·44	1·42	1·40	1·38	1·36	1·34	1·33	1·31	1·29	1·28	1·26	1·25	1·24	47
48	1·57	1·54	1·52	1·49	1·47	1·45	1·43	1·41	1·39	1·37	1·36	1·34	1·32	1·31	1·30	1·28	48
49	1·63	1·60	1·57	1·55	1·52	1·50	1·48	1·46	1·44	1·42	1·40	1·39	1·37	1·36	1·34	1·33	49
50	1·69	1·66	1·63	1·60	1·58	1·56	1·53	1·51	1·49	1·47	1·45	1·44	1·42	1·41	1·39	1·38	50
51	1·75	1·72	1·69	1·66	1·64	1·61	1·59	1·57	1·55	1·53	1·51	1·49	1·47	1·46	1·44	1·43	51
52	1·81	1·78	1·75	1·72	1·70	1·67	1·65	1·62	1·60	1·58	1·56	1·54	1·53	1·51	1·49	1·48	52
53	1·88	1·84	1·81	1·79	1·76	1·73	1·71	1·68	1·66	1·64	1·62	1·60	1·58	1·56	1·55	1·53	53
54	1·95	1·91	1·88	1·85	1·82	1·80	1·77	1·75	1·72	1·70	1·68	1·66	1·64	1·62	1·61	1·59	54
55	2·02	1·99	1·95	1·92	1·89	1·86	1·84	1·81	1·79	1·77	1·74	1·72	1·70	1·68	1·67	1·65	55
56	2·10	2·06	2·03	2·00	1·96	1·94	1·91	1·88	1·86	1·83	1·81	1·79	1·77	1·75	1·73	1·71	56
57	2·18	2·14	2·11	2·07	2·04	2·01	1·98	1·95	1·93	1·90	1·88	1·86	1·84	1·82	1·80	1·78	57
58	2·27	2·22	2·19	2·15	2·12	2·09	2·06	2·03	2·00	1·98	1·95	1·93	1·91	1·89	1·87	1·85	58
59	2·36	2·31	2·28	2·24	2·21	2·17	2·14	2·11	2·08	2·06	2·03	2·01	1·98	1·96	1·94	1·92	59
60	2·45	2·41	2·37	2·33	2·29	2·26	2·23	2·20	2·17	2·14	2·11	2·09	2·07	2·04	2·02	2·00	60
Dec.	135° 225°	134° 226°	133° 227°	132° 228°	131° 229°	130° 230°	129° 231°	128° 232°	127° 233°	126° 234°	125° 235°	124° 236°	123° 237°	122° 238°	121° 239°	120° 240°	Dec.

B HOUR ANGLE. **B**

B—Always named the same as Declination.

B—Always named the same as Declination.

TABLE A — HOUR ANGLE.

A—Named opposite to Latitude, **except** when Hour Angle is between 90° and 270°

Lat.°	60° / 300°	61° / 299°	62° / 298°	63° / 297°	64° / 296°	65° / 295°	66° / 294°	67° / 293°	68° / 292°	69° / 291°	70° / 290°	71° / 289°	72° / 288°	73° / 287°	74° / 286°	75° / 285°	Lat.°
0	·00	·00	·00	·00	·00	·00	·00	·00	·00	·00	·00	·00	·00	·00	·00	·00	0
1	·01	·01	·01	·01	·01	·01	·01	·01	·01	·01	·01	·01	·01	·01	·01	·01	1
2	·02	·02	·02	·02	·02	·02	·02	·02	·02	·02	·01	·01	·01	·01	·01	·01	2
3	·03	·03	·03	·03	·03	·02	·02	·02	·02	·02	·02	·02	·02	·02	·02	·01	3
4	·04	·04	·04	·04	·03	·03	·03	·03	·03	·03	·03	·02	·02	·02	·02	·02	4
5	·05	·05	·05	·04	·04	·04	·04	·04	·04	·04	·03	·03	·03	·03	·03	·02	5
6	·06	·06	·06	·05	·05	·05	·05	·05	·04	·04	·04	·04	·03	·03	·03	·03	6
7	·07	·07	·07	·06	·06	·06	·06	·05	·05	·05	·05	·04	·04	·04	·04	·03	7
8	·08	·08	·07	·07	·07	·07	:06	·06	·06	·05	·05	·05	·05	·04	·04	·04	8
9	·09	·09	·08	·08	·08	·07	·07	·07	·06	·06	·06	·06	·05	·05	·05	·04	9
10	·10	·10	·09	·09	·09	·08	·08	·08	·08	·07	·07	·06	·06	·06	·05	·05	10
11	·11	·11	·10	·10	·10	·09	·09	·08	·08	·08	·07	·07	·06	·06	·06	·05	11
12	·12	·12	·11	·11	·10	·10	·10	·09	·09	·08	·08	·07	·07	·07	·06	·06	12
13	·13	·13	·12	·12	·11	·11	·10	·10	·09	·09	·08	·08	·08	·07	·07	·06	13
14	·14	·14	·13	·13	·12	·12	·11	·11	·10	·10	·09	·09	·08	·08	·07	·07	14
15	·15	·15	·14	·14	·13	·12	·12	·11	·11	·10	·10	·09	·09	·08	·08	·07	15
16	·17	·16	·15	·15	·14	·13	·13	·12	·12	·11	·10	·10	·09	·09	·08	·08	16
17	·18	·17	·16	·16	·15	·14	·14	·13	·12	·12	·11	·11	·10	·09	·09	·08	17
18	·19	·18	·17	·17	·16	·15	·15	·14	·13	·13	·12	·11	·11	·10	·09	·09	18
19	·20	·19	·18	·18	·17	·16	·15	·15	·14	·13	·13	·12	·11	·11	·10	·09	19
20	·21	·20	·19	·19	·18	·17	·16	·15	·15	·14	·13	·13	·12	·11	·10	·10	20
21	·22	·21	·20	·20	·19	·18	·17	·16	·16	·15	·14	·13	·13	·12	·11	·10	21
22	·23	·22	·21	·21	·20	·19	·18	·17	·16	·16	·15	·14	·13	·12	·12	·11	22
23	·25	·24	·23	·22	·21	·20	·19	·18	·17	·16	·15	·15	·14	·13	·12	·11	23
24	·26	·25	·24	·23	·22	·21	·20	·19	·18	·17	·16	·15	·15	·14	·13	·12	24
25	·27	·26	·25	·24	·23	·22	·21	·20	·19	·18	·17	·16	·15	·14	·13	·13	25
26	·28	·27	·26	·25	·24	·23	·22	·21	·20	·19	·18	·17	·16	·15	·14	·13	26
27	·29	·28	·27	·26	·25	·24	·23	·22	·21	·20	·19	·18	·17	·16	·15	·14	27
28	·31	·29	·28	·27	·26	·25	·24	·23	·22	·20	·19	·18	·17	·16	·15	·14	28
29	·32	·31	·29	·28	·27	·26	·25	·24	·22	·21	·20	·19	·18	·17	·16	·15	29
30	·33	·32	·31	·29	·28	·27	·26	·25	·23	·22	·21	·20	·19	·18	·17	·16	30
31	·35	·33	·32	·31	·29	·28	·27	·26	·24	·23	·22	·21	·20	·18	·17	16	31
32	·36	·35	·33	·32	·31	·29	·28	·27	·25	·24	·23	·22	·20	·19	·18	·17	32
33	·37	·36	·35	·33	·32	·30	·29	·28	·26	·25	·24	·22	·21	·20	·19	·17	33
34	·39	·37	·36	·34	·33	·31	·30	·29	·27	·26	·25	·23	·22	·21	·19	·18	34
35	·40	·39	·37	·36	·34	·33	·31	·30	·28	·27	·26	·24	·23	·21	·20	·19	35
36	·42	·40	·39	·37	·35	·34	·32	·31	·29	·28	·26	·25	·24	·22	·21	·20	36
37	·44	·42	·40	·38	·37	·35	·34	·32	·30	·29	·27	·26	·25	·23	·22	·20	37
38	·45	·43	·42	·40	·38	·36	·35	·33	·32	·30	·28	·27	·25	·24	·22	·21	38
39	·47	·45	·43	·41	·40	·38	·36	·34	·33	·31	·30	·28	·26	·25	·23	·22	39
40	·48	·47	·45	·43	·41	·39	·37	·36	·34	·32	·31	·29	·27	·26	·24	·23	40
41	·50	·48	·46	·44	·42	·41	·39	·37	·35	·33	·32	·30	·28	·27	·25	·23	41
42	·52	·50	·48	·46	·44	·42	·40	·38	·36	·35	·33	·31	·29	·28	·26	·24	42
43	·54	·52	·50	·48	·46	·43	·42	·40	·38	·36	·34	·32	·30	·29	·27	·25	43
44	·56	·54	·51	·49	·47	·45	·43	·41	·39	·37	·35	·33	·31	·30	·28	·26	44
45	·58	·55	·53	·51	·49	·47	·45	·42	·40	·38	·36	·34	·33	·31	·29	·27	45
46	·60	·57	·55	·53	·51	·48	·46	·44	·42	·40	·38	·36	·34	·32	·30	·28	46
47	·62	·59	·57	·55	·52	·50	·48	·45	·43	·41	·39	·37	·35	·33	·31	·29	47
48	·64	·62	·59	·57	·54	·52	·49	·47	·45	·43	·40	·38	·36	·34	·32	·30	48
49	·66	·64	·61	·59	·56	·54	·51	·49	·47	·44	·42	·40	·37	·35	·33	·31	49
50	·69	·66	·63	·61	·58	·56	·53	·51	·48	·46	·43	·41	·39	·36	·34	·32	50
51	·71	·68	·66	·63	·60	·58	·55	·52	·50	·47	·45	·43	·40	·38	·35	·33	51
52	·74	·71	·68	·65	·62	·60	·57	·54	·52	·49	·47	·44	·42	·39	·37	·34	52
53	·77	·74	·71	·68	·65	·62	·59	·56	·54	·51	·48	·46	·43	·41	·38	·36	53
54	·79	·76	·73	·70	·67	·64	·61	·58	·56	·53	·50	·47	·45	·42	·40	·37	54
55	·82	·79	·76	·73	·70	·67	·64	·61	·58	·55	·52	·49	·46	·44	·41	·38	55
56	·86	·82	·79	·76	·72	·69	·66	·63	·60	·57	·54	·51	·48	·45	·43	·40	56
57	·89	·85	·82	·78	·75	·72	·69	·65	·62	·59	·56	·53	·50	·47	·44	·41	57
58	·92	·89	·85	·81	·78	·75	·71	·68	·65	·61	·58	·55	·52	·49	·46	·43	58
59	·96	·92	·88	·85	·81	·78	·74	·71	·67	·64	·61	·57	·54	·51	·48	·45	59
60	1·00	·96	·92	·88	·85	·81	·77	·74	·70	·67	·63	·60	·56	·53	·50	·46	60
Lat.	120° / 240°	119° / 241°	118° / 242°	117° / 243°	116° / 244°	115° / 245°	114° / 246°	113° / 247°	112° / 248°	111° / 249°	110° / 250°	109° / 251°	108° / 252°	107° / 253°	106° / 254°	105° / 255°	Lat.

A — HOUR ANGLE.

B																	**TABLE B — HOUR ANGLE.**	B

Dec. °	60° 300°	61° 299°	62° 298°	63° 297°	64° 296°	65° 295°	66° 294°	67° 293°	68° 292°	69° 291°	70° 290°	71° 289°	72° 288°	73° 287°	74° 286°	75° 285°	Dec. °
0	·00	·00	·00	·00	·00	·00	·00	·00	·00	·00	·00	·00	·00	·00	·00	·00	0
1	·02	·02	·02	·02	·02	·02	·02	·02	·02	·02	·02	·02	·02	·02	·02	·02	1
2	·04	·04	·04	·04	·04	·04	·04	·04	·04	·04	·04	·04	·04	·04	·04	·04	2
3	·06	·06	·06	·06	·06	·06	·06	·06	·06	·06	·06	·06	·06	·06	·06	·05	3
4	·08	·08	·08	·08	·08	·08	·08	·08	·08	·08	·07	·07	·07	·07	·07	·07	4
5	·10	·10	·10	·10	·10	·10	·10	·10	·09	·09	·09	·09	·09	·09	·09	·09	5
6	·12	·12	·12	·12	·12	·12	·12	·11	·11	·11	·11	·11	·11	·11	·11	·11	6
7	·14	·14	·14	·14	·14	·14	·13	·13	·13	·13	·13	·13	·13	·13	·13	·13	7
8	·16	·16	·16	·16	·16	·16	·15	·15	·15	·15	·15	·15	·15	·15	·15	·15	8
9	·18	·18	·18	·18	·18	·17	·17	·17	·17	·17	·17	·17	·17	·17	·16	·16	9
10	·20	·20	·20	·20	·20	·19	·19	·19	·19	·19	·19	·19	·19	·18	·18	·18	10
11	·22	·22	·22	·22	·22	·21	·21	·21	·21	·21	·21	·21	·20	·20	·20	·20	11
12	·25	·24	·24	·24	·24	·23	·23	·23	·23	·23	·23	·23	·22	·22	·22	·22	12
13	·27	·26	·26	·26	·26	·25	·25	·25	·25	·25	·25	·24	·24	·24	·24	·24	13
14	·29	·29	·28	·28	·28	·28	·27	·27	·27	·27	·27	·26	·26	·26	·26	·26	14
15	·31	·31	·30	·30	·30	·30	·29	·29	·29	·29	·29	·28	·28	·28	·28	·28	15
16	·33	·33	·32	·32	·32	·32	·31	·31	·31	·31	·31	·30	·30	·30	·30	·30	16
17	·35	·35	·35	·34	·34	·34	·33	·33	·33	·33	·33	·32	·32	·32	·32	·32	17
18	·38	·37	·37	·36	·36	·36	·36	·35	·35	·35	·35	·34	·34	·34	·34	·34	18
19	·40	·39	·39	·39	·38	·38	·38	·38	·37	·37	·37	·36	·36	·36	·36	·36	19
20	·42	·42	·41	·41	·41	·40	·40	·40	·39	·39	·39	·39	·38	·38	·38	·38	20
21	·44	·44	·43	·43	·43	·42	·42	·42	·41	·41	·41	·41	·40	·40	·40	·40	21
22	·47	·46	·46	·45	·45	·45	·44	·44	·44	·43	·43	·43	·43	·42	·42	·42	22
23	·49	·49	·48	·48	·47	·47	·46	·46	·46	·46	·45	·45	·45	·44	·44	·44	23
24	·51	·51	·50	·50	·50	·49	·49	·48	·48	·48	·47	·47	·47	·47	·46	·46	24
25	·54	·53	·53	·52	·52	·52	·51	·51	·50	·50	·49	·49	·49	·49	·49	·48	25
26	·56	·56	·55	·55	·54	·54	·53	·53	·53	·52	·52	·52	·51	·51	·51	·50	26
27	·59	·58	·58	·57	·57	·56	·56	·55	·55	·55	·54	·54	·54	·53	·53	·53	27
28	·61	·61	·60	·60	·59	·59	·58	·58	·57	·57	·57	·56	·56	·56	·55	·55	28
29	·64	·63	·63	·62	·62	·61	·61	·60	·60	·59	·59	·59	·58	·58	·58	·57	29
30'	·67	·66	·65	·65	·64	·64	·63	·63	·62	·62	·62	·61	·61	·60	·60	·60	30
31	·69	·69	·68	·67	·67	·66	·66	·65	·65	·64	·64	·64	·63	·63	·63	·62	31
32	·72	·71	·71	·70	·70	·69	·68	·68	·67	·67	·67	·66	·66	·65	·65	·65	32
33	·75	·74	·74	·73	·72	·72	·71	·71	·70	·70	·69	·69	·68	·68	·68	·67	33
34	·78	·77	·76	·76	·75	·75	·74	·73	·73	·72	·72	·71	·71	·71	·70	·70	34
35	·81	·80	·79	·79	·78	·77	·77	·76	·76	·75	·74	·74	·74	·73	·73	·72	35
36	·84	·83	·82	·82	·81	·80	·80	·79	·78	·78	·77	·77	·76	·76	·76	·75	36
37	·87	·86	·85	·85	·84	·83	·82	·82	·81	·81	·80	·80	·79	·79	·78	·78	37
38	·90	·89	·88	·88	·87	·86	·86	·85	·84	·84	·83	·83	·82	·82	·81	·81	38
39	·94	·93	·92	·91	·90	·89	·89	·88	·87	·87	·86	·86	·85	·85	·84	·84	39
40	·97	·96	·95	·94	·93	·93	·92	·91	·91	·90	·89	·89	·88	·88	·87	·87	40
41	1·00	·99	·98	·98	·97	·96	·95	·94	·94	·93	·92	·92	·91	·91	·90	·90	41
42	1·04	1·03	1·02	1·01	1·00	·99	·99	·98	·97	·96	·96	·95	·95	·94	·94	·93	42
43	1·08	1·07	1·06	1·05	1·04	1·03	1·02	1·01	1·01	1·00	·99	·99	·98	·98	·97	·97	43
44	1·11	1·10	1·09	1·08	1·07	1·07	1·06	1·05	1·04	1·03	1·03	1·02	1·02	1·01	1·00	1·00	44
45	1·15	1·14	1·13	1·12	1·11	1·10	1·10	1·09	1·08	1·07	1·06	1·06	1·05	1·05	1·04	1·04	45
46	1·20	1·18	1·17	1·16	1·15	1·14	1·13	1·12	1·12	1·11	1·10	1·10	1·09	1·08	1·08	1·07	46
47	1·24	1·23	1·22	1·20	1·19	1·18	1·17	1·16	1·16	1·15	1·14	1·13	1·13	1·12	1·12	1·11	47
48	1·28	1·27	1·26	1·25	1·24	1·22	1·22	1·21	1·20	1·19	1·18	1·17	1·17	1·16	1·16	1·15	48
49	1·33	1·32	1·30	1·29	1·28	1·27	1·26	1·25	1·24	1·23	1·22	1·22	1·20	1·20	1·20	1·19	49
50	1·38	1·36	1·35	1·34	1·33	1·31	1·31	1·29	1·29	1·28	1·27	1·26	1·25	1·25	1·24	1·23	50
51	1·43	1·41	1·40	1·39	1·37	1·36	1·35	1·34	1·33	1·32	1·31	1·31	1·30	1·29	1·28	1·28	51
52	1·48	1·46	1·45	1·44	1·42	1·41	1·40	1·39	1·38	1·37	1·36	1·35	1·35	1·34	1·33	1·33	52
53	1·53	1·52	1·50	1·49	1·48	1·46	1·45	1·44	1·43	1·42	1·41	1·40	1·40	1·39	1·38	1·37	53
54	1·59	1·57	1·56	1·55	1·53	1·52	1·51	1·50	1·48	1·47	1·46	1·46	1·45	1·44	1·43	1·42	54
55	1·65	1·63	1·62	1·60	1·59	1·58	1·56	1·55	1·54	1·53	1·52	1·51	1·50	1·49	1·49	1·48	55
56	1·71	1·70	1·68	1·66	1·65	1·64	1·62	1·61	1·60	1·59	1·58	1·57	1·56	1·55	1·54	1·54	56
57	1·78	1·76	1·74	1·73	1·71	1·70	1·69	1·67	1·66	1·65	1·64	1·63	1·62	1·61	1·60	1·60	57
58	1·85	1·83	1·81	1·80	1·78	1·77	1·75	1·74	1·73	1·71	1·71	1·69	1·68	1·67	1·66	1·66	58
59	1·92	1·90	1·89	1·87	1·85	1·84	1·82	1·81	1·79	1·78	1·77	1·76	1·75	1·74	1·73	1·72	59
60	2·00	1·98	1·96	1·94	1·93	1·91	1·89	1·88	1·87	1·86	1·84	1·83	1·82	1·81	1·80	1·79	60
Dec.	120° 240°	119° 241°	118° 242°	117° 243°	116° 244°	115° 245°	114° 246°	113° 247°	112° 248°	111° 249°	110° 250°	109° 251°	108° 252°	107° 253°	106° 254°	105° 255°	Dec.

B—Always named the **same** as Declination.

B	**HOUR ANGLE.**	B

TABLE A — HOUR ANGLE.

A—Named opposite to Latitude, **except** when Hour Angle is between 90° and 270°

Lat.°	75° 285°	76° 284°	77° 283°	78° 282°	79° 281°	80° 280°	81° 279°	82° 278°	83° 277°	84° 276°	85° 275°	86° 274°	87° 273°	88° 272°	89° 271°	90° 270°	Lat.°
0	·00	·00	·00	·00	·00	·00	·00	·00	·00	·00	·00	·00	·00	·00	·00	·00	0
1	·01	·00	·00	·00	·00	·00	·00	·00	·00	·00	·00	·00	·00	·00	·00	·00	1
2	·01	·01	·01	·01	·01	·01	·01	·01	·00	·00	·00	·00	·00	·00	·00	·00	2
3	·01	·01	·01	·01	·01	·01	·01	·01	·01	·01	·01	·00	·00	·00	·00	·00	3
4	·02	·02	·02	·02	·01	·01	·01	·01	·01	·01	·01	·01	·00	·00	·00	·00	4
5	·02	·02	·02	·02	·02	·02	·01	·01	·01	·01	·01	·01	·01	·00	·00	·00	5
6	·03	·03	·02	·02	·02	·02	·02	·02	·01	·01	·01	·01	·01	·00	·00	·00	6
7	·03	·03	·03	·03	·02	·02	·02	·02	·02	·01	·01	·01	·01	·00	·00	·00	7
8	·04	·04	·03	·03	·03	·03	·02	·02	·02	·02	·01	·01	·01	·01	·00	·00	8
9	·04	·04	·04	·03	·03	·03	·03	·02	·02	·02	·01	·01	·01	·01	·00	·00	9
10	·05	·04	·04	·04	·03	·03	·03	·03	·02	·02	·02	·01	·01	·01	·00	·00	10
11	·05	·05	·05	·04	·04	·03	·03	·03	·02	·02	·02	·01	·01	·01	·00	·00	11
12	·06	·05	·05	·05	·04	·04	·03	·03	·03	·02	·02	·02	·01	·01	·00	·00	12
13	·06	·06	·05	·05	·05	·04	·04	·03	·03	·02	·02	·02	·01	·01	·00	·00	13
14	·07	·06	·06	·05	·05	·04	·04	·04	·03	·03	·02	·02	·01	·01	·00	·00	14
15	·07	·07	·06	·06	·05	·05	·04	·04	·03	·03	·02	·02	·01	·01	·01	·00	15
16	·08	·07	·07	·06	·06	·05	·05	·04	·04	·03	·03	·02	·02	·01	·01	·00	16
17	·08	·08	·07	·07	·06	·05	·05	·04	·04	·03	·03	·02	·02	·01	·01	·00	17
18	·09	·08	·08	·07	·06	·06	·05	·05	·04	·04	·03	·03	·02	·01	·01	·00	18
19	·09	·09	·08	·07	·07	·06	·06	·05	·04	·04	·03	·02	·02	·01	·01	·00	19
20	·10	·09	·08	·08	·07	·06	·06	·05	·05	·04	·03	·03	·02	·01	·01	·00	20
21	·10	·10	·09	·08	·07	·07	·06	·05	·05	·04	·03	·03	·02	·01	·01	·00	21
22	·11	·10	·09	·09	·08	·07	·06	·06	·05	·04	·04	·03	·02	·01	·01	·00	22
23	·11	·11	·10	·09	·08	·08	·07	·06	·05	·05	·04	·03	·02	·02	·01	·00	23
24	·12	·11	·10	·10	·09	·08	·07	·06	·06	·05	·04	·03	·02	·02	·01	·00	24
25	·13	·12	·11	·10	·09	·08	·07	·07	·06	·05	·04	·03	·02	·02	·01	·00	25
26	·13	·12	·11	·10	·10	·09	·08	·07	·06	·05	·04	·03	·03	·02	·01	·00	26
27	·14	·13	·12	·11	·10	·09	·08	·07	·06	·05	·05	·04	·03	·02	·01	·00	27
28	·14	·13	·12	·11	·10	·09	·08	·08	·07	·06	·05	·04	·03	·02	·01	·00	28
29	·15	·14	·13	·12	·11	·10	·09	·08	·07	·06	·05	·04	·03	·02	·01	·00	29
30	·16	·14	·13	·12	·11	·10	·09	·08	·07	·06	·05	·04	·03	·02	·01	·00	30
31	·16	·15	·14	·13	·12	·11	·10	·08	·07	·06	·05	·04	·03	·02	·01	·00	31
32	·17	·16	·14	·13	·12	·11	·10	·09	·08	·07	·06	·04	·03	·02	·01	·00	32
33	·17	·16	·15	·14	·13	·12	·10	·09	·08	·07	·06	·05	·03	·02	·01	·00	33
34	·18	·17	·16	·14	·13	·12	·11	·10	·08	·07	·06	·05	·04	·02	·01	·00	34
35	·19	·18	·16	·15	·14	·12	·11	·10	·09	·07	·06	·05	·04	·02	·01	·00	35
36	·20	·18	·17	·15	·14	·13	·12	·10	·09	·08	·06	·05	·04	·03	·01	·00	36
37	·20	·19	·17	·16	·15	·13	·12	·11	·09	·08	·07	·05	·04	·03	·01	·00	37
38	·21	·20	·18	·17	·15	·14	·12	·11	·10	·08	·07	·06	·04	·03	·01	·00	38
39	·22	·20	·19	·17	·16	·14	·13	·11	·10	·09	·07	·06	·04	·03	·01	·00	39
40	·23	·21	·19	·18	·16	·15	·13	·12	·10	·09	·07	·06	·04	·03	·02	·00	40
41	·23	·22	·20	·19	·17	·15	·14	·12	·11	·09	·08	·06	·05	·03	·02	·00	41
42	·24	·22	·21	·19	·18	·16	·14	·13	·11	·10	·08	·06	·05	·03	·02	·00	42
43	·25	·23	·22	·20	·18	·16	·15	·13	·11	·10	·08	·07	·05	·03	·02	·00	43
44	·26	·24	·22	·21	·19	·17	·15	·14	·12	·10	·09	·07	·05	·03	·02	·00	44
45	·27	·25	·23	·21	·19	·18	·16	·14	·12	·11	·09	·07	·05	·04	·02	·00	45
46	·28	·26	·24	·22	·20	·18	·16	·15	·13	·11	·09	·07	·05	·04	·02	·00	46
47	·29	·27	·25	·23	·21	·19	·17	·15	·13	·11	·09	·08	·06	·04	·02	·00	47
48	·30	·28	·26	·24	·22	·20	·18	·16	·14	·12	·10	·08	·06	·04	·02	·00	48
49	·31	·29	·27	·25	·22	·20	·18	·16	·14	·12	·10	·08	·06	·04	·02	·00	49
50	·32	·30	·28	·25	·23	·21	·19	·17	·15	·13	·10	·08	·06	·04	·02	·00	50
51	·33	·31	·29	·26	·24	·22	·20	·17	·15	·13	·11	·09	·07	·04	·02	·00	51
52	·34	·32	·30	·27	·25	·23	·20	·18	·16	·14	·11	·09	·07	·05	·02	·00	52
53	·36	·33	·31	·28	·26	·23	·21	·19	·16	·14	·12	·09	·07	·05	·02	·00	53
54	·37	·34	·32	·29	·27	·24	·22	·19	·17	·15	·12	·10	·07	·05	·02	·00	54
55	·38	·36	·33	·30	·28	·25	·23	·20	·18	·15	·13	·10	·08	·05	·03	·00	55
56	·40	·37	·34	·32	·29	·26	·24	·21	·18	·16	·13	·10	·08	·05	·03	·00	56
57	·41	·38	·36	·33	·30	·27	·24	·22	·19	·16	·14	·11	·08	·05	·03	·00	57
58	·43	·40	·37	·34	·31	·28	·25	·23	·20	·17	·14	·11	·08	·06	·03	·00	58
59	·45	·42	·38	·35	·32	·29	·26	·23	·20	·18	·15	·12	·09	·06	·03	·00	59
60	·46	·43	·40	·37	·34	·31	·27	·24	·21	·18	·15	·12	·09	·06	·03	·00	60
Lat.	105° 255°	104° 256°	103° 257°	102° 258°	101° 259°	100° 260°	99° 261°	98° 262°	97° 263°	96° 264°	95° 265°	94° 266°	93° 267°	92° 268°	91° 269°	90° 270°	Lat.

A   HOUR ANGLE.   A

TABLE B — HOUR ANGLE.

Dec. °	75° 285°	76° 284°	77° 283°	78° 282°	79° 281°	80° 280°	81° 279°	82° 278°	83° 277°	84° 276°	85° 275°	86° 274°	87° 273°	88° 272°	89° 271°	90° 270°	Dec. °
0	·00	·00	·00	·00	·00	·00	·00	·00	·00	·00	·00	·00	·00	·00	·00	·00	0
1	·02	·02	·02	·02	·02	·02	·02	·02	·02	·02	·02	·02	·02	·02	·02	·02	1
2	·04	·04	·04	·04	·04	·04	·04	·04	·04	·04	·04	·04	·04	·04	·03	·03	2
3	·05	·05	·05	·05	·05	·05	·05	·05	·05	·05	·05	·05	·05	·05	·05	·05	3
4	·07	·07	·07	·07	·07	·07	·07	·07	·07	·07	·07	·07	·07	·07	·07	·07	4
5	·09	·09	·09	·09	·09	·09	·09	·09	·09	·09	·09	·09	·09	·09	·09	·09	5
6	·11	·11	·11	·11	·11	·11	·11	·11	·11	·11	·11	·11	·11	·11	·11	·11	6
7	·13	·13	·13	·13	·13	·12	·12	·12	·12	·12	·12	·12	·12	·12	·12	·12	7
8	·15	·15	·14	·14	·14	·14	·14	·14	·14	·14	·14	·14	·14	·14	·14	·14	8
9	·16	·16	·16	·16	·16	·16	·16	·16	·16	·16	·16	·16	·16	·16	·16	·16	9
10	·18	·18	·18	·18	·18	·18	·18	·18	·18	·18	·18	·18	·18	·18	·18	·18	10
11	·20	·20	·20	·20	·20	·20	·20	·20	·20	·20	·20	·20	·20	·19	·19	·19	11
12	·22	·22	·22	·22	·22	·22	·22	·22	·21	·21	·21	·21	·21	·21	·21	·21	12
13	·24	·24	·24	·24	·24	·23	·23	·23	·23	·23	·23	·23	·23	·23	·23	·23	13
14	·26	·26	·26	·26	·26	·25	·25	·25	·25	·25	·25	·25	·25	·25	·25	·25	14
15	·28	·28	·28	·28	·27	·27	·27	·27	·27	·27	·27	·27	·27	·27	·27	·27	15
16	·30	·30	·29	·29	·29	·29	·29	·29	·29	·29	·29	·29	·29	·29	·29	·29	16
17	·32	·32	·31	·31	·31	·31	·31	·31	·31	·31	·31	·31	·31	·31	·31	·31	17
18	·34	·34	·33	·33	·33	·33	·33	·33	·33	·33	·33	·33	·33	·33	·33	·32	18
19	·36	·36	·35	·35	·35	·35	·35	·35	·35	·35	·35	·35	·35	·35	·34	·34	19
20	·38	·38	·37	·37	·37	·37	·37	·37	·37	·37	·37	·37	·36	·36	·36	·36	20
21	·40	·40	·39	·39	·39	·39	·39	·39	·39	·39	·39	·39	·38	·38	·38	·38	21
22	·42	·42	·42	·41	·41	·41	·41	·41	·41	·41	·41	·41	·41	·40	·40	·40	22
23	·44	·44	·44	·43	·43	·43	·43	·43	·43	·43	·43	·43	·43	·43	·43	·42	23
24	·46	·46	·46	·46	·45	·45	·45	·45	·45	·45	·45	·45	·45	·45	·45	·45	24
25	·48	·48	·48	·48	·48	·47	·47	·47	·47	·47	·47	·47	·47	·47	·47	·47	25
26	·50	·50	·50	·50	·50	·50	·49	·49	·49	·49	·49	·49	·49	·49	·49	·49	26
27	·53	·53	·52	·52	·52	·52	·52	·52	·51	·51	·51	·51	·51	·51	·51	·51	27
28	·55	·55	·55	·54	·54	·54	·54	·54	·54	·54	·53	·53	·53	·53	·53	·53	28
29	·57	·57	·57	·57	·57	·56	·56	·56	·56	·56	·56	·56	·56	·56	·55	·55	29
30	·60	·60	·59	·59	·59	·59	·59	·58	·58	·58	·58	·58	·58	·58	·58	·58	30
31	·62	·62	·62	·61	·61	·61	·61	·61	·61	·60	·60	·60	·60	·60	·60	·60	31
32	·65	·64	·64	·64	·64	·63	·63	·63	·63	·63	·63	·63	·63	·63	·63	·62	32
33	·67	·67	·67	·66	·66	·66	·66	·66	·65	·65	·65	·65	·65	·65	·65	·65	33
34	·70	·70	·69	·69	·69	·68	·68	·68	·68	·68	·68	·68	·68	·68	·68	·67	34
35	·72	·72	·72	·72	·71	·71	·71	·71	·71	·70	·70	·70	·70	·70	·70	·70	35
36	·75	·75	·75	·74	·74	·74	·74	·73	·73	·73	·73	·73	·73	·73	·73	·73	36
37	·78	·78	·77	·77	·77	·77	·76	·76	·76	·76	·75	·76	·76	·75	·75	·75	37
38	·81	·81	·80	·80	·80	·79	·79	·79	·79	·79	·78	·78	·78	·78	·78	·78	38
39	·84	·84	·83	·83	·83	·82	·82	·82	·82	·81	·81	·81	·81	·81	·81	·81	39
40	·87	·87	·86	·86	·86	·85	·85	·85	·85	·85	·84	·84	·84	·84	·84	·84	40
41	·90	·90	·89	·89	·89	·88	·88	·88	·88	·87	·87	87	·87	·87	·87	·87	41
42	·93	·93	·92	·92	·92	·91	·91	·91	·91	·91	·90	·90	·90	·90	·90	·90	42
43	·97	·96	·96	·95	·95	·95	·94	·94	·94	·94	·94	·94	·93	·93	·93	·93	43
44	1·00	1·00	·99	·99	·99	·98	·98	·98	·97	·97	·97	·97	·97	·97	·97	·97	44
45	1·04	1·03	1·03	1·02	1·02	1·02	1·01	1·01	1·01	1·01	1·00	1·00	1·00	1·00	1·00	1·00	45
46	1·07	1·07	1·06	1·06	1·05	1·05	1·05	1·05	1·04	1·04	1·04	1·04	1·04	1·04	1·04	1·04	46
47	1·11	1·11	1·10	1·10	1·09	1·09	1·09	1·08	1·08	1·08	1·08	1·07	1·07	1·07	1·07	1·07	47
48	1·15	1·14	1·14	1·14	1·13	1·13	1·12	1·12	1·12	1·12	1·11	1·11	1·11	1·11	1·11	1·11	48
49	1·19	1·19	1·18	1·18	1·17	1·17	1·16	1·16	1·16	1·16	1·15	1·15	1·15	1·15	1·15	1·15	49
50	1·23	1·23	1·22	1·22	1·21	1·21	1·21	1·20	1·20	1·20	1·20	1·19	1·19	1·19	1·19	1·19	50
51	1·28	1·27	1·27	1·26	1·26	1·25	1·25	1·25	1·24	1·24	1·24	1·24	1·24	1·24	1·24	1·23	51
52	1·33	1·32	1·31	1·31	1·30	1·30	1·30	1·29	1·29	1·29	1·28	1·28	1·28	1·28	1·28	1·28	52
53	1·37	1·37	1·36	1·36	1·35	1·35	1·34	1·34	1·34	1·33	1·33	1·33	1·33	1·33	1·33	1·33	53
54	1·42	1·42	1·41	1·41	1·40	1·40	1·39	1·39	1·39	1·38	1·38	1·38	1·38	1·38	1·38	1·38	54
55	1·48	1·47	1·47	1·46	1·46	1·45	1·45	1·44	1·44	1·44	1·43	1·43	1·43	1·43	1·43	1·43	55
56	1·54	1·53	1·52	1·52	1·51	1·51	1·50	1·50	1·49	1·49	1·49	1·49	1·48	1·48	1·48	1·48	56
57	1·60	1·59	1·58	1·57	1·57	1·57	1·56	1·55	1·55	1·55	1·55	1·54	1·54	1·54	1·54	1·54	57
58	1·66	1·65	1·64	1·64	1·63	1·63	1·62	1·62	1·61	1·61	1·61	1·60	1·60	1·60	1·60	1·60	58
59	1·72	1·71	1·71	1·70	1·70	1·69	1·69	1·68	1·68	1·67	1·67	1·67	1·67	1·67	1·66	1·66	59
60	1·79	1·79	1·78	1·77	1·76	1·76	1·75	1·75	1·75	1·74	1·74	1·74	1·73	1·73	1·73	1·73	60
Dec.	105° 255°	104° 256°	103° 257°	102° 258°	101° 259°	100° 260°	99° 261°	98° 262°	97° 263°	96° 264°	95° 265°	94° 266°	93° 267°	92° 268°	91° 269°	90° 270°	Dec.

B—Always named the same as Declination.

HOUR ANGLE.

A — Named opposite to Latitude, **except** when Hour Angle is between 90° and 270°

A TABLE A — HOUR ANGLE. A

Lat. °	Diff.	0° 15′ 359° 45′	0° 30′ 359° 30′	0° 45′ 359° 15′	1° 00′ 359° 00′	1° 15′ 358° 45′	1° 30′ 358° 30′	1° 45′ 358° 15′	2° 00′ 358° 00′	2° 15′ 357° 45′	2° 30′ 357° 30′	2° 45′ 357° 15′	3° 00′ 357° 00′	3° 15′ 356° 45′	3° 30′ 356° 30′	3° 45′ 356° 15′	Lat. °
60	*16·0*	397	198	132	99·2	79·4	66·1	56·7	49·6	44·1	39·7	36·1	33·0	30·5	28·3	26·4	**60**
61	*17·0*	414	207	138	103	82·7	68·9	59·0	51·7	45·9	41·3	37·6	34·4	31·8	29·5	27·5	**61**
62	*18·2*	431	216	144	108	86·2	71·8	61·6	53·9	47·9	43·1	39·2	35·9	33·1	30·8	28·7	**62**
63	*19·4*	450	225	150	112	89·9	74·9	64·2	56·2	50·0	45·0	40·9	37·4	34·6	32·1	29·9	**63**
64	*20·8*	470	235	157	117	94·0	78·3	67·1	58·7	52·2	47·0	42·7	39·1	36·1	33·5	31·3	**64**
65	*22·4*	491	246	164	123	98·3	81·9	70·2	61·4	54·6	49·1	44·6	40·9	37·8	35·1	32·7	**65**
66	*24·2*	515	257	172	129	103	85·8	73·5	64·3	57·2	51·4	46·8	42·9	39·6	36·7	34·3	**66**
67	*26·2*	540	270	180	135	108	90·0	77·1	67·5	60·0	54·0	49·0	45·0	41·5	38·5	35·9	**67**
68	*28·5*	567	284	189	142	113	94·5	81·0	70·9	63·0	56·7	51·5	47·2	43·6	40·5	37·8	**68**
69	*29·8*	597	299	199	149	119	99·5	85·3	74·6	66·3	59·7	54·2	49·7	45·9	42·6	39·7	**69**
70	*32·6*	630	315	210	157	126	105	89·9	78·7	69·9	62·9	57·2	52·4	48·4	44·9	41·9	**70**
71	*35·9*	666	333	222	166	133	111	95·1	83·2	73·9	66·5	60·5	55·4	51·2	47·5	44·3	**71**
72	*39·8*	705	353	235	176	141	118	101	88·1	78·3	70·5	64·1	58·7	54·2	50·3	47·0	**72**
73	*44·2*	750	375	250	187	150	125	107	93·7	83·3	74·9	68·1	62·4	57·6	53·5	49·9	**73**
74	*49·6*	799	400	266	200	160	133	114	99·9	88·8	79·9	72·6	66·6	61·4	57·0	53·2	**74**
75	*56·1*	855	428	285	215	171	143	122	107	95·0	85·5	77·7	71·3	65·7	61·0	56·9	**75**
76	*63·9*	919	460	306	230	184	153	131	115	102	91·9	83·5	76·5	70·6	65·6	61·2	**76**
77	*73·5*	993	496	331	248	199	165	142	124	110	99·2	90·2	82·7	76·3	70·8	66·1	**77**
78	*85·3*	1078	539	359	270	216	180	154	135	120	108	97·9	89·7	82·9	76·9	71·8	**78**
79	*101*	1179	590	393	295	236	197	168	147	131	118	107	98·2	90·6	84·1	78·5	**79**
80	*121*	1300	650	433	325	260	217	186	162	144	130	118	108	100	92·7	86·5	**80**
81	*147*	1447	724	482	362	289	241	207	181	161	145	131	121	111	103	96·3	**81**
82	*184*	1631	815	544	408	326	272	233	204	181	163	148	136	125	116	109	**82**
83	*236*	1867	933	622	467	373	311	267	233	207	187	170	155	143	133	124	**83**
Lat.	Diff.	179° 45′ 180° 15′	179° 30′ 180° 30′	179° 15′ 180° 45′	179° 00′ 181° 00′	178° 45′ 181° 15′	178° 30′ 181° 30′	178° 15′ 181° 45′	178° 00′ 182° 00′	177° 45′ 182° 15′	177° 30′ 182° 30′	177° 15′ 182° 45′	177° 00′ 183° 00′	176° 45′ 183° 15′	176° 30′ 183° 30′	176° 15′ 183° 45′	Lat.

A HOUR ANGLE. A

A TABLE A — HOUR ANGLE. A

Lat. °	3° 45′ 356° 15′	4° 00′ 356° 00′	4° 15′ 355° 45′	4° 30′ 355° 30′	4° 45′ 355° 15′	5° 00′ 355° 00′	5° 15′ 354° 45′	5° 30′ 354° 30′	5° 45′ 354° 15′	6° 00′ 354° 00′	6° 15′ 353° 45′	6° 30′ 353° 30′	6° 45′ 353° 15′	7° 00′ 353° 00′	7° 15′ 352° 45′	7° 30′ 352° 30′	Lat. °
60	26·4	24·8	23·3	22·0	20·8	19·8	18·8	18·0	17·2	16·5	15·8	15·2	14·6	14·1	13·6	13·2	**60**
61	27·5	25·8	24·3	22·9	21·7	20·6	19·6	18·7	17·9	17·2	16·5	15·8	15·2	14·7	14·2	13·7	**61**
62	28·7	26·9	25·3	23·9	22·6	21·5	20·5	19·5	18·7	17·9	17·2	16·5	15·9	15·3	14·8	14·3	**62**
63	29·9	28·1	26·4	24·9	23·6	22·4	21·4	20·4	19·5	18·7	17·9	17·2	16·6	16·0	15·4	14·9	**63**
64	31·3	29·3	27·6	26·1	24·7	23·4	22·3	21·3	20·4	19·5	18·7	18·0	17·3	16·7	16·1	15·6	**64**
65	32·7	30·7	28·9	27·2	25·8	24·5	23·3	22·3	21·3	20·4	19·6	18·8	18·1	17·5	16·9	16·3	**65**
66	34·3	32·1	30·2	28·5	27·0	25·7	24·4	23·3	22·3	21·4	20·5	19·7	19·0	18·3	17·7	17·1	**66**
67	35·9	33·7	31·7	29·9	28·4	26·9	25·6	24·5	23·4	22·4	21·5	20·7	19·9	19·2	18·5	17·9	**67**
68	37·8	35·4	33·3	31·4	29·8	28·3	26·9	25·7	24·6	23·5	22·6	21·7	20·9	20·2	19·5	18·8	**68**
69	39·7	37·3	35·1	33·1	31·4	29·8	28·4	27·1	25·9	24·8	23·8	22·9	22·0	21·2	20·5	19·8	**69**
70	41·9	39·3	37·0	34·9	33·1	31·4	29·9	28·5	27·3	26·1	25·1	24·1	23·2	22·4	21·6	20·9	**70**
71	44·3	41·5	39·1	36·9	35·0	33·2	31·6	30·2	28·9	27·6	26·5	25·5	24·5	23·7	22·8	22·1	**71**
72	47·0	44·0	41·4	39·1	37·0	35·2	33·5	32·0	30·6	29·3	28·1	27·0	26·0	25·1	24·2	23·4	**72**
73	49·9	46·8	44·0	41·6	39·4	37·4	35·6	34·0	32·5	31·1	30·0	28·7	27·6	26·6	25·7	24·8	**73**
74	53·2	50·0	46·9	44·3	42·0	39·9	38·0	36·2	34·6	33·2	31·9	30·6	29·5	28·4	27·4	26·5	**74**
75	56·9	53·4	50·2	47·4	44·9	42·7	40·6	38·8	37·1	35·5	34·1	32·8	31·5	30·4	29·3	28·4	**75**
76	61·2	57·4	54·0	51·0	48·3	45·8	43·7	41·7	39·8	38·2	36·6	35·2	33·9	32·7	31·5	30·5	**76**
77	66·1	62·0	58·3	55·0	52·1	49·5	47·1	45·0	43·0	41·2	39·6	38·0	36·6	35·2	34·1	32·9	**77**
78	71·8	67·3	63·3	59·8	56·6	53·8	51·2	48·9	46·7	44·8	43·0	41·3	39·8	38·3	37·0	35·7	**78**
79	78·5	73·6	69·2	65·4	61·9	58·8	56·0	53·4	51·1	49·0	47·0	45·2	43·5	41·9	40·0	39·1	**79**
80	86·5	81·1	76·3	72·1	68·3	64·8	61·7	58·9	56·3	54·0	51·8	49·8	47·9	46·2	44·6	43·1	**80**
81	96·3	90·3	85·0	80·2	76·0	72·2	68·7	65·6	62·7	60·1	57·7	55·4	53·3	51·4	49·6	48·0	**81**
82	109	102	96·0	90·4	85·6	81·3	77·4	73·9	70·7	67·7	65·0	62·5	60·1	58·0	55·9	54·0	**82**
83	124	117	110	104	98·0	93·1	88·6	84·6	80·9	77·5	74·4	71·5	68·8	66·3	64·0	61·9	**83**
Lat	176° 15′ 183° 45′	176° 00′ 184° 00′	175° 45′ 184° 15′	175° 30′ 184° 30′	175° 15′ 184° 45′	175° 00′ 185° 00′	174° 45′ 185° 15′	174° 30′ 185° 30′	174° 15′ 185° 45′	174° 00′ 186° 00′	173° 45′ 186° 15′	173° 30′ 186° 30′	173° 15′ 186° 45′	173° 00′ 187° 00′	172° 45′ 187° 15′	172° 30′ 187° 30′	Lat.

A HOUR ANGLE. A

TABLE B — HOUR ANGLE.

Dec. °	Diff.	0° 15' / 359° 45'	0° 30' / 359° 30'	0° 45' / 359° 15'	1° 00' / 359° 00'	1° 15' / 358° 45'	1° 30' / 358° 30'	1° 45' / 358° 15'	2° 00' / 358° 00'	2° 15' / 357° 45'	2° 30' / 357° 30'	2° 45' / 357° 15'	3° 00' / 357° 00'	3° 15' / 356° 45'	3° 30' / 356° 30'	3° 45' / 356° 15'	Dec. °
60	16·0	397	198	132	99·2	79·4	66·2	56·7	49·6	44·1	39·7	36·1	33·1	30·6	28·4	26·5	60
61	17·0	414	207	138	103	82·7	68·9	59·1	51·7	46·0	41·4	37·6	34·5	31·8	29·6	27·6	61
62	18·2	431	216	144	108	86·2	71·8	61·6	53·9	47·9	43·1	39·2	35·9	33·2	30·8	28·8	62
63	19·4	450	225	150	112	90·0	75·0	64·3	56·2	50·0	45·0	40·9	37·5	34·6	32·1	30·0	63
64	20·8	470	235	157	117	94·0	78·3	67·1	58·7	52·2	47·0	42·7	39·2	36·2	33·6	31·4	64
65	22·4	491	246	164	123	98·3	81·9	70·2	61·4	54·6	49·2	44·7	41·0	37·8	35·1	32·8	65
66	24·2	515	257	172	129	103	85·8	73·5	64·4	57·2	51·5	46·8	42·9	39·6	36·8	34·3	66
67	26·2	540	270	180	135	108	90·0	77·1	67·5	60·0	54·0	49·1	45·0	41·6	38·6	36·0	67
68	28·5	567	284	189	142	113	94·6	81·0	70·9	63·0	56·7	51·6	47·3	43·7	40·5	37·8	68
69	29·8	597	299	199	149	119	99·5	85·3	74·6	66·4	59·7	54·3	49·8	46·0	42·7	39·8	69
70	32·6	630	315	210	157	126	105	90·0	78·7	70·0	63·0	57·3	52·5	48·5	45·0	42·0	70
71	35·9	666	333	222	166	133	111	95·1	83·2	74·0	66·6	60·5	55·5	51·2	47·6	44·4	71
72	39·8	705	353	235	176	141	118	101	88·2	78·4	70·6	64·2	58·8	54·3	50·4	47·1	72
73	44·2	750	375	250	187	150	125	107	93·7	83·3	75·0	68·2	62·5	57·7	53·6	50·0	73
74	49·6	799	400	266	200	160	133	114	100	88·8	80·0	72·7	66·6	61·5	57·1	53·3	74
75	56·1	855	428	285	214	171	143	122	107	95·1	85·6	77·8	71·3	65·8	61·1	57·1	75
Dec.	Diff.	179° 45' / 180° 15'	179° 30' / 180° 30'	179° 15' / 180° 45'	179° 00' / 181° 00'	178° 45' / 181° 15'	178° 30' / 181° 30'	178° 15' / 181° 45'	178° 00' / 182° 00'	177° 45' / 182° 15'	177° 30' / 182° 30'	177° 15' / 182° 45'	177° 00' / 183° 00'	176° 45' / 183° 15'	176° 30' / 183° 30'	176° 15' / 183° 45'	Dec.

B — HOUR ANGLE. — B

TABLE B — HOUR ANGLE.

Dec. °	3° 45' / 356° 15'	4° 00' / 356° 00'	4° 15' / 355° 45'	4° 30' / 355° 30'	4° 45' / 355° 15'	5° 00' / 355° 00'	5° 15' / 354° 45'	5° 30' / 354° 30'	5° 45' / 354° 15'	6° 00' / 354° 00'	6° 15' / 353° 45'	6° 30' / 353° 30'	6° 45' / 353° 15'	7° 00' / 353° 00'	7° 15' / 352° 45'	7° 30' / 352° 30'	Dec. °
60	26·5	24·8	23·4	22·1	20·9	19·9	18·9	18·1	17·3	16·6	15·9	15·3	14·7	14·2	13·7	13·3	60
61	27·6	25·9	24·3	23·0	21·8	20·7	19·7	18·8	18·0	17·3	16·6	15·9	15·3	14·8	14·3	13·8	61
62	28·8	27·0	25·4	24·0	22·7	21·6	20·6	19·6	18·8	18·0	17·3	16·6	16·0	15·4	14·9	14·4	62
63	30·0	28·1	26·5	25·0	23·7	22·5	21·4	20·5	19·6	18·8	18·0	17·3	16·7	16·1	15·6	15·0	63
64	31·3	29·4	27·7	26·1	24·8	23·5	22·4	21·4	20·5	19·6	18·8	18·1	17·4	16·8	16·2	15·7	64
65	32·8	30·7	28·9	27·3	25·9	24·6	23·4	22·4	21·4	20·5	19·7	18·9	18·2	17·6	17·0	16·4	65
66	34·3	32·2	30·3	28·6	27·1	25·8	24·5	23·4	22·4	21·5	20·6	19·8	19·1	18·4	17·8	17·2	66
67	36·0	33·8	31·8	30·0	28·4	27·0	25·7	24·6	23·5	22·5	21·6	20·8	20·1	19·3	18·7	18·0	67
68	37·8	35·5	33·4	31·5	29·9	28·4	27·0	25·8	24·7	23·7	22·7	21·9	21·1	20·3	19·6	19·0	68
69	39·8	37·3	35·2	33·2	31·5	29·9	28·5	27·2	26·0	24·9	23·9	23·0	22·2	21·4	20·6	20·0	69
70	42·0	39·4	37·1	35·0	33·2	31·5	30·0	28·7	27·4	26·3	25·2	24·3	23·4	22·5	21·8	21·1	70
71	44·4	41·6	39·2	37·0	35·1	33·3	31·7	30·3	29·0	27·8	26·7	25·7	24·7	23·8	23·0	22·6	71
72	47·1	44·1	41·5	39·2	37·2	35·3	33·6	32·1	30·7	29·4	28·3	27·2	26·2	25·3	24·4	23·3	72
73	50·0	46·9	44·1	41·7	39·5	37·5	35·8	34·1	32·7	31·3	30·0	28·9	27·8	26·8	25·9	25·1	73
74	53·3	50·0	47·1	44·5	42·1	40·0	38·1	36·4	34·8	33·4	32·0	30·8	29·7	28·6	27·6	26·7	74
75	57·1	53·5	50·4	47·6	45·1	42·8	40·8	38·9	37·3	35·7	34·3	33·0	31·8	30·6	29·6	28·6	75
Dec.	176° 15' / 183° 45'	176° 00' / 184° 00'	175° 45' / 184° 15'	175° 30' / 184° 30'	175° 15' / 184° 45'	175° 00' / 185° 00'	174° 45' / 185° 15'	174° 30' / 185° 30'	174° 15' / 185° 45'	174° 00' / 186° 00'	173° 45' / 186° 15'	173° 30' / 186° 30'	173° 15' / 186° 45'	173° 00' / 187° 00'	172° 45' / 187° 15'	172° 30' / 187° 30'	Dec.

B — HOUR ANGLE. — B

TABLE A. — HOUR ANGLE.

A—Named opposite to Latitude, **except** when Hour Angle is between 90° and 270°

Lat.	7°30'	7°45'	8°00'	8°15'	8°30'	8°45'	9°00'	9°15'	9°30'	9°45'	10°00'	10°15'	10°30'	10°45'	11°00'	11°15'	Lat.
°	352°30'	352°15'	352°00'	351°45'	351°30'	351°15'	351°00'	350°45'	350°30'	350°15'	350°00'	349°45'	349°30'	349°15'	349°00'	348°45'	°
60	13·2	12·7	12·3	11·9	11·6	11·3	10·9	10·6	10·4	10·1	9·82	9·58	9·35	9·12	8·91	8·71	60
61	13·7	13·3	12·8	12·4	12·1	11·7	11·4	11·1	10·8	10·5	10·2	9·98	9·73	9·50	9·28	9·07	61
62	14·3	13·8	13·4	13·0	12·6	12·2	11·9	11·5	11·2	10·9	10·7	10·4	10·1	9·91	9·68	9·46	62
63	14·9	14·4	14·0	13·5	13·1	12·8	12·4	12·1	11·7	11·4	11·1	10·9	10·6	10·3	10·1	9·87	63
64	15·6	15·1	14·6	14·1	13·7	13·3	12·9	12·6	12·3	11·9	11·6	11·3	11·1	10·8	10·5	10·3	64
65	16·3	15·8	15·3	14·8	14·3	13·9	13·5	13·2	12·8	12·5	12·2	11·9	11·6	11·3	11·0	10·8	65
66	17·1	16·5	16·0	15·5	15·0	14·6	14·2	13·8	13·4	13·1	12·7	12·4	12·1	11·8	11·6	11·3	66
67	17·9	17·3	16·8	16·2	15·8	15·3	14·9	14·5	14·1	13·7	13·4	13·0	12·7	12·4	12·1	11·8	67
68	18·8	18·2	17·6	17·1	16·6	16·1	15·6	15·2	14·8	14·4	14·0	13·7	13·4	13·0	12·7	12·4	68
69	19·8	19·1	18·5	18·0	17·4	16·9	16·5	16·0	15·6	15·2	14·8	14·4	14·1	13·7	13·4	13·1	69
70	20·9	20·2	19·6	19·0	18·4	17·9	17·4	16·9	16·4	16·0	15·6	15·2	14·8	14·5	14·1	13·8	70
71	22·1	21·3	20·7	20·0	19·4	18·9	18·3	17·8	17·4	16·9	16·5	16·1	15·7	15·3	14·9	14·6	71
72	23·4	22·6	21·9	21·2	20·6	20·0	19·4	18·9	18·4	17·9	17·5	17·0	16·6	16·2	15·8	15·5	72
73	24·8	24·0	23·3	22·6	21·9	21·3	20·6	20·1	19·5	19·0	18·6	18·1	17·7	17·2	16·8	16·4	73
74	26·5	25·6	24·8	24·1	23·3	22·7	22·0	21·4	20·8	20·3	19·8	19·3	18·8	18·4	17·9	17·5	74
75	28·4	27·4	26·6	25·7	25·0	24·3	23·6	22·9	22·3	21·7	21·2	20·6	20·1	19·7	19·2	18·8	75
76	30·5	29·5	28·5	27·7	26·8	26·1	25·3	24·6	24·0	23·3	22·8	22·2	21·6	21·1	20·6	20·2	76
77	32·9	31·8	30·8	29·9	29·0	28·1	27·4	26·6	25·9	25·2	24·6	24·0	23·4	22·8	22·3	21·8	77
78	35·7	34·6	33·5	32·5	31·5	30·6	29·7	28·9	28·1	27·4	26·7	26·0	25·4	24·8	24·2	23·7	78
79	39·1	37·8	36·6	35·5	34·4	33·4	32·5	31·6	30·7	29·9	29·2	28·5	27·8	27·1	26·5	25·9	79
80	43·1	41·7	40·4	39·1	38·0	36·9	35·8	34·8	33·9	33·0	32·2	31·4	30·6	29·9	29·2	28·5	80
81	48·0	46·4	44·9	43·6	42·2	41·0	39·9	38·8	37·7	36·7	35·8	34·9	34·1	33·3	32·5	31·7	81
82	54·0	52·3	50·6	49·1	47·6	46·2	44·9	43·7	42·5	41·4	40·4	39·4	38·4	37·5	36·6	35·8	82
83	61·9	59·9	58·0	56·2	54·5	52·9	51·4	50·0	48·7	47·4	46·2	45·0	43·9	42·9	41·9	41·0	83
Lat.	172°30'	172°15'	172°00'	171°45'	171°30'	171°15'	171°00'	170°45'	170°30'	170°15'	170°00'	169°45'	169°30'	169°15'	169°00'	168°45'	Lat.
	187°30'	187°45'	188°00'	188°15'	188°30'	188°45'	189°00'	189°15'	189°30'	189°45'	190°00'	190°15'	190°30'	190°45'	191°00'	191°15'	

HOUR ANGLE.

TABLE A — HOUR ANGLE.

Lat.	11°15'	11°30'	11°45'	12°00'	12°15'	12°30'	12°45'	13°00'	13°15'	13°30'	13°45'	14°00'	14°15'	14°30'	14°45'	15°00'	Lat.
°	348°45'	348°30'	348°15'	348°00'	347°45'	347°30'	347°15'	347°00'	346°45'	346°30'	346°15'	346°00'	345°45'	345°30'	345°15'	345°00'	°
60	8·71	8·51	8·33	8·15	7·98	7·81	7·65	7·50	7·36	7·21	7·08	6·95	6·82	6·70	6·58	6·46	60
61	9·07	8·87	8·67	8·49	8·31	8·14	7·97	7·81	7·66	7·51	7·37	7·24	7·10	6·98	6·85	6·73	61
62	9·46	9·24	9·04	8·85	8·66	8·48	8·31	8·15	7·99	7·83	7·69	7·54	7·41	7·27	7·14	7·02	62
63	9·87	9·65	9·44	9·23	9·04	8·85	8·67	8·50	8·33	8·17	8·02	7·87	7·73	7·59	7·45	7·32	63
64	10·3	10·1	9·86	9·65	9·44	9·25	9·06	8·88	8·71	8·54	8·38	8·22	8·07	7·93	7·79	7·65	64
65	10·8	10·5	10·3	10·1	9·88	9·67	9·48	9·29	9·11	8·93	8·76	8·60	8·44	8·29	8·15	8·00	65
66	11·3	11·0	10·8	10·6	10·3	10·1	9·93	9·73	9·54	9·36	9·18	9·01	8·84	8·68	8·53	8·38	66
67	11·8	11·6	11·3	11·1	10·9	10·6	10·4	10·2	10·0	9·81	9·63	9·45	9·28	9·11	8·95	8·79	67
68	12·4	12·2	11·9	11·6	11·4	11·2	10·9	10·7	10·5	10·3	10·1	9·93	9·75	9·57	9·40	9·24	68
69	13·1	12·8	12·5	12·3	12·0	11·8	11·5	11·3	11·1	10·9	10·7	10·5	10·3	10·1	9·90	9·72	69
70	13·8	13·5	13·2	12·9	12·7	12·4	12·1	11·9	11·7	11·4	11·2	11·0	10·8	10·6	10·4	10·3	70
71	14·6	14·3	14·0	13·7	13·4	13·1	12·8	12·6	12·3	12·1	11·9	11·7	11·4	11·2	11·0	10·8	71
72	15·5	15·1	14·8	14·5	14·2	13·9	13·6	13·3	13·1	12·8	12·6	12·3	12·1	11·9	11·7	11·5	72
73	16·4	16·1	15·7	15·4	15·1	14·8	14·5	14·2	13·9	13·6	13·4	13·1	12·9	12·7	12·4	12·2	73
74	17·5	17·1	16·8	16·4	16·1	15·7	15·4	15·1	14·8	14·5	14·3	14·0	13·7	13·5	13·3	13·0	74
75	18·8	18·3	17·9	17·6	17·2	16·8	16·5	16·2	15·9	15·6	15·3	15·0	14·7	14·4	14·2	13·9	75
76	20·2	19·7	19·3	18·9	18·5	18·1	17·7	17·4	17·0	16·7	16·4	16·1	15·8	15·5	15·2	15·0	76
77	21·8	21·3	20·8	20·4	20·0	19·5	19·2	18·8	18·4	18·0	17·7	17·4	17·1	16·8	16·5	16·2	77
78	23·7	23·1	22·6	22·1	21·7	21·2	20·8	20·4	20·0	19·6	19·2	18·9	18·5	18·2	17·9	17·6	78
79	25·9	25·3	24·7	24·2	23·7	23·2	22·7	22·3	21·9	21·4	21·0	20·6	20·3	19·9	19·5	19·2	79
80	28·5	27·9	27·3	26·7	26·1	25·6	25·1	24·5	24·1	23·6	23·2	22·8	22·3	21·9	21·5	21·2	80
81	31·7	31·0	30·4	29·7	29·1	28·5	27·9	27·4	26·8	26·3	25·8	25·3	24·9	24·4	24·0	23·6	81
82	35·8	35·0	34·2	33·5	32·8	32·1	31·5	30·8	30·2	29·6	29·1	28·5	28·0	27·5	27·0	26·6	82
83	41·0	40·0	39·2	38·3	37·5	36·7	36·0	35·3	34·6	33·9	33·3	32·7	32·1	31·5	30·9	30·4	83
Lat.	168°45'	168°30'	168°15'	168°00'	167°45'	167°30'	167°15'	167°00'	166°45'	166°30'	166°15'	166°00'	165°45'	165°30'	165°15'	165°00'	Lat.
	191°15'	191°30'	191°45'	192°00'	192°15'	192°30'	192°45'	193°00'	193°15'	193°30'	193°45'	194°00'	194°15'	194°30'	194°45'	195°00'	

HOUR ANGLE.

B					TABLE B — HOUR ANGLE.										B

Dec.°	7° 30'	7° 45'	8° 00'	8° 15'	8° 30'	8° 45'	9° 00'	9° 15'	9° 30'	9° 45'	10° 00'	10° 15'	10° 30'	10° 45'	11° 00'	11° 15'	Dec.°
	352° 30'	352° 15'	352° 00'	351° 45'	351° 30'	351° 15'	351° 00'	350° 45'	350° 30'	350° 15'	350° 00'	349° 45'	349° 30'	349° 15'	349° 00'	348° 45'	
60	13·3	12·8	12·4	12·1	11·7	11·4	11·1	10·8	10·5	10·2	9·97	9·73	9·50	9·29	9·08	8·88	60
61	13·8	13·4	13·0	12·6	12·2	11·9	11·5	11·2	10·9	10·7	10·4	10·1	9·90	9·67	9·45	9·25	61
62	14·4	13·9	13·5	13·1	12·7	12·4	12·0	11·7	11·4	11·1	10·8	10·6	10·3	10·1	9·86	9·64	62
63	15·0	14·6	14·1	13·7	13·3	12·9	12·5	12·2	11·9	11·6	11·3	11·0	10·8	10·5	10·3	10·1	63
64	15·7	15·2	14·7	14·3	13·9	13·5	13·1	12·8	12·4	12·1	11·8	11·5	11·3	11·0	10·7	10·5	64
65	16·4	15·9	15·4	14·9	14·5	14·1	13·7	13·3	13·0	12·7	12·3	12·1	11·8	11·5	11·2	11·0	65
66	17·2	16·7	16·1	15·7	15·2	14·8	14·4	14·0	13·6	13·3	12·9	12·6	12·3	12·0	11·8	11·5	66
67	18·0	17·5	16·9	16·4	15·9	15·5	15·1	14·7	14·3	13·9	13·6	13·2	12·9	12·6	12·3	12·1	67
68	19·0	18·4	17·8	17·2	16·7	16·3	15·8	15·4	15·0	14·6	14·3	13·9	13·6	13·3	13·0	12·7	68
69	20·0	19·3	18·7	18·2	17·6	17·1	16·7	16·2	15·8	15·4	15·0	14·6	14·3	14·0	13·7	13·4	69
70	21·1	20·4	19·7	19·1	18·6	18·1	17·6	17·1	16·6	16·2	15·8	15·4	15·1	14·7	14·4	14·1	70
71	22·3	21·5	20·9	20·2	19·6	19·1	18·6	18·1	17·6	17·1	16·7	16·3	15·9	15·6	15·2	14·9	71
72	23·6	22·8	22·1	21·4	20·8	20·2	19·7	19·1	18·6	18·2	17·7	17·3	16·9	16·5	16·1	15·8	72
73	25·1	24·3	23·5	22·8	22·1	21·5	20·9	20·3	19·8	19·3	18·8	18·4	18·0	17·5	17·1	16·8	73
74	26·7	25·9	25·1	24·3	23·6	22·9	22·3	21·7	21·1	20·6	20·1	19·6	19·1	18·7	18·3	17·9	74
75	28·6	27·7	26·8	26·0	25·3	24·5	23·9	23·2	22·6	22·0	21·5	21·0	20·5	20·0	19·6	19·1	75
Dec.	172° 30'	172° 15'	172° 00'	171° 45'	171° 30'	171° 15'	171° 00'	170° 45'	170° 30'	170° 15'	170° 00'	169° 45'	169° 30'	169° 15'	169° 00'	168° 45'	Dec.
	187° 30'	187° 45'	188° 00'	188° 15'	188° 30'	188° 45'	189° 00'	189° 15'	189° 30'	189° 45'	190° 00'	190° 15'	190° 30'	190° 45'	191° 00'	191° 15'	

B		HOUR ANGLE.	B

B					TABLE B — HOUR ANGLE.										B

Dec.°	11° 15'	11° 30'	11° 45'	12° 00'	12° 15'	12° 30	12° 45'	13° 00'	13° 15'	13° 30'	13° 45'	14° 00'	14° 15'	14° 30'	14° 45'	15° 00'	Dec.°
	348° 45'	348° 30'	348° 15'	348° 00'	347° 45'	347° 30'	347° 15'	347° 00'	346° 45'	346° 30'	346° 15'	346° 00'	345° 45'	345° 30'	345° 15'	345° 00'	
60	8·88	8·69	8·51	8·33	8·16	8·00	7·85	7·70	7·56	7·42	7·29	7·16	7·04	6·92	6·80	6·69	60
61	9·25	9·05	8·86	8·68	8·50	8·34	8·17	8·02	7·87	7·73	7·59	7·46	7·33	7·21	7·09	6·97	61
62	9·64	9·43	9·24	9·05	8·86	8·69	8·52	8·36	8·21	8·06	7·91	7·77	7·64	7·51	7·39	7·27	62
63	10·1	9·84	9·64	9·44	9·25	9·07	8·89	8·72	8·56	8·41	8·26	8·11	7·97	7·84	7·71	7·58	63
64	10·5	10·3	10·1	9·86	9·66	9·47	9·29	9·11	8·95	8·78	8·63	8·48	8·33	8·19	8·05	7·92	64
65	11·0	10·8	10·5	10·3	10·1	9·91	9·72	9·53	9·36	9·19	9·02	8·86	8·71	8·57	8·42	8·29	65
66	11·5	11·3	11·0	10·8	10·6	10·4	10·2	9·98	9·80	9·62	9·45	9·28	9·12	8·97	8·82	8·68	66
67	12·1	11·8	11·6	11·3	11·1	10·9	10·7	10·5	10·3	10·1	9·91	9·74	9·57	9·41	9·25	9·10	67
68	12·7	12·4	12·2	11·9	11·7	11·4	11·2	11·0	10·8	10·6	10·4	10·2	10·1	9·89	9·72	9·56	68
69	13·4	13·1	12·8	12·5	12·3	12·0	11·8	11·6	11·4	11·2	11·0	10·8	10·6	10·4	10·2	10·1	69
70	14·1	13·8	13·5	13·2	12·9	12·7	12·4	12·2	12·0	11·8	11·6	11·4	11·2	11·0	10·8	10·6	70
71	14·9	14·6	14·3	14·0	13·7	13·4	13·2	12·9	12·7	12·4	12·2	12·0	11·8	11·6	11·4	11·2	71
72	15·8	15·4	15·1	14·8	14·5	14·2	13·9	13·7	13·4	13·2	12·9	12·7	12·5	12·3	12·1	11·9	72
73	16·8	16·4	16·1	15·7	15·4	15·1	14·8	14·5	14·3	14·0	13·8	13·5	13·3	13·1	12·8	12·6	73
74	17·9	17·5	17·1	16·8	16·4	16·1	15·8	15·5	15·2	14·9	14·7	14·4	14·2	13·9	13·7	13·5	74
75	19·1	18·7	18·3	18·0	17·6	17·2	16·9	16·6	16·3	16·0	15·7	15·4	15·2	14·9	14·7	14·4	75
Dec.	168° 45'	168° 30'	168° 15'	168° 00'	167° 45'	167° 30'	167° 15'	167° 00'	166° 45'	166° 30'	166° 15'	166° 00'	165° 45'	165° 30'	165° 15'	165° 00'	Dec.
	191° 15'	191° 30'	191° 45'	192° 00'	192° 15'	192° 30'	192° 45'	193° 00'	193° 15'	193° 30'	193° 45'	194° 00'	194° 15'	194° 30'	194° 45'	195° 00'	

B		HOUR ANGLE.	B

B—Always named the **same** as Declination

TABLE A — HOUR ANGLE.

Lat.°	15°00' 345°00'	15°30' 344°30'	16°00' 344°00'	16°30' 343°30'	17°00' 343°00'	17°30' 342°30'	18°00' 342°00'	18°30' 341°30'	19°00' 341°00'	19°30' 340°30'	20°00' 340°00'	20°30' 339°30'	21°00' 339°00'	21°30' 338°30'	22°00' 338°00'	22°30' 337°30'	Lat.°
60	6·46	6·25	6·04	5·85	5·67	5·49	5·33	5·18	5·03	4·89	4·76	4·63	4·51	4·40	4·29	4·18	60
61	6·73	6·51	6·29	6·09	5·90	5·72	5·55	5·39	5·24	5·09	4·96	4·83	4·70	4·58	4·47	4·36	61
62	7·02	6·78	6·56	6·35	6·15	5·96	5·79	5·62	5·46	5·31	5·17	5·03	4·90	4·77	4·65	4·54	62
63	7·32	7·08	6·84	6·63	6·42	6·23	6·04	5·87	5·70	5·54	5·39	5·25	5·11	4·98	4·86	4·74	63
64	7·65	7·39	7·15	6·92	6·71	6·50	6·31	6·13	5·95	5·79	5·63	5·48	5·34	5·21	5·08	4·95	64
65	8·00	7·73	7·48	7·24	7·01	6·80	6·60	6·41	6·23	6·06	5·89	5·74	5·59	5·44	5·31	5·18	65
66	8·38	8·10	7·83	7·58	7·35	7·12	6·91	6·71	6·52	6·34	6·17	6·01	5·85	5·70	5·56	5·42	66
67	8·79	8·50	8·22	7·95	7·71	7·47	7·25	7·04	6·84	6·65	6·47	6·30	6·14	5·98	5·83	5·69	67
68	9·24	8·92	8·63	8·36	8·10	7·85	7·62	7·40	7·19	6·99	6·80	6·62	6·45	6·28	6·13	5·98	68
69	9·72	9·39	9·09	8·80	8·52	8·26	8·02	7·79	7·57	7·36	7·16	6·97	6·79	6·61	6·45	6·29	69
70	10·3	9·91	9·58	9·28	8·99	8·71	8·46	8·21	7·98	7·76	7·55	7·35	7·16	6·98	6·80	6·63	70
71	10·8	10·5	10·1	9·81	9·50	9·21	8·94	8·68	8·43	8·20	7·98	7·77	7·57	7·37	7·19	7·01	71
72	11·5	11·1	10·7	10·4	10·1	9·76	9·47	9·20	8·94	8·69	8·46	8·23	8·02	7·81	7·62	7·43	72
73	12·2	11·8	11·4	11·1	10·7	10·4	10·1	9·78	9·50	9·24	8·99	8·75	8·52	8·30	8·10	7·90	73
74	13·0	12·6	12·2	11·8	11·4	11·1	10·7	10·4	10·1	9·85	9·58	9·33	9·09	8·85	8·63	8·42	74
75	13·9	13·5	13·0	12·6	12·2	11·8	11·5	11·2	10·8	10·5	10·3	9·98	9·72	9·47	9·24	9·01	75
76	15·0	14·5	14·0	13·5	13·1	12·7	12·4	12·0	11·7	11·3	11·1	10·7	10·5	10·2	9·93	9·68	76
77	16·2	15·6	15·1	14·6	14·2	13·7	13·3	13·0	12·6	12·2	11·9	11·6	11·3	11·0	10·7	10·5	77
78	17·6	17·0	16·4	15·9	15·4	14·9	14·5	14·1	13·7	13·3	12·9	12·6	12·3	11·9	11·7	11·4	78
79	19·2	18·6	17·9	17·4	16·8	16·3	15·8	15·4	14·9	14·5	14·1	13·8	13·4	13·1	12·7	12·4	79
80	21·2	20·5	19·8	19·2	18·6	18·0	17·5	17·0	16·5	16·0	15·6	15·2	14·8	14·4	14·0	13·7	80
81	23·6	22·8	22·0	21·3	20·7	20·0	19·4	18·9	18·3	17·8	17·4	16·9	16·5	16·0	15·6	15·2	81
82	26·6	25·7	24·8	24·0	23·3	22·6	21·9	21·3	20·7	20·1	19·6	19·0	18·5	18·1	17·6	17·2	82
83	30·4	29·4	28·4	27·5	26·6	25·8	25·1	24·3	23·7	23·0	22·4	21·8	21·2	20·7	20·2	19·7	83
Lat.	165°00' 195°00'	164°30' 195°30'	164°00' 196°00'	163°30' 196°30'	163°00' 197°00'	162°30' 197°30'	162°00' 198°00'	161°30' 198°30'	161°00' 199°00'	160°30' 199°30'	160°00' 200°00'	159°30' 200°30'	159°00' 201°00'	158°30' 201°30'	158°00' 202°00'	157°30' 202°30'	Lat.

A   HOUR ANGLE.   A

TABLE A — HOUR ANGLE.

Lat.°	22°30' 337°30'	23°00' 337°00'	23°30' 336°30'	24°00' 336°00'	24°30' 335°30'	25°00' 335°00'	25°30' 334°30'	26°00' 334°00'	26°30' 333°30'	27°00' 333°00'	27°30' 332°30'	28°00' 332°00'	28°30' 331°30'	29°00' 331°00'	29°30' 330°30'	30°00' 330°00'	Lat.°
60	4·18	4·08	3·98	3·89	3·80	3·71	3·63	3·55	3·47	3·40	3·33	3·26	3·19	3·12	3·06	3·00	60
61	4·36	4·25	4·15	4·05	3·96	3·87	3·78	3·70	3·62	3·54	3·47	3·39	3·32	3·25	3·18	3·12	61
62	4·54	4·43	4·33	4·22	4·13	4·03	3·94	3·86	3·77	3·69	3·61	3·54	3·46	3·39	3·32	3·26	62
63	4·74	4·62	4·51	4·41	4·31	4·21	4·11	4·02	3·94	3·85	3·77	3·69	3·61	3·54	3·47	3·40	63
64	4·95	4·83	4·72	4·61	4·50	4·40	4·30	4·20	4·11	4·02	3·94	3·86	3·78	3·70	3·62	3·55	64
65	5·18	5·05	4·93	4·82	4·71	4·60	4·50	4·40	4·30	4·21	4·12	4·03	3·95	3·87	3·79	3·71	65
66	5·42	5·29	5·17	5·04	4·93	4·82	4·71	4·61	4·51	4·41	4·31	4·22	4·13	4·05	3·97	3·89	66
67	5·69	5·55	5·42	5·29	5·17	5·05	4·94	4·83	4·73	4·62	4·53	4·43	4·34	4·25	4·16	4·08	67
68	5·98	5·83	5·69	5·56	5·43	5·31	5·19	5·07	4·97	4·86	4·75	4·65	4·56	4·47	4·38	4·29	68
69	6·29	6·14	5·99	5·85	5·72	5·59	5·46	5·34	5·23	5·11	5·00	4·90	4·80	4·70	4·60	4·51	69
70	6·63	6·47	6·32	6·17	6·03	5·89	5·76	5·63	5·51	5·39	5·28	5·17	5·06	4·96	4·86	4·76	70
71	7·01	6·84	6·68	6·52	6·37	6·23	6·09	5·96	5·83	5·70	5·58	5·46	5·35	5·24	5·13	5·03	71
72	7·43	7·25	7·08	6·91	6·75	6·60	6·45	6·31	6·17	6·04	5·91	5·79	5·67	5·55	5·44	5·33	72
73	7·90	7·71	7·52	7·35	7·18	7·01	6·86	6·71	6·56	6·42	6·28	6·15	6·03	5·90	5·78	5·67	73
74	8·42	8·22	8·02	7·83	7·65	7·48	7·31	7·15	6·99	6·84	6·70	6·56	6·42	6·29	6·16	6·04	74
75	9·01	8·79	8·58	8·38	8·19	8·00	7·82	7·65	7·49	7·32	7·17	7·02	6·87	6·73	6·60	6·46	75
76	9·68	9·45	9·22	9·01	8·80	8·60	8·41	8·22	8·04	7·87	7·71	7·54	7·39	7·24	7·09	6·95	76
77	10·5	10·2	9·96	9·73	9·51	9·29	9·08	8·88	8·69	8·50	8·32	8·15	7·98	7·81	7·66	7·50	77
78	11·4	11·1	10·8	10·6	10·3	10·1	9·86	9·65	9·44	9·23	9·04	8·85	8·67	8·49	8·32	8·15	78
79	12·4	12·1	11·8	11·6	11·3	11·0	10·8	10·6	10·3	10·1	9·88	9·68	9·48	9·28	9·09	8·91	79
80	13·7	13·4	13·0	12·7	12·4	12·2	11·9	11·6	11·4	11·1	10·9	10·7	10·4	10·3	10·0	9·82	80
81	15·2	14·9	14·5	14·2	13·9	13·5	13·2	13·0	12·7	12·4	12·1	11·9	11·6	11·4	11·2	10·9	81
82	17·2	16·8	16·4	16·0	15·6	15·3	14·9	14·6	14·3	14·0	13·7	13·4	13·1	12·8	12·6	12·3	82
83	19·7	19·2	18·7	18·3	17·9	17·5	17·1	16·7	16·3	16·0	15·7	15·3	15·0	14·7	14·4	14·1	83
Lat.	157°30' 202°30'	157°00' 203°00'	156°30' 203°30'	156°00' 204°00'	155°30' 204°30'	155°00' 205°00'	154°30' 205°30'	154°00' 206°00'	153°30' 206°30'	153°00' 207°00'	152°30' 207°30'	152°00' 208°00'	151°30' 208°30'	151°00' 209°00'	150°30' 209°30'	150°00' 210°00'	Lat.

A   HOUR ANGLE.   A

B — TABLE B — HOUR ANGLE. — B

Dec.	15° 00'	15° 30'	16° 00'	16° 30'	17° 00'	17° 30'	18° 00'	18° 30'	19° 00'	19° 30'	20° 00'	20° 30'	21° 00'	21° 30'	22° 00'	22° 30'	Dec.
°	345° 00'	344° 30'	344° 00'	343° 30'	343° 00'	342° 30'	342° 00'	341° 30'	341° 00'	340° 30'	340° 00'	339° 30'	339° 00'	338° 30'	338° 00'	337° 30'	°
60	6·69	6·48	6·28	6·10	5·92	5·76	5·61	5·46	5·32	5·19	5·06	4·95	4·83	4·73	4·62	4·53	60
61	6·97	6·75	6·55	6·35	6·17	6·00	5·84	5·69	5·54	5·40	5·27	5·15	5·03	4·92	4·82	4·71	61
62	7·27	7·04	6·82	6·62	6·43	6·25	6·09	5·93	5·78	5·63	5·50	5·37	5·25	5·13	5·02	4·91	62
63	7·58	7·34	7·12	6·91	6·71	6·53	6·35	6·19	6·03	5·88	5·74	5·60	5·48	5·36	5·24	5·13	63
64	7·92	7·67	7·44	7·22	7·01	6·82	6·63	6·46	6·30	6·14	5·99	5·85	5·72	5·59	5·47	5·36	64
65	8·29	8·02	7·78	7·55	7·33	7·13	6·94	6·76	6·59	6·42	6·27	6·12	5·98	5·85	5·72	5·60	65
66	8·68	8·40	8·15	7·91	7·68	7·47	7·27	7·08	6·90	6·73	6·57	6·41	6·27	6·13	6·00	5·87	66
67	9·10	8·82	8·55	8·29	8·06	7·83	7·62	7·42	7·24	7·06	6·89	6·73	6·57	6·43	6·29	6·16	67
68	9·56	9·26	8·98	8·72	8·47	8·23	8·01	7·80	7·60	7·41	7·24	7·07	6·91	6·75	6·61	6·47	68
69	10·1	9·75	9·45	9·17	8·91	8·66	8·43	8·21	8·00	7·80	7·62	7·44	7·27	7·11	6·95	6·81	69
70	10·6	10·3	9·97	9·67	9·40	9·14	8·89	8·66	8·44	8·23	8·03	7·85	7·67	7·50	7·33	7·18	70
71	11·2	10·9	10·5	10·2	9·93	9·66	9·40	9·15	8·92	8·70	8·49	8·29	8·10	7·92	7·75	7·59	71
72	11·9	11·5	11·2	10·8	10·5	10·2	9·96	9·70	9·45	9·22	9·00	8·79	8·59	8·40	8·22	8·04	72
73	12·6	12·2	11·9	11·5	11·2	10·9	10·6	10·3	10·0	9·80	9·57	9·34	9·13	8·93	8·73	8·55	73
74	13·5	13·1	12·7	12·3	11·9	11·6	11·3	11·0	10·7	10·4	10·2	9·96	9·73	9·52	9·31	9·11	74
75	14·4	14·0	13·5	13·1	12·8	12·4	12·1	11·8	11·5	11·2	10·9	10·7	10·4	10·2	9·96	9·75	75
Dec.	165° 00'	164° 30'	164° 00'	163° 30'	163° 00'	162° 30'	162° 00'	161° 30'	161° 00'	160° 30'	160° 00'	159° 30'	159° 00'	158° 30'	158° 00'	157° 30'	Dec.
	195° 00'	195° 30'	196° 00'	196° 30'	197° 00'	197° 30'	198° 00'	198° 30'	199° 00'	199° 30'	200° 00'	200° 30'	201° 00'	201° 30'	202° 00'	202° 30'	

B — HOUR ANGLE. — B

B — TABLE B — HOUR ANGLE. — B

Dec.	22° 30'	23° 00'	23° 30'	24° 00'	24° 30'	25° 00'	25° 30'	26° 00'	26° 30'	27° 00'	27° 30'	28° 00'	28° 30'	29° 00'	29° 30'	30° 00'	Dec.
°	337° 30'	337° 00'	336° 30'	336° 00'	335° 30'	335° 00'	334° 30'	334° 00'	333° 30'	333° 00'	332° 30'	332° 00'	331° 30'	331° 00'	330° 30'	330° 00'	°
60	4·53	4·43	4·34	4·26	4·18	4·10	4·02	3·95	3·88	3·82	3·75	3·69	3·63	3·57	3·52	3·46	60
61	4·71	4·62	4·52	4·44	4·35	4·27	4·19	4·12	4·04	3·97	3·91	3·84	3·78	3·72	3·66	3·61	61
62	4·91	4·81	4·72	4·62	4·54	4·45	4·37	4·29	4·22	4·14	4·07	4·01	3·94	3·88	3·82	3·76	62
63	5·13	5·02	4·92	4·83	4·73	4·64	4·56	4·48	4·40	4·32	4·25	4·18	4·11	4·05	3·99	3·93	63
64	5·36	5·25	5·14	5·04	4·94	4·85	4·76	4·68	4·60	4·52	4·44	4·37	4·30	4·23	4·16	4·10	64
65	5·60	5·49	5·38	5·27	5·17	5·07	4·98	4·89	4·81	4·72	4·64	4·57	4·49	4·42	4·35	4·29	65
66	5·87	5·75	5·63	5·52	5·42	5·31	5·22	5·12	5·03	4·95	4·86	4·78	4·71	4·63	4·56	4·49	66
67	6·16	6·03	5·91	5·79	5·68	5·57	5·47	5·37	5·28	5·19	5·10	5·02	4·94	4·86	4·78	4·71	67
68	6·47	6·33	6·21	6·09	5·97	5·86	5·75	5·65	5·55	5·45	5·36	5·27	5·19	5·11	5·03	4·95	68
69	6·81	6·67	6·53	6·40	6·28	6·16	6·05	5·94	5·84	5·74	5·64	5·55	5·46	5·37	5·29	5·21	69
70	7·18	7·03	6·89	6·75	6·63	6·50	6·38	6·27	6·16	6·05	5·95	5·85	5·76	5·67	5·58	5·49	70
71	7·59	7·43	7·28	7·14	7·00	6·87	6·75	6·63	6·51	6·40	6·29	6·19	6·08	5·99	5·90	5·81	71
72	8·04	7·88	7·72	7·57	7·42	7·28	7·15	7·02	6·90	6·78	6·67	6·56	6·45	6·35	6·25	6·15	72
73	8·55	8·37	8·20	8·04	7·89	7·74	7·60	7·46	7·33	7·21	7·08	6·97	6·86	6·75	6·64	6·54	73
74	9·11	8·93	8·75	8·57	8·41	8·25	8·10	7·96	7·82	7·68	7·55	7·43	7·31	7·19	7·08	6·97	74
75	9·75	9·55	9·36	9·18	9·00	8·83	8·67	8·52	8·37	8·22	8·08	7·95	7·82	7·70	7·58	7·46	75
Dec.	157° 30'	157° 00'	156° 30'	156° 00'	155° 30'	155° 00'	154° 30'	154° 00'	153° 30'	153° 00'	152° 30'	152° 00'	151° 30'	151° 00'	150° 30'	150° 00'	Dec.
	202° 30'	203° 00'	203° 30'	204° 00'	204° 30'	205° 00'	205° 30'	206° 00'	206° 30'	207° 00'	207° 30'	208° 00'	208° 30'	209° 00'	209° 30'	210° 00'	

B — HOUR ANGLE. — B

B—Always named the **same** as Declination.

A — TABLE A — HOUR ANGLE.

Lat.	30°	31°	32°	33°	34°	35°	36°	37°	38°	39°	40°	41°	42°	43°	44°	45°	Lat.
°	330°	329°	328°	327°	326°	325°	324°	323°	322°	321°	320°	319°	318°	317°	316°	315°	°
60	3·00	2·88	2·77	2·67	2·57	2·47	2·38	2·30	2·22	2·14	2·06	1·99	1·92	1·86	1·79	1·73	60
61	3·12	3·00	2·89	2·78	2·67	2·58	2·48	2·39	2·31	2·23	2·15	2·08	2·00	1·93	1·87	1·80	61
62	3·26	3·13	3·01	2·90	2·79	2·69	2·59	2·50	2·41	2·32	2·24	2·16	2·09	2·02	1·95	1·88	62
63	3·40	3·27	3·14	3·02	2·91	2·80	2·70	2·60	2·51	2·42	2·34	2·26	2·18	2·10	2·03	1·96	63
64	3·55	3·41	3·28	3·16	3·04	2·93	2·82	2·72	2·62	2·53	2·44	2·36	2·28	2·20	2·12	2·05	64
65	3·71	3·57	3·43	3·30	3·18	3·06	2·95	2·85	2·74	2·65	2·56	2·47	2·38	2·30	2·22	2·14	65
66	3·89	3·74	3·59	3·46	3·33	3·21	3·09	2·98	2·87	2·77	2·68	2·58	2·49	2·41	2·33	2·25	66
67	4·08	3·92	3·77	3·63	3·49	3·36	3·24	3·13	3·02	2·91	2·81	2·71	2·62	2·53	2·44	2·36	67
68	4·29	4·12	3·96	3·81	3·67	3·53	3·41	3·28	3·17	3·06	2·95	2·85	2·75	2·65	2·56	2·48	68
69	4·51	4·34	4·17	4·01	3·86	3·72	3·59	3·46	3·34	3·22	3·11	3·00	2·89	2·79	2·70	2·61	69
70	4·76	4·57	4·40	4·23	4·07	3·92	3·78	3·65	3·52	3·39	3·27	3·16	3·04	2·95	2·85	2·75	70
71	5·03	4·83	4·65	4·47	4·31	4·15	4·00	3·85	3·72	3·59	3·46	3·34	3·23	3·11	3·01	2·90	71
72	5·33	5·12	4·93	4·74	4·56	4·40	4·24	4·08	3·94	3·80	3·67	3·54	3·42	3·30	3·19	3·08	72
73	5·67	5·44	5·23	5·04	4·85	4·67	4·50	4·34	4·19	4·04	3·90	3·76	3·63	3·51	3·39	3·27	73
74	6·04	5·80	5·58	5·37	5·17	4·98	4·80	4·63	4·46	4·31	4·16	4·01	3·87	3·74	3·61	3·49	74
75	6·46	6·21	5·97	5·75	5·53	5·33	5·14	4·95	4·78	4·61	4·45	4·29	4·15	4·00	3·87	3·73	75
76	6·95	6·68	6·42	6·18	5·95	5·73	5·52	5·32	5·13	4·95	4·78	4·61	4·45	4·30	4·15	4·01	76
77	7·50	7·21	6·93	6·67	6·42	6·19	5·96	5·75	5·54	5·35	5·16	4·98	4·81	4·65	4·49	4·33	77
78	8·15	7·83	7·53	7·25	6·98	6·72	6·48	6·24	6·02	5·81	5·61	5·41	5·23	5·05	4·87	4·70	78
79	8·91	8·56	8·23	7·92	7·63	7·35	7·08	6·83	6·59	6·35	6·13	5·92	5·71	5·52	5·33	5·14	79
80	9·82	9·44	9·08	8·73	8·41	8·10	7·81	7·53	7·26	7·00	6·76	6·52	6·30	6·08	5·87	5·67	80
81	10·9	10·5	10·1	9·72	9·36	9·02	8·69	8·38	8·08	7·80	7·52	7·26	7·01	6·77	6·54	6·31	81
82	12·3	11·8	11·4	11·0	10·6	10·2	9·79	9·44	9·11	8·79	8·48	8·19	7·90	7·63	7·37	7·12	82
83	14·1	13·6	13·0	12·5	12·1	11·6	11·2	10·8	10·4	10·1	9·71	9·37	9·05	8·73	8·43	8·14	83
Lat.	150°	149°	148°	147°	146°	145°	144°	143°	142°	141°	140°	139°	138°	137°	136°	135°	Lat.
	210°	211°	212°	213°	214°	215°	216°	217°	218°	219°	220°	221°	222°	223°	224°	225°	

A — HOUR ANGLE. — A

A — TABLE A — HOUR ANGLE. — A

Lat.	45°	46°	47°	48°	49°	50°	51°	52°	53°	54°	55°	56°	57°	58°	59°	60°	Lat.
°	315°	314°	313°	312°	311°	310°	309°	308°	307°	306°	305°	304°	303°	302°	301°	300°	°
60	1·73	1·67	1·62	1·56	1·51	1·45	1·40	1·35	1·31	1·26	1·21	1·17	1·12	1·08	1·04	1·00	60
61	1·80	1·74	1·68	1·62	1·57	1·51	1·46	1·41	1·36	1·31	1·26	1·22	1·17	1·13	1·08	1·04	61
62	1·88	1·82	1·75	1·69	1·63	1·58	1·52	1·47	1·42	1·37	1·32	1·27	1·22	1·18	1·13	1·09	62
63	1·96	1·90	1·83	1·77	1·71	1·65	1·59	1·53	1·48	1·43	1·37	1·32	1·27	1·23	1·18	1·13	63
64	2·05	1·98	1·91	1·85	1·78	1·72	1·66	1·60	1·55	1·49	1·44	1·38	1·33	1·28	1·23	1·18	64
65	2·14	2·07	2·00	1·93	1·86	1·80	1·74	1·68	1·62	1·56	1·50	1·45	1·39	1·34	1·29	1·24	65
66	2·25	2·17	2·09	2·02	1·95	1·88	1·82	1·75	1·69	1·63	1·57	1·52	1·46	1·40	1·35	1·30	66
67	2·36	2·28	2·20	2·12	2·05	1·98	1·91	1·84	1·78	1·71	1·65	1·59	1·53	1·47	1·42	1·36	67
68	2·48	2·39	2·31	2·23	2·15	2·08	2·00	1·93	1·87	1·80	1·73	1·67	1·61	1·55	1·49	1·43	68
69	2·61	2·52	2·43	2·35	2·26	2·19	2·11	2·04	1·96	1·89	1·82	1·76	1·69	1·63	1·57	1·50	69
70	2·75	2·65	2·56	2·47	2·39	2·31	2·23	2·15	2·07	2·00	1·92	1·85	1·78	1·72	1·65	1·59	70
71	2·90	2·80	2·71	2·62	2·52	2·44	2·35	2·27	2·19	2·11	2·03	1·96	1·89	1·82	1·75	1·68	71
72	3·08	2·97	2·87	2·77	2·68	2·58	2·49	2·41	2·32	2·24	2·16	2·08	2·00	1·92	1·85	1·78	72
73	3·27	3·16	3·05	2·95	2·84	2·74	2·65	2·56	2·47	2·38	2·29	2·21	2·12	2·04	1·97	1·89	73
74	3·49	3·37	3·25	3·14	3·03	2·93	2·82	2·73	2·63	2·53	2·44	2·35	2·27	2·18	2·10	2·01	74
75	3·73	3·60	3·48	3·36	3·24	3·13	3·02	2·92	2·81	2·71	2·61	2·52	2·42	2·33	2·24	2·16	75
76	4·01	3·87	3·74	3·61	3·49	3·37	3·25	3·13	3·02	2·91	2·81	2·71	2·61	2·51	2·41	2·32	76
77	4·33	4·18	4·04	3·90	3·77	3·64	3·51	3·38	3·26	3·15	3·03	2·92	2·81	2·71	2·60	2·50	77
78	4·70	4·54	4·39	4·24	4·09	3·95	3·81	3·68	3·55	3·42	3·29	3·17	3·06	2·94	2·83	2·72	78
79	5·14	4·97	4·80	4·63	4·47	4·32	4·17	4·02	3·88	3·74	3·60	3·47	3·34	3·22	3·09	2·97	79
80	5·67	5·48	5·29	5·11	4·93	4·76	4·59	4·43	4·27	4·12	3·97	3·83	3·68	3·54	3·41	3·27	80
81	6·31	6·10	5·89	5·69	5·49	5·30	5·11	4·93	4·76	4·59	4·42	4·26	4·10	3·95	3·79	3·65	81
82	7·12	6·87	6·64	6·41	6·19	5·97	5·76	5·56	5·36	5·17	4·98	4·80	4·62	4·45	4·28	4·11	82
83	8·14	7·86	7·60	7·33	7·08	6·83	6·60	6·36	6·14	5·92	5·70	5·49	5·29	5·09	4·89	4·70	83
Lat.	135°	134°	133°	132°	131°	130°	129°	128°	127°	126°	125°	124°	123°	122°	121°	120°	Lat.
	225°	226°	227°	228°	229°	230°	231°	232°	233°	234°	235°	236°	237°	238°	239°	240°	

A — HOUR ANGLE. — A

| B | | | | | | TABLE B — HOUR ANGLE. | | | | | | | | | B |

Dec.	30°	31°	32°	33°	34°	35°	36°	37°	38°	39°	40°	41°	42°	43°	44°	45°	Dec.
°	330°	329°	328°	327°	326°	325°	324°	323°	322°	321°	320°	319°	318°	317°	316°	315°	°
60	3·46	3·36	3·27	3·18	3·10	3·02	2·95	2·88	2·81	2·75	2·69	2·64	2·59	2·54	2·49	2·45	60
61	3·61	3·50	3·40	3·31	3·23	3·15	3·07	3·00	2·93	2·87	2·81	2·75	2·70	2·65	2·60	2·55	61
62	3·76	3·65	3·55	3·45	3·36	3·28	3·20	3·13	3·05	2·99	2·93	2·87	2·81	2·76	2·71	2·66	62
63	3·93	3·81	3·70	3·60	3·51	3·42	3·34	3·26	3·19	3·12	3·05	2·99	2·93	2·88	2·83	2·78	63
64	4·10	3·98	3·87	3·76	3·67	3·57	3·49	3·41	3·33	3·26	3·19	3·13	3·06	3·01	2·95	2·90	64
65	4·29	4·16	4·05	3·94	3·84	3·74	3·65	3·56	3·48	3·41	3·34	3·27	3·20	3·14	3·09	3·03	65
66	4·49	4·36	4·24	4·12	4·02	3·92	3·82	3·73	3·65	3·57	3·49	3·42	3·36	3·29	3·23	3·18	66
67	4·71	4·57	4·45	4·33	4·21	4·11	4·01	3·91	3·83	3·74	3·67	3·59	3·52	3·45	3·39	3·33	67
68	4·95	4·81	4·67	4·54	4·43	4·32	4·21	4·11	4·02	3·93	3·85	3·77	3·70	3·63	3·56	3·50	68
69	5·21	5·06	4·92	4·78	4·66	4·54	4·43	4·33	4·23	4·14	4·05	3·97	3·89	3·82	3·75	3·68	69
70	5·49	5·33	5·18	5·04	4·91	4·79	4·67	4·56	4·46	4·37	4·27	4·19	4·11	4·03	3·95	3·89	70
71	5·81	5·64	5·48	5·33	5·19	5·06	4·94	4·83	4·72	4·61	4·52	4·43	4·34	4·26	4·18	4·11	71
72	6·15	5·98	5·81	5·65	5·50	5·37	5·24	5·11	5·00	4·89	4·79	4·69	4·60	4·51	4·43	4·35	72
73	6·54	6·35	6·17	6·01	5·85	5·70	5·57	5·44	5·31	5·20	5·09	4·99	4·89	4·80	4·71	4·63	73
74	6·97	6·77	6·58	6·40	6·24	6·08	5·93	5·79	5·67	5·54	5·43	5·32	5·21	5·11	5·02	4·93	74
75	7·46	7·25	7·04	6·85	6·67	6·51	6·35	6·20	6·06	5·93	5·81	5·69	5·58	5·47	5·37	5·28	75
Dec.	150°	149°	148°	147°	146°	145°	144°	143°	142°	141°	140°	139°	138°	137°	136°	135°	Dec.
	210°	211°	212°	213°	214°	215°	216°	217°	218°	219°	220°	221°	222°	223°	224°	225°	

| B | HOUR ANGLE. | B |

| B | | | | | | TABLE B — HOUR ANGLE. | | | | | | | | | B |

Dec.	45°	46°	47°	48°	49°	50°	51°	52°	53°	54°	55°	56°	57°	58°	59°	60°	Dec.
°	315°	314°	313°	312°	311°	310°	309°	308°	307°	306°	305°	304°	303°	302°	301°	300°	°
60	2·45	2·41	2·37	2·33	2·29	2·26	2·23	2·20	2·17	2·14	2·11	2·09	2·07	2·04	2·02	2·00	60
61	2·55	2·51	2·47	2·43	2·39	2·36	2·32	2·29	2·26	2·23	2·20	2·18	2·15	2·13	2·10	2·08	61
62	2·66	2·62	2·57	2·53	2·49	2·46	2·42	2·39	2·35	2·32	2·30	2·27	2·24	2·22	2·19	2·17	62
63	2·78	2·73	2·68	2·64	2·60	2·56	2·53	2·49	2·46	2·43	2·40	2·36	2·34	2·31	2·29	2·27	63
64	2·90	2·85	2·80	2·76	2·72	2·68	2·64	2·60	2·57	2·53	2·50	2·47	2·44	2·42	2·39	2·37	64
65	3·03	2·98	2·93	2·89	2·84	2·80	2·76	2·72	2·69	2·65	2·62	2·59	2·56	2·53	2·50	2·48	65
66	3·18	3·12	3·07	3·02	2·98	2·93	2·89	2·85	2·81	2·78	2·74	2·71	2·68	2·65	2·62	2·59	66
67	3·33	3·28	3·22	3·17	3·12	3·08	3·03	2·99	2·95	2·91	2·88	2·84	2·81	2·78	2·75	2·72	67
68	3·50	3·44	3·38	3·33	3·28	3·23	3·18	3·14	3·10	3·06	3·02	2·99	2·95	2·92	2·89	2·86	68
69	3·68	3·62	3·56	3·51	3·45	3·40	3·35	3·31	3·26	3·22	3·18	3·14	3·11	3·07	3·04	3·01	69
70	3·89	3·82	3·76	3·70	3·64	3·59	3·54	3·49	3·44	3·40	3·35	3·31	3·28	3·24	3·20	3·17	70
71	4·11	4·04	3·97	3·91	3·85	3·79	3·74	3·68	3·64	3·59	3·54	3·50	3·46	3·42	3·39	3·35	71
72	4·35	4·28	4·21	4·14	4·08	4·02	3·96	3·91	3·85	3·80	3·76	3·71	3·67	3·63	3·59	3·55	72
73	4·63	4·55	4·47	4·40	4·33	4·27	4·21	4·15	4·10	4·04	3·99	3·95	3·90	3·86	3·82	3·78	73
74	4·93	4·85	4·77	4·69	4·62	4·55	4·49	4·43	4·37	4·31	4·26	4·21	4·16	4·11	4·07	4·03	74
75	5·28	5·19	5·10	5·02	4·95	4·87	4·80	4·74	4·67	4·61	4·56	4·50	4·45	4·40	4·35	4·31	75
Dec.	135°	134°	133°	132°	131°	130°	129°	128°	127°	126°	125°	124°	123°	122°	121°	120°	Dec.
	225°	226°	227°	228°	229°	230°	231°	232°	233°	234°	235°	236°	237°	238°	239°	240°	

| B | HOUR ANGLE. | B |

B—Always named the same as Declination.

TABLE A — HOUR ANGLE.

Lat. °	60° 300°	61° 299°	62° 298°	63° 297°	64° 296°	65° 295°	66° 294°	67° 293°	68° 292°	69° 291°	70° 290°	71° 289°	72° 288°	73° 287°	74° 286°	75° 285°	Lat. °
60	1·00	·96	·92	·88	·85	·81	·77	·74	·70	·67	·63	·60	·56	·53	·50	·46	60
61	1·04	1·00	·96	·92	·88	·84	·80	·77	·73	·69	·66	·62	·59	·55	·52	·48	61
62	1·09	1·04	1·00	·96	·92	·88	·84	·80	·76	·72	·68	·65	·61	·58	·54	·50	62
63	1·13	1·09	1·04	1·00	·96	·92	·87	·83	·79	·75	·71	·68	·64	·60	·56	·53	63
64	1·19	1·14	1·09	1·05	1·00	·96	·91	·87	·83	·79	·75	·71	·67	·63	·59	·55	64
65	1·24	1·19	1·14	1·09	1·05	1·00	·96	·91	·87	·82	·78	·74	·70	·66	·62	·57	65
66	1·30	1·25	1·19	1·14	1·10	1·05	1·00	·95	·91	·86	·82	·77	·73	·69	·64	·60	66
67	1·36	1·31	1·25	1·20	1·15	1·10	1·05	1·00	·95	·90	·86	·81	·77	·72	·68	·63	67
68	1·43	1·37	1·32	1·26	1·21	1·15	1·09	1·05	1·00	·95	·90	·85	·80	·76	·71	·66	68
69	1·50	1·44	1·39	1·33	1·27	1·22	1·16	1·11	1·05	1·00	·95	·90	·85	·80	·75	·70	69
70	1·59	1·52	1·46	1·40	1·34	1·28	1·22	1·17	1·11	1·05	1·00	·95	·89	·84	·79	·74	70
71	1·68	1·61	1·54	1·48	1·42	1·35	1·29	1·23	1·17	1·11	1·06	1·00	·94	·89	·83	·78	71
72	1·78	1·71	1·64	1·57	1·50	1·44	1·37	1·31	1·24	1·18	1·12	1·06	1·00	·94	·88	·82	72
73	1·89	1·81	1·74	1·67	1·60	1·53	1·46	1·39	1·32	1·26	1·19	1·13	1·06	1·00	·94	·88	73
74	2·01	1·93	1·85	1·78	1·70	1·63	1·55	1·48	1·41	1·34	1·27	1·20	1·13	1·07	1·00	·93	74
75	2·16	2·07	1·98	1·90	1·82	1·74	1·66	1·58	1·51	1·43	1·36	1·29	1·21	1·14	1·07	1·00	75
76	2·32	2·22	2·13	2·04	1·96	1·87	1·79	1·70	1·62	1·54	1·46	1·38	1·30	1·23	1·15	1·07	76
77	2·50	2·40	2·30	2·21	2·11	2·02	1·93	1·84	1·75	1·66	1·58	1·49	1·41	1·32	1·24	1·16	77
78	2·72	2·61	2·50	2·40	2·29	2·19	2·09	2·00	1·90	1·81	1·71	1·62	1·53	1·44	1·35	1·26	78
79	2·97	2·85	2·74	2·62	2·51	2·40	2·29	2·18	2·08	1·97	1·87	1·77	1·67	1·57	1·48	1·37	79
80	3·27	3·14	3·02	2·89	2·77	2·64	2·53	2·41	2·29	2·18	2·06	1·95	1·84	1·73	1·63	1·52	80
81	3·65	3·50	3·36	3·22	3·08	2·94	2·81	2·68	2·55	2·42	2·30	2·17	2·05	1·93	1·81	1·69	81
82	4·11	3·94	3·78	3·63	3·47	3·32	3·17	3·02	2·87	2·73	2·59	2·45	2·31	2·18	2·04	1·91	82
83	4·70	4·51	4·33	4·15	3·97	3·80	3·63	3·46	3·29	3·13	2·96	2·80	2·65	2·49	2·34	2·18	83
Lat.	120° 240°	119° 241°	118° 242°	117° 243°	116° 244°	115° 245°	114° 246°	113° 247°	112° 248°	111° 249°	110° 250°	109° 251°	108° 252°	107° 253°	106° 254°	105° 255°	Lat.

A   HOUR ANGLE.   A

TABLE A — HOUR ANGLE.

Lat. °	75° 285°	76° 284°	77° 283°	78° 282°	79° 281°	80° 280°	81° 279°	82° 278°	83° 277°	84° 276°	85° 275°	86° 274°	87° 273°	88° 272°	89° 271°	90° 270°	Lat. °
60	·46	·43	·40	·37	·34	·31	·27	·24	·21	·18	·15	·12	·09	·06	·03	·00	60
61	·48	·45	·42	·38	·35	·32	·29	·25	·22	·19	·16	·13	·10	·06	·03	·00	61
62	·50	·47	·43	·40	·37	·33	·30	·26	·23	·20	·16	·13	·10	·07	·03	·00	62
63	·53	·49	·45	·42	·38	·35	·31	·28	·24	·21	·17	·14	·10	·07	·03	·00	63
64	·55	·51	·47	·44	·40	·36	·33	·29	·25	·22	·18	·14	·11	·07	·04	·00	64
65	·57	·54	·50	·46	·42	·38	·34	·30	·26	·23	·19	·15	·11	·08	·04	·00	65
66	·60	·56	·52	·48	·44	·40	·36	·32	·28	·24	·20	·16	·12	·08	·04	·00	66
67	·63	·59	·54	·50	·46	·42	·37	·33	·29	·25	·21	·17	·12	·08	·04	·00	67
68	·66	·62	·57	·53	·48	·44	·39	·35	·30	·26	·22	·17	·13	·09	·04	·00	68
69	·70	·65	·60	·55	·51	·46	·41	·37	·32	·27	·23	·18	·14	·09	·05	·00	69
70	·74	·69	·63	·58	·53	·48	·44	·39	·34	·29	·24	·19	·14	·10	·05	·00	70
71	·78	·72	·67	·62	·57	·51	·46	·41	·36	·31	·25	·20	·15	·10	·05	·00	71
72	·82	·77	·71	·65	·60	·54	·49	·43	·38	·32	·27	·22	·16	·11	·05	·00	72
73	·88	·82	·76	·70	·64	·58	·52	·46	·40	·34	·29	·23	·17	·11	·06	·00	73
74	·93	·87	·81	·74	·68	·61	·55	·49	·43	·37	·31	·24	·18	·12	·06	·00	74
75	1·00	·93	·86	·79	·73	·66	·59	·53	·46	·39	·33	·26	·20	·13	·07	·00	75
76	1·07	1·00	·93	·85	·78	·71	·64	·56	·49	·42	·35	·28	·21	·14	·07	·00	76
77	1·16	1·08	1·00	·92	·84	·76	·69	·61	·53	·46	·38	·30	·23	·15	·08	·00	77
78	1·26	1·17	1·09	1·00	·91	·83	·75	·66	·58	·49	·41	·33	·25	·16	·08	·00	78
79	1·37	1·28	1·19	1·09	1·00	·91	·82	·72	·63	·54	·45	·36	·27	·18	·09	·00	79
80	1·52	1·41	1·31	1·21	1·10	1·00	·90	·80	·70	·60	·50	·40	·30	·20	·10	·00	80
81	1·69	1·57	1·46	1·34	1·23	1·11	1·00	·89	·78	·66	·55	·44	·33	·22	·11	·00	81
82	1·91	1·77	1·64	1·51	1·38	1·26	1·13	1·00	·87	·75	·62	·50	·37	·25	·12	·00	82
83	2·18	2·03	1·88	1·73	1·58	1·44	1·29	1·15	1·00	·86	·71	·57	·43	·28	·14	·00	83
Lat.	105° 255°	104° 256°	103° 257°	102° 258°	101° 259°	100° 260°	99° 261°	98° 262°	97° 263°	96° 264°	95° 265°	94° 266°	93° 267°	92° 268°	91° 269°	90° 270°	Lat.

A   HOUR ANGLE.   A

| B | TABLE B — HOUR ANGLE. | | | | | | | | | | | | | | B |

Dec.	60°	61°	62°	63°	64°	65°	66°	67°	68°	69°	70°	71°	72°	73°	74°	75°	Dec.
°	300°	299°	298°	297°	296°	295°	294°	293°	292°	291°	290°	289°	288°	287°	286°	285°	°
60	2·00	1·98	1·96	1·94	1·93	1·91	1·89	1·88	1·87	1·86	1·84	1·83	1·82	1·81	1·80	1·79	60
61	2·08	2·06	2·04	2·03	2·01	1·99	1·98	1·96	1·95	1·93	1·92	1·91	1·90	1·89	1·88	1·87	61
62	2·17	2·15	2·13	2·11	2·09	2·08	2·06	2·04	2·03	2·01	2·00	1·99	1·98	1·97	1·96	1·95	62
63	2·27	2·24	2·22	2·20	2·18	2·17	2·15	2·13	2·12	2·10	2·09	2·08	2·06	2·05	2·04	2·03	63
64	2·37	2·34	2·32	2·30	2·28	2·26	2·24	2·23	2·21	2·20	2·18	2·16	2·16	2·14	2·13	2·12	64
65	2·48	2·45	2·43	2·41	2·39	2·37	2·35	2·33	2·31	2·30	2·28	2·27	2·25	2·24	2·23	2·22	65
66	2·59	2·57	2·54	2·52	2·50	2·48	2·46	2·44	2·42	2·41	2·39	2·38	2·36	2·35	2·34	2·33	66
67	2·72	2·69	2·67	2·64	2·62	2·60	2·58	2·56	2·54	2·52	2·51	2·49	2·48	2·46	2·45	2·44	67
68	2·86	2·83	2·80	2·78	2·75	2·73	2·71	2·69	2·67	2·65	2·63	2·62	2·60	2·59	2·57	2·56	68
69	3·01	2·98	2·95	2·92	2·90	2·87	2·85	2·83	2·81	2·79	2·77	2·76	2·74	2·72	2·71	2·70	69
70	3·17	3·14	3·11	3·08	3·06	3·03	3·01	2·98	2·96	2·94	2·92	2·91	2·89	2·87	2·86	2·84	70
71	3·35	3·32	3·29	3·26	3·23	3·20	3·18	3·16	3·13	3·11	3·09	3·07	3·05	3·04	3·02	3·01	71
72	3·55	3·52	3·49	3·45	3·42	3·40	3·37	3·34	3·32	3·30	3·28	3·26	3·24	3·22	3·20	3·19	72
73	3·78	3·74	3·70	3·67	3·64	3·61	3·58	3·55	3·53	3·50	3·48	3·46	3·44	3·42	3·40	3·39	73
74	4·03	3·99	3·95	3·91	3·88	3·85	3·82	3·79	3·76	3·74	3·71	3·69	3·67	3·65	3·63	3·61	74
75	4·31	4·27	4·23	4·19	4·15	4·12	4·09	4·05	4·03	4·00	3·97	3·95	3·92	3·90	3·88	3·86	75
Dec.	120°	119°	118°	117°	116°	115°	114°	113°	112°	111°	110°	109°	108°	107°	106°	105°	Dec.
	240°	241°	242°	243°	244°	245°	246°	247°	248°	249°	250°	251°	252°	253°	254°	255°	

| B | HOUR ANGLE. | | | | | | | | | | | | | | B |

B—Always named the **same** as Declination.

| B | TABLE B — HOUR ANGLE. | | | | | | | | | | | | | | B |

Dec.	75°	76°	77°	78°	79°	80°	81°	82°	83°	84°	85°	86°	87°	88°	89°	90°	Dec.
°	285°	284°	283°	282°	281°	280°	279°	278°	277°	276°	275°	274°	273°	272°	271°	270°	°
60	1·79	1·79	1·78	1·77	1·76	1·76	1·75	1·75	1·75	1·74	1·74	1·74	1·73	1·73	1·73	1·73	60
61	1·87	1·86	1·85	1·84	1·84	1·83	1·83	1·82	1·82	1·81	1·81	1·81	1·81	1·80	1·80	1·80	61
62	1·95	1·94	1·93	1·92	1·92	1·91	1·90	1·90	1·89	1·89	1·89	1·89	1·88	1·88	1·88	1·88	62
63	2·03	2·02	2·01	2·01	2·00	1·99	1·99	1·98	1·98	1·97	1·97	1·97	1·97	1·96	1·96	1·96	63
64	2·12	2·11	2·10	2·10	2·09	2·08	2·08	2·07	2·07	2·06	2·06	2·06	2·05	2·05	2·05	2·05	64
65	2·22	2·21	2·20	2·19	2·18	2·18	2·17	2·17	2·16	2·16	2·15	2·15	2·15	2·14	2·14	2·14	65
66	2·33	2·31	2·31	2·30	2·29	2·28	2·27	2·27	2·26	2·26	2·25	2·25	2·25	2·25	2·25	2·25	66
67	2·44	2·43	2·42	2·41	2·40	2·39	2·39	2·38	2·37	2·37	2·36	2·36	2·36	2·36	2·36	2·36	67
68	2·56	2·55	2·54	2·53	2·52	2·51	2·51	2·50	2·49	2·49	2·48	2·48	2·48	2·48	2·48	2·48	68
69	2·70	2·68	2·67	2·66	2·65	2·65	2·64	2·63	2·62	2·62	2·62	2·61	2·61	2·61	2·61	2·61	69
70	2·84	2·83	2·82	2·81	2·80	2·79	2·78	2·78	2·77	2·76	2·76	2·75	2·75	2·75	2·75	2·75	70
71	3·01	2·99	2·98	2·97	2·96	2·95	2·94	2·93	2·93	2·92	2·92	2·91	2·91	2·90	2·90	2·90	71
72	3·19	3·17	3·16	3·15	3·14	3·13	3·12	3·11	3·10	3·09	3·09	3·09	3·08	3·08	3·08	3·08	72
73	3·39	3·37	3·36	3·34	3·33	3·32	3·31	3·30	3·30	3·29	3·28	3·28	3·28	3·27	3·27	3·27	73
74	3·61	3·59	3·58	3·57	3·55	3·54	3·53	3·52	3·51	3·51	3·50	3·50	3·49	3·49	3·49	3·49	74
75	3·86	3·85	3·83	3·82	3·80	3·79	3·78	3·77	3·76	3·75	3·75	3·74	3·74	3·73	3·73	3·73	75
Dec.	105°	104°	103°	102°	101°	100°	99°	98°	97°	96°	95°	94°	93°	92°	91°	90°	Dec.
	255°	256°	257°	258°	259°	260°	261°	262°	263°	264°	265°	266°	267°	268°	269°	270°	

| B | HOUR ANGLE. | | | | | | | | | | | | | | B |

B—Always named the **same** as Declination.

TABLE C

C   *C + Lat F*   TABLE C   *Cl + Latt T*   C

A & B CORRECTION.

AZIMUTHS.

Lat.	'00'	'01'	'02'	'03'	'04'	'05'	'06'	'07'	'08'	'09'	'10'	'11'	'12'	'13'	'14'	'15'	Lat.
0	90·0	89·4	88·9	88·3	87·7	87·1	86·6	86·0	85·4	84·9	84·3	83·7	83·2	82·6	82·0	81·5	0
5	90·0	89·4	88·9	88·3	87·7	87·1	86·6	86·0	85·4	84·9	84·3	83·7	83·2	82·6	82·1	81·5	5
10	90·0	89·4	88·9	88·3	87·7	87·2	86·6	86·1	85·5	84·9	84·4	83·8	83·3	82·7	82·1	81·6	10
14	90·0	89·5	88·9	88·3	87·8	87·2	86·7	86·1	85·6	85·0	84·4	83·9	83·4	82·8	82·3	81·7	14
18	90·0	89·5	88·9	88·4	87·8	87·3	86·7	86·2	85·6	85·1	84·5	84·0	83·5	83·0	82·4	81·9	18
20	90·0	89·5	88·9	88·4	87·8	87·3	86·8	86·2	85·7	85·2	84·6	84·1	83·6	83·0	82·5	82·0	20
22	90·0	89·5	88·9	88·4	87·9	87·3	86·8	86·3	85·8	85·2	84·7	84·2	83·7	83·1	82·6	82·1	22
24	90·0	89·5	89·0	88·4	87·9	87·4	86·9	86·3	85·8	85·3	84·8	84·3	83·7	83·2	82·7	82·2	24
26	90·0	89·5	89·0	88·5	87·9	87·4	86·9	86·4	85·9	85·4	84·9	84·4	83·8	83·3	82·8	82·3	26
28	90·0	89·5	89·0	88·5	88·0	87·5	87·0	86·5	86·0	85·5	85·0	84·5	84·0	83·5	83·0	82·5	28
30	90·0	89·5	89·0	88·5	88·0	87·5	87·0	86·5	86·0	85·5	85·1	84·6	84·1	83·6	83·1	82·6	30
31	90·0	89·5	89·0	88·5	88·0	87·5	87·1	86·6	86·1	85·6	85·1	84·6	84·1	83·6	83·2	82·7	31
32	90·0	89·5	89·0	88·5	88·1	87·6	87·1	86·6	86·1	85·6	85·2	84·7	84·2	83·7	83·2	82·8	32
33	90·0	89·5	89·0	88·6	88·1	87·6	87·1	86·6	86·2	85·7	85·2	84·7	84·3	83·8	83·3	82·8	33
34	90·0	89·5	89·1	88·6	88·1	87·6	87·2	86·7	86·2	85·7	85·3	84·8	84·3	83·8	83·4	82·9	34
35	90·0	89·5	89·1	88·6	88·1	87·7	87·2	86·7	86·3	85·8	85·3	84·9	84·4	83·9	83·5	83·0	35
36	90·0	89·5	89·1	88·6	88·1	87·7	87·2	86·8	86·3	85·8	85·4	84·9	84·5	84·0	83·5	83·1	36
37	90·0	89·5	89·1	88·6	88·2	87·7	87·3	86·8	86·3	85·9	85·4	85·0	84·5	84·1	83·6	83·2	37
38	90·0	89·5	89·1	88·6	88·2	87·7	87·3	86·8	86·4	85·9	85·5	85·0	84·6	84·2	83·7	83·3	38
39	90·0	89·6	89·1	88·7	88·2	87·8	87·3	86·9	86·4	86·0	85·6	85·1	84·7	84·2	83·8	83·4	39
40	90·0	89·6	89·1	88·7	88·2	87·8	87·4	86·9	86·5	86·1	85·6	85·2	84·7	84·3	83·9	83·4	40
41	90·0	89·6	89·1	88·7	88·3	87·8	87·4	87·0	86·5	86·1	85·7	85·3	84·8	84·4	84·0	83·5	41
42	90·0	89·6	89·1	88·7	88·3	87·9	87·4	87·0	86·6	86·2	85·7	85·3	84·9	84·5	84·1	83·6	42
43	90·0	89·6	89·2	88·7	88·3	87·9	87·5	87·1	86·7	86·2	85·8	85·4	85·0	84·6	84·2	83·7	43
44	90·0	89·6	89·2	88·8	88·4	87·9	87·5	87·1	86·7	86·3	85·9	85·5	85·1	84·7	84·2	83·8	44
45	90·0	89·6	89·2	88·8	88·4	88·0	87·6	87·2	86·8	86·4	86·0	85·6	85·1	84·7	84·3	83·9	45
46	90·0	89·6	89·2	88·8	88·4	88·0	87·6	87·2	86·8	86·4	86·0	85·6	85·2	84·8	84·4	84·1	46
47	90·0	89·6	89·2	88·8	88·4	88·0	87·7	87·3	86·9	86·5	86·1	85·7	85·3	84·9	84·5	84·2	47
48	90·0	89·6	89·2	88·9	88·5	88·1	87·7	87·3	86·9	86·6	86·2	85·8	85·4	85·0	84·6	84·3	48
49	90·0	89·6	89·2	88·9	88·5	88·1	87·7	87·4	87·0	86·6	86·2	85·9	85·5	85·1	84·8	84·4	49
50	90·0	89·6	89·3	88·9	88·5	88·2	87·8	87·4	87·1	86·7	86·3	86·0	85·6	85·2	84·9	84·5	50
51	90·0	89·6	89·3	88·9	88·6	88·2	87·8	87·5	87·1	86·8	86·4	86·0	85·7	85·3	85·0	84·6	51
52	90·0	89·6	89·3	88·9	88·6	88·2	87·9	87·5	87·2	86·8	86·5	86·1	85·8	85·4	85·1	84·7	52
53	90·0	89·7	89·3	89·0	88·6	88·3	87·9	87·6	87·2	86·9	86·6	86·2	85·9	85·5	85·2	84·8	53
54	90·0	89·7	89·3	89·0	88·7	88·3	88·0	87·6	87·3	87·0	86·6	86·3	86·0	85·6	85·3	85·0	54
55	90·0	89·7	89·3	89·0	88·7	88·4	88·0	87·7	87·4	87·0	86·7	86·4	86·1	85·7	85·4	85·1	55
56	90·0	89·7	89·4	89·0	88·7	88·4	88·1	87·8	87·4	87·1	86·8	86·5	86·2	85·8	85·5	85·2	56
57	90·0	89·7	89·4	89·1	88·8	88·4	88·1	87·8	87·5	87·2	86·9	86·6	86·3	86·0	85·6	85·3	57
58	90·0	89·7	89·4	89·1	88·8	88·5	88·2	87·9	87·6	87·3	87·0	86·7	86·4	86·1	85·8	85·5	58
59	90·0	89·7	89·4	89·1	88·8	88·5	88·2	87·9	87·6	87·3	87·1	86·8	86·5	86·2	85·9	85·6	59
60	90·0	89·7	89·4	89·1	88·9	88·6	88·3	88·0	87·7	87·4	87·1	86·9	86·6	86·3	86·0	85·7	60
61	90·0	89·7	89·4	89·2	88·9	88·6	88·3	88·1	87·8	87·5	87·2	86·9	86·7	86·4	86·1	85·8	61
62	90·0	89·7	89·5	89·2	88·9	88·7	88·4	88·1	87·8	87·6	87·3	87·0	86·8	86·5	86·2	86·0	62
63	90·0	89·7	89·5	89·2	89·0	88·7	88·4	88·2	87·9	87·7	87·4	87·1	86·9	86·6	86·4	86·1	63
64	90·0	89·7	89·5	89·2	89·0	88·7	88·5	88·2	88·0	87·7	87·5	87·2	87·0	86·7	86·5	86·2	64
65	90·0	89·8	89·5	89·3	89·0	88·8	88·5	88·3	88·1	87·8	87·6	87·3	87·1	86·9	86·6	86·4	65
66	90·0	89·8	89·5	89·3	89·1	88·8	88·6	88·4	88·1	87·9	87·7	87·4	87·2	87·0	86·7	86·5	66
67	90·0	89·8	89·6	89·3	89·1	88·9	88·7	88·4	88·2	88·0	87·8	87·5	87·3	87·1	86·9	86·6	67
68	90·0	89·8	89·6	89·4	89·1	88·9	88·7	88·5	88·3	88·1	87·9	87·6	87·4	87·2	87·0	86·8	68

AZIMUTHS.

A±B= '00' '01' '02' '03' '04' '05' '06' '07' '08' '09' '10' '11' '12' '13' '14' '15' =A±B

A & B Same Names, } RULE TO FIND { A & B Different Names,
   take Sum, (add) } C CORRECTION { take Difference. (sub.)

C CORRECTION, (A±B) is named the same as the greater of these quantities.

AZIMUTH takes combined names of C Correction and Hour Angle.

C   C

TABLE C

A & B CORRECTION.

AZIMUTHS.

Lat.	·15′	·16′	·17′	·18′	·19′	·20′	·21′	·22′	·23′	·24′	·25′	·26′	·27′	·28′	·29′	·30′	Lat.
0	81.5	80.9	80.4	79.8	79.2	78.7	78.1	77.6	77.0	76.5	76.0	75.4	74.9	74.4	73.8	73.3	0
5	81.5	80.9	80.4	79.8	79.3	78.7	78.2	77.6	77.1	76.6	76.0	75.5	74.9	74.4	73.9	73.4	5
10	81.6	81.0	80.5	79.9	79.4	78.9	78.3	77.8	77.2	76.7	76.2	75.6	75.1	74.6	74.1	73.5	10
14	81.7	81.2	80.6	80.1	79.6	79.1	78.5	78.0	77.4	76.9	76.4	75.8	75.3	74.8	74.3	73.7	14
18	81.9	81.3	80.8	80.3	79.8	79.2	78.7	78.2	77.7	77.1	76.6	76.1	75.6	75.1	74.6	74.1	18
20	82.0	81.4	80.9	80.4	79.9	79.4	78.8	78.3	77.8	77.3	76.8	76.3	75.8	75.3	74.8	74.3	20
22	82.1	81.6	81.0	80.5	80.0	79.5	79.0	78.5	78.0	77.5	76.9	76.4	75.9	75.4	75.0	74.5	22
24	82.2	81.7	81.2	80.7	80.2	79.6	79.1	78.6	78.1	77.6	77.1	76.6	76.1	75.7	75.2	74.7	24
26	82.3	81.8	81.3	80.8	80.3	79.8	79.3	78.8	78.3	77.8	77.3	76.8	76.4	75.9	75.4	74.9	26
28	82.5	82.0	81.5	81.0	80.5	80.0	79.5	79.0	78.5	78.0	77.6	77.1	76.6	76.1	75.6	75.2	28
30	82.6	82.1	81.6	81.1	80.7	80.2	79.7	79.2	78.7	78.3	77.8	77.3	76.8	76.4	75.9	75.4	30
31	82.7	82.2	81.7	81.2	80.7	80.3	79.8	79.3	78.8	78.4	77.9	77.4	77.0	76.5	76.0	75.6	31
32	82.8	82.3	81.8	81.3	80.8	80.4	79.9	79.4	79.0	78.5	78.0	77.6	77.1	76.6	76.2	75.7	32
33	82.8	82.4	81.9	81.4	80.9	80.5	80.0	79.5	79.1	78.6	78.2	77.7	77.2	76.8	76.3	75.9	33
34	82.9	82.4	82.0	81.5	81.0	80.6	80.1	79.7	79.2	78.7	78.3	77.8	77.4	76.9	76.5	76.0	34
35	83.0	82.5	82.1	81.6	81.2	80.7	80.2	79.8	79.3	78.9	78.4	78.0	77.5	77.1	76.6	76.2	35
36	83.1	82.6	82.2	81.7	81.3	80.8	80.4	79.9	79.5	79.0	78.6	78.1	77.7	77.2	76.8	76.4	36
37	83.2	82.7	82.3	81.8	81.4	80.9	80.5	80.0	79.6	79.1	78.7	78.3	77.8	77.4	77.0	76.5	37
38	83.3	82.8	82.4	81.9	81.5	81.0	80.6	80.2	79.7	79.3	78.9	78.4	78.0	77.6	77.1	76.7	38
39	83.4	82.9	82.5	82.0	81.6	81.2	80.7	80.3	79.9	79.4	79.0	78.6	78.1	77.7	77.3	76.9	39
40	83.4	83.0	82.6	82.1	81.7	81.3	80.9	80.4	80.0	79.6	79.2	78.7	78.3	77.9	77.5	77.1	40
41	83.5	83.1	82.7	82.3	81.8	81.4	81.0	80.6	80.2	79.7	79.3	78.9	78.5	78.1	77.7	77.2	41
42	83.6	83.2	82.8	82.4	82.0	81.5	81.1	80.7	80.3	79.9	79.5	79.1	78.7	78.2	77.8	77.4	42
43	83.7	83.3	82.9	82.5	82.1	81.7	81.3	80.9	80.5	80.0	79.6	79.2	78.8	78.4	78.0	77.6	43
44	83.8	83.4	83.0	82.6	82.2	81.8	81.4	81.0	80.6	80.2	79.8	79.4	79.0	78.6	78.2	77.8	44
45	83.9	83.5	83.1	82.7	82.3	82.0	81.6	81.2	80.8	80.4	80.0	79.6	79.2	78.8	78.4	78.0	45
46	84.1	83.7	83.3	82.9	82.5	82.1	81.7	81.3	80.9	80.5	80.1	79.8	79.4	79.0	78.6	78.2	46
47	84.2	83.8	83.4	83.0	82.6	82.2	81.8	81.5	81.1	80.7	80.3	79.9	79.6	79.2	78.8	78.4	47
48	84.3	83.9	83.5	83.1	82.8	82.4	82.0	81.6	81.3	80.9	80.5	80.1	79.8	79.4	79.0	78.6	48
49	84.4	84.0	83.6	83.3	82.9	82.5	82.2	81.8	81.4	81.1	80.7	80.3	80.0	79.6	79.2	78.9	49
50	84.5	84.1	83.8	83.4	83.0	82.7	82.3	82.0	81.6	81.2	80.9	80.5	80.2	79.8	79.4	79.1	50
51	84.6	84.3	83.9	83.5	83.2	82.8	82.5	82.1	81.8	81.4	81.1	80.7	80.4	80.0	79.7	79.3	51
52	84.7	84.4	84.0	83.7	83.3	83.0	82.6	82.3	81.9	81.6	81.2	80.9	80.6	80.2	79.9	79.5	52
53	84.8	84.5	84.2	83.8	83.5	83.1	82.8	82.5	82.1	81.8	81.4	81.1	80.8	80.4	80.1	79.8	53
54	85.0	84.6	84.3	84.0	83.6	83.3	83.0	82.6	82.3	82.0	81.6	81.3	81.0	80.7	80.3	80.0	54
55	85.1	84.8	84.4	84.1	83.8	83.5	83.1	82.8	82.5	82.2	81.8	81.5	81.2	80.9	80.6	80.2	55
56	85.2	84.9	84.6	84.3	83.9	83.6	83.3	83.0	82.7	82.4	82.0	81.7	81.4	81.1	80.8	80.5	56
57	85.3	85.0	84.7	84.4	84.1	83.8	83.5	83.2	82.9	82.6	82.2	81.9	81.6	81.3	81.0	80.7	57
58	85.5	85.2	84.9	84.6	84.3	84.0	83.7	83.4	83.1	82.8	82.5	82.2	81.9	81.6	81.3	81.0	58
59	85.6	85.3	85.0	84.7	84.4	84.1	83.8	83.5	83.2	83.0	82.7	82.4	82.1	81.8	81.5	81.2	59
60	85.7	85.4	85.1	84.9	84.6	84.3	84.0	83.7	83.4	83.2	82.9	82.6	82.3	82.0	81.7	81.5	60
61	85.8	85.5	85.3	85.0	84.7	84.5	84.2	83.9	83.6	83.4	83.1	82.8	82.5	82.3	82.0	81.7	61
62	86.0	85.7	85.4	85.2	84.9	84.6	84.4	84.1	83.8	83.6	83.3	83.0	82.8	82.5	82.2	82.0	62
63	86.1	85.8	85.6	85.3	85.1	84.8	84.6	84.3	84.0	83.8	83.5	83.3	83.0	82.8	82.5	82.2	63
64	86.2	86.0	85.7	85.5	85.2	85.0	84.7	84.5	84.2	84.0	83.7	83.5	83.2	83.0	82.8	82.5	64
65	86.4	86.1	85.9	85.6	85.4	85.2	84.9	84.7	84.4	84.2	84.0	83.7	83.5	83.3	83.0	82.8	65
66	86.5	86.3	86.0	85.8	85.6	85.3	85.1	84.9	84.7	84.4	84.2	84.0	83.7	83.5	83.3	83.0	66
67	86.6	86.4	86.2	86.0	85.8	85.5	85.3	85.1	84.9	84.6	84.4	84.2	84.0	83.8	83.5	83.3	67
68	86.8	86.6	86.4	86.1	85.9	85.7	85.5	85.3	85.1	84.9	84.6	84.4	84.2	84.0	83.8	83.6	68

AZIMUTHS.

A±B = ·15′ ·16′ ·17′ ·18′ ·19′ ·20′ ·21′ ·22′ ·23′ ·24′ ·25′ ·26′ ·27′ ·28′ ·29′ ·30′ = A±B

A & B Same names, take Sum, (add.) } RULE TO FIND C CORRECTION { A & B Different names, take Difference, (sub.)

C CORRECTION, (A±B) is named the same as the greater of these quantities.

AZIMUTH takes combined names of C Correction and Hour Angle.

C TABLE C C

A. & B. CORRECTION.

A±B=	·30′	·31′	·32′	·33′	·34′	·35′	·36′	·37′	·38′	·39′	·40′	·41′	·42′	·43′	·44′	·45′=A±B
Lat.							AZIMUTHS.									Lat.
°	°	°	°	°	°	°	°	°	°	°	°	°	°	°	°	°	°
0	73·3	72·8	72·3	71·7	71·2	70·7	70·2	69·7	69·2	68·7	68·2	67·7	67·2	66·7	66·3	65·8	0
5	73·4	72·8	72·3	71·8	71·3	70·8	70·3	69·8	69·3	68·8	68·3	67·3	66·8	66·4	65·9		5
10	73·5	73·0	72·5	72·0	71·5	71·0	70·5	70·0	69·5	69·0	68·5	68·0	67·5	67·0	66·6	66·1	10
14	73·7	73·2	72·7	72·2	71·7	71·2	70·7	70·2	69·7	69·3	68·8	68·3	67·8	67·4	66·9	66·4	14
18	74·1	73·6	73·1	72·6	72·1	71·6	71·1	70·6	70·1	69·6	69·2	68·7	68·2	67·8	67·3	66·8	18
20	74·3	73·8	73·3	72·8	72·3	71·8	71·3	70·8	70·3	69·9	69·4	68·9	68·5	68·0	67·5	67·1	20
22	74·5	74·0	73·5	73·0	72·5	72·0	71·5	71·1	70·6	70·1	69·7	69·2	68·7	68·3	67·8	67·4	22
24	74·7	74·2	73·7	73·2	72·7	72·3	71·8	71·3	70·9	70·4	69·9	69·5	69·0	68·6	68·1	67·7	24
26	74·9	74·4	74·0	73·5	73·0	72·5	72·1	71·6	71·1	70·7	70·2	69·8	69·3	68·9	68·4	68·0	26
28	75·2	74·7	74·2	73·8	73·3	72·8	72·4	71·9	71·5	71·0	70·5	70·1	69·7	69·2	68·8	68·3	28
30	75·4	75·0	74·5	74·1	73·6	73·1	72·7	72·2	71·8	71·3	70·9	70·5	70·0	69·6	69·1	68·7	30
31	75·6	75·1	74·7	74·2	73·8	73·3	72·9	72·4	72·0	71·5	71·1	70·6	70·2	69·8	69·3	68·9	31
32	75·7	75·3	74·8	74·4	73·9	73·5	73·0	72·6	72·1	71·7	71·3	70·8	70·4	70·0	69·5	69·1	32
33	75·9	75·4	75·0	74·5	74·1	73·6	73·2	72·8	72·3	71·9	71·5	71·0	70·6	70·2	69·7	69·3	33
34	76·0	75·6	75·1	74·7	74·3	73·8	73·4	72·9	72·5	72·1	71·7	71·2	70·8	70·4	70·0	69·5	34
35	76·2	75·8	75·3	74·9	74·4	74·0	73·6	73·1	72·7	72·3	71·9	71·4	71·0	70·6	70·2	69·8	35
36	76·4	75·9	75·5	75·1	74·6	74·2	73·8	73·3	72·9	72·5	72·1	71·6	71·2	70·8	70·4	70·0	36
37	76·5	76·1	75·7	75·2	74·8	74·4	74·0	73·5	73·1	72·7	72·3	71·9	71·5	71·0	70·6	70·2	37
38	76·7	76·3	75·8	75·4	75·0	74·6	74·2	73·7	73·3	72·9	72·5	72·1	71·7	71·3	70·9	70·5	38
39	76·9	76·5	76·0	75·6	75·2	74·8	74·4	74·0	73·5	73·1	72·7	72·3	71·9	71·5	71·1	70·7	39
40	77·1	76·6	76·2	75·8	75·4	75·0	74·6	74·2	73·8	73·4	73·0	72·6	72·2	71·8	71·4	71·0	40
41	77·2	76·8	76·4	76·0	75·6	75·2	74·8	74·4	74·0	73·6	73·2	72·8	72·4	72·0	71·6	71·2	41
42	77·4	77·0	76·6	76·2	75·8	75·4	75·0	74·6	74·2	73·8	73·4	73·1	72·7	72·3	71·9	71·5	42
43	77·6	77·2	76·8	76·4	76·0	75·6	75·2	74·9	74·5	74·1	73·7	73·3	72·9	72·5	72·2	71·8	43
44	77·8	77·4	77·0	76·6	76·3	75·9	75·5	75·1	74·7	74·3	73·9	73·6	73·2	72·8	72·4	72·1	44
45	78·0	77·6	77·3	76·9	76·5	76·1	75·7	75·3	75·0	74·6	74·2	73·8	73·5	73·1	72·7	72·3	45
46	78·2	77·8	77·5	77·1	76·7	76·3	76·0	75·6	75·2	74·8	74·5	74·1	73·7	73·4	73·0	72·6	46
47	78·4	78·1	77·7	77·3	76·9	76·6	76·2	75·8	75·5	75·1	74·7	74·4	74·0	73·7	73·3	72·9	47
48	78·6	78·3	77·9	77·5	77·2	76·8	76·5	76·1	75·7	75·4	75·0	74·7	74·3	73·9	73·6	73·2	48
49	78·9	78·5	78·1	77·8	77·4	77·1	76·7	76·4	76·0	75·6	75·3	74·9	74·6	74·2	73·9	73·6	49
50	79·1	78·7	78·4	78·0	77·7	77·3	77·0	76·6	76·3	75·9	75·6	75·2	74·9	74·5	74·2	73·9	50
51	79·3	79·0	78·6	78·3	77·9	77·6	77·2	76·9	76·6	76·2	75·9	75·5	75·2	74·9	74·5	74·2	51
52	79·5	79·2	78·9	78·5	78·2	77·8	77·5	77·2	76·8	76·5	76·2	75·8	75·5	75·2	74·8	74·5	52
53	79·8	79·4	79·1	78·8	78·4	78·1	77·8	77·4	77·1	76·8	76·5	76·1	75·8	75·5	75·2	74·8	53
54	80·0	79·7	79·3	79·0	78·7	78·4	78·1	77·7	77·4	77·1	76·8	76·5	76·1	75·8	75·5	75·2	54
55	80·2	79·9	79·6	79·3	79·0	78·6	78·3	78·0	77·7	77·4	77·1	76·8	76·5	76·1	75·8	75·5	55
56	80·5	80·2	79·9	79·5	79·2	78·9	78·6	78·3	78·0	77·7	77·4	77·1	76·8	76·5	76·2	75·9	56
57	80·7	80·4	80·1	79·8	79·5	79·2	78·9	78·6	78·3	78·0	77·7	77·4	77·1	76·8	76·5	76·2	57
58	81·0	80·7	80·4	80·1	79·8	79·5	79·2	78·9	78·6	78·3	78·0	77·7	77·5	77·2	76·9	76·6	58
59	81·2	80·9	80·6	80·4	80·1	79·8	79·5	79·2	78·9	78·6	78·4	78·1	77·8	77·5	77·2	77·0	59
60	81·5	81·2	80·9	80·6	80·4	80·1	79·8	79·5	79·2	79·0	78·7	78·4	78·1	77·9	77·6	77·3	60
61	81·7	81·5	81·2	80·9	80·6	80·4	80·1	79·8	79·6	79·3	79·0	78·8	78·5	78·2	78·0	77·7	61
62	82·0	81·7	81·5	81·2	80·9	80·7	80·4	80·1	79·9	79·6	79·4	79·1	78·8	78·6	78·3	78·1	62
63	82·2	82·0	81·7	81·5	81·2	81·0	80·7	80·5	80·2	80·0	79·7	79·5	79·2	79·0	78·7	78·5	63
64	82·5	82·3	82·0	81·8	81·5	81·3	81·0	80·8	80·5	80·3	80·1	79·8	79·6	79·3	79·1	78·8	64
65	82·8	82·5	82·3	82·1	81·8	81·6	81·3	81·1	80·9	80·6	80·4	80·2	79·9	79·7	79·5	79·2	65
66	83·0	82·8	82·6	82·4	82·1	81·9	81·7	81·4	81·2	81·0	80·8	80·5	80·3	80·1	79·9	79·6	66
67	83·3	83·1	82·9	82·7	82·4	82·2	82·0	81·8	81·6	81·3	81·1	80·9	80·7	80·5	80·2	80·0	67
68	83·6	83·4	83·2	83·0	82·7	82·5	82·3	82·1	81·9	81·7	81·5	81·3	81·1	80·8	80·6	80·4	68
Lat.							AZIMUTHS.									Lat.
A±B=	·30′	·31′	·32′	·33′	·34′	·35′	·36′	·37′	·38′	·39′	·40′	·41′	·42′	·43′	·44′	·45′=A±B

A & B Same names } RULE TO FIND { A & B Different names,
take Sum, (add.) } C CORRECTION { take Difference, (sub.)

C CORRECTION, (A±B) is named the same as the greater of these quantities.

AZIMUTH takes combined names of C Correction and Hour Angle.

C C

TABLE C

A & B CORRECTION.

A±B=	·45′	·46′	·47′	·48′	·49′	·50′	·51′	·52′	·53′	·54′	·55′	·56′	·57′	·58′	·59′	·60′=A±B
Lat.							**AZIMUTHS.**									**Lat.**
0	65·8	65·3	64·8	64·4	63·9	63·4	63·0	62·5	62·1	61·6	61·2	60·8	60·3	59·9	59·5	59·0	**0**
5	65·9	65·4	64·9	64·4	64·0	63·5	63·1	62·6	62·2	61·7	61·3	60·8	60·4	60·0	59·6	59·1	**5**
10	66·1	65·6	65·2	64·7	64·2	63·8	63·3	62·9	62·4	62·0	61·6	61·1	60·7	60·3	59·8	59·4	**10**
14	66·4	65·9	65·5	65·0	64·6	64·1	63·7	63·2	62·8	62·3	61·9	61·5	61·1	60·6	60·2	59·8	**14**
18	66·8	66·4	65·9	65·5	65·0	64·6	64·1	63·7	63·2	62·8	62·4	62·0	61·5	61·1	60·7	60·3	**18**
20	67·1	66·6	66·2	65·7	65·3	64·8	64·4	64·0	63·5	63·1	62·7	62·2	61·8	61·4	61·0	60·6	**20**
22	67·4	66·9	66·5	66·0	65·6	65·1	64·7	64·3	63·8	63·4	63·0	62·6	62·1	61·7	61·3	60·9	**22**
24	67·7	67·2	66·8	66·3	65·9	65·5	65·0	64·6	64·2	63·7	63·3	62·9	62·5	62·1	61·7	61·3	**24**
26	68·0	67·5	67·1	66·7	66·2	65·8	65·4	64·9	64·5	64·1	63·7	63·3	62·9	62·5	62·1	61·7	**26**
28	68·3	67·9	67·5	67·0	66·6	66·2	65·8	65·3	64·9	64·5	64·1	63·7	63·3	62·9	62·5	62·1	**28**
30	68·7	68·3	67·9	67·4	67·0	66·6	66·2	65·8	65·3	64·9	64·5	64·1	63·7	63·3	62·9	62·5	**30**
31	68·9	68·5	68·1	67·6	67·2	66·8	66·4	66·0	65·6	65·2	64·8	64·4	64·0	63·6	63·2	62·8	**31**
32	69·1	68·7	68·3	67·9	67·4	67·0	66·6	66·2	65·8	65·4	65·0	64·6	64·2	63·8	63·4	63·0	**32**
33	69·3	68·9	68·5	68·1	67·7	67·2	66·8	66·4	66·0	65·6	65·2	64·8	64·5	64·1	63·7	63·3	**33**
34	69·5	69·1	68·7	68·3	67·9	67·5	67·1	66·7	66·3	65·9	65·5	65·1	64·7	64·3	63·9	63·6	**34**
35	69·8	69·4	68·9	68·5	68·1	67·7	67·3	66·9	66·5	66·1	65·7	65·4	65·0	64·6	64·2	63·8	**35**
36	70·0	69·6	69·2	68·8	68·4	68·0	67·6	67·2	66·8	66·4	66·0	65·6	65·2	64·9	64·5	64·1	**36**
37	70·2	69·8	69·4	69·0	68·6	68·2	67·8	67·4	67·1	66·7	66·3	65·9	65·5	65·1	64·8	64·4	**37**
38	70·5	70·1	69·7	69·3	68·9	68·5	68·1	67·7	67·3	66·9	66·6	66·2	65·8	65·4	65·1	64·7	**38**
39	70·7	70·3	69·9	69·5	69·2	68·8	68·4	68·0	67·6	67·2	66·9	66·5	66·1	65·7	65·4	65·0	**39**
40	71·0	70·6	70·2	69·8	69·4	69·0	68·7	68·3	67·9	67·5	67·2	66·8	66·4	66·0	65·7	65·3	**40**
41	71·2	70·9	70·5	70·1	69·7	69·3	68·9	68·6	68·2	67·8	67·5	67·1	66·7	66·4	66·0	65·6	**41**
42	71·5	71·1	70·7	70·4	70·0	69·6	69·2	68·9	68·5	68·1	67·8	67·4	67·0	66·7	66·3	66·0	**42**
43	71·8	71·4	71·0	70·7	70·3	69·9	69·5	69·2	68·8	68·4	68·1	67·7	67·4	67·0	66·7	66·3	**43**
44	72·1	71·7	71·3	71·0	70·6	70·2	69·9	69·5	69·1	68·8	68·4	68·1	67·7	67·4	67·0	66·7	**44**
45	72·3	72·0	71·6	71·3	70·9	70·5	70·2	69·8	69·5	69·1	68·7	68·4	68·0	67·7	67·4	67·0	**45**
46	72·6	72·3	71·9	71·6	71·2	70·8	70·5	70·1	69·8	69·4	69·1	68·7	68·4	68·1	67·7	67·4	**46**
47	72·9	72·6	72·2	71·9	71·5	71·2	70·8	70·5	70·1	69·8	69·4	69·1	68·8	68·4	68·1	67·7	**47**
48	73·2	72·9	72·5	72·2	71·8	71·5	71·2	70·8	70·5	70·1	69·8	69·5	69·1	68·8	68·5	68·1	**48**
49	73·6	73·2	72·9	72·5	72·2	71·8	71·5	71·2	70·8	70·5	70·2	69·8	69·5	69·2	68·8	68·5	**49**
50	73·9	73·5	73·2	72·9	72·5	72·2	71·8	71·5	71·2	70·9	70·5	70·2	69·9	69·6	69·2	68·9	**50**
51	74·2	73·9	73·5	73·2	72·9	72·5	72·2	71·9	71·6	71·2	70·9	70·6	70·3	69·9	69·6	69·3	**51**
52	74·5	74·2	73·9	73·5	73·2	72·9	72·6	72·2	71·9	71·6	71·3	71·0	70·7	70·3	70·0	69·7	**52**
53	74·8	74·5	74·2	73·9	73·6	73·3	72·9	72·6	72·3	72·0	71·7	71·4	71·1	70·8	70·5	70·1	**53**
54	75·2	74·9	74·6	74·2	73·9	73·6	73·3	73·0	72·7	72·4	72·1	71·8	71·5	71·2	70·9	70·6	**54**
55	75·5	75·2	74·9	74·6	74·3	74·0	73·7	73·4	73·1	72·8	72·5	72·2	71·9	71·6	71·3	71·0	**55**
56	75·9	75·6	75·3	75·0	74·7	74·4	74·1	73·8	73·5	73·2	72·9	72·6	72·3	72·0	71·7	71·5	**56**
57	76·2	75·9	75·6	75·3	75·1	74·8	74·5	74·2	73·9	73·6	73·3	73·0	72·8	72·5	72·2	71·9	**57**
58	76·6	76·3	76·0	75·7	75·4	75·2	74·9	74·6	74·3	74·0	73·8	73·5	73·2	72·9	72·6	72·4	**58**
59	77·0	76·7	76·4	76·1	75·8	75·6	75·3	75·0	74·7	74·5	74·2	73·9	73·6	73·4	73·1	72·8	**59**
60	77·3	77·0	76·8	76·5	76·2	76·0	75·7	75·4	75·2	74·9	74·6	74·4	74·1	73·8	73·6	73·3	**60**
61	77·7	77·4	77·2	76·9	76·6	76·4	76·1	75·9	75·6	75·3	75·1	74·8	74·6	74·3	74·0	73·8	**61**
62	78·1	77·8	77·6	77·3	77·0	76·8	76·5	76·3	76·0	75·8	75·5	75·3	75·0	74·8	74·5	74·3	**62**
63	78·5	78·2	78·0	77·7	77·5	77·2	77·0	76·7	76·5	76·2	76·0	75·7	75·5	75·2	75·0	74·8	**63**
64	78·8	78·6	78·4	78·1	77·9	77·6	77·4	77·2	76·9	76·7	76·4	76·2	76·0	75·7	75·5	75·3	**64**
65	79·2	79·0	78·8	78·5	78·3	78·1	77·8	77·6	77·4	77·1	76·9	76·7	76·5	76·2	76·0	75·8	**65**
66	79·6	79·4	79·2	79·0	78·7	78·5	78·3	78·1	77·8	77·6	77·4	77·2	76·9	76·7	76·5	76·3	**66**
67	80·0	79·8	79·6	79·4	79·2	78·9	78·7	78·5	78·3	78·1	77·9	77·7	77·4	77·2	77·0	76·8	**67**
68	80·4	80·2	80·0	79·8	79·6	79·4	79·2	79·0	78·8	78·6	78·4	78·2	77·9	77·7	77·5	77·3	**68**
Lat.							**AZIMUTHS.**									**Lat.**
A±B=	·45′	·46′	·47′	·48′	·49′	·50′	·51′	·52′	·53′	·54′	·55′	·56′	·57′	·58′	·59′	·60′=A±B

A & B **S**ame names, take **S**um, (add.) } RULE TO FIND { **C** CORRECTION { **A & B** **D**ifferent names, take **D**ifference, (sub.)

C CORRECTION, (A±B) is named the same as the greater of these quantities.

AZIMUTH takes combined names of **C** Correction and Hour Angle.

C	TABLE C															C

A & B CORRECTION.

A±B=·60′	·62′	·64′	·66′	·68′	·70′	·72′	·74′	·76′	·78′	·80′	·82′	·84′	·86′	·88′	·90′=A±B	Lat.

AZIMUTHS.

Lat.	·60′	·62′	·64′	·66′	·68′	·70′	·72′	·74′	·76′	·78′	·80′	·82′	·84′	·86′	·88′	·90′	Lat.
0	59·0	58·2	57·4	56·6	55·8	55·0	54·2	53·5	52·8	52·0	51·3	50·6	50·0	49·3	48·7	48·0	0
5	59·1	58·3	57·5	56·7	55·9	55·1	54·3	53·6	52·9	52·2	51·5	50·8	50·1	49·4	48·8	48·1	5
10	59·4	58·6	57·8	57·0	56·2	55·4	54·7	53·9	53·2	52·5	51·8	51·1	50·4	49·7	49·1	48·4	10
14	59·8	59·0	58·2	57·4	56·6	55·8	55·1	54·3	53·6	52·9	52·2	51·5	50·8	50·2	49·5	48·9	14
18	60·3	59·5	58·7	57·9	57·1	56·3	55·6	54·9	54·1	53·4	52·7	52·1	51·4	50·7	50·1	49·4	18
20	60·6	59·8	59·0	58·2	57·4	56·7	55·9	55·2	54·5	53·8	53·1	52·4	51·7	51·1	50·4	49·8	20
22	60·9	60·1	59·3	58·5	57·8	57·0	56·3	55·5	54·8	54·1	53·4	52·8	52·1	51·4	50·8	50·2	22
24	61·3	60·5	59·7	58·9	58·2	57·4	56·7	55·9	55·2	54·5	53·8	53·2	52·5	51·8	51·2	50·6	24
26	61·7	60·9	60·1	59·3	58·6	57·8	57·1	56·4	55·7	55·0	54·3	53·6	52·9	52·3	51·7	51·0	26
28	62·1	61·3	60·5	59·8	59·0	58·3	57·6	56·8	56·1	55·4	54·8	54·1	53·4	52·8	52·2	51·5	28
30	62·5	61·8	61·0	60·2	59·5	58·8	58·1	57·3	56·6	56·0	55·3	54·6	54·0	53·3	52·7	52·1	30
31	62·8	62·0	61·3	60·5	59·8	59·0	58·3	57·6	56·9	56·2	55·6	54·9	54·2	53·6	53·0	52·4	31
32	63·0	62·3	61·5	60·8	60·0	59·3	58·6	57·9	57·2	56·5	55·8	55·2	54·5	53·9	53·3	52·7	32
33	63·3	62·5	61·8	61·0	60·3	59·6	58·9	58·2	57·5	56·8	56·1	55·5	54·8	54·2	53·6	53·0	33
34	63·6	62·8	62·1	61·3	60·6	59·9	59·2	58·5	57·8	57·1	56·4	55·8	55·1	54·5	53·9	53·3	34
35	63·8	63·1	62·3	61·6	60·9	60·2	59·5	58·8	58·1	57·4	56·8	56·1	55·5	54·8	54·2	53·6	35
36	64·1	63·4	62·6	61·9	61·2	60·5	59·8	59·1	58·4	57·7	57·1	56·4	55·8	55·2	54·6	53·9	36
37	64·4	63·7	62·9	62·2	61·5	60·8	60·1	59·4	58·7	58·1	57·4	56·8	56·1	55·5	54·9	54·3	37
38	64·7	64·0	63·2	62·5	61·8	61·1	60·4	59·8	59·1	58·4	57·8	57·1	56·5	55·9	55·3	54·7	38
39	65·0	64·3	63·6	62·8	62·1	61·5	60·8	60·1	59·4	58·8	58·1	57·5	56·9	56·2	55·6	55·0	39
40	65·3	64·6	63·9	63·2	62·5	61·8	61·1	60·5	59·8	59·1	58·5	57·9	57·2	56·6	56·0	55·4	40
41	65·6	64·9	64·2	63·5	62·8	62·2	61·5	60·8	60·2	59·5	58·9	58·2	57·6	57·0	56·4	55·8	41
42	66·0	65·3	64·6	63·9	63·2	62·5	61·9	61·2	60·5	59·9	59·3	58·6	58·0	57·4	56·8	56·2	42
43	66·3	65·6	64·9	64·2	63·6	62·9	62·2	61·6	60·9	60·3	59·7	59·0	58·4	57·8	57·2	56·6	43
44	66·7	66·0	65·3	64·6	63·9	63·3	62·6	62·0	61·3	60·7	60·1	59·5	58·9	58·3	57·7	57·1	44
45	67·0	66·3	65·7	65·0	64·3	63·7	63·0	62·4	61·7	61·1	60·5	59·9	59·3	58·7	58·1	57·6	45
46	67·4	66·7	66·0	65·4	64·7	64·1	63·4	62·8	62·2	61·5	60·9	60·3	59·7	59·1	58·6	58·0	46
47	67·7	67·1	66·4	65·8	65·1	64·5	63·8	63·2	62·6	62·0	61·4	60·8	60·2	59·6	59·0	58·5	47
48	68·1	67·5	66·8	66·2	65·5	64·9	64·3	63·7	63·0	62·4	61·8	61·2	60·7	60·1	59·5	58·9	48
49	68·5	67·9	67·2	66·6	66·0	65·3	64·7	64·1	63·5	62·9	62·3	61·7	61·1	60·6	60·0	59·4	49
50	68·9	68·3	67·6	67·0	66·4	65·8	65·2	64·6	64·0	63·4	62·8	62·2	61·6	61·1	60·5	60·0	50
51	69·3	68·7	68·1	67·4	66·8	66·2	65·6	65·0	64·4	63·8	63·3	62·7	62·1	61·6	61·0	60·5	51
52	69·7	69·1	68·5	67·9	67·3	66·7	66·1	65·5	64·9	64·4	63·8	63·2	62·7	62·1	61·6	61·0	52
53	70·1	69·5	68·9	68·3	67·7	67·2	66·6	66·0	65·4	64·9	64·3	63·7	63·2	62·6	62·1	61·6	53
54	70·6	70·0	69·4	68·8	68·2	67·6	67·1	66·5	65·9	65·4	64·8	64·3	63·7	63·2	62·6	62·1	54
55	71·0	70·4	69·8	69·3	68·7	68·1	67·6	67·0	66·4	65·9	65·4	64·8	64·3	63·7	63·2	62·7	55
56	71·5	70·9	70·3	69·7	69·2	68·6	68·1	67·5	67·0	66·4	65·9	65·4	64·9	64·3	63·8	63·3	56
57	71·9	71·3	70·8	70·2	69·7	69·1	68·6	68·0	67·5	67·0	66·5	65·9	65·4	64·9	64·4	63·9	57
58	72·4	71·8	71·3	70·7	70·2	69·6	69·1	68·6	68·1	67·5	67·0	66·5	66·0	65·5	65·0	64·5	58
59	72·8	72·3	71·8	71·2	70·7	70·2	69·7	69·1	68·6	68·1	67·6	67·1	66·6	66·1	65·6	65·1	59
60	73·3	72·8	72·3	71·7	71·2	70·7	70·2	69·7	69·2	68·7	68·2	67·7	67·2	66·7	66·3	65·8	60
61	73·8	73·3	72·8	72·3	71·8	71·3	70·8	70·3	69·8	69·3	68·8	68·3	67·8	67·4	66·9	66·4	61
62	74·3	73·8	73·3	72·8	72·3	71·8	71·3	70·8	70·4	69·9	69·4	68·9	68·5	68·0	67·6	67·1	62
63	74·8	74·3	73·8	73·3	72·8	72·4	71·9	71·4	71·0	70·5	70·0	69·6	69·1	68·7	68·2	67·8	63
64	75·3	74·8	74·3	73·9	73·4	72·9	72·5	72·0	71·6	71·1	70·7	70·2	69·8	69·3	68·9	68·5	64
65	75·8	75·3	74·9	74·4	74·0	73·5	73·1	72·6	72·2	71·8	71·3	70·9	70·5	70·0	69·6	69·2	65
66	76·3	75·8	75·4	75·0	74·5	74·1	73·7	73·2	72·8	72·4	72·0	71·6	71·1	70·7	70·3	69·9	66
67	76·8	76·4	76·0	75·5	75·1	74·7	74·3	73·9	73·5	73·1	72·6	72·2	71·8	71·4	71·0	70·6	67
68	77·3	76·9	76·5	76·1	75·7	75·3	74·9	74·5	74·1	73·7	73·3	72·9	72·5	72·1	71·8	71·4	68

AZIMUTHS.

A±B=·60′	·62′	·64′	·66′	·68′	·70′	·72′	·74′	·76′	·78′	·80′	·82′	·84′	·86′	·88′	·90′=A±B	Lat.

A & B **S**ame names take **S**um, (add.) } RULE TO FIND **C** CORRECTION { A & B **D**ifferent names, take **D**ifference, (sub.)

C CORRECTION, (A±B) is named the same as the greater of these quantities.

AZIMUTH takes combined names of **C** Correction and Hour Angle.

C

| C | | | | | | TABLE C | | | | | | | | | | C |

A & B CORRECTION.

A±B=·90'	·92'	·94'	·96'	·98'	1·00'	1·02'	1·04'	1·06'	1·08'	1·10'	1·12'	1·14'	1·16'	1·18'	1·20'=A±B
Lat.						AZIMUTHS.									**Lat.**
0	48·0	47·4	46·8	46·2	45·6	45·0	44·4	43·9	43·3	42·8	42·3	41·8	41·3	40·8	40·3	39·8	**0**
5	48·1	47·5	46·9	46·3	45·7	45·1	44·5	44·0	43·4	42·9	42·4	41·9	41·4	40·9	40·4	39·9	**5**
10	48·4	47·8	47·2	46·6	46·0	45·4	44·9	44·3	43·8	43·2	42·7	42·2	41·7	41·2	40·7	40·2	**10**
14	48·9	48·3	47·7	47·1	46·5	45·9	45·3	44·7	44·2	43·7	43·1	42·6	42·1	41·6	41·2	40·7	**14**
18	49·4	48·8	48·2	47·6	47·0	46·4	45·9	45·3	44·8	44·2	43·7	43·2	42·7	42·2	41·7	41·2	**18**
20	49·8	49·2	48·5	47·9	47·4	46·8	46·2	45·7	45·1	44·6	44·1	43·5	43·0	42·5	42·0	41·6	**20**
22	50·2	49·5	48·9	48·3	47·7	47·2	46·6	46·0	45·5	45·0	44·4	43·9	43·4	42·9	42·4	41·9	**22**
24	50·6	50·0	49·3	48·7	48·2	47·6	47·0	46·5	45·9	45·4	44·9	44·3	43·8	43·3	42·9	42·4	**24**
26	51·0	50·4	49·8	49·2	48·6	48·1	47·5	46·9	46·4	45·9	45·3	44·8	44·3	43·8	43·3	42·8	**26**
28	51·5	50·9	50·3	49·7	49·1	48·6	48·0	47·4	46·9	46·4	45·8	45·3	44·8	44·3	43·8	43·3	**28**
30	52·1	51·5	50·9	50·3	49·7	49·1	48·5	48·0	47·4	46·9	46·4	45·9	45·4	44·9	44·4	43·9	**30**
31	52·4	51·7	51·1	50·6	50·0	49·4	48·8	48·3	47·7	47·2	46·7	46·2	45·7	45·2	44·7	44·2	**31**
32	52·7	52·0	51·4	50·9	50·3	49·7	49·1	48·6	48·0	47·5	47·0	46·5	46·0	45·5	45·0	44·5	**32**
33	53·0	52·3	51·7	51·2	50·6	50·0	49·5	48·9	48·4	47·8	47·3	46·8	46·3	45·8	45·3	44·8	**33**
34	53·3	52·7	52·1	51·5	50·9	50·3	49·8	49·2	48·7	48·2	47·6	47·1	46·6	46·1	45·6	45·1	**34**
35	53·6	53·0	52·4	51·8	51·2	50·7	50·1	49·6	49·0	48·5	48·0	47·5	47·0	46·5	46·0	45·5	**35**
36	53·9	53·3	52·7	52·2	51·6	51·0	50·5	49·9	49·4	48·9	48·3	47·8	47·3	46·8	46·3	45·8	**36**
37	54·3	53·7	53·1	52·5	52·0	51·4	50·8	50·3	49·8	49·2	48·7	48·2	47·7	47·2	46·7	46·2	**37**
38	54·7	54·1	53·5	52·9	52·3	51·8	51·2	50·7	50·1	49·6	49·1	48·6	48·1	47·6	47·1	46·6	**38**
39	55·0	54·4	53·9	53·3	52·7	52·1	51·6	51·1	50·5	50·0	49·5	49·0	48·5	48·0	47·5	47·0	**39**
40	55·4	54·8	54·2	53·7	53·1	52·5	52·0	51·5	50·9	50·4	49·9	49·4	48·9	48·4	47·9	47·4	**40**
41	55·8	55·2	54·6	54·1	53·5	53·0	52·4	51·9	51·3	50·8	50·3	49·8	49·3	48·8	48·3	47·8	**41**
42	56·2	55·6	55·1	54·5	53·9	53·4	52·8	52·3	51·8	51·2	50·7	50·2	49·7	49·2	48·8	48·3	**42**
43	56·6	56·1	55·5	54·9	54·4	53·8	53·3	52·7	52·2	51·7	51·2	50·7	50·2	49·7	49·2	48·7	**43**
44	57·1	56·5	55·9	55·4	54·8	54·3	53·7	53·2	52·7	52·2	51·6	51·1	50·6	50·2	49·7	49·2	**44**
45	57·5	57·0	56·4	55·8	55·3	54·7	54·2	53·7	53·1	52·6	52·1	51·6	51·1	50·6	50·2	49·7	**45**
46	58·0	57·4	56·9	56·3	55·8	55·2	54·7	54·2	53·6	53·1	52·6	52·1	51·6	51·1	50·7	50·2	**46**
47	58·5	57·9	57·3	56·8	56·2	55·7	55·2	54·7	54·1	53·6	53·1	52·6	52·1	51·7	51·2	50·7	**47**
48	58·9	58·4	57·8	57·3	56·7	56·2	55·7	55·2	54·7	54·1	53·6	53·2	52·7	52·2	51·7	51·2	**48**
49	59·4	58·9	58·3	57·8	57·3	56·7	56·2	55·7	55·2	54·7	54·2	53·7	53·2	52·7	52·3	51·8	**49**
50	60·0	59·4	58·9	58·3	57·8	57·3	56·7	56·2	55·7	55·2	54·7	54·2	53·8	53·3	52·8	52·4	**50**
51	60·5	59·9	59·4	58·9	58·3	57·8	57·3	56·8	56·3	55·8	55·3	54·8	54·3	53·9	53·4	52·9	**51**
52	61·0	60·5	59·9	59·4	58·9	58·4	57·9	57·4	56·9	56·4	55·9	55·4	54·9	54·5	54·0	53·5	**52**
53	61·6	61·0	60·5	60·0	59·5	59·0	58·5	58·0	57·5	57·0	56·5	56·0	55·5	55·1	54·6	54·2	**53**
54	62·1	61·6	61·1	60·6	60·1	59·6	59·1	58·6	58·1	57·6	57·1	56·6	56·2	55·7	55·3	54·8	**54**
55	62·7	62·2	61·7	61·2	60·7	60·2	59·7	59·2	58·7	58·2	57·8	57·3	56·8	56·4	55·9	55·5	**55**
56	63·3	62·8	62·3	61·8	61·3	60·8	60·3	59·8	59·3	58·9	58·4	57·9	57·5	57·0	56·6	56·1	**56**
57	63·9	63·4	62·9	62·4	61·9	61·4	60·9	60·5	60·0	59·5	59·1	58·6	58·2	57·7	57·3	56·8	**57**
58	64·5	64·0	63·5	63·0	62·6	62·1	61·6	61·1	60·7	60·2	59·8	59·3	58·9	58·4	58·0	57·5	**58**
59	65·1	64·6	64·2	63·7	63·2	62·7	62·3	61·8	61·4	60·9	60·5	60·0	59·6	59·1	58·7	58·3	**59**
60	65·8	65·3	64·8	64·4	63·9	63·4	63·0	62·5	62·1	61·6	61·2	60·8	60·3	59·9	59·5	59·0	**60**
61	66·4	66·0	65·5	65·0	64·6	64·1	63·7	63·2	62·8	62·4	61·9	61·5	61·1	60·6	60·2	59·8	**61**
62	67·1	66·6	66·2	65·7	65·3	64·9	64·4	64·0	63·5	63·1	62·7	62·3	61·8	61·4	61·0	60·6	**62**
63	67·8	67·3	66·9	66·5	66·0	65·6	65·2	64·7	64·3	63·9	63·5	63·0	62·6	62·2	61·8	61·4	**63**
64	68·5	68·0	67·6	67·2	66·8	66·3	65·9	65·5	65·1	64·7	64·3	63·9	63·4	63·0	62·6	62·3	**64**
65	69·2	68·8	68·3	67·9	67·5	67·1	66·7	66·3	65·9	65·5	65·1	64·7	64·3	63·9	63·5	63·1	**65**
66	69·9	69·5	69·1	68·7	68·3	67·9	67·5	67·1	66·7	66·3	65·9	65·5	65·1	64·7	64·4	64·0	**66**
67	70·6	70·2	69·8	69·4	69·0	68·7	68·3	67·9	67·5	67·1	66·7	66·4	66·0	65·6	65·2	64·9	**67**
68	71·4	71·0	70·6	70·2	69·8	69·5	69·1	68·7	68·3	68·0	67·6	67·2	66·9	66·5	66·2	65·8	**68**
Lat.						AZIMUTHS.									**Lat.**
A±B=·90'	·92'	·94'	·96'	·98'	1·00'	1·02'	1·04'	1·06'	1·08'	1·10'	1·12'	1·14'	1·16'	1·18'	1·20'=A±B

A & B **S**ame names } RULE TO FIND { **A & B D**ifferent names,
take **S**um, (add.) } **C** CORRECTION { take **D**ifference, (sub.)

C CORRECTION, (A±B) is named the same as the greater of these quantities.

AZIMUTH takes combined names of **C** Correction and Hour Angle.

C   **C**

TABLE C

A & B CORRECTION.

A±B=	1·20'	1·24'	1·28'	1·32'	1·36'	1·40'	1·44'	1·48'	1·52'	1·56'	1·60'	1·64'	1·68'	1·72'	1·76'	1·80'=A±B	
Lat.							AZIMUTHS.										Lat.
0	39·8	38·9	38·0	37·1	36·3	35·5	34·8	34·0	33·3	32·7	32·0	31·4	30·8	30·2	29·6	29·1	0
5	39·9	39·0	38·1	37·3	36·4	35·6	34·9	34·1	33·4	32·8	32·1	31·5	30·9	30·3	29·7	29·2	5
10	40·2	39·3	38·4	37·6	36·8	36·0	35·2	34·5	33·8	33·1	32·4	31·8	31·2	30·6	30·0	29·4	10
14	40·7	39·7	38·8	38·0	37·2	36·4	35·6	34·9	34·2	33·5	32·8	32·1	31·5	30·9	30·4	29·8	14
18	41·2	40·3	39·4	38·5	37·7	36·9	36·1	35·4	34·7	34·0	33·3	32·7	32·0	31·4	30·9	30·3	18
20	41·6	40·6	39·7	38·9	38·0	37·2	36·5	35·7	35·0	34·3	33·6	33·0	32·4	31·8	31·2	30·6	20
22	42·0	41·0	40·1	39·3	38·4	37·6	36·8	36·1	35·4	34·7	34·0	33·3	32·7	32·1	31·5	30·9	22
24	42·4	41·4	40·5	39·7	38·8	38·0	37·2	36·5	35·8	35·1	34·4	33·7	33·1	32·5	31·9	31·3	24
26	42·8	41·9	41·0	40·1	39·3	38·5	37·7	36·9	36·2	35·5	34·8	34·2	33·5	32·9	32·3	31·7	26
28	43·3	42·4	41·5	40·6	39·8	39·0	38·2	37·4	36·7	36·0	35·3	34·6	34·0	33·4	32·8	32·2	28
30	43·9	43·0	42·1	41·2	40·3	39·5	38·7	38·0	37·2	36·5	35·8	35·1	34·5	33·9	33·3	32·7	30
31	44·2	43·3	42·4	41·5	40·6	39·8	39·0	38·2	37·5	36·8	36·1	35·4	34·8	34·1	33·5	32·9	31
32	44·5	43·6	42·7	41·8	40·9	40·1	39·3	38·5	37·8	37·1	36·4	35·7	35·1	34·4	33·8	33·2	32
33	44·8	43·9	43·0	42·1	41·2	40·4	39·6	38·9	38·1	37·4	36·7	36·0	35·4	34·7	34·1	33·5	33
34	45·1	44·2	43·3	42·4	41·6	40·7	39·9	39·2	38·4	37·7	37·0	36·3	35·7	35·0	34·4	33·8	34
35	45·5	44·6	43·7	42·8	41·9	41·1	40·3	39·5	38·8	38·0	37·3	36·7	36·0	35·4	34·7	34·1	35
36	45·8	44·9	44·0	43·1	42·3	41·4	40·6	39·9	39·1	38·4	37·7	37·0	36·3	35·7	35·1	34·5	36
37	46·2	45·3	44·4	43·5	42·6	41·8	41·0	40·2	39·5	38·8	38·1	37·4	36·7	36·1	35·4	34·8	37
38	46·6	45·7	44·8	43·9	43·0	42·2	41·4	40·6	39·9	39·1	38·4	37·7	37·1	36·4	35·8	35·2	38
39	47·0	46·1	45·2	44·3	43·4	42·6	41·8	41·0	40·3	39·5	38·8	38·1	37·4	36·8	36·2	35·6	39
40	47·4	46·5	45·6	44·7	43·8	43·0	42·2	41·4	40·7	39·9	39·2	38·5	37·8	37·2	36·6	36·0	40
41	47·8	46·9	46·0	45·1	44·3	43·4	42·6	41·8	41·1	40·3	39·6	38·9	38·3	37·6	37·0	36·4	41
42	48·3	47·3	46·4	45·6	44·7	43·9	43·1	42·3	41·5	40·8	40·1	39·4	38·7	38·0	37·4	36·8	42
43	48·7	47·8	46·9	46·0	45·2	44·3	43·5	42·7	42·0	41·2	40·5	39·8	39·1	38·5	37·8	37·2	43
44	49·2	48·3	47·4	46·5	45·6	44·8	44·0	43·2	42·4	41·7	41·0	40·3	39·6	38·9	38·3	37·7	44
45	49·7	48·8	47·9	47·0	46·1	45·3	44·5	43·7	42·9	42·2	41·5	40·8	40·1	39·4	38·8	38·2	45
46	50·2	49·3	48·4	47·5	46·6	45·8	45·0	44·2	43·4	42·7	42·0	41·3	40·6	39·9	39·3	38·7	46
47	50·7	49·8	48·9	48·0	47·2	46·3	45·5	44·7	44·0	43·2	42·5	41·8	41·1	40·4	39·8	39·2	47
48	51·2	50·3	49·4	48·5	47·7	46·9	46·1	45·3	44·5	43·8	43·0	42·3	41·7	41·0	40·3	39·7	48
49	51·8	50·9	50·0	49·1	48·3	47·4	46·6	45·8	45·1	44·3	43·6	42·9	42·2	41·5	40·9	40·3	49
50	52·4	51·5	50·6	49·7	48·8	48·0	47·2	46·4	45·7	44·9	44·2	43·5	42·8	42·1	41·5	40·8	50
51	52·9	52·0	51·2	50·3	49·4	48·6	47·8	47·0	46·3	45·5	44·8	44·1	43·4	42·7	42·1	41·4	51
52	53·5	52·6	51·8	50·9	50·1	49·2	48·4	47·7	46·9	46·2	45·4	44·7	44·0	43·4	42·7	42·1	52
53	54·2	53·3	52·4	51·5	50·7	49·9	49·1	48·3	47·5	46·8	46·1	45·4	44·7	44·0	43·4	42·7	53
54	54·8	53·9	53·0	52·2	51·4	50·6	49·8	49·0	48·2	47·5	46·8	46·1	45·4	44·7	44·0	43·4	54
55	55·5	54·6	53·7	52·9	52·0	51·2	50·5	49·7	48·9	48·2	47·5	46·8	46·1	45·4	44·7	44·1	55
56	56·1	55·3	54·4	53·6	52·7	51·9	51·2	50·4	49·6	48·9	48·2	47·5	46·8	46·1	45·5	44·8	56
57	56·8	56·0	55·1	54·3	53·5	52·7	51·9	51·1	50·4	49·7	48·9	48·2	47·5	46·9	46·2	45·6	57
58	57·5	56·7	55·9	55·0	54·2	53·4	52·7	51·9	51·1	50·4	49·7	49·0	48·3	47·7	47·0	46·4	58
59	58·3	57·4	56·6	55·8	55·0	54·2	53·4	52·7	51·9	51·2	50·5	49·8	49·1	48·5	47·8	47·2	59
60	59·0	58·2	57·4	56·6	55·8	55·0	54·2	53·5	52·8	52·0	51·3	50·6	50·0	49·3	48·7	48·0	60
61	59·8	59·0	58·2	57·4	56·6	55·8	55·1	54·3	53·6	52·9	52·2	51·5	50·8	50·2	49·5	48·9	61
62	60·6	59·8	59·0	58·2	57·4	56·7	55·9	55·2	54·5	53·8	53·1	52·4	51·7	51·1	50·4	49·8	62
63	61·4	60·6	59·8	59·1	58·3	57·6	56·8	56·1	55·4	54·7	54·0	53·3	52·7	52·0	51·4	50·7	63
64	62·3	61·5	60·7	59·9	59·2	58·5	57·7	57·0	56·3	55·6	55·0	54·3	53·6	53·0	52·4	51·7	64
65	63·1	62·3	61·6	60·8	60·1	59·4	58·7	58·0	57·3	56·6	55·9	55·3	54·6	54·0	53·4	52·7	65
66	64·0	63·2	62·5	61·8	61·1	60·3	59·6	59·0	58·3	57·6	56·9	56·3	55·7	55·0	54·4	53·8	66
67	64·9	64·1	63·4	62·7	62·0	61·3	60·6	60·0	59·3	58·6	58·0	57·3	56·7	56·1	55·5	54·9	67
68	65·8	65·1	64·4	63·7	63·0	62·3	61·7	61·0	60·3	59·7	59·1	58·4	57·8	57·2	56·6	56·0	68
Lat.							AZIMUTHS.										Lat.
A±B=	1·20'	1·24'	1·28'	1·32'	1·36'	1·40'	1·44'	1·48'	1·52'	1·56'	1·60'	1·64'	1·68'	1·72'	1·76'	1·80'=A±B	

A & B Same Names, ⎫ RULE TO FIND ⎰ A & B Different Names,
take Sum, (add) ⎭ C CORRECTION ⎱ take Difference. (sub.)

C CORRECTION, (A±B) is named the same as greater of these quantities.

AZIMUTH is named same as C Correction and the Hour Angle.

| C | | | | | | | TABLE C | | | | | | | | | C |

A & B CORRECTION.

AZIMUTHS.

Lat.	A±B=1·80′	1·84′	1·88′	1·92′	1·96′	2·00′	2·04′	2·08′	2·12′	2·16′	2·20′	2·24′	2·28′	2·32′	2·36′	2·40′=A±B	Lat.
0	29·1	28·5	28·0	27·5	27·0	26·6	26·1	25·7	25·2	24·8	24·4	24·0	23·7	23·3	23·0	22·6	0
5	29·1	28·6	28·1	27·6	27·1	26·7	26·2	25·8	25·3	24·9	24·5	24·1	23·8	23·4	23·0	22·7	5
10	29·4	28·9	28·4	27·9	27·4	26·9	26·5	26·0	25·6	25·2	24·8	24·4	24·0	23·6	23·3	22·9	10
14	29·8	29·3	28·7	28·2	27·7	27·3	26·8	26·4	25·9	25·5	25·1	24·7	24·3	24·0	23·6	23·2	14
18	30·3	29·7	29·2	28·7	28·2	27·7	27·3	26·8	26·4	26·0	25·6	25·2	24·8	24·4	24·0	23·7	18
20	30·6	30·0	29·5	29·0	28·5	28·0	27·6	27·1	26·7	26·2	25·8	25·4	25·0	24·6	24·3	23·9	20
22	30·9	30·4	29·8	29·3	28·8	28·3	27·9	27·4	27·0	26·5	26·1	25·7	25·3	24·9	24·6	24·2	22
24	31·3	30·8	30·2	29·7	29·2	28·7	28·2	27·8	27·3	26·9	26·5	26·1	25·7	25·3	24·9	24·5	24
26	31·7	31·2	30·6	30·1	29·6	29·1	28·6	28·2	27·7	27·3	26·8	26·4	26·0	25·6	25·2	24·9	26
28	32·2	31·6	31·1	30·5	30·0	29·5	29·0	28·6	28·1	27·7	27·2	26·8	26·4	26·0	25·6	25·3	28
30	32·7	32·1	31·6	31·0	30·5	30·0	29·5	29·0	28·6	28·1	27·7	27·3	26·9	26·5	26·1	25·7	30
31	32·9	32·4	31·8	31·3	30·8	30·3	29·8	29·3	28·8	28·4	27·9	27·5	27·1	26·7	26·3	25·9	31
32	33·2	32·7	32·1	31·6	31·0	30·5	30·0	29·6	29·1	28·6	28·2	27·8	27·3	26·9	26·5	26·2	32
33	33·5	32·9	32·4	31·8	31·3	30·8	30·3	29·8	29·4	28·9	28·5	28·0	27·6	27·2	26·8	26·4	33
34	33·8	33·2	32·7	32·1	31·6	31·1	30·6	30·1	29·6	29·2	28·7	28·3	27·9	27·5	27·1	26·7	34
35	34·1	33·6	33·0	32·4	31·9	31·4	30·9	30·4	29·9	29·5	29·0	28·6	28·2	27·8	27·4	27·0	35
36	34·5	33·9	33·3	32·8	32·2	31·7	31·2	30·7	30·3	29·8	29·3	28·9	28·5	28·1	27·7	27·3	36
37	34·8	34·2	33·7	33·1	32·6	32·1	31·6	31·1	30·6	30·1	29·6	29·2	28·8	28·4	28·0	27·6	37
38	35·2	34·6	34·0	33·5	32·9	32·4	31·9	31·4	30·9	30·4	30·0	29·5	29·1	28·7	28·3	27·9	38
39	35·6	35·0	34·4	33·8	33·3	32·8	32·3	31·8	31·3	30·8	30·3	29·9	29·4	29·0	28·6	28·2	39
40	36·0	35·4	34·8	34·2	33·7	33·1	32·6	32·1	31·6	31·2	30·7	30·2	29·8	29·4	28·9	28·5	40
41	36·4	35·8	35·2	34·6	34·1	33·5	33·0	32·5	32·0	31·5	31·1	30·6	30·2	29·7	29·3	28·9	41
42	36·8	36·2	35·6	35·0	34·5	33·9	33·4	32·9	32·4	31·9	31·5	31·0	30·5	30·1	29·7	29·3	42
43	37·2	36·6	36·0	35·5	34·9	34·4	33·8	33·3	32·8	32·3	31·9	31·4	31·0	30·5	30·1	29·7	43
44	37·7	37·1	36·5	35·9	35·3	34·8	34·3	33·8	33·3	32·8	32·3	31·8	31·4	30·9	30·5	30·1	44
45	38·2	37·6	37·0	36·4	35·8	35·3	34·7	34·2	33·7	33·2	32·7	32·3	31·8	31·4	30·9	30·5	45
46	38·7	38·1	37·5	36·9	36·3	35·7	35·2	34·7	34·2	33·7	33·2	32·7	32·3	31·8	31·4	31·0	46
47	39·2	38·6	38·0	37·4	36·8	36·2	35·7	35·2	34·7	34·2	33·7	33·2	32·7	32·3	31·9	31·4	47
48	39·7	39·1	38·5	37·9	37·3	36·8	36·2	35·7	35·2	34·7	34·2	33·7	33·2	32·8	32·4	31·9	48
49	40·3	39·6	39·0	38·4	37·9	37·3	36·8	36·2	35·7	35·2	34·7	34·2	33·8	33·3	32·9	32·4	49
50	40·8	40·2	39·6	39·0	38·4	37·9	37·3	36·8	36·3	35·8	35·3	34·8	34·3	33·8	33·4	33·0	50
51	41·4	40·8	40·2	39·6	39·0	38·5	37·9	37·4	36·9	36·3	35·8	35·3	34·9	34·4	34·0	33·5	51
52	42·1	41·4	40·8	40·2	39·6	39·1	38·5	38·0	37·5	36·9	36·4	35·9	35·5	35·0	34·5	34·1	52
53	42·7	42·1	41·5	40·9	40·3	39·7	39·2	38·6	38·1	37·6	37·1	36·6	36·1	35·6	35·1	34·7	53
54	43·4	42·8	42·1	41·5	41·0	40·4	39·8	39·3	38·7	38·2	37·7	37·2	36·7	36·3	35·8	35·3	54
55	44·1	43·5	42·8	42·2	41·7	41·1	40·5	40·0	39·4	38·9	38·4	37·9	37·4	36·9	36·5	36·0	55
56	44·8	44·2	43·6	43·0	42·4	41·8	41·2	40·7	40·1	39·6	39·1	38·6	38·1	37·6	37·2	36·7	56
57	45·6	44·9	44·3	43·7	43·1	42·6	42·0	41·4	40·9	40·4	39·8	39·3	38·8	38·4	37·9	37·4	57
58	46·4	45·7	45·1	44·5	43·9	43·4	42·8	42·2	41·7	41·2	40·6	40·1	39·6	39·1	38·6	38·2	58
59	47·2	46·5	45·9	45·3	44·7	44·2	43·6	43·0	42·5	42·0	41·4	40·9	40·4	39·9	39·4	39·0	59
60	48·0	47·4	46·8	46·2	45·6	45·0	44·4	43·9	43·3	42·8	42·3	41·8	41·3	40·8	40·3	39·8	60
61	48·9	48·3	47·7	47·1	46·5	45·9	45·3	44·8	44·2	43·7	43·2	42·7	42·2	41·7	41·2	40·7	61
62	49·8	49·2	48·6	48·0	47·4	46·8	46·2	45·7	45·1	44·6	44·1	43·6	43·1	42·6	42·1	41·6	62
63	50·7	50·1	49·5	48·9	48·3	47·8	47·2	46·6	46·1	45·6	45·0	44·5	44·0	43·5	43·0	42·5	63
64	51·7	51·1	50·5	49·9	49·3	48·8	48·2	47·6	47·1	46·6	46·0	45·5	45·0	44·5	44·0	43·5	64
65	52·7	52·1	51·5	50·9	50·4	49·8	49·2	48·7	48·1	47·6	47·1	46·6	46·1	45·6	45·1	44·6	65
66	53·8	53·2	52·6	52·0	51·4	50·9	50·3	49·8	49·2	48·7	48·2	47·7	47·2	46·7	46·2	45·7	66
67	54·9	54·3	53·7	53·1	52·6	52·0	51·4	50·9	50·4	49·8	49·3	48·8	48·3	47·8	47·3	46·8	67
68	56·0	55·4	54·8	54·3	53·7	53·2	52·6	52·1	51·5	51·0	50·5	50·0	49·5	49·0	48·5	48·0	68

AZIMUTHS.

| Lat. | A±B=1·80′ | 1·84′ | 1·88′ | 1·92′ | 1·96′ | 2·00′ | 2·04′ | 2·08′ | 2·12′ | 2·16′ | 2·20′ | 2·24′ | 2·28′ | 2·32′ | 2·36′ | 2·40′=A±B | Lat. |

A & B **S**ame Names, ⎫ RULE TO FIND ⎰ A & B **D**ifferent Names,
take **S**um, (add) ⎰ **C** CORRECTION ⎱ take **D**ifference. (sub.)

C CORRECTION, (A±B) is named the same as greater of these quantities.

AZIMUTH is named same as **C** Correction and the Hour Angle.

C							TABLE C									C

A & B CORRECTION.

A±B=	2·40'	2·45'	2·50'	2·55'	2·60'	2·65'	2·70'	2·75'	2·80'	2·90'	3·00'	3·10'	3·20'	3·30'	3·40'	3·50'=A±B
Lat.							AZIMUTHS.									Lat.
0	22·6	22·2	21·8	21·4	21·0	20·7	20·3	20·0	19·7	19·0	18·4	17·9	17·4	16·9	16·4	15·9	0
5	22·7	22·3	21·9	21·5	21·1	20·8	20·4	20·1	19·7	19·1	18·5	17·9	17·4	16·9	16·4	16·0	5
10	22·9	22·5	22·1	21·7	21·3	21·0	20·6	20·3	19·9	19·3	18·7	18·1	17·6	17·1	16·6	16·2	10
14	23·2	22·8	22·4	22·0	21·6	21·3	20·9	20·5	20·2	19·6	19·0	18·4	17·9	17·4	16·9	16·4	14
18	23·7	23·2	22·8	22·4	22·0	21·6	21·3	20·9	20·6	19·9	19·3	18·7	18·2	17·7	17·2	16·7	18
20	23·9	23·5	23·1	22·7	22·3	21·9	21·5	21·2	20·8	20·2	19·5	18·9	18·4	17·9	17·4	16·9	20
22	24·2	23·8	23·3	22·9	22·5	22·2	21·8	21·4	21·1	20·4	19·8	19·2	18·6	18·1	17·6	17·1	22
24	24·5	24·1	23·6	23·2	22·8	22·5	22·1	21·7	21·4	20·7	20·0	19·4	18·9	18·4	17·8	17·4	24
26	24·9	24·4	24·0	23·6	23·2	22·8	22·4	22·0	21·7	21·0	20·3	19·7	19·2	18·6	18·1	17·6	26
28	25·3	24·8	24·4	23·9	23·5	23·1	22·8	22·4	22·0	21·3	20·7	20·1	19·5	18·9	18·4	17·9	28
30	25·7	25·2	24·8	24·4	23·9	23·5	23·2	22·8	22·4	21·7	21·1	20·4	19·8	19·3	18·8	18·3	30
31	25·9	25·5	25·0	24·6	24·2	23·8	23·4	23·0	22·6	21·9	21·2	20·6	20·0	19·5	18·9	18·4	31
32	26·2	25·7	25·3	24·8	24·4	24·0	23·6	23·2	22·8	22·1	21·5	20·8	20·2	19·7	19·1	18·6	32
33	26·4	26·0	25·5	25·1	24·6	24·2	23·8	23·4	23·1	22·4	21·7	21·0	20·4	19·9	19·3	18·8	33
34	26·7	26·2	25·8	25·3	24·9	24·5	24·1	23·7	23·3	22·6	21·9	21·3	20·7	20·1	19·5	19·0	34
35	27·0	26·5	26·0	25·6	25·2	24·7	24·3	23·9	23·6	22·8	22·1	21·5	20·9	20·3	19·8	19·2	35
36	27·3	26·8	26·3	25·9	25·4	25·0	24·6	24·2	23·8	23·1	22·4	21·7	21·1	20·5	20·0	19·5	36
37	27·6	27·1	26·6	26·2	25·7	25·3	24·9	24·5	24·1	23·4	22·7	22·0	21·4	20·8	20·2	19·7	37
38	27·9	27·4	26·9	26·5	26·0	25·6	25·2	24·8	24·4	23·6	22·9	22·3	21·6	21·0	20·5	19·9	38
39	28·2	27·7	27·2	26·8	26·3	25·9	25·5	25·1	24·7	23·9	23·2	22·5	21·9	21·3	20·7	20·2	39
40	28·5	28·1	27·6	27·1	26·7	26·2	25·8	25·4	25·0	24·2	23·5	22·8	22·2	21·6	21·0	20·5	40
41	28·9	28·4	27·9	27·5	27·0	26·6	26·1	25·7	25·3	24·6	23·8	23·1	22·5	21·9	21·3	20·7	41
42	29·3	28·8	28·3	27·8	27·4	26·9	26·5	26·1	25·7	24·9	24·2	23·5	22·8	22·2	21·6	21·0	42
43	29·7	29·2	28·7	28·2	27·7	27·3	26·9	26·4	26·0	25·2	24·5	23·8	23·1	22·5	21·9	21·3	43
44	30·1	29·6	29·1	28·6	28·1	27·7	27·2	26·8	26·4	25·6	24·9	24·2	23·5	22·8	22·2	21·7	44
45	30·5	30·0	29·5	29·0	28·5	28·1	27·6	27·2	26·8	26·0	25·2	24·5	23·8	23·2	22·6	22·0	45
46	31·0	30·4	29·9	29·5	29·0	28·6	28·1	27·6	27·2	26·4	25·6	24·9	24·2	23·6	22·9	22·4	46
47	31·4	30·9	30·4	29·9	29·4	29·0	28·5	28·1	27·6	26·8	26·0	25·3	24·6	24·0	23·3	22·7	47
48	31·9	31·4	30·9	30·4	29·9	29·4	29·0	28·5	28·1	27·3	26·5	25·7	25·0	24·4	23·7	23·1	48
49	32·4	31·9	31·4	30·9	30·4	29·9	29·4	29·0	28·6	27·7	26·9	26·2	25·5	24·8	24·1	23·5	49
50	33·0	32·4	31·9	31·4	30·9	30·4	30·0	29·5	29·1	28·2	27·4	26·6	25·9	25·2	24·6	24·0	50
51	33·5	33·0	32·4	31·9	31·4	30·9	30·5	30·0	29·6	28·7	27·9	27·1	26·4	25·7	25·0	24·4	51
52	34·1	33·5	33·0	32·5	32·0	31·5	31·0	30·6	30·1	29·3	28·5	27·7	26·9	26·2	25·5	24·9	52
53	34·7	34·1	33·6	33·1	32·6	32·1	31·6	31·1	30·7	29·8	29·0	28·2	27·4	26·7	26·0	25·4	53
54	35·3	34·8	34·2	33·7	33·2	32·7	32·2	31·7	31·3	30·4	29·6	28·8	28·0	27·3	26·6	25·9	54
55	36·0	35·4	34·9	34·4	33·8	33·4	32·9	32·4	31·9	31·0	30·2	29·4	28·6	27·9	27·1	26·5	55
56	36·7	36·1	35·6	35·1	34·5	34·0	33·5	33·1	32·6	31·7	30·8	30·0	29·2	28·5	27·7	27·1	56
57	37·4	36·8	36·3	35·8	35·2	34·7	34·2	33·8	33·3	32·4	31·5	30·6	29·8	29·1	28·4	27·7	57
58	38·2	37·6	37·0	36·5	36·0	35·5	35·0	34·5	34·0	33·1	32·2	31·3	30·5	29·8	29·0	28·3	58
59	39·0	38·4	37·8	37·3	36·8	36·3	35·7	35·2	34·7	33·8	32·9	32·1	31·2	30·5	29·7	29·0	59
60	39·8	39·2	38·7	38·1	37·6	37·1	36·5	36·0	35·5	34·6	33·7	32·8	32·0	31·2	30·5	29·8	60
61	40·7	40·1	39·5	39·0	38·4	37·9	37·4	36·9	36·4	35·4	34·5	33·6	32·8	32·0	31·2	30·5	61
62	41·6	41·0	40·4	39·9	39·3	38·8	38·3	37·8	37·3	36·3	35·4	34·5	33·6	32·8	32·1	31·3	62
63	42·5	42·0	41·4	40·8	40·3	39·7	39·2	38·7	38·2	37·2	36·3	35·4	34·5	33·7	32·9	32·2	63
64	43·5	43·0	42·4	41·8	41·3	40·7	40·2	39·7	39·2	38·2	37·3	36·4	35·5	34·7	33·9	33·1	64
65	44·6	44·0	43·4	42·9	42·3	41·8	41·2	40·7	40·2	39·2	38·3	37·4	36·5	35·7	34·9	34·1	65
66	45·7	45·1	44·5	44·0	43·4	42·9	42·3	41·8	41·3	40·3	39·3	38·4	37·6	36·7	35·9	35·1	66
67	46·8	46·2	45·7	45·1	44·6	44·0	43·5	42·9	42·4	41·4	40·5	39·5	38·7	37·8	37·0	36·2	67
68	48·0	47·4	46·9	46·3	45·8	45·2	44·7	44·1	43·6	42·6	41·7	40·7	39·8	39·0	38·1	37·3	68
Lat.							AZIMUTHS.									Lat.
A±B=	2·40'	2·45'	2·50'	2·55'	2·60'	2·65'	2·70'	2·75'	2·80'	2·90'	3·00'	3·10'	3·20'	3·30'	3·40'	3·50'=A±B

A & B **S**ame Names,⎫ RULE TO FIND ⎰ A & B **D**ifferent Names,
take **S**um, (add) ⎰ C CORRECTION ⎰ take **D**ifference. (sub.)

C CORRECTION, (A±B) is named the same as greater of these quantities.
AZIMUTH is named same as C Correction and the Hour Angle.

C

C TABLE C C

A & B CORRECTION.

AZIMUTHS.

A±B=3·50′	3·60′	3·70′	3·80′	3·90′	4·00′	4·10′	4·20′	4·30′	4·40′	4·50′	4·60′	4·70′	4·80′	4·90′	5·00′=A±B
Lat.															Lat.
0 15·9	15·5	15·1	14·7	14·4	14·0	13·7	13·4	13·1	12·8	12·5	12·3	12·0	11·8	11·5	11·3 0
5 16·0	15·6	15·2	14·8	14·4	14·1	13·8	13·5	13·2	12·9	12·6	12·3	12·1	11·8	11·6	11·4 5
10 16·2	15·8	15·4	15·0	14·6	14·2	13·9	13·6	13·3	13·0	12·7	12·4	12·2	11·9	11·7	11·5 10
14 16·4	16·0	15·6	15·2	14·8	14·4	14·1	13·8	13·5	13·2	12·9	12·6	12·4	12·1	11·9	11·6 14
18 16·7	16·3	15·9	15·5	15·1	14·7	14·4	14·1	13·7	13·4	13·1	12·9	12·6	12·4	12·1	11·9 18
20 16·9	16·5	16·0	15·6	15·3	14·9	14·6	14·2	13·9	13·6	13·3	13·0	12·8	12·5	12·3	12·0 20
22 17·1	16·7	16·3	15·8	15·5	15·1	14·7	14·4	14·1	13·8	13·5	13·2	12·9	12·7	12·4	12·2 22
24 17·4	16·9	16·5	16·1	15·7	15·3	14·9	14·6	14·3	14·0	13·7	13·4	13·1	12·8	12·6	12·3 24
26 17·6	17·2	16·7	16·3	15·9	15·5	15·2	14·8	14·5	14·2	13·9	13·6	13·3	13·1	12·8	12·5 26
28 17·9	17·5	17·0	16·6	16·2	15·8	15·4	15·1	14·8	14·4	14·1	13·8	13·5	13·3	13·0	12·8 28
30 18·3	17·8	17·3	16·9	16·5	16·1	15·7	15·4	15·0	14·7	14·4	14·1	13·8	13·5	13·3	13·0 30
31 18·4	18·0	17·5	17·1	16·7	16·3	15·9	15·5	15·2	14·9	14·5	14·2	13·9	13·7	13·4	13·1 31
32 18·6	18·1	17·7	17·2	16·8	16·4	16·0	15·7	15·3	15·0	14·7	14·4	14·1	13·8	13·6	13·3 32
33 18·8	18·3	17·9	17·4	17·0	16·6	16·2	15·8	15·5	15·2	14·9	14·5	14·2	14·0	13·7	13·4 33
34 19·0	18·5	18·1	17·6	17·2	16·8	16·4	16·0	15·7	15·3	15·0	14·7	14·4	14·1	13·8	13·6 34
35 19·2	18·7	18·3	17·8	17·4	17·0	16·6	16·2	15·9	15·5	15·2	14·9	14·6	14·3	14·0	13·7 35
36 19·5	19·0	18·5	18·0	17·6	17·2	16·8	16·4	16·1	15·7	15·4	15·0	14·7	14·4	14·2	13·9 36
37 19·7	19·2	18·7	18·2	17·8	17·4	17·0	16·6	16·3	15·9	15·6	15·2	14·9	14·6	14·3	14·1 37
38 19·9	19·4	18·9	18·5	18·0	17·6	17·2	16·8	16·5	16·1	15·8	15·4	15·1	14·8	14·5	14·2 38
39 20·2	19·7	19·2	18·7	18·3	17·8	17·4	17·0	16·7	16·3	16·0	15·6	15·3	15·0	14·7	14·4 39
40 20·5	19·9	19·4	19·0	18·5	18·1	17·7	17·3	16·9	16·5	16·2	15·8	15·5	15·2	14·9	14·6 40
41 20·7	20·2	19·7	19·2	18·8	18·3	17·9	17·5	17·1	16·8	16·4	16·1	15·7	15·4	15·1	14·8 41
42 21·0	20·5	20·0	19·5	19·0	18·6	18·2	17·8	17·4	17·0	16·7	16·3	16·0	15·7	15·4	15·1 42
43 21·3	20·8	20·3	19·8	19·3	18·9	18·4	18·0	17·6	17·3	16·9	16·6	16·2	15·9	15·6	15·3 43
44 21·7	21·1	20·6	20·1	19·6	19·2	18·7	18·3	17·9	17·5	17·2	16·8	16·5	16·2	15·8	15·5 44
45 22·0	21·4	20·9	20·4	19·9	19·5	19·0	18·6	18·2	17·8	17·5	17·1	16·8	16·4	16·1	15·8 45
46 22·4	21·8	21·3	20·8	20·3	19·8	19·3	18·9	18·5	18·1	17·8	17·4	17·1	16·7	16·4	16·1 46
47 22·7	22·2	21·6	21·1	20·6	20·1	19·7	19·2	18·8	18·4	18·1	17·7	17·4	17·0	16·7	16·3 47
48 23·1	22·5	22·0	21·5	21·0	20·5	20·0	19·6	19·2	18·8	18·4	18·0	17·7	17·3	17·0	16·6 48
49 23·5	22·9	22·4	21·9	21·4	20·9	20·4	19·9	19·5	19·1	18·7	18·3	18·0	17·6	17·3	17·0 49
50 24·0	23·4	22·8	22·3	21·8	21·3	20·8	20·3	19·9	19·5	19·1	18·7	18·3	18·0	17·6	17·3 50
51 24·4	23·8	23·3	22·7	22·2	21·7	21·2	20·7	20·3	19·9	19·5	19·1	18·7	18·3	18·0	17·6 51
52 24·9	24·3	23·7	23·2	22·6	22·1	21·6	21·2	20·7	20·3	19·9	19·5	19·1	18·7	18·3	18·0 52
53 25·4	24·8	24·2	23·6	23·1	22·6	22·1	21·6	21·2	20·7	20·3	19·9	19·5	19·1	18·7	18·4 53
54 25·9	25·3	24·7	24·1	23·6	23·1	22·6	22·1	21·6	21·1	20·7	20·3	19·9	19·5	19·1	18·8 54
55 26·5	25·8	25·2	24·6	24·1	23·6	23·1	22·6	22·1	21·6	21·2	20·8	20·4	20·0	19·6	19·2 55
56 27·1	26·4	25·8	25·2	24·6	24·1	23·6	23·1	22·6	22·1	21·7	21·3	20·8	20·4	20·1	19·7 56
57 27·7	27·0	26·4	25·8	25·2	24·7	24·1	23·6	23·1	22·7	22·2	21·8	21·3	20·9	20·6	20·2 57
58 28·3	27·7	27·0	26·4	25·8	25·3	24·7	24·2	23·7	23·2	22·8	22·3	21·9	21·5	21·1	20·7 58
59 29·0	28·4	27·7	27·1	26·5	25·9	25·3	24·8	24·3	23·8	23·4	22·9	22·5	22·0	21·6	21·2 59
60 29·8	29·1	28·4	27·8	27·2	26·6	26·0	25·5	24·9	24·4	24·0	23·5	23·1	22·6	22·2	21·8 60
61 30·5	29·8	29·1	28·5	27·9	27·3	26·7	26·2	25·6	25·1	24·6	24·2	23·7	23·3	22·8	22·4 61
62 31·3	30·5	29·9	29·3	28·7	28·0	27·5	26·9	26·4	25·8	25·3	24·8	24·4	23·9	23·5	23·1 62
63 32·2	31·5	30·8	30·1	29·5	28·8	28·2	27·7	27·1	26·6	26·1	25·6	25·1	24·6	24·2	23·8 63
64 33·1	32·4	31·7	31·0	30·3	29·7	29·1	28·5	27·9	27·4	26·9	26·4	25·9	25·4	25·0	24·5 64
65 34·1	33·3	32·6	31·9	31·2	30·6	30·0	29·4	28·8	28·3	27·8	27·2	26·7	26·2	25·8	25·3 65
66 35·1	34·3	33·6	32·9	32·2	31·6	30·9	30·3	29·8	29·2	28·7	28·1	27·6	27·1	26·7	26·2 66
67 36·2	35·4	34·7	34·0	33·3	32·6	32·0	31·4	30·8	30·2	29·6	29·1	28·6	28·1	27·6	27·1 67
68 37·3	36·6	35·8	35·1	34·4	33·7	33·1	32·4	31·8	31·2	30·7	30·1	29·6	29·1	28·6	28·1 68
Lat.															Lat.

AZIMUTHS.

A±B=3·50′	3·60′	3·70′	3·80′	3·90′	4·00′	4·10′	4·20′	4·30′	4·40′	4·50′	4·60′	4·70′	4·80′	4·90′	5·00′=A±B

A & B Same Names, ⎫ RULE TO FIND ⎰ A & B Different Names,
take Sum, (add) ⎰ C CORRECTION ⎱ take Difference. (sub.)

C CORRECTION, (A±B) is named the same as greater of these quantities.

AZIMUTH is named same as C Correction and the Hour Angle.

C   C

TABLE C

C   **C**

A & B CORRECTION.

Lat.	A±B=5·00′	5·20′	5·40′	5·60′	5·80′	6·00′	6·20′	6·40′	6·60′	6·80′	7·00′	7·20′	7·40′	7·60′	7·80′	8·00′=A±B	Lat.
							AZIMUTHS.										
0	11·3	10·9	10·5	10·1	9·8	9·5	9·2	8·9	8·6	8·4	8·1	7·9	7·7	7·5	7·3	7·1	0
5	11·4	10·9	10·5	10·2	9·8	9·5	9·2	8·9	8·6	8·4	8·2	7·9	7·7	7·5	7·3	7·2	5
10	11·5	11·0	10·6	10·3	9·9	9·6	9·3	9·0	8·7	8·5	8·3	8·0	7·8	7·6	7·4	7·2	10
14	11·6	11·2	10·8	10·4	10·1	9·7	9·4	9·1	8·9	8·7	8·4	8·1	7·9	7·7	7·5	7·3	14
18	11·9	11·4	11·0	10·6	10·3	9·9	9·6	9·3	9·1	8·8	8·5	8·3	8·1	7·9	7·7	7·5	18
20	12·0	11·6	11·1	10·8	10·4	10·1	9·7	9·4	9·2	8·9	8·6	8·4	8·2	8·0	7·8	7·6	20
22	12·2	11·7	11·3	10·9	10·5	10·2	9·9	9·6	9·3	9·0	8·8	8·5	8·3	8·1	7·9	7·7	22
24	12·3	11·9	11·5	11·1	10·7	10·3	10·0	9·7	9·4	9·1	8·9	8·6	8·4	8·2	8·0	7·8	24
26	12·5	12·1	11·6	11·2	10·9	10·5	10·2	9·9	9·6	9·3	9·0	8·8	8·6	8·3	8·1	7·9	26
28	12·8	12·3	11·8	11·4	11·0	10·7	10·4	10·0	9·7	9·5	9·2	8·9	8·7	8·5	8·3	8·1	28
30	13·0	12·5	12·1	11·7	11·3	10·9	10·5	10·2	9·9	9·6	9·4	9·1	8·9	8·6	8·4	8·2	30
31	13·1	12·6	12·2	11·8	11·4	11·0	10·7	10·3	10·0	9·7	9·5	9·2	9·0	8·7	8·5	8·3	31
32	13·3	12·8	12·3	11·9	11·5	11·1	10·8	10·4	10·1	9·8	9·6	9·3	9·1	8·8	8·6	8·4	32
33	13·4	12·9	12·5	12·0	11·6	11·2	10·9	10·6	10·2	9·9	9·7	9·4	9·2	8·9	8·7	8·5	33
34	13·6	13·1	12·6	12·2	11·7	11·4	11·0	10·7	10·4	10·1	9·8	9·5	9·3	9·0	8·8	8·6	34
35	13·7	13·2	12·7	12·3	11·9	11·5	11·1	10·8	10·5	10·2	9·9	9·6	9·4	9·1	8·9	8·7	35
36	13·9	13·4	12·9	12·4	12·0	11·6	11·3	10·9	10·6	10·3	10·0	9·7	9·5	9·2	9·0	8·8	36
37	14·1	13·5	13·1	12·6	12·2	11·8	11·4	11·1	10·7	10·4	10·1	9·9	9·6	9·4	9·1	8·9	37
38	14·2	13·7	13·2	12·8	12·3	11·9	11·6	11·2	10·9	10·6	10·3	10·0	9·7	9·5	9·2	9·0	38
39	14·4	13·9	13·4	12·9	12·5	12·1	11·7	11·4	11·0	10·7	10·4	10·1	9·9	9·6	9·4	9·1	39
40	14·6	14·1	13·6	13·1	12·7	12·3	11·9	11·5	11·2	10·9	10·6	10·3	10·0	9·7	9·5	9·3	40
41	14·8	14·3	13·8	13·3	12·9	12·5	12·1	11·7	11·4	11·0	10·7	10·4	10·2	9·9	9·6	9·4	41
42	15·1	14·5	14·0	13·5	13·1	12·6	12·2	11·9	11·5	11·2	10·9	10·6	10·3	10·0	9·8	9·5	42
43	15·3	14·7	14·2	13·7	13·3	12·8	12·4	12·1	11·7	11·4	11·1	10·8	10·5	10·2	9·9	9·7	43
44	15·5	15·0	14·4	13·9	13·5	13·0	12·6	12·3	11·9	11·6	11·2	10·9	10·6	10·4	10·1	9·9	44
45	15·8	15·2	14·7	14·2	13·7	13·3	12·8	12·5	12·1	11·7	11·4	11·1	10·8	10·5	10·3	10·0	45
46	16·1	15·5	14·9	14·4	13·9	13·5	13·1	12·7	12·3	12·0	11·6	11·3	11·0	10·7	10·5	10·2	46
47	16·3	15·7	15·2	14·7	14·2	13·7	13·3	12·9	12·5	12·2	11·8	11·5	11·2	10·9	10·6	10·4	47
48	16·6	16·0	15·5	14·9	14·4	14·0	13·6	13·1	12·8	12·4	12·1	11·7	11·4	11·1	10·8	10·6	48
49	17·0	16·3	15·8	15·2	14·7	14·3	13·8	13·4	13·0	12·6	12·3	12·0	11·6	11·3	11·1	10·8	49
50	17·3	16·7	16·1	15·5	15·0	14·5	14·1	13·7	13·3	12·9	12·5	12·2	11·9	11·6	11·3	11·0	50
51	17·6	17·0	16·4	15·8	15·3	14·8	14·4	13·9	13·5	13·2	12·8	12·4	12·1	11·8	11·5	11·2	51
52	18·0	17·3	16·7	16·2	15·6	15·1	14·7	14·2	13·8	13·4	13·1	12·7	12·4	12·1	11·8	11·5	52
53	18·4	17·7	17·1	16·5	16·0	15·5	15·0	14·6	14·1	13·7	13·4	13·0	12·7	12·3	12·0	11·7	53
54	18·8	18·1	17·5	16·9	16·3	15·8	15·3	14·9	14·5	14·1	13·7	13·3	13·0	12·6	12·3	12·0	54
55	19·2	18·5	17·9	17·3	16·7	16·2	15·7	15·2	14·8	14·4	14·0	13·6	13·3	12·9	12·6	12·3	55
56	19·7	19·0	18·3	17·7	17·1	16·6	16·1	15·6	15·2	14·7	14·3	13·9	13·6	13·2	12·9	12·6	56
57	20·2	19·5	18·8	18·2	17·6	17·0	16·5	16·0	15·5	15·1	14·7	14·3	13·9	13·6	13·2	12·9	57
58	20·7	20·0	19·3	18·6	18·0	17·5	16·9	16·4	16·0	15·5	15·1	14·7	14·3	13·9	13·6	13·3	58
59	21·2	20·5	19·8	19·1	18·5	17·9	17·4	16·9	16·4	15·9	15·5	15·1	14·7	14·3	14·0	13·6	59
60	21·8	21·0	20·3	19·7	19·0	18·4	17·9	17·4	16·9	16·4	15·9	15·5	15·1	14·7	14·4	14·0	60
61	22·4	21·6	20·9	20·2	19·6	19·0	18·4	17·9	17·4	16·9	16·4	16·0	15·6	15·2	14·8	14·5	61
62	23·1	22·3	21·5	20·8	20·2	19·6	19·0	18·4	17·9	17·4	16·9	16·5	16·1	15·7	15·3	14·9	62
63	23·8	23·0	22·2	21·5	20·8	20·2	19·6	19·0	18·5	18·0	17·5	17·0	16·6	16·2	15·8	15·4	63
64	24·5	23·7	22·9	22·2	21·5	20·8	20·2	19·6	19·1	18·6	18·1	17·6	17·1	16·7	16·3	15·9	64
65	25·3	24·5	23·7	23·0	22·2	21·5	20·9	20·3	19·7	19·2	18·7	18·2	17·7	17·3	16·9	16·5	65
66	26·2	25·3	24·5	23·7	23·0	22·3	21·6	21·0	20·4	19·9	19·4	18·9	18·4	17·9	17·5	17·1	66
67	27·1	26·2	25·4	24·6	23·8	23·1	22·4	21·8	21·2	20·6	20·1	19·6	19·1	18·6	18·2	17·8	67
68	28·1	27·2	26·3	25·5	24·7	24·0	23·3	22·6	22·0	21·4	20·9	20·3	19·8	19·4	18·9	18·5	68
Lat.							AZIMUTHS.										Lat.
	A±B=5·00′	5·20′	5·40′	5·60′	5·80′	6·00′	6·20′	6·40′	6·60′	6·80′	7·00′	7·20′	7·40′	7·60′	7·80′	8·00′=A±B	

A & B **S**ame Names, take **S**um, (add)   RULE TO FIND **C** CORRECTION   A & B **D**ifferent Names, take **D**ifference. (sub.)

C CORRECTION, (A±B) is named the same as greater of these quantities.

AZIMUTH is named same as **C** Correction and the Hour Angle.

C   **C**

| C | | | | | | | TABLE C | | | | | | | | | | C |

A & B CORRECTION.

A±B=8·00'	8·20'	8·40'	8·60'	8·80'	9·00'	9·20'	9·40'	9·60'	9·80'	10·0'	10·3'	10·6'	11·0'	11·5'	12·0'=A±B
Lat.						**AZIMUTHS.**									**Lat.**
0	7·1	7·0	6·8	6·6	6·5	6·3	6·2	6·1	5·9	5·8	5·7	5·5	5·4	5·2	5·0	4·8	**0**
5	7·2	7·0	6·8	6·7	6·5	6·4	6·2	6·1	6·0	5·8	5·7	5·6	5·4	5·2	5·0	4·8	**5**
10	7·2	7·1	6·9	6·7	6·6	6·4	6·3	6·2	6·0	5·9	5·8	5·6	5·5	5·3	5·0	4·8	**10**
14	7·3	7·2	7·0	6·8	6·7	6·5	6·4	6·3	6·1	6·0	5·9	5·7	5·6	5·4	5·1	4·9	**14**
18	7·5	7·3	7·1	7·0	6·8	6·7	6·5	6·4	6·3	6·1	6·0	5·8	5·7	5·5	5·2	5·0	**18**
20	7·6	7·4	7·2	7·1	6·9	6·7	6·6	6·5	6·3	6·2	6·1	5·9	5·7	5·5	5·3	5·1	**20**
22	7·7	7·5	7·3	7·1	7·0	6·8	6·7	6·5	6·4	6·3	6·2	6·0	5·8	5·6	5·4	5·1	**22**
24	7·8	7·6	7·4	7·3	7·1	6·9	6·8	6·6	6·5	6·4	6·2	6·1	5·9	5·7	5·4	5·2	**24**
26	7·9	7·7	7·5	7·4	7·2	7·0	6·9	6·7	6·6	6·5	6·3	6·2	6·0	5·8	5·5	5·3	**26**
28	8·1	7·9	7·7	7·5	7·3	7·2	7·0	6·9	6·7	6·6	6·5	6·3	6·1	5·9	5·6	5·4	**28**
30	8·2	8·0	7·8	7·6	7·5	7·3	7·2	7·0	6·9	6·7	6·6	6·4	6·2	6·0	5·7	5·5	**30**
31	8·3	8·1	7·9	7·7	7·6	7·4	7·2	7·1	6·9	6·8	6·7	6·5	6·3	6·1	5·8	5·6	**31**
32	8·4	8·2	8·0	7·8	7·6	7·5	7·3	7·2	7·0	6·9	3·7	6·5	6·3	6·1	5·9	5·6	**32**
33	8·5	8·3	8·1	7·9	7·7	7·5	7·4	7·2	7·1	6·9	6·8	6·6	6·4	6·2	5·9	5·7	**33**
34	8·6	8·4	8·2	8·0	7·8	7·6	7·5	7·3	7·2	7·0	6·9	6·7	6·5	6·3	6·0	5·7	**34**
35	8·7	8·5	8·3	8·1	7·9	7·7	7·6	7·4	7·2	7·1	7·0	6·8	6·6	6·3	6·1	5·8	**35**
36	8·8	8·6	8·4	8·2	8·0	7·8	7·7	7·5	7·3	7·2	7·0	6·8	6·7	6·4	6·1	5·9	**36**
37	8·9	8·7	8·5	8·3	8·1	7·9	7·8	7·6	7·4	7·3	7·1	6·9	6·7	6·5	6·2	6·0	**37**
38	9·0	8·8	8·6	8·4	8·2	8·0	7·9	7·7	7·5	7·4	7·2	7·0	6·9	6·6	6·3	6·0	**38**
39	9·1	8·9	8·7	8·5	8·3	8·1	8·0	7·8	7·6	7·5	7·3	7·1	6·9	6·7	6·4	6·1	**39**
40	9·3	9·0	8·8	8·6	8·4	8·3	8·1	7·9	7·7	7·6	7·4	7·2	7·0	6·8	6·5	6·2	**40**
41	9·4	9·2	9·0	8·8	8·6	8·4	8·2	8·0	7·9	7·7	7·5	7·3	7·1	6·9	6·6	6·3	**41**
42	9·5	9·3	9·1	8·9	8·7	8·5	8·3	8·1	8·0	7·8	7·7	7·4	7·2	7·0	6·7	6·4	**42**
43	9·7	9·5	9·2	9·0	8·8	8·6	8·5	8·3	8·1	7·9	7·8	7·6	7·4	7·1	6·8	6·5	**43**
44	9·9	9·6	9·4	9·2	8·9	8·8	8·6	8·4	8·2	8·1	7·9	7·7	7·5	7·2	6·9	6·6	**44**
45	10·0	9·8	9·6	9·3	9·1	8·9	8·7	8·6	8·4	8·2	8·0	7·8	7·6	7·3	7·0	6·7	**45**
46	10·2	10·0	9·7	9·5	9·3	9·1	8·9	8·7	8·5	8·4	8·2	8·0	7·7	7·5	7·1	6·8	**46**
47	10·4	10·1	9·9	9·7	9·5	9·3	9·1	8·9	8·7	8·5	8·3	8·1	7·9	7·6	7·3	7·0	**47**
48	10·6	10·3	10·1	9·9	9·6	9·4	9·2	9·0	8·8	8·7	8·5	8·3	8·0	7·7	7·4	7·1	**48**
49	10·8	10·5	10·3	10·1	9·8	9·6	9·4	9·2	9·0	8·8	8·7	8·4	8·2	7·9	7·6	7·2	**49**
50	11·0	10·7	10·5	10·3	10·0	9·8	9·6	9·4	9·2	9·0	8·8	8·6	8·3	8·0	7·7	7·4	**50**
51	11·2	11·0	10·7	10·5	10·2	10·0	9·8	9·6	9·4	9·2	9·0	8·8	8·5	8·2	7·9	7·5	**51**
52	11·5	11·2	10·9	10·7	10·5	10·2	10·0	9·8	9·6	9·4	9·2	9·0	8·7	8·4	8·0	7·7	**52**
53	11·7	11·5	11·2	10·9	10·7	10·5	10·2	10·0	9·8	9·6	9·4	9·2	8·9	8·6	8·2	7·9	**53**
54	12·0	11·7	11·4	11·2	10·9	10·7	10·5	10·3	10·0	9·8	9·7	9·4	9·1	8·8	8·4	8·1	**54**
55	12·3	12·0	11·7	11·5	11·2	11·0	10·7	10·5	10·3	10·1	9·9	9·6	9·3	9·0	8·6	8·3	**55**
56	12·6	12·3	12·0	11·7	11·5	11·2	11·0	10·8	10·6	10·3	10·1	9·8	9·6	9·2	8·8	8·5	**56**
57	12·9	12·6	12·3	12·1	11·8	11·5	11·3	11·1	10·8	10·6	10·4	10·1	9·8	9·5	9·1	8·7	**57**
58	13·3	13·0	12·7	12·4	12·1	11·8	11·6	11·4	11·1	10·9	10·7	10·4	10·1	9·7	9·3	8·9	**58**
59	13·6	13·3	13·0	12·7	12·4	12·2	11·9	11·7	11·4	11·2	11·0	10·7	10·4	10·0	9·6	9·2	**59**
60	14·0	13·7	13·4	13·1	12·8	12·5	12·3	12·0	11·8	11·5	11·3	11·0	10·7	10·3	9·9	9·5	**60**
61	14·5	14·1	13·8	13·5	13·2	12·9	12·6	12·4	12·1	11·9	11·7	11·3	11·0	10·6	10·2	9·8	**61**
62	14·9	14·6	14·2	13·9	13·6	13·3	13·0	12·8	12·5	12·3	12·0	11·7	11·4	11·0	10·5	10·1	**62**
63	15·4	15·0	14·7	14·4	14·1	13·8	13·5	13·2	12·9	12·7	12·4	12·1	11·7	11·3	10·8	10·4	**63**
64	15·9	15·5	15·2	14·9	14·5	14·2	13·9	13·6	13·4	13·1	12·9	12·5	12·1	11·7	11·2	10·8	**64**
65	16·5	16·1	15·7	15·4	15·1	14·7	14·4	14·1	13·8	13·6	13·3	12·9	12·6	12·1	11·6	11·2	**65**
66	17·1	16·7	16·3	16·0	15·6	15·3	15·0	14·7	14·4	14·1	13·8	13·4	13·1	12·6	12·1	11·6	**66**
67	17·8	17·3	16·9	16·6	16·2	15·9	15·5	15·2	14·9	14·6	14·4	14·0	13·6	13·1	12·5	12·0	**67**
68	18·5	18·0	17·6	17·2	16·9	16·5	16·2	15·9	15·5	15·2	14·9	14·6	14·2	13·6	13·1	12·5	**68**
Lat.						**AZIMUTHS.**									**Lat.**
A±B=8·60'	8·20'	8·40'	8·60'	8·80'	9·00'	9·20'	9·40'	9·60'	9·80'	10·0'	10·3'	10·6'	11·0'	11·5'	12·0'=A±B

A & B **S**ame Name, take **S**um, (add.) **⎱RULE TO FIND⎰** A & B **D**ifferent Name, take **D**ifference. (sub). **⎰C CORRECTION⎱**

C CORRECTION, (A±B) is named the same as greater of these quantities.

AZIMUTH is named same as **C** Correction and the Hour Angle.

| C | | | | | | | | | | | | | | | | | C |

C   TABLE C   C

A & B CORRECTION.

AZIMUTHS.

Lat.	12·0′	12·5′	13·0′	13·5′	14·0′	14·5′	15·0′	16·0′	17·0′	18·0′	19·0′	20·0′	21·0′	22·0′	23·0′	25·0′	Lat.
0	4·8	4·6	4·4	4·2	4·1	3·9	3·8	3·6	3·4	3·2	3·0	2·9	2·7	2·6	2·5	2·3	0
5	4·8	4·6	4·4	4·3	4·1	4·0	3·8	3·6	3·4	3·2	3·0	2·9	2·7	2·6	2·5	2·3	5
10	4·8	4·6	4·5	4·3	4·1	4·0	3·9	3·6	3·4	3·2	3·1	2·9	2·8	2·6	2·5	2·3	10
14	4·9	4·7	4·5	4·4	4·2	4·1	3·9	3·7	3·5	3·3	3·1	2·9	2·8	2·7	2·6	2·4	14
18	5·0	4·8	4·6	4·4	4·3	4·1	4·0	3·8	3·5	3·3	3·2	3·0	2·9	2·7	2·6	2·4	18
20	5·1	4·9	4·7	4·5	4·3	4·2	4·1	3·8	3·6	3·4	3·2	3·0	2·9	2·8	2·6	2·4	20
22	5·1	4·9	4·7	4·6	4·4	4·3	4·1	3·9	3·6	3·4	3·3	3·1	2·9	2·8	2·7	2·5	22
24	5·2	5·0	4·8	4·6	4·5	4·3	4·2	3·9	3·7	3·5	3·3	3·1	3·0	2·8	2·7	2·5	24
26	5·3	5·1	4·9	4·7	4·5	4·4	4·2	4·0	3·7	3·5	3·4	3·2	3·0	2·9	2·8	2·5	26
28	5·4	5·2	5·0	4·8	4·6	4·5	4·3	4·0	3·8	3·6	3·4	3·2	3·1	2·9	2·8	2·6	28
30	5·5	5·3	5·1	4·9	4·7	4·6	4·4	4·1	3·9	3·7	3·5	3·3	3·1	3·0	2·9	2·6	30
31	5·6	5·3	5·1	4·9	4·8	4·6	4·4	4·2	3·9	3·7	3·5	3·3	3·2	3·0	2·9	2·7	31
32	5·6	5·4	5·2	5·0	4·8	4·6	4·5	4·2	4·0	3·7	3·6	3·4	3·2	3·1	2·9	2·7	32
33	5·7	5·4	5·2	5·0	4·9	4·7	4·5	4·3	4·0	3·8	3·6	3·4	3·2	3·1	3·0	2·7	33
34	5·7	5·5	5·3	5·1	4·9	4·8	4·6	4·3	4·1	3·8	3·6	3·5	3·3	3·1	3·0	2·8	34
35	5·8	5·6	5·4	5·2	5·0	4·8	4·7	4·4	4·1	3·9	3·7	3·5	3·3	3·2	3·0	2·8	35
36	5·9	5·7	5·4	5·2	5·0	4·9	4·7	4·4	4·2	3·9	3·7	3·5	3·4	3·2	3·1	2·8	36
37	6·0	5·7	5·5	5·3	5·1	4·9	4·8	4·5	4·2	4·0	3·8	3·6	3·4	3·3	3·1	2·9	37
38	6·0	5·8	5·6	5·4	5·2	5·0	4·8	4·5	4·3	4·0	3·8	3·6	3·5	3·3	3·2	2·9	38
39	6·1	5·9	5·7	5·5	5·3	5·1	4·9	4·6	4·3	4·1	3·9	3·7	3·5	3·3	3·2	2·9	39
40	6·2	6·0	5·7	5·5	5·3	5·2	5·0	4·7	4·4	4·1	3·9	3·7	3·6	3·4	3·2	3·0	40
41	6·3	6·1	5·8	5·6	5·4	5·2	5·0	4·7	4·5	4·2	4·0	3·8	3·6	3·4	3·3	3·0	41
42	6·4	6·1	5·9	5·7	5·5	5·3	5·1	4·8	4·5	4·3	4·1	3·8	3·7	3·5	3·3	3·1	42
43	6·5	6·2	6·0	5·8	5·6	5·4	5·2	4·9	4·6	4·3	4·1	3·9	3·7	3·6	3·4	3·1	43
44	6·6	6·3	6·1	5·9	5·7	5·5	5·3	5·0	4·7	4·4	4·2	4·0	3·8	3·6	3·5	3·2	44
45	6·7	6·5	6·2	6·0	5·8	5·6	5·4	5·1	4·8	4·5	4·3	4·0	3·9	3·7	3·5	3·2	45
46	6·8	6·6	6·3	6·1	5·9	5·7	5·5	5·1	4·8	4·6	4·3	4·1	3·9	3·7	3·6	3·3	46
47	7·0	6·7	6·4	6·2	6·0	5·8	5·6	5·2	4·9	4·7	4·4	4·2	4·0	3·8	3·6	3·4	47
48	7·1	6·8	6·6	6·3	6·1	5·9	5·7	5·3	5·0	4·7	4·5	4·3	4·1	3·9	3·7	3·4	48
49	7·2	7·0	6·7	6·4	6·2	6·0	5·8	5·4	5·1	4·8	4·6	4·4	4·2	4·0	3·8	3·5	49
50	7·4	7·1	6·8	6·6	6·3	6·1	5·9	5·6	5·2	4·9	4·7	4·4	4·2	4·0	3·9	3·6	50
51	7·5	7·2	7·0	6·7	6·5	6·2	6·0	5·7	5·3	5·0	4·8	4·5	4·3	4·1	4·0	3·6	51
52	7·7	7·4	7·1	6·9	6·6	6·4	6·2	5·8	5·5	5·2	4·9	4·6	4·4	4·2	4·0	3·7	52
53	7·9	7·6	7·3	7·0	6·8	6·5	6·3	5·9	5·6	5·3	5·0	4·7	4·5	4·3	4·1	3·8	53
54	8·1	7·8	7·5	7·2	6·9	6·7	6·5	6·1	5·7	5·4	5·1	4·9	4·6	4·4	4·2	3·9	54
55	8·3	7·9	7·6	7·4	7·1	6·9	6·6	6·2	5·9	5·5	5·2	5·0	4·7	4·5	4·3	4·0	55
56	8·5	8·1	7·8	7·5	7·3	7·0	6·8	6·4	6·0	5·7	5·4	5·1	4·9	4·6	4·4	4·1	56
57	8·7	8·4	8·0	7·7	7·5	7·2	7·0	6·5	6·2	5·8	5·5	5·2	5·0	4·8	4·6	4·2	57
58	8·9	8·6	8·3	8·0	7·7	7·4	7·2	6·7	6·3	6·0	5·7	5·4	5·1	4·9	4·7	4·3	58
59	9·2	8·8	8·5	8·2	7·9	7·6	7·4	6·9	6·5	6·2	5·8	5·5	5·3	5·0	4·8	4·4	59
60	9·5	9·1	8·7	8·4	8·1	7·9	7·6	7·1	6·7	6·3	6·0	5·7	5·4	5·2	5·0	4·6	60
61	9·8	9·4	9·0	8·7	8·4	8·1	7·8	7·3	6·9	6·5	6·2	5·9	5·6	5·4	5·1	4·7	61
62	10·1	9·7	9·3	9·0	8·7	8·4	8·1	7·6	7·1	6·7	6·4	6·1	5·8	5·5	5·3	4·9	62
63	10·4	10·0	9·6	9·3	8·9	8·6	8·4	7·8	7·4	7·0	6·6	6·3	6·0	5·7	5·5	5·0	63
64	10·8	10·3	10·0	9·6	9·3	8·9	8·6	8·1	7·6	7·2	6·8	6·5	6·2	5·9	5·7	5·2	64
65	11·2	10·7	10·3	9·9	9·6	9·3	9·0	8·4	7·9	7·5	7·1	6·7	6·4	6·1	5·9	5·4	65
66	11·6	11·1	10·7	10·3	10·0	9·6	9·3	8·7	8·2	7·8	7·4	7·0	6·7	6·4	6·1	5·6	66
67	12·0	11·6	11·1	10·7	10·4	10·0	9·7	9·1	8·6	8·1	7·7	7·3	6·9	6·6	6·3	5·8	67
68	12·5	12·1	11·6	11·2	10·8	10·4	10·1	9·5	8·9	8·4	8·0	7·6	7·2	6·9	6·6	6·1	68

AZIMUTHS.

A & B Same Name, take Sum, (add.) } RULE TO FIND C CORRECTION { A & B Different Name, take Difference, (sub.)

C CORRECTION, (A±B) is named the same as greater of these quantities.

AZIMUTH is named same as C Correction and the Hour Angle.

TABLE C

C   C

Lat.	A±B=25·0'	27·0'	30·0'	33·0'	36·0'	40·0'	45·0'	50·0'	60·0'	70·0'	80·0'	100'	150'	200'	400'	800'=A±B	Lat.
							AZIMUTHS.										
0	2·3	2·1	1·9	1·7	1·6	1·4	1·3	1·1	1·0	0·8	0·7	0·6	0·4	0·3	0·1	0·1	0
5	2·3	2·1	1·9	1·7	1·6	1·4	1·3	1·2	1·0	0·8	0·7	0·6	0·4	0·3	0·1	0·1	5
10	2·3	2·2	1·9	1·8	1·6	1·5	1·3	1·2	1·0	0·8	0·7	0·6	0·4	0·3	0·1	0·1	10
14	2·4	2·2	2·0	1·8	1·6	1·5	1·3	1·2	1·0	0·8	0·7	0·6	0·4	0·3	0·1	0·1	14
18	2·4	2·2	2·0	1·8	1·7	1·5	1·3	1·2	1·0	0·9	0·8	0·6	0·4	0·3	0·2	0·1	18
20	2·4	2·2	2·0	1·8	1·7	1·5	1·4	1·2	1·0	0·9	0·8	0·6	0·4	0·3	0·2	0·1	20
22	2·5	2·3	2·1	1·9	1·7	1·5	1·4	1·2	1·0	0·9	0·8	0·6	0·4	0·3	0·2	0·1	22
24	2·5	2·3	2·1	1·9	1·7	1·6	1·4	1·3	1·0	0·9	0·8	0·6	0·4	0·3	0·2	0·1	24
26	2·5	2·4	2·1	1·9	1·8	1·6	1·4	1·3	1·1	0·9	0·8	0·6	0·4	0·3	0·2	0·1	26
28	2·6	2·4	2·2	2·0	1·8	1·6	1·4	1·3	1·1	0·9	0·8	0·6	0·4	0·3	0·2	0·1	28
30	2·6	2·4	2·2	2·0	1·8	1·7	1·5	1·3	1·1	0·9	0·8	0·7	0·4	0·3	0·2	0·1	30
31	2·7	2·5	2·2	2·0	1·9	1·7	1·5	1·3	1·1	1·0	0·8	0·7	0·4	0·3	0·2	0·1	31
32	2·7	2·5	2·3	2·0	1·9	1·7	1·5	1·4	1·1	1·0	0·8	0·7	0·5	0·3	0·2	0·1	32
33	2·7	2·5	2·3	2·1	1·9	1·7	1·5	1·4	1·1	1·0	0·9	0·7	0·5	0·3	0·2	0·1	33
34	2·8	2·6	2·3	2·1	1·9	1·7	1·5	1·4	1·2	1·0	0·9	0·7	0·5	0·3	0·2	0·1	34
35	2·8	2·6	2·3	2·1	2·0	1·7	1·6	1·4	1·2	1·0	0·9	0·7	0·5	0·3	0·2	0·1	35
36	2·8	2·6	2·4	2·1	2·0	1·8	1·6	1·4	1·2	1·0	0·9	0·7	0·5	0·4	0·2	0·1	36
37	2·9	2·7	2·4	2·2	2·0	1·8	1·6	1·4	1·2	1·0	0·9	0·7	0·5	0·4	0·2	0·1	37
38	2·9	2·7	2·4	2·2	2·0	1·8	1·6	1·5	1·2	1·0	0·9	0·7	0·5	0·4	0·2	0·1	38
39	2·9	2·7	2·5	2·2	2·0	1·8	1·6	1·5	1·2	1·0	0·9	0·7	0·5	0·4	0·2	0·1	39
40	3·0	2·8	2·5	2·3	2·1	1·9	1·7	1·5	1·2	1·1	0·9	0·7	0·5	0·4	0·2	0·1	40
41	3·0	2·8	2·5	2·3	2·1	1·9	1·7	1·5	1·3	1·1	1·0	0·8	0·5	0·4	0·2	0·1	41
42	3·1	2·9	2·6	2·3	2·1	1·9	1·7	1·5	1·3	1·1	1·0	0·8	0·5	0·4	0·2	0·1	42
43	3·1	2·9	2·6	2·4	2·2	2·0	1·7	1·6	1·3	1·1	1·0	0·8	0·5	0·4	0·2	0·1	43
44	3·2	2·9	2·7	2·4	2·2	2·0	1·8	1·6	1·3	1·1	1·0	0·8	0·5	0·4	0·2	0·1	44
45	3·2	3·0	2·7	2·5	2·2	2·0	1·8	1·6	1·4	1·2	1·0	0·8	0·5	0·4	0·2	0·1	45
46	3·3	3·1	2·7	2·5	2·3	2·1	1·8	1·6	1·4	1·2	1·0	0·8	0·5	0·4	0·2	0·1	46
47	3·4	3·1	2·8	2·5	2·3	2·1	1·9	1·7	1·4	1·2	1·1	0·8	0·6	0·4	0·2	0·1	47
48	3·4	3·2	2·9	2·6	2·4	2·1	1·9	1·7	1·4	1·2	1·1	0·9	0·6	0·4	0·2	0·1	48
49	3·5	3·2	2·9	2·6	2·4	2·2	1·9	1·7	1·5	1·2	1·1	0·9	0·6	0·4	0·2	0·1	49
50	3·6	3·3	3·0	2·7	2·5	2·2	2·0	1·8	1·5	1·3	1·1	0·9	0·6	0·4	0·2	0·1	50
51	3·6	3·4	3·0	2·8	2·5	2·3	2·0	1·8	1·5	1·3	1·2	0·9	0·6	0·5	0·2	0·1	51
52	3·7	3·4	3·1	2·8	2·6	2·3	2·1	1·9	1·6	1·3	1·2	0·9	0·6	0·5	0·2	0·1	52
53	3·8	3·5	3·2	2·9	2·6	2·4	2·1	1·9	1·6	1·4	1·2	1·0	0·6	0·5	0·2	0·1	53
54	3·9	3·6	3·2	3·0	2·7	2·4	2·2	1·9	1·6	1·4	1·2	1·0	0·6	0·5	0·2	0·1	54
55	4·0	3·7	3·3	3·0	2·8	2·5	2·2	2·0	1·7	1·4	1·2	1·0	0·7	0·5	0·2	0·1	55
56	4·1	3·8	3·4	3·1	2·8	2·6	2·3	2·0	1·7	1·5	1·3	1·0	0·7	0·5	0·3	0·1	56
57	4·2	3·9	3·5	3·2	2·9	2·6	2·3	2·1	1·8	1·5	1·3	1·1	0·7	0·5	0·3	0·1	57
58	4·3	4·0	3·6	3·3	3·0	2·7	2·4	2·2	1·8	1·5	1·4	1·1	0·7	0·5	0·3	0·1	58
59	4·4	4·1	3·7	3·4	3·1	2·8	2·5	2·2	1·9	1·6	1·4	1·1	0·7	0·6	0·3	0·1	59
60	4·6	4·2	3·8	3·5	3·2	2·9	2·5	2·3	1·9	1·6	1·4	1·1	0·8	0·6	0·3	0·1	60
61	4·7	4·4	3·9	3·6	3·3	3·0	2·6	2·4	2·0	1·7	1·5	1·2	0·8	0·6	0·3	0·1	61
62	4·9	4·5	4·1	3·7	3·4	3·0	2·7	2·4	2·0	1·7	1·5	1·2	0·8	0·6	0·3	0·2	62
63	5·0	4·7	4·2	3·8	3·5	3·2	2·8	2·5	2·1	1·8	1·6	1·3	0·8	0·6	0·3	0·2	63
64	5·2	4·8	4·3	4·0	3·6	3·3	2·9	2·6	2·2	1·9	1·6	1·3	0·9	0·7	0·3	0·2	64
65	5·4	5·0	4·5	4·1	3·8	3·4	3·0	2·7	2·3	1·9	1·7	1·4	0·9	0·7	0·3	0·2	65
66	5·6	5·2	4·7	4·3	3·9	3·5	3·1	2·8	2·3	2·0	1·8	1·4	0·9	0·7	0·4	0·2	66
67	5·8	5·4	4·9	4·4	4·1	3·7	3·3	2·9	2·4	2·1	1·8	1·5	1·0	0·7	0·4	0·2	67
68	6·1	5·6	5·1	4·6	4·2	3·8	3·4	3·1	2·5	2·2	1·9	1·5	1·0	0·8	0·4	0·2	68
Lat.							**AZIMUTHS.**										Lat.
A±B=25·0'	27·0'	30·0'	33·0'	36·0'	40·0'	45·0'	50·0'	60·0'	70·0'	80·0'	100'	150'	200'	400'	800'=A±B		

A & B **S**ame Name,   }RULE TO FIND{ A & B **D**ifferent Name,
take **S**um, (add.)   } C CORRECTION { take **D**ifference, (sub.)

C CORRECTION, (A±B) is named the same as greater of these quantities.
AZIMUTH is named same as C Correction and the Hour Angle.

C   C

TABLE C.

C   **A & B CORRECTION**   **C**

AZIMUTHS.

A±B=	·00′	·01′	·02′	·03′	·04′	·05′	·06′	·07′	·08′	·09′	·10′	·11′	·12′	·13′	·14′	·15′=A±B
68	90·0	89·8	89·6	89·4	89·1	88·9	88·7	88·5	88·3	88·1	87·9	87·6	87·4	87·2	87·0	86·8
69	90·0	89·8	89·6	89·4	89·2	89·0	88·8	88·6	88·4	88·2	88·0	87·7	87·5	87·3	87·1	86·9
70	90·0	89·8	89·6	89·4	89·2	89·0	88·8	88·6	88·4	88·2	88·0	87·8	87·7	87·5	87·3	87·1
71	90·0	89·8	89·6	89·4	89·3	89·1	88·9	88·7	88·5	88·3	88·1	87·9	87·8	87·6	87·4	87·2
72	90·0	89·8	89·6	89·5	89·3	89·1	88·9	88·8	88·6	88·4	88·2	88·1	87·9	87·7	87·5	87·3
73	90·0	89·8	89·7	89·5	89·3	89·2	89·0	88·8	88·7	88·5	88·3	88·2	88·0	87·8	87·7	87·5
74	90·0	89·8	89·7	89·5	89·4	89·2	89·0	88·9	88·7	88·6	88·4	88·3	88·1	87·9	87·8	87·6
75	90·0	89·9	89·7	89·6	89·4	89·3	89·1	89·0	88·8	88·7	88·5	88·4	88·2	88·1	87·9	87·8
76	90·0	89·9	89·7	89·6	89·4	89·3	89·2	89·0	88·9	88·8	88·6	88·5	88·3	88·2	88·1	87·9
77	90·0	89·9	89·7	89·6	89·5	89·4	89·2	89·1	89·0	88·8	88·7	88·6	88·5	88·3	88·2	88·1
78	90·0	89·9	89·8	89·6	89·5	89·4	89·3	89·2	89·0	88·9	88·8	88·7	88·6	88·5	88·3	88·2
79	90·0	89·9	89·8	89·7	89·6	89·5	89·4	89·2	89·1	89·0	88·9	88·8	88·7	88·6	88·5	88·4
80	90·0	89·9	89·8	89·7	89·6	89·5	89·4	89·3	89·2	89·1	89·0	88·9	88·8	88·7	88·6	88·5
81	90·0	89·9	89·8	89·7	89·6	89·6	89·5	89·4	89·3	89·2	89·1	89·0	88·9	88·8	88·7	88·7
82	90·0	89·9	89·8	89·8	89·7	89·6	89·5	89·5	89·4	89·3	89·2	89·1	89·0	89·0	88·9	88·8
83	90·0	89·9	89·9	89·8	89·7	89·7	89·6	89·5	89·4	89·4	89·3	89·2	89·2	89·1	89·0	89·0

TABLE C.

C   **A & B CORRECTION**   **C**

AZIMUTHS.

A±B=	·15′	16′	17′	18′	19′	20′	·21′	·22′	·23′	·24′	25′	26′	27′	28′	29′	30′=A±B
68	86·8	86·6	86·4	86·1	85·9	85·7	85·5	85·3	85·1	84·9	84·6	84·4	84·2	84·0	83·8	83·6
69	86·9	86·7	86·5	86·3	86·1	85·9	85·7	85·5	85·3	85·1	84·9	84·7	84·5	84·3	84·1	83·9
70	87·1	86·9	86·7	86·5	86·3	86·1	85·9	85·7	85·5	85·3	85·1	84·9	84·7	84·5	84·3	84·1
71	87·2	87·0	86·8	86·6	86·5	86·3	86·1	85·9	85·7	85·5	85·3	85·2	85·0	84·8	84·6	84·4
72	87·3	87·2	87·0	86·8	86·6	86·5	86·3	86·1	85·9	85·8	85·6	85·4	85·2	85·1	84·9	84·7
73	87·5	87·3	87·2	87·0	86·8	86·7	86·5	86·3	86·2	86·0	85·8	85·6	85·5	85·3	85·2	85·0
74	87·6	87·5	87·3	87·2	87·0	86·8	86·7	86·5	86·4	86·2	86·1	85·9	85·7	85·6	85·4	85·3
75	87·8	87·6	87·5	87·3	87·2	87·0	86·9	86·7	86·6	86·4	86·3	86·2	86·0	85·9	85·7	85·6
76	87·9	87·8	87·6	87·5	87·4	87·2	87·1	87·0	86·8	86·7	86·5	86·4	86·3	86·1	86·0	85·9
77	88·1	87·9	87·8	87·7	87·6	87·4	87·3	87·2	87·0	86·9	86·8	86·7	86·5	86·4	86·3	86·1
78	88·2	88·1	88·0	87·9	87·7	87·6	87·5	87·4	87·3	87·1	87·0	86·9	86·8	86·7	86·6	86·4
79	88·4	88·3	88·1	88·0	87·9	87·8	87·7	87·6	87·5	87·4	87·3	87·2	87·1	86·9	86·8	86·7
80	88·5	88·4	88·3	88·2	88·1	88·0	87·9	87·8	87·7	87·6	87·5	87·4	87·3	87·2	87·1	87·0
81	88·7	88·6	88·5	88·4	88·3	88·2	88·1	88·0	87·9	87·9	87·8	87·7	87·6	87·5	87·4	87·3
82	88·8	88·7	88·6	88·6	88·5	88·4	88·3	88·2	88·2	88·1	88·0	87·9	87·8	87·8	87·7	87·6
83	89·0	88·9	88·8	88·7	88·7	88·6	88·5	88·5	88·4	88·3	88·3	88·2	88·1	88·0	88·0	87·9

TABLE C

C   **A & B CORRECTION**   **C**

AZIMUTHS.

A±B=	·30′	·31′	32′	33′	34′	35′	36′	37′	38′	39′	40′	41′	42′	43′	44′	·45′=A±B
68	83·6	83·4	83·2	83·0	82·7	82·5	82·3	82·1	81·9	81·7	81·5	81·3	81·1	80·8	80·6	80·4
69	83·9	83·7	83·5	83·3	83·1	82·9	82·7	82·5	82·3	82·0	81·8	81·6	81·4	81·2	81·0	80·8
70	84·1	83·9	83·7	83·6	83·4	83·2	83·0	82·8	82·6	82·4	82·2	82·0	81·8	81·6	81·4	81·3
71	84·4	84·2	84·0	83·9	83·7	83·5	83·3	83·1	83·0	82·8	82·6	82·4	82·2	81·9	81·9	81·7
72	84·7	84·5	84·4	84·2	84·0	83·8	83·7	83·5	83·3	83·1	83·0	82·8	82·6	82·4	82·3	82·1
73	85·0	84·8	84·7	84·5	84·3	84·2	84·0	83·8	83·7	83·5	83·3	83·2	83·0	82·8	82·7	82·5
74	85·3	85·1	85·0	84·8	84·6	84·5	84·3	84·2	84·0	83·9	83·7	83·6	83·4	83·2	83·1	82·9
75	85·6	85·4	85·3	85·1	85·0	84·8	84·7	84·5	84·4	84·2	84·1	83·9	83·8	83·7	83·5	83·4
76	85·9	85·7	85·6	85·4	85·3	85·2	85·0	84·9	84·8	84·6	84·5	84·3	84·2	84·1	83·9	83·8
77	86·1	86·0	85·9	85·8	85·6	85·5	85·4	85·2	85·1	85·0	84·9	84·7	84·6	84·5	84·4	84·2
78	86·4	86·3	86·2	86·1	86·0	85·8	85·7	85·6	85·5	85·4	85·3	85·1	85·0	84·9	84·8	84·7
79	86·7	86·6	86·5	86·4	86·3	86·2	86·1	86·0	85·9	85·7	85·6	85·5	85·4	85·3	85·2	85·1
80	87·0	86·9	86·8	86·7	86·6	86·5	86·4	86·3	86·2	86·1	86·0	85·9	85·8	85·7	85·6	85·5
81	87·3	87·2	87·1	87·0	87·0	86·9	86·8	86·7	86·6	86·5	86·4	86·3	86·2	86·2	86·1	86·0
82	87·6	87·5	87·5	87·4	87·3	87·2	87·1	87·1	87·0	86·9	86·8	86·7	86·7	86·6	86·5	86·4
83	87·9	87·8	87·8	87·7	87·6	87·6	87·5	87·4	87·4	87·3	87·2	87·1	87·1	87·0	86·9	86·9

TABLE C.

A & B CORRECTION

A±B =	·45′	·46′	·47′	·48′	·49′	·50′	·51′	·52′	·53′	·54′	·55′	·56′	·57′	·58′	·59′	·60′ =A±B
Lat.							AZIMUTHS.									Lat.
68	80·4	80·2	80·0	79·8	79·6	79·4	79·2	79·0	78·8	78·6	78·4	78·2	77·9	77·7	77·5	77·3	68
69	80·8	80·6	80·4	80·2	80·0	79·8	79·6	79·5	79·3	79·1	78·9	78·7	78·5	78·3	78·1	77·9	69
70	81·3	81·1	80·9	80·7	80·5	80·3	80·1	79·9	79·7	79·5	79·4	79·2	79·0	78·8	78·6	78·4	70
71	81·7	81·5	81·3	81·1	80·9	80·8	80·6	80·4	80·2	80·0	79·9	79·7	79·5	79·3	79·1	79·0	71
72	82·1	81·9	81·7	81·6	81·4	81·2	81·0	80·9	80·7	80·5	80·4	80·2	80·0	79·8	79·7	79·5	72
73	82·5	82·3	82·2	82·0	81·9	81·7	81·5	81·4	81·2	81·0	80·9	80·7	80·5	80·4	80·2	80·1	73
74	82·9	82·8	82·6	82·5	82·3	82·2	82·0	81·9	81·7	81·5	81·4	81·2	81·1	80·9	80·8	80·6	74
75	83·4	83·2	83·1	82·9	82·8	82·6	82·5	82·3	82·2	82·1	81·9	81·8	81·6	81·5	81·3	81·2	75
76	83·8	83·7	83·5	83·4	83·2	83·1	83·0	82·8	82·7	82·6	82·4	82·3	82·2	82·0	81·9	81·8	76
77	84·2	84·1	84·0	83·8	83·7	83·6	83·5	83·3	83·2	83·1	83·0	82·8	82·7	82·6	82·4	82·3	77
78	84·7	84·5	84·4	84·3	84·2	84·1	84·0	83·8	83·7	83·6	83·5	83·4	83·2	83·1	83·0	82·9	78
79	85·1	85·0	84·9	84·8	84·7	84·6	84·5	84·3	84·2	84·1	84·0	83·9	83·7	83·6	83·5	83·5	79
80	85·5	85·4	85·3	85·2	85·1	85·0	84·9	84·8	84·7	84·6	84·5	84·5	84·4	84·3	84·2	84·1	80
81	86·0	85·9	85·8	85·7	85·6	85·5	85·4	85·3	85·3	85·2	85·1	85·0	84·9	84·8	84·7	84·6	81
82	86·4	86·3	86·3	86·2	86·1	86·0	85·9	85·9	85·8	85·7	85·6	85·5	85·5	85·4	85·3	85·2	82
83	86·9	86·8	86·7	86·7	86·6	86·5	86·4	86·4	86·3	86·2	86·2	86·1	86·0	85·9	85·8	85·8	83

TABLE C.

A & B CORRECTION

A±B =	·60′	·62′	·64′	·66′	·68′	·70′	·72′	·74′	·76′	·78′	·80′	8 2′	·84′	·86′	·88′	·90′ =A±B
Lat.							AZIMUTHS.									Lat.
68	77·3	76·9	76·5	76·1	75·7	75·3	74·9	74·5	74·1	73·7	73·3	72·9	72·5	72·1	71·8	71·4	68
69	77·9	77·5	77·1	76·7	76·3	75·9	75·5	75·2	74·8	74·4	74·0	73·6	73·3	72·9	72·5	72·1	69
70	78·4	78·0	77·7	77·3	76·9	76·5	76·2	75·8	75·4	75·1	74·7	74·3	74·0	73·6	73·3	72·9	70
71	79·0	78·6	78·2	77·9	77·5	77·2	76·8	76·5	76·1	75·8	75·4	75·1	74·7	74·4	74·0	73·7	71
72	79·5	79·2	78·8	78·5	78·1	77·8	77·5	77·1	76·8	76·5	76·1	75·8	75·5	75·1	74·8	74·5	72
73	80·1	79·7	79·4	79·1	78·8	78·4	78·1	77·8	77·5	77·2	76·8	76·5	76·2	75·9	75·6	75·3	73
74	80·6	80·3	80·0	79·7	79·4	79·1	78·8	78·5	78·2	77·9	77·6	77·3	77·0	76·7	76·4	76·1	74
75	81·2	80·9	80·6	80·3	80·0	79·7	79·4	79·2	78·9	78·6	78·3	78·0	77·7	77·5	77·2	76·9	75
76	81·8	81·5	81·2	80·9	80·7	80·4	80·1	79·9	79·6	79·3	79·0	78·8	78·5	78·3	78·0	77·7	76
77	82·3	82·1	81·8	81·6	81·3	81·1	80·8	80·6	80·3	80·1	79·8	79·6	79·3	79·1	78·8	78·5	77
78	82·9	82·7	82·4	82·2	82·0	81·7	81·5	81·3	81·0	80·8	80·6	80·3	80·1	79·9	79·6	79·4	78
79	83·5	83·3	83·0	82·8	82·6	82·4	82·2	82·0	81·8	81·5	81·3	81·1	80·9	80·7	80·5	80·3	79
80	84·1	83·9	83·7	83·5	83·3	83·1	82·9	82·7	82·5	82·3	82·1	81·9	81·7	81·5	81·3	81·1	80
81	84·6	84·5	84·3	84·1	83·9	83·7	83·6	83·4	83·2	83·1	82·9	82·7	82·5	82·3	82·2	82·0	81
82	85·2	85·1	84·9	84·8	84·6	84·4	84·3	84·1	84·0	83·8	83·6	83·5	83·3	83·2	83·0	82·9	82
83	85·8	85·7	85·5	85·4	85·3	85·1	85·0	84·8	84·7	84·6	84·4	84·3	84·2	84·0	83·9	83·7	83

TABLE C

A & B CORRECTION

A±B =	·90′	·92′	·94′	·96′	·98′	1·00′	1·02′	1·04′	1·06′	1·08′	1·10′	1·12′	1·14′	1·16′	1·18′	1·20′ =A±B
Lat.							AZIMUTHS.									Lat.
68	71·4	71·0	70·6	70·2	69·8	69·5	69·1	68·7	68·3	68·0	67·6	67·2	66·9	66·5	66·2	65·8	68
69	72·1	71·8	71·4	71·0	70·6	70·3	69·9	69·5	69·2	68·9	68·5	68·1	67·8	67·4	67·1	66·7	69
70	72·9	72·5	72·2	71·8	71·5	71·1	70·8	70·4	70·1	69·7	69·4	69·0	68·7	68·4	68·0	67·7	70
71	73·7	73·3	73·0	72·6	72·3	72·0	71·6	71·3	71·0	70·6	70·3	70·0	69·6	69·3	69·0	68·7	71
72	74·5	74·1	73·8	73·5	73·2	72·8	72·5	72·2	71·9	71·5	71·2	70·9	70·6	70·3	70·0	69·7	72
73	75·3	74·9	74·6	74·3	74·0	73·7	73·4	73·1	72·8	72·5	72·2	71·9	71·6	71·3	71·0	70·7	73
74	76·1	75·8	75·5	75·2	74·9	74·6	74·3	74·0	73·7	73·4	73·1	72·8	72·6	72·3	72·0	71·7	74
75	76·9	76·6	76·3	76·0	75·8	75·5	75·2	74·9	74·7	74·4	74·1	73·8	73·6	73·3	73·0	72·7	75
76	77·7	77·5	77·2	76·9	76·7	76·4	76·1	75·9	75·6	75·4	75·1	74·8	74·6	74·3	74·1	73·8	76
77	78·5	78·3	78·1	77·8	77·6	77·3	77·1	76·8	76·6	76·4	76·1	75·9	75·6	75·4	75·1	74·9	77
78	79·4	79·2	78·9	78·7	78·5	78·3	78·0	77·8	77·6	77·4	77·1	76·9	76·7	76·4	76·2	76·0	78
79	80·3	80·0	79·8	79·6	79·4	79·2	79·0	78·8	78·6	78·4	78·1	77·9	77·7	77·5	77·3	77·1	79
80	81·1	80·9	80·7	80·5	80·3	80·1	80·0	79·8	79·6	79·4	79·2	79·0	78·8	78·6	78·4	78·2	80
81	82·0	81·8	81·6	81·5	81·3	81·1	80·9	80·8	80·6	80·4	80·2	80·1	79·9	79·7	79·5	79·4	81
82	82·9	82·7	82·5	82·4	82·2	82·1	81·9	81·8	81·6	81·4	81·3	81·1	81·0	80·8	80·7	80·5	82
83	83·7	83·6	83·5	83·3	83·2	83·0	82·9	82·8	82·6	82·5	82·4	82·2	82·1	81·9	81·8	81·7	83

TABLE C.

A & B CORRECTION — AZIMUTHS

A±B=	1·20'	1·24'	1·28'	1·32'	1·36'	1·40'	1·44'	1·48'	1·52'	1·56'	1·60'	1·64'	1·68'	1·72'	1·76'	1·80'=A±B
Lat.																**Lat.**
68	65·8	65·1	64·4	63·7	63·0	62·3	61·7	61·0	60·3	59·7	59·1	58·4	57·8	57·2	56·6	56·0	68
69	66·7	66·0	65·4	64·7	64·0	63·4	62·7	62·1	61·4	60·8	60·2	59·6	59·0	58·4	57·8	57·2	69
70	67·7	67·0	66·4	65·7	65·0	64·4	63·8	63·2	62·5	61·9	61·3	60·7	60·1	59·5	58·9	58·4	70
71	68·7	68·0	67·1	66·7	66·1	65·5	64·9	64·3	63·7	63·1	62·5	61·9	61·3	60·7	60·2	59·6	71
72	69·7	69·0	68·4	67·8	67·2	66·6	66·0	65·4	64·9	64·3	63·7	63·1	62·6	62·0	61·4	60·9	72
73	70·7	70·1	69·5	68·9	68·3	67·7	67·2	66·6	66·0	65·5	64·9	64·4	63·8	63·3	62·8	62·2	73
74	71·7	71·1	70·6	70·0	69·5	68·9	68·4	67·8	67·2	66·7	66·2	65·7	65·2	64·6	64·1	63·6	74
75	72·7	72·2	71·7	71·1	70·6	70·1	69·6	69·0	68·5	68·0	67·5	67·0	66·5	66·0	65·5	65·0	75
76	73·8	73·3	72·8	72·3	71·8	71·3	70·8	70·3	69·8	69·3	68·8	68·4	67·9	67·4	66·9	66·5	76
77	74·9	74·4	73·9	73·5	73·0	72·5	72·0	71·6	71·1	70·7	70·2	69·8	69·3	68·8	68·4	68·0	77
78	76·0	75·5	75·1	74·7	74·2	73·8	73·3	72·9	72·5	72·0	71·6	71·2	70·7	70·3	69·9	69·5	78
79	77·1	76·7	76·3	75·9	75·5	75·1	74·6	74·2	73·8	73·4	73·0	72·6	72·2	71·8	71·4	71·0	79
80	78·2	77·9	77·5	77·1	76·7	76·4	75·9	75·6	75·2	74·8	74·5	74·1	73·7	73·4	73·0	72·6	80
81	79·4	79·0	78·7	78·4	78·0	77·7	77·3	77·0	76·6	76·3	76·0	75·6	75·3	75·0	74·6	74·3	81
82	80·5	80·2	79·9	79·6	79·3	79·0	78·7	78·4	78·0	77·8	77·5	77·1	76·8	76·5	76·2	75·9	82
83	81·7	81·4	81·1	80·9	80·6	80·3	80·0	79·8	79·5	79·2	79·0	78·7	78·4	78·2	77·9	77·6	83

TABLE C — A & B CORRECTION — AZIMUTHS

A±B=	1·80'	1·84'	1·88'	1·92'	1·96'	2·00'	2·04'	2·08'	2·12'	2·16'	2·20'	2·24'	2·28'	2·32'	2·36'	2·40'=A±B
Lat.																**Lat.**
68	56·0	55·4	54·8	54·3	53·7	53·2	52·6	52·1	51·5	51·0	50·5	50·0	49·5	49·0	48·5	48·0	68
69	57·2	56·6	56·0	55·5	54·9	54·4	53·8	53·3	52·8	52·2	51·8	51·3	50·8	50·3	49·8	49·3	69
70	58·4	57·8	57·2	56·7	56·2	55·6	55·1	54·6	54·0	53·5	53·0	52·6	52·1	51·6	51·1	50·5	70
71	59·6	59·1	58·5	58·0	57·5	56·9	56·4	55·9	55·4	54·9	54·4	53·9	53·4	52·9	52·5	52·0	71
72	60·9	60·4	59·8	59·3	58·8	58·3	57·8	57·3	56·8	56·3	55·8	55·3	54·8	54·4	53·9	53·4	72
73	62·2	61·7	61·2	60·7	60·2	59·7	59·2	58·7	58·2	57·7	57·3	56·8	56·3	55·9	55·4	54·9	73
74	63·6	63·1	62·6	62·1	61·6	61·1	60·6	60·2	59·7	59·2	58·8	58·3	57·9	57·4	56·9	56·5	74
75	65·0	64·5	64·0	63·6	63·1	62·6	62·2	61·7	61·2	60·8	60·3	59·9	59·5	59·0	58 6	58·1	75
76	66·5	66·0	65·6	65·1	64·6	64·2	63·7	63·3	62·8	62·4	62·0	61·6	61·1	60·7	60·3	59·8	76
77	68·0	67·5	67·1	66·6	66·2	65·8	65·4	64·9	64·5	64·1	63·7	63·2	62·8	62·5	62·0	61·6	77
78	69·5	69·1	68·7	68·2	67·8	67·4	67·0	66·6	66·2	65·8	65·4	65·0	64·6	64·3	63·9	63·5	78
79	71·0	70·7	70·3	69·9	69·5	69·1	68·7	68·4	68·0	67·6	67·2	66·9	66·5	66·1	65·8	65·4	79
80	72·6	72·3	71·9	71·6	71·2	70·8	70·5	70·1	69·8	69·4	69·1	68·8	68·4	68·1	67·7	67·4	80
81	74·3	73·9	73·6	73·3	72·9	72·6	72·3	72·0	71·7	71·3	71·0	70·7	70·4	70·1	69·7	69·4	81
82	75·9	75·6	75·3	75·0	74·7	74·5	74·2	73·9	73·6	73·3	73·0	72·7	72·4	72·1	71·8	71·5	82
83	77·6	77·3	77·1	76·8	76·6	76·3	76·0	75·8	75·5	75·2	74·9	74·7	74·5	74·2	73·9	73·7	83

TABLE C. — A & B CORRECTION — AZIMUTHS

A±B=	2·40'	2·45'	2·50'	2·55'	2·60'	2·65'	2·70'	2·75'	2·80'	2·90'	3·00'	3·10'	3·20'	3·30'	3·40'	3·50'=A±B
Lat.																**Lat.**
68	48·0	47·4	46·9	46·3	45·8	45·2	44·7	44·1	43·6	42·6	41·7	40·7	39·8	39·0	38·1	37·3	68
69	49·3	48·7	48·2	47·6	47·0	46·5	45·9	45·4	44·9	43·9	42·9	42·0	41·1	40·2	39·4	38·6	69
70	50·5	50·0	49·5	48·9	48·4	47·8	47·3	46·8	46·2	45·2	44·3	43·3	42·4	41·5	40·7	39·9	70
71	52·0	51·4	50·9	50·3	49·8	49·2	48·8	48·3	47·7	46·7	45·7	44·7	43·8	42·9	42·1	41·3	71
72	53·4	52·9	52·4	51·8	51·2	50·7	50·2	49·7	49·2	48·2	47·2	46·2	45·3	44·4	43·6	42·7	72
73	54·9	54·4	53·9	53·3	52·8	52·2	51·7	51·2	50·7	49·7	48·8	47·8	46·9	46·0	45·2	44·3	73
74	56·5	56·0	55·5	54·9	54·4	53·9	53·4	52·8	52·3	51·4	50·4	49·5	48·6	47·7	46·9	46·0	74
75	58·1	57·6	57·1	56·6	56·1	55·6	55·1	54·6	54·1	53·1	52·2	51·3	50·4	49·5	48·7	47·8	75
76	59·8	59·3	58·8	58·3	57·8	57·3	56·8	56·4	55·9	54·9	54·0	53·1	52·3	51·4	50·6	49·8	76
77	61·6	61·1	60·7	60·2	59·7	59·2	58·7	58·3	57·8	56·9	56·0	55·1	54·3	53·4	52·6	51·8	77
78	63·5	63·0	62·5	62·1	61·6	61·1	60·7	60·3	59·8	58·9	58·1	57·2	56·4	55·5	54·7	53·9	78
79	65·4	64·9	64·5	64·1	63·6	63·2	62·7	62·3	61·9	61·0	60·2	59·4	58·6	57·8	57·0	56·3	79
80	67·4	67·0	66·6	66·1	65·7	65·3	64·9	64·5	64·1	63·3	62·5	61·7	60·9	60·2	59·5	58·8	80
81	69·4	69·0	68·7	68·3	67·9	67·5	67·1	66·7	66·4	65·6	64·9	64·1	63·4	62·7	61·9	61·3	81
82	71·5	71·2	70·8	70·5	70·1	69·8	69·4	69·1	68·7	68·0	67·4	66·7	66·0	65·3	64·7	64·0	82
83	73·7	73·4	73·0	72·7	72·4	72·1	71·8	71·5	71·1	70·5	69·9	69·3	68·7	68·1	67·5	66·9	83

TABLE C.

A & B CORRECTION

A±B=3·50'	3·60'	3·70'	3·80'	3·90'	4·00'	4·10'	4·20'	4·30'	4·40'	4·50'	4·60'	4·70'	4·80'	4·90'	5·00'=A±B
Lat.						AZIMUTHS.									**Lat.**
68	37·3	36·6	35·8	35·1	34·4	33·7	33·1	32·4	31·8	31·2	30·7	30·1	29·6	29·1	28·6	28·1	68
69	38·6	37·8	37·0	36·3	35·6	34·9	34·2	33·6	32·9	32·4	31·8	31·2	30·7	30·2	29·7	29·2	69
70	39·9	39·1	38·3	37·6	36·8	36·2	35·5	34·8	34·2	33·6	33·0	32·4	31·8	31·3	30·8	30·3	70
71	41·3	40·5	39·8	38·9	38·2	37·5	36·8	36·2	35·5	34·9	34·3	33·7	33·2	32·6	32·1	31·5	71
72	42·7	41·9	41·2	40·4	39·7	38·9	38·3	37·6	36·9	36·3	35·7	35·1	34·5	33·9	33·4	32·9	72
73	44·3	43·4	42·7	41·9	41·2	40·5	39·8	39·2	38·5	37·8	37·2	36·6	36·0	35·5	34·9	34·4	73
74	46·0	45·2	44·4	43·7	42·9	42·2	41·5	40·8	40·1	39·5	38·9	38·3	37·7	37·1	36·5	35·9	74
75	47·8	47·0	46·2	45·5	44·8	44·0	43·3	42·6	41·9	41·3	40·6	40·0	39·4	38·8	38·2	37·7	75
76	49·8	48·9	48·2	47·4	46·7	45·9	45·2	44·6	43·9	43·2	42·6	41·9	41·3	40·7	40·1	39·6	76
77	51·8	51·0	50·2	49·5	48·7	48·0	47·3	46·6	45·9	45·3	44·6	44·0	43·4	42·8	42·2	41·6	77
78	53·9	53·2	52·4	51·7	50·9	50·2	49·5	48·9	48·2	47·6	46·9	46·3	45·7	45·1	44·5	43·9	78
79	56·3	55·5	54·8	54·1	53·3	52·6	51·9	51·3	50·6	49·9	49·3	48·7	48·1	47·5	46·9	46·3	79
80	58·8	58·0	57·3	56·6	55·9	55·2	54·5	53·9	53·2	52·6	52·0	51·4	50·8	50·2	49·6	49·0	80
81	61·3	60·6	59·9	59·3	58·6	57·9	57·3	56·7	56·1	55·5	54·8	54·3	53·7	53·1	52·5	51·9	81
82	64·0	63·4	62·7	62·1	61·5	60·9	60·3	59·7	59·1	58·5	57·9	57·3	56·8	56·2	55·7	55·2	82
83	66·9	66·3	65·7	65·1	64·6	64·0	63·5	62·9	62·3	61·8	61·2	60·7	60·2	59·7	59·2	58·6	83

TABLE C

A & B CORRECTION

A±B=5·00'	5·20'	5·40'	5·60'	5·80'	6·00'	6·20'	6·40'	6·60'	6·80'	7·00'	7·20'	7·40'	7·60'	7·80'	8·00'=A±B
Lat.						AZIMUTHS.									**Lat.**
68	28·1	27·2	26·3	25·5	24·7	24·0	23·3	22·6	22·0	21·4	20·9	20·3	19·8	19·4	18·9	18·5	68
69	29·2	28·2	27·3	26·5	25·7	24·9	24·2	23·5	22·9	22·3	21·7	21·2	20·7	20·2	19·7	19·2	69
70	30·3	29·3	28·4	27·6	26·8	25·9	25·2	24·5	23·9	23·3	22·7	22·1	21·6	21·0	20·5	20·1	70
71	31·5	30·6	29·6	28·7	27·9	27·1	26·3	25·6	24·9	24·3	23·7	23·1	22·5	22·0	21·5	21·0	71
72	32·9	31·9	30·9	30·0	29·1	28·3	27·5	26·8	26·1	25·4	24·8	24·2	23·6	23·1	22·6	22·0	72
73	34·4	33·3	32·3	31·4	30·5	29·7	28·9	28·1	27·4	26·7	26·1	25·4	24·8	24·2	23·7	23·1	73
74	35·9	34·9	33·9	32·9	32·0	31·2	30·3	29·6	28·8	28·1	27·4	26·7	26·1	25·5	25·0	24·4	74
75	37·7	36·6	35·6	34·6	33·7	32·8	31·9	31·1	30·3	29·6	28·9	28·2	27·6	27·0	26·4	25·8	75
76	39·6	38·5	37·4	36·4	35·5	34·6	33·7	32·8	32·1	31·3	30·6	29·8	29·2	28·5	27·9	27·3	76
77	41·6	40·5	39·4	38·4	37·5	36·5	35·6	34·8	33·9	33·2	32·4	31·7	30·9	30·3	29·7	29·0	77
78	43·9	42·8	41·7	40·7	39·7	38·7	37·8	36·9	36·1	35·3	34·5	33·7	33·0	32·3	31·6	31·0	78
79	46·3	45·2	44·1	43·1	42·1	41·1	40·2	39·3	38·4	37·6	36·8	36·1	35·3	34·6	33·9	33·2	79
80	49·0	47·9	46·8	45·8	44·8	43·8	42·9	41·9	41·1	40·3	39·4	38·6	37·9	37·2	36·5	35·8	80
81	51·9	50·8	49·8	48·8	47·8	46·8	45·9	44·9	44·1	43·2	42·4	41·6	40·8	40·1	39·4	38·6	81
82	55·2	54·1	53·1	52·1	51·1	50·1	49·2	48·3	47·4	46·6	45·7	44·9	44·2	43·4	42·6	41·9	82
83	58·6	57·6	56·6	55·7	54·7	53·8	52·9	52·0	51·2	50·3	49·5	48·7	47·9	47·2	46·4	45·7	83

TABLE C

A & B CORRECTION

A±B=8·00'	8·20'	8·40'	8·60'	8·80'	9·00'	9·20'	9·40'	9·60'	9·80'	10·0'	10·3'	10·6'	11·0'	11·5'	12·0'=A±B
Lat.						AZIMUTHS.									**Lat.**
68	18·5	18·0	17·6	17·2	16·9	16·5	16·2	15·9	15·5	15·2	14·9	14·6	14·2	13·6	13·1	12·5	68
69	19·2	18·8	18·4	18·0	17·6	17·2	16·9	16·5	16·2	15·8	15·5	15·2	14·8	14·2	13·6	13·1	69
70	20·1	19·6	19·2	18·8	18·4	18·0	17·6	17·2	16·9	16·6	16·3	15·9	15·5	14·9	14·2	13·7	70
71	21·0	20·5	20·1	19·6	19·2	18·8	18·5	18·1	17·8	17·4	17·1	16·6	16·1	15·6	15·0	14·4	71
72	22·0	21·5	21·1	20·6	20·2	19·8	19·4	19·0	18·6	18·2	17·9	17·5	17·0	16·4	15·7	15·1	72
73	23·1	22·6	22·2	21·7	21·2	20·8	20·4	20·0	19·6	19·2	18·9	18·4	17·9	17·3	16·6	15·9	73
74	24·4	23·9	23·4	22·9	22·4	21·9	21·5	21·1	20·7	20·3	19·9	19·4	18·9	18·3	17·5	16·8	74
75	25·8	25·2	24·7	24·2	23·7	23·2	22·8	22·3	21·9	21·5	21·1	20·6	20·1	19·4	18·6	17·9	75
76	27·3	26·7	26·2	25·7	25·2	24·7	24·2	23·7	23·3	22·9	22·5	21·9	21·3	20·6	19·8	19·0	76
77	29·0	28·4	27·9	27·3	26·8	26·3	25·8	25·3	24·8	24·4	24·0	23·4	22·8	22·0	21·1	20·3	77
78	31·0	30·4	29·8	29·2	28·7	28·1	27·6	27·1	26·6	26·1	25·7	25·0	24·3	23·6	22·7	21·9	78
79	33·2	32·5	31·9	31·3	30·8	30·2	29·7	29·1	28·6	28·1	27·7	27·0	26·3	25·5	24·5	23·7	79
80	35·8	35·1	34·4	33·8	33·2	32·6	32·0	31·5	31·0	30·4	29·9	29·2	28·5	27·6	26·6	25·7	80
81	38·6	37·9	37·3	36·6	36·0	35·4	34·8	34·2	33·7	33·1	32·6	31·8	31·0	30·2	29·1	28·1	81
82	41·9	41·2	40·5	39·8	39·2	38·6	38·0	37·4	36·8	36·2	35·7	34·9	34·1	33·2	32·0	31·0	82
83	45·7	45·0	44·3	43·6	43·0	42·3	41·7	41·1	40·5	39·9	39·4	38·5	37·6	36·7	35·5	34·4	83

TABLE C.

C   **C**

A & B CORRECTION

A=B±	12·0'	12·5'	13·0'	13·5'	14·0'	14·5'	15·0'	16·0'	17·0'	18·0'	19·0'	20·0'	21·0'	22·0'	23·0'	25·0'=A±B
Lat.							AZIMUTHS.									**Lat.**
°	°	°	°	°	°	°	°	°	°	°	°	°	°	°	°	°
68	12·5	12·1	11·6	11·2	10·8	10·4	10·1	9·5	8·9	8·4	8·0	7·6	7·2	6·9	6·6	6·1	68
69	13·1	12·6	12·1	11·7	11·3	10·9	10·5	9·9	9·4	8·8	8·4	7·9	7·6	7·2	6·9	6·4	69
70	13·7	13·2	12·7	12·2	11·8	11·4	11·0	10·4	9·8	9·2	8·7	8·3	7·9	7·6	7·3	6·7	70
71	14·4	13·8	13·3	12·8	12·4	12·0	11·6	10·9	10·2	9·7	9·2	8·7	8·3	7·9	7·6	7·0	71
72	15·1	14·5	14·0	13·5	13·0	12·6	12·2	11·4	10·8	10·2	9·7	9·2	8·8	8·4	8·0	7·4	72
73	15·9	15·3	14·7	14·2	13·7	13·3	12·8	12·1	11·4	10·8	10·2	9·7	9·3	8·8	8·5	7·8	73
74	16·8	16·2	15·6	15·0	14·5	14·1	13·6	12·8	12·1	11·4	10·8	10·3	9·8	9·4	9·0	8·3	74
75	17·9	17·2	16·6	16·0	15·4	14·9	14·5	13·6	12·8	12·1	11·5	10·9	10·4	10·0	9·6	8·8	75
76	19·0	18·3	17·6	17·0	16·3	15·6	15·4	14·5	13·7	12·9	12·3	11·7	11·1	10·7	10·2	9·4	76
77	20·3	19·6	18·9	18·2	17·6	17·0	16·5	15·5	14·7	13·9	13·2	12·5	11·9	11·4	10·9	10·1	77
78	21·9	21·1	20·3	19·6	19·0	18·4	17·8	16·7	15·8	15·0	14·2	13·5	12·9	12·3	11·9	10·9	78
79	23·7	22·8	22·0	21·2	20·5	19·9	19·7	18·1	17·1	16·2	15·4	14·7	14·0	13·4	12·9	11·8	79
80	25·7	24·7	23·9	23·1	22·4	21·7	21·0	19·8	18·7	17·8	16·9	16·1	15·3	14·7	14·0	13·0	80
81	28·1	27·1	26·2	25·3	24·5	23·8	23·1	21·8	20·6	19·6	18·6	17·7	16·9	16·2	15·5	14·3	81
82	31·0	29·9	29·0	28·0	27·2	26·4	25·6	24·2	22·9	21·8	20·7	19·8	18·9	18·1	17·4	16·0	82
83	34·4	33·3	32·3	31·3	30·4	29·5	28·7	27·2	25·8	24·5	23·4	22·3	21·1	20·5	19·6	18·2	83

C   TABLE C   **C**

A & B CORRECTION

A±B=	25·0'	27·0'	30·0'	33·0'	36'0	40·0'	45·0'	50·0'	60·0'	70·0'	80·0'	100'	150'	200'	400'	800'=A±B
Lat.							AZIMUTHS.									**Lat.**
°	°	°	°	°	°	°	°	°	°	°	°	°	°	°	°	°
68	6·1	5·6	5·1	4·6	4·2	3·8	3·4	3·1	2·5	2·2	1·9	1·5	1·0	0·8	0·4	0·2	68
69	6·4	5·9	5·3	4·8	4·4	4·0	3·5	3·2	2·7	2·3	2·0	1·6	1·1	0·8	0·4	0·2	69
70	6·7	6·2	5·6	5·1	4·6	4·2	3·7	3·4	2·8	2·4	2·1	1·7	1·1	0·8	0·4	0·2	70
71	7·0	6·5	5·9	5·3	4·9	4·4	3·9	3·5	2·9	2·5	2·2	1·8	1·2	0·9	0·4	0·2	71
72	7·4	6·8	6·2	5·6	5·1	4·6	4·1	3·7	3·1	2·6	2·3	1·9	1·2	0·9	0·5	0·2	72
73	7·8	7·3	6·5	5·9	5·4	4·9	4·4	3·9	3·3	2·8	2·4	2·0	1·3	1·0	0·5	0·2	73
74	8·3	7·7	6·9	6·3	5·8	5·2	4·6	4·2	3·5	3·0	2·6	2·1	1·4	1·0	0·5	0·3	74
75	8·8	8·1	7·3	6·7	6·1	5·5	4·9	4·4	3·7	3·2	2·8	2·2	1·5	1·1	0·6	0·3	75
76	9·4	8·7	7·8	7·1	6·6	5·9	5·3	4·7	3·9	3·4	3·0	2·4	1·6	1·2	0·6	0·3	76
77	10·1	9·4	8·4	7·7	7·0	6·3	5·6	5·1	4·2	3·6	3·2	2·6	1·7	1·3	0·6	0·3	77
78	10·9	10·1	9·1	8·3	7·6	6·9	6·1	5·5	4·6	3·9	3·4	2·8	1·8	1·4	0·7	0·3	78
79	11·8	11·0	9·9	9·0	8·3	7·5	6·7	6·0	5·0	4·3	3·8	3·0	2·0	1·5	0·7	0·4	79
80	13·0	12·0	10·9	9·9	9·1	8·2	7·3	6·6	5·5	4·7	4·1	3·3	2·2	1·7	0·8	0·4	80
81	14·3	13·3	12·0	11·0	10·1	9·1	8·1	7·3	6·1	5·2	4·6	3·7	2·4	1·8	0·9	0·5	81
82	16·0	14·9	13·5	12·3	11·3	10·2	9·1	8·2	6·8	5·9	5·1	4·1	2·7	2·1	1·0	0·5	82
83	18·2	16·9	15·3	14·0	12·8	11·6	10·3	9·3	7·8	6·7	5·9	4·7	3·1	2·3	1·2	0·6	83

CONVERSION OF ARC TO TIME

Arc °	Time h.m. m.s.	Arc °	Time h.m.	Arc °	Time h.m.	Arc °	Time h.m.	Arc °	Time h.m.	Arc °	Time h.m.	Parts of 1'	Time s.	Arc ″	Time s.
0	0 00	60	4 00	120	8 00	180	12 00	240	16 00	300	20 00	0·1	0·4	0	0·00
1	0 04	61	4 04	121	8 04	181	12 04	241	16 04	301	20 04			1	0·07
2	0 08	62	4 08	122	8 08	182	12 08	242	16 08	302	20 08	0·2	0·8	2	0·13
3	0 12	63	4 12	123	8 12	183	12 12	243	16 12	303	20 12			3	0·20
4	0 16	64	4 16	124	8 16	184	12 16	244	16 16	304	20 16	0·3	1·2	4	0·27
5	0 20	65	4 20	125	8 20	185	12 20	245	16 20	305	20 20			5	0·33
6	0 24	66	4 24	126	8 24	186	12 24	246	16 24	306	20 24	0·4	1·6	6	0·40
7	0 28	67	4 28	127	8 28	187	12 28	247	16 28	307	20 28			7	0·47
8	0 32	68	4 32	128	8 32	188	12 32	248	16 32	308	20 32	0·5	2·0	8	0·53
9	0 36	69	4 36	129	8 36	189	12 36	249	16 36	309	20 36			9	0·60
10	0 40	70	4 40	130	8 40	190	12 40	250	16 40	310	20 40	0·6	2·4	10	0·67
11	0 44	71	4 44	131	8 44	191	12 44	251	16 44	311	20 44			11	0·73
12	0 48	72	4 48	132	8 48	192	12 48	252	16 48	312	20 48	0·7	2·8	12	0·80
13	0 52	73	4 52	133	8 52	193	12 52	253	16 52	313	20 52			13	0·87
14	0 56	74	4 56	134	8 56	194	12 56	254	16 56	314	20 56	0·8	3·2	14	0·93
15	1 00	75	5 00	135	9 00	195	13 00	255	17 00	315	21 00			15	1·00
16	1 04	76	5 04	136	9 04	196	13 04	256	17 04	316	21 04	0·9	3·6	16	1·07
17	1 08	77	5 08	137	9 08	197	13 08	257	17 08	317	21 08			17	1·13
18	1 12	78	5 12	138	9 12	198	13 12	258	17 12	318	21 12			18	1·20
19	1 16	79	5 16	139	9 16	199	13 16	259	17 16	319	21 16			19	1·27
20	1 20	80	5 20	140	9 20	200	13 20	260	17 20	320	21 20			20	1·33
21	1 24	81	5 24	141	9 24	201	13 24	261	17 24	321	21 24			21	1·40
22	1 28	82	5 28	142	9 28	202	13 28	262	17 28	322	21 28			22	1·47
23	1 32	83	5 32	143	9 32	203	13 32	263	17 32	323	21 32			23	1·53
24	1 36	84	5 36	144	9 36	204	13 36	264	17 36	324	21 36			24	1·60
25	1 40	85	5 40	145	9 40	205	13 40	265	17 40	325	21 40			25	1·67
26	1 44	86	5 44	146	9 44	206	13 44	266	17 44	326	21 44			26	1·73
27	1 48	87	5 48	147	9 48	207	13 48	267	17 48	327	21 48			27	1·80
28	1 52	88	5 52	148	9 52	208	13 52	268	17 52	328	21 52			28	1·87
29	1 56	89	5 56	149	9 56	209	13 56	269	17 56	329	21 56			29	1·93
30	2 00	90	6 00	150	10 00	210	14 00	270	18 00	330	22 00			30	2·00
31	2 04	91	6 04	151	10 04	211	14 04	271	18 04	331	22 04			31	2·07
32	2 08	92	6 08	152	10 08	212	14 08	272	18 08	332	22 08			32	2·13
33	2 12	93	6 12	153	10 12	213	14 12	273	18 12	333	22 12			33	2·20
34	2 16	94	6 16	154	10 16	214	14 16	274	18 16	334	22 16			34	2·27
35	2 20	95	6 20	155	10 20	215	14 20	275	18 20	335	22 20			35	2·33
36	2 24	96	6 24	156	10 24	216	14 24	276	18 24	336	22 24			36	2·40
37	2 28	97	6 28	157	10 28	217	14 28	277	18 28	337	22 28			37	2·47
38	2 32	98	6 32	158	10 32	218	14 32	278	18 32	338	22 32			38	2·53
39	2 36	99	6 36	159	10 36	219	14 36	279	18 36	339	22 36			39	2·60
40	2 40	100	6 40	160	10 40	220	14 40	280	18 40	340	22 40			40	2·67
41	2 44	101	6 44	161	10 44	221	14 44	281	18 44	341	22 44			41	2·73
42	2 48	102	6 48	162	10 48	222	14 48	282	18 48	342	22 48			42	2·80
43	2 52	103	6 52	163	10 52	223	14 52	283	18 52	343	22 52			43	2·87
44	2 56	104	6 56	164	10 56	224	14 56	284	18 56	344	22 56			44	2·93
45	3 00	105	7 00	165	11 00	225	15 00	285	19 00	345	23 00			45	3·00
46	3 04	106	7 04	166	11 04	226	15 04	286	19 04	346	23 04			46	3·07
47	3 08	107	7 08	167	11 08	227	15 08	287	19 08	347	23 08			47	3·13
48	3 12	108	7 12	168	11 12	228	15 12	288	19 12	348	23 12			48	3·20
49	3 16	109	7 16	169	11 16	229	15 16	289	19 16	349	23 16			49	3·27
50	3 20	110	7 20	170	11 20	230	15 20	290	19 20	350	23 20			50	3·33
51	3 24	111	7 24	171	11 24	231	15 24	291	19 24	351	23 24			51	3·40
52	3 28	112	7 28	172	11 28	232	15 28	292	19 28	352	23 28			52	3·47
53	3 32	113	7 32	173	11 32	233	15 32	293	19 32	353	23 32			53	3·53
54	3 36	114	7 36	174	11 36	234	15 36	294	19 36	354	23 36			54	3·60
55	3 40	115	7 40	175	11 40	235	15 40	295	19 40	355	23 40			55	3·67
56	3 44	116	7 44	176	11 44	236	15 44	296	19 44	356	23 44			56	3·73
57	3 48	117	7 48	177	11 48	237	15 48	297	19 48	357	23 48			57	3·80
58	3 52	118	7 52	178	11 52	238	15 52	298	19 52	358	23 52			58	3·87
59	3 56	119	7 56	179	11 56	239	15 56	299	19 56	359	23 56			59	3·93
60	4 00	120	8 00	180	12 00	240	16 00	300	20 00	360	24 00			60	4·00

CONVERSION OF TIME TO ARC

Time h. m. (m. s.)	Arc ° (′)	Time h. m.	Arc °	Time h. m.	Arc °	Time h. m.	Arc °	Time h. m.	Arc °	Time h. m.	Arc °	Time m. s.	Arc ° ′ (′ ″)	Critical Table * s.	′
0 00	0	4 00	60	8 00	120	12 00	180	16 00	240	20 00	300	0	0 00		
0 04	1	4 04	61	8 04	121	12 04	181	16 04	241	20 04	301	1	0 15	0·0	
0 08	2	4 08	62	8 08	122	12 08	182	16 08	242	20 08	302	2	0 30		0·0
0 12	3	4 12	63	8 12	123	12 12	183	16 12	243	20 12	303	3	0 45	0·2	
0 16	4	4 16	64	8 16	124	12 16	184	16 16	244	20 16	304	4	1 00		0·1
														0·5	
0 20	5	4 20	65	8 20	125	12 20	185	16 20	245	20 20	305	5	1 15		0·2
0 24	6	4 24	66	8 24	126	12 24	186	16 24	246	20 24	306	6	1 30	1·0	
0 28	7	4 28	67	8 28	127	12 28	187	16 28	247	20 28	307	7	1 45		0·3
0 32	8	4 32	68	8 32	128	12 32	188	16 32	248	20 32	308	8	2 00	1·3	
0 36	9	4 36	69	8 36	129	12 36	189	16 36	249	20 36	309	9	2 15		0·4
														1·8	
0 40	10	4 40	70	8 40	130	12 40	190	16 40	250	20 40	310	10	2 30		0·5
0 44	11	4 44	71	8 44	131	12 44	191	16 44	251	20 44	311	11	2 45	2·1	
0 48	12	4 48	72	8 48	132	12 48	192	16 48	252	20 48	312	12	3 00		0·6
0 52	13	4 52	73	8 52	133	12 52	193	16 52	253	20 52	313	13	3 15	2·6	
0 56	14	4 56	74	8 56	134	12 56	194	16 56	254	20 56	314	14	3 30		0·7
														2·9	
1 00	15	5 00	75	9 00	135	13 00	195	17 00	255	21 00	315	15	3 45		0·8
1 04	16	5 04	76	9 04	136	13 04	196	17 04	256	21 04	316	16	4 00	3·4	
1 08	17	5 08	77	9 08	137	13 08	197	17 08	257	21 08	317	17	4 15		0·9
1 12	18	5 12	78	9 12	138	13 12	198	17 12	258	21 12	318	18	4 30	3·7	
1 16	19	5 16	79	9 16	139	13 16	199	17 16	259	21 16	319	19	4 45		1·0
														4·0	
1 20	20	5 20	80	9 20	140	13 20	200	17 20	260	21 20	320	20	5 00		
1 24	21	5 24	81	9 24	141	13 24	201	17 24	261	21 24	321	21	5 15		
1 28	22	5 28	82	9 28	142	13 28	202	17 28	262	21 28	322	22	5 30		
1 32	23	5 32	83	9 32	143	13 32	203	17 32	263	21 32	323	23	5 45		
1 36	24	5 36	84	9 36	144	13 36	204	17 36	264	21 36	324	24	6 00		
1 40	25	5 40	85	9 40	145	13 40	205	17 40	265	21 40	325	25	6 15		
1 44	26	5 44	86	9 44	146	13 44	206	17 44	266	21 44	326	26	6 30		
1 48	27	5 48	87	9 48	147	13 48	207	17 48	267	21 48	327	27	6 45		
1 52	28	5 52	88	9 52	148	13 52	208	17 52	268	21 52	328	28	7 00		
1 56	29	5 56	89	9 56	149	13 56	209	17 56	269	21 56	329	29	7 15		
2 00	30	6 00	90	10 00	150	14 00	210	18 00	270	22 00	330	30	7 30		
2 04	31	6 04	91	10 04	151	14 04	211	18 04	271	22 04	331	31	7 45		
2 08	32	6 08	92	10 08	152	14 08	212	18 08	272	22 08	332	32	8 00		
2 12	33	6 12	93	10 12	153	14 12	213	18 12	273	22 12	333	33	8 15		
2 16	34	6 16	94	10 16	154	14 16	214	18 16	274	22 16	334	34	8 30		
2 20	35	6 20	95	10 20	155	14 20	215	18 20	275	22 20	335	35	8 45		
2 24	36	6 24	96	10 24	156	14 24	216	18 24	276	22 24	336	36	9 00		
2 28	37	6 28	97	10 28	157	14 28	217	18 28	277	22 28	337	37	9 15		
2 32	38	6 32	98	10 32	158	14 32	218	18 32	278	22 32	338	38	9 30		
2 36	39	6 36	99	10 36	159	14 36	219	18 36	279	22 36	339	39	9 45		
2 40	40	6 40	100	10 40	160	14 40	220	18 40	280	22 40	340	40	10 00		
2 44	41	6 44	101	10 44	161	14 44	221	18 44	281	22 44	341	41	10 15		
2 48	42	6 48	102	10 48	162	14 48	222	18 48	282	22 48	342	42	10 30		
2 52	43	6 52	103	10 52	163	14 52	223	18 52	283	22 52	343	43	10 45		
2 56	44	6 56	104	10 56	164	14 56	224	18 56	284	22 56	344	44	11 00		
3 00	45	7 00	105	11 00	165	15 00	225	19 00	285	23 00	345	45	11 15		
3 04	46	7 04	106	11 04	166	15 04	226	19 04	286	23 04	346	46	11 30		
3 08	47	7 08	107	11 08	167	15 08	227	19 08	287	23 08	347	47	11 45		
3 12	48	7 12	108	11 12	168	15 12	228	19 12	288	23 12	348	48	12 00		
3 16	49	7 16	109	11 16	169	15 16	229	19 16	289	23 16	349	49	12 15		
3 20	50	7 20	110	11 20	170	15 20	230	19 20	290	23 20	350	50	12 30		
3 24	51	7 24	111	11 24	171	15 24	231	19 24	291	23 24	351	51	12 45		
3 28	52	7 28	112	11 28	172	15 28	232	19 28	292	23 28	352	52	13 00		
3 32	53	7 32	113	11 32	173	15 32	233	19 32	293	23 32	353	53	13 15		
3 36	54	7 36	114	11 36	174	15 36	234	19 36	294	23 36	354	54	13 30		
3 40	55	7 40	115	11 40	175	15 40	235	19 40	295	23 40	355	55	13 45		
3 44	56	7 44	116	11 44	176	15 44	236	19 44	296	23 44	356	56	14 00		
3 48	57	7 48	117	11 48	177	15 48	237	19 48	297	23 48	357	57	14 15		
3 52	58	7 52	118	11 52	178	15 52	238	19 52	298	23 52	358	58	14 30		
3 56	59	7 56	119	11 56	179	15 56	239	19 56	299	23 56	359	59	14 45		
4 00	60	8 00	120	12 00	180	16 00	240	20 00	300	24 00	360	60	15 00		

* For use when arc quantity is required only to nearest 0′·1. In critical cases use upper value.

HOURS AND MINUTES TO DECIMAL OF A DAY

Min.	HOURS											
	0	1	2	3	4	5	6	7	8	9	10	11
0	·0000	·0417	·0833	·1250	·1667	·2083	·2500	·2917	·3333	·3750	·4167	·4583
1	·0007	·0424	·0840	·1257	·1674	·2090	·2507	·2924	·3340	·3757	·4174	·4590
2	·0014	·0431	·0847	·1264	·1681	·2097	·2514	·2931	·3347	·3764	·4181	·4597
3	·0021	·0438	·0854	·1271	·1688	·2104	·2521	·2938	·3354	·3771	·4188	·4604
4	·0028	·0444	·0861	·1278	·1694	·2111	·2528	·2944	·3361	·3778	·4194	·4611
5	·0035	·0451	·0868	·1285	·1701	·2118	·2535	·2951	·3368	·3785	·4201	·4618
6	·0042	·0458	·0875	·1292	·1708	·2125	·2542	·2958	·3375	·3792	·4208	·4625
7	·0049	·0465	·0882	·1299	·1715	·2132	·2549	·2965	·3382	·3799	·4215	·4632
8	·0056	·0472	·0889	·1306	·1722	·2139	·2556	·2972	·3389	·3806	·4222	·4639
9	·0063	·0479	·0896	·1313	·1729	·2146	·2563	·2979	·3396	·3813	·4229	·4646
10	·0069	·0486	·0903	·1319	·1736	·2153	·2569	·2986	·3403	·3819	·4236	·4653
11	·0076	·0493	·0910	·1326	·1743	·2160	·2576	·2993	·3410	·3826	·4243	·4660
12	·0083	·0500	·0917	·1333	·1750	·2167	·2583	·3000	·3417	·3833	·4250	·4667
13	·0090	·0507	·0924	·1340	·1757	·2174	·2590	·3007	·3424	·3840	·4257	·4674
14	·0097	·0514	·0931	·1347	·1764	·2181	·2597	·3014	·3431	·3847	·4264	·4681
15	·0104	·0521	·0938	·1354	·1771	·2188	·2604	·3021	·3438	·3854	·4271	·4688
16	·0111	·0528	·0944	·1361	·1778	·2194	·2611	·3028	·3444	·3861	·4278	·4694
17	·0118	·0535	·0951	·1368	·1785	·2201	·2618	·3035	·3451	·3868	·4285	·4701
18	·0125	·0542	·0958	·1375	·1792	·2208	·2625	·3042	·3458	·3875	·4292	·4708
19	·0132	·0549	·0965	·1382	·1799	·2215	·2632	·3049	·3465	·3882	·4299	·4715
20	·0139	·0556	·0972	·1389	·1806	·2222	·2639	·3056	·3472	·3889	·4306	·4722
21	·0146	·0563	·0979	·1396	·1813	·2229	·2646	·3063	·3479	·3896	·4313	·4729
22	·0153	·0569	·0986	·1403	·1819	·2236	·2653	·3069	·3486	·3903	·4319	·4736
23	·0160	·0576	·0993	·1410	·1826	·2243	·2660	·3076	·3493	·3910	·4326	·4743
24	·0167	·0583	·1000	·1417	·1833	·2250	·2667	·3083	·3500	·3917	·4333	·4750
25	·0174	·0590	·1007	·1424	·1840	·2257	·2674	·3090	·3507	·3924	·4340	·4757
26	·0181	·0597	·1014	·1431	·1847	·2264	·2681	·3097	·3514	·3931	·4347	·4764
27	·0188	·0604	·1021	·1438	·1854	·2271	·2688	·3104	·3521	·3938	·4354	·4771
28	·0194	·0611	·1028	·1444	·1861	·2278	·2694	·3111	·3528	·3944	·4361	·4778
29	·0201	·0618	·1035	·1451	·1868	·2285	·2701	·3118	·3535	·3951	·4368	·4785
30	·0208	·0625	·1042	·1458	·1875	·2292	·2708	·3125	·3542	·3958	·4375	·4792
31	·0215	·0632	·1049	·1465	·1882	·2299	·2715	·3132	·3549	·3965	·4382	·4799
32	·0222	·0639	·1056	·1472	·1889	·2306	·2722	·3139	·3556	·3972	·4389	·4806
33	·0229	·0646	·1063	·1479	·1896	·2313	·2729	·3146	·3563	·3979	·4396	·4813
34	·0236	·0653	·1069	·1486	·1903	·2319	·2736	·3153	·3569	·3986	·4403	·4819
35	·0243	·0660	·1076	·1493	·1910	·2326	·2743	·3160	·3576	·3993	·4410	·4826
36	·0250	·0667	·1083	·1500	·1917	·2333	·2750	·3167	·3583	·4000	·4417	·4833
37	·0257	·0674	·1090	·1507	·1924	·2340	·2757	·3174	·3590	·4007	·4424	·4840
38	·0264	·0681	·1097	·1514	·1931	·2347	·2764	·3181	·3597	·4014	·4431	·4847
39	·0271	·0688	·1104	·1521	·1938	·2354	·2771	·3188	·3604	·4021	·4438	·4854
40	·0278	·0694	·1111	·1528	·1944	·2361	·2778	·3194	·3611	·4028	·4444	·4861
41	·0285	·0701	·1118	·1535	·1951	·2368	·2785	·3201	·3618	4035	·4451	·4868
42	·0292	·0708	·1125	·1542	·1958	·2375	·2792	·3208	·3625	·4042	·4458	·4875
43	·0299	·0715	·1132	·1549	·1965	·2382	·2799	·3215	·3632	·4049	·4465	·4882
44	·0306	·0722	·1139	·1556	·1972	·2389	·2806	·3222	·3639	·4056	·4472	·4889
45	·0313	·0729	·1146	·1563	·1979	·2396	·2813	·3229	·3646	·4063	·4479	·4896
46	·0319	·0736	·1153	·1569	·1986	·2403	·2819	·3236	·3653	·4069	·4486	·4903
47	·0326	·0743	·1160	·1576	·1993	·2410	·2826	·3243	·3660	·4076	·4493	·4910
48	·0333	·0750	·1167	·1583	·2000	·2417	·2833	·3250	·3667	·4083	·4500	·4917
49	·0340	·0757	·1174	·1590	·2007	·2424	·2840	·3257	·3674	·4090	·4507	·4924
50	·0347	·0764	·1181	·1597	·2014	·2431	·2847	·3264	·3681	·4097	·4514	·4931
51	·0354	·0771	·1188	·1604	·2021	·2438	·2854	·3271	·3688	·4104	·4521	·4938
52	·0361	·0778	·1194	·1611	·2028	·2444	·2861	·3278	·3694	·4111	·4528	·4944
53	·0368	·0785	·1201	·1618	·2035	·2451	·2868	·3285	·3701	·4118	·4535	·4951
54	·0375	·0792	·1208	·1625	·2042	·2458	·2875	·3292	·3708	·4125	·4542	·4958
55	·0382	·0799	·1215	·1632	·2049	·2465	·2882	·3299	·3715	·4132	·4549	·4965
56	·0389	·0806	·1222	·1639	·2056	·2472	·2889	·3306	·3722	·4139	·4556	·4972
57	·0396	·0813	·1229	·1646	·2063	·2479	·2896	·3313	·3729	·4146	·4563	·4979
58	·0403	·0819	·1236	·1653	·2069	·2486	·2903	·3319	·3736	·4153	·4569	·4986
59	·0410	·0826	·1243	·1660	·2076	·2493	·2910	·3326	·3743	·4160	·4576	·4993
60	·0417	·0833	·1250	·1667	·2083	·2500	·2917	·3333	·3750	·4167	·4583	·5000

For an additional 12 hours, increase the tabulated value by 0·5.,
e.g., 14 h. 38 m. = 12 h. + 2 h. 38 m. = 0·5 + ·1097 = 0·6097 of a day.
Similarly, ·8972 of a day = 0·5 + ·3972 = 12 h. + 9 h. 32 m. = 21 h. 32 m.

TRUE AMPLITUDES

Lat.	Declination														
	1°	2°	3°	4°	5°	6°	7°	8°	9°	10°	11°	12°	13°	14°	15°
°	°	°	°	°	°	°	°	°	°	°	°	°	°	°	°
2	1·0	2·0	3·0	4·0	5·0	6·0	7·0	8·0	9·0	10·0	11·0	12·0	13·0	14·0	15·0
4	1·0	2·0	3·0	4·0	5·0	6·0	7·0	8·0	9·0	10·0	11·0	12·0	13·0	14·0	15·0
6	1·0	2·0	3·0	4·0	5·0	6·0	7·1	8·1	9·1	10·1	11·1	12·1	13·1	14·1	15·1
8	1·0	2·0	3·0	4·1	5·1	6·1	7·1	8·1	9·1	10·1	11·1	12·1	13·1	14·1	15·2
10	1·0	2·0	3·1	4·1	5·1	6·1	7·1	8·1	9·2	10·2	11·2	12·2	13·2	14·2	15·3
12	1·0	2·1	3·1	4·1	5·1	6·1	7·2	8·2	9·2	10·2	11·3	12·3	13·3	14·3	15·4
14	1·0	2·1	3·1	4·1	5·2	6·2	7·2	8·3	9·3	10·3	11·3	12·4	13·4	14·4	15·5
16	1·1	2·1	3·1	4·2	5·2	6·2	7·3	8·3	9·4	10·4	11·5	12·5	13·5	14·6	15·6
18	1·1	2·1	3·2	4·2	5·3	6·3	7·4	8·4	9·5	10·5	11·6	12·6	13·7	14·7	15·8
20	1·1	2·1	3·2	4·3	5·3	6·4	7·5	8·5	9·6	10·7	11·7	12·8	13·9	14·9	15·9
22	1·1	2·2	3·2	4·3	5·4	6·5	7·6	8·6	9·7	10·8	11·9	13·0	14·1	15·1	16·2
24	1·1	2·2	3·3	4·4	5·5	6·6	7·7	8·8	9·9	11·0	12·1	13·2	14·3	15·4	16·5
26	1·1	2·2	3·4	4·5	5·6	6·7	7·8	8·9	10·0	11·2	12·3	13·4	14·5	15·6	16·8
28	1·1	2·3	3·4	4·5	5·7	6·8	7·9	9·1	10·2	11·4	12·5	13·6	14·8	15·9	17·1
30	1·2	2·3	3·5	4·6	5·8	6·9	8·1	9·3	10·4	11·6	12·7	13·9	15·1	16·2	17·4
31	1·2	2·3	3·5	4·7	5·8	7·0	8·2	9·4	10·5	11·7	12·9	14·0	15·2	16·4	17·6
32	1·2	2·4	3·6	4·7	5·9	7·1	8·3	9·5	10·6	11·8	13·0	14·2	15·4	16·6	17·8
33	1·2	2·4	3·6	4·8	6·0	7·2	8·4	9·6	10·8	12·0	13·2	14·4	15·6	16·8	18·0
34	1·2	2·4	3·6	4·8	6·0	7·3	8·5	9·7	10·9	12·1	13·3	14·5	15·8	17·0	18·2
35	1·2	2·5	3·7	4·9	6·1	7·3	8·6	9·8	11·0	12·2	13·5	14·7	16·0	17·2	18·4
36	1·2	2·5	3·7	5·0	6·2	7·4	8·7	9·9	11·2	12·4	13·7	14·9	16·2	17·4	18·7
37	1·3	2·5	3·8	5·0	6·3	7·5	8·8	10·0	11·3	12·6	13·8	15·1	16·4	17·6	18·9
38	1·3	2·5	3·8	5·1	6·4	7·6	8·9	10·2	11·4	12·7	14·0	15·3	16·6	17·9	19·2
39	1·3	2·6	3·9	5·2	6·4	7·7	9·0	10·3	11·6	12·9	14·2	15·5	16·8	18·1	19·5
40	1·3	2·6	3·9	5·2	6·5	7·9	9·2	10·5	11·8	13·1	14·4	15·8	17·1	18·4	19·8
41	1·3	2·7	4·0	5·3	6·6	8·0	9·3	10·6	12·0	13·3	14·7	16·0	17·4	18·7	20·1
42	1·4	2·7	4·0	5·4	6·7	8·1	9·4	10·8	12·2	13·5	14·9	16·3	17·6	19·0	20·4
43	1·4	2·7	4·1	5·5	6·9	8·2	9·6	11·0	12·4	13·7	15·1	16·5	17·9	19·3	20·7
44	1·4	2·8	4·2	5·6	7·0	8·4	9·8	11·2	12·6	14·0	15·4	16·8	18·2	19·7	21·1
45	1·4	2·8	4·3	5·7	7·1	8·5	9·9	11·4	12·8	14·2	15·7	17·1	18·6	20·0	21·5
46	1·4	2·9	4·3	5·8	7·2	8·7	10·1	11·6	13·0	14·5	16·0	17·4	18·9	20·4	21·9
47	1·5	2·9	4·4	5·9	7·4	8·8	10·3	11·8	13·3	14·8	16·3	17·8	19·3	20·8	22·3
48	1·5	3·0	4·5	6·0	7·5	9·0	10·5	12·0	13·5	15·1	16·6	18·1	19·7	21·2	22·8
49	1·5	3·1	4·6	6·1	7·6	9·2	10·7	12·3	13·8	15·4	16·9	18·5	20·1	21·6	23·2
50	1·6	3·1	4·7	6·2	7·8	9·4	10·9	12·5	14·1	15·7	17·3	18·9	20·5	22·1	23·8
50½	1·6	3·1	4·7	6·3	7·9	9·5	11·0	12·6	14·2	15·8	17·5	19·1	20·7	22·4	24·0
51	1·6	3·1	4·8	6·4	8·0	9·6	11·2	12·8	14·4	16·0	17·7	19·3	21·0	22·6	24·3
51½	1·6	3·2	4·8	6·4	8·0	9·7	11·3	12·9	14·6	16·2	17·8	19·5	21·2	22·9	24·6
52	1·6	3·3	4·9	6·5	8·1	9·8	11·4	13·1	14·7	16·4	18·1	19·7	21·4	23·2	24·9
52½	1·6	3·3	4·9	6·6	8·2	9·9	11·5	13·2	14·9	16·6	18·3	20·0	21·7	23·4	25·2
53	1·7	3·3	5·0	6·7	8·3	10·0	11·7	13·4	15·1	16·8	18·5	20·2	22·0	23·7	25·5
53½	1·7	3·4	5·0	6·7	8·4	10·1	11·8	13·5	15·2	17·0	18·7	20·5	22·2	24·0	25·8
54	1·7	3·4	5·1	6·8	8·5	10·3	12·0	13·7	15·4	17·2	19·0	20·7	22·5	24·3	26·1
54½	1·7	3·4	5·2	6·9	8·6	10·4	12·2	13·9	15·6	17·4	19·2	21·0	22·8	24·6	26·5
55	1·8	3·5	5·2	7·0	8·7	10·5	12·3	14·1	15·8	17·6	19·4	21·3	23·1	25·0	26·8
55½	1·8	3·5	5·3	7·1	8·9	10·6	12·4	14·2	16·0	17·9	19·7	21·5	23·4	25·3	27·2
56	1·8	3·6	5·4	7·2	9·0	10·8	12·6	14·4	16·3	18·1	20·0	21·8	23·7	25·6	27·6
56½	1·8	3·6	5·4	7·3	9·1	10·9	12·8	14·6	16·5	18·3	20·2	22·1	24·1	26·0	28·0
57	1·8	3·7	5·5	7·4	9·2	11·1	12·9	14·8	16·7	18·6	20·5	22·4	24·4	26·4	28·4
57½	1·9	3·7	5·6	7·5	9·3	11·2	13·1	15·0	16·9	18·9	20·8	22·8	24·8	26·8	28·9
58	1·9	3·8	5·7	7·6	9·5	11·4	13·3	15·2	17·2	19·1	21·1	23·1	25·1	27·2	29·2
58½	1·9	3·8	5·7	7·7	9·6	11·5	13·5	15·4	17·4	19·4	21·4	23·4	25·5	27·6	29·7
59	2·0	3·9	5·8	7·8	9·8	11·7	13·7	15·7	17·7	19·7	21·8	23·8	25·9	28·0	30·2
59½	2·0	3·9	5·9	7·9	9·9	11·9	13·9	15·9	18·0	20·0	22·1	24·2	26·3	28·5	30·7
60	2·0	4·0	6·0	8·0	10·0	12·1	14·1	16·2	18·2	20·3	22·4	24·6	26·7	28·9	31·2
60½	2·0	4·1	6·1	8·1	10·2	12·3	14·3	16·4	18·5	20·6	22·8	25·0	27·2	29·4	31·7
61	2·1	4·1	6·2	8·3	10·4	12·5	14·6	16·7	18·8	21·0	23·2	25·4	27·7	29·9	32·3
61½	2·1	4·2	6·3	8·4	10·5	12·7	14·8	17·0	19·1	21·3	23·6	25·8	28·1	30·5	32·8
62	2·1	4·3	6·4	8·6	10·7	12·9	15·1	17·3	19·5	21·7	24·0	26·3	28·6	31·0	33·5
62½	2·2	4·3	6·5	8·7	10·9	13·1	15·3	17·5	19·8	22·1	24·4	26·8	29·2	31·6	34·1

TRUE AMPLITUDES

Lat.	Declination														
	16°	17°	18°	19°	20°	20½°	21°	21½°	22°	22½°	23°	23½°	24°	24½°	25°
2	16·0	17·0	18·0	19·0	20·0	20·5	21·0	21·5	22·0	22·5	23·0	23·5	24·0	24·5	25·0
4	16·0	17·1	18·1	19·1	20·1	20·6	21·1	21·6	22·1	22·6	23·1	23·6	24·1	24·6	25·1
6	16·1	17·1	18·1	19·1	20·1	20·6	21·1	21·6	22·1	22·6	23·1	23·6	24·1	24·6	25·1
8	16·2	17·2	18·2	19·2	20·2	20·7	21·2	21·7	22·2	22·7	23·2	23·7	24·3	24·8	25·3
10	16·3	17·3	18·3	19·3	20·3	20·8	21·4	21·8	22·4	22·9	23·4	23·9	24·4	24·9	25·4
12	16·4	17·4	18·4	19·4	20·5	21·0	21·5	22·0	22·5	23·0	23·6	24·1	24·6	25·1	25·6
14	16·5	17·5	18·6	19·6	20·6	21·2	21·7	22·2	22·7	23·2	23·8	24·3	24·8	25·3	25·8
16	16·7	17·7	18·8	19·8	20·9	21·4	21·9	22·4	22·9	23·5	24·0	24·5	25·0	25·6	26·1
18	16·9	17·9	19·0	20·0	21·1	21·6	22·1	22·7	23·2	23·7	24·3	24·8	25·3	25·9	26·4
20	17·1	18·1	19·2	20·3	21·4	21·9	22·4	23·0	23·5	24·0	24·6	25·1	25·7	26·2	26·7
22	17·3	18·4	19·5	20·6	21·7	22·2	22·7	23·3	23·8	24·4	24·9	25·5	26·0	26·6	27·1
24	17·6	18·7	19·8	20·9	22·0	22·5	23·1	23·7	24·2	24·8	25·3	25·9	26·4	27·0	27·6
26	17·9	19·0	20·1	21·2	22·3	22·9	23·5	24·1	24·6	25·2	25·8	26·3	26·9	27·5	28·0
28	18·2	19·3	20·5	21·6	22·8	23·4	24·0	24·5	25·1	25·7	26·3	26·8	27·4	28·0	28·6
30	18·6	19·7	20·9	22·1	23·3	23·9	24·5	25·0	25·6	26·2	26·8	27·4	28·0	28·6	29·2
31	18·8	20·0	21·1	22·3	23·5	24·1	24·7	25·3	25·9	26·5	27·1	27·7	28·3	28·9	29·5
32	19·0	20·2	21·4	22·6	23·8	24·4	25·0	25·6	26·2	26·8	27·4	28·0	28·7	29·3	29·9
33	19·2	20·4	21·6	22·9	24·1	24·7	25·3	25·9	26·5	27·1	27·8	28·4	29·0	29·6	30·3
34	19·4	20·6	21·9	23·1	24·4	25·0	25·6	26·2	26·9	27·5	28·1	28·7	29·4	29·9	30·7
35	19·7	20·9	22·2	23·4	24·7	25·3	26·0	26·6	27·2	27·9	28·5	29·1	29·8	30·4	31·1
36	19·9	21·2	22·5	23·7	25·0	25·7	26·3	26·9	27·6	28·2	28·9	29·5	30·2	30·8	31·5
37	20·2	21·5	22·8	24·1	25·4	26·0	26·7	27·3	28·0	28·6	29·3	30·0	30·6	31·3	31·9
38	20·5	21·8	23·1	24·4	25·7	26·4	27·1	27·7	28·4	29·1	29·7	30·4	31·1	31·8	32·4
39	20·8	22·1	23·4	24·8	26·1	26·8	27·5	28·1	28·8	29·5	30·2	30·9	31·6	32·3	32·9
40	21·1	22·4	23·8	25·2	26·5	27·2	27·9	28·6	29·3	30·0	30·7	31·4	32·1	32·8	33·5
41	21·4	22·8	24·2	25·6	27·0	27·6	28·4	29·1	29·8	30·5	31·2	31·9	32·6	33·3	34·1
42	21·8	23·2	24·6	26·0	27·4	28·1	28·8	29·6	30·3	31·0	31·7	32·5	33·2	33·9	34·7
43	22·1	23·6	25·0	26·4	27·9	28·6	29·3	30·1	30·8	31·6	32·3	33·0	33·8	34·5	35·3
44	22·5	24·0	25·4	26·9	28·4	29·1	29·9	30·6	31·4	32·1	32·9	33·7	34·4	35·2	36·0
45	23·0	24·4	25·9	27·4	28·9	29·7	30·5	31·3	32·0	32·8	33·6	34·3	35·1	35·9	36·7
46	23·4	24·9	26·4	28·0	29·5	30·3	31·1	31·8	32·6	33·4	34·2	35·0	35·8	36·7	37·5
47	23·8	25·4	27·0	28·5	30·1	30·9	31·7	32·5	33·3	34·1	35·0	35·8	36·6	37·5	38·3
48	24·3	25·9	27·5	29·1	30·7	31·6	32·4	33·2	34·1	34·9	35·7	36·6	37·4	38·3	39·2
49	24·9	26·5	28·1	29·8	31·4	32·3	33·1	34·0	34·8	35·7	36·6	37·4	38·3	39·2	40·1
50	25·4	27·1	28·7	30·4	32·2	33·0	33·9	34·8	35·6	36·5	37·4	38·3	39·3	40·2	41·1
50½	25·7	27·4	29·1	30·8	32·5	33·4	34·3	35·2	36·1	37·0	37·9	38·8	39·8	40·7	41·6
51	26·0	27·7	29·4	31·2	32·9	33·8	34·7	35·6	36·5	37·5	38·4	39·3	40·3	41·2	42·2
51½	26·3	28·0	29·8	31·5	33·3	34·2	35·1	36·1	37·0	37·9	38·9	39·8	40·8	41·8	42·8
52	26·6	28·4	30·1	31·9	33·8	34·7	35·6	36·5	37·5	38·4	39·4	40·4	41·4	42·3	43·4
52½	26·9	28·7	30·5	32·3	34·2	35·1	36·1	37·0	38·0	38·9	39·9	40·9	41·9	42·9	44·0
53	27·3	29·1	30·9	32·8	34·6	35·6	36·6	37·5	38·5	39·5	40·5	41·5	42·5	43·6	44·6
53½	27·6	29·4	31·3	33·2	35·1	36·1	37·0	38·0	39·0	40·0	41·1	42·1	43·1	44·2	45·3
54	28·0	29·8	31·7	33·6	35·6	36·6	37·6	38·6	39·6	40·6	41·7	42·7	43·8	44·9	46·0
54½	28·3	30·2	32·2	34·1	36·1	37·1	38·1	39·1	40·1	41·2	42·3	43·4	44·5	45·6	46·7
55	28·7	30·7	32·6	34·6	36·6	37·6	38·7	39·7	40·8	41·9	42·9	44·0	45·2	46·3	47·5
55½	29·1	31·1	33·1	35·1	37·1	38·2	39·3	40·3	41·4	42·5	43·6	44·7	45·9	46·9	48·3
56	29·5	31·5	33·6	35·6	37·7	38·8	39·9	41·0	42·1	43·2	44·3	45·5	46·7	47·9	49·1
56½	30·0	32·0	34·0	36·1	38·3	39·4	40·5	41·6	42·8	43·9	45·1	46·3	47·5	48·7	50·0
57	30·4	32·5	34·6	36·7	38·9	40·0	41·2	42·3	43·5	44·6	45·9	47·1	48·3	49·6	50·9
57½	30·9	33·0	35·1	37·3	39·5	40·7	41·8	43·0	44·2	45·4	46·7	47·9	49·2	50·5	51·9
58	31·3	33·5	35·7	37·9	40·2	41·4	42·6	43·8	45·0	46·2	47·5	48·8	50·1	51·5	52·9
58½	31·8	34·0	36·3	38·5	40·9	42·1	43·3	44·5	45·8	47·1	48·4	49·7	51·1	52·5	54·0
59	32·4	34·6	36·9	39·2	41·6	42·8	44·1	45·4	46·7	48·0	49·4	50·7	52·2	53·6	55·1
59½	32·9	35·2	37·5	39·9	42·4	43·6	44·9	46·2	47·6	48·9	50·3	51·8	53·3	54·8	56·4
60	33·5	35·8	38·2	40·6	43·2	44·5	45·8	47·1	48·5	49·9	51·4	52·9	54·4	56·0	57·7
60½	34·0	36·4	38·8	41·4	44·0	45·3	46·7	48·1	49·5	51·0	52·5	54·1	55·7	57·4	59·1
61	34·7	37·1	39·6	42·2	44·9	46·3	47·7	49·1	50·6	52·1	53·7	55·3	57·0	58·8	60·6
61½	35·3	37·8	40·4	43·0	45·8	47·2	48·7	50·2	51·7	53·3	55·0	56·7	58·5	60·4	62·3
62	36·0	38·5	41·2	43·9	46·8	48·2	49·8	51·3	52·9	54·6	56·3	58·1	60·0	62·0	64·2
62½	36·7	39·3	42·0	44·8	47·8	49·3	50·9	52·5	54·2	56·0	57·8	59·7	61·7	63·9	66·2

TRUE AMPLITUDES

Lat.	Declination							
	25½°	26°	26½°	27°	27½°	28°	28½°	29°
2	25.5	26.0	26.5	27.0	27.5	28.0	28.5	29.0
4	25.6	26.1	26.6	27.1	27.6	28.1	28.6	29.1
6	25.7	26.2	26.7	27.2	27.7	28.2	28.7	29.2
8	25.8	26.3	26.8	27.3	27.8	28.3	28.8	29.3
10	25.9	26.4	26.9	27.5	28.0	28.5	29.0	29.5
12	26.1	26.6	27.1	27.7	28.2	28.7	29.2	29.7
14	26.3	26.9	27.4	27.9	28.4	28.9	29.5	30.0
16	26.6	27.1	27.7	28.2	28.7	29.2	29.8	30.3
18	26.9	27.5	28.0	28.5	29.0	29.6	30.1	30.7
20	27.3	27.8	28.3	28.9	29.4	30.0	30.5	31.1
22	27.7	28.2	28.8	29.3	29.9	30.4	31.0	31.5
24	28.1	28.7	29.2	29.8	30.4	30.9	31.5	32.1
26	28.6	29.2	29.8	30.3	30.9	31.5	32.1	32.6
28	29.2	29.8	30.4	30.9	31.5	32.1	32.7	33.3
30	29.8	30.4	31.0	31.6	32.2	32.8	33.4	34.1
31	30.1	30.8	31.4	32.0	32.6	33.2	33.8	34.5
32	30.5	31.1	31.7	32.4	33.0	33.6	34.2	34.9
33	30.9	31.5	32.1	32.8	33.4	34.0	34.7	35.3
34	31.3	31.9	32.6	33.2	33.8	34.5	35.1	35.8
35	31.7	32.4	33.0	33.7	34.3	35.0	35.6	36.3
36	32.2	32.8	33.5	34.1	34.8	35.5	36.1	36.8
37	32.6	33.3	34.0	34.6	35.3	36.0	36.7	37.4
38	33.1	33.8	34.5	35.2	35.9	36.6	37.3	38.0
39	33.6	34.3	35.0	35.8	36.5	37.2	37.9	38.6
40	34.2	34.9	35.6	36.4	37.1	37.8	38.5	39.3
41	34.8	35.5	36.2	37.0	37.7	38.5	39.2	40.0
42	35.4	36.2	36.9	37.7	38.4	39.2	39.9	40.7
43	36.1	36.8	37.6	38.4	39.2	39.9	40.7	41.5
44	36.8	37.5	38.3	39.2	39.9	40.7	41.6	42.4
45	37.5	38.3	39.1	39.9	40.8	41.6	42.4	43.3
46	38.3	39.1	40.0	40.8	41.7	42.5	43.4	44.3
47	39.1	40.0	40.9	41.7	42.6	43.5	44.4	45.3
48	40.0	40.9	41.8	42.7	43.6	44.5	45.5	46.4
49	41.0	41.9	42.9	43.8	44.7	45.7	46.7	47.7
50	42.0	43.0	44.0	44.9	45.9	46.9	47.9	49.0
50½	42.6	43.6	44.5	45.5	46.5	47.6	48.6	49.7
51	43.2	44.2	45.2	46.2	47.2	48.2	49.3	50.4
51½	43.8	44.8	45.8	46.8	47.9	49.0	50.0	51.2
52	44.4	45.4	46.4	47.5	48.6	49.7	50.8	52.0
52½	45.0	46.0	47.1	48.2	49.3	50.5	51.6	52.8
53	45.7	46.7	47.9	49.0	50.1	51.3	52.5	53.7
53½	46.4	47.5	48.6	49.8	50.9	52.1	53.3	54.6
54	47.1	48.2	49.4	50.6	51.8	53.0	54.3	55.6
54½	47.8	49.0	50.2	51.4	52.7	53.9	55.3	56.6
55	48.6	49.8	51.1	52.3	53.6	54.9	56.3	57.7
55½	49.5	50.7	52.0	53.3	54.6	56.0	57.4	58.9
56	50.3	51.6	52.9	54.3	55.7	57.1	58.6	60.1
56½	51.3	52.6	53.9	55.4	56.8	58.3	59.8	61.4
57	52.2	53.6	55.0	56.5	58.0	59.6	61.2	62.9
57½	53.2	54.7	56.1	57.7	59.2	60.7	62.6	65.0
58	54.3	55.8	57.4	58.9	60.6	62.4	64.2	66.2
58½	55.5	57.0	58.6	60.3	62.1	64.0	66.0	68.1
59	56.7	58.3	60.0	61.8	63.7	65.7	67.9	70.3
59½	58.0	59.7	61.5	63.4	65.5	67.7	70.1	72.8
60	59.4	61.2	63.2	65.2	67.4	69.9	72.6	75.8
60½	61.0	62.9	65.0	67.2	69.7	72.4	75.7	79.9
61	62.6	64.7	67.0	69.4	72.3	75.5	79.8	90.0
61½	64.5	66.7	69.3	72.1	75.4	79.7	90.0	—
62	66.2	69.0	71.9	75.2	79.6	90.0	—	—
62½	68.8	71.7	75.1	79.5	90.0	—	—	—

Amplitude Corrections

Lat.	Declination					
	0°	5°	10°	15°	20°	25°
0	0.0	0.0	0.0	0.0	0.0	0.0
5	0.1	0.1	0.1	0.1	0.1	0.1
10	0.1	0.1	0.1	0.1	0.1	0.1
15	0.2	0.2	0.2	0.2	0.2	0.2
20	0.2	0.2	0.2	0.2	0.2	0.3
25	0.3	0.3	0.3	0.3	0.3	0.3
30	0.4	0.4	0.4	0.4	0.4	0.4
35	0.4	0.5	0.5	0.5	0.5	0.5
40	0.5	0.6	0.6	0.6	0.6	0.7
42	0.6	0.6	0.6	0.6	0.7	0.7
44	0.6	0.6	0.7	0.7	0.7	0.7
46	0.7	0.7	0.7	0.7	0.8	0.8
48	0.7	0.8	0.8	0.8	0.9	0.9
50	0.8	0.8	0.8	0.9	0.9	1.0
52	0.8	0.9	0.9	0.9	1.0	1.1
54	0.9	0.9	1.0	1.0	1.1	1.3
56	0.9	0.9	1.0	1.0	1.2	1.5
58	1.0	1.0	1.1	1.2	1.3	1.7
60	1.1	1.2	1.2	1.3	1.5	2.1
62	1.2	1.2	1.3	1.4	1.8	2.9

COMPASS ERROR BY AMPLITUDE

The true amplitudes given in the main table are calculated for the instant when the true altitude of the body is precisely 0° 00′. In the case of the sun (owing to the effects of dip, refraction and parallax) the lower limb at this instant will appear to be approximately half a diameter above the visible horizon. If the compass bearing is taken at that moment there will be no need to apply any correction.

However, should the bearing be observed when the sun's centre appears to be in the visible horizon, the correction obtained from the subsidiary table should be applied by being added to the observed azimuth reckoned from the elevated pole as shown in the example below. (Lat. 62° N., decl. 20° S.).

Obs'd. Azi.	S. 41°.5 E.
From elev. pole	N. 138°.5 E.
Corr'n.	+ 1°.8
Sum	N. 140°.3 E.
Corr'd. obs'd. Amp.	E. 50°.3 S.
T. Amp. from table	E. 46°.8 S.
Comp. Error	3°.5 W.

Observations of rising or setting stars and planets are seldom practicable but, if obtained, should be treated in the same way as those of the sun's centre.

In the case of the moon that body will be approximately one-third of a degree below the horizon at the moment when its true altitude is 0° 00′. If observed when its centre appears in the visible horizon, two-thirds of the correction from the subsidiary table should be subtracted from the observed azimuth reckoned from the elevated pole.

EQUIVALENTS of THERMOMETER SCALES

TABLE FOR CONVERTING TEMPERATURE READINGS ON THE FAHRENHEIT AND CENTIGRADE SCALES TO THE ABSOLUTE SCALE

Fahr.	Cent.	Abs.	Fahr.	Cent.	Abs.	Fahr.	Cent.	Abs.	Fahr.	Cent.	Abs.
−99	−72.8	200.2	−44	−42.2	230.8	+11	−11.7	261.3	+66	+18.9	291.9
98	72.2	200.8	43	41.7	231.3	12	11.1	261.9	67	19.4	292.4
97	71.7	201.3	42	41.1	231.9	13	10.6	262.4	68	20.0	293.0
96	71.1	201.9	41	40.6	232.4	14	10.0	263.0	69	20.6	293.6
95	70.6	202.4	40	40.0	233.0	15	9.4	263.6	70	21.1	294.1
94	70.0	203.0	39	39.4	233.6	16	8.9	264.1	71	21.7	294.7
93	69.4	203.6	38	38.9	234.1	17	8.3	264.7	72	22.2	295.2
92	68.9	204.1	37	38.3	234.7	18	7.8	265.2	73	22.8	295.8
91	68.3	204.7	36	37.8	235.2	19	7.2	265.8	74	23.3	296.3
90	67.8	205.2	35	37.2	235.8	20	6.7	266.3	75	23.9	296.9
89	67.2	205.8	34	36.7	236.3	21	6.1	266.9	76	24.4	297.4
88	66.7	206.3	33	36.1	236.9	22	5.6	267.4	77	25.0	298.0
87	66.1	206.9	32	35.6	237.4	23	5.0	268.0	78	25.6	298.6
86	65.6	207.4	31	35.0	238.0	24	4.4	268.6	79	26.1	299.1
85	65.0	208.0	30	34.4	238.6	25	3.9	269.1	80	26.7	299.7
84	64.4	208.6	29	33.9	239.1	26	3.3	269.7	81	27.2	300.2
83	63.9	209.1	28	33.3	239.7	27	2.8	270.2	82	27.8	300.8
82	63.3	209.7	27	32.8	240.2	28	2.2	270.8	83	28.3	301.3
81	62.8	210.2	26	32.2	240.8	29	1.7	271.3	84	28.9	301.9
80	62.2	210.8	25	31.7	241.3	30	1.1	271.9	85	29.4	302.4
79	61.7	211.3	24	31.1	241.9	31	−0.6	272.4	86	30.0	303.0
78	61.1	211.9	23	30.6	242.4	32	0.0	273.0	87	30.6	303.6
77	60.6	212.4	22	30.0	243.0	33	+0.6	273.6	88	31.1	304.1
76	60.0	213.0	21	29.4	243.6	34	1.1	274.1	89	31.7	304.7
75	59.4	213.6	20	28.9	244.1	35	1.7	274.7	90	32.2	305.2
74	58.9	214.1	19	28.3	244.7	36	2.2	275.2	91	32.8	305.8
73	58.3	214.7	18	27.8	245.2	37	2.8	275.8	92	33.3	306.3
72	57.8	215.2	17	27.2	245.8	38	3.3	276.3	93	33.9	306.9
71	57.2	215.8	16	26.7	246.3	39	3.9	276.9	94	34.4	307.4
70	56.7	216.3	15	26.1	246.9	40	4.4	277.4	95	35.0	308.0
69	56.1	216.9	14	25.6	247.4	41	5.0	278.0	96	35.6	308.6
68	55.6	217.4	13	25.0	248.0	42	5.6	278.6	97	36.1	309.1
67	55.0	218.0	12	24.4	248.6	43	6.1	279.1	98	36.7	309.7
66	54.4	218.6	11	23.9	249.1	44	6.7	279.7	99	37.2	310.2
65	53.9	219.1	10	23.3	249.7	45	7.2	280.2	100	37.8	310.8
64	53.3	219.7	9	22.8	250.2	46	7.8	280.8	101	38.3	311.3
63	52.8	220.2	8	22.2	250.8	47	8.3	281.3	102	38.9	311.9
62	52.2	220.8	7	21.7	251.3	48	8.9	281.9	103	39.4	312.4
61	51.7	221.3	6	21.1	251.9	49	9.4	282.4	104	40.0	313.0
60	51.1	221.9	5	20.6	252.4	50	10.0	283.0	105	40.6	313.6
59	50.6	222.4	4	20.0	253.0	51	10.6	283.6	106	41.1	314.1
58	50.0	223.0	3	19.4	253.6	52	11.1	284.1	107	41.7	314.7
57	49.4	223.6	2	18.9	254.1	53	11.7	284.7	108	42.2	315.2
56	48.9	224.1	−1	18.3	254.7	54	12.2	285.2	109	42.8	315.8
55	48.3	224.7	0	17.8	255.2	55	12.8	285.8	110	43.3	316.3
54	47.8	225.2	+1	17.2	255.8	56	13.3	286.3	111	43.9	316.9
53	47.2	225.8	2	16.7	256.3	57	13.9	286.9	112	44.4	317.4
52	46.7	226.3	3	16.1	256.9	58	14.4	287.4	113	45.0	318.0
51	46.1	226.9	4	15.6	257.4	59	15.0	288.0	114	45.6	318.6
50	45.6	227.4	5	15.0	258.0	60	15.6	288.6	115	46.1	319.1
49	45.0	228.0	6	14.4	258.6	61	16.1	289.1	116	46.7	319.7
48	44.4	228.6	7	13.9	259.1	62	16.7	289.7	117	47.2	320.2
47	43.9	229.1	8	13.3	259.7	63	17.2	290.2	118	47.8	320.8
46	43.3	229.7	9	12.8	260.2	64	17.8	290.8	+119	+48.3	321.3
−45	−42.8	230.2	+10	−12.2	260.8	+65	+18.3	291.3			

BRITISH & METRIC WEIGHTS & MEASURES

Conversion factors	British to Metric		Metric to British	
LENGTH	1 inch	= 25·399972 millimetres	1 millimetre	= 0·03937 inches
	1 foot	= 30·479972 centimetres	1 centimetre	= 0·39370 inches
	1 yard	= 0·91439916 metres	1 metre	=39·3701130 inches
	1 rod	= 5·02919 metres	1 ,,	= 3·2808429 feet
	1 chain	= 20·1168 metres	1 ,,	= 1·0936143 yards
	1 furlong	= 0·2011678 kilometres	1 kilometre	= 0·621372 statute miles
	1 statute mile	= 1·60934252 kilometres	1 ,,	= 0·539612 nautical miles
	1 nautical mile	= 1·85318230 kilometres		
WEIGHT	1 grain	= 0·0647989 grammes	1 gramme	=15·4323487 grains
	1 ounce	= 28·34953 grammes	1 kilogramme	= 2·20462124 pounds
	1 pound	= 0·45359265 kilos.	1 tonne (metric ton)	= 0·98420591 tons
	1 stone	= 6·350296 kilogrammes		
	1 hundredweight	= 50·8023772 kilos.		
	1 ton	=1016·04754 kilogrammes		
CAPACITY	1 cubic inch	= 16·38702 cubic cms.	1 cubic centimetre	= 0·0610239 cu. inches
	1 cubic foot	= 28·31677 cu. decimetres	1 cubic metre	=35·31476 cubic feet
	1 cubic yard	= 0·7645527 cu. metres	1 ,, ,,	= 1·307954 cubic yards
	1 pint	= 0·5682454 litres	1 litre	=61·02562 cubic inches
	1 quart	= 1·136490 litres	1 ,,	= 1·759803 pints
	1 gallon	= 4·545963 litres		

BRITISH & METRIC WEIGHTS & MEASURES

Tables for converting British quantities into their Metric equivalents, & vice versa

Units	Inches to Milli-metres	Feet to Metres	Yards to Metres	Fathoms to Metres	Ounces to Grammes	Pounds to Kilo-grammes	Hundred-weights to Kilo-grammes	Tons to Kilo-grammes	Tons to Metric Tons
1	25·4	0·3047997	0·9143992	1·8287983	28·3495	0·453593	50·80238	1016·0475	1·0160475
2	50·8	0·6095994	1·8287983	3·6575966	56·6991	0·907185	101·60475	2032·0951	2·0320951
3	76·2	0·9143992	2·7431975	5·4863950	85·0486	1·360778	152·40713	3048·1426	3·0481426
4	101·6	1·2191989	3·6575966	7·3151933	113·3982	1·814371	203·20951	4064·1902	4·0641902
5	127·0	1·5239986	4·5719958	9·1439916	141·7477	2·267963	254·01189	5080·2377	5·0802377
6	152·4	1·8287983	5·4863950	10·9727899	170·0972	2·721556	304·81426	6096·2852	6·0962852
7	177·8	2·1335980	6·4007941	12·8015882	198·4468	3·175149	355·61664	7112·3328	7·1123328
8	203·2	2·4383978	7·3151933	14·6303866	226·7963	3·628741	406·41902	8128·3803	8·1283803
9	228·6	2·7431975	8·2295924	16·4591849	255·1459	4·082334	457·22139	9144·4279	9·1444279
10	254·0	3·0479972	9·1439916	18·2879832	283·4954	4·535926	508·02377	10160·4754	10·1604754

Units	Cubic Ins. to Cu. Centimetres	Fluid Ozs. to Cu. Centimetres	Pints to Litres	Gallons to Litres	Lbs. perSq. In. to Kgs. per Sq.Cm.	Centimetres to Inches	Metres to Feet	Metres to Yards	Metres to Fathoms
1	16·38702	28·413	0·568245	4·54596	0·07031	0·3937011	3·280843	1·093614	0·54681
2	32·77404	56·826	1·136491	9·09193	0·14061	0·7874023	6·561686	2·187229	1·09361
3	49·16106	85·239	1·704736	13·63789	0·21092	1·1811034	9·842529	3·280843	1·64042
4	65·54808	113·652	2·272982	18·18385	0·28123	1·5748045	13·123372	4·374457	2·18723
5	81·93510	142·065	2·841227	22·72982	0·35154	1·9685057	16·404215	5·468072	2·73404
6	98·32212	170·478	3·409472	27·27578	0·42184	2·3622068	19·685057	6·561686	3·28084
7	114·70914	198·891	3·977718	31·82174	0·49215	2·7559079	22·965900	7·655300	3·82765
8	131·09616	227·304	4·545963	36·36770	0·56246	3·1496090	26·246743	8·748914	4·37446
9	147·48318	255·717	5·114209	40·91367	0·63276	3·5433102	29·527586	9·842529	4·92126
10	163·87020	284·130	5·682454	45·45963	0·70307	3·9370113	32·808429	10·936143	5·46807

Units	Grammes to Grains	Kgs. to Ounces	Kgs. to Pounds	Metric Tons to Tons	Cubic Centimetres to Cu. Ins.	Litres to Fl. Ozs.	Litres to Pints	Litres to Gallons	Kgs. per Sq. Cm. to Lbs. per Sq. In.
1	15·43235	35·27394	2·204621	0·9842059	0·061024	35·1961	1·75980	0·219975	14·22
2	30·86470	70·54788	4·409242	1·9684118	0·122048	70·3921	3·51961	0·439951	28·45
3	46·29705	105·82182	6·613864	2·9526177	0·183072	105·5882	5·27941	0·659926	42·67
4	61·72939	141·09576	8·818485	3·9368236	0·244096	140·7842	7·03921	0·879902	56·89
5	77·16174	176·36970	11·023106	4·9210296	0·305120	175·9803	8·79902	1·099877	71·12
6	92·59409	211·64364	13·227727	5·9052355	0·366143	211·1764	10·55882	1·319852	85·34
7	108·02644	246·91758	15·432349	6·8894414	0·427167	246·3724	12·31862	1·539828	99·56
8	123·45879	282·19152	17·636970	7·8736473	0·488191	281·5685	14·07842	1·759803	113·79
9	138·89114	317·46546	19·841591	8·8578532	0·549215	316·7646	15·83823	1·979779	128·01
10	154·32349	352·73940	22·046212	9·8420591	0·610239	351·9606	17·59803	2·199754	142·23

For comparative values of Statute Miles, Nautical Miles & Kilometres—See page 564

TANK TONNAGE TABLE.

For Calculating Quantity of Liquid in Tons.

RULE.—To the log of the number of cubic feet add the specific gravity constant and the resultant log will give the number of tons.

S.G.	Constant.	S.G.	Constant.	S.G.	Constant.	S.G.	Constant.	S.G.	Constant.	S.G.	Constant.
·600	8·222242	·667	8·268217	·734	8·309787	·800	8·347181	·867	8·382110	·934	8·414438
·601	8·222965	·668	8·268867	·735	8·310378	·801	8·347723	·868	8·382611	·935	8·414903
·602	8·223687	·669	8·269517	·736	8·310969	·802	8·348265	·869	8·383111	·936	8·415367
·603	8·224408	·670	8·270166	·737	8·311558	·803	8·348806	·870	8·383610	·937	8·415831
·604	8·225128	·671	8·270814	·738	8·312147	·804	8·349347	·871	8·384109	·938	8·416294
·605	8·225846	·672	8·271460	·739	8·312735	·805	8·349887	·872	8·384607	·939	8·416757
·606	8·226564	·673	8·272106	·740	8·313323	·806	8·350426	·873	8·385105	·940	8·417219
·607	8·227280	·674	8·272751	·741	8·313909	·807	8·350964	·874	8·385602	·941	8·417681
·608	8·227995	·675	8·273395	·742	8·314495	·808	8·351502	·875	8·386099	·942	8·418142
·609	8·228708	·676	8·274038	·743	8·315080	·809	8·352039	·876	8·386595	·943	8·418603
·610	8·229421	·677	8·274680	·744	8·315664	·810	8·352576	·877	8·387091	·944	8·419063
·611	8·230132	·678	8·275321	·745	8·316247	·811	8·353112	·878	8·387585	·945	8·419523
·612	8·230842	·679	8·275961	·746	8·316830	·812	8·353647	·879	8·388080	·946	8·419982
·613	8·231551	·680	8·276600	·747	8·317412	·813	8·354181	·880	8·388574	·947	8·420441
·614	8·232259	·681	8·277238	·748	8·317993	·814	8·354715	·881	8·389067	·948	8·420899
·615	8·232966	·682	8·277875	·749	8·318573	·815	8·355249	·882	8·389560	·949	8·421357
·616	8·233672	·683	8·278512	·750	8·319152	·816	8·355781	·883	8·390052	·950	8·421815
·617	8·234376	·684	8·279147	·751	8·319731	·817	8·356313	·884	8·390543	·951	8·422271
·618	8·235079	·685	8·279782	·752	8·320309	·818	8·356844	·885	8·391034	·952	8·422728
·619	8·235782	·686	8·280415	·753	8·320886	·819	8·357375	·886	8·391525	·953	8·423184
·620	8·236483	·687	8·281048	·754	8·321462	·820	8·357905	·887	8·392015	·954	8·423639
·621	8·237183	·688	8·281679	·755	8·322038	·821	8·358434	·888	8·392504	·955	8·424094
·622	8·237881	·689	8·282310	·756	8·322613	·822	8·358963	·889	8·392993	·956	8·424549
·623	8·238579	·690	8·282940	·757	8·323187	·823	8·359491	·890	8·393481	·957	8·425003
·624	8·239276	·691	8·283569	·758	8·323760	·824	8·360018	·891	8·393969	·958	8·425456
·625	8·239971	·692	8·284197	·759	8·324333	·825	8·360545	·892	8·394456	·959	8·425910
·626	8·240665	·693	8·284824	·760	8·324905	·826	8·361071	·893	8·394942	·960	8·426362
·627	8·241359	·694	8·285450	·761	8·325476	·827	8·361596	·894	8·395428	·961	8·426814
·628	8·242051	·695	8·286076	·762	8·326046	·828	8·362121	·895	8·395914	·962	8·427266
·629	8·242742	·696	8·286700	·763	8·326615	·829	8·362646	·896	8·396399	·963	8·427717
·630	8·243432	·697	8·287324	·764	8·327184	·830	8·363169	·897	8·396883	·964	8·428168
·631	8·244120	·698	8·287946	·765	8·327752	·831	8·363692	·898	8·397367	·965	8·428618
·632	8·244808	·699	8·288568	·766	8·328320	·832	8·364214	·899	8·397851	·966	8·429068
·633	8·245495	·700	8·289189	·767	8·328886	·833	8·364736	·900	8·398334	·967	8·429517
·634	8·246180	·701	8·289809	·768	8·329452	·834	8·365257	·901	8·398816	·968	8·429966
·635	8·246865	·702	8·290428	·769	8·330017	·835	8·365777	·902	8·399297	·969	8·430415
·636	8·247548	·703	8·291046	·770	8·330582	·836	8·366297	·903	8·399779	·970	8·430863
·637	8·248230	·704	8·291664	·771	8·331145	·837	8·366816	·904	8·400259	·971	8·431310
·638	8·248912	·705	8·292280	·772	8·331708	·838	8·367335	·905	8·400740	·972	8·431757
·639	8·249592	·706	8·292896	·773	8·332270	·839	8·367853	·906	8·401219	·973	8·432204
·640	8·250271	·707	8·293510	·774	8·332832	·840	8·368370	·907	8·401698	·974	8·432650
·641	8·250949	·708	8·294124	·775	8·333393	·841	8·368887	·908	8·402177	·975	8·433096
·642	8·251626	·709	8·294737	·776	8·333953	·842	8·369403	·909	8·402655	·976	8·433541
·643	8·252302	·710	8·295349	·777	8·334512	·843	8·369919	·910	8·403132	·977	8·433986
·644	8·252977	·711	8·295961	·778	8·335071	·844	8·370433	·911	8·403609	·978	8·434430
·645	8·253651	·712	8·296571	·779	8·335628	·845	8·370948	·912	8·404086	·979	8·434874
·646	8·254324	·713	8·297181	·780	8·336186	·846	8·371461	·913	8·404562	·980	8·435317
·647	8·254995	·714	8·297789	·781	8·336742	·847	8·371974	·914	8·405037	·981	8·435760
·648	8·255666	·715	8·298397	·782	8·337298	·848	8·372487	·915	8·405512	·982	8·436202
·649	8·256336	·716	8·299004	·783	8·337853	·849	8·372999	·916	8·405986	·983	8·436644
·650	8·257004	·717	8·299610	·784	8·338407	·850	8·373510	·917	8·406460	·984	8·437086
·651	8·257672	·718	8·300215	·785	8·338961	·851	8·374021	·918	8·406934	·985	8·437527
·652	8·258339	·719	8·300820	·786	8·339514	·852	8·374531	·919	8·407406	·986	8·437968
·653	8·259004	·720	8·301423	·787	8·340066	·853	8·375040	·920	8·407879	·987	8·438408
·654	8·259669	·721	8·302026	·788	8·340617	·854	8·375549	·921	8·408351	·988	8·438848
·655	8·260332	·722	8·302628	·789	8·341168	·855	8·376057	·922	8·408822	·989	8·439287
·656	8·260995	·723	8·303229	·790	8·341718	·856	8·376565	·923	8·409293	·990	8·439726
·657	8·261656	·724	8·303830	·791	8·342267	·857	8·377072	·924	8·409763	·991	8·440166
·658	8·262317	·725	8·304429	·792	8·342816	·858	8·377578	·925	8·410233	·992	8·440603
·659	8·262976	·726	8·305028	·793	8·343364	·859	8·378084	·926	8·410702	·993	8·441040
·660	8·263635	·727	8·305625	·794	8·343911	·860	8·378589	·927	8·411171	·994	8·441477
·661	8·264292	·728	8·306222	·795	8·344458	·861	8·379094	·928	8·411639	·995	8·441914
·662	8·264949	·729	8·306819	·796	8·345004	·862	8·379598	·929	8·412107	·996	8·442350
·663	8·265605	·730	8·307414	·797	8·345549	·863	8·380102	·930	8·412574	·997	8·442786
·664	8·266259	·731	8·308008	·798	8·346094	·864	8·380605	·931	8·413041	·998	8·443221
·665	8·266913	·732	8·308602	·799	8·346638	·865	8·381107	·932	8·413507	·999	8·443656
·666	8·267565	·733	8·309195	·800	8·347181	·866	8·381609	·933	8·413973	1·00	8·444091

IMPERIAL GALLONS, U.S.A. GALLONS, LITRES.

Imperial Gallons	U.S.A. Gallons	Litres	U.S.A. Gallons	Imperial Gallons	Litres	Litres	Imperial Gallons	U.S.A. Gallons
1	1·20	4·55	1	0·83	3·79	1	0·22	0·26
2	2·40	9·09	2	1·67	7·57	2	0·44	0·53
3	3·60	13·64	3	2·50	11·36	3	0·66	0·79
4	4·80	18·18	4	3·33	15·14	4	0·88	1·06
5	6·00	22·73	5	4·16	18·93	5	1·10	1·32
6	7·21	27·28	6	5·00	22·71	6	1·32	1·59
7	8·41	31·82	7	5·83	26·50	7	1·54	1·85
8	9·61	36·37	8	6·66	30·28	8	1·76	2·11
9	10·81	40·91	9	7·49	34·07	9	1·98	2·38
10	12·01	45·46	10	8·33	37·85	10	2·20	2·64
11	13·20	50·01	11	9·16	41·64	11	2·42	2·91
12	14·41	54·55	12	9·99	45·42	12	2·64	3·17
13	15·61	59·10	13	10·82	49·21	13	2·86	3·43
14	16·81	63·64	14	11·66	52·99	14	3·08	3·70
15	18·01	68·19	15	12·49	56·78	15	3·30	3·96
16	19·22	72·74	16	13·32	60·56	16	3·52	4·23
17	20·42	77·28	17	14·16	64·35	17	3·74	4·49
18	21·62	81·83	18	14·99	68·14	18	3·96	4·76
19	22·82	86·37	19	15·82	71·92	19	4·18	5·02
20	24·02	90·92	20	16·65	75·71	20	4·40	5·28
21	25·22	95·47	21	17·49	79·49	21	4·62	5·55
22	26·42	100·01	22	18·32	83·28	22	4·84	5·81
23	27·62	104·56	23	19·15	87·06	23	5·06	6·08
24	28·82	109·10	24	19·98	90·85	24	5·28	6·34
25	30·02	113·65	25	20·82	94·63	25	5·50	6·60
26	31·22	118·19	26	21·65	98·42	26	5·72	6·87
27	32·43	122·74	27	22·48	102·20	27	5·94	7·13
28	33·63	127·29	28	23·31	105·99	28	6·16	7·40
29	34·83	131·83	29	24·15	109·77	29	6·38	7·66
30	36·03	136·38	30	24·98	113·56	30	6·60	7·93
31	37·23	140·92	31	25·81	117·34	31	6·82	8·19
32	38·43	145·47	32	26·65	121·13	32	7·04	8·45
33	39·63	150·02	33	27·48	124·91	33	7·26	8·72
34	40·83	154·56	34	28·31	128·70	34	7·48	8·98
35	42·03	159·11	35	29·14	132·49	35	7·70	9·25
36	43·23	163·65	36	29·98	136·27	36	7·92	9·51
37	44·44	168·20	37	30·81	140·06	37	8·14	9·77
38	45·64	172·75	38	31·64	143·84	38	8·36	10·04
39	46·84	177·29	39	32·47	147·63	39	8·58	10·30
40	48·04	181·84	40	33·31	151·41	40	8·80	10·57
41	49·24	186·38	41	34·14	155·20	41	9·02	10·83
42	50·44	190·93	42	34·97	158·98	42	9·24	11·10
43	51·64	195·48	43	35·80	162·77	43	9·46	11·36
44	52·84	200·02	44	36·64	166·55	44	9·68	11·62
45	54·04	204·57	45	37·47	170·34	45	9·90	11·89
46	55·24	209·11	46	38·30	174·12	46	10·12	12·15
47	56·44	213·66	47	39·14	177·91	47	10·34	12·42
48	57·65	218·21	48	39·97	181·69	48	10·56	12·68
49	58·85	222·75	49	40·80	185·48	49	10·78	12·94
50	60·05	227·30	50	41·63	189·26	50	11·00	13·21
60	72·06	272·76	60	49·96	227·12	60	13·20	15·85
70	84·07	318·22	70	58·29	264·97	70	15·40	18·49
80	96·08	363·68	80	66·61	302·82	80	17·60	21·13
90	108·09	409·14	90	74·94	340·68	90	19·80	23·78
100	120·10	454·60	100	83·27	378·53	100	22·00	26·42
200	240·19	909·19	200	166·53	757·06	200	43·99	52·84
300	360·29	1363·79	300	249·80	1135·59	300	65·99	79·25
400	480·38	1818·38	400	333·07	1514·11	400	87·99	105·67
500	600·48	2272·98	500	416·34	1892·64	500	109·99	132·09
600	720·57	2727·58	600	499·60	2271·17	600	131·98	158·51
700	840·67	3182·17	700	582·87	2649·70	700	153·98	184·93
800	960·76	3636·77	800	666·14	3028·23	800	175·98	211·34
900	1080·86	4091·36	900	749·40	3406·76	900	197·97	237·76
1000	1200·95	4545·96	1000	832·67	3785·29	1000	219·97	264·18

NAUTICAL MILES, STATUTE MILES, KILOMETRES.

Nautical Miles	Statute Miles	Kilo-Metres	Statute Miles	Nautical Miles	Kilo-Metres	Kilo-Metres	Nautical Miles	Statute Miles
1	1·15	1·85	1	0·87	1·61	1	0·54	0·62
2	2·30	3·71	2	1·74	3·22	2	1·08	1·24
3	3·45	5·56	3	2·61	4·83	3	1·62	1·86
4	4·61	7·41	4	3·47	6·44	4	2·16	2·49
5	5·76	9·27	5	4·34	8·05	5	2·70	3·11
6	6·91	11·12	6	5·21	9·66	6	3·24	3·73
7	8·06	12·97	7	6·08	11·27	7	3·78	4·35
8	9·21	14·83	8	6·95	12·87	8	4·32	4·97
9	10·36	16·68	9	7·82	14·48	9	4·86	5·59
10	11·52	18·53	10	8·68	16·09	10	5·40	6·21
11	12·67	20·39	11	9·55	17·70	11	5·94	6·84
12	13·82	22·24	12	10·42	19·31	12	6·48	7·46
13	14·97	24·09	13	11·29	20·92	13	7·01	8·08
14	16·12	25·94	14	12·16	22·53	14	7·55	8·70
15	17·27	27·80	15	13·03	24·14	15	8·09	9·32
16	18·42	29·65	16	13·89	25·75	16	8·63	9·94
17	19·58	31·50	17	14·76	27·36	17	9·17	10·56
18	20·73	33·36	18	15·63	28·97	18	9·71	11·19
19	21·88	35·21	19	16·50	30·58	19	10·25	11·81
20	23·03	37·06	20	17·37	32·19	20	10·79	12·43
21	24·18	38·92	21	18·24	33·80	21	11·33	13·05
22	25·33	40·77	22	19·10	35·40	22	11·87	13·67
23	26·48	42·62	23	19·97	37·01	23	12·41	14·29
24	27·64	44·48	24	20·84	38·62	24	12·95	14·91
25	28·79	46·33	25	21·71	40·23	25	13·49	15·53
26	29·94	48·18	26	22·58	41·84	26	14·03	16·16
27	31·09	50·04	27	23·45	43·45	27	14·57	16·78
28	32·24	51·89	28	24·32	45·06	28	15·11	17·40
29	33·39	53·74	29	25·18	46·67	29	15·65	18·02
30	34·55	55·60	30	26·05	48·28	30	16·19	18·64
31	35·70	57·45	31	26·92	49·89	31	16·73	19·26
32	36·85	59·30	32	27·79	51·50	32	17·27	19·88
33	38·00	61·16	33	28·66	53·11	33	17·81	20·51
34	39·15	63·01	34	29·53	54·72	34	18·35	21·13
35	40·30	64·86	35	30·39	56·33	35	18·89	21·75
36	41·45	66·71	36	31·26	57·93	36	19·43	22·37
37	42·61	68·57	37	32·13	59·54	37	19·97	22·99
38	43·76	70·42	38	33·00	61·15	38	20·50	23·61
39	44·91	72·27	39	33·87	62·76	39	21·04	24·23
40	46·06	74·13	40	34·74	64·37	40	21·58	24·85
41	47·21	75·98	41	35·60	65·98	41	22·12	25·48
42	48·36	77·83	42	36·47	67·59	42	22·66	26·10
43	49·51	79·69	43	37·34	69·20	43	23·20	26·72
44	50·67	81·54	44	38·21	70·81	44	23·74	27·34
45	51·82	83·39	45	39·08	72·42	45	24·28	27·96
46	52·97	85·25	46	39·95	74·03	46	24·82	28·58
47	54·12	87·10	47	40·81	75·64	47	25·36	29·21
48	55·27	88·95	48	41·68	77·25	48	25·90	29·83
49	56·42	90·81	49	42·55	78·86	49	26·44	30·45
50	57·58	92·66	50	43·42	80·47	50	26·98	31·07
60	69·09	111·19	60	52·11	96·56	60	32·38	37·28
70	80·61	129·72	70	60·79	112·65	70	37·77	43·50
80	92·12	148·25	80	69·47	128·75	80	43·17	49·71
90	103·64	166·79	90	78·16	144·84	90	48·57	55·92
100	115·15	185·32	100	86·84	160·93	100	53·96	62·14
200	230·30	370·64	200	173·68	321·87	200	107·92	124·28
300	345·45	555·95	300	260·53	482·80	300	161·88	186·41
400	460·61	741·27	400	347·37	643·74	400	215·84	248·55
500	575·76	926·59	500	434·21	804·67	500	269·81	310·69
600	690·91	1111·91	600	521·05	965·61	600	323·77	372·83
700	806·06	1297·23	700	607·89	1126·54	700	377·73	434·96
800	921·21	1482·55	800	694·74	1287·47	800	431·69	497·10
900	1036·36	1667·86	900	781·58	1448·41	900	485·65	559·23
1000	1151·52	1853·18	1000	868·42	1609·34	1000	539·61	621·37

PORTS OF THE WORLD

PORTS OF THE WORLD
Alphabetical List
with
INDEX

Name.	Index No.	Name.	Index No.	Name.	Index No.
Brunei	1476	Canakkale (*See Chanak*)		Chalna	1382
Brunsbuttel	699	Canay	1500	Champerico	2118
Brunswick	2580	Candia	1060	Chanak	989
Brussels	753	Candle	1936	Chanaral	2184
Buchupureo	2206	Canea	1057	Chancay	2160
Buckie	230	Cannanore	1351	Chandarly	1038
Bucksport	2650	Cannes	887	Charleston	2053
Buctouche	2734	Cantley	283	Charlestown, *England*	57
Bude	67	Canton	1554	Charlestown, *Nevis*	2557
Budrum (*See Bodrum*)		Canvey Island	13	Charlotte (*See Skidegate*)	
Buenaventuro	2134	Cape Coast	1166	Chartlottetown	2823
Buenos Aires	2246	Cape Cove	2753	Chatham, *Cape Cod*	2639
Buffalo	2787	Capelle a/d Yssel	735	Chatham, *England*	18
Buff Bay	2493	Cape North, Cape Breton Id.		Chefoo	1583
Bulangan	1486	(*Near Newhaven*)		Chemainus	2015
Buleleng (*See Beliling*)		Cape Palmas	1160	Chemulpo (*See Inchon*)	
Bullen Bay	2361	Capetown	1224	Cherbourg	779
Bull's River	2583	Cap Haitien	2510	Chesme	1041
Bunbury	1862	Carabelle	2450	Chester, *Nova Scotia*	2699
Buncrana	210	Caraga	1498	Chesterfield	2870
Bundaberg	1810	Caragues	2137	Chetican	2716
Burea	494	Caraquet (*See Carraquette*)		Chiba	1657
Burgas	995	Caravelles	2294	Chicago	2810
Burnham-on-Crouch	293	Carbonear	2833	Chicoutimi	2768
Burketown	1878	Cardenas	2460	Chignik	1948
Burnie	1888	Cardiff	79	Chiltepec	2408
Burntisland	244	Cardigan	86	Chimbote	2153
Burton Port	207	Cardwell	1800	Chinde	1237
Barriana	873	Carentan	777	Chinkiang	1572
Burutu	1181	Caripito	2346	Chinnanpo	1592
Bushire	1324	Carleton, *N.B.*	2263	Chinwangtao	1587
Busselton	1861	Carleton, *Quebec*	2745	Chioggia	947
Butaritari Makin	1760	Carloforte	900	Chittagong	1383
Butedale	1999	Carlsborg (*See Karlsborg*)		Chorak	1055
Buton	1456	Carmen	2405	Christmas Island,	
Butuan	1503	Carmen de Patagones	2239	E. *Indies*	1425
		Carnarvon, *Australia*	1866	Christmas Island, *Oceania*	1773
Cabadello	2306	Carnarvon, *Wales*	94	Christianshaab	2875
Cabanas	2486	Carraquette	2740	Christiansted	2550
Cabarette	2514	Carrara	907	Chuanchowfu	1558
Cabimas (*Maracaibo Lake port*)		Carrizal Bajo	2186	Chungking	1580
Cachew	1147	Cartagena, *Spain*	860	Churchill	2869
Cadiz	849	Cartagena, *Colombia*	2371	Cienfuegos	2480
Caen	776	Cartwright	2853	Ciudad Bolivar	2342
Cagayan	1505	Carupano	2348	Ciudad del Carmen (*See	
Cagliari	895	Casablanca	1122	Carmen*)	
Cahirciveen	191	Cascumpique	2819	Ciudad Trujillo	2519
Caibarien	2462	Casilda	2479	Civitavechia	912
Caimanera	2471	Casma	2155	Clare Castle	197
Cairn Ryan (*Head of Loch Ryan*)		Castellamare	919	Claremont	2598
Cairns	1798	Castellon	874	Clarke City	2851
Calabar	1187	Castro Alen	819	Clarke Harbour	2691
Calais	764	Castro Urdiales (*See Castro Alen*)		Claveria	1522
Calcutta	1381	Catania	933	Claxton	1989
Caldera, *Chile*	2185	Cavite	1518	Clayoquot	2025
Caldera, *Costa Rica*	2128	Caybarien (*See Caibarien*)		Clementsport	2683
Caleta Buena d'Sur	2175	Cayenne	2328	Cleveland	2790
Caleta Colosa	2182	Cayo Mambi (*Near Sagua*)		Clifden	200
Calicut	1355	Cebu	1509	Cliffe	8
Callao	2161	Cedar Keys	2452	Clode Sound	2830
Calvi	892	Celestun	2401	Clonakilty	186
Camana	2169	Cerro de Azul	2164	Clo-oose	2022
Camocim	2311	Cette	881	Clovelly	68
Campana	2247	Ceuta	1115	Clydebank	150
Campbell River	2010	Chagres	2375	Coal Harbour	1946
Campbellton	2744	Chahbar	1329	Coatzacoalcos	2411
Campbeltown	142	Chala	2168	Cobh	182
Campeche	2402	Chalkis	981	Cobija	2178

PORTS OF THE WORLD

Name.	Index No.	Name.	Index No.	Name.	Index No.
Cocagne	2733	Dalsbruk	559	Dulcigno	959
Cochin	1359	Daman	1346	Duluth	2815
Coconada (*See Kakinada*)		Dammam	1311	Dumaquete	1508
Codrington	2554	Danmarks Havn	2892	Dumbarton	149
Coff's Harbour	1821a	Dar-es-Salaam	1277	Duncannon	177
Colastine	2260	Darien (*See Sapelo*)		Dundalk	168
Colchester	290	Darlowo	598	Dundee	239
Coleraine	213	Dartmouth, *England*	51	Dunedin	1924
Collingwood	1935	Dartmouth, *N. Scotia*	2701	Dungarvan	180
Collo	1100	Darwin	1875	Dunglass	156
Colombo	1364	Davao	1497	Dunkirk	762
Colon	2374	Davisville (*Near Toronto*)		Dunlaoghaire	171
Colonia	2272	Dawes Island (*Near Lagos,*		Dunoon	145
Comeau Bay	2849	*Nigeria*)		Durazzo	960
Comox	2011	Deauville	775	Durban	1233
Conakry	1151	Dedeagach	986	Dutch Harbour	1944
Concarneau	794	Deering	1937	Dwarka	1340
Concepcion, *Chile*	2209	Degerhamn	412	Dzaudzi (*See Mayotte Island*)	
Concepcion del Uruguay	2274	Delft	739		
Condon	1870	Delfzyl	716	Easter Island	1774
Connah's Quay	100	Deli, *Timor*	1445	Eastham	102
Constanza	997	Dellys	1099	East London	1231
Coos Bay	2071	Demarara (*See Georgetown,*		East Main Fort	2862
Copenhagen	628	*Br. Guiana*)		Easton	2609
Coquimbo	2191	Den Helder	724	Eastport	2657
Corcubion	829	Denia	865	Ebeltoft	667
Cordova	1960	Derby	1872	Eccles	113
Corfu	962	Derince	1028	Eckerneforde	614
Corinth	970	Derna	1045	Eckero	556
Corinto	2125	Detroit	1795	Eda	1671
Cork	184	Devonport, *England*	54	Eden	1829
Corner Brook	2846	Devonport, *Tasmania*	1889	Edenton	2593
Cornwall	2778	Diamante	2258	Edgartown	2637
Cornwallis	1911	Diamond Harbour	1380	Egedsminde	2876
Coronel	2210	Diego Garcia	1263	Egersund	351
Corosal	2394	Diego Suarez	1242	Ekenas	561
Corpus Christi	2419	Dielette	780	El Akaba (*See Aqaba*)	
Corral	2215	Dieppe	768	El Arrish (*See Larache*)	
Corrientes	2268	Digby	2684	Elbing	593
Corunna	828	Dingwall, Cape Breton Id.		El Bluff (*See Bluefields*)	
Coryton	11	(*Near Newhaven*)		Eleusis (*Near Corinth*)	
Covenas	2372	Djakarta	1429	Eling	40
Coverack	59	Djibouti	1288	Elizabeth	2594
Cowes	36	Djupivogur	2902	Ellesmere Port	103
Cowichan Harbour	2017	Djupvik	486	Ellington	1746
Cox's Bazaar	1384	Docksta	475	Ellsworth	2652
Crab Island (*See Porto Mulas*)		Doha	1308	Elsfleth	708
Crapaud	2822	Dokkum (*Near Leeuwarden*)		Elsinore (*See Helsingor*)	
Crescent City	2074	Domsjo (*Near Ornskoldsvik*)		El Wej	1299
Crofton	2016	Donaghadee	162	Emden	713
Crotone (*Toe of Italy*)		Donges	798	Empedrado	2267
Cruz Grande	2189	Donggala	1461	Enanger	452
Cuddalore	1369	Dordrecht	743	English Harbour	2559
Cullera	867	Doreh	1722	Eregli	1026
Cumana	2350	Douarnenez	790	Erie	2789
Cumarebo	2358	Douglas	128	Erith	3
Curacao (*See Willemstad*)		Douglastown	2756	Esashi	1709
Curanipe	2205	Dover	29	Esbjerg	691
Cutaco (*See Cutuco*)		Drammen	374	Escombreras Harbour	861
Cutuco	2122	Drewin (*Near Sassandra*)		Eskifjordur	2901
Cuxhaven	701	Drobak	376	Esmeraldas	2136
		Drogheda	169	Esperance Bay	1859
Dadens	1978	Duala	1190	Esquimalt	2019
Dagami	1511	Dubai	1306	Esquina	2264
Dairen	1590	Dublin	170	Essviken	453
Daiquiri	2472	Dubrovnic	956	Esteros Bay	2090
Dakar	1142	Dugirat (*Near Split*)		Eten	2417
Dalhousie	2743	Duisburg (*Rhine port*)		Eucla	1858

PORTS OF THE WORLD

Name.	Index No.	Name.	Index No.	Name.	Index No.
Grimsby	277	Hankmeri	542	Hoyanger	339
Grimstad	359	Harlingen	718	Hoyer	693
Groningen	717	Harrang	435	Hsinkong (See Tongku New	
Groton	2627	Harstad	320	Harbour)	
Grouw (Near Harlingen)		Harta Point	1319	Huanillo	2179
Grundvik (See Soderhamn)		Hartlepool	260	Huarmei	2156
Guam	1766	Harvey	2667	Huasco	2187
Guanape Islands	2152	Harwich	288	Hudiksvall	457
Guane	2484	Haugesund	345	Hue	1544
Guanica	2547	Haukipudas	519	Huelva	847
Guanta	2351	Havana	2458	Hukow	1576
Guantanamo Bay	2470a	Havre (See Le Havre)		Hull	268
Guayacan	2192	Havre St. Pierre	2848	Hundested	632b
Guayama	2542	Hawk's Nest Anchorage	2509	Hunnesbostrand	382
Guayanilla	2546	Hayle	65	Husavik	2896
Guayaquil	2140	Heart's Content	2832	Husum	694
Guaymas	2106	Heilingenhafen	611	Hythe	39
Guiria	2347	Helgenas	419		
Gulfport	2442	Hellville	1241		
Gullsmedvik	324	Helmsdale	226	Ibicuy	2249
Gumboda	491	Helsingborg	399	I-Chang	1579
Guysboro	2704	Helsingor	630	Ifni	1133
Gwadar	1331	Helsinki	564	Igarka	298
Gwattar	1330	Helvoet	736	Iggesund	455
Gweek	61	Hermopolis (See Syra)		Ileus	2297
Gythion	975	Hernosand	466	Ilico	2203
		Heroy	335	Iligan	1507
		Herring Cove	2666	Ilo	2172
Haapai Group	1793	Hesquiat Harbour	2026	Iloilo	1510
Haarlem	727	Heysham	124	Imabaru	1695
Haast	1932	Hillsboro	2668	Imbitiba	2292
Hachinohe	1647	Hilo	1778	Imbituba	2281
Hadersleben	660	Himanko	529	Immingham	276
Haderslev	658	Hirakata	1654	Ince	105
Hadsund	672	Hirohata	1679	Inchon	1593
Hafnarfjordur	2907	Hirtshals	680	Independencia	2273
Haifa	1076	Hjo	391	Indianola	2421
Haiphong	1545	Hobart	1883	Ineboli	1022
Hakata	1618	Hobro	671	Inganish	2713
Hakodate	1704	Hodeida	1303	Inhambane	1235
Halden	380	Hoganas	398	Inverary	140
Halifax	2700	Hoihow	1549	Invercargill	1931
Hallsta	436	Holbak	635	Invergordon	227
Hallstavik (Near Stockholm)		Holehaven	12	Inverness, B.C.	1988
Halmstad	397	Hollandia	1724	Inverness, N. Scotia	2717
Haltenau	615	Holmestrand	373	Inverness, Scotland	228
Hamada	1633	Holstenborg	2877	Ipswich, Australia	1815
Hamble	37	Holyhead	95	Ipswich, England	287
Hamburg	697	Holyrood	2835	Iquique	2176
Hamilton, Bermuda	2570	Honfleur	773	Iquitos	2326
Hamilton, Ontario	2785	Hongay	1546	Iraklion (See Candia)	
Hamina (See Frederikshamn)		Hongkong	1555	Irlam	111
Hammarby (Near Sandviken)		Honningsvaag (N)	314	Irvine	143
Hammerfest	315	Honningsvaag (S)	336	Irwell	116
Hamnholmen	548	Honolulu	1787	Isabela	1495
Hampton Roads	2595	Honuapo	1779	Isabeli	2388
Hanasaki	1714	Hook	270	Isafjordur	2894
Hane	1634	Hoonah	1964	Isigny-Sur-Mer	778
Hango	562	Hopedale	2855	Iskenderun	1070
Hankmo	535	Hornafjordur	2903	Isle of Grain	15
Hankow	1578	Hornbak	631	Ismailia	1079
Hantsport	2678	Hornefors	483	Istanbul	993
Haparanda	514	Horsens	664	Itajahy	2283
Hapsal	585	Horten	372	Itea	968
Harbour Grace	2834	Horton	2679	Itozaki	1683
Harbour Island	2506	Houston	2428	Ivigut	2885
Harbour Springs	2813	Hov	665	Iviza	869
Hargshamn	437	Howth	172	Izmir	1039

PORTS OF THE WORLD

Name.	Index No.	Name.	Index No.	Name.	Index No.
Jacksonville	2577	Karrebaksminde (See Karrebak)		Kobe	1678
Jacmel	2524	Karskar	440	Koche	1700
Jacobshaven	2874	Karumba	1879	Kodiak	1950
Jacobstad	532	Kaska	540	Koge (See Kjoge)	
Jaluit	1761	Kastron	1040	Kogo	1194
Jamlosund	509	Kastrup	627	Kogon River	1150
Jask	1328	Katakolon	973	Koh-Sichang	1537
Jeddah	1301	Katalla	1961	Kohukohu	1914
Jedway	1995	Kavalla	984	Koivoluoto	517
Jerba	1087	Kaveli	1452	Kokkola (See Gamla Karleby)	
Jeremie	2526	Kavieng Harbour	1741	Kokura	1620
Jesselton	1478	Kawanoishi	1692	Kolaka	1466
Jideh	1023	Kawasaki	1660	Kolbergermunde	599
Jijelli	1101	Kawhia	1908	Kolding	661
Jobos	2544	Keadby	272	Kolo	1784
Joggins Wharf	2669	Keelung	1564	Kolobrzeg (See Kolbergermunde)	
Johnston Island	1769	Keflavik	2908	Komatsushima	1699
Jolo (See Sulu)		Keil	612	Konakry (See Conakry)	
Jonesport	2655	Kem	303	Kongsmoen	326a
Jonkoping	392	Kemi	516	Konigsburg	591
Jordan River	2693	Kempen a/d Yssel	722	Kopervik	346
Jose Panganiban	1515	Kerasund (See Gireson)		Koping	428
Juan Fernandez	2200	Kerch	1003	Kopmanholmen	477
Jucaro	2477	Kerguelen Island	1257	Kopparverkshamn (Near	
Julianhaab	2886	Kesennuma	1651	Helsingborg)	
Juneau	1970	Keta	1171	Korlovassi	1044
Juniskar	460	Ketchikan	1983	Korsakovsk	1602
Juutas	533	Key West	2457	Korsnas	538
		Kherson	1000	Korsor	639
Kabinda	1205	Khorramshahr	1320	Kortgene	748
Kagoshima	1605	Kieta	1742	Kos	1046
Kahului	1777	Kildala	2006	Kosseir	1295
Kakinada	1374	Kilia	1027	Kota Baru	1493
Kakkis	576	Kilindini (See Mombasa)		Kota Raja	1409
Kalajoki	528	Killala	202	Kotka	571
Kalama	2062	Killeany Bay	198	Kotlik	1942
Kalamaki Bay	1051	Killingholme	273	Kotonu	1175
Kalamata	974	Killybegs	205	Kouilou (See Pointe Noire)	
Kalamuti Bay	985	Kilmarnock, Virginia	2602	Kovda	307
Kalapan	1517	Kilrush	196	Kovic	2859
Kalingapatam	1378	Kilwa Kivinje	1276	Kozhikode (See Calicut)	
Kaliningrad (See Konigsburg)		Kilwa Masoko	1275	Kragero	364
Kalix	510	Kingscote	1847	Kralendjik	2359
Kallundborg	637	Kingsgate	27	Kramfors	470
Kallviken	492	King's Lynn	280	Kribi	1192
Kalmar	413	Kingsnorth	19	Krik	687
Kalyvia (Near Corinth)		Kingsport	2681	Kristiansand	357
Kamaishi	1650	Kingston, Australia	1840	Kristiansund	331
Kamalo Harbour	1783	Kingston, Jamaica	2496	Kristinehamn	386
Kandalacksha	304	Kingston, New York	2619	Kristinestad	541
Kandla	1337	Kingston, Ontario	2780	Kronstadt	580
Kankesanturai (Palk Strait,		Kingston, St. Vincent	2566	Kuala Dungun	1532
Ceylon)		Kingwa	2860	Kuantan	1531
Kaohsiung	1561	Kinlochleven	159	Kubikenborg	463
Kaolock	1143	Kirkaldy	243	Kuching	1473
Karabane	1145	Kirkehavn	354	Kuchinotsu	1612
Karabogha	1033	Kirkenes	311	Kudat	1479
Karachi	1333	Kirkwall	222	Kuivaniemi	518
Karatsu	1616	Kismayu	1283	Kumai	1469
Kardeljevo (Near Ragusa)		Kitimat	1998	Kum Kale	988
Karenko	1566	Kiukiang	1577	Kunda	582
Karikal	1368	Kjerteminde	651	Kupang	1446
Karlsborg, Gulf of Bothnia	511	Kjoge	626	Kure	1684
Karlsborg, Lake Vetter	390	Klagsvig	2909	Kuruk	1326
Karlshamn	409	Klaipeda	590	Kusadasi (See Scalanova)	
Karlskrona	411	Klintehamn	434	Kusaie	1762
Karlstad	387	Knik	1952	Kushiro	1716
Karrebak	642	Koba	1418	Kuwait	1313

Name.	Index No.	Name.	Index No.	Name.	Index No.
Necochea	2242	North Sydney	2711	Owen Sound	2803
Neder over Heembeeck		Norwich	284	Oxelosund	422
(*Near Brussels*)		Noshiro	1645		
Negapatam	1367	Nossi Be (*See Hellville*)		Pacasmayo	2148
Neguac	2738	Noumea	1745	Pacofi	1993
Nelson	1916	Novo Redondo	1213	Padang	1423
Nemoro	1713	Novorossisk	1007	Pago Pago	1754
Nemours	1111	Nuevitas	2463	Paimboeuf	799
Neuhaus-Oste	700	Nukualofa	1794	Paimpol	784
Neustadt	610	Nuske	476	Paisley	152
New Amsterdam (*See Berbice*)		Nyborg	650	Paita	2144
Newark, *N.J.*	2612	Nykjobing, *Falster*	621	Pakhoi	1548
New Bedford	2634	Nykjobing, *Mors*	683	Paknam	1538
Newbern	2589	Nykjobing, *Zealand*	636	Palamos	878
Newburg	2617	Nykoping, *Sweden*	423	Palapo	1465
Newcastle, *N.B.*	2737	Nyland	472	Palembang	1415
Newcastle, *N.S.W.*	1824	Nynashamn	425	Palermo	924
Newcastle-on-Tyne	256	Nystad	553	Palma	870
Newchwang	1588	Nysted	619	Palmyra Island	1770
New Glasgow	2724	Nyuchotsk	302	Pampatar	2349
Newhaven, C. Breton Id.				Panama	2133
(*N. of Inganish*)		Oakland	2081	Panama City	2447
New Haven, *Conn.*	2623	Oamaru	1923	Panarukan	1437
Newhaven, *England*	31	Obbola	484	Panderma (*See Perama*)	
Newhaven, *N. Scotia*	2714	Obidos	2323	Pangani	1279
Newhaven, *N. Zealand*	1926	Obligado	2252	Pango Pango (*See Pago Pago*)	
New London	2626	Ocean Falls	2003	Panjim	1348
New Mills	2742	Ocean Island	1758	Papeete	1755
New Orleans	2435	Odda	343	Papenburg	715
New Plymouth	1906	Odense	652	Paphos	1066
Newport, *B.C.*	2031	Odessa	990	Papudo	2196
Newport, *I.O.Wight*	35	Oginohama	1653	Par	56
Newport, *Mon.*	78	Old Harbour	2499	Para (*See Belem*)	
Newport, *N.B.*	2750	Old Kilpatrick	154	Parahiba (*See Parahyba*)	
Newport News	2600	Oleh Leh	1408	Parahyba	2305
Newport, *Oregon*	2067	Olhao	844	Paramaraibo	2330
Newport, *Rhode Island*	2629	Olympia	2045	Parana	2259
New Richmond	2746	Omoa	2386	Paranagua	2285
New Ross	179	Omura	1614	Pare Pare	1463
Newry	167	Onega	301	Parga	964
New Westminster	2038	Onehunga	1910	Pargas	555
New York	2613	Oplo (*Near Narvik*)		Parika	2335
Nhatrang	1542	Opobo	1185	Parnahyba	2312
Nicaro (*Near Preston, Cuba*)		Oporto	835	Parnaiba (*See Parnahyba*)	
Nice	888	Opotiki	1898	Paris	772
Nickerie	2332	Opua	1893	Parrsboro	2671
Niigata	1642	Oran	1108	Parry Sound	2801
Niihama (*Near Matsuyama*)		Oranjestad	2362	Partington	110
Nikolaiev	1001	Orange	2432	Partipique	2672
Nikolaievsk	1600	Orange Town	2555	Pasajes	817
Ningpo	1569	Oravais	534	Pascagoula	2444
Nisao	2520	Ordu	1017	Paskallavic	414
Nizampatam	1372	Oregrund	438	Pasni	1332
Nome	1940	Ornskoldsvik	478	Paspebiac	2748
Nomuka	1792	Orviken (*Near Skelleftea*)		Passage West	185
Nonopapa	1789	Osaka	1677	Pass Christian	2441
Nordby	690	Oskarshamn	415	Passir	1492
Nordenham	710	Oslo	375	Pasuruan	1435
Nordfjordur	2900	Ostend	761	Patani, *Halmahera*	1454
Nordmaling	480	Osterby	676	Patani, *Thailand*	1534
Norfolk, *Va.*	2596	Ostermoor (*Keil Canal*)		Patea	1905
Normanton	1880	Ostrand (*Near Sundsvall*)		Pateniemi	520
Norre Sundby	674	Oswego	2781	Patras	971
Norris Arm	2828	Otago Harbour (*See Dunedin*)		Pauillac	810
Norrkoping	421	Otaru	1706	Payo Oosipo	2395
Norrsundet	442	Otterbacken	388	Peekskill	2615
Norrtelje	427	Oulu	522	Peel	129
Northport	2727	Owendo	1196	Pekalongan	1432

PORTS OF THE WORLD
Geographical List
with
LATITUDES and LONGITUDES

Index No.	Name.	Lat.	Long.
	SECTION 1. BRITISH ISLES AND EIRE.		
		° ′	° ′
1	London, *Upper Pool*	51 30 N.	0 05 W.
2	Woolwich . . .	51 29 N.	0 04 E.
3	Erith	51 28 N.	0 06 E.
4	Purfleet . . .	51 29 N.	0 15 E.
5	Greenhithe . .	51 27 N.	0 17 E.
6	Gravesend . .	51 27 N.	0 22 E.
7	Tilbury . . .	51 28 N.	0 22 E.
8	Cliffe	51 28 N.	0 29 E.
9	Thames Haven .	51 32 N.	0 31 E.
10	Shell Haven .	51 32 N.	0 32 E.
11	Coryton . . .	51 31 N.	0 34 E.
12	Holehaven .	51 30 N.	0 32 E.
13	Canvey Island .	51 32 N.	0 35 E.
14	Southend . .	51 31 N.	0 45 E.
15	Isle of Grain .	51 27 N.	0 42 E.
16	Strood	51 23 N.	0 29 E.
17	Rochester . .	51 23 N.	0 30 E.
18	Chatham . .	51 24 N.	0 34 E.
19	Kingsnorth .	51 07 N.	0 52 E.
20	Queenborough .	51 25 N.	0 45 E.
21	Sheerness . .	51 27 N.	0 45 E.
22	Faversham . .	51 19 N.	0 54 E.
23	Whitstable .	51 22 N.	1 02 E.
24	Margate . .	51 24 N.	1 23 E.
25	Ramsgate . .	51 20 N.	1 25 E.
26	Richborough .	51 18 N.	1 21 E.
27	Kingsgate . .	51 23 N.	1 27 E.
28	Sandwich . .	51 17 N.	1 20 E.
29	Dover	51 07 N.	1 19 E.
30	Folkestone .	51 05 N.	1 12 E.
31	Newhaven . .	50 47 N.	0 03 E.
32	Shoreham . .	50 49 N.	0 14 W.
33	Littlehampton .	50 47 N.	0 32 W.
34	Portsmouth . .	50 48 N.	1 07 W.
35	Newport, *I.O.W.* .	50 43 N.	1 18 W.
36	Cowes	50 46 N.	1 17 W.
37	Hamble . . .	50 53 N.	1 18 W.
38	Southampton . .	50 55 N.	1 24 W.
39	Hythe	50 55 N.	1 25 W.
40	Eling	50 55 N.	1 29 W.
41	Fawley . . .	50 49 N.	1 20 W.
42	Poole	50 40 N.	1 56 W.
43	Weymouth . .	50 37 N.	2 27 W.
44	Bridport . . .	50 43 N.	2 46 W.
45	Exmouth . .	50 37 N.	3 25 W.
46	Exeter	50 43 N.	3 34 W.
47	Teignmouth .	50 33 N.	3 30 W.
48	Tor Bay . . .	50 25 N.	3 33 W.
49	Brixham . . .	50 23 N.	3 31 W.
50	Berry Head . .	50 24 N.	3 29 W.
51	Dartmouth . .	50 21 N.	3 35 W.
52	Totnes . . .	50 26 N.	3 42 W.
53	Plymouth . . .	50 20 N.	4 09 W.
54	Devonport . .	50 22 N.	4 11 W.
55	Fowey	50 20 N.	4 38 W.
56	Par	50 21 N.	4 43 W.
57	Charlestown .	50 20 N.	4 45 W.
58	Truro	50 16 N.	5 03 W.
59	Coverack . . .	50 01 N.	5 05 W.
60	Falmouth . . .	50 09 N.	5 03 W.
61	Gweek	50 06 N.	5 13 W.
62	Porthleven . .	50 05 N.	5 19 W.
63	Penzance . . .	50 06 N.	5 33 W.
64	St. Ives . . .	50 12 N.	5 28 W.
65	Hayle	50 11 N.	5 25 W.
66	Portreath . . .	50 16 N.	5 17 W.
67	Bude	50 50 N.	4 33 W.
68	Clovelly . . .	51 00 N.	4 24 W.
69	Appledore . . .	51 03 N.	4 12 W.
70	Bideford . . .	51 02 N.	4 12 W.
71	Fremington . .	51 04 N.	4 07 W.
71a	Barnstaple . .	51 04 N.	4 03 W.
72	Watchet . . .	51 11 N.	3 20 W.
73	Bridgwater . .	51 08 N.	3 00 W.
74	Portishead . .	51 29 N.	2 46 W.
75	Avonmouth . .	51 30 N.	2 41 W.
76	Bristol	51 27 N.	2 35 W.
77	Sharpness . .	51 43 N.	2 28 W.
78	Newport, *Mon.* .	51 34 N.	2 59 W.
79	Cardiff	51 27 N.	3 10 W.
80	Barry	51 23 N.	3 16 W.
81	Port Talbot . .	51 35 N.	3 49 W.
82	Swansea . . .	51 34 N.	3 58 W.
83	Llanelly . . .	51 40 N.	4 10 W.
84	Pembroke Dock, *Milford Haven*	51 42 N.	4 57 W.
85	Fishguard . . .	52 00 N.	4 58 W.
86	Cardigan . . .	52 05 N.	4 39 W.
87	Aberystwith . .	52 25 N.	4 05 W.
88	Aberdovey . .	52 33 N.	4 03 W.
89	Barmouth . . .	52 43 N.	4 03 W.
90	Portmadoc . .	52 55 N.	4 08 W.
91	Pwllheli . . .	52 54 N.	4 24 W.
92	Trevor	53 00 N.	4 25 W.
93	Port Dinorwic .	53 11 N.	4 13 W.
94	Carnarvon . . .	53 08 N.	4 15 W.
95	Holyhead . . .	53 19 N.	4 37 W.
96	Penmaenmawr .	53 15 N.	4 00 W.
97	Llanddulas . .	53 17 N.	3 38 W.
98	Rhyl	53 18 N.	3 29 W.
99	Mostyn . . .	53 19 N.	3 16 W.
100	Connah's Quay .	53 13 N.	3 03 W.
101	Port Sunlight .	53 21 N.	2 59 W.
102	Eastham . . .	53 18 N.	2 58 W.
103	Ellesmere Port .	53 17 N.	2 53 W.
104	Stanlow . . .	53 18 N.	2 52 W.
105	Ince	53 17 N.	2 47 W.
106	Weston Point .	53 19 N.	2 44 W.
107	Runcorn . . .	53 20 N.	2 43 W.
108	Warrington . .	53 21 N.	2 37 W.
109	Latchford . . .	53 23 N.	2 34 W.
110	Partington . .	53 27 N.	2 22 W.
111	Irlam	53 28 N.	2 21 W.
112	Barton	53 29 N.	2 20 W.
113	Eccles	53 29 N.	2 19 W.
114	Weaste . . .	53 29 N.	2 18 W.
115	Manchester . .	53 28 N.	2 14 W.

Index No.	Name.	Lat.	Long.	Index No.	Name.	Lat.	Long.
		° ′	° ′			° ′	° ′
116	Irwell	53 32 N.	2 18 W.	182	Cobh	51 50 N.	8 18 W.
117	Garston	53 21 N.	2 55 W.	183	Rushbrooke Dock	51 51 N.	8 19 W.
118	Liverpool	53 25 N.	3 00 W.	184	Cork	51 54 N.	8 27 W.
119	Birkenhead	53 23 N.	3 02 W.	185	Passage West	51 52 N.	8 22 W.
120	Preston	53 45 N.	2 43 W.	186	Clonakilty	51 35 N.	8 50 W.
121	Fleetwood	53 56 N.	3 00 W.	187	Baltimore	51 27 N.	9 16 W.
122	Glasson Dock	54 00 N.	2 51 W.	188	Bantry	51 41 N.	9 28 W.
123	Lancaster	54 03 N.	2 47 W.	189	Glengariff	51 45 N.	9 34 W.
124	Heysham	54 02 N.	2 55 W.	190	Valentia Harbour	51 56 N.	10 18 W.
125	Barrow	54 06 N.	3 12 W.	191	Cahirciveen	51 57 N.	10 14 W.
126	Millom	54 13 N.	3 16 W.	192	Ballykissane	52 07 N.	9 47 W.
127	Ramsey, *I.O.M.*	54 19 N.	4 22 W.	193	Fenit	52 18 N.	9 52 W.
128	Douglas, *I.O.M.*	54 09 N.	4 28 W.	194	Foynes	52 37 N.	9 07 W.
129	Peel, *I.O.M.*	54 14 N.	4 42 W.	195	Limerick	52 40 N.	8 38 W.
130	Whitehaven	54 33 N.	3 36 W.	196	Kilrush	52 38 N.	9 30 W.
131	Workington	54 39 N.	3 34 W.	197	Clare Castle	52 49 N.	8 57 W.
132	Maryport	54 43 N.	3 30 W.	198	Killeany Bay	53 07 N.	9 38 W.
133	Silloth	54 52 N.	3 24 W.	199	Galway	53 16 N.	9 03 W.
134	Port Carlisle	54 57 N.	3 10 W.	200	Clifden	53 30 N.	10 00 W.
135	Stranraer	55 00 N.	5 03 W.	201	Westport	53 48 N.	9 31 W.
136	Ayr	55 28 N.	4 39 W.	202	Killala	54 14 N.	9 12 W.
137	Troon	55 33 N.	4 41 W.	203	Ballina	54 07 N.	9 10 W.
138	Scallasaig, *Colonsay*	56 03 N.	6 11 W.	204	Sligo	54 18 N.	8 34 W.
139	Rothsay	55 50 N.	5 02 W.	205	Killybegs	54 34 N.	8 27 W.
140	Inveraray	56 14 N.	5 05 W.	206	Ballyshannon	54 31 N.	8 12 W.
141	Port Ellen, *Islay*	55 37 N.	6 12 W.	207	Burton Port	55 00 N.	8 45 W.
142	Campbeltown	55 25 N.	5 37 W.	208	Mulroy	55 15 N.	7 47 W.
143	Irvine	55 36 N.	4 41 W.	209	Letterkenny	54 57 N.	7 44 W.
144	Ardrossan	55 38 N.	5 40 W.	210	Buncrana	55 08 N.	7 27 W.
145	Dunoon	55 57 N.	4 55 W.	211	Moville	55 12 N.	7 03 W.
146	Gourock	55 58 N.	4 50 W.	212	Londonderry	55 00 N.	7 19 W.
147	Greenock	55 57 N.	4 46 W.	213	Coleraine	55 08 N.	6 40 W.
148	Port Glasgow	55 56 N.	4 41 W.	214	Port Stewart	55 11 N.	6 43 W.
149	Dumbarton	55 57 N.	4 33 W.	215	Portrush	55 12 N.	6 39 W.
150	Clydebank	55 55 N.	4 24 W.	216	Ullapool	57 53 N.	5 10 W.
151	Renfrew	55 53 N.	4 25 W.	217	Stornaway, *Outer Hebrides*	58 12 N.	6 23 W.
152	Paisley	55 51 N.	4 26 W.	218	Thurso	58 36 N.	3 33 W.
153	Glasgow	55 51 N.	4 17 W.	219	Lerwick, *Shetland Is.*	60 09 N.	1 08 W.
154	Old Kilpatrick	55 56 N.	4 26 W.	220	Scalloway, *Shetland Is.*	60 08 N.	1 16 W.
155	Bowling	55 56 N.	4 29 W.	221	Stromness, *Orkney Is.*	58 58 N.	3 18 W.
156	Dunglass	55 55 N.	4 30 W.	222	Kirkwall *Orkney Is.*	58 59 N.	2 59 W.
157	Faslane	**56 04 N.**	**4 49 W.**	223	Stronsay, *Orkney Is.*	59 05 N.	2 36 W.
158	Finnart	**56 07 N.**	**4 50 W.**	224	Scapa, *Orkney Is.*	58 58 N.	2 59 W.
159	Kinlochleven	56 43 N.	4 58 W.	225	Wick	58 26 N.	3 05 W.
160	Larne	54 51 N.	5 47 W.	226	Helmsdale	58 08 N.	3 38 W.
161	Belfast	54 36 N.	5 55 W.	227	Invergordon	57 41 N.	4 10 W.
162	Donaghadee	54 38 N.	5 32 W.	228	Inverness	57 30 N.	4 15 W.
163	Ardglass	54 16 N.	5 37 W.	229	Lossiemouth	57 43 N.	3 18 W.
164	Greenore	54 02 N.	6 08 W.	230	Buckie	57 40 N.	2 58 W.
165	Giles Quay	54 05 N.	6 10 W.	231	Macduff	57 40 N.	2 29 W.
166	Warrenpoint, *L. Carlingford*	54 06 N.	6 15 W.	232	Fraserburgh	57 41 N.	2 00 W.
167	Newry	54 11 N.	6 19 W.	233	Peterhead	57 30 N.	1 46 W.
168	Dundalk	54 00 N.	6 21 W.	234	Aberdeen	57 09 N.	2 05 W.
169	Drogheda	53 44 N.	6 20 W.	235	Stonehaven	56 58 N.	2 12 W.
170	Dublin	53 21 N.	6 13 W.	236	Montrose	56 42 N.	2 27 W.
171	Dunlaoghaire	53 18 N.	6 08 W.	237	Abroath	56 33 N.	2 27 W.
172	Howth	53 23 N.	6 04 W.	238	Broughty Ferry	56 29 N.	2 52 W.
173	Wicklow	52 59 N.	6 02 W.	239	Dundee	56 27 N.	2 58 W.
174	Arklow	52 47 N.	6 08 W.	240	Perth	56 24 N.	3 27 W.
175	Wexford	52 20 N.	6 27 W.	241	Tayport	56 27 N.	2 54 W.
176	Rosslare	52 15 N.	6 21 W.	242	Methil	56 11 N.	3 00 W.
177	Duncannon	52 13 N.	6 55 W.	243	Kirkaldy	56 07 N.	3 10 W.
178	Waterford	52 15 N.	7 07 W.				
179	New Ross	52 23 N.	6 56 W.				
180	Dungarvan	52 05 N.	7 38 W.				
181	Youghal	51 57 N.	7 50 W.				

Index No.	Name.	Lat.	Long.	Index No.	Name.	Lat.	Long.
		° ′	° ′			° ′	° ′
244	Burntisland	56 03 N.	3 14 W.	302	Nyuchotsk	64 01 N.	36 03 E.
245	Rosyth	56 01 N.	3 27 W.	303	Kem	64 58 N.	34 45 E.
246	Alloa	56 07 N.	3 47 W.	304	Kandalacksha	67 08 N.	32 34 E.
247	Grangemouth	56 02 N.	3 39 W.	305	Ponoi	67 00 N.	41 00 E
248	Bo'ness	56 02 N.	3 36 W.	306	Murmansk	68 55 N.	33 10 E.
249	Granton	55 59 N.	3 13 W.	307	Kovda	66 42 N.	32 52 E.
250	Leith	55 59 N.	3 10 W.	308	Umba	66 43 N.	34 10 E.
251	Berwick	55 47 N.	2 00 W.	309	Advent Bay, *Spitzbergen*	78 18 N.	15 42 E.
252	Warkworth	55 21 N.	1 36 W.	310	Green Harbour, *Spitzbergen*	78 03 N.	14 30 E.
253	Blyth	55 07 N.	1 29 W.	311	Kirkenes	69 54 N.	30 03 E.
254	Tynemouth	55 01 N.	1 26 W.	312	Vadso	70 04 N.	29 45 E.
255	Wallsend	55 00 N.	1 31 W.	313	Syltefjord	70 34 N.	30 14 E.
256	Newcastle-on-Tyne	54 58 N.	1 36 W.	314	Honningsvaag	70 59 N.	25 59 E.
257	South Shields	55 01 N.	1 25 W.	315	Hammerfest	70 40 N.	23 40 E.
258	Sunderland	54 55 N.	1 21 W.	316	Alta	69 57 N.	23 10 E.
259	Seaham	54 50 N.	1 19 W.	317	Tromso	69 38 N.	18 58 E.
260	Hartlepool	54 41 N.	1 11 W.	318	Narvik	68 26 N.	17 25 E.
261	West Hartlepool	54 41 N.	1 13 W.	319	Svolvaer	68 25 N.	14 35 E.
262	Billingham	54 36 N.	1 17 W.	320	Harstad	68 48 N.	16 33 E.
263	Middlesbrough	54 35 N.	1 13 W.	321	Ballengan	68 21 N.	16 50 E.
264	Stockton	54 34 N.	1 18 W.	322	Bodo	67 17 N.	14 25 E.
265	Whitby	54 29 N.	0 37 W.	323	Glomfjord	66 49 N.	13 59 E.
266	Bridlington	54 05 N.	0 11 W.	324	Gullsmedvik	66 20 N.	14 09 E.
267	Salt End	53 44 N.	0 14 W.	325	Mo, *Norway*	66 19 N.	14 08 E.
268	Hull	53 45 N.	0 19 W.	326	Velsen	65 51 N.	13 13 E.
269	Goole	53 42 N.	0 52 W.	326a	Kongsmoen	64 53 N.	12 26 E.
270	Hook	53 43 N.	0 51 W.	327	Namsos	64 28 N.	11 30 E.
271	Selby	53 47 N.	1 04 W.	328	Steinkjer	64 01 N.	11 30 E.
272	Keadby, *R. Trent*	53 35 N.	0 44 W.	329	Trondheim	63 27 N.	9 44 E.
273	Killingholme, *R. Trent*	53 38 N.	0 14 W.	330	Thamshamn	63 18 N.	9 53 E.
274	Flixborough, *R. Trent*	53 37 N.	0 41 W.	331	Kristiansund	63 07 N.	7 45 E.
275	Gainsborough, *R. Trent*	53 24 N.	1 46 W.	332	Sundalsoren	62 41 N.	8 34 E.
276	Immingham	53 37 N.	0 11 W.	333	Molde	62 45 N.	7 09 E.
277	Grimsby	53 35 N.	0 04 W.	334	Aalesund	62 29 N.	6 10 E.
278	Boston	52 58 N.	0 01 W.	335	Heroy	62 18 N.	5 45 E.
279	Wisbech	52 48 N.	0 13 E.	336	Honningsvaag	62 12 N.	5 13 E.
280	King's Lynn	52 45 N.	0 24 E.	337	Ardalstangen	61 14 N.	7 42 E.
281	Wells	52 57 N.	0 51 E.	338	Floro	61 36 N.	5 03 E.
282	Great Yarmouth	52 34 N.	1 44 E.	339	Hoyanger	61 10 N.	6 04 E.
283	Cantley	52 34 N.	1 32 E.	340	Tofto	60 28 N.	4 57 E.
284	Norwich	52 38 N.	1 17 E.	341	Bergen	60 24 N.	5 19 E.
285	Lowestoft	52 29 N.	1 46 E.	342	Aalvik	60 26 N.	6 29 E.
286	Felixstowe	51 58 N.	1 21 E.	343	Odda	60 03 N.	6 32 E.
287	Ipswich	52 05 N.	1 09 E.	344	Sauda	59 38 N.	6 23 E.
288	Harwich	51 57 N.	1 17 E.	345	Haugesund	59 25 N.	5 12 E.
289	Wivenhoe	51 51 N.	0 58 E.	346	Kopervik	59 17 N.	5 14 E.
290	Colchester	51 53 N.	0 53 E.	347	Skudesneshavn	59 08 N.	5 16 E.
291	Stansgate	51 43 N.	0 48 E.	348	Stavanger	58 58 N.	5 40 E.
292	River Blackwater	51 43 N.	0 46 E.	349	Sandnes	58 52 N.	5 43 E.
293	Burnham-on-Crouch	51 38 N.	0 49 E.	350	Tou	59 04 N.	5 55 E.
294	St. Mary's, *Scilly Is.*	49 55 N.	6 19 W.	351	Egersund	58 29 N.	6 00 E.
295	St. Helier, *Jersey*	49 11 N.	2 07 W.	352	Sogndal	58 21 N.	6 17 E.
296	St. Peter's Port, *Guernsey*	49 27 N.	2 31 W.	353	Flekkefjord	58 20 N.	6 38 E.
297	Braye, *Alderney*	49 43 N.	2 12 W.	354	Kirkehavn	58 14 N.	6 32 E.
				355	Farsund	58 05 N.	6 44 E.
	SECTION 2. ARCTIC AND NORTH-EASTERN EUROPE.			356	Mandal	58 02 N.	7 28 E.
				357	Kristiansand	58 09 N.	7 57 E.
				358	Lillesand	58 15 N.	8 20 E.
298	Igarka	67 28 N.	86 33 E.	359	Grimstad	58 21 N.	8 34 E.
299	Mezane	65 51 N.	44 17 E.	360	Arendal	58 27 N.	8 46 E.
300	Archangel	64 55 N.	40 17 E.	361	Tvedestrand	58 38 N.	8 56 E.
301	Onega	63 55 N.	38 12 E.	362	Lyngoer	58 38 N.	9 10 E.
				363	Risor	58 42 N.	9 13 E.
				364	Kragero	58 52 N.	9 22 E.
				365	Langesund	59 00 N.	9 42 E.
				366	Porsgrunn	59 08 N.	9 39 E.

Index No.	Name.	Lat.	Long.	Index No.	Name.	Lat.	Long.
		° ′	° ′			° ′	° ′
367	Skien . . .	59 14 N.	9 34 E.	429	Vesteras . .	59 37 N.	16 31 E.
368	Brevik. . .	59 03 N.	9 39 E.	430	Stockholm . .	59 20 N.	18 03 E.
369	Larvik. . .	59 03 N.	10 05 E.	431	Stockvik, Gotland	56 59 N.	18 21 E.
370	Sandefjord .	59 08 N.	10 12 E.	432	Slite, Gotland .	57 41 N.	18 49 E.
371	Tonsberg . .	59 16 N.	10 25 E.	433	Visby, Gotland	57 37 N.	18 19 E.
372	Horten . .	59 25 N.	10 30 E.	434	Klintehamn,		
373	Holmestrand .	59 29 N.	10 19 E.		Gotland . .	57 23 N.	18 13 E.
374	Drammen . .	59 43 N.	10 11 E.	435	Harrang . .	60 07 N.	18 40 E.
375	Oslo . . .	59 55 N.	10 43 E.	436	Hallsta . .	60 03 N.	18 36 E.
376	Drobak . .	59 40 N.	10 39 E.	437	Hargshamn .	60 12 N.	18 24 E.
377	Moss . . .	59 27 N.	10 42 E.	438	Oregrund . .	60 19 N.	18 25 E.
378	Sarpsborg . .	59 17 N.	11 07 E.	439	Skutskar . .	60 39 N.	17 23 E.
379	Frederikstad .	59 12 N.	11 00 E.	440	Karskar . .	60 40 N.	17 15 E.
380	Halden . .	59 08 N.	11 22 E.	441	Gefle . . .	60 40 N.	17 09 E.
381	Stromstad. .	58 58 N.	11 11 E.	442	Norrsundet .	60 56 N.	17 07 E.
382	Hunnesbostrand .	58 28 N.	11 20 E.	443	Axmar. . .	61 01 N.	17 08 E.
383	Lysekil . .	58 18 N.	11 28 E.	444	Wallvik . .	61 03 N.	17 08 E.
384	Uddevalla. .	58 21 N.	11 58 E.	445	Ljusne. . .	61 12 N.	17 08 E.
385	Gothenburg .	57 42 N.	11 58 E.	446	Sandarne . .	61 15 N.	17 11 E.
386	Kristinehamn,			447	Stugsund . .	61 17 N.	17 08 E.
	L. Vener .	59 20 N.	14 06 E.	448	Soderhamn .	61 18 N.	17 04 E.
387	Karlstad,			449	Langplagan .	61 26 N.	17 08 E.
	L. Vener .	59 23 N.	13 30 E.	450	Langvind . .	61 28 N.	17 08 E.
388	Otterbacken,			451	Fjalviken . .	61 31 N.	17 08 E.
	L. Vener .	58 58 N.	14 02 E.	452	Enanger . .	61 33 N.	17 02 E.
389	Mariestad,			453	Essviken . .	61 33 N.	17 08 E.
	L. Vener .	58 40 N.	13 50 E.	454	Snaikmor . .	61 37 N.	17 05 E.
390	Karlsborg,			455	Iggesund . .	61 41 N.	17 03 E.
	L. Vetter .	58 31 N.	14 30 E.	456	Basfjord . .	61 42 N.	17 12 E.
391	Hjo, L. Vetter.	58 19 N.	14 18 E.	457	Hudiksvall . .	61 44 N.	17 09 E.
392	Jonkoping,			458	Lingaro . .	61 43 N.	17 17 E.
	L. Vetter .	57 46 N.	14 10 E.	459	Stockviken . .	61 54 N.	17 20 E.
393	Motala, L. Vetter.	58 32 N.	15 03 E.	460	Juniskar . .	62 17 N.	17 25 E.
394	Lidkoping. .	58 30 N.	13 11 E.	461	Svartvik . .	62 19 N.	17 21 E.
395	Varberg . .	57 08 N.	12 07 E.	462	Sundsvall . .	62 23 N.	17 16 E.
396	Falkenberg .	56 55 N.	12 30 E.	463	Kubikenborg .	62 22 N.	17 20 E.
397	Halmstad . .	56 40 N.	12 51 E.	464	Soraka. . .	62 31 N.	17 29 E.
398	Hoganas . .	56 13 N.	12 33 E.	465	Aviken . .	62 31 N.	17 41 E.
399	Helsingborg .	56 03 N.	12 43 E.	466	Hernosand .	62 38 N.	17 58 E.
400	Landskrona .	55 53 N.	12 50 E.	467	Utansjo . .	62 46 N.	17 54 E.
401	Limhamn . .	55 35 N.	12 58 E.	468	Ramvik . .	62 49 N.	17 51 E.
402	Malmo. . .	55 36 N.	13 02 E.	469	Bjorknas . .	62 56 N.	17 48 E.
403	Skarelage . .	55 24 N.	13 04 E.	470	Kramfors . .	62 57 N.	17 46 E.
404	Trelleborg. .	55 22 N.	13 09 E.	471	Bollsta . .	62 59 N.	17 44 E.
405	Ystad . . .	55 25 N.	13 48 E.	472	Nyland . .	63 00 N.	17 42 E.
406	Simrishamn .	55 35 N.	14 20 E.	473	Sando . . .	62 53 N.	17 54 E.
407	Ahus . . .	55 56 N.	14 19 E.	474	Maviken . .	62 58 N.	18 23 E.
408	Solvesborg .	56 04 N.	14 35 E.	475	Docksta . .	63 03 N.	18 18 E.
409	Karlshamn .	56 11 N.	14 52 E.	476	Nuske . . .	63 08 N.	18 29 E.
410	Ronneby . .	56 12 N.	15 20 E.	477	Kopmanholmen	63 10 N.	18 33 E.
411	Karlskrona .	56 12 N.	15 37 E.	478	Ornskoldsvik .	63 18 N.	18 42 E.
412	Degerhamn .	56 21 N.	16 25 E.	479	Rundvik . .	63 34 N.	19 23 E.
413	Kalmar . .	56 40 N.	16 21 E.	480	Nordmaling .	63 35 N.	19 29 E.
414	Paskallavik .	57 10 N.	16 29 E.	481	Kylorn . .	63 34 N.	19 48 E.
415	Oskarshamn .	57 16 N.	16 28 E.	482	Mo, Sweden .	63 37 N.	19 53 E.
416	Blankaholm .	57 35 N.	16 30 E.	483	Hornefors . .	63 38 N.	19 55 E.
417	Westervik. .	57 46 N.	16 37 E.	484	Obbola . .	63 43 N.	20 19 E.
418	Gamleby . .	57 53 N.	16 25 E.	485	Umea . . .	63 50 N.	20 21 E.
419	Helgenas . .	58 00 N.	16 32 E.	486	Djupvik . .	63 46 N.	20 22 E.
420	Valdesmarsvik	58 12 N.	16 38 E.	487	Sandviken . .	63 43 N.	20 23 E.
421	Norkoping .	58 34 N.	16 12 E.	488	Ratan . . .	64 00 N.	20 50 E.
422	Oxelosund . .	58 40 N.	17 08 E.	489	Marieberg . .	63 03 N.	20 52 E.
423	Nykoping, Sweden	58 45 N.	17 02 E.	490	Sikea . . .	64 11 N.	20 59 E.
424	Trosa . . .	58 55 N.	17 35 E.	491	Gumboda . .	64 14 N.	21 03 E.
425	Nynashamn .	58 55 N.	17 55 E.	492	Kallviken . .	64 20 N.	21 24 E.
426	Sodertelje . .	59 13 N.	17 37 E.	493	Bjuro . . .	64 25 N.	21 30 E.
427	Norrtelje . .	59 45 N.	18 41 E.	494	Burea . . .	64 36 N.	21 15 E.
428	Koping . . .	59 31 N.	15 48 E.	495	Yttrevik . .	64 40 N.	21 10 E.

Index No.	Name.	Lat.	Long.	Index No.	Name.	Lat.	Long.
		° ′	° ′			° ′	° ′
496	Bjornsholm . .	64 42 N.	21 12 E.	563	Fagervik . . .	60 02 N.	23 50 E.
497	Skelleftea . .	64 45 N.	20 58 E.	564	Helsinki . . .	60 10 N.	24 58 E.
498	Ursvik. . .	64 42 N.	21 14 E.	565	Sornas. . . .	60 13 N.	25 00 E.
499	Feruogrund .	64 54 N.	21 14 E.	566	Tolkis	60 18 N.	25 31 E.
500	Brannfors . .	65 02 N.	21 23 E.	567	Borga	60 24 N.	25 40 E.
501	Ronnskar . .	65 03 N.	21 34 E.	568	Pernoviken .	60 23 N.	25 56 E.
502	Munksund . .	65 18 N.	21 31 E.	569	Walkom . . .	60 25 N.	26 16 E.
503	Skuthamn . .	65 16 N.	21 31 E.	570	Lovisa . . .	60 27 N.	26 16 E.
504	Storfors . .	65 17 N.	21 29 E.	571	Kotka	60 28 N.	26 57 E.
505	Pitea . . .	65 19 N.	21 29 E.	572	Frederikshamn	60 34 N.	27 14 E.
506	Lulea . . .	65 35 N.	22 08 E.	573	Pyterlahti. .	60 35 N.	27 42 E.
507	Robertsvik .	65 36 N.	22 10 E.	574	Wederlax . .	60 30 N.	27 42 E.
508	Ranea . . .	65 52 N.	22 18 E.	575	Viborg. . . .	60 43 N.	28 45 E.
509	Jamlosund . .	65 52 N.	22 28 E.	576	Kakkis . . .	60 31 N.	28 42 E.
510	Kalix . . .	65 52 N.	23 08 E.	577	Bjorko. . . .	60 22 N.	28 40 E.
511	Karlsborg, *G. of Bothnia* . .	65 48 N.	23 16 E.	578	Systerbak . .	60 08 N.	29 59 E.
				579	Leningrad. . .	59 57 N.	30 20 E.
512	Axelvik . .	65 46 N.	23 12 E.	580	Kronstadt. . .	60 00 N.	29 46 E.
513	Sandvik . .	65 43 N.	23 45 E.	581	Narva. . . .	59 22 N.	28 11 E.
514	Haparanda .	65 50 N.	24 08 E.	582	Kunda. . . .	59 30 N.	27 35 E.
515	Tornea. . .	65 51 N.	24 09 E.	583	Tallinn . . .	59 26 N.	24 45 E.
516	Kemi . . .	65 49 N.	24 32 E.	584	Baltiski . . .	59 20 N.	24 05 E.
517	Koivoluoto .	65 40 N.	24 54 E.	585	Hapsal . . .	58 57 N.	23 35 E.
518	Kuivaniemi .	65 34 N.	25 09 E.	586	Pernau . . .	58 23 N.	24 31 E.
519	Haukipudas .	65 12 N.	25 17 E.	587	Riga	56 58 N.	24 08 E.
520	Pateniemi . .	65 05 N.	25 24 E.	588	Ventspils . .	57 22 N.	21 36 E.
521	Toppila . .	65 04 N.	25 27 E.	589	Libau . . .	56 33 N.	21 02 E.
522	Oulu . . .	65 01 N.	25 30 E.	590	Klaipeda . .	55 43 N′	21 09 E.
523	Warjakka . .	64 58 N.	25 17 E.	591	Konigsberg .	54 42 N.	20 30 E.
524	Sukajoki . .	64 51 N.	24 43 E.	592	Pillau . . .	54 38 N.	19 54 E.
525	Brahestad. .	64 42 N.	24 28 E.	593	Elbing . . .	54 09 N.	19 24 E.
526	Siniluoto . .	64 39 N.	24 26 E.	594	Gdansk . . .	54 21 N.	18 41 E.
527	Pyhajoki . .	64 28 N.	24 18 E.	595	Gdynia . . .	54 32 N.	18 33 E.
528	Kalajoki . .	64 16 N.	23 58 E.	596	Wladyslawowo	54 47 N.	18 26 E.
529	Himanko . .	64 03 N.	23 41 E.	597	Stolpmunde .	54 34 N.	16 52 E.
530	Yxpila. . .	63 54 N.	23 07 E.	598	Darlowo . .	54 25 N.	16 25 E.
531	Gamla Karleby	63 50 N.	23 08 E.	599	Kolbergermunde .	54 11 N.	15 32 E.
532	Jacobstad .	63 41 N.	22 41 E.	600	Szczecin . .	53 23 N.	14 32 E.
533	Juutas . . .	63 29 N.	22 27 E.	601	Swinemunde .	53 55 N.	14 15 E.
534	Oravais . .	63 18 N.	22 22 E.	602	Wolgast . .	54 04 N.	13 52 E.
535	Hankmo . .	63 13 N.	21 55 E.	603	Greifswald .	54 05 N.	13 21 E.
536	Vasa . . .	63 04 N.	21 44 E.	604	Stralsund . .	54 19 N.	13 05 E.
537	Gamla Vasa .	63 05 N.	21 44 E.	605	Rostock . .	54 04 N.	12 09 E.
538	Korsnas . .	62 49 N.	21 12 E.	606	Warnemunde .	54 09 N.	12 03 E.
539	Toiby . . .	62 41 N.	21 12 E.	607	Wismar . . .	53 54 N.	11 28 E.
540	Kasko . . .	62 23 N.	21 14 E.	608	Lubeck . . .	53 53 N.	10 40 E.
541	Kristinestad .	62 17 N.	21 22 E.	609	Travemunde .	53 58 N.	10 53 E.
542	Harkmeri . .	62 12 N.	21 27 E.	610	Neustadt . .	54 09 N.	10 52 E.
543	Boberg . . .	62 10 N.	21 23 E.	611	Heilingenhafen	54 22 N.	10 59 E.
544	Skaftung . .	62 08 N.	21 22 E.	612	Keil	54 20 N.	10 09 E.
545	Fladahamn .	62 00 N.	21 20 E.	613	Rendsburg, *Keil Canal* . .	54 17 N.	9 40 E.
546	Risby . . .	61 57 N.	21 28 E.	614	Eckernforde .	54 28 N.	9 48 E.
547	Brandohamn .	61 50 N.	21 30 E.	615	Haltenau . .	54 27 N.	10 07 E.
548	Hamnholmen .	61 45 N.	21 34 E.	616	Flensburg . .	54 47 N.	9 25 E.
549	Rafso . . .	61 37 N.	21 27 E.	617	Nakskov . .	54 50 N.	11 08 E.
550	Pori	61 27 N.	21 49 E.	618	Bandholm . .	54 50 N.	11 28 E.
551	Mantyluoto .	61 35 N.	21 35 E.	619	Nysted . . .	54 40 N.	11 43 E.
552	Raumo . . .	61 08 N.	21 31 E.	620	Rodby . . .	54 41 N.	11 23 E.
553	Nystad . . .	60 48 N.	21 25 E.	621	Nykjobing, *Falster*	54 47 N.	11 52 E.
554	Abo	60 27 N.	22 17 E.	622	Stubbekjobing .	54 53 N.	12 03 E.
555	Pargas . . .	60 18 N.	22 18 E.	623	Masnedsund .	55 02 N.	11 53 E.
556	Eckero. . .	60 12 N.	19 37 E.	624	Faxo	55 13 N.	12 09 E.
557	Mariehamn .	60 07 N.	19 56 E.	625	Rodvig . . .	55 16 N.	12 22 E.
558	Skinnarvik .	60 08 N.	22 30 E.	626	Kjoge . . .	55 28 N.	12 10 E.
559	Dalsbruk . .	60 02 N.	22 30 E.	627	Kastrup . . .	55 39 N.	12 38 E.
560	Lappvik . .	59 55 N.	23 12 E.	628	Copenhagen .	55 42 N.	12 33 E.
561	Ekenas . . .	60 00 N.	23 31 E.	629	Vedbak . . .	55 50 N.	12 33 E.
562	Hango . . .	59 46 N.	22 57 E.				

Index No.	Name.	Lat.	Long.
		° ′	° ′
630	Helsingor . . .	56 02 N.	12 38 E.
631	Hornbak . . .	56 06 N.	12 27 E.
632	Frederiksvaerk .	55 58 N.	12 01 E.
632a	Lynaes . . .	55 57 N.	11 53 E.
632b	Hundested . .	55 58 N.	11 53 E.
633	Frederiksund . .	55 50 N.	12 05 E.
634	Roskilde . . .	55 39 N.	12 05 E.
635	Holbak . . .	55 42 N.	11 43 E.
636	Nykjobing, *Zealand* .	55 56 N.	11 42 E.
637	Kallundborg .	55 41 N.	11 08 E.
638	Mullerup . . .	55 29 N.	11 11 E.
639	Korsor . . .	55 09 N.	11 09 E.
640	Skjelskor . . .	55 15 N.	11 18 E.
641	Bisserup . . .	55 13 N.	11 31 E.
642	Karrebak . . .	55 12 N.	11 39 E.
643	Naestved . . .	55 15 N.	11 47 E.
644	Vordingborg . .	55 02 N.	11 53 E.
645	Marstal . . .	54 52 N.	10 30 E.
646	Aroskjobing . .	54 53 N.	10 26 E.
647	Rudkjobing . .	54 57 N.	10 45 E.
648	Faaborg . . .	55 07 N.	10 14 E.
649	Svendborg . .	55 04 N.	10 37 E.
650	Nyborg . . .	55 19 N.	10 48 E.
651	Kjerteminde . .	55 28 N.	10 38 E.
652	Odense . . .	55 23 N.	10 23 E.
653	Bogense . . .	55 33 N.	10 04 E.
654	Middelfart . . .	55 31 N.	9 44 E.
655	Assens . . .	55 17 N.	9 53 E.
656	Sonderburg . .	54 55 N.	9 47 E.
657	Aabenraa . . .	55 03 N.	9 25 E.
658	Haderslev . . .	55 16 N.	9 30 E.
659	Aarosund . . .	55 16 N.	9 40 E.
660	Hadersleben . .	55 17 N.	9 28 E.
661	Kolding . . .	55 29 N.	9 27 E.
662	Fredericia . . .	55 35 N.	9 44 E.
663	Vejle . . .	55 42 N.	9 31 E.
664	Horsens . . .	55 52 N.	9 48 E.
665	Hov . . .	55 56 N.	10 13 E.
666	Aarhus . . .	56 09 N.	10 11 E.
667	Ebeltoft . . .	56 12 N.	10 42 E.
668	Grenaa . . .	56 26 N.	10 53 E.
669	Randers . . .	56 30 N.	10 02 E.
670	Marieager . . .	56 39 N.	9 59 E.
671	Hobro . . .	56 39 N.	9 47 E.
672	Hadsund . . .	56 44 N.	10 07 E.
673	Aalborg . . .	57 03 N.	9 55 E.
674	Norre Sundby .	57 05 N.	9 55 E.
675	Saby . . .	57 19 N.	10 31 E.
676	Osterby . . .	57 18 N.	11 08 E.
677	Frederikshavn .	57 26 N.	10 32 E.
678	Albak . . .	57 36 N.	10 26 E.
679	Skagen . . .	57 43 N.	10 35 E.
680	Hirtshals . . .	57 36 N.	9 57 E.
681	Fur Id. . . .	56 50 N.	9 00 E.
682	Thisted . . .	56 58 N.	8 40 E.
683	Nykjobing, *Mors* .	56 48 N.	8 50 E.
684	Skive . . .	56 35 N.	9 00 E.
685	Struer . . .	56 29 N.	8 37 E.
686	Lemvig . . .	56 33 N.	8 17 E.
687	Krik . . .	56 47 N.	8 18 E.
688	Sondervig . .	56 08 N.	8 04 E.
689	Ringkjobing . .	56 07 N.	8 14 E.
690	Nordby . . .	55 27 N.	8 22 E.
691	Esbjerg . . .	55 29 N.	8 28 E.
692	Skaerback . . .	55 10 N.	8 47 E.
693	Hoyer . . .	54 58 N.	8 42 E.
694	Husum . . .	54 29 N.	9 02 E.

Index No.	Name.	Lat.	Long.
		° ′	° ′
695	Tonning . . .	54 19 N.	8 57 E.
696	Vollerwick . .	54 18 N.	8 48 E.
	SECTION 3. N.W. EUROPE—ELBE TO GIBRALTAR.		
697	Hamburg . . .	53 33 N.	9 58 E.
698	Glueckstadt . .	53 42 N.	9 25 E.
699	Brunsbuttel . .	53 54 N.	9 09 E.
700	Neuhaus-Oste .	53 48 N.	9 02 E.
701	Cuxhaven . . .	53 52 N.	8 43 E.
702	Bremerhaven . .	53 33 N.	8 35 E.
703	Blumenthal . .	53 12 N.	8 33 E.
704	Farge . . .	53 12 N.	8 32 E.
705	Vegesack . . .	53 10 N.	8 38 E.
706	Lemwerder . .	53 10 N.	8 36 E.
707	Bremen . . .	53 05 N.	8 48 E.
708	Elsfleth . . .	53 15 N.	8 28 E.
709	Brake . . .	53 20 N.	8 29 E.
710	Nordenham . .	53 30 N.	8 30 E.
711	Blexen . . .	53 32 N.	8 31 E.
712	Wilhelmshaven .	53 31 N.	8 09 E.
713	Emden . . .	53 21 N.	7 12 E.
714	Leer . . .	53 13 N.	7 28 E.
715	Papenburg . .	53 06 N.	7 23 E.
716	Delfzyl . . .	53 20 N.	6 56 E.
717	Groningen . . .	53 13 N.	6 35 E.
718	Harlingen . . .	53 10 N.	5 25 E.
719	Leeuwarden . .	53 12 N.	5 50 E.
720	Sneek . . .	53 02 N.	5 40 E.
721	Zwolle . . .	52 31 N.	6 07 E.
722	Kempen a/d Yssel	52 33 N.	5 55 E.
723	Terschelling . .	53 22 N.	5 13 E.
724	Den Helder . .	52 58 N.	4 45 E.
725	Ymuiden . . .	52 28 N.	4 34 E.
726	Velsen . . .	52 27 N.	4 35 E.
727	Haarlem . . .	52 23 N.	4 38 E.
728	Zaandam . . .	52 26 N.	4 49 E.
729	Wormerveer . .	52 30 N.	4 18 E.
730	Amsterdam . .	52 22 N.	4 53 E.
731	Alkmaar . . .	52 38 N.	4 43 E.
732	Utrecht . . .	52 05 N.	5 07 E.
733	Scheveningen . .	52 06 N.	4 16 E.
734	Maasluis . . .	51 55 N.	4 15 E.
735	Capelle a/d Yssel	51 51 N.	4 36 E.
736	Helvoet . . .	51 49 N.	4 09 E.
737	Vlaardingen . .	51 54 N.	4 21 E.
738	Schiedam . . .	51 54 N.	4 24 E.
739	Delft . . .	52 01 N.	4 22 E.
740	Leiden . . .	52 09 N.	4 30 E.
741	Rotterdam . . .	51 55 N.	4 29 E.
742	Gouda . . .	52 01 N.	4 42 E.
743	Dordrecht . . .	51 49 N.	4 40 E.
744	Alblasserdam . .	51 51 N.	4 40 E.
745	's Hertogenbosch .	51 42 N.	5 18 E.
746	Brouwershaven .	51 44 N.	3 55 E.
747	Zierikzee . . .	51 39 N.	3 56 E.
748	Kortgene . . .	51 33 N.	3 48 E.
749	Flushing . . .	51 27 N.	3 36 E.
750	Antwerp . . .	51 14 N.	4 25 E.
751	Zwyndrecht . .	51 14 N.	4 19 E.
752	Vilvorde . . .	50 55 N.	4 22 E.
753	Brussels . . .	50 50 N.	4 20 E.
754	Middelburg . .	51 32 N.	3 39 E.
755	Terneuzen . . .	51 20 N.	3 50 E.
756	Sas Van Ghent .	51 14 N.	3 47 E.
757	Selzaete . . .	51 12 N.	3 48 E.
758	Ghent . . .	51 03 N.	3 42 E.

Index No.	Name.	Lat.	Long.	Index No.	Name.	Lat.	Long.
		° ′	° ′			° ′	° ′
759	Zeebrugge.	51 21 N.	3 12 E.	823	Musel	43 34 N.	5 43 W.
760	Bruges.	51 13 N.	3 13 E.	824	Aviles	43 35 N.	5 58 W.
761	Ostend.	51 14 N.	2 55 E.	825	San Esteban de		
762	Dunkirk	51 03 N.	2 22 E.		Pravia	43 34 N.	6 08 W.
763	Gravelines.	51 01 N.	2 06 E.	826	Ribadeo	43 32 N.	7 03 W.
764	Calais	50 58 N.	1 51 E.	827	Ferrol	43 28 N.	8 10 W.
765	Boulogne	50 44 N.	1 35 E.	828	Corunna	43 23 N.	8 22 W.
766	St. Valery sur			829	Corcubion	42 57 N.	9 11 W.
	Somme	50 11 N.	1 38 E.	830	Villagarcia	42 36 N.	8 46 W.
767	Treport	50 04 N.	1 22 E.	831	Marin	42 22 N.	8 43 W.
768	Dieppe.	49 56 N.	1 06 E.	832	Vigo	42 15 N.	8 45 W.
769	Fecamp	49 46 N.	0 21 E.	833	Vianna	41 41 N.	8 50 W.
770	Le Havre	49 26 N.	0 06 E.	834	Leixoes	41 11 N.	8 42 W.
771	Rouen.	49 27 N.	1 05 E.	835	Oporto.	41 08 N.	8 37 W.
772	Paris	48 52 N.	2 20 E.	836	Figueira	40 50 N.	9 10 W.
773	Honfleur	49 25 N.	0 14 E.	837	Aveiro.	40 38 N.	8 45 W.
774	Gonfreville			838	Peniche	39 00 N.	9 22 W.
	L'Orcher	49 29 N.	0 14 E.	839	Lisbon.	38 42 N.	9 08 W.
775	Trouville—			840	Setubal	38 31 N.	8 54 W.
	Deauville	49 22 N.	0 04 E.	841	Lagos, *Portugal*	37 06 N.	8 40 W.
776	Caen	49 11 N.	0 23 W.	842	Villa Nova de		
777	Carentan	49 18 N.	1 15 W.		Portimao	37 08 N.	8 31 W.
778	Isigny-sur-Mer	49 19 N.	1 17 W.	843	Faro	37 01 N.	7 53 W.
779	Cherbourg.	49 39 N.	1 37 W.	844	Olhao	37 02 N.	7 49 W.
780	Dielette	49 33 N.	1 52 W.	845	Vila Real	37 10 N.	7 26 W.
781	Granville	48 50 N.	1 37 W.	846	Ayamonte.	37 12 N.	7 25 W.
782	St. Malo	48 38 N.	2 02 W.	847	Huelva	37 15 N.	6 55 W.
783	St. Brieuc.	48 31 N.	2 46 W.	848	Seville	37 22 N.	6 00 W.
784	Paimpol	48 47 N.	3 02 W.	849	Cadiz	36 31 N.	6 18 W.
785	Treguier	48 47 N.	3 12 W.	850	Tarifa	36 00 N.	5 37 W.
786	Morlaix	48 35 N.	3 50 W.	851	Algeciras	36 07 N.	5 27 W.
787	Roscoff	48 43 N.	3 58 W.	852	Gibraltar	36 07 N.	5 21 W.
788	Brest	48 23 N.	4 29 W.				
789	Landerneau	48 27 N.	4 15 W.		**SECTION 4. MEDITERRANEAN,**		
790	Douarnenez	48 06 N.	4 20 W.		**BLACK SEA AND SEA OF AZOV.**		
791	Audierne	48 01 N.	4 33 W.	853	Malaga	36 43 N.	4 25 W.
792	Loctudy	47 50 N.	4 10 W.	854	Motril	36 45 N.	3 34 W.
793	Quimper	48 00 N.	4 06 W.	855	Adra	36 43 N.	3 00 W.
794	Concarneau	47 52 N.	3 54 W.	856	Almeria	36 50 N.	2 31 W.
795	Lorient	47 45 N.	3 21 W.	857	Garrucha	37 10 N.	1 50 W.
796	Le Croisic	47 18 N.	2 31 W.	858	Aguilas	37 24 N.	1 39 W.
797	St. Nazaire	47 16 N.	2 12 W.	859	Mazarron	37 38 N.	1 16 W.
798	Donges	47 18 N.	2 03 W.	860	Cartagena.	37 37 N.	0 59 W.
799	Paimboeuf	47 17 N.	2 02 W.	861	Escombreras		
800	Nantes.	47 13 N.	1 35 W.		Harbour	37 33 N.	0 58 W.
801	Basse-Indres	47 15 N.	1 45 W.	862	Torrevieja.	37 59 N.	0 41 W.
802	Les Sables			863	Santa Pola	38 13 N.	0 32 W.
	d'Olonne	46 30 N.	1 47 W.	864	Alicante	38 20 N.	0 29 W.
803	La Pallice.	46 10 N.	1 13 W.	865	Denia	38 54 N.	0 03 E.
804	Marans	46 18 N.	1 02 W.	866	Gandia	38 58 N.	0 10 W.
805	La Rochelle	46 09 N.	1 09 W.	867	Cullera.	39 11 N.	0 15 W.
806	Rochefort	45 57 N.	0 58 W.	868	Valencia	39 29 N.	0 23 W.
807	Tonnay-Charente.	45 57 N.	0 51 W.	869	Iviza, *Iviza Id.*	38 55 N.	1 30 E.
808	Verdon	45 33 N.	1 01 W.	870	Palma, *Majorca*	39 35 N.	2 43 E.
809	Marennes	45 49 N.	1 09 W.	871	Mahon, *Minorca*	39 52 N.	4 13 E.
810	Pauillac	45 11 N.	0 45 W.	872	Sagunto	39 01 N.	0 17 W.
811	Blaye	45 08 N.	0 38 W.	873	Burriana	39 54 N.	0 06 W.
812	Bordeaux	44 50 N.	0 35 W.	874	Castellon	40 01 N.	0 03 W.
813	Libourne	44 55 N.	0 15 W.	875	Tarragona.	41 09 N.	1 16 E.
814	Arachon	44 40 N.	1 10 W.	876	Barcelona.	41 22 N.	2 10 E.
815	Boucau	43 32 N.	1 29 W.	877	San Feliu de		
816	Bayonne	43 30 N.	1 29 W.		Guixols.	41 47 N.	3 04 E.
817	Pasajes	43 20 N.	1 55 W.	878	Palamos	41 53 N.	3 08 E.
818	Bilbao.	43 16 N.	2 58 W.	879	Port Vendres	42 29 N.	3 09 E.
819	Castro Alen	43 24 N.	3 16 W.	880	La Nouvelle	42 57 N.	3 01 E.
820	Santander	43 30 N.	3 47 W.	881	Cette	43 24 N.	3 40 E.
821	Ribadesella	43 30 N.	5 04 W.	882	St. Louis, *Rhone*	43 23 N.	4 49 E.
822	Gijon	43 35 N.	5 40 W.	883	Port de Bouc.	43 25 N.	4 58 E.

Index No.	Name.	Lat.	Long.	Index No.	Name.	Lat.	Long.
		° ′	° ′			° ′	° ′
884	Marseilles . . .	43 18 N.	5 22 E.	945	Rimini. . .	44 02 N.	12 32 E.
885	La Seyne . . .	43 05 N.	5 22 E.	946	Ravenna . .	44 23 N.	12 11 E.
886	Toulon. . .	43 07 N.	5 56 E.	947	Chioggia . .	45 13 N.	12 17 E.
887	Cannes. . .	43 31 N.	7 01 E.	948	Venice . .	45 26 N.	12 20 E.
888	Nice . . .	43 42 N.	7 15 E.	949	Trieste. . .	45 39 N.	13 45 E.
889	Villefranche .	43 42 N.	7 19 E.	950	Pola . .	44 52 N.	13 50 E.
890	Monaco . .	43 44 N.	7 26 E.	951	Rijeka . .	45 20 N.	14 26 E.
891	Ajaccio, Corsica .	41 56 N.	8 43 E.	952	Zara . . .	44 08 N.	15 12 E.
892	Calvi, Corsica.	42 33 N.	8 44 E.	953	Sibenic . .	43 44 N.	15 52 E.
893	Bastia, Corsica	42 41 N.	9 25 E.	954	Solin . .	43 33 N.	16 29 E.
894	Porto Vecchio, Corsica . .	41 34 N.	9 17 E.	955	Split . . .	43 31 N.	16 26 E.
895	Cagliari, Sardinia	39 13 N.	9 08 E.	956	Dubrovnic .	42 39 N.	18 09 E.
896	Alghero, Sardinia	40 32 N.	8 22 E.	957	Zelenika .	42 27 N.	18 37 E.
897	Porto Torres, Sardinia .	40 50 N.	8 24 E.	958	Bar . . .	42 05 N.	19 04 E.
898	Golfo Aranci, Sardinia .	40 55 N.	9 35 E.	959	Dulcigno . .	41 48 N.	19 14 E.
899	Tortoli, Sardinia .	39 55 N.	9 38 E.	960	Durazzo . .	41 19 N.	19 27 E.
900	Carloforte, Sardinia .	39 08 N.	8 19 E.	961	Valona. . .	40 28 N.	19 29 E.
901	Portovesme, Sardinia .	39 12 N.	8 23 E.	962	Corfu . .	39 37 N.	19 53 E.
902	Vado . . .	44 17 N.	8 26 E.	963	Santi Quaranta .	39 52 N.	20 02 E.
903	Porto Maurizio .	43 52 N.	8 01 E.	964	Parga . . .	39 18 N.	20 22 E.
904	Savona . .	44 20 N.	8 27 E.	965	Leukas . .	38 48 N.	20 42 E.
905	Genoa . . .	44 23 N.	8 56 E.	966	Samos, Cephalonia	38 15 N.	20 40 E.
906	Spezia . . .	44 07 N.	9 47 E.	967	Lixuri, Cephalonia	38 12 N.	20 27 E.
907	Carrara . .	44 03 N.	10 06 E.	968	Itea . . .	38 26 N.	22 25 E.
908	Leghorn . .	43 33 N.	10 18 E.	969	Aegion. . .	38 14 N.	22 07 E.
909	Piombino . .	42 58 N.	10 31 E.	970	Corinth . .	37 55 N.	23 00 E.
910	Follonica . .	42 55 N.	10 44 E.	971	Patras . . .	38 13 N.	21 45 E.
911	Porto San Stefano . .	42 25 N.	11 06 E.	972	Zante . . .	37 46 N.	20 54 E.
912	Civitavecchia .	42 04 N.	11 48 E.	973	Katakolon .	37 40 N.	21 20 E.
913	Fiumicino. . .	41 46 N.	12 15 E.	974	Kalamata . .	37 03 N.	22 10 E.
914	Anzio . . .	41 28 N.	12 38 E.	975	Gythion . .	36 46 N.	22 37 E.
915	Pozzuoli . .	40 51 N.	14 07 E.	976	Nauplia . .	37 35 N.	22 49 E.
916	Bagnoli . .	40 50 N.	14 10 E.	977	Xyli Bay . .	36 40 N.	22 52 E.
917	Naples. . .	40 50 N.	14 15 E.	978	Falconera . .	36 50 N.	23 53 E.
918	Torre Annunziata	40 46 N.	14 26 E.	979	Piræus. . .	37 58 N.	23 40 E.
919	Castellamare di Stabia .	40 40 N.	14 29 E.	980	Laurium . .	37 42 N.	24 04 E.
920	Salerno . .	40 39 N.	14 46 E.	981	Chalkis . .	38 27 N.	23 37 E.
921	Reggio. . .	38 06 N.	15 39 E.	982	Volo . . .	39 24 N.	22 59 E.
922	Messina, Sicily .	38 12 N.	15 34 E.	983	Salonika . .	40 38 N.	22 57 E.
923	Milazzo, Sicily .	38 13 N.	15 15 E.	984	Kavalla . .	40 56 N.	24 26 E.
924	Palermo, Sicily .	38 08 N.	13 22 E.	985	Kalamuti Bay .	40 50 N.	24 30 E.
925	Trapani, Sicily .	38 00 N.	12 34 E.	986	Dedeagach . .	40 49 N.	25 53 E.
926	Marsala, Sicily .	37 54 N.	12 24 E.	987	Alexandropoulis .	40 50 N.	25 55 E.
927	Mazzara, Sicily .	37 39 N.	12 35 E.	988	Kum Kale . .	39 59 N.	26 12 E.
928	Sciacca, Sicily .	37 28 N.	13 08 E.	989	Chanak . .	40 04 N.	26 14 E.
929	Licata, Sicily .	37 06 N.	13 56 E.	990	Gallipoli, Turkey .	40 26 N.	26 38 E.
930	Gela, Sicily .	37 03 N.	14 16 E.	991	Rodosto . .	40 59 N.	27 28 E.
931	Syracuse, Sicily .	37 03 N.	15 18 E.	992	Scutari . .	41 02 N.	29 02 E.
932	Augusta, Sicily .	37 14 N.	15 13 E.	993	Istanbul . .	41 04 N.	28 57 E.
933	Catania, Sicily .	37 29 N.	15 06 E.	994	Vassiliko Bay.	42 10 N.	27 52 E.
934	Avola . .	36 54 N.	15 09 E.	995	Burgas. . .	42 30 N.	27 29 E.
935	Valetta, Malta .	35 53 N.	14 31 E.	996	Varna . . .	43 13 N.	27 57 E.
936	Taranto . .	40 28 N.	17 13 E.	997	Constanza. .	44 10 N.	28 39 E.
937	Gallipoli, Italy .	40 02 N.	18 02 E.	998	Sulina . .	45 07 N.	29 41 E.
938	Brindisi . .	40 39 N.	17 58 E.	999	Odessa. . .	46 29 N.	30 43 E.
939	Bari . . .	41 07 N.	16 49 E.	1000	Kherson . .	46 42 N.	32 35 E.
940	Molfetta . .	41 12 N.	16 34 E.	1001	Nikolaiev . .	46 49 N.	31 58 E.
941	Barletta . .	41 19 N.	16 15 E.	1002	Sebastapol .	44 36 N.	33 29 E.
942	Manfredonia .	41 38 N.	15 56 E.	1003	Kerch . . .	45 20 N.	36 28 E.
943	Vasto . . .	42 07 N.	14 42 E.	1004	Mariupol . .	47 10 N.	37 30 E.
944	Ancona . .	43 37 N.	13 30 E.	1005	Taganrog . .	47 13 N.	38 23 E.
				1006	Rostov . .	47 30 N.	39 50 E.
				1007	Novorossisk .	44 42 N.	37 46 E.
				1008	Tuapse . .	44 21 N.	39 02 E.
				1009	Poti . . .	42 07 N.	41 40 E.
				1010	Batum . .	41 38 N.	41 40 E.
				1011	Rize . . .	40 03 N.	40 32 E.
				1012	Trabzon . .	41 00 N.	39 42 E.

Index No.	Name.	Lat.	Long.	Index No.	Name.	Lat.	Long.
		° ′	° ′			° ′	° ′
1013	Akcaabat	41 02 N.	39 35 E.	1078	Port Said	31 14 N.	32 18 E.
1014	Gorele	41 04 N.	39 21 E.	1079	Ismailia, *Suez*		
1015	Tireboli	41 01 N.	38 49 E.		*Canal*	30 42 N.	32 15 E.
1016	Gireson	40 54 N.	38 23 E.	1080	Port Tewfik, *Suez*		
1017	Ordu	41 01 N.	37 51 E.		*Canal*	29 55 N.	32 35 E.
1018	Fatsa	41 02 N.	37 29 E.	1081	Alexandria	31 12 N.	29 52 E.
1019	Unye	41 07 N.	37 17 E.	1082	Mersa Matruh	31 23 N.	27 13 E.
1020	Samsun	41 17 N.	36 20 E.	1083	Tobruk	32 05 N.	24 00 E.
1021	Sinop	42 01 N.	35 09 E.	1084	Derna	32 45 N.	22 40 E.
1022	Ineboli	41 57 N.	33 46 E.	1085	Benghazi	32 07 N.	20 03 E.
1023	Jideh	41 53 N.	33 00 E.	1086	Tripoli, *Libya*	32 54 N.	13 11 E.
1024	Amasra	41 43 N.	32 23 E.	1087	Jerba	33 45 N.	10 50 E.
1025	Zonguldak	41 28 N.	31 48 E.	1088	Gabes	33 53 N.	10 07 E.
1026	Eregli	41 17 N.	31 28 E.	1089	Sfax	34 44 N.	10 46 E.
1027	Kilia	41 09 N.	29 39 E.	1090	Mahedia	35 30 N.	11 05 E.
1028	Derince	40 48 N.	29 56 E.	1091	Monastir	35 45 N.	10 51 E.
1029	Yalova	40 39 N.	29 17 E.	1092	Susa	35 49 N.	10 38 E.
1030	Gemlik	40 27 N.	29 09 E.	1093	Tunis	36 48 N.	10 10 E.
1031	Mudania	40 22 N.	28 51 E.	1094	La Goulette	36 49 N.	10 19 E.
1032	Perama	40 20 N.	28 00 E.	1095	Bizerta	37 16 N.	9 53 E.
1033	Karabogha	40 25 N.	27 17 E.	1096	Tabarca	36 58 N.	8 46 E.
1034	Lapsaki	40 19 N.	26 39 E.	1097	Bona	36 54 N.	7 46 E.
1035	Mudros, *Lemnos*	39 51 N.	25 17 E.	1098	Philippville	36 53 N.	6 54 E.
1036	Aivaly	39 19 N.	26 42 E.	1099	Dellys	36 55 N.	6 34 E.
1037	Mytilene	39 04 N.	26 34 E.	1100	Collo	37 00 N.	6 34 E.
1038	Chandarly	38 56 N.	26 55 E.	1101	Jijelli	36 46 N.	5 45 E.
1039	Izmir	38 24 N.	27 10 E.	1102	Les Falaises	36 39 N.	5 25 E.
1040	Kastron, *Chios*	38 24 N.	26 08 E.	1103	Bougie	36 46 N.	5 06 E.
1041	Chesme	38 19 N.	26 19 E.	1104	Algiers	36 47 N.	3 04 E.
1042	Scalanova	37 52 N.	27 19 E.	1105	Port Briera	36 32 N.	1 35 E.
1043	Vathy, *Samos*	37 47 N.	27 00 E.	1106	Mostanganem	35 56 N.	0 06 E.
1044	Karlovassi, *Samos*	37 46 N.	26 42 E.	1107	Arzew	35 53 N.	0 17 W.
1045	Syra, *Hermopolis*	37 28 N.	24 58 E.	1108	Oran	35 44 N.	0 42 W.
1046	Kos	36 50 N.	27 10 E.	1109	Mers el Kebir	35 45 N.	0 44 W.
1047	Marmaras	36 53 N.	28 20 E.	1110	Benisaf	35 20 N.	1 23 W.
1048	Bodrum	37 03 N.	27 27 E.	1111	Nemours	35 06 N.	1 52 W.
1049	Makri	36 38 N.	29 12 E.	1112	Port Kelah	35 06 N.	2 09 W.
1050	Fethiye	36 39 N.	29 09 E.	1113	Melilla	35 18 N.	2 57 W.
1051	Kalamaki Bay	36 11 N.	29 26 E.	1114	Tetuan	35 37 N.	5 17 W.
1052	Fineka	36 19 N.	30 10 E.	1115	Ceuta	35 54 N.	5 17 W.
1053	Adalia	36 52 N.	30 45 E.				
1054	Alanza	36 35 N.	32 03 E.		SECTION 5. WEST, SOUTH		
1055	Chorak	36 04 N.	32 54 E.		AND EAST AFRICA.		
1056	Mersin	36 58 N.	34 37 E.	1116	Tangier	35 45 N.	5 50 W.
1057	Canea, *Crete*	35 29 N.	24 01 E.	1117	Larrache	35 13 N.	6 07 W.
1058	Suda Bay, *Crete*	35 30 N.	24 10 E.	1118	Port Lyautey	34 17 N.	6 35 W.
1059	Rethymno, *Crete*	35 20 N.	24 28 E.	1119	Mehediya	34 18 N.	6 40 W.
1060	Candia, *Crete*	35 18 N.	25 09 E.	1120	Rabat	34 04 N.	6 49 W.
1061	Rhodes	36 25 N.	28 17 E.	1121	Fedalah	33 45 N.	7 22 W.
1062	Lindos, *Rhodes*	36 05 N.	28 08 E.	1122	Casablanca	33 36 N.	7 37 W.
1063	Famagusta,			1123	Mazagan	33 15 N.	8 31 W.
	Cyprus	35 05 N.	33 58 E.	1124	Saffi	32 20 N.	9 17 W.
1064	Larnarca, *Cyprus*	34 57 N.	33 38 E.	1125	Mogador	31 31 N.	9 47 W.
1065	Limassol, *Cyprus*	34 42 N.	33 05 E.	1126	Agadir	30 26 N.	9 37 W.
1066	Paphos, *Cyprus*	34 48 N.	32 25 E.	1127	Fayal, *Azores*	38 32 N.	28 37 W.
1067	Port Latchi,			1128	Graciosa, *Azores*	39 05 N.	28 00 W.
	Cyprus	35 02 N.	32 25 E.	1129	Terceira, *Azores*	38 39 N.	27 13 W.
1068	Morphou Bay,			1130	St. Michael's,		
	Cyprus	35 15 N.	32 50 E.		*Azores*	37 44 N.	25 40 W.
1069	Vassiliko, *Cyprus*	34 43 N.	33 19 E.	1131	Santa Maria,		
1070	Iskenderun	36 36 N.	36 10 E.		*Azores*	36 56 N.	25 10 W.
1071	Lattakia	35 32 N.	35 47 E.	1132	Funchal, *Madeira*	32 38 N.	16 55 W.
1072	Banias	35 11 N.	35 58 E.	1133	Ifni	29 29 N.	10 10 W.
1073	Tripoli, *Lebanon*	34 27 N.	35 50 E.	1134	Teneriffe, *Canary*		
1074	Beirut	33 53 N.	35 31 E.		*Is.*	28 29 N.	16 14 W.
1075	Sidon	33 33 N.	35 22 E.	1135	Las Palmas,		
1076	Haifa	32 49 N.	34 59 E.		*Canary Is.*	28 08 N.	15 40 W.
1077	Tel-Aviv	32 03 N.	34 48 E.	1136	Villa Cisneros	23 29 N.	15 58 W.

Index No.	Name.	Lat.	Long.	Index No.	Name.	Lat.	Long.
		° ′	° ′			° ′	° ′
1137	Port Etienne .	21 00 N.	17 10 W.	1201	Mayumba . . .	3 23 S.	10 38 E.
1138	Portendik . .	18 08 N.	16 08 W.	1202	Pointe Noire . .	4 48 S.	11 50 E.
1139	St. Louis, *Senegal*	16 05 N.	16 30 W.	1203	Loango . . .	4 42 S.	11 50 E.
1140	St. Vincent, *Cape*			1204	Landana . . .	5 20 S.	12 15 E.
	Verde Is. . .	16 53 N.	25 00 W.	1205	Kabinda . . .	5 33 S.	12 12 E.
1141	Porto Praia,			1206	Banana . . .	6 00 S.	12 23 E.
	Cape Verde Is. .	14 55 N.	23 31 W.	1207	Boma	5 51 S.	13 03 F.
1142	Dakar	14 40 N.	17 25 W.	1208	Matadi. . .	5 40 S.	13 30 E.
1143	Kaolock . . .	14 09 N.	16 06 W.	1209	Ambrizete. . .	7 15 S.	12 54 E.
1144	Bathurst,			1210	Port Ambriz . .	7 52 S.	13 08 E.
	R. Gambia . .	13 27 N.	16 34 W.	1211	Luanda . . .	8 47 S.	13 14 E.
1145	Karabane . . .	12 33 N.	16 42 W.	1212	Porto Amboim .	10 45 S.	13 48 E.
1146	Zighinkov . . .	12 34 N.	16 20 W.	1213	Novo Redondo .	11 07 S.	13 54 E.
1147	Cachew . . .	12 18 N.	16 10 W.	1214	Lobito. . .	12 20 S.	13 35 E.
1148	Bissao	11 52 N.	15 35 W.	1215	Benguela . . .	12 34 S.	13 24 E.
1149	Bolama . . .	11 35 N.	15 29 W.	1216	Mossamedes . .	15 12 S.	12 09 E.
1150	Kogon River . .	11 20 N.	14 30 W.	1217	Porto Alexandre .	15 48 S.	11 51 E.
1151	Conakry . . .	9 31 N.	13 43 W.	1218	Ascension. . .	7 55 S.	14 25 W.
1152	Pepel	8 35 N.	13 04 W.	1219	St. Helena . .	15 55 S.	5 42 W.
1153	Freetown . . .	8 30 N.	13 14 W.	1220	Tristan da Cunha .	37 02 S.	12 18 W.
1154	Sherbro . . .	7 32 N.	12 30 W.	1221	Walvis Bay . .	22 57 S.	14 30 E.
1155	Robertsport . .	6 40 N.	11 25 W.	1222	Luderitz . . .	26 38 S.	15 08 E.
1156	Monrovia . . .	6 19 N.	10 48 W.	1223	Port Nolloth . .	29 15 S.	16 52 E.
1157	Marshall . . .	6 05 N.	10 20 W.	1224	Capetown . . .	33 54 S.	18 25 E.
1158	Grand Bassa . .	5 54 N.	10 04 W.	1225	Simonstown . .	34 10 S.	18 25 E.
1159	Tabou . . .	4 28 N.	7 20 W.	1226	Port Beaufort . .	34 24 S.	20 49 E.
1160	Cape Palmas . .	4 22 N.	7 44 W.	1227	Mossel Bay . .	34 11 S.	22 09 E.
1161	Sassandra . . .	4 57 N.	6 03 W.	1228	Knysna . . .	34 04 S.	23 03 E.
1162	Abidjan . . .	5 16 N.	4 10 W.	1229	Port Elizabeth . .	33 58 S.	25 38 E.
1163	Axim	4 52 N.	2 12 W.	1230	Port Alfred . .	33 36 S.	26 54 E.
1164	Takoradi . . .	4 54 N.	1 45 W.	1231	East London . .	33 02 S.	27 55 E.
1165	Sekondi . . .	4 56 N.	1 42 W.	1232	Port St. John's .	31 37 S.	29 33 E.
1166	Cape Coast . .	5 06 N.	1 14 W.	1233	Durban . . .	29 52 S.	31 03 E.
1167	Salt Pond . . .	5 09 N.	1 05 W.	1234	Lourenco Marques	25 58 S.	32 34 E.
1168	Winneba . . .	5 23 N.	0 35 W.	1235	Inhambane . .	23 52 S.	35 23 E.
1169	Accra	5 35 N.	0 12 W.	1236	Beira	19 50 S.	34 50 E.
1170	Ada	5 46 N.	0 41 E.	1237	Chinde. . .	18 34 S.	36 30 E.
1171	Keta	5 54 N.	1 00 E.	1238	Quelimane . .	18 00 S.	36 54 E.
1172	Lome	6 07 N.	1 13 E.	1239	Majunga,		
1173	Great Popo . .	6 12 N.	1 33 E.		*Madagascar*	15 44 S.	46 19 E.
1174	Whydah . . .	6 22 N.	2 02 E.	1240	Analalave,		
1175	Kotonu . . .	6 21 N.	2 26 E.		*Madagascar*	14 39 S.	47 45 F.
1176	Porto Novo . .	6 30 N.	2 34 E.	1241	Hellville,		
1177	Lagos	6 27 N.	3 23 E.		*Madagascar* .	13 24 S.	48 18 E.
1178	Sapele	5 55 N.	5 42 E.	1242	Diego Suarez,		
1179	Warri	5 31 N.	5 45 E.		*Madagascar* .	12 15 S.	49 19 E.
1180	Forcados . . .	5 22 N.	5 26 E.	1243	Port Leven,		
1181	Burutu . . .	5 21 N.	5 31 E.		*Madagascar* .	13 00 S.	49 47 E.
1182	Akassa. . . .	4 19 N.	6 04 E.	1244	Vohemar,		
1183	Brass	4 19 N.	6 15 E.		*Madagascar* .	13 10 S.	49 50 E.
1184	Bonny	4 27 N.	7 10 E.	1245	Antalaha,		
1185	Opobo	4 23 N.	7 35 E.		*Madagascar* .	14 53 S.	50 18 E.
1186	Port Harcourt .	4 46 N.	7 00 E.	1246	Maroantsetra,		
1187	Calabar . . .	4 58 N.	8 19 E.		*Madagascar* .	15 25 S.	49 48 E.
1188	Victoria, *Nigeria*.	4 00 N.	9 12 E.	1247	Mananara,		
1189	Tiko	4 04 N.	9 24 E.		*Madagascar* .	16 14 S.	49 40 E.
1190	Duala	4 03 N.	9 43 E.	1248	Port St. Mary,		
1191	Santa Isabel . .	3 30 N.	8 40 E.		*Madagascar* .	17 00 S.	49 50 E.
1192	Kribi	2 56 N.	9 55 E.	1249	Tamatave,		
1193	Bata, *Spanish*				*Madagascar* .	18 09 S.	49 26 E.
	Guinea . . .	1 51 N.	9 48 E.	1250	Mananjara,		
1194	Kogo	1 05 N.	9 42 E.		*Madagascar* .	21 17 S.	48 20 E.
1195	Libreville . . .	0 21 N.	9 30 E.	1251	Manakara,		
1196	Owendo . . .	0 17 N.	9 32 E.		*Madagascar* .	22 00 S.	48 10 E.
1197	Princes Id. . .	1 39 N.	7 26 E.	1252	Farafangana,		
1198	St. Thomas Id. .	0 21 N.	6 44 E.		*Madagascar* .	22 55 S.	47 48 E.
1199	Port Gentil . .	0 43 S.	8 48 E.	1253	Fort Dauphin,		
1200	Sette Cama . .	2 32 S.	9 44 E.		*Madagascar* .	25 00 S.	46 50 E.

Index No.	Name.	Lat.	Long.	Index No.	Name.	Lat.	Long.
		° ′	° ′			° ′	° ′
1254	Tulear, Madagascar	23 20 S.	43 45 E.	1310	Sitra, Bahrein	26 10 N.	50 37 E.
1255	Morondava, Madagascar	20 18 S.	44 28 E.	1311	Dammam	26 27 N.	50 06 E.
				1312	Ras Tanura	26 38 N.	50 10 E.
1256	Maintirano, Madagascar	18 11 S.	44 00 E.	1313	Kuwait	29 21 N.	47 55 E.
				1314	Mena al Ahmadi	29 04 N.	48 10 E.
1257	Kerguelen Id.	48 30 S.	69 40 E.	1315	Shatt al Arab (Outer Bar)	29 50 N.	48 43 E.
1258	Port Louis, Mauritius	20 09 S.	57 29 E.	1316	Fao	29 58 N.	48 29 E.
1259	Rodriguez Id.	19 40 S.	63 26 E.	1317	Abadan	30 20 N.	48 16 E.
1260	St. Denis, Reunion	20 52 S.	55 27 E.	1318	Abu Flus	30 27 N.	48 02 E.
				1319	Harta Point	30 22 N.	48 11 E.
1261	Pte. de Galets, Reunion	20 55 S.	55 17 E.	1320	Khorramshahr	30 26 N.	48 10 E.
				1321	Bandar Mashur	30 28 N.	49 11 E.
1262	St. Pierre, Reunion	21 20 S.	55 29 E.	1322	Bandar Shapur	30 27 N.	49 05 E.
				1323	Basrah	30 30 N.	47 53 E.
1263	Diego Garcia, Chagos Arch.	7 14 S.	72 24 E.	1324	Bushire	28 54 N.	50 45 E.
				1325	Lingah	26 30 N.	54 50 E.
1264	Mahe, Seychelles	4 45 S.	55 30 E.	1326	Kuruk	28 52 N.	58 27 E.
1265	Farquhar Id.	10 10 S.	51 05 E.	1327	Bandar Abbas	27 12 N.	56 15 E.
1266	Maroni, Comoro Is.	11 40 S.	43 05 E.	1328	Jask	26 45 N.	57 12 E.
				1329	Chahbar	25 16 N.	60 37 E.
1267	Mayotte Id.	12 50 S.	45 00 E.	1330	Gwattar	25 09 N.	61 30 E.
1268	Mozambique	15 02 S.	40 48 E.	1331	Gwadar	25 05 N.	62 18 E.
1269	Nacala	14 29 S.	40 40 E.	1332	Pasni	25 12 N.	63 30 E.
1270	Port Simuko	13 59 S.	40 36 E.	1333	Karachi	24 48 N.	66 58 E.
1271	Porto Amelia	12 58 S.	40 29 E.	1334	Mandvi	22 50 N.	69 20 E.
1272	Mikindani	10 16 S.	40 02 E.	1335	Navalakhi	22 58 N.	70 24 E.
1273	Mtwara	10 17 S.	40 05 E.	1336	Bedi-Bunder	22 31 N.	70 09 E.
1274	Lindi	10 00 S.	39 44 E.	1337	Kandla	23 00 N.	70 12 E.
1275	Kilwa Masoko	8 58 S.	39 30 E.	1338	Navanagar	22 28 N.	70 04 E.
1276	Kilwa Kivinje	8 45 S.	39 25 E.	1339	Port Okha	22 28 N.	69 05 E.
1277	Dar-es-Salaam	6 49 S.	39 19 E.	1340	Dwarka	22 13 N.	69 01 E.
1278	Zanzibar	6 09 S.	39 11 E.	1341	Porbandar	21 38 N.	69 37 E.
1279	Pangani	5 25 S.	38 59 E.	1342	Veraval	20 53 N.	70 22 E.
1280	Tanga	5 04 S.	39 07 E.	1343	Mhowa	21 05 N.	71 45 E.
1281	Mombasa	4 04 S.	39 39 E.	1344	Bhaunagar	21 48 N.	72 09 E.
1282	Lamu	2 16 S.	40 55 E.	1345	Surat	21 11 N.	72 48 E.
1283	Kismayu	0 22 S.	42 33 E.	1346	Daman	20 25 N.	72 49 E.
1284	Barawa	1 10 N.	44 05 E.	1347	Bombay	18 55 N.	72 50 E.
1285	Merka	1 43 N.	44 47 E.	1348	Panjim	15 30 N.	73 49 E.
1286	Mogadiscio	2 02 N.	45 21 E.	1349	Mormugao	15 25 N.	73 48 E.
1287	Berbera	10 26 N.	45 01 E.	1350	Mangalore	12 51 N.	74 50 E.
1288	Djibouti	11 35 N.	43 09 E.	1351	Cannanore	11 51 N.	75 22 E.
1289	Perim	12 38 N.	43 24 E.	1352	Tellicherry	11 45 N.	75 29 E.
1290	Assab	13 00 N.	42 44 E.	1353	Badagara	11 35 N.	75 35 E.
1291	Rasa	14 56 N.	40 24 E.	1354	Quiland	11 28 N.	75 39 E.
1292	Massowah	15 37 N.	39 28 E.	1355	Calicut	11 15 N.	75 46 E.
1293	Suakin	19 02 N.	37 18 E.	1356	Beypore	11 10 N.	75 48 E.
1294	Port Sudan	19 36 N.	37 15 E.	1357	Ponnani	10 47 N.	75 54 E.
1295	Kosseir	26 08 N.	34 10 E.	1358	Malipuram	10 01 N.	76 15 E.
1296	Safaja Id.	26 46 N.	34 00 E.	1359	Cochin	9 58 N.	76 15 E.
1297	Abu Zenima	29 02 N.	33 06 E.	1360	Alleppey	9 35 N.	76 16 E.
				1361	Quilon	8 53 N.	76 34 E.
				1362	Trivandrum	8 32 N.	76 57 E.
	SECTION 6. ASIA.			1363	Tuticorin	8 48 N.	78 00 E.
1297a	Aqaba	29 29 N.	35 01 E.	1364	Colombo	6 57 N.	79 51 E.
1298	Adabiya	29 52 N.	32 28 E.	1365	Galle	6 01 N.	80 13 E.
1299	El Wej	26 15 N.	36 18 E.	1366	Trincomali	8 33 N.	81 13 E.
1300	Yanbo el Bahr	24 10 N.	37 50 E.	1367	Negapatam	10 46 N.	79 51 E.
1301	Jeddah	21 28 N.	39 11 E.	1368	Karikal	10 58 N.	79 46 E.
1302	Gizan	16 54 N.	42 31 E.	1369	Cuddalore	11 43 N.	79 47 E.
1303	Hodeida	14 49 N.	42 55 E.	1370	Pondicherry	11 55 N.	79 49 E.
1304	Aden	12 47 N.	44 59 E.	1371	Madras	13 06 N.	80 18 E.
1305	Muscat	23 37 N.	58 36 E.	1372	Nizampatam	15 53 N.	80 38 E.
1306	Dubai	25 16 N.	55 18 E.	1373	Masulipatam	16 10 N.	81 11 E.
1307	Umm Said	24 57 N.	51 35 E.	1374	Kakinada	16 56 N.	82 15 E.
1308	Doha	25 21 N.	51 34 E.	1375	Vizagapatam	17 41 N.	83 17 E.
1309	Manama, Bahrein	26 14 N.	50 35 E.	1376	Bheemunipatam	17 53 N.	83 27 E.

Index No.	Name.	Lat.	Long.	Index No.	Name.	Lat.	Long.
		° ′	° ′			° ′	° ′
1377	Bimlipatam	17 55 N.	83 31 E.	1442	Sape	8 36 S.	119 00 E.
1378	Kalingapatam	18 20 N.	84 07 E.	1443	Reo	8 16 S.	120 31 E.
1379	Gopalpur	19 16 N.	84 55 E.	1444	Maumeri	8 35 S.	122 12 E.
1380	Diamond Harbour	22 11 N.	88 11 E.	1445	Deli, *Timor*	8 31 S.	125 36 E.
1381	Calcutta	22 33 N.	88 19 E.	1446	Kupang	10 10 S.	123 40 E.
1382	Chalna	22 35 N.	89 32 E.	1447	Ambogaga	8 46 S.	121 40 E.
1383	Chittagong	22 20 N.	91 50 E.	1448	Waingapu	9 35 S.	120 12 E.
1384	Cox's Bazar	21 26 N.	91 59 E.	1449	Beliling	8 05 S.	115 05 E.
1385	Akyab	20 08 N.	92 54 E.	1450	Ampenan	8 34 S.	116 04 E.
1386	Kyaukpyu	19 26 N.	93 33 E.	1451	Amboina	3 40 S.	128 14 E.
1387	Sandoway	18 30 N.	94 24 E.	1452	Kaveli	3 15 S.	127 10 E.
1388	Bassein	16 47 N.	94 47 E.	1453	Bachian	0 39 S.	127 29 E.
1389	Rangoon	16 46 N.	96 10 E.	1454	Patani	0 15 N.	128 45 E.
1390	Maulmain	16 29 N.	97 37 E.	1455	Ternate	0 47 N.	127 22 E.
1391	Tavoy	14 04 N.	98 02 E.	1456	Buton	5 28 S.	122 37 E.
1392	Mergui	12 26 N.	98 36 E.	1457	Gorontalo	0 30 N.	123 03 E.
1393	Maya Bandar, *Andaman Is.*	12 56 N.	92 54 E.	1458	Menado	1 53 N.	124 50 E.
				1459	Kwandang	0 54 N.	122 50 E.
1394	Port Blair, *Andaman Is.*	11 41 N.	92 46 E.	1460	Tolitoli	1 02 N.	120 49 E.
				1461	Donggala	0 37 S.	119 44 E.
1395	Penang	5 25 N.	100 21 E.	1462	Majene	3 33 S.	118 59 E.
1396	Prai	5 04 N.	100 05 E.	1463	Pare Pare	4 00 S.	119 30 E.
1397	Telok Anson	4 01 N.	101 01 E.	1464	Macassar	5 09 S.	119 24 E.
1398	Port Swettenham	3 00 N.	101 23 E.	1465	Palapo	3 02 S.	120 18 E.
1399	Port Dickson	2 31 N.	101 47 E.	1466	Kolaka	4 00 S.	121 30 E.
1400	Malacca	2 11 N.	102 15 E.	1467	Banjermasini	3 20 S.	114 36 E.
1401	Pulo Bukom	1 14 N.	103 46 E.	1468	Sampit	3 10 S.	113 15 E.
1402	Singapore	1 17 N.	103 51 E.	1469	Kumai	2 40 S.	111 55 E.
1403	Pulo Sebarok	1 12 N.	103 48 E.	1470	Sukudana	1 14 S.	109 57 E.
1404	Pulo Sambu	1 09 N.	103 54 E.	1471	Pontianak	0 01 S.	109 20 E.
1405	Tanjong Uban	1 03 N.	104 13 E.	1472	Singkawang	0 59 N.	109 02 E.
1406	Segli	5 18 N.	96 05 E.	1473	Kuching	1 34 N.	110 21 E.
1407	Sabang	5 50 N.	95 20 E.	1474	Sungei Rajang	2 09 N.	111 15 E.
1408	Oleh Leh	5 35 N.	95 18 E.	1475	Miri	4 23 N.	113 59 E.
1409	Kota Raja	5 25 N.	95 20 E.	1476	Brunei	4 54 N.	114 59 E.
1410	Susu	4 07 N.	98 13 E.	1477	Labuan	5 17 N.	115 15 E.
1411	Belawan Deli	3 47 N.	98 40 E.	1478	Jesselton	5 59 N.	116 04 E.
1412	Bengkalis	1 30 N.	102 10 E.	1479	Kudat	6 53 N.	116 51 E.
1413	Tanjong Pinang	0 55 N.	104 26 E.	1480	Sandakan	5 50 N.	118 07 E.
1414	Muntok	2 05 S.	105 15 E.	1481	Lahad Datu	5 02 N.	118 20 E.
1415	Palembang	2 55 S.	104 50 E.	1482	Semporna	4 29 N.	118 37 E.
1416	Sungei Gerong	2 59 S.	104 51 E.	1483	Tawao	4 14 N.	117 53 E.
1417	Pladju	3 00 S.	104 50 E.	1484	Wallace Bay	4 12 N.	117 54 E.
1418	Koba	2 30 S.	106 30 E.	1485	Tarakan	3 25 N.	117 40 E.
1419	Toboali	3 00 S.	106 30 E.	1486	Bulangan	2 49 N.	117 30 E.
1420	Tanjong Pandan	2 45 S.	107 38 E.	1487	Sambiliong	2 02 N.	117 45 E.
1421	Telok Betong	5 20 S.	105 48 E.	1488	Telok Seliman	1 24 N.	118 34 E.
1422	Benkulen	3 45 S.	102 19 E.	1489	Sangkulirang	0 45 N.	118 15 E.
1423	Padang	0 58 S.	100 22 E.	1490	Samarinda	0 36 S.	117 15 E.
1424	Siboga	1 45 N.	98 50 E.	1491	Balik Papan	1 15 S.	116 45 E.
1425	Christmas Island	10 24 S.	105 43 E.	1492	Passir	2 05 S.	116 10 E.
1426	Port Refuge, *Cocos Is.*	12 05 S.	96 53 E.	1493	Kota Baru	3 20 S.	116 10 E.
				1494	Sulu	6 04 N.	121 00 E.
1427	Anjer Kidoel	6 00 S.	106 05 E.	1495	Isabela	6 40 N.	122 00 E.
1428	Tanjong Priok	6 06 S.	106 52 E.	1496	Zamboanga	7 10 N.	121 55 E.
1429	Djakarta	6 10 S.	106 50 E.	1497	Davao	7 05 N.	125 28 E.
1430	Tjirebon	6 42 S.	108 34 E.	1498	Caraga	7 43 N.	126 25 E.
1431	Tegal	6 55 S.	109 08 E.	1499	Bislig	8 07 N.	126 20 E.
1432	Pekalongan	6 58 S.	109 24 E.	1500	Canay	8 49 N.	126 12 E.
1433	Semerang	6 58 S.	110 25 E.	1501	Surigao	9 48 N.	125 29 E.
1434	Surabaya	7 13 S.	112 44 E.	1502	Nasipit	8 59 N.	125 20 E.
1435	Pasuruan	7 40 S.	112 52 E.	1503	Butuan	8 52 N.	125 31 E.
1436	Probolinggo	7 46 S.	113 13 E.	1504	Medina	8 54 N.	125 01 E.
1437	Panarukan	7 44 S.	113 59 E.	1505	Cagayan	8 28 N.	124 34 E.
1438	Banjuwangi	8 12 S.	114 20 E.	1506	Misamis	8 09 N.	123 51 E.
1439	Tjilitjap	7 41 S.	109 05 E.	1507	Iligan	8 14 N.	124 14 E.
1440	Pelaboehan	7 02 S.	106 37 E.	1508	Dumaguete	9 22 N.	123 18 E.
1441	Bima	8 27 S.	118 43 E.	1509	Cebu	10 18 N.	123 54 E.

Index No.	Name.	Lat.	Long.	Index No.	Name.	Lat.	Long.
		° ′	° ′			° ′	° ′
1510	Iloilo	10 42 N.	122 34 E.	1572	Chinkiang,		
1511	Dagami . . .	11 00 N.	124 50 E.		*Yangtse River* .	32 13 N.	119 26 E.
1512	Tacloban . . .	11 20 N.	124 55 E.	1573	Nanking, *Yangtse*		
1513	Mangarin . . .	12 21 N.	121 05 E.		*River* . . .	32 06 N.	118 45 E.
1514	Sablayan . . .	12 50 N.	120 45 E.	1574	Wuhu, *Yangtse*		
1515	Jose Panganiban .	14 17 N.	122 42 E.		*River* . . .	31 20 N.	118 21 E.
1516	Siain	13 58 N.	122 00 E.	1575	Anking, *Yangtse*		
1517	Kalapan . . .	13 26 N.	121 11 E.		*River* . . .	30 35 N.	117 00 E.
1518	Cavite	14 30 N.	120 53 E.	1576	Hukow, ,,	29 43 N.	116 16 E.
1519	Manila	14 35 N.	120 58 E.	1577	Kiukiang, ,,	29 40 N.	116 05 E.
1520	Lingayen . . .	15 55 N.	120 15 E.	1578	Hankow, ,,	30 33 N.	114 22 E.
1521	San Fernando .	16 35 N.	120 25 E.	1579	I-Chang, ,,	31 02 N.	111 00 E.
1522	Claveria . . .	18 37 N.	121 06 E.	1580	Chungking, ,,	29 40 N.	106 05 E.
1523	Aparri	18 22 N.	121 38 E.	1581	Tsingtao . .	36 05 N.	120 19 E.
1524	Port Dimalansan .	17 20 N.	122 22 E.	1582	Wei-Hei-Wei . .	37 30 N.	122 10 E.
1525	Baler	15 45 N.	121 35 E.	1583	Chefoo . . .	37 33 N.	121 33 E.
1526	Port Lampon . .	14 40 N.	121 37 E.	1584	Taku Bar .	38 56 N.	117 50 E.
1527	Tabako . . .	13 22 N.	123 44 E.	1585	Tongku New		
1528	Lamitan . , .	13 58 N.	123 32 E.		Harbour . .	39 00 N.	117 44 E.
1529	Legaspi . . .	13 09 N.	123 45 E.	1586	Tientsin . . .	39 09 N.	117 12 E.
1530	Masbate . . .	12 22 N.	123 44 E.	1587	Chinwangtao . .	39 54 N.	119 38 E.
1531	Kuantan . . .	3 50 N.	103 20 E.	1588	Newchwang . .	40 41 N.	122 16 E.
1532	Kuala Dungun .	4 47 N.	103 26 E.	1589	Ryojun . . .	38 48 N.	121 15 E.
1533	Trengganu . .	5 21 N.	103 08 E.	1590	Dairen . . .	38 56 N.	121 39 E.
1534	Patani	6 57 N.	101 17 E.	1591	Antung . . .	40 07 N.	124 24 E.
1535	Singora . . .	7 13 N.	100 38 E.	1592	Chinnanpo . .	38 43 N.	125 24 E.
1536	Bandon . . .	9 05 N.	99 25 E.	1593	Inchon . . .	37 29 N.	126 37 E.
1537	Koh Sichang . .	13 10 N.	100 47 E.	1594	Mokpo . . .	34 54 N.	126 26 E.
1538	Paknam . . .	13 35 N.	100 40 E.	1595	Pusan . . .	35 06 N.	129 02 E.
1539	Bangkok . . .	13 44 N.	100 30 E.	1596	Wonsan . . .	39 11 N.	127 21 E.
1540	Saigon . . .	10 46 N.	106 41 E.	1597	Songchin . . .	40 41 N.	129 09 E.
1541	Pnom Penh . .	11 40 N.	104 55 E.	1598	Vladivostock . .	43 07 N.	131 54 E.
1542	Nhatrang . . .	12 12 N.	109 12 E.	1599	Alexandrovsk,		
1543	Tourane . . .	16 05 N.	108 11 E.		*Siberia* . .	51 45 N.	140 40 E.
1544	Hue	16 33 N.	107 39 E.	1600	Nikolaievsk . .	52 08 N.	140 40 E.
1545	Haiphong . . .	20 52 N.	106 42 E.	1601	Alexandrovski,		
1546	Hongay . . .	20 57 N.	107 04 E.		*Sakhalin* .	51 00 N.	142 00 E.
1547	Port Campha . .	21 02 N.	107 22 E.	1602	Korsakovsk,		
1548	Pakhoi . . .	21 29 N.	109 04 E.		*Sakhalin* .	46 43 N.	142 44 E.
1549	Hoihow . . .	20 03 N.	110 19 E.	1603	Petropavlovsk,		
1550	Yulin	18 13 N.	109 34 E.		*Kamchatka* .	53 00 N.	158 20 E.
1551	Kwang Chow			1604	Tamano . . .	32 38 N.	128 36 E.
	Wan . . .	21 10 N.	110 26 E.	1605	Kagoshima . .	31 36 N.	130 34 E.
1552	Macao	22 11 N.	113 33 E.	1606	Yamagawa . .	31 12 N.	130 38 E.
1553	Whampoa . .	23 05 N.	113 26 E.	1607	Yatsushiro . .	32 30 N.	130 35 E.
1554	Canton . . .	23 10 N.	113 20 E.	1608	Misumi . . .	32 38 N.	130 28 E.
1555	Hongkong . . .	22 18 N.	114 10 E.	1609	Miike	33 00 N.	130 23 E.
1556	Swatow . . .	23 20 N.	116 45 E.	1610	Suminaye . .	33 12 N.	130 12 E.
1557	Amoy	24 27 N.	118 04 E.	1611	Shimabara . .	32 47 N.	130 22 E.
1558	Chuanchowfu .	24 53 N.	118 33 E.	1612	Kuchinotsu . .	32 36 N.	130 11 E.
1559	Shajo, *Formosa* .	22 04 N.	120 41 E.	1613	Nagasaki . . .	32 43 N.	129 51 E.
1560	Toko, *Formosa* .	22 28 N.	120 27 E.	1614	Omura . . .	32 54 N.	129 57 E.
1561	Kaohsiung,			1615	Sasebo . . .	33 10 N.	129 43 E.
	Formosa .	22 37 N.	120 16 E.	1616	Karatsu . . .	33 27 N.	129 59 E.
1562	Anping, *Formosa*.	23 01 N.	120 10 E.	1617	Fukuoka . . .	33 36 N.	130 23 E.
1563	Tamsui, *Formosa*.	25 11 N.	121 26 E.	1618	Hakata . . .	33 35 N.	130 25 E.
1564	Keelung,			1619	Wakamatsu,		
	Formosa .	25 08 N.	121 41 E.		*Kyushu* . .	33 55 N.	130 49 E.
1565	Suo Wan,			1620	Kokura . . .	33 53 N.	130 54 E.
	Formosa .	24 35 N.	121 52 E.	1621	Yawata, *Kiushiu*.	33 52 N.	130 49 E.
1566	Karenko,			1622	Moji	33 57 N.	131 00 E.
	Formosa .	24 59 N.	121 30 E.	1623	Nakatsu . . .	33 37 N.	131 10 E.
1567	Pinan, *Formosa* .	24 59 N.	121 05 E.	1624	Beppu . . .	33 16 N.	131 30 E.
1568	Foochow (*Pagoda*			1625	Usuki . . .	33 06 N.	131 47 E.
	Anchorage) . .	25 59 N.	119 27 E.	1626	Tsukumi . . .	33 05 N.	131 52 E.
1569	Ningpo . . .	29 53 N.	121 34 E.	1627	Saegi	32 59 N.	131 51 E.
1570	Woosung . . .	31 24 N.	121 30 E.	1628	Mimitsu . . .	32 20 N.	131 38 E.
1571	Shanghai . . .	31 15 N.	121 29 E.	1629	Takanabe . . .	32 10 N.	131 32 E.

Index No.	Name.	Lat.	Long.	Index No.	Name.	Lat.	Long.
		° ′	° ′			° ′	° ′
1630	Miyosake . . .	31 56 N.	131 24 E.	1694a	Matsuyama . .	33 50 N.	132 46 E.
1631	Shibushi . . .	31 29 N.	131 06 E.	1695	Imabaru . . .	34 04 N.	132 59 E.
1632	Shimonoseki . .	33 58 N.	130 57 E.	1696	Tadotsu . . .	34 14 N.	133 44 E.
1633	Hamada . . .	34 55 N.	132 04 E.	1697	Takamatsu . .	34 20 N.	134 04 E.
1634	Hane	35 12 N.	132 30 E.	1698	Tokushima . .	34 04 N.	134 34 E.
1635	Miyazu . . .	35 32 N.	135 11 E.	1699	Komatsushima .	34 00 N.	134 36 E.
1636	Tsuruga . . .	35 39 N.	136 08 E.	1700	Kochc	33 32 N.	133 33 E.
1637	Wajima . . .	37 24 N.	136 51 E.	1701	Susaki	33 24 N.	133 17 E.
1638	Nanao . . .	37 03 N.	136 59 E.	1702	Masaki . . .	32 58 N.	132 59 E.
1639	Fushiki . . .	36 47 N.	137 04 E.	1703	Sukumo . . .	32 54 N.	132 42 E.
1640	Naoetsu . . .	37 11 N.	138 11 E.	1704	Hakodate. . .	41 47 N.	140 43 E.
1641	Minato. . . .	38 03 N.	138 26 E.	1705	Fukuyama . .	41 26 N.	140 06 E.
1642	Niigata . . .	37 57 N.	139 03 E.	1706	Otaru	43 13 N.	141 01 E.
1643	Sakata. . . .	38 55 N.	139 53 E.	1707	Mashike . . .	43 47 N.	141 30 E.
1644	Akita	39 41 N.	140 07 E.	1708	Soya	45 30 N.	142 00 E.
1645	Noshiro . . .	40 03 N.	140 01 E.	1709	Esashi	44 58 N.	142 29 E.
1646	Aomori . . .	40 51 N.	140 44 E.	1710	Mombetsu. . .	44 23 N.	143 19 E.
1647	Hachinohe . .	40 32 N.	141 33 E.	1711	Abashiri . . .	44 00 N.	144 18 E.
1648	Miyako . . .	39 38 N.	141 58 E.	1712	Shibetsu . . .	43 38 N.	145 06 E.
1649	Yamada,			1713	Nemoro . . .	43 20 N.	145 35 E.
	N. Honshu .	39 28 N.	141 58 E.	1714	Hanasaki. . .	43 17 N.	145 35 E.
1650	Kamaishi . . .	39 16 N.	141 54 E.	1715	Akishi	43 02 N.	144 52 E.
1651	Kesennuma .	38 53 N.	141 37 E.	1716	Kushiro . . .	42 58 N.	144 22 E.
1652	Motoyoshi. . .	38 40 N.	141 27 E.	1717	Muroran . . .	42 19 N.	140 58 E.
1653	Oginohama . .	38 23 N.	141 28 E.	1718	Naha, Okinawa .	26 12 N.	127 40 E.
1654	Hirakata . . .	36 51 N.	140 48 E.				
1655	Tateyama. . .	35 02 N.	139 51 E.				
1656	Yawata, Tokio B.	35 30 N.	140 05 E.		**SECTION 7. AUSTRALASIA AND**		
1657	Chiba	35 36 N.	140 07 E.		**PACIFIC ISLANDS.**		
1658	Funabashi. . .	35 42 N.	140 00 E.	1719	Saonek . . .	0 27 S.	130 46 E.
1659	Tokio	35 40 N.	139 46 E.	1720	Sorong. . . .	0 53 S.	131 14 E.
1660	Kawasaki. . .	35 32 N.	139 42 E.	1721	Saukris . . .	0 35 S.	133 00 E.
1661	Yokohama . .	35 27 N.	139 38 E.	1722	Doreh	0 53 S.	134 01 E.
1662	Yokosuka . .	35 17 N.	139 40 E.	1723	Mokmer . . .	1 10 S.	136 06 E.
1663	Shimizu, Honshu.	35 00 N.	138 30 E.	1724	Hollandia . . .	2 32 S.	140 44 E.
1664	Nagoya . . .	35 05 N.	136 51 E.	1725	Vanemo . . .	2 40 S.	141 10 E.
1665	Yokkaichi. . .	34 58 N.	136 36 E.	1726	Aitape. . . .	3 09 S.	142 31 E.
1666	Tsu	34 43 N.	136 32 E.	1727	Wewak . . .	3 30 S.	143 35 E.
1667	Yamada, Owari			1728	MonumboHarbour	4 22 S.	144 55 E.
	Bay . . .	34 29 N.	136 42 E.	1729	Madang . . .	5 12 S.	145 49 E.
1668	Gokasho . . .	34 19 N.	136 40 E.	1730	Astrolabe Bay .	5 28 S.	145 48 E.
1669	Nagashima . .	34 12 N.	136 30 E.	1731	Finsch Harbour .	6 33 S.	147 52 E.
1670	Atashika . . .	33 55 N.	136 09 E.	1732	Samarai . . .	10 37 S.	150 40 E.
1671	Eda	33 30 N.	135 40 E.	1733	Tufi Harbour .	9 05 S.	149 19 E.
1672	Susami . . .	33 34 N.	135 30 E.	1734	Port Harvey . .	8 55 S.	148 35 E.
1673	Tanabe . . .	33 43 N.	135 22 E.	1735	Morobe . . .	7 56 S.	147 44 E.
1674	Satomura . . .	33 59 N.	135 09 E.	1736	Port Moresby. .	9 26 S.	147 06 E.
1675	Wakayama . .	34 13 N.	135 12 E.	1737	Port Romilly . .	7 41 S.	144 49 E.
1676	Sakai	34 37 N.	135 25 E.	1738	Rabaul . . .	4 16 S.	152 11 E.
1677	Osaka	34 39 N.	135 26 E.	1739	Gasmata . . .	6 20 S.	150 30 E.
1678	Kobe	34 41 N.	135 11 E.	1740	Talasea . . .	5 10 S.	150 00 E.
1679	Hirohata . . .	34 46 N.	134 37 E.	1741	Kavieng Harbour	2 40 S.	150 50 E.
1680	Tomotsu . . .	34 20 N.	134 23 E.	1742	Kieta	6 15 S.	155 35 E.
1681	Sakaide . . .	34 19 N.	133 51 E.	1743	Port Vila . . .	17 41 S.	168 19 E.
1682	Shimotsu . . .	34 26 N.	133 48 E.	1744	Bourai. . . .	21 37 S.	165 26 E.
1683	Itozaki . . .	34 23 N.	133 06 E.	1745	Noumea . . .	22 16 S.	166 27 E.
1684	Kure	34 14 N.	132 33 E.	1746	Ellington, Fiji	17 20 S.	178 14 E.
1684a	Fukuyama Honshu	34 26 N.	133 27 E.	1747	Suva	18 09 S.	178 26 E.
1685	Ujina	34 22 N.	132 27 E.	1748	Levuka . . .	17 41 S.	178 51 E.
1686	Miyajima . . .	34 17 N.	132 15 E.	1749	Savu Savu . .	16 45 S.	179 20 E.
1687	Murotsu . . .	33 51 N.	132 08 E.	1750	Lautoka . . .	17 36 S.	177 26 E.
1688	Tokuyama . .	34 05 N.	131 50 E.	1751	Labasa . . .	16 20 S.	179 24 E.
1689	Mitajiri . . .	34 03 N.	131 34 E.	1752	Tongatabu . .	21 10 S.	175 15 W.
1690	Uwajima . . .	33 13 N.	132 35 E.	1753	Apia, Samoa . .	13 49 S.	171 46 W.
1691	Takayama . .	33 20 N.	132 26 E.	1754	Pago Pago . .	14 17 S.	170 41 W.
1692	Kawanoishi . .	33 29 N.	132 35 E.	1755	Papeete, Tahiti	17 32 S.	149 34 W.
1693	Nagahama . .	33 36 N.	132 31 E.	1756	Rarotonga . .	21 12 S.	159 46 W.
1694	Mitsuhama . .	33 52 N.	132 42 E.	1757	Nauru	0 33 S.	166 55 E.
				1758	Ocean Island . .	0 50 S.	169 35 E.

Index No.	Name.	Lat.	Long.	Index No.	Name.	Lat.	Long.
		° ′	° ′			° ′	° ′
1759	Tarawa . . .	1 25 N.	173 00 E.	1822	Port Macquarie .	31 26 S.	152 56 E.
1760	Butaritari Makin.	3 00 N.	173 00 E.	1823	Port Stephens .	32 43 S.	152 09 E.
1761	Jaluit . . .	6 00 N.	169 30 E.	1824	Newcastle, N.S.W	32 56 S.	151 47 E.
1762	Kusaie. . .	5 15 N.	163 05 E.	1825	Sydney . .	33 52 S.	151 13 E.
1763	Ponapi. . .	6 55 N.	158 10 E.	1826	Port Hacking. .	34 04 S.	151 10 E.
1764	Truk . . .	7 20 N.	151 50 E.	1827	Port Kembla . .	34 29 S.	150 55 E.
1765	Yap . . .	9 30 N.	138 05 E.	1828	Ulladulla . . .	35 22 S.	150 30 E.
1766	Guam . . .	13 30 N.	144 45 E.	1829	Eden . . .	37 05 S.	149 59 E.
1767	Saipan. . .	15 10 N.	145 40 E.	1830	Bairnsdale . .	37 49 S.	147 40 E.
1768	Wake Island .	19 10 N.	166 32 E.	1831	Port Albert . .	38 40 S.	146 45 E.
1769	Johnston Island .	16 45 N.	169 32 W.	1832	Melbourne. . .	37 45 S.	144 58 E.
1770	Palmyra Island .	5 52 N.	162 06 W.	1833	Williamstown. .	37 52 S.	144 55 E.
1771	Washington Island . .	4 42 N.	160 25 W.	1834	Port Phillip . .	38 18 S.	144 38 E.
				1835	Geelong . .	38 06 S.	144 18 E.
1772	Fanning Island .	3 51 N.	159 22 W.	1836	Warrnambool. .	38 24 S.	142 29 E.
1773	Christmas Island.	1 59 N.	157 28 W.	1837	Port Fairy . .	38 18 S.	142 13 E.
1774	Easter Island . .	27 09 S.	109 17 W.	1838	Portland . .	38 21 S.	141 37 E.
1775	Welles Harbour .	28 13 N.	177 22 W.	1839	Port McDonnell .	38 04 S.	140 42 E.
1776	Pitcairn Island .	25 07 S.	130 20 W.	1840	Kingston . .	36 50 S.	139 51 E.
1777	Kahului . .	20 54 N.	156 29 W.	1841	Port Elliott .	35 29 S.	138 48 E.
1778	Hilo . . .	19 44 N.	155 04 W.	1842	Port Victor . .	35 31 S.	138 32 E.
1779	Honuapo . .	19 05 N.	155 33 W.	1843	Adelaide . .	34 51 S.	138 30 E.
1780	Napupu . .	19 28 N.	155 55 W.	1844	Port Wakefield .	34 16 S.	138 06 E.
1781	Mahukona. .	20 11 N.	155 54 W.	1845	Ardrossan. . .	34 24 S.	137 52 E.
1782	Pukoo Harbour .	21 04 N.	156 47 W.	1846	Stansbury. . .	34 55 S.	137 42 E.
1783	Kamalo Harbour.	21 03 N.	156 53 W.	1847	Kingscote . .	35 37 S.	137 38 E.
1784	Kolo . . .	21 06 N.	157 12 W.	1848	Port Pirie . .	33 10 S.	138 01 E.
1785	Laie . . .	21 39 N.	157 56 W.	1849	Port Germein. .	32 59 S.	138 05 E.
1786	Waikane . .	21 30 N.	157 51 W.	1850	Whyalla . .	33 02 S.	137 36 E.
1787	Honolulu . .	21 18 N.	157 52 W.	1851	Port Augusta. .	32 30 S.	137 46 E.
1788	Nawiliwili. .	21 57 N.	159 21 W.	1852	Port Lincoln . .	34 43 S.	135 52 E.
1789	Nonopapa. .	21 52 N.	160 14 W.	1853	Port Victoria . .	34 30 S.	137 27 E.
1790	Majuro . .	7 08 N.	171 21 E.	1854	Franklin Harbour	33 42 S.	136 57 E.
1791	Vavua Group. .	18 40 S.	174 00 W.	1855	Wallaroo . .	33 54 S.	137 36 E.
1792	Nomuka, *Tonga Is.* . .	20 16 S.	174 48 W.	1856	Thevenard . .	32 09 S.	133 39 E.
				1857	Port Eyre. . .	32 00 S.	132 27 E.
1793	Haapai Group, *Tonga Is.* . .	19 50 S.	174 30 W.	1858	Eucla . . .	31 44 S.	128 53 E.
1794	Nukualofa. .	21 08 S.	175 12 W.	1859	Esperance Bay .	33 52 S.	121 54 E.
1795	Thursday Island .	10 35 S.	142 13 E.	1860	Albany . . .	35 02 S.	117 53 E.
1796	Cooktown . .	15 28 S.	145 15 E.	1861	Busselton . .	33 39 S.	115 22 E.
1797	Port Douglas . .	16 29 S.	145 28 E.	1862	Bunbury . .	33 17 S.	115 38 E.
1798	Cairns . . .	16 55 S.	145 47 E.	1863	Fremantle. .	32 03 S.	115 44 E.
1799	Mourilyan Harbour .	17 36 S.	146 07 E.	1864	Port Gregory . .	28 12 S.	114 15 E.
				1865	Geraldton . .	28 48 S.	114 34 E.
1800	Cardwell . .	18 15 S.	146 02 E.	1866	Carnarvon. . .	24 52 S.	113 39 E.
1801	Townsville . .	19 16 S.	146 50 E.	1867	Ashburton . .	21 43 S.	114 50 E.
1802	Ayr . . .	19 37 S.	147 28 E.	1868	Port Walcott . .	20 39 S.	117 13 E.
1803	Bowen. . .	20 01 S.	148 15 E.	1869	Port Hedland. .	20 18 S.	118 35 E.
1804	Proserpine . .	20 31 S.	148 45 E.	1870	Condon . .	20 00 S.	119 30 E.
1805	Mackay . .	21 10 S.	149 15 E.	1871	Broome . .	18 00 S.	122 13 E.
1806	St. Lawrence . .	22 20 S.	149 36 E.	1872	Derby . . .	17 22 S.	123 45 E.
1807	Port Clinton . .	22 32 S.	150 45 E.	1873	Wyndham. .	15 27 S.	128 06 E.
1808	Rockhampton .	23 23 S.	150 31 E.	1874	Port Keats . .	14 03 S.	129 34 E.
1809	Gladstone . .	23 50 S.	151 15 E.	1875	Darwin . .	12 28 S.	130 51 E.
1810	Bundaberg . .	24 50 S.	152 20 E.	1876	Port Roper . .	14 46 S.	135 30 E.
1811	Urangan . .	25 16 S.	152 54 E.	1877	Port McArthur .	15 58 S.	136 45 E.
1812	Pialba . .	25 18 S.	152 54 E	1878	Burketown . .	17 45 S.	139 34 E.
1813	Maryborough . .	25 53 S.	153 07 E.	1879	Karumba . .	17 30 S.	140 52 E.
1814	Brisbane . .	27 28 S.	153 03 E.	1880	Normanton . .	17 52 S.	141 05 E.
1815	Ipswich . .	27 35 S.	152 45 E.	1881	Launceston . .	41 26 S.	147 08 E.
1816	Southport. . .	27 57 S.	153 26 E.	1882	Georgetown . .	41 07 S.	146 49 E.
1817	Narang . .	27 58 S.	153 21 E.	1883	Hobart . . .	42 53 S.	147 20 E.
1818	Ballina . .	28 52 S.	153 35 E.	1884	Franklin . .	43 06 S.	147 00 E.
1819	Wardell . .	28 57 S.	153 29 E.	1885	Port Davey . .	43 19 S.	146 00 E.
1820	Yamba . . .	29 25 S.	153 25 E.	1886	Pillinger . .	42 15 S.	145 29 E.
1821	Grafton . .	29 42 S.	152 57 E.	1887	Stanley . . .	40 42 S.	145 23 E.
1821a	Coff's Harbour .	30 18 S.	153 10 E.	1888	Burnie. . .	41 02 S.	145 53 E.
				1889	Devonport . .	41 09 S.	146 23 E.

Index No.	Name.	Lat.	Long.	Index No.	Name.	Lat.	Long.
		° ′	° ′			° ′	° ′
1890	Beaconsfield . .	41 11 S.	146 48 E.	1955	Port Bainbridge .	60 04 N.	148 12 W.
1891	Mangonui . . .	35 00 S.	173 33 E.	1956	Macleod Harbour	59 53 N.	147 46 W.
1892	Whangaroa . .	35 02 S.	173 47 E.	1957	Port Chalmers,		
1893	Opua	35 20 S.	174 01 E.		*Alaska* . . .	60 14 N.	147 14 W.
1894	Port Russell . .	35 16 S.	174 07 E.	1958	Port Etches . .	60 20 N.	146 33 W.
1895	Whangerai . .	35 45 S.	174 20 E.	1959	Valdez. . . .	61 07 N.	146 17 W.
1896	Auckland . . .	36 51 S.	174 46 E.	1960	Cordova . . .	60 33 N.	145 46 W.
1897	Tauranga . . .	37 41 S.	176 11 E.	1961	Katalla . . .	60 10 N.	144 32 W.
1898	Opotiki . . .	38 02 S.	177 18 E.	1962	Yakutat . . .	59 33 N.	139 40 W.
1899	Gisborne . . .	38 41 S.	178 02 E.	1963	Port Althorp . .	58 07 N.	136 17 W.
1900	Wairoa . . .	39 03 S.	177 26 E.	1964	Hoonah . . .	58 07 N.	135 27 W.
1901	Napier. . . .	39 28 S.	176 54 E.	1965	Swanson Harbour	58 13 N.	135 07 W.
1902	Wellington . .	41 17 S.	174 47 E.	1966	Sitka	57 03 N.	135 20 W.
1903	Foxton . . .	40 28 S.	175 17 E.	1967	Port Banks . .	56 34 N.	134 59 W.
1904	Wanganui. . .	39 54 S.	175 03 E.	1968	Port Walter . .	56 23 N.	134 40 W.
1905	Patea . . .	39 47 S.	174 31 E.	1969	Skagway . . .	59 27 N.	135 18 W.
1906	New Plymouth .	39 04 S.	174 02 E.	1970	Juneau . . .	58 18 N.	134 24 W.
1907	Waitara . . .	39 01 S.	174 13 E.	1971	Taku Harbour .	58 04 N.	134 00 W.
1908	Kawhia . . .	38 05 S.	174 49 E.	1972	Port McArthur .	56 04 N.	134 07 W.
1909	Raglan . . .	37 46 S.	174 55 E.	1973	Port Protection .	56 19 N.	133 36 W.
1910	Onehunga. . .	36 56 S.	174 47 E.	1974	St. John Harbour	56 26 N.	132 57 W.
1911	Cornwallis. . .	37 00 S.	174 36 E.	1975	Petersburg . .	56 49 N.	132 58 W.
1912	Aotea	36 34 S.	174 23 E.	1976	Port Alice. . .	55 48 N.	133 36 W.
1913	Tekopuru . . .	36 02 S.	173 56 E.	1977	Port Santa Cruz .	55 16 N.	133 24 W.
1914	Kohukohu . .	35 22 S.	173 32 E.	1978	Dadens . . .	54 11 N.	132 59 W.
1915	Ahipara . . .	35 12 S.	173 08 E.	1979	Naden Harbour .	54 02 N.	132 34 W.
1916	Nelson. . . .	41 17 S.	173 16 E.	1980	Port Chester . .	55 08 N.	131 34 W.
1917	Port Hardy . .	40 49 S.	173 54 E.	1981	Ratz Harbour .	55 53 N.	132 36 W.
1918	Picton. . . .	41 17 S.	174 00 E.	1982	Wrangell . . .	56 28 N.	132 22 W.
1919	Port Underwood .	41 19 S.	174 10 E.	1983	Ketchikan. . .	55 21 N.	131 39 W.
1920	Port Lyttleton .	43 36 S.	172 44 E.	1984	Stewart . . .	55 56 N.	130 00 W.
1921	Akaroa . . .	43 48 S.	172 58 E.	1985	Port Nelson . .	54 56 N.	130 00 W.
1922	Timaru . . .	44 24 S.	171 15 E.	1986	Port Simpson . .	54 34 N.	130 26 W.
1923	Oamaru . . .	45 07 S.	170 59 E.	1987	Prince Rupert .	54 19 N.	130 20 W.
1924	Dunedin . . .	45 53 S.	170 33 E.	1988	Inverness . . .	54 12 N.	130 15 W.
1925	Port Chalmers .	45 49 S.	170 39 E.	1989	Claxton . . .	54 05 N.	130 06 W.
1926	Newhaven. . .	46 30 S.	169 44 E.	1990	Port Essington .	54 10 N.	129 57 W.
1927	Waikawa . . .	46 39 S.	169 08 E.	1991	Port Louis . .	53 42 N.	132 57 W.
1928	Bluff	46 36 S.	168 20 E.	1992	Skidegate . .	53 16 N.	132 04 W.
1929	Port Pegasus .	47 13 S.	167 43 E.	1993	Pacofi	52 52 N.	131 52 W.
1930	Port William . .	46 51 S.	168 06 E.	1994	Lockeport. . .	52 42 N.	131 58 W.
1931	Invercargil . .	46 25 S.	168 22 E.	1995	Jedway . . .	52 15 N.	131 14 W.
1932	Haast. . . .	43 50 S.	169 20 E.	1996	Port Canaveral .	53 35 N.	130 09 W.
1933	Greymouth . .	42 26 S.	171 13 E.	1997	Port Stephens .	53 21 N.	129 43 W.
1934	Westport . . .	41 44 S.	171 36 E.	1998	Kitimat . . .	53 58 N.	128 42 W.
1935	Collingwood . .	40 32 S.	172 41 E.	1999	Butedale. . .	53 10 N.	128 45 W.
				2000	Swanson Bay .	53 01 N.	128 30 W.
	SECTION 8. NORTH AND SOUTH			2001	Port Blackney .	52 19 N.	128 21 W.
	AMERICA, PACIFIC COAST.			2002	Bella Bella . .	52 09 N.	128 07 W.
1936	Candle. . . .	66 02 N.	161 45 W.	2003	Ocean Falls . .	52 21 N.	127 42 W.
1937	Deering . . .	62 02 N.	162 35 W.	2004	Namu . . .	51 51 N.	127 51 W.
1938	York	65 31 N.	167 34 W.	2005	Wadhams . . .	51 31 N.	127 31 W.
1939	Port Clarence. .	65 13 N.	166 28 W.	2006	Kildala . . .	51 42 N.	127 21 W.
1940	Nome	64 30 N.	165 26 W.	2007	Port Hardy . .	50 43 N.	127 28 W.
1941	St. Michael . .	63 29 N.	162 02 W.	2008	Alert Bay . . .	50 35 N.	126 56 W.
1942	Kotlik. . . .	62 55 N.	163 15 W.	2009	Port Kusum . .	50 22 N.	125 59 W.
1943	Apokak . . .	60 08 N.	162 10 W.	2010	Campbell River .	50 01 N.	125 18 W.
1944	Dutch Harbour .	53 54 N.	166 32 W.	2011	Comox. . . .	49 40 N.	124 55 W.
1945	Bellcofski . .	55 07 N.	162 01 W.	2012	Nanoose . . .	49 16 N.	124 10 W.
1946	Coal Harbour. .	55 15 N.	160 40 W.	2013	Nanaimo . . .	49 10 N.	123 57 W.
1947	Unga . . .	55 10 N.	160 35 W.	2014	Ladysmith . .	48 59 N.	123 49 W.
1948	Chignik . . .	56 15 N.	158 20 W.	2015	Chemainus . .	48 56 N.	123 42 W.
1949	Uyak . . .	57 40 N.	154 00 W.	2016	Crofton . . .	48 52 N.	123 39 W.
1950	Kodiak . . .	57 52 N.	152 35 W.	2017	Cowichan		
1951	Afognak . . .	58 02 N.	152 54 W.		Harbour . .	48 45 N.	123 37 W.
1952	Knik	61 30 N.	149 28 W.	2018	Victoria . . .	48 26 N.	123 23 W.
1953	Seldovia . . .	59 26 N.	151 43 W.	2019	Esquimalt. . .	48 26 N.	123 26 W.
1954	Seward . . .	60 00 N.	149 20 W.	2020	Port Renfrew. .	48 34 N.	124 24 W.

Index No.	Name.	Lat.	Long.	Index No.	Name.	Lat.	Long.
		° ′	° ′			° ′	° ′
2021	Sooke . . .	48 22 N.	123 44 W.	2078	Fort Bragg . .	39 26 N.	123 49 W.
2022	Clo-oose . .	48 41 N.	124 51 W.	2079	Markham . .	38 28 N.	123 08 W.
2023	Banfield . .	48 50 N.	125 08 W.	2080	San Francisco .	37 48 N.	122 27 W.
2024	Port Alberni .	49 14 N.	124 49 W.	2081	Oakland . .	37 46 N.	122 14 W.
2025	Clayoquot. .	49 09 N.	125 55 W.	2082	Vallejo. . .	38 10 N.	122 15 W.
2026	Hesquiat Harbour .	49 22 N.	126 31 W.	2083	Port Chicago .	38 04 N.	122 02 W.
2027	Port Alice. .	50 23 N.	127 28 W.	2084	Antioch, *California* . .	38 01 N.	121 51 W.
2028	Quatsino . .	50 32 N.	127 37 W.	2085	Martinez . .	38 01 N.	122 08 W.
2029	Powell River .	50 00 N.	124 40 W.	2086	Stockton . .	37 57 N.	121 17 W.
2030	Stillwater . .	49 48 N.	124 24 W.	2087	Sacramento .	38 33 N.	121 30 W.
2031	Newport, *B.C.*	49 44 N.	123 11 W.	2088	Santa Cruz .	36 58 N.	122 01 W.
2032	Britannia . .	49 36 N.	123 10 W.	2089	Monterey . .	36 36 N.	121 52 W.
2033	Vancouver, *B.C.*	49 18 N.	123 07 W.	2090	Esteros Bay .	35 25 N.	120 48 W.
2034	Port Moody .	49 17 N.	122 52 W.	2091	Port San Luis .	35 10 N.	120 45 W.
2035	Sumas . . .	49 08 N.	122 05 W.	2092	Santa Barbara .	34 25 N.	119 41 W.
2036	Port Haney .	49 13 N.	122 36 W.	2093	Ventura . .	34 20 N.	119 18 W.
2037	Port Coquitlam	49 14 N.	122 46 W.	2094	Port Hueneme .	34 09 N.	119 12 W.
2038	New Westminster	49 12 N.	122 55 W.	2095	Santa Monica. .	34 00 N.	118 30 W.
2039	Blaine . . .	49 00 N.	122 46 W.	2096	Los Angeles Harbour .	33 43 N.	118 16 W.
2040	Bellingham .	48 45 N.	122 30 W.	2097	San Pedro, *California* .	33 44 N.	118 19 W.
2041	Anacortes . .	48 31 N.	122 37 W.	2098	Avalon . . .	33 21 N.	118 20 W.
2042	Everett, *Washington.*	47 59 N.	122 13 W.	2099	San Diego. .	32 43 N.	117 10 W.
2043	Seattle. . .	47 37 N.	122 20 W.	2100	Magdalena Bay .	24 38 N.	112 09 W.
2044	Tacoma . .	47 17 N.	122 25 W.	2101	San Jose del Cabo	23 03 N.	109 40 W.
2045	Olympia . .	47 03 N.	122 54 W.	2102	La Paz, *Mexico* .	24 10 N.	110 20 W.
2046	Arcadia . .	47 12 N.	122 57 W.	2103	Puerto Loreto .	26 00 N.	111 22 W.
2047	Union . . .	47 22 N.	123 06 W.	2104	Santa Rosalia. .	27 20 N.	112 16 W.
2048	Quilcene . .	47 49 N.	122 53 W.	2105	Puerto Isabel. .	31 46 N.	114 44 W.
2049	Port Orchard .	47 30 N.	122 39 W.	2106	Guaymas . .	27 55 N.	110 54 W.
2050	Bremerton .	47 33 N.	122 38 W.	2107	Topolobampo. .	25 36 N.	109 03 W.
2051	Port Madison .	47 40 N.	122 31 W.	2108	Altata . . .	24 50 N.	107 51 W.
2052	Port Gamble .	47 51 N.	122 36 W.	2109	Mazatlan . .	23 11 N.	106 26 W.
2053	Charleston .	47 32 N.	122 42 W.	2110	Puerto San Blas .	21 32 N.	105 19 W.
2054	Port Townsend .	48 08 N.	122 46 W.	2111	Manzanillo .	19 03 N.	104 20 W.
2055	Port Angeles .	48 07 N.	123 26 W.	2112	Acapulco . .	16 50 N.	99 56 W.
2056	Twin Rivers .	48 09 N.	123 57 W.	2113	Puerto Angel .	15 44 N.	96 41 W.
2057	Aberdeen, *Gray's Harbour.* .	46 58 N.	123 51 W.	2114	Salina Cruz .	16 10 N.	95 12 W.
				2115	La Puerta. .	15 56 N.	93 50 W.
2058	South Bend, *Gray's Harbour*	46 38 N.	123 49 W.	2116	Soconusco. .	15 09 N.	92 55 W.
2059	Bay City, *Oregon.*	46 52 N.	124 02 W.	2117	San Benito .	14 48 N.	92 16 W.
2060	Westport, *Washington.*	46 52 N.	124 08 W.	2118	Champerico .	14 18 N.	91 56 W.
2061	Astoria, *Columbia River.*	46 13 N.	123 46 W.	2119	San Jose . .	13 55 N.	90 50 W.
				2120	Acajutla . .	13 35 N.	89 51 W.
2062	Kalama, *Columbia River.*	46 01 N.	122 49 W.	2121	La Libertad .	13 29 N.	89 19 W.
2063	St. Helens, *Columbia River.*	45 52 N.	122 48 W.	2122	Cutuco. . .	13 21 N.	89 50 W.
				2123	La Union . .	13 20 N.	87 49 W.
2064	Vancouver, *Washington.*	45 41 N.	122 41 W.	2124	Ampala . .	13 18 N.	87 39 W.
2065	Portland, *Oregon.*	45 31 N.	122 40 W.	2125	Corinto . .	12 28 N.	87 12 W.
2066	Tillamook . .	45 29 N.	123 49 W.	2126	San Juan Del Sur	11 15 N.	85 53 W.
2067	Newport, *Oregon.*	44 38 N.	124 03 W.	2127	Puntarenas . .	9 58 N.	84 50 W.
2068	Yaquina . .	44 36 N.	124 02 W.	2128	Caldera, *Costa Rica*. . .	9 56 N.	84 44 W.
2069	Florence . .	43 58 N.	124 06 W.	2129	Quepos . .	9 24 N.	84 10 W.
2070	Gardiner . .	43 44 N.	124 07 W.	2130	Golfito. . .	8 38 N.	83 13 W.
2071	Coos Bay . .	43 23 N.	124 13 W.	2131	Armuelles. .	8 16 N.	82 51 W.
2072	Bandon . .	43 07 N.	124 25 W.	2132	Balboa. . .	8 57 N.	79 34 W.
2073	Wedderburn .	42 26 N.	124 25 W.	2133	Panama . .	9 00 N.	79 34 W.
2074	Crescent City .	41 45 N.	124 12 W.	2134	Buenaventura .	3 54 N.	77 05 W.
2075	Trinidad, *California* .	41 03 N.	124 09 W.	2135	Tumaco . .	1 50 N.	78 44 W.
				2136	Esmeraldas . .	0 57 N.	79 40 W.
				2137	Caraques . .	0 43 S.	80 28 W.
2076	Eureka . . .	40 48 N.	124 10 W.	2138	Manta . . .	0 57 S.	80 44 W.
2077	Westport, *California* .	39 38 N.	123 45 W.	2139	Santa Elena . .	2 12 S.	80 56 W.
				2140	Guayaquil. .	2 12 S.	79 52 W.
				2141	Puerto Bolivar .	2 40 S.	79 45 W.

Index No.	Name.	Lat.	Long.	Index No.	Name.	Lat.	Long.
		° ′	° ′			° ′	° ′
2142	Lobitos	4 27 S.	81 17 W.	2209	Concepcion, *Chile*	36 58 S.	73 05 W.
2143	Talara	4 34 S.	81 16 W.	2210	Coronel	37 02 S.	73 10 W.
2144	Paita	5 05 S.	81 07 W.	2211	Lota	37 05 S.	73 12 W.
2145	Puerto Bayovar	5 48 S.	81 02 W.	2212	Laraquete	37 20 S.	73 15 W.
2146	Pimentel	6 50 S.	79 58 W.	2213	Lebu	37 43 S.	73 40 W.
2147	Eten	6 57 S.	79 52 W.	2214	Valdivia	39 49 S.	73 15 W.
2148	Pacasmayo	7 30 S.	79 38 W.	2215	Corral	39 53 S.	73 28 W.
2149	Puerto Chicama	7 42 S.	79 25 W.	2216	Puerto Montt	41 29 S.	72 58 W.
2150	Trujillo, *Peru*.	8 10 S.	79 10 W.	2217	Ancud	41 52 S.	73 50 W.
2151	Salaverry	8 20 S.	78 55 W.	2218	Punta Arenas	53 10 S.	70 54 W.
2152	Guanape Islands	8 34 S.	78 47 W.	2219	Ushuaia	54 49 S.	68 13 W.
2153	Chimbote	9 05 S.	78 39 W.				
2154	Samanco	9 14 S.	78 35 W.		SECTION 9. SOUTH AMERICA,		
2155	Casma	9 25 S.	78 22 W.		ATLANTIC COAST.		
2156	Huarmei	10 05 S.	78 09 W.	2220	Moltke Harbour,		
2157	Puerto Bermejo	10 33 S.	77 53 W.		*S. Georgia*	54 31 S.	36 01 W.
2158	Supe	10 38 S.	77 47 W.	2221	Leith Harbour,		
2159	Puerto Huacho	11 08 S.	77 36 W.		*S. Georgia*	54 08 S.	36 41 W.
2160	Chancay	11 32 S.	77 25 W.	2222	Stanley Harbour,		
2161	Callao	12 04 S.	77 10 W.		*Falkland Is.*	51 42 S.	57 51 W.
2162	Puerto de Chilka	12 30 S.	76 50 W.	2223	Port Louis,		
2163	Tambo de Moro	13 19 S.	76 12 W.		*Falkland Is.*	51 33 S.	58 09 W.
2164	Cerro de Azul	13 03 S.	76 31 W.	2224	Port Salvador,		
2165	Pisco	13 43 S.	76 14 W.		*Falkland Is.*	51 26 S.	58 17 W.
2166	San Juan, *Peru*	15 20 S.	75 10 W.	2225	Port Albemarle,		
2167	Lomas	15 33 S.	74 52 W.		*Falkland Is.*	52 09 S.	60 30 W.
2168	Chala	15 43 S.	74 25 W.	2226	Chatham Harbour		
2169	Camana	16 25 S.	72 45 W.		*Falkland Is.*	51 50 S.	60 56 W.
2170	Port Matarani	17 00 S.	72 07 W.	2227	Port Egmont,		
2171	Mollendo	17 05 S.	71 53 W.		*Falkland Is.*	51 21 S.	60 03 W.
2172	Ilo	17 40 S.	71 22 W.	2228	St. John Harbour,		
2173	Arica	18 29 S.	70 20 W.		*Staten Id.*	54 40 S.	63 50 W.
2174	Pisagua	19 35 S.	70 13 W.	2229	Port Vancouver,		
2175	Caleta Buena				*Staten Id.*	54 47 S.	64 05 W.
	d'Sur	19 55 S.	70 10 W.	2230	Punta Loyola	51 36 S.	69 01 W.
2176	Iquique	20 13 S.	70 10 W.	2231	Puerto Santa Cruz	50 11 S.	68 13 W.
2177	Tocopilla	22 06 S.	70 14 W.	2232	Puerto Deseado	47 45 S.	65 55 W.
2178	Cobija	22 34 S.	70 18 W.	2233	Rivadavia	45 52 S.	67 29 W.
2179	Huanillo	22 39 S.	70 12 W.	2234	Rawson	43 11 S.	65 00 W.
2180	Mejillones	23 06 S.	70 28 W.	2235	Puerto Madryn	42 46 S.	65 02 W.
2181	Antofagasta	23 39 S.	70 25 W.	2236	Puerto Piramides	42 35 S.	64 17 W.
2182	Caleta Colosa	23 45 S.	70 26 W.	2237	San Antonio	40 48 S.	64 52 W.
2183	Taltal	25 26 S.	70 31 W.	2238	Viedma	40 50 S.	63 00 W.
2184	Chanaral	26 21 S.	70 38 W.	2239	Carmen de		
2185	Caldera, *Chile*.	27 04 S.	70 52 W.		Patagones	40 46 S.	62 50 W.
2186	Carrizal Bajo	28 04 S.	71 11 W.	2240	Bahia Blanca	38 49 S.	62 17 W.
2187	Huasco	28 28 S.	71 13 W.	2241	Puerto Militar	39 00 S.	61 55 W.
2188	Pena Blanca	28 50 S.	71 30 W.	2242	Necochea	38 42 S.	58 40 W.
2189	Cruz Grande	29 10 S.	71 20 W.	2243	Mar Del Plata	38 03 S.	57 33 W.
2190	Totoralillo	29 28 S.	71 20 W.	2244	San Clemente,		
2191	Coquimbo	29 56 S.	71 20 W.		*Argentina*	36 33 S.	56 48 W.
2192	Guayacan	30 05 S.	71 30 W.	2245	La Plata (*Eva Peron*)	34 50 S.	57 53 W.
2193	Tongoy	30 15 S.	71 32 W.	2246	Buenos Aires	34 36 S.	58 22 W.
2194	Los Vilos	31 54 S.	71 32 W.	2247	Campana	34 20 S.	58 58 W.
2195	Pichidanqui	32 05 S.	71 30 W.	2248	Zarate	34 15 S.	59 10 W.
2196	Papudo	32 30 S.	71 28 W.	2249	Ibicuy	33 38 S.	59 17 W.
2197	Quintero Bay	32 45 S.	71 31 W.	2250	Baradero	33 58 S.	59 35 W.
2198	Valparaiso	33 01 S.	71 37 W.	2251	San Pedro,		
2199	San Antonio	33 36 S.	71 38 W.		*Argentina*	33 40 S.	59 40 W.
2200	Juan Fernandez	33 38 S.	78 50 W.	2252	Obligado	33 30 S.	59 50 W.
2201	Matanzas, *Chile*.	34 00 S.	71 50 W.	2253	San Nicolas	33 30 S.	60 12 W.
2202	Pichilemu	34 30 S.	72 00 W.	2254	Villa Constitucion	33 18 S.	60 20 W.
2203	Ilico	34 46 S.	72 06 W.	2255	Rosario	33 05 S.	60 40 W.
2204	Constitucion	35 18 S.	72 25 W.	2256	San Lorenzo	32 40 S.	60 40 W.
2205	Curanipe	35 49 S.	72 37 W.	2257	Puerto Gomez	32 24 S.	60 38 W.
2206	Buchupureo	36 04 S.	72 50 W.	2258	Diamante	32 08 S.	60 35 W.
2207	Talcahuano	36 41 S.	73 06 W.	2259	Parana	31 45 S.	60 30 W.
2208	San Vicente	36 47 S.	73 17 W.	2260	Colastine	31 37 S.	60 30 W.

Index No.	Name.	Lat.	Long.	Index No.	Name.	Lat.	Long.
		° ′	° ′			° ′	° ′
2261	Santa Fe . . .	31 35 S.	60 40 W.	2325	Tabatinga. . .	4 15 S.	69 50 W.
2262	Santa Elena,			2326	Iquitos . . .	3 47 S.	73 19 W.
	Argentina . .	31 00 S.	59 42 W.	2327	Nauta	4 35 S.	73 45 W.
2263	La Paz, *Uruguay*.	30 51 S.	59 35 W.	2328	Cayenne . . .	4 55 N.	52 20 W.
2264	Esquina . . .	29 58 S.	59 30 W.	2329	Sinnamari. . .	5 19 N.	52 54 W.
2265	Goya	29 20 S.	59 20 W.	2330	Paramaribo . .	5 49 N.	55 09 W.
2266	Bella Vista . .	28 35 S.	59 10 W.	2331	Batavia, *Dutch*		
2267	Empedrado . .	28 08 S.	58 50 W.		*Guiana* . . .	5 47 N.	55 55 W.
2268	Corrientes. . .	27 35 S.	58 48 W.	2332	Nickerie . . .	5 55 N.	56 50 W.
2269	Villa del Pilar .	26 55 S.	58 17 W.	2333	Berbice . . .	6 17 N.	57 29 W.
2270	Formosa,			2334	Georgetown,		
	Argentina .	26 12 S.	58 12 W.		*Demarara* . .	6 48 N.	58 10 W.
2271	Asuncion . . .	25 25 S.	57 38 W.	2335	Parika,		
2272	Colonia . . .	34 28 S.	57 51 W.		*Essequibo River*	6 52 N.	58 25 W.
2273	Independencia .	33 15 S.	58 10 W.	2336	Makauria,		
2274	Concepcion del				*Essequibo River*	6 28 N.	58 35 W.
	Uruguay .	32 13 S.	58 12 W.	2337	Barima . . .	8 35 N.	60 18 W.
2275	Montevideo . .	34 55 S.	56 13 W.	2338	Boca Grande . .	8 39 N.	60 35 W.
2276	Maldonado . .	34 54 S.	54 52 W.	2339	Barrancas . .	8 29 N.	62 29 W.
2277	Rio Grande do Sul	32 02 S.	52 06 W.	2340	Puerto Tablas .	8 14 N.	63 03 W.
2278	Pelotas . . .	31 45 S.	52 30 W.	2341	San Felix . .	8 18 N.	63 08 W.
2279	Porto Alegre . .	30 02 S.	51 15 W.	2342	Ciudad Bolivar .	8 09 N.	63 59 W.
2280	Laguna . . .	28 30 S.	48 46 W.				
2281	Imbituba . . .	28 14 S.	48 40 W.		SECTION 10. WEST INDIES, GULF OF		
2282	Florianopolis . .	27 36 S.	48 34 W.		MEXICO AND CARIBBEAN SEA.		
2283	Itajahy . . .	26 53 S.	48 45 W.	2343	Brighton,		
2284	Sao Francisco do				*Trinidad* .	10 15 N.	61 38 W.
	Sul . . .	26 14 S.	48 38 W.	2344	Port of Spain,		
2285	Paranagua . .	25 31 S.	48 27 W.		*Trinidad* .	10 39 N.	61 31 W.
2286	Antonina . . .	25 20 S.	48 35 W.	2345	San Fernando,		
2287	Santos. . . .	23 57 S.	46 24 W.		*Trinidad* .	10 12 N.	61 25 W.
2288	Angra dos Reis .	23 01 S.	44 19 W.	2346	Caripito . . .	10 09 N.	63 05 W.
2289	Rio de Janeiro .	22 54 S.	43 10 W.	2347	Guiria	10 32 N.	62 28 W.
2290	Saquarema . .	22 55 S.	42 30 W.	2348	Carupano . . .	10 35 N.	63 14 W.
2291	Macahe . . .	22 30 S.	41 50 W.	2349	Pampatar . . .	10 58 N.	63 53 W.
2292	Imbitiba . . .	22 23 S.	41 45 W.	2350	Cumana . . .	10 18 N.	64 07 W.
2293	Vitoria. . . .	20 19 S.	40 17 W.	2351	Guanta . . .	10 07 N.	64 36 W.
2294	Caravellas. . .	17 50 S.	39 23 W.	2352	Puerto la Cruz .	10 14 N.	64 38 W.
2295	Porto Seguro . .	16 25 S.	39 07 W.	2353	Barcelona,		
2296	Belmonte . . .	15 50 S.	39 00 W.		*Venezuela* .	10 05 N.	64 45 W.
2297	Ilheus	14 50 S.	39 06 W.	2354	La Guaira. . .	10 37 N.	66 56 W.
2298	Marahu . . .	14 10 S.	39 05 W.	2355	Puerto Cabello .	10 19 N.	68 02 W.
2299	Bahia	12 58 S.	38 31 W.	2356	Tucacas . . .	10 40 N.	68 24 W.
2300	Aracaju . . .	10 56 S.	37 07 W.	2357	San Miguel,		
2301	Penedo . . .	10 10 S.	36 35 W.		*Venezuela* .	10 56 N.	68 22 W.
2302	Maceio. . . .	9 40 S.	35 43 W.	2358	Cumareto . . .	11 21 N.	69 29 W.
2303	Tamandare . .	8 44 S.	35 05 W.	2359	Kralendjik,		
2304	Recife	8 04 S.	34 53 W.		*Bonaire* .	12 09 N.	68 17 W.
2305	Parahyba . . .	7 10 S.	34 53 W.	2360	Willemstad,		
2306	Cabadello . . .	6 59 S.	34 19 W.		*Curacao* .	12 05 N.	68 59 W.
2307	Natal, *Brazil* . .	5 47 S.	35 12 W.	2361	Bullen Bay,		
2308	Macau	5 15 S.	36 39 W.		*Curacao* .	12 11 N.	69 01 W.
2309	Aracaty . . .	4 24 S.	37 45 W.	2361a	La Vela . . .	11 19 N.	69 39 W.
2310	Fortaleza . . .	3 41 S.	38 34 W.	2362	Oranjestad, *Aruba*	12 31 N.	70 03 W.
2311	Camocim . . .	2 53 S.	40 52 W.	2363	Las Piedras . .	11 42 N.	70 13 W.
2312	Parnahyba . .	3 06 S.	41 50 W.	2364	Amuay Bay . .	11 46 N.	70 15 W.
2313	Tutoya . . .	2 41 S.	42 14 W.	2365	Punta Cardon .	11 38 N.	70 15 W.
2314	Sao Luiz . . .	2 32 S.	44 18 W.	2366	Altagracia. . .	10 38 N.	71 35 W.
2315	Alcantara . . .	2 25 S.	44 32 W.	2367	Maracaibo. . .	10 34 N.	71 48 W.
2316	Vizeu	1 10 S.	46 06 W.	2368	Santa Marta . .	11 11 N.	74 14 W.
2317	Braganca . . .	1 00 S.	47 08 W.	2369	Puerto Colombia	10 58 N.	75 10 W.
2318	Belem	1 27 S.	48 30 W.	2370	Barranquilla . .	10 56 N.	74 51 W.
2319	Breves. . . .	1 40 S.	50 29 W.	2371	Cartagena,		
2320	Porto de Moz. .	2 00 S.	52 00 W.		*Colombia* .	10 26 N.	75 34 W.
2321	Prainha . . .	1 35 S.	53 40 W.	2372	Covenas . . .	9 25 N.	75 41 W.
2322	Santarem . . .	2 30 S.	54 45 W.	2373	Porto Bello,		
2323	Obidos. . . .	1 50 S.	55 53 W.		*Panama* . .	9 29 N.	79 37 W.
2324	Manaos . . .	3 00 S.	60 05 W.	2374	Colon	9 21 N.	79 55 W.

Index No.	Name.	Lat.	Long.	Index No.	Name.	Lat.	Long.
		° ′	° ′			° ′	° ′
2375	Chagres	9 17 N.	79 59 W.	2437	Baton Rouge	30 27 N.	91 06 W.
2376	Bocas del Toro	9 20 N.	82 12 W.	2438	Natchez	31 29 N.	91 26 W.
2377	Almirante	9 07 N.	82 28 W.	2439	Vicksburg	32 17 N.	90 54 W.
2378	Port Limon	10 00 N.	83 02 W.	2440	Bay St. Louis	30 21 N.	89 29 W.
2379	Greytown	10 55 N.	83 42 W.	2441	Pass Christian	30 22 N.	89 14 W.
2380	Blewfields	12 00 N.	83 40 W.	2442	Gulfport	30 25 N.	89 03 W.
2380a	Puerto Cabezas	14 02 N.	83 25 W.	2443	Biloxi	30 24 N.	88 51 W.
2381	Gracios a Dios	14 58 N.	83 14 W.	2444	Pascagoula	30 25 N.	88 35 W.
2382	Truxillo	15 51 N.	86 00 W.	2445	Mobile	30 41 N.	88 02 W.
2383	La Ceiba	15 43 N.	86 56 W.	2446	Pensacola	30 24 N.	87 13 W.
2384	Tela	15 55 N.	87 35 W.	2447	Panama City	30 10 N.	85 40 W.
2385	Puerto Cortez	15 50 N.	87 57 W.	2448	Port St. Joe	29 55 N.	85 23 W.
2386	Omoa	15 43 N.	88 01 W.	2449	Apalachicola	29 44 N.	85 06 W.
2387	Puerto Barrios	15 43 N.	88 34 W.	2450	Carabelle	29 55 N.	84 37 W.
2388	Isabel, *Guatamala*	15 21 N.	89 11 W.	2451	St. Mark	30 12 N.	84 11 W.
2389	Livingston	15 49 N.	88 52 W.	2452	Cedar Keys	29 08 N.	83 02 W.
2390	Punta Gorda, *Brit. Honduras*	15 05 N.	88 55 W.	2453	St. Petersburg	27 46 N.	82 38 W.
				2454	Tampa	27 58 N.	82 28 W.
2391	All Pines	16 50 N.	88 16 W.	2455	Punta Gorda, *Florida*	26 53 N.	82 01 W.
2392	Stann Creek	16 58 N.	88 14 W.	2456	Punta Rassa	26 29 N.	81 59 W.
2393	Belize	17 29 N.	88 13 W.	2457	Key West	24 33 N.	81 48 W.
2394	Corosal	18 23 N.	88 33 W.	2458	Havana, *Cuba*	23 09 N.	82 20 W.
2395	Payo Ooispo	18 31 N.	88 25 W.	2459	Matanzas, *Cuba*	23 04 N.	81 32 W.
2396	Xcalak	18 18 N.	87 50 W.	2460	Cardenas, *Cuba*	23 04 N.	81 12 W.
2397	Vigia	19 40 N.	87 40 W.	2461	Sagua, *Cuba*	22 50 N.	80 11 W.
2398	Silan	21 24 N.	88 54 W.	2462	Caibarien, *Cuba*	22 28 N.	79 32 W.
2399	Progreso	21 17 N.	89 40 W.	2463	Nuevitas, *Cuba*	21 35 N.	77 15 W.
2400	Sisal	21 03 N.	90 12 W.	2464	Manati, *Cuba*	21 19 N.	76 50 W.
2401	Celestun	20 47 N.	90 26 W.	2465	Puerto Padre, *Cuba*	21 12 N.	76 36 W.
2402	Campeche	19 50 N.	90 32 W.	2466	Puerto Gibara, *Cuba*	21 06 N.	76 08 W.
2403	Lerma	19 46 N.	90 40 W.	2467	Puerto Banes, *Cuba*	20 54 N.	75 44 W.
2404	Sabancuy	19 03 N.	91 05 W.	2468	Antilla, *Cuba*	20 50 N.	75 44 W.
2405	Carmen	18 39 N.	91 51 W.	2469	Preston, *Cuba*	20 43 N.	75 36 W.
2406	Xicalango	19 37 N.	91 50 W.	2470	Baracoa, *Cuba*	20 19 N.	74 30 W.
2407	Frontera	18 33 N.	92 42 W.	2471	Caimanera, *Cuba*	19 58 N.	75 10 W.
2408	Chiltepec	18 25 N.	93 05 W.	2472	Daiquiri, *Cuba*	19 52 N.	75 36 W.
2409	Tupilco	18 24 N.	93 27 W.	2473	Santiago, *Cuba*	20 00 N.	75 50 W.
2410	Santa Ana, *Mexico*	18 12 N.	93 56 W.	2474	Puerto Pilon, *Cuba*	19 54 N.	77 19 W.
2411	Coatzacoalcos	18 09 N.	94 25 W.	2475	Manzanillo, *Cuba*	20 20 N.	77 07 W.
2412	Alvarado	18 52 N.	95 50 W.	2476	Santa Cruz del Sur, *Cuba*	20 48 N.	78 09 W.
2413	Vera Cruz	19 12 N.	96 08 W.	2477	Jucaro, *Cuba*	21 35 N.	78 58 W.
2414	Tuxpan	20 59 N.	97 30 W.	2478	Tunas de Zara, *Cuba*	21 39 N.	79 38 W.
2415	Tampico	22 15 N.	97 51 W.	2479	Casilda, *Cuba*	21 45 N.	79 58 W.
2416	Matamoros, *Mexico*	25 53 N.	97 28 W.	2480	Cienfuegos, *Cuba*	22 09 N.	80 27 W.
2417	Brownsville	25 59 N.	97 28 W.	2481	Xagua, *Cuba*	22 16 N.	81 02 W.
2418	Port Isabel, *Texas*	26 04 N.	97 13 W.	2482	Batabano, *Cuba*	22 43 N.	82 19 W.
2419	Corpus Christi	27 45 N.	97 25 W.	2483	La Coloma, *Cuba*	22 17 N.	83 38 W.
2420	Rockport, *Texas*	28 03 N.	97 04 W.	2484	Guane, *Cuba*	22 19 N.	83 56 W.
2321	Indianola	28 28 N.	96 37 W.	2485	Bahia Honda, *Cuba*	22 55 N.	83 06 W.
2422	Matagorda	28 40 N.	95 59 W.	2486	Cabanas, *Cuba*	22 59 N.	82 46 W.
2423	Freeport	28 56 N.	95 18 W.	2487	Mariel, *Cuba*	23 01 N.	82 46 W.
2424	Velasco	29 01 N.	95 20 W.	2488	Montego Bay, *Jamaica*	18 30 N.	78 00 W.
2425	Texas City	29 28 N.	95 05 W.	2489	Falmouth, *Jamaica*	18 30 N.	77 42 W.
2426	Galveston	29 19 N.	94 48 W.	2490	St. Anns Bay, *Jamaica*	18 27 N.	77 14 W.
2427	Baytown	29 43 N.	95 01 W.	2491	Port Maria, *Jamaica*	18 24 N.	76 55 W.
2428	Houston	29 16 N.	94 50 W.	2492	Annatto Bay, *Jamaica*	18 18 N.	76 49 W.
2429	Port Arthur	29 55 N.	93 57 W.				
2430	Sabine	29 43 N.	93 51 W.				
2431	Beaumont	30 08 N.	94 10 W.				
2432	Orange	30 06 N.	93 47 W.				
2433	Lake Charles	30 09 N.	93 10 W.				
2434	Port Eads	29 01 N.	89 10 W.				
2435	New Orleans	29 58 N.	90 02 W.				
2436	Port Allen	30 26 N.	91 10 W.				

Index No.	Name.	Lat.	Long.	Index No.	Name.	Lat.	Long.
		° ′	° ′			° ′	° ′
2493	Buff Bay, Jamaica	18 14 N.	76 40 W.	2533	Port de Paix, Haiti	19 58 N.	72 50 W.
2494	Port Antonio, Jamaica	18 12 N.	76 27 W.	2534	Le Borgne, Haiti	19 51 N.	72 32 W.
2495	Port Morant, Jamaica	17 55 N.	76 21 W.	2535	Acul, Haiti	19 40 N.	72 20 W.
2496	Kingston, Jamaica	18 00 N.	76 49 W.	2536	Mayaguez, Porto Rico	18 09 N.	67 09 W.
2497	Port Royal, Jamaica	17 57 N.	76 50 W.	2537	Aguadilla, Porto Rico	18 28 N.	67 12 W.
2498	Port Esquivel, Jamaica	17 53 N.	77 08 W.	2538	Arecibo, Porto Rico	18 29 N.	66 42 W.
2499	Old Harbour, Jamaica	17 58 N.	77 13 W.	2539	San Juan, Porto Rico	18 29 N.	66 07 W.
2500	Salt River, Jamaica	17 50 N.	77 16 W.	2540	Fajardo, Porto Rico	18 20 N.	65 40 W.
2501	Milk River, Jamaica	17 53 N.	77 22 W.	2541	Porto Mulas, Porto Rico	18 10 N.	65 30 W.
2502	Alligator Pond, Jamaica	17 52 N.	77 37 W.	2542	Guayama, Porto Rico	18 01 N.	66 04 W.
2503	Black River, Jamaica	18 06 N.	77 52 W.	2543	Arroyo, Porto Rico	17 56 N.	66 04 W.
2504	Bluefields, Jamaica	18 08 N.	78 02 W.	2544	Jobos, Porto Rico.	17 59 N.	66 11 W.
2505	Savanna la Mar, Jamaica	18 14 N.	78 10 W.	2545	Ponce, Porto Rico.	17 58 N.	66 37 W.
2506	Harbour Island, Bahama Is..	25 32 N.	76 43 W.	2546	Guayanilla, Porto Rico	18 00 N.	66 48 W.
2507	Nassau, Bahama Is..	25 05 N.	77 21 W.	2547	Guanica, Porto Rico	18 00 N.	66 58 W.
2508	Matthew Town, Great Inagua	20 57 N.	73 41 W.	2548	St. Thomas, Virgin Is.	18 20 N.	64 56 W.
2509	Hawk's Nest Anchorage, Turks Is.	21 26 N.	71 07 W.	2549	Fredericksted, St. Croix	17 44 N.	64 52 W.
2510	Cap Haitien, Haiti	19 45 N.	72 12 W.	2550	Christiansted, St. Croix	17 47 N.	64 46 W.
2511	Fort Liberte, Haiti	19 38 N.	71 50 W.	2551	Marigot, St. Martin, L.I.	18 04 N.	63 05 W.
2512	Monte Cristi, Haiti	19 50 N.	71 40 W.	2552	Philipsburg, St. Martin, L.I.	18 01 N.	63 03 W.
2513	Puerto del Plata, Haiti	19 47 N.	70 45 W.	2553	Gastavia, St. Barthelmy, L.I.	17 54 N.	62 51 W.
2514	Cabarette, Haiti.	19 44 N.	70 24 W.	2554	Codrington, Barbuda, L.I.	17 39 N.	61 50 W.
2515	Sanchez, Haiti	19 12 N.	69 41 W.	2555	Orange Town, St. Eustache, L.I.	17 27 N.	62 58 W.
2516	Santa Barbara de Samana, Haiti.	19 11 N.	69 23 W.	2556	Basse Terre, St. Kitts, L.I.	17 18 N.	62 45 W.
2517	La Romana, Haiti	18 23 N.	68 59 W.	2557	Charlestown, Nevis, L.I..	17 08 N.	62 36 W.
2518	San Pedro de Macoris, Haiti.	18 30 N.	69 21 W.	2558	St. John's, Antigua, L.I.	17 06 N.	61 50 W.
2519	Ciudad Trujillo, Haiti	18 28 N.	69 53 W.	2559	English Harbour, Antigua, L.I.	17 00 N.	61 46 W.
2520	Nisao, Haiti	18 20 N.	70 15 W.	2560	Plymouth, Montserrat, L.I.	16 42 N.	62 13 W.
2521	Bani, Haiti	18 19 N.	70 20 W.	2561	Basse Terre, Guadaloupe, L.I.	16 00 N.	61 44 W.
2522	Azua, Haiti	18 26 N.	70 38 W.	2562	Pointe a Pitre, Guadaloupe, L.I.	16 14 N.	61 33 W.
2523	Barahona, Haiti	18 10 N.	71 05 W.	2563	Roseau, Dominica, L.I..	15 17 N.	61 24 W.
2524	Jacmel, Haiti.	18 13 N.	72 34 W.				
2525	Aux Cayes, Haiti.	18 16 N.	73 45 W.	2564	Fort de France, Martinique, W.I.	14 36 N.	61 04 W.
2526	Jeremie, Haiti	18 38 N.	74 09 W.				
2527	Miragoane, Haiti.	18 28 N.	73 06 W.	2565	Port Castries, St. Lucia, W.I.	14 01 N.	61 00 W.
2528	Petit Goave, Haiti	18 24 N.	72 47 W.	2566	Kingstown, St. Vincent, W.I.	13 09 N.	61 14 W.
2529	Port au Prince, Haiti	18 34 N.	72 22 W.	2567	St. George, Grenada, W.I.	12 03 N.	61 45 W.
2530	St. Marc, Haiti	19 05 N.	72 45 W.				
2531	Gonaives, Haiti	19 26 N.	72 42 W.	2568	Bridgetown, Barbados, W.I..	13 06 N.	59 37 W.
2532	Mole St. Nicholas, Haiti	19 49 N.	73 24 W.				

Index No.	Name.	Lat.	Long.	Index No.	Name.	Lat.	Long.
		° ′	° ′			° ′	° ′
2569	Scarborough,			2621	Albany, *New York*	42 39 N.	73 45 W.
	Tobago, W.I. .	11 11 N.	60 43 W.	2622	Bridgeport . .	41 10 N.	73 11 W.
2570	Hamilton,			2623	New Haven,		
	Bermuda . .	32 15 N.	64 50 W.		*Connecticut* .	41 14 N.	72 55 W.
				2624	Saybrook . . .	41 16 N.	72 21 W.
	SECTION 11. N. AMERICA—			2625	Greenport . . .	41 06 N.	72 22 W.
	ATLANTIC COAST, GREENLAND,			2626	New London . .	41 21 N.	72 06 W.
	ICELAND, Etc.			2627	Groton. . . .	41 21 N.	72 04 W.
2571	Miami	25 46 N.	80 12 W.	2628	Stonington . .	41 22 N.	71 53 W.
2572	Port Everglades .	26 06 N.	80 11 W.	2629	Newport,		
2573	West Palm Beach	26 43 N.	80 04 W.		*Rhode Island* .	41 30 N.	71 20 W.
2574	Fort Pierce . .	27 28 N.	80 19 W.	2630	Bristol, *Rhode*		
2575	St. Augustine. .	29 53 N.	81 18 W.		*Island* . .	41 40 N.	71 16 W.
2576	Mayport . . .	30 24 N.	81 26 W.	2631	Wickford . .	41 34 N.	71 29 W.
2577	Jacksonville . .	30 19 N.	81 40 W.	2632	Providence . .	41 49 N.	71 24 W.
2578	Fernandina . .	30 41 N.	81 28 W.	2633	Fall River. . .	41 42 N.	71 09 W.
2579	Satilla River . .	30 54 N.	81 32 W.	2634	New Bedford . .	41 36 N.	70 54 W.
2580	Brunswick . . .	31 09 N.	81 20 W.	2635	Wareham, *Mass.*.	41 47 N.	70 44 W.
2581	Sapelo	31 33 N.	81 26 W.	2636	Woods Hole . .	41 31 N.	70 40 W.
2582	Savannah . . .	32 05 N.	81 05 W.	2637	Edgartown . .	41 24 N.	70 32 W.
2583	Bull's River . .	32 20 N.	80 40 W.	2638	Nantucket . .	41 16 N.	70 07 W.
2584	Beaufort . . .	32 26 N.	80 40 W.	2639	Chatham, *Cape*		
2585	Charleston . .	32 47 N.	79 55 W.		*Cod* . .	41 40 N.	69 57 W.
2586	Georgetown . .	33 22 N.	79 17 W.	2640	Plymouth, *Mass.*.	41 58 N.	70 42 W.
2587	Wilmington . .	34 14 N.	77 57 W.	2641	Boston, *Mass.* .	42 21 N.	71 03 W.
2588	Morehead City .	34 44 N.	76 43 W.	2642	Salem . . .	42 32 N.	70 53 W.
2589	Newbern . . .	35 05 N.	77 05 W.	2643	Gloucester, *Mass.*	42 38 N.	70 30 W.
2590	Bayboro . . .	35 09 N.	76 42 W.	2644	Portsmouth,		
2591	Washington,				*New Hampshire*	43 05 N.	70 45 W.
	N. Carolina .	35 32 N.	77 02 W.	2645	Portland, *Maine* .	43 40 N.	70 15 W.
2592	Belhaven . . .	35 22 N.	76 34 W.	2646	Bath, *Maine* . .	43 55 N.	69 49 W.
2593	Edenton . . .	36 03 N.	76 33 W.	2647	Richmond, *Maine*	44 05 N.	69 48 W.
2594	Elizabeth,			2648	Rockland . . .	44 06 N.	69 06 W.
	N. Carolina .	36 19 N.	76 13 W.	2649	Belfast, *Maine* .	44 25 N.	69 00 W.
2595	Hampton Roads .	36 57 N.	76 20 W.	2650	Bucksport. . .	44 36 N.	68 50 W.
2596	Norfolk, *Virginia*	36 51 N.	76 16 W.	2651	Bangor, *Maine* .	44 48 N.	68 46 W.
2597	Portsmouth,			2652	Ellsworth . . .	44 32 N.	68 27 W.
	Virginia . .	36 49 N.	76 18 W.	2653	Mount Desert		
2598	Claremont. . .	37 12 N.	76 57 W.		Ferry . .	44 30 N.	68 20 W.
2599	Richmond,			2654	Bar Harbour . .	44 23 N.	68 12 W.
	Virginia . .	37 32 N.	77 25 W.	2655	Jonesport . . .	44 32 N.	67 36 W.
2600	Newport News .	36 59 N.	76 22 W.	2656	Machias . .	44 43 N.	67 28 W.
2601	West Point,			2657	Eastport . . .	44 54 N.	66 59 W.
	Virginia . .	37 32 N.	76 47 W.	2658	Welshpool. . .	44 53 N.	66 57 W.
2602	Kilmarnock,			2659	St. George, *N.B.* .	45 06 N.	66 54 W.
	Virginia . .	37 40 N.	76 21 W.	2660	St. Stephens . .	45 11 N.	67 17 W.
2603	Port Tobacco . .	38 31 N.	77 13 W.	2661	St. Andrew, *N.B.*	45 04 N.	67 05 W.
2604	Potomac . . .	38 31 N.	77 18 W.	2662	Musquash . . .	45 12 N.	66 18 W.
2605	Fort Washington.	38 41 N.	77 00 W.	2663	Carleton, *N.B.* .	45 15 N.	66 06 W.
2606	Annapolis . . .	38 59 N.	76 29 W.	2664	St. John, *N.B.* .	45 15 N.	66 04 W.
2607	Alexandria,			2665	Quaco . . .	45 20 N.	65 32 W.
	Virginia . .	38 48 N.	77 02 W.	2666	Herring Cove . .	45 35 N.	64 58 W.
2608	Baltimore . . .	39 16 N.	76 35 W.	2667	Harvey . . .	45 42 N.	64 44 W.
2609	Easton. . . .	38 44 N.	76 03 W.	2668	Hillsboro . . .	45 56 N.	64 42 W.
2610	Marcus Hook . .	39 48 N.	75 30 W.	2669	Joggins Wharf .	45 42 N.	64 27 W.
2611	Philadelphia . .	39 57 N.	75 08 W.	2670	Advocate		
2612	Newark, *New*				Harbour .	45 20 N.	64 50 W.
	Jersey . .	40 44 N.	74 10 W.	2671	Parrsboro . . .	45 23 N.	64 22 W.
2613	New York. . .	40 40 N.	74 00 W.	2672	Partipique . .	45 25 N.	63 46 W.
2614	Yonkers . . .	40 56 N.	73 54 W.	2673	Truro, *Nova*		
2615	Peekskill . . .	41 17 N.	73 56 W.		*Scotia* . .	45 22 N.	63 20 W.
2616	West Point,			2674	Walton . . .	45 11 N.	64 25 W.
	New York . .	41 24 N.	73 57 W.	2675	Summerville . .	45 08 N.	64 11 W.
2617	Newburg . . .	41 30 N.	74 01 W.	2676	Windsor, *Nova*		
2618	Poughkeepsie . .	41 42 N.	73 56 W.		*Scotia* . .	45 00 N.	64 08 W.
2619	Kingston, *New*			2677	Avondale . . .	44 58 N.	64 08 W.
	York . . .	51 55 N.	74 00 W.	2678	Hantsport. . .	45 02 N.	64 12 W.
2620	Tivoli	42 04 N.	73 56 W.	2679	Horton . . .	45 06 N.	64 13 W.

Index No.	Name.	Lat.	Long.	Index No.	Name.	Lat.	Long.
		° ′	° ′			° ′	° ′
2680	Wolfville . . .	45 10 N.	64 22 W.	2732	Shediac . . .	46 15 N.	64 32 W.
2681	Kingsport . .	45 09 N.	64 24 W.	2733	Cocagne . . .	46 18 N.	64 37 W.
2682	Annapolis, *Nova Scotia* . .	44 45 N.	65 30 W.	2734	Buctouche . .	46 28 N.	64 44 W.
				2735	Richibucto . .	46 43 N.	64 48 W.
2683	Clementsport .	44 41 N.	65 36 W.	2736	Chatham, *N.B.* .	47 00 N.	65 30 W.
2684	Digby	44 38 N.	65 45 W.	2737	Newcastle, *N.B.* . .	47 00 N.	65 34 W.
2685	Weymouth, *Nova Scotia* . .	44 27 N.	66 01 W.	2738	Neguac . . .	47 17 N.	65 04 W.
2686	Bellivean Cove .	44 20 N.	66 10 W.	2739	Shippigan . .	47 43 N.	64 44 W.
2687	Tiverton, *Nova Scotia* . .	44 24 N.	66 14 W.	2740	Caraquette . .	47 51 N.	64 54 W.
				2741	Bathurst, *N.B.* .	47 36 N.	65 39 W.
2688	Yarmouth, *Nova Scotia* . .	43 48 N.	66 08 W.	2742	New Mills . .	47 07 N.	66 10 W.
				2743	Dalhousie, *N.B.* .	48 05 N.	66 22 W.
2689	Tusket . .	43 52 N.	65 57 W.	2744	Campbellton . .	47 59 N.	66 41 W.
2690	Barrington .	43 31 N.	65 36 W.	2745	Carleton, *Quebec* .	48 08 N.	66 10 W.
2691	Clarke Harbour .	43 26 N.	65 38 W.	2746	New Richmond .	48 10 N.	65 51 W.
2692	Shelburne . .	43 45 N.	65 20 W.	2747	Bonaventuro .	48 04 N.	65 30 W.
2693	Jordan River .	43 40 N.	65 12 W.	2748	Paspebiac . .	48 04 N.	65 13 W.
2694	Lockeport . .	43 44 N.	65 05 W.	2749	Port Daniel .	48 12 N.	64 57 W.
2695	Port Jolie . .	43 54 N.	64 55 W.	2750	Newport, *N.B.* .	48 17 N.	64 47 W.
2696	Liverpool, *Nova Scotia* . .	44 02 N.	64 42 W.	2751	Grand Pabos .	48 22 Nl	64 43 W.
				2752	Grand River .	48 25 N.	64 31 W.
2697	Bridgewater, *Nova Scotia*	44 23 N.	64 31 W.	2753	Cape Cove . .	48 26 N.	64 22 W.
				2754	Perce . . .	48 31 N.	64 15 W.
2698	Lunenburg .	44 23 N.	64 19 W.	2755	Port St. Peter .	48 37 N.	64 13 W.
2699	Chester, *Nova Scotia* . .	44 34 N.	64 18 W.	2756	Douglastown .	48 44 N.	64 24 W.
				2757	Gaspe . . .	48 50 N.	64 32 W.
2700	Halifax, *Nova Scotia* . .	44 40 N.	63 34 W.	2758	St. Anne des Monts . .	49 07 N.	66 32 W.
2701	Dartmouth, *Nova Scotia* .	44 41 N.	63 33 W.	2759	Matane . .	48 52 N.	67 32 W.
				2760	Blanche River .	48 46 N.	67 42 W.
2702	Sheet Harbour .	44 47 N.	62 33 W.	2761	Metis . . .	48 37 N.	68 09 W.
2703	Liscomb . .	45 00 N.	62 00 W.	2762	Rimouski . .	48 26 N.	68 32 W.
2704	Guysboro . .	45 24 N.	61 30 W.	2763	Tadoussac . .	48 08 N.	69 43 W.
2705	Port Mulgrave .	45 37 N.	61 24 W.	2764	Anse St. Jean.	48 14 N.	70 04 W.
2706	Arichat . .	45 31 N.	61 02 W.	2765	Bagotville . .	48 20 N.	71 53 W.
2707	Louisburg . .	45 54 N.	60 00 W.	2766	Port Alfred .	48 20 N.	70 53 W.
2708	Port Morien .	46 08 N.	59 50 W.	2767	St. Alexis . .	48 22 N.	70 49 W.
2709	Bridgeport, *Nova Scotia* .	46 12 N.	59 57 W.	2768	Chicoutimi . .	48 25 N.	71 06 W.
				2769	Fraserville . .	47 51 N.	69 31 W.
2710	Sydney, *Nova Scotia* . .	46 09 N.	60 12 W.	2770	Murray Bay .	47 39 N.	70 09 W.
				2771	Quebec . . .	46 49 N.	71 11 W.
2711	North Sydney .	46 10 N.	60 16 W.	2772	Portneuf . .	46 42 N.	71 56 W.
2712	St. Ann's Harbour . .	46 15 N.	60 34 W.	2773	Lotbiniere. .	46 37 N.	71 57 W.
2713	Inganish . .	46 41 N.	60 20 W.	2774	Three Rivers .	46 22 N.	72 34 W.
2714	Newhaven, *Nova Scotia* . .	46 50 N.	60 19 W.	2775	Sorel . . .	46 02 N.	73 08 W.
				2776	Montreal . .	45 30 N.	73 34 W.
2715	Pleasant Bay .	46 49 N.	60 48 W.	2777	Valleyfield . .	45 15 N.	74 08 W.
2716	Chetican . .	46 39 N.	61 00 W.	2778	Cornwall, *Ontario*.	45 03 N.	74 43 W.
2717	Inverness, *Nova Scotia* . .	46 13 N.	61 17 W.	2779	Prescott . .	44 45 N.	75 30 W.
				2780	Kingston, *Ontario*	44 16 N.	76 30 W.
2718	Mabou . . .	46 05 N.	61 22 W.	2781	Oswego . . .	43 27 N.	76 30 W.
2719	Port Hood . .	46 01 N.	61 30 W.	2782	Rochester, *New York* . .	42 02 N.	77 35 W.
2720	Port Hastings .	45 42 N.	61 25 W.				
2721	Port Hawkesbury	45 37 N.	61 18 W.	2783	Port Colborne .	42 52 N.	79 14 W.
2722	Antigonish . .	45 42 N.	61 53 W.	2784	Port Weller .	43 14 N.	79 13 W.
2723	Merigonish Harbour . .	45 35 N.	62 25 W.	2785	Hamilton, *Ontario*	43 25 N.	79 50 W.
				2786	Toronto . .	43 40 N.	79 25 W.
2724	New Glasgow .	45 35 N.	62 38 W.	2787	Buffalo . .	42 52 N.	78 50 W.
2725	Pictou . . .	45 41 N.	62 42 W.	2788	Welland . .	43 00 N.	79 15 W.
2726	Pugwash . .	45 53 N.	63 41 W.	2789	Erie . . .	42 08 N.	80 07 W.
2727	Northport, *Nova Scotia* . .	45 56 N.	63 52 W.	2790	Cleveland . .	41 30 N.	81 45 W.
				2791	Sandusky . .	41 28 N.	82 48 W.
2728	Tidnish . .	45 59 N.	64 01 W.	2792	Port Clinton .	41 30 N.	82 56 W.
2729	Port Elgin . .	46 04 N.	64 05 W.	2793	Toledo. . .	41 45 N.	83 35 W.
2730	Moncton . .	46 05 N.	64 46 W.	2794	Monroe . .	41 58 N.	83 27 W.
2731	Bayfield . .	46 08 N.	63 48 W.	2795	Detroit . .	42 25 N.	83 07 W.
				2796	Sarnia . .	43 00 N.	82 24 W.
				2797	Port Huron .	43 00 N.	82 35 W.

Index No.	Name.	Lat.	Long.	Index No.	Name.	Lat.	Long.
		° ′	° ′			° ′	° ′
2798	Bay City, *Michigan*	43 35 N.	83 51 W.	2839	Baine Harbour,		
2799	Saginaw . . .	43 26 N.	83 58 W.		*Newfoundland* .	47 26 N.	54 56 W.
2800	Alpena. . . .	45 04 N.	83 27 W.	2840	Marytown,		
2801	Parry Sound .	45 19 N.	80 03 W.		*Newfoundland* .	47 10 N.	55 09 W.
2302	Midland . . .	44 45 N.	79 53 W.	2841	Fortune,		
2803	Owen Sound .	44 35 N.	80 56 W.		*Newfoundland* .	47 05 N.	55 56 W.
2804	Gladstone . .	45 52 N.	87 02 W.	2842	St. Pierre,		
2805	Menominee .	45 08 N.	87 38 W.		*Newfoundland* .	46 49 N.	56 19 W.
2806	Green Bay .	45 00 N.	87 30 W.	2843	Port au Basque,		
2807	Manitowoc .	44 06 N.	87 40 W.		*Newfoundland* .	47 45 N.	59 15 W.
2808	Sheboygan .	43 47 N.	87 45 W.	2844	Flat Bay,		
2809	Milwaukee .	43 05 N.	87 55 W.		*Newfoundland* .	48 26 N.	58 36 W.
2810	Chicago . .	41 50 N.	87 40 W.	2845	Bay of Islands,		
2811	Manistee . .	44 15 N.	86 18 W.		*Newfoundland* .	48 55 N.	57 54 W.
2812	Traverse City. .	44 38 N.	86 13 W.	2846	Corner Brook,		
2813	Harbour Springs .	45 27 N.	84 58 W.		*Newfoundland* .	48 58 N.	57 57 W.
2814	Sault Ste Marie .	46 28 N.	84 22 W.	2847	Port Saunders,		
2815	Duluth . .	46 48 N.	92 14 W.		*Newfoundland* .	50 39 N.	57 16 W.
2816	Port Arthur,			2848	Havre St. Pierre .	50 14 N.	63 36 W.
	Ontario . . .	48 33 N.	89 07 W.	2849	Comeau Bay .	49 15 N.	68 07 W.
2817	Tignish, *Prince*			2850	Seven Islands.	50 08 N.	66 20 W.
	Edward Id. .	46 58 N.	64 00 W.	2851	Clarke City .	50 12 N.	66 41 W.
2818	Alberton, *Prince*			2852	Port Marnham .	52 23 N.	55 44 W.
	Edward Id. .	46 48 N.	64 04 W.	2853	Cartwright . .	53 42 N.	57 02 W.
2819	Cascumpique,			2854	Rigolet . .	54 11 N.	58 26 W.
	Prince Edward			2855	Hopedale . .	55 27 N.	60 13 W.
	Id. . . .	46 40 N.	63 03 W.	2856	Nain . . .	56 33 N.	61 45 W.
2820	Souris, *Prince*			2857	Port Manvers.	57 00 N.	61 17 W.
	Edward Id. .	46 21 N.	62 15 W.	2858	Fort Chimo .	58 00 N.	68 16 W.
2821	Georgetown,			2859	Kovic . . .	61 40 N.	78 00 W.
	Prince Edward			2860	Kingwa . .	60 50 N.	78 10 W.
	Id. . . .	46 10 N.	62 29 W.	2861	Fort George .	53 52 N.	79 00 W.
2822	Crapaud, *Prince*			2862	East Main Fort .	52 05 N.	78 20 W.
	Edward Id. .	46 12 N.	63 30 W.	2863	Ruperts House .	51 30 N.	78 43 W.
2823	Charlottetown,			2864	Moosonee . .	51 19 N.	80 34 W.
	Prince Edward			2865	Moose Factory .	51 14 N.	80 42 W.
	Id. . . .	46 13 N.	63 08 W.	2866	Fort Albany .	52 13 N.	81 30 W.
2824	Summerside,			2867	Fort Severn .	56 03 N.	89 30 W.
	Prince Edward			2868	Port Nelson,		
	Id. . . .	46 23 N.	63 48 W.		*Hudson Bay* .	57 05 N.	92 36 W.
2825	Tilt Cove,			2869	Churchill . .	58 47 N.	94 12 W.
	Newfoundland .	49 54 N.	55 46 W.	2870	Chesterfield .	63 00 N.	91 25 W.
2826	Betts Cove,			2871	Thule . . .	76 32 N.	68 54 W.
	Newfoundland .	49 50 N.	55 51 W.	2872	Upernivik .	72 47 N.	56 03 W.
2827	Botwood,			2873	Godhavn . .	69 15 N.	53 33 W.
	Newfoundland .	49 02 N.	55 30 W.	2874	Jakobshavn .	69 15 N.	51 00 W.
2828	Norris Arm,			2875	Christianshaab .	68 30 N.	51 00 W.
	Newfoundland .	49 00 N.	55 20 W.	2876	Egedsminde .	68 30 N.	53 00 W.
2829	Gambo,			2877	Holstenborg .	66 56 N.	53 42 W.
	Newfoundland .	48 42 N.	54 18 W.	2878	Sukkertoppen.	65 23 N.	52 55 W.
2830	Clode Sound,			2879	Tovkussak .	64 51 N.	52 10 W.
	Newfoundland .	48 18 N.	54 12 W.	2880	Godthaab . .	64 11 N.	51 45 W.
2831	Goose Bay,			2881	Narsarssuak .	63 59 N.	51 36 W.
	Newfoundland .	48 25 N.	53 52 W.	2882	Faeringehavn .	63 42 N.	51 33 W.
2832	Heart's Content,			2883	Lichtenfels .	63 03 N.	50 47 W.
	Newfoundland .	47 53 N.	53 23 W.	2884	Frederikshaab .	62 00 N.	49 43 W.
2833	Carbonear,			2885	Ivigtut . .	61 12 N.	48 10 W.
	Newfoundland .	47 49 N.	53 06 W.	2886	Julianehaab .	60 43 N.	46 02 W.
2834	Harbour Grace,			2887	Nanortalik .	60 10 N.	45 13 W.
	Newfoundland .	47 41 N.	53 12 W.	2888	Frederiksdal .	60 00 N.	44 37 W.
2835	Holyrood,			2889	Prins Christians		
	Newfoundland .	47 25 N.	53 07 W.		Sund . .	60 03 N.	43 12 W.
2836	Wabana,			2890	Finnsbu . .	63 24 N.	41 17 W.
	Newfoundland .	47 40 N.	53 00 W.	2891	Angmagssalik. .	65 35 N.	37 30 W.
2837	St. John's,			2892	Danmarks Havn .	76 46 N.	18 46 W.
	Newfoundland .	47 34 N.	52 41 W.	2893	Reykjavik, *Iceland*	64 09 N.	21 56 W.
2838	Placentia Harbour			2894	Isafjordur,		
	Newfoundland .	47 17 N.	54 01 W.		*Iceland* . .	66 04 N.	23 08 W.

Index No.	Name.	Lat.	Long.	Index No.	Name.	Lat.	Long.
		° ′	° ′			° ′	° ′
2895	Siglufjordur, Iceland . . .	66 09 N.	18 54 W.	2904	Vik, *Iceland* . .	63 25 N.	19 00 W.
2896	Husavik, *Iceland*.	66 02 N.	19 09 W.	2905	Vestmann Island, *Iceland* . . .	63 27 N.	20 15 W.
2897	Akureyri, *Iceland*	65 41 N.	18 03 W.	2906	Eyrarbakki, *Iceland* . . .	63 52 N.	21 08 W.
2898	Vopnafjordur, *Iceland* . . .	65 45 N.	14 50 W.	2907	Hafnarfjordur, *Iceland* . . .	64 04 N.	21 57 W.
2899	Seydisfjordur, *Iceland* . . .	65 13 N.	13 44 W.	2908	Keflavik, *Iceland*.	64 00 N.	22 33 W.
2900	Nordfjordur, *Iceland* . . .	65 09 N.	13 42 W.	2909	Klaksvig, *Faroes*.	62 14 N.	6 35 W.
2901	Eskifjordur, *Iceland* . . .	65 04 N.	14 01 W.	2910	Thorshavn, *Faroes*	62 00 N.	6 46 W.
2902	Djupivogur, *Iceland* . . .	64 39 N.	14 18 W.	2911	Fuglefjord, *Faroes*	62 15 N.	6 49 W.
2903	Hornafjordur, *Iceland* . . .	64 15 N.	15 12 W.	2912	Trangisvaag, *Faroes* . . .	61 33 N.	6 50 W.
				2913	Vestmanhavn, *Faroes* . . .	62 09 N.	7 10 W.

GENERAL INDEX